COLORADO Breeding Bird ATLAS

HUGH E. KINGERY, EDITOR

Illustrated by Radeaux

COLORADO BIRD ATLAS PARTNERSHIP

Published by Colorado Bird Atlas Partnership

COLORADO DIVISION OF WILDLIFE

Copublished by Colorado Division of Wildlife

Library of Congress Catalog Number: 98-074577

ISBN Number: 0-9668506-0-2

Printed in the United States of America

PUBLISHER
Colorado Bird Atlas Partnership
Copublished by Colorado Division of Wildlife

EDITOR
Hugh E. Kingery

ILLUSTRATIONS
Radeaux, Pueblo, Colorado

MAPS
Don Schrupp, Colorado Division of Wildlife, Wildlife Resource Information System

DESIGN, LAYOUT, AND TYPOGRAPHY
Cathy Holtz, Blonde Ambition, Denver, Colorado

PUBLICATION MANAGEMENT
Bob Berman, Colorado Springs, Colorado

DISTRIBUTED BY
Colorado Wildlife Heritage Foundation
P.O. Box 211512
Denver, Colorado 80221-0394
FAX: 303-291-7106

THE FOLLOWING ORGANIZATIONS PROVIDED FUNDING FOR THIS BOOK:

COLORADO BIRD ATLAS PARTNERSHIP

National Fish and Wildlife Foundation

GREAT OUTDOORS COLORADO

THESE ORGANIZATIONS QUALIFIED AS SPONSORS OF THE ATLAS BY MAKING FINANCIAL CONTRIBUTIONS PROPORTIONATE TO THEIR MEMBERSHIPS:

Aiken Audubon Society
Arkansas Valley Audubon Society
Audubon Society of Greater Denver
Boulder Bird Club
Boulder County Audubon Society
Colorado Field Ornithologists
Denver Field Ornithologists
Durango Bird Club
Evergreen Naturalists Audubon Society
Foothills Audubon Club
Fort Collins Audubon Society
Grand Valley Audubon Society
Heart of the Rockies Audubon Society
Rabbit Ears Audubon Society
Tuesday Birders

THESE ORGANIZATIONS PROVIDED FUNDING AND IN-KIND SERVICES FOR ATLAS FIELD WORK:

Colorado Audubon Council
Colorado Chapter, The Wildlife Society
Colorado Division of Wildlife
Colorado Wildlife Federation
Colorado Wildlife Heritage Foundation
Denver Museum of Natural History
National Audubon Society
Owl Mountain Partnership
U.S. Bureau of Land Management
U.S. Fish and Wildlife Service
U.S. Forest Service

THESE INDIVIDUALS MADE SUBSTANTIAL FINANCIAL CONTRIBUTIONS TO ATLAS FIELD WORK:

Kathy Bollhoefer
Toni Brevillier
Mrs. George Milton
Helen Traylor
Scott Yarberry
Bartlesville Audubon Society, *in memory of Sophia Mery*

Dedicated to the field workers who made this project and book possible.

FIELD WORKERS WHO WORKED SEVEN OR MORE BLOCKS:

John Barber . . . . .Fort Collins
Norman Barrett . . . . .Kremmling
Jerry Besser . . . . .Lakewood
Virginia Bleck . . . . .Denver
Steve Boyle . . . . .Montrose
Toni Brevillier . . . . .Colorado Springs
Dan Bridges . . . . .Aurora
Alex Brown . . . . .Boulder
Gillian Brown . . . . .Boulder
Jim Chace . . . . .Boulder
Marilyn Colyer . . . . .Mesa Verde
Alex Cringan . . . . .Fort Collins
Vernon Dayhoff . . . . . . .Colorado Springs
Coen Dexter . . . . .Palisade
Beth Dillon . . . . .Fort Collins
Mike Figgs . . . . .Boulder
Bill Funk . . . . .Littleton
Gerald Fyler . . . . .Dolores
Maxine Fyler . . . . .Dolores
Marina Graves . . . . .Lakewood
J.R. Guadagno . . . . .Paonia
David Gulbenkian . . . . .Lakewood
Martha Gulbenkian . . . . .Lakewood
Dave Hallock . . . . .Nederland
Flora Harrison . . . . .Salida
Dave Hawksworth . . . . .Socorro, NM
Philip Hayes . . . . .Lakewood
Jon Horn . . . . .Montrose
Dave Johnson . . . . .Pueblo
Stephanie Jones . . . . .Denver
Stephen Jones . . . . .Boulder
Bill Kaempfer . . . . .Boulder
Mike Ketchen . . . . .Pueblo West
Clara King . . . . .Ordway
Hugh Kingery . . . . .Franktown
Urling Kingery . . . . .Franktown
Ruth Kuenning . . . . . . .Colorado Springs
Walt Kuenning . . . . . . .Colorado Springs
Ron Lambeth . . . . .Grand Junction
David Leatherman . . . . .Fort Collins
Randy Lentz . . . . .Aurora
Richard Levad . . . . .Grand Junction
Jean Maguire . . . . .Denver
Jack Merchant . . . . .Eagle
Ronald Meyer . . . . .Gunnison
Wayne Mollhoff . . . . .Ashland, NE
Duane Nelson . . . . .Golden
Paul Opler . . . . .Loveland
Sandra Opler . . . . .Fort Collins
David Pantle . . . . .Cañon City
Sherril Pantle . . . . .Cañon City
Kim Potter . . . . .Rifle
Dick Pratt . . . . .Longmont
Bill Pulliam . . . . .Fort Collins
John Rawinski . . . . .Monte Vista
Bob Righter . . . . .Denver
Dick Roth . . . . .Pueblo
Ronald A. Ryder . . . . .Fort Collins
David Silverman . . . . .Rye
Kent Simon . . . . .Evergreen
Robert Skalka . . . . .Trinidad
Todd Soderberg . . . . .Littleton
Mort Staatz . . . . .Golden
Leon Stigen . . . . .Grand Junction
Art Stiles-Wainwright . . . . .Boulder
Helen Stiles-Wainwright . . . . .Boulder
Janeal Thompson . . . . .Walsh
John Toolen . . . . .Grand Junction
Van Truan . . . . .Pueblo
Alan Versaw . . . . .Colorado Springs
Alan Wallace . . . . .Reno, NV
Judy Ward . . . . .Sheridan, WY
Roberta Winn . . . . .Woodland Park
Barbara Winternitz . . . . .Colorado Springs
Dave Woodward . . . . .Grand Junction
Ronda Woodward . . . . . . .Grand Junction
Brenda Wright . . . . .Clifton
Mark Yaeger . . . . .Pueblo
Don Youkey . . . . .Seward, AK
Vic Zerbi . . . . .Glenwood Springs

FIELD WORKERS WHO WORKED TWO TO SIX BLOCKS:

Kathy Abramson	Grand Junction
Paul Adamus	Corvallis, OR
Helen Allen	Grand Junction
John Alves	Monte Vista
Catherine Anderson	Englewood
Elaine Anderson	Denver
Idabelle Arndt	Berthoud
Jeff Bachant	Boulder
Bev Baker	Boulder
Cecilia Barr	Grand Junction
Bonnie Barton	Fort Collins
Gifford Beaton II	Marietta, GA
R.G. Beaton	Isle of Palms, SC
Donald Beaver	East Lansing, MI
Jim Berry	Ipswich, MA
Mary Jane Black	Denver
Ron Blaj	Colorado Springs
Bev Blank	Gunnison
Timothy Blank	Gunnison
Ann Bonnell	Littleton
Steve Borichevsky	Boulder
Steve Bouricius	Palisade
Jeff Brigham	Grand Junction
Leon Bright	Pueblo
Bill Brockner	Evergreen
Sylvia Brockner	Evergreen
Diane Brown	Nederland
Jeffery Burkhard	Enid, OK
Jerry Cairo	Cheyenne, WY
Mary Calkum	Limon
Lawrence Campbell	Monument
Dave Carey	Paonia
Mike Carter	Brighton
John Chrisafis	Loveland
David Clark	Carbondale
Douglas Clark	Alamosa
Jeff Connor	Estes Park
Gary Conover	Colorado Springs
Jim Conway	Glenwood Springs
Fred Crowley	Knox, PA
DeWitt Dagget	Meredith
Anne Dakin	Glenwood Springs.
Robert Darnell	Alamosa
Joshua Dickinson	Fort Collins
Bob Dickson	Pueblo
Johnny Dickson	Pueblo
Ferd Dirckx	Broomfield
Jo Dirckx	Broomfield
Ruby Ebright	Salida
Aaron Ellingson	Fort Collins
Fran Enright	Evergreen
Norman Erthal	Arvada
Gail Evans	Denver
Margaret Ewing	McCoy
Fred Facer	Grand Junction
Jack Fagan	Moab, UT
Linda Fagan	Moab, UT
Roxanne Falise	Craig
Gerald Fedrizzi	Eagle
Linda Ferguson	Colorado Springs
Ray Ferguson	Colorado Springs
Tammy Ferraira	
Warren Finch	Lakewood
Judy Fisher	Boulder
Ray Fisher	Boulder
Donna Flageolle	Westminster
Terry Flageolle	Westminster
Elva Fox	Durango
Linda Frederick	Rolla, MO
David Freed	Colorado Springs
Bob Frye	Pagosa Springs
Mark Fuller	Carbondale
Dolores Gable	Arvada
Jim Gable	Arvada
Tolline Gallagher	Aurora
Rosamond Garcia	Hayden
Caye Geer	Durango
Ken Giesen	Fort Collins
William Gillespie	Fort Collins
Scott Gillihan	Fort Collins
Eric Gordon	Arvada
Laurel Goulding	Grand Junction
Jay Graham	Rangely
Van Graham	Grand Junction
Larry Green	Glenwood Springs
Julie Grode	Rifle
Jim Grohs	Grand Junction
Craig Grother	Norwood
David Guertin	LaPorte
Dan Guthrie	Claremont, CA
Dorothy Hahn	Palisade
Helen Hale	Denver
John Hale	Denver
Jon Halverson	Steamboat Springs
Karen Harbaugh	Cortez
Joe Harrison	Longmont
Frank Hawksworth	Fort Collins
Scott Haywood	Austin, TX
Dale Hein	Fort Collins
Dick Heller	Mission, TX
Mike Henwood	Grand Junction
Robert Hernbrode	Denver
Jim Herold	Casper, WY
Verna Herold	Casper, WY
Elaine Hill	Boulder
Ed Hollowed	Meeker
Bill Howe	Albuquerque, NM
Tom Hyden	Windsor
Terry Ireland	Grand Junction
Mark Janos	Pueblo
Dave Jasper	Portal, AZ
Linda Johnson	Pueblo
Jan Justice-Waddington	Golden
Kevin Kempf	Longmont
Dave Kenvin	South Fork
Kathleen Kenvin	South Fork
Nyla Kladder	Grand Junction
Doris Kneuer	Florissant
Frances Krehbiel	Grand Junction
Chuck La Rue	Kayenta, AZ
Tom Laurion	Lander, WY
Jim Liewer	Kremmling
Steve Litz	Boulder
Jean William Lockett	Leavenworth, KS
Forrest Luke	Craig
Larry Malone	Aurora
Earl Markley	Homelake
Steve Marks	Claremont, CA
Kathy Martin	Wellington
Kent Martin	Rocky Ford
Nancy Martin	Rocky Ford
Steve Martin	Wellington
Mary Massey	Meeker
John Maynard	Manitou Springs
Virginia Maynard	Manitou Springs
Connie McDonald	Englewood
Kathleen McGinley	Grand Junction
Linda McMenamy	Breckenridge
Doug McVean	Grand Junction
Ann Means	Loveland
Cynthia Melcher	Fort Collins
Ed Merritt	McGrath, AK
David Michael	Aspen
Gary Miller	Colorado Springs
Mike Monahan	Denver
John Moody	Boulder
Thomas Moran	Grand Junction
Anne Morkill	Alamosa
Diana Mullineaux	Arvada
Don Mullineaux	Arvada
Naseem Munshi	Lafayette
Kathleen Munthe	Fort Collins
Kirk Navo	Monte Vista
Kathleen Nelson	Yampa
Reed Noss	Moscow, ID
Jean Parker	Greeley
Jim Parker	Tucson, AZ
Marianne Parker	Tucson, AZ
Bud Pearson	Fruita
Juliene Pimental	Denver
Pam Piombino	Boulder
Bob Plaska	Alamosa

FIELD WORKERS WHO WORKED TWO TO SIX BLOCKS (continued):

Jo Ann Potter . . . . Avon
Bill Prather . . . . Longmont
Glenn Pritchard . . . . Craig
Dan Purrington . . . . Metairie, LA
Phyllis Radice . . . . Grand Junction
Ray Rauch . . . . Commerce City
Claudia Rector . . . . Mancos
Chuck Reichert . . . . Meeker
Mike Reid . . . . Pagosa Springs
Lonnie Renner . . . . Grand Junction
Marjorie Rhoades . . . . Fort Collins
R. Rilling . . . .
Laura Roberts . . . . Evergreen
Steve Roberts . . . . Evergreen
Andrea Robinson . . . . Montrose
Jerry Robinson . . . . Golden
Julie Roederer . . . . Estes Park
Scott Roederer . . . . Estes Park
Joe Roller . . . . Golden
Jan Roth . . . . Craig
Maurice Rowe . . . . Hotchkiss
Pearle Sandstrom-Smith . . . . Pueblo
Rick Schnaderbeck . . . . Alamosa
Mary Jane Schock . . . . Denver
Glenna Schwalbe . . . . Port Jefferson, PA
Paul Schwalbe . . . . . . Port Jefferson, PA
Jim Sedgwick . . . . Fort Collins
William Shuster . . . . Delta
Julie Simpson . . . . Winchester, VA
Lynelle Skorkowsky . . . . Craig
Robert Skorkowsky . . . . Craig
Jennie Slater . . . . Lamar
Roger Sleeper . . . . Holly
Michael Smith . . . . Golden
Ben Sorensen . . . . Colorado Springs
Sally Sorensen . . . . Colorado Springs
Bob Spencer . . . . Golden
Marj Swies . . . . Cedar Park, TX
Nancy Taggart . . . . Colorado Springs
Joe Ten Brink . . . . Denver
William Theimer . . . . Gunnison
Mike Tindall . . . . Grand Junction
Don Traver . . . . Webster, NY
Donna Traver . . . . Webster, NY
Helen Traylor . . . . Grand Junction
Dave Valasek . . . . Golden
Else Van Erp . . . . Morrison
Ray Varney . . . . Walden
Linda Vidal . . . . Aspen
Scott Wait . . . . Saguache
Doug Ward . . . . Denver
Barbara Waters . . . . Durango
Jim Watts . . . . Penrose
Rosie Watts . . . . Penrose
Betsy Webb . . . . Denver
Carl Wells . . . . Denver
Julie Wells . . . . Denver
Anne Whitfield . . . . Beulah
Nathaniel Whitney . . . . . . Rapid City, SD
Jack Wieland . . . . Holyoke
Bob Wilkinson . . . . Fort Collins
Earl Williams . . . . Grand Junction
Bob Wing . . . . Boulder
Ru Wing . . . . Boulder
John Yaeger . . . . Pueblo
Jerry Zaninelli . . . . Estes Park

Boulder Bird Club . . . . Boulder
Boulder County Nature Assoc. . . . Boulder
Colorado Field Ornithologists . . . .
NERC Terrestrial Systems . . . Fort Collins
Yampa Ranger District, U.S. Forest Service . . . . Yampa
Zoology Dept., Denver Museum of Natural History . . . . Denver

FIELD WORKERS WHO WORKED ONE BLOCK:

Chuck Aid . . . . Brighton
Jim Allison . . . . Littleton
Kimberly Anderson . . . . Fort Collins
Randy Anderson . . . . Akron
Bob Andrews . . . . Denver
LaVonne Axford . . . . Littleton
Jamie Bailey . . . . Silverthorne
Sarah M.E. Bailey . . . . Paonia
Bruce Baker . . . . Fort Collins
Kelly Baker . . . . Woodland Park
Brett Bannon . . . . Folsom, NM
Dave Barber . . . . Pueblo
Fred Blackburn . . . . Cortez
Sarah Bogart . . . . Hotchkiss
Woody Brady . . . . Denver
John Brandt . . . . Alamosa
Ruth Breckon . . . . Denver
Erik Brekke . . . . Cañon City
Whitney Brevillier . . . . Colorado Springs
Barbara Bridges . . . . Lakewood
Merlynn Brown . . . . Evergreen
Virginia Carlson . . . . . . Colorado Springs
Kay Cheney . . . . Stoneham
Gloria Childress . . . . Durango
Jim Childress . . . . Durango
Dave Christian . . . . Denver
Sarah Christian . . . . Denver
Dina Clark . . . . Denver
Mike Claussner . . . . Edwards
Amanda Clements . . . . Mesa
Mary Coffey . . . . Pueblo
Walter Collins . . . . Boulder
Jack Coss . . . . Longmont
Theressa Coss . . . . Longmont
Paul Creedon . . . . Fruita
Don Cunningham . . . . Gateway
Thom Curdts . . . . Fort Collins
Carol Cushman . . . . Boulder
Raymond Davis . . . . Lyons
Harley Delafield . . . . Dinosaur
Howard Delfiner . . . . . . Elkins Park, PA
John Demmitt . . . . Chesterfield, MO
Mary Demmitt . . . . . . Chesterfield, MO
Jim Dennis . . . . Fort Collins
Marty Dick . . . . Boulder
Sharon Dooley . . . . Boulder
Marj Dunmire . . . . Estes Park
Alice Dunnick . . . . Palisade
Leanne Egeland . . . . St. Paul, MN
Gregory Everhart . . . . Lamar
Suzanne Fellows . . . . Maybell
Jim Ferguson . . . . Montrose
Fern Ford . . . . Boulder
Mike Foster . . . . Denver
Tom Fratt . . . . Steamboat Springs
Mel Gabel . . . . Evergreen
Ron Garcia . . . . Alamosa
B.J. Gard . . . . Evergreen
Tom Giffen . . . . Yampa
Walt Graul . . . . Fort Collins
Brian Gray . . . . Grand Junction
Paul Gray . . . . Grand Junction
Bob Gustafson . . . . Palisade
Nancy Gustafson . . . . Palisade
Joe Hall . . . . Grand Junction
Evan Hannay . . . . Evergreen
Ives Hannay . . . . Evergreen
Wendy Hanophy . . . . Denver
Paula Hansley . . . . Golden
Ron Harden . . . . Loveland
Joedy Hartman . . . . Burlington
Susan Hartman . . . . Burlington
Betty Harwood . . . . Morrison
Stan Harwood . . . . Morrison
Alan Hay . . . . Arvada

Peg Hayden . . . . Evergreen
J. B. Hayes . . . . Littleton
Louise Hector . . . . Denver
Helen Hernandez . . . . Greeley
Ann Higgins . . . . Colorado Springs
Jim Hill . . . . Boulder
Judy Hilty . . . . Salida
Joan Hoffman . . . . . . Steamboat Springs
Steve Hokenson . . . . Alamosa
John Hollingsworth . . . . Bellevue
Karen Hollingsworth . . . . Bellevue
Bill Householder . . . . Fort Collins
Jim Howell . . . . Lafayette
William Iko . . . . Fort Collins
Audrey Jensen . . . . Nucla
Bob Jickling . . . . Boulder
Bill Johnson . . . . Carbondale
Eric Johnson . . . . Albuquerque, NM
Peggy Jordan . . . . Glenwood Springs
Rod Jordan . . . . Glenwood Springs
Marjorie Joy . . . . Menasha, WI
Richard Joy . . . . Menasha, WI
Beth Kaeding . . . . Yellowstone Park, WY
Joe Kaplan . . . .
Aurel Kelly . . . . Arvada
D. W. King . . . . Lyons
Jackie King . . . . Commerce City
Connie Knapp . . . . Alamosa
Chris Knight . . . . Pueblo West
JoAnn Kozan . . . . Greeley
David Kramer . . . . Evergreen
Richard Kramer . . . . Loveland
Rose Kreher . . . . Holyoke
Mary Kronvall . . . . Grand Junction
Naomi Kuhlman . . . . . Colorado Springs
Paul S. Kunkel . . . . Eagle
Joe LaFleur . . . . Fort Collins
Ann Lambeth . . . . Grand Junction
Ken Lane . . . . Denver
Diana Laughlin . . . . Fort Collins
Robert Lewis . . . . Farmington, MO
Gene Ligon . . . . New Castle
Cindy Lippincott . . . . . Colorado Springs
Bill Lisowsky . . . . Fort Collins
Paula Lisowsky . . . . Fort Collins
Linda Litteral . . . . Steamboat Springs
Thomas Litteral . . . . . Steamboat Springs
Peggy Locke . . . . La Junta
Jackie Loughridge . . . Steamboat Springs
Keith Madsen . . . . Turin, IA
Jo Marchand . . . . Fort Collins
Joe Margoshes . . . . Denver
Scott Marsh . . . . Craig
David Martin . . . . Westminster
Ginny Martin . . . . Colorado Springs
Jim Massey . . . . Whitewater
Ed McConkey . . . . Boulder
Van McCutcheon . . . . Boulder
Kay McGinnis . . . . Breckenridge
George McKinnon . . . . Pueblo
Linda McKinnon . . . . Pueblo
Dorothy Mierow . . . Green Mountain Falls
Betty Minges . . . . Lakewood
Martha Mohan . . . . Hotchkiss
Jim Moore . . . . Buena Vista
Mary Moore . . . . Crawford
Karen Morris . . . . Bellevue
Sue Moyer . . . . Grand Junction
Marcia Murdock . . . . Conifer
Ron Murdock . . . . Conifer
Kenneth Nanney . . . . Longview, TX
Larry Nell . . . . Fort Collins
Tom Nicholls . . . . St. Paul, MN
LeRoy Nickelson . . . . Lamar
Mona Nielsen . . . . Denver
Carmen Nue . . . . Lakewood
Molly Ohlheiser . . . . Crawford
Betty Oliver . . . . Denver
Harold Oliver . . . . Denver
Brad Parks . . . . Atlanta, GA
Bruce Paton . . . . Denver
Brandon Percival . . . . Pueblo
Kathleen Phelps . . . . Silverthorne
Polly Phillips . . . . Evergreen
David Powell . . . . Kalamazoo, MI
John Prasuhn . . . . Estes Park
Francie Pusateri . . . . Fort Collins
Helen Randal . . . . Cascade
Elizabeth Rawinski . . . . Monte Vista
Mark Rebelo . . . . Gustine, CA
Jenny Rechel . . . . Riverside, CA
Jack Reddall . . . . Englewood
David Rees . . . . Bloomfield, NM
Jan Rees . . . . Bloomfield, NM
John Rees . . . . Bloomfield, NM
Ann Rich . . . . Steamboat Springs
Sue Ridd . . . . Glen Allen, VA
Marjorie Roberts . . . . Wheat Ridge
Alice Robinson . . . . Craig
Molly Robinson . . . . Glade Park
Wayne Robinson . . . . Craig
Mike Rogers . . . . . . Mountain View, CA
Louise Roloff . . . . Denver
Will Russell . . . . Tucson, AZ
John Schenck . . . . Littleton
Karleen Schofield . . . . Denver
Patty Schrader . . . . Grand Junction
Bill Schreier . . . . Lakewood
Cheryl Schreier . . . . Lakewood
Andrew Schroeder . . . . Alamosa
Don Schrupp . . . . Denver
Beverly Sears . . . . Boulder
Wes Sears . . . . Boulder
Aaron Sell . . . . Littleton
Ruth Shields . . . . Durango
Carol Shryock . . . . Ignacio
Clif Smith . . . . Pueblo
Kate Smith . . . . Eagle
Lynn Solomon . . . . Denver
Anne Southcott . . . . Littleton
Ray Sperger . . . . Morrison
Larry Springston . . . . Parker
Steve Stachowiak . . . . Aurora
Kip Stransky . . . . Durango
Skip Swies . . . . Cedar Park, TX
Dean Swift . . . . Jaroso
Pattie Swift . . . . Jaroso
Beverly Tomberlin . . . . Durango
Guy Tomberlin . . . . Durango
Ken Toy . . . . Aspen
Mike Trujillo . . . . Yuma
Timothy Tucey . . . . Pueblo
Tom Tustison . . . . Grand Junction
Candy Vallado . . . . Littleton
Jean Van Loan . . . . Denver
Larry Vincent . . . . Crownsville, MD
Mary Kay Waddington . . . . . Englewood
Dick Wagner . . . . Dolores
Evelyn Wagner . . . . Dolores
Melissa Walker . . . . Colorado Springs
Dick Walton . . . . Concord, MA
Andrea Wang . . . . Paonia
Mary Washburne . . . . Englewood
Lorraine Weber . . . . Clark
Dorothy Wells . . . . Sparks, NV
Barbara Whipple . . . . Buena Vista
Tony Wiegant . . . . Grand Junction
Marilyn Willard . . . . Mayville, MI
Myles Willard . . . . Mayville, MI
Dorothea Williams . . . . Colorado Springs
Robert Wilson . . . . Broomfield
X. Wilson . . . . Fort Collins
Marcella Wirth . . . . New Raymer
Bob Witt . . . . Grand Junction
Joyce Wolff . . . . Los Alamos, NM
Chris Wood . . . . Morrison
Marvin Woolf . . . . Boulder
Bill Wuerthele . . . . Denver
Suzanne Wuerthele . . . . Denver
Michael Wunder . . . . Boulder
Scott Yarberry . . . . Littleton
Barbara Zinn . . . . Green River, UT
Ed Zipser . . . . College Station, TX

Piñon Mesa 4-H Club . . . . . . Glade Park
Tuesday Birders . . . . Littleton
Del Norte District,
U.S. Forest Service . . . . Del Norte

FIELD HELPERS:

Diane Adams
Jane Adams
Mary Alcock
Christie Allan
Bill Alldredge
Jan Alldredge
Jim Aragon
George Armbrust
Donald Arndt
Jim Atkinson
Stu Auchincloss
Thomas Bailey
Jim Baird
David Balogh
Mary Ellen Barber
Sondra Barnett
Todd Bauman
Tim Baumann
Rebecca Beaton
Gay Beatty
Brenda Bechter
Ted Bennett
Joe Bens
Darlene Benson
Cynthia Bergen
Sandra Besseghini
Blake Besser
Linda Bessette
Ian Billick
Lin Bird
Mel Bird
John Black
Susan Blackshaw
Marcie Blanchard
X. Blasoz
Hartley Bloomfield
Barbara Boland
Andy Bondsha
Susan Bonfield
Jessica Bowles
Joyce Bowles
Walter Bowles
Betty Boyd
Jean Boylan
Mike Braal
Sandra Braasch
Silja Brandt
Mary Ann Brenner
Kate Brewer
Sreva Bright
Doris Brown
Joanna Brown
Evi Buckner
Rich Bunn
Carol Burns
Robert Burns
Jean Callahan
Min Canavan
Ruth Carlson
Fred Carney
Ken Carpenter
Duane Carr
Amy Cervens
Sherry Chapman
Janet Chu
Ray Chu
Dan Clarkson
Sue Cleintaur
Charles Clifton
Theo Colburn
Leonard Coleman
Chris Collins
Jenny Kate Collins
Mary Comfitt
Deb Conroy
Kevin Cook
Robin Corcoran
Wayne Cornils
Nancy Crafton
Jerry Craig
June Cringan
Randy Crook
Rich Crook
Michael Crosby
Gretchen Cross
Doris Cruze
Ed Curry
Martha Curry
Mike Daigneault
Elizabeth Darnell
Nathan Davidson
Nina Davis
Georgia Dayhoff
Rob De Baca
Tylan Dean
Russ DeFusco
Linda Delair
Lisa DeMoss
Irma Demshki
Rod DeWeese
Doug Dickman
Virginia Dionigi
Barbara Doehring
Lora Domingo
David Dominick
Gladys Donohue
Catherine Doyle
Dan Doyle
Robert Doyle
Lois Drury
Sam Duerkson
Hugh Duffy
Peggy Durant
Will Durant
Dave Dyson
Sonda Eastlock
Patty Echelmeyer
Mike Edgington
John Edwards
Marlene Eggerling
Justin Ellenberger
Dan Ellison
David Ely
Paul Ermigiotti
Vera Evenson
Clark Ewing
Denise Falzone-Schirm
Barbara Fay
Don Fiene
Matt Fisher
Marguerite Flanagan
Edward Floyd
Ann Ford
Bill Ford
Barbara Foster
Jan Fowler
Matthew Frank
Tari Frank
Jeanne Frazier
John Frederick
Myriam Friggens
Howard Frisch
Steven Frye
Wilber Fulker
Jim Fuller
Barbara Furniss
Todd Furniss
Yale Fyler
David Galinat
David Gallagher
Diana Gansemer
Jeff Garcia
Dolly Gardner
Richard Geer
Darin Geiger
Peter Gent
Dan George
Janet George
Mark Gershman
Randy Gietzen
Keith Giezentanner
Bob Girvin
Anne Goiran
John Goss
Les Goss
Gail Goulding
Courtney Graves
Candy Gray
Brandon Grebence
Dan Green
Wendy Green
Richard Greenlee
Mike Grode
Ann Groshek
X. Gruenberg
Gunner Guinan
Tony Gurzick
Glenn Hageman
Jeane Hageman
Lisa Hahner
Bruce Hale
Linda Hale
Sandra Hamm
Becky Hammond
Katie Hamrick
Carla Hansen
Rita Harden
Susan Harris
Randy Harrison
Jonathon Hart
Phil Hart
Susan Hart
Jim Hasart
Marilyn Hasart
Helen Hassemer
John Hawkins
Rick Hayes
Barry Heath
Jennifer Heintzelman
Arian Hemphill
Mindy Hetrick
Susan Hiebert
Mona Hill
Tony Hoag
Lyn Hoffman
Julie Hofmann
Eric Hollan
Tom Holland
Victoria Holland
Harold Holt
Karen Holzer
Karen Horak
Cal Howe
Emily Howe
Penny Howe
Russell Howe

Claire Hultgrove
Jonathan Hunt
Tory Hurst
Scott Hutchings
Tom Ingersoll
Roger Jakonbek
Susan Jenkins
Tina Jones
Joe Kamby
Norma Kamby
Amy Keatley
Ted Keatley
Reed Kelley
Joey Kellner
Lisa Kerman
Rhonda Ketchen
Steven Ketler
Ann Ketterman
Bill Ketterman
Mary King
Pauline King
Steve King
Anne Kingery
Kate Kingery
Katie Kinney
Bill Kleiman
Kent Kleman
Tasha Kotliar
Tom Kozan
Rich Kraeger
Marcy Kraetz
Heidi Krapfl
Heidi Krause
Barbara Krekeler
Dorothy Krimm
Hans Krimm
David LaLiberte
Brent Lambeth
Rachel Lambeth
Kelley Lawler
Carol Laycock
Nan Lederer
Barbara Lester
Tony Leukering
Karen Levad
Kenny Levad
Jerry Ligon
Lori Lilly
Emma Locke
Marvin Locke
Melvin Locklear
Chuck Loeffler
Bobbie Lohr
Cherie Long
Chuck Lopeshaisky
Liv Loria
Bruce Lowry
David Luke
Lou Maddis
Jeff Madison
George Maentz
Joe Mammoser
Linda Martin
Sue Martin
Dean Mason
Linda Mason
Terry Matheson
Mike Maura
Mark Maurogaines
Steve Maybon
Marilyn McBirney
Neal McCaskill
Priscilla McClain
Tracy McCoy
Peg McElwain
Kevin McGarigal
Alexie McKechnie
James McKinley
Sarah McKinley
Carron Meaney
Charles Meier
Monica Melloci
Cecilia Meyer
John Miley
Bob Miller
Gwyn Miller
Jim Miller
Judy Miller
Martin Miller
Palo Milliken
Charles Mills
Brenda Mitchell
Lori Miyasato
Pat Monaco
Mel Moody
William Moody
Muriel Morley
Arthur Morris
Heidi Morris
Jim Morris
Marianne Morris
Mark Morris
Natasha Munro
Daniel Murphy
Doug Muscanell
Don Myers
Adrienne Myser
Jane Nagel
Christina Nealson
Jack Neiman
Judy Neiman
Carl Nelson
John Nelson
Peter Nelson
Sally Nelson
Geoff Nemnick
Lori Nielsen
Mark Nikas
Carolyn Norblom
Jim Nowak
Lee O'Brien
John O'Rourke
Dan Ohn
Sharon Oldt
Thea Oldt
Pat Oppelt
Ted Oppelt
Catherine Orgish
Barb Osmundson
Marilyn Overly
Scott Overly
Chris Pague
Sandy Paige
Barb Palmer
Teresa Paquet
Ron Parkinen
Jared Parks
Greg Pasquaniello
John Paynter
Brad Petch
Brett Peterson
Dale Peterson
Judy Peterson
Mark Peterson
Cynthia Phillips
Eleanor Pickering
Robert Pike
Vivian Pliler
Jerry Poe
Carl Poenisch
Carla Poenisch
P. Pollock
Mike Post
Ray Potter
Inez Prather
Charles Preston
Tara Prince
Kay Quann
Horace Quick
Colleen Quinn
Don Radovich
Scott Rashid
Bettie Rauch
Andrea Reichert
Ellen Reichert
Ellen J. Reichert
K. Remmick
Jack Renee
Jim Rhette
Sara Rhette
Rick Richards
Sandy Righter
J. C. Rigli
James Ringelman
Maurice Ritter
Pam Rizor
Barb Roberts
Catherine Robertson
John Robertson
Mark Robertson
Leslie Robinette
Larry Rogstad
Ardis Rohwer
Chuck Romero
Jo Romero
Clyde Rose
Virginia Rose
Milton Ross
B. Rowe
W. O. Royall, Jr.
Bob Rozinski
Greg Ruff
Helen Ryan
Karin Sable
Catherine Sadler
Mary Safford
Miguel Angel Salas Paez
Bill Sanborn
Bob Sanders
Scott Sarner
Mark Schaefer
Judy Scherpelz
Debbi Schmuck
Karen Schneller-McDonald
X. Schomaker
Dick Schottler
Cheryl Schreier
Shirley Schroeder
Katie Schulte
Chris Schultz
Diane Seltman
Scott Seltman
Stan Senner
Scott Severs
Alice Jane Shane
Sara Shane
Tom Shane
Neil Sherwood
Susan Shields
Janet Shin
Gary Skiba
Bill Smith
Dusty Smith
Jim Smith

Linda Smith
Terri Snyder
Richard Sojda
Robert Spahn
Frances Sperl
John Sporleder
Carol Spurrier
Peter Stangel
Mark Stark
Harry Starkey
Gertrude Stattlemyer
Janis Steenberg
Rick Steenberg
Carol Steingraeber
David Steingraeber
Myrna Steinkamp
Maggie Stevens
Mike Stevens
Carol Stewart
Leslie Stewart
S. Stroup
Rosemary Sucee
Judy Sugden
Stephanie Swift
Mike Szymczak
Kerri Tashiro
Doug Tedrick
Christi Terry
Andy Thompson
Bliss Thompson
Dale Thompson
Dan Thompson
Rick Thompson
M. Thorp
Bert Tignor
Bridget Tisthammer
Ryan Toolen
Charles Traylor
Robin Troup
Kristyn Truan
Diane Truelson
Mary Tucey
Nichola Tucey
Richard Tucey
Mike Tupper
Jim Turley
Evelyn Upham
Lee Upham
Maureen Vallenta
Tom Van Erp
Marv Vanderkolk
Jon Verner
Dan Versaw
Raymond Versaw
Ruth Ann Versaw
Mike Villa
Joshua Voorhis
Dave Waddington
Joannie Wagner
Norman Walker
Neil Ward
Ellen Warren
John Waters
Bruce Webb
Marilyn Weber
Mark Weber
Thomas Weber
William Weber
Mary Ann Weisenen
Ron Wemple
Stephen Wenger
Judy Wertenbaker
Steve West
Sylvia Wheelock
Claude White
Ben Whitfield
Debbie Whitmore
John Whitmore
Jodi Whittier
Anne Wilbar
Lynn Willcockson
Eva Willenbuccher
Hans Willenbuccher
Ilse Willenbuccher
Erin Williams
Neal Williams
Virgil Williams
Betti Willow
Mark Willow
Dale Wills
Ann Wilson
James Wilson
Mike Wilson
Susan Wilson
Kathy Winternitz
Roy Wittenberg
Russ Wood
Daniella Wooddell
Robert Woodworth
Emily Wortman
Cliff Wright
Dale Wright
Karen Wright
Ann Yaeger
Barbara Yaeger
Robin Yates
Merle Yoston
Barbara Zaninelli
Louise Zemaitis
Sheri Zimmer
David Zook

FOREWORD

We live in a dynamic world—and at a time when, measured in human terms, impacts on our wildlife are occurring more rapidly than ever before documented. Bird ranges expand and contract, measured not in centuries but in decades. We are concerned not only with changes in range, but also in global populations of individual species and whether they are stable, declining, or increasing.

One hundred years ago Wells W. Cooke (1897) published the first statewide list of Colorado birds. Professor Cooke categorized Colorado birds by their seasonal occurrence and determined that 228 of Colorado's 360 "species and varieties" were known to breed in the state. He boasted that Nebraska was the only state that exceeded Colorado in the number of bird species recorded! In the next three years the Colorado totals were raised to 387 species and varieties of which 243 were recognized as breeding (Cooke 1900).

No attempt was made to map the ranges of Colorado birds until Kingery and Graul (1978) published the Colorado Bird Distribution Latilong Study, the precursor of the present Atlas. Those maps, crude by today's atlas standards, gave a general impression of the statewide distribution of each species and defined the principal breeding habitats. The summer maps in Andrews and Righter (1992) pinpointed known localities of rarities and estimated general ranges of the more common species.

This first statewide breeding bird atlas in the Rockies represents a giant leap forward in documenting the breeding distribution of North American avifauna. Following the standard format and field procedures adopted by the European and North American Ornithological Atlas Committees, the Colorado Bird Atlas Partnership has successfully designed and completed one of the most challenging ornithological field projects ever undertaken on this continent.

Professor Cooke (1897), in his Introduction to *Birds of Colorado*, stated, "There is no State in the Union that offers a more difficult field for thorough work." Unlike smaller and less rugged states, Colorado has many priority atlas blocks that are not only roadless, but also lie many kilometers from the nearest public access over extremely rugged terrain. The successful completion of the enjoyable but vigorous field work is a credit to the dedication of the large corps of volunteers and blockbusters who worked so tirelessly to accomplish their goal.

Ever since publication of the first grid-based breeding bird atlas in 1970, *Atlas of the Breeding Birds of the West Midlands* by Lord and Munns, I have been fascinated by not only the success of each atlas project—each one exceeding expectations—but also by the ingenuity of the organizers and editors. Although every atlas follows the same standard field procedures, which is important for maintaining comparability and ensuring credibility, each one incorporates ingenious novelties of presentation that add to the value and interest of the publication. Unique in the Colorado atlas are the habitat charts, the detailed breeding phenology charts, and the summary of species totals within each latilong.

Completion of this book represents not an end but a beginning. Having established a statewide base for comparison, the atlas can be repeated many times in the future, demonstrating changes in bird distribution and relative abundance. Comparisons can be made with other states and provinces as similar studies are completed elsewhere. I predict that atlas data from many states, provinces, and countries will be shared on the Internet until, ultimately, species distributions can be viewed and analyzed at a continent-wide or even a worldwide scale. This will present unlimited possibilities for international collaboration in protection of critical habitats. I compliment the Colorado organizers for their foresight in describing in the Results section how interested researchers may obtain atlas data either in computer format or on paper. It is important that the results of this project be readily available for research and conservation purposes.

Atlas results from Colorado are already being used locally and statewide by resource managers, wildlife biologists, and administrators. Perhaps the percentage of atlas blocks in which species are recorded will be used, together with other information, to prioritize conservation efforts, as is taking place in some other states. In Colorado, this Atlas will stimulate further research on rare and local species that were poorly sampled by the atlas effort. The success of this Atlas will stimulate observers in other mountain states to undertake similar projects that they may have considered too awesome to contemplate in the past. And finally, an indirect benefit of this project was the training of hundreds of field observers who can now tackle other exciting partnership projects, leading to a better understanding of our natural resources and their conservation.

Chandler S. Robbins

Chandler S. Robbins
Biological Resources Division
U.S. Geological Survey

BIRD ILLUSTRATION IDENTIFICATION

In Front Matter

Cover, title page	Mountain Plover
Back cover	American Dipper
Pages iv and x	White-throated Swifts
Page 1	Cliff Swallows
Pages 2 and 3	Common Nighthawks
Page 36	Broad-tailed Hummingbird
Page 38	Acorn Woodpeckers

In Supplemental Species Accounts

Page 546	Least Flycatcher
Page 548	Chestnut-sided Warbler
Page 550	Black Rail
Page 551	Hooded Warbler

TABLE OF CONTENTS

INTRODUCTION

Colorado birdwatchers undertook an audacious venture when they launched a breeding bird atlas. An atlas project faced formidable obstacles: a state with a large land area, rugged topography, relatively small human population, limited financial resources. The project had one important asset: a receptive community of birdwatchers, amateur and professional. By the time of completion, 1,295 people did field work, contributed data, worked at administrative chores, or provided financial support.

Project organizers dreamed of providing a "snapshot in time" of the breeding birds in Colorado on a scale never before attempted. That the resulting body of data provides such a picture reflects the perseverance and dedication of everyone involved in the project. By surveying (one-sixth of) 99% of the state, atlasers produced an extraordinary body of information about Colorado's breeding birds.

In sending observers across the entire state, the Atlas surveyed places that birdwatchers had never explored. A few highlights of Atlas results provide a general, and probably unfair, sense of the body of information provided by the Atlas effort.

In non-riparian zones of the high plains—a principal beneficiary of the random, statewide, assignments—atlasers found Rock Wrens, Brown Thrashers, and Orchard Orioles nesting more widely than previously realized and they found House Finches spreading. They discovered that more Mountain Plovers breed in southern Colorado than previously thought—and that over half the world's Mountain Plovers breed in Colorado. Atlasers also discovered that two species thought to breed commonly across the High Plains—Black-capped Chickadee and Song Sparrow—actually do not breed there.

In the mountains atlasers filled in geographical gaps in breeding ranges of dozens of species from Boreal and Flammulated owls to Evening Grosbeak. They documented recent range expansions by Blue-gray Gnatcatchers up the Front Range and Great-tailed Grackles all over the state. They found new habitat data for Cassin's Finch and expanded the elevational range for Williamson's Sapsucker, House Wren, and Green-tailed Towhee.

Collectively, atlasers determined that more Horned Larks breed in Colorado than any other species and that American Robins breed in more places in the state than any other species. Colorado hosts a major proportion of the breeding population of several species besides Mountain Plover and the almost endemic Brown-capped Rosy Finch: Black Swift, Gray Vireo, Virginia's Warbler, Green-tailed Towhee, and the state bird, the Lark Bunting.

Atlasers confirmed the diversity of breeding species within the state—264 species—even as they also affirmed, generally, less diversity within the confines of single Atlas blocks than found by atlas projects in the eastern half of the continent. The list of 264 breeding species exceeds that found by any other U.S. atlas project published so far (New York had the most, 242 species), although Texas, Florida, New Mexico, Arizona, and California have either not published or not conducted atlas projects. Both Ontario and Quebec reported 292 breeding species. We salute our volunteers for their unparalleled contribution to Colorado ornithology.

EDITOR'S NOTE

With this book the editor and authors have tried to present not only the results but also the spirit of Colorado atlasing. Participants in the project—the atlasers—derived many rewards. They had memorable experiences in the field, enjoyed the distinctive atlasing techniques, observed facets of bird behavior new to them, explored new places, and met welcoming Coloradans.

As Jean Maguire (1990) wrote in an Atlas newsletter, "We've found Atlas people to be the best part of atlasing. Get to know the people in your block. You'll be well fed with delicious food, wonderful birds, interesting conversation, and warm friendships." This book aims to depict the rewards and fun of atlas-style birdwatching, and to acknowledge the contributions of the many field workers.

The collated results from their field efforts have produced an extensive database about Colorado breeding bird distribution, with new details about locations, habitats, and breeding cycles.

We have also tried to convey the magic of learning about the lives of birds. Each species account includes facets of breeding biology. Readers can learn about the differences and similarities of various species. You can learn about behavior to look for in the field and how your own observations might add to understanding bird behavior as well as distribution. The *Journal of the Colorado Field Ornithologists* welcomes short and long articles about bird behavior, distribution, habitat use, and breeding habits.

ORGANIZATION OF THE BOOK

The Organization and Methods section describes the organization of the Atlas project, and the Results section discusses the results of the project. The Colorado Environment section relates species distribution to ecosystems. The section on Post-settlement Changes to Colorado Habitats outlines a few dramatic changes to Colorado birdlife associated with the settlement that began in the mid-1800s. The Colorado Ornithologists section describes some of the people who have contributed to Colorado ornithology.

The Species Accounts section contains species accounts for 253 of the species breeding in Colorado. The Short Species Accounts section provides short accounts about three categories of birds: 11 additional species that bred during the Atlas period but in no more than three places, species not confirmed as breeders but which exhibited breeding behavior (such as territorial conduct), and species recorded breeding in the state before or after the period but not during the Atlas years.

Appendices include Block Statistics, Abundance Statistics derived from Atlas estimates, Cowbird parasitism data, a sample field card, a Gazetteer of Colorado place names used in the text, and the Scientific Names for plants and animals other than birds covered in the Species Accounts. The list of citations contains 1,338 references used by the authors.

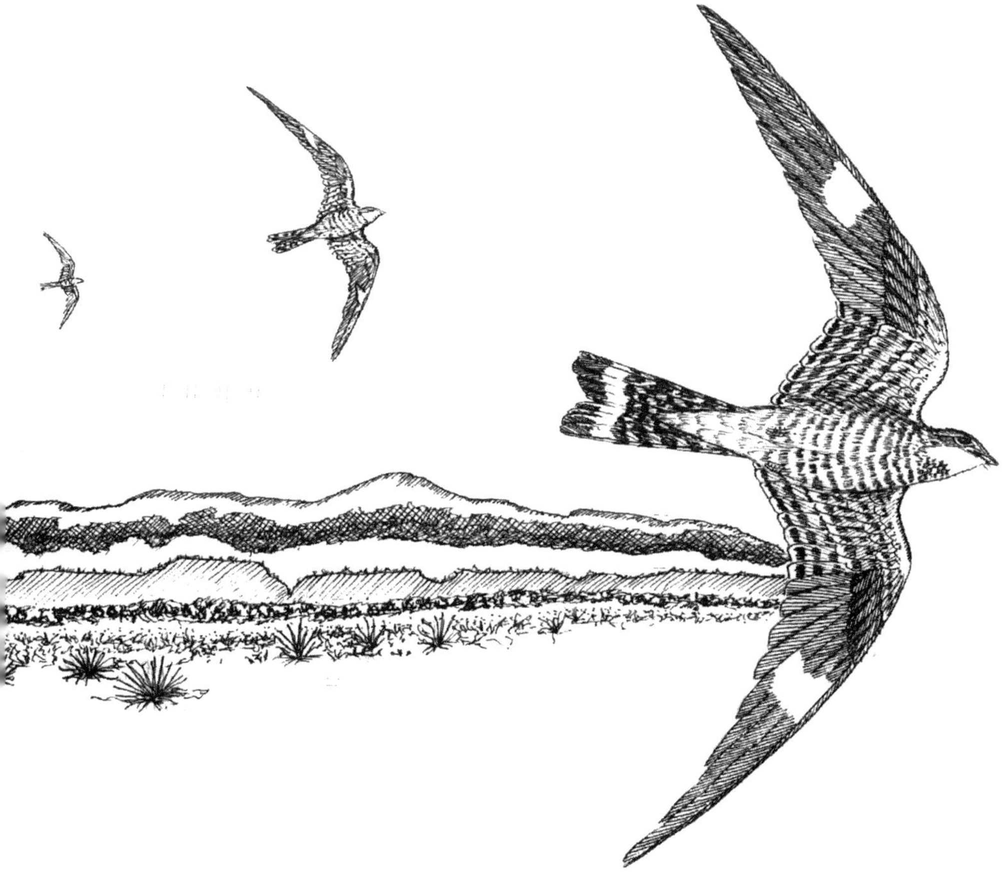

ATLAS ORGANIZATION AND METHODS

ORGANIZATION

Volunteers sparked the Colorado Breeding Bird Atlas Project (CBAP). Their enthusiasm, dedication, and diligence enabled the project to accomplish its objectives. CBAP operated without paid staff; volunteers made it happen.

The project began with a proposal to the Board of the Colorado Field Ornithologists. The directors accepted with enthusiasm. In 1986 an organizing committee established an independent organization, the Colorado Bird Atlas Partnership, to administer the project. A Steering Committee governed Atlas administration and progress throughout the Project. The Steering Committee set up three operating committees: Regional Coordinators, Technical Committee, and Publications Committee. Table 1 lists the members of these committees.

Regional Coordinators, after dividing the state into geographic regions (most with an urban center that could provide a source of field workers), took responsibility for supervising field work. They recruited field workers, issued topographic maps and Atlas materials, received and reviewed field cards, and reviewed Proof Listings derived from the field cards.

A Technical Committee made decisions about Atlas protocol and processed Verification Forms that documented unusual occurrences. The Publications Committee supervised preparation of this book.

President of the Steering Committee, Alex T. Cringan, maintained regular contact with the Project Director, Hugh E. Kingery. The Project Director communicated with the Regional Coordinators and, through them, the field workers. He developed Atlas materials, prepared the newsletter, directed field worker recruitment, supervised paid field workers, supervised data processing, and did a lot of field work.

FINANCING

CBAP proceeded with many partners. Volunteers administered the project and volunteers did most of the field work. Several government agencies, especially the Colorado Division of Wildlife, U.S. Forest Service, and the Denver Museum of Natural History, contributed substantial in-kind services. The Atlas proceeded with a minimal budget.

The Colorado Field Ornithologists contributed the initial seed money and supported the project with generous annual donations. Ten Audubon chapters and five bird clubs in the state became Atlas Sponsors, as did the Bartlesville, Oklahoma, Audubon chapter through a memorial fund for Colorado bird bander Sophia Mery. Over 150 individuals became Partnership members by providing financial support.

At the outset the U.S. Bureau of Land Management made it feasible to inaugurate field work by providing CBAP with a mountain of maps—two entire sets of topographic maps for Colorado. The Colorado Division of Wildlife provided field workers and Atlas administrators and, along with Great Outdoors Colorado, the U.S. Forest Service, U.S. Bureau of Land Management, and U.S. Fish and Wildlife Service made grants for completion of field work.

This publication received funding through a grant from the National Fish and Wildlife Foundation, matched by Great Outdoors Colorado through the Colorado Division of Wildlife.

TABLE 1

ATLAS COMMITTEES

REGIONAL COORDINATORS

David Blue	Colorado Springs	1987
Toni Brevillier	Colorado Springs	1987–1989
Coen Dexter	Palisade	1989–1994
Beth Dillon	Fort Collins	1987–1994
Mike Figgs	Boulder	1987
Gerald Fyler	Dolores	1992–1994
Ron Garcia	Alamosa	1993–1994
Philip Hayes	Lakewood	1987–1994
Mark Janos	Delta	1987–1988
Dave Johnson	Pueblo	1987–1994
Bill Kaempfer	Boulder	1988–1994
Connie Knapp	Alamosa	1992
Ruth Kuenning	Colorado Springs	1989–1994
Walt Kuenning	Colorado Springs	1989–1994
Ron Lambeth	Grand Junction	1987–1994
Tom Light	Colorado Springs	1987
David Martin	Westminster	1987–1990
Jack Merchant	Eagle	1987–1994
Pat Monaco	Wetmore	1989–1991
Anne Morkill	Alamosa	1991
Paul Opler	Loveland	1987–1994
Dick Roth	Pueblo	1992–1994
Dick Schottler	Golden	1987–1994
Jim Sedgwick	Fort Collins	1987–1994
Gary Skiba	Durango	1990
Jennie Slater	Lamar	1987–1994
John Toolen	Grand Junction	1987–1994
Alan Versaw	Cortez	1991–1992
Judy Ward	Lafayette	1987
Jim Watts	Penrose	1987–1988
Rosie Watts	Penrose	1987–1988
Jeanne Willetto	Durango	1987–1989

STEERING COMMITTEE

Mike Carter . . . . .Brighton
Alex Cringan . . . . . . .Fort Collins
Jo Dirckx . . . . .Broomfield
Norm Erthal . . . . .Arvada
Hugh Kingery . . . . .Franktown
David Martin . . . . . . .Westminster
Paul Opler . . . . .Loveland
David Pantle . . . . .Cañon City
Charles Preston . . . . .Denver
Bob Righter . . . . .Denver
Don Schrupp . . . . .Denver
Judy Sheppard . . . . .Denver
Judy Ward . . . . .Lafayette
Betsy Webb . . . . .Denver
Dale Wills . . . . .Arvada
Vic Zerbi . . . . .Glenwood Springs

TECHNICAL COMMITTEE

Jim Dennis . . . . .Fort Collins
Robert Hernbrode . . . . . .Denver
Stephen Jones . . . . .Boulder
David Leatherman . . .Fort Collins
Duane Nelson . . . . .Golden
Ronald A. Ryder . . . . .Fort Collins
Jim Sedgwick . . . . .Fort Collins
Alan Versaw . . . .Colorado Springs

ADMINISTRATIVE VOLUNTEERS

Virginia Bleck . . . . .Denver
Ann Bonnell . . . . .Littleton
Ruth Breckon . . . . .Denver
Sarah Christian . . . . .Denver
Beth Dillon . . . . .Fort Collins
Gail Evans . . . . .Denver
Marilyn Hackett . . . . .Denver
Helen Hayes . . . . .Lakewood
Allan Lavery . . . . .Denver
David Martin . . . . . . .Westminster
John Moody . . . . .Boulder
Julene Pimental . . . . .Denver
Dick Pratt . . . . .Longmont
Jeanette Scherbath . Commerce City
Ray Scherbath . . .Commerce City
Ellen Swanson . . . . .Denver
Jean Van Loan . . . . .Denver
Judy Ward . . . . .Denver

COMPUTER INPUT

Gail Evans . . . . .Denver
Barbara Fox . . . . .Boulder
Phil Hayes . . . . .Lakewood
Rich Levad . . . . . . .Grand Junction
Marjorie Rhoades . . . .Fort Collins
Mike Smith . . . . .Golden
Alan Versaw . . . .Colorado Springs
Doug Ward . . . . .Lakewood
Dale Wills . . . . .Arvada
Robin Yates . . . . .Denver

PUBLICATIONS COMMITTEE

Alex Cringan . . . . . . .Fort Collins
Robert Hernbrode . . . . . .Denver
Hugh Kingery . . . . .Franktown
Paul Opler . . . . .Loveland
David Pantle . . . . .Cañon City

PUBLICATION VOLUNTEERS

Bev Baker . . . . .Boulder
Cindy Lippincott . Colorado Springs
Karen Metz . . . . .Franktown
Alan Wallace . . . . .Reno, NV
Scott Yarberry . . . . .Denver

PEER REVIEWERS

Craig Benkman . . .Las Cruces, NM
Carl Bock . . . . .Boulder
Greg Butcher . . .Colorado Springs
Jerry Craig . . . . .Fort Collins
Alex Cruz . . . . .Boulder
Bob Dorn . . . . .Cheyenne, WY
Jane Dorn . . . . . . .Cheyenne, WY
John Emerick . . . . .Golden
Ken Giesen . . . . .Fort Collins
Jon Greenlaw . . . . .Cape Coral, FL
Dave Hallock . . . . .Nederland
Bill Howe . . . . . .Albuquerque, NM
Don Kroodsma . . . .Amherst, MA
Ron Lambeth . . . . .Grand Junction
Richard Levad . . . .Grand Junction
Steve Martin . . . . .Wellington
Chris Pague . . . . .Boulder
Bruce Peterjohn . . . . . .Laurel, MD
Van Remsen . . . .Baton Rouge, LA
Chan Robbins . . . . .Laurel, MD
Scott Seltman . . . . . . .Nekoma, KS
Tom Shane . . . . . .Garden City, KS
Dave Shuford . .Stinson Beach, CA
Mike Szymczak . . . . . .Fort Collins
James Travis . . . .Albuquerque, NM
Alan Versaw . . . .Colorado Springs
S. O. Williams III . . . .Santa Fe, NM
John L. Zimmerman . Manhattan, KS

METHODS

The Colorado Atlas followed the protocol recommended by the North American Ornithological Atlas Committee (NORAC) at its second conference, held in Ithaca, New York, April 25–27, 1986 (Sutcliffe et al. 1986).

BLOCKS

NORAC recommends that atlas blocks contain 10 square miles (25 km^2). CBAP used topographic maps published by the U.S. Geological Survey. Dividing the maps in half vertically and in thirds horizontally resulted in six blocks per map that measure approximately 10 square miles. Colorado contains 10,848 such blocks wholly or partially within the state.

Because of the formidable coverage difficulties, CBAP elected to survey one block in each topographic map—a total of 1,760 blocks. For administrative simplicity the Steering Committee decided to survey the same block in each topographic map. By random selection it picked the South East sector as the "Priority Block." This random selection theoretically provides a statistically valid sample of representative habitats throughout the state. (See the section on Biases and Limitations for comments on the disadvantages of this block selection.) The committee randomly selected the South West and Central East blocks, in order, as substitute Priority blocks when access to the South East block proved impossible.

Atlas plans omitted from coverage the 32 Priority blocks on the eastern border and the 32 Priority blocks on the western border because they do not lie entirely within Colorado. The Steering Committee determined that adding those 64 blocks to the other 1,760 would make completion of the project more formidable and would not add materially to the project's results.

Colorado blocks measure 2.9 miles (4.7 km) north/south and 3.52 to 3.34 miles (5.7 to 5.4 km) east/west (larger on the New Mexico border, smaller to the north as longitudinal lines converge). On a flat map surface, Atlas blocks contain 9.7 to 10.2 square miles (25.1–26.4 km^2). Over half the blocks, however, contain far more surface area than the ostensible 10 square miles; mountain blocks can contain 50% more surface area. Colorado chauvinists declare that if you flattened Colorado it would be bigger than Texas.

Observers nonetheless conducted some field work in 12 blocks on the west border and six blocks on the east border; the map includes results from these blocks but block statistics omit them. The Atlas also received data from special areas in Non-priority blocks, and obtained information on a number of species with limited distribution, much of it in Non-priority blocks.

The Atlas originally announced a field work period of 1987 to 1991. Not surprisingly, completion of field work required a three-year extension to 1994. In 1995 an informal network of field workers worked in 11 previously unworked blocks (usually one-day efforts) and 24 partially completed blocks.

Prior to the 1993 season, Atlas organizers determined that, even in eight years, volunteers alone could not survey all 1,760 blocks. They obtained funding to hire five field workers in 1993 and ten in 1994, all of whom had previous experience as volunteers for the Atlas. The field workers completed 328 new blocks and finished 162 others. The Colorado Division of Wildlife, Great Outdoors Colorado, U.S. Forest Service, U.S. Bureau of Land Management, and U.S. Fish and Wildlife Service contributed funding, and the Colorado Bird Observatory administered the payroll.

FIGURE 1

MAP BLOCK SYSTEM

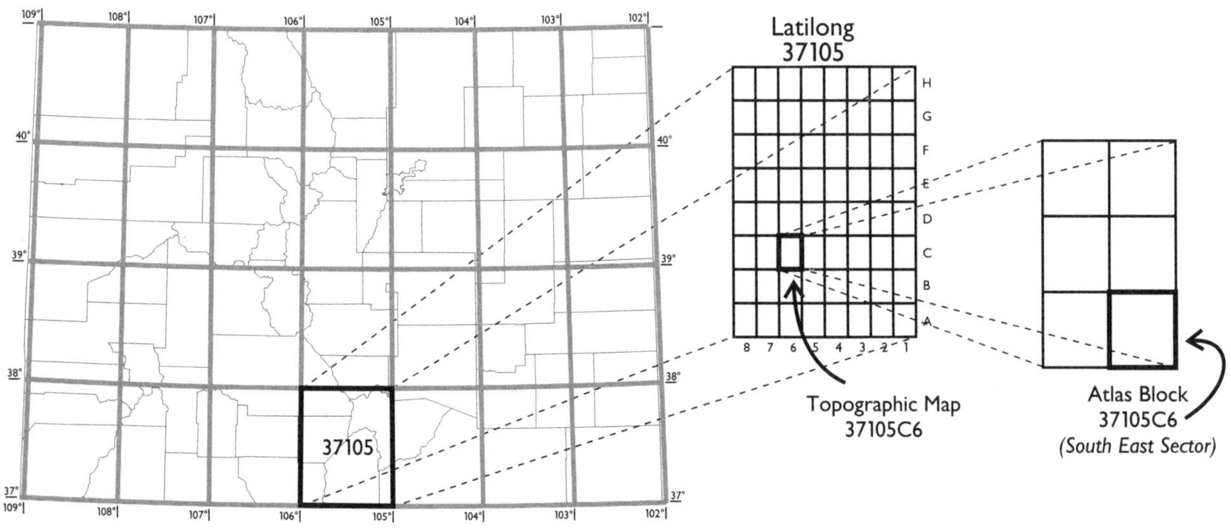
109°
108°
107°
106°
105°
104°
103°
102°
40°
39°
38°
37°
37105
Latilong
37105
H
G
F
E
D
C
B
A
8 7 6 5 4 3 2 1
Topographic Map
37105C6
Atlas Block
37105C6
(South East Sector)

RENDEZVOUS

To cover blocks with difficult access and to train field workers, the Atlas sponsored 29 "Rendezvous" from 1990 to 1994. Conducted by Regional Coordinators and other volunteers, these weekend excursions helped to survey 78 blocks from Baca County in the southeast to the Ute Mountain Ute Reservation in the southwest to Moffat County in the northwest.

PROTOCOL

The Atlas Handbook directed field workers to make four or more half-day trips, at least a week apart, to each block (depending on diversity and complexity of habitats), and one night trip for owls and poorwills if the habitat warranted it. If the block had only one or two simple habitats, then two trips of two to four hours sufficed for adequate coverage. This latter criterion applied to plains blocks with only prairie, cropland, and an occasional farmhouse, to a few blocks in the San Luis Valley that have only shrublands, and to sagebrush-covered blocks in the northwestern corner of the state.

FIELD CARDS (APPENDIX D)

The handbook instructed the observers to submit field cards for each block after each year of field work, even if the block was not complete. Field workers reported Breeding, Habitat, and Abundance Codes, and the Date of observation of the highest breeding code. They also recorded their Atlas effort in hours and miles and identified other participants. Appendix D presents a sample field card filled out for one year of field work.

The inside back cover of this book contains abbreviated summaries of the meanings of the Breeding, Habitat, and Abundance Codes for the reader to use in interpreting the Species Accounts. The explanations that follow provide detailed descriptions of the various codes.

BREEDING CRITERIA

Breeding Codes (Table 2)

Atlas instructions categorized breeding according to the type of breeding behavior observed. Field workers recorded the species as Possible, Probable, or Confirmed only if they saw the species (a) in suitable breeding habitat, (b) during the breeding season. Therefore all blocks depicted on Species Maps, whether Possible, Probable, or Confirmed, represent blocks where Atlas protocol deems the bird a breeding species. In the Species Accounts, because of the dictionary meaning of the word "confirmed," this Atlas capitalizes Confirmed when the term refers to Atlas-defined confirmations. The book also capitalizes Atlas references to Probable and Possible categories, and their constituent codes (e.g., Territorial, Agitated).

CBAP used the breeding codes specified at the NORAC conference. Starting in 1992 some field workers subdivided the confirmation code "FY" (Feeding Young) into "CF" for Carrying Food and "FF" for Feeding Fledgling. This division permitted distinguishing observations of actual feeding of fledglings from observations of food-carrying for either nestlings or fledglings. This also clarified the FY code when used for raptors and corvids (which carry food for their own consumption) and helped in developing the phenology of breeding for all species.

TABLE 2

BREEDING CODES AND THEIR INTERPRETATION

All the following codes apply to a species seen or heard *during its breeding season*:

OBSERVED

O Migrants and non-breeding species observed in block during breeding season.

POSSIBLE

SPECIES FOUND or breeding calls heard in suitable nesting habitat.

X SINGING MALE: present in suitable nesting habitat during its breeding season.

PROBABLE

M MULTIPLE MALES *seven* different *singing* males heard on one day in suitable nesting habitat.

P PAIR observed in suitable nesting habitat.

T Permanent TERRITORY presumed through territorial behavior: defense, chasing other birds, or song in the same location on at least two occasions a week or more apart.

C COURTSHIP behavior between a male and a female or COPULATION. Includes display or food exchange.

V VISITING probable nest site, but no further evidence obtained.

A AGITATED behavior or anxiety calls of adult, indicating nest site or young in vicinity.

N NEST BUILDING by Bald Eagles or excavation of nest hole by wrens and woodpeckers, and construction of dummy nests.

CONFIRMED

NB NEST BUILDING or adult carrying nesting material.

PE PHYSIOLOGICAL EVIDENCE: refers to cloacal protuberances and brood patches. For use only by banders.

DD DISTRACTION DISPLAY or injury feigning.

UN USED NEST or eggshells found.

FL Recently FLEDGED young (or downy young of ducks, grebes, grouse, and sandpipers still dependent on adults), with limited mobility, including young incapable of sustained flight.

ON OCCUPIED NEST indicated by adult entering or leaving nest site in circumstances indicating occupied nest.

FY FEEDING YOUNG: adult seen carrying food for young, or feeding recently fledged young. Also subdivided into:
- **CF** Adult CARRYING FOOD for young.
- **FF** FEEDING FLEDGLING: adult feeding fledgling.

FS Adult carrying FECAL SAC away from nest for disposal.

NE NEST with EGGS.

NY NEST with YOUNG seen or heard.

ADDITIONAL DATA

CBAP requested, in addition to the breeding code, three other pieces of information. Field workers recorded the Date of the highest breeding code: these data provide the basis for the species phenology charts. Dates also flagged inappropriate codes and helped to identify migrants.

The Division of Wildlife asked CBAP to record Habitats used for nesting (see the following chapter on the Colorado Environment). A group of ornithologists and botanists developed the habitat codes. These codes provide the source for the habitat charts on the Species Account pages. Table 3 explains the Atlas Habitat codes.

ABUNDANCE CODES (TABLE 4)

Observers also provided estimates of Abundance for each species in the block. They made broad estimates based on a logarithmic progression of abundance.

BLOCK PHOTOGRAPHS

Many field workers provided photographs of the habitats in their blocks. The pictures illustrate the diversity of habitats in the state, and the diversity of structure that habitats which have the same description. The file of these photographs (prints and slides) will lodge with the Archives Department of the Denver Museum of Natural History.

TABLE 3

HABITAT CODES

Code	Classification	Description	Dominant Vegetation
GRASSLANDS			
TTR	Tallgrass–Riparian	Meadow associated with cottonwood woodlands; often old agricultural areas, foothill remnants	
TTS	Tallgrass–Sandsage	Sandsage and remnant tallgrass; NE Colorado	
TMGP	Midgrass Prairie	Eastern Slope grasslands	Wheatgrass, sideoats grama, little bluestem
TMGB	Western Intermountain Grassland	Western Slope grasslands	Bluebunch wheatgrass, galleta, wild rye, needle grass
TMXP	Mixed grasses of habitat alteration	Plains plantings and introduced grasses	Crested wheatgrass, sideoats grama, little bluestem, thistle
TSG	Shortgrass Prairie	Shortgrass prairie and cactus lands	Buffalo grass, blue grama, walkingstick cholla
TMG	Montane Grasslands	Grasslands in middle and high elevations	Mountain muhly, fescues, danthonias, etc.
TUNDRA			
TAL	Alpine Tundra	Meadows, grasses, and shrubs above timberline	
SHRUBLANDS			
SBR	Sagebrush	Sagebrush, all elevations	
SLS	Lowland Sagebrush	Sagebrush at lower elevations (SE Colorado, western tablelands)	
SMS	Mountain Sagebrush	Middle to higher elevations	Mountain big sagebrush
SMT	Mat Saltbush	Low mat like desert shrubs	Saltbush
SOS	Oak Scrub	Oak brush, tree-like oak woodlands	Gambel oak, New Mexico locust

Code	Classification	Description	Dominant Vegetation
SMM	Mountain Shrublands	Middle elevation, mixed species	Mountain mahogany, serviceberry, antelope bitterbrush, snowberry, bush cinquefoil, chokecherry, squawapple
SLC	Lowland Carr	Lowland willow, riparian skunkbrush thickets	
SMC	Montane Carr	Middle-elevation willow and alder thickets	
SBC	Willow Carr	High-elevation willow thickets	
SKR	Evergreen Shrublands	Shrubby timberline conifers, prostrate or nearly so	Engelmann spruce, subalpine fir
STD	Tall Desert Shrub	Desert shrublands of medium to tall height; also low-growing shadscale with small and medium sparse greasewood	Fourwing saltbush, shadscale, black greasewood, blackbrush, saltbush sp.
SSL #	Logged-over areas	Logged areas with or without regenerated growth, but prior to succession to habitats that fall into another code	Specify surrounding forest type
SBL #	Burned-over areas	Burned areas similar to SSL	
WOODLANDS			
LRD	Lowland Riparian	Cottonwood woodlands along streams or lakes	Plains cottonwood, tamarisk (salt cedar)
WPJ	Pinyon/Juniper	All P/J, singly or in associations	
WPP	Ponderosa Pine	Typical open-canopy ponderosa	
WMO #	Montane Woodland	Open-canopy woodland, mountains	
FOREST			
FUC #	Upland Coniferous	Mountain conifer forests with closed canopies	Engelmann spruce, subalpine fir, lodgepole pine, white fir, Douglas-fir, ponderosa pine
FUD	Upland Deciduous	Aspen, pure or mixed stands, no more than 50% conifer	
FRD	Riparian Deciduous	Mountains/foothills/canyon streams	Narrowleaf cottonwood, boxelder
FRC	Riparian Coniferous	More or less pure stands along stream banks	Usually Colorado blue spruce
WETLANDS			
ASG	Salt Meadows	Salt grass meadows	Miscellaneous alkali grasses
AEM	Emergent Wetlands, Marshes	Saturated soil, emergent vegetation	Cattails, sedges, rushes, bulrushes, arrowroot

Code	Classification	Description	Dominant Vegetation
ASU	Submergent or Floating	Floating mats	Pondweed, duckweed, water lilies
OWL	Open Water–Lakes	Lacks rooted vegetation	
OWS	Open Water–Streams	Lacks rooted vegetation	
URBAN/AGRICULTURAL			
URB	Residential	Ornamental plants, yards, feeders	
UPK	Parks	City parks planted to grass and trees	
UIN	Industrial	Downtowns, commercial districts, warehouses, little vegetation	
RRL	Rural	Scattered farm buildings and shelterbelts	
CWD	Cultivated Woodlands	Orchards, tree farms	
CPL	Croplands	Hay meadows, corn fields, wheat fields	
TSU	Barren Ground	Plowed fields, blowouts, sand dunes	
MCL	Cliffs	No significant vegetation	
MSC	Other	Talus, scree, rocks, hanging gardens, caves, manmade structures not above	

#: For WMO, FUC, SBL, and SSL, use these subcodes; e.g., a thick lodgepole pine forest = FUCL.

S = Engelmann spruce/subalpine fir
L = Lodgepole pine
D = Douglas-fir
B = Bristlecone pine
R = Limber pine
M = Mixed conifer/aspen woodland, less than 50% aspen

Low elevation: Plains or Western Slope river bottoms west of the mountains
Middle elevation: Foothills, mesas, up to and including elevations that support lodgepole pine or white fir
High elevation: The elevation at or above which spruce/fir forests grow

TABLE 4
ABUNDANCE CODES

Code	Explanation
A1	One breeding pair in block
A2	2–10 breeding pairs in block
A3	11–100 breeding pairs in block
A4	101–1,000 breeding pairs in block
A5	More than 1,000 breeding pairs in block

BIASES AND LIMITATIONS

For all bird atlases, the experience of the field worker and the time spent on field work in particular blocks affect the completeness of the information. The data demonstrate presence of a species, but do not necessarily denote absence. The block selection and uneven block coverage prevent an unequivocal statement that a species does not breed in a particular geographic area where atlasers did not locate it.

Habitats

The block selection system sampled Colorado habitats well. Nonetheless it resulted in some anomalies or limitations.

*1. **Riparian Habitats.*** Plains riparian habitats comprise only 3% of Colorado's land area, but they have a disproportionately high species richness and density of breeding birds. Because riparian corridors run like ribbons across the landscape—and topographic maps—the Atlas maps may show spotty distribution of species restricted to such habitats.

The random selection missed the Arkansas River from Bent's Fort to Holly (9 blocks, 63 miles [100 km]). North of the river, irrigated farmland provides rural habitat that simulates, to a degree, the continuous cottonwood forests along the Arkansas. This woodland, supplied by rural shelterbelts and roadside plantings, does not provide good habitat for some riparian species that require larger expanses of riparian forest. Upstream from Bent's Fort to Pueblo, 6 of 11 Priority blocks included the river.

Along the South Platte the Priority block selection included riverine stretches in 12 of 21 blocks. Block selection on the Western Slope and the San Luis Valley picked up a higher proportion of riparian sections along major rivers—Yampa, White, Colorado, Uncompahgre, Gunnison, and Rio Grande. This irregular sampling of plains riparian habitats (Longitudes 102, 103, and 104) affected results for birds such as Yellow Warbler, Red-headed Woodpecker, flickers, and screech-owls. However, because riparian habitats comprise such a small proportion of Colorado's land area, the block statistics most likely represent well the distribution frequency of riparian species.

*2. **Isolated and Discontinuous Habitats.*** In the San Luis Valley, only one Priority block contained a substantial part of any of the major refuges and major waterfowl nesting areas. In the rest of the state, Priority blocks did not include many waterbird specialty spots or well-known special sites like Roxborough, Castlewood Canyon, and Barr Lake state parks. It did include representative parts of such areas as Arapaho and Alamosa NWRs, and Bonny and Chatfield state parks. The block selection system, however, did sample bird populations away from traditional birdwatching spots—one reason the Atlas provides so much new information.

*3. **Alpine Habitats.*** A special form of discontinuous habitat, or habitat that Priority blocks skip, the alpine also presents formidable access challenges. In some blocks, field workers did not climb above timberline. In addition, each alpine species favors a different niche and so presents a different detection challenge. American Pipits and Horned Larks sing, court, and perch over different parts of the tundra; secretive ptarmigans hide until almost stepped on and nest where rocks complement their cryptic coloration. Many blocks lacked suitable habitat for Brown-capped Rosy-Finches, which nest in cliffs, but some with proper habitat presented major access problems.

Species Detection

*1. **Confirmations.*** The nature and protocol of Atlas field work does not allow sufficient field time "to Confirm" breeding by all species in a block. In 20–40 hours field workers could not Confirm every species found. Obviously, Confirmed observations give greater assurance of breeding in the block, but according to atlasing theory each block shown as occupied on a species map represents a breeding location.

For a few species this assumption does not hold. Atlasers recorded a few far-ranging species in many blocks where they likely do not breed. For example, Turkey Vultures and Common Ravens, conspicuous birds, undoubtedly do not breed in every place with adequate breeding habitat.

Seeking Confirmations had two objectives. First, Confirmations do provide more likely indicators of breeding. Second, they defined goals for field work in two ways: by encouraging and motivating field workers and by setting a standard for block completion. The Atlas urged that field workers Confirm 50% of the species they found in each block. If field workers met that goal, they probably detected most species nesting in the block.

*2. **Night Birds.*** The Atlas underrepresents nocturnal species like Common Poorwill and most owls (perhaps excepting Great Horned and Burrowing)—both in recording them at all and in Confirming them. The essential night surveys offer both mental difficulty, due to observers' daytime clocks, and major physical difficulties. The peak calling time for small mountain owls—March to May—occurs when several feet of snow cover the mountainsides as well as backcountry roads and trails. Spring access to blocks, at night, in the dark, in rugged mountain terrain usually deep in snow, is not only difficult but also dangerous due to avalanches.

*3. **Scarce, Secretive Species.*** Other secretive and scarce species likewise escaped detection simply because of their habits or limited numbers. These include accipiters, rails, and cuckoos including Greater Roadrunner.

*4. **Early and Late Breeders.*** Some species conduct the bulk of their breeding prior to the season when atlasers did the bulk of their field work, and a few nest after most other species. Atlas data probably depict these species less accurately. Examples are Clark's Nutcracker, Gray and Pinyon jays, and Common Raven (early) and American Goldfinch (late).

*5. **Atlaser Ability.*** Atlasers' abilities vary as to mobility and identification skills. The former meant that, physically, some observers could not thoroughly cover all habitats within their blocks. The latter probably resulted in fewer reports of some hard-to-identify groups such as *Empidonax* flycatchers and plains and sagebrush sparrows. Lack of familiarity with songs and calls could limit species detection by some observers.

*6. **Dates and Habitats.*** The Species Accounts present date and habitat data in charts and text. The protocol, which instructed field workers to provide only the highest breeding code, meant that the field cards usually provided only that code and date. Entry of higher codes entailed dropping lower-level codes, with their dates. For example, an observation of a nest with eggs would supplant the NB code, and if the observer returned and found young in the nest, NY would replace NE.

Not all atlasers recorded dates and habitats, and many recorded two or three habitats. Subsequent analysis suggests that some reported habitats were not breeding habitats at all—e.g., lakes and streams for tree-nesting birds. A substantial number of Confirmations of waterfowl involved flotillas of fledglings observed in open water—not particularly informative as to nesting habitat.

RESPONSES TO LIMITATIONS

To fill in map gaps, the Atlas solicited field work in Non-priority blocks in some sites with special habitats, largely state wildlife areas. These included sites in the San Luis and Arkansas valleys, lower South Platte (Tamarack State Wildlife Area), and North Park (Walden Reservoir and Lake John).

Biologists at Arapaho National Wildlife Refuge and Mesa Verde National Park provided coverage of all blocks—Priority and Non-priority—touched by their property. Researchers with the National Biological Service (now the Biological Resource Division of the U.S. Geological Survey) and the National Park Service contributed data from their studies, particularly in the San Luis Valley and Rocky Mountain National Park. Volunteers surveyed all preserves of The Nature Conservancy. The Colorado Division of Wildlife shared raptor and grouse data. Banding stations conducted by the Colorado Bird Observatory in Summit County and the U.S. Army at Fort Carson (El Paso County) and Pinyon Canyon (Las Animas County) contributed information.

The U.S. Forest Service provided results of owl surveys on the Rio Grande, San Juan, and Grand Mesa/Uncompahgre/Gunnison national forests and of statewide Spotted Owl surveys. As a result Flammulated and Northern Saw-whet owls and Northern Pygmy-Owl have a good representation in southwestern Colorado (Latilongs 37105, 37106, 37107, and 37108), perhaps disproportionately to the rest of the state.

ATLAS ARCHIVES

CBAP will deposit the Atlas database and other original materials with agencies that will make them available to anyone interested in them. Researchers, environmental consultants, and any interested persons may contact these agencies to review or use the materials.

Copies of the database reside in computers of several organizations concerned with wildlife. Interested persons may obtain data in computer format or on paper. By spring 1999, the following organizations will have the database and make it available: Colorado Division of Wildlife in Denver, Colorado Bird Observatory at Barr Lake State Park, Colorado Natural Heritage Program in Fort Collins, and Audubon Society of Greater Denver, soon to move to Chatfield State Park.

The Archives Department of the Denver Museum of Natural History will keep the principal Atlas records. These include: (a) the sources of the database: original field cards and proof listings that field workers reviewed; (b) photographs of block habitats that field workers contributed with their field cards; (c) complete file of Atlas instructions, newsletters, and other records.

RESULTS

The final database contains 86,499 bird records, 80,930 of them CF, PR, and PO reports (34,470, 17,495, and 28,965 respectively). This massive file consists of one record for each species recorded in each block. From it come all the Atlas data in the Species Accounts.

The database provides many kinds of source data with the information that atlasers provided on breeding codes, habitats, dates, and abundance. Researchers can extract information about single species, Atlas blocks, or particular habitats. They can calculate abundance or study breeding phenology, by species, location, habitats, or combinations of some or all of these. Species Account authors use these techniques in their discussions.

PEOPLE

FIELD WORKERS

The 1,170 field workers turned in 4,515 field cards. They recorded a nine-year total of 73,486 hours of field work plus 26,766 hours to get to and from the blocks. In traveling in, to, and from the blocks they covered 649,433 miles. They also spent uncounted hours filling out field cards and verification forms, studying maps, and reviewing proof listings of their block observations.

They reported some exceptional field experiences: a Black-throated Gray Warbler mistaking a field worker for its mother; a long-tailed weasel running up a steep bank with a baby Brewer's Blackbird; a brilliance of Mountain Bluebirds bathing after a thunderstorm; a Loggerhead Shrike shredding nesting material out of a nest belonging to frantic Western Kingbirds; innumerable aspen trees with two to five species nesting in them.

TABLE 5
FIELD WORK STATISTICS

	PRIORITY BLOCKS					NON-PRIORITY BLOCKS	FIELD WORKER EFFORTS		
Year	Field Cards	Cumulative Blocks Worked		Cumulative Blocks Completed		Field Cards	Field Work Hours	Travel Hours	Travel Miles
1987	225	225	13%	50	3%	42	4,723	1,533	38,400
1988	297	418	24%	91	5%	61	8,672	3,000	32,678
1989	313	583	33%	177	10%	82	6,886	2,072	51,064
1990	417	806	46%	273	16%	105	6,882	2,748	83,059
1991	443	976	55%	442	25%	158	8,802	3,559	89,230
1992	451	1,207	69%	614	35%	54	7,630	3,934	101,042
1993	535	1,427	81%	927	53%	126	11,651	4,200	111,186
1994	772	1,734	99%	1,645	93%	204	12,396	4,640	122,045
1995	35	1,745	99%	1,650	94%	174	415	65	2,544
TOTALS									
PRIORITY BLOCKS, 1987–1995	3,509					1,006	68,057	25,751	631,248
NON-PRIORITY BLOCKS, 1987–1995	1,006						5,429	1,015	18,185
TOTAL, ALL BLOCKS	**4,515**						**73,486**	**26,766**	**649,433**

LANDOWNERS

Hundreds of Colorado ranchers, farmers, and other landowners contributed to the Atlas by welcoming field workers and directing them to fruitful places. In order to survey most blocks in the eastern third of the state and many blocks elsewhere, atlasers needed permission to work on private land. Almost uniformly, landowners both permitted access and shared their observations. (A substantial number of pheasants, quail, owls, and swallows owe their black blocks on the map to observant landowners.) Some ferried field workers over their lands, and others allowed camping and entertained atlasers with historical and family tales; one offered a bedroom on a night punctuated by thunderstorms.

Only a few landowners refused to permit access—some concerned about uncaring trespassers, others fearful that we might find something that would jeopardize their property rights. ("We don't want you to find information that can be used against us," one said.) One plains farmer lacked sympathy with the idea of studying birds: "Besides," he said, "I don't have anything except doves and the state bird." Unfortunately these people don't understand that as we learn more about wildlife distribution, rare birds tend to become more common.

BLOCKS

COVERAGE

The Atlas project gathered momentum as it progressed: from covering 226 blocks in 1987 to 776 blocks in 1994, and from 50 completed in 1987 to 1,650 completed by 1995.

The Atlas contains data from 1,745 of the 1,760 Priority blocks (including alternate Priority blocks), with 1,650 (92%) deemed "complete." "Completed" blocks include those with a 40% Confirmation rate (down from the originally recommended 50%) and some with a lower rate but which the Project Director deemed had received adequate coverage. The Atlas also contains data from 18 border blocks and 1,006 Non-priority blocks. Major landowners denied access to 42 blocks, for which atlasers substituted alternates. Atlasers backpacked into 67 blocks and in fact relished the challenge of finding birds in rugged and scenic wildernesses.

Fifteen blocks received no coverage. Landowners refused permission for access to nine of them, field workers failed to cover four, and CBAP did not assign two.

TABLE 6

BLOCKS WITH HIGHEST SPECIES COUNTS

Quadcode	Block	County	# of Species	# of Habitats	Elevation Min.	Elevation Max.	Elevation Diff.	Field Work Party-Hours
40105A3	Boulder	Boulder	101	15	5,270	6,979	1,809	197
39105H3	Eldorado Springs	Jefferson	97	14	6,000	8,945	2,045	25
40107E2	Hooker Mountain	Routt	97	12	6,380	7,285	905	150
38106E8	Gunnison	Gunnison	96	12	7,630	8,624	1,899	44
38107H5	Bowie	Delta	94	11	5,700	8,234	2,534	27
40105D6	Trail Ridge	Larimer	93	9	8,526	11,722	3,196	406
37106D8	Saddle Mountain	Mineral/Archuleta	92	14	7,600	10,000	2,400	30
40107E3	Rock Spring Gulch	Routt	92	13	6,320	7,320	1,000	53
39106F7	Eagle	Eagle	90	14	6,600	7,840	1,240	90
39107F7	Horse Mountain	Garfield	87	14	5,860	7,840	1,980	38
39104C7	Castle Rock South	Douglas	87	11	6,600	7,030	1,670	87
40105B5	Allens Park	Boulder	87	15	8,520	10,810	2,290	100+
40105E2	Horsetooth Res.	Larimer	85	12	5,140	9,712	4,562	52
40106A4	Kremmling	Grand	85	8	7,320	9,320	2,000	42
39107C8	Hawxhurst Creek	Mesa	84	9	6,337	8,048	1,711	89
40105C3	Pinewood Lake	Boulder	84	14	5,600	6,700	1,100	84
40107F3	Slide Mountain	Routt	84	11	6,400	7,400	1,000	17
37107C8	Durango West	La Plata	83	10	6,465	8,320	1,855	97
40105B3	Lyons	Boulder	83	13	5,350	6,954	1,604	53
37108D3	Millwood	Montezuma	80	10	7,160	8,102	600	50+

SPECIES PER BLOCK

The average block tallied 42 species. The number of species ranged from 101 in Boulder (40105A1) to six in two treeless blocks: Alkali Lake (38102D6), in the dry grasslands of Kiowa County, and Mesito Reservoir (37105B6), an expanse of desert shrubland in the San Luis Valley. Map 1 shows the geographical spread of species per block, and Figure 2 graphs blocks according to the total number of species.

The number of species per block increases from south to north. Of the blocks that recorded more than 75 species, Latitudes 37 and 38 have eight each, Latitude 39 has 13, and Latitude 40 has 19.

Plains blocks without riparian habitat rarely have as many as 35 species. Those with fairly extensive riparian habitat yielded up to 79 species. At least 38 blocks have no trees at all (28 grassland and cropland blocks on the plains, four shrubland blocks in the San Luis Valley, and six sagebrush blocks in Moffat County) and 18 more had fewer than ten trees.

Blocks with the most species (Table 6) have a variety of habitats, usually high-quality riparian and diverse forest types. Many of these blocks lie in the foothills or mountains, have altitudinal ranges up to 4,000 feet (1220 m) contributing to habitat diversity, and have urban settlements in which human alterations add new habitats.

SPECIES PER LATILONG

One degree of latitude and longitude outlines a "latilong." In Colorado a latilong measures about 70 by 50–55 miles (122 by 80–88 km)—3,500–3,800 square miles (9,065–9,850 km^2), equal to about three-quarters of the area of Connecticut. Latilongs provide standard-sized geographical units that permit comparison. Using only Priority blocks within the latilong, Figure 3 shows the total number of species recorded in the latilong, the highest and lowest block counts, and the arithmetic means for blocks in the latilong.

MAP 1

SPECIES PER BLOCK

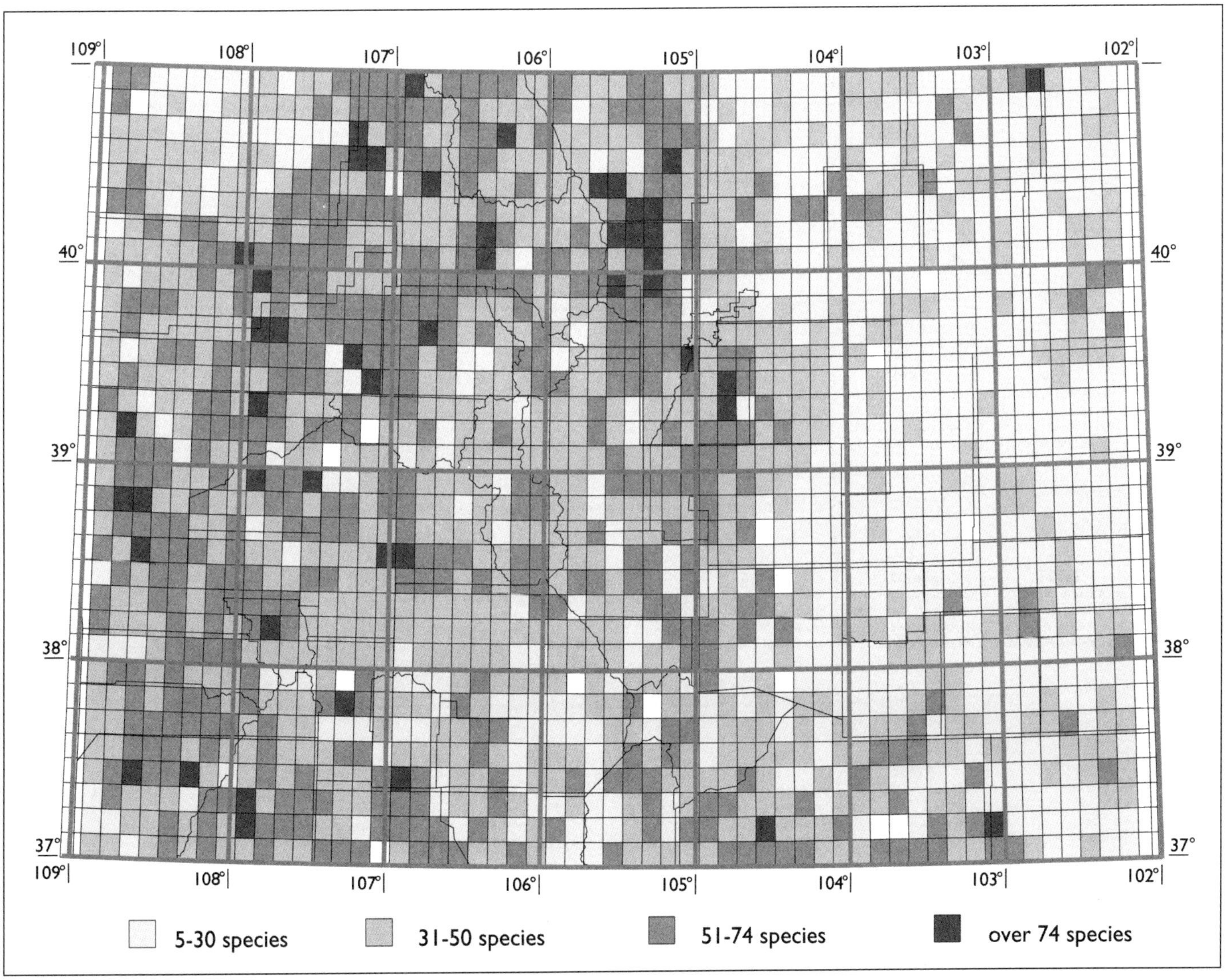
109°
108°
107°
106°
105°
104°
103°
102°
40°
39°
38°
37°
5-30 species
31-50 species
51-74 species
over 74 species

FIGURE 2

SPECIES PER BLOCK DISTRIBUTION

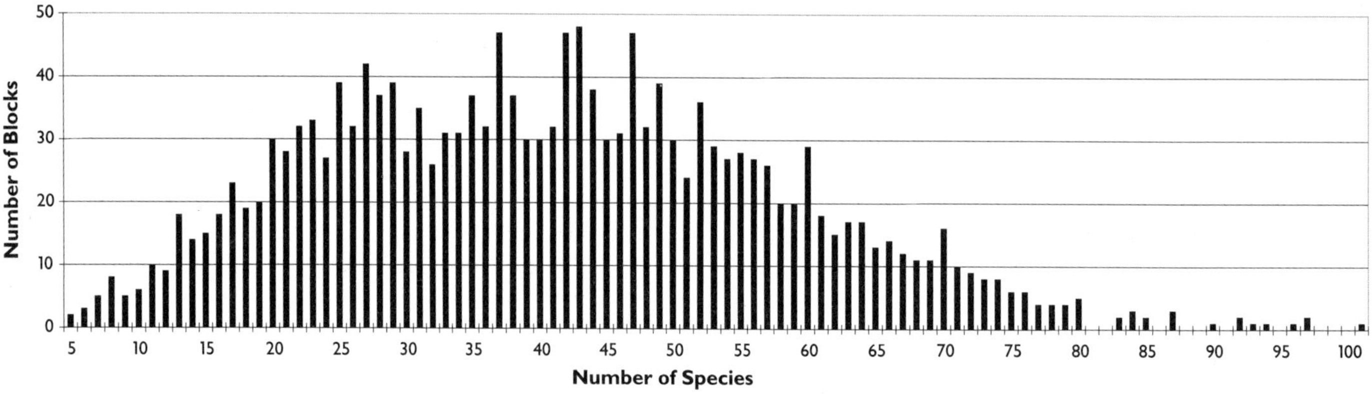

Number of Blocks
50
40
30
20
10
0
5
10
15
20
25
30
35
40
45
50
55
60
65
70
75
80
85
90
95
100
Number of Species

FIGURE 3

SPECIES PER LATILONG

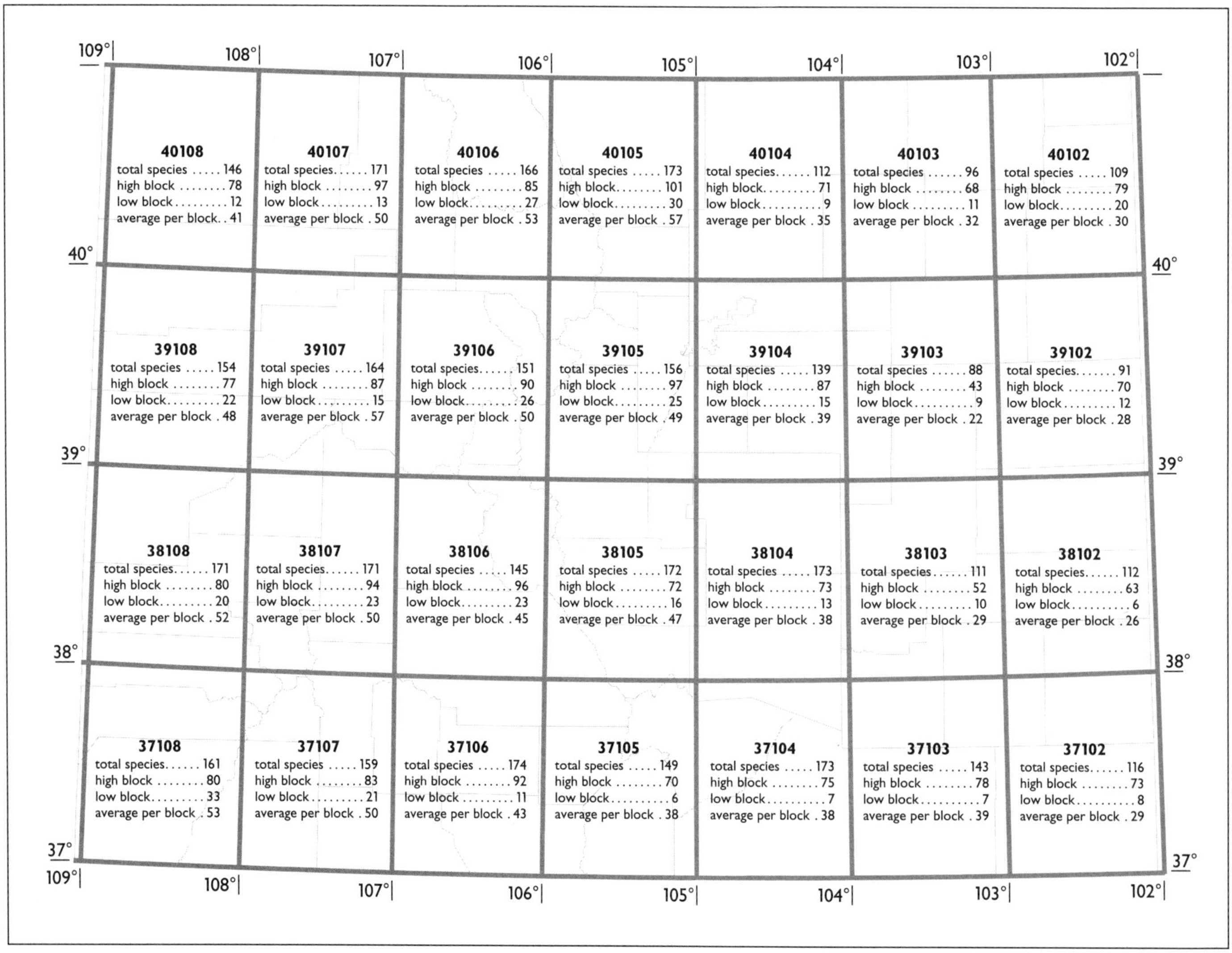

109°
108°
107°
106°
105°
104°
103°
102°
40°
39°
38°
37°
40108
total species 146
high block 78
low block 12
average per block . . 41
40107
total species 171
high block 97
low block 13
average per block . 50
40106
total species 166
high block 85
low block 27
average per block . 53
40105
total species 173
high block 101
low block 30
average per block . 57
40104
total species 112
high block 71
low block 9
average per block . 35
40103
total species 96
high block 68
low block 11
average per block . 32
40102
total species 109
high block 79
low block 20
average per block . 30
39108
total species 154
high block 77
low block 22
average per block . 48
39107
total species 164
high block 87
low block 15
average per block . 57
39106
total species 151
high block 90
low block 26
average per block . 50
39105
total species 156
high block 97
low block 25
average per block . 49
39104
total species 139
high block 87
low block 15
average per block . 39
39103
total species 88
high block 43
low block 9
average per block . 22
39102
total species 91
high block 70
low block 12
average per block . 28
38108
total species 171
high block 80
low block 20
average per block . 52
38107
total species 171
high block 94
low block 23
average per block . 50
38106
total species 145
high block 96
low block 23
average per block . 45
38105
total species 172
high block 72
low block 16
average per block . 47
38104
total species 173
high block 73
low block 13
average per block . 38
38103
total species 111
high block 52
low block 10
average per block . 29
38102
total species 112
high block 63
low block 6
average per block . 26
37108
total species 161
high block 80
low block 33
average per block . 53
37107
total species 159
high block 83
low block 21
average per block . 50
37106
total species 174
high block 92
low block 11
average per block . 43
37105
total species 149
high block 70
low block 6
average per block . 38
37104
total species 173
high block 75
low block 7
average per block . 38
37103
total species 143
high block 78
low block 7
average per block . 39
37102
total species 116
high block 73
low block 8
average per block . 29

BIRDS

The Atlas database contains 264 Confirmed breeding species and 14 Possible breeders. Atlasers did not Confirm seven species previously, and two subsequently, recorded as breeders in Colorado.

During the eight years of the Atlas the state recorded 13 new breeding species. Two, Least Flycatcher and Great Crested Flycatcher, probably represent permanent range expansions. Five others probably represent temporary, or one-time, range extensions: Little Blue Heron, Acorn Woodpecker, Carolina Wren, White-eyed Vireo, and Golden-winged Warbler. Six hard-to-find species probably nested in the state all along: Bufflehead, Spotted Owl, Northern Waterthrush, Field Sparrow, Eastern Meadowlark, and White-winged Crossbill.

Prior to the Atlas, birdwatchers tended to do their field work on public lands—state parks and wildlife areas and federal refuges, forests, and parks. Particularly in eastern Colorado, Atlas data offer significant new information because the Atlas protocol forced often shy birdwatchers to summon their nerve and ask permission for access to private land. This private-land effort documented range extensions, contradicted some assumptions about presence or absence of some species, and showed some species as more common than previously thought. It should alleviate concerns, temporarily anyway, about some species needing conservation efforts in order to survive into the twenty-first century.

Significant observations for a number of species emerged from the project. The Atlas documented that Mountain Plovers, although much diminished from 1900, nest more widely than previously thought. Atlas workers found clumps of McCown's and Chestnut-collared longspurs breeding south of their well-documented sites in the Pawnee Grassland. They discovered Gray Vireo and Hepatic Tanager nests in eastern Las Animas County and verified breeding of such rarities as Black Phoebe and Great Crested Flycatcher. They found that Rock Wrens, Brown Thrashers, and Orchard Orioles nest widely across the plains, but that, surprisingly, Black-capped Chickadees and Song, Vesper, and Chipping sparrows do not. They located what may be the first two nests of Rufous-crowned Sparrows actually found in Colorado. The project demonstrated that Eastern Phoebes and Purple Martins have more extensive ranges than previously thought, and that in Colorado, contrary to their habits in the eastern U.S., both species use natural nest sites.

In the mountains volunteers found two cavity-nesting ducks—Barrow's Goldeneye and Bufflehead—not previously known as established breeders. They filled in gaps in the ranges of species such as Three-toed Woodpecker, Brown Creeper, Golden-crowned Kinglet, Gray Vireo, Orange-crowned Warbler, Fox Sparrow, and Evening Grosbeak. And they found Cassin's Finches, generally described as conifer obligates, nesting in deciduous riparian habitats. Atlasers established high-altitude nesting records for several species including House Wren (11,800 feet, 3597 m), American Dipper (11,820 feet, 3602 m), and Brown-capped Rosy-Finch (14,200 feet, 4328 m).

The statewide nature of the survey led to new discoveries about the Colorado distribution of some common birds. House Finches have expanded across the plains, and Blue-gray Gnatcatchers now breed up and down the Front Range. Boreal and Flammulated owls have a widespread distribution in the mountains, Grace's Warblers nest, although sparsely, north of the San Juan Mountains, and Common Poorwills nest at the east edge of the Black Forest, maybe as far east as Limon. Gray Catbirds nest in as many blocks on the Western Slope as on the Eastern Slope. During the Atlas period American White Pelicans and California Gulls each established two new breeding sites in the state. Common and Great-tailed grackles expanded their ranges.

For several species, the Atlas shows a range more limited than that documented by earlier observers: Snowy Egret, Northern Harrier, American Avocet, Wilson's Phalarope, Forster's and Black terns, Short-eared Owl, Bank Swallow, and Chihuahuan Raven. A problematic Red-backed Hawk, whose origin Colorado ornithologists debate, probably bred with a Swainson's Hawk for two or more years.

Several species, mostly Neotropical migrants, maintained territories for at least two weeks, but atlasers did not detect additional nesting behavior. The Black Rail, discovered in 1991 in the lower Arkansas Valley and extremely hard to confirm breeding, probably does breed in a few marshes.

SPECIES DISTRIBUTION

Due to the diverse nature and dissimilarities of Colorado habitats, the Colorado Atlas recorded no species uniformly across the state—unlike eastern states that recorded several species in 90–99% of their Atlas blocks. No Colorado species nests on the prairies and in every kind of coniferous forest. Table 7 lists the 20 most frequently recorded species—all of them natives.

The most frequently recorded, American Robin, nests in 1,428 blocks (81%). The ubiquitous robins occupy every habitat with trees. Mourning Doves, second in frequency, occur in a variety of habitat niches. They avoid lodgepole pine and spruce/fir forests, and do not often breed above 8,000 feet (2440 m). Unlike robins, they breed on the plains in small shelterbelts with only a few trees or with only one or two kinds of trees; in grasslands they nest on the ground. The third most frequent species, Western Meadowlark, has persisted on the plains despite expanding human settlement. These superb songsters inhabit grasslands all over the state. Atlas observers found that Northern Flickers breed in any habitat with trees old enough to provide wood for drilling holes. As primary cavity-nesters, flickers contribute an essential ingredient to forest and woodland ecosystems: nest holes for species ranging from swallows to House Wren to Bufflehead. Colorado's state bird, the Lark Bunting, thickly populates the plains; Atlas figures suggest that it ranks as the fifth most abundant bird in the state even though 37 species occurred in more blocks. Red-tailed Hawks top the raptors in distribution and ranked sixth in number of blocks, but abundance estimates suggest a greater total abundance of breeding American Kestrels (13th in number of blocks).

The results, naturally, reflect the impact of humans on Colorado birdlife. Among the top 20 species, most of them habitat generalists, half have benefited from landscape changes since settlement began in 1860. A list of the 20 most frequent species in 1860 would probably look much different and include more grassland specialists.

SPECIES CONFIRMATIONS

The conspicuous American Robins exhibit conspicuous breeding behavior throughout the breeding cycle and have higher densities next to trails than away from them (Miller and Knight 1995). Field workers Confirmed robins in 84% of the blocks where they found them (Table 8). In contrast, they Confirmed Northern Harriers (secretive about nest sites), Common Nighthawks and Common Poorwills (cryptic nesters), American Goldfinches (late nesters), and swifts and Common Yellowthroats (inaccessible nest sites) in very few blocks. Despite the size and conspicuousness of Turkey Vultures, finding their nests proved almost impossible.

ABUNDANCE ESTIMATES

Appendix B lists estimates, based on Atlas abundance codes, of the number of breeding pairs of most species that breed in the state. These abundance numbers are only "first approximations," numbers that readers should use with caution.

This admonition applies particularly to species with isolated, colonial breeding sites (e.g., pelicans, gulls, and terns) or whose habitat Priority blocks tend to miss (e.g., some waterfowl and shorebirds). The introduction to Appendix B provides more details about the derivation and use of these numbers. This volume reports these because they at least provide a beginning point for evaluating relative abundances of species breeding in Colorado. Some species accounts also refer to these calculations.

In total, the estimates claim 40,057,000 breeding pairs in the state, or one pair per 1.67 acres. The actual number of breeding birds in Colorado must exceed that number substantially, but any certainty about which species have greater numbers must await a study using better statistical sampling methods. Therefore, although we report these estimates, users should not construe these numbers as true population figures.

Residents and non-residents think of Colorado as synonymous with the high Rockies, yet plains cover a third of the state. Atlas abundance estimates (Table 9) reflect that the plains had most of the most abundant species, Horned Lark. In fact, three of the five most abundant species nest mainly on the plains (the other two are habitat generalists). Three of the second five most abundant species breed primarily in conifers.

Despite the shortcomings of Atlas abundance estimates, Horned Larks most likely are the most abundant Colorado bird. Hal Borland (1956) frequently referred to them in his autobiographical account of homesteading on the prairie south of Brush:

> *And our horned larks . . . sang again, spiraling like*
> *the storied skylarks that sang so beautifully in English poetry.*
> *Our larks sang beautifully, too, and in reality, not in poems*
> *that use verbs like wert and wingest.*

TABLE 7

SPECIES REPORTED IN THE MOST BLOCKS

Rank/Species		Number of Blocks	% of Blocks in Which Recorded
1	American Robin	1,428	82%
2	Mourning Dove	1,386	79%
3	Western Meadowlark	1,187	68%
4	Northern (Red-shafted) Flicker	1,162	67%
5	Brown-headed Cowbird	1,119	64%
6	Red-tailed Hawk	1,105	63%
7	Killdeer	1,049	60%
8	Red-winged Blackbird	1,046	60%
9	Barn Swallow	1,039	60%
10	Common Nighthawk	1,020	58%
11	Horned Lark	982	56%
12	House Wren	976	56%
13	American Kestrel	973	56%
14	Black-billed Magpie	905	52%
15	Western Kingbird	893	51%
16	Mountain Bluebird	891	51%
17	Broad-tailed Hummingbird	865	50%
18	Mallard	862	49%
19	Western Wood-Pewee	861	49%
20	Common Raven	837	48%
21	Violet-green Swallow	815	47%
22	Mountain Chickadee	785	45%
23	European Starling	782	45%
24	Chipping Sparrow	777	45%
25	Great Horned Owl	775	44%
26	Lark Sparrow	771	44%
27	Bullock's Oriole	762	44%
28	Yellow Warbler	735	42%
29	Pine Siskin	732	42%
30	Cliff Swallow	715	41%

TABLE 8

SPECIES WITH HIGHEST CONFIRMATION RATES*

Rank/Species		# of Confirm.	Total Blocks in Which Recorded	Confirm. %
1	Bald Eagle	17	19	89.5
2	American Robin	1,205	1,427	84.4
3	Eastern Phoebe	19	23	82.6
4	House Sparrow	545	692	78.8
5	Barn Swallow	818	1039	78.7
6	Black-billed Magpie	673	905	74.4
7	Dark-eyed (Gray-headed) Junco	451	613	73.6
8	American Pipit	90	123	73.2
9	Mountain Bluebird	652	891	73.2
10	Cliff Swallow	520	715	72.7
11	Bullock's Oriole	553	762	72.6
12	Western Kingbird	627	893	70.2
13	White-crowned Sparrow	547	782	69.9
14	Canada Goose	141	202	69.8
15	Red-naped Sapsucker	321	468	68.6
16	House Wren	655	976	67.1
17	Common Grackle	428	652	65.6
18	Sandhill Crane	15	23	65.2
19	Mountain Chickadee	500	785	63.7
20	Eared Grebe	12	19	63.2
21	Lincoln's Sparrow	261	413	63.2
22	Osprey	10	16	62.5
23	Horned Lark	612	982	62.3
24	Western Meadowlark	740	1,187	62.3
25	Eastern Bluebird	12	20	60.0

Recorded in at least 15 blocks

TABLE 9

MOST ABUNDANT SPECIES

NUMBER OF BREEDING PAIRS

Rank*/Species		Estimate Low	Estimate Mid	Estimate High
1	Horned Lark	938,608	2,246,503	4,199,890
2	American Robin	590,168	1,903,131	3,511,966
3	Western Meadowlark	528,006	1,900,006	3,493,099
4	Mourning Dove	392,020	1,490,767	2,735,244
5	Lark Bunting	540,470	1,308,072	2,444,009
6	Mountain Chickadee	326,633	1,173,645	2,157,763
7	Dark-eyed (Gray-headed) Junco	343,084	1,103,625	2,037,118
8	Pine Siskin	318,127	1,050,534	1,937,056
9	House Wren	327,548	1,032,685	1,906,956
10	Ruby-crowned Kinglet	418,100	1,011,162	1,889,318
11	Yellow-rumped Warbler	354,152	971,564	1,804,674
12	Green-tailed Towhee	310,386	856,377	1,590,112
13	Warbling Vireo	345,820	842,146	1,572,584
14	Broad-tailed Hummingbird	205,362	828,075	1,516,471
15	Chipping Sparrow	198,874	827,502	1,514,120
16	Lark Sparrow	190,752	755,420	1,384,157
17	Cliff Swallow	184,183	747,493	1,368,671
18	Red-winged Blackbird	171,429	671,944	1,231,090
19	House Sparrow	222,154	661,838	1,224,766
20	Violet-green Swallow	132,049	639,093	1,164,358
21	Vesper Sparrow	176,985	618,099	1,137,179
22	Green-tailed Towhee	208,626	588,180	1,091,107
23	White-crowned Sparrow	232,371	580,760	1,083,311
24	Hermit Thrush	154,432	555,209	1,020,385
25	Western Wood-Pewee	106,939	529,392	963,514
26	Brewer's Sparrow	176,591	520,484	963,749
27	European Starling	144,767	501,100	921,734
28	Lincoln's Sparrow	149,826	487,635	899,516
29	Mountain Bluebird	114,527	481,178	879,572
30	Blue-gray Gnatcatcher	104,875	473,560	864,641
31	Tree Swallow	93,345	432,585	788,923
32	Barn Swallow	104,891	424,216	775,924

Rank based on Mid Estimate

HABITATS AND THEIR Breeding Birds

Colorado boasts a colorful diversity of habitats, and this section portrays most of those included in the Atlas protocol. Most depict Atlas Priority blocks. By linking the codes in the Habitat charts to the codes in the picture captions, readers may visualize the habitats a particular species uses. This section also includes photographs of nests, fledglings, and nesting activity, shown side-by-side with the appropriate habitats.

Vegetative type and structure exert a major influence on bird distribution. Atlas field workers developed an awareness of how changes in habitat, even within short distances, cause changes in the birdlife. These pictures demonstrate differences among habitats and within ecosystems, e.g., grasslands, shrublands, and forests.

The captions refer to the applicable Atlas Habitat Code or Breeding Code, and identify the Atlas block in which the picture was taken.

TTS: TALLGRASS, SANDSAGE
Sedgwick County
Atlas block 40102H5, Sedgwick

STEPHEN R. JONES

TMGP: MID-GRASS
Yuma County
Atlas block 39102G4, Vernon

HUGH KINGERY

HUGH KINGERY

NE: NEST WITH EGGS
LARK BUNTING
Washington County
Atlas block 39103F7, Cottonwood Valley North

STEPHEN R. JONES

TSG: SHORTGRASS
Lincoln County
Atlas block 39103C7, River Bend

DAVID LEATHERMAN

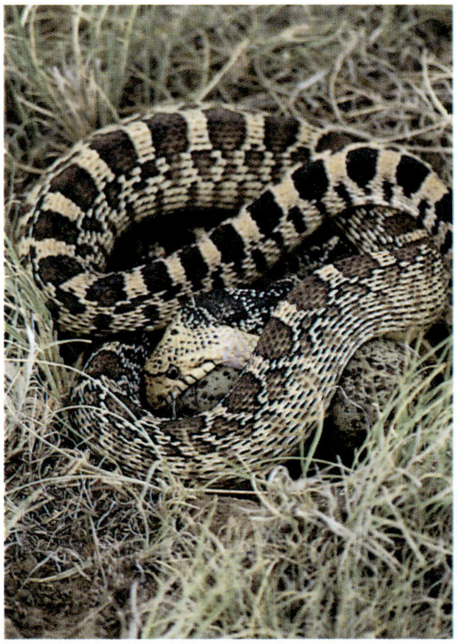

HUGH KINGERY

Left—
FL: FLEDGED YOUNG
MOUNTAIN PLOVER
Pawnee National Grassland, Weld County
Atlas block #40104E3, Briggsdale

Right—
NE: NEST WITH EGGS
LONG-BILLED CURLEW
Bullsnake raiding nest
Las Animas County
Atlas block 37103C8, Painted Canyon

NE: NEST WITH EGGS
HORNED LARK
Lincoln County
Atlas block #38103G4, McKenzie Draw

ALAN VERSAW

GRASSLAND HABITATS

HUGH KINGERY

TMXP: MIXED INTRODUCED AND PLANTED GRASSES
La Plata County
Atlas block 37108A1, Pinkerton Mesa

HUGH KINGERY

TMG: MONTANE GRASSLAND
Mt. Meeker and Longs Peak, Larimer County
Atlas block 40105D5, Estes Park

TUNDRA HABITAT

TAL: ALPINE GRASSLAND
La Plata Mountains,
Montezuma and La Plata Counties
Atlas block 37108E1, Orphan Butte

HUGH KINGERY

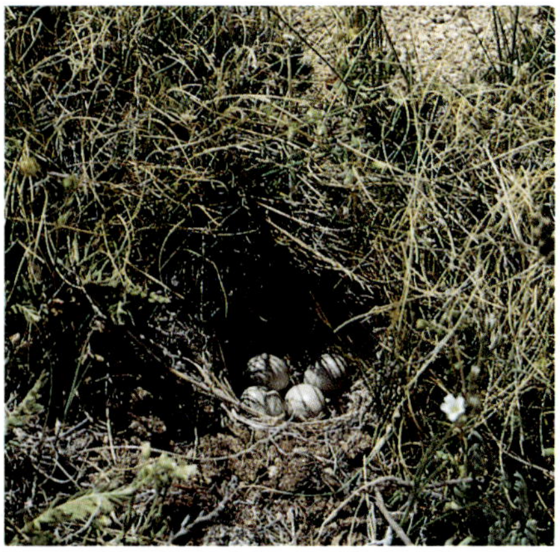

ALAN VERSAW

NE: NEST WITH EGGS
AMERICAN PIPIT
Sawatch Range, Chaffee County
Atlas block 38106F4, Cumberland Pass

SMS: MOUNTAIN SAGEBRUSH
Costilla County
Atlas block 37105C4, Fort Garland

HUGH KINGERY

SLS: LOWLAND SAGEBRUSH
Lincoln County
Atlas block 38103H3, Boyero

DAVID LEATHERMAN

STD: TALL DESERT SHRUB
Dolores River, San Miguel County
Atlas block 38108B8, Anderson Mesa

ALAN VERSAW

FL: FLEDGED YOUNG
LOGGERHEAD SHRIKE
Montezuma County
Atlas block 37108B7, Mariano Wash East

HUGH KINGERY

NE: NEST WITH EGGS
COMMON NIGHTHAWK
Saguache County
Atlas block 37106G4, Twin Mountains

HUGH KINGERY

SSLS: LOGGED-OVER AREA
Wolf Creek Pass, Archuleta County
Atlas block 37106D7, Wolf Creek Pass

HUGH KINGERY

SMT: MAT SALTBUSH
San Luis Valley, Costilla County
Atlas block 37105A7, Kiowa Hill

HUGH KINGERY

HUGH KINGERY

SMM: MONTANE SHRUBLANDS
Deer Creek Canyon County Park,
Jefferson County
Atlas block 39105E2, Indian Hills

HUGH KINGERY

SMC: MONTANE CARR

Mt. Elbert, highest point in Colorado (14,443 feet; 4399 m), Lake County

Atlas block 39106B4, Mt. Massive

HUGH KINGERY

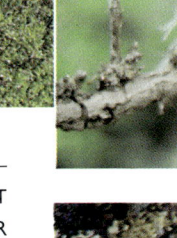

Top Right—
ON: OCCUPIED NEST
DUSKY FLYCATCHER
Montezuma County
Atlas block 37108C7, Battle Rock

HUGH KINGERY

Middle Right—
FL: FLEDGED YOUNG
CANYON WREN

BEVERLY TOMBERLIN

Lower Right—
NY: NEST WITH YOUNG
TURKEY VULTURE
San Juan National Forest

ALAN VERSAW

SLC: LOWLAND CARR

MCL: CLIFFS

Dolores River, San Miguel and Montrose Counties

Atlas block 38108B8, Anderson Mesa

SOS: OAK SCRUB
Chimney Peak and Courthouse Mountain,
Ouray County
Atlas block 38107B6, Dallas

HUGH KINGERY

ANN HIGGINS

FL: FLEDGED YOUNG
BROWN-HEADED COWBIRD FLEDGLING
FED BY VIRGINIA'S WARBLER
Manitou Springs, El Paso County
Atlas block 38104G8, Manitou Springs

SBC: WILLOW CARR
Colorado Trail, San Juan County
Atlas block 37107G6, Silverton

HUGH KINGERY

DAVID LEATHERMAN

SKR: KRUMMHOLZ–SHRUBBY TIMBERLINE CONIFERS
Loveland Pass, Summit County
Atlas block 39105F8, Loveland Pass

HUGH KINGERY

WPJ: PINYON/JUNIPER
Sawatch Range (Mt. Shavano on left),
Chaffee County
Atlas block 38106E1, Salida West

DAVID LEATHERMAN

ON: OCCUPIED NEST, GRAY VIREO
Colorado National Monument, Mesa County
Atlas block 39108A6, Colorado National Monument

WPP: PONDEROSA PINE
Lake Dorothey State Wildlife Area, Las Animas County
Atlas block 37104A3, Barela

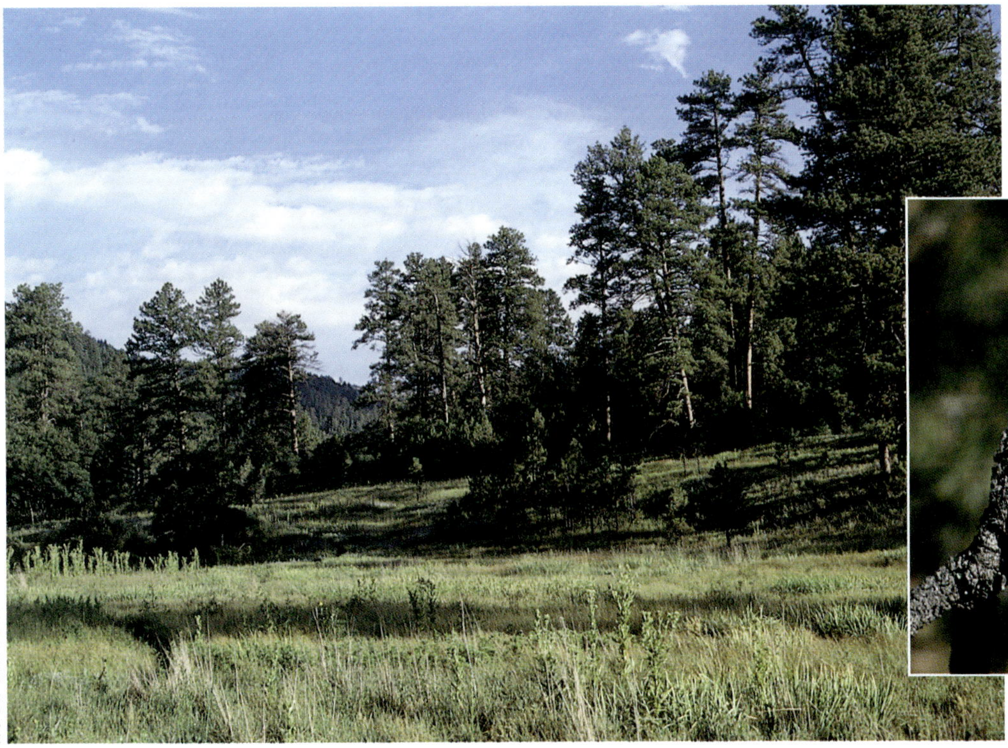

HUGH KINGERY

NE: NEST WITH EGGS
BROAD-TAILED HUMMINGBIRD
Big Thompson Canyon, Larimer County
Atlas block 40105D4, Glen Haven

DAVID LEATHERMAN

FL: FLEDGED YOUNG
GREAT HORNED OWL
Crow Valley Campground, Weld County
Atlas block 40104E3, Briggsdale

DAVID LEATHERMAN

LRD: LOWLAND RIPARIAN
Chatfield State Park, Jefferson County
Atlas block 39105E1, Littleton

DAVID LEATHERMAN

HUGH KINGERY

UN: USED NEST
YELLOW-BREASTED CHAT
Unaweep Seep
Atlas block 38108G7, Fish Creek

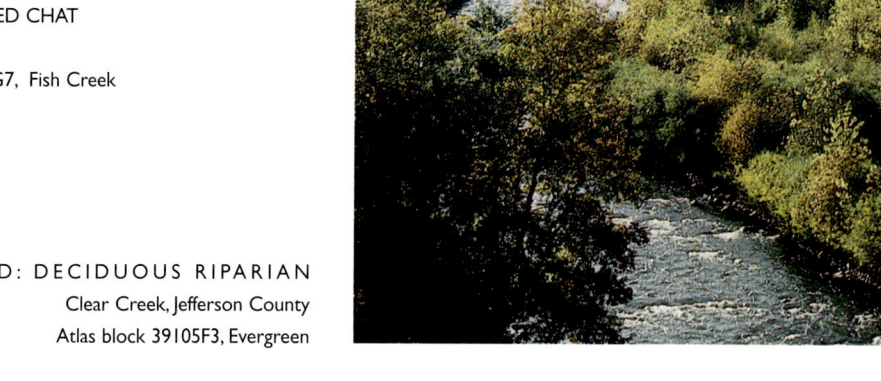

DAVID LEATHERMAN

FRD: DECIDUOUS RIPARIAN
Clear Creek, Jefferson County
Atlas block 39105F3, Evergreen

HUGH KINGERY

WMOS: MONTANE WOODLAND
Colorado Trail, San Juan County
Atlas block 37107G7, Ophir

HUGH KINGERY

NB: NEST BUILDING
EVENING GROSBEAK
GATHERING NESTING MATERIAL
Chaffee County
Atlas block 38106F1, Buena Vista East

HUGH KINGERY

FRC: CONIFEROUS RIPARIAN
West Fork, San Juan River, Archuleta County
Atlas block 37106D8, Saddle Mountain

HUGH KINGERY

FL: FLEDGED YOUNG
HAMMOND'S FLYCATCHER FAMILY
Routt National Forest, Routt County
Atlas block 40106G7, Farwell Mountain

HUGH KINGERY

CF: CARRYING FOOD
PINE GROSBEAK WITH SEEDS IN THROAT
Archuleta County
Atlas block 37106B7, Harris Lake

FUC: UPLAND CONIFEROUS
South San Juan Wilderness, Conejos County
Atlas block 37106C6, Summit Peak

HUGH KINGERY

STEPHEN R. JONES

FUD: UPLAND DECIDUOUS
Monarch Lake, Grand County
Atlas block 40105A6, Monarch Lake

HUGH KINGERY

ON: OCCUPIED NEST
RED-NAPED SAPSUCKER
Arapaho National Forest, Summit County
Atlas block 39106F1, Dillon

FUCS: SPRUCE/FIR FOREST
Cameron Pass, Larimer County
Atlas block 40105E8, Clark Peak

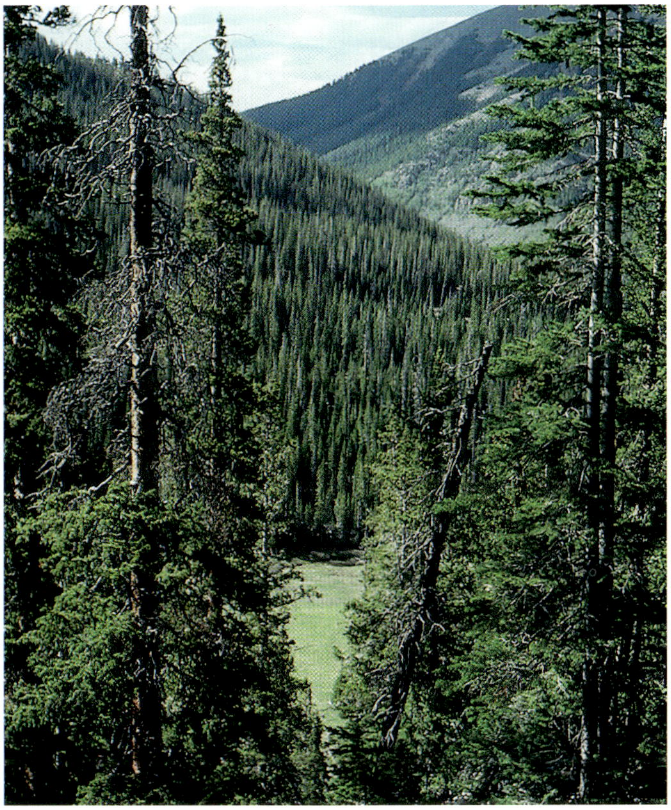

DAVID LEATHERMAN

HUGH KINGERY

OWS: OPEN WATER, STREAM
South San Juan Wilderness,
Conejos County
Atlas block 37106C6, Summit Peak

Inset—
ON: OCCUPIED NEST
CORDILLERAN FLYCATCHER
Conejos County
Atlas block 37106C6, Summit Peak

HUGH KINGERY

FUCL: UPLAND CONIFEROUS, LODGEPOLE PINE
Lake County
Atlas block 39106B4,
Mt. Massive

DAVID LEATHERMAN

NE: NEST WITH EGGS
AMERICAN ROBIN
Cameron Pass, Larimer County
Atlas block 40105E8, Clark Peak

AEM: EMERGENT WETLANDS
Rocky Mountain Arsenal National Wildlife Refuge, Adams County
Atlas block 39104G8, Commerce City

HUGH KINGERY

DAVID LEATHERMAN

NE: NEST WITH EGGS
AMERICAN AVOCET
Fossil Creek Slough
Atlas block 40105E1, Fort Collins

OWS: OPEN WATER, STREAM
SLC: LOWLAND CARR
Arkansas River as it leaves Colorado; lowest point in the state (3,350 feet; 1021 m); Prowers County
Atlas block 38102A1, Holly East

HUGH KINGERY

FL: FLEDGED YOUNG
BUFFLEHEADS
Big Creek Lake, Jackson County
Atlas block 40106H5, Pearl

BOB RIGHTER

OWL: OPEN WATER, LAKE
Quien Sabe Lake, Archuleta County
Atlas block 37107D1, Pagosa Peak

HUGH KINGERY

BOB DARNELL

DAVID LEATHERMAN

CPL: CROPLANDS
Alamosa County
Atlas block 37105D8, Alamosa West

Inset—
FL: FLEDGED YOUNG
KILLDEER
Moffat County

THE
CHICAGO

ALAN VERSAW

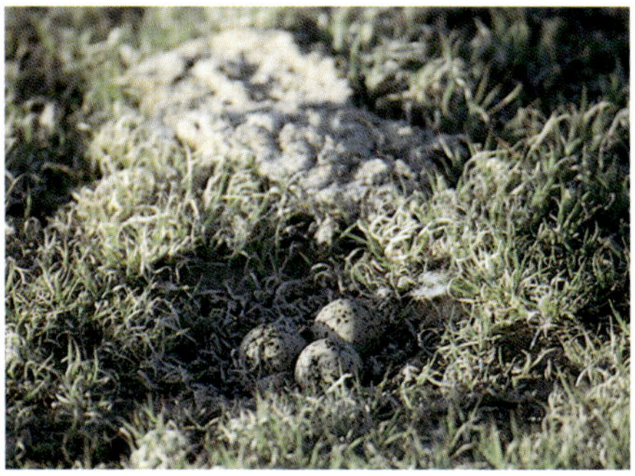

HUGH KINGERY

NE: NEST WITH EGGS
MOUNTAIN PLOVER
Cheyenne County
Atlas block 38102E3, Cheyenne Wells 3 SE

ON: OCCUPIED NEST
CHIHUAHUAN RAVEN
Baca County
Atlas block 37103C1, Utleyville

TSU: BARREN GROUND
CPL: CROPLAND
Logan County
Atlas block 40102E6, Rockland

N. R. WHITNEY

URB: URBAN, RESIDENTIAL
Denver
Atlas block 39104F8, Englewood

HUGH KINGERY

HUGH KINGERY

FL: FLEDGED YOUNG
AMERICAN ROBIN
Denver
Atlas block 39104F8, Englewood

HUGH KINGERY

RRL: RURAL
Last Chance, Washington County
Atlas block 39103G5,
Woodlin School

WALTER COLLINS

NE: NEST WITH EGGS
YELLOW WARBLER AND
BROWN-HEADED COWBIRD
Grand County
Atlas block 40105A8, Granby

DAVID LEATHERMAN

FL: FLEDGED YOUNG
BLACK-BILLED MAGPIE
Garden of the Gods Park, El Paso County
Atlas block 38104G8, Manitou Springs

THE COLORADO ENVIRONMENT

By Charles R. Preston and Hugh E. Kingery

Colorado offers a spectacular variety of potential home sites for the prospective avian resident. From the fragile carpet of cushion plants covering the alpine tundra to the tall, dense, evergreen fabric of the subalpine conifer forest to the scattered, remnant islands of shortgrass prairie on the eastern plains, Colorado provides for a plethora of avian lifestyles. Indeed, our state boasts one of the largest bird lists (460 species) of any inland state, and the Atlas project recorded 264 species of breeding birds between 1987 and 1994.

Where a bird lives depends on the suite of challenges and opportunities offered by the places available. These challenges and opportunities are often linked directly to the character of the vegetation, which, in turn, is influenced by physical parameters such as topography, climate, and soil texture and chemistry. Thus, to understand the avifaunal diversity of Colorado, it is first essential to consider the complex and dynamic character of the physical environment.

OVERVIEW: GEOGRAPHICAL SYSTEMS

Colorado encompasses an area of 104,247 square miles (270,000 km^2), and includes an elevational relief of more than 11,000 feet (3350 m). In the center of the state, the highest point, the summit of Mount Elbert, thrusts 14,433 feet (4398 m) above sea level, and the lowest point (3,350 feet, 1020 m) lies about 250 miles (400 km) to the southeast, where the Arkansas River leaves Colorado east of Holly.

Physical geographers segregate Colorado into five physiographic provinces, or geographic regions (Fenneman 1931, Mutel and Emerick 1992). Together, the Great Plains, Colorado Plateau, and Southern Rocky Mountains provinces comprise most of the state. The Middle Rocky Mountains (Uinta Mountains) and Wyoming Basin intrude in the far northwestern corner. Although biologists often use physiographic provinces to help understand species distribution (e.g., BBS data), these designations stem from physical geography; they do not consistently reflect floral and faunal characteristics.

The Great Plains Province occupies about 40% of Colorado, eastward from the foothills of the Southern Rocky Mountains. The terrain varies from level to gently rolling, but also contains some shallow canyons and impressive escarpments, buttes, mesas, and volcanic remnants. These elements not only add relief to the landscape, but provide nesting and feeding opportunities for many bird species that do not occur over most of the province, especially raptors, swifts, and swallows. The pinyon/juniper woodlands of Mesa de Maya, the red sandstone canyons carved by the Purgatoire River, and the

MAP 2

PHYSIOGRAPHIC REGIONS

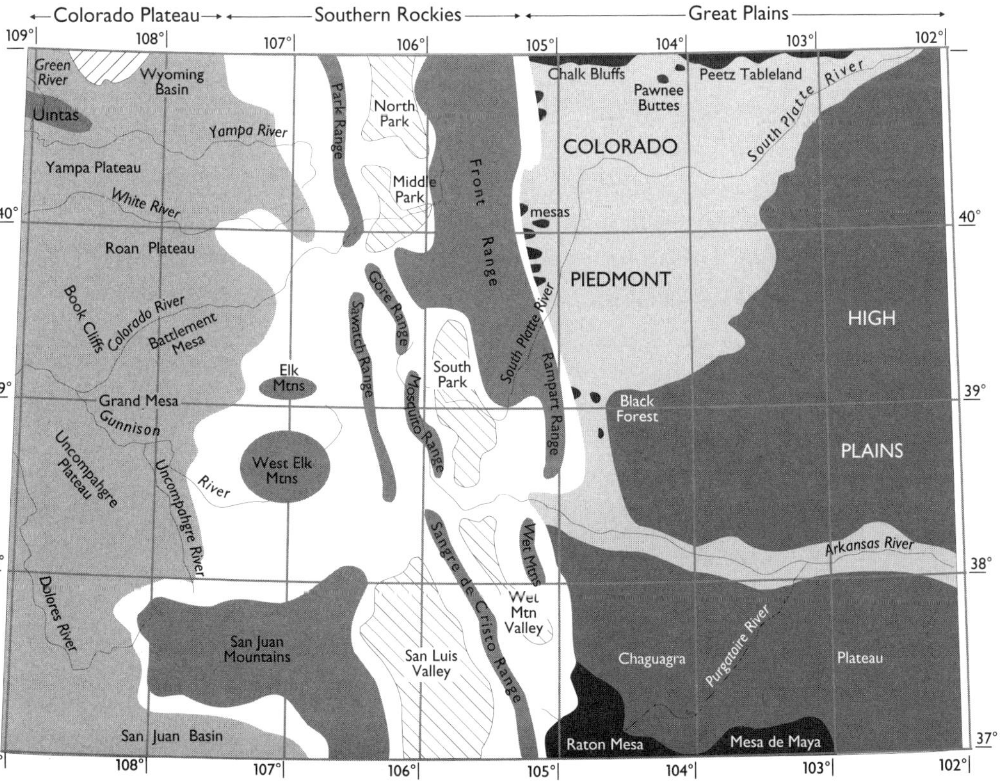
Colorado Plateau
Southern Rockies
Great Plains
109°
108°
107°
106°
105°
104°
103°
102°
Green River
Wyoming Basin
Uintas
Yampa River
Park Range
North Park
Chalk Bluffs
Pawnee Buttes
Peetz Tableland
South Platte River
COLORADO
Yampa Plateau
White River
Middle Park
Front Range
mesas
40°
Roan Plateau
PIEDMONT
Gore Range
HIGH
Book Cliffs
Colorado River
Battlement Mesa
Sawatch Range
South Park
South Platte River
Rampart Range
Elk Mtns
39°
Grand Mesa
Gunnison
Mosquito Range
Black Forest
West Elk Mtns
Uncompahgre Plateau
Uncompahgre River
River
PLAINS
Sangre de Cristo Range
Wet Mtns
Arkansas River
38°
Dolores River
Wet Mtn Valley
San Juan Mountains
San Luis Valley
Chaguagra
Purgatoire River
Plateau
37°
San Juan Basin
Raton Mesa
Mesa de Maya

Map adapted with permission from *Atlas of Colorado* by Kenneth Erikson and Albert Smith, University Press of Colorado, 1985, pp.6–7.

ponderosa-clad Black Forest add non-plains topography and vegetation to the more widespread grasslands and stream bottoms. They also introduce western and southwestern species of hummingbirds, flycatchers, wrens, bluebirds, sparrows, and others.

Typically, the climate is windy and dry. Annual precipitation here in the shadow of the Rockies varies from about 12 inches (30.5 cm) near the Rocky Mountain foothills to 18 inches (46 cm) near the Kansas border. Summer temperatures often rise above 100°F (38°C), and winter temperatures may drop to well below 0°F (-18°C).

Traversing the Great Plains like giant, life-supporting arteries, the South Platte River in northeastern Colorado and the Arkansas River to the southeast drain much of the water from the Eastern Slope of the Southern Rockies.

The gently rolling terrain of the Great Plains is a mere prologue to the dramatic relief displayed in the Southern Rocky Mountains Province. In broad outline, the Colorado Rockies consists of two magnificent ridges, running north–south between the Great Plains to the east and the Colorado Plateau to the west, but is actually a cordillera comprised of many, more or less parallel, ridges. The Continental Divide denotes the separation between the Eastern Slope and the Western Slope of the Rockies (i.e., the line between the Atlantic and Pacific drainages).

Fifty-four peaks on these rugged ridges rise above 14,000 feet (4270 m), presiding over their foothills 8,000 feet (2440 m) below. The transition between the Great Plains and Southern Rocky Mountains is especially dramatic in the foothills along the Front Range Corridor, from Fort Collins to Colorado Springs.

Climate varies widely along the steep elevational gradient from foothills to mountain peaks. Ambient temperature typically decreases with increasing elevation at the rate of about 3°F per 1,000 feet (1.8°C/300 m) elevation. Precipitation generally increases with elevation. Alpine peaks may receive more than twice as much annual precipitation as the Great Plains; most falls as snow. Because wind also tends to increase with elevation, much of the snow that falls on the mountain peaks blows away and accumulates on the mountainsides below. Deep snow typically covers the subalpine zones from October through May each year. Frost may occur on any night of the year above 9,000 feet (2745 m) elevation (Fitzgerald et al. 1994).

Between the two great ridges of the Southern Rocky Mountains Province lie large, treeless basins, North Park, Middle Park, South Park, and the San Luis and Wet Mountain valleys. Dense, cold air sinks into these areas and, especially in winter, temperatures often drop much lower than in adjacent uplands. Because the great peaks surrounding them block moisture from the Pacific Ocean and the Gulf of Mexico, these parks and valleys receive significantly less precipitation than the surrounding uplands.

West of the Southern Rocky Mountains Province lies the Colorado Plateau Province, a place of often breathtaking beauty and surprising variety. The region includes numerous mesas and plateaus fractured by deep, richly hued canyons cut by the Colorado River and its tributaries. Grand Mesa and Battlement Mesa each exceed 10,000 feet (3048 m). Other prominent, yet somewhat lower highlands in this province are the Book Cliffs and the Roan and Uncompahgre plateaus. The varied topography greatly influences local climate, with warm semi-arid lowlands contrasting markedly with moist canyon bottoms and cooler, more mesic uplands.

Map 3 and Figure 5 offer two ways to view Colorado geography. In Map 3, shades of gray, darker with increasing elevation, dramatically depict Colorado landforms. The river systems and mountain parks

FIGURE 4

COLORADO PROFILE

This schematic presentation represents a cross-section of Colorado, along the 39th Parallel. Symbols represent an approximation of the zones of various types of vegetation.

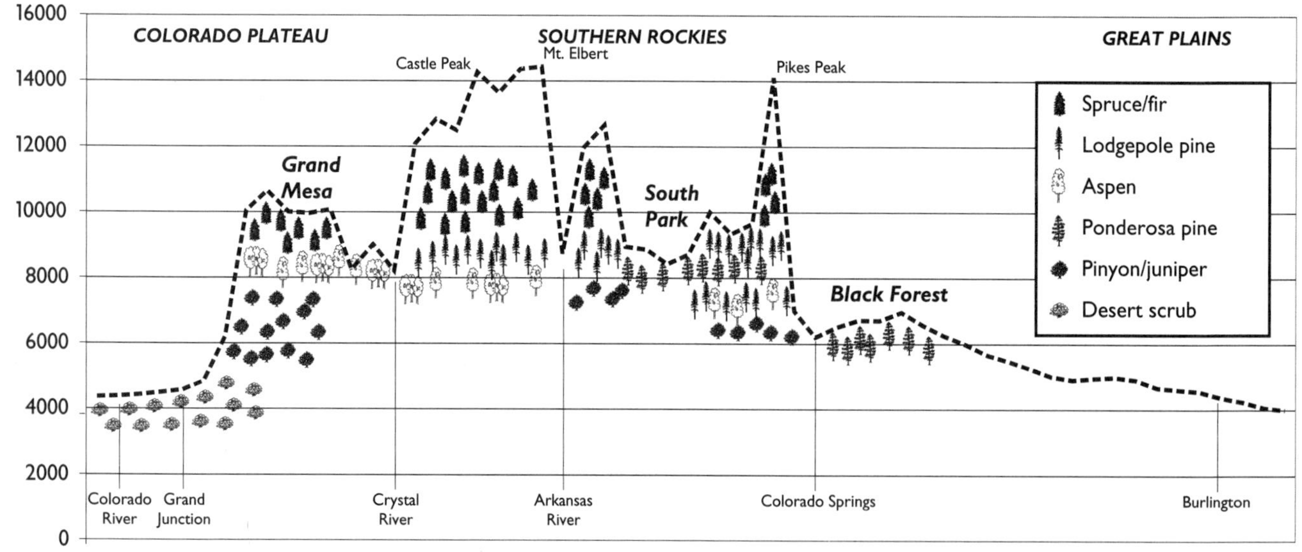
16000
14000
12000
10000
8000
6000
4000
2000
0
COLORADO PLATEAU
SOUTHERN ROCKIES
GREAT PLAINS
Castle Peak
Mt. Elbert
Pikes Peak
Grand Mesa
South Park
Black Forest
Spruce/fir
Lodgepole pine
Aspen
Ponderosa pine
Pinyon/juniper
Desert scrub
Colorado River
Grand Junction
Crystal River
Arkansas River
Colorado Springs
Burlington

MAP 3

COLORADO ELEVATIONS

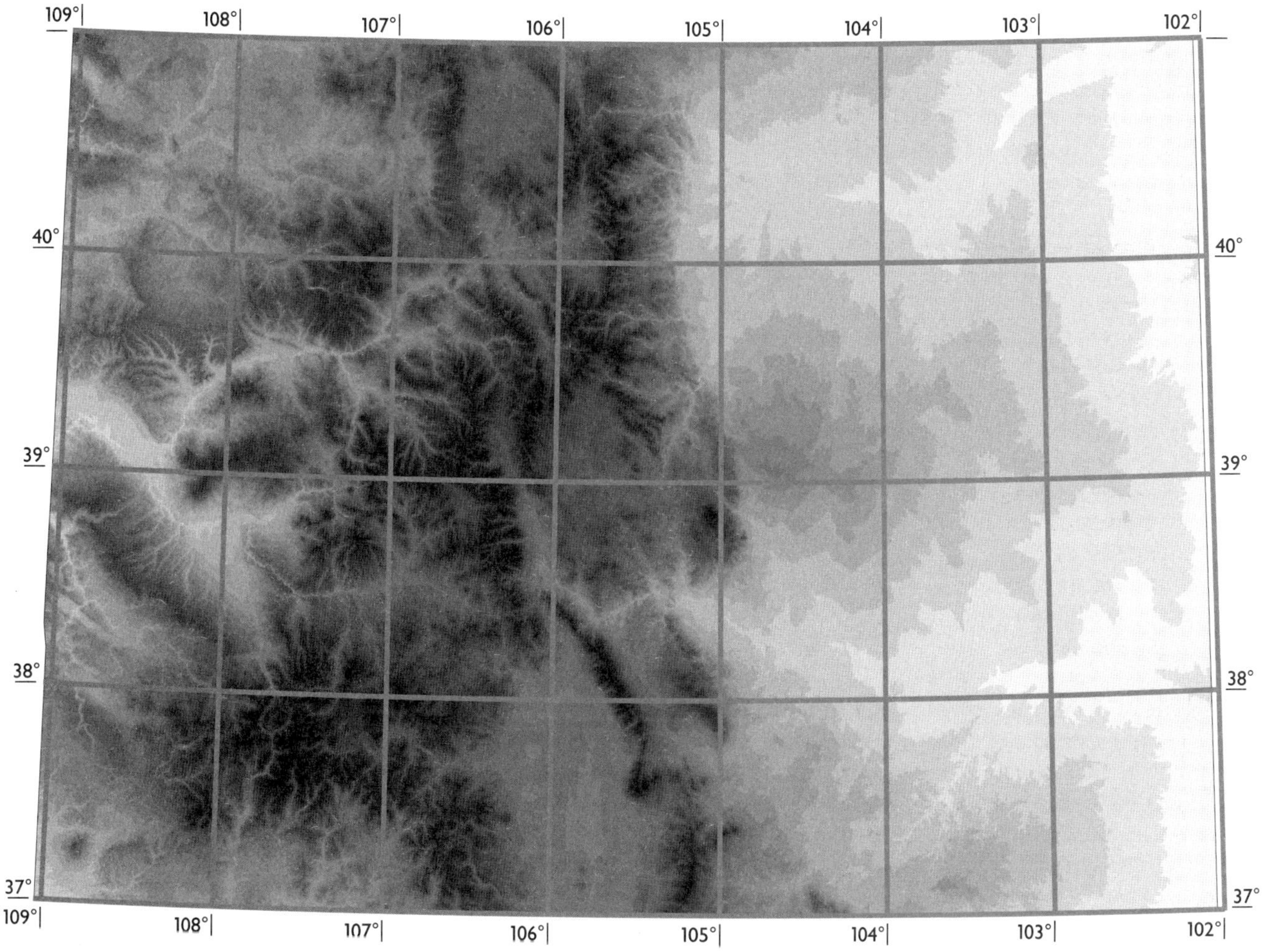
109°
108°
107°
106°
105°
104°
103°
102°
40°
39°
38°
37°

show distinctly on this map. The Colorado cross-section in Figure 5 slices the state approximately along the 39th Parallel. Its exaggerated profile shows how much the state changes from east to west.

CLASSIFYING ECOSYSTEMS

Physical attributes vary greatly within and between physiographic provinces in Colorado, as described above. Geological history, local topography, and elevation all have their effects. Human land use, especially activities associated with farming, ranching, and urbanization, has profoundly altered the native landscape and added to the already extraordinary environmental mosaic. The great diversity of intrinsic physical attributes and of human land-use patterns sets the stage for an equally stunning diversity of plant and animal assemblages across Colorado.

Ecologists refer to the physical environment, together with all the organisms living in a given area, as an *ecosystem*. An ecosystem can be of any size, from the gut of an insect, to a discrete patch of landscape, to the entire life-support system of the planet. Typically, a landscape-based ecosystem is named for dominant plant species or physical characteristics found within it. Ecosystems, of course, typically blend into one another, forming transition zones, or *ecotones*. Scientists have offered numerous approaches to classifying broad, landscape-based ecosystems in Colorado (e.g., Cary 1911, Gregg 1963, Armstrong 1972, Bailey 1978, Erickson and Smith 1985, Benedict 1991, Mutel and Emerick 1992, Fitzgerald et al. 1994). Each of these schemes represents a human construct developed from a slightly different perspective. All are useful in gaining an understanding of the environment. Here, we identify 21 distinct ecosystems, each presenting a unique suite of challenges and opportunities for resident breeding birds.

The Atlas provides a unique source to assess the avian inhabitants of each ecosystem. Linked with the ecosystem descriptions below, Habitat Use tables show the number of Priority blocks that reported the habitat and the species reported most frequently in that habitat. For instance, atlasers reported 736 Priority blocks with at least one species in TSG (shortgrass prairie), and they located Western Meadowlarks in 651 of those blocks. The tables list only the most frequently reported species for the particular habitats.

GRASSLANDS

Grasslands typically cover broad stretches of flat or gently rolling landscape. The simple structure of the vegetation provides relatively few opportunities for nesting birds. Consequently, the breeding bird fauna in grassland ecosystems is less diverse than in others.

SHORTGRASS PRAIRIE ECOSYSTEM

TSG: Shortgrass Prairie
736 blocks
Includes Atlas codes **TSG, TMGP, TMXP, TTS**

# of Blocks	Species
651	Western Meadowlark
582	Horned Lark
550	Lark Bunting
381	Lark Sparrow
305	Grasshopper Sparrow
278	Mourning Dove
278	Cassin's Sparrow
240	Killdeer
225	Common Nighthawk
214	Burrowing Owl
211	Swainson's Hawk

The shortgrass prairie once covered most of the Great Plains. Since European settlement, farming, grazing, and urbanization have greatly altered the vegetation and character of the shortgrass prairie. Consequently, many bird species associated with this system have exhibited population declines and reduced distributions during this century. Comanche National Grassland, in southeastern Colorado, and Pawnee National Grassland, in the northeast, preserve some of the best remnants of native shortgrass prairie.

The harsh, desiccating climate on the plains shapes the character of the native vegetation. Two grasses, buffalo grass and blue grama, combine to form a short, thick turf typical of native shortgrass prairie. On drier sites, the turf is broken by frequent patches of bare ground and prickly pear or, in southeastern Colorado, cholla cactus. Alternatively, taller native grasses, such as needle-and-thread, little bluestem, western wheatgrass, and side-oats grama, or exotic species such as cheatgrass and crested wheatgrass, may accompany or replace buffalo grass and blue grama, depending on minor differences in local moisture, soil characteristics, and degree of disturbance. In sandy areas, yucca and sandsage dominate, together with red three-awn, sand dropseed, sand bluestem, and other grasses. Tall grasses, such as big bluestem and switchgrass, occupy localized, moist sites, especially along stream courses and near the foothills. The Atlas adopted codes to describe short, tall, and mixed grasslands, but their application proved difficult. The Atlas classified sandsage as a shrubland (SLS).

Insects and other arthropods are abundant in the prairie during much of the breeding season. But the hot, desiccating summer conditions and the structural simplicity of the shortgrass prairie keep avian breeding diversity relatively low.

The numbers of the most abundant breeding species vary with grass species and other site characteristics. The distribution and abundance of prairie bird species also fluctuate greatly with yearly variation in precipitation and plant species composition. Colonies of black-tailed prairie dogs provide centers of bird activity in the shortgrass prairie. Horned Larks, Mountain Plovers, and Burrowing Owls are especially linked to sites with short, sparse vegetation and patches of bare ground in and around prairie dog towns. Abundant mice, voles, ground squirrels, and jackrabbits in the shortgrass prairie attract several species of raptors during the breeding season. The distribution of these birds within the prairie is linked to topography and vegetation as well as prey availability. For example, large bluffs provide nesting and perching sites for Golden Eagles and Prairie Falcons, and scattered trees or large shrubs provide the same for Swainson's Hawks during summer. Ferruginous Hawks may use either bluffs or trees but may also nest or perch on slightly elevated ground that provides some view of the surrounding prairie.

As human activities continue to encroach on the shortgrass prairie ecosystem, several species merit close monitoring: BBS data suggest alarming declines by at least half the breeding birds on the shortgrass prairie. (See Species Accounts for Mountain Plover, Long-billed Curlew, Loggerhead Shrike, Cassin's Sparrow, the longspurs, and even the Colorado state bird, the Lark Bunting.)

MOUNTAIN GRASSLAND ECOSYSTEM

TMG: Mountain Grassland
373 blocks

# of Blocks	Species
140	Vesper Sparrow
109	Mountain Bluebird
97	Western Meadowlark
63	Horned Lark
62	Brewer's Blackbird

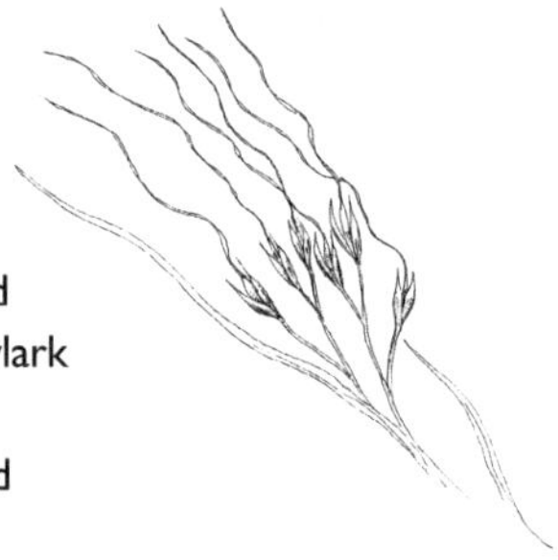

Similar in many respects to the shortgrass prairie, the mountain grassland ecosystem occurs chiefly in the mountain parks and San Luis Valley. Mountain grasslands have varying climates, but always cooler and moister regimes than the shortgrass prairie. Plant composition varies greatly, but typical grasses include Junegrass, mountain muhly, and needle-and-thread. Thurber fescue grows commonly at higher elevations. In optimum conditions numerous plants intermingle with the grasses to form luxuriant growth.

Shrubs may include shrubby cinquefoil, greasewood, and golden rabbitbrush. Insects abound in warm months, and the dense ground cover provides nesting sites for ground nesters. Species that breed in adjacent forested ecosystems but forage in the open areas enhance bird diversity.

ALPINE TUNDRA

ALPINE TUNDRA ECOSYSTEM

TAL: Alpine Tundra
119 blocks

# of Blocks	Species
111	American Pipit
61	White-tailed Ptarmigan
58	Horned Lark
49	Brown-capped Rosy-Finch

Sitting atop the Southern Rocky Mountains is the treeless alpine tundra ecosystem. Structurally like a grassland but actually composed of a rich mix of diminutive forbs, low-growing grasses, and dwarf shrubs, tundra supports only a handful of bird species. Rugged topography further characterizes this most scenic of Colorado ecosystems. Each patch of tundra exists as a virtual island; vast tracts of lower-elevation ecosystems separate patches from each other. Winds here often exceed 100 mph (160 kph), and the mean annual temperature is below freezing. The harsh conditions and short growing season offer an inhospitable habitat for most birds. Atlasers tended to lump all above-timberline habitats—true tundra, cliffs, scree slopes, talus piles, and rocky peak tops—into the tundra habitat.

SHRUBLANDS

Short (usually less than 10 feet; 3 m), woody vegetation dominates shrubland ecosystems. Some shrublands occupy vast, contiguous areas, but often they occur in small patches among grasslands and dry woodlands. Structurally less complex than forests and woodlands but more complex than grasslands, shrublands typically support an intermediate diversity of breeding bird species.

SEMIDESERT SHRUBLAND ECOSYSTEM

Semidesert Shrubland
168 blocks
Includes Atlas codes **STD, SMT**

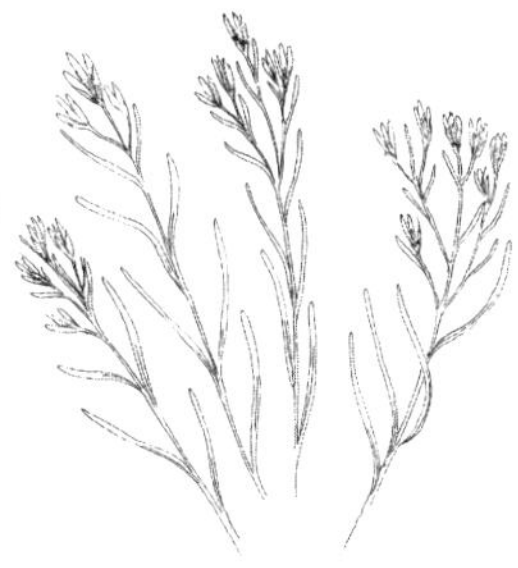

# of Blocks	Species
61	Western Meadowlark
55	Horned Lark
54	Brewer's Sparrow
54	Lark Sparrow
49	Sage Thrasher
39	Mourning Dove
39	Vesper Sparrow

The semidesert shrubland ecosystem covers large parts of the Colorado Plateau, mostly between 4,000 and 6,500 feet (1220–1980 m), although in the San Luis Valley the system spreads from 7,500 to 8,000 feet (2290–2440 m). Species composition varies with soil characteristics; prominent species include greasewood, rabbitbrush, sagebrush, shadscale, and four-winged saltbush. Greasewood typically dominates in areas with alkaline soils. Taller shrubs often grow along watercourses, with low shrubs and desert grasses covering the uplands. Semidesert shrublands have lower avian diversity and density than other shrublands.

SAGEBRUSH SHRUBLAND ECOSYSTEM

SGBR: Sagebrush Shrubland
574 blocks
Includes Atlas codes **SLS, SMS**

# of Blocks	Species
294	Brewer's Sparrow
269	Vesper Sparrow
212	Green-tailed Towhee
163	Western Meadowlark
147	Lark Sparrow
136	Sage Thrasher
126	Mourning Dove

The sagebrush shrubland ecosystem occurs primarily in the Colorado Plateau, mountain parks (especially North Park and Middle Park), and portions of the San Luis and Gunnison valleys. Scattered patches exist in the foothills along the Eastern Slope and large tracts cover prairie rangeland. Sagebrush spans a large elevation range—from 4,000 to 9,500 feet (1220–2900 m) elevation. Big sagebrush predominates in some areas (e.g., Gunnison Basin, North Park) and mountain sagebrush in others; sand sagebrush covers tracts on the plains, from the sandhills of Yuma County south to the Comanche National Grassland.

Scattered junipers, a wide variety of grasses, and other shrubs may intermingle with the two primary sagebrush species. This ecosystem attracts slightly more bird diversity and density than the semidesert.

SCRUB OAK ECOSYSTEM

SOS: Scrub Oak
326 blocks

# of Blocks	Species
150	Rufous-sided Towhee
141	Virginia's Warbler
138	Green-tailed Towhee
98	Dusky Flycatcher
87	Western Scrub-Jay
83	Blue-gray Gnatcatcher
83	Black-headed Grosbeak
78	Orange-crowned Warbler

FOOTHILLS SHRUBLAND ECOSYSTEM

SMM: Foothills Shrublands
414 blocks

# of Blocks	Species
246	Green-tailed Towhee
129	Broad-tailed Hummingbird
97	Rufous-sided Towhee
97	MacGillivray's Warbler
77	Virginia's Warbler
75	Dusky Flycatcher
75	Blue-gray Gnatcatcher

The foothills shrubland ecosystem commands hillsides between 5,000 and 8,500 feet (1525–2590 m) and occurs most expansively in western and southern Colorado. Atlasers subdivided it into two types: scrub oak and mountain shrublands.

Scrub (Gambel) oaks typically form a patchwork of dense thickets, sometimes intermingled with other shrubs, and in southern Colorado they grow within ponderosa pine woodlands. On the Eastern Slope, scrub oaks grow north only to the South Platte River. They grow to taller heights than other shrubs (up to 25 feet, 8 m), almost tree-like in some places. The Colorado state insect, the Colorado Hairstreak butterfly, emerges in midsummer for one to two weeks to mate in scrub oak thickets.

Structurally similar but lower in height, foothills shrublands consist of mixtures of mountain mahogany, skunkbrush, chokecherry, squaw currant, bitterbrush, buckbrush, serviceberry, scrub oak, and a variety of other less frequent shrubs. A tree-sized mountain mahogany grows on the western portion of the Colorado Plateau and on the south-facing escarpment of Cold Springs Mountain, in the state's far northwestern corner.

Scrub oak supports lower bird species diversity than the more varied foothills shrublands, and both exceed that of semidesert and sagebrush.

WILLOW CARR ECOSYSTEM

Willow Carrs
520 blocks
Includes Atlas codes **SMC, SBC**

# of Blocks	Species
325	Lincoln's Sparrow
262	Wilson's Warbler
260	White-crowned Sparrow
231	Song Sparrow
182	Yellow Warbler
155	MacGillivray's Warbler
136	Broad-tailed Hummingbird
134	Dusky Flycatcher
122	Fox Sparrow

The willow carr ecosystem, especially well developed in broad, glaciated valleys, occupies streamsides and other wetlands from 8,000 to 12,000 feet (2440–3660 m). It features a relatively low plant species diversity—mostly various willow species plus shrubby cinquefoil and bog birch—and a short growing season. Willow thickets provide protected nest sites, and insects are fairly plentiful for a limited time. Avian species diversity is relatively low but density can reach very high levels (Hallock 1984a).

FORESTS AND WOODLANDS—CONIFEROUS

The ecosystems described under this heading may vary widely from one another in geographic location, climate, dominant tree species, foraging opportunities, and even growth form. However, they offer relatively complex structures. Thus, avian species diversity in forests and woodlands typically exceeds that of other ecosystems. Trees provide opportunities for many birds that require cavities for nesting or elevated sites for singing, perching, or nest placement.

Conifers cover far more of Colorado than do deciduous trees. The relatively few conifer species form five significant coniferous ecosystems, largely separated by elevation—pinyon/juniper, ponderosa pine, Douglas-fir, lodepole pine, and spruce/fir—plus a limited one, bristlecone pine.

PINYON/JUNIPER WOODLAND ECOSYSTEM

WPJ: Pinyon/juniper Woodland
470 blocks

# of Blocks	Species
241	Pinyon Jay
241	Chipping Sparrow
232	Mourning Dove
231	Mountain Bluebird
230	Blue-gray Gnatcatcher
226	Western Scrub-Jay
219	Plain Titmouse
218	Common Nighthawk
213	Ash-throated Flycatcher
184	Black-throated Gray Warbler
178	Northern (Red-shafted) Flicker
173	Mountain Chickadee
171	Gray Flycatcher
161	Bushtit
160	Bewick's Wren
160	Spotted Towhee
153	Red-tailed Hawk

Occupying warm, dry sites on mountain slopes, mesas, and plateaus, the pinyon/juniper woodland ecosystem features two characteristic trees: pinyon pines and one of three species of junipers: one-seed juniper in southern Colorado, Utah juniper on the Western Slope, and, above 6,500 feet (1980 m), Rocky Mountain juniper. At the upper edges only pinyon pines grow, and at lower edges the pines thin out in favor of junipers. The most extensive stands lie between 5,000 and 7,000 feet (1525–2135 m) on the Colorado Plateau. These woodlands also fringe the San Luis Valley and cover the foothills from Colorado Springs and Buena Vista south. Patches extend east and south of Walsenburg into the canyonlands of

Huerfano and Las Animas counties.

Trees in this "pygmy forest" are short (usually less than 20 feet; 6 m) and widely spaced. A sparse ground cover may include blue grama, Indian ricegrass, Junegrass, and paintbrush and a few other flowering plants. Semidesert, sagebrush, and foothills shrublands may mix into pinyon/juniper woodlands, depending on site characteristics and disturbance.

This ecosystem supports several breeding bird species essentially restricted to it (at least in Colorado): Gray Flycatcher, Pinyon Jay, Juniper Titmouse, Bushtit, and Black-throated Gray Warbler (Sedgwick 1987). Gray Vireos and Scott's Orioles breed only in the low-elevation stands of pure junipers.

PONDEROSA PINE FOREST ECOSYSTEM

WPP: Ponderosa Pine Woodland

284 blocks

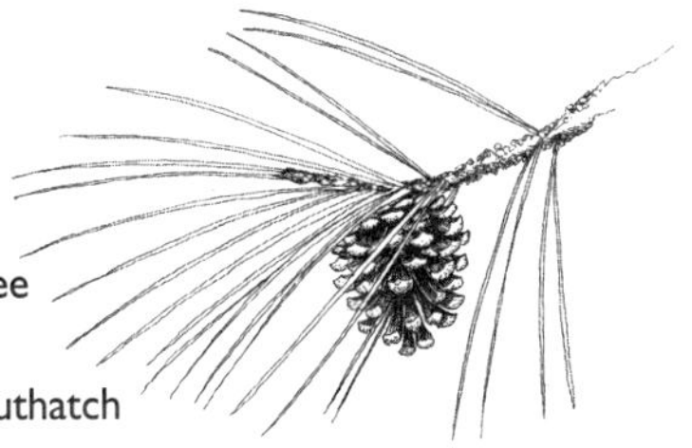

# of Blocks	Species
164	Pygmy Nuthatch
139	Western Tanager
130	Steller's Jay
120	Mountain Chickadee
113	Chipping Sparrow
112	White-breasted Nuthatch
107	Western Wood-Pewee
106	Northern (Red-shafted) Flicker

The ponderosa pine forest ecosystem grows above the pinyon/juniper ecosystem in southern Colorado and, along the Front Range in northern Colorado, above the foothills shrubland ecosystem. Ponderosa pine forests occur primarily between 5,500 and 8,500 feet (1675–2600 m), although South Park has stands at 10,000 feet (3048 m). This ecosystem supports more bird species than any other coniferous forest ecosystem in Colorado. Pygmy Nuthatches, Western Bluebirds, and Grace's Warblers breed almost exclusively in ponderosa pines.

Plant species and general structural characteristics of ponderosa pine forests vary greatly with elevation, slope exposure, soil characteristics, age, and disturbance. Large, broad-crowned, widely spaced trees and luxuriant ground-cover characterize old-growth ponderosa pine forests. To occur naturally, old growth needs frequent, low-intensity fires (Benedict 1991). Due to extensive logging and fire suppression through much of this century, few of these parklike, old-growth stands remain in Colorado. Some extensive, mature stands remain in the foothills west of Denver and in the rolling hills between Pagosa Springs and Durango (Mutel and Emerick 1992).

UPLAND CONIFER FORESTS

(All types combined)

622 blocks

Includes Atlas codes **FUC, FUCD, FUCL, FUCM, WMO, WMOB, WMOD, WMOL, WMOM**

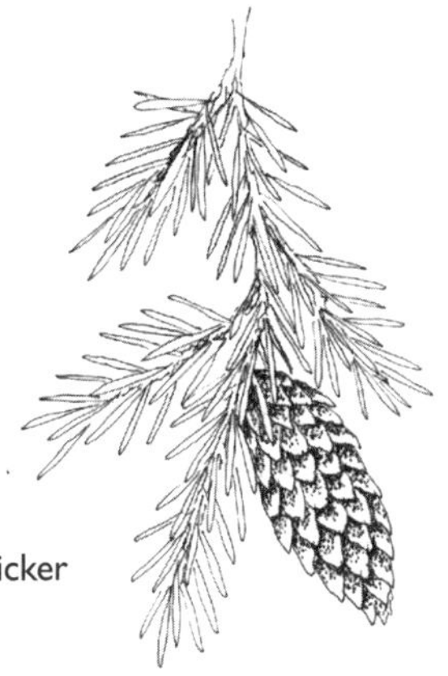

# of Blocks	Species
476	Ruby-crowned Kinglet
474	Hermit Thrush
469	Mountain Chickadee
451	Gray-headed Junco
433	Yellow-rumped Warbler
425	Pine Siskin
373	Red-breasted Nuthatch
364	Steller's Jay
343	American Robin
338	Clark's Nutcracker
320	Gray Jay
309	Northern (Red-shafted) Flicker
290	Cassin's Finch
277	Townsend's Solitaire
267	Olive-sided Flycatcher
267	Red Crossbill
261	Pine Grosbeak
245	Western Tanager
227	Common Raven
219	Broad-tailed Hummingbird
217	Brown Creeper
213	Hairy Woodpecker
211	Red-tailed Hawk
206	Chipping Sparrow

Atlasers used codes to distinguish between forests and woodlands (based on canopy cover), but this subjective distinction proved an enigmatic concept. The Upland Conifer forests table lists the most frequently observed species in all the Upland Conifer forests (atlasers often lumped spruce/fir, lodgepole pine, Douglas-fir, and mixed conifer forests and woodlands together). The following tables present the data for specific conifer systems.

DOUGLAS-FIR FOREST ECOSYSTEM

FUCD: Douglas-fir Forest

7 blocks

Includes Atlas codes **FUCD, WMOD**

# of Blocks	Species
32	Steller's Jay
30	Red-breasted Nuthatch
27	Mountain Chickadee
24	Hermit Thrush
24	Western Tanager
23	Pine Siskin
21	Townsend's Solitaire

Sharing roughly the same elevation with ponderosa pine forests, the Douglas-fir forest ecosystem occupies relatively cool, moist sites, especially on shady, north-facing slopes. Douglas-fir forests grow much more densely than ponderosa pine forests, and little sunlight penetrates the foliage to the forest floor. Few mature, single-species stands of Douglas-fir forest remain in Colorado today, due primarily to extensive logging, but these stands were never abundant. Current stands usually include significant numbers of white fir (southern Colorado), ponderosa pine, Colorado blue spruce, lodgepole pine, or other conifers. Douglas-firs in the Pacific Northwest may tower well over 200 feet (61 m), but the short growing season and comparatively arid conditions limit tree height in Colorado to under 100 feet (30 m). Shrubs, herbs, and grasses are absent or sparse. Mosses and lichens grow abundantly in the shadier locations.

Foraging opportunities include plentiful seed-bearing cones and abundant insects; major outbreaks of the western spruce budworm are a key feature of these forests. Older forests offer abundant nest cavities. Atlasers confirmed that the avifauna occupying this ecosystem comprises representatives of wide-ranging species from the adjacent woodlands; none is restricted to Douglas-fir (Winternitz 1976, A&R).

LODGEPOLE PINE FOREST ECOSYSTEM

Lodgepole Pine Forest

64 blocks

Includes Atlas codes **FUCL, WMOL**

# of Blocks	Species
24	Ruby-crowned Kinglet
21	Yellow-rumped Warbler
20	Red Crossbill
19	Pine Siskin
18	Hermit Thrush
17	Mountain Chickadee

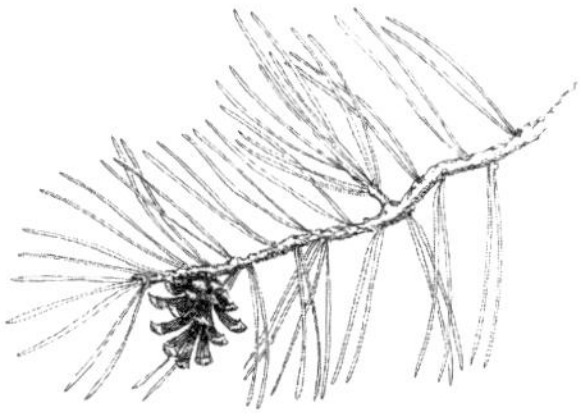

Lodgepole pines grow widely in the Southern Rocky Mountains Province. Although they may intermingle with other conifer species anywhere above 6,500 feet (1982 m), they occur primarily in extensive, closely packed, dense, evenly spaced, and evenly aged stands (especially between 8,000 and 10,000 feet; 2440–3048 m). The vast monocultures of this species represent the lodgepole pine forest ecosystem.

Ecologists sometimes call this ecosystem a "fire forest," because lodgepole pines thrive in the aftermath of fire. Some lodgepole pines produce serotinous cones. Serotinous cones open with extreme heat, releasing a profusion of seeds. These long-lived cones may wait decades for a fire to release their seeds. The seedlings flourish in open, sunny areas. The number of serotinous trees in a stand can be quite variable, perhaps reflecting the fire history of the stand. Stands that colonize an area after a windfall or an insect epidemic generally contain a smaller proportion of serotinous trees than stands that colonize an area after fire (Muir and Lotan 1985). Shrub and herbaceous layers are poorly developed in lodgepole pine forests, and plant species diversity is low.

These forests often are eerily quiet, virtually devoid of birds. The serotinous cones provide formidable challenges for seedeaters. In comparison, stands that contain higher proportions of non-serotinous cones can suffer more intensive infestations by mountain pine beetles and other wood-boring insects. Thus, these stands may provide increased food opportunities for birds. Despite the wide tracts of this monoculture in central and northern Colorado, atlasers applied this code infrequently (probably because of the paucity of birds in this habitat).

SPRUCE/FIR FOREST ECOSYSTEM

Spruce/fir Forest

191 blocks

Includes Atlas codes **FUCS, WMOS**

# of Blocks	Species
130	Ruby-crowned Kinglet
124	Hermit Thrush
113	Mountain Chickadee
111	Yellow-rumped Warbler
103	Pine Siskin
99	Gray-headed Junco
99	Pine Grosbeak
92	Gray Jay
92	Red-breasted Nuthatch

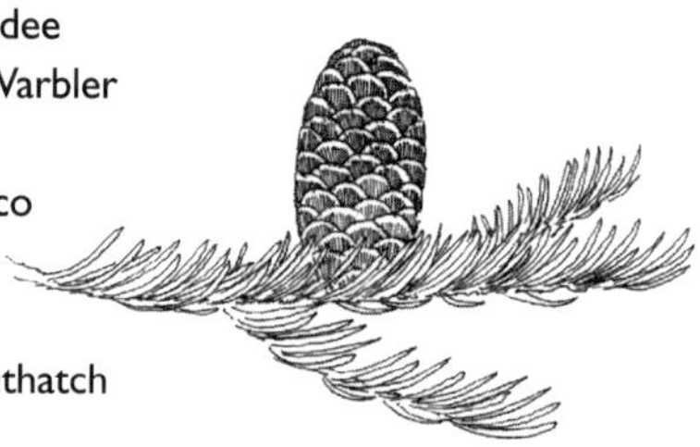

Broad tracts of the Engelmann spruce/subalpine fir forest ecosystem blanket mountainsides from 9,000 to 11,500 feet (2745–3500 m). This ecosystem receives several feet of snow during the winter and remains cool and moist during the summer. The snowpack endures on the shady forest floor well into the summer. The frost-free period is often less than two months. The two dominant tree species (usually with far more spruce than firs) form dense stands with a closed canopy. Where windfalls or stream courses produce gaps in the dense growth, sunlight penetrates to the forest floor and supports lush understory and ground cover, which creates a patchy pattern of undergrowth.

Despite the cold, harsh climate, spruce/fir forests provide a wealth of opportunities for breeding birds. The forests usually have a profusion of snags and cavities. The densely packed conifers produce an abundance of readily available seed-bearing cones, and insects (especially bark-dwellers) are generally abundant. A few species, including Boreal Owl, Three-toed Woodpecker, Gray Jay, and Pine Grosbeak, breed almost exclusively in this ecosystem.

BRISTLECONE PINE ECOSYSTEM

Occupying a few, usually exposed, windy sites with coarse, rocky soils, the bristlecone pine–limber pine woodland ecosystem contains few plant species. The broad-crowned bristlecone and limber pines seldom reach 30 feet (9.1 m) in height. This open woodland occurs from foothills to timberline, most frequently above 8,500 feet (2590 m). Because of the sparse understory, severe local climate, and scarce resources, few breeding birds occupy this ecosystem in the few places where it occurs. Atlasers coded this ecosystem for only 12 species in three blocks.

Not all Colorado forests fit into neat ecosystem classifications. Aspen often intermingle with conifers and various species of conifers grow together, creating woodlands and forests with diverse characteristics and diverse species of birds. For example, north of Pagosa Springs at 8,000–9,000 feet (2440–2745 m), a combination of aspen, ponderosa pine, Douglas-fir, white fir, and Engelmann spruce interposes a distinctly different forest between the ponderosa pine woodlands and the spruce/fir and aspen forests. Scattered limber pines grow in association with all these systems. Foothills riparian systems often include both deciduous and coniferous trees. Small patches of conifers intermingle with large tracts of aspen and vice versa, and by natural succession, spruce and fir gradually supplant lodgepole monocultures and aspen woodlands.

MIXED CONIFER FORESTS

SKR: Krummholz
57 blocks

# of Blocks	Species
35	White-crowned Sparrow
11	Mountain Bluebird
10	Clark's Nutcracker

The ecotone marking the transition between the Engelmann spruce–subalpine fir ecosystems and alpine tundra is termed "krummholz" (i.e., crooked wood). These bands of tortured, dwarf conifers sprawl at timberline between 11,000 and 12,000 feet (3500–3660 m). The windy, harsh climate, sparse vegetation, and limited food supply provide little support for breeding birds, but White-crowned Sparrows commonly breed here. Other species, notably White-tailed Ptarmigan, Clark's Nutcracker, and American Pipit, common in the adjacent ecosystems, use krummholz for foraging and shelter.

FORESTS AND WOODLANDS—DECIDUOUS

LOWLAND RIPARIAN ECOSYSTEM

Lowland Riparian
536 blocks
Includes Atlas codes **LRD, SLC, TTR**

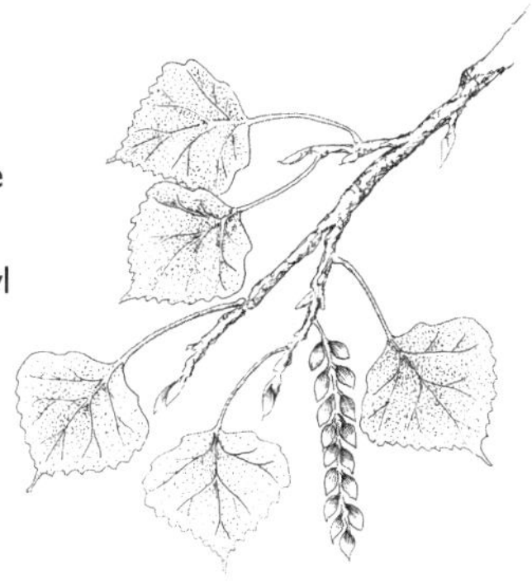

# of Blocks	Species
281	Bullock's Oriole
225	Yellow Warbler
207	Western Kingbird
197	Black-billed Magpie
196	Northern Flicker
186	Great Horned Owl
183	Mourning Dove
177	House Wren
177	American Robin
159	Eastern Kingbird
157	American Kestrel
146	Blue Grosbeak
146	Western Wood-Pewee

The rivers that drain the mountains braid green ribbons through the eastern plains and western deserts—the lowland riparian forest ecosystem. Plains cottonwoods line the plains rivers and Fremont cottonwoods follow the watercourses on the Western Slope and in the San Luis Valley. Humans have altered this ecosystem, although ecologists debate about how much. Exotic species, such as Russian-olive along the South Platte and tamarisks on the other river systems, mix with the native cottonwoods and usurp the position of native willows and other shrubs.

Thickets of various shrubs, usually including sandbar willows, intervene to vary the uniformity of the cottonwood woodlands. In some draws near the Kansas and Nebraska state lines these consist of skunkbrush interspersed with hackberry trees; near the Front Range foothills thickets of wild plums, hackberries, hawthorns, chokecherries, boxelders, and willows line the streamsides. In southeastern Colorado, New Mexico locusts mix with cottonwoods to line some stream courses such as the Purgatoire River. On the Western Slope willows mix with red-osier dogwood, buffaloberry, greasewood, and other shrubs, and often the alien tamarisk forms a monoculture, there and on the lower Arkansas River.

This system provides dispersal corridors for woodland birds across otherwise treeless terrain. The lowland riparian ecosystem typically supports many more species of native birds than surrounding grassland or shrubland ecosystems (Bottorff 1974, Knopf 1985). Indeed, it contains the richest avian diversity of any ecosystem in Colorado and the highest densities, except perhaps for marshes (Bottorff et al. 1971–1984). Riparian systems occupy only a small proportion of Colorado's land area (about 3%); therefore they house only a small proportion of the total number of breeding birds in the state. From a global perspective, Colorado lowland riparian habitat does not host a significant population of any of these species.

FOOTHILLS RIPARIAN FOREST ECOSYSTEM

FRD: Foothills Riparian Deciduous Forest

498 blocks

# of Blocks	Species
204	Yellow Warbler
188	American Robin
158	Northern (Red-shafted) Flicker
158	House Wren
155	Warbling Vireo
155	Song Sparrow
142	Western Wood-Pewee
127	Broad-tailed Hummingbird

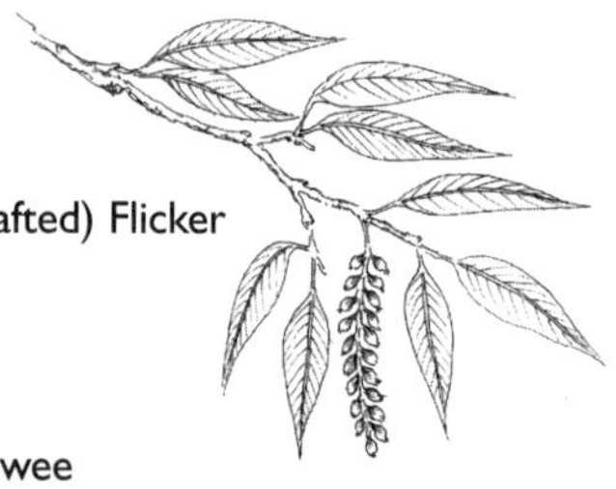

FRC: Foothills Riparian Conifer Forest

94 blocks

# of Blocks	Species
66	Cordilleran Flycatcher
35	Broad-tailed Hummingbird
29	Ruby-crowned Kinglet
29	American Robin
27	Golden-crowned Kinglet
26	Swainson's Thrush
23	Mountain Chickadee
23	Yellow-rumped Warbler
22	Western Tanager

The foothills riparian forest ecosystem traces watercourses from the foothills into the mountains (5,500–10,000 feet; 1675–3048 m). Narrowleaf cottonwood replaces the plains cottonwood of the lowlands, and peach-leaved willow, boxelder, mountain alder, and river birch form deciduous systems, whereas Colorado blue spruce and other conifers create coniferous riparian zones. The understory is characteristically lush with a wide variety of shrubs (e.g., chokecherry, red-osier dogwood, shrubby cinquefoil) and a high diversity of herbaceous plants.

ASPEN FOREST ECOSYSTEM

Aspen Forest

533 blocks

# of Blocks	Species
396	Warbling Vireo
335	House Wren
302	Red-naped Sapsucker
297	Northern (Red-shafted) Flicker
262	Tree Swallow
260	Western Wood-Pewee
228	Violet-green Swallow
228	American Robin
189	Mountain Bluebird
182	Yellow-rumped Warbler
170	Gray-headed Junco

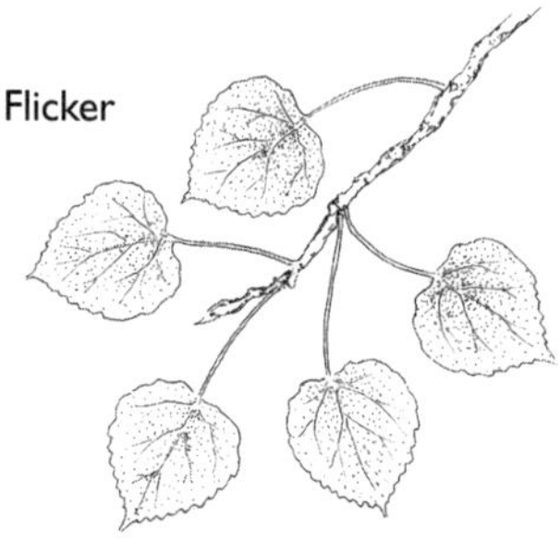

The only widespread deciduous species that grows in the Southern Rocky Mountains Province, quaking aspen occur throughout Colorado from 5,600 feet to tree line. Large, luxuriant groves, especially between 8,000 feet and 10,000 feet (2440–3048 m) compose the aspen forest ecosystem. Typically, sites supporting well-developed aspen groves have deeper, less rocky soils than those sites dominated by conifer woodlands (Mutel and Emerick 1992). Aspens can tolerate a wide variety of soil and local climate conditions as long as they do not suffer prolonged periods of high temperatures or drought. This ecosystem typically contains a profuse, diverse understory of shrubs, grasses, and herbaceous plants, including the Colorado columbine. Foliage-dwelling insects are plentiful in the aspen forest ecosystem, and the structure provides openings for insectivores that feed on the wing. The thick ground cover provides opportunities for ground nesters, and older forest stands, depending on their condition, provide many cavities. Consequently, aspen forests typically support higher avian diversity than the surrounding conifer-dominated forests. Only Red-naped Sapsuckers are virtually restricted to breeding here (Flack 1976, A&R, Hejl et al. 1995) although Warbling Vireos offer the quintessential birdsong of the aspen groves.

AQUATIC

Aquatic environments, both natural and manmade, support large numbers of breeding bird species and individuals. Food and cover generally abound in these habitats, and they often provide islands of sanctuary in the midst of otherwise inhospitable regions. The size and depth of a wetland, as well as its geographic location, elevation, and the characteristics of adjacent habitats, help determine the composition of its avifauna. Birds may locate nests in vegetation or bare substrate found on islands or along shorelines of these aquatic ecosystems.

OPEN WATER ECOSYSTEM

OWL: Open Water, Lakes

490 blocks

# of Blocks	Species
322	Mallard
128	Green-winged Teal
114	Spotted Sandpiper

OWS: Open Water, Streams

467 blocks

# of Blocks	Species
179	American Dipper
179	Spotted Sandpiper
142	Mallard
83	Belted Kingfisher
58	Killdeer
53	Common Merganser

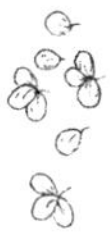

The open water ecosystem, which occurs statewide, consists of streams and rivers, natural lakes and beaver ponds, and reservoirs, along with associated islands and shorelines. Specific characteristics vary greatly with degree of openness, elevation, and surrounding vegetation. Only one species uses this ecosystem exclusively, the American Dipper.

EMERGENT WETLAND ECOSYSTEM

Emergent Wetlands

706 blocks

Includes Atlas codes **AEM, ASG, ASU**

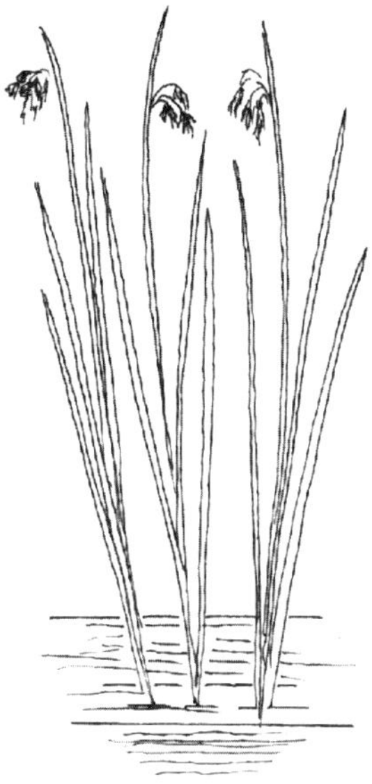

# of Blocks	Species
	Waterbirds
256	Mallard
81	Green-winged Teal
68	Cinnamon Teal
63	American Coot
62	Blue-winged Teal
	Shorebirds
196	Killdeer
161	Common Snipe
94	Sora
87	Spotted Sandpiper
50	Wilson's Phalarope
43	American Avocet
40	Virginia Rail
	Landbirds
478	Red-winged Blackbird
123	Yellow-headed Blackbird
100	Common Yellowthroat
65	Song Sparrow
56	Brewer's Blackbird
37	Savannah Sparrow
29	Northern Harrier

The emergent wetland ecosystem includes a wide variety of wet meadows and marshes. Specific vegetation and characteristics vary, especially with soil chemistry and duration of water inundation. Dominant emergent vegetation usually includes cattails, bulrushes, and various sedges and rushes. The usually dense vegetation provides cover for nesting birds, and typically hosts a plentiful supply of insects, other invertebrates, and small fish. A large number of species uses this ecosystem, most of which use this and open water exclusively, e.g., grebes, ducks, rails, and terns. The density of breeding birds in these marshes frequently exceeds that in other habitats.

HUMAN DOMAIN

Humans have effected major alterations upon native vegetation and other landscape components by, among other activities, introducing exotic species and manmade structures. Human uses overpower the original nature of many landscapes; these uses control the ecosystems, which now exist to service human demands.

URBAN/RURAL ECOSYSTEM

Urban Areas
162 blocks
Includes Atlas codes **URB, UIN, UPK**

# of Blocks	Species
90	House Sparrow
70	House Finch
70	American Robin
62	European Starling
50	Common Grackle
37	Rock Dove
30	Western Kingbird

Rural Areas
1007 blocks
Includes Atlas codes **RRL, CWD**

# of Blocks	Species
667	Barn Swallow
544	House Sparrow
487	Western Kingbird
443	European Starling
427	American Robin
391	Common Grackle
389	Mourning Dove
331	Say's Phoebe
325	Bullock's Oriole
239	Black-billed Magpie
238	Rock Dove
233	Loggerhead Shrike

Unlike any ecosystem discussed thus far, the urban/rural ecosystem is rapidly expanding. (When atlasers surveyed the Louisville Atlas block [39105H2] in 1989, about 200 people lived in it; 20,000 people lived there in 1998 [Bill Kaempfer, pers. comm.[1]].) The urban/rural ecosystem includes major cities, small towns, and suburban residential areas, and farms and ranches surrounded by croplands, riparian systems, or forests. The structural diversity of human dwellings and other buildings found in the urban/rural ecosystem defies coherent description here. Vegetation is equally diverse, including remnant patches of native flora and a mind-boggling array of ornamental trees, shrubs, grasses, and flowering plants, most of them non-native. Any Atlas block with human habitation contained a sample of this ecosystem, and that included 65% of all the blocks.

Serious challenges to breeding birds include chemical pesticides, frequent human activity, and high densities of predators such as Blue Jays, Common Grackles, American Crows, fox squirrels, and domestic and feral cats. Despite the structural diversity and abundant foraging opportunities offered by urban and rural areas, bird species diversity is generally low. Presumably, few species can successfully meet the challenges of constant human activity.

1. Atlasers surveyed the Central East sector of the Louisville quadrangle because of denial of access to both the South East and South West sectors.

CROPLAND ECOSYSTEM

Croplands
652 blocks
Includes Atlas codes **CPL, TSU**

# of Blocks	Species
358	Western Meadowlark
274	Killdeer
269	Horned Lark
224	Red-winged Blackbird
206	Ring-necked Pheasant
163	Lark Bunting
161	Mourning Dove

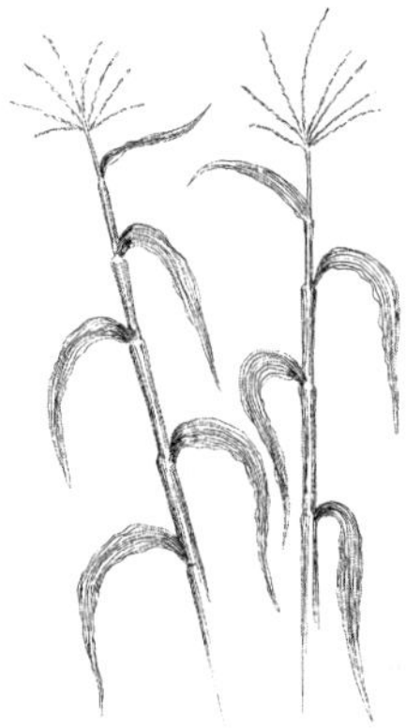

The cropland ecosystem includes irrigated and non-irrigated croplands across the Great Plains and in the San Luis Valley and western valleys. Winter wheat is the main non-irrigated crop; alfalfa, corn, and sugar beets are the main irrigated crops. Apples, peaches, and cherries are important orchard crops, especially in the western valleys. Ranchers have converted many mountain valleys into hayfields. Croplands provide limited nesting sites and low bird diversity.

The species that commonly use this ecosystem either occupy less trammeled portions (Western Meadowlark, Horned Lark) or take advantage of the exotic plants for their habitat requirements (Ring-necked Pheasant, Lark Bunting, Red-winged Blackbird). A different contingent of species exploits the mountain hayfields: Sora, Common Snipe, Savannah Sparrow, Red-winged Blackbird, and some ducks.

ECOSYSTEMS OF COLORADO: EPILOGUE

The environment in Colorado has undergone profound changes through the millennia. Great mountains arose, eroded, uplifted, and eroded again. Deposition of marine shales and limestones helped to form soils. Volcanoes and glaciers alternately left their marks on the land. Marine fish, dinosaurs, mastodons, dire wolves, and giant vultures lived in various incarnations of the Colorado landscape. But the rapid changes in flora and fauna that occurred during the past 150 years have little to do with glaciers, volcanoes, or mountain uplifts, or the ebb and flow of natural extinction. They are the products of human behavior: farming, ranching, mining, logging, urbanization, and recreation. As the human population of Colorado grows, and our hunger for fossil fuels, lumber, shopping malls, manicured lawns, ski slopes, golf courses, hiking trails, and other mainstays of modern civilization increases, our native ecosystems will continue to change. Bird species composition and abundance will no doubt reflect these changes. And, for better or worse, subsequent descriptions of Colorado ecosystems and breeding bird populations written 25, 50, or 100 years hence will continue to reflect the values and behavior of our civilization.

POST-SETTLEMENT CHANGES TO COLORADO HABITATS

By Hugh E. Kingery

The Colorado landscape has undergone many visible and subtle changes since settlement began in earnest in 1859. This chapter, not a scholarly examination, offers a brief summary of conspicuous changes in an effort to provide a context for the current habitats and birdlife.

PLAINS

RIVERS

On the ground, the Colorado plains today look like the map: ribbons of trees flanking thin lines of blue, which string through a vast landscape of prairies and croplands. Ecologists debate whether or not the prairie rivers, pre-settlement, had trees, and if so, how extensively and whether or not it really made a difference to the birds.

In historical accounts by early explorers and early settlers, trees along the river elicited specific comments, suggesting that trees grew as exceptions rather than as the rule. For example, Zebulon Pike, headed up the Arkansas River and approaching present-day Lamar in 1806, wrote, "The river banks begin to be entirely covered by woods on both sides, but no other specie than cotton wood" (Hart and Hulbert 1932, 117). That site today bears the name "Big Timbers."

One hypothesis says that, prior to settlement, annual spring floods used to scour the streams and wash away any sprouting cottonwoods. Bison found a food source in the surviving tender young trees and, later, so did cattle (Hamil 1976). The combination left a tree-deprived stream, a shallow waterway with numerous sandbars. The old cliché, "a mile wide and an inch deep," described well the South Platte and Arkansas rivers, along with the Republican, Arikaree, Smoky Hill, and Big Sandy rivers.

On the South Platte, "Old-timers recalled that before there were fences to keep cattle out of the river, only a few scattered trees had been visible. They supported the story with reference to a battered old cottonwood between Iliff and Proctor that everybody called 'Lone Tree.' . . . It was in the annals of the neighborhood that the buffalo had chewed away at young trees and sprouts exactly as the cattle had, and that no white man had ever seen as many trees along the South Platte as there were in the opening decades of the twentieth century" (Hamil 1976).

Cottonwoods reproduce through seeds and root runners. Successful seeding requires the perfect flood—one that occurs at the right time of year (June, when the trees produce their seeds), with flooding that leaves bare, disturbed mudflats, and with slow regression. This combination provides ideal germination conditions for cottonwood seeds. During early settlement, changed conditions promoted cottonwood establishment: hunters wiped out the bison, then settlers built fences. Without hungry bison and cattle, cottonwoods seeding during floods grew without impediments (Hamil 1976). The current cottonwood forests host tree-dwelling orioles, vireos, warblers, flycatchers, and screech-owls, and the continuity of woodlands has allowed a meeting of eastern and western birds.

Some ornithologists think that the Great Plains formed a barrier that separated eastern and western species. Settlement's trees, they postulate, created suitable habitat for 14 pairs of closely related, ecologically similar, geographically separated species to encounter each other. Nine of these non-identical-twin species and subspecies now meet in Colorado: Red- and Yellow-shafted flickers, Steller's and Blue jays, Mountain and Eastern bluebirds, Spotted and Eastern towhees, the two meadowlarks, Bullock's and Baltimore orioles, Black-headed and Rose-breasted grosbeaks, and perhaps Western and Eastern screech-owls. The Atlas shows that the ninth pair of species included in this group, Lazuli and Indigo buntings, share most of Colorado in their range. Atlasers also demonstrated that the bluebird twins are really triplets: Mountain, Eastern, and Western bluebirds nest together in the Black Forest (Elbert, Douglas, and El Paso counties). Except for the flickers, these twins (according to current ornithological opinion) do not display "free interbreeding," so that the AOU maintains them as separate species. The hybrid zones show stable locations that apparently predate settlement and support classifying most species as separate (Rising 1983).

In the twentieth century came upstream dams, built for flood control, urban water supplies, and irrigation. Flow manipulation for these purposes limits the flood events that cottonwoods need to seed. As a result, riparian areas today contain miles of mature cottonwoods but have few younger, replacement trees. As the mature cottonwoods die, river bottoms open up and habitat for woodland birds disappears. Conceivably, flood control could re-create the Great Plains barrier, partially anyway, as a decadent habitat degenerates back to a treeless watercourse, and again isolate these bird species. On the other hand, urbanization produces a countertrend by imposing (urban) forests of mainly eastern trees on former grasslands.

SHORTGRASS PRAIRIE

Settlement introduced two major impacts on the prairie: crops and cattle. Settlers transformed most of the Colorado prairie either into wheat fields and irrigated farmland or into rangeland grazed principally by cattle. Animal control efforts altered interspecies relationships, and reservoirs altered stream flow and created new habitat.

AGRICULTURE: Large sections of the once vast prairie now grow crops. A few birds use the croplands: Red-winged Blackbirds, Ring-necked Pheasants, some Lark Buntings. Settlements connected to crops and grazing—farms with their shelterbelts, small towns—provide another new habitat. Trees surround the farms that dot the plains and provide opportunity for birds that used to find the plains a barrier: orioles, kingbirds, grackles, House Sparrows.

GRAZING: Crops make obvious changes, but grazing produces subtler ones. Grass species change; often non-natives take over and the native species exhibit differing growth patterns. Sometimes the result is a structurally similar but botanically changed landscape.

ANIMAL CONTROL: Settlers actively eliminated various animals they did not like, from "chicken hawks" to grizzly bears. Their actions affected birdlife, directly and indirectly. Prairie dog towns used to stretch across the plains; ranchers have an aversion to prairie dogs, and croplands cover many former colonies. As a result, the prairie dog population plummeted, and with it so have the numbers of associated breeding birds such as Burrowing Owls and Mountain Plovers. Coyotes, despite control programs, thrive throughout the state (probably because, in part, successful "control" programs exterminated larger predators that imposed natural controls on smaller predators [coyotes] and partly because activities connected with settlement provided new habitat for coyote food: mice and rats).

Wildlife advocates engage in affirmative animal control. Ranchers and city bird-lovers put up bird boxes for bluebirds and other small birds; the Forest Service erects boxes for small owls. Birdhouses, bird feeders, and hummingbird spas dot mountain landscapes. The Colorado Division of Wildlife, acting under federal and state endangered species statutes, brought back Peregrine Falcons and Bald Eagles from extirpation in the state. A much longer and very successful Division commitment—to big game animals—abets their impact upon mountain habitats.

RESERVOIRS: Dotting the landscape, especially a 50-mile strip from Denver north at the edge of the foothills, reservoirs have created a new breeding habitat for species of birds that before could not have nested there. Usually off the main stem of the river, strings of irrigation reservoirs follow east along the South Platte and Arkansas rivers. Away from the mountains only three reservoirs dam the actual riverbeds—Pueblo and John Martin reservoirs on the Arkansas and Chatfield on the Platte. Irrigation diversion structures siphon water out of the rivers to reservoirs and crops.

The green ribbons provided highways for expansionist species such as Mississippi Kite, Blue Jay, Common Grackle, and Eastern Screech-Owl. Croplands render prairies unsuitable for many native species (Mountain Plover, Burrowing Owl, Common Nighthawk, longspurs) yet farms create habitat for kingbirds, Barn Swallows, Red-winged Blackbirds, grackles, and House Sparrows. Reservoirs created habitat for water and marsh birds—not only ducks but also grebes, rails, yellowthroats, and blackbirds.

MOUNTAIN FORESTS

Several factors changed the pre-settlement mountain landscape. Early nineteenth-century fashion dispatched mountain men to Colorado: they trapped as many beaver as they could, and eliminated them from many streams. Removal of that mammalian engineer changed the character of riparian areas all across Colorado (Dick Roth pers. comm.). In the twentieth century, ranchers converted willow carrs and many of the remaining beaver ponds into hay meadows. This metamorphosis reduced habitat for many flycatcher, thrush, warbler, and sparrow species. In sites with enough moisture, it also created habitat for Soras and Common Snipes (John Emerick pers. comm.).

Early miners cut down forests for firewood and buildings and to shore up mine tunnels, and man-caused fires burned huge segments of forest (see, e.g., Fiester 1973, 22–23). More than a century later, at many sites close to timberline, new forests have yet to regenerate.

After the miners left their mark, fire suppression became a management mantra. Conventional forestry practices also attempt to eliminate insects and disease from forest ecosystems. These timbering practices tend to create an homogenized forest. Today vast stretches of "dark timber" contrast with pictures of a nineteenth-century hodgepodge of landscape types, which mixed forests, woodlands, meadows, and streamside vegetation (Dick Roth, pers. comm.). Under the more modern and controversial ecosystem management, biologists recognize fire, insects, disease, and the species tied to them as an integral part of forest health (Norm Barrett pers. comm.).

Changes brought on by fire suppression, timber management, and human settlement have significantly altered many of our ecosystems and affect numbers and distribution of many species (Hejl et al. 1995). During the 1990s settlement has imposed a new change to western ranchland birds. Ranchlands transformed into housing developments attract such human commensals as Black-billed Magpies, European Starlings, grackles, Brown-headed Cowbirds, House Finches, and House Sparrows (John Emerick pers. comm.).

GRAZING: Grazing has affected the landscape in the mountains as well as the plains. Cattle and sheep often damage riparian habitats. Overgrazing changes plant associations and deprives streambanks of soil-holding roots; gully erosion results. Above timberline, sheep have changed the vegetation on their allotments. The monotonous tone of heavily grazed sheep allotments contrasts with the lush rainbow colors of tundra flowers protected in Rocky Mountain National Park and Yankee Boy Basin.

ASPEN FORESTS

Contributed by Norman Barrett

Colorado's extensive aspen forests provide the only large expanses of deciduous trees in the state. Aspens guarantee nesting habitat not only for deciduous obligates such as Warbling Vireos and House Wrens, but also for hole-nesting species. With continued fire prevention aspen forests mature and decay, thereby creating additional nest sites. In many areas conifer forests will gradually replace many of these aspen stands through natural succession. Aspen birds such as House Wrens, Warbling Vireos, and Red-naped Sapsuckers will give way to conifer birds like Ruby-crowned Kinglets, Pine Grosbeaks, and crossbills.

However, aspen acreage probably will not decline significantly over the long run for three reasons, two man-caused and one natural. Timber harvest of conifers opens up habitat for aspen regeneration. Fire suppression creates higher fuel levels, raising the possibility of catastrophic fires eventually removing the conifer cover, as happened in Yellowstone. The natural cause, avalanches, opens up narrow swatches within the forest (Chuck Preston pers. comm.), particularly at higher elevations and on steep slopes. Within these small patches the sprouting aspens rarely grow into tall trees, because of poor soil and subsequent avalanches.

An increasingly common forestry practice requires leaving snags and replacement trees inside clear-cut units. As long as suitable nest sites remain, Red-naped Sapsuckers and other woodpeckers will use aspen forests with no reduction in breeding success (Tobalske 1992) and continue to create prospects for secondary cavity-nesters such as owls, bluebirds, and swallows. However, the minimum number of snags required by U.S. Forest Service policy—0.6–2 per acre (1.5–5/ha; NMB)—leaves minimal habitat for a thriving community of snag nesters. Olive-sided Flycatchers use such snags as feeding posts in small logged clearings—2 acres or less (Scott and Crouch 1987)—and even nest in live conifers left within small clear-cuts (Dick Roth pers. comm.), yet BBS data report this species in decline. However, hole nesters such as woodpeckers, small owls, swallows, chickadees, nuthatches, and Brown Creepers will find fewer opportunities in logged areas with few snags, and forest-interior species will not find the necessary habitat continuity for successful nesting.

PINYON/JUNIPER WOODLANDS

Contributed by Ron Lambeth

Originally, throughout the West, pinyon/juniper woodlands (the "pygmy forest") grew mostly on steeper slopes. Sagebrush and grasses grew on flatter ground and deeper soil. Fire kept the trees at bay: on flat ground, grass and brush provided "fuel ladders," but on the steep slopes the brush dropped out—no understory. As the pinyons and junipers advanced onto flat ground, fires would sweep through and burn them up, but fire usually didn't go up the hillsides.

When cattle came in, they cropped off the grass under the sagebrush, and therefore reduced the flammability of the flats. That, plus fire suppression, meant that nothing stopped the advancing trees.

After World War II, ranchers and government agencies instituted chaining to fight off the advance of the pygmy forest. Nowadays chaining, but not fire suppression, has dropped out of favor, so that the pinyons and junipers again enjoy an expanding range. Probably the West now has more pinyon/juniper woodland than it did 150 years ago.

OTHER HUMAN-CAUSED CHANGES

Roads and trails create edge habitats; robins and cowbirds thrive, but most other species avoid these corridors (Miller and Knight 1995). High-country dams flood narrow canyons and destroy habitat for cliff-nesting raptors, swifts, swallows, Canyon Wrens, and American Dippers and riparian passerines.

The rural and urban population booms in the 1900s joined to convert prairie Sharp-tailed Grouse habitat into farmlands, suburban developments, ranchettes, and other habitat types (Braun et al. 1994). Along the Front Range, development means a net loss of meadowlarks and Horned Larks, and a net gain of American Robins, House Finches, and Common Grackles.

Non-native plants also arrived and changed landscapes. Some culprits came by accident, such as knapweed and various thistles that plague many grasslands. Nurseries sold tamarisks and purple loosestrife (and some still sell them), plants that escape from cultivation to infest stream bottoms and wetlands. For planting in plains shelterbelts, wildlife agencies once recommended Russian-olive trees, which, encountering a congenial climate and lacking a native insect constituency, developed into aggressive, invasive, non-beneficial interlopers.

Settlement also brought non-native species, which fall into three categories: those that arrived in North America and Colorado without human help (Cattle Egret); those introduced into the eastern U.S. and which spread to Colorado on their own (European Starling and House Sparrow); and game species (Ring-necked Pheasant, Chukar) that the Division of Wildlife introduced for hunters.

Human settlement and its habitat manipulation increased the abundance of many species, from Barn Swallows to Wood Ducks to Brown-headed Cowbirds, and perhaps attracted other species that, pre-settlement, occurred rarely if at all (Blue Jay, grackles). Some of the interlopers have negative effects on native birds: starlings usurp flicker nest holes and Common Grackles apparently displaced Brewer's Blackbirds from the cityscapes that attracted the blackbirds in the first place. The same changes caused substantial decreases in the numbers of other species such as Mountain Plover, Sharp-tailed Grouse, Peregrine Falcon, and Yellow-billed Cuckoo. Settlement also caused less conspicuous declines by displacing species that specialize in converted habitats: longspurs on the plains, riparian species in the mountains. Even though meadowlarks rank as the state's second most abundant species, the Front Range metropolitan areas now occupy hundreds of square miles where meadowlarks once sang.

The Species Accounts that follow include short discussions about how the changes affected particular species. A definitive discussion would requires a different research design and a bigger book than this one.

Today as we travel across Colorado, most of us can barely imagine what the landscape looked like before settlement recast it. If, in 1600, a squirrel could tree-hop from the Atlantic Ocean to the Mississippi River, could a Burrowing Owl in 1800 have winged across the Colorado plains always in sight of a prairie dog town? In 1860 could a traveler have traversed the prairie on a June morning always within sight and sound of larking longspurs and Lark Buntings? Did park-like ponderosa pine woodlands span the foothills and support healthy populations of Plumbeous Vireos without cowbirds to sink their populations? Could Three-toed Woodpeckers have found, in every spruce/fir forest, insect-rich trees killed by beetles and fires?

Settlement changed the landscape and its inhabitants, but we don't have a baseline from times past. One hundred and fifty years after humans launched this transformation, the Atlas provides the first baseline, a profile of Colorado birdlife over one decade. It provides a starting point for future studies and comparisons.

COLORADO ORNITHOLOGISTS

Atlas volunteers perpetuated a long tradition of an eclectic mix of citizens, explorers, ornithologists, and government employees who contributed to Colorado ornithology. This section reviews, briefly, the principal publications on which our present knowledge relies. A much more thorough review of early Colorado ornithologists appears in B&N. Thompson Marsh, an avid Denver birder, completed his master's thesis at the University of Denver (1931) by documenting the first 60 species of birds reported from the state by the first ornithological explorers.

Appropriately for the Centennial State, the expedition of Dominguez and Escalante reported the first bird from the state in 1776: a grouse, presumably a Blue Grouse, which they shot and ate on 26 August and found "exceedingly palatable" (Marsh 1931).

Zebulon Pike came to Colorado in 1806; he kept a journal that reported on five conspicuous bird species (Wild Turkey, Common Raven, Black-billed Magpie, Canada Goose, and, presumably, Greater Roadrunner). The description of the roadrunner is a little odd (greenish color) and the location a place where they do not occur today (Salida).

The Long expedition of 1820 had two scientists, Edwin James and Thomas Say, who collected and described 23 species. Over the rest of the nineteenth century many people explored the state, mainly along the Front Range, for birds; many were egg collectors. They included Denis Gale (Boulder County), Edwin Carter (Summit, Park, and Grand counties), Charles Aiken (El Paso), Horace Smith (Denver area), and Ed Andrews (Estes Park area).

Wells W. Cooke produced a state compendium and three supplements (1897, 1989, 1900, 1909). In 1912 W.L. Sclater published the 576-page *A History of the Birds of Colorado,* the first book devoted to Colorado birds. William Bergtold published the second one, in 1928.

Several ornithologists studied specific locations, such as H.W. Henshaw (1875) of the Wheeler Survey, who reported on birds in the Denver area and the San Luis Valley. Other ornithologists produced local lists. They include La Plata County (Morrison 1888); Mesa County (Rockwell 1908); El Paso County (Aiken and Warren 1914); three for Boulder County, by Junius Henderson (1908) Norman Betts (1913), and University of Colorado ornithologist Gordon Alexander (1937); and Gunnison County, by Western State College ornithologist Sidney Hyde (1979).

L.C. Keyser (1902) wrote a travel book in which he described birds he observed, particularly in the Colorado Springs area. Two books by Hal Borland, not ornithological tomes, vividly and literately describe growing up as a homesteaders' boy on the Colorado plains south of Fort Morgan and later as a newspaperman's son in Flagler. These books, *High, Wide and Lonesome* (1956) and *Country Editor's Boy* (1970), weave natural history with autobiography.

Birds of Denver and Mountain Parks, by Robert J. Niedrach and Robert Rockwell (1939), drew on extensive field experience and the collections of the Denver Museum of Natural History.

The two-volume *Birds of Colorado,* by Alfred M. Bailey and Robert J. Niedrach (1965), ranks as a classic presentation not only of Colorado birds, but also of birds in any state. It towers as the most authoritative reference to Colorado birds. This Atlas refers to it so frequently that we cite it simply as "B&N."

Bob Niedrach began his career at the museum in 1913. He not only conducted pioneering field work throughout the state, but also inspired many amateur ornithologists over the years (including the editor of this Atlas). The Colorado Bird Club, now the Denver Field Ornithologists, formed after a class taught by Mr. Niedrach wanted to continue their association.

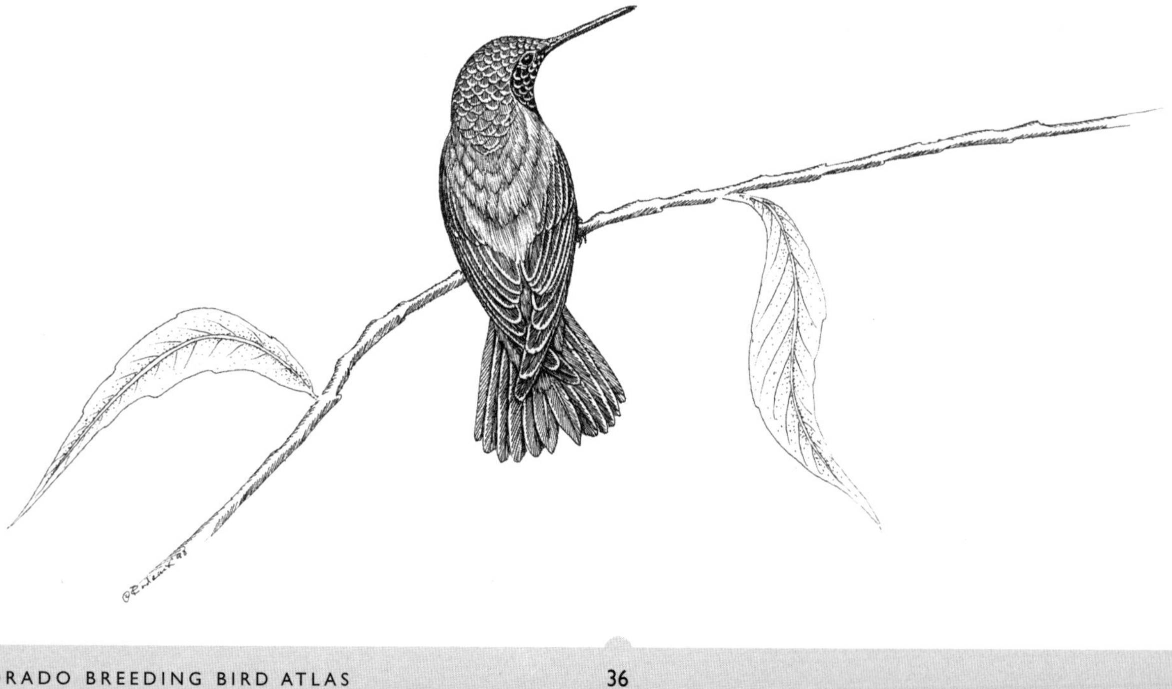

The Denver Field Ornithologists began to publish *Colorado Bird Notes,* when Don Thatcher initiated seasonal notes of observations by bird club members. Harold Holt, who succeeded him, revamped the bird notes (now renamed the *Lark Bunting*) so that it reports all observations on club field trips and additional sightings reported by members and others.

Harold subsequently compiled summaries of arrival and departure dates for Denver-area birds. In 1972 he incorporated these data as bar graphs in the first bird-finding book on Colorado, *A Birder's Guide to Eastern Colorado.*

In 1992 Bob Andrews and Bob Righter published an authoritative book, *Colorado Birds,* an update of Bailey and Niedrach's 1965 classic. They meticulously researched as much material as they could possibly find, including unpublished private field notes of many field and professional ornithologists and CFO records. The book has distribution maps, seasonal and altitudinal graphs, and a brief text describing the status of each species recorded in the state. The Atlas refers to this essential text so often that we cite it as "A&R."

The Colorado Field Ornithologists and Colorado Division of Wildlife published the first Colorado "atlas," a short publication that portrayed Colorado bird distribution by degrees of latitude and longitude—"latilongs." It collected data from as many field and professional ornithologists as would respond and distilled the data into graph-type maps with terse occurrence and habitat codes. The three editions (1978, 1982, and 1988) built on each other; this Atlas refers to them (mainly the 1988 version) as the "Latilong Study."

Since its beginning in 1967, the journal published by the Colorado Field Ornithologists has contained articles about Colorado birds, their habitats, behavior, distribution, and occurrences.

During his 40 years at Colorado State University, Professor Ronald A. Ryder has contributed more ornithological research than any contemporary. Practically every wildlife professional in the state has studied under him or enjoyed a professional association with him. He has encouraged amateur ornithology in Colorado through the Colorado Field Ornithologists, Fort Collins Audubon Society, Denver Field Ornithologists, and every other group that he could find or which sought him, including the Atlas organization. His enthusiasm and quiet support permeate Colorado ornithology in the second half of the twentieth century. The CFO journal devoted its April 1995 issue as a well-deserved tribute to him (Leatherman 1995).

Biologists with the Colorado Division of Wildlife, funded mainly by income from hunters, conduct research on waterfowl, upland game- birds, raptors, and endangered species. Their voluminous publications document the biology and status of these well-studied species. The Division protects and creates habitat for these birds and, as an important by-product, for many other wildlife species.

Citizen scientists have become prominent contributors to Colorado ornithology during the past 40 years. Many Colorado field workers conducted Breeding Bird Censuses and Winter Bird Censuses, sponsored by the National Audubon Society and, later, the Cornell Laboratory of Ornithology. *Audubon Field Notes, American Birds,* and the *Journal of Field Ornithology* published them in turn. These censuses provide data on density and territories of breeding birds in particular habitats. Coloradans have submitted over 50 censuses, covering most habitat types in the state. Starting in 1954, Louise Hering conducted the most protracted, consistent Breeding Bird Census in the state, a 26-year study of a ponderosa pine plot on Enchanted Mesa near Boulder.

Breeding Bird Surveys, sponsored by the U.S. Fish and Wildlife Service and Canadian Wildlife Service, began in Colorado in 1968 under the direction of Ron Ryder. The state originally had 28 routes (one in each latilong). USF&W added a second set of routes in 1987, a third set in 1992, a fourth in 1994, and, in 1998, a fifth set of routes only in the Southern Rockies and Colorado Plateau.

These roadside routes, run once a year over a 24.5-mile course selected randomly, involve 50 stops where the observer counts all the birds seen and heard in three minutes. Collectively these routes contribute data from which scientists derive population trends species by species. Because Colorado had so few routes for the first 20 years, most statewide trend data lack statistical significance, although the plains routes provide fairly dependable data. After a few more years with five routes per latilong, the data should prove worthwhile for mountain and Colorado Plateau species. Atlas Species Accounts discuss BBS data that provide useful trends and information.

Atlasers have made the most recent and the most substantial citizen-science contribution to Colorado ornithology. By combining together to study the entire state, they consummated a unique endeavor.

Citizen science composes an essential part of ornithology. Both intellectual and emotional reasons motivate those who participate. The contributors to the Atlas venture, and the readers of this book, share a sense of wonder about the wild creatures that share the earth with us. The knowledge secured by these citizen-science contributions fuels wildlife preservation today and, thereby, affects wildlife tomorrow. Aldo Leopold (1949) caught the spirit:

> *What one remembers is the invisible hermit thrush*
> *pouring silver chords from impenetrable shadows.*

SPECIES ACCOUNTS

INTRODUCTION

The bulk of the Atlas consists of species accounts for each species found breeding during the Atlas period, 1987–1995. Thirty atlasers wrote these accounts. Each account summarizes Atlas information, discusses the status of the bird in Colorado, and provides information about the breeding biology of the species. Each Species Account has a suite of visual presentations to explain the status of the species as a Colorado breeder.

ILLUSTRATION

Using, as one source, his own observations during eight years as an Atlas field worker, artist Radeaux drew all the illustrations in the book. Each sketch depicts the species in some sort of breeding behavior—with all the Atlas breeding codes represented in his sketches. In a poster at an exhibit of his paintings at the Sangre de Cristo Art Center in Pueblo, he described how he linked his art with atlasing:

> Patterns are a common element in all my art, as they are in all of nature. . . . In 1987 I got involved with the Atlas, initially designed to use only volunteer field workers. In order to complete the survey a number of us had to turn professional and work intensively to finish it. As a paid birdwatcher I accessed areas of southern Colorado that were unknown as far as ornithology is concerned. I climbed mountains. I climbed canyons. I lost weight (I gained it back) and I discovered many things. I found dinosaur bones, irate landowners, cooperative landowners, archeological sites, a hunter's wallet with $300 cash, bears, many other animals, and many, many birds. I found birds that were rare or unknown in southern Colorado. I discovered and documented an entirely new species for Colorado (Acorn Woodpecker). And I survived.
>
> In that final year of the project I went several times into 32 blocks (I went into 50 blocks over the years of the project). The block maps have interesting and varied names. These names are the titles of my paintings [in the exhibit at the Sangre de Cristo Art Center]. The subjects of the paintings are things that are found in these areas, all in southeastern Colorado—and just a fraction of the variety of life that is found here.

MAP

Each block shown on species maps represents a breeding location for that species. On the map, each square depicts a whole topographic map but displays data from one Atlas block.

Maps depict data from both Priority and Non-priority blocks. If the Atlas database includes records from both Priority and Non-priority blocks for a particular topographic map, the Species Map shows the highest code reported for any block in that map. If a species did not occur in the Priority block but did occur in a Non-priority block, the species appears on the map with the breeding code from the Non-priority block. On each map an overlay of latitudes and longitudes defines the latilongs and provides a frame of reference for comparing species distribution with other maps in the book, such as those that show altitudes, habitats, and geographical locations.

Computer specialist Don Schrupp of the Habitat Section of the Colorado Division of Wildlife designed the database and supervised preparation of the maps.

BLOCK STATISTICS

A statistical analysis, based on Priority blocks only, appears above each map. It charts the number and percentage of Priority blocks in which observers reported the species; this appraises the relative distribution of each species. It also shows the number and percentage of each level of confirmation—Possible, Probable, and Confirmed.

The Priority block system ensures that the Atlas project reports a statistically valid sample of the occurrence and distribution of each breeding species in Colorado (except, perhaps, those with few nesting sites, mainly water birds such as pelicans, herons, some ducks, shorebirds with restricted ranges, gulls, and terns).

Block Statistics use only data from Priority blocks. The map shows blocks along the east and west borders (field workers provided data for a few of these blocks). Block Statistics do not include data from these blocks or from Non-priority blocks.

HABITAT USE

Most Species Accounts include charts showing habitats used by the species. The charts include data from both Priority and Non-priority blocks. The number of habitats in the charts does not correspond to the number of blocks from which field workers recorded the species. Observers often reported more than one habitat for a species, and some reported none. The data also include Non-priority blocks. Habitat charts omit infrequently used habitats, and most combine related habitats. The habitat graphs collate 83,737 habitat entries. The color folio of Habitat Photographs shows examples of most Colorado habitats.

NESTING PHENOLOGY

Using the dates provided on the field cards (both Priority and Non-priority blocks), a Phenology chart shows the timing of the breeding cycle for each species. A few species had too few dates for a chart.

TEXT

The text, divided into three parts on Habitat, Breeding, and Distribution, uses Atlas data to discuss each of these subjects, particularly distribution discovered by the Atlas statewide canvas. The authors present basic information about the breeding biology of each species, both to put Atlas results into context and to stimulate the interest of readers in the diverse habits of Colorado's diverse collection of breeding birds. The Distribution sections report on historical accounts of distribution and compare them with the Atlas maps.

Many accounts employ Atlas data to report on geographical centers of abundance and habitat preferences. Computer calculations that combined abundance estimates with block location or habitat codes provided this information. Researchers have already used Atlas data in various ways, from simple block and species lists to abundance/habitat relationships.

TERMS AND ABBREVIATIONS

In the text, a variety of abbreviations and symbols permits presentation of more information in less space. The inside back cover also explains them.

TERMS

Breeding Codes, Habitat Codes, Abundance Codes: The inside back cover has keys to the Breeding, Habitat, and Abundance codes, and Tables 2, 3, and 4 define them.

Priority blocks: Blocks, roughly 3 by 3.5 miles on a side, in the South East sector of each topographic map.

Non-priority blocks: Similarly sized blocks in other sectors of a topographic map.

Block codes: CBAP used the code system applied by the U.S. Geological Survey for topographic maps. This consists of three parts: two numbers define the latitude and three identify the longitude; these specify a "latilong." Each latilong contains 64 maps—eight rows of eight maps. The last two units of the code define the location of the map within the latilong, first a letter (A–H, corresponding to the numbers 1–8) that locates it along the north/south axis, and second a number (1–8) that places it east and west. The numbering system starts from the southeast corner (of the latilong or the state). For example, the Mount Elbert quadrangle, with a code of 39106A4, lies in Latilong 39106, i.e., north of Latitude 39 and west of Longitude 106. "A4" places it as the first map north and fourth west of the southeast corner of the latilong. In the computer listings used in Appendix A, numbers replace the letters (A–H = 1–8). See Figure 1 on page 6.

CERTAIN STANDARD REFERENCES

The text uses shorthand for some seminal references to Colorado birds:

A&R: Andrews, Robert A., and Robert Righter. 1992. *Colorado Birds.* Denver Museum of Natural History, Denver. The most recent standard reference book on Colorado bird distribution; meticulously researched. Maps, elevation charts, seasonal occurrences.

B&N: Bailey, A.M., and R.J. Niedrach. 1965. *Birds of Colorado.* Denver Museum of Natural History, Denver. The basic reference on Colorado birds. Two volumes of descriptions, distribution, records, and discussion.

BBS: 1966–1996. Breeding Bird Survey Trend Analysis. Unpublished data furnished by the Breeding Bird Survey Office, National Biological Service, Patuxent Wildlife Research Center, Laurel, Maryland.

Latilong Study: Kingery, H.E. 1988. *Colorado Bird Distribution Study.* Colorado Field Ornithologists, in cooperation with the Colorado Division of Wildlife, Denver. Maps occurrences, seasonally, by latilong, from 1965 to 1987, derived from contributions by most Colorado birdwatchers.

Priorities 1995: Colorado Bird Observatory and Colorado Division of Wildlife 1995. *Priorities for Bird Conservation and Monitoring in Colorado.* Categorizes bird species, using Partners in Flight criteria, for developing priorities for bird conservation and monitoring.

ABBREVIATIONS

A&R	Andrews and Righter 1992
AOU	American Ornithologists' Union
B&N	Bailey and Niedrach 1965
BBS	Breeding Bird Survey
BLM	U.S. Bureau of Land Management
CBAP	Colorado Bird Atlas Partnership
CDOW	Colorado Division of Wildlife
CFO	Colorado Field Ornithologists
DFO	Denver Field Ornithologists
NWR	National Wildlife Refuge
RBF	CFO Rare Bird form
SRA	State Recreation Area
SWA	State Wildlife Area
USF&WS	U.S. Fish & Wildlife Service
USFS	U.S. Forest Service
VF	CBAP Verification form

Pied-billed Grebes distinguish themselves from other, more gregarious family members by exhibiting secretive and solitary breeding behaviors. Belonging to one of the oldest groups of birds, Pied-billeds combine the submarine abilities of diving ducks with the underwater expertise of loons.

HABITAT

Strictly aquatic like all grebes, Pied-billeds inhabit the fresh waters of sluggish streams, shallow lakes, and ponds at lower elevations. They prefer still waters with abundant vegetation, both emergent and shoreline (Palmer 1976). Three habitat codes describe over 90% of all reported Pied-billed locations: OWL (lakes), AEM (emergent wetlands and marshes), and ASU (submergent floating vegetation). Frequently, a single pair nests on a small pond (B&N) but several will nest in larger bodies of water. They remain on their home marsh to breed and to forage until colder weather and ice force southward migration. Migratory birds more readily use exposed, larger reservoirs that lack rotted vegetation.

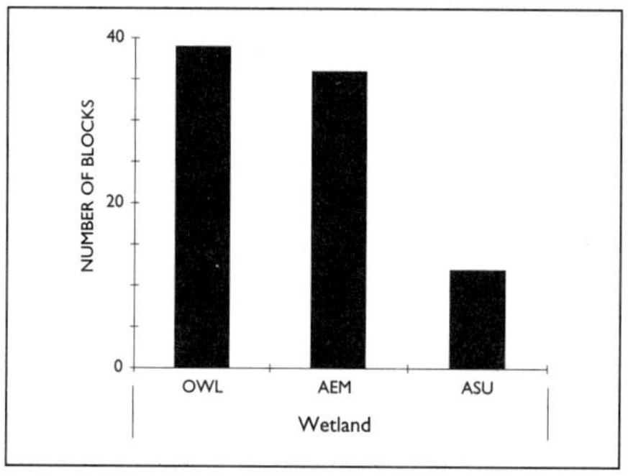

40
20
0
NUMBER OF BLOCKS
OWL
AEM
ASU
Wetland

Despite proficient diving abilities, Pied-billeds tend to forage in the top 10–15 feet (3–4 m) of water in search of slow-moving prey. Opportunistic in their choices, Pied-billeds eat aquatic insects, beetles, dragonfly larvae, crustaceans, mollusks, amphibians, and fish (del Hoyo et al. 1992).

BREEDING

When constructing nests, Pied-billed Grebes build platforms of decaying aquatic vegetation brought up from the bottom of a shallow pond, lake, or marsh. Because of the fresh condition of the green stems of water plants used in nest construction, the nest platform floats with sufficient buoyancy to bear the weight of the nest platform, the usual 4–7 eggs, and the brooding adult. A pair may build more than one platform, but only one becomes the actual nest. Pied-billeds usually nest off-shore in association with emergent plants. The surrounding emergent vegetation anchors the nest, acts as a wave break, and conceals the nest from predators (Forbes and Ankney 1988).

Both males and females develop brood patches and share incubating duties (Forbes and Ankney 1988). The adaptable Pied-billeds will nest two or three times in a season depending on weather conditions (Terres 1980).

Precocial young leave the nest within two days of hatching, either by swimming or by crawling upon the back, clinging to the tail, or clamping their bills over the feathers of a parent.

A time-line of Pied-billeds' breeding activities reveals a wide range of variability. Birds begin arriving in Colorado in late April or early May, with courtship beginning soon after arrival. Their secretive nature and nest-concealment tactics probably contributed to Atlas workers making only two nest-building observations and finding only two nests with eggs. The only observation of a nest with young occurred on a notably late date of 4 September 1995 (Rifle, 39107E7; KMP). Detection of boldly striped fledglings led to two-thirds of all breeding confirmations, the dates ranging over two and one-half months.

BREEDING PHENOLOGY

CODE		# OF RECORDS	DATE RANGE
NB	Nest Building	2	29 Apr–4 Jul
ON	Occupied Nest	5	7 Jun–13 Jul
NE	Nest with Eggs	2	9 Jun–12 Jul
NY	Nest with Young	1	4 Sep
FY	Feeding Young	1	26 Jul
FL	Fledged Young	21	16 May–1 Aug

DISTRIBUTION

Distributed widely throughout the Americas, Pied-billed Grebes rank as the most widespread grebes in North America. Birds breeding in the upper latitudes move south in winter from Canada and the northern U.S. to the southern U.S. and Central America.

In 1897 Colorado ornithologists knew of only one Pied-billed Grebe nest site, near Loveland, although they expected to find more breeding. In southern Colorado, only migrants occurred (Cooke 1897). By 1965 the species had become more abundant on the eastern plains than on the Western Slope (B&N).

Atlas workers found Pied-billeds present statewide in suitable habitat, with equal numbers in the south half and the north half. They mapped Pied-billeds in all but two latilongs. Breeding birds concentrated on plains marshes along the edge of the Front Range, in the San Luis Valley, and in the San Juan Basin. The largest concentration of breeders, in the San Luis Valley, received the only A3 (11 to 100 breeding pairs) abundance codes. The biggest voids on the map come from the plains and from northwestern Colorado, probably because of the paucity of sluggish streams and shallow lakes and ponds.

Information gathered by Atlas workers shows that the upward trend of Pied-billed breeding continues and depicts wider distribution than previously recorded. Due to manmade farm ponds and irrigation reservoirs, Colorado probably has more appropriate Pied-billed habitat than at any time in the past.

BREEDING EVIDENCE

REPORTED IN 63 (4%) OF 1745 PRIORITY BLOCKS

□	Possible	34 blocks	(54%)
▩	Probable	6 blocks	(10%)
■	Confirmed	23 blocks	(36%)

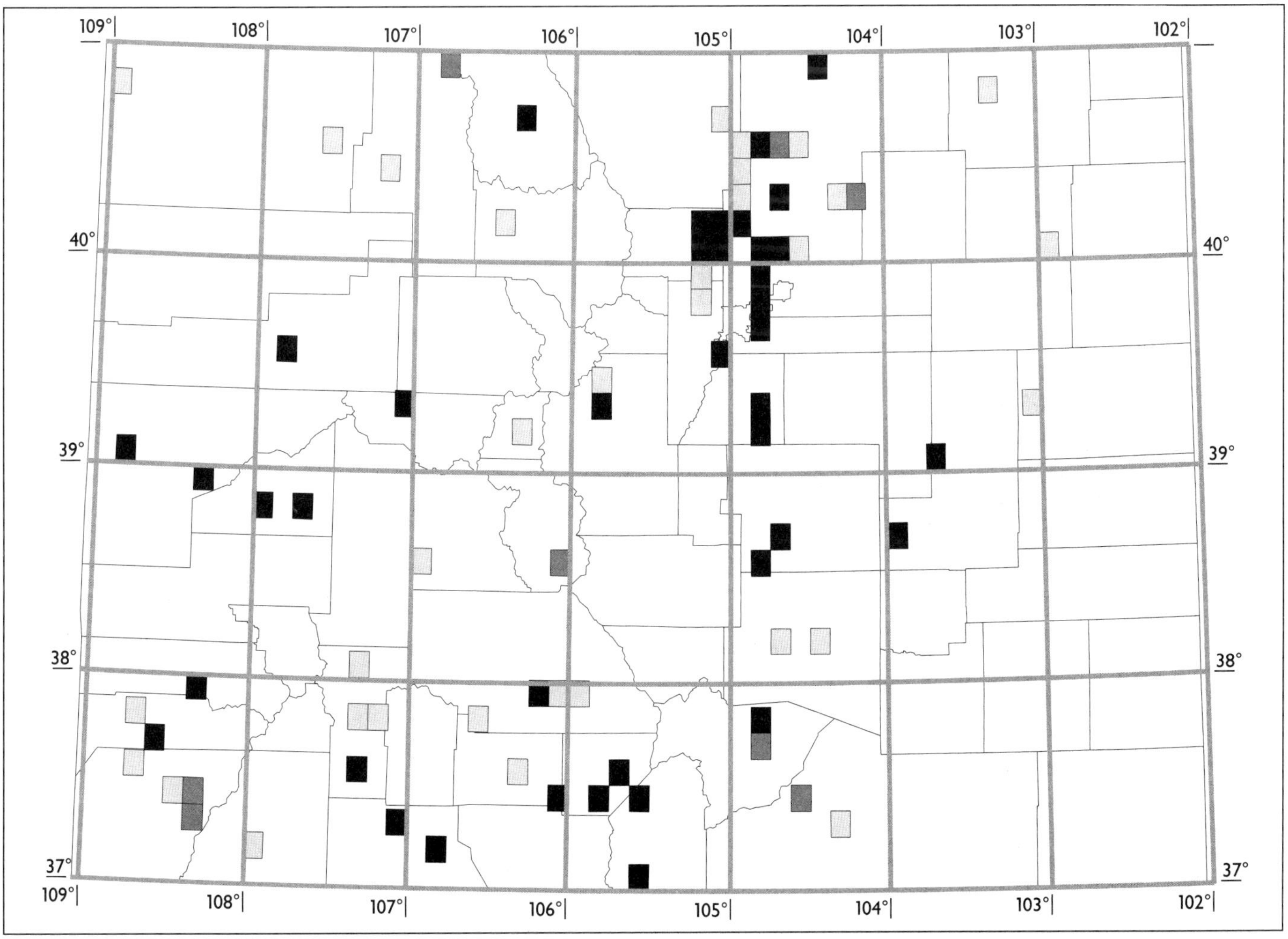
109°
108°
107°
106°
105°
104°
103°
102°
40°
39°
38°
37°
109°
108°
107°
106°
105°
104°
103°
102°
40°
39°
38°
37°

COLORADO DIVISION OF WILDLIFE / WILDLIFE RESOURCE INFORMATION SYSTEM

Breeding in dense colonies, Eared Grebes seem to gain security from the sheer strength of numbers in the ponds and lakes where they nest. In favorable sites, they place their nests as close as one foot apart. Other Colorado grebe species nest in loose colonies or alone; when agitated they slink away from nests to watch intruders from great distances. Only Eared Grebes, when disturbed, slowly swim away, gather to watch the colony from a close vantage until they perceive (sometimes wrongly) that danger has passed, and return quickly to their nests.

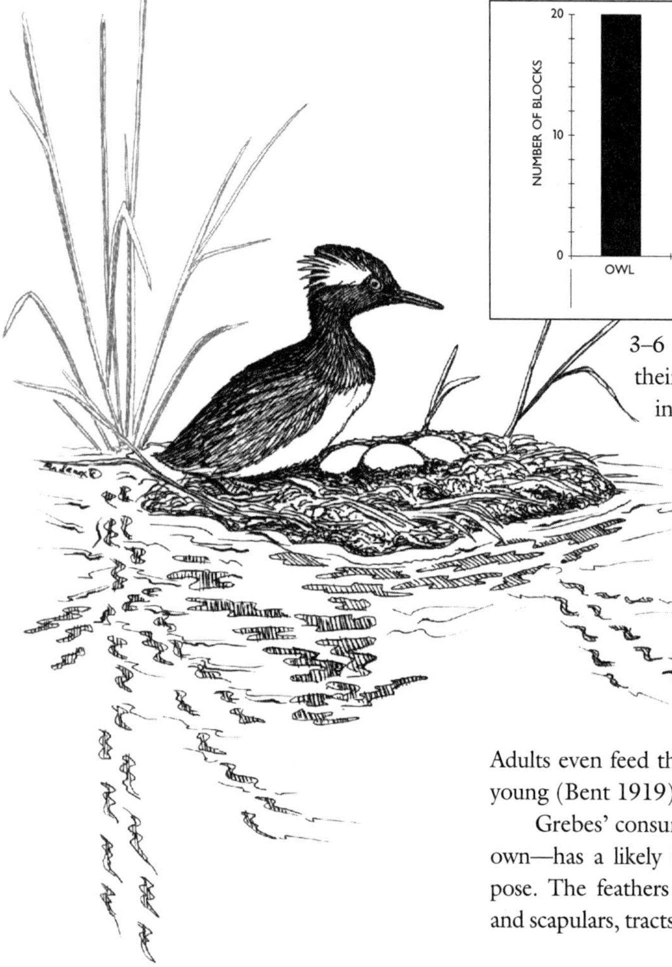

HABITAT

Colonial nesters, Eared Grebes require shallow water with emergent vegetation to which they anchor their floating nests. Emergent wetlands on the margins of intermountain lakes and ponds provide most of the suitable habitat in Colorado, with low-altitude wetlands used less regularly.

Decaying reeds dredged from nearby water provide construction material for the notably sloppy nests. Anchored by live reeds, the large nest cups sway with the wind. The bottom of the nest cup rarely clears the waterline, and the bottoms of the eggs are usually wet (Rockwell *in* Bent 1919). The open emergent vegetation used to anchor the nests typically grows in water 18–24 inches (45–60 cm) deep.

Eared Grebes used only three Atlas habitat types: they situated nests in emergent wetlands (AEM) and floating mats (ASU), and fledglings swam in open water, sometimes far from nests.

Feeding almost exclusively on aquatic organisms, Eared Grebes have little need to leave the water except while incubating their 3–6 egg clutches. Most of their diet consists of aquatic insects and larvae (Bent 1919), although they also will eat brine shrimp, tadpoles, fish, and a small amount of plant matter. Grebes as a group are the only birds that regularly consume feathers as part of their diet. Adults even feed their own feathers to their young (Bent 1919).

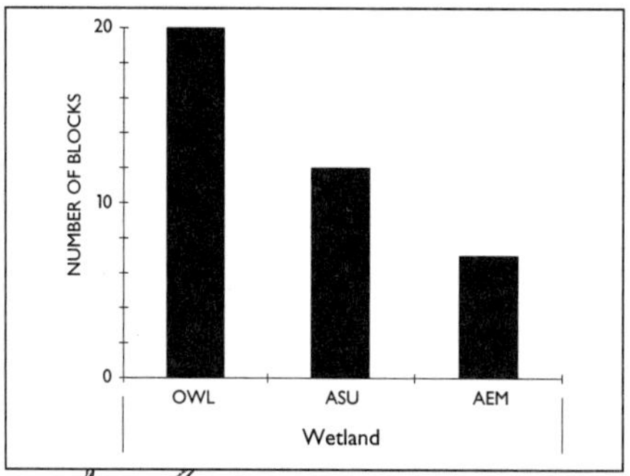
NUMBER OF BLOCKS
20
10
0
OWL
ASU
AEM
Wetland

Grebes' consumption of feathers—their own—has a likely though not proven purpose. The feathers (mainly from the flanks and scapulars, tracts in almost constant molt) form a ball in the stomach. Researchers theorize that the feather ball keeps fish bones from injuring the stomach lining and/or helps form pellets for oral expulsion (Storer and Nuechterlein 1992).

BREEDING

Eared Grebes have water-level dances comparable to those of Western and Clark's grebes, but unlike those grebes they perform only on breeding lakes. The performances include head shaking, a "cat attitude" posture addressed to an underwater companion, a "penguin dance" involving a race, and a pivoting display (Palmer 1962). Courtship also involves a noisy chorus of mellow *poo-eep* sounds; in a large colony the chorus rivals a "first-class frog pond" (Dawson 1909).

Nest-building is an active affair. During construction, the nests sometimes collapse when the birds climb onto them; birds may change nests and abandon old ones (including an egg or two) up to six times before finally settling down to one nest. Incubation begins with the first egg so that the young hatch a day or more apart from each other. The first-hatched youngster broods on the back of a parent, and the free parent (the non-brooder) feeds the young as it rides. The parents split the brood when the chicks become a week old and thereafter shun, or even act antagonistically to, each other. At this stage the chicks, immediately after a parent feeds them, flee from the parent (Palmer 1962).

Wetland habitats used by nesting colonies do not allow easy viewer access. The occupied nest (ON) code, when used for this usually non-secretive species, implies breeding activity too distant to determine nest contents.

Most Confirmations involved dependent young getting fed by adults, hitching rides on their parents' backs, or swimming near adults. Observations of fledglings seldom occurred near nest sites.

BREEDING PHENOLOGY

CODE		# OF RECORDS	DATE RANGE
NB	Nest Building	1	13 May
ON	Occupied Nest	5	1 Jun–7 Aug
NE	Nest with Eggs	3	9 Jul–20 Jul
NY	Nest with Young	1	9 Jul
FY	Feeding Young	1	9 Jul
FL	Fledged Young	14	15 Jun–17 Aug

DISTRIBUTION

Eared Grebes nest in the interior of western North America from Canada's prairie provinces to Minnesota and Iowa, and south to Baja California and Texas. They winter along the coast from British Columbia to northern South America. A few winter in the interior U.S., sometimes on large reservoirs in Colorado (A&R). Eared Grebes also nest in Eurasia, Africa, and Colombia.

Historic nesting in Colorado occurred in the San Luis Valley and Front Range piedmont. Unstable water levels at nesting reservoir sites eliminated nesting colonies at places like Barr Lake and Yuma County (B&N). By 1992 nesting had expanded to Middle Park, Browns Park, and east-central Colorado (A&R).

Atlasers found that these grebes continue to nest in their previously known colonies in North Park and the San Luis Valley. Away from these two locations, breeders scatter across wet areas at low elevations on the eastern plains and western valleys. Even as the San Luis Valley holds the most breeders, atlasers assigned A3 codes to colonies in Hinsdale, San Miguel, Gunnison, Park, Weld, and Moffat counties. The Atlas map substantially tracks the A&R map of "summer residents," a term that may include nonbreeders. The Atlas shows fewer grebes along the Arkansas and South Platte rivers (three blocks each), none in Middle Park, and only one block in South Park. During the Atlas, North Park as a whole apparently had fewer than 100 nesting pairs; Lake John, where 50–100 pairs formerly nested (HEK), rated only an A2 abundance code. However, by 1998 Walden Reservoir had more than 400 nests (Rich Levad pers. comm.).

Historically, Eared Grebe numbers dropped when trappers collected their breast feathers for the turn-of-the-century millinery trade in New York. Their pickled eggs were also sold on the open markets of that time (Bent 1919). Populations rebounded as persecution stopped.

BREEDING EVIDENCE

REPORTED IN 19 (1%) OF 1745 PRIORITY BLOCKS

Possible	4 blocks	(21%)
Probable	3 blocks	(16%)
Confirmed	12 blocks	(63%)

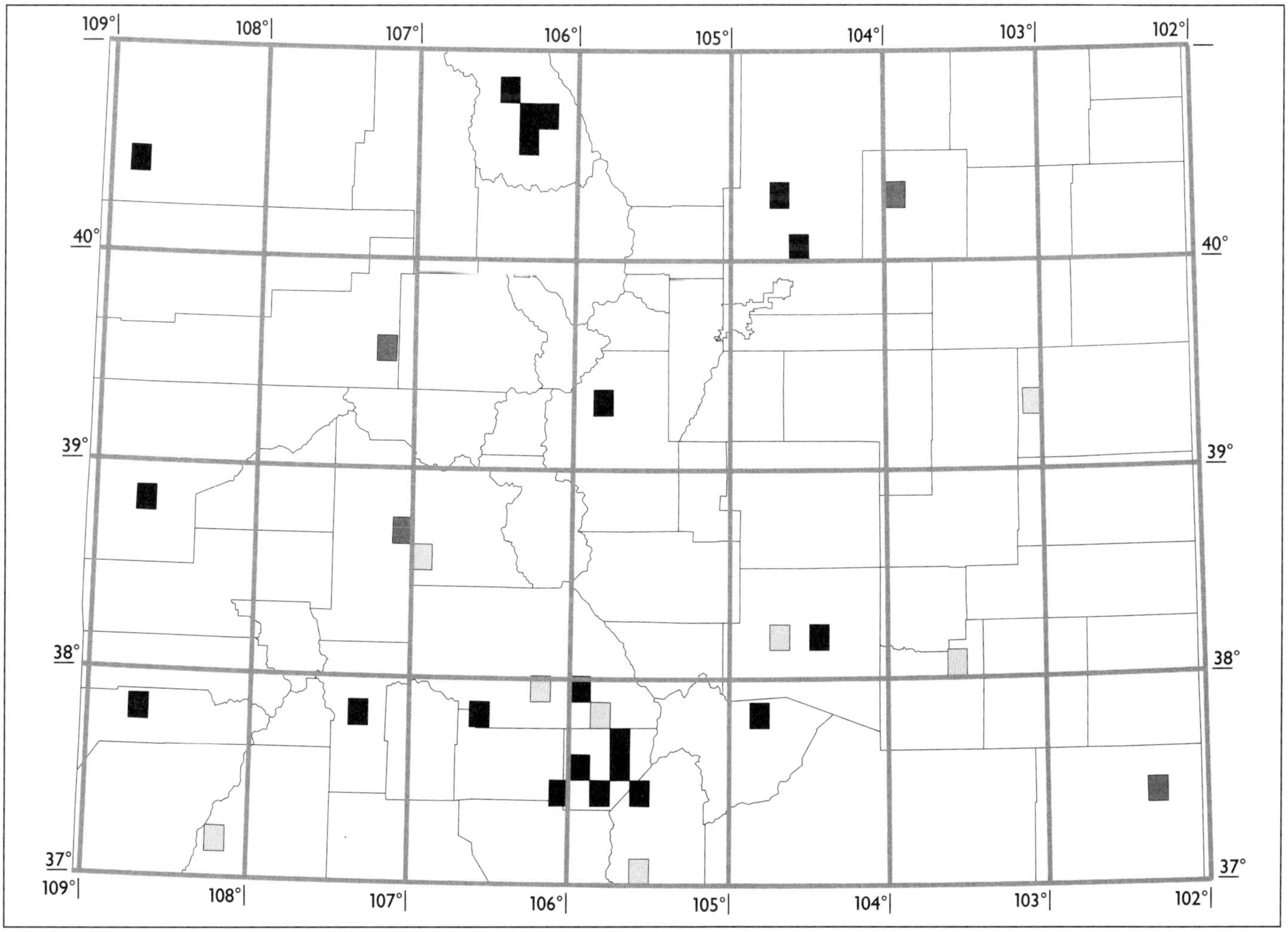
109°
108°
107°
106°
105°
104°
103°
102°
40°
39°
38°
37°

COLORADO DIVISION OF WILDLIFE / WILDLIFE RESOURCE INFORMATION SYSTEM

The synchronized courtship display of Western Grebes stands out as one of nature's true marvels. From a side-by-side swimming position, necks uniformly outstretched, weeds draped in their bills, both male and female spring to their rear-set, webbed feet and excitedly tap-dance, tiptoe fashion, across a section of water.

HABITAT

All grebes have their legs and feet attached at the rear of the body. This architectural adaptation, though perfect for aquatic habitats, makes locomotion on land awkward; therefore, they favor habitats with extensive open water—several square kilometers—bordered by flooded vegetation such as cattails, smartweed, or willows. The lakes where atlasers found grebes conform to the requirement of marshes linked to large bodies of water—from Barr Lake to the San Luis Valley. All Colorado breeding sites except Russell and San Luis lakes in the San Luis Valley are reservoirs or natural lakes with dam-enhanced water levels. All have several acres of open water and marshy shorelines.

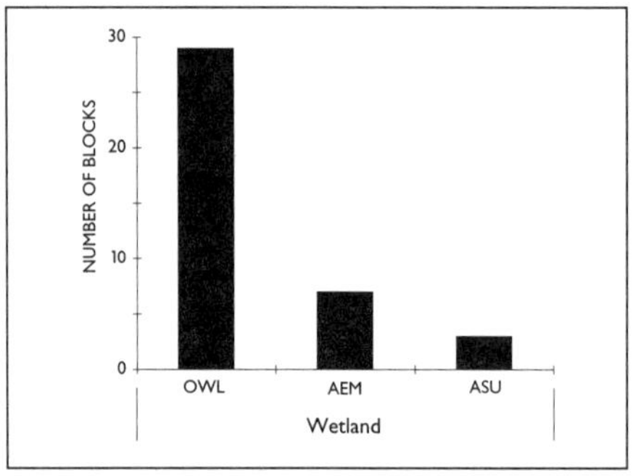
30
20
10
0
NUMBER OF BLOCKS
OWL
AEM
ASU
Wetland

Shortly after Western Grebes arrive on their breeding grounds, their flight muscles atrophy. They cannot fly until a complete molt in late summer (Storer and Nuechterlein 1992). This means that they must select large, food-rich lakes in which to breed and to spend the summer.

A unique characteristic of *Aechmophorus* grebes, including the Western Grebe, is that the neck, like a heron's, can spring forward, so that the grebe can spear prey with its bill while the bird is under water. Western Grebes consume a variety of protein-rich prey, but fish remain the dinner of choice.

BREEDING

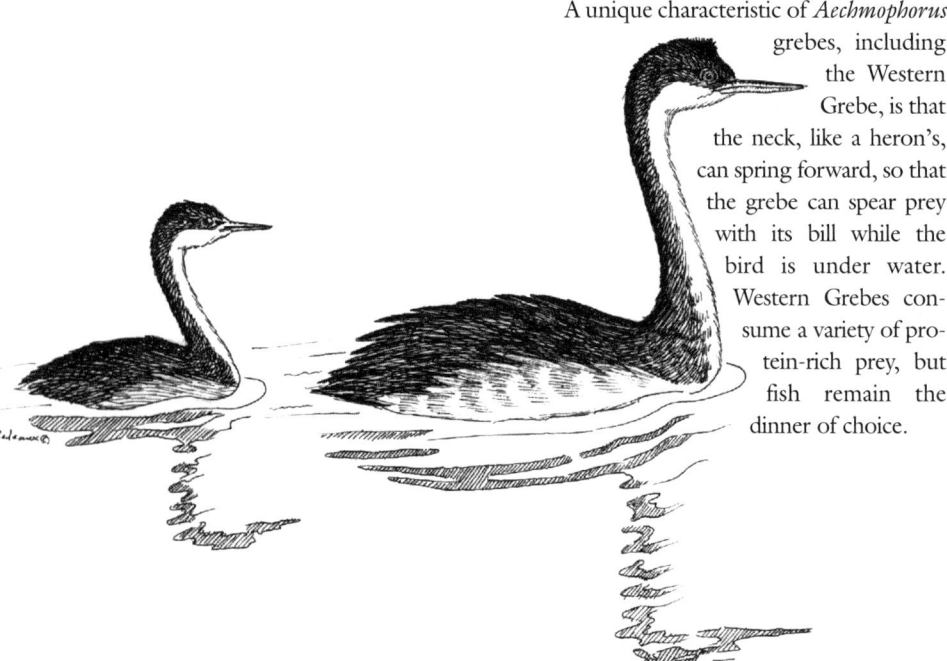

Breeding occurs in Colorado only in a few places and succeeds only in years when the right water levels continue throughout the breeding season. If water levels rise too high, breeding fails because the vegetation used for nests becomes scarce. If water levels drop, nest failures also occur: available food may decline, and increased, wind-caused, wave action washes out nests. Low water levels create land bridges that predators use (Storer and Nuechterlein 1992). Reservoir drawdowns for irrigation pose major problems. Many grebes spend the summer but do not breed because of fluctuating water levels. In addition, Western Grebes usually do not breed until they are two years old.

Together the pair constructs a nest of dry and sodden vegetation. Most frequently they use bulrushes, *phragmites,* and cattails. They build a solid mound, typically in floating, emergent vegetation, with a depression for the eggs. They may build up from the bottom of the lake, attach the nest to a snag, or float it anchored to emergent or floating plants. They choose anchor plants not by plant species but according to water depth; they require water more than 10 inches (25 cm) and less than 10 feet (3 m) deep (Storer and Nuechterlein 1992).

Both sexes incubate, for 24 days. Within minutes of hatching, the downy young scramble onto a parent's back for transportation to water, where brooding takes place. At the end of its shift, the brooding parent rises up with flapping wings so that the young tumble onto the water's surface like plastic bathtub ducks. The alternate parent then assists the young onto its back by holding out a foot as a ramp (Storer and Nuechterlein 1992). On the rare occasion when they nest on land, parents move the young to water by carrying the nestlings under their wings (Palmer 1962).

Parents feed fledglings for eight weeks; the diet parallels that of adults—mainly fish—but smaller items and a greater proportion of insect larvae. When the parent with food approaches its young, a bare patch on the youngster's head turns red (Storer and Nuechterlein 1992). In 1908 Frank Chapman reported finding 238 and 331 adult feathers in the stomachs of two young less than three days old (see Eared Grebe account).

Atlasers Confirmed breeding from 23 June on, later than the early May date cited by B&N, to 12 August.

BREEDING PHENOLOGY

CODE		# OF RECORDS	DATE RANGE
C	Courtship	2	29 Apr–9 Jun
ON	Occupied Nest	1	2 Jul
NE	Nest with Eggs	4	23 Jun–20 Jul
NY	Nest with Young	1	30 Jul
FY	Feeding Young	3	9 Jul–12 Aug
FL	Fledged Young	5	10 Jul–1 Aug

DISTRIBUTION

As the name suggests, Western Grebes breed principally in western North America, south into central Mexico. Virtually the entire population winters along the Pacific Coast, which suggests an inland, basically east–west, migration pattern (Johnsgard 1987). Presumably, Colorado grebes follow this pattern.

The movement west by European settlers started the conversion of shortgrass prairie to croplands. Because the Colorado plains receive an average annual rainfall of only 10–20 inches (25–50 cm), the settlers augmented their water supply with water impoundments—which greatly benefited Western Grebes. In the early 1900s Western Grebes occurred in Colorado only as rare fall migrants (Sclater 1912). Niedrach and Rockwell (1939) reported "never seeing more than four in any year at Barr Lake until 1937 when suddenly 52 were recorded." The first documentation of breeding came in 1940 at Trites Lake in the San Luis Valley (Bailey and Brandenberg 1941). By 1992 A&R reported 17 breeding locations.

Atlasers found Western Grebes in 35 quads, encompassing the western valleys, mountain parks, and eastern plains, although in only 19 Priority blocks. Because several breeding sites lie outside Priority blocks, the map does not reflect all current breeding sites (e.g., Lake Meredith and Browns Park).

Although Non-priority blocks had 12 of the 15 Confirmations, atlasers discovered breeding in nine new reservoirs, making a total of about 25 sites. During the Atlas, Western Grebes started breeding near Pagosa Springs, the first San Juan Basin site (Bob Frye pers. comm.). Atlasers also found new breeding at three San Luis Valley and two Arkansas Valley lakes, two near Denver, and one in North Park. Total breeders, derived from Atlas abundance codes, number less than 1,000 pairs for this species in Appendix B.

Increasing human population and shifting agricultural practices, both in Colorado and downstream states, will alter water management patterns. This will no doubt modify the status and distribution of these grebes within the state; whether the changes benefit them or not depends upon the nature of the changes.

BREEDING EVIDENCE

REPORTED IN 19 (1%) OF 1745 PRIORITY BLOCKS

Possible	11 blocks	(58%)
Probable	5 blocks	(26%)
Confirmed	3 blocks	(16%)

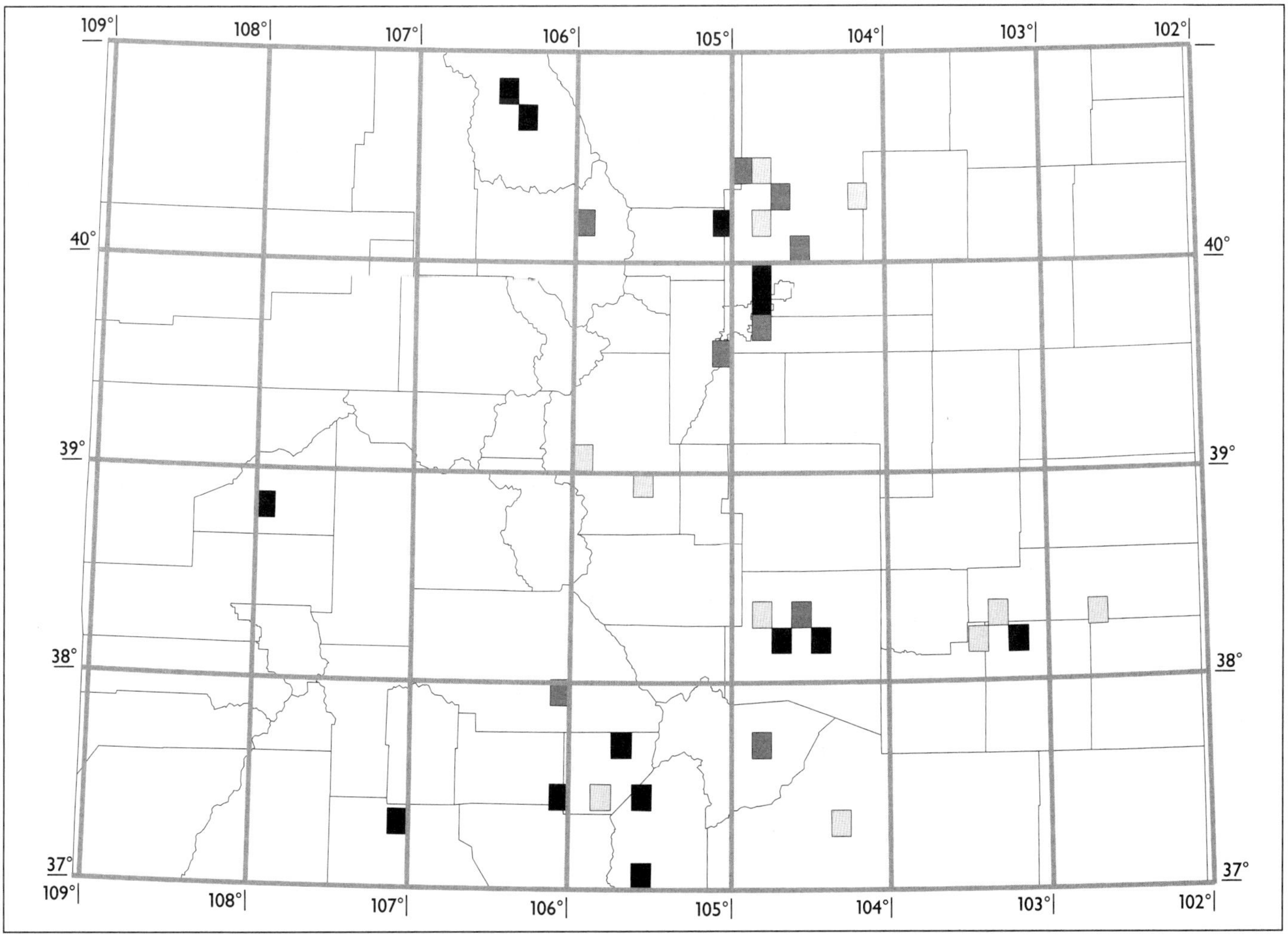
109°
108°
107°
106°
105°
104°
103°
102°
40°
39°
38°
37°

COLORADO DIVISION OF WILDLIFE / WILDLIFE RESOURCE INFORMATION SYSTEM

CLARK'S GREBE

Aechmophorus clarkii

In 1985 a very important event occurred in the life history of the Clark's Grebe: the AOU elevated it to full species status. Previously, it existed as just a humble color morph of the Western. Clark's Grebes possess a unique call, mate only (or mainly) with other Clark's Grebes, possess subtly different morphological features, and display a distinct DNA profile (Storer and Nuechterlein 1992). Combined, this evidence sufficed to promote it up the biological ladder to its own identity.

HABITAT

Because for so long the odd head pattern on some populations of Western Grebes was accepted as just a color morph rather than a feature associated with a subspecies or species, ornithologists apparently had little incentive to investigate the aberrant form. Consequently, almost no knowledge exists about the biological differences between these two similar grebes. The opportunity to discover completely new biology about a bird species is rarely encountered in North America these days. Both *Aechmophorus* grebes reside in Colorado during spring, summer, and fall, providing an excellent opportunity to learn more about the individual characteristics of each of these two grebes.

Clark's Grebes apparently have habitat requirements similar to Western Grebes, inasmuch as the two frequent the same marshes and often nest together in the same places. Atlasers found Clark's Grebes, like Westerns, in large lakes with open water for feeding and marshy edges for breeding. The recent AOU monograph on the two species noted only one habitat difference: Clark's may tend to forage farther from shore and in deeper water than do Westerns (Storer and Nuechterlein 1992).

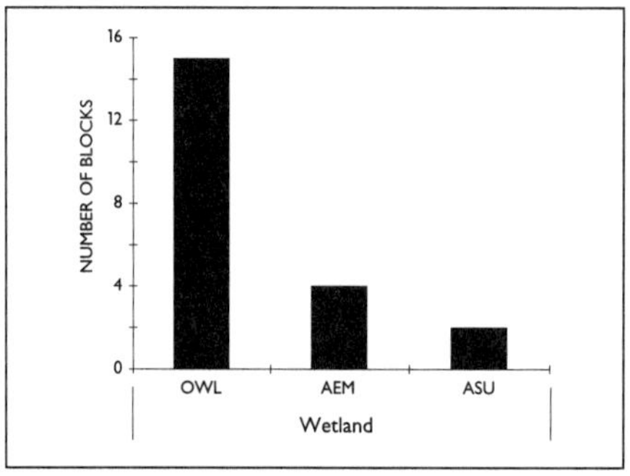
NUMBER OF BLOCKS
16
12
8
4
0
OWL
AEM
ASU
Wetland

John Henry Clark, namesake of this grebe, shares the grebe's main distinction: very little has been recorded of his life history. No one even knows for sure when Clark was born or when he died. Clark belonged to an army of collectors that Spencer Baird of the Smithsonian Institution sent far afield to collect whatever Baird wanted: mammals, insects, reptiles, amphibians, birds, and anything else. Clark, assigned to the U.S. and Mexican Boundary Survey from 1851 to 1855, focused on collecting in the region contiguous with the border. He collected the type specimen of this grebe in the Mexican state of Chihuahua. In 1858 George Lawrence, one of the major authors of the voluminous "Pacific Railroad Survey," first described, named, and gave full species status to this grebe in volume 9 of the survey's report. In 1886 the AOU demoted Clark's Grebe to a mere color morph of the Western Grebe (Means 1992).

Recent genetic analysis has shown that the DNAs of Western and Clark's grebes differ at least as much as do DNA sequences between other closely related species that belong to the same genera. This information, along with the other distinctions mentioned above, led the AOU (1985) to recant and to acknowledge Mr. Clark's grebe.

BREEDING

Clark's and Western grebes have different "advertising calls," which both males and females give. These calls differentiate the two species during courtship and breeding. Westerns have a harsh *cree creet*, quite variable among individuals, whereas Clark's have only a single note, also individually variable. Where the two species nest together, males of one species respond only to calls of their own species. In single species marshes, the males discriminate less consistently between taped calls of the two (Storer and Nuechterlein 1992).

As Colorado had no breeding records of Western Grebes until 1940, Clark's presumably also did not breed here at least until then. Atlas data show only 21 blocks with Clark's Grebes, with seven breeding Confirmations. One possibly informative result is that of the 21 sightings, five (24%) occurred on bodies of water where Western Grebes do not breed. Birds sat on nests at Banner Lakes SWA (Keenesburg, 40104AS) on 6 June, a month before any Atlas dates for Western Grebes; Western Grebes were on the same lakes.

BREEDING PHENOLOGY

CODE		# OF RECORDS	DATE RANGE
C	Courtship	1	5 Aug
ON	Occupied Nest	1	6 Jun
NE	Nest with Eggs	3	23 Jun–19 Jul
NY	Nest with Young	1	30 Jul
FL	Fledged Young	3	14 Jul–30 Jul

DISTRIBUTION

To date no definitive knowledge exists of where, if anywhere, in North America and Mexico Clark's reside and Westerns do not. Consequently, at this time ornithologists consider both species to occupy the same geographical range. In the Mexican states of Michoacan and Guerrero, the dilemma is further exacerbated as 30–33% of the combined population of these two grebes is thought to be intermediate morphs (Feerer 1977).

Clark's Grebes probably arrived in Colorado after Westerns started to breed. At San Luis and Head lakes in the San Luis Valley, all grebes picked up dead in the 1940s and 1950s were Westerns. By the 1980s Clark's and Westerns occurred together at these lakes (Ron Ryder pers. comm.).

A&R reported Clark's at 12 breeding sites; atlasers Confirmed breeding at four more (two north of Denver and two in the San Luis Valley). The map depicts all the A&R sites, but atlasing may not have sampled other sites where the species breeds. The map also shows Probable or Possible breeding in five other new places, all but one (Lake Granby) near previously known breeding locations.

Year-to-year fluctuations of the water levels at known *Aechmophorus* grebe breeding sites in Colorado lead both species to utilize these sites in irregular and unpredictable patterns. Consequently, no one knows with certainty where Clark's and Westerns will appear or how their patterns of distribution may differ. Generally, Clark's occur less often in the western valleys and northeastern plains and more often in the San Luis Valley (A&R). Because atlasers sampled one-sixth of the state, they could not find all nesting sites for all species. Maps for birds such as Clark's and Western grebes, which have specialized breeding requirements and limited suitable sites, do not portray the whole breeding distribution.

BREEDING EVIDENCE

REPORTED IN 13 (1%) OF 1745 PRIORITY BLOCKS

☐	Possible	8 blocks	(62%)
▩	Probable	3 blocks	(23%)
■	Confirmed	2 blocks	(15%)

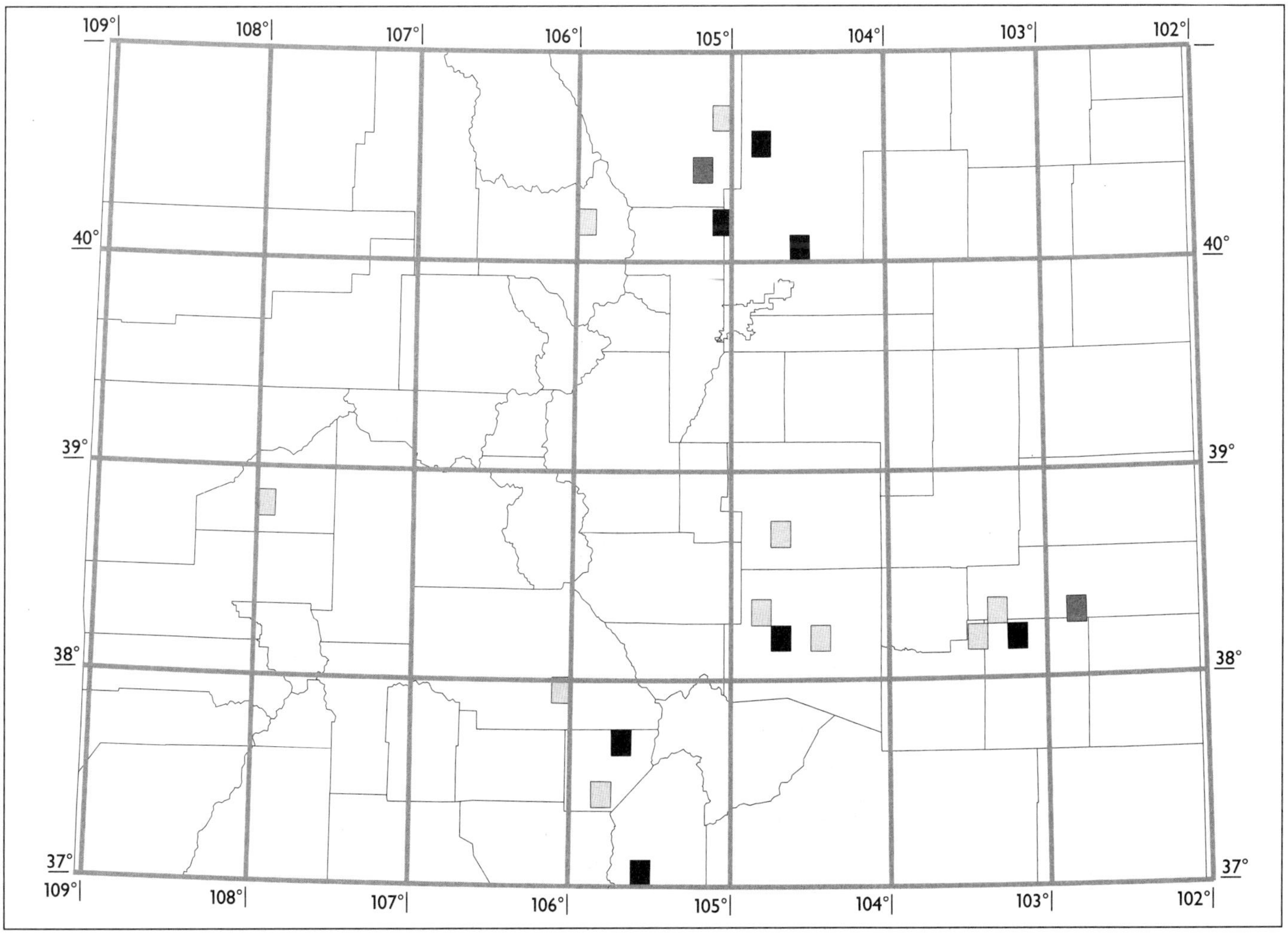
109°
108°
107°
106°
105°
104°
103°
102°
40°
39°
38°
37°

COLORADO DIVISION OF WILDLIFE / WILDLIFE RESOURCE INFORMATION SYSTEM

AMERICAN WHITE PELICAN

Pelecanus erythrorhynchos

What a treat to watch the graceful elegance of the synchronized, gliding flight of a precision of white pelicans! Only three decades ago did state wildlife biologists (on a routine waterfowl survey) discover the first Colorado breeding colony of American White Pelicans. Since 1962 white pelican numbers have soared; annual production of young has exceeded 2,000.

HABITAT

Across the West, American White Pelicans breed on islands in large bodies of water, such as Yellowstone Lake, Great Salt Lake, and Pyramid Lake. In response to losses of breeding sites and feeding habitat they adapted to human settlement. In Colorado during the Atlas they bred in three sites, each on an island in an irrigation reservoir. Colony elevations range from 4,829 feet (1472 m) at Riverside Reservoir on the northeastern plains, to 8,100 feet (2470 m) at MacFarlane Reservoir in North Park, to 8,940 feet (2725 m) at Antero Reservoir in South Park.

White pelicans forage up to 30 miles (50 km) away from breeding sites (Evans and Knopf 1993) in marshes, lakes, and rivers. They consume fish, crayfish, and salamanders. Scooping prey into their two-gallon pouches, these pelicans feed individually or in cooperatives. Foraging parties spread out and drive fish toward shallow shores for easier capture.

BREEDING

In the three Colorado colonies the gregarious American White Pelicans nest in association with up to seven species of colonial waterbirds: Double-crested Cormorants, Great, Snowy, and Cattle egrets, Black-crowned Night-Herons, Great Blue Herons, and California Gulls (Miller 1978, Atlas).

In late April and early May the pelicans begin to arrive in Colorado and by the first week of June only breeding pelicans remain on the islands. Courtship commences upon arrival on the breeding grounds and continues into late June (Schaller 1994).

White pelicans nest on the ground. They construct mounded nests; while sitting they haul in materials within reach of their bills, from all directions. Males and females incubate by holding the eggs, usually two, under their large foot webs (Evans and Knopf 1993).

During the 30-day incubation one adult guards the nest at all times; adults stay on the nest for stretches of 1–3 days while the mate feeds, often far from the colony (Evans and Knopf 1993). Adults brood the nestlings for 15–18 days, during which time the second chick usually dies from starvation due to harassment by its older sibling. After the brooding period young birds cluster into pods and adults begin foraging full time. Altricial at birth, the young require 10–11 weeks of care before their first flights.

It appears that only adult pelicans return to, and remain at, the breeding grounds. Subadults wander widely: although some do not return north until their third year, when they begin breeding, many others return sooner. Throughout the summer hundreds of non-breeding birds (they can amount to as much as 22% of the number of breeders; Evans and Knopf 1993) throng many reservoirs in northeastern Colorado.

BREEDING PHENOLOGY

CODE		# OF RECORDS	DATE RANGE
NE	Nest with Eggs	1	6 Jun
NY	Nest with Young	2	14 Jul–16 Jul

DISTRIBUTION

American White Pelicans breed in widely scattered inland colonies in western North America from northern Alberta to western Ontario and northeastern California to Utah and Colorado. From 1979 to 1981 the United States had 14–17 colonies (del Hoyo et al. 1992); twice as many birds breed in Canada. In winter these pelicans migrate to coastal Texas and Mexico, especially Veracruz (Ryder 1981); a few even breed in Mexico in the winter (Evans and Knopf 1993).

About 2.5% of the total population breeds in Colorado (Evans and Knopf 1993; 5–9% according to Priorities 1995). American White Pelicans reportedly nested in Colorado before 1900 (Cooke 1900), but B&N could not track down any confirming details. Definite nesting was confirmed in 1962, at Riverside Reservoir (Ryder 1981). Band returns suggest that the Riverside pelicans originally came from Chase Lake, North Dakota (Ryder 1981).

During the Atlas they nested only at the three colonies mentioned above and shown on the map. In 1996 engineers drained Antero Reservoir to make structural repairs. No nesting by colonial waterbirds occurred in 1996 and none is expected until the lake refills in 1998 (Chuck Loeffler pers. comm.). Also in 1996 pelicans attempted, unsuccessfully, to nest on the island at Adobe Creek Reservoir SWA (Chuck Loeffler pers. comm.).

In 1994 American White Pelicans produced approximately 1,800 young at Riverside, 150 young at MacFarlane, and 125 young at Antero. Ron Ryder has led a banding program for pelicans since the first discovery of breeding at Riverside in 1962.

Pelicans seem to have adapted to habitat losses of the past 50 years (e.g., Antero Reservoir), a major threat to pelicans across their breeding range. Human intrusion can cause desertions in breeding colonies, especially during courtship and early incubation. Critical factors affect nesting success: cold air temperatures, hailstorms, and especially varying water levels.

BBS data show a significant, moderately large increase—2.65% per year. The number of pelicans nesting in Colorado continues to increase, suggesting that local production and recruitment from other colonies over the past 35 years have more than replaced losses (Ryder 1981) and also implying that the species has a capacity to rebound from developments adverse to nesting habitat.

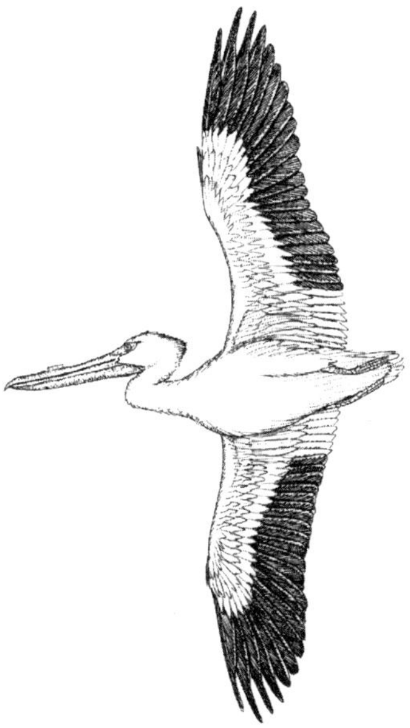

BREEDING EVIDENCE

REPORTED IN 1 (0%) OF 1745 PRIORITY BLOCKS

☐	Possible	0 blocks	(0%)
▩	Probable	0 blocks	(0%)
■	Confirmed	1 block	(100%)

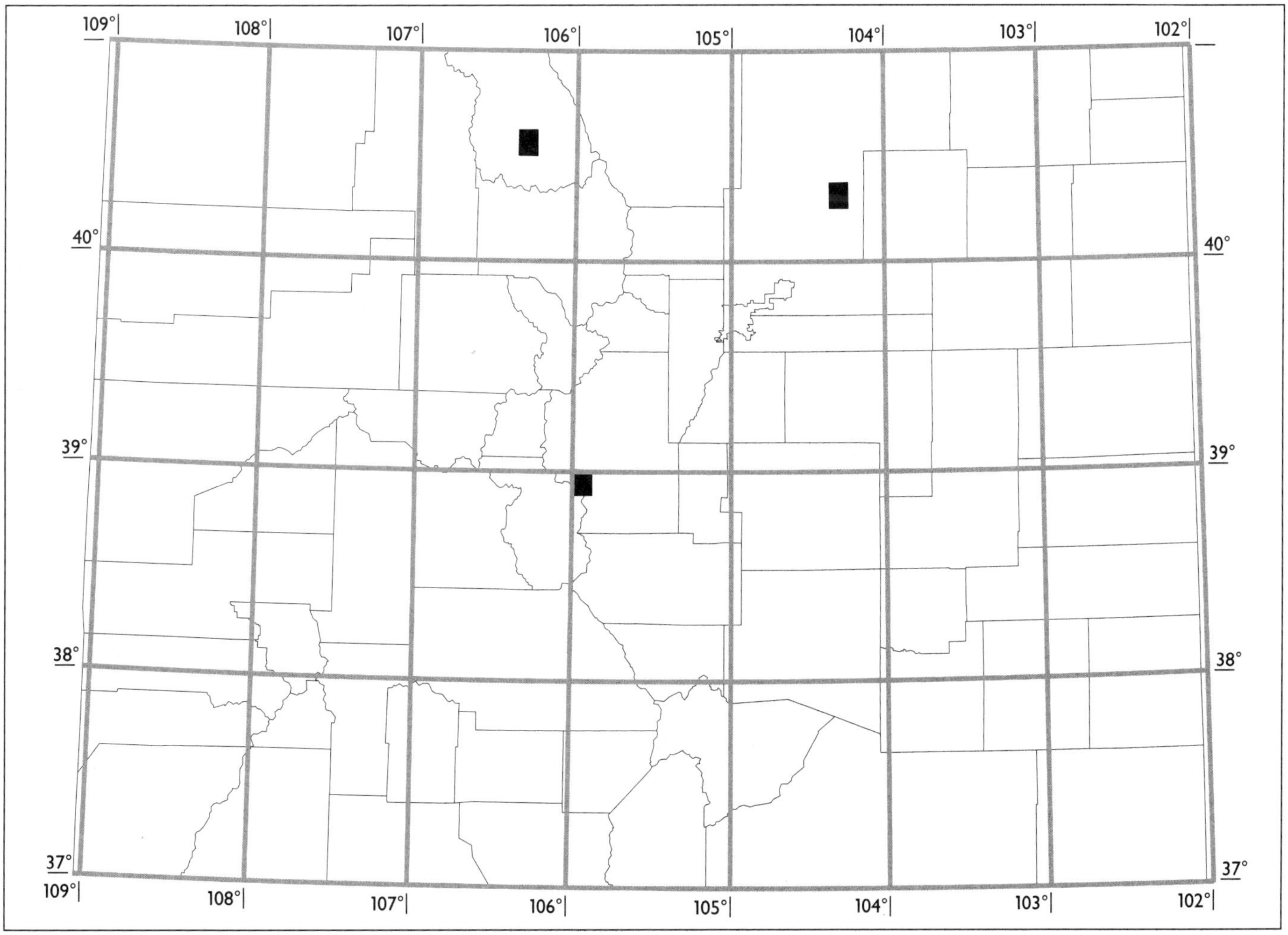
109°
108°
107°
106°
105°
104°
103°
102°
40°
39°
38°
37°

COLORADO DIVISION OF WILDLIFE / WILDLIFE RESOURCE INFORMATION SYSTEM

DOUBLE-CRESTED CORMORANT *Phalacrocorax auritus*

As settlers advanced across the continent, the numbers of Double-crested Cormorants dropped and their colonies disappeared. From a low in 1925 they started to rebound; they came to nest in Colorado in 1931. Since then they have not only increased dramatically but have also started to displace other species with which they share their nesting colonies (Palmer 1962, Ryder 1996).

HABITAT

Double-crested Cormorants in Colorado use rivers and reservoirs for breeding-colony sites and nest in cottonwood groves or on the ground near water with abundant fish populations. Most nest on the plains, although a few cormorants go to higher elevations to breed: during the Atlas, 8,100 feet (2470 m) at MacFarlane Reservoir in North Park, 8,940 feet (2725 m) at Antero Reservoir in South Park, and 7,600 feet (2315 m) at Eastdale Reservoir in the San Luis Valley (Garcia, 37105A5, Dean Swift pers. comm.). All breeding Confirmations reported by Atlas workers occurred within two habitat types: cottonwoods or other trees along streams or lakes and islands in reservoirs.

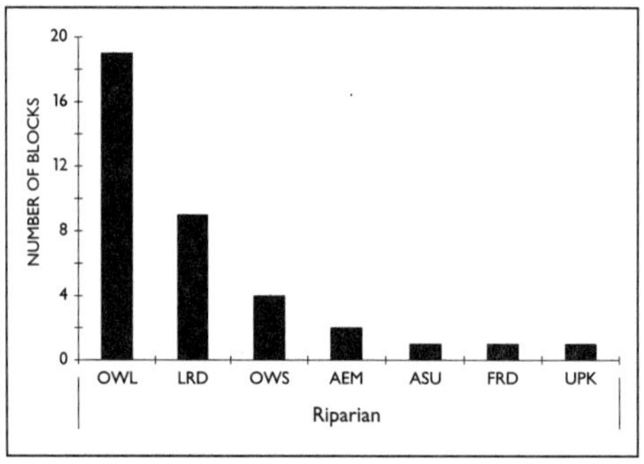
NUMBER OF BLOCKS
0
4
8
12
16
20
OWL
LRD
OWS
AEM
ASU
FRD
UPK
Riparian

Eyes adapted for aerial as well as underwater vision help cormorants to locate their prey. From a surface dive, using feet and wings to swim, they pursue small fish, principally carp (B&N), but also crappies, perch, sticklebacks, sunfish, salamanders, and crayfish (del Hoyo 1992). Cormorants forage gregariously in close proximity to the nesting site. In fact an adequate nest site requires an adequate nearby food supply (Palmer 1962). Some believe that cormorants compete for game fish and warrant control efforts (Ryder 1996). Most references explain that cormorants pursue the most abundant small fish available and usually consume non-sport fish (Pearson 1936, Terres 1980, Campo et al. 1993).

BREEDING

Double-crested Cormorants in Colorado nest with other colonial waterbirds. They share colonies variously with American White Pelicans, Great, Snowy, and Cattle egrets, Black-crowned Night-Herons, Great Blue Herons, and California Gulls (Miller 1978).

Cormorants most frequently nest in live or dead trees at any height, as far out as the limbs will support them. On an island at Riverside Reservoir they nest on the ground (B&N).

These cormorants normally first breed at three years of age. Older, more experienced birds breed earliest and occupy nest sites in the center of the colony. Pre-breeders engage in both courtship displays and some nest-building, mainly on the edge of the colony; even pre-flight young may engage in a little nest-building (Palmer 1962). They raise one brood per season. Both sexes incubate usually 3–4 eggs, which hatch asynchronously.

Cormorants begin nesting later than Great Blue Herons, as the Atlas phenology chart shows, which does not deter them from crowding into heronries and building nests among the heron nests.

BREEDING PHENOLOGY

CODE		# OF RECORDS	DATE RANGE
NB	Nest Building	1	6 Jun
NY	Nest with Young	4	16 Jun–14 Jul

DISTRIBUTION

Double-crested Cormorants breed along the eastern and western coasts of North America and the Bahamas. Inland they breed mainly at lakes and reservoirs in the Great Basin and the northern Great Plains. Banded Colorado birds have been recovered at wintering grounds in coastal Texas and Mexico (Ryder 1995). Very few remain in the state during winter.

Sclater (1912) reported the species as rare throughout Colorado, with only four records verified up to 1912, and Davis (1969) considered them very rare visitors in western Colorado. The first Colorado breeding record came in 1931 when Bailey and Niedrach found eight nests at Barr Lake; the colony had 30 pairs by 1939 (B&N) and 248 nests in 1995 (George 1996). By the 1980s the state population increased to an estimated

1,000 pairs (Ryder 1996), a number similar to Atlas figures. Proliferating Double-crested Cormorants now breed at more than a dozen sites on the Eastern Slope; three Denver-area colonies contain over 100 nests each. Cormorants expanded into North and South parks during the Atlas (Ryder 1995), and the Atlas recorded the first San Luis Valley breeding site, at Eastdale Reservoir.

BBS data show a very large increase across the continent. Various sources, including 26 years of Jack Reddall's field trips, show a trend of increasing cormorants and decreasing herons in the South Platte River valley. Cormorants may be usurping traditional heron nesting areas (Carter et al. 1992).

At Chatfield State Park, a colony in which Great Blue Herons had nine nests in 1971 increased to 171 heron nests by 1986. Cormorants joined the colony in 1979; by 1986 cormorants had more nests than the herons, whose numbers started to drop. The colony peaked in 1987 with 190 cormorant and 157 heron nests (Bottorff et al. 1971–1984, HEK). By 1995 the colony consisted of 68 Great Blue Heron nests and 171 Double-crested Cormorant nests (Nettleship et al. 1995). City Park in Denver exhibits a similar pattern with cormorants and Black-crowned Night-Herons (HEK).

Atlas data confirm breeding of cormorants in the big mountain parks, along the Arkansas and Platte rivers, and on the northeastern plains. The data indicate no Confirmed breeding on the Western Slope, but do show cormorants present during the breeding season in one block each along the Yampa, White, Colorado, and Gunnison river courses. Because cormorants breed on the Eastern Slope and in Utah, and given their increasing population trends, we can expect cormorants to breed on the Western Slope in the near future. Legal protection of the species since 1972, the DDT ban, and reservoir construction account for the increases in numbers and range expansion.

BREEDING EVIDENCE

REPORTED IN 27 (2%) OF 1745 PRIORITY BLOCKS

▢	Possible	24 blocks	(89%)
■	Probable	2 blocks	(7%)
■	Confirmed	1 blocks	(4%)

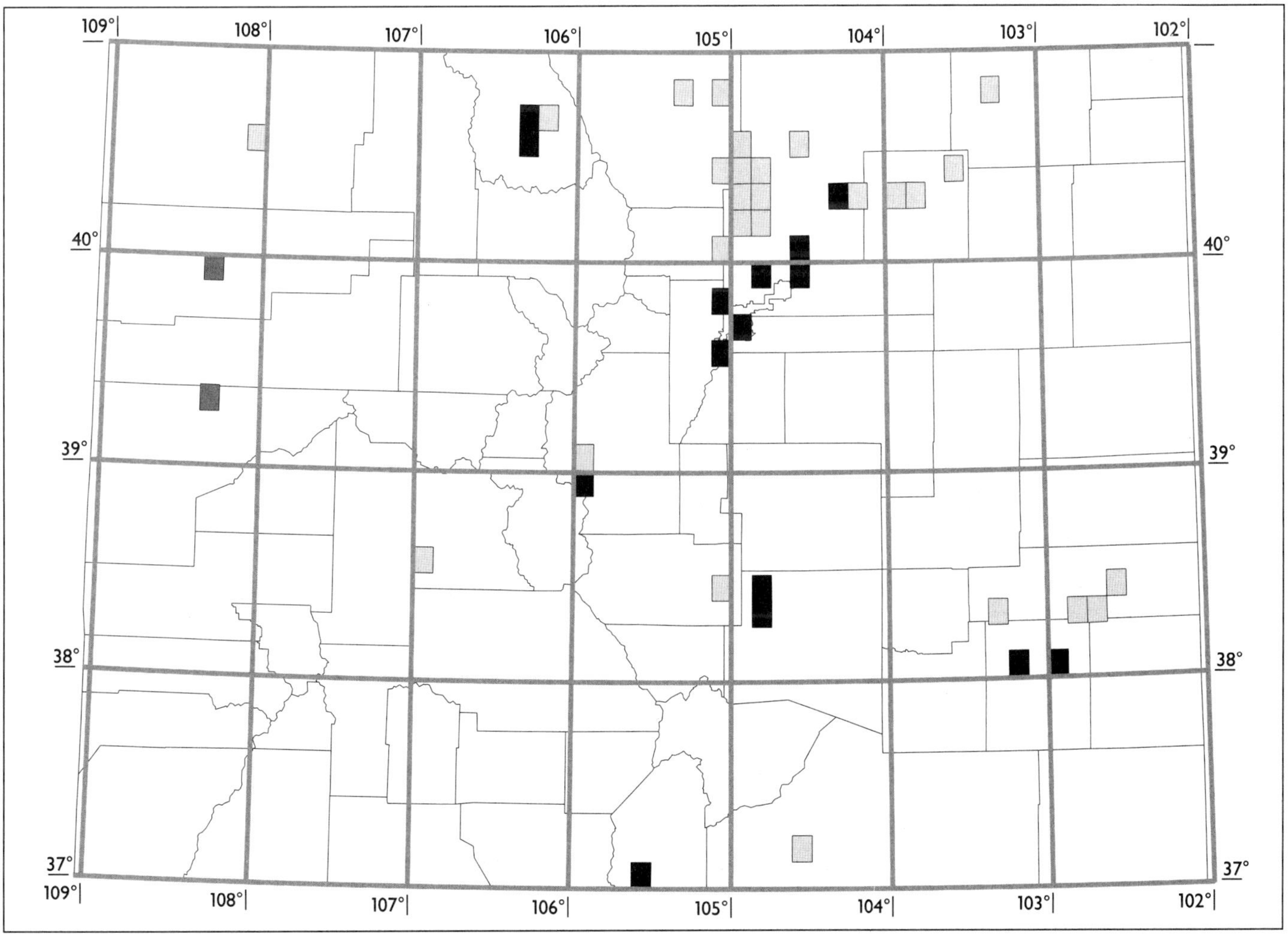
109°
108°
107°
106°
105°
104°
103°
102°
40°
39°
38°
37°

COLORADO DIVISION OF WILDLIFE / WILDLIFE RESOURCE INFORMATION SYSTEM

A remarkable sound, resonating from a cattail marsh and something akin to a bullfrog on steroids, signals the presence of an American Bittern. The sound, described by some as *plunk-er-lunk*, can carry up to half a mile and may give the only clue to this secretive species, a master of remaining still and invisible. In my only encounter with a nesting American Bittern, the bird stayed completely still with its head erect until it realized we had discovered it. Instead of flushing and abandoning the young on the nest, it convinced my canoe party to retreat by hissing at us and then regurgitating tadpoles in our direction.

HABITAT

American Bitterns seldom wander far from a marsh, swamp, or bog. They reside in cattails, rushes, grasses, or sedges of wet meadows or marshes. Less frequently they inhabit dry fields where the grass grows tall. All Atlas blocks showed bitterns in marshes except two that reported lowland riparian habitat.

American Bitterns stalk wetlands in search of their favorite foods, which consist of any small animal found in a marsh: insects, frogs, fish, snakes, meadow mice, etc. The birds swallow these animals whole but will crush larger vertebrates and crustaceans before swallowing (Bent 1926).

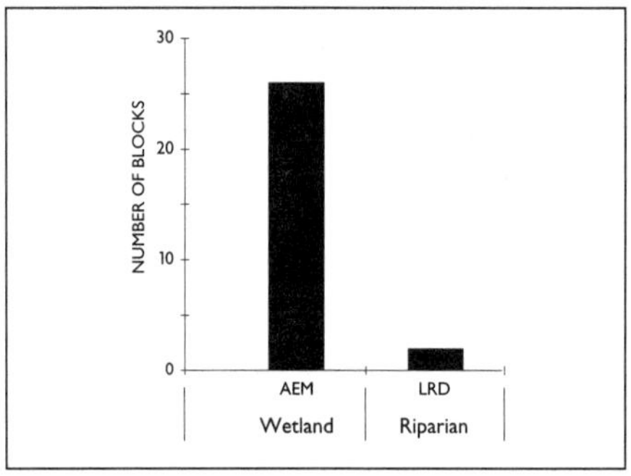
30
20
10
0
NUMBER OF BLOCKS
AEM
LRD
Wetland
Riparian

BREEDING

Primarily solitary nesters, American Bitterns move onto the breeding grounds in mid April (B&N). Nests consist of a loose platform of dead reeds, cattails, and grasses built a few inches above the water. Sometimes bitterns build nests on the ground. Occasional groupings of nests in loose colonies hint at possible polygynous behavior (Palmer 1962). Bitterns build a path leading to and from the nest so that the bird never directly alights on the nest (Bent 1926). The nest, usually completed by late May, contains 4–5 eggs (range, 3–7) colored brown to olive-buff. The female apparently does all of the nest-building (Palmer 1962). As incubation progresses, new plant growth conceals the nest completely by the time the eggs hatch (Bent 1926). Harrison (1979) found a nest in Colorado with five bittern eggs and one Redhead egg.

Atlas observations range from birds in suitable habitat in early May to fledglings on 1 August. Bent (1926) describes a remarkable courtship display in which birds show an extensive white plume, or ruff, above their shoulders. Atlasers saw courtship in two blocks on 25 May in Bent County. B&N note young birds on 12 July. Only four blocks have Confirmations of nesting—one nest with eggs, one nest with young, one used nest, and one block with fledglings. The nest-with-young Confirmation on 30 July certainly adds to the scarce phenology data.

BREEDING PHENOLOGY

CODE		# OF RECORDS	DATE RANGE
C	Courtship	2	25 May
NY	Nest with Young	1	30 Jul
FL	Fledged Young	1	1 Aug

DISTRIBUTION

The breeding territory of American Bitterns ranges from south-central British Columbia and southern Mackenzie to Newfoundland. In the U.S. the species nests in all western and northern states. They winter from southern British Columbia, Utah, and southern New England to the Gulf states, Cuba, and Central America.

American Bitterns "breed in suitable locations throughout Colorado, but few nests have been found" (B&N). A&R showed them mainly along the lower Arkansas and South Platte valleys, north-central Colorado, San Luis Valley, and North Park. The sparse Atlas findings follow this pattern with a few surprises (Possible breeding on the headwaters of the Rio Grande in Hinsdale County and on the Yampa River in eastern Moffat County). Priority blocks did not include the requisite marshes to find them along the South Platte, in Browns Park NWR, or at Bonny Reservoir. Their absence in these places and their scarcity in the Arkansas Valley reflect Atlas block selection, but it also could signify cause for concern. A&R showed secondary populations (less common) in Middle Park, South Park, and the Wet Mountain Valley, also places where atlasers did not find any bitterns.

Only 16 Priority and 15 Non-priority blocks recorded American Bitterns. Two blocks, at Alamosa and Monte Vista NWRs, recorded more than 10 (A3) breeding pairs. Eight blocks specified A2 abundance (2–10 breeding pairs), but three of those blocks (all on refuges in the San Luis Valley) used the multiple male (M) breeding code (seven singing males), which suggests a strong probability of A3 codes—more than 10 pairs in each. The remaining blocks show only one breeding pair. This still does not add up to many American Bitterns.

Bitterns seem most active at dusk and at night (Terres 1980). This, as well as a possible reluctance by atlasers to muck through swamps, could factor into the low number found during the Atlas. If one takes into account how loud bitterns call, the more likely evidence points to an absence of American Bitterns.

Previous authors described this species as "common" in Colorado (Niedrach and Rockwell 1939, B&N) and A&R believed that bitterns may occur in greater numbers than most observers perceive. In 1985 Colorado listed the species as one of special concern (ranked fourth from the top) due to marsh disturbance (Webb 1986).

Continentally, BBS routes record a large decrease of 2.2% per year. For the 31 years of the survey this computes to a potentially disastrous drop of half the entire bittern population. The WatchList categorizes American Bittern as a Moderate Priority species (Carter et al. 1996).

BREEDING EVIDENCE

REPORTED IN 16 (1%) OF 1745 PRIORITY BLOCKS

Possible	10 blocks	(63%)
Probable	3 blocks	(19%)
Confirmed	3 blocks	(19%)

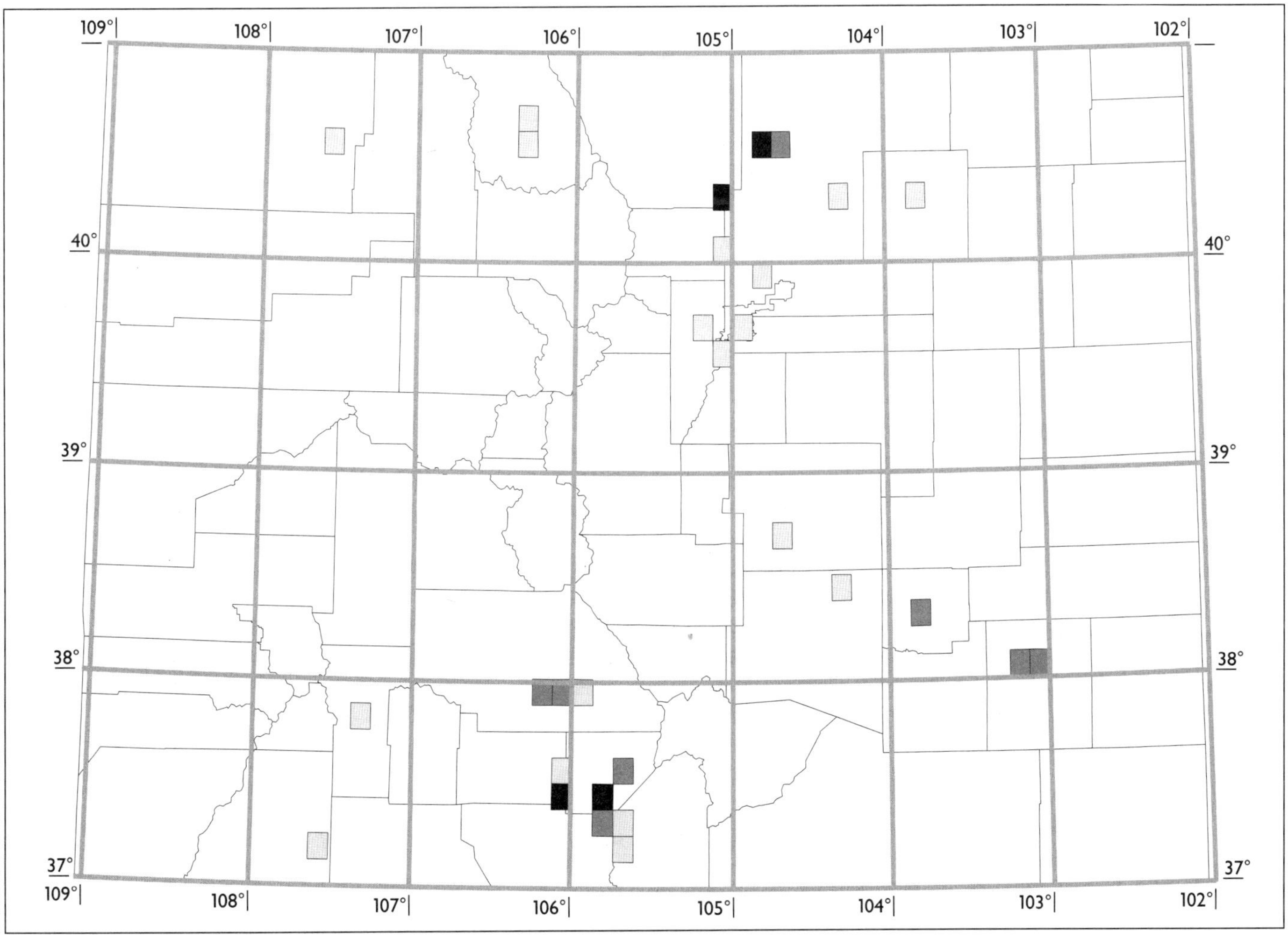
109°
108°
107°
106°
105°
104°
103°
102°
40°
39°
38°
37°

Proudly sported on T-shirts and pictured on travel and wildlife brochures, postcards, calendars, and other showy locations, Great Blue Herons appear in more places and more frequently than most other birds. Man had a much different relationship with herons in the nineteenth century, when plumes from slaughtered herons and egrets decorated spectacular ladies' hats. Conservation and wetlands laws now protect them; although some species have yet to recover, Great Blue Herons now prosper in Colorado and nationwide.

HABITAT

The most widespread of all herons in North America, adaptable Great Blue Herons find food in aquatic zones of all types (Terres 1980). Atlas field workers found them feeding mainly along streams and rivers at low and middle elevations but also in shallow lakes and marshes. They found them nesting in groves of very tall, live or dead, trees, usually located close to water. As members of a sociable species that nests in close, congested communities, these herons require a large grove rather than an isolated large tree. The tree species is not as important as size, number, and location of the trees.

Atlas field workers found nests mostly in cottonwood trees but also in other species of trees including ponderosa pines in Castlewood Canyon State Park. The "blue cranes" (as Great Blues are often miscalled) prefer the tallest nest sites available, but at

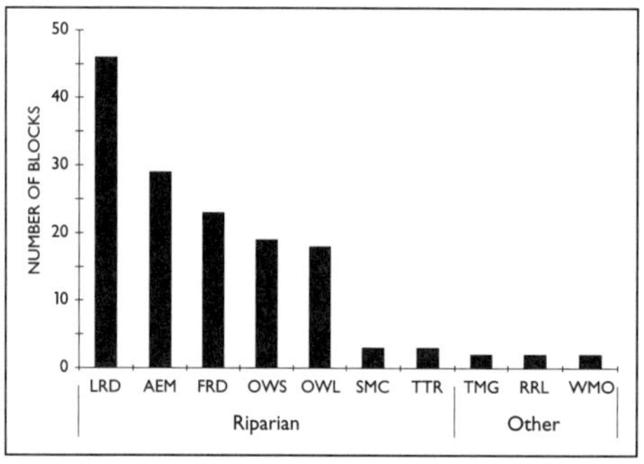

NUMBER OF BLOCKS
0
10
20
30
40
50
LRD
AEM
FRD
OWS
OWL
SMC
TTR
TMG
RRL
WMO
Riparian
Other

sites without trees they nest on the ground (as at Riverside Reservoir) or in shrubs (Ryser 1985). Site selection also depends on food availability, remoteness, and inaccessibility (Bent 1926).

Many Colorado heronries do not meet Bent's three criteria. Choice nesting habitat in Colorado follows the major river systems and older reservoirs; large groves of big trees that provide nesting habitat often surround reservoirs and line irrigation canals. Private ownership makes many inaccessible but state parks such as Barr Lake and Chatfield support large colonies, although in sections removed from the centers of human activity. Herons fly several miles to feed (Walt Graul pers. comm., atlasers), and some Possible records on the Atlas map may represent feeding birds that traveled as far as 25 miles from their nesting colonies.

BREEDING

Atlas field workers found 52 breeding colonies (21 in Priority blocks, 31 in Non-priority blocks). This compares with 63 colonies found in a six-year survey (1978–1983; Miller and Graul 1987); however, the theory of the Atlas protocol assumes that, by surveying one-sixth of each topographic map, atlasers would find in Priority blocks one-sixth of the birds of a particular species. Applying this theory to heron colonies suggests that the state has 126 colonies—twice the 1987 estimate. Great Blue Herons switch or abandon nest sites periodically, and the total number of colonies in use at any one time can be smaller than all the sites used over several years. On the other hand, atlasers had no predetermined notions of where to look for heronries, so they uncovered some new sites not surveyed by the earlier study.

Field workers reported 2–10 nests in half the colonies and over 10 nests in the other half. Only heronries at Barr Lake and Boulder had over 100 nests. The 1978–1983 study found an average of 24 nests per colony, including six colonies with more than 100 nesting pairs (Miller and Graul 1987).

Other species often nest among the herons, most prominently Double-crested Cormorants, which may out-compete Great Blues for nest sites. Black-crowned Night-Herons and Snowy Egrets share some colonies in Colorado, and in one or two colonies Great and Cattle egrets nest. Frequently a pair of Great Horned Owls nests among the herons, with the two species tolerating each other (Bottorff et al. 1971–1984), and in at least one colony Red-tailed Hawks raised a family (Bottorff et al. 1971).

Herons normally use sticks to form a large nest platform, which they place in treetops. Birds lay 3–5 eggs. Both sexes incubate, for a total of 25–29 days. Both parents attend the young, who do not leave the nest for two to three months after hatching (Harrison 1978). Parents feed their young regurgitated fish, amphibians, reptiles, mammals, birds, crustaceans, and even insects (Ryser 1985). The long in-nest period, coupled with staggered egg-laying dates, causes a long breeding season; Atlas dates span four and one-half months.

BREEDING PHENOLOGY			
CODE		# OF RECORDS	DATE RANGE
NB	Nest Building	2	20 Mar–3 Jul
ON	Occupied Nest	14	1 Apr–6 Jun
NY	Nest with Young	16	8 May–30 Jul

DISTRIBUTION

Great Blue Herons range from southeastern Alaska and southeastern Canada throughout the U.S., Mexico, the West Indies, and northwestern South America. Most herons from Colorado and north migrate; to the south many birds are resident. Some herons winter in Colorado along major rivers and at lower elevations (A&R).

Great Blues historically bred chiefly in northeastern Colorado, with some breeding up to 8,000 feet (2440 m) in the mountain parks (Cooke 1900, Sclater 1912). Atlasers found them throughout the state, with most colonies along major rivers and their tributaries. Most western Colorado heronries are on the Yampa, White, Gunnison, and Colorado rivers. East of the Continental Divide most lie along the South Platte and Arkansas rivers and their numerous water-diversion reservoirs.

Atlas workers discovered many new nesting colonies, especially in the Arkansas and Colorado river drainages. Mesa County supported only four known sites in 1983, but had seven in 1995. Montrose and Pitkin counties now have breeders. Southeastern Colorado has new colonies along various Arkansas River drainages in Las Animas, Kiowa, Prowers, and Bent counties, and other new colonies in El Paso, Baca, and Douglas counties.

The BBS shows this heron increasing in North America at a healthy 2.1%/year. Comparisons of the A&R map and known sites listed by Miller and Graul (1987) with the Confirmed breeding locations on the Atlas map give support for a higher nesting population than previously thought.

BREEDING EVIDENCE		
REPORTED IN 105 (6%) OF 1745 PRIORITY BLOCKS		
Possible	80 blocks	(76%)
Probable	3 blocks	(3%)
Confirmed	22 blocks	(21%)

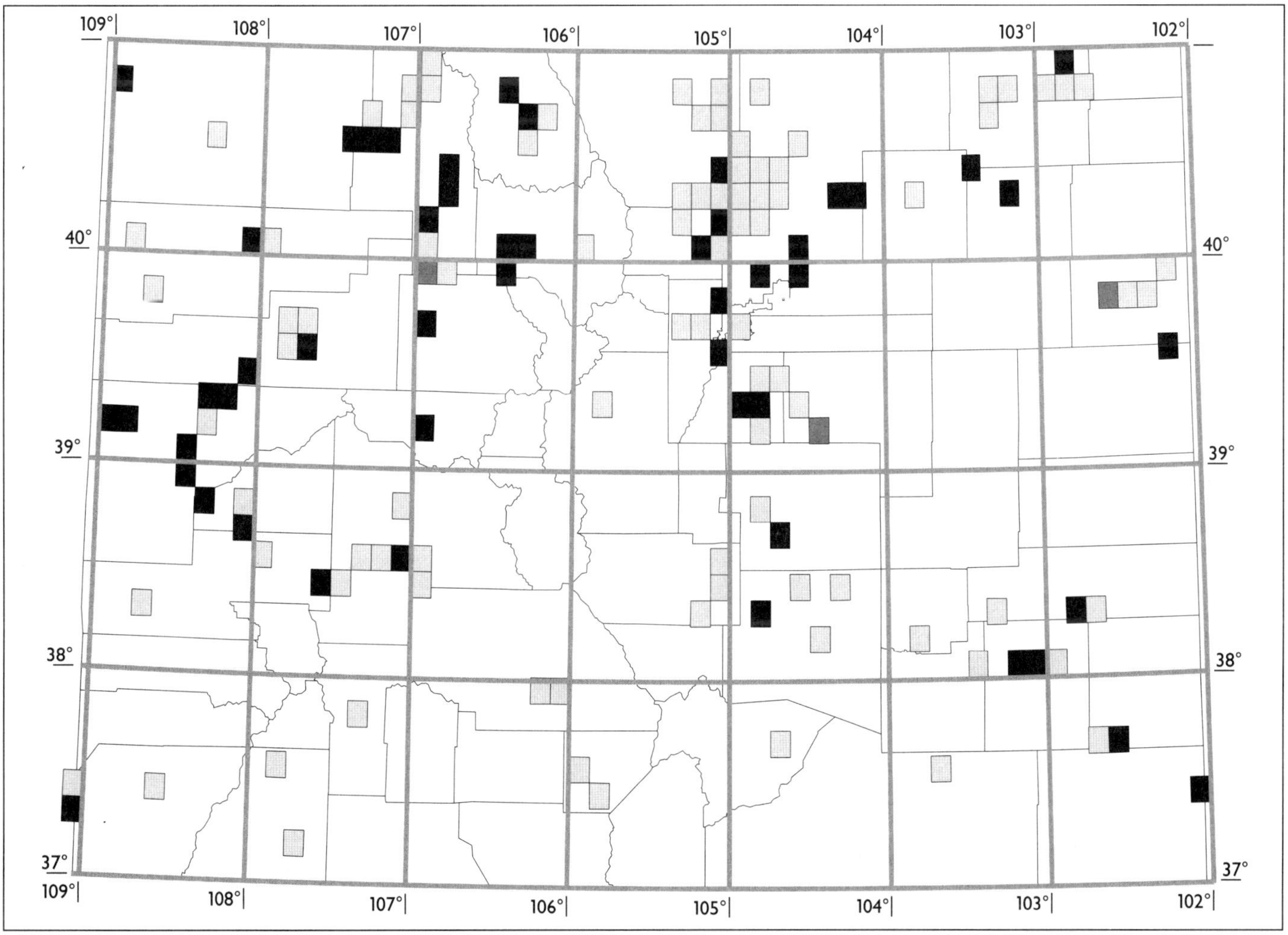
109°
108°
107°
106°
105°
104°
103°
102°
40°
39°
38°
37°

COLORADO DIVISION OF WILDLIFE / WILDLIFE RESOURCE INFORMATION SYSTEM

Snowy Egrets are the most abundant and widespread white heron in Colorado; they far outnumber the Great Egrets and Cattle Egrets that may nest in the same colonies. Snowies have nested regularly in Colorado since 1937 (B&N) and probably earlier (RAR).

HABITAT

For feeding, Snowy Egrets frequent mainly marshes, wet meadows, streams, rivers, and shores of shallow ponds and reservoirs. They nest in colonies in trees (usually willows or cottonwoods) and in tall emergents, especially cattails and bulrushes (RAR). Atlasers reported all nesting colonies in marshes except two that use cottonwood woodlands, at Barr Lake and Riverside Reservoir.

Post-breeding or non-breeding egrets wander into the high mountain parks and to the plains and valley marshes (A&R).

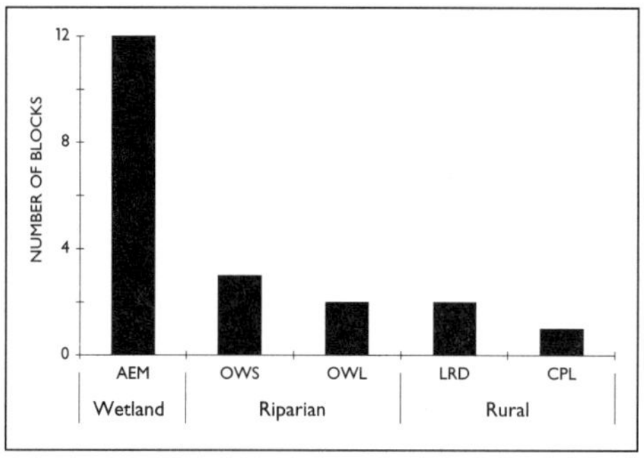
NUMBER OF BLOCKS
12
8
4
0
AEM
OWS
OWL
LRD
CPL
Wetland
Riparian
Rural

BREEDING

Early accounts (Sclater, Cooke, Bergtold) listed Snowy Egrets as only accidental summer visitors to Colorado. First nesting was noted 8 July 1937 in cottonwoods at Barr Lake north of Denver (Niedrach and Rockwell 1939). They may have nested at Russell Lakes in the San Luis Valley in 1937 (Phillips 1951), and they definitely did nest there in 1940 (Bailey and Brandenburg 1941) and in 1949 and 1950 (Ryder 1950). Egrets now nest annually in the San Luis Valley and irregularly at Riverside Reservoir and near Walden. Sporadic nesting has occurred near Greeley and Fort Collins (Ryder et al. 1979). During the Atlas, possibly breeding Snowy Egrets occurred in nine Priority blocks and 13 Non-priority blocks. Atlasers reported a long nesting season, from nests with eggs on 19 May to nests with young on 3 August.

In Colorado Snowy Egrets nest mainly in mixed colonies with Black-crowned Night-Herons and White-faced Ibises. Cattle Egrets and Great Egrets have nested in Snowy Egret colonies both in the San Luis Valley and at Riverside Reservoir. In 1988 one colony at the Monte Vista NWR in 1988 contained a nesting pair of Little Blue Herons (Ryder et al. 1989).

In the San Luis Valley, Snowy Egrets arrive in April from their wintering areas in Texas, California, and Mexico (Ryder 1978). They commence nest construction in mid April and early May (Schreur 1987), later than Black-crowned Night-Herons, about the same time as White-faced Ibises, and earlier than Cattle Egrets in the same colonies. Schreur (1987) noted Snowy Egret eggs as early as 16 April, in 1985. In all, she counted 135 nests in three colonies in 1984 and 139 nests in six colonies in 1985, more each year than either ibises or night-herons.

In San Luis Valley marshes, 48 nests were 10.8–24.7 inches (27.4–62.7 cm) above August water levels and in water 4.7–15.7 inches (11.9–39.8 cm) deep (Schreur 1987). Ninety-three egret clutches ranged from 2 to 6 eggs with a mean of 4.6 and mode of 5 (i.e., most frequent clutch size). Two to five young/nest reached at least 10 days of age (mode 3). Nesting success in different colonies varied from 10 to 82%. The low came at Russell Lakes in 1984 when two photographers disturbed the colony during early incubation (George et al. 1991). Birds move among colonies: two adults found dead in a colony at Monte Vista NWR had been banded four years earlier as flightless young at Russell Lakes (RAR). Eggs are lost mainly to ravens and magpies, adults and nestlings to raccoons and Great Horned Owls (RAR).

BREEDING PHENOLOGY

CODE		# OF RECORDS	DATE RANGE
ON	Occupied Nest	1	19 May
NE	Nest with Eggs	2	21 May–15 Jul
NY	Nest with Young	6	22 Jun–3 Aug

DISTRIBUTION

Snowy Egrets nest most abundantly in the Great Basin, Texas, Louisiana, and Florida, but also in high densities in the San Luis Valley of Colorado. B&N credited their relatively recent arrival as Colorado breeders to the development of prairie reservoirs like Barr Lake, where small numbers continue to nest (B&N, Atlas).

Statewide surveys of Snowy Egrets are less complete than for Great Blue Herons, but in 1965, 1973, and 1979 Colorado State University and CDOW personnel conducted mail surveys and inventoried most known colonies. A high of 600 pairs nested in the San Luis Valley in 1976 (Ryder et al. 1979). Since then egret numbers have fluctuated widely both in the San Luis Valley and at Riverside Reservoir (BBS, RAR). Complete failures occur some years, probably caused by fluctuating water levels, predators, and human disturbance (RAR).

The Atlas map reflects little change from those older surveys. The San Luis Valley has the most Snowy Egrets, but the birds still nest at Riverside Reservoir and Barr Lake. The isolated observations constitute Possible breeders, but produced no Confirmations during the Atlas.

Breeding season observations by atlasers were rather widespread; the map omits non-breeding reports. Long-distance, post-nesting-season wandering is not as pronounced in Snowy Egrets as in other southern herons (Ryder 1978).

Although not considered a species of special concern in Colorado, many Snowy Egrets unfortunately were shot or trapped by fish hatchery personnel in the past. With federal protection since 1972 and improved law enforcement, this loss has lessened considerably. Improved water management on state and federally owned wetlands would benefit the species.

BREEDING EVIDENCE

REPORTED IN 9 (1%) OF 1745 PRIORITY BLOCKS

Possible	7 blocks	(78%)
Probable	1 block	(11%)
Confirmed	1 block	(11%)

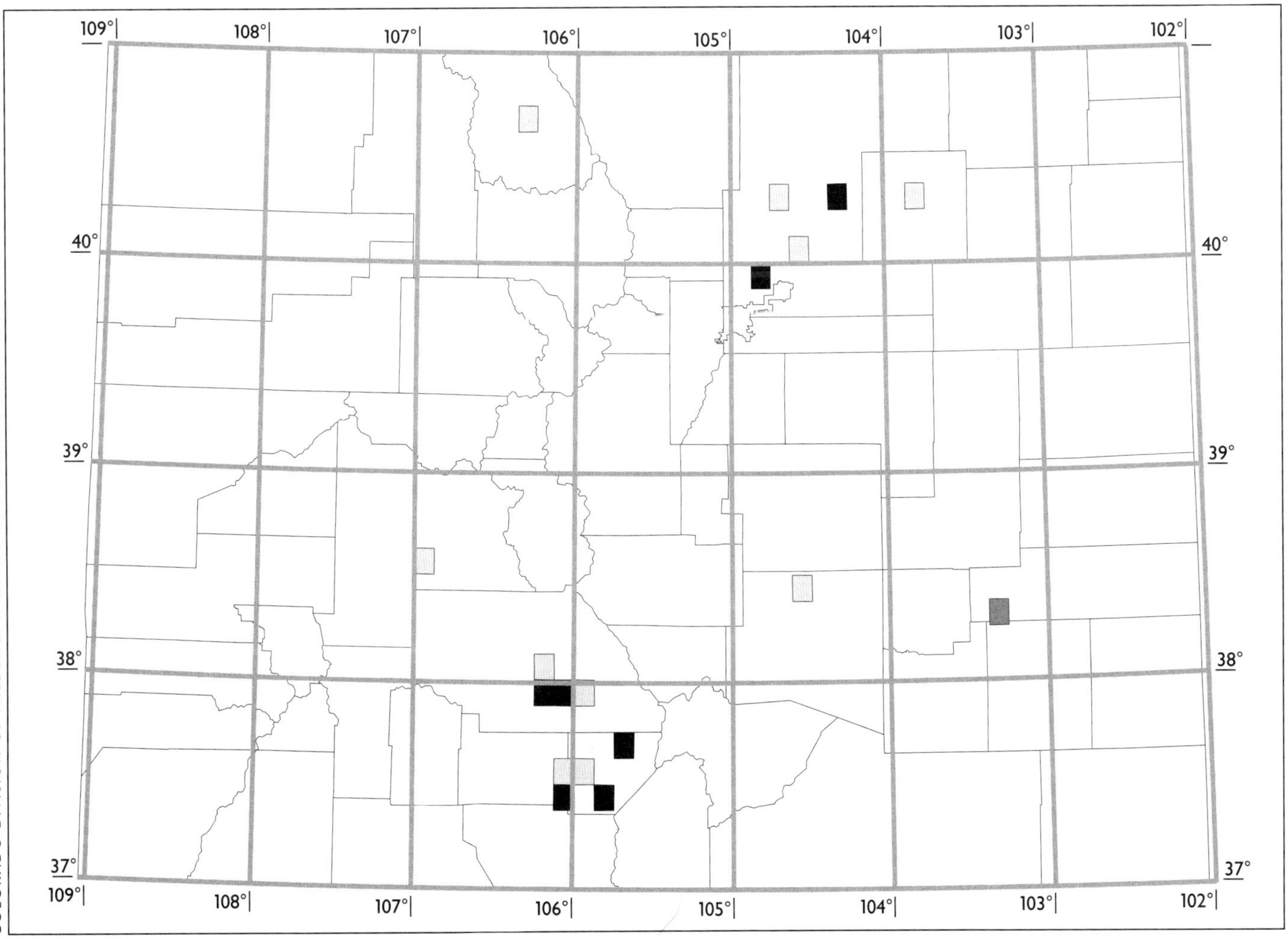
109°
108°
107°
106°
105°
104°
103°
102°
40°
39°
38°
37°

COLORADO DIVISION OF WILDLIFE / WILDLIFE RESOURCE INFORMATION SYSTEM

Newcomers to Colorado, Cattle Egrets now seem as much at home among North American cattle as with the antelope herds of their native Africa. This gregarious white egret, quite unlike its North American relatives, typically hunts for insects and other prey at the feet of grazing animals. First seen in Colorado in 1964 (B&N), Cattle Egrets remain rare breeders in the state.

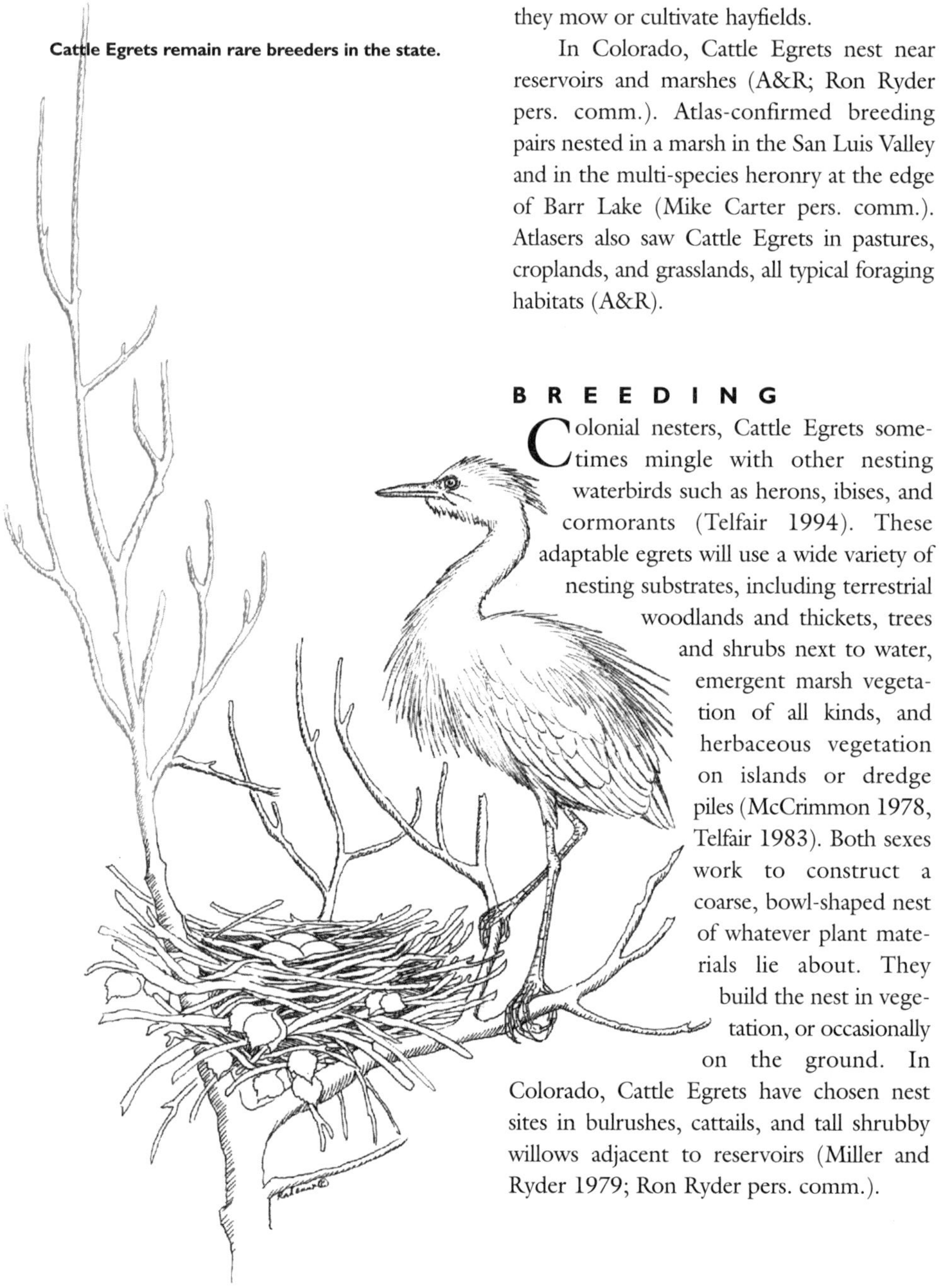

HABITAT

Throughout North America Cattle Egrets occur in many habitats, both terrestrial and aquatic. Adapted to disturbed areas, Cattle Egrets feel at home in converted landscapes such as agricultural lands and towns; true to their name, they often forage in close association with cattle (Telfair 1994). Ever adaptable, Cattle Egrets may also join more conventional wading birds at wetland edges, or seek prey on road margins, parks, sports fields, or city dumps. In fact, farm equipment can serve as livestock surrogates—Cattle Egrets sometimes feed alongside tractors as they mow or cultivate hayfields.

In Colorado, Cattle Egrets nest near reservoirs and marshes (A&R; Ron Ryder pers. comm.). Atlas-confirmed breeding pairs nested in a marsh in the San Luis Valley and in the multi-species heronry at the edge of Barr Lake (Mike Carter pers. comm.). Atlasers also saw Cattle Egrets in pastures, croplands, and grasslands, all typical foraging habitats (A&R).

BREEDING

Colonial nesters, Cattle Egrets sometimes mingle with other nesting waterbirds such as herons, ibises, and cormorants (Telfair 1994). These adaptable egrets will use a wide variety of nesting substrates, including terrestrial woodlands and thickets, trees and shrubs next to water, emergent marsh vegetation of all kinds, and herbaceous vegetation on islands or dredge piles (McCrimmon 1978, Telfair 1983). Both sexes work to construct a coarse, bowl-shaped nest of whatever plant materials lie about. They build the nest in vegetation, or occasionally on the ground. In Colorado, Cattle Egrets have chosen nest sites in bulrushes, cattails, and tall shrubby willows adjacent to reservoirs (Miller and Ryder 1979; Ron Ryder pers. comm.).

In North America, Cattle Egrets usually lay 3–4 eggs, and renest only if the clutch is lost. The semi-altricial young hatch after about 24 days.

BREEDING PHENOLOGY

CODE		# OF RECORDS	DATE RANGE
NE	Nest with Eggs	1	16 Aug

DISTRIBUTION

Cattle Egrets originated in Africa, probably alongside the immense herds of large grazing mammals. Since about 1800 they have colonized most of the world's tropical and temperate regions from their original strongholds in tropical Africa and Asia—one of the animal kingdom's most remarkable natural range expansions during historic times. Cattle Egrets reached South America in the late 1800s, presumably by transatlantic flight, and spread throughout most of South and North America in the past 40 years (Telfair 1994). Cattle Egrets are such good colonizers because of the strong penchant among juveniles for long-distance wandering after the breeding season, their adaptability to many habitats, and the increasing conversion of much of the world to farm- and pastureland (Siegfried 1978).

In Colorado, known Cattle Egret breeding sites include marshes and reservoirs in the San Luis Valley, Barr Lake, and several reservoirs in Weld County (A&R; Ron Ryder pers. comm.). In any given year, only a handful of pairs nest. The few nest sites known lie outside of Priority blocks, and atlasers Confirmed breeding Cattle Egrets only in two Non-priority blocks, at Barr Lake and San Luis Lake SWA.

Possible breeding records came from San Luis Valley wetlands and a reservoir margin near LaSalle. However, vagrants visit Colorado regularly from April through September (especially in August), most commonly in the low valleys of the Gunnison, Colorado, South Platte, and Arkansas rivers, and in the San Luis Valley (A&R). Many of these birds probably represent juveniles exhibiting nondirectional wandering. Atlas results reflect this pattern, with six of the 10 total observations recorded as non-breeders between 10 May and 24 June. Two new county records came from Grand County (presumably a summer vagrant), and Gerald Fyler's backyard in Montezuma County, where it seems that every bird species in Colorado must appear at least once.

Although rare in Colorado, these unusual and adaptable herons would seem to find here all that they need to breed. Given their remarkable talents for colonization, Cattle Egrets would seem destined to become more common residents of Colorado's waterbird colonies wherever wetlands and pastures abound. Surprisingly, however, North American populations have apparently peaked in the 1970s and 1980s; many peripheral populations have declined or disappeared in the last decade. Therefore an expansion in Colorado seems unlikely (Bruce Peterjohn pers. comm.).

BREEDING EVIDENCE

REPORTED IN 1 (0%) OF 1745 PRIORITY BLOCKS

Possible	1 block	(100%)
Probable	0 blocks	(0%)
Confirmed	0 blocks	(0%)

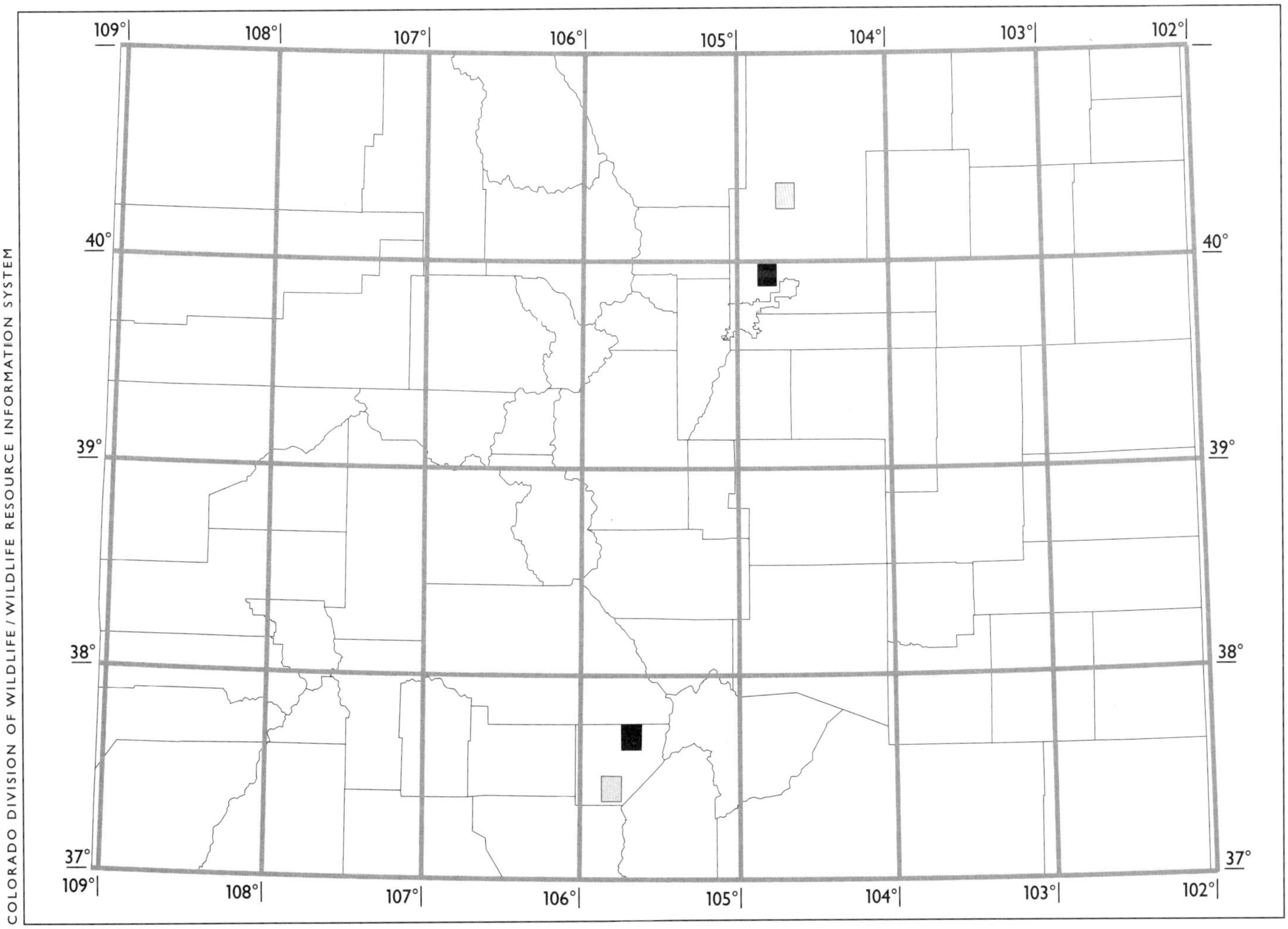
109°
108°
107°
106°
105°
104°
103°
102°
40°
39°
38°
37°

COLORADO DIVISION OF WILDLIFE / WILDLIFE RESOURCE INFORMATION SYSTEM

An arresting sight: a Green Heron, taut and motionless, as it waits for prey to blunder within striking distance. As the bird crouches on a low-slung branch, just inches above a quiet pool, its body and spear-like bill align perfectly with its gaze, aimed obliquely at the water. In a flash, the bird finally releases all that potential energy as its bill shoots out, unreeling a surprisingly long neck, to snare a wriggling minnow.

HABITAT

Throughout their range, Green Herons display adaptable and cosmopolitan habitat choices for herons (Hancock and Kushlan 1984). Although they rarely venture far from water, the birds make themselves at home in many wetland types, including the shorelines of rivers, lakes, reservoirs, and park ponds, as well as mangrove stands and fresh or saltwater marshes and swamps (Davis and Kushlan 1994). They find the cover of dense, woody vegetation along the water's edge especially enticing, but occasionally they breed in more open habitats, such as pastures, orchards, and willows along irrigation ditches.

Few Green Herons wander into Colorado, and Atlas workers found them in habitat associations typical of their species. All selected lowland riparian sites along Colorado's major rivers and reservoir shorelines at less than 6,000 feet (1830 meters) in elevation. In these habitats, the birds found their preferred cover—deciduous trees (mostly cottonwoods), willow shrubs, and other thicket-forming, woody vegetation.

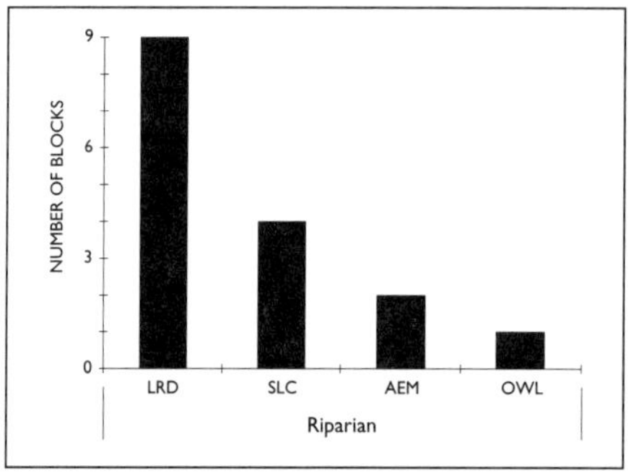
9
6
3
0
NUMBER OF BLOCKS
LRD
SLC
AEM
OWL
Riparian

Typically, Green Herons tuck their nests well out of view in shrubby thickets or among low-hanging tree branches, especially those that reach out over water. Some nest on the ground, but more often they build their nests between 10 and 20 feet (3–7 m) above the ground (Davis and Kushlan 1984). In Fort Collins, a pair nested 18 feet (5.5 m) above the ground in a box-elder tree near the Cache la Poudre River. Green Herons occasionally nest in dry orchards or woodlands (Bent 1926) if productive feeding habitat can be found nearby (Davis and Kushlan 1994). In Colorado, Atlas workers Confirmed breeding only in lowland riparian habitat.

BREEDING

The secretive nesting habits and rare occurrences of Green Herons made it difficult to document breeding, yet all four Atlas Confirmations superseded the previous earliest-known nesting date (18 May, A&R). On 7 May 1994 in Fort Collins, an Atlas worker discovered an adult bird clipping dead twigs from a tree (Dave Leatherman, nest card); the next day the bird began egg-laying and chicks hatched on or about 26 May. The next year, two pairs nested in the same area; one of the two pairs began nest-building 23 April (Dave Leatherman pers. comm.). In Pueblo an Atlas worker observed a Green Heron building a nest on 15 May 1993 (Dave Johnson pers. comm.). These nests provided indisputable Atlas breeding records.

Stick-carrying behavior constituted the basis for the other two Confirmations (both in the Denver area on 12 May), leaving some uncertainty as to their validity because Green Herons use sticks to lure fish as well as to build nests! They will dangle sticks in the water to entice fish or drop pieces of stick into the water and then seize fish that come to inspect the "tidbits" (Higuchi 1986, 1988). The birds also fish with mayflies (Keenan 1981), bread (Lovell 1958), worms, berries, feathers, and plastic foam (Higuchi 1986, 1988). To date, no one has reported Green Herons using stick bait in North America; nonetheless, unless a bird carries a stick to its nest, other behaviors may provide more reliable breeding confirmations.

BREEDING PHENOLOGY

CODE		# OF RECORDS	DATE RANGE
NB	Nest Building	3	12 May–15 May
NY	Nest with Young	1	11 Jun

DISTRIBUTION

Green Herons continue to ride a merry-go-round of taxonomic debate. Earlier this century, authorities distinguished Green-backed Heron (*Butorides striatus*) as a species separate from Green Heron (*B. virescens*; Bent 1926, AOU 1957). The AOU (1983) subsequently lumped both species into a superspecies comprising 18–30 subspecies distributed throughout much of the world (Hancock and Kushlan 1984). More recent changes, however, once again distinguished the Green Heron (*B. virescens*) as a full species (AOU 1993) comprising four subspecies. The subspecies in Colorado, *B. v. virescens*, ranges from the Dakotas south to Panama, east to the northeast coast of the U.S., and south through the Caribbean islands.

Green Herons attain their greatest breeding densities in mangrove thickets along the humid, southeastern U.S. coast. Colorado's arid climate defines the northwestern edge of their (*B. v. virescens*) breeding range, explaining the species' infrequent occurrence here. Birds that do wander across the state line apparently travel along the major lowland riparian corridors.

Green Herons find suitable nesting habitat over a greater portion of Colorado than initially believed. At one of the Denver sites, observers from the Denver Audubon Society have observed Green Herons throughout the breeding seasons of 1993–1998. Generally, these herons inhabit eastern Colorado, but atlasers reaffirmed that a few appear west of the Continental Divide, primarily in the Colorado River drainage system during migration periods (Latilong Study, A&R). Ironically, the only three winter records of Green Herons in Colorado all occurred west of the Divide, and an agitated bird, recorded by an Atlas worker on 25 May in the vicinity of Grand Junction, expanded the species' previously known breeding range to the Western Slope.

Historical records show that observations of these birds in Colorado have increased since the early 1900s (Cooke 1909, Bent 1926, B&N, Denver Field Ornithologists 1967, Johnsgard 1986b, A&R). B&N attribute this increase to the species' westward range expansion, possibly in response to the development of reservoirs and other wetland habitat changes associated with European settlement. The first breeding record for Colorado occurred in 1973, when a pair nested at Rocky Ford SWA (A&R). Prior to the Atlas, Tamarack Ranch hosted the only other Green Herons known to have nested in Colorado (A&R).

BREEDING EVIDENCE

REPORTED IN 7 (0%) OF 1745 PRIORITY BLOCKS

☐	Possible	6 blocks	(86%)
■	Probable	0 blocks	(0%)
■	Confirmed	1 block	(14%)

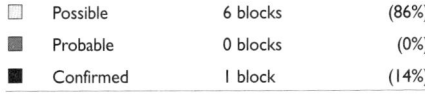

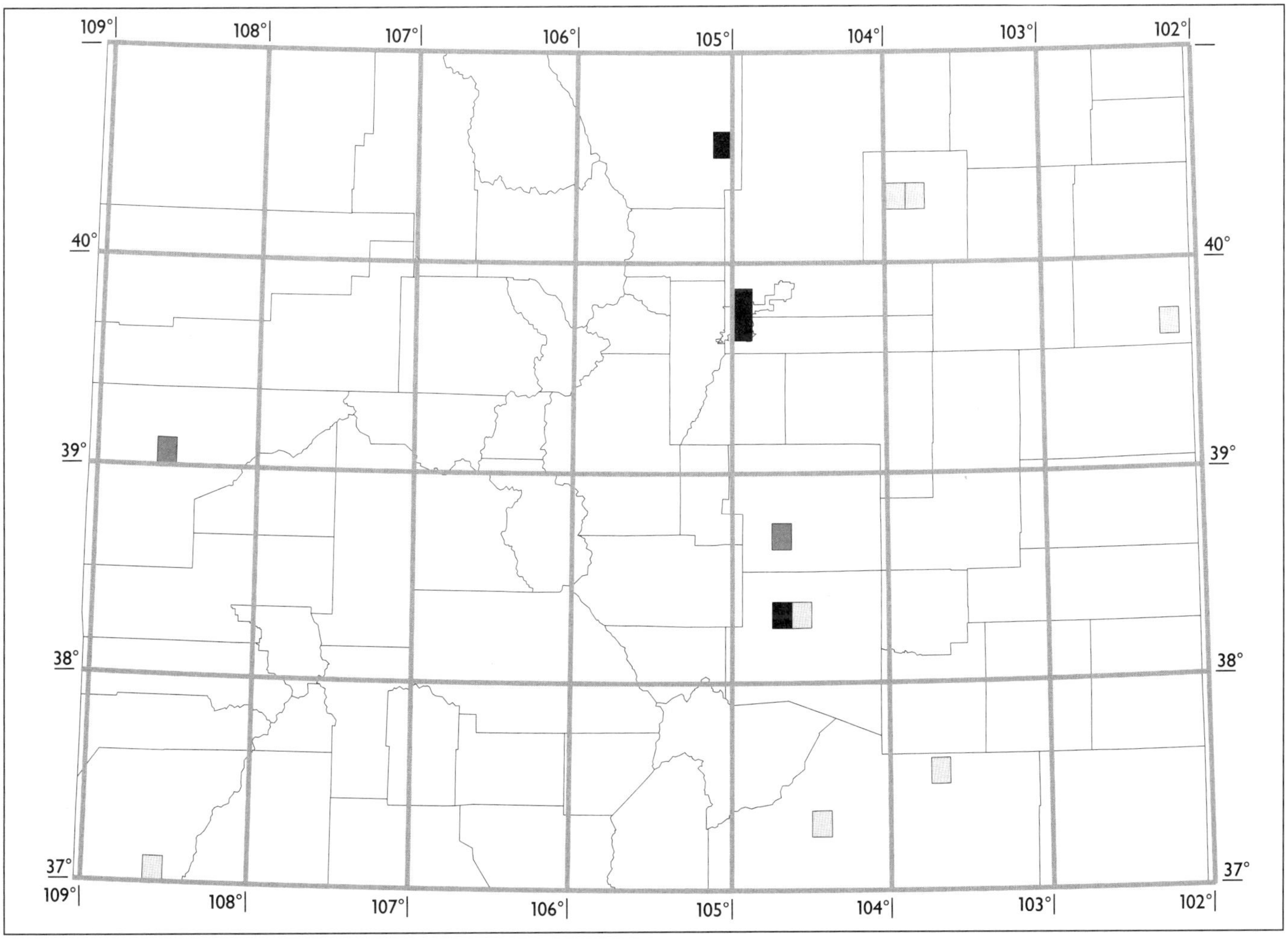
109°
108°
107°
106°
105°
104°
103°
102°
40°
39°
38°
37°

BLACK-CROWNED NIGHT-HERON *Nycticorax nycticorax*

Feeding from dusk until dawn, Black-crowned Night-Herons go unnoticed except for the few honks or squawks uttered during the commute to and from home or when the busy breeding season forces them into overtime day shifts. Although cosmopolitan in their breeding range, their behavior renders them relatively inconspicuous. This species nests in colonies with Great Blue Herons, but night-herons receive only a fraction of the publicity of their conspicuous day-shift neighbors.

HABITAT

Black-crowned Night-Herons forage in shallow water bodies and along water edges in all types of wet habitats, even in cities and at fish hatcheries. They eat worms, mollusks, insects, fish, amphibians, reptiles, birds, small mammals, plant material, eggs, carrion, and garbage. As solitary foragers, they defend feeding territories and may chase conspecifics from the area (Davis 1993).

When selecting nesting sites night-herons consider primarily areas near water, in close proximity to foraging habitat. In Colorado, they nest in single-species colonies and also in colonies variously with Snowy, Great, and Cattle egrets, Double-crested Cormorants, California Gulls, American White Pelicans, and Great Blue Herons (Miller 1978). Crepuscular and nocturnal feeding habits probably accommodate harmonious colonization with daytime foraging species. Colony sites include reservoirs, rivers, marshes,

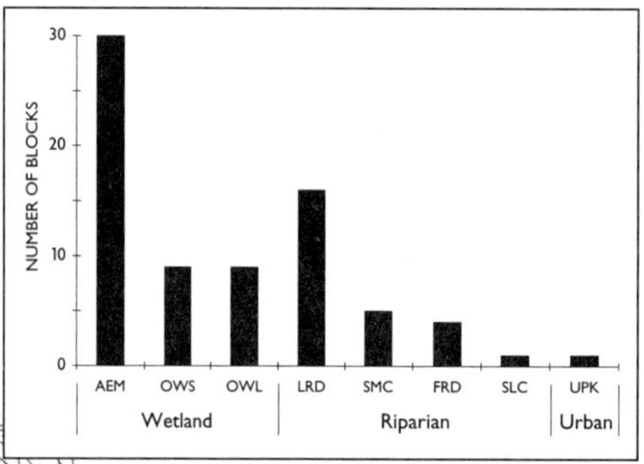
NUMBER OF BLOCKS
30
20
10
0
AEM
OWS
OWL
LRD
SMC
FRD
SLC
UPK
Wetland
Riparian
Urban

ponds, islands, and irrigation ditches. The highest colony reported by Atlas workers, at over 10,000 feet (3048 m) at the headwaters of the Rio Grande, uses emergent wetlands.

BREEDING

As nest-structure generalists, night-herons may assemble flimsy twig nests in the lower branches of cottonwoods, willows, or shrubs. Sometimes they build structures of weed stalks on or above water among cattails. Night-herons locate their small nests in dense cover, making nests in mixed colonies difficult to distinguish.

Night-herons begin arriving at nest sites in late March, lay eggs in late April to early May, and fledge young in mid July to mid August. Clutch sizes average 3–4 eggs in one brood per season.

Atlas workers observed adult night-herons on nests (ON) 1 May through 15 July, and young in the nest (NY) 15 June through 3 August. The young fly to feeding areas with adults and beg for food at 6–7 weeks (Palmer 1962). Because night-herons may hunt several miles from their nest, the dependent fledgling (FL) code has more significance for identifying foraging areas than breeding sites. Observations of night-heron feeding territories recorded as "possible" breeding may have resulted in inflated breeding data for this species.

In the San Luis Valley and on the plains, Black-crowned Night-Heron colonies exceed 100 breeding pairs. City Park in Denver hosts one of the largest colonies, with over 300 nests (George 1995). Initially a single-species colony established in the 1970s, the City Park colony attracted Double-crested Cormorants to join the night-herons in the early 1990s.

BREEDING PHENOLOGY

CODE		# OF RECORDS	DATE RANGE
NB	Nest Building	1	14 May
ON	Occupied Nest	3	1 May–5 Jul
NE	Nest with Eggs	2	22 Jun–15 Jul
NY	Nest with Young	11	15 Jun–3 Aug
FL	Fledged Young	7	21 Jun–1 Aug

DISTRIBUTION

Black-crowned Night-Herons breed on every continent except Australia and Antarctica (Davis 1993). Early accounts of the Colorado distribution by Sclater (1912) and others appear similar to that of A&R, which reads, "locally common summer residents of the water areas of eastern Colorado and less common on the Western Slope." Migration toward Mexico begins in late September or October (Davis 1993), but a few night-herons regularly winter along the South Platte River near Denver. Night-herons migrate mainly at night, flying singly or in small groups. Extensive post-breeding dispersal in all directions accounts for many records of northward movement and vagrancy by night-herons (del Hoyo et al. 1992).

At some breeding sites numbers have declined dramatically, whereas at others numbers have held steady or increased; the herons also shift locations and colonize new sites (A&R). At Gunnison, over a span of 40 years, the birds moved to at least three sites (Cook 1978); the second site, which had 20 nests in 1954 (Hyde 1979), no longer exists (A&R), but night-herons now have about six nests below the Roaring Judy Fish Hatchery north of Gunnison (KMP).

Atlas work located several new colonies and emphasized concentrations of night-herons in North Park, the northern plains, and the San Luis Valley. New colony sites include the Colorado River at Grand Junction, Hermit Lakes in Hinsdale County, and Fountain Creek south of Colorado Springs. Non-priority blocks represent about one-third of the blocks shown on the map. In the South Platte River Valley night-herons showed a significant declining trend (8.7% per year) over the past 26 years (Carter et al. 1992). Habitat alterations that effect changes in food availability and human disturbances at colony nesting sites pose the greatest threats to these night-shift birds.

BREEDING EVIDENCE

REPORTED IN 41 (2%) OF 1745 PRIORITY BLOCKS

☐	Possible	30 blocks	(73%)
▪	Probable	3 blocks	(7%)
■	Confirmed	8 blocks	(20%)

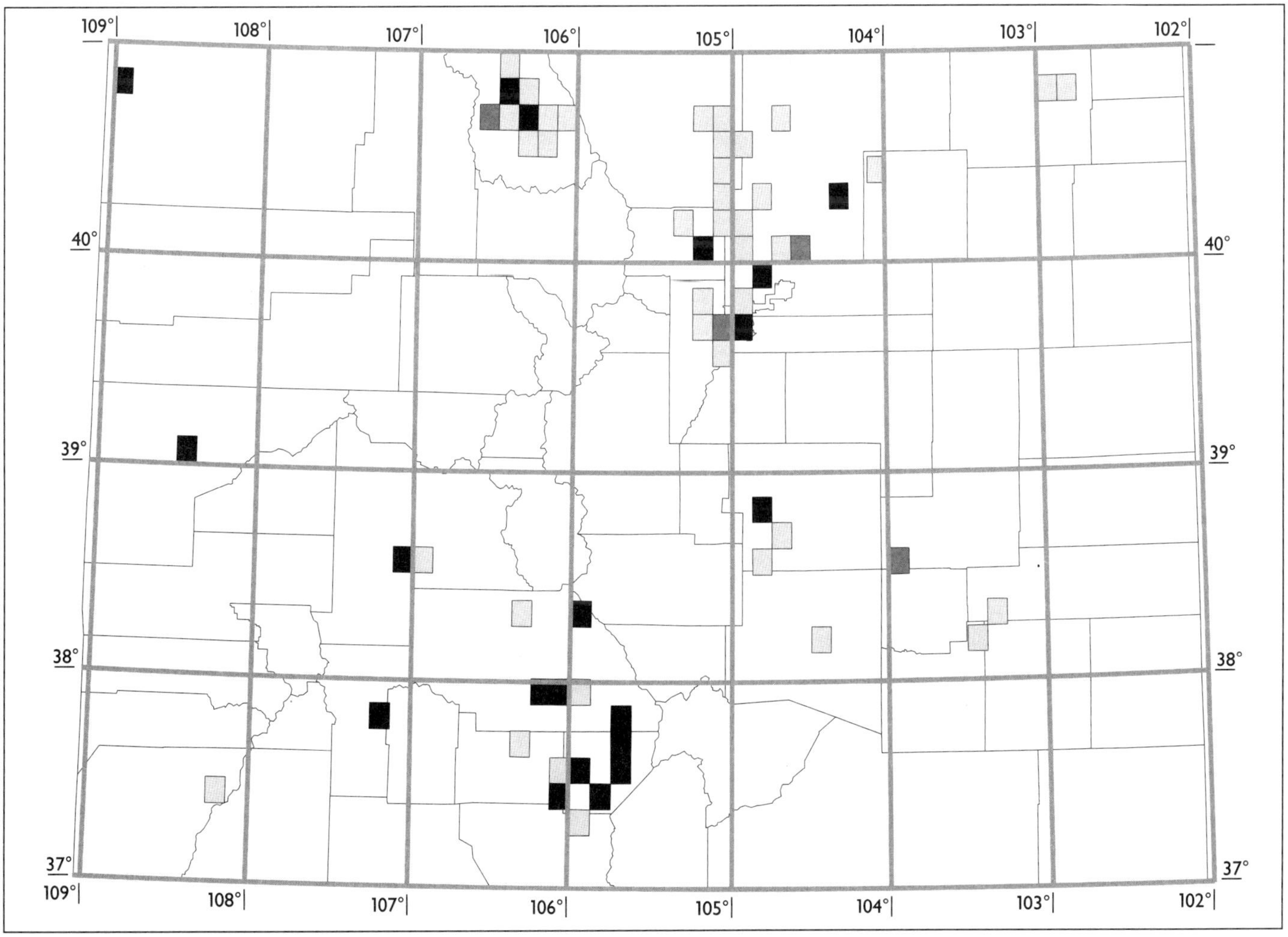
109°
108°
107°
106°
105°
104°
103°
102°
40°
39°
38°
37°

COLORADO DIVISION OF WILDLIFE / WILDLIFE RESOURCE INFORMATION SYSTEM

Each spring flocks of tropical-appearing, blackish, curlew-like birds fly in loose Vs over Colorado wetlands. Wading in shallow water, they appear coppery-brown with purplish, metallic-brown wings. Adults have red eyes, and distinctive white ovals surround the base of their long decurved beaks. New World counterparts to the ibises that decorate the tombs of the Pharaohs, White-faced Ibises lend an exotic aura to Colorado wetlands.

HABITAT

White-faced Ibises feed in wet hay meadows and flooded agricultural croplands as well as in marshes and the shallow water of ponds, lakes, and reservoirs (Ryder and Manry 1994). Most ibises nesting in Colorado favor tall emergents such as bulrushes and cattails growing as "islands" surrounded by water more than 18 inches (45 cm) deep. The dominance of marsh habitat shows up in Atlas habitat codes, over two-thirds of which are emergent wetlands (AEM). Twice, however, I observed nests about 6 feet (2 m) above water level in flooded willows at Riverside Reservoir. Ibises (migrants and "off-duty" nesters) usually roost in flocks, mainly in flooded emergents like those preferred for nesting (Ryder and Manry 1994).

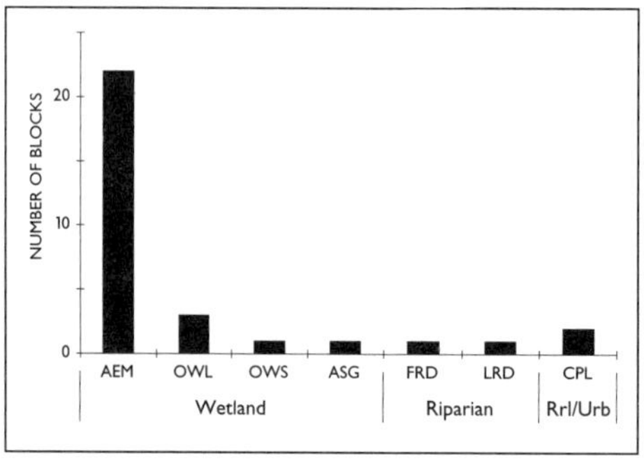
NUMBER OF BLOCKS
0
10
20
AEM
OWL
OWS
ASG
FRD
LRD
CPL
Wetland
Riparian
Rrl/Urb

BREEDING

As early as 1872, Aiken found White-faced Ibises nesting in the San Luis Valley (Sclater 1912), where in 1993 they still nested in at least 11 colonies. Ibises usually arrive in the San Luis Valley by mid April and nest in May, but dates vary from year to year. Both sexes take part in nest-building. Courtship occurs mainly in the nesting colonies, with some calling and bickering between rivals. Nests consist of bulrush and cattail stems and average about 3 feet (1 m) above water. Colonies may consist of fewer than 5 pairs up to 100 pairs, usually intermixed with nesting Snowy Egrets, Black-crowned Night-Herons, and in recent years Cattle Egrets.

In the San Luis Valley incubation and hatching in 1984 extended from early June to late July, yet from early May to late July in 1985 (Schreur 1987). Young usually fly by late July to mid August. Atlas dates, covering observations over several years, show an intermediate period from incubation to fledging, from mid June to early August.

BREEDING PHENOLOGY

CODE		# OF RECORDS	DATE RANGE
NB	Nest Building	1	8 Jul
ON	Occupied Nest	1	14 May
NE	Nest with Eggs	5	15 Jun–15 Jul
NY	Nest with Young	8	15 Jun–3 Aug
FL	Fledged Young	1	5 Aug

DISTRIBUTION

In North America, White-faced Ibises nest from central Mexico to Louisiana and Texas (mainly coastal) and through the Great Basin, with isolated colonies in Alberta, New Mexico, California, Montana, North Dakota, Iowa, and Kansas (Ryder and Manry 1994). A large population of mainly resident White-faced Ibises nests in South America (especially in Argentina) with no known mixing with North American birds (Hancock et al. 1992).

The Atlas confirms earlier reports (B&N, Ryder et al. 1979) that in Colorado most ibises continue to nest in the San Luis Valley, especially in marshes around Russell Lakes, between Head and San Luis lakes, at Adams Lake, and on Alamosa and Monte Vista NWRs. In 1982 and 1986, I found single nests at Riverside Reservoir, and in 1970 and 1976, five nests at Lower Latham Reservoir, south of Greeley. Post-Atlas, in 1996, two pairs nested on the Arapaho NWR. One nest contained two eggs, the second two eggs and three young on 24 July. Also in 1996 at least 11 pairs pioneered a new nesting site in Kiowa County, in a tangle of shrubs and trees at the outlet to Nee Noshe Reservoir—the tangle so thick that the precise number of nests could not be determined (Duane Nelson pers. comm.). The Atlas map omits these 1996 reports, which occurred after completion of Atlas field work.

Suspected nesting has been reported for the Mile High Gun Club near Brighton (before Atlas) and near Gunnison and Cortez (both Atlas fledgling reports). In northwestern Colorado, the species nests at

Browns Park NWR. Migrant ibises, reported from all 28 latilongs (Latilong Study), occur most often on the eastern plains and western plateaus. Most ibises leave Colorado in September, some as late as October. Based on band recoveries, most ibises from Colorado winter in Texas and Mexico (Ryder 1993, 1994).

A "boom or bust" species, breeding populations vary considerably from year to year, depending on water levels in favored marshes (Ryder 1967). In 1949 at least 12 pairs nested at Russell Lakes (Ryder 1950); in 1965 only 10 pairs bred in the whole San Luis Valley (Ryder 1967). During the Atlas an estimated 355 pairs nested on Alamosa NWR (Ron Garcia pers. comm.), and over 470 pairs in the Valley in 1993 (Rilling and Falzone-Schrim 1993).

Nests, eggs, and young suffer from human disturbance, overgrazing by livestock, and, especially in the San Luis Valley, heavy predation from magpies, ravens, and raccoons. Breeding populations in Texas, Utah, Nevada, and Oregon declined in the 1970s, probably because of pesticides, but increased in the 1980s and 1990s.

The U.S. Fish and Wildlife Service has at various times designated the Great Basin population of White-faced Ibises a "sensitive species," a "species of management concern," and a "category 2 candidate" for possible listing as a threatened species, based on the birds' limited number of consistent breeding sites and uncertain status (Sharp 1985). Subsequently Ehrlich et al. (1992) concurred with possible designation as "threatened" in the U.S. but said that listing required further study. In 1994 the CDOW considered it a species of "undetermined status." Refuge operators need to adopt management controls to keep this strange copper-colored marsh-dweller flying in the Colorado skies.

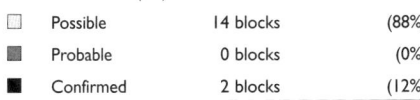

BREEDING EVIDENCE

REPORTED IN 16 (1%) OF 1745 PRIORITY BLOCKS

☐	Possible	14 blocks	(88%)
▩	Probable	0 blocks	(0%)
■	Confirmed	2 blocks	(12%)

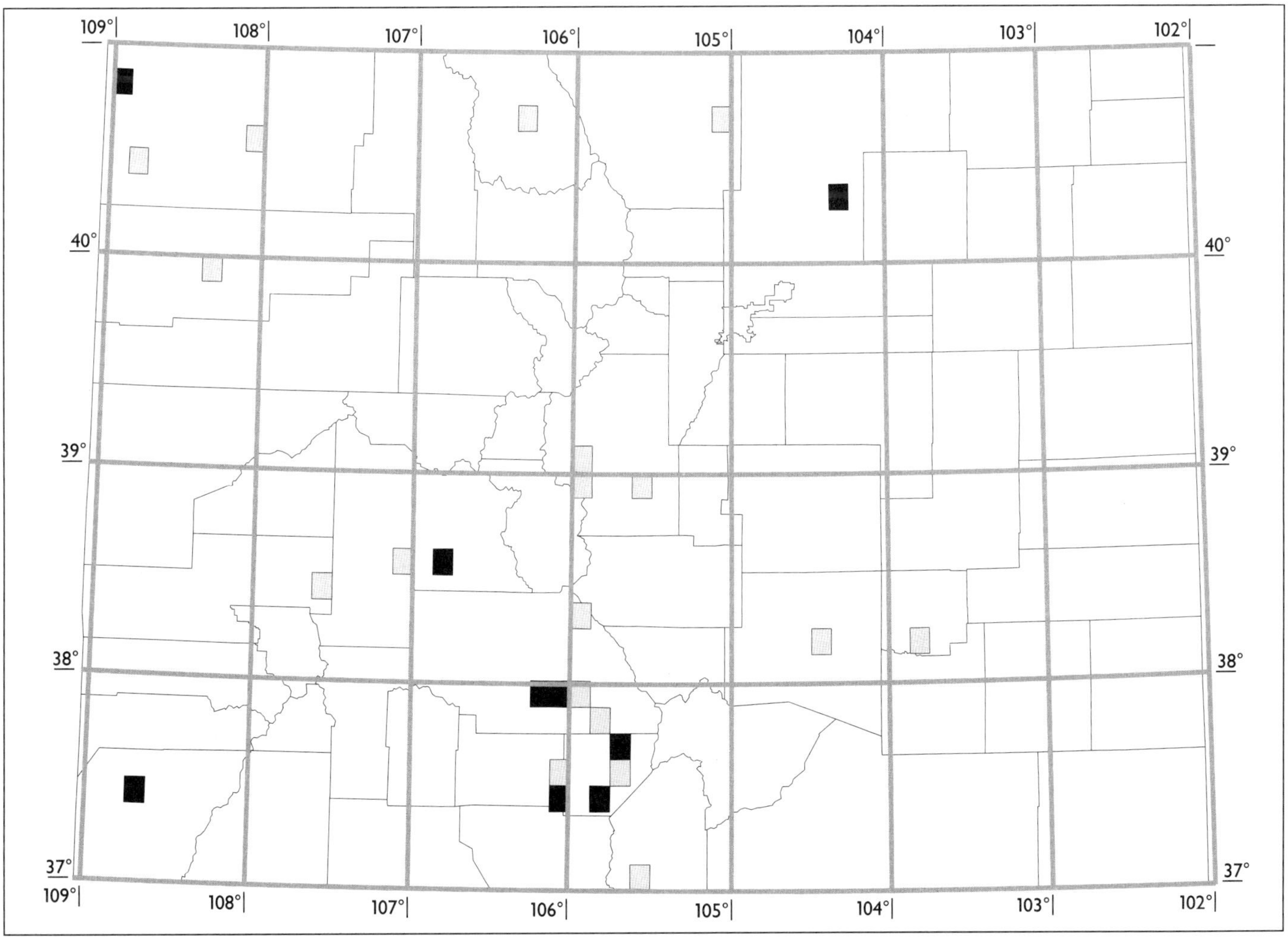
109°
108°
107°
106°
105°
104°
103°
102°
40°
39°
38°
37°

COLORADO DIVISION OF WILDLIFE / WILDLIFE RESOURCE INFORMATION SYSTEM

The chief avian scavengers of Colorado seek the dead by smell, not sight (Bierregaard *in* del Hoyo et al. 1994), yet they use sight, like a detective, to locate other vultures that have found carrion (Jerry Craig pers. comm.). Even as Turkey Vultures exemplify the perfection of static, soaring flight, secretiveness pervades their breeding behavior. To add to the contradictions about these magnificent carrion-eaters, current ornithological thinking posits that these birds are more closely related to storks than to raptors.

HABITAT

Extremely dry deserts, grasslands, temperate forests, and even dense tropical rain forests provide breeding habitat for Turkey Vultures. Encouraged by the abundance of animals killed on highways, vultures have expanded northward into new regions (Houston *in* del Hoyo et al. 1994).

Vultures cruise over most of Colorado's habitats except the high mountain valleys and peaks and the croplands and grasslands north of the Arkansas divide. The habitat reports, for the most part, represent habitats underneath foraging and soaring vultures. With that qualification, Atlas habitat codes identify the vultures' favorite haunts as the pinyon/juniper woodlands of the canyons, mesas, and plateaus across the state, particularly below 8,000 feet (2440 m)—one-fifth of all reports. The next most frequently reported category, shrublands, composed one-sixth of the reports, distributed among all the major shrub types. Atlasers also reported 15% soaring over various deciduous habitats, 8% over ponderosa woodlands, and 10% over other assorted conifers. Grasslands, dominated by shortgrass prairies, amounted to one-tenth of the total habitat reports.

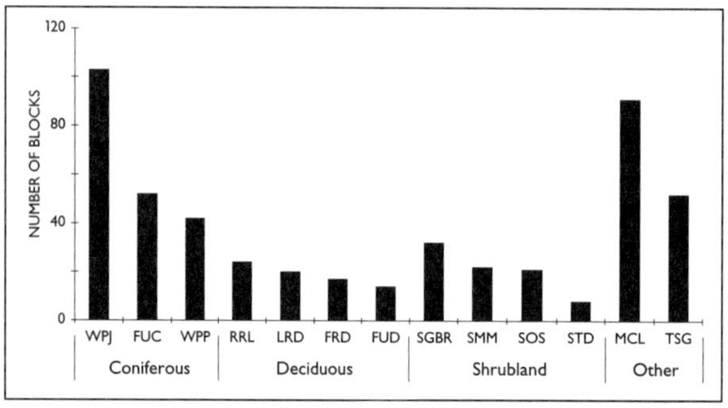
NUMBER OF BLOCKS
120
80
40
0
WPJ
FUC
WPP
RRL
LRD
FRD
FUD
SGBR
SMM
SOS
STD
MCL
TSG
Coniferous
Deciduous
Shrubland
Other

Soaring Turkey Vultures depend on thermal upwellings. They often need to cover many miles to find food. Foothills, low-elevation mountains, and canyons that receive excellent solar heating and exposure to westerly winds create abundant thermals. The conditions for flying may be just as important to the bird as the habitat type that it soars above (Ryser 1985).

Nest sites of Turkey Vultures often do not relate directly to habitat. Rock outcrops, tree cavities, and man-made structures offer nesting opportunities. Vultures even nest on the ground. All Colorado nests that observers have found have been in caves or recesses in cliffs and mesa walls. Watchers find them by studying vulture movements and by using their sense of smell to detect the pungent odors peculiar to food left at nest sites.

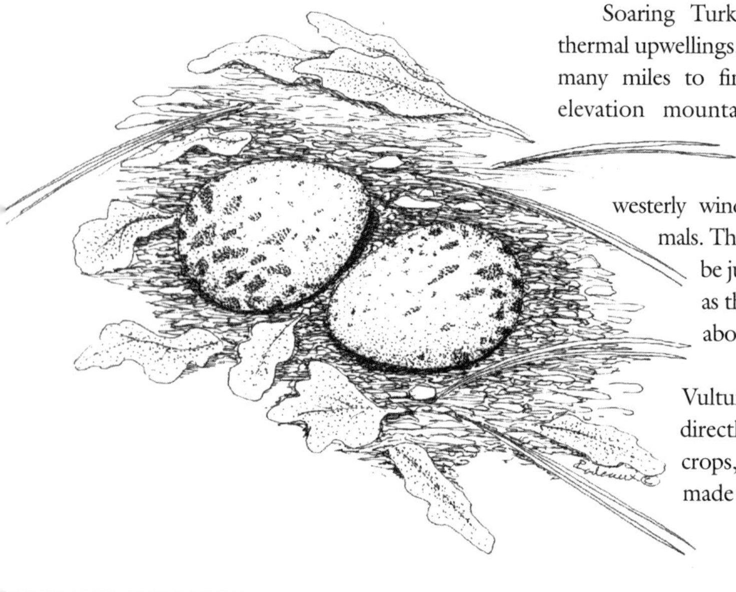

BREEDING

Atlasers found little breeding evidence for Turkey Vultures. They Confirmed nesting in only 10 of the 482 Priority blocks where vultures summer and may breed, and found only 12 nests (6 in Non-priority blocks). Nest sites reported by atlasers include rocky outcrops, Indian ruins, an aspen tree, and a cave. Vultures occasionally use abandoned nests of other species—e.g., an old hawk nest in a cottonwood in Nevada. In Mesa Verde National Park breeding vultures used Anasazi ruins nearly 100 years ago, and today vultures still use the same sites (Marilyn Colyer, Mesa Verde archives). Montezuma County had the most Atlas Confirmations, in five blocks.

The breeding cycle takes more than three months. Vultures lay eggs from April to June (Terres 1980). The eggs hatch in 38–41 days. The two young stay in the nest a lengthy 66–88 days—two to three months (Ehrlich et al. 1988).

Vultures are widespread, conspicuous summer residents in Colorado. Why should the massive Atlas effort glean so little breeding information? The answer involves several factors. Vultures do not reach sexual maturity until five years of age, and breeding may not occur every year thereafter. They nest in inaccessible locations. Feeding by regurgitation decreases movement to and from nest sites. Hawks and eagles advertise their nest sites by feeding their young freshly caught prey and making many trips daily to and from nests; in contrast vultures leave their nests in the morning and do not return until evening. As a result of all of these factors, nest sites often go undetected (Bierregaard *in* del Hoyo et al. 1994).

BREEDING PHENOLOGY

CODE		# OF RECORDS	DATE RANGE
C	Courtship	2	27 Jun–6 Jul
NE	Nest with Eggs	3	16 May–25 May
NY	Nest with Young	4	10 Jul–25 Jul
FL	Fledged Young	3	9 Jul–4 Aug

DISTRIBUTION

Turkey Vultures nest throughout southern Canada and the U.S. to Argentina, yet are the only truly migratory New World vultures. An average of 1.2 million vultures migrated over Veracruz, Mexico, in the falls of 1992–1996 (Ruelas Inzunza 1998). In North America, BBS data show vultures increasing at 1% per year.

Vultures find three-fourths of Colorado suitable for summer living. They decrease from southwest to northeast; on the plains north of the Arkansas Valley only 4% of the blocks reported them. The high mountain valleys—San Luis, Gunnison, Middle and North parks (which lack nest sites)—and the high peaks that surround them also had few vultures.

The canyons and mesas of southern and western Colorado and the Front Range contain the greatest densities. These habitats have cliffs or outcroppings where vultures can roost and, presumably, nest. In the Front Range foothills and in the two latilongs east of the Continental Divide on the New Mexico line, 60% of the blocks had vultures; in the San Juan Basin of southwestern Colorado they occurred in 44%. Only about 40 other Colorado species occurred in as many blocks as vultures did.

Caveat on the Atlas map: The map shows ***all*** observations of Turkey Vultures during the breeding season (1 May–31 July). It shows vultures as Possible breeders in 165 blocks where observers reported them as merely Observed. The nature of Atlas field work makes it difficult for field workers to distinguish between breeders and non-breeders due to three components of Turkey Vulture biology: they cover large areas in their feeding excursions—far greater than an Atlas block; they use such diverse sites for their nests that "suitable" breeding habitat is an anomaly; and many non-breeders populate the state.

Also, at the request of the atlaser, one Confirmed report on the plains has its location "masked" on the map in order to keep the site safe from disturbance.

BREEDING EVIDENCE

REPORTED IN 482 (28%) OF 1745 PRIORITY BLOCKS

Possible	444 blocks	(92%)
Probable	28 blocks	(6%)
Confirmed	10 blocks	(2%)

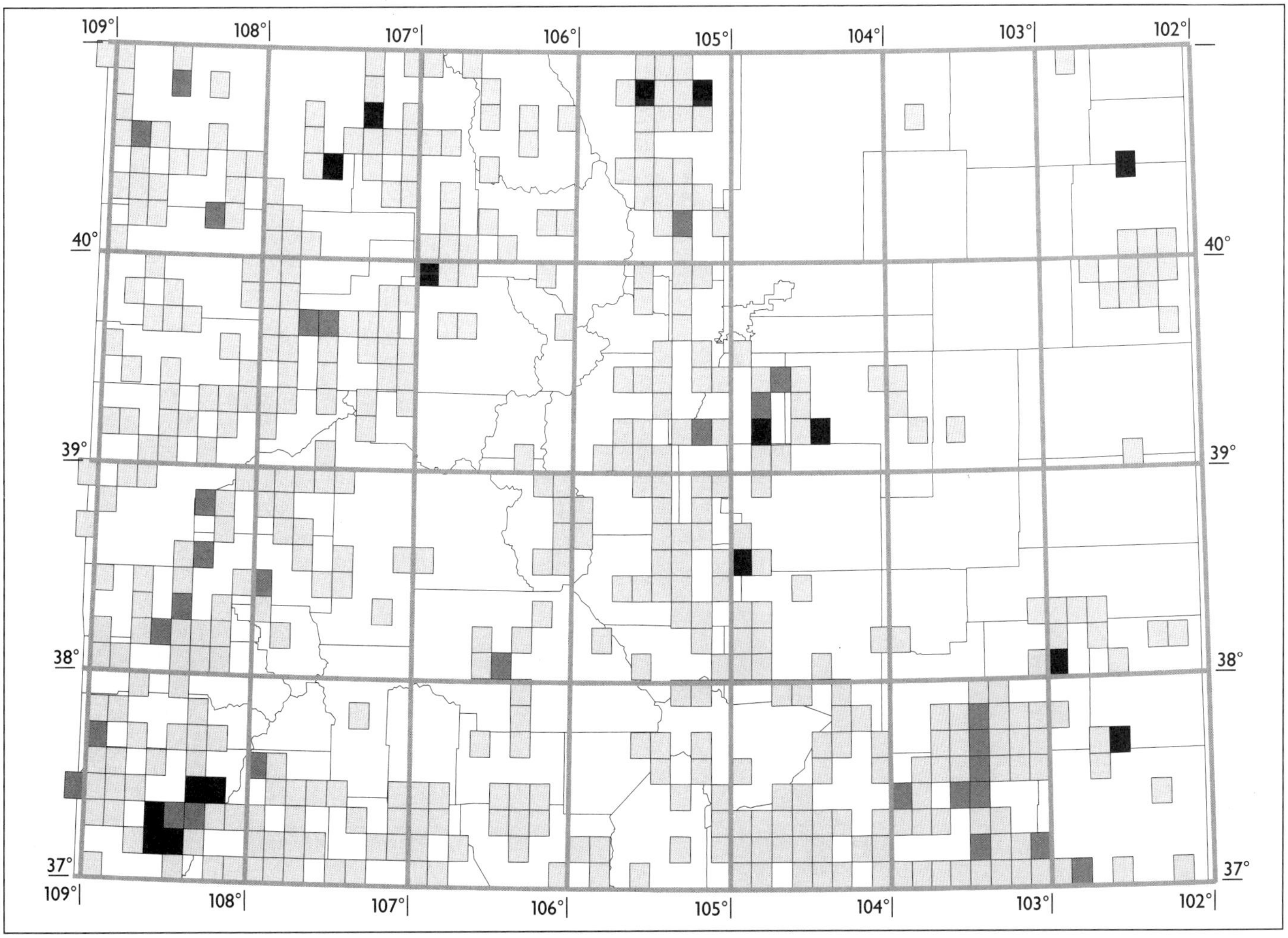
109°
108°
107°
106°
105°
104°
103°
102°
40°
39°
38°
37°

COLORADO DIVISION OF WILDLIFE / WILDLIFE RESOURCE INFORMATION SYSTEM

Never had resident Canada Geese lived along the northern Front Range until, in 1957, the Division of Wildlife released 40 geese near Fort Collins to try to develop a breeding flock. The project succeeded too well. Now resident birds number about 10,000, and in winter they entice another 50,000 migrating geese to stop instead of going on to New Mexico (Carty 1986). Hunting increased, but the geese increased more, and so did conflicts with the burgeoning numbers of people from Denver to Fort Collins—geese became pests (Szymczak 1975). CDOW has since transplanted geese elsewhere in Colorado and given them to other states as well.

HABITAT

For their breeding areas geese need water, secure nest sites, and nearby feeding areas. Resident geese need year-round open water. Geese prefer to nest on islands, if available, or in other protected sites where the gander maintains a clear lookout to watch for predators. Wherever they find a pond, marsh, or reservoir, geese may breed. They accept artificial nesting platforms (Carty 1986).

Atlasers located geese in wetlands, lakes, streams, and riparian (both lowland and foothills) habitats. Agricultural products benefit the geese, which graze particularly on corn, millet, and oats and on the green forage of lawns, golf courses, winter wheat, and native riparian plants (Bellrose 1980). Because geese graze and feed in croplands and grasslands, atlasers also reported rural, cropland, park, and urban habitats. Geese have grown so numerous on golf courses and city parks that they have become problems.

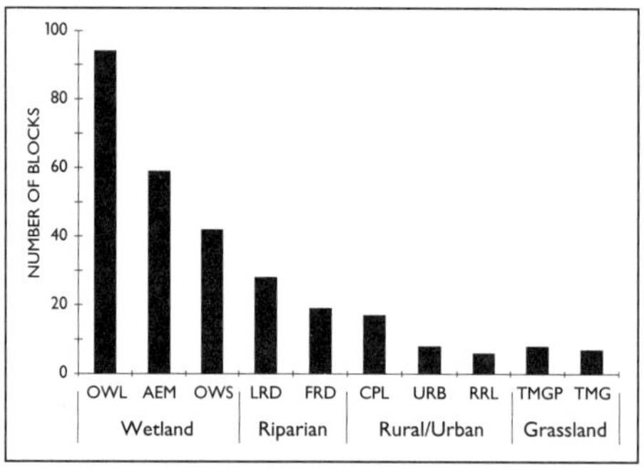
NUMBER OF BLOCKS
100
80
60
40
20
0
OWL AEM OWS
Wetland
LRD FRD
Riparian
CPL URB RRL
Rural/Urban
TMGP TMG
Grassland

BREEDING

Canada Geese mate for life (Kortright 1942) and return to breed in the place where they were raised. Atlasers found nests with eggs from 30 March into May, with a couple of late stragglers. Incubation, by the female alone, takes 28 to 30 days (Bent 1925). Despite young that atlasers found as early as 20 April, the majority hatch in May and June. Adults do not feed the goslings, but indirectly provide their young with food by defending a feeding site and protecting offspring against other geese (Friedl 1993).

From 63 to 86 days after hatching and before the young fly, the adults molt and are flightless (Bent 1925). During this period, the families may join other geese at larger, more secure reservoirs; adults and young may walk there, sometimes several miles. Families stay together until the following year. Then the young, because they do not usually breed until they are three years of age, form roaming flocks (Terres 1980).

BREEDING PHENOLOGY

CODE		# OF RECORDS	DATE RANGE
ON	Occupied Nest	8	31 Mar–6 Jun
NE	Nest with Eggs	10	30 Mar–16 Jul
NY	Nest with Young	8	29 Apr–13 Jul
FY	Feeding Young	2	23 May–20 Jul
FL	Fledged Young	133	20 Apr–4 Aug

DISTRIBUTION

Canada Geese migrate through and spend the winter in much of the southern U.S. Historically, most of them bred in the far north of Canada and Alaska. Each flock was distinct and several subspecies of different sizes developed. In Colorado, Canada Geese historically nested "at higher secluded lakes" in the mountains, especially in Middle and North parks ("in large numbers"—Coues *in* Sclater 1912), and also on the plains at Niwot (Cooke 1897, 1898). Apparently both flocks diminished and nesting birds disappeared from those places during the early twentieth century (B&N). The natural breeding flock in northwestern Colorado along the Yampa River, first reported in 1895 (B&N), forms part of the Great Basin Flock in Utah and Wyoming. Now wildlife biologists have created breeding flocks all over Colorado. The geese that atlasers saw (other than the Yampa River and Browns Park birds in Moffat County) have descended from transplanted geese.

This development of breeding flocks stems from CDOW's Canada Goose restoration program. It began in 1955 in the San Luis Valley with a few pinioned geese pairs. Goslings from these birds returned to the area the following years. In 1957, the goose-spreaders released 40 geese near Fort Collins (Szymczak 1975). They live-trapped some wild goslings. They took eggs from a captive flock southwest of Denver and used bantam hens as foster mothers. They erected nesting platforms. The project started slowly, but eventually ballooned—by 1967, they could move goslings from the flock in northeastern Colorado to other parts of the state. Several years ago CDOW initiated an annual gosling roundup at City Park in Denver to collect birds to move within and outside the state (Jo Marchant pers. comm.). By 1992 they had moved over 5,100 goslings to 36 other locations in Colorado (Mike Szymczak pers. comm.).

Atlas observations show that the geese transplanted to the mountains now nest to 10,000 feet (3048 m). They nest close to the headwaters of the Rio Grande in Hinsdale County and the Colorado River in Grand County. Those that breed at higher altitudes migrate to lower altitudes or farther south to find open water in winter. The greatest number, however, breeds along the urban Front Range corridor from Denver north. This population, protected by the dense human presence, faces little pressure from hunting.

Canada Geese now live as year-round residents in ponds and wetlands throughout Colorado except in the high mountains and dry eastern plains. Augmented by migrants from the north, Canada Geese winter on the plains, and up to 7,500 feet (2300 m) in the San Luis Valley (A&R).

The purpose of the breeding and stocking of Canada Geese was to provide hunting in the state, and it did obtain that objective. By learning to stay close to homes, parks, and golf courses—areas necessarily closed to hunting—geese have multiplied, uncontrolled. Sadly, these magnificent symbols of the wild North have become too numerous and, in many places, a nuisance.

BREEDING EVIDENCE

REPORTED IN 202 (12%) OF 1745 PRIORITY BLOCKS

☐	Possible	39 blocks	(19%)
▩	Probable	22 blocks	(11%)
■	Confirmed	141 blocks	(70%)

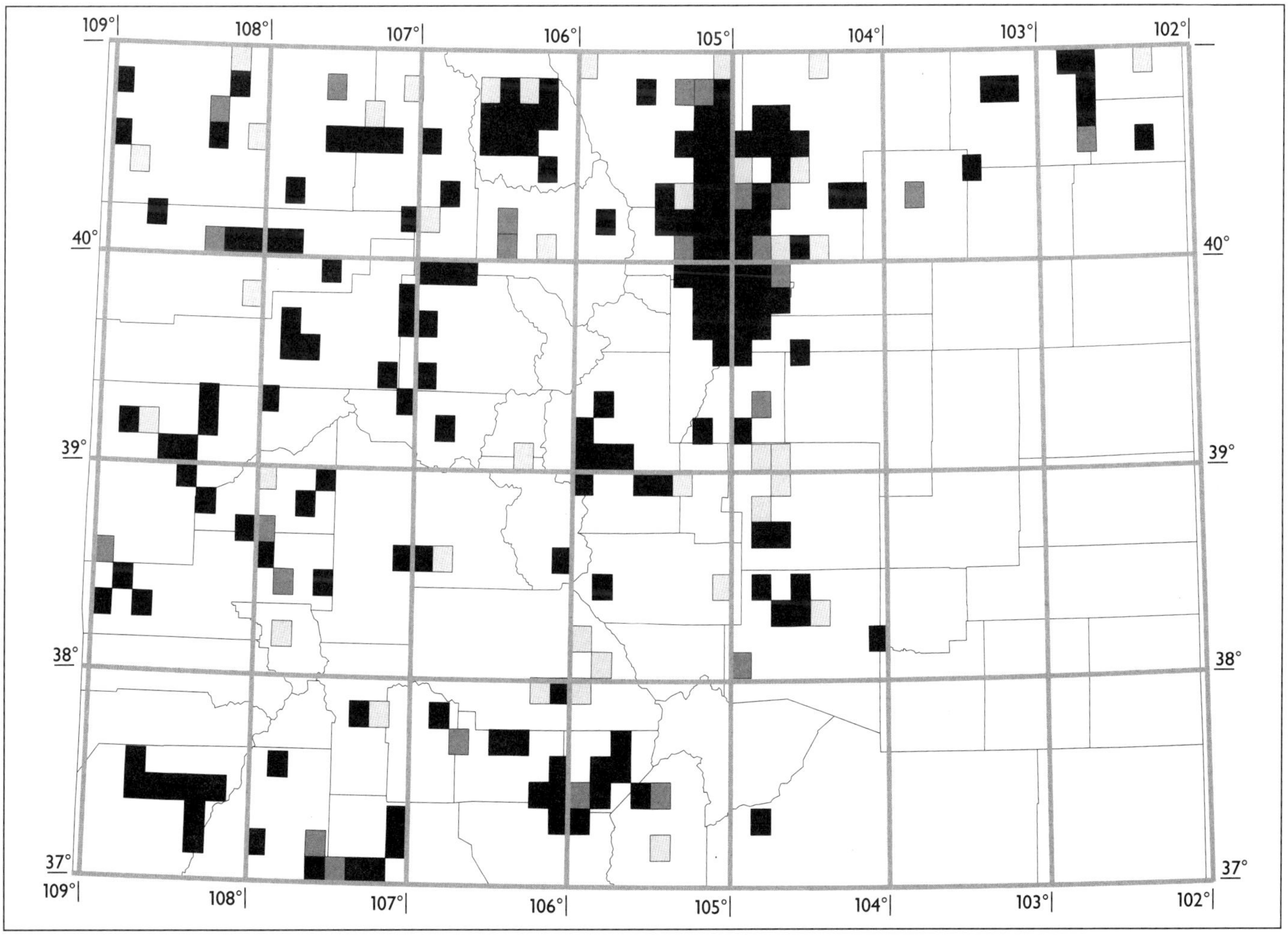
109°
108°
107°
106°
105°
104°
103°
102°
40°
39°
38°
37°

COLORADO DIVISION OF WILDLIFE / WILDLIFE RESOURCE INFORMATION SYSTEM

A remarkable spectacle occurs each March in the cottonwoods along the Colorado River near Grand Junction. The tree branches come alive, not yet with leaves, but with amorous Wood Ducks. Nearly unknown there as recently as 1970, these brilliantly colored ducks have established a thriving population in one of Colorado's most arid regions.

HABITAT

Wood Ducks seldom venture far from wooded country along watercourses. They fly through these woods with an ease befitting smaller woodland passerines. Although Colorado lacks the densely wooded riversides and backwaters that Wood Ducks prefer in other regions, the ducks have found homes in the state's lowland riparian habitats. Nest boxes may have reduced the need for trees with large cavities that Wood Ducks historically used, but they seem to shun nest boxes in many areas. In one case, a hole in a dirt bank provided an apparent nest site (Coen Dexter pers. comm.). Part of the success of Wood Ducks stems from their ability to nest in areas with extensive human activity (Hester and Dermid 1973).

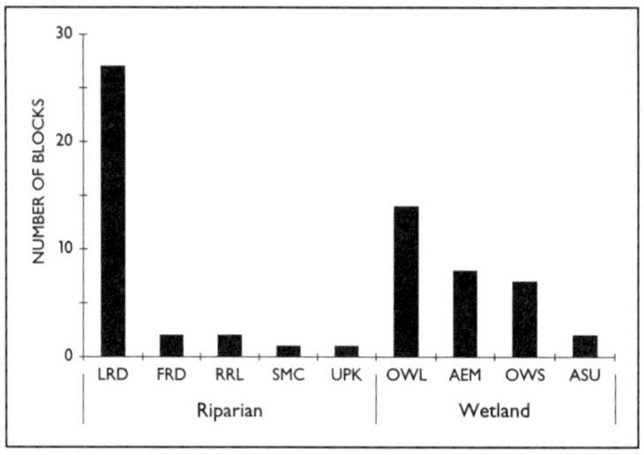
NUMBER OF BLOCKS
30
20
10
0
LRD
FRD
RRL
SMC
UPK
OWL
AEM
OWS
ASU
Riparian
Wetland

Colorado's primary Wood Duck concentrations fall along the lower reaches of three river systems—the Colorado, South Platte, and Arkansas. Populations concentrate almost entirely below 5,500 feet (1700 m) elevation, although a few Wood Ducks breed at higher elevations.

The cavities preferred by Wood Ducks measure at least 3.5 inches (9 cm) wide and have an internal diameter of about 8 inches (20 cm) or more; they prefer sites located directly above water (Johnsgard 1978).

BREEDING

Atlasers Confirmed breeding in 16 Priority blocks and several Non-priority blocks. Typical of duck species, fledglings provided the overwhelming majority of Confirmations. No Atlas records used the Courtship code because of the timing and places of pairing, which begins in the fall and continues into spring. Display sequences associated with pair formation occur primarily in early fall and last nearly twice as long as in the spring. Pair formation may take place at sites far removed from breeding sites. Once paired, Wood Ducks stay together through incubation (Armbruster 1982).

The early pair formation of Wood Ducks affords them a chronological advantage in selecting nest cavities. Because Wood Ducks do not appropriate already occupied sites (Hester and Dermid 1973), they must select nest sites before their competitors (both avian and mammalian).

When multiple pairs breed in an area, Wood Duck hens frequently engage in egg-dumping, i.e., laying eggs in the nest of another hen. Semel and Sherman (1986) noted dump nests in which the complement of eggs increased by as many as 8 eggs in a single day and, in one case, a nest that accumulated 37 eggs. As a defense against such nest parasitism, Wood Duck females behave surreptitiously near their own nest sites, which they do not approach in the presence of other females. They aggressively defend their nests from other females; sometimes they even expel intruders in the process of egg-dumping. Nest parasitism increases when females use boxes rather than natural sites (Semel and Sherman 1986). In 1996 a nest box in Englewood contained 2 unhatched Wood Duck eggs and 2 unhatched Hooded Merganser eggs (Cat Anderson pers. comm., Kingery in press).

On the positive side, well-designed nest boxes can provide greater protection from predators than natural cavities. Principal nest predators of Wood Ducks include squirrels and raccoons (Bellrose 1976).

BREEDING PHENOLOGY

CODE		# OF RECORDS	DATE RANGE
ON	Occupied Nest	2	6 May–8 Jun
NE	Nest with Eggs	1	6 May
FL	Fledged Young	10	24 May–18 Jul

DISTRIBUTION

In 1976 Wood Ducks nested regularly in all states east of the Mississippi River and along the Pacific Coast. By 1990 they had expanded into Idaho and Montana, as well as Colorado.

Reports from the early years of Colorado statehood accorded Wood Ducks "rare" status (or did not even mention them), with one dubious exception. Morrison (1888) reported the Wood Duck as common along the La Plata River "headwaters" near Durango. No population has existed there in this century, and Wood Ducks have only rarely visited southwestern Colorado since that time. Yet the only breeding record cited by B&N came from Dolores County, 25–50 miles across the San Juan Mountains from the La Plata but far removed from any current concentrations. Atlasers found none in either area.

The Atlas map depicts thriving Wood Duck populations along the South Platte River from Waterton Canyon to Fort Morgan. It also reveals smaller, yet substantial, pockets of breeding Wood Ducks between Boulder and Fort Collins, in the vicinity of Bonny Reservoir, near Pueblo, and a surprising group scattered in several blocks across southeastern Colorado, as well as the amorous Grand Junction flock.

Although the current distribution may reflect the species' extensive range expansion across the Great Plains during the 1950s and 1960s (Bellrose 1976), introductions also contributed. The Pueblo population, in particular, received considerable impetus from introductions (David Silverman pers. comm.).

Atlas field work documented breeding in eight new latilongs, including one in southeastern Colorado without any prior observations. Although most Wood Ducks leave the state by November, the river valleys that support the largest populations host a few birds year round.

Wood Ducks proved especially susceptible to unregulated hunting in the nineteenth and early twentieth centuries. By 1900, their once-plentiful numbers had plummeted to the point that extirpation across much of their range appeared likely. Regulation of hunting and the Migratory Bird Treaty Act of 1918 helped to bring them back from potential disaster (Bellrose 1976). Today, the greatest threats come from cutting dead or decaying trees and from clearing, draining, and filling of swamplands for development and agriculture. Despite all this, Wood Duck populations show a general upward trend. BBS data reveal a substantial increase across the continent of 7% per year with increases on three-fifths of the routes.

BREEDING EVIDENCE

REPORTED IN 46 (3%) OF 1745 PRIORITY BLOCKS

Possible	13 blocks	(28%)
Probable	17 blocks	(37%)
Confirmed	16 blocks	(35%)

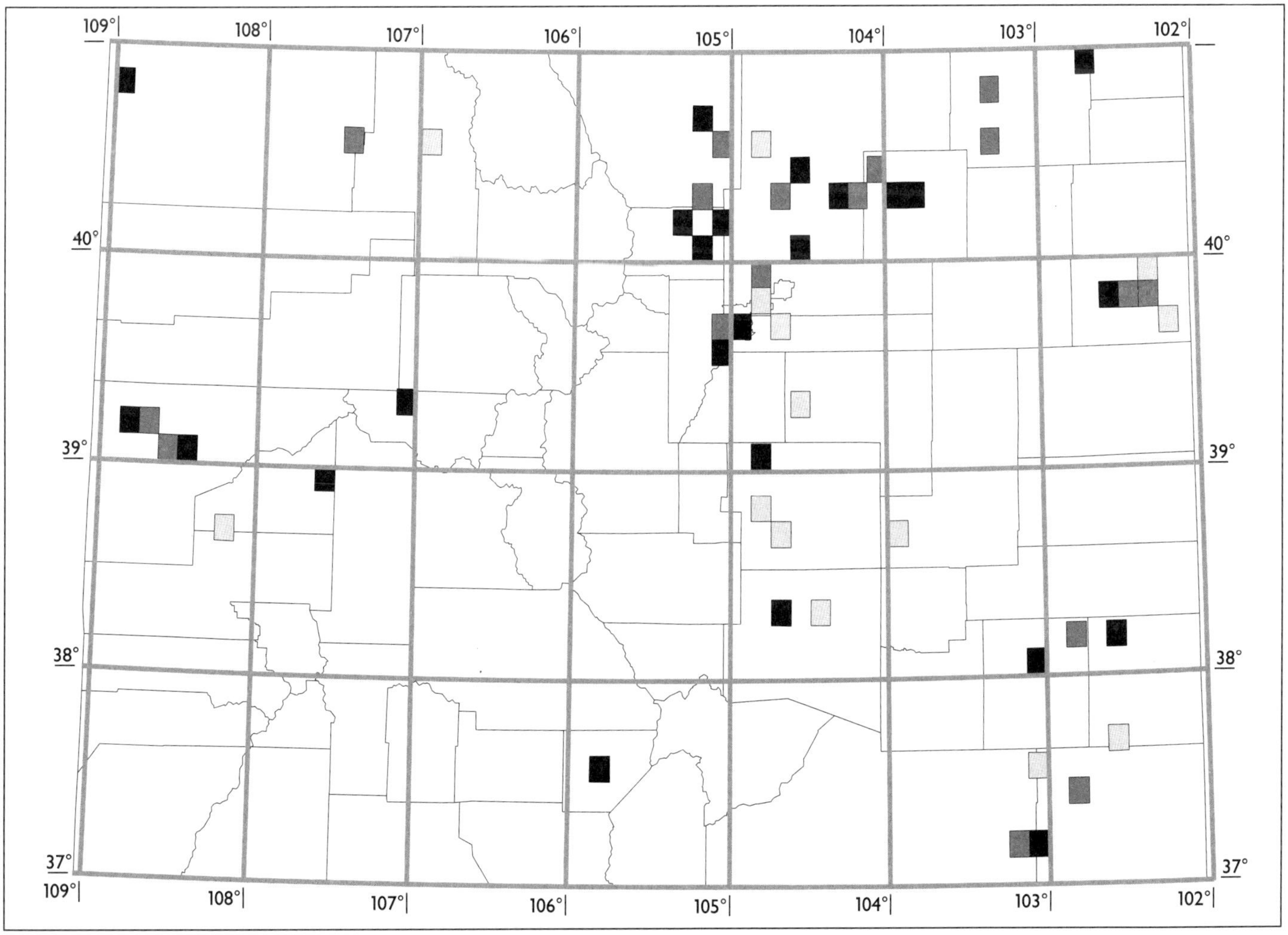
109°
108°
107°
106°
105°
104°
103°
102°
40°
39°
38°
37°

COLORADO DIVISION OF WILDLIFE / WILDLIFE RESOURCE INFORMATION SYSTEM

Plain country cousins among the typically more colorful puddle ducks, Gadwall drakes and hens both wear somber, cryptic plumage, similar to female Mallards and American Wigeons. Gadwalls inhabit a vast range across the Northern Hemisphere, but seem to be common nowhere. North American population counts in this century show wide fluctuations as mysterious as the origins of this duck's common name (Terres 1980).

HABITAT

Gadwalls breed in more stable, permanent wetlands than most other puddle ducks (Johnson and Grier 1988). They prefer shallow marshes and ponds with plenty of wetland vegetation (Bellrose 1980). They feed almost entirely on plants, often underwater leaves and stems rather than the hard seeds preferred by other dabblers, and they seldom venture into upland grain fields.

Gadwalls typically nest in upland sites, within 300 feet (90 m) of water, in dense vegetation including weed patches, nettles, and tall rank grass of ditchbanks and dikes. At Monte Vista NWR Gadwalls choose the densest vegetation available fairly close to water (Gilbert et al. 1996). Atlasers found few nests; they nearly always established breeding by observing ducklings or adult pairs. Wetland vegetation or water made up 89% of habitat observations (lakes six times as often as streams), with another 6% from riparian forest or shrubland, and the rest mainly open rural and field habitats. In three Atlas nest reports, nest vegetation included montane carr and grassland, saltgrass meadow, emergent vegetation, and upland deciduous forest—the last a Routt County nest placed in tall grass and wildflowers at the base of an aspen tree.

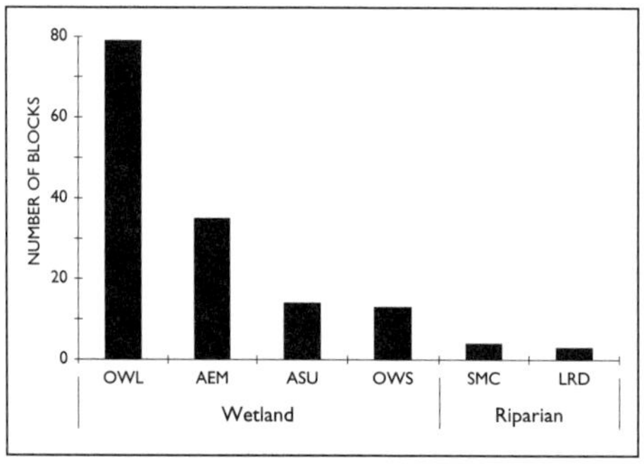
80
60
40
20
0
NUMBER OF BLOCKS
OWL
AEM
ASU
OWS
SMC
LRD
Wetland
Riparian

Like other dabblers, Gadwalls move to expansive marshes for the wing molt in late summer, and during fall migration and winter they use many kinds of wetlands including rivers, large lakes, and impoundments (Bellrose 1980).

BREEDING

Gadwalls, like other puddle ducks, form pair bonds in winter (Oring 1969). On the breeding grounds in spring, pairs nest in loosely defined territories that seem to center on the hen's location (Duebbert 1966). Atlas records show that Gadwalls start to nest at about the same time as most Coloado dabblers. Records for nests with eggs and ducklings suggest nest initiation mainly in late May and the first half of June, as reported throughout their far-flung range (Bellrose 1980). However, a few occasionally break ranks to nest early, as shown by three Atlas observations of ducklings by 1 June, and B&N's report of downy young on 18 May near Longmont.

Gadwall nest success averages over 67%, higher than any other dabbling duck (Bellrose 1980). Their late-nesting habits and choice of dense nesting cover presumably reduce nest losses to predators and harsh weather. Broods average about 50 days to fledge (Oring 1968), leaving hens barely enough time to undergo wing molt in late August and September before fall migration begins.

BREEDING PHENOLOGY

CODE		# OF RECORDS	DATE RANGE
ON	Occupied Nest	2	30 May–16 Jun
NE	Nest with Eggs	3	21 Jun–30 Jun
FL	Fledged Young	39	29 May–5 Aug

DISTRIBUTION

Gadwalls roam the entire Northern Hemisphere, mostly breeding north of and wintering south of 40 degrees latitude. North American breeding grounds center on the northern prairies, but unusually large numbers also breed in western mountains and basins (Bellrose 1980). Colorado hosts large numbers of Gadwalls as spring and fall migrants in the Great Plains, western valleys, and mountain parks (A&R, B&N). The Atlas map shows breeding concentrations in the mountain parks and reservoirs on the plains along the Front Range. It includes scattered breeding records throughout the rest of the state, in the mountains up to about 9,000 feet (2745 m). Banding recoveries show that most Colorado breeders winter in Texas or Mexico (Szymczak 1986).

Gadwall populations in the northern prairies cycle enigmatically (Bellrose 1980), but during droughts in the 1980s when most other dabblers declined, Gadwall counts increased. BBS data show large average annual increases of almost 6% continent-wide. In Colorado breeding Gadwalls seem uncommon generally (Cooke 1897, B&N); a few concentrations in mountain parks account for most Colorado production. In the 1980s Gadwalls made up 15% of the nesting ducks in North Park, and 10% in the San Luis Valley (Szymczak 1986), as they did in the 1940s (Ryder 1951, Wampole 1951). Gilbert et al. (1996) reported nest densities at Monte Vista NWR of $51/mi^2$ ($20/km^2$), more than the highest estimates for the prairie-pothole region (Bellrose 1980).

Atlas data reflect the concentration of breeding Gadwalls in mountain parks; atlasers usually rated block abundance as code A2, uncommonly A3, and only two blocks as A4, both near Alamosa.

Protecting Gadwall breeding populations in Colorado requires conservation of existing wetlands, especially the extensive marshes in the mountain parks. Livestock grazing and other land uses that reduce upland nesting cover reduce Gadwall nest production (Kirsh 1969). Maintaining water quality and hydrology of wetlands will also help to ensure that Gadwalls flock to Colorado marshes for years to come.

BREEDING EVIDENCE

REPORTED IN 124 (7%) OF 1745 PRIORITY BLOCKS

Possible	35 blocks	(28%)
Probable	51 blocks	(41%)
Confirmed	38 blocks	(31%)

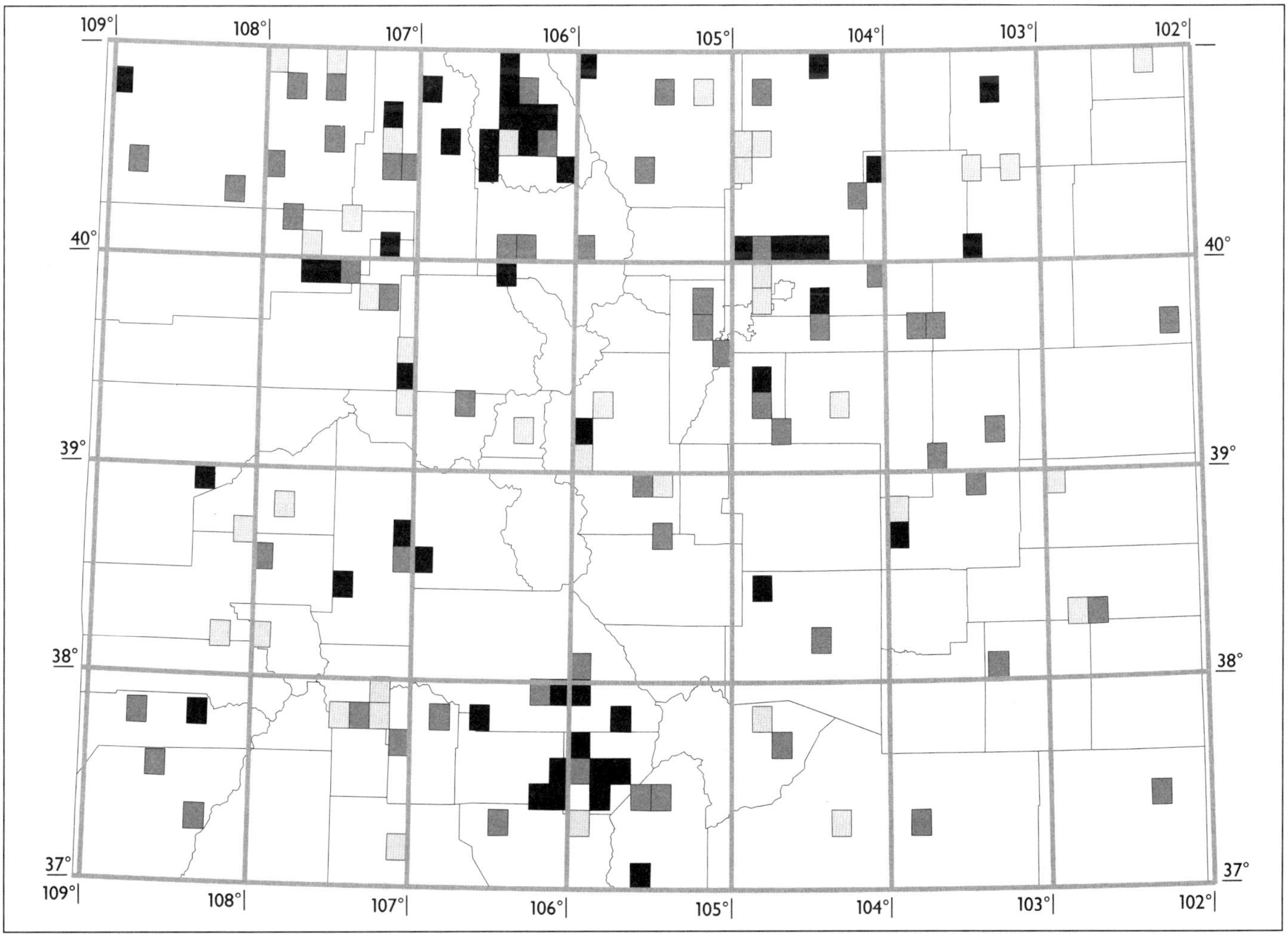
109°
108°
107°
106°
105°
104°
103°
102°
40°
39°
38°
37°

COLORADO DIVISION OF WILDLIFE / WILDLIFE RESOURCE INFORMATION SYSTEM

This duck kept its good looks but not its name through 40 years of nomenclatural turmoil. First the AOU (1957) took away its distinctive, descriptive, hunter's name, "Baldpate," and then, in 1983, took the "d" out of "Widgeon." The AOU could not change the flashy white cap—the "bald pate"—nor substitute a mundane quack for the unique whistle.

HABITAT

American Wigeons utilize many watery habitats in Colorado during much of the year, but retreat to shallow, permanent or semi-permanent ponds and marshes for nesting. For nesting they like marshes with open waterways and exposed shorelines; they do not use small temporary ponds as do Mallards and pintails (Palmer 1976a), nor do they like fast-flowing streams. Of 14 Atlas OWS (streams) records, eight came from the meandering streams of North Park.

Wigeons, both more terrestrial and more inclined toward deep water than any other *Anas* ducks, spend much of their time grazing on the shore (Palmer 1976a). Yet they feed extensively in open water as well, in company with Redheads, Canvasbacks, and other diving ducks. Here they become pirates—when a diving duck brings up a beakful of wild celery or eelgrass, wigeons often rob them of the bounty (Kortright 1942, Palmer 1976a). Even ducklings, shortly after hatching, engage in piracy (Palmer 1976a).

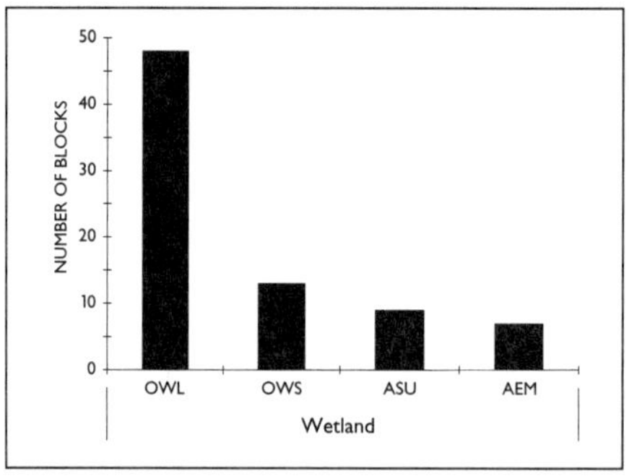
NUMBER OF BLOCKS
50
40
30
20
10
0
OWL
OWS
ASU
AEM
Wetland

Female wigeons often place their nests a "considerable distance from the water." They choose dry upland sites from a few yards to as many as 400 yards from water (Bellrose 1976) and along the Yukon River, as far as one-half mile (Bent 1923).

BREEDING

Like most waterfowl, American Wigeons begin pair formation on the winter grounds, although as many as 20% may delay pairing until reaching the breeding grounds (Bellrose 1976). The courtship flight begins by two males approaching a female on the water. When the hen takes flight, the males follow, with an individual male darting ahead of the female, setting his wings into a glide, and throwing his head up, thus displaying his attributes (white crown and wing patches) to the female (Wetmore 1920). The pair bond has a short duration; most males abandon their mates shortly after initiation of incubation and head off to safe areas such as large marshes. There they undergo prebasic molt, during which they remain flightless (about a month; Bellrose 1976).

Females incubate the eggs and raise the young by themselves. Females often leave to molt before their broods can fly, though usually not until the ducklings have advanced fairly far through the preflight stage. Often several broods will join together under one or several females (Palmer 1976a). Abandonment by the mother hen may generate the habit of duckling piracy.

The nests resemble those of other ducks, and the hens hide them effectively. When they leave the nest they often cover the eggs with down, making them invisible (Bent 1923). The average clutch size for the species is 8.5 eggs. The incubation period ranges from 23 to 25 days (Hochbaum 1944, Delacour 1956); the time between hatching and first flight ranges from 37 to 48 days (Hochbaum 1944, Hooper 1951). Authorities disagree about the timing of the breeding cycle. Bellrose (1976) says that first clutches in much of the species' range (Alberta to Alaska) come in May, whereas Bent (1923) says that wigeons observe a later nesting cycle than many ducks and most frequently do not start incubation until June. Atlas data support the latter for Colorado.

Atlasers recorded many pairs in June through 11 July and found only two nests, one on 28 June, the other on 18 July. Atlasers found most broods between 7 and 27 July, with one late brood reported on 18 August—a span of 42 days. More southerly-breeding birds possibly start nesting earlier than those breeding in North Park.

The fall and spring migrations of this species in Colorado occur mainly from late September to early November and mid March to mid May, respectively (A&R). Most recoveries outside Colorado of wigeons banded in the state have come from New Mexico, California, Canada, and Mexico, but a hunter shot one as far east as Maryland (B&N).

BY TONY LEUKERING AND HUGH E. KINGERY

BREEDING PHENOLOGY

CODE		# OF RECORDS	DATE RANGE
C	Courtship	1	2 Jun
NE	Nest with Eggs	2	28 Jun–18 Jul
FL	Fledged Young	16	7 Jul–18 Aug

DISTRIBUTION

American Wigeons breed from coast to coast through Canada (although eastern Canadian populations are small) and Alaska. In the lower 48 states, they nest in the prairie-pothole region west to the Cascades and south in the Rockies to northern New Mexico. A&R depicted this species as occurring in summer in most of the Colorado mountains and the two main drainages in eastern Colorado, but the state actually has few breeding records away from Browns Park NWR and North and Middle parks (Ron Garcia, Mike Szymczak pers. comm.).

Atlas data place the center of abundance in North and Middle parks; Jackson and Grand counties supplied more than one-third of the blocks that recorded wigeons and all but four of the Confirmations. Wigeons occasionally breed anywhere within the summer range, as is shown by two Confirmations by the Atlas in lowland areas, one near Carbondale and one on the plains at Jackson Reservoir.

BREEDING EVIDENCE

REPORTED IN 62 (4%) OF 1745 PRIORITY BLOCKS

☐	Possible	19 blocks	(31%)
■	Probable	30 blocks	(48%)
■	Confirmed	13 blocks	(21%)

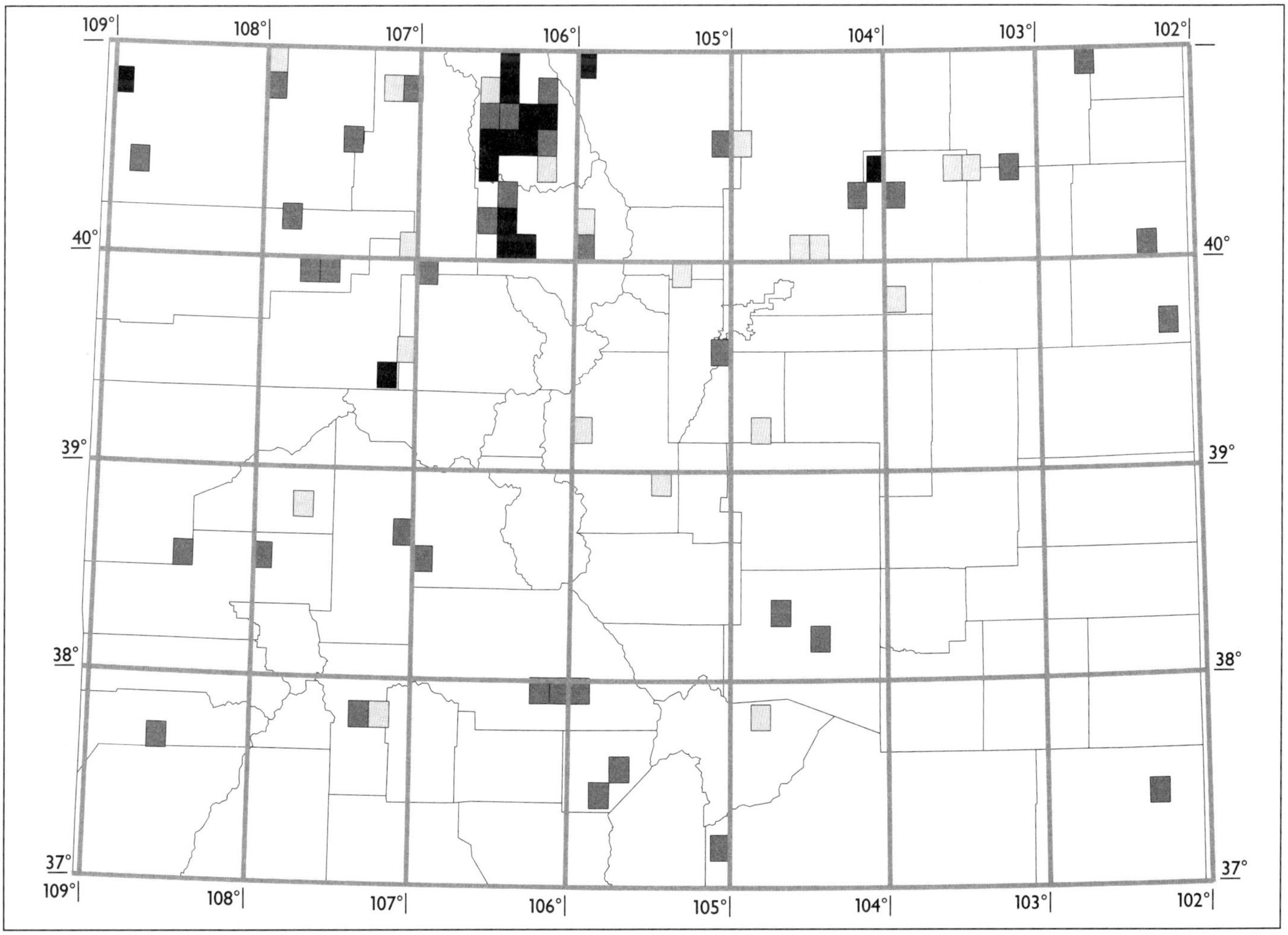
109°
108°
107°
106°
105°
104°
103°
102°
40°
39°
38°
37°

COLORADO DIVISION OF WILDLIFE / WILDLIFE RESOURCE INFORMATION SYSTEM

For naturalists, hunters, and other people who have business in marshes, the sight and sound of Mallards conjure thoughts of the wildness and bounty of wetlands and the wonder of migration with the changing seasons. Tamed since antiquity, Mallards have fathered all races of domestic ducks except the Muscovy (Terres 1980). Drakes, or "greenheads," wear their showy plumage to impress hens, who choose a mate based partly on his good appearance (Weidmann 1990).

HABITAT

Although wetlands figure prominently in Mallard life, these generalists (as ducks go) use a tremendous variety of habitats at different times of year. Atlas workers recorded Mallards on water or in aquatic vegetation in 72% of habitat observations, with another 12% in riparian vegetation; the remaining habitat types included mostly open upland habitats.

Mallards choose various forested, open, wetland, or upland situations for nesting. The 63 Atlas records that combined nest observations with vegetation type divided evenly between aquatic vegetation (35%), riparian situations including forest, willow, and meadow (33%), and upland sites including cropland, shrubland, and grassland (32%). Although Mallards sometimes nest as far as a mile from water, they tend to nest in denser vegetation and nearer to water than other dabblers (Lokemoen et al. 1990, Gilbert et al. 1996), and often exploit

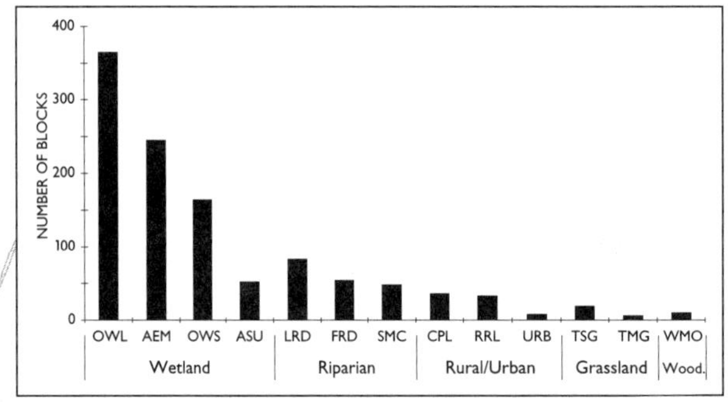
400
300
200
100
0
NUMBER OF BLOCKS
OWL AEM OWS ASU LRD FRD SMC CPL RRL URB TSG TMG WMO
Wetland
Riparian
Rural/Urban
Grassland
Wood.

ephemeral or semi-permanent wetlands (Johnson and Grier 1988, Rotella and Ratti 1992). Nesting hens feed on invertebrates to increase dietary protein; shallow ephemeral wetlands, which warm quickly in spring, produce rich blooms of invertebrates sooner than do deeper, permanent waters (Ringelman 1991). Hens usually take broods to wetlands with ample emergent vegetation for security against predators and harsh weather (Ringelman 1991). Outside of the breeding season, most Mallards move to larger marshes and rivers and often forage on waste grain in farm fields.

BREEDING

Bellrose (1980) observed, "Mallards and Pintails vie with each other in the northward race back to the breeding grounds," and in Colorado Mallards nest as early as winter's retreat allows. Birds pair in fall and early winter, and then migrate to breeding grounds where pairs select territories. In Colorado, Mallards stake territories from mid March through May, depending on elevation (Ringelman 1991). Hens build a ground nest of coarse plant matter, well hidden in dense vegetation. Atlasers found relatively few nests, and most often Confirmed breeding by observing ducklings. Hens lay a clutch of usually nine eggs, then incubate about 28 days (Bellrose 1980). Frequency of Atlas NE records peaked in the last half of May; late records extending into July probably reflect renests.

Hatchlings huddle in the nest for only 12–24 hours before the hen walks the brood to water, where they stay until fledging. About this time drakes also abandon the nest territory, moving to large wetlands to molt. After ducklings fledge, hens also move to molting areas, but later and often in different areas than drakes. The four-month span of Atlas fledgling observations delineates the longest breeding season of any Colorado duck.

BREEDING PHENOLOGY

CODE		# OF RECORDS	DATE RANGE
C	Courtship	2	15 May–17 May
NB	Nest Building	3	23 May–14 Jun
ON	Occupied Nest	14	5 May–30 Jun
NE	Nest with Eggs	27	27 Apr–8 Jul
NY	Nest with Young	9	25 May–24 Jul
FL	Fledged Young	421	7 May–31 Aug

DISTRIBUTION

Mallards occur throughout the Northern Hemisphere, perhaps the most common wild duck in the world (Bellrose 1980). About 10 million breed in the North American heartland, and substantial numbers breed in the western mountains from Alaska to Arizona.

In Colorado Mallards may breed in any small wetland, but they concentrate on rivers, small reservoirs of the South Platte and Cache la Poudre watersheds, and wetlands of the mountain parks (B&N, Szymczak 1986). Atlas breeding records came from virtually all parts of the state except alpine tundra and completely waterless prairies.

Relying on censuses of only the main breeding centers, CDOW estimates that Colorado hosts 30,000 breeding pairs (CDOW 1989); Atlas data suggest a population at least twice that by dispatching atlasers throughout the state, the Atlas detected more Mallards than CDOW's surveys conducted only at core waterfowl concentrations. Atlasers found Mallards breeding in every county, and one or two pairs in small ponds throughout the state.

At Monte Vista NWR, where breeding ducks reach densities as high as any in North America, Mallards make up half of the breeding ducks, with 427 pairs/mi^2 (165 pairs/km^2); Ryder 1951, Gilbert et al. 1996). In North Park, Mallards make up one-quarter of the breeding ducks (Szymczak 1986).

Banding returns show that most mountain park nesters remain in Colorado in winter, when they move to large, lower-elevation wetlands and reservoirs (Szymczak 1986); migrants from the north more than double their ranks.

The ability to use ephemeral wetlands created by agriculture has probably helped Mallards overcome natural wetland losses. BBS data show large average annual increases continentally and in Colorado, especially on western Colorado routes.

Mallards hold tremendous appeal as game animals, and Colorado hunters took about 87,000 in the 1995–1996 season (Jim Gammonly pers. comm.). Much public and private effort goes toward management and enhancement of wetlands for waterfowl, but such gains may only offset ongoing habitat losses. Livestock grazing in wetlands and riparian areas reduces vegetation density and decreases nest density and success rate of Mallards (Kirsch 1969, Gilbert et al. 1996). Without question, Mallards have benefited tremendously from waste grain left in fields (Bellrose 1980), but more efficient farming coupled with losses of farm ground and wetlands to urban development continues to reduce the value of rural landscapes as Mallard habitat.

BREEDING EVIDENCE

REPORTED IN 862 (49%) OF 1745 PRIORITY BLOCKS

□	Possible	173 blocks	(20%)
▩	Probable	232 blocks	(27%)
■	Confirmed	457 blocks	(53%)

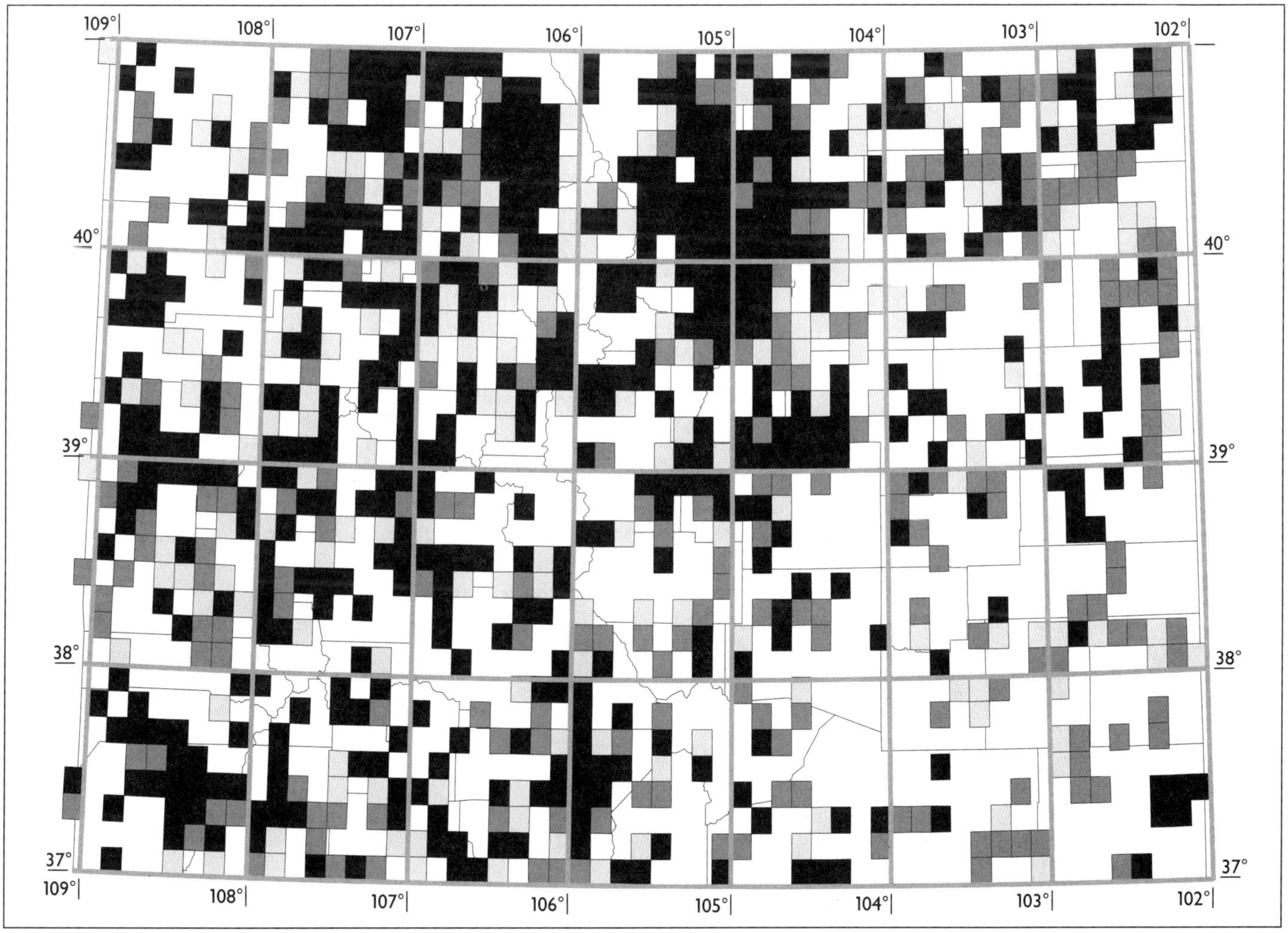
109°
108°
107°
106°
105°
104°
103°
102°
40°
39°
38°
37°

COLORADO DIVISION OF WILDLIFE / WILDLIFE RESOURCE INFORMATION SYSTEM

These small, fast-flying teal make a handsome sight darting across a marsh with blue wing patches "glistening like polished steel," as Audubon described them. Blue-wings arrive late to breeding areas and leave early, for good reason: they make the longest migrations of any North American duck, mostly to South America, up to 4,000 miles (6400 km) from breeding sites.

HABITAT

Phillips (1922) concluded that "bluewings never met a surface water they did not like"—that they use almost any marsh, lake, pond, ditch, or even rain puddle. An omnivorous diet helps this very adaptable duck to inhabit diverse wetlands and to adapt to changing conditions within those wetlands (Botero and Rusch 1994, DuBowy 1985). More than most ducks, they thoroughly exploit temporary or seasonal wetlands for breeding (Johnson and Grier 1988). Atlas workers found Blue-wings widely distributed across the state in wetland habitats (82% of habitat observations, with riparian vegetation accounting for another 9%).

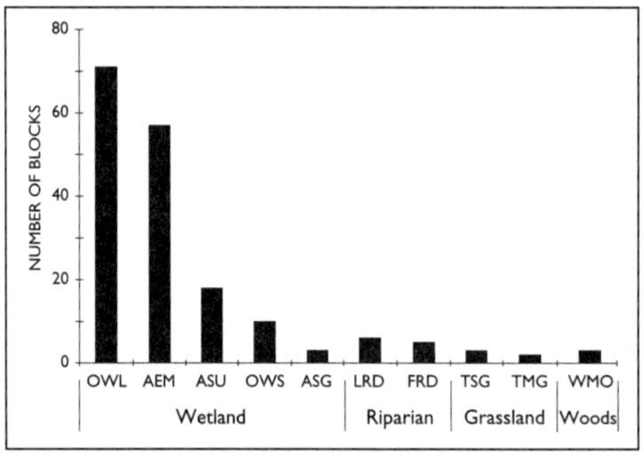
NUMBER OF BLOCKS
0
20
40
60
80
OWL AEM ASU OWS ASG LRD FRD TSG TMG WMO
Wetland
Riparian
Grassland
Woods

Hens usually conceal nests in dense and fairly tall cover, including rank grass along ditch banks and roadsides (Lokemoen et al. 1990) or damp sites adjacent to marshes including grass and sedge meadows, bulrushes, and grass mixed with cattails (Bellrose 1980, B&N). Of nests reported by atlasers who included vegetation information, four of five occurred in emergent vegetation. Gilbert et al. (1996) reported grass/sedge nest sites most commonly for all teal species at Monte Vista NWR.

BREEDING

Monogamous pairs form anew each year on winter ranges, and much courtship takes place during spring migration and the first few weeks on the breeding ranges (B&N). Birds arrive in Colorado from distant winter ranges in April and begin nesting later than Mallards and other teal but earlier than most other ducks. Egg-laying begins in late May or early June (B&N), confirmed by Atlas dates for nest-building and nests with eggs. Observations of hens with broods did not peak until mid July. During nesting drakes defend a small area around the hen; near the end of incubation they abandon the family and make sometimes extensive molt migrations to large marshes with plenty of cattails or other emergent vegetation, where they remain flightless for about a month (McHenry 1971).

Blue-winged Teal average the poorest nest-success rate of any dabbler, usually well under 50% (Bellrose 1980), but, strangely, are the least likely to renest following nest destruction (Keith 1961).

BREEDING PHENOLOGY

CODE		# OF RECORDS	DATE RANGE
C	Courtship	1	10 Jun
NB	Nest Building	1	22 May
ON	Occupied Nest	1	7 May
NE	Nest with Eggs	4	16 May–20 Jun
FL	Fledged Young	32	21 May–5 Aug

DISTRIBUTION

Blue-winged Teal breed only in North America; they concentrate in the parklands and prairie-pothole country of the northern Great Plains where they exceed all other ducks in number. They can dramatically alter their breeding distribution from year to year in response to changing water conditions; for example, after massive floods in Illinois and Louisiana, thousands of pairs bred in places that had not seen a Blue-wing in years (Bellrose 1980).

Nineteenth-century observers found them rare on the Great Plains (Cooke 1897). Atlasers found them breeding widely on the plains, where they obviously have taken advantage of water impoundments and canals. They found fewer in the moun-

tain parks and fewer still in the western river valleys. Across the continent Cinnamon Teal largely replace Blue-wings west of the Continental Divide; Colorado's mountain parks represent a transition zone, with the two species about even in North Park and Cinnamon Teal leading three to one in the San Luis Valley (Szymczak 1986).

Nationwide, Blue-wing numbers have fluctuated unpredictably around an average of about 5 million birds (Bellrose 1980). On the Colorado plains, where they trail only Mallards as the most common breeding duck, "every little pond has one or two breeding pairs" (B&N). Atlasers estimated abundance conservatively, ranking them code A2 in 67% of the blocks, and code A1 in a surprisingly large 25%. Surveys of major waterfowl breeding areas by CDOW in May 1996 found 2,135 pairs, 67% in the South Platte River drainage and very few in western valleys (Jim Gammonly pers. comm.).

Blue-winged Teal play a big part in the waterfowl hunting season, with about 8,400 birds (nearly all migrants) taken in Colorado in the 1995–1996 season (Jim Gammonly pers. comm.). Based on banding data, they suffer the highest annual (non-hunting) mortality rates of any North American duck (Bellrose 1980), perhaps because of their grueling and hazardous intercontinental migrations. Although highly adaptable in choice of wetlands, Blue-wings nevertheless need healthy wetlands with ample vegetation and good water quality. Like other Neotropical migrants, the survival of Blue-winged Teal depends on how well mankind can protect critical wildlife habitat on not one, but two continents.

BREEDING EVIDENCE

REPORTED IN 156 (9%) OF 1745 PRIORITY BLOCKS

Possible	44 blocks	(28%)
Probable	80 blocks	(51%)
Confirmed	32 blocks	(21%)

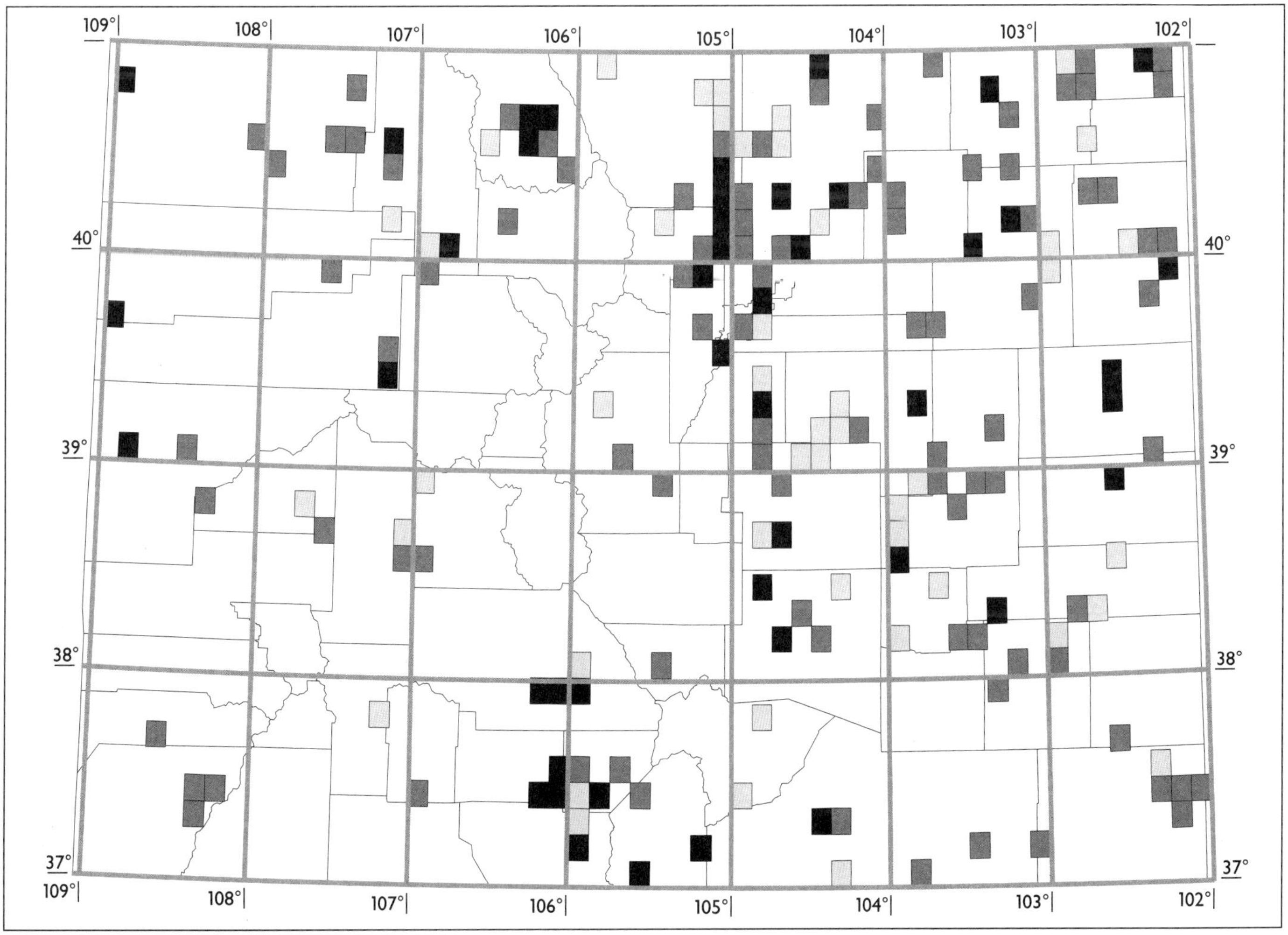
109°
108°
107°
106°
105°
104°
103°
102°
40°
39°
38°
37°

COLORADO DIVISION OF WILDLIFE / WILDLIFE RESOURCE INFORMATION SYSTEM

CINNAMON TEAL

Anas cyanoptera

The coppery-red body and glowing red eyes of male Cinnamon Teal in nuptial plumage delight the beholder and make these teal unique among North American ducks. Except in eclipse plumage, their red eyes and cinnamon color distinguish them from their relatives, the Blue-winged Teal. The drab females of these two teal cannot be identified in the field except by the company they keep.

HABITAT

Shallow tule-bordered lakes, freshwater ponds, sluggish creeks, reservoirs, irrigation ditches, marshes, and wet meadows find Cinnamon Teal at home (Johnsgard 1986a). Atlasers found them almost exclusively in ponds, lakes, and associated marshes. Like Gadwalls, but unlike Blue-winged Teal, they may seek alkaline waters (Johnsgard 1975). Their food requirements of bulrush seeds (particularly alkali bulrush), pondweed seeds and leaves, and saltgrass seeds dictate their choice of areas rich in submerged aquatic vegetation surrounded by marsh plants. They consume very little animal food, and that consists mostly of mollusks and insects (Bellrose 1980).

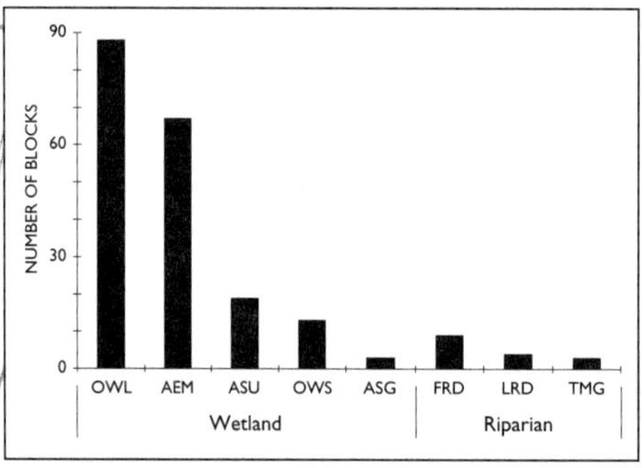
NUMBER OF BLOCKS
90
60
30
0
OWL
AEM
ASU
OWS
ASG
FRD
LRD
TMG
Wetland
Riparian

Rockwell (1911) found four Cinnamon Teal nests in the Barr Lake region of Colorado: two in very wet locations and two in dry meadows. They nest in vegetation that also provides much of their food—bulrushes and saltgrass. Atlas volunteers found these teal nesting at 8,860 feet (2700 m) in wet meadows with grass so short that it seems impossible for ducks to disappear, but disappear they do (RRK).

BREEDING

Cinnamon Teal renew the pair bond each year, if both have survived the winter (McKinney 1970). Yearlings breed (Bellrose 1976). Some teal arrive at the breeding grounds already mated, but mate selection continues through May into June (Oring 1964). Both Bent (1923) and Johnsgard (1986) give nesting dates as 3 May through 27 June, but Nelson (1993) lists Colorado breeding dates as 21 April through 20 August. Atlas workers found fledglings as early as 5 May and as late as 18 August, confirming the accuracy of Nelson's extended dates and depicting a longer breeding season than any ducks except Mallards.

Males usually desert the females during incubation, but about 12% of the males remain in attendance until the eggs hatch (Oring 1964). Although Tyler (*in* Bent 1923) said, "It is unusual to find the young not accompanied by both parents," more recent studies have found no evidence of this behavior. He may have based this misconception on the fact that males often remain in the rearing wetlands prior to wing molt, and feed and loaf in areas also occupied by females with their broods (Gammonly 1996).

Females usually select nest sites near water, but they have chosen places as much as 220 yards (200 m) from water, even in a dry meadow (Bellrose 1976). They make shallow scrapes under dense cover of vegetation or, if that is not available, a deeper bowl in the earth. They always line it with dry grass and down (Bellrose 1976). Spencer (1953) considered female Cinnamon Teal excellent mothers. They keep their young near cover and feign injury to divert predators.

BREEDING PHENOLOGY

CODE		# OF RECORDS	DATE RANGE
C	Courtship	3	29 Apr–21 Jun
NB	Nest Building	1	7 May
ON	Occupied Nest	4	7 May–2 Jun
NE	Nest with Eggs	6	17 May–21 Jun
NY	Nest with Young	1	21 May
FL	Fledged Young	45	5 May–18 Aug

DISTRIBUTION

Unique in their range, as well as their color, these teal are the only ducks breeding in North America that do not normally occur in the eastern part of the continent (Bent 1923). Of all the ducks that breed in North America, only Cinnamon Teal and Ruddy Ducks have a breeding range in South America as well (Terres 1980). The North American subspecies of the Cinnamon Teal, however, migrates no farther south than Guatemala (Johnsgard 1975), and some in the southern part of the U.S. are year-round residents (Peterson 1990).

Over half of the North American population breeds in Utah (Spencer 1953), and in 1976 about 5,000 pairs called Colorado home (Bellrose 1976). CDOW in 1996 estimated a pair count of 6,277 (Jim Gammonly pers. comm.), but this count included no data from the Front Range west, except the San Luis Valley, North Park, and the Yampa River.

Atlas work expands these 1976 and 1996 abundance estimates. These teal nest in the state wherever they find suitable habitat, with 30% breeding in the San Luis Valley. Atlas data show them nesting in widely scattered locations, but most breed west of the Front Range. Despite observations of single and paired teal on the plains during the nesting season, atlasers did not Confirm any nesting there. Atlasers found them in at least as many locations outside the CDOW survey area as within it. Estimates from Atlas figures suggest a population twice Bellrose's estimate.

Cold, wet weather and predators such as skunks and gulls may cause nest failure. Redheads and even Mallards parasitize their nests, but Cinnamon Teal are persistent breeders and can thrive in spite of these disasters. Only humans make a significant difference in their breeding success, by draining, cultivating, and destroying their habitat.

BREEDING EVIDENCE

REPORTED IN 154 (9%) OF 1745 PRIORITY BLOCKS

☐	Possible	38 blocks	(25%)
▩	Probable	61 blocks	(40%)
■	Confirmed	55 blocks	(35%)

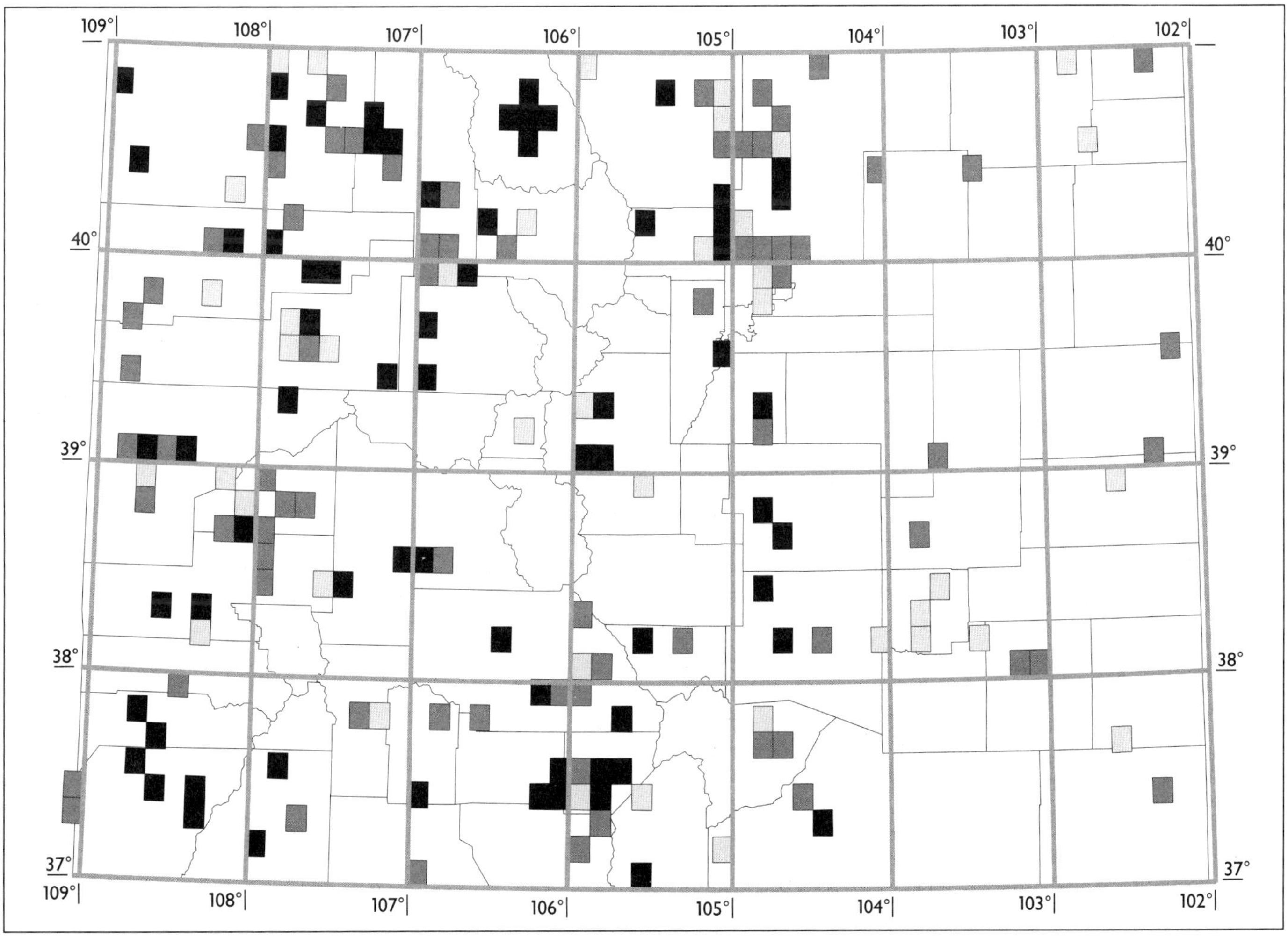
109°
108°
107°
106°
105°
104°
103°
102°
40°
39°
38°
37°

COLORADO DIVISION OF WILDLIFE / WILDLIFE RESOURCE INFORMATION SYSTEM

This puddle duck's wide, curving bill makes it easy to recognize and has inspired many interesting names, from "spoonbill" to its original genus name, *Spatula.* "Spoonies" seem comically burdened by their oversized bills, which they hold sloping downward about 30 degrees; on the water, as Bellrose (1980) says, they look "down in the bow."

HABITAT

Shoveler bills relate to their dietary specialty: straining water for tiny invertebrates. Their special diet, in turn, explains their preference for shallow, muddy, even stagnant wetlands with plenty of bottom ooze—wetlands that provide a bounty of free-swimming invertebrates (Bellrose 1980, DuBowy 1996). Not surprisingly, Northern Shovelers do especially well on sewage settling ponds (Belanger and Couture 1988). Breeding wetlands typically encompass margins of open, shallow marshes and ponds among prairie, sagebrush, or aspen parklands. Preferred breeding wetlands possess plenty of open water compared to emergent

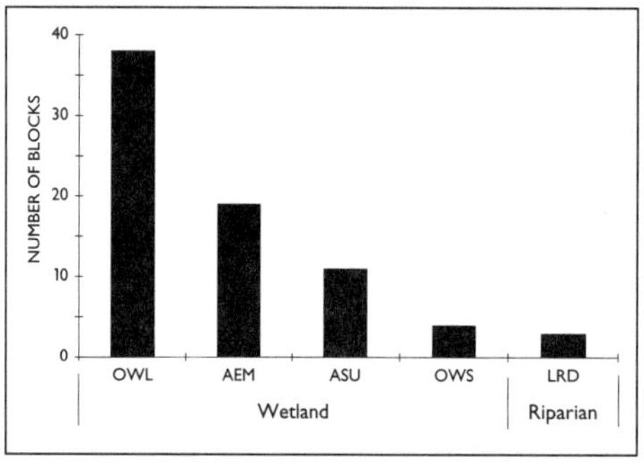
NUMBER OF BLOCKS
40
30
20
10
0
OWL
AEM
ASU
OWS
LRD
Wetland
Riparian

vegetation (at least 50/50), and much submergent vegetation. Atlasers nearly always found shovelers on or around water, lakes more than five times as often as streams; shovelers do not forage for seeds in upland fields the way many other dabblers do.

Atlasers found very few shoveler nests, and usually discerned breeding status by observing paired adults or broods. Hens usually select nest sites in low grassy or weedy vegetation within 160 feet (50 m) of water (DuBowy 1996). Most nests in Monte Vista NWR (Gilbert et al. 1996) and one in North Park (B&N) lay among grasses or sedges. In saline playas shovelers will nest in saltgrass meadows (Williams and Marshall 1938).

Hens may lead their broods to diverse wetlands including roadside ditches and irrigation canals (Poston 1969). Unlike most dabblers, shovelers do not make discernible late-summer movements to molting areas, perhaps because of their late-nesting habits; typically, adults hide in breeding-area marshes during the flightless period (DuBowy 1996).

BREEDING

Shovelers pair mostly on winter ranges, then make a leisurely, protracted spring migration to breeding areas. They move into and through Colorado from mid April to early June (B&N, A&R). As a result, shovelers initiate nesting when early nesters like Mallards already have half-grown broods. Nearly all Atlas brood sightings came after 22 June, and B&N reported Colorado nests with eggs in late June.

Nesting pairs staunchly defend a "mobile, conceptual space" centered on the hen (Seymour 1974); drakes drive away interlopers by lunges on the water and short aerial chases. As usual for ducks, nesting territories do not last past hatching, and by late August drakes begin the fall migration, followed later in September by hens and young.

BREEDING PHENOLOGY

CODE		# OF RECORDS	DATE RANGE
ON	Occupied Nest	1	3 Jul
NE	Nest with Eggs	1	15 May
FL	Fledged Young	15	1 Jun–12 Aug

DISTRIBUTION

Northern Shovelers breed throughout the temperate Northern Hemisphere, with the greatest North American share in the mid-continent aspen parkland north of the prairie-pothole region. Altitude duplicates that latitudinal trend in Colorado, where greater numbers breed in the mountains than on the plains. Atlas records cluster in North Park and the San Luis Valley, the same centers as reported before (Sclater 1912, B&N, A&R), with scattered breeding in plains reservoirs and river bottoms, especially in Larimer and Weld counties. B&N thought that many plains shovelers might be non-breeding residents. Atlas data show a higher breeding Confirmation rate in the mountain parks, but many of these records resulted from waterfowl nesting inventories on national wildlife refuges where nest discovery rates would be understandably higher.

Shoveler numbers have remained relatively stable nationally for the past 45 years (DuBowy 1996). Assessments of abundance in Colorado (A&R), mostly unchanged since the earliest comprehensive reports (Sclater 1912), agree that Northern Shovelers breed commonly in mountain parks, and uncommonly to rarely elsewhere. Densities at Monte Vista NWR reach 26 nests/mi^2 (10/km^2; Gilbert et al. 1996). Ponds in Alberta less than 1.5 acres (0.6 ha) typically harbor only one nesting pair, whereas larger ponds hold more (Poston 1969). During waterfowl surveys in May 1996, CDOW counted 3,700 pairs in selected count areas (Jim Gammonly pers. comm.).

Fall migration greatly swells the ranks of Colorado shovelers, and hunters took about 4,100 statewide in the 1995–1996 season (Jim Gammonly pers. comm.). Waterfowlers regard "spoonies" as too small and gamey (due to their diet) to make first-class table fare, and thus game biologists have not lavished management and research efforts on this species. Shovelers, like other wetland birds, have suffered much habitat loss from pollution, grazing, land-filling, and water depletion. Shovelers seem less susceptible than many ducks to poisoning from agricultural pesticides and lead shot, because they do not feed in uplands and seldom filter bottom sediments (DuBowy 1996). These wonderfully odd ducks should continue to thrive as long as we leave them enough high-quality wetlands.

BREEDING EVIDENCE

REPORTED IN 63 (4%) OF 1745 PRIORITY BLOCKS

☐	Possible	25 blocks	(40%)
■	Probable	28 blocks	(44%)
■	Confirmed	10 blocks	(16%)

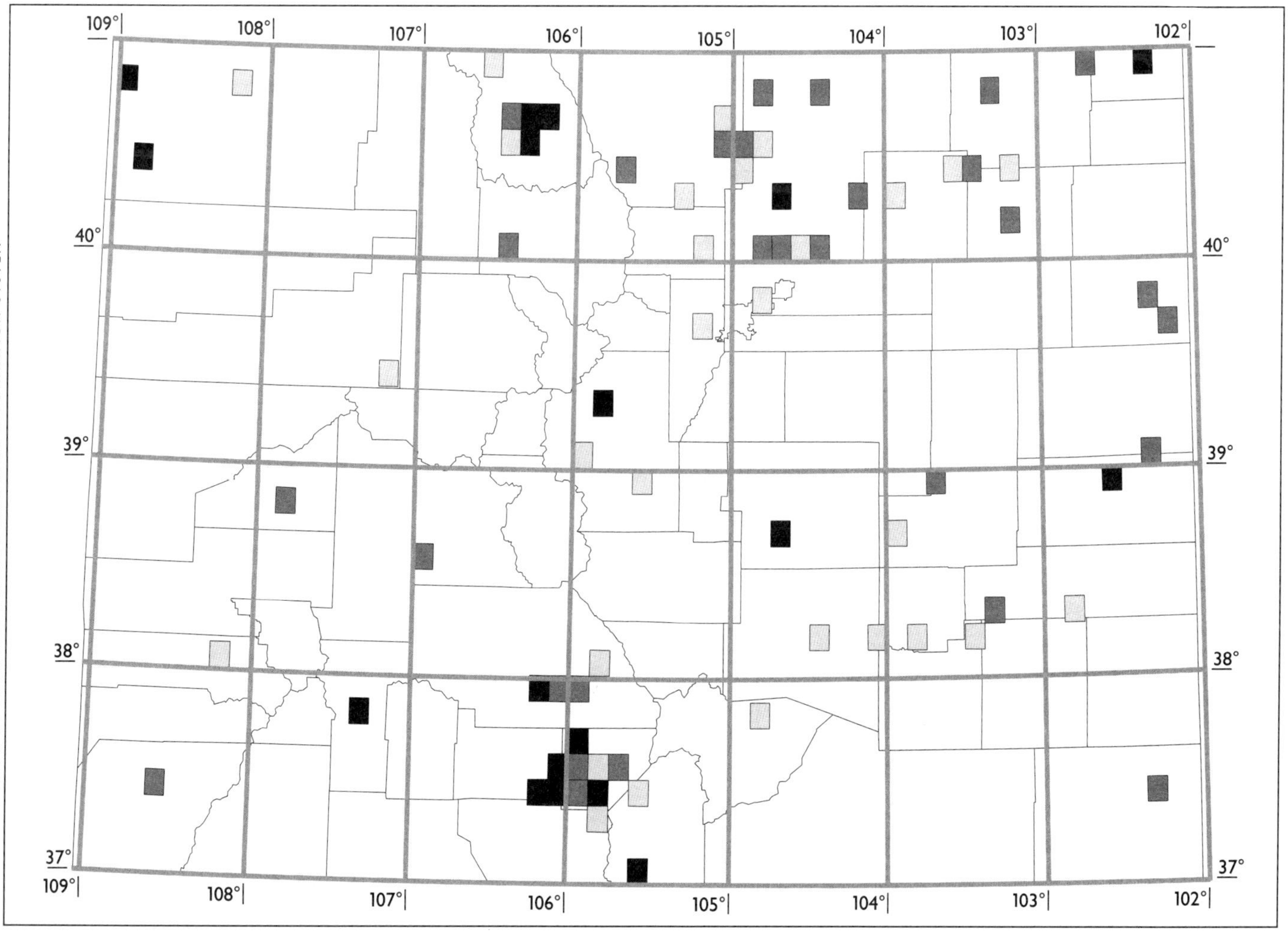
109°
108°
107°
106°
105°
104°
103°
102°
40°
39°
38°
37°

Not the showy type, Northern Pintail drakes wear conservative brown and white plumage and vocalize far less than chatty ducks like the Mallard. Yet a streamlined body, pointed tail feathers, and slender wings more like a gull than a duck lend to pintails a remarkable elegance in flight and on the water. Modern naturalists and sportsmen are not the first to admire these handsome wildfowl—artisans of monumental architecture in ancient Egypt pictured Northern Pintails more than any other duck (Phillips 1922).

HABITAT

Pintails use a variety of wetlands to meet the demands of life in different seasons. In spring and summer they nest on upland sites near shallow wetlands (Austin and Miller 1995). Later in the summer they tend to move to larger marshes for greater security during the wing molt, and in fall and winter they use various wetlands including large lakes and reservoirs (Bellrose 1980).

Pintails breed more often in seasonal or semi-permanent wetlands rather than around deep permanent waters such as lakes, reservoirs, and rivers (Stewart and Kantrud 1973, Austin and Miller 1995). Of all North American dabblers, these ducks make the most use of ephemeral wetlands for breeding and, consequently, show the weakest homing tendency to natal or breeding sites (Johnson and Grier 1988).

Pintails choose the most open nest sites of any dabbler. They usually nest in low upland cover well away from water (Gilbert et al. 1996), occasionally with the sitting hen in full view (B&N).

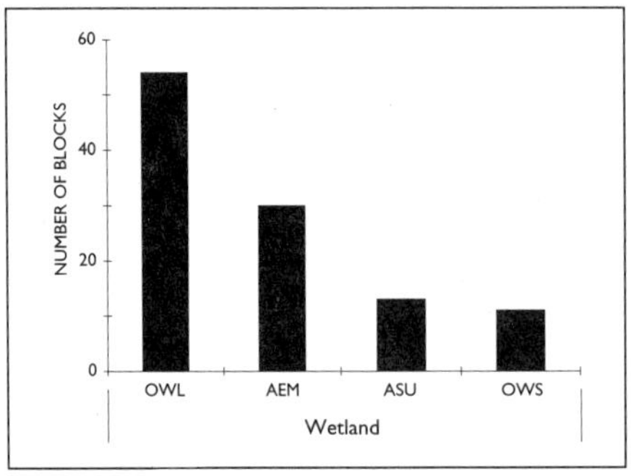
60
40
20
0
NUMBER OF BLOCKS
OWL
AEM
ASU
OWS
Wetland

Atlasers found very few nests; they mainly Confirmed breeding by observing ducklings. Consequently 86% of habitat observations describe water or wetland vegetation; the remainder include mostly open rural or field habitats, and forests comprise only 3%.

Hens with broods retire to seasonal or semi-permanent wetlands with abundant emergent vegetation; these habitats provide the best mix of invertebrate and plant foods with escape cover (Mack and Flake 1980, Kaminski and Prince 1984).

BREEDING

Pair bonds form each year by early winter and persist until the incubation stage (Derrickson 1977). Pintails nest early; most pairs arrive in Colorado by April (Bellrose 1980). Older hens nest right away, but yearlings tend to delay a week or so (Duncan 1987). Atlas dates for NE and FL (ducklings) indicate nest initiation a little later than reported by B&N, A&R, and Ringelman (1991). The July nest date shown on the Atlas phenology chart is later than expected (Mike Szymczak pers. comm.).

Sociable ducks, pintails do not defend breeding territories, except perhaps a few yards around the nest (Austin and Miller 1995). Hens incubate for 22–24 days while drakes begin to drift off to molting areas. The precocial ducklings must undergo a hike to water of up to a mile (1.6 km) during their first day of life. When the young fledge in four to six weeks, hens usually depart to a molting marsh, where the loss of flight feathers grounds them for about 36 days (Miller et al. 1992).

BREEDING PHENOLOGY

CODE		# OF RECORDS	DATE RANGE
C	Courtship	1	13 Jul
ON	Occupied Nest	2	11 Jun–3 Jul
NE	Nest with Eggs	2	15 May–16 Jun
FL	Fledged Young	25	3 Jun–1 Aug

DISTRIBUTION

Northern Pintails breed around the world north of the equator. Their New World strongholds lie in Alaska and the northern prairies; fewer breed in the mountains from Alaska to New Mexico (Bellrose 1980). Colorado's greatest counts by far occur during migrations, most in the Great Plains, western valleys, and mountain parks (B&N, A&R). Some winter in the lowest river valleys, the San Luis Valley, and large plains reservoirs (A&R).

Atlas records show breeding concentrations in North Park, the San Luis Valley, and the northeastern Great Plains, especially in the South Platte and Cache la Poudre drainages. Mountain parks produce the most pintails in Colorado by far (Szymczak 1986); North Park and the San Luis Valley held 86 percent of the 2,500 pairs of pintails counted by the CDOW survey in May 1996 (Jim Gammonly pers. comm.). This CDOW count covered only six areas; by covering the whole state, atlasers found additional concentrations on the plains in the South Platte drainage. Atlas data also show that scattered pintails breed in the western valleys and lower mountains. Some of the Possible records, however, may show birds on molt migration, more pronounced in pintails than other ducks (Salomonsen 1968).

Nest densities at Monte Vista NWR reach 93 nests/mi^2 (36 nests/km^2; Gilbert et al. 1996), probably the highest density in Colorado. Pintails compose about 9% of the breeding ducks in the San Luis Valley and North Park (Szymczak 1986), but only about 3% in the mountains west of the San Luis Valley (Rutherford and Hayes 1976).

Although Northern Pintails are still abundant continent-wide, counts dropped in half from 1975 to 1992 (USFWS and CWS 1994). BBS data also show a major continent-wide decline, 4.8% annually, which calculates to a drop of over two-thirds in 31 years. Sharp declines in North American counts correlate with prairie-region droughts (Bellrose 1980). Farm practices also inadvertently destroy many nests in croplands and hayfields, and wetland habitats suffer from pollution, water depletion, and conversion to other land uses.

Wildlife agencies have adopted measures to try to regain the former numbers. Lower populations and hunting restrictions have reduced mid 1990s pintail harvests to less than 30% of 1985 levels. Reducing hunters' harvest rates, even drastically, has only a small impact on the total population. Only protecting wetlands from land use changes that reduce breeding habitat and worsen the effects of drought, especially in the prairie strongholds, can bring back the elegant pintail and its relatives. Cooperative government programs improve waterfowl habitat on thousands of public and private acres each year. Pintails can continue to flock in Colorado marshes and fields only if habitats receive better protection and wildlife agencies continue to limit hunting.

BREEDING EVIDENCE

REPORTED IN 97 (6%) OF 1745 PRIORITY BLOCKS

□	Possible	35 blocks	(36%)
▒	Probable	36 blocks	(37%)
■	Confirmed	26 blocks	(27%)

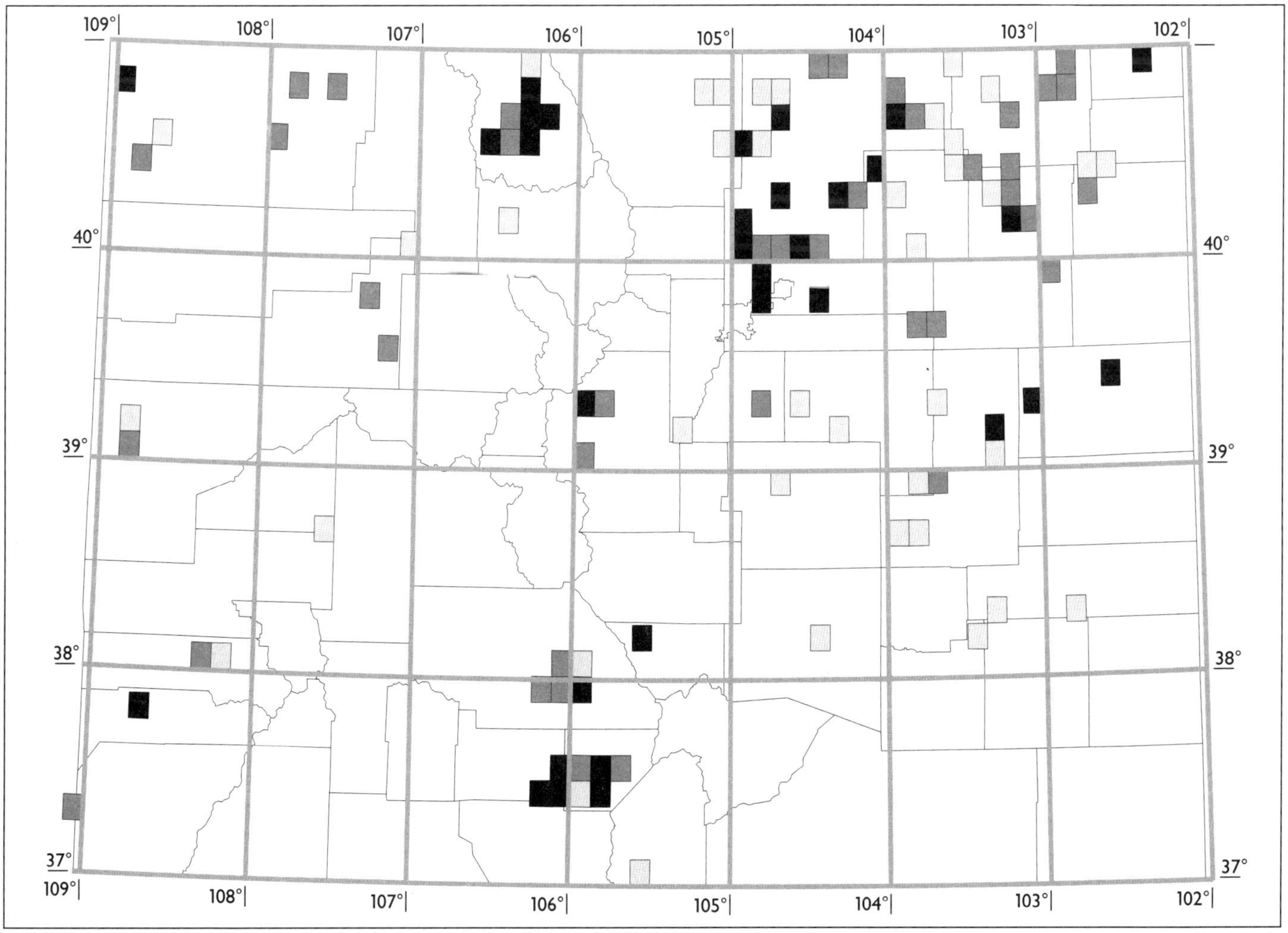
109°
108°
107°
106°
105°
104°
103°
102°
40°
39°
38°
37°

COLORADO DIVISION OF WILDLIFE / WILDLIFE RESOURCE INFORMATION SYSTEM

The smallest North American dabbling duck, the Green-winged Teal belongs to a widespread "superspecies" that includes the South American and Baikal teals. Tiny and quick, these pigeon-sized dabblers dart and swoop in flight like shorebirds, and their erratic flight makes them seem even faster. Although common breeders in mountain wetlands, Green-wings can escape attention because of their tiny size and preference for emergent vegetation.

HABITAT

In North America, the highest breeding densities occur in deciduous parklands, boreal forests, and prairies mixed with brushlands (Bellrose 1980). Typical breeding sites include grass/sedge meadows where shrub thickets or woodlands lie adjacent to small ponds and marshes (Johnson 1995). Green-winged Teal commonly use beaver ponds and artificial wetlands (Baldassarre and Bolen 1994). Atlasers recorded mostly wetland habitats (lakes about three times as often as streams); the upland habitats recorded indicate this teal's preference for wetlands within shrub or forest environments.

Hens like plenty of cover for nesting. They usually conceal the nest in dense grass or sedges against or within shrub thickets or woodlands not far from water (Keith 1961). Of nine Atlas records reporting nest vegeta-

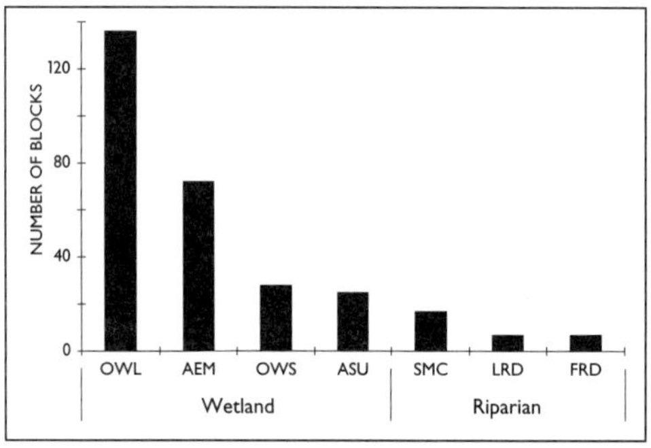

NUMBER OF BLOCKS
0
40
80
120
OWL
AEM
OWS
ASU
SMC
LRD
FRD
Wetland
Riparian

tion codes with nests (NB, NE, ON), five listed emergents, two listed shrubland sites, and one listed riparian forest. Two nest record cards described nests on the ground in grass among shrubs. At Monte Vista NWR most Green-winged Teal conceal their nests in grasses or sedges (Gilbert et al. 1996).

BREEDING

Like most dabblers, pairs form in winter, and courtship continues through spring migration and onto the breeding grounds (McKinney 1965). Green-wings pass into and through Colorado in March and April, but they take their time initiating breeding, not surprising inasmuch as most nest at relatively high elevations. Nesting pairs do not defend a breeding territory, but paired drakes stay busy attempting forced copulations with hens from other pairs and driving other drakes away from their own mates (McKinney and Stolen 1982). One atlaser found nest-building underway on 21 April, but typically laying does not begin until at least mid May (Johnson 1995). Atlasers found nests with eggs mostly after 13 June, similar to reports by B&N. Nearly all Atlas sightings of broods came after mid June, and observations peaked late in that month. Replacement clutches may be initiated as late as mid July (Munro 1949). Green-winged Teal nest over a wide altitudinal range in Colorado, which undoubtedly causes nesting dates to vary.

BREEDING PHENOLOGY

CODE		# OF RECORDS	DATE RANGE
NB	Nest Building	1	21 Apr
ON	Occupied Nest	1	18 May
NE	Nest with Eggs	6	29 May–10 Jul
FL	Fledged Young	87	24 May–12 Aug

DISTRIBUTION

Found throughout the Northern Hemisphere, most of the North American contingent breeds in the aspen woodlands, boreal forests, and Arctic shrublands of northwest Canada and Alaska (Johnson 1995). In Colorado, Green-wings breed rarely on the plains but commonly in the mountain parks and other high country to nearly 11,000 feet (3350 m; B&N, A&R), as reported by early observers (Cooke 1897). Atlas records confirm high numbers in the mountain parks, widespread breeding elsewhere in the mountains, and more scattered breeding in the Great Plains, mostly near the Front Range. Mountain parks also serve as staging grounds from which spring arrivals disperse to smaller wetlands in surrounding mountains (Szymczak 1986).

Green-wings trail only Mallards as the most common breeding duck in Colorado (Szymczak 1986); in the mountains they reach densities of 0.47 pairs/mi^2 (0.18/km^2; Rutherford and Hayes 1976). Atlasers found Green-wings in more blocks than any other duck except Mallards, and usually recorded abundance code A2.

Nationwide population estimates have remained at or above 1.9 million birds for many years, despite prairie droughts (Chu et al. 1995). Waterfowl surveys by CDOW in May 1996 counted only 5,685 pairs, 72% of them in North Park (Jim Gammonly pers. comm.). The survey focused only six areas of concentration and missed a great many Green-wings in small wetlands throughout the state. The Atlas map shows the wide distribution in small wetlands throughout the mountains and Colorado Plateau, and Atlas data suggest a higher population, around 8,000 pairs.

Green-winged Teal usually rank second in nationwide duck-hunter bags (Chu et al. 1995), and Colorado hunters take about 13,000 (mostly migrants) each year (Jim Gammonly pers. comm.). Like all waterfowl, teal face problems of wetland loss and degradation. Their breeding wetlands, both continent-wide and in Colorado, tend to be fairly remote, so that they have not suffered as much as many North American ducks from habitat loss and drought. However, as Colorado's growing human population puts more pressure on wildlife habitat, only wetland conservation practices will ensure that Green-wings continue to dart and flash through Colorado marshes.

BREEDING EVIDENCE

REPORTED IN 247 (14%) OF 1745 PRIORITY BLOCKS

□	Possible	84 blocks	(34%)
■ (grey)	Probable	73 blocks	(30%)
■	Confirmed	90 blocks	(36%)

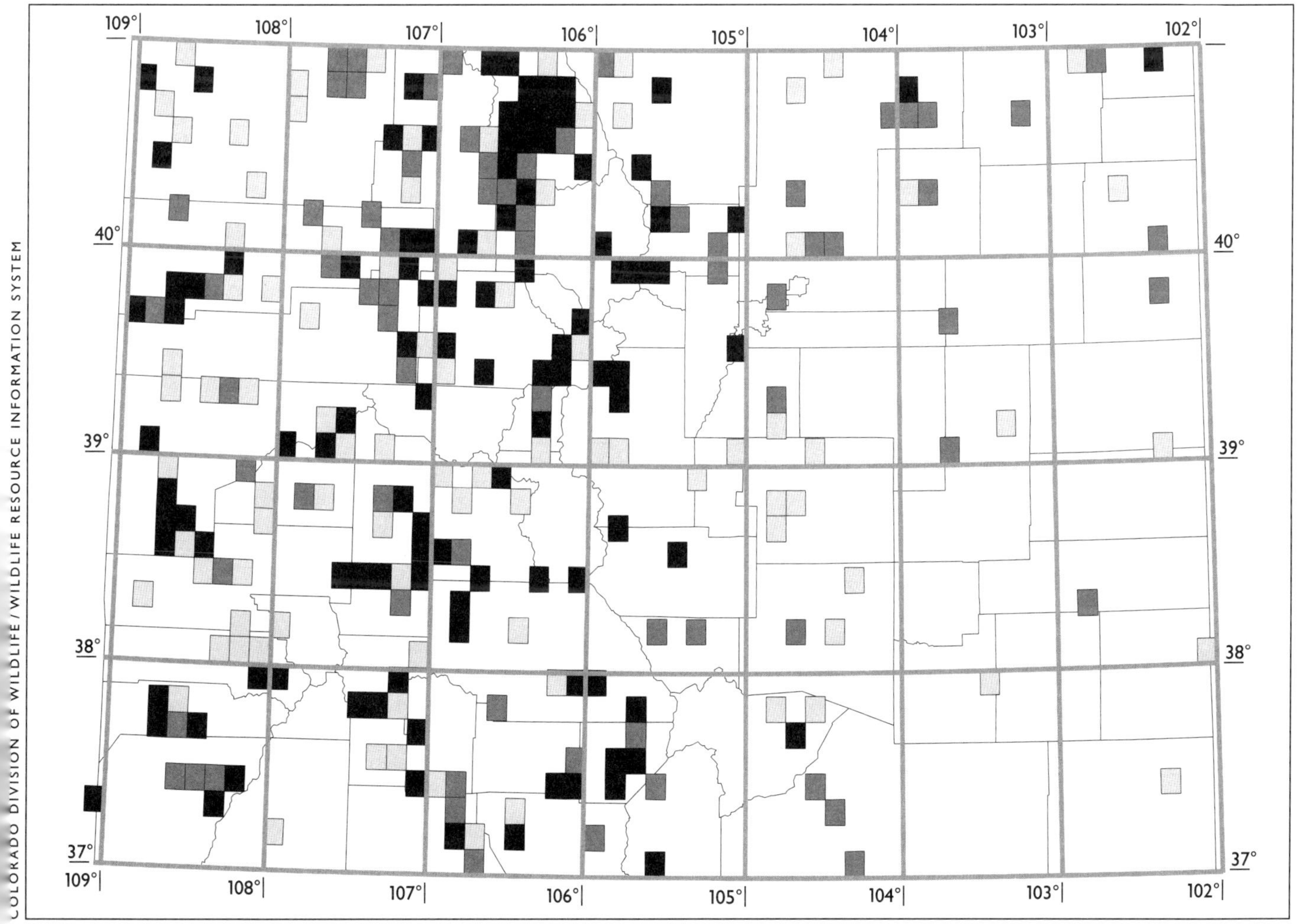
109°
108°
107°
106°
105°
104°
103°
102°
40°
39°
38°
37°
COLORADO DIVISION OF WILDLIFE / WILDLIFE RESOURCE INFORMATION SYSTEM

Waterfowl texts from the last century often crowned the Canvasback the "King of Ducks," in deference to its beauty, wariness, and perhaps also because "Cans" fetched more than five times the price of "ordinary" ducks in the game meat trade of those days. This elegant diver's common name comes from its pale gray back and white sides, which are finely flecked with gray in wavy patterns suggesting canvas fabric (Terres 1980).

HABITAT

Canvasbacks use various ponds, sloughs, and marshes in summer. They prefer smaller ponds than their relatives, the Redheads, which usually outnumber them on large marshes (Bellrose 1980). Within their home range, they use deeper ponds for feeding, resting, and courting, and shallower waters for nesting (Trauger and Stoudt 1974). Nest ponds, completely encircled by bulrushes or cattails, usually cover less than an acre (0.4 ha). Brood ponds vary more, but typically hold plenty of emergent vegetation. Atlas workers nearly always recorded wetlands or wetland vegetation with Canvasback sightings, and noted lakes rather than streams for all breeding birds except two May sightings on rivers (possibly migrants).

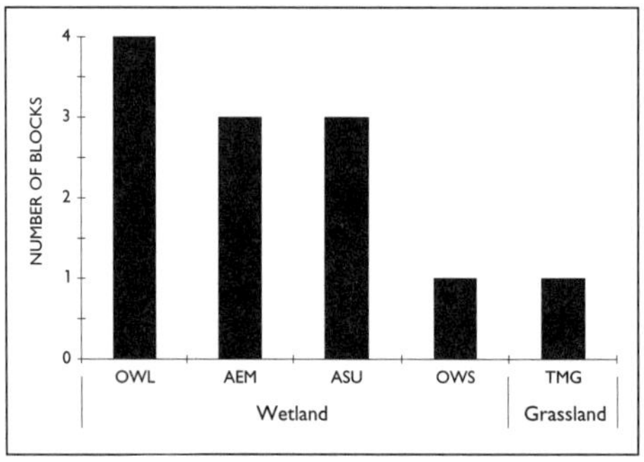

NUMBER OF BLOCKS
0
1
2
3
4
OWL
AEM
ASU
OWS
TMG
Wetland
Grassland

Like most divers, Canvasbacks nest in dense marsh vegetation over shallow water; the nest rests on the mud bottom or floats anchored to emergent growth. Bulrushes, sometimes mixed with cattails or sedges, seem preferred as attaching plants (Erickson 1948), although dense cattails, flooded willows, and common reeds will also serve (Bellrose 1980). Sometimes hens will nest on small islands, including the heaped vegetation mounds of muskrat houses (Hochbaum 1944, B&N).

BREEDING

Canvasbacks leave most courtship business for the spring migration and first few weeks on breeding ranges. During nesting, males show little inclination for establishing territories, at best defending a small space around the hen. Spring migrants pass through Colorado mostly in March and April (A&R), with some stragglers "well into May" (B&N). Most Canvasbacks fly on through to more northerly breeding places, but a few remain to breed in Colorado's larger marshes. They nest later than most other ducks, and Atlas dates for eggs and broods confirm earlier reports of egg-laying mostly in June (B&N) and later (Bellrose 1980). Brood parasitism plays a regular part in Canvasback nesting, on both the giving and receiving ends. Redheads in some areas parasitize over 50% of Canvasback nests, and can add three or more Redhead eggs per Canvasback nest (Sorenson 1991). Adding to the nesting melee themselves, Canvasbacks frequently dump eggs in each other's nests (and rarely those of other species, and then, as far as known, only Redheads). In a high-density Manitoba population, time-lapse photographs revealed that at least 36% of Canvasback clutches were parasitized by other Canvasback hens, and that Cans laid about 10% of their eggs parasitically. Nest success rates demonstrate that nest parasitism amounts to a relatively unsuccessful strategy used more as a last resort after failed clutches or during bad-weather years (Sorenson 1993).

BREEDING PHENOLOGY

CODE		# OF RECORDS	DATE RANGE
FL	Fledged Young	3	29 Jun–18 Jul

DISTRIBUTION

Canvasbacks breed predominantly in the northern parts of the prairie provinces of Canada (Bellrose 1980). They make less use of intermountain marshes than many other ducks, resulting in very small breeding numbers in Colorado. Cooke (1897) mentioned them only as uncommon migrants. B&N reported small but regular numbers breeding in North Park, and rare irregular breeding at Denver-area marshes and unspecified reservoirs on the plains. Atlasers found small, widely scattered groups breeding in Browns Park, North Park, South Park, the San Luis Valley, and the Yampa Valley, places where a few have nested regularly for many years (Wampole 1951, A&R). Two Atlas records in Weld and Garfield counties may represent migrants or non-breeding summer residents—in each case a May sighting on a stream without a breeding pair noted. In the northern prairies, Canvasbacks sometimes make significant molt migrations during July and August, but no records of such movements exist for Colorado Cans.

Atlas records agree with A&R's observation of low Colorado breeding numbers.

Waterfowl surveys by CDOW during May 1996 counted just 21 breeding pairs, one in the San Luis Valley and the rest in Browns Park (Jim Gammonly pers. comm.). Calculations from Atlas data suggest that the state has three times as many breeders, with breeders in North and South parks and possible breeders in four other sites.

Canvasbacks have averaged about 560,000 birds continent-wide for the past 40 years. Serious drought-period declines of Canvasbacks in this century have been followed by increasingly poor recoveries (Madson 1990). Wild celery, once a staple food in the central and eastern U.S. (the plant's genus, *Vallisneria,* inspired the species name—though its namer, Alexander Wilson, misspelled it; Gruson 1972), has been all but eliminated by stream pollution, and the effect on Canvasback numbers has yet to be reckoned. Canvasbacks seem especially sensitive to drought, grazing, and human-caused water depletions, which worsen predation losses and increase Redhead parasitism (Madson 1990). Hunting restrictions imposed since the 1960s will not restore Canvasback populations by themselves. Ensuring the survival of North America's "King of Ducks" will require conservation and restoration of wetlands in the northern Great Plains, where so much habitat already has been lost.

BREEDING EVIDENCE

REPORTED IN 7 (0%) OF 1745 PRIORITY BLOCKS

☐	Possible	2 blocks	(29%)
▩	Probable	3 blocks	(42%)
■	Confirmed	2 blocks	(29%)

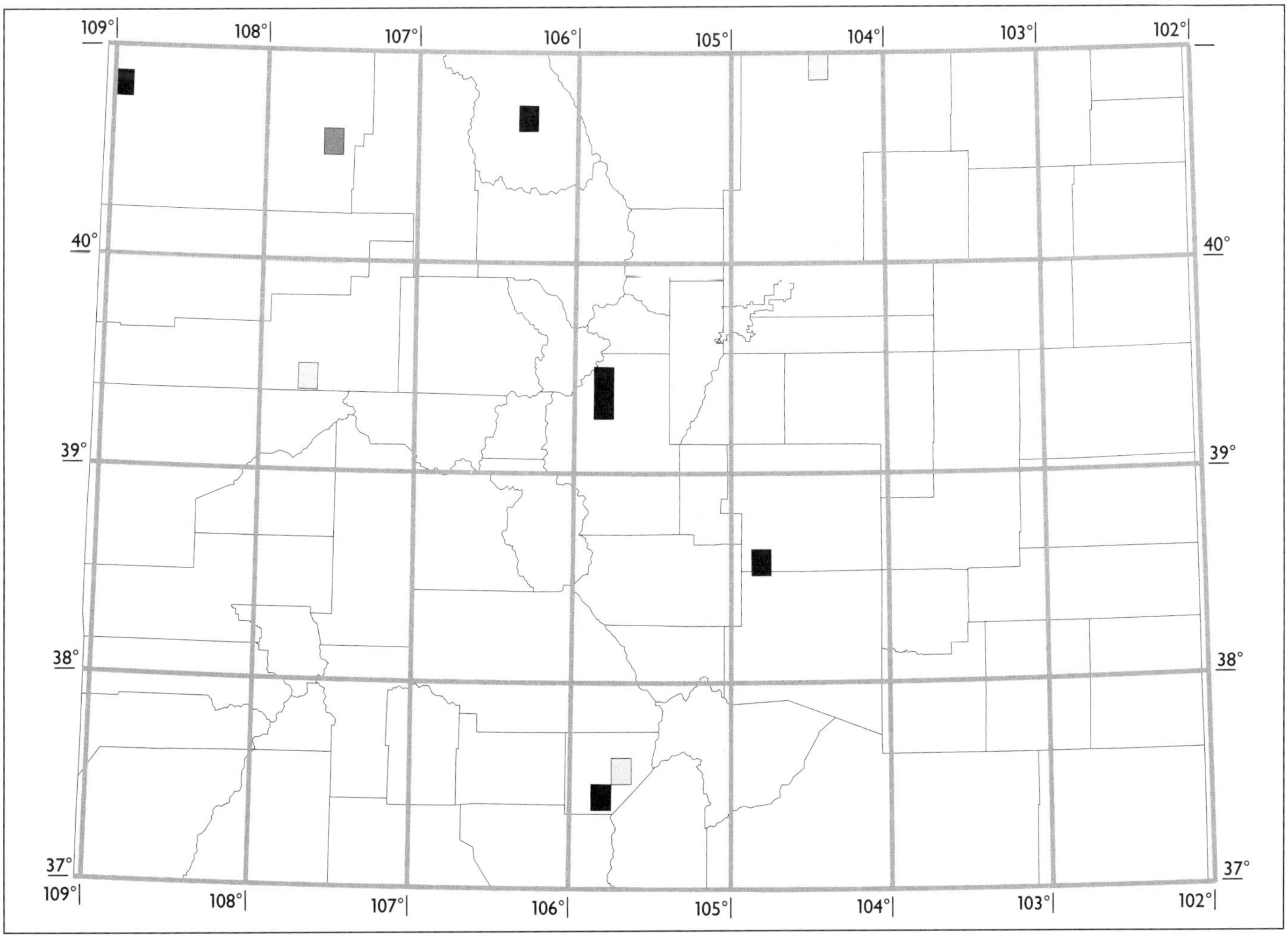
109°
108°
107°
106°
105°
104°
103°
102°
40°
39°
38°
37°

COLORADO DIVISION OF WILDLIFE / WILDLIFE RESOURCE INFORMATION SYSTEM

Not exactly typical diving ducks, Redheads spend more time tipping and dabbling than most divers would think proper. Redhead hens practice "facultative" brood parasitism by laying some of their eggs in nests of other Redheads and various other duck species as well as nesting the conventional way. Regardless of their idiosyncrasies, Redheads arouse the aesthetic instincts of duck watchers. Waterfowl expert J.C. Phillips (1922) described flocks in flight as dark ribbons against the sky, undulating as the trailing ducks followed the leader's minor course corrections.

HABITAT

For breeding, Redheads prefer stable, permanent wetlands with large expanses of open water (Johnson and Grier 1988). Seeds of shallow marsh emergents form a staple in their diet, and Redheads forage in shallow lake margins, ponds, and sloughs more than most diving ducks (Bellrose 1980). Atlas results suggest that Colorado birds prefer large wetland complexes, primarily in the mountain parks, and wetlands along the large Great Plains rivers. Atlas workers nearly always found breeding birds associated with open water or wetland vegetation, and recorded lakes more than 20 times as often as streams.

Redheads usually place their nests in emergent vegetation over water (Ringelman 1991), and Bellrose (1980) ranked hardstem bulrush, then cattails, and finally sedges as the emergents of choice. In some places significant numbers of Redheads nest on dryland sites (McKnight 1974, Keith 1961), but nests usually lie within a few feet of

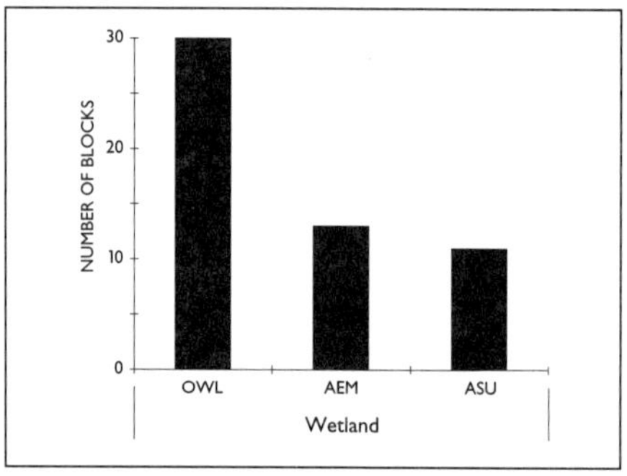
30
20
10
0
NUMBER OF BLOCKS
OWL
AEM
ASU
Wetland

water. Within hours of hatching, hens lead their broods to favored surface waters, occasionally some distance away. Drakes in some areas make a significant midsummer molt migration to favored marshes that have excellent security cover for the flightless period.

BREEDING

Pair formation begins in late winter and continues during spring migration (Weller 1965). Redheads arrive in Colorado from late February through April. Many pass through, but some remain for a protracted breeding season that stretches egg-laying well into July. Atlasers found the earliest broods on 11 June.

Phillips (1922) thought, a little unfairly, that "Redheads show a certain irresponsibility in depositing their eggs in the nests of other species." Moral judgments aside, population ecologists view brood parasitism as an interesting strategy for accomplishing something every species does: maximizing reproduction with minimum energy cost (Weller 1959, Sorenson 1991). Redhead hens lay up to 75% of their eggs in nests other than their own (victims include Redheads plus other duck species), and up to 50% of Redhead ducklings grow up in "foster" broods (Sorenson 1991). Redhead hens stay flexible regarding parasitism: in good water and food years, they may lay a parasitic clutch and then a typical one; in bad years, they tend to choose one or the other, but not both. When parasitizing other species, Redheads choose Canvasbacks most often, followed by Ruddy Ducks, pintails, Mallards, Lesser Scaup, and Cinnamon and Blue-winged teal (Sorenson 1991). Redheads provide the best example of facultative parasitism known among birds, although closer study is showing more and more of this behavior in many birds, not only ducks but also passerines.

Females take a leisurely five minutes to lay eggs, and when they try this in another duck's nest, the owner often interrupts the process, resulting in broken eggs scattered around. Many parasitic females do not even develop brood patches, so do not incubate their own eggs. When they do brood their own eggs, their interest in the ducklings wanes and they desert them before the young can fly (Ryser 1985). Nevertheless their reproductive system succeeds. Redheads have one of the highest average annual increases on BBS surveys (3.8%).

BREEDING PHENOLOGY

CODE		# OF RECORDS	DATE RANGE
C	Courtship	1	12 Jun
FL	Fledged Young	12	11 Jun–1 Aug

DISTRIBUTION

Redheads breed only in North America, primarily in the pothole region of the northern Great Plains and the intermountain marshes of the West. The greatest massing of breeding Redheads occurs in the Great Salt Lake marshes of Utah, where densities reach an astounding 335 birds/mi^2 (137/km^2) (Bellrose 1980). In Colorado Redheads breed regularly in about 18 of 28 latilongs (Ryder 1991b)—and the Atlas map shows them in 18 latilongs—but they concentrate in the big marshes of the mountain parks. B&N listed the Barr Lake drainage as a "regular" breeding area, but 25 years later A&R regarded Redheads as rare breeders on the plains. This may indicate a decline in breeding numbers on the Colorado plains, mirroring the serious decline of Redhead numbers in the northern Great Plains during the droughty 1980s (Ryder 1991b). Atlas records that Confirmed Redheads concentrated in North Park, Browns Park, and the San Luis Valley. Several blocks in the South Platte drainage from Barr Lake to Fort Collins reported them, including breeding at Rocky Mountain Arsenal NWR.

Surveys at major waterfowl breeding areas in May 1996 by CDOW (Jim Gammonly pers. comm.) counted 3,383 pairs, 52% in the South Platte drainage and 22% in North Park. Atlasers rarely regarded Redheads as common; they almost always assigned A2 abundance codes, which implies a maximum density of one pair/mi^2 (0.4/km^2).

Redheads provide choice hunting for waterfowlers, but Colorado hunters took only an estimated 967 birds in the 1995–1996 hunting season (Jim Gammonly pers. comm.), down from over 4,000 in some recent years (Ryder 1991b). Waterfowl managers continue to regard northern prairie populations as precarious, and hunting restrictions, including full-season closures in some years, have been the rule since the early 1980s. Although Redheads can quickly pioneer new areas of suitable habitat (McKnight 1974), they also remain very vulnerable to drought and manmade water depletions, because of breeding habits so closely tied to permanent and stable surface waters (Bellrose 1980). Only with dedicated conservation of wetlands will Redheads continue to trace undulating ribbons across Colorado skies.

BREEDING EVIDENCE

REPORTED IN 40 (2%) OF 1745 PRIORITY BLOCKS

Possible	12 blocks	(30%)
Probable	22 blocks	(55%)
Confirmed	6 blocks	(15%)

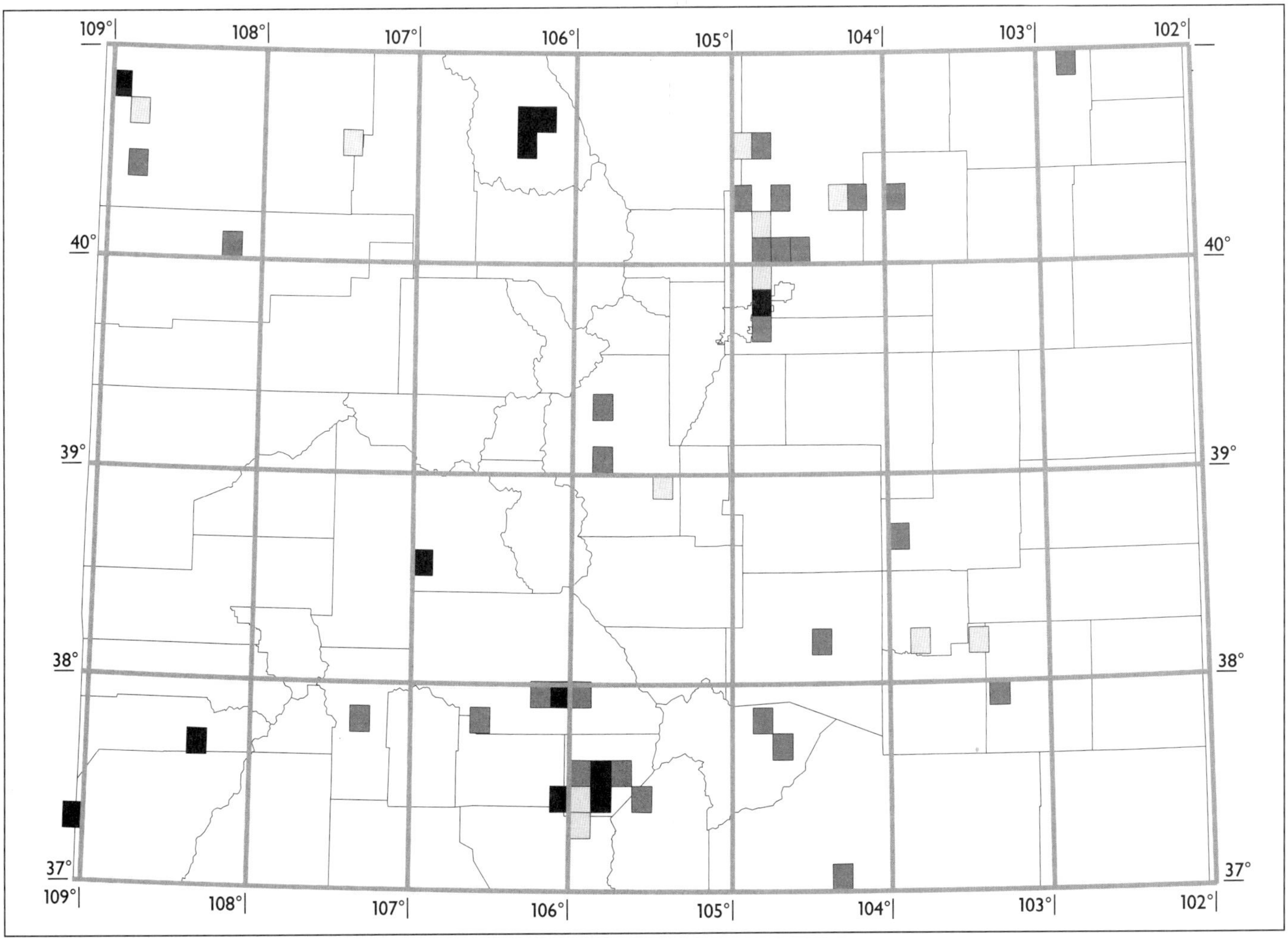
109°
108°
107°
106°
105°
104°
103°
102°
40°
39°
38°
37°

COLORADO DIVISION OF WILDLIFE / WILDLIFE RESOURCE INFORMATION SYSTEM

Shallow ponds in mountain forests harbor Colorado's nesting Ring-necked Ducks. Although classified as diving ducks, they share some characteristics with dabblers: they can spring up from the water at a steep angle, they take young to thick vegetation for safety, and they often feed on surface vegetation (Palmer 1976).

HABITAT

Known chiefly as ducks of Canada's north country, Ring-necked Ducks find a few parcels of sufficiently similar habitat among the ponds and beaver-dammed creeks in Colorado's mountain forests. Here they breed at elevations ranging from 7,000 to 10,000 feet (2135–3048 m), almost exclusively west of the Continental Divide.

In contrast to other North American diving ducks, Ring-necks readily take to ponds with shallow water and submerged vegetation. They normally feed in water less than 6 feet (2 m) deep (Bellrose 1976).

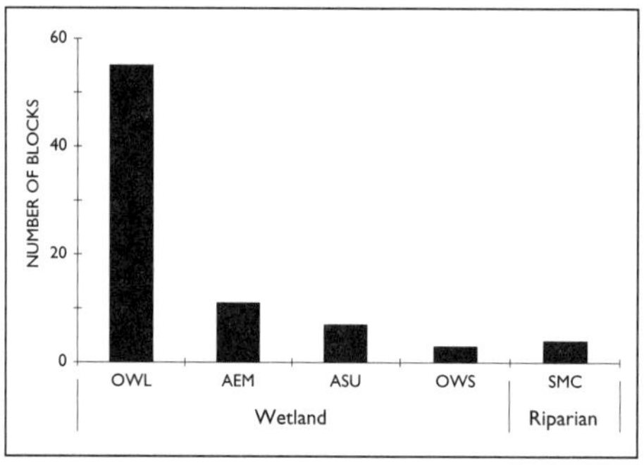
60
40
20
0
NUMBER OF BLOCKS
OWL
AEM
ASU
OWS
SMC
Wetland
Riparian

Hummocks, waterside grasses, and sedge meadows serve as nesting areas for the hens (Palmer 1976b, Hohman 1986). They seldom nest in emergent vegetation over water (Johnsgard 1978), but the majority of their nest sites fall within 15 yards (14 m) of permanent open water. Nest sites in sedge meadows, although usually not as close to feeding areas as waterside sites, provide added protection from raccoons, a principal predator of Ring-necked Duck nests (Hohman 1986).

Atlasers discovered a meager two nests, both close to open water. At both locations birds built nests in marshy areas near open water; one nest sat among cattails, the other among bulrushes (Randy Lentz pers. comm., HEK). Although this provides too small a sample from which to establish a clear pattern, neither does it indicate any great departure from observations in the northern part of their range. Observations of fledglings in the company of a hen provided most breeding confirmations. Such Confirmations came mostly from open water and provide few clues to the locations of nests. Nest ponds varied in size from less than one acre to 10 acres (0.4–4 ha).

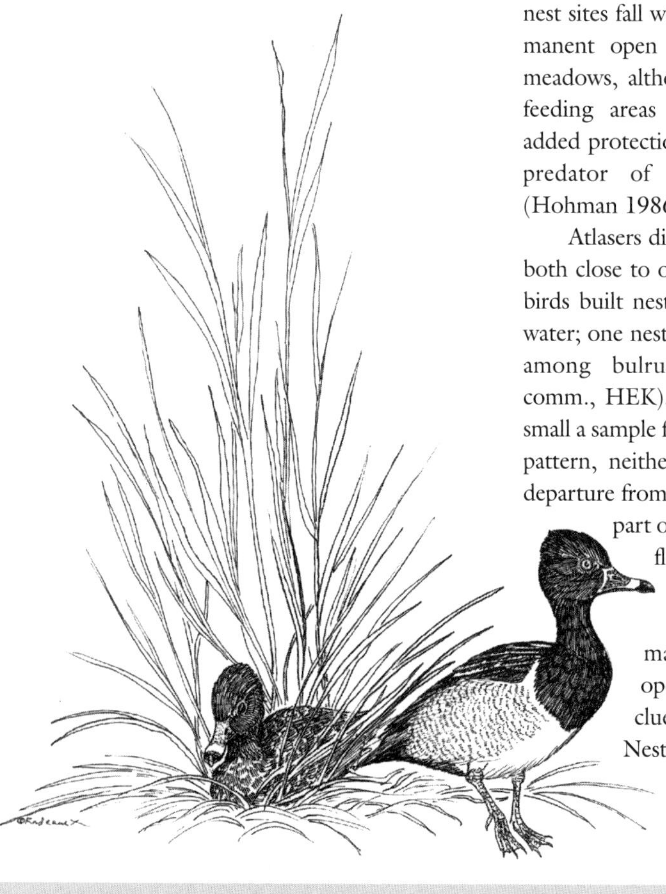

BREEDING

Pair formation by Ring-necked Ducks begins on their wintering grounds and continues at stopover points during migration. In some cases pair formation takes place after arrival on the breeding grounds, where the birds arrive after periods of ice- and snowmelt (Hohman 1986, Palmer 1976b). In some parts of their range, they arrive as early as the middle of April. Colorado's higher lakes and beaver ponds often remain icebound until June, a factor that delays the onset of nesting here. As of 1976, 1 May stood as the earliest recorded date of egg-laying for this species; the latest records ran into the third week of July (Palmer 1976b). Atlas and other Colorado observations of fledglings that run into late August and September suggest that egg-laying frequently extends at least as late as the late dates given by Palmer. Some broods hatch so late that the ducklings must race against freeze-up to fledge (Bob Sanders pers. comm.).

Ring-necked Ducks maintain pair bonds longer than most duck species. Typically, drakes remain with their mates until the four-week incubation period has nearly ended. The persistence of this pair bond accounts for the regular attempts at renesting among hens that have lost nests (Palmer 1976b). Some of the late broods reported by Atlas field workers may represent renesting attempts. Others may simply reflect the late arrival of summer and consequent delayed breeding in Colorado's high country.

Within a day after hatching, the hen leads her brood to water. During the first few days, the hen broods the young for the greater part of the day. She performs a distraction display in the face of threats to her young. By five days the young begin diving for food, primarily animal matter for the first few weeks. Adults and more developed ducklings consume a diet of mostly vegetable matter (Palmer 1976b).

BREEDING PHENOLOGY

CODE		# OF RECORDS	DATE RANGE
NE	Nest with Eggs	2	4 Jul–15 Jul
FL	Fledged Young	30	13 Jun–23 Aug

DISTRIBUTION

The bulk of the breeding range of Ring-necked Ducks falls within Canada, New England, the Great Lakes region, and the Pacific Northwest. Scattered breeding occurs within the Rocky Mountains. Winter takes them to open water, chiefly in the southern half of the contiguous states.

Unless early Colorado ornithologists simply overlooked Ring-necks, these small but striking ducks have become much more numerous in recent years. Cooke (1897), Sclater (1912), and Bergtold (1928) each regarded this species as one of Colorado's rarest ducks. Similarly, Bailey (1928) described them as "very rare" in New Mexico. Ryder (1950) found the first Colorado nesting in 1949—19 nests and some broods near Creede. The first hint of something different came in 1965 when B&N, cognizant that the numbers reported "in the literature" lagged behind the numbers implied by their own observations, suggested that hunters, and perhaps others as well, routinely misidentified these birds as Lesser Scaups. Nevertheless, they could cite only three nesting records of Ring-necked Ducks within Colorado.

By the 1980s breeding records existed for most of Colorado's mountainous latilongs (Latilong Study). Atlasers confirmed breeding in three more latilongs and upgraded their status to that of probable breeder in a fourth. All this provides strong evidence that Ring-necked Ducks have substantially expanded both their range and population within Colorado. If so, this expansion more or less echoes a similar expansion in the northeastern United States starting in the 1930s (Bellrose 1976).

BREEDING EVIDENCE

REPORTED IN 64 (4%) OF 1745 PRIORITY BLOCKS

Possible	10 blocks	(16%)
Probable	26 blocks	(40%)
Confirmed	28 blocks	(44%)

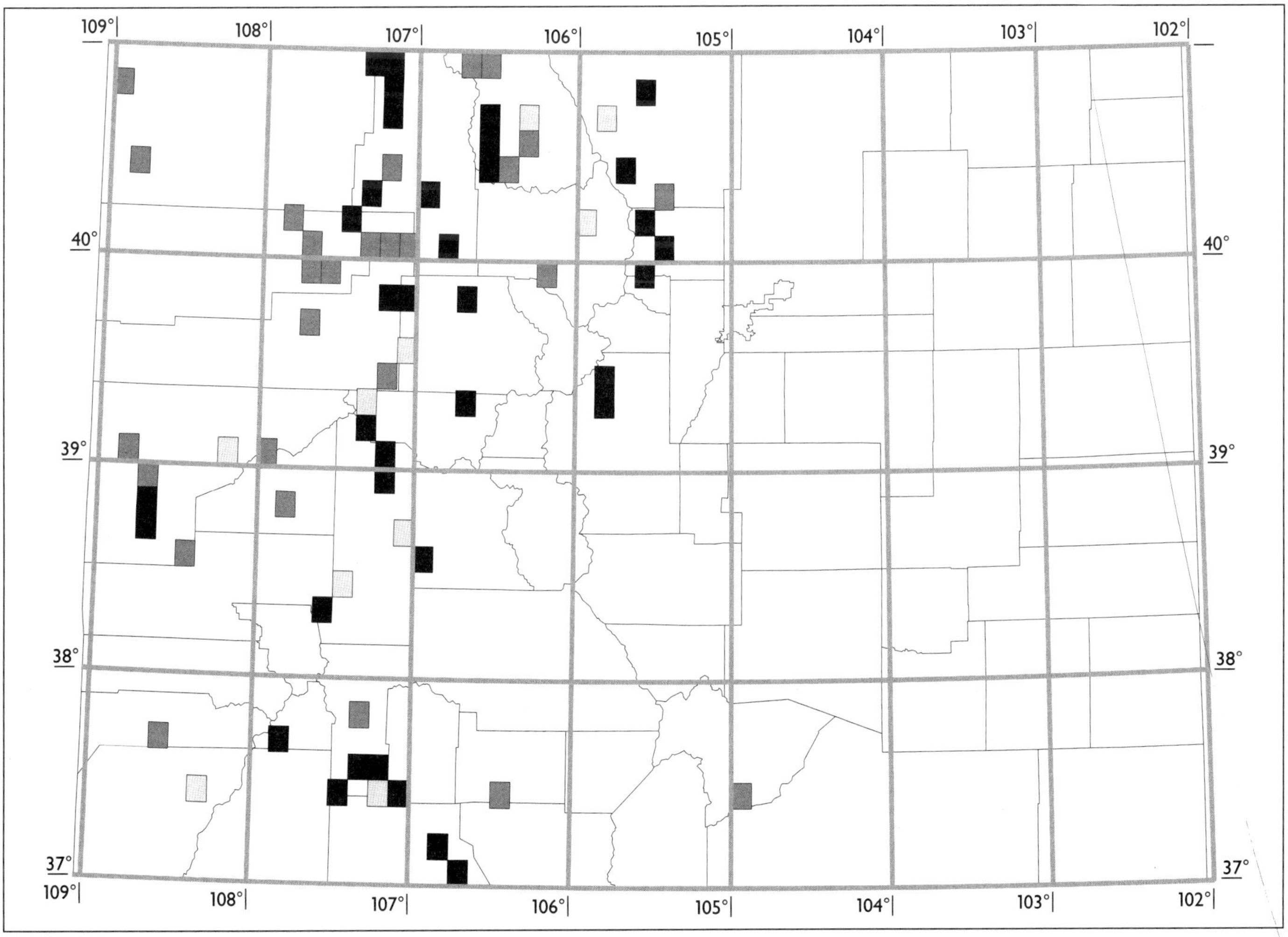
109°
108°
107°
106°
105°
104°
103°
102°
40°
39°
38°
37°

COLORADO DIVISION OF WILDLIFE / WILDLIFE RESOURCE INFORMATION SYSTEM

Secluded potholes in the forests and prairies of northern and western Canada provide summer homes for most of North America's Lesser Scaups (Terres 1980), but a few find a little bit of Canada in Colorado's Rocky Mountains.

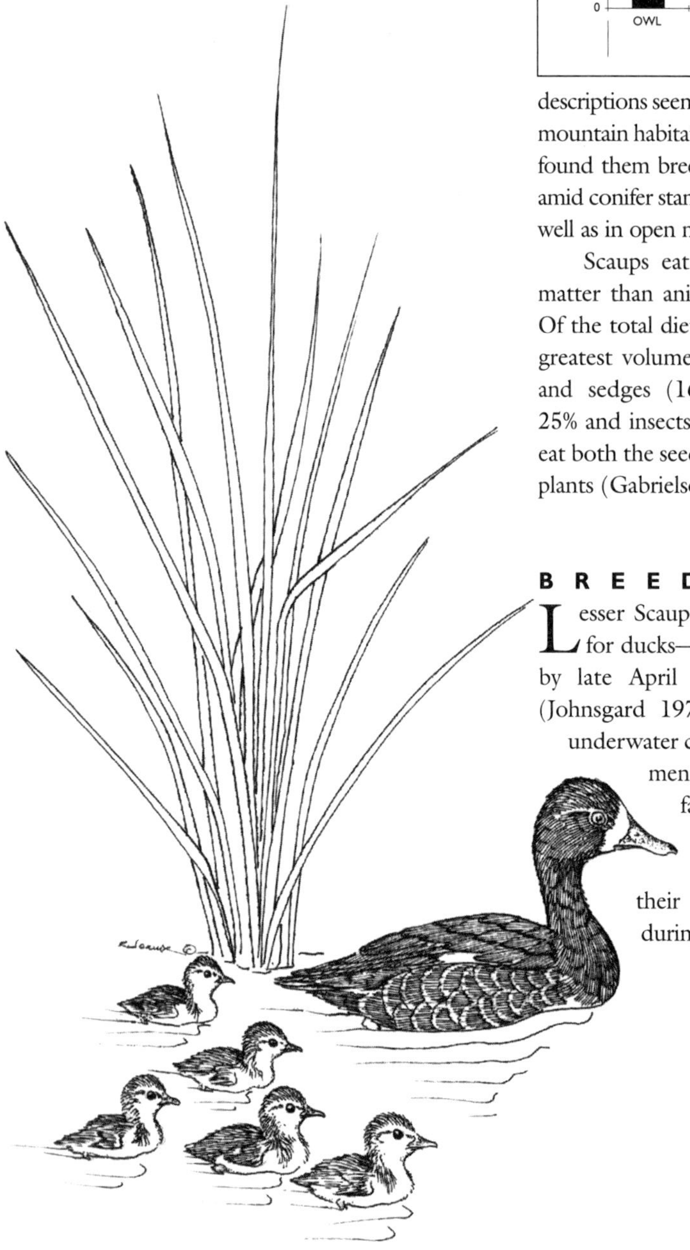

HABITAT

According to Johnsgard (1978), "Lesser Scaup typically breed in the vicinity of interior lakes and ponds with associated low islands and moist sedge meadows and surrounding environments of prairies or partially wooded parklands." In Canada, they breed most abundantly in wetlands of boreal forests (Bellrose 1976), especially grass-margined ponds and lakes (Palmer 1975). These

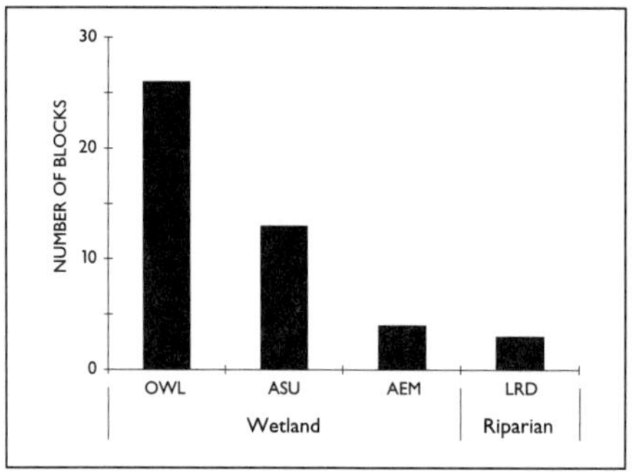
NUMBER OF BLOCKS
30
20
10
0
OWL
ASU
AEM
LRD
Wetland
Riparian

descriptions seem a perfect fit for the Colorado mountain habitats used by this species. Atlasers found them breeding in high mountain lakes amid conifer stands and mountain meadows as well as in open mountain parks.

Scaups eat somewhat more vegetable matter than animal material (60% to 40%). Of the total diet, pondweed contributes the greatest volume (19%), followed by grasses and sedges (16%); gastropods contribute 25% and insects 12% (Palmer 1976b). They eat both the seeds and vegetative parts of the plants (Gabrielson and Lincoln 1959).

BREEDING

Lesser Scaups form their pair bonds late for ducks—in March during migration; by late April most females have mates (Johnsgard 1978). Their displays include underwater components and rapid movements such as head-shaking so fast that the human eye sees it as a blur. Presumably many Colorado scaups form their pairs on prairie reservoirs during spring migration.

Members of this species breed relatively late throughout their range (Kortright 1943). Breeding females tend to return to the previous year's nesting pond in succeeding years. They also tend to nest in a semicolonial fashion rather than scatter their nests (Palmer 1976b). The females conceal their nests in grasses in a hollow, not far from water but on dry ground, and line them with grass and feathers. The 8–12 eggs hatch in 21–22 days. Like many ducks, scaups suffer from and engage in nest parasitism. Ruddy Ducks lay eggs in scaup nests, and Lesser Scaups lay eggs in the nests of wigeons, shovelers, and Redheads (Palmer 1976b).

Most nesting takes place at relatively high altitudes in Colorado, so nesting generally does not begin until mid or late June. Atlas workers reported only one Lesser Scaup nest, at Arapaho NWR on 28 June, and a distraction display in South Park on 10 July. The remainder of the Atlas Confirmations reported fledged young. Nine of 12 fledgling reports with dates came within a 10-day period—15–26 July. The other three came on an early 26 June from Gunnison and 20 and 26 August from two Flat Tops Wilderness blocks.

BREEDING PHENOLOGY

CODE		# OF RECORDS	DATE RANGE
C	Courtship	1	5 Jul
NY	Nest with Young	1	28 Jun
FL	Fledged Young	12	26 Jun–26 Aug

DISTRIBUTION

Lesser Scaups breed inland from Alaska to Manitoba and in the northern U.S. Colorado lies at the southernmost edge of the breeding range. B&N mention only North Park as a location of summering scaups in Colorado. A&R described this species as a "fairly common" resident in North Park and South Park, and they reported other scattered locations where breeding had occurred. Atlasers confirmed North Park as a center of Colorado abundance

(11 Priority blocks), but they reported this species in only one, Non-priority, block in South Park. Atlas work revealed a second center for this species in the Flat Tops Wilderness on the White River Plateau (9 blocks). Atlasers made three breeding Confirmations outside of these two centers, one each in South Park, near Gunnison, and on Grand Mesa. Atlas workers Confirmed breeding in 10 of 26 Priority and Non-priority blocks in the northwestern quarter of the state, yet in only one of 13 in the southern half. Most of the southern reports and the scattered reports from other mountain park and plains areas probably represent non-breeding birds and late migrants.

According to Johnsgard (1978), Lesser Scaups' use of relatively undisturbed habitats in Alaska and Canada has enabled the species to maintain a fairly stable population. The Colorado population probably also benefits from these habits. In fact, Atlas work seems to suggest a larger Colorado breeding population than previously known.

Arapaho NWR personnel in North Park have estimated the number of young raised in the refuge each year since 1972. The numbers range from a low of 77 in 1981 to a high of 1,638 in 1986. Over the past decade the refuge has produced approximately 1,000 young per year, nearly double that of the previous decade. A limited census of ducks in the Flat Tops Wilderness since 1992 has found about a dozen ducklings and a couple of dozen adults each year (Mary Massey pers. comm.). Atlas workers covering additional areas of the White River Plateau Confirmed breeding in several spots not covered by the Forest Service Survey.

BREEDING EVIDENCE

REPORTED IN 33 (2%) OF 1745 PRIORITY BLOCKS

Possible	6 blocks	(18%)
Probable	20 blocks	(61%)
Confirmed	7 blocks	(21%)

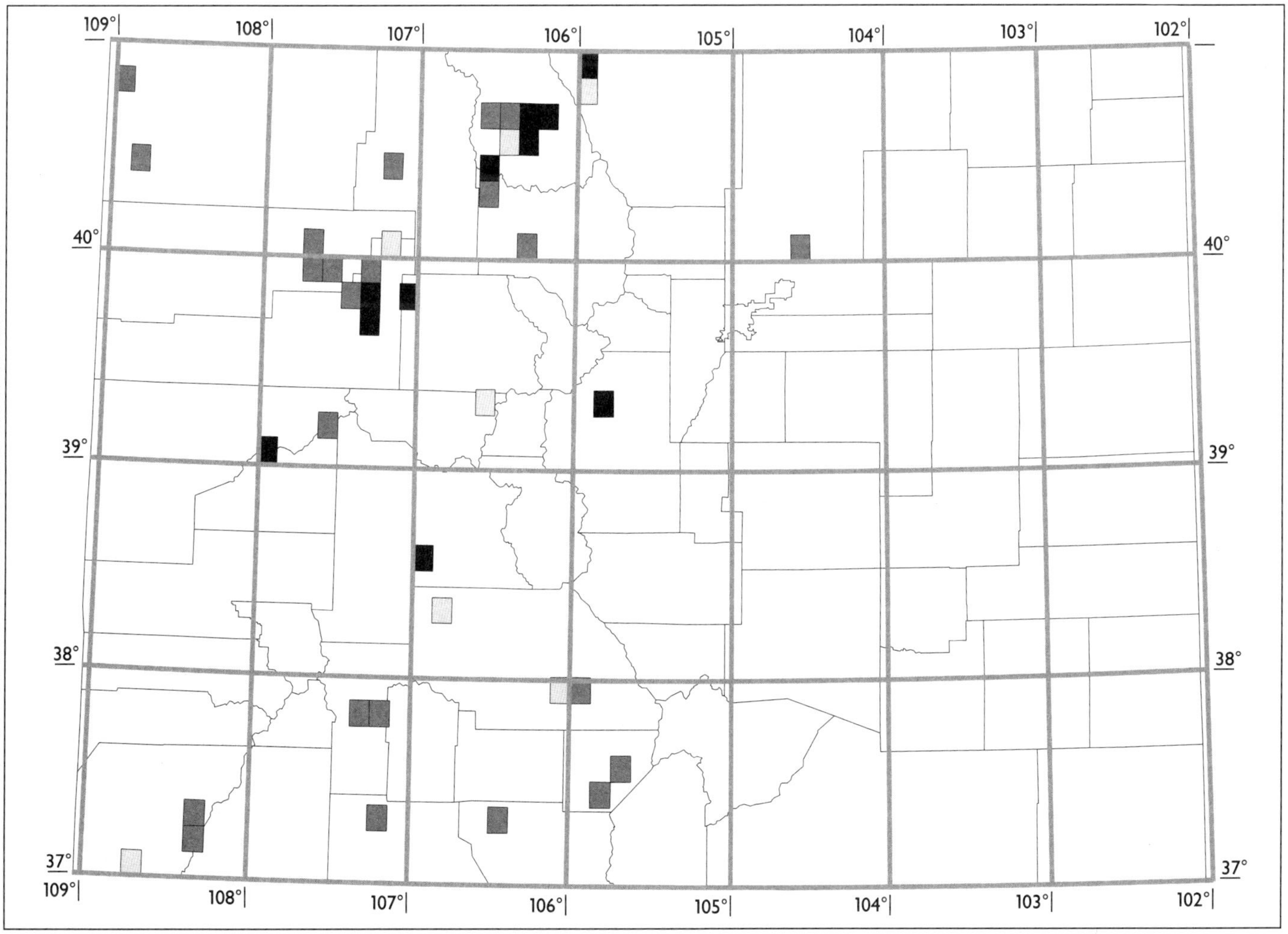
109°
108°
107°
106°
105°
104°
103°
102°
40°
39°
38°
37°
109°
108°
107°
106°
105°
104°
103°
102°
40°
39°
38°
37°

COLORADO DIVISION OF WILDLIFE / WILDLIFE RESOURCE INFORMATION SYSTEM

The Bufflehead hen circled the pond twice before flying toward the flicker hole. Her target, 22 feet up the aspen, measured barely two inches across. How could a duck that big fit through a hole that small, while in full flight? I watched her go in and still don't know.

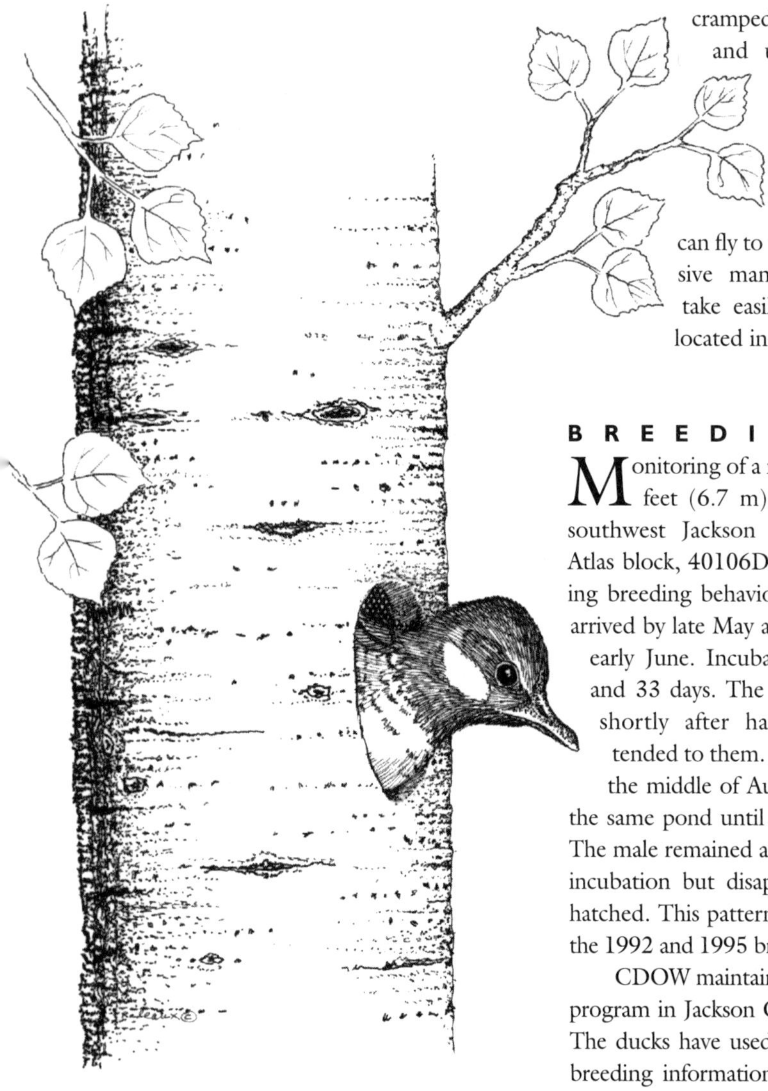

HABITAT

Buffleheads breed primarily in the boreal forests of the far North. In northern Colorado a few breed in the Park Range beside forested ponds similar to those in that northern habitat. Because they dive for food, Buffleheads require deep ponds and small lakes that have a large supply of insects and other invertebrates for the adults and young to eat.

Buffleheads prefer aspen stands where large dead or dying trees provide nest sites adjacent to a pond, although they also use conifer sites with suitable cavities. The nest tree can be as far as 220 yards (185 m) from the shore (Terres 1980). Old flicker nest holes and other large cavities provide cramped quarters for the hen and up to a dozen eggs (Ringelman and Kehmeier 1990). The entrance hole must face a clear flight path so that the hen can fly to the hole without extensive maneuvering. Buffleheads take easily to large nest boxes located in suitable habitats.

BREEDING

Monitoring of a nest, in a flicker hole 22 feet (6.7 m) up in a live aspen in southwest Jackson County (Spicer Peak Atlas block, 40106D4), showed the following breeding behavior. The Bufflehead pair arrived by late May and egg-laying began in early June. Incubation took between 28 and 33 days. The ducklings left the nest shortly after hatching, and the hen tended to them. The young could fly by the middle of August, but remained on the same pond until the end of the month. The male remained at the pond site through incubation but disappeared once the eggs hatched. This pattern occurred during both the 1992 and 1995 breeding seasons (NMB).

CDOW maintains an extensive nest-box program in Jackson County for Buffleheads. The ducks have used a few boxes, but host breeding information comes from sightings of broods after nesting is complete. Pairs often reuse successful breeding sites in following years, typically using the same nest hole or nest box if they bred successfully the previous year (Gauthier 1990).

BREEDING PHENOLOGY

CODE		# OF RECORDS	DATE RANGE
NY	Nest with Young	1	20 Jun
FL	Fledged Young	1	1 Aug

DISTRIBUTION

Buffleheads occur only in North America, with the vast majority breeding in the boreal forests of Canada and Alaska. The first breeding record for Colorado came in 1987 when atlaser Jennifer Rechel located a brood in the Pearl block (40106H5) in northwestern Jackson County. CDOW surveys since then have located broods on several ponds on the east side of the Park Range scattered from the Wyoming border south to Muddy Pass (Ringelman and Kehmeier 1990).

In 1991 the nest tree located in the Spicer Peak Atlas block extended the known breeding range farther south than previously known, and it is the only natural nest reported in the state. The Atlas map shows the CDOW brood-finds and atlaser discoveries, all within this restricted zone. Although suitable breeding habitat occurs in other parts of the state, Buffleheads apparently do not breed there (Jim Ringelmann pers. comm.). In 1992 and 1994 Buffleheads, observed in suitable habitat immediately south of the Continental Divide in Grand County, showed no indications of breeding (NMB).

An estimated 50 to 100 pairs of Buffleheads nest in Colorado, all in western Jackson County. The remote nature of many potential breeding sites in the high mountains makes a precise estimate of numbers difficult. Colorado's breeding Buffleheads may represent a new population or one simply undetected in the past.

Buffleheads may be sensitive to disturbance. By breeding on remote lakes with little human activity, they typically avoid lakes with extensive recreation use. The history of the Spicer Peak breeding illustrates the results of people impact. The pair fledged young for three consecutive years, but in 1994 road repair work took place adjacent to the pond during the courtship and nesting period. The adults stayed on the pond but did not attempt to nest. A gate greatly reduced traffic past the nest site. Reintroduced beaver repaired the failing dam and maintained deep water levels. Wrapping the trunk of the nest tree with fencing wire protected it from the beaver. The Buffleheads nested again in 1995, after closure of the road and in 1995 and 1996 fledging success increased from 3–4 young/year to 9–10 per year.

Buffleheads readily nest in boxes placed close to suitable ponds (Gauthier 1988). In addition, other management actions can improve Bufflehead breeding habitat: protection of suitable nesting trees, creation of snags, beaver management, and road closures. For safety reasons, land managers remove dead and diseased trees around recreation sites. Because many recreation areas occur near open water, this practice eliminates many suitable Bufflehead nesting sites.

Buffleheads receive no special protection at this time, but they have limited distribution in the state. Because almost all nest habitat falls on U.S. Forest Service lands, the Forest Service and the CDOW monitor known breeding sites and have undertaken an active management program to increase the numbers of breeding pairs in the state, in hopes of maintaining long-term viability of the local population.

BREEDING EVIDENCE

REPORTED IN 3 (0%) OF 1745 PRIORITY BLOCKS

☐	Possible	2 blocks	(67%)
▩	Probable	0 blocks	(0%)
■	Confirmed	1 block	(33%)

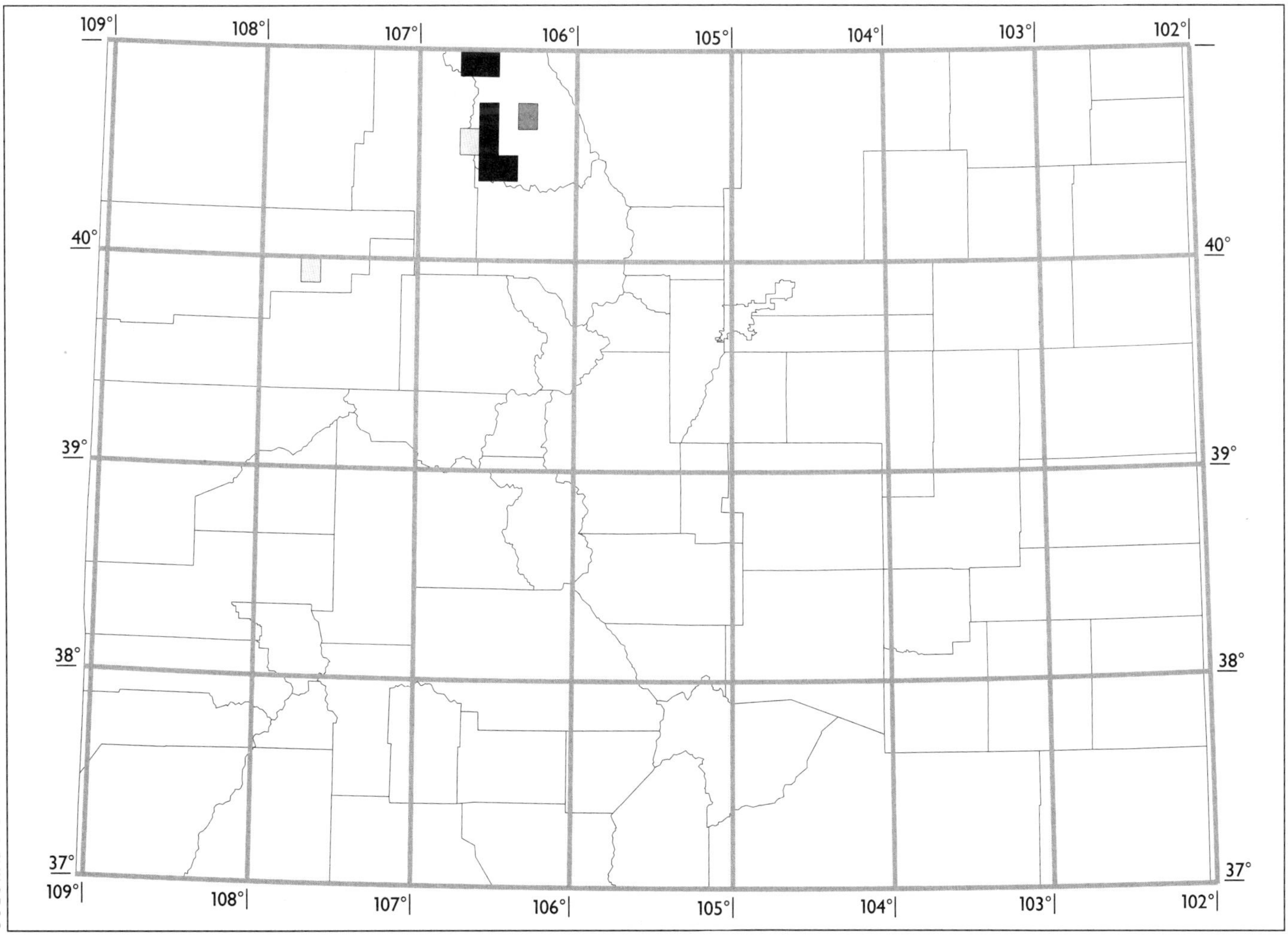
109°
108°
107°
106°
105°
104°
103°
102°
40°
39°
38°
37°

COLORADO DIVISION OF WILDLIFE / WILDLIFE RESOURCE INFORMATION SYSTEM

The Flat Tops Wilderness produced an exciting Atlas discovery: Confirmed breeding of Barrow's Goldeneyes in Colorado for the first time in this century. Finding these birds, the southernmost of the Rocky Mountain population, added more exhilaration to the striking scenery seen on wilderness backpacks.

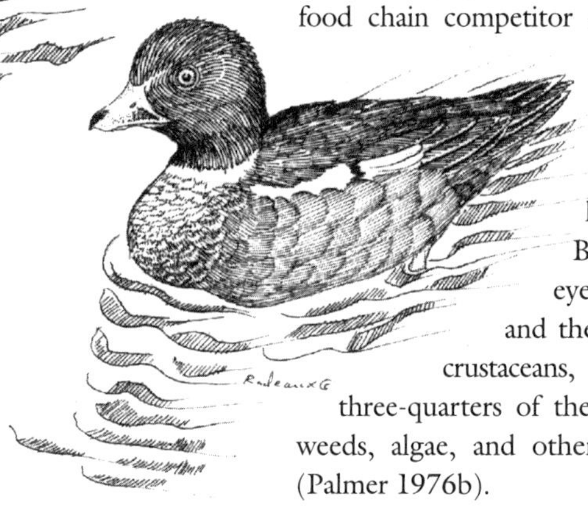

HABITAT

As cavity-nesters, these diving ducks find nest holes among the thousands of beetle-killed trees that blanket the Flat Tops Wilderness. Broods of Barrow's Goldeneyes in the Flat Tops inhabit shallow lakes at elevations near 10,500 feet (3200 m). Here hundreds of small lakes form a landscape mosaic within spruce/fir stands, willow carrs, grasslands, and lava rock fields.

Two prehistoric bark beetle outbreaks, which began in 1716 and 1827 (Veblen et al. 1994), plus the most recent epidemic in the 1940s, affected extensive stands of mature spruce. Each of these cyclic infestations left thousands of acres of standing dead trees. The beetle-killed spruce undoubtedly attracted large influxes of woodpeckers, which drilled many cavities. As new trees grew up among the spruce skeletons and matured, the old cavity trees may have disappeared and left the goldeneyes without nest sites. The 100-year gap in goldeneyes may parallel the 110-year beetle cycle.

Many small lakes on the Flat Tops may have high lake-edge insect populations because they lack continuous replenishment from flowing mountain streams. Oxygen depletion during winter, which commonly causes fish populations to die, eliminates a food chain competitor of the goldeneyes. (British Columbia goldeneyes breed in alkaline lakes in which fish cannot live; Palmer 1976b.) Barrow's Goldeneyes eat aquatic insects and their larvae, mollusks, crustaceans, amphipods (about three-quarters of the diet), plus pondweeds, algae, and other vegetable matter (Palmer 1976b).

BREEDING

Atlasers identified fledged young or broods on three shallow lakes in the Flat Tops on 6, 10, and 23 July, at elevations of 10,800–11,000 feet (3292–3350 m). During the second week of July 1991 a female Barrow's Goldeneye flew into a cavity of a broken spruce snag approximately 1,000 feet (305 m) west of one small lake. Brood records at this lake date continuously from 1988 to 1991 (Chuck Reichert pers. comm.), vouching for strong nest-site fidelity. These goldeneyes typically return to their natal lakes to breed and they reuse nest sites as long as the cavities remain functional (Palmer 1976b).

Barrow's Goldeneyes will nest up to 1.3 miles (2 km) away from water, depending on nest site availability (Savard et al. 1991), but usually they pick sites within 100 feet (30 m) of water. They nest in cavities in trees and stumps, in abandoned woodpecker holes, and, rarely, in rock crevices. The clutch size averages 10–13 eggs, and the incubation period lasts approximately 33 days. The precocial young jump out of the nest within 24–36 hours of hatching. At hatching time bad weather conditions and predators pose hazards to ducklings.

Males remain on territories until 7–10 days after incubation begins, then gather in small flocks (or in the Flat Tops, singly) to move to molt lakes (Bellrose 1976, Atlas).

Upright and fallen snags amid dense growths of young spruce hinder observer access to the lakes scattered in valleys and benches throughout the forest. Hundreds of square miles of suitable habitat exist within the formal wilderness and the adjoining national forest. These goldeneyes probably breed at many more sites, and their abundance could range from 11 to 100 breeding pairs per block.

A high degree of sensitivity to alterations in breeding habitat renders this species vulnerable to logging impacts (del Hoyo et al. 1992). Potentially, timber salvage sales pose the greatest threat.

BREEDING PHENOLOGY

CODE		# OF RECORDS	DATE RANGE
C	Courtship	1	30 May
FL	Fledged Young	3	6 Jul–23 Jul

DISTRIBUTION

Approximately 150,000 Barrow's Goldeneyes breed in the western mountains from Alaska to central California, 90% of the total population. The other 10% breed in Iceland, Greenland, northern Labrador, and possibly northern Quebec (Gauthier and Aubry 1996). During the winter most goldeneyes move down to nearby brackish coastal waters. Taxonomists recognize no subspecies, but the goldeneyes of Colorado belong to a unique population that breeds and winters inland on freshwater lakes, reservoirs, and rivers in Idaho, Montana, Wyoming, and Colorado. The Colorado contingent probably does not exceed 100 pairs, according to Atlas computations.

By mid November, Colorado birds breeding at higher elevations apparently drop into the lower valleys. Migrants and local wintering birds use open waters in northern and central Colorado west of the Continental Divide, especially reservoirs and fish hatcheries. They feed in conspecific groups, although sometimes in close proximity to flocks of Common Goldeneyes.

Early collectors considered the species numerous and resident in central and southern Colorado mountain areas (B&N). Cooke (1897) reported, "In the mountains it has been found breeding throughout the whole western half of Colorado, usually at about 8,000 feet." Edwin Carter found six nests along the Colorado River in 1876–1877, including the first North American breeding record, at the mouth of the Blue River (B&N). Two other nineteenth-century breeding records came from Summit and Boulder counties, and a dubious one from La Plata County (B&N).

More than a century passed before anyone again found Barrow's Goldeneyes breeding in Colorado, the result of Atlas field workers backpacking into the Flat Tops Wilderness. (The putative breeding record of six sleeping goldeneyes, photographed at treeless Walden Reservoir on 10 August 1982, lacks the details to substantiate breeding in Colorado; Charles Preston pers. comm., HEK.)

Atlas work discovered Probable breeding goldeneyes in 1988 and Confirmed breeding in 1990, 1991, and 1994. The records come from scattered lakes in four Priority and three Non-priority blocks within the Flat Tops mosaic of beetle-killed spruce. Atlas backpackers contributed greatly to the limited knowledge of Colorado's breeding Barrow's Goldeneyes and closed a century-long gap in history.

BREEDING EVIDENCE

REPORTED IN 4 (0%) OF 1745 PRIORITY BLOCKS

☐	Possible	1 block	(25%)
▩	Probable	1 block	(25%)
■	Confirmed	2 blocks	(50%)

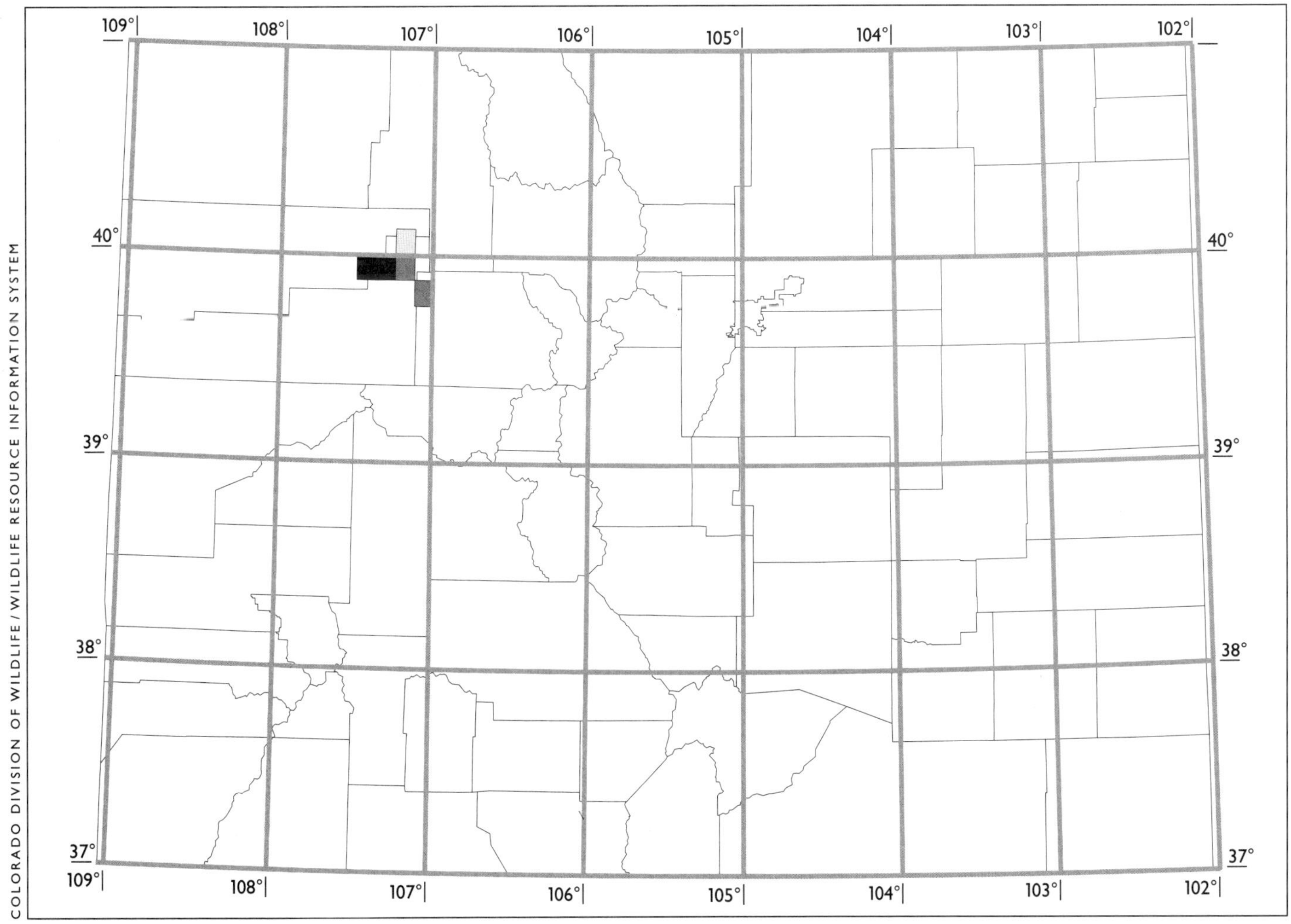
109°
108°
107°
106°
105°
104°
103°
102°
40°
39°
38°
37°
COLORADO DIVISION OF WILDLIFE / WILDLIFE RESOURCE INFORMATION SYSTEM

With "hairstyles" only their mother could fully appreciate, a clutch of young Common Mergansers will spend their first weeks close by her side. Infrequently, the young hitch a ride atop her back as she cruises along a river. This idyllic scene erupts into comedy when a youngster loses its perch and must scramble, with no sympathy from Mom, to regain its place.

HABITAT

Common Mergansers require clear water for feeding. They abandon sites with muddy water or that have grown dense with aquatic vegetation (Palmer 1976b). Atlasers saw most mergansers (in 90% of the blocks) in streams and streamside riparian habitats. The rest they reported on lakes, reservoirs, and beaver ponds.

In most places, tree cavities provide the favored nest sites for Common Mergansers (Bellrose 1976). B&N described Colorado nest sites as "holes in trees, old stubs, depressions in cliffs, and occasionally on the ground under low bushes." Although no atlaser had the good fortune of looking in on a nest of eggs, some nest sites did turn up—three in streamside deciduous trees, one in willow brush, and two on cliffs. One site, set high above the Dolores River on a cliff, provided an exciting Confirmation (Gerald Fyler pers. comm.).

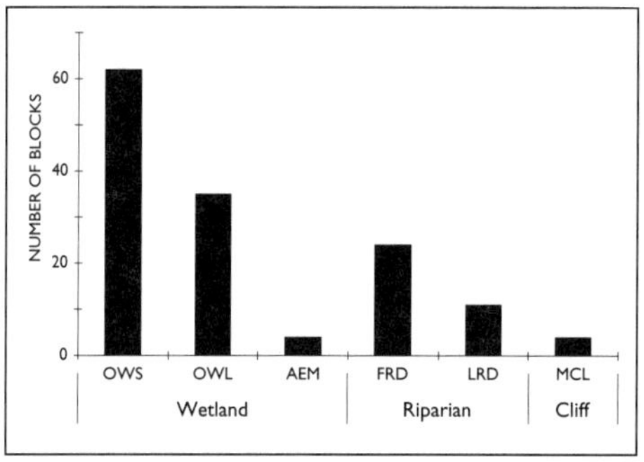
NUMBER OF BLOCKS
60
40
20
0
OWS
OWL
AEM
FRD
LRD
MCL
Wetland
Riparian
Cliff

Such breeding in cliff sites probably happens more frequently than records suggest. As evidenced by the number of fledgling broods observed along waterways, Common Mergansers breed with regularity along Western Slope creeks and rivers. In many of these sites, such as those on the Colorado Plateau, suitable trees never seem plentiful, but cliffs and rock faces abound. Although mergansers will accept nest boxes (Palmer 1976b), these birds lack the popular appeal of Wood Ducks and have had few, if any, boxes constructed on their behalf in Colorado.

Common Mergansers usually place their nests close to water but records exist of nests as far as 200 yards (180 m) from shoreline (Bellrose 1976). Whether nesting in cliffs, trees, or some other site, they invariably select sites concealed from above; darkness usually surrounds a nest. When nesting in trees, they prefer cavities with an entrance about 5 inches (12 cm) wide and an internal diameter of 10 inches (25 cm; Johnsgard 1978).

BREEDING

Because nearly all Atlas Confirmations came by observation of fledglings, assigning specific dates for egg-laying and incubation remains speculative. Although reports of fledglings ranged from late May to early August, most came from late June to mid July. A long fledgling period of 60–70 days (Johnsgard 1978) helps to account for the wide range of dates.

In the southern part of their range Common Mergansers start breeding as early as April (Palmer 1976b), although probably not in Colorado. Lingering cold temperatures at higher elevations may delay the onset of nesting here.

Clutch sizes vary widely; reports range from seven to the midteens (Palmer 1976b). Incubation averages about 30 days (Bellrose 1976). Fledglings exhibit well-developed swimming and diving skills and show greater facility walking on land than do adult birds. The young birds eat mostly insects until they gain proficiency at catching fish (Johnsgard 1978). As broods grow older, they tend to move downstream from their nest sites, in some cases perhaps due to depletion of prey in the upper reaches of the stream (Palmer 1976b). Before the young reach full growth, the mother departs to begin her molt.

Although Common Mergansers nest in sites with little accessibility to predators, and females cover their eggs before departing the nest during the incubation period (Palmer 1976b), some nests still succumb to predation, often by raccoons.

BREEDING PHENOLOGY

CODE		# OF RECORDS	DATE RANGE
ON	Occupied Nest	6	9 May–13 Jul
FL	Fledged Young	42	21 May–10 Aug

DISTRIBUTION

Common Mergansers breed throughout the boreal zones of northern Asia, Europe, and North America. Winter pushes their populations southward toward open water.

Early accounts of Colorado's birdlife recognized Common Mergansers as a breeding species (Cooke 1897, Sclater 1912). B&N

relegated their breeding chiefly to the southwestern mountains, and Davis (1969) regarded them as rare summer residents in the western Colorado mountains. Now Atlas data reflect a broad distribution there: of the counties west of Longitude 106, only Lake and Pitkin yielded no records. Atlas data, paralleling the A&R map, show roughly equal distribution in northwestern and southwestern Colorado. To the A&R map the Atlas adds breeding in the plateau country north of the Colorado River and in the river canyons west and north of the San Juan Mountains. A&R, however, indicate a sizable population along the lower Yampa River not reflected in Atlas data. This disparity may owe to a generally poor match between the locations of Priority blocks and the river.

These mergansers also breed more widely in the Front Range than previously realized. Larimer County produced the largest share, and other reports came from the foothills in Jefferson, Teller, and Park counties. Surprisingly, Confirmations came from the plains in Weld County and along the South Platte from Denver to the foothills. Atlasers provided first confirmations of breeding in latilongs 37105, 37108, and 40105.

The fish-eating habits give a flavor to their flesh that makes them less-than-favored table fare. Consequently, Common Mergansers face minimal pressure from legal hunting. They occasionally raise the ire of fishermen; although mergansers consume large numbers of rough fish, they also consume sport fish where available (Palmer 1976b).

Powerboats and other waterborne recreational vehicles can disturb Common Mergansers, occasionally in nesting areas. Destruction of nest trees, though, probably creates more disturbance to their breeding. Common Mergansers often use the same nest site year after year (Palmer 1976b). In places with limited nest sites, destruction of stubs and hollow trees can severely restrict breeding possibilities for these birds.

BREEDING EVIDENCE

REPORTED IN 124 (7%) OF 1745 PRIORITY BLOCKS

☐	Possible	45 blocks	(36%)
▪	Probable	34 blocks	(27%)
■	Confirmed	45 blocks	(36%)

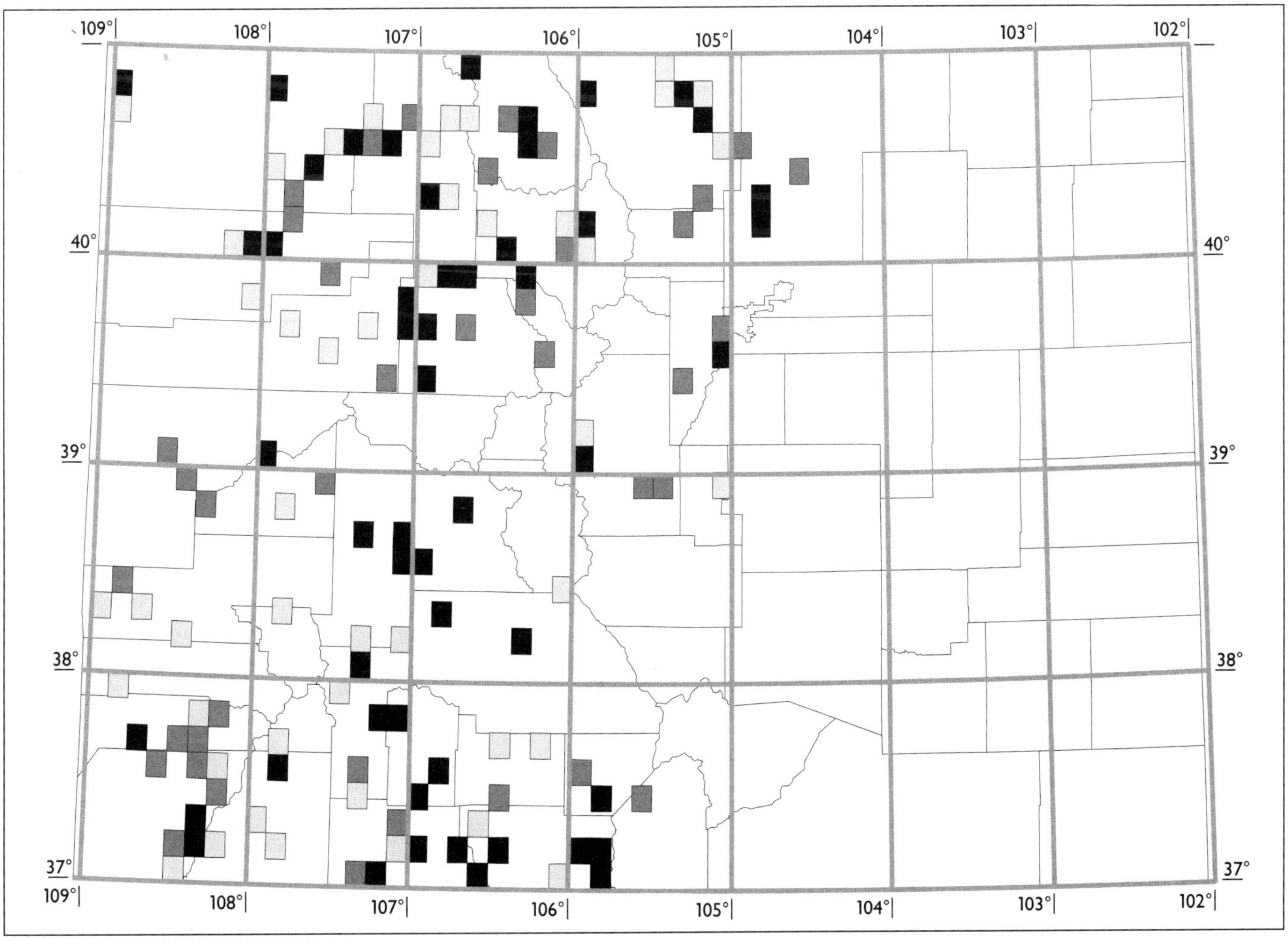

109°
108°
107°
106°
105°
104°
103°
102°
40°
39°
38°
37°

COLORADO DIVISION OF WILDLIFE / WILDLIFE RESOURCE INFORMATION SYSTEM

RUDDY DUCK *Oxyura jamaicensis*

Although he spends the entire winter wearing the dress of an avian pauper, the drake Ruddy Duck in late spring changes to princely dress. Sporting a dapper black cap, sky-blue bill, and white cheek patch, he assumes an entirely different bearing than his nondescript alter ego.

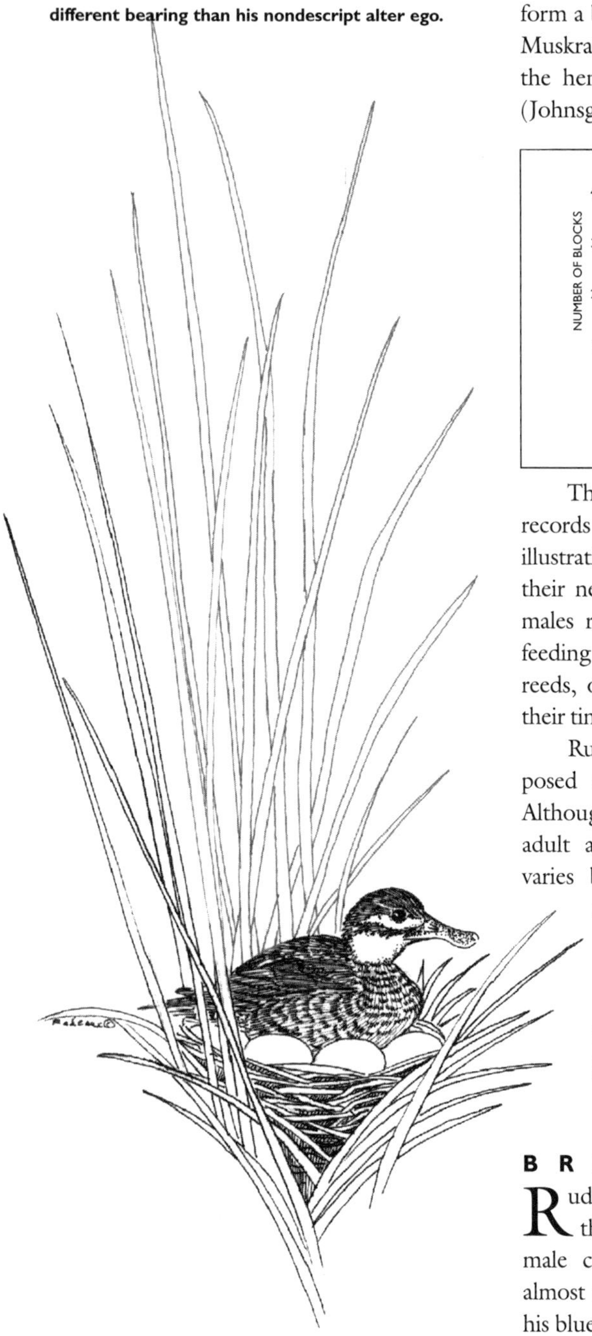

HABITAT

Ruddy Ducks conceal their nests in fresh water and alkaline marshes that have stable water levels. They prefer shallow marshes with extensive emergent vegetation and enough open water to facilitate takeoffs and landings (Johnsgard 1978). The female constructs a nest either by building up from the marsh bottom or by amassing a floating platform that she anchors to emergent vegetation. Later, she adds a rim to the nest to form a bowl (Bellrose 1976, Palmer 1976a). Muskrat paths through the vegetation allow the hen easy access to and from her nest (Johnsgard 1978).

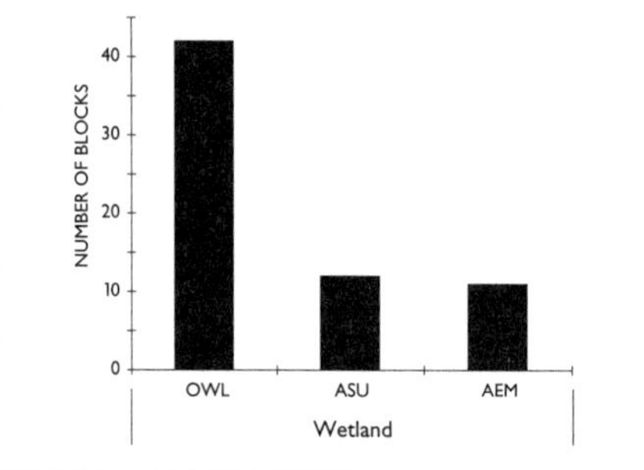
NUMBER OF BLOCKS
0
10
20
30
40
OWL
ASU
AEM
Wetland

The overwhelming majority of Atlas records came from open water on ponds, illustrating how well Ruddy Ducks conceal their nests from human observation. While males remain highly visible swimming and feeding in open water adjacent to cattails, reeds, or bulrushes, females spend most of their time under the cloak of deep vegetation.

Ruddy Ducks consume a diet composed of both animal and plant matter. Although the ratio of animal matter (mostly adult and larval insects) to plant matter varies by location, the diet never seems strongly skewed toward either (Palmer 1976a). Ruddys apparently prefer to feed in water just deep enough that the bottom mud is beyond the reach of dabbling ducks (Gooders and Boyer 1986).

BREEDING

Ruddy Ducks form pairs after arriving on their breeding grounds. In courtship a male circles the female with tail cocked almost to his head, slaps his red chest with his blue bill, and puffs out his chest by inflating an air sac (unique among ducks). He calls continually and almost stands as he scoots over the water surface (Kortright 1942). Once formed, the pair bond lasts at least until the hen begins to incubate the eggs, and often longer. Drakes occasionally accompany the hens well into the summer, an unusual behavior among duck species. A hen so accompanied drives away all drakes but one (presumably the one with which she paired), indicating an extension of the original pair bond (Joyner 1977).

Clutch sizes for Ruddy Ducks average 8 eggs. Hens, however, routinely lay many more eggs in the nests of other females or in dump nests, presumably to ensure a higher level of breeding success. (A dump nest receives eggs from multiple females, and sometimes contains two to three times the number of eggs in a normal clutch.) In all, a female may lay as many as 60 eggs in a single season (Bellrose 1976). Ruddy Ducks also parasitize nests of birds that share their marshy home, including grebes, coots, and other ducks (Palmer 1976a). Where Redheads coexist with Ruddy Ducks—as in the San Luis Valley—Ruddys themselves occasionally suffer nest parasitism from female Redheads (Joyner 1983, Ron Ryder pers. comm.).

So large are the eggs of Ruddy Ducks that a single clutch may exceed the weight of the hen. Incubation lasts 23 days. After hatching, the ducklings remain on the nest for almost one day. The ducklings stay with the hen for three to four weeks, but she usually abandons her young before they have finished fledging (Palmer 1976a, Johnsgard 1978).

Although the three Atlas reports of Ruddy Duck nests with eggs fell in July and August, the brood seen on 13 June signifies May egg-laying. The 3 August egg date betokens unusually late nest initiation for most species, although probably not for Ruddy Ducks (Ron Ryder pers. comm.). Fledglings seen in September provide additional evidence that these birds rear young at late dates with regularity. Because very little Atlas work took place during the months of August and September, other similarly late broods may have escaped the eyes of the atlasers.

BREEDING PHENOLOGY

CODE		# OF RECORDS	DATE RANGE
C	Courtship	3	4 Jun–16 Jul
NE	Nest with Eggs	3	9 Jul–3 Aug
NY	Nest with Young	1	28 Jun
FL	Fledged Young	13	13 Jun–1 Aug

DISTRIBUTION

Ruddy Ducks breed primarily in the prairie lakes and potholes of the western U.S. and Canada. Most migrate south into Texas and Mexico to spend the winter; flocks of 50–100 stop over in Colorado during spring and fall migration. The Atlas map shows that the San Juan Basin, San Luis Valley, and North Park provide the primary breeding grounds within the state, with lesser breeding numbers in several blocks in the South Platte River basin northeast of Denver.

Atlas field work documented breeding in one northeastern Colorado latilong (39103) that previously lacked confirmed breeding. Five latilongs with prior breeding records produced few or no reports.

Cooke (1897) accorded Ruddys the status of common summer resident in Colorado—a status that he extended, without qualification, to only two other duck species. Bergtold (1928) also regarded this species as a common summer resident and labeled it Colorado's third most common duck, with an estimated 10% of the state's total duck population. Atlas abundance estimates show fewer than 2,000 breeding pairs.

Even allowing for a substantial margin of error in Cooke's claims, and in the Atlas estimates, and despite the fact that Ruddy Ducks still appear in most parts of the state, this species has undergone a considerable decline. This drop in Colorado parallels similar decreases across North America. Periodic losses due to oil spills in the winter range have taken their toll. More importantly, continuing loss of breeding habitat, whether due to drought conditions or "reclamation" of shallow wetlands for agricultural purposes, has reduced the Ruddy Duck population to a fraction of its earlier numbers (Johnsgard 1978). Continued destruction of shallow marshland habitat will only exacerbate this already dismal trend.

BREEDING EVIDENCE

REPORTED IN 44 (3%) OF 1745 PRIORITY BLOCKS

☐	Possible	14 blocks	(32%)
■	Probable	19 blocks	(43%)
■	Confirmed	11 blocks	(25%)

COLORADO DIVISION OF WILDLIFE / WILDLIFE RESOURCE INFORMATION SYSTEM

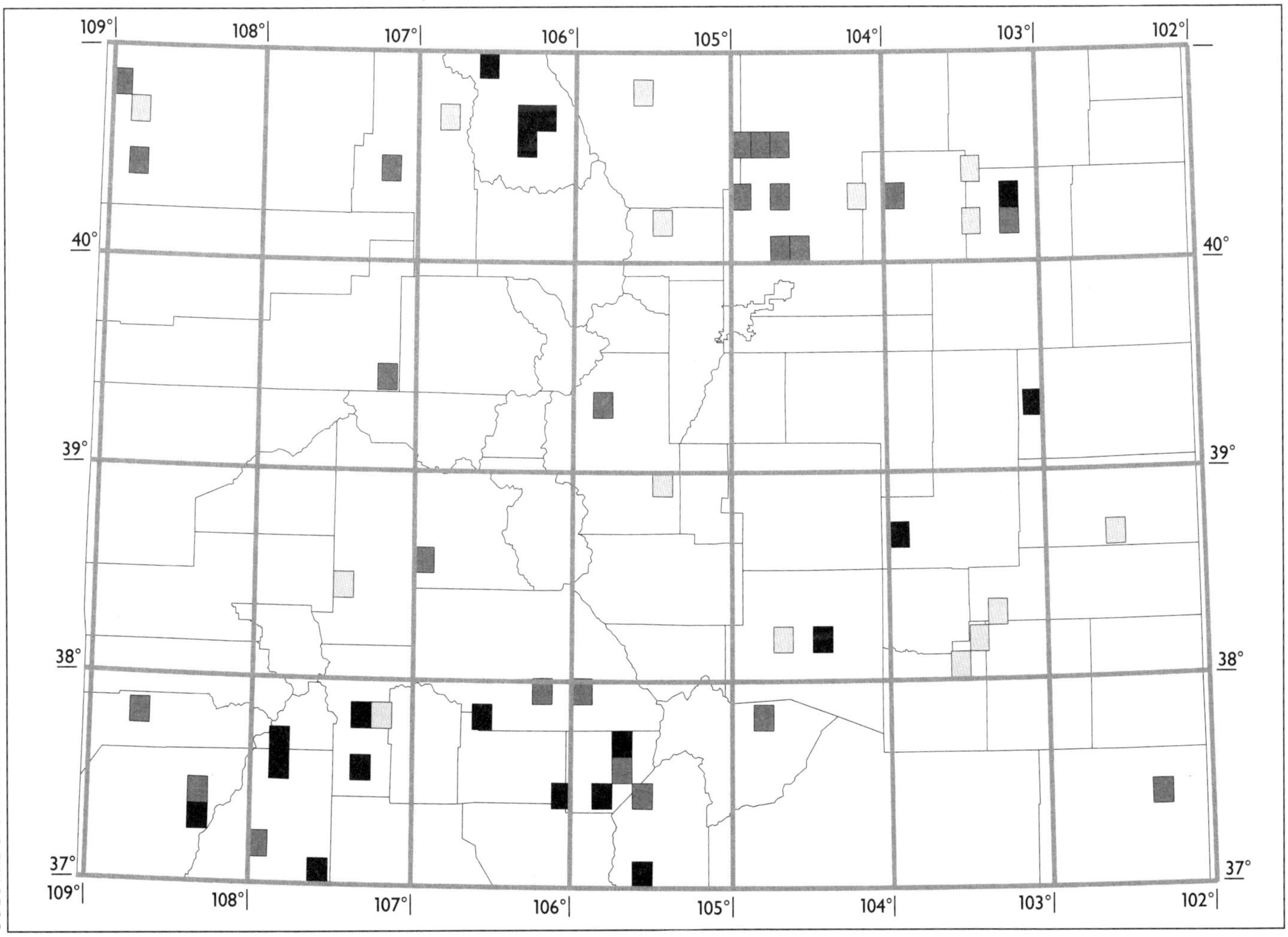
109°
108°
107°
106°
105°
104°
103°
102°
40°
39°
38°
37°

When it comes to nest-building, Ospreys combine the best skills of architects with the talents of high-wire artists. Although cranes find stability building on the ground and Bald Eagles use sturdy tree branches, Ospreys build their high nests delicately balanced atop dead trees and telephone poles. If a nest blows over, they simply start again, and again, and again.

HABITAT

Ospreys nest in a wide variety of places, making "suitable habitat" for them a broader choice than for species limited to a single specialized niche. They require two major components in their habitat: a large body of water with fish large enough to catch, and suitable nest sites. Tall dead trees, dead broken-off treetops, power poles, goose nest platforms—all serve equally well. Even floating platforms at water level work if other suitable sites are unavailable (NMB).

In Colorado they often nest in flooded groves or trees on islands. Thus 9 of 13 Atlas block habitat reports listed lakes and 2 more listed streams. Atlasers described four nests, all

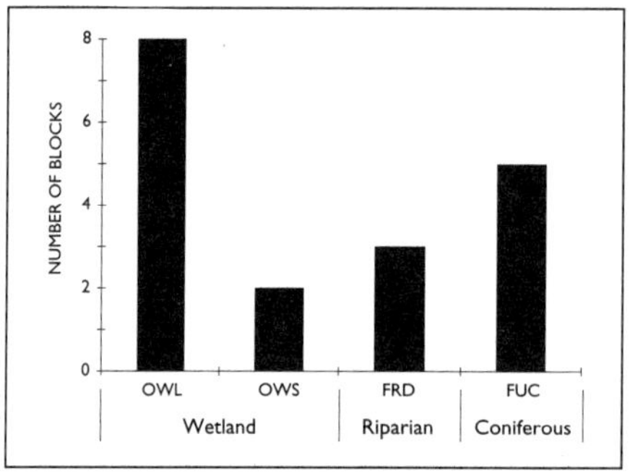
NUMBER OF BLOCKS
0
2
4
6
8
OWL
OWS
FRD
FUC
Wetland
Riparian
Coniferous

balanced 30–100 feet up in the tops of live conifers or dead snags. Many of Colorado's active nest sites do not lie within Priority blocks. Several Osprey pairs use nest platforms put up by CDOW in an effort to recover failing populations across the state.

BREEDING

Numbers of breeding Ospreys over the last three decades have gradually climbed across the U.S., and populations in Colorado follow the same pattern. Ospreys nest in single pairs or in loose colonies, but tend to concentrate in familiar breeding areas and do not readily expand to new locations (Jerry Craig pers. comm.). Birds return to the same nest or territory each year, sometimes for decades.

Ospreys engage in a protracted nesting cycle—four months from nest-building to fledging. A new nest may take three weeks to build, and the birds add sticks throughout the incubation and nestling phases, exemplified by the Atlas nest-building date of 11 August. In Colorado Ospreys lay eggs around mid April and typically raise two young. Incubation lasts 38 days and nestlings fledge after about 50 days (Palmer 1988a). Females brood the nestlings almost constantly for a month; the young can succumb to direct sunlight or rain. Atlas nest dates span two months—three by including the one (August) NB report. Most atlasing did not commence in the mountains early enough to detect nest-building, and even if it did, on-nest codes would have superseded them according to Atlas protocol.

The large nests sometimes provide nesting sites for smaller birds without conflict. Common Grackles, House Sparrows, Tree Swallows, and Northern Flickers all have nested within Osprey nests or in the supporting structures while the Ospreys used the nest (Evans et al. 1994). This can prove dangerous, as Ospreys occasionally capture songbirds (NMB).

BREEDING PHENOLOGY

CODE		# OF RECORDS	DATE RANGE
NB	Nest Building	1	11 Aug
ON	Occupied Nest	3	5 May–15 Jul
NY	Nest with Young	5	7 Jun–15 Jul
FY	Feeding Young	1	19 Jul

DISTRIBUTION

Ospreys occur on all continents around the world except Antarctica. In North America they breed across the U.S. and southern Canada. Historical records agree that they have always nested in Colorado, but offer no enlightenment as to population size. Cooke (1897) described them as "not uncommon, locally," and Sclater (1912) and B&N recounted only scattered nest records. Sclater mentioned nests at Sweetwater Lake and Summit County (to which they returned during Atlas) and Twin Lakes.

In the 1960s Ospreys disappeared from most of Colorado due to DDT and other pesticide-related problems. One small, remnant population survived in Jackson County, in the vicinity of Big Creek Lakes (Jerry Craig pers. comm.). Since then a population expansion aided by the DDT ban, along with one reintroduction effort by CDOW, has established local populations around lakes and rivers scattered across the state. The Atlas map shows the small concentrations that now occur around Shadow Mountain and Dillon reservoirs, Fort Collins, and the San Juan Basin. A pair began nesting at Pueblo Reservoir in 1990 (Mark and John Yaeger pers. comm.; Swallows block, 38104C7) and from 1996 to 1998 pairs pioneered new sites near Boulder, Fort Collins, and Denver.

Ospreys migrate out of the state for the winter. Band returns indicate that most winter on the Gulf Coast from Texas to Central America, and on the Pacific coast of Central America.

Although local populations now seem well established, numbers remain low. A local problem could have catastrophic impacts on the state's Ospreys. Potential problems include renewed exposure to pesticides and other chemicals, loss of suitable nest sites, and disturbance of established nest sites. By establishing new population centers, local threats to one population may not affect the others.

Pesticides remain an international problem because Ospreys migrate and they occupy the top of the food chain. DDT use in Mexico and Central America continues as a problem for several bird species. Any efforts to relax current water-quality laws could reverse the current population trend and again deplete numbers of Ospreys and of many other species.

Nest-site loss revolves around snag (dead tree) management. Most prime nesting habitat occurs around reservoirs and large rivers, all popular recreation sites. Safety requirements for campgrounds and parks call for removal of most snags as hazards. Development of marinas and vacation housing also results in a loss of suitable nest trees. Preservation of known nest trees and creation of artificial nest sites help to maintain and to bolster current populations of Ospreys, because they take readily to man-made nest platforms. The development of recreation at these prime nest sites also creates several disturbance problems. Boating around nest sites disturbs breeding birds; boating over the entire water body may interfere with fishing possibilities for the Ospreys. Current conservation measures include seasonal closure of areas around known nests.

BREEDING EVIDENCE

REPORTED IN 16 (1%) OF 1745 PRIORITY BLOCKS

Possible	6 blocks	(37%)
Probable	0 blocks	(0%)
Confirmed	10 blocks	(63%)

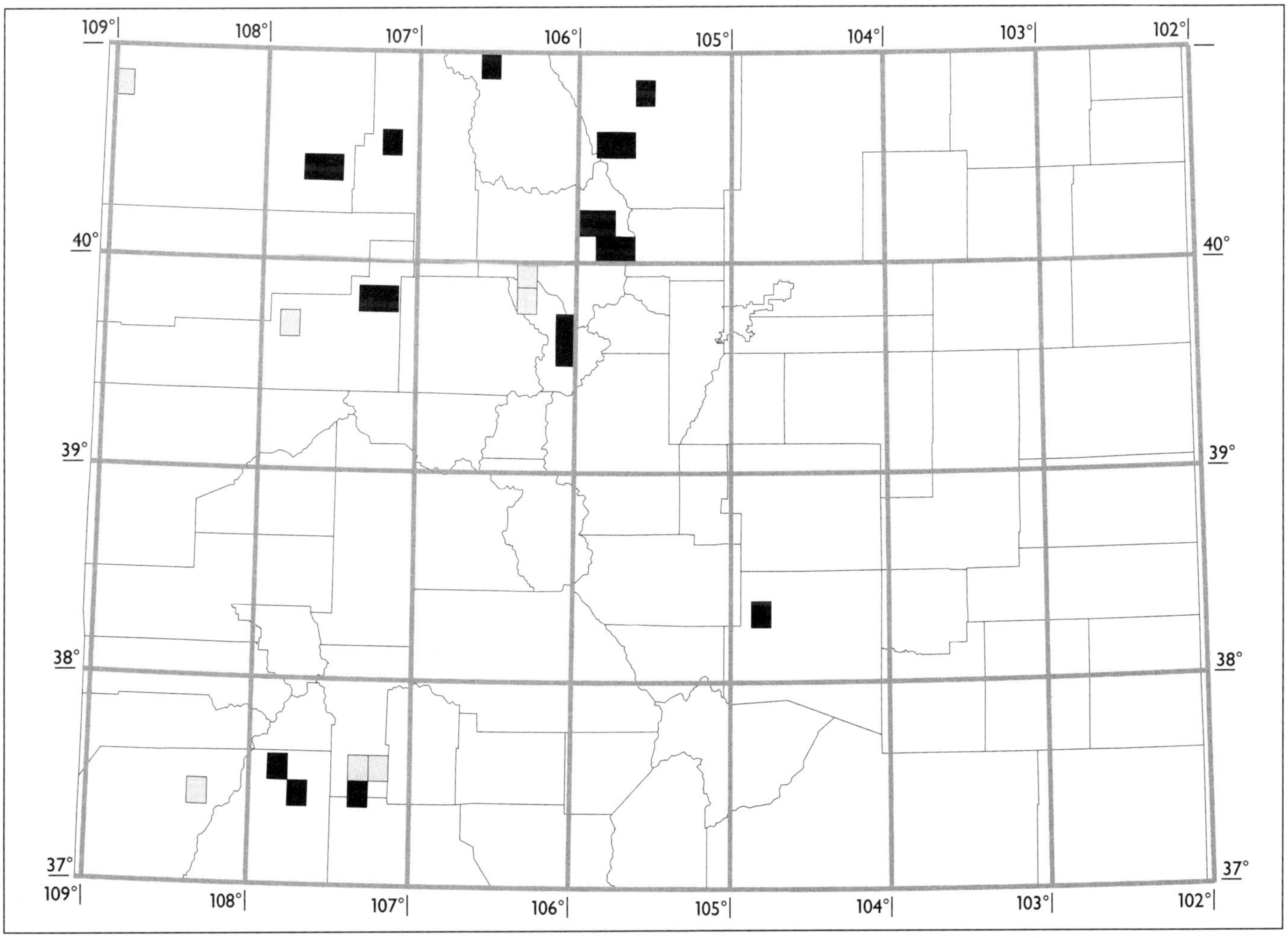
109°
108°
107°
106°
105°
104°
103°
102°
40°
39°
38°
37°

COLORADO DIVISION OF WILDLIFE / WILDLIFE RESOURCE INFORMATION SYSTEM

Whether gliding silently over open country, swooping down to snatch a grasshopper from midair, or hanging suspended in the wind, these graceful raptors make flying seem joyous and effortless. In the last half century over 100 breeding pairs of Mississippi Kites have settled in southeastern Colorado, where they nest colonially in cottonwood groves and city parks.

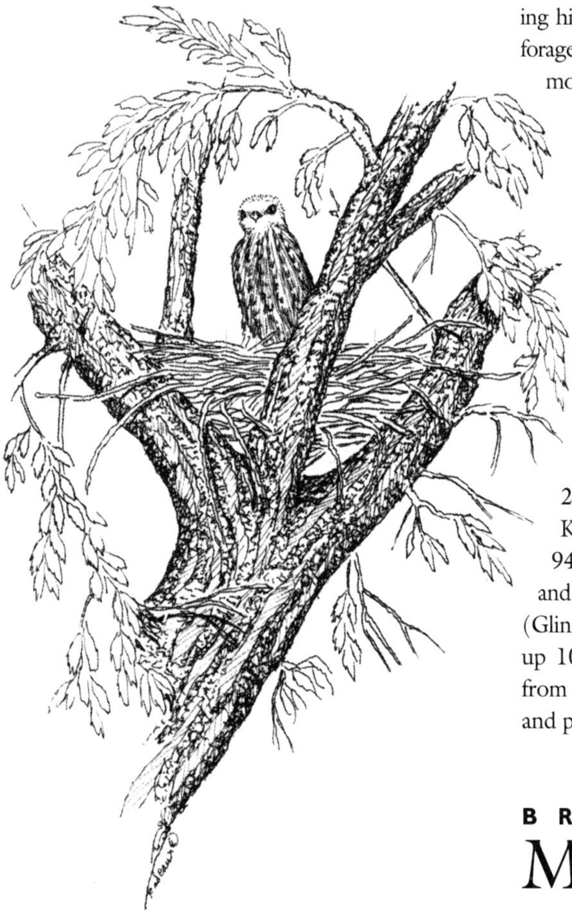

HABITAT

Mississippi Kites soar over the rimrock canyons, mesas, river valleys, and cities of southeastern Colorado. Atlasers saw them principally along cottonwood stream bottoms and over city parks and adjacent urban areas. Throughout their breeding range they typically nest in woodlots, riparian woodlands, and forest edges (Johnsgard 1990).

Atlasers found them in both very urban and very wild situations—from city parks in Lamar and Pueblo to cottonwood groves in the canyons of Las Animas and Baca counties. Highly social, these kites choose tall trees and frequently nest in tightly packed or loose colonies.

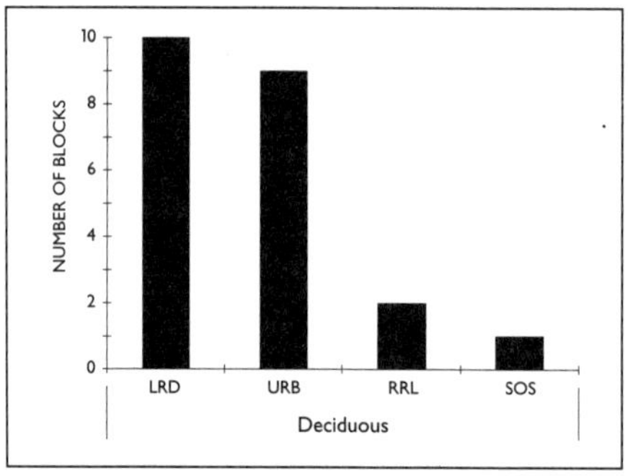
NUMBER OF BLOCKS
0
2
4
6
8
10
LRD
URB
RRL
SOS
Deciduous

Their ability to capture insects while soaring high overhead enables Mississippi Kites to forage in a variety of habitats, ranging from moist deciduous forests to arid grasslands. In the southern Great Plains they hunt over grasslands, rangelands, croplands, pinyon/juniper woodlands, oak woodlands, shrublands, and riparian woodlands (Parker 1974).

During prolonged foraging flights, Mississippi Kites capture grasshoppers, cicadas, beetles, and, occasionally, small bats (Johnsgard 1990). Infrequently they also prey on songbirds and scavenge a variety of road-killed invertebrates. In a study of 2,636 prey deliveries to three Mississippi Kite nests in Arizona, insects composed 94% of prey items; bats, toads, lizards, and frogs made up the remaining 6% (Glinski and Ohmart 1983). Insects made up 100% of identified prey items collected from 205 pellets deposited beneath nests and perches in Kansas (Fitch 1963).

BREEDING

Mississippi Kites arrive in southeastern Colorado in May (A&R) and begin nest-building in late May or early June. Atlas field workers observed nest-building in Cottonwood Canyon in southern Las Animas County on 22 May. A few miles to the south in Sheep Pen Canyon (Furnish Canyon West, 37103A2) a pair of kites had completed nest construction and begun incubation by 16 June (SRJ). A dependent fledgling in Holyoke initiated awkward experimental flights from the nest tree on 1 August (Jack Wieland, VF). A pair of Mississippi Kites nesting in an Atlas field worker's yard in Pueblo began nest construction in late June and fledged one young in late August (Mark Yaeger pers. comm.).

Because Mississippi Kites nest in the same location year after year, often reusing or reconstructing old nests, nesting colonies may become magnets for predators. Parker (1974) attributed a majority of nest failures in Kansas to predation. In Colorado Great Horned Owls, American Crows, raccoons, skunks, and various snakes probably prey on eggs and nestlings.

These kites breed in one of the hottest parts of Colorado and begin nesting after the young of many other species have already fledged, so that Atlas data may understate nesting densities. Most Atlas field work in southeastern Colorado took place in May or early June, before many kites had begun nesting. Nevertheless, Atlas volunteers found four previously unreported nesting sites.

BREEDING PHENOLOGY

CODE		# OF RECORDS	DATE RANGE
NB	Nest Building	2	22 May–20 Jun
ON	Occupied Nest	1	16 Jun
NY	Nest with Young	3	1 Jun–22 Aug
FL	Fledged Young	1	25 Jul

DISTRIBUTION

Southeastern Colorado lies at the northwestern corner of the Mississippi Kite's breeding range, which extends in a band 100–700 miles wide from central Arizona eastward across Texas, western Oklahoma, and western Kansas, and across the South to central North Carolina and northern Florida. These kites also breed northward along the Mississippi River to southern Illinois. They winter from central Texas to Argentina (Johnsgard 1990); an average of 34,900 passed by the Veracruz, Mexico, hawk-watch in 1992–1996 (Ruelas Inzunza 1998).

Although Mississippi Kites may have nested in Colorado prior to European settlement (Johnsgard 1990), Colorado had no breeding records until 1971, when a pair nested in La Junta (Parker and Ogden 1979). Early authors characterized the species as a "straggler" or "casual" in Colorado (Sclater 1912, Bent 1942, B&N).

Atlasers found the largest colonies nesting in the Lamar City Park (10–15 pairs) and in parks, residential areas, and riverine cottonwood groves in Pueblo (50–100 pairs). Smaller colonies (1–3 pairs) inhabit other Arkansas Valley towns and canyon-bottom cottonwood groves in southern Baca and southern Las Animas counties. A single pair nested in an urban park in Holyoke, in northeastern Colorado, from 1992 to 1994, and two pairs nested there in 1995.

The recent range expansion into eastern Colorado probably stems from the spread of cottonwoods along High Plains rivers and streams, the planting of shelterbelts around farm and ranch houses, the growth of Siberian elms in prairie communities, and reduced mortality from shooting. Parker and Ogden (1979) concluded that both the proliferation of deciduous trees on the Great Plains and the clearing of deciduous forests in the Southeast have benefited Mississippi Kite populations. However, destruction of riverine forest habitat in Illinois and Tennessee has probably led to the species' decline in that region (Kalla and Alsop 1983).

Because Mississippi Kites range widely and can breed as yearlings, they can rapidly colonize new nesting areas (Parker and Ogden 1979). They also adapt well to urban settings. In Kansas, where they often nest in urban parks, breeding pairs have begun to colonize golf courses (Parker 1988). On some courses golfers wear "kite-proof" hats or detour around certain holes to avoid the parental fury of nesting birds.

BREEDING EVIDENCE

REPORTED IN 13 (1%) OF 1745 PRIORITY BLOCKS

☐	Possible	10 blocks	(77%)
▩	Probable	0 blocks	(0%)
■	Confirmed	3 blocks	(23%)

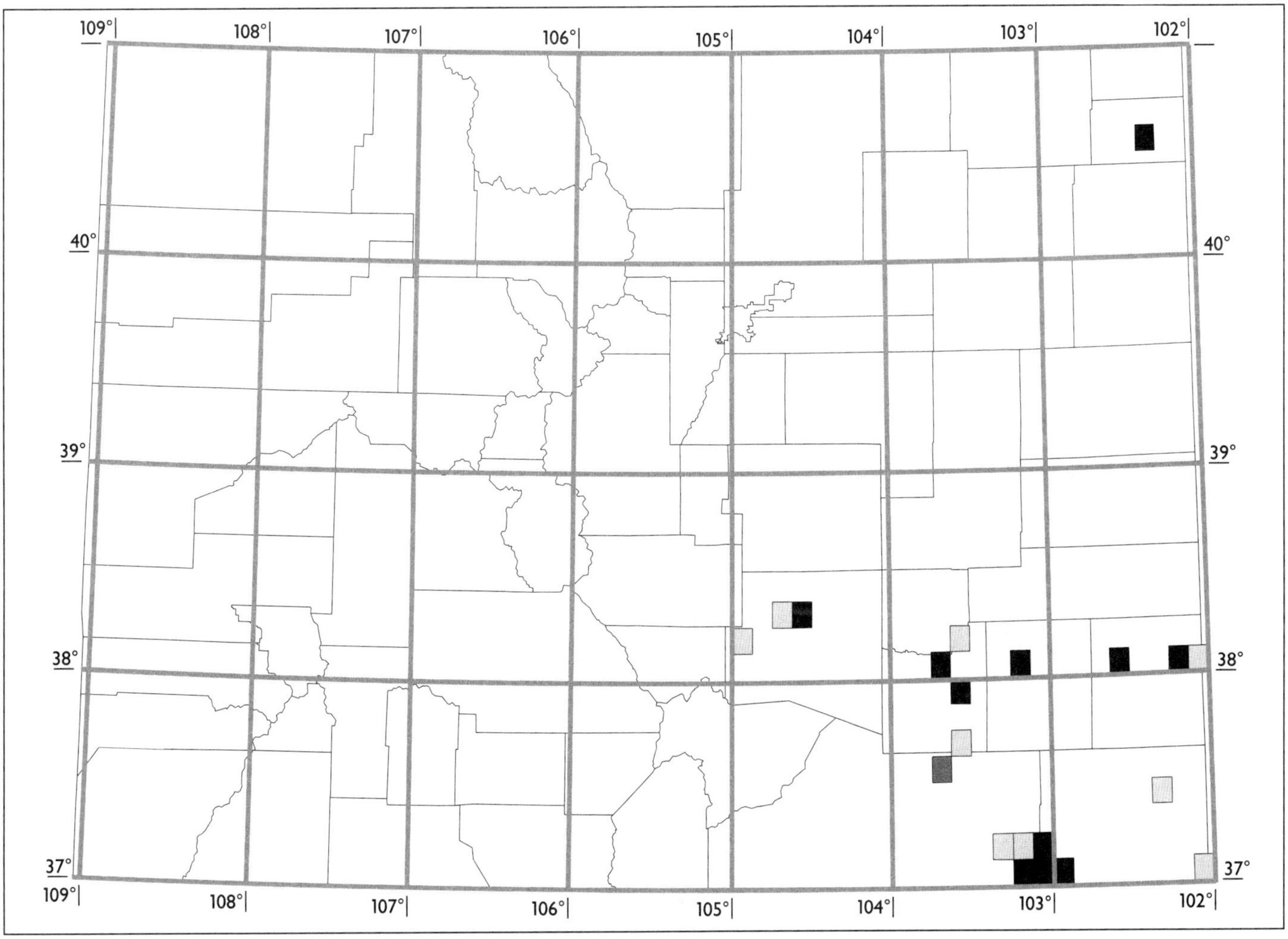
109°
108°
107°
106°
105°
104°
103°
102°
40°
39°
38°
37°

COLORADO DIVISION OF WILDLIFE / WILDLIFE RESOURCE INFORMATION SYSTEM

BALD EAGLE *Haliaeetus leucocephalus*

What a beautiful, impressive bird! Our national symbol! When you see it perched or in flight you know that you behold something important. Its large size, broad wingspan, white head and tail contrasting with dark body, fierce eyes and beak, and threatening cries all tell you to beware of and to respect this presence.

HABITAT

Those pairs of Bald Eagles that breed in Colorado use large, mature cottonwoods or pines to hold their heavy nests. Of the nests on the Atlas map, eagles built three-quarters in cottonwoods—both plains and narrowleaf—and one-quarter in conifers. In the Denver area they use trees on reservoir edges. Western Slope birds most often pick cottonwoods along rivers such as the Yampa, Little Snake, White, and Colorado. San Juan Basin eagles use conifers that grow near either lakes or streams, although one pair selected a spreading cottonwood in a horse pasture. Human disturbance creates stress for nesting Bald Eagles. Hence pairs persist at sites well protected from people traffic (usually—one pair nested adjacent to the golf course in Craig).

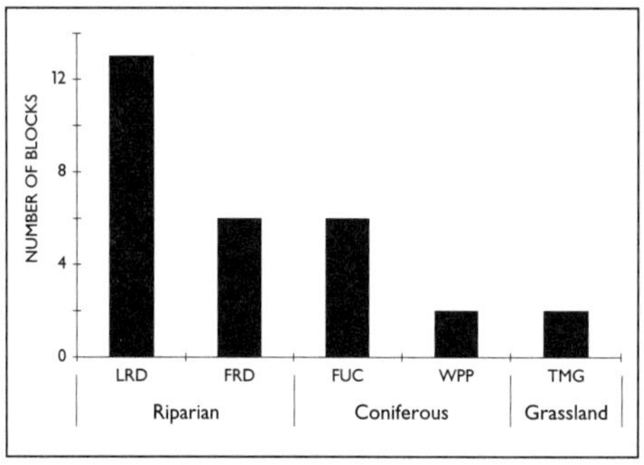

NUMBER OF BLOCKS
0
4
8
12
LRD
FRD
FUC
WPP
TMG
Riparian
Coniferous
Grassland

Colorado's Bald Eagle population soars in winter, when they occur most often on plains, western river systems, and mountain parks (A&R). Scavenging along reservoir shores and rivers, they look for stranded fish or crippled waterfowl (B&N). The Rocky Mountain Arsenal NWR has established a blind for viewing a winter roost; often 30 or more Bald Eagles fly in to spend the night.

Bald Eagles relish fish and often steal them from Ospreys, who sometimes return the favor (Steve Stachowiak pers. comm.). They also drive Ferruginous Hawks off prairie dog kills. Sometimes they use the behavior of other eagles and hawks as a clue to locate food (Bent 1937, Root 1988).

BREEDING

Bald Eagles build a nest, or eyrie, in a fork near the crown of a large tree (Harrison 1979). An immense pile of sticks, the nest has a lining of grasses, moss, twigs, sod, and forbs. The female lays 1–3 eggs, commonly 2, at intervals of 3–4 days. Both parents incubate for 34–35 days (Harrison 1979). Although early observers said that often only one young fledges (because the first hatched is stronger and larger and the adults ignore the needs of the punier infant; Bent 1937), Colorado eagles frequently fledge two and sometimes three young (Jerry Craig pers. comm.)

Atlas nest cards record five Bald Eagle nests in cottonwoods at an average of 49 feet (15 m) above the ground and one in a ponderosa pine 60 feet (18.5 m) high. Ground elevations of the nests ranged from 4,460 to 8,475 feet (1370–2600 m). The four successful nests had young on 19 April, 20 April, 26 June, and 8 July.

They return to old sites—a pair has nested at Barr Lake since 1986. Older nest areas exist: eagles first used those at Electra Lake and on the Little Snake River in 1974 and 1975 respectively (Jerry Craig pers. comm.). Eagles breed every year and sometimes shift nest sites; a pair's presence year after year in the same area indicates that the birds have nested somewhere nearby (Jerry Craig pers. comm.).

BREEDING PHENOLOGY

CODE		# OF RECORDS	DATE RANGE
ON	Occupied Nest	3	27 Feb–20 May
NY	Nest with Young	6	20 Apr–29 Jun
FY	Feeding Young	1	26 Jun

DISTRIBUTION

Bald Eagles live throughout North America—from Alaska to Newfoundland, and from the tip of Florida to southern California—but with settlement they suffered a catastrophic decline. Shooting, nest disturbance, loss of nest trees and nesting habitat, plus contamination of their food by pesticides, ravaged the population (Terres 1980). The decline started in the last century (Aiken and Warren 1914). B&N attributed the drop to man and gun as well as DDT. Both the federal and Colorado governments designated our national symbol as an Endangered Species.

Historically, Bald Eagles were first found nesting in Colorado in 1889 (B&N), although Oregon-bound settlers saw them in 1839 on the Blue River in Grand County (Marsh 1931). B&N cited several mountain and one plains (Bent County) nesting sites, but how many pairs nested historically is apparently unknown.

They now nest across Colorado. Over the last 10 years raptor biologist Jerry Craig has supervised a steady increase. From 1987 to 1995 Atlas and CDOW workers Confirmed 38 nests (or occupied nest sites), which represent 33 different breeding pairs. Sites sort into four clumps, mainly on the Western Slope.

After official protection increased public attention through the DDT ban and the Endangered Species listing, Bald Eagle populations rebounded. Diligent recovery efforts in the U.S. and Canada rescued them. In 1995 the birds graduated from Endangered to Threatened status. We still need to worry about their living conditions and population health, but we have stemmed their downward slide toward extinction. Bald Eagles are magnificent to watch, appropriate symbols of freedom, and a prime example of how the Endangered Species Act has worked.

CAVEAT ON THE ATLAS MAP: The Atlas map reflects both Atlas data and information provided by the CDOW study team, but at the request of that agency the Atlas map masks the specific locations in order to protect the birds from human predation. The map clumps breeding sites in the center of each latilong rather than showing the specific blocks.

BREEDING EVIDENCE

REPORTED IN 19 (1%) OF 1745 PRIORITY BLOCKS

□	Possible	2 blocks	(11%)
■	Probable	0 blocks	(0%)
■	Confirmed	17 blocks	(89%)

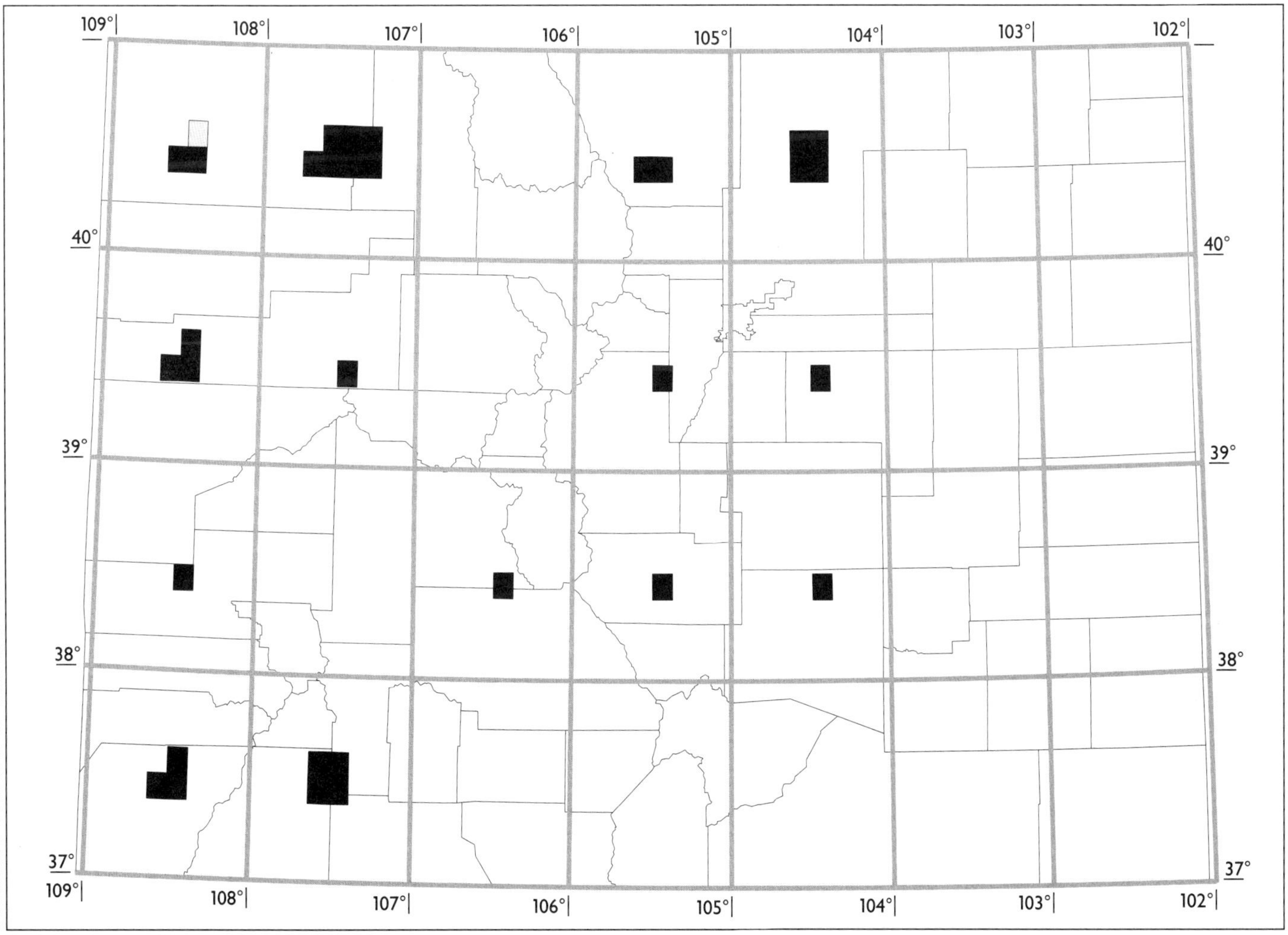
109°
108°
107°
106°
105°
104°
103°
102°
40°
39°
38°
37°

COLORADO DIVISION OF WILDLIFE / WILDLIFE RESOURCE INFORMATION SYSTEM

Who hasn't watched, spellbound, the foraging flights of this hawk, which is more like an owl than a hawk? Northern Harriers hunt and find their prey on the wing by listening with ears asymmetrically embedded in an owl-like face. Even more impressive than the foraging flights are the male's roller-coaster courtship flights and the aerial prey transfers between males and females. Paradoxically, nesting, roosting, and resting all take place on the ground, relegating this accomplished flyer to a life spent mostly afoot rather than aloft.

HABITAT

The main habitats of Northern Harriers include native and non-native grasslands, agricultural lands, and marshes, but in fall they also range up to the alpine tundra (A&R). Harriers nest in a variety of habitats from grasslands to marshes, with the only requirement being abundant cover such as that provided by tall reeds, cattails, and grasses (Johnsgard 1979). Atlas workers confirmed this diverse habitat use by finding harriers in 30 habitats. Atlas workers most commonly saw harriers in croplands, shortgrass prairie, emergent wetland marshes, and mountain sagebrush. These habitats, which seem largely dissimilar, composed 52% of all sightings. For breeding and hunting in these diverse habitats the birds select the parts with dense cover such as swales, draws, fencerows, and canal banks.

The frequency of habitats in which harriers were Confirmed breeders followed the general pattern of habitats in which they were observed. The most commonly listed Confirmed breeding habitats (31% of all Confirmations) were emergent wetlands, croplands, and tall desert shrublands. No Atlas worker recorded an abundance above A2 (2–10 pairs) in any block—consistent with harrier territory sizes, 1–1.5 mi² (2.6–3.9 km²), although territories often have linear shapes of up to one mile long (Craighead and Craighead 1956).

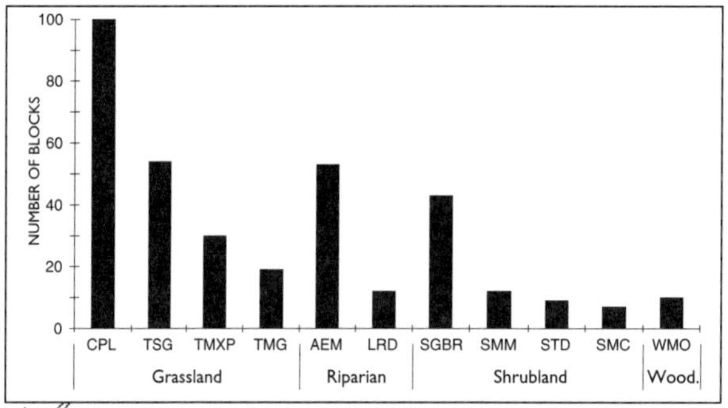
NUMBER OF BLOCKS
100
80
60
40
20
0
CPL
TSG
TMXP
TMG
AEM
LRD
SGBR
SMM
STD
SMC
WMO
Grassland
Riparian
Shrubland
Wood.

Due to their well-developed auditory capability, harriers can find prey in dense vegetation. They fly low over a field listening as well as watching for prey. Upon hearing something, harriers flap their wings in a way that accelerates them toward the ground and a potential meal. In total darkness, using only auditory cues, harriers can regularly strike a target the size of a baseball (Rice 1982). However, their strategy is to pounce on everything that even acts or sounds like food; consequently, they miss more times than they succeed. After a failure they take flight again, continuing to course over what seems like every inch of a field.

BREEDING

In the spring, males migrate before females and arrive first on breeding sites to establish territories that eventually may hold one or more females (Hamerstrom et al. 1985). In mid March males begin displaying in a spectacular series of roller-coaster undulations. Females later join in the aerobatics. Harriers exhibit semi-social and polygynous mating systems. Polygyny is strongly tied to food availability; its incidence increases in years of abundant prey (Hamerstrom et al. 1985). Females construct the nest, and lay 4–6 eggs. They begin incubation before laying the last egg, thereby ensuring an asynchronous hatch. Incubation lasts 24–30 days and the young fledge in approximately 5 weeks (Johnsgard 1979). Males feed the females during incubation and brooding and provide food for the young through aerial prey transfers. The male initiates prey transfers after catching a prey item. He calls loudly to the female, swoops in near her, and drops the prey. As the prey tumbles through the air, it seems impossible that this non-falcon will ever catch it—but the female swoops up and catches the food in her talons.

Atlas workers Confirmed harriers in only 19 Priority blocks for a Confirmation rate of 7%. This low percentage reflects the secrecy with which harriers nest. In addition, because their breeding territories are so large, actually finding a nest is very difficult—and venturing too near a harrier nest poses real hazards. Both adults vigorously defend the nest area with low dives, and occasionally make contact with the intruder. The annoying raucous calling by both male and female compels one to leave the immediate area even without the strafing attacks. With the conspicuous prey transfers between male and female, feedings of the young can be easily observed. However, Atlas workers observed this behavior only twice.

BREEDING PHENOLOGY			
CODE		# OF RECORDS	DATE RANGE
C	Courtship	9	23 Apr–27 Jun
NB	Nest Building	1	20 Jun
NE	Nest with Eggs	5	19 Apr–3 Jul
NY	Nest with Young	7	12 Jun–15 Jul
FY	Feeding Young	5	3 May–1 Aug
FL	Fledged Young	4	3 Jun–10 Jul

DISTRIBUTION

Northern Harriers breed throughout North America and parts of Eurasia. In North America they reach their highest densities in the prairie-pothole region of the U.S. and Canada (Price et al. 1995). The Atlas map shows a shotgun distribution, with fewer blocks with harriers at high elevations and in dry areas (most of the central Rockies and southeastern Colorado). Of Priority blocks reporting harriers, 57% were on the plains, proportionately more in the northern half.

Cooke (1897) listed the Northern Harrier as one of the most common hawks of the plains. The Atlas shows that today, even though the plains had the most harriers, only a quarter of plains blocks reported them. Other hawks now substantially outnumber harriers on the plains (Red-tailed and Swainson's, also kestrels). Harriers declined in the 1970s due to DDT, as did most raptors, but they continue to decline due to habitat loss (Ehrlich et al. 1992). In Colorado, loss of wetland habitats probably poses the greatest threat.

BREEDING EVIDENCE			
REPORTED IN 283 (16%) OF 1745 PRIORITY BLOCKS			
□	Possible	212 blocks	(75%)
■ (grey)	Probable	52 blocks	(18%)
■ (black)	Confirmed	19 blocks	(7%)

109°
108°
107°
106°
105°
104°
103°
102°
40°
39°
38°
37°

COLORADO DIVISION OF WILDLIFE / WILDLIFE RESOURCE INFORMATION SYSTEM

SHARP-SHINNED HAWK *Accipiter striatus*

Few developments better illustrate the recent triumph of ecology than the changing popular attitudes toward Sharp-shinned Hawks. Earlier in this century, sober ornithological literature referred to this bird as an "audacious murderer," a "feathered bandit" (Bent 1937), and a "murderous little villain" (Pearson 1936). Now, all but the most sentimental bird fanciers consider this hawk's forays at the backyard feeder an exciting spectacle of nature in action rather than an outrage.

HABITAT

Sharp-shins forage in the forest canopy for small birds and utilize a variety of woodland habitats (Reynolds 1989). Atlas workers reported 26 different habitat codes, eight in double figures. The lowest, most basic breeding code (#), for birds seen in appropriate habitat, accounted for more than 75% of the reports. Most of these observations probably indicate hunting birds, perhaps some distance from the immediate nesting territory, and they depict a highly generalized woodland raptor. Atlasers reported conifer codes somewhat more frequently than deciduous codes (115 to 90), and they found "Sharpies" in each Colorado life zone in roughly similar numbers (53 Hudsonian zone blocks, 62 Canadian, 44 Transition, and 44 Upper Sonoran). The reported codes virtually duplicate those reported for Cooper's Hawks. For each, the ten habitat

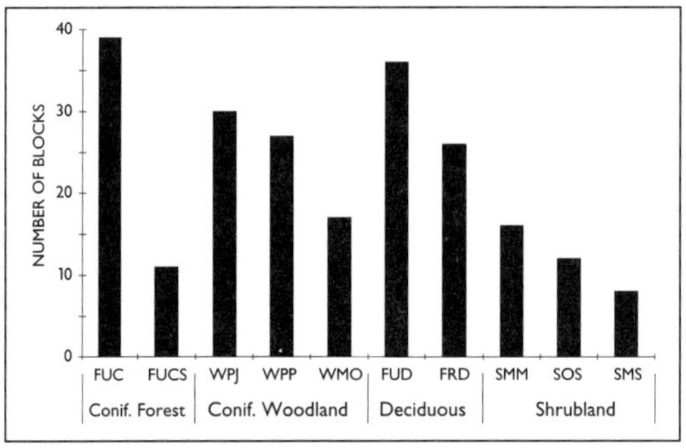

NUMBER OF BLOCKS
40
30
20
10
0
FUC
FUCS
WPJ
WPP
WMO
FUD
FRD
SMM
SOS
SMS
Conif. Forest
Conif. Woodland
Deciduous
Shrubland

codes most frequently reported were the same and atlasers reported these codes in very nearly the same order of frequency.

Pairing the habitat codes with breeding status codes reveals that Sharp-shins require high foliage density for nest sites (Reynolds 1989). Young or even-aged stands of conifers most effectively provide this density (Bent 1937, Palmer 1988a). Atlasers found nine nests, eight in coniferous trees. Most Sharp-shins breed in boreal forests (Palmer 1988a), in both broken forest and large continuous stands of conifers (Reynolds 1989). In Colorado small stands of conifers within mature aspen forests seem particularly significant (Joy et al. 1994). Scrub oak also can provide the required dense foliage, and B&N reported several nests in oak brush.

Ninety percent of the diet of this hawk consists of small birds; in summer 60% consists of nestlings or fledglings. Most frequent prey items in west-central Colorado are Yellow-rumped Warblers, White-crowned Sparrows, and juncos (Joy et al. 1989).

BREEDING

Sharp-shins arrive at breeding sites in April and early May (Palmer 1988a). Atlasers did not report any observations of courtship behavior, which remains largely unknown (Stokes 1989), or of nest-building. The birds generally nest in the same area for several years but build a new nest each year (Stokes 1989). Atlas workers Confirmed breeding in two blocks by finding used nests. The female begins incubating after laying the last of 4–5 eggs (Palmer 1988a). Nine Atlas observations of birds on the nest ranged from 20 May to 16 July; one dated 8 August seems late. During the 30-day incubation period, the female rarely leaves the nest, and the smaller male provides food (Ehrlich et al. 1988). The great size difference between the sexes—the greatest of any North American bird (Palmer 1988a)—may enable Sharp-shins to specialize on different prey and to increase their efficiency in providing for youngsters (Cade 1982).

Adults cackle noisily and may attack intruders in the nesting area. Atlasers reported agitated behavior in a dozen blocks and Confirmed breeding by distraction displays in four more. The young spend 24–27 days on the nest (the smaller males fledge earlier) and depend on the adults for another 3–4 weeks after fledging (Palmer 1988a). Atlas fledgling dates, which fall between 3 July and 19 August, fit within that time period.

BREEDING PHENOLOGY

CODE		# OF RECORDS	DATE RANGE
ON	Occupied Nest	5	20 May–8 Aug
NY	Nest with Young	5	15 Jun–16 Jul
FL	Fledged Young	6	3 Jul–19 Aug

DISTRIBUTION

Sharp-shinned Hawks, like the other accipiters, breed in woodland habitats throughout the United States and Canada. The ranges of the accipiters overlap nearly completely (Palmer 1988a), but differing prey preferences prevent direct competition (Reynolds 1989). Atlas data indicate that Sharp-shins range to somewhat higher elevations than Cooper's Hawks in Colorado; they reported a few more Sharp-shins in spruce/fir forests and a few more Cooper's in lowland riparian forests. Atlasers reported Sharp-shins in each latilong almost precisely in proportion to the extent of that latilong's

BY RICHARD LEVAD

wooded area; extensive blank areas on the Atlas distribution map correspond to treeless areas, except that the plains stream bottoms do not support breeding Sharp-shins.

A&R described Sharp-shins as rare to uncommon summer residents, but the birds' secretive behavior may lead observers to overlook them (Viverette et al. 1996). As a result their population in Colorado may well exceed current estimates (A&R). Atlasers reported codes of A2 and A1 at approximately a 3:2 ratio. Most of the A2 blocks probably held only two or three pairs because their rather large territories, with nests 2 or 3 miles apart (Stokes 1989), would preclude more than three or four pairs nesting in any Atlas block.

Sharp-shin numbers seem relatively stable. In the East several hawk-watch migration counts show a decline, but Christmas Counts north of these points indicate an increase, suggesting perhaps a change in behavior rather than a change in numbers (Viverette et al. 1996). In the 1960s and 1970s, DDT adversely affected reproduction of this species, and it remains a problem for those birds that winter in Mexico and Central America (Reynolds 1989). The practice of thinning large areas of forest may threaten the nesting habitat of Sharp-shins.

BREEDING EVIDENCE

REPORTED IN 198 (11%) OF 1745 PRIORITY BLOCKS

☐	Possible	157 blocks	(79%)
▩	Probable	21 blocks	(11%)
■	Confirmed	20 blocks	(10%)

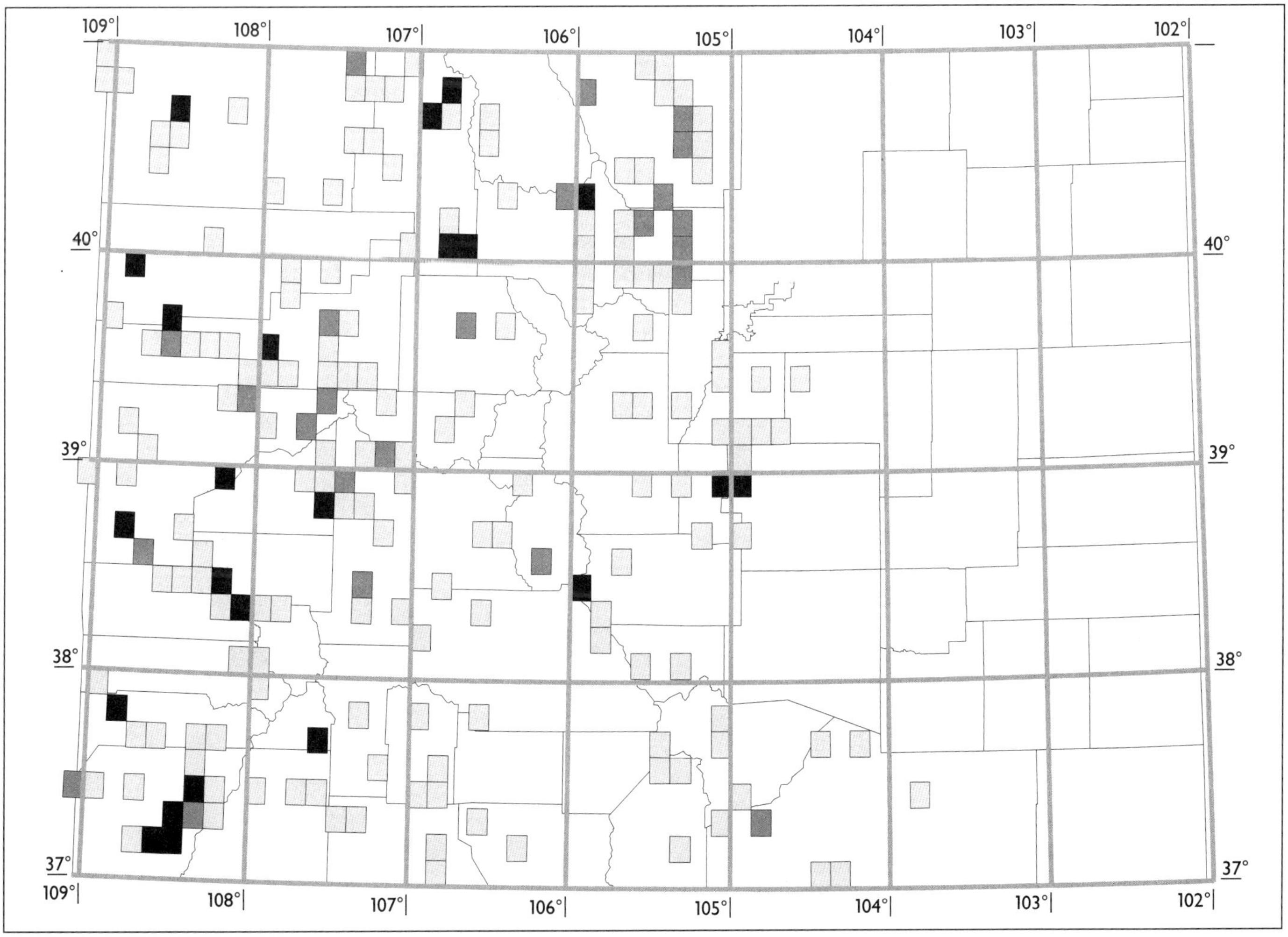
109°
108°
107°
106°
105°
104°
103°
102°
40°
39°
38°
37°

COLORADO DIVISION OF WILDLIFE / WILDLIFE RESOURCE INFORMATION SYSTEM

Female Cooper's Hawks average one-third larger than males, among the greatest size dimorphisms of any of the world's hawks. Presumably the larger females can, and do, catch larger prey than the comparatively diminutive males. Females do all the incubation and most brooding. Males feed incubating females and provision the young via the home-abiding female—the size of what you eat probably doesn't matter if somebody else catches it for you.

HABITAT

Cooper's Hawks nest in most Colorado forest types, from cottonwood riparian to spruce/fir at 10,000 feet (3048 m). Most nest in ponderosa pine, Douglas-fir, lodgepole pine, and aspen (A&R), but Atlas data confirm A&R's speculation: the species also breeds in spruce/fir, riparian forests, and pinyon/juniper woodlands. In fact pinyon/juniper turned up more than any other habitat code. Atlasers also found 11 instances of breeding activity in scrub oak, where the hawks place their nests in older, tall, mature stands of oak (JFT). Lowland riparian records come primarily from the southern and western parts of the state. Although atlasers' habitat codes for Sharp-shinned and Cooper's hawks echoed each other, Cooper's Hawks are more abundant in lowland riparian than Sharp-shins, and less so in spruce/fir. Goshawks tend to nest at higher elevations, but the two species nested within 200 yards (180 m) of each other in a canyon south of Mesa Verde (HEK).

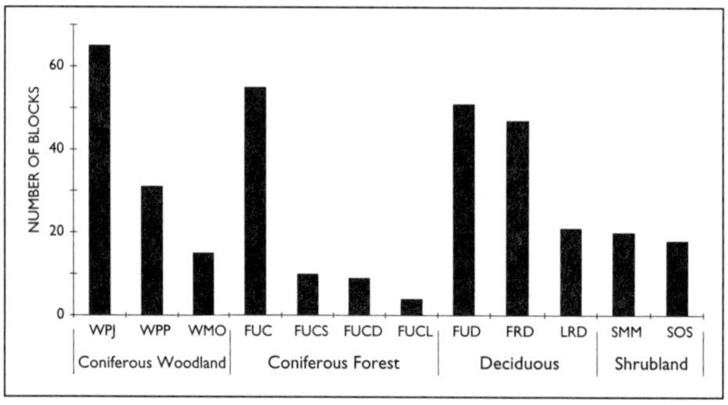
NUMBER OF BLOCKS
0
20
40
60
WPJ
WPP
WMO
FUC
FUCS
FUCD
FUCL
FUD
FRD
LRD
SMM
SOS
Coniferous Woodland
Coniferous Forest
Deciduous
Shrubland

Ornithologists have little information on the cause or function of the size dimorphism. They have not settled on how size affects the diet of the two sexes, or vice versa (Rosenfield *in* Palmer 1988a). One monograph on Cooper's Hawks (Rosenfield and Bielefeldt 1993) does not mention food differences between the sexes, and another work has only two sentences on the subject (Palmer 1988a).

In one study captive females consumed 50% more food than the males; both males and females ate 16–17% of their weight each day (Craighead and Craighead 1956). In wild populations, 60–80% of the summer diet consists of small- and medium-sized birds; sparrows compose one-quarter (Palmer 1998a). Other studies report, without specifying the sex of the consumer, principal prey items as medium-sized birds—robins and starlings—and Northern Bobwhites (Rosenfield and Bielefeldt 1993). Their bird diet consists mainly of ground-feeding birds. Cooper's also eat rabbits, hares, ground squirrels, and tree squirrels. Individual Cooper's Hawks may specialize on one prey item: specific birds have concentrated on bobwhites in Georgia, lizards in California, and even bats in Texas (Palmer 1988a).

When going to or from the nest or when hunting, Cooper's drop down from the nest or perch, then burst along within 10 feet (3 m) of the ground. While hunting they use cover—usually bushes—to shield their approach toward prey (Palmer 1988a). This hunting technique probably facilitates catching birds on the ground.

BREEDING

Cooper's Hawks place their nests under dense canopy cover—64–95% cover in five studies. In Utah and Oregon, and very likely Colorado, they orient their nests on the presumably cooler north or east slopes of the terrain (Palmer 1988a).

Historical accounts shed little light on the abundance of Cooper's Hawks. As a Colorado breeder, this midsized version of the three North American accipiters outnumbers its relatives. Abundance codes assigned by atlasers suggest that, like other top-of-the-food-chain species, they are uncommon when compared to passerines with similar distributions; atlasers gave abundance codes of only A1 and A2.

Elevations on six nest cards ranged from 4,600 feet (1400 m) in Montrose County to 7,920 feet (2415 m) in San Miguel County. In over 60 blocks atlasers reported Cooper's in aspen, spruce/fir, lodgepole pines, and Douglas-fir, most well over 8,000 feet (2440 m) in elevation. Dates for breeding activity vary widely, without any geographic or altitudinal pattern.

BREEDING PHENOLOGY

CODE		# OF RECORDS	DATE RANGE
C	Courtship	1	9 Jun
ON	Occupied Nest	22	5 May–12 Jul
NE	Nest with Eggs	5	16 May–5 Jul
NY	Nest with Young	16	29 May–31 Jul
FY	Feeding Young	3	30 May–4 Jul
FL	Fledged Young	18	8 Jun–1 Sep

DISTRIBUTION

Cooper's Hawks breed across southern Canada, all of the U.S., and the mountainous areas of northwestern Mexico. Canadian and northern U.S. birds are described as migratory (Palmer 1988a) or partially migratory (Rosenfield and Bielefeldt 1993); the hawk-watch west of Denver counts several hundred each spring (254–569, average 400). Winter range includes most of Mexico, Guatemala, and Belize.

Atlas data, matching the A&R map, show that Colorado birds breed primarily in the mountainous western two-thirds of the state. They also inhabit southeastern Colorado—the Purgatoire River, Comanche National Grassland, and other parts of Las Animas County—and the Black Forest northeast of Colorado Springs. Cooper's do not nest in the high mountains, the large mountain parks, or the sagebrush/grasslands of Moffat County.

The historical distribution of Cooper's Hawks in Colorado differs little from current distribution. No historic records exist for plains riparian areas as they do for similar areas on the Western Slope. Considering the extent to which riparian woodlands in eastern Colorado have expanded since settlement, one might expect Cooper's Hawks to expand into them, but atlasers did not find them breeding on the plains. In other states they have adapted to urbanization (Stewart et al. 1996), but in Colorado only in winter.

In the northeastern U.S., residues of DDT and other organochlorines that accumulated in prey species impacted this species and numbers dipped very low in the 1940s. The hawks picked up residues on both breeding and wintering grounds. Restrictions on DDT have diminished but not eliminated its effects because the poison persists in the environment and because countries south of the U.S. continue to allow its use (Gauthier and Aubry 1996). Western raptors probably have similar problems, although no available data document this. Within Colorado the known breeding range of Cooper's Hawks has remained consistent, and the Atlas found them fairly widespread across the western two-thirds of the state.

BREEDING EVIDENCE

REPORTED IN 290 (17%) OF 1745 PRIORITY BLOCKS

Possible	179 blocks	(62%)
Probable	32 blocks	(11%)
Confirmed	79 blocks	(27%)

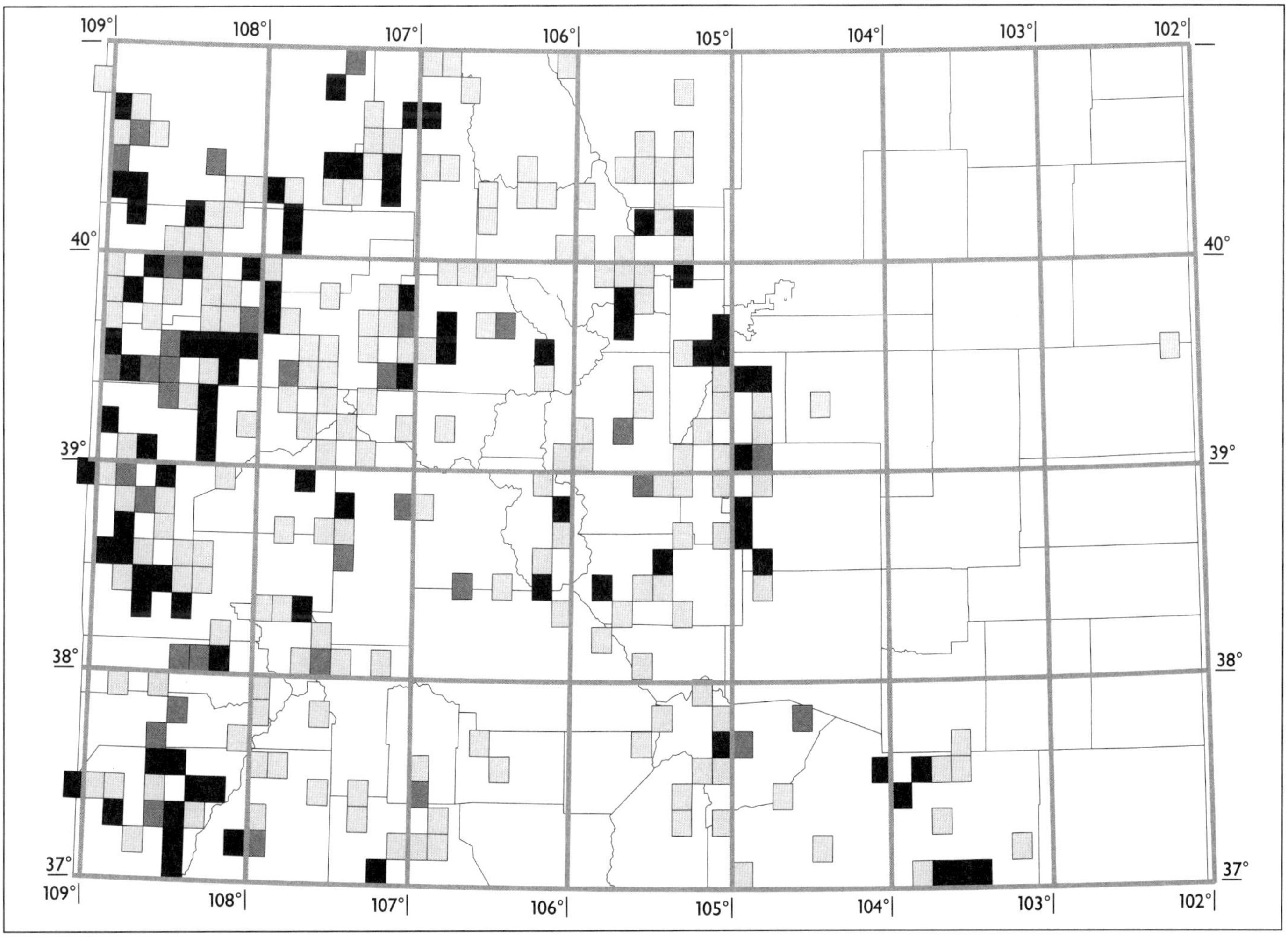

109°
108°
107°
106°
105°
104°
103°
102°
40°
39°
38°
37°

COLORADO DIVISION OF WILDLIFE / WILDLIFE RESOURCE INFORMATION SYSTEM

Completely fearless, Northern Goshawks in their nest space take on intruders of any size—including humans. Goshawks defending a nest site swoop within inches of the interloper's head and occasionally strike observers who come too close. What a wonderful opportunity to observe talons and undertail coverts at close range!

HABITAT

Habitat use by Northern Goshawks varies by region, so the question of what constitutes suitable nesting habitat greatly concerns land managers across the country (Braun et al. 1996). Most Colorado goshawks breed in conifer forests (64% of Atlas blocks), and apparently do not discriminate among tree species. Many breed in ponderosa pines in southern Colorado and along much of the Front Range, as they do in New Mexico and Arizona (Fletcher and Sheppard 1994).

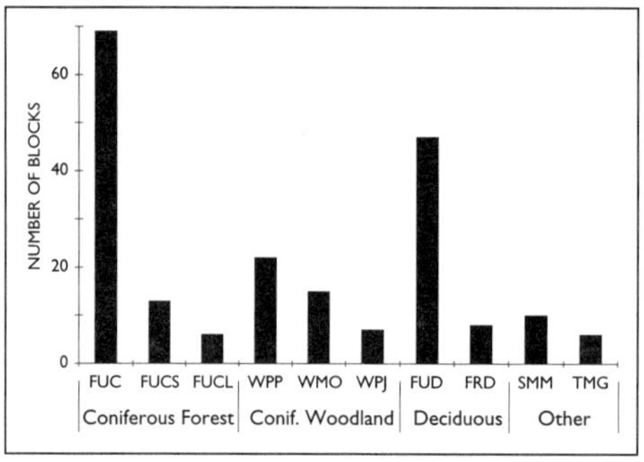
NUMBER OF BLOCKS
0
20
40
60
FUC FUCS FUCL WPP WMO WPJ FUD FRD SMM TMG
Coniferous Forest
Conif. Woodland
Deciduous
Other

In southern Wyoming Northern Goshawks typically nest up to 9,000 feet (2950 m) in open conifer stands dominated by large, older trees. Just across the border in northwestern Colorado, however, they often nest in aspen, sometimes in conifer stands less than 100 years old, and up to 10,000 feet (3048 m). They even nest occasionally in cottonwoods. Atlasers reported goshawks in aspen in 32% of the blocks in the northern three-quarters of Colorado, but in fewer in the southwest—the San Juan and Sangre de Cristo mountains (only 11%).

No matter the habitat type, this species requires large blocks of forest for nesting and foraging. Forest Service biologists often report small openings (less than one acre) associated with nest and forage sites. Foothills populations thrive in mosaic landscapes. Goshawks tend to choose nest trees on shallow slopes, flat benches in steep country, and fluvial pans on small stream junctions. A fairly open understory allows the birds to maneuver under the canopy although they forage over open meadows and sagebrush as well, even to timberline.

Fierce hunters, goshawks hunt for small birds and mammals from perches in the lower forest canopy. They attack suddenly, at tremendous speed, over short distances. They fearlessly crash into brush and will walk or run after prey (Palmer 1988a). Over half the breeding diet in Moose, Wyoming, consists of mammals (38% squirrels [red and ground], 9% rodents and hares), and 45% comprises birds (38% small and medium-sized birds, 7% grouse; Craighead and Craighead 1956). In the Sierra Nevadas a nesting male feeding his mate and young preyed mainly on robins and Steller's Jays (Schnell 1958). The larger females can catch larger prey.

BREEDING

An 11 May nest represents the earliest Atlas Confirmation, but a fledgling seen on 3 June in Mesa Verde National Park suggests an even earlier nesting effort. With a 36–38-day incubation and 45-day nestling period (Terres 1980) egg-laying would have occurred in late April. Breeding does not correlate with latitude: northern birds nested as early (or late) as southern ones.

Goshawks reuse the same territory year after year and sometimes reuse the same nest. Pairs typically have one or more alternate nests within the territory and may desert one nest and then return to it in a later year.

Because of the size of their nests and the nature of their preferred prey base, breeding Northern Goshawks require mature forests—the same forests needed for economic lumbering. Environmental groups attribute declining goshawk populations across the continent to this classic confrontation between commodity production and wildlife needs. In 1991 this led to a petition for listing the southwestern U.S. population as Threatened under the Endangered Species Act. Although the U.S. Fish and Wildlife Service determined that insufficient data exist to warrant listing, the U.S. Forest Service placed the Northern Goshawk on its Sensitive Species lists due to these declining populations. This requires the agency to survey for the species and to monitor management activities near nest sites.

Logging activities that fragment forests have resulted in nest abandonment on national forest lands. From 1991 to 1995 on the Routt National Forest, goshawks experienced a severe reduction in breeding success, with nest failures both on logged and on unmanaged forest (those parts not managed by the Forest Service for timber cutting, other commodity uses, and recreation). Inasmuch as goshawk numbers show cyclical variations in some parts of the country, the decline on the Routt may reflect a natural fluctuation (unpubl. USFS data).

BREEDING PHENOLOGY

CODE		# OF RECORDS	DATE RANGE
C	Courtship	1	26 Jun
ON	Occupied Nest	7	11 May–13 Jul
NY	Nest with Young	16	5 Jun–7 Aug
FY	Feeding Young	2	12 Jul–14 Aug
FL	Fledged Young	13	3 Jun–11 Sep

DISTRIBUTION

Colorado's goshawk records date from 1873, with reports from western counties and the Front Range. Northern Goshawks currently breed across the same range, throughout the forested mountains of Colorado. The Atlas map shows them well distributed in the San Juan Mountains and Uncompahgre Plateau and across the northern mountain ranges. The map depicts an inexplicable absence from the Elk, Sawatch, Wet, northern Sange de Cristo, and northeastern San Juan mountains. Some of these absences may stem from gaps in coverage (e.g., the Sawatch; Alan Wallace, pers. comm.), but in part they probably represent actual holes in distribution.

On a larger scale, Northern Goshawks occur on all continents in the Northern Hemisphere. In North America they breed throughout Canada and the northern and western U.S., south into Mexico. Western mountain ranges provide habitats and temperatures similar to those found farther north. Some individual goshawks follow regular migratory routes yet others wander randomly in the fall and winter but remain in the same general region as their breeding territory.

Although some pairs tolerate limited disturbance around their nests, human activities in the vicinity of nests upsets more. Conservation for Northern Goshawks requires preservation of large blocks of habitat, not only for the breeding birds but also for dispersing juveniles, and for their prey species.

BREEDING EVIDENCE

REPORTED IN 155 (9%) OF 1745 PRIORITY BLOCKS

☐	Possible	109 blocks	(70%)
▩	Probable	13 blocks	(8%)
■	Confirmed	33 blocks	(21%)

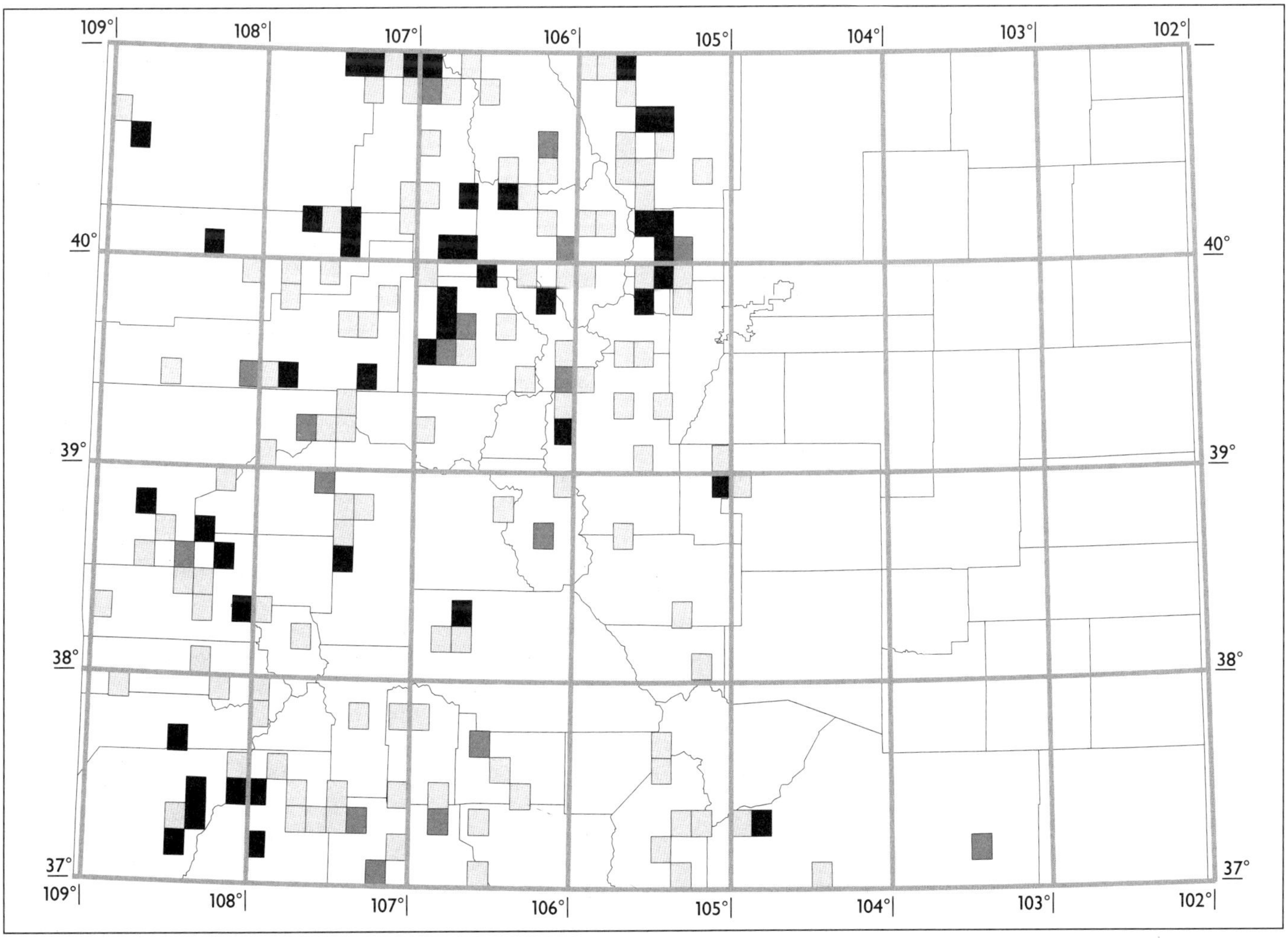
109°
108°
107°
106°
105°
104°
103°
102°
40°
39°
38°
37°

COLORADO DIVISION OF WILDLIFE / WILDLIFE RESOURCE INFORMATION SYSTEM

On powerful wings, surprisingly long and pointed for a buteo, Swainson's Hawks make an annual round-trip odyssey that may exceed 14,000 miles (22,540 km) between their North American breeding grounds and South American wintering grounds. Seasonally versatile, Swainson's Hawks prey on small birds and mammals through the nesting period, but shift to a diet of grasshoppers and other insects for the rest of the year. During late summer and early fall, flocks of several hundred birds assemble on croplands in eastern Colorado to harvest the bounty of grasshoppers saturating the region.

HABITAT

Overlapping to some degree with Red-tailed and Ferruginous hawks, Swainson's Hawks typically inhabit sites in arid grassland, desert, and agricultural areas with scattered trees and shrubs (B&N, Olendorff 1972, Andersen 1991, A&R). Optimum habitat probably includes somewhat less woodland than that used by Red-tailed Hawks, but more than that used by Ferruginous Hawks. On the Pawnee National Grassland Olendorff (1972) reported 61% of occupied Swainson's Hawk nests in creek bottoms, 25% in pure grassland, and 14% in cultivated lands. Atlas results reflect this association: roughly 70% of all reports originated in blocks with rural, shortgrass prairie, lowland riparian woodland, and cropland habitats. Atlasers reported nests placed in a variety of structures, including Siberian elm, Colorado blue spruce, lodgepole pine, a utility pole, a particularly thick and tall serviceberry, and, most commonly, plains cottonwoods.

Because of the dramatic post-breeding shift by Swainson's Hawks to an insect diet, farmers in eastern Colorado call them "Grasshopper Hawks." In fall these locals direct visiting birders to flocks by pointing out fields scheduled for plowing that day.

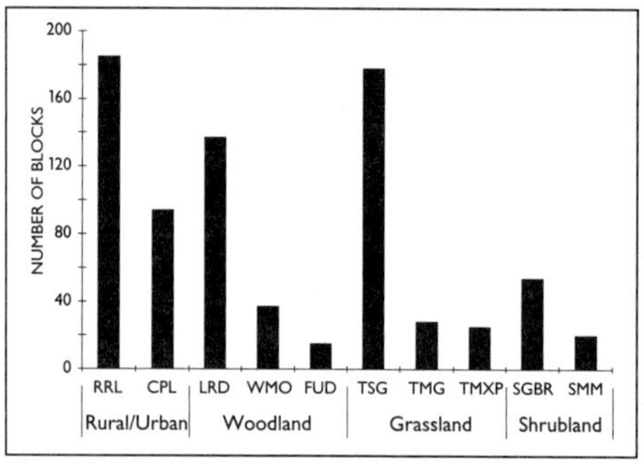
200
160
120
80
40
0
NUMBER OF BLOCKS
RRL
CPL
LRD
WMO
FUD
TSG
TMG
TMXP
SGBR
SMM
Rural/Urban
Woodland
Grassland
Shrubland

BREEDING

Nesting generally begins later for Swainson's Hawks than for either Red-tailed or Ferruginous hawks (see Palmer 1988b for review). Swainson's Hawks build their own stick nests, usually near the tops of isolated trees, or use nests built by magpies, crows, ravens, or other buteos. The egg-laying period varies widely across the breeding range, but typically occurs between mid May and mid June in Colorado (Olendorff 1972, Palmer 1988b). Hatching, from mid June through mid July, often coincides with the appearance of prey such as fledgling birds and young ground squirrels (Olendorff 1972, Palmer 1988b). Probably due to conspicuous nest placement by Swainson's Hawks, Atlas workers Confirmed breeding in 45% of reporting blocks, the second highest Confirmation rate among diurnal raptors (behind only Peregrine Falcons at 56%). Atlasers found Swainson's on nests in 243 blocks, from mid April to early August, and noted the first fledglings on 29 May. The dates do not correlate to altitude or latitude.

BREEDING PHENOLOGY

CODE		# OF RECORDS	DATE RANGE
C	Courtship	8	28 Apr–29 Jun
NB	Nest Building	11	19 Apr–4 Jun
ON	Occupied Nest	192	14 Apr–30 Jul
NE	Nest with Eggs	2	12 Jun
NY	Nest with Young	49	19 May–3 Aug
FL	Fledged Young	25	29 May–17 Aug

DISTRIBUTION

Colorado lies nearly in the center of the Swainson's Hawk breeding range. The species breeds locally in east-central Alaska, Yukon, and Mackenzie, and from central British Columbia east to southern Manitoba and southern Minnesota, south to southern Texas and northern Mexico. The western boundary extends from central Washington through northeastern California, western Nevada, central Arizona, south to Sonora and Chihuahua. A local breeding population occupies California's Central Valley.

The Colorado Latilong Study reported Swainson's Hawk as a definite breeder in all but one latilong, and atlasers Confirmed breeding in all but four. Atlasers recorded Swainson's Hawks in 670 Priority blocks, with 77% of them lying east of Longitude 105. Greatest densities occur in the Great Plains, North Park, and the San Luis Valley, reflecting the association of these birds with wide, open spaces containing sparse vegetation and scattered trees; only about 15% of Atlas reports originated outside these locations. Atlas workers reported only one pair in 53% of all blocks occupied by Swainson's Hawks. Published breeding densities vary from one nest/1.6 mi^2 (2.56 km^2) in North Dakota (Gilmer and Stewart 1984) to one pair/4.0 mi^2 (6.7 km^2) in Wyoming (Dunkle 1977). In his 414-mi^2 (1073-km^2) study area on Pawnee National Grassland,

Olendorff (1972) estimated their breeding density as one pair/17.2 mi^2 (44.6 km^2).

Primarily through tree-plantings and agriculture, humans have altered much of the Swainson's Hawk stronghold in the Great Plains during the past 150 years. Some of these changes appear to have benefited the species. Indeed, Olendorff (1972) noted that one-third of the Swainson's Hawks he studied in northeastern Colorado nested in situations created by humans (e.g., abandoned farmsteads, windmills). It remains unclear whether human activities in Colorado have caused a summary increase or decrease in Swainson's Hawk populations.

Early reports list Swainson's Hawk as "common" or "fairly common" through eastern Colorado (Cooke 1897, Beidleman 1949). B&N referred to the species as the most abundant of the large hawks in the state. They also suggested that it had declined with human settlement but provided no supporting data. Although the Swainson's Hawk has declined in many parts of its range, notably California (Bloom 1980), Oregon (Janes 1987), Nevada (Herron et al. 1985), Saskatchewan, and Alberta (Houston and Schmutz 1995), its numbers have increased significantly in North Dakota and Montana (Dobkin 1994). Harlow and Bloom (1987) considered it common and stable in Colorado. Atlas data depict far more Red-tailed Hawks statewide, but more Swainsons on the plains.

This species ranks high on several recent conservation priority lists for Colorado breeding birds. These lists are based in part on BBS results, however, and the BBS may not adequately sample wide-ranging, sparsely distributed, open-country raptors. Recently, concern for Swainson's Hawk populations increased with the report of several hundred insecticide-caused deaths (including some Colorado breeders) on wintering grounds in Argentina (Woodbridge et al. 1995, Goldstein et al. 1996). In light of these reports and the rapidly changing face of arid grassland habitats in North America, Swainson's Hawks warrant systematic, long-term study and monitoring in Colorado.

BREEDING EVIDENCE

REPORTED IN 670 (38%) OF 1745 PRIORITY BLOCKS

□	Possible	279 blocks	(42%)
▩	Probable	89 blocks	(13%)
■	Confirmed	302 blocks	(45%)

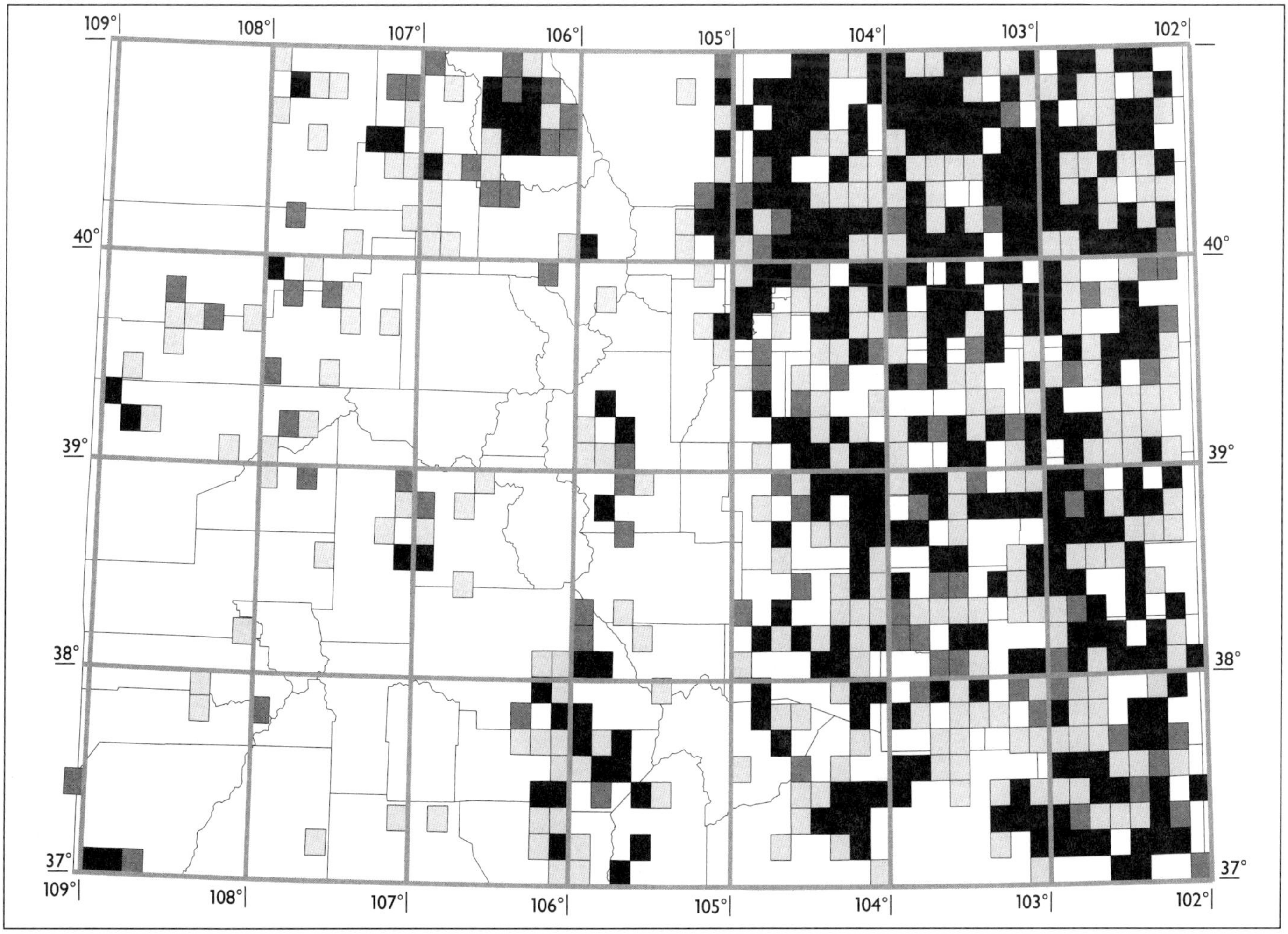
109°
108°
107°
106°
105°
104°
103°
102°
40°
39°
38°
37°

COLORADO DIVISION OF WILDLIFE / WILDLIFE RESOURCE INFORMATION SYSTEM

Few wild scenes in North America seem complete without the distinctive *kee-eeee-arrr* of the Red-tailed Hawk exploding overhead. This amazingly adaptable predator, represented by as many as 16 different races across its realm, hunts rodents, rabbits, reptiles, and Ring-necked Pheasants with equal mastery. It exploits a broad range of environmental opportunities, and atlasers encountered this species in a great variety of situations.

HABITAT

Across their range, Red-tailed Hawks inhabit open areas interspersed with patches of trees or other elevated perch and nest sites that provide a commanding view of the surrounding countryside (Preston and Beane 1993). All regions of Colorado provide opportunities for nesting Red-tails, and atlasers recorded these birds in virtually every habitat searched. Pinyon/juniper woodland appears in more reports than any other habitat (nearly 13% of all reports), but lowland riparian woodland, upland deciduous forest, upland coniferous forest, ponderosa pine woodland, and shortgrass prairie each appear in 6% or more of the reports. Together, these six habitats comprise 52% of all reports. Atlasers documented nests placed on cliff ledges, on utility poles, and in several tree species, including plains and narrowleaf cottonwoods, Utah juniper, quaking aspen, ponderosa pine, and Colorado blue spruce.

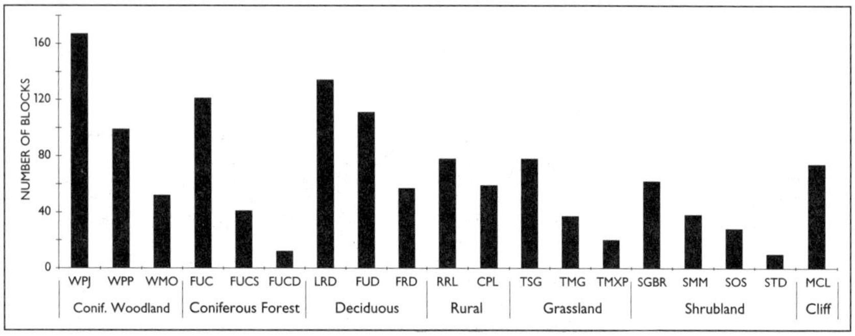
NUMBER OF BLOCKS
160
120
80
40
0
WPJ WPP WMO FUC FUCS FUCD LRD FUD FRD RRL CPL TSG TMG TMXP SGBR SMM SOS STD MCL
Conif. Woodland
Coniferous Forest
Deciduous
Rural
Grassland
Shrubland
Cliff

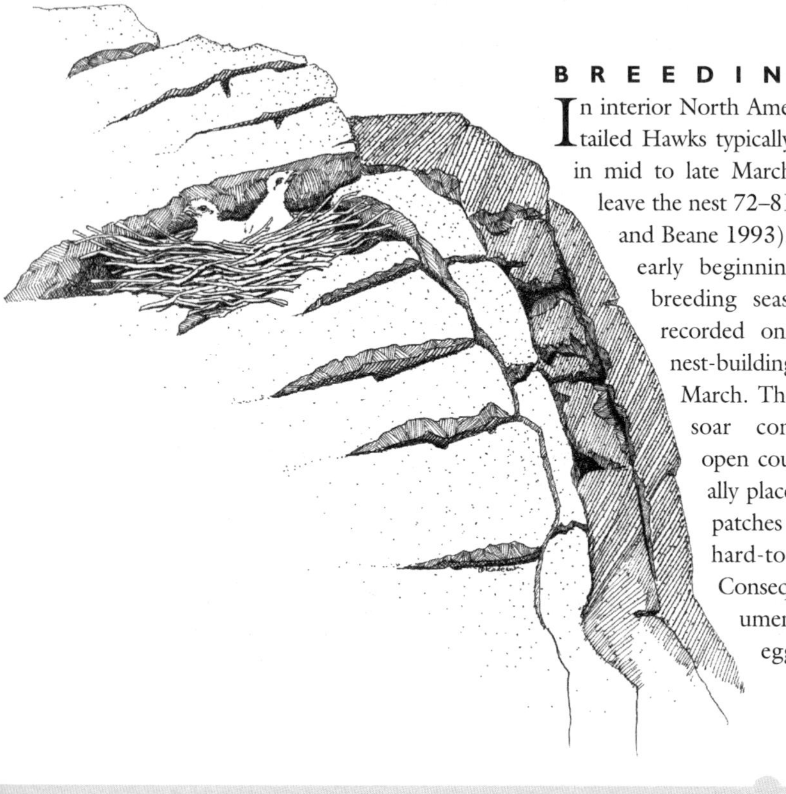

BREEDING

In interior North America, breeding Red-tailed Hawks typically lay their first eggs in mid to late March, and young birds leave the nest 72–81 days later (Preston and Beane 1993). Perhaps due to the early beginning of the Red-tail breeding season, Atlas workers recorded only six instances of nest-building, the earliest on 30 March. These birds perch and soar conspicuously above open country, but they usually place their nests high in patches of tall trees or on hard-to-reach cliff faces. Consequently, atlasers documented no nests with eggs and Confirmed attempted breeding in only 29% of the blocks. Atlasers found Red-tails on nests over a four-month period, from 1 April to 1 August. They reported nests with young in 89 blocks beginning on 7 May, but the later ON codes undoubtedly involve nests with young also. Timing of reproduction showed no consistent trends related to geographic features such as elevation, habitat, or latitude.

BREEDING PHENOLOGY

CODE		# OF RECORDS	DATE RANGE
C	Courtship	15	16 Apr–10 Jul
NB	Nest Building	6	30 Mar–18 Jun
ON	Occupied Nest	86	1 Apr–1 Aug
NY	Nest with Young	89	7 May–19 Jul
FY	Feeding Young	7	23 May–13 Jul
FL	Fledged Young	95	5 May–21 Sep

DISTRIBUTION

Red-tailed Hawks range from central Alaska south to Panama and east to Nova Scotia and the Virgin Islands. The vast majority of birds breeding in Colorado belong to the race *B. j. calurus.* A few individuals of the eastern race, *B. j. borealis* or *calurus* x *borealis* intergrades, may breed along the eastern border. Individuals from several races, especially *calurus, harlani,* and *kriderii,* overwinter across the state, although there remains some question whether *kriderii* merits subspecific status (Preston and Beane 1993).

The Latilong Study reported the Red-tail as a resident breeder in every latilong. Atlas results reaffirm the Latilong Study and report Red-tails in more Atlas blocks (1,105) than any other raptor, and the sixth most frequently recorded of all species. Atlasers found them least frequently in treeless regions of the Great Plains and in large, unbroken patches of subalpine forest.

In Garfield County, McGovern and McNurney (1986) reported a breeding density of one nest/2.2 mi^2 (5.7 km^2) and one nest/0.8 m^2 (2.0 km^2) in two disjunct areas of aspen woodland. Estimated breeding densities from throughout the species' range vary from one pair/19.3 mi^2 (50.0 km^2) in Ohio (Shelton 1971) to one pair/0.5 mi^2 (1.3 km^2) in California (Fitch et al. 1946). Nesting density generally depends on prey density and distribution of available nest sites (Preston and Beane 1993).

During the last 100 years various authors (Cooke 1897, Beidleman 1949, B&N, A&R) have characterized Red-tailed Hawks as "common" or "abundant" from plains to high mountains in Colorado. These designations apply only in the context of raptor abundance—abundance codes of A1 and A2 (the latter, in most blocks, meaning two nesting pairs) appear with nearly equal frequency in Atlas reports. Despite decades of persecution (Beidleman 1949, B&N) and the precipitous decline of some prey species, especially prairie dogs, Red-tails continue to thrive in Colorado. Indeed, the overall range and abundance of Red-tailed Hawks have increased in recent decades as deforestation in eastern North America and fire suppression in western North America have created more of the open woodland habitat it favors (Houston and Bechard 1983).

In much of eastern Colorado, especially near the Front Range corridor, patchy woodland—characteristic of agricultural, rural, and suburban development—has largely replaced native grassland habitats. These changes favor Red-tailed Hawks at the expense of Swainson's and, especially, Ferruginous hawks. Changes in the relative abundance and distribution of these three species thus reflect human land-use changes to some extent. Atlas data may provide the best basis, to date, for future assessment of these changes.

BREEDING EVIDENCE

REPORTED IN 1105 (63%) OF 1745 PRIORITY BLOCKS

□	Possible	575 blocks	(52%)
▩	Probable	215 blocks	(19%)
■	Confirmed	315 blocks	(29%)

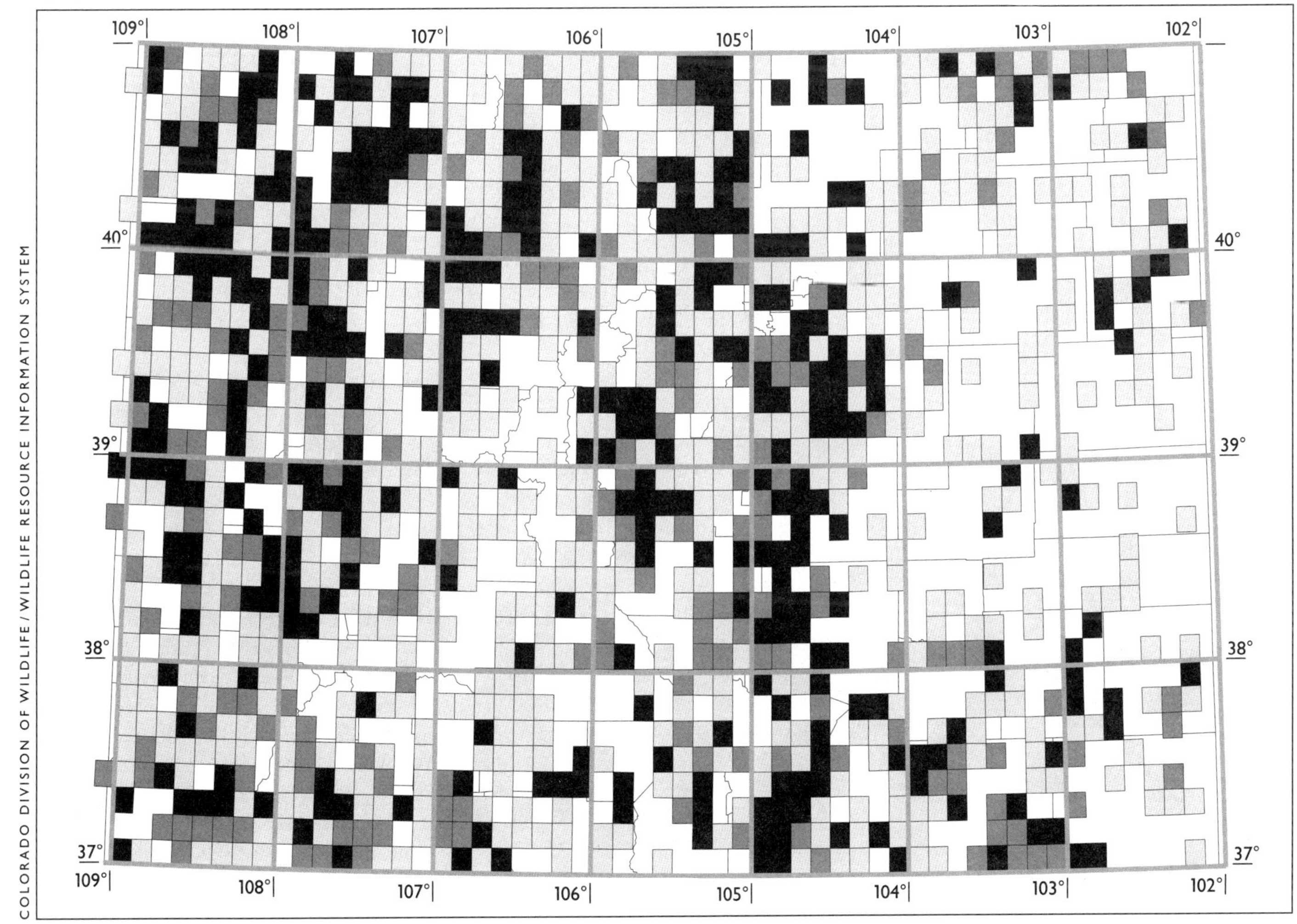
109°
108°
107°
106°
105°
104°
103°
102°
40°
39°
38°
37°
COLORADO DIVISION OF WILDLIFE / WILDLIFE RESOURCE INFORMATION SYSTEM

Like a royal specter, this great predator glides high above its grassy kingdom in search of prey. The largest of all North American buteos, the Ferruginous Hawk feeds nearly exclusively on medium-sized mammals such as rabbits, prairie dogs, and ground squirrels. Unlike Red-tailed and Swainson's hawks, this buteo does not require trees or similar elevated structures for hunting or nesting. When not soaring on high, Ferruginous Hawks regularly rest on the ground, even where elevated perches abound. Consequently, even seasoned observers often underestimate the abundance of this splendid raptor.

HABITAT

Optimum Ferruginous Hawk habitat consists of vast expanses of ungrazed or lightly grazed grassland and shrubland with varied topography, including hills, ridges, and valleys (Ensign 1983). For nest sites Ferruginous Hawks use trees or similar structures when available, but they readily nest on the ground. They typically place ground nests on hilltops that provide a panoramic view. Not surprisingly, shortgrass prairie appears in 44% of Atlas reports for this species.

Food availability probably exerts a strong influence on habitat use. Typically more than 90% of the diet (by biomass) consists of small to medium-sized mammals. The main prey west of the Continental Divide is jackrabbits and cottontails, whereas ground squirrels and prairie dogs predominate in diets east of the Divide. Colorado's Ferruginous Hawks prey heavily on prairie dogs, especially in winter (e.g., Preston and Beane 1996). Surprisingly, in the summary of 12 food studies on Ferruginous Hawks, Palmer (1988b) did not mention prairie dogs, even for one involving Colorado birds.

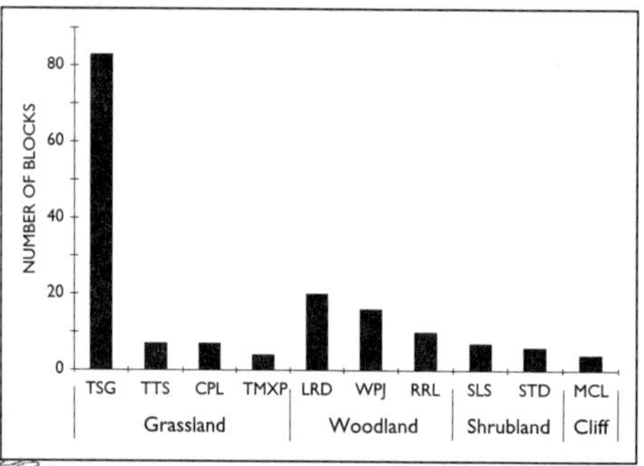
NUMBER OF BLOCKS
0
20
40
60
80
TSG
TTS
CPL
TMXP
LRD
WPJ
RRL
SLS
STD
MCL
Grassland
Woodland
Shrubland
Cliff

Like other buteos, "Ferrugs" hunt from the air and from elevated perches, but they also exhibit a curious ground- feeding technique in prairie dog colonies. A hawk crouches next to a prairie dog burrow and even puts its head in the hole to listen; when it hears a prairie dog come close to the surface, it leaps up, extends its talons down the hole to grab the animal, and flies up with it. Often the hawks have dirty breasts and necklaces of dirt from this hunting technique (Urling Kingery pers. comm.). Occasionally in winter two or more hawks surround a burrow, ready to pounce on unwary prey. If successful, Ferrugs must sometimes contend with Bald Eagles, who readily pirate their prey.

BREEDING

Depending on prey availability and weather, Ferruginous Hawks typically begin laying eggs from mid March to early April (Palmer 1988b, Bechard and Schmutz 1995). Probably owing to the early nesting season and the hawks' proclivity for nesting on the ground, atlasers Confirmed breeding in only 25% of the Priority blocks that reported Ferrugs. They reported only two nests with eggs, although they first saw one on a nest on 17 April. Seventeen nests had young between 22 May and 15 July. Fledging occurs 38–50 days after hatching, and birds as young as 52 days can take live prey (Angell 1969). Atlasers recorded fledglings in three blocks between 26 June and 22 July. None of these dates displayed a chronological/geographical relationship.

In 90% of blocks where they found Ferruginous Hawks atlasers reported only one pair. Typically, these hawks occur in low densities throughout their breeding range. Published estimates range from one nest/772 mi^2 (2000 km^2) in a Washington sector with a large proportion of unsuitable habitat (Fitzner et al. 1977) to one nest/3.3 mi^2 (8.6 km^2) in Alberta (Schmutz et al. 1980). In his 414-mi^2 (1073 km^2) study area in the Pawnee National Grassland, Olendorff (1972) estimated breeding density as one pair/41.6 mi^2 (108 km^2).

BREEDING PHENOLOGY

CODE		# OF RECORDS	DATE RANGE
C	Courtship	2	12 Jun–17 Jun
NB	Nest Building	1	16 Jun
ON	Occupied Nest	14	17 Apr–15 Jul
NE	Nest with Eggs	2	4 May–1 Jun
NY	Nest with Young	20	22 May–15 Jul
FL	Fledged Young	3	26 Jun–22 Jul

DISTRIBUTION

Endemic to the Great Plains and other grassland and shrub-steppe areas of western North America, Ferruginous Hawks occupy the smallest breeding range of any North American buteo. They breed from Saskatchewan, southern Alberta, eastern Washington, and south to northwestern Texas, central New Mexico, and northern Arizona (Bechard and Schmutz 1995). Some individuals winter as far east as eastern Nebraska, Kansas, Oklahoma, and Texas, and as far south as central Mexico. Atlasers found this species in 153 blocks, three-fourths of these broadly scattered across the plains. Virtually all other reports originated in the San Luis Valley and on the Colorado Plateau of northwestern Colorado.

The breeding distribution of Ferruginous Hawks in southern Canada has declined to about 50% of its historic range, probably due to cultivation of native grassland and concomitant loss of ground squirrels and other prey populations (Houston and Bechard 1984, Schmutz 1984, Schmutz 1987, Schmutz et al. 1992). Environment Canada listed the Ferruginous Hawk as Threatened in 1980 and Vulnerable in 1995, but the U.S. Fish and Wildlife Service (1992) rejected a petition to list it for protection under the Endangered Species Act (Ure et al. 1991). Population increases have been reported in Canada (Schmutz and Hungle 1989), California (Warkentin and James 1988), Montana, and North Dakota (Dobkin 1994). Olendorff (1993) confirmed population declines during the past ten years only in eastern Nevada and northern Utah. He deemed Ferruginous Hawk numbers "stable" in Colorado between 1979 and 1992.

Priorities 1995 ranked this bird 14th among 264 species on its conservation priority list of Colorado breeding birds. Because the list relies in part on BBS surveys, which may not effectively sample wide-ranging, sparsely distributed birds (e.g., open-country raptors), readers should use it cautiously. Also, BBS surveys cover only the last 31 years; changes to the prairie breeding range of Ferruginous Hawks have occurred over the past 150 years of human settlement. Not only conversion of prairie to cropland but also the war on prairie dog towns conducted in eastern Colorado undoubtedly have affected Ferrug numbers. It seems clear from studies in Canada (Houston and Bechard 1984, Schmutz 1984) and Utah (Woofinden and Murphy 1989) that conversion of grassland and moderately grazed rangeland to cropland and urban development poses significant threats to Ferruginous Hawk populations through much of its range.

BREEDING EVIDENCE

REPORTED IN 153 (9%) OF 1745 PRIORITY BLOCKS

□	Possible	103 blocks	(67%)
■	Probable	11 blocks	(7%)
■	Confirmed	39 blocks	(25%)

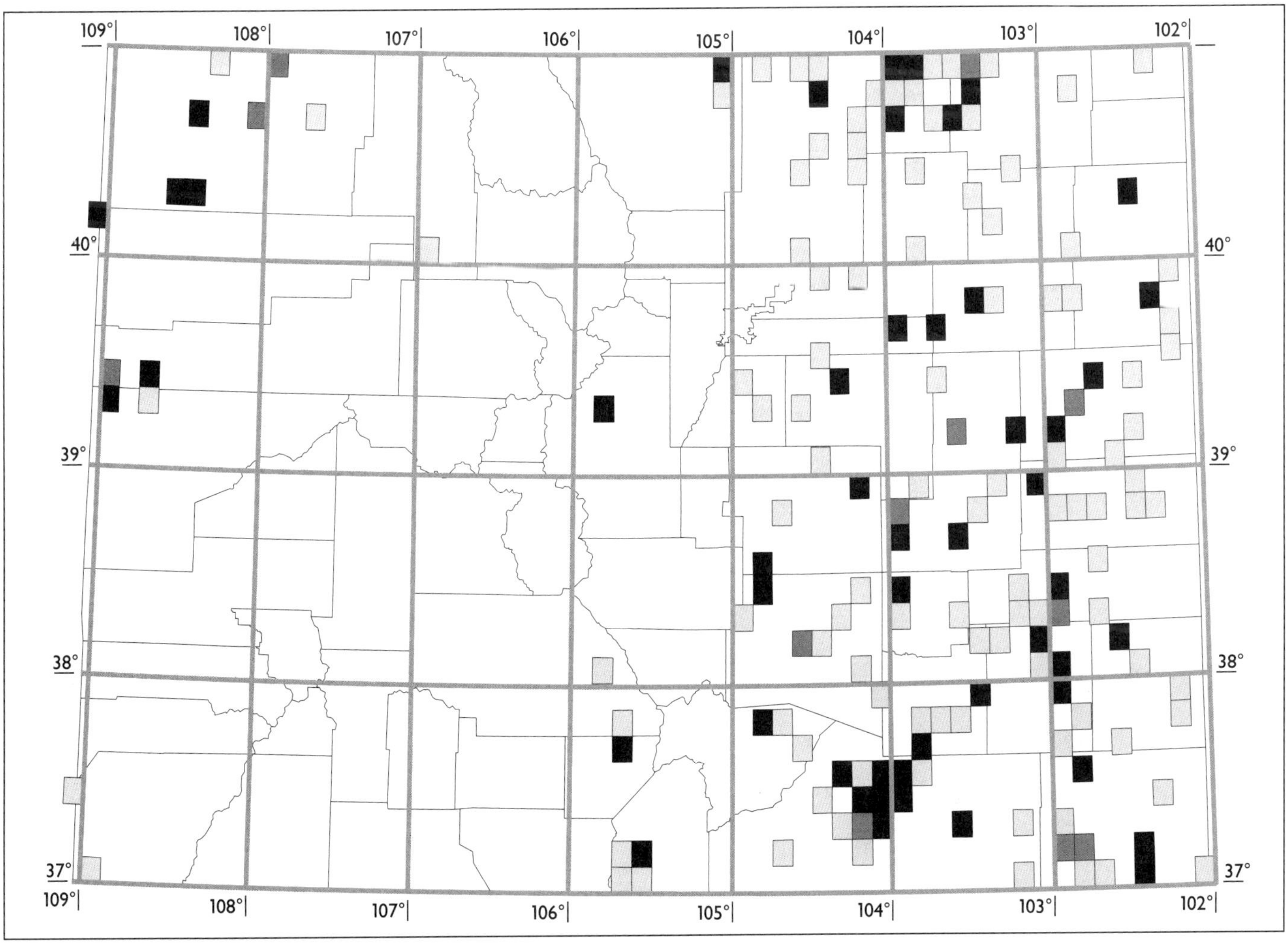
109°
108°
107°
106°
105°
104°
103°
102°
40°
39°
38°
37°

COLORADO DIVISION OF WILDLIFE / WILDLIFE RESOURCE INFORMATION SYSTEM

GOLDEN EAGLE

Aquila chrysaetos

Jackrabbits and ground squirrels, although they don't know it, exist to satisfy the appetites of Golden Eagles and to sustain their effortless sailing flight. Eagles have survived despite a century of persecution by western ranchers. Golden Eagles function as winter vultures of the high Rockies—vultures with style. Thriving on roadkills, they may actually end the winter in better shape than they started it. As snow drives deer and elk closer to Colorado's highways, car-killed carcasses feed hundreds of eagles, and even a smudge of what was once a rabbit often has two or three eagles standing around it.

HABITAT

Golden Eagles nest in a variety of habitats; they most often nest on cliffs, sometimes in trees. Nest sites, however, represent only a small part of the habitat the birds require to raise young successfully. Because of their size Golden Eagles can, and because of their prey base must, forage widely over open habitats, including grasslands, sagebrush, farmlands, and even tundra. A suitable territory needs vast expanses of hunting space, and a large supply of rabbits helps (Bates and Moretti 1994). Suitable mixes of sagebrush and cliffs can support high concentrations. For example, eagles have two cliff nests in the Kremmling block (40106A4), which is 80% sagebrush, and within 5 miles of the block have four more eyries. Atlas habitat data reflect both parts of the territory—nest sites and hunting habitats.

Pairs maintain large territories—average size in four states ranged from 22 to 66 mi^2 (50–170 km^2). Terrain affects territory shape and dimensions; in areas with varied elevations (canyons and valleys) eagles hold smaller territories and on flatter ground claim larger ones (Palmer 1988b).

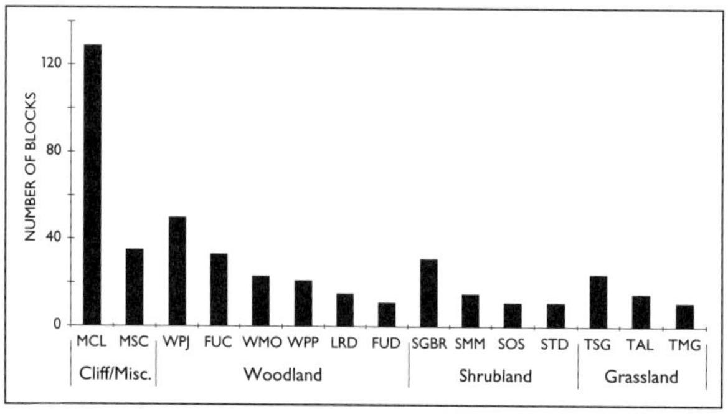
NUMBER OF BLOCKS
0
40
80
120
MCL MSC WPJ FUC WMO WPP LRD FUD SGBR SMM SOS STD TSG TAL TMG
Cliff/Misc.
Woodland
Shrubland
Grassland

Golden Eagles dine mainly on small rodents, hares, and rabbits, and in winter, carrion. Overall breeding success correlates positively with jackrabbit numbers (Ryser 1985). Eagles also like ground squirrels, marmots, and prairie dogs (Palmer 1988b). A massive analysis of 7,094 prey items divulged a diet of 84% mammals, 15% birds, and 1% reptiles and fish (Olendorff 1976).

BREEDING

Golden Eagles need a long breeding season: incubation takes 43–45 days (Terres 1980) and young fledge about 65 days after hatching (Palmer 1988b). As residents with an early season food source of rabbits, Goldens begin courtship early in March (Atlas data). By mid April some nests have young. Other atlasers reported young in nests as late as 20 July, connoting nest initiation during May or June. Adults stock up on food for nestlings—eyries can hold such stashes as 22 ground squirrels or 12 jackrabbits surrounding the eaglets (Palmer 1988b).

Eagles apparently mate for life. They renew the pair bond in late winter with spectacular "high-circling" courtship displays in which the two may rise 2,000 feet (600 m) above the ground.

Two essentials define the nest site: space so that an adult burdened with prey can fly in without hindrance, and shelter from heat and cold. Eagles exhibit site loyalty—they have occupied one New England site since 1689 (Palmer 1988b).

Returning pairs may reuse the previous year's nest or move to another one. Technically the pair using the Kremmling cliffs rotates among three nests, though after ten years the nests slowly grew into each other and now form a giant mass of sticks on the cliff face. The nest site now represents an avian split-level condo with Cliff Swallows in the attic and White-throated Swifts in the basement.

Eagles add material to the nest, both before and during breeding. Beyond ordinary sticks, leaves, and grass, they may add cow bones, antlers, barbed wire, burlap, newspaper, stockings, and other rubbish. Like many raptors, they add greenery during the season.

BREEDING PHENOLOGY

CODE		# OF RECORDS	DATE RANGE
C	Courtship	3	12 Apr–19 Jun
NB	Nest Building	2	19 Jan–4 May
ON	Occupied Nest	11	22 Apr–25 Jul
NY	Nest with Young	43	15 Apr–20 Jul
FY	Feeding Young	5	9 May–17 Jun
FL	Fledged Young	10	8 May–12 Aug

DISTRIBUTION

In North America the circumpolar Golden Eagles range from the Arctic to Mexico and coast to coast except in the Southeast. Palmer (1988b) estimated a North American winter population of 70,000 Golden Eagles, with 10% in Colorado; Colorado's breeding population is 5–10% of the winter numbers.

Colorado's population has changed little in 100 years (cf. Cooke 1897). Although eagles occur throughout the state, the Atlas shows breeding concentrated in the western two-thirds of the state. The map differs from Craig's report in 1979 (A&R), which showed more breeding in northeastern Colorado and less in the southeast. Craig reported that Goldens have 500 nest sites in Colorado but in any year they use less than half. The Atlas, spanning eight years, did not detect shifting site use, although atlasers did find some new sites.

Many Possible and Probable observations on the Atlas map likely represent breeders foraging far from their nests or nesting outside Priority blocks, or non-breeders, immatures and adults. Even using only Confirmed nests, Atlas data double Craig's estimate of 191 Colorado breeding pairs.

Dietary differences protected Golden Eagles from the drastic decline suffered by Bald Eagles in the 1950s and 1960s. They also appear less sensitive to human disturbance, because they forage and sometimes nest near towns and farmhouses. Showing great adaptability, the pair nesting on the Kremmling cliffs inevitably fledge their young a week before the fourth of July fireworks are fired from atop the cliff (NMB).

Nonetheless humans are a major cause of death for this species. Winter roadkills contribute significantly to eagle deaths. Eagles that find a tasty carcass often gorge themselves so that takeoffs prove difficult when cars approach, and they also become smears on the highway.

In the past illegal shooting and intentional poisoning by ranchers severely impacted numbers in parts of the West, notably Texas and Wyoming. Other human-caused deaths result from electrocution on power lines and accidental poisoning during predator-control efforts. Each of these threats kills some birds, but cumulatively the species as a whole seems able to overcome them.

BREEDING EVIDENCE

REPORTED IN 408 (23%) OF 1745 PRIORITY BLOCKS

☐	Possible	250 blocks	(61%)
▩	Probable	47 blocks	(12%)
■	Confirmed	111 blocks	(27%)

109°
108°
107°
106°
105°
104°
103°
102°
40°
39°
38°
37°

COLORADO DIVISION OF WILDLIFE / WILDLIFE RESOURCE INFORMATION SYSTEM

AMERICAN KESTREL *Falco sparverius*

Like a miniature helicopter, an American Kestrel hovers in midair, then drops suddenly to the ground, plunging after a grasshopper or mouse. If successful the bird flies to a perch to eat. If the prey escapes, the kestrel returns to hover or perches to watch for the movement or sound that denotes food. Even though kestrels masqueraded as "Sparrow Hawks" for a century (the same name used for a small European accipiter), birds constitute only 5% of their food.

HABITAT

Ecologically versatile, American Kestrels need elevated perches (such as dead tree stubs, rock outcrops, and often utility poles and wires), open terrain for hunting insects and small vertebrates, and nesting cavities (tree cavities, crevices, cliffs, or nest boxes; Johnsgard 1990). In forests they seek edges and openings. They breed up to 10,000 feet (3084 m; Sclater 1912). The highest Atlas report, north of Craig (Freeman Res. 40107G4), was a nest with young at 9,200 feet (2800 m; HEK, nest card). Atlas observers saw about one-fourth of the kestrels near habitations—in rural and urban areas—with the remainder divided among the mountains, grasslands, and shrublands.

Kestrels maintain the smallest territories of any Colorado hawks; the size varies with prey density. California birds defended home ranges estimated at 250 acres (109 ha; Balgooyen 1976), yet in Utah five pairs located nests within 200 feet (61 m) of at least one other nest, with two within 40 feet (12 m) of each other (see Palmer 1988b).

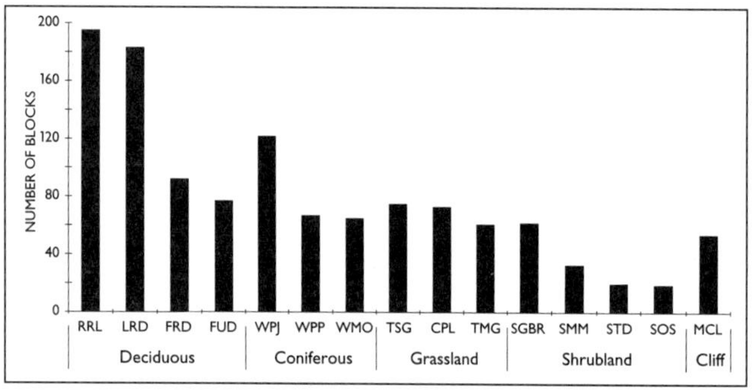

NUMBER OF BLOCKS
200
160
120
80
40
0
RRL LRD FRD FUD
Deciduous
WPJ WPP WMO
Coniferous
TSG CPL TMG
Grassland
SGBR SMM STD SOS
Shrubland
MCL
Cliff

Birds compose only 5% of their prey. A multi-year study of prey items in 703 stomachs found grasshoppers most frequently (in 491 stomachs, 70%). Other principal prey items included beetles and moths in 229 (33%); small mammals, especially voles, in 192 (27%); spiders in 153 (22%); lizards and snakes in 62 (9%); and birds in only 38 stomachs (Palmer 1988b).

BREEDING

In the fall, most kestrels migrate, although some males maintain territories through the winter to use the following year. Returning males arrive on the breeding grounds in March or early April and females follow about a week later. A male establishes his territory, then attracts the female with aerial displays, courtship feeding, nest-site inspection, and vocal signals (Balgooyen 1976). They use tree cavities when available, and may evict flickers or other birds. In order of importance, kestrels pick as nest sites flicker and other woodpecker holes, natural cavities, cavities in buildings, magpie nests, holes in cliffs, and dirt banks. In contests over nest holes, kestrels usually dominate woodpeckers, starlings, flickers, chipmunks, and squirrels, although sometimes the starlings prevail (Palmer 1988b). Kestrels in a Utah study had 28 of 41 nests in trees, with 19 in old flicker holes and seven in natural cavities; 11 used buildings (Smith et al. 1972). Atlas observers reported one nest in a cavity in the side of a wooden house (Steve Boyle, nest card), as well as several in nest boxes.

Atlasers reported courtship as early as March, and four active nests in April. The first young kestrels out of the nest, discovered 7 and 8 May, had fledged in low, hot valleys (Grand Junction, 39108A5, and Northeast Pueblo, 38104C5). Atlasers dated the majority of nests with young from late May to mid July, with most fledglings reported in July. Last dates for occupied nests progressed chronologically from south to north: in latitudes 37–40 respectively 11, 19, 22, and 28 July. Renesting may replace a lost clutch, and second nestings were thought possible in El Paso County (Stahlecker and Griese 1977).

The female lays 4–5 eggs and incubates them 29–31 days. The male feeds the female on the nest, and some males incubate at times (it varies with individual males). The young leave the nest 30–31 days after hatching (Willoughby and Cade 1964).

BREEDING PHENOLOGY

CODE		# OF RECORDS	DATE RANGE
C	Courtship	22	8 Apr–10 Jul
NB	Nest Building	1	10 Jun
ON	Occupied Nest	74	17 Apr–28 Jul
NE	Nest with Eggs	4	18 May–12 Jun
NY	Nest with Young	38	12 May–21 Jul
FY	Feeding Young	49	17 Apr–4 Aug
FL	Fledged Young	110	7 May–20 Sep

DISTRIBUTION

The North American races of American Kestrel breed nearly as far north as the limits of the boreal forests in northern Canada and Alaska south through the U.S.

to Panama. Other subspecies occupy the West Indies and South America to Tierra del Fuego (Tyler *in* Bent 1938).

Edwin James first reported the kestrel in Colorado on 15 July 1820 near Colorado Springs, as he returned from his first attempt to climb Pikes Peak (B&N). Cooke and Sclater regarded them as the commonest of all Colorado raptors.

Atlas data show that kestrels breed throughout most of Colorado except the high mountains and some of the plains, especially those sections devoted mainly to crops. Kestrels avoid the croplands of the east-central plains and the high-altitude spines of the mountain cordilleras. They follow all the stream bottoms on the plains and foothills and occupy most mountain blocks below 8,000 feet (2440 m). On the plains they prefer riparian to rural habitats: atlasers recorded them in equal numbers of each, but, because rural far outnumbers riparian, in a much higher percentage of riparian. Many rural settings lack either nest sites or suitable nearby hunting terrain. The presence of pinyon/juniper, ponderosa, riparian, and rural habitats determine which mountain blocks support kestrels, and the western stream bottoms sustain continuous populations.

American Kestrel was on the National Audubon Society Blue List between 1971 and 1979 because of eggshell thinning from pesticides (Tate 1986). However, BBS surveys from 1966 to 19 79 suggested that continental numbers of American Kestrels had increased (Fuller et al. 1987) and the 1966–1996 trend analysis suggests a stable population. From the number of birds found by the Atlas, it appears that American Kestrels maintain a healthy population in Colorado.

BREEDING EVIDENCE

REPORTED IN 973 (56%) OF 1745 PRIORITY BLOCKS

☐	Possible	436 blocks	(45%)
▩	Probable	264 blocks	(27%)
■	Confirmed	273 blocks	(28%)

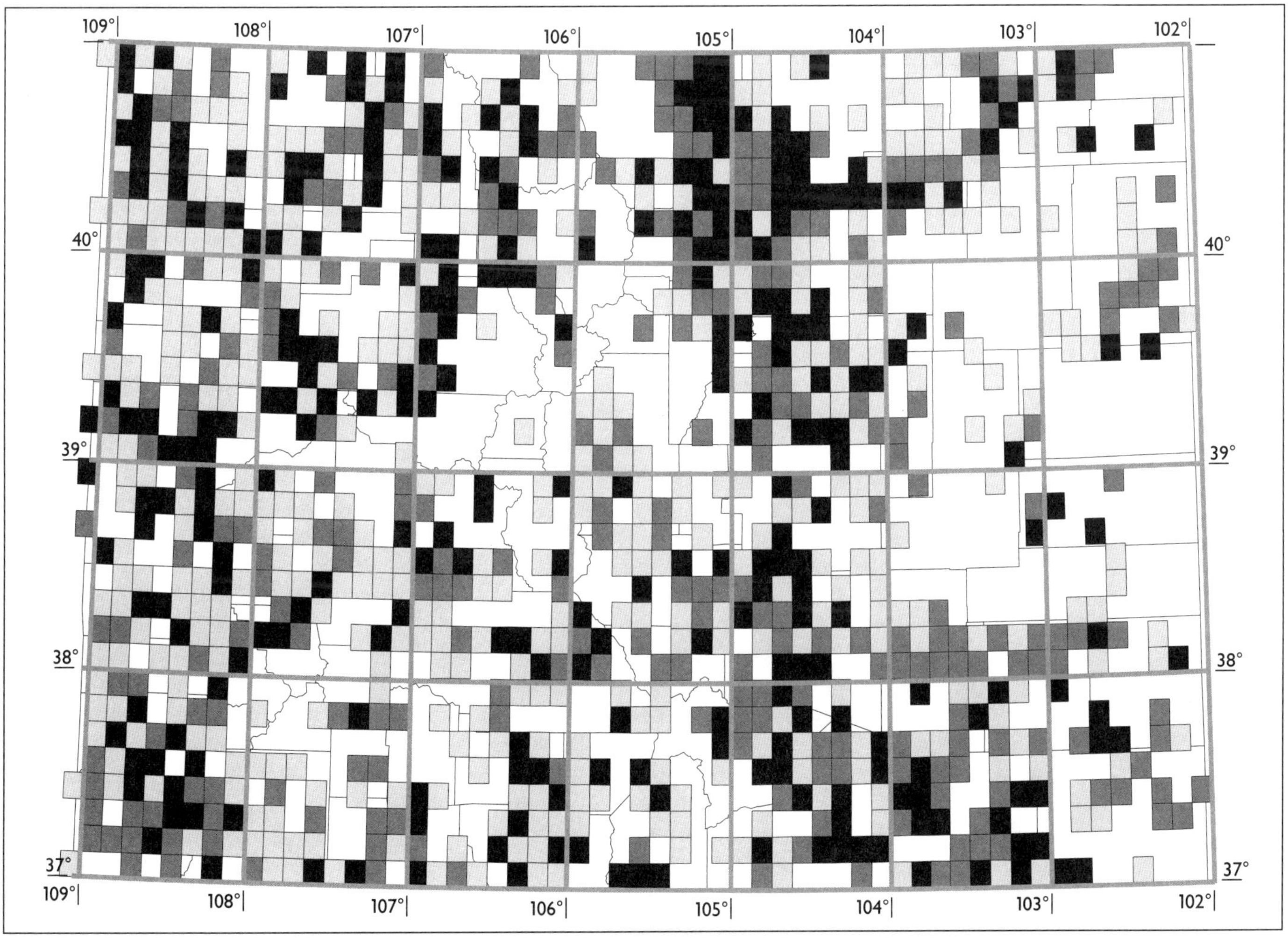
109°
108°
107°
106°
105°
104°
103°
102°
40°
39°
38°
37°

COLORADO DIVISION OF WILDLIFE / WILDLIFE RESOURCE INFORMATION SYSTEM

In the 1950s and 1960s DDT hammered the American Peregrine Falcon population down to almost nil. Use of that falconicide has since declined, and wildlife agencies have launched intensive restoration efforts. Colorado's Peregrine Falcon population has rebounded, and Rock Doves, White-throated Swifts, and other favored prey rest a bit less easily with every passing year.

HABITAT

Peregrines usually nest on ledges of high cliffs. When Peregrine populations hit their nadir, cliffs higher than 200 feet (60 m) sustained the last remaining eyries (Rocky Mtn./SW Peregrine Recovery Team 1977). Early records described Peregrines nesting in much more accessible sites, even on low dikes (Porter and White 1973) and in marshes and mud banks (Coues 1874, Bendire 1892), but only more remote sites have withstood increasing human disturbance.

Peregrines nest on foothills and mountain cliffs from 4,500 to over 9,000 feet (1370–2745 m), although most locate their eyries near the lower end of this range

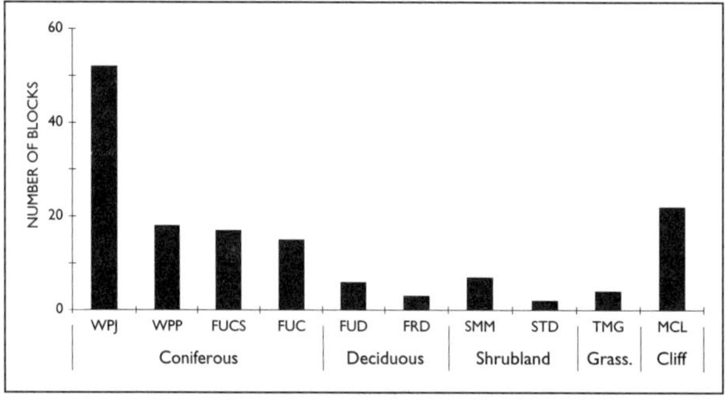
60
40
20
0
NUMBER OF BLOCKS
WPJ
WPP
FUCS
FUC
FUD
FRD
SMM
STD
TMG
MCL
Coniferous
Deciduous
Shrubland
Grass.
Cliff

(Rocky Mtn./SW Peregrine Recovery Team 1977). Pinyon/juniper grows in the vicinity of about half of all nest sites, and ponderosa pines at about one-quarter of the sites; a variety of other vegetation makes up the remainder (Jerry Craig pers. comm.). Atlas workers reported cliffs at all sites where they located nests. Omitting cliffs, habitat reports specified pinyon/juniper most frequently, and ponderosa and spruce/fir nearly as frequently. Atlasers saw these falcons frequently in pinyon/juniper and ponderosa primarily because these trees grow at the same elevations as the best cliffs, and the birds hunt in these habitats.

BREEDING

The Rocky Mountain/Southwestern Recovery Plan (1977) indicates that although some Peregrines remain in the vicinity of their eyries throughout the year, most winter south of Colorado. Pairs generally arrive at the nesting area in March, and most begin incubating in April.

An ideal eyrie has a wide view and plentiful prey in the vicinity, is near water, and receives little disturbance. Pairs defend a small area of about 100 yards (90 m) around their eyries; they do not patrol the boundaries but rather announce the territory with a sky dance and a high circling display (Palmer 1988b). A level site, at least 2 feet (60 cm) in diameter (to make space for the nestlings), suits them if it has a sheltering overhang and some debris (sand, sticks) in which to make a scrape for the eggs.

The female lays 3–4 eggs (usually); incubation lasts 32–35 days for each egg. The young remain in the nest for 39–46 days. Both adults share in incubating and caring for the young, although males do more of the hunting. After fledging, young birds may remain in the nesting area for several weeks and adults will continue to provide food. Adults pass food to fledged young on perches and in the air; air transfers enable the parents to avoid aggressive attacks from the young (Palmer 1988b).

BREEDING PHENOLOGY

CODE		# OF RECORDS	DATE RANGE
C	Courtship	3	30 Mar–31 May
ON	Occupied Nest	2	10 Jun–13 Jun
NY	Nest with Young	15	10 May–11 Jul
FL	Fledged Young	3	23 Jul–8 Aug

DISTRIBUTION

Peregrines have a more extensive worldwide range than any other bird: they occur on all continents except Antarctica, and on many islands (Ehrlich et al. 1988).

Cooke (1897) considered the Peregrine Falcon a "locally common" bird in Colorado, and reported Peregrines as breeding up to 10,000 feet (3048 m) in the mountains. Sclater (1912), however, stated that Peregrines nested only rarely in Colorado, although suitable locations did harbor some. He added that the southern and western parts of the state had provided no reports of this species, although Rockwell (1908) earlier had indicated that it bred "in different parts of western Colorado."

A survey of 15 known Peregrine nesting sites found only six occupied in 1964 (Enderson 1965). By 1973, researchers had located 23 sites, 11 of them occupied that year (Enderson and Craig 1974). As Peregrine numbers crashed throughout the country, the Colorado population in 1977 dropped to a low of four nesting pairs (Gray 1995). During the next 10 years wildlife agencies released more than 500 captive-reared Peregrine Falcons in Colorado alone (Gray 1995). This intensive effort has built up the population to the point that in 1995 Peregrines occupied 71 sites, and 68 pairs attempted to nest (Jerry Craig pers. comm.).

The 71 eyries are scattered throughout the mountains. Peregrines breed in several blocks along the foothills of the Front Range, but higher concentrations nest in the river valleys and canyons of the Western Slope, which has had nearly three-fourths of the eyries (Craig 1991, 1993, 1994). The highest concentrations occur along the Dolores and Colorado river canyons in Mesa and Montrose counties and in Dinosaur National Monument.

The prime objective of the 1977 Recovery Plan called for increasing Peregrine populations in the Rocky Mountain/Southwestern region to a minimum of 100 breeding pairs by 1995. Glinski and Ambrose (1990) reported that the Colorado Plateau of the southwestern U.S. has "a tremendous density of Peregrines." The Grand Canyon alone sustains nearly 100 breeding pairs. Craig reckons that Colorado will have 100 to 120 nest sites when the population has fully recovered (Gray 1995). The effort in restoring the Peregrine Falcon has clearly paid off, and barring any major relapse, this falcon will harass other cliff-dwelling birds for years to come.

Caveat on the Atlas map: The Atlas map reflects the information provided by the CDOW study team, but at the request of that agency the Atlas map masks the specific locations in order to protect the birds from human predation. The map clumps breeding sites in the center of each latilong rather than showing the specific blocks.

BREEDING EVIDENCE

REPORTED IN 36 (2%) OF 1745 PRIORITY BLOCKS

Possible	10 blocks	(28%)
Probable	6 blocks	(17%)
Confirmed	20 blocks	(56%)

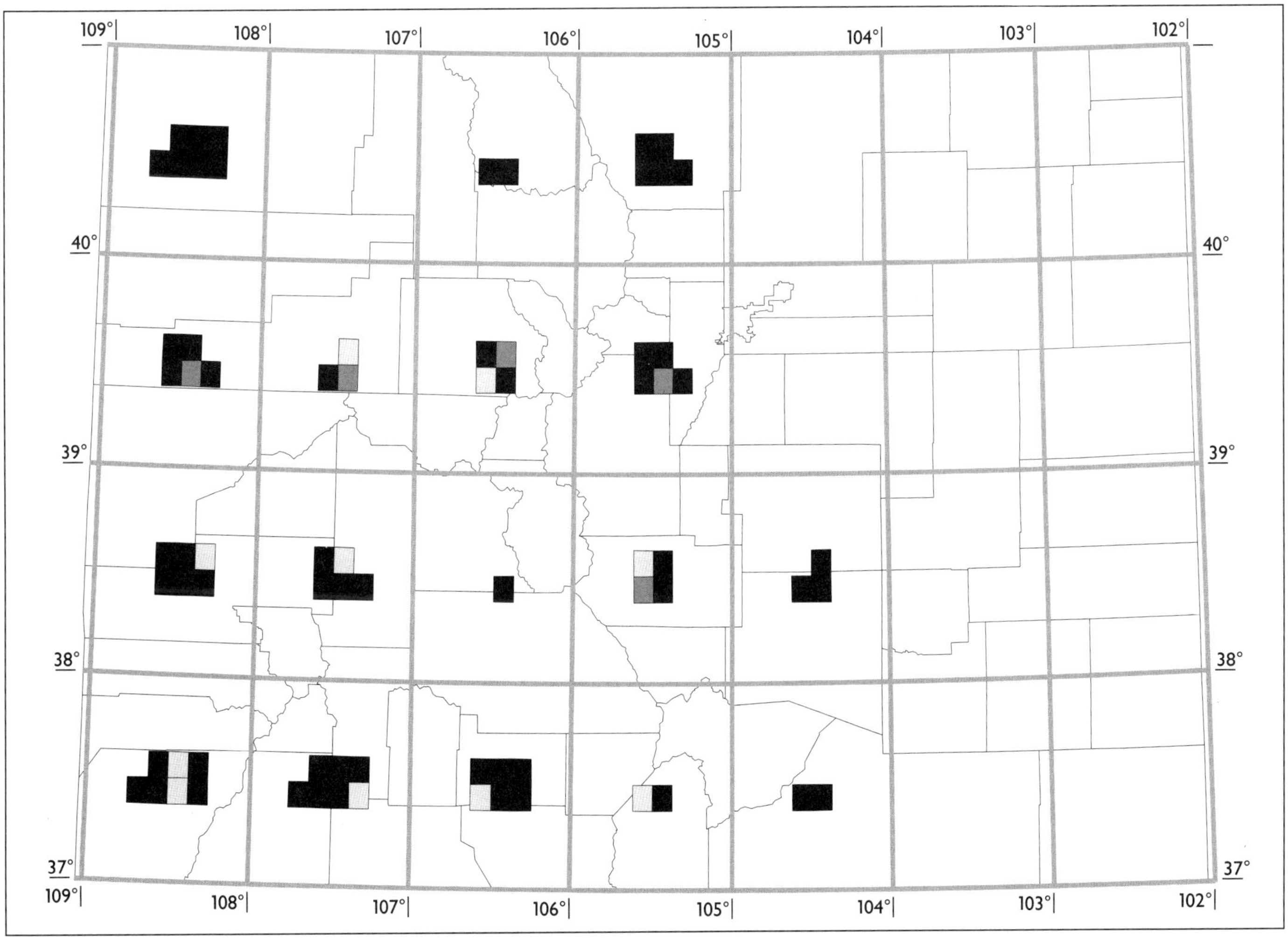
109°
108°
107°
106°
105°
104°
103°
102°
40°
39°
38°
37°

COLORADO DIVISION OF WILDLIFE / WILDLIFE RESOURCE INFORMATION SYSTEM

The screams of nesting Prairie Falcons echo off cliff faces and canyon walls throughout Colorado from March to July. Nesting locations range from chalk bluffs in the Pawnee National Grassland to conglomerate cliffs in Castlewood State Park to red sandstone monoliths in Colorado National Monument.

HABITAT

These swift, acrobatic falcons typically nest on cliff faces in open country below 10,000 feet (3048 m), but they occasionally breed above timberline. An eyrie on a volcanic headwall in Rocky Mountain National Park at 12,140 feet (3700 m) had three young in 1968 (Collister 1970, Marti and Braun 1975); this site remained active in 1994. In suitable habitat at lower elevations, breeding densities may approach one pair per square mile, with some nests only 200 yards (183 m) apart (Enderson 1964). Webster (1944) counted 23 nesting pairs along 16 miles of cliffs in Weld County. Atlasers found relatively high nesting densities in the canyons of Baca and Las Animas counties and in the Front Range foothills.

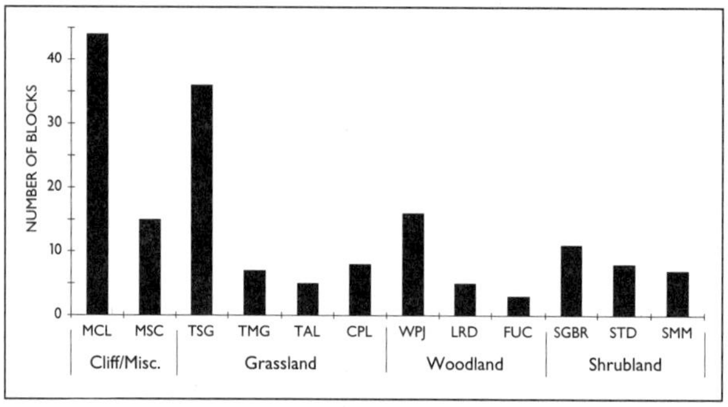
NUMBER OF BLOCKS
0
10
20
30
40
MCL MSC TSG TMG TAL CPL WPJ LRD FUC SGBR STD SMM
Cliff/Misc.
Grassland
Woodland
Shrubland

Prairie Falcons hunt birds and small mammals from perches or while soaring, flapping, or swooping low over open country (Sherrod 1978). In 13 studies of Prairie Falcon diets summarized by Sherrod (1978), cottontails, ground squirrels, and other small mammals composed as little as 8% and as much as 66% of prey items. During winter, when the falcons range over grasslands, fields, and deserts far from summer nest sites, they may prey almost exclusively on songbirds, especially Horned Larks (Enderson 1964).

BREEDING

Pairs arrive on breeding territories in February or March (Enderson 1964, B&N). Incubation starts as early as 17 April on the Boulder Flatirons (SRJ), and Colorado atlasers recorded them on nests on a northeastern escarpment and a southeastern canyon on 20 and 29 May. Brooding dates (nest with young) ranged from 31 May to 13 July.

Females usually lay five eggs (range 2–6) in a shallow depression on a ledge or in a pothole (B&N, Millsap 1981). Occasionally they select abandoned raptor nests, such as Golden Eagle nests at Pawnee Buttes where Prairie Falcons nest in the accumulated dirt and gravel in the nest bowl.

Prairie Falcons compete for nest sites with Peregrine Falcons, Golden Eagles, Great Horned Owls, and Common Ravens. When Peregrines declined in Colorado, Prairie Falcons moved into several territories, but as Peregrines came back they repossessed their old sites (Jerry Craig pers. comm.). This occurred in Rocky Mountain National Park (Jeff Conner pers. comm.) and the pothole eyries in the Flatirons near Boulder. (Common Ravens also displaced Prairies there; Lederer and Armstead 1995.) Peregrines also usurped several historic Prairie Falcon nesting sites along the Front Range foothills. Great Horned Owls frequently prey on Prairie Falcon nestlings, and the falcons will attack and sometimes kill the owls (Webster 1944, Enderson 1964).

Some nests may produce five or more young (Johnsgard 1990). In early June atlasers observed four scruffy-looking chicks huddled on a narrow, whitewash-splattered ledge in a Las Animas County canyon. In mid June a pothole eyrie south of Boulder harbored five nearly full-grown young.

BREEDING PHENOLOGY

CODE		# OF RECORDS	DATE RANGE
C	Courtship	2	18 Apr–2 Jun
ON	Occupied Nest	15	21 Apr–15 Jul
NY	Nest with Young	11	31 May–13 Jul
FL	Fledged Young	6	3 Jun–24 Jul

DISTRIBUTION

Prairie Falcons range over the western half of North America, from southern Canada to central Mexico. Wintering falcons fly as far east as Illinois and as far south as Oaxaca (Enderson 1964, Schmutz et al. 1991).

Early ornithologists called Prairie Falcons "quite numerous" in western Colorado (Cooke 1897), "moderately common" in the state (Sclater 1912), and an "infrequent summer resident" in Boulder County (Betts 1913). The Atlas map shows major nesting clusters in Las Animas and Baca counties, and along the Front Range.

Most of the somewhat scattered blocks in western Colorado contain either deep river canyons or imposing cliffs.

Early estimates of the Colorado breeding population varied from 500 pairs in 1944 (Webster 1944), to 300 pairs in 1964 (Enderson 1964), to 400–500 pairs in 1994 (Jerry Craig, CDOW unpubl. data). Cade (1982) estimated the total North American breeding population at 5,000–6,000 pairs.

Prairie Falcons forage over wide areas and often nest in remote, inaccessible sites. By including blocks with Possible and Probable observations, Atlas data probably overestimate the number of blocks in which these wide-ranging falcons actually nest, even as the map does not depict Confirmations in all Priority blocks where they breed. Many previously documented nests lie outside of Priority blocks, especially in the Chalk Bluffs area of northern Weld County. A calculation of the state population using all Atlas Priority blocks projects a rather high count of about 1,000 nesting pairs; a calculation using only blocks with Confirmations results in 300 pairs. The actual number probably lies between the two estimates, closer to Craig's 1994 figure.

Illegal taking of nestlings, nest sabotage, and illegal hunting have impacted Colorado Prairie Falcon populations throughout this century. Webster (1944) cited shooting as the leading cause of mortality of young falcons. Enderson (1964) observed frequent nest depredations by falconers on an escarpment in northern Colorado: between 1960 and 1962, 26 nests on the escarpment fledged a total of only three young.

Shooting of perched birds appears to have declined in recent years (Jerry Craig pers. comm.), but human disturbance of nests and destruction of nesting habitat still threaten these falcons. A 13-year (1982–1994) study of nesting success in nine Jefferson and Boulder County territories noted no instances of nest failure due to taking of nestlings or shooting. However, eyries located in popular rock-climbing and hiking areas often failed to produce young (Lederer and Armstead 1995). During the 1980s Prairie Falcons abandoned several eyries in the Front Range foothills as urban growth gobbled up foraging habitat (Jerry Craig pers. comm.). Colorado wildlife agencies and many conservation groups have initiated nest-monitoring programs to protect existing eyries.

BREEDING EVIDENCE

REPORTED IN 149 (9%) OF 1745 PRIORITY BLOCKS

☐	Possible	86 blocks	(58%)
■	Probable	17 blocks	(11%)
■	Confirmed	46 blocks	(31%)

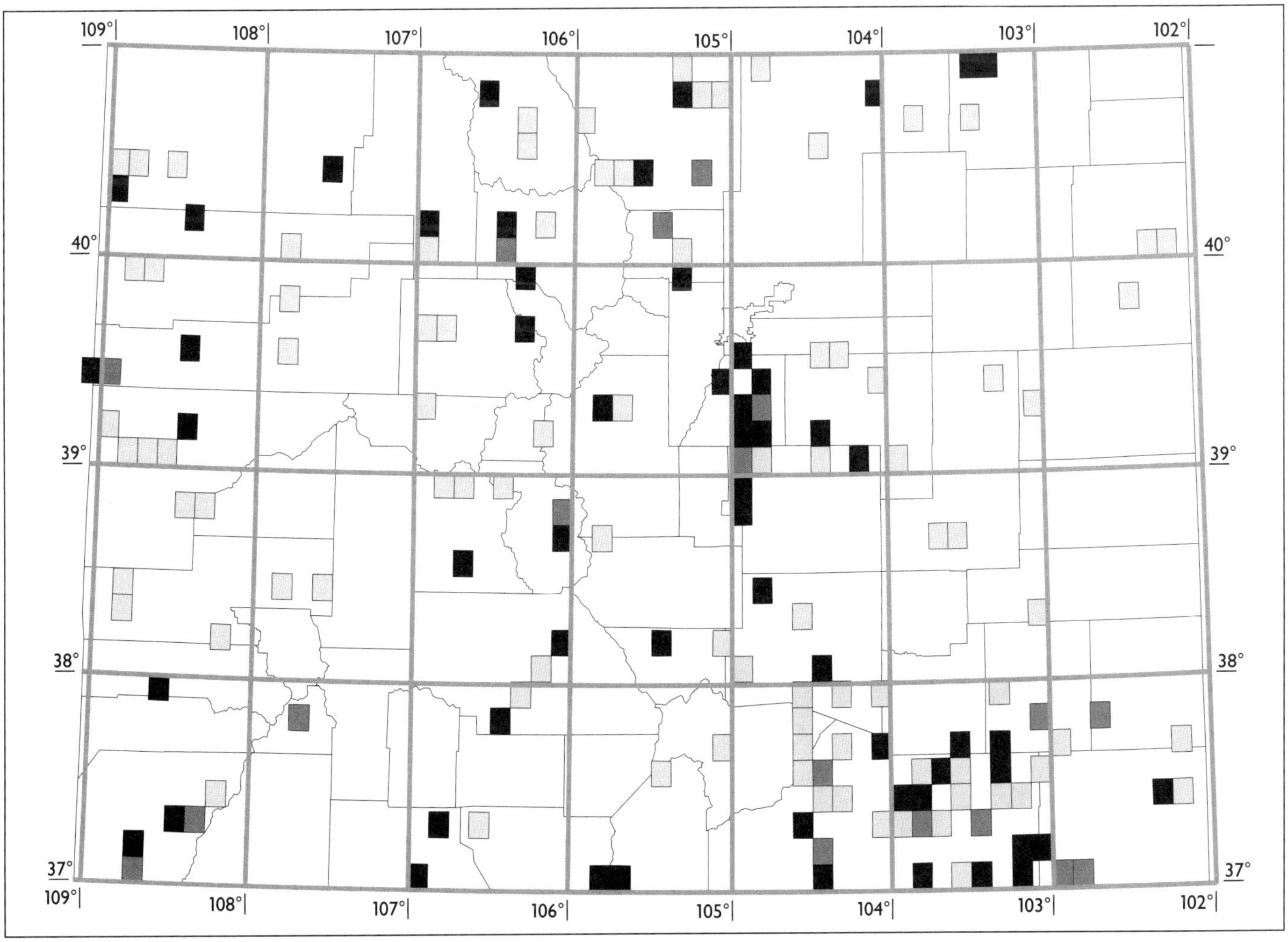
109°
108°
107°
106°
105°
104°
103°
102°
40°
39°
38°
37°

COLORADO DIVISION OF WILDLIFE / WILDLIFE RESOURCE INFORMATION SYSTEM

Hunters, because of their love affair with chicken birds, bear total responsibility for the successful introduction of Chukars into the vast, arid landscape of the Great Basin. Where similar landscapes extend eastward into western Colorado, a few of these colorful, well-camouflaged Asian birds ramble over the rocky slopes.

HABITAT

Desert areas with rocky canyons, steep hillsides, scattered bushes, and blankets of cheatgrass meet the needs of Chukars. Short hot summers and moderately cold winters characterize the climate. Seeds, leaves, and grass represent the primary food items, plus occasional insects (Christensen 1996). Rabbitbrush, saltbush, and greasewood make shade during summer heat and afford protection from avian predators (Christensen 1970).

Atlasers cited desert shrub communities as primary Chukar habitat—medium to tall shrublands of greasewood, fourwing saltbush,

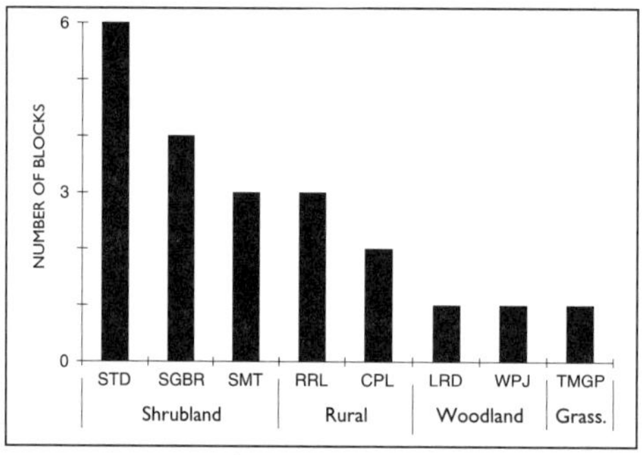
NUMBER OF BLOCKS
0
3
6
STD
SGBR
SMT
RRL
CPL
LRD
WPJ
TMGP
Shrubland
Rural
Woodland
Grass.

blackbrush, and shadscale; some Chukars use sagebrush and the lower-growing species of mat saltbush.

Availability of water limits their range during hot, dry summer months; Chukars, especially the young, require water (Galbreath and Moreland 1953). CDOW initiated a program of building guzzlers, a type of water collector, to expand Chukar range, but it had limited success (Braun et al. 1994). Birds use numerous natural springs located in many of the canyons, as well as streams and rivers, as their primary water sources.

In winter Chukars need south-facing slopes, free of deep snow. Low-elevation canyons in western Colorado receive little snow—which sunny skies quickly melt—keeping food available for the birds. Deep snow results in winterkill (Galbreath and Moreland 1953).

Heard more often than seen, Chukars frequently call from rocky outcrops on steep slopes. In this habitat they prove a difficult bird for the hunter. Their effective camouflage, reluctance to fly, and habit of running uphill when pursued make them a difficult quarry (Bohl 1957).

BREEDING

Research by CDOW provides evidence that only a few Chukars live more than two years (Ron Basagartia pers. comm.). Breeding birds constitute only a small part of the total population. This leads to big swings in abundance. In successful breeding years a few breeding adults produce many young, and birds remain plentiful into the fall. In poor years breeding pairs produce few young and, with little carryover from prior years, numbers crash.

During mid March Chukars pair off for mating. Soon after, the birds establish territories and by the first of May the mated pairs complete nest construction. The nests are notoriously hard to find (Christensen 1996). The birds typically place them on rocky hillsides and hide them under shrubs or beside rocks in shallow depressions. Chukars scratch a depression in the ground and line it with dry grasses and feathers; hens lay 10–21 eggs (Christensen 1996). Upon completion of the clutch, incubation begins, taking about 24 days (Bohl 1957). Broods come off the nest from late May to mid June. Upon hatching, young follow the female, who shows them food but does not feed them. They can fly in two to three weeks. Males, on occasion, remain with the family (Molini 1976). The number of surviving chicks in a brood drops to seven or eight by July or August (Sandfort 1952). Pairs produce only one brood per year, but persistently renest to succceed at producing a brood (Christensen 1996).

Atlasers found no nests and instead made their few Confirmations by finding fledglings. The dates spanned two months starting at the end of May and continuing until fledglings had achieved high mobility in late July, but young remain with the adults until late winter (Christensen 1996).

BREEDING PHENOLOGY

CODE		# OF RECORDS	DATE RANGE
FL	Fledged Young	8	30 May–23 Jul

DISTRIBUTION

Native from Greece to northern India and Mongolia, Chukars attracted the attention of hunters in the 1930s. Wildlife agencies introduced them into 42 states and Canada. Game farms can easily raise Chukars in captivity; their massive propagation efforts supplied state agencies with large numbers (936,000 according to Christensen 1996). Most stocking attempts failed. At present only ten western states and British Columbia have self-sustaining populations (Christensen 1996).

Initial Chukar plantings in Colorado occurred in 1937. By 1950, after plantings in 52 Colorado counties, most in unsuitable habitat, only 15 colonies remained. They succeeded best in the canyons and hills of west-central Colorado in Montrose, Delta, Mesa, and Garfield counties. By 1988 the CDOW discontinued most large-scale plantings.

As the Atlas map shows, today the population lives mostly below 6,600 feet (2000 m), mainly in latilong 39108, and reaches its greatest abundance along the Colorado and Gunnison river drainages. The numerous canyons and mesas along permanent water sources in this arid landscape provide the best home the state has to offer this Asian transplant.

In the fall of 1958 Colorado opened its first hunting season on Chukars—182 hunters bagged 400 birds (Sandfort 1967). Presently 500–1,000 hunters kill 1,000–3,000 Chukars annually (Braun et al. 1994).

Because game farms and recreational bird enthusiasts raise and release them annually (Fergus 1993), Chukars can turn up almost anywhere in the state. The fledglings that Atlasers found in northeastern Colorado and Chaffee County probably represent escaped captive birds or hunters' introductions unlikely to succeed. For many years a gun club made annual releases at Rocky Mountain Arsenal. Many birders between flights at Stapleton Airport drove along the south fence to add Chukar to their life lists (probably inappropriately, as the population there could not sustain itself). The releases have stopped, and the Chukars have disappeared.

CDOW reports that it has inadequate knowledge of the distribution, density, habitat requirements, and biology of Chukars in Colorado (Braun et al. 1994). The species has a low priority for research and management because of its limited distribution. Although this status probably will not change significantly, strategies to improve assessment of distribution and abundance may be necessary to justify hunting seasons (Braun et al. 1994).

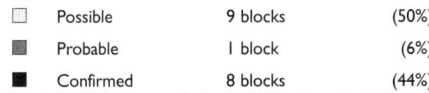

BREEDING EVIDENCE

REPORTED IN 18 (1%) OF 1745 PRIORITY BLOCKS

☐	Possible	9 blocks	(50%)
▩	Probable	1 block	(6%)
■	Confirmed	8 blocks	(44%)

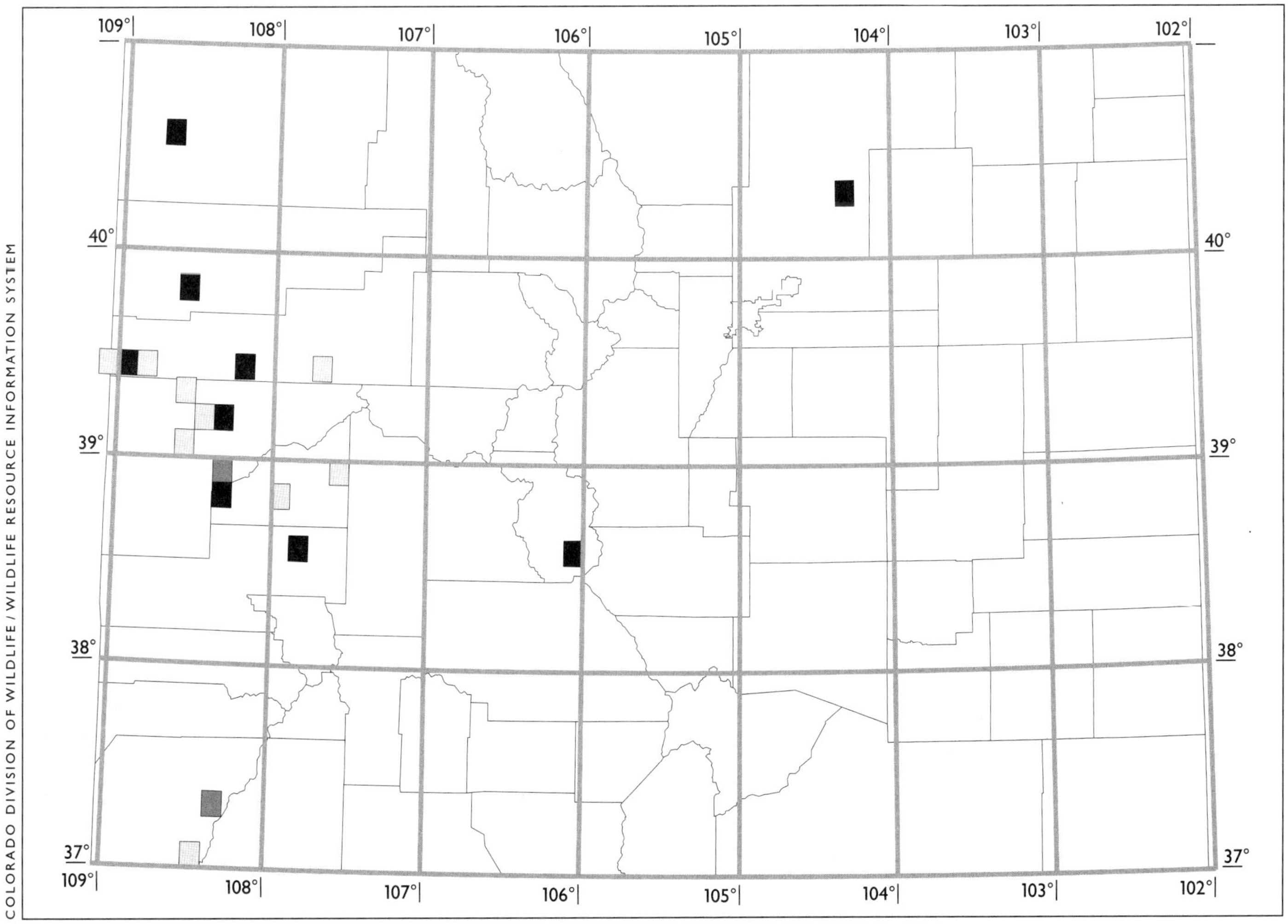
109°
108°
107°
106°
105°
104°
103°
102°
40°
39°
38°
37°
COLORADO DIVISION OF WILDLIFE / WILDLIFE RESOURCE INFORMATION SYSTEM

RING-NECKED PHEASANT *Phasianus colchicus*

For sheer beauty and good eating the Ring-necked Pheasant has no close rival among the upland game birds of Colorado. Its lineage goes back many centuries to wild ancestors in China, and to more recent ancestors in England, where it was introduced in the eleventh century or earlier.

HABITAT

Ring-necked Pheasants prefer agricultural areas, croplands, and hayfields. They do not use open plains and grasslands because they must have cover for nest protection, concealment from predators, and shelter from rough weather. This cover may consist of shelterbelts, thickets, weedy fencerows, edges of marshes, stubble, brush heaps, or even abandoned farm machinery (Terres 1980).

In Colorado, as well as in many other states, the pheasant population has declined

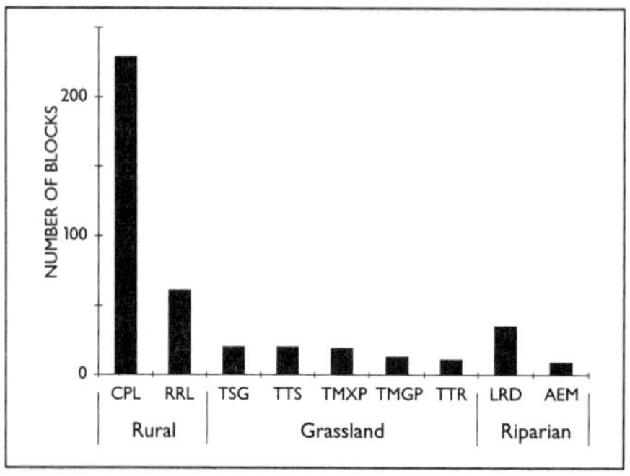
NUMBER OF BLOCKS
200
100
0
CPL
RRL
TSG
TTS
TMXP
TMGP
TTR
LRD
AEM
Rural
Grassland
Riparian

since the 1950s because of changes in habitat that come with modern farming methods (Remington and Snyder 1991). Small farms with their patchiness and variety have given way to large agricultural businesses. Larger fields and cultivation right up to the fencerows leave little weedy cover. Plowing fields immediately after harvesting the crop destroys the protective stubble. Even highway department mowing and weed spraying have reduced the cover needed by the pheasant (Remington and Snyder 1991).

Despite the economics of efficient farming, in more than two-thirds of the Atlas blocks where atlasers found pheasants, the birds occupied croplands and rural settings. In other blocks they used a variety of habitats from riparian stream bottoms and marshes to sagebrush and scrub oak. Croplands in northeastern and eastern Colorado continue to be the center of abundance because of pheasant dietary needs. Waste from cultivated grains supplies the major part of their food. They also eat acorns, pine seeds, wild berries, and green plants (in spring), and grasshoppers, caterpillars, and other invertebrates (Terres 1980).

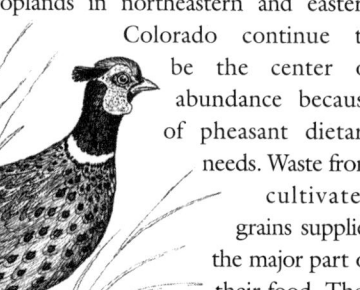

BREEDING

Ring-necked Pheasants are polygynous. Strutting with raised ear tufts and the bare skin about the eyes engorged and brilliant red (Bent 1932), cocks begin their spring courting in Colorado in March. Crowing announces the cocks' territories and attracts the females. One cock will copulate with two to four hens or more and fight other cocks to keep his territory and his harem (Bent 1932).

Nesting may go on from 1 April to 10 September (Nelson 1993). Atlasers found fledged young as early as 15 May. Hens make their shallow scrapes in one cock's territory or crowing area. They hide their nests under weeds or brush, and their protective color and pattern enable them to "sit tight" until almost stepped on (Bent 1932). Neither the females nor their newly hatched precocial young have an odor, a help for escaping marauding dogs, coyotes, and foxes (Terres 1980). Tegetmeier (1911) reported instances of cocks incubating as well as guarding the nest. In the fall, families gather to form flocks of 30–40 birds. These flocks break up in the spring, although hens still remain somewhat gregarious.

The hens' success in concealing their nests accounts for the vast majority of Atlas Confirmations, which consist of fledged young. Atlasers found very few nests.

BREEDING PHENOLOGY

CODE		# OF RECORDS	DATE RANGE
C	Courtship	2	22 Apr–18 Jun
ON	Occupied Nest	4	22 May–28 Jun
NE	Nest with Eggs	4	19 May–13 Jun
FL	Fledged Young	70	15 May–24 Aug

DISTRIBUTION

Widespread in their native China, Ring-necked Pheasants have been introduced worldwide. In North America the first introductions were from China into Oregon in 1881, and from England into the eastern U.S. in 1887 (Bent 1932). W.F. Kendrick brought pheasants to Colorado in

1894 from England, China, and Japan (B&N). They soon flourished in the agricultural land east of the mountains and west of the Continental Divide wherever they found suitable habitat. Although the Latilong Study showed the pheasant as a breeding resident in all but four of the 28 latilongs, newer Atlas data suggest that five latilongs have no pheasants and six have only scattered breeders. This finding coincides with the nationwide population decline that began in the 1950s (Remington and Snyder 1991). BBS data show that the decline continues: a continent-wide drop of 1%/year.

The Atlas map shows that pheasants persist throughout northeastern Colorado except in the shortgrass prairie of the Pawnee National Grassland. They occupy farmlands south to the state line, and west along the Arkansas valley to Pueblo. Agricultural areas in the Grand and Uncompahgre valleys of Mesa, Delta, and Montrose counties and in Montezuma County still have fair numbers. Except at the Rocky Mountain Arsenal NWR, pheasants have largely disappeared from the Denver area as it becomes more solidly urban.

CDOW has initiated determined efforts to increase the pheasant population, not only for hunting but also because of the pleasure non-hunters derive from seeing this lovely bird. Increasing the production and release of pen-raised birds does not work, because these birds fall victim to predators more easily than wild birds, and without improved habitat they would soon disappear (Remington and Snyder 1991). Habitat improvement requires the cooperation of farmers and state agencies. Plowing fields before or after the nesting season and raising the cutter bar to keep small patches of taller stubble will help retain cover. The farmers can establish thickets and deliberately leave uncultivated edges. Spot control of weeds, instead of burning or spraying, can do much to restore the cover these pheasants must have (Remington and Snyder 1991). Whether this necessary cooperation can be achieved between the farmers and the state remains questionable. What is good for the pheasants is not necessarily good for the farmer when he weighs his own economic gain against the Ring-necked Pheasant's habitat needs. Many farmers do value wildlife and make space on their land for pheasants (and for native wildlife from quail to owls to buntings to pronghorns).

BREEDING EVIDENCE

REPORTED IN 345 (20%) OF 1745 PRIORITY BLOCKS

□	Possible	172 blocks	(50%)
▩	Probable	76 blocks	(22%)
■	Confirmed	97 blocks	(28%)

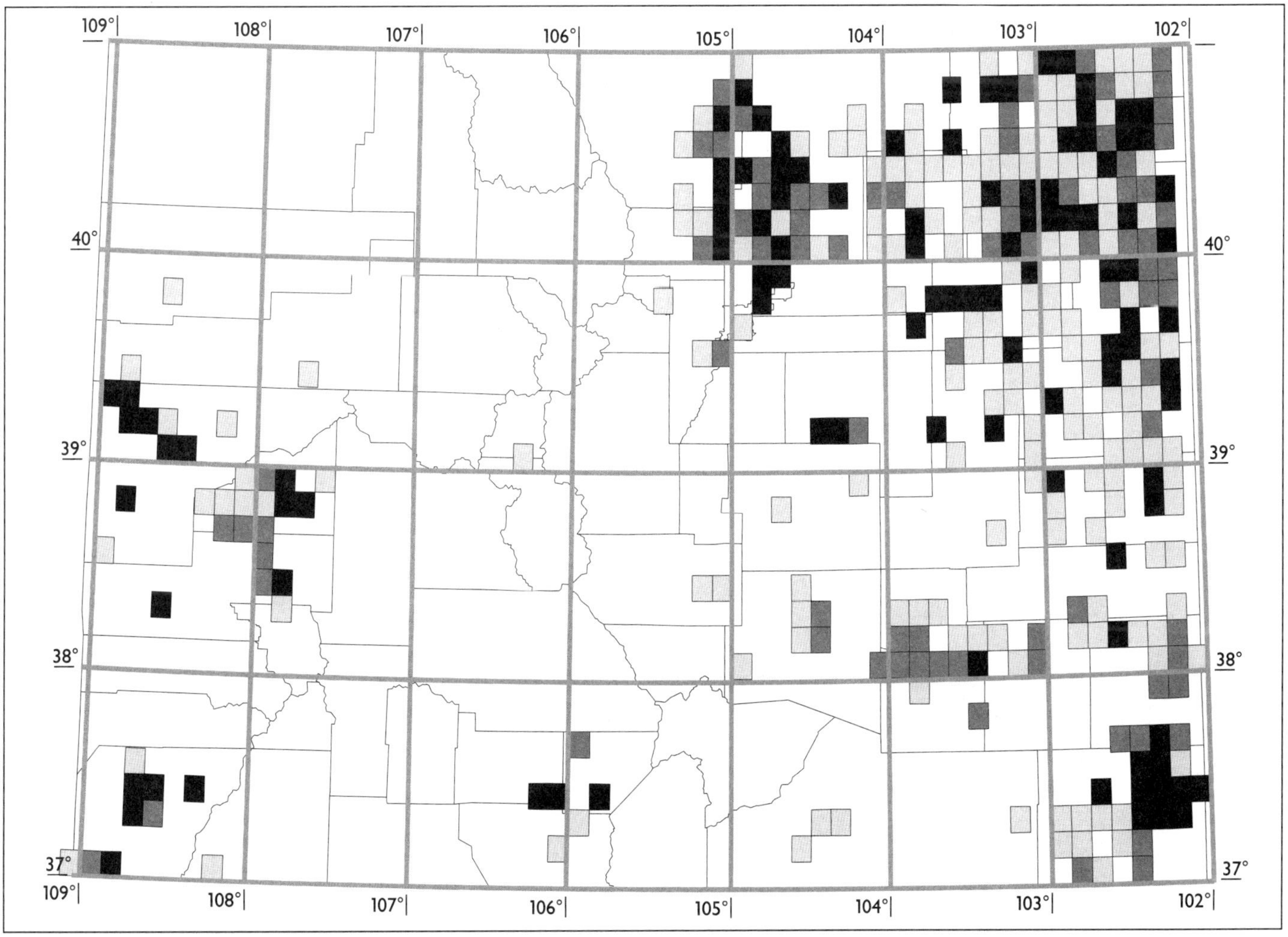
109°
108°
107°
106°
105°
104°
103°
102°
40°
39°
38°
37°

COLORADO DIVISION OF WILDLIFE / WILDLIFE RESOURCE INFORMATION SYSTEM

SAGE GROUSE *Centrocercus urophasianus*

As Lewis and Clark crossed the rolling sagebrush hills of the American West, they encountered Sage Grouse in flocks of thousands (Terres 1980). Hordes of settlers followed close on the explorers' heels, their plows and cows in tow. They converted much of the sagebrush to field and pasture, and the flocks of thousands soon became flocks of hundreds, then of dozens, and now, in many places, they have become merely memories.

HABITAT

Sage Grouse depend totally upon sagebrush-dominated habitats (Ellis et al. 1989, Benson et al. 1993) and prefer large contiguous areas of sagebrush on flat or gently rolling terrain. They avoid very steep slopes and sagebrush intermixed with pinyon/juniper (Rogers 1964). Although Atlas workers reported habitats other than sagebrush in 14 blocks, these habitats all typically intermingle with or stand adjacent to sagebrush. Sage Grouse use all of the three subspecies of big sagebrush (*Artemesia tridentata tridentata*, basin; *A. t. vaseyana*, mountain; and *A. t. wyomingensis*, Wyoming), but seem to have some preference for

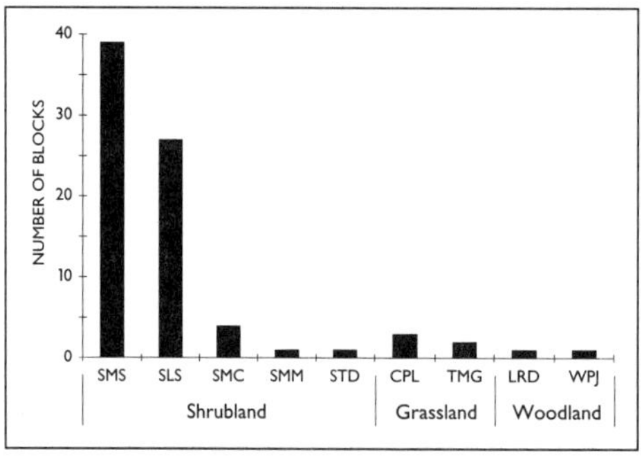

40
30
20
10
0
NUMBER OF BLOCKS
SMS
SLS
SMC
SMM
STD
CPL
TMG
LRD
WPJ
Shrubland
Grassland
Woodland

mountain big sagebrush (Welch et al. 1988, 1991). Atlas workers generally did not discriminate between sage subspecies, and they reported mountain (SMS) and lowland (SLS) sagebrush with almost equal frequency.

At least three habitat factors seem important in maintaining Sage Grouse numbers. Males require tall, mature sagebrush near their leks for day use, and damage to stands of these plants often leads to abandonment of the lek (Rogers 1964, Ellis et al. 1989). The grouse also require large contiguous areas of big sagebrush. When land management practices such as burning, spraying with herbicides, and overgrazing fragment these large contiguous areas of big sagebrush, Sage Grouse populations decline (Braun et al. 1977, Benson et al. 1993, Braun et al. 1994). Finally, they require the presence of tall grass within the sagebrush communities to protect nests from predators, and reducing such cover can negatively affect Sage Grouse nesting success (DeLong et al. 1995).

BREEDING

The spectacular courtship rituals of the Sage Grouse extend from mid March to mid May (B&N). Although dozens of cocks may strut on a lek, only a few dominant ones actually copulate with visiting females (B&N, Wiley 1978). After copulation, each female lays 6–8 eggs at a site as much as 2.5 miles (4 km) from the lek (Wiley 1978). The females incubate for 25–27 days on a nest in a shallow depression beneath a sagebrush (Terres 1980). The precocial young leave the nest soon after hatching, fly at one to two weeks of age, and soon become independent.

The combination of early nesting, well-hidden nests, and precocial young makes Sage Grouse nests difficult to find, and atlasers found none. Sightings of fledged young provided the only breeding Confirmations of this species. Nearly 80% of these sightings occurred between 20 June and 20 July.

BREEDING PHENOLOGY

CODE		# OF RECORDS	DATE RANGE
C	Courtship	11	5 Apr–16 Jun
FL	Fledged Young	34	27 May–5 Aug

DISTRIBUTION

Sage Grouse range from the Canadian border southward through the intermountain west to central Nevada, Utah, and Colorado (Terres 1980). The original distribution of Sage Grouse in Colorado apparently extended to the southern border of the state with populations in at least 23 counties and probably in 27 (Braun 1995). Except in the northwestern corner of the state, forested regions interrupt sage habitats, breaking Sage Grouse range into several discontinuous patches. These birds fare better in broad, continuous stretches of sagebrush, and from our earliest knowledge of them, their population density has decreased from north to south in Colorado as the habitat becomes less homogeneous (Rogers 1964).

Overgrazing and sagebrush removal for irrigated and dryland agriculture have greatly reduced Sage Grouse numbers in the last century (Rogers 1964, Braun 1995, Buchanan 1995). Presently only five counties have stable populations (over 500 breeding pairs), and the species has been greatly reduced in 15 counties and extirpated in several more (Braun 1995). Even in those counties that sustain the highest populations, hunting results and counts of males on leks suggest a precipitous decline in populations—approximately 50% between 1980 and 1993 (Braun 1995).

Articles scheduled for publication in the next year will propose full species status for the "Gunnison" Sage Grouse, which live south of the Colorado River (Clait Braun, Ken Giesen pers. comm.). Most of these birds reside in the Gunnison Basin, with small scattered populations in Montrose, San Miguel, and Mesa counties (Braun 1995). These birds, significantly smaller than the nominate race, differ in feather morphology and in both visual and acoustical mating behavior patterns (Hupp and Braun 1991, Young et al. 1994).

Inasmuch as lek activity ceases by mid May, and most Atlas field work took place in June and July, Atlas work did not census this species very effectively. The Atlas map represents much more data than that gathered by Atlas field workers. Atlasers encountered Sage Grouse in only 64 Priority blocks, and they directly observed courtship, presumably lek activity, in only nine blocks. A survey of lek records of the CDOW Northwest Region's database added records for 23 Priority and 56 Non-priority blocks in 61 additional topographic maps.

BREEDING EVIDENCE

REPORTED IN 87 (5%) OF 1745 PRIORITY BLOCKS

☐	Possible	22 blocks	(25%)
▩	Probable	35 blocks	(40%)
■	Confirmed	30 blocks	(34%)

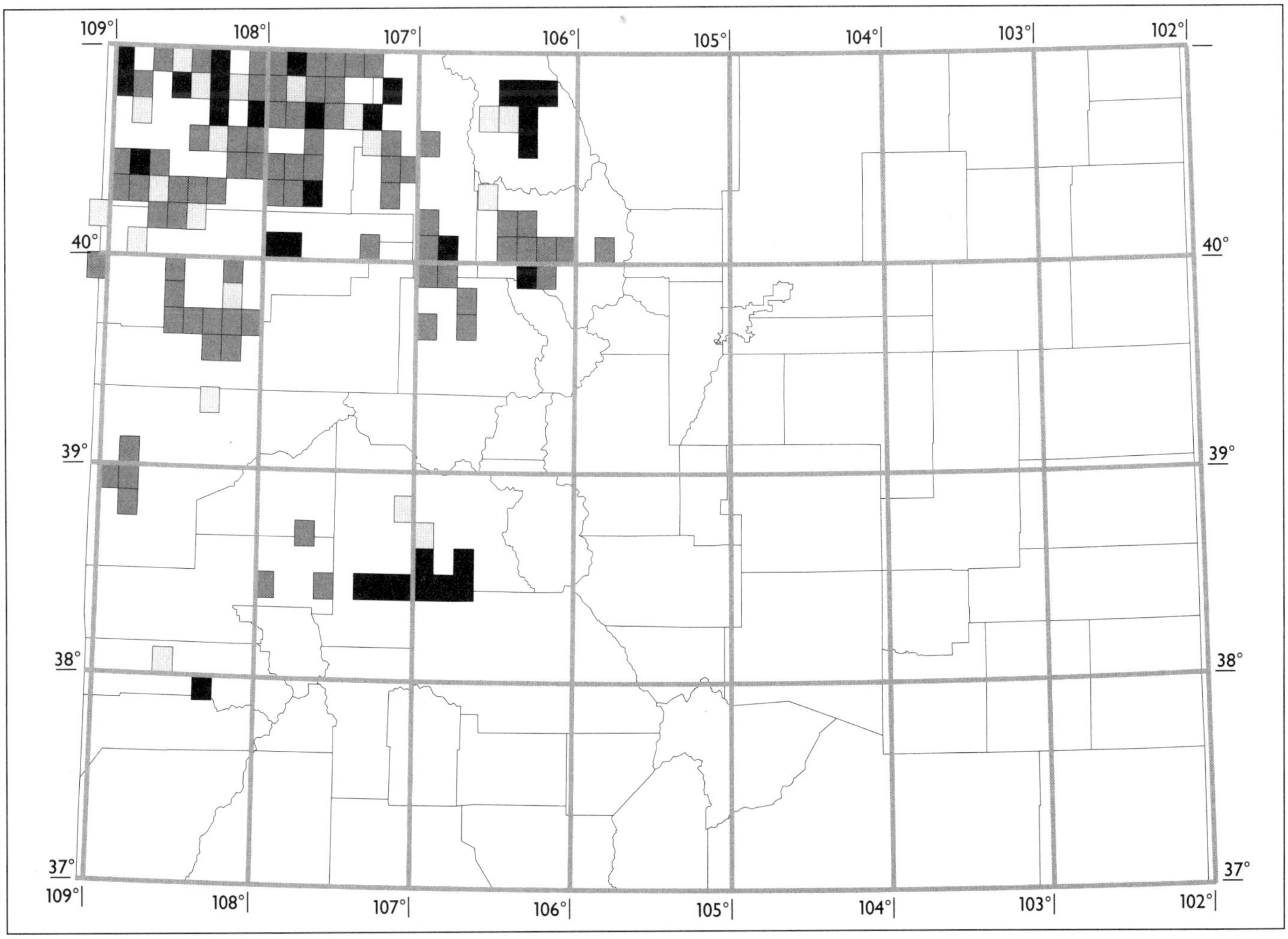
109°
108°
107°
106°
105°
104°
103°
102°
40°
39°
38°
37°

COLORADO DIVISION OF WILDLIFE / WILDLIFE RESOURCE INFORMATION SYSTEM

WHITE-TAILED PTARMIGAN *Lagopus leucurus*

Some people climb Colorado's high peaks "because they are there." Pity these folks! There exists an infinitely better reason to endure windburn, strained muscles, blisters, and oxygen deprivation: catching a glimpse of the elusive White-tailed Ptarmigan.

HABITAT

No other Colorado bird can lay a more conclusive claim to the alpine tundra as home. As the pipits and rosy-finches leave the high country when the brief summer season gives way to winter, ptarmigan, usually, retreat only a few hundred vertical feet to forage among willows, either above timberline or in subalpine sites near tundra regions (Hoffman and Braun 1977). Under extreme winter conditions, they may venture as low as 8,000 feet (2400 m) along streams lined by willows or alders. Although both males and females spend the winter feeding on the tips and buds of willows, males winter closer to the breeding grounds and return there sooner than do females (Braun et al. 1976b).

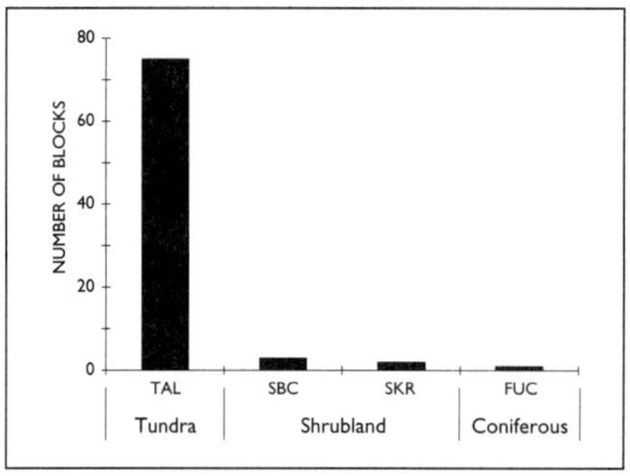
NUMBER OF BLOCKS
80
60
40
20
0
TAL
SBC
SKR
FUC
Tundra
Shrubland
Coniferous

When ptarmigan return to breeding grounds and select a nest site, most locate nests among rock fields or tundra grasses adjacent to sheltering rocks. A small percentage situate their nests among willows or krummholz. Natural depressions, sometimes enlarged by the female, serve as the nest site. To these depressions, they add feathers, leaves, grasses, sedges, and lichens. Ptarmigan tolerate a good deal of slope around the nest site but appear reluctant to use sites with surrounding slopes with a gradient over 40% (Giesen et al. 1980, Braun et al. 1993).

Although Atlas habitat data, at first glance, appear somewhat inconsistent with the discussion above, the anomaly disappears under closer inspection. The TAL code, denominated as grasslands and meadows above timberline, subsumed the rock fields where many ptarmigan nest. This code dominated ptarmigan habitat reports. Also, habitat codes often reflect where birds forage rather than where they nest.

BREEDING

Pair formation among White-tailed Ptarmigan begins in late April when females return to the breeding grounds. Courtship displays continue through the pre-incubation period. Egg-laying begins, in Colorado, in early June (Braun et al. 1993). During the incubation period, the hen spends almost all of her time on the nest and allows humans, and presumably other predators as well, to approach very closely. In some cases, hens refuse to flush from the nest until touched (Giesen and Braun 1979). This reluctance to flush from the nest, combined with a plumage that blends extraordinarily well into the surrounding rocks, makes finding a ptarmigan nest a difficult affair. Atlasers located only five nests.

A typical clutch consists of 4–8 eggs with an average of 6; older females produce somewhat larger clutches than yearlings. Usually during the first two weeks of July, the young emerge from the eggs in a precocial state (Giesen et al. 1980). They leave the nest within 6–12 hours of hatching to forage on their own (Braun et al. 1993).

The chicks grow rapidly but remain with the hen through the remainder of the summer. Atlas fledgling observations extend through early September; hikers in Colorado's high country find hens accompanying their broods throughout September.

BREEDING PHENOLOGY

CODE		# OF RECORDS	DATE RANGE
C	Courtship	1	28 Jun
ON	Occupied Nest	2	6 Jul–13 Jul
NE	Nest with Eggs	3	23 Jun–28 Jul
FL	Fledgling Young	37	1 Jul–28 Aug

DISTRIBUTION

White-tailed Ptarmigan occupy alpine areas from north-central New Mexico northward into the Yukon and southern Alaska. They inhabit all alpine regions of Colorado except the Wet Mountains and the Spanish Peaks. Their presence on Pikes Peak, however, owes to a 1975 transplant project (Hoffman and Giesen 1983). The Atlas map shows them in all these high ranges, but because suitable alpine habitat tends to lie in relatively narrow corridors along mountain ridges, Priority blocks sometimes bypassed extended sections of ptarmigan habitat. Hence, ptarmigan in fact occur in some places where Atlas data depict discontinuous distribution, as in the Park, Sangre de Cristo, and Sawatch ranges.

Most of Colorado's alpine areas remain relatively inaccessible—to atlasers and to serious human impacts. Wilderness designation applies to many of these regions. Yet, even where human impact reaches its highest levels (e.g., Trail Ridge Road and Mt. Evans), ptarmigan maintain a presence. Congregations of corvids attracted to these centers of human activity may pose a greater threat than the human activity per se. An armada of voracious corvids might easily wipe out an entire season's nesting activity within a localized area.

Unfavorable developments on wintering grounds exact a greater impact on ptarmigan populations than do human intrusions onto breeding grounds. Demand for water has increased dramatically in Colorado over the past few generations. This, in turn, has led to the construction of high-altitude reservoirs that have inundated large expanses of willows previously used by ptarmigan for winter forage. Examples include Dillon and Homestake reservoirs and Turquoise Lake (Braun et al. 1976). In some places, such as Rocky Mountain National Park, artificially large populations of elk have damaged extensive stretches of willow habitat on ptarmigan wintering grounds—and, to some extent, on breeding grounds as well (Cynthia Melcher pers. comm.). Other factors that can reduce the availability of winter forage include livestock grazing, road construction along stream courses, and outdoor recreation such as ski-area development and snowmobile activity.

Pressure to expand activities of the sort discussed above will continue as Colorado's population and affluence continue to grow. Without careful planning and management the ptarmigan will face a future plagued by adversity.

BREEDING EVIDENCE

REPORTED IN 66 (4%) OF 1745 PRIORITY BLOCKS

Possible	22 blocks	(33%)
Probable	8 blocks	(12%)
Confirmed	36 blocks	(55%)

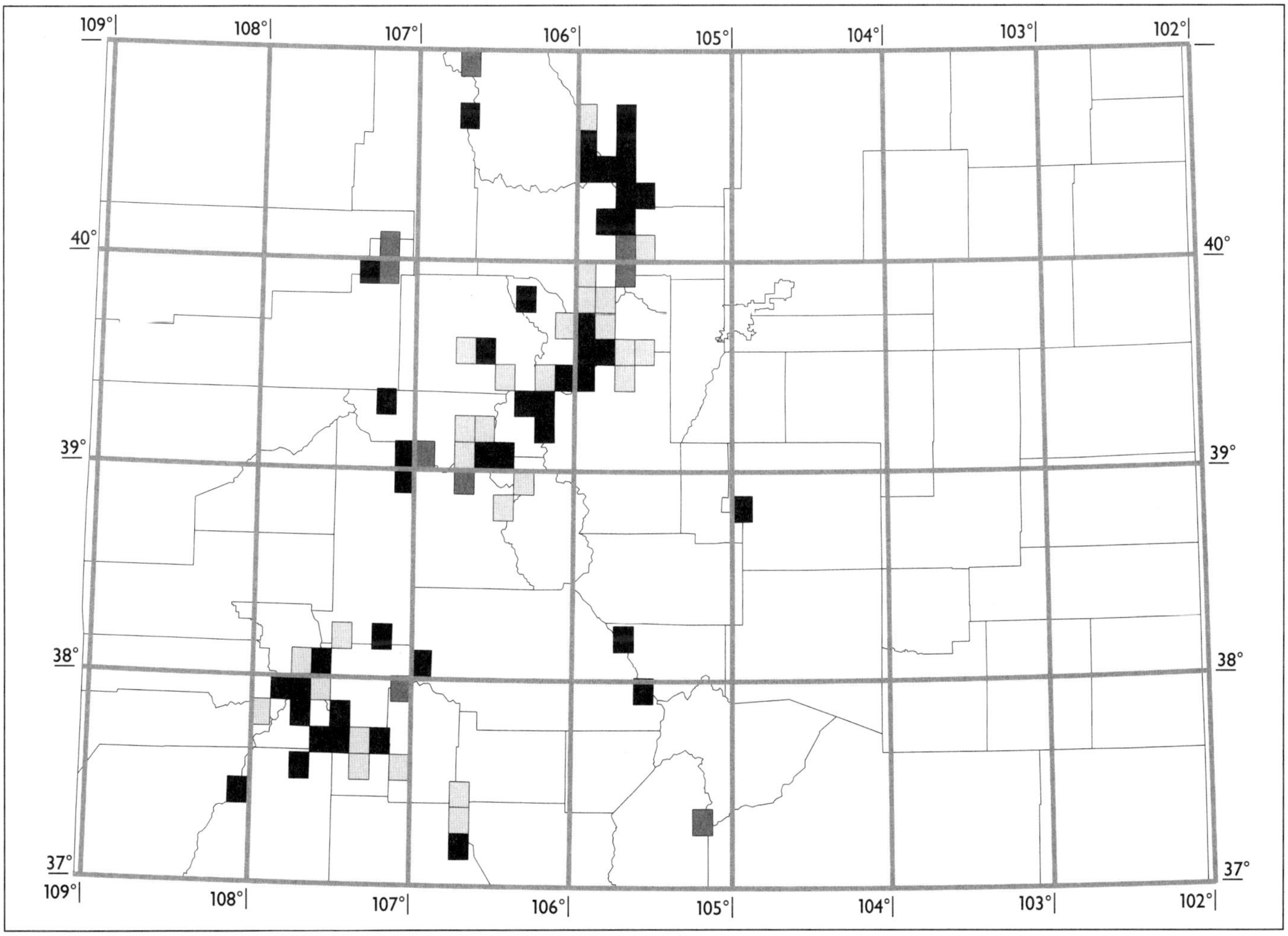
109°
108°
107°
106°
105°
104°
103°
102°
40°
39°
38°
37°

COLORADO DIVISION OF WILDLIFE / WILDLIFE RESOURCE INFORMATION SYSTEM

In winter, Blue Grouse switch the downward movement of many wildlife species for a "reverse migration" pattern; most move to higher elevations in order to find their preferred winter food of conifer needles. In summer, supremely confident in their camouflage, Blue Grouse tend to sidle slowly away from intruders or to explode noisily into the air when approached too closely. When not disturbed they act unwary of humans and provide viewers with leisurely looks.

HABITAT

Blue Grouse breed from the foothills to timberline. They breed in, or along the edge of, nearly all types of forest with relatively open canopies and a shrub understory (Zwickel 1992). Half the Atlas records came from conifer habitats—all types above the pinyon/juniper zone. This includes lodgepole pine, ponderosa pine, Douglas-fir, Engelmann spruce, subalpine fir, and aspen, and any combination of these six species. A quarter of the records came from shrubby habitats, 22% from aspen, and the rest from grasslands. Mountain shrub communities (also important for food and cover) provide ideal breeding habitat; they include scrub oak, serviceberry, chokecherry, mountain mahogany, bitterbrush, and sagebrush, frequently interspersed with aspen. Shrub communities associated with pinyon/juniper woodlands also provide breeding habitat in some areas.

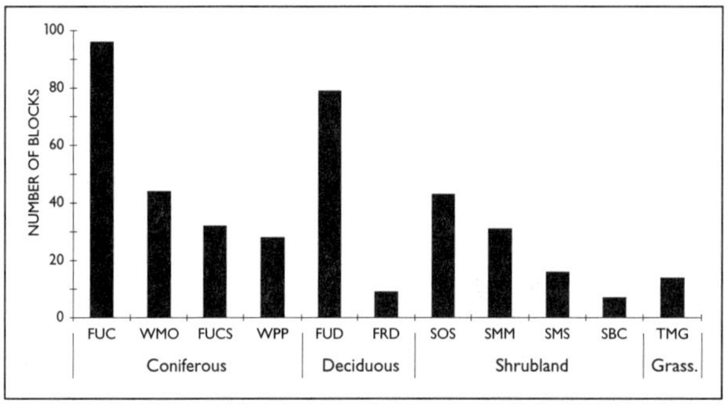
100
80
60
40
20
0
NUMBER OF BLOCKS
FUC
WMO
FUCS
WPP
FUD
FRD
SOS
SMM
SMS
SBC
TMG
Coniferous
Deciduous
Shrubland
Grass.

Primarily ground-dwellers, Blue Grouse feed on a broad variety of plants and insects in summer and conifer needles in winter. They winter almost exclusively in conifer forests, and Douglas-fir needles are their preferred food (Remington and Hoffman 1996).

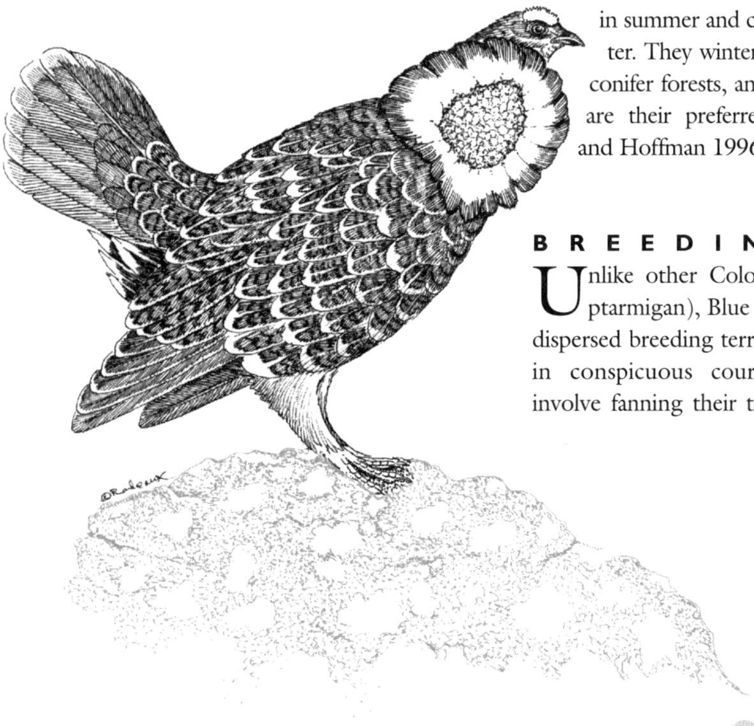

BREEDING

Unlike other Colorado grouse (except ptarmigan), Blue Grouse males defend dispersed breeding territories. Males engage in conspicuous courtship displays that involve fanning their tails, puffing up their air sacs, and "singing." They perform wing-fluttering displays audible for one-quarter mile. Their "blowing-in-an-empty-bottle" hoots, perceptible only within 50 yards (45 m), are heard less frequently. They use these displays and a variety of threat postures and vocalizations to attract females and to discourage neighboring males from entering their territories. Males promiscuously direct their courtship toward any female who comes close. Despite this known male philandering, researchers do not know whether females mate with more than one male (Zwickel 1992). However, a female's home range may include the territories of as many as three males (JFT).

Courtship peaks from mid April to late May, although atlasers reported courtship six times in June and three in July. Daylight length apparently stimulates the males to display; occasionally in October they echo the spring displays (Clait Braun pers. comm., HEK).

The peak hatch in Colorado occurs between 18 June and 12 July (Hoffman 1981). Atlas data conform to this, with 76% of fledgling codes (FL) reported in July. This code represented more than 90% of all Confirmation codes reported—atlasers found only one nest with eggs, and that report lacked a date. Females lay their eggs in inconspicuous scrape nests, well concealed and difficult to find.

Young leave the nest the morning after the entire brood has hatched. They feed themselves, but for a week stay within 5–15 yards (5–15 m) of the supervising hen. They peck at leaves for insects and test different plant and animal foods (Zwickel 1992). Family units disband at the end of summer.

BREEDING PHENOLOGY

CODE		# OF RECORDS	DATE RANGE
C	Courtship	15	19 Apr–29 Jul
NY	Nest with Young	2	2 Jul–6 Jul
FY	Feeding Young	2	28 Jun–9 Jul
FL	Fledged Young	147	24 May–26 Aug

DISTRIBUTION

Blue Grouse are residents in all the major mountains of the western U.S. and Canada, north to the southern Yukon. Most abundant and widespread of Colorado's seven grouse species, they inhabit 43 of the 63 counties (Rogers 1968, Atlas). Recent maps of Blue Grouse distribution in Colorado show a wider distribution than previously thought. Atlas information extends the range beyond Zwickel (1992) and closely matches A&R, which included early Atlas information.

The Atlas map shows that they occur in all the Colorado foothills and mountains west of Longitude 105 and also in all the smaller, isolated ranges, mesas, and plateaus. Western Slope populations tend toward higher densities and more continuous distribution than Eastern Slope populations.

Their tameness, or reluctance to fly, makes Blue Grouse hard to detect. Like ptarmigan, they probably occur somewhat more widely and abundantly than Atlas records depict. Grouse probably occur more contiguously than the map shows in western Moffat County, the Bookcliffs/Roan Plateau area, the Sawatch and Sangre de Cristo mountains, and between the Gunnison and San Luis valleys. Extensive suitable habitat exists to connect these to each other and to the Uinta Mountains in Utah.

Using various definitions, previous authors (Cooke, B&N, A&R) regarded them as "common." Atlasers' abundance codes ranged from A1 to A4, with the vast majority of codes A2s or A3s.

Data from CDOW show that hunter kills of Blue Grouse have dropped in recent decades. Since 1979 Blue Grouse wings collected from hunters have shown a steady downward trend. Despite some upward movement in the late 1980s, the decline resumed by 1995. Numbers may fluctuate over periods of 3–5 years due to weather conditions, but the declines could also result from fewer hunters or fewer birds or both (Hoffman 1995).

Wildlife managers have a poor understanding of, and little success at predicting, population levels, and trends (Zwickel 1992). Hunting apparently has a minor impact on population levels although other activities such as grazing, fire suppression, and habitat alterations impact them. Their numbers and distribution have probably changed little in the twentieth century, although they may have diminished in the nineteenth century due to subsistence hunting by miners, ranchers, and other settlers.

BREEDING EVIDENCE

REPORTED IN 313 (18%) OF 1745 PRIORITY BLOCKS

□	Possible	134 blocks	(43%)
▩	Probable	28 blocks	(9%)
■	Confirmed	151 blocks	(48%)

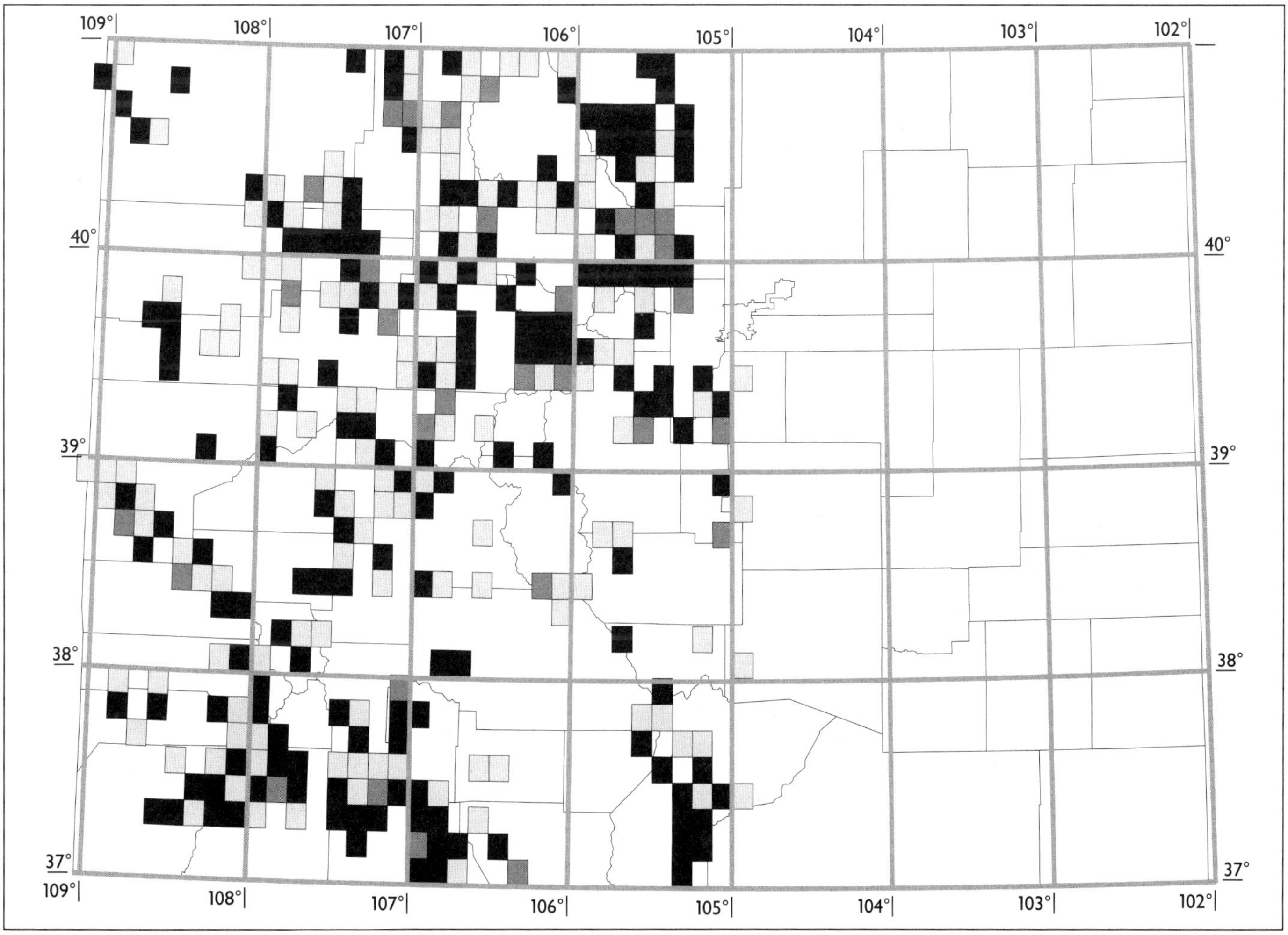
109°
108°
107°
106°
105°
104°
103°
102°
40°
39°
38°
37°

COLORADO DIVISION OF WILDLIFE / WILDLIFE RESOURCE INFORMATION SYSTEM

Prairie Chicken. Willow Grouse. Sage Chicken. Partridge. Early settlers used these and other names for Sharp-tailed Grouse, Sage Grouse, both prairie-chickens, and other species—indiscriminately. The name confusion makes a clear history of Sharp-tailed Grouse hard to track. As it has both a plains and a mountain ("Columbian") subspecies that occupy different habitats, the problem becomes even harder (Rogers 1969). Many Colorado bird species select habitats different from those their counterparts use east of the Rockies, but only one species—the Sharp-tailed Grouse—has two subspecies that, within the state, inhabit such divergent habitats.

HABITAT

Researchers consistently write about shrubs and brush as the dominant habitat component for Sharp-tailed Grouse (Evans 1968, Rogers 1969, Braun et al. 1992). For the Columbian race (*T. p. columbianus*), all Atlas blocks specified sagebrush, and one placed it also in mountain shrublands.

The Plains race (*T. p. jamesi*) historically occupied a narrow strip from El Paso County north to Larimer County, an area "dominated by mixed tall and short grasses with abundant deciduous shrubs" (Braun et al. 1994). Douglas County birds—the only ones of that group left today—use rolling hills with scrub oak thickets and grassy glades. As an equivalent to sagebrush, they use scrub oaks, serviceberries, and willows. These brushy sites provide critical winter shelter and food sources. Douglas County grouse may occupy an atypical habitat; typically the Plains race occupies medium to tall grasslands for courtship and nesting (Ken Giesen pers. comm.). In the northern Weld County blocks the grouse occur in mixtures of native and introduced tall grasses planted through the Conservation Reserve Program (Beth Dillon pers. comm.) and the field card from Logan County specified a grass/sandsage mixture.

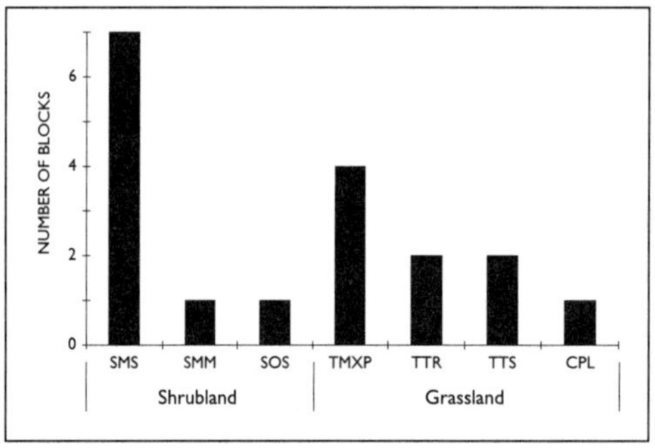
NUMBER OF BLOCKS
0
2
4
6
SMS
SMM
SOS
TMXP
TTR
TTS
CPL
Shrubland
Grassland

Birds of the Columbian subspecies strut on leks in sagebrush; Plains birds strut in grasslands. Patches of shrubs of some kind form common components on these breeding grounds.

BREEDING

Male grouse begin lek dancing in March on the plains and in early April in the mountains; both continue into June. Females give them plenty of time to sort out dominance before they show up in mid to late April to select mates. Typical of lek species, dominant males mate with the most females. Nesting generally occurs fairly close to the lek; few birds travel more than a mile from the site (Evans 1968). The female locates her nest under or near overhanging vegetation. While growing up, the young remain within a half-mile of the nest (Evans 1968).

Populations of the Columbian race fluctuate in numbers considerably from year to year, but over the long term appear stable, based on data from lek counts (Giesen and Braun 1993).

Native Plains Sharp-tail numbers have declined so severely that the population could soon become extirpated. Several factors led to this decline. In the 1800s cattle grazing converted much of their habitat to grass and less suitable brush patterns. In the twentieth century Colorado's rural and urban population booms joined to convert Sharp-tail habitat into farmland, suburban developments, ranchettes, and other altered habitats. Fire suppression and urban sprawl in recent years exacerbated the habitat loss. The Plains Sharp-tail decline has accelerated as Denver's sprawl moves southward (Braun et al. 1994).

CDOW lists the Plains subspecies as endangered, but this does not provide additional protection or funding for recovery. CDOW initiated transplant efforts in northeastern Colorado, but an unexpected problem arose: Greater Prairie-Chickens planted at Tamarack Ranch attracted Sharp-tails from Nebraska and the two started to breed with each other. Interbreeding dilutes the gene pool, accelerating the loss of the pure Plains subspecies.

BREEDING PHENOLOGY

CODE		# OF RECORDS	DATE RANGE
FL	Fledged Young	3	23 May–15 Aug

DISTRIBUTION

Prior to 1880, plentiful Sharp-tails occupied northern Colorado (Cooke 1897) as well as several other states north and west to British Columbia. In 1839 Illinois emigrants en route to Oregon provided the first description of a Sharp-tail in Colorado, a bird along the Blue River near Kremmling (they mentioned the feathered legs; Marsh 1931). The two main units on the Atlas map represent Colorado's remnant populations of the two subspecies.

The Columbian subspecies now occurs only in isolated pockets across its former range. They once occurred across the Western Slope; B&N, presumably in the 1930s, found them numerous on the Uncompahgre Plateau. Today they inhabit only a few spots in five Western Slope counties, a sharp reduction from the historic range. As the Atlas map shows, Routt and Moffat counties have the most. Mesa and Montrose counties still have some birds (Giesen and Braun 1993), but atlasers found them in only one block there.

Plains Sharp-tailed Grouse once roamed the plains north from El Paso and Kit Carson counties, but they "declined dramatically between 1877 and 1887." Douglas County holds the only vestige of the population that originally inhabited the edge of the Front Range. In northeastern Colorado the map shows the Tamarack Ranch transplants and, in two blocks, birds that entered the state from Nebraska and Wyoming. CDOW also transplanted birds to Las Animas County near Raton Mesa (Clait Braun pers. comm.), not on the Atlas map. The isolation of these new populations could prove critical for species survival in a catastrophe, such as a poor breeding season or disease hitting one population.

The isolation of the remaining pockets of Columbian Sharp-tailed Grouse across the continent led to a 1995 proposal to list this subspecies as Threatened under the federal Endangered Species Act (Biodiversity Legal Foundation 1995). This petition remains under consideration. CDOW maintains a limited hunting season on Columbian Sharp-tails.

BREEDING EVIDENCE

REPORTED IN 11 (1%) OF 1745 PRIORITY BLOCKS

Possible	2 blocks	(18%)
Probable	5 blocks	(45%)
Confirmed	4 blocks	(36%)

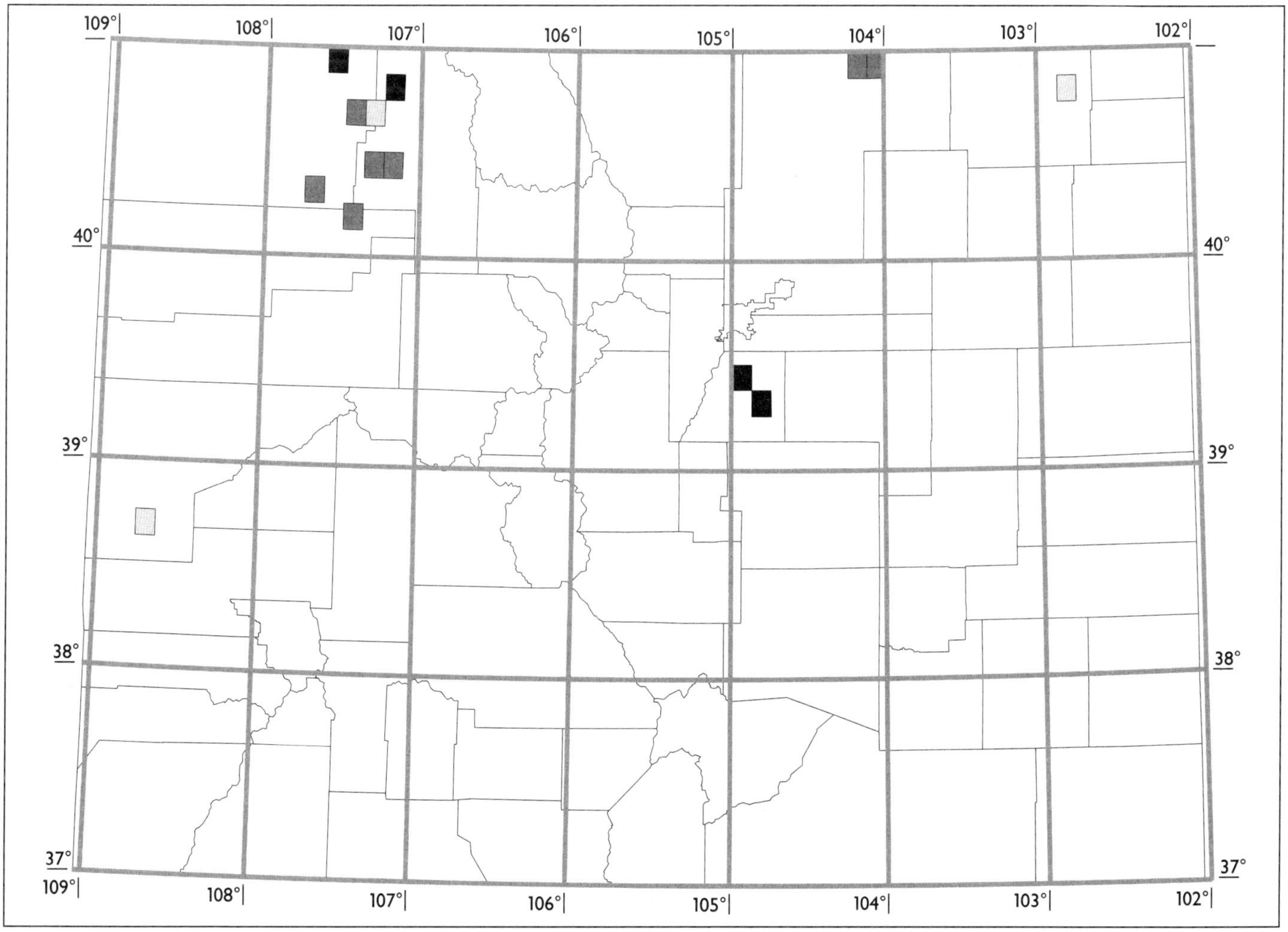
109°
108°
107°
106°
105°
104°
103°
102°
40°
39°
38°
37°

COLORADO DIVISION OF WILDLIFE / WILDLIFE RESOURCE INFORMATION SYSTEM

An August 1863 article in the Sioux City, Iowa, *Register* proclaimed, "never . . . have prairie-chickens been as numerous as the present season," and added that 36 men had recently killed 1,269 in a single day. Within a few decades, cultivation of tallgrass prairies and overhunting had driven Greater Prairie-Chickens from most of their historical range. Meanwhile, cultivation of cereal grains in Colorado enabled these grouse to expand westward into relatively arid and seed-poor sandsage prairies.

HABITAT

Greater Prairie-Chickens breed in midgrass and tallgrass prairies, often near food sources such as scrub oak woodlands, oak forests, and croplands. In northeastern Colorado they nest in sandsage prairie and frequently forage in corn and wheat fields (Van Sant and Braun 1990). About 80% of Colorado Atlas observations occurred in sandsage.

From early March into late May males gather at booming grounds (leks) where they conduct elaborate displays. They raise their neck feathers, stamp their feet, spread their wings, jump into the air, and expand esophageal air sacs to produce low-pitched "booming" vocalizations. Leks often lie on rises or hilltops where relatively short grasses give displaying males and hens a clear view of their surroundings (Schroeder and Robb 1993). After mating, females disperse into the neighboring grasslands to nest.

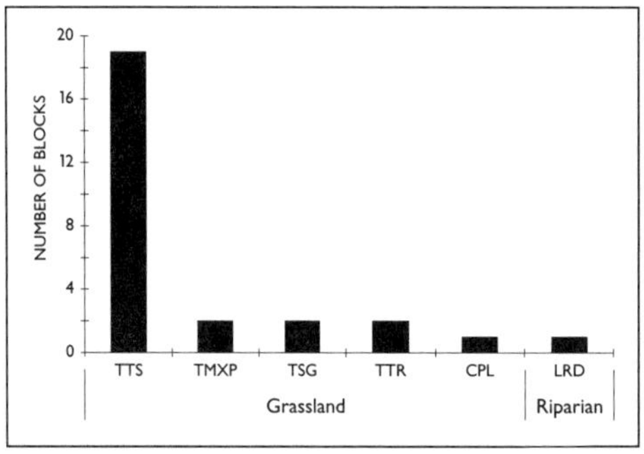
NUMBER OF BLOCKS
20
16
12
8
4
0
TTS
TMXP
TSG
TTR
CPL
LRD
Grassland
Riparian

In late summer both males and hens migrate short distances (usually less than 25 miles [40 km]) to wintering areas. Current late-summer dispersal patterns in northeastern Colorado appear random and show no clear relationship to food availability or weather conditions (Schroeder and Braun 1993).

Highest nesting success occurs in habitats with thick vegetation and few predators (Schroeder and Robb 1993). Predators of adults and chicks include hawks, owls, coyotes, foxes, weasels, badgers, dogs, and cats. Ring-necked Pheasants occasionally parasitize nests; in an Illinois study, parasitized nests failed twice as often as those not parasitized (Vance and Westemeier 1979).

In addition to seeds, Greater Prairie-Chickens feed on leaves, buds, fruits, and insects. Prior to European settlement of the Great Plains, acorn mast comprised a major portion of their winter diet. Consumption of insects increases during the breeding season (Schroeder and Robb 1993).

BREEDING

Hens lay from six to more than 20 eggs in a bowl-shaped depression lined with feathers, dried grass, leaves, and small twigs. After an incubation period of 23–25 days, the chicks hatch and soon leave the nest with the hen. Greater Prairie-Chickens produce only one brood, but the hen may lay a second clutch if predators destroy the first one.

Schroeder and Robb (1993) reported average copulation dates of 20–30 April and nest initiation dates of mid April to mid June throughout the Great Plains region. Schroeder and Braun (1993) observed that most hens nesting in northeastern Colorado disperse from breeding sites by mid July. Atlas field workers observed only one Confirmation, a nest with eggs on 20 May. Landowners supplied the other Confirmations.

BREEDING PHENOLOGY

CODE		# OF RECORDS	DATE RANGE
C	Courtship	5	17 Apr–20 Jun
NE	Nest with Eggs	1	20 May

DISTRIBUTION

The Greater Prairie-Chicken's historical range extended from central Alberta eastward through southern Saskatchewan, southern Manitoba, and parts of southern Ontario; southward through the Ohio and Mississippi valleys to southern Texas and northern Tamaulipas; but west only to central Texas, Oklahoma, Kansas, Nebraska, and the Dakotas. A separate population nested along the eastern seaboard from northern Virginia to Massachusetts.

As European settlers moved west, Greater Prairie-Chickens quickly disappeared from the northeastern states and most of the Ohio and Mississippi valleys. Today, scattered populations hold on in portions of the Midwest from Wisconsin, Michigan, and Illinois west to the Dakotas and northeastern Colorado, and south to Texas.

No evidence exists that Greater Prairie-Chickens occurred historically in Colorado; no reliable records existed until 1897, when J.S. Robertson reported them near Julesburg (Sclater 1912). In 1900 Cooke described this species as "not an uncommon breeder" near Wray. Hersey and Rockwell (1909) observed a nest and collected a specimen at Barr Lake in 1907.

The advent of small-grain cultivation during the late nineteenth century along the South Platte, Republican, and Arikaree river drainages probably enabled these grouse to expand westward into northeastern Colorado. Populations increased into the 1920s, when local residents of Yuma talked of "prairie-chickens by the hundreds, maybe the thousands," and then decreased from the mid 1930s to the mid 1950s. By 1963 CDOW estimated that the statewide population had plummeted to 700–800 birds, and in 1974 the Division declared the Greater Prairie-Chicken endangered within the state. Lek counts conducted by the Division from 1981 to 1983 indicated that the Colorado population had increased to 3,000–6,000 birds (Van Sant and Braun 1990). Reintroduction programs and improved management of livestock grazing have contributed to this species' recovery in northeastern Colorado (Van Sant and Braun 1990).

Colorado Atlas observations correspond with the concentration of current breeding populations in the sandsage prairies of southern Phillips and eastern Washington and Yuma counties. A trap-and-release program undertaken by the CDOW has established separate nesting populations at Tamarack Ranch (two blocks on the map) and Morgan County (one block) and several locations in Weld and Washington counties not shown on the map.

BREEDING EVIDENCE

REPORTED IN 21 (1%) OF 1745 PRIORITY BLOCKS

☐	Possible	12 blocks	(57%)
■	Probable	7 blocks	(33%)
■	Confirmed	2 blocks	(10%)

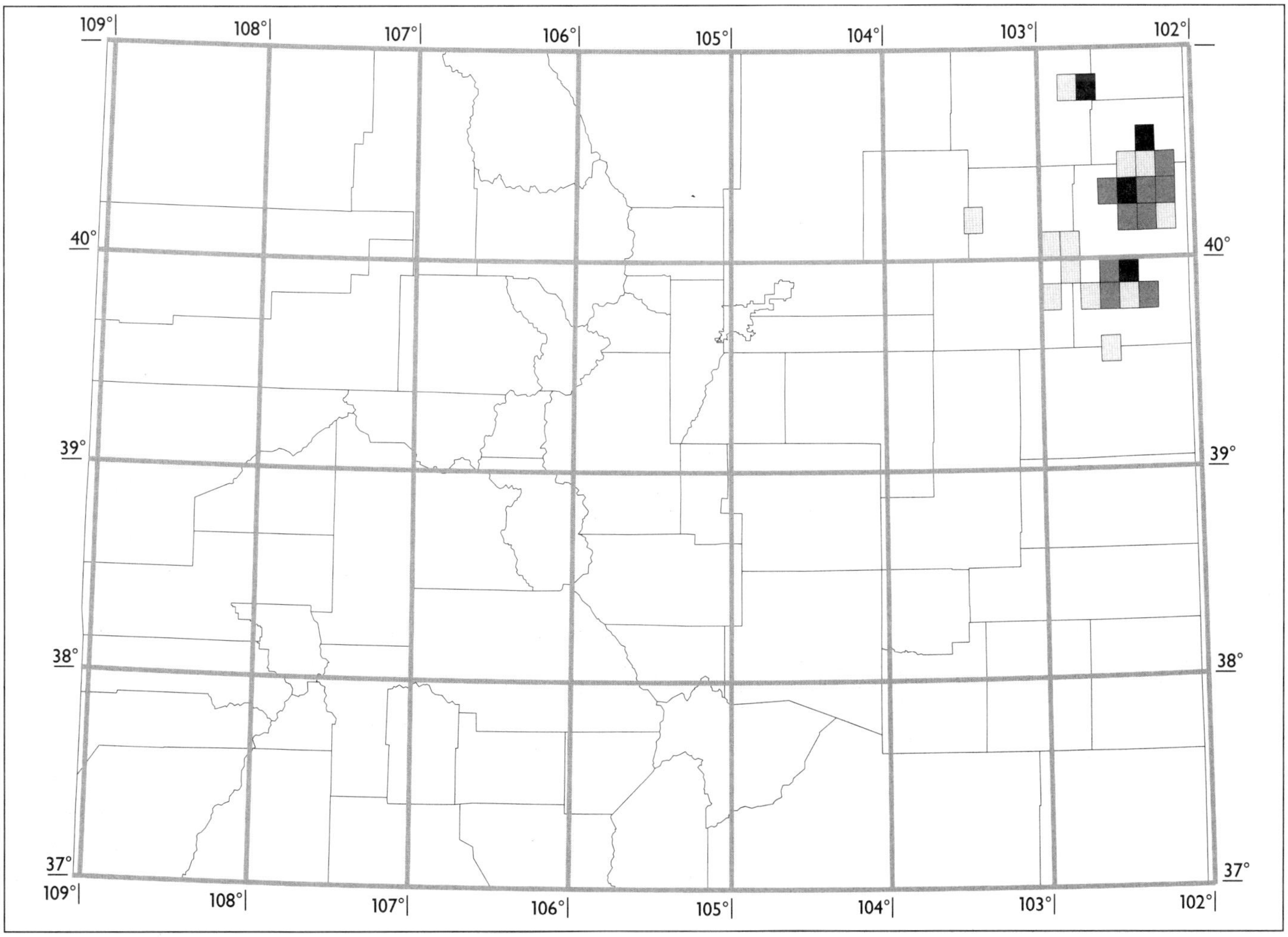
109°
108°
107°
106°
105°
104°
103°
102°
40°
39°
38°
37°

COLORADO DIVISION OF WILDLIFE / WILDLIFE RESOURCE INFORMATION SYSTEM

Unless you already know the location of a lek of Lesser Prairie-Chickens in southeastern Colorado, you will have a difficult time finding one. Nevertheless, they display on 50 or so hilltops and their courting dances provide fascinating watching at daybreak in April and May.

HABITAT

In Colorado Lesser Prairie-Chickens depend on the sandsage/grassland community (A&R). Optimal habitat consists of midgrass to tallgrass prairies for nest and winter cover. The forb and shrub component of these rangelands provide foraging substrate and nesting habitat (Giesen 1994b). In New Mexico and Texas, shinnery oak is also part of their habitat. In southeastern Colorado grazing, plowing, and drought (especially the Dust Bowl of the 1930s)

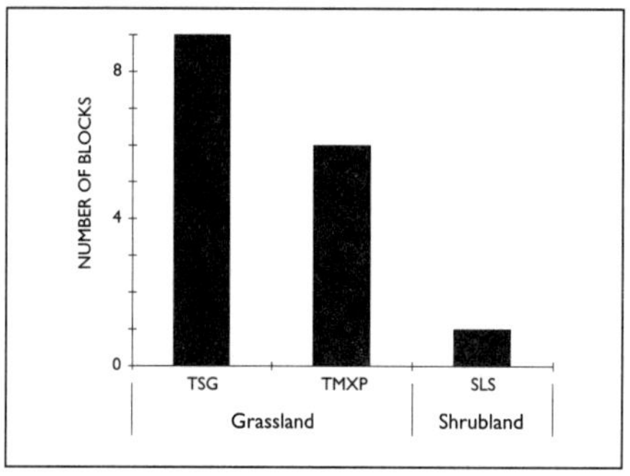
NUMBER OF BLOCKS
8
4
0
TSG
TMXP
SLS
Grassland
Shrubland

changed much of the nineteenth-century tallgrass and midgrass habitat to a shortgrass dis-climax. In the last 50 years, better management of private lands and the protection of the grasslands by the establishment of the Comanche National Grassland (managed by the U.S. Forest Service) have provided habitat that again includes mixed grasses with the shortgrass prairie (Hoffman 1963).

Some grazing can occur without adverse effects if overgrazing is prevented (Rakestraw 1995). Research in Kansas determined that a mix of 75% grassland and 25% cropland provided optimum habitat for Lesser Prairie-Chickens (Horak 1984). Leks (display grounds) are located within a short distance of nesting habitat on slightly higher ground with shorter grass or on a disturbed area. Atlas observers indicated both shortgrass and mixed-grass habitat in leks censused, with low sagebrush in one area.

Grasshoppers and insects provide the primary food during the summer, whereas winter food consists largely of plant materials such as seeds and leaves, and grain and milo from farmers' fields.

BREEDING

Breeding Lesser Prairie-Chickens, like other grouse, are polygynous. The males attract the females to the leks with elaborate displays. They inflate their reddish-purple air sacs and yellow combs as they crouch and spread their wings, stamp their feet, and leap into the air. All but one block on the Atlas map represent lek observations, with most data provided by CDOW (Ken Giesen pers. comm.).

The nest consists of a scrape in the sandy soil; the hen lines it with grass and hides it well under a grass clump or small shrub. Clutches average 12 eggs, which hatch in 24–26 days (Rakestraw 1995). The precocial young stay with the female for six to eight weeks (Copelin 1963). Their home range occupies an estimated 160–256 acres (65–104 ha). In winter flocks wander out of the home range in search of food. During the adult molting period, the juveniles form separate flocks (Taylor and Guthery 1980). Later the older males reestablish territories and leks, with juveniles as observers.

BREEDING PHENOLOGY

CODE		# OF RECORDS	DATE RANGE
C	Courtship	1	15 Apr
NE	Nest with Eggs	1	19 May

DISTRIBUTION

Where the habitat meets their specific needs, Lesser Prairie-Chickens range over western Kansas, southeastern Colorado, the Oklahoma Panhandle, western Texas, and eastern New Mexico. Their numbers have plunged; in 1904 one observer claimed that he saw 15,000–20,000 in one southwestern Kansas field (Bent 1932), but 75 years later the total population had plummeted to 40,000 (Terres 1980).

Although Cooke (1909) reported prairie-chickens in Baca County, presumably Lessers, the first confirmed record came from a bird collected in Baca County by Frederick Lincoln in 1914. When the grasslands were unbroken they were probably "fairly common" residents in southeastern Colorado (B&N). Then, they probably covered their present range in Baca and Bent counties and, north of the Arkansas River, went into Crowley, Kiowa, Cheyenne, and

Lincoln counties (Hoffman 1963). In the Dust Bowl years of the 1930s their range decreased about 65% from pre-settlement days and populations declined by an estimated 97% (Taylor and Guthery 1980).

CDOW listed them as "Threatened in Colorado" in 1973 (Giesen 1994a). Since the listing in 1973 populations have slowly increased, due to better habitat management and habitat protection on Comanche National Grassland (Braun et al. 1994). The Colorado population in the late 1980s was estimated at 1,000 to 2,000 individuals (Giesen 1994a). The Atlas map depicts their present range in Baca, Prowers, and Kiowa counties. A study by CDOW from 1986 to 1990 found 58 active display grounds—40 of them in Baca County and most near Campo on Comanche National Grassland (Giesen 1994a).

On this grassland in southern Baca County the U.S. Forest Service provides special management for prairie-chickens, with a goal of maintaining the sandy rangeland in good to excellent condition with a diversity of plant life-forms—largely a function of grazing management (Taylor and Guthery 1980). CDOW has attempted to transplant flocks into similar non-federal habitat in southeastern Colorado with so-far unknown success; the Atlas map does not include those locations. CDOW proposes to continue research into the habitat requirements and to plan for possible additional transplants (Braun et al. 1994).

BREEDING EVIDENCE

REPORTED IN 9 (1%) OF 1745 PRIORITY BLOCKS

☐	Possible	0 blocks	(0%)
■	Probable	8 blocks	(89%)
■	Confirmed	1 block	(11%)

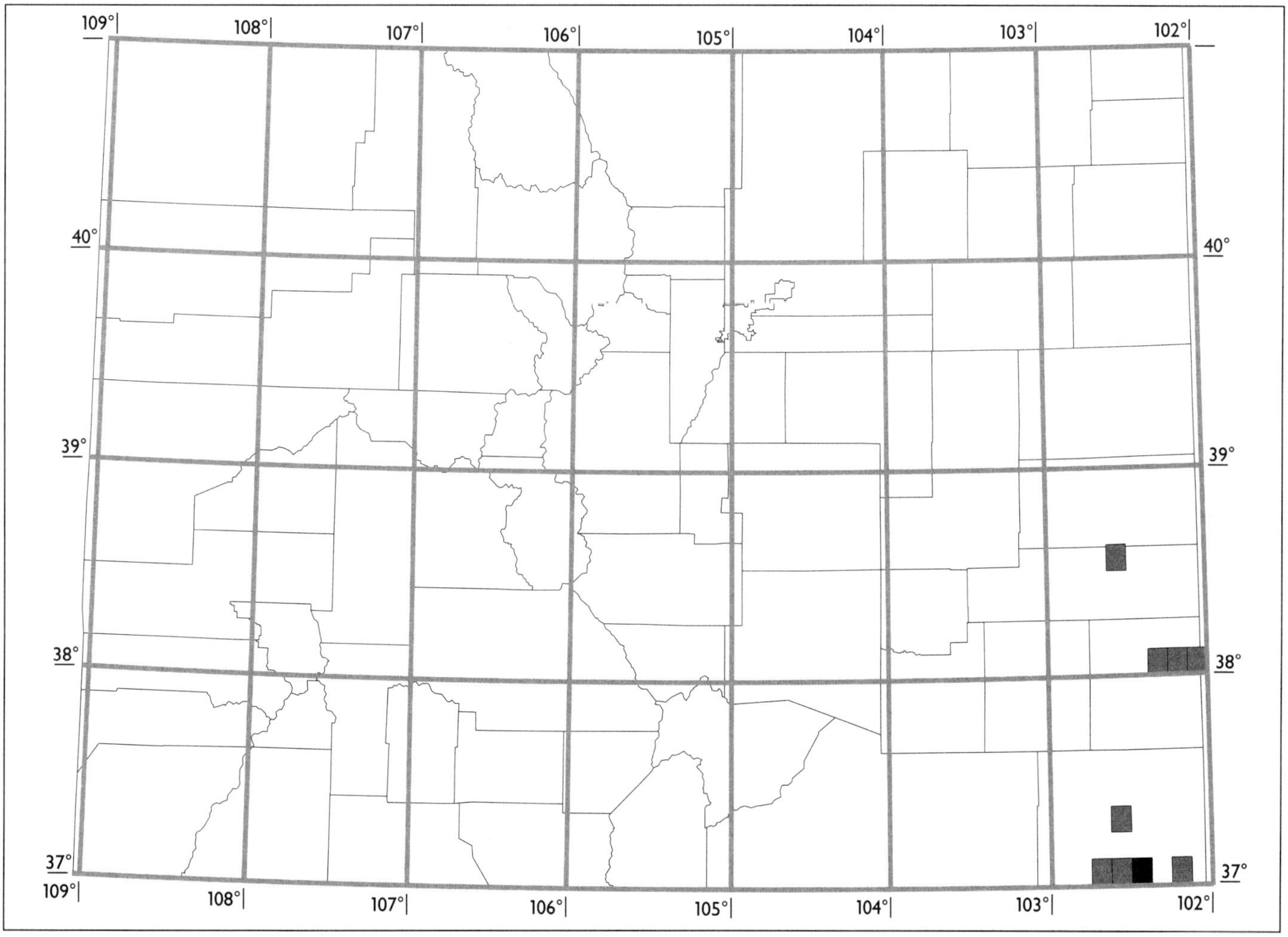
109°
108°
107°
106°
105°
104°
103°
102°
40°
39°
38°
37°

COLORADO DIVISION OF WILDLIFE / WILDLIFE RESOURCE INFORMATION SYSTEM

An encounter with Wild Turkeys impresses almost everyone, even people who don't ordinarily notice birds. Turkeys evoke powerful images—a bountiful Thanksgiving feast, or gobblers heard in a misty spring dawn. Thousands of Colorado hunters pursue turkeys each year; Native Americans also hunted turkeys and used their feathers and bones for clothing, tools, and jewelry (Dickson 1992).

HABITAT

Merriam's Turkeys (*M. g. merriami*), the subspecies native to Colorado, range primarily in dry forests of broken, mountainous terrain to about 8,000 feet (2440 m; Hoffman et al. 1993, A&R). Atlasers found Merriam's Turkeys most often in forested habitats (85% of observations), primarily lower-elevation conifers and oakbrush. In southwest Colorado scrub oak commonly grows as an understory shrub or patchy co-dominant of ponderosa pine, providing both important cover and mast. Riparian deciduous forests, usually cottonwoods, account for 16% of Merriam's records. Turkeys forage rather than breed in the shrub, cropland, and grassland habitats listed by field workers. Merriam's Turkeys search in forest openings or agricultural lands for leaves, seeds, and invertebrates (Hoffman 1962). Use of unforested areas or pinyon/juniper woodlands depends in all seasons on proximity to stands of tall roost trees, usually ponderosa pines or cottonwoods (Ligon 1946, Burget 1957, Hoffman 1968). They seldom breed in pinyon/juniper unless it adjoins high-altitude ponderosa areas (Eaton 1992).

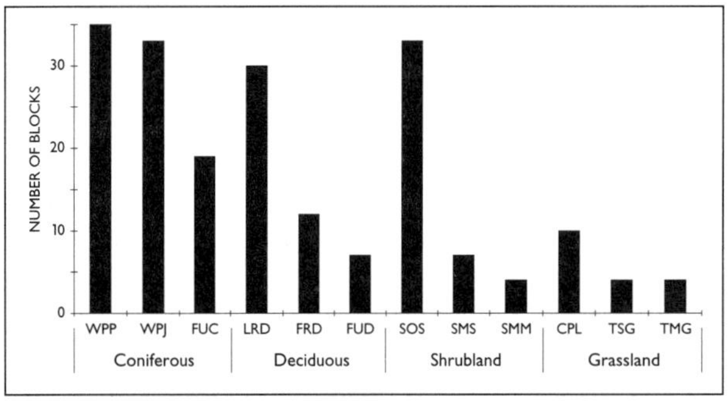
NUMBER OF BLOCKS
0
10
20
30
WPP
WPJ
FUC
LRD
FRD
FUD
SOS
SMS
SMM
CPL
TSG
TMG
Coniferous
Deciduous
Shrubland
Grassland

To increase hunting opportunities, CDOW transplanted Rio Grande Turkeys (*M. g. intermedia*), native to Kansas, Oklahoma, and Texas, to forests that now line major rivers of Colorado's eastern plains. Atlas records confirm the birds' close association with riparian cottonwoods (68% of records), where flocks roost year round. Rio Grande Turkeys forage in river bottoms but also visit adjacent crop fields and rangeland (Schmutz 1988), as Atlas records in open habitats confirm.

BREEDING

In spring mixed-sex winter flocks break up into smaller all-male or all-female groups (Dickson 1992). Turkeys do not defend territories—males fight for breeding opportunities rather than real estate. Courtship, including the well-known gobbling call, begins in April (Hoffman 1990). Hens lay 9–12 eggs in a shallow scrape and frequently dump eggs in each other's nests. They incubate for 26–28 days, and the precocial poults leave the nest within a day of hatching (Dickson 1992). For the first few days the hen feeds the poults and they roost close to and under her outspread wings (Eaton 1992). Turkeys conceal the nest well—atlasers found just two nests with eggs, both in late May. Atlas workers most often Confirmed breeding by observing the conspicuous poults in feeding flocks throughout summer.

BREEDING PHENOLOGY

CODE		# OF RECORDS	DATE RANGE
C	Courtship	4	10 May–1 Jul
ON	Occupied Nest	1	28 Jun
NE	Nest with Eggs	2	23 May–31 May
FL	Fledged Young	43	28 May–8 Oct

DISTRIBUTION

Following transplants by wildlife agencies Wild Turkeys now inhabit all the conterminous United States, southern Canada, and north-central Mexico (Dickson 1992), including previously unoccupied range in the western states and Canada. As early as A.D. 500 in southwest Colorado, pueblo dwellers kept turkeys in pens; in fact Colorado's "native" Merriam's Turkeys may be feral descendants of domestic strains kept by pre-Columbian Indians (McKusick 1980). Colorado's European explorers and early settlers found Merriam's Turkeys on the Arkansas River near Lamar, on the Front Range, in the Uncompahgre and San Juan river valleys, and elsewhere (Burget 1957, B&N). Unregulated hunting, grazing, and diseases took a heavy toll and by 1930 few turkeys survived in Colorado. Transplants by CDOW beginning in the 1930s restored Merriam's Turkeys to approximately their historic distribution, and the agency introduced Rio Grande Turkeys to the Great Plains (Braun et al. 1994).

Game managers have thoroughly documented the Wild Turkey's Colorado distribution (Braun et al. 1994), and atlasers found turkeys in all major regions where known populations exist (Clait Braun pers. comm.). Atlasers reported the most Merriam's Turkeys in southern Colorado and the Front Range, particularly in habitats with scrub oak mixed with ponderosa pines; scattered blocks north of Grand Junction also had them. Field workers located the introduced Rio Grande Turkeys in a smattering of riparian woodlands in plains blocks along the South Platte, Republican, and Arkansas rivers.

However, turkeys occur more widely than Atlas records indicate in Dolores County, the San Juan Basin, and the Uncompahgre Plateau. Because turkeys are rather conspicuous, low densities partially account for discontinuity of Atlas records. Furthermore, Atlas Priority blocks on the plains often fell outside of narrow riparian corridors, causing discontinuous distribution records for Rio Grande Turkeys on the Arkansas and South Platte rivers.

Although distribution of Merriam's Turkeys in Colorado has not changed much in the past 120 years, density has fluctuated markedly. CDOW and hunters' groups have reversed sharp declines of the past by transplants and habitat management (Braun et al. 1994), and turkeys now occur in virtually all suitable habitats in Colorado. On the northern Front Range and Great Plains, where most birds inhabit private lands, grazing and development erode Wild Turkey habitat (Culotta and Hoffman 1989, Braun et al. 1994). Intensive grazing depletes food supplies and reduces herbaceous and shrub structure necessary for cover. Turkeys thrive elsewhere, generally on public lands, although managers suspect that grazing and disease magnify the effects of harsh winters on some flocks (Hoffman et al. 1993). In the long run, their widespread appeal and economic value will encourage the conservation of Colorado's Wild Turkeys.

BREEDING EVIDENCE

REPORTED IN 142 (8%) OF 1745 PRIORITY BLOCKS

☐	Possible	81 blocks	(57%)
▩	Probable	14 blocks	(10%)
■	Confirmed	47 blocks	(33%)

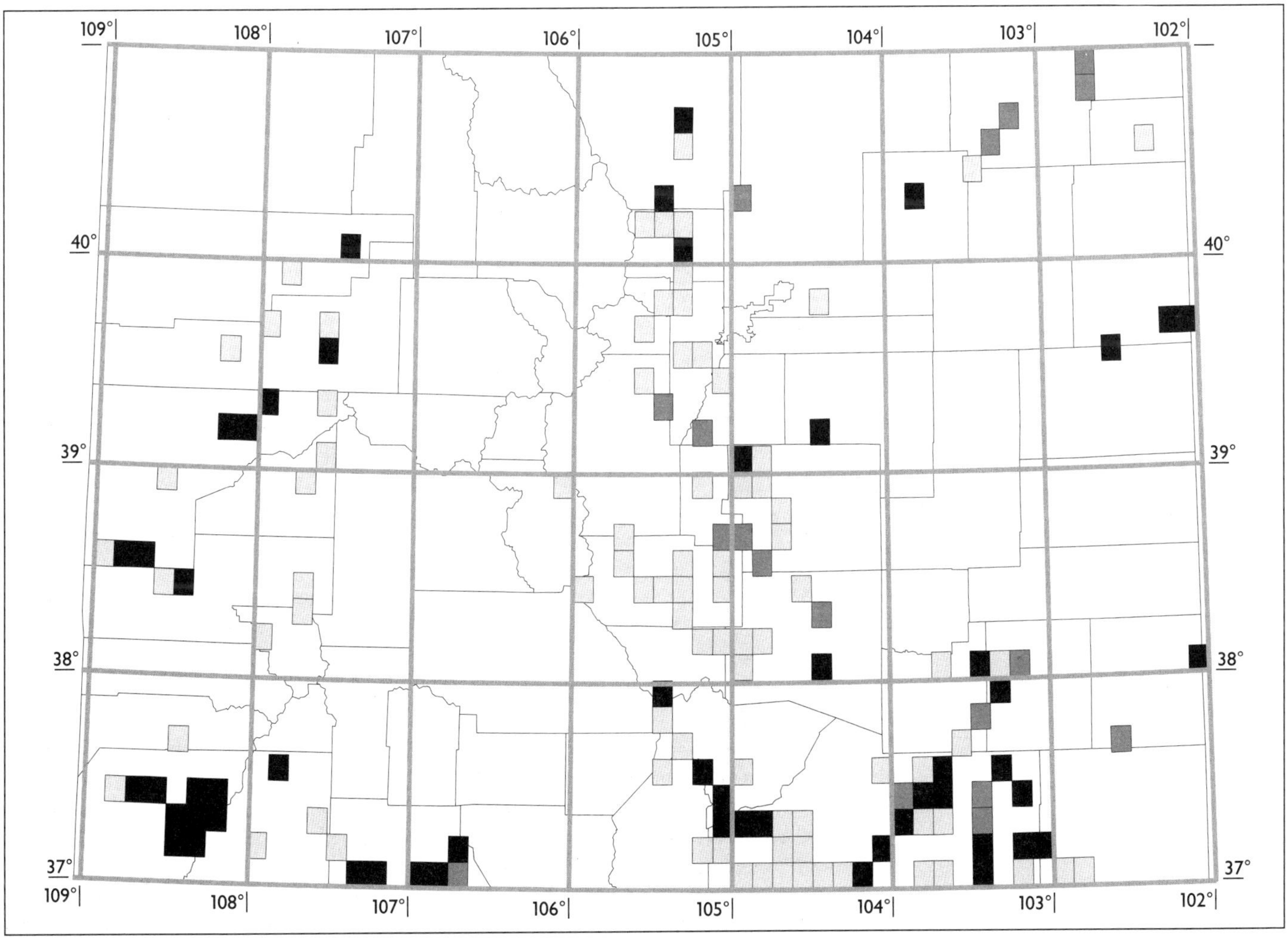
109°
108°
107°
106°
105°
104°
103°
102°
40°
39°
38°
37°

COLORADO DIVISION OF WILDLIFE / WILDLIFE RESOURCE INFORMATION SYSTEM

Like windup toys, their feet a twinkling blur, Scaled Quail escape their enemies by running rapid zigzags across the barren ground. These blue-gray quail, familiarly known as "cottontops" for their up-standing white or creamy crests, seldom burst into flight. Although game birds, they often become tame around farms and ranches.

HABITAT

Scaled Quail center in the Chihuahuan Desert, a place characterized by creosote bushes, scattered cacti, and yuccas—a habitat that extends from Mexico into New Mexico and western Texas (Schemnitz 1994). As their second choice in habitat they pick desert-like grasslands with scattered cacti and shrubs. The most important Colorado habitats are sagebrush in sandy soils, followed by dense cholla cactus and yucca stands in short grasslands and open pinyon/juniper woodlands (Hoffman 1965). Although Atlas workers Confirmed breeding

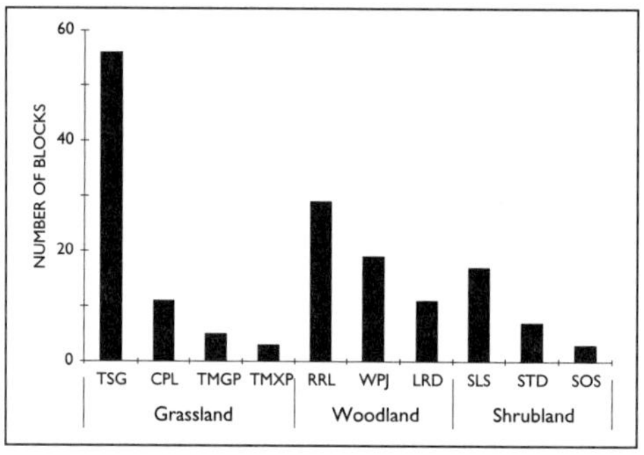
NUMBER OF BLOCKS
60
40
20
0
TSG
CPL
TMGP
TMXP
RRL
WPJ
LRD
SLS
STD
SOS
Grassland
Woodland
Shrubland

in all of these habitats, their observations showed that these quail use short grasslands with cholla cactus three times as often as sandsage and pinyon/juniper woodlands. "Cottontops" also use yucca, sage, and stubblefields for winter feeding. Brush piles, manmade structures, trash piles, and even abandoned vehicles and machinery provide shelter for their nests.

About 30% of the Scaled Quail diet consists of insects; seeds, grain, and leafy material complete the menu. These quail usually forage on the ground, but they will perch in small trees and bushes to feed on seeds and leaves when snow covers the ground (Schemnitz 1961). They usually feed in the early morning and late afternoon until dusk and remain quiet during the middle of the day.

BREEDING

At the end of the summer, Scaled Quail gather in flocks of as many as a hundred birds (Schemnitz 1961). In Colorado the flock stays together until pairing begins in March or early April. A pair will stay together monogamously during the breeding season, which may last from mid April through September in the southeastern corner of the state. Atlas observations verify these dates, with fledged young discovered as early as 15 April and as late as 14 September.

Hens make a shallow scrape lined with grasses under any available protection. The well-camouflaged nests defy detection, so the FL code provided 93% of the Confirmed Atlas observations. The cocks often act as lookouts by perching on fence posts or trash piles; they give an alarm "chip" call to warn the hens brooding or leading the chicks (Johnsgard 1973). Hens often nest again while chicks of the first broods gather in coveys. The cocks rarely share in incubation, but in one instance, at least, a male tended the first brood after hatching while the female began a second nesting (Wallmo 1956). The coveys of young quail gradually merge with other young quail until the larger winter flocks of adults and young form.

BREEDING PHENOLOGY

CODE		# OF RECORDS	DATE RANGE
NE	Nest with Eggs	1	28 Jun
NY	Nest with Young	1	15 Jun
FL	Fledged Young	37	15 Apr–14 Sep

DISTRIBUTION

Primarily Mexican birds, Scaled Quail live year round on the high desert plateaus of the Chihuahuan Desert as far south as the Valley of Mexico near Mexico City. Their range extends northward to southeastern Arizona and across New Mexico to southwestern Kansas.

In 1897 Scaled Quail, then named Scaled Partridges, were listed as rare in Colorado, with only one specimen then existing (Cooke 1897). Now they occupy the southeastern quadrant of the state. The Atlas map shows them solidly entrenched as far north as Cheyenne, Lincoln, and El Paso counties and west up the Arkansas drainage to the Cañon City area. Their Colorado range reportedly extends as high as 7,000 feet (2134 m; Hoffman 1965), although the Atlas map does not reflect that. Atlasers found a few in Montezuma County, introduced according to A&R, but that remains an unverified assumption; Scaled Quail also occur around Farmington and Shiprock, New Mexico (Jacobs 1986).

Hunters periodically attempt to introduce these quail to other parts of the state; the

Washington County Atlas record probably represents such an attempt. Attempts to increase the range of these quail in Colorado generally fail. B&N thought them poor game birds because of their habit of running away at high speed rather than hiding and then flushing. They thought that the hunter had little chance of getting his quarry.

BBS data show a significant downward trend across the Scaled Quail's range: a continent-wide 3.4%/year drop, which translates to a decline of 62% from 1966 to 1996. A study in New Mexico indicates that hunting has little effect on population numbers (Campbell et al. 1973). Lack of cover and food caused by overgrazing may contribute more to this decline (Snyder 1967). Guidance for more effective management practices awaits more detailed, thorough, and careful studies (Schemnitz 1994).

BREEDING EVIDENCE

REPORTED IN 139 (8%) OF 1745 PRIORITY BLOCKS

☐	Possible	40 blocks	(29%)
■	Probable	52 blocks	(37%)
■	Confirmed	47 blocks	(34%)

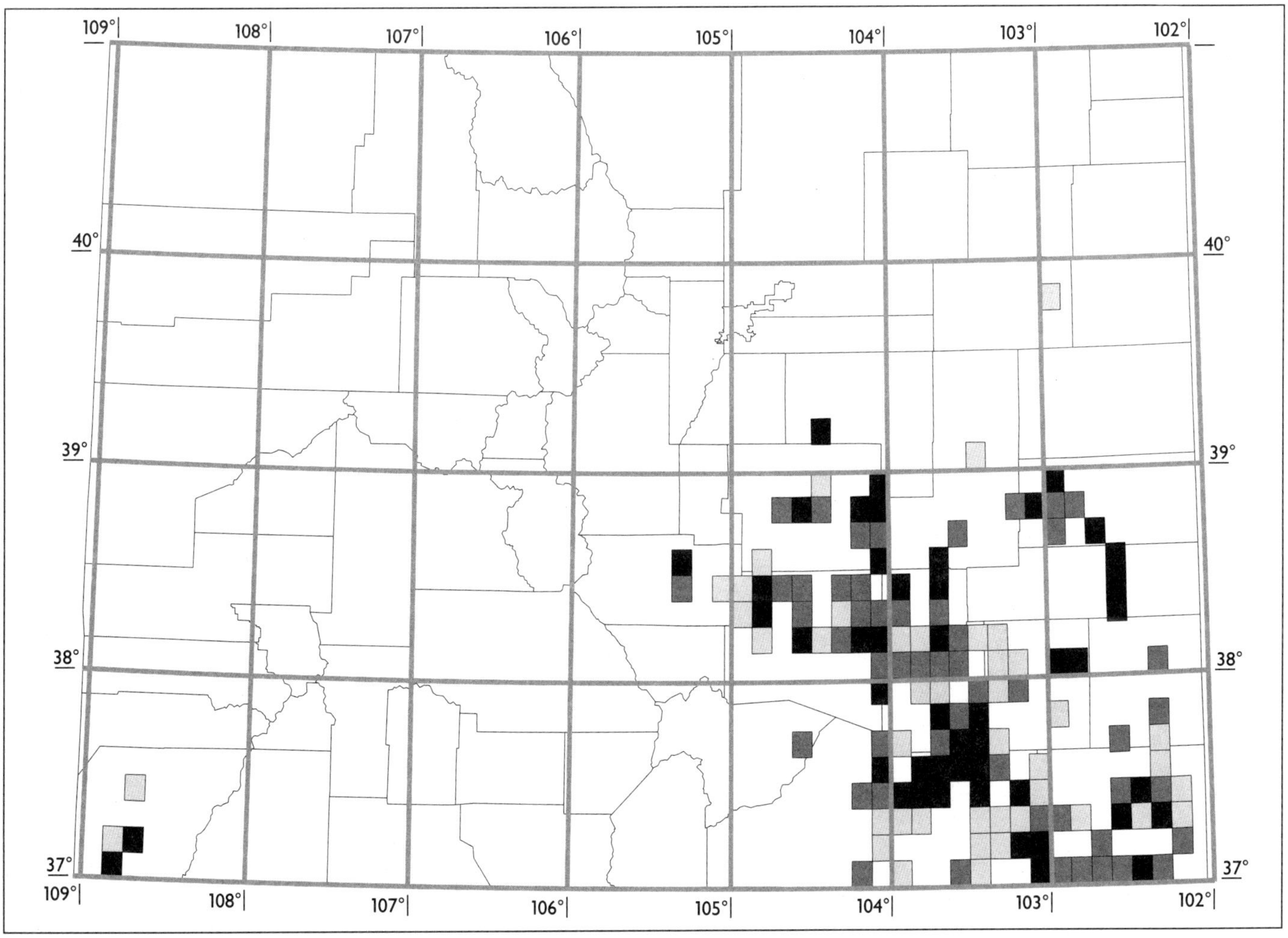
109°
108°
107°
106°
105°
104°
103°
102°
40°
39°
38°
37°

COLORADO DIVISION OF WILDLIFE / WILDLIFE RESOURCE INFORMATION SYSTEM

The Uncompahgre and Grand valleys opened for settlement in 1880 and 1881. The desert valleys supported no game birds, but by 1885 hunters had introduced a sporting target, Gambel's Quail, which they called the "California Partridge." The handsome quarry liked its new territory, became fruitful, and multiplied.

HABITAT

Gambel's Quail require tall shrubs near water (Gullion 1960), and they seem to find the skunkbrush and greasewood thickets along the streams, rivers, and irrigation canals of western Colorado ideal (Figgins 1913). Although Atlas workers reported lowland riparian deciduous forest (LRD) most frequently of all habitats (one-third of all reports), the understory shrubs associated with cottonwoods, especially skunkbrush, provide the required cover for this species (A&R). The second most frequently reported habitat, tall desert shrub (STD), usually greasewood, also supports good numbers of these quail. Workers reported seeing Gambel's Quail in several other grassland, brushland, and agricultural habitats. These habitats generally lie adjacent to the required cover, and Gambel's Quail use them for foraging.

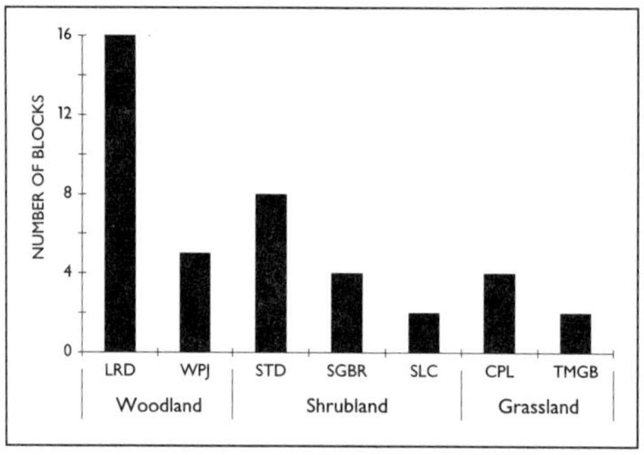
NUMBER OF BLOCKS
16
12
8
4
0
LRD
WPJ
STD
SGBR
SLC
CPL
TMGB
Woodland
Shrubland
Grassland

BREEDING

In May, June, and July males sing conspicuously from bush tops and fence posts, marking out their territories and putting on a splendid show. Females lay 10–12 eggs in a depression in the ground hidden under brush that provides shade from the desert sun and protection from raptors (Ehrlich et al. 1988). Apparently this strategy protects the nest from discovery by atlasers as well; observers found no Gambel's Quail nests. The precocial young leave the nest almost immediately after hatching. Atlas workers Confirmed breeding only by finding fledged young.

Most accounts of Gambel's Quail indicate that they generally rear a single brood (Ehrlich et al. 1988, Gorsuch 1934). The wide range of fledgling reports by Atlas workers (1 June to 5 August) tends to support earlier indications of frequent double-clutching in Colorado (Figgins 1914), although late nest dates of other quail arise from renesting after nest failures (Thompson and Ely 1989). In the Grand Junction area observers see downy young Gambel's scurrying across roads and along trails from early June to early September (RL).

BREEDING PHENOLOGY

CODE		# OF RECORDS	DATE RANGE
FL	Fledged Young	6	1 Jun–5 Aug

DISTRIBUTION

Originally, Gambel's Quail lived only in northwest Mexico and the southwestern U.S. from southern New Mexico to southern California and Nevada. Introductions across the U.S. succeeded on a large scale only in Colorado ("a remarkable achievement"—Bent 1932). The first plantings in Colorado apparently occurred in Montrose County in 1885, and others soon followed in Mesa and Montezuma counties (Figgins 1913, B&N, Braun et al. 1994). Within the very limited areas of introductions, the birds soon became abundant.

The 1,000 birds introduced into Montrose County came from southern California; hence early reports called them California Quail, *Lophortyx californicus*, and questioned reports of Gambel's Quail southwest of Durango (Cooke 1897, 1909, Rockwell 1908, Sclater 1912). Rockwell (1908) reported that since their introduction near Grand Junction, "California Partridge" had multiplied rapidly and had become "so plentiful as to become a nuisance" and that "they are common at least as far up the Grand River as Debeque." Cooke (1909) and Sclater (1912) both alluded to Rockwell's report, listing the birds as *L. californicus*. Figgins (1913) established that the introduced birds were *L. gambelii* and debunked the reports of native Gambel's Quail in southwestern Colorado. B&N regarded this species as "plentiful in the southwestern quarter of the state," and A&R described it as "uncommon to fairly common . . . in valleys and low foothills

from Garfield County south to Montrose County and in Montezuma County."

Atlas field workers affirmed that the lower Uncompahgre, Gunnison, and Colorado river valleys remain the center of this bird's abundance, as all but six of the Atlas reports came from these valleys and their tributaries. The report from western San Miguel County reflects a small introduction along the Dolores River. Field workers found the birds in only one block in Montezuma County, but added four blocks in La Plata County, all along tributaries of the San Juan River. These birds apparently spread northward from introductions in the San Juan Valley in New Mexico where they have become well established (Ligon 1961).

Atlas workers reported an abundance code of A3 for only four of the 21 Priority blocks in which they found this species. However, they assigned A3 codes to eight of 12 Non-priority blocks. Prime habitat can maintain fairly dense populations; in Grand Junction in May, one cannot walk a mile along the trail through the Connected Lakes unit of Colorado River State Park without hearing a dozen cocks calling (RL). Block workers may have tended to underestimate their numbers.

Although Gambel's Quail thrived after their introduction into Colorado, their future seems insecure. Growing urban areas and intensified farming eliminate many acres of greasewood and skunkbrush each year. CDOW upland game bird specialists predict that its habitat will gradually deteriorate, and "population declines will continue at a slow pace" (Braun et al. 1994).

BREEDING EVIDENCE

REPORTED IN 21 (1%) OF 1745 PRIORITY BLOCKS

Possible	10 blocks	(48%)
Probable	6 blocks	(28%)
Confirmed	5 blocks	(24%)

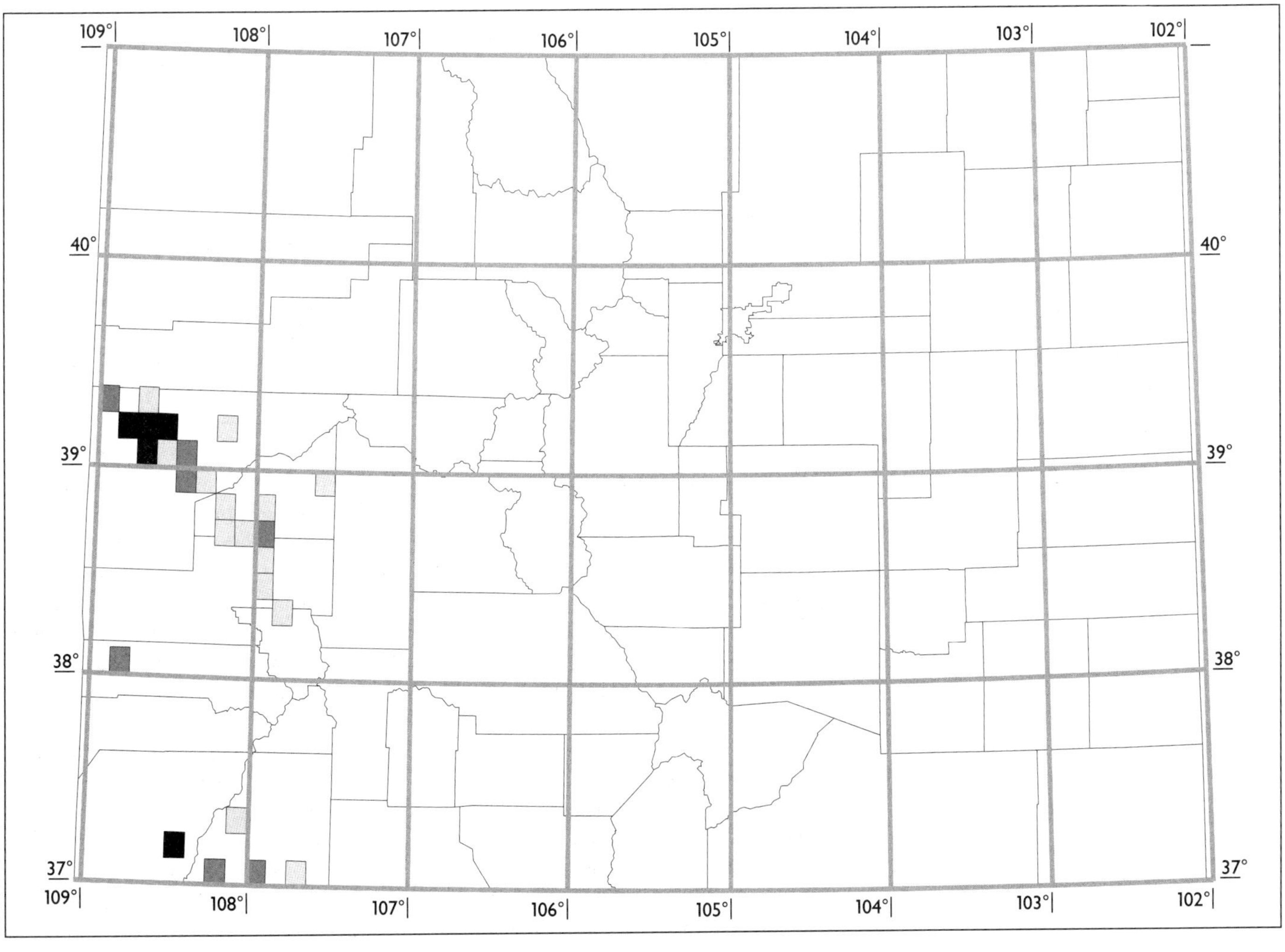
109°
108°
107°
106°
105°
104°
103°
102°
40°
39°
38°
37°

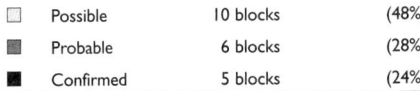

COLORADO DIVISION OF WILDLIFE / WILDLIFE RESOURCE INFORMATION SYSTEM

NORTHERN BOBWHITE *Colinus virginianus*

In 1846 Lieutenant J.W. Abert spent time at Bents Fort by the Arkansas River near present-day La Junta. In a later report he stated, "A quail like those of the eastern United States were then to be found in that neighborhood." The discovery of this report in 1909 (Cooke) settled a dispute as to whether Northern Bobwhites inhabit Colorado as a native or introduced species. Introductions definitely took place in various parts of the state, starting in 1870 and extending to 1936 (Braun et al. 1994). Despite the many introductions, bobwhites remained a protected species until 1952, when they became legal game.

HABITAT

Tall dense shrubs or brushy cover near a food source fulfill the basic habitat requirements. Food sources vary from grass seeds, wild and cultivated grains, and legumes to wild fruits, pine seeds, and acorns. In spring the quail supplement their diet with tender green plants and, in summer, with insects.

The aforementioned Lt. Abert observed that this quail met its upper limit in New Mexico where the wild plum thickets dwindled (B&N). In Colorado they commonly dwell in lowland riparian habitat and less commonly in sandsage grasslands and agricultural areas (A&R). The range has expanded with settlement and farming (Braun et al. 1994). Several habitats recorded by atlasers have an agricultural connection, from cropland to rural shelterbelts to a cultivated woodland. In southeastern Colorado atlasers also found them in shortgrass prairie and sandsage.

Bobwhites are "the least mobile of American game birds; some spend their entire lives within one-quarter mile of their birthplace, and most others stay within the square mile where they are hatched" (Murphey and Baskett 1952). Judging by this, most Atlas observations occurred in their breeding places.

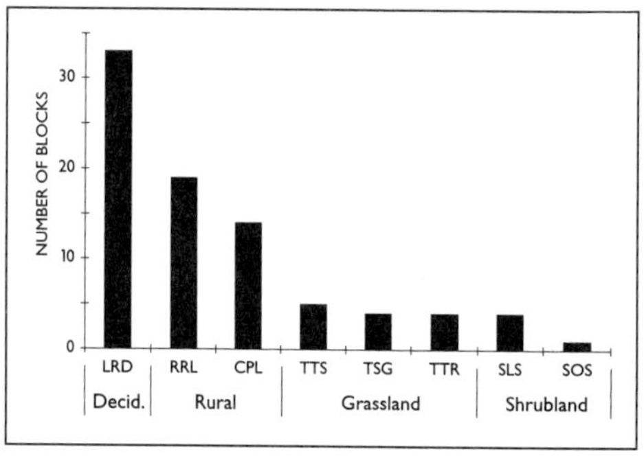
NUMBER OF BLOCKS
30
20
10
0
LRD
RRL
CPL
TTS
TSG
TTR
SLS
SOS
Decid.
Rural
Grassland
Shrubland

BREEDING

"Males call from fence posts and will combat for their territory" (B&N). Nests consist of a well-concealed depression in the ground lined with grasses. Weeds or grasses woven into an arch over the nest with a small opening in the side help to conceal it. Both sexes build the nest and both incubate the eggs (the male only a quarter of the time; Thompson and Ely 1989). Pearson (1936) reports some evidence of polygyny. Bobwhites sit so closely that observers find nests only by accident (Harrison 1979). Atlas results bear this out, as no observers found nests.

Colorado nests have dates ranging from 17 May to 24 July (B&N). B&N mentioned a mixed set of two bobwhite and eight Ring-necked Pheasant eggs found on 19 May 1919 in Jefferson County. Bobwhites lay between 12 and 18, usually 14–16, white or cream-white eggs. More than 18 eggs means two or more hens used the same nest (Terres 1980). All Atlas Confirmations came from fledglings, found over a two-month period.

BREEDING PHENOLOGY

CODE		# OF RECORDS	DATE RANGE
FL	Fledged Young	10	12 Jun–23 Aug

DISTRIBUTION

Northern Bobwhites range from Massachusetts to the southeastern U.S., west to South Dakota and eastern New Mexico. Historically, the thickest populations in Colorado occurred along the Arkansas and South Platte rivers (A&R). Less commonly they occur in sandsage rangelands and agricultural areas of northeastern Colorado. Low-density populations also inhabit agricultural and shortgrass prairie areas of Baca County (A&R). Atlasers found birds in all these areas, with equivalent abundance estimates. A&R showed populations along the South Platte as far west as Denver and in farmland north almost to Wyoming. Atlasers did find scattered populations in this area but not as far as Denver.

Areas of previous bobwhite introductions include the foothills from Pueblo to Boulder, Estes Park, the Wet Mountain Valley, and Grand Junction (Cooke 1900, B&N). Both A&R and CDOW showed them along Big Sandy Creek in Kiowa and Cheyenne counties but atlasers found no remnants of these populations (but they did record Scaled Quail on the Big Sandy). Field workers did find other unmapped birds in Phillips, Kit Carson, and Cheyenne counties, possibly recent introductions.

Atlasers reported bobwhites in 67 blocks. They recorded A3 abundance estimates (over 10 breeding pairs) in 42% of these blocks; of the A3 codes, 71% occurred in riparian habitat, the rest in agricultural areas (shelterbelts and croplands).

Populations of bobwhites on the fringes of their range fluctuate with the severity of winter. Blizzards or freezing rains occasionally deplete populations of Colorado's bobwhites, particularly in agricultural areas (Braun et al. 1994).

Overhunting of bobwhites may occur on CDOW properties but habitat improvement may mitigate this impact and increase populations. CDOW researchers stated that "late season hunting during severe weather may decrease populations" and that "the extent of the problem is unknown and needs to be evaluated further" (Braun et al. 1994). Introductions of bobwhites into new areas or to bolster existing populations do not seem currently in vogue (MY). Pough (1951) believed that artificial stocking weakens the genetic makeup of native bobwhite populations, affecting their ability to survive in fringe areas.

BBS trends (1994) show a 2.5% decline per year continentally, representing a substantial reduction in numbers. In Kansas hunters bagged almost 2 million birds in 1986, making the species the prime game bird in that state. Young of the year composed 85% of the bag, which Thompson and Ely (1989) say shows that most bobwhites breed only once during their lifetime.

Colorado's fringe population probably will disappear where housing developments gobble up farmland and continue to fluctuate due to weather conditions, but will maintain stable numbers in riparian areas that remain protected.

BREEDING EVIDENCE

REPORTED IN 67 (4%) OF 1745 PRIORITY BLOCKS

☐	Possible	25 blocks	(37%)
■	Probable	27 blocks	(40%)
■	Confirmed	15 blocks	(22%)

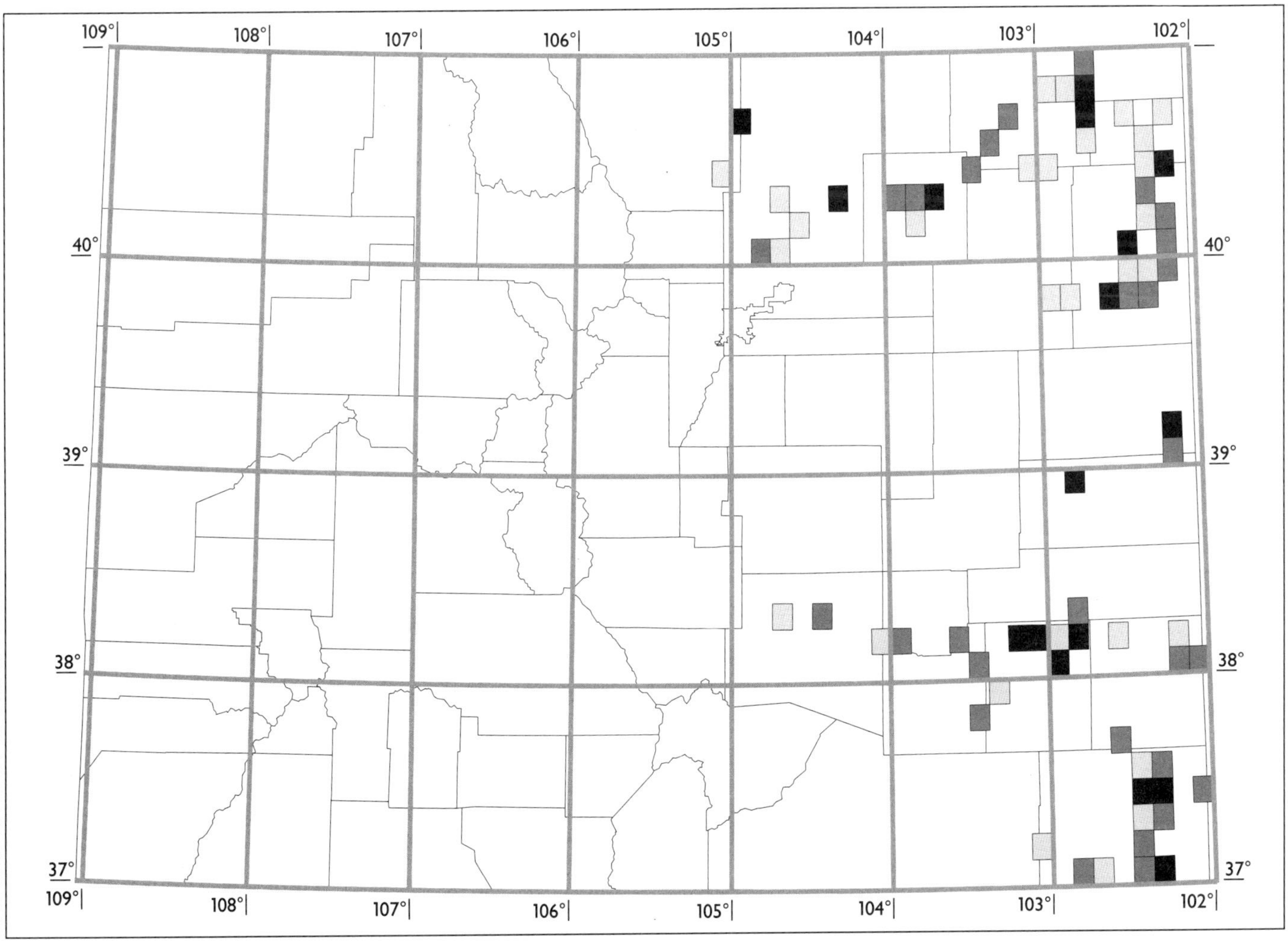
109°
108°
107°
106°
105°
104°
103°
102°
40°
39°
38°
37°

COLORADO DIVISION OF WILDLIFE / WILDLIFE RESOURCE INFORMATION SYSTEM

For most of the year these shy marsh birds creep furtively through the cattails. Whenever they sense movement, they dart back into the heavy, concealing vegetation. Yet when breeding season starts in the spring, males sing *ki-dick′ ki-dick′* from the cattails and boldly run into the open.

HABITAT

Virginia Rails need shallow water, emergent cover, and high invertebrate abundance for their habitat; they avoid dry stands of emergents (Conway and Eddleman 1994). Changing water levels, especially flooding of nests, disrupt their breeding. Although Virginia Rails live primarily in cattail marshes, they also breed in wet meadows and irrigated hayfields, especially at higher elevations (A&R).

Virginia Rails and Soras often breed in the same marshes; differences in diet permit coexistence (Horak 1970). Soras eat mainly seeds and Virginia Rails consume a diet consisting of 85–97% animal food—small aquatic invertebrates such as beetles, snails, spiders, true bugs, and diptera larvae (Conway 1995).

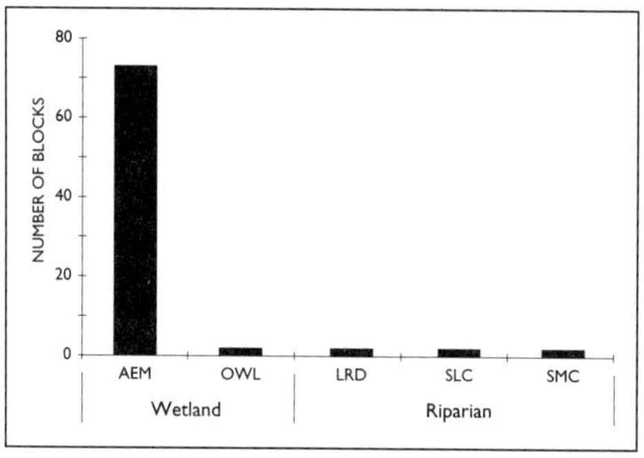
80
60
40
20
0
NUMBER OF BLOCKS
AEM
OWL
LRD
SLC
SMC
Wetland
Riparian

Virginia Rails use drier areas of the marsh than Soras, usually parts of the marsh in the early stages of succession—those without high stem densities. The most important features of the habitat are standing water, emergent plants such as cattails or bulrushes, and a substrate with high invertebrate abundance. They feed in shallow water (0–6 inches, 0–15 cm), with a muddy or silty bottom. Rails as a group have strong legs and the highest leg-muscle to flight-muscle proportion of any birds, a 25:15% ratio of body weight (Conway 1995).

Eighty-nine percent of Atlas reports come from marshes with cattails or sedges. Suitable marshes exist near rivers and in mountain parks. Irrigation reservoirs and ditches and runoff from irrigation create many of these wetlands. Springs, even far out on the plains, supply a few marshes. Occasionally the rails feed in adjacent upland habitats (Conway 1995), which may explain aberrant Atlas habitat codes (shrubs and mountain meadows).

BREEDING

Virginia Rails reside in Colorado primarily from early April to mid October (Conway 1995) and breed from May through August (B&N). Males voice most of their courtship calls from mid to late April (Glahn 1974). Nest-building begins with the laying of the first egg, or slightly before. The birds conceal their nests well in robust marsh vegetation; they place them touching, slightly below, or up to 6 inches (15 cm) above the water surface (Conway 1995).

Atlasers obtained only one report of nest with eggs and one of nest with young. Atlasers based most Confirmed breeding on observing fledglings over a long period, 21 May through 12 August, although 8 of the 12 dates came in July. This span of dates accommodates the production of double broods.

BREEDING PHENOLOGY

CODE		# OF RECORDS	DATE RANGE
C	Courtship	2	19 May–29 May
FL	Fledged Young	12	21 May–12 Aug

DISTRIBUTION

Virginia Rails nest widely across the northern half of the lower 48 states, across southern Canada, and also in Mexico and South America. In winter most withdraw to the southern edge of the breeding range, including eastern Colorado.

In Colorado, Virginia Rails now probably nest more widely and in greater numbers than before European settlement. Construction of reservoirs for irrigation and recreation, and creation of ponds from irrigation runoff and for livestock watering have produced additional and more widely scattered marshes.

The density of breeding rails depends on habitat quality (Conway 1995). A survey of 252 Colorado townships (over 50% of those with marshes larger than 2.5 acres [1 ha] detected Virginia Rails in 100 townships (Griese et al. 1980). This survey found rail densities higher in eastern than in western Colorado, and as great as 1.9 pairs per acre (4.7/ha) of marsh in southeastern Colorado. It also found, within these large wetlands, a higher breeding density of Virginia Rails than of Soras, but located Soras in more places (125:100). Atlasers surveyed small marshes as well as large ones;

Atlas results depict Soras in over twice as many blocks and more than twice as numerous as Virginia Rails, in part because Soras venture to higher altitudes than Virginia Rails.

Atlasers reported Virginia Rails throughout Colorado, widely scattered in those blocks where suitable marshes exist (a typical pattern for the species). The only A4 abundance estimates (between 101 and 1,000 pairs) came from blocks in the big marshes around John Martin Reservoir in the Arkansas River Valley.

Populations of Virginia Rails nationwide have declined in the last ten years, a trend most evident in the central U.S. where severe loss and degradation of wetlands have occurred (Conway and Eddleman 1994). BBS trends lack precision, but they do provide the best data available and may even underestimate negative trends (Conway 1995). Wetland loss is the greatest continent-wide threat to rail populations; agricultural development has caused 87% of recent national wetland losses (Eddleman et al. 1988). Livestock grazing of marshes is usually detrimental to marsh habitat (Eddleman et al. 1988). Many wetlands used by rails during migration and in winter occur on national wildlife refuges.

An early Colorado observer considered Virginia Rails "seldom seen [but] fairly common" on the plains in summer (Sclater 1912), a status largely unchanged today. However, continued human population growth threatens their future. Draining wetlands for agriculture, roads, shopping centers, housing, and the like reduces their habitat and therefore their numbers. Selling agricultural water to cities dries up wetlands otherwise replenished by irrigation. The latter wetlands did not exist before settlement, but development has obliterated many presettlement wetlands on the plains and especially along the Front Range. Loss of wetlands affects not only Virginia Rails, but numerous other species, a few on the brink of disappearing from the state.

BREEDING EVIDENCE

REPORTED IN 48 (3%) OF 1745 PRIORITY BLOCKS

Possible	22 blocks	(46%)
Probable	13 blocks	(27%)
Confirmed	13 blocks	(27%)

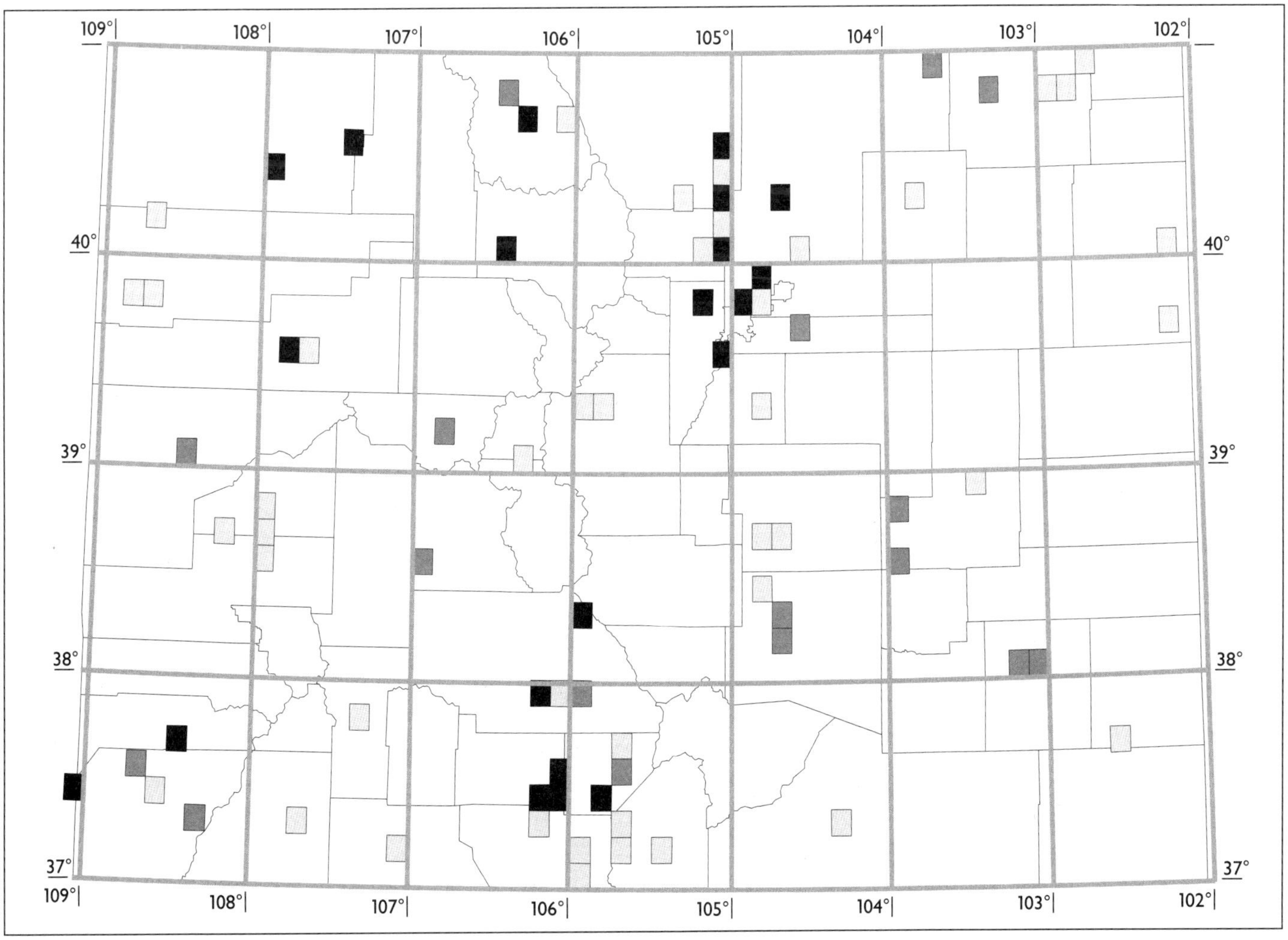

109°
108°
107°
106°
105°
104°
103°
102°
40°
39°
38°
37°

COLORADO DIVISION OF WILDLIFE / WILDLIFE RESOURCE INFORMATION SYSTEM

In soggy mountain swales and prairie cattail marshes, the secretive and rarely detected Soras spend most of their time skulking around in heavy vegetation. Infrequently one may venture into open shallows next to the cattails, and sometimes they even swim in open water. Early one morning in June, as I sat beside a North Park beaver pond where moose frequently browse, an adult Sora swam out into the open water and called a sharp *keck, keck.* Three fluffy black chicks swam out toward her—a rare observation and a satisfying Atlas Confirmation.

HABITAT

Soras breed in a wide variety of wetlands, primarily cattail marshes, and also in grassy or sedge marshes, wet meadows, and irrigated hayfields especially in mountain parks (A&R). Soras prefer cattails with shallow water for breeding and escape, and 80% of Atlas reports come from marshes, with cattails usually the dominant vegetation. Atlasers reported much less use of other habitats, such as hay meadows, open water, willow carrs, lowland riparian thickets, and salt meadows. Although on the plains Soras and Virginia Rails may share the same wetlands, Soras also nest at higher, cooler elevations than do Virginia Rails. Atlasers recorded Soras in at least four blocks above 9,000 feet (2745 m), including one at a 9,988-foot (3044-m) pond on Kenosha Pass (Jefferson, 39105D7).

Soras frequently breed in cattail marshes near Virginia Rails. The latter, with long, slender, decurved bills, eat mainly insects and other animal life; Soras, which have short, heavy beaks, eat mainly seeds. Especially in the spring, they eat some invertebrates such as beetles and snails. These differences in food habits between the two species allow them to live together without serious competition for food (Horak 1970). The two species shared 27 Atlas blocks, and most shared the same marshes.

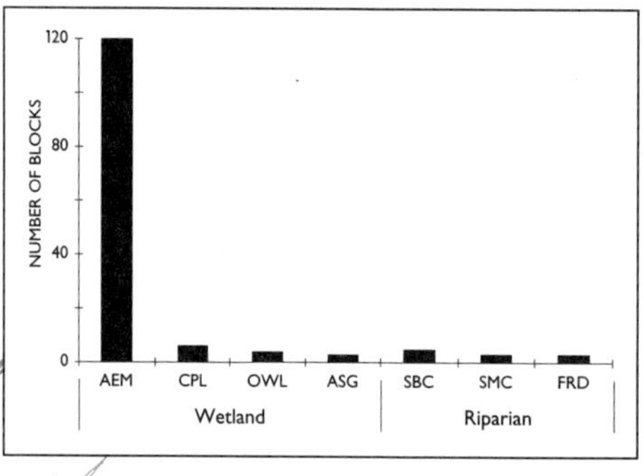

NUMBER OF BLOCKS
120
80
40
0
AEM
CPL
OWL
ASG
SBC
SMC
FRD
Wetland
Riparian

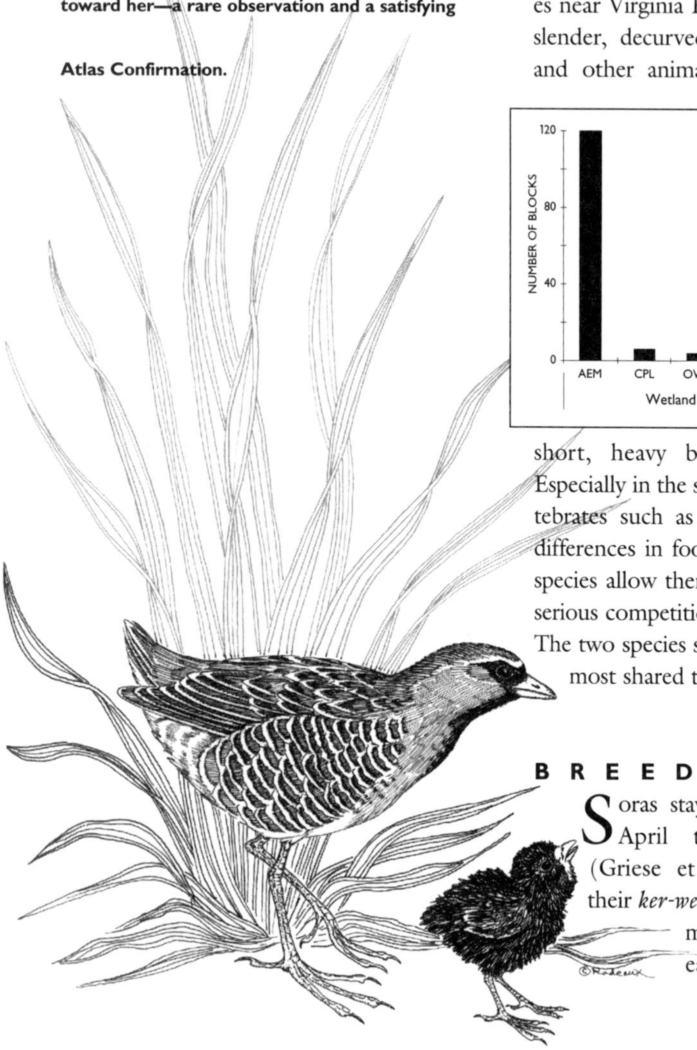

BREEDING

Soras stay in Colorado from early April through early October (Griese et al. 1980). Males voice their *ker-wee* call to form territories primarily from mid April to early May (Glahn 1974). They lay eggs from late May through early July (B&N). They have small territories. Six pairs of Soras (and 10–18 pairs of Virginia Rails) nested in a 25-acre (10-ha) marsh near Fort Collins, with only 35 feet (10 m) separating the closest Sora nests (Glahn 1974).

Soras lay a substantial 8–12 eggs; the black downy young leave the nest almost immediately upon hatching—as soon as their down dries (B&N). Soras carefully conceal their nests, and atlasers based most Confirmations on observations of fledglings. Atlasers also frequently reported breeding based on territorial behavior, multiple calling males, and agitated adults.

BREEDING PHENOLOGY

CODE		# OF RECORDS	DATE RANGE
ON	Occupied Nest	2	12 May–10 Jun
NE	Nest with Eggs	1	14 Jul
FY	Feeding Young	1	5 Aug
FL	Fledged Young	14	27 May–15 Aug

DISTRIBUTION

Soras breed from coast to coast across southern Canada and the northern and central U.S. to as far south as Colorado. Rarely, a very few remain in Colorado in winter, but most winter from the southern edge of the United States to South America.

Atlasers reported Soras breeding throughout the state where suitable wetlands exist. Buttressing the Latilong Study, they confirmed breeding in one additional latilong (38103) and reported probable breeding in another (40103). They did not observe breeding in a few counties, mainly in the eastern plains, that have few to no wetlands in Priority blocks. The Atlas map both expands and contracts the previously mapped range in Colorado. It shows little breeding on the plains, including the major river valleys. Excluding North Park and the San Luis Valley, half again as many Priority blocks recorded Soras on the Western Slope as on the Eastern Slope, and most of the latter came along the Front Range corridor and west to the Continental Divide.

Secretive Soras live in dense habitats, making it difficult to estimate their population density. An early observer termed them a common summer resident from the plains to about 9,000 feet (2745 m; Sclater 1912).

Soras are the most abundant and widely distributed rail in North America (Melvin and Gibbs 1994). Some researchers report that the Sora population plunged in central North America during the past 30 years, largely the result of natural loss of wetlands from a long drought from which both wetlands and rail populations may yet recover (Conway et al. 1994). Others report that, although habitat is declining, Sora population trends are uncertain (Melvin and Gibbs 1994).

Soras presently need little conservation assistance in our state. Privately owned marshes preserved for duck hunting, wildlife refuges, and state parks and wildlife areas provide much of the breeding habitat for Soras. Loss of habitat would threaten their future, however, such as that caused by draining and filling marshes or drying them up by transferring water from agricultural irrigation to municipal and other uses.

BREEDING EVIDENCE

REPORTED IN 105 (6%) OF 1745 PRIORITY BLOCKS

☐	Possible	62 blocks	(59%)
▩	Probable	27 blocks	(26%)
■	Confirmed	16 blocks	(15%)

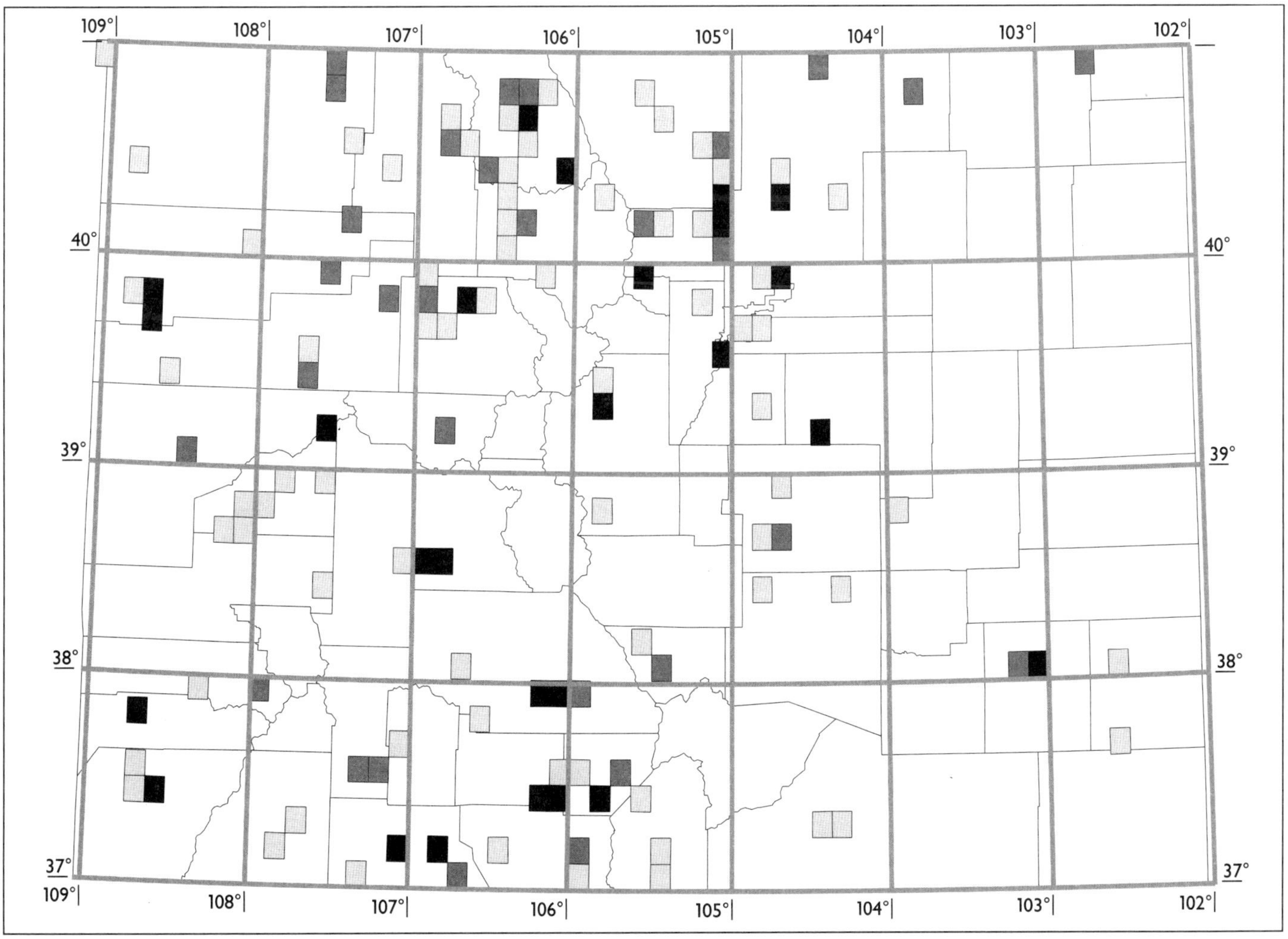
109°
108°
107°
106°
105°
104°
103°
102°
40°
39°
38°
37°

COLORADO DIVISION OF WILDLIFE / WILDLIFE RESOURCE INFORMATION SYSTEM

Often confused with ducks, coots more resemble black chickens with white beaks. Their fiery red eyes indicate their pugnacious nature (Ryder 1959). Their huge, lobe-toed feet enable coots to swim well and to dive to moderate depths (more than six feet) to pull up the aquatic vegetation on which they feed.

HABITAT

Although coots frequent a wide variety of wetlands, from wet meadows to large reservoirs (Manci 1986), their favorite breeding habitats are marshes with tall emergents, especially cattails and bulrushes, ideally about 50% open water and 50% flooded emergents (Ryder 1961).

During spring migration and breeding they need water levels of 11.75 inches (30 cm) or greater (sufficient to flood emergents) and aquatic food species (e.g., *Potamogeton* sp. and *Chara* sp.).

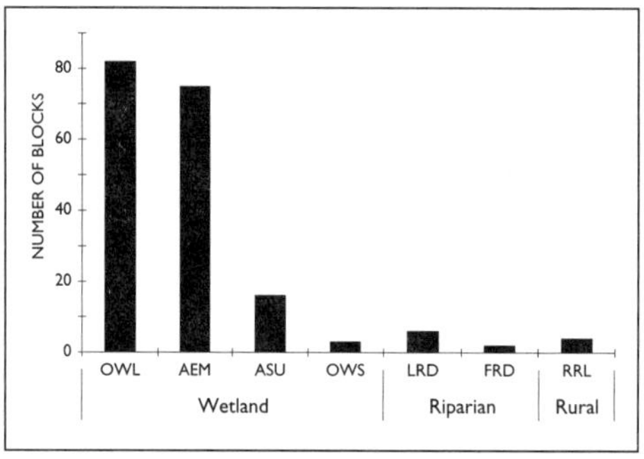
NUMBER OF BLOCKS
80
60
40
20
0
OWL
AEM
ASU
OWS
LRD
FRD
RRL
Wetland
Riparian
Rural

Most coot nests are platforms over water in cattail and bulrushes (Gorenzel et al. 1983). The birds build nests of local materials (cattails and bulrushes), which often are floating but anchored to emergents. Occasionally coots nest in sedges, tamarisks, or willows, such as along ditches and irrigation canals. A small colony regularly nests in sedges around a small lake on Kenosha Pass at 9,988 feet (3044 m) in the Jefferson Atlas block (39105D7)—the same pond where atlasers Confirmed Soras (Evan and Ives Hannay pers. comm.). Generally, coot marshes are one or more acres in size (RAR).

BREEDING

In 1897 coots were a common summer resident breeding throughout Colorado in suitable places on the plains and in the mountain parks up to 8,000 feet (2440 m; Cooke 1897). Sclater (1912) reported nesting above 10,000 feet (3048 m). Gorenzel et al. (1982, 1983) had summer records in 32 of the 63 counties in Colorado, with highest densities in Browns Park, North Park, and the San Luis Valley. Atlasers reported coots in 134 Priority blocks, with nesting Confirmed in 54% of those blocks, mainly by observing the peculiar-looking fledglings.

Gorenzel (1979) and Gorenzel et al. (1981, 1982) reported on the fate of 354 coot nests that they followed. Nests were started mid April to mid July, with the peak 23 April–12 June. Nest success averaged 80.2% (and ranged from 33.3 to 97.3%) on four study areas. Clutches averaged 8.4 eggs (range, 6.5–9.2). Predation was the main cause of nest failure, but desertion was also important. Atlas observers also found that nest-building and incubation extended to mid July and saw fledglings, at various stages of development, through August.

Gorenzel (1979) reported densities of 0.4–14.0 successful nests per acre (1.0 to 35.1 per ha) of emergents on his four study areas. The lowest was at Beebe Draw near Greeley following a burn in 1978 and highest at Ice Pond near Buena Vista. Landowners burned the emergent area at Beebe Draw in an effort to "control" cattails. The coots at Beebe Draw bounced back to a near-normal density one year following burning (1977, 77 nests; 1978, 6 nests; 1979, 56 nests; Gorenzel et al. 1981).

BREEDING PHENOLOGY

CODE		# OF RECORDS	DATE RANGE
C	Courtship	2	30 Apr–21 Jun
NB	Nest Building	5	25 May–19 Jul
ON	Occupied Nest	4	28 May–13 Jun
NE	Nest with Eggs	8	30 May–10 Jul
NY	Nest with Young	3	2 Jun–28 Jun
FY	Feeding Young	4	3 Jun–23 Jul
FL	Fledged Young	64	29 May–23 Aug

DISTRIBUTION

American Coots breed across the continent from southern Canada south to Costa Rica. Even early observers (Cooke, Sclater, and Bergtold) commented on the widespread range and abundance of coots in Colorado, especially in migration.

Gorenzel et al. (1981) documented nesting in all but two of the 28 latilong blocks. Atlasers Confirmed nesting in 43 counties (72 Priority blocks) but did not confirm breeding in those two latilongs. Highest densities were in North Park, the San Luis Valley, and plains areas from Wyoming to New Mexico. BBS routes in

Colorado that record coots average 0.7 coots/route, above that for Wyoming and New Mexico, but well below that for the Prairie Provinces in Canada and the northern plains states (Alisauskas and Arnold 1994). Most coots nesting in Colorado winter in Mexico, mainly in the central highlands and the west coast states (Ryder 1963).

Although considered a game bird in Colorado, coots are not regularly counted in waterfowl surveys, summer or winter. Coots are dark-fleshed, very palatable, and favored over ducks by Cajuns but not by Colorado hunters; hence CDOW does not count them regularly. Frederickson et al. (1977) estimated 8,000 pairs of coots nesting in the state, somewhat lower than the Atlas estimate of 12,060 pairs (see Appendix B).

Estimates of coot numbers in the prairie-parkland region of the U.S. and Canada have increased in the last three decades (Alisauskas and Arnold 1994), but comparable trend data are not available for Colorado, because CDOW does not count coots. The sequence of events at Beebe Draw (see above) indicates that coot populations can be easily controlled by management activities—water manipulation, burning, and dredging done between November and February, when coots are absent (Gorenzel et al. 1981).

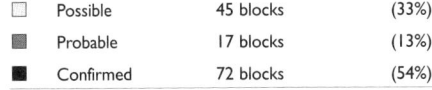

BREEDING EVIDENCE

REPORTED IN 134 (8%) OF 1745 PRIORITY BLOCKS

Possible	45 blocks	(33%)
Probable	17 blocks	(13%)
Confirmed	72 blocks	(54%)

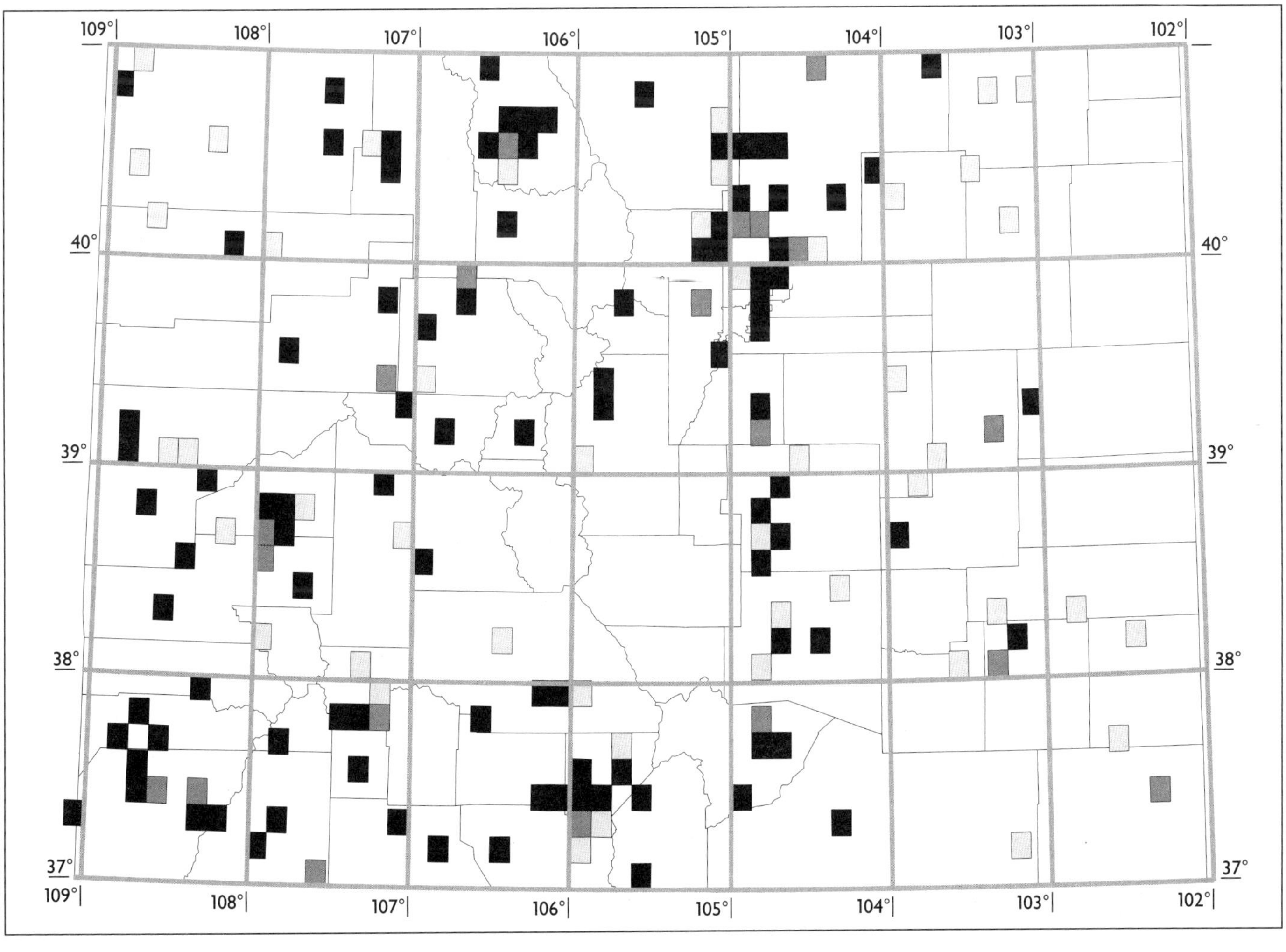
109°
108°
107°
106°
105°
104°
103°
102°
40°
39°
38°
37°

COLORADO DIVISION OF WILDLIFE / WILDLIFE RESOURCE INFORMATION SYSTEM

Courting in the San Luis Valley and nesting in northwestern Colorado, Sandhill Cranes signal the start of spring to many Coloradans. Haunting calls issue from the western Colorado skies as the cranes migrate from the San Luis Valley to northwestern Colorado and points beyond. Monte Vista honors them each March with a Crane Festival for crane fanciers.

HABITAT

Two subspecies of Sandhill Cranes occur in Colorado. This account discusses Greater Sandhill Cranes (*G. c. tabida*), the subspecies that breeds in northwestern Colorado.

Atlasers found Sandhill Cranes breeding in a variety of wetland habitats, particularly flooded fields and beaver ponds, and also marshes and wet meadows. The birds require a suitable nest site and enough open space to become airborne. Disturbance is one factor in the decline of crane numbers (Enriquez 1979), but if the potential nest site seems suitable, matters such as highway traffic and human activity sometimes become secondary. Atlasers found one crane on her nest in a flooded pasture west of Steamboat Springs within 100 feet of U.S. Highway 40. Another nest contained a single egg atop a beaver den in the backyard of a summer home (NMB).

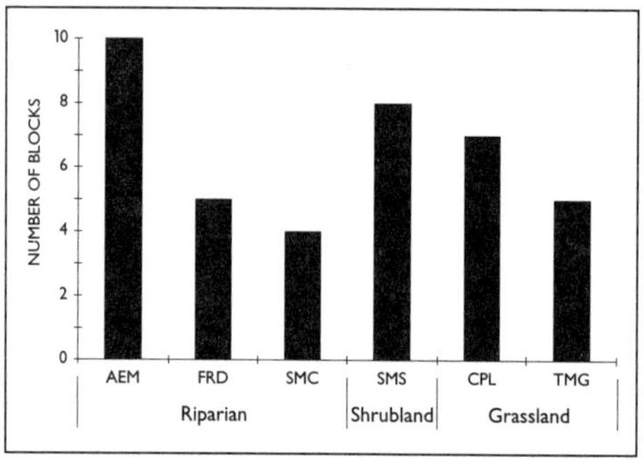
NUMBER OF BLOCKS
10
8
6
4
2
0
AEM
FRD
SMC
SMS
CPL
TMG
Riparian
Shrubland
Grassland

Omnivorous feeders, Sandhills consume cultivated grains, roots, many small mammals, frogs, toads, snakes, crayfish, and numerous insects. They dig up earthworms with their bills, and apparently prey on nestling Red-winged Blackbirds.

After the young reach the flight stage, Colorado's cranes gather in pastures and meadows along the Yampa River. They use ranch hayfields as staging areas preparatory to migration south. Researchers discuss spring staging sites like the San Luis Valley as places to build up energy reserves prior to breeding (Tacha et al. 1992), but have not explained the function of fall staging sites. With the relatively short flight distance between the Yampa Valley, late fall staging in the San Luis Valley, and their wintering grounds in New Mexico, energy storage for migration does not seem to explain the behavior. Possibly it serves a social function by concentrating isolated family groups into larger migratory flocks, or it may represent an evolutionary holdover from a period when the flocks migrated greater distances.

BREEDING

Greater Sandhill Cranes begin to arrive in the San Luis Valley in late February, and begin their courtship in March. Some migrate through northwestern Colorado as early as the middle weeks in March; later ones fly through well into May.

Local breeders must wait in the lowlands for the snow to melt off the high-country breeding habitat before nesting. Thus birds that breed at lower elevations begin considerably earlier than those in the higher mountains, explaining the two-month range of incubation dates reported by atlasers.

Nests consist of piles of grasses, reeds, and other aquatic vegetation; they sometimes sit in shallow water but more typically are built on raised land surrounded by water, such as beaver or muskrat huts. Here the cranes lay and incubate the clutch of one or two eggs. Once the eggs hatch, the light-mahogany-colored crane chicks spend their time wading in the shallow grasses or in the aspen stands typically found around beaver pond complexes.

BREEDING PHENOLOGY

CODE		# OF RECORDS	DATE RANGE
ON	Occupied Nest	3	1 May–30 May
NE	Nest with Eggs	5	5 May–30 Jun
NY	Nest with Young	1	1 Jun
FL	Fledged Young	6	6 Jun–20 Jul

DISTRIBUTION

Greater Sandhill Cranes nest west of the Continental Divide in the Great Basin and north into southern British Columbia and Alberta. Historic records in Colorado date back to 1822 when beaver trappers killed one in the San Juan Mountains at the headwaters of the Rio Grande (Marsh 1931). Breeding cranes in early reports (La Plata, Gunnison, and Routt counties; Cooke 1897, Sclater 1912) probably belonged to the Greater Sandhill Crane subspecies, which once bred from Idaho to Arizona and probably, therefore, in western Colorado.

By the 1940s hunting, habitat change, and disturbance reduced the Colorado population to isolated nests in Routt and Jackson counties (Bieniasz 1978). Since then numbers have slowly increased and distribution expanded. The Atlas shows that the vast majority of the birds breed in Routt County and adjacent Moffat County, with a few in Rio Blanco, Jackson, and Grand counties. Atlasers also found pairs nesting in possible historic sites in Mesa County.

Crane population dynamics involve beaver pond complexes and riverfront grasslands. Due to reduced beaver trapping, suitable nesting habitat now exists throughout the mountainous regions of the state. Each year valley hayfields become scarcer as river-bottom developments such as ski areas, subdivisions, golf courses, reservoirs, and even airports transform them. A substantial loss of fall staging habitat in the Yampa Valley—usually on private land—may start to limit crane numbers in Routt County.

The potentially damaging habitat changes could alter migratory patterns or increase mortality due to inadequate foraging opportunities prior to migration. These changes might force the cranes to find new breeding sites (if they can) or may wipe out this breeding population.

CDOW considers the Greater Sandhill Cranes that breed in Colorado as threatened because of the few pairs thinly scattered across northwestern Colorado. CDOW closed the hunting season on Greater Sandhills on the Western Slope and in the San Luis Valley.

The U.S. Forest Service, which manages most of the suitable breeding habitat, in 1994 listed the Greater Sandhill Crane as a sensitive species. This requires seasonal restrictions and other mitigation to reduce the impacts of timber harvest, road construction, and grazing. In addition, positive actions such as aspen manipulation, beaver reintroduction, and riparian protection will benefit the species.

BREEDING EVIDENCE

REPORTED IN 23 (1%) OF 1745 PRIORITY BLOCKS

☐	Possible	4 blocks	(17%)
▦	Probable	4 blocks	(17%)
■	Confirmed	15 blocks	(65%)

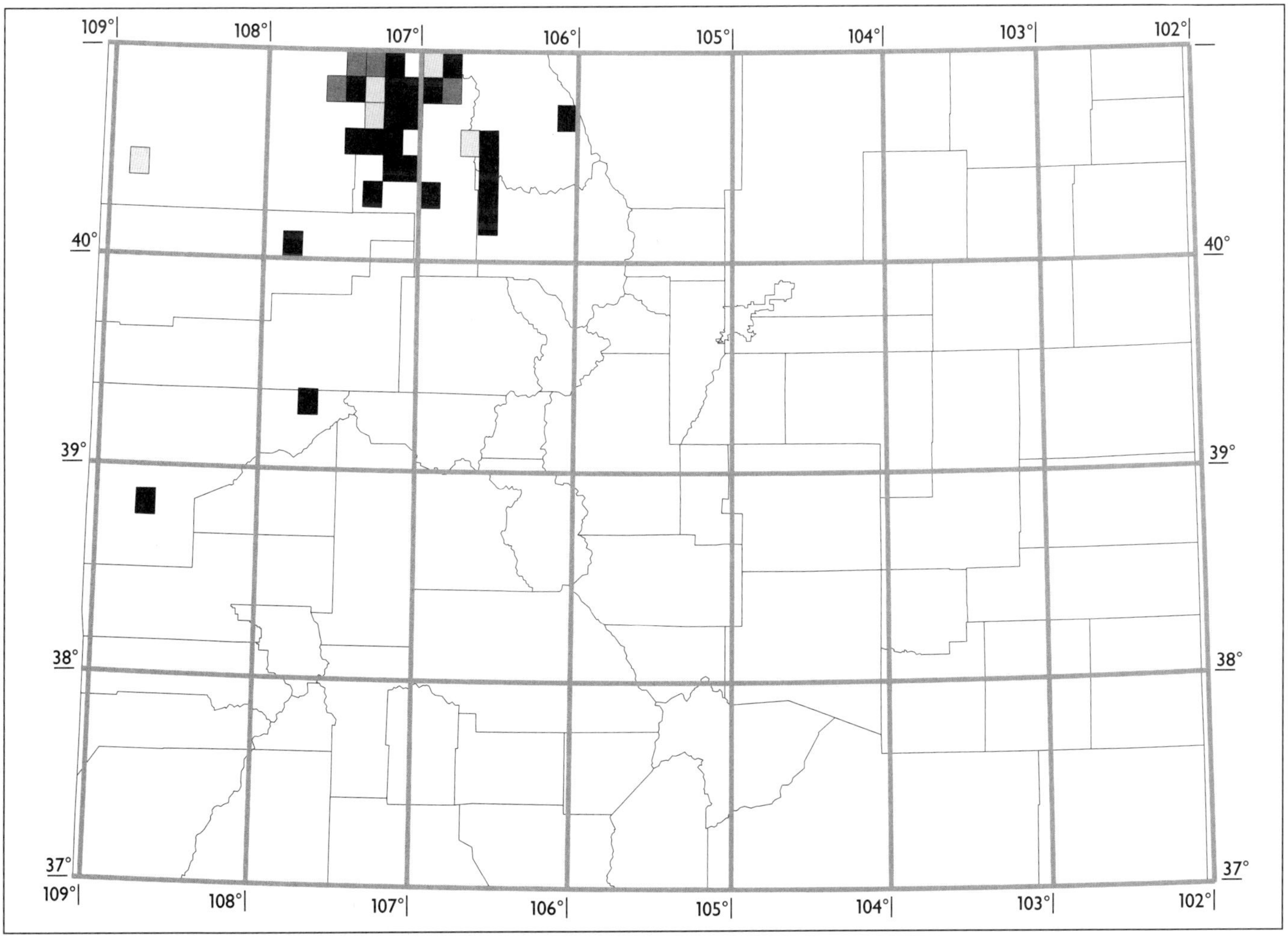
109°
108°
107°
106°
105°
104°
103°
102°
40°
39°
38°
37°

COLORADO DIVISION OF WILDLIFE / WILDLIFE RESOURCE INFORMATION SYSTEM

Snowy Plover

The plaintive *ter-wee-it* call of a Snowy Plover announces its territory on the salt-encrusted shorelines it needs for nesting. Shy and skittish, these ghosts of the salt flats rarely betray the location of their nests, even though they typically place them in very exposed beach locations.

©Rubeaux

HABITAT

Snowy Plovers breed in sandy ocean beaches, dry salt flats, dredge spoils, and river bars. Inland they breed in natural habitats such as ephemeral alkali playas and alongside manmade sewage and evaporation ponds (Page et al. 1995). In southeastern Colorado they breed only in manmade habitats: reservoir edges. Atlasers found them skittering along the sandy shores of several large reservoirs. In the San Luis Valley the plovers formerly used alkali-covered playas near San Luis Lake, but abandoned them when reservoir levels rose and covered them.

Necessarily, Snowy Plovers adapted to changing habitat quality by colonizing new sites when old ones became unusable. Snowies may have moved into Colorado because of disturbance to nest colonies elsewhere, perhaps in the panhandle of Oklahoma or southwestern Kansas (DLN).

Atlas nesting records from the large prairie reservoirs, especially the Great Plains Reservoirs in Kiowa County, came in years when those reservoirs were shallow and close to drying up. In years with marginal habitat, birds may even use weedy beaches. They place their nests up to 500 feet (150 m) from water. They feed on terrestrial and aquatic invertebrates although they rarely feed in the water, and then only in water "1–2 cm deep" (Page et al. 1995).

They share the alkali-bordered reservoirs along the Arkansas River with Piping Plovers, Killdeer, Spotted Sandpipers, American Avocets, Black-necked Stilts, Wilson's Phalaropes, and Least Terns. In June 1993 a male Snowy Plover attempted copulation with a female Piping Plover at Neenoshe Reservoir (DLN), unusual because in their entire ranges only at the Great Plains Reservoirs do the two species come into contact.

BREEDING

Early migrants, male Snowy Plovers arrive on territory the first week of April. Females choose from displaying males almost as soon as they return a few days later. They place their nests up to 500 feet (150 m) from water. Even from very close distances the nests—simple scrapes in the ground—are practically invisible. During courtship males construct "dummy" nests; the female chooses the nest site from several prepared for her. She often cradles the nest in the crotch of driftwood or nestles it against rocks. This may protect it from trampling or inhibit detection by predators. Atlas Confirmation codes bear out the near invisibility of nests: nests with eggs were found in only one block. Young plovers with adults provided the other Confirmations.

Clutches almost always contain three eggs when complete. Snowy Plovers lay their eggs an average of 2.5 days apart; incubation begins upon clutch completion. Both sexes share incubation; Great Plains males typically incubate overnight. Temperatures on the seemingly inhospitable beaches can lead to overheating of the eggs. The parents exchange places frequently. The off-duty parent may bathe in water before replacing its mate, and thus cool the eggs (Ryser 1985).

An extended hatching process starts eight days before the chicks emerge, when fine cracks appear in the eggs. Starting 3–4 days before hatching, the chicks start tapping on the eggshells, and 1–2 days before hatching they start to peep. Incubation on the Great Plains takes 24–26 days; usually all eggs hatch near dawn on the same day (Page et al. 1995).

As soon as they hatch, the precocial young feed themselves under the careful eye of their parents. The males stay with the young through fledging (29–47 days). Meanwhile the females (on the Pacific coast and Great Salt Lake) may attempt to find another mate (polyandry), but Great Plains females stay with their chicks until fledging. Atlas dates show a short breeding season. On the Great Plains (including Colorado) the short nesting season apparently precludes polyandry (Page et al. 1995).

BREEDING PHENOLOGY

CODE		# OF RECORDS	DATE RANGE
ON	Occupied Nest	1	23 May
NE	Nest with Eggs	1	11 Jun
FL	Fledged Young	4	26 Jun–7 Aug

DISTRIBUTION

Almost cosmopolitan in distribution, Snowy Plovers nest in temperate regions of all continents except Antarctica. Across the world, subspecies vary significantly in appearance, breeding biology, and vocalizations (Terres 1980). In the United States they nest patchily along the Gulf and Pacific coasts, and in the Great Basin and Great Plains.

The first Colorado record came from Denver in 1939. A month later Brandenburg and Bailey (1940) found them nesting in Kiowa County—along the same reservoir shorelines where they precariously persist today. In 1965 Ryder first found them nesting in the San Luis Valley when he banded two downy young (B&N).

Atlas field work found them in the same two locations—the lower Arkansas River Valley and the San Luis Valley. They nested at Antero Reservoir in South Park, probably for about 12 years (1971–1983), but not during the Atlas period. This species likely numbers fewer than 100 pairs statewide.

People and their pets, using the reservoirs for various types of recreation from fishing to off-road vehicles to birdwatching, can impact these little plovers. Some plovers nest at remote alkali playas, but many pick state wildlife areas that attract heavy recreational use. Changing demands for water from the prairie reservoirs may pose even greater danger to the nesting regime. In 1996 the Great Plains Reservoirs brimmed over with water stored for irrigation and for delivery to down-river users in Kansas, which meant flooded and therefore unsuitable nest sites. However, in years when Snowy Plovers nest with endangered Least Terns and Piping Plovers, they benefit from state efforts to foster those endangered species.

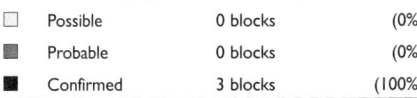

BREEDING EVIDENCE

REPORTED IN 3 (0%) OF 1745 PRIORITY BLOCKS

☐	Possible	0 blocks	(0%)
▩	Probable	0 blocks	(0%)
■	Confirmed	3 blocks	(100%)

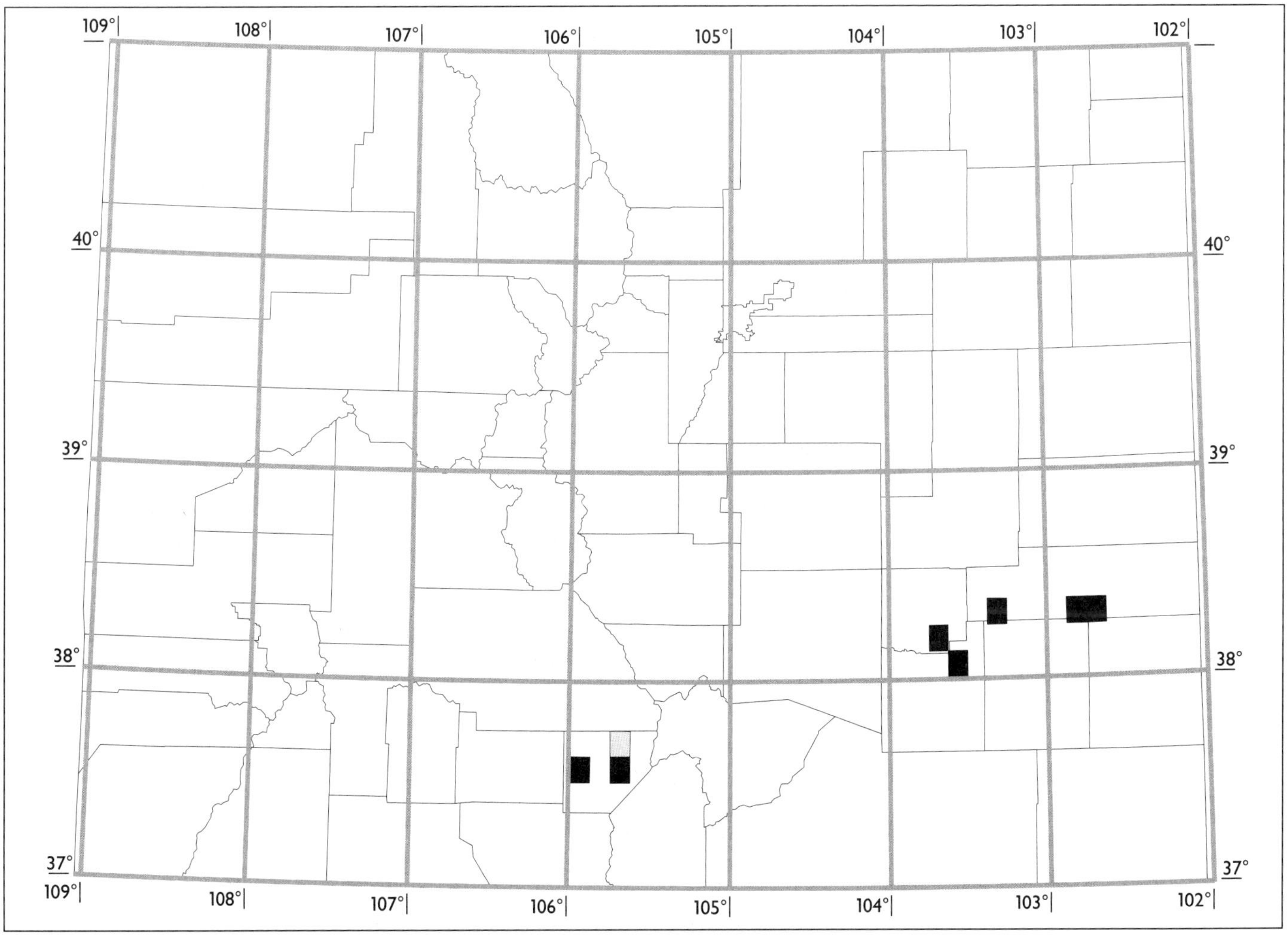
109°
108°
107°
106°
105°
104°
103°
102°
40°
39°
38°
37°

COLORADO DIVISION OF WILDLIFE / WILDLIFE RESOURCE INFORMATION SYSTEM

An explosive, loud, and melancholy *peep*, unlike any other sound in nature, announces the breeding territory of one of Colorado's rarest nesting species, the Piping Plover. From 1990 to 1995 between three and eight pairs nested annually in the state.

HABITAT

Piping Plovers in Colorado nest on broad, sandy beaches, preferably on islands, but water-level fluctuations from year to year force movement between sites. In wet years when water covers island beaches, they have nested on gravel bars, sandstone benches between bands of cliffs, and even a boat ramp (DLN).

Piping Plovers successfully adapted to nesting on the shores of off-stream irrigation storage reservoirs. In most years, they initiate nesting as water levels begin to drop due to irrigation withdrawals and evaporation. Reservoir water levels vary because the lakes receive water at irregular intervals—up to ten years; suitable habitat does not exist annually. Complicating the cycle, cottonwoods, tamarisks, and rank, exotic weeds such as kochia and white sweet-clover have invaded the shores of the reservoirs.

A few species—Snowy Plovers, Killdeer, Spotted Sandpipers, and American Avocets—nest with them in the same habitats. Endangered Least Terns also nest on sandy beaches, but show a stronger affinity to sites on islands. California Gulls, a threat because they eat eggs and young, nest in the same habitat but in segregated colonies. Whereas Piping Plovers depend on cryptic coloration, the other species, especially Least Terns, provide indirect protection to the plovers by aggressively defending their territories and driving away gulls and other predators.

BREEDING

Adults return from their winter grounds along the Gulf Coast of Texas in late April. The first pairs begin to nest in early May. They raise only one brood per year, but if nests fail, renesting occurs most years, especially if habitat quality remains high throughout the summer. Renesting may occur as late as mid July.

Males make the nests, simple scrapes in the ground, usually far from cover. They line the nests by tossing pebbles, and sometimes twigs, into the nest cup. Females ultimately choose one from among the several nest sites the male has constructed (Haig et al. 1992).

Complete clutches contain 3–4 cryptically colored eggs, although three is usual for renesting attempts (Haig et al. 1992, DLN). They lay eggs every other day, and begin to incubate after the clutch is complete. The sexes share incubation; at changeover, the incubating bird stands and performs a tilt display as the mate walks underneath to perch on the eggs (Haig et al. 1992). Incubation in Colorado averages 26 days. All eggs usually hatch on the same day. Fledging (for precocial birds like these, the date of first flight) takes 28–31 days.

Piping Plover nests blend well into the sand, silt, and gravel. Even with the camouflaged eggs, predators frequently succeed in marauding the nests and young. Practically invisible while incubating eggs, a female will quietly slip off her nest and distract predators with loud peeps to draw them away from the nest site. Adults may engage in distraction displays to protect young, but only on or near the day of hatching: the young "freeze" motionless while the adults try to lure predators away from the young birds.

BREEDING PHENOLOGY

CODE		# OF RECORDS	DATE RANGE
NY	Nest with Young	2	26 Jun–6 Jul
FL	Fledged Young	1	28 Jun

DISTRIBUTION

Piping Plovers nest in three separate regions of temperate North America. One population, currently only a few pairs and classified as Endangered, now nests only on the shores of northern Lake Michigan. Another, classified as Threatened, is strictly coastal, from the Maritime Provinces of Canada south to South Carolina. The third, also classified as Threatened, now nests in the prairies from southern Canada south to Kansas and Colorado. Only 2,100 pairs remain in the three populations, which continue to decline (Haig et al. 1992).

Except for one failed nest at Prewitt Reservoir in 1949 (B&N), all Colorado nesting records are from southeastern Colorado from 1989 and later. Piping Plovers probably colonized Colorado in response to droughts elsewhere (Sidle and Kirsch 1993) or after high water flooded habitat at the nearest known nesting site in the panhandle of Oklahoma (Nelson 1993b) or both.

The Atlas map shows the four sites where Piping Plovers nested at some time between 1987 and 1995. In 1990 and 1991, they nested only at the four Great Plains Reservoirs in Kiowa County. In 1992 they colonized Adobe Creek and John Martin reservoirs (Nelson and Aid 1993). In 1995 John Martin became the primary location.

Many factors limit their populations: detrimental human activities on recreational beaches (people, dogs, vehicles), destruction of beach habitat, water control policies, and predation on eggs and chicks (Haig et al. 1992). CDOW has implemented a recovery plan that includes habitat improvement and vegetation control, agreements with irrigation companies, predator exclosures, and closures of nest areas to human activity. Habitat recovery includes manually removing cottonwood saplings of all sizes that envelop nesting beaches at three reservoirs. This habitat repair has now reclaimed the island on Adobe Creek Reservoir for use by terns and plovers. CDOW closes beaches wherever the plovers nest in order to keep out people and livestock. Of questionable effectiveness, fences built around some especially vulnerable nests keep large mammals and gulls at bay but do not work as well with snakes and small predators such as weasels (Nelson and Aid 1993, Estelle and Mabee 1994).

Without human intervention, Piping Plovers would likely not survive for long in Colorado. The peril of all populations of Piping Plovers justifies the intensive management efforts necessary for their survival in Colorado and elsewhere—a survival by no means assured.

BREEDING EVIDENCE

REPORTED IN 3 (0%) OF 1745 PRIORITY BLOCKS

Possible	0 blocks	(0%)
Probable	1 block	(33%)
Confirmed	2 blocks	(67%)

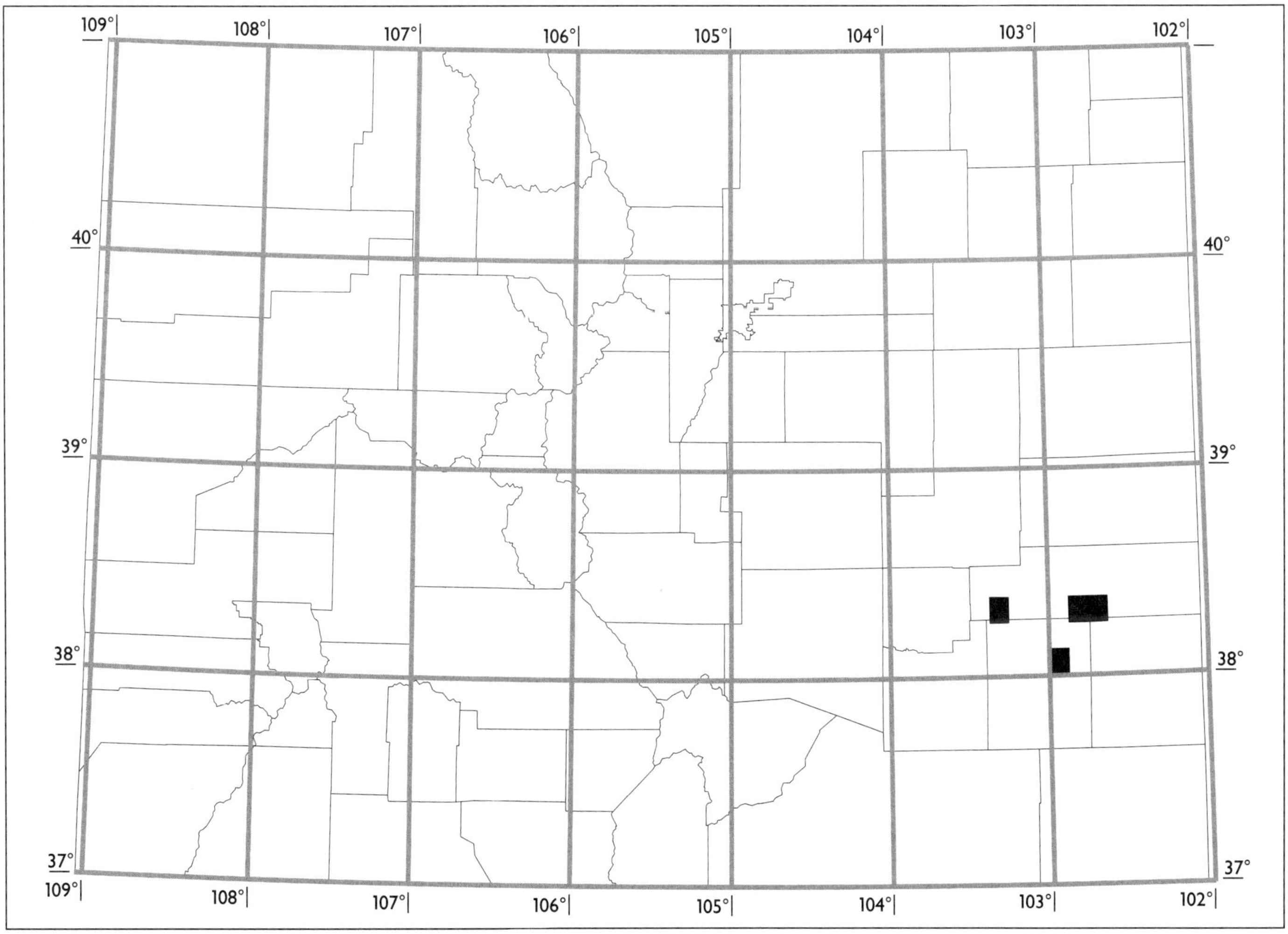
109°
108°
107°
106°
105°
104°
103°
102°
40°
39°
38°
37°

COLORADO DIVISION OF WILDLIFE / WILDLIFE RESOURCE INFORMATION SYSTEM

Any attempt to describe the lifestyle of Killdeer requires an overflowing supply of superlatives. When a Killdeer circles an intruding hunter or biologist, its *vociferus* (sic) calling of its name demands attention and means that other birds will soon notice and take wing. The broken-wing distraction display has few rivals in the bird world. In 1840 Alexander Wilson described Killdeer as "known to almost every inhabitant of the U.S," (Santner *in* Brauning 1992). As valuable as any bird species to farmers and ranchers, Killdeer eat massive quantities of destructive insects (Bent 1927). The hard-to-overlook Killdeer merits more appreciation by birdwatchers.

HABITAT

Unique in their widespread distribution, Killdeer stake out their nesting territories in habitats that vary from shortgrass prairies to fallow fields, and from mountain meadows to riverbeds and lakeshores. Although noticeably more abundant near the shores of lakes and streams, they can nest many miles from water (Bent 1927).

Like other plovers, Killdeer nest on barren ground, but they feed in a multitude of habitats: wet places, dry grasslands, gravel roads, etc. They often associate with wild animals or livestock at stock ponds, corrals, and farms; on 29 June 1988, above timberline in a Flat Tops Wilderness Atlas block (Crescent Lake, 39107H2), one group of four scurried around an elk wallow (HEK).

Higher precipitation at higher elevations means taller vegetation. Consequently, those nesting in the mountains utilize mainly cobbly riverbanks.

Killdeer use such diverse habitats as heavily grazed pastures, newly plowed fields, agricultural lands, disturbed areas, and old fields. They can exploit patches of open country, smaller than those used by other grassland nesters such as Mountain Plovers, Grasshopper Sparrows, and longspurs, which need more extensive stretches of habitat.

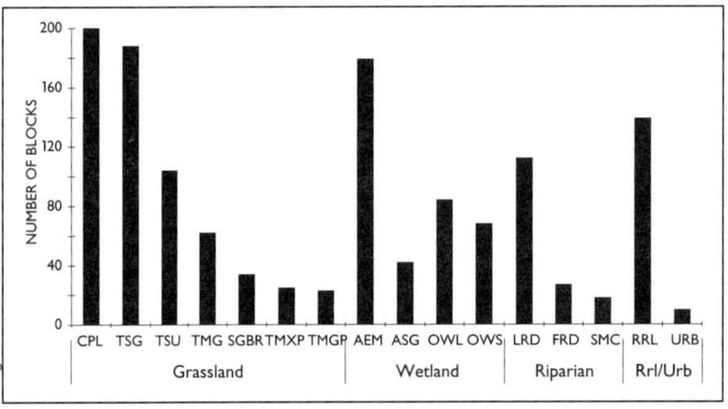

200
160
120
80
40
0
NUMBER OF BLOCKS
CPL TSG TSU TMG SGBR TMXP TMGP AEM ASG OWL OWS LRD FRD SMC RRL URB
Grassland
Wetland
Riparian
Rrl/Urb

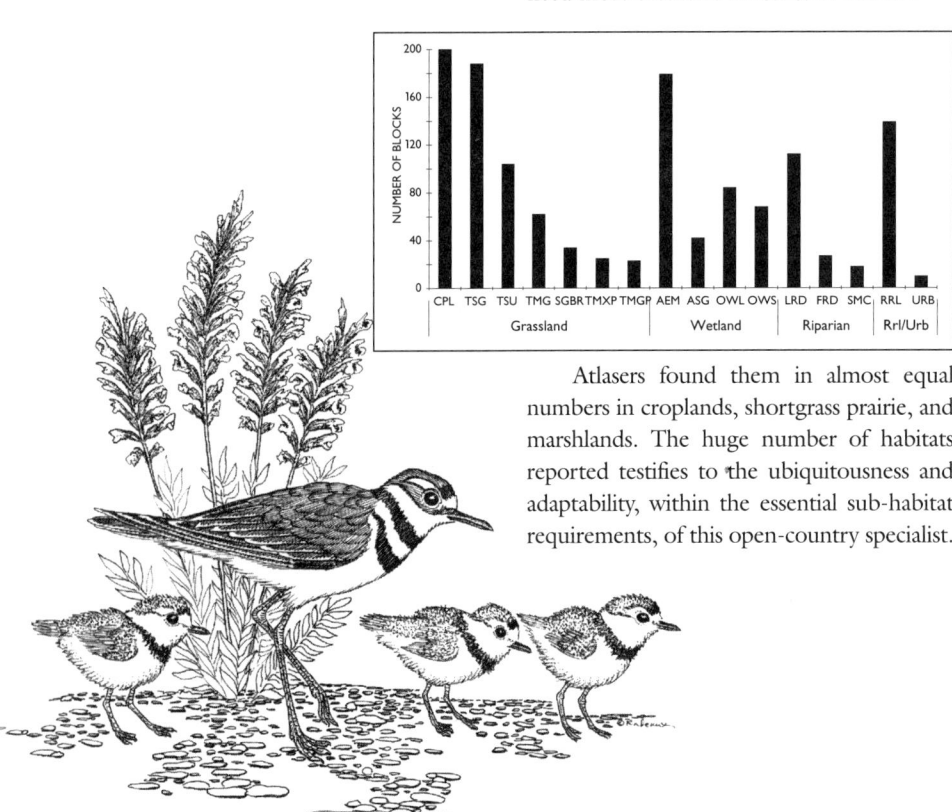

Atlasers found them in almost equal numbers in croplands, shortgrass prairie, and marshlands. The huge number of habitats reported testifies to the ubiquitousness and adaptability, within the essential sub-habitat requirements, of this open-country specialist.

BREEDING

Nest sites always have an extended view of the surrounding area (Bent 1927). They have certain consistent features: sparse cover, a raised area for the nest, nearby large objects such as dead wood or mounds of earth (perhaps for concealment), and ground soft enough for scrapes. The birds need, within a yard of the nest, a supply of stems, wood chips, or small pebbles; these they toss into the nest with their bills (Eaton *in* Andrle and Carroll 1988, Chu *in* Brewer et al. 1991, Palmer-Ball 1996).

Pebbles usually line the nests, which are simple scrapes on the ground. The nest lining can also contain twigs, weeds, seeds, or other materials. The inconsequential nest and inconspicuous eggs serve to disguise the nest effectively.

Complete clutches almost always contain four, surprisingly large, eggs. Like the Sandgrouse of the Sahara Desert, Killdeer incubating in hot weather may wet their bellies to cool their eggs (Ehrlich et al. 1988). Killdeer need some open water near their nesting territories for drinking and bathing, but it can be the smallest of puddles.

The precocial young literally "hit the ground running." The adults keep an incredibly vigilant eye on their young until they fledge, but the young feed themselves.

Able to exploit many habitats, Killdeer nest as early as April in Colorado (DLN) and may have up to three broods a season (Bent 1927). Their propensity to perform distraction displays yielded over half of the Confirmations. Very likely some atlaser DD codes referred to mere "agitated behavior" —perhaps comparable to real distraction displays of other, less excitable, species. Only 40 Confirmations came through nest observations, not surprising considering the great lengths the birds employ to divert eyes away from the nests.

A few Killdeer winter in Colorado along rivers and streams that do not freeze. Migrating Killdeer are noisy and can act agitated, making the true nesting status, especially in late summer, difficult to determine.

BREEDING PHENOLOGY

CODE		# OF RECORDS	DATE RANGE
C	Courtship	7	14 May–18 Jun
NB	Nest Building	1	23 May
ON	Occupied Nest	11	29 Apr–5 Jul
NE	Nest with Eggs	22	30 Apr–12 Jul
NY	Nest with Young	6	27 May–4 Jul
FL	Fledged Young	218	12 May–15 Aug

DISTRIBUTION

Killdeer occur only in the Western Hemisphere, where they nest below the treeline in Canada and the U.S. In Colorado they nest statewide in open habitat, up to 10,000 feet (3048 m) in South Park. Well over half of all Atlas blocks in the state recorded Killdeers; atlasers found them in all 63 counties and Confirmed breeding in all but one. Their distraction displays and diminutive, downy fledglings enabled observers to Confirm breeding in 54% of the blocks where they found these constantly agitated sentinels.

Killdeer are so conspicuous that they are easy to take for granted, but in the late eighteenth century, market hunting and habitat disturbance caused their near extirpation in New England (Bent 1927, Ehrlich et al. 1988). They rebounded as persecution abated.

Priorities 1995 ranked the Killdeer as the species least in need of conservation measures or monitoring of all species nesting in Colorado (264th out of 264 species). These beneficial, native species are worthy citizens of our state. They do not negatively impact other species and they even provide protection to other shorebirds (including diminishing Snowy and Piping plovers) by their vigilant defense of nesting territory.

BREEDING EVIDENCE

REPORTED IN 1049 (60%) OF 1745 PRIORITY BLOCKS

☐	Possible	270 blocks	(26%)
▩	Probable	214 blocks	(20%)
■	Confirmed	565 blocks	(54%)

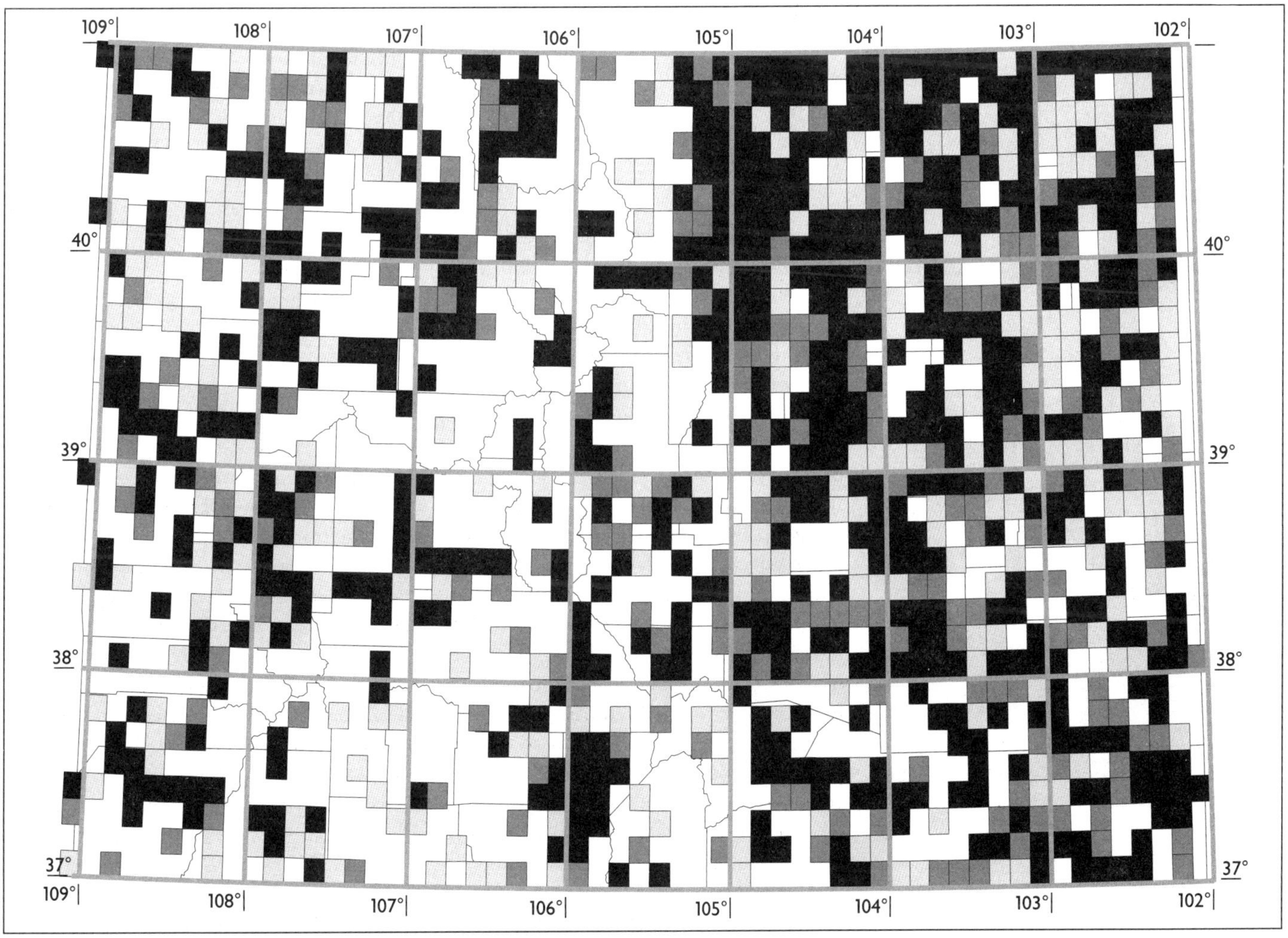
109°
108°
107°
106°
105°
104°
103°
102°
40°
39°
38°
37°

COLORADO DIVISION OF WILDLIFE / WILDLIFE RESOURCE INFORMATION SYSTEM

MOUNTAIN PLOVER *Charadrius montanus*

More than half of the world's Mountain Plovers breed in Colorado. Despite their name and classification, they breed neither in the mountains nor at the shore—we might more appropriately call them "Prairie" or "Bison" Plovers. Unlike their noisy relatives, the Killdeer, they seldom call attention to themselves; on the shortgrass prairies where they breed, observers can easily overlook these neat, slim plovers.

HABITAT

In 1839 Townsend collected the first prairie plovers, in central Wyoming; he called the species "Rocky Mountain Plover," later shortened to its present, inappropriate, name. Atlasers found them primarily in the arid grasslands of the Great Plains, where the grass grows no taller than 3 inches. In South Park and similar mountain parks, shortgrass habitat (short structure, different grasses) attracts breeders. In the San Luis Valley they occupy stunted shrublands of widely spaced dwarf rabbitbrush (HEK).

Nesting plovers choose shortgrass prairie grazed by prairie dogs, bison, or (now) cattle, overgrazed tallgrass, and (on fragmented prairie) fallow fields (Knopf 1996). Atlasers discovered nests (only five) in shortgrass only, and found fledglings in both grasslands and plowed fields as well as mat shrublands.

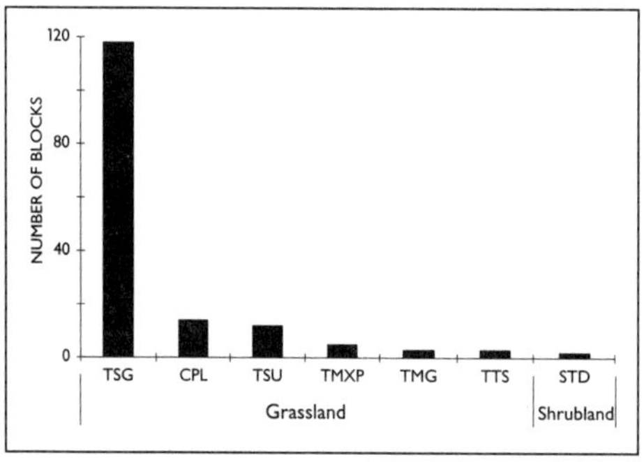
NUMBER OF BLOCKS
120
80
40
0
TSG
CPL
TSU
TMXP
TMG
TTS
STD
Grassland
Shrubland

Prior to settlement, these plovers shared their territories with roaming bison. They even have a distraction display in which they fly up in the face of an intruding bison—or cow—to divert the animal from stepping on nest or young (Fritz Knopf pers. comm.).

BREEDING

Mountain Plovers arrive at their Colorado breeding grounds in March. Males attract hens with a "falling leaf" flight-display: rocking back and forth with wings held in a sharp V, they drop from 15–30 feet in the air. Then they perform a "butterfly display" with slow, deep wing-beats, and finally conduct a ritualized nest scraping for the females. They situate their nests on hill-tops and in dry swales, sometimes in the shelter of manure piles (Graul 1975).

Hens often lay the first eggs on bare ground, but they add a small quantity of rootlets and grass by the time they complete the clutch of three (Bradbury 1918b). All Colorado plover species share this bare-ground affinity (Knopf and Miller 1994). The pair bond is fleeting; females frequently leave their first nest to the males so they can lay and incubate a second set of eggs (often with a different male), which the female incubates (Graul 1975).

The young, led away from the nest site soon after hatching, need shade; taller plants, like those along roadsides or field edges, or even dirt clods in plowed fields, provide this survival element. Their use of road edges explains why 84% of Atlas Confirmations involve fledglings.

BBS trends and population surveys by various researchers indicate a major decline of this bison/cow associate. The BBS shows a drop of two-thirds in 25 years (Knopf 1996). The population totals 8,000–10,000 birds (Knopf 1996).

An estimate using Atlas abundance codes suggests that Colorado has 3,600 breeding "pairs" of Mountain Plovers—at least three-fourths of all Mountain Plovers. Atlas observers, in making their estimates, did not take into account the female-dominant breeding biology. They also made conservative estimates, as they did for most comparatively unusual species.

BREEDING PHENOLOGY

CODE		# OF RECORDS	DATE RANGE
C	Courtship	3	21 May–25 May
NB	Nest Building	1	7 Apr
NE	Nest with Eggs	1	7 Jun
NY	Nest with Young	2	1 Jun–15 Jun
FL	Fledged Young	44	20 May–8 Aug

DISTRIBUTION

Mountain Plovers breed only in western North America, from southern Alberta and Montana to New Mexico and western Texas. They winter in habitats similar to their summer haunts, but in two geographically separate places: southern California, and southwest Texas and northern Mexico. To which winter haven Colorado plovers go is undetermined.

An important population breeds in Phillips County, Montana, but the major strongholds are the Pawnee National Grassland (10–20% of the total breeding population) and southeastern Colorado (40–50%; Graul and Webster 1976, Knopf 1996, Atlas).

As cultivation and irrigation consumed their preferred habitat, Mountain Plovers abandoned large areas where they formerly flourished (Graul and Webster 1976). By 1900 Colorado Springs already had "built over" plover habitat (Aiken and Warren 1914). Farther east, the plow turned much grassland to cropland.

By sending observers to all parts of the state, the Atlas found more widespread breeding than expected. The significant population south of Weld County comprises 75% of the state's plovers—and therefore almost half of the world breeders. South Park also has a significant number, in the area from Fairplay south to the Park County line.

Atlasers found them in three counties where they formerly nested and in two parts of the state without recent reports: North Park and the San Luis Valley. In 1873 Henshaw, stationed in Fort Garland, collected three—one ready to deposit an egg—on the "dry plains near the Rio Grande."

The Atlas map almost matches that of another declining species—the swift fox. Plovers tagged with radio telemetry devices proved a fox/plover connection: in the Pawnee every radio tag recovered was three feet down in a swift fox burrow (Jim Fitzgerald pers. comm.).

Population estimates by researchers project twice as many of the rapidly declining Mountain Plovers as of the endangered Piping Plover, a species spread out over half the continent (Haig et al. 1992, Knopf 1996). "Preservation" of the core population of Mountain Plovers—with a more compact distribution—requires efforts in only a few states. In March 1996, Denver Audubon Society initiated coordination of conservation efforts for this Great Plains endemic.

Forest Service biologists intend to ensure that these plovers do not join the ranks of endangered species—but Colorado's two national grasslands alone cannot preserve the species. Biologists must spread their efforts to private ranchland. Because Mountain Plovers have the unusual characteristic of a biology compatible with the grazing economy of eastern Colorado, recovery may prove less contentious than for some other species.

BREEDING EVIDENCE

REPORTED IN 116 (7%) OF 1745 PRIORITY BLOCKS

☐	Possible	42 blocks	(36%)
▩	Probable	28 blocks	(24%)
■	Confirmed	46 blocks	(40%)

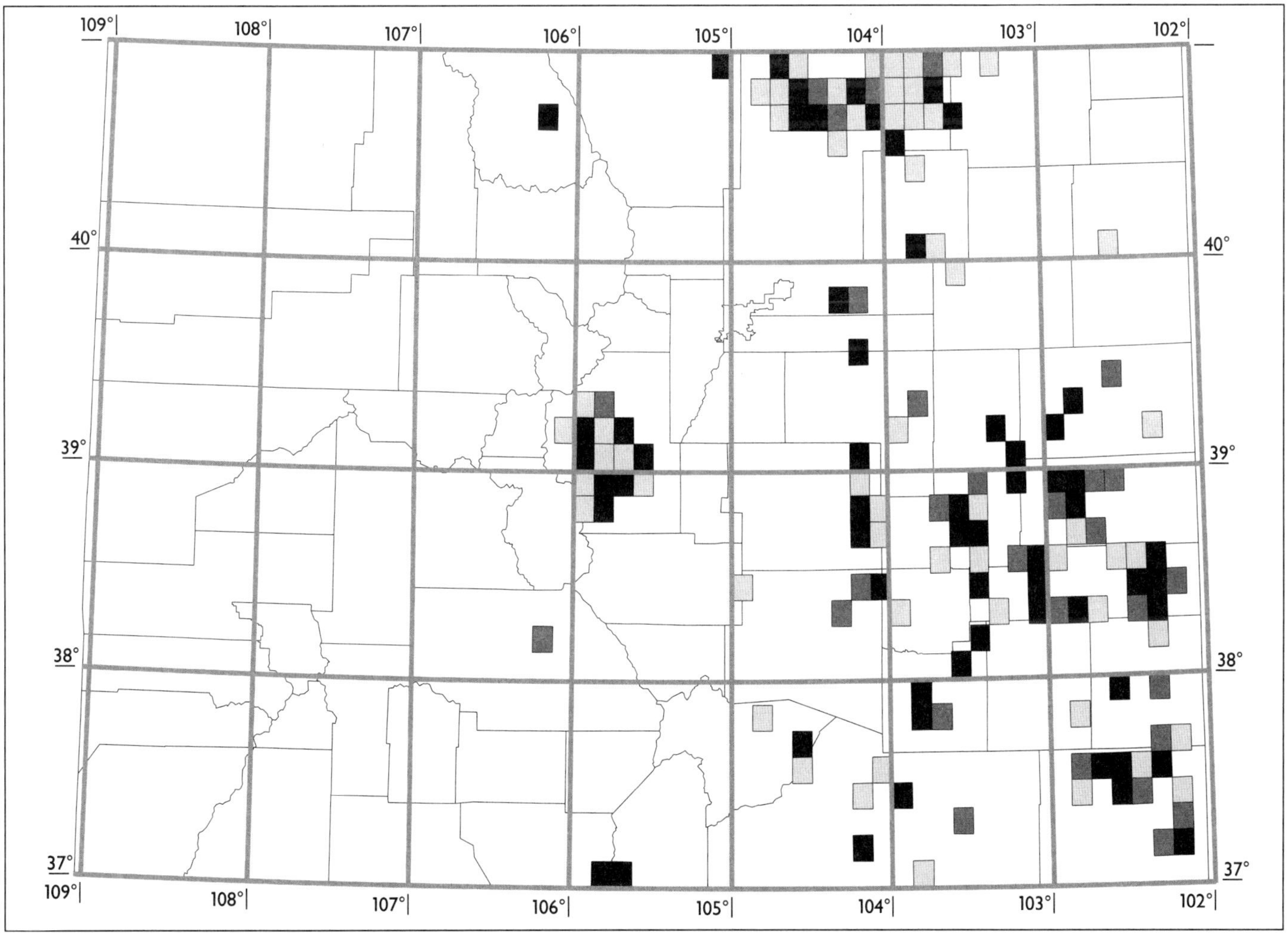
109°
108°
107°
106°
105°
104°
103°
102°
40°
39°
38°
37°

COLORADO DIVISION OF WILDLIFE / WILDLIFE RESOURCE INFORMATION SYSTEM

BLACK-NECKED STILT

Himantopus mexicanus

Slender elegance in black and white describes this dainty summer resident of Colorado. They dignify Colorado wetlands—a few of them anyway—as they wade with long, slim, pink-red legs and seize insects with long, needle-pointed bills. First cousins of the web-footed American Avocet, Black-necked Stilts have feet only partially webbed, and therefore swim less proficiently than avocets (Bent 1927).

HABITAT

In the West stilts prefer freshwater lakes and ponds, wet meadows, and irrigated fields (Ryser 1985, A&R). They often associate with avocets but their habitat selection differs somewhat. Stilts prefer fresher water, marshes, and flooded grassy areas; they do not venture as far out from shore as do avocets (Ryser 1985). Their major food is insects, taken in or on water or on land, and they peck with their bills rather than sweep them through the water like avocets (Ryser 1985). Atlas workers found them in marshes and on lake edges.

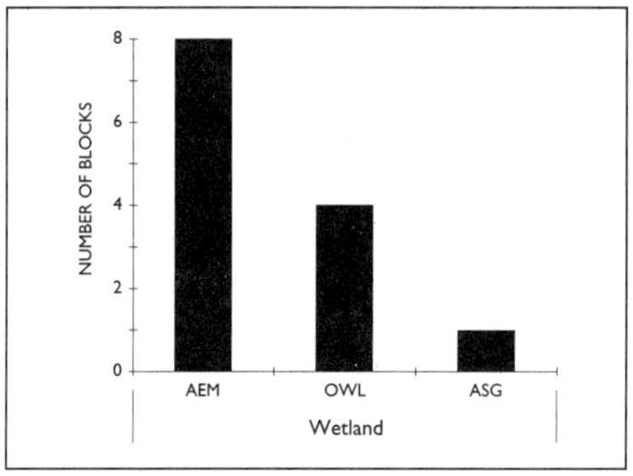

NUMBER OF BLOCKS
0
2
4
6
8
AEM
OWL
ASG
Wetland

BREEDING

Stilts nest in small colonies of a few to several dozen pairs. They pick dry sites near water or build a mound in shallow water (Harrison 1979). The colonies may include avocets, and the two species even lay eggs in each other's nests (Ryser 1985). Nests vary from mere scrapes lined with shells to an elaborate mound of mud with sticks, shells, and other adornment (Harrison 1979). Like avocets, stilts may shove additional nesting material under the eggs if the water level rises (Ryser 1985). They produce single broods from April through July (Harrison 1979). Atlas workers' nest data indicate June and July nesting here.

The female lays 3–5 eggs; both parents incubate for about 25 days (Bent 1927, B&N). Stilts vociferously defend their nests (B&N), and sometimes the whole colony cooperates to distract intruders (Harrison 1979). Stilts differ from avocets in their distraction displays in that they fly above and scream at intruders, but don't dive on them (Ryser 1985). They also perform distraction incubation (crouching as if on eggs), wing flagging (raising one wing at a time), and injury feigning. One other interesting display, the "Butterfly Flight," may have territorial meaning (Ryser 1985). In it the bird hovers 15–30 feet (5–10 m) over the water, neck in, tail spread, and long red legs dangling. It sinks slowly to the water's surface, then rises and repeats the process nearby.

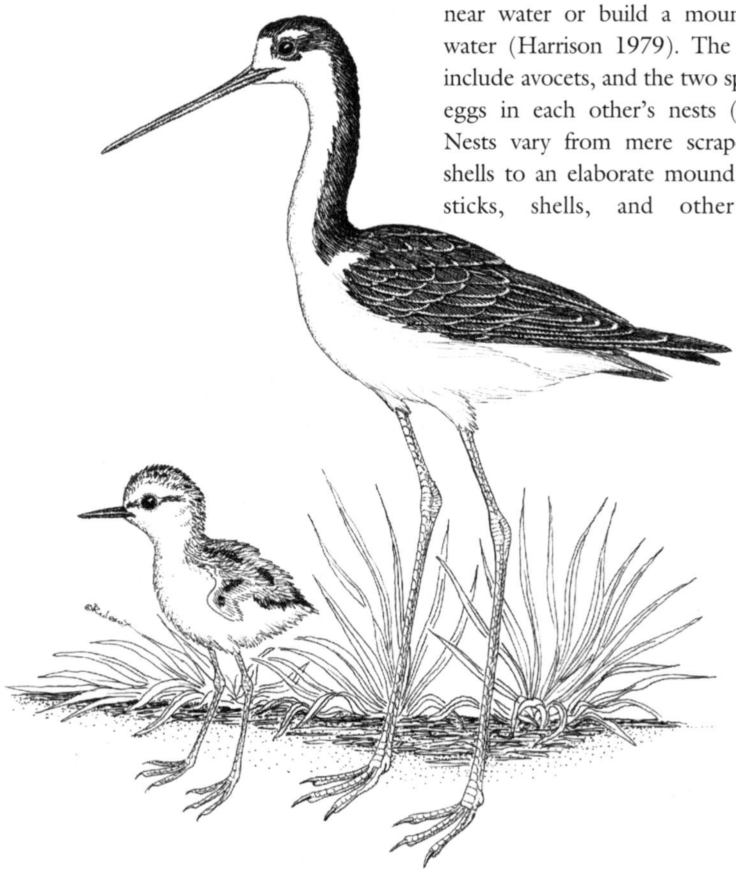

BREEDING PHENOLOGY

CODE		# OF RECORDS	DATE RANGE
ON	Occupied Nest	4	23 May–20 Jun
NE	Nest with Eggs	1	5 Jul
FL	Fledged Young	1	3 Jul

DISTRIBUTION

Stilts breed in scattered sites across the western United States and the Atlantic and Gulf coasts, south into South America. In winter they retreat southward to the coasts of south Texas and Mexico.

In Colorado stilts have their longest known history in the San Luis Valley; Charles Aiken collected a female at San Luis Lake on 7 May 1871. The earliest record of nesting also came from the valley, at Fort Garland in 1875 (Henshaw of the Wheeler Survey). Stilts have bred in southeastern Colorado since the late 1970s, and subsequently a few pairs have nested on the northeastern plains and in North Park (A&R).

Atlas workers found the same breeding pattern: none in western Colorado, but breeding Confirmed in the San Luis Valley and in southeastern and northeastern Colorado, and Probable breeding in North Park. Fewer than 500 pairs, probably fewer than 250 pairs, nest in Colorado. These elegant birds may be increasing their breeding range, and they may reflect this increase in Colorado.

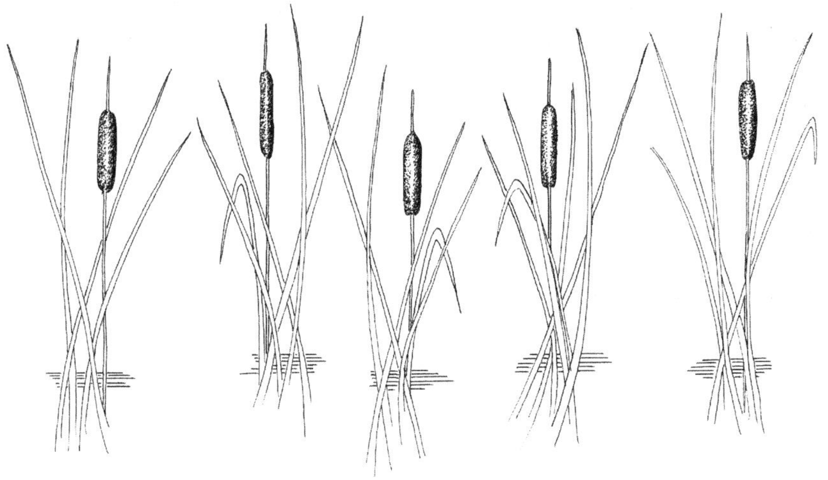

BREEDING EVIDENCE

REPORTED IN 5 (0%) OF 1745 PRIORITY BLOCKS

☐	Possible	1 block	(20%)
▨	Probable	1 block	(20%)
■	Confirmed	3 blocks	(60%)

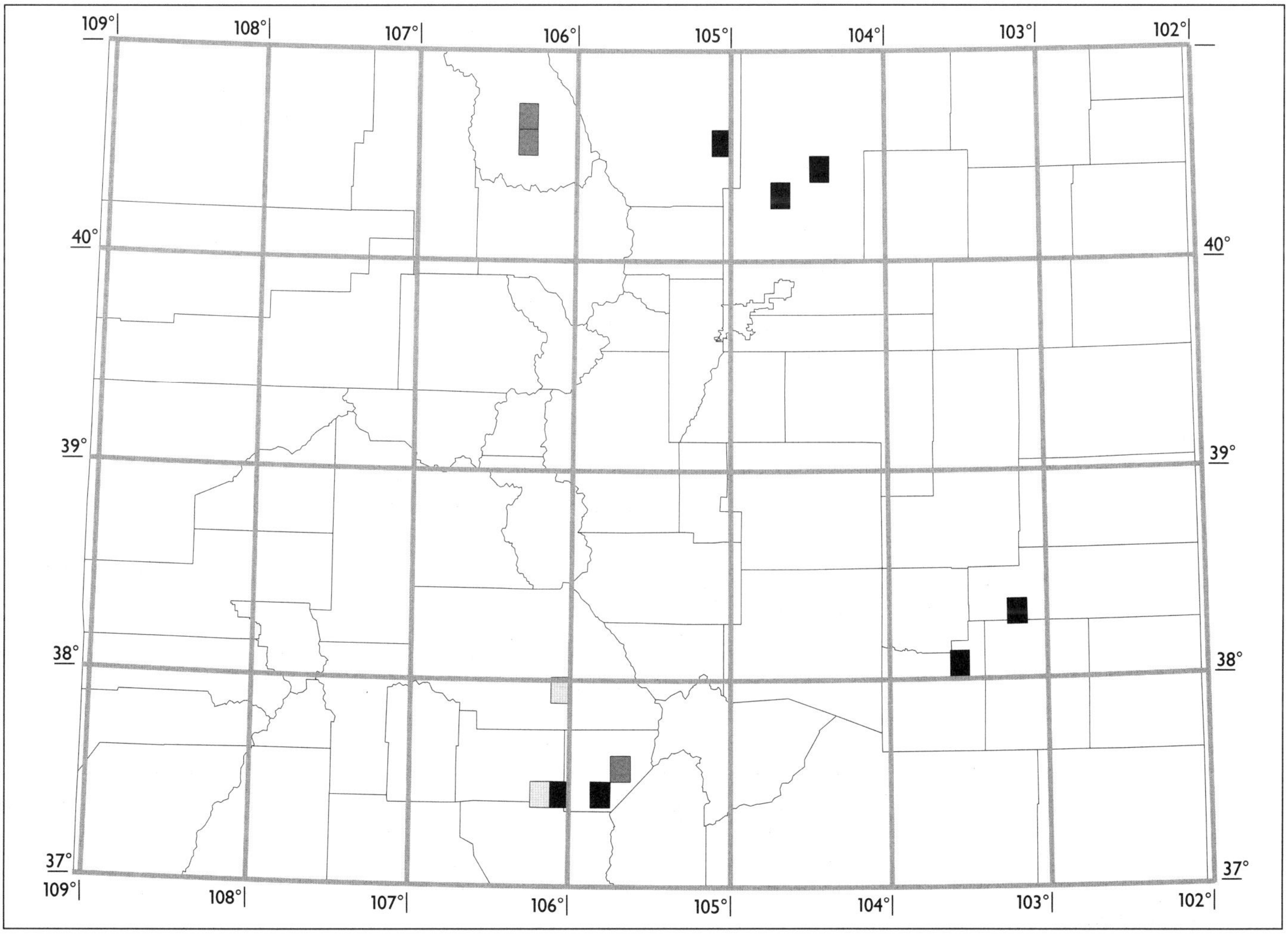
109°
108°
107°
106°
105°
104°
103°
102°
40°
39°
38°
37°

COLORADO DIVISION OF WILDLIFE / WILDLIFE RESOURCE INFORMATION SYSTEM

With long blue legs, the "Blueshanks" wades the shallows and, in search of food, cuts its bill through the water like a scythe. Its bright cinnamon-colored head and neck and black-and-white body make it one of our most conspicuous and beautiful shorebirds.

HABITAT

Avocets frequent "shallow, foul bodies of alkaline or brackish water and their fringing flats of mud, alkali, or salt" (Ryser 1985). They favor saline and semi-saline lakes with little human disturbance and mudflats or otherwise barren edges (Hamilton 1975). Of the Atlas avocet reports, 74% occurred in emergent wetland marshes and lakes.

Their long legs, with the three front toes encased in a web, make wading on the mud and swimming easy. Their upcurved sharp bills sweep from side to side and find food by touch in the bottom mud. Groups of them sometimes form a phalanx scything in unison as the birds wade forward together. They eat more animal (mostly aquatic insects and crustaceans) than plant food (Ryser 1985).

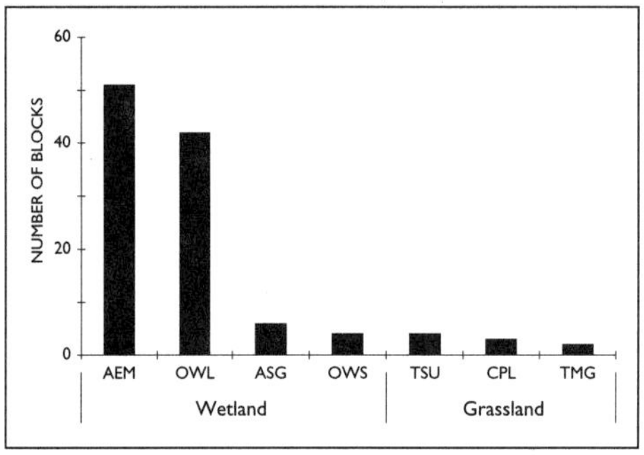
60
40
20
0
NUMBER OF BLOCKS
AEM
OWL
ASG
OWS
TSU
CPL
TMG
Wetland
Grassland

BREEDING

The female initiates pair bonding by repeatedly approaching the male in a nonaggressive posture. She even initiates copulation, but afterwards both male and female cross bills and run together, sometimes with his wing spread out over her back (Ryser 1985).

With no attempt at concealment, they place nests in simple scrapes on the ground or on elevated piles of debris (B&N, Harrison 1979). The 3–4 eggs blend in well with the surroundings (Harrison 1979, Terres 1980) and both parents incubate. The length of incubation drops as the weather warms: 27 days for a nest started in April, down to 23 for one started in June (Gibson 1971). Atlas workers found nests with eggs from late May through June.

Breeding avocets maintain three territories. Before egg-laying, a pair has one territory centered on a feeding area that the pair keeps during the entire nesting period; the nest site lies within a few hundred yards. Adults, singly or as a pair, defend this area, which has definite boundaries respected by other avocets. The second territory, a small one around the nest, the pair maintains during incubation, but defends only when both mates are present; often one parent is away defending the feeding territory. The third territory develops after hatching. Chick-centered, it has somewhat mobile boundaries with a diameter of 20–100 yards (20–100 m). The birds defend it both intra-specifically and inter-specifically—e.g., against blackbirds and ducks (Gibson 1971).

American Avocets often nest in loose colonies of 15–20 pairs. They disperse their nests for better concealment: mean distances between them vary from 100 to 260 feet (30–80 m; Gibson 1971). The birds defend their nests in a group distraction display: all adults leave their eggs, gather together, and run at the intruder with shrill cries, often with outstretched neck and wings (Harrison 1979, Ryser 1985, B&N). Aiken and Warren (1914) noted a "milling dance" on an island in Utah, maybe the group circling display. In this display the adults, all facing inward, form a rough circle and lean forward with necks extended and bill near and parallel to the water. Trumpeting loudly, they then circle around first this way and then that way (Hamilton 1975). Arrival of a potential new nesting pair may instigate this display (Robinson et al. 1997). Young defend themselves by diving underwater; using their wings and feet, they can swim underwater as far as 20 feet (7 m; Robinson et al. 1997).

Aiken and Warren (1914) also noted that avocets don't taste good—they have a fishy flavor—and most of the Utah birds they saw had a heavy infestation of worms in their digestive tracts, abdominal cavities, and even eye sockets. Another singular characteristic: unlike other shorebirds they leave the water to defecate! (Hamilton 1975).

Atlasers found the earliest nest in mid April (comparable to the earliest date in Utah), but the remainder they found from mid May to 30 June. They recorded the precocial young only from mid June to mid July. Atlas workers found most nests and young in marshes (42%), around lakeshores (30%), and in salt meadows (8%).

BY BARBARA L. WINTERNITZ

BREEDING PHENOLOGY

CODE		# OF RECORDS	DATE RANGE
ON	Occupied Nest	13	14 Apr–30 Jun
NE	Nest with Eggs	3	20 May–28 Jun
FL	Fledged Young	17	15 Jun–17 Jul

DISTRIBUTION

American Avocets breed commonly throughout the West in the Great Basin and Great Plains south to California and Texas. They formerly bred along the Atlantic coast and farther north in Canada (Robinson et al. 1997). They winter on the Pacific and Gulf coasts from southern California and Florida into Mexico.

In 1875 the San Luis Valley yielded the first Colorado avocet nest, found by Henshaw of the Wheeler Survey in Costilla County (B&N). Although large numbers nest around the Great Salt Lake in Utah, a gap used to exist between there and breeding avocets in the Colorado mountain parks. However, since 1986 they have extended their breeding range to the Western Slope (A&R).

The Atlas map shows substantial breeding in northeastern Colorado, both along the South Platte corridor and the edge of the Front Range from Denver north. The cluster in east-central Colorado consists of birds scattered at reservoirs and at permanent and ephemeral ponds from the Arkansas River north to the Smoky Hill River in Cheyenne County. The San Luis Valley and North Park have concentrations, but atlasers found more in South Park than expected and a surprising cluster in four blocks east of I-25 in Huerfano and Las Animas counties. On the Western Slope they located only a few breeders from Delta to Browns Park and Confirmed breeding in only two blocks in the Grand Junction area. The Atlas map suggests a gradual expansion by these blue-shanked waders.

BREEDING EVIDENCE

REPORTED IN 97 (6%) OF 1745 PRIORITY BLOCKS

☐	Possible	41 blocks	(42%)
■	Probable	21 blocks	(22%)
■	Confirmed	35 blocks	(36%)

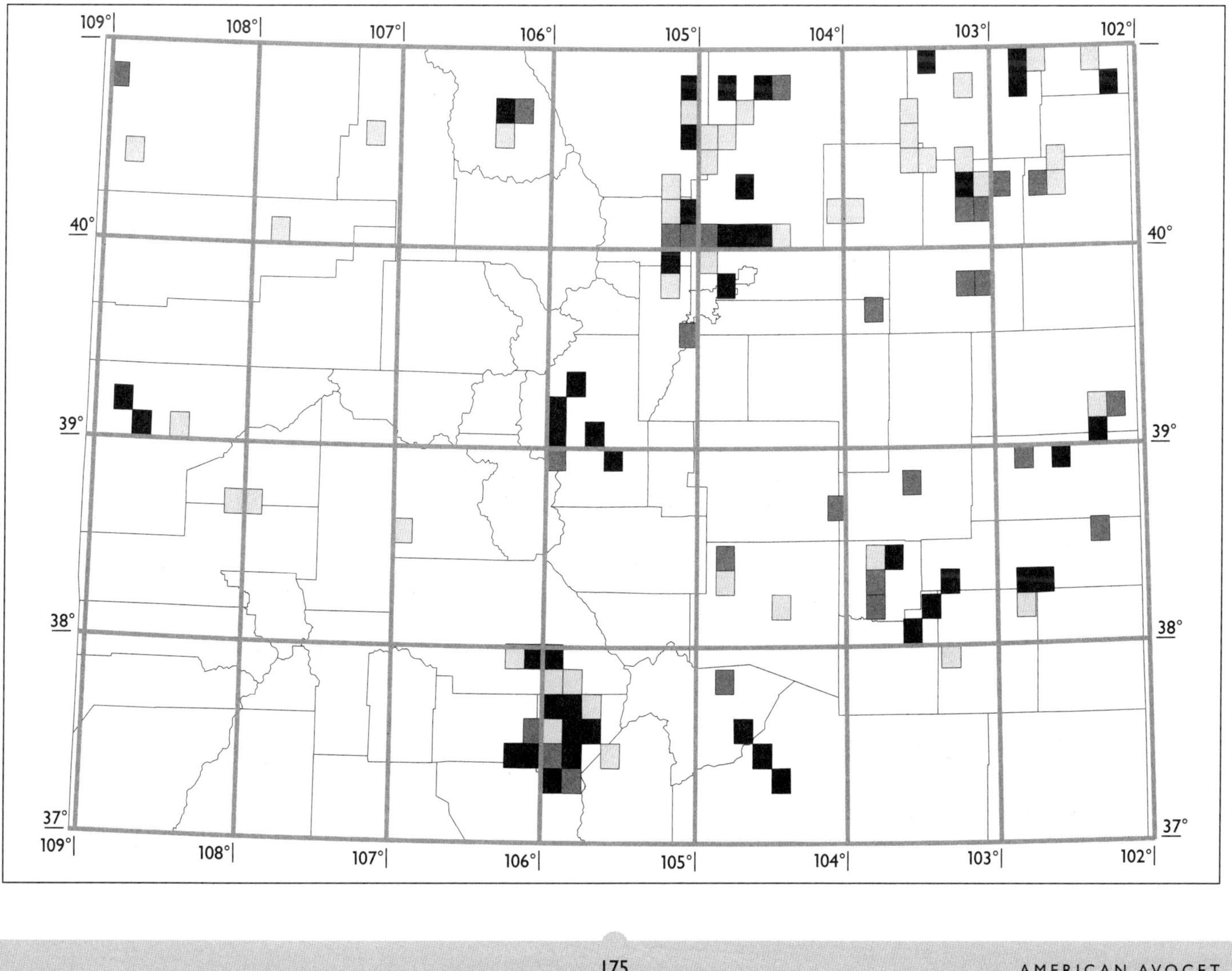

109°
108°
107°
106°
105°
104°
103°
102°
40°
39°
38°
37°

COLORADO DIVISION OF WILDLIFE / WILDLIFE RESOURCE INFORMATION SYSTEM

The wide-open spaces and harsh climate of Wyoming spill south dramatically into Colorado in North Park. Bitterly cold winters come early and extend far into what, on the plains, is springtime. Cool summers and a short growing season limit agriculture to haying operations. The spell of the North extends to the wildlife inhabiting North Park. Transplanted moose thrive in the valley and adjacent forests. Birds with northern affinities such as Buffleheads, Northern Waterthrushes, and Willets reach their southern breeding limit here.

HABITAT

Willets require nesting situations with proximity to open water for foraging, drinking, and bathing, which concentrates them near small ponds. They nest in nearby prairies or close to beaches on alkali flats (Bent 1929).

Drab breeding plumage renders silent Willets inconspicuous. However, a territorial Willet makes a Killdeer seem practically mellow by comparison. Their habit of "flying a long distance to meet the intruder and making a great fuss everywhere but near the nest" (Bent 1929) makes it extremely difficult to locate nests.

They wade in deeper water than most shorebirds (Terres 1980), and use this specialized feeding habitat to feed mainly on aquatic prey, including aquatic insects, worms, and fish (Ehrlich et al. 1988).

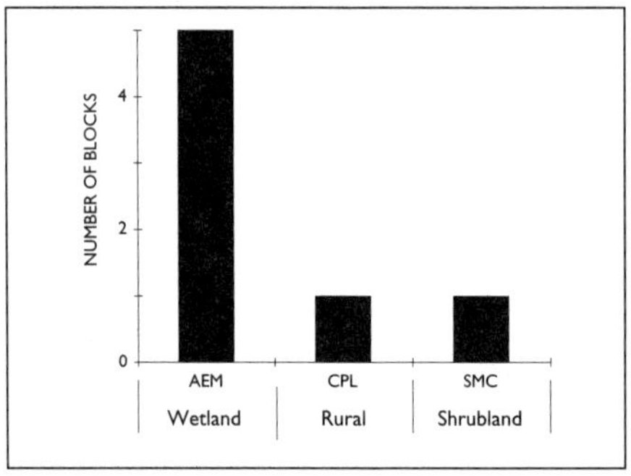
NUMBER OF BLOCKS
0
2
4
AEM
Wetland
CPL
Rural
SMC
Shrubland

BREEDING

Due to the difficulty of approaching nests and observing incubation, all Confirmations involved either distraction displays (two) or observations of fledgling birds (five). Six of the Confirmations came from emergent wetlands, the other from cropland.

Nests occur in any number of settings and vary in type of construction. Sometimes Willets place them far from water in prairie uplands; if the territory has perches where they can keep a vigilant watch for predators, so much the better. They will perch and perform distraction displays from any high point. Nest-cup construction ranges from well built and concealed by vegetation to simple nests in remarkably open situations.

North Park Willets lay single clutches and complete them in late May. Nearly all North Park Willets synchronize their breeding; the young hatch in mid to late June. This explains the fledgling Confirmations concentrated in the narrow window in the first week of July.

Atlas abundance estimates suggest a population of 25–50 nesting pairs in the state.

BREEDING PHENOLOGY

CODE		# OF RECORDS	DATE RANGE
FL	Fledged Young	2	2 Jul–3 Jul

DISTRIBUTION

Two subspecies of Willets nest in North America. One nests in saltwater marshes from the Maritime Provinces of Canada south along the Atlantic and Gulf coasts into northern Mexico. The other subspecies inhabits the Great Basin and the prairies of western Canada and the U.S. The southern limit of the prairie population extends from northern California, northern Utah, and northern Colorado east to central Nebraska. The prairie subspecies differs from the coastal birds by having paler plumage and larger size (Hayman et al. 1986). Interior Willets winter on the Atlantic, Gulf, and Pacific coasts (Terres 1980). They migrate north early in the season, and usually return to their breeding grounds by 1 May (Bent 1929).

Only three Priority and seven Nonpriority blocks recorded breeding activity. The seven blocks with Confirmations of nesting came from only three contiguous topographic maps, all in Jackson County.

Before their young take their first flights, adult females depart for the south. Female Willets are one of the first migrants to head south in the "fall." As early as the last week of June they arrive at prairie reservoirs, along with southbound Greater Yellowlegs and Marbled Godwits. Atlasers took care not to assign Possible breeding to southbound transients.

Outside of North Park, Colorado has four breeding records in two places (Barr Lake and Antero Reservoir; A&R). In North Park most nests are on Arapaho NWR. The refuge protects the habitat from grazing and other habitat disturbance. Away from protected areas Willets survive because they have adapted to limited human intrusion. In other states they have accepted mowing and burns in their habitat without deserting nest sites (Ehrlich et al. 1988).

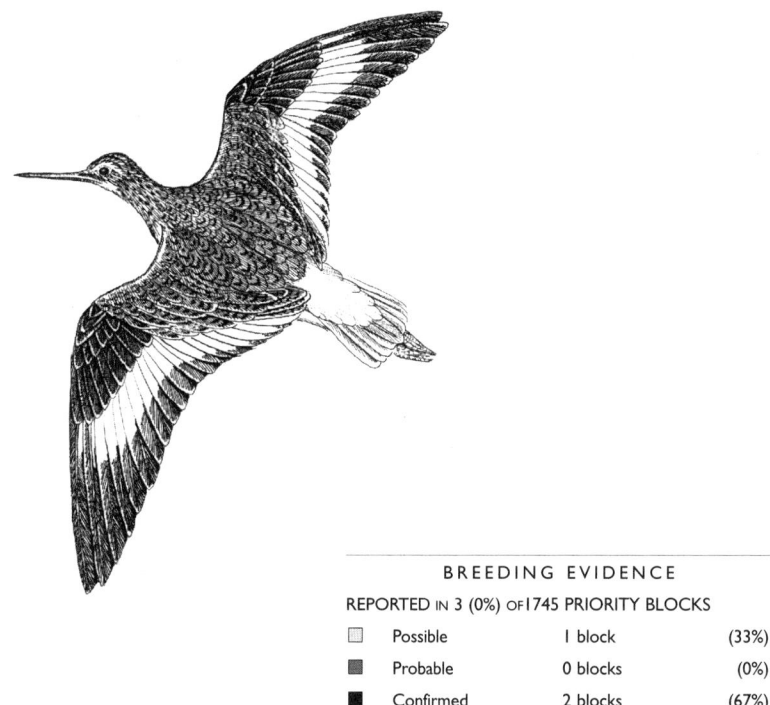

BREEDING EVIDENCE

REPORTED IN 3 (0%) OF 1745 PRIORITY BLOCKS

☐	Possible	1 block	(33%)
■	Probable	0 blocks	(0%)
■	Confirmed	2 blocks	(67%)

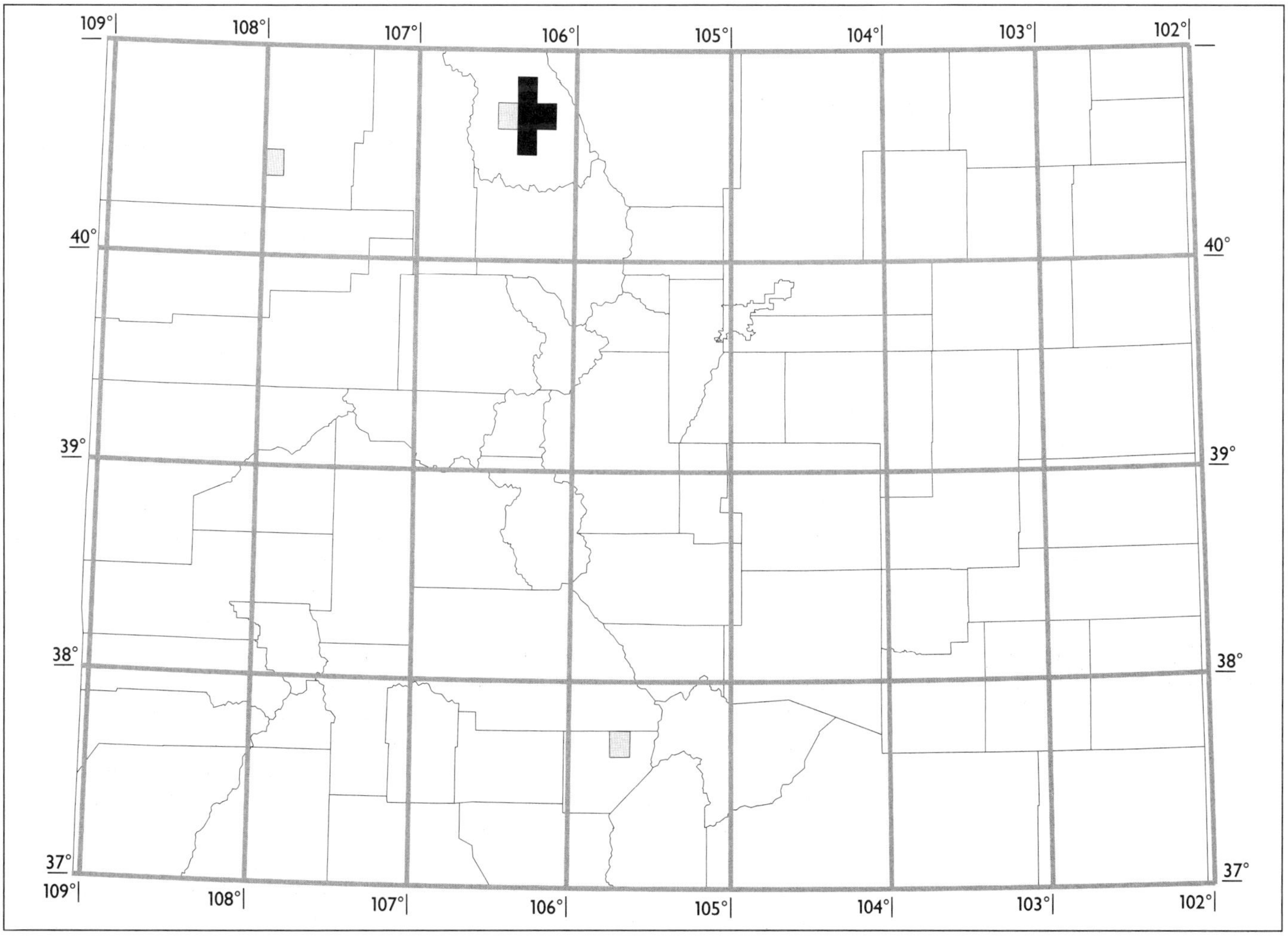
109°
108°
107°
106°
105°
104°
103°
102°
40°
39°
38°
37°

COLORADO DIVISION OF WILDLIFE / WILDLIFE RESOURCE INFORMATION SYSTEM

Teetering along the shoreline of a stream or lake, Spotted Sandpipers give no clue to their tumultuous mating system. Male Spotted Sandpipers fit the dictionary definition of "gigolos"—although the female's attachment to a male lasts only long enough for her to lay her eggs before she switches to another male for a similarly short tête-à-tête. The female may acquire as many as five mates during the breeding season.

HABITAT

The only sandpipers to breed extensively in temperate North America, Spotted Sandpipers nest in a vast array of waterside habitats. The common thread seems to be the availability of some sort of protective cover along the edge of the water (although sometimes they nest far from water).

In Colorado, Spotted Sandpipers nest mainly in the mountains and in the canyons and mesas of the Colorado Plateau. Atlasers found more along streams than on lake edges and in marshes. High-country streams where they found "Spotties" typically feature shallow water and cobblestones, and shores lined with shrubs—willows, alders, birches, and maples. These streams often flow through montane meadows covered with grasses, forbs, and shrubs (especially shrubby cinquefoil). Beaver ponds, the mountain lakes they most often frequent, often have dams covered with thick grass and edges choked with willows.

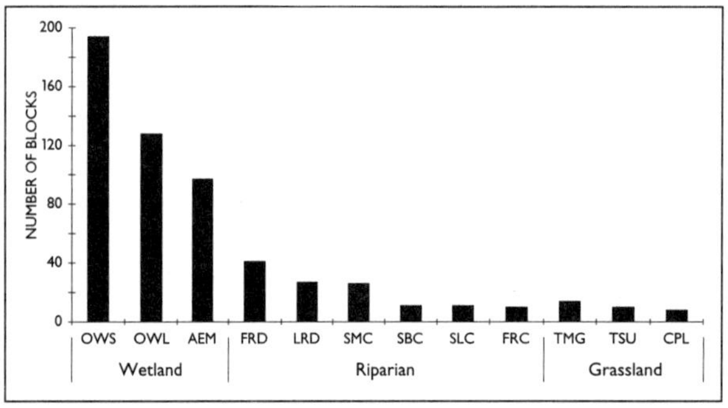
NUMBER OF BLOCKS
200
160
120
80
40
0
OWS OWL AEM
Wetland
FRD LRD SMC SBC SLC FRC
Riparian
TMG TSU CPL
Grassland

These sandpipers seem not to breed on the edges of large reservoirs, most of which lack the protective cover the birds find in the high country. On the plains and in intermountain valleys, Spotties share nesting habitat with several shorebird species, but except in the large mountain parks they are usually the only shorebird in the montane and subalpine zones.

Spotted Sandpipers walk or perch on rocks, tree limbs, beaches, weedstalks, or even utility wires, their bodies bobbing constantly. Even the structure of their toes suits their shoreline habits. The three forward toes are spaced closely together and the hind toe is strong and flexible, allowing them to clasp perches better (Bent 1927).

Spotted Sandpipers feed on invertebrates, mainly insects, along shorelines and wade only occasionally. They also catch prey on the wing. Where croplands abut wetlands, they feed in those as well. They eat a diverse selection, from aquatic insects to shoreline beetles and larvae to cropland grasshoppers and crickets (Johnsgard 1981).

BREEDING

Spotted Sandpipers place their nests in a variety of sites—in clumps of grass, among streamside rocks, under decaying logs, on beaver dams, along driftwood lining streams and lakes, on rock ledges, and in thick, rank vegetation.

Incubating males often reveal nest sites when flushed from the well-hidden nest cups. Grass lines the well-built nest cups, thicker in the northern part of the range (Bent 1927), perhaps to insulate eggs from the cold. Could nests at high altitudes in Colorado have correspondingly thicker linings?

In order to exploit the long temperate breeding season and raise more broods in a season, they have adopted polyandry (where one female mates with one or more males). Early accounts of breeding behavior assumed that aggressive males claimed territories, but autopsies of presumed males invariably identified aggressive birds as females (Bent 1927). This led to nearly a century of uncertainty about their life history. However, recent studies in northern Minnesota (Oring et al. 1983) have answered early questions with most unusual answers.

Females fight among each other to breed with males, always in short supply. They form their pair bonds within minutes, and stay monogamous until the female completes laying the clutch, almost invariably of four eggs. She cuts off their association and seeks another mate. Depending on the availability of males, females in northern Minnesota lay up to 20 eggs (five 4-egg clutches) during a five-week egg-laying season. Males must do all the incubation in order for the female to have enough time to mate and to produce eggs, although a female may assist with incubating and brooding her last clutch. Females wear themselves out; their life spans average only 3.7 years, the shortest of all North American shorebirds. Remarkably precocial, the young leave the nest within a half hour of hatching (Johnsgard 1981, Oring et al. 1983).

Colorado Atlas dates for nests with eggs came in a seven-week period between 28 May and 20 July. The dates showed two patterns:

after the median date of 21 June, 75% of NE codes (Nest with Eggs) were in the north half of the state (and all north of Latitude 38); and the later dates tended to come from high-mountain blocks.

BREEDING PHENOLOGY

CODE		# OF RECORDS	DATE RANGE
C	Courtship	8	6 May–29 Jun
NB	Nest Building	1	29 May
ON	Occupied Nest	2	30 May–18 Jun
NE	Nest with Eggs	23	27 May–30 Jul
NY	Nest with Young	3	13 Jul–20 Jul
FL	Fledged Young	110	8 Jun–14 Aug

DISTRIBUTION

One of the most widespread of all Nearctic shorebirds, Spotted Sandpipers nest from the northern edge of the taiga in Alaska and Canada south to California and North Carolina. They winter mostly south of the U.S., to central South America. A few winter along the coast in California and the southeastern U.S.

The Atlas map shows them widespread throughout the mountains and in the streams of the western canyons. They persist even above timberline: a family of fledglings found atop Independence Pass at 12,093 feet (3686 m) provides the highest Confirmed nesting record (Randy Lentz, nest card). South Park and the San Luis Valley have fewer Spotties, probably because the streams lack cover. Although the map shows only a small contingent on the eastern plains, mainly along the South Platte plus some scattered on other ponds and streams, low-altitude streams on the Western Slope like the San Juan, Uncompahgre, Gunnison, White, and Yampa (and their tributaries) support solid populations.

The conservation priority schemes rank Spotted Sandpiper as one of the most secure nesting species in the state. Their unique nesting strategy serves them well, and their continued breeding in Colorado seems assured.

BREEDING EVIDENCE

REPORTED IN 467 (27%) OF 1745 PRIORITY BLOCKS

☐	Possible	215 blocks	(46%)
■	Probable	98 blocks	(21%)
■	Confirmed	154 blocks	(33%)

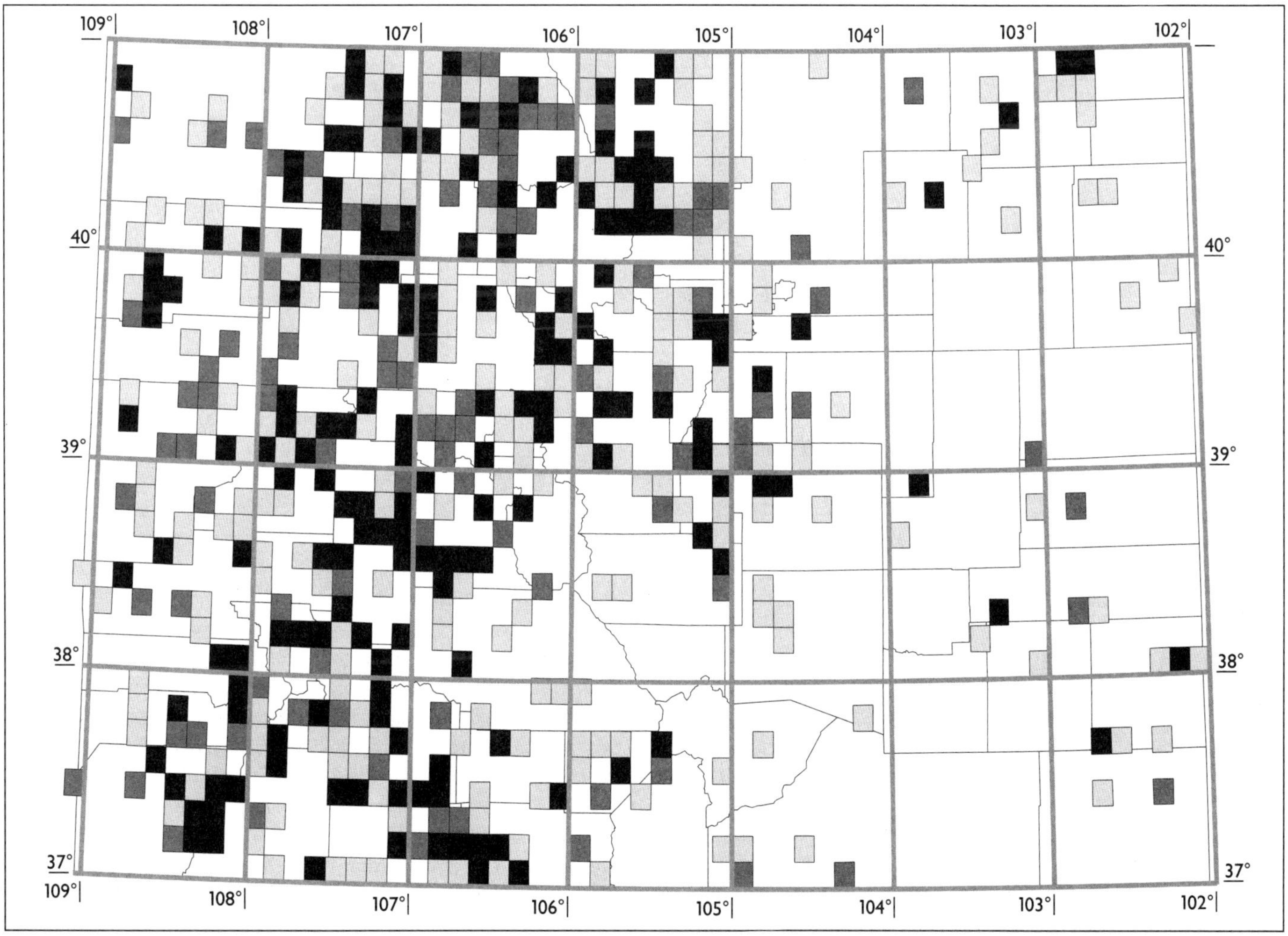
109°
108°
107°
106°
105°
104°
103°
102°
40°
39°
38°
37°

COLORADO DIVISION OF WILDLIFE / WILDLIFE RESOURCE INFORMATION SYSTEM

Although Upland Sandpipers by placement on the phylogenetic tree belong to the sandpiper family, by no means do they match the stereotypes. They acquired their former moniker, "Upland Plover," because they run, then stop, then run, like plovers. They do not wade, and rarely frequent the margins of lakes. In flight, the disproportionately long tail makes them look more like falcons than shorebirds. The small bill and head give them a gentle appearance. They perch on posts like meadowlarks and they call while skylarking, like longspurs or Lark Buntings.

HABITAT

Well named, Upland Sandpipers usually nest far from water, although atlasers found that in their northeast Colorado stronghold, the mid-height grasslands that they utilize abut river bottoms. Midgrass prairies far from water also provide suitable habitat, especially where atlasers found them north of the Arikaree River in Yuma County. The usual proximity to river valleys probably relates more to a wetter microhabitat supporting taller grass than to the importance of the stream course.

Upland Sandpipers seek elevated perches such as fence posts to use as lookout posts, because thick midgrass prairies render viewing approaching danger difficult.

Vast numbers of grasshoppers, locusts, weevils, and other injurious insects make up the bulk of their diet (Bent 1927). One of the few shorebirds that consume seeds, they include grain as about 3% of their diet. True to the beneficial nature of their insect diet, seeds of invasive weeds and spilled grain predominate, not those of valuable crops (Bent 1927).

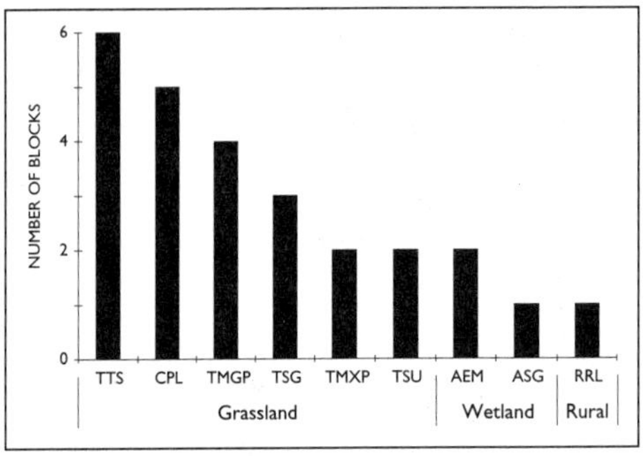
NUMBER OF BLOCKS
0
2
4
6
TTS
CPL
TMGP
TSG
TMXP
TSU
AEM
ASG
RRL
Grassland
Wetland
Rural

BREEDING

Thick grasslands provide nest sites where predators have difficulty locating nests. Tall grass usually arches over the nest cups (Bent 1927), rendering the location nearly invisible. Meadowlarks sharing the same habitat employ the same strategy.

Agitated behavior and distraction displays make finding nests of almost all shorebird species difficult. With Upland Sandpipers the impediments to access into thick midgrass prairies, along with the concealed nests, made Confirmations even more difficult. The only Atlas-period Confirmation involved a family group seen foraging in croplands in Logan County. The young had already hatched by 13 June.

They vacate the nesting grounds by mid July. Unlike some shorebirds, Upland Sandpipers on migration call frequently from high in the sky. The ringing *whi-brr-brr* call, audible for several hundred yards, draws attention both on the nesting grounds and in migration. Their vocal repertoire on the breeding grounds far exceeds the migration tune.

BREEDING PHENOLOGY

CODE		# OF RECORDS	DATE RANGE
C	Courtship	1	4 Jul
FL	Fledged Young	1	13 Jun

DISTRIBUTION

Once abundant on the prairies and fields that covered much of the midwestern United States, Upland Sandpiper populations have fluctuated wildly since European settlement. They moved east as farmers converted forests to fields (Anderle and Carroll 1988, Peterjohn and Rice 1991), but these "dough-birds" of the plains nearly joined the Passenger Pigeon in extinction. Easily shot by hunters, they too migrated in huge flocks that 1880s market hunters could "harvest" so as to procure much meat in a short time for city stores. Upland Sandpipers came close to taking the place of the Passenger Pigeons at markets as the pigeons became increasingly rare (Bent 1927, Ehrlich et al. 1988). On the wintering grounds on the pampas of Argentina, they also appeared on dinner tables, when the once-abundant Eskimo Curlew disappeared there. The Migratory Bird Act saved them (Bent 1927).

Currently Upland Sandpipers nest in scattered locales from Alaska to Maine, south to northwestern Oklahoma and the mid-Atlantic states. The breeding range is shrinking in the East as pastures grow back to woodlands and suburbs replace farmland (Ehrlich et al. 1988).

The range has shrunk in Colorado as well. Markets in Colorado Springs had "many" in April 1882 (Sclater 1912). They continued to migrate and breed north and east of Denver through 1960 (B&N), but by the Atlas period they had become scarce breeders, and most at least 100 miles from Denver.

Colorado atlasers found breeding only in the northeastern quarter of the state, in mid-height grasslands and adjacent croplands. These sandpipers occupy grasslands, hayfields, and croplands along the South Platte, and the lush tallgrass and midgrass prairies of the sandhills north of the Arikaree River in Yuma and Washington counties.

The two observations in southeastern Colorado came on 6 and 25 June; these fall within the window of breeding dates (Nelson 1993) but easily could have referred to migrants. In Kansas Upland Sandpipers do not breed within 90 miles of Colorado, and become commonplace only 150 miles east of the line. June birds seen in western Kansas have not exhibited breeding behavior (Scott Seltman pers. comm.).

Because of concern over loss of habitat, Upland Sandpipers have appeared on both state and national lists of species deserving special attention (Webb 1985, Tate 1986). Low numbers in the state will direct management attention to other species, but because so many grassland species exhibit population drops, these sandpipers may benefit by association. An air of optimism surrounds the future for Upland Sandpipers. Fortunately, they have adapted to cropland habitat as grassland habitat has diminished, although nowhere have the numbers approached former levels (Terres 1980).

BREEDING EVIDENCE

REPORTED IN 20 (1%) OF 1745 PRIORITY BLOCKS

Possible	13 blocks	(65%)
Probable	6 blocks	(30%)
Confirmed	1 block	(5%)

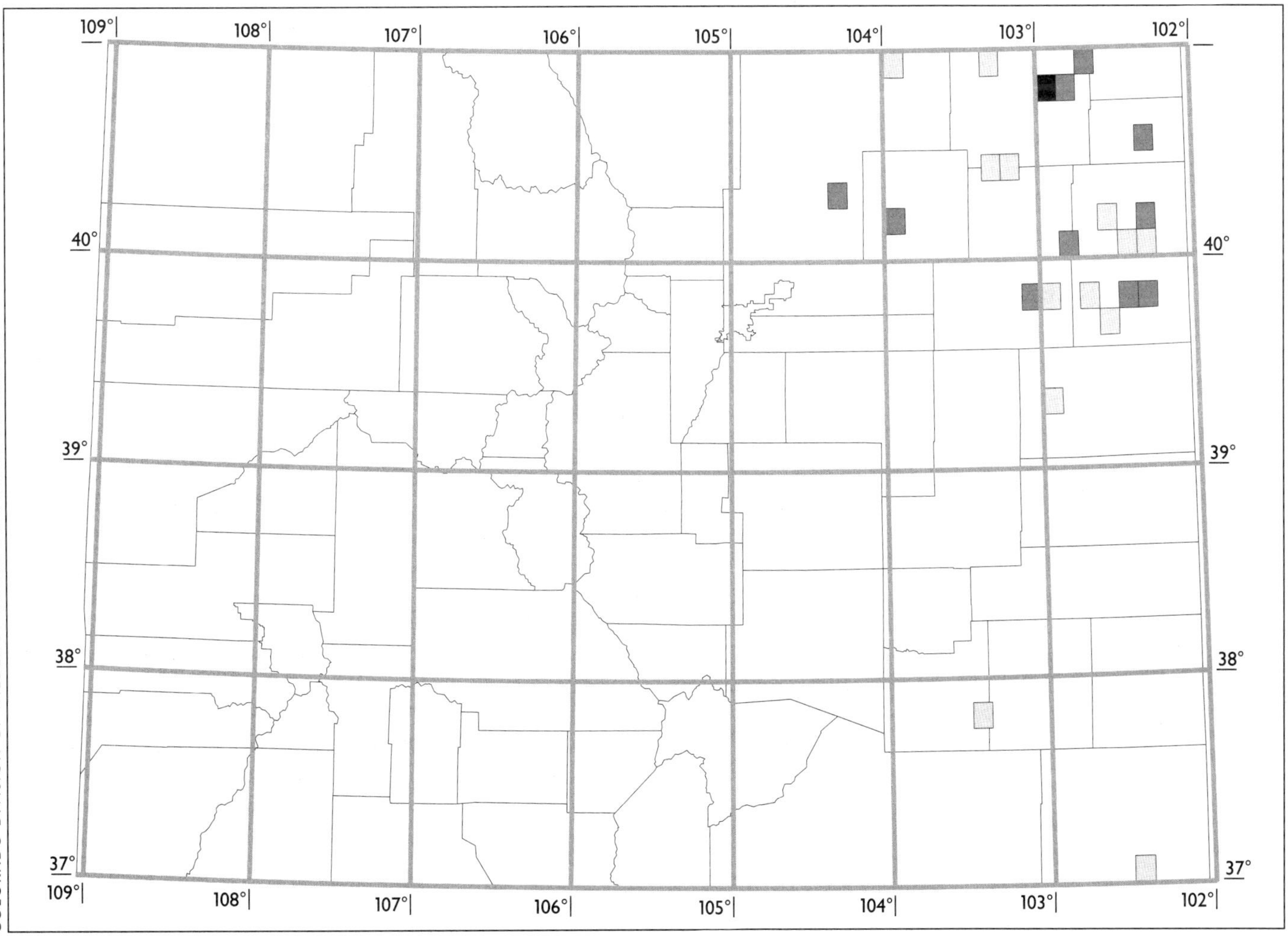
109°
108°
107°
106°
105°
104°
103°
102°
40°
39°
38°
37°

COLORADO DIVISION OF WILDLIFE / WILDLIFE RESOURCE INFORMATION SYSTEM

LONG-BILLED CURLEW *Numenius americanus*

An artist with an active imagination would have a hard time designing a bird as outrageous as a Long-billed Curlew. The huge size (largest of all the North American shorebirds) and flashy cinnamon underwings make it hard to overlook. And then—that incredible decurved bill: at up to 8 3/4 inches, the bill alone almost equals the entire length of a Mountain Plover, its frequent prairie companions. As if its appearance weren't enough, the ringing territorial calls draw attention from a mile away.

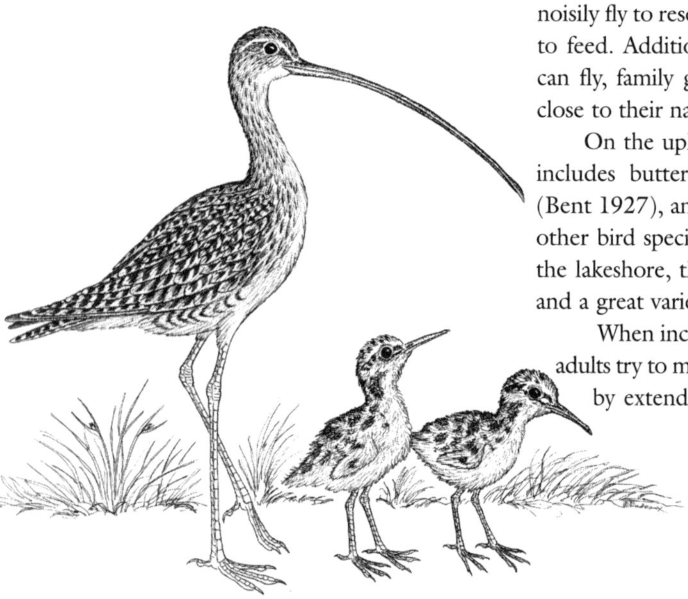

HABITAT

An indicator species for healthy native grasslands, Long-billed Curlews nest mostly on shortgrass prairies. Only a few scattered prickly-pear cactus plants or small shrubs break up the expansive uniformity of their grasslands.

Colorado has vast areas of apparently suitable habitat unoccupied by curlews. For example, in Kiowa and Bent counties, atlasing found that occupied territories abut all the reservoirs, yet much of the intervening prairie remains barren of these dramatic birds. An explanation may lie in a dependency on nearby ponds, playas, or lakes for feeding, bathing, or drinking. A substantial number found in a 1974–1975 Baca County study were observed near standing water (42% within 100 yards, 68% within one-quarter mile). The presence of water may influence initiation of nesting the first year. Site fidelity may induce them to return even if the nearby water has dried up (McCallum et al. 1977).

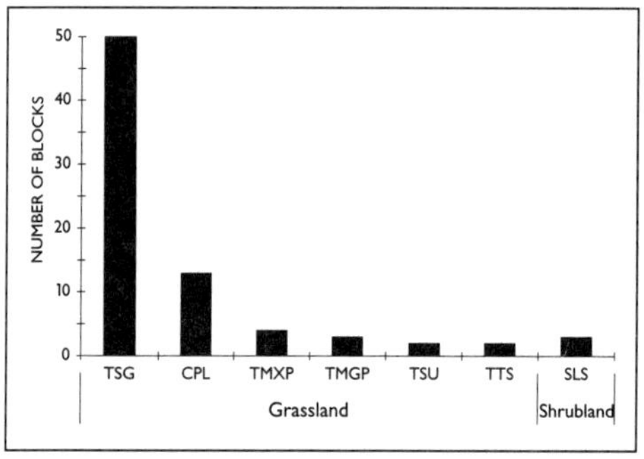
NUMBER OF BLOCKS
50
40
30
20
10
0
TSG
CPL
TMXP
TMGP
TSU
TTS
SLS
Grassland
Shrubland

During migration, especially post-breeding, Long-billed Curlews feed along the shorelines of prairie reservoirs. At the height of nesting in May and June, adults noisily fly to reservoir shorelines to drink and to feed. Additionally, as soon as the young can fly, family groups gather on lake edges close to their natal sites.

On the upland breeding grounds, prey includes butterflies, insects, toads, berries (Bent 1927), and even eggs and nestlings of other bird species (Ehrlich et al. 1988). On the lakeshore, they probe for worms, snails, and a great variety of aquatic invertebrates.

When incubating their 4-egg clutches, adults try to make themselves inconspicuous by extending their necks forward and laying their heads on the ground. Coyotes are egg predators (Bent 1927), and in Colorado, so are bullsnakes (Painted Canyon, 37103C8; Kingery and Kingery 1995).

BREEDING

Like most shorebird species, adult Long-billed Curlews expertly distract observers away from the actual nest sites. Nest-building activity revealed one of the seven nests found by atlasers. Three nests with eggs and three nests with young rounded out observations of actual nests. Most Confirmations involved the ubiquitous shorebird strategy of distraction displays or adults seen with dependent fledglings.

Curlews raise only one brood per year, and nesting fits into a compact time period. Atlas dates show that adults arrive on the breeding grounds in April. Most clutches, laid in May, hatch from early to mid June. The 21 June Atlas egg date is later than most (and the one that involved the bullsnake coiled in the nest). Most of the precocial young can fly by the first of July. Only eight Atlas records came from after 1 July. Birds seen after 15 July could either come from locally produced clutches or be southbound migrants.

BREEDING PHENOLOGY

CODE		# OF RECORDS	DATE RANGE
C	Courtship	2	19 Apr–18 Jun
NB	Nest Building	1	29 May
NE	Nest with Eggs	2	22 May–21 Jun
NY	Nest with Young	1	13 Jun
FL	Fledged Young	10	11 Jun–15 Jul

DISTRIBUTION

The current range of the Long-billed Curlew has contracted from historic times. The historical range extended from southern British Columbia to Manitoba, southeast to Wisconsin, Illinois, and Kansas, south to northern California and northern Texas, with a disjunct population on the coastal Texas prairies.

Even by Bent's time (1927), nesting no longer occurred east of the Mississippi River. Curlews once migrated along the entire Atlantic coast, but by the early 1900s had disappeared north of South Carolina (Bent

1927). Even in the western strongholds, numbers declined dramatically. Habitat loss due to conversion of prairies to agriculture caused much of this decline, and their size and taste made them a popular main dish on the dinner tables of the time.

Colorado's first record, two birds shot by Mr. Peale and Edwin James, graced the tables of the Long Expedition on 5 July 1820 (B&N). Curlews once nested in Middle and South parks (Sclater 1912), and Niedrach photographed nesting curlews 20 miles from Denver in 1956 (B&N). McCallum et al. (1977) surveyed Colorado birdwatchers and wildlife professionals in 1974–1975; from 773 observations they plotted a distribution map (essentially reproduced in A&R) that showed a much diminished range, with most curlews in Baca County, a few north of Lamar, and a few others scattered to the north.

The greater and longer Atlas effort substantially enlarged the breeding range. Three separate prairie regions now contain almost the entire statewide population. The heaviest concentration extends from Baca County west in Las Animas County to the Purgatoire River. A second population breeds north of the Arkansas River from eastern El Paso County to Kansas. Many nest on large private ranches. In northeastern Colorado the third cluster, with lower densities and wider spaces between blocks, affirms that breeding curlews have not entirely deserted the northern plains. These birds nest in select prairies bordering the South Platte River and, sporadically, on the Pawnee National Grassland.

A small contingent apparently nests on the Western Slope, definitely in Mesa and probably in Moffat County, although atlasers did not Confirm breeding during the period in either county.

Declines in range and population have led to concern about its status. Long-billed Curlews share an unfortunate bond with other shortgrass prairie specialists because of threats to remaining shortgrass habitat. Almost all species—songbirds, raptors, and shorebirds—are declining. A paramount biological challenge of the 1990s is to stabilize populations of these declining grassland bird species.

BREEDING EVIDENCE

REPORTED IN 78 (4%) OF 1745 PRIORITY BLOCKS

▢	Possible	33 blocks	(42%)
▩	Probable	21 blocks	(27%)
■	Confirmed	24 blocks	(31%)

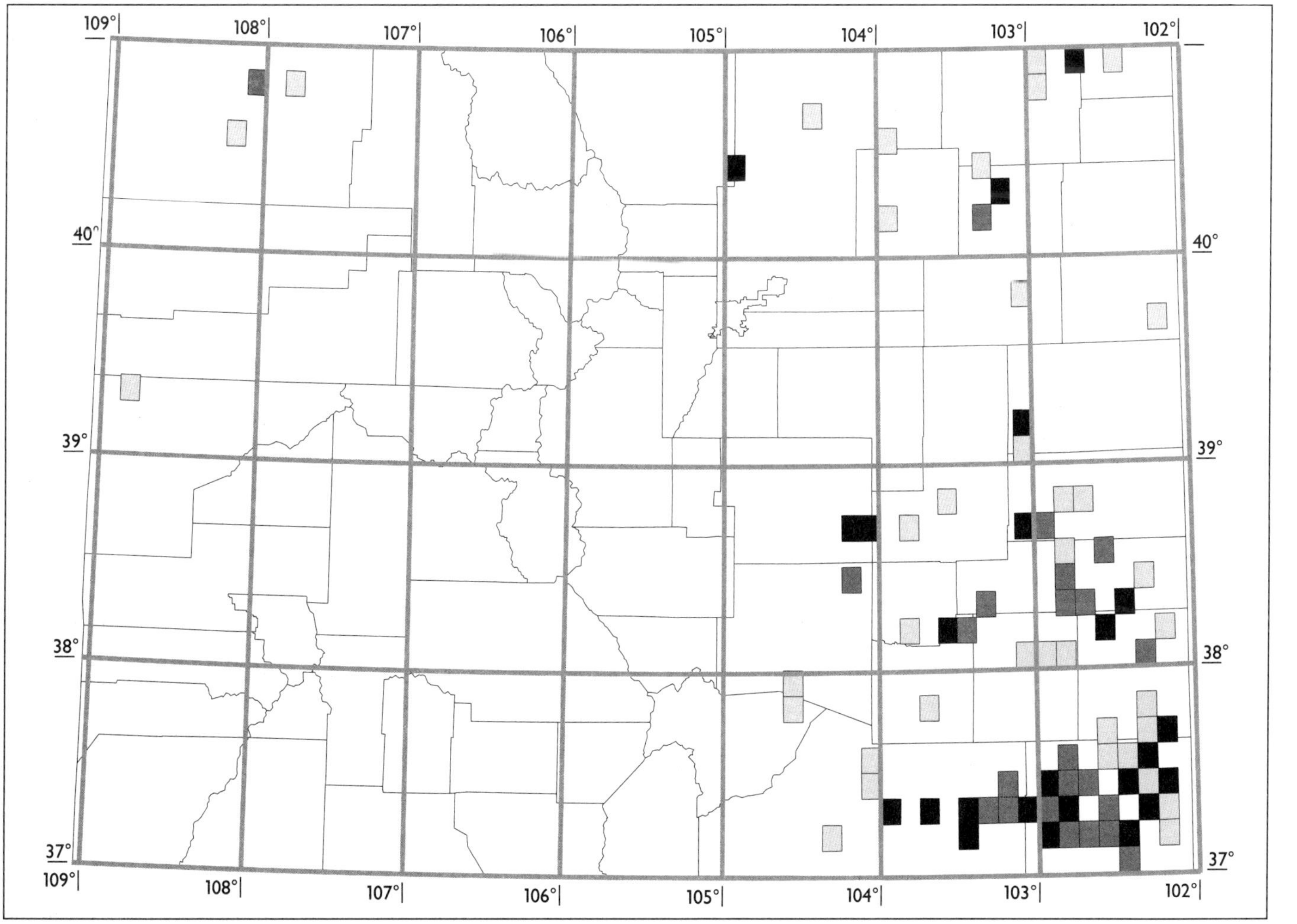

109°
108°
107°
106°
105°
104°
103°
102°
40°
39°
38°
37°

COLORADO DIVISION OF WILDLIFE / WILDLIFE RESOURCE INFORMATION SYSTEM

The strange sounds produced by the Common Snipe, both its haunting, disembodied winnowing from high in the air and its persistent protestations from hiding places in bogs and fens, have led the uninitiated to confuse this species with walruses, goats, owls, and even spirits (Chadwick 1989). For the Atlas worker, however, the same sounds made this species rather easy to locate and to census.

HABITAT

Typically, Common Snipes nest in wet grass or sedge meadows—moist places beside streams and ponds. They sometimes pick wet areas with fairly dense, low woody growth, ideally with fairly open terrain close by (B&N, A&R, Stout 1967, Johnsgard 1981). The most suitable sites contain shallow, stable, discontinuous water levels (Johnson and Ryder 1977). Snipes place their nests in fairly dry places in these wet surroundings (Terres 1980).

Atlas results affirm prior work, with two-thirds of all habitat reports denoting either emergent vegetation or pasturelands and hay meadows. Although Common Snipes most frequently nest in low grass and sedge, they also occasionally nest among taller willows (Johnson and Ryder 1977). Willow carrs (SBC, SMC) appeared regularly in Atlas reports, accounting for 13% of all habitat reports.

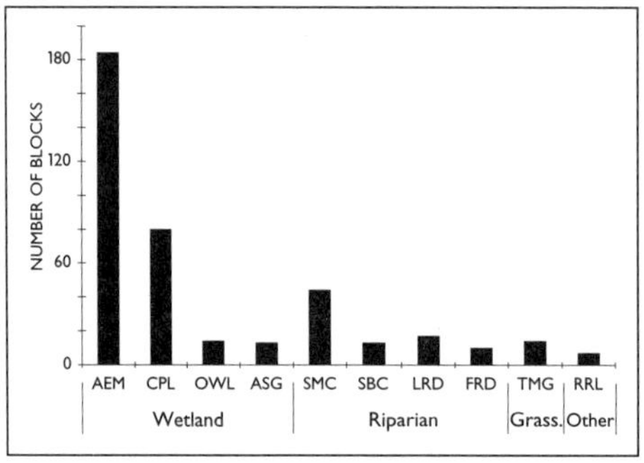
180
120
60
0
NUMBER OF BLOCKS
AEM
CPL
OWL
ASG
SMC
SBC
LRD
FRD
TMG
RRL
Wetland
Riparian
Grass.
Other

BREEDING

By vibrating tail feathers, males produce the eerie winnowing sound as they fly high over the breeding territory, sometimes as high as 300 feet (100 m). These courtship flights made Common Snipes relatively easy for atlasers to detect. Field workers reported the courtship code in one-third (103) of the Priority blocks in which they found this species. A number of the 37 reports of territorial birds probably also involved observations of this winnowing flight.

A wide range of dates for courtship, nesting, and fledglings reflects the Common Snipe's use of habitats at a wide range of elevations. Some males begin displaying on their breeding territories by April (Tuck 1972), and Atlas workers recorded courtship activity in five blocks in April, one as early as 12 April. Courtship flights continued well into the summer, the latest recorded on 9 August.

The males have strong territorial tendencies and initially defend a fairly large area—perhaps 25 acres (10 ha)—but during nesting their territories shrink to a small section immediately surrounding the nest. Snipes place their nests close to water, often adjacent to shallow channels among sedges or grasses. The nests perch only slightly above the water—amid the grasses, on a tussock, in a sedgy tuft, or on a floating island.

Confirming breeding of Common Snipe proved much more difficult than detecting its presence; the Confirmation rate barely reached 15%. Atlasers found only 13 of the well-concealed nests, the dates ranging from 21 May in the Carbondale block (39107D2) at 6,000 feet (1830 m) and at San Luis Lakes SWA at 7,500 feet (2285 m) to 21 July in the Flat Tops Wilderness (Blair Mtn., 39107G4), above 10,000 feet (3048 m).

Eggs hatch after an 18–20-day incubation period (Ehrlich et al. 1988). The parents lead the precocial young from the nest as soon as they are dry; they divide the brood between them. The bills of the chicks grow fairly slowly, so the parents must feed them. The parent pokes around in the bottom and edge mud in their home ground; as it pulls out its bill the chicks crowd close around to feed (Stout 1967). Chicks reach self-sufficiency in 10 days and can maintain sustained flight at 20 days (Terres 1980). Atlas workers found fledglings in 13 blocks, with a two-month range of dates from 6 June to 5 August.

BREEDING PHENOLOGY

CODE		# OF RECORDS	DATE RANGE
C	Courtship	116	12 Apr–9 Aug
NB	Nest Building	2	20 May–12 Jun
ON	Occupied Nest	6	5 Jun–5 Jul
NE	Nest with Eggs	11	21 May–21 Jul
FY	Feeding Young	1	9 Jul
FL	Fledged Young	11	6 Jun–5 Aug

DISTRIBUTION

Circumpolar in distribution, Common Snipes nest across the northern half of Europe and Asia and South America east of the Andes from Colombia to the Straits of Magellan. Colorado lies at the southern edge of the North American breeding range, which extends from the arctic tundra south to (barely) northern New Mexico.

In Colorado the population thins noticeably from the northern border to the southern. Atlas workers found Common Snipe breeding in 126 Priority blocks lying above 40 degrees and 70, 62, and 47 blocks above 39, 38, and 37 degrees respectively. Five of the six blocks reported with an abundance code of A4 (101–1,000) lie north of the 40th parallel.

Atlasers found that some Common Snipes breed in wetlands on the plains near the Front Range and in western valleys. Higher elevations as well as higher latitudes seem more hospitable, these hosting populations in a greater percentage of blocks and in greater densities. The highest densities occur in wet mountain parks and meadows between 7,500 and 10,000 feet (2285–3048 m) above sea level.

Atlas workers reported abundance codes of A2 most frequently (58% of all reports) followed by A3 (26%). In the market-hunting days of the late nineteenth century, overshooting greatly diminished this bird's populations. Although many states, including Colorado, still manage the Common Snipe as a game bird, hunting probably has little effect on their population today (Gooch 1993). Wetlands have been reduced nationwide, but the number of snipes that atlasers found in Colorado pastures and hayfields indicates that irrigation has provided significant new habitat for this bird. It appears that human activity, because it no longer includes wholesale slaughter, now has more positive than negative effects on this species (Chadwick 1989).

BREEDING EVIDENCE

REPORTED IN 305 (17%) OF 1745 PRIORITY BLOCKS

☐	Possible	97 blocks	(32%)
▩	Probable	163 blocks	(53%)
■	Confirmed	45 blocks	(15%)

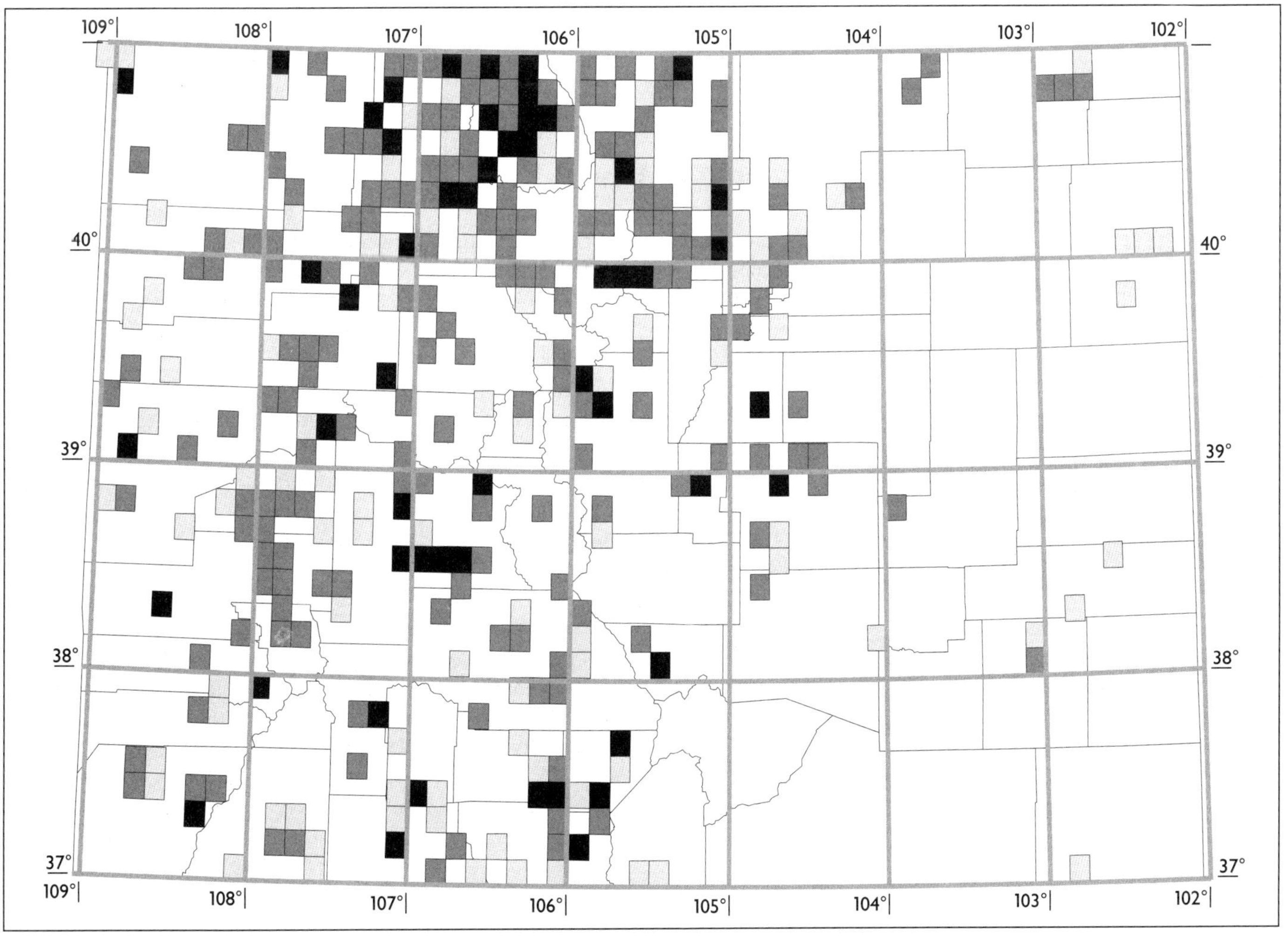
109°
108°
107°
106°
105°
104°
103°
102°
40°
39°
38°
37°

COLORADO DIVISION OF WILDLIFE / WILDLIFE RESOURCE INFORMATION SYSTEM

Quiet and unassuming, Wilson's Phalaropes live by different rules than the other shorebirds nesting in Colorado. They mix with the other species while feeding on shorelines, but are the only "locals" that swim just to feed. Many shorebirds have polyandrous habits, but only the female phalaropes both assume traditional male roles and wear brighter breeding plumage. They fight each other for the privilege of breeding with males. Conversely, males perform most of the nesting duties.

HABITAT

Wilson's Phalaropes summer across much of Colorado, although their breeding status is not always easy to determine. They prefer to nest in moist sedge and rush meadows characterized by low plant height. Low sedges may grow in highly alkaline soil. Adjacent open water rounds out the habitat requirements (Bent 1927). Colorado nesting habitat centers on intermountain valleys at 7,000–10,000 feet (2135–3048 m). Suitable habitat also occurs at various Western Slope and prairie marshes.

Males and females use somewhat different habitats during breeding. Males feed in a greater variety of aquatic and terrestrial (flooded meadows, beaches) habitats whereas females feed more in water (Colwell and Jehl 1994).

Phalaropes employ a unique feeding technique: spinning. A bird spins like a top—as fast as 60 revolutions per minute. This creates a whirlpool that either sucks in insects from the surface or stirs them up from the bottom of the pond. The phalarope picks insects out of its self-created food source (Ryser 1985, HEK). They also feed in more ordinary ways, such as pecking prey from the surface, lunging at passing flies, and probing soft mud (Colwell and Jehl 1994).

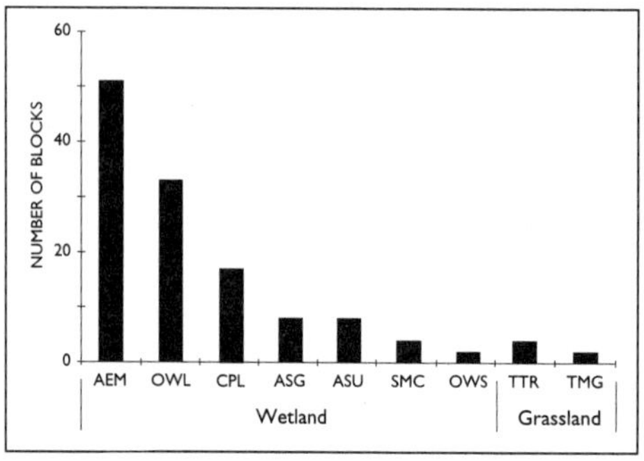
NUMBER OF BLOCKS
0
20
40
60
AEM
OWL
CPL
ASG
ASU
SMC
OWS
TTR
TMG
Wetland
Grassland

The diet of the Wilson's Phalarope consists of typical aquatic larvae, especially mosquitoes and crane flies, true bugs (Heteroptera), and beetles including predaceous diving beetles that prey on young fish and minnows (Bent 1927, Colwell and Jehl 1994).

BREEDING

Female Wilson's Phalaropes compete aggressively with each other for the privilege of mating with unattached males. Up to 11% practice polyandry (mating with more than one male), but she of the brighter plumage assumes none of the familial duties. The male's drab plumage suits his domestic job description. He alone constructs the nest. He develops two brood patches so as to provide all incubation and accompanies the young to fledging. After the female lays the eggs, she serves as a sentry to protect nests. She may even provide lookouts, briefly, for two closely spaced nests simultaneously incubated by her different mates—but not for long. A pair associates closely as long as the pair bond lasts. Females desert males one to 18 days after completion of the clutch (Colwell and Jehl 1994).

Bent (1927) considered finding nests "exasperating" work. Males sit tight when incubating, and nests avoid human detection unless the male flushes at the last possible instant. Additionally, few humans (atlasers included) venture into soggy sedge meadows. The recently hatched young do not cooperate much better. Young stay in heavy vegetation until old enough to fly, and, shortly after fledging, depart the breeding grounds. Still, most Confirmations were of fledglings. Ten Confirmations came from Non-priority blocks, six of them from Alamosa, Monte Vista, and Arapaho NWRs; these account for having more Confirmations on the Phenology chart and map than in the Breeding Evidence table.

BREEDING PHENOLOGY

CODE		# OF RECORDS	DATE RANGE
C	Courtship	3	18 May–22 Jun
ON	Occupied Nest	3	30 May–29 Jun
NE	Nest with Eggs	5	8 Jun–10 Jul
FL	Fledged Young	14	16 Jun–5 Aug

DISTRIBUTION

Wilson's Phalaropes nest mainly in western North America from northern Alberta to central Nevada, Colorado, and western Minnesota; an expanding population breeds around the Great Lakes and recently reached southwest Quebec (first recorded breeding in 1974; Gauthier and

Aubry 1996). During migration huge flocks descend on four major staging areas in saline wetlands where they undergo molt and gain weight for migration; the Great Salt Lake attracts the most—600,000 (Colwell and Jehl 1994). The bulk of the population winters in saline lakes in the altiplano of Bolivia, Chile, and Argentina.

In Colorado, the known range in the late 1800s restricted it to the northeast quarter of the state (Cooke 1897), but by 1965 B&N regarded them as common nesters not only in prairie marshes but also in mountain valleys. That situation has changed; the Atlas map shows that currently they nest in scattered places across much of the state, but have subsided on the plains, where many marshes have turned into wheat fields, corn patches, and subdivisions.

Nesting strongholds during the Atlas included North Park, the San Luis Valley, the Gunnison Valley, and the Yampa watershed. Although numerous records came from the plains, that area yielded only one Confirmation. Atlas results of current distribution may reflect comprehensive coverage of habitat more than any changes in range, population, or distribution, but the Atlas protocol may have hampered obtaining "Confirmations," especially on the plains.

With the recent conservation emphasis on maintaining wetlands, Wilson's Phalaropes are likely to maintain, or even to increase, their numbers. Unlike other shorebirds migrating to South America, they escaped hunting pressure. Thus, they avoided the population declines so evident in other long-distance migrant shorebirds.

BREEDING EVIDENCE

REPORTED IN 100 (6%) OF 1745 PRIORITY BLOCKS

Possible	41 blocks	(41%)
Probable	44 blocks	(44%)
Confirmed	15 blocks	(15%)

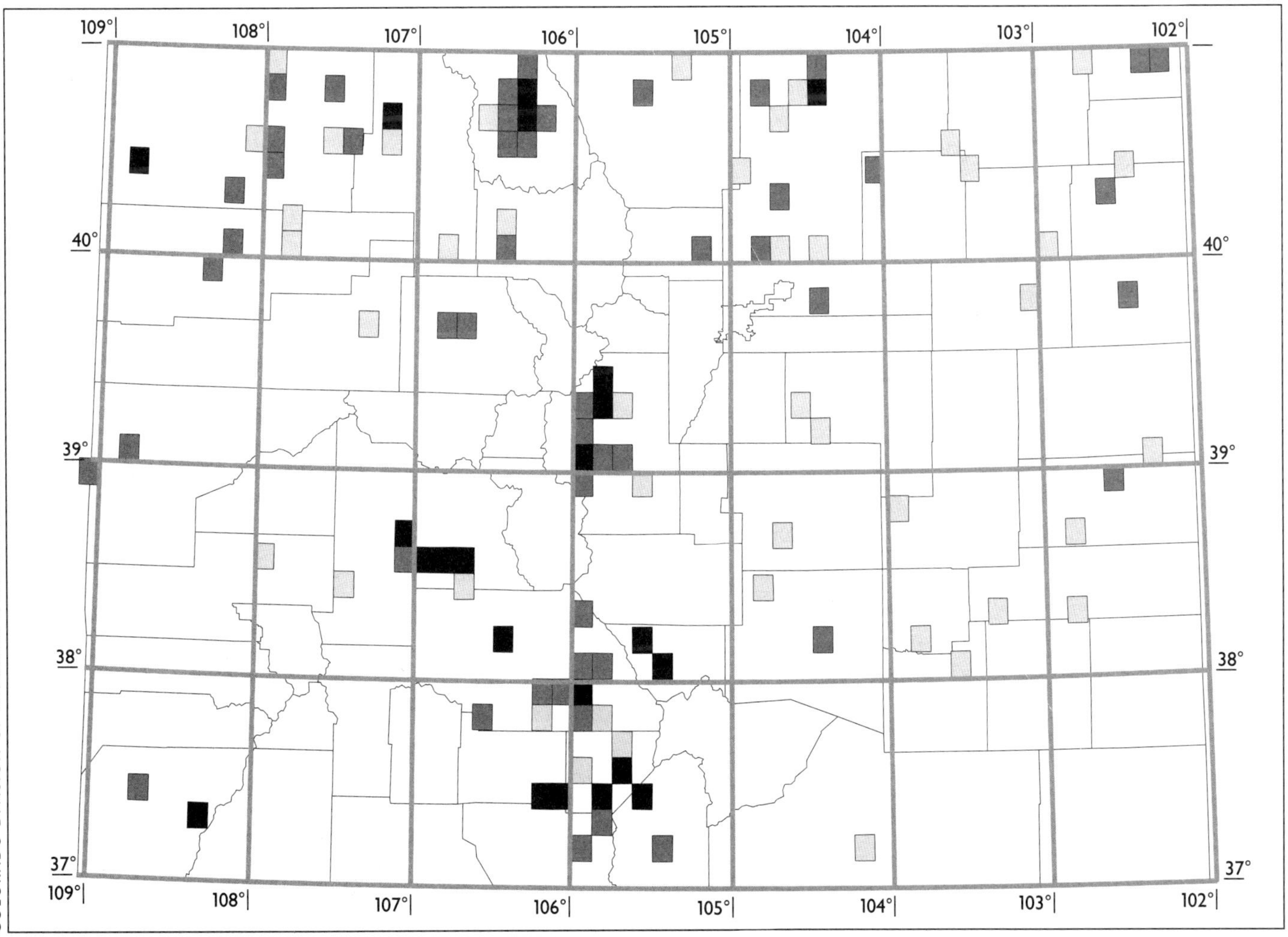
109°
108°
107°
106°
105°
104°
103°
102°
40°
39°
38°
37°

COLORADO DIVISION OF WILDLIFE / WILDLIFE RESOURCE INFORMATION SYSTEM

California Gulls have not nested in Colorado for long, but like humans flocking here from California and other faraway points, they find it appealing. Like their human counterparts, the gulls rapidly change the character of every neighborhood in which they settle.

HABITAT

Several species of white-headed gulls loaf and feed on reservoirs on the plains and in intermountain valleys. The presence of gulls does not necessarily indicate nesting nearby. During the Atlas only one gull species, California Gull, nested in Colorado, and at only five reservoirs.

The most important component of site selection and colonization is the availability of an island. Vegetation on their Colorado nesting islands varies—from barren ground at Antero Reservoir to heavy vegetation at Adobe Creek Reservoir. Correspondingly, nests vary from eggs laid directly on sand at Antero (Chuck Loeffler pers. comm.) to bulky nests made of grass, cockleburs, sticks, feathers, and trash at Adobe Creek (DLN, 1990–1996).

California Gulls feed on a wide range of food—fish, crayfish, shorebird eggs and young, insects, plants, and garbage (Winkler 1996). Breeding or non-breeding, California Gulls may travel far to feed; Antero birds often cruise around the landfill between Buena Vista and Salida, 25 miles (40 km) away (HEK).

BREEDING

California Gulls nest in colonies. In 1931 the Salt Lake basin held 80,000 birds and in 1980 at Mono Lake, California, an estimated 50,000 birds nested (Ryser 1985, Young 1981, respectively). Those numbers have burgeoned to 130,000 at Great Salt Lake and 45,000–65,000 in various years at Mono Lake (Dave Shuford pers. comm.). In Colorado they have not achieved such high numbers: the largest colonies of 1,000–1,500 nest at Riverside and Antero reservoirs, but colonies vary in size from year to year. Between 1975 and 1995, breeding pairs peaked at 1,500 at Riverside, 1,200 at Antero, 25–30 at Elevenmile (not 2,000 as reported in A&R; Chuck Loeffler pers. comm.), 65 at Adobe Creek, and 100 in North Park (A&R, Ron Ryder pers. comm.).

Normally California Gulls do not breed until their fourth year, when they achieve adult plumage. They build nests on the ground from dead plants, rubbish, and straw; bones and feathers from birds that died in prior years constitute a substantial part of the nest (Winkler 1996). The gulls place their nests as close together as 2–3 feet (Bent 1921). Nests vary in size, but typically measure 14–18 inches (35–46 cm) in diameter with rims 2–5 inches (5–12 cm) above ground or higher at Adobe Creek (DLN). A few days after hatching, young can run about and even swim.

Adults feed the young until fledging (at about 40 days). They swallow food to carry it to the nest and then regurgitate it as the nestlings beg, peck at the adults' bills, and poke their heads into the parents' gullets (Winkler 1996).

Occasionally, California Gulls hybridize with Herring Gulls (*Larus argentatus*). This happened at Antero Reservoir in three different years (Chase 1984) and at Adobe Creek Reservoir in 1996. There, a mixed pair produced a nest with three eggs that eventually failed when rising water flooded the nest (DLN).

BREEDING PHENOLOGY

CODE		# OF RECORDS	DATE RANGE
NE	Nest with Eggs	1	16 May
NY	Nest with Young	3	20 Jun–14 Jul

DISTRIBUTION

Named for their usual wintering sites along the coast of California, California Gulls did not breed along the coast until 1980, at San Francisco Bay (Dave Shuford pers. comm.). During the breeding season, they visit large interior lakes from the east slope of the Sierra Nevadas north to the Northwest Territories, east to North Dakota, and south to Colorado. The Colorado sites in Bent and Kiowa counties mark, for the moment, the southeastern limits of their nesting range. In 1939, California Gulls, considered "rare stragglers" in Colorado (Niedrach and Rockwell 1939), nested no closer than Utah and Yellowstone National Park, Wyoming (Bent 1921).

The first nest documented in Colorado came in 1963 at Riverside Reservoir (Ryder 1964). By 1965 the gulls had colonized Antero in South Park and, by 1977, nearby Elevenmile Reservoir (soon abandoned). Nesting expanded to Adobe Creek in 1988 and North Park in 1990 (A&R). In 1995 a small colony of 11 pairs pioneered an isolated gravel bar at John Martin Reservoir, the first non-island breeding site in Colorado (Nelson 1995). The Denver Water Board drained Antero in 1995, and a few pairs moved back to Elevenmile (Chuck Loeffler pers. comm.). The Atlas map shows all those colonies except Elevenmile, not used during the period.

The impact of California Gulls ranges from possibly beneficial to potentially devastating, depending on the part of Colorado they inhabit. Economically, some nest colonies have a positive impact as the gulls consume grasshoppers in nearby fields (Ehrlich et al. 1988). To early Mormon pioneers in Utah, California Gulls symbolized divine intervention; Salt Lake City has a gilded gull statue to commemorate the gulls for saving Mormon crops from grasshoppers.

Biologically, colonies in southeastern Colorado have negative impacts on the nesting success of endangered Least Terns and Piping Plovers. The most benign product of gull colonization comes when they usurp prime island habitat from the plovers and terns (DLN). More direct consequences result from the gulls' appetites for tern and plover eggs and chicks, and when they rob the terns of food. A possible, purely speculative example: about the time that California Gulls arrived at Antero, Snowy Plovers disappeared. Because no one monitored the Snowies, no data exist even to pinpoint the date of the Snowies' departure. California Gulls do prey on Snowy Plover eggs at Mono Lake, California, as well as on grebe chicks and various ducklings in Manitoba and Utah (Winkler 1996).

High densities of gulls assault a variety of victims across their range (USFWS 1988). The ominous shadow of these recent colonists looms over southeastern Colorado and poses tough management questions for the people responsible for recovery of the gulls' endangered neighbors.

BREEDING EVIDENCE

REPORTED IN 1 (0%) OF 1745 PRIORITY BLOCKS

□	Possible	0 blocks	(0%)
▪	Probable	0 blocks	(0%)
■	Confirmed	1 block	(100%)

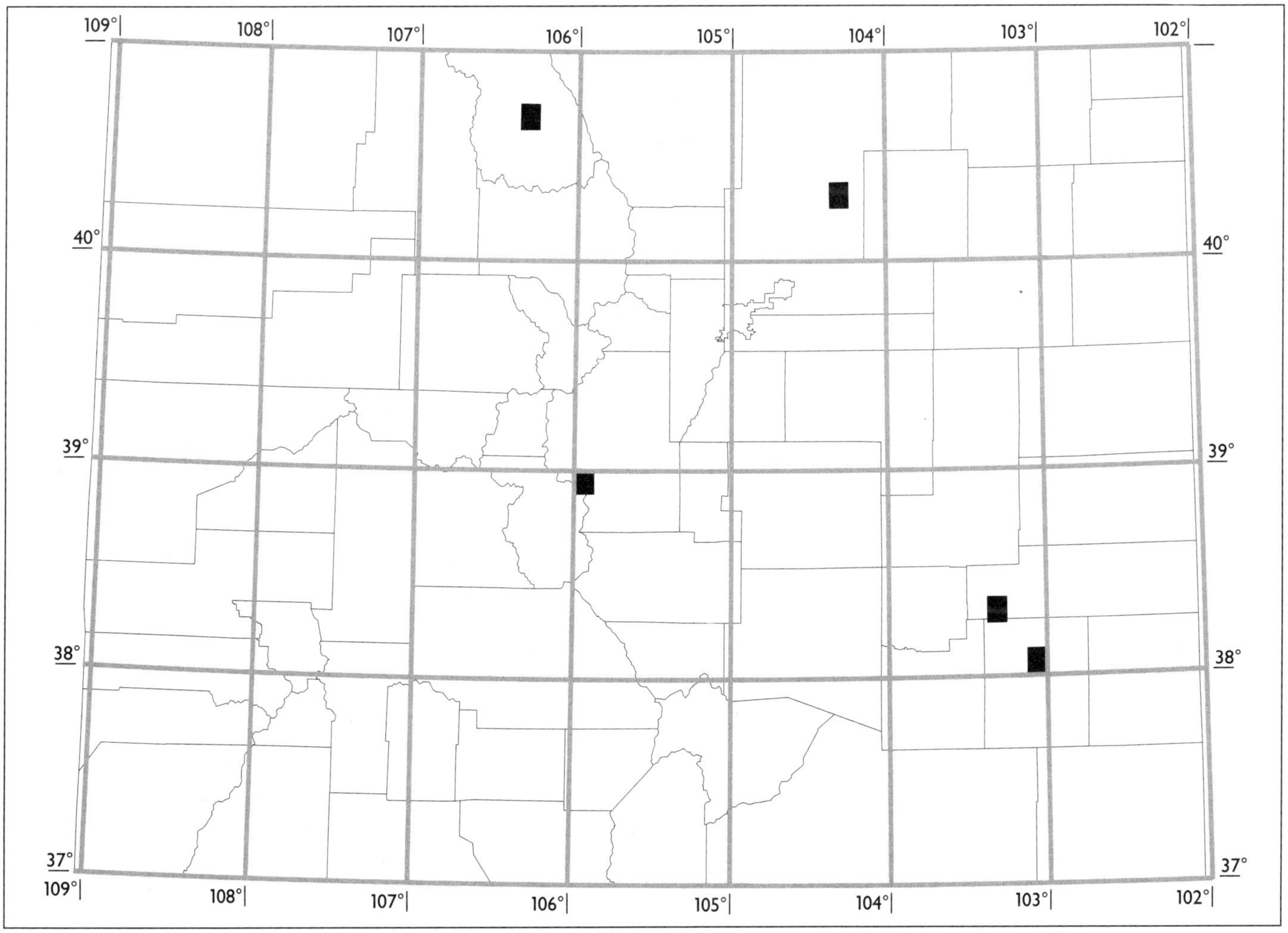
109°
108°
107°
106°
105°
104°
103°
102°
40°
39°
38°
37°

COLORADO DIVISION OF WILDLIFE / WILDLIFE RESOURCE INFORMATION SYSTEM

Forster's Tern mystified early ornithologists; they thought it conspecific with the Common Tern. Not until 1840 did Audubon describe it as a new species, based only on the distinctive winter plumage. Not surprisingly, even the breeding biology was misunderstood. Coues's 1877 account identified breeding only in "British America" (probably the Prairie Provinces) instead of the continental United States (Bent 1921). When research finally confirmed their nesting affinity for emergent wetlands, Bent (1921) proffered that this species deserved the common name "Marsh Tern."

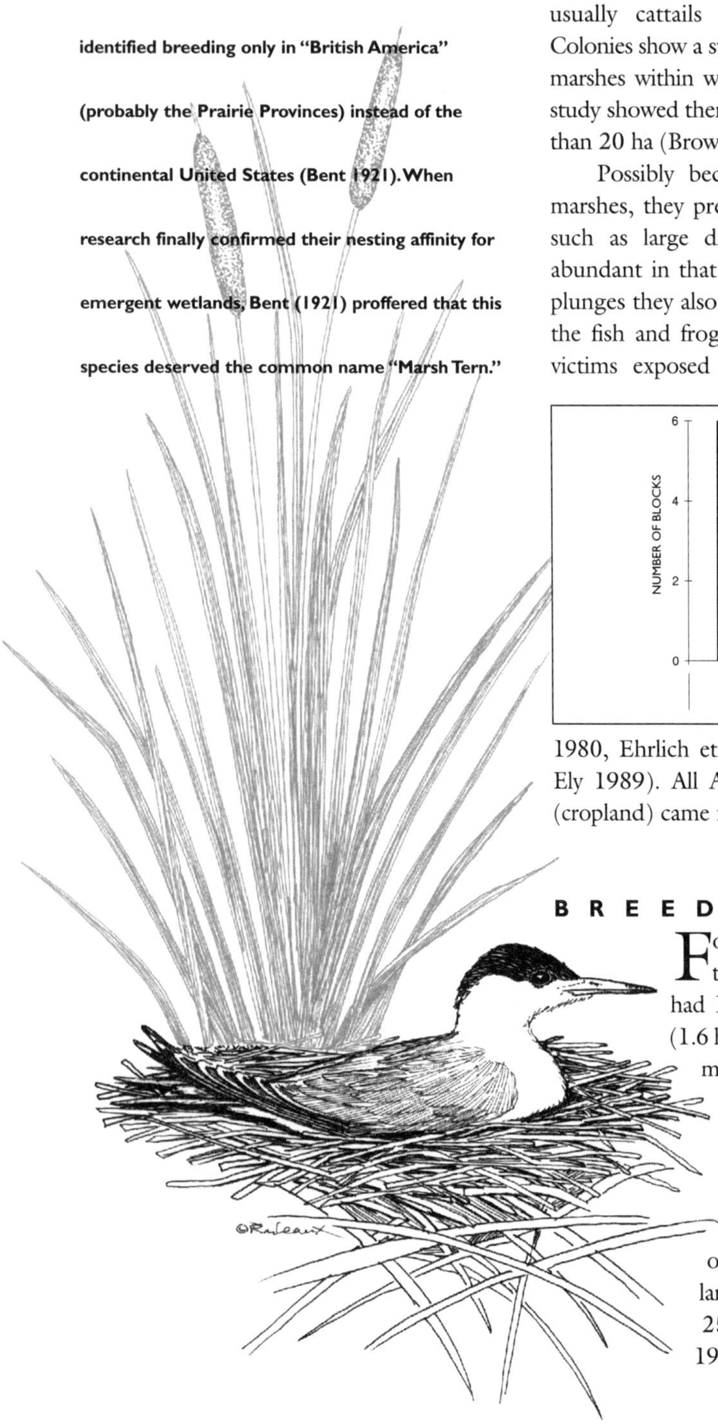

HABITAT

Every large lake on the plains or in the intermountain valleys of Colorado seems to have flocks of Forster's Terns flying over at some time during the summer. They look at home foraging over the water and landing on sandbars with the legions of gulls, other terns, and shorebirds, but they breed at few of these lakes.

"Marsh tern" fits: they breed and feed in marshes. For breeding they need large marshes with extensive emergent vegetation, usually cattails or sedges in Colorado. Colonies show a strong tendency to use larger marshes within wetland complexes; an Iowa study showed them using only marshes larger than 20 ha (Brown and Dinsmore 1986).

Possibly because of their affinity for marshes, they prey mostly on flying insects such as large dragonflies, a food source abundant in that habitat. With their diving plunges they also catch small fish. In spring, the fish and frogs they eat are often dead, victims exposed as the ice melts (Terres 1980, Ehrlich et al. 1988, Thompson and Ely 1989). All Atlas observations but one (cropland) came from marshes.

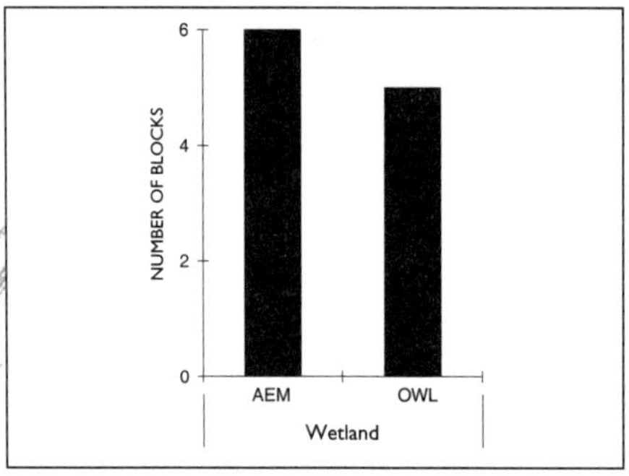
6
4
2
0
NUMBER OF BLOCKS
AEM
OWL
Wetland

BREEDING

Forster's Terns nest close together; an Iowa marsh had 150 nests within four acres (1.6 ha) of a 1,000-acre (405-ha) marsh. They build conical mounds of plant detritus about a foot across. Relatively social, they may place nests so close together as to touch each other, up to eight on one large muskrat house of only 25 square feet (Provost 1947, Burleigh 1971).

Nests usually sit in emergent reedy wetlands, although other wetland types may suffice. Unlike Black Terns, Forster's Terns situate their nests in association with open water (Provost 1947). Almost always elaborately constructed and attached to floating cattail root stalks, the nests "looked like works of art compared with the slovenly nests built by other species of terns" (Bent 1921). In addition to reedy wetlands, muskrat huts and even abandoned nests of Western Grebes have served as nest sites (Bent 1921). Unlike many wetland nesters, these terns like their eggs to remain dry. The final nest cup usually rises at least 2–6 inches (5–16 cm) above the water level (Bergman et al. 1970).

BREEDING PHENOLOGY

CODE		# OF RECORDS	DATE RANGE
NE	Nest with Eggs	1	9 Jul
NY	Nest with Young	1	8 Jul

DISTRIBUTION

Forster's Terns breed mainly from the Prairie Provinces of western Canada south to northwestern Iowa, and in scattered sites from southern California to Colorado. A disjunct population nests in coastal marshes on the mid Atlantic and from Louisiana to Tamaulipas, Mexico. They winter from coastal California and Mexico to the southern Gulf Coast and Puerto Rico.

In the early 1900s Barr Lake had colonies numbering up to 100 pairs (Sclater 1912, B&N). In 1965 B&N knew only of that site and one at Riverside Reservoir. In the 15-year period between 1967 and 1982, breeding was reported in five eastern plains latilongs (Latilong Study), but the reports lack documentation and details. During the Atlas observers did not Confirm nesting anywhere in eastern Colorado. Atlas Confirmations, confined to wildlife refuges in the San Luis Valley and North Park, reflect the present location of most, if not all, of Colorado's breeding pairs. The number of breeding sites in Colorado has dropped, and probably the number of breeding pairs has as well, to less than 100 pairs.

The North American population is shrinking for reasons not yet well known, but no agencies have yet categorized the species as endangered or threatened. In Minnesota a two-year study showed an average fledging rate of 0.46 per pair, below the rate necessary to sustain Common Tern populations and presumably below that for Forster's (Cuthbert 1993). The Iowa studies monitored success rates as low as 4% (during a year of heavy rain and rising water levels) to 36%. Factors such as wind and wave action, muskrat activity, predation, and intraspecific strife caused the failures (Bergman et al. 1970). Colorado terns apparently mirror these trends.

If the species faces plummeting numbers throughout its range as serious as those in Colorado, Minnesota, and Iowa, ornithologists have cause for concern. Forster's and Black terns are "marsh terns," and the wetlands where they nest often abut agricultural land. Pesticides running off into wetlands can kill the invertebrates on which the terns feed, and consumption of surviving insects may cause toxic buildups in the terns. Draining marshes for agriculture and development has reduced available habitat (Webb 1985) and could cause some of the decline. Wetland recovery efforts and control of agricultural runoff should allow for population stabilization and recovery.

BREEDING EVIDENCE

REPORTED IN 6 (0%) OF 1745 PRIORITY BLOCKS

☐	Possible	6 blocks	(100%)
■	Probable	0 blocks	(0%)
■	Confirmed	0 blocks	(0%)

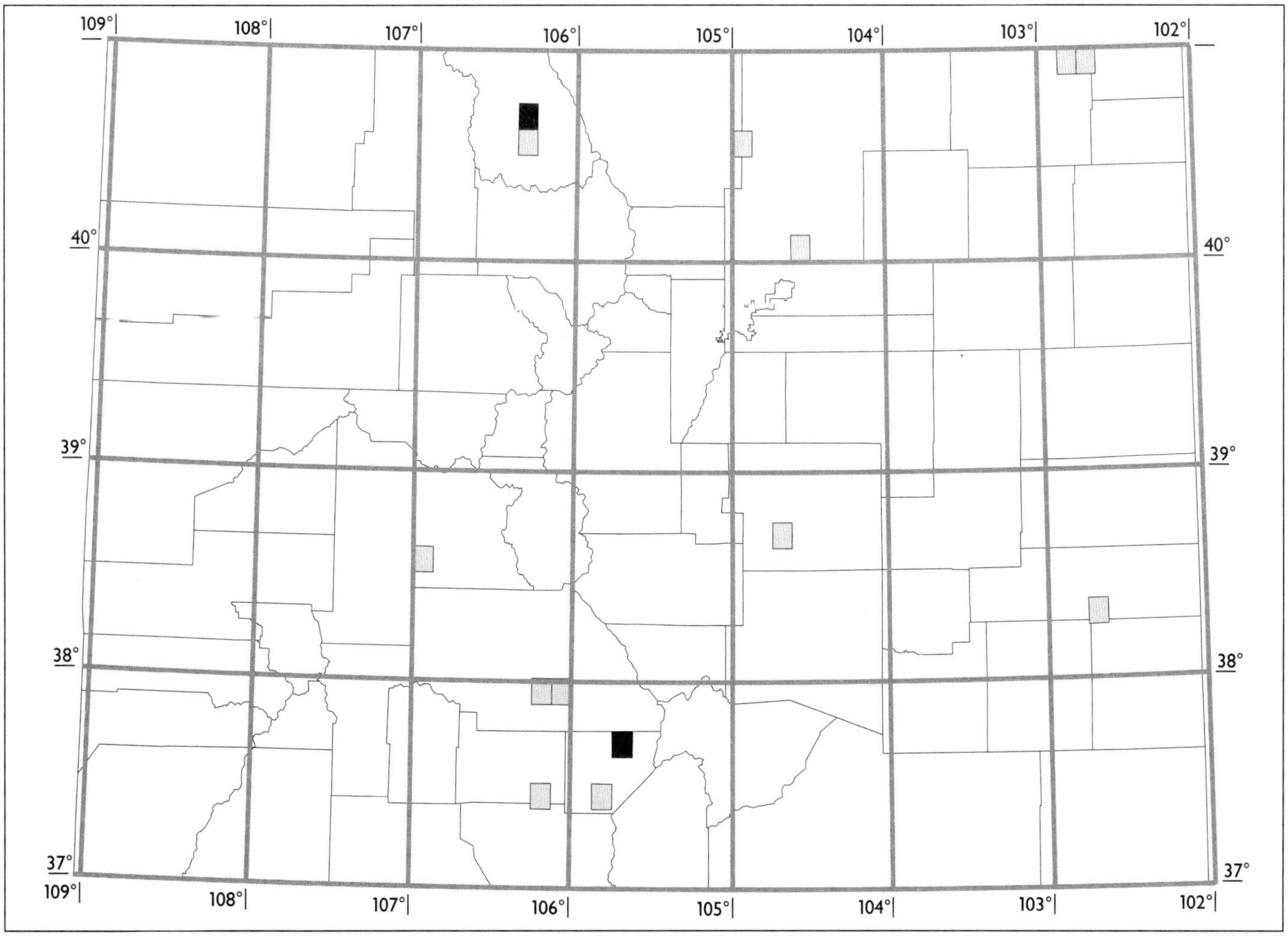
109°
108°
107°
106°
105°
104°
103°
102°
40°
39°
38°
37°

COLORADO DIVISION OF WILDLIFE / WILDLIFE RESOURCE INFORMATION SYSTEM

Tenacious defenders of their nesting territories, Least Terns dive-bomb colony trespassers with vocal assaults and often plaster human and animal invaders with copious quantities of well-aimed excrement. Sadly, this aggressiveness has not succeeded in keeping this graceful little tern from becoming endangered.

HABITAT

Historically, interior Least Terns in the central Great Plains nested on sandbars scoured by heavy runoff from spring rains combined with mountain snowmelt. In this current age of irrigation and river channelization, dams and canals have altered historic flooding cycles, rendering the former habitat changed and unusable.

In Colorado Least Terns have adapted to using the shores of irrigation reservoirs built to hold spring runoff. Although the terns may nest on mainland beaches, they are especially drawn to, and experience much greater success nesting on, islands. Islands, especially when periodically flooded, eliminate the threat of terrestrial predators.

Sparsely vegetated sandy, gravelly, or silty beaches provide nest sites. Often, nest cups involve only simple scrapes on the ground, but occasionally they have a pebble or twig lining. The tan, speckled eggs match their surroundings perfectly, rendering nests nearly invisible to predators.

By returning later in spring than most migrants, Least Terns synchronize nesting with receding water levels. In years with heavy runoff, traditionally used reservoirs, beaches, or islands may have no habitat. Whole colonies may relocate, wait for water to recede, or leave without nesting.

They feed exclusively on small fish, which limits nesting to lakes with healthy fish populations and excludes alkali playas. When fish die during the nesting season, as at Neeskah Reservoir in 1993, nests fail.

BREEDING

Adults return from their winter grounds after 15 May, with numbers continuing to build through June. Least Terns incorporate an aerial glide into their courtship, and they have a distinctive fish-flight call that males and females use during courtship, incubation, and nestling feeding (Massey 1976).

Pioneer Least Terns begin laying eggs the last few days of May, and colonies often recruit more pairs almost daily. Nests typically contain two or three eggs when complete, with larger clutches predominating early in the season. Clutch size also varies with habitat quality. Nests on islands with good nesting habitat, in lakes with abundant fish, average more eggs than nests on marginal silty or muddy sites characterized by intruding weedy growth.

Both sexes share incubation. Nationwide, incubation takes 19–25 days (Thompson et al. 1997); in Colorado 19 days is typical (DLN). The eggs often hatch asynchronously. First flight takes place at 19–21 days. Groups of recently fledged terns remain for a few days on the edges of the nesting colony (DLN).

BREEDING PHENOLOGY

CODE		# OF RECORDS	DATE RANGE
NY	Nest with Young	1	26 Jun
FL	Fledged Young	1	19 Aug

DISTRIBUTION

The Least Tern was once considered conspecific with the Little Tern of the Old World, but a study on vocalizations, behavior, and fine points of morphology (Massey 1976) led to splitting them into two species (AOU 1983). They breed in the West Indies, Central America, and in three groups in North America: East Coast, Pacific Coast, and along the major midwestern river systems. They winter on the coasts of Mexico, Central America, and northern South America. The USFWS has classified both the Interior and California Least Terns as endangered.

Interior Least Terns nest along rivers and lakes of the Mississippi, Ohio, Missouri, and Arkansas river drainages, and several Texas rivers. Population counts for this form have varied from 1,400–1,800 terns at the time of listing in 1985 (Whitman 1988) to 4,932 individuals in a 1988 survey (Sidle and Harrison 1990).

Colorado had only three records of the species before the first nesting confirmation in 1949 at two sites near Prewitt Reservoir (B&N). No further nesting was found until 1978 when nests in the Arkansas Valley at Horse Creek and Adobe Creek reservoirs re-established them as Colorado breeders (Chase 1979). Systematic monitoring for

CDOW confirmed nesting at Adobe Creek and Neenoshe reservoirs in 1990 (Nelson and Carter 1990). The Atlas map shows the sites discovered during research sponsored by CDOW, but only those used in 1987–1995 (DLN). Statewide nest counts during the Atlas period ranged from a high of 23 nests in 1991 to a low of 12 in 1995. In post-Atlas 1996 Least Terns rebounded to a record 37 nests, including 19 nests at Adobe Creek, where in 1994 and 1995 water covered the island (DLN).

The tiny, vocal Least Tern is a federally listed Endangered Species (Whitman 1988), the same status accorded it by the CDOW. The primary limiting factor for the terns, as for Piping Plovers, is availability of nesting habitat. A CDOW statewide recovery plan aims to restore and maintain a healthy population, primarily by recovery and creation of habitat. Managing Tern Island at Adobe Creek Reservoir for the benefit of Least Terns is the cornerstone of the recovery efforts. By 1996 CDOW had cut and burned all of the approximately 100,000 saplings in a belt of vegetation around the island. This created proper habitat at any water level. Additionally, trees cut on an ephemeral island at John Martin Reservoir in May 1996 lured Least Terns to nest there for the first time. Over 30 young eventually fledged from this small island. Other corollary activities such as beach closures and law enforcement continue.

Least Terns currently need human intervention in order to survive, mainly because degradation of habitat poses such a threat. Even with intensive management efforts, the future of Least Terns in Colorado and elsewhere remains uncertain. However their rebound in 1996 signals cause for great optimism.

BREEDING EVIDENCE

REPORTED IN 4 (0%) OF 1745 PRIORITY BLOCKS

□	Possible	1 block	(25%)
■	Probable	2 blocks	(50%)
■	Confirmed	1 block	(25%)

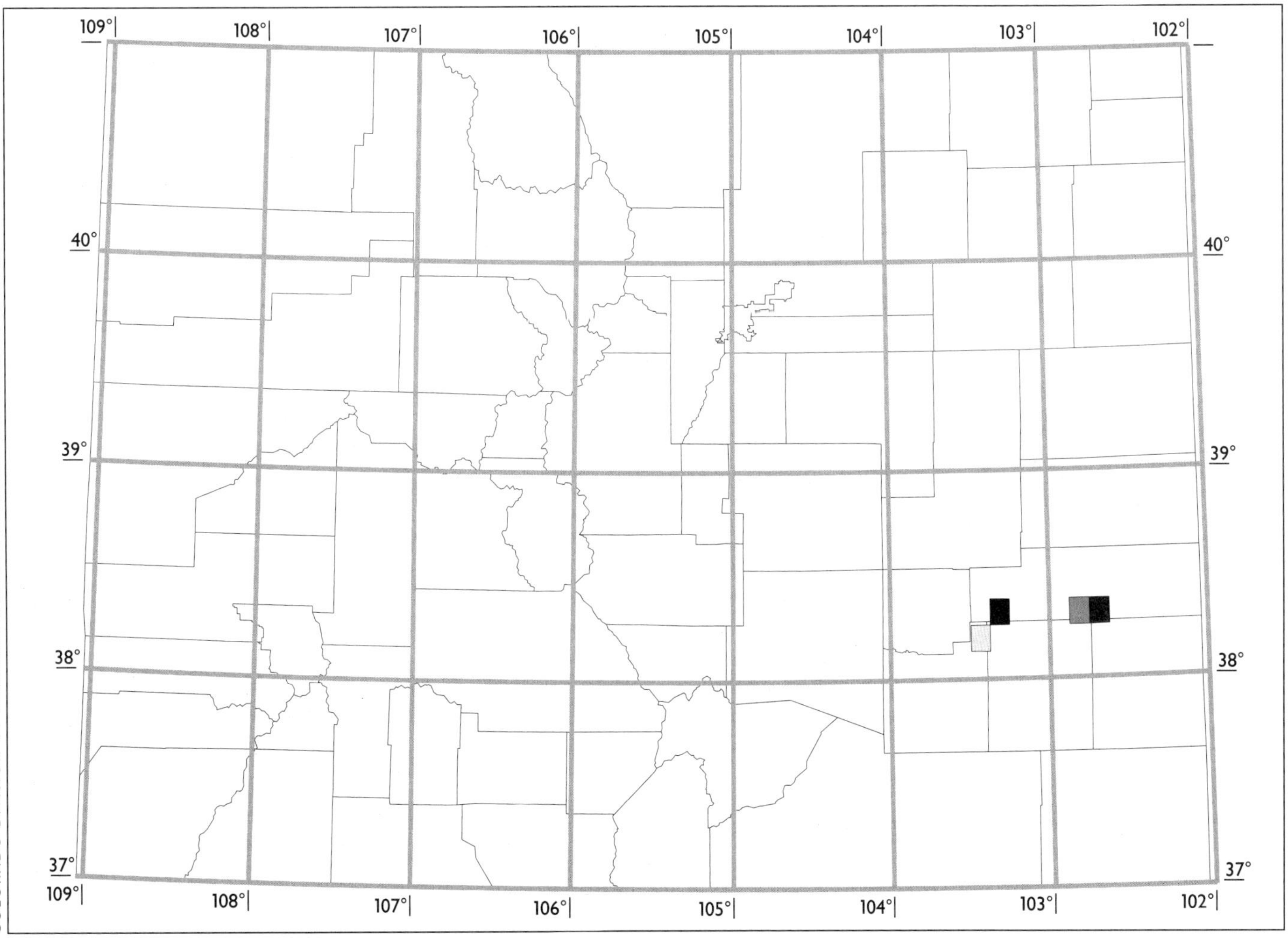
109°
108°
107°
106°
105°
104°
103°
102°
40°
39°
38°
37°

COLORADO DIVISION OF WILDLIFE / WILDLIFE RESOURCE INFORMATION SYSTEM

Cloaked in mystery, Black Terns appear abruptly as dark apparitions over lakes or wetlands. Sometimes flocks numbering in the hundreds appear, only to disappear by the next day, or the next hour. Their customary silence, unusual for a tern species, only adds to the enigma. Furthermore, they have drastically dropped in numbers during the twentieth century, to one-quarter or less of their numbers in 1900 (Dunn and Agro 1995).

HABITAT

Black Terns lead a double life. They breed in freshwater marshes and winter along marine coasts. During the nesting season they eat only insects, which they pick from the air and from the surface of the water. During the winter they feed mainly on small fish (Dunn and Agro 1995). Restlessly they fly like swallows above the surface, and rarely dive into the water like other terns (Burleigh 1972, Terres 1980, Ehrlich et al. 1988).

True prairie terns, they nest only at the edges of ponds with emergent reedy wetlands, especially prairie sloughs and pseudo-prairie of the high Colorado mountain parks. They also nest in forested areas with marshes and ponds interspersed (Dunn and Agro 1995).

Black and Forster's terns will nest sympatrically. Forster's pick higher and drier sites; Blacks choose lower sites on wet substrates. They prefer marsh complexes of at least 50 acres (20 ha; Brown and Dinsmore 1986). Colorado has few marshes that large, although smaller marshes suffice—the smallest reported wetland, in Iowa, is 12 acres (5 ha; Brown and Dinsmore 1986)—but with the marsh the terns need open water and fields for feeding (Bergman et al. 1970). Black Terns spend much time in the air winging over marshes, water, and fields hawking for insects.

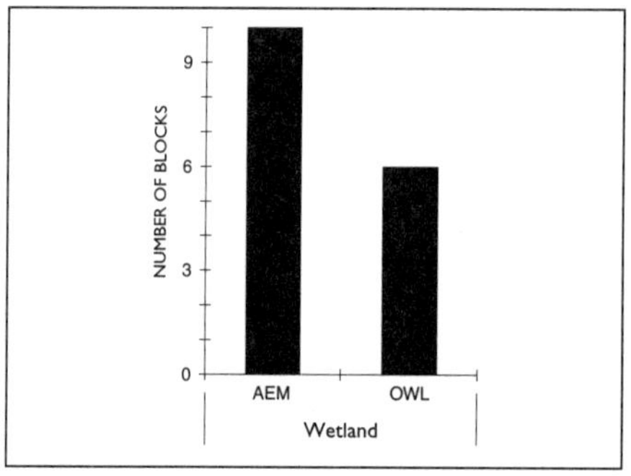
NUMBER OF BLOCKS
0
3
6
9
AEM
OWL
Wetland

Although Colorado has many marshes, their small size and lack of sufficient open water nearby limit these terns to a few places. Yet most apparently suitable habitat remains unclaimed today by Black Terns. Atlasers found them mainly at wildlife refuges, with many former sites unoccupied.

BREEDING

Black Terns have two types of display flights, one called a "fish flight," but although catching fish may be part of this ritual, the birds normally carry insects rather than fish. The other courtship flight, "flock flight," involves most of the terns nesting in the colony (Johnsgard 1979).

Black Terns place their nests in a variety of conditions: on floating, matted vegetative growth from a previous year, on top of muskrat houses, and sometimes over water with growing emergents. They distribute their nests in groups of 10–20 scattered across the marsh. Most nest sites lie in water at least 2 feet deep, but with a variety of surrounding vegetation from dense cattails to open water (Bergman et al. 1970).

Nests vary from very simple, or nonexistent, "an apology for a nest," to elaborate. Black Terns often put the nest on algal mats buoyed by plant remains. They build the nest cup, a moist and spongy, sloppy, conical mound of plant detritus (Provost 1974). They use fragments of reeds and cattails, usually gathered in the immediate vicinity of the nest rather than carried in from other parts of the marsh (Burleigh 1971, Johnsgard 1979, Spahn *in* Andrle and Carroll 1988). The three eggs lie barely above the water level and even tolerate being wet (Davis and Ackerman 1985). Atlas data provide little information on the nesting cycle. The only Atlas Confirmation dates were nests found in two blocks on 7–8 July, plus fledglings in late July.

Although highly sensitive to human disturbance, they defend their nests aggressively, even to the point of striking intruders (Bent 1921).

BREEDING PHENOLOGY

CODE		# OF RECORDS	DATE RANGE
ON	Occupied Nest	1	8 Jul
NY	Nest with Young	1	7 Jul
FL	Fledged Young	1	28 Jul

DISTRIBUTION

Black Terns nest—or did nest—in temperate parts of Eurasia and North America. North American birds breed from southern Canada to northern California, southern Colorado, and southern New England, and they winter along the Pacific and Caribbean coasts from Mexico to northern South America.

They have absented many historic sites throughout their range, although their overall range probably remains unchanged (Dunn and Agro 1995). More may once have nested in Colorado; their breeding and migrating numbers have dwindled.

A "large" colony nested in a "marshy tract along the Arkansas River" at Fort Lyon in 1864 (Coues 1874). Eggs presented to the Colorado College Museum documented breeding near Greeley in 1902 (Sclater 1912). In 1907–1908 Hersey and Rockwell found two nests along the chain of marshes north of Barr Lake (Niedrach and Rockwell 1939), and a few nested there regularly at least through 1938 (B&N).

They no longer nest in these places. The Latilong Study indicates confirmed or probable nesting status in all but two of the eastern Colorado latilongs, not accurate today and perhaps not when published (HEK). During the Atlas no Confirmations of nesting came from anywhere in eastern Colorado. Definite breeding came only from San Luis Lake SWA and Alamosa and Arapaho NWRs, although these mysterious terns could hide their nests in the prairie marshes on the map with Possible and Probable breeding.

Three factors contribute to the Black Tern's current diminishing status. Habitat degradation, the most obvious, has followed drainage of wetlands for agriculture or development (Webb 1985, A&R, Dunn and Agro 1995). On the wintering grounds, stocks of small fish collapsed in 1972 in the Pacific Ocean off Central America, and overfishing has kept the stocks depleted; this has the potential to devastate Black Terns. Pesticides may reduce insect foods; Black Terns suffered from eggshell thinning in the 1970s, although they have since recovered. They also achieve nest success varying from 13 to 72%: exceptionally low (Dunn and Agro 1995) and perhaps below levels necessary to maintain populations (Dinsmore *in* Jackson et al. 1996). The reasons for their current decline remain as mysterious as the birds themselves.

BREEDING EVIDENCE

REPORTED IN 11 (1%) OF 1745 PRIORITY BLOCKS

Possible	10 blocks	(91%)
Probable	0 blocks	(0%)
Confirmed	1 block	(9%)

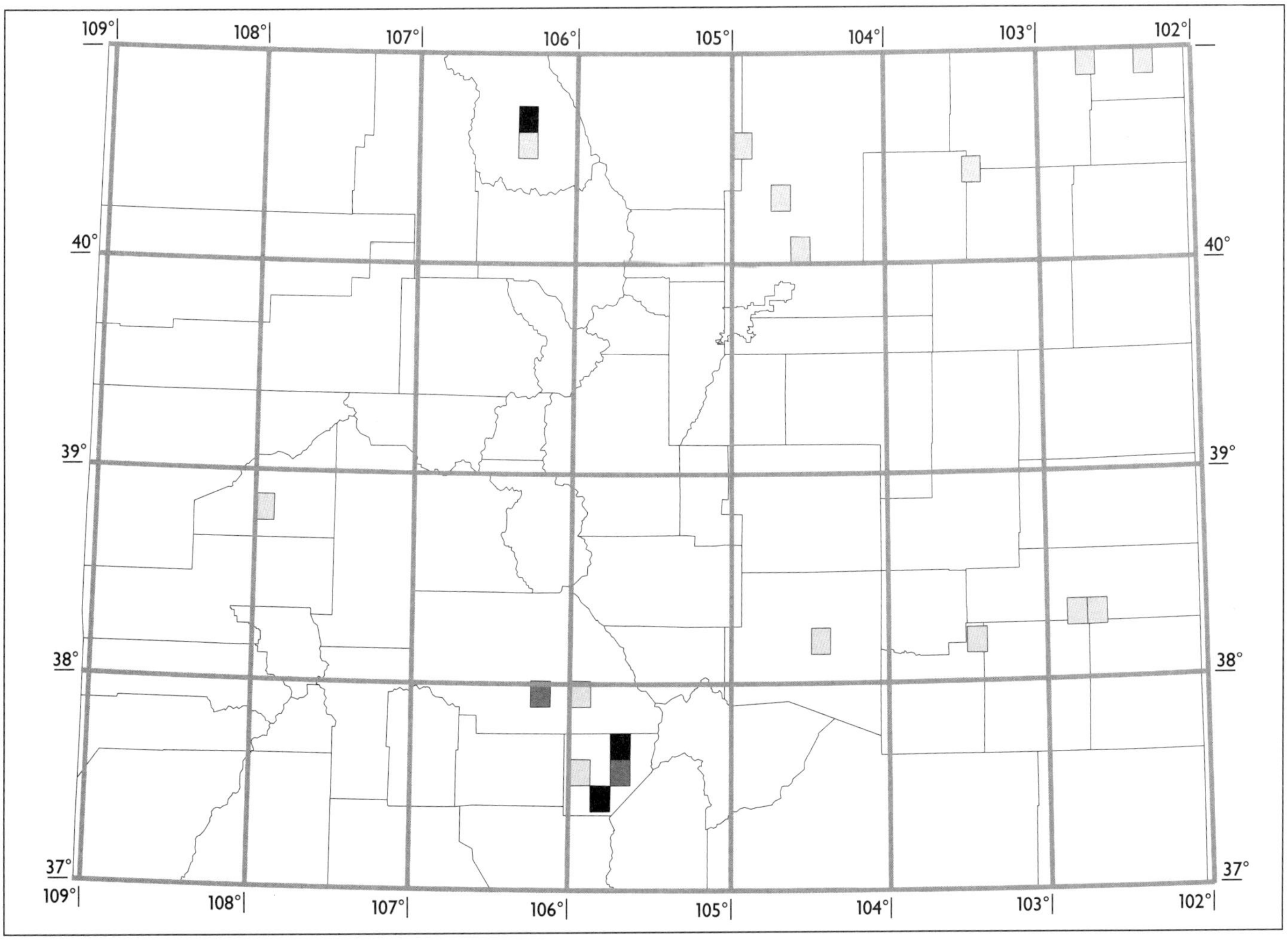
109°
108°
107°
106°
105°
104°
103°
102°
40°
39°
38°
37°

COLORADO DIVISION OF WILDLIFE / WILDLIFE RESOURCE INFORMATION SYSTEM

ROCK DOVE

Columba livia

With a link to humans that predates the recorded history of man (who raised them for food over 6,500 years ago), the ubiquitous Rock Doves were the first domesticated birds (Zeuner 1963). Eurasian natives, they entered the New World with the first settlers, and these pigeons soon became feral and part of the American avifauna.

HABITAT

Following the habits of their wild ancestors, feral Rock Doves still breed in holes and caves in rocky cliffs. Ten percent of Atlas reports specified cliffs; generally these birds centered in Baca/Las Animas counties, the Front Range north of Denver, and the mesa/canyon country west of Grand Junction.

About three-fourths of the Atlas reports came from manmade habitats. Rock Doves readily substitute structures, such as buildings and bridges, for natural nesting sites. Therefore they easily find suitable places to live and to breed in throughout most of Colorado. Characteristic city dwellers, they gather in large numbers on rooftops and ledges, and feed in parks, lawns, and streets. Urban downtowns reproduce the natural canyons from which they began their odyssey.

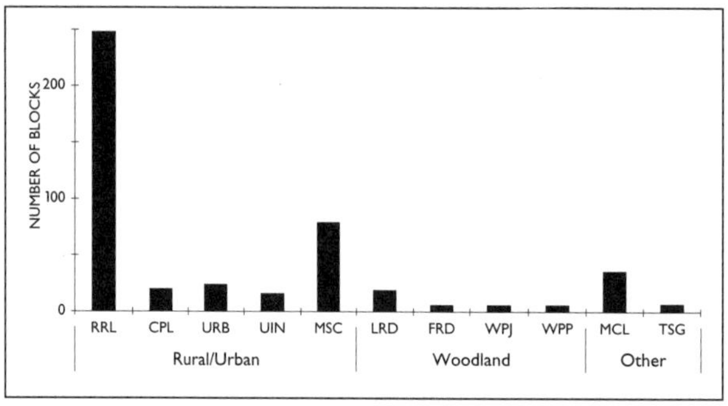
NUMBER OF BLOCKS
0
100
200
RRL
CPL
URB
UIN
MSC
LRD
FRD
WPJ
WPP
MCL
TSG
Rural/Urban
Woodland
Other

Atlas workers recorded them more often in rural habitats—in over half of the blocks—but only because such a small part of Colorado is urban. In rural areas Rock Doves procure their food of seeds and grain in the fields and feedlots, and the birds breed around the barns and outbuildings of farms and under road bridges.

BREEDING

In Colorado, as in most of the Temperate Zone, these birds nest in every month of the year (Goodwin 1983). Although they lay only two eggs, they achieve high breeding success for many reasons. Monogamous, both parents care for the young and produce the rich crop milk with which they feed the young by regurgitation.

Usually colonial nesters, Rock Doves can guard against predators more easily than birds that nest singly. By nesting year round, they can raise as many as five broods of youngsters in a year, and these youngsters themselves can breed at six months. They make a shallow nest of twigs and straw in the chosen niche and expend little energy at the task. Atlas workers recorded NY (nest with young) as early as 20 February and NE (nest with eggs) as late as 22 November. One pair of these doves made an unusual nest on a small shelf inside a totem pole in the middle of the courtyard at the Fine Arts Center in Colorado Springs (RRK). This nest held eggs being brooded in December.

BREEDING PHENOLOGY

CODE		# OF RECORDS	DATE RANGE
C	Courtship	10	27 Mar–14 Jul
NB	Nest Building	7	7 May–1 Jul
ON	Occupied Nest	112	24 Feb–24 Jul
NE	Nest with Eggs	10	28 May–22 Nov
NY	Nest with Young	14	20 Feb–12 Jul
FY	Feeding Young	6	8 May–15 Aug
FL	Fledged Young	12	20 Apr–10 Aug

DISTRIBUTION

British ornithologists distinguish between wild Rock Doves and "Feral Pigeons" (Robbins et al. 1993). In 1983 wild Rock Doves still inhabited the Faeroe, Shetland, and Hebridian islands, the coasts of Scotland, Ireland, the Iberian Peninsula, and northwestern Africa, and the northern shores of the Mediterranean (Goodwin 1983, Robbins et al. 1993). However, by 1993 they existed as wild birds in Europe only in southwestern Portugal (Johnston 1994). The common feral Rock Doves or "city pigeons" descended from these wild doves by way of the domesticated pigeon. They now live worldwide wherever people have settled.

The wild ancestors of Rock Doves never inhabited the Americas. The French introduced the birds to this continent at Port Royal, Nova Scotia, in 1607, but by

1750 these Canadian introductions had failed (Erskine 1992). The English brought them to Virginia as food supplies in 1620 (Schorger 1952). In North America they are now absent only from wilderness areas, high mountains, dense forests, and barren deserts. They thrive in every Colorado latilong (Latilong Study 1987), but Atlas volunteers did not report them from sagebrush habitats or high mountains (although they did Confirm nesting at 10,000 feet [3048 m] in downtown Leadville). Atlasers missed these birds in South Park and the Gunnison Valley, but at least in Gunnison they probably breed around ranches (Hyde 1979).

Many people think of Rock Doves as nuisances in cities because their droppings soil buildings and statues and cause chemical deterioration of stones and paint. Others, especially the elderly and children, derive much pleasure from "feeding the pigeons" and are delighted when the birds show that they recognize their benefactors. Rock Doves in turn become unwitting benefactors of the Peregrine Falcons recently reintroduced into some North American cities. Pigeons are the main food for these urban birds of prey. Near Colorado Springs, the Prairie Falcons nesting on Gateway Rocks in the Garden of the Gods and the Golden Eagles nesting in the cliffs at Glen Eyrie make many a meal of their neighbors, the Rock Doves.

BREEDING EVIDENCE

REPORTED IN 423 (24%) OF 1745 PRIORITY BLOCKS

□	Possible	161 blocks	(38%)
▩	Probable	90 blocks	(21%)
■	Confirmed	172 blocks	(41%)

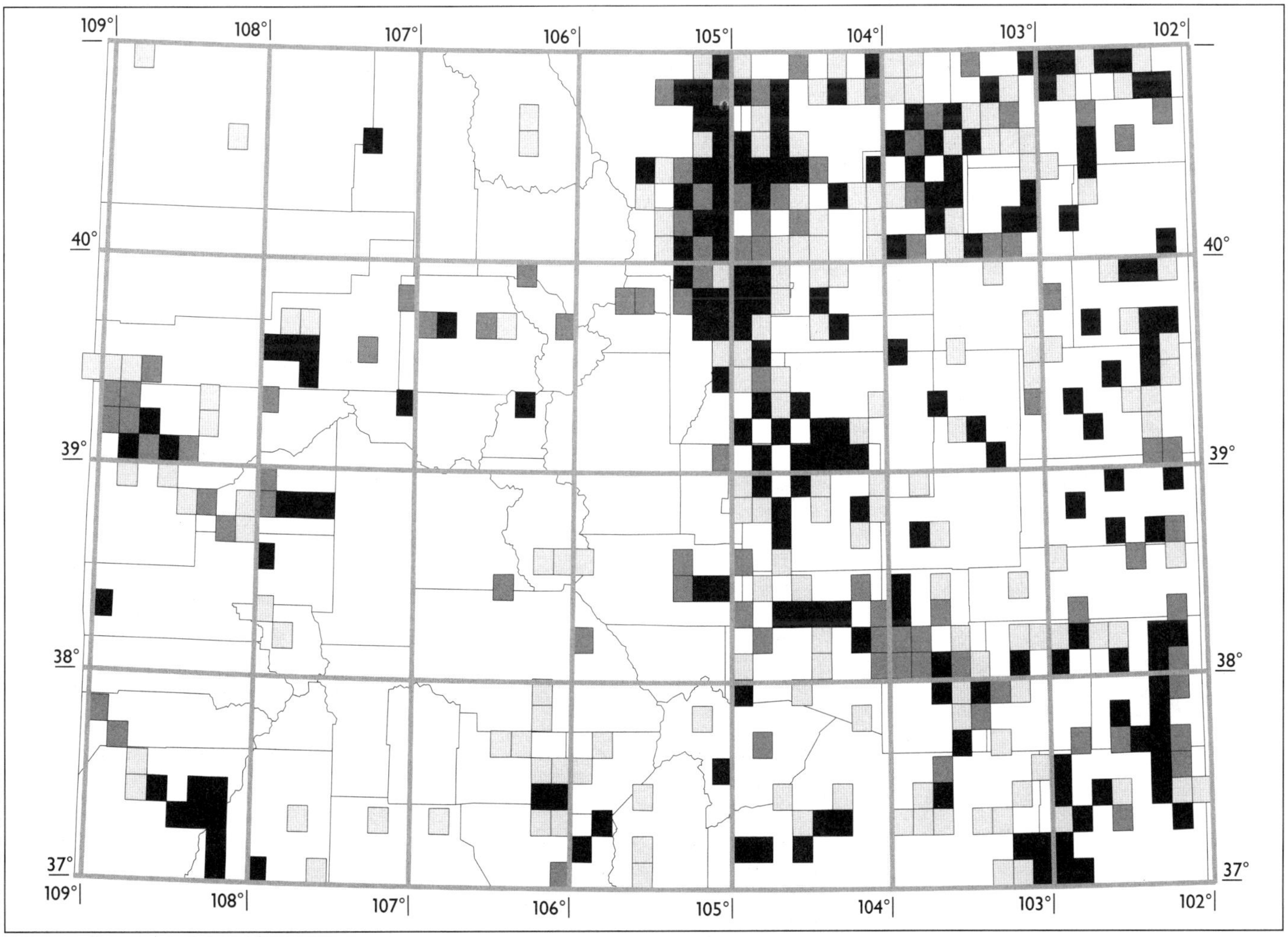
109°
108°
107°
106°
105°
104°
103°
102°
40°
39°
38°
37°

COLORADO DIVISION OF WILDLIFE / WILDLIFE RESOURCE INFORMATION SYSTEM

BAND-TAILED PIGEON *Columba fasciata*

Only in hand or through binoculars can the exquisite coloration of Colorado's only true native wild pigeon be appreciated. The feathers refract the light into brilliant colors ranging from mauve to metallic bronze to green. At the end of the broad, dark-gray tail a wide, pale-gray band gives the bird its common name.

HABITAT

Atlasers found that Band-tailed Pigeons use mostly ponderosa pine, other types of coniferous and mixed forest, and scrub oak shrublands. They reach their highest density between 6,000 and 9,000 feet (1830–2745 m) in mountainous and foothills terrain dominated by scrub oaks and ponderosa and pinyon pines, which produce their favorite food—acorns and pinyon nuts (Braun 1976, A&R). Nesting also occurs in many other habitat types and foraging frequently extends to cultivated fields, stream courses, and livestock feeding areas (Braun 1976).

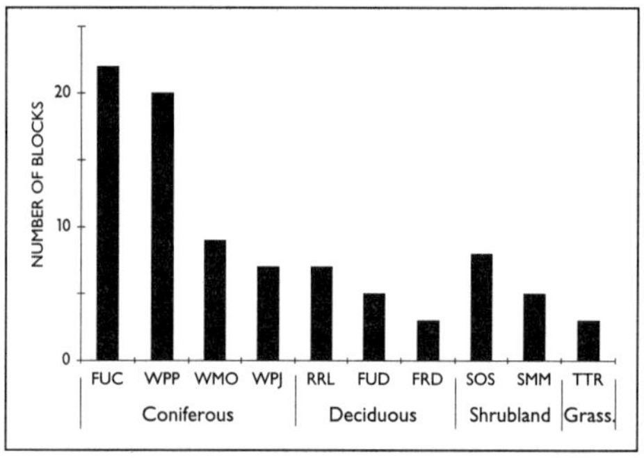
NUMBER OF BLOCKS
0
10
20
FUC
WPP
WMO
WPJ
RRL
FUD
FRD
SOS
SMM
TTR
Coniferous
Deciduous
Shrubland
Grass.

Band-taileds may fly long distances to feed and to satisfy their daily need for fresh water (Shuford 1993). Recent research indicates that during times of limited food supply they may seek food almost anywhere (Schroeder and Braun 1993). Atlas field workers used the "Observed" code in 11 blocks because they concluded that the birds used the particular habitat as a food source but did not breed in the block. In 12 other blocks field workers found them in a habitat where they probably do not breed.

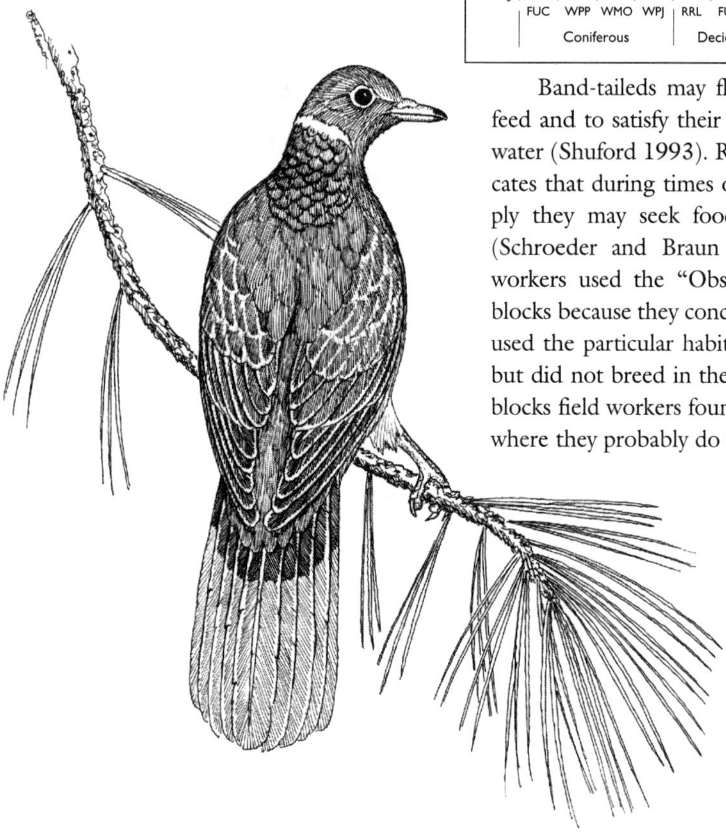

BREEDING

Thorough studies have addressed most game species, including Band-tailed Pigeons. For three decades CDOW biologists trapped birds at 103 stations in nearly half the counties in Colorado, and banded nearly 27,000 birds (Schroeder and Braun 1993). In contrast to such well-funded efforts, the Atlas added little new information.

Pigeons begin arriving in late March and continue to increase until breeding season, which begins in May and can continue into September (A&R). Depending upon the food supply, a pair may nest two or more times during a season.

Pigeons usually fabricate a poorly constructed stick nest placed 12–40 feet (3.5–12 m) high in a conifer or deciduous tree. In 1888 near Durango, Morrison found nests in scrub oaks and on the ground (Sclater 1912). The birds usually lay one egg, rarely two. Both parents take turns incubating the clutch for 19–20 days and both feed the nestling, which fledges 25–28 days after hatching. For their first 4–5 days the young receive regurgitated pigeon's milk, a fatty, curd-like substance produced in the adults' crops. Gradually their diet shifts to the acorn mast, grains, seeds, and fruits eaten by adults (Shuford 1993).

Field workers Confirmed breeding of this species at a rate near the very bottom of all of Colorado's birds. The pigeons hide their nests in heavy cover. When feeding, young adults tend to visit nest sites infrequently, making nest detection difficult (Braun 1994). Atlasers reported that the birds become very quiet when they approached nest sites.

BREEDING PHENOLOGY

CODE		# OF RECORDS	DATE RANGE
C	Courtship	1	3 Jun
FL	Fledged Young	7	17 Jun–5 Aug

DISTRIBUTION

Band-tailed Pigeons breed in western mountain ranges from northwestern North America to Argentina. Presently, two subspecies are recognized north of Mexico. The coastal *C. f. montilla* ranges west of the Sierra Nevadas from southeastern Alaska to Baja California. The interior *C. f. fasciata* breeds from northern Colorado and central Utah south through Arizona and New Mexico into Mexico (Braun 1994).

In 1820 the Long Expedition found these pigeons, then new to science, near Sedalia. Thomas Say described the bird lavishly—"brilliant golden-green" neck and sides, "purplish cinerous" head, "bluish ash" rump and tail (Marsh 1931). According to

Cooke (1897 et seq.) and Sclater (1912) they had an irregular distribution, chiefly south from Cañon City and Glenwood Springs.

The southwestern quarter of Colorado may harbor the most pigeons (Braun 1994, A&R) but the Atlas map, surprisingly, shows them in equal numbers of blocks in the northern and southern halves of the state. Field workers found Band-tailed Pigeons in all but two of the latilongs from the foothills west. Many areas where the CDOW trapped and banded feeding pigeons from 1969 to 1975 (Braun 1976) did not produce any Atlas reports. Most of these locations are in southern Colorado adjacent to New Mexico.

Based on banding results, 53% of the young survive (Schroeder and Braun 1993). Recapture rates of banded birds in 1972 provided an estimate of a fall population of about 70,000 pigeons in Colorado (Braun 1994).

Adjusting for birds of the year, Atlas abundance estimates compute to a population half that of CDOW's 1972 fall count of 70,000. Atlas field work may have missed the birds, in part because of their quiet demeanor around their nests. On the other hand, BBS data show a major continental decline of 2.7%/year, so that the difference between the Atlas observations and CDOW's data from 25 years ago may reflect a real change.

BREEDING EVIDENCE

REPORTED IN 83 (5%) OF 1745 PRIORITY BLOCKS

□	Possible	61 blocks	(74%)
■	Probable	16 blocks	(19%)
■	Confirmed	6 blocks	(7%)

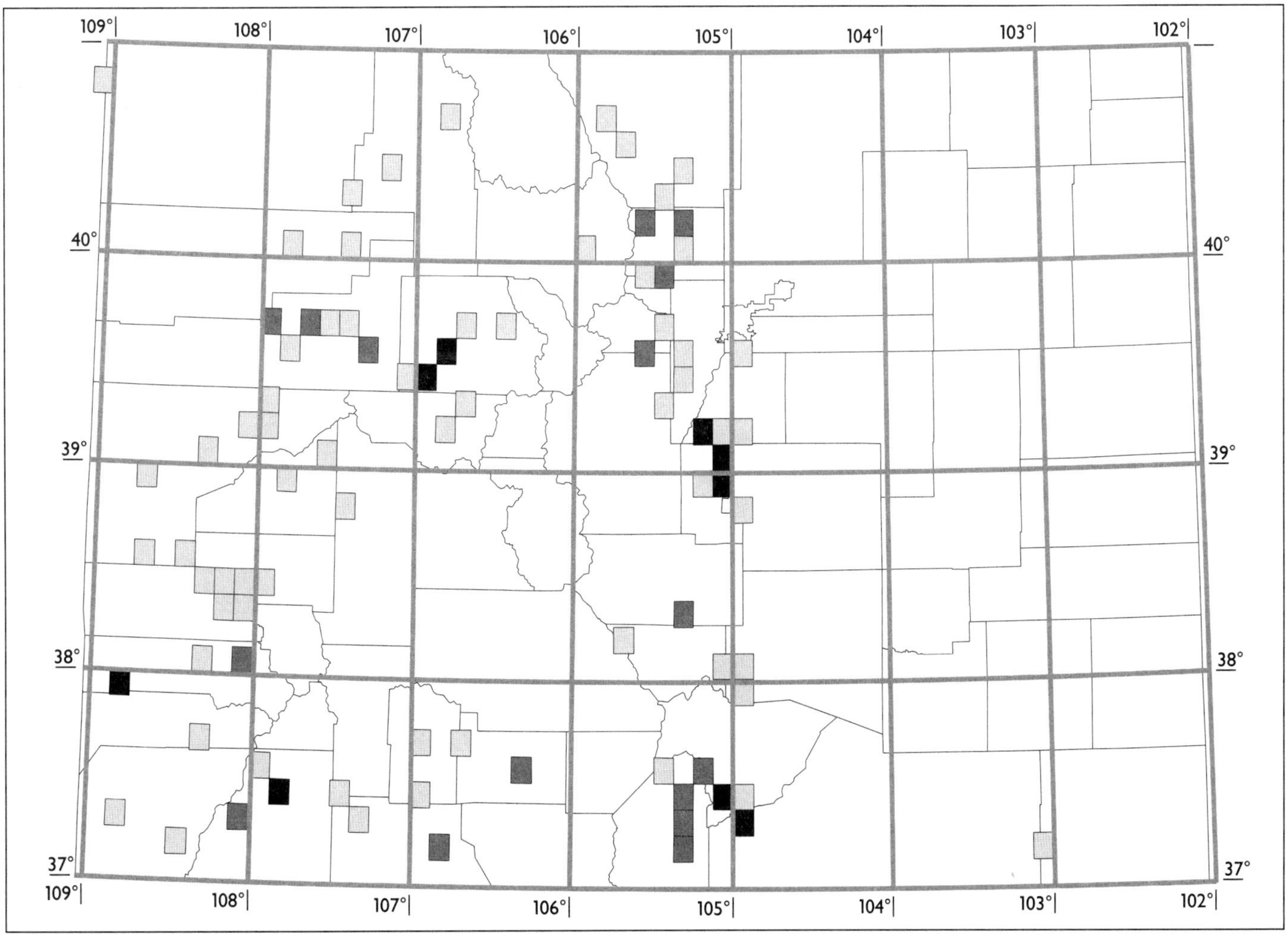
109°
108°
107°
106°
105°
104°
103°
102°
40°
39°
38°
37°

COLORADO DIVISION OF WILDLIFE / WILDLIFE RESOURCE INFORMATION SYSTEM

Feathered inconspicuously in Quaker drab and cooing in soft, low tones, Mourning Doves invite the adjective "gentle." However, like many other members of their pigeon tribe, these adaptable survivors are "tough" both physically and psychologically (Goodwin 1983). Their Colorado range and abundance attest to their tenacity.

HABITAT

The extremely wide choice of habitats by Mourning Doves emphasizes their adaptability. Doves reside from sea level to 9,000 feet (2750 m) and occasionally even to 10,000 feet (3048 m; B&N, Atlas). Only San Juan and Lake, the two highest-elevation counties in the state, lacked Atlas records. In the open lands of South Park, atlasers found them nesting up to 10,000 feet (3048 m).

Mourning Doves have no decided habitat preference. Habitats dominated by humans dominated Atlas dove reports: almost one-third specified rural settings, croplands, and shelterbelts, and 8% named urban residential and park areas. Equal proportions specified deciduous habitats (mainly lowland riparian), grasslands (mainly shortgrass prairie), shrublands (various categories), and conifers. The latter, though, consist essentially of only two types: pinyon/juniper (the second most frequent single habitat reported overall) and ponderosa pine. Altitude, closed canopies, or food requirements probably prevent doves from nesting in the higher conifer forests.

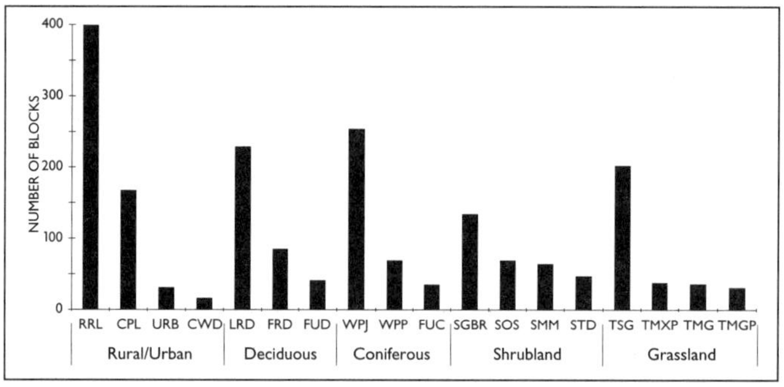

NUMBER OF BLOCKS
400
300
200
100
0
RRL CPL URB CWD LRD FRD FUD WPJ WPP FUC SGBR SOS SMM STD TSG TMXP TMG TMGP
Rural/Urban
Deciduous
Coniferous
Shrubland
Grassland

Seed-eaters, especially of weeds and wild grasses but also of cultivated grains, Mourning Doves have no difficulty finding the food they need. Although classed as ground feeders, they can change their habits when they find a good crop of conifer seeds. They often resort to feeding in trees by extracting the seeds of various pine cones (Tyler *in* Bent 1932).

BREEDING

These migratory doves return to Colorado in March; atlasers found nest-building as early as 14 April. The almost continuous cooing of the males as they woo their mates makes it seem as if they arrive before the females, but possibly only because their cooing makes them more conspicuous (Gutierrez 1971). Males also court with billing and charging displays toward the females, and spectacular "flap-glide" flights. In this, they arise in noisy flapping flights from perches and suddenly descend in long spiraling dives, resembling the flights of Sharp-shinned Hawks (Stokes 1983).

After mating, males select nest sites, which the females accept or reject. Males gather all the nesting materials—small twigs—which the females arrange until a shallow ramshackle nest results. They choose sites on horizontal branches in trees, on the ground, on ledges of cliffs or buildings, on top of used nests, or even, as one atlaser found, on a pile of tumbleweeds.

Although they lay only two eggs, doves brood so diligently that they never leave the nest exposed, and both young usually survive. Males brood during the day and females at night (Nice 1922, 1923). Both parents feed the young with crop milk. Champions at multiple brooding—two broods at least in Colorado and up to four in the southern states—these tough birds ensure the survival of their species (Ehrlich et al. 1988). Atlasers found fledglings from 4 May through 26 August.

Tyler (1932) claimed that Mourning Doves, in contrast to Passenger Pigeons, nested well away from each other. However, Stockard (1905) reported them nesting in small colonies in Mississippi. Atlasers also found them turning to colonial nesting in the prime habitat offered by shelterbelts on the eastern plains, where a single shelterbelt may harbor a nest in every fourth tree.

BREEDING PHENOLOGY

CODE		# OF RECORDS	DATE RANGE
C	Courtship	54	21 Apr–22 Jul
NB	Nest Building	24	14 Apr–11 Jul
ON	Occupied Nest	183	16 Apr–29 Jul
NE	Nest with Eggs	164	29 Apr–28 Jul
NY	Nest with Young	62	7 May–27 Jul
FY	Feeding Young	18	20 May–18 Jul
FL	Fledged Young	235	30 Apr–26 Aug

DISTRIBUTION

Their breeding range includes southern Canada, the U.S., Central America, and the West Indies. The first settlers on the eastern seaboard appreciated these doves as abundant game birds (Tyler 1932). Hunters still shoot them in Colorado, but some northern states protect them as songbirds.

Edwin James first described Mourning Doves in Colorado on 15 July 1820 (Marsh 1931), and every Colorado ornithologist since then has chronicled their ubiquity on the plains and foothills. Most limit breeding to altitudes below 8,000–8,500 feet (2440–2600 m), although Carter found them at Breckenridge at 9,500 feet (2900 m; Sclater 1912).

Atlas results show that these doves almost solidly blanket the eastern plains and thrive abundantly in all other parts of the state except the high mountains. Atlas workers found only American Robins breeding in more blocks than Mourning Doves. They recorded doves in almost every block with elevations below 9,000 feet (2745 m).

Hunting and even man-caused habitat changes have no appreciable effect on these efficient breeders who change their ways to fit the environment in which they find themselves. They even tend to increase in numbers in Colorado, as well as other states, when farmers destroy original vegetation and replace it with secondary growth, farmyard cultivation including shelterbelts, croplands, and pasture (Goodwin 1983).

BREEDING EVIDENCE

REPORTED IN 1386 (79%) OF 1745 PRIORITY BLOCKS

Possible	352 blocks	(25%)
Probable	309 blocks	(22%)
Confirmed	725 blocks	(52%)

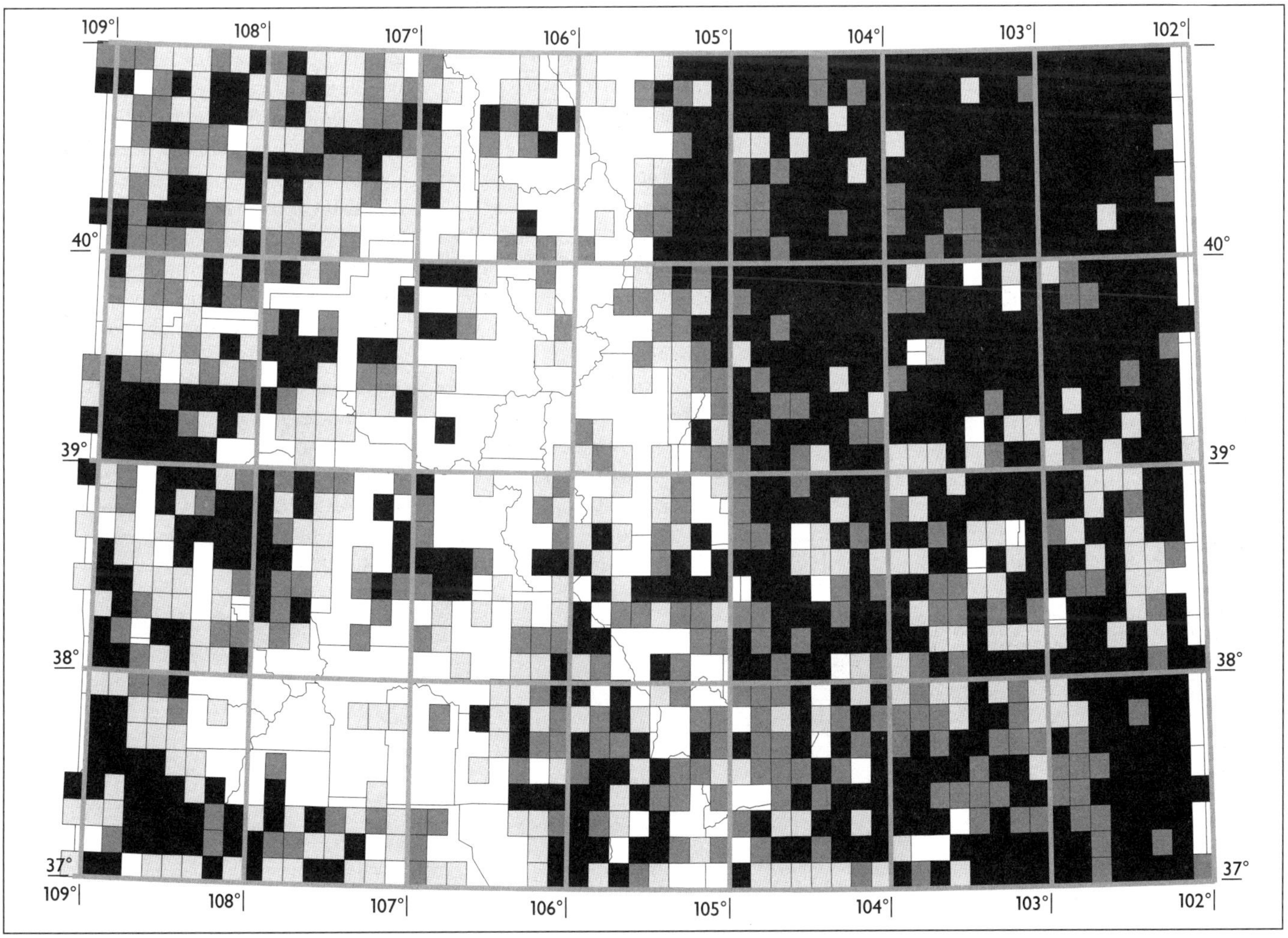
109°
108°
107°
106°
105°
104°
103°
102°
40°
39°
38°
37°

COLORADO DIVISION OF WILDLIFE / WILDLIFE RESOURCE INFORMATION SYSTEM

BLACK-BILLED CUCKOO

Coccyzus erythropthalmus

North American cuckoos, unlike their European counterparts, build their own nests (usually), although they possess only rudimentary nest-building skills. They occasionally exhibit a latent streak of parasitism; observers in the eastern United States have found cuckoo eggs in the nests of several common, cuckoo-sized species.

HABITAT

These secretive birds live in lowland riparian forests and in tall urban trees (B&N, A&R). Groves of trees, moist thickets, and forest edges or overgrown pastures attract them (Harrison 1979). Very similar in habits, behavior, and appearance, Yellow-billed and Black-billed cuckoos often inhabit the same woodlands (Bent 1940). No studies so far separate the niches of these two cuckoos, although some observe that Black-billeds stick more to wooded edges than to dense canopied habitats and act even more retiring than their Yellow-billed cousins (Bent 1940, Johnsgard 1979, Terres 1980).

The Latilong Study lists Colorado habitat as lowland riparian. Atlas workers found Black-billed Cuckoos in only four Non-priority blocks, with all the birds in cottonwoods. The limited number of cuckoos in Colorado precludes distinguishing habitat choices by the two species, but both seem to occur in mature cottonwood forests.

These cuckoos feed on "hairy" caterpillars, and their numbers cycle in unison with outbreaks of these caterpillars—in the East often those of tent caterpillars, webworms, mourning cloaks, and tussock moths including gypsy moths. Among the few birds that will touch these creatures, cuckoos have stomachs that become "so felted with a mass of hairs and spines that it obstructs digestion"—which supposedly causes them to shed their stomach lining and grow a new one. This predilection to consume destructive insects earned praise from the early "economic ornithologists" who evaluated, and indeed defended, birds according to their capacity for destroying pest insects (Pistorius *in* Laughlin and Kibbe 1985). Colorado has few hosts for these insects in cuckoo habitat, which may explain the infrequency with which we see either cuckoo species.

BREEDING

Black-billeds place their own nests in either deciduous or evergreen trees. The nests range from 2 to 20 feet high (0.6–6 m), with an average height of 6 feet (1.8 m) (Harrison 1979). They line the nest with cottony fibers, catkins, and leaves (Harrison 1979), and Black-billeds have a reputation of building better, stronger nests than those of Yellow-billeds (Bent 1940). A 1951 Colorado nest was reported in a lilac bush (B&N). The female lays two to four small green-blue eggs, and both parents incubate them for up to 14 days. They leave the eggshells in the nest after hatching (Harrison 1979). Atlas results list one observation of nest-building, on 1 June 1994.

Both Black-billed and Yellow-billed cuckoos lay eggs in each other's nests. Black-billeds also parasitize nests of Yellow Warblers, Chipping Sparrows, Eastern Wood-Pewees, Northern Cardinals, Cedar Waxwings, Gray Catbirds, Wood Thrushes, and Mourning Doves; the list of hosts parasitized by Yellow-billeds is almost the same (Bent 1940).

This cuckoo's newly hatched altricial young are vigorous and very strong; they quickly assume a very upright posture in the nest (Bent 1940). Like Yellow-billeds, the young leave the nest before they can fly and clamber around the nest tree (Johnsgard 1979). The years in which tent caterpillars cycle into great numbers coincide with increases in Black-billed Cuckoo numbers, increases in clutch size, and earlier nesting (Sealy 1978).

BREEDING PHENOLOGY			
CODE		# OF RECORDS	DATE RANGE
NB	Nest Building	1	1 Jun

DISTRIBUTION

These cuckoos breed from central Alberta and western Montana south to Colorado and east to the mid Atlantic. Colorado has never had many; Cooke (1897) reported the first ones, from Fort Collins and Loveland. Most Colorado records involve strays, although B&N reported three nesting records from Denver to Colorado Springs 30–40 years ago.

The Latilong Study showed them breeding in seven latilongs, likely breeding in two others, and migrant in eight. Three Atlas observations, with the only Confirmed nest, came from contiguous Non-priority blocks in Tamarack SWA; a singing bird near Wray provided the fourth report. Other sites along the South Platte may host a few breeding pairs but the Latilong Study seems unduly optimistic, at least for the Atlas period. Yet the three nests of the past known to B&N, in Colorado Springs, Denver, and Golden, indicate a wider breeding range—perhaps at times of caterpillar infestations? Never numerous in Colorado, Black-billed Cuckoos may have reduced their range from earlier in the twentieth century.

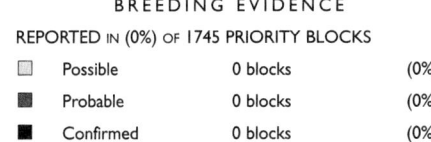

BREEDING EVIDENCE			
REPORTED IN (0%) OF 1745 PRIORITY BLOCKS			
	Possible	0 blocks	(0%)
	Probable	0 blocks	(0%)
	Confirmed	0 blocks	(0%)

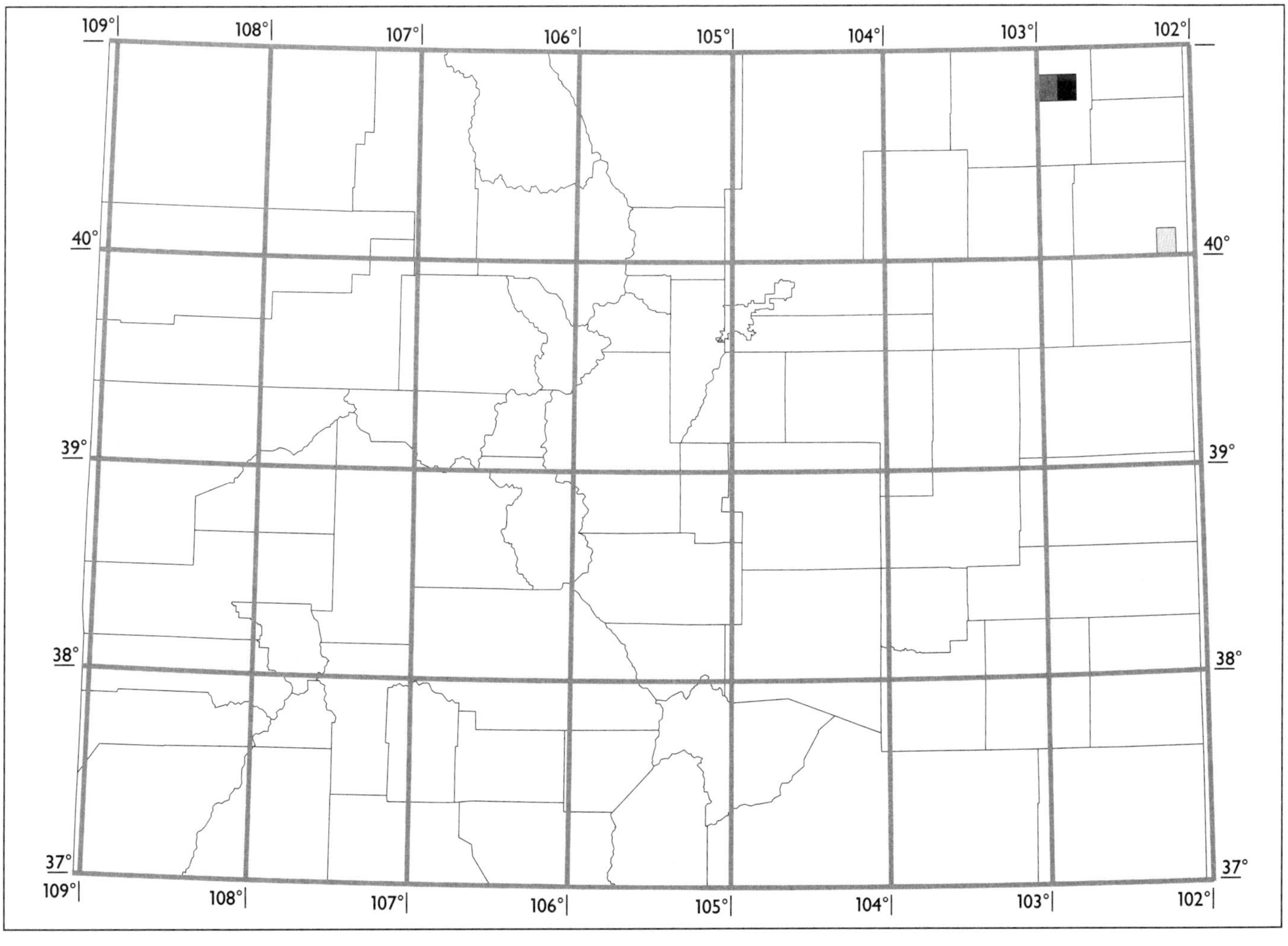
109°
108°
107°
106°
105°
104°
103°
102°
40°
39°
38°
37°

Because of its propensity to sing on cloudy or rainy days, and often at night, the Yellow-billed Cuckoo has earned the nickname "Rain Crow." Its secretive nature adds to its mystery, but not to the fact that, in much of the western United States, cuckoos are extirpated. Once common in the eastern United States, there they have become less common. Colorado serves as a microcosm of the nation: the Eastern Slope still has modest numbers of cuckoos (but probably fewer than it did historically) while the Western Slope now has only a few pairs.

HABITAT

The Rocky Mountains separate the two subspecies of Yellow-billed Cuckoo (AOU 1957). The western (*C. a. occidentalis*) and eastern (*C. a. americanus*) subspecies exemplify the significant ecological differences between these areas (Ehrlich et al. 1992). In the West, Yellow-billed Cuckoos depend on old-growth riparian woodlands with dense understories. In other areas of the country, more open woodlands with thick undergrowth seem to suffice.

The three Western Slope records came from cottonwoods on the Yampa, Colorado, and Uncompahgre rivers. In eastern Colorado atlasers found cuckoos primarily in cottonwood woodlands (71% of 52 Priority blocks) but also in five other habitats. This mixture of habitats agrees with A&R, who assign the species to lowland riparian forests and urban areas with tall trees. The eastern subspecies has more catholic habitat preferences than the western one, which confines itself mainly to riparian habitats (in Grand Junction they once nested in orchards; Sclater 1912, B&N).

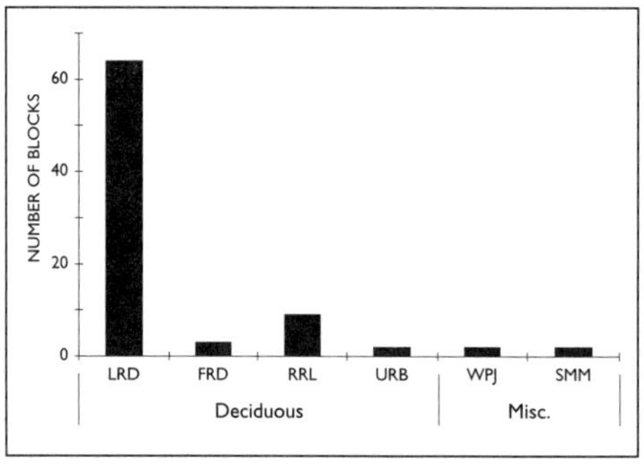
NUMBER OF BLOCKS
0
20
40
60
LRD
FRD
RRL
URB
WPJ
SMM
Deciduous
Misc.

BREEDING

B&N documented cuckoos in Colorado from 8 May to 22 September, with eggs recorded from 18 June to 5 August and one exceptionally late record, a dead fledgling found 10 September 1963. This long nesting season suggests possible double-brooding in Colorado as in southern states (Johnsgard 1979, 1986). Atlas dates show only a two-month breeding season, reflecting perhaps a small number of birds, little field work on the plains in August, or only one brood cycle.

Cuckoos hide their nests, usually in shrubs under riparian canopy but sometimes high in trees, on horizontal limbs 2–20 feet (0.5–6 m) above ground (Johnsgard 1986). B&N described a nest in Denver in a blue spruce. Cuckoos lay 2–5 eggs at irregular intervals often several days apart. Eggs hatch asynchronously because incubation, estimated at 14 days, begins during egg-laying. Both adults tend the nest but will leave late-hatching young behind to perish in the nest.

Young leave the nest still flightless—10 days before they can fly—but are agile climbers and clamber around in trees. Young grow extraordinarily fast—in their first four days they gain half their adult weight (Thompson and Ely 1989). Developing feathers remain shrouded in sheaths (rather than the usual process in which tips of sheaths slough off as the feather grows), which leaves the young resembling tiny porcupines. Later, the sheaths split and chicks appear well feathered (Thompson and Ely 1989).

The few Atlas observations, with only one Confirmation on the Western Slope and 10 on the plains (including four in Non-priority blocks in Tamarack SWA), reflect the birds' skulking nature as well as their rarity. The one Western Slope Confirmation, adults carrying food, occurred in cottonwoods along the Yampa River near Hayden (40107E2).

BREEDING PHENOLOGY

CODE		# OF RECORDS	DATE RANGE
C	Courtship	1	21 Jun
NB	Nest Building	4	1 Jun–14 Jun
NE	Nest with Eggs	2	15 Jun–26 Jun
FY	Feeding Young	2	28 Jun–26 Jul
FL	Fledged Young	3	13 Jun–26 Jul

DISTRIBUTION

Yellow-billed Cuckoos once ranged throughout much of the U.S., southern Canada, and Mexico. The western subspecies historically occurred from British Columbia into Mexico. That subspecies was extirpated from British Columbia in the 1920s, Washington in the 1930s, and Oregon in the 1940s. By 1987 only 85 pairs remained in California and western Arizona (Ehrlich et al. 1992). Howell and Webb (1995) list them as "fairly common to common" breeders in Mexico but offer no comment on declines.

Yellow-billed Cuckoos winter in northern South America south to Bolivia and northern Argentina. No cuckoos banded in

Colorado have been recovered; however, one banded in Cowley County, Kansas, was found dead in Brazil.

In all likelihood, more cuckoos used to nest on both the Eastern and Western slopes than do so today. Various observers found cuckoos plentiful in 1914–1915 in Prowers and Yuma counties on the Kansas line, and located nests from 1894 to 1963 along the Front Range from Fort Collins to Sedalia (presumably the eastern subspecies). Grand Junction has two nesting reports, probably the western subspecies (Rockwell 1908, B&N).

The Atlas map depicts the distribution described by Bent (1940), but probably consists of fewer breeders than formerly. To a degree it also resembles the A&R map: atlasers found the most cuckoos along the two major plains rivers, the South Platte and Arkansas. The Atlas adds a center in the east-central plains (Republican and Big Sandy rivers) including fledglings that Virginia Bleck found at Arriba (39103C3), the first breeding record in that latilong. The map depicts another center in Las Animas and Baca counties. On the Western Slope atlasers found them in only three blocks, fewer than A&R reported. Although the Western Slope probably never supported many cuckoos, the birds have now become extremely rare.

In general, their Colorado status mimics the rest of the continent—in the West nearly extirpated and in the East a once-common species that has become uncommon to rare. Although still widespread, cuckoos in other areas of the country seem to be declining. Statistically significant BBS data indicate a continent-wide decline of 1.6% per year with drops on 59% of the routes.

Ehrlich et al. (1992) cited three primary reasons for the declines: loss of riparian woodlands, prey scarcity (especially the loss of sphinx moth caterpillars to pesticides), and, in the West, drought; plus direct pesticide stress on breeding, migration, and wintering grounds.

BREEDING EVIDENCE

REPORTED IN 53 (3%) OF 1745 PRIORITY BLOCKS

☐	Possible	26 blocks	(49%)
■	Probable	20 blocks	(38%)
■	Confirmed	7 blocks	(13%)

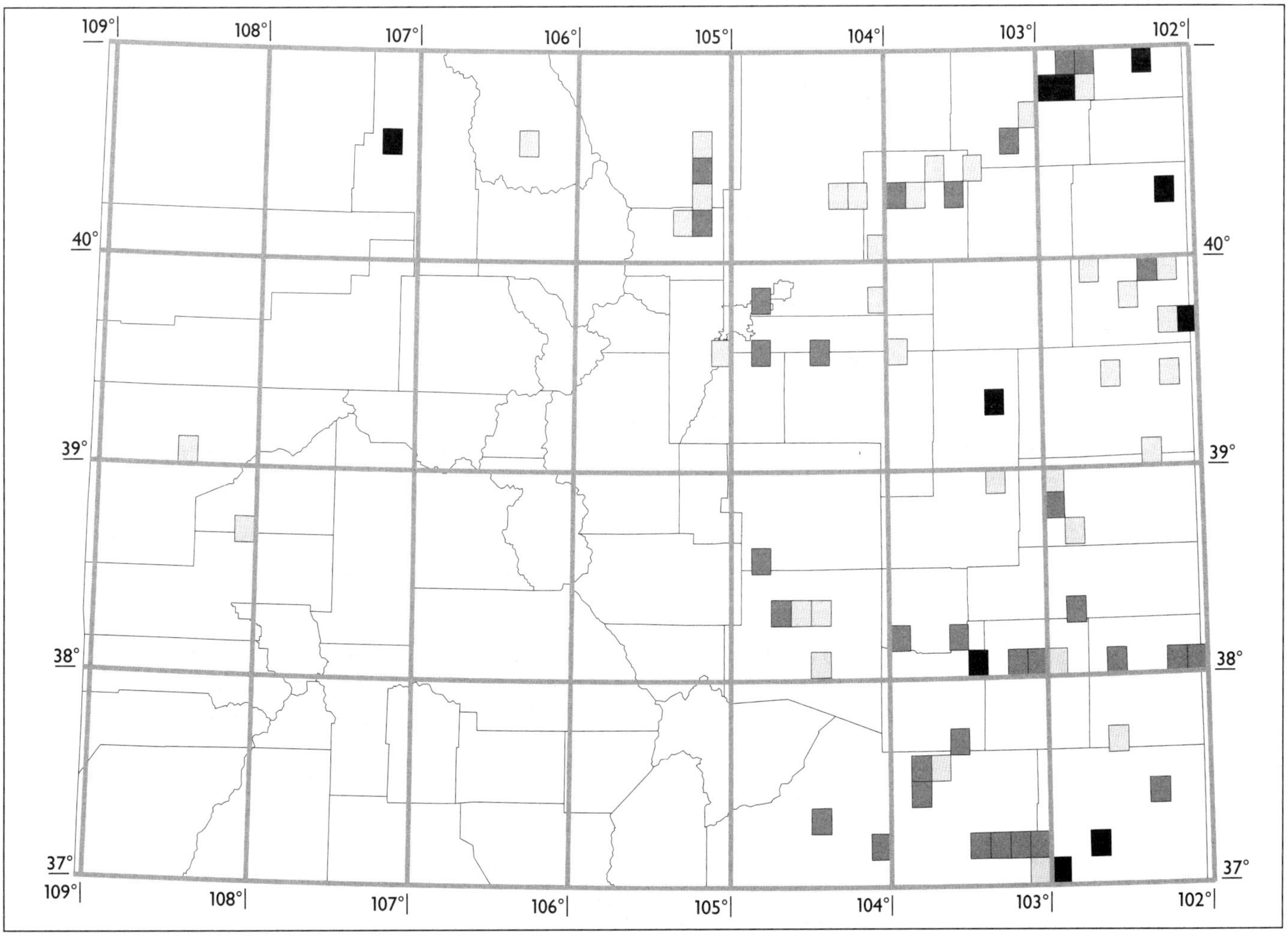
109°
108°
107°
106°
105°
104°
103°
102°
40°
39°
38°
37°

COLORADO DIVISION OF WILDLIFE / WILDLIFE RESOURCE INFORMATION SYSTEM

Imagine a tall, skinny chicken with long blue legs, a frowsy head, and blue-black eyeshadow. Then stick three long feathers in its rear end and watch it run like 60! Everything about roadrunners invites caricature. Picture a "tall thin tramp in a swallow-tailed coat" or a "long striped snake on two legs" (Sutton *in* Bent 1940). All agree the roadrunner is unique, singular in both appearance and personality.

HABITAT

Roadrunners live in the arid and semi-arid Southwest. Root (1988) correlates abundance with high evaporation rates (low humidity), and says that their range coincides with regions having at least 140 full days of sunshine per year. The Latilong Study described them as resident in desert scrub, pinyon/juniper, sagebrush/rabbitbrush, sandsage, and cactus grassland. Atlas workers found them in 14 different habitats, most often in pinyon/juniper (44%), then shortgrass prairie (18%), and plains riparian (12%). Eight of the 11 Confirmed nestings were in pinyon/juniper, one in shortgrass prairie, one in a ranch yard amid shortgrass prairie, and one in desert shrub.

Carnivory is the way of life for roadrunners, and carnivores can never be as common as their prey. Roadrunners are voracious predators capable of running at least 15–20 mph (24–30 kph; Sutton *in* Bent 1940, Terres 1980) but reluctant and deficient flyers (Hughes 1996). They chase snakes, lizards, frogs, scorpions, tarantulas, small mammals, birds, and insects at high speed over the landscape. The diet includes a minimum amount of vertebrates (1–6% birds and 1–4% mammals) and a few more reptiles (4–16%). Mostly it consists of insects: 72–86% (Hughes 1996).

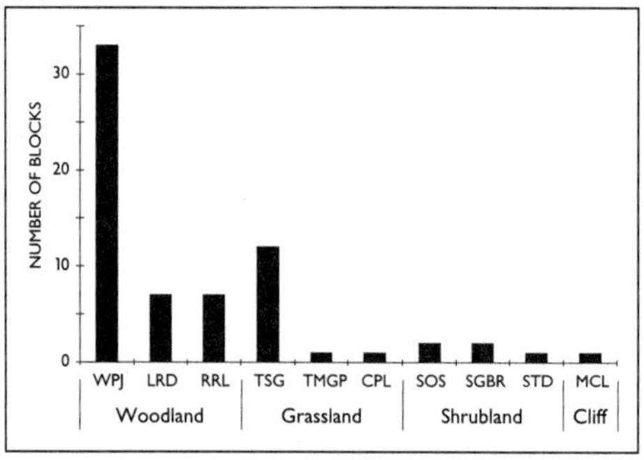

NUMBER OF BLOCKS
0
10
20
30
WPJ
LRD
RRL
TSG
TMGP
CPL
SOS
SGBR
STD
MCL
Woodland
Grassland
Shrubland
Cliff

They allegedly eat quail eggs and chicks, and in Colorado their range coincides with that of the Scaled Quail (Sutton *in* Bent 1940, BLW). Day after day a roadrunner may follow a covey as the quail feed and move through brush—but they trail quail in order to feed on insects flushed by the moving covey (Gorsuch 1934, Hughes 1996). Their penchant for quail-following has caused humans to persecute them on the assumption that they prey on eggs and chicks. Yet coyotes, hawks, ravens, and crows prey on roadrunners, too (Sutton *in* Bent 1940).

Roadrunners have a patch of orange skin behind the eye that they expose during disturbance. When a captive roadrunner, used to humans, saw its keeper in orange clothing, it fled (Hughes 1996).

BREEDING

George Miksch Sutton as a boy in Texas raised a pair of roadrunners (Bent 1940). Particularly eloquent about their sounds, he described a spring song given at sunrise, a love song, he said: a long series of *coo*s in descending pitch given while the roadrunner's head rose slightly with each *coo*. Roadrunners also make a chuckling crowing noise, as well as a purring noise (perhaps by rapidly rolling their mandibles together).

Roadrunners form permanent pair bonds (Terres 1980). They become year-round residents in suitable habitat, but only in small numbers (B&N, A&R). The Atlas map omits some out-of-season observations that could represent breeding birds after the season. The few Atlas reports add little to the breeding phenology, reported to run 1 May–10 August (Nelson 1993). Atlasers reports start with pairs observed 10 May and run to territorial behavior 11 July.

The breeding pair builds a nest in a low tree or shrub; they create a foundation of sticks lined with leaves, grass, roots, feathers, snakeskin, and manure (Harrison 1979). They incubate 3–6 eggs for 18 days, starting with the first laid, so the hatched young can differ in size by up to six days' growth (B&N). They may raise two broods per year (Harrison 1979).

The male incubates at night. Although the body temperature of females and non-breeding males drops at night, that of incubating males (which expend 36% more energy) stays constant. Breeding males, with conspicuous fat deposits, weigh significantly more than non-breeders (Hughes 1996).

An atlaser observed a nest near Springfield in a juniper, 2.5 feet (0.8 m) above the ground in a rural yard planted with junipers and buffalo grass. On 4 July, one egg had fallen on the ground and three eggs remained in the nest, presumably the result of a predation event (Janeal Thompson, nest card).

BREEDING PHENOLOGY

CODE		# OF RECORDS	DATE RANGE
NB	Nest Building	1	22 May
FY	Feeding Young	1	11 Jun
FL	Fledged Young	4	27 May–13 Jun

DISTRIBUTION

Greater Roadrunners breed and winter from central California and Oklahoma south into central Mexico. They roam the foothills and mesas, and some plains ranchyards, in southeastern Colorado, all through the year. Some stray north on the plains and into the mountains. Zebulon Pike trapped two of them on Christmas Day in 1806 in the pinyon/juniper woodlands between Buena Vista and Salida (Aiken and Warren 1914; see also Marsh 1931). (Marsh tagged this duo as roadrunners, although Pike's description leaves some doubt as to their identity: "of green color, almost the size of a quail, had a small tuft on its head like a pheasant . . . carnivorous.") In El Paso County Charles Aiken found them most often in cholla cactus and scrub oak (Aiken and Warren 1914). The highest elevation record for the state came in 1907 from Marshall Pass near Salida, at 10,000 feet (3048 m; B&N).

Atlas work found roadrunners in 52 blocks and Confirmed breeding in 11, all in southeastern Colorado. The map tracks closely with the A&R map, although it shows a few breeding farther north into Bent and Otero counties. Although atlasers found them in four blocks from Pueblo to the Cañon City area, they did not duplicate Pike's Salida birds of 1806.

Although the birds are not common here, the possibility of seeing a roadrunner makes field work exciting, finding one a real delight, and memories unforgettable.

BREEDING EVIDENCE

REPORTED IN 52 (3%) OF 1745 PRIORITY BLOCKS

☐	Possible	29 blocks	(56%)
■	Probable	12 blocks	(23%)
■	Confirmed	11 blocks	(21%)

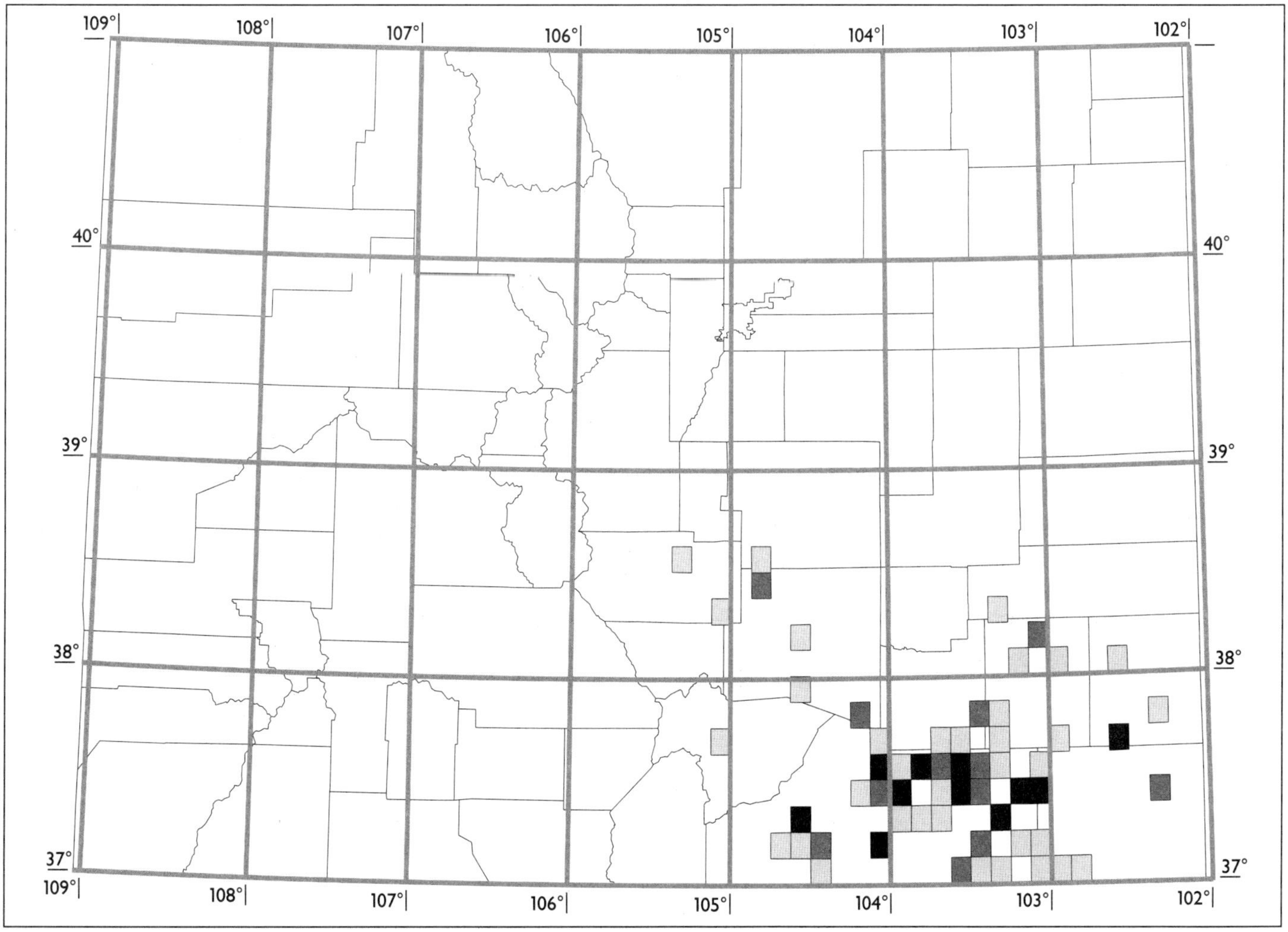
109°
108°
107°
106°
105°
104°
103°
102°
40°
39°
38°
37°

COLORADO DIVISION OF WILDLIFE / WILDLIFE RESOURCE INFORMATION SYSTEM

BARN OWL *Tyto alba*

Rats and mice have followed human settlement around the globe; fortunately for humans, but less so for the vermin, one of nature's best mousetraps, the Barn Owl, also prospers amid human activity.

HABITAT

Before the arrival of ranchers and farmers, Colorado's Barn Owls made their homes in natural cavities of trees and cliffs in riparian and grassland areas (B&N). Most modern owls, however, seem to favor human-built structures—churches, ruins, lofts, and barns (Terres 1980, Prestt and Wagstaffe 1973)—and the first Colorado report of this species described a bird "caught in the town hall of South Denver" (Cooke 1897). Atlas workers reported Barn Owls more frequently (32 reports) in rural buildings than in any other habitat, and they found only one nest in a tree cavity.

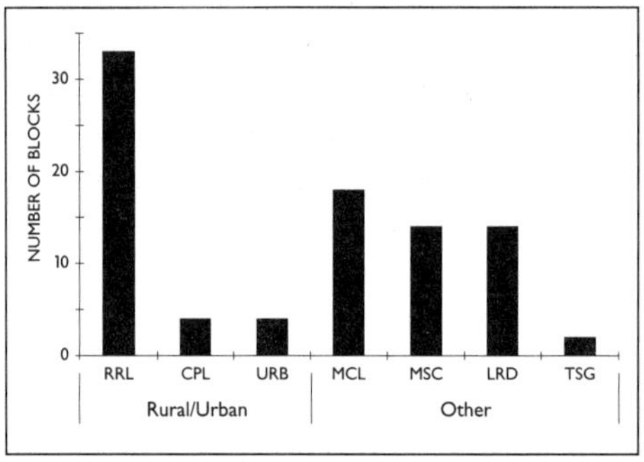
NUMBER OF BLOCKS
30
20
10
0
RRL
CPL
URB
MCL
MSC
LRD
TSG
Rural/Urban
Other

From earliest times, Barn Owls probably also constructed burrows in the dirt banks of arroyos and rivers, but researchers have only recently discovered this habit (Martin 1973b, Millsap and Millsap 1987). B&N reported a nest found in a small cavity in an arroyo bank on Cherry Creek south of Denver and owls nesting in a cutbank along Wild Cat Creek in Morgan County. In west-central Colorado, all Barn Owls reported by atlasers resided in these burrows. The location of these burrows within or adjacent to irrigated agriculture suggests human influence here as well (Levad 1994). Atlas data include one such nest in southeastern Colorado (McClave 38102B5). Studies in the early 1970s indicated that in Weld and Larimer counties Barn Owls preferred burrows (Millsap and Millsap 1987), and several owls still used the same burrows in 1991 (Brian Millsap pers. comm.). Atlas work in that area, however, reported nests only in rural buildings.

BREEDING

Atlas workers found only a limited number of Barn Owl nests, and the scope of the Atlas did not include nest monitoring; the Atlas data, therefore, shed little light on the breeding cycle of Colorado Barn Owls. The dates probably fall within the limits of the dates determined in Utah. There, pairs often begin roosting together in the future nest site in November (Marti 1992). Courtship typically begins in January (Smith and Marti 1976), and egg laying usually begins in early March (Marti 1994). Incubation runs 30–35 days, hatching begins in early April, and the fledgling period lasts 52–56 days (Ehrlich et al. 1988).

Atlasers reported birds on nests (28 reports) from 2 April to 25 September, the latter at Rocky Mountain Arsenal NWR (Mindy Hetrick pers. comm.). An observer may not detect young in a nest occupied by adults, and adult owls at a roost may present an appearance similar to nesting owls; these reports, therefore, offer little help in determining actual incubation periods. Barn Owls seldom produce second clutches in temperate areas (Marti 1994), but the late date (16 June) of the only nest with eggs found by Atlas workers suggests either a second clutch or renesting after a nest failure. A single report had adults feeding young on 8 June, and atlasers saw fledgling birds as early as 28 May and as late as 18 September.

BREEDING PHENOLOGY

CODE		# OF RECORDS	DATE RANGE
ON	Occupied Nest	18	2 Apr–25 Jul
NE	Nest with Eggs	1	16 Jun
NY	Nest with Young	9	21 May–25 Sep
FY	Feeding Young	1	8 Jun
FL	Fledged Young	5	28 May–18 Sep

DISTRIBUTION

Not well adapted to cold weather (Marti 1994), Barn Owls reside around the world in warm climates, generally from 40 degrees north to 40 degrees south of the equator (Prestt and Wagstaffe 1973). In Colorado they inhabit only lower, warmer regions of the state, and seldom venture above 6,000 feet (1830m).

Cooke's first report of a Barn Owl in the South Denver Town Hall prefigured a pattern of the birds inhabiting urban and agricultural area. The availability of nest cavities and rodents seems to limit their breeding numbers (Marti 1992), and human activities have tended to increase both. Accordingly Atlas workers found most of these birds in agricultural areas.

BY RICHARD LEVAD

The farm- and ranchlands of north-central, southeast, and west-central Colorado sustain the greatest populations.

Early reports emphasized the relative rarity of Barn Owls in Colorado. Sclater (1912) called it a rare bird, "probably a resident, though not known to breed." B&N also termed it rare, and the only Western Slope records that they reported came from the southwestern corner of the state early in the century. A&R also called them "rare, probably locally uncommon," and their map indicates breeding in only 16 locales.

Atlas work suggests that the owl's populations may significantly exceed previous estimates. Workers found them in 72 Priority blocks and reported abundance codes that indicate that more than 1,000 pairs reside in Colorado. The birds' nocturnal habits and somewhat secretive nature, combined with atlasers' limited awareness of their proclivity for arroyo banks, undoubtedly caused the workers to overlook them in many blocks, and the actual population may significantly exceed this number.

BREEDING EVIDENCE

REPORTED IN 75 (4%) OF 1745 PRIORITY BLOCKS

Possible	34 blocks	(45%)
Probable	6 blocks	(8%)
Confirmed	35 blocks	(47%)

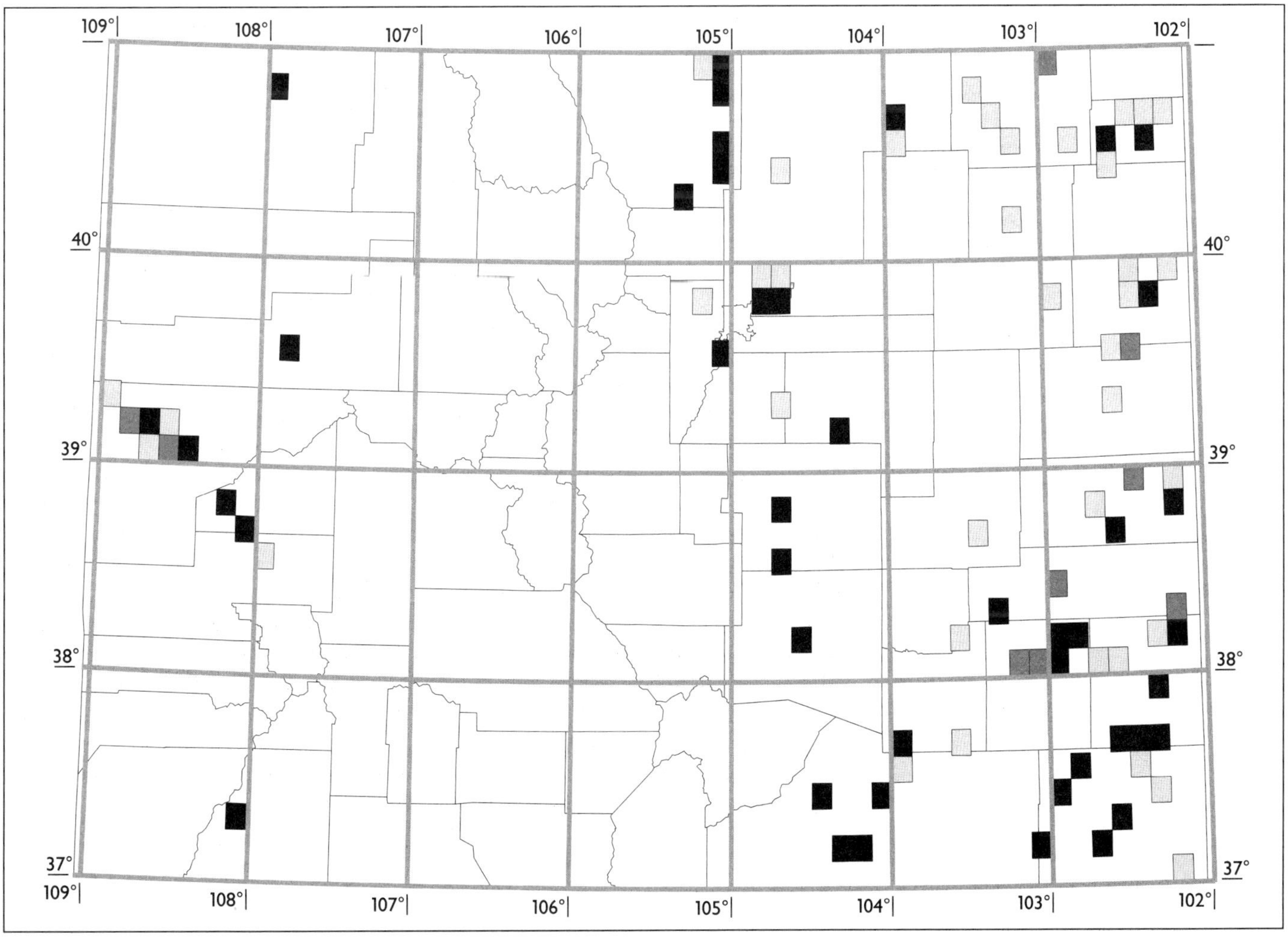
109°
108°
107°
106°
105°
104°
103°
102°
40°
39°
38°
37°

COLORADO DIVISION OF WILDLIFE / WILDLIFE RESOURCE INFORMATION SYSTEM

FLAMMULATED OWL *Otus flammeolus*

The deep and ventriloquial hoot of the Flammulated Owl belies its small size. Even the quavering begging calls of the young are hard to trace. This quality of their calls, their protective coloration and dark eyes, and their use of the forest understory help protect them as they hawk moths and other insects from dusk to dawn.

HABITAT

Researchers, hampered by the quiet, elusive, nocturnal habits of Flammulated Owls, nevertheless continue to generate new information about them. All agree that the owls depend on cavities for nesting, open forests for catching insects, and brush or dense foliage for roosting. Near Woodland Park researchers identified old-growth ponderosa pines as key habitat (Reynolds and Linkhart 1992). Another study found them in aspen-dominated habitats, but of four nests, all in aspen, three were in ponderosa/aspen ecotones and only one in pure aspen (Webb 1982a). The latest summary says that they use primarily "open ponderosa pine or forests with similar features, e.g., dry montane conifer or aspen . . . forests, often with dense saplings." They roost in sprawling ponderosa pines and Douglas-firs (McCallum 1994b). A small-owl survey in Boulder County elicited a habitat preference for ponderosa/Douglas-fir forests and dense shrubs along small streams that had larger trees and higher snag densities than average (Jones 1991).

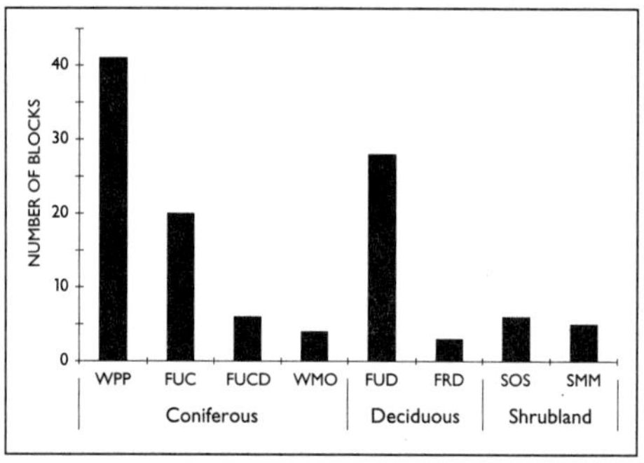
NUMBER OF BLOCKS
0
10
20
30
40
WPP
FUC
FUCD
WMO
FUD
FRD
SOS
SMM
Coniferous
Deciduous
Shrubland

Because "Flams" depend on existing cavities for nest sites, the attraction to aspen may arise from plentiful flicker holes, and possibly from greater prey abundance (Webb 1982a).

Atlas observations strike a balance: ponderosa pine and aspen had the most records, 40% and 28% of the total. Almost as many occurred in mixed forests of ponderosa, Douglas-fir, aspen, lodgepole pine, and/or spruces; a few records originated in shrubby habitats. Several of the aspen tracts did not include nearby conifers. The sum of Atlas records suggests that these little owls are definitely not ponderosa obligates. Reportedly they also use old-growth pinyon/juniper woodlands (A&R), an observation not borne out by Atlas records or the recent AOU monograph (McCallum 1994c).

These owls occur regularly from 6,000 to 10,000 feet (1830–3048 m; B&N). The two lowest Atlas records were 5,850 feet (1918 m) near Mesa Verde National Park—a calling owl (HEK)—and 6,800 feet (2230 m)—a nesting cavity near Golden (Dan George, nest card). Atlasers reported most between 7,650 and 9,500 feet (2508–3114 m.).

BREEDING

Flams start to return to Colorado in late April. Late snowstorms can kill migrants and cold spring weather may eliminate insects, causing the owls to die from starvation.

Pairs usually re-mate each year, usually in their old territories. They normally pick a flicker hole or natural cavity but also use holes made by other woodpeckers (Reynolds and Linkhart 1992). In some areas, shortages cause competition for nesting cavities, and the owls may evict other cavity-nesting birds (McCallum 1994b).

Territories apparently range from 90 to 140 acres (36–55 ha). They may nest in loose colonies and leave intervening stretches of suitable habitat unoccupied (McCallum 1994c).

In June, the female lays 2–3 eggs, occasionally four, which she incubates 21–24 days (McCallum 1994c). Courtship involves food solicitation by the female and the male feeding her. Incubating females, fed by the males, enlarge in size up to 68% and have trouble flying. Not until 10 days after the young hatch do females resume feeding themselves (McCallum 1994b). Males may feed the young directly or relay food via the female (Reynolds and Linkhart 1987b). Both adults feed older nestlings. Young fledge at 23–24 days (usually the last half of July) over two nights. By the third night after fledging, adults divide the fledglings into two groups (Linkhart and Reynolds 1987). The young disperse 30–35 days after fledging (Linkhart and Reynolds 1987).

Most Atlas records involved owls that responded to taped calls. Atlasers added little to the nesting phenology as they Confirmed this species in only nine blocks and found only five nesting cavities.

BREEDING PHENOLOGY

CODE		# OF RECORDS	DATE RANGE
ON	Occupied Nest	2	15 Jun–8 Jul
NY	Nest with Young	2	25 Jun–23 Jul
FY	Feeding Young	2	29 Jun–17 Jul
FL	Fledged Young	3	6 Jul–30 Aug

DISTRIBUTION

Flammulated Owls breed in montane forests of the West from southern British Columbia to the highlands of Mexico and Guatemala. Winter range is probably in southern Mexico and northern Central America (McCallum 1994b).

They occur throughout Colorado montane forests, with more in the south. Cooke (1897) called this little owl the rarest in Colorado and perhaps in the U.S. He listed 11 Colorado specimens, taken between Mosca Pass and Estes Park, and only six from the rest of the U.S. In 1875, Aiken found the "first nest known to science" near Poncha Pass (Cooke 1897).

Recent studies suggest that these owls are among the most abundant birds of prey in some areas (McCallum 1994c, Atlas). Whether early naturalists mistakenly considered them rare because of their secrecy (Rich Levad pers. comm.), or the owls declined and rebounded (RW), or they really have increased only recently, we do not know.

Like most owls, Flammulated Owls probably occur more widely than the Atlas map shows. Difficulty of access, rigors of nighttime owling, non-cooperative owls, and luck all affected the detection rate. Data from Forest Service surveys in the Rio Grande and San Juan national forests (where workers found Flammulated Owls in 46, mostly Non-priority, blocks) weight the south half of the Atlas map.

The map in Hayward and Verner (1994) showed no Flams in northwestern Colorado where atlasers found them in several blocks on the Roan Plateau, in western Rio Blanco and Garfield counties, and in Routt National Forest.

Although CDOW gives it no special status, the U.S. Forest Service listing as a "sensitive species" dictates development of management plans (Verner 1994). Logging of old-growth forests and use of pesticides may be detrimental to this species—whether or not it has increased over the past 100 years.

BREEDING EVIDENCE

REPORTED IN 71 (4%) OF 1745 PRIORITY BLOCKS

▢	Possible	49 blocks	(69%)
■	Probable	17 blocks	(24%)
■	Confirmed	5 blocks	(7%)

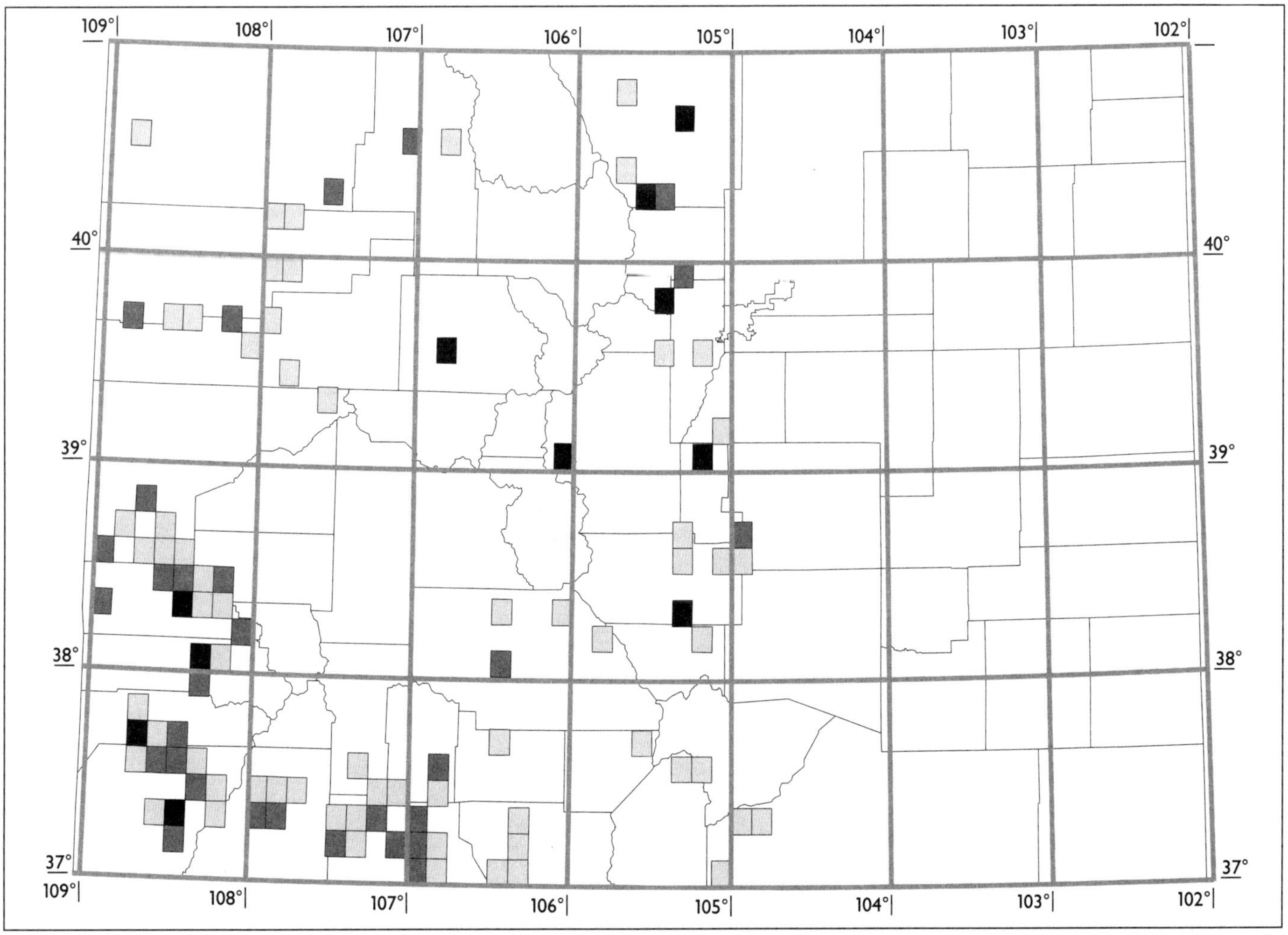
109°
108°
107°
106°
105°
104°
103°
102°
40°
39°
38°
37°
COLORADO DIVISION OF WILDLIFE / WILDLIFE RESOURCE INFORMATION SYSTEM

Good neighbors conduct themselves so unobtrusively that they often go unnoticed. Such demeanor characterizes Western Screech-Owls, who reside in many backyards completely unknown to the homeowners.

HABITAT

The Western Screech-Owl belongs to a group of closely related species of screech-owls that live in open woodlands from Alaska to Argentina (Hekstra 1973). In the irrigated western United States, screech-owls have extended their usual domain from lowland riparian woodlands to farmstead woodlots and urban shade trees (Marti 1979). Atlasers found Western Screech-Owls

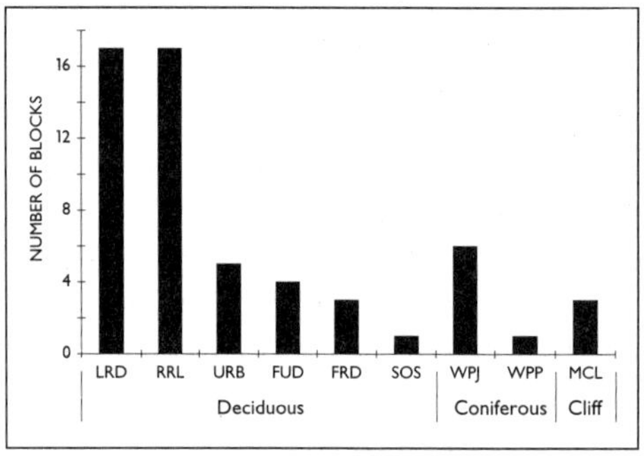
NUMBER OF BLOCKS
16
12
8
4
0
LRD
RRL
URB
FUD
FRD
SOS
WPJ
WPP
MCL
Deciduous
Coniferous
Cliff

most frequently in developed areas (RRL, URB), followed closely by broad-leaved cottonwoods along river bottoms. Observers reported Western Screech-Owls in pinyon/juniper in four blocks, and Van Truan found a nest with young in an old-growth pinyon/juniper woodland. Atlasers may have missed this species a number of times in this habitat, for New Mexico birders expect to find Western Screech-Owls in pinyon/juniper (Sandy Williams pers. comm.).

Several writers mention foothills riparian woodlands (narrowleaf cottonwood) and aspen as Western Screech-Owl habitat (B&N, Fitton 1973, A&R), but atlasers submitted few records from these habitats and little breeding evidence. Observations in these higher-elevation settings usually occur after the breeding season and perhaps indicate dispersing young. At least some offspring of many owl species, including Eastern Screech-Owls, disperse considerable distances after fledging (VanCamp and Henry 1975); research has not yet established this pattern for Western Screech-Owls, but it seems likely.

BREEDING

Most male Western Screech-Owls maintain territories through most of the year and begin calling in search of a mate in February. Although pairs sometimes remain on territory throughout the winter, generally the females move in shortly after calling begins. The pair then advertises its presence with a duet, the male's low tones answered by the female's higher and slightly faster response (RL). The female incubates a clutch of 2–8 eggs (usually four) for 26–28 days, generally beginning in late March or early April (Terres 1980, Ehrlich et al. 1988). Both parents feed the young on the nest for four weeks. For Eastern Screech-Owls the nestling period in May coincides with the height of passerine migration and passerines then become the owls' staple diet (VanCamp and Henry 1975), but thus far observers have reported Western Screech-Owls consuming only moths and other insects during this period (RL). Around the end of May the young become visible as they peer out of nest holes, and they leave the nest soon afterward (RL). Through June and into July the parents continue to feed the young in and around the nest. As summer progresses, insects become a large portion of the diet, providing easy prey for inexperienced young owls as they disperse. Small mammals replace the insects as cold weather sets in (VanCamp and Henry 1975).

BREEDING PHENOLOGY

CODE		# OF RECORDS	DATE RANGE
ON	Occupied Nest	1	13 Jun
NY	Nest with Young	3	13 May–15 Jun
FY	Feeding Young	1	19 Jun
FL	Fledged Young	6	15 Apr–25 Jun

DISTRIBUTION

Western Screech-Owls reside from coastal British Columbia to Central America. The AOU established Eastern and Western as separate species in 1983, but research has not yet conclusively defined their respective ranges. The dividing line apparently runs from eastern New Mexico through central Colorado to central Wyoming; north of there, the line of separation seems uncertain. Marshall speculated that a zone of overlap between Western and Eastern screech-owls might be found along the Cimarron and Arkansas rivers in southeastern Colorado (1967). Recent observations by atlasers and others suggest that along the Arkansas Western Screech-Owls give way to Easterns in Bent County (Dan Bridges pers. comm.), where the two species overlap somewhat. Observers reported seeing both species singing in the same tree at Fort Lyon wildlife area in May (Percival et al. 1994).

Atlasers found Western Screech-Owls breeding below 6,000 feet (1830 m) in river drainages in the southern, west-central, and southwestern parts of the state. The Atlas map shows the heaviest concentration of blocks in the lower Colorado and Gunnison drainages in west-central Colorado. Rather dense populations occur in appropriate habitat; during one winter Levad and Moran located 102 Western Screech-Owls in the Grand Valley (Levad 1989), and the Grand Junction Christmas Count recorded 34 in 1995 and 38 in 1996.

This concentration may represent the densest population or else the most thorough canvassing by atlasers. Some observers from the Pueblo area believe that Western Screech-Owl populations along the Arkansas may rival those of western riparian areas; however, the best habitat there did not fall into Priority blocks, so atlasers there spent less time searching for them (Van Truan pers. comm.).

Atlas data suggest a more extensive population in southwestern Colorado than previous accounts have indicated. Workers found no Western Screech-Owls along the Colorado River upstream from Palisade even though extensive broad-leaved cottonwood stands extend another 60 miles. A&R mapped it as secondary range. Debeque Canyon, a 15-mile-long gorge with few mature trees, separates the populated from the apparently unpopulated areas.

Atlasers also found no screech-owls in the White and Yampa river valleys, although in northeastern Utah this species does reside in these valleys (Cook 1984) and some Colorado habitat seems appropriate (Ed Hollowed pers. comm.). A&R also mapped these valleys as secondary range, and the Latilong Study designates this species as a resident in the latilongs through which these rivers flow.

Atlas workers most commonly reported A2 for the abundance code. An estimate using abundance codes provided by Atlas field workers suggests that Colorado has 1,100 breeding pairs of Western Screech-Owls. Because owling coverage lacked completeness, the population may exceed this estimate.

BREEDING EVIDENCE

REPORTED IN 37 (2%) OF 1745 PRIORITY BLOCKS

□	Possible	17 blocks	(46%)
■	Probable	13 blocks	(35%)
■	Confirmed	7 blocks	(19%)

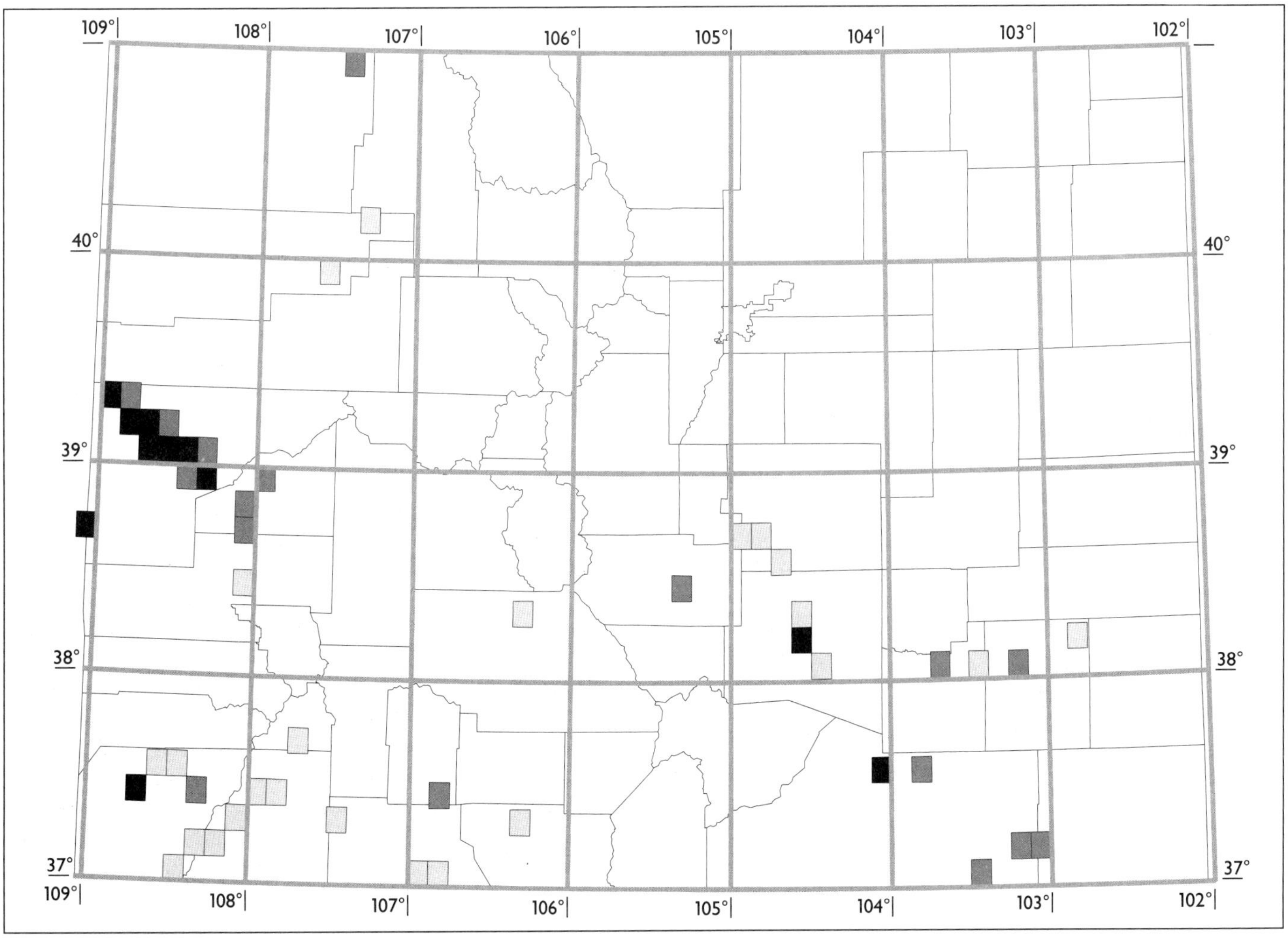
109°
108°
107°
106°
105°
104°
103°
102°
40°
39°
38°
37°

COLORADO DIVISION OF WILDLIFE / WILDLIFE RESOURCE INFORMATION SYSTEM

Dark groves, where the bare limbs of half-dead cottonwoods, maples, or elms frame the night sky, provide the perfect backdrop for the Eastern Screech-Owl's ghostly wails. These habitats occur throughout northeastern Colorado, but the secretive and perfectly camouflaged screech-owls avoid detection with haunting regularity. As a result, many questions about their distribution and population dynamics in Colorado remain unanswered.

HABITAT

Eastern Screech-Owls inhabit woodlots, open woods, shady urban parks, and residential areas, often near water, fields, or forest clearings (Marshall 1967, Gehlbach 1995). In Colorado and Wyoming, highest nesting densities appear to occur in groves containing large-diameter plains cottonwoods (Rockwell 1907, Dorn and Dorn 1994). They usually nest in natural cavities of deciduous trees but will also use artificial nest boxes. In the eastern United States, where they inhabit a wide range of forest types, population densities correspond more strongly with forest structure than with tree species composition (Lynch and Smith 1984, Smith et al. 1987). The cities of Denver, Boulder, and Fort Collins have complements of nesting Eastern Screech-Owls in their older residential sections—especially those with old, big cottonwoods and maples.

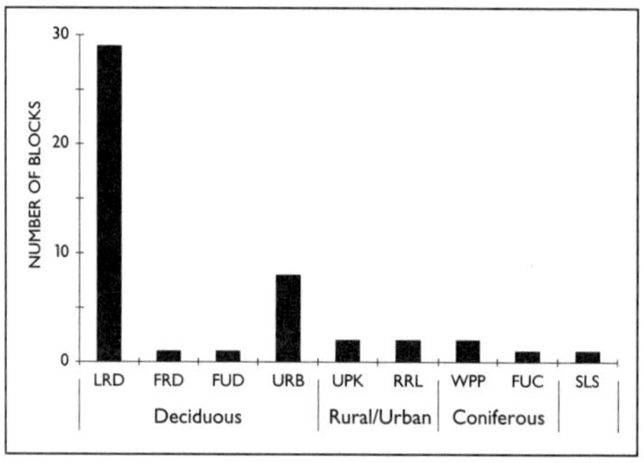
30
20
10
0
NUMBER OF BLOCKS
LRD
FRD
FUD
URB
UPK
RRL
WPP
FUC
SLS
Deciduous
Rural/Urban
Coniferous

Of 47 observations reported by Atlas field workers, 29 occurred in lowland riparian woodland. Small numbers of Eastern Screech-Owls may nest along the fingers of riparian growth that extend into the Front Range foothills. During May and June, atlasers heard calling owls, but could not Confirm nesting, in coniferous forest near Golden Gate State Park, at 9,000 feet (2743 m), in ponderosa pine forest near Estes Park at 7,500 feet (2286 m), and in aspen near Red Feather Lakes at 8,500 feet (2591 m). These nocturnal owls feed opportunistically on small mammals, birds, reptiles, amphibians, small fish, and a variety of invertebrates (Bent 1938). A single nest observed by Allen (1924) contained the remains of 24 bird species. Eastern Screech-Owls typically hunt along forest edges, in meadows, and along streams. They take their prey during short flights from an exposed perch (Marshall 1967).

BREEDING

Establishment of nesting territories begins in January or February. During the mating period, the incidence of the "monotonic trill" (a series of rapid whistles on the same pitch) increases, and the incidence of the "whinnying" call decreases (Bent 1938, Smith et al. 1987, Gehlbach 1994). The female lays 2–6 eggs (typically 4–5) in a tree cavity, usually 10–40 feet (3–12 m) off the ground. Incubation takes 26–34 days but because the female starts incubating with the first egg laid, the young hatch asynchronously. Nestlings fledge after 28 days, so that the entire in-nest process requires about two months. Egg dates for Colorado range from 3 April to 12 May (B&N). However, Atlas field workers reported seeing fledged young on 5, 17, and 19 May, suggesting that egg-laying may often begin in March.

Atlasers Confirmed breeding in only 16 of 47 (Priority and Non-priority) blocks. This low Confirmation rate attests to the difficulty of locating an active nest without making repeated visits to a suspected nest site. Moreover, frequency of vocalization decreases during incubation and brooding, making the owls difficult to detect during the height of the breeding season (Bent 1938, Smith et al. 1987, Gehlbach 1994).

The young fledge about 28 days after hatching. A second peak in vocal activity (of both the "monotonic trill" and the "whinnying" call) occurs around the time of fledging. Adults also emit a variety of alarm calls, including hoots, barks, slow trills, and screeches, around the nest. Young disperse from their parents' home range about 10 weeks after fledging (Bent 1938, Gehlbach 1994). In a central Texas study, pairs nesting in urban areas consistently fledged more young than did pairs nesting in rural areas, presumably because the urban areas supported more prey, more suitable nesting trees, and fewer nest predators (Gehlbach 1994).

BREEDING PHENOLOGY

CODE		# OF RECORDS	DATE RANGE
ON	Occupied Nest	2	13 Apr–5 May
NY	Nest with Young	1	27 May
FL	Fledged Young	10	15 May–22 Jul

DISTRIBUTION

Eastern Screech-Owls breed east of the Rocky Mountains, from southern Canada to Florida and central Mexico. They appear to be non-migratory throughout their breeding range (Johnsgard 1988). In Wyoming higher-elevation breeders (6,000–7,000 feet, 1830–2135 m) may move to lower elevations for the winter. In Colorado these screech-owls breed primarily in the northeast; of the 47 Atlas observations, 44 occurred north of the Palmer Divide. Atlas observations clump into three main places: Front Range cities with old residential sections and trees over 50 years old, the South Platte River valley downstream from Denver, and a complex around Bonny Reservoir. They also may inhabit riparian lowlands in the eastern Arkansas Valley (A&R).

Dorn and Dorn (1994) used tape playbacks to census Eastern and Western screech-owl populations throughout Wyoming. They reported Eastern Screech-Owl responses at 21 calling stations, all east of the Continental Divide and below 6,200 feet (1890 m). All confirmed Western Screech-Owl responses came from west of the Continental Divide. In Montana, the Continental Divide also appears to separate these two species (Marti and Marks 1989).

Population densities on the plains close to the mountains remain largely undetermined. In Boulder County nesting populations approach one pair per linear mile along stretches of Boulder and South Boulder creeks (Boulder Co. Nature Assoc., unpubl. data). Adam (1987) documented eight territories along a 15-mile (24-km) stretch of the Souris River in southeastern Saskatchewan.

Cooke (1897) described screech-owls as "common" in the eastern foothills and around Denver, Boulder, and Loveland; B&N described them as "uncommon residents" in northeastern Colorado. Betts (1913) said the screech-owl was "common" along the creeks of Boulder County from the foothills eastward, and Alexander (1937) said it was "a common resident on (the) plains and at moderate elevations in the mountains" of Boulder County. Fort Collins, Boulder, and Denver Christmas Count data reveal no clear directional trends in Eastern Screech-Owl populations from 1950 to 1994.

BREEDING EVIDENCE

REPORTED IN 20 (1%) OF 1745 PRIORITY BLOCKS

☐	Possible	12 blocks	(60%)
■	Probable	2 blocks	(10%)
■	Confirmed	6 blocks	(30%)

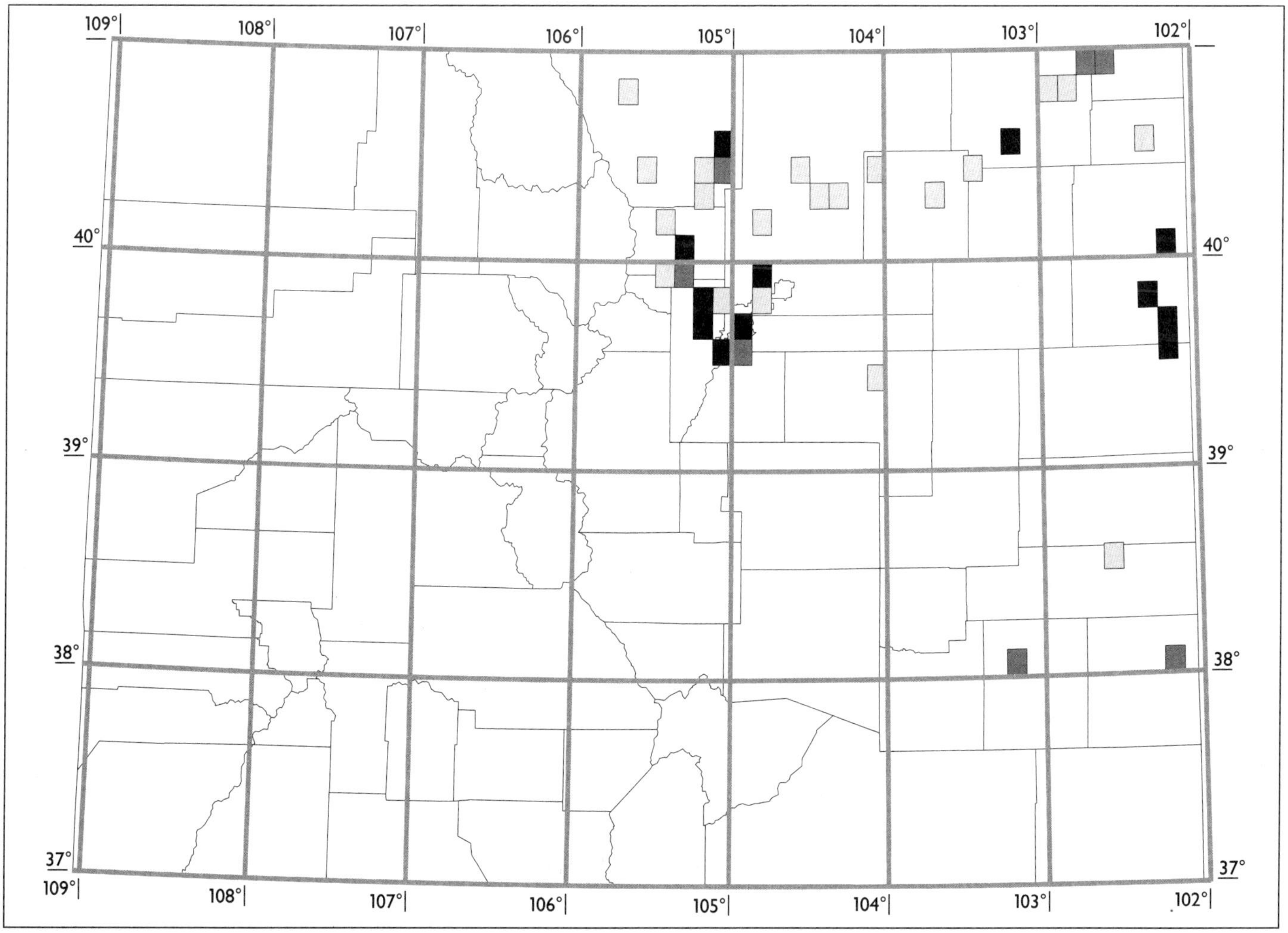
109°
108°
107°
106°
105°
104°
103°
102°
40°
39°
38°
37°

COLORADO DIVISION OF WILDLIFE / WILDLIFE RESOURCE INFORMATION SYSTEM

GREAT HORNED OWL *Bubo virginianus*

When night blankets Colorado's farms and forests with darkness, the familiar yet strangely eerie hoots of Great Horned Owls evoke the mystery of owls—Shakespeare's "fatal bellman, Which gives the stern'st goodnight." This powerful nocturnal hunter, Colorado's largest owl, uses marvelous powers of sight, hearing, and silent flight to capture prey ranging from insects to mammals as large as porcupines.

HABITAT

No other owl in North America makes use of so many diverse habitats and climatic situations. Great Horned Owls inhabit most terrain that combines at least a few trees with open areas rich in rodents and other prey (Johnsgard 1988). Atlasers found them in a diverse range of habitats. Shelterbelts and lowland riparian forest ranked highest, attesting to the abundance of owls in farm and ranch country of both the Great Plains and western valleys. Forest types account for the top seven habitats reported (equating RRL with shelterbelts in this context), and all forest types account for nine-tenths of all habitat observations. One-third of the owl habitat reports came from deciduous habitats, slightly more than a quarter from rural/urban (mostly deciduous), and another quarter from conifers.

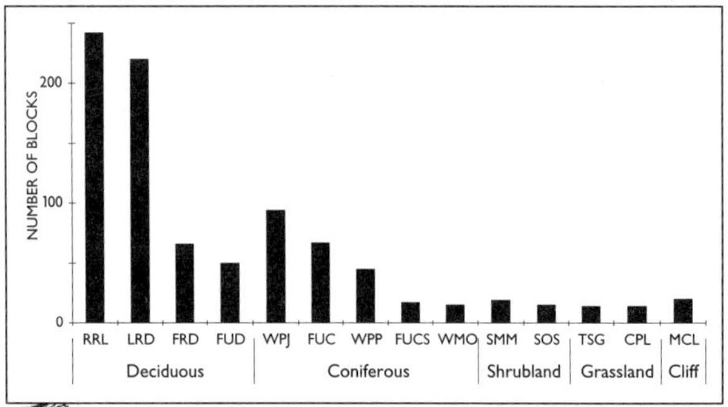
NUMBER OF BLOCKS
200
100
0
RRL
LRD
FRD
FUD
WPJ
FUC
WPP
FUCS
WMO
SMM
SOS
TSG
CPL
MCL
Deciduous
Coniferous
Shrubland
Grassland
Cliff

Sightings from shrublands, grasslands, and farm fields reflect the owls' preference for open hunting sites with scattered trees or forest edges (Bent 1938, Austing and Holt 1966, Craighead and Craighead 1956). Great Horned Owls studied in Larimer and Weld counties usually hunted in open terrain by making short sallies from perches such as trees, fences, or windmills (Marti 1974).

BREEDING

Large obvious nests in rural settings helped Atlas workers achieve a high Confirmation rate. A higher Confirmed rate in Great Plains blocks (51%) compared to Rocky Mountain and Colorado Plateau blocks (29%) probably reflects easier field worker access to eastern owls, especially during the early spring nest stages. Great Horned Owls typically nest on stick platforms built by other species, often magpies (Johnsgard 1988). However, they sometimes choose a rocky cliff or dirt bank (B&N) or, rarely, nest on the ground (Truslow 1966). Among Atlas nest finds, shelterbelts or deciduous riparian woodlands comprised 80% of habitats recorded, with 6% in other forest habitats. Habitat distribution for all observations differed substantially—57% shelterbelt/riparian and 31% forest; finding nests in forests, especially conifer forests, is substantially harder, and access more difficult in spring.

The unevenness of search effort by atlasers from late winter through summer complicates the estimation of nesting cycle dates for these early-nesting owls. However, the ratio of NY to FL observations, which should be independent of search effort, indicates that fledging peaks in the last half of May. Inasmuch as incubation and nestling periods last about 30 and 35 days respectively (Johnsgard 1988), egg-laying peaks in mid March and hatching in late April.

BREEDING PHENOLOGY

CODE		# OF RECORDS	DATE RANGE
C	Courtship	2	12 May–18 Aug
ON	Occupied Nest	42	11 Feb–1 Jul
NE	Nest with Eggs	1	20 Apr
NY	Nest with Young	77	22 Feb–27 Aug
FY	Feeding Young	8	24 Apr–26 Aug
FL	Fledged Young	157	11 Apr–18 Sep

DISTRIBUTION

Great Horned Owls occupy a vast region, from the northern edge of boreal forest in Alaska and Canada to the southern tip of South America. The closely related Eurasian Eagle Owl (*Bubo bubo*) inhabits most of Eurasia; this "superspecies" therefore covers a large part of the world (Johnsgard 1988). Atlasers found Great Horned Owls at least possibly breeding in 44% of blocks statewide, the most widespread owl species by far. Atlas records, in close agreement with former distribution assessments (B&N, A&R), indicate that Great Horned Owls breed commonly on the Great Plains and on the Colorado Plateau, especially the valleys. The owls decrease in abundance with elevation, and become rare in subalpine forest and absent from alpine tundra.

Atlas workers usually assigned abundance codes A2 or A1, and recorded A3 in only a few Great Plains riparian woodlands and Western Slope uplands around Grand Mesa.

Great Horned Owls' ability to coexist with humans in altered landscapes helps them thrive in rural and suburban areas, as well as in forested wildlands. With few natural enemies, these top predators still face habitat loss and human-caused mortality as threats. Indiscriminate shooters take their toll, including farmers and small-pet owners who fail to comprehend these owls' tremendous contribution to controlling rodent pests. Of 374 Great Horned Owls banded in the U.S., Stewart (1969) found that 52% were eventually shot, and estimated, from indirect evidence, that 86% may have been. Despite this persecution, Great Horned Owls seem to fare well. Minor et al. (1993) reported that nesting populations in New York suburban and urban habitats were as dense, and as successful at fledging young, as populations in wild landscapes. As long as open space for nesting and foraging persists, Colorado's Great Horned Owls should continue to deliver "the stern'st goodnight."

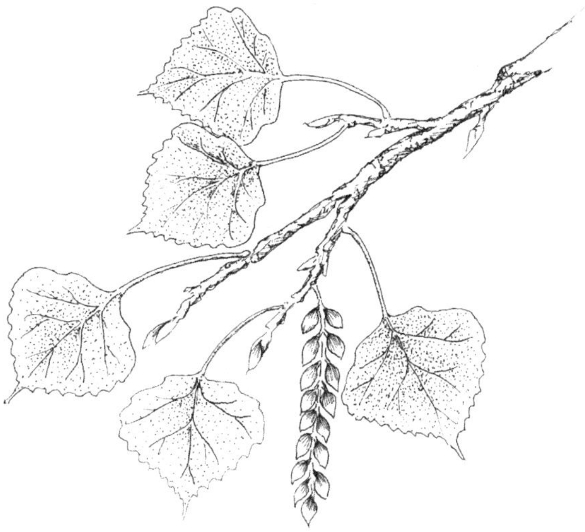

BREEDING EVIDENCE

REPORTED IN 775 (44%) OF 1745 PRIORITY BLOCKS

□	Possible	352 blocks	(45%)
■ (grey)	Probable	99 blocks	(13%)
■	Confirmed	324 blocks	(42%)

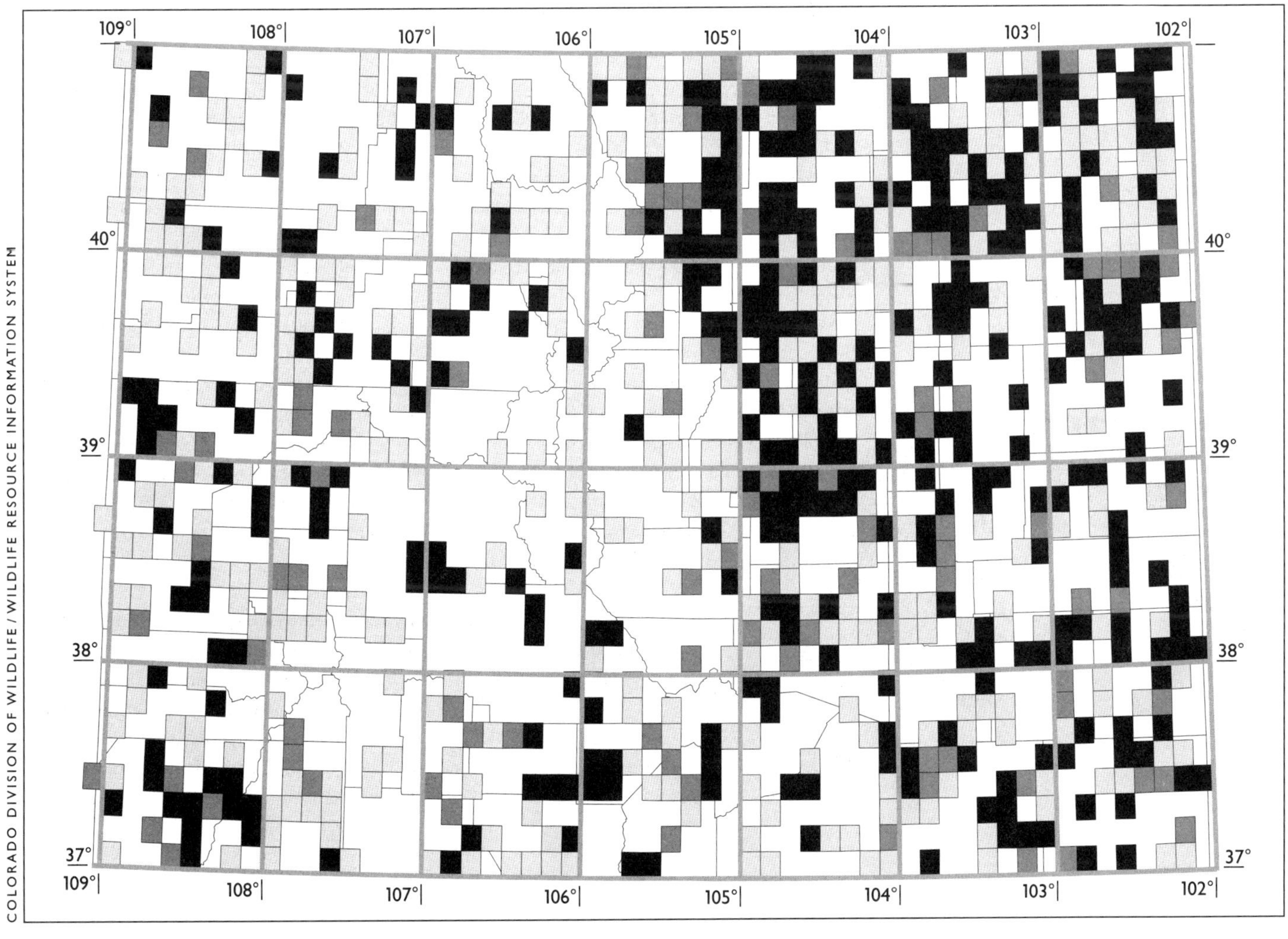
109°
108°
107°
106°
105°
104°
103°
102°
40°
39°
38°
37°
COLORADO DIVISION OF WILDLIFE / WILDLIFE RESOURCE INFORMATION SYSTEM

Few predators can match the ounce-for-ounce ferocity of these tenacious little owls. Despite measuring less than seven inches from head to tail tip, Northern Pygmy-Owls can kill jays, doves, and an occasional small grouse (Bent 1938). No wonder they attract so much attention from songbirds, whose mobbing and querulous chattering sometimes betray a roost or nest site.

HABITAT

Northern Pygmy-Owls inhabit coniferous forests and deciduous woodlands throughout much of western North America. In Colorado atlasers found them breeding primarily in ponderosa pine forests, pinyon/juniper woodlands, and aspen woodlands from the lower foothills to the upper montane life zones. Pygmy-owls nest in woodpecker holes or natural cavities (Reynolds et al. 1989).

Front Range nest sites documented by atlasers and others during the Atlas include a natural cavity in a cottonwood near Lyons at 5,400 feet (1645 m; D.W. King pers. comm.); natural cavities and woodpecker holes in ponderosa pines near Boulder at 5,950 feet (1810 m), 6,495 feet (1980 m), and 7,200 feet (2195 m; Jones 1991); and a woodpecker hole in an aspen near Estes Park at 8,600 feet (2621 m; Ron Ryder pers. comm.). In an aspen grove on an Atlas Rendezvous south of Craig (Slide Creek, 40107B4), observers watched pygmy-owls feeding voles to three fledglings perched in the treetops: the parents tore the voles apart and parceled out the pieces. They also watched one of the youngsters lose one big morsel, which dropped to the ground (Roxanne Falice pers. comm.).

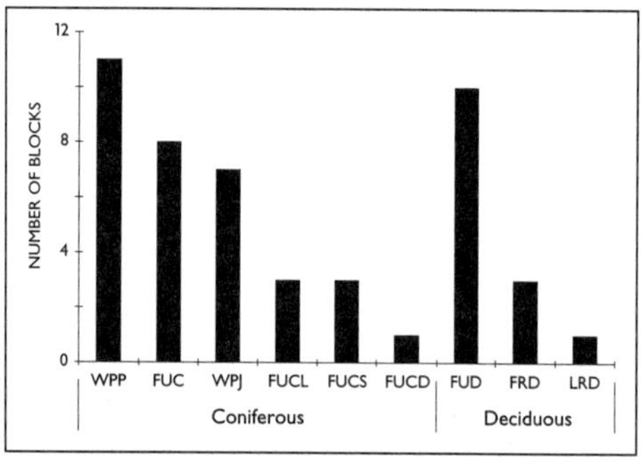
12
8
4
0
NUMBER OF BLOCKS
WPP
FUC
WPJ
FUCL
FUCS
FUCD
FUD
FRD
LRD
Coniferous
Deciduous

Atlas field workers Confirmed nesting in ten locations, at elevations ranging from 5,400 to 10,100 feet (1645–3078 m). The majority of nesting Confirmations and sightings occurred between 6,000 and 9,000 feet (1829–2743 m) in open coniferous forests, aspen groves, and riparian woodlands. All conifers totaled 70% of habitat reports, including one-fifth in ponderosa pine woodlands. Aspen groves accounted for one-fifth of the habitat reports.

Northern Pygmy-Owls prey on birds, small mammals, reptiles, amphibians, and insects (Bent 1938). They ambush their prey from a favored perch, usually located 15 feet (4.6 m) or more off the ground. Hunting activity peaks around sunrise and sunset (Johnsgard 1988).

BREEDING

Males begin to establish nesting territories in February or early March (Jones 1991). Their territorial song, a monotonous series of breathy, hollow whistles given at the rate of 1–2 whistles per second, often ceases abruptly after the onset of incubation, in April or early May (Johnsgard 1988, Jones 1991).

The female lays 3–5 eggs in a tree cavity located 7–65 feet (2–20 m) off the ground (Johnsgard 1988, Jones 1991). Incubation requires about 29 days, with the young fledging about 30 days later (Johnsgard 1988). Colorado egg dates range from 17 May to 22 June (B&N).

Prior to the Atlas, only two dated Colorado records existed for nests with young: 15 June in Boulder (Jones 1991) and 7 July near Estes Park (Ron Ryder pers. comm.). Atlas field workers added a third: a nest with young near Lyons on 10 June. Field workers observed fledged young in eight blocks throughout the state from 15 May to 30 August.

Around the nest, Northern Pygmy-Owls emit a variety of vocalizations, including a high, soft twitter, a loud ascending whistle of alarm, and a flicker-like series of very fast, staccato whistles. These pugnacious owls vigorously defend their nesting territories against all comers. In a Boulder Mountain Park a male pygmy-owl repeatedly swooped over the head of an Atlas volunteer who sat on a rock watching the nest 50 yards away. Before the volunteer departed, the owl had taken out his frustration by knocking a magpie off its perch in a nearby Douglas-fir.

BREEDING PHENOLOGY

CODE		# OF RECORDS	DATE RANGE
NY	Nest with Young	1	10 Jun
FY	Feeding Young	1	24 Jun
FL	Fledged Young	7	15 May–30 Aug

DISTRIBUTION

Northern Pygmy-Owls range from the east slope of the Rocky Mountains westward to British Columbia, Washington, Oregon, and coastal California; and south through the interior of Mexico to Guatemala and central Honduras. In winter they undertake short elevational migrations but probably do not stray far from summer breeding areas (B&N, Johnsgard 1988).

The highest nesting densities in Colorado probably occur in the foothills and lower mountains. From 1985 to 1989 in a Boulder/Gilpin county small-owl survey (sponsored by Boulder Co. Nature Assoc.), volunteers used tape playbacks to locate 31 singing Northern Pygmy-Owls. Of these 31 contacts, 29 occurred between 5,500 and 8,600 feet (1700–2600 m), with the highest densities occurring in foothills canyons from 6,500–7,600 feet (1981–2316 m; Jones 1991).

Atlas data suggest that Northern Pygmy-Owls nest throughout the mountains of Colorado. The few records center on the Front Range, Flat Tops, Gore, Elk, and Sangre de Cristo mountains, Uncompahgre Plateau, and Mesa Verde. The scarcity of Atlas records stems from the near invisibility of these lilliputian owls and the timing of most Atlas field work long after the March–April period of peak vocal activity.

Insufficient data exist to assess population trends in Colorado. In the foothills of Boulder County, nesting population densities appear to fluctuate from year to year. Factors that may affect nesting densities include snag availability, vole population cycles, and abundance of avian prey (Jones 1991). Although the Atlas and other recent research projects have added to our knowledge of the Northern Pygmy-Owl's life history, much remains to be learned about this, one of the most elusive of Colorado birds.

BREEDING EVIDENCE

REPORTED IN 37 (2%) OF 1745 PRIORITY BLOCKS

☐	Possible	28 blocks	(76%)
■	Probable	4 blocks	(11%)
■	Confirmed	5 blocks	(13%)

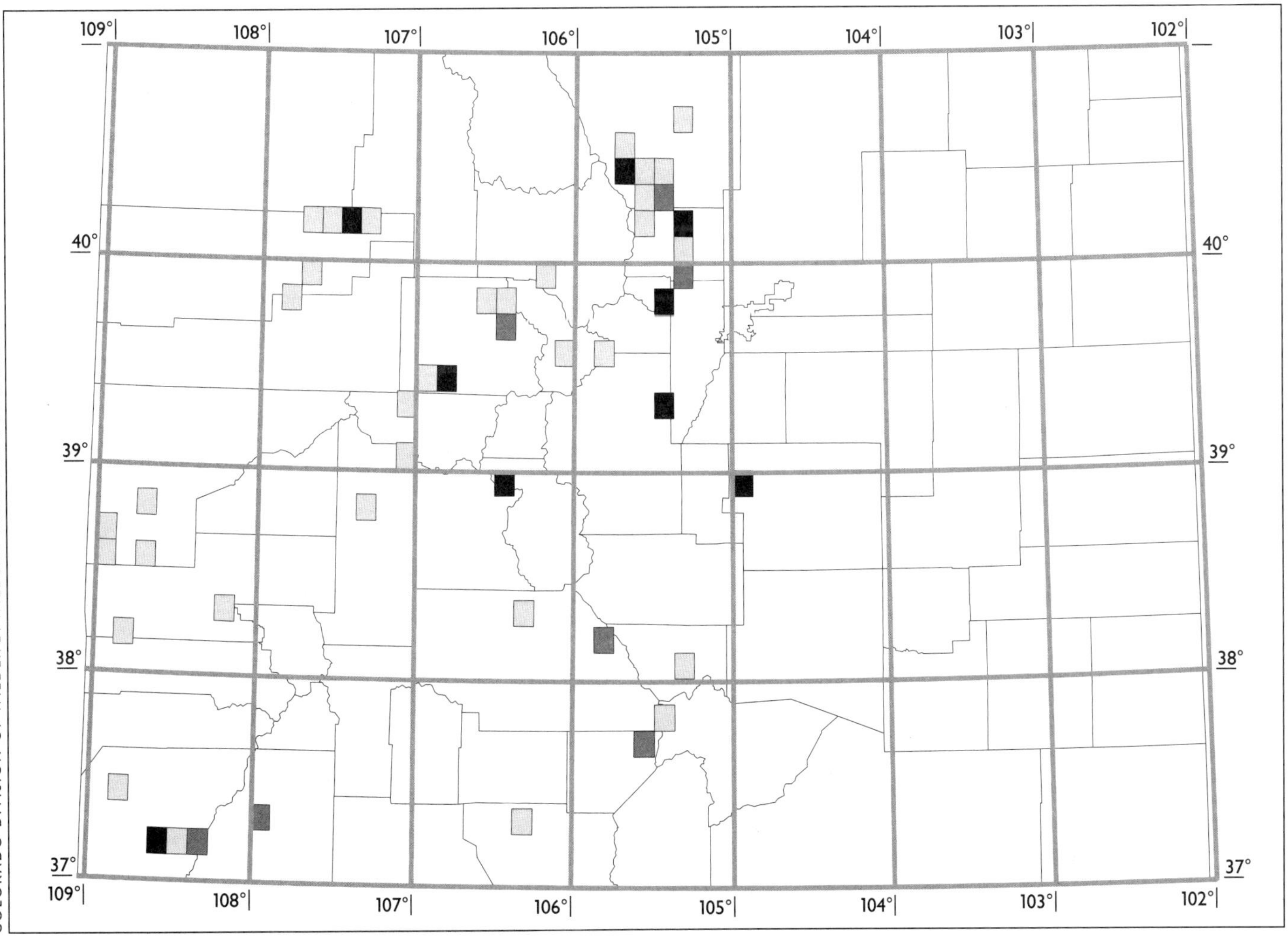
109°
108°
107°
106°
105°
104°
103°
102°
40°
39°
38°
37°

COLORADO DIVISION OF WILDLIFE / WILDLIFE RESOURCE INFORMATION SYSTEM

BURROWING OWL *Athene cunicularia*

Perching on prairie dog mounds, these little owls bob up and down on spindly legs, turn their round heads nearly upside down to peer at potential predators, and coo seductively at each other while rubbing beaks. Their winsomeness makes their decline, brought on by poisoning of prey and destruction of habitat, seem all the more tragic.

HABITAT

Burrowing Owls nest primarily in rodent burrows—in grasslands, shrublands, deserts, and grassy urban areas such as golf courses and airports. They can excavate their own burrows in sandy soils but usually use ready-made holes dug by burrowing animals. In eastern Colorado they favor prairie dog colonies, which provide well-maintained burrows for nesting, mounds for perching, and close-cropped vegetation that affords a clear view of terrestrial predators. In western Colorado, they use burrows of prairie dogs, Wyoming ground squirrels, rock squirrels, and other ground squirrels. When plague or poisoning kills off a majority of the rodents in a particular colony and vegetation grows more than ankle-high, Burrowing Owls abandon their nest burrows (MacCracken et al. 1985, Plumpton and Lutz 1993b).

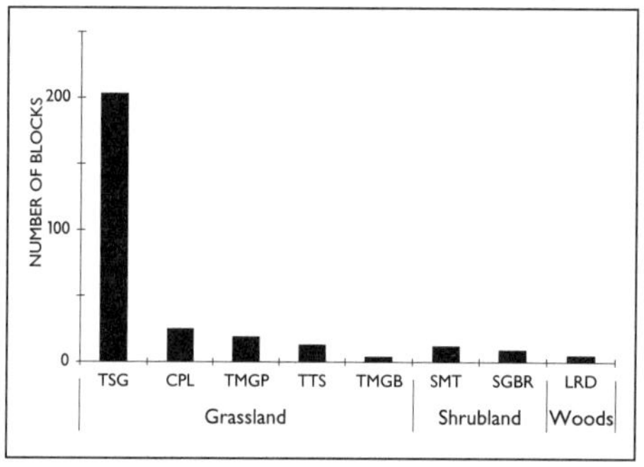
NUMBER OF BLOCKS
200
100
0
TSG
CPL
TMGP
TTS
TMGB
SMT
SGBR
LRD
Grassland
Shrubland
Woods

Colorado Atlas observations reflect this owl's propensity for nesting in sparsely vegetated habitat. More than 70% of the sightings occurred in shortgrass prairie. Most of the remaining sightings also occurred in habitats with low or scattered vegetation.

Burrowing Owls feed on insects, small rodents, and, occasionally, small songbirds. During the breeding season they typically forage for insects throughout the day and then hunt voles and mice at night (Bent 1938, Plumpton and Lutz 1993a, Haug et al. 1993). Insects predominate numerically in Burrowing Owl diets, but mammals usually compose a higher percentage of the total ingested biomass (Gleason and Craig 1979, Plumpton and Lutz 1993a).

BREEDING

The first individuals arrive in Colorado in late March or early April, and nesting begins within a few weeks. Egg dates reported by Bent (1938) for Colorado and Kansas ranged from 29 March to 1 July. During the incubation period of 27–30 days, males bring food to the females, who rarely leave the nest (Bent 1938, Plumpton and Lutz 1993a). The young appear above ground about 14 days after hatching (Johnsgard 1988). Consistent with Bent's data, Colorado Atlas dates for nests with young ranged from 7 May to 5 August.

Mean reported fledge rates for various populations range from 2 to 5 fledglings per burrow (Johnsgard 1988) although sometimes families are much bigger. Ten stood around one burrow on 13 June 1991 in a block near Burlington (Alpine Ranch NW, 39102B6). These numbers, high relative to those of other small owls, may reflect the evolutionary advantages of nesting underground as well as a need to compensate for high post-fledging mortality rates (Johnsgard 1988). Of 326 adults and juveniles banded at the Rocky Mountain Arsenal in 1991 and 1992, only 28 (9%) returned to nest in 1992 or 1993 (Pezzolesi 1994).

The appearance of colonial nesting in this species may reflect a scarcity of nest sites as much as a lack of territoriality among breeding pairs. In northeastern Colorado, pairs breeding in larger prairie dog colonies tend to nest farther apart than do pairs breeding in smaller colonies (Hughes 1993). Atlasers reported approximately equal numbers of blocks containing only one nesting pair (15%) and blocks with 11 or more pairs (14%). Competition for food among pairs nesting close together in Idaho's Snake River Plain often resulted in nest abandonment (Green and Anthony 1989).

BREEDING PHENOLOGY

CODE		# OF RECORDS	DATE RANGE
ON	Occupied Nest	75	2 May–19 Jul
NY	Nest with Young	21	7 May–5 Aug
FY	Feeding Young	9	28 May–19 Aug
FL	Fledged Young	49	8 May–8 Aug

DISTRIBUTION

Burrowing Owls breed from south-central British Columbia eastward to southern Saskatchewan and south through most of the western U.S., Mexico, Central America, and South America to southern Chile. They also breed in Florida, the Bahamas, and Hispaniola. Individuals that nest more than 35 degrees north or south of the equator migrate to warmer regions during fall and winter, but little is known about their migratory movements (Haug et al. 1993).

Populations have declined since the turn of the century, when many observers considered Burrowing Owls common on the high plains (Henderson 1908, Sclater 1912, Bent 1938, Tate 1986). From 1976 to 1987 the number of breeding pairs in southern Saskatchewan declined by 50%, and from 1982 to 1994 the number of known breeding pairs in Manitoba declined from 76 to fewer than 10 (Dundas 1995). Along the Front Range urban corridor these owls have disappeared from much of their historic range (Ron Ryder pers. comm., SRJ).

Atlas field workers Confirmed breeding in only 18 Colorado latilongs and only four of the ten Western Slope latilongs within which these owls have occurred historically (B&N, Latilong Study). Burrowing Owls continue to breed across the plains despite the problems facing their prairie dog burrow-diggers. On the plains, field workers found Burrowing Owls in 40% of the Priority blocks.

On the plains the number of blocks with owls gradually decreases from east to west and from south to north. In Longitude 102, 46% of the plains Priority blocks had owls; in 103, 35% and in 104, 33% had them. In latilong 37102, 57% had owls; going north in Latitudes 38, 39, and 40, owls occurred in, respectively, 42%, 29%, and 32% of the plains blocks.

Atlasers also documented breeding continuity in the Grand Valley from Grand Junction to the Utah line. They did not find owls in northwestern Colorado or South and Middle parks, and recorded them in only three blocks in southwestern Colorado and four in the San Luis Valley.

Habitat loss, habitat fragmentation, pesticide poisoning of insect populations, and collisions with vehicles have contributed to the decline in North American populations (Haug et al. 1993). In rural areas, poisoning of rodent colonies, plague outbreaks in rodent colonies, and conversion of rangeland to farmland have eliminated nest sites (Griess 1995, Boulder Co. Aud. Soc. 1978–1997). In urban areas, habitat fragmentation has isolated existing populations, possibly exposing them to greater predation risk from Great Horned Owls, red foxes, and other urban-edge predators. In addition, continued use of DDT and other toxic pesticides in Mexico may threaten wintering populations (Johnsgard 1988).

BREEDING EVIDENCE

REPORTED IN 259 (15%) OF 1745 PRIORITY BLOCKS

□	Possible	53 blocks	(20%)
▦	Probable	54 blocks	(21%)
■	Confirmed	152 blocks	(59%)

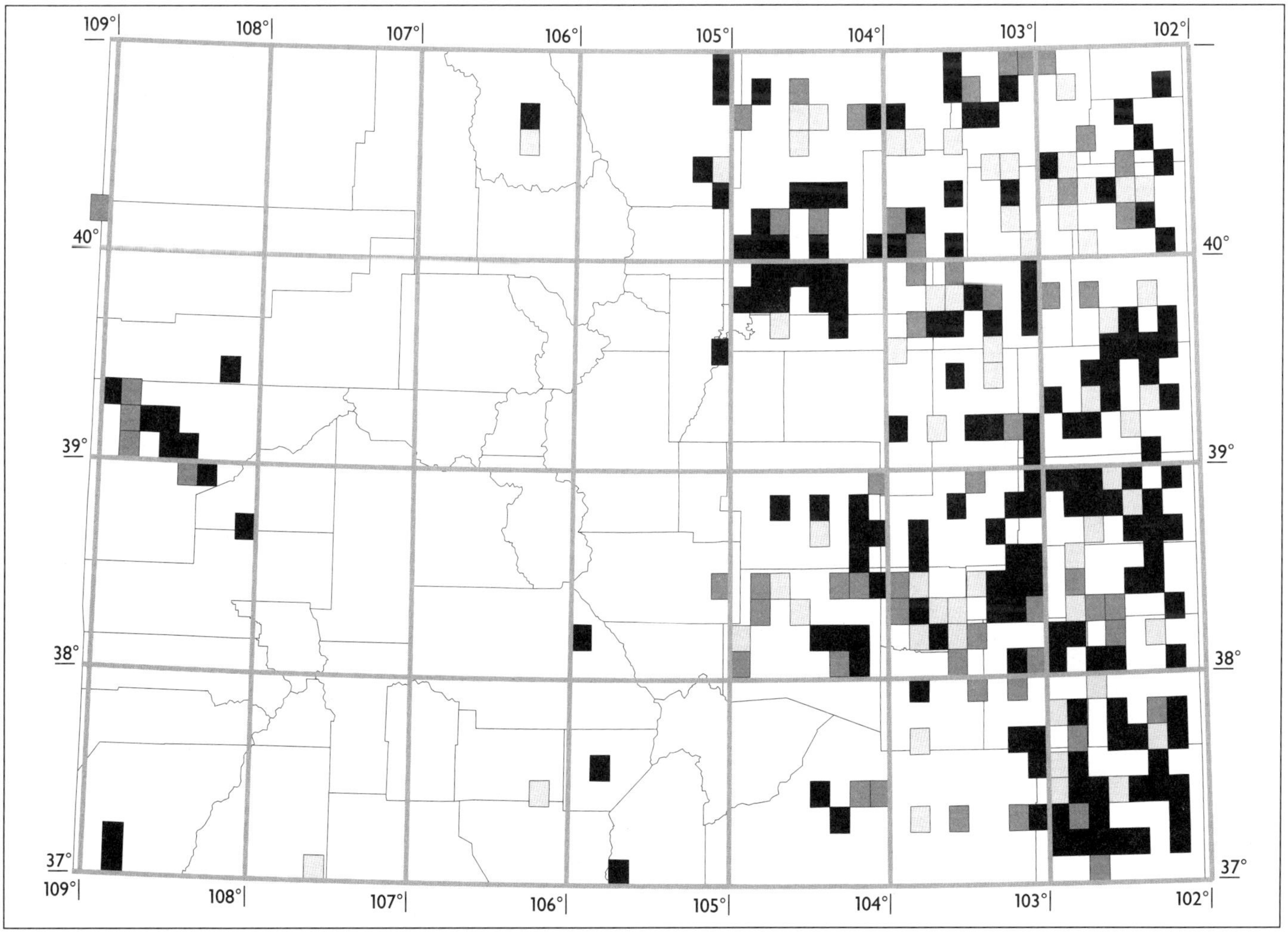
109°
108°
107°
106°
105°
104°
103°
102°
40°
39°
38°
37°

Few observers have heard, and fewer still have seen, the rarest resident owl in Colorado. Spotted Owls, western counterparts of the closely related Barred Owl, roost quietly by day in coniferous foliage or in caves and rock crevices of shady canyons. They emerge after dusk to hunt the dark, lonely canyons for rodents and other prey. Dark eyes, medium size, and spotted plumage help distinguish Spotted Owls from other Colorado owls, along with their gruff, bark-like call.

HABITAT

The Mexican Spotted Owl (*S. o. lucida*) differs from the two West Coast subspecies in habitat preferences and subtle plumage differences. Mexican Spotted Owls are well isolated geographically from other subspecies (Johnsgard 1988), and are possibly a separate species according to genetic studies (Barrowclough and Gutierrez 1990).

Throughout their range, Spotted Owls choose rocky canyons or forested mountains below 9,500 feet (2970 m). Like the West Coast subspecies, Mexican Spotted Owls favor old-growth coniferous forest (Ganey and Balda 1994, Zwank et al. 1994), units of forest that have more complex structure than surrounding forests (Gutierrez et al. 1995). They also inhabit sparsely forested canyons (David Willey pers. comm.). In central Colorado Spotted Owls seclude themselves

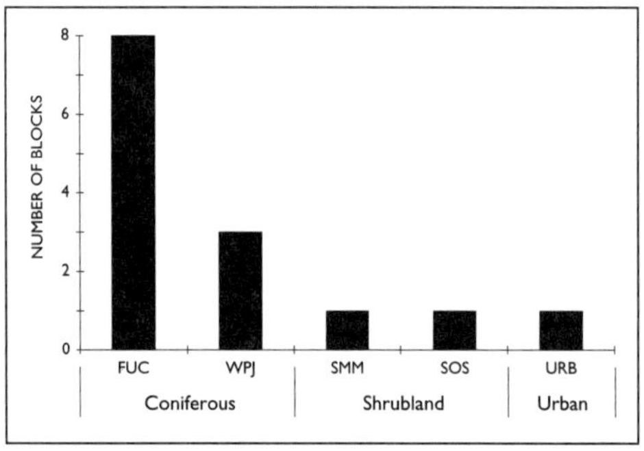
NUMBER OF BLOCKS
0
2
4
6
8
FUC
WPJ
SMM
SOS
URB
Coniferous
Shrubland
Urban

in deep rocky canyons with tall conifers in the canyon bottoms or between bands of cliffs (Charles Johnson pers. comm.). In southwestern Colorado, they occur in narrow slickrock canyons that slice through pinyon/juniper woodlands and contain occasional small stands of dense mixed conifers in shaded corners (Marilyn Colyer pers. comm.). Old-growth conifers and slickrock canyons not only furnish the necessary nesting and foraging substrates, but they also provide cool, shady microclimates, presumably important for Spotted Owls because they possess less ability than some other owls to dissipate body heat (Ganey et al. 1993). In Mesa Verde National Park Spotted Owls find coolness in the recesses of Anasazi ruins; the Urban entry in the Habitat Chart refers to Cliff Palace in the park.

BREEDING

Atlasers found no Spotted Owls during regular field work, as might be expected due to their extreme rarity in Colorado (even though many atlasers searched for them). The Atlas map shows results from ongoing research begun in 1989 (Charles Johnson pers. comm.) and surveys for Spotted Owls conducted in southwestern and south-central Colorado by Forest Service and BLM biologists. The few nest sites found in the state include crevices or caves in rocky cliffs, tall conifers, and the stump of a broken-topped tree. Spotted Owls in Colorado build their stick platform nests by March and lay eggs in mid to late April; eggs hatch in May, and young fledge by June.

Low nest productivity seems to characterize Spotted Owl breeding. Although pairs generally stay together year after year, they produce just one clutch per season of usually two or three eggs, and do not breed every season. Ten nesting attempts known in Colorado since 1989 produced an average of 1.4 fledglings. In the unusually cold and wet spring of 1995, none of the Spotted Owl pairs under observation nested successfully (Charles Johnson pers. comm.).

BREEDING PHENOLOGY

CODE		# OF RECORDS	DATE RANGE
NE	Nest with Eggs	1	9 Apr
NY	Nest with Young	1	7 Apr
FL	Fledged Young	2	28 Jul–6 Aug

DISTRIBUTION

Mexican Spotted Owls range from the Colorado Rockies and southern Utah south to central Mexico, but occur only in widely scattered areas within this vast range (Ganey et al. 1988). Biologists estimate the total population in the U.S. as between 777 and 1,554 birds (Gutierrez et al. 1995).

Colorado's first Confirmed nesting arose from a fledgling observed and photographed at Cliff Palace ruin in Mesa Verde National Park (Kingery 1988b). Before the fledgling perched on Cliff Palace, Colorado had a paltry eight verified and nine other possible records—yet only one each from the two now-confirmed nesting regions in the state (Kingery 1991). Atlas records adequately reflect the two known breeding populations of Spotted Owls in Colorado: three adult birds at Mesa Verde, and 20–30 in 22 sites in the south-central mountains—the larger portion on the southern massif of Pikes Peak and smaller numbers in the Wet Mountains (Charles Johnson pers. comm.). Systematic searches of appropriate habitat in southwestern and south-central Colorado, begun in 1989, have produced only a few records of apparently transient birds; it seems unlikely that other breeding populations of significant size exist in Colorado. Former distribution is difficult to infer from the 20 or so historic records. Many historic records may represent vagrants wandering at the northern limit of the species' range (Webb 1982b). Except for unmated transients and young birds dispersing, Spotted Owls usually reside year round near their breeding territories, or make short-distance altitudinal migrations (Ganey and Balda 1989, Charles Johnson pers. comm.).

Spotted Owls face an insecure future throughout their range. In 1993 low population numbers and habitat loss from timber harvesting prompted federal listing of the Mexican subspecies as Threatened (McDonald et al. 1991). As with the northern subspecies in the Pacific Northwest, conservation efforts to protect Mexican Spotted Owls and their habitats generate controversy because of the commercial value of old-growth forests for timber harvest (Bonnett and Zimmerman 1991). In Colorado, the highest count ever (in 1993) tallied only 20 owls, with seven breeding pairs. Research in Colorado has not yet identified population trends, but the extremely low number of owls, coupled with exacting habitat requirements and low nesting productivity, make Colorado's Spotted Owls very vulnerable to extirpation.

Caveat on Atlas Map: The Atlas map reflects both Atlas data and information from Spotted Owl surveys. Because of this owl's status as a Threatened species under the Endangered Species Act, the Atlas map masks specific breeding locations. The map clumps breeding sites in the center of each latilong rather than showing specific blocks.

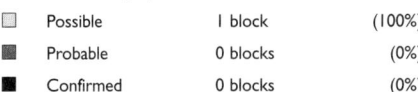

BREEDING EVIDENCE

REPORTED IN 1 (0%) OF 1745 PRIORITY BLOCKS

☐	Possible	1 block	(100%)
▪	Probable	0 blocks	(0%)
■	Confirmed	0 blocks	(0%)

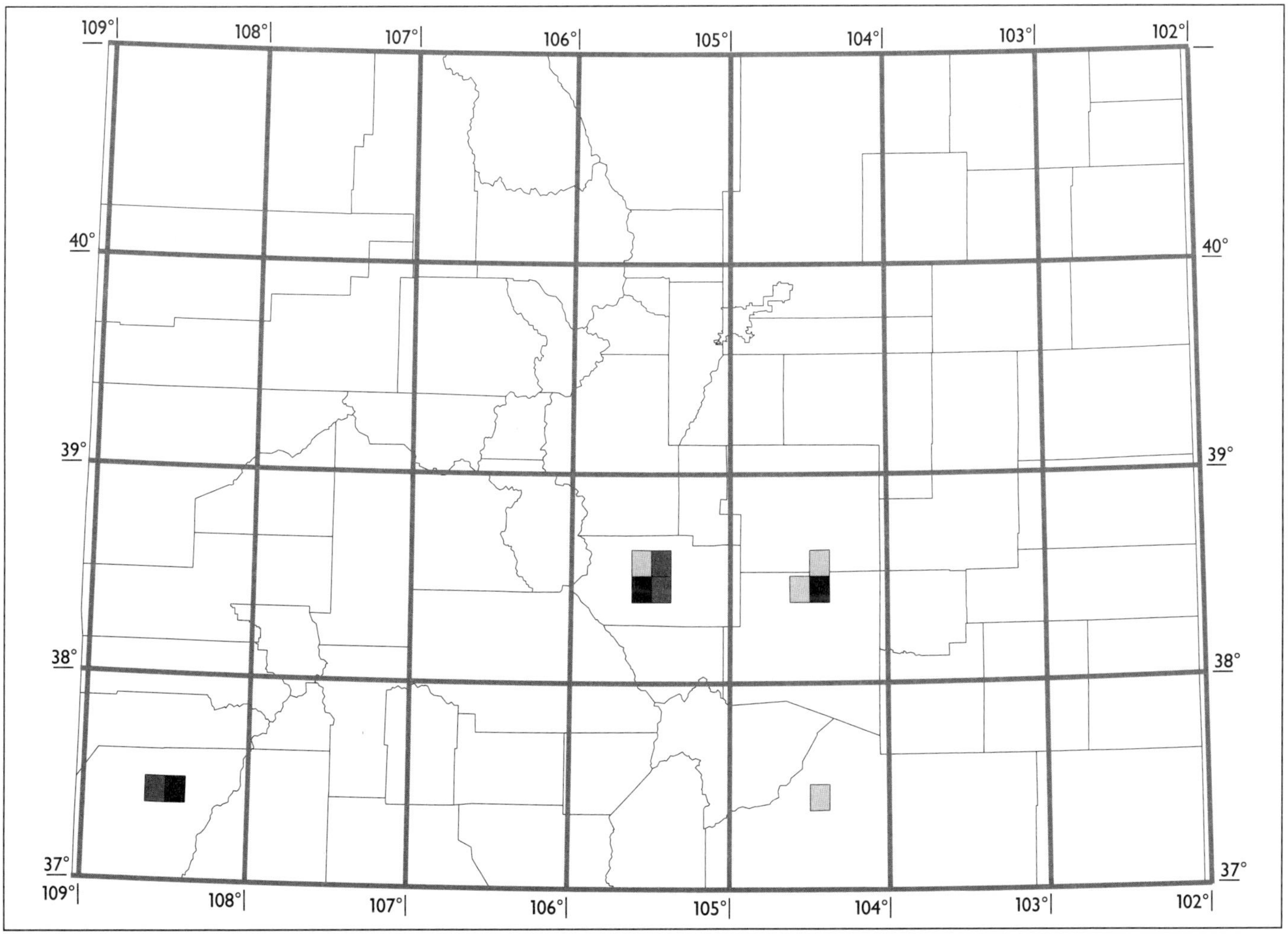
109°
108°
107°
106°
105°
104°
103°
102°
40°
39°
38°
37°

COLORADO DIVISION OF WILDLIFE / WILDLIFE RESOURCE INFORMATION SYSTEM

It takes persistence, and maybe a little luck, to find Long-eared Owls. These secretive owls retire by day in dense tree foliage, then emerge at nightfall to hunt for rodents in open country or forest clearings. The feather tufts that give this owl its name, not really ears, play an important role in the complex body language of owl communication. Much of this interplay takes place at night, and the tufts work like signal flags in silhouette, visible to other owls in dim light (Johnsgard 1988).

HABITAT

Long-eared Owls make their living at the edge between forests and open country (Johnsgard 1988, Marks et al. 1994). They sometimes inhabit dense forests, but more often mixed shrublands, prairie, and rural areas where riparian stringers or planted shelterbelts provide roost sites (B&N, Marti 1974, Marks 1986). A majority (88%) of Atlas records came from coniferous and deciduous forest or riparian shrub habitats, and the rest from shrublands or grasslands typically associated with forest or shrub patches. Nest and roost sites reported include stands of young, dense cottonwoods, willow patches, scrub oak, junipers, and dense forest of mixed conifers and aspen. These owls favor close-packed, but not necessarily tall, vegetation for nesting and roosting, often a clump of shrubs or trees surrounded by more open country (Moulton et al. 1976). Atlasers found them in sprouting cottonwoods at Chatfield State Park, tamarisk thickets in the Grand Valley, scrub oak thickets near Mesa Verde, and a young spruce/lodgepole/aspen stand at 10,000 feet (3048 m) near Fairplay.

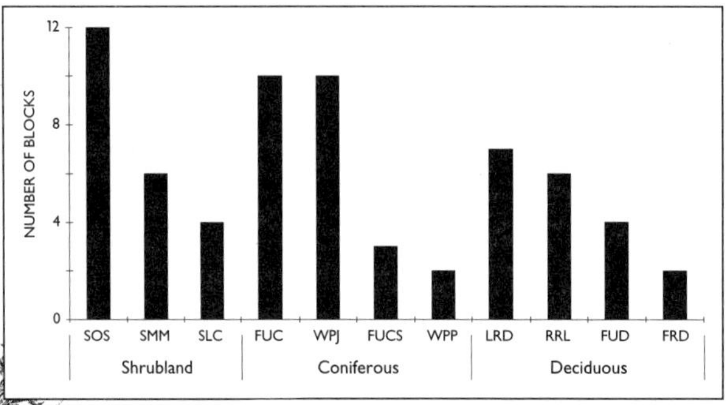
NUMBER OF BLOCKS
12
8
4
0
SOS
SMM
SLC
FUC
WPJ
FUCS
WPP
LRD
RRL
FUD
FRD
Shrubland
Coniferous
Deciduous

BREEDING

Like most owls in Colorado, Long-eared Owls begin to nest in early spring, often before snowmelt. The earliest Atlas nest report fell on 9 March in Prowers County, and Levad (1993) found a pair nesting on 13 March in Mesa County. Field workers recorded eggs or nestlings mostly before 15 June and fledglings after that date, similar to published dates (B&N). Among Colorado owl species, Long-eared Owls tied for second place in Confirmation percentage (46%). Nesting makes these otherwise retiring owls more conspicuous, and in many blocks atlasers probably located the owls by first spotting the nest. Old corvid nests, often placed conspicuously in trees or brushy tangles, provide most nest sites in Colorado (B&N, Moulton et al. 1976) and elsewhere in the western U.S. (Marks et al. 1994).

BREEDING PHENOLOGY

CODE		# OF RECORDS	DATE RANGE
ON	Occupied Nest	4	9 Mar–30 May
NE	Nest with Eggs	1	4 Apr
NY	Nest with Young	7	21 Apr–15 Jun
FY	Feeding Young	1	29 Jun
FL	Fledged Young	11	28 May–22 Jul

DISTRIBUTION

One of several Colorado owl species that Eurasian birders would recognize, Long-eared Owls inhabit middle latitudes of the entire Northern Hemisphere. Broad overlap exists between winter and summer ranges, and Colorado hosts birds in all seasons. Breeding adults often do not nest in the same areas in consecutive years (Marti 1974). Fluctuating density of rodent prey plays a role in such nomadism in Europe, and presumably in North America as well (Marks et al. 1994).

Atlas workers found Long-eared Owls in few blocks but widely scattered through the Great Plains, mountain parks and valleys, and valleys of the Colorado Plateau, affirming previous reports that the species occurs widely at lower elevations (B&N). However, numbers have dropped in Colorado. Historic reports (Cooke 1897, Sclater 1912, B&N) described them as common breeders, at least in the Great Plains adjacent to the Front Range, but recently A&R described them as rare and declining. Field workers found them in less than 3% of the Atlas blocks, with conspicuous gaps in some areas such as Boulder County, where they once bred as the most common owl (Gale 1884, Henderson 1908). Other longtime observers affirm this species' decline in the northern Front Range (Ron Ryder, D. Conner pers. comm.). Atlasers usually assigned abundance code A1, less often A2. Potential for high densities exists in good habitat; Moulton et al. (1976) found four active nests within about 50 feet (16 m) of each other in Cherry Creek Reservoir SRA, in a willow strand surrounded by prairie. At Rocky Mountain Arsenal NWR Long-eareds had three, none, and nine nests in

1994–1996 respectively. In 1996 they fledged 4.1 young/successful nest (Mindy Hetrick pers. comm.)

Continent-wide, Long-eared Owls have declined in California and some eastern and midwestern states, presumably from habitat loss (Marks et al. 1994), whereas researchers postulate stable numbers elsewhere (Marti and Marks 1989). The apparent recent decline of Long-eared Owls in northeastern Colorado has coincided with an increase in Great Horned Owl density (Steve Jones, Ron Ryder pers. comm.). Competition with the bigger Great Horneds for limited prey resources or interspecific intolerance at nest sites may partly explain declines of the much smaller Long-eared Owl.

Ultimately, land-use changes accelerating along the Front Range may correlate with population declines. The demise of shelterbelts for more efficient farming, degradation of riparian woodlands, and loss of rural lands to urban development all pose problems for Long-eared Owls. Such habitat changes remove nesting and roost sites, reduce foraging areas or rodent prey, and may increase populations of nest predators such as raccoons and of potentially competing Great Horned Owls. More research on population trends, habitat use, and competitive interactions would clarify the status of these fascinating and beneficial nighttime hunters, and help ensure that Long-eared Owls remain in Colorado.

BREEDING EVIDENCE

REPORTED IN 37 (2%) OF 1745 PRIORITY BLOCKS

Possible	14 blocks	(38%)
Probable	6 blocks	(16%)
Confirmed	17 blocks	(46%)

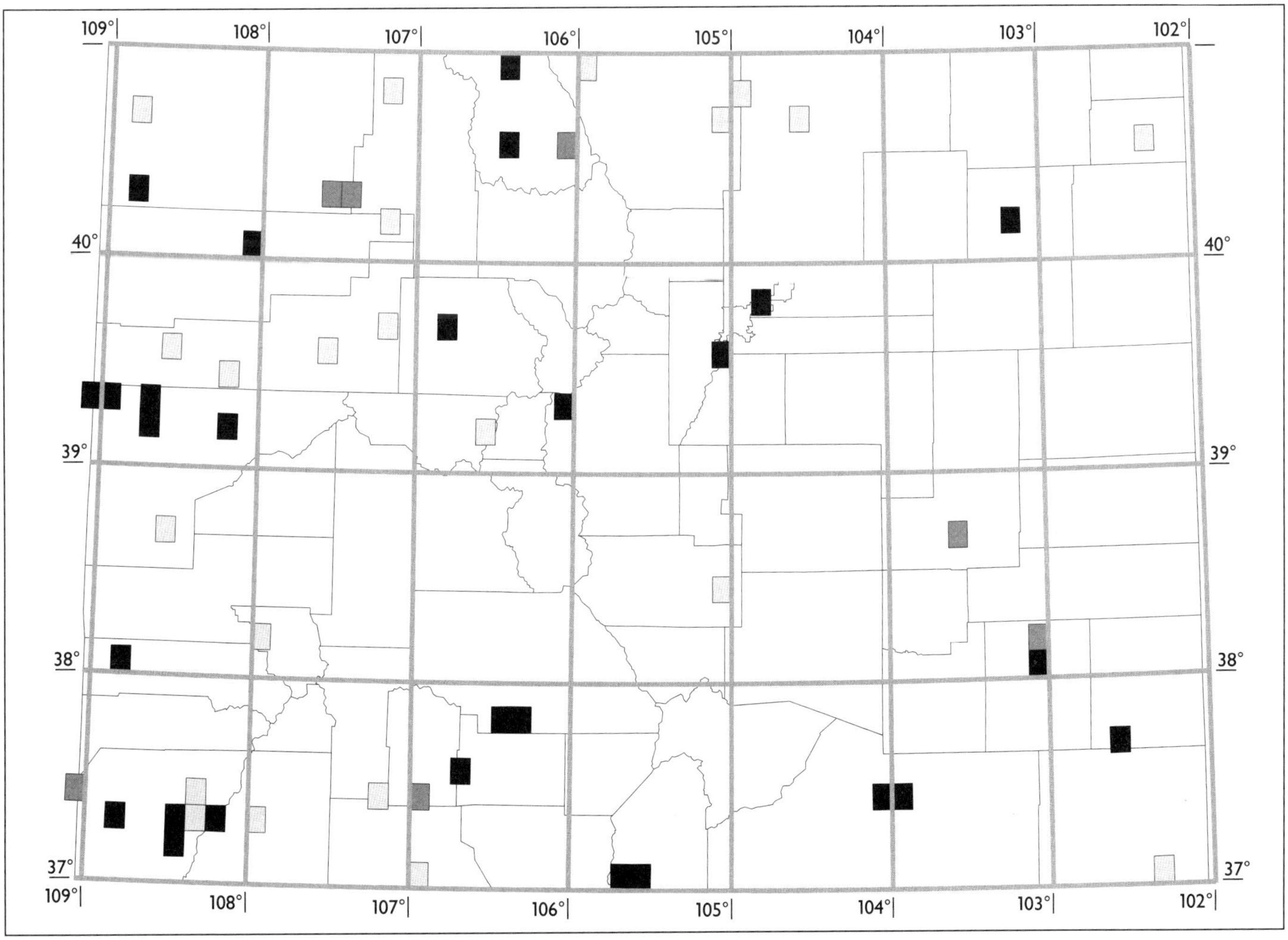
109°
108°
107°
106°
105°
104°
103°
102°
40°
39°
38°
37°

COLORADO DIVISION OF WILDLIFE / WILDLIFE RESOURCE INFORMATION SYSTEM

Open and windswept country suits prairie-loving Short-eared Owls. On broad wings with distinctive black wrist marks these medium-sized owls course low over prairies, hayfields, and marshes. One of the easier owls to observe, they often take wing or in late afternoon stand sentinel on fence posts.

HABITAT

With anatomy and behavior well designed for life in open country, Short-eared Owls range over a variety of open habitats including grasslands, marsh edges, shrub-steppes, and agricultural lands (A&R, Holt and Leasure 1993). They nest on the ground amid vegetation tall and dense enough to conceal the incubating female (Clark 1975). Atlasers found two nests in montane grasslands, one nest in a rural shelterbelt, and one in a marsh. They roost on the ground in similar environments throughout the year, except in snowy winter conditions when they may flock to tree roosts, sometimes with Long-eared Owls (Bosakowski 1986).

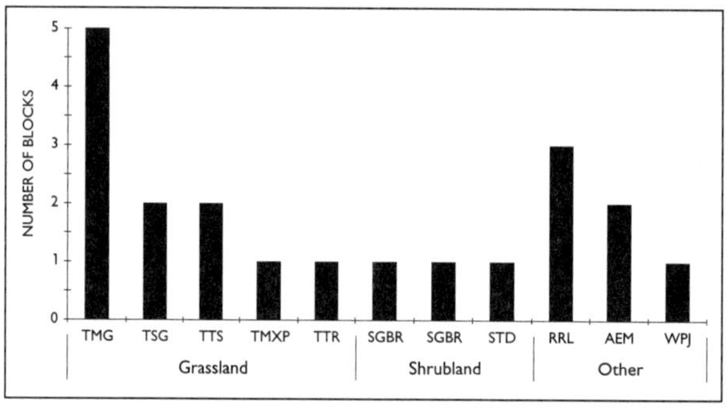
NUMBER OF BLOCKS
0
1
2
3
4
5
TMG
TSG
TTS
TMXP
TTR
SGBR
SGBR
STD
RRL
AEM
WPJ
Grassland
Shrubland
Other

Atlas workers recorded Short-eared Owls most often in grasslands, prairies, and wetlands (68% of habitat observations), less commonly in shrublands, and rarely in scrub oak or pinyon/juniper woodlands. These observations reflect consistent use of mostly treeless terrain for foraging as well as breeding.

BREEDING

One of the few owls to build their own nests, Short-eared Owl females make a shallow scrape on the ground and construct a usually sizable nest of grass blades, forbs, and downy feathers (Clark 1975, Holt and Leasure 1993). Owls may transport vegetation from some distance, making it possible to Confirm breeding by observing a Short-eared Owl carrying nest materials. Because ground-nesting makes females and young vulnerable to terrestrial predators, incubating females often sit tight until intruders approach very closely, making nests hard to find by flushing nesting females. Young are more precocial than those of most owls; they gain weight very rapidly and often leave the nest on foot when only 14–17 days old, well before fledging. These ground-nesting adaptations, along with general rarity in Colorado, help explain why Atlas workers found just three nests and Confirmed breeding in only 19% of the few blocks where they found Short-eared Owls. Dates recorded for dependent young lie within the May–August range reported for Colorado (B&N, Nelson 1993).

BREEDING PHENOLOGY

CODE		# OF RECORDS	DATE RANGE
NY	Nest with Young	1	31 May
FL	Fledged Young	1	12 Jun

DISTRIBUTION

Short-eared Owls breed in temperate parts of North and South America, and throughout most of the Northern Hemisphere as well (Johnsgard 1988). In cooler parts of their range, including Colorado, they migrate seasonally, and Colorado hosts more birds in winter than in summer (A&R, B&N). These far-ranging owls seem prone to nomadic wanderings as well, which has resulted in far-flung island populations and occasionally the odd sight of a Short-eared Owl landing on a ship at sea (Gray 1945). Migrants and nomads may flock in large numbers in response to concentrations of or changes to prey populations (Bent 1938); 60 owls seen at Monte Vista NWR in mid August 1958 (B&N) probably represented a post-breeding flock.

Atlasers verified previous reports that Short-eared Owls are scarce and local breeders in Colorado (B&N, A&R). Breeding populations occur in North Park, the San Luis Valley, and the northern Great Plains; the few records from southeastern Colorado and the Four Corners area suggest very scattered and perhaps impermanent breeding populations in those areas. On the Western Slope, migrants or wintering birds comprise nearly all published records, but a specimen taken in Gunnison County in June 1903 (Warren 1910) suggests at least one possible breeding record.

Serious declines in northeastern states put this species on the Partners in Flight WatchList of threatened birds (Carter et al. 1996), and BBS data suggest recent declines in Oregon, Washington, and Idaho (Holt and Leasure 1993). Short-eared Owls probably never bred commonly in Colorado, as early observers reported these rather conspicuous owls uncommonly (Cooke 1897, Sclater 1912). More recent observations suggest the possibility of a decline in Colorado as well. Atlasers failed to find Short-eared Owls in several Great Plains breeding locations reported by B&N, including Barr Lake, Arapahoe County near Denver, and the Arkansas River in Bent County.

Loss of habitat due to more intensive agriculture and urbanization, including the greening of the formerly treeless Great Plains with shelterbelts and riparian forests, may partly explain the apparent decline in Colorado, especially near the Front Range. However, much seemingly suitable habitat remains unoccupied throughout North America (Holt and Leasure 1993), suggesting that other factors are also at work. For example, nest predation may increase when nest-destroying feral dogs and cats, foxes, and skunks proliferate with human settlement. Conservation of these fascinating grassland wanderers in Colorado will require protection of large open spaces for owls to hunt, and maintenance of nesting cover in grasslands and shrublands.

BREEDING EVIDENCE

REPORTED IN 16 (1%) OF 1745 PRIORITY BLOCKS

□	Possible	10 blocks	(62%)
■	Probable	3 blocks	(19%)
■	Confirmed	3 blocks	(19%)

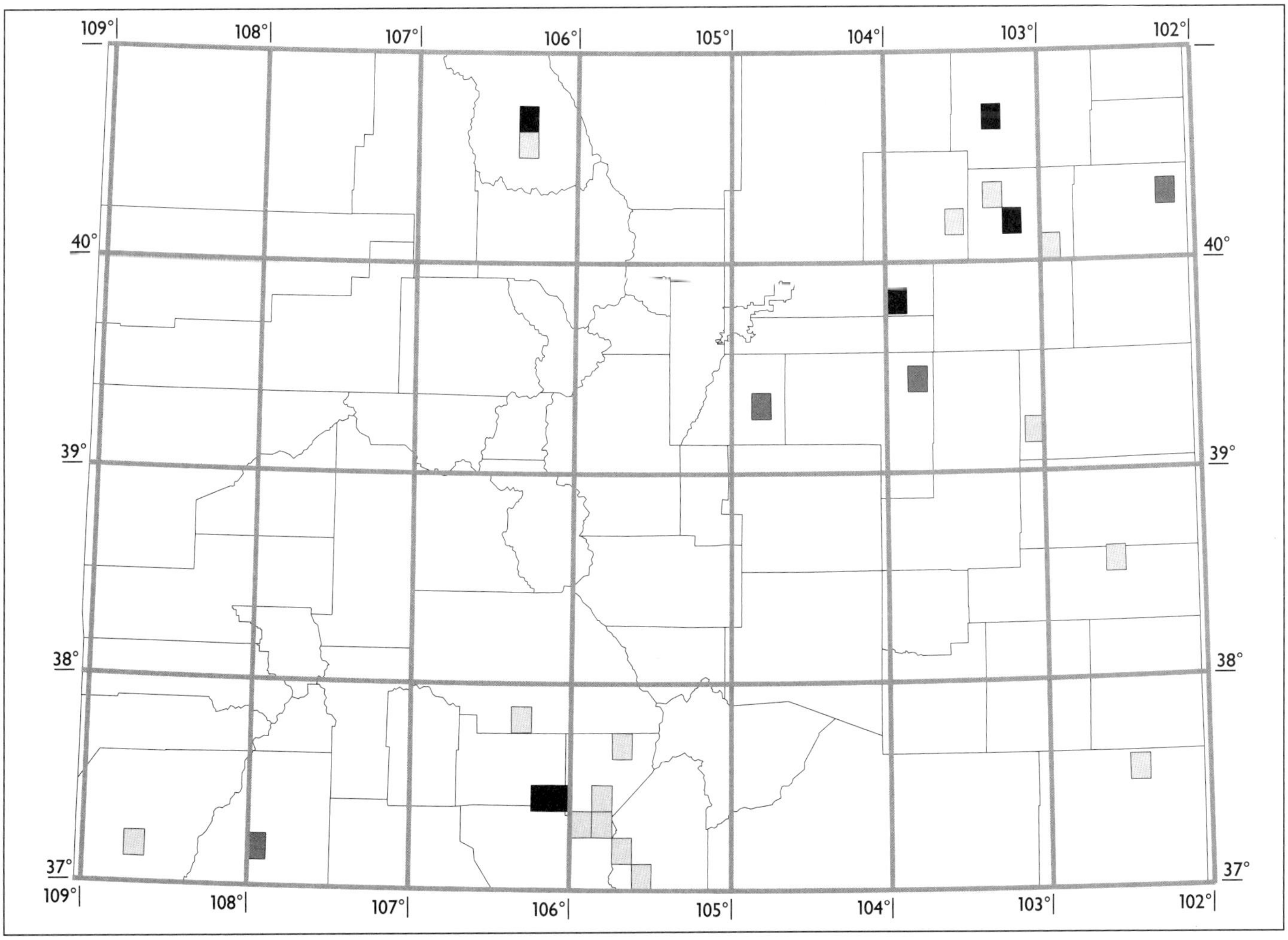
109°
108°
107°
106°
105°
104°
103°
102°
40°
39°
38°
37°

The Colorado high country has two nighttime whinnies to tantalize birdwatchers. Boreal Owls and Common Snipe make similar sounds—owls with their vocal chords from February to May, snipes with their tails, May to July. The owls' clear six- to eight-syllable calls issue from perches in subalpine forests above 9,000 feet (2745 m). Intrigued by this elusive owl, birders have even put "How to find a Boreal Owl" on a computer internet web page (Pulliam 1995)!

HABITAT

In Colorado Boreal Owls occur mainly in mature to old-age (over 150 years) Engelmann spruce and subalpine fir above 9,000 feet (2745 m), but they also frequent higher-elevation lodgepole pine and aspen (Palmer 1986, Hayward and Hayward 1993, Holland and Schultz in press). They prefer wet situations near streams or bogs, because these often have good populations of small rodents that the owls can catch even under soft snow cover. After nesting they may wander to open ponderosa pine (RAR) and even pinyon/juniper (Tom Holland pers. comm.).

Atlas habitat reports cited spruce (mainly) or lodgepole forests. The general nature of the Atlas habitat codes did not permit analysis of forest-age categories.

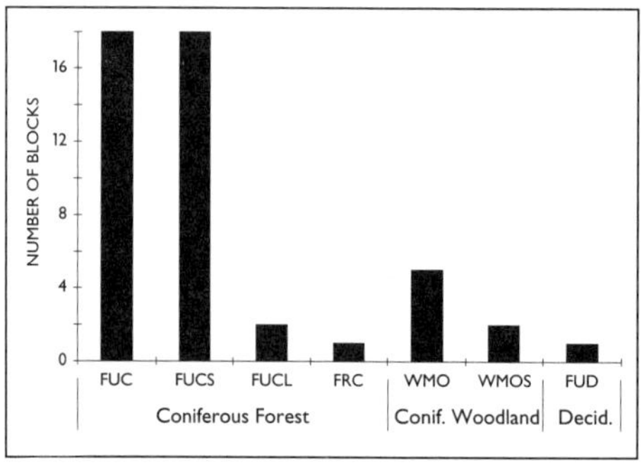
NUMBER OF BLOCKS
0
4
8
12
16
FUC
FUCS
FUCL
FRC
WMO
WMOS
FUD
Coniferous Forest
Conif. Woodland
Decid.

BREEDING

Early accounts (Sclater, Cooke, Bergtold) listed Boreal Owls only as winter visitors. However, recent studies document widespread nesting from Wyoming to New Mexico (Webb 1982a, Palmer and Ryder 1984, Ryder et al. 1987, Reynolds et al. 1990, Ryder 1991a, Jones 1991, Rawinski et al. 1993, Stahlecker and Duncan 1996). Atlas data verify this broad range but with low numbers. Boreal Owls occurred in only 27 of 1,745 Atlas Priority blocks and in 25 Non-priority blocks. Observers searching for them at the peak of their March-to-May singing period risk strong winds, deep snow, and avalanches.

Atlasers most frequently Confirmed breeding with eggs or young in nests (8 observations) and fledged young (4 observations). They saw Probable breeding most often by territorial behavior (7).

In Colorado nest initiations (31 reports of egg-laying) range from 22 May to 30 June (median = 7 June). Colorado observers have reported only five nests in natural cavities (mainly flicker holes), plus 26 in nest boxes (Ryder 1991a; Holland and Schultz in press). The natural sites include a Larimer County lodgepole pine cavity in which Art Wolfe photographed adults and young on 24–25 July 1982 (Bergman 1985).

One female nesting in a box near Cameron Pass was observed 22 June 1990 by more than 200 participants from an American Birding Association convention (Kingery 1990b; photograph by Rick Bowers in *American Birds* 44:1211). Despite birders' attention all summer long, three young fledged from this box, but owls did not use the box the next year. Colorado State University wildlife students and CDOW, State Forest, and U.S. Forest Service personnel have installed over 60 boxes on the Roosevelt National Forest, 100 on the Colorado State Forest (RAR), and 400 on the Grand Mesa/Uncompahgre/Gunnison National Forest (Rich Levad pers. comm.).

Summer home ranges of three radio-marked adults near Cameron Pass varied from 593 to 869 acres (240–352 ha) from 18 May to 31 July, and winter ranges covered 1,961 to 3,631 acres (794–1470 ha), from 1 August to 12 January. Year-round totals for the two males encompassed 3,447 and 3,894 acres (1395 and 1576 ha) and overlapped one another by more than 90% (Palmer 1986).

BREEDING PHENOLOGY

CODE		# OF RECORDS	DATE RANGE
C	Courtship	2	13 Mar–29 Apr
ON	Occupied Nest	3	21 Apr–21 Jun
NE	Nest with Eggs	1	20 Jun
NY	Nest with Young	4	22 Jun–9 Jul
FL	Fledged Young	4	23 Jul–2 Sep

DISTRIBUTION

A circumpolar species, Boreal Owls are resident across North America and northern Eurasia. They range through Canada's boreal forests from Yukon to Newfoundland (Hayward and Hayward 1993). Scattered populations occur in northern Minnesota, the Cascade ranges, and south in the Rocky Mountains into north-central New Mexico (Stahlecker and Duncan 1996). Most reside year round in their range, but some, especially in Canada and Scandinavia, show nomadic/migratory/irruptive movements in years of food scarcity.

Colorado birds are primarily year-round residents in subalpine spruce/fir forests (Ryder et al. 1987). One radio-marked juvenile, however, moved from spruce/fir on Grand Mesa north about 30 miles (18 km) to scrub oak and pinyon/juniper on Roan Creek (Tom Holland pers. comm.).

Boreal Owls have a much greater range in Colorado than early authors thought (Cooke, Sclater, B&N). The only three specimens known before 1963 include an 1896 breeding-season bird (Ryder et al. 1987). In August 1963, Baldwin and Koplin (1966) collected a juvenile female on Deadman Hill, near Red Feather Lakes.

Now, field workers have found Boreal Owls in most Colorado mountain ranges (Bridges 1992b, Atlas). Atlasers located the first Boreals in the Elk and San Juan mountains and the Forest Service lured them to Grand Mesa nest boxes. Bridges, searching all the high ranges, located calling Boreals in Atlas blocks in the Flat Tops, Park Range, and southern San Juans but elicited no responses in the rather linear Sangre de Cristos and Wet Mountains. Boreal Owls have probably been longtime Colorado residents rather than recent arrivals (Stahlecker and Duncan 1996).

Bridges (1992a) ranks the Boreal Owl as the tenth most abundant owl species in Colorado (more abundant than only Short-eared, Spotted, Snowy, and Barred owls). During 16 years of annual roadside surveys near Cameron Pass, the number of calling males fluctuated widely (0–27), probably due mainly to availability of prey, red-backed voles (RAR).

Although considered a "Species of Concern" in Idaho and Montana, Boreal Owls have no special protection in Colorado. The U.S. Forest Service designates the Boreal Owl as a "sensitive species" requiring special management programs. Nest boxes erected in some national forests provide nest sites, but no sound data demonstrate that natural cavities are limited. Annual surveys on several forests monitor trends in calling males. More detailed studies could help to evaluate the effects of various land-use practices (especially logging) on the species. Hayward (1997) observes that clear-cut areas remain unsuitable for a century or more, and recommends partial cutting to maintain productive mature and older spruce/fir forests. Boreal Owls currently are well distributed across a large geographic range in a largely unexploited forest type and not in any immediate peril (Hayward and Verner 1994).

BREEDING EVIDENCE

REPORTED IN 27 (2%) OF 1745 PRIORITY BLOCKS

Possible	16 blocks	(59%)
Probable	6 blocks	(22%)
Confirmed	5 blocks	(19%)

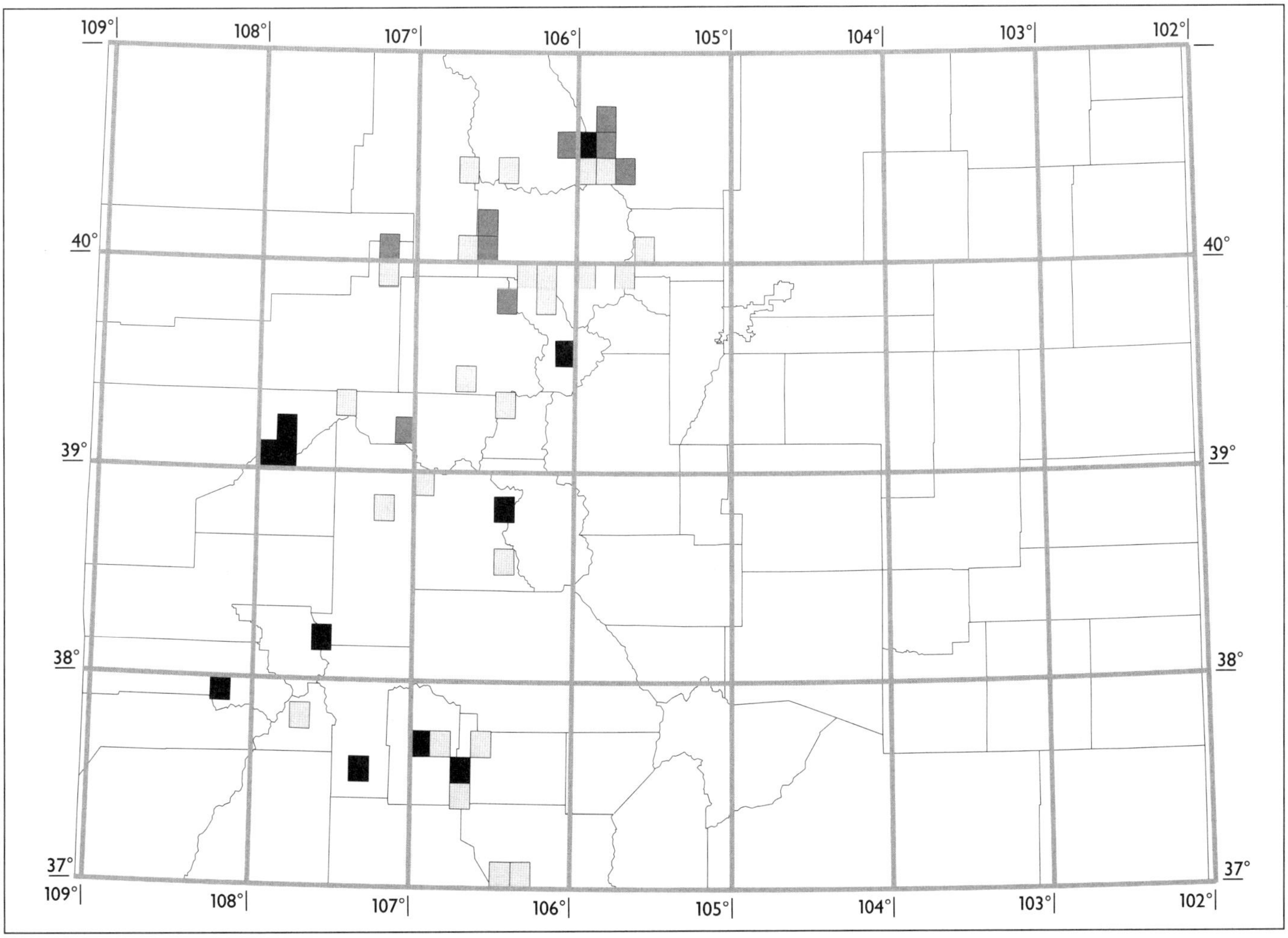
109°
108°
107°
106°
105°
104°
103°
102°
40°
39°
38°
37°
COLORADO DIVISION OF WILDLIFE / WILDLIFE RESOURCE INFORMATION SYSTEM

NORTHERN SAW-WHET OWL *Aegolius acadicus*

This voracious predator of field mice acts much bigger than its size. Northern Saw-whet Owls sit quietly in thick conifers by day, so tame they often allow humans within arm's length. Males call for hours on end in early spring with monotonous whistles and quickly respond to imitated calls (Webb 1982b, Jones 1991). From low perches, saw-whet owls spot their prey with superb hearing and low-light vision, then dart forth on silent wings to make the catch.

HABITAT

Mountain and foothills forests and canyon country woodlands provide homes for these owls in Colorado. Atlasers almost always recorded them in forest habitats (92% of records). Previous workers in Colorado observed that saw-whets dwell more in ponderosa pine woodlands or ponderosa pine mixed with Douglas-fir than in in other forest types (Webb 1982b, Jones 1991). Atlas records do indicate many owls in ponderosa pine and unspecified upland conifers, but they also show significant use of pinyon/juniper woodland (a habitat not mentioned in other analyses; e.g., Cannings 1993). The prevalence of Atlas records from pinyon/juniper makes it surprising that other observers have not mentioned the importance of this habitat. Atlasers observed that saw-whets chose old-growth over younger pinyon/juniper woodlands.

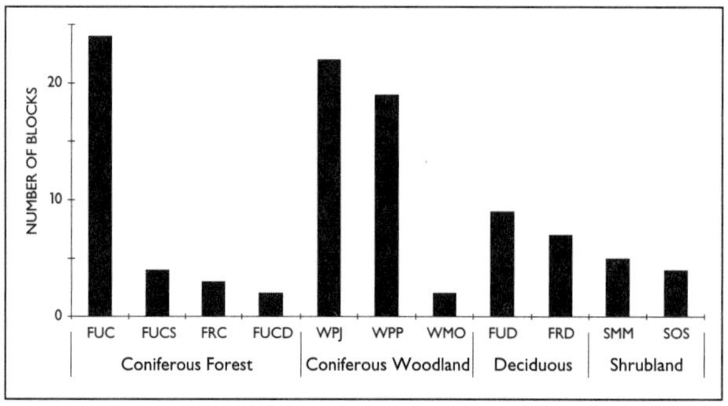

NUMBER OF BLOCKS
0
10
20
FUC
FUCS
FRC
FUCD
WPJ
WPP
WMO
FUD
FRD
SMM
SOS
Coniferous Forest
Coniferous Woodland
Deciduous
Shrubland

A&R speculated that Northern Saw-whet Owls are rare in densely forested high mountains, and Atlas results suggest that may be so, with many more records from mid-elevation habitats than spruce/fir forest.

Saw-whets select fairly open woodlands over dense forest (Palmer 1986), and they will even use narrow riparian forests among wide-open shrublands (Marks and Doremus 1988, Hayward and Garton 1988). In Oregon conifer forests, saw-whets use sites with more open understory than typical of the forest as a whole (Boula 1982).

BREEDING

These forest owls nest in tree cavities, usually woodpecker holes in conifers (Webb 1982b, Jones 1991) or aspen (2 Atlas nest cards). Like most Colorado owls, Northern Saw-whet Owls start nesting long before winter has yielded its icy grip. Palmer (1986) and Jones (1991) reported that some males begin to stake breeding territories in February; calling activity peaks in April and May. Atlas records of nests with young in mid May and fledglings by late May confirm early nesting. A late observation of nestlings on 13 July, and published reports of June nests in Colorado (Webb 1981, B&N), strongly suggest second clutches. Although unconfirmed for this species, second broods seem likely from continent-wide reports of late nesting dates (Cannings 1993). Females often leave the nest permanently after nestlings are fully feathered; in such cases the male tends the nestlings of the first brood while females may seek a second mate and nest again, a mating system called sequential polyandry.

Atlasers Confirmed breeding in only 13% of the blocks with saw-whets. This percentage, third lowest among Colorado's 12 owl species, reflects the difficulty of finding cavity nests compared to the relative ease of detecting these highly vocal owls by night calling.

BREEDING PHENOLOGY

CODE		# OF RECORDS	DATE RANGE
C	Courtship	1	6 May
ON	Occupied Nest	2	21 May–27 Jun
NY	Nest with Young	4	13 May–13 Jul
FY	Feeding Young	2	11 Jul–21 Jul
FL	Fledged Young	5	26 May–20 Jul

DISTRIBUTION

Northern Saw-whet Owls range across the timbered stretches of northern North America and forested mountains of the western U.S. In Colorado, Atlas workers recorded them widely in the Rocky Mountains and Colorado Plateau except the mountain parks, lower river valleys, and the Gunnison Basin, all treeless areas. (A high rate of occurrence in southwestern Colorado and the Uncompahgre Plateau resulted from Northern Saw-whet Owl sightings incidental to Spotted Owl surveys independent of the Atlas.) This distribution seems unchanged from earlier reports by B&N, who listed records from most western counties. A&R rightly regarded the few Great Plains records as accidentals.

Historically, observers described Northern Saw-whet Owls as uncommon in Colorado, based primarily on anecdotal information (B&N). More recent systematic searches using taped calls found that saw-whets occur widely in suitable habitats (Webb 1982b, Palmer 1986, Jones 1991). Atlasers usually rated them in higher abundance categories than usual for birds of prey, assigning A2 (2–10 pairs) in 75% of blocks reporting the owls, and A3 (11–100) in 15%. Cannings (1993) regarded 2.6 breeding pairs per square mile (1 pair/km^2) as a reasonable maximum density in good habitat. At that density, an Atlas block of continuous suitable habitat could hold 25 nesting pairs, so abundance ranks of A2 and A3 seem reasonable.

Availability of secondary tree cavities for nesting may limit populations of Northern Saw-whet Owls (Webb 1982b), and forest management practices that eliminate snags or older-age trees could reduce owl populations. Furthermore, young regenerating stands usually lack the edge habitat and open understory Northern Saw-whet Owls require for foraging (Boula 1982). Forest management for snag retention and diverse stands of uneven-aged trees will help ensure the survival of Northern Saw-whet Owls and other members of the cavity-nesting guild.

BREEDING EVIDENCE

REPORTED IN 72 (4%) OF 1745 PRIORITY BLOCKS

☐	Possible	50 blocks	(69%)
▩	Probable	13 blocks	(18%)
■	Confirmed	9 blocks	(13%)

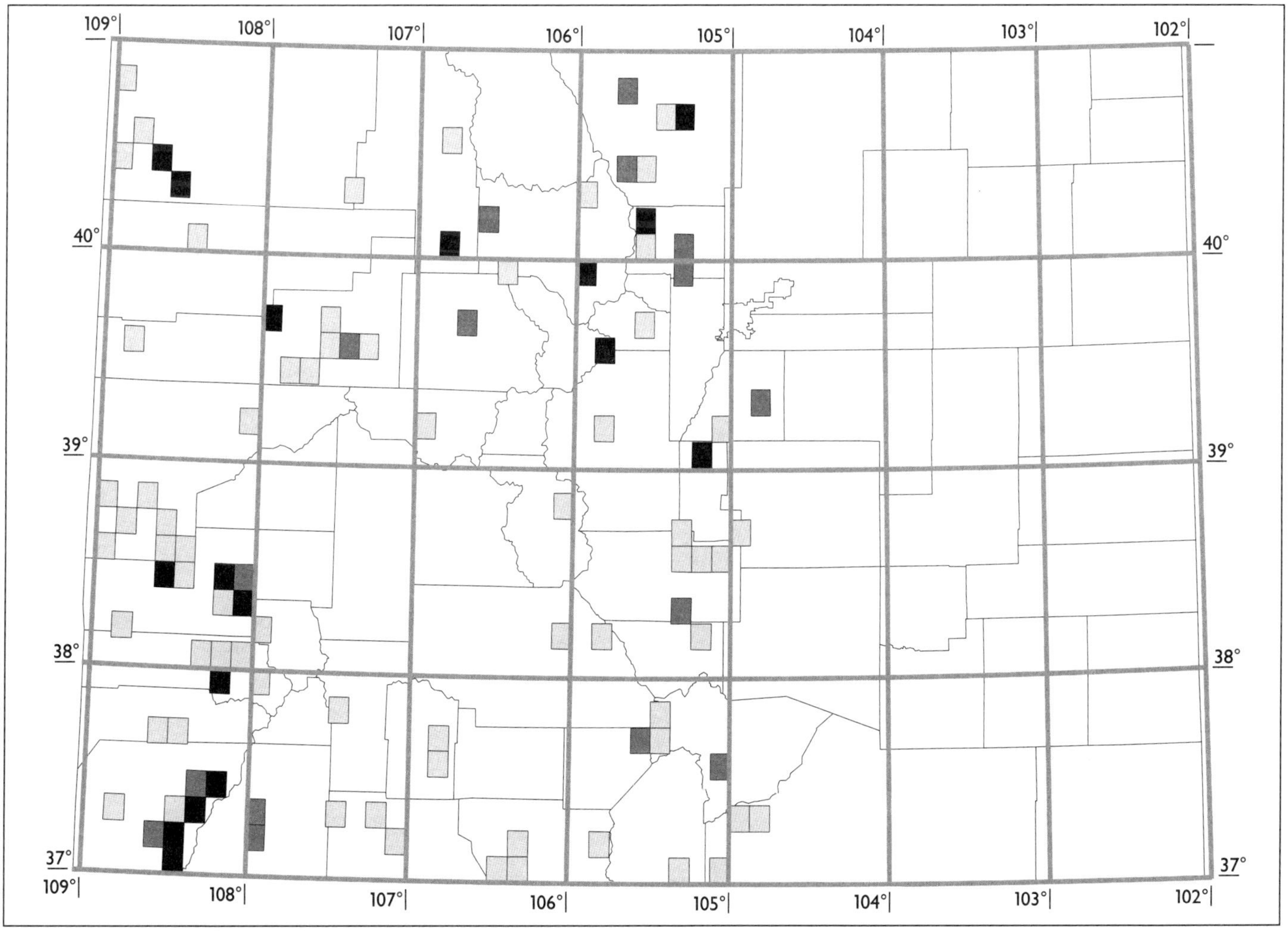
109°
108°
107°
106°
105°
104°
103°
102°
40°
39°
38°
37°

COLORADO DIVISION OF WILDLIFE / WILDLIFE RESOURCE INFORMATION SYSTEM

Casual observers and ornithologists alike marvel at the buoyant flight and aerodynamic prowess of the Common Nighthawk. Better known to many as the "bullbat," its skills go on display on summer evenings as the heat of the day wanes and insects take flight.

HABITAT

Although the variety of habitat types atlasers listed for Common Nighthawks rivals that of any bird in the state, the birds exercise somewhat greater selectivity in breeding habitat than the length of the list would indicate. Many of the recorded habitat codes come from places visited by nighthawks on evening feeding flights rather than from breeding territories.

If one looks only at habitat codes associated with verified nest sites (ON, NE, and NY codes), the nesting requirements of Common Nighthawks become evident. Open pinyon/juniper and ponderosa pine woodlands, sometimes associated with scrub oak or sagebrush, and shortgrass prairie provide two-thirds of the nest sites.

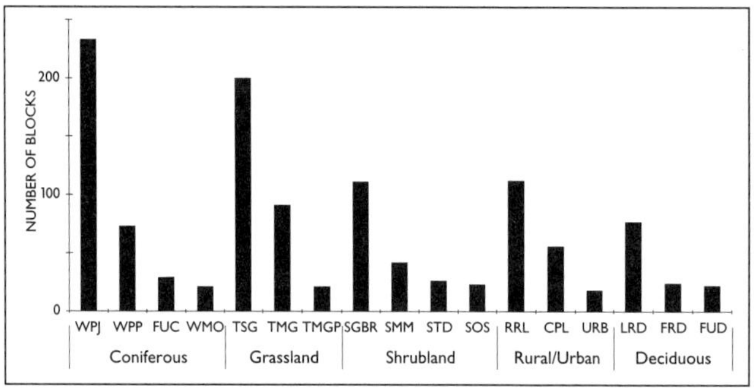
NUMBER OF BLOCKS
200
100
0
WPJ WPP FUC WMO TSG TMG TMGP SGBR SMM STD SOS RRL CPL URB LRD FRD FUD
Coniferous
Grassland
Shrubland
Rural/Urban
Deciduous

Dry conditions prevail on the shortgrass prairie and in the open pinyon/juniper and ponderosa woodlands preferred by nighthawks. These habitats contain an abundance of the bare, often gravel-strewn, spots on which nighthawks lay and incubate their eggs. Some nest cards specifically mention a slight southern exposure at the nest site. A southern exposure would, of course, accentuate any xeric conditions already associated with a site. Occasionally, burned-over areas leave conditions suitable for a nest site (Bent 1940).

In the western U.S. the coloration of nighthawks inhabiting a particular geographic area tends to change with the general characteristics of the soil in that area. They appear lighter in the Great Basin where grayish and light buff soils predominate and they appear tawny-colored where soils are reddish, in the southwestern U.S. and northern Mexico (Selander 1954). This would appear to afford them some protection from predators.

BREEDING

Common Nighthawks arrive in Colorado well after most other migrants. Many parts of the state do not see their first nighthawks of the year until late May or early June. Apparently, however, this species loses little time before laying eggs. Nighthawks require no investment of energy in building a nest and may thus begin laying eggs shortly after arriving on their breeding grounds. Almost invariably, clutches consist of two eggs. Apparently, males play a relatively minor role, and sometimes no role whatsoever, in the incubation of the eggs (Bent 1940, B&N). The incubation period lasts almost three weeks.

Although atlasers recorded a few "nests" in late May and early June, the overwhelming majority of observations of nests with eggs came from mid June through mid July. Reports of nests with young came primarily during July. The nestling period lasts for about three weeks. During this time both parents feed the young (Bent 1940, B&N). Between feedings the young birds sit quietly in the area of the nest and become all but impossible to distinguish from the surrounding ground clutter.

The late arrival date combined with a relatively long incubation and period prior to fledging (40 days) precludes any possibility of double-brooding. By mid August nighthawks have begun their southward migration, although some birds remain in the state well into October.

A matter of special interest in the breeding activities of Common Nighthawks concerns the interpretation of the "booming" flights of the male. Although many regard this simply as courtship behavior, these flights continue, though somewhat abated, well after the female has laid her eggs—sometimes even into migration (Bent 1940). The frequency of these booming flights elicited an enormous number of "courtship" records in the Atlas database, including ones with dates as late as the end of July and August. One must wonder about alternate interpretations of the meaning of booming flights.

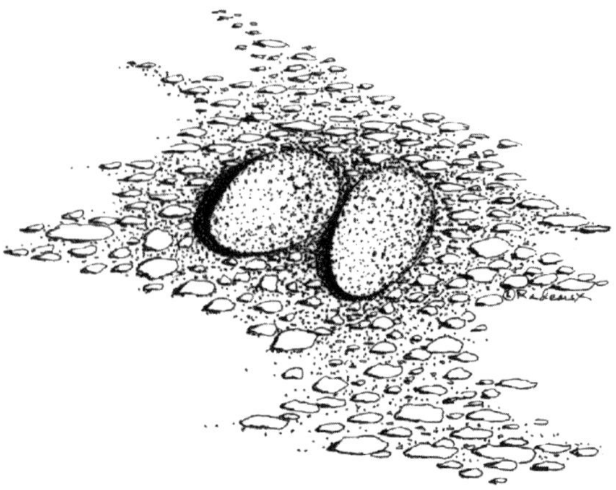

BREEDING PHENOLOGY			
CODE		# OF RECORDS	DATE RANGE
C	Courtship	256	3 May–5 Aug
ON	Occupied Nest	11	2 Jun–30 Jul
NE	Nest with Eggs	39	31 May–26 Jul
NY	Nest with Young	13	30 Jun–6 Aug
FL	Fledged Young	24	6 Jun–30 Aug

DISTRIBUTION

Common Nighthawks breed across most of the lower 48 states and Canada. Their fall migration takes them all the way to South America where they spend the winter.

Common Nighthawks appear throughout Colorado up to timberline. Sightings of this species have become increasingly infrequent in the larger cities, however. Whether due to predation (largely from cats, dogs, and corvids), loss of prey base, loss of suitable habitat, or, most likely, some combination of these factors, they no longer grace the urban skies on summer evenings as they once did. Their strongholds remain in the southeastern plains and around pinyon/juniper and ponderosa woodlands.

Atlas field work Confirmed breeding in two of the three remaining Colorado latilongs where no records of breeding existed. Only in the latilong surrounding Burlington (39102) have breeding Common Nighthawks escaped documentation.

Numbers of Common Nighthawks decrease as elevation increases. The only two counties to yield no Atlas records of nighthawks—Lake and San Juan—both lack any low-elevation areas within their boundaries. In addition to this gap at higher elevations, Common Nighthawks show a considerable thinning in their populations across the northeastern plains of Colorado. Although older records of nighthawks nesting in croplands exist (Bent 1940), the extensive agricultural development in this area seems to offer a plausible explanation for the relative scarcity of this species in many parts of the northeastern plains.

Increasing suburbanization of many areas once favored by Common Nighthawks similarly threatens their numbers within the state. The manicured lawns and pets that attend the growth of the suburbs leave no quarter for nighthawks.

BREEDING EVIDENCE			
REPORTED IN 1020 (58%) OF 1745 PRIORITY BLOCKS			
☐	Possible	533 blocks	(52%)
▩	Probable	384 blocks	(38%)
■	Confirmed	103 blocks	(10%)

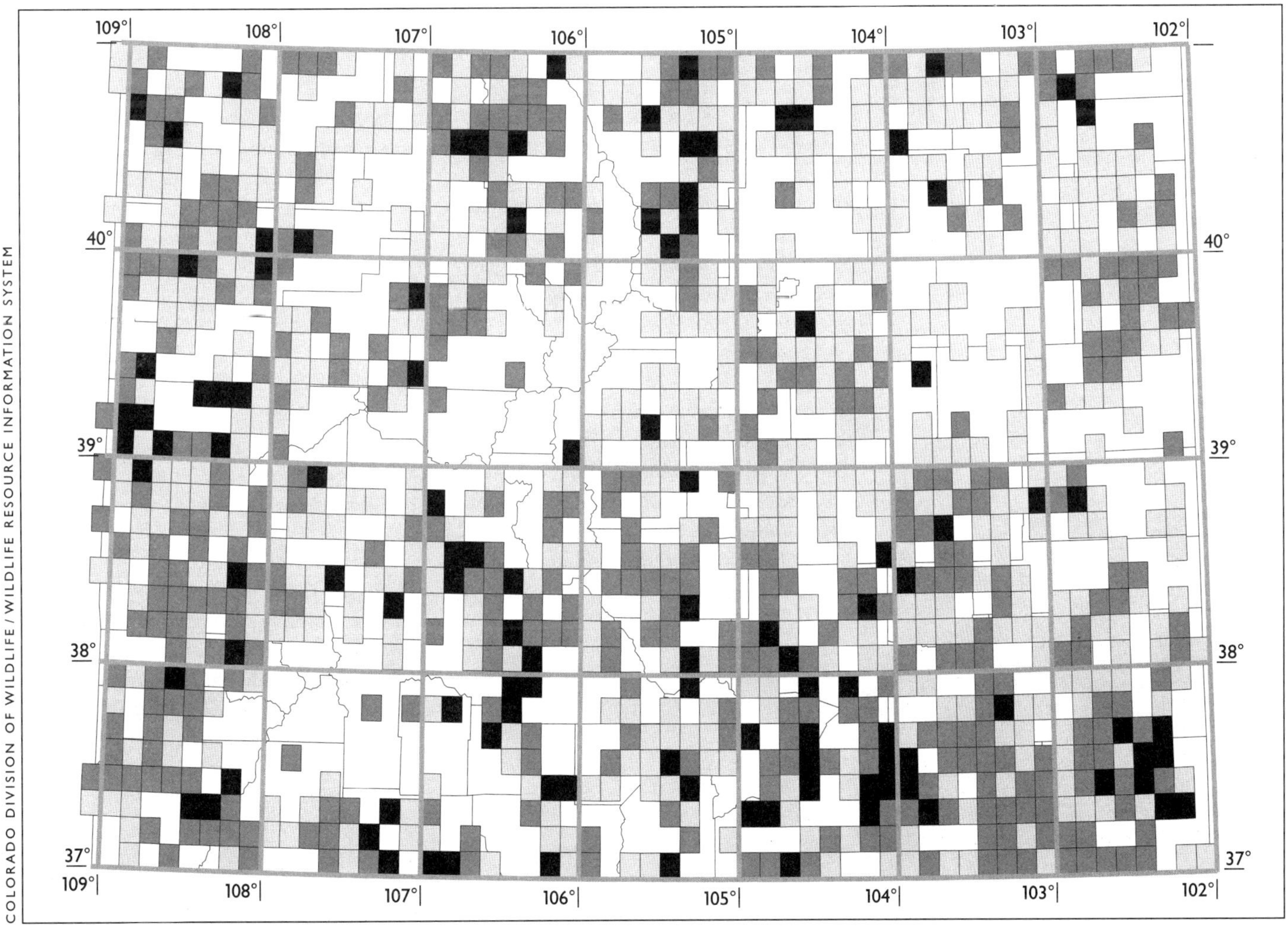
109°
108°
107°
106°
105°
104°
103°
102°
40°
39°
38°
37°
COLORADO DIVISION OF WILDLIFE / WILDLIFE RESOURCE INFORMATION SYSTEM

Cryptic coloration and nocturnal habits, along with observers' distaste for wandering about after dark in rattlesnake habitat, combine to surround Common Poorwills in an aura of mystery. The adjective "common" would seem well chosen for the Common Poorwill only in western Colorado, where it has widespread distribution and plentiful numbers, but the description confounds daytime-oriented birdwatchers who rarely encounter this wraith of the evening.

HABITAT

Common Poorwills breed in dry, open grassy or shrubby areas in western North America (Csada and Brigham 1992). In Colorado, they nest in "foothill shrublands (scrub oak, mountain mahogany, serviceberry, etc.), pinyon/juniper woodlands, and ponderosa pine forests" (A&R), but not in the prairie grasslands that they use farther north (Csada and Brigham 1992). Common Nighthawks, which do use the grasslands, also nest in the same habitats as Common Poorwills, and observers have found nests of the two separated by as little as 16 feet (5 m; Csada and Brigham 1992)

Atlasers reported pinyon/juniper woodlands most frequently, in nearly 40% of the reported blocks. Sagebrush and mountain shrubs (SMM—serviceberry, chokecherry, mountain mahogany, scrub oak) accounted for another 34%. Previous reports did not emphasize sagebrush as important habitat for this species (A&R, Csada and Brigham 1992), but atlasers regularly reported them from this habitat (25 blocks).

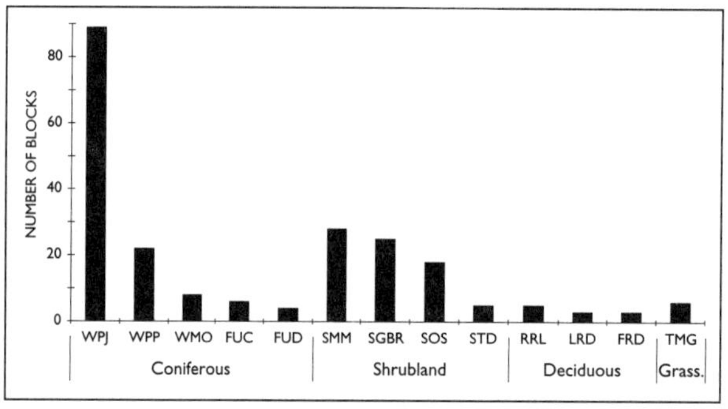
NUMBER OF BLOCKS
0
20
40
60
80
WPJ
WPP
WMO
FUC
FUD
SMM
SGBR
SOS
STD
RRL
LRD
FRD
TMG
Coniferous
Shrubland
Deciduous
Grass.

Poorwills feed by sallying for insects from a low perch or an open patch of ground. They require some light to catch their prey and feed most actively at dusk, at dawn, and during moonlit nights (Csada and Brigham 1992). Roadbeds qualify as open ground, and Atlas field workers often found this bird in their headlights (RL, HEK).

BREEDING

In their monograph of this species, Csada and Brigham (1992) nearly two dozen times used phrases indicating that we know little or nothing about some aspect of the Common Poorwill's breeding activity; atlasers found very few nests, and so added little to the scanty knowledge base. Although poorwills nest out in the open (Ryser 1985), observers seldom find the nests because the birds have such excellent protective coloration that they sit very tightly and seldom flush from the nest (Ligon 1961). Of 194 Atlas reports of Common Poorwills, only ten recorded nests, and the Confirmation rate for this species did not reach 10%.

Common Poorwills apparently begin singing on territory in mid May (earliest Atlas date: 12 May). In Canada calling males maintain rather large territories, spacing themselves an average of 1,640 feet (500 m) apart (Csada and Brigham 1994). Colorado's warmer climate may permit smaller territories, for Atlas workers often would have several respond at once to a tape recording of their call (Coen Dexter pers. comm., RL). Egg-laying begins by the end of May: the first Atlas NE (nest with eggs) recorded 28 May. Csada and Brigham (1992) report that Common Poorwills regularly double-clutch, and the other Atlas NE dates, spread about two weeks apart from 21 June to 25 July, suggest that pattern for Colorado as well.

BREEDING PHENOLOGY

CODE		# OF RECORDS	DATE RANGE
C	Courtship	1	26 May
ON	Occupied Nest	2	10 Jun–1 Jul
NE	Nest with Eggs	5	28 May–25 Jul
NY	Nest with Young	3	15 Jun–2 Jul
FL	Fledged Young	3	15 Jun–12 Aug

DISTRIBUTION

The breeding range of the Common Poorwill extends from southern British Columbia through the western U.S. into Mexico. Early ornithologists described Common Poorwills as common but inconspicuous summer birds on the plains and mountains to 8,000 feet (2440 m). Even allowing for their nocturnal and cryptic behavior, that description seems a bit expansive.

The Atlas map depicts a wide distribution below 8,000 feet (2440 m) on both sides of the Divide, but not on the plains. Atlasers found Common Poorwills regularly in the pinyon/juniper woodlands of southeastern Colorado and the Arkansas Valley. Northward, the Front Range foothills support a significant population; there the birds primarily use ponderosa woodlands and scrub oak. A nest found in central Elbert County (Joe TenBrink VF; TenBrink 1995, 1997) confirmed that poorwills breed in the scattered stands of ponderosas in the Black Forest, east of the previously mapped breeding limit.

Each of the latilongs along the western border of the state had records in more than ten blocks, making that area the center of abundance in Colorado. In northwestern Colorado workers found Common Poorwills in several sagebrush country blocks that had few previous records (A&R).

Even with Atlas reports in more than 200 quads, the birds' nocturnal habits probably have led to underreporting, perhaps by as much as half. Atlasers also seemed somewhat conservative in designating the abundance of this species, although they did report an abundance code of A3 (11–100) in 50% of the blocks.

BREEDING EVIDENCE

REPORTED IN 188 (11%) OF 1745 PRIORITY BLOCKS

Possible	148 blocks	(79%)
Probable	25 blocks	(13%)
Confirmed	15 blocks	(8%)

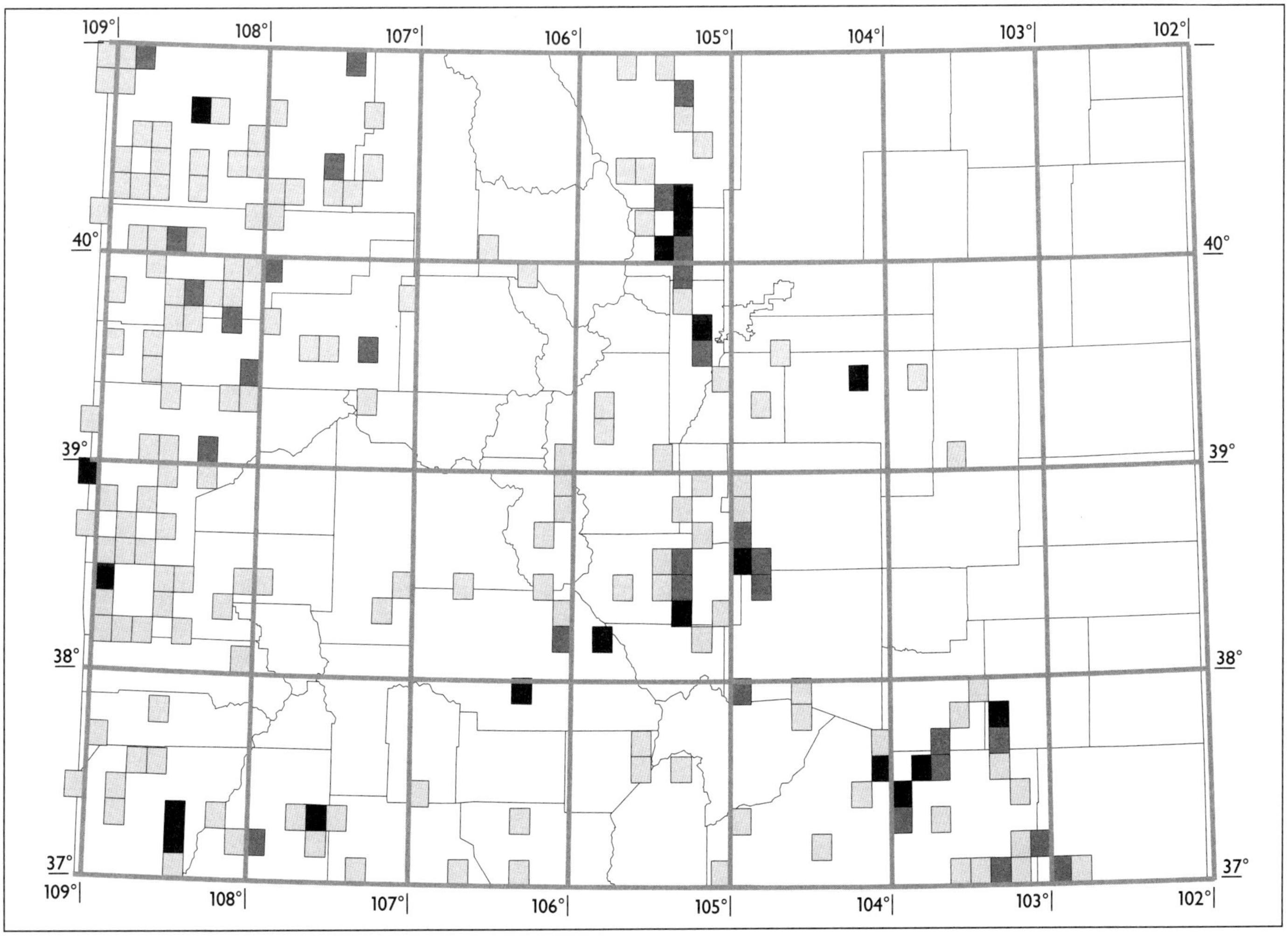
109°
108°
107°
106°
105°
104°
103°
102°
40°
39°
38°
37°

COLORADO DIVISION OF WILDLIFE / WILDLIFE RESOURCE INFORMATION SYSTEM

Black Swift

Cypseloides niger

In a contest for Colorado's most eccentric bird, the Black Swift would rank as a front-runner. Black Swifts stand out for their quirky breeding biology, including a penchant for nesting on rock faces in the coldest, dampest spots they can find. Because of the inaccessibility of nest cliffs, and the difficulty of studying or even properly identifying these high-flying birds, much remains unknown about their distribution and habits (Stiles and Negret 1994).

HABITAT

Black Swifts invariably nest on vertical rock faces, near waterfalls or in dripping caves (Lack 1956). Beyond that requirement, they inhabit a variety of landscapes, from seacoasts to the high fastness of the Rocky Mountains. Black Swifts spend most of the daylight hours on the wing pursuing aerial insects (Lack 1956), often ranging far from nesting areas in search of this abundant but patchy resource (Holroyd 1993). Waterfalls pour down sheer cliffs in mountain canyons close to all nesting colonies found by atlasers, a situation the birds almost always choose in Colorado (Knorr 1961, B&N).

Several Atlas records of Black Swifts flying in blocks without suitable nesting habitat (HEK) substantiate earlier observations (Knorr 1961, B&N) that Black Swifts in Colorado forage over most montane and lowland habitats. They sometimes cruise over the summits of 14,000-foot peaks and over croplands or deserts 25 miles (40 km) from nesting colonies.

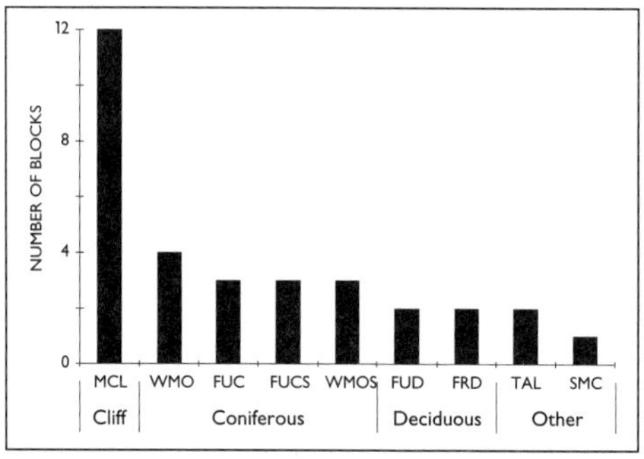
NUMBER OF BLOCKS
0
4
8
12
MCL
WMO
FUC
FUCS
WMOS
FUD
FRD
TAL
SMC
Cliff
Coniferous
Deciduous
Other

BREEDING

Atlasers found Black Swifts nesting in expected circumstances. Nests in a colony near Silver Plume (39105F6; Thom Curdts VF) closely resembled those described by previous observers (Knorr 1961): well-built cups of moss and mud in a niche on sheer rock cliffs, well shaded and bathed in mist from the frigid splash of a nearby cascade.

After arriving in Colorado in June, Black Swifts take all summer to raise a single nestling. The Silver Plume nests held eggs on 19 July, and in Ouray (38107A6) nests held young on 11 September. Earliest atlas egg records came in early July, although diligent observations in 1996 at the Ouray colony ascertained that many eggs appeared there the last week of June, one on 19 June (Sue Hirshman pers. comm; see also Hirshman 1998). The earliest known hatching date is 17 August, and most nestling records fall in early September (B&N, Prather 1991).

Nestlings have weak, tiny feet. With no place to walk on the precipitous cliffs anyway, they stay in the nest until they fledge 45 to 49 days after hatching (Ehrlich et al. 1988, Marin 1997), much longer than other swift species. They probably start on their migration as soon as the young fledge (Marin 1997).

The curious nesting behavior of Black Swifts seems closely tied to their feeding strategy. They seek widely scattered "blooms" of certain aerial insects, particularly flying ants, which leads foraging adults to travel widely. Nestlings spend the day alone, so these swifts nest on cliffs beyond the reach of terrestrial predators. Young also must endure a long day without food: typically, adults return only in the evening with a meal of partly digested insects. Infrared techniques show that they feed the young all night long (Owen Knorr pers. comm.). Faced with day-long fasts, young grow extremely slowly, and may even exhibit torpor, a slowing of metabolism that cold temperatures would aid. Nestling torpor could explain why Black Swifts invariably choose cold, damp nest sites even when dry ledges exist nearby (Holroyd 1993). Another theory postulates that two other factors dictate the choice of nest sites: constant temperatures to ameliorate outside changes and high humidity to aid in attaching the nest to the cliff (Marin 1997).

BREEDING PHENOLOGY

CODE		# OF RECORDS	DATE RANGE
C	Courtship	1	6 Jul
ON	Occupied Nest	1	4 Aug
NE	Nest with Eggs	2	5 Jul–19 Jul
NY	Nest with Young	1	11 Sep

DISTRIBUTION

Black Swifts breed in scattered colonies in western North America, from southeast Alaska to central Mexico, and they migrate to the Neotropics in winter (B&N, Stiles and Negret 1994). In Colorado, Atlas records agree with previous assessments that Black Swifts breed most commonly in the San Juan Mountains (23 of 32 blocks reporting breeding evidence), with scattered colonies in four other mountain ranges—Sangre de Cristo, Flat Tops, Gore, and Front. Knorr (1961) found 27 nesting colonies in ten counties during his thorough study in the 1950s; atlasers saw Black Swifts in all the same areas except one west of Aspen. In view of the challenging (to atlasers) nest sites, most blocks on the map represent breeding colonies. Still, some Possible records may not pertain to breeders because Black Swifts forage so far from nest sites.

Due to their exacting nesting requirements, Black Swifts probably never have been numerous in Colorado (Knorr 1961). Although Black Swifts always congregate to nest, colonies usually hold fewer than ten pairs (Knorr 1961). Atlasers agreed: they estimated abundance at most colonies as 2–10 pairs, and as A3 (11–100 pairs) only at Ouray and Little Bear Peak in the Sangre de Cristo Mountains (37105E4). The statewide total probably does not exceed a few hundred pairs; Atlas abundance codes suggest a population of 700–800 pairs. (See Appendix B for comments on abundance estimates for this species.)

The recent WatchList ranked Black Swift nationally as a High Priority species (Carter et al. 1996). Priorities 1995, setting conservation priorities for Colorado bird species, ranked Black Swifts fifth in the state. Seven factors contributed to this ranking, including the 26-year BBS trend and the hypothesis that at least 20% of all Black Swifts breed in Colorado. The latter theory combined with Atlas data intimates that the total continental population could not exceed 10,000 birds. The map in Price, Droege, and Price (1995) insinuates a lesser number in Colorado, but even if the state has only 10%, then the total number of breeding Black Swifts does not exceed 15,000 and Colorado's colonies constitute a crucial component of the population of these fascinating and unusual birds.

BREEDING EVIDENCE

REPORTED IN 25 (1%) OF 1745 PRIORITY BLOCKS

Possible	19 blocks	(76%)
Probable	4 blocks	(16%)
Confirmed	2 blocks	(8%)

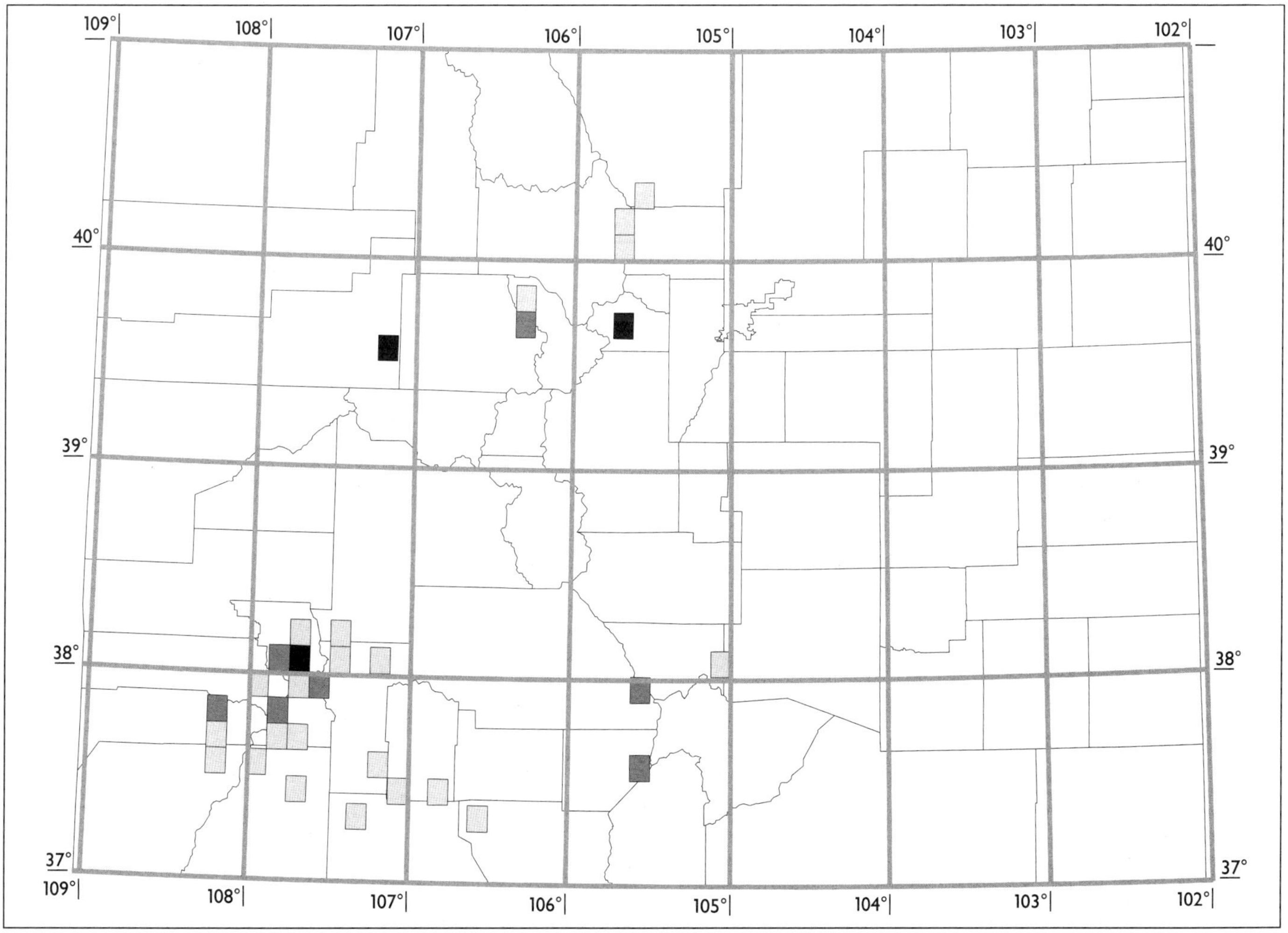
109°
108°
107°
106°
105°
104°
103°
102°
40°
39°
38°
37°

COLORADO DIVISION OF WILDLIFE / WILDLIFE RESOURCE INFORMATION SYSTEM

CHIMNEY SWIFT *Chaetura pelagica*

For a bird associated exclusively with humans, Chimney Swifts keep their nesting biology well concealed from their habitat-providers. Originally these swifts nested in hollow trees, but in Colorado, at least, they inhabit only plains towns, and nest only in manmade structures that mimic hollow trees—mainly chimneys and smokestacks.

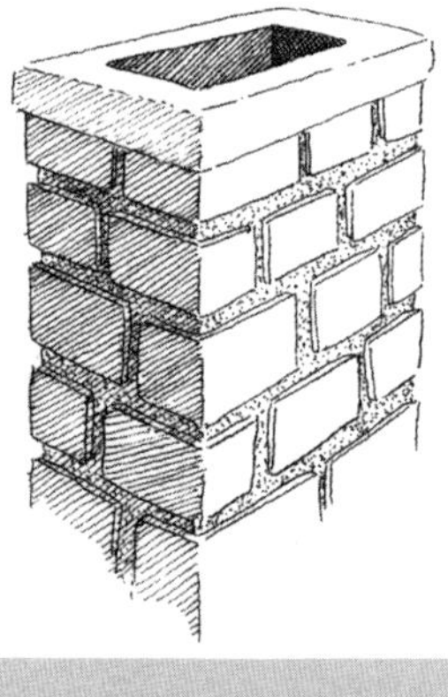

HABITAT

By 1672 Chimney Swifts had already identified chimneys as satisfactory nest sites; yet in 1840, in Kentucky, Indiana, and Illinois, John James Audubon found them roosting in hollow trees. By 1870 they made extensive use of chimneys and buildings, and since 1920 few observers have referred to tree-nesting (Fischer 1958).

In Colorado they rarely stray from urban areas. They inhabit plains towns and Front Range cities, where atlasers often saw them feeding high above streets, homes, and business districts. Chimney Swifts forage widely, and atlasers occasionally saw them feeding over croplands and cottonwood stream bottoms.

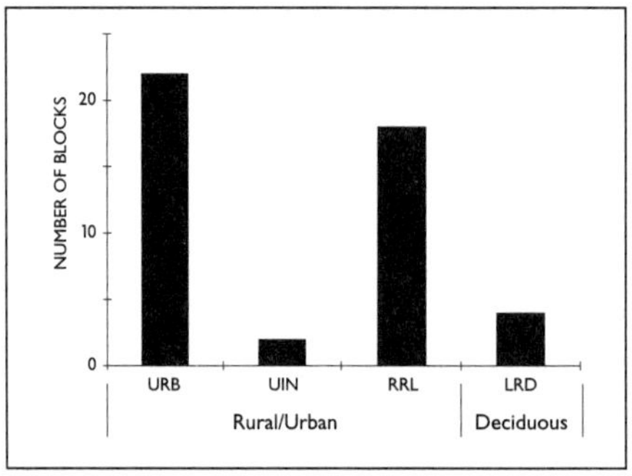
NUMBER OF BLOCKS
0
10
20
URB
UIN
RRL
LRD
Rural/Urban
Deciduous

Their choice of nest site dictates their choice of air space. Because they do not find many hollow trees in Colorado, Chimney Swifts occur exclusively in or near chimney-rich urban areas. In fact Chimney Swift habitat in Colorado may more accurately be described as "the air above our urban areas." As Tyler said, they "never, except by accident, set . . . foot upon one inch of this vast land" (Bent 1940). Chimney Swifts feed, court, and mate in the air. They tend to ignore, or not to notice, humans.

BREEDING

As aerial specialists, Chimney Swifts pair in the air. First a group flies in association; two birds separate for a while, then rejoin the group. They break into a trio (usually two males chasing a female), which cruises around for up to five minutes. Chipping loudly, the three rise to 500 feet (out of sight of binoculars) and execute sudden, angular turns in unison. After pairing they use flight to maintain their bond. The two fly in unison; the rear bird snaps its wings into a V; the leader, who can't see the rear bird, does the same; the follower snaps back to normal sailing position; then the leader follows suit (Fischer 1958).

These aerialists put their nests in dark places. They collect sticks while in flight. Picking sticks off trees with their feet as they circle the tree, they use saliva to plaster the sticks into a small saucer on the vertical nest surface (Fischer 1958).

At the nest Chimney Swifts sometimes have helpers—the third bird usually an offspring of the parents from a previous year. The female lays 3–5 eggs, which only the parents incubate. After the young hatch, both parents and helper feed them by regurgitation. The helper also broods the young, at least after dark. The number of young determines when they leave the nest—at 14 days for broods of 4–5, 19 days for smaller broods. Nestlings don't depart very far—they cling to a wall within a foot of the nest until ready to fly, about 30 days after hatching (Fischer 1958).

Finding nests demands a combination of patience, detective work, and mountain-climbing or chimney-sweep skills. Although they do nest in abandoned buildings, swifts mostly select dark sites typified by their namesake chimneys. Two of the nests atlasers found offered something of an adventure.

For years Janeal Thompson watched swifts swoop in and out of a 12-foot chimney on an old school building in Walsh. Finally she persuaded a roof repair crew to allow her to inspect the chimney. A crew member put a long ladder against the shaky chimney (one brick fell into the abyss). With the aid of a spotlight she detected an old nest structure several feet down from the top of the chimney, thus verifying breeding for the first time in southeastern Colorado.

In June 1995 a homeowner called Jack Wieland, a Division of Wildlife manager in Holyoke, to investigate raccoon or squirrel noises in a stovepipe. He discovered instead several baby Chimney Swifts clinging to the inside of the stovepipe, next to a nest. He took them to a rehabilitator whose treatment fledged two of the six; she fed them on puréed house flies.

BREEDING PHENOLOGY

CODE		# OF RECORDS	DATE RANGE
C	Courtship	1	15 Jun
NY	Nest with Young	1	9 May

DISTRIBUTION

Chimney Swifts nest from New Brunswick west to southeastern Alberta and south to Texas and Florida. Their wintering grounds, unknown until recently, lie in the Amazon Basin of Brazil and Peru, south to northern Chile.

The first Colorado records came from one cruising over City Park in Denver in 1917 and a second in Denver in 1930 (B&N). In 1965, the state had so few records that B&N listed them all and cited only three nesting records, from Denver and Boulder. By the time of the Atlas, however, swifts had become regular on the plains and in Front Range cities. Nevertheless, the two nests found by atlasers are the only ones reported since 1965. The swifts' selection of chimneys as nest sites has brought them to Colorado towns and cities, but only relatively recently. In contrast to the apparent increase in Colorado, BBS trends show a continental decline of 1% per year, a drop of about 25% in 31 years.

Atlasers, by fanning out over the plains, found Chimney Swifts in more plains towns than previously mapped. Swifts undoubtedly nest in all these towns, but observers have not had the fortitude to search for nests.

As urban newcomers to the state, Chimney Swifts presumably have adapted to urban conditions in Colorado, although they have become less conspicuous in recent years (HEK). As adaptable birds they probably will continue to maintain their successful association with the air over human-provided nest sites.

BREEDING EVIDENCE

REPORTED IN 37 (2%) OF 1745 PRIORITY BLOCKS

☐	Possible	31 blocks	(84%)
▩	Probable	3 blocks	(8%)
■	Confirmed	3 blocks	(8%)

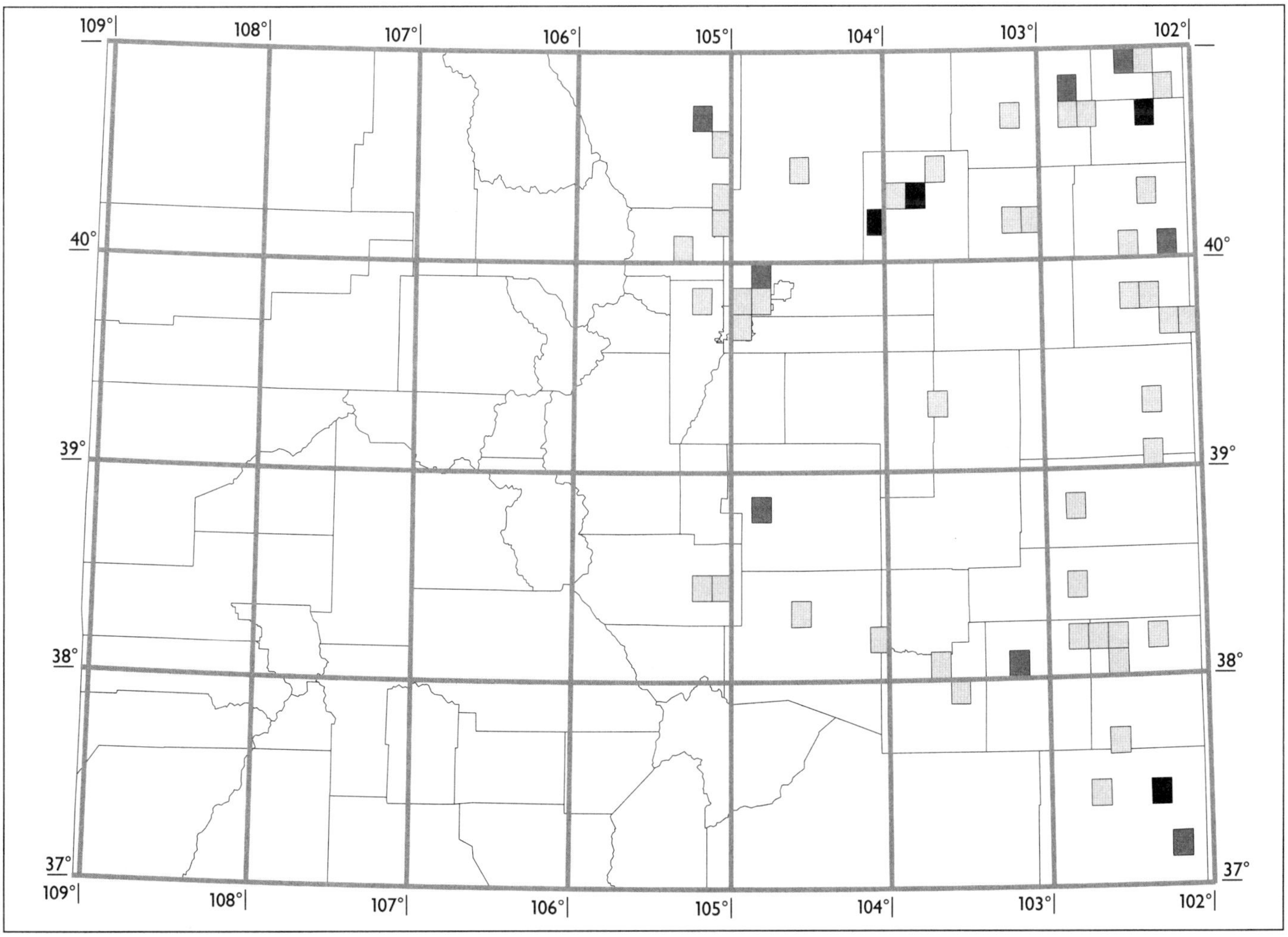
109°
108°
107°
106°
105°
104°
103°
102°
40°
39°
38°
37°

COLORADO DIVISION OF WILDLIFE / WILDLIFE RESOURCE INFORMATION SYSTEM

WHITE-THROATED SWIFT

Aeronautes saxatalis

This bird's Latin name, meaning "airy sailor of the rocks," describes it well. White-throated Swifts, perhaps the fastest North American birds (faster than Peregrine Falcons; Terres 1980), provide a spectacular aerial show against a backdrop of some of Colorado's most dramatic canyon scenery.

HABITAT

White-throated Swifts occupy rocky canyon country, where they spend the day in spectacular flight—dashing, swooping, and cornering as though unbound by gravity. Rock cliffs for nesting determine where White-throated Swifts breed, although they forage over any forested or open country adjacent to nest sites. Atlasers recorded cliffs, rock, or talus for 65% of all observations, and 85% of Confirmed observations that involved sighting nests. Other

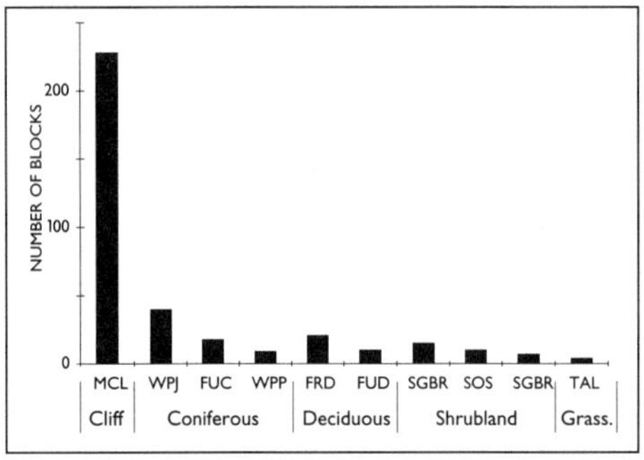
NUMBER OF BLOCKS
200
100
0
MCL
WPJ
FUC
WPP
FRD
FUD
SGBR
SOS
SGBR
TAL
Cliff
Coniferous
Deciduous
Shrubland
Grass.

habitat types recorded by atlasers included a variety of mostly mid- to low-elevation vegetation types characteristic of canyon and foothill regions. These observations indicate the wide range of vegetation types White-throated Swifts will occupy if nesting cliffs exist. Five Atlas records in alpine tundra, including two Confirmed nest sites, verify earlier reports that White-throated Swifts occasionally forage above tree line (B&N), and extend the known Colorado breeding range up to the alpine zone.

BREEDING

White-throated Swifts return to Colorado from their wintering grounds, often to nest sites used previously, in April and May. These dedicated aerialists court and even copulate on the wing, mates pressed together as they pinwheel down through space. White-throated Swifts nest in cracks on tall rock faces, in small colonies of up to about a dozen pairs. Like most swifts, they use saliva to glue feathers, and sometimes a little vegetation, into a rigid nest cup, firmly cemented to the rock surface in a narrow crevice (Bent 1940). Atlasers often saw White-throated Swifts but usually could not Confirm breeding, not surprising because as Bradbury (1918a) noted, the birds usually nest "beyond the depredations of the most enthusiastic oological crank unless the life of the latter is insured for twice its value!" Atlasers more often inferred Probable nesting, especially by watching birds visiting nest sites.

Although White-throated Swifts arrive in Colorado in April and May, they apparently do not nest until summer. Atlas observers saw nest-building in June, incubation in late June, and nestlings in July and August, corroborating reports by B&N of late-June incubation. Atlas nestling records for 10 and 16 August suggest the possibility of second clutches, an observation not yet recorded for this species (Ehrlich et al. 1988).

BREEDING PHENOLOGY

CODE		# OF RECORDS	DATE RANGE
C	Courtship	31	22 May–15 Jul
ON	Occupied Nest	28	9 May–26 Jul
NB	Nest Building	2	13 Jun–23 Jun
NE	Nest with Eggs	1	24 Jun
NY	Nest with Young	9	2 Jul–16 Aug
FY	Feeding Young	1	23 Jun
FL	Fledged Young	1	3 Jul

DISTRIBUTION

White-throated Swifts breed from the mountains of the interior U.S. and coastal California south to El Salvador, and they winter in Mexico and Central America. Historical records blanket the Colorado mountain and plateau regions (Bradbury 1918a, B&N, A&R), and Atlas records indicate that White-throated Swifts remain widespread in those areas, especially on the Colorado Plateau. Atlasers found breeding birds commonly between 5,500 feet (1720 m) and 8,200 feet (2560 m), as reported by A&R, occasionally in the higher mountains to 10,000 feet (3120 m), and rarely above timberline.

White-throated Swifts rarely venture more than a few miles from nest sites during breeding season (A&R), and with few exceptions, the eastern limit of Atlas records closely follows the boundary of the Rocky Mountains, including the canyon country of Las Animas County. Historical records extend southeast to western Baca County (B&N), and atlasers found them nearly that far east. A few scattered breeding populations occur around bluffs and buttes at the north edge of Colorado's Great Plains. A single Atlas record for Kirchnavy Buttes

(40103H3) lies well east of the breeding range described by A&R or B&N, and breeding records exist for Pawnee Buttes (HEK).

Atlasers usually estimated abundance as 11–100 pairs per block, and slightly less often 2–10 pairs, suggesting that many blocks contained more than one nesting colony. A calculation using the Atlas abundance estimates suggests that the state has about 70,000 breeding pairs. Their abundance in Colorado notwithstanding, BBS surveys indicate a continent-wide decline of 2.5% annually. Such a decline could result from pesticide intake or land-use changes in winter range habitat. Aerial insect-feeders can quickly ingest some kinds of pesticides with their food; B&N described a rapid die-off of over 200 White-throated Swifts near Colorado Springs in 1964, probably due to pesticide poisoning.

BREEDING EVIDENCE

REPORTED IN 303 (17%) OF 1745 PRIORITY BLOCKS

☐	Possible	199 blocks	(66%)
▪	Probable	66 blocks	(22%)
■	Confirmed	38 blocks	(12%)

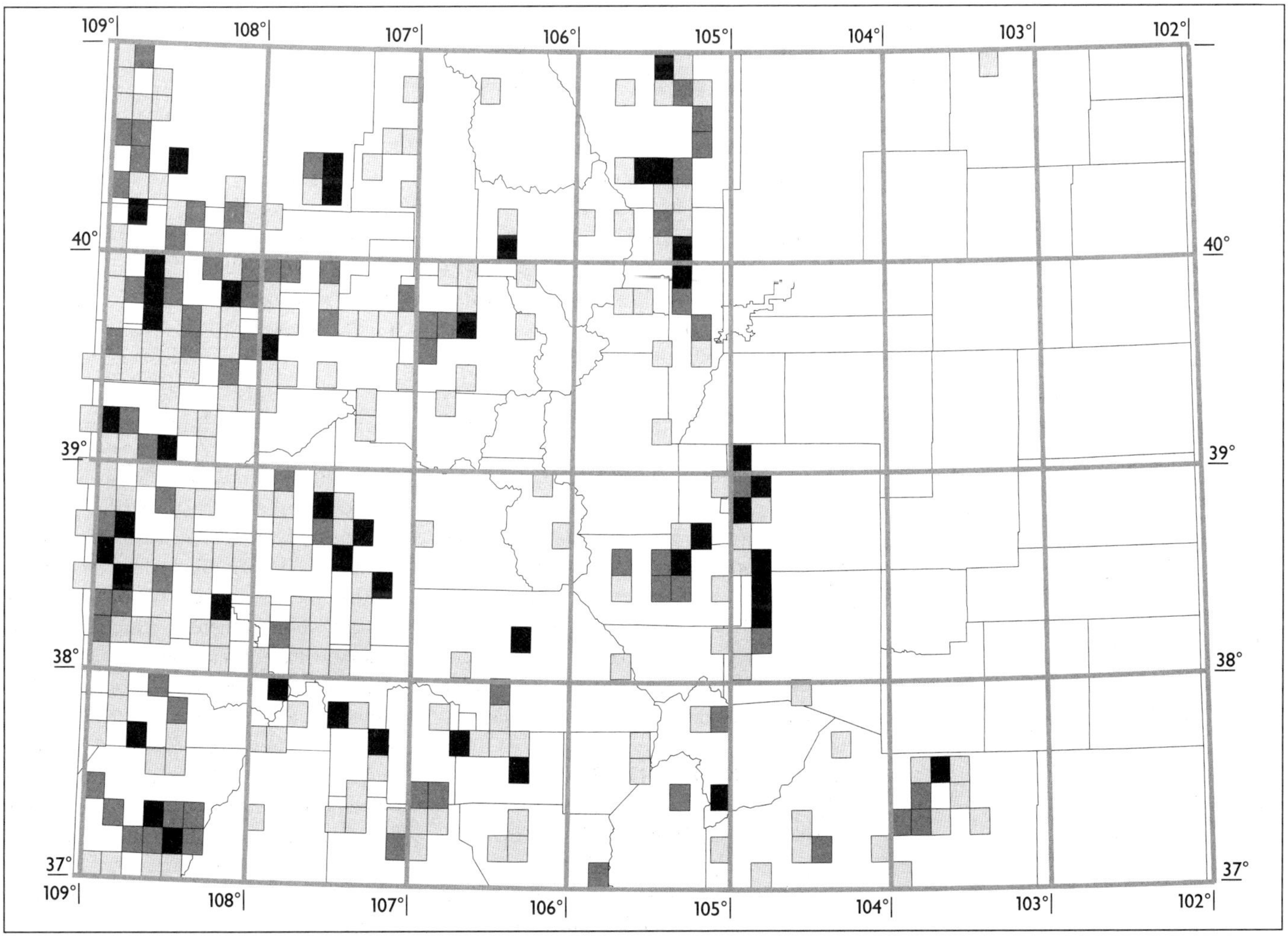
109°
108°
107°
106°
105°
104°
103°
102°
40°
39°
38°
37°

COLORADO DIVISION OF WILDLIFE / WILDLIFE RESOURCE INFORMATION SYSTEM

Hummingbirds can hardly be described without superlatives. Turn-of-the-century ornithologist Robert Ridgway found no other group of birds "so varied on form, so brilliant in plumage, and so different in their way of life." Black-chinned Hummingbirds hover at feeders or garden flowers, green iridescent feathers flashing in the sun, before they dash away to a favorite perch or to tend young at the nest.

HABITAT

Hummingbirds intrigue scientists and naturalists with their unusual adaptations for a nectar-feeding lifestyle. The ability to hover, thanks to unusual shoulder anatomy and wing beats reaching 80 per second, permits hummingbirds to feed from flowers they otherwise could not reach. Hummingbirds also co-evolved with plants in intricate evolutionary dances. Many familiar wildflowers specialize in attracting hummingbirds; they reward hummers with nutritious nectar in exchange for carrying out crucial pollination.

Throughout the desert Southwest, Black-chinned Hummingbirds favor arid and semi-arid lowlands and mid elevations, especially deciduous vegetation of riparian forests, shelterbelts, orchards, and suburban greenways (Bent 1940, Johnsgard 1983).

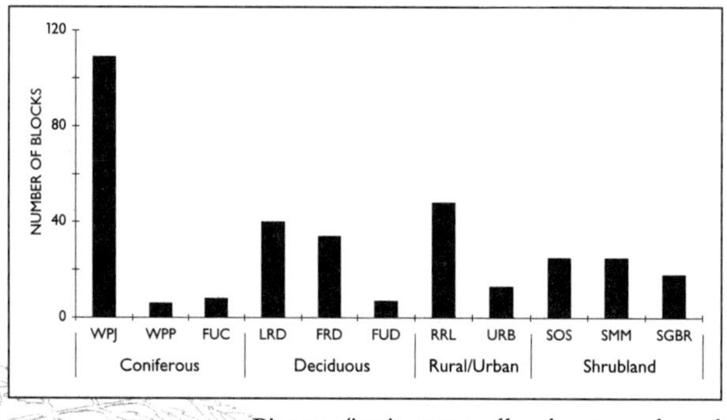

NUMBER OF BLOCKS
120
80
40
0
WPJ
WPP
FUC
LRD
FRD
FUD
RRL
URB
SOS
SMM
SGBR
Coniferous
Deciduous
Rural/Urban
Shrubland

Pinyon/juniper woodland outnumbered any other habitats reported by atlasers, although they reported riparian (low plus mid-elevation) almost as frequently. As a whole, deciduous habitats, including riparian, urban, and rural situations (including shelterbelts, orchards, and manmade structures), account for 64% of the habitats; these plus lower-elevation brushy and urban habitats total 84%. Atlasers did not find Black-chins in large areas of mostly treeless brush such as the Wyoming Basin. Thus, Atlas habitat records suggest that open brush, meadows, and fields serve as foraging areas, but that the birds require trees or other cover for nest sites. Brown (1992) reported similar findings from the Grand Canyon, Arizona, where Black-chinned Hummingbirds nested in tamarisk but foraged in adjacent desert brushlands.

BREEDING

Migrants begin returning in the last few days of April (B&N). Males arrive first, to set up territories that they defend vigorously (Bene 1947). Beginning in late May, atlasers observed the notable courtship display—the male executes a spectacular pendulum-like swooping flight over the perched female—although the hummingbirds probably began their courtship before many atlasers went afield.

The tiny nests, made of tightly felted plant down and spiderwebs, resemble a globe with the top third missing (Bent 1940). Although atlasers recorded nest-building primarily in June, state records exist as early as 6 May (B&N), perhaps again reflecting less Atlas field work in May. Black-chins usually place the nest 3–10 feet (1–3 m) high under cover of dense foliage, over a streambed, or beneath an overhanging rock or porch roof (Johnsgard 1983, nest cards).

Two (rarely three) eggs appear starting in late May. Females often produce more than one clutch per season (Ehrlich et al. 1988), and Cogswell (1949) found one busy female tending two nests at once. Atlasers found evidence of multiple clutches, noting young in nests throughout the summer months and observing courtship flights, which continue through egg-laying (Bene 1947), into midsummer.

BREEDING PHENOLOGY

CODE		# OF RECORDS	DATE RANGE
C	Courtship	18	21 May–18 Jul
NB	Nest Building	6	28 May–25 Jun
ON	Occupied Nest	18	16 May–14 Jul
NE	Nest with Eggs	3	23 May–24 Jun
NY	Nest with Young	6	7 Jun–9 Aug
FY	Feeding Young	5	9 Jun–3 Aug
FL	Fledged Young	12	9 Jun–8 Aug

DISTRIBUTION

The breeding range covers western North America from British Columbia to northern Mexico, but the greatest densities occur in the desert Southwest (Johnsgard 1983). Black-chins migrate to Mexico for the winter. Within Colorado, atlasers found them breeding in the Colorado Plateau and tributary valleys, San Luis Valley, and the southern foothills of the Front Range southeast to the canyons and plateaus of Las Animas County. They come to feeders in many Western Slope cities, as well as Pueblo and vicinity. Atlasers did not find Black-chins in the higher mountain ranges or Great Plains.

B&N suggested that they occurred mostly on the Western Slope even while mentioning several specimens taken from the southern Front Range. Cooke (1897) reported that this hummingbird occurred in western Colorado, but early naturalists probably missed it in southeastern Colorado. Blocks with the highest abundance estimates lie primarily in the Grand Valley (Mesa County) and the lowlands of the southwestern counties, but also from Eastern Slope Las Animas County.

Reported nest densities in southwest riparian sites range from 0.2/acre (0.4/ha; Baltosser 1986) to up to 1.6/acre (3.5/ha; Brown 1992). Atlas workers most often assigned abundance category A3 but half as often chose A2, suggesting a perception of Black-chins as fairly common but not abundant. Previous works (Cooke 1897, B&N) echo this assessment, suggesting that no substantial changes in abundance have occurred during this century.

Because Black-chins use plantings and structures in farms, gardens, and cities for foraging and nesting, they should be less vulnerable than some other species to habitat changes, although pesticides or other toxins could cause serious harm. The splendid beauty and fascinating behavior of hummingbirds provide ample reasons to keep rural and suburban environments safe for wildlife.

BREEDING EVIDENCE

REPORTED IN 242 (14%) OF 1745 PRIORITY BLOCKS

□	Possible	142 blocks	(59%)
■	Probable	52 blocks	(21%)
■	Confirmed	48 blocks	(20%)

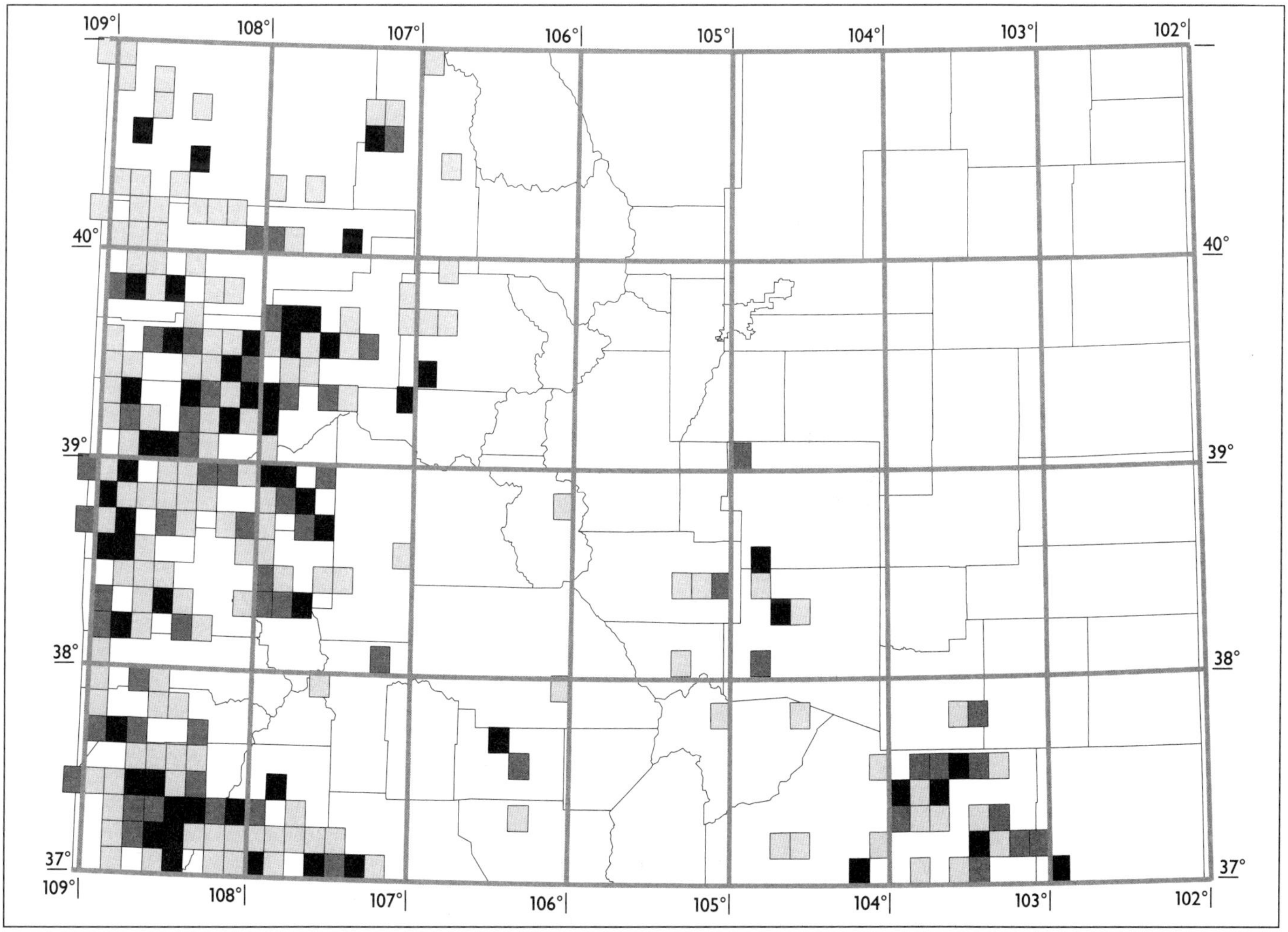
109°
108°
107°
106°
105°
104°
103°
102°
40°
39°
38°
37°

BROAD-TAILED HUMMINGBIRD *Selasphorus platycercus*

When hummingbirds return each spring, thousands of Coloradans refill nectar feeders and wait for the show to start. These tiny but audacious birds provide endless entertainment in flower-decked meadows or at mountain feeders with their dazzling iridescent plumage and pugnacious behavior. This common breeding hummingbird of the Colorado Rockies uses decidedly uncommon adaptations to survive. Hummingbirds consume food equal to their weight each day to feed their roaring metabolic furnaces, and they sometimes use hypothermic torpor, a controlled drop in body temperature, to conserve energy (Hainsworth et al. 1977, Calder 1994). Highly aggressive and territorial, hummingbirds compete intensely for rich but patchy and unpredictable nectar sources.

HABITAT

Broad-tails breed in montane forests, and shrublands provided some forest is nearby (Calder and Calder 1992). Atlasers recorded 37 habitat types, indicating a wide diversity of sites used for breeding and foraging. The 18 categories most frequently listed (93% of observations) include foothills and montane forest and shrub habitats, especially lower-elevation and more open habitats. Conifers and shrublands each composed one-third of the habitat reports and deciduous types totaled one-quarter.

B&N and A&R said that Broad-tailed Hummingbirds reach their greatest breeding densities in ponderosa pine, but atlasers recorded aspen most frequently, twice as often as ponderosa pine, and with greater densities. Field workers noted A4 and A5 abundance codes three times as often in aspen as in ponderosa (53:15). Foothills riparian, mountain shrublands, and pinyon/juniper woodlands also outnumbered ponderosa reports.

Atlasers listed rural and urban habitats in 6% of their reports, indicating only in part the broad use hummingbirds make of gardens, parks, and ranch yards. A few Broad-tailed Hummingbirds nest in eastern Colorado cottonwood river bottoms, close to the foothills. Sixteen Atlas nest cards reported elevations from 5,200 to 10,720 feet (1620–3350 m), with half between 7,000 and 8,500 feet (2180–2650 m).

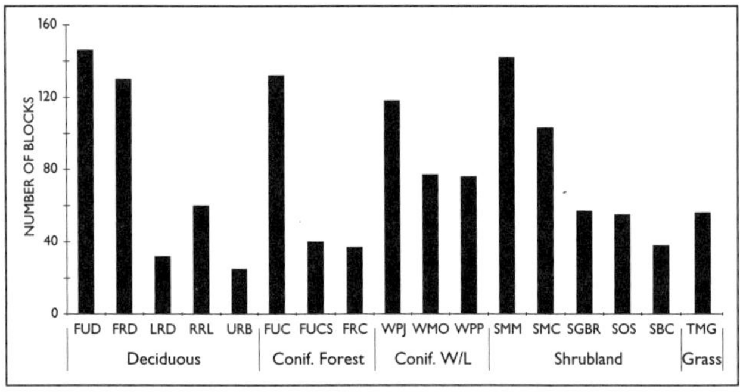

NUMBER OF BLOCKS
160
120
80
40
0
FUD FRD LRD RRL URB FUC FUCS FRC WPJ WMO WPP SMM SMC SGBR SOS SBC TMG
Deciduous
Conif. Forest
Conif. W/L
Shrubland
Grass

BREEDING

During the nesting season, females face many difficulties in meeting energy needs in an often harsh climate, and this dictates much of the species' breeding biology. Males arrive in early May, before the females, but do not firmly establish territories at higher elevations until enough wildflowers bloom (Calder and Calder 1992). Breeding and nest initiation coincide closely with peak flowering of nectar sources (Waser 1976) and consequently vary from year to year with weather (Calder 1981).

Males engage in spectacular courtship flights over perched females. Males mate promiscuously and vigorously defend their territory from other males until mating opportunities decline in mid July. Male territory size averaged 0.9 acre (0.4 ha) over several years near Gothic, and varied with food availability (Miller and Inouye 1983).

Of courtship observations recorded by atlasers, 90% occurred between 20 May and 14 July. Some of these records, however, may represent observations of territorial defense flights, which resemble courtship displays.

Females and males pursue different strategies during the breeding season: females attempt to raise young, whereas males try to mate as often as possible and participate not at all in the nesting process. Consequently, male and female energy requirements and habitat selection differ.

With increased nutritional needs from nest-building and care of young, females tend to nest in valleys where they find nectar sources more dependable than on higher mountain slopes (Calder and Calder 1992). Males roam to higher slopes where they can often roost at warmer temperatures above nighttime cold-air sinks in valleys. Females select nest sites carefully to minimize nighttime heat loss; generally they choose a branch under another branch or overhanging foliage (Calder 1973). Atlasers found 11 nests on conifer branches, usually spruce or fir, and five in deciduous bushes, vines, or trees.

Females provide the sole care for their clutch of two; they contend with freezing temperatures, food shortages from storms and drought, high predation, and their own considerable energy demands. Despite the adults' nectar-centered diet, females feed nestlings insects (e.g., gnats, flies, aphids), and observations of chick droppings suggest that the nestling diet includes little (if any) nectar (Calder and Calder 1992). This explains how these birds can breed in the flower-poor plains stream bottoms close to the mountains.

Atlasers' observations of nestlings peaked in the last third of July before dropping sharply, probably reflecting reduced field work in August. At higher elevations, the brief summer limits Broad-tailed Hummingbirds to a single clutch, but Bailey

(1974) observed a second clutch near Denver and a female fed a fledgling in Chatfield State Park (39105E1) on 9 September 1992 (HEK).

BREEDING PHENOLOGY

CODE		# OF RECORDS	DATE RANGE
C	Courtship	183	6 May–30 Jul
NB	Nest Building	27	6 May–18 Jul
ON	Occupied Nest	45	22 May–24 Jul
NE	Nest with Eggs	10	6 Jun–4 Jul
NY	Nest with Young	27	15 May–10 Aug
FY	Feeding Young	13	9 Jun–23 Jul
FL	Fledged Young	52	1 Jun–9 Sep

DISTRIBUTION

Broad-tailed Hummingbirds breed in the continental cordilleras from northern Wyoming south to Guatemala, and winter from central Mexico south (Calder and Calder 1992). Colorado atlasers found them breeding throughout the foothills and mountains up to about 10,300 feet (3230 m), as reported by earlier authors (B&N, A&R). (The range map in Calder and Calder [1992] omits half the Colorado breeding range.) The showy, fearless birds usually make themselves obvious, and Atlas workers found them in most mountain blocks. Atlasers recorded abundance category A3 most often (and A4 as often as A2), confirming previous descriptions of Broad-tails as common in Colorado (Cooke 1897, Sclater 1912). Long-term studies at Gothic yielded a density estimate of one pair/5 acres (1 pair/2 ha; Calder and Calder 1992).

Observers have noted population fluctuations from year to year. Feeder observers frequently notice changes that may occur because of fluctuations in natural food sources or activity at other feeders. No evidence exists of historic changes in numbers or distribution (Calder and Calder 1992). The popularity of hummingbird feeders may maintain unnaturally high hummingbird populations during times of food shortage such as droughts, and it may extend the distribution of hummingbirds early and late in the breeding season.

BREEDING EVIDENCE

REPORTED IN 865 (50%) OF 1745 PRIORITY BLOCKS

☐	Possible	368 blocks	(43%)
▩	Probable	314 blocks	(36%)
■	Confirmed	183 blocks	(21%)

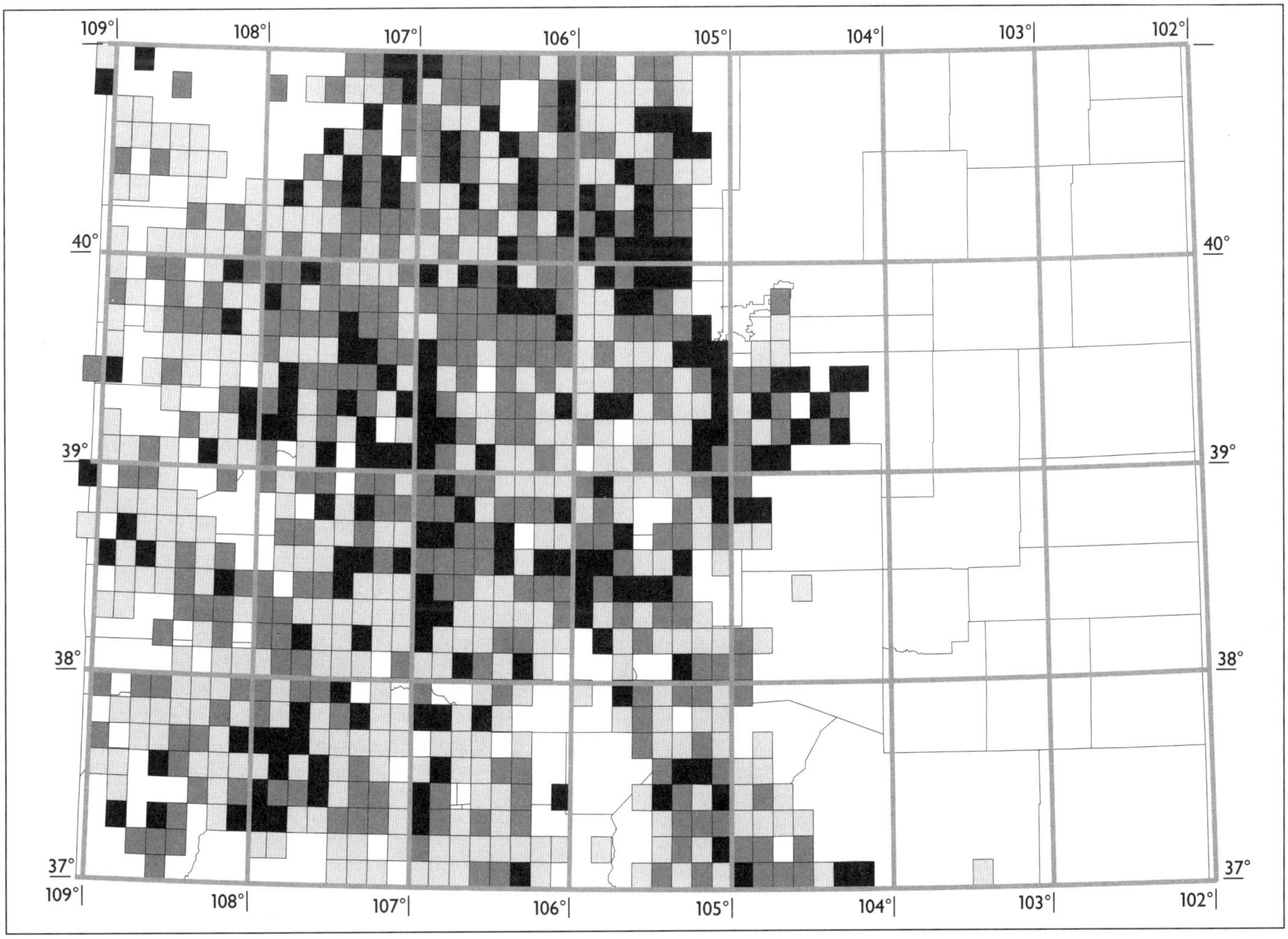
109°
108°
107°
106°
105°
104°
103°
102°
40°
39°
38°
37°

COLORADO DIVISION OF WILDLIFE / WILDLIFE RESOURCE INFORMATION SYSTEM

Like raptors, Belted Kingfishers regurgitate pellets, which accumulate under favorite perches. Examining bones and fish scales in pellets reveals more than the birds' dinner menu—even the ages of fish prey can be read from scale rings. The loud rattling calls of kingfishers are one of the characteristic sounds of Colorado waterways. Intently scanning for their primary prey, fish, they hunt from conspicuous perches over water. Always alert, kingfishers usually call loudly at the first sign of human intruders, making the birds (or intruders) easy to detect.

HABITAT

Riparian habitats composed 92% of habitat observations. Field workers found kingfishers on streams twice as often as on lakes, and found nests on streams more than four times as often. Streams apparently provide more concentrated food resources, because kingfishers hold smaller territories on streams than on lakes (Hamas 1994). The birds require clear, slow-moving water for hunting. Riffles provide the most dependable fishing. In Ohio, smaller territories have proportionately more riffles than larger territories (Davis 1982). On lakes, kingfishers prefer shallow coves (Hamas 1994). Availability of nest sites and perch cover may limit the birds' use of many reservoirs in unforested parts of Colorado.

Kingfishers excavate a nest burrow in a steep earth bank, usually close to water, but occasionally at some distance away (as much as half a mile) in a ditch bank, gravel pit, or roadcut (Hamas 1994), e.g., a cutbank along U.S. 285 west of Bailey (39105D5, Shawnee).

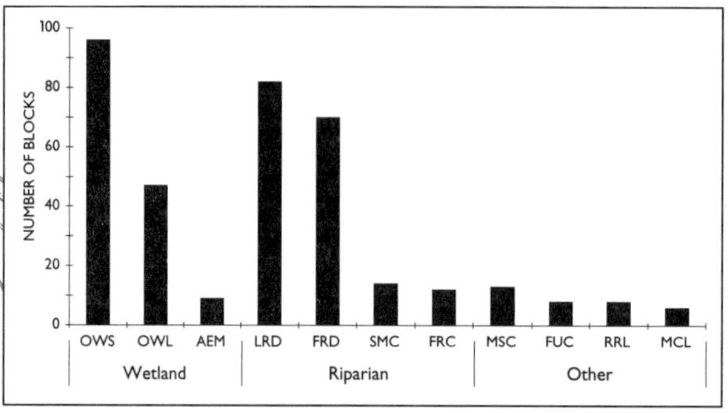
NUMBER OF BLOCKS
100
80
60
40
20
0
OWS OWL AEM LRD FRD SMC FRC MSC FUC RRL MCL
Wetland
Riparian
Other

Kingfishers eat mainly small fish, less than 6 inches long. Along their nesting streams and lakes they must use all their expert skills at fish catching, although to the dismay of some hatchery managers they sometimes zero in on vulnerable trout-rearing ponds (Bent 1940). They sometimes prey on other food as well: the list ranges from other aquatic denizens to snakes, grasshoppers, butterflies and other insects, mice, and even berries. In 1979 one spent a week (with unknown success) trying to spear Pine Siskins in Ron Ryder's backyard in Fort Collins (*Am. Birds* 34:295).

BREEDING

After staking out a breeding territory in spring, males attract and court a mate, then the pair defends the territory against other kingfishers with noisy chases (Davis 1980). Courtship includes ritual feeding of the female by the male, and both partners begin digging the nest burrow soon after. Atlas sightings of courtship and nest-building show that some Colorado kingfishers begin nesting in early May. Understandably, atlasers did not record any nests with eggs, which usually cannot be seen easily in the 3–7-foot (1–2 m) deep burrows. Two furrows, like inverted railroad tracks, characterize nest burrows. The birds make the tracks as they enter and exit the burrow.

Kingfishers need about 50 days to nest (22 days to incubate and 28 days to fledge; Hamas 1975), which limits Colorado birds to a single clutch of 6–7 eggs per season. The young hatch, according to Bent (1940), "naked, blind, and helpless, a shapeless mass of reddish flesh with a huge conical bill." Dates that atlasers reported for nests with young, fledglings, and feeding young (NY, FL, CF, FY) suggest that most Colorado nestlings fledge by the first half of July. Parents sporadically feed fledglings for about three weeks, then parents and young tend to disperse to a solitary existence until the next breeding season (Hamas 1994).

BREEDING PHENOLOGY

CODE		# OF RECORDS	DATE RANGE
C	Courtship	6	4 May–13 Jun
NB	Nest Building	1	1 May
ON	Occupied Nest	10	23 May–30 Jun
NY	Nest with Young	3	9 Jun–27 Jun
FY	Feeding Young	33	28 May–17 Aug
FL	Fledged Young	12	5 Jun–4 Aug

DISTRIBUTION

Belted Kingfishers breed throughout most of North America, and winter from the U.S.-Canada border to South America (Hamas 1994). Colorado harbors birds in all seasons; some breeders may remain in winter, but they retreat to the lowest elevations where open water persists reliably (A&R). B&N reported that Colorado receives migrants from the north in winter, and fewer kingfishers occur in the state in winter than in summer (A&R).

Atlas results agree with B&N and A&R that kingfishers occur primarily in western valleys and along Great Plains watercourses. Atlasers did not find kingfishers in the Wyoming Basin, high mountains, dry parts of the Four Corners, the Great Plains away from principal rivers, or most of South Park and the San Luis Valley, all of which lack either water or nest sites or both.

B&N stated that kingfishers "seem more numerous" in eastern counties than on the Western Slope. Atlasers, however, found them breeding in more blocks west of 106 degrees longitude than east (151 to 89), and gave slightly higher abundance estimates to western blocks (mean abundance codes 1.56 cf. 1.47). Statewide, atlasers rated abundance codes A1 and A2 equally and rarely used A3. Published densities range from 0.3 pair/mile of stream (0.5/km) in Ohio to 2 pairs/mile (3/km) in New Brunswick (White 1953, Davis 1982).

Conservation of kingfishers depends on maintenance of water quality, perching substrates, and adequate nest sites (Prose 1985). Toxicity studies have found little evidence that water-borne pollutants or DDT have affected kingfisher productivity (Fox 1974, Hamas 1994). Nevertheless, BBS data indicate a substantial continent-wide decline of 2.2% annually, which suggests that in 31 years the population has dropped in half. In the past, kingfishers often drew persecution from anglers and fishery managers who viewed them as competitors for trout stocks (Salyer and Lagler 1946). In the future, human population pressures may provide the greatest threat, through habitat degradation and inadvertent disturbance of nesting birds.

BREEDING EVIDENCE

REPORTED IN 284 (16%) OF 1745 PRIORITY BLOCKS

Possible	165 blocks	(58%)
Probable	60 blocks	(21%)
Confirmed	59 blocks	(21%)

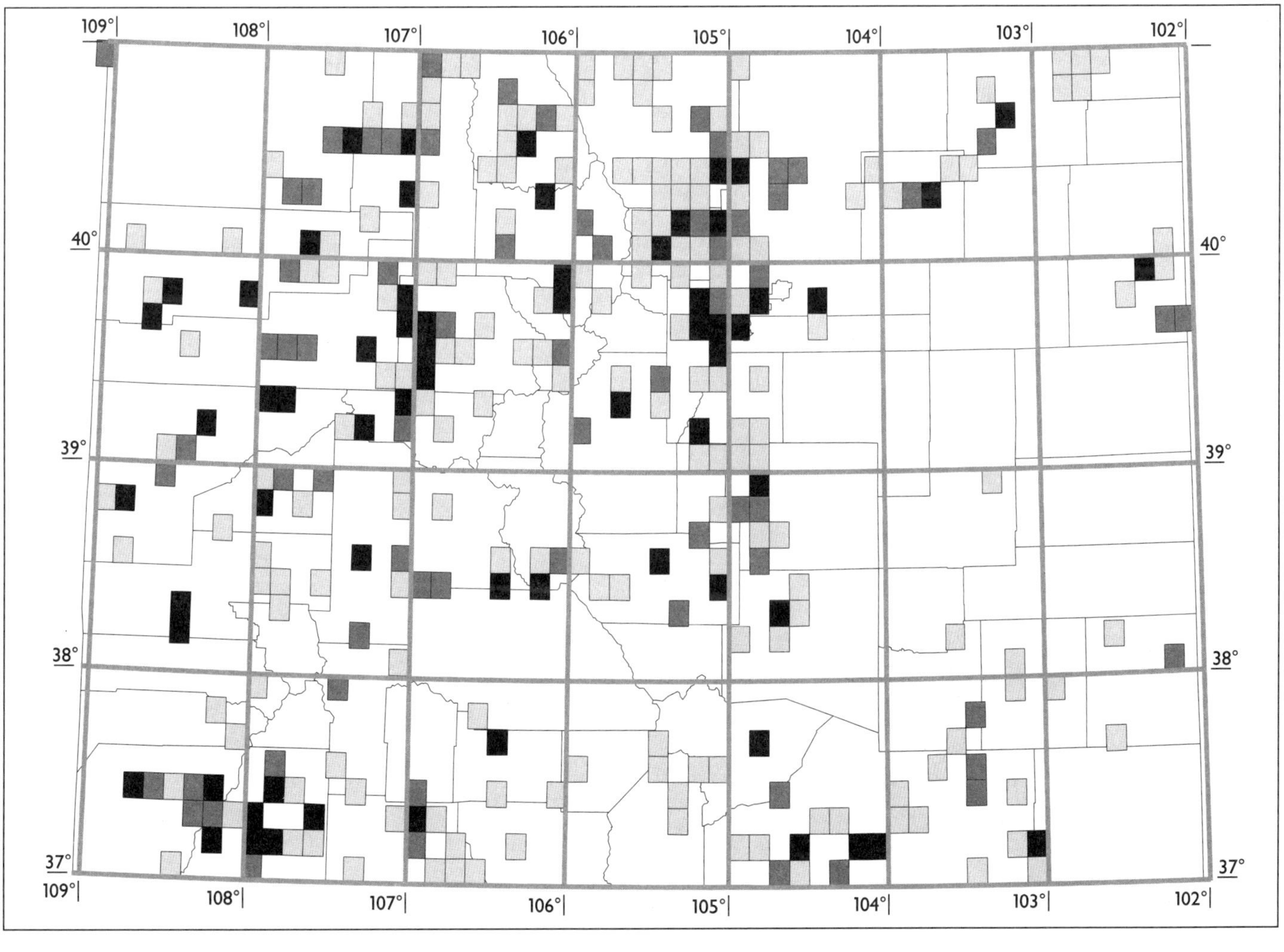
109°
108°
107°
106°
105°
104°
103°
102°
40°
39°
38°
37°

COLORADO DIVISION OF WILDLIFE / WILDLIFE RESOURCE INFORMATION SYSTEM

"Different" describes the rosy-breasted Lewis's Woodpecker. Taxonomists formerly classified it in a genus of its own, *Asyndesmus*, because the melanistic plumage obscured its taxonomic affinities. Crow-like flight, aerial acrobatics, expert flycatching, and weaker skull and bill structure distinguish Lewis's from most other woodpeckers. The only woodpeckers to perch on wires (Ehrlich et al. 1988), they sometimes even soar with swallows to catch insect meals.

HABITAT

During the breeding season Lewis's Woodpeckers feed almost exclusively on emergent insects rather than on the grubs like those that other woodpeckers dig from trees (Bock 1970). As woodpeckers that specialize in flycatching, they need open habitats for their foraging methods to succeed. They prefer open pine forests, burnt-over areas with abundant snags and stumps, riparian and rural cottonwoods, and pinyon/juniper woodlands (Johnsgard 1986b). In 1914 in Colorado they were primarily mountain birds preferring ponderosa pine habitat for breeding (Aiken and Warren 1914). By 1959 they had expanded their nesting onto the plains as cottonwoods matured in stream bottoms and around farms (Knorr 1959). On the plains they avoid riparian habitats if Red-headed Woodpeckers—fly-catching competitors—already occupy them (Bock et al. 1971).

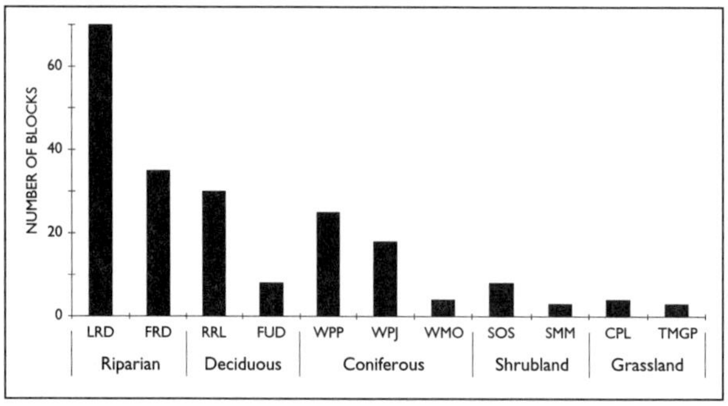
NUMBER OF BLOCKS
0
20
40
60
LRD
FRD
RRL
FUD
WPP
WPJ
WMO
SOS
SMM
CPL
TMGP
Riparian
Deciduous
Coniferous
Shrubland
Grassland

By 1996, data from Atlas work show them breeding preponderantly in riparian habitats where they nest in old decadent cottonwoods. They still use ponderosa pine and pinyon/juniper woodlands in the foothills, but not nearly as much as they use riparian areas. Yet, most of the riparian areas they pick lie within sight of pinyon/juniper or ponderosa zones. Two exceptions to this proximity exist: a small enclave in the southern San Luis Valley and a large population in the Arkansas River Valley downstream from Pueblo. This latter group apparently has moved down-valley in the past 30–80 years as the cottonwoods have matured (B&N, Hadow 1973) and extensive corn plantings provided winter mast (Vierling 1997).

B&N cited several high-elevation records in places where it would surprise today's birdwatchers to find them. Lewis's Woodpeckers wander in fall in search of fruits and berries before settling into winter habitat; these high-elevation birds probably were post-breeding nomads (Carl Bock pers. comm.).

BREEDING

The comparatively weak skull and bill dictate these woodpeckers' choice of dead wood, trees weakened by fire, and natural cavities for their nesting sites (Sclater 1912). They often use previously excavated holes of other woodpeckers, but they drill their own holes when they find suitably soft woods. They form permanent pair bonds and show strong nest fidelity (Bock 1970). Pairs sometimes use the same holes from year to year (Short 1982). After the nesting season, when insects are no longer available, many Lewis's do not migrate to warmer climates but rather move to different localities. They feed on berries, corn, and acorns, which they cache in holes and crevices. Some Arkansas Valley birds move upstream after breeding, to the foothills of the Wet Mountains. Others stay near their nest sites where they store corn mast in old trees (Vierling 1997).

Males and females use separate sleeping holes. When the breeding season begins, in mid April according to Atlas dates, the female may choose the male's hole for the nest if that hole is sounder (Skutch 1985). Males incubate and care for the young as much as females do. The birds become extremely agitated when danger approaches and will desert the nest if under observation too long. However, the noisy young made it fairly easy for atlasers to Confirm breeding using the NY code. Young woodpeckers venture from the nest and climb about on nearby limbs just prior to flying; rarely, a kestrel may pick off one (Bock 1970).

BREEDING PHENOLOGY

CODE		# OF RECORDS	DATE RANGE
C	Courtship	1	17 May
N	Nest Building	4	16 Apr–27 Jun
ON	Occupied Nest	25	15 May–15 Jul
NE	Nest with Eggs	1	13 Jun
NY	Nest with Young	18	8 Jun–16 Jul
FY	Feeding Young	42	22 May–30 Jul
FL	Fledged Young	10	6 Jun–4 Aug

DISTRIBUTION

Lewis and Clark collected the first specimen of Lewis's Woodpecker in 1806 in Idaho. Strictly a species of western North America, these woodpeckers breed from Colorado west to the Pacific and from southern British Columbia to Arizona and New Mexico.

In 1900 the Colorado range differed significantly from that of today. Lewis's Woodpeckers have changed their range as man has changed the landscape. Back then they did not breed east of the foothills in the Arkansas Valley or in Baca County. They did breed on the Western Slope north to Steamboat Springs. After 1910 they started colonizing eastward onto the plains along the Arkansas River.

The Atlas map, like A&R, shows the southern Colorado birds concentrated in three main clumps: the Arkansas River watershed, the pinyon/juniper country of Las Animas and Huerfano counties, and the San Juan Basin. North of the San Juans they maintain significant breeding areas up to Grand Junction, in the Black Forest northeast of Colorado Springs, and along the edge of the Front Range from Denver to Wyoming.

Twice as many BBS routes show decreases as increases, but the continental BBS trend, which hints at a decline, lacks statistical validity. Relying on the BBS data and threats to the breeding grounds, Partners in Flight included Lewis's Woodpecker on its WatchList (Carter et al. 1996). The short duration of the BBS (31 years) may fail to account for the longer-term expansion, and Colorado BBS routes necessarily follow roads that largely traverse, at best, only the edges of the new riparian haunts of these woodpeckers.

Lewis's Woodpeckers do well where insects, especially grasshoppers, abound, so that the use of insecticides affects their well-being. These woodpeckers need open areas in which to forage; fire suppression, especially in ponderosa woodlands, may well have removed the essential openness of the ponderosa woodlands that these woodpeckers favor. This, along with their discovery of mature cottonwoods on the plains, may have pushed this species to new breeding areas and new wintering sites.

BREEDING EVIDENCE

REPORTED IN 167 (10%) OF 1745 PRIORITY BLOCKS

Possible	49 blocks	(29%)
Probable	27 blocks	(16%)
Confirmed	91 blocks	(54%)

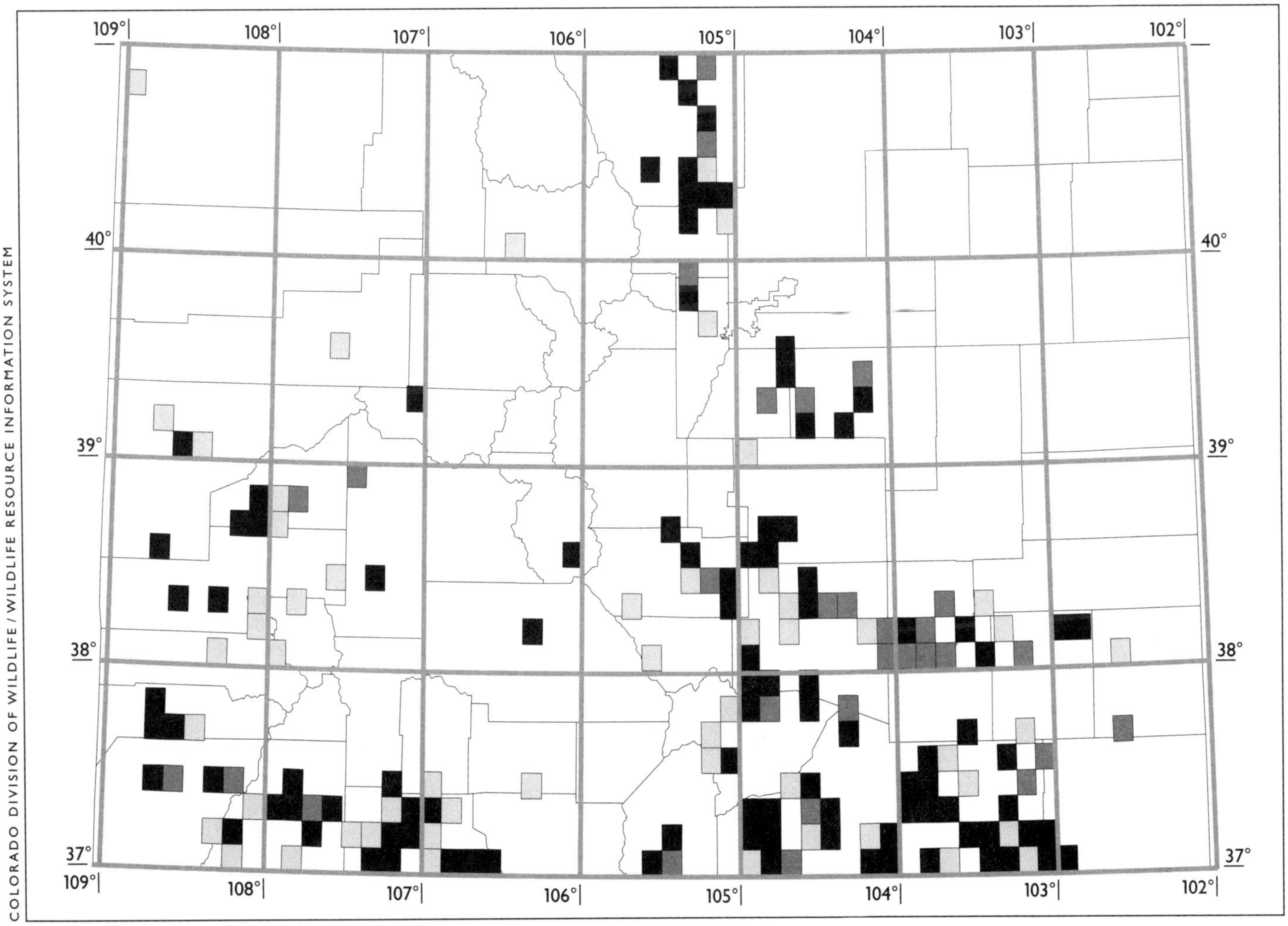
109°
108°
107°
106°
105°
104°
103°
102°
40°
39°
38°
37°

COLORADO DIVISION OF WILDLIFE / WILDLIFE RESOURCE INFORMATION SYSTEM

RED-HEADED WOODPECKER *Melanerpes erythrocephalus*

As if dressed for their own coronations, Red-headed Woodpeckers flash about their sparsely wooded habitats with all the pomp and circumstance of royalty. Their regalia includes a blood-red hood seemingly textured like velvet, and a white suit accented smartly with a short, black cape and black "tails." Mate chases and territorial battles among these birds delight observers not only with their color, but with the sheer lack of diplomacy that makes royalty all the more visible and entertaining.

HABITAT

Red-headed Woodpeckers inhabit open woodlands, riparian lowlands, wooded urban parks and suburbs, and old burns (Bent 1939, Johnsgard 1979, 1986a). In the northern Rockies they breed in aspen parks, but in Colorado aspens grow above the species' elevational breeding range. The birds' choice of open habitats undoubtedly facilitates their foraging strategy, which entails gleaning insects from the ground and hawking them from the air (Bock et al. 1971). During the Atlas, lowland riparian and rural habitats composed most habitats reported for this species (59% and 25% respectively).

These birds prefer to nest in deciduous trees, including cottonwoods, elms, oaks, and beeches, but occasionally they use pine trees (Johnsgard 1986a). They typically select dead trees from which bark has sloughed away and limbs have broken off (Jackson 1976, Kilham 1977b). Atlas nest cards indicated that two of four nesting trees were dead (two not described), and nesting cavity heights ranged from 25 to 35 feet (7.6–10.7 m) above the ground, well within the typical range of 8–80 feet (2.5–24 m) (Bent 1939). Mature cottonwoods constitute the most important Colorado habitat, perhaps because of rotten cores that allow the birds to drill nest holes (Carl Bock pers. comm., Atlas).

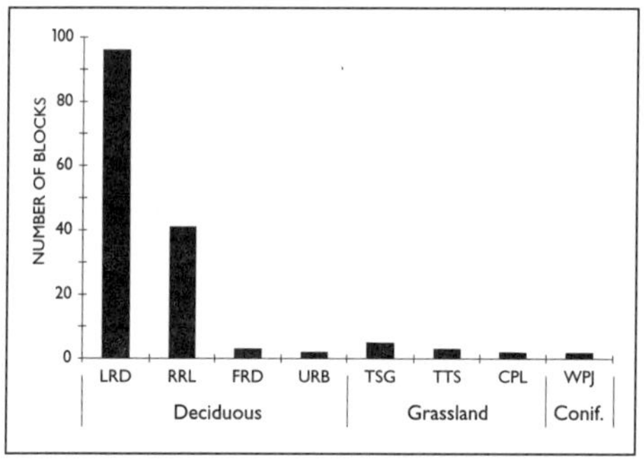
NUMBER OF BLOCKS
0
20
40
60
80
100
LRD
RRL
FRD
URB
TSG
TTS
CPL
WPJ
Deciduous
Grassland
Conif.

BREEDING

The earliest date that Atlas workers first recorded Red-headed Woodpeckers—7 May—reaffirmed that most arrive in early to mid May (A&R). Courtship includes "mutual tapping" (Kilham 1959, 1977a, Reller 1972), during which the male calls to the female until she comes to inspect a nesting cavity he has prepared. He then taps from within his cavity, at which point she indicates her level of interest by tapping from the outside, or by flying off, leaving the male to make another nest. Although this and other courtship behaviors could hardly escape observer detection, Atlas workers recorded courtship just twice. Perhaps other codes used during the courtship period (#, P, T, N, V) represent misinterpreted courtship behaviors; also, Confirmation codes sometimes superseded the Courtship code.

The lack of records for NB (invalid code for woodpeckers; Nelson 1993) and NE (inaccessibility of nests) contributed to the low breeding Confirmation rate of 37%. Other codes (C, N), however, indicated that nesting activities probably begin between 10 May and 20 June, corresponding with egg dates reported from locations throughout the species' range (20 April–23 July; B&N, Jackson 1976, Bent 1939), the earliest date coming from Alabama. Although Atlas workers recorded no second broods, these birds do double-brood on occasion (Ingold 1987). They are also persistent renesters, as demonstrated when one pair tried six times and laid 32 eggs in one season (Bent 1939).

BREEDING PHENOLOGY

CODE		# OF RECORDS	DATE RANGE
C	Courtship	3	10 May–21 Jun
N	Nest Building	7	28 May–20 Jun
ON	Occupied Nest	23	24 May–17 Jul
NY	Nest with Young	7	11 Jun–16 Jul
FY	Feeding Young	20	2 Jun–14 Jul
FL	Fledged Young	2	16 Jun–15 Jul

DISTRIBUTION

The Red-headed Woodpecker's breeding range extends from eastern Saskatchewan to southern New England, south to the Gulf Coast, west to central New Mexico. Bent (1939), AOU (1983), and A&R suggested that, historically, they occupied a broader range at greater densities, but these discussions did not quantify the degree of change. Many authors have reported on population declines and cyclical fluctuations in the East (e.g., Spahn *in* Andrle and Carroll 1988, Pitcher *in* Brewer et al. 1991).

Atlas results depict a widespread though not continuous distribution of Red-headed Woodpeckers in eastern Colorado. They concentrate along most of the lowland rivers and major creeks lined with cottonwoods, at altitudes of 5,500 feet (1830 m) or less. Cooke (1897) made the unsubstantiated and unlikely statement that they breed up to 10,000 feet

(3048 m); A&R probably explained this when they cited scattered records of migrants and non-breeding birds from areas well west of the Continental Divide, including foothills and mountain parks.

An abundance of food resources, including mast, corn, and insects, as well as supplies in their own caches (Kilham 1958), delimits the species' winter range. In most years, they winter in the southern and eastern two-thirds of their breeding range, but in winters preceded by excellent mast crops they may winter casually to the northern and western limits of their breeding range. In winter Red-headed Woodpeckers depend upon stored mast (acorns or corn). Although they could find this food in Colorado they normally do not stay the winter, likely because of competition with Lewis's Woodpecker for this winter food (Carl Bock pers. comm.). A&R plotted locations of four birds that overwintered in Colorado, and an early winter record came from Loveland on 11 December 1926, where an observer found a bird caching supplies of corn in holes and crevices (B&N).

BBS coverage in Colorado does not capture this species adequately enough to show any population trends, but the data show significant continent-wide annual declines of 2%, suggesting that half the population has disappeared in 31 years. Among the problems facing these birds today, habitat loss from logging, removing snags, and channeling rivers and streams undoubtedly outrank all others. Creosote in utility poles the birds use for nesting, competition with starlings for nest sites, population booms among Corvids that steal winter food caches, collisions with cars, and intentional eradications by utility companies and farmers (despite the species' appetite for injurious insects) also contribute to their troubles (Bent 1939, B&N, Rumsey 1970, Ingold 1989).

BREEDING EVIDENCE

REPORTED IN 137 (8%) OF 1745 PRIORITY BLOCKS

☐	Possible	54 blocks	(39%)
▩	Probable	33 blocks	(24%)
■	Confirmed	50 blocks	(36%)

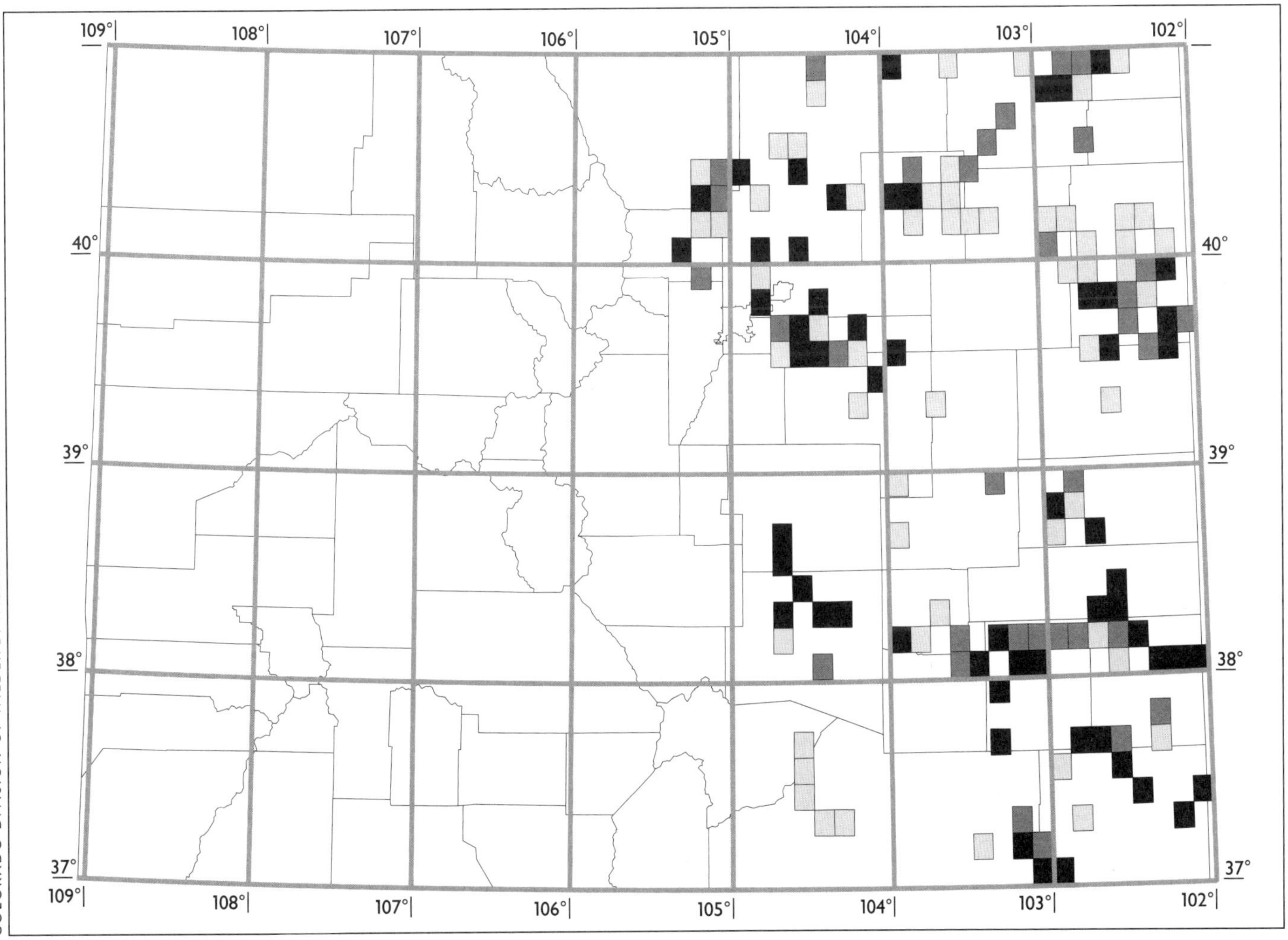
109°
108°
107°
106°
105°
104°
103°
102°
40°
39°
38°
37°

COLORADO DIVISION OF WILDLIFE / WILDLIFE RESOURCE INFORMATION SYSTEM

RED-BELLIED WOODPECKER *Melanerpes carolinus*

These eye-catching woodpeckers win first place among birds with the wrong English name! The red on the belly, sometimes just a yellowish tinge, is always hard to see. More showy distinguishing marks are the bright red nape on both sexes and the flashy red scalp on the male, plus the black-and-white striped back. The name "Ladderback" is already taken, although people in eastern Colorado call them that anyway. Southerners call them "Zebrabird."

HABITAT

A bird of open woodlands, suburbs, and parks in the eastern United States, these beautiful woodpeckers inhabit the deciduous trees in a few cottonwood/willow river bottoms and Siberian elm woodlots on Colorado's eastern border (B&N, A&R). The first Colorado birds inhabited a woodlot full of dying Siberian elms at Bonny State Park (HEK). Atlasers found them still at Bonny, but now they inhabit big, mature, cottonwoods and planted woodlots. Along the South Platte they also use old cottonwood groves. In Wray they inhabit the older residential sections.

Red-bellied Woodpeckers feed in upper stories of live and dead deciduous trees. They choose to feed in dead trees (depending on availability) one-third to one-half of the time. During breeding they glean; at other times of the year they probe tree surfaces (Winkler et al. 1995).

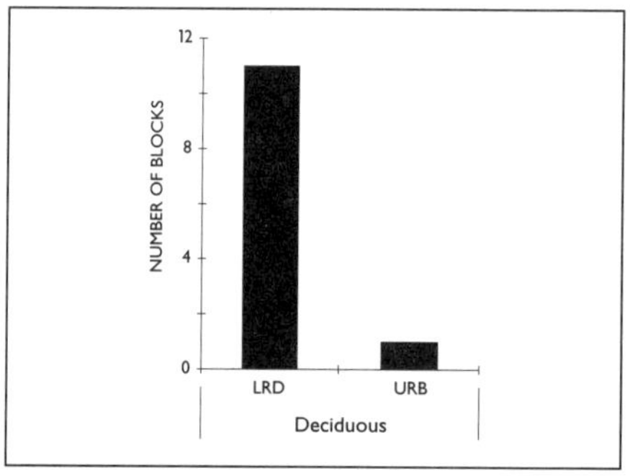

12
8
4
0
NUMBER OF BLOCKS
LRD
URB
Deciduous

The ecological niche of Red-bellieds approaches that of Lewis's Woodpeckers; the two have mutually exclusive ranges (Root 1988). The two species have similar diets: over 60% plant foods, mostly acorn mast and other fruits (Beal 1911), although Lewis's Woodpeckers switch to insects in summer. Red-bellieds store acorns, other nuts, and seeds, as do Lewis's Woodpeckers (Winkler et al. 1995).

Red-bellieds do occur with Red-headed Woodpeckers, but Red-bellieds tend to use denser woodlands and those with more understory. Red-headeds tend to nest in trees without bark whereas Red-bellieds do not show that preference (Winkler et al. 1995). Some aggressive interactions have been seen between Red-bellied and Red-headed woodpeckers (Root 1988). Red-headeds dominate Red-bellieds, and the latter will move to other sites rather than compete or use different strata in their habitat, yet sometimes they overlap without aggression (Winkler et al. 1995). With similar habitats, these species may exhibit avoidance behaviors to lower competition—a subject for future research. Bonny Reservoir, where both occur, would be the spot for exploring this (see Atlas maps).

BREEDING

Both sexes of Red-bellieds take part in digging nest holes in living trees (B&N), from 5 to 70 feet (1.5–21.4 m), usually under 40 feet (12.2 m; Harrison 1979). They may use a deserted nest hole of another woodpecker or a birdhouse, but starlings sometimes force them out of their excavated or adopted hole. Eggs number 3–8, generally 4–5, and are pure white. Incubation lasts 14 days; like most woodpeckers, both parents participate (B&N, Harrison 1979). In the southern part of their range they may raise two broods, but northern birds raise only one.

Atlas workers found five sets of fledged young, all in late June to late July, and all in cottonwood/willow riparian habitat. They reported no dates for nest-building or incubation, but saw courtship in mid May.

BREEDING PHENOLOGY

CODE		# OF RECORDS	DATE RANGE
C	Courtship	1	12 May
FY	Feeding Young	4	21 Jun–26 Jul
FL	Fledged Young	1	15 Jul

DISTRIBUTION

As a southeastern species, Red-bellieds barely enter Colorado on the eastern border. Charles Aiken collected one in Fountain in 1873, the earliest record for the state (Aiken and Warren 1914). They now stay all year along the eastern border, and a few—mainly post-breeding wanderers—follow the Platte and Arkansas river courses toward the foothills (A&R). These woodpeckers recently have expanded into the northeastern and north-central U.S., and they may now have started to expand their range in a westerly direction (B&N, A&R). Audubon Christmas Counts from 1965 to 1982 found them regularly at Bonny State Park, with occasional individuals straying to Front Range locations from Fort Collins to Pueblo.

Breeding was confirmed at Springfield in Baca County in 1937 but not since (B&N).

Observers documented breeding by a small colony at Bonny starting in the 1960s and along the South Platte at Ovid and Tamarack Ranch SWA. Atlasers Confirmed breeding in both places. Red-bellied Woodpeckers undoubtedly also breed in Wray (40102A2), but residents in that Non-priority block, who know the "Ladderbacks" well, could not Confirm breeding.

Atlas workers consistently estimated their abundance as low, even when the birds were raising young. Using those abundance estimates suggests that the state has fewer than 50 nesting pairs. Nevertheless, because the birds are so conspicuous and the young somewhat noisy, Colorado really must have very few.

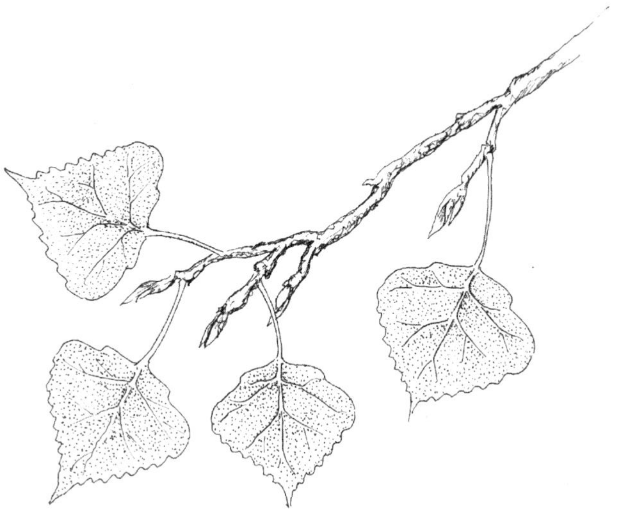

BREEDING EVIDENCE

REPORTED IN 4 (0%) OF 1745 PRIORITY BLOCKS

☐	Possible	2 blocks	(50%)
■	Probable	2 blocks	(50%)
■	Confirmed	0 blocks	(0%)

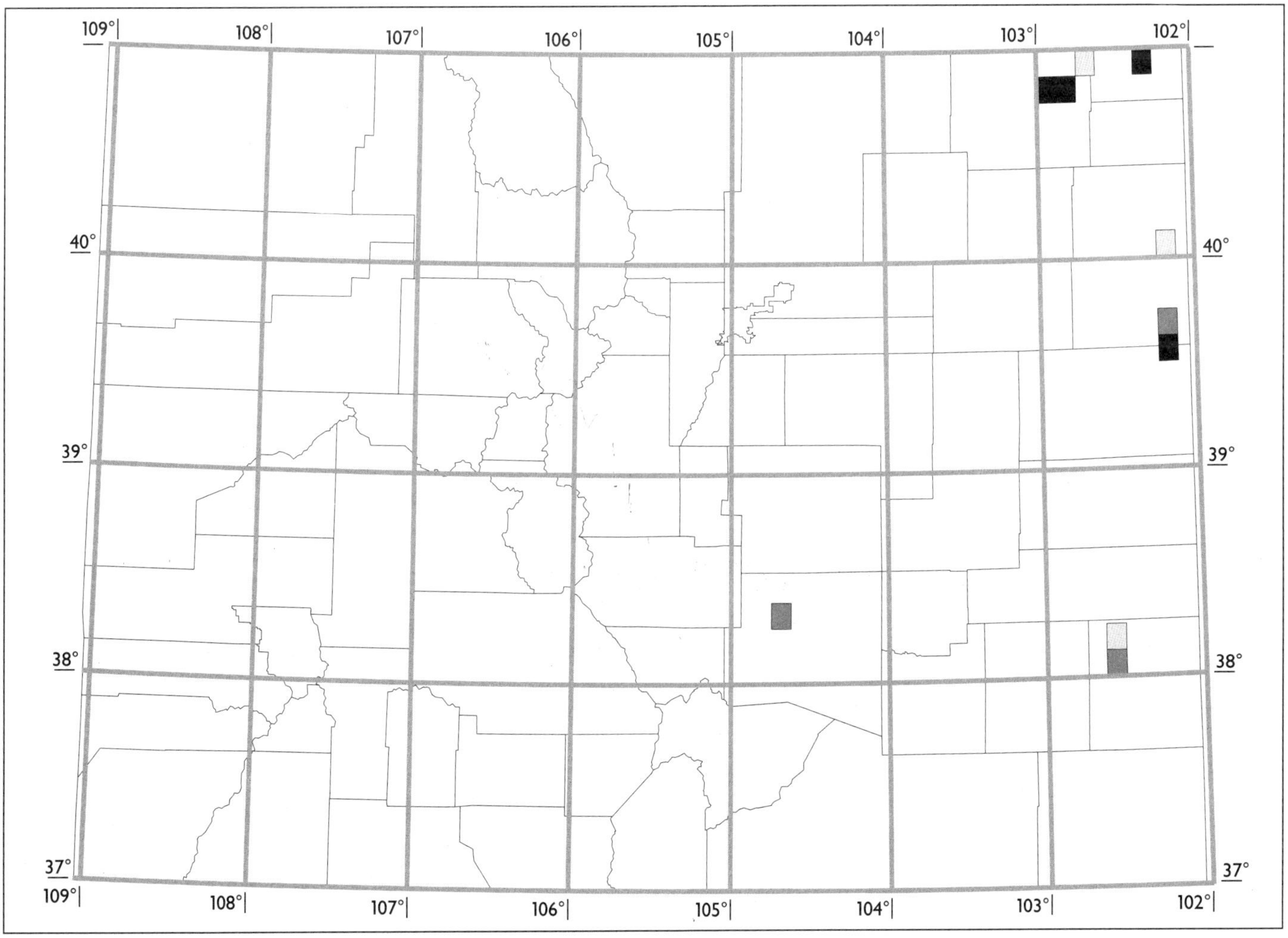
109°
108°
107°
106°
105°
104°
103°
102°
40°
39°
38°
37°

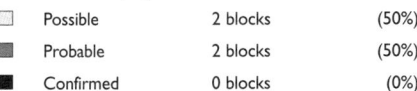

COLORADO DIVISION OF WILDLIFE / WILDLIFE RESOURCE INFORMATION SYSTEM

WILLIAMSON'S SAPSUCKER

Sphyrapicus thyroideus

The sexes of Williamson's Sapsuckers look so different that in 1852 and 1855, two ornithologists described the male and female as separate species. Twenty years later, near Fort Garland, Henry Henshaw (1875), a member of the Wheeler Survey, collected both a male and a female coming from one nest hole and thus proved that the two actually are the same species. The vast difference in plumage, unique among woodpeckers (Henshaw 1875), may amount simply to a reversion to the juvenile pattern by the female (Short and Morony 1970).

HABITAT

Williamson's Sapsuckers occupy conifer forests, often mixed with aspen, from 7,000 to 10,700 feet (2300–3260 m). Over two-thirds of Atlas reports came from coniferous habitats, mainly ponderosa pine, and conifer habitats housed two-thirds of the nest sites. Aspens composed only 26% of the habitat reports and 34% of the Confirmation habitats.

Aspen compose an essential element of the nesting habitat because the birds most often use aspens as a nest tree (Crockett and Hadow 1975). Atlas nest cards reflect the mix: in nine of the 14 cards, sapsuckers used aspens for nest trees even though seven cards listed the habitat as a mixed aspen/conifer forest. The cards showed four nests in ponderosas and one in a dead spruce.

Even though many Williamson's nest in aspens, they feed almost exclusively in coniferous trees, where they forage for insects and make sap wells (Crockett and Hadow 1975). Males and females have different feeding habits as well as plumage. Females forage for insects by gleaning on tree trunks; males glean mainly on limbs and the ground (Winkler et al. 1995).

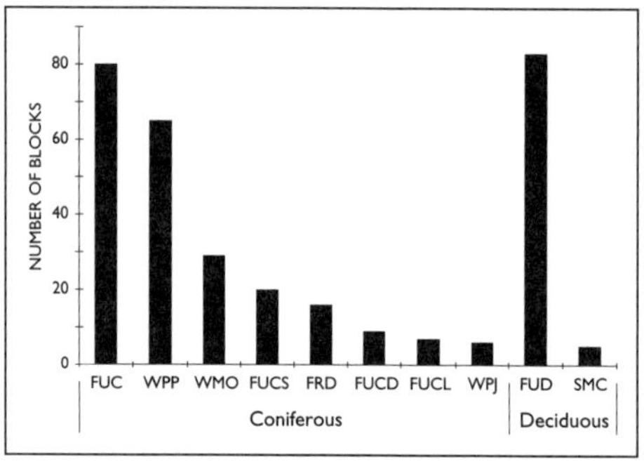

NUMBER OF BLOCKS
80
60
40
20
0
FUC
WPP
WMO
FUCS
FRD
FUCD
FUCL
WPJ
FUD
SMC
Coniferous
Deciduous

Although their habitat selection involves a combination of conifers and aspens, it can overlap with Red-naped Sapsuckers' preference for aspen. In some places Williamson's maintain interspecific territories to exclude other sapsucker species (Winkler et al. 1995). The exclusionary tactics do not operate uniformly, as the two species occasionally nest in the same tree (RW). At several sites in Rocky Mountain National Park Red-naped and Williamson's frequently nest in the same aspen stand, sometimes within 50 feet (15 m) of each other (Crockett and Hadow 1975).

BREEDING

Males arrive in Colorado in mid April and drum to claim territories and to attract females, who arrive a week or two later. The drumming differs from that of woodpeckers (except other sapsuckers) by its irregular cadence. Courting includes posturing and noisy chases up the pine trees (Winn 1981). In Rocky Mountain National Park, territories measured 2.5 acres (6.75 ha; Crockett 1975b). Individual pairs of Williamson's nest no closer together than 575 feet (175 m).

The sapsuckers drill cavities in fungus-infected aspens within coniferous forests, or in partially dead pines, Douglas-firs, or other conifers (Crockett and Hadow 1975). In some places Williamson's Sapsuckers build most of their nests in aspens: five of six in a small Rocky Mountain National Park study area and 97% in an Arizona site (Crockett 1975b, Li and Martin 1991). Eight Atlas nest cards reported nesting above 9,000 feet (2745 m), and two above 10,000 feet (10,200 and 10,680; 3110 and 3255 m) in subalpine aspen/spruce and lodgepole forests.

Males excavate nesting cavities—usually a new hole each year, often in the same tree as previous years. This creates "apartment house trees" where other cavity-nesting birds take residence.

Both adults incubate, brood the young, and feed them. Males perform somewhat more of the nursery duties: they incubate at night and part of the day (Winkler et al. 1995). The females lay 4–6 eggs. Young hatch within two weeks and fledge about four weeks after hatching (Crockett 1975a). They make little sound at first, but the begging noise gradually increases in volume and eventually becomes almost constant during the day.

Nests with young provided over half the Atlas Confirmations. Atlasers found some nests with young in early June, but most from 20 June to 13 July. The clamor may betray the nests to unmated males who sometimes intrude to feed larger nestlings, over the objections of the parents (Winkler et al. 1995). The racket also may attract predators (Skutch 1985), as evidenced by trees near Woodland Park with sapsucker holes torn out, apparently by bears (RW). Family units break down quickly, with young soon foraging for themselves. More drumming, limited courtship, and excavation occur in July, but second nestings have not been recorded (Crockett and Hansley 1977).

BREEDING PHENOLOGY

CODE		# OF RECORDS	DATE RANGE
N	Nest Building	3	26 May–24 Jun
ON	Occupied Nest	15	20 May–15 Jul
NY	Nest with Young	73	2 Jun–18 Jul
FY	Feeding Young	24	9 Jun–17 Jul
FL	Fledged Young	27	7 Jul–18 Aug

DISTRIBUTION

Williamson's Sapsuckers breed from southeastern British Columbia to central Arizona and New Mexico. They winter primarily in Arizona and Mexico.

In Colorado, as the Atlas map shows, the east flank of the Rockies from Wyoming to New Mexico provides optimal ponderosa pine habitat. Williamson's do not breed in the Black Forest, which lacks aspen and willows close to its ponderosa groves. The sapsuckers breed throughout the mixed forests of South Park, the Wet and Sangre de Cristo Mountains, and Cochetopa Hills. Atlasers confirmed breeding throughout the lower slopes of the San Juan Mountains, wherever ponderosas grow in pure stands or mixed with aspen, Douglas-fir, white fir, subalpine fir, and Engelmann spruce. The birds also breed around Grand Mesa, the Flat Tops, the mountains surrounding North and Middle parks, and the Park Range. They do not breed in Moffat County.

A comparison with Atlas maps of two other ponderosa specialists, Pygmy Nuthatch and Western Bluebird, demonstrates broader habitat use by these sapsuckers. In outline, the Williamson's range tracks that of the Red-naped, but Red-napeds, with their aspen-based affinities, occur more widely.

Williamson's Sapsuckers depend upon short-lived, diseased, dying, or dead trees, scattered through the forests. Sapsuckers may increase in numbers during outbreaks of forest insects. With commercial logging for firewood and for residential construction and amenities, the number of trees suitable for breeding is decreasing, especially along the Front Range. The winter range in Mexico, subject to heavy timber cutting, may also become a critical factor in their long-range population trends.

BREEDING EVIDENCE

REPORTED IN 266 (15%) OF 1745 PRIORITY BLOCKS

☐	Possible	83 blocks	(31%)
▩	Probable	47 blocks	(18%)
■	Confirmed	136 blocks	(51%)

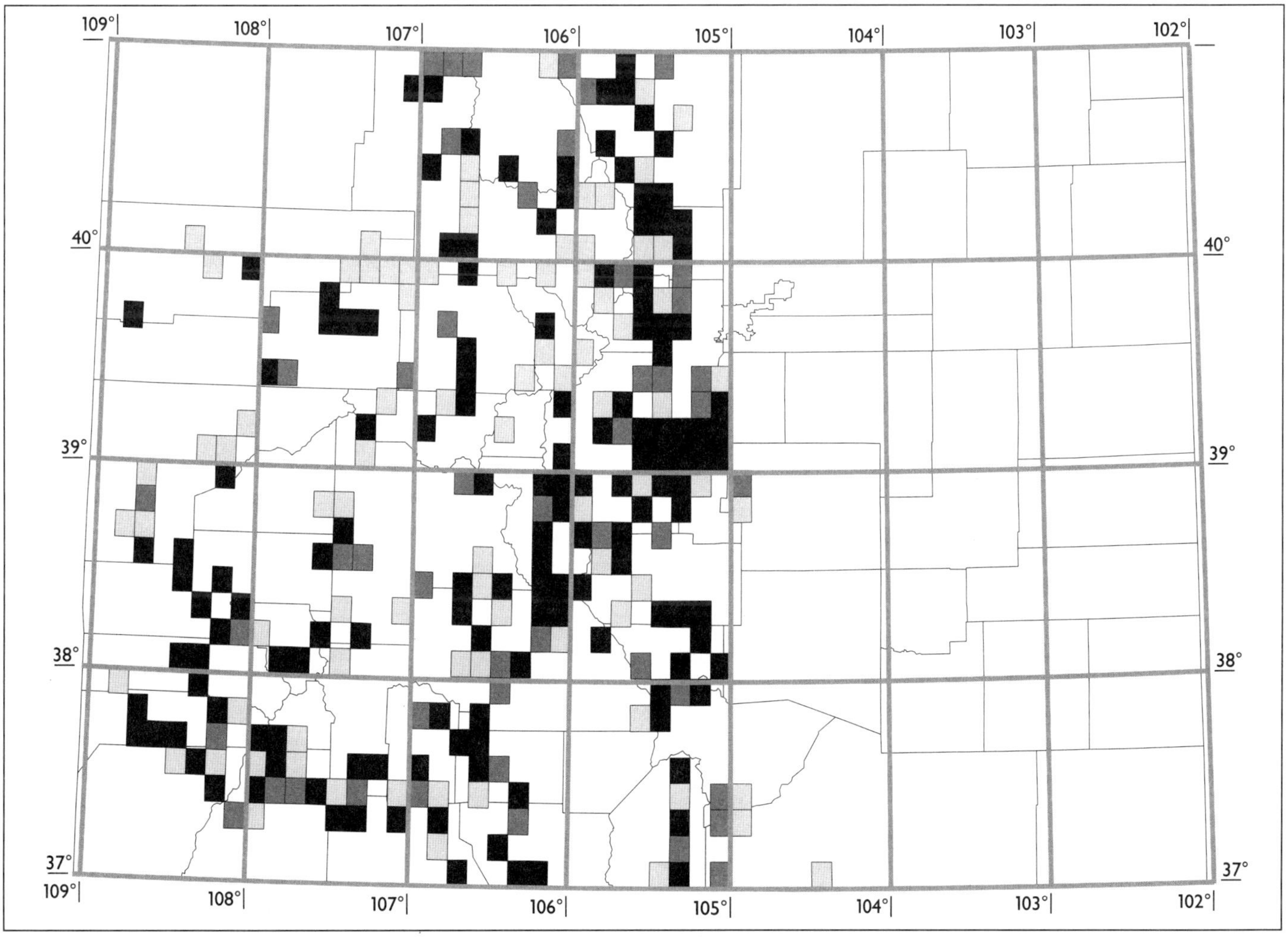
109°
108°
107°
106°
105°
104°
103°
102°
40°
39°
38°
37°

COLORADO DIVISION OF WILDLIFE / WILDLIFE RESOURCE INFORMATION SYSTEM

The theory of Indicator Species states that certain species serve as indicators of the health of entire ecosystems. As aspen obligates, Red-naped Sapsuckers serve as indicators, but their connection to aspens does not translate into a connection with a pure "healthy" grove of aspen trees. Robust aspen stands have few sapsuckers. Rather, because the birds require disease and rot for digging cavities, the presence of these sapsuckers acts as an indicator of the maturity of the ecosystem, not simply of the vigor of one specific stand of aspen.

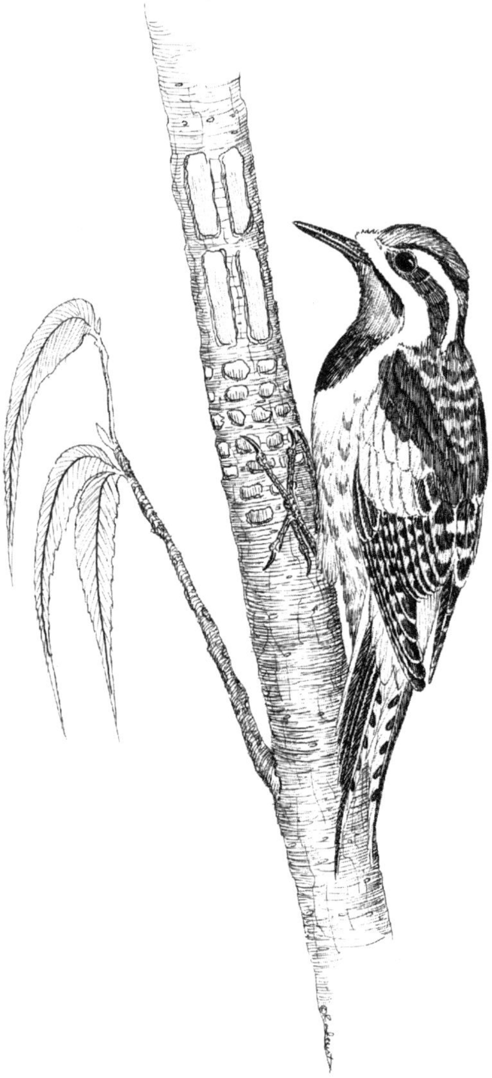

HABITAT

Red-naped Sapsuckers forage in aspen, willows, and cottonwoods close to their nest site (Crockett and Hadow 1975) but breed almost exclusively in mature aspen stands. Of 21 nest cards submitted to the Atlas, 20 reported nests in aspen. If the Red-naped Sapsucker serves as an indicator species for aspen, its abundance would seem to demonstrate that Colorado has healthy aspen ecosystems, but to sapsuckers not all aspen groves are equal. Typical nest stands, dominated by large aspen, have a variety of diseases that create the heart rot needed for suitable cavity excavation.

Nesting Red-naped Sapsuckers require aspen groves with two characteristics: aspen trees infected with shelf or heartwood fungus (for drilling nest holes) and nearby willow carrs (for drilling sap wells). They reject aspen groves that lack nearby willow riparian habitat (Ehrlich and Daily 1993). In

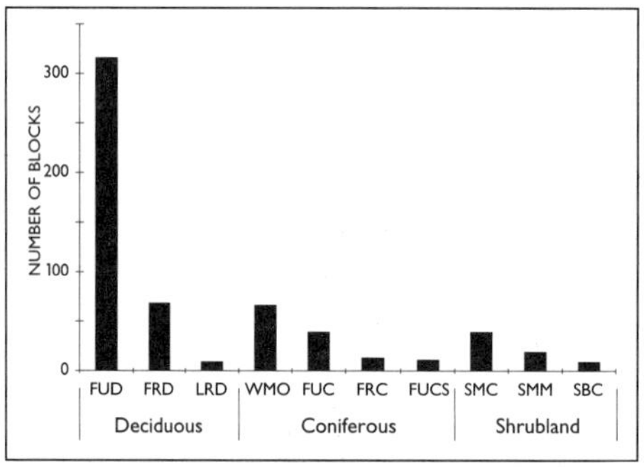
300
200
100
0
NUMBER OF BLOCKS
FUD
FRD
LRD
WMO
FUC
FRC
FUCS
SMC
SMM
SBC
Deciduous
Coniferous
Shrubland

younger, thin-barked trees (not willows exclusively) they drill sap wells, an important food source. For the most part atlasers disregarded this as a discrete component of nesting habitat (11% of habitat codes).

Just as a variety of species use sapsucker nest cavities, a variety use their sap wells—hummingbirds, insects, and other wildlife flock to the sugary sap. Sap wells at a study area near Gothic had at least one Orange-crowned Warbler in attendance almost all day, and female and immature Broad-tailed and Rufous hummingbirds about one-fifth of the day. Chipmunks, a squirrel, and a squadron of vespid wasps visited the wells regularly (Ehrlich and Daily 1988).

Red-naped Sapsuckers share their habitat with Williamson's Sapsuckers but Red-napeds do not exhibit the conifer connection of Williamson's. Whereas Red-napeds require nearby willows, Williamson's require nearby conifers. The aspen stands that Williamson's pick either contain a large mix of conifers or grow near conifer forests (Crockett and Hadow 1975).

BREEDING

Red-napeds choose, when available, aspen nest trees infected with fungus. The two sapsucker species choose similar sites in similar-sized trees (Crockett and Hadow 1975). Red-naped Sapsuckers occasionally reuse a nest hole in following years, but more often they excavate new nests, and the old ones become homes for House Wrens, swallows, and other small, cavity-nesting birds.

Red-naped Sapsuckers willingly share their nest trees with other species. From 1992 to 1996, Red-naped Sapsuckers nested in the same aspen in the Spicer Peak Atlas block (40106D4) with four other species at the same time. Over the years Northern Flickers, House Wrens, Mountain Bluebirds, Buffleheads, Tree Swallows, and Violet-green Swallows used the tree—the non-woodpeckers in old flicker and sapsucker holes. In the high mountains Tree and Violet-green swallows depend almost entirely on sapsucker cavities for nesting. Other cavity-nesting species also nest more frequently in areas where Red-naped Sapsuckers breed (Ehrlich and Daily 1993).

The noisy chatter of young sapsuckers in nests made Confirmations easier than for most species (69% rate). Atlas reports of nests with young started early, on 29 May, and extended into early August. Peak breeding occurred between 14 June and 20 July. The cavity-nesting habits of Red-naped Sapsuckers make NE confirmations difficult; determining the initiation of egg-laying requires backdating from the earliest detection of young in nests. With incubation taking two weeks (Terres 1980), egg-laying must start in mid May.

BREEDING PHENOLOGY

CODE		# OF RECORDS	DATE RANGE
C	Courtship	5	6 Jun–29 Jun
N	Nest Building	4	20 May–1 Jul
ON	Occupied Nest	41	17 May–19 Jul
NY	Nest with Young	141	29 May–4 Aug
FY	Feeding Young	73	7 Jun–31 Jul
FL	Fledged Young	68	22 Jun–25 Aug

BY NORMAN M. BARRETT

DISTRIBUTION

Lumping (AOU 1957) and splitting (AOU 1983) Red-naped and Yellow-bellied sapsuckers confused distribution records (A&R), although not for Colorado breeders. Red-napeds breed from the Rockies west to eastern California and Oregon, and from southern Canada to Arizona and New Mexico. Aspens do not occur uniformly across this range, and Red-naped Sapsuckers have adapted to other tree species. In parts of Oregon they nest in alders (NMB) and in other parts of the range they use larch and Engelmann spruce (Terres 1980).

Early records list Red-naped Sapsuckers as common summer residents in western Colorado (Cooke 1897). Atlas data verify that the eastern foothills of the Rockies still represent the eastern boundary for Red-napeds.

Red-naped Sapsuckers breed throughout the Colorado mountains. Their strong affinity for aspen results in a pattern on the Atlas map that closely duplicates the forested mountains of the state. The birds breed in an altitudinal pattern where they follow aspen to the upper limits of the mountain slopes but do not nest regularly on valley floors. The Atlas map provides new distribution information only in Moffat County, where atlasers located Red-napeds in aspen stands in three blocks north of the Yampa River but found few south of the Yampa. The Sangre de Cristos and other mountains south of Pikes Peak have fewer than expected, partly due to Atlas block selection.

The need of this species (and other woodpeckers) for mature, diseased trees conflicts with conventional forestry practices that attempt to eliminate insects and disease from forest ecosystems. Under the more modern and controversial ecosystem management, biologists recognize the roles of insects, disease, and the species tied to them as an integral part of forest health. If aspen forests give way through natural succession to conifers, in another 50 years or so sapsucker numbers could drop dramatically.

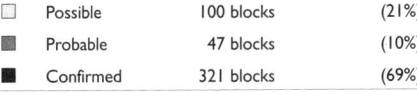

BREEDING EVIDENCE

REPORTED IN 468 (27%) OF 1745 PRIORITY BLOCKS

Possible	100 blocks	(21%)
Probable	47 blocks	(10%)
Confirmed	321 blocks	(69%)

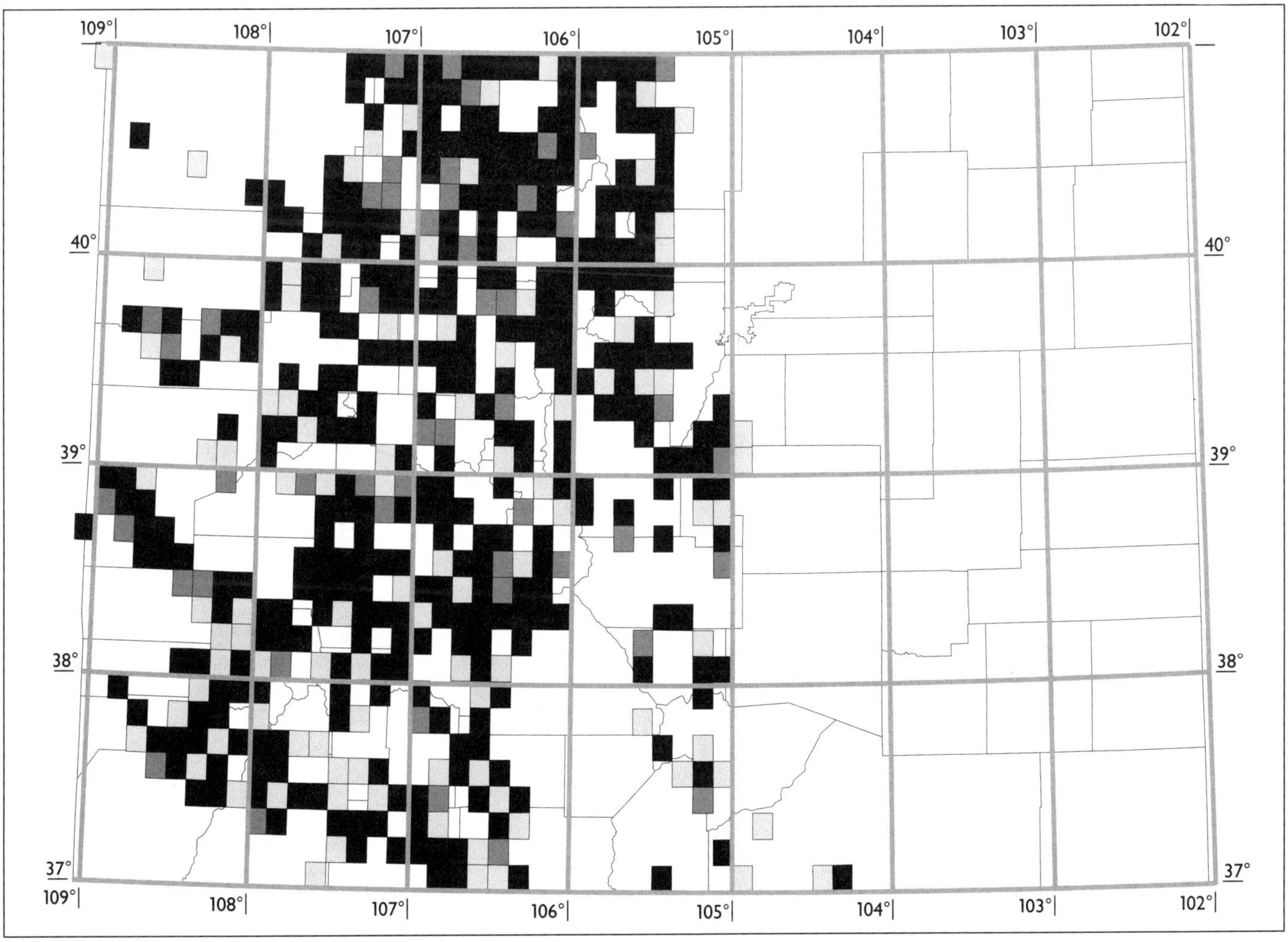
109°
108°
107°
106°
105°
104°
103°
102°
40°
39°
38°
37°

COLORADO DIVISION OF WILDLIFE / WILDLIFE RESOURCE INFORMATION SYSTEM

Several species occupy different habitats in Colorado than they do in other parts of their range. The Great Plains generally serve as a divide between eastern and western habitat selections. Colorado's Ladder-backed Woodpeckers use different habitats than distant Ladder-backs, but in this case the habitat partition lies to the south in New Mexico. Cooke and Sclater refer to these birds as "Texan Woodpeckers," and Texas is the center of their range—for Colorado-type birds.

HABITAT

Ladder-backs in Arizona and New Mexico live year round in cactus- and yucca-studded deserts and other arid country. They also occupy woodlands with deciduous trees (such as sycamores) along seasonally dry streams (Winkler et al. 1995). Birds in Los Alamos County, New Mexico, choose juniper/yucca/cholla grasslands in canyon bottoms (Travis 1992). In Mexico they occupy deserts and deciduous and pine/oak woodlands (Howell and Webb 1995).

The AOU in 1957 recognized birds of Colorado and Texas as a distinct subspecies, but this form will probably be subsumed into the Arizona/New Mexico subspecies (Winkler et al. 1995). Birds in the easterly range have broader habitat tastes than those in Arizona and New Mexico. They frequent open riparian woods (cottonwoods and oaks), mesquite woodlands, and shade trees in towns and rural areas (Bent 1939).

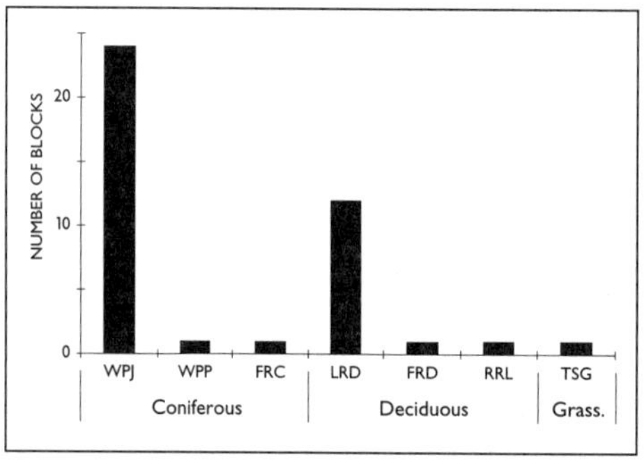
NUMBER OF BLOCKS
0
10
20
WPJ
WPP
FRC
LRD
FRD
RRL
TSG
Coniferous
Deciduous
Grass.

Even males and females tend to use different habitats, by foraging in different substrates. In Arizona, males forage mainly in cholla, females in mesquite. Males use more kinds of plants, and during post-breeding season, when both feed in mesquites, the males feed on the trunks and larger branches; females feed on small and peripheral branches, and at greater heights than the males (Austin 1976).

Bent and others list nearby water as a habitat essential, but an analysis of Christmas Count data postulates that they inhabit areas with very dry air—with a high annual evaporation rate. Because this theory limits them to desert and semi-arid oak woodlands, (unspecified) pine forests, shrub savanna, and cactus/grassland (Root 1988), it seems not to apply to Colorado birds.

Colorado habitat reports exhibit a dichotomy: B&N reported Ladder-backs in "dry pinyon and juniper country" and Atlas workers located the most birds, 55%, in pinyon/juniper, followed by 26% in riparian deciduous areas. In contrast, most nests are reported in deciduous trees. Niedrach found Baca County pairs with nests in a cottonwood in 1914 and a scrub oak in 1938 (B&N). Atlas workers found only one nest in pinyon/juniper and five others in riparian cottonwood river bottoms. Atlasers also observed two birds drilling nest holes—one in pinyon/juniper and one in riparian.

The closely related Ladder-backed and Downy woodpeckers occupy similar niches in their woodland habitats, and barely overlap in Colorado; atlasers found Downies and Ladder-backs in only five of the same blocks. The two woodpeckers occupied different habitats in the common blocks except in one with nesting by both species Confirmed in cottonwoods (Hasser Ranch 37102F6).

BREEDING

Ladder-backed Woodpeckers nest in rotten stubs or dead, decayed branches of oak and willow alongside water, or in fenceposts or utility poles, from 4 to 25 feet (1–7.5 m) above ground (B&N). The males drill the holes, with some help from the females, in dead parts of trees or the other sites. Males aggressively enforce their domination of feeding sites and the sexes probably have somewhat different diets (Winkler et al. 1995). Clutches contain 2–7 eggs, usually three or four (Winkler et al. 1995).

Atlas workers found six active nests from late May to early July, and fledged young on 25 July. A pair near Kim had a nest in a fence post, amid juniper and cottonwood vegetation, with adults entering the hole on 27 May and tending young on 4 July (Marina Graves, nest card).

Highly carnivorous (a diet of 92% animal matter), they eat primarily wood-boring beetles as well as caterpillars and ants (Beal 1911).

BREEDING PHENOLOGY

CODE		# OF RECORDS	DATE RANGE
N	Nest Building	2	29 May–24 Jun
ON	Occupied Nest	5	22 May–5 Jul
NY	Nest with Young	1	15 Jun
FL	Fledged Young	1	25 Jul

DISTRIBUTION

Ladder-backs live, year round, in the southwestern U.S. and most of Mexico south into Nicaragua. In Colorado Cooke (1897) called the species, at the edge of its range, strangely absent from early accounts because birds then occurred regularly in Pueblo and Huerfano counties. A year later he reported the state's first breeding record southwest of Pueblo. Niedrach found them nesting in cottonwoods in Furnish Canyon from 1914 on, and Knorr found a nest near Colorado Springs in 1950 (B&N). The Latilong Study lists them as breeding along the Arkansas River in Otero and Pueblo counties and as wanderers downstream along the river; one strayed to Boulder County in 1976 (A&R).

Atlasers found Ladder-backs in only 33 Priority blocks scattered across southeastern Colorado; Las Animas and Pueblo counties produced the most breeding birds. They found none in El Paso County despite coverage of two Non-priority blocks along Fountain Creek. Atlas block selection may have caused field work to miss some birds in riparian strips, e.g., Huerfano County.

BBS trends show a large continental population decline of 2%. The average number of birds per route is less than one (connoting a small sample size), but 60% of the routes show decreases. The numbers in Colorado are not high enough to show any statistical trend. It seems these woodpeckers have lived in Colorado always in modest numbers, as befits a species on the periphery of its range.

BREEDING EVIDENCE

REPORTED IN 34 (2%) OF 1745 PRIORITY BLOCKS

☐	Possible	21 blocks	(62%)
▩	Probable	7 blocks	(20%)
■	Confirmed	6 blocks	(18%)

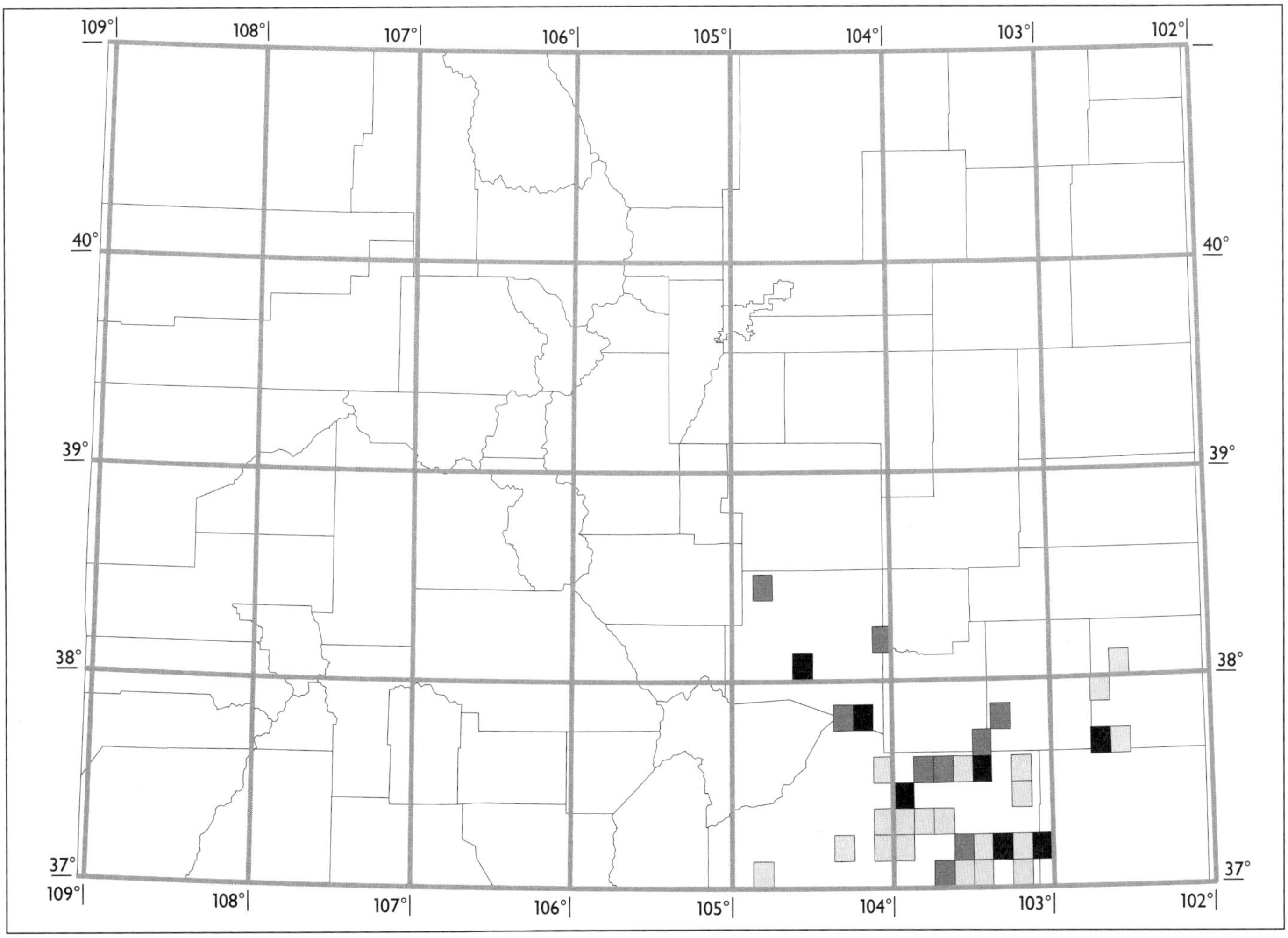
109°
108°
107°
106°
105°
104°
103°
102°
40°
39°
38°
37°

COLORADO DIVISION OF WILDLIFE / WILDLIFE RESOURCE INFORMATION SYSTEM

Sportscasters who gush over a gymnast's ability "to stick" her landings might benefit from an introduction to the Downy Woodpecker. Downies long ago perfected the skill humans find so elusive—hurtling through the air and then suddenly, somehow, alighting cleanly in an upright posture. And, one might add, tree bark provides a landing surface much less forgiving of error than foam mats.

HABITAT

Across their range Downy Woodpeckers choose to live in deciduous habitats (Winkler et al. 1995)—forests, river groves, orchards, and urban areas (Peterson 1990). Although they inhabit all the deciduous forest types in New York (Confer *in* Andrle and Carroll 1988) and even hemlock forests and pine plantations in Ohio (Peterjohn and Rice 1991), they achieve their highest densities in woodlands that include small or young trees with low canopy heights (Lemieux *in* Gauthier and Aubry 1996, Foss 1994).

In Colorado they maintain the partiality to deciduous woodlands—cottonwoods on the plains and in the western river valleys, aspens in the mountains. Atlas habitat codes highlight this favoritism—71% of the blocks listed deciduous habitats, and aspen commanded the most block habitat reports. Atlas abundance codes also emphasized aspen: half of the blocks with aspen (17% of all blocks) reported A3 and A4 codes. Although blocks with riparian habitats outnumbered those with aspen, only half as many claimed A3 (and no A4) codes. Available trees provide the key to these abundance codes; blocks with riparian areas (which cross the landscape in narrow strips) do not have as many trees as blocks blanketed with aspen forests. Blocks with conifers that had A3 and A4 codes composed a mere 3% of blocks with Downies.

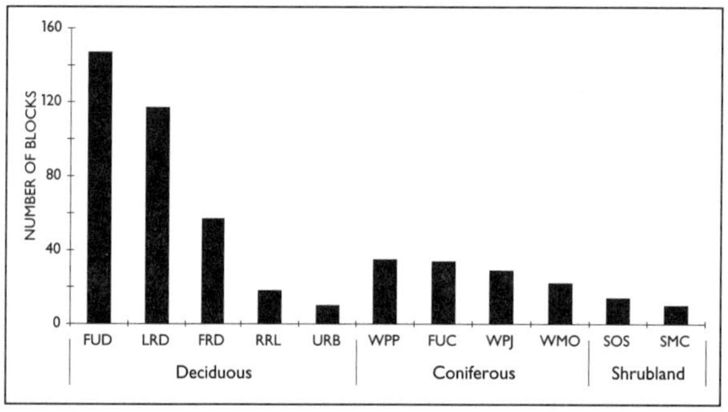
160
120
80
40
0
NUMBER OF BLOCKS
FUD
LRD
FRD
RRL
URB
WPP
FUC
WPJ
WMO
SOS
SMC
Deciduous
Coniferous
Shrubland

BREEDING

Male and female Downies maintain separate territories through the winter. The males excavate roost holes more diligently, so that when pairs come together to breed, the female most often picks a hole dug by the male. During breeding they zealously defend a small territory around the nest tree and, less vigorously, a larger home feeding range. After breeding the pair separates into individual winter territories, then the same birds re-mate in the spring (Lawrence 1966). The attraction appears to be not the mate but the territory, so that the "lifetime" pair bond in fact consists of "renewed pairing of erstwhile partners rather than permanent mating," and instigated by real estate (Skutch 1985).

Both sexes incubate the 3–6 eggs for a total of 12 days and the young fledge after three weeks. They raise one brood per year. The opening to the nest hole is a perfect circle 1.26 inches (3.2 cm) in diameter (Harrison 1979)—so precise that you can measure an old nest hole and tell if a Downy made it. Many other bird species use old Downy nest holes, among them chickadees, wrens, and swallows.

The birds excavate nest holes in live and dead trees, stumps, stubs, and fence posts (Harrison 1979). Henshaw in 1875 reported the first Downy nest in Colorado, near Fort Garland (B&N). B&N said that Downies tend to drill in dead trees for nest holes, and Winkler et al. (1995) say they "inevitably" excavate nests in dead or partially dead trees but, of nine Atlas nest cards, five were in live aspen and four in dead aspen. Live trees that most woodpeckers use usually have some kind of inner decay like heart rot (Norm Barrett pers. comm.). Ground elevation for these nest trees extended from 8,000 to 9,400 feet (2460–2890 m). Nest hole heights above the ground ranged from 4 to 40 feet (1.2–12 m) and averaged 18.1 feet (5.6 m). Downies probably drill holes higher than this in the bigger cottonwoods of plains and Western Slope riparian systems.

Like abundance codes, habitats in blocks with Confirmations differed from overall block habitat reports. Aspen had a higher proportion—47% of all Confirmations (i.e., in blocks with Downies, field workers Confirmed breeding in almost half the ones with aspen). Plains and foothills riparian Confirmations composed 24% and 12% respectively. The remainder came from conifers.

Downies found by atlasers manifested a lengthy breeding season, from nests occupied in mid May to fledglings, still attended by adults feeding them, in mid August. Confirmations showed a definite altitudinal progression; after 21 June all but three of feeding young (FY) codes and all but four of 30 fledgling (FL) codes came from mountain blocks.

Downy Woodpeckers resolutely defend their nests; one pair contended for five hours to repel a red squirrel. After the squirrel finally left the birds were so agitated that they attacked each other (Skutch 1985).

BREEDING PHENOLOGY

CODE		# OF RECORDS	DATE RANGE
C	Courtship	1	3 Jun
N	Nest Building	11	7 May–21 Jul
ON	Occupied Nest	17	16 May–31 Jul
NY	Nest with Young	32	2 Jun–15 Jul
FY	Feeding Young	51	23 May–15 Aug
FL	Fledged Young	33	6 Jun–10 Aug

DISTRIBUTION

Downy Woodpeckers live coast to coast, from central Alaska almost to Mexico. Two subspecies breed in Colorado (B&N). Recognizable in the field, these two differ visibly in coloration and subtly in size. Most Colorado Downies belong to the subspecies that breeds in the mountains; they have black wing coverts with very few or no white spots on them. An eastern race breeds on the plains in eastern Colorado. These birds have extensive white spotting on the wing coverts, and the outer white tail feathers show conspicuous black barring (B&N).

The Atlas map shows a greater distribution of Downies in the western two-thirds of the state, with large gaps only in the northwest and San Luis Valley. Their presence in the eastern half of the state, linked to the major river systems, is much more spotty—widespread but lots of open country in between. Still, more Downy breeding occurred in eastern Colorado than predicted by A&R's summer map. Possibly increased habitat along the prairie streams and in now-forested urban areas has induced a gradual, subtle increase in distribution on the Colorado plains.

BREEDING EVIDENCE

REPORTED IN 420 (24%) OF 1745 PRIORITY BLOCKS

☐	Possible	201 blocks	(48%)
▪	Probable	89 blocks	(21%)
■	Confirmed	130 blocks	(31%)

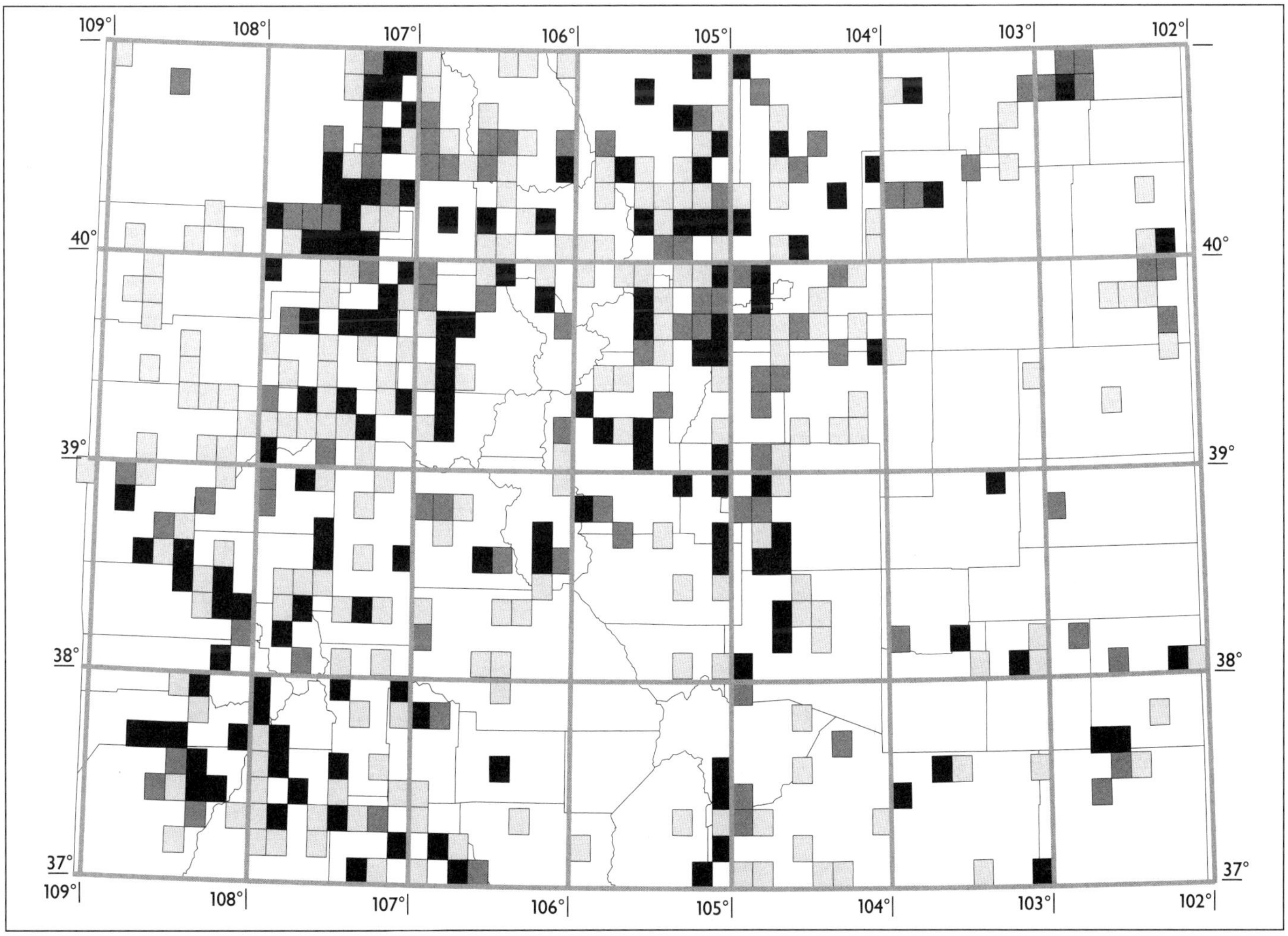
109°
108°
107°
106°
105°
104°
103°
102°
40°
39°
38°
37°

COLORADO DIVISION OF WILDLIFE / WILDLIFE RESOURCE INFORMATION SYSTEM

The sounds of their penetrating calls and vigorous hammering carry long distances through the conifer forests where Hairy Woodpeckers breed. These loud, black-and-white tree-huggers search noisily for tree borers and other wood-inhabiting insects as they explore the bark of pines, spruces, and firs in the forests in which they live. They call loudly and excitably as they swoop from tree to tree, then settle on tree trunks to probe for grubs.

HABITAT

Hairy Woodpeckers live throughout Colorado from timberline to the plains, more common in the mountains in summer, and at lower elevations in winter (A&R, B&N). They choose ponderosa pine, Douglas-fir, lodgepole, spruce/fir, and aspen forests, as well as riparian forests, pinyon/juniper woodland, and urban areas with tall trees (A&R).

Across the country Hairy Woodpeckers breed in extensive, mature forests of deciduous, coniferous, and mixed trees. They seek woodlands with dense canopies formed by large trees and a complement of dead stubs and fallen logs. In the East, when they nest in coniferous habitat, they prefer places with at least some deciduous component (Lemieux *in* Gauthier and Aubry 1996, Sibley *in* Andrle and Carroll 1988).

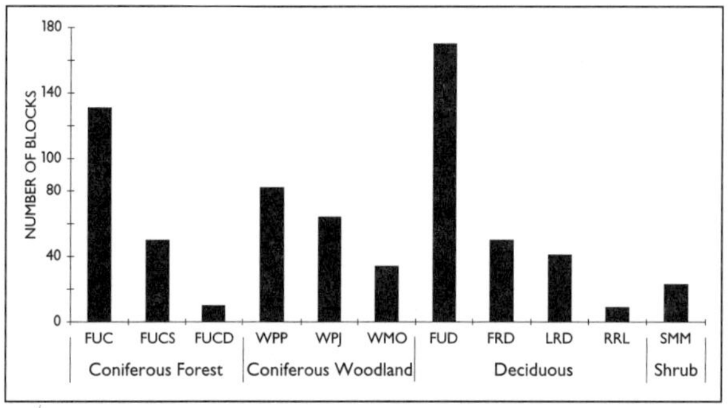
NUMBER OF BLOCKS
180
140
100
80
40
0
FUC
FUCS
FUCD
WPP
WPJ
WMO
FUD
FRD
LRD
RRL
SMM
Coniferous Forest
Coniferous Woodland
Deciduous
Shrub

B&N said that the various subspecies (12 in number) have different preferred habitats. Atlas results appear to support this in Colorado. Most plains-nesting Hairies belong to the deciduously inclined eastern subspecies (*P. v. villosus*), which has far more white spotting in the wings than the mountain form. They breed in LRD habitats, "the great cottonwoods of the prairie streams," as B&N said. Not many Hairies nest there: all the Atlas plains blocks recorded A1 and A2 abundance codes. The prairie stream bottoms do not support the extensive closed-canopy woodlands that eastern Hairies prefer.

Mountain Hairy Woodpeckers belong to another subspecies, *P. v. monticola* (B&N; included [apparently] in subspecies *P. v. orius* according to Winkler et al. 1995); they occur predominantly in conifers (61% of Atlas records), although a substantial number (36%) use aspen forests and deciduous foothills streams (which usually have some streamside or mountainside conifers). The distribution of A4 abundance codes suggests that the highest numbers occur in the San Juan Mountains.

Like other woodpeckers, the nest holes Hairies drill each year benefit a suite of other bird species that need the holes for breeding but lack the ability to make them. These other birds recycle the holes over and over. Forty percent of montane breeding birds require nest holes, and only 8% of these are woodpeckers able to drill the holes (Winternitz 1976).

BREEDING

Charles Aiken discovered the first Hairy Woodpecker nest in Colorado in El Paso County in 1872 (B&N). Both sexes excavate the nest hole, but the male does the most, taking two to three weeks (Harrison 1979). They favor aspen or trees with decayed heartwood, and dig a new hole each year. The entrance hole has a somewhat elongated shape, measuring 2.5 by 2 inches (6.4 x 5.1 cm). One brood of 3–6 eggs hatches in 12 days, incubated in the day by the female and at night by the male (Harrison 1979).

Females are aggressive in pair formation, which starts in November or December. They indulge in intense drumming—400–500 times a day, up to 10 times a minute—to attract a mate; the male may respond with a desultory four or five bursts in a minute (Kilham 1960).

In the 196 blocks where Atlas workers Confirmed breeding, they found them at nearly the same percentages as the bird's general occurrence in Colorado habitats—31% in aspen forest, 20% in upland coniferous forest, 14% in ponderosa pine, 11% in pinyon/juniper. Nine Atlas nest cards for the mountains listed seven nests in aspen, one in a narrowleaf cottonwood, and one in a pinyon pine, and on the plains one each in a willow and a cottonwood. Altitudes ranged from 3,700 to 10,200 feet (1140–3140 m). The height of nest holes varied from 5–60 feet (1.5–18 m) and averaged 22.4 feet (7 m). Atlas workers found nests with young from early June to mid July (most during June) and fledged young from mid May to 10 August (most during July).

BREEDING PHENOLOGY			
CODE		# OF RECORDS	DATE RANGE
C	Courtship	4	20 May–30 Jun
N	Nest Building	7	23 May–4 Jul
ON	Occupied Nest	15	20 May–15 Jul
NE	Nest with Eggs	1	17 Jun
NY	Nest with Young	49	20 May–17 Jul
FY	Feeding Young	74	26 May–28 Aug
FL	Fledged Young	66	9 Jun–10 Aug

DISTRIBUTION

Like Downy Woodpeckers, Hairy Woodpeckers live coast to coast in the U.S. and Canada, north into west-central Alaska, but unlike Downy Woodpeckers, they breed south to Panama.

In Colorado they breed throughout the mountains and in scattered locations on the plains. B&N recorded an upward altitudinal migration after fledging, and a downward movement into the foothills by September.

The Atlas map shows many more Hairies than Downies nesting in the western part of the state, and far fewer Hairies in eastern Colorado. In the west they breed fairly uniformly throughout the mountains. Mountain populations fluctuate with insect infestations and the occurrence of forest fires (A&R). Atlasers verified breeding in northwestern Colorado for the first time by visiting conifers that birders had not explored and, surprisingly, in the Black Forest area of Elbert County. Hairies avoid the treeless mountain parks and San Luis Valley. To the sparse assortment of Hairies on the plains the Atlas adds scattered breeders on the Arikaree, Republican, and Big Sandy rivers.

The map does not distinguish between the plains and mountain subspecies, as atlasers did not make the distinction. The birds nesting on the plains in latilong 37102 and in the two easternmost latilongs from the Arkansas River north probably belong to the eastern subspecies. Those nesting at the edge of the Front Range foothills could belong to either subspecies or could be intermediate forms. Birds in the mountains, Black Forest, and Las Animas County probably belong to the mountain subspecies.

BREEDING EVIDENCE		
REPORTED IN 545 (31%) OF 1745 PRIORITY BLOCKS		
Possible	249 blocks	(46%)
Probable	100 blocks	(18%)
Confirmed	196 blocks	(36%)

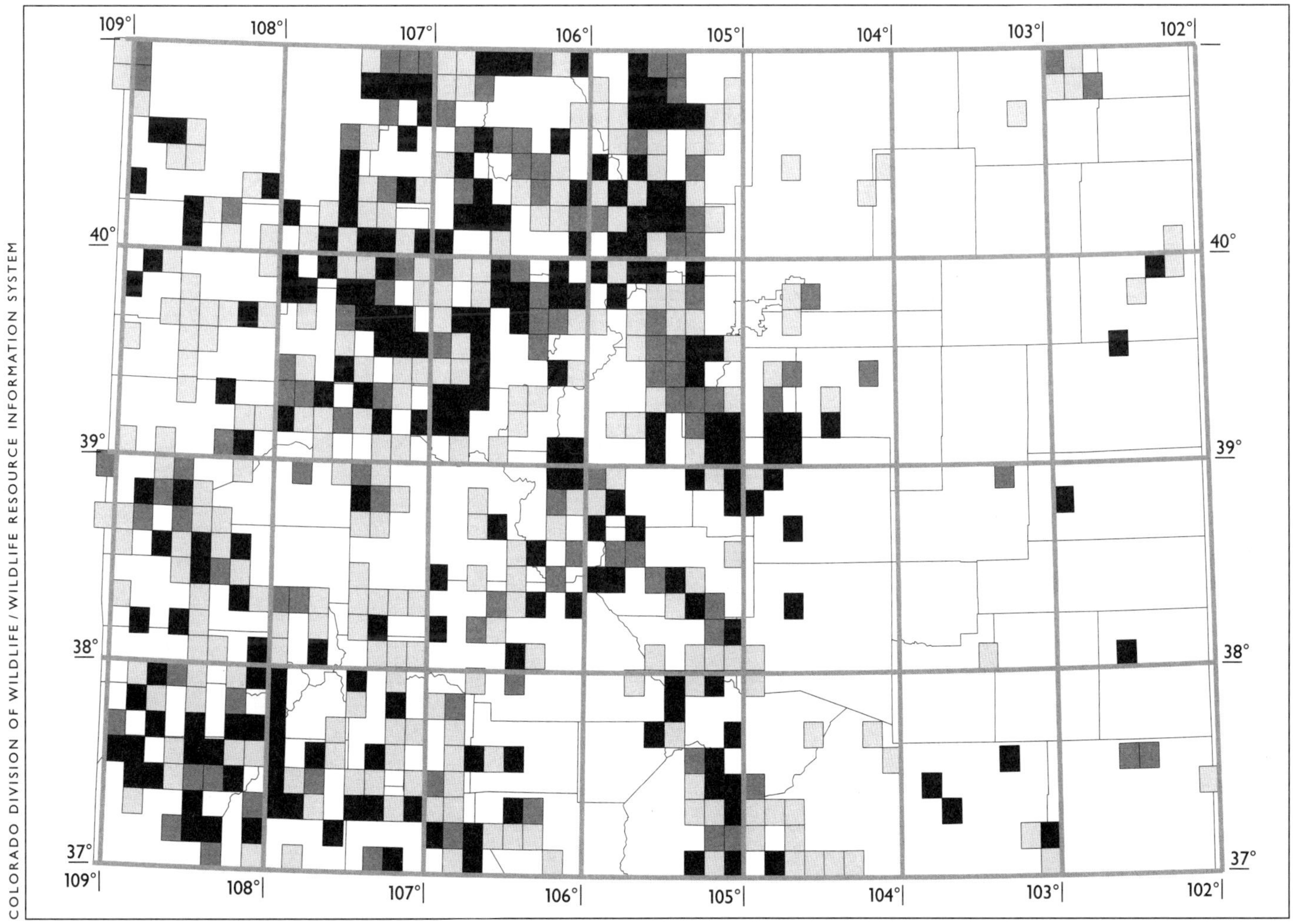
109°
108°
107°
106°
105°
104°
103°
102°
40°
39°
38°
37°
COLORADO DIVISION OF WILDLIFE / WILDLIFE RESOURCE INFORMATION SYSTEM

THREE-TOED WOODPECKER *Picoides tridactylus*

This woodpecker perches high atop the "most wanted species" list of many birders. Its quiet ways and narrow choice of habitats, mostly far from the convenience of paved roads, lend it an aura that enchants birdwatchers.

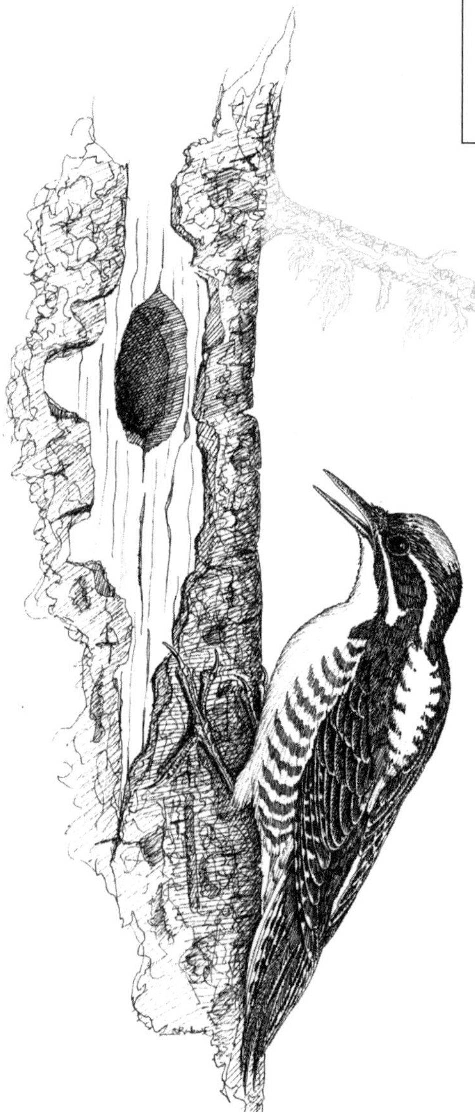

HABITAT

Although regarded by many as denizens of areas recently logged or ravaged by fire (Bent 1964), Three-toed Woodpeckers exhibit more adaptability than their reputation suggests. Indeed, the Atlas habitat codes contain only a single reference to a recently burned habitat and none to logged areas. If a preference for burned areas does not stand out, a decided preference for subalpine coniferous forests certainly does. Spruce/fir habitats claim, by far, the largest

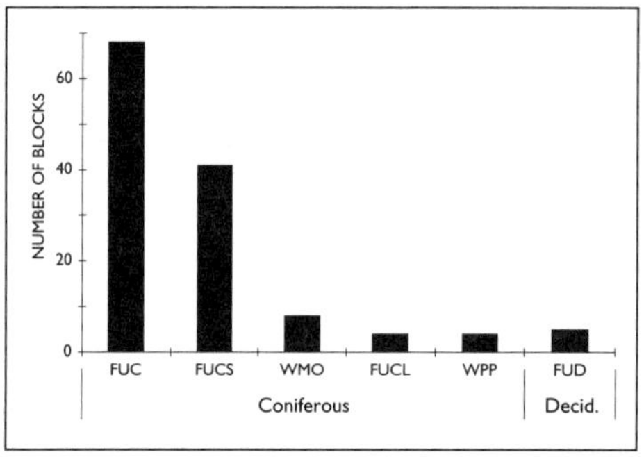
NUMBER OF BLOCKS
0
20
40
60
FUC
FUCS
WMO
FUCL
WPP
FUD
Coniferous
Decid.

share of observations, with only a handful of records from ponderosa, lodgepole, and aspen habitats. Unfortunately, these data do not address two other crucial factors in the site preferences of this species—insect populations and diseased trees. Wood-boring insects descend upon dead and dying trees. As the numbers of both insects and larvae increase, so do the numbers of Three-toed Woodpeckers that come to enjoy a feast of their own.

Records emerging from "lesser" habitats provide evidence of occasional non-loyalty to spruce/fir and burns. Atlas field workers recorded three nest sites in ponderosa pines (at least one of which had scorched bark but was still alive—AV) and two in aspen trees. Both habitats represent a deviation from typical nest locations, particularly the latter. Although records of nesting in aspen and other deciduous trees exist, such records in Colorado appear rarely in the literature (Bent [1939] cites one example). Atlas observations of Three-toed Woodpeckers ranged from about 7,000 to 12,000 feet (2100–3700 m) in elevation, with most above 9,000 feet (2700 m).

Regardless of the lack of Atlas data referring to burned areas, such sites figure heavily in the natural history of Three-toed Woodpeckers. Their densities reach much higher levels in burned areas than elsewhere. In the years immediately following a fire, these birds come to forage on the numerous wood-borers that chew away at the wood of fire-killed trees. The populations of wood-borers and woodpeckers gradually diminish over time. By five years after a fire, the Three-toed Woodpeckers have left for other areas, leaving behind a legacy of nest cavities for rebounding populations of bluebirds, swallows, and other secondary cavity-nesters (Taylor and Barmore 1980).

BREEDING

Courtship and breeding of Three-toed Woodpeckers occurs in April and May over much of their range (Winkler et al. 1995). Atlas data show, however, that breeding dates in Colorado fall later than this. Disregarding an enigmatic "courtship" record from 3 August 1987 (a male and female drumming and chasing each other as they spiraled up the drumming tree; Mike Carter pers. comm.), the data show that most young leave their holes in July. Fledglings exit wearing adult plumage—juvenile males even have the yellow crown (Bent 1939). Observations of dependent fledglings continued until well into August. Such dates point to the latter half of May and June for courtship and egg-laying.

Four eggs represent a normal clutch size. Both parents incubate, and the male attends both eggs and chicks during the night. The incubation period lasts 14 days. The nestling period lasts 22–26 days, after which the fledglings remain dependent on their parents for another month (Bent 1939, Winkler et al. 1995). Adults feed nestlings and fledglings a diet similar to their own. Wood-boring insect larvae and pupae extracted from beneath the bark of trees constitute the main part of the diet. Like other woodpeckers, Three-toeds raise only a single brood.

BREEDING PHENOLOGY

CODE		# OF RECORDS	DATE RANGE
C	Courtship	1	3 Aug
ON	Occupied Nest	5	26 Jun–9 Aug
NE	Nest with Eggs	2	30 May–6 Jul
NY	Nest with Young	8	20 Jun–13 Jul
FY	Feeding Young	9	8 Jun–14 Aug
FL	Fledged Young	15	2 Jul–19 Aug

DISTRIBUTION

Three-toed Woodpeckers occupy boreal forests in Scandinavia, northern Europe, Asia, and North America, always closely, but not exclusively, tied to spruce trees (Bock and Bock 1974). Colorado stands near the southern terminus of the species' distribution in North America, and atlasers found them occupying all the higher mountain ranges in the state. In the Rockies they occasionally move en masse to burns and insect infestations, but generally they stay in the spruce/fir.

Cooke (1897) described this bird as "not common anywhere but . . . scattered quite generally through the mountains." This seems as current today as when originally penned. Regions that provided food and shelter 100 years ago still do so today: the Atlas map shows Three-toed Woodpeckers throughout the mountains. The scattered nature of the distribution probably reflects the scattered nature of mature spruce/fir forests with decadent trees. Never abundant, Three-toeds rated an A4 abundance code in only one block and A3 in 28 blocks; these codes clustered in the Front and Park ranges, Flat Tops, Grand Mesa, and San Juan Mountains.

Atlas field work produced first Confirmations of breeding in five western Colorado latilongs, including the first documentation of the species' presence, as well as Confirmed breeding, in latilongs 37108 and 40107. This underscores that much of our information regarding the range of these woodpeckers in Colorado has come only recently.

One change has taken place within the world of Three-toed Woodpeckers: fire suppression has doubtless led to fewer available burned-over areas. Conversely, fire suppression has led to highly favorable conditions for infestations of the wood-boring insects that this species likes (David Leatherman pers. comm.). This suggests a population that remains fairly near historically high densities in unburned portions of the forest but which largely lacks the local concentrations that exploit the aftermath of fires.

BREEDING EVIDENCE

REPORTED IN 118 (7%) OF 1745 PRIORITY BLOCKS

☐	Possible	61 blocks	(52%)
▩	Probable	22 blocks	(18%)
■	Confirmed	35 blocks	(30%)

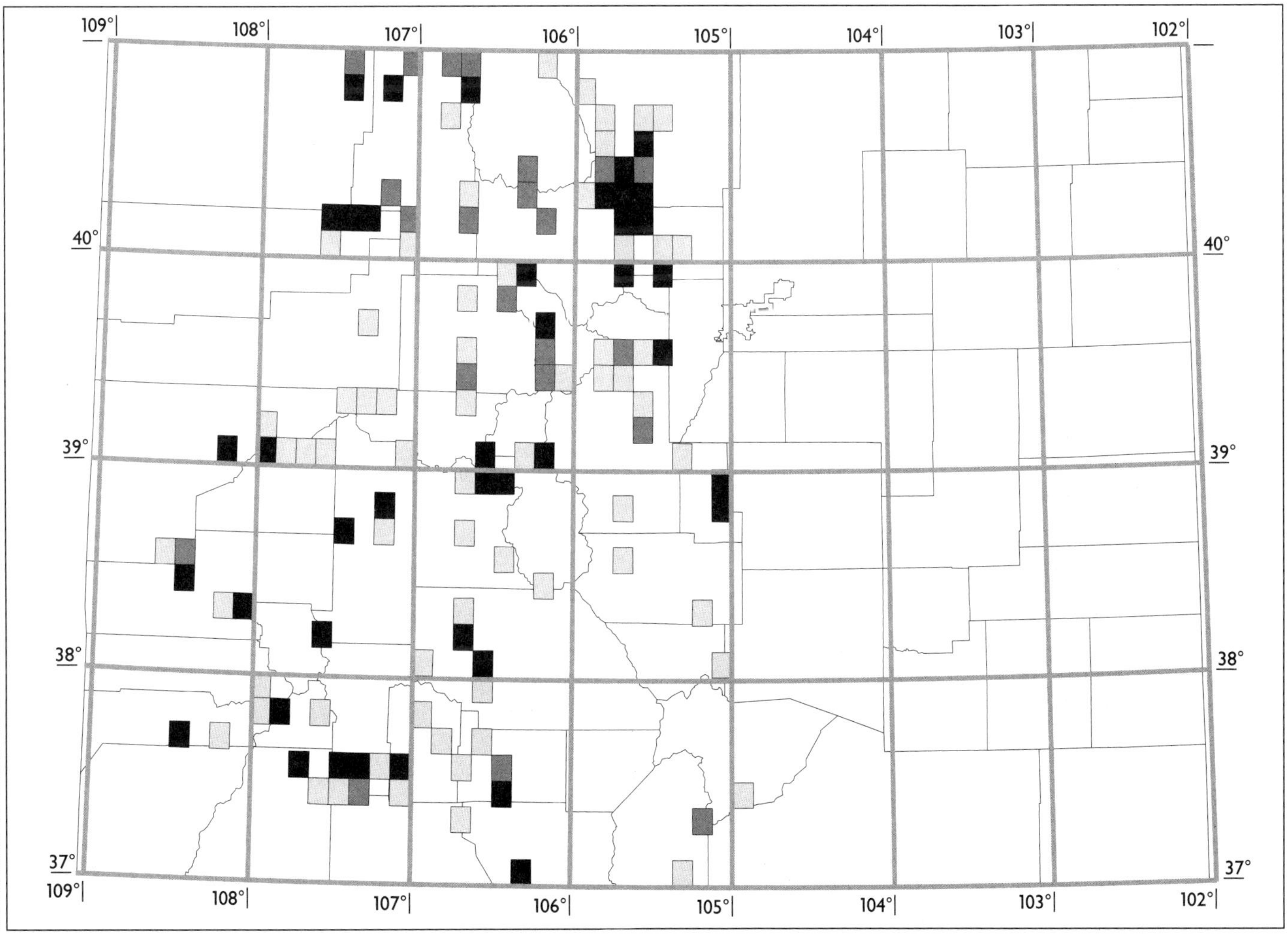
109°
108°
107°
106°
105°
104°
103°
102°
40°
39°
38°
37°

COLORADO DIVISION OF WILDLIFE / WILDLIFE RESOURCE INFORMATION SYSTEM

Flickers rank fourth on the list of the most frequently recorded species by the Colorado Atlas—with many good reasons for such high visibility. These large woodpeckers display a bold demeanor, they have bright colors, they fly in undulating swoops that show off the white rump patch, and they frequently broadcast the *flicka-flicka* vocalization that gave them the name "flicker."

HABITAT

In Colorado, Northern Flickers range from the plains to the high mountains, from prairie streams to cities to timberline (B&N). They feed on the ground more than most woodpeckers and eat more plant foods than most woodpeckers, although ants form about 50% of their diet (Moore 1995).

Ornithologists formerly classified Yellow-shafted and Red-shafted Flickers as separate species. Now they lump both forms together into one species, Northern Flicker (AOU 1983). Even when ornithologists considered them separate species, they noted that the two forms had very similar habits, habitats, and breeding behaviors (Beal 1911, Bent 1939, B&N, Terres 1980). The major difference between them is in coloration of the faces and feather shafts.

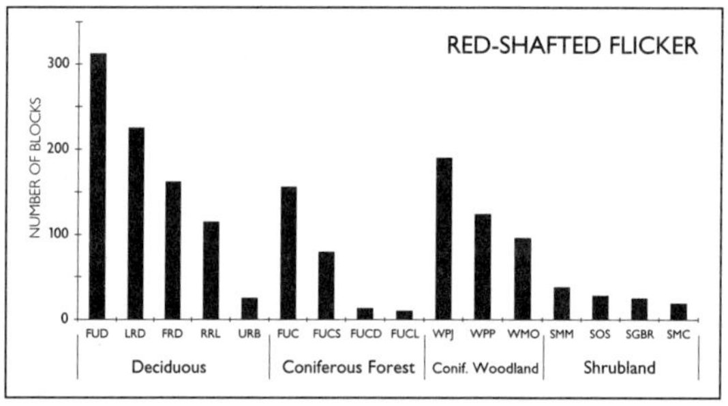
RED-SHAFTED FLICKER
NUMBER OF BLOCKS
0
100
200
300
FUD LRD FRD RRL URB FUC FUCS FUCD FUCL WPJ WPP WMO SMM SOS SGBR SMC
Deciduous
Coniferous Forest
Conif. Woodland
Shrubland

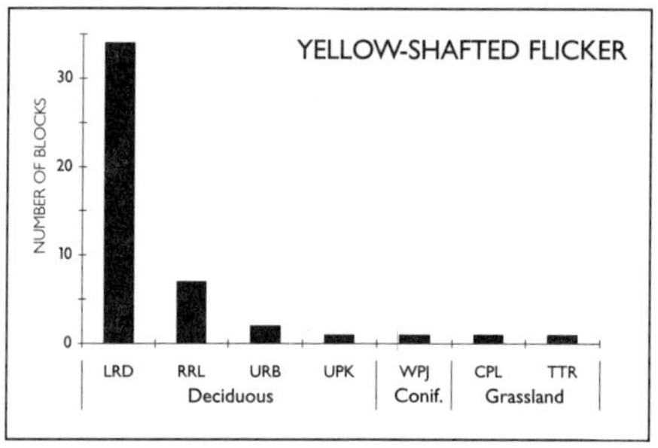
YELLOW-SHAFTED FLICKER
NUMBER OF BLOCKS
0
10
20
30
LRD RRL URB UPK WPJ CPL TTR
Deciduous
Conif.
Grassland

Atlas workers found the Red-shafted subspecies widespread, in habitats from aspen forest to rural settings. Atlasers found Red-shafted Flickers almost equally in deciduous (50% of reports) and coniferous (41%) habitats. Aspen topped the habitat frequency, although plains and foothills riparian combined exceeded aspen. Yellow-shafted birds occurred mostly in cottonwood river bottoms. The chart divides the habitats between deciduous and coniferous; an alternative analysis, contrasting plains and mountains, parallels the map, and shows a substantial majority in mountain habitats.

Conflicts arise between flickers and starlings, and other woodpeckers, for suitable nest-hole trees (B&N). Starlings can evict flickers from their holes. In a Denver cemetery, different starling pairs evicted a Red-shafted male from five nest holes in succession after he had excavated them. He and his mate finally bred in peace in the sixth one (Tina Jones pers. comm.). Flickers often clash with suburban homeowners when they drill holes in the wood siding or under the eaves or hammer on drain pipes. Old nest holes are in high demand by other birds besides starlings; the large size of the hole often attracts larger species such as American Kestrels and small owls (Moore 1995).

BREEDING

Flickers nest in holes drilled in trees, fence posts, utility poles, houses, in natural cavities, and in nest boxes if the size is right (B&N, Harrison 1979). They may reuse old nest cavities (Moore 1995). Both sexes excavate for one to two weeks. The entrance hole has a more irregular shape than that of other woodpeckers; it can measure 2–4 inches (5–10 cm) wide or high (Harrison 1979, Moore 1995). They prefer large trees (mean diameter at breast height 15.7 inches [40 cm]; Flack 1976). Nest holes tend to have an orientation toward the east or south (Moore 1995). The female lays one egg per day, with clutches ranging from 3 to 12, an average of 6.5, and both sexes incubate the eggs for 11 days (Moore 1995).

In 30 Atlas nest cards, aspen composed 80% of the nest trees (4 dead and 20 live). The ground elevation of the nests ranged from 5,400 to 11,480 feet (1660–3530 m). The height of nest holes above the ground ranged from 4 to 50 feet (1–15 m), with an average of 21 feet (6.5 m). The only Yellow-shafted nest card reported a cottonwood at 3,600 feet (1107 m) elevation.

In blocks in which atlasers found nests, they specified habitats in 274; 38% indicated aspen habitat—much higher than the 26% shown for all Atlas flicker sightings. This may illustrate a habitat difference or may show that atlasers could find nests more easily in aspen groves. Other habitats in blocks with Confirmations matched more closely the birds' general habitat use. That 8% of all Confirmations occurred in rural and urban areas documents that flickers tolerate humans during breeding.

The number of Atlas NY dates (from mid May to late July), compared with the number of nests with eggs, illustrates the ease of locating nests because of noisy nestlings. The young call for, greet, and solicit the parents returning to feed them.

BREEDING PHENOLOGY

RED-SHAFTED FLICKER

CODE		# OF RECORDS	DATE RANGE
C	Courtship	27	13 Apr–20 Jul
NB	Nest Building	11	16 Apr–13 Jul
ON	Occupied Nest	143	27 Apr–15 Aug
NE	Nest with Eggs	4	29 May–10 Jun
NY	Nest with Young	129	14 May–22 Jul
FY	Feeding Young	116	21 Apr–10 Aug
FL	Fledged Young	201	21 May–25 Aug

YELLOW-SHAFTED FLICKER

CODE		# OF RECORDS	DATE RANGE
C	Courtship	2	23 Apr–1 Jun
NB	Nest Building	1	2 Jun
ON	Occupied Nest	13	6 May–22 Jun
NY	Nest with Young	1	10 Jun
FY	Feeding Young	3	3 Jun–23 Jun
FL	Fledged Young	3	22 Jun–17 Jul

DISTRIBUTION

Northern Flickers live over most of North America, south into Guatemala. Of the two Colorado forms, the Yellow-shafted has by far the largest range, across the eastern U.S., and into a large part of Alaska, with the Red-shafted in the western U.S. except the Sonoran desert. Intergrades occur mainly in a broad band (the "rain shadow" of the Rockies) along the west edge of the Great Plains and the eastern foothills transition zone (Flack 1976, Moore 1995).

The two Atlas maps demonstrate the range division of the two forms. Yellow-shafteds breed on the eastern edge of Colorado. (See map for Yellow-shafted on page 544.) Red-shafteds occur all over the state but more widely in the west. Intergrades with traits of both parents can occur anywhere, but more commonly on the plains between the strongholds of the two subspecies. Atlas workers found only a few spots in the state without flickers—parts of the San Luis Valley and some empty spots on the eastern plains, due to lack of adequate trees. Flickers breed along the plains rivers and in some plains cities but the high plains lack the trees that flickers need.

BBS data show a major decline for Yellow-shafted Flickers, but no trend for Red-shafteds. Terres (1980) indicates that the decline since 1950 may be due to increased competition from starlings. Moore (1995) attributes it to starlings and snag removal.

BREEDING EVIDENCE

(RED-SHAFTED FLICKER)

REPORTED IN 1162 (67%) OF 1745 PRIORITY BLOCKS

Possible	340 blocks	(29%)
Probable	221 blocks	(19%)
Confirmed	601 blocks	(52%)

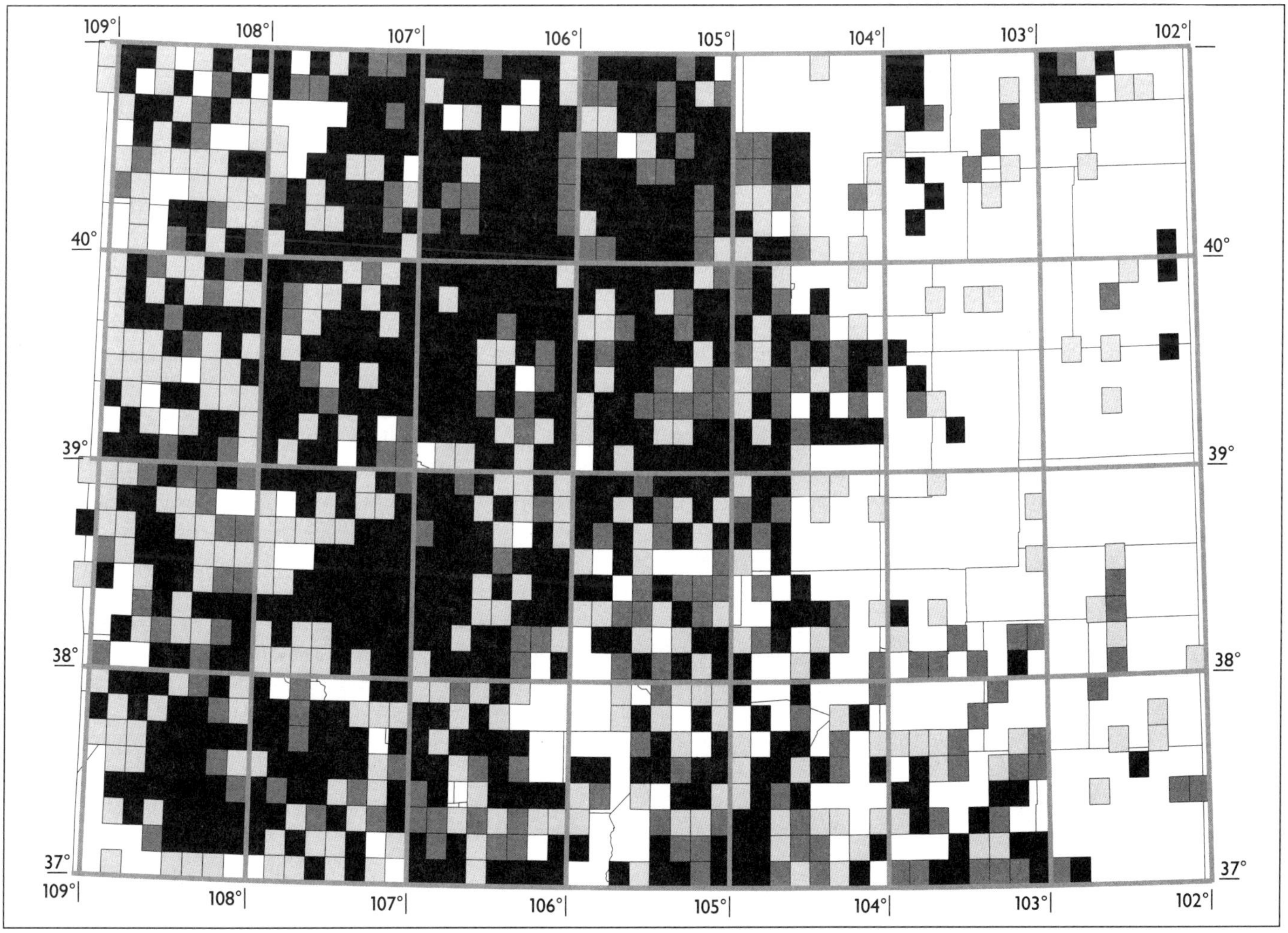
109°
108°
107°
106°
105°
104°
103°
102°
40°
39°
38°
37°

COLORADO DIVISION OF WILDLIFE / WILDLIFE RESOURCE INFORMATION SYSTEM

One of the most distinctive sounds heard in the high-country forests is the loud, strongly accented, three-part, whistled song, *Quick, three beers,* sung from the highest treetops by Olive-sided Flycatchers. They are birds of the forests, their favorite haunts the open coniferous woodlands where they look over some of Colorado's best mountain scenery.

HABITAT

Olive-sided Flycatchers commonly breed in the solitude of the forests where their breeding habitat has two basic components: snags and conifers. They most often inhabit parts of the forest with natural clearings, bogs, stream and lake shores with water-killed trees, forest burns, and logged areas with standing dead trees (Godfrey 1966).

In early successional forests Olive-sided Flycatchers may depend upon snags or residual tall trees for foraging and singing perches. In the Rockies of the northern U.S. they are one of the most abundant species in early post-fire communities (Hutto 1995) and they use clear-cuts provided that the loggers leave snags (Scott et al. 1982, Hutto 1995). Two percent of Atlas habitat codes referred to logged or burned areas.

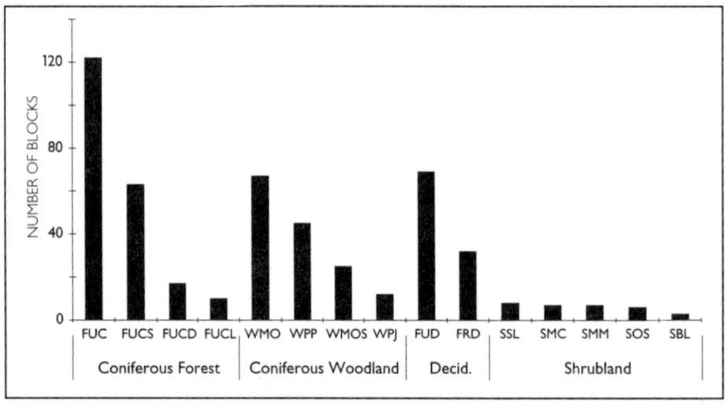
NUMBER OF BLOCKS
120
80
40
0
FUC FUCS FUCD FUCL WMO WPP WMOS WPJ FUD FRD SSL SMC SMM SOS SBL
Coniferous Forest
Coniferous Woodland
Decid.
Shrubland

Feeding technique dictates the snag association. Most flycatchers consume some non-flying insects (Murphy 1989) but Olive-sideds prey almost exclusively on flying insects, particularly bees, flies, moths, grasshoppers, and dragonflies (Bent 1942). They perch on snags to spot their prey.

In much of their range these flycatchers breed in old-growth coniferous forests (Raphael 1984, Carey et al. 1991) with nearby water (Bent 1942). Colorado birds do not always adhere to the water requirement as the flycatchers call their distinctive refrains from streamside to ridge top.

They occur less regularly and less abundantly in deciduous or mixed aspen/conifer forests (Finch and Reynolds 1988, Scott and Crouch 1988). As part of their breeding territory they use aspen forests clear-cut in patches with snags and remnant spruce trees (Scott and Crouch 1987).

Atlas habitat codes, which lack the detail to delineate snag densities or the presence or dimensions of forest openings, accentuate conifers. Atlasers recorded habitats dominated by conifers in 72% of the blocks with the flycatchers. Those blocks contained 84% of the birds (calculated from Atlas abundance codes). Atlasers recorded aspen in 20% of the blocks, but with only 13% of the birds.

Although many think of the Olive-sided as a high mountain species, atlasers found them in ponderosa pine (9% of habitat reports) and even Confirmed breeding in pinyon/juniper. Observers cited the latter habitat in a surprising 12 blocks—2% of the habitat reports.

Summer (boreal forests) and winter (tropical rain forests) habitats have similar physical structures—tall exposed perches near openings—suggesting that Olive-sideds depend more on forest structure than on tree species composition (Bruce Peterjohn pers. comm.).

BREEDING

Olive-sided Flycatchers aggressively defend their breeding territory. The male uses his treetop perch not only for singing and foraging but also to warn his mate of approaching intruders (Bent 1942).

In the West the bird's generally place their nests high in the trees, 15–70 feet up (4.5–21 m), and hide them in clusters of needles on horizontal branches. The birds use conifer twigs to build a loosely woven, quite shallow nest, and commonly line it with lichens, rootlets, grasses, and mosses (Bent 1942).

The many Atlas observations of breeding behavior in late May, earlier than "Blue Book" dates (Nelson 1993), suggest that these flycatchers begin to nest earlier than previously assumed. Incubation takes 16–17 days (Godfrey 1966) and young fledge in 21–23 days (Bent 1942). The Atlas provides the first phenology data from Colorado. Atlasers observed late nest-building activity in mid July when birds in most nests were feeding young, which implies either renesting or second broods.

BREEDING PHENOLOGY

CODE		# OF RECORDS	DATE RANGE
C	Courtship	1	25 Jun
NB	Nest Building	9	5 Jun–25 Jul
ON	Occupied Nest	8	2 Jul–26 Jul
NE	Nest with Eggs	1	5 Jul
NY	Nest with Young	11	15 Jul–31 Jul
FY	Feeding Young	24	23 Jun–4 Aug
FL	Fledged Young	10	23 Jun–2 Aug

DISTRIBUTION

Olive-sideds breed in the boreal forests from Alaska to Newfoundland and in the mountains of the western U.S. They winter from Mexico south to Peru. In Colorado they breed in the mountains from 7,000 to 11,000 feet (2135–3350 m; A&R).

The Atlas map depicts these flycatchers with their principal range in the high mountains and coniferous forests. The blocks cover all the high-country spruce/fir forests, and most other blocks that reported them lie adjacent to high-country blocks. The map, which tracks A&R almost precisely, also shows them absent from the intermountain parks and northwestern Colorado away from the high country.

BBS trends across the continent show a steady, statistically significant, negative trend of 3.9%/year. This implies a population drop of three-quarters in 31 years.

Interpretation of this trend is hampered by the lack of natural history information and an absence of data detailing specific factors that adversely affect the species. Changes in their winter habitats may have caused at least some of the decline (Altman 1997).

Some logging practices increase forest openings and edge habitats, which would seem to increase breeding habitat. The contrast of increased forest edge with declining populations may connote logged areas as "ecological traps" where the habitat appears suitable but in which the birds reproduce poorly and achieve low survival rates (Hutto 1995).

Demographic studies that compare productivity in human-induced forest-edge habitats with that in natural, mature forests would provide data essential for assessing population health and for evaluating BBS data (Hutto 1995, Altman 1997). Conservation of remaining mature forests becomes important to the future of this species.

Because of these trends and concerns, the U.S. Fish and Wildlife Service listed Olive-sided Flycatcher as a species of management concern in 1987 and 1995 (USFWS 1995). This attention may presage a hopeful future for these flycatchers with their ethereal treetop calls.

BREEDING EVIDENCE

REPORTED IN 411 (24%) OF 1745 PRIORITY BLOCKS

Possible	258 blocks	(63%)
Probable	93 blocks	(23%)
Confirmed	60 blocks	(14%)

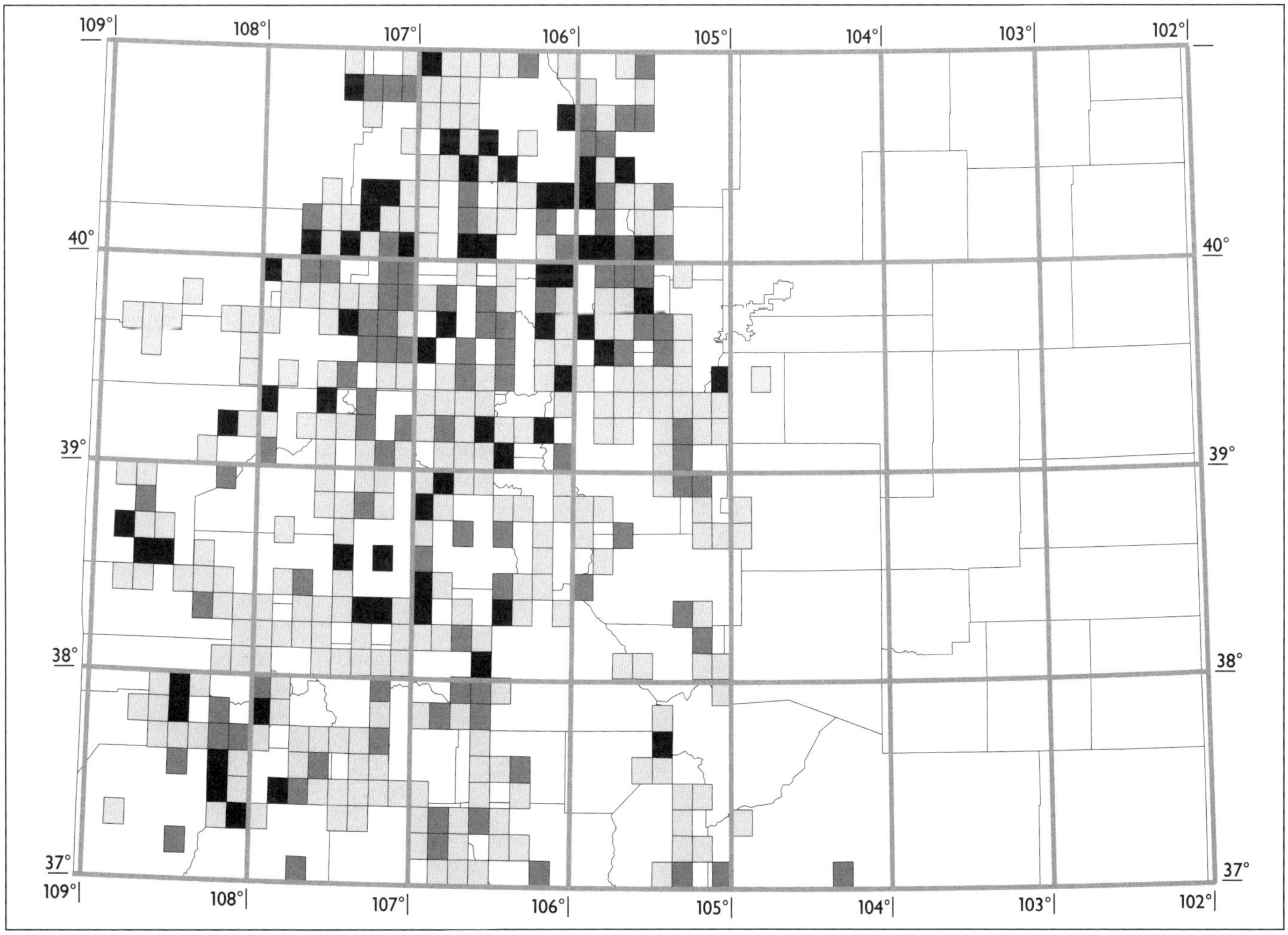
109°
108°
107°
106°
105°
104°
103°
102°
40°
39°
38°
37°

COLORADO DIVISION OF WILDLIFE / WILDLIFE RESOURCE INFORMATION SYSTEM

WESTERN WOOD-PEWEE *Contopus sordidulus*

Perhaps exasperated from attempts to distinguish between this species and its look-alike, the Eastern Wood-Pewee, a systematist gave this bird a scientific name that means "dirty little one" (Choate 1985). Bent (1942) described its unique song as possessing a "tinge of melancholy." Could it be lamenting its sordid name with its sad song?

HABITAT

Western Wood-Pewees breed in a wide variety of habitats, most commonly aspen forests, riparian deciduous areas, and ponderosa pine woodlands, and they range in elevation from 3,800 to 10,000 feet (1150–3000 m). Atlasers found them in almost every available forest, woodland, and shrub habitat. In any of these habitats, pewees require trees with dead tops or exposed branches, from 20 to 75 feet (6–23 m) above the ground, for foraging and singing posts (USDA Forest Service 1994). Pewees avoid nesting near trails in a Boulder open-space area with intense recreational pressure (Miller and Knight 1996, unpubl. rept.).

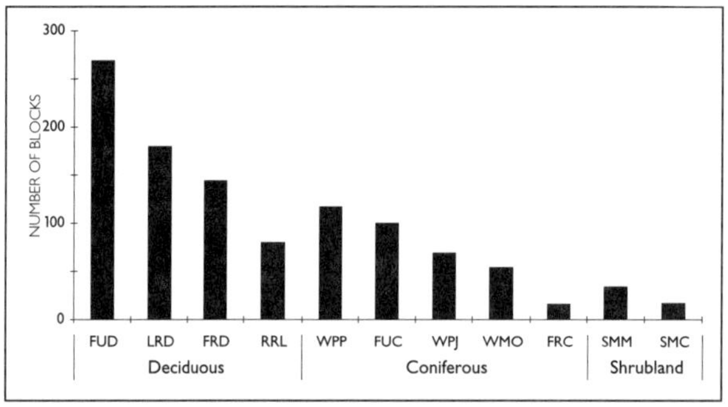

NUMBER OF BLOCKS
300
200
100
0
FUD
LRD
FRD
RRL
WPP
FUC
WPJ
WMO
FRC
SMM
SMC
Deciduous
Coniferous
Shrubland

The diet of pewees consists of 99% insects, about half of them flies (Bent 1942), which they capture by hawking their prey from high perches. They dart quickly into open areas, usually aiming their sallies horizontally or downward (Verbeek 1975, Cruz and Sanders 1996). After a successful sally, they usually return to the same perch but do not display perch loyalty after unsuccessful forays (Cruz and Sanders 1996).

Research suggests that pewees avoid competition with similar species through foraging techniques and habitat selection. Western Wood-Pewees differ from Cordilleran Flycatchers in foraging tactics and in choice of nest and perch sites. They use perch sites similar to those selected by Olive-sided Flycatchers (Verbeek 1975), but prefer different nesting habitats (A&R). The smaller Dusky and Hammond's flycatchers also use different habitats. The most likely competitor, the closely related Eastern Wood-Pewee, does not nest in Colorado.

Pewees attain their greatest abundance in the western two-thirds of Colorado, particularly in forested, mountainous areas. Field workers estimated over 100 nesting pairs in more than 100 blocks.

BREEDING

Arriving in Colorado in early May (Sclater 1912, Niedrach and Rockwell 1939), pewees begin courting as early as mid May. Males defend territories of three to four acres (7.5–8 ha; Eckhardt 1976). From nest-building through fledgling independence, their nesting cycle lasts 44–55 days (Ehrlich et al. 1988, Terres 1980, Lu Bainbridge, nest card).

Pewees place their cup nests from 8 to 80 feet (2–24 m) above ground in Colorado, most commonly 20–30 feet (6–9 m) high. They saddle the nest on a horizontal limb, often a dead branch. Females usually build the nest, made of twigs, plant fibers, leaves, bark, chrysalids, cocoons, and spiderwebs and lined with grass (Niedrach and Rockwell 1939, Bent 1942, B&N, Ehrlich et al. 1988). Off to a slow start, one pair of pewees in Jackson County spent four hours flying to and from a nest site rearranging one fiber. Apparently they got back on track, as two days later they had built about half the nest (Norm Barrett, nest card).

Normally single-brooded, pewees will quickly replace a lost nest (Bent 1942). The wide range of dates within each Confirmation code, from 44 to 68 days, may partially result from the broad elevational range. This disparity may further suggest replacement of nests lost to storms or other causes.

Pewees in the Front Range of Colorado exhibit low rates of nest predation (Miller and Knight, unpubl. data). Males attack jays and other nest robbers (Bent 1942), a behavior that could contribute to low predation rates. Displaying an unusual trait for a non-colonial passerine, pewees reuse nest sites between years, including, in one instance, a pair that successfully renested after depredation in the same year. The depredated nest site was also used successfully the following year; whether the same individuals reused the site is not known (Andrews et al. 1996). Pewees rarely host cowbirds (Ehrlich et al. 1988), with only two of 99 nests parasitized in a Boulder County study, neither of which succeeded

(Miller and Knight 1996 unpubl. data). In another Boulder County study, cowbirds parasitized none of 26 pewee nests at the same time they parasitized 47% of 132 Plumbeous Vireo nests (Chace et al. 1997).

Polygyny was observed in one instance in Colorado, in Rocky Mountain National Park. On three days in late July 1974, one male pewee fed nestlings in two nests, constructed by two different females in separate stands of trees within his territory (Eckhardt 1976).

BREEDING PHENOLOGY

CODE		# OF RECORDS	DATE RANGE
C	Courtship	5	17 May–30 Jun
NB	Nest Building	28	31 May–14 Jul
ON	Occupied Nest	59	25 May–24 Jul
NE	Nest with Eggs	6	30 Jun–17 Jul
NY	Nest with Young	35	13 Jun–12 Aug
FY	Feeding Young	70	24 May–7 Aug
FL	Fledged Young	39	15 Jun–9 Aug

DISTRIBUTION

Western Wood-Pewees breed throughout western North America, from eastern Alaska east to North Dakota, and south to Mexico and Central America. They spend the winter from Panama south to Bolivia and Peru. In Colorado, historical accounts report them as a common summer resident from the plains to about 10,000 feet (3048 m) in the mountains. Edwin Carter reported the first state nesting record on 8 July 1876, a nest with two eggs in Summit County (Sclater 1912, B&N).

The Atlas map differs from the A&R distribution map in details but not in overall distribution. Atlasers found pewees in a few blank spots on the A&R map: mainly scattered blocks on the plains and including a concentration in Elbert and Lincoln counties, plus the first nests in Latilong 37103—in the canyon country of eastern Las Animas County. They found pewees less frequently in Moffat County and western Rio Blanco, Garfield, and Mesa counties, and absent from much of the San Luis Valley. They also Confirmed breeding in four latilongs in which the Latilong Study listed them as only probable breeders.

BBS data show Western Wood-Pewees exhibiting a 1.5% per year continental decline, which signals a major population problem. The decline, however, has not attracted the attention of any Neotropical bird ranking systems.

BREEDING EVIDENCE

REPORTED IN 861 (49%) OF 1745 PRIORITY BLOCKS

▢	Possible	345 blocks	(40%)
▣	Probable	270 blocks	(31%)
■	Confirmed	246 blocks	(29%)

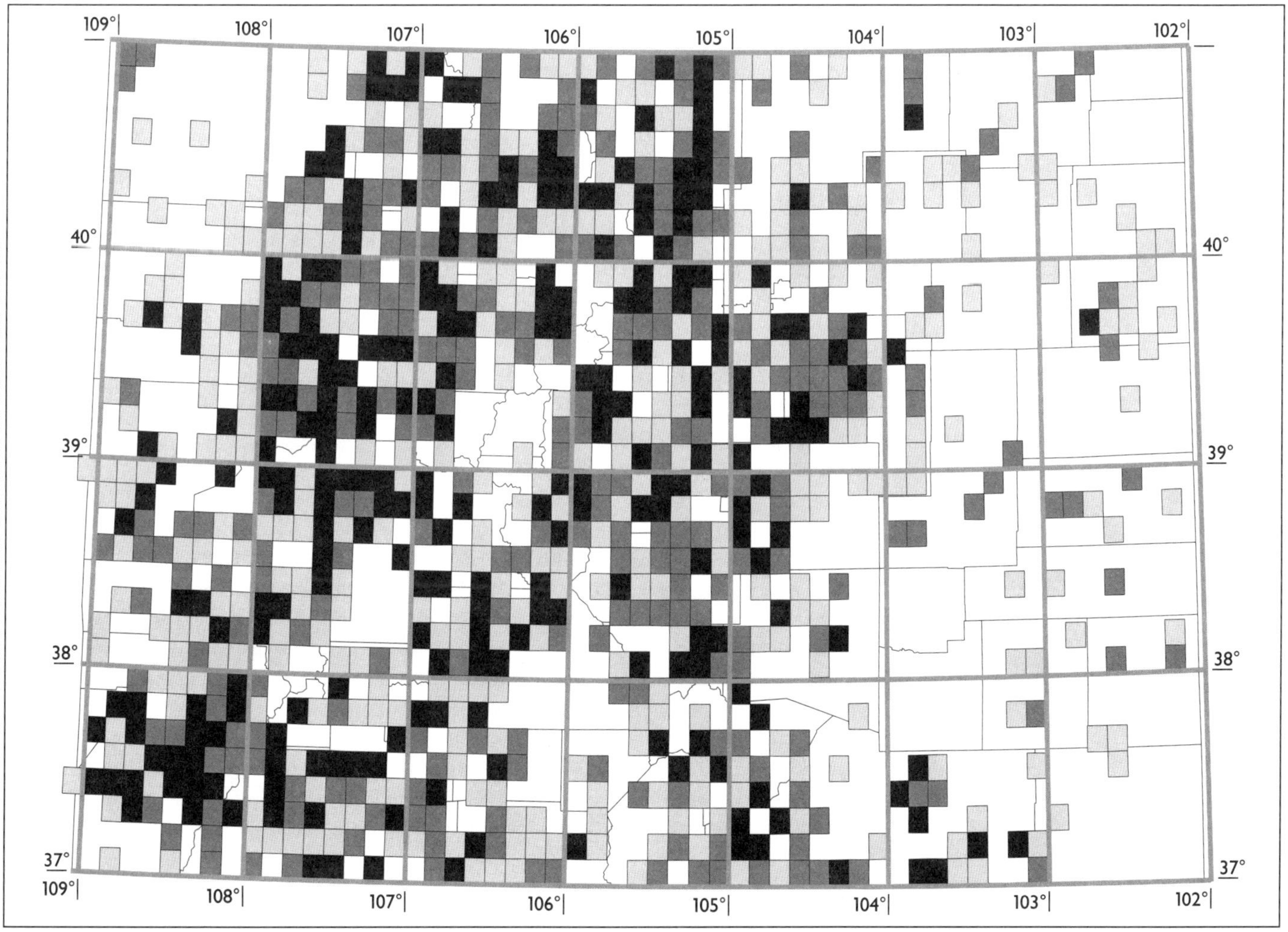
109°
108°
107°
106°
105°
104°
103°
102°
40°
39°
38°
37°

COLORADO DIVISION OF WILDLIFE / WILDLIFE RESOURCE INFORMATION SYSTEM

The diagnostic *fitz-bew* song identifies Willow Flycatchers as they sing along willow-lined streams and montane willow thickets. They have other distinctive traits such as diagnostic messy nests, which they sometimes build in the same fork of the same willow as the previous year.

HABITAT

In Colorado, Willow Flycatchers breed primarily in willows along foothills streams (B&N) and in middle- and high-altitude willow and alder carrs (JAS). They favor riparian thickets in the foothills and montane zones and willow-dominated open valleys and mountain parks—usually distant from trees. They often pick shrubbery with two or three layers of shrub height. The presence of water around the willows increases the forage basis by producing an

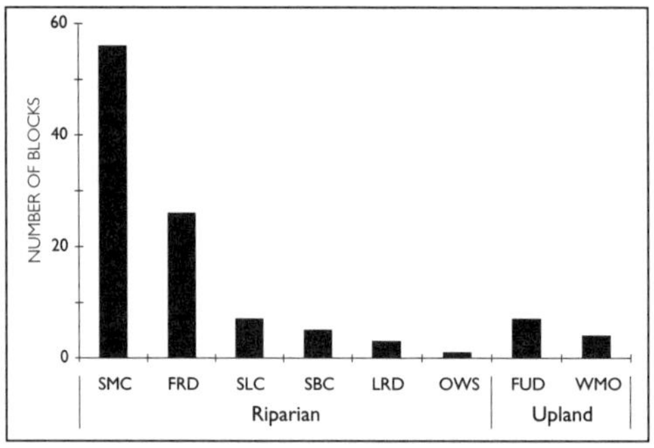
60
40
20
0
NUMBER OF BLOCKS
SMC
FRD
SLC
SBC
LRD
OWS
FUD
WMO
Riparian
Upland

abundance of insects. Atlas results substantiate this, with the predominant habitat—willow carrs at various elevations—reported in 61% of the blocks, and streamside riparian habitats reported in another 26%.

Dusky Flycatchers sometimes replace Willow Flycatchers in montane willow thickets adjacent to forests and in smaller patches of willows (A&R). Willow Flycatchers also occur in western Colorado where tamarisk has invaded riparian corridors.

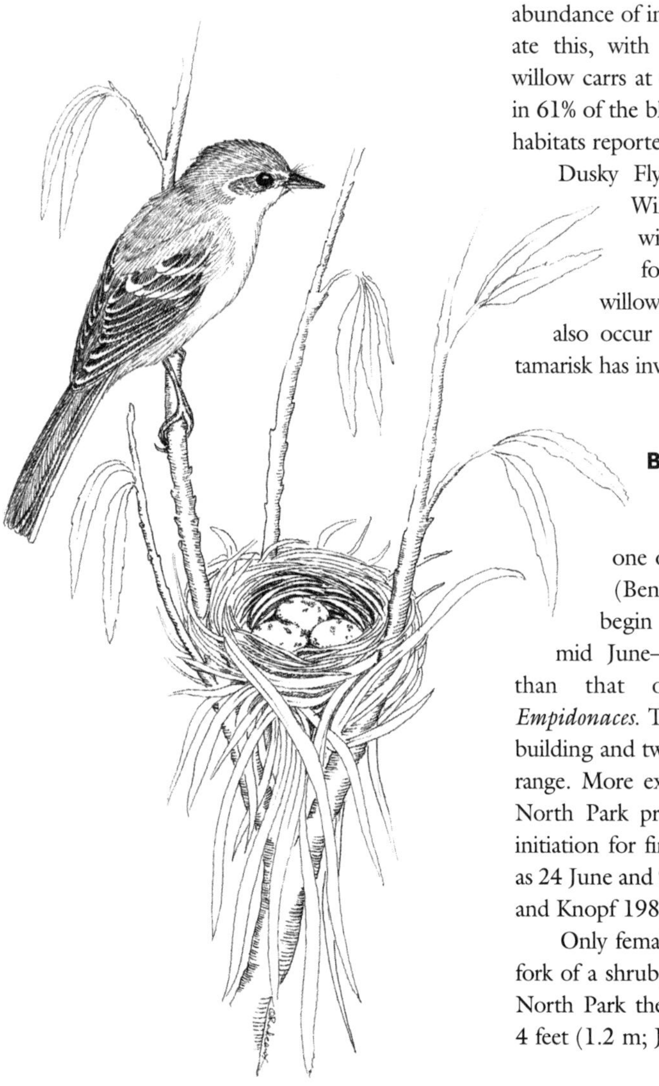

BREEDING

Late spring migrants, Willow Flycatchers are one of the latest passerine nesters (Bent 1942). Nesting does not begin until June—often not until mid June—but usually ends no later than that of any other Colorado *Empidonaces*. The four Atlas dates for nest-building and two for nests spanned a narrow range. More extensive data for 27 nests in North Park provide mean dates of clutch initiation for first nests and renest attempts as 24 June and 9 July, respectively (Sedgwick and Knopf 1988).

Only females build nests, usually in the fork of a shrub, especially willow or rose; in North Park the mean height of 33 nests is 4 feet (1.2 m; JAS). Nests resemble those of Yellow Warblers but are usually more loosely constructed (messier) and often have streamers of grass or shredded bark hanging down from the bottom. They practice polygyny in Colorado and elsewhere (Sedgwick and Knopf 1989)—at Malheur NWR in Oregon up to 10% of the males in a given year are polygynous (JAS unpubl. data).

Willow Flycatchers often sing while they have fledglings, and unpaired males sing all summer long (JAS). Atlas records for feeding young came mostly in July and records for fledglings came in late July and early August. Although nests are fairly easy to find and males are relatively vocal and conspicuous, breeding was Confirmed in only 19% of the blocks, lowest of any *Empidonax*.

Once a Willow Flycatcher finds suitable nesting habitat it tends to return year after year. Both sexes display notable site tenacity, especially after nesting successfully the previous year. In Oregon more than 50% return to the same general area, many to the same territory, and not uncommonly, they place the nest in the same bush, or even the same fork, as in previous years (JAS).

Cowbirds parasitize this species heavily —41% of the nests at Arapaho NWR. The narrow, linear nature of willow riparian habitats (such as that at the refuge and along many montane streams) favored by Willow Flycatchers provides edge and ideal opportunities for cowbird parasitism (Sedgwick and Knopf 1988).

BREEDING PHENOLOGY

CODE		# OF RECORDS	DATE RANGE
NB	Nest Building	4	4 Jun–18 Jun
NE	Nest with Eggs	2	18 Jun–28 Jun
FY	Feeding Young	9	15 Jun–9 Aug
FL	Fledged Young	5	17 Jul–18 Aug

DISTRIBUTION

Willow Flycatchers breed from central British Columbia east to Nova Scotia and south to northern Georgia and southern California. Winter range extends from southern Mexico to Panama.

The Atlas map shows a patchy distribution, like that of their favored habitats. They breed at elevations ranging from 6,000 feet to 10,000 feet (1830–3048 m), largely west of the Continental Divide except for North Park, which has a large breeding population. Atlasers found Willow Flycatchers in only 96

Atlas blocks, the fewest of any Colorado *Empidonax,* presumably due to the patchy and limited distribution of suitable willows. Some suitable habitat exists along rivers and streams on the plains but the flycatchers apparently do not breed there.

In the willow thickets preferred by Willow Flycatchers, atlasers reported abundance most commonly as 2–10 pairs per block. In the middle elevation willow thickets of Arapaho NWR, densities ranged from 0.17–0.48 *birds*/acre (42.5–117.5 *birds*/km^2) between 1980 and 1989 (Knopf and Stanley unpubl. data). BBS data show a moderate continental decrease of 1.2% per year; too few Colorado routes detect them to provide a valid trend for the state. Willow Flycatchers may have dropped in numbers, especially at lower elevations in western Colorado, because of degradation and modification of riparian habitats from overgrazing by livestock.

A subspecies—the Southwestern Willow Flycatcher (*E. t. extimus*)—has recently been listed as Endangered; it may have occurred historically in western Colorado. Status surveys and taxonomic studies currently under way will determine if populations of this endangered subspecies occur in the lowland willows and tamarisks of western Colorado (JAS).

Because Willow Flycatchers breed in riparian habitats, they are especially vulnerable to impacts associated with modification of riparian habitats such as channelization, recreational development, grazing, and agricultural conversions. Populations of Willow Flycatchers increase with the reduction of cattle grazing in shrub willow habitats and with the cessation of poisoning and removal of riparian willows (Taylor and Littlefield 1986).

BREEDING EVIDENCE

REPORTED IN 96 (6%) OF 1745 PRIORITY BLOCKS

☐	Possible	47 blocks	(49%)
▩	Probable	31 blocks	(32%)
■	Confirmed	18 blocks	(19%)

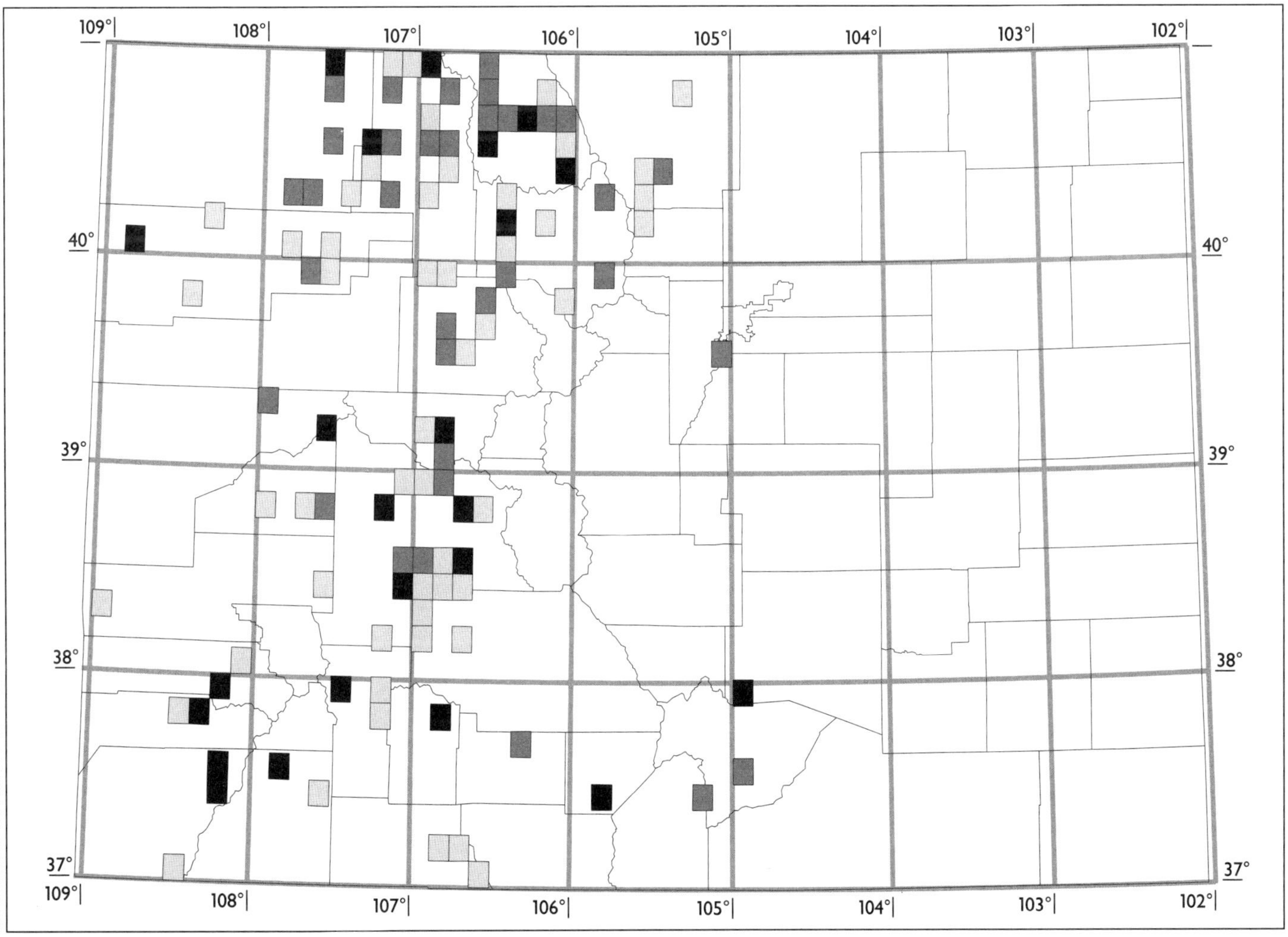
109°
108°
107°
106°
105°
104°
103°
102°
40°
39°
38°
37°

COLORADO DIVISION OF WILDLIFE / WILDLIFE RESOURCE INFORMATION SYSTEM

*E*mpidonax **flycatchers have long confused amateur and professional ornithologists alike. Without vocal, habitat, or nest-site clues, the identity of small flycatchers is often best left as "an Empid." But in a coniferous forest in Colorado, if you see a small, drab flycatcher, hear the burry, unmusical song notes of the male from high in the canopy, or hear the occasional, spaced, Pygmy Nuthatch–like *pip* notes, you are probably in the territory of a pair of Hammond's Flycatchers. The architecture of Hammond's nests also helps to distinguish them: they saddle their nests on limbs, as wood-pewees do.**

HABITAT

Each *Empidonax* has a few distinctive traits. Hammond's like conifer forests with little ground cover, and they sometimes feed on the ground. In Colorado, they breed from 7,000 to 10,000 feet (2135–3048 m; B&N). Some type of coniferous forest composed the habitat in nearly 85% of the blocks where atlasers recorded Hammond's Flycatchers. Hammond's prefer mature forests of spruce/fir, ponderosa pine, Douglas-fir, or mixed coniferous/aspen, particularly those with limited understory. They also occur in Colorado in aspen and riparian deciduous habitats, but these compose only a small portion (12%) of blocks where Hammond's were recorded. In mature, closed-canopy coniferous forests, they often are the only *Empidonax* (A&R). Habitat can overlap with that of Cordilleran, Dusky, and, occasionally, Gray flycatchers, but each has distinctive songs and calls, nests, and nest placement. In migration Hammond's Flycatchers occur in all wooded habitats (A&R).

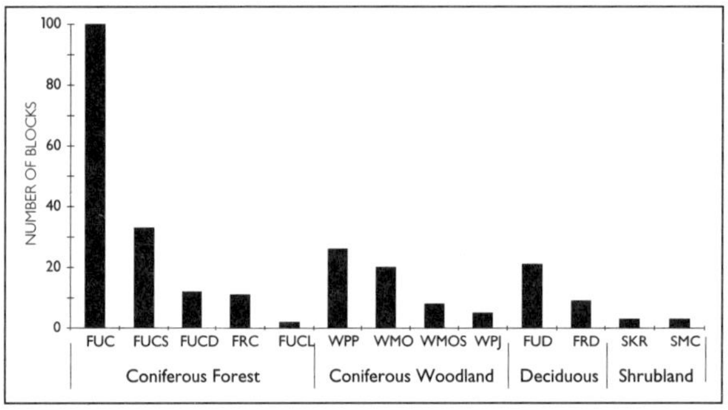
NUMBER OF BLOCKS
100
80
60
40
20
0
FUC
FUCS
FUCD
FRC
FUCL
WPP
WMO
WMOS
WPJ
FUD
FRD
SKR
SMC
Coniferous Forest
Coniferous Woodland
Deciduous
Shrubland

Another trait distinguishes Hammond's from other Empids: they spend an appreciable amount of time feeding on or near the ground—despite feeding mostly in "shaded airways" 20–100 feet (6–30 m) above the ground (Sedgwick 1994).

BREEDING

Hammond's Flycatchers nest from late May to early August in Colorado, comparable to their nesting phenology throughout the West (Sedgwick 1994). Atlas nest-building dates ranged from late May to mid June, and nestlings and fledglings were observed primarily in July. Attesting to the difficulty in finding nests of this species, atlasers reported only five instances of nest-building, two nests with eggs, and only one nest with young. They Confirmed breeding in only 19% of the blocks with Hammond's, the lowest, with Willow Flycatcher, of the Empids. Nests, usually saddled on a horizontal conifer branch or next to the main stem of an aspen, are constructed only by the female. Nests more closely resemble those of wood-pewees in shape (short, squat) and location (on horizontal limbs 10–100 feet [3–31 m] high) than those of other *Empidonaces* (Mannan 1984, Sakai 1987, Sedgwick 1975, 1994).

Calls of this species are distinctive as well. The distinctive *pip* alarm notes (given by both sexes) distinguish them from all other Empids. Also, irregularly throughout the breeding season, Hammond's males give a soft, whistle-like *k-lear* call, which superficially sounds like the *du* element of the Dusky Flycatcher *du-hic* sequence (Sedgwick 1994). This call, in combination with buzzier *wheezee* notes, is sometimes given during the pair-formation period as birds swoop and chase in a series of slow flights among the trees (Sedgwick 1994).

Neither Atlas nor other observers have recorded instances of cowbird parasitism in Colorado (Chace and Cruz 1996), but Hammond's have served as cowbird hosts elsewhere (Friedmann and Kiff 1985).

BREEDING PHENOLOGY

CODE		# OF RECORDS	DATE RANGE
NB	Nest Building	5	31 May–24 Jun
NE	Nest with Eggs	2	4 Jul–22 Jul
NY	Nest with Young	2	9 Jul–18 Jul
FY	Feeding Young	16	16 Jun–5 Aug
FL	Fledged Young	16	6 Jun–6 Aug

DISTRIBUTION

Hammond's Flycatchers breed in mature and old-growth coniferous habitats from Alaska to northern New Mexico and south-central California. Winter range extends from southern Arizona through the highlands of Mexico to north-central Nicaragua. In Colorado, Hammond's Flycatchers occur in conifers throughout the mountains, above 6,000 feet (1830 m). Atlasers recorded them in only 199 blocks, and historical nesting records (late 1800s, early 1900s) are few as well (from only four mountain counties; B&N). Both historical records and the apparent discontinuities in distribution shown on the Atlas map reflect more the furtive nature of Hammond's Flycatchers and, especially, identification problems, rather than actual abundance and distribution.

In coniferous habitats in Colorado, atlasers most commonly estimated their abundance as 2–10 or 11–100 breeding pairs/block. Densities of 0.06 to 0.19 *birds*/acre (14–47 *birds*/km²) in aspen and 0.06 to 0.11 *birds*/acre (14–28 *birds*/km²) in aspen/coniferous habitat in Colorado were reported by Beaver and Baldwin (1975). Densities in other western states (California, Oregon, Montana) range from 0.06 to 0.33 pairs/acre (14.5–80.4 pairs/km²; Sedgwick 1994).

On the BBS, the northwest Wyoming mountains recorded the highest number of birds/route (5.14); Colorado had 1.06 birds/route. Continentally, significantly more routes showed increases (49.8%) than decreases (43.1%) in the 1966–1991 period.

Because they like mature and old-growth forests, logging of such stands can adversely affect this species. They need tracts larger than 25 acres (10 ha) and a minimum of 80–90 years of age to sustain viable populations (Sakai 1987). Intact older stands of more than 37 acres (15 ha) benefit them more than those with openings and scattered large trees (Sakai and Noon 1991).

BREEDING EVIDENCE

REPORTED IN 200 (11%) OF 1745 PRIORITY BLOCKS

☐	Possible	125 blocks	(62%)
▩	Probable	36 blocks	(18%)
■	Confirmed	39 blocks	(20%)

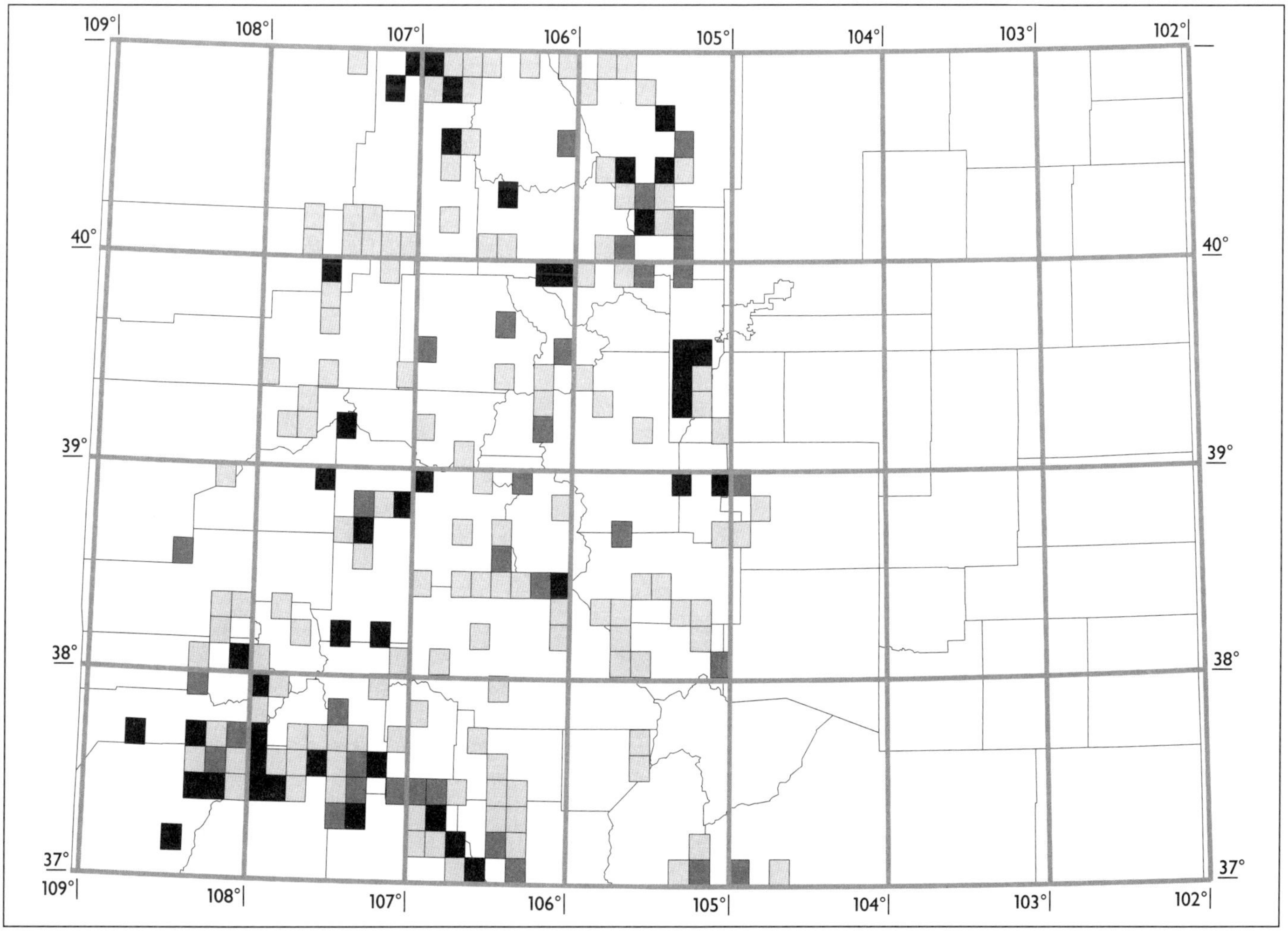
109°
108°
107°
106°
105°
104°
103°
102°
40°
39°
38°
37°

COLORADO DIVISION OF WILDLIFE / WILDLIFE RESOURCE INFORMATION SYSTEM

Because they dip their tails downward slowly and evenly, like a subdued phoebe, and breed only in pinyon/juniper associations, Gray Flycatchers are one of the easiest of the Colorado *Empidonax* species to identify. They inhabit an ecosystem (most of which lies far from the populated Front Range of eastern Colorado) largely ignored by avian ecologists (Sedgwick 1987) and infrequently visited by amateur ornithologists. In fact, until 1924, Gray Flycatchers were considered only as stragglers in the state; the first nesting record for this bird in Colorado was not secured until 31 May 1924 (B&N).

HABITAT

Gray Flycatchers prefer dry habitats, particularly open pinyon/juniper woodlands mixed with sagebrush. In Colorado they occur almost exclusively in pinyon/juniper; atlasers reported this habitat type in 83% of the blocks in which they recorded these birds. Sagebrush and tall desert shrub composed an additional 7% of reported habitats—not nesting habitat but rather part of the nesting territory. In other parts of the West, Gray Flycatchers also occur in open pine/oak associations, and open ponderosa pine and Jeffrey pine forests with sagebrush in the understory (Johnson 1963, AOU 1983). An *Empidonax* in any other habitat during the summer months most certainly is not a Gray Flycatcher.

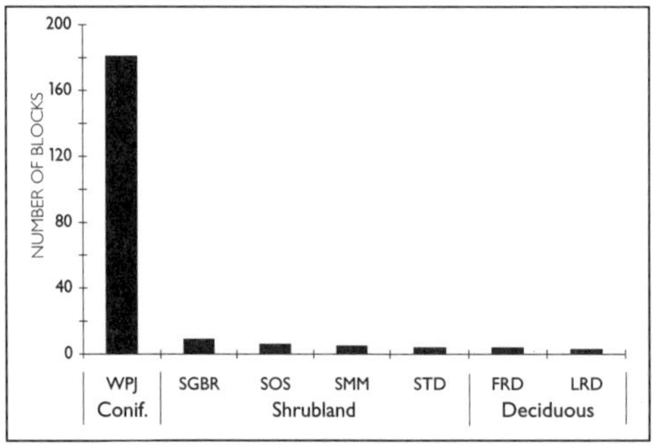
200
160
120
80
40
0
NUMBER OF BLOCKS
WPJ
Conif.
SGBR
SOS
SMM
STD
Shrubland
FRD
LRD
Deciduous

Only two species of *Empidonax* are likely to occur in pinyon/juniper during the breeding season in Colorado: Gray and Dusky. When these two species do come into contact, territories usually do not overlap (Johnson 1963). In migration, Gray Flycatchers occur in sagebrush, semidesert shrublands, and riparian areas (A&R).

BREEDING

Of the five commonly breeding Colorado *Empidonaces,* Gray Flycatchers arrive in spring the earliest, often by late April (B&N). Atlas records reflect this, with dates for nest-building as early as 13 May, nests with eggs by 4 June, and nests with young as early as 9 June. The nesting season continues through early to mid August.

Usually only the female builds the nest, but occasionally the male helps (Johnson 1963). Typically she places the nest, somewhat bulky and loosely constructed, in the crotch of a juniper, pine, or sagebrush. She constructs it of bark, plant down and stems, pine needles, and grass, and lines it with feathers and hair (Bent 1942, Johnson 1963). By placing their nests in the forks of juniper trees and building them mainly of gray, weathered strands of juniper bark, the birds make the nests surprisingly difficult to see. Males defend territories—intraspecifically against other Gray Flycatchers and interspecifically in areas of overlap with Dusky Flycatchers (Johnson 1963, Sedgwick 1993). Brown-headed Cowbirds occasionally parasitize Gray Flycatchers in Colorado (two Atlas records; Chace and Cruz 1996; Appendix C) and elsewhere (Friedmann and Kiff 1985).

BREEDING PHENOLOGY

CODE		# OF RECORDS	DATE RANGE
C	Courtship	2	29 May–9 Jul
NB	Nest Building	6	13 May–24 Jun
ON	Occupied Nest	1	2 Jun
NE	Nest with Eggs	3	4 Jun–3 Jul
NY	Nest with Young	5	9 Jun–14 Jul
FY	Feeding Young	21	23 May–31 Jul
FL	Fledged Young	16	22 Jun–1 Aug

DISTRIBUTION

Gray Flycatchers breed in all western states except Montana and winter from southern California and central Arizona south into Mexico (AOU 1983). They have the most limited distribution of any Colorado *Empidonax*; they occur only on mesas and foothills from 5,000 to 7,000 feet (1525–2135 m; A&R). Atlas records come mostly from the western third of the state, and more in the northwest than the southwest.

East of the Continental Divide, Fremont County has the most breeding records, and the flycatchers follow the pinyon/juniper woodlands north to Buena Vista. Surprisingly, very few follow those woodlands east to the mesas of Las Animas County (included in the A&R range). Atlas data stretch the altitudinal range up to 8,400 feet (2560 m) east of Buena Vista and 8,700 feet (2650 m) in the San Luis Valley.

Atlas workers most commonly reported Gray Flycatcher abundance as 11–100 breeding pairs per block in pinyon/juniper woodlands. The birds achieve densities of 13 birds/mi^2 (5/km^2) in pinyon/juniper habitat in Utah (Graham and Nettell 1992). Breeding Bird Surveys from the 1968–1994 period show a significantly increasing trend for Gray Flycatchers in both the pinyon/juniper stratum (12.4%/year, n=14 routes, 0.73 birds/route) and the U.S. (6.4%/year, n = 67 routes, 1.49 birds/route).

Two conflicting trends—one long-term, the other more recent—affect Gray Flycatcher habitat. Over the long term, grazing has caused pinyon/juniper woodlands to replace grasslands in much of western Colorado. More recently, because of the low commercial value of pinyon and juniper, vast areas of pinyon/juniper woodland have been converted to grazing lands, especially between 1945 and 1965. The most widely used conversion technique, chaining, knocks down the trees by dragging an anchor chain between two bulldozers (Sedgwick 1987). Chaining dramatically alters vegetation, and Gray Flycatchers, among other species, drop to significantly lower densities in chained plots (Sedgwick and Ryder 1986).

Gray Flycatchers, as well as most members of other nesting and foraging guilds of birds dependent on mature pinyon/juniper—the foliage-gleaning, live-bark-foraging, foliage-nesting, and cavity-nesting guilds—cannot tolerate removal of the pinyon/juniper overstory canopy. Because Gray Flycatchers nest almost exclusively in pinyon/juniper woodlands in Colorado, they are especially vulnerable to alterations of these woodlands.

Colorado hosts 10–19% of the entire breeding population of Gray Flycatchers (Priorities 1995); the Atlas estimate of 167,000 breeding pairs gives support to the importance of Colorado to this Great Basin species. Accordingly, Priorities 1995 ranked it in the top 20 species needing special attention. Two positive factors address the species' status. BBS routes show the species increasing in the pinyon/juniper physiographic region (5.7%/year) and a stable trend in Colorado. Second, as chaining has subsided, the long-term trend replacement of grasslands by pinyon/juniper woodlands has resumed. This would seem to indicate improving conditions for Gray Flycatchers, but the species bears watching because of its limited range.

BREEDING EVIDENCE

REPORTED IN 186 (11%) OF 1745 PRIORITY BLOCKS

☐	Possible	98 blocks	(53%)
▩	Probable	39 blocks	(21%)
■	Confirmed	49 blocks	(26%)

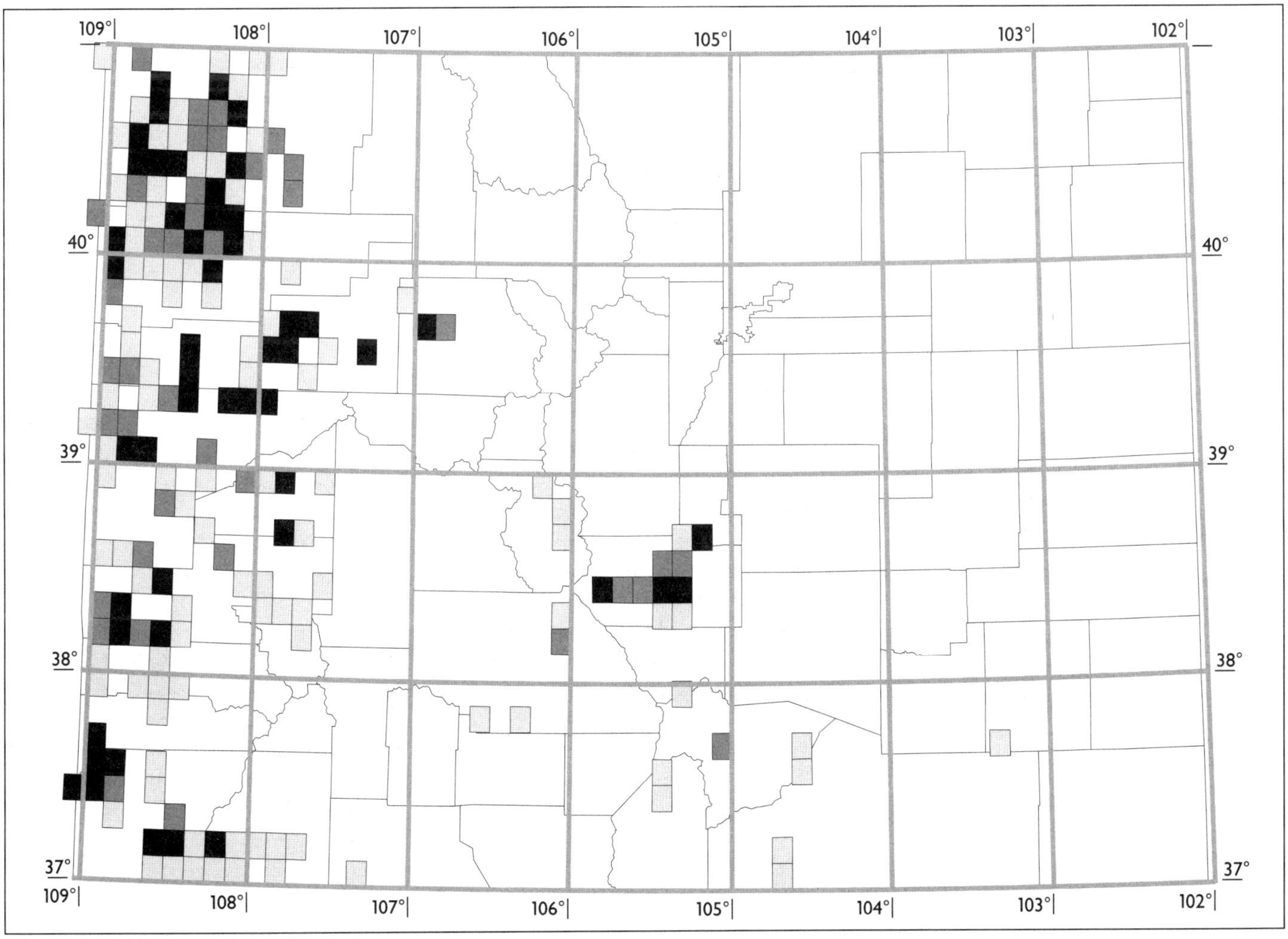
109°
108°
107°
106°
105°
104°
103°
102°
40°
39°
38°
37°

"Notoriously difficult to identify" characterizes Dusky Flycatchers and indeed, many of the *Empidonax* flycatchers. Habitat generalists compared to other Empids, Duskys select a variety of habitats, with shrubs the most consistent component. Their nest architecture contrasts with Hammond's: Duskys build their nests in the forks of shrubs instead of saddled on a tree limb. Perhaps the most distinguishing characteristic of Dusky Flycatchers is the plaintive, whistle-like, *du-hic* call, given by the males sporadically throughout the day but especially around dusk, from high atop a conifer or other large tree (Sedgwick 1993a).

HABITAT

Dusky Flycatchers occur in a wide diversity of habitats, including oak shrublands, willow riparian, aspen groves, open coniferous forest, mountain chaparral, and open brushy areas, often with scattered trees (Grinnell et al. 1930, Sedgwick 1975, AOU 1983). Colorado Atlas workers found them most commonly in areas dominated by shrubs (scrub oak, willow thickets, and mountain shrublands) and in aspen stands. Duskys also occur regularly in mountain and foothills riparian habitats, in open-canopy ponderosa pine forests, and in openings in pinyon/juniper woodland (Sedgwick 1987, 1993a). Dusky Flycatchers are scarce in forests with little shrubby understory. Although habitat can overlap with that of both Gray (pinyon/juniper) and Hammond's (open coniferous forests) flycatchers, each of the three has distinctive nests and nest placement.

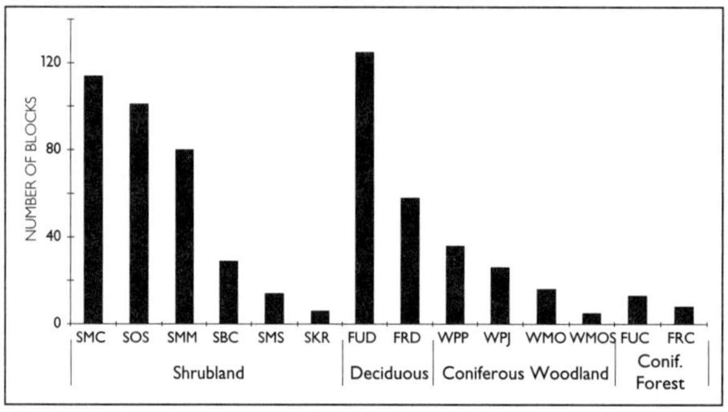
NUMBER OF BLOCKS
0
40
80
120
SMC SOS SMM SBC SMS SKR FUD FRD WPP WPJ WMO WMOS FUC FRC
Shrubland
Deciduous
Coniferous Woodland
Conif. Forest

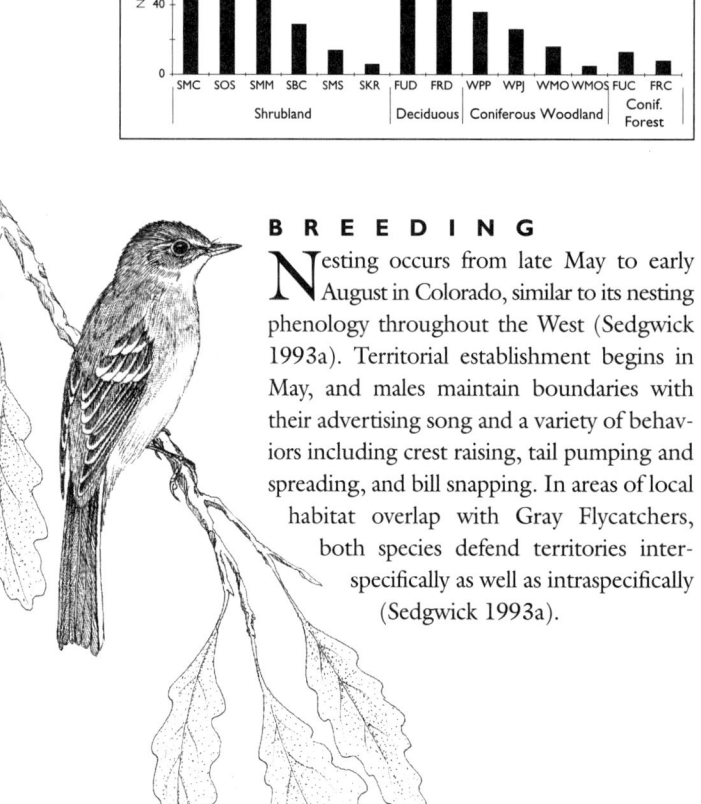

BREEDING

Nesting occurs from late May to early August in Colorado, similar to its nesting phenology throughout the West (Sedgwick 1993a). Territorial establishment begins in May, and males maintain boundaries with their advertising song and a variety of behaviors including crest raising, tail pumping and spreading, and bill snapping. In areas of local habitat overlap with Gray Flycatchers, both species defend territories interspecifically as well as intraspecifically (Sedgwick 1993a).

Like most *Empidonax* flycatchers, only the female builds the nest, a soft, neatly woven, cup-shaped structure constructed largely of grasses and finely shredded plant material. They generally place nests in the upright crotches of shrubs, but occasionally use aspens, cottonwoods, and conifers as nest substrates. The species typically nests within a few meters of the ground.

Incubation feeding, the delivery of food by males to incubating females, occurs fairly frequently during incubation, and mate feeding continues into the early stages of the brooding phase (Sedgwick 1993b). Incubation feeding, which occurs among 54% of western Montana birds, distinguishes this species from most other members of the genus (Sedgwick 1993b).

Atlas records suggest that June is the primary period of nest-building in Colorado, but one block had an occupied nest on 17 May. Atlasers observed nearly all nests with young during July and most adults carrying food for young (FY, CF) in July as well. Late July NB and NY and August FY and CF codes probably pertain to pairs whose earlier nests failed. Most Confirmations of this furtive species were fledglings or adults carrying food.

Brown-headed Cowbirds occasionally parasitize Dusky Flycatchers in Colorado (four Atlas records; Chace and Cruz 1996; Appendix C) and elsewhere (Friedmann and Kiff 1985, JAS).

BREEDING PHENOLOGY

CODE		# OF RECORDS	DATE RANGE
C	Courtship	3	5 Jun–20 Jun
NB	Nest Building	20	23 May–16 Jul
ON	Occupied Nest	23	17 May–16 Jul
NE	Nest with Eggs	11	15 Jun–20 Jul
NY	Nest with Young	15	11 Jun–5 Aug
FY	Feeding Young	69	7 Jun–7 Aug
FL	Fledged Young	28	2 Jul–9 Aug

DISTRIBUTION

Dusky Flycatchers breed from southwestern Yukon, southwestern Alberta, and southwestern Saskatchewan south throughout much of mountainous western North America. They winter from southern Arizona south through Mexico to northwestern Guatemala. They breed throughout western Colorado and east of the Continental Divide to the edge of the foothills. Atlasers recorded them in more blocks than any other *Empidonax* except for Cordilleran Flycatcher and Confirmed breeding equally with Cordillerans (35%). Historical nesting records (late 1800s, early 1900s) come from only four Front Range and four Western Slope counties (B&N). Despite a relative paucity of historical records, their distribution today probably resembles historical distribution.

In the three habitats where Dusky Flycatchers were most commonly reported (aspen, willow/alder, and scrub oak), Atlas observers estimated their abundance most frequently as 11–100 breeding pairs/block. Densities from a variety of habitats and regions in the West ranged from 0.08–0.22 territories/acre (20–54 territories/km^2; Sedgwick 1993a). In a montane mesic willow carr in Boulder County, Hallock (1992) reported an astounding 0.96 territories/acre (237.5 territories/km^2).

Dusky Flycatchers are relatively common in western North America and appear to be faring well with no special management efforts. Association with shrubby habitats and logged-over lands suggests this species probably benefits from forestry practices that thin dense coniferous stands or leave small openings in the forest. Populations occurring in streamside habitats (e.g., willow thickets) are vulnerable to the array of potential impacts associated with riparian habitats, such as channelization, recreational development, grazing, and agricultural conversions.

BREEDING EVIDENCE

REPORTED IN 466 (27%) OF 1745 PRIORITY BLOCKS

☐	Possible	208 blocks	(45%)
■	Probable	93 blocks	(20%)
■	Confirmed	165 blocks	(35%)

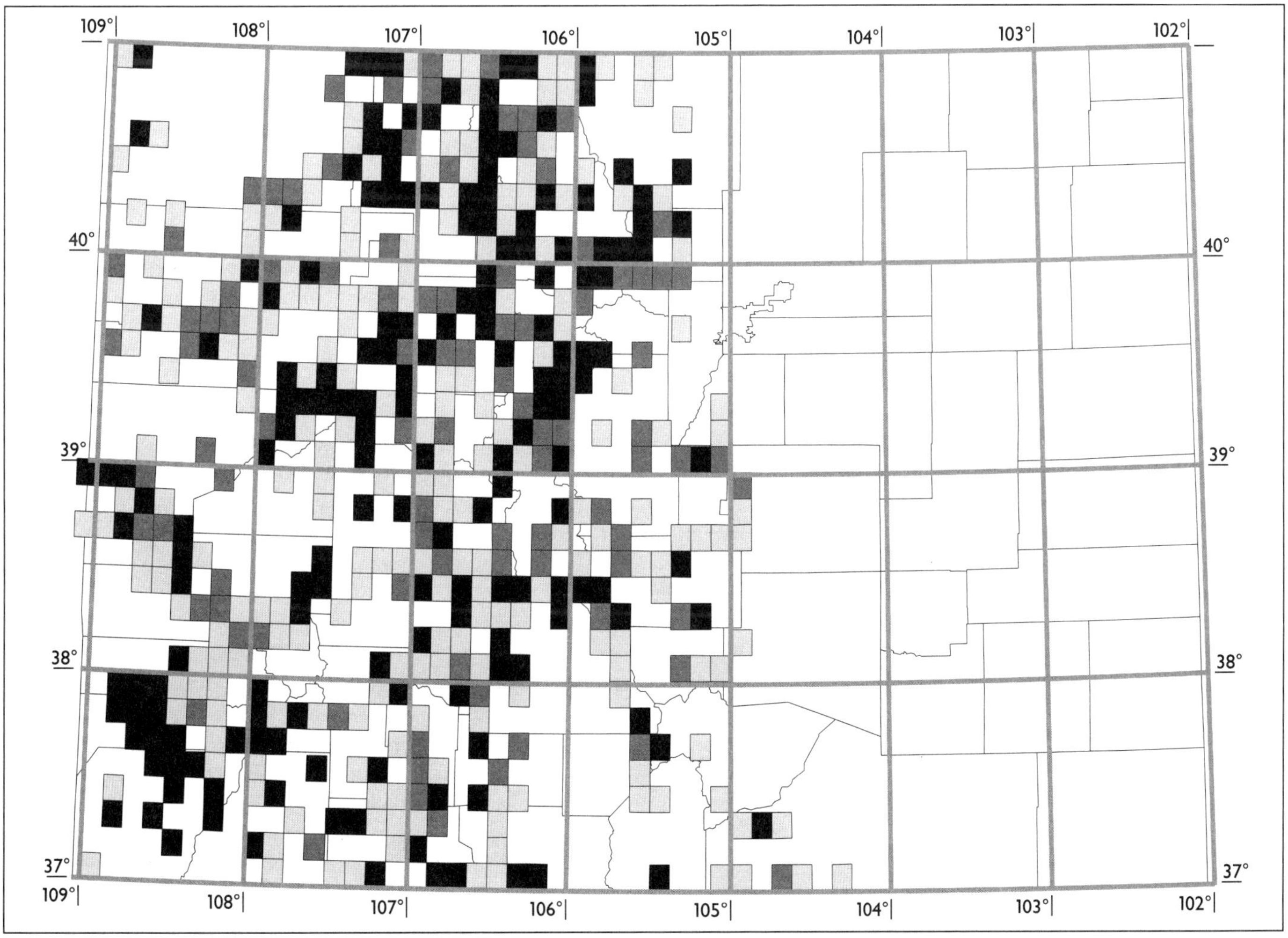
109°
108°
107°
106°
105°
104°
103°
102°
40°
39°
38°
37°

COLORADO DIVISION OF WILDLIFE/WILDLIFE RESOURCE INFORMATION SYSTEM

Coloradans lucky enough to own a summer cabin in the mountains often enjoy a nesting pair of Cordilleran Flycatchers. These birds have discovered that porch beams, rafters, window ledges, open garages, and similar structures serve as nest sites just as well as natural sites like cliffs and stream banks.

HABITAT

This distinctively plumaged Colorado Empid changed its name during the Atlas period. The Atlas field card listed it as "Western Flycatcher," its name until the American Ornithologists' Union split the species into Pacific-slope and Cordilleran flycatchers (AOU 1989).

Cordilleran Flycatchers in Colorado inhabit primarily montane and subalpine forests. Although reportedly more common from 6,000 to 8,000 feet (1830–2040 m) in the foothills and lower mountains, they occur up to 10,000 feet (3048 m; A&R, B&N). They frequent streams and moist ravines with vertical cliffs, outcrops, or steep banks, and they often occur around cabins and mines and where they find dirt cutbanks or rock ledges (A&R). Atlasers most often reported riparian deciduous (narrowleaf cottonwood), riparian coniferous (blue spruce), upland coniferous (spruce/fir, Douglas-fir, ponderosa pine), and aspen forests as habitat types (55%). They also found them as commonly from 8,000 to 10,000 feet (2440–3048 m) as they did below those elevations.

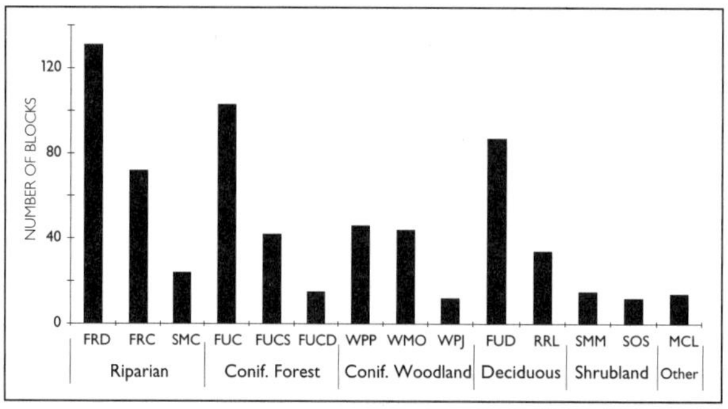
NUMBER OF BLOCKS
0
40
80
120
FRD
FRC
SMC
FUC
FUCS
FUCD
WPP
WMO
WPJ
FUD
RRL
SMM
SOS
MCL
Riparian
Conif. Forest
Conif. Woodland
Deciduous
Shrubland
Other

The habitat choices of Cordilleran Flycatchers more closely resemble those of Hammond's Flycatchers than of any other Colorado *Empidonax;* territories of the two species are frequently adjacent or overlapping. Although territories of Hammond's and Cordilleran flycatchers overlapped in aspen/conifer habitat in one Colorado study, the two species exhibited only a few aggressive interactions (Beaver and Baldwin 1975). Competition for resources likely occurs between the two, but coexistence is apparently maintained by partial vertical separation of foraging niches, diet differences, and differences in nest sites. Cordilleran Flycatchers take primarily Lepidopterans (moths and butterflies) and Dipterans (flies and mosquitoes) and feed mostly from ground level to 30 feet (9 m), whereas Hammond's take mostly Coleopterans (beetles) and Lepidopterans and feed most often from 20 to 40 feet (6–12 m) high (Beaver and Baldwin 1975).

BREEDING

Cordilleran Flycatchers normally arrive in Colorado in early May, begin nesting in late May to early June, and complete the breeding cycle by early to mid August. Seven Atlas dates for nest-building before 10 June suggest that this species begins its nesting cycle somewhat earlier than other Colorado *Empidonaces* except Gray Flycatchers. Although Cordillerans normally nest only once in a season except in cases of predation, renesting may extend the nesting season well into August (Atlas records on 5, 5, 8, and 17 August for NE, NY, FL, and FY respectively).

Nest-site preferences of this species are as varied as they are unusual. Rock faces, dirt cutbanks, and depressions in banks of streams are among the more common nest sites. The birds also place nests in or on roots of upturned trees, in tree cavities, and on flaps of tree bark. Mountain settlement expanded their choices: they build nests on cabin beams, porches, and window ledges, on rafters in buildings, in old mine shafts, in tires, and under bridges (Bent 1942, Davis et al. 1963, Beaver and Baldwin 1975, Atlas records). Of 15 Atlas nest cards, seven describe nests on manmade structures, six on rock faces or in cutbanks, and two in trees.

Only the female builds the nest; she uses moss, lichen, rootlets, grass, leaves, and bark and lines it with shredded bark, hair, and feathers. They often use old nests as the foundation for new material in subsequent years (B&N, Davis et al. 1963). Brown-headed Cowbirds rarely parasitize Cordilleran Flycatchers—the Atlas record of adults attending a cowbird fledgling (Ben and Sally Sorensen; Chace and Cruz 1996; Appendix C) is the only record of parasitism of the species.

BREEDING PHENOLOGY

CODE		# OF RECORDS	DATE RANGE
C	Courtship	1	30 Jun
NB	Nest Building	20	19 May–11 Jul
ON	Occupied Nest	30	23 May–19 Jul
NE	Nest with Eggs	18	10 Jun–5 Aug
NY	Nest with Young	29	20 Jun–5 Aug
FY	Feeding Young	41	19 Jun–17 Aug
FL	Fledged Young	26	26 Jun–8 Aug

DISTRIBUTION

Ornithologists have not worked out a definitive range for this newly defined species, especially its western edges. Cordilleran Flycatchers do breed in the mountains of western North America from southern Alberta south to Oaxaca in the Mexican highlands. Winter range extends from Baja California and northern Mexico south to the Isthmus of Tehuantepec (AOU 1983). In Colorado they occur throughout the foothills and mountains. Atlasers recorded the species in more blocks and had a greater Confirmation percentage (36%) than any other *Empidonax.* The few historical breeding season (June–August) records (late 1800s, early 1900s) come from Larimer, Boulder, Summit, Park, Chaffee, Alamosa, and Conejos counties (B&N).

In the flycatchers' preferred habitats, atlasers most commonly reported abundance of Cordilleran Flycatchers as 2–10 and 11–100 breeding pairs/block. These flycatchers occurred at densities of 0.28 and 0.05 *birds*/acre (69 and 12 *birds*/km^2) in aspen/conifer habitat in the Wet Mountains of southern Colorado in two different years (Beaver and Baldwin 1975). BBS data from the 1968–1993 period show no significant trend, either in Colorado or the U.S. Although only a small amount of mature forest land is logged annually in Colorado, the association of Cordilleran Flycatchers with a wide range of deciduous and coniferous forested habitats makes them potentially vulnerable to habitat destruction by logging.

BREEDING EVIDENCE

REPORTED IN 489 (28%) OF 1745 PRIORITY BLOCKS

☐	Possible	211 blocks	(43%)
■	Probable	103 blocks	(21%)
■	Confirmed	175 blocks	(36%)

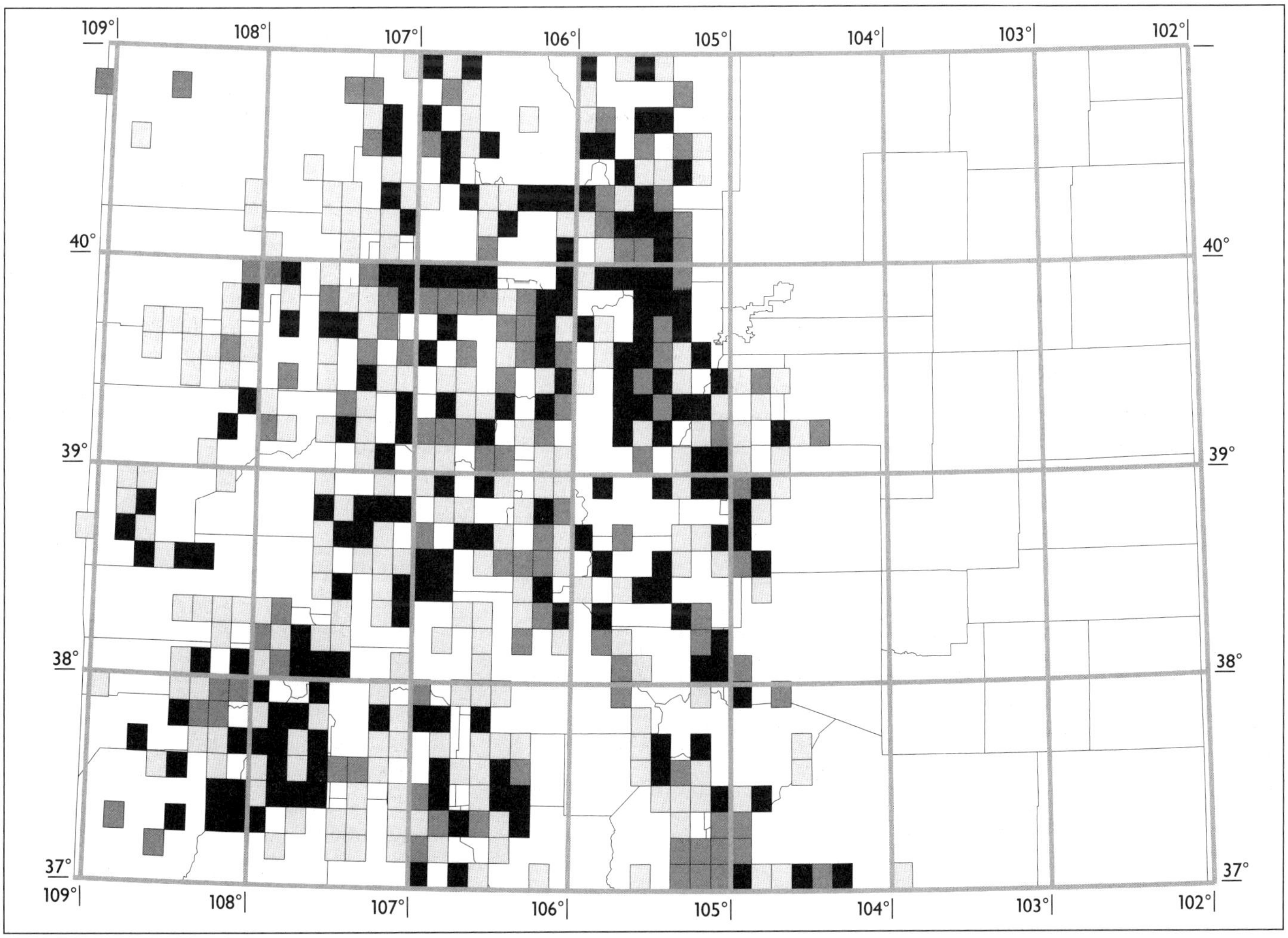
109°
108°
107°
106°
105°
104°
103°
102°
40°
39°
38°
37°

Atlasers poking into remote, underbirded parts of Colorado found the second through fourth Colorado breeding sites of the elegant Black Phoebe. These disjunct discoveries hint that a few more Black Phoebes may breed in undiscovered grottos.

HABITAT

No other flycatcher displays such partiality to water as this mud-nest specialist. Invariably associated with water, Black Phoebes occur along streams and canyons and in residential and farmland areas that have scattered trees and water (Bent 1942). Atlasers recorded the habitats as juniper, lowland and foothills riparian, and cliffs. These phoebes use areas that have both water and suitable nest sites, regardless of the surrounding vegetation.

The birds plaster their mud nests against rock faces, under bridges, in niches or on ledges, and, occasionally, in trees. The nests always have some type of protective overhang. The site must be within carrying distance of the mud used in nest construction. The nest is usually a mud cup or half cup with grass and fibers, lined with fine soft fibers, wool, hair, and feathers (Bent 1942).

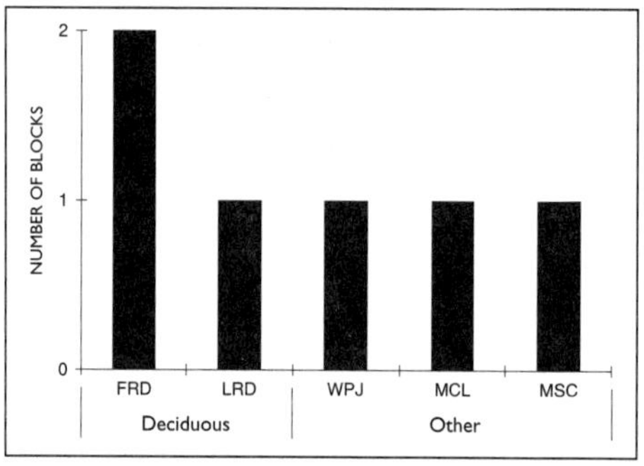
NUMBER OF BLOCKS
0
1
2
FRD
LRD
WPJ
MCL
MSC
Deciduous
Other

The birds display no known correlation between type of water (fast-moving or still) and nesting sites. Of the three nests found by atlasers, the birds placed one along the fast-flowing San Miguel River, one on a bank of Pueblo Reservoir, and one on a cliff above a shaded pool in an ephemeral stream.

BREEDING

Of the three nests atlasers found, Mark Yaeger found two. The first year of the Atlas, on 7 June 1987, he and Bob Doyle canoed along Pueblo Reservoir in the Swallows block (38104C7). Gliding into a small cove, they saw a Black Phoebe sitting on a rock pumping its tail. Bob saw it fly off a nest on the cliff 3.5 feet (1 m) above the water. Two weeks later it dipped into the nest as if feeding young. The next week the nest, now abandoned, held two dead nestlings and two unhatched eggs (Yaeger 1987). The other pair, in the Lambing Spring block (37104C1), had a nest in a deep cut in a bank, above a pool in a stream that flows during spring melt and seasonal rains. Rocky Mountain junipers surrounded the streambed (Mark Yaeger VF).

At the San Miguel River, phoebes, first seen on 5 May 1993, returned the next four years. The first 1993 observation involved two adults carrying food into a niche 4 feet above the water in a 12-foot (3.6-m) vertical sandstone cliff (which had recesses created by flaking chunks of rock). By 6 June three fledglings perched on a nearby willow branch (Coen Dexter pers. comm.).

Black Phoebes demonstrate strong nest-site tenacity, probably due to a limited supply of natural nest sites. Birds frequently reuse the same site or the same nest year after year. Courtship includes a nest-site-showing display in which the male hovers in front of a nest site. The female builds the nest from mud pellets that she gathers within 250 feet (75 m) of the nest. She shapes the nest into a half-hemisphere, with the top usually within 3 inches (7 cm) of a protective ceiling (Wolf 1997).

BREEDING PHENOLOGY

CODE		# OF RECORDS	DATE RANGE
ON	Occupied Nest	1	7 Jun
NE	Nest with Eggs	1	8 Jun
NY	Nest with Young	2	5 May–23 Jun
FL	Fledged Young	2	6 Jun–12 Jun

DISTRIBUTION

Black Phoebes occur on the Pacific coast from southern Oregon to the tip of Baja California, and from southern Nevada to southwestern Texas and south to northern

Argentina. In New Mexico they breed north to Santa Fe (Sandy Williams pers. comm.) and probably along the Rio Grande in Los Alamos County (Travis 1992). Upstream in Colorado in July 1996, one actively skimmed the Rio Grande near Antonito (Urling Kingery pers. comm., RBR).

All Colorado records (about 15) are scattered across the southern part of the state, in seven different latilongs with no discernible pattern except for the connection with water. The phoebes' affinity for water may partially account for the lack of any other pattern. The first Colorado nesting record involved a pair that nested for three years along the St. Charles River 20 miles (32 km) south of Pueblo Reservoir (Ligon and Griffiths 1972).

Colorado now has four confirmed breeding records. The Pueblo Reservoir birds did not return (Yaeger saw only one bird there), but the St. Charles and San Miguel river birds did return in subsequent years. Atlasers could not return to the Las Animas County site to determine the nest-site fidelity.

Categorized as non-migratory by the AOU (1957), Black Phoebes in midsummer occur in Arizona north and east of the normal breeding range. The two non-Confirmations on the Atlas map, both July observations, may depict breeders or may represent dispersal either from an adjacent breeding site or a longer distance. Black Phoebes are resident over much of their range but partly migratory in the northern part; many make seasonal movements (Wolf 1997, Phillips et al. 1964). Winter survival may explain why they have not colonized Colorado, although one spent the 1995–1996 winter along the Arkansas River in Cañon City (Ely 1996).

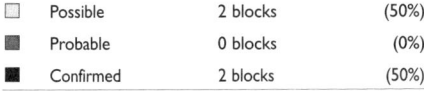

BREEDING EVIDENCE

REPORTED IN 4 (0%) OF 1745 PRIORITY BLOCKS

☐	Possible	2 blocks	(50%)
▪	Probable	0 blocks	(0%)
■	Confirmed	2 blocks	(50%)

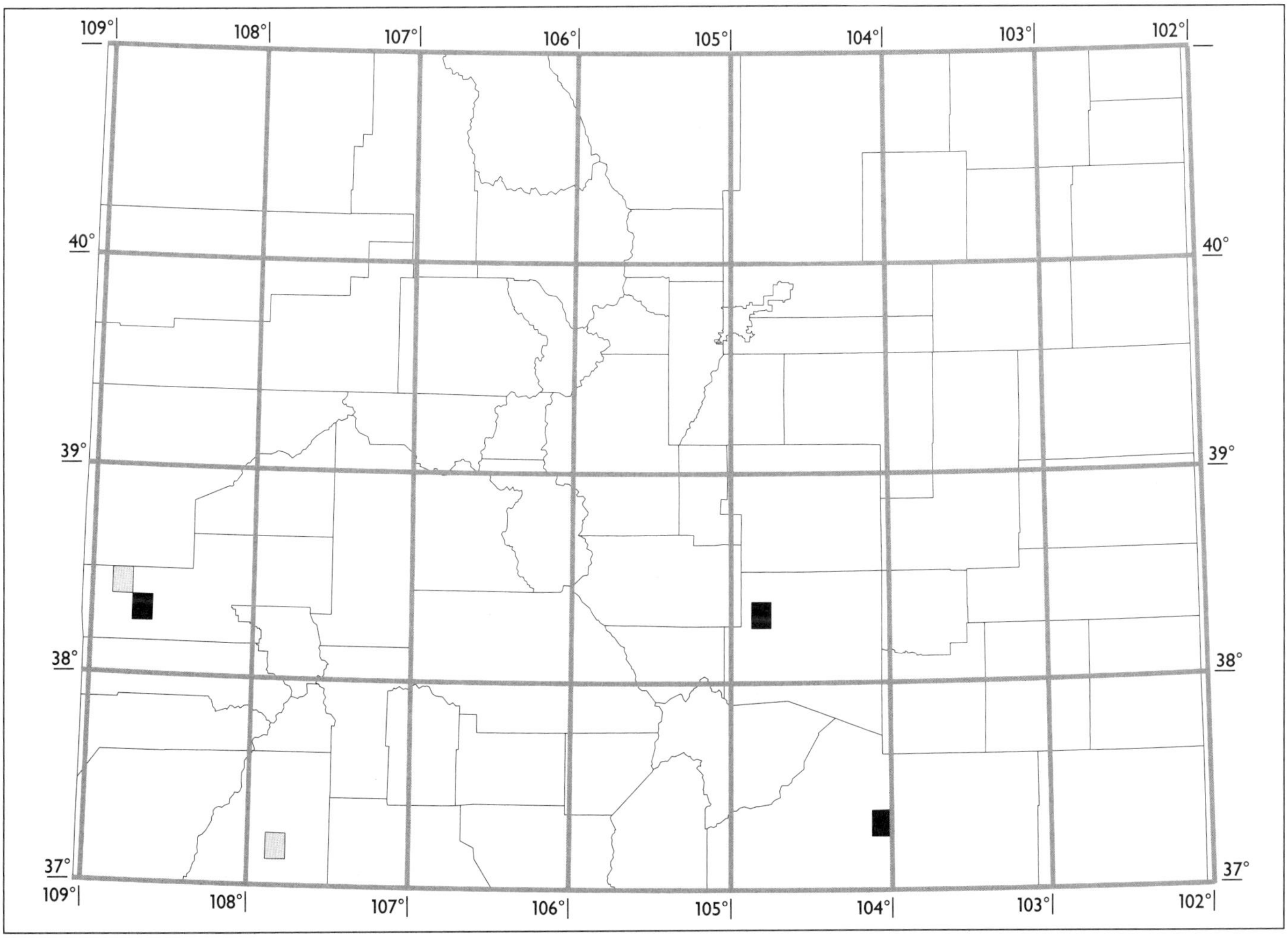
109°
108°
107°
106°
105°
104°
103°
102°
40°
39°
38°
37°

COLORADO DIVISION OF WILDLIFE / WILDLIFE RESOURCE INFORMATION SYSTEM

New England naturalist Winsor Marrett Tyler described this flycatcher as "a member of the happy community that makes up rural life—the pigs in their sty, the hens in their coops, the horses and cows in the barn, and the Phoebe in the back shed" (Bent 1942). In Colorado, on the western frontier of their North American range, Eastern Phoebes display their wilder side, nesting primarily in secluded canyons far from farms and towns.

HABITAT

Eastern Phoebes inhabit plains grassland, plains riparian woodland, pinyon/juniper woodland, rangeland, and cropland, mainly in southeastern Colorado. They require trees or other woody vegetation for perching and hiding from predators (Weeks 1994). Highest Colorado nesting densities (2–5 pairs per Priority block) occur in the canyons of Las Animas County, where they construct their mud and moss nests under overhanging cliffs and boulders in moist, shady grottos.

A nest in Pat Canyon (Big Hole Canyon, 37103A8) clung to a mossy overhang at the head of a narrow side canyon choked with juniper, wild grape, and poison ivy. A few yards away, a Canyon Wren sat on her nest in a narrow crevice, and an incubating Ash-throated Flycatcher poked her head out of a hole in a tall, half-dead juniper. Meanwhile, under the shady overhang, the adult phoebes busily delivered insects to their two oversized nestlings, both cowbirds.

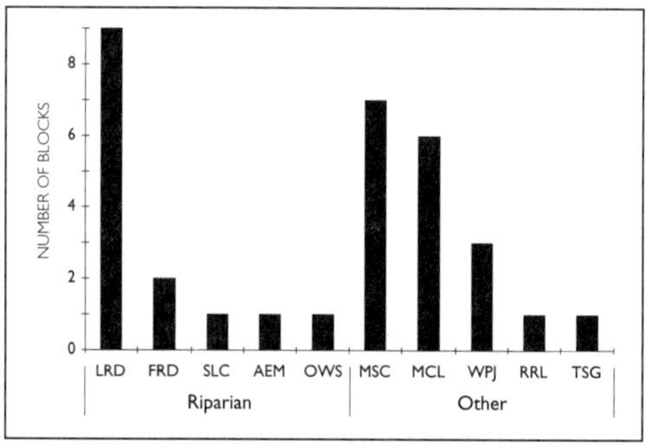
NUMBER OF BLOCKS
0
2
4
6
8
LRD
FRD
SLC
AEM
OWS
MSC
MCL
WPJ
RRL
TSG
Riparian
Other

Atlas volunteers found only four nests outside the canyons of Las Animas and western Baca counties. All of these nests were attached to human structures. Eastern Phoebes originally nested mostly in canyons and ravines but after European settlement adapted readily to human structures (Tyler in Bent 1942). Of 148 nests in southern Indiana, the phoebes built only 13 (9%) on natural surfaces and constructed the remainder on bridges (47%), culverts (31%), buildings (7%), and miscellaneous structures (7%; Weeks 1979).

These agile flycatchers prey primarily on insects, but they also consume small quantities of fruit and seeds, especially during fall, winter, and early spring (Beal 1912). They typically catch their prey in the air during short sallies from an exposed perch or during more prolonged flitting and hovering flights over water (Bent 1942).

BREEDING

Breeding pairs arrive in southeastern Colorado by late March (A&R) and probably begin nest-building by mid April (Weeks 1994). By fledging the first brood in late May or early June, they can lay a second clutch, often in the same nest (Weeks 1978). A single pair frequently uses the same nest from one year to the next.

Eastern Phoebes seem particularly susceptible to nest predation and nest parasitism. Weeks (1979) attributed almost half of the nest failures in his Indiana study area to predators, including snakes, raccoons, and white-footed mice. He observed no instances of nest loss due to human or livestock interference. In three midwestern studies of nesting success, rates of nest parasitism by cowbirds ranged from 5% of nests in southern Indiana to 24% of nests in Kansas and 30% of nests in southern Illinois (Klaas 1970, Graber et al. 1974, Weeks 1979). Weeks reported much higher cowbird parasitism rates for nests situated in natural sites than for those situated on bridges or buildings. He cited cowbird parasitism as a selective factor favoring Eastern Phoebes that nest on human structures.

Assuming that cowbirds are at least as observant as Atlas field workers, they should have little trouble locating active nests in Colorado. Field workers managed to Confirm nesting in 76% of the blocks in which Eastern Phoebes occurred, the third highest Confirmation rate for any Colorado species.

BREEDING PHENOLOGY

CODE		# OF RECORDS	DATE RANGE
NB	Nest Building	2	15 May–26 May
ON	Occupied Nest	7	29 May–18 Jun
NE	Nest with Eggs	4	11 Jun–22 Jun
NY	Nest with Young	5	29 May–20 Jun
FY	Feeding Young	4	8 Jun–15 Jul

DISTRIBUTION

The Eastern and Say's phoebes' breeding ranges fit together like pieces of a jigsaw puzzle. The ranges overlap along a narrow north–south band extending from western Kansas, eastern Colorado, western Nebraska, and the Dakotas to Alberta and the Northwest Territories. The canyons of Las Animas County and the Black Hills in South Dakota constitute the westernmost extent of the Eastern Phoebe's breeding range in the United States.

The isolated southeastern Colorado breeding population survives in cool, shady micro-habitats within an arid environment. Few breeding records exist for these sparsely populated canyons, making it impossible to assess long-term population trends.

B&N, who reported nesting in western Baca County as early as 1914 and found three nests there in 1938, characterized the Eastern Phoebe as an "uncommon summer resident" in that region. A&R reported nesting in the canyons of Las Animas and western Baca counties as well as at Tamarack Ranch, Logan County, and Bonny Reservoir, Yuma County. Results of the Atlas survey extend this species' known Colorado breeding range in natural sites westward in central Las Animas County and northward into Otero County. The Atlas also adds bridge and barn sites in Bent, Cheyenne, and Yuma counties.

Populations increased throughout the United States from 1966 to 1995. In Colorado an inability to adapt to dry, sparsely vegetated environments may continue to restrict nesting populations to a few favored micro-habitats. Future fluctuations in the High Plains aquifer and flow rates of natural springs could disrupt the delicate environmental balance that has enabled Eastern Phoebes to thrive in the shady canyons of southeastern Colorado.

BREEDING EVIDENCE

REPORTED IN 23 (1%) OF 1745 PRIORITY BLOCKS

☐	Possible	4 blocks	(17%)
■	Probable	0 blocks	(0%)
■	Confirmed	19 blocks	(83%)

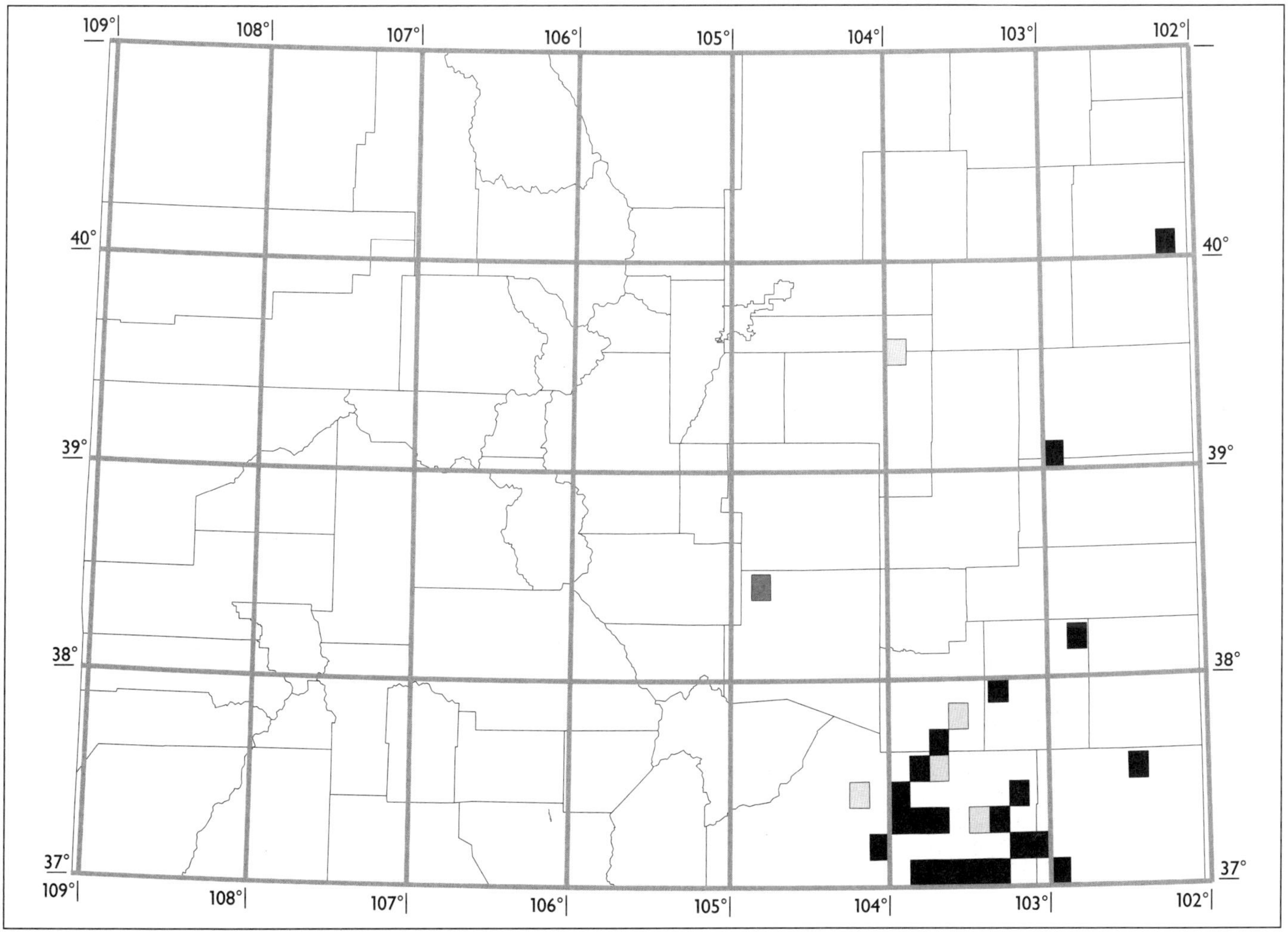

109°
108°
107°
106°
105°
104°
103°
102°
40°
39°
38°
37°

The plaintive two-note call of the Say's Phoebe evokes images of pinyon pines frosted with spring snow, remote rimrock canyons, and abandoned ranch buildings on the empty prairie. Although these hardy and resourceful flycatchers typically inhabit open, sparsely populated country, they also make good use of human artifacts by nesting on buildings, under bridges, and even in wells and mailboxes.

HABITAT

Say's Phoebes breed in grassland, riparian woodland, rural shelterbelts, rangeland, pinyon/juniper and oak woodlands, and shrublands, usually below 9,000 feet (2745 m). Nearly 50% of breeding season sightings reported by Atlas volunteers occurred in rural habitats. Atlasers used the MSC code for 11% of their observations, probably referring mainly to bridges. Nest sites included a townhouse eave in Boulder, a bucket nailed to the inside wall of a shed in Moffat County, a metal culvert in Elbert County, and abandoned farmhouses throughout the eastern plains and Western Slope.

In the canyons of Baca and Las Animas counties, Say's and Eastern phoebes occasionally nest near each other; atlasers found some nests of the two species separated by fewer than 50 yards (46 m; SRJ). In this region Say's Phoebes occupy a variety of sites, including cliff crevices, boulders, recycled Cliff Swallow nests, and abandoned ranch buildings, whereas Eastern Phoebes nest predominantly in shady cliff faces near springs or standing water.

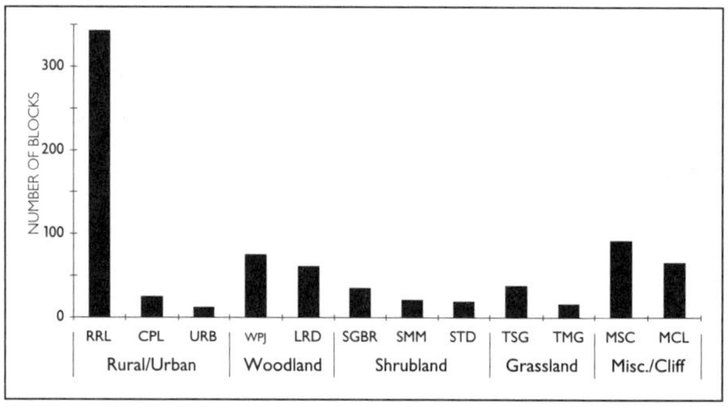

300
200
100
0
NUMBER OF BLOCKS
RRL
CPL
URB
WPJ
LRD
SGBR
SMM
STD
TSG
TMG
MSC
MCL
Rural/Urban
Woodland
Shrubland
Grassland
Misc./Cliff

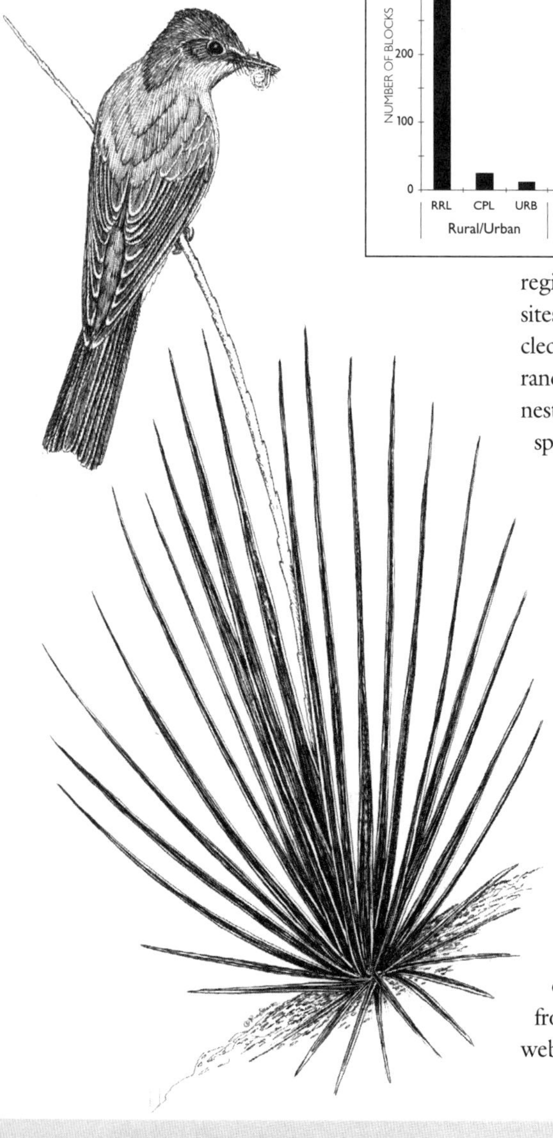

Say's Phoebes forage for insects from low perches on rocks, bushes, fence posts, and small trees. They catch their prey in midair or by hovering and dropping to the ground. Common prey includes bees, wasps, grasshoppers, crickets, flies, beetles, moths, caterpillars, and true bugs (Bent 1942, Ohlendorf 1976).

BREEDING

Pairs arrive in Colorado with the first wave of Neotropical migrants in late March or early April (A&R, B&N). They begin nest construction by mid April and incubation by early May. They construct their nests from grasses, forbs, mosses, wool, spiderwebs, and pieces of cloth and lint (Bent 1942). During a single nesting season, pairs frequently produce two and sometimes three broods in the same nest, and they often reuse the nest in subsequent years (Ohlendorf 1976). Late summer sightings of fledglings by Atlas volunteers reflect this species' tendency to produce multiple broods.

Say's Phoebes fledge 2.0–2.1 young per nest, and their eggs have a 45–57% chance of hatching and fledging. Losses of eggs or young occur most frequently from predation and weather-related causes, especially high winds (Shukman 1993). Although these phoebes typically nest in conspicuous locations, often on or near human habitations, cowbirds rarely parasitize their nests (Ehrlich et al. 1988). None of 46 clutches observed by Ohlendorf (1976) in southwest Texas or of 24 in west-central Kansas (Shukman 1993) contained cowbird eggs. Colorado Atlas volunteers found no cowbird eggs in 26 observed clutches.

BREEDING PHENOLOGY

CODE		# OF RECORDS	DATE RANGE
C	Courtship	6	15 Apr–21 Jun
NB	Nest Building	13	15 Apr–19 Jun
ON	Occupied Nest	89	4 May–7 Jul
NE	Nest with Eggs	24	8 May–15 Jul
NY	Nest with Young	67	21 May–24 Jul
FY	Feeding Young	90	19 May–8 Aug
FL	Fledged Young	46	13 May–10 Aug

DISTRIBUTION

Colorado lies within the eastern one-third of the Say's Phoebe's breeding range, which extends from southern Saskatchewan, Mackenzie, and Yukon to north-central Alaska, south through the western half of the U.S. to central Mexico. Individuals winter southward from central California, southern Arizona, and southern New Mexico to south-central Mexico.

Atlas volunteers observed Say's Phoebes in approximately half of low- to mid-elevation blocks on the eastern plains and Western Slope. Pairs also nest in mountain valleys and parks, including South Park and the Wet Mountain and San Luis valleys. Highest nesting densities (11–100 pairs per block) occurred on the Western Slope below 8,000 feet (2438 m). Relatively low nesting densities (2–10 pairs per block) occurred on the eastern plains and in mountain valleys. Low availability of nest sites in these sparsely populated regions may restrict nesting populations. A Costilla County block with no trees within 10 miles (16 km), and in which desert shrubs comprised the only vegetation taller than 6 inches, demonstrated such a limitation. Here five young crowded into a nest in a wooden crate dropped next to a corral (Mesito Res. 37105B6).

Colorado populations appear to have remained fairly stable throughout this century. Sclater (1912) characterized this species as a "common summer resident, chiefly along the eastern foothills, but also throughout the mountain valleys up to 9,500 feet." B&N described it as a "common migrant [and] summer resident in the upper Sonoran and Transition zones."

Urbanization and human disturbance of nests probably pose the greatest threats to breeding populations. Ohlendorf (1976) blamed nearly 50% of nest failures in his southwestern Texas study area on human destruction of active nests. In Boulder County the nesting population declined from 1907 to 1994 as housing developments replaced rural landscapes (Henderson 1907, Betts 1913, Alexander 1937, Boulder County Audubon Society 1978–1994).

BREEDING EVIDENCE

REPORTED IN 687 (39%) OF 1745 PRIORITY BLOCKS

Possible	230 blocks	(33%)
Probable	97 blocks	(14%)
Confirmed	360 blocks	(52%)

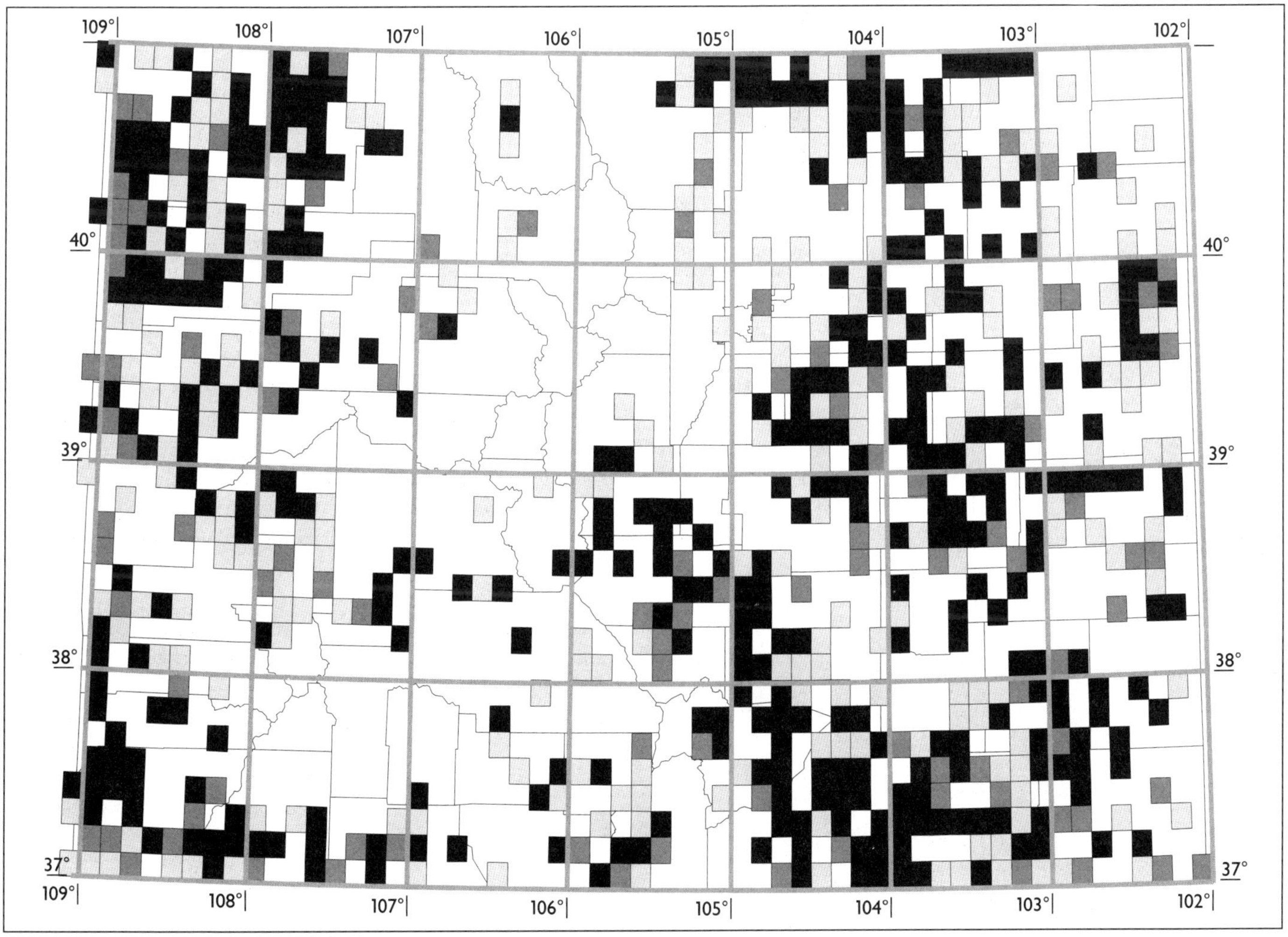
109°
108°
107°
106°
105°
104°
103°
102°
40°
39°
38°
37°

COLORADO DIVISION OF WILDLIFE / WILDLIFE RESOURCE INFORMATION SYSTEM

ASH-THROATED FLYCATCHER *Myiarchus cinerascens*

Ash-throated Flycatchers differ from most of the flycatcher tribe in several respects. They nest in cavities (in Colorado, only the extralimital Great Crested shares that characteristic). They usually do not return to the same perch after capturing an insect, but instead move on to a new perch where they may remain in an upright posture with tail drooped for several minutes. The aggressive, belligerent actions that characterize kingbirds do not exemplify Ash-throateds, which attend to the business of catching insects in a quiet, inconspicuous manner (Bent 1942).

HABITAT

Ash-throated Flycatchers in Colorado exhibit narrow habitat selection. They breed exclusively in pinyon/juniper woodlands and riparian woodlands. Atlas field workers found two-thirds in pinyon/juniper and 16% in riparian woodlands. In coastal California they display wider habitat tolerance, by breeding in various types of open woodlands: broad-leaved evergreens, mixed evergreens/conifers, oak savannah woodlands, and chaparral (Roberson and Tenney 1993, Shuford 1993). Everywhere they avoid densely forested areas (Ryser 1985).

Ash-throated Flycatchers achieve their greatest abundance between 4,500 and 7,000 feet (1400–2100 m) elevation; above this elevation their population decreases rapidly. Atlas data show that in riparian habitats they favor lowland cottonwoods over foothills deciduous by a two-to-one margin, consistent with a preference for low elevations. Atlasers estimated over 100 breeding pairs in 12 blocks, all of them along the low-elevation pinyon/juniper and riparian woodlands adjacent to New Mexico and Utah.

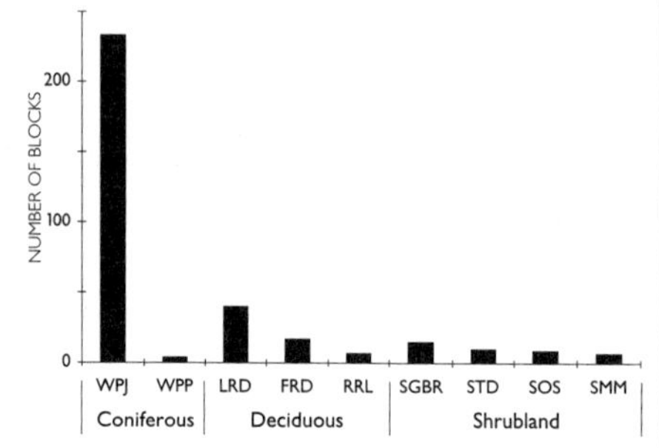

NUMBER OF BLOCKS
200
100
0
WPJ
WPP
LRD
FRD
RRL
SGBR
STD
SOS
SMM
Coniferous
Deciduous
Shrubland

In 1991 the Atlas Confirmed breeding in the San Luis Valley for the first time. A pair had a nest located at 7,500 feet (2285 m) in southern Saguache County, in a dead tree in an abandoned farmyard surrounded by greasewood and farm fields (HEK, nest card). This expansion into a different habitat, attributable to a man-altered landscape, came in an area that must offer a suitable climate and food supply but lacks natural nest sites. Yet, like Plain Titmice, Ash-throateds do not nest in the pinyon belt that rings the valley—a high-elevation pinyon woodland that must lack some essential component for the flycatchers. Perhaps they miss old junipers because pinyon pines do not make good cavity trees (Rich Levad pers. comm.)

BREEDING

Ash-throateds nest in a wide variety of natural and artificial cavities. They nest in natural cavities and old woodpecker holes in junipers, pinyons, and cottonwoods. Their selection of nest sites creates many interesting situations. They use abandoned nests of other birds, drain pipes, old tin cans, empty mailboxes, and even a pair of overalls left too long on the clothesline (Bent 1942). Manmade cavities in nest boxes and in gates and fences made from pipe have expanded available nest sites. When they choose metal pipes as a nest site, however, the young often die because of excessive heat (Dunning and Bowers 1990).

Ash-throated Flycatchers rarely place their nests higher than 20 feet (6 m) above the ground. The cup nest varies considerably depending upon the size of the hole the birds have selected. Nest-building takes as little time as one day; materials used include weed stems, rootlets, grass, hair, fur, and occasionally a snake or lizard skin.

Atlas field workers found evidence of Ash-throated Flycatchers breeding from mid May to the end of July. The four or five eggs laid take 13–15 days to hatch and are incubated by the female. Both parents join in feeding the young, which leave the nest in 14–16 days (Ehrlich et al. 1988). Most birds arrive after 1 May and depart before the end of August (A&R). Although the breeding period seems long, atlasers supplied no evidence to support double-brooding in Colorado.

BREEDING PHENOLOGY

CODE		# OF RECORDS	DATE RANGE
C	Courtship	3	6 Jun–10 Jul
NB	Nest Building	6	13 May–15 Jul
ON	Occupied Nest	10	11 May–5 Jul
NY	Nest with Young	12	23 May–5 Jul
FY	Feeding Young	37	27 May–24 Jul
FL	Fledged Young	25	31 May–30 Jul

DISTRIBUTION

Ash-throated Flycatchers breed from southeastern Washington and southern Idaho south into most of the arid areas of California, Nevada, Utah, Arizona, New Mexico, west Texas, and Mexico. Early authors declared these flycatchers "rare summer residents" in the state, specifically in Mesa County, though "not uncommon" in Fremont County (Sclater 1912). Cooke (1909) did note a "great extension eastward" into Baca County, and they now occur rarely in southwestern Kansas (Thompson and Ely 1992).

Atlasers found that nearly all the Ash-throated Flycatchers in Colorado breed southwest of a line from the northwestern to the southeastern corners of the state. The greatest abundance found by field workers inhabits the western tier of latilongs next to Utah. This area contains 50% of the blocks where field workers found them. A second substantial breeding population (22% of the blocks with flycatchers) exists in southeastern Colorado. These two populations, broadly separated by the San Juan Mountains and mountain parks, have a geographical connection via New Mexico (Ligon 1961). The Atlas data do not show any difference in habitat requirements or breeding phenology for these two separate populations.

Ash-throateds show a statistically significant continental increase on BBS routes of 1.8% per year; Colorado routes do not show any trend. Other pinyon/juniper birds show declining trends; one wonders why this retiring flycatcher has had successful breeding in recent years.

BREEDING EVIDENCE

REPORTED IN 282 (16%) OF 1745 PRIORITY BLOCKS

☐	Possible	116 blocks	(41%)
▩	Probable	80 blocks	(28%)
■	Confirmed	86 blocks	(30%)

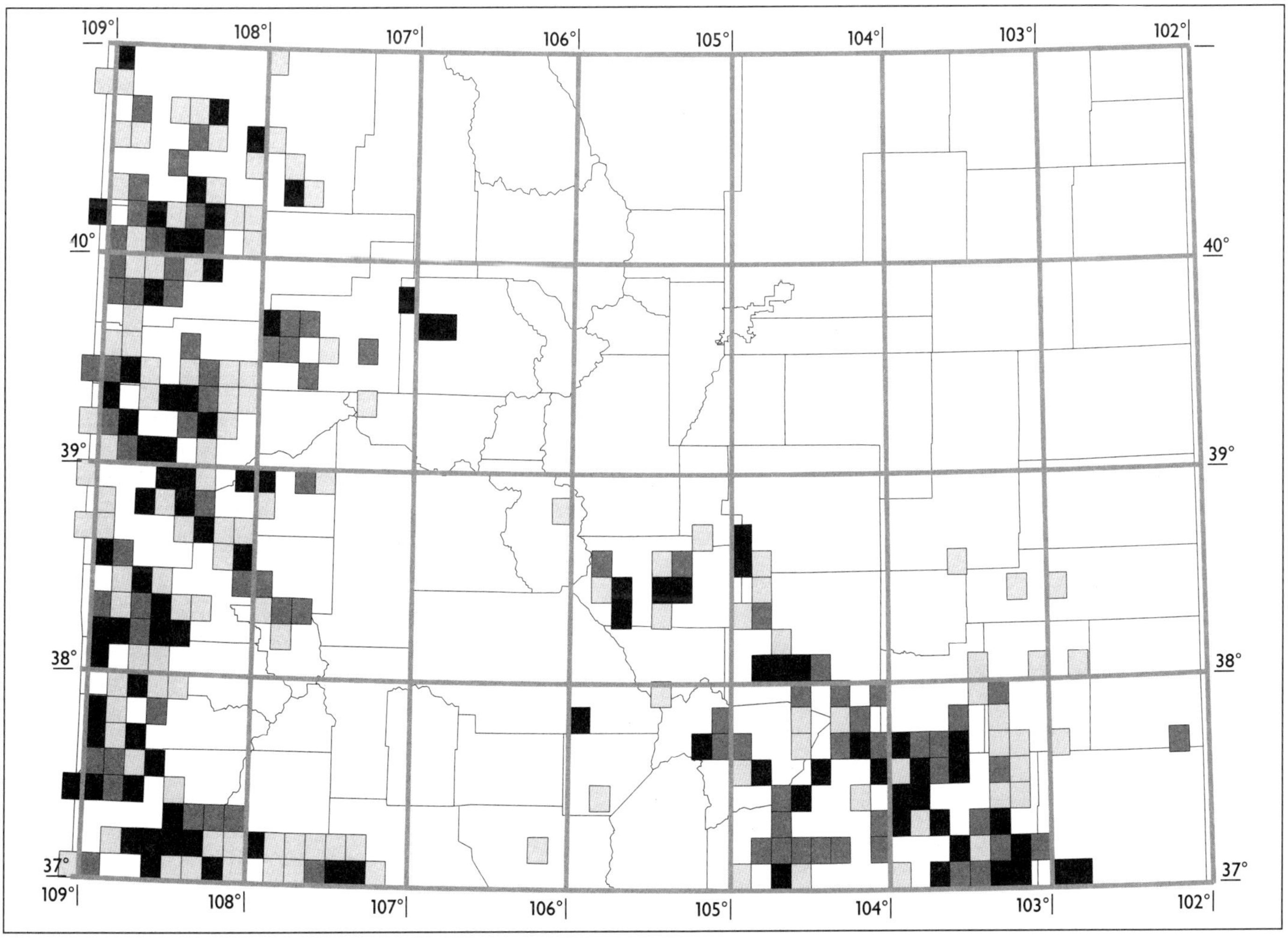
109°
108°
107°
106°
105°
104°
103°
102°
40°
39°
38°
37°

COLORADO DIVISION OF WILDLIFE / WILDLIFE RESOURCE INFORMATION SYSTEM

GREAT CRESTED FLYCATCHER *Myiarchus crinitus*

Thanks to the growth of cottonwood corridors along the South Platte and Arkansas rivers, Great Crested Flycatchers have recently extended their breeding range into eastern Colorado. Occasionally they appear in other parts of the state during migration. More often heard than seen, they call a harsh rising *wheep* from riparian woods.

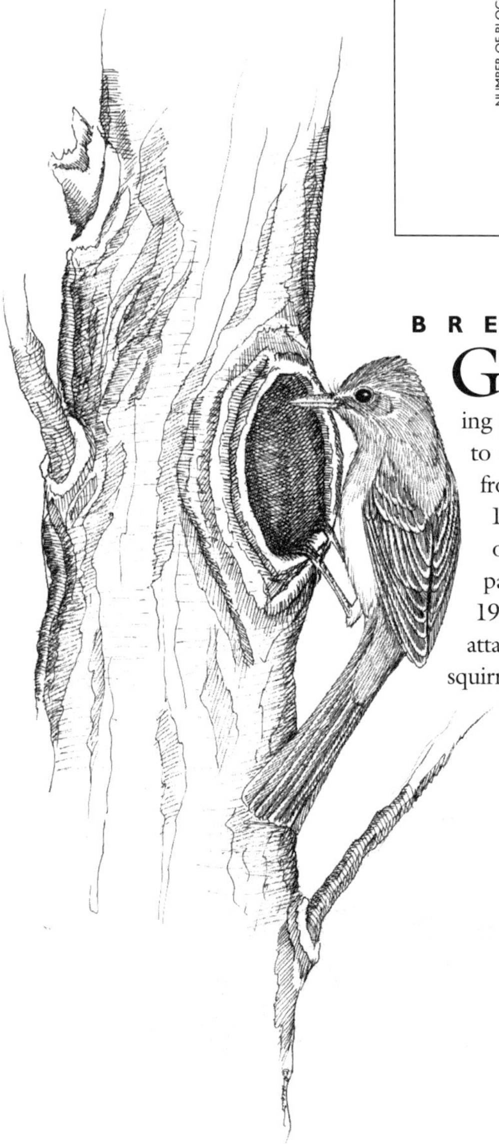

HABITAT

Great Crested Flycatchers place their bulky nests in natural cavities in trees or holes drilled by woodpeckers (Bent 1942). East of Colorado, they breed in a wide array of deciduous woods, both open and less open. They also nest in parks, orchards, and around residences. They prefer nest sites at the borders of woods, and apparently would rather nest in natural cavities than in woodpecker holes (Hebert and Elkins *in* Foss 1994). In eastern Colorado, at the extreme western edge of their range, atlasers and other observers have found them breeding only in riparian cottonwoods, but the potential exists for them to use residential areas with large trees, as suggested by the lone rural habitat report.

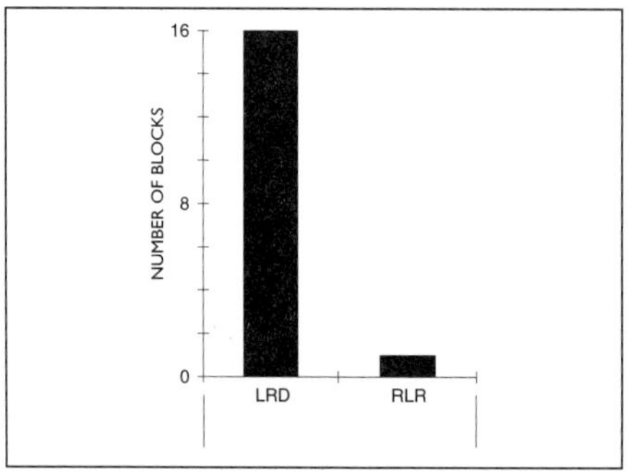
16
8
0
NUMBER OF BLOCKS
LRD
RLR

BREEDING

Great Crested Flycatchers probably return to Colorado from wintering grounds in early May and remain to early September. They lay eggs from May to early July (Johnsgard 1979). They usually produce only one brood, and cowbirds rarely parasitize them (Ehrlich et al. 1988). Pugnacious and fearless, they attack predators such as snakes and squirrels that threaten their nests or young (Bent 1942). Nevertheless, starlings, as they continue to increase in numbers and distribution, usurp many available nesting cavities (Bent 1942).

During territory formation these flamboyant flycatchers are quite noisy, but they become silent near the nest. They feed within the canopy so observers hear their *wheep* more often than they see the birds (Hebert and Elkins *in* Foss). Atlasers collected very limited breeding data in Colorado. With the limited Atlas observations, the few dates from the end of May to mid June for nest-building and carrying food do not reflect the span of dates for all breeding activities.

BREEDING PHENOLOGY

CODE		# OF RECORDS	DATE RANGE
NB	Nest Building	4	30 May–1 Jun
FY	Feeding Young	1	15 Jun
FL	Fledged Young	1	31 Jul

DISTRIBUTION

Great Crested Flycatchers breed throughout the eastern U.S. and southeastern Canada, as far west as the Great Plains. They winter in southern Florida, and central Mexico south to northern South America. They breed commonly in eastern and central Kansas, but in western Kansas they breed very locally, and then only in riparian habitats (Thompson and Ely 1992). In western Nebraska the Platte River supports some breeding birds up to the Colorado line and the Colorado breeders represent their westernmost outpost.

Some thirty years ago, B&N termed them stragglers on the eastern prairies, but quite likely to breed in the area of Bonny Reservoir, where atlasers in 1987 did report Possible breeding. In 1988 the first

Colorado nest was found by Glenn and Jeane Hageman near Prewitt Reservoir in the Messex block (40103D4). Atlasers later reported evidence of breeding in nine other blocks along the South Platte between Greeley and Julesburg. Atlasers also extended the known or possible breeding range into southeastern Colorado along the Arkansas and Cimarron rivers and Big Sandy Creek.

These increasing observations stem from two factors: maturing cottonwoods along the stream bottoms provide newly suitable nest sites, and intensive searching by atlasers and others found these flycatchers in places not previously explored for birds. So long as mature cottonwood stands survive, Great Cresteds can find potential nest sites along the eastern Colorado streams.

BREEDING EVIDENCE

REPORTED IN 6 (0%) OF 1745 PRIORITY BLOCKS

□	Possible	3 blocks	(50%)
■	Probable	2 blocks	(33%)
■	Confirmed	1 block	(17%)

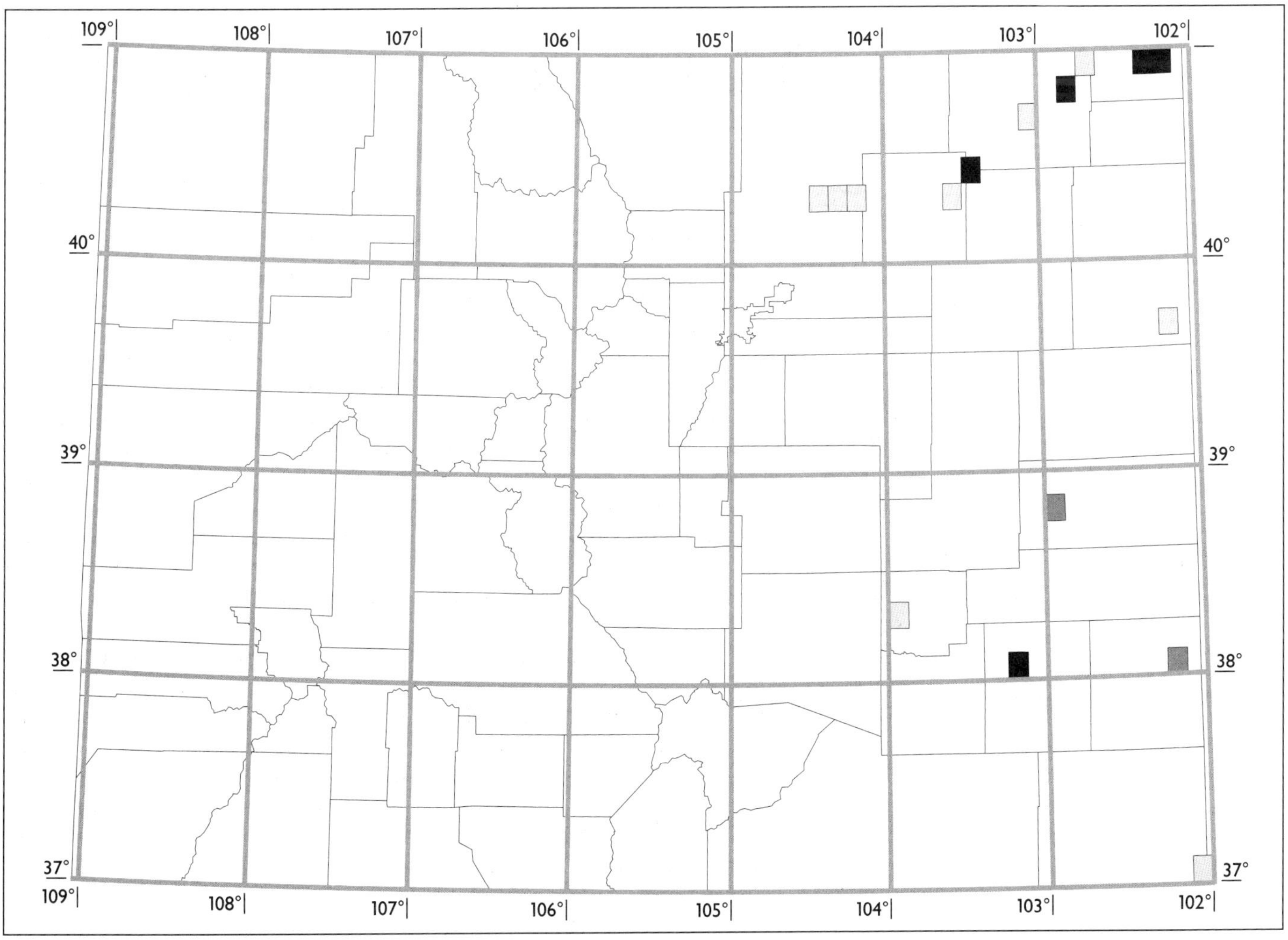
109°
108°
107°
106°
105°
104°
103°
102°
40°
39°
38°
37°

COLORADO DIVISION OF WILDLIFE / WILDLIFE RESOURCE INFORMATION SYSTEM

The Latin name *vociferans* aptly describes this pugnacious flycatcher, whose harsh cries explode from prairie canyons and pinyon/juniper uplands well before daybreak. Although they look and behave much like Western Kingbirds, Cassin's occupy a slightly different niche; they often inhabit rougher, more heavily vegetated country and range to higher elevations.

HABITAT

Atlasers found Cassin's Kingbirds nesting in pinyon/juniper woodland, plains riparian woodland, rural shelterbelts, and grasslands, primarily in southeastern and western Colorado. Highest nesting densities occur on the mesas and in the canyons bordering the Apishapa and Purgatoire rivers, east of Trinidad and Walsenburg. Here they nest predominantly in pinyon/juniper, whereas Western Kingbirds show a decided preference for lowland riparian woodland and rural shelterbelts. In western Colorado, Cassin's Kingbirds nest predominantly in pinyon/juniper (12 of 17 occurrences). In northeastern Colorado, where Western Kingbirds abound, a few scattered pairs of Cassin's nest in cottonwoods and elms around farmhouses.

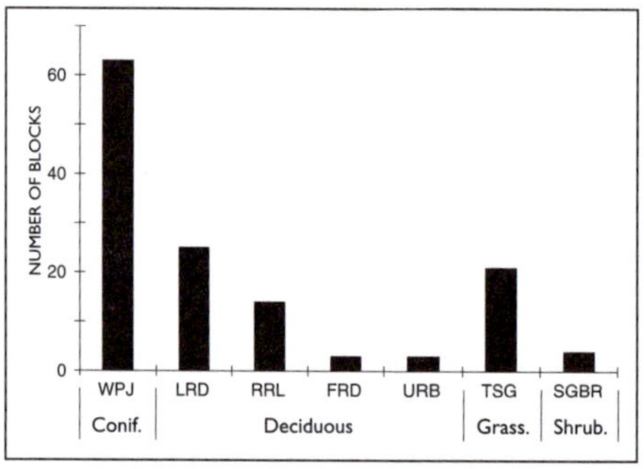
NUMBER OF BLOCKS
0
20
40
60
WPJ
LRD
RRL
FRD
URB
TSG
SGBR
Conif.
Deciduous
Grass.
Shrub.

Colorado nesting patterns for these two kingbirds roughly parallel those observed in southeastern Arizona, where Cassin's Kingbirds tend to nest in dense riparian creek habitat, whereas Westerns nest in more open riparian and desert habitat (Blancher and Robertson 1984). In southwestern Texas, Cassin's also tend to nest at higher elevations and in denser vegetation than Westerns (Ohlendorf 1974). However, throughout their breeding ranges, these two species commonly share nesting habitat in lowland riparian woodlands.

Cassin's Kingbirds catch a variety of insects by sallying from an exposed perch or by dropping down to pick their prey off the ground. They supplement their insect diet with small amounts of spiders, seeds, and fruits (Ohlendorf 1974). Researchers have noted only subtle differences between diets and foraging strategies of Cassin's and Western kingbirds. In fact, biologists have reported greater differences in diets and foraging strategies of Cassin's Kingbirds from one habitat to another than in those of Cassin's and Western kingbirds nesting in the same habitat (Goldberg 1979, Blancher and Robertson 1984).

BREEDING

Pairs arrive in eastern Colorado by mid April (A&R) and begin nest-building by early May. The female lays 2–5 eggs in a robust nest usually constructed on a horizontal tree branch in the upper third of the tree canopy (Bent 1942, Ohlendorf 1974, Blancher and Robertson 1984). The young hatch after 12–14 days of incubation and leave the nest about two weeks later (Bent 1942).

Bent (1942) reported that this species "is said to raise two broods" in the southern part of its range. Subsequent research has not confirmed this behavior (Ohlendorf 1974, Blancher and Robertson 1984). The nesting chronology reported by Colorado Atlas volunteers, with only about 30 days' difference between the earliest and latest reported dates for most nesting categories (NB, NE, NY, and FL), suggests that Colorado's Cassin's Kingbirds typically raise only one brood.

The relatively low rate of nest Confirmations for Cassin's Kingbirds (32%) compared to Western Kingbirds (72%) probably reflects the Cassin's tendency to nest higher off the ground, in denser vegetation, and later in the season (Ohlendorf 1974, Blancher and Robertson 1984) as well as their lower numbers. Neither species exhibits particular guile or secretiveness around the nest. When feeding their young, Cassin's emit a high buzzing sound audible from at least 50 yards away. When defending nesting territories, their shrill ***ki-dear ki-dear ki-dear ki-dear*** holds its own against most other avian and human sounds, including jet-airplane noise.

BREEDING PHENOLOGY

CODE		# OF RECORDS	DATE RANGE
C	Courtship	3	16 Jun–27 Jun
NB	Nest Building	6	10 May–23 Jun
ON	Occupied Nest	11	25 May–23 Jun
NE	Nest with Eggs	5	26 May–27 Jun
NY	Nest with Young	5	13 Jun–14 Jul
FY	Feeding Young	15	29 May–25 Jul
FL	Fledged Young	5	11 Jun–11 Jul

BY STEPHEN R. JONES

DISTRIBUTION

Cassin's Kingbirds breed from southeastern Montana south through eastern Wyoming and western Texas to central Mexico, and west to southern Nevada and central California. Throughout the U.S. and northern Mexico their breeding range lies within the more extensive breeding range of Western Kingbirds.

Early ornithologists reported these birds as breeding along the foothills, plains, and western valleys, although less common north of the Palmer Divide than to the south (Sclater 1912). B&N reported only small numbers, and "surprisingly few" reports.

Colorado Atlas results generally confirm previous assessments of this kingbird's distribution within the state (B&N, A&R, Latilong Study). However, atlasers confirmed breeding in five new counties: Montezuma, Dolores, and Mesa in western Colorado and Elbert and Weld in northeastern Colorado. Atlasers reported relatively high nesting densities in northern Las Animas and southern Pueblo counties and a broader presence on the Western Slope than suggested by A&R.

Mean numbers reported on BBS routes throughout the continent declined from 1966 to 1996, with the sharpest declines occurring on routes traversing pinyon/ juniper woodland, but none of the BBS data categories provides an adequate statistical sample. The future of Cassin's Kingbirds in the Rocky Mountain region probably depends largely on the degree of protection afforded this critical but often neglected ecosystem.

BREEDING EVIDENCE

REPORTED IN 121 (7%) OF 1745 PRIORITY BLOCKS

☐	Possible	45 blocks	(37%)
▩	Probable	30 blocks	(25%)
■	Confirmed	46 blocks	(38%)

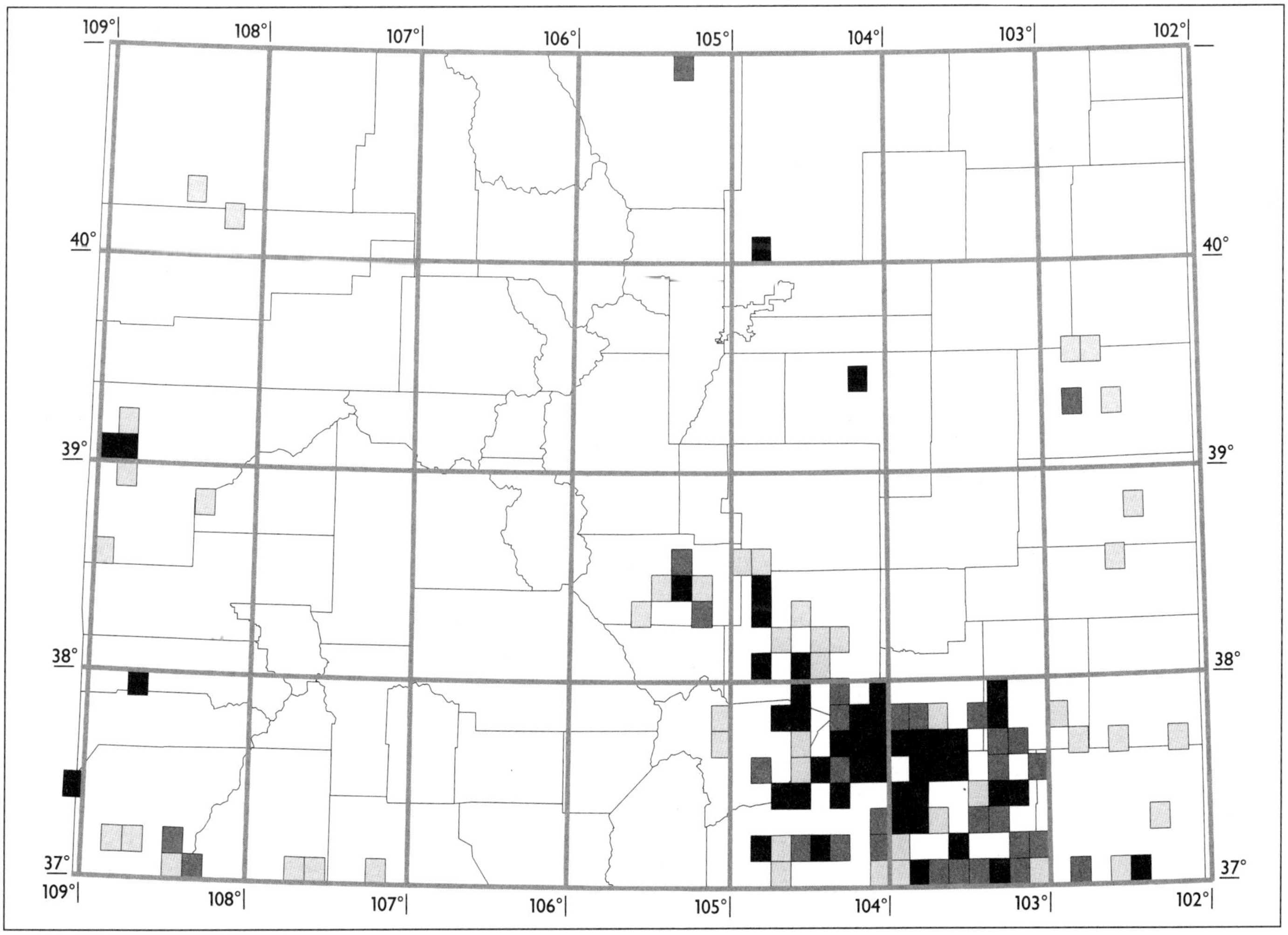

109°
108°
107°
106°
105°
104°
103°
102°
40°
39°
38°
37°

COLORADO DIVISION OF WILDLIFE / WILDLIFE RESOURCE INFORMATION SYSTEM

Insomniacs who have slept out on the prairie should have no trouble recognizing the Western Kingbird's discordant song. The clamor usually begins an hour or two before dawn and continues well into the evening as these aggressive flycatchers advertise and defend nesting sites. Their constant vigilance enables them to breed successfully in open country where predators can easily find and climb into active nests.

HABITAT

Western Kingbirds breed in grasslands, deserts, shrublands, riparian woodlands, and rural shelterbelts, usually below 7,000 feet (2135 m). Nearly two-thirds of the Atlas sightings occurred in either rural residential areas or lowland riparian woodlands. These colorful kingbirds typically construct their nests in trees, but they also nest on building ledges, telephone poles, fence posts, windmills, stumps, or in tree cavities (Bent 1937). Bergin (1992) reported that Western Kingbirds nesting in western Nebraska consistently selected large, widely spaced cottonwoods surrounded by grass cover. When their primary nesting territories lack suitable foraging habitat they feed in open country nearby (Pleasants 1979), a habit exemplified by the many kingbirds that atlasers saw sallying for insects from fence posts up to one-quarter mile from a farmhouse or shelterbelt.

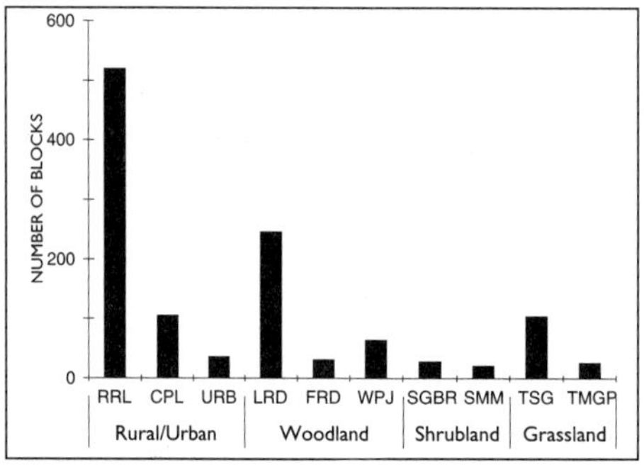
600
400
200
0
NUMBER OF BLOCKS
RRL
CPL
URB
LRD
FRD
WPJ
SGBR
SMM
TSG
TMGP
Rural/Urban
Woodland
Shrubland
Grassland

Nest predators include Swainson's Hawks, Common Ravens, Black-billed Magpies, Blue Jays, Common Grackles, small rodents, and raccoons (Bent 1937, Bergin 1993). In Bergin's western Nebraska study, 31% of nests suffered some predation.

Western Kingbirds prey primarily on large insects, including bees and wasps, crickets, and grasshoppers (Beal 1912). Ridgway (1877) watched a captive male consume 120 grasshoppers in a single day. Wheelock (1904) observed adults teaching their young to flycatch. The parents brought wounded insects to the nest tree and dropped them, so that their fledglings had to fly out and snatch the prey from midair. Kingbirds learn at an early age to distinguish between potentially dangerous worker bees and stingerless drones (Wheelock 1904).

BREEDING

Breeding pairs arrive in Colorado in April and begin nest-building in May. Females lay 3–7 eggs in a robust nest often placed in a prominent fork of a tree trunk or limb. Bent (1937) reported Colorado egg dates of 1–17 June; Atlas field workers reported a much wider range of nests with eggs, 20 May–16 July, and reported nests with young as late as 6 August. Observers have suspected that this species may raise more than one brood, but no one has documented double-brooding (Bent 1937, Ehrlich et al. 1988).

Kingbirds readily recognize and remove cowbird eggs, and cowbirds rarely parasitize their nests (Ehrlich et al. 1988, Sealy and Bazin 1995). Chace and Cruz (1996) found no Colorado records of cowbirds parasitizing Western Kingbird nests prior to 1994, when Atlas volunteers working a block east of Longmont observed two instances of Western Kingbird adults feeding Brown-headed Cowbird fledglings (Appendix C).

Studies undertaken by the U.S. Fish and Wildlife Service in New Mexico and Texas during the early 1980s revealed high concentrations of DDT residues in Western Kingbird nestlings (White and Krynitsky 1986). Other studies have shown that egg production and nestling survival correspond closely to availability of insect prey (Blancher 1982).

BREEDING PHENOLOGY

CODE		# OF RECORDS	DATE RANGE
C	Courtship	13	13 May–4 Jul
NB	Nest Building	55	8 May–1 Aug
ON	Occupied Nest	265	9 May–21 Jul
NE	Nest with Eggs	17	20 May–16 Jul
NY	Nest with Young	76	12 May–6 Aug
FY	Feeding Young	142	20 May–19 Aug
FL	Fledged Young	56	22 May–28 Aug

DISTRIBUTION

Western Kingbirds breed throughout much of western North America, from southern British Columbia and southern Manitoba to southern Texas and southern Baja. They winter from central Mexico south to Costa Rica. Summer and winter ranges have expanded since 1900 as deforestation has created new habitat (Ehrlich et al. 1988).

Atlas data depict the highest Colorado populations on the eastern plains, where about 10% of the blocks supported 100 or more nesting pairs. Throughout much of eastern Colorado, nearly every grove of trees around a well or farmhouse contains a nest; 94% of the plains blocks had them. The highest Western Slope nesting densities occur at lower elevations, especially in the Colorado River drainage.

Both Cooke (1897) and Sclater (1907) considered this kingbird common on the eastern plains and in western valleys. The Colorado distribution reported by Atlas volunteers corresponds closely with that reported by B&N and A&R. BBS estimates indicate that populations have increased in Colorado (1.4% per year) and continentally (0.6% per year) since 1966. This increase may stem in part from the planting and maturing of deciduous trees along rivers and around farmhouses on the High Plains.

BREEDING EVIDENCE

REPORTED IN 893 (51%) OF 1745 PRIORITY BLOCKS

Possible	127 blocks	(14%)
Probable	139 blocks	(16%)
Confirmed	627 blocks	(70%)

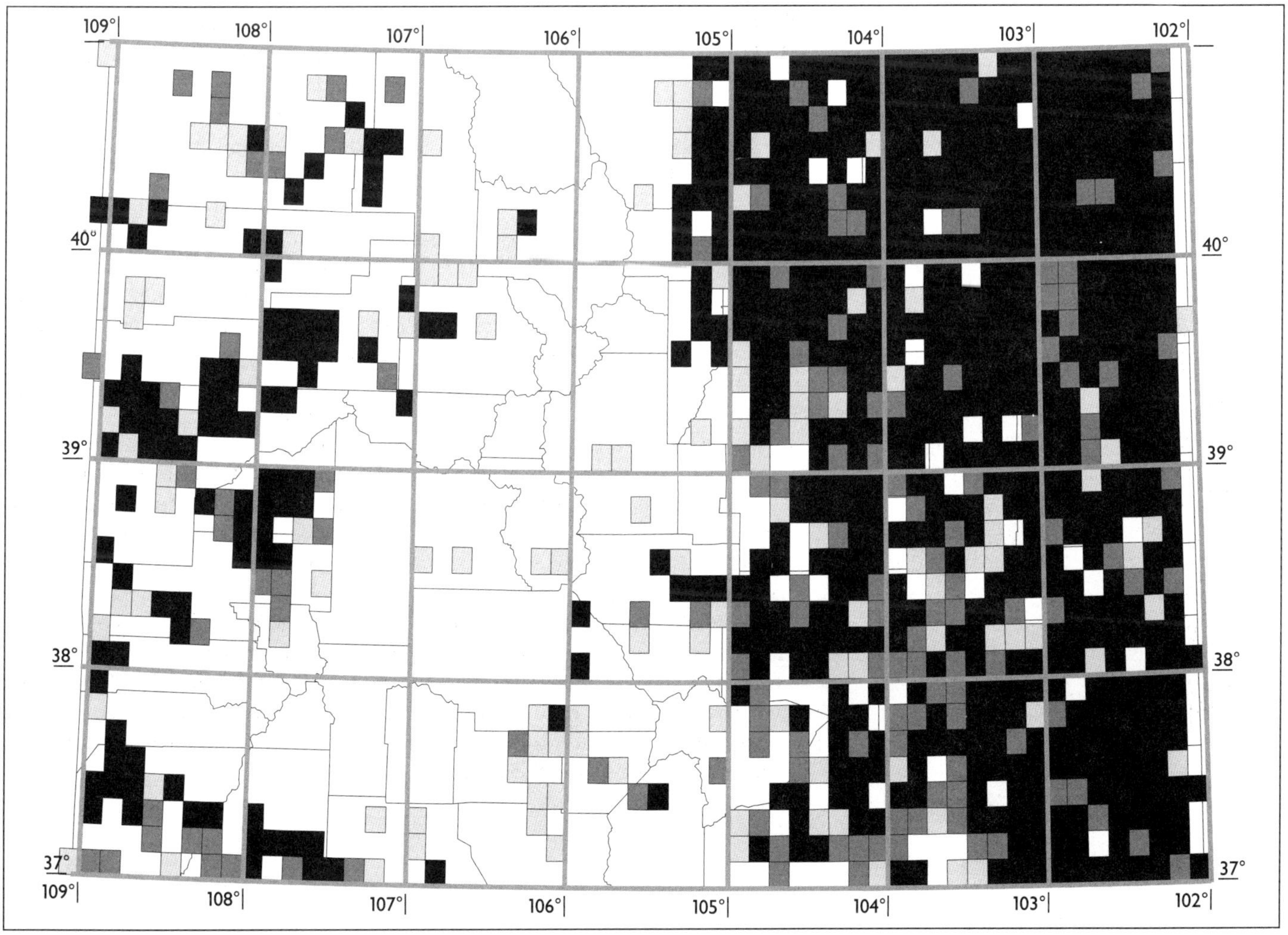
109°
108°
107°
106°
105°
104°
103°
102°
40°
39°
38°
37°

COLORADO DIVISION OF WILDLIFE / WILDLIFE RESOURCE INFORMATION SYSTEM

Most passerines deal with danger by avoiding it—lying low or retreating deep into the foliage usually keeps them from harm's way. Eastern Kingbirds, however, specialize in confrontation. Many a would-be predator has discovered that this bird knows no fear when it comes to removing suspicious characters from the nest area.

HABITAT

The sheer number of Eastern Kingbirds seen as one travels across the midwestern states testifies to the variety of wooded habitats that this species will call home. Atlasers found them in single trees, shelterbelts, and, somewhat more frequently, in riparian woodlands, each of which provides suitable habitat for this species. Like their congeners, Western Kingbirds, they inhabit riparian and deciduous woodlands as well as agricultural areas. Eastern Kingbirds, however, tend to associate much more closely with woodlands adjacent to open water.

Cottonwoods and elms provide most of the nesting sites for Eastern Kingbirds in Colorado. Frequently, orioles, Yellow Warblers, or other passerines nest in the same tree as a pair of feisty kingbirds. It seems likely that the kingbirds' aggressive behavior affords these less contentious passerines an added measure of security.

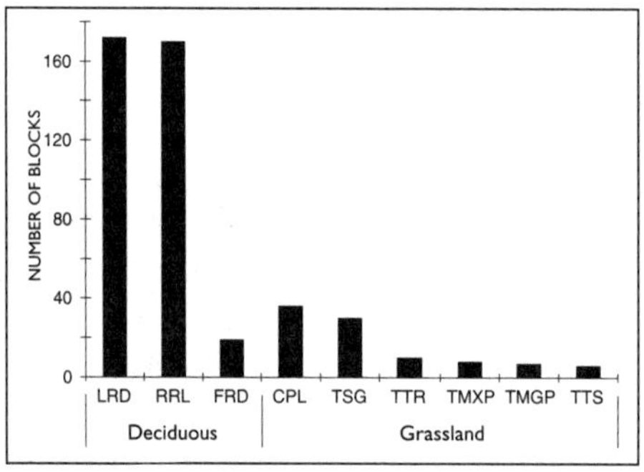
NUMBER OF BLOCKS
0
40
80
120
160
LRD
RRL
FRD
CPL
TSG
TTR
TMXP
TMGP
TTS
Deciduous
Grassland

BREEDING

Eastern Kingbirds arrive in Colorado by the first or second week in May. Soon thereafter, they begin nest construction and egg-laying. Heights for their nests range from 6 to 80 feet (2–25 m) and beyond. Eastern Kingbirds usually position their nests a bit more than halfway out on horizontal or diagonal branches of a large, deciduous tree. On the whole, the lower the site selected for the nest, the farther, proportionately, its location will fall from the main trunk of the tree. Such placement helps to deter predators, which can climb to the lower branches more easily. Site selection figures heavily in the incidence of nest predation. Nests placed at lower levels suffer consistently higher levels of predation than nests placed higher in trees. Conversely, nests placed at higher levels tend to experience greater losses from weather-related causes (Murphy 1983).

Egg-laying occurs earlier in years with higher insect populations (Murphy 1986). Atlasers' observations of nests with eggs span the first three weeks of June; nestling records begin in mid June and continue through mid July. During this period, the female performs all of the incubation duties. The male perches near the nest when the female leaves the nest but does not incubate (Morehouse and Brewer 1968).

The entire period from egg-laying to fledging spans just over one month (Murphy 1983). Clutch sizes vary from two to four eggs (Morehouse and Brewer 1968). Atlas data suggest that this species normally begins its nesting cycle in Colorado anywhere from mid May to mid June, but nine late June records of nest-building probably indicate failed first attempts.

Adult kingbirds feed their nestlings a diet very similar to their own. Nestlings consume mainly adult insects, including grasshoppers, dragonflies, flies, butterflies, and moths. Some nestlings receive occasional fruit items. Once able to forage on their own, young birds often continue to eat moderate amounts of fruit, where available. In addition to flycatching, fledglings may also glean vertebrates off the foliage of plants. Adult birds continue to feed their young for a little over a month after fledging, longer for Eastern Kingbirds than for most other birds of comparable size (Morehouse and Brewer 1968). This period of dependence may help to explain why this species raises only one brood per season.

Predators of kingbird nests and young include kestrels, squirrels, and corvids (Murphy 1983). Colorado has no records of Eastern Kingbirds raising cowbird young, although the species' habit of ejecting cowbird eggs doubtless conceals the frequency with which cowbirds attempt to parasitize their nests.

BREEDING PHENOLOGY

CODE		# OF RECORDS	DATE RANGE
C	Courtship	1	4 Jul
NB	Nest Building	32	21 May–30 Jun
ON	Occupied Nest	53	17 May–17 Jul
NE	Nest with Eggs	4	4 Jun–22 Jun
NY	Nest with Young	7	11 Jun–15 Jul
FY	Feeding Young	45	6 Jun–29 Aug
FL	Fledged Young	19	18 Jun–19 Aug

DISTRIBUTION

Although the modifier "eastern" suggests a species whose range might, at best, just reach the western states, the range of this bird extends well into all western states except California and Arizona. Only by way of comparison with Western Kingbirds, whose range barely crosses to the eastern side of the Mississippi River, do Eastern Kingbirds qualify as an eastern species. Eastern Kingbirds spend their winters in the northern half of western South America.

The Atlas map shows that Eastern Kingbirds, although much more common on the plains in eastern Colorado, breed over much of western Colorado, especially the northern half. This appears little changed from earlier periods; Warren (1908) reported a handful of Eastern Kingbirds on an early bird survey in northwestern Colorado. Perhaps the most notable expansion of the species' range in Colorado has taken place in Mesa and Delta counties. Rockwell (1908) could cite no Mesa County records, but atlasers found these birds in nine blocks across irrigated areas of the Grand Valley and the Gunnison Valley below Delta.

Although their numbers diminish dramatically west of the Front Range, kingbirds exhibit another smaller-scale decline that appears as one travels southward. In general, the northern parts of Colorado receive greater precipitation and hold more open water than the southern parts. This, in turn, leads to smaller numbers of Eastern Kingbirds in the southern regions of the state.

Atlas field work provided first Confirmations of breeding in two northwestern Colorado latilongs. A fledgling near Kremmling filled in one latilong (40106). An adult carrying food west of Meeker and a bird on a nest near Rangely provided Confirmations in the other (40108).

Human development in the West, typically accompanied by irrigation and planting of trees, has provided new and more widespread breeding opportunities for Eastern Kingbirds.

BREEDING EVIDENCE

REPORTED IN 404 (23%) OF 1745 PRIORITY BLOCKS

☐	Possible	136 blocks	(34%)
▩	Probable	109 blocks	(27%)
■	Confirmed	159 blocks	(39%)

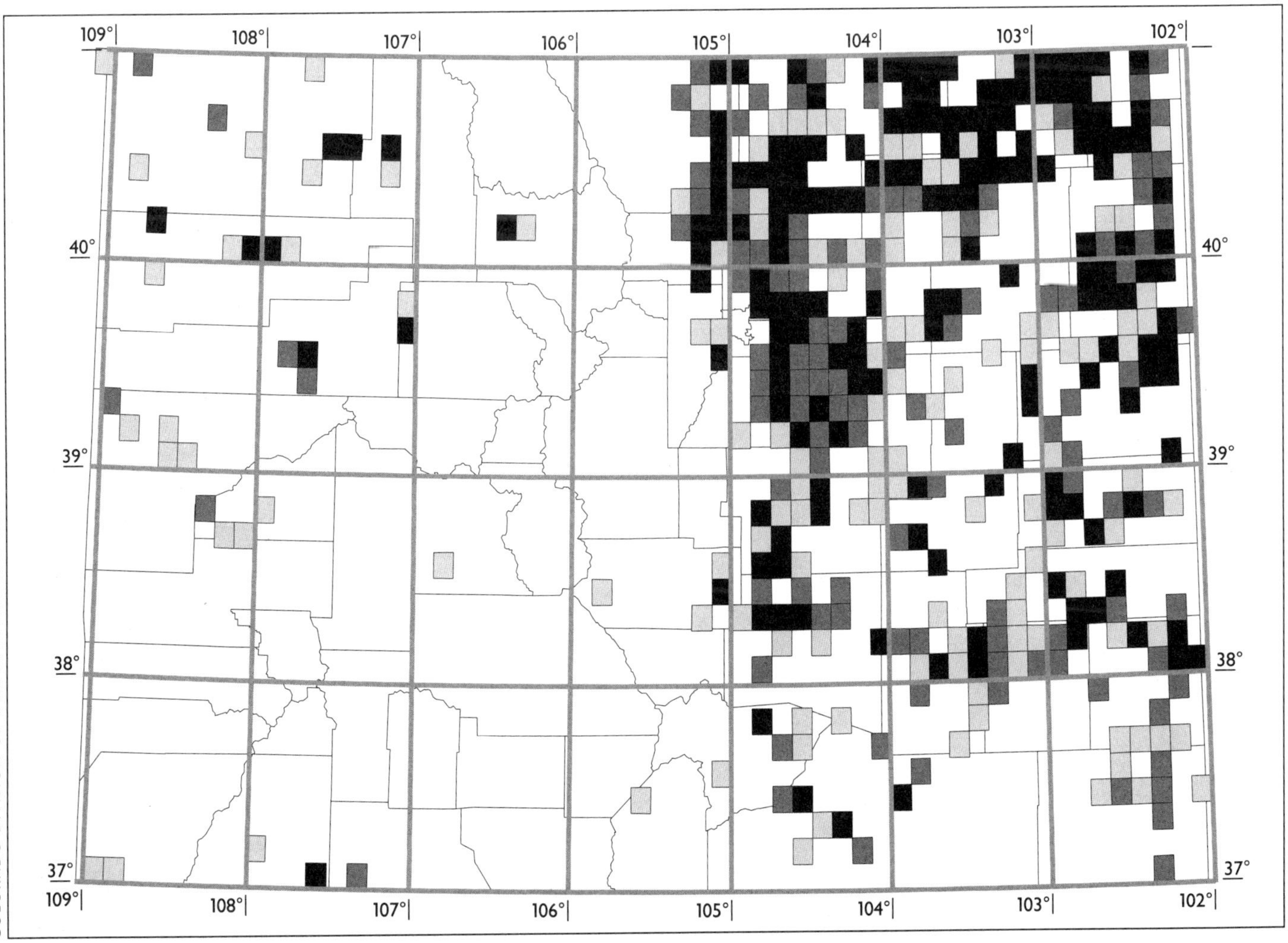
109°
108°
107°
106°
105°
104°
103°
102°
40°
39°
38°
37°

COLORADO DIVISION OF WILDLIFE / WILDLIFE RESOURCE INFORMATION SYSTEM

Like the Cimarron River, Scissor-tailed Flycatchers reside only briefly in far southeastern Colorado. From New Mexico the Cimarron River flows through Baca County for 15 miles (24 km) before moving onward into the heart of real Scissor-tail country. Along this sandy, usually dry, quarter-mile-wide streambed filled with cottonwoods and tamarisks, these flycatchers build their nests, spend two or three months, and then move on.

HABITAT

The Cimarron River provides a corridor of nesting habitat that extends the Scissor-tailed Flycatchers' breeding range into Colorado. Atlas workers observed these birds, more typical of open chaparral, defending territories and nesting in cottonwood trees along the Cimarron at elevations between 3,500 and 4,000 feet (1070–1220 m). The flycatchers set up their territories and build nests in the smaller, younger trees at the open edges of the woodland.

Insectivorous Scissor-tailed Flycatchers forage almost exclusively within their territories during breeding season (Teather 1992) by flycatching from perches. Their diet consists largely of grasshoppers, crickets, locusts, beetles, and bees.

BREEDING

Unique among Colorado kingbirds, Scissor-tailed Flycatcher males and females exhibit a striking dimorphism in their tail length: males have greatly elongated and forked tails. Only males engage in singing and perform the "tumble-flight" display described by Bent (1942). The bonded pair exhibits a high degree of site fidelity between years. They will return year after year to the same site and use the same tree, even if the tree is dead (e.g., killed by herbicides). Of eight reused nests found in dead mesquite trees in south Texas, all failed (Nolte and Fulbright 1996).

Males arrive earliest to claim territories. Females begin nest-building almost immediately upon their arrival. In Oklahoma recorded dates of nest construction range from 31 May to 29 June; in Colorado from 16 May to 23 June (B&N, HEK).

In nest-site selection they make a tradeoff, providing open air space around the nest with less cover as a defense against predators, at the expense of increased exposure to the vagaries of weather—wind, rain, snow, and solar radiation (Nolte and Fulbright 1996).

Atlas workers saw nest construction on 29 May 1988. Males do not participate in nest-building, incubation, or brooding of young. Both sexes feed nestlings and respond aggressively to potential predators near the nest. These flycatchers normally raise one brood, but sometimes two (Regosin 1995).

No records of parasitism by Brown-headed Cowbirds exist for Colorado (Chace and Cruz 1996), due to the small size of this fringe population, but Oklahoma has had three cases. In all three the flycatchers ejected the uninvited eggs from the nest, a trait shared with only 11 other species, which include Eastern and Western kingbirds (Regosin 1994).

BREEDING PHENOLOGY

CODE		# OF RECORDS	DATE RANGE
NB	Nest Building	1	29 May
NY	Nest with Young	1	28 Jul

DISTRIBUTION

Scissor-tailed Flycatchers breed from southeastern Colorado and southeastern Nebraska south to Texas, western Arkansas, and western Louisiana. They winter from southern Mexico to Panama. The Cimarron River lies on the western fringe of this species' normal breeding range and comprises their only predictable breeding area within the state. On 31 May 1923 Niedrach noted two Scissor-tailed Flycatcher nests under construction along the river, the first breeding records in Colorado (B&N). These flycatchers wander throughout much of North America; Colorado has over 30 records of spring vagrancy and at least 10 records for fall (A&R).

No Atlas blocks included the Cimarron; the one Confirmation from Baca County came from a Non-priority block. Ten miles (16 m) south of Kit Carson, and 100 miles (160 km) north of the Baca County population, a male Scissor-tailed Flycatcher, apparently unmated, exhibited territorial behavior by staying for two weeks in June 1992 in trees that outline a gas transmission station (Lewis Lake, 38103F7, Dave Leatherman pers. comm.).

The most unusual breeding observation occurred in Garfield County, more than 330 miles (530 km) northwest of Baca County. A female Scissor-tailed Flycatcher unsuccessfully attempted to nest in 1993, and she returned in 1994 to raise three young. White outer tail-feathers and yellow coloration on the breast and abdomen of one of the juveniles suggest that the female had mated with a Western Kingbird (Dexter 1995). Tyler collected a hybrid Scissor-tailed Flycatcher x Western Kingbird in southwestern Oklahoma in May 1988 and reported only two other documented cases of this combination before 1988 (Tyler and Parkes 1992).

Colorado's Scissor-tailed Flycatcher population, clearly a fringe group, has remained steady in numbers since first recorded in 1923. They have no state or federal agency rankings because the species, as a whole, ranks as globally stable. Based on their rarity they could easily become extirpated from the state; species in trouble may first withdraw from the edges of their ranges. In this capacity, Colorado's breeding population could act as an indicator for the overall health of the species (Colo. Nat. Herit. Prog. 1995).

BREEDING EVIDENCE

REPORTED IN 3 (0%) OF 1745 PRIORITY BLOCKS

□	Possible	2 blocks	(67%)
■	Probable	1 block	(33%)
■	Confirmed	0 blocks	(0%)

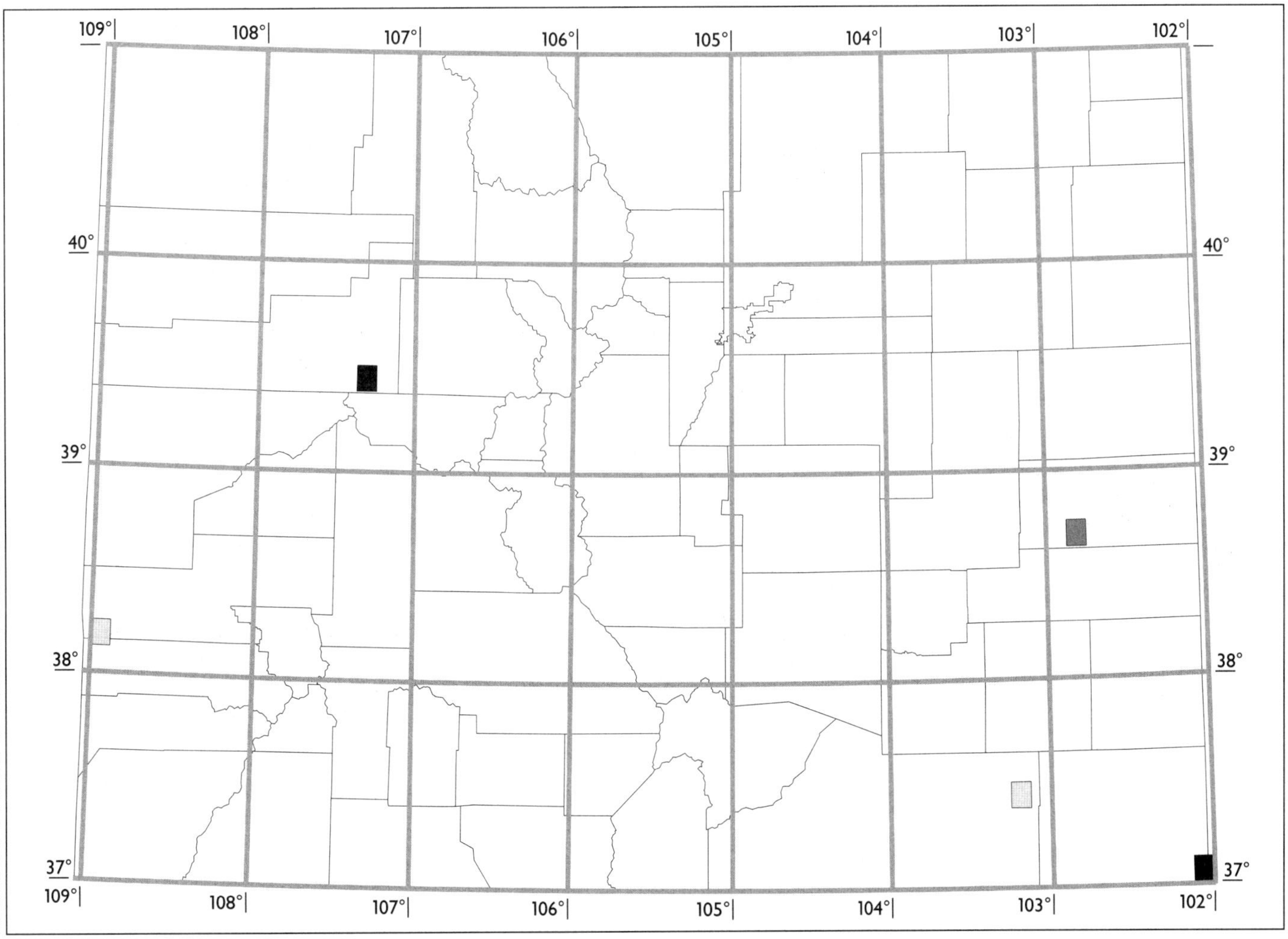
109°
108°
107°
106°
105°
104°
103°
102°
40°
39°
38°
37°

COLORADO DIVISION OF WILDLIFE / WILDLIFE RESOURCE INFORMATION SYSTEM

One of the fiercest passerine birds around is the "butcher bird." Its modus operandi includes harrying hapless sparrows, subduing mice with raptorial skill, impaling insect victims on barbed wire, and bludgeoning prey to death with powerful blows from its massive bill. In spite of these aggressive traits, the Loggerhead Shrike is experiencing population declines, probably because of habitat loss.

HABITAT

Early Colorado ornithologists (Cooke 1897, Sclater 1912) described Loggerhead Shrikes as quite common, chiefly on the plains—at that time dominated by shortgrass prairie. Over 60% of the shrikes seen by atlasers occupied only two habitats, but shortgrass prairie (15% of all records) came in a distant second to rural (47%). Certainly in 1897 suitable rural habitat barely existed and Loggerheads depended almost entirely on shortgrass prairies, at least in eastern Colorado. Because shrikes nest in trees, the prairie places that they pick always have a component of a few small, scattered trees (4–10 trees will do). These little copses often grow in shallow draws on the plains. Shrikes commonly pick hackberries or small cottonwoods under 15 feet (5 m). The rural habitats the shrikes use today often consist of abandoned farmyards with untended trees (often Siberian elms).

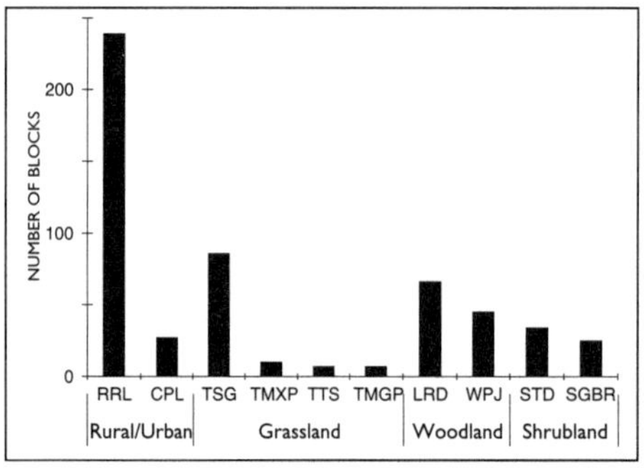
NUMBER OF BLOCKS
200
100
0
RRL
CPL
TSG
TMXP
TTS
TMGR
LRD
WPJ
STD
SGBR
Rural/Urban
Grassland
Woodland
Shrubland

Atlas workers also found Loggerheads in lowland riparian (10%) and pinyon/juniper woodlands (8%). Shrubby habitats—mainly greasewood, saltbush, and sagebrush with thorny bushes for nesting—dominate shrike habitat in the San Luis Valley and Western Slope. Essentially, Loggerhead Shrikes prefer open country, and select areas with scattered trees and shrubs.

Shrikes eat mostly insects—two-thirds of their diet. They also eat vertebrates (24–28%), about half of them birds, and reptiles (e.g., lizards), frogs, and toads. Because shrikes have passerine feet, they impale large prey on thorns (or barbed wire) for plucking. Some prey protected by noxious chemicals (monarch butterfly, certain toads) they leave impaled for up to three days, presumably waiting for the toxins to degrade (Yosef 1996). As a top-of-the-food-chain passerine raptor, Loggerheads never densely pack their habitat—only 34 blocks claimed more than ten breeding pairs, and these blocks probably had 15 breeding pairs at best.

BREEDING

Atlas workers easily Confirmed shrikes as breeders—in 57% of all blocks. Because Loggerhead Shrikes nest in open habitats, their nest trees are often obvious. So are the nests, which the birds often place below the crown of the tree. The bulky nest is a well-built woven affair of twigs and bark strips, with finer materials for lining. Loggerhead Shrikes lay 4–7 eggs that they incubate 16–17 days. The young fledge in 17–21 days and adults often feed the young for 3–4 weeks after fledging (Yosef 1996). Incubation probably starts with the first or second egg: atlasers in Montezuma County found a fledgling family of stair-step-aged young (Kingery 1989).

On the Pawnee National Grassland, shrikes arrive in early April and establish territories by early May. With this early start they are among the earliest returning passerines (Porter et al. 1975), yet atlasers found nest-building on an average date of 1 June (only six records)—later than expected and comparable to most other plains-nesting passerines.

Most in-nest activity occurs 16 May–28 June. Average dates for nesting activities reported by atlasers followed a logical sequence that matches usual durations for incubation and fledging: nests with eggs (2 June), nests with young (12 June), carrying food for young (21 June), and fledged young (29 June). Loggerheads still completed their in-nest activities by early July, earlier than most passerines.

BREEDING PHENOLOGY

CODE		# OF RECORDS	DATE RANGE
C	Courtship	2	30 Apr–6 Jun
NB	Nest Building	6	23 May–16 Jun
ON	Occupied Nest	53	4 May–5 Jul
NE	Nest with Eggs	18	16 May–27 Jun
NY	Nest with Young	25	21 May–6 Jul
FY	Feeding Young	55	17 May–19 Aug
FL	Fledged Young	106	26 May–24 Aug

DISTRIBUTION

Loggerhead Shrikes occur across the U.S., north to central Washington, the Canadian prairies, and Virginia. The southern states and the central plains, not including eastern Colorado, support the highest densities (Price et al. 1995). They winter from Nevada and Virginia to southern Mexico.

The Atlas map shows a distinctly eastern Colorado concentration—88% of the blocks reporting shrikes. Atlasers found shrikes in 60% of the plains blocks, and 85% of the blocks with more than ten breeding pairs were on the plains. Of the A3 codes, 38% were rural, and 16–22% in each of lowland riparian, shortgrass prairie, and pinyon/juniper. Atlas data Confirmed breeding at 7,900 feet (2410 m) on the Grand/Summit county line (King Creek, 39106H3), not far from the highest locality known to Sclater (1912). The map also shows Possible breeding at higher elevations, 8,500–8,900 feet (2590–2715 m), in Chaffee, western Boulder, and Saguache counties.

Continental BBS trends indicate a significant decline of 3.5% per year (with 63% of the routes reporting drops), which computes to a loss, in 31 years, of two-thirds of the total population. Causes of the continental decline seem to center on breeding-ground problems, ranging from habitat loss to pesticide exposure to car kills (Ehrlich et al. 1992). Others propose that, in the southeastern U.S. wintering grounds, land-use changes and fire ant infestations limit shrikes (Lymn and Temple 1991).

In vivid contrast, routes in the High Plains region (including eastern Colorado) manifest a large increase of 3.8% per year, a strong indication of healthy populations in the Great Plains. On the Western Slope spotty distribution and (inconclusive) BBS trends pointing downward may indicate a decline there. This contrasts strongly with the A&R view that shrikes are doing fine on the Western Slope and have been extirpated from parts of eastern Colorado. Historical context may put this into perspective. In the nineteenth century Loggerhead Shrikes once lived largely in the shortgrass prairie, much of it now replaced by rural habitats. Atlas records showing heavy use of rural habitats and BBS data bolster the view that Loggerhead Shrikes have a healthy Eastern Slope presence, and that they currently do better in rural habitats than in shortgrass prairie.

BREEDING EVIDENCE

REPORTED IN 489 (28%) OF 1745 PRIORITY BLOCKS

Possible	145 blocks	(30%)
Probable	64 blocks	(13%)
Confirmed	280 blocks	(57%)

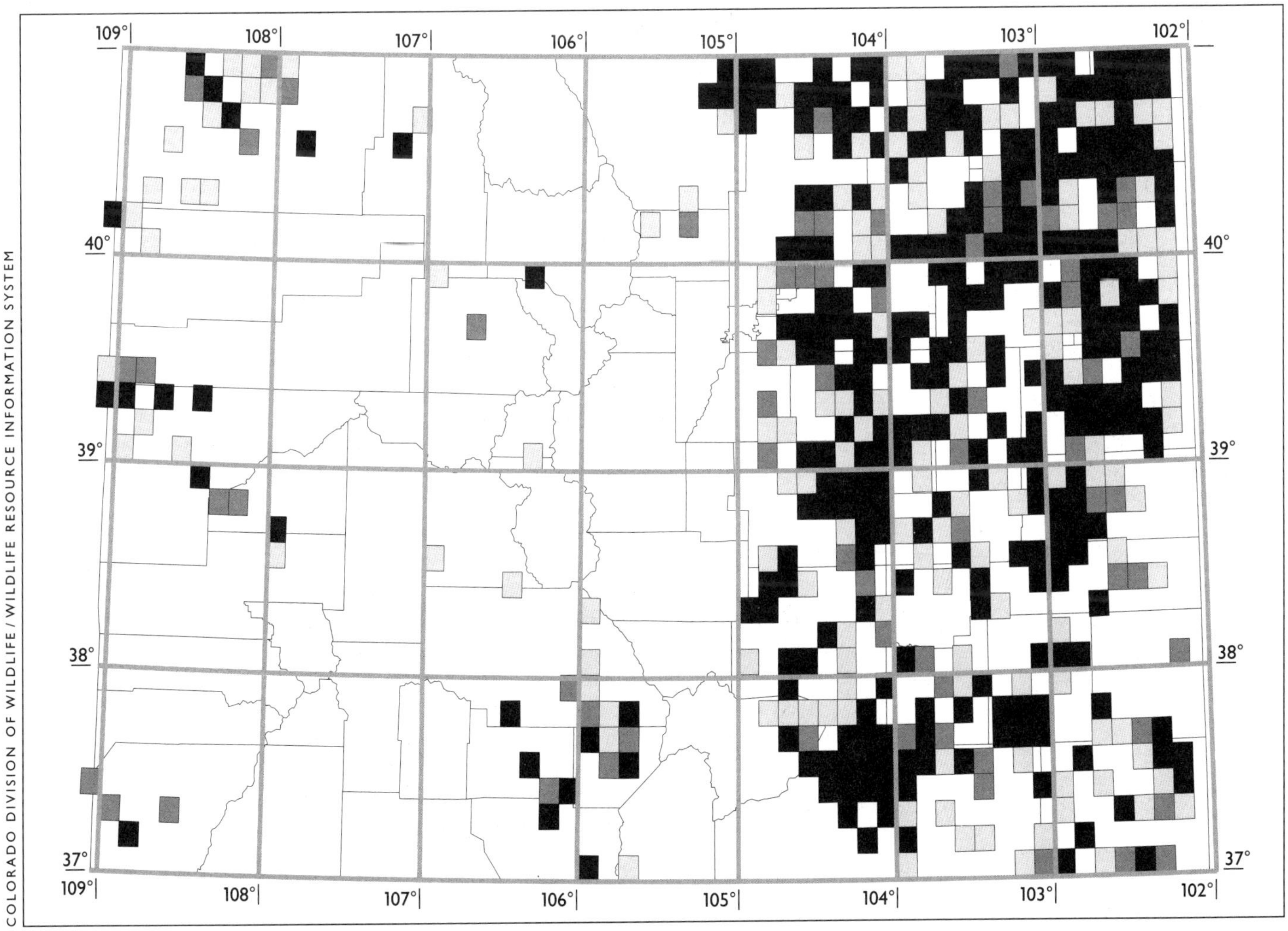
109°
108°
107°
106°
105°
104°
103°
102°
40°
39°
38°
37°
COLORADO DIVISION OF WILDLIFE / WILDLIFE RESOURCE INFORMATION SYSTEM

A quick buzzy scold deep in a tangle of riparian vegetation signals the presence of a Bell's Vireo. Males sing regularly during the breeding season, although apparently not from the nest as Warbling and Plumbeous vireos do. Throughout their range lost breeding habitat, the depredations of cats, and Brown-headed Cowbird brood parasitism have brought on a disastrous population decline.

HABITAT

The key habitat component for breeding and wintering Bell's Vireos is dense, often impenetrable, shrubby vegetation (Barlow 1980, Brown 1993). Atlasers found them only in plains riparian and shelterbelt thickets. Typical shrubs included skunkbush, wild plum, and hackberry. During the breeding season Bell's Vireos restlessly glean insects off the foliage and branches of deciduous shrubs for a diet that consists almost entirely of insects (99.3%; Chapin 1925, Salata 1983).

Food preference, foraging space, and nest-site selection may reduce competition between Bell's Vireos and other foliage-gleaning insectivorous birds that nest within

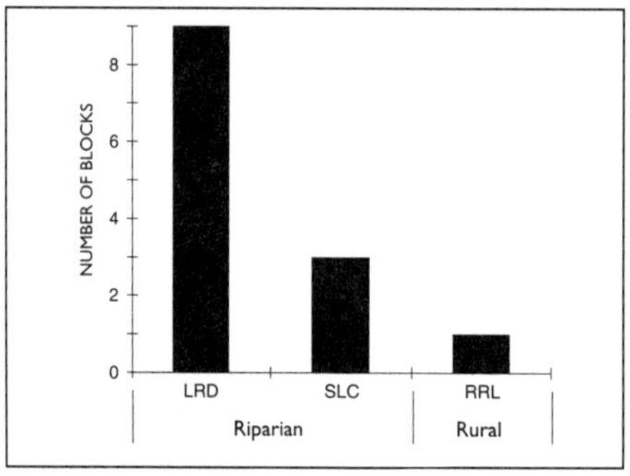
NUMBER OF BLOCKS
0
2
4
6
8
LRD
SLC
RRL
Riparian
Rural

the riparian habitats of eastern Colorado. Bell's Vireos have a narrow niche breadth; they forage among the dense branches of the understory, whereas other thicket-linked species (e.g., Yellow-breasted Chat, Northern Mockingbird, Brown Thrasher, Eastern Towhee) forage on the ground or have a broader diet (Ehrlich et al. 1988). Unlike this shrub-nesting vireo, other insectivorous foliage-gleaners (e.g., Warbling Vireo, House Wren) prefer cottonwood woodlands to the dense shrubby vegetation used by Bell's Vireos (Ehrlich et al. 1988).

BREEDING

Bell's Vireos arrive on the breeding grounds in early May. Males court females by rapid chases punctuated with bursts of fluttering, feather spreading, and song (Ehrlich et al. 1988). Once the pair forms, they build a small cup-shaped nest in about four days (Nolan 1960). Males maintain small breeding territories of 1.2 acres ± 1 acre (0.5 ± 0.4 ha) with constant singing interspersed with bouts of chasing and physical contact between neighboring territorial males (Barlow 1962).

Nests often have a cone-shaped profile (Bent 1950). The birds place it in a shrub 1–5 feet (0.5–1.5 m) high, usually in the fork of a lateral twig halfway between the stem and tip of the branch (Nolan 1960). Vireos construct nests of dried leaves and shredded bark, hold them together with spider silk, and line them with grasses, hair, and down (Harrison 1979). Bell's Vireos lay three to five oval white eggs with brown specks. Both sexes incubate, for 14 days, and young fledge from the nest about 11 days after hatching (Harrison 1979, Ehrlich et al. 1988). Both adults feed and protect post-fledging young.

Brown-headed Cowbirds have parasitized Bell's Vireo nests in Kansas, Arizona, and Oklahoma. Colorado has no such records, probably due to the low number of vireos and of field ornithologists in far eastern Colorado (Chace and Cruz 1996). These vireos vigorously attack cowbirds and predators near the nest site, and they sometimes construct a new nest over an earlier nest that contains foreign eggs (Nolan 1960, Brown 1993). These defensive adaptations against brood parasitism probably evolved in concert with a long-term host relationship with Brown-headed Cowbirds. However, the mechanisms do not work well in regions of high cowbird densities because often the cowbirds parasitize renesting attempts too. Thus, a strategy that probably succeeded with transient cowbirds that followed nomadic bison now fails because packs of these once-itinerant parasites have become stationary throughout the breeding season.

BREEDING PHENOLOGY

CODE		# OF RECORDS	DATE RANGE
ON	Occupied Nest	1	23 Jun
FY	Feeding Young	2	21 Jun–22 Jun
FL	Fledged Young	1	12 Jul

DISTRIBUTION

Bell's Vireos range over the Midwest from the Dakotas and Indiana south to central Texas and northern Mexico, and from Mexico into southern New Mexico and Arizona; an isolated (and threatened) population breeds in southern California. In the winter they forage among dense thickets along the western coast of Mexico south to Nicaragua.

Sclater (1912) considered them "a regular summer bird in the eastern half" of Colorado and B&N thought them "fairly common summer residents" along the eastern border. Smith collected 26 specimens from Julesburg, Wray, and Holly (B&N), but his Holly specimens bear dates of 9–23 May and probably were migrants. Atlasers found them only near Julesburg and Wray, and not near Holly. Atlas data track the range depicted by A&R. Colorado lies at the western fringe of the midwestern Bell's Vireo range; despite the early reports, they probably have never nested in numbers in eastern Colorado, which has too few shrubby drainages.

From 1966 to 1996 Bell's Vireos declined significantly on BBS routes, and several authors and agencies have designated them a species of special concern (Robbins et al. 1986, Tate 1986, Sauer and Droege 1992, Carter et al. 1996). Nationally, Bell's Vireos have declined 3.4% per year, calculating to a population drop of half in 31 years. Their decline stems from a now ineffective, although aggressive, strategy to counter cowbird parasitism coupled with the loss of breeding habitat caused by draining wetlands in the midwestern center of their range.

BREEDING EVIDENCE

REPORTED IN 4 (0%) OF 1745 PRIORITY BLOCKS

Possible	1 block	(25%)
Probable	1 block	(25%)
Confirmed	2 blocks	(50%)

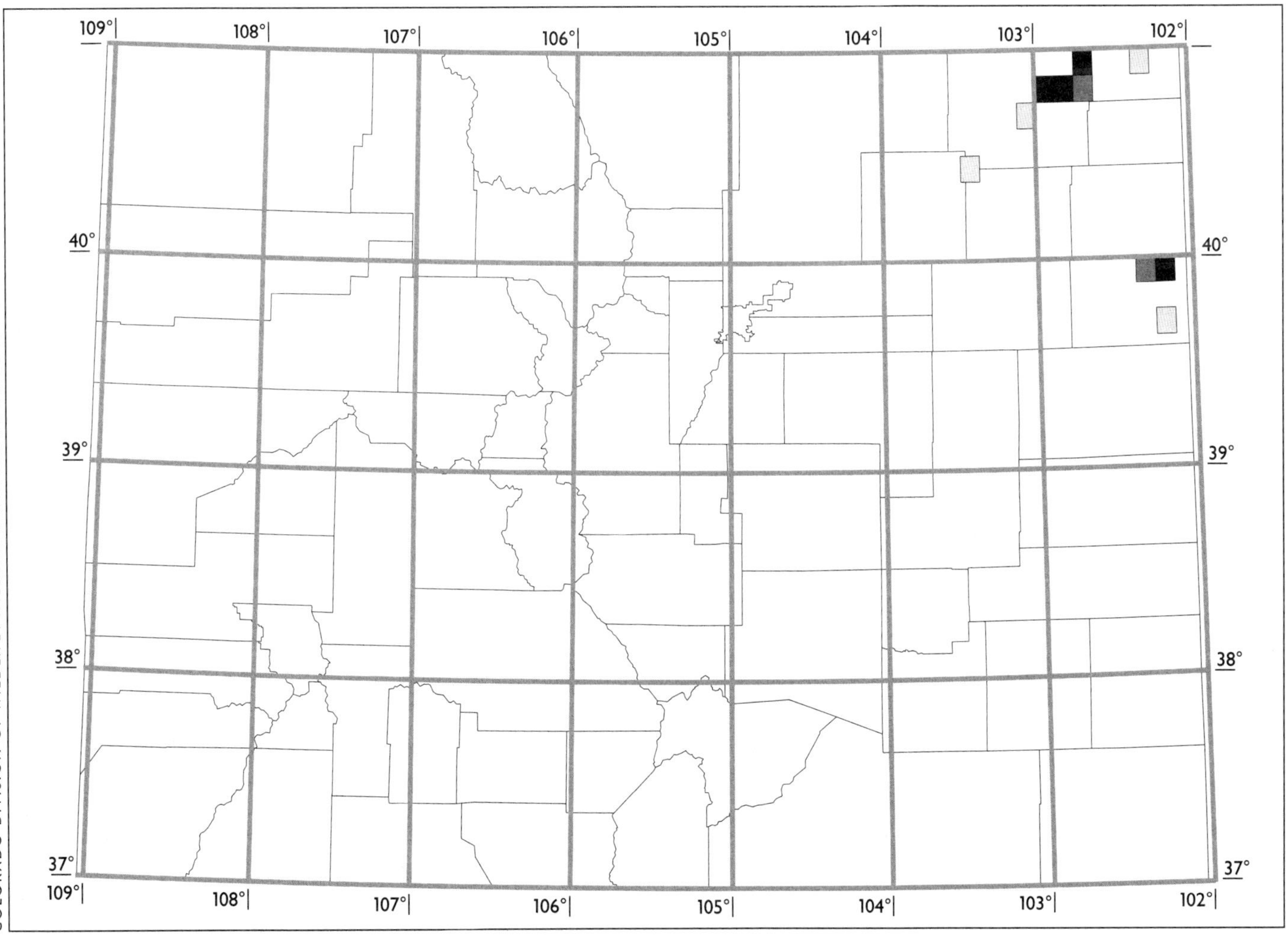
109°
108°
107°
106°
105°
104°
103°
102°
40°
39°
38°
37°

COLORADO DIVISION OF WILDLIFE / WILDLIFE RESOURCE INFORMATION SYSTEM

From a hot, dry, inhospitable Colorado environment comes, arguably, the sweetest melody any vireo can sing (Bent 1950). With tails twitching, Gray Vireos hop and dart among the junipers as they forage all day long, except for the short breaks they take to sing. The grays and greens of juniper, sagebrush, and desert scrub camouflage these tail-wagging and little-known vireos. Inconspicuous inhabitants of an obscure habitat—scattered junipers—they did not arrive on the list of Colorado birds until 1907.

HABITAT

On lightly forested mesas, steep hillsides, canyons, and wide valleys where scattered juniper trees grow spaced apart, Atlas field workers located Gray Vireos. The juniper habitat that Gray Vireos prefer allows room for grasses, sagebrush, and desert scrub to flourish. They avoid denser sectors of pinyon/juniper "pygmy forests."

Las Animas County field workers found Gray Vireos in junipers between 4,500 and 6,000 feet (1400–1800 m) elevation. There pinyon pines cover the mesa tops; downslope, junipers mix with pinyons and, at the bottom, only sparsely spaced junipers grow. Among the scattered junipers, atlaser Randy Lentz found two nests at 4,700 feet (1430 m).

In western Colorado atlasers found a greater elevational range from 4,400 to 6,400 feet (1200–2000 m). Gray Vireos rarely range higher into the taller, mixed

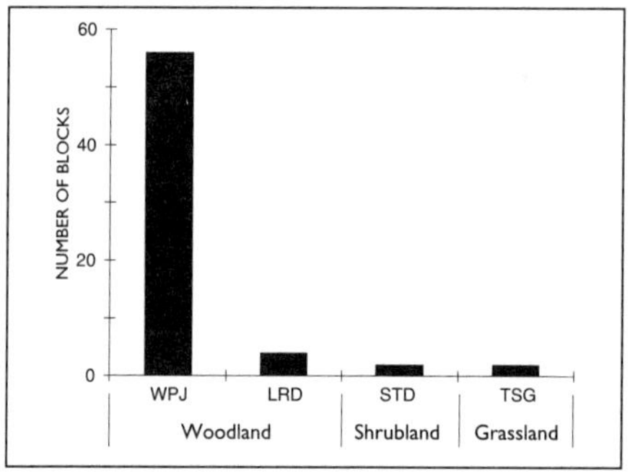
60
40
20
0
NUMBER OF BLOCKS
WPJ
LRD
STD
TSG
Woodland
Shrubland
Grassland

pinyon/juniper woodlands where they might compete with the bigger and more common Plumbeous Vireo (A&R, Scott Hutchings pers. comm.).

In all but two of the 51 blocks (Priority and Non-priority) for which field workers reported habitat codes, they recorded pinyon/juniper habitat; Atlas habitat codes did not distinguish between pinyon/juniper mix and junipers only. Atlasers reported six other habitats used by Gray Vireos, all associated with the perimeter of juniper stands—habitats the vireos use to forage for food. In three blocks field workers reported birds in riparian but listed them as only Possible breeders. Large expanses of seemingly suitable habitat in western Colorado lack these vireos—similar to New Mexico where they display a spotty or "island" distribution (Ligon 1961).

BREEDING

Gray Vireos arrive from Mexico by early May, the second week in May in Colorado National Monument; the nesting cycle starts soon after their arrival. Each pair needs a large territory (40 acres, 16 ha). They place their nests 2–8 feet (0.6–2.5 m) above the ground in a thorny or twiggy shrub or, most frequently, in a juniper tree with a snag protruding from the top (Scott Hutchings pers. comm.).

In three to four days they bind the nest at its rim to twigs and construct it from dry grasses, plant fibers, stems, and hair. They often camouflage (and perhaps fumigate; Ron Lambeth pers. comm.) the nest by adding sagebrush leaves to the exterior. The female vireo lays three to five eggs. Both parents incubate and the eggs hatch in 13–14 days. In another 13–14 days the young leave the nest with both parents still in attendance.

Although reportedly a frequent host to Brown-headed Cowbirds (Terres 1980; contra, Chace and Cruz 1996), Colorado has but two records, including one from the 1995 CBO study (Chace and Cruz 1996). A bigger threat appears to be nest predation. The suspects—jays, rock squirrels, and chipmunks—destroyed one-half of the nests found in 1995 in Colorado National Monument (Scott Hutchings pers. comm.).

BREEDING PHENOLOGY

CODE		# OF RECORDS	DATE RANGE
C	Courtship	2	23 Jun–24 Jun
NB	Nest Building	3	22 May–9 Jun
ON	Occupied Nest	2	23 May–17 Jun
NE	Nest with Eggs	1	11 Jun
NY	Nest with Young	1	29 Jun
FY	Feeding Young	8	7 Jun–11 Jul
FL	Fledged Young	5	20 Jun–27 Jul

DISTRIBUTION

Gray Vireos breed from southern California, Nevada, and Utah into western Colorado. The range extends eastward across New Mexico, dips upward into southeastern Colorado (known from Atlas field work), then south to western Texas. Short-range migrants, most winter on the arid northwest coast of Mexico.

The first Colorado records came from Lamar in 1907 (Cooke 1909) and Baca County in 1914 (B&N). Gray Vireos were not even recorded in western Colorado until 1944. This observation, birds that nested in front of the Mesa Verde National Park museum, also constituted the first state breeding record (B&N). By 1992 A&R could show only five breeding sites. An Atlas Rendezvous discovered one of those: two nests found in Brown Canyon (37103E3) on 3 June 1990 confirmed for the first time breeding in southeastern Colorado (Randy Lentz, nest cards).

Expanding the previously known range, the Atlas map depicts a fairly widespread distribution of this species in western Colorado, from Rangely to the Four Corners. It also depicts the small population in Las Animas County. Most blocks that recorded the species (all except Mesa Verde and Colorado National Monument) lie in places never previously surveyed for birds. West of Rangely at low elevations along the White River, field workers added Confirmed breeding in northwestern Colorado. Transformation of juniper woodlands into farmland in Dolores, San Miguel, and Montezuma counties may have reduced the population there. In the summers of 1995 and 1996, CBO found 21 Gray Vireo nests in Colorado National Monument.

Three Colorado conservation priority lists place Gray Vireos among the ten most threatened Colorado breeders. The main breeding range is confined to the Great Basin. If, as postulated by Priorities 1995, Colorado has more than 40% of the breeding distribution, the total number of these obscure vireos must be tiny—30,000 breeding pairs, calculated by projecting Atlas abundance estimates. One report lists them as globally secure, but vulnerable in Colorado (Colo. Nat. Herit. Prog. 1995). The birds winter in a very small area (Priorities 1995), which contributes to a perilous status for the species. An effective conservation program first needs more research on Gray Vireo status and habitat requirements (Scott Hutchings pers. comm.).

BREEDING EVIDENCE

REPORTED IN 45 (3%) OF 1745 PRIORITY BLOCKS

☐	Possible	16 blocks	(35%)
■	Probable	12 blocks	(27%)
■	Confirmed	17 blocks	(38%)

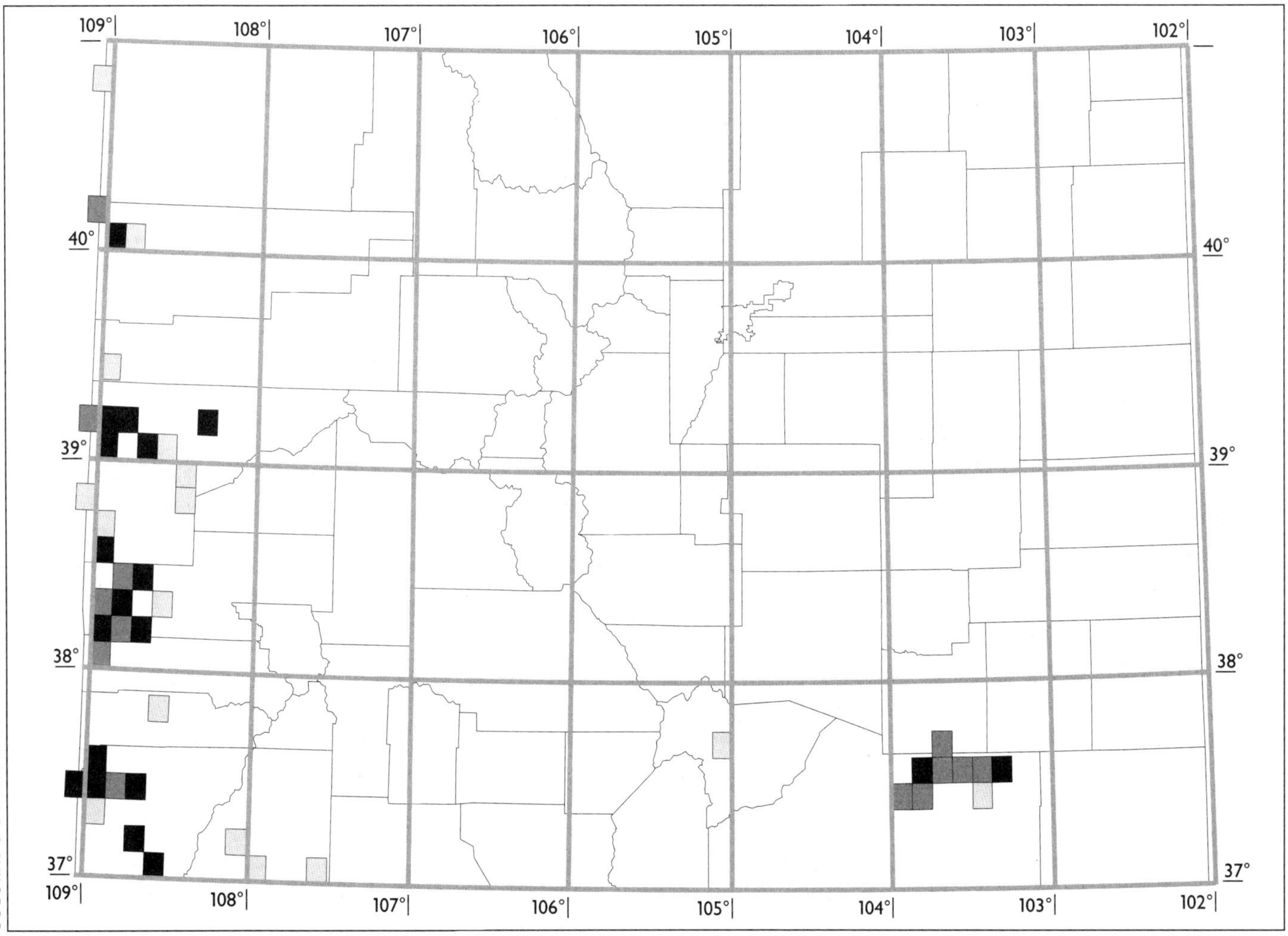
109°
108°
107°
106°
105°
104°
103°
102°
40°
39°
38°
37°

COLORADO DIVISION OF WILDLIFE / WILDLIFE RESOURCE INFORMATION SYSTEM

The loud and continuous song of the spectacled Plumbeous Vireo rings out throughout the pinyon/juniper and ponderosa pine forests of Colorado. These vireos pick coniferous habitats, forage in tree interiors, and are bigger than other vireos. Sometimes, males exhibit the peculiar behavior of singing while on the nest.

HABITAT

Colorado atlasers found Plumbeous Vireos breeding in the pygmy forests of pinyon pines and junipers and the park-like forests of ponderosa pine. To a lesser extent they found the birds in deciduous trees along foothills streams (which most often lie next to coniferous hillsides) and aspen (also often bordered by conifers). They found these vireos as well in riparian woodlands in the western valleys, and to some extent in scrub oak and montane shrubs. Within these habitats vireos glean insects from tree trunks and branches and often hawk insects. During migration and winter, like other vireos, they become frugivorous.

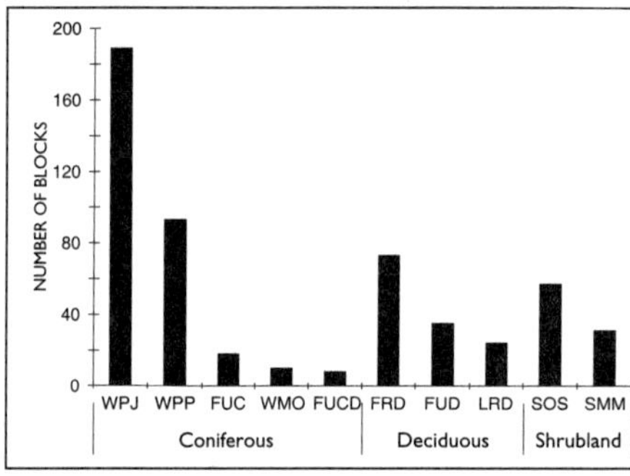
NUMBER OF BLOCKS
0
40
80
120
160
200
WPJ WPP FUC WMO FUCD FRD FUD LRD SOS SMM
Coniferous
Deciduous
Shrubland

Colorado vireo species may reduce competitive interactions by habitat selection, foraging locations, and size differences. Warbling Vireos typically breed in the foothills and montane riparian zones adjacent to, and in higher elevations than, the pine forests where Plumbeous Vireos breed, yet atlasers recorded Warbling Vireos in 61% of the blocks with Plumbeous. When Plumbeous Vireos inhabit deciduous foothills and aspen forests they reduce food competition with Warbling Vireos by foraging more centrally in the vegetative structure. Where the two coexist, Plumbeous may displace Warbling Vireos (22% smaller) from foraging areas (Barlow 1980). Size differences may reduce competition by affecting food-size selection.

Plumbeous Vireos inhabit 78% of the Atlas blocks in which Gray Vireos breed. Gray Vireos are about 25% smaller by weight; the size difference may enable the larger Plumbeous to exploit different resources in pinyon pine and juniper forests where they coexist. They also exercise differential habitat selection: Gray Vireos use the drier sections of pinyon/juniper woodlands dominated by junipers, usually at lower elevations than Plumbeous Vireos.

BREEDING

Male Plumbeous Vireos arrive on the breeding grounds in early May. Courtship and establishment of territories continue until late May when nest-building begins (the Atlas shows this 2 as 6 May–24 June).

The female chooses the nest site and does most nest-building while the singing male follows. She places the nest in a distal fork of a low branch usually 6–15 feet (2–4.5 m) high, but up to 32 feet (10 m). The nests contain grasses, forbs, and bark; insect cocoons and spiderweb material hold the materials together. The birds attach lichens and mosses to the outer surface and sometimes white cocoon material hangs well below the nest.

Females lay 3–5 eggs on consecutive days (Chace et al. in press). The Atlas dates of 21 May–30 June show the peak of incubation, although sometimes incubation continues into mid July. Eggs hatch after 14 days and young fledge 14 days later. Males develop a partial brood patch and both adults incubate, feed the young, and care for fledglings. During incubation the males often sing from the nest, thus making nest finding easier for Atlas field workers. Although single-brooded, Plumbeous Vireos will, like most birds, rebuild and initiate a new clutch following nest failure.

Cowbird parasitism severely impacts Plumbeous Vireos (Marvil and Cruz 1989). Atlasers reported this in four blocks and on 22 nest cards (Chace and Cruz 1996; Appendix C). In Boulder County Plumbeous Vireos have sustained nearly 50% parasitism for at least ten years (Chace et al. in press). Cowbird parasitism, coupled with high rates of nest predation, reduces productivity so severely that the Boulder population would soon disappear without immigration by birds from more productive regions (Chace 1995). When these vireos nest near canopy openings and edge created by roads and residential areas, they suffer more parasitism by cowbirds and more predation by jays, magpies, rock squirrels, and other predators (Chace 1995, Ron Lambeth pers. comm.).

BREEDING PHENOLOGY

CODE		# OF RECORDS	DATE RANGE
C	Courtship	4	20 May–4 Jul
NB	Nest Building	13	26 May–24 Jun
ON	Occupied Nest	17	26 May–21 Jun
NE	Nest with Eggs	8	21 May–30 Jun
NY	Nest with Young	10	7 Jun–7 Jul
FY	Feeding Young	41	19 May–30 Jul
FL	Fledged Young	40	14 Jun–1 Aug

DISTRIBUTION

Plumbeous Vireos breed throughout the U.S. Rocky Mountains and Great Basin ranges. Like many western breeding birds, Plumbeous Vireos have a large breeding distribution but a relatively small wintering range in western Mexico (Terborgh 1980). Early ornithologists found them in Colorado's pinyon/juniper and ponderosa pine forests (Henshaw 1875, Rockwell 1908), where Atlas workers found high abundances.

The Atlas shows these vireos with a strong presence in ponderosas along the Front Range from Wyoming to New Mexico and in the pinyon/juniper woodlands upstream from Pueblo to Buena Vista. Relatively few occur in the more scattered pinyon/juniper woodlands of eastern Las Animas County and in the pinyon/juniper band around the San Luis Valley. On the Western Slope, where both ponderosas and pinyon/junipers grow extensively, Plumbeous are more abundant.

Populations in the central Rockies have shown a small decline since 1966, but a greater decline in Colorado. Plumbeous Vireos respond positively to low-level disturbance on the breeding grounds (Bock and Bock 1983, Hejl and Woods 1991), yet they stick to undisturbed habitats in their winter range (Hutto 1992).

Further research needs to determine what factors affect reproduction in open canopy forests of pinyon pine and juniper, and in what regions Plumbeous Vireo populations achieve a net production of young. First-year birds probably tend to breed in less productive places, thereby stabilizing populations that otherwise lack the capability to replace themselves. Conservation strategies for this species should focus on two things: preserving winter habitat and identifying and protecting breeding populations where Plumbeous Vireos can breed successfully without major cowbird parasitism and predation (Chace 1995).

BREEDING EVIDENCE

REPORTED IN 407 (23%) OF 1745 PRIORITY BLOCKS

☐	Possible	174 blocks	(43%)
▩	Probable	108 blocks	(26%)
■	Confirmed	125 blocks	(31%)

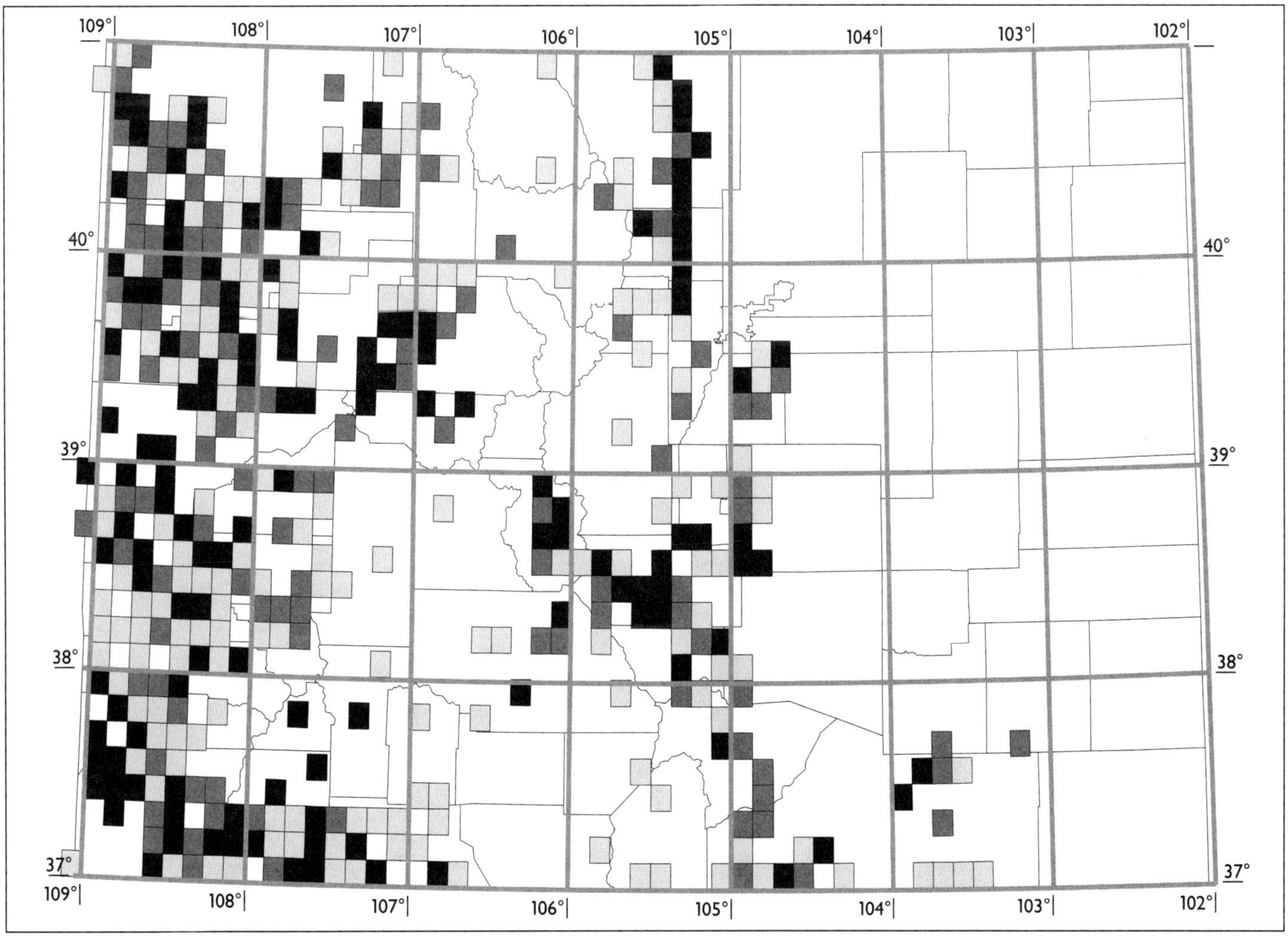
109°
108°
107°
106°
105°
104°
103°
102°
40°
39°
38°
37°

COLORADO DIVISION OF WILDLIFE / WILDLIFE RESOURCE INFORMATION SYSTEM

Classic survival theory states that a breeding bird that draws attention to its nest invites disaster and will never pass on its genes. Distraction displays, camouflage, construction in inaccessible locations, and other behaviors have evolved to avoid attracting predators, cowbirds, and other unpleasant creatures. Contravening this concept, the male Warbling Vireo sings while on the nest, making nest discovery by atlasers a lot easier.

HABITAT

Warbling Vireos forage and breed almost exclusively in deciduous habitats. In the East they breed in woodland edges, villages lined with shade trees, and open woodlands with trees widely spaced and little undergrowth (Hebert and Elkins *in* Foss 1994, Villard *in* Gauthier and Aubry 1996). Colorado birds occupy two main habitat types: riparian stream bottoms and aspen forests, each of which has a relatively open canopy equivalent to the eastern habitats.

Along the prairie streams small numbers of these vireos breed in cottonwoods; Breeding Bird Censuses reported in *American Birds* in the 1970s showed densities ranging from territories of 1.7–3.2 acres (0.7–1.3 ha) in a Chatfield State Park cottonwood grove to 17 acres (7 ha) along the edge of Barr Lake. Atlasers reported them in about twice as many foothills riparian as plains riparian blocks.

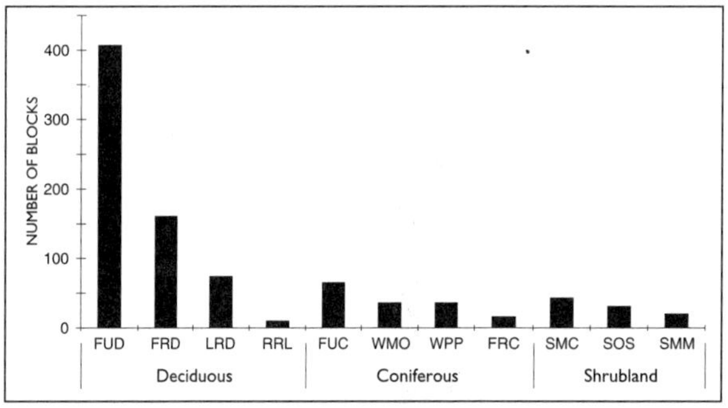
NUMBER OF BLOCKS
400
300
200
100
0
FUD
FRD
LRD
RRL
FUC
WMO
WPP
FRC
SMC
SOS
SMM
Deciduous
Coniferous
Shrubland

The bulk of Colorado's breeders inhabit aspen woodlands (407 of 729 blocks reporting habitat), where they sound omnipresent. Although they seem to have smaller territories in aspen, censuses reported in *American Birds* (1972–1980) show low densities of one pair per 3.4–8.0 acres (1.4–3.4 ha). The vireos also use a third habitat—willow carrs—in which Boulder County drainages packed them more tightly, with territories ranging from 2.4 to 4.3 acres (1–1.7 ha; Hallock 1984a, 1984b, Figgs 1984).

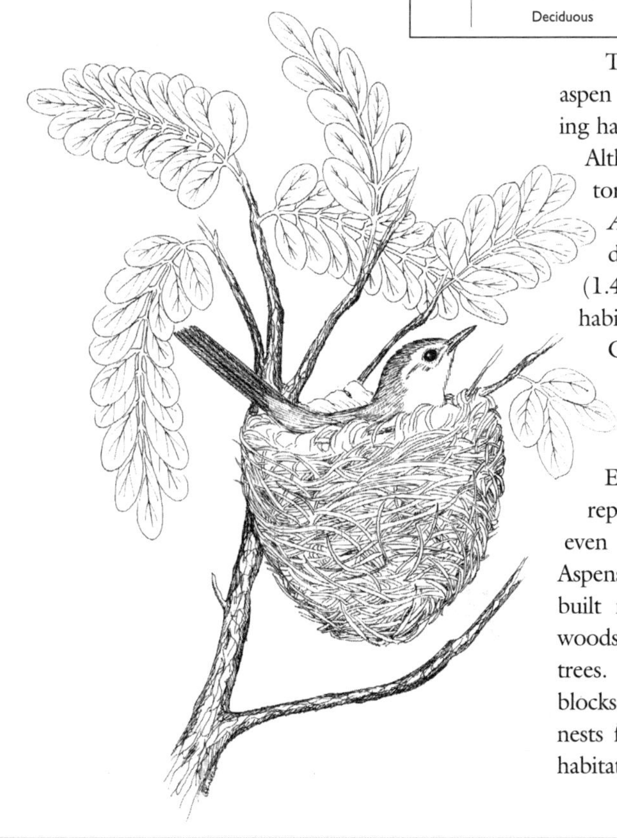

Every nest card submitted by atlasers reported the nest in a deciduous plant, even when the habitat included conifers. Aspens dominated, but Warbling Vireos also built nests in willows, narrowleaf cottonwoods, and a variety of other deciduous trees. Nearly 10% of all Warbling Vireo blocks contained conifers, but the lack of nests found in conifers indicates that these habitats also contained deciduous trees or that the birds used the blocks for foraging instead of breeding.

BREEDING

This species nests in everything from low brush to mature trees. Warbling Vireos in the West put their nests in forks well away from the trunk, in bushes or aspens reputedly within 12 feet (4 m) of the ground (B&N, Harrison 1979), but up to 50 feet (16 m) according to Atlas observations (Steve Boyle, Lu Bainbridge, nest cards). This flexibility increases the number of suitable nest sites, improving the chance for successful nesting.

As the map indicates, a relatively high proportion of breeding blocks with Warbling Vireos had Confirmed nesting. The male's habit of singing on the nest may explain this. Atlas nest-building dates began the last week of May; almost all pairs complete building by the end of June. Three records of nests with young from 2–14 June and two fledgling dates of 8 and 14 June demonstrate earlier starts than the first, 23 May, nest-building date.

Males singing from their nests makes nest location easier for cowbirds as well as for atlasers. Chace and Cruz (1996) report that Brown-headed Cowbirds parasitized 48% of 29 Warbling Vireo nests in a Boulder area. With the potential to attract predators or cowbirds, vireos need to have benefits to offset losses produced by this behavior. Observations of pairs show that vireos use song to maintain contact between each other and to coordinate movements on and off the nest. The complexity of the songs indicates that the singing may serve other purposes as well (Howes-Jones 1985).

BREEDING PHENOLOGY

CODE		# OF RECORDS	DATE RANGE
C	Courtship	4	27 May–27 Jun
NB	Nest Building	25	23 May–19 Jul
ON	Occupied Nest	70	30 May–21 Jul
NE	Nest with Eggs	5	24 Jun–28 Jun
NY	Nest with Young	26	2 Jun–24 Jul
FY	Feeding Young	106	10 Jun–14 Aug
FL	Fledged Young	71	8 Jun–23 Aug

DISTRIBUTION

Warbling Vireos occur across most of the U.S. and southern Canada. They breed throughout Colorado but concentrate in the western mountains. Habitat needs, rather than geographic or geological patterns, determine distribution within the state. The scattered birds on the eastern plains concentrate along the rivers, where cottonwoods supply suitable habitat. Most plains blocks had low abundance codes; only one rated an A4 code and the rest had lower estimates. The western two-thirds of the state harbors the most; aspen woodlands garnered the biggest proportion of high abundance codes—47% A4s and A5s. Atlas abundance calculations rank this vireo as the thirteenth most abundant species in the state, with over 842,000 breeding pairs.

Ornithologists may declare the two groups of Warbling Vireos, now designated as Eastern (*V. g. gilvus*) and Western (*V. g. swainsonii* and *V. g. leucopolius*), as separate species. A singing vireo near Holly (38102B2) that sounded like the eastern subspecies (Duane Nelson pers. comm.) was Confirmed with an ON code (on nest), the only Atlas report. Like records for Warbling Vireos in the rest of the state, the habitat contained deciduous trees—in this case plains cottonwoods.

Possible long-term threats to Warbling Vireos include fire suppression and pesticides. Fire suppression allows conifer forests to replace aspen gradually. It seems unlikely, however, that aspen acreage will significantly decline over the long run for two reasons. Timber harvest of conifers opens up habitat for aspen regeneration. Fire suppression creates higher fuel levels, increasing the probability of catastrophic fires eventually removing the conifer cover.

Pesticide use poses a second potential threat, not so much in Colorado as in other parts of the summer and winter range. All vireos eat insects, so increased use of insecticides in the U.S. and in their Central American wintering grounds could threaten birds through loss of food or by secondary poisoning.

BBS trends show Warbling Vireos increasing at a healthy rate of 1.2% per year. If habitat changes, pesticides, and even singing on the nest have negatively impacted this species, some countering influences must help them to survive.

BREEDING EVIDENCE

REPORTED IN 683 (39%) OF 1745 PRIORITY BLOCKS

Possible	166 blocks	(24%)
Probable	206 blocks	(30%)
Confirmed	311 blocks	(46%)

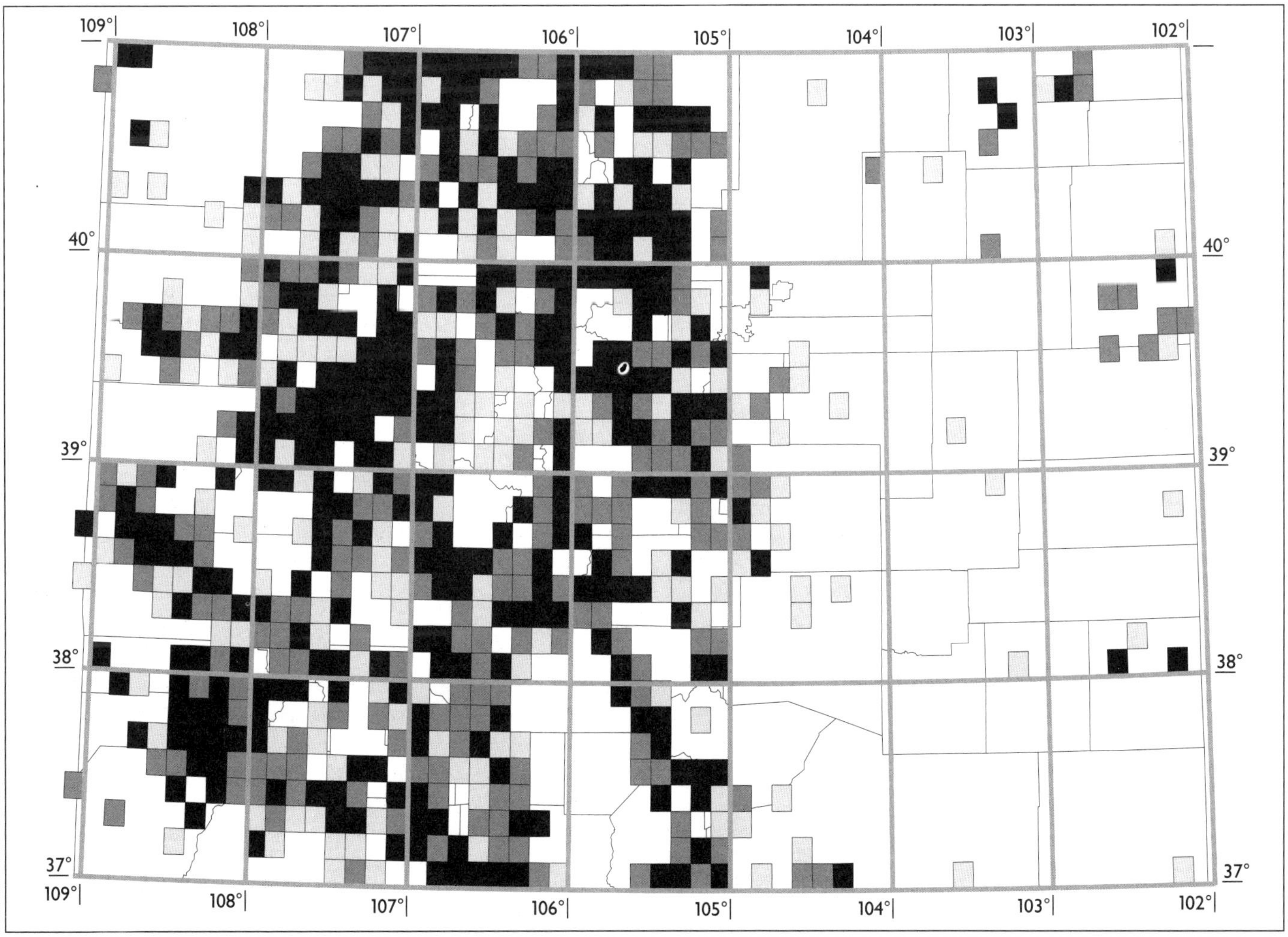
109°
108°
107°
106°
105°
104°
103°
102°
40°
39°
38°
37°

COLORADO DIVISION OF WILDLIFE / WILDLIFE RESOURCE INFORMATION SYSTEM

Prior to European settlement, squirrels supposedly could travel from the Atlantic to the Mississippi by leaping from tree to tree without touching the ground and always be within earshot of singing Red-eyed Vireos. These forest-canopy birds found little suitable habitat west of the Mississippi, where grasslands replaced the vast eastern deciduous forest and where rivers like the Platte had few cottonwood groves for these forest habitués.

HABITAT

With settlement, the river regimes on the Great Plains changed and fostered the growth of cottonwoods. Now the South Platte and Arkansas rivers nurture giant cottonwoods that provide nest sites that Red-eyed Vireos could not find 150 years ago. The vireos also have found breeding territory among the relatively new deciduous forests associated with Front Range metropolitan areas. Atlasers found them in South Platte and Front Range cottonwoods and in a few urban/rural trees.

During the breeding season Red-eyed Vireos relentlessly glean insects from the high-canopy deciduous foliage; sometimes they hover to pick off prey from the underside of a leaf. Males typically forage higher in the vertical structure of the foliage than do females (Williamson 1971). They limit their insectivorous habits to the breeding season. During the temperate winter Red-eyed Vireos eat primarily fruit: they inhabit low-elevation sites in northern Amazonia with abundant sources of fruit (Barlow 1980).

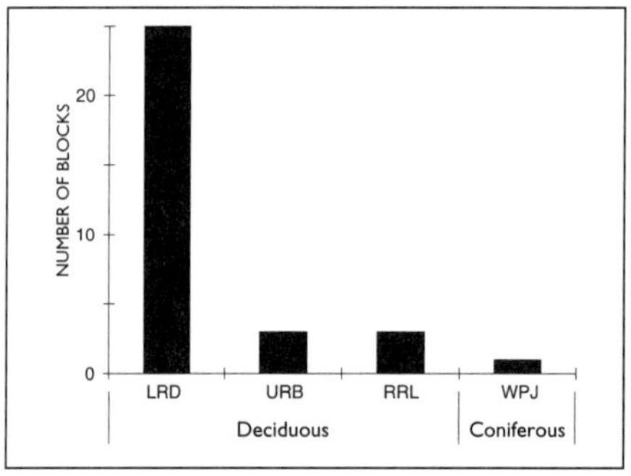
NUMBER OF BLOCKS
0
10
20
LRD
URB
RRL
WPJ
Deciduous
Coniferous

BREEDING

In early May when the males arrive at the breeding grounds, they establish territories of 0.8 to 1.7 acres (0.3–0.7 ha; Williamson 1971, Rice 1978). Their characteristic high-canopy habits hindered atlasers trying to Confirm breeding. The few Atlas Confirmations provide little data on the nesting phenology, but from the two dates it appears that the few Colorado Red-eyed Vireos may have a later cycle than eastern vireos.

Males patrol territories and sing incessantly and monotonously until the young fledge. The constant singing reassures the female, who becomes alert when the singing stops, and alarmed when the male breaks into a chattering rattle call (Stokes 1979).

Females select the nest site and build a small basket-like nest made of bark, fine grass, forbs, spiderwebs, and cocoons in the prong of a forked twig (Harrison 1979). Females incubate three to five oval white eggs with brown specks for 11–14 days. Young fledge 10–12 days after hatching and both sexes tend the young. Brown-headed Cowbirds commonly parasitize Red-eyed Vireo nests throughout their range; two records of this exist for Colorado (Friedmann and Kiff 1985).

BREEDING PHENOLOGY

CODE		# OF RECORDS	DATE RANGE
NE	Nest with Eggs	1	27 Jun
FY	Feeding Young	1	12 Jul

DISTRIBUTION

As long-distance Neotropical migrants, Red-eyed Vireos breed throughout southern Canada and the U.S. except in the Southwest, but they spend seven to eight months—October through April—in northern South America (Barlow 1980). Most writers considered the Red-eyed Vireo a "rare summer bird in the plains country of eastern Colorado [that] does not extend into the foothills or mountains" (Sclater 1912; accord, A&R).

The Atlas map shows more Red-eyed Vireos than that, although still not a large number. Atlasers recorded these vireos in 16 Priority and 15 Non-priority blocks but Confirmed them in only three. They recorded the vireos in ten blocks along the South Platte from Greeley downstream. Red-eyed Vireos definitely breed at the edge of the foothills in the Pueblo and Denver/Boulder areas. Atlasers found a few in four blocks in southeastern Colorado from 25 May to 3 June. The birds may have been migrants; observers either did not return to these blocks later in June or when they did they did not find any vireos. Two reports dated 29 May and 24 July, from west of the Divide along the Yampa River (omitted from the map), apparently represent migrants rather than breeders; in Canada these vireos breed all the way to the Pacific coast.

The increase of cottonwood corridors across the Midwest and the introduction of eastern trees in many Front Range communities probably facilitated this change. Despite the increase in range, they breed in low densities across Colorado, due, in part, to the linear and discontinuous structure of Colorado riparian forests.

At the turn of the century, habitat fragmentation had sent populations of Red-eyed Vireos into serious decline. Clearing eastern forests for crops and dairy herds both reduced available habitat for vireos and increased suitable habitat for Brown-headed Cowbirds. Increased cowbird parasitism dramatically reduces the reproductive success of Red-eyed Vireos. Even though in many areas of the East forests have come back in abandoned croplands, another problem plagues the vireos. The increase in insecticide use in recent years hampers this foliage-gleaning insectivore by lowering insect densities; this in turn forces the vireos to use larger, and thus to have fewer, territories (Cooper et al. 1990). In addition, Red-eyed Vireos are extremely vulnerable to tropical deforestation (Morton 1992).

Despite tropical deforestation and insecticides, BBS results show a significant population increase of Red-eyed Vireos across the continent. Because the survey did not begin until 1966, it does not measure the comeback compared to the days of squirrels crossing half the continent, but it does reflect the nascent return of eastern deciduous forest. Due to low abundance, no Colorado routes record them, but comparison of the historic records with the Atlas suggests little change or perhaps a modest range expansion here.

BREEDING EVIDENCE

REPORTED IN 16 (1%) OF 1745 PRIORITY BLOCKS

☐	Possible	8 blocks	(50%)
■	Probable	6 blocks	(38%)
■	Confirmed	2 blocks	(12%)

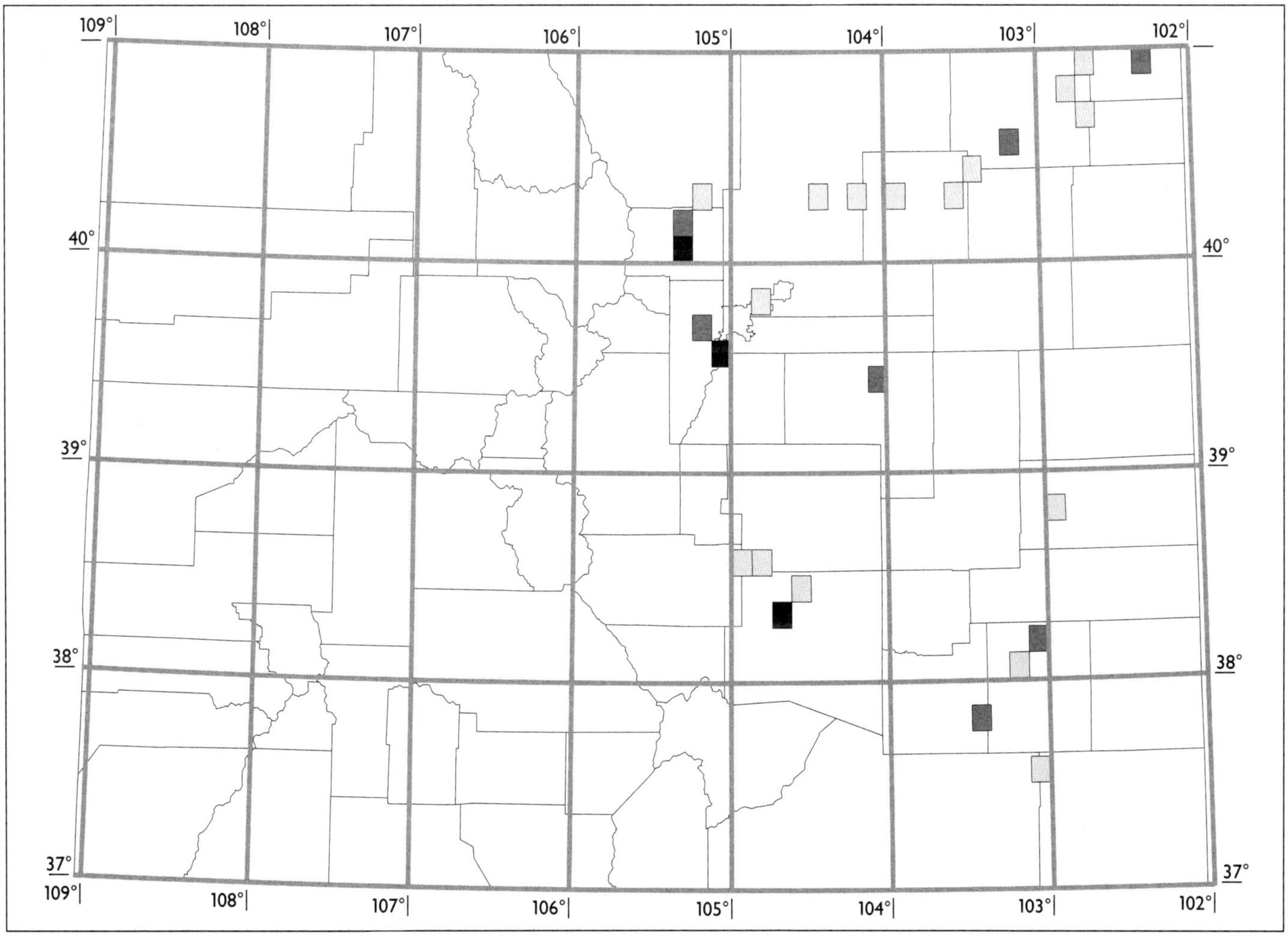
109°
108°
107°
106°
105°
104°
103°
102°
40°
39°
38°
37°

COLORADO DIVISION OF WILDLIFE / WILDLIFE RESOURCE INFORMATION SYSTEM

Curious, bold, silent, and deceptively gentle, "Camp Robbers" can take the sandwich from your fingers or even from your lips, when you picnic in the high country. No foods are safe, even in the cookpot. They like cheese and baked beans best (Goodwin 1976). The Cree tribe called them "wiss ca chon" (Bent 1946), a name soon anglicized to another alias for the Gray Jay—the "Whiskey Jack."

HABITAT

These mountain birds live from near 8,500 feet (2615 m) to timberline (B&N). They choose to be in conifers, usually spruce/fir forests or lodgepole stands, even timberline tree islands (A&R). Less common in lower mountain parks and foothills, they descend rarely to the eastern and western plains in fall and winter (A&R, B&N). Atlas workers observed them most often in upland coniferous closed-canopy forests of spruce and lodgepole, and open canopy montane woodlands. Spruce trees particularly attract them (Strickland and Ouellet 1993).

They eat anything edible, and to sandwiches, cheese, and beans add conifer seeds, fir needles, oats, fruits, mushrooms, lichens, insects, ticks, small birds and mammals, and carrion (Bent 1946, Root 1988, Goodwin 1976, Strickland and Ouellet 1993). They prey on nestling Cassin's Finches (Mt. View Crest Atlas block, 37107E6; Kingery and Kingery 1995), and take fish, ptarmigan, Pine Grosbeaks, crossbills, and other Gray Jays as well (Strickland and Ouellet 1993).

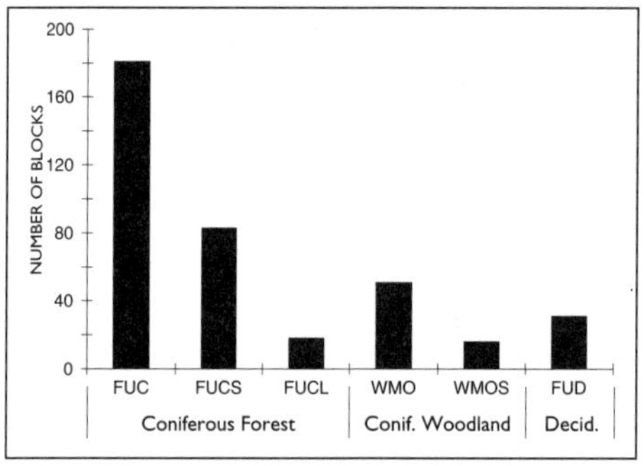
NUMBER OF BLOCKS
200
160
120
80
40
0
FUC
FUCS
FUCL
WMO
WMOS
FUD
Coniferous Forest
Conif. Woodland
Decid.

They store food for later use in a sticky bolus of food mixed with saliva. They place this on a limb, tucked under bark or a lichen. This habit allows them to survive times of low food supply (Strickland and Ouellet 1993).

They are attracted to humans—hunters, trappers, and campers (B&N, Goodwin 1976). When attempting to steal such human foods at a campsite, they fly in and perch for a while looking things over, then swoop down and try it, first carrying the food off to eat, then staying and eating it on the spot. Their habit of holding large food items in their feet instead of their bill allows them to carry heavier loads than usual (Strickland and Ouellet 1993). In addition to food, they carry off matches, tobacco, pencils, soap, and candles (Terres 1980) and even try to rob traps of bait or eat carcasses in the traps (Rennicke 1990). Various hawks and owls prey upon them and red squirrels kill the nestlings (Strickland and Ouellet 1993).

BREEDING

Gray Jays form long-term pair bonds, and the birds keep in constant visual and vocal contact with each other. The breeding pair starts very early for the high country—in February, March (B&N, Goodwin 1976), and April (Strickland and Ouellet 1993). They occupy and defend a large, permanent, all-purpose territory (average size, 148 acres, 60 ha). Together the pair constructs a new nest each year, in a spruce or fir tree, touching or within a nest-width of the trunk, and often on the south side of the tree. At the start, the male does most of the work, but the female contributes an equal amount by the time they have installed the nest lining. First they make a donut of twigs and cocoons (particularly tent caterpillars in Ontario), and then fill it with finer twigs, bark, feathers, lichens, down, and rabbit fur. Two to five (usually 3–4) young form the clutch that the female incubates for 18.5 days. She begins incubation with the first egg laid, and raises only one brood (Strickland and Ouellet 1993).

Atlas workers found one nest with young in June, and recently fledged young from June through August. Adults fed young throughout June and July. Most Confirmed nests were in conifer forests, especially spruce/fir (51%), and 97% of the Atlas observations were in conifers. After the young fledge, the most dominant fledgling expels its siblings from the parental territory—keeping the territory (and its stored food) for itself and the parent birds, until the next spring breeding season when the parents expel it (Strickland and Ouellet 1993).

BREEDING PHENOLOGY

CODE		# OF RECORDS	DATE RANGE
C	Courtship	1	6 Jul
NY	Nest with Young	1	9 Jun
FY	Feeding Young	20	6 Jun–6 Aug
FL	Fledged Young	100	10 Jun–23 Aug

DISTRIBUTION

Gray Jays live in the boreal forests of North America from Alaska across most of Canada to the Atlantic shores. They inhabit similar habitat on the northwestern coast of the United States south into California, and the coniferous forests of the Rocky Mountains to Arizona and New Mexico. In Colorado they reside in coniferous forests year round, although they may wander more widely in fall and winter down into lower elevations (Aiken and Warren 1914, Bent 1946), to Denver and Colorado Springs and even Nebraska (B&N). On a trek to Oregon, Thomas Jefferson Farnham first recorded these jays in Colorado—he called them "Jack-daws"—where the Blue River empties into the Colorado River south of Kremmling (Marsh 1931). B&N recorded them in ten other counties before 1900. The Atlas map shows their Colorado range throughout the mountains, with large gaps visible in conifer-poor North and South parks and the San Luis Valley.

Atlas workers found them in 312 blocks, 80% of them in closed-canopy coniferous forest. Atlasers Confirmed breeding solidly across the expected high-altitude forests but also in some outlying lower-altitude locations. They Confirmed or found Probable breeding in several blocks in the Rampart Range between Denver and Colorado Springs, the Wet Mountains, the Uncompahgre Plateau, and west of the Park Range on the Wyoming line.

Their numbers appear stable or nearly so in Colorado, and BBS analysis shows no significant trends (Strickland and Ouellet 1993). Their absence would disappoint many high-altitude hikers, campers, and skiers—sure to miss the companionship and funny antics of this opportunistic bird.

BREEDING EVIDENCE

REPORTED IN 312 (18%) OF 1745 PRIORITY BLOCKS

Possible	129 blocks	(41%)
Probable	60 blocks	(19%)
Confirmed	123 blocks	(39%)

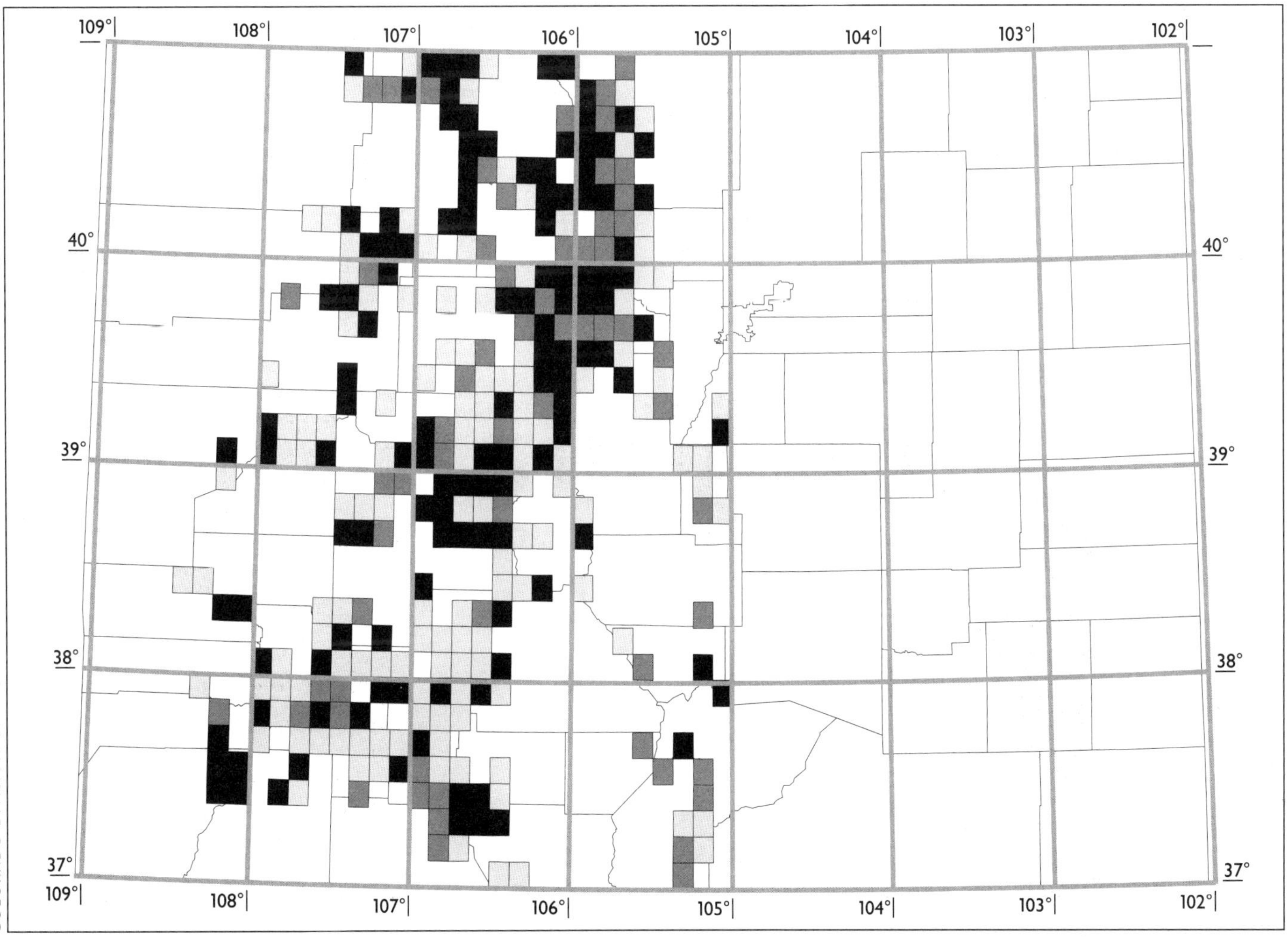
109°
108°
107°
106°
105°
104°
103°
102°
40°
39°
38°
37°

COLORADO DIVISION OF WILDLIFE / WILDLIFE RESOURCE INFORMATION SYSTEM

STELLER'S JAY *Cyanocitta stelleri*

In the forests of the West, Steller's Jays make themselves very visible and noisy—except during the nesting season. Any time of the year one of these jays could win the high-jumping championship of the bird world as it climbs a pine tree "as if it were a spiral staircase, starting down low and hopping upwards from branch to branch around the trunk" (Peterson 1941b). Although frequently misspelled (appropriately) as "Stellar," the name of this jay actually honors Georg Steller, who collected it during Vitus Bering's Arctic expedition of 1841 (Gruson 1972). Elliott Coues described Steller's Jays with lavish prose: "a stranger to modesty and forebearance.... [prone] to fillibuster [sic] ... an elegant dashing fellow" (Bent 1946).

HABITAT

Foothills and lower mountains from 6,000 to 9,000 feet (1830–2745 m; A&R) comprise this jay's major breeding habitat: conifer forests. Atlas field workers saw the most in ponderosa pine and significant numbers in Douglas-fir, lodgepole, spruce, and pinyon/juniper. Atlasers found a surprising number in aspen woodlands as well. Atlas codes show them proportionately more abundant in ponderosa woodlands, with somewhat lower densities in coniferous forests (FUC, WMO) and aspen forests. According to the abundance codes, pinyon/juniper woodlands have half the densities of the other principal habitat types.

The 12 breeding Confirmations involving nests occurred only in ponderosa pine and unspecified conifers, except for one in pinyon/juniper/mountain shrub habitat. Steller's occupy the niche above the plains and the territories of Blue Jays and Western Scrub-Jays, and below the spruce/fir forests of Gray Jays, with overlap on both edges.

Steller's live year round in the mountains; bird feeders may provide a considerable portion of the winter diet for some individuals. Limited banding returns suggest that most are sedentary and breed within 6–9 miles (10–15 km) of the place where they hatched (Morrison and Yoder-Williams 1984). In a California study, all the banded young dispersed in the late fall (Brown 1963). In the fall in Colorado, some birds move temporarily to the scrub oaks of the lower foothills or the cities at the foothills' edge (B&N).

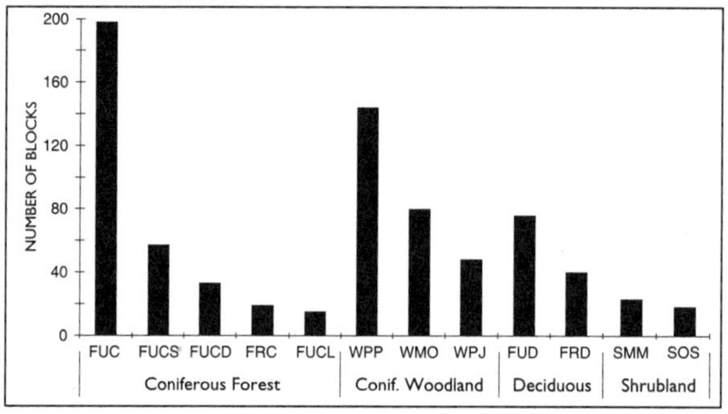

NUMBER OF BLOCKS
200
160
120
80
40
0
FUC
FUCS
FUCD
FRC
FUCL
WPP
WMO
WPJ
FUD
FRD
SMM
SOS
Coniferous Forest
Conif. Woodland
Deciduous
Shrubland

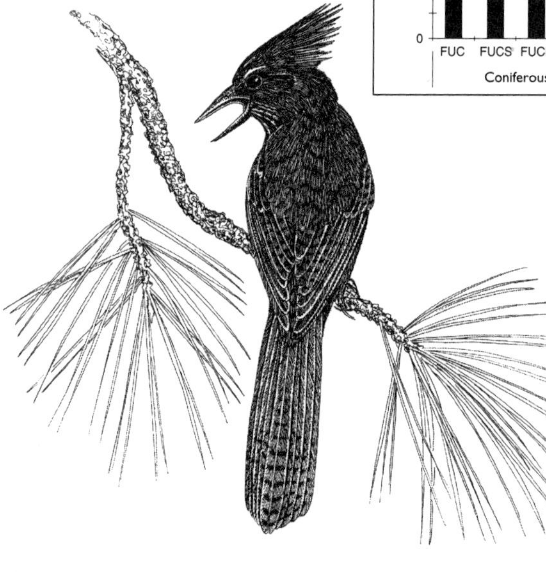

Insects and other animal matter compose about 43% of their summer diet; in winter their food consists primarily of plant material such as seeds and berries (Martin et al. 1951). Surprisingly, they feed mainly on the ground, by tossing loose materials like leaf litter and soil to find food underneath. The stair-step hops up trees lead to other food items such as fruits, nuts, caterpillars, and other insects. Occasionally they rob other birds' nests of eggs and young—in June their diet swells with animal matter. They also readily take foods supplied by man. They store food within their home ranges all year long, mainly in the ground although sometimes in trees. A bird will dig a hole with its bill, insert the food item—corn, sunflower seeds, e.g.—and cover it with litter (Shuford 1993). Later they dig up the cache with unerring accuracy, their bills serving as "miniature mattocks" (Abbott 1929).

BREEDING

During the nesting season, these jays become quiet and shy, and difficult to find. They build a bulky nest, usually in evergreens, located from a few feet to 25 feet (7.5 m) in height (Terres 1980). Although males are aggressive, with a hierarchy for feeding, they apparently do not guard a territory except the area around the nesting site. Once paired, they rarely change mates, unless one permanently disappears (Brown 1963). The female lays 3–5 eggs, and only she incubates.

An atlaser found a nest with eggs on an early date, 25 March (Kent Simon, nest card), but otherwise Atlas nest-building records commenced in May, with two nests with young in June. Other Colorado egg records run from 23 April to 3 June (Johnsgard 1986a). Raucous, begging, fledged young, which led to most Atlas Confirmations, came throughout June, July, and August. Many Confirmations came in habitats where atlasers did not find nests; the families may have roamed from their nesting habitats. Juveniles lack the white eye-stripe, which makes the young easy to identify.

BREEDING PHENOLOGY

CODE		# OF RECORDS	DATE RANGE
C	Courtship	4	1 Apr–13 Jul
NB	Nest Building	4	10 May–24 Jun
ON	Occupied Nest	5	8 May–25 Jun
NE	Nest with Eggs	1	25 Mar
NY	Nest with Young	2	3 Jun–24 Jun
FY	Feeding Young	22	10 May–24 Aug
FL	Fledged Young	88	19 May–30 Aug

DISTRIBUTION

Ornithologists divide Steller's Jays into several subspecies, some with a black crest and some blue-crested. Coloring, size, and crest vary slightly. *C. s. macrolopha*, the form in Colorado, has a black crest with distinct whitish streaks in the forehead and a white patch above the eye. Only the two Rocky Mountain subspecies (which range from Alberta into central Mexico) have the white eye-patch. Blue-crested jays without eye spots inhabit southern Mexico and Central America (Madge and Burn 1994).

Steller's Jays range from the Rocky Mountains to the Pacific coast, and from southeastern Alaska to Mexico and Nicaragua. Lieutenant J.W. Abert, on Fremont's third exploration, first described this conspicuous jay in Colorado on 22 August 1845, on the headwaters of the Purgatoire River in Las Animas County. He wrote of it as "a jay whose plumage partook of the color of the darkest blue of a clear sky" (Marsh 1931).

The Atlas map shows them throughout the mountain forests; the gaps in the San Luis Valley, North Park, and Colorado Plateau mirror the lack of pines in those places. The jays do not extend east with the pinyon/juniper woodlands of Las Animas County. This range follows the pattern shown in A&R, although the Atlas shows larger gaps in the places mentioned above and in Delta and Montrose counties. The map adds a contingent in the ponderosas of western Moffat County.

People have moved into the Steller's Jay ponderosa country; with the largesse from their feeders, they guarantee the jays their ponderosa staircases.

BREEDING EVIDENCE

REPORTED IN 583 (33%) OF 1745 PRIORITY BLOCKS

☐	Possible	353 blocks	(61%)
▩	Probable	105 blocks	(18%)
■	Confirmed	125 blocks	(21%)

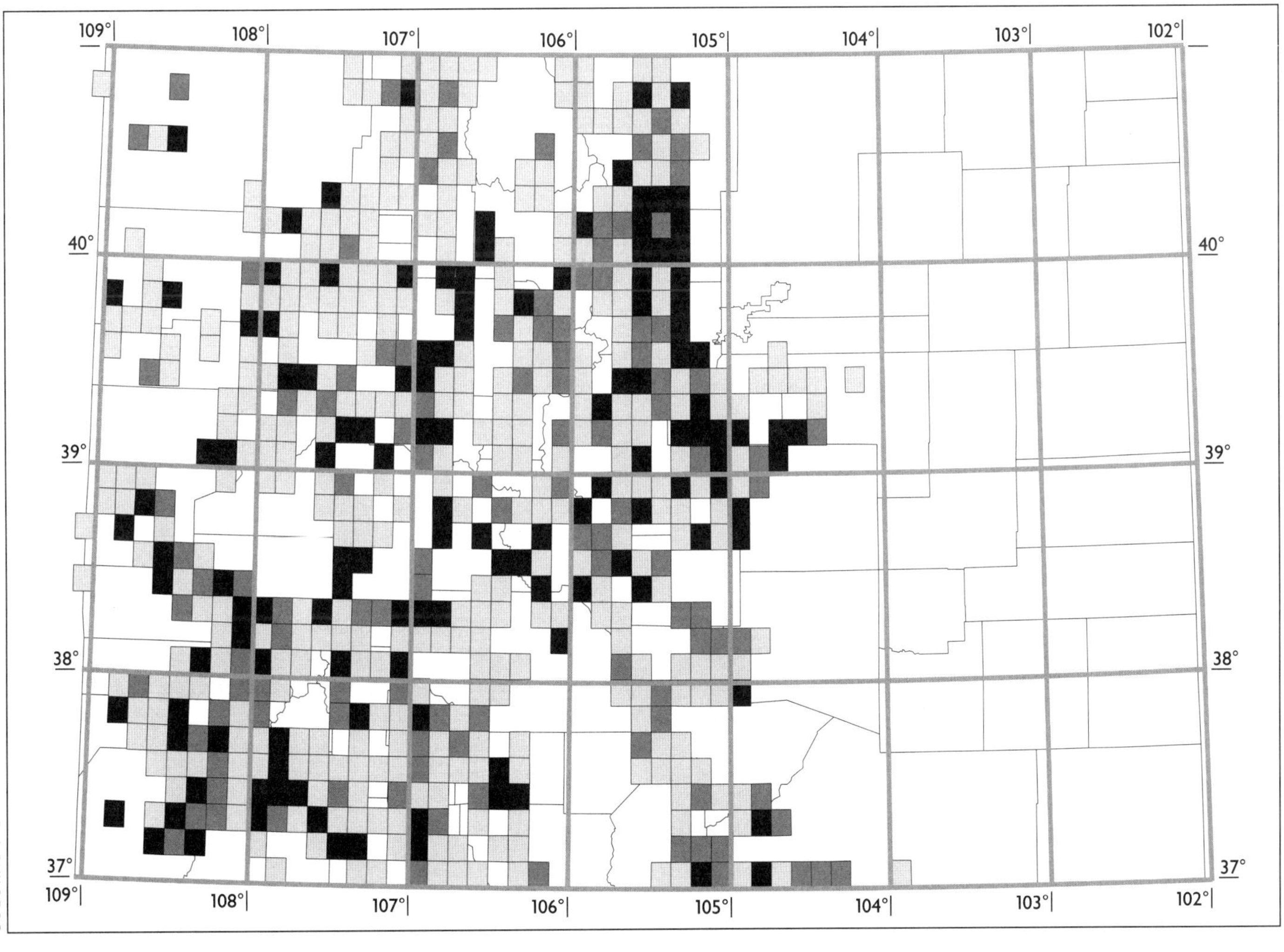
109°
108°
107°
106°
105°
104°
103°
102°
40°
39°
38°
37°

COLORADO DIVISION OF WILDLIFE / WILDLIFE RESOURCE INFORMATION SYSTEM

BLUE JAY *Cyanocitta cristata*

By raucous squawks and flashes of blue bouncing up escalators of tree branches, Blue Jays proclaim their presence to the world. In an abrupt change of personality the same birds become quiet, shy, and extremely circumspect during nesting time. Then, some even call this noisy bird "furtive" (Madge and Burn 1994).

HABITAT

When European settlers arrived in North America Blue Jays exclusively inhabited mixed deciduous forests, especially beech and oak, and coniferous woodlands (Tyler *in* Bent 1946). Although they still use forests widely, Blue Jays have adapted to well-treed residential neighborhoods, gardens, urban parks, small towns, and farmyards.

In Colorado, where these jays have lived for fewer than 100 years, they also use linear forests along streams and rivers on the plains. Atlas statistics show 43% of the habitat records are for lowland riparian. Urban and rural habitats take second and third place, but together they outnumber riparian habitats.

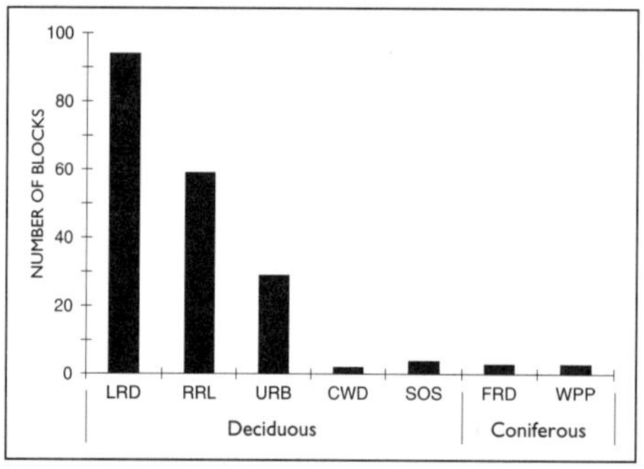
NUMBER OF BLOCKS
0
20
40
60
80
100
LRD
RRL
URB
CWD
SOS
FRD
WPP
Deciduous
Coniferous

The food choices of these omnivorous birds dictate their choice of habitat. Over 75% of their food consists of vegetable matter, primarily nuts and acorns, but they also eat seed, grain, berries, and some fruit. The remaining 25% animal matter includes insects and their larvae, eggs, small birds, and rodents. However, examinations of stomach contents belie the oft-repeated charge that Blue Jays are inveterate nest robbers (Beal 1897). They have not earned their bad reputation and economically do more good than harm (Tyler *in* Bent 1946).

BREEDING

In April or early May a group of five to eight Blue Jays, noisily chasing each other and bobbing up and down as they all suddenly alight in a tree, signifies courtship. Soon after this display, a pair separates from the mob and begins nest-building. Both males and females build the nests, usually well hidden in crotches of evergreens. They may also use thickets, deciduous trees, and even arbors or vines close to buildings as nest sites (Tyler *in* Bent 1946). They build bulky nests of sticks, rags, paper, Styrofoam, leaves, moss, and any other interesting materials they can find, neatly line them with rootlets, and carefully cement them with mud (Madge and Burn 1994).

The partners share incubation of the four to five eggs until the hatchlings emerge after 17–18 days. The pair also shares feeding duties for the nestlings and the fledglings. Atlas observers report NB (nest-building) as early as 21 April and as late as 1 June. Atlas records also show FL (fledged young) as early as 10 May and as late as 13 August, indicating a long breeding season in Colorado. Second nests are almost unknown (Tyler *in* Bent 1946). Secretive behavior at the nest probably accounts for the fact that 45% of the Atlas Confirmations were of fledglings.

BREEDING PHENOLOGY

CODE		# OF RECORDS	DATE RANGE
NB	Nest Building	5	13 Apr–1 Jun
ON	Occupied Nest	7	31 May–27 Jun
NE	Nest with Eggs	1	4 Jun
NY	Nest with Young	2	23 Jun–12 Jul
FY	Feeding Young	19	19 May–3 Jul
FL	Fledged Young	23	10 May–13 Aug

DISTRIBUTION

These jays reside in eastern North America from Newfoundland to Florida, west through Quebec to central Alberta and Manitoba (Madge and Burn 1994). In 1970 their western limit was the foothills of the Rockies in eastern Wyoming and eastern Colorado to west Texas. Since the 1970s their range has gradually advanced westward (Smith 1978). Smith believes that "city hopping," facilitated by the planting of gardens and trees in urban areas and the landscaping of city parks, contributed to this western expansion by providing suitable habitat. Atlas data suggest that the jays' use of riparian corridors also contributes to this expansion.

Blue Jays reached Colorado in 1903 and first nested in 1905 at Yuma. By 1917 they had reached Denver, where Niedrach and Rockwell still considered them rare stragglers in 1939. They bred in Pueblo in 1970 (A&R) and had reached Salida by 1987 (Salida West Atlas block, 38106E1). The Atlas map shows them well entrenched along the South Platte corridor, the Front Range from Wyoming to Pueblo, the Republican drainage in Yuma County, and the Arkansas from Kansas to Pueblo. Some have moved into a few towns across the plains. Despite their advance to states west of Colorado, no mountain towns except Salida reported breeding during the Atlas.

Colorado observers have noted two previously unreported behaviors of these jays. The first recorded hybridization between Blue Jays and Steller's Jays (*Cyanocitta stelleri*) occurred from 1969 to 1971 in Boulder (Williams and Wheat 1971). Other hybrids visited feeders and yards in Grand Lake 1978–1980 (*Am. Birds*), Coal Creek Canyon near Golden in 1995 (Dave and Jan Waddington pers. comm.), and Denver, May–August 1996 (Norm Lewis, Moras Shubert pers. comm.).

In Colorado Springs in 1993, Blue Jays engaged in cooperative breeding. A third jay assisted a pair in nest-building and feeding of nestlings and fledglings from May through mid July (Kuenning 1994). Ornithologists had documented this behavior in the United States in several other species of jays, but not in Blue Jays (Angell 1978, Skutch 1987).

BREEDING EVIDENCE

REPORTED IN 153 (9%) OF 1745 PRIORITY BLOCKS

☐	Possible	70 blocks	(46%)
▩	Probable	31 blocks	(20%)
■	Confirmed	52 blocks	(34%)

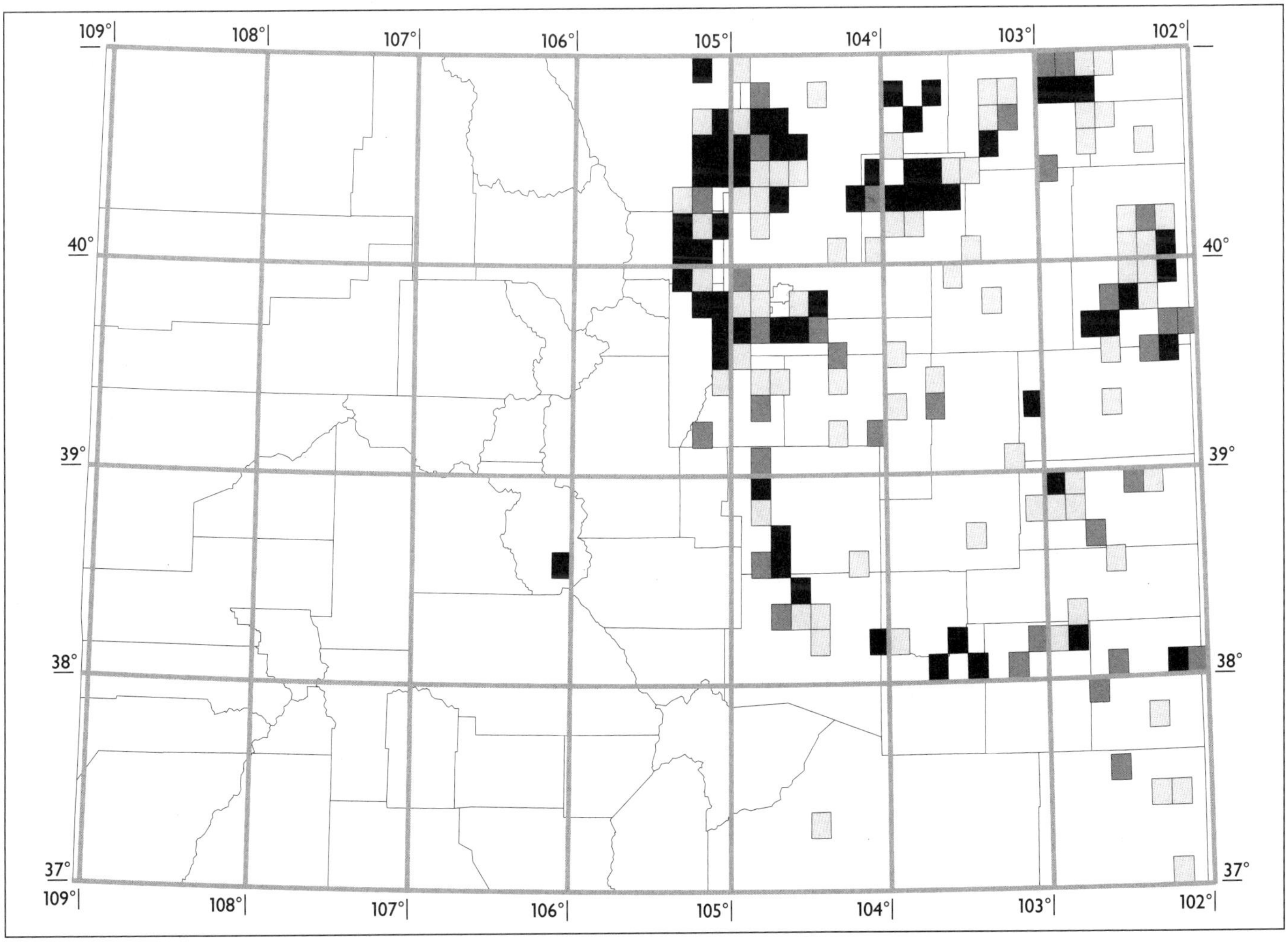
109°
108°
107°
106°
105°
104°
103°
102°
40°
39°
38°
37°

COLORADO DIVISION OF WILDLIFE / WILDLIFE RESOURCE INFORMATION SYSTEM

A sort of Jekyll-and-Hyde bird, the Western Scrub-Jay most of the year conducts itself loudly and aggressively, but during that period maintains a benign vegetarian diet. During the nesting season, however, it becomes a shy, retiring recluse and eats baby birds.

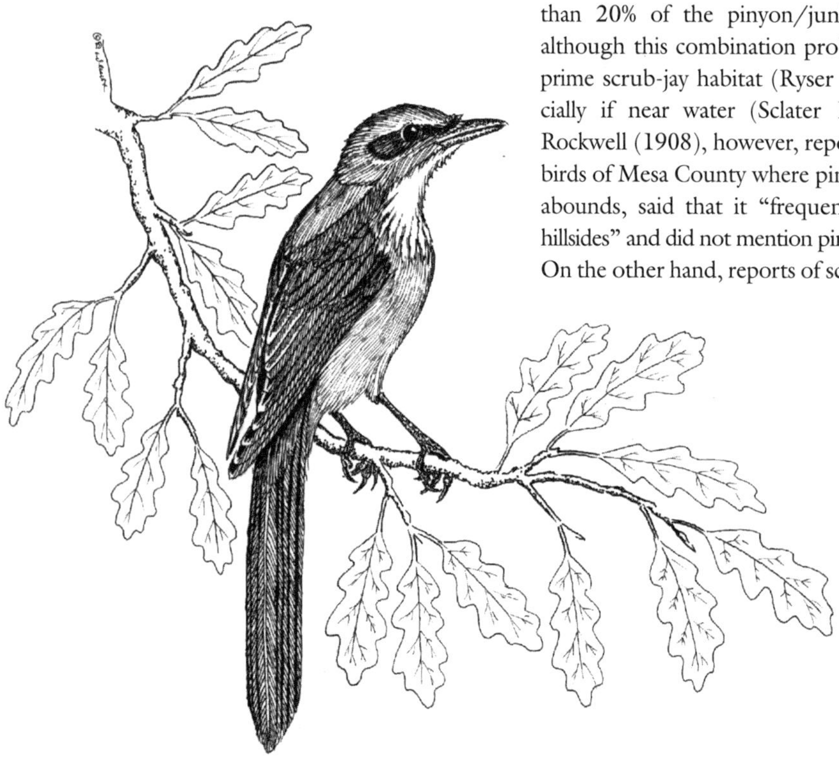

HABITAT

The primary references on Colorado birds emphasize scrub oak as the most important habitat for Western Scrub-Jays. A&R described the scrub-jay's habitat as "primarily scrub oak shrublands," and add mountain mahogany and pinyon/juniper as other habitats. B&N described the scrub-jay as a "common species in scrub oak associations of the eastern foothills" and add sagebrush, serviceberry, and pinyon/juniper in the western and southern parts of the state.

In contrast, Atlas reports of pinyon/juniper outnumbered scrub oak reports by three to one. This discrepancy may result

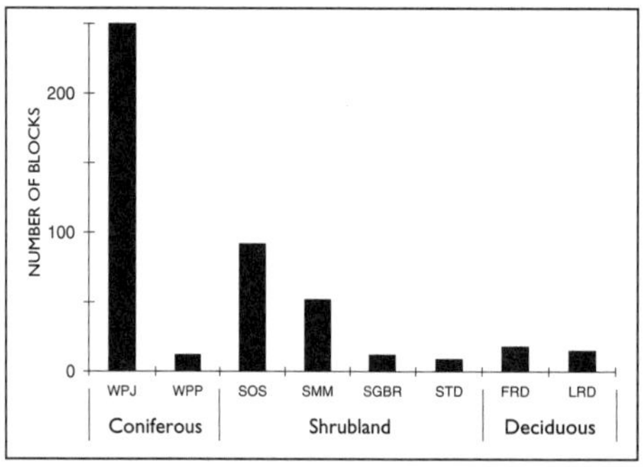
NUMBER OF BLOCKS
0
100
200
WPJ
WPP
SOS
SMM
SGBR
STD
FRD
LRD
Coniferous
Shrubland
Deciduous

from the fact that numerous scrub-jays and numerous birders live within 50 miles of Denver, where no significant pinyon/juniper grows. Or it may result from Atlas workers not regularly reporting a second habitat. They reported riparian or scrub habitats in association with pinyon/juniper on fewer than 20% of the pinyon/juniper reports, although this combination probably defines prime scrub-jay habitat (Ryser 1985), especially if near water (Sclater 1912). Even Rockwell (1908), however, reporting on the birds of Mesa County where pinyon/juniper abounds, said that it "frequents scrub-oak hillsides" and did not mention pinyon/juniper. On the other hand, reports of scrub-jay nests found in scrub oak by atlasers outnumbered the pinyon/juniper reports by the same ratio—six in scrub oak, two in pinyon/juniper. (Two additional nest reports did not specify a habitat.)

BREEDING

Scrub-jays build extremely well-hidden nests (Rockwell 1908, B&N), and Atlas workers found only ten nests for this rather common bird. B&N reported that "the average date for fresh eggs seems to be about the first week of May in the southern half of Colorado, and a week later in the vicinity of Denver." Ron Lambeth found the Atlas project's only nest with eggs on 28 April, only a couple of days earlier than B&N's average. After an incubation period of 16 days, the hatchlings remain in the nest for about 18 days before fledging (Ehrlich et al. 1988).

For most of the year scrub-jays subsist on acorns, pinyon nuts, and other fruits and seeds, but their diet during the nesting season shifts to softer fare more easily handled by young birds—insects and small animals, especially bird's eggs and nestlings (Ehrlich et al. 1988). A two-year study of Black-headed Grosbeaks in New Mexico found that Steller's Jays and scrub-jays had raided 60% of their nests (Hill 1995).

Atlas dates for other nesting activities cover a broad span of time; Atlas workers observed nest-building in mid June, weeks after the earliest reports of adults feeding fledglings. Either many birds delay nesting until much later than the earliest nesters or second attempts or double-clutching occurs regularly. Previous accounts consider the scrub-jay single-brooded with second broods rare (Terres 1980, Ehrlich et al. 1988).

BREEDING PHENOLOGY

CODE		# OF RECORDS	DATE RANGE
C	Courtship	2	13 Apr–15 May
NB	Nest Building	5	27 Apr–12 Jun
ON	Occupied Nest	4	4 Jun–22 Jun
NE	Nest with Eggs	1	28 Apr
NY	Nest with Young	4	27 May–25 Jun
FY	Feeding Young	20	28 May–21 Aug
FL	Fledged Young	95	10 May–24 Aug

BY RICHARD LEVAD

DISTRIBUTION

Western Scrub-Jays range from Oregon, southern Idaho, and Wyoming to southern Mexico. Cooke (1897) described them as common Colorado residents, "most common along the base of the foothills and the lower wooded mountains," but did not mention their presence on the Western Slope. Sclater (1912), though, described them as common residents "throughout the western half of the state" as well as all of the eastern foothills counties. B&N termed them "common . . . in preferred habitat in the southern two-thirds of the state."

The Atlas map conforms closely to that in A&R. On the Eastern Slope, scrub-jays reside fairly commonly in the foothills from Denver to New Mexico. They extend eastward to Baca County in the pinyon/juniper-clad hills south of the Arkansas River and reach upstream along the Arkansas Valley to Buena Vista. Scrub-jays also breed north of Denver to Fort Collins but in scant numbers. No Eastern Slope block north of Denver received an abundance code higher than A2 (2–10 pairs).

On the Western Slope, scrub-jays reside in appropriate habitat from Moffat County to New Mexico. Atlas workers found them along the Colorado River and its tributaries east as far as State Bridge, Wolcott, and Aspen. They found scrub-jays on the southern slope of the San Juan Mountains and along the San Juan River from Navajo Reservoir upstream to Pagosa Springs. A few reports of small numbers of breeding scrub-jays also came from the San Luis Valley, probably representing a northward extension from New Mexico.

Scrub-jay numbers seem fairly uniform throughout most of its Colorado range; Atlas workers reported an abundance code of A4 (101–1,000) from blocks in ten different latilongs. They reported an A3 code most frequently, in 46% of all blocks. Colorado jay species divide habitats among them, and scrub-jays face little competition in their pinyon/juniper and scrub oak habitats, which grow plentifully throughout Colorado.

BREEDING EVIDENCE

REPORTED IN 358 (21%) OF 1745 PRIORITY BLOCKS

☐	Possible	168 blocks	(47%)
▩	Probable	58 blocks	(16%)
■	Confirmed	132 blocks	(37%)

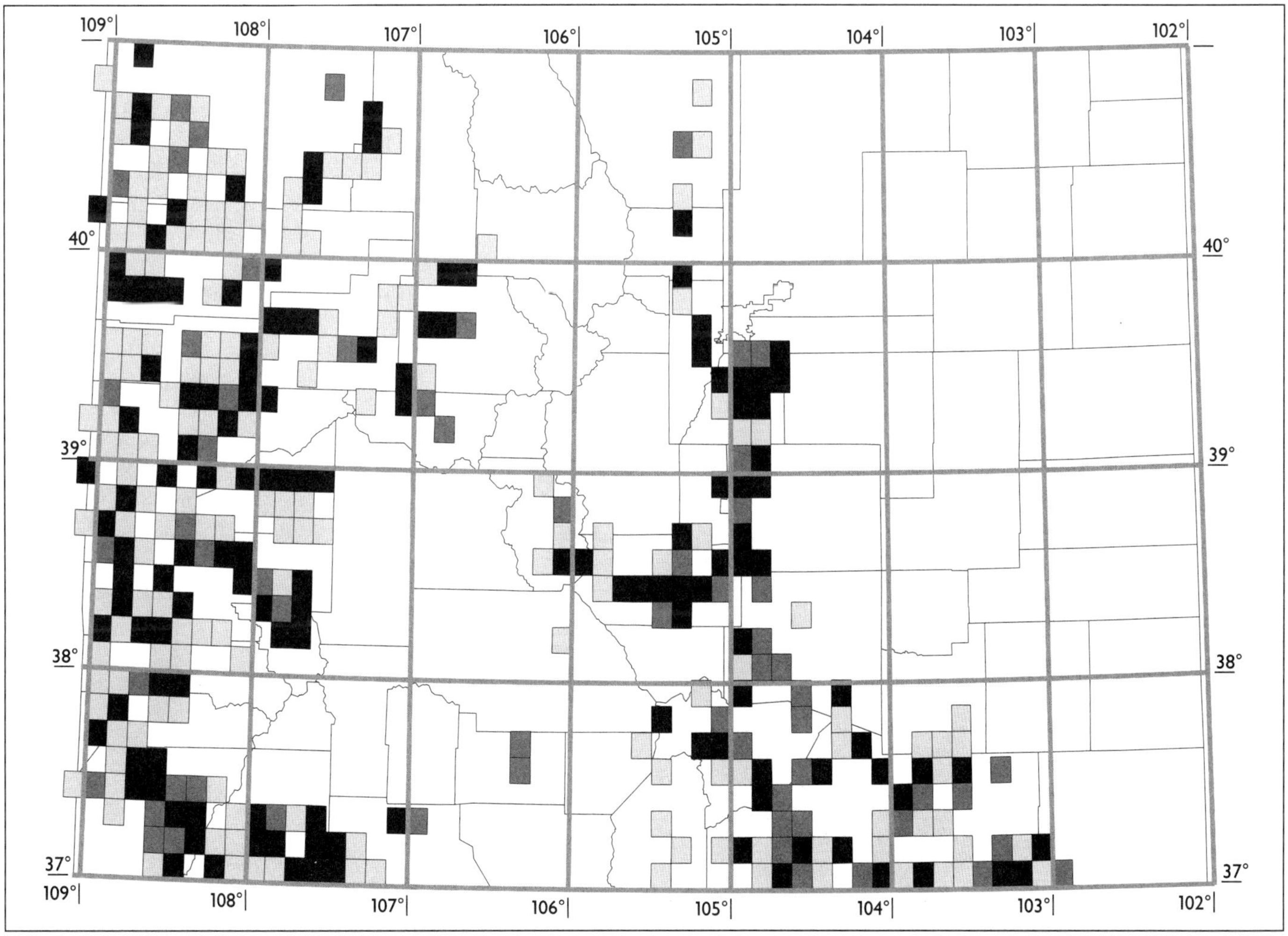
109°
108°
107°
106°
105°
104°
103°
102°
40°
39°
38°
37°

COLORADO DIVISION OF WILDLIFE / WILDLIFE RESOURCE INFORMATION SYSTEM

The marvel of Pinyon Jays uttering their querulous notes as they wander restlessly from mesa to mesa over canyons, ridges, and hills in pursuit of food often rewarded atlasers during a hot day of work in the pinyons. These flocking, cooperatively breeding corvids, which some call "blue crows," breed exclusively in pinyon/juniper and ponderosa woodlands of the West.

HABITAT

Pinyon Jays thrive in Colorado wherever pinyon/juniper woodlands grow. In the fringes of their range that lack pinyons, such as southeastern and northwestern Colorado, they use juniper habitat. Here the juniper berry replaces the pinyon nut as the primary food item (Ryser 1985). The limited area in which field workers found jays using ponderosa pines lies along the foothills on the east side of the Continental Divide. In southwestern Colorado atlasers failed to find any jays using the extensive ponderosa pine forest. Of Atlas Confirmations 96% were in pinyon/juniper woodlands, with ponderosa pines a very distant second at 2%. Few Colorado species have as close a tie to one breeding habitat type as do Pinyon Jays.

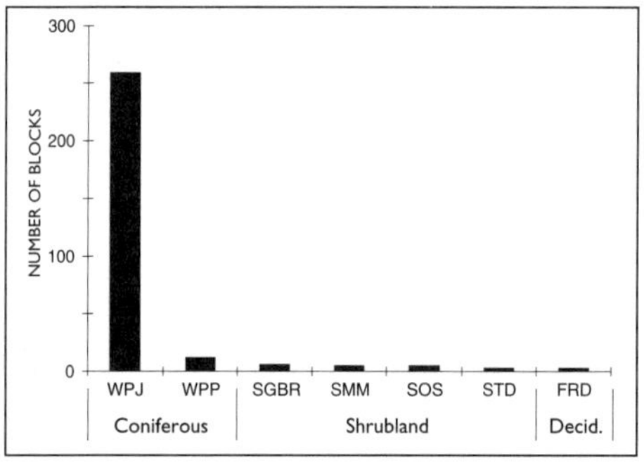
300
200
100
0
NUMBER OF BLOCKS
WPJ
WPP
SGBR
SMM
SOS
STD
FRD
Coniferous
Shrubland
Decid.

Field workers recorded jays using several other habitats. They often feed in adjacent, open areas of sagebrush that lie next to the pinyon/juniper woodlands. Jays eat berries, insects, and seeds found in these nearby habitats but usually do not stray far from the pinyon/juniper forest (Ehrlich et al. 1988).

Over the ages Pinyon Jays and pinyon pines co-evolved. Pines provide a food source, nesting sites, and protection. Pines become benefactors in years of plentiful nut crops. The jays harvest tremendous numbers of seeds and cache them by storing them in the ground; they retrieve many but also leave many to grow into trees (Ryser 1985).

BREEDING

Pinyon Jays begin their breeding cycle in winter with courtship behavior. Adults, some of which (like most corvids) stay paired year round, may start nesting as early as February (Harrison 1978). They build a deep, bulky nest with a framework of twigs and shreds of bark supporting the well-felted cup. Usually pairs build their nests in a pinyon or juniper from 3 to 18 feet (1–6 m) up. These gregarious birds breed in colonies of up to several dozen pairs. Sometimes as many as three pairs may nest in the same tree (Bent 1946). Incubation of the 3–5 eggs requires 16–17 days and young remain in the nest for 21 days before fledging (Ehrlich et al. 1988).

During the last 4–5 days that the young remain in the nest, adults from the flock feed them communally—adults feed young not their own; as many as seven adults may attend one nest. After fledging, groups of families form nursery groups: guarded by some adults, the young perch quietly until the foraging adults return with food. Adults feed the young communally until they reach six weeks of age; then only the actual parents feed them (Ryser 1985).

Atlasers located Pinyon Jays in 267 blocks, but Confirmed breeding in only 68 blocks (26%). Nesting jays avoided detection by field workers because of early spring nesting. The bulk of Atlas work started in May and by this time the jays were already bringing young off the nest. Prior to the Atlas nests with eggs were reported in Colorado from 23 March to 19 May (Sclater 1912, Bent 1946, B&N). By starting in early May, atlasers caught only the second half of the breeding period for this species. Of the 68 Confirmations, 59 (87%) represent young birds out of the nest (FF and FL).

BREEDING PHENOLOGY

CODE		# OF RECORDS	DATE RANGE
C	Courtship	2	25 Mar–16 Apr
NB	Nest Building	1	21 May
ON	Occupied Nest	2	21 May–10 Jun
NY	Nest with Young	1	27 May
FY	Feeding Young	6	6 May–3 Jul
FL	Fledged Young	58	8 May–12 Aug

DISTRIBUTION

Pinyon Jays range from central Oregon, east-central Montana, and the Black Hills of South Dakota south to northern Baja California, Arizona, and New Mexico. The center of their range lies within the Great Basin and on the Colorado Plateau; Colorado hosts 10–20% of the population (Priorities 1995).

Colorado's jays live mainly southwest of the diagonal from the northwest to the southeast corners of the state. The highest population density lies between 5,000 and 8,000 feet (1525–2440 m) elevation (A&R). Cooke (1897, 1909) noted them in this same area and east into Baca County by 1909, but also commented that they continually changed locations depending upon food supply.

The western tier of latilongs has the most pinyon pines and the most Pinyon Jays. Adjacent to Utah, 115 of 256 blocks (45%) sustain the jays, including ten of the 16 blocks where atlasers reported an abundance of over 100 breeding pairs. Atlasers defined another large area of high density for Pinyon Jays in southeastern Colorado. Two latilongs, 37103 and 37104, which lie adjacent to New Mexico, recorded 56 blocks with jays (44%).

On the west and north sides of the San Luis Valley in Rio Grande and Saguache counties, Pinyon Jays occur in fewer blocks but atlasers found high densities; field workers estimated that 20% had more than 100 pairs. Less densely populated areas lie along the Arkansas River drainage from Buena Vista to Cañon City, and in Eagle County near the Colorado River and its tributaries. A flock that cruised through ponderosa pines in Douglas County in April 1989 engaged in courtship but apparently did not nest (David Martin pers. comm.).

Clark's Nutcrackers, scrub-jays, and Steller' Jays all join Pinyon Jays (and rock squirrels) in a free-for-all for nuts when the trees produce a big crop. How the species allot the crop, population indices, harvesting, caching, and retrieval of nut crops and other foods offer intriguing subjects for future research, both formal and informal (Ron Lambeth pers. comm.).

BREEDING EVIDENCE

REPORTED IN 267 (15%) OF 1745 PRIORITY BLOCKS

☐	Possible	168 blocks	(63%)
■ (gray)	Probable	31 blocks	(12%)
■	Confirmed	68 blocks	(25%)

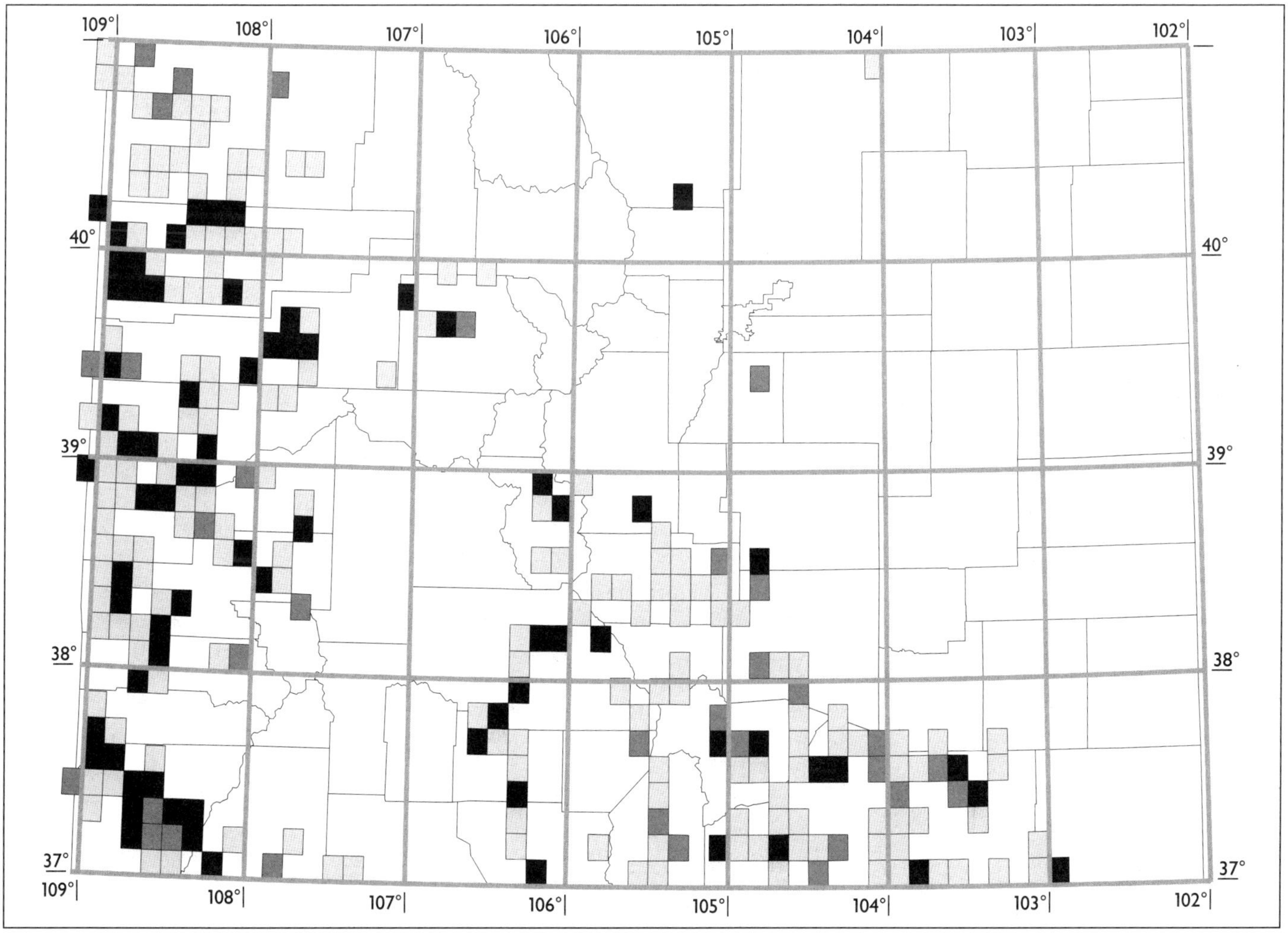
109°
108°
107°
106°
105°
104°
103°
102°
40°
39°
38°
37°

COLORADO DIVISION OF WILDLIFE / WILDLIFE RESOURCE INFORMATION SYSTEM

The Lewis and Clark expedition of 1803–1806 discovered this bird and named it after Captain William Clark. Some call it "Clark's Crow" due to its crow-like walk, querulous calls, and flocking habits. But nutcracking is its specialty, and pinyon nuts its favorite food.

HABITAT

High-country birds like their Gray Jay relatives, Clark's Nutcrackers live in the coniferous forests of the West, all the way to timberline (B&N). The vast majority of Atlas habitat codes cover forests, especially spruce/fir and mixed conifer forests. Nutcrackers also inhabit pinyon/juniper woodlands throughout the breeding season; B&N reported several nests in pinyon pines in the San Luis Valley. Because the birds rely extensively on pinyon seeds, this habitat has great importance to them all year, whether they nest in the pinyons or collect the seeds and cache them at higher elevations.

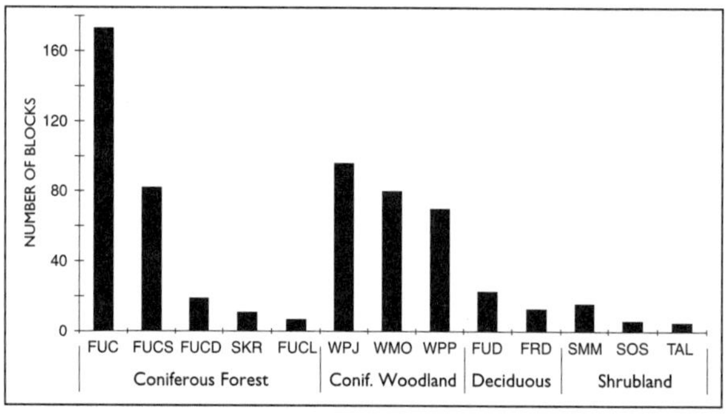
NUMBER OF BLOCKS
0
40
80
120
160
FUC FUCS FUCD SKR FUCL WPJ WMO WPP FUD FRD SMM SOS TAL
Coniferous Forest
Conif. Woodland
Deciduous
Shrubland

Atlas records show nutcrackers making substantial use of pinyon/juniper habitat from April through July, particularly in June, but many Atlas pinyon/juniper Confirmations may involve families far from their nest sites. Adults continue to feed fledglings capable of extended flight.

In the fall they travel down to pinyon/juniper zones and harvest large quantities of pinyon nuts, which they collect in their throats. A nutcracker in Yosemite Park carried 70 good-sized seeds in its sublingual pouch (Grinnell and Storer 1924) and another California study found them carrying 30–150 nuts at once (Tomback 1980). The birds carry the seeds up to their higher breeding grounds and store them in underground caches, often on south-facing slopes. Then, during the spring breeding season, they feed the seeds to the young (Bent 1946, B&N, Goodwin 1976, Tomback 1980).

Omnivorous, nutcrackers eat seeds, berries, sprouting wheat, insects such as beetles, ants, grasshoppers, spiders, young birds and eggs, small mammals, carrion, and human picnic foods (Bent 1946, Goodwin 1976). They also flycatch in the air, and, like a woodpecker, feed on bark insects (Goodwin 1976).

BREEDING

Clark's Nutcrackers begin breeding in early spring when deep snows keep the breeding grounds inaccessible to possible predators as well as to curious birdwatchers looking for nests. They breed in conifers; the prime requisite for a nest site seems not to be warmth but rather shelter—from the winds that howl over high-country ridges in March and April. Nutcrackers often pick sites cold in temperature and deep in snow but which provide protection from wind (Mewaldt 1956; Ryser 1985).

Colorado naturalists universally assume that nutcrackers breed in spruce/fir because many spend the early part of the year there, but no records of nests in that habitat seem to exist. Nests have been found in ponderosa pines, junipers, pinyon pines, and (Douglas?) firs (B&N). Availability of seed storehouses enables nutcrackers to initiate nesting in late winter or early spring (Tomback 1977). Colorado's earliest nest date of 5 March comes from a Boulder County ponderosa pine, where in 1888 Dennis Gale collected a nest with three eggs (B&N).

Both sexes gather nest materials of twigs, bark, and grass, but the female builds most of it over 5–8 days (B&N, Harrison 1979). The birds build in a conifer, usually well out on a horizontal limb, 8–45 feet (2.4–14 m) high (Harrison 1979). Both parents incubate the two to four eggs for 16–18 days; they raise only one brood (Harrison 1979).

An Atlas nest card from the Cochetopa Hills of Saguache County (Chester, 38106C3; Scott Wait) described a nest in Douglas-fir/grassland habitat, at 9,600 feet (2950 m), midway up a Douglas-fir on a branch 18 feet (5.5 m) above the ground. On 29 June 1992 the female was in the midst of egg-laying. Pinyon/juniper forests grow within 10 miles (16 km) of the site. Except for that nest, atlasers Confirmed breeding only by finding fledglings and parents feeding young, from mid May to mid August.

The young often hatch while snow still covers the nesting territory, but the birds can still find their nut caches. An adult lands at the cache site and probes into the ground. When the prod succeeds, the bird digs with sidesweeps, hulls the nut, and leaves a pile of seed coats. Prodding produced a 67% success rate (far greater than chance) in one study in the Sierra Nevadas (Tomback

1980). Caches covered by snow add to the mystery of whether or not nutcrackers breed in spruce/fir: digging through several feet of snow in March and April seems beyond even their unerring ability to find caches.

BREEDING PHENOLOGY

CODE		# OF RECORDS	DATE RANGE
C	Courtship	2	1 Jul–2 Jul
NE	Nest with Eggs	1	29 Jun
FY	Feeding Young	11	2 Jun–3 Aug
FL	Fledged Young	62	18 May–21 Aug

DISTRIBUTION

Nutcrackers are widespread in western North America in the mountain conifers from British Columbia to Baja California and New Mexico. In Colorado they range in summer from 8,000 feet (2460 m) to timberline (B&N). In winter, particularly when food sources dip (Goodwin 1976), they wander widely down to lower elevations and even onto the northeastern plains (A&R).

Atlasers Confirmed breeding in only 14% of the blocks where they found nutcrackers. The early-season nesting schedule and the inaccessibility of high-country terrain in early spring explain this low Confirmation rate. The map shows their decidedly western Colorado breeding range—almost entirely west of Longitude 105—with a small eastward extension in Mesa de Maya pinyon/juniper on the southern border. Several gaps in distribution are visible in the mountain parks and the sagebrush hills of the northwestern corner of the state.

Abundance is hard to judge, for although they are territorial and well spaced when breeding, afterwards they can form large flocks when wandering or traveling to good nut crops. Then the air is filled with the beautiful black, white, and gray of the birds and their grating calls of *kraaa, kraaa.*

BREEDING EVIDENCE

REPORTED IN 508 (29%) OF 1745 PRIORITY BLOCKS

☐	Possible	376 blocks	(74%)
■	Probable	59 blocks	(12%)
■	Confirmed	73 blocks	(14%)

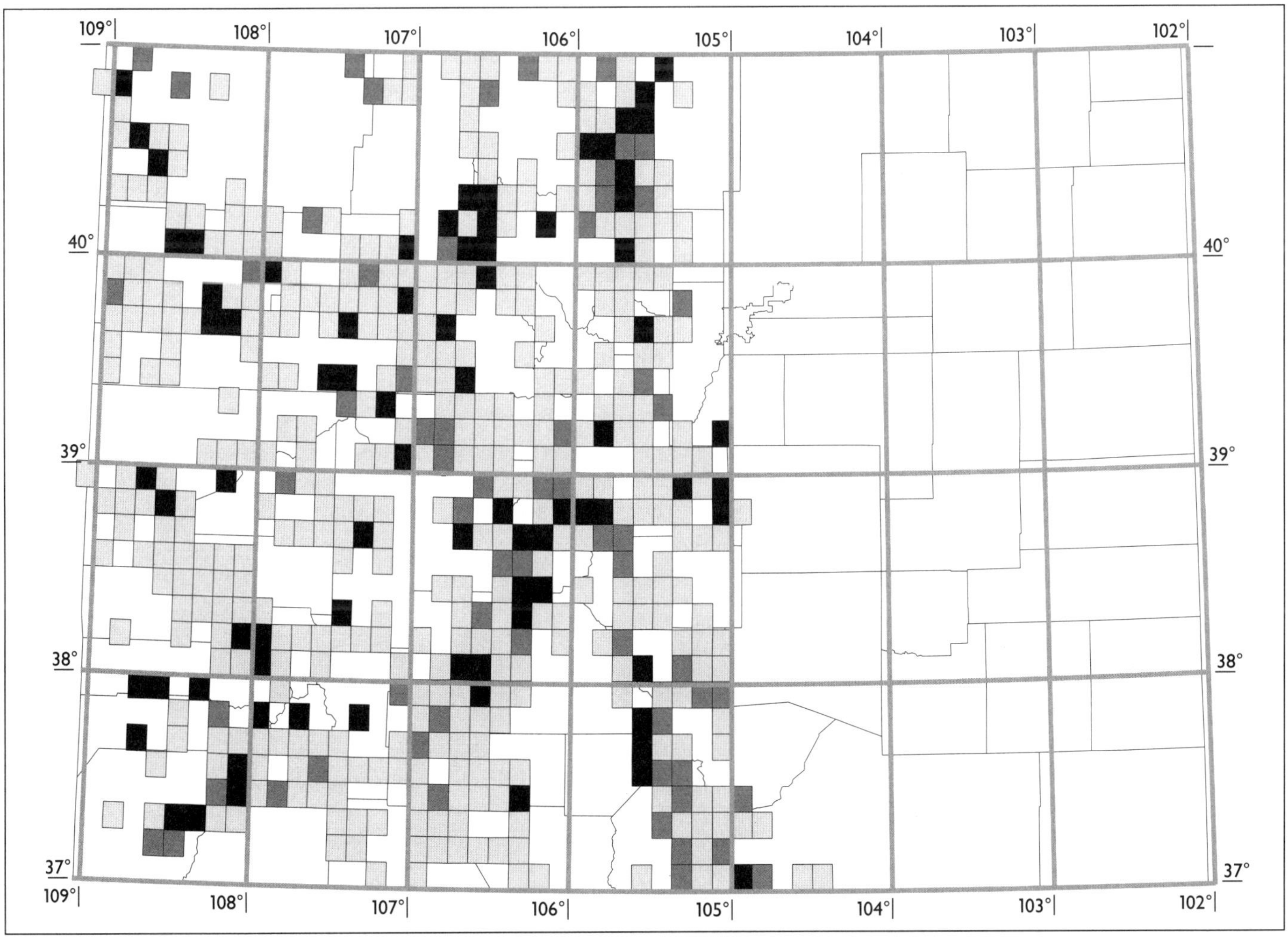
109°
108°
107°
106°
105°
104°
103°
102°
40°
39°
38°
37°

COLORADO DIVISION OF WILDLIFE / WILDLIFE RESOURCE INFORMATION SYSTEM

Black-billed Magpie

The flamboyant black-and-white magpie with its very long tail attracts the attention of newcomers and natives alike, so it is often the first bird seen and identified by birdwatchers in Colorado. The social habits and noisy behavior of the birds make them easy to see and fun to watch. They have both good and bad reputations—smart and cocky, raucous, carrion-eaters, and opportunistic birds that can injure livestock; therefore, state law until recently classified them as "varmints."

HABITAT

Magpies range statewide at all elevations from plains throughout the mountains, occasionally past timberline (B&N, Latilong Study, A&R). The birds frequent streamside thickets and scattered trees in open country, and range over sagebrush, croplands, and pastures (Terres 1980). They avoid treeless grasslands, deserts, and dense forests (A&R).

Atlas workers found them breeding most often in riparian, rural, and pinyon/juniper areas, three to five times as often as in cropland or urban habitats. The magpie map shows substantial distribution in the mountains, but, except for ponderosa woodlands, the habitat codes demonstrate an almost complete avoidance of conifers.

Magpies, like woodpeckers, contribute their old nests to other creatures. Many species of birds depend on old magpie nests for nesting or for storm shelters. American Kestrels use the nests with the roofs intact. Many hawks, owls, and herons flatten the domed nests to form a stable platform for their own nests (Rockwell 1909, Bent 1946, B&N, Terres 1980). Feral house cats and even gray foxes may use these nests (Bent 1946).

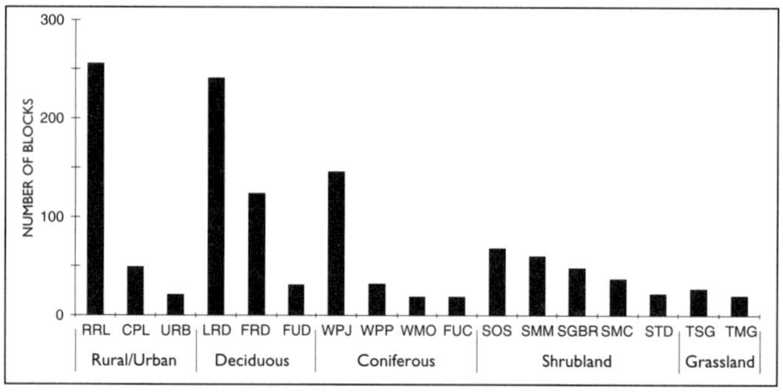
300
200
100
0
NUMBER OF BLOCKS
RRL CPL URB LRD FRD FUD WPJ WPP WMO FUC SOS SMM SGBR SMC STD TSG TMG
Rural/Urban
Deciduous
Coniferous
Shrubland
Grassland

Magpies definitely have traits that led to the legislative hostility. They prey on birds' eggs and young (B&N, Bill Howe pers. comm.), and harm livestock by pecking flesh from open sores (Coues 1895, Terres 1980). On 1 December 1806 Pike's journal reports a violent blizzard that dumped a foot of snow while his party camped on Turkey Creek southwest of present-day Colorado Springs. "Our horses were obliged to scrape [the snow] away to obtain their miserable pittance; and to increase their misfortune our animals were attacked by the magpies, which, attracted by the scent of their sore backs, alighted on them, and in defiance of their wincing and kicking, picked many places quite raw. The difficulty of procuring food rendered these birds so bold as to alight on our men's arms and eat meat out of their hands" (Coues 1895, Marsh 1931, Hart and Hulbert 1932).

Magpies also scavenge roadkills and other carrion, and pick and eat ticks from the backs of elk, deer, sheep, and other livestock (Terres 1980).

BREEDING

Magpies breed from March through July, from our lowest elevations to 10,000 feet (3048 m). They build nests in cottonwoods, boxelders, junipers, pinyons, and willows, from 6 to 60 feet (1.8–18.5 m) high. Nests consist of bulky 2 x 4 foot (0.6 x 1.2 m) masses of sticks. Mud holds the nest base together and forms the nest cup, which the birds line with rootlets and horsehair. Other sticks roof over the nest cup in a canopy or dome; magpies usually leave two openings in the sides for access (Terres 1980).

Nest-building is probably a part of courtship (Austin 1961). In Colorado courtship begins in mid March, and Atlas results show most nest-building in April and early May. Seven to 12 eggs form the clutch (B&N, Terres 1980), and incubation lasts 16–18 days (Terres 1980). Colorado's earliest eggs were found on 31 March at 6,500 feet (1980 m; B&N). Atlas workers report nests with eggs from 18 April until 4 June. Magpies raise only one brood each year although the pair may renest two or three times if they lose the first set (B&N). Atlasers saw nests with young from the end of April to the end of July, and bobtailed young from mid May through August, with most observed in June.

BREEDING PHENOLOGY

CODE		# OF RECORDS	DATE RANGE
C	Courtship	1	22 Jun
NB	Nest Building	16	17 Mar–1 Jun
ON	Occupied Nest	103	10 Mar–22 Jul
NE	Nest with Eggs	9	18 Apr–4 Jun
NY	Nest with Young	31	25 Apr–30 Jul
FY	Feeding Young	26	10 May–31 Jul
FL	Fledged Young	227	11 May–24 Aug

DISTRIBUTION

The worldwide range of this species includes western North America, Europe, and parts of Asia. In North America they range from Alaska to California, eastward to the 100th Meridian. Magpies in the U.S. do not tolerate high heat or high humidity (Hayworth and Weathers 1984), or high rainfall levels (Root 1988); this restricts them to the western U.S. The birds do not migrate with the change in season, but they may wander in winter (Robbins et al. 1983).

With records from 886 blocks, magpies rank as the 14th most widely distributed species in Colorado. They rank sixth for Confirmations, due to their very conspicuous nests and noisy fledglings. The Atlas map shows the birds breeding all over the state with a surprising hole in the central eastern counties. Here the heat levels may be too high and not ameliorated by streamside vegetation, two essential habitat requirements (Hayworth and Weathers 1984). Also here, and in the Pawnee National Grassland, few trees grow to present them with nest sites. These holes hint at an expansion of magpie range concurrently with European settlement and surging plains riparian forests.

Surprisingly, BBS trends (1994) show them decreasing in numbers across the continent by 1.2%/year. In the High Plains Physiographic Region, the data exhibit a drastic drop of 4.5% per year with two-thirds of the routes showing declines. In Colorado the data suggest a stable trend, but magpies have also dropped in number on Colorado Christmas counts. Riparian habitat does not appear to have decreased, and livestock numbers and roadkills (another favorite food) have definitely not done so. A decline—its actual existence and any causes—remains unexplained.

BREEDING EVIDENCE

REPORTED IN 905 (52%) OF 1745 PRIORITY BLOCKS

☐	Possible	189 blocks	(21%)
▩	Probable	43 blocks	(5%)
■	Confirmed	673 blocks	(74%)

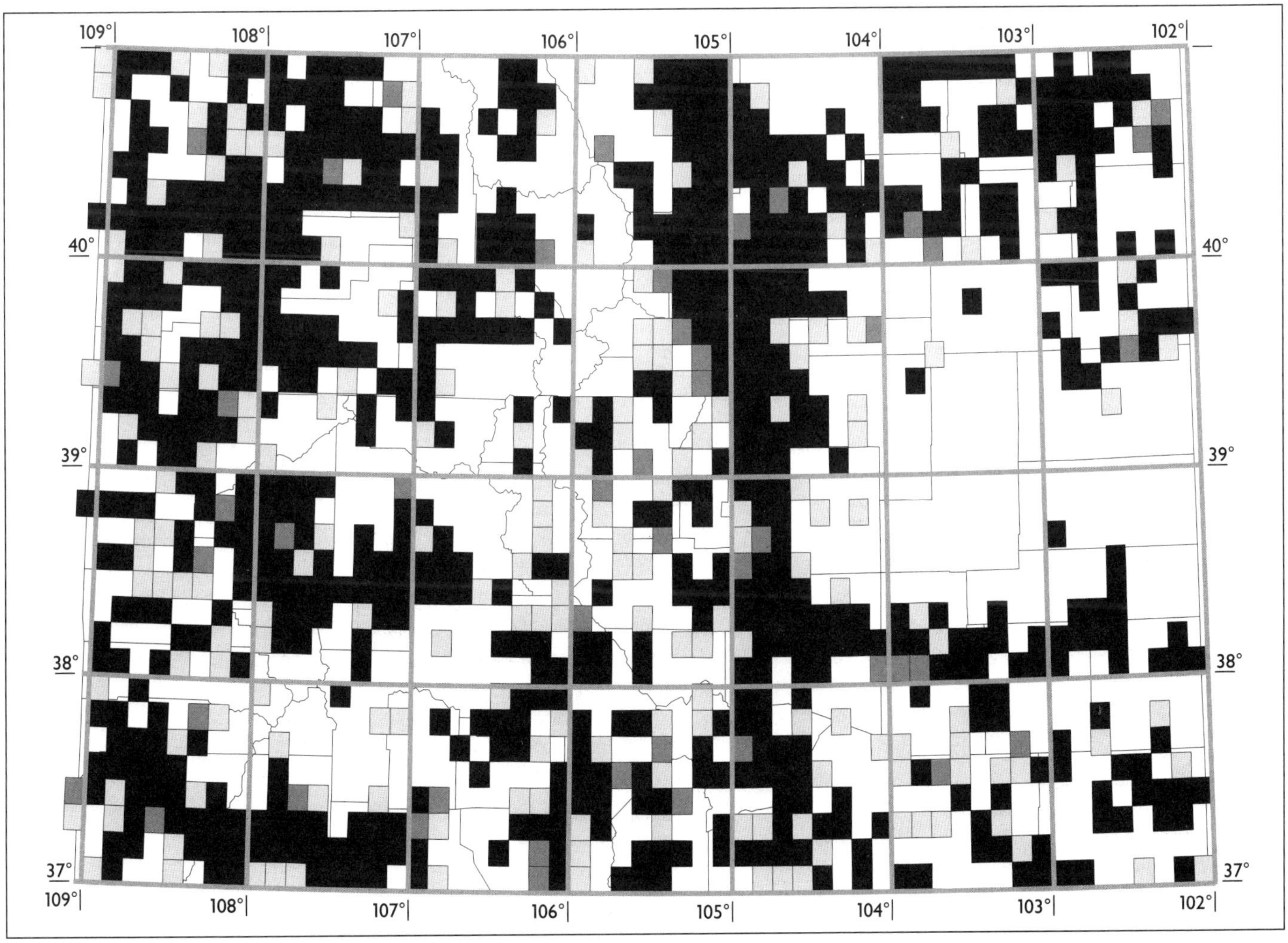
109°
108°
107°
106°
105°
104°
103°
102°
40°
39°
38°
37°

COLORADO DIVISION OF WILDLIFE / WILDLIFE RESOURCE INFORMATION SYSTEM

Here is a bird with a bad reputation and many charges of bad behavior—we have all heard terms such as "eat crow" and "crowbait." In the East farmers and others shoot crows as agricultural pests and devilers of other smaller creatures. Root (1988) calls them persecuted birds. But they are very smart birds—can count to three or four, solve many puzzles, and have a language all their own (Terres 1980).

HABITAT

In the eastern U.S. American Crows generally inhabit open country, farms, forest edges, open woodlands, and parks, and they form communal roosts in winter (Goodwin 1976). In New Jersey they also have a penchant for visiting landfills to feed, and one winter roost had 5,000 crows (Stouffer and Caccamise 1991).

Colorado birds reside up to 10,000 feet (3048 m). They form big fall flocks here and big winter roosts in some localities, but they do not always do so, and then not nearly as large as in the East. Atlas data show that here they breed mainly in riparian, agricultural, and urban areas.

Atlasers found them most frequently in rural areas and croplands, lowland and montane riparian areas, pinyon/juniper and ponderosa pine woodlands, and aspen stands. They reported a few in lodgepole pine stands, shrublands, and even cactus grassland. Unlike magpies, crows use conifers, both those artificially added to urban areas and various forest and woodland types. In residential Denver crows pick tall spruces and pines for nest sites.

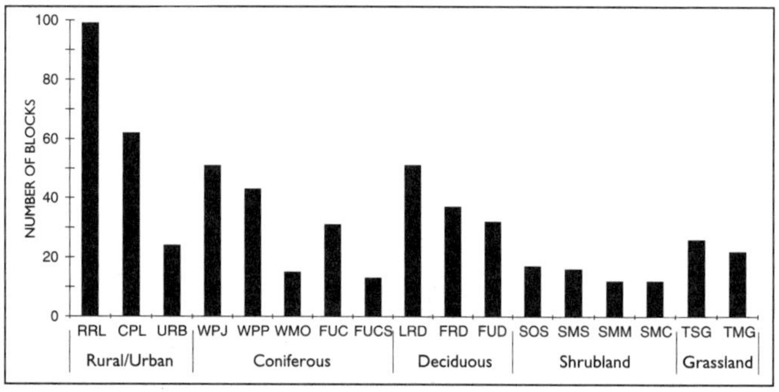
NUMBER OF BLOCKS
0
20
40
60
80
100
RRL CPL URB WPJ WPP WMO FUC FUCS LRD FRD FUD SOS SMS SMM SMC TSG TMG
Rural/Urban
Coniferous
Deciduous
Shrubland
Grassland

BREEDING

In the East breeding crows apparently behave differently, in some ways, than they do in the West. Eastern flocks defend breeding territories year round. They sometimes breed cooperatively—where young (up to eight) from a previous year's brood return to help their parents raise a new brood, and they remain as helpers for up to four years (Kilham 1984). These young helpers participate in all aspects of nesting (except egg-laying). In California, only a third of the breeding pairs have any helpers, and then often only one female yearling giving minimal help to the parents (Caffrey 1992).

Cooperative breeding in Colorado has not yet been studied. Young crows in California may form a non-breeding flock of 20–30 birds, using an area of 10 acres or so, but they freely wander into the territories of breeding pairs (Caffrey 1992).

Crows build nests in trees, shrubs, or on utility poles, 10–70 feet (3–20 m) high, averaging 25 feet (7.6 m; Harrison 1979). Both male and female cooperate to build the nest in about 12 days and to incubate the clutch for 18 days. The female lays 4–7 eggs, 5–6 most commonly, and the pair tends the nestlings for four to five weeks (Harrison 1979, Ehrlich 1988).

The long span of nest dates found by atlasers—from nest-building in March to fledglings still dependent in August—makes a very long breeding season. Nest-building reports spanned three months; although the first such observation came in March, the first on-nest report bears a 17 April date. Atlasers found them on nests over a two-and-one-half-month period.

BREEDING PHENOLOGY

CODE		# OF RECORDS	DATE RANGE
C	Courtship	3	13 Jun–4 Jul
NB	Nest Building	14	8 Mar–20 Jun
ON	Occupied Nest	20	17 Apr–2 Jul
NY	Nest with Young	5	21 May–5 Jul
FY	Feeding Young	13	29 May–4 Aug
FL	Fledged Young	52	31 May–28 Jul

DISTRIBUTION

True North American birds, American Crows are widespread across the continent except in the arid Southwest. They do not migrate except when large flocks form and roost together in fall and winter (A&R, B&N).

Their Colorado distribution has changed over the years. Cooke (1897), Sclater (1912), and Aiken and Warren (1914) called them common only in the northeast, particularly the Greeley/Fort Collins area and South Platte valley, and rare elsewhere. Sclater added La Plata County. B&N provided a similar though ambivalent analysis, but Davis (1969) called them common in western Colorado. A&R first noted a different statewide distribution.

Compared with historic reports, the Atlas map shows puzzling differences. It shows crows most widespread from 5,000 to 9,000 feet (1525–2750 m). In eastern Colorado far more breed in and near the foothills, including the Front Range urban areas, than on the plains to the northeast. In recent years crows have become noisy, visible residents of Colorado cities and towns along the Front Range from Wyoming to New Mexico. They throng into the mountains as high as Woodland Park (8,500 feet, 2590 m), Dillon (9,100 feet, 2775 m) and Leadville (10,200 feet, 3110 m).

Colorado crows nest in the pines of the Black Forest between Denver and Colorado Springs, in the pinyon/juniper woodlands of Las Animas County, and in the cottonwoods of rivers as far apart as the South Platte, Rio Grande, and San Juan. In southwestern Colorado breeding areas cover the San Juan Basin and fringe the Uncompahgre Plateau. Recently crows moved into the Grand Valley in the Grand Junction area, but inasmuch as the map shows a blank there, apparently after 1990 when atlasers completed most of their field work in that area (Rich Levad pers. comm.).

Atlasers found the highest numbers (A4 abundance) in rural cropland and lowland riparian habitats. Crows do feed in cropland but they nest in nearby drainages. They are surprisingly absent from several areas dominated by agriculture—the High Plains of eastern Colorado and the San Luis and Gunnison valleys, areas deficient in nest sites.

Atlas abundance estimates also contradict the historical reports. They show very few crows breeding on the plains—fewer than 3,000 pairs. The greatest numbers occur in north-central Colorado, the San Juan Basin, and the Front Range plains and foothills, each region with over 10,000 breeding pairs.

BBS trends show a slight increase nationally (0.8% per year). In Colorado, although the BBS shows an indefinite trend, other observers see them increasing in both rural and urban settings (A&R). We need more problem solvers, and this bird has a language all its own! Now if we could just figure out what they are saying . . .

BREEDING EVIDENCE

REPORTED IN 479 (27%) OF 1745 PRIORITY BLOCKS

☐	Possible	291 blocks	(61%)
■	Probable	77 blocks	(16%)
■	Confirmed	111 blocks	(23%)

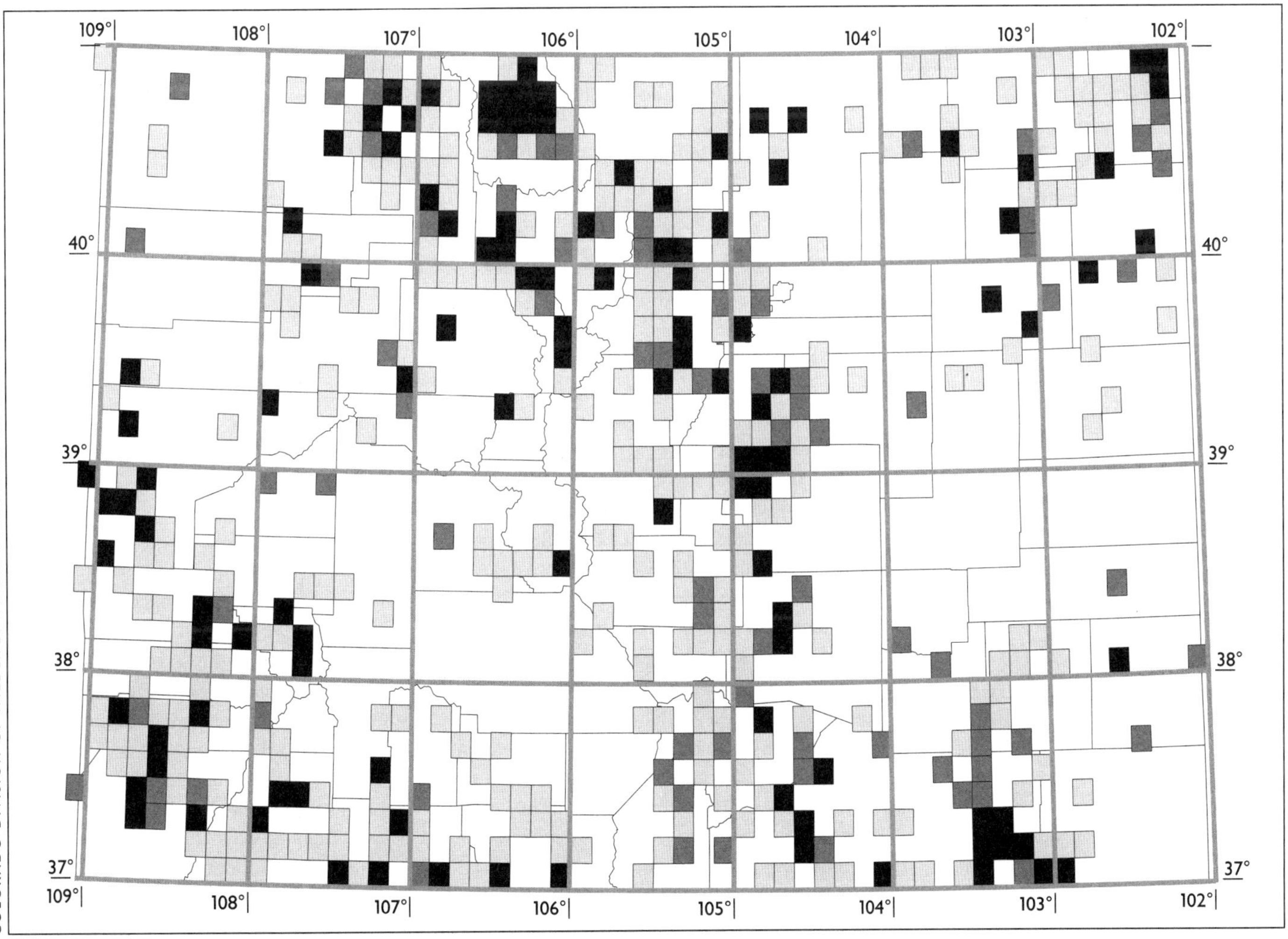
109°
108°
107°
106°
105°
104°
103°
102°
40°
39°
38°
37°

COLORADO DIVISION OF WILDLIFE / WILDLIFE RESOURCE INFORMATION SYSTEM

CHIHUAHUAN RAVEN

Corvus cryptoleucus

This raven's Latin name, *cryptoleucus,* translates to "hidden white" and describes these birds better than the old common name, "White-necked Raven." The feathers around the neck have white bases, rarely visible and then only in optimal viewing conditions.

HABITAT

Many know that Chihuahuan Ravens used to occur regularly far to the north of their current range (Henshaw 1875, Marsh 1931). Few realize that before a brief period of prosperity in the latter half of the nineteenth century, these ravens probably had the same restricted range as now. The wanton slaughter of bison, by producing a never-ending supply of fresh meat, seduced the ravens into a range expansion (Aiken and Warren 1914).

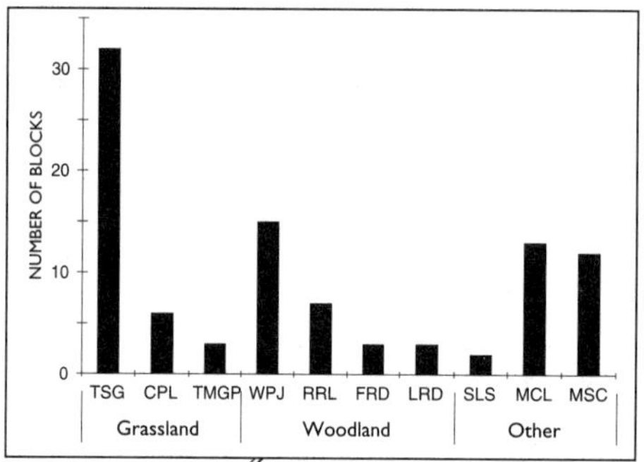
NUMBER OF BLOCKS
0
10
20
30
TSG
CPL
TMGP
WPJ
RRL
FRD
LRD
SLS
MCL
MSC
Grassland
Woodland
Other

As settlement progressed on the plains, tidy homesteads replaced the smorgasbord left by hunters. When the ravens lost their major food source, they withdrew to their former domain.

Before the bison glut they probably nested only along stream courses with trees and in the canyon country where the larger Common Ravens also nest. The shortgrass prairie lacked nest sites (which settlement now provides—trees, windmills, and utility poles).

Today, as the Atlas habitat codes illustrate, Chihuahuan Ravens inhabit semi-arid shortgrass prairies and cholla cactus grasslands, adjacent riparian corridors, pinyon/juniper woodlands, and nearby croplands. They often nest in isolated trees, especially near abandoned homesteads. They also place nests on windmills, atop the struts of telephone poles and, rarely, under railroad or highway bridges. These human artifacts do not show up in Atlas habitat codes. The ravens feed on whatever they can find, from grain to living or dead birds and mammals to roadkill.

During the breeding season, nest sites and food sources limit their distribution. Opportunistic colonists, they nested on the prairies wherever they found trees (Aiken and Warren 1914). Noticeably sociable, these ravens sometimes cluster many nests close together in suitable places with a good food supply—in "huge colonies along water-courses" (Henshaw 1875). In 1837 Rufus Sage encountered thousands, many in groups nesting on tributaries of the South Platte in Morgan County, and he feasted on their eggs (Marsh 1931, B&N).

BREEDING

Because Chihuahuan Ravens typically nest in exposed locations and act conspicuously around the nest, atlasers Confirmed a high percentage, and found nests in 29 Priority blocks and 14 Non-priority blocks.

In the Sonoran desert of Arizona these ravens nest in the midsummer wet season, to take advantage of the bounty it provides—e.g., bird eggs and young, and weak mammals of the year. However, in southeastern Colorado, nesting occurs earlier in the year, to benefit from our wetter spring. The March-to-June Atlas nest dates reflect that; nests contained young by 15 May. Perversely, the only Atlas record for a nest with eggs came on 17 June.

Any account of Chihuahuan Ravens should mention the difficulty of separating them from Common Ravens. Common Ravens nest on cliffs in the southeastern Colorado canyonlands, where unheard ravens should remain unidentified as to species.

BREEDING PHENOLOGY

CODE		# OF RECORDS	DATE RANGE
C	Courtship	1	13 May
NB	Nest Building	2	21 Mar–27 May
ON	Occupied Nest	12	29 Apr–8 Jul
NE	Nest with Eggs	1	17 Jun
NY	Nest with Young	20	15 May–15 Jul
FY	Feeding Young	2	27 May–12 Jun
FL	Fledged Young	5	28 May–19 Jun

DISTRIBUTION

In the boom cycle of the late 1800s, Chihuahuan Ravens ranged north along the edge of the Front Range foothills into southern Wyoming, south-central Nebraska, and much of Kansas. Now they range from southwestern Kansas and central Texas to southeastern Arizona and central Mexico.

The Colorado range prior to the bison boom probably coincided with the current range (Bent 1946). They retreated from northeastern Colorado in the 1900s and gradually decreased in the southeast after the 1930s (A&R). By 1946 the northern margin of their range had retreated south to Hugo, in Lincoln County (Bent 1946), and by 1962 they no longer occurred commonly at Kit Carson (B&N).

All Atlas Confirmations but one came from the nine counties in southeastern Colorado, from Pueblo County east to Kiowa County and south to the state line, with most south of the Arkansas River. The exception expanded the known range to southwestern Colorado, with fledglings documented in a block along the San Juan River (HEK VF).

Atlas reviewers cautiously handled a few extralimital records in marginal habitat or farther north in the former range. Very possibly some outlying but undocumented reports actually pertained to Chihuahuan Ravens, but without documentation the Atlas map omits such observations.

The historic decline in population and range led to concern about the species (Webb 1985). The decline continues. For example, between 1990 and 1995 one colony in Kiowa County fell from more than ten nesting pairs to only one (DLN). In the last 30 years Kansas breeders have declined dramatically, with perhaps only 25 nesting pairs left (Scott Seltman pers. comm.).

American Crows have proliferated in Colorado. In southeastern Colorado, where the Chihuahuan Raven previously was the only resident corvid, the Atlas map shows some overlap between the two in Las Animas and Otero counties, but not much elsewhere. Chase (1980) postulated that the recent increase of crows could have heightened competition for food and nest sites. Although displacement could have occurred in northeastern Colorado, historic and Atlas data do not support this thesis. First, the withdrawal of Chihuahuan Ravens from the northeast probably predates any influx of crows there (even assuming crows occurred in greater numbers than today). Second, the two species occupy different habitats; Chihuahuan Raven colonies in the High Plains occupy open-country habitat not suitable for crows. Common Ravens make more formidable competitors for both breeding sites and food sources, and the two raven species do in fact meet in southeastern Colorado. Atlasers reported 27 blocks with both ravens, and in 13 both species occupied the same habitats.

Persecution by man still occurs. Utility companies may remove nests placed on telephone poles. Lack of nest sites probably serves as a strong limiting factor. Finally, perhaps the prairies no longer provide the bounty needed to support these birds.

BREEDING EVIDENCE

REPORTED IN 80 (5%) OF 1745 PRIORITY BLOCKS

☐	Possible	25 blocks	(31%)
▪	Probable	19 blocks	(24%)
■	Confirmed	36 blocks	(45%)

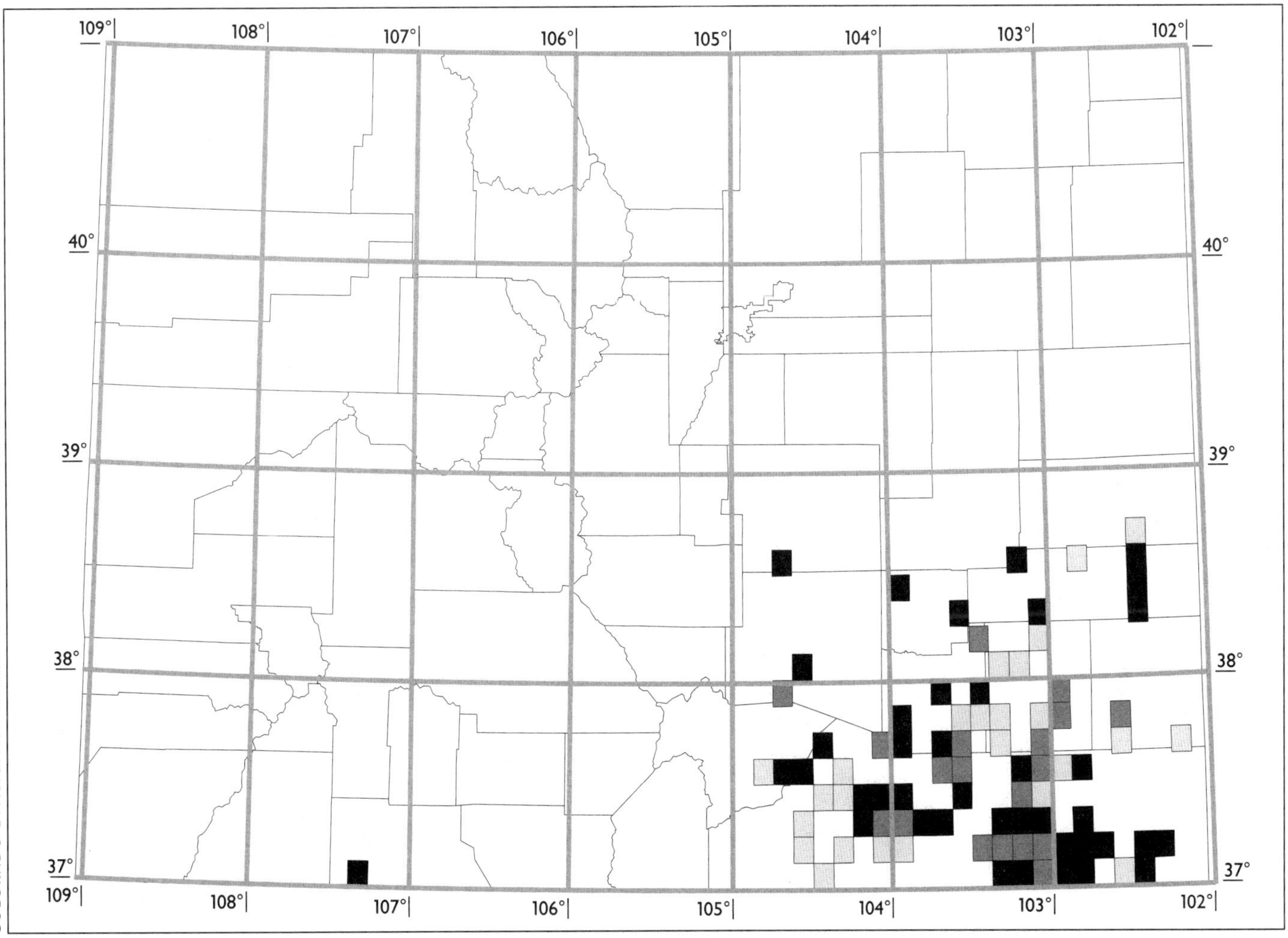
109°
108°
107°
106°
105°
104°
103°
102°
40°
39°
38°
37°

COLORADO DIVISION OF WILDLIFE / WILDLIFE RESOURCE INFORMATION SYSTEM

Almost every desert canyon and mountain valley in Colorado has its pair of ravens. Depending on topography, an Atlas block can have one to three pairs. Pike first mentioned them in Colorado when, after his unsuccessful assault on his peak, the party ate "a piece of deer's ribs the ravens had left us, being the first we had eaten in that forty-eight hours" (Marsh 1931). Most early naturalists reported ravens, and they played prominent parts in Indian legends (Goodwin 1976). Fossils of the crow family, species unknown, have been found in sedimentary deposits in Colorado laid down 12 million years ago (Brodkorb 1978).

HABITAT

The mountains and forests provide the major habitat for ravens. Atlas observers found them in a variety of habitats, from the foothills to the alpine, primarily in the montane forests, woodlands, shrublands, and associated mountain areas. Cliffs and rocky outcroppings provide most of the breeding sites. Although ravens also nest in large evergreen trees and aspens (Bent 1946, Dorn 1972), atlasers reported no such nests. Atlasers found pairs in the San Luis Valley and the mountain parks nesting on power poles in lieu of cliffs (e.g., Whale Hill, 38106C1).

The omnivorous ravens eat almost anything that they can catch, kill, or find, including carrion, small rodents, insects, and grain (Bent 1946), and the contents of dumps and trash bins. Foraging territory may cover several square miles, and usually seems to encompass one valley, ridge to ridge. Other raptors, especially Great Horned Owls and Prairie Falcons, sometimes use old raven nests or live in close proximity without conflict (Stiehl 1985), although ravens will defend against some other birds.

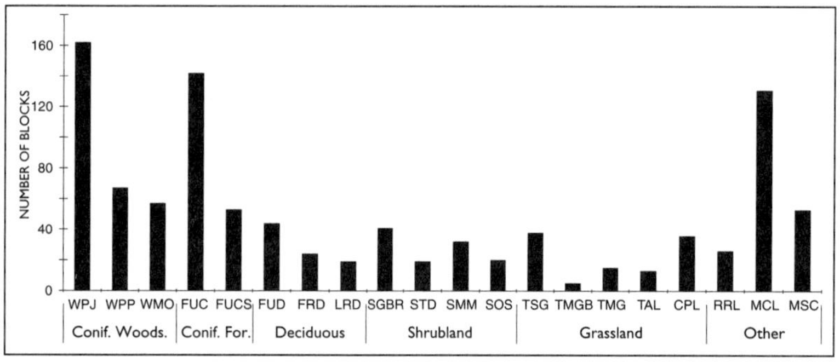

NUMBER OF BLOCKS
0
40
80
120
160
WPJ WPP WMO FUC FUCS FUD FRD LRD SGBR STD SMM SOS TSG TMGB TMG TAL CPL RRL MCL MSC
Conif. Woods.
Conif. For.
Deciduous
Shrubland
Grassland
Other

BREEDING

Ravens mate for life (Heinrich 1989). As year-round residents, they show high fidelity to their nesting sites and use the same area for a number of years (Ratcliffe 1962). Large winter roosts may consist largely of immature birds (up to four years of age) and non-breeding adults.

Although some may initiate courtship at six months of age, they do not breed until 3–4 years of age. Territory size depends on the food supply; it can vary from 3.5 mi^2 (9.5 km^2) in Wales (where they find lots of sheep carcasses) to 18.6 mi^2 (48 km^2) in Germany (Gibbons et al. 1993, Ratcliffe 1962). In Utah, Smith and Murphy (1973) measured raven territories as 2.5 mi^2 (6.5 km^2), and as close together as one mile (1.6 km). In Maine 1.2 miles (2 km) ordinarily separates nests (Heinrich 1989).

Although ravens begin to court in the fall, atlasers saw aerial courtship from March to July; the ritual includes billing, cooing, and allopreening. Nesting begins in very early spring (Goodwin 1976). A pair builds a bulky nest of sticks in a well-protected crevice of a cliff or rock outcrop, or well hidden in a tree. Sclater (1912) noted a tendency for Colorado ravens to nest on cliffs having a south aspect. Nests are difficult to find; atlasers observed only seven instances of nest-building and 25 nests with young. Most of the few Confirmed breeding observations (17%) stemmed from finding fledged young still dependent on their parents.

Clutches typically consist of 4–5 eggs, probably incubated only by the female, who depends on the male to bring her food for 20–21 days (Stiehl 1985, Heinrich 1989). The parent birds can become very agitated and defensive while the young are vulnerable. Immatures remain with their parents for nearly six months before dispersing in the fall, never to return to their natal home (Heinrich 1989).

BREEDING PHENOLOGY

CODE		# OF RECORDS	DATE RANGE
C	Courtship	18	5 Mar–9 Jul
NB	Nest Building	8	21 Mar–13 May
ON	Occupied Nest	18	15 Apr–6 Jul
NY	Nest with Young	25	1 May–15 Jul
FY	Feeding Young	19	22 May–22 Jul
FL	Fledged Young	60	24 May–2 Aug

DISTRIBUTION

Common Ravens soar from the Arctic to the top of Pikes Peak, and over the forests, mountains, wildernesses, and sea cliffs of the Northern Hemisphere—Great Britain, northern Europe, north Africa, Asia south to the Himalayas, and arctic and forested areas across Canada south to Nicaragua. They occur widely in the western U.S., and in scattered areas in the Appalachians and forests of the eastern states.

Ravens inhabit the Colorado mountains wherever cliffs provide good nesting habitat, and they range southeast to Las Animas and Baca counties, where their range overlaps with that of Chihuahuan Ravens. The map depicts them as blanketing the mountains, mesa and canyon country, and the Colorado Plateau, including most of the mountain parks.

The Atlas map shows few Confirmations, but a pair or two nest in almost every block shown on the map. Pairs cruise most mountain valleys and desert canyons, from Carrizo Canyon in Baca to Mesa Verde up to the peaks of all the mountain ranges.

After a drastic decrease in the early 1900s Common Ravens have achieved a dramatic recovery. In the East atlasers found them in 17% of Pennsylvania, 26% of West Virginia, and even in six blocks in the Kentucky Appalachians (Brauning 1992, Buckelew and Hall 1994, Palmer-Ball 1996). BBS statistics delineate a large increase across the continent, over 3% per year, with increases on 60% of the routes, as well as on 59% of the Colorado routes.

People are their major enemy, through shooting, harassment of nesting areas, and poison baits. The present laws against shooting and the vast extent of open land protect them from major depredations. Ravens have the potential to become nuisances. If they multiply, already large winter roosts would burgeon and wandering flocks of immatures could become a greater aggravation. Too many birds could generate calls for controls, like those perceived as necessary for crows and blackbirds.

BREEDING EVIDENCE

REPORTED IN 837 (48%) OF 1745 PRIORITY BLOCKS

☐	Possible	554 blocks	(66%)
■	Probable	141 blocks	(17%)
■	Confirmed	142 blocks	(17%)

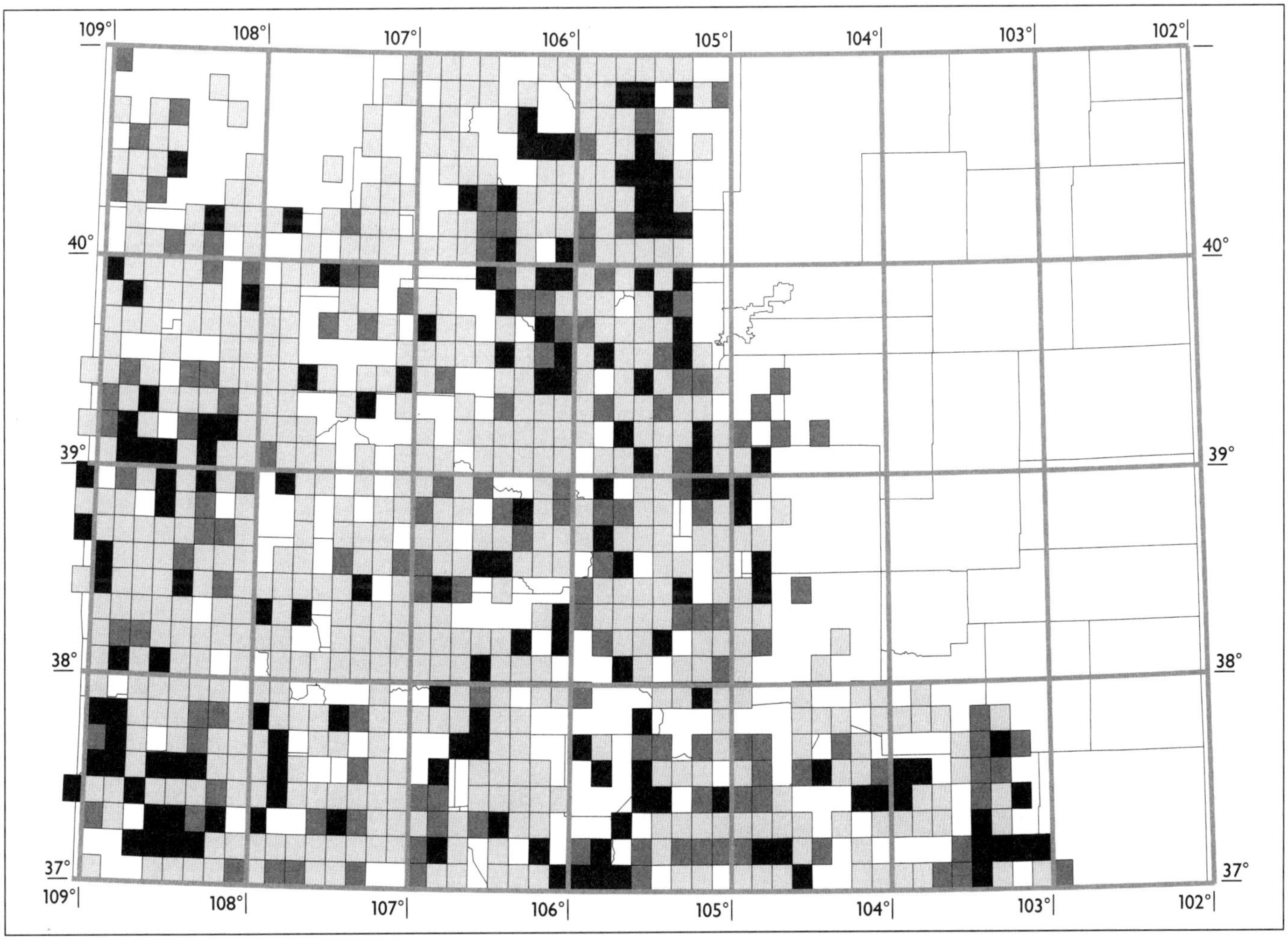
109°
108°
107°
106°
105°
104°
103°
102°
40°
39°
38°
37°

COLORADO DIVISION OF WILDLIFE / WILDLIFE RESOURCE INFORMATION SYSTEM

HORNED LARK *Eremophila alpestris*

Horned Larks are probably Colorado's most abundant and one of the most widespread nesters. They feed along roadsides, and the sparrow-sized birds that dash in front of your car showing black tails with narrow white outer feathers are usually Horned Larks.

HABITAT

Horned Larks breed in open, generally barren country. They prefer shortgrass prairie with considerable bare ground and grasses no taller than a few centimeters (Beason 1995). In agricultural areas, they inhabit bare ground and fields of row crops. In areas grazed by livestock, population densities are highest in the most heavily grazed areas (Boyd 1976, Creighton and Baldwin 1974, Giezentanner 1970, Porter and Ryder 1974, Ryder 1980, Strong and Ryder 1971, With and Webb 1993). Atlasers found breeding Horned Larks mainly in shortgrass prairie (447 blocks) and croplands (218 blocks), but also in 21 other habitat types. In arid parts of the San Luis Valley atlasers found Horned Larks where shrubs grow widely spaced and no higher than a foot. One of the four passerines that regularly nest above timberline, Horned Larks avoid snow-accumulation areas for nesting (Braun 1980, Conry 1978, Melcher 1992). Horned Larks occurred in 63 alpine blocks, where atlasers typically found them in dryer, shorter grass components of the tundra. The highest record came from 13,300 feet, of several birds singing on a shoulder of Mount Columbia in the Sawatch Range (Alan Versaw pers. comm.).

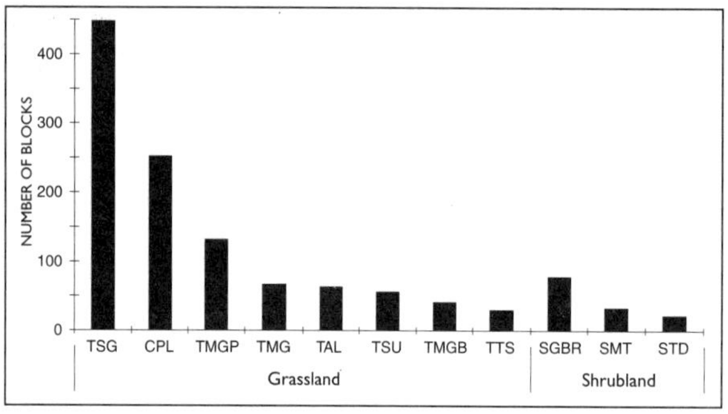

NUMBER OF BLOCKS
400
300
200
100
0
TSG
CPL
TMGP
TMG
TAL
TSU
TMGB
TTS
SGBR
SMT
STD
Grassland
Shrubland

Nests on the Central Plains Experimental Range (CPER) in Weld County were mainly in shortgrass (64.7%); the rest 1.6% in midgrass, 5.8% in sedge, 2.3% in prickly pear cactus, 0.8% in shrubs (saltbush), 16.9% on bare ground, and 1.2% in rock (Creighton and Baldwin 1974).

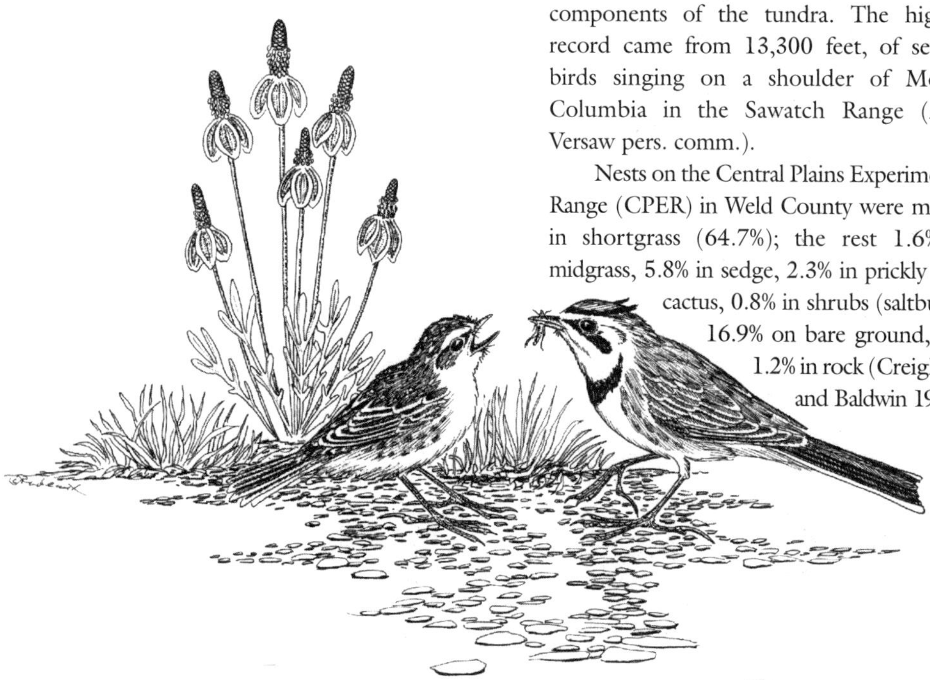

BREEDING

Strong and Ryder (1971) followed the fate of 84 nests on the CPER. Nesting success ranged from 36% in 1970 to 41% in 1971. In 1970, 51% of 78 eggs hatched; 60% of 140 eggs in 1971. Clutches averaged 3.0 eggs (range 2–4). Predation was the main cause of nest failure. An average of 1.7 eggs hatched per nest, and 1.0 young fledged per nest. Boyd (1976) found that Horned Larks double-brood on the CPER. He followed 62 nests in 1975 with an average clutch of 2.6. Only females incubated, but males shared in feeding the young. Fledglings "normally" depart the nest 9–10 days after hatching (Boyd 1976).

Creighton (1974) followed 50 nests on the Pawnee National Grassland. Earliest nest initiation was mid April, and it continued until late June. Nest-building reported by atlasers fits within those dates, but they found 14 nests with eggs at later dates, 1 July to 5 August, eight of them above timberline.

The oldest known Horned Lark was 7 years and 11 months old. Van Truan banded it west of Pueblo on 2 August 1976 and recaptured it on 26 May 1983 (Klimkiewicz and Futcher 1989).

Atlasers noted two instances of cowbird parasitism: Scott Gillihan observed a fledgling cowbird fed by a Horned Lark in Saguache County, as did Jack Wieland in Phillips County (Chace and Cruz 1996; Appendix C).

BREEDING PHENOLOGY

CODE		# OF RECORDS	DATE RANGE
C	Courtship	21	12 Mar–25 Jul
NB	Nest Building	3	21 Apr–12 Jun
ON	Occupied Nest	12	22 May–19 Jul
NE	Nest with Eggs	34	8 Apr–5 Aug
NY	Nest with Young	9	27 Apr–8 Jul
FY	Feeding Young	121	13 May–19 Aug
FL	Fledged Young	423	7 Apr–22 Aug

DISTRIBUTION

This circumpolar species breeds across Europe, Asia, and northern Africa, and in North America from the Canadian and Alaskan tundra and grasslands south to Colombia and the Andes. One of Colorado's hardiest and most widespread birds, Horned Larks occur border to border, from the lowest point, about 3,390 feet (1050 m) near

Holly, up to 13,500 feet in the higher mountains. Early observers reported them as abundant nesters and winter residents (Cooke 1897, Sclater 1912, Bergtold 1928). Atlasers Confirmed Horned Larks in 611 of the 982 Priority blocks in which they found them—a very high rate of 62.2%. Horned Larks ranked eleventh in the number of blocks in which atlasers found them.

On 20 July 1820, Edwin James, a member of Major Long's expedition, saw three Horned Larks on the plains in what is now Baca County (Marsh 1931). Larks have been observed in every county since then, and although atlasers Confirmed breeding in all 28 latilongs, they found the most occurrences on the eastern plains.

Colorado has more Horned Larks than any other species, over 2.7 million breeding pairs of Horned Larks, according to estimates using abundance codes provided by Atlas field workers. This far exceeds the 1.98 million breeding pairs calculated for American Robins, the second most numerous species.

Horned Larks constitute the most abundant species on many Colorado BBS routes. In 1993 Colorado BBS routes tallied 7,300 Horned Larks, also making the species the most abundant BBS species counted in the state. In that year 51 routes detected them; five reported them at all 50 stops, and 13 more recorded them at more than 40 stops. The Pinneo route in Washington County (run by Peter Gent since 1982) has an average abundance of 534.9, the highest in the country (Price et al. 1995). BBS trends from 1966–1996 indicate a moderate decline across the continent (1.34%/year) but do not show a trend in Colorado, even with a mean of 93 birds/route on 65 routes.

BREEDING EVIDENCE

REPORTED IN 982 (56%) OF 1745 PRIORITY BLOCKS

☐	Possible	174 blocks	(18%)
▩	Probable	196 blocks	(20%)
■	Confirmed	612 blocks	(62%)

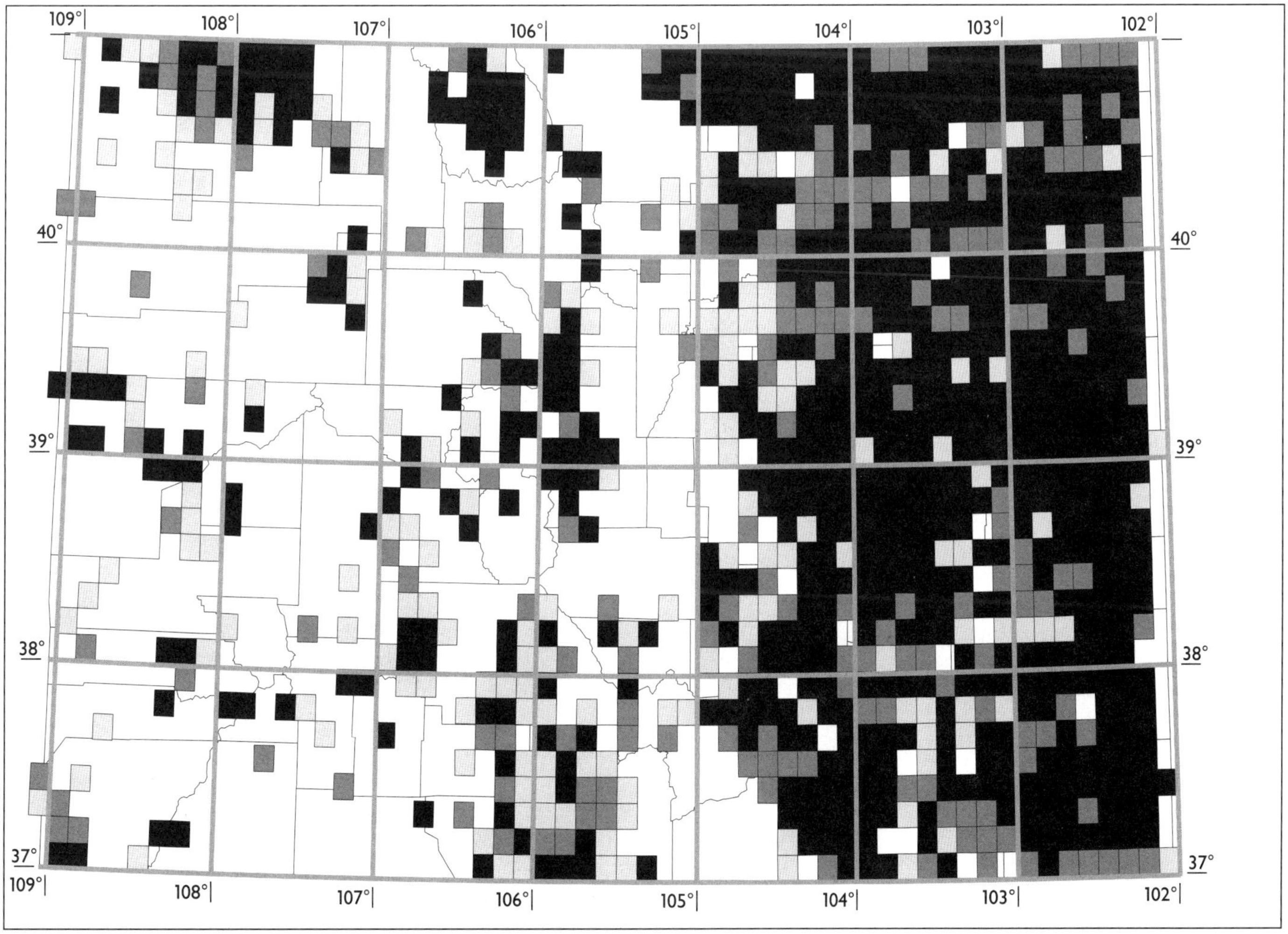
109°
108°
107°
106°
105°
104°
103°
102°
40°
39°
38°
37°

COLORADO DIVISION OF WILDLIFE / WILDLIFE RESOURCE INFORMATION SYSTEM

Crowded apartment dwellers in the East often dream of escaping to reclusive mountain retreats in the Colorado Rockies; a few lucky Purple Martins actually have made the switch.

HABITAT

Purple Martins occupy a variety of habitats in other states—cavities in cacti, cliffs, trees, gourds, and especially manmade houses (Terres 1980). In Colorado, however, their preference seems very specific: the edges of old-growth aspen stands, usually near a stream, spring, or pond (Reynolds et al.

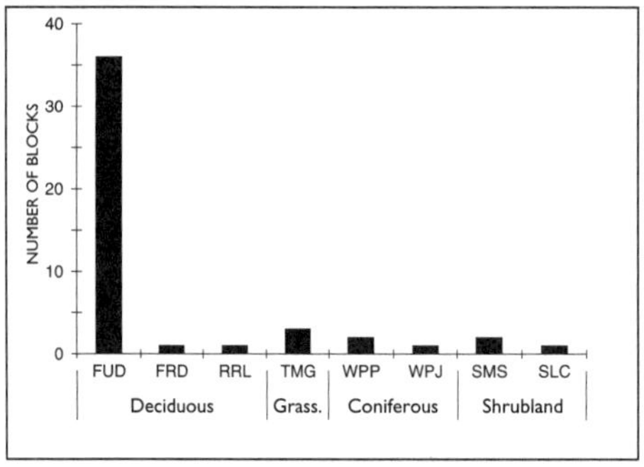
NUMBER OF BLOCKS
0
10
20
30
40
FUD
FRD
RRL
TMG
WPP
WPJ
SMS
SLC
Deciduous
Grass.
Coniferous
Shrubland

1991). Atlasers found them in aspen stands from 8,000 to 9,000 feet (2440–2745 m). These aspen groves often stand adjacent to ponderosa pine or spruce/fir woodlands and to open brushlands—snowberry, sage, serviceberry, mountain mahogany, and scrub oak. Except for a single nest in ponderosa pine, habitats other than aspen reported for this species described birds feeding in flight rather than birds at nests. The martins' preferred habitat also hosts flickers and Red-naped Sapsuckers (which provide nest holes for the martins), and other cavity-nesting species such as Flammulated Owl, House Wrens, Mountain Bluebird, Tree Swallows, and Violet-green Swallows.

Ecologists describe aspen as a successional species, one gradually giving way to conifers; in parts of Colorado's Western Slope, however, aspen qualifies as climax vegetation (Flack 1976). The range of the Purple Martin in Colorado rather closely coincides with the range of climax aspen (RL).

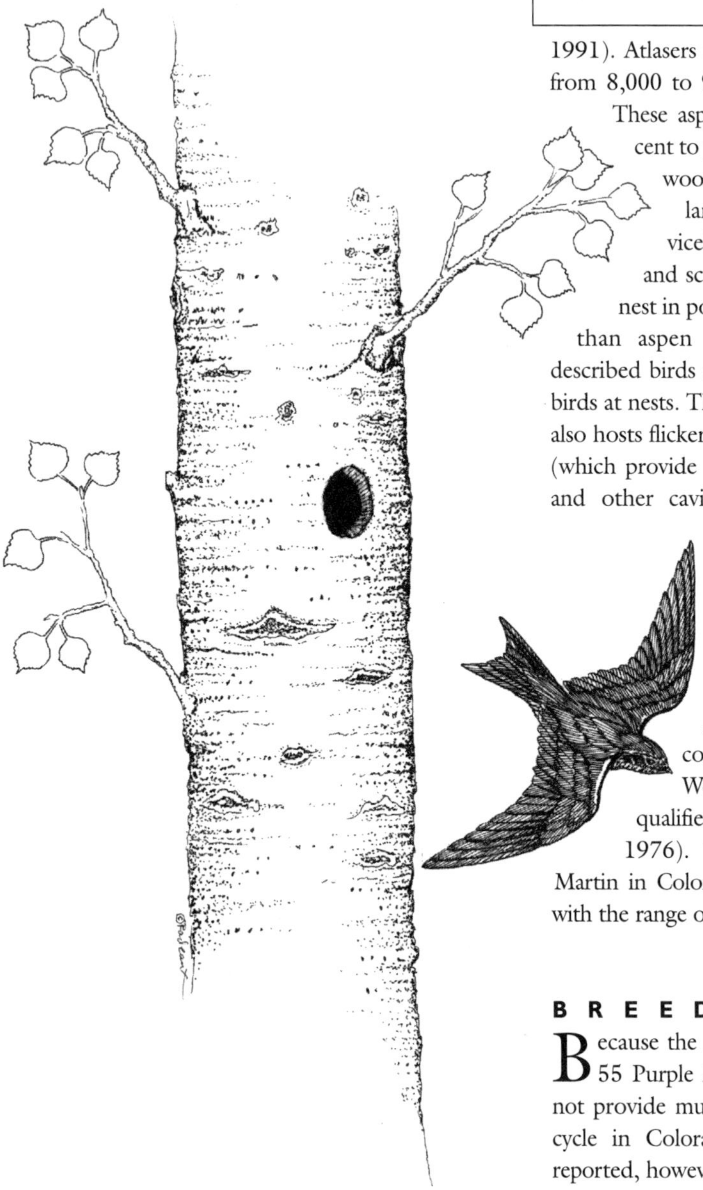

BREEDING

Because the Atlas project produced only 55 Purple Martin records, the data do not provide much insight into its breeding cycle in Colorado. The dates and codes reported, however, do seem to follow closely those reported for Alberta (Finlay 1971), at the northern limits of the bird's range. The habitat—aspen parkland—and climate of that region resemble the habitat and climate where martins nest in western Colorado. Most Purple Martins probably first arrive in western Colorado in mid May (earliest Atlas record: 24 May) and immediately begin to engage in nest-cavity defense and pair-bond formation (earliest pair report: 24 May). By the first of June, incubating has begun (16–17 days), and hatching generally begins in mid June (earliest nest with young: 14 June). The 28-day nestling stage lasts through most of July and, for later nests, into early August (earliest fledged young date: 17 July).

BREEDING PHENOLOGY

CODE		# OF RECORDS	DATE RANGE
ON	Occupied Nest	10	6 Jun–21 Jul
NY	Nest with Young	4	14 Jun–7 Jul
FY	Feeding Young	5	6 Jul–4 Aug
FL	Fledged Young	4	17 Jul–31 Jul

DISTRIBUTION

Purple Martins breed locally throughout the eastern U.S. from the Atlantic to the Great Plains, across the Southwest, and up the Pacific coast from south-central California to British Columbia. They winter in South America.

In Colorado observers first confirmed breeding in 1872 in western Las Animas County (B&N). Despite several sightings in the interim, a century passed without further nesting records. Finally, in 1978, Stoner Mesa in Montezuma County produced another nesting record (Svoboda et al. 1980), and in 1984 observers began to find nests annually on McClure Pass in Gunnison County (Zerbi 1985).

The paucity of records led to omission of the Purple Martin from the Atlas field card. However, Tom Moran and I found a nest on the Uncompahgre Plateau in 1987, the first year of field work, and many more write-ins followed as Atlas workers reported Purple Martins as at least Possible breeders in 39 Priority blocks and Confirmed breeding in 22. They also Confirmed them in two of the 16 Non-priority blocks where they found them. Atlas work has demonstrated that a thinly scattered population of Purple Martins breeds throughout the plateaus of western Colorado, but within a narrow elevation range, 8,000–9,000 feet (2440–2745 m).

An Atlas report from Prowers County in southeastern Colorado may represent an outpost of the midwestern population. The date (6 June) and the habitat (RRL) fit that population's breeding pattern, and B&N reported two early records for that area, one a nest record. Kansas has breeding records in three fairly close counties (Thompson and Ely 1992). Another report of Observed birds, on 7 June from Baca County (not on the map), also suggests that a few Purple Martins may occasionally breed in this part of the state.

Purple Martins do not form large colonies in Colorado; only four or five pairs inhabited most Priority blocks. Field workers reported an abundance code of A2 (2–10 pairs) for three-fourths of the blocks for which they gave abundance figures; each of the six Priority blocks with a reported abundance code of A3 (11–100 pairs) holds only a dozen or so pairs at most (Coen Dexter pers. comm.).

The practice of removing standing dead trees has reduced nest availability in other areas (Ehrlich et al. 1988), so the relatively small size of the population, combined with the Purple Martin's strong site fidelity (Allen and Nice 1952), indicates that the clear-cutting of aspen in its territory could disturb its nesting.

In the Midwest and East, European Starlings compete vigorously for nest cavities (Allen and Nice 1952), and in Colorado they have begun to use the sort of open aspen parkland that Purple Martins favor. These neighbors may soon threaten Purple Martins in remote mountain retreats as they do in the more crowded East.

BREEDING EVIDENCE

REPORTED IN 39 (2%) OF 1745 PRIORITY BLOCKS

Possible	10 blocks	(26%)
Probable	7 blocks	(18%)
Confirmed	22 blocks	(56%)

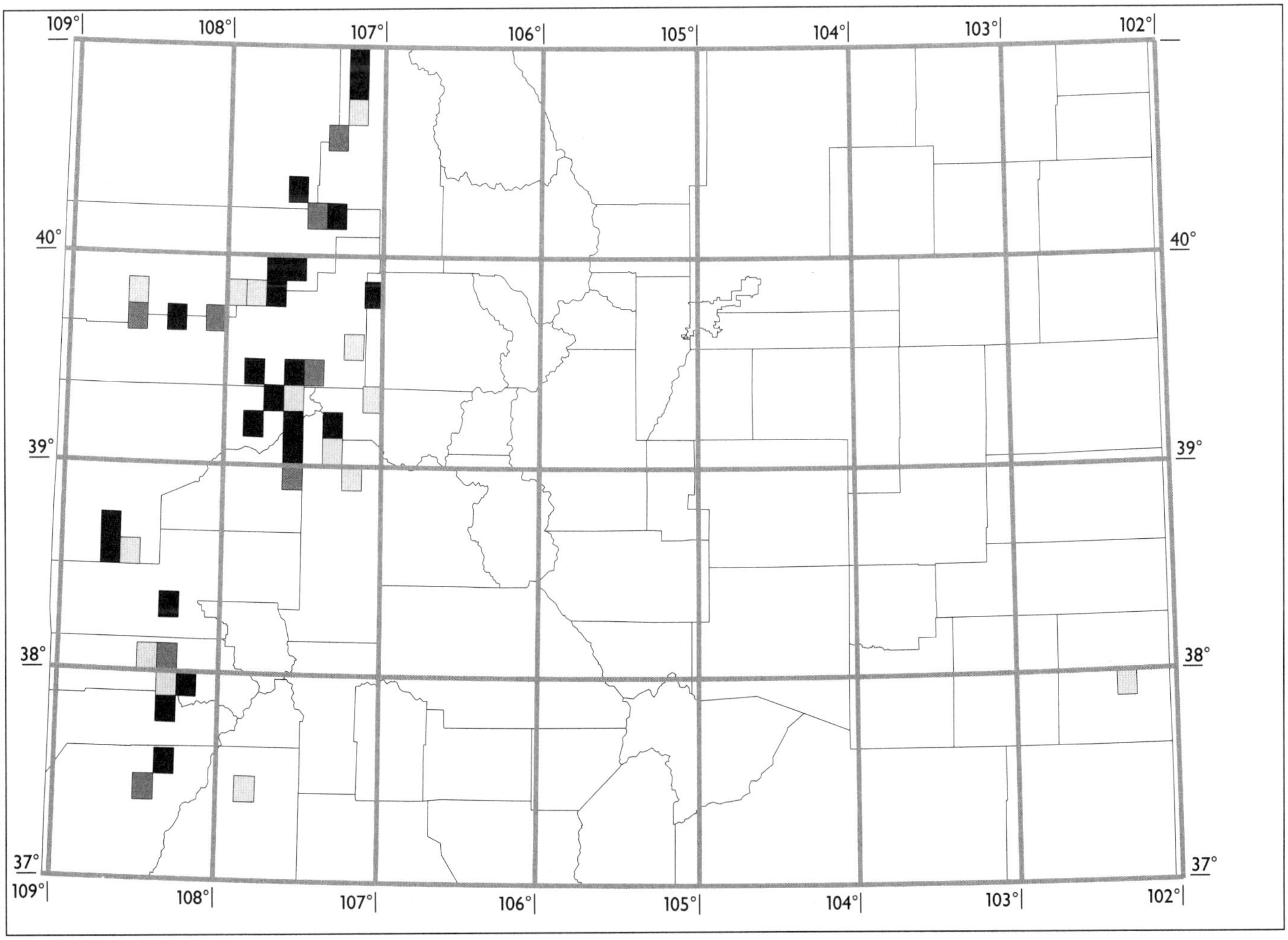
109°
108°
107°
106°
105°
104°
103°
102°
40°
39°
38°
37°

COLORADO DIVISION OF WILDLIFE / WILDLIFE RESOURCE INFORMATION SYSTEM

Hardy, graceful, acrobatic, aggressive, playful Tree Swallows, obsessed with feathers, show up in the Colorado mountains—Winter Park, Vail Pass—when snow still covers the ground. Ornithologists say that they play with feathers; they engage in aerial chases and games over possession of feathers.

HABITAT

Tree Swallows require open areas for breeding and feeding but need trees for nest cavities. They frequently nest near water, especially beaver ponds and streams. Atlas habitat codes show them primarily in deciduous trees, especially aspens, by far the most frequent habitat reported. These swallows do not nest in the middle of extensive aspen forests; rather they pick the edges, open glades, and small groves. Riparian cottonwood groves, the other main deciduous habitat both in the foothills and on the plains, by nature are linear; their growth pattern provides nest site access and open feeding areas. In recent years Tree Swallows have colonized a few cottonwood groves adjacent to reservoirs and streams on the plains.

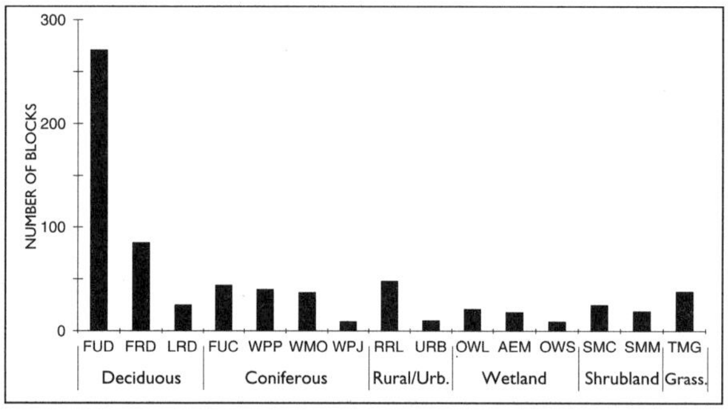
NUMBER OF BLOCKS
300
200
100
0
FUD FRD LRD FUC WPP WMO WPJ RRL URB OWL AEM OWS SMC SMM TMG
Deciduous
Coniferous
Rural/Urb.
Wetland
Shrubland
Grass.

Atlasers also reported Tree Swallows in coniferous habitats; most were woodlands of various tree species, with open canopies. Rural, shrubland, and grassland habitats represent either feeding areas or the location of bluebird houses that these swallows usurp.

Although nest strategies distinguish swallows, feeding locations do not. Tree Swallows range over many habitats to feed but availability of nest sites limits breeding populations (Robertson et al. 1992). They compete for nest cavities with House Wrens, bluebirds, starlings, and House Sparrows. An earlier observer reported them formerly more common; a few even tried to nest in the outskirts of Denver, for example, but House Sparrows prevented them (Bergtold 1928). Tree Swallows, rather quarrelsome themselves, frequently oust Violet-green Swallows from nesting places (Sclater 1912).

Although removal of dead trees reduces available nest sites, Tree Swallows readily breed in nest boxes and frequently use nest boxes of bluebird trails. Bluebird trail proprietors have learned to put two houses together so that the swallows will leave one house for the bluebirds.

Atlasers found Tree Swallow nests mainly in aspens; 17 nest holes ranged 5–60 feet (1.5–18 m) above the ground. Atlasers located one nest in a parking shed in Leadville (Pam Piombino, nest card) and two nests in dead spruce stubs. Tree Swallows nested north of Kremmling in 1994 in an aspen cavity that Buffleheads had used the two previous years (Norman Barrett, nest card).

BREEDING

These hardy swallows migrate back to Colorado in early spring, when they risk cold weather and snow. They breed 1 May–16 August, and depart in September (Sclater 1912, Nelson 1993, Atlas).

They frequently arrive at mountain breeding grounds while weather remains unsettled—in Winter Park by early April when snow still covers the ground. If the temperature turns too cold to catch insects in the breeding area around Rollinsville, the swallows may fly 20 miles (32 km) to feed on the plains (Bob Cohen pers. comm.). On cloudy days they travel to ponds where the air over the water is warmer and therefore have more insect activity (Robertson et al. 1992).

They produce single broods in Colorado; of several thousand nestings in Gilpin County, Bob Cohen observed no second broods (Robertson et al. 1992), and regularly monitored nest boxes in Eagle County have never produced second broods (Jack Merchant pers. comm.). Atlasers reported occupied nests from 9 May to 28 July, consistent with raising only one brood.

Mainly monogamous, Tree Swallows are also promiscuous. In some populations 50% of the nests contain offspring not descended from the resident male (Robertson et al. 1992).

Tree Swallows have a romance with feathers, but this idiosyncrasy apparently has a practical effect. Young grow better in well-feathered nests, which, compared to nests with feathers experimentally removed, have lower parasite infestations (Robertson et al. 1992).

Competition over feathers "may contain an element of play behavior," identify the best breeding males, or fulfill some other social function. Tree Swallows engage in active and aggressive chases after feathers. Sometimes one possessing a feather that others ignore will drop it, a ploy that attracts birds who begin to chase (Robertson et al. 1992).

BREEDING PHENOLOGY

CODE		# OF RECORDS	DATE RANGE
C	Courtship	1	19 Jun
NB	Nest Building	17	20 May–1 Jul
ON	Occupied Nest	162	9 May–28 Jul
NE	Nest with Eggs	2	12 Jun–24 Jun
NY	Nest with Young	63	27 May–1 Aug
FY	Feeding Young	55	7 Jun–5 Aug
FL	Fledged Young	31	9 Jun–16 Aug

DISTRIBUTION

Tree Swallows breed across the continent from the northern tree line to Georgia, Arkansas, and south-central California, except in the Great Plains and Great Basin. They winter mainly along the Gulf and Caribbean coasts from Florida to Central America.

Atlasers found them mainly in the foothills and lower mountains. They added Confirmed breeding in two new latilongs (the Latilong Study had positive breeding in 21 of 28 latilongs). They Confirmed breeding by Tree Swallows in approximately twice as many blocks on the Western Slope as on the Eastern Slope and in greater abundance on the Western Slope. Approximately 25% of the Confirmed blocks received a population estimate of A4 (101–1,000 breeding pairs), including 30% of the blocks on the Western Slope but only 14% on the Eastern Slope.

Breeding along the plains rivers appears to have developed only recently; 30 years ago B&N could cite only one such event, at Cherry Creek Reservoir near Denver. Atlasers found them in a dozen blocks along the South Platte River from the foothills to Nebraska.

Tree Swallows have increased throughout the continent in recent years, although some local populations have declined because of competition for nest holes from starlings and House Sparrows (Turner and Rose 1989). In recent years, their breeding range has expanded into the southeastern states (Robertson et al. 1992).

BREEDING EVIDENCE

REPORTED IN 566 (32%) OF 1745 PRIORITY BLOCKS

☐	Possible	197 blocks	(35%)
▩	Probable	48 blocks	(8%)
■	Confirmed	321 blocks	(57%)

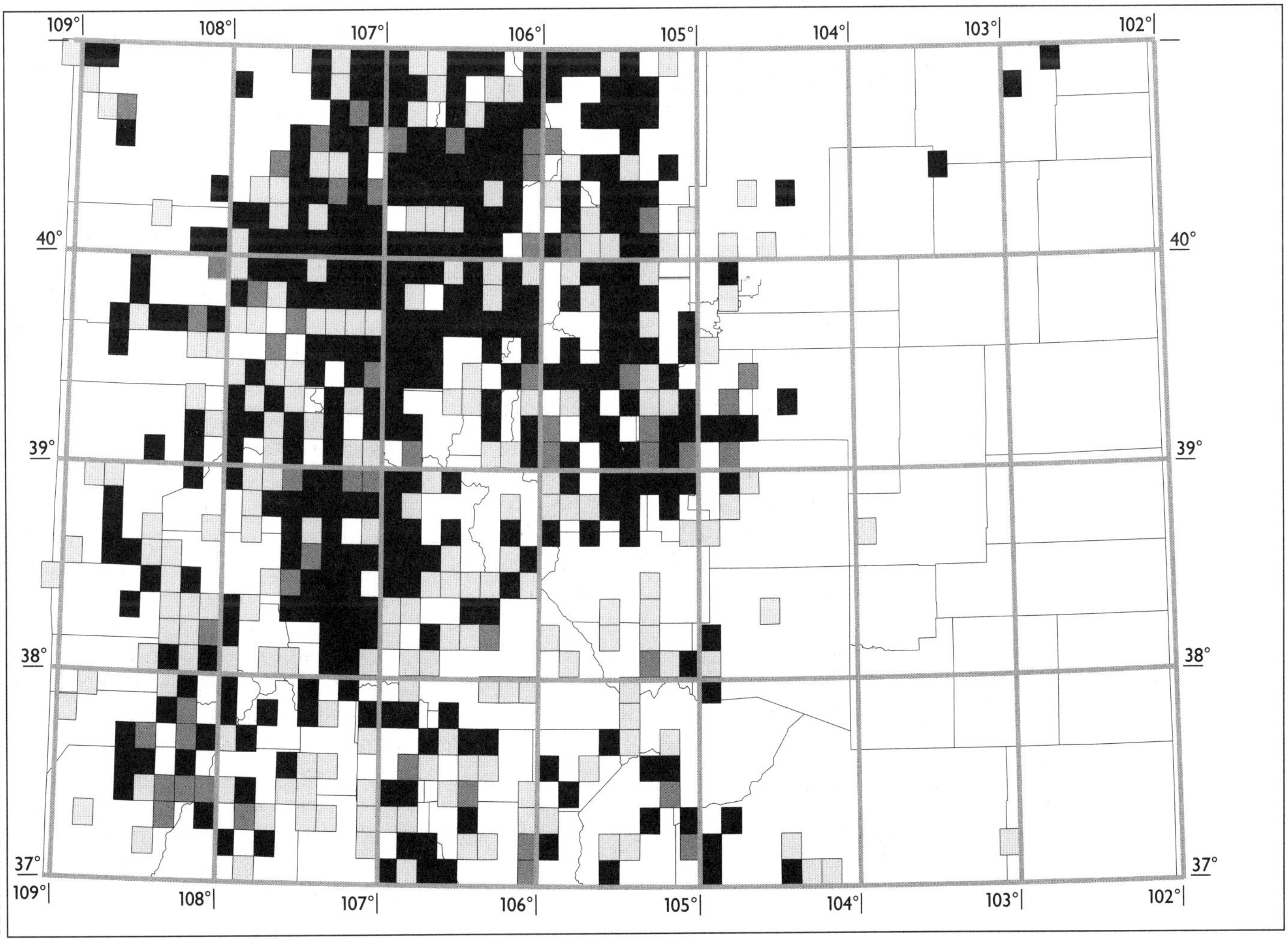
109°
108°
107°
106°
105°
104°
103°
102°
40°
39°
38°
37°

COLORADO DIVISION OF WILDLIFE / WILDLIFE RESOURCE INFORMATION SYS-EM

These handsome swallows face keen competition for nesting cavities and therefore select a variety of nesting sites. Atlasers found nests in such diverse locations as an opening in ski-lift machinery at Winter Park, a hole in the wall of a mountain residence, ventilation openings in a mountain condominium, and ledges of cliffs, as well as more mundane sites like holes in cottonwood and aspen trees.

HABITAT

Violet-green Swallows breed in a variety of habitats that offer cavities, from trees to cliffs to condos. They breed in open wooded areas in the foothills and mountains. B&N cited the center of abundance as the ponderosa pine belt, but Atlas habitat reports show three times as many nesting in aspens. By nesting in crevices in cliffs, situated in all kinds of habitats from aspen and ponderosa to pinyon/juniper and desert shrublands, Violet-greens have adapted to a source of nest cavities that other swallows do not use. Below the aspen woods, they breed mainly in deciduous stream bottoms, and also in pinyon/ juniper and ponderosa woodlands.

Cliffs, which atlasers reported as the second most common habitat, draw Violet-greens to some of Colorado's most spectacular scenery. They nest among the cliffs, canyons, and striking rock formations throughout the state. Sites along the Front Range include Red Rocks, Roxborough, Castlewood Canyon, and Garden of the Gods parks; farther west they inhabit the spectacular river canyons of the Gunnison in Black Canyon, the Yampa and Green in Dinosaur, and the Arkansas, Rio Grande, Colorado, and Dolores.

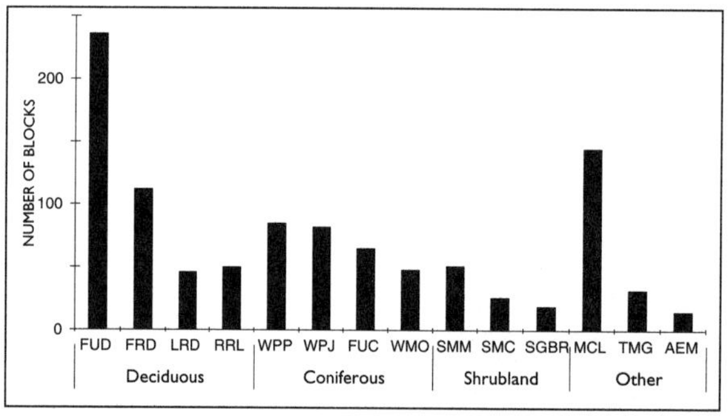

NUMBER OF BLOCKS
200
100
0
FUD FRD LRD RRL
Deciduous
WPP WPJ FUC WMO
Coniferous
SMM SMC SGBR
Shrubland
MCL TMG AEM
Other

They find nesting sites in conifer woodlands as well; atlasers reported them in five different conifer associations. Many of the shrubland and grassland habitats where atlasers found them include enough trees to provide nest sites, and others have bluebird trails where both Violet-green and Tree swallows find nest sites. Other blocks may have represented feeding strata rather than nesting sites.

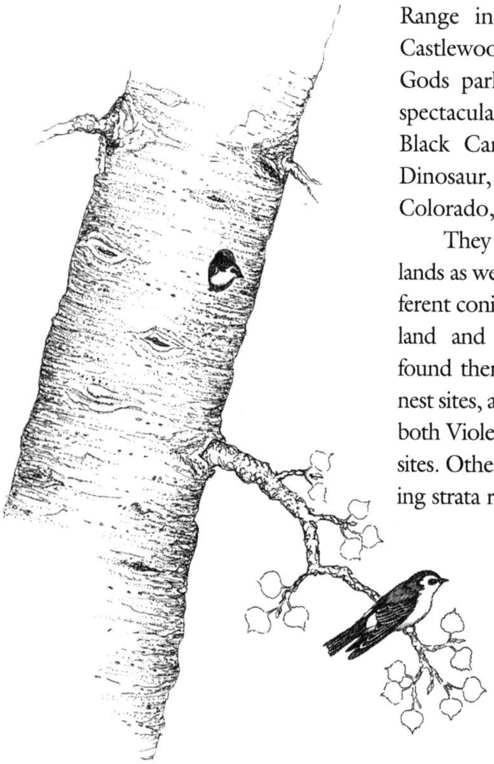

BREEDING

Biologists have very little information about the life history of these strictly western swallows. The recent AOU monograph cited 48 items about which the authors could find little or no information (Brown et al. 1992).

Violet-green Swallows arrive in Colorado in late April and leave by early September (A&R). Females commonly return to the same breeding site in successive years (Brown et al. 1992). Atlasers found occupied nests from late May until mid July. Egg-laying occurs from mid June through mid July (B&N). They nest both solitarily and in small colonies and most raise only one brood (Brown et al. 1992).

Atlasers submitted more than 20 nest cards. Most nests found by atlasers were in holes in aspens 4–50 feet (1–15 m) above the ground. One card lists a nest 300 feet (91 m) up on a cliff in Colorado National Monument (Ronda Woodward, nest card).

Violet-green Swallows often compete with other hole-nesting species (Brown et al. 1992). They frequently share apartment trees with other species, each occupying an old woodpecker hole. In a block east of Meeker, the swallows used a nest hole 40 feet (12 m) high in a dead aspen located in mixed woods near a small pond. Purple Martins nested in a hole 2 feet below, Red-naped Sapsuckers nested in a hole 4 feet above. At Winter Park they shared openings in ski-lift machinery with Mountain Bluebirds. Atlasers reported Violet-green Swallows sharing apartment trees with Mountain Bluebirds, White-breasted Nut-hatches, Tree Swallows, House Wrens, sapsuckers, and flickers. One atlaser reported Violet-green Swallows as one of five species nesting in the same aspen in Jackson County (Norman Barrett, nest card). They sometimes nest colonially; B&N cited an 1888 report of 20 pairs nesting in one ancient, dead pine.

Violet-green Swallows share many habits of their close relatives, Tree Swallows. One shared attribute seems to be the fixation with feathers. In Eagle, where Tree Swallows nest in backyard birdhouses, atlaser Jack Merchant threw a feather in the air. Before it could float to the ground a Violet-green Swallow grabbed it and fled, with Tree Swallows in pursuit; the Violet-green retained custody.

BREEDING PHENOLOGY

CODE		# OF RECORDS	DATE RANGE
C	Courtship	13	28 May–14 Jul
NB	Nest Building	17	24 May–6 Jul
ON	Occupied Nest	196	12 May–3 Aug
NY	Nest with Young	68	30 May–30 Jul
FY	Feeding Young	66	2 Jun–25 Aug
FL	Fledged Young	49	26 May–12 Aug

DISTRIBUTION

Violet-green Swallows breed in western North America, from Alaska to Mexico. They winter primarily in Mexico south through Central America. In Colorado, more breed in the foothills and lower mountains than in the higher mountains (A&R).

Atlasers easily located these conspicuous swallows. Recorded in 815 blocks, they rank 21st in the number of blocks; they occurred in 76% of the blocks west of the 105th Meridian. Over 25% of the Western Slope blocks with breeding Confirmed received abundance estimates of A4 (101–1,000 breeding pairs). These blocks are scattered throughout the Western Slope with no apparent center of abundance. Two widely separated Western Slope blocks received estimates of A5, more than 1,000 breeding pairs. Atlas abundance estimates suggest that Violet-greens are the second most common swallow breeding in Colorado, after Cliff Swallows.

In some parts of the West, House Sparrows and European Starlings nest earlier than Violet-green Swallows, and aggressively preclude these swallows from occupying tree cavities in residential areas at lower elevations (Bent 1942). However, only a few mountain towns in Colorado, such as Eagle, Winter Park, and Vail, harbor many House Sparrows. Cutting dead standing timber reduces available nesting cavities, but this may be offset by additional nesting sites in buildings and other structures provided by human activity.

BREEDING EVIDENCE

REPORTED IN 815 (47%) OF 1745 PRIORITY BLOCKS

☐	Possible	310 blocks	(38%)
▩	Probable	119 blocks	(15%)
■	Confirmed	386 blocks	(47%)

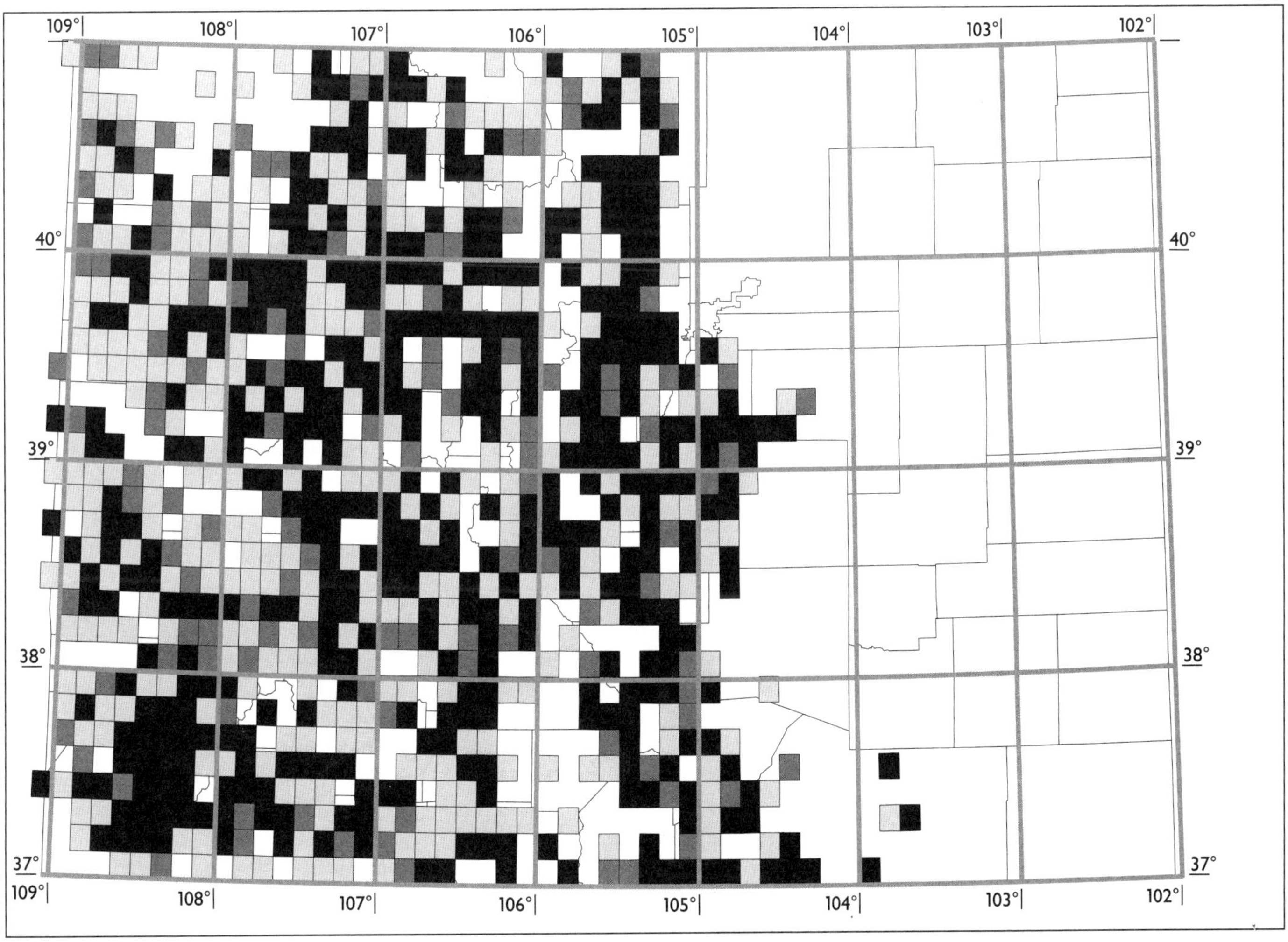
109°
108°
107°
106°
105°
104°
103°
102°
40°
39°
38°
37°

COLORADO DIVISION OF WILDLIFE / WILDLIFE RESOURCE INFORMATION SYSTEM

NORTHERN ROUGH-WINGED SWALLOW *Stelgidopteryx serripennis*

In 1819 John James Audubon, fatigued after an unsuccessful pursuit of too-wary ibises, noticed a flock of small swallows. He "gazed upon the Swallows, wishing that I could travel with as much ease and rapidity as they, and thus return to my family as readily as they could to their winter quarters." He recognized the birds he collected that day as different from the European Sand Martins (our Bank Swallows) but he did not get around to describing them for ornithology until 1838 (Bent 1942). Serrations on the outer edges of the primary feathers give the wings a rough edge and the swallow its English and Latin, Audubon-assigned, names.

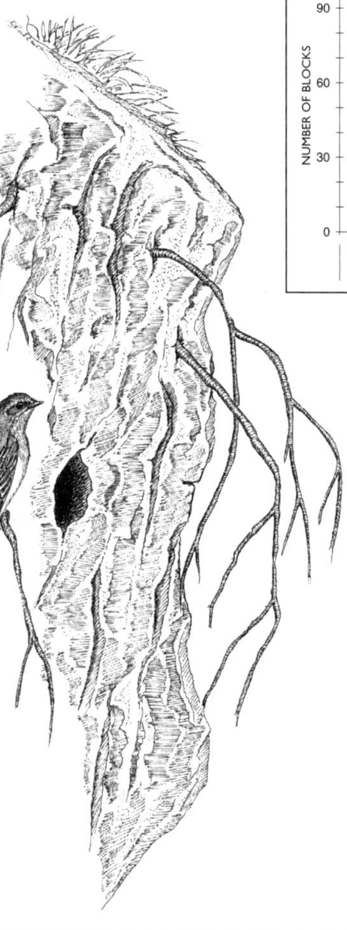

HABITAT

Rough-winged swallows prefer to nest in open areas with bodies of water that have banks or in roadcuts. Atlasers reported them from a wide variety of streamside, open habitats. Unlike other Colorado swallows, rough-winged swallows do not nest in colonies. Availability of suitable nesting sites limits their distribution. One atlaser described a location near Steamboat Springs as a marsh in a meadow with a 3-foot-wide stream cut about 5 ½ feet deep. The nest hole was 18 inches from the top of the bank and 4 feet above the stream (Kent Simon, nest card).

Authors disagree whether the swallows dig their own nest holes, but it appears that generally they rely on holes dug by others such as rodents, kingfishers, or Bank Swallows (Lunk 1962). They eat insects, caught primarily on the wing, with a languid flight (compared to other swallows). They feed close to the ground, usually along the nesting streams, and sometimes they even land on the ground to catch food. During the nesting season they frequently perch close to the nest burrow on branches of small trees along streams.

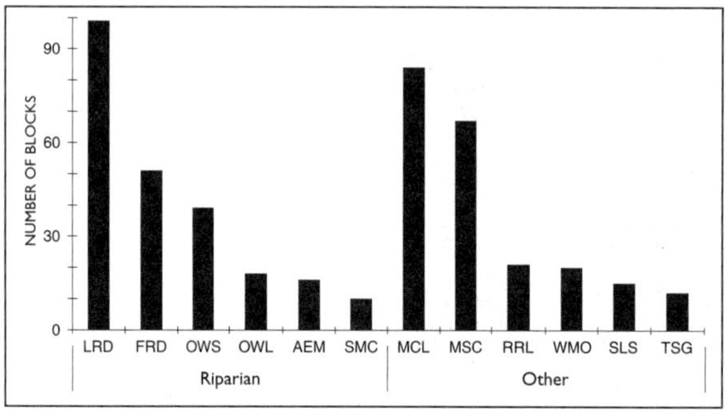
NUMBER OF BLOCKS
0
30
60
90
LRD
FRD
OWS
OWL
AEM
SMC
MCL
MSC
RRL
WMO
SLS
TSG
Riparian
Other

BREEDING

Rough-winged swallows typically nest in dirt banks cut by streams. They will use manmade imitations of stream banks—roadcuts and gravel pits. They nest not only in dirt banks but also in a variety of other situations. Examples include fissures in large rocks, crevices in masonry, crannies under bridges, drainpipes and other structures, and an occasional tree (Levine *in* Andrle and Carroll 1988, Shuford 1993). In Tennessee one pair nested in a steamboat that traveled 24 miles per day, each way, leaving at 10 A.M. and returning at 6 P.M. The adults followed the boat all the way to feed their young (Howell *in* Bent 1942). In Pennsylvania swallow enthusiasts have established for rough-wings "analogs" of bluebird trails by attaching short lengths of tubing to bridge abutments in order to provide nest sites (Schwalbe *in* Brauning 1992).

Atlasers usually found these quiet birds nesting in banks cut by prairie, foothills, and desert streams. Northern Rough-winged Swallows breed in solitary pairs, although several pairs may nest near one another in favorable locations (Bent 1942). Field workers reported only nine blocks with more than 100 pairs. Although they nest solitarily with respect to others of their own species, a pair will sometimes share a bank with Bank Swallows (Levine *in* Andrle and Carroll 1988).

Rough-winged swallows return to Colorado in late April and leave by the middle of September (A&R). Atlasers found occupied nests from mid May through mid July, both on the eastern plains and on the Western Slope. They may delay nesting to wait for favorable weather and food availability. These swallows produce only a single brood (Lunk 1962).

BREEDING PHENOLOGY

CODE		# OF RECORDS	DATE RANGE
C	Courtship	4	16 May–15 Jul
NB	Nest Building	15	14 Apr–18 Jun
ON	Occupied Nest	73	13 May–17 Jul
NE	Nest with Eggs	1	30 Jun
NY	Nest with Young	6	4 Jun–14 Jul
FY	Feeding Young	11	6 Jun–23 Jul
FL	Fledged Young	12	6 Jun–23 Jul

DISTRIBUTION

Northern Rough-winged Swallows breed throughout the United States and southern Canada, south through Central America. They winter in Mexico and Central America. The closely related Southern Rough-winged Swallows breed from Central America south to the middle of South America.

Early observers reported that Rough-wings breed throughout Colorado up to about 7,500 feet (2285 m), wherever the birds can find suitable nesting banks (Bergtold 1928). Later researchers considered them more common on the eastern plains than in western valleys and occasional in the mountain parks (A&R). The Latilong Study reported them as breeders or likely breeders in all 28 latilongs.

Atlasers found them breeding in about as many blocks on the Western Slope as on the eastern plains, with no significant difference in abundance between the two regions. Field workers found them up to 9,800 feet (3000 m) in the San Juan Mountains. The Atlas has added confirmed breeding in three latilongs in which the Latilong Study classified the species as only a likely breeder. Atlasers also found them numerous in three of the four big intermountain valleys (North and Middle parks, San Luis Valley). South Park, with smaller streams, probably lacks suitable dirt banks for nests that streams in the others provide; it has only a few Rough-wings.

The range of Northern Rough-winged Swallows expanded in the northeastern U.S. in recent decades, and their population increased in the eastern and central U.S. (Turner and Rose 1989). BBS trends appear stable. In Colorado, atlasers found them to be fairly common, but widely scattered in locally small numbers.

BREEDING EVIDENCE

REPORTED IN 374 (21%) OF 1745 PRIORITY BLOCKS

☐	Possible	196 blocks	(52%)
▩	Probable	55 blocks	(15%)
■	Confirmed	123 blocks	(33%)

109°
108°
107°
106°
105°
104°
103°
102°
40°
39°
38°
37°

COLORADO DIVISION OF WILDLIFE / WILDLIFE RESOURCE INFORMATION SYSTEM

Bank Swallows, smallest of the Colorado swallows, breed in colonies, in burrows that they dig into the faces of steep sandy banks. Their scientific name suits them well, as *riparia* refers to the bank of a stream, where these swallows find the most nest sites. They have one of the most extensive world ranges of any swallow. Across the Atlantic, the English call them "Sand Martins."

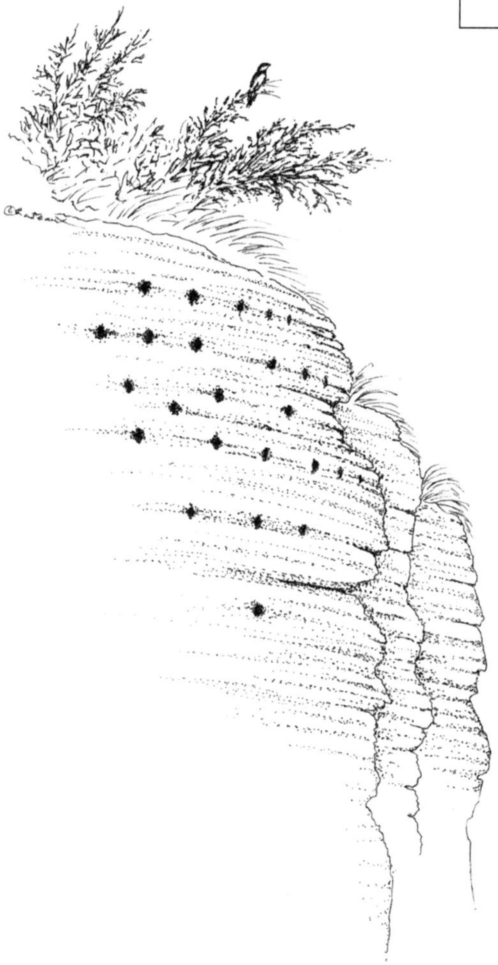

HABITAT

Vertical banks of soft sand, gravel, clay, or mud comprise the nest stratum for Bank Swallows. They have a spotty distribution due to the spottiness of suitable nesting sites (Armistead 1983). They usually nest near water, perhaps because nest sites occur around streams and lakes; but they also use manmade dirt cliffs in gravel pits and roadcuts (Burrill and Bonney *in* Andrle and Carroll 1988, Schwalbe *in* Brauning 1992).

The transient colonies come and go as the banks with suitable material become available and then lose their essential characteristics. Human activities and erosion affect the topography of their dirt cliffs and send these gregarious birds to alternate sites.

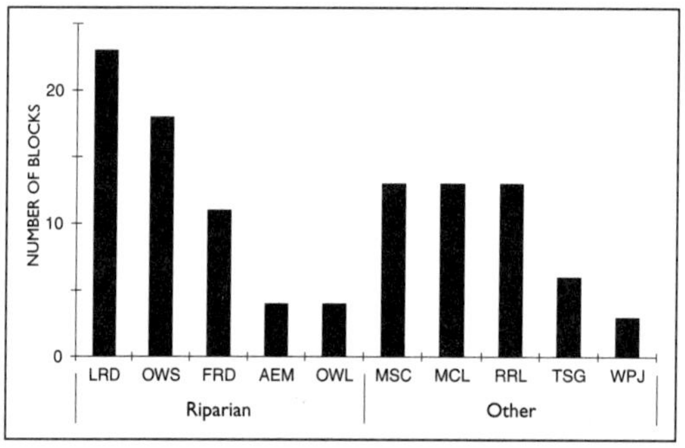
NUMBER OF BLOCKS
0
10
20
LRD
OWS
FRD
AEM
OWL
MSC
MCL
RRL
TSG
WPJ
Riparian
Other

Atlasers found them in a variety of open areas, most frequently near streams. Colonies breed in such locations as the Colorado River near Grand Junction, Fountain Creek north of Pueblo, and the South Platte near Fort Morgan. For habitats, atlasers sometimes specified general codes (riparian, rural), sometimes the specific nest site (cliffs), and sometimes feeding sites (lakes and streams).

BREEDING

These swallows arrive in Colorado in late April and remain until the middle of September (A&R). The colony excavates in burrows near the top of a high bank, with openings as close as 6 inches from one another (Petersen 1955). Colonies may shift from year to year depending on erosion or other damage to nest sites. During the Atlas period, a colony along the South Platte in Denver nested in a 50-foot pile of dirt in a brickyard; the next year the birds lost their nest site because the owners turned the dirt into bricks (Dick Pratt pers. comm.).

These industrious swallows, one of the few North American birds to do so, dig their own burrows. Although they usually do their own digging, they sometimes may refurbish an old burrow (Petersen 1955), or use one dug by a kingfisher (Nelson 1993). The burrows, flat on the bottom and arched over the top, end in an enlarged hollow where the females build their nests. The male starts the digging, and then attracts a female who helps complete it. She builds a nest of dry vegetation in a hollow at the end where she, like a Tree Swallow, lines it with feathers. The pair completes excavation by the middle of May and incubates in June (B&N).

Atlasers reported occupied nests from mid May to mid July, but observed adults feeding young only to the end of July. In this part of their range, Bank Swallows raise only one brood (Petersen 1955), although in other areas they may raise two broods (Turner and Rose 1989). Cowbirds parasitize them only by accident (Bent 1942).

BREEDING PHENOLOGY

CODE		# OF RECORDS	DATE RANGE
C	Courtship	3	20 May–16 Jun
NB	Nest Building	1	21 Jun
ON	Occupied Nest	29	10 May–10 Jul
NY	Nest with Young	1	3 Jul
FY	Feeding Young	7	24 May–30 Jul

DISTRIBUTION

Bank Swallows range nearly worldwide. They breed in North America, Europe, and Asia. North American breeders spend the winter in South America.

In Colorado Bank Swallows may have increased in abundance from early in the century. Sclater (1912) considered them either quite rare in Colorado or else overlooked by observers. A later writer termed them a rare summer resident, more common toward the eastern border, but breeding on the plains to the foothills (Bergtold 1928). More recent authors classified the Bank Swallow as a "fairly common" breeder on the eastern plains and in some western counties (B&N), or as a "common to abundant" breeder in western valleys, the San Luis Valley, and eastern plains, but perhaps more common in western Colorado than in the east (A&R). The Latilong Study reported breeding or probable breeding in 25 of 28 latilongs.

Atlas results do not bear out the more recent estimates of abundance. Atlasers reported breeding in only 86 blocks, fewer than 5% of the total Atlas blocks, and these are widely scattered throughout the state. They found Bank Swallows in approximately 50% more blocks on the Eastern Slope than on the Western. Abundance estimates per block did not differ materially between Eastern and Western slopes. Atlasers did Confirm breeding in seven latilongs that previously had only reported them as likely breeders.

Two facets of Bank Swallow biology may lead observers to overestimate their abundance. First, of course, the shifting nature of their nesting dirt banks causes field workers, from one year to the next, to add sites to the inventory without subtracting abandoned sites. Secondly, the swallows forage several miles from their nests, so that birders may record them where they do not actually nest. On the other hand, because of the spotty distribution of nest sties, Atlas block selection may have skipped over some colonies.

The BBS provides imprecise data, although the 1965–1979 trends indicated decreases in the eastern and central regions and an increase in the west (Turner and Rose 1989).

Growth of cities and other development destroyed some historic nesting sites in Colorado (B&N). Bank Swallows compete for nest sites with city and state highway departments and construction businesses (such as brickyards) that use sand and gravel for roads and buildings. Sometimes they benefit from the sand-and-gravel business, but available sites change from year to year. The human uses prevail, so Bank Swallows have to settle for transitory nest sites, most of them away from the centers of human population.

BREEDING EVIDENCE

REPORTED IN 86 (5%) OF 1745 PRIORITY BLOCKS

▢	Possible	35 blocks	(41%)
▩	Probable	8 blocks	(9%)
■	Confirmed	43 blocks	(50%)

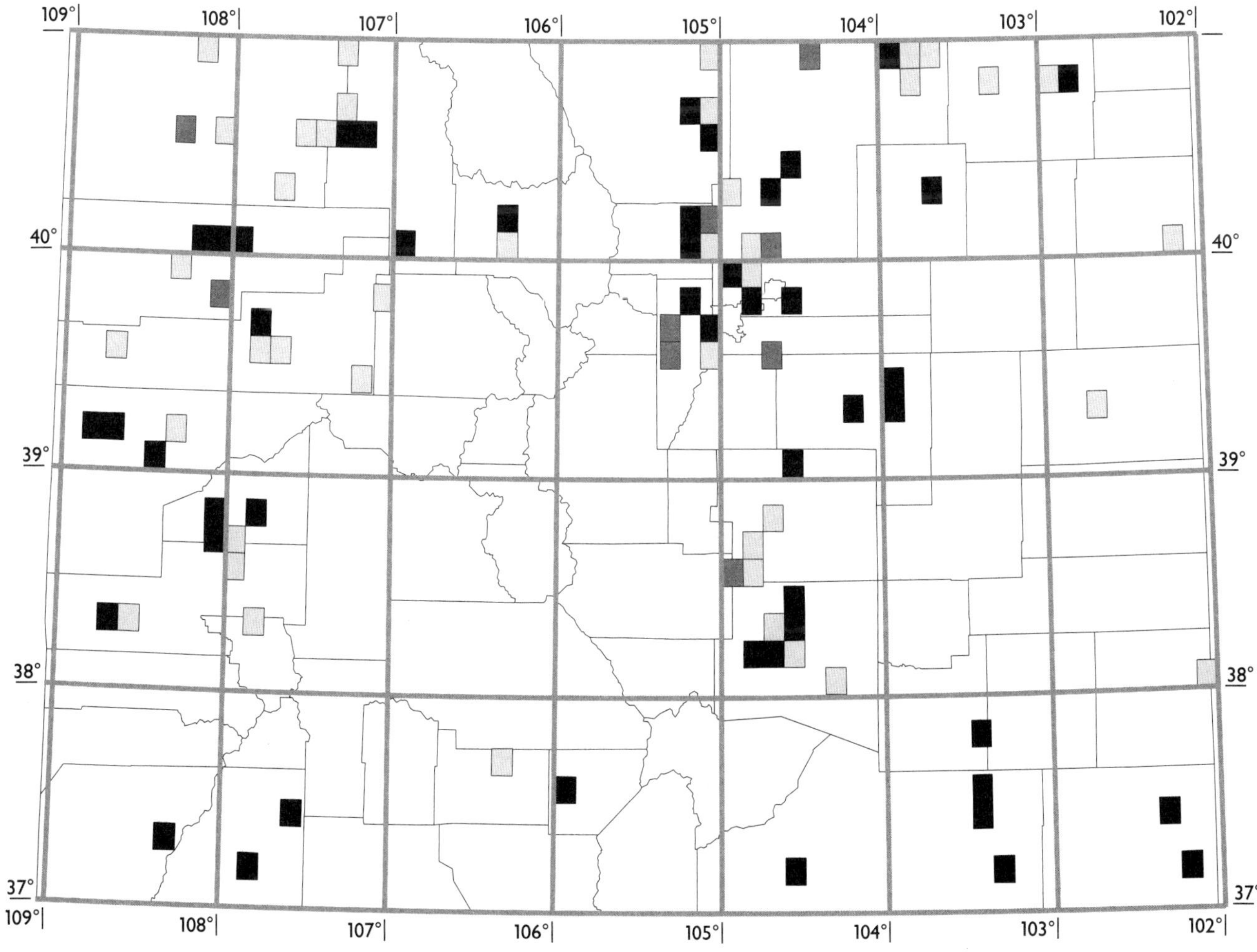
109°
108°
107°
106°
105°
104°
103°
102°
40°
39°
38°
37°

COLORADO DIVISION OF WILDLIFE / WILDLIFE RESOURCE INFORMATION SYSTEM

CLIFF SWALLOW *Petrochelidon pyrrhonota*

These handsome swallows breed in large colonies, their domed mud nests so close that they touch one another. They exhibit the highest degree of colonial breeding of any swallow in the world. As their name implies, Cliff Swallows historically nested on the sides of steep cliffs. Many colonies still use cliffs, but many others now use manmade structures, such as buildings and bridges. By adapting to these new sites, they have increased their population and expanded their range.

HABITAT

Originally Cliff Swallows inhabited open canyons, escarpments, and river valleys that offered a vertical cliff face and an overhang (Brown and Brown 1995). They require open foraging areas, a vertical substrate for nest attachment, and a supply of mud for nest construction (Emlen 1954). They may travel up to several kilometers for nest mud (Brown and Brown 1995). Colonial nesting, rare for North American passerines other than swallows, may have evolved because of the scarcity of nest sites before European settlement (Emlen 1952).

Colorado Cliff Swallows breed in open country, mostly at lower elevations, across the state in the western valleys, mountain parks, and eastern plains. They also range to the alpine zone: a colony in the Ophir Atlas block (37107G7) nests on glaciated headwater cliffs

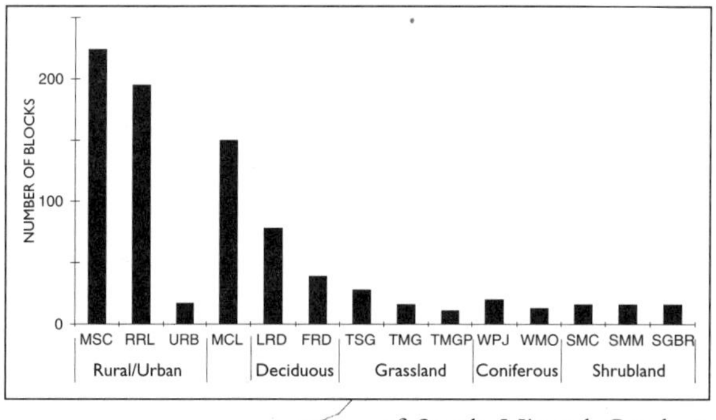

NUMBER OF BLOCKS
200
100
0
MSC
RRL
URB
MCL
LRD
FRD
TSG
TMG
TMGP
WPJ
WMO
SMC
SMM
SGBR
Rural/Urban
Deciduous
Grassland
Coniferous
Shrubland

of South Mineral Creek at 12,000 feet (3655 m)—the species' highest known breeding site. Colonies found by atlasers occupied a variety of nesting sites: cliffs 4–250 feet (1–76 m) above the ground, bridges, dams, buildings, and barns.

They place their gourd-shaped nests where heavy rains will least affect them, e.g., under overhanging cliffs or the eaves of buildings. As protection against predators, they ordinarily place their nests at least 8 feet above land or 3 feet above water (Emlen 1954). However, in a large culvert, one colony of 40 nests was only 5 feet above the bottom of the culvert, in a sagebrush area of northern Routt County (Norm Barrett, nest card).

Until a revision in 1990, Atlas habitat codes did not include cliffs, which observers included under MSC (miscellaneous); extrapolation suggests that at least one-third of the colonies nest on cliffs. Another large batch of MSC codes pertains to bridges (thus the inclusion of MSC under Rural/Urban in the Habitat graph). Many other habitat codes represent feeding areas or habitats surrounding the nests rather than nest sites, so that a higher proportion actually nests on cliffs.

BREEDING

Cliff Swallows arrive in mid April and remain until late September (A&R). Many adults return each year to breed in the same colony (Turner and Rose 1989). They reoccupy, and repair if necessary, old nests from prior years, or build new nests. Atlasers reported occupied nests from mid May through early August. In some parts of their range they may raise two broods (Bent 1942), although Brown and Brown (1995) state unequivocally that they produce only a single brood but lay replacement clutches when a nest fails in the early part of the breeding season. In Colorado they produce only one brood (B&N).

Barely territorial, they defend only the nest space (or, before building the nest, the wall space where they will build it). Their territorial defense does not succeed. Females lay eggs in the nests of other pairs, both adults actively engage in extra-pair copulations, and some even transfer their eggs, in their bills, to other nests. Up to 43% of the nests may contain at least one nestling unrelated to either parent (Brown and Brown 1995).

The diet depends on available food; the main objective seems to be swarming insects—concentrations of easily captured food. They seek thermals that transport insects into the air. Colonies within a few miles of each other may consume entirely different groups of insects (Brown and Brown 1995).

BREEDING PHENOLOGY

CODE		# OF RECORDS	DATE RANGE
C	Courtship	1	27 May
NB	Nest Building	47	11 May–16 Jul
ON	Occupied Nest	304	14 May–9 Aug
NE	Nest with Eggs	3	17 Jun–3 Jul
NY	Nest with Young	56	12 May–30 Jul
FY	Feeding Young	45	24 May–29 Aug
FL	Fledged Young	10	10 Jun–25 Jul

DISTRIBUTION

Cliff Swallows breed throughout most of North America and Mexico and winter in southeastern South America. Historically, they inhabited mainly the mountainous western part of North America where cliffs offered nesting sites (Brown and Brown 1995). Construction of human structures provided many more nesting sites; in the nineteenth century Cliff Swallows spread across much of the country (Turner and Rose 1989) and more recently into the Southeast (Grant and Quay 1977).

Atlasers Confirmed breeding throughout Colorado, in roughly the same number of blocks in the western as in the eastern half of the state. Cliff Swallow ranked 30th in number of reporting Atlas blocks and had the tenth highest Confirmation percentage. Approximately 60% of Confirmations were based on locating their durable old nests.

Large numbers of Cliff Swallows breed in Colorado, and their numbers have increased since settlement. Already common in the early 1900s, they bred from the plains to 10,000 feet (3048 m; Sclater 1912). House Sparrows usurp their nests and their activities caused a drop in Cliff Swallow numbers in the early twentieth century (Bergtold 1928). Populations rebounded as highway bridges, dams, houses, and other structures produced more nest sites (Turner and Rose 1989), and as House Sparrow numbers declined (Bent 1942).

Eight Atlas blocks, scattered around the state, had over 1,000 breeding pairs. Atlaser estimates suggest that Cliff Swallows are the most abundant Colorado swallow. BBS trends indicate a continent-wide increase of 1.4% per year.

Some property owners destroy swallow nests built on residences because they cannot tolerate the noise and unsightly excrement produced by masses of nesting swallows. In contrast, a colony thrives at a Winter Park ski lodge, where the manager waits to hose off the mud nests until after the nesting season. Destruction of nests after fledging may actually benefit the swallows by preventing injury to young from parasites in reused nests (Turner and Rose 1989). Many other species, such as Say's Phoebes, Canyon Wrens, Mountain Bluebirds, and rosy-finches, use the old nests for breeding or winter roosts (B&N, Atlas).

BREEDING EVIDENCE

REPORTED IN 715 (41%) OF 1745 PRIORITY BLOCKS

□	Possible	179 blocks	(25%)
▩	Probable	16 blocks	(2%)
■	Confirmed	520 blocks	(73%)

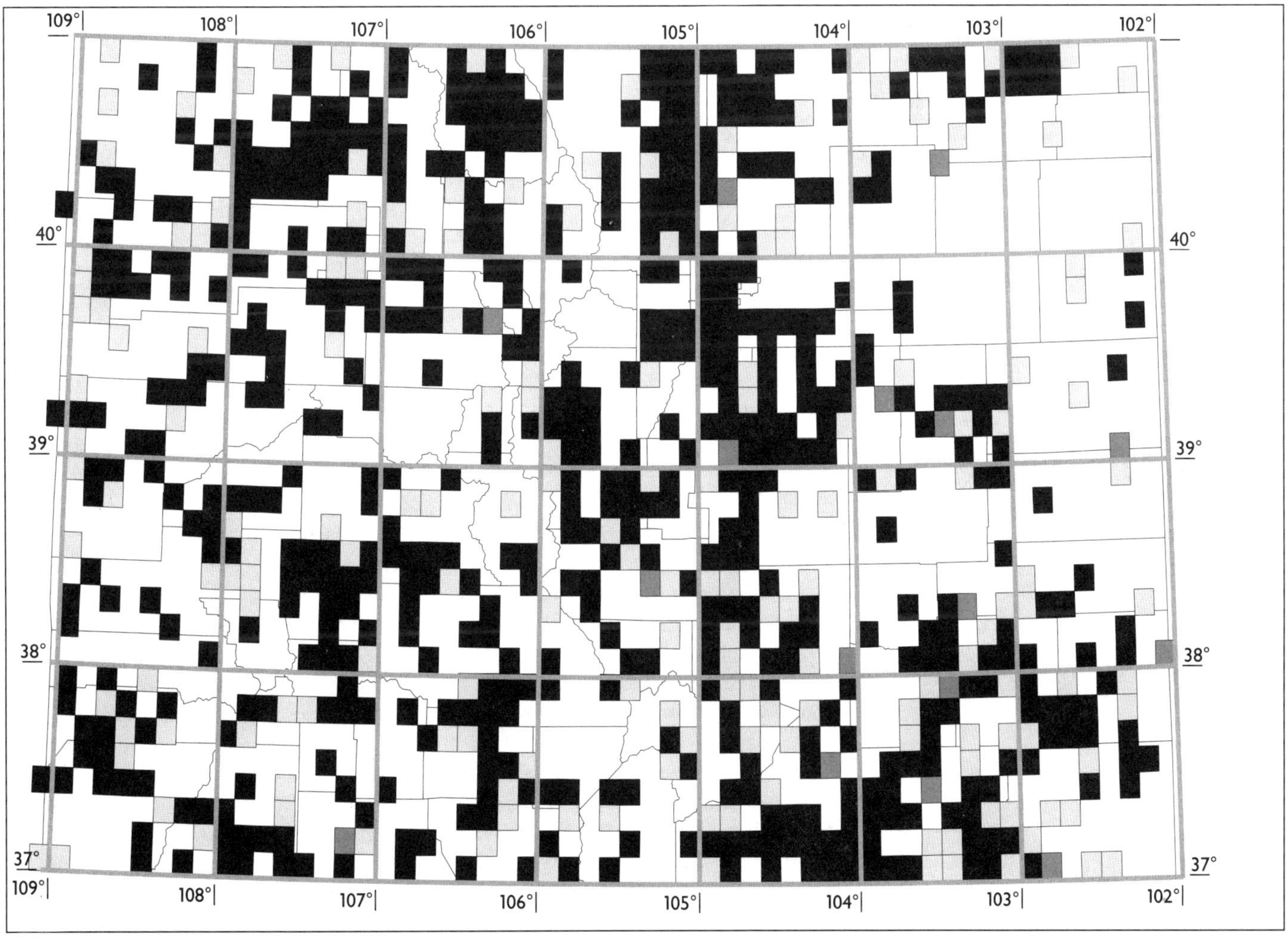
109°
108°
107°
106°
105°
104°
103°
102°
40°
39°
38°
37°

COLORADO DIVISION OF WILDLIFE / WILDLIFE RESOURCE INFORMATION SYSTEM

Perhaps better than any other North American species, Barn Swallows have adapted to European settlement. It seems likely that even before the Pilgrim and Jamestown settlements these voracious insect consumers had discovered the permanent Native American villages as sources of both food and shelter, and that Americans have long appreciated their capacious appetite for insects. They seem to flourish around people, and few place their nests in natural sites.

HABITAT

Barn Swallows prefer open country, such as meadows and fields for feeding, and now almost exclusively use nearby manmade structures for nest sites. Open areas, including bodies of water, provide the flying insects for which they have such capacious appetites. Prior to European settlement, they nested mainly or entirely in caves and on cliffs and so inhabited more local and restricted areas, mainly around coasts and uplands (Turner and Rose 1989). They now plaster their nests inside buildings, under bridges, on houses and barns, and in similar locations, and they have greatly expanded both their habitat and their numbers. In over half the blocks atlasers reported code RRL—referring to farmhouses, barns, and sheds.

Constrained by suitable natural nest sites, Barn Swallows have readily adapted to the opportunities that human settlement provides. They congregate around farms and farmhouses, and occasionally around suburban homes. Nesting pairs swoop around intersections in Denver and other cities, usually close to streams where bridges provide nest sites. The habitat code MSC signifies bridges (urban and rural), as well as other structures that the swallows find inviting.

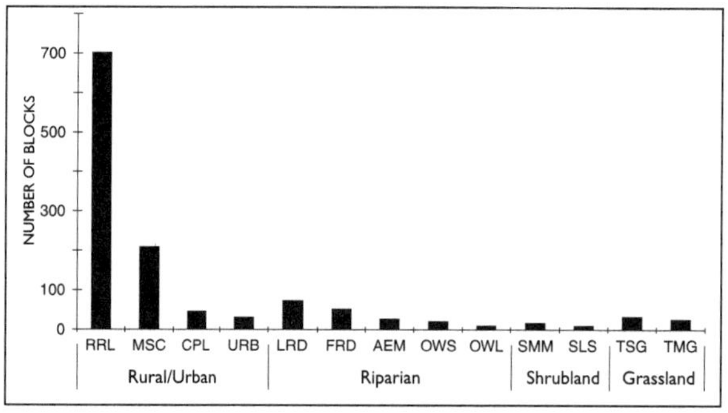
NUMBER OF BLOCKS
700
500
300
100
0
RRL
MSC
CPL
URB
LRD
FRD
AEM
OWS
OWL
SMM
SLS
TSG
TMG
Rural/Urban
Riparian
Shrubland
Grassland

BREEDING

These abundant summer residents arrive in mid April and a few remain until mid October (A&R). They nest mainly during June and July, but the breeding season extends over a very long period—from April to September. Atlas field workers observed nest-building as early as 22 April near Cortez, and nests with young as late as 12 September, also near Cortez. Many pairs raise two broods (B&N), the first brood in June, the second in mid July or early August (Sclater 1912). Cowbirds rarely deposit eggs in their nests (Bent 1942).

Adults return to breed at the same nest site used in previous years (B&N), and frequently reuse old nests. They breed in isolated pairs and in small colonies. These swallows build nests on ledges they find in a wide variety of locations. Atlasers found nests under the deck of a ski resort, inside abandoned ranch houses or homestead cabins, under the eaves of a country store, inside cattle sheds and barns, and under bridges and culverts. A pair even built a nest on a Forest Service signboard (HEK nest card). When they nest inside manmade structures they prefer those that have large openings such as barns and sheds (Snapp 1976).

BREEDING PHENOLOGY

CODE		# OF RECORDS	DATE RANGE
C	Courtship	3	26 May–22 Jun
NB	Nest Building	49	22 Apr–10 Jul
ON	Occupied Nest	376	3 May–1 Aug
NE	Nest with Eggs	46	17 May–17 Jul
NY	Nest with Young	110	12 May–12 Sep
FY	Feeding Young	59	5 May–22 Aug
FL	Fledged Young	34	6 Jun–30 Aug

DISTRIBUTION

The most widespread of all swallow species, Barn Swallows nest in North America, Europe, Africa, and Asia. Those that breed in North America migrate far to the south and winter from Panama through much of South America. In Colorado they breed in western valleys, in mountain parks up to 10,000 feet (3048 m), and on the eastern plains (A&R), in both rural and urban settings. They breed almost solidly across the eastern plains.

Barn Swallows rank ninth in number of reporting blocks. Atlasers Confirmed breeding in 79% of these blocks, the fifth highest Confirmation rate. Atlas workers easily located active nests, and they Confirmed breeding in 145 blocks by finding used nests.

Barn Swallows breed in arid areas that have abandoned buildings; one abandoned building in a sea of wheat fields was enough to attract a breeding pair (HEK). Atlasers reported only one or two used or occupied nests in blocks with few trees, such as in central North Park or in the San Luis Valley near the New Mexico line. They found many more swallows in blocks with more moisture and more manmade structures. Four widely scattered blocks had more than 1,000 breeding pairs (A5), and 57 blocks, concentrated in northeastern Colorado but a few scattered throughout the state, had more than 100 nesting pairs (A4).

An early Colorado naturalist reported fewer Barn Swallows than other swallows, although he still called them "far from rare" (Sclater 1912). Earlier in this century, their Colorado population declined because of displacement by House Sparrows (Bergtold 1928). Recently they have increased in all regions of the U.S. (Turner and Rose 1989) including Colorado; Atlas abundance estimates suggest that they have now become the state's third most abundant swallow.

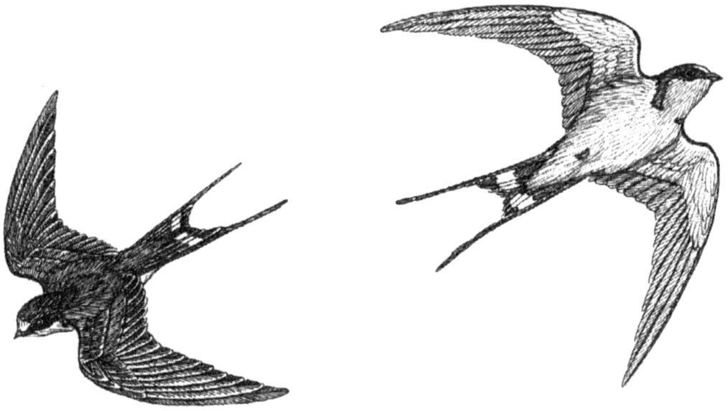

BREEDING EVIDENCE

REPORTED IN 1039 (60%) OF 1745 PRIORITY BLOCKS

☐	Possible	176 blocks	(17%)
▦	Probable	45 blocks	(4%)
■	Confirmed	818 blocks	(79%)

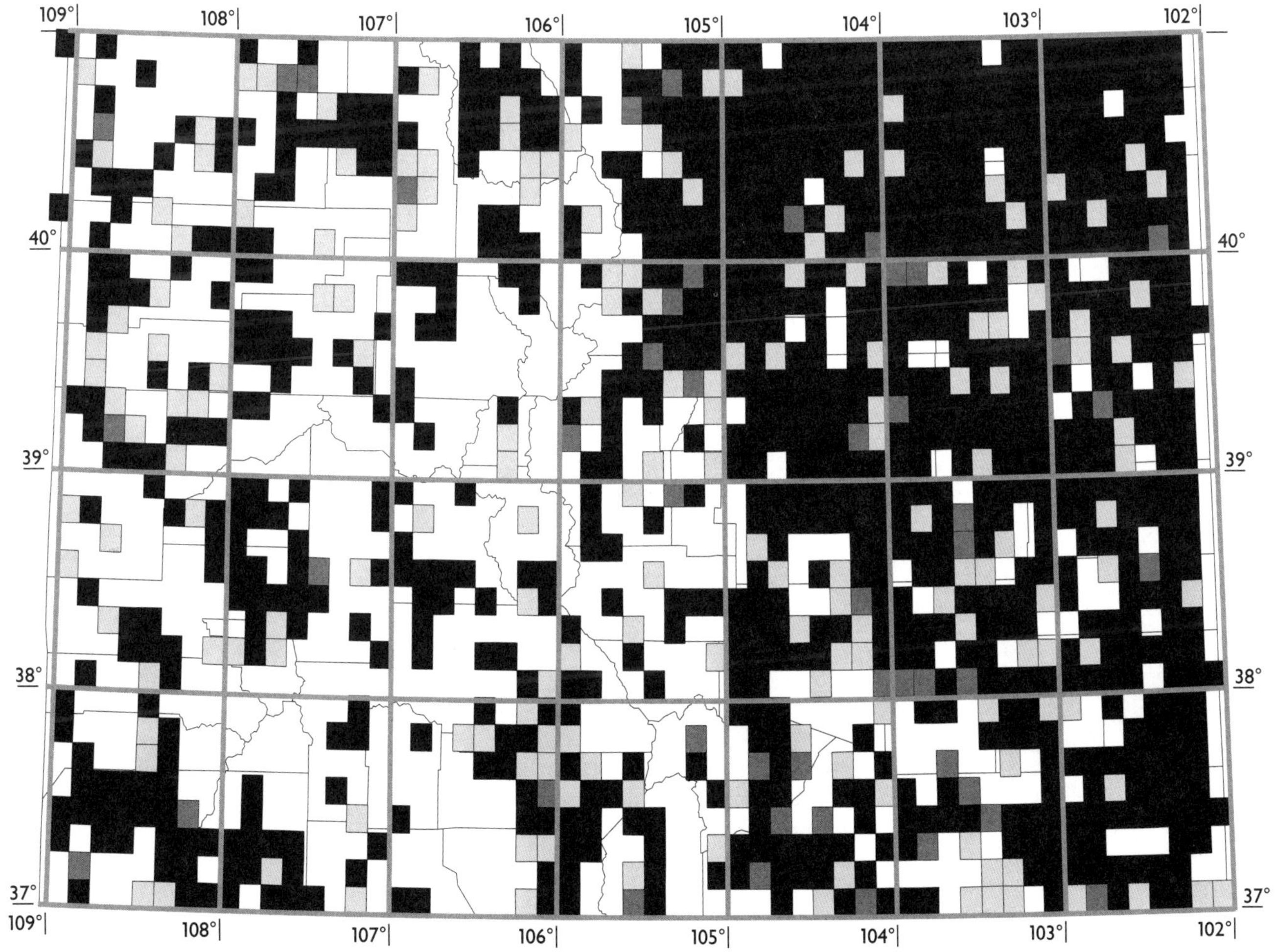
109°
108°
107°
106°
105°
104°
103°
102°
40°
39°
38°
37°

COLORADO DIVISION OF WILDLIFE / WILDLIFE RESOURCE INFORMATION SYSTEM

In the chickadee world, Colorado has only two parts: deciduous and coniferous. Black-capped Chickadees choose deciduous.

HABITAT

The nesting strategy of Black-capped Chickadees dictates their Colorado distribution: they excavate their own cavities (Odum 1941b). A suitable nest tree has a firm outer shell and rotten inner wood (Ryser 1985); old aspens and old cottonwoods suffice perfectly.

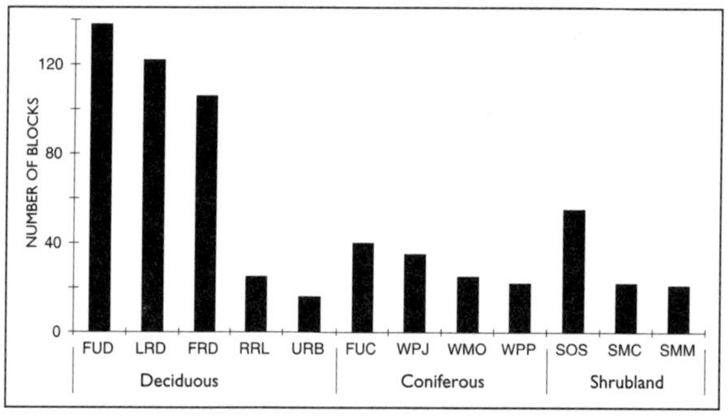
NUMBER OF BLOCKS
0
40
80
120
FUD
LRD
FRD
RRL
URB
FUC
WPJ
WMO
WPP
SOS
SMC
SMM
Deciduous
Coniferous
Shrubland

They achieve their greatest abundance at mountain elevations of 5,000 to 9,000 feet (1525–2745 m), especially in the northwestern quarter of Colorado. Atlasers estimated over 100 nesting pairs in 33 blocks, 27 of which form a triangle from Vail southwest to Aspen and north to Steamboat Springs, a definitive pattern in Latilongs 39106, 39107, and 40107. Large stands of aspens in these areas provide the key to this abundance; both there and throughout the mountains atlasers reported these chickadees in aspens and narrowleaf cottonwoods.

On both sides of the Continental Divide they occupy the second main habitat reported, lowland riparian—the broad-leaved cottonwoods of the piedmont streams and western river corridors. They also nest sparsely along a few prairie streams and in urban parks and residential areas. They nest more commonly near edges of wooded areas, but will also breed within them (Smith 1993).

These chickadees formerly reached their greatest abundance in ponderosa/aspen (B&N), but this has, apparently, changed, perhaps because of manmade alterations to habitat. Fire suppression, post-settlement logging, and subdivisions have caused ponderosa woodlands to grow more dense, with more and smaller trees. The Black-cap population now concentrates in aspens, although Atlas habitat reports show that some still use conifers.

BREEDING

Chickadees maintain both breeding and non-breeding territories. In winter flocks defend territories of 25 acres (10.1 ha; Odum 1941a). Flocks usually consist of a stable group of one or more pairs of adults, plus unrelated juveniles and single adults. The flock circulates through its territory and stops to feed at productive sites (Stokes 1979).

As spring approaches, the dominant male becomes less tolerant of other members of the flock. He and his mate gradually form a territory and start to defend it a few weeks before nest-building. During nest excavation, skirmishes with neighbors may last for 45 minutes. The defended breeding area—initially about 10 acres—gradually decreases in size during the nesting phase, and disappears entirely during the fledgling phase (Odum 1941a).

During nest excavation Black-capped Chickadees have the "marvelous habit" of carrying excavated wood chips away from the nest and dropping them from a nearby perch (Stokes 1979). Both members excavate the hole, and then the female lines the bottom of the cavity with fiber, moss, cocoons, feathers, wool, and hair. "Every furry denizen of the woods and some domestic animals may contribute hair" to the nests (Bent 1946). During incubation, when the female leaves the nest, she, like a duck, covers the eggs with the nest lining (Bent 1946).

Their nesting cycle lasts about 40 days from nest-building to fledging (Smith 1993). Atlas nest-building dates show that they start early, long before most migrant species arrive. Chickadees rarely have second broods (Smith 1993), so that the later starts (NB codes spanned 74 days) probably represent replacement broods.

The BBS analysis shows their continental population increasing at 1.4% per year, but too few Colorado routes record them to measure a trend here. Because of their more limited habitat opportunities in Colorado, one-sixth as many Black-caps as Mountain Chickadees breed in Colorado, according to Atlas abundance estimates.

BY **HUGH E. KINGERY**

BREEDING PHENOLOGY			
CODE		# OF RECORDS	DATE RANGE
C	Courtship	8	19 May–11 Jun
NB	Nest Building	12	20 Apr–2 Jul
ON	Occupied Nest	15	18 May–2 Aug
NY	Nest with Young	11	21 May–27 Jun
FY	Feeding Young	79	12 May–11 Aug
FL	Fledged Young	89	2 Jun–12 Aug

DISTRIBUTION

Black-capped Chickadees range coast to coast across the northern U.S. and southern Canada, north into the aspen and birch zone of central Alaska. In Colorado they breed most widely in the lower mountains and stream corridors where they find aspens or cottonwoods.

The Atlas map shows a somewhat different pattern from the range described in standard Colorado reference books (B&N, A&R), which show them occurring statewide except in the southeast. A&R identified them as most abundant on the plains river corridors and as uncommon to rare in aspen, pinyon/juniper woodland, and coniferous forests above 6,000 feet (1830 m).

Atlas results show a different pattern. Black-capped Chickadees breed on the Republican and South Platte drainages east to Nebraska. The random Atlas block selection disguises their continuity along the South Platte, but they breed sparingly, if at all, on the lower Arkansas below Manzanola, although a pair nested at Lamar in 1996 (Duane Nelson pers. comm.). They bypass the High Plains: rural shelterbelts and the High Plains towns do not harbor chickadees in their introduced hardwood Siberian elms and Russian-olives.

In the mountains their distribution corresponds to that of aspen woodlands. In the western valleys they use the cottonwood river bottoms. They spurn the conifer forests of the high country and the sagebrush and greasewood flats of the west and San Luis Valley.

Our predilection for cutting dead and dying trees and logging mature ones affects the populations of many cavity-nesting birds, chickadees among them. Black-capped Chickadees occasionally use birdhouses, but substituting nest boxes for natural holes will not have much impact on their populations if their aspen woodlands and foothills riparian lose older, decadent trees. Interference with natural succession does not benefit them: they need their old, deciduous, world.

BREEDING EVIDENCE			
REPORTED IN 510 (29%) OF 1745 PRIORITY BLOCKS			
☐	Possible	227 blocks	(44%)
▩	Probable	86 blocks	(17%)
■	Confirmed	197 blocks	(39%)

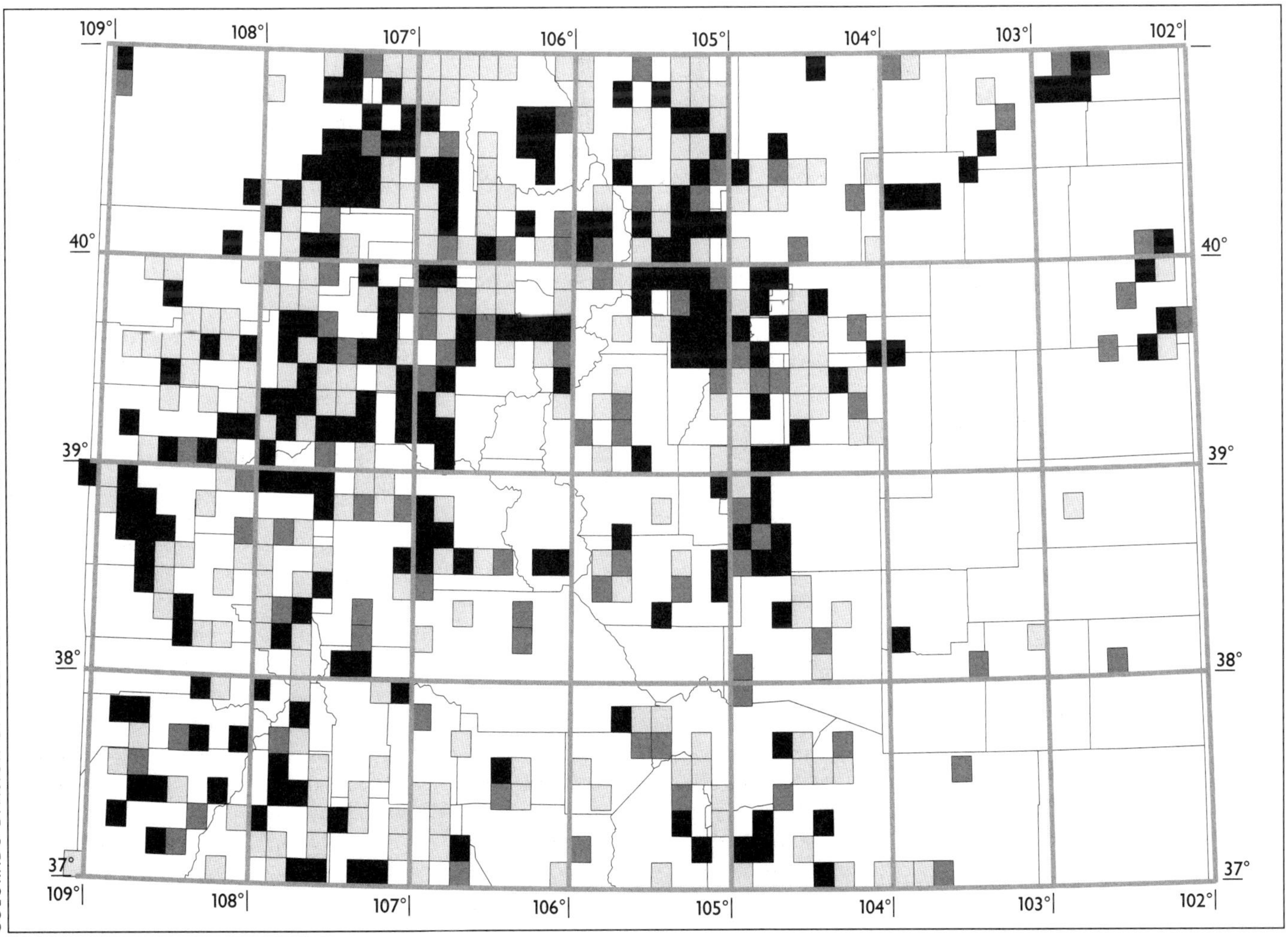
109°
108°
107°
106°
105°
104°
103°
102°
40°
39°
38°
37°

COLORADO DIVISION OF WILDLIFE / WILDLIFE RESOURCE INFORMATION SYSTEM

Populating exactly the territory their name connotes, Mountain Chickadees breed from 6,500 to 12,000 feet (1960–3600 m). As one of the commonest conifer birds, they add a lively component to deep forests. Also, like other chickadee species, their garrulous alertness leads observers to other less conspicuous forest inhabitants, such as, in Colorado conifers, Boreal, Northern Pygmy-, and Northern Saw-whet owls.

HABITAT

Mountain Chickadees epitomize mountain conifer forests; 78% of Atlas habitat records come from conifer forest types. Atlasers found them in more blocks than any other conifer-specialist species. They reach their highest densities in old-growth spruce/fir, lodgepole pine, and ponderosa pine forests, and breed less frequently in pinyon/juniper (Hallock 1990, Atlas, BBS). Although more *blocks* reported them in pinyon/juniper, more birds bred in ponderosas.

Despite their coniferous preference they also nest in aspen, where they coexist with Black-capped Chickadees. Up to about 9,000 feet (2745 m) atlasers found both in aspen woodlands. Aspens have lots of holes: sapsuckers, woodpeckers, and flickers like their soft wood, and Mountain Chickadees use the old holes.

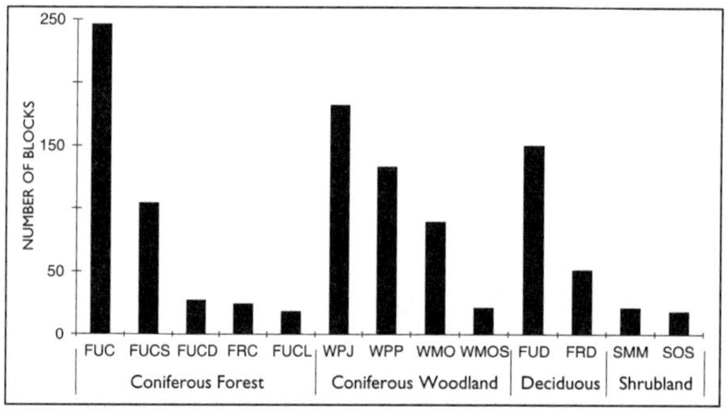
250
150
50
0
NUMBER OF BLOCKS
FUC
FUCS
FUCD
FRC
FUCL
WPJ
WPP
WMO
WMOS
FUD
FRD
SMM
SOS
Coniferous Forest
Coniferous Woodland
Deciduous
Shrubland

Mountain Chickadees find their insect prey high in the trees—usually 50 feet (15 m) or more above ground. They glean insects, larvae, and insect eggs from tree branch tips and dead twigs, typically circling around the tree as they flit from branch to branch (Ryser 1985).

BREEDING

Mountain Chickadees use old woodpecker holes or natural cavities; unlike Black-capped Chickadees, they do not carve their own holes. A few move into their cavity nests by mid May, but the greatest activity occurs in June and July. The two-and-one-half month continuum of on-nest Atlas dates (NE, NY, ON), adjusted for later nesting at higher altitudes and by allowing for replacement broods, supports the premise that chickadees have second broods only rarely (Smith 1993).

Calculations that combine Atlas habitat codes with abundance codes corroborate the strong preference for conifers. Mountain Chickadees rank as the seventh most abundant bird in the state, and second most abundant of the conifer specialists (after Ruby-crowned Kinglet). Of an estimated 1,200,000 breeding pairs, 80% breed in conifers and 20% in aspen woodlands. Of the 80% in conifers, 64% use spruce/fir; smaller proportions use Douglas-fir and lodgepole pine (8% and 4% respectively). Ponderosa pine and pinyon/ juniper each hold 12%. Although the state has less acreage in ponderosa woodlands than in pinyon/ juniper, atlasers estimated higher densities in ponderosas.

The relations and interactions between Mountain and Black-capped chickadees are mostly unstudied, although Mountains evicted Black-cappeds from a nest box in Jefferson County (Tina Jones pers. comm.). Mountain and Black-capped chickadees hybridize occasionally—in 1994 a mixed pair on the plains in Larimer County fledged six nestlings (Martin and Martin 1996). Mountain Chickadees sometimes winter on the plains, though they rarely stay to breed. The Larimer pair had formed a pair bond during the previous fall.

Mountain Chickadees stuff the nest cavity with bark, soft moss, and animal fur, then lay 7–12 white eggs. Incubation lasts 14 days. The female sits closely; instead of fleeing from a predator when threatened, she hisses loudly and flutters her wings. Even the young hiss—e.g., when someone reflects light into the nest. For the first four days after the young hatch, the male collects food that he feeds the young by regurgitation; the female tends the young at the nest during that time. Young fledge in 15–16 days (Bent 1946), a relatively short time for cavity-nesters.

The state has one record of cowbird predation—a nestling in a box with four chickadee nestlings, apparently the first such record for this species. It occurred in a year with high cowbird numbers and a delayed nesting season (Brockner 1984).

BREEDING PHENOLOGY

CODE		# OF RECORDS	DATE RANGE
C	Courtship	4	17 May–3 Jul
NB	Nest Building	8	2 May–24 Jun
ON	Occupied Nest	33	19 May–26 Jul
NE	Nest with Eggs	1	20 May
NY	Nest with Young	78	1 Jun–3 Aug
FY	Feeding Young	200	20 May–22 Aug
FL	Fledged Young	180	27 May–24 Aug

DISTRIBUTION

Mountain Chickadees nest in the western mountains from northern British Columbia to Arizona and New Mexico. In Colorado they breed in coniferous forests and avoid the lower valleys and open parks. The Atlas map tracks closely with the one in A&R.

The breeding distribution blackens in the Atlas map in western Colorado—the mountain and plateau country—wherever even a few conifers grow. Although absent from the treeless parts of North Park and the San Luis Valley, these chickadees nest in the woodlands scattered across Middle and South parks. Atlas abundance estimates show them most abundant in the San Juan Mountains; they also describe high numbers in the Front, Gore, and Park ranges.

Through the winter, adults apparently stay in pairs on the breeding territories they occupy during their first breeding season. Altitudinal migrants—both those that move up to timberline and those that drop down to the plains and valleys—are usually first-year birds (Dixon and Gilbert 1964).

Removal of old and rotten trees and logging old-growth forests directly impact this chickadee, just as these practices affect all hole-nesting birds. Tree removal also has indirect effects on these and other bird species by altering forest ecosystems and associated food chains.

BREEDING EVIDENCE

REPORTED IN 785 (45%) OF 1745 PRIORITY BLOCKS

Possible	182 blocks	(23%)
Probable	103 blocks	(13%)
Confirmed	500 blocks	(64%)

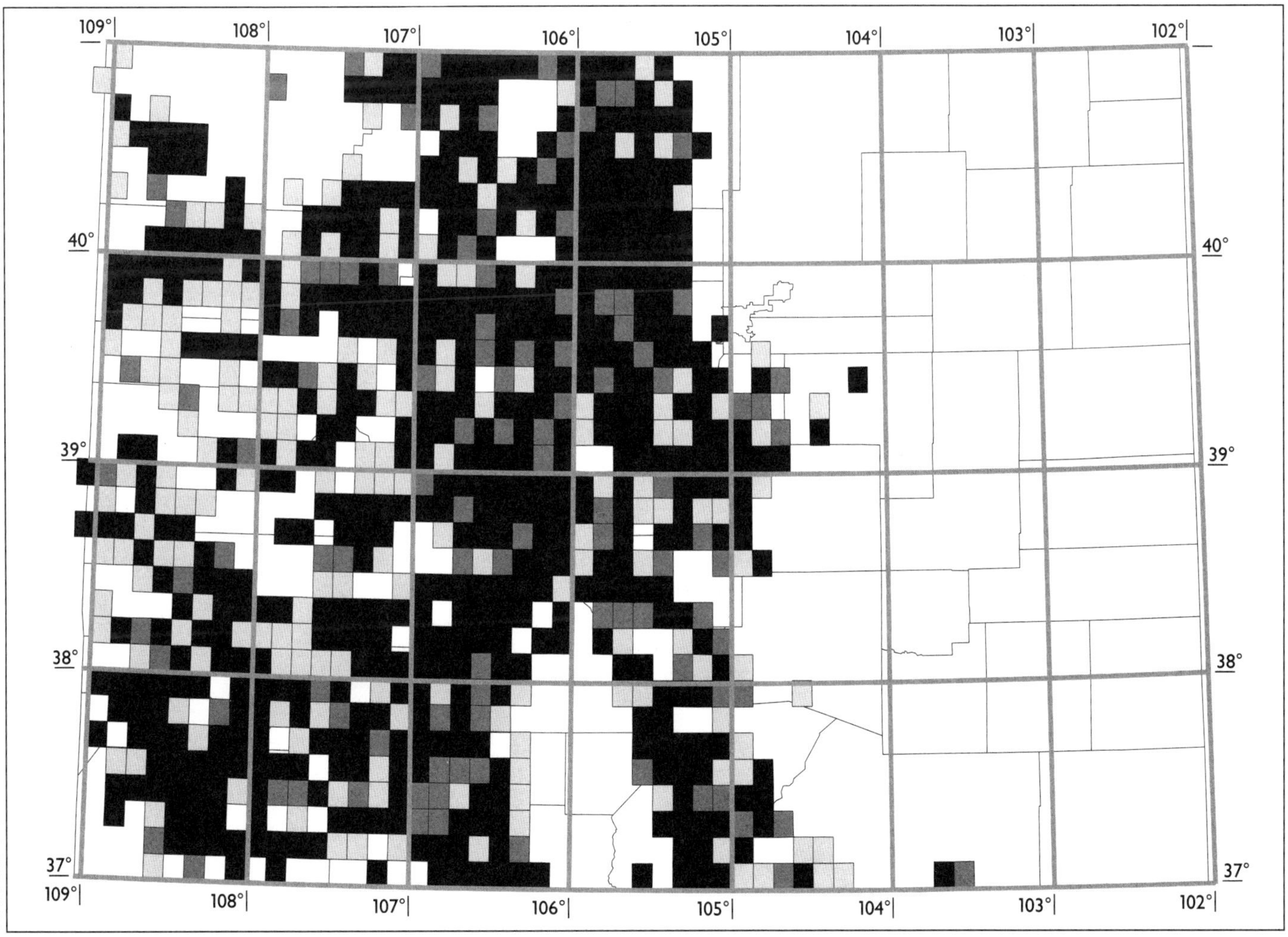
109°
108°
107°
106°
105°
104°
103°
102°
40°
39°
38°
37°

COLORADO DIVISION OF WILDLIFE / WILDLIFE RESOURCE INFORMATION SYSTEM

The gray wardrobe of the Juniper Titmouse and its careful adherence to dense cover seem appropriate for a secretive, reclusive hermit that observers might overlook. Titmice are curious hermits, however, and their compulsion to investigate any activity in their territories made these birds relatively easy for squeaking-and-pishing atlasers to census.

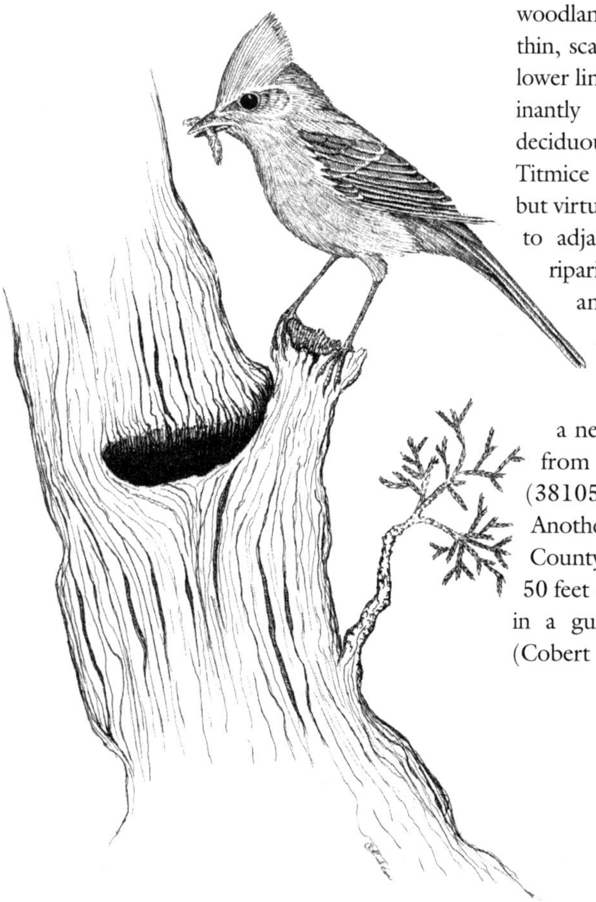

HABITAT

Atlas workers found only a few species as closely tied to a single habitat as the Juniper Titmouse. They reported pinyon/juniper in over 96% of the blocks that harbored this species. The pattern corroborates earlier reports that the titmice found in the Great Basin and Rocky Mountain foothills live almost exclusively in pinyon/juniper woodlands (Gabrielson and Jewett 1940, A&R). Juniper Titmice share a preference for dense canopy cover with their congeners, Oak Titmice (Ryser 1985). (These two species comprised a single species, the Plain Titmouse, until after completion of Atlas field work [AOU 1997].) Pinyon/juniper woodlands vary greatly in composition, from thin, scattered stands of pure juniper at the lower limits to very dense stands of predominantly pinyon pine liberally mixed with deciduous shrubs at the upper limits. Juniper Titmice seem at home in all these varieties, but virtually nowhere else. The birds wander to adjacent habitats—various shrubs and riparian cottonwoods—to gather food, and these excursions account for nearly all of the other habitat reports.

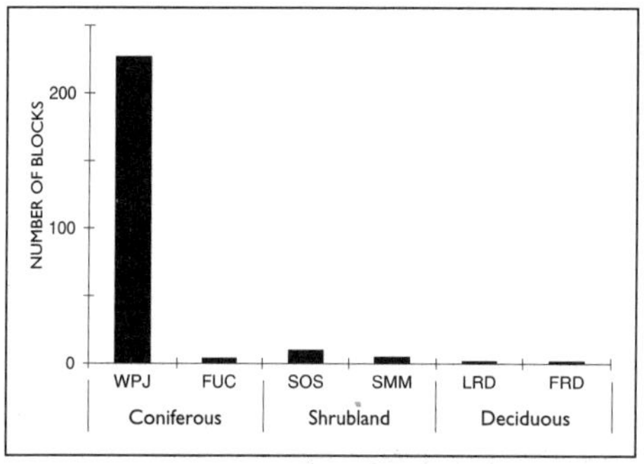
NUMBER OF BLOCKS
200
100
0
WPJ
FUC
SOS
SMM
LRD
FRD
Coniferous
Shrubland
Deciduous

One clear exception to Colorado's pinyon/juniper pattern, a nest with young in a scrub oak, came from the Phantom Canyon block (38105E1) northeast of Cañon City. Another exception occurred in Las Animas County where adults fed young in a cavity 50 feet (15 m) up in a cottonwood growing in a gully below pinyon/juniper hillsides (Cobert Mesa North, 37103A5).

BREEDING

Juniper Titmice nest in knotholes and other natural cavities (Bent 1946), which occur abundantly in junipers. They begin nesting at least by early May, before most field work began; atlasers reported a single instance of nest-building, on 5 May. The females do all the incubating and seldom flush from the nest (Bent 1946). One female sat so tightly that she was discovered in a nest box when the landlady took it down for cleaning (Phyllis Radice pers. comm.) After an incubation period of 14–16 days, newly hatched young remain on the nest for another 16–21 days (Ehrlich et al. 1988). This combination of early nest-building, well-concealed nest sites, and tight-sitting females led to atlasers finding few nests and none until June, when nestlings became vocal, field workers became active, and Confirmations became abundant. Atlas workers observed adults carrying food in 29 blocks, 24 of these before 1 July. Aggressive and noisy begging for food made the new fledglings quite conspicuous in July, and atlasers Confirmed breeding of Juniper Titmice in 56 blocks by observing these little beggars. The uniform pattern of dates suggests that this species, like other Parids, rears single broods and that second clutches rarely, if ever, occur.

Atlas workers reported an abundance code of A3 (11–100 pairs) in more than 60% of the 200 blocks with abundance codes recorded. The aridity of pinyon/juniper woodlands precludes high populations, and atlasers recorded only 24 codes of A4 (and no A5s)—still fairly dense for an arid habitat.

Colorado's densest populations, according to the distribution of A4s, occur in the southwest part of the state. Atlasers reported A4 in 25% of the Western Slope blocks south of 39 degrees, compared to 8% in the southeast and only 4% in the northwest.

BREEDING PHENOLOGY

CODE		# OF RECORDS	DATE RANGE
NB	Nest Building	1	8 May
ON	Occupied Nest	1	8 Jun
NY	Nest with Young	6	27 May–24 Jun
FY	Feeding Young	41	11 May–31 Jul
FL	Fledged Young	56	3 May–10 Aug

DISTRIBUTION

The range of the Juniper Titmouse extends from southern Oregon, Idaho, and Wyoming south to the Mexican border.

Cooke reported it in the pinyons of southern Colorado north to Mesa, Fremont, and El Paso counties, as well as in the Escalante Hills of Routt County (Cooke 1897, 1909, Sclater 1912, B&N). A&R refined those accounts, with a detailed map following the foothills and mesas east to southwestern Baca County.

Atlas data bear out these descriptions precisely. The map generated by the Atlas data generally coincides with that drawn by A&R. A&R listed the species as rare on the periphery of the San Luis Valley, and Atlas workers reported none from that area. The Latilong Study shows this species as a migrant in the San Luis latilong and absent from the Monte Vista latilong; one long-time resident of the valley has never seen any there (John Rawinski pers. comm).

Colorado supports a significant proportion (10–19%, Priorities 1995) of the Juniper Titmouse population. That the species depends almost exclusively on a single habitat, pinyon/juniper, would seem to make it vulnerable, but this habitat has probably expanded since settlement (West 1984). Thus, this bird's population in Colorado has probably changed little in historical times, and it perhaps has even grown somewhat.

BREEDING EVIDENCE

REPORTED IN 233 (13%) OF 1745 PRIORITY BLOCKS

Possible	94 blocks	(40%)
Probable	33 blocks	(14%)
Confirmed	106 blocks	(45%)

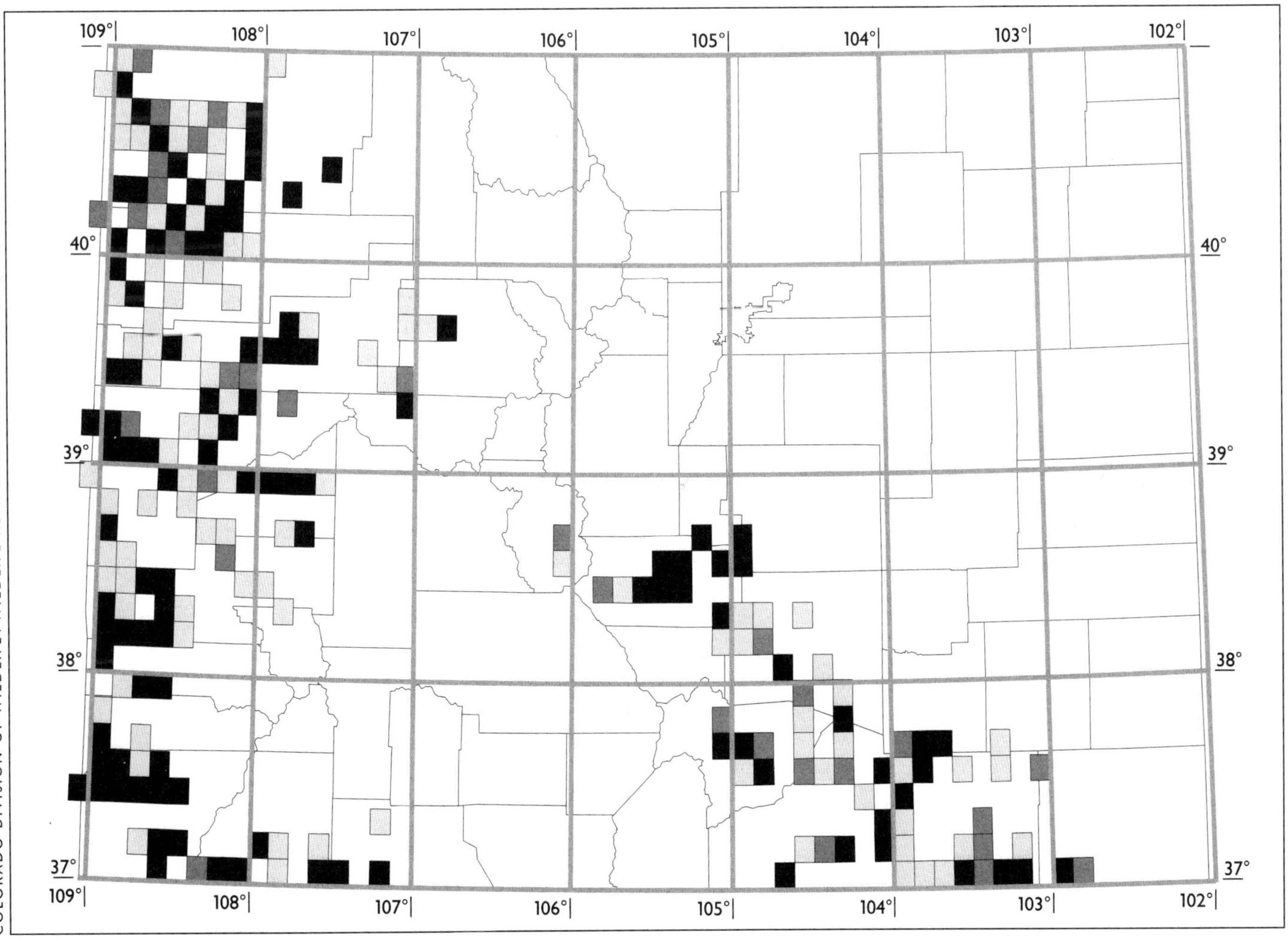
109°
108°
107°
106°
105°
104°
103°
102°
40°
39°
38°
37°

COLORADO DIVISION OF WILDLIFE / WILDLIFE RESOURCE INFORMATION SYSTEM

"Bushtits flock in small bands, flitting nervously through trees and bushes, hanging, prying, picking, and gleaning, and keeping contact through a constant banter of soft chirps. They pervade a small area, then vanish, and reappear a couple of hundred yards away" (Udvardy 1977). "Flocks sound like animated, squeaky toys, uttering a variety of sibilant squeaks and excited twitters" (Farrand 1988). They manifest their nervous energy in an extended nest-building phase, which can last almost two months, to produce a huge pendent nest that can accommodate an entire Bushtit family.

HABITAT

Bushtits breed primarily below 8,000 feet (2500 m), in pinyon/juniper woodlands and, to a lesser degree, in shrublands, both upland and riparian (A&R). Coastal subspecies breed in a variety of habitats such as live oaks and chaparral—most of them evergreen; the Colorado subspecies (*P. m. plumbeus*) finds this requirement for year-round foliage in pinyon/juniper woodlands.

In 85% of the blocks where they Confirmed breeding Bushtits, atlasers found them in pinyon/juniper woodlands. Field workers reported many other habitats, often those adjacent to pinyon/juniper, such as riparian deciduous trees, mountain shrub with mountain mahogany, and tall sagebrush. They may use other habitats during post-nest wandering (Terres 1980).

Year-round residents, Bushtits glean insects: aphids, treehoppers, leafhoppers, scale insects, beetles, wasps, ants, caterpillars, and pupae of codling moths. Highly gregarious, family units join together by July to create flocks of 10–75 Bushtits. Bushtits often associate with other species: kinglets, wrens, titmice, warblers, and chickadees (Ehrlich et al. 1988). Movement by the group may dictate the habitats in use rather than choice by the Bushtits.

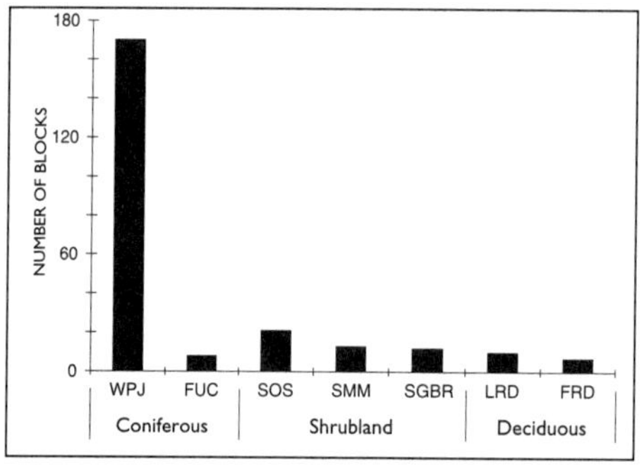

NUMBER OF BLOCKS
180
120
60
0
WPJ
FUC
SOS
SMM
SGBR
LRD
FRD
Coniferous
Shrubland
Deciduous

BREEDING

Bushtits pair off and begin nest-building as early as late March. They create a huge, gourd-shaped, hanging pocket, woven around and supported by twigs, moss, lichen, leaves, cocoons, grass, and flowers. They reinforce it with spiderwebs and line it with plant down, hair, and feathers. Construction of this palace can take 13–51 days (Bent 1946, Ehrlich et al. 1988). The big gap in time between commencement of nest-building and the first sign of young happens because of the large amount of time it takes to build the nest.

Both parents and, with some frequency, other adults may roost together in the nest. Early spring nesting in cold weather by such a small bird may require drastic measures to conserve energy: they eat prodigious amounts and, particularly on cold nights, roost with several birds huddled close together (Ryser 1985). Bushtits sometimes leave a nest during construction only to come back in a few days to finish it. They place their nests 5–20 feet (1.5–6 m) high in shrubs or trees (Bent 1946). They hide the nests well: atlasers found only ten.

Nests usually contain five to eight eggs and both birds incubate, although only the female has a brood patch (B&N). Young hatch in 12–13 days and remain in the nest 12–14 days. The slightest disturbance will cause the entire brood, when nearly full grown, to "pop" with a flurry of wings and to disperse into the nearby vegetation (B&N). If something disturbs the nest during building, egg-laying, or incubation, the birds often desert, even change mates, and nest again (Ehrlich et al. 1988).

In the Chiricahua Mountains of Arizona breeding pairs of Bushtits, in all stages of the nesting process, accept the help of other Bushtits. The helpers can be unmated males, juveniles, or birds from failed nesting attempts. Those that arrive during the nest-building or egg-laying phase may have unsuccessfully courted the resident female (or she may have elected to have two mates). Those helpers that arrive later assist with feeding young. Bushtits normally lay a full clutch of six eggs; nests with male helpers never had more than six but nests with more eggs had female helpers (Sloane 1996).

Once the young left the nest, field workers found Bushtits easy to confirm. Fledged young make lots of sounds as they beg and receive food from parents. Over half of the Confirmations involved young out of their nests.

BREEDING PHENOLOGY

CODE		# OF RECORDS	DATE RANGE
C	Courtship	1	29 May
NB	Nest Building	6	26 Mar–20 Jun
ON	Occupied Nest	2	1 Apr–30 May
NY	Nest with Young	6	11 May–20 Jun
FY	Feeding Young	32	23 May–31 Jul
FL	Fledged Young	67	19 May–6 Aug

DISTRIBUTION

Bushtits find suitable places to live from the southern tip of British Columbia, southern Idaho, and southwestern Wyoming south to Guatemala. Most breeding populations exist southwest of the diagonal line from the northwestern to southeastern corners of the state, as do pinyon/juniper woodlands. Colorado hosts less than 10% of the total population of this species (Priorities 1995), but a substantial proportion of the Great Basin subspecies nests in the state.

Early Colorado naturalists knew that Bushtits bred in western Colorado, but because they did little field work in the western pinyon/juniper habitats they regarded the species as uncommon and scattered (Cooke 1897). By 1965 B&N knew that they occurred commonly in pinyon/juniper north as far as Routt County (where atlasers did not find any). The Atlas map resembles that of A&R, except that atlasers did not find any nesting along the Front Range north of Colorado Springs.

Atlas records show that Bushtits concentrate in the western third of Colorado, where 65% of the blocks reporting them are located. The southeastern section—essentially the pinyon woodlands on tributaries to the Arkansas River, from Baca County to Buena Vista—hosts a quarter of the blocks. Most of the pinyon/juniper woodland that fringes the San Luis Valley has Bushtits, even though for the most part the valley lacks some species typical of that habitat, such as Ash-throated Flycatcher, Gray Flycatcher, Juniper Titmouse, Bewick's Wren, and Black-throated Gray Warbler.

BREEDING EVIDENCE

REPORTED IN 202 (12%) OF 1745 PRIORITY BLOCKS

Possible	68 blocks	(34%)
Probable	24 blocks	(12%)
Confirmed	110 blocks	(54%)

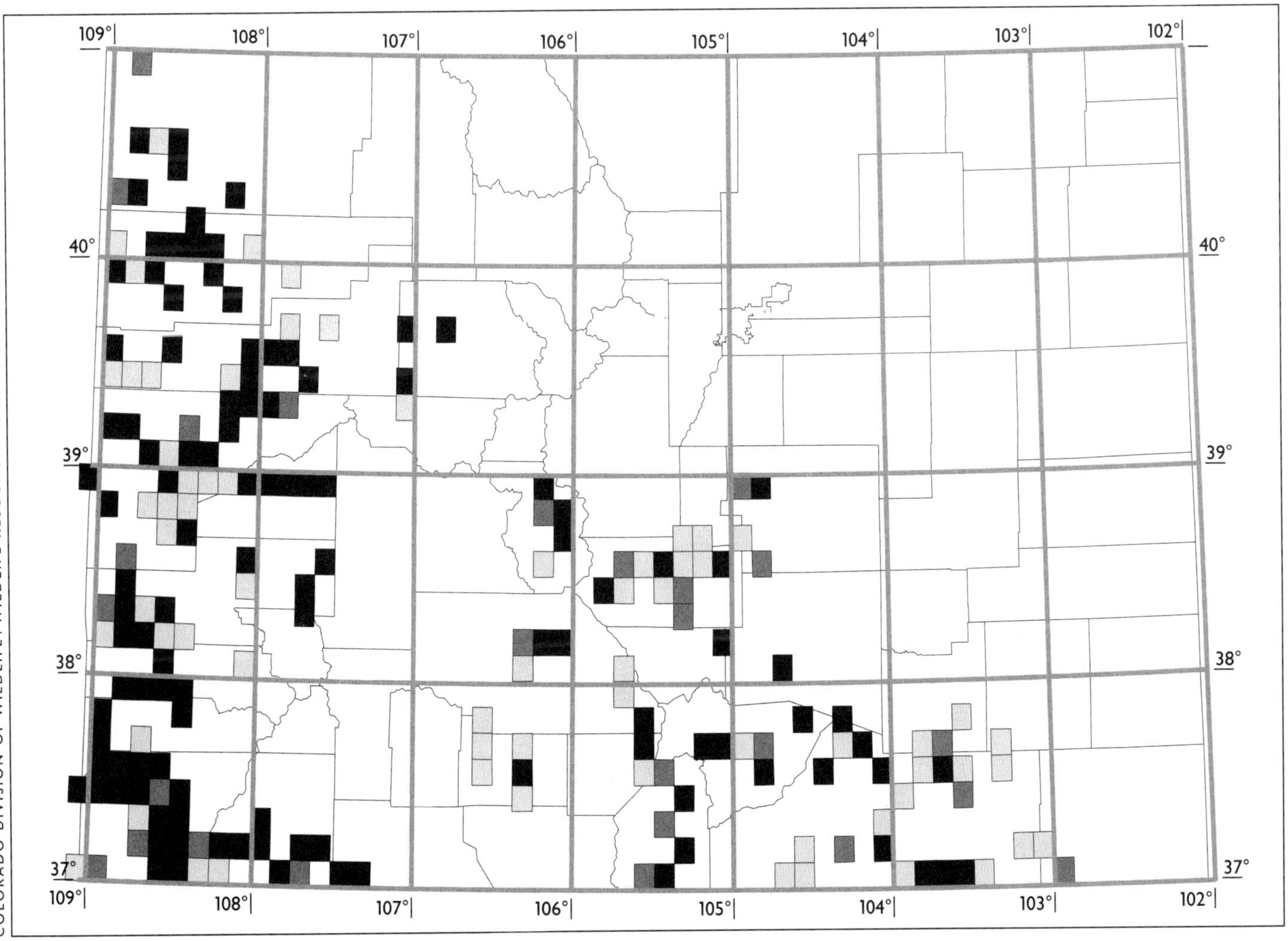
109°
108°
107°
106°
105°
104°
103°
102°
40°
39°
38°
37°

COLORADO DIVISION OF WILDLIFE / WILDLIFE RESOURCE INFORMATION SYSTEM

RED-BREASTED NUTHATCH *Sitta canadensis*

Red-breasted Nuthatches fill a niche between woodpeckers and secondary cavity-nesting species. Where suitably sized cavities commonly occur, the nuthatches readily use old nests. At sites with few cavities and sufficiently soft, rotten trees, they pound away, excavating their own nests.

HABITAT

Although occasionally they inhabit deciduous habitats, Red-breasted Nuthatches predominantly occupy conifer forests. The Atlas habitat data skew strongly to conifers (83% cf. 11% deciduous). The strong conifer emphasis may mislead those unfamiliar with Colorado mountain forests. Many conifer stands contain a substantial aspen component or may neighbor aspen stands. Recognizing the importance of deciduous trees for this species, B&N described Red-breasted Nuthatch habitat as the evergreen/quaking aspen association.

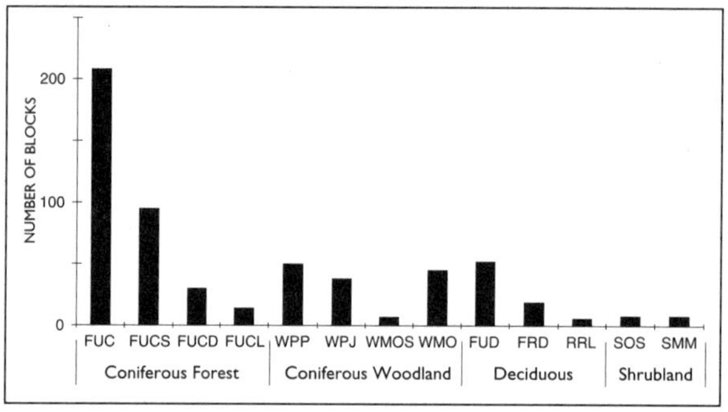
NUMBER OF BLOCKS
200
100
0
FUC
FUCS
FUCD
FUCL
WPP
WPJ
WMOS
WMO
FUD
FRD
RRL
SOS
SMM
Coniferous Forest
Coniferous Woodland
Deciduous
Shrubland

The value of aspen and other deciduous trees lies in their softer wood and their patterns of rotting. Aspens develop heart rot quickly through dead branches, dead or broken tops, and other injuries. Nuthatches find this soft, rotten wood easier to excavate than the hard wood shell of typical conifers.

Atlasers found only 21 nest-type Confirmations and none submitted nest cards. One nuthatch excavated a nest in a small, dead aspen in a Non-priority block (NMB), and in three of the four historic records in B&N that mentioned tree species, the birds used aspens.

Conifer forests provide the preferred foods for Red-breasted Nuthatches in the form of seeds in cones and a variety of insects that also feed on the conifers. Conifers also provide suitable nest sites in the form of broken-off branches and old woodpecker holes. Not any conifer forest will do, however. Red-breasted Nuthatches show a strong correlation with large trees, resulting in few birds in lodgepole pine forests and many more in spruce/fir stands. Timber harvest practices that fragment mature forests lead to lower numbers of these nuthatches in the remaining blocks of forest (Keller 1987).

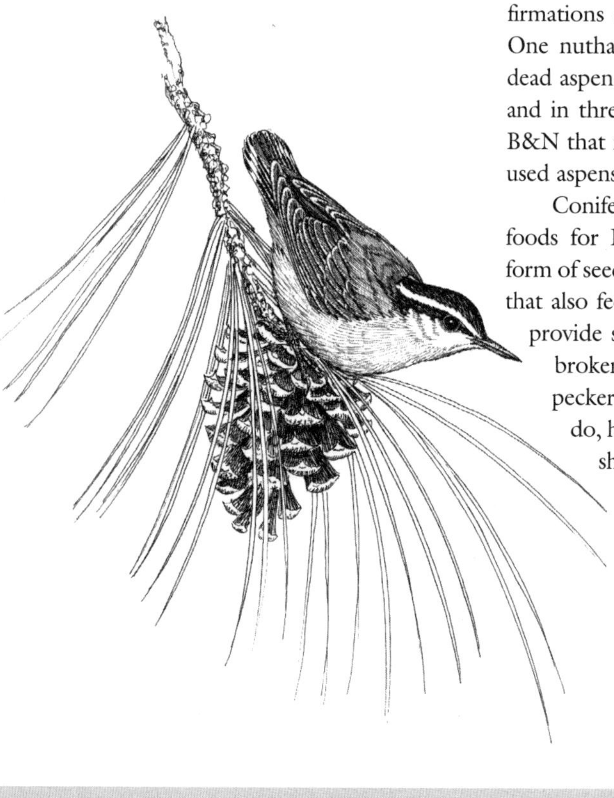

BREEDING

Breeding records for this species begin later than for many resident species in Colorado. Backdating from the earliest NY report (using an incubation period of 12 days), the first egg-laying for an Atlas-observed nest occurred in late May. The vast majority of breeding activity reported by atlasers occurred in July, ranging from nest-building in the first two weeks to fledged young found on almost every day in July.

The close tie to conifer forests includes a close tie to higher elevations; Red-breasted Nuthatches use habitats up to timberline. Later snowmelts at these high elevations, which lead to delayed springtime plant growth, seed development, and insect hatches needed for feeding young, may explain the relatively late starting date for breeding.

BREEDING PHENOLOGY

CODE		# OF RECORDS	DATE RANGE
C	Courtship	1	20 May
NB	Nest Building	5	14 May–10 Jul
ON	Occupied Nest	11	19 May–20 Jul
NY	Nest with Young	5	11 Jun–15 Jul
FY	Feeding Young	56	28 May–30 Jul
FL	Fledged Young	79	13 Jun–21 Aug

DISTRIBUTION

The species nests from south-coastal Alaska across southern Canada to the northern U.S. and western mountains. The Colorado Atlas map closely resembles a map of forested lands (excluding pinyon/juniper and aspen). The scale of the Atlas blocks prevents determining any altitude limits from the basic report data. Some early records placed an upper limit of 8,000 feet (2440 m) with occasional birds up to 10,000 feet (3048 m; Cooke 1897), but other authors' early accounts showed their primary habitats ranged 8,500–11,000 feet (2600–3350 m; Betts 1913). A&R reported them from 6,000 to 11,500 feet (1830–3500 m), which more closely reflects the elevation band of the conifer habitats in which atlasers reported these nuthatches.

BY NORMAN M. BARRETT

The difference may indicate a lack of data in the early reports or an actual change in range. Forest management by fire suppression after extensive nineteenth-century fires has changed the dominant tree types. Now, after a century, mature aspen forests have conifers invading them, resulting in thousands of acres of mixed aspen/conifer forests that provide the food and nest sites needed by these nuthatches.

Red-breasted Nuthatches share this transition zone with White-breasted and, to a degree, Pygmy nuthatches. All three species have the ability to excavate cavities to some extent. In the spruce/fir/aspen intermix Red-breasted Nuthatches occur more frequently, and the other two species dominate other habitat mixes. B&N considered Red-breasted Nuthatches less common than the other nuthatches in Colorado. Atlas numbers suggest that the state has roughly equal numbers of White-breasted and Pygmy nuthatches, and about two-thirds as many Red-breasteds.

You can hear the nasal *neep neep neep* of the Red-breasted Nuthatch in any of the high conifer forests of Colorado's Rockies. Even in the seemingly sterile, pure lodgepole pine forests, one can find these small birds probing among the needles and cones of the trees and constantly calling to one another.

BREEDING EVIDENCE

REPORTED IN 477 (27%) OF 1745 PRIORITY BLOCKS

☐	Possible	233 blocks	(49%)
■	Probable	87 blocks	(18%)
■	Confirmed	157 blocks	(33%)

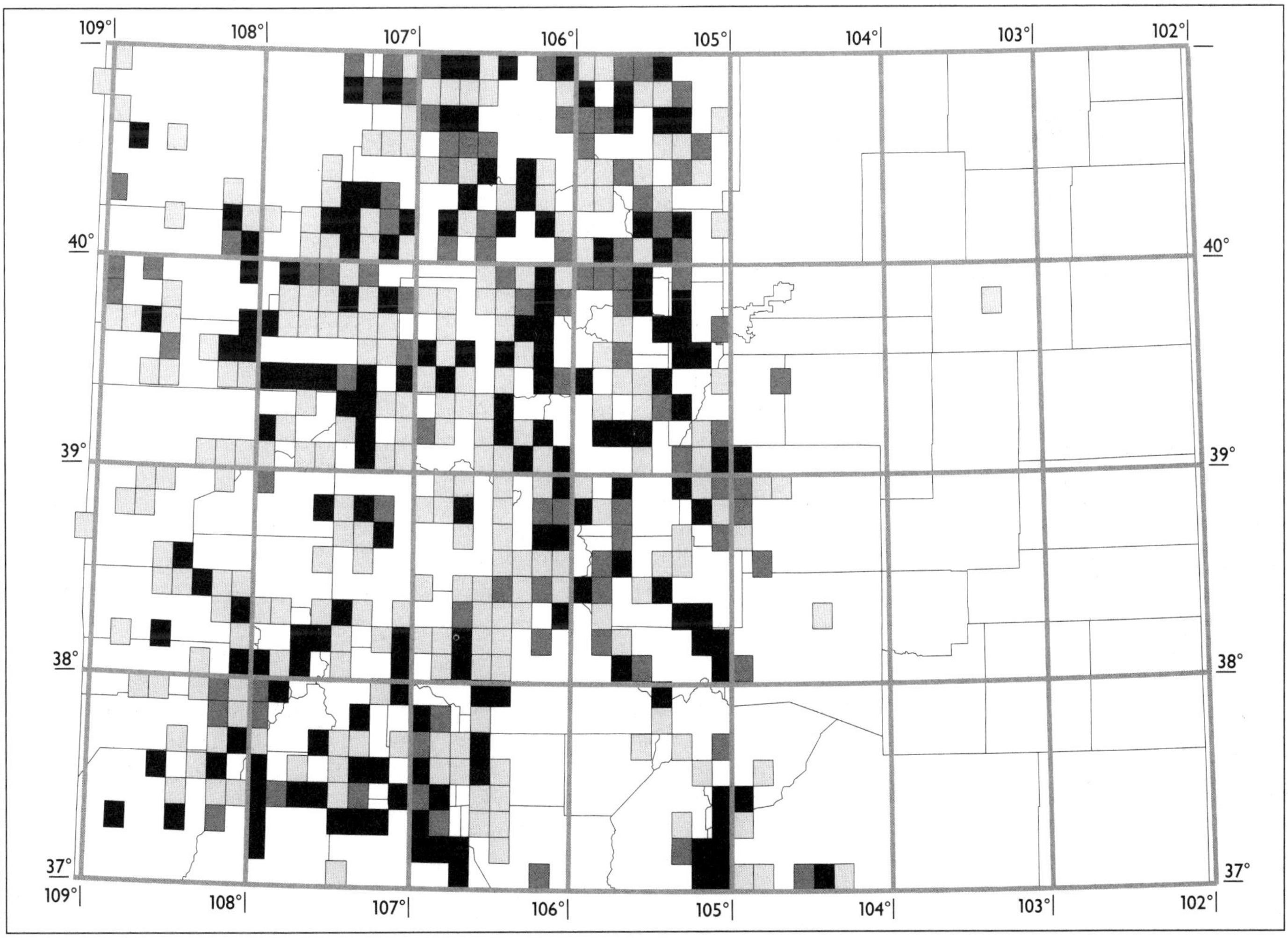

109°
108°
107°
106°
105°
104°
103°
102°
40°
39°
38°
37°

COLORADO DIVISION OF WILDLIFE / WILDLIFE RESOURCE INFORMATION SYSTEM

This neat, gray, black, and white acrobat causes an astonished double-take when the beginning birder sees one for the first time. Although many other species forage the tree trunks for insects, larvae, and eggs, only the nuthatches habitually feed head-downward, or in human concept, "upside down."

HABITAT

In all parts of their range White-breasted Nuthatches reside in wooded areas, but the preferred trees differ due to geography. Eastern breeding bird atlases emphasize their occurrence primarily in deciduous forests (Bonney *in* Andrle and Carroll 1988, Hamas *in* Brewer 1991, Palmer-Ball 1996), and in California coastal counties they choose oak woodlands (Roberson and Tenney 1993, Shuford 1993). In the interior West they favor pines (Idaho, Burleigh 1971; New Mexico, Travis 1992). Colorado birds favor conifers 4:1, almost exclusively the lower-

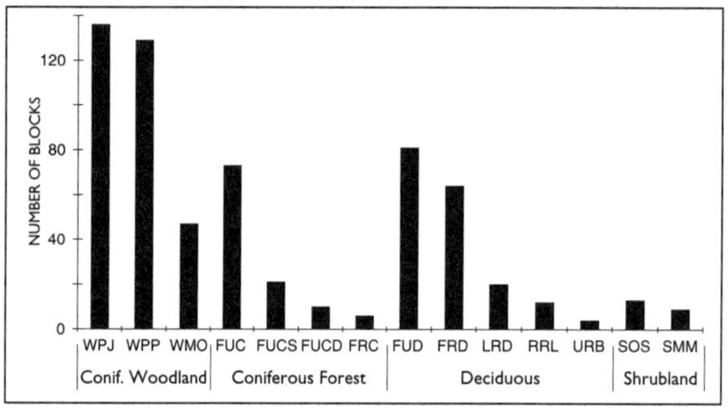
NUMBER OF BLOCKS
120
80
40
0
WPJ WPP WMO FUC FUCS FUCD FRC FUD FRD LRD RRL URB SOS SMM
Conif. Woodland
Coniferous Forest
Deciduous
Shrubland

elevation coniferous forests, with secondary use of aspen and foothills riparian habitats. Atlasers recorded them most frequently in pinyon/juniper woodlands and open ponderosa pine forests.

As tree-trunk gleaners, in summer they eat insect eggs, larvae, and adults, which they find by meticulous probing of the bark, up, down, and around the trunks and larger branches of the trees (Pravosudov and Grubb 1993). In winter their diet shifts to 60% seeds and nuts. They also use crevices in the bark to hold large seeds and nuts for cracking as well as for storage of these food items. In Colorado males and females forage at different levels in ponderosa pine, with females foraging higher and on smaller branches (McEllin 1996). Feeding techniques also differ between the sexes: females tend to peer and poke, males to scale off flakes of bark (Shuford 1993).

BREEDING

The monogamous pairs of these nuthatches remain in the same area all winter, but apparently pay little attention to each other (Tyler *in* Bent 1948). Males will even treat females as rivals at a feeding tray. However, as early as the end of January, males begin singing in earnest and bow and strut before the females (Stokes 1983). Their courtships increase in tempo in April and nesting goes on from 10 May through 20 August in Colorado (Nelson 1993). Atlas data expand these dates to 28 April for nest-building and to 24 August for still feeding young.

They always nest in cavities, often natural knotholes, woodpecker holes, or nest boxes. B&N mentioned finding a strange nest site in a crack in a boulder at Castlewood Dam (now Castlewood State Park). Atlasers watched a pair of these birds using a similar rock cranny south of Elbert (RRK). They often nest close to Western Bluebirds, Violet-green Swallows, and Pygmy Nuthatches (B&N). For at least 20 years White-breasted Nuthatches and Violet-green Swallows—undoubtedly different individuals—have used the same nest box near Green Mountain Falls. The nuthatches always nest first, followed by the swallows, who sometimes begin to harass the nuthatches in an effort to hurry them up (RRK, Walt Kuenning).

Inside the cavity the nest consists of a layer of bark strips, then a layer of dried earth or mud, and lastly some animal fur. Tyler (*in* Bent 1948) reported that to get the desired material White-breasted Nuthatches will even pluck hairs from live squirrels. Only females build nests and incubate the eggs, but both parents feed the young (Bent 1948).

BREEDING PHENOLOGY

CODE		# OF RECORDS	DATE RANGE
C	Courtship	3	9 Apr–7 Jun
NB	Nest Building	7	28 Apr–10 Jul
ON	Occupied Nest	15	24 May–26 Jul
NE	Nest with Eggs	1	25 Jun
NY	Nest with Young	18	18 May–24 Jul
FY	Feeding Young	77	20 May–24 Aug
FL	Fledged Young	71	27 May–18 Aug

DISTRIBUTION

Across southern Canada from British Columbia to Nova Scotia, in each of the lower 48 states, and as far as southern Mexico, White-breasted Nuthatches make their homes. Only treeless plains, deserts, and mountains above 10,000 feet (3048 m) have none. The map clearly shows how distribution in Colorado is determined by the location of trees, which in turn is determined by the mountains and the watercourses. The paucity of trees on the High

Plains, the mountain parks, and the desert areas accounts for the very few Atlas observations from these parts of Colorado. Atlas observations coincide very well with the A&R map with a few exceptions. Atlasers did not find them breeding in eastern Las Animas County, even in the pinyon/juniper habitat like that in which they found the nuthatches most frequently in the rest of the state. They did not find them in the San Luis Valley or central Moffat County, which lack the essential trees.

Very few nest in eastern Colorado and those that do, on the South Platte and Republican rivers, seem disconnected from both eastern populations and the nuthatches in the Colorado mountains. The Atlas map shows a 75-mile gap between the plains breeders and those in the rest of the state. They do not nest along the Platte in Nebraska, or in Kansas except on the eastern edge (Johnsgard 1979, Thompson and Ely 1992).

White-breasted Nuthatches help control insect pests in Colorado forests. They only suffer from man's activities if forestry management practices include the removal of old dead trees, thus eliminating nesting cavities (Pravosudov and Grubb 1993). This practice is unlikely to have much impact in Colorado on this widespread species. This bird's friendly *yank, yank, yank* should persist in our woodlands.

BREEDING EVIDENCE

REPORTED IN 515 (30%) OF 1745 PRIORITY BLOCKS

☐	Possible	288 blocks	(56%)
■	Probable	48 blocks	(9%)
■	Confirmed	179 blocks	(35%)

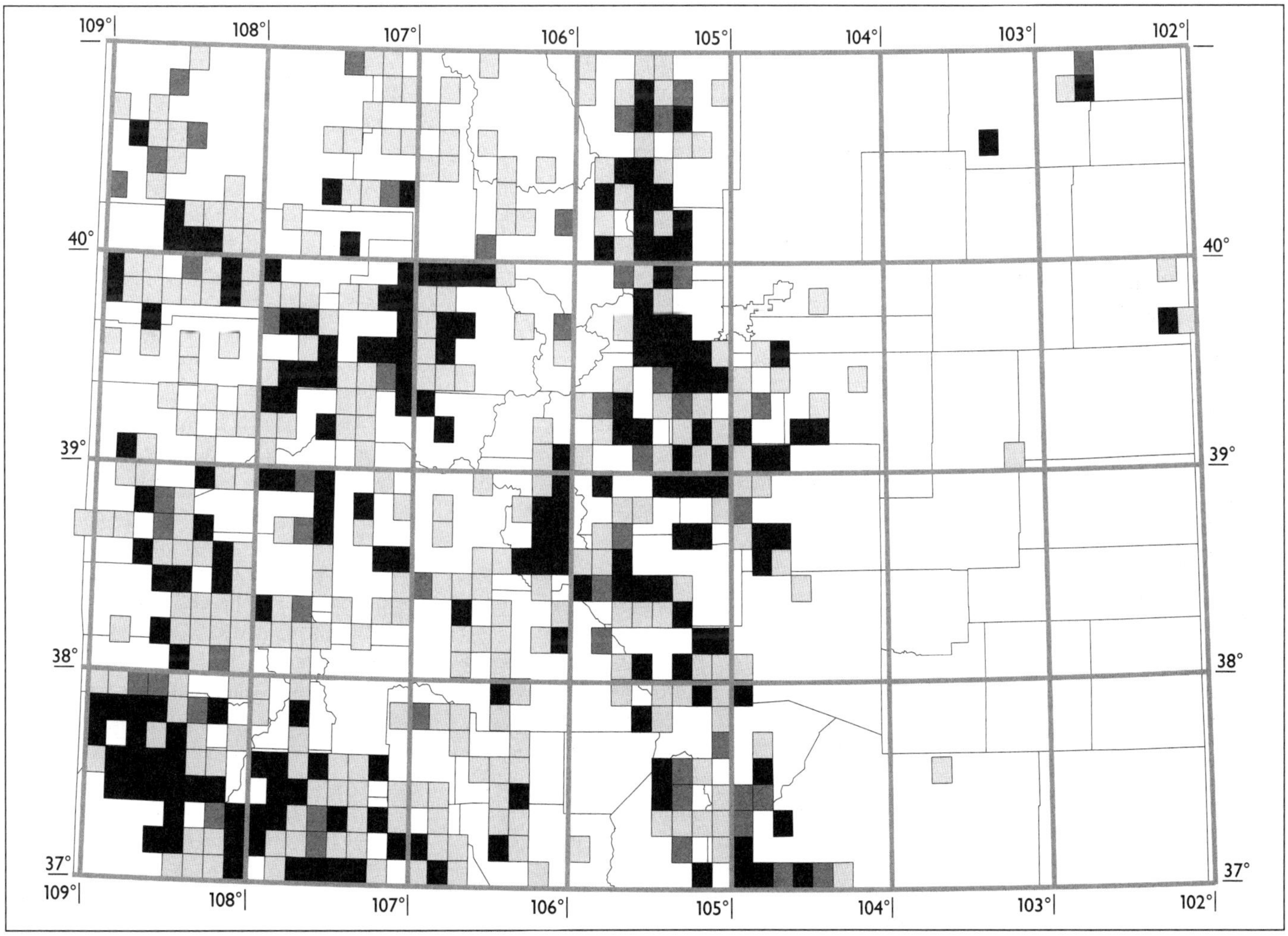
109°
108°
107°
106°
105°
104°
103°
102°
40°
39°
38°
37°

COLORADO DIVISION OF WILDLIFE / WILDLIFE RESOURCE INFORMATION SYSTEM

PYGMY NUTHATCH *Sitta pygmaea*

Pygmy Nuthatches, unable to suppress their querulous chatter for long, travel in babbling flocks, piping high-pitched notes as they flit from pine to pine. The sociability extends to the nesting process, as a third of the breeding pairs enlist "helpers" to assist during the incubation and nestling periods.

HABITAT

Pygmy Nuthatches typify Colorado ponderosa pine forests. Sixty-six percent of Atlas observations came from ponderosa woodlands or habitats with a ponderosa component, and another 27% came from other coniferous forest habitats. Of 269 habitat entries, atlasers reported Pygmys using spruce/fir in only two blocks, lodgepole pine in three, pinyon/juniper in five, Douglas-fir in eight, aspen in eleven, and shrublands in four—but in half these blocks the birds also used ponderosa.

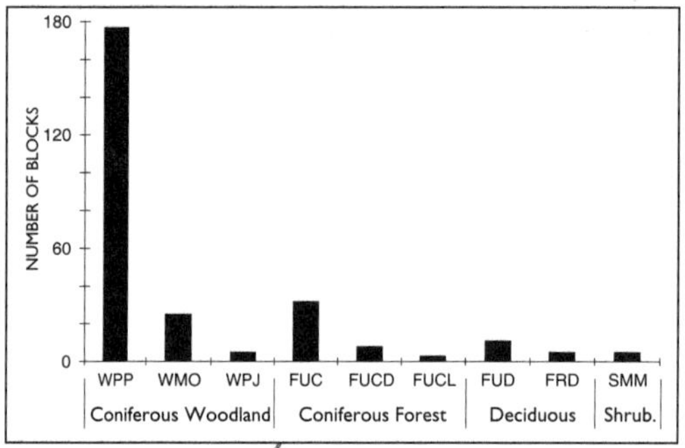

NUMBER OF BLOCKS
180
120
60
0
WPP
WMO
WPJ
FUC
FUCD
FUCL
FUD
FRD
SMM
Coniferous Woodland
Coniferous Forest
Deciduous
Shrub.

Because these nuthatches usually drill their own cavities, they need mature ponderosas, with old or decayed wood. B&N described nests in a "half rotted pine [and in] towering pine stubs." The ideal habitat consists of park-like, open forests of tall ponderosas where the pines have broken-off stubs of branches or treetops (Bent 1948). They depend largely on healthy, mature ponderosas, and occur in lower densities in logged tracts (Ewell and Cruz 1998).

Colorado's three species of nuthatches breed in different habitats. Although all three favor conifers, only Pygmys demonstrate a strong affinity to ponderosas. White-breasted Nuthatches also nest in pines but do not dig their own cavities, and Red-breasted Nuthatches exploit boreal forests.

In fall these year-round residents form noisy flocks, often traveling in loose flocks of chickadees, other nuthatches, creepers, warblers, and kinglets, and flocking persists into winter. These flocks often consist of hatching-year birds, resulting in the noisy active juvenile gangs of the coniferous forest. Pygmys roost communally in nest cavities; near Colorado Springs one evening Knorr (1959) watched a flock of 100–150 converge into a hole in a ponderosa. In another instance near Evergreen, a family of seven roosted for five nights in a nest box until a pair of House Wrens stuffed the box with sticks and dislodged the roosting nuthatches (Jones 1930).

The bulk of the Pygmys' summer diet consists of insects associated with pine twigs and foliage (Anderson 1976). During breeding season they eat insects and spiders and in winter also consume conifer seeds that they cache in pine trees. The three nuthatch species show considerable dietary overlap. In winter flocks the three species feed in different parts of the trees: Pygmys tend to forage in the crowns of ponderosas, out on smaller and terminal branches, twigs, and needles, especially on cone-needle clusters (Ewell and Cruz 1998); White-breasted Nuthatches habitually cruise the trunks and larger branches; and Red-breasteds station themselves in the top third of the tree but in the inner branches and trunks (Ryser 1985). Mountain Chickadees also feed with Pygmys, but they may dissipate direct competition by foraging at different heights within the tree (Manolis 1977).

BREEDING

Although Pygmy Nuthatches usually excavate their own nest cavities, they occasionally use deserted woodpecker holes. The entrance, 10–25 feet (3–7.5 m) up, has an irregular shape, 1–1.25 inches (2.5–3 cm) in diameter. The bottom of the cavity is usually 8 inches (20 cm) below the entrance and the birds add a scanty bed of plant down, cocoons, wool, bark shreds, hair, and feathers (Bent 1948).

Pygmy Nuthatches maintain their territories all year but defend only the immediate vicinity of the nest cavity (Anderson 1976); the larger forest becomes a feeding area for flocks.

Like many cavity-nesters, Pygmy Nuthatches increase in numbers in plots experimentally enhanced by nest boxes—13-fold in ponderosa pines and six-fold in higher-elevation mixed conifer forests. Their numbers increased more where the trees had fewer snags, indicating that a lack of snags could limit their numbers (Bock and Fleck 1995).

A unique aspect of these nuthatches' biology is their cooperative breeding, i.e., use of "helpers" at the nests. Although adults form long-term pair bonds, the breeding unit consists of 2–5 birds: two breeding adults and helpers. In an Arizona study area, about 30% of the nests had helpers. Most helpers, usually unmated males related to the adult territory holders, are second-year young or siblings of the adults. Helpers assist in nest maintenance, feed the female on the nest, and feed nestlings and fledglings (Sydeman et al. 1988).

BREEDING PHENOLOGY

CODE		# OF RECORDS	DATE RANGE
C	Courtship	2	17 May–6 Jun
NB	Nest Building	4	7 May–12 Jun
ON	Occupied Nest	14	24 May–1 Jul
NY	Nest with Young	19	3 Jun–22 Jul
FY	Feeding Young	48	5 Jun–25 Aug
FL	Fledged Young	42	23 May–25 Aug

DISTRIBUTION

Pygmy Nuthatches are resident, principally in pine forests, from British Columbia to southern Mexico. In Colorado, the Atlas map shows that the ranges of Pygmys and ponderosa pines coincide almost exactly. By targeting the ponderosas of the Douglas Mountains in Moffat County atlasers verified breeding in northwestern Colorado for the first time (Coen Dexter pers. comm.). Disjunct records in Elbert County reflect the discontinuity of ponderosa pines in the Black Forest.

The year-round range map in A&R included a large area in western Colorado where the Atlas map shows no occurrences. These areas, apparently covered with spruce/fir and lodgepole pine, may represent winter records. The Atlas covered that section of the state well and the inappropriate habitats in these areas would seem to exclude breeding nuthatches. A&R also reported these nuthatches as "common" in the lodgepole pine forests in Summit and Grand counties, but Atlas records do not support that. About 20 years ago Pygmys colonized some mature lodgepole forest stands in Summit County but only a few, and in only a few places (HEK).

Their habitat linkage means that Pygmy Nuthatch populations will probably wax and wane in proportion to the health of mature ponderosa pines.

BREEDING EVIDENCE

REPORTED IN 226 (13%) OF 1745 PRIORITY BLOCKS

▢	Possible	79 blocks	(35%)
▦	Probable	20 blocks	(9%)
■	Confirmed	127 blocks	(56%)

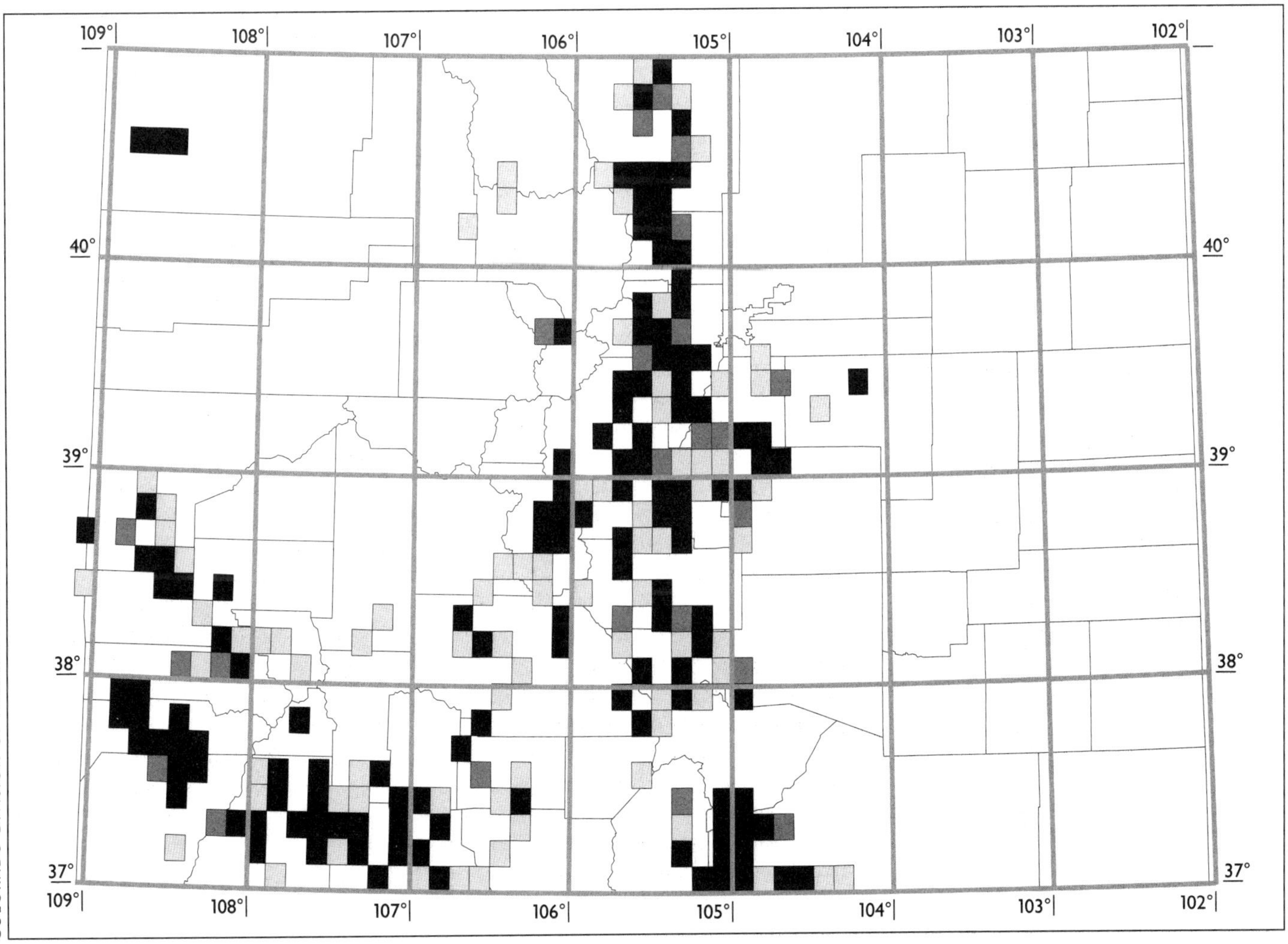
109°
108°
107°
106°
105°
104°
103°
102°
40°
39°
38°
37°

COLORADO DIVISION OF WILDLIFE / WILDLIFE RESOURCE INFORMATION SYSTEM

Although many North American animals ascend a tree with greater abandon, Brown Creepers outclimb them all over the course of their lifetimes. What they lack in speed they overcome with persistence as they spend the greater part of their existence hitching their way up tree trunks.

HABITAT

Brown Creepers occupy all varieties of coniferous forest and woodland habitats in Colorado, ranging from ponderosa pine to subalpine spruce/fir. Most frequently, though, they breed among mature spruce/fir and lodgepole communities from 9,000 to 11,500 feet (2700–3500 m) in elevation (A&R, B&N). They take to the biologically deprived lodgepole communities as readily as any Colorado species. Spruce/fir dominated Atlas habitat reports, over half the blocks in which atlasers specified forest types. Atlasers indicated ponderosa pine in 16% of those blocks and aspen, Douglas-fir, and lodgepole each in 7–9%.

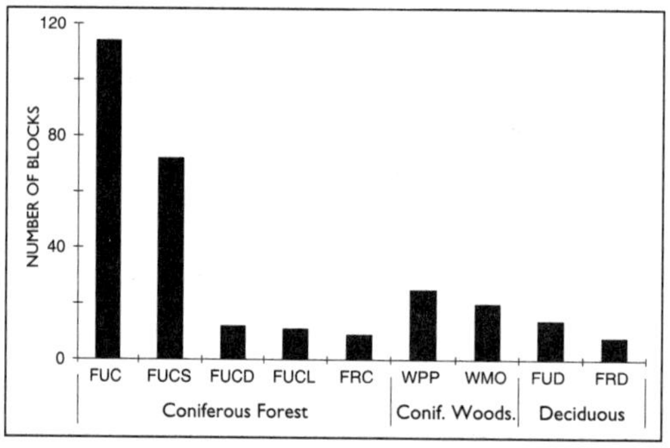
120
80
40
0
NUMBER OF BLOCKS
FUC
FUCS
FUCD
FUCL
FRC
WPP
WMO
FUD
FRD
Coniferous Forest
Conif. Woods.
Deciduous

Within these habitats Brown Creepers consume a diet of spiders, beetles, moths, and other insects (Davis 1978). Adult birds start foraging near the base of a tree and continue working upward, often until branch density makes further maneuvering too difficult (Franzreb 1985). Having reached this point, the bird will descend to the base of a nearby tree and repeat the process.

Creepers suspend their hammock-like nest behind a piece of bark peeling away from a tree. The nest usually has a foundation of twigs, leaves, and smaller pieces of bark with a lining of grasses, feathers, and, occasionally, moss (Davis 1978).

Nuthatches and woodpeckers exhibit foraging habits similar to creepers, but the curved bill of creepers makes them better suited than either nuthatches or woodpeckers for probing into crevices and behind pieces of bark. Due to the adaptability of the species, they live in close proximity to all three of Colorado's nuthatches and most of Colorado's woodpeckers. Thus, Brown Creepers may compete for food with both woodpeckers and nuthatches, although habitat overlap probably reaches its greatest level with Red-breasted Nuthatches in subalpine spruce/fir forests. Even here, however, the different method of gleaning minimizes direct competition.

Creepers avoid recently logged areas, even where a few old trees remain standing (Franzreb and Ohmart 1978). They conserve energy—at least during the breeding season—by restricting their foraging activities to dense stands of mature trees (Franzreb 1985). Accordingly they are among the very last species to return to a regenerating forest.

BREEDING

Courtship feeding plays an important role in the breeding cycle of Brown Creepers. Wing-fluttering reminiscent of that shown by fledglings often precedes courtship feeding. This, accompanied by the fact that courtship feeding may continue until eggs hatch and males begin feeding young, provides ample opportunity to mistake courtship activity for the feeding of young. Therefore, early Atlas Confirmations of food-carrying require cautious treatment.

Atlasers found nests within a brief, month-long period from 3 June to 4 July, plus one late nest on 29 July. The nesting period may extend longer because many of the birds carrying food could have had nests with young as their destinations. Confirmation dates showed no geographical pattern: atlasers found fledglings in August from the San Juan Mountains in the south up to the Front Range in the north.

Both male and female gather nesting materials, but the female assumes the primary role in nest-building. A typical clutch consists of five or six eggs, and incubation takes just over two weeks. Females perform most, if not all, of the incubation duties. During this time males bring food items to their mates on the nest (Davis 1978).

Both parents feed the young after hatching. Fledglings begin to forage on their own approximately nine days after fledging, but continue to receive food from parents for several more days (Davis 1978).

At night fledglings characteristically cluster together to roost. This cluster may be either in a circular pattern with the heads grouped together in the center or with two rows of birds, in which case the birds in the lower row situate their heads between the flanks of the birds in the upper row. Adult birds do not join these roosting clusters (Davis 1978).

Predators of creeper nests and fledglings include squirrels, small owls, and accipiters. Rarely do cowbirds parasitize the nests (Davis 1978) and no records of such activity exist from Colorado.

BREEDING PHENOLOGY

CODE		# OF RECORDS	DATE RANGE
C	Courtship	2	13 Jun–21 Jun
NB	Nest Building	1	23 May
ON	Occupied Nest	3	3 Jun–22 Jun
NY	Nest with Young	5	12 Jun–29 Jul
FY	Feeding Young	30	29 May–1 Aug
FL	Fledged Young	33	25 Jun–17 Aug

DISTRIBUTION

Brown Creepers breed throughout the Northern Hemisphere. Although widely distributed across the mountainous regions of Colorado and the western states, they never occur in large numbers. The lack of reports from several Priority blocks with suitable lodgepole or spruce/fir habitat further indicates that their numbers rarely attain a point where they become difficult to miss.

Most breeding Confirmations came from the mountains of central and southwestern Colorado. Atlasers produced first-ever records of breeding in four western Colorado latilongs. Most noteworthy among these is the discovery of Brown Creepers breeding in an isolated pocket of ponderosa pine near Dinosaur National Monument.

In much of the western U.S., Brown Creepers live as more or less permanent residents, retreating short distances to lower elevations during the colder months of winter. Creepers bring their charm to stream bottoms, parks, and residential areas during these months. Something closer to full-scale migration occurs in the eastern states and Canada. There, they venture far north to breed in the coniferous forests of Canada and the Great Lakes region. During winter some retreat as far as the woodlands of the southern tier of states.

The Colorado range shows little evidence of change from pre-settlement days. Most changes in range take place in relatively small parcels where logging has occurred. In these places, Brown Creepers will remain absent until the forests regenerate adequately to support their nesting and foraging requirements.

BREEDING EVIDENCE

REPORTED IN 263 (15%) OF 1745 PRIORITY BLOCKS

☐	Possible	164 blocks	(62%)
▩	Probable	29 blocks	(11%)
■	Confirmed	70 blocks	(27%)

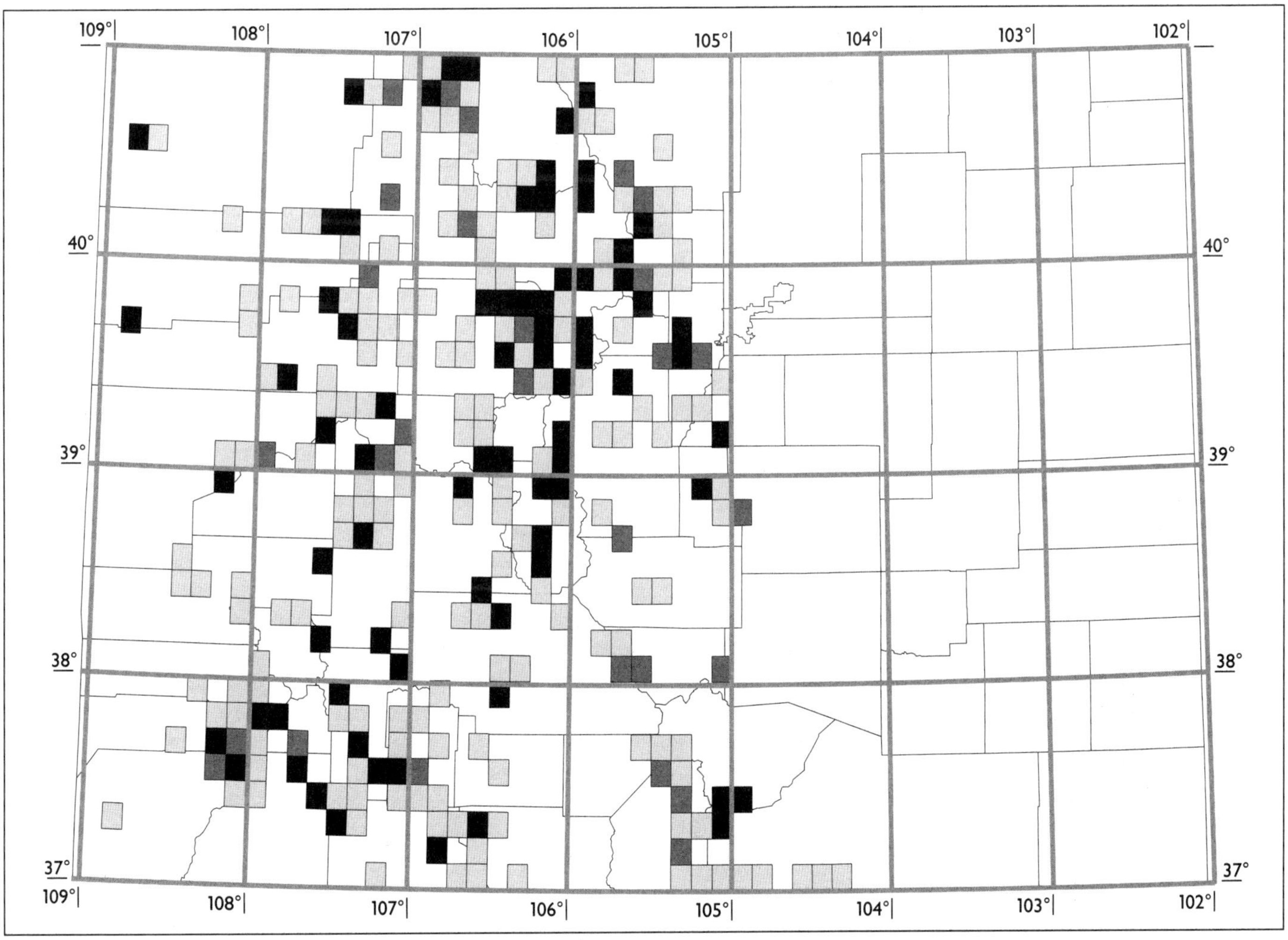
109°
108°
107°
106°
105°
104°
103°
102°
40°
39°
38°
37°

COLORADO DIVISION OF WILDLIFE / WILDLIFE RESOURCE INFORMATION SYSTEM

As Colorado's leading avian landscape architects, Rock Wrens have the distinctive breeding behavior of building runways of small flat stones leading to their nest cavities. Visually inconspicuous but active and bouncy, they breed in rocky habitats from prairies to peak-tops. Their harsh yet melodious songs repeat variable series of two notes, and repeat each sequence three or four times as they bob up and down.

HABITAT

These wrens largely confine their activities to open arid habitats dominated by rocks, including talus slopes, boulders, scree, and cliffs. They consistently use bare, open, windswept, sunny areas, with piles of broken rocks, scattered boulders, or rocky outcrops. Many of these habitats have little vegetation. In Colorado, they breed from the prairies (3,900 feet, 1190 m) to above timberline (13,000 feet, 3660 m; A&R, atlasers).

Atlasers used the MSC (Miscellaneous) code in 27% of the blocks—more than any other habitat code. This code applied to dirt banks and cliffs before the Atlas adopted the MCL code in 1990. On the plains, these wrens still maintain their affinity to rocks by using riprap on dams, small rock piles, individual boulders, eroded slopes, badlands (Johnsgard 1979), and dirt banks (Johnsgard 1979, atlasers).

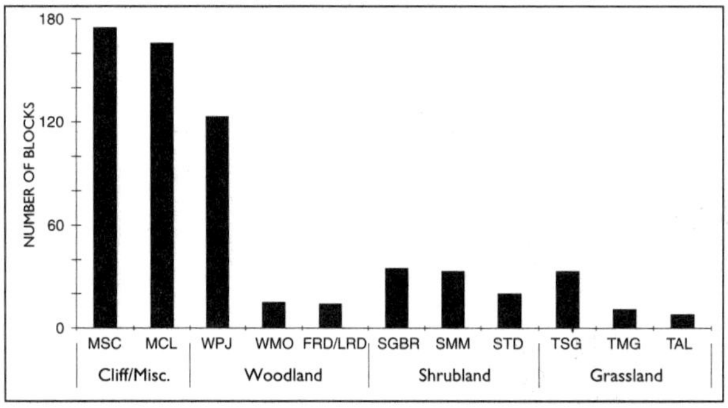

180
120
60
0
NUMBER OF BLOCKS
MSC
MCL
WPJ
WMO
FRD/LRD
SGBR
SMM
STD
TSG
TMG
TAL
Cliff/Misc.
Woodland
Shrubland
Grassland

Their habitat tolerance for small as well as big rocks contrasts with that of Canyon Wrens, which use a more limited habitat of larger rocks and steep canyons. The Rock Wrens' larger selection of breeding habitats allows them to breed farther out on the prairies, yet they still occur only in arid areas west of the Missouri River.

Field workers reported them in 84% of the vegetation divisions used by the Atlas; this does not demonstrate an affinity for any particular vegetation, but rather shows that suitable rock habitat occurs in many different vegetation types. The only published data for Colorado report them breeding within a pinyon/juniper woodland (Sedgwick 1987). Of Atlas-reported vegetative codes, pinyon/juniper composed 41% (four times the next most frequent, shortgrass prairie). Nevertheless Rock Wren occurrence shows no correlation with the surrounding vegetation type (Rumble 1987).

BREEDING

Rock Wrens typically nest among rocks, in crevices of rocky piles and canyon walls (Bent 1948), and occasionally even in tree holes or dirt banks (Johnsgard 1979, atlasers). These wrens are one of the least studied of the North American birds. Although a recent article describes the breeding ecology in New Mexico (Merola 1995) no detailed data are available from Colorado.

They commonly build a small runway of stones leading to the nest—sometimes quite an elaborate one. Atlaser-artist Radeaux observed one nest with 40–50 small rocks leading to it! They use more stones when the nest has earthen floors; nests in steep areas may have no walkways (Bent 1948).

The birds construct the nest of twigs and grasses with a soft lining of wool, feathers, hair, and plant material (Bent 1948). They ordinarily are single-brooded but occasionally double-brood in Colorado (SLJ). The extended Atlas breeding dates allow ample time for double-brooding.

Two observers have documented cowbird parasitism in Colorado (Chace and Cruz 1996) and recent data indicate a high percentage of parasitized nests in Kansas (G. Farley pers. comm.).

BREEDING PHENOLOGY

CODE		# OF RECORDS	DATE RANGE
C	Courtship	2	9 May–16 Jul
N	Nest Building	13	27 Apr–1 Jul
ON	Occupied Nest	9	12 May–4 Jul
NE	Nest with Eggs	3	27 May–29 Jun
NY	Nest with Young	9	9 Jun–18 Jul
FY	Feeding Young	116	16 May–6 Aug
FL	Fledged Young	83	23 May–12 Aug

DISTRIBUTION

Rock Wrens occur throughout western North America from the western edge of the Great Plains to the Pacific slope, north just into Canada and south through Mexico into Central America. They are short-distance migrants in parts of their range but resident in others; most migrate out of Colorado in early September.

Thomas Say on the 1820 Long expedition collected the first Rock Wren described to science, where the South Platte comes out of the mountains (Marsh 1931), probably Waterton Canyon. Atlasers corroborated that Rock Wrens breed throughout the state. The distribution of these wrens can be patchy and uneven yet they breed in most counties in Colorado; atlasers found them in all but eight.

Rock Wrens occur on the rugged plateaus of the Western Slope and Las Animas County and on the steep rocky slopes of the Arkansas River upstream from Pueblo. They breed along the dirt draws of the greasewood flats in western Colorado and among the jumbles of rocks that surround the San Luis Valley. Only eight blocks reported them with the tundra vegetative code but in several other blocks atlasers found them on talus slopes above timberline. On the rocky slopes along the Barr Trail on Pikes Peak, they outnumber American Pipits to over 13,000 feet (3660 m; Alan Versaw pers. comm.).

Atlasers found these rock-loving wrens breeding in many places on the prairie (sometimes in dirt banks instead of rock piles). This conforms both to Cooke's report (1897) and to their status in western Kansas (Thompson and Ely 1992). Most Atlas observations from the plains came from the Pawnee National Grassland, the breaks and draws of Yuma County, and the mesas and canyons of Las Animas County.

These breeding records indicate a more widespread use of the plains in Colorado than recently reported (B&N, A&R). Topography explains the patchy plains distribution. The wrens concentrate in rangelands—rougher country with more topographic relief than the broad stretches of croplands where they do not occur.

Negative population trends apply throughout their range according to BBS data: a 1.4% decline per year across the continent (a total drop of one-third of the population over 31 years). BBS data also show negative trends for shorter periods (15-year increments) and for various states and physiographic regions but they lack statistical significance (Sauer et al. 1996). Consistent with the paucity of life history information, no credible theory or data explain these declines.

BREEDING EVIDENCE

REPORTED IN 527 (30%) OF 1745 PRIORITY BLOCKS

☐	Possible	178 blocks	(34%)
▩	Probable	122 blocks	(23%)
■	Confirmed	227 blocks	(43%)

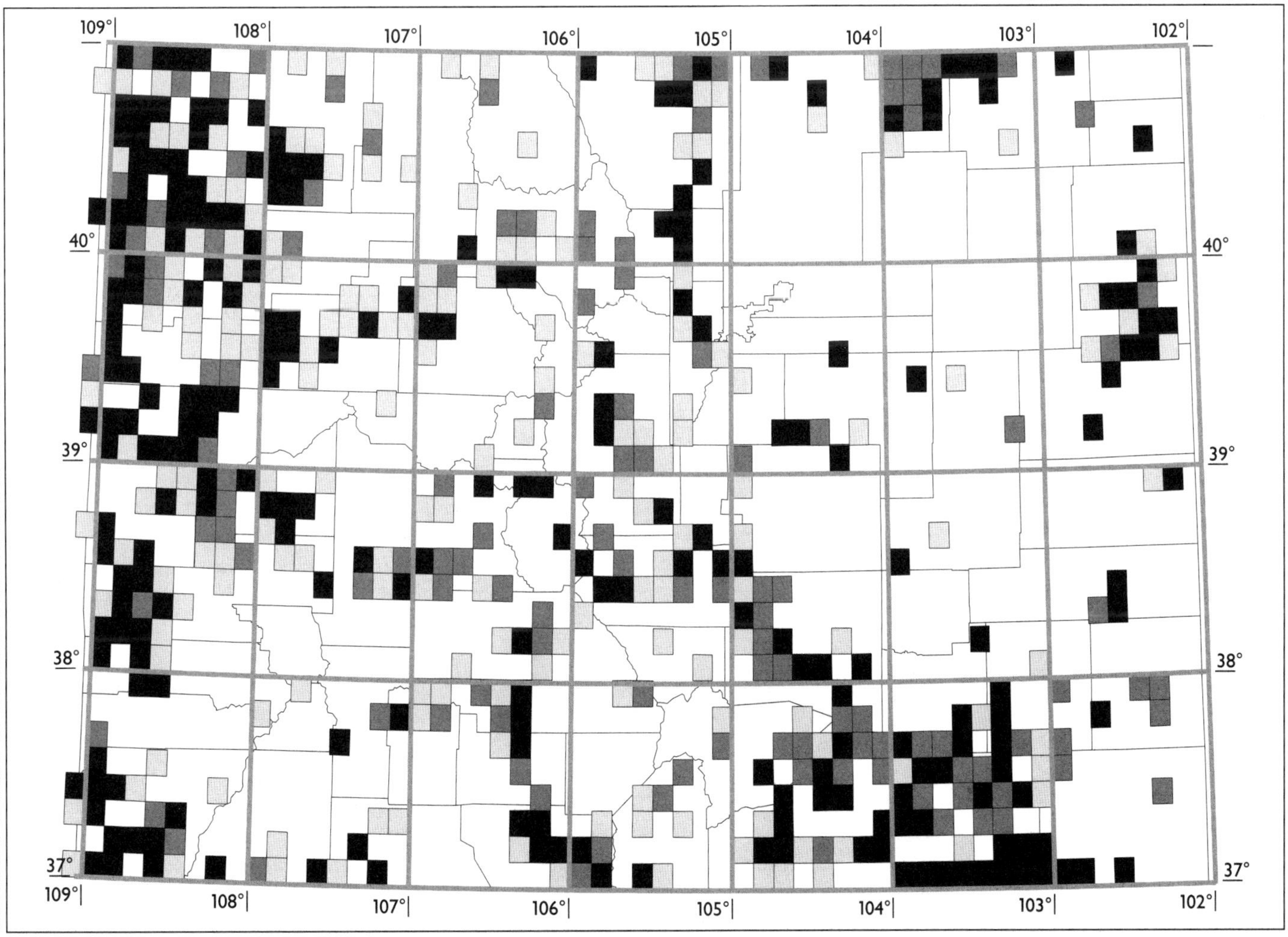

109°
108°
107°
106°
105°
104°
103°
102°
40°
39°
38°
37°

COLORADO DIVISION OF WILDLIFE / WILDLIFE RESOURCE INFORMATION SYSTEM

One of the most distinctive sounds in the West is the Canyon Wren's cascading song, which echoes throughout Colorado's canyonlands. These wrens are elusive and difficult to spot, but for many people their unforgettable song defines the canyon country.

HABITAT

Canyon Wrens creep and hop over rocks and into crevices. They nest, feed, roost, and live entirely among rocks and cliffs. Typically they inhabit steep-sided canyons. Although their habitat often includes running streams, the wrens also occur in dry areas. Water may simply be the force that created the vertical component, rather than a limiting factor (Jones and Dieni 1995). These habitats provide protective shade and cool temperatures during intense summer heat. Typical sites include the jutting sandstone outcrops of Roxborough Park, rimrocks of Mesa Verde, and vertical canyon cliffs of Black Canyon and Dinosaur national monuments.

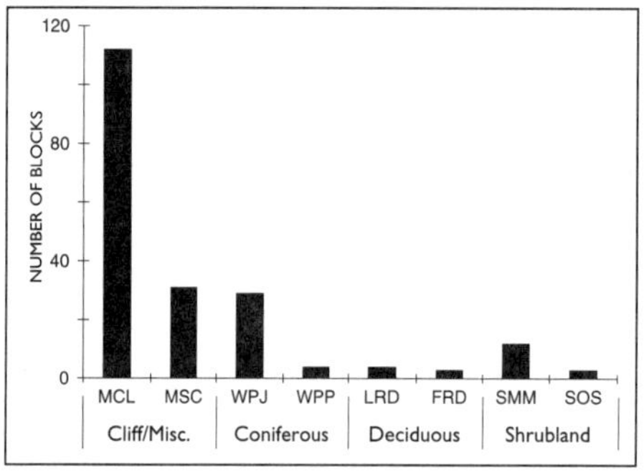
120
80
40
0
NUMBER OF BLOCKS
MCL
MSC
WPJ
WPP
LRD
FRD
SMM
SOS
Cliff/Misc.
Coniferous
Deciduous
Shrubland

Many vegetative types can surround their cliffy habitats. Atlasers reported them in ponderosa pine, shrublands, pinyon/juniper, scrub oak, and grasslands. Individual wrens will roam into riparian habitats but never stray far from rocks.

Canyon Wrens use the same habitat in winter, and pairs can remain together and maintain winter territories. Other individuals may wander the canyons and re-establish territories in February (Jones and Dieni 1995).

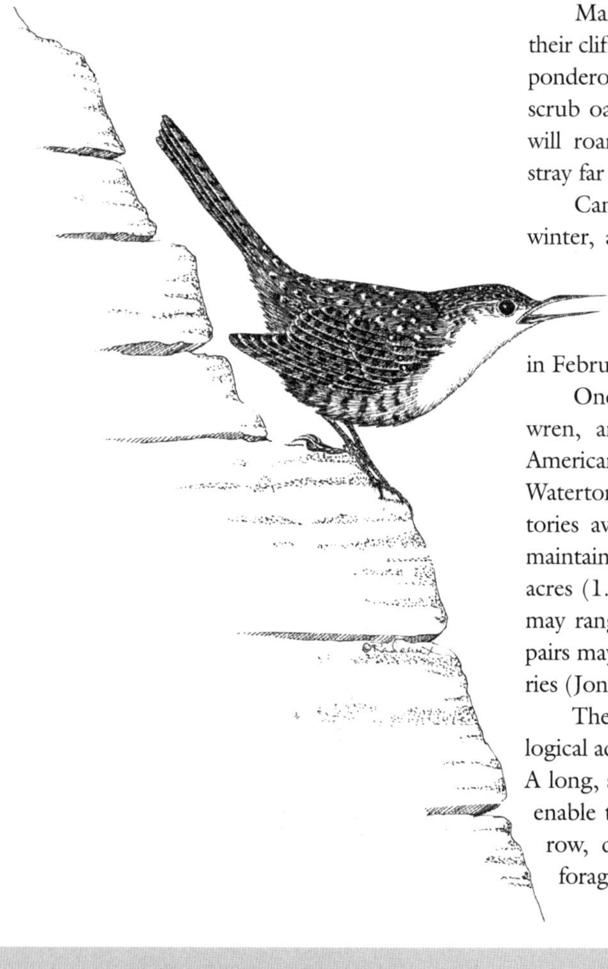

One major study on the biology of this wren, among the least known of North American birds, occurred in Colorado. In Waterton Canyon, the size of breeding territories averages 2.25 acres (0.91 ha). Pairs maintain winter territories that average 3.5 acres (1.44 ha), a 58% increase. Individuals may range far during the day and multiple pairs may overlap into larger feeding territories (Jones and Dieni 1995).

The Canyon Wren has several morphological adaptations for living in rock crevices. A long, slender bill and a flattened cranium enable the wren to probe deeply into narrow, deep crevices. Short tarsi facilitate foraging in rocks by lowering the bird's height at the shoulders without affecting locomotion (Mirsky 1976).

They glean spiders and insects from rocky surfaces. Occasionally they attempt (ineffectual) flycatching (Tramontano 1964, SLJ) and steal cached spiders from trypoxylid wasp nests (Martin 1971).

BREEDING

Canyon Wrens are not known to build "dummy" nests like other wrens. They approach their nests evasively and defend territories with more than one suitable nest site (such as ones used for previous broods or in prior years). They occasionally reuse old nests, both for second broods and in subsequent years (B&N). Atlas nesting dates correspond well to the known phenology.

The birds attach their nests to rock faces in caves or crevices. Frequently an overhead ledge protects the nest, or the wrens may squeeze it into a small cranny or hole (SLJ), and occasionally they build nests on abandoned buildings, roads, and fences. They fasten the nest by a stick and twig base to which they affix a cup made of moss, twigs, grasses, and dead leaves and line it with lichens, plant down, wool, cobwebs, feathers, or other soft material (Jones and Dieni 1995).

Breeding dates in Colorado probably correlate with spring weather patterns. Egg-laying takes 4–6 days for an average clutch size of five, and incubation averages 16 days. The nestling period apparently averages 16 days. Fledglings receive adult care for up to two weeks, and the young may remain with the adults on the natal territory all winter (Jones and Dieni 1995). Canyon Wrens initiate second clutches in Colorado; Atlas dates seem to support that.

Canyon and Rock wrens often live in close association. Their territories commonly overlap but the micro-habitats differ. Atlasers found Rock Wrens in 130 of the 177 blocks in which they found Canyon Wrens, although this does not necessarily mean that the two species shared territories. Each has slightly different, specialized nesting and foraging habitats. Rock Wrens use piles of broken rocks and scattered boulders in areas with little vegetation (Bent 1948) as well as the cliffs that Canyon Wrens choose. Canyon Wrens of both sexes try to repulse Rock Wrens from breeding territories by flying toward, singing at, and chasing the intruders. Rock Wrens respond to recordings of Canyon Wren songs by singing loudly and repeatedly. Rock Wrens will, however, successfully nest within successful Canyon Wren territories.

White-throated Swifts nest in cliffs used by Canyon Wrens. Responding to Canyon Wren song recordings, swifts consistently leave their nests and vocalize aloft.

BREEDING PHENOLOGY

CODE		# OF RECORDS	DATE RANGE
C	Courtship	3	28 Apr–13 Jul
N	Nest Building	1	13 Jun
ON	Occupied Nest	3	28 May–15 Jul
NE	Nest with Eggs	2	20 May–21 Jun
NY	Nest with Young	4	3 Jun–18 Jul
FY	Feeding Young	17	21 May–26 Jul
FL	Fledged Young	15	8 Jun–12 Aug

DISTRIBUTION

Canyon Wrens occur from the Rocky Mountain foothills to British Columbia and south down the Pacific slope to Chiapas, Mexico. They do not migrate, but some individuals may make short seasonal movements. In Colorado, Cooke (1897) first termed the species rare but subsequently (1909) noted them in southwestern Colorado. Sclater (1912) and B&N restricted their range to the southern two-thirds of the state. Finally, A&R showed wider distribution similar to, yet more extensive than, the Atlas map.

Atlas records show that in Colorado they breed from 3,900 feet (1150 m) to 8,500 feet (2500 m). They nest away from the foothills and southeastern canyon country only at John Martin (A&R) and Two Buttes (Atlas) reservoirs. Particularly along the Front Range, the Atlas map reflects the discontinuity of proper habitat, with clusters from Denver north, from Pikes Peak to the Arkansas River gorge, and in Las Animas County canyons. The map presents less exclusive clusters in Western Slope mesas and canyons, which offer more continuous cliffs and canyons. Atlasers did not find the wrens anywhere in the San Luis Valley, but in 1996 three birds sang in the Rio Grande canyon near New Mexico (HEK).

With apparently stable populations, they face no known threats although rock climbers destroyed one nest west of Denver (SLJ). This beautiful and distinctive song should continue to echo throughout the canyons of Colorado and delight all who hear it.

BREEDING EVIDENCE

REPORTED IN 160 (9%) OF 1745 PRIORITY BLOCKS

☐	Possible	76 blocks	(47%)
▪	Probable	41 blocks	(26%)
■	Confirmed	43 blocks	(27%)

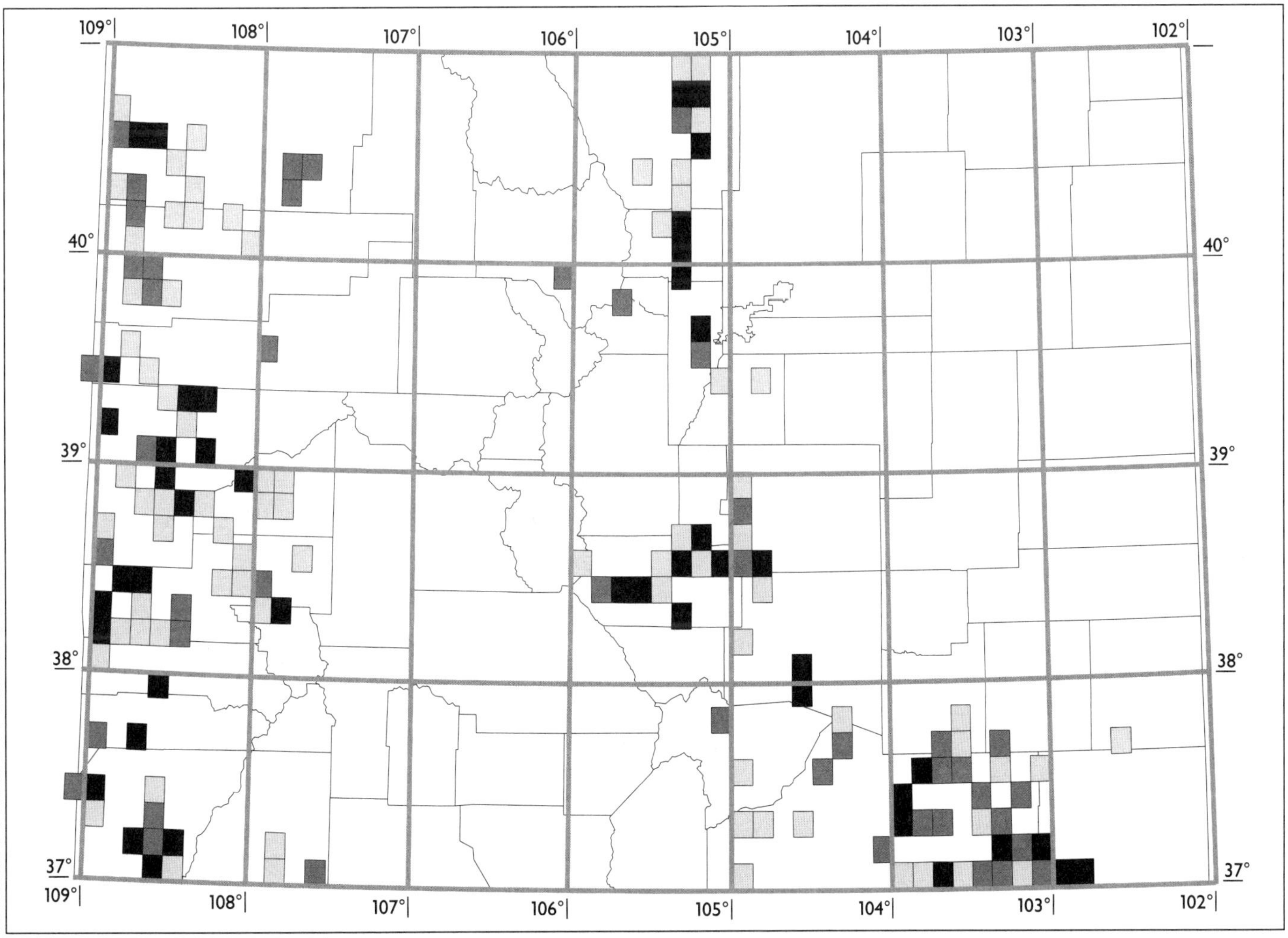
109°
108°
107°
106°
105°
104°
103°
102°
40°
39°
38°
37°

COLORADO DIVISION OF WILDLIFE / WILDLIFE RESOURCE INFORMATION SYSTEM

Only a featherweight, and absolutely incapable of inflicting physical harm, the feisty Bewick's Wren nonetheless rings out its song forcefully and berates intruders on its premises like a pugnacious bouncer.

HABITAT

Throughout their range, Bewick's Wrens occupy a variety of habitats that contain dense brush with some openings (Shuford 1993). B&N indicated that in Colorado they inhabit "dry canyon and pinyon/juniper forest." According to A&R, in southeastern Colorado this species also nests, in addition to arid shrublands and pnyon/juniper woodlands, in a wetter habitat: lowland riparian

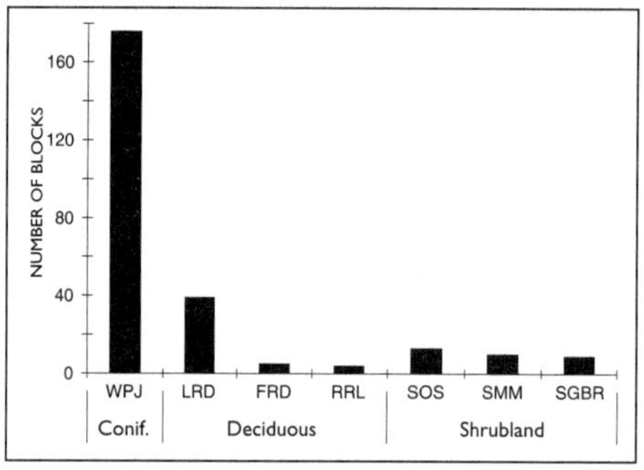
NUMBER OF BLOCKS
160
120
80
40
0
WPJ
LRD
FRD
RRL
SOS
SMM
SGBR
Conif.
Deciduous
Shrubland

forests. Atlasers found them in all of these places and in the riparian forests along the lower Colorado and Gunnison valleys in western Colorado as well. These seemingly disparate habitats all support dense brush that these birds require—juniper in the dry uplands, greasewood and threeleaf sumac along the Western Slope watercourses. On the Eastern Slope, riparian areas that they select incline toward rough, rock-strewn creek bottoms with a variety of shrubby growth.

Atlas work confirmed a strong preference by Bewick's Wrens in Colorado for pinyon/juniper, this preference stronger in western Colorado. Atlasers listed pinyon/juniper in 81% of the Western Slope Priority blocks compared to 67% of those on the Eastern Slope. Riparian habitats accounted for most of the rest—14% on the Western Slope, 28% on the Eastern Slope. The small stands of juniper scattered throughout the sagebrush hills of northwestern Colorado nearly always hosted a few pairs. Semi-desert brushland (STD) appeared only three times as a habitat for this species, and sagebrush only seven times, all in the northwestern corner of the state and generally in association with pinyon/juniper.

BREEDING

Bewick's Wrens carefully conceal their nests in woodpecker holes, knotholes, and other natural cavities, all abundant in juniper and cottonwood trees. They will use virtually any sort of hole or cranny close to the ground (Ryser 1985). B&N noted that nests of this species usually elude detection, and they listed only five records of nests with eggs from Colorado. Atlas workers also experienced difficulty in finding nests; during the project, they found only seven nests—including one birdhouse nest reported by a landowner in Las Animas County.

Atlas dates reflect a lengthy breeding season for this species. Field workers reported multiple males singing on territory as early as 30 March and as late as 4 July. Of 22 multiple-singing-male records, nearly half occurred in May. Unfortunately, Atlas workers did not find any nests with eggs or with young; dates reported in B&N range from 13 May to 10 June. Incubation and nestling periods of 14 days each (Terres 1980) indicate that some birds begin incubating in late April, because Atlas workers first found adults carrying food on 1 May. Atlasers observed adults carrying food as late as 19 July—an 80-day period—and they found fledglings from 29 May to 6 August—a 70-day period. They made only two observations of nest-building and three of birds on nests; these all occurred in June, weeks after the earliest reports of fledglings and adults carrying food. These dates strongly suggest that at least some Bewick's Wrens raise two broods.

BREEDING PHENOLOGY

CODE		# OF RECORDS	DATE RANGE
N	Nest Building	2	8 Jun–19 Jun
ON	Occupied Nest	3	16 Jun–25 Jun
FY	Feeding Young	39	1 May–19 Jul
FL	Fledged Young	27	29 May–6 Aug

DISTRIBUTION

Bewick's Wrens breed from the Appalachians west to the Pacific coast and south into Mexico, and winter in the southern U.S. to south-central Mexico. The Atlas map generally supports earlier descriptions of Bewick's Wren distribution in Colorado (B&N, A&R). It outlines a Colorado range that rather closely follows that of the pinyon/juniper woodlands found in the foothills and mesas of western and southeastern parts of the state. These wrens also inhabit the riparian lowlands along the lower Colorado River and the Arkansas River bottomlands, but only as far east as Crowley County (Mark Janos pers. comm.).

The Atlas range map generally coincides with the map in A&R, although the Atlas depicts a more extensive range in Moffat County. House Wrens replace Bewick's at approximately 7,000 feet (2135 m) and also replace them in lowland riparian habitat in most of the state. Whether differences in habitat structure or dominance by House Wrens accounts for this displacement remains uncertain (Kroodsma 1973, Ryser 1985, Roberson and Tenney 1993).

When atlasers found Bewick's Wrens, they usually encountered dozens of them, reporting A3 (11–100) in slightly more than half the blocks. They reported A5 in two blocks, Brown Canyon and Sheep Canyon (37103E3, 37103F7), both in the southeastern pinyon/juniper hills, a center of abundance for this bird in Colorado.

As cavity-nesters, these wrens face little threat from cowbird parasitism, although the Atlas contributed two records (Chace and Cruz 1996; Appendix C). As pinyon/juniper residents, they face little threat from habitat destruction. Livestock grazing has reduced herbaceous plants—particularly native grasses—and has led to less frequent fires, allowing pinyon/juniper to expand significantly since settlement (West 1984).

BREEDING EVIDENCE

REPORTED IN 212 (12%) OF 1745 PRIORITY BLOCKS

☐	Possible	69 blocks	(33%)
■ (grey)	Probable	68 blocks	(32%)
■	Confirmed	75 blocks	(35%)

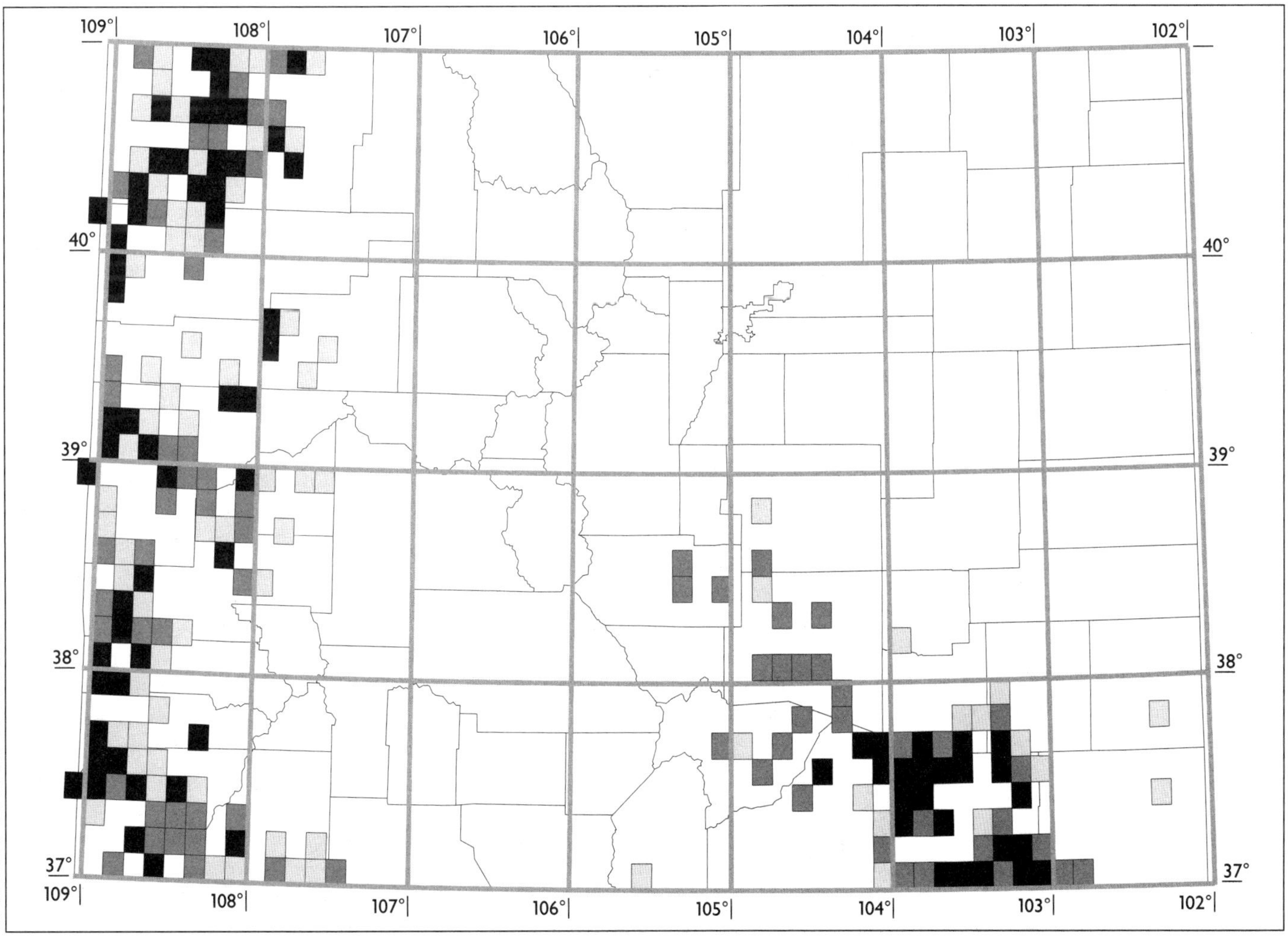
109°
108°
107°
106°
105°
104°
103°
102°
40°
39°
38°
37°

COLORADO DIVISION OF WILDLIFE / WILDLIFE RESOURCE INFORMATION SYSTEM

How much can you pack into the smallest space? Take the lungs and skills of the finest opera soprano. Add the strength and endurance of Paul Bunyan. Throw in the courage of a Doberman (and the foolhardiness of a Chihuahua). Squeeze it all into a half-ounce ball (less than 10 grams) and you have a House Wren. You have one of the best singers in the Rockies; you have a miniature tornado that readily and audaciously takes on all who offend it. Even the size, teeth, and appetite of a pine marten don't faze this tiny whirlwind.

HABITAT

The Atlas's twelfth most frequently recorded species, House Wrens display catholic habitat tastes, although Atlas reports tilt toward the deciduous. In half the Atlas blocks House Wrens used aspen or deciduous riparian woodlands (plains and mountains). On the plains wrens congregate densely along the stream bottoms, yet they avoid urban areas. The deciduous presence becomes stronger by adding rural, another 10%; largely deciduous, farms and ranches often have cottonwoods and Siberian elms planted around them.

Forested parts of western Colorado hold the bulk of Atlas records. In the mountains, wrens occupy most habitats with trees (especially aspens and streamside cottonwoods) to the upper limits of the aspen groves. Aspens housed most of the nests found: 21 of 30 nest cards identified aspen. Wrens will occupy small patches of aspens within lodgepole pine or spruce/fir forests. The map has empty spaces in croplands and grazing lands, along the Continental Divide and other high mountain ridges, and in the open greasewood and sagebrush parks.

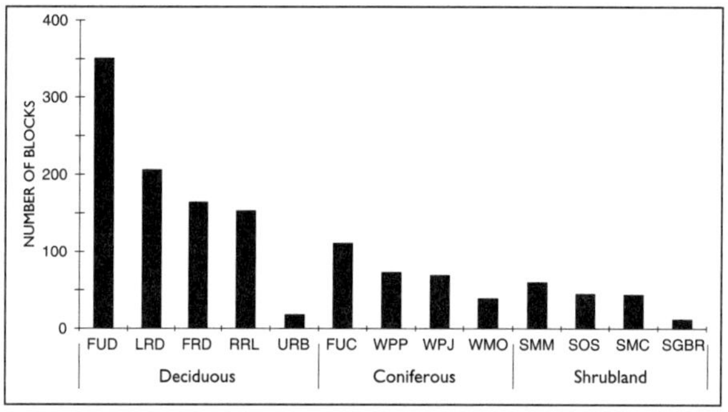
NUMBER OF BLOCKS
400
300
200
100
0
FUD LRD FRD RRL URB FUC WPP WPJ WMO SMM SOS SMC SGBR
Deciduous
Coniferous
Shrubland

In the canyon country of Las Animas County and the sagebrush/pinyon/juniper country of Moffat County, few blocks reported them; many without House Wrens had Bewick's Wrens, a species partial to pinyon/juniper woodlands. In Latilongs 37102–37104, Bewick's or House wrens occurred in 91 blocks but shared only one-sixth. Habitat separated them: House wrens selected riparian habitats and shunned pinyon/junipers.

BREEDING

House Wrens are not particular about what they nest in, as long as it has a cavity and an entrance hole. Nest sites range from old trees and birdhouses to gutter pipes, old boots, abandoned autos, and stranger places. Atlasers found nests in the brain cavity of a cow skull hanging on the outside wall of a shed (NMB) and on a light fixture (Marina Graves, nest card). More typically, though, trees with old woodpecker holes and broken-off limbs provide the cavities prized by these birds.

The high number of blocks with Confirmed breeding reflects the high visibility of House Wrens. Only 15 species had higher Confirmation percentages. Of course, with their noisy, aggressive tactics, they proved hard to miss in the woods.

Male House Wrens seeking mates usually build one or more dummy nests, so observations of nest-building do not necessarily represent actual breeding. Extra nests may function to make males more attractive to females (Terres 1980). Building multiple nests requires energy, but these bird dynamos have plenty of that. The males need extra energy because they raise the young with little help from the female. Females often mate with more than one male at the same time (Johnson and Searcy 1993, Johnson 1993).

The practice of building ancillary nests signals the beginning of breeding activities. Colorado wrens start in May, with some birds still at it in July. Reports of young in nests began in late May, and one nest had young in early August. Atlas data do not show a real peak in breeding activity; instead atlasers consistently found nests from 11 May through 2 August.

House Wrens have tiny territories—one-quarter to one-half acre (0.1–0.5 ha; Sydlik *in* Brewer et al.). Density in a cottonwood grove at Chatfield State Park varied over nine years from 255 to 360 territorial males/100 acres (630–890/100 ha; Bottorff et al. 1971–1979).

Atlas abundance codes show the highest aggregations of House Wrens in forested blocks: 40% of the blocks with aspen and 30% of the blocks with conifers earned A4s and A5s. Ponderosa blocks had many of the two higher abundance codes (35%) but pinyon/juniper had only 14%. Although House Wrens crowd close together along riparian stream bottoms, the prairie and plateau rivers run like green ribbons across the landscape and provide only narrow corridors for wrens—only 18% of these blocks had A4 or A5 abundance codes. Rural habitats had primarily A2 and A3 codes.

In another activity that seems to spend energy without benefit, House Wrens nesting near Rocky Mountain National Park fed baby bluebirds in a neighboring hole (Pollock 1991). House Wrens are excellent neighbors in other ways. One physically attacked a pine marten attempting to climb an aspen with five active bird nests while the other species simply flew around the area, squawking and chipping. The marten fled (NMB).

BREEDING PHENOLOGY

CODE		# OF RECORDS	DATE RANGE
C	Courtship	6	19 May–30 Jun
N	Nest Building	31	5 May–25 Jul
ON	Occupied Nest	131	11 May–20 Jul
NE	Nest with Eggs	4	10 Jun–14 Jul
NY	Nest with Young	141	27 May–2 Aug
FY	Feeding Young	232	16 May–25 Aug
FL	Fledged Young	159	13 May–18 Aug

DISTRIBUTION

House Wrens breed throughout the Americas; they winter from the southern U.S. south. Throughout Colorado their distribution mirrors the presence or absence of trees.

Several nest cards recorded nests in the 9,000–10,000-foot band (2745–3048-m). Most authors place an upper breeding limit of 10,000 feet (3048 m; Cooke 1897, B&N) or 11,000 feet (3350 m; A&R). Atlas observations in two blocks push higher—a singing male (Bowers Peak, 37106H5) and nest-building (Ojito Peak, 37105C3), both at 11,800 feet (3597 m). The habitats, at timberline, consisted of old burns with snags providing potential nesting cavities (Righter et al. 1989, Alan Wallace pers. comm.).

Various types of habitat manipulation (fire suppression, logging) and water management (dams, flood control) affect the aspen groves and prairie streams where House Wrens nest. These little whirlwinds, with their adaptable and aggressive nature, use both deciduous and conifer habitats and can persist in the face of the changes man makes to the natural world.

BREEDING EVIDENCE

REPORTED IN 976 (56%) OF 1745 PRIORITY BLOCKS

☐	Possible	154 blocks	(16%)
▩	Probable	167 blocks	(17%)
■	Confirmed	655 blocks	(67%)

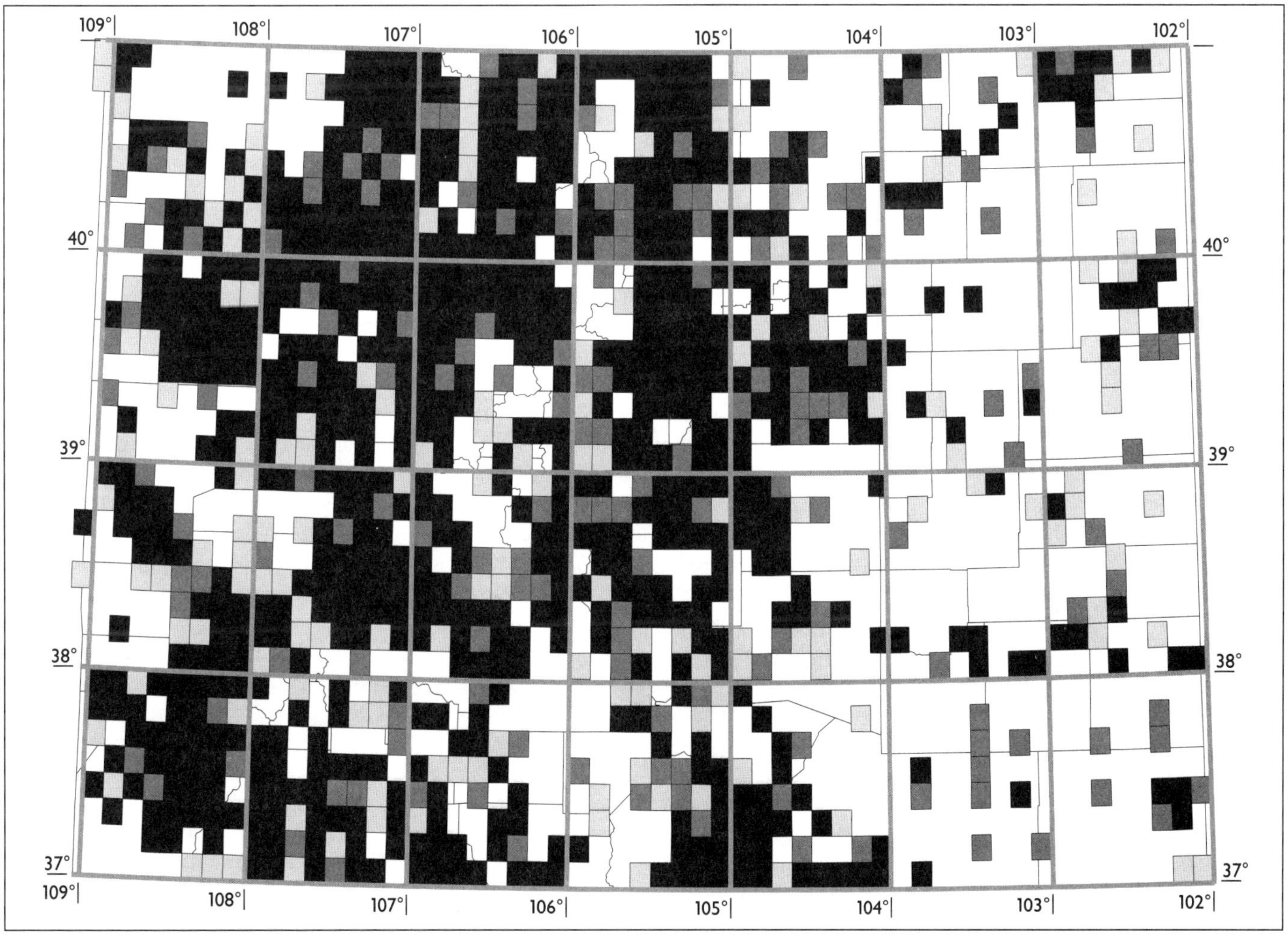
109°
108°
107°
106°
105°
104°
103°
102°
40°
39°
38°
37°

COLORADO DIVISION OF WILDLIFE / WILDLIFE RESOURCE INFORMATION SYSTEM

The reedy, bubbling song of the Marsh Wren belies the aggressive character of this species. Males dart invisibly through the cattails and sing much of the time, day and night, throughout the breeding season. They also display by singing in flight (Kroodsma and Verner 1978). All is not idyllic, however: after coupling, males sometimes attack their own mates' nests and puncture the eggs or kill the nestlings (Picman 1977a, 1977b).

HABITAT

Marsh Wrens breed only in wetland swamps and marshes with an abundance of emergent vegetation such as cattails and reeds. Across their range various subspecies use a variety of nest sites, although Colorado birds almost invariably build in cattails.

They build characteristic domed nests in cattail marshes ("emergent wetlands" in the language of the Atlas habitat code). The Colorado marshes they use are fairly large—usually. In one block in the San Luis Valley (Swede Corners, 37106H2) a pair attended fledglings in a tenth-of-a-mile-long willow and cattail marsh between a fence line and a county road (HEK). Only one block reported a different habitat—a male singing from a patch of willows along the White River

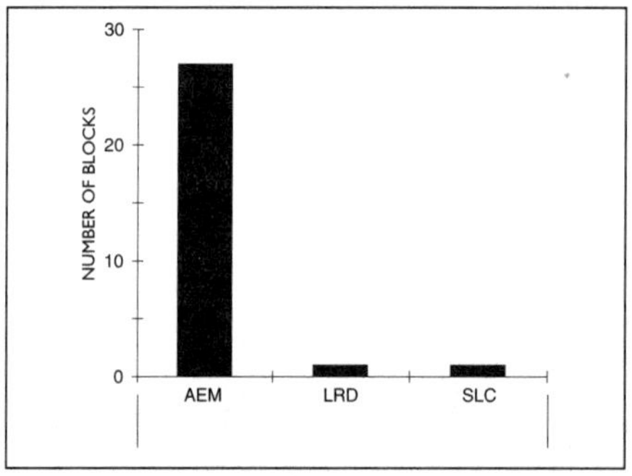

NUMBER OF BLOCKS
30
20
10
0
AEM
LRD
SLC

upstream from Rangely—and Marsh Wrens nest in willows at Arapaho NWR (Stephanie Jones pers. comm.).

A major predator in these marshes, these wrens persistently destroy the nests and eggs of other marsh birds. Blackbirds attack Marsh Wrens when they encounter them and often nest in marsh units segregated from the gangs of wrens. In one marsh, Marsh Wrens destroyed or disrupted 5% of Yellow-headed Blackbird nests and nesting attempts (Bump 1986) and, in another, up to 52% of Red-winged Blackbird nests (Picman 1977a, 1977b). Most baffling, Marsh Wrens also attack eggs and nestlings of other Marsh Wrens—even those of their own mates. Female Marsh Wrens defend their nests vigorously against all wrens who dare to come too close. A male may puncture the eggs and kill his mate's nestlings if she leaves them unguarded, and even females attack neighbors' nestlings (Kroodsma and Verner 1997). The soft, shock-absorbing nest lining cushions, and thicker-than-average shells protect, the eggs from the depredations of invading wrens trying to puncture the eggs (Picman 1977b). As a climax to this eccentric behavior, males assist in feeding fledged young, at least in some populations (Don Kroodsma pers. comm.).

BREEDING

Individual males may build an extraordinary number of nests. The average number of nests correlates with the success of the male in attracting a mate(s). Successful males build more nests. In a Washington study 13 bachelor males *averaged* 17.4 nests apiece, monogamous males averaged 22.1 nests, and bigamous males averaged 24.9 nests (Ryser 1985, Verner and Engelsen 1970).

The selective advantage to this behavior remains a mystery; the dummy nests could function as decoys to avoid predation or they may cue the females in selecting the most vigorous mates (Leonard and Picman 1987). Males may use unlined dummy nests as courting sites (Kroodsma and Verner 1997). After pairing the female either builds a new brood nest or may line one of the dummy nests with grasses and other vegetation (Verner 1965). Marsh Wrens also use dummy nests as winter and night roosting sites (Bent 1948).

The birds attach the nests to cattails or other plants, 1–3 feet (0.3–0.9 m) above water (Kroodsma and Verner 1997). The nest, globular in shape with a 5-inch (13-cm) diameter (Sclater 1912), is built in layers, with a side-entrance usually below the "equator" (Don Kroodsma pers. comm.), commonly facing south or east (Verner 1965).

Throughout the state atlasers found Marsh Wrens singing in June. Nest-building proceeded sequentially northward: 15 May in the San Luis Valley and, north of the 40th parallel, 10 June at Craig and 19 July in North Park. Atlas observations of nests and fledglings did not show the same geographical progression: all such Confirmations, San Luis Valley and north of Latitude 40, came in July.

BREEDING PHENOLOGY

CODE		# OF RECORDS	DATE RANGE
N	Nest Building	3	15 May–15 Jul
ON	Occupied Nest	3	7 Jul–30 Jul
FY	Feeding Young	1	31 Jul
FL	Fledged Young	1	20 Jul

DISTRIBUTION

Marsh Wrens breed in a belt that crosses the northern U.S. and southern Canada. Resident on both coasts, they also winter in the southern U.S. to central Mexico.

Cooke (1897) thought that more summered in southern than in northern Colorado, and more at the base of the foothills than farther east, a pattern to which the Atlas map seems to conform. They do not breed in all suitable marshes (A&R). Like the reasons for many facets of their bizarre breeding behavior, the factors limiting their breeding and distribution in Colorado are undetermined.

Despite many apparently suitable marshes across the state, Marsh Wrens breed only in the scattered places shown on the Atlas map. The San Luis Valley has the only extensive breeding population; cattail marshes from Denver north hold the only other Colorado concentration. The wrens do not breed in many marshes where they migrate or in plains wetlands where they spend the winter.

Cooke (1900) reported that Carter had found them breeding in South Park, but B&N cited no later records from that high-altitude expanse (which has few, if any, cattail marshes), nor did atlasers find them there.

Apparently always infrequent in Colorado, Marsh Wrens have benefited from the cattail marshes that burgeon around irrigation reservoirs and their down-dam seepage sloughs. Habitat destruction, including the draining and filling of wetlands, may affect Colorado Marsh Wrens, but their low numbers or absence from many suitable marshes also constrains their numbers. The state has a population probably more or less equivalent to peak historic numbers.

BREEDING EVIDENCE

REPORTED IN 19 (1%) OF 1745 PRIORITY BLOCKS

Possible	8 blocks	(42%)
Probable	5 blocks	(26%)
Confirmed	6 blocks	(32%)

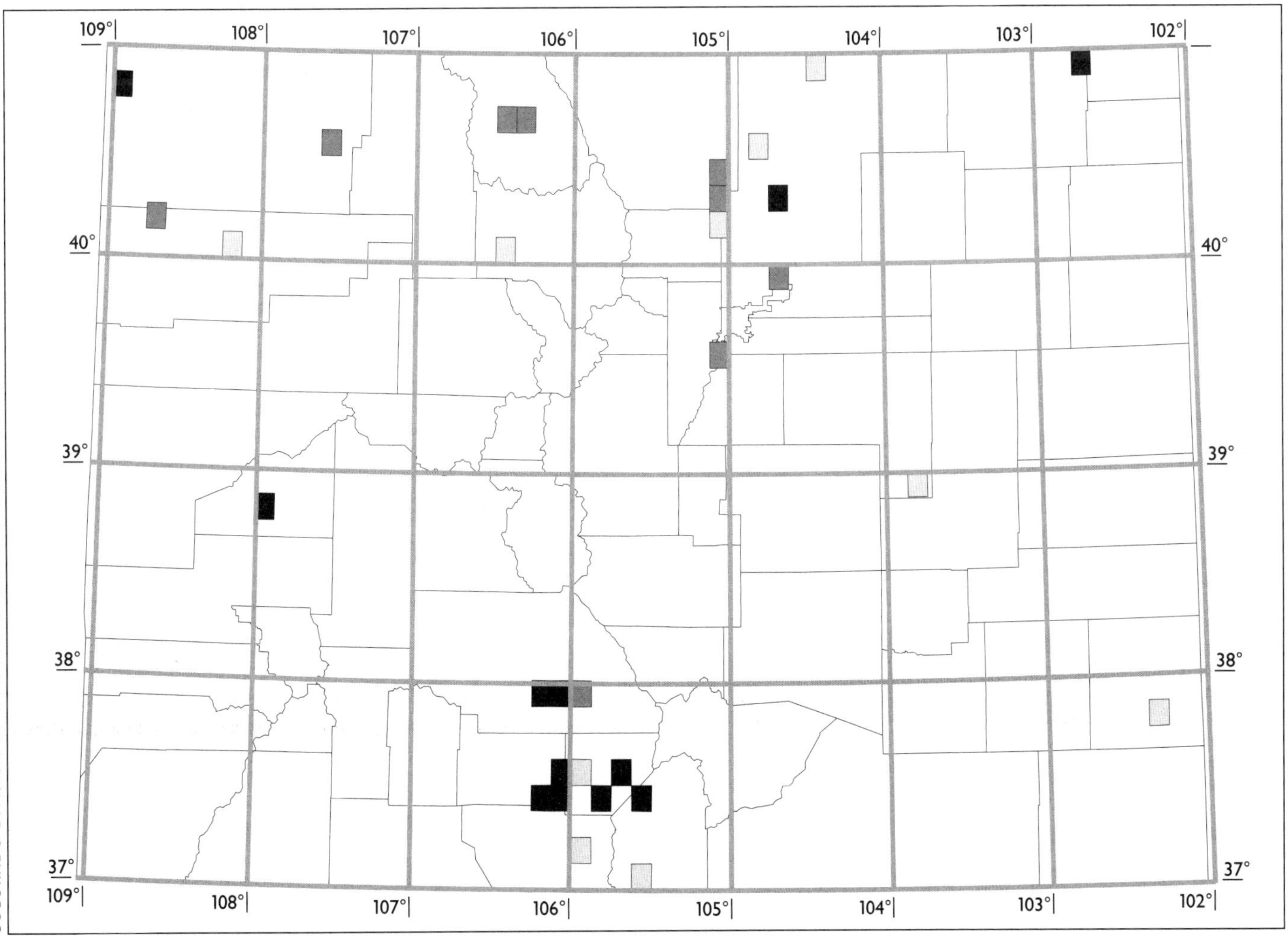
109°
108°
107°
106°
105°
104°
103°
102°
40°
39°
38°
37°

COLORADO DIVISION OF WILDLIFE / WILDLIFE RESOURCE INFORMATION SYSTEM

AMERICAN DIPPER *Cinclus mexicanus*

Dippers sing the music of the streams and act like the tumbling cataracts. Moving constantly, they chase upstream and down, and chatter louder than waterfalls. Their song proclaims "the deep booming notes of the falls, the trills of the rapids, the gurgling of margin eddies, the low whispering of level reaches, and the sweet tinkle of drops oozing from mosses" (Muir 1894).

HABITAT

As Colorado's only aquatic passerines, American Dippers spend their lives along fast-moving streams. Here they find food among the noisy cascades, riffles, and waterfalls, and find nest sites and shelter in adjacent cliffs, rocks, and stream banks.

Nest sites limit distribution. Dippers use natural and manmade nest sites that share three basic requirements: the obvious one, proximity to water, usually fast water, usually very noisy; security from predators and floods; and a ledge or crevice for support (Ealey 1977).

Dippers eat every type and stage of stream insect (Mitchell 1968), which they

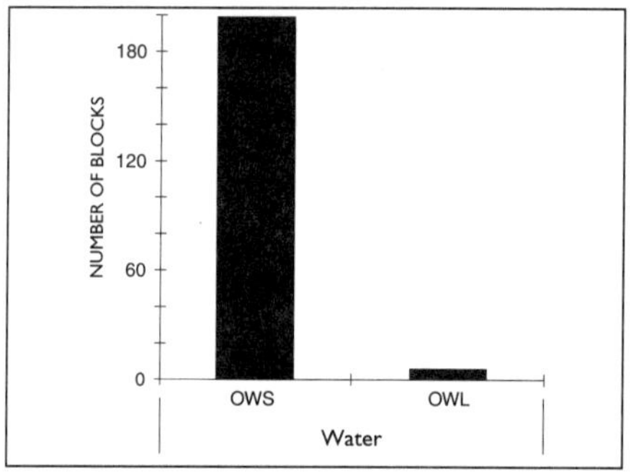

180
120
60
0
NUMBER OF BLOCKS
OWS
OWL
Water

catch while diving, swimming, or standing. When diving, dippers plunge into the stream head first, legs extended, propelled by their wings (Goodge 1959). To subdue large prey (large larvae or an occasional fish) a dipper may shake or slam it on a rock for as long as five minutes.

Atlasers found dippers in streams throughout the mountains, regardless of streamside habitat. Dipper streams run through anything from ponderosa pine to aspen woodlands to dense spruce/fir forests. On a 1994 Atlas Rendezvous in the San Juan Mountains, atlasers found, at 11,820 feet (3602 m) in Hinsdale County, the highest dipper nest ever recorded (Jack Neiman, nest card; see photograph in the Color Folio).

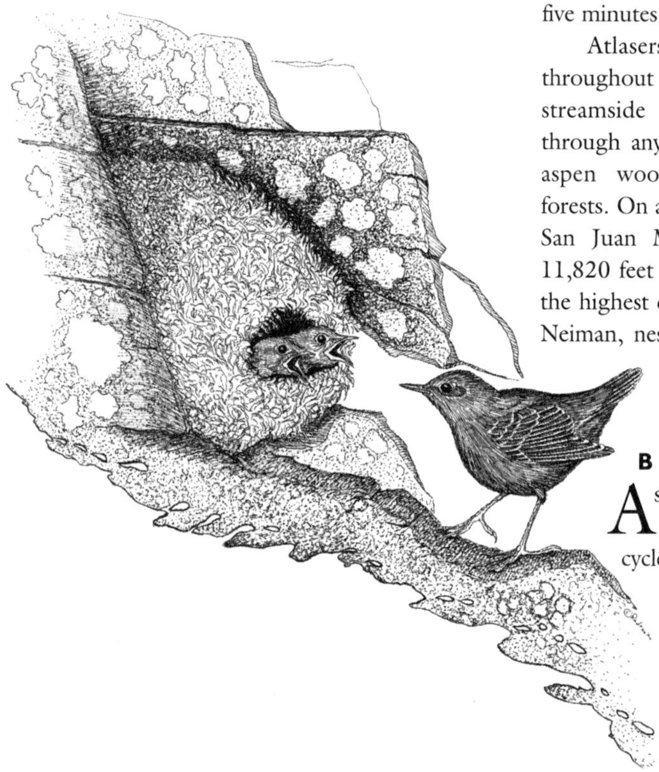

BREEDING

As ice leaves the winter streams, dippers begin their nesting cycle. Prolonged chases along the stream and ten-minute singing bouts confirm pair bonding and set territory boundaries. The start of breeding varies from March to May depending on snowmelt (Price and Bock 1983). Early and high spring runoff delays breeding, but if the snow melts slowly, breeding begins early. On Bear Creek at the foothills' edge, a pair started rebuilding a previous year's nest on 21 February 1996 (HEK). In early seasons, at lower elevations, they may start second broods. None tried second broods above 6,000 feet (1830 m) in Boulder County (Price and Bock 1983) or at 9,500 feet (2900 m) in Gunnison County (Hann 1950). Atlas dates show nesting in March and April, early enough to permit double broods, below 7,600 feet (2300 m) in Boulder and Eagle counties.

Dippers do not need a misty nest site—only about half build where spray keeps the nest wet (Ealey 1977). Yet they dip moss in the stream before stuffing it into the nest (Kingery 1996). Females build the nests by standing inside the nest and forcing the material "into interstices of an already constructed and partially dry wall . . . as a shoemaker uses his awl" (Gale *in* Henderson 1908). The nest has an outside shell made of moss and grass and an inside globular cup of grass and leaves of streamside deciduous shrubs.

Polygyny by dippers depends on nest-site availability: a male with several nest sites in his territory can accommodate a second mate. Four of 31 pairs on site-poor Boulder Creek (Boulder County) practiced polygyny (Price and Bock 1983).

When the young fledge, the parents may divide their stream territory and their brood. Each parent feeds and tends its coterie, and both teach dipper manners by singing to the fledglings after feeding them (Fite 1984).

Few predators catch dippers—an occasional *Accipiter*, probably some mammals, and, rarely, a heron or a trout (Kingery 1996). Dippers can flatten themselves on the water surface so that a hawk seeking a meal sees only a stick-like form floating downstream (Sullivan 1973).

BREEDING PHENOLOGY

CODE		# OF RECORDS	DATE RANGE
NB	Nest Building	6	21 Feb–2 Jun
ON	Occupied Nest	18	14 May–23 Jul
NE	Nest with Eggs	1	22 Apr
NY	Nest with Young	23	15 May–4 Aug
FY	Feeding Young	38	20 Apr–18 Aug
FL	Fledged Young	29	15 May–17 Aug

DISTRIBUTION

American Dippers inhabit western mountain, coastal, and desert streams from Alaska to Panama, and they migrate, if at all, only far enough to escape freeze-up.

Rattles louder than stream babble, extraordinary songs, flamboyant chases, and hungry squalling fledglings helped atlasers find nesting dippers. Dippers do not nest in several areas shown on the A&R map (which combines summer and winter ranges). The mountain parks, lower stretches of the big western rivers, and the western Colorado plateaus lack streams with appropriate habitat, as do the valleys and foothills east of the Sangre de Cristo Range from South Park to New Mexico.

In the heart of the range, the Atlas map shows discontinuities. This reflects both reality and artifacts of Atlas block selection. Streams that cross broad mountain valleys lack adequate nest sites. Where some Priority blocks lack suitable nest sites although contiguous Non-priority blocks have them, the map also shows gaps. This occurred most noticeably in the narrow Sangre de Cristo Range. In sum, the map accurately outlines the breeding range in Colorado but lacks the continuity of their actual range.

Man's effect on dippers parallels man's treatment of mountain streams. Dippers benefit from some human intrusions into their wilderness: road and foot bridges—and recently, nest boxes—have expanded dipper habitat to streams deficient in natural nest sites (Sullivan 1973, Steve Bouricius pers. comm.). They do not benefit from dams that flood mountain valleys that once welcomed their music. Streams polluted by sediment from poorly managed lands or by acidity (from natural sources or from mine and chemical wastes) poison dipper food supply. Overall, dippers have both gained and lost from human settlement, and their songs will sparkle on Colorado high-country streams for as long as we protect the high-country water quality.

BREEDING EVIDENCE

REPORTED IN 251 (14%) OF 1745 PRIORITY BLOCKS

☐	Possible	91 blocks	(36%)
▩	Probable	17 blocks	(7%)
■	Confirmed	143 blocks	(57%)

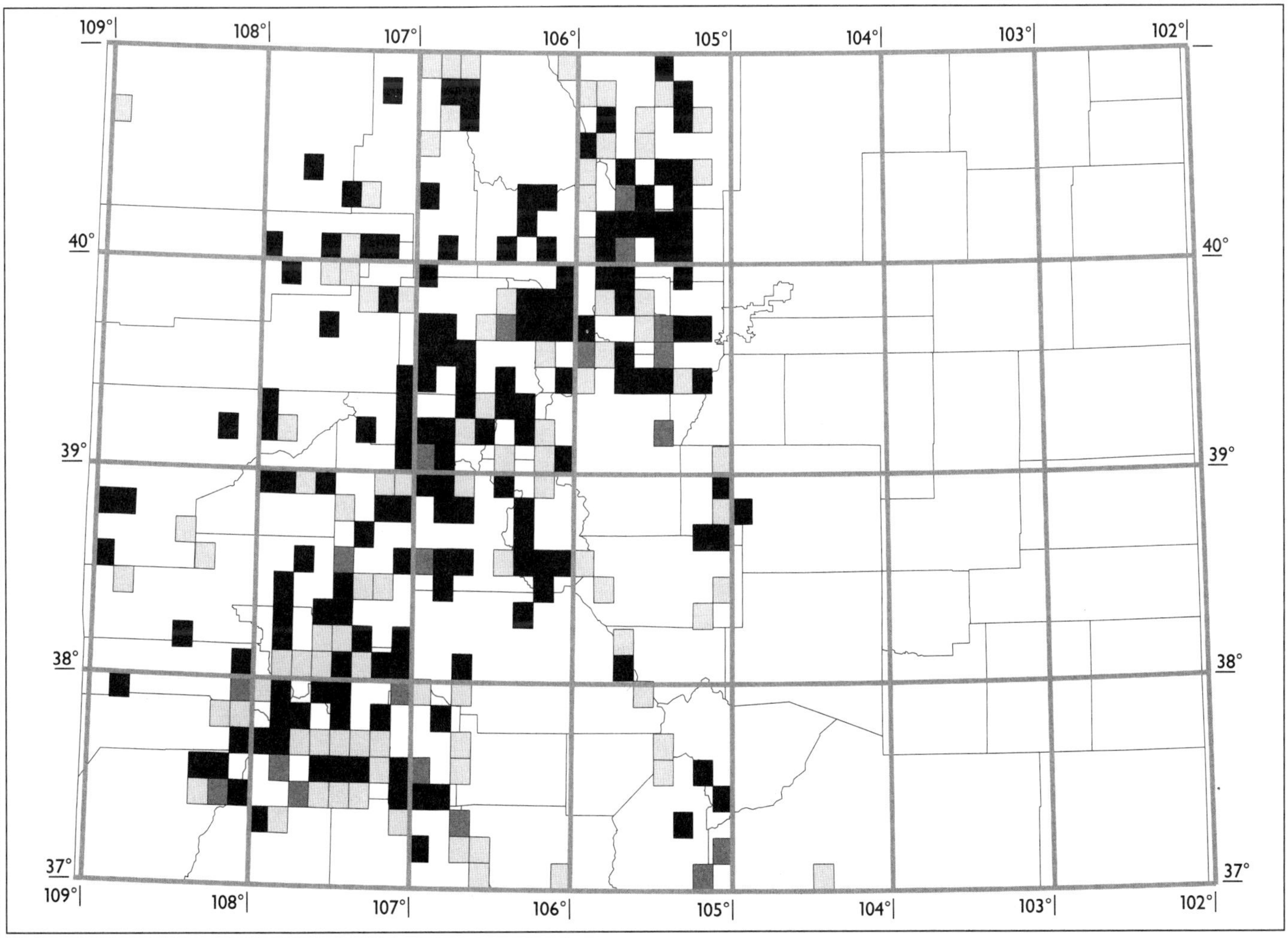

109°
108°
107°
106°
105°
104°
103°
102°
40°
39°
38°
37°

COLORADO DIVISION OF WILDLIFE / WILDLIFE RESOURCE INFORMATION SYSTEM

Among the smallest passerines in the world at two ounces, Golden-crowned Kinglets weigh in only slightly heavier than most hummingbirds (Heinrich 1993). Surprisingly these diminutive insectivores do not claim Neotropical migratory status; they remain in Colorado as year-round conifer specialists.

HABITAT

Birds of the forest interior, Golden-crowned Kinglets use old-growth coniferous forests or stands with old-growth characteristics (Rick Thompson unpubl. data). Most abundant in spruce/fir, kinglets occur from 7,600 to 11,600 feet (2316–3536 m). Where atlasers designated tree type, they specified spruce/fir almost exclusively. Atlas data show a very limited breeding presence in Colorado blue spruce along foothills streams, lodgepole pine forests, Douglas-fir, and ponderosa pine.

These kinglets forage for insects in tall, dense conifers. Concentrating at medium

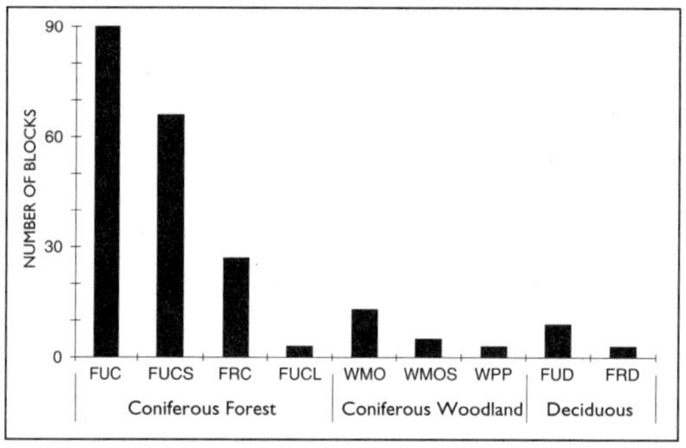

NUMBER OF BLOCKS
90
60
30
0
FUC
FUCS
FRC
FUCL
WMO
WMOS
WPP
FUD
FRD
Coniferous Forest
Coniferous Woodland
Deciduous

heights, they capture prey primarily by gleaning on foliage (Franzreb 1984, Airola and Barrett 1985). With a specialized foot structure adapted for clinging, they can grasp and cling at a pronounced downward angle from conifer needles. Secondarily, kinglets capture prey by hovering, making short flights to hang in front of twig tips to pick off insects (Keast and Saunders 1991). They expend most foraging efforts at the tips of boughs in foliage, on small twigs and branches, and only rarely on trunks, cones, shrubs, logs, or on the ground (Shuford 1993). Kinglets flick their wings nervously while moving amid foliage. Fidgety movements, sibilant songs, high-pitched call notes, and small size convey an impression of juvenility (Jewett et al. 1953).

Their diet consists predominantly of arthropods—adults, larvae, and eggs of wasps, ants, true bugs, flies, beetles, moths, butterflies, caterpillars, spiders, pseudo-scorpions—and very rarely fruits and seeds (Keast and Saunders 1991).

More particular than Ruby-crowned Kinglets, Golden-crowned Kinglets generally occupy mature, dense old-growth trees. They often share their habitat with Brown Creepers, and both species have similar, high-pitched *see* calls. Atlas workers may have missed some kinglets because of the difficulty of hearing the songs and calls in dense spruce/fir.

BREEDING

Kinglets weave a pensile nest, resembling those of bushtits and orioles, made of fine bark, lichens, mosses, webs, and hair. The materials, interwoven over the entrance, conceal both nest and female from observers below, while long conifer branches conceal the nest from weather and predators above. They line the nests with feathers of ground birds such as grouse and Hermit Thrushes. The usually lofty nest placement ranges from 6 to 60 feet (1.8–18.3 m) high. Females, by themselves, build nests, usually in spruces. The nest has an arched, 1-inch diameter entrance with an inside diameter of 1.75 inches (Tufts 1961, Harrison 1979).

The nest is so tiny and the female lays so many eggs (8–9) that she has to deposit the eggs in two layers (Bent 1949). She incubates for 14–15 days while the male patrols the territory within a radius of 50 yards (46 m). The female leaves the nest about every 15 minutes and joins the male, who offers her food that she accepts with fluttering wings, suggestive of fledgling behavior (Tufts 1961). Exceptional productivity may counterbalance heavy winter loss; most years a pair produces 7–12 young from two broods (Heinrich 1993).

Atlas data show a breeding season beginning 28 May with nest construction and ending in late August with fledgling activities. This range of dates conforms to a two-brood nesting pattern. Of the total breeding Confirmations, 92% occurred during the second half of the breeding season. May and June access to the high country, restricted by lingering deep snow, probably caused this bias. Fledgling-related observations—FY, FF, FL—accounted for 84% of the Confirmations.

BREEDING PHENOLOGY

CODE		# OF RECORDS	DATE RANGE
C	Courtship	2	28 Jun–25 Jul
NB	Nest Building	3	28 May–22 Jun
ON	Occupied Nest	1	28 Jun
NY	Nest with Young	2	23 Jul–25 Jul
FY	Feeding Young	30	18 Jun–24 Aug
FL	Fledged Young	37	30 Jun–30 Aug

BY RICHARD ROTH AND KIM M. POTTER

DISTRIBUTION

Golden-crowned Kinglets inhabit coniferous forests year round across the northern U.S. and Canada; in winter they withdraw from the northern edge of their breeding range. Atlas data show that Colorado Golden-crowneds, although never abundant, occur mostly west of the Continental Divide.

Historically, ornithologists thought that Golden-crowned Kinglets bred rarely in the state, and only near timberline (Cooke 1897). Rockwell (1908) wrote that Mesa County had no place "of sufficient altitude to attract these birds during the breeding season." The abundance and elevation ranges of breeding kinglets remained little understood until publication of A&R in 1992.

The Atlas map presents the fullest picture yet of Golden-crowned Kinglets in Colorado. They occupy spruce/fir forests in most of the high country. This includes the Flat Tops, Gore Range, Grand Mesa, West Elks, and northern San Juans, sites not previously mapped for them. Atlas work found them breeding in Mesa County and the first Confirmation of breeding in Latilong 37108. Atlasers also discovered Golden-crowneds in montane habitats on the Uncompahgre Plateau and Battlement Mesa. They did not find them along the Continental Divide in Saguache County (the Cochetopa Hills), where the dry spruce/fir forests may lack the attributes of old growth that these kinglets require.

Kinglet numbers vary from year to year, and in some localities they become exceedingly rare for an entire breeding season (Jewett et al. 1953). Populations of New Hampshire breeders apparently fluctuate because of severe weather on their wintering grounds south of New England (Sabo 1980), but their Colorado movements seem tied to some factor other than winter weather.

Populations of kinglets probably experienced declines at the turn of the century because of intense logging and mining activities. Even small clear-cuts cause a significant drop in their numbers (Scott et al. 1982). Populations subsequently recovered with regeneration of dense forest due to fire suppression, changes in logging techniques, and, more recently, wilderness designations (Shuford 1993, Kit Buell pers. comm.). Many of the new sites atlasers found lie in or near wilderness areas—Flat Tops, Eagles Nest, West Elks, and Big Blue.

BREEDING EVIDENCE

REPORTED IN 188 (11%) OF 1745 PRIORITY BLOCKS

☐	Possible	79 blocks	(42%)
■	Probable	42 blocks	(22%)
■	Confirmed	67 blocks	(36%)

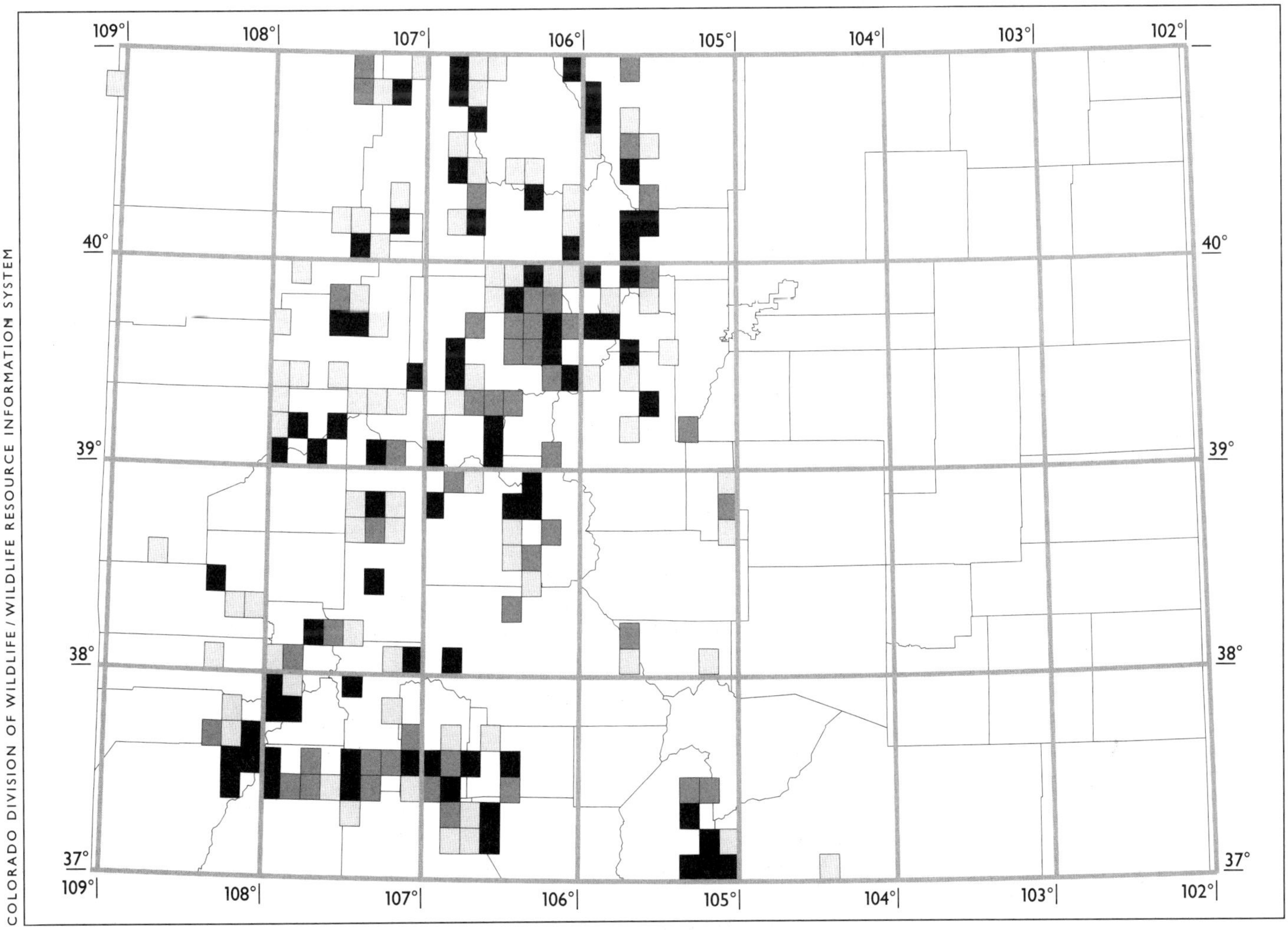
109°
108°
107°
106°
105°
104°
103°
102°
40°
39°
38°
37°

COLORADO DIVISION OF WILDLIFE / WILDLIFE RESOURCE INFORMATION SYSTEM

One of the loudest songs in the coniferous forests comes belting out of one of the smallest packages in the forest. Ruby-crowned Kinglets seem to perch and sing from every other tree and even became a standing joke at the Atlas Rendezvous in Routt and Grand counties: atlasers would code seven singing males before they even arrived in the block, and they wanted a code to indicate 20 or 30 singing males.

HABITAT

Both Ruby- and Golden-crowned kinglets show strong ties to coniferous forests, but in some parts of the country Ruby-crowneds forage more readily in deciduous trees and brush than do Golden-crowneds (in Ontario, Keast and Saunders 1991; and Oregon, NMB). They do not maintain as close a tie to dense conifers as the Golden-crowneds, and they often use open, sunlit parts of the forest (Ryser 1985).

Colorado Ruby-crowneds display an exclusive affinity for conifers; only 2% of block habitat codes specified aspen without associated conifers. Atlas field workers could find kinglets in any little patch of conifers, even those surrounded by acres of aspen. Thirty block Confirmations used nest-related codes (ON, NB, NE, NY); 24 identified only conifer as the associated habitat and three identified mixed conifer/deciduous. Only three reported pure deciduous habitats. All five nest card reports specified conifers. Engelmann spruce held three nests; lodgepole pine and ponderosa pine claimed the remaining two.

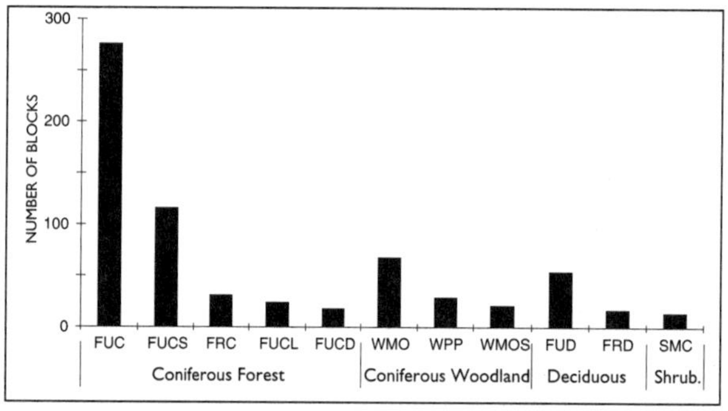
NUMBER OF BLOCKS
300
200
100
0
FUC
FUCS
FRC
FUCL
FUCD
WMO
WPP
WMOS
FUD
FRD
SMC
Coniferous Forest
Coniferous Woodland
Deciduous
Shrub.

Small territories permit Ruby-crowneds to pack closely together; in two spruce/fir forests in northern Utah territories averaged 1.5 and 1.8 acres (0.6 and 0.7 ha). An aspen forest studied at the Beaver Creek Ski Area (Grouse Mt. 39106E5) had larger territories—one pair/6.25 acres (2.5 ha); each of the five kinglet territories contained one of the five conifers in the study area (Bottorff 1972). The deciduous habitats that these birds use add to the supply of insects that serve as the major portion of the Ruby-crowned Kinglet's diet.

BREEDING

Although extremely common breeders across western Colorado, Ruby-crowned Kinglets hide their nests very well. The nest sits snugly tucked near the trunk of the conifer, with its mosses, lichens, and other plant materials blending in with the bark.

Atlasers saw Ruby-crowned Kinglets migrating in April and arriving on their nesting grounds in early May. The birds immediately settle in to breed. Multiple-singing-male records commenced on 5 May and the first Atlas nest-building record came on 15 May. Atlasers found nests with young as late as 5 August and parents feeding fledged young throughout the summer, showing a wide range of nest initiation dates.

Breeding Ruby-crowned Kinglets face a variety of challenges in nesting. Two of the five nest cards mention problems with jays. In one the adults defended their nest with young from a Steller's Jay (Ronda Woodward pers. comm.). In the second case both adults attacked a Gray Jay that completely ignored them while it dismantled their nest (NMB). Cowbirds also occasionally take their toll. Chase and Cruz (1996) reported three records of Brown-headed Cowbird parasitism on Ruby-crowned Kinglets (Appendix C). These natural controls prevent Ruby-crowned Kinglets from overwhelming us. A single pair can lay up to 12 eggs (Ingold and Wallace 1994). If every one of those eggs hatched for every pair in the forest, kinglets would become more common than the trees in which they sing.

BREEDING PHENOLOGY

CODE		# OF RECORDS	DATE RANGE
C	Courtship	3	8 Jun–25 Jul
NB	Nest Building	11	15 May–20 Jul
ON	Occupied Nest	11	29 May–9 Jul
NE	Nest with Eggs	1	16 Jul
NY	Nest with Young	7	11 Jul–5 Aug
FY	Feeding Young	116	26 May–9 Aug
FL	Fledged Young	85	14 Jun–19 Aug

DISTRIBUTION

Ruby-crowned Kinglets breed in forests from Alaska across Canada to northern New England and south in the western mountains. In Colorado the breeding range reflects the strong tie to conifer forests, with nesting limited to the forested mountain slopes. The large elevational variation within Atlas blocks makes inferences about altitude limits on a species difficult, but the high number of Confirmations along the high mountain divides indicate that Ruby-crowned Kinglets can breed at very high elevations. A&R reported them above 11,000 feet (3380 m) during breeding season.

The Atlas map shows a surprisingly broader distribution than the A&R map. The differences expand the breeding range to middle elevations of the mountains, particularly along the Front Range, Uncompahgre Plateau, and west of the Flat Tops on the Colorado Plateau. The expansion links with the limited numbers that breed in ponderosa pines, Douglas-firs, and lodgepole pines.

The most common breeder in Colorado's mountains according to Atlas abundance estimates, Ruby-crowneds would seem secure, but reason for concern exists. Timber harvests on kinglet habitat may impact them negatively, but several studies provide divergent answers. One study of the effects of 12 small clear-cuts in a 100-acre (40.5 ha) spruce/fir forest in the Fraser Experimental Forest provided the predictable result: although the kinglets continued to nest in the remaining forest fragments, their numbers dropped significantly (Scott et al. 1982). Two other studies, one in spruce/fir and one in aspen, found little response by Ruby-crowned populations to habitat fragmentation from timber harvests (Keller 1987, Scott and Crouch 1987).

Comparing Cooke's "common" of 1897 with current numbers cannot be done, but the species continues to dominate the avian chorus of Colorado's conifer forests.

BREEDING EVIDENCE

REPORTED IN 507 (29%) OF 1745 PRIORITY BLOCKS

□	Possible	116 blocks	(23%)
▩	Probable	161 blocks	(32%)
■	Confirmed	230 blocks	(45%)

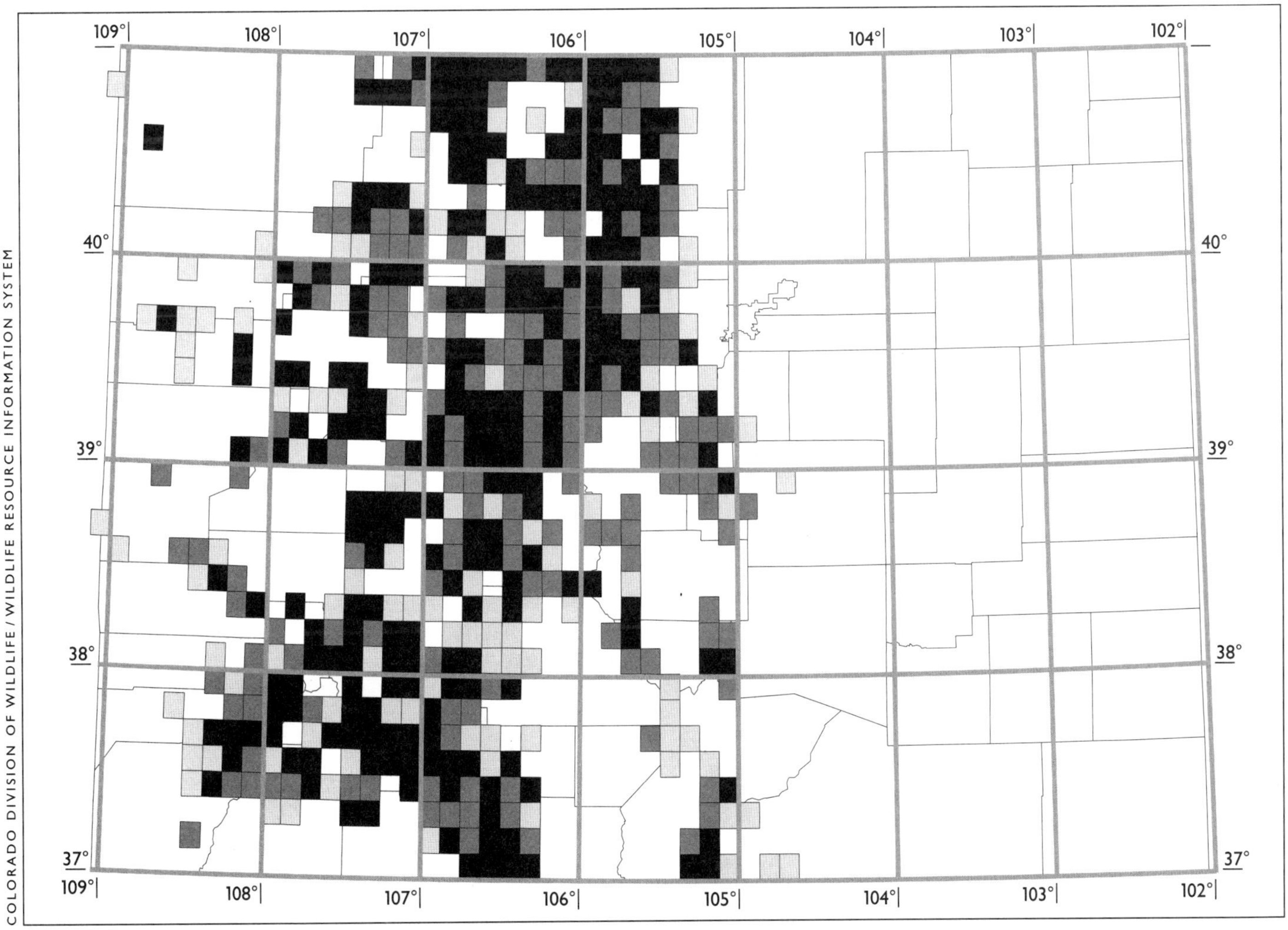
109°
108°
107°
106°
105°
104°
103°
102°
40°
39°
38°
37°

COLORADO DIVISION OF WILDLIFE / WILDLIFE RESOURCE INFORMATION SYSTEM

BLUE-GRAY GNATCATCHER *Polioptila caerulea*

Although at somewhat of a disadvantage in both size and song, Blue-gray Gnatcatchers rarely go undetected. A delightful blend of perpetual motion, an expressive tail, and a nasal scolding call makes these birds difficult to miss.

HABITAT

Across the U.S. Blue-gray Gnatcatchers breed in a wide array of wooded and brushy habitats (Weston *in* Bent 1949). In Colorado, however, they exercise greater selectivity. Here, according to Atlas data, they rely on three principal habitat types: pinyon/juniper woodlands, scrub oak thickets, and mountain mahogany or serviceberry shrublands. Of these, pinyon/juniper

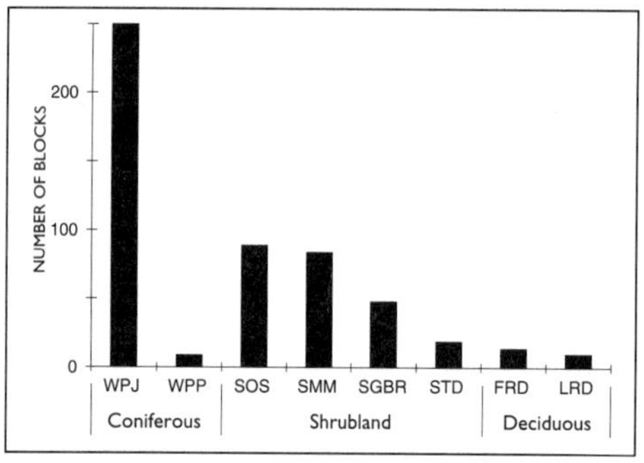
NUMBER OF BLOCKS
0
100
200
WPJ
WPP
SOS
SMM
SGBR
STD
FRD
LRD
Coniferous
Shrubland
Deciduous

woodlands support the largest numbers. Where gnatcatchers occupy scrub oak or other shrubland habitats, pinyon/juniper woodlands usually lie nearby. Although riparian woodlands provide favored habitat for this species in some parts of the country (Ellison 1992), a relatively small number of Atlas records come from these habitats. Despite the claim of Ellison (1992) that Blue-gray Gnatcatchers are "rare or absent from habitats dominated by needle-leaved conifers," these birds rely heavily on such habitat in Colorado.

Other sites used less frequently by Blue-gray Gnatcatchers include sagebrush and ponderosa pine habitats. Many of the sagebrush records come from sites where the brush grows in association with pinyon/juniper woodlands. When atlasers found them in ponderosa pine woodlands, a scrub oak component usually grew nearby. Gnatcatchers involved in an expansion along the northern Front Range seem more inclined to use scrub oak and mountain mahogany hillsides without the conifer components than they do in the rest of the state.

BREEDING

Few migrant passerines precede Blue-gray Gnatcatchers in starting the breeding cycle. One atlaser witnessed nest-building activity near Grand Junction as early as 30 April. Even around Colorado Springs, where cool springtime temperatures prevail, males defend territories by the end of April.

These gnatcatchers select and defend a territory immediately after arriving on breeding grounds (Root 1969). The male actively patrols the perimeter of his territory. He frequently engages in boundary disputes with nearby males. Many of these boundary disputes culminate in chasing and actual combat. By the time nesting has begun, well-defined boundaries exist and males patrol them less frequently.

Both sexes contribute to building the nest (Root 1969). They construct a cup-shaped nest in the fork of two small branches or saddled on a horizontal limb (Weston *in* Bent 1949). Typically they adorn the outside of their nest with crustose lichens similar to, and probably appropriated from, those found on the bark of the trees or brush that support their nests (Ellison 1992). This appears to help camouflage the nest but, in some cases, affixed lichens actually make the nest more visible (Root 1969).

Whether due to failure of the first nest, disturbance, or simply when starting a second brood, Blue-gray Gnatcatchers reuse material from their first nest to construct a second. They regularly double-brood in warmer parts of their range and occasionally do so as far north as Vermont (Ellison 1992). The range of dates for nest-building and fledglings implies that double-brooding occurs in Colorado. Typical clutches contain four or five eggs (Root 1969).

Blue-gray Gnatcatchers appear prone to cowbird parasitism. Ellison (1992) and Chace and Cruz (1996) cite numerous examples, and atlasers contributed five records (Appendix C). Cowbirds, however, constitute but one hazard to gnatcatcher broods. Western Scrub-Jays evoke extreme agitation among adult birds and regularly take gnatcatcher eggs or young. Other predators include accipiters, small owls, magpies, shrikes, snakes, and small mammals (Root 1969).

BREEDING PHENOLOGY

CODE		# OF RECORDS	DATE RANGE
C	Courtship	4	8 May–4 Jun
NB	Nest Building	27	30 Apr–11 Jul
ON	Occupied Nest	21	7 May–4 Jul
NE	Nest with Eggs	10	25 May–26 Jun
NY	Nest with Young	15	27 May–6 Aug
FY	Feeding Young	76	21 May–20 Aug
FL	Fledged Young	33	2 Jun–15 Aug

BY ALAN E. VERSAW

DISTRIBUTION

Blue-gray Gnatcatchers breed across the contiguous states except the Pacific Northwest and northern Great Plains, and in much of Mexico. They winter across the southern tier of states and through Mexico into Guatemala. They occasionally show up on lower-elevation Christmas Bird Counts in western Colorado.

The Colorado range tracks the pinyon/juniper woodlands. On the Western Slope, where pinyons and junipers dominate wooded regions up to elevations of about 7,000 feet (2135 m), Blue-gray Gnatcatchers sustain both large populations and a broad distribution. On the other hand, the perimeter of the San Luis Valley, an area dominated by pinyon/juniper woodlands, hosts few gnatcatchers; elevation may provide a partial clue to their absence here. Around the arid San Luis Valley, pinyon/juniper woodlands only begin at about 8,000 feet (2440 m). On the eastern side of the mountains most gnatcatchers breed from Colorado Springs southward in close association with stands of pinyon/juniper.

North of Colorado Springs the smaller breeding populations rely on different habitats—scrub oak, ponderosa pine woodlands, mountain mahogany hillsides, and shrubby riparian thickets. Blue-gray Gnatcatchers have not always occupied the northern part of the Front Range. Cooke (1897) regarded Blue-gray Gnatcatchers as unknown north of El Paso County. Both Sclater (1912) and Bergtold (1928) cited Boulder-area records, but they mentioned no other records north of Cañon City. Niedrach and Rockwell (1939) said the Blue-gray Gnatcatcher "rarely wanders" into Denver and the adjacent mountains. Even in 1987, Holt and Lane stated (a bit mistakenly) that these birds "barely make it to the eastern slope."

This has changed in the last ten years. On 13 May 1986 the Tuesday Birders found the first Denver-area nest (Ann Bonnell pers. comm.). The Atlas documents a major range expansion into Douglas, Jefferson, and Boulder counties. Isolated reports from Larimer County may hint at a continuing northward range expansion. This movement fits nicely within a century-long pattern of northward expansion across most of the continent (Ellison 1992). Whereas many migratory species have experienced serious declines, Blue-gray Gnatcatchers have increased, in both range and numbers.

BREEDING EVIDENCE

REPORTED IN 369 (21%) OF 1745 PRIORITY BLOCKS

Possible	109 blocks	(29%)
Probable	87 blocks	(24%)
Confirmed	173 blocks	(47%)

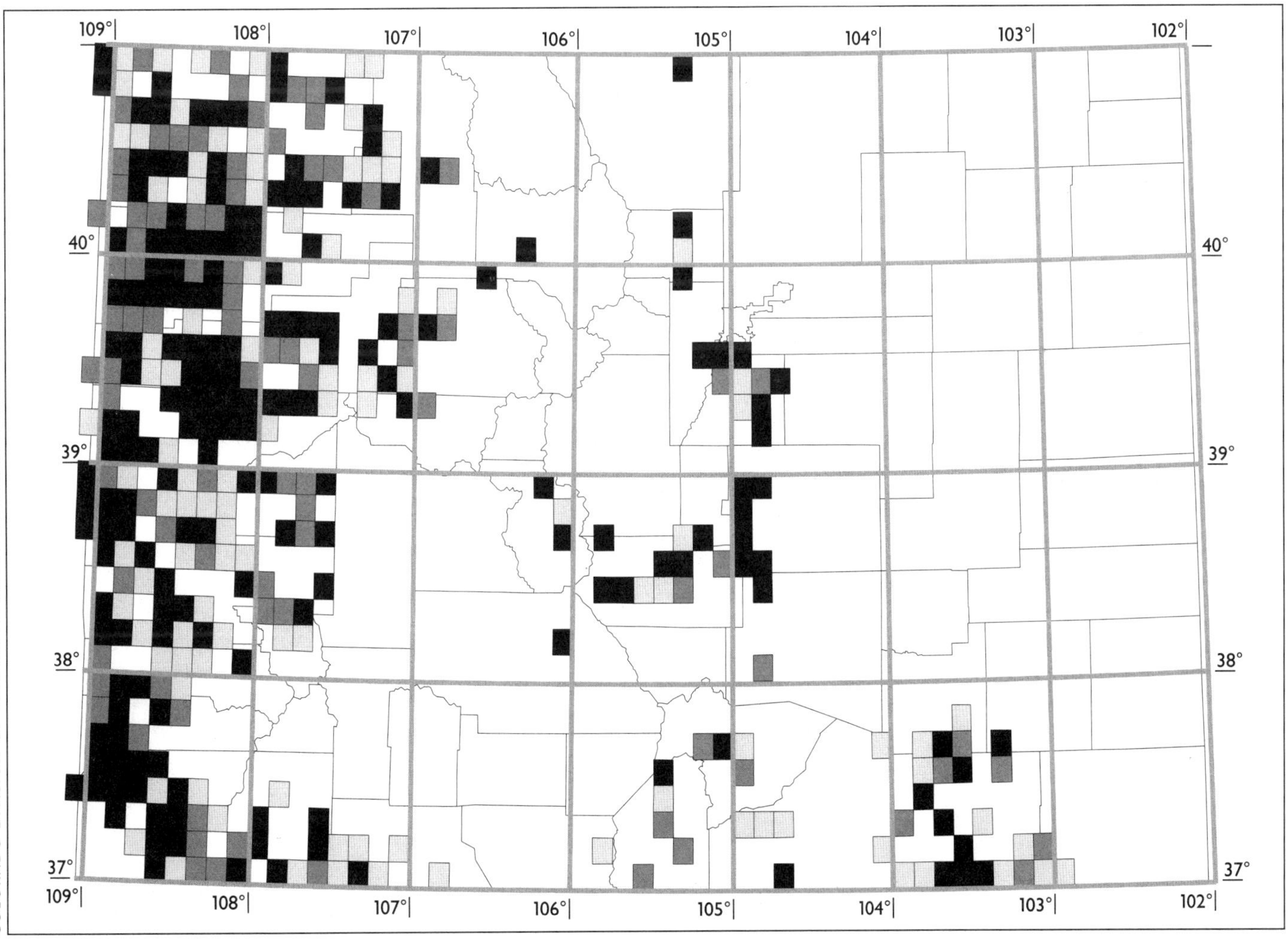

109°
108°
107°
106°
105°
104°
103°
102°
40°
39°
38°
37°

COLORADO DIVISION OF WILDLIFE / WILDLIFE RESOURCE INFORMATION SYSTEM

Eastern Bluebird

"Who has ever seen the bluest sky as blue as the bluebird's back!" (Bent 1949). This exclamation refers to the Eastern Bluebird, but the description and admiration belong to all three species. The early English settlers on the eastern seaboard of North America called it the "blue robin" and loved it as a sign of spring and a reminder of home.

HABITAT

In eastern North America Eastern Bluebirds choose open country, farmyards, orchards, and fencerows rather than forests. Early colonists, by turning forests into fields and planting orchards, enlarged bluebird habitat and attracted them close to their homes. Bluebirds nest in cavities. Orchards, allowed to age naturally, attracted woodpeckers and the bluebirds that use woodpecker holes.

Several species accustomed in the East to birdhouses, bridges, and barns for nesting—Eastern Phoebes, Purple Martins, and Eastern Bluebirds—still use natural nest sites in Colorado. Knorr (1959) believed that Colorado bluebirds shun farmyards in order to elude the House Sparrows and starlings that monopolize nesting holes around farms.

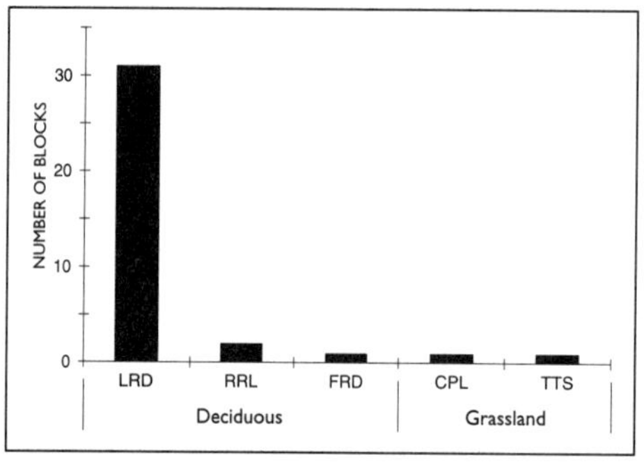
NUMBER OF BLOCKS
0
10
20
30
LRD
RRL
FRD
CPL
TTS
Deciduous
Grassland

Atlasers found Eastern Bluebirds in cottonwoods along streams and on the edges of ponderosa pine forests. Over 88% of the Atlas habitat reports specify lowland riparian, and all six Atlas nest cards identify a cottonwood nest tree.

Insects (adults and larvae), spiders, and other small invertebrates form the bulk of their food. Bluebirds also consume wild fruit and berries (Bent 1949). Wintering birds or early returnees sometimes perish when ice storms cover their food supply of wild berries (Zeleny 1976).

BREEDING

Eastern Bluebirds begin to nest early. Birds in Atlas blocks along the Kansas line already had young in their nests by early May. Most often atlasers found them using old woodpecker holes in old or dead cottonwoods. Males choose the nesting cavities and coax the females with soft warbles to enter the chosen holes. Courtships may last an hour, a day, or a week, but as soon as the females accept the cavities, they start to make the nests—loose cups lined with grass, stems, twigs, needles, and feathers (Zeleny 1976).

The female lays 4–5 eggs (she sometimes does not start laying for a week after completion of the nest), then incubates them for 13–14 days. The adults, mainly males during the first few days, feed the fledglings for 3–4 weeks (Stokes and Stokes 1989). Often young of the first brood or from a prior year assist in feeding a second brood. Double-brooding occurs commonly in Colorado (B&N).

BREEDING PHENOLOGY

CODE		# OF RECORDS	DATE RANGE
NB	Nest Building	1	29 May
ON	Occupied Nest	3	5 May–13 Jun
NY	Nest with Young	7	5 May–6 Jun
FY	Feeding Young	8	8 May–28 Jul
FL	Fledged Young	1	31 Jul

DISTRIBUTION

Eastern Bluebirds breed east of the Rocky Mountains from southern Canada to Central America. The western edge of their range coincides with the eastern edge of the ranges of both Western and Mountain bluebirds (Stokes and Stokes 1991). Colorado, happily, has all three bluebird species breeding. Rarely the three nest in close proximity; in 1931, in the same aspen grove at 8,000 feet (2440 m) near Evergreen, all three nested (B&N). The state has two similar instances. In 1952 the three species nested close to each other south of Colorado Springs (Knorr 1959). In 1989 we (RRK, Walt Kuenning) found Eastern, Western, and Mountain bluebirds nesting in the same Atlas block in Elbert County, a block with meadows, ponderosa pine woodlands, and willow-lined stream bottoms (39104B5). On the same ranch in 1997 the three species attempted to nest, although the singing Eastern Bluebird could not attract a mate (Karen Metz pers. comm.).

Aiken procured Colorado's first Eastern Bluebird specimen in El Paso County in 1872 (Holden and Aiken 1872). Eastern Bluebirds spread across the plains in the early twentieth century, probably helped by the developing cottonwood stream corridors that provided suitable habitat. In the 1920s they nested regularly in residential Denver and often used birdhouses (B&N). When their numbers plummeted in the eastern U.S., their Colorado range retracted east to the state line, but Atlas results show a recovery; again they are moving west.

The 37 Atlas observations (including Non-priority blocks) occurred in scattered locations on the plains. The area along the Republican River near the Kansas line in Yuma County has the highest concentration, with 150–200 nesting pairs. Another group nests along the South Platte, restricted to the cottonwood river bottom. The map shows this group discontinuously in eight blocks from Greeley downstream because present-day Colorado bluebirds have not adapted to manmade habitats, and Priority blocks do not coincide with the course of the river. In 1995 one pair attempted, unsuccessfully, to nest in a birdhouse at the edge of the Boulder foothills (Virginia Dionigi pers. comm.).

All bluebirds, but especially Eastern Bluebirds, suffered severe population declines from the early 1900s to 1978 (Stokes and Stokes 1991). Introduced House Sparrows and European Starlings and even native House Wrens usurped their nesting holes. The spreading use of pesticides, commercial agriculture that left no undisturbed edges, and substitution of metal fence posts for wood contributed to loss of habitat and food (Stokes and Stokes 1991). People began bluebird-box trails in the 1930s but appreciable population recovery occurred only after the DDT ban and the formation of the North American Bluebird Society by Lawrence Zeleny in 1978.

Bluebird trails, consisting of a few to hundreds of boxes, had an important impact on the recovery. In 1995 Denver Audubon Society and CDOW inaugurated a statewide bluebird trail, aimed at helping all three species. Boxes in farmyards and roadsides along stream courses like the South Platte might lure the bluebirds away from their cottonwood stream bottoms.

Verifying the success of the bluebird campaigns, BBS statistics document the recovery from the mid-century decline—an increase of over 100% since 1966 when the BBS started. Increasing reports in Colorado reflect this triumph by bird-lovers.

BREEDING EVIDENCE

REPORTED IN 20 (1%) OF 1745 PRIORITY BLOCKS

Possible	3 blocks	(15%)
Probable	5 blocks	(25%)
Confirmed	12 blocks	(60%)

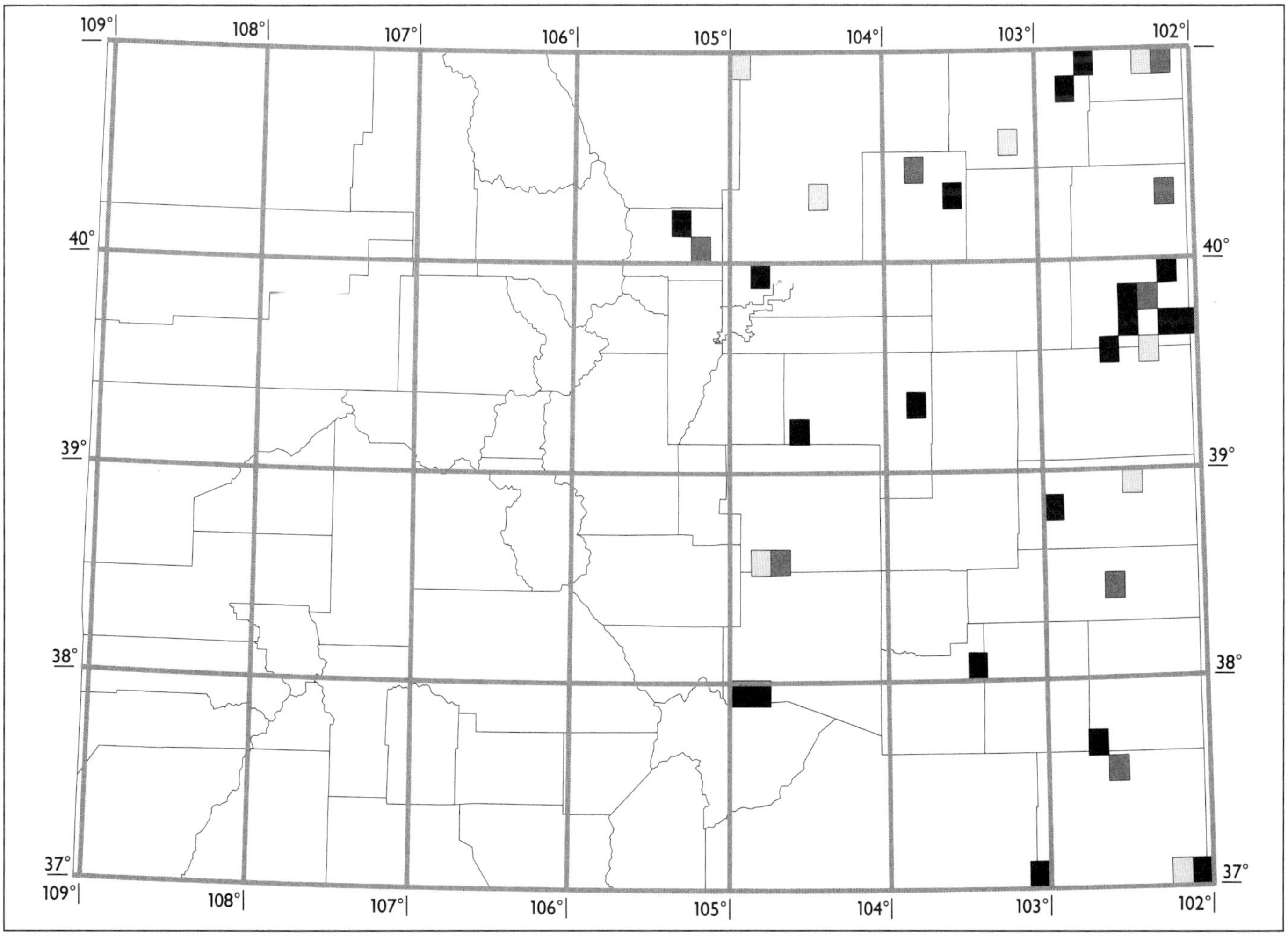
109°
108°
107°
106°
105°
104°
103°
102°
40°
39°
38°
37°

COLORADO DIVISION OF WILDLIFE / WILDLIFE RESOURCE INFORMATION SYSTEM

Other states may match Colorado by attracting all three bluebird species, but Colorado ups them all. Only here do all three breed in close proximity; all three nested in one Atlas block (Elbert, 39104B5). Western Bluebirds, the jewels of the ponderosa pines, in Colorado display another virtue: they carry their breast color on their backs. Thus they rate distinction as a subspecies identifiable in the field, the Chestnut-backed Bluebird (*S. m. bairdi*).

HABITAT

Two regions separated by high mountains make up most of Colorado's ponderosa pine habitat. Pine woodlands cover much of the foothills along the Front Range from Wyoming to New Mexico. Similar habitat abounds in a triangle south of the Colorado River from Grand Junction south to Cortez and east to Pagosa Springs. These two zones hold Colorado's greatest Western Bluebird populations. They achieve their greatest breeding density in ponderosa pine woodlands at elevations of 5,000 to 8,000 feet (1525–2440 m; A&R).

Pines compose 65% of the habitat reports for these bluebirds, and ponderosas compose two-thirds of those. Deciduous habitats (scrub oak, aspen, cottonwood riparian zones), grasslands, and shrublands compose the rest. Like Mountain Bluebirds,

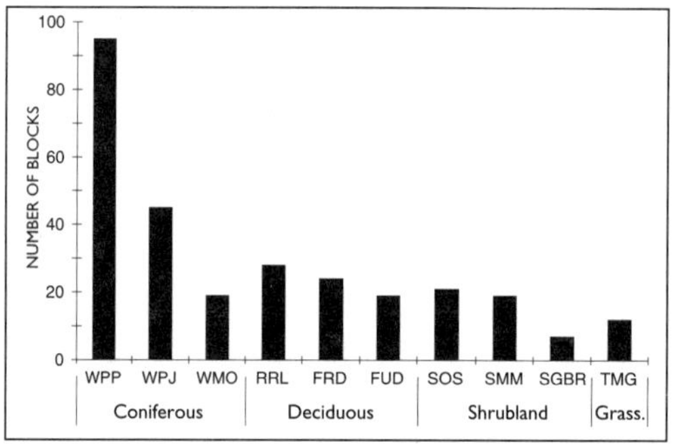
100
80
60
40
20
0
NUMBER OF BLOCKS
WPP
WPJ
WMO
RRL
FRD
FUD
SOS
SMM
SGBR
TMG
Coniferous
Deciduous
Shrubland
Grass.

Westerns use nest boxes, which bird enthusiasts often place in habitats such as grasslands and shrublands where the birds would not otherwise nest. Mountain Bluebirds breed, atlasers found, in more habitats, with a wider altitudinal range, and to higher elevations, than Westerns. Where habitat overlap between the two species occurs, Mountain Bluebirds use the edge zones and Western Bluebirds inhabit open areas in woodland interiors (Stokes and Stokes 1991).

Nest site availability may limit density. By supplying nest boxes, which the bluebirds used readily, Arizona researchers achieved an artificially high breeding density of one pair per 2.5 acres (1 pair/ha; With and Balda 1990). Bluebird trails increase nesting opportunities and densities, as atlasers found especially on the Eastern Slope.

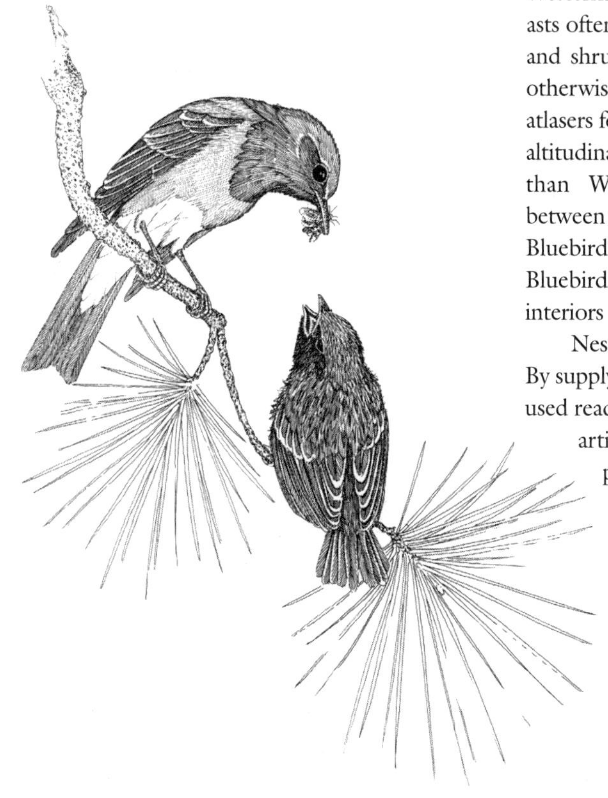

BREEDING

Western Bluebirds arrive on their breeding territories in mid April (Nelson 1993). By mid May atlasers found pairs in courtship and others already with young in the nest. Because of the ease in Confirming this conspicuous species (58% of the blocks) the Atlas missed much of the early breeding phenology. Field workers did most of their work in June and July, too late to observe the courtship and nest-building phases. Also, the Atlas protocol, which directed atlasers to report the highest Confirmation codes, removed early dates of lower codes such as courtship and nest-building.

Western Bluebirds nest in old tree cavities carved by woodpeckers or in nest boxes. They often nest in gigantic trees riddled with old woodpecker holes, where they associate and live amicably with sapsuckers, other woodpeckers, swallows, chickadees, wrens, and nuthatches.

Preparing the cup nest, made of fine grasses, weed stalks, pine needles, and occasionally fine rootlets, takes about four to five days (Stokes and Stokes 1991). Clutch size in Western Bluebirds ranges from four to six eggs. Incubation takes 14 days and upon hatching the nestlings receive food from both parents for about 20 days. The parents continue to feed their fledged young for about two weeks (With and Balda 1990).

Some pairs of Western Bluebirds attempt a second nest. Arthropod abundance, moisture, and nest location affect whether they make a second nest attempt. Second nest attempts have a lower success rate (Brawn 1991).

BREEDING PHENOLOGY

CODE		# OF RECORDS	DATE RANGE
C	Courtship	2	12 May–6 Jun
NB	Nest Building	5	3 May–23 Jun
ON	Occupied Nest	17	11 May–23 Jul
NE	Nest with Eggs	3	28 May–18 Jul
NY	Nest with Young	26	14 May–14 Jul
FY	Feeding Young	46	15 May–23 Jul
FL	Fledged Young	40	19 Jun–26 Aug

DISTRIBUTION

Western Bluebirds breed from southern British Columbia into Mexico and in the southern Rockies from Wyoming south, wherever ponderosa pines thrive. Although it was thought that in Colorado this species reaches its highest density south of the Colorado River and the Arkansas-Platte divide (A&R), Atlas data show that substantial numbers breed along the Front Range from the Arkansas River north almost to Wyoming.

Atlas workers found these bluebirds concentrated in two main sections of the state: the San Juan Basin and the Front Range from Fort Collins south to the Wet Mountains, especially in the Black Forest. In the San Juan Basin, 38 of 109 blocks had more than 11 pairs; 56% of 23 blocks in the Elbert/Douglas/Park county area had more than 11 pairs.

Western Bluebirds breed over the entire area shown on the CDOW map of ponderosa pine, but they also breed in pinyon/juniper. In the San Juan Basin atlasers reported them in ponderosa in 19 blocks and pinyon/juniper in 16, and in northwestern Colorado found them in seven pinyon/juniper and two ponderosa (the Douglas Mountains) blocks. The Atlas map tracks that of A&R fairly closely—at least those sections rated as having "common to abundant" breeding. Atlasers found some differences: fewer in northwestern Colorado and the Gunnison area, a few more in Las Animas County.

Reports from several other locations came from isolated sites and formed no particular pattern in relationship to habitat; they may represent migrant or wandering, nonbreeding birds.

Human impacts can have deleterious effects on these and many other hole-nesting species. Leaving one snag per acre during logging (U.S. Forest Service policy) can trigger nest-site competition among a host of hole-nesters, and fire suppression that results in less open ponderosa pine woodlands deprives them of suitable habitat. Cutting down dead trees and trimming dead limbs reduce nest holes and nesting densities by depriving these and other cavity-nesting species of nest sites (Ehrlich et al. 1988).

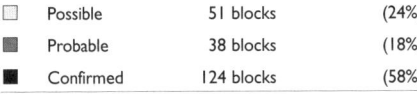

BREEDING EVIDENCE

REPORTED IN 213 (12%) OF 1745 PRIORITY BLOCKS

Possible	51 blocks	(24%)
Probable	38 blocks	(18%)
Confirmed	124 blocks	(58%)

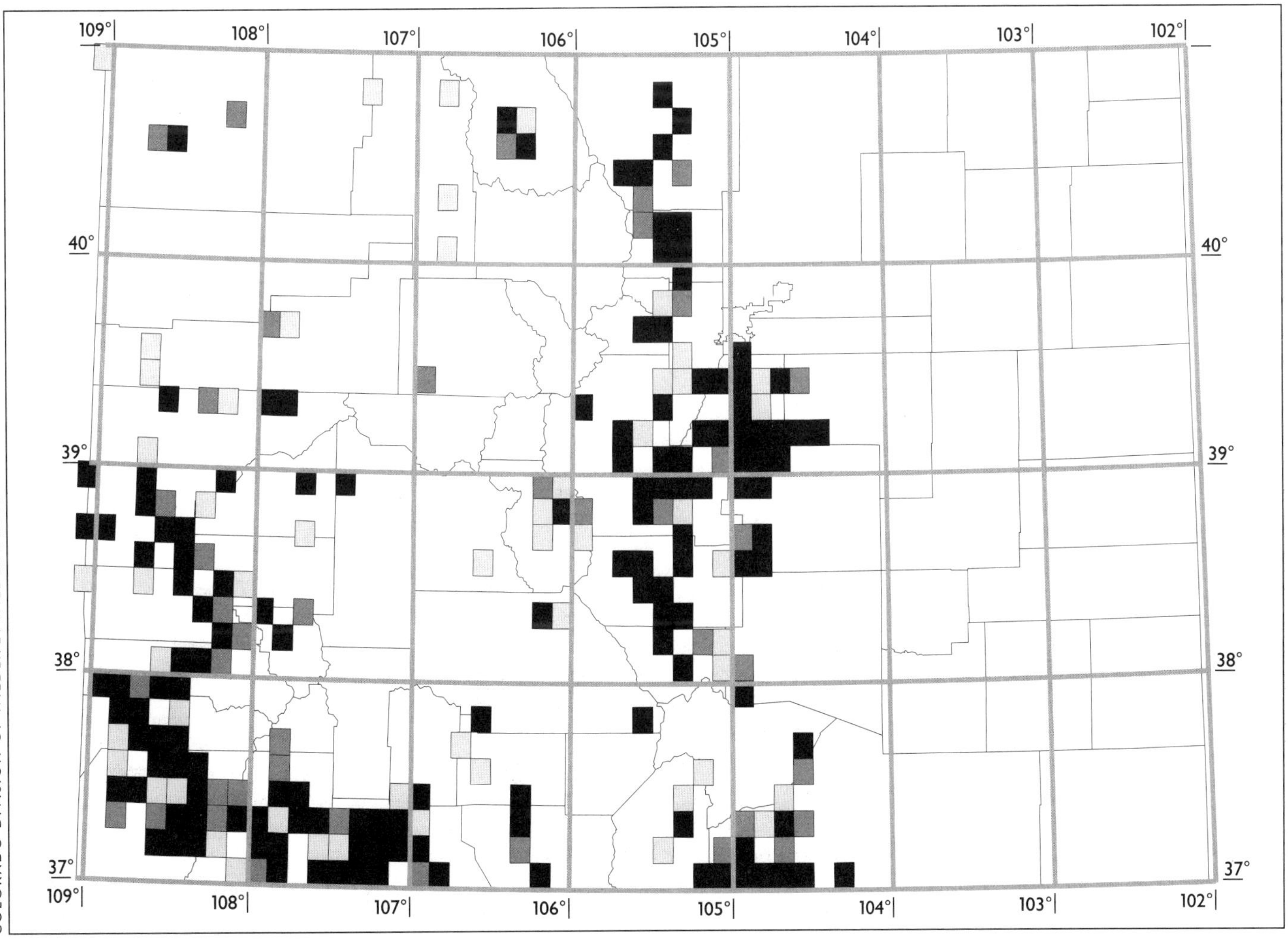
109°
108°
107°
106°
105°
104°
103°
102°
40°
39°
38°
37°

COLORADO DIVISION OF WILDLIFE / WILDLIFE RESOURCE INFORMATION SYSTEM

Colorado's Mountain Bluebirds feed and breed from the Great Plains to the tundra. Nesting in dead and live trees, on old and new buildings, on cliffs and ski lifts, from 4 feet to 200 feet (1–60 m) above the ground, these sky-colored birds show an adaptability not displayed by the other bluebird species.

HABITAT

Mountain Bluebirds use pinyon/juniper woodlands more frequently than any other recorded habitat—in 20% of the blocks. Bluebirds occurred in aspen forests in almost 200 blocks (16%) and in mountain grasslands in 100 (9%). Twenty habitats had them in over ten blocks. Atlas abundance estimates portray an even greater use of pinyon/juniper—30% of the total Colorado population of Mountain Bluebirds—with 18% in aspen and 8% in mountain grasslands. Although they need trees for nesting, Mountain Bluebirds do not breed in heavily forested habitats; they prefer forest edges and mixtures of woodland and open habitats.

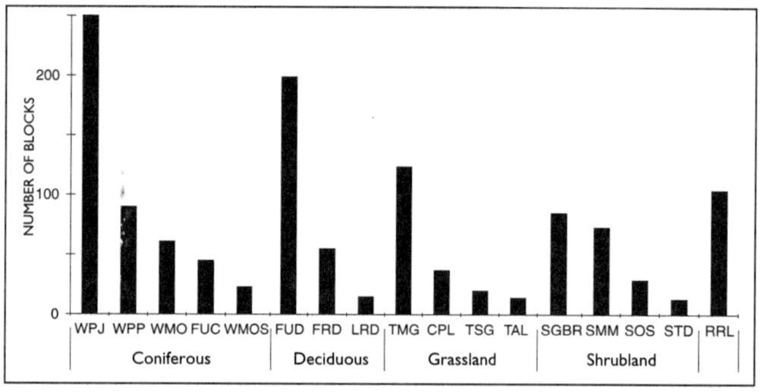
NUMBER OF BLOCKS
200
100
0
WPJ WPP WMO FUC WMOS FUD FRD LRD TMG CPL TSG TAL SGBR SMM SOS STD RRL
Coniferous
Deciduous
Grassland
Shrubland

As cavity-nesters, they find their habitat choices restricted by the availability of nest sites. Old woodpecker holes and natural cavities provide natural nest sites. Nest boxes have expanded their opportunities, and when placed in open fields these open up a habitat previously unavailable. An extreme example of opening up new habitat: in June 1976 a pair fed young in a nest in an abandoned miner's cabin packed full of snow, on a shoulder of Mount Bross at 13,500 feet (4115 m; Louise Roloff, nest card).

Old woodpecker holes and nest boxes accounted for most nests reported on Atlas nest cards. One-third of the cards reported more unexpected locations. These included a campground information box, an abandoned Cliff Swallow nest, a cattle chute, a ski-lift building, and one higher-class bird using a hole in the wall of a condominium. A tendency to submit nest cards on atypical nest locations probably overemphasizes these odd sites but makes the overall variety of sites no less surprising. This penchant for recording unusual nesting sites dates back to 1902 when a pair attempted to nest in the coupler of a train (Keyser 1902). Use of non-tree nest sites characterizes an adaptation to unforested habitats: 12 of the 13 atypical nest reports came from grasslands, tundra, desert, and other open habitats.

The presence of cavities may determine breeding density. Where adequate numbers of cavities occur, pairs may breed within 100 yards of each other. In situations with limited cavities they compete vigorously with other birds (Power and Lombardo 1996), yet they often share nest trees with other species—the Colorado list includes Bufflehead, Red-naped Sapsucker, Violet-green and Tree swallows, Pygmy Nuthatch, and House Wren.

BREEDING

Their affinity for open country and their vivid electric-blue color make Mountain Bluebirds among the easiest birds to locate during the breeding season. The Confirmation rate exceeded 72%, eighth highest for the Atlas.

The earliest nest-building recorded by an atlaser occurred on 11 April, three weeks before the second report. Several reports of fledged young in mid May affirm that this first date represents the true start of nesting. Nest-building commenced as late as 5 July, at both mid and higher elevations, and some parents still feed young in mid August. Double-brooding occurs frequently.

Inside the cavity the female builds a nest of fine plant materials including grasses, shreds of bark, and pine needles. She lays 4–6 eggs and incubates for 13 days. Only she broods newly hatched young, and only the male provides food for the first few days after hatching. The female obstructs the male from delivering food directly to the young—she takes the food from him and feeds them herself. After a few days she starts to catch food for the young and she allows the male to feed them directly (Power and Lombardo 1996). A nest box starts out 4–6 inches deep (10–15 cm); the nestlings, by the time they fledge, mash it down to half that size (Urling Kingery pers. comm.).

Occasionally another female will lay an egg in the nest (Pollock and Pollock 1993). Some parents receive help feeding young. Single females sometimes assist a pair, or young from a previous brood chip in to help (Power and Lombardo 1996). The Pollocks saw a House Wren serve as helper, over the objections of the bluebird parents.

Both male and female Mountain Bluebirds show strong site fidelity. This occasionally results in the same birds pairing up in subsequent years without having pair bonds that last through the year (Power and Lombardo 1996).

BREEDING PHENOLOGY

CODE		# OF RECORDS	DATE RANGE
C	Courtship	5	9 May–21 Jul
NB	Nest Building	17	11 Apr–5 Jul
ON	Occupied Nest	84	19 May–21 Jul
NE	Nest with Eggs	9	14 May–23 Jun
NY	Nest with Young	110	17 May–24 Jul
FY	Feeding Young	239	13 May–15 Aug
FL	Fledged Young	213	18 May–26 Aug

DISTRIBUTION

Mountain Bluebirds breed in western North America from east-central Alaska to New Mexico. The Atlas map shows nesting throughout the western two-thirds of Colorado. Only 15 species occurred in more Priority blocks. The only sizable gap occurs in the agricultural fields of the San Luis Valley, which offer no nest sites. The single plains record near Idalia (Vernon NW, 39102H4; Beth Dillon VF) documents a location over 100 miles (160 km) beyond the previous easternmost nesting site.

Mountain Bluebird populations have increased at the state, national, and continental levels. In Colorado conservation of suitable breeding habitat means preservation of dead (snags) and live trees with suitable cavities for nesting. Two major threats face these trees in western Colorado. Firewood collectors chop down roadside snags and those left for wildlife in logged areas, and safety concerns result in removal of trees that might fall on homes, campgrounds, roads, and other high-use areas. On the other hand, poor lumber prices for aspen and a lack of market for pinyon pine unintentionally protect some of the best bluebird habitat.

Like Eastern and Western bluebirds, Mountain Bluebirds take readily to nest boxes and willingly nest adjacent to homes and offices. Bluebird trails and isolated boxes help to ensure that generations to come will continue to enjoy the visions of electric blue.

BREEDING EVIDENCE

REPORTED IN 891 (51%) OF 1745 PRIORITY BLOCKS

□	Possible	112 blocks	(13%)
▩	Probable	127 blocks	(14%)
■	Confirmed	652 blocks	(73%)

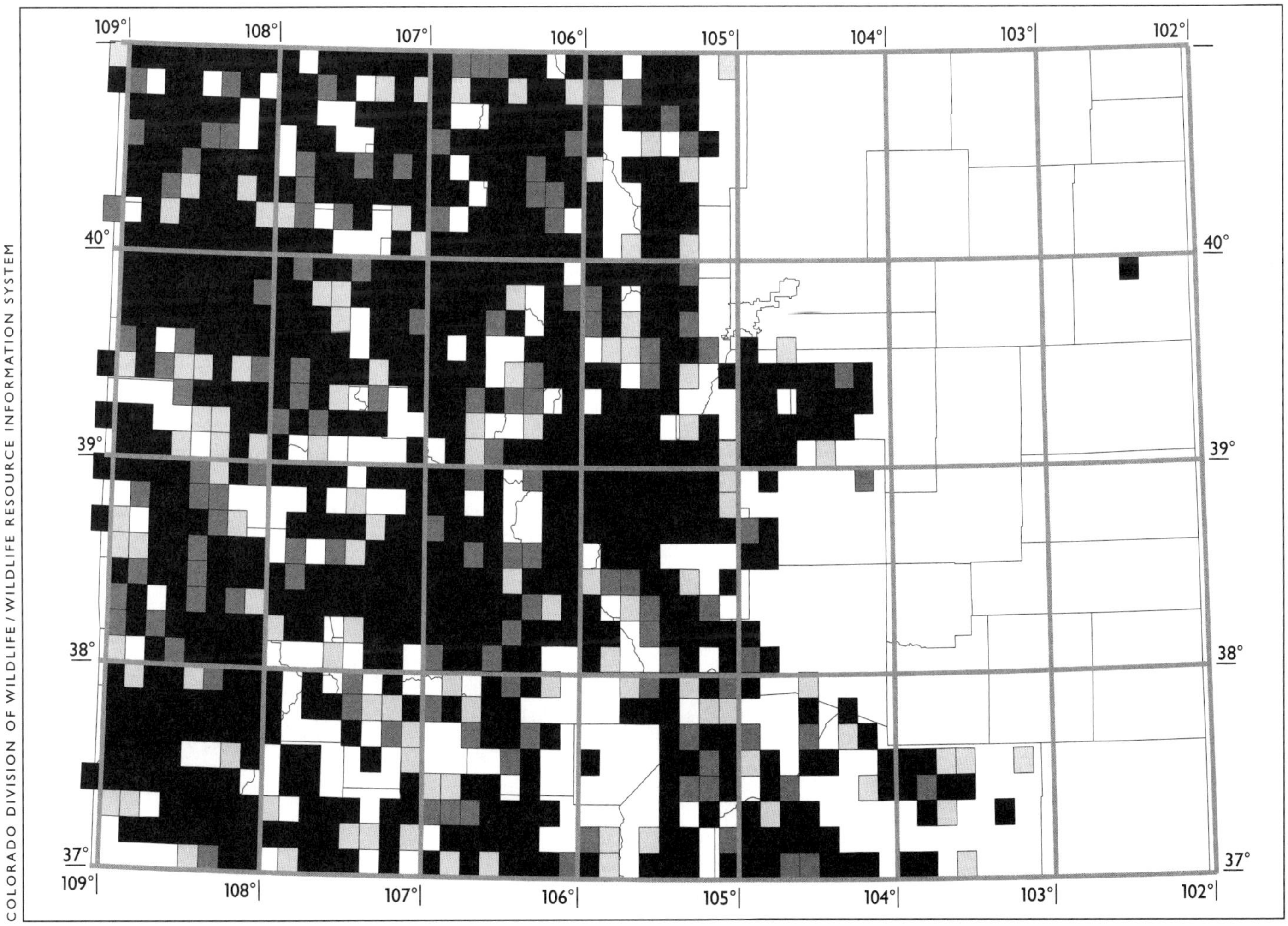
109°
108°
107°
106°
105°
104°
103°
102°
40°
39°
38°
37°
COLORADO DIVISION OF WILDLIFE / WILDLIFE RESOURCE INFORMATION SYSTEM

A shy, retiring bird, cloaked in unassuming gray, the Townsend's Solitaire sings a flight song that often seems disembodied, a song that graces June mornings, summer evenings, and winter days alike. This soft warbling song, a ventriloquial call note, and an ethereal gray presence give the solitaire a blend of melancholy and mystery beloved by romantic poets. Had Keats lived in Colorado, we surely would have an "Ode to a Solitaire."

HABITAT

Accounts of Townsend's Solitaires typically describe their habitat as open coniferous forest (Keyser 1902, Ehrlich et al. 1988, A&R, Bowen 1997). Atlas habitat reports of closed-canopy forest outnumbered those of open woodland. Atlas records, however, suggest that dominant vegetation forms may not particularly interest solitaires: atlasers reported nearly all montane habitats approximately in the same proportion as they occur. They reported 27 different habitat codes associated with breeding, including substantial numbers in deciduous forests. The wide variety of codes suggests that elements other than vegetation type determine solitaires' choice of nesting sites.

For nest placement solitaires scout cliffs, steep dirt banks, ditch sides, and roadcuts. They build their nests in hollows and nooks on the ground, always beneath an overhanging ledge, rock, stump, or root that shelters the nest from above (B&N, Bowen 1997). These conditions may occur in virtually any habitat between 7,000 and 11,000 feet (3135–3500 m). One pair near a Larimer County ghost town tended young in a nest on a ledge 8 feet below ground level in a mine shaft (Swanson 1971).

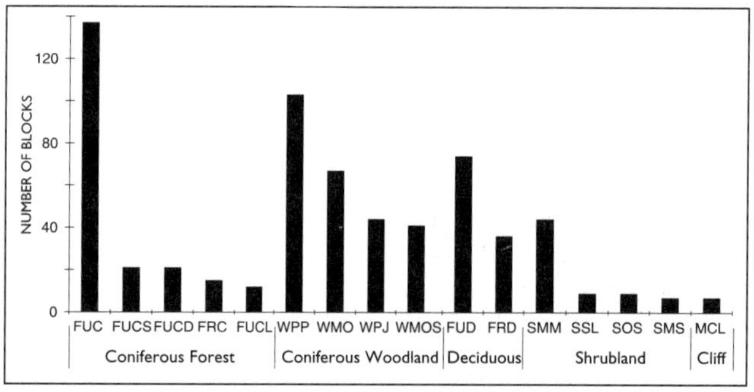
NUMBER OF BLOCKS
0
40
80
120
FUC
FUCS
FUCD
FRC
FUCL
WPP
WMO
WPJ
WMOS
FUD
FRD
SMM
SSL
SOS
SMS
MCL
Coniferous Forest
Coniferous Woodland
Deciduous
Shrubland
Cliff

Atlas workers reported solitaires in pinyon/juniper in 54 blocks, suggesting a significant extension of its breeding range down into that habitat. Most of these reports, however, came very early in the breeding season, or they described nonbreeding birds, second habitats, or July fledglings. It seems likely that solitaires do not breed in this habitat as regularly as the number of reports might suggest.

The birds catch their insect diet in several ways. During the breeding season they forage mostly in open areas of the forest understory and also feed by pouncing from low perches for insects on the ground or on tree trunks. In late summer and sometimes in winter they hawk insects, flycatcher fashion, above the canopy (Bowen 1997).

BREEDING

During the winter, Townsend's Solitaires feed primarily on fruit, especially juniper berries, from the plains up into the foothills and lower mesas. Individual birds often defend winter territories, where they sing to advertise their presence and aggressively chase intruders. Early in May, they leave their winter haunts and begin to move up into the higher mountains. Males perform elaborate courtship flight-songs that have enthralled many observers; they often soar virtually out of sight in performing this ritual (Keyser 1902, Ligon 1961, Ryser 1985). Atlas workers, however, seldom reported courtship of any type (only two records).

Nesting begins in May at the lowest elevations and stretches nearly into August at higher elevations. Atlasers reported nests with eggs or young in early May in ponderosa pine woodlands, and they found nests with eggs or young in spruce/fir into the middle of July. Only the female incubates, for 11–14 days, while the male feeds her. Nestlings fledge 10–14 days after hatching; usually all depart together. Parents divide the fledglings between them and each becomes sole provider for one or more (Bowen 1997). Solitaires normally produce a single brood unless nestlings from the first brood fledge before 26 June (in California; Bowen 1997). Atlas workers found nests with eggs over a 50-day period, but the dates seem to show an altitudinal progression rather than multiple broods.

BREEDING PHENOLOGY

CODE		# OF RECORDS	DATE RANGE
C	Courtship	12	1 Apr–6 Jul
NB	Nest Building	6	5 May–9 Jul
ON	Occupied Nest	3	9 Jun–31 Jul
NE	Nest with Eggs	8	3 Jun–22 Jul
NY	Nest with Young	8	9 Jun–14 Jul
FY	Feeding Young	23	8 Jun–10 Aug
FL	Fledged Young	96	20 Jun–4 Sep

DISTRIBUTION

Townsend's Solitaires' breeding range spans the mountainous regions of western North America from Alaska to New

Mexico. Early writers described the solitaire in Colorado as breeding from 7,000 feet to timberline (2135–3500 m) throughout the state (Cooke 1897, Sclater 1912). Little has been added since then to modify their interpretation, although B&N reported an above-timberline nest at 12,000 feet (3660 m) on Loveland Pass.

The Atlas map for this species comes close to defining the wooded areas of the state above 7,000 feet. Blank spots, besides indicating alpine tundra and mountain parks, may have resulted from atlasers occasionally missing this elusive bird; its solitary nature, its demure bearing, and a song easily mistaken for that of a distant robin or a warbling Cassin's Finch make it rather easy to overlook.

Despite the solitaire's widespread distribution, the total population probably does not rival many other species with similar distributions. In more than half the blocks in which they recorded abundance codes, atlasers used A2 (2–10 pairs). They estimated more than 100 pairs (A4) in only seven blocks. Solitaires, because of their wide range of breeding habitats, seem destined to a secure future in which to continue their warbling and to wait for an appreciative poet.

BREEDING EVIDENCE

REPORTED IN 508 (29%) OF 1745 PRIORITY BLOCKS

Possible	266 blocks	(52%)
Probable	101 blocks	(20%)
Confirmed	141 blocks	(28%)

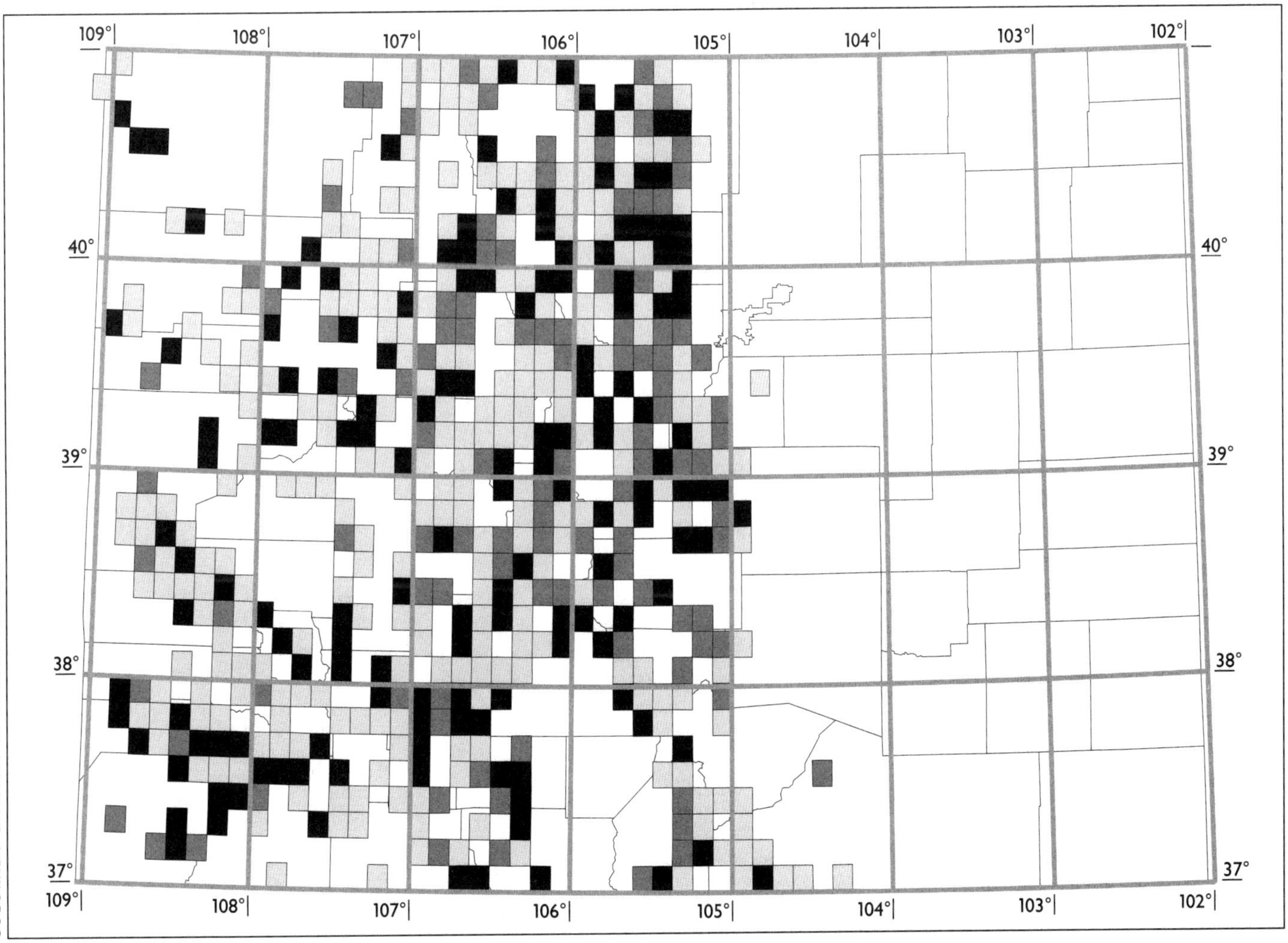
109°
108°
107°
106°
105°
104°
103°
102°
40°
39°
38°
37°

COLORADO DIVISION OF WILDLIFE / WILDLIFE RESOURCE INFORMATION SYSTEM

Veeries mate after several days of song duels between the members of the prospective pair. Particularly at dawn and dusk, their distinctive song, a downward spiraling *veer, veer, veer,* announces their presence during the short summer breeding season. "Veery," an onomatopoeic rendition of the memorable song, fits the species; the English and Latin names of the Colorado subspecies—Willow Thrush (*C. f. salicicola*)—identify the habitat the birds select in Colorado.

HABITAT

Habitat selection for all spot-breasted thrushes seems to vary somewhat depending upon competition with other thrushes. When other thrushes are present in the vicinity, Veeries select a broader range of habitats or different successional stages (Moskoff 1995).

Across their range Veeries select damp habitats with thick understory. In the eastern U.S. Veeries nest in damp, deciduous forests, and sometimes in mixed deciduous/conifer forests; on the Great Plains they pick riparian areas. The nesting habitats have the common component of dense shrubs (Moskoff 1995).

As the Atlas habitat chart shows, Colorado's few Veeries breed not in forests but rather in moist, dense riparian thickets, such as willow carrs or cottonwood saplings. They also nest in tangles of alders and willows in secluded ravines in the Front Range foothills (B&N). Occasionally a few deviate to drier hillside thickets near streams or in aspen forests (A&R); the Atlas had one such record.

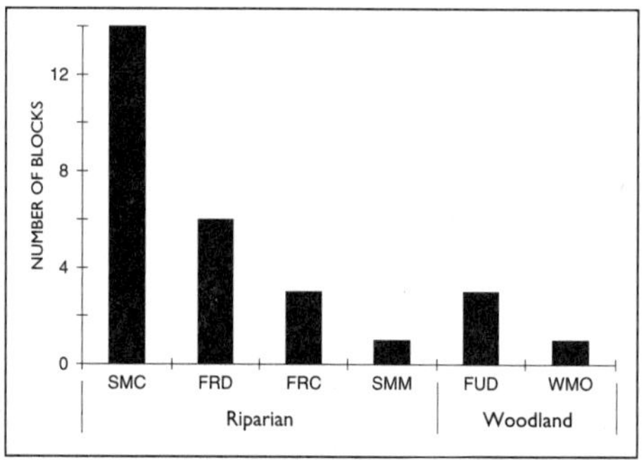
NUMBER OF BLOCKS
0
4
8
12
SMC
FRD
FRC
SMM
FUD
WMO
Riparian
Woodland

Atlasers found Veeries along montane rivers and streams. They found Swainson's Thrushes in similar but higher-altitude and less dense willow bottoms than Veeries, and Hermit Thrushes mainly in drier conifer forests away from streams. Atlasers located Veeries primarily in willows in North Park, in shrub-choked cottonwood stream bottoms along the Yampa River, and in willow/alder tangles southwest of Denver.

During the breeding season Veeries forage mainly on the ground and eat mostly insects. They switch to a fruit diet in late summer and fall, with a winter diet of insects and fruit (Moskoff 1995).

BREEDING

Veeries begin to arrive on the plains in early May, and most pass through only as migrants. Those few that remain to nest in Colorado have a short breeding season, from 11 June through early August (Nelson 1993, Atlas).

Using song duels, particularly at dawn and dusk, they take several days to form their pair bond, during which both sexes sing back and forth to each other. Veeries can recognize individual songs of their neighbors, and they act aggressively when a stranger intrudes into the territory (Moskoff 1995).

Colorado atlasers did not find any nests. The female Veery, alone, builds the nest and incubates. Veeries hide their nests effectively on the ground (about 40% of the nests) or in vegetation within 5 feet (1.5 m) of the ground (Moskoff 1995). Double brooding, although unlikely and not reported in Colorado with its short breeding season, has been reported by observers elsewhere (Bent 1949).

Even though Veeries hide their nests well from humans, they do not always succeed as well with cowbirds, which often parasitize their nests: from 19% of the nests on an Ontario site to 87% in Alberta and Manitoba. Veeries lack defenses against cowbirds, perhaps unnecessary in the East because of their predilection to nest in forest interiors (Moskoff 1995). The Colorado subspecies nests in a linear stream habitat with thickets interspersed in grassland openings, and the highly victimized Alberta and Manitoba birds belong to the same subspecies. That, plus the presence of Colorado Veery habitat on cattle ranches, implies that here also Veeries suffer from heavy cowbird predation, but the state has no confirming records.

Atlasers Confirmed breeding in only four blocks, all Confirmations based on adults carrying food for young. Other breeding evidence includes territorial behavior, agitated behavior by adults, a pair together, and singing males.

BREEDING PHENOLOGY

CODE		# OF RECORDS	DATE RANGE
FY	Feeding Young	4	1 Jul–5 Aug

DISTRIBUTION

Veeries breed across southern Canada and the northern and central U.S., and winter in South America. In northern and central Colorado they breed in a handful of sites in the foothills, lower mountains, and mountain parks (A&R). Atlasers reported breeding scattered over a wide area within that range, plus one isolated breeding record from a historical site in Gunnison County. The latter fits with reports from other observers of occasional breeding in Arizona. Although the Latilong Study listed definite breeding in seven latilongs and likely breeding in five more, atlasers found them in only seven. They did Confirm breeding in one latilong previously listed as likely (40105), but did not locate any in fringe areas to the east and south. Previously thought to breed up to 8,000 feet (2440 m), Veeries bred in one Atlas block (Boreas Pass 39105D8) at 10,000 feet (3048 m), among tangled willows and beaver ponds.

In 1882, Henshaw found these thrushes plentiful at Fort Garland and collected two nests and the type specimen of the Colorado subspecies (Sclater 1912, AOU 1957). Even though two Atlas blocks near Fort Garland contain probably suitable Veery habitat (and possibly Henshaw's site), neither there nor anywhere else in the San Luis Valley did atlasers discover them. It seems probable that this population has disappeared, despite the birds that breed occasionally in Arizona (Moskoff 1995). Atlasers also found none in other historic sites such as Colorado Springs and Boulder.

Atlasers estimated low abundances of breeding pairs; probably fewer than 1,000 pairs breed in Colorado. Veeries remain an infrequent breeder, apparently unchanged from their early status as scarce summer residents (Sclater 1912). Continent-wide, however, the BBS shows a decline of 1.4% per year from 1966 to 1996, which suggests a drop in population of one-third in 31 years.

BREEDING EVIDENCE

REPORTED IN 22 (1%) OF 1745 PRIORITY BLOCKS

☐	Possible	9 blocks	(41%)
■	Probable	9 blocks	(41%)
■	Confirmed	4 blocks	(18%)

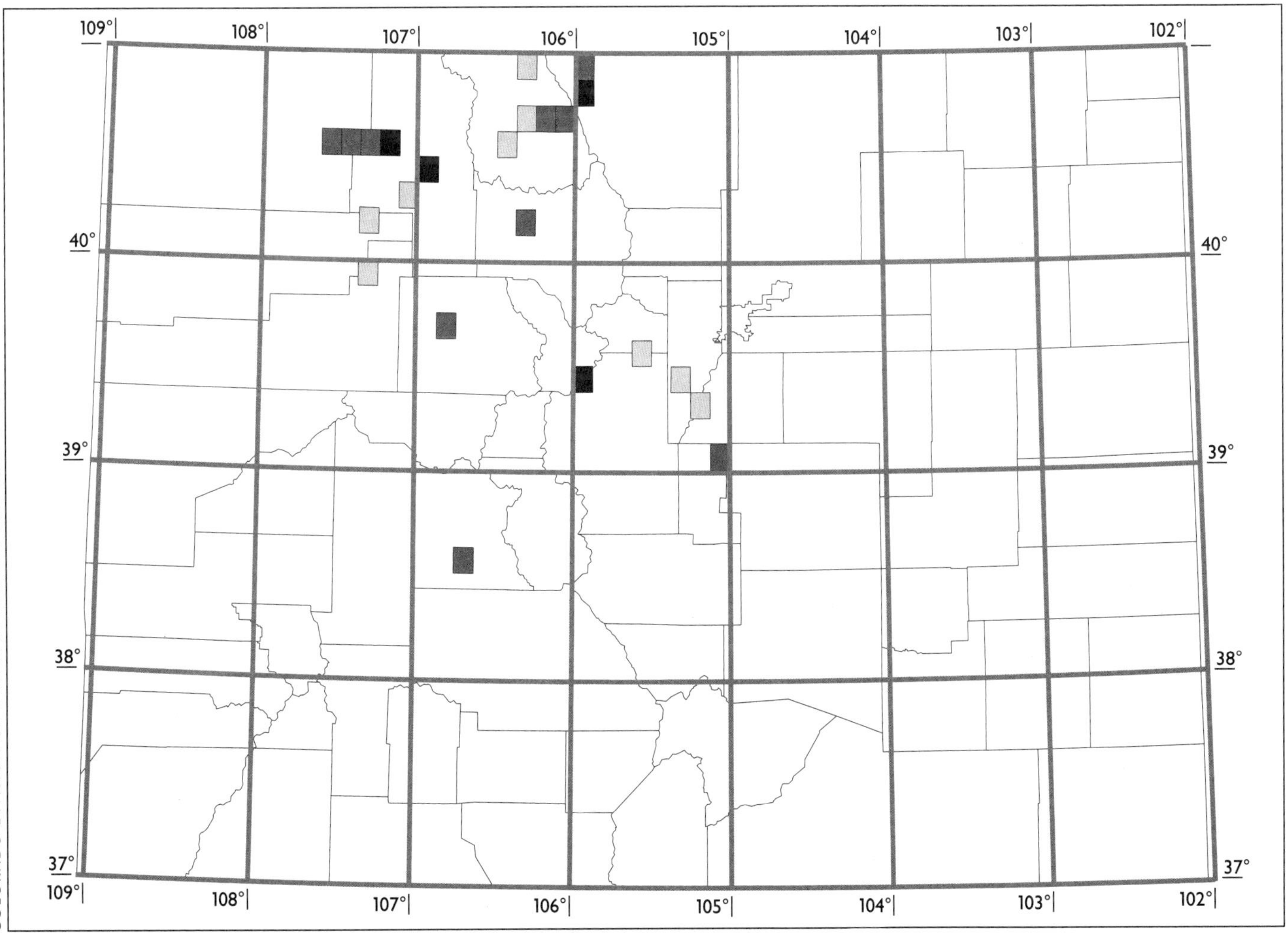
109°
108°
107°
106°
105°
104°
103°
102°
40°
39°
38°
37°
109°
108°
107°
106°
105°
104°
103°
102°
40°
39°
38°
37°

COLORADO DIVISION OF WILDLIFE / WILDLIFE RESOURCE INFORMATION SYSTEM

With flute-like upward-spiraling songs, male Swainson's Thrushes begin in mid May to announce their territories in the Colorado high country. Yet in late May, the Colorado plains teem with migrating Swainson's Thrushes. The plains migrants probably do not make a left turn into the mountains but rather continue northward. The singing Colorado birds occupy different habitats than more northern breeders.

HABITAT

In the East Swainson's Thrushes live in conifer forests. As arboreal foragers, they also have adept flycatching skills although most thrushes feed on the ground. To facilitate this skill of hawking insects, within the forests they occupy openings and gaps in the canopy (Laughlin and Kibbe 1985, Murray *in* Brewer et al. 1991, Elkins *in* Foss 1994).

In Colorado, in contrast, Atlas habitat codes show that Swainson's Thrushes nest in moist mountain valleys, usually in riparian thickets of alder and willow. Atlasers reported over 40% of breeding records from montane carrs, that is, middle-elevation willow and alder thickets; all streamside habitats contributed over 65%. A substantial number of Swainson's Thrushes breed in aspen

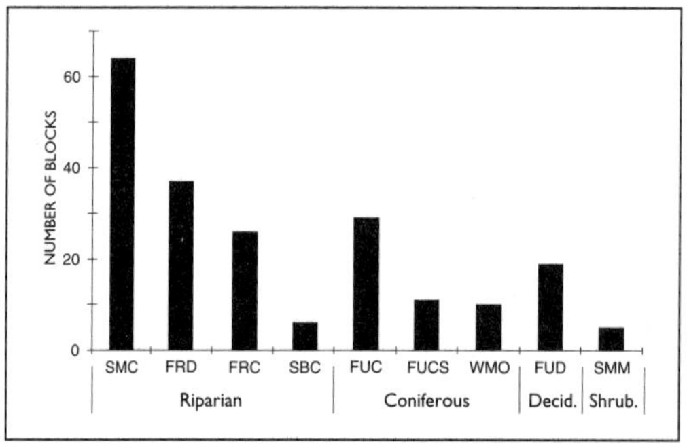
NUMBER OF BLOCKS
0
20
40
60
SMC
FRD
FRC
SBC
FUC
FUCS
WMO
FUD
SMM
Riparian
Coniferous
Decid.
Shrub.

woodlands and mountain shrubland mixes (11% of Atlas habitat codes). One-quarter breed in conifer forests (as in the East), especially in spruce/fir forests. A different analysis shows 62% breeding in deciduous habitats (streamside shrubs and aspens) and 38% in conifer-dominated habitats. Swainson's frequently breed near Hermit Thrushes, but Swainson's prefer damp willow-lined stream bottoms whereas Hermit Thrushes inhabit the drier conifer-covered hillsides. In the East, Swainson's breed at higher elevations than Hermit Thrushes (Elkins *in* Foss 1994); in Colorado the more abundant Hermits nest both lower and higher.

Swainson's share the willows with Fox, Song, Lincoln's, and White-crowned sparrows, Yellow and Wilson's warblers, and Willow and Dusky flycatchers, but Swainson's seem to require larger willows growing in bigger patches than those smaller birds. The thrushes that sing in the conifer forests may use the shrubby understory for breeding.

BREEDING

Although Swainson's Thrushes reportedly breed from mid May to mid August (A&R), Atlas reports fit within the dates reported by B&N, 22 June to 1 August. These thrushes may not start nesting as early as mid May; Nelson (1993) showed nesting dates of 11 June to 10 August. If so, perhaps some of the throngs of thrushes on the plains actually do turn into the high country.

Swainson's Thrushes build hard-to-locate nests in heavy vegetation near the ground. Atlasers found only four nests of this, as B&N discreetly put it, "modestly dressed," thrush of equally modest behavior. The nests described for Colorado (B&N) all were 3–5 feet (1–1.5 m) from the ground in willows, alders, and a tree stump. In the East these thrushes build their nests as high as 20 feet (6 m).

The rolling, spiraling songs of territorial males produced many of the Atlas breeding records; otherwise these thrushes go about their breeding process secretively. Atlasers had to search to find the birds after the brief singing season. Even their flycatching habit does not make them easier to find, at least in Colorado.

They breed during a short summer season, and probably produce only one brood. Cowbirds seldom breed in Colorado's high mountain valleys, but in other states Swainson's Thrushes occasionally fall victim to cowbird parasitism (Bent 1949). With the spread of cowbirds into the mountains, their depredations may become more frequent.

BREEDING PHENOLOGY

CODE		# OF RECORDS	DATE RANGE
C	Courtship	1	9 Jun
NB	Nest Building	1	10 Jun
ON	Occupied Nest	2	26 Jun–17 Jul
FY	Feeding Young	16	24 Jun–31 Jul
FL	Fledged Young	9	13 Jul–11 Aug

DISTRIBUTION

Swainson's Thrushes breed in wooded areas across Canada and Alaska, and in the U.S. in wooded mountains in the Northeast and the West. They winter from Mexico to South America. A staggering quantity migrate across the Colorado plains; each shelterbelt and stream bottom has its complement during late May and early June. Atlasers observed them widely as spring migrants on the eastern plains; in 87 plains blocks where they reported migrants, the median date was 25 May. These birds probably continue on a northward course, perhaps toward conifers like eastern Swainson's Thrushes. In northern Idaho Swainson's nest in the thick fir and spruce woods, whereas a few in southern Idaho nest in riparian thickets similar to those they occupy in Colorado (Burleigh 1972). Colorado breeders and migrants probably belong to different subspecies (B&N).

Atlasers reported breeding throughout the mountains in damp, wooded habitat. Swainson's breed around beaver ponds in such locations as the Illinois River in North Park (Jack Creek Ranch, 40106D1) and South Cottonwood Creek below Mount Yale (38106G3), stretches of willow-clogged streams along the Eagle River and Homestake Creek upstream from Red Cliff (Mt. Holy Cross, 39106D4), and the Blue River downstream from Silverthorne (Squaw Creek, 39106G2).

An early observer thought them rare summer residents (Bergtold 1928), whereas later observers reported larger breeding numbers. Recent authors considered them "uncommon to fairly common" in the summer, but expressed concern that they had recently declined (A&R). Atlasers reported widely scattered breeders in substantial numbers; they estimated abundance code A3 (11–100 breeding pairs) for 63 blocks, and A2 (2–10 pairs) for 71 blocks.

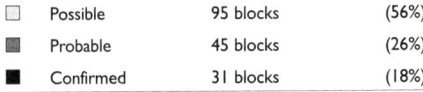

BREEDING EVIDENCE

REPORTED IN 171 (10%) OF 1745 PRIORITY BLOCKS

Possible	95 blocks	(56%)
Probable	45 blocks	(26%)
Confirmed	31 blocks	(18%)

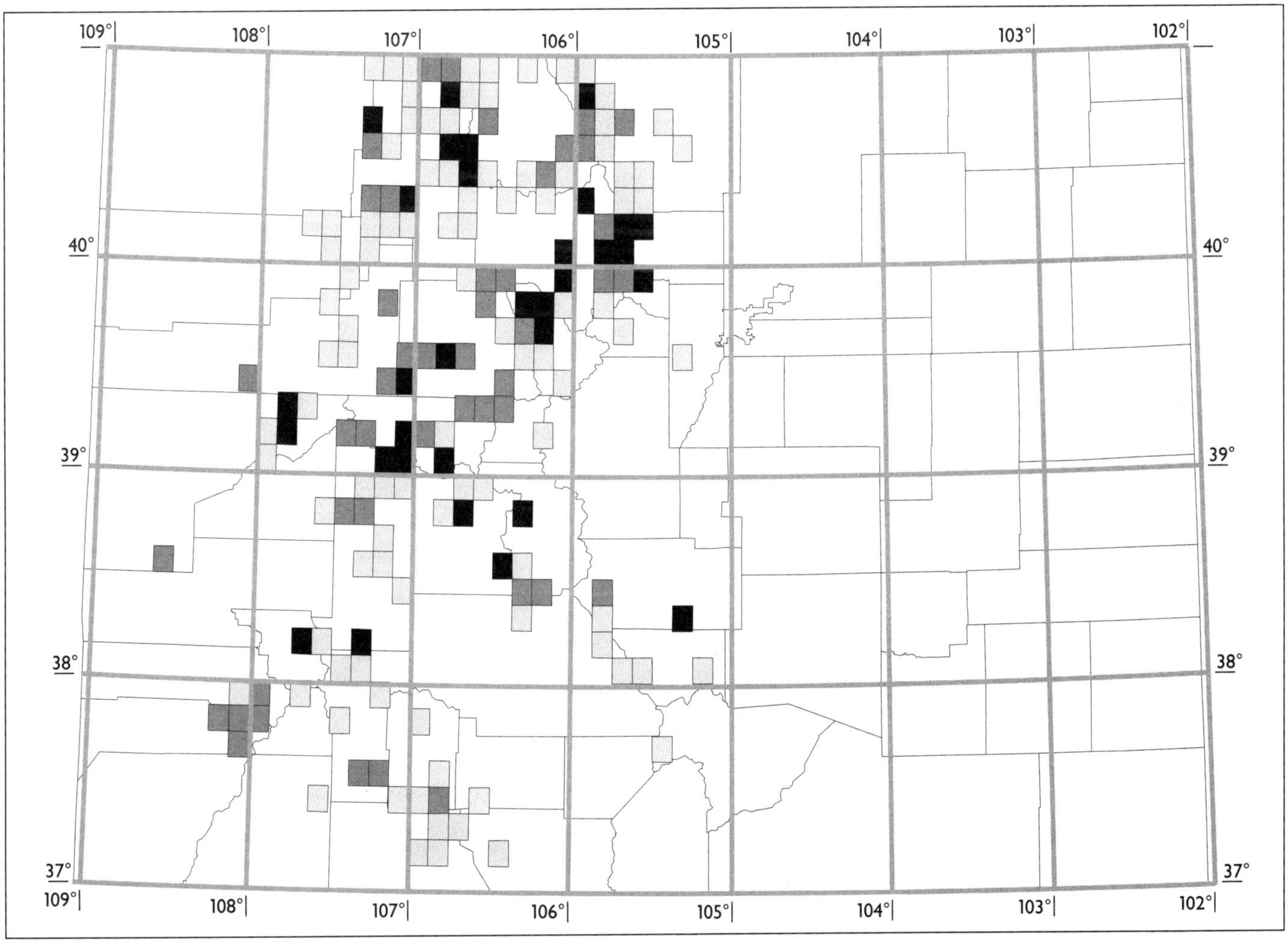
109°
108°
107°
106°
105°
104°
103°
102°
40°
39°
38°
37°

COLORADO DIVISION OF WILDLIFE / WILDLIFE RESOURCE INFORMATION SYSTEM

HERMIT THRUSH *Catharus guttatus*

A woodland concerto by one singer, perhaps comparable to the harmonic and alternating notes of a flute, oboe, and bell, identifies the presence of the "American Nightingale," the Hermit Thrush. Often the song emanates from a distance as the introverted bird seldom performs knowingly before a human audience. Poets have waxed extravagantly about the song, like Henry Van Dyke (1911): "the molten gold of that ethereal tone floating and falling through the wood."

HABITAT

Across their range Hermit Thrushes dwell in coniferous and hardwood forests, wooded canyons, bogs, and pine barrens, habitats that contain leaf litter for foraging as a common component. In Colorado the species "breeds in the mountains from 8,000 feet to timber-line and occasionally to the lower foothills" (Cooke 1897).

Atlas results exhibit a decided bias toward conifers; 82% of the blocks (539) reported Hermit Thrushes in conifer habitats (and, by extrapolation, half in spruce/fir). Ponderosa pine accounted for 8% and pinyon/juniper for 5%. Hermits occurred exclusively in deciduous habitats in only 13% of the blocks (9% of them aspen), and in shrublands in 4%. A&R showed them as fairly common in dense upper-elevation pinyon/juniper woodlands, and occasional in scrub oak shrublands with scattered conifers; both are minor habitats according to Atlas records.

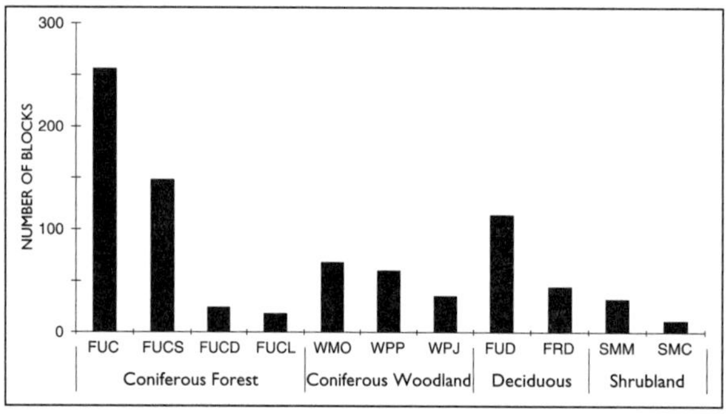

300
200
100
0
NUMBER OF BLOCKS
FUC
FUCS
FUCD
FUCL
WMO
WPP
WPJ
FUD
FRD
SMM
SMC
Coniferous Forest
Coniferous Woodland
Deciduous
Shrubland

Hermit Thrushes eat insects, spiders, earthworms, sow bugs, and snails. The animal portion of their diet shifts from 93–99% in summer to 40–56% in winter; wild fruits, berries, and some seeds make up the balance (Shuford 1993). In comparison, a photographer studying a Colorado nest saw nestlings fed a balanced diet of insects and currants (B&N).

Little comes to light on the foraging habits of western birds. Eastern birds forage mostly on the ground (Holmes and Robinson 1988), but in one study they spent more time in saplings and the midstory of trees (Paszowski 1984). Except when the birds sang from treetops, atlasers saw Hermits most frequently on or within 6 feet of the ground.

BREEDING

In the East Hermit Thrushes build nests almost exclusively on the ground (Bonney *in* Andrle and Carroll 1988, Peterjohn and Rice 1991, Vernon *in* Foss 1994). Colorado birds nest in conifers or other trees or bushes 3–10 feet (1–3 m) above the ground, usually near a tree trunk, sometimes farther out in the dense branches (B&N, Atlas nest cards). Bent (1949) notes no record of the western subspecies nesting on the ground but one atlaser found a nest built on the ground in leaf litter next to an aspen (Randy Lentz, nest card). These glorious singers conceal their nests well—atlasers found only 27 nests.

Nests consist of twigs, bark fiber, rotten wood, leaves, dried grasses, and mosses lined with pine needles, fine grasses, horse or porcupine hair, and plant fiber. Some authors, referring to eastern birds, describe a small cup of mud beneath the lining; Harrison (1979) states that they contain no mud. None of the eight Colorado nests in the Denver Museum of Natural History collection (all Front Range) contain mud (HEK). Perhaps the Colorado subspecies never uses mud, or perhaps eastern Hermit Thrushes, nesting on the ground, use mud to hold the nest together.

Females build the nest and incubate the eggs (Harrison 1979). Both adults of Lentz's ground-nesting birds brooded and fed naked nestlings on 6 July and both fed fully feathered nestlings on 13 July. At another nest, in the split bark of an aspen tree, adults traded places to brood young on 3 July. Although most Confirmations involved adults carrying food or feeding young, or fledglings, three field workers saw distraction displays.

Cowbirds uncommonly parasitize these birds. Atlasers detected two instances (Chace and Cruz 1996; Appendix C).

BREEDING PHENOLOGY

CODE		# OF RECORDS	DATE RANGE
C	Courtship	2	28 Jun–4 Jul
NB	Nest Building	7	27 May–12 Jul
ON	Occupied Nest	11	3 Jun–12 Jul
NE	Nest with Eggs	8	9 Jun–29 Jul
NY	Nest with Young	8	8 Jun–23 Jul
FY	Feeding Young	103	30 May–13 Aug
FL	Fledged Young	65	22 Jun–23 Aug

DISTRIBUTION

Hermit Thrushes breed from Newfoundland to central Alaska, south to the northern Great Lakes states, and in the western mountains to New Mexico and California. They winter in the U.S. along Pacific coast and southern states, south to Guatemala.

The Atlas map closely resembles a map of Colorado's mountain forest lands. For atlasers, the birds' secrecy superseded their ethereal tones—only a 34% Confirmation rate.

Altitudes on nest cards range from 8,600 to 11,100 feet (2620–3383 m), consistent with A&R but not with Bent (1949) who said they breed up to 12,000 feet (3660 m), at least in New Mexico. A sighting of fledglings constitutes the lone Atlas record above timberline. Atlasers Confirmed breeding down to 6,200 feet (1990 m). A&R showed the range extending farther east and northeast in Las Animas County (even into Baca County) than atlasers found. Atlasers also did not uncover as extensive a range in the western corners of the state as A&R mapped.

Atlasers submitted generous abundance estimates—nine blocks with over 1,000 breeding pairs and 100 blocks with 100–1,000. BBS trend analysis shows a continental increase of 1.5% per year.

Researchers differ on the impact of logging. Obviously, large clear-cuts cannot support forest obligates like Hermit Thrushes, but studies of selective logging offer differing conclusions (see discussion in Jones and Donovan 1996). Two studies in Colorado conducted by U.S. Forest Service researchers in ponderosa pine and aspen show similar populations on selectively logged and unlogged tracts (Burgoyne 1980, Scott and Crouch 1987). In contrast, a study in a Wyoming old-growth ponderosa pine forest identifies Brown Creepers and Hermit Thrushes as the species "most negatively influenced" by clear-cuts (Keller and Anderson 1992). In Colorado spruce/fir, Hermits dropped 14% in a selectively clear-cut forest while, in the unlogged control area, they increased by 12% (Scott et al. 1982). In Arizona conifers, the number of breeding territories dropped almost in half after logging (Franzreb and Ohmart 1978).

Regardless of the response to logging, the mountain wildernesses offer adequate habitat to please mountain hikers with the liquid harmonies of these spirits of the wilderness.

BREEDING EVIDENCE

REPORTED IN 625 (36%) OF 1745 PRIORITY BLOCKS

Possible	207 blocks	(33%)
Probable	208 blocks	(33%)
Confirmed	210 blocks	(34%)

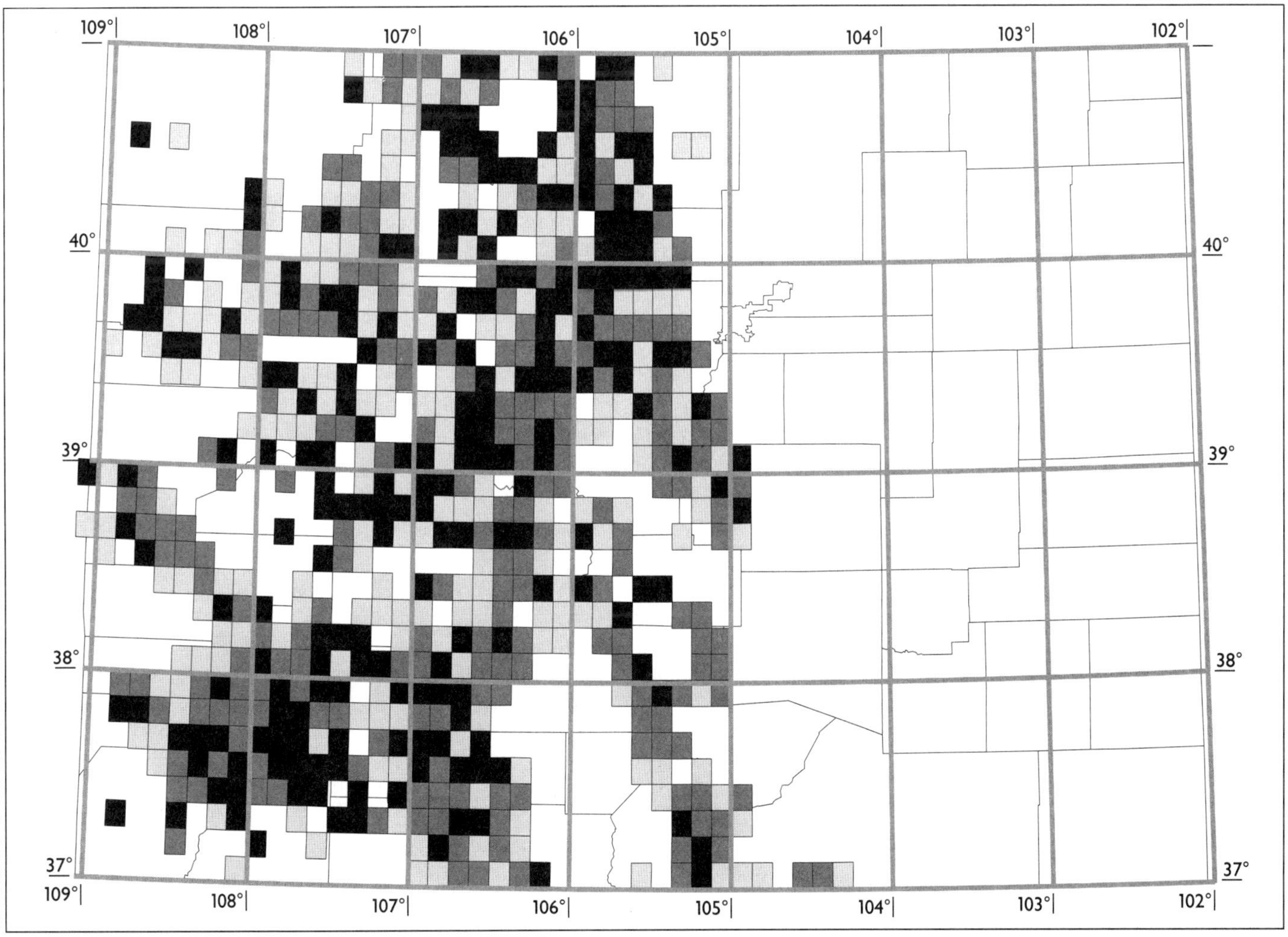
109°
108°
107°
106°
105°
104°
103°
102°
40°
39°
38°
37°

COLORADO DIVISION OF WILDLIFE / WILDLIFE RESOURCE INFORMATION SYSTEM

Everyone enjoys watching robins as they hop tamely across mowed lawns and cock their heads, listening for earthworms. Even children easily locate their nests, in plain sight on tree limbs, and examine their blue eggshells lying in halves on the ground. These abundant and conspicuous thrushes flourish in close association with people and in almost every habitat in Colorado.

HABITAT

Colorado atlasers found American Robins in more blocks and more habitats, and Confirmed them at a higher rate, than any other species. Robins breed in urban areas, around farmhouses and windbreaks, in riparian, coniferous, and aspen forests, and in krummholz. Atlas volunteers recorded this habitat generalist in 47 habitat codes and subcodes. Rural farmsteads and shelterbelts received the largest count, but often with low abundance—ten or fewer breeding pairs. Greater numbers breed in mountain conifers and aspen woodlands. Atlasers estimated A5 abundance—more than 1,000 breeding pairs—in 15 mainly coniferous blocks and five mainly aspen blocks, in both northwestern and southwestern Colorado.

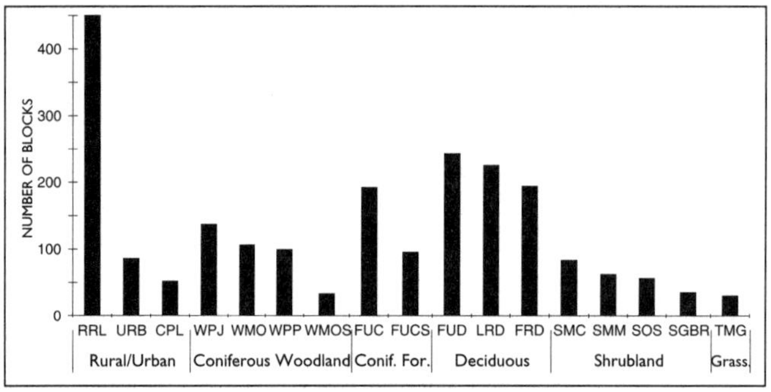
NUMBER OF BLOCKS
400
300
200
100
0
RRL URB CPL WPJ WMO WPP WMOS FUC FUCS FUD LRD FRD SMC SMM SOS SGBR TMG
Rural/Urban
Coniferous Woodland
Conif. For.
Deciduous
Shrubland
Grass

Robins' conspicuous habits draw attention and they tend to use edge habitats along roads and trails, which may lead observers to overestimate their abundance. On a well-used Boulder Open Space trail through ponderosa pine, robins were the only woodland species with greater abundance close to the trail than away from it (Miller and Knight 1995). In counterpoint, on a different Boulder Open Space trail, a riparian one with heavy people, dog, and bicycle traffic, robins experienced a 100% nest failure (Carl Bock pers. comm.).

Much of the breeding success of robins results from their ability to forage in different ways and on a wide variety of foods, including insects, tiny vertebrates, fruits (Paszkowski 1982), and, of course, earthworms.

Their predilection for earthworms contributed a major element to Rachel Carson's case against DDT. From 1954 to 1958 the Michigan State University campus turned into a graveyard for robins. To combat Dutch elm disease powerful sprayers each year had squirted a stream of poison to all parts of the tallest trees, killing not only the target bark beetles but all other insects within the zone of spraying. In autumn the leaves, retaining the oily residue of the pesticide, fell to the ground; over the winter earthworms ingested the leaves. The next spring robins died in dramatic numbers—11 earthworms can transfer a lethal dose to a robin. The large campus population of robins died out in four years; no fledglings at all appeared in 1958 (Carson 1962). After the DDT ban, the robins, proving their adaptability, recovered normal numbers on the campus by 1979 (Brewer et al. 1991). In New York, however, pesticide mortalities continued into the 1980s in suburbs where applicators applied diazinon and organophosphates to lawns (Bonney *in* Andrle and Carroll 1988).

BREEDING

The robin's Confirmation percentage exceeded that of any other species. Abundant breeding pairs, obvious nests, multiple broods, and visible fledglings enhanced the ease of confirmations.

Atlas dates for nest-building, nests with eggs, and nests with young document a very long breeding season. Fledglings, easily identified because of their breast spots, short tails, and attendant adults, provide the most obvious indicia of breeding. Robins raise two, sometimes three, broods (Bent 1949). Grackles and jays frequently pillage their eggs and young, but cowbirds rarely parasitize them (Ehrlich et al. 1988). Robins loudly scold intruders, which often led atlasers to find and to Confirm breeding of predators like hawks and owls.

Robins usually nest on tree limbs where they can find platforms for their bulky nests—often near the trunk but sometimes well out on a branch. Near human habitations they place nests on almost anything that will support them (porches, window ledges, under eaves, etc.). The female, usually by herself, builds the nest of mud and grasses with a lining of fine grass. They will fly as far as a quarter of a mile to obtain mud, and they make their own mud by wetting their feathers and then shaking to moisten dust. One even took dry soil and wet it in a bird bath (Shuford 1993).

Atlasers located nests in a wide variety of supporting plants or structures, including blue spruces, subalpine firs, aspens, and cottonwoods. Robins even built nests in big sagebrush (Lu Bainbridge, nest card), a lilac bush (Norm Barrett, nest card), and under a bridge on a support beam (Ronda Woodward, nest card).

BREEDING PHENOLOGY

CODE		# OF RECORDS	DATE RANGE
C	Courtship	1	20 Apr
NB	Nest Building	43	29 Mar–7 Jul
ON	Occupied Nest	137	5 May–4 Aug
NE	Nest with Eggs	57	8 Apr–22 Jul
NY	Nest with Young	151	2 May–2 Aug
FY	Feeding Young	497	21 Apr–23 Aug
FL	Fledged Young	327	7 May–26 Aug

DISTRIBUTION

American Robins breed throughout North America, from the northern limit of trees in Alaska and Canada south into Mexico. They winter throughout the southern half of their breeding range and south to Guatemala. Originally a resident of open woods, their range in Colorado and elsewhere expanded with the spread of European settlement (Bent 1949). Robins moved into the Great Plains as settlers planted trees and built houses, which provide nest sites, and as lawn watering and other irrigation augmented sources for more of their soft food.

Robins breed very commonly throughout Colorado. A few blocks without trees for nesting lack breeding records—some devoted to croplands and grazing on the plains, shrublands and agricultural fields in the San Luis Valley, and sagebrush-covered hills along the western border. The Latilong Study cited breeding in 27 of the 28 latilongs; atlasers Confirmed breeding in the remaining one (39103). BBS data suggest that robins have increased their numbers by one-quarter in 31 years, with a stable trend in Colorado.

We can expect American Robins, with their ability to adapt to any habitat that has a few trees, to continue to thrive all across Colorado. With the enactment of pesticide controls, these birds which signaled a problem still make their noisy and persistent contributions to the dawn (and predawn) chorus.

BREEDING EVIDENCE

REPORTED IN 1428 (82%) OF 1745 PRIORITY BLOCKS

☐	Possible	139 blocks	(10%)
▩	Probable	83 blocks	(6%)
■	Confirmed	1206 blocks	(84%)

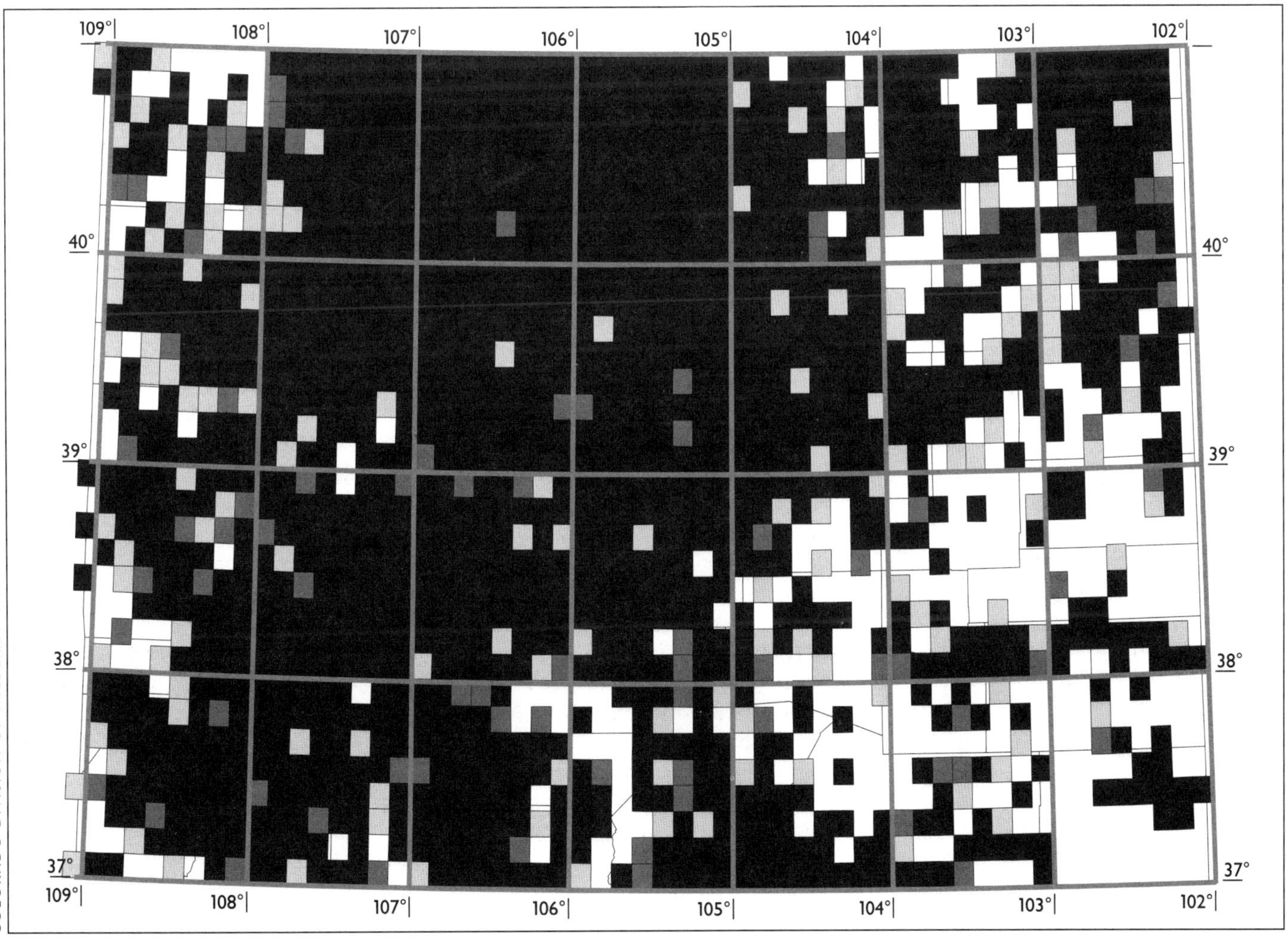
109°
108°
107°
106°
105°
104°
103°
102°
40°
39°
38°
37°

COLORADO DIVISION OF WILDLIFE / WILDLIFE RESOURCE INFORMATION SYSTEM

Finding Colorado's Gray Catbirds in late summer usually involves no more effort than poking around among ripening crops of chokecherries along our riparian corridors. Here, these rather shy mimids gorge themselves, building stores of fat for their impending journey to warmer climes.

HABITAT

Throughout their breeding range, Gray Catbirds prefer dense shrubs and vines (Cimprich and Moore 1995). Atlas habitat data reveal a bird with a decided proclivity for streamsides; few other places in this arid state offer sufficiently dense vegetation and adequate quantities of small fruits in season. Montane shrublands and willow carrs do, however, provide secondary breeding habitat.

Although catbirds typically breed in areas of dense foliage, the location of the nest site provides adult birds a good deal of visibility. They usually construct their nests within 3 feet (1 m) of the top or side of the supporting shrub—sometimes right on the edge. Nest heights can range from ground level to more than 50 feet (15 m), but most nests stand about 5 feet (1.3 m) above the ground (Nickell 1965, Cimprich and Moore 1995).

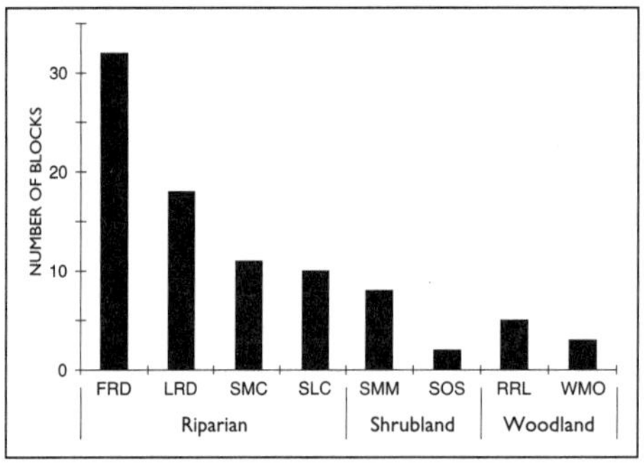
NUMBER OF BLOCKS
0
10
20
30
FRD
LRD
SMC
SLC
SMM
SOS
RRL
WMO
Riparian
Shrubland
Woodland

The closely related Brown Thrashers breed in similar habitats. Some competition for nest sites exists (Nickell 1965), but within Colorado habitat choices by the two species help to restrict competition. In Colorado and Nebraska along the South and mainstem Platte rivers, catbirds nest in sites significantly different from the typical riparian habitat. They select sites with denser foliage and, presumably, greater shrub heights (Andrews 1987, A&R). Atlas data illustrate that Brown Thrashers have broader habitat preferences than Gray Catbirds.

BREEDING

Pair formation and nest construction take place shortly after the birds return to their breeding grounds in May. Although males usually help to gather nesting material, females do most of the nest construction (Nickell 1965). A successful pair will frequently mate together in subsequent years (Darley et al. 1977).

Typical clutches contain three or four eggs (rarely six). Clutch sizes decrease later in the season and with second nestings (Nickell 1965). The range of the few Atlas Confirmation dates (especially nest dates) suggests double-brooding in Colorado. Atlasers found that the most activity involving food carrying and fledglings happened in July—half of the FY, CF, FF, and FL reports—with only one such report (CF) in August.

Only the female incubates, but the male guards the nest area during periods while the female takes a recess from incubation. During these periods, the pair maintains contact through wing-flipping and soft singing. Such coordinated efforts at nest-guarding apparently help to reduce the incidence of predation and nest parasitism (Slack 1976). When cowbirds lay eggs in their nests, catbirds promptly remove them (Scott 1977). No records of cowbird parasitism exist from Colorado.

Incubation normally lasts 11–15 days, followed by a nestling period of 9–12 days (Nickell 1965). Early in the nestling period, the female spends much of her time brooding and shading the young while the male brings most of the food. As the nestlings develop, the female spends less time at the nest and assumes a greater portion of the feeding responsibilities, especially when shade protects the nestlings from direct sunlight (Johnson and Best 1982).

First broods receive a diet of mostly small insects and spiders. As fruits and berries ripen later in the summer, second broods receive greater quantities of fruit (Nickell 1965). Parents continue to feed young birds for two weeks after they leave the nest (Cimprich and Moore 1995). Predators of nests and young include crows, rodents, and raccoons (Nickell 1965).

BREEDING PHENOLOGY

CODE		# OF RECORDS	DATE RANGE
NB	Nest Building	2	2 Jun–22 Jun
ON	Occupied Nest	2	29 May–22 Jul
NE	Nest with Eggs	1	18 May
FY	Feeding Young	8	7 Jun–12 Aug
FL	Fledged Young	4	10 Jun–25 Jul

DISTRIBUTION

The breeding range extends from British Columbia to Nova Scotia in the north and from eastern Texas to Georgia in the south. States south and west of Colorado support few catbirds. The primary winter range lies along the Atlantic coast and Gulf of Mexico from New England to Central America.

The Atlas map shows that Gray Catbirds concentrate in and near the foothills from Trinidad to Fort Collins and in a few mid-elevation riparian corridors of western Colorado. Compared to A&R, Atlas field workers detected catbirds in somewhat greater numbers from Colorado City to Trinidad. Atlas data contain one major surprise: this "eastern" species turned up in as many blocks on the Western Slope as on the eastern side of the mountains.

Cooke (1897) regarded the species as "fairly common on the plains" and "rare in western Colorado," the latter status echoed by Davis (1969). Much of the apparent increase in western Colorado—particularly northwestern Colorado—appears to stem from a dearth of observers until recent years.

Surely, however, the species never boasted a sizable breeding population on the plains. Many catbirds migrate across the plains in May, but suitable dense shrubbery exists only in isolated pockets on the plains. Nearly all of the suitable habitat in eastern Colorado falls along the Front Range and from Cañon City to Trinidad. Atlasers found no breeders along the South Platte from Greeley to Nebraska (11 Priority blocks) nor along the Arkansas from Pueblo to Kansas (nine Priority blocks). These stream bottoms largely lack dense shrubbery and, during the breeding season, lack catbirds (Bill Prather, Mark Janos pers. comm.). Field workers reported no breeding in three latilongs (38103, 40103, and 40104) where the Latilong Study had, but did provide the first Confirmations in Latilong 37105.

Destruction of winter habitat, due to development accompanying rapid growth of human populations in coastal areas, represents a significant threat to this species (Cimprich and Moore 1995). BBS data show a minor decline across the continent. Human habitat alterations, both favorable and unfavorable, in breeding areas appear to have had little net effect on the numbers of this species.

BREEDING EVIDENCE

REPORTED IN 68 (4%) OF 1745 PRIORITY BLOCKS

Possible	30 blocks	(44%)
Probable	20 blocks	(29%)
Confirmed	18 blocks	(26%)

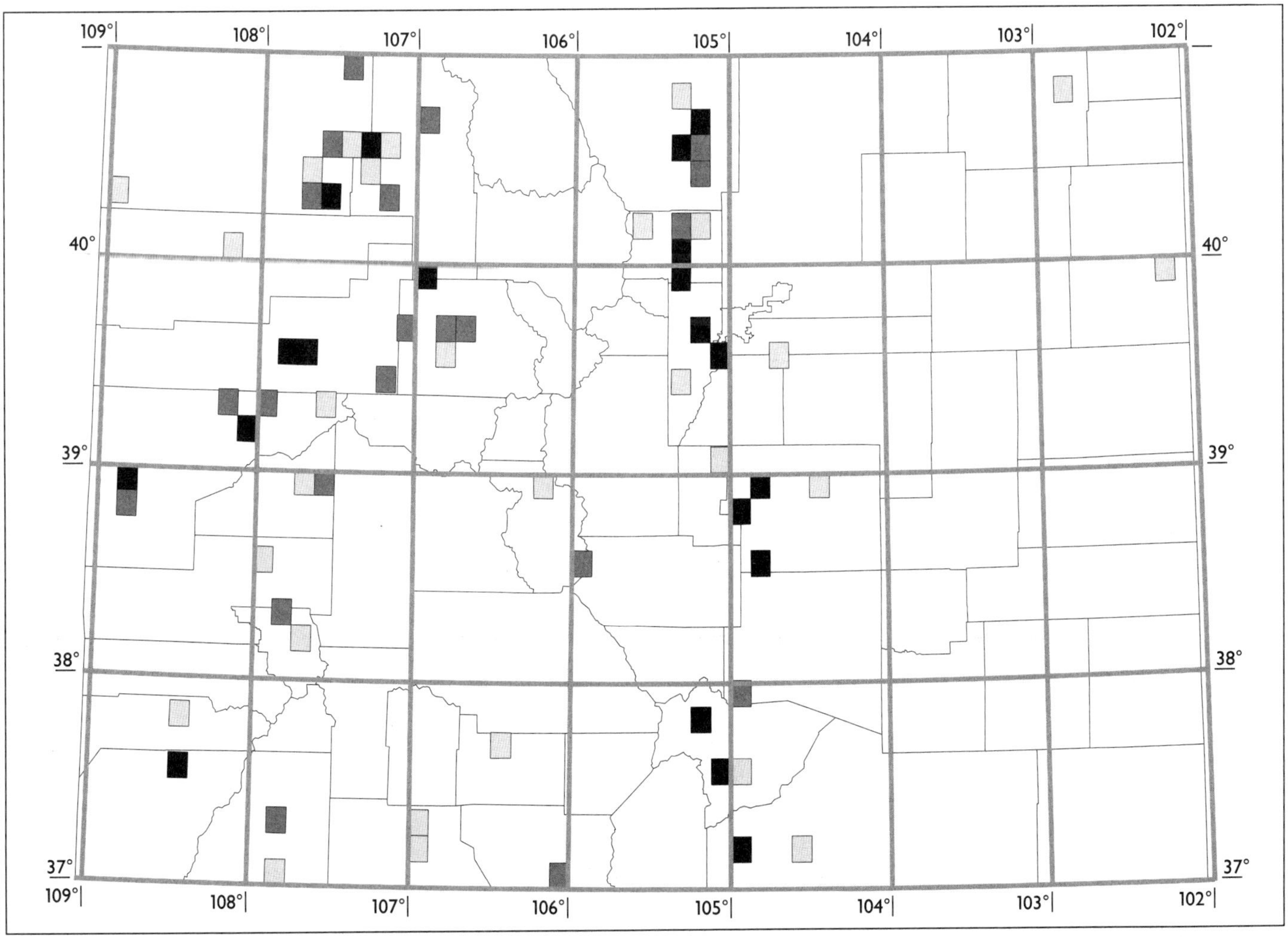
109°
108°
107°
106°
105°
104°
103°
102°
40°
39°
38°
37°

COLORADO DIVISION OF WILDLIFE / WILDLIFE RESOURCE INFORMATION SYSTEM

No other grayish bird adds so much color to our lives. At once both a prince and a rogue, the Northern Mockingbird seldom ventures far from the spotlight of human attention.

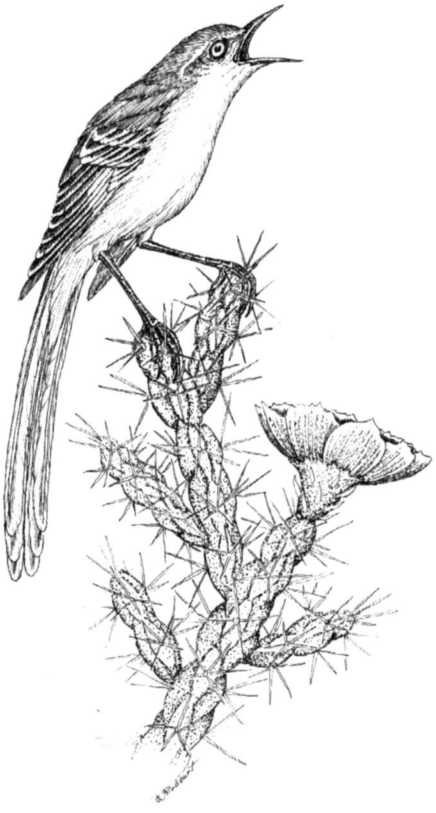

HABITAT

Although common in parks and suburban yards throughout much of their range, Northern Mockingbirds in Colorado use fewer artificial habitats. Only a trickle of Atlas reports came from parks and residential areas. In much of eastern Colorado this species finds suitable habitat in remnant windbreaks, shrubs, and scattered trees. Such sites occur mainly in association with agricultural areas and in expanses of relatively undisturbed prairie. Many atlasers reported isolated cottonwood trees on the prairie as "lowland riparian" habitat due to the lack of a clearly defined category for cottonwoods growing away from open water. As a result, the mockingbirds' usage of lowland riparian areas appears somewhat inflated.

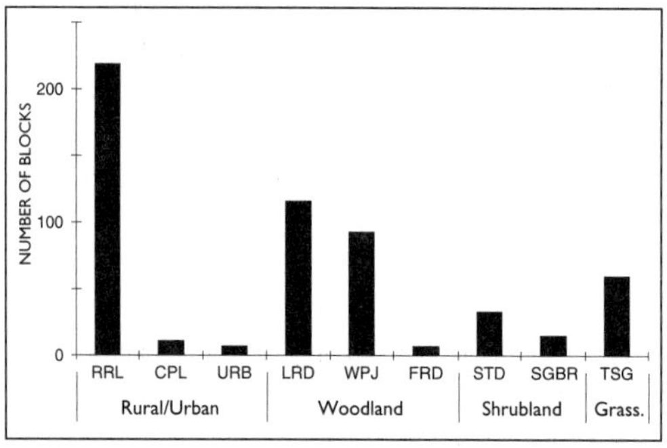
NUMBER OF BLOCKS
0
100
200
RRL
CPL
URB
LRD
WPJ
FRD
STD
SGBR
TSG
Rural/Urban
Woodland
Shrubland
Grass.

In much of western Colorado and Las Animas County, pinyon/juniper woodlands attract sizable populations of breeders. There, these relatively short trees replace the shrubs and plantings the birds use in eastern Colorado.

BREEDING

Mockingbirds maintain a small year-round population in Colorado in most years. Their numbers receive an enormous boost from migration, however, beginning in April and continuing into early May (A&R, Dexter and Levad 1995).

Shortly after arriving, the birds establish territories and begin the breeding cycle. Male mockingbirds sing most frequently during the nest-building stage of reproduction. This singing appears to help trigger a response in the reproductive system of the female (Logan 1983). Clutch sizes typically range from 3–5 eggs (Laskey 1962).

Mockingbirds build nearly all their nests 3–10 feet (1–3 m) above the ground (Derrickson and Breitwisch 1992). Nest heights do, however, tend to increase later in the season as the density of the foliage of the trees and shrubs used for nesting increases (Laskey 1962). Nesting success correlates positively with the degree of isolation of the tree or shrub supporting the nest (Joern and Jackson 1983).

Especially in southeastern Colorado, nestlings and fledglings begin to appear during the second half of May. Males and females bear roughly equal portions of the feeding of nestlings, although females perform almost all of the brooding. Nestlings receive both animal and fruit food items, but animal prey constitute a much larger part of their overall diet. The amount of fruit consumed increases with the age of the nestlings (Breitwisch et al. 1986).

Nestlings leave the nest about 12 days after hatching. Both parents continue to feed fledglings, although both may devote time (not simultaneously) to the construction of a nest for a subsequent brood during this time. Parental feeding continues for approximately three weeks after fledging (Derrickson and Breitwisch 1992).

Atlas phenology depicts a relatively short breeding period. The economy of starting a second nest while tending to a fledged brood may lessen the time required for two broods.

Mockingbirds sometimes maintain a pair bond for life; more frequently the pair bond spans the length of the breeding season. Both sexes reach sexual maturity at one year of age, but year-old males often fail to attract a mate because males outnumber females (Derrickson and Breitwisch 1992).

Predators of nests and fledglings include corvids, snakes, and squirrels. Cowbirds infrequently parasitize the nests of this species (Derrickson and Breitwisch 1992), and no records of cowbird parasitism exist from Colorado.

BREEDING PHENOLOGY

CODE		# OF RECORDS	DATE RANGE
C	Courtship	6	9 May–2 Jul
NB	Nest Building	13	23 May–3 Jul
ON	Occupied Nest	27	22 May–4 Jul
NE	Nest with Eggs	13	22 May–5 Jul
NY	Nest with Young	12	3 Jun–13 Jul
FY	Feeding Young	44	15 May–23 Jul
FL	Fledged Young	25	26 May–29 Jul

DISTRIBUTION

Northern Mockingbirds range through the southern half of the U.S., north to New England, the Great Lakes, and northern California. Although the winter range nearly coincides with the breeding range, substantial numbers of birds withdraw from northern parts of the breeding range during winter. Most of the birds that breed in Colorado have departed by the end of October (A&R).

Cooke (1897, 1900) regarded this species as present east of the mountains and all across southern Colorado. Then, as now, the species' stronghold fell in the southeastern quadrant of the state; along the Arkansas east of Pueblo he termed it "as abundant as at any place in the south." The distribution described by Cooke resembles the Atlas map, except for a continuing movement away from the Front Range cities and a small population in northwestern Colorado that Cooke did not mention. The latter population has, in all likelihood, existed all along but escaped notice until human activity increased in the region.

Atlas work provided a first Confirmation of breeding in Latilong 39103 but did not even find them in the Denver metro area—probably due to suburbanization of old farmsteads. Except in the San Luis Valley, where mockingbirds nest up to 8,500 feet (2600 m), and single blocks in Gunnison, Douglas, Teller, and Huerfano counties, atlasers found them breeding only up to about 6,000 feet (1830 m).

Changing agricultural practices have reduced available habitat for mockingbirds within Colorado. The small trees and bushes that once grew abundantly at the edges of fields and around farmhouses have started to disappear as fewer and fewer houses dot the rural plains and as fencerows continue to succumb to the practice of planting right up to the edges of county roads.

BREEDING EVIDENCE

REPORTED IN 465 (27%) OF 1745 PRIORITY BLOCKS

☐	Possible	178 blocks	(38%)
▩	Probable	137 blocks	(29%)
■	Confirmed	150 blocks	(32%)

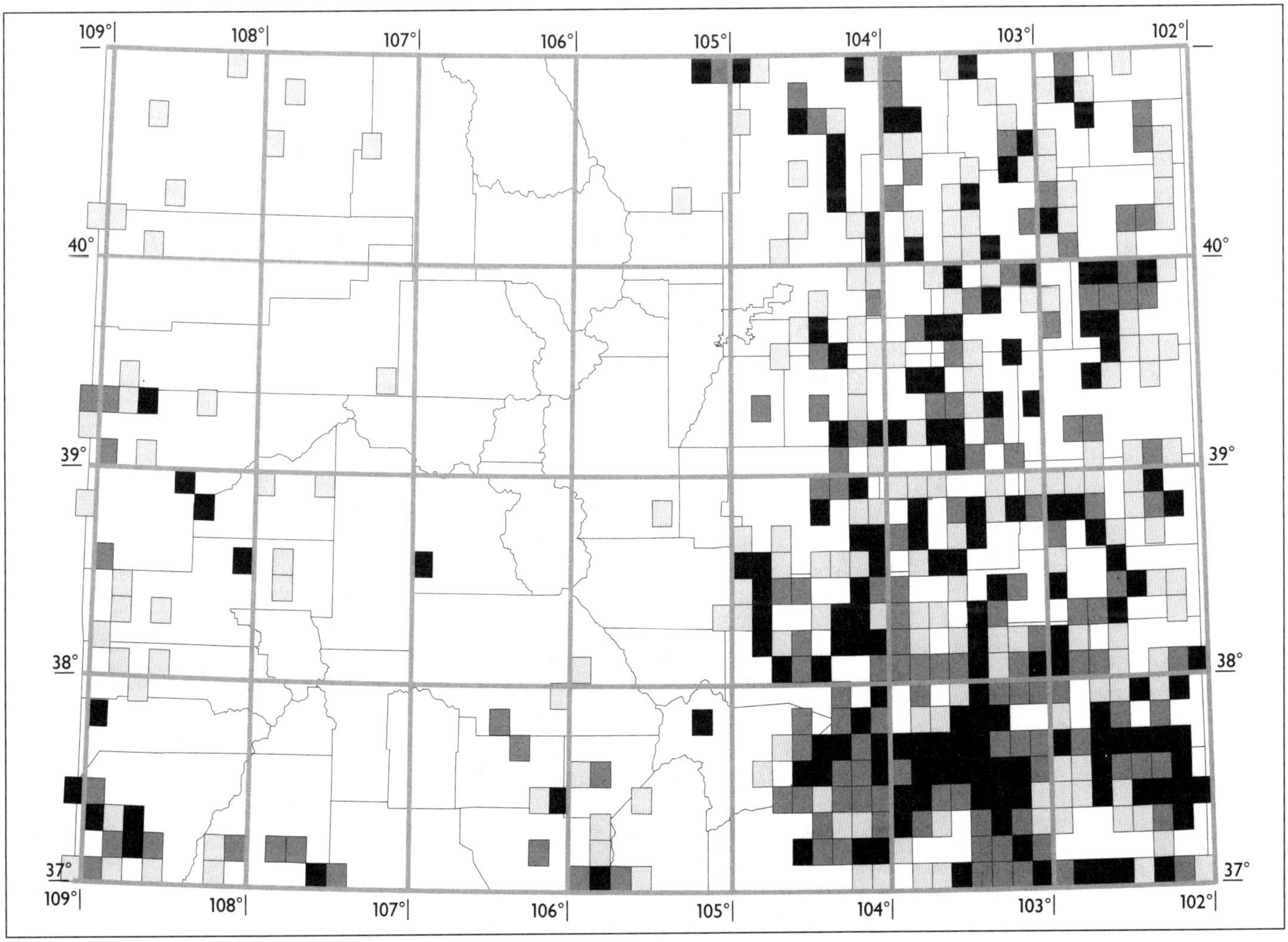
109°
108°
107°
106°
105°
104°
103°
102°
40°
39°
38°
37°

COLORADO DIVISION OF WILDLIFE / WILDLIFE RESOURCE INFORMATION SYSTEM

Few avian species claim the vast expanse of desert shrublands in the arid West. Perhaps the most conspicuous birds one encounters amidst the gray-green sea of shrubs are Sage Thrashers. Heralded as the "poet of the lonesome sagebrush plain" by Bent (1948), the thrasher relieves the monotony of its stark landscape through its sweet warbling songs.

HABITAT

Bestowed with a fitting name, Sage Thrashers inhabit the sagebrush-dominated rangelands of the western U. S. Only a few other bird species—Sage Grouse, Sage Sparrow, and Brewer's Sparrow—regularly occupy this habitat. These species maintain such a strong association with sagebrush that they have been called "obligates" (Braun et al. 1976a). Atlas records however, show that Sage Thrashers also use other habitats.

Besides a propensity for sagebrush, thrashers inhabit and nest in other desert shrubland communities including greasewood, rabbitbrush, shadscale, and saltbush, occasionally cholla cactus, and in mountain shrublands (A&R). Over 80% of all Atlas observations occurred in shrubland habitats, and sagebrush alone accounted for over half of all habitats recorded. Reports of thrashers using desert shrubland vegetation occurred most frequently in the San Luis Valley.

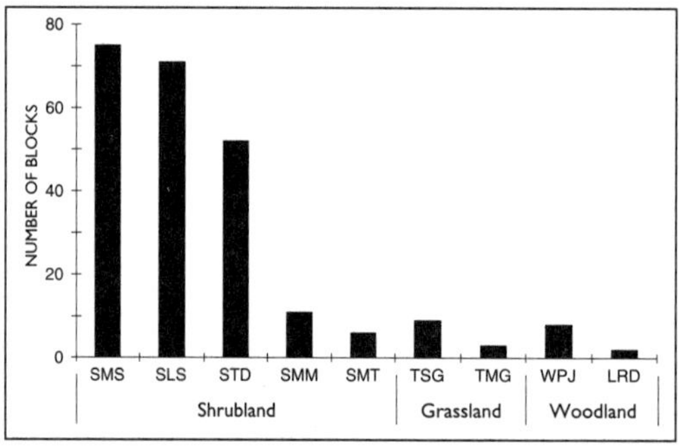
NUMBER OF BLOCKS
80
60
40
20
0
SMS
SLS
STD
SMM
SMT
TSG
TMG
WPJ
LRD
Shrubland
Grassland
Woodland

BREEDING

Visually and vocally conspicuous during breeding, Sage Thrashers often perch atop shrubs, quietly perusing their surroundings or singing cheerily. This behavior may account for the large number of Possible Atlas records. In the latter part of the breeding season, Confirmations surpassed all observations. Fledged young accounted for over half of all Atlas Confirmations. At the nest site, thrashers become more secretive, as shown by the very few nests detected during Atlas work.

Thrashers begin arriving in Colorado in mid March, with peak numbers in April (A&R). The earliest nesting date in Colorado is a nest with young found on 13 May 1931 near Cañon City (B&N). The latest dates, nests with young found during the Atlas, came on 10 July 1988 in the San Luis Valley and 13 July 1990 in North Park.

Thrashers conceal their bulky nests in or beneath the tallest and densest clumps of sagebrush or other shrubs (Rich 1980, Petersen and Best 1991). Vegetation that shades the nest from the afternoon sun appears important in nest-site selection (Reynolds and Rich 1978). Some birds even construct twig platforms above their nests to provide shade (Bent 1948, Rich 1985). Sage Thrashers usually rear one brood, although occasionally they may produce two. Cowbirds rarely parasitize Sage Thrashers, which will eject cowbird eggs (Rich and Rothstein 1985).

BREEDING PHENOLOGY

CODE		# OF RECORDS	DATE RANGE
C	Courtship	1	23 Jun
NB	Nest Building	4	2 Jun–21 Jun
NE	Nest with Eggs	2	18 Jun–29 Jun
NY	Nest with Young	3	19 Jun–13 Jul
FY	Feeding Young	34	21 May–27 Jul
FL	Fledged Young	56	2 Jun–6 Aug

DISTRIBUTION

During the breeding season, Sage Thrashers occupy shrub-steppe habitats throughout most of the western states, from southern British Columbia to northern Arizona and New Mexico. In winter months they retreat to southern Arizona and New Mexico, western Texas, and northern Mexico.

Within Colorado, Sage Thrashers most commonly occur in North Park, the San Luis Valley, the Gunnison Basin, and the plateaus and mesas in the northwestern corner of the state (A&R). They less commonly inhabit western mesas and valleys, Middle Park, and the Wet Mountain Valley. Once parental duties have ceased, the birds, either singly or in family units, spread out from their nesting areas into adjacent habitats. Atlasers reported many observations of thrashers from mid to late July on the eastern plains and at timberline. By early to mid October most birds leave the state. Only rarely do thrashers remain during the winter in Colorado.

Atlas records generally concur with A&R, with the greatest complexes of blocks and highest densities reported in North Park, the San Luis Valley, and especially Moffat County. Field workers reported high contiguity of occupied blocks but lower densities in the Gunnison Basin and the Uncompahgre Valley, and low density and spotty distribution in the Colorado River basin from Middle Park to the Utah line. Atlasers found no conclusive evidence of nesting in northeastern Colorado where A&R described them as "uncommon" local inhabitants. Although these thrashers occasionally nest on the Pawnee National Grassland (CFO, DFO), their occurrence and distribution there remain spotty. The Atlas deemed the many July reports from northeastern Colorado (not shown on the map) as birds on migration or dispersal.

Several reports of thrashers came from southeastern Colorado. Although scattered geographically, the dates and behaviors observed confirm some local breeding in this region as well. Both Kansas and Oklahoma have reported one record each of Sage Thrashers breeding in counties adjoining Colorado (Sutton 1967, Thompson and Ely 1992). In the high-altitude sagebrush near Leadville (39106B3) atlasers found a few Sage Thrashers on 5 July (and saw Brewer's Sparrows carrying food in the same block), suggesting that Sage Thrashers may nest up to 10,000 feet.

These thrashers show a large increase on Colorado BBS routes, over 4% per year. This may explain the high numbers and new nesting sites found by the Atlas.

BREEDING EVIDENCE

REPORTED IN 212 (12%) OF 1745 PRIORITY BLOCKS

Possible	72 blocks	(34%)
Probable	44 blocks	(21%)
Confirmed	96 blocks	(45%)

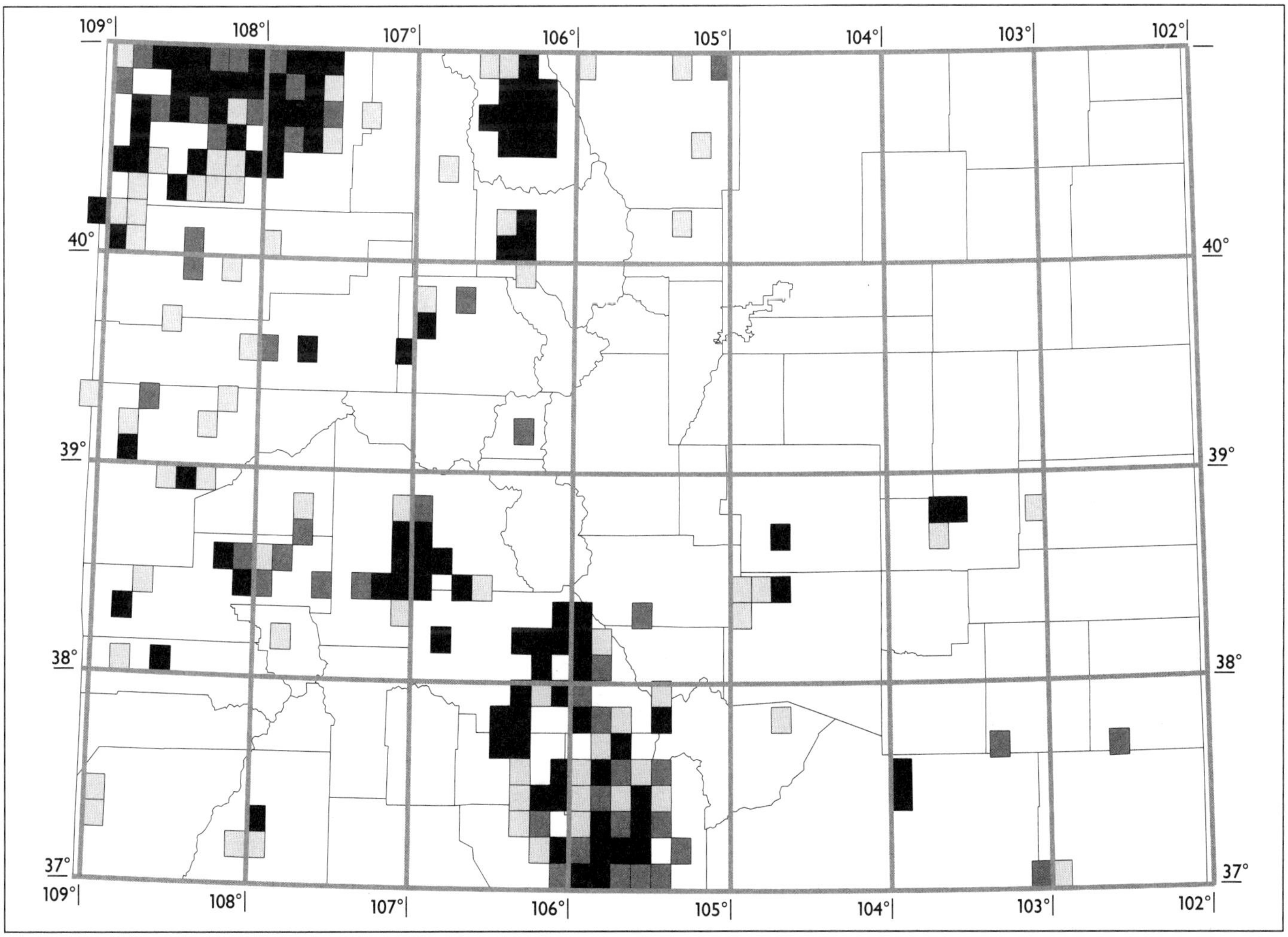
109°
108°
107°
106°
105°
104°
103°
102°
40°
39°
38°
37°

COLORADO DIVISION OF WILDLIFE / WILDLIFE RESOURCE INFORMATION SYSTEM

For vocal virtuosity and flashy appeal few species rival the Brown Thrasher. Whether singing from the highest vantage or dashing into cover, thrashers enliven the stream bottoms and farmsteads that dot the eastern plains. Our desire to enhance our surroundings with trees and shrubs has promoted the expansion of this species in Colorado.

HABITAT

The dense shrubbery and thickets that Brown Thrashers favor in the East and Midwest become increasingly limited on the western edge of their range. Here they find their favored habitat only in riparian corridors, brushy draws, and manmade habitats such as shelterbelts, woodlots, shrub thickets, and plantings around rural and suburban houses. These shy, retiring birds relish almost any dense vegetation that provides escape cover and a place to conceal their nests. Along the South and mainstem Platte rivers in Colorado and Nebraska, Brown Thrashers select sites similar to, or at least not different from, random sites, with regard to numbers of trees and shrubs, shrub height, and foliage density (Andrews 1987).

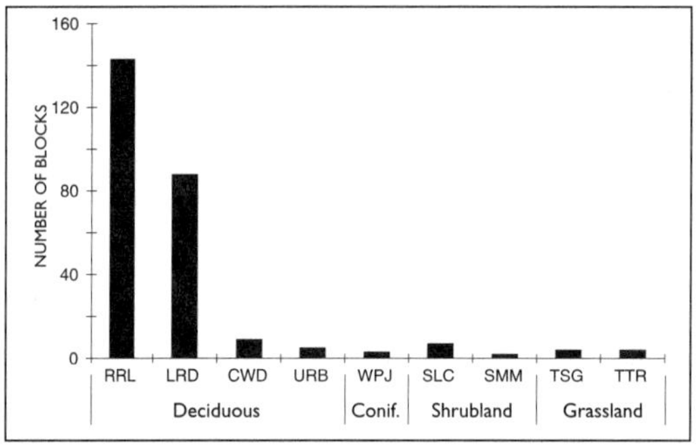

NUMBER OF BLOCKS
160
120
80
40
0
RRL
LRD
CWD
URB
WPJ
SLC
SMM
TSG
TTR
Deciduous
Conif.
Shrubland
Grassland

Brown Thrashers in Colorado have keyed in on introduced plantings in rural environments and shrubby riparian vegetation. The predominant Atlas habitat consists of vegetation associated with rural settings (RRL). This habitat typically contains linear or block plantings of trees and shrubs around houses and outbuildings. Combining this and the very similar cultivated woodland habitat (CWD) makes up over half of all Atlas habitats reported. The next most-reported habitat, lowland riparian, provides suitable cover when it contains a shrub component. All Atlas nests and nest-building activity occurred within these three habitats.

BREEDING

Arriving in early May, male thrashers herald the breeding season by singing exuberantly from a prominent perch. From this vantage point they pour forth their repertoire of twice-repeated phrases. This is often the best time to see and to hear these accomplished songsters because they otherwise exhibit a decided shyness in their daily routine.

This shyness, and a proclivity for dense cover, serve well to conceal their bulky nests in shrubs, trees, or tangles of vines. With a bit of knowledgeable searching, however, atlasers frequently discovered their nests. In the West, thrashers usually nest in shrubs and trees, whereas in the Northeast they often nest on the ground (Bent 1948). A few ground nests have been documented in Kansas and Oklahoma (Snow 1994, Tyler 1994, Facemire 1978), so the possibility exists that some thrashers may nest on the ground in Colorado.

Brown Thrashers bear the unfortunate distinction as the largest known passerine parasitized by Brown-headed Cowbirds (Bent 1948). The Atlas documented Colorado's first record of a cowbird parasitizing a thrasher nest. The nest contained one large cowbird and several smaller thrasher nestlings (MBD; Appendix C).

BREEDING PHENOLOGY

CODE		# OF RECORDS	DATE RANGE
C	Courtship	1	18 Jun
NB	Nest Building	10	7 May–9 Jun
ON	Occupied Nest	12	23 May–4 Jul
NE	Nest with Eggs	5	20 May–22 Jun
NY	Nest with Young	8	8 Jun–18 Jul
FY	Feeding Young	41	20 May–7 Aug
FL	Fledged Young	15	7 Jun–31 Jul

DISTRIBUTION

Brown Thrashers breed throughout the eastern U.S. and southern Canada west to Alberta and eastern Texas. During winter months they withdraw to the southeastern states. At the western edge of their range in

Colorado, thrashers historically bred only on the plains along the forested river drainages and appeared sporadically in the mountains and western valleys. Most observations away from the plains represent non-breeding birds, although the state has two mountain nesting reports, in Estes Park and La Plata County (B&N, A&R). In winter, thrashers occur rarely along the Front Range and accidentally on the Western Slope.

Atlas results substantially (and surprisingly) expand prior concepts of Brown Thrasher distribution in Colorado. Atlas records generally follow those in the Latilong Study but cover a far broader area than the range outlined in A&R. Those authors portrayed thrashers as occurring primarily along the South Platte and Arkansas rivers, but this distribution may also reflect a bias in the preferences of birdwatchers concentrating on public lands in riparian areas.

Atlas records show that Brown Thrashers breed solidly across northeastern Colorado and south along the Arkansas River into Baca County. As well as riparian corridors, they occupy shelterbelts and farmyards on the High Plains. Observers most often reported an A2 abundance (2–10 pairs/block), with twice as many in rural habitat as in riparian. Higher densities of A3 (11–100 pairs/block) occurred most often in riparian habitat along the South Platte and Arkansas rivers, with a few A3 reports in rural habitat along the eastern border.

This broader distribution stems from two factors. First, the systematic Atlas coverage of the eastern plains facilitated documentation of thrashers in areas not previously surveyed; and second, thrashers clearly have embraced artificial habitats. Since the settlement of eastern Colorado, thrashers have expanded their range as manmade habitats have increased and matured over the years.

Brown Thrashers display a continuing drop on BBS routes, 1.2%/year across the continent (a drop of one-third of the population in 31 years). Too few routes in Colorado count them to detect a trend here, but Atlas data emphasize a greater presence than previously thought.

BREEDING EVIDENCE

REPORTED IN 238 (14%) OF 1745 PRIORITY BLOCKS

Possible	89 blocks	(37%)
Probable	58 blocks	(24%)
Confirmed	91 blocks	(38%)

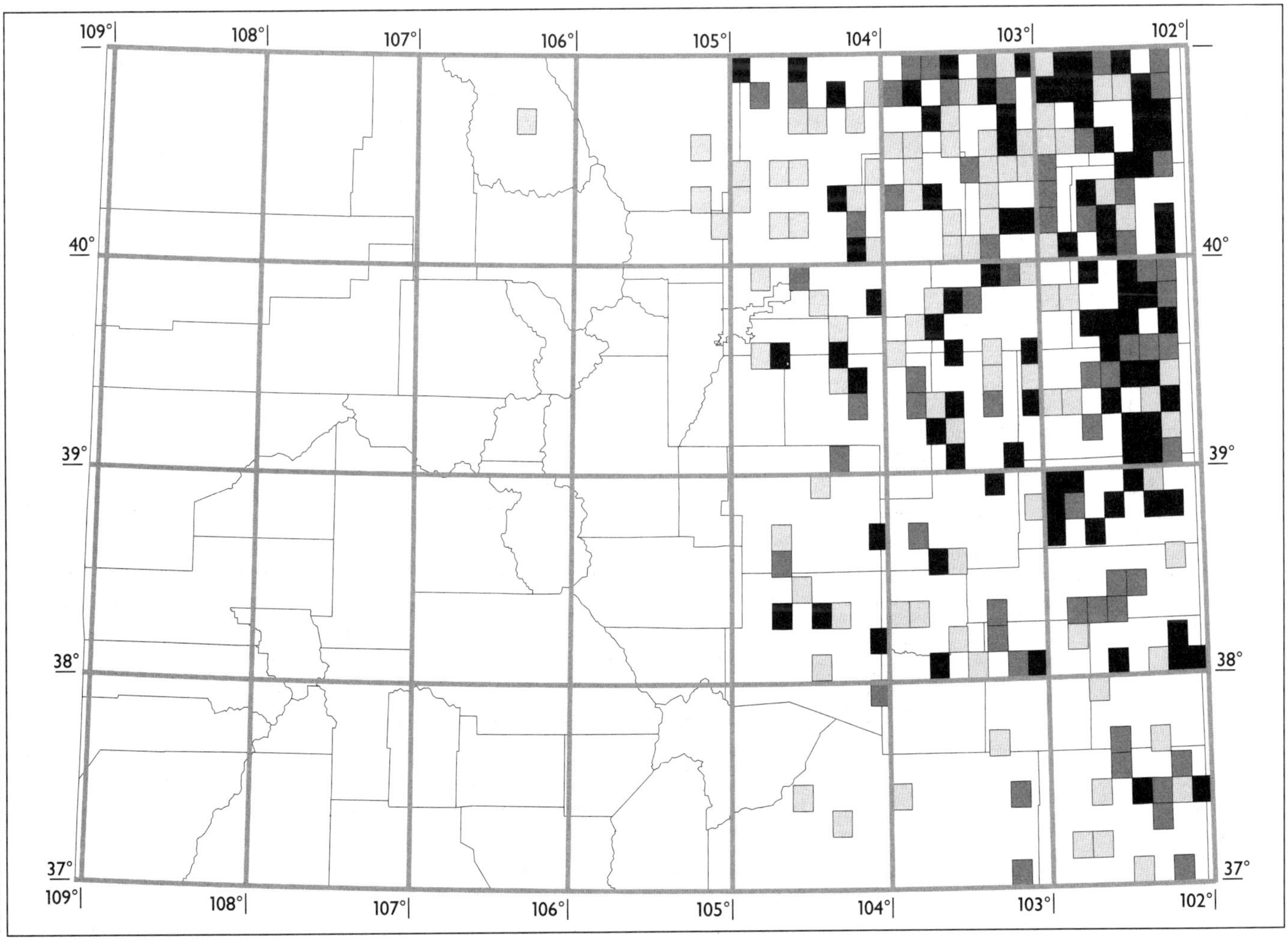
109°
108°
107°
106°
105°
104°
103°
102°
40°
39°
38°
37°

COLORADO DIVISION OF WILDLIFE / WILDLIFE RESOURCE INFORMATION SYSTEM

Denizens of cactus-strewn landscapes, Curve-billed Thrashers thrive amidst forests of tall cacti. Making the most of their prickly neighbors, these thrashers regularly employ them in their annual housekeeping duties. Only recently in ornithological history did these desert dwellers discover Colorado's cactus-dotted lands.

HABITAT

Throughout the Southwest, Curve-billed Thrashers show an affinity for arid shrublands and grasslands dotted with tall cactus or other thorny plants. Their preferred range in Colorado matches that preference, with cholla cactus grasslands of southeastern Colorado serving as their primary habitat. Nearly half of all the Atlas observations of this species occurred in shortgrass prairie containing tall cacti. To a lesser extent, they inhabit pinyon/juniper, riparian woodlands, rural shelterbelts, and shrublands. Atlas results reflect the intermittent use of these other habitats with rural (RRL) the second most reported habitat.

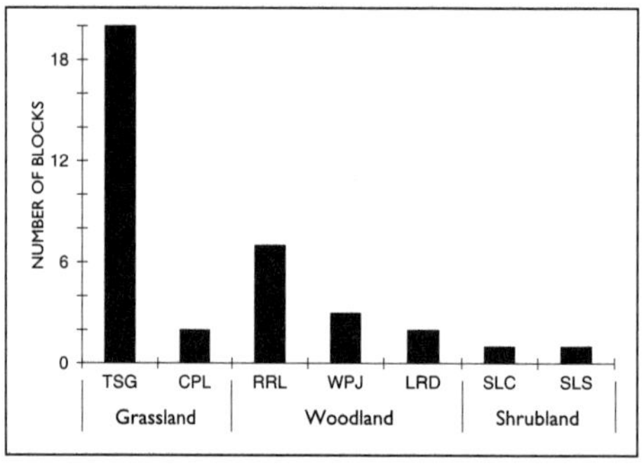

NUMBER OF BLOCKS
18
12
6
0
TSG
CPL
RRL
WPJ
LRD
SLC
SLS
Grassland
Woodland
Shrubland

In winter in the Oklahoma panhandle, Curve-billed Thrashers shift from grasslands to shrubby thickets in canyons and on the lower slopes of mesas (Sutton 1967). This species tames easily and ventures around rural yards and buildings; they occasionally drink from water troughs and dripping faucets (Bent 1948).

BREEDING

In the broad expanse of shortgrass prairie in southeastern Colorado, cholla cactus provides the primary structure (or substrate) for nest sites. Curve-billed Thrashers (here and in similar habitat in Oklahoma and Kansas) nest almost exclusively in this spiny fortress (Sutton 1967, Thompson and Ely 1992). All nests reported by atlasers, and all from prior years, have occurred in cholla cactus (CFO, DFO, var.) except one Atlas record: one pair near Fowler used an abandoned magpie nest 10 feet (3 m) up in a shrub-like tree (Robert Dickson pers. comm.). In much of their range, Curve-billed Thrashers maintain their territory and pair bond throughout the year (Fischer 1980). Within this territory, several older nest sites may be scattered among cacti. Some birds build roosting platforms during the winter and occasionally use them as foundations for future nests (Anderson and Anderson 1973).

Nesting phenology for Colorado remains spotty because few nests have been reported by the Atlas or in the Colorado literature. For the few nests recorded, the dates suggest a fairly lengthy breeding season. The first state breeding record described a nest with three eggs on 7 May 1972 in Baca County (Kingery 1973). In southwestern Kansas, nest-building begins by mid April (Thompson and Ely 1992). In Oklahoma, nesting ranged from a nest with eggs on 24 April to recently fledged young on 2 October (Sutton 1967). In southern parts of their range Curve-billed Thrashers regularly raise two broods (Tweit 1996), and probably do so in Oklahoma (Sutton 1967) and Kansas (Thompson and Ely 1992). Atlas data show a shorter span of dates, about six weeks, partly due to the small number of observations.

Curve-billed Thrashers generally incur little to no cowbird parasitism, and Colorado has no reports. In Texas, thrashers repeatedly chased cowbirds from nest sites (Fischer 1980), and Bronzed Cowbird eggs placed in four nests were ejected within 48 hours (Carter 1986).

BREEDING PHENOLOGY

CODE		# OF RECORDS	DATE RANGE
NB	Nest Building	1	18 Jun
NY	Nest with Young	6	21 May–4 Jul
FY	Feeding Young	2	3 Jun–9 Jul
FL	Fledged Young	2	10 Jun–11 Jul

DISTRIBUTION

Historically, Curve-billed Thrashers ranged from central Arizona to western Texas, and south into Mexico. They first appeared in Colorado in 1951, near Granada, Prowers County (B&N), and gradually have extended their range in southeastern Colorado. By 1970 Colorado had records from Baca to Pueblo counties; the Atlas map shows them in the same area. The species now lives year round in dispersed pockets from Baca County north to the Arkansas Valley and west to Huerfano and Fremont counties (A&R, Atlas). Prior to their discovery in Colorado, an isolated population of thrashers occurred near Kenton, Oklahoma (Sutton 1967). In 1968 they arrived in northeastern New Mexico, their expansion possibly tied to the spread of cholla caused by grazing (Darling 1970). They continue to expand their range in Oklahoma and a few breed in southwestern Kansas (Thompson and Ely 1992).

The Atlas reports these thrashers in only 26 blocks, basically within their known range. Their distribution may be more widespread than Atlas records show; because of the thrashers' spotty occurrence and low density, atlasers may have missed them. Most observers reported an abundance of 2–10 pairs/block. This density parallels that reported by A&R and the Latilong Study.

Vagrant non-breeding thrashers occasionally stray into northeastern Colorado; the Atlas had two such records, in Douglas and Larimer counties. Dispersal by juveniles may explain some of these vagrant sightings. For example, in Texas, pairs showed nest-site fidelity, whereas banded nestlings did not return to their natal areas the following year (Fischer 1980).

Long-term BBS data show a major decline for Curve-billed Thrashers across their range, but Texas (the only state showing a decline) dominates the drop because it has most of the U.S. population. The clearing of southern Texas brushlands may explain this decline (Tweit 1996). Too few routes record Curve-billeds in Colorado to determine the rate here, although the species has maintained the same distribution in the state since the first nest found in 1972. Until they incorporate some other suitable substrate for their nesting needs, the Curve-billed Thrasher range in Colorado will remain closely linked to the distribution of cholla cactus.

BREEDING EVIDENCE

REPORTED IN 26 (1%) OF 1745 PRIORITY BLOCKS

Possible	15 blocks	(58%)
Probable	2 blocks	(8%)
Confirmed	9 blocks	(34%)

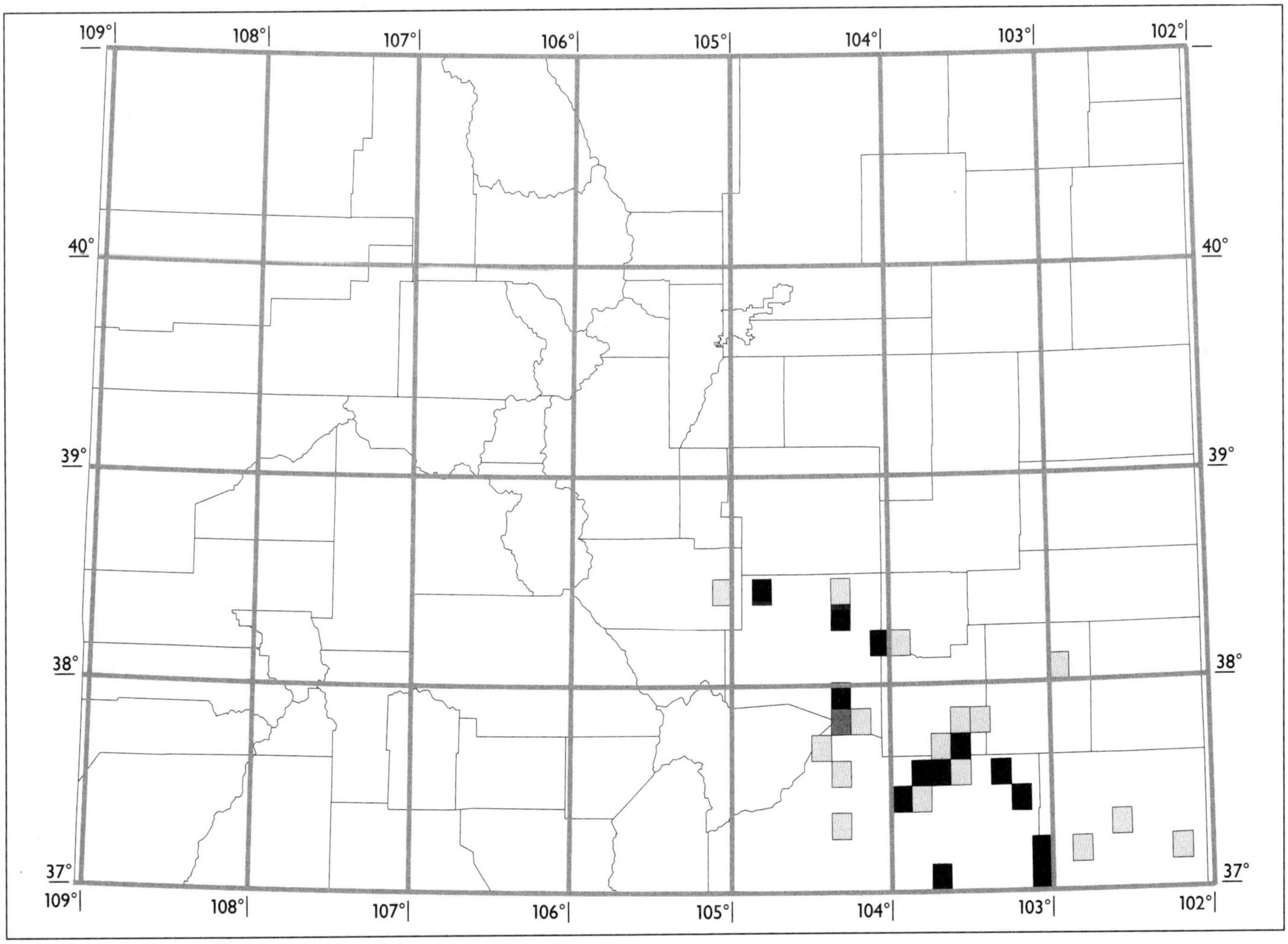
109°
108°
107°
106°
105°
104°
103°
102°
40°
39°
38°
37°

COLORADO DIVISION OF WILDLIFE / WILDLIFE RESOURCE INFORMATION SYSTEM

Castigated in North America for usurping the holes of native cavity-nesting birds, and disliked for their large, loud, and dirty roosts, European Starlings earn praise elsewhere for their economic value. Stock-growers in New Zealand put up nesting boxes along the fence lines to attract these introduced birds, which they consider beneficial in consuming the grubs that ravage the pastures. B&N verified starlings' usefulness in insect control.

HABITAT

Historical records by Pliny and Aristotle attest that European Starlings have associated closely with people since the advent of agriculture (Feare 1984). In Colorado, they use urban as well as rural areas, and Atlas records show them now breeding up to 10,000 feet (3048 m). Only dense forests, treeless plains, and deserts have scant starling populations. Wherever cavities exist, whether natural, bird-made, or in manmade structures, starlings make their nests. They forage for invertebrates principally in shortgrass such as mowed lawns and grazed pastures. Where these feeding areas lie close to nesting cavities and roost sites such as trees or buildings, their abundance increases (Cabe 1993).

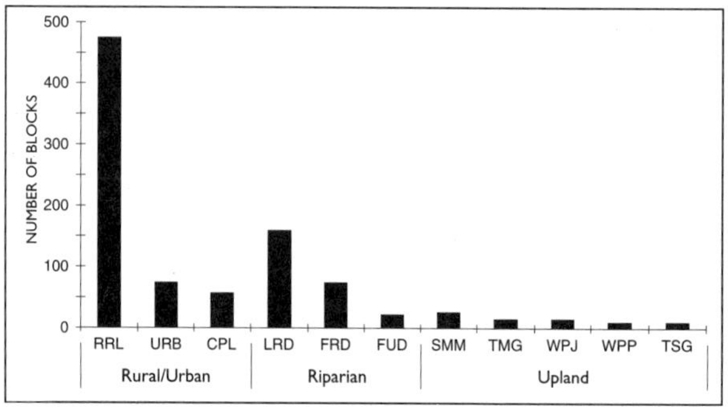
NUMBER OF BLOCKS
500
400
300
200
100
0
RRL URB CPL LRD FRD FUD SMM TMG WPJ WPP TSG
Rural/Urban
Riparian
Upland

Atlas results affirm the ancient association of starlings with people. Atlasers found them in three times as many blocks with rural habitats (including farm buildings and yards as well as shelterbelts) as in the next most used habitat, riparian woodlands. Urban areas replace riparian as second most used habitat by applying Atlas abundance estimates. These portray urban areas with higher densities than other habitats (A5 estimates in seven urban blocks, and only one each in rural and riparian blocks). Calculations from these estimates allot 45% of the nesting pairs to rural habitats, 26% to urban areas, and 22% to cottonwood stream bottoms. The lower abundance estimates in riparian blocks stem from the linear nature of forested stream bottoms.

Nevertheless starlings pack tightly into mature cottonwood groves where flickers and woodpeckers have drilled nest holes. Here many pairs can nest in close proximity; atlasers frequently Confirmed breeding by seeing busy parents flying back and forth from open fields to stands of tall cottonwoods. Conspicuous throughout their nesting cycle, starlings yielded a 70% Confirmation rate; 43% of the Confirmations involved these obvious trips carrying food for young.

BREEDING

The rapid population growth of European Starlings springs from many factors. Nelson (1993) gives breeding dates for Colorado as 1 February through 31 July. Atlas data record the earliest nest-building as 4 March on the plains, with fledglings noted till late August in the same area. The early start allows them to commandeer nest holes before native cavity-nesters start breeding. The long breeding season gives time for double-brooding (Cabe 1993). They nest colonially if many nest sites are available. Cooperative breeding adds to their success, as first-year males may feed the young and guard the nest (Kessel 1957).

The nest, loosely constructed of twigs and other vegetable matter within the cavity, often becomes filthy and befouled with mites and other parasites. Although starlings often use the same nest for more than one brood and more than one year, they acquire protection from disease-carrying pests by adding fresh green vegetation that acts as a natural fumigant (Clark and Mason 1985).

Starlings persistently eject bluebirds, woodpeckers, and other cavity-nesters from holes and appropriate them—even from birds bigger than they, such as flickers and Wood Ducks (Cabe 1993). In Fairmount Cemetery in Denver five successive pairs of starlings evicted the same pair of flickers from a series of newly drilled holes before the beleaguered flickers finally bred in peace in the sixth one (Tina Jones pers. comm.).

BREEDING PHENOLOGY

CODE		# OF RECORDS	DATE RANGE
C	Courtship	1	28 May
NB	Nest Building	39	4 Mar–30 Jun
ON	Occupied Nest	63	29 Apr–4 Jul
NE	Nest with Eggs	1	14 Jun
NY	Nest with Young	58	7 May–7 Jul
FY	Feeding Young	253	7 Apr–6 Aug
FL	Fledged Young	144	19 Apr–22 Aug

DISTRIBUTION

Europeans have introduced starlings wherever they have settled, and these adaptable birds now flourish worldwide

(Clements 1981). Attempts to establish them in North America began in the nineteenth century in places as widely separated as Quebec (1875) and Oregon (1889). Only the introductions in New York City (1890–1891) succeeded. After six years starlings began an expansion in every direction until their still-growing range now includes North America south of the treeline through northern Mexico (Godfrey 1986, Howell and Webb 1995). *The Denver Post* reported the first Colorado sighting, 16 February 1937 (B&N), and Breiding (1943) found the first nest at Lowry Field in Denver, 16 May 1943. By 1987 the Latilong Study reported them breeding in every latilong. Atlasers recorded them in 782 blocks (starlings ranked 23rd in number of blocks) and in every county.

The map shows a solid presence along the populated Front Range, and almost as solid breeding on the High Plains and cottonwood-lined stream bottoms from the South Platte to the San Juan. Generally starlings do not occur at high altitudes, although atlasers found a colony breeding in downtown Leadville at 10,200 feet (3110 m); Leadville also hosts the highest recorded nesting sites for Rock Doves and House Sparrows.

Atlas abundance estimates suggest that over 500,000 pairs of European Starlings nest in Colorado. This places them as the 27th most numerous species in the state, surprisingly low to observers who think of them as ubiquitous.

Also surprising to city dwellers and birdwatchers along the stream bottoms, BBS statistics show a continental decline of 1%/year; the Colorado trend is undetermined. Starling numbers are also declining in parts of Europe and the United Kingdom (Feare 1994).

As a feeding platoon of starlings moves across a lawn or pasture with machine-like precision, scarcely a grub or cutworm escapes detection and capture. Numerous studies (Kalmbach and Gabrielson 1921, Russell 1971, Feare 1984, Fischl and Caccamise 1987) prove that starlings' diet consists of 44–66% invertebrates, depending on the season, and that they consume more wild fruits and berries than crops. Their economic value is beyond dispute. Only esthetics dictates that we appreciate bluebirds more than starlings.

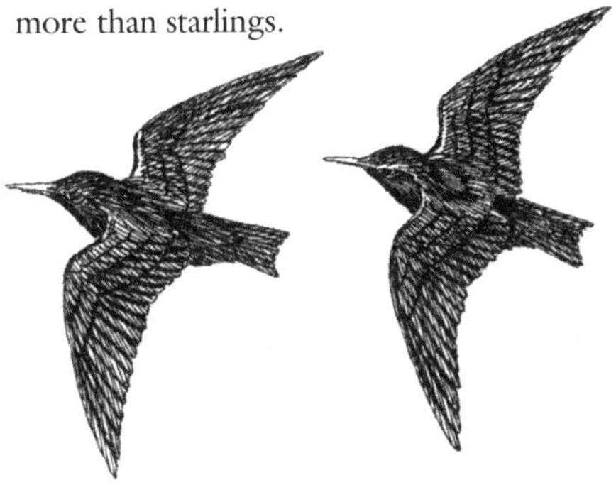

BREEDING EVIDENCE

REPORTED IN 782 (45%) OF 1745 PRIORITY BLOCKS

☐	Possible	173 blocks	(22%)
▩	Probable	62 blocks	(8%)
■	Confirmed	547 blocks	(70%)

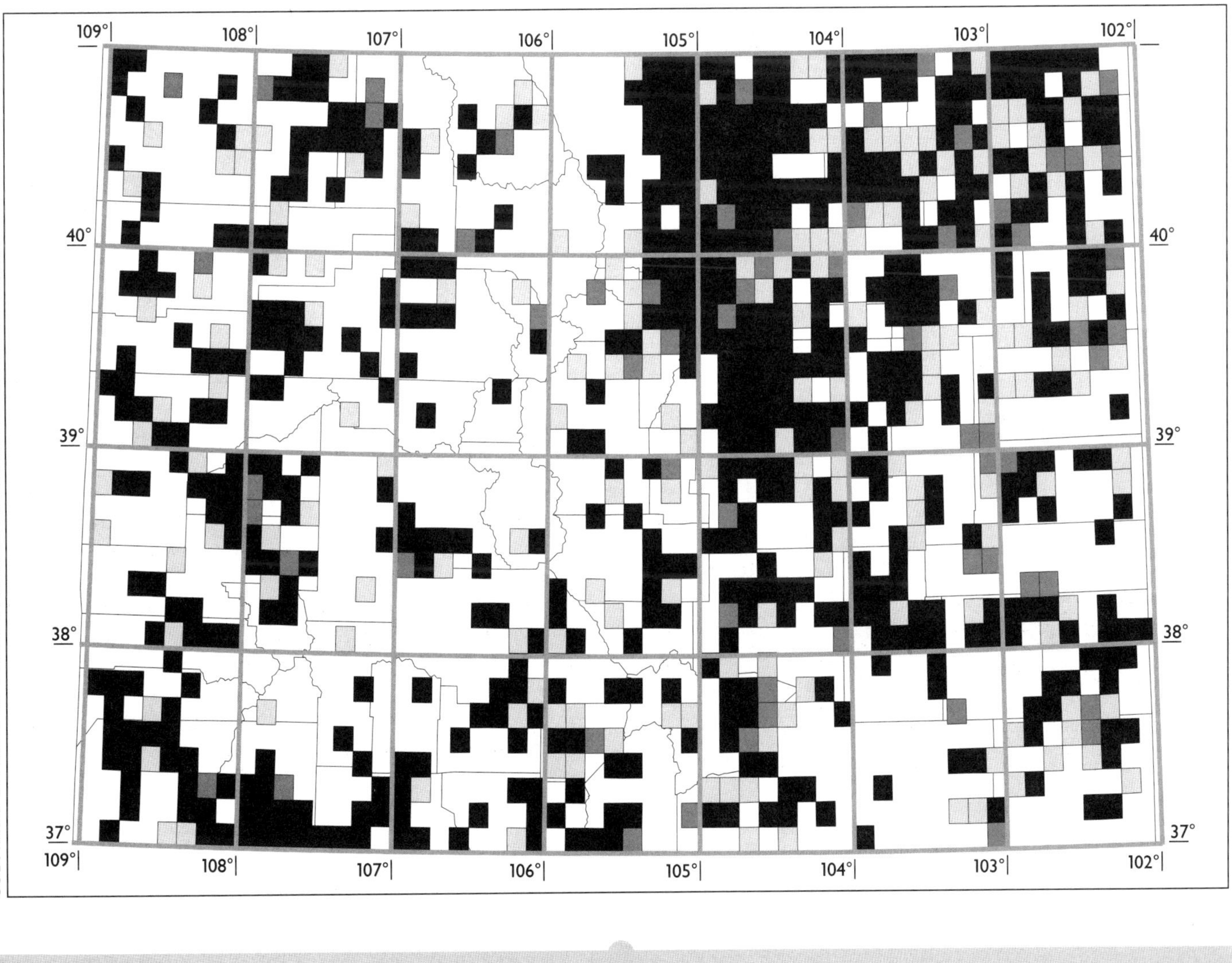
109°
108°
107°
106°
105°
104°
103°
102°
40°
39°
38°
37°

COLORADO DIVISION OF WILDLIFE / WILDLIFE RESOURCE INFORMATION SYSTEM

None would dare dispute the hardiness of American Pipits. In March, they endure bitter winds while foraging along shores of half-frozen lakes. During May they arrive on the alpine tundra—Colorado's most forbidding natural habitat. Here they spend the summer raising their young while waging a ceaseless battle against onslaughts of frost, wind, hail, and even snow.

HABITAT

Only a handful of avian species manage to eke out an existence on Colorado's alpine tundra. And, within this brutal environment, no species rivals the numbers of American Pipits. Atlas figures suggest that pipits compose three-quarters of the breeding birds above timberline.

Rosy-finches, ptarmigans, Rock Wrens, White-crowned Sparrows, and Horned Larks all breed nearby, but the pipits' most direct competition probably comes from Horned Larks. The diets of these two species overlap extensively. Adults of both species feed their young diets composed entirely of insects. Adult Horned Larks do consume some plant material, not usually present in the diet of adult pipits (Verbeek 1967, 1970).

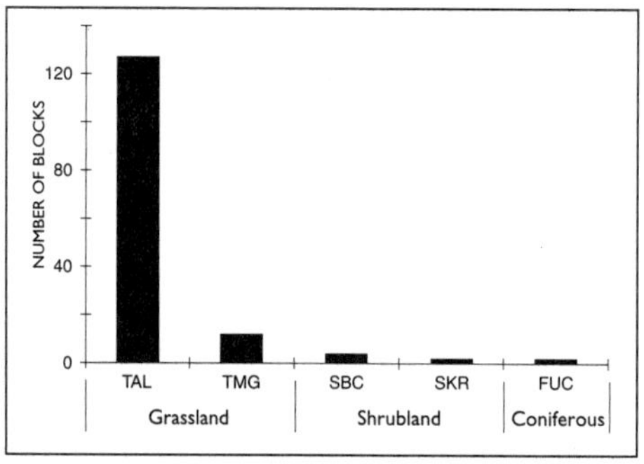
NUMBER OF BLOCKS
0
40
80
120
TAL
TMG
SBC
SKR
FUC
Grassland
Shrubland
Coniferous

Despite the similarity of diets, habitat preferences limit somewhat the extent of competition between the species. Verbeek (1970) noted that Horned Larks prefer drier less dense, tundra for nest sites and feeding. American Pipits adapt readily to both wet and dry conditions on the tundra. Atlas data do not always show this distinction, in part because tundra conditions may vary widely within even a single Priority block. Nonetheless many atlasers observed that alpine Horned Larks tend to use drier sections than pipits.

On rare occasions, American Pipits venture below timberline into montane meadows to breed (Ryser 1985). Colorado's first record of subalpine breeding resulted from Atlas work done on Buffalo Pass (40106E6). Atlas workers Confirmed subalpine breeding in montane grasslands in eight blocks and reported Possible breeding in four others. The 12 included four blocks in the San Juans, three in the Flat Tops, two in the Elks, and one each in the Front Range, Park Range, and Grand Mesa.

BREEDING

Snow conditions figure heavily in the onset of the pipit breeding cycle. Birds take up residence and may begin nesting activity as soon as the snow recedes from a suitable site. Intervals of snow and poor weather early in the season, however, may prompt them to abandon territories until more favorable weather returns (Verbeek 1970). As a result, nesting often does not occur until July.

Although it would seem difficult to hide a nest on the alpine tundra, American Pipits perform the feat remarkably well. Neatly tucked away beneath a rock, tufts of grass, or taller plants, the small bowl of grasses, stems, and, occasionally, hair eludes all but the most careful or fortunate of observers. The birds select nest sites in areas ranging from bare ground to dense, rock-studded tundra, but talus and boulder fields do not meet their nesting requirements.

Atlasers located nests with eggs as early as 22 June, but this figure probably better marks the first arrival of atlasers on the tundra than the onset of egg-laying. Incubation, the responsibility solely of the female, lasts approximately two weeks (Verbeek 1970). Clutch sizes reported on nest cards ranged from three to five eggs.

Newly hatched pipits remain in the nest for nearly two weeks. Once leaving the nest, the well-camouflaged nestlings can fly short distances but tend to remain hidden unless predators approach very closely. It takes about two weeks after leaving the nest for fledglings to become independent of their parents (Verbeek 1970).

Atlas data show sizable densities of pipits on their breeding grounds. Verbeek (1970) reported an average territory size of approximately 1.5 acres (3.7 ha). Thus, in favorable conditions, the breeding densities reported often exceeded 100 pairs per block.

BREEDING PHENOLOGY

CODE		# OF RECORDS	DATE RANGE
C	Courtship	4	19 Jun–7 Jul
NB	Nest Building	1	5 Jul
ON	Occupied Nest	3	22 Jun–21 Jul
NE	Nest with Eggs	16	22 Jun–23 Jul
NY	Nest with Young	9	30 Jun–3 Aug
FY	Feeding Young	44	20 Jun–13 Aug
FL	Fledged Young	23	8 Jul–22 Aug

DISTRIBUTION

American Pipits breed widely throughout alpine areas of North America. The subspecies to which Colorado's birds belong breeds in the Rocky Mountains from central New Mexico into Alberta and British Columbia. Perhaps as recently as the 1970s, this subspecies established a toehold in the Sierra Nevada as well (Miller and Green 1987). The winter range of this subspecies includes Mexico and the southwestern states. American Pipits appear in small numbers (1–30 individuals) on up to four Colorado Christmas Bird Counts in a year and a few occasionally winter over within Colorado.

During the breeding season, American Pipits occupy nearly every patch of tundra from eastern Montezuma to western Las Animas counties in the south to Routt, Jackson, and Larimer counties in the north. The Atlas map shows this well except in the linear Sangre de Cristo Range, where Atlas blocks either missed the tundra or offered major access difficulties to atlasers. (Only one of its eight 14,000-foot peaks has even part of its massif within a Priority block, and three Priority blocks have at least a 4,000-foot [1220 m] elevation differential and no trails to the alpine zone.)

Most nesting areas have remained relatively undisturbed by human development over the last century. On the other hand, recreational hiking in Colorado's high country has exploded in popularity over the last 20 years. Thus far, however, human intrusion onto the tundra has not visibly reduced pipit abundance. Additionally, relatively few hikers in the alpine areas bother to venture beyond narrow corridors along trails to alpine lakes, across passes, and to the summits of 14,000-foot peaks. Lower summits remain largely unvisited. These circumstances may change in years to come as congestion grows worse on the highest peaks and as hikers migrate to less-peopled routes.

BREEDING EVIDENCE

REPORTED IN 123 (7%) OF 1745 PRIORITY BLOCKS

☐	Possible	16 blocks	(13%)
▩	Probable	17 blocks	(14%)
■	Confirmed	90 blocks	(73%)

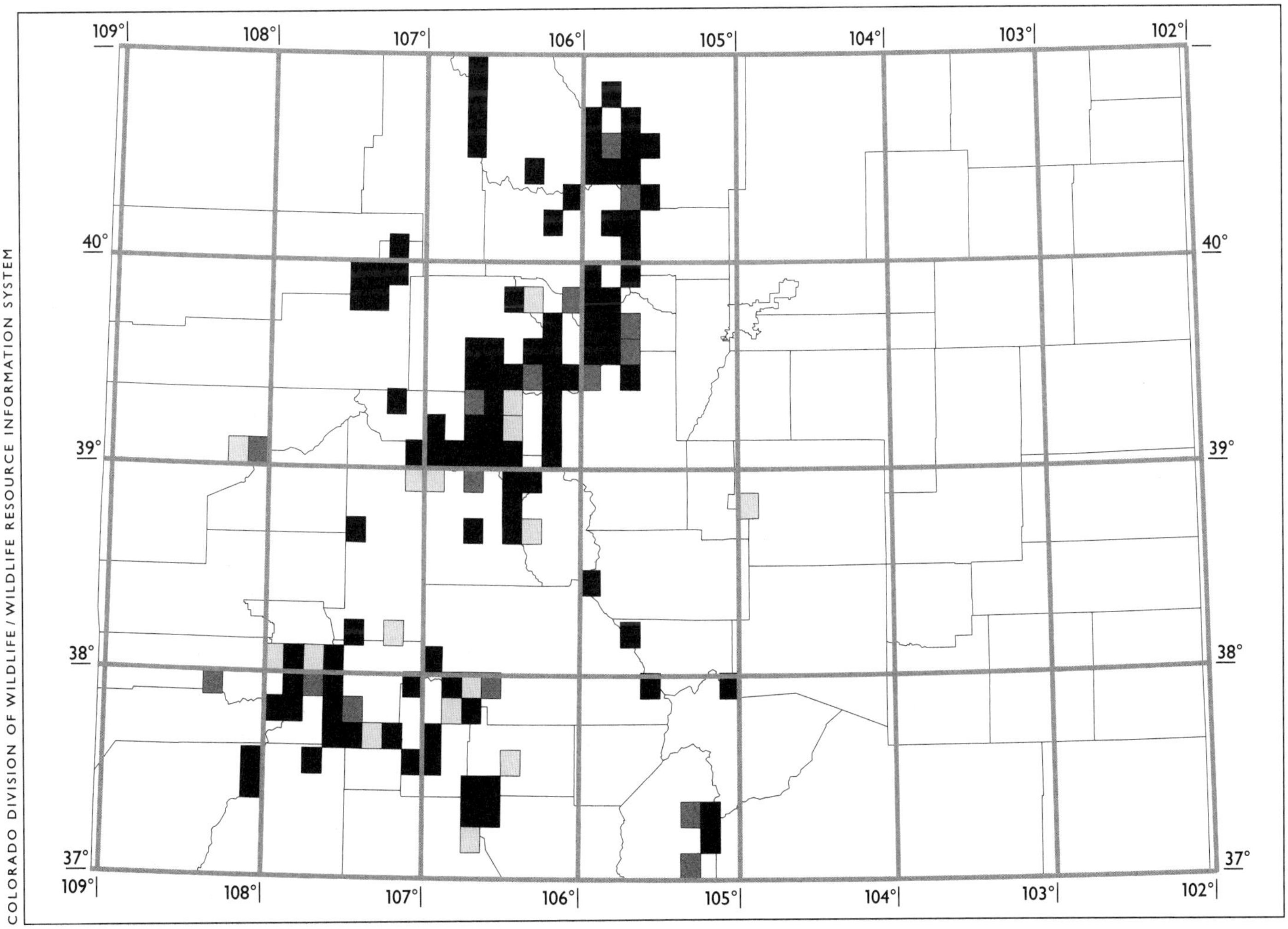
109°
108°
107°
106°
105°
104°
103°
102°
40°
39°
38°
37°

COLORADO DIVISION OF WILDLIFE / WILDLIFE RESOURCE INFORMATION SYSTEM

Unlike the majority of North American wildlife, Cedar Waxwings profit from man's seemingly insatiable desire to plant exotic species in his yard. Feasting on the fruits of pyracantha, hollies, flowering fruit trees, and other ornamental plants, waxwings readily take to the cities and towns for food. After feeding and partial digestion they drop seeds primed for germination, spreading these exotic plants to new locations (Robinson 1986).

HABITAT

Cedar Waxwings in Colorado, at least when nesting, do not show this affiliation with exotics. Here they show a strong preference for deciduous riparian habitats during the breeding season. This limits them to lower elevations, with no Atlas Confirmations above 7,500 feet (2285 m). A&R specifically mentioned urban habitats, as do many researchers across the range of the species (McPherson 1987, Robinson 1986). Atlas data indicate that breeding Colorado waxwings do not exhibit this strong preference (only three urban reports—in Loveland, La Veta, and Gunnison).

Only six of Colorado's Atlas records involved nests and no Atlas submissions contained nest cards, making analysis of suitable nesting habitat characteristics difficult beyond the tie to deciduous riparian areas. In Oregon, waxwings build their nests from 2 to 40 feet (0.6–12 m) off the ground in blackberry bushes and deciduous trees (NMB).

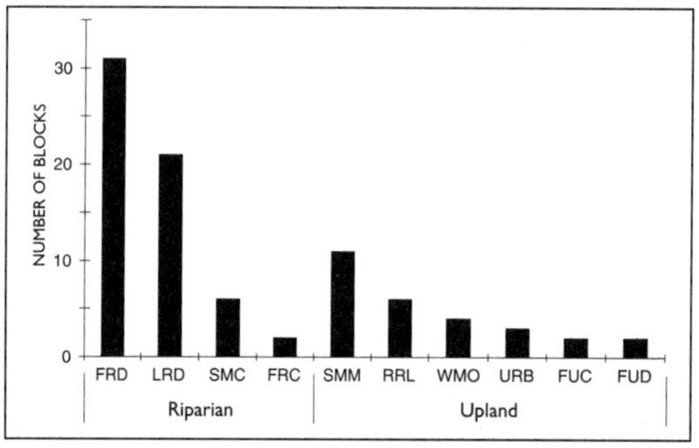
NUMBER OF BLOCKS
0
10
20
30
FRD
LRD
SMC
FRC
SMM
RRL
WMO
URB
FUC
FUD
Riparian
Upland

BREEDING

Cedar Waxwings engage in an extended breeding season. A report of fledged young on 2 June at Cedaredge, in west-central Colorado, implies a starting date for nest-building in early May. At the other extreme, waxwings fed fledglings at Waterton on 26 August 1992 and a pair was constructing a nest as late as 16 August in Colorado Springs. Cedar Waxwings occasionally have second clutches (Terres 1980), and this late nest may represent one of these. The latest date for Colorado fledglings came from Chatfield State Park on 7 September 1985 (HEK). Atlas dates show earlier nesting on the Western Slope—a median date of 27 June compared to 16 July on the Eastern Slope, and twice as many Confirmations before 31 July (8:4) as on the Eastern Slope.

Fruit and other food sources strongly drive waxwing behavior. As food sources wax and wane in one region, breeding flocks move to new nest sites. A site supporting several nests one year might have none for several years later if the food source declines. With strong flocking instincts, Cedar Waxwings often nest in loose groups (Terres 1980). Shared food sources and flock ties keep the birds together, although the nesting pairs scatter too widely to qualify as colony nesters. Individuals defend the immediate nest but do not defend actual territories.

BREEDING PHENOLOGY

CODE		# OF RECORDS	DATE RANGE
C	Courtship	1	6 Jun
NB	Nest Building	3	23 Jun–16 Aug
ON	Occupied Nest	2	12 May–23 Jun
NY	Nest with Young	1	16 Jul
FY	Feeding Young	6	13 Jun–14 Aug
FL	Fledged Young	7	2 Jun–26 Aug

DISTRIBUTION

Cedar Waxwings breed across southern Canada and the northern U.S. and many remain through the winter. In Colorado the species has always been uncommon and irregular in its distribution (Cooke 1897). Sclater could cite only three summer records, and B&N cited several more that reinforced their vagabond distribution. The Atlas map appears random at first. Atlas blocks across the state had waxwings, but elevation actually appears to guide distribution, with breeding limited to the lower elevations and most below 7,000 feet (2135 m).

The Atlas map shows clumps of records along the edge of the Front Range from Wyoming to New Mexico, the upper Yampa drainage in Moffat and Routt counties, the lower reaches of Grand Mesa, and the Fryingpan and Roaring Fork rivers from Basalt to Glenwood Springs. Although a few waxwings were present in some of the higher valleys during the breeding season (NMB), no breeding records came from high-elevation blocks. This distribution agrees with that found on the A&R map.

The highly nomadic behavior of this species makes determining population trends

difficult. Nationally, BBS data indicate a positive trend in bird numbers but Colorado has too few waxwings to determine any trends. Numbers may fluctuate widely in some areas yet remain stable across the entire range. Possible threats include use of pesticides on the fruits on which waxwings feed and the perceived threat of Cedar Waxwings to crops (on a small scale). Waxwing flocks occasionally damage small home fruit plots and larger winter flocks have caused damage to commercial crops in Florida (Brugger et al. 1994).

Cowbirds parasitize Cedar Waxwing nests but waxwings recognize the eggs and often eject them (Neudorf 1992). They do this by piercing the eggs and removing them from the nest. No one reported cowbird parasitism for Cedar Waxwings during the Atlas. With the few records of nests and fledged young, evidence of parasitism could easily be missed—but few nests also offer meager opportunities to cowbirds.

With few threats to their survival, an ability to adapt to some of man's habitats, and the ability to counter cowbird parasitism, Cedar Waxwing populations appear safe. Their flocking and nomadic behavior recall a quote of a now-anonymous birder. When asked if Cedar Waxwings were common he replied, "Cedar Waxwings are extremely common where found in large numbers."

BREEDING EVIDENCE

REPORTED IN 70 (4%) OF 1745 PRIORITY BLOCKS

☐	Possible	48 blocks	(68%)
■	Probable	6 blocks	(9%)
■	Confirmed	16 blocks	(23%)

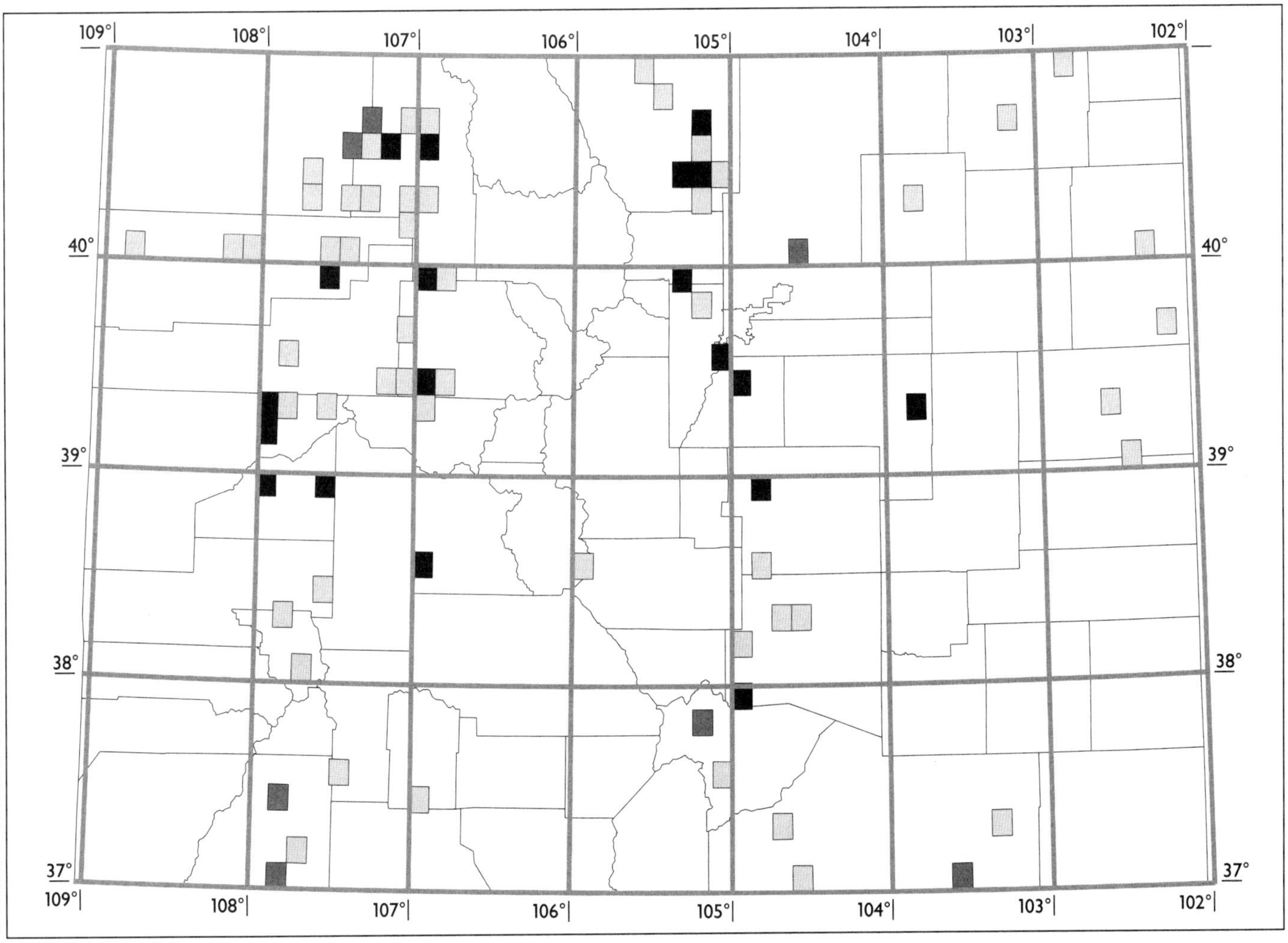
109°
108°
107°
106°
105°
104°
103°
102°
40°
39°
38°
37°

COLORADO DIVISION OF WILDLIFE / WILDLIFE RESOURCE INFORMATION SYSTEM

ORANGE-CROWNED WARBLER *Vermivora celata*

Among North American wood-warblers, Orange-crowned Warblers carry the least glamorous credentials. Nothing about their song, habits, or plumage attracts much attention. They do, however, have the redeeming quality of hanging around our state well after most warblers have departed for Mexico.

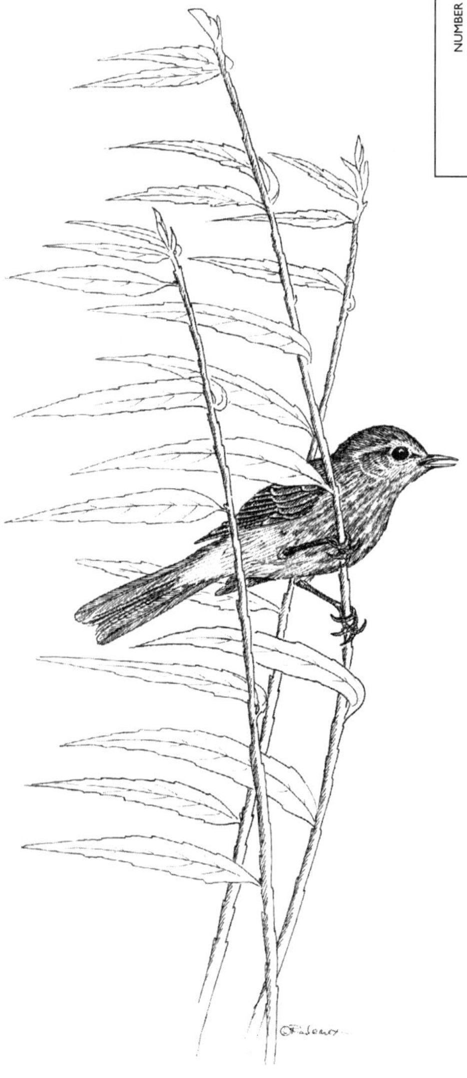

HABITAT

Although widespread across the state during migration, Orange-crowned Warblers breed only in the montane and subalpine zones. Within these zones, atlasers found them breeding in scrub oak, aspen, mountain shrubs such as serviceberry and mountain mahogany, and willow carrs. A few records came from spruce/fir forests, but pure stands of spruce/fir lack the understory required for nesting. Nearly all Atlas records came from 6,500 to 9,500 feet (2000–2900 m).

These warblers occupy habitats that often place them in proximity with Virginia's and MacGillivray's warblers. Orange-crowneds shared 121 blocks with Virginia's and 171

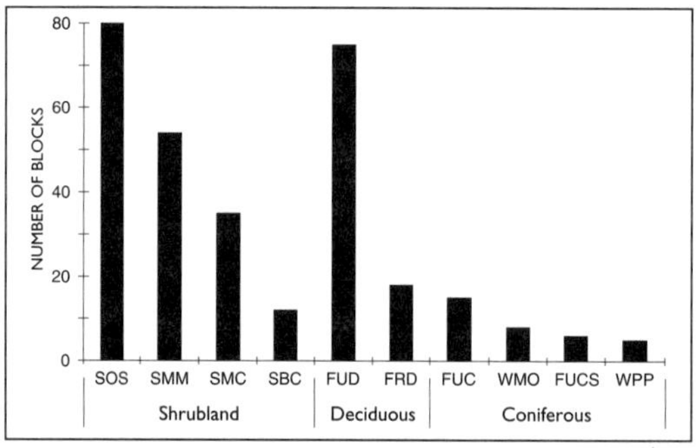

NUMBER OF BLOCKS
80
60
40
20
0
SOS
SMM
SMC
SBC
FUD
FRD
FUC
WMO
FUCS
WPP
Shrubland
Deciduous
Coniferous

with MacGillivray's (48% and 68% respectively of Orange-crowned blocks), but closely related species occurring in close proximity usually do not use the same habitat. Orange-crowneds frequently occur with Virginia's Warblers where scrub oak grows in association with ponderosa pines. In this habitat Orange-crowneds tend to use higher-elevation and less xeric sites, although some regions of overlap exist. MacGillivray's Warblers tend to select sites closer to water than do Orange-crowned Warblers, but, similarly, overlap exists. The foraging patterns of Orange-crowneds show more versatility, however, than do those of MacGillivray's Warblers. The former forage at all levels of the vegetation whereas the latter stay near to the ground (Curson et al. 1994).

In almost any habitat Orange-crowneds could suffer from under-reporting. Aside from an occasional spell of singing from an exposed perch, the birds do little to attract notice. No flashes of brilliant yellow attend their movements through the brush, nor do they venture far from cover when foraging for insects and spiders. Even if deliberately pursued, they seem to possess an uncanny ability to disappear into the foliage. Thus, unless detected by song, many Orange-crowned Warblers could escape notice.

BREEDING

Although the presence of Orange-crowned Warblers in Colorado until late fall testifies to their hardiness, they still take their time getting the breeding process under way. Atlasers reported them carrying food as early as 10 June but found the overwhelming majority of food carrying and fledglings in late June and July.

Although Orange-crowned Warblers do not rush the breeding cycle, representatives arrive in Colorado earlier than most other warblers, in late April (B&N, Dexter and Levad 1992). Migrants pass through lower elevations until late May. Whether these birds typically go to the Colorado mountains to breed or continue moving north to breed elsewhere remains unknown. A still more basic uncertainty surrounds the question of what percentage of these May birds belong to the subspecies *orestera* that breeds in the state. During both spring and fall migration, representatives of both the West Coast *lutescens* (Tony Leukering pers. comm.) and the northern *V. c. celata* subspecies (B&N) move through the state.

Atlas field workers managed to locate only two nests. Orange-crowned Warblers construct their nests on the ground, setting the rim flush with the ground. Although they build fairly large nests for so small a bird, they conceal them well among the ground clutter associated with shrubs and thickets (Harrison 1984). A typical nest contains 4–5 eggs (Curson et al. 1994).

The first record of cowbird parasitism in Colorado did not come until 1996, when an adult Orange-crowned fed a cowbird fledgling near Meeker on 9 July (Kim Potter pers. comm.). Other Colorado ground-nesting species using brushy areas for nest sites (e.g., Spotted and Green-tailed towhees) experience some cowbird parasitism, and it seems likely that Colorado's Orange-crowneds also suffer some from it as they do in other parts of the country (Friedmann and Kiff 1985).

BREEDING PHENOLOGY

CODE		# OF RECORDS	DATE RANGE
NB	Nest Building	1	30 May
ON	Occupied Nest	3	22 Jun–9 Jul
NE	Nest with Eggs	2	6 Jun–24 Jun
FY	Feeding Young	27	10 Jun–7 Aug
FL	Fledged Young	18	29 Jun–10 Aug

DISTRIBUTION

Orange-crowned Warblers breed throughout much of Canada and Alaska and most of the western United States. Few warblers rival the distribution of the Orange-crowned in the western states. The Colorado breeding subspecies *orestera* breeds from the Yukon and British Columbia south to the mountains of Arizona and New Mexico and winters in southern California, Arizona, and Mexico.

Early accounts of birdlife in Colorado uniformly recognized the presence of Orange-crowned Warblers in Colorado's montane regions. Sclater (1912) doubted that they bred here but B&N said that the species "probably breeds in the mountains." At some time after 1965 someone confirmed breeding; the Latilong Study accepted breeding in nine latilongs and A&R mapped an extensive summer range. Atlas data verify breeding throughout western Colorado and add breeding in six more latilongs. New territory includes extensive breeding in the San Juan Mountains (four latilongs in southwestern Colorado); Confirmed breeding in the Sangre de Cristos, Gores, and Mosquitos; and Possible breeding in the Sawatch and Wet mountains. Atlasers established Confirmed or Probable records in Rio Grande, Saguache, Custer, and Chaffee counties, entirely excluded (or nearly so) from the summer range by A&R.

Probably the range of this species in the state has changed little over the last 100 years. Only recently has development accelerated into areas favored by this bird. Although the impact of this development remains uncertain, much of the habitat this species uses is in national forests and remains relatively protected.

Although Orange-crowned Warblers occupy large parts of Colorado's Western Slope, their distribution and numbers peak in the San Juan, Uncompahgre, White River, and Routt national forests. Lands encompassed by the Rio Grande, Grand Mesa, and Gunnison national forests show spottier distributions and lesser densities. Other national forests in the state, such as the Eastern Slope San Isabel, Arapaho, and Roosevelt, host comparatively few of these birds.

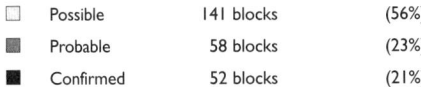

BREEDING EVIDENCE

REPORTED IN 251 (14%) OF 1745 PRIORITY BLOCKS

□	Possible	141 blocks	(56%)
▨	Probable	58 blocks	(23%)
■	Confirmed	52 blocks	(21%)

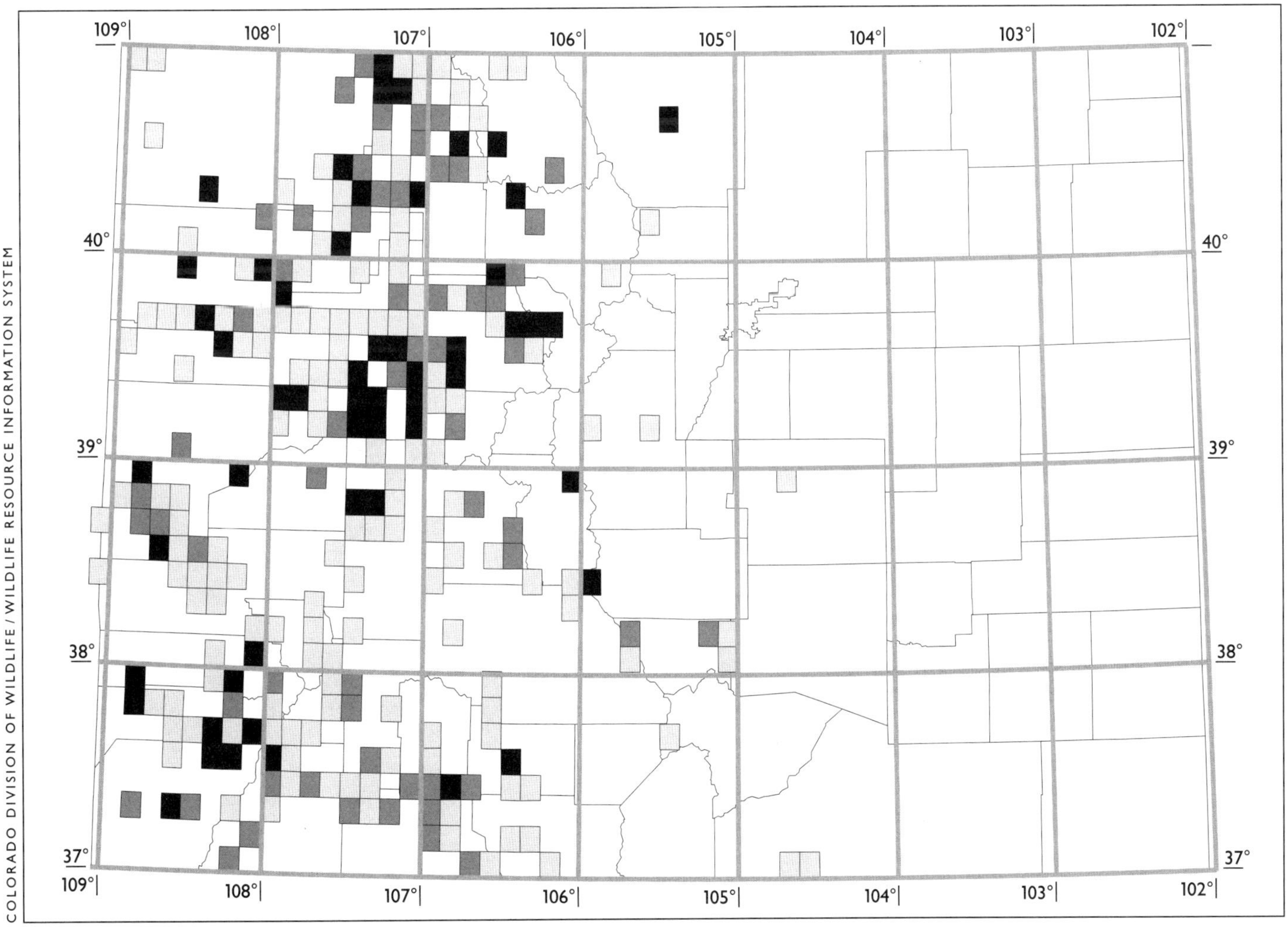

109°
108°
107°
106°
105°
104°
103°
102°
40°
39°
38°
37°

COLORADO DIVISION OF WILDLIFE / WILDLIFE RESOURCE INFORMATION SYSTEM

Virginia's Warblers play hard to get, easily frustrating the most seasoned birders. They maintain low profiles as they dart furtively through the dense cover of their brushy hillside habitats. Discovering their nests proves even more challenging; Hal Harrison (1984), warbler biologist extraordinaire, ranked them among the hardest warbler nests to find. Their common name honors Virginia Anderson, wife of Dr. William Anderson, an army surgeon who sent bird specimens to Spencer Baird (of the Smithsonian) from New Mexico and who first discovered the species (Gruson 1972).

HABITAT

The dense shrublands and scrub forests adorning the slopes of mesas, foothills, open ravines, and mountain valleys in semi-arid country compose the favored haunts of Virginia's Warblers (Bent 1953, A&R, Curson et al. 1994). Although Atlas workers found more in pygmy forests of oak (30% of all habitat records) than in any other habitat, they also recorded many in pinyon/juniper woodlands (18%) and the brushy cover of foothills and montane streamsides (17%). A variety of mountain shrubland and upland forest types, particularly mountain mahogany and ponderosa pine, respectively, provided suitable habitat as well.

Virginia's Warblers sneak like "little gray mice" (Bent 1953) through the dense understory of their habitats, where their "quiet colors" (Bailey and Niedrach 1938) work well as camouflage. The males reveal their presence, however, by singing continually; on occasion they pop into view when they tee up on tall shrubs or trees (Bailey and Niedrach 1938). They nest on the ground, often in little hollows with the nest rim at ground level. They may tuck their nests among dead leaves or hide them under rocky ledges, logs, tree roots, or overhanging grassy tussocks (Bailey and Niedrach 1938, Harrison 1978).

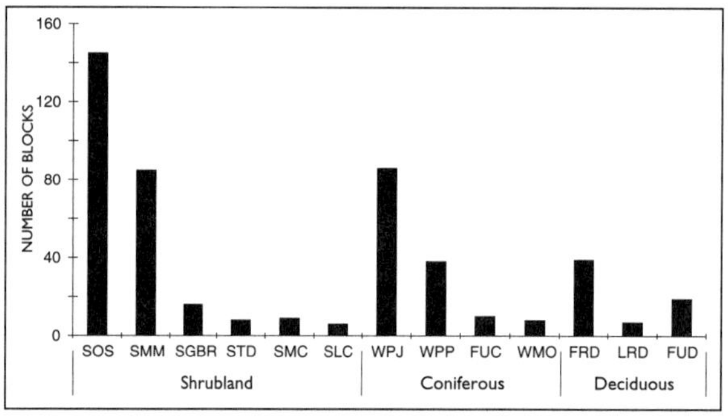
NUMBER OF BLOCKS
160
120
80
40
0
SOS SMM SGBR STD SMC SLC WPJ WPP FUC WMO FRD LRD FUD
Shrubland
Coniferous
Deciduous

BREEDING

Atlas workers Confirmed breeding in only 28% of the blocks in which they reported Virginia's Warblers. Apparently, the birds' strategy for nest concealment and their rough nesting terrain conspired to make breeding Confirmations difficult. Almost all Confirmations (88%) were made after the eggs hatched, when behaviors associated with attending nestlings or fledglings tipped off the observers. Atlas workers also used the DD code a few times, affirming that these birds try to lure potential predators away from their offspring by feigning injuries (Bailey and Niedrach 1938).

Early Atlas dates (20 and 28 April) conform to reports that Virginia's Warblers begin to arrive in Colorado during the last third of April (A&R). The three egg dates, ranging from 28 May to 4 July, expanded the known range of egg dates in Colorado (1 to 26 June; Bent 1939, Johnsgard 1986). Atlas workers found no nests with young, but Johnsgard (1986) reported a 5 June date for nestlings in Colorado, and Bailey and Niedrach (1938) mentioned that hatching coincides with the hatch of larvae that "denude nearby trees of their leaves." Family groups move downslope to lowland riparian areas after the young fledge and disperse by mid July (AV), thus implying single-brooding (Bailey and Niedrach 1938). However, Atlas records—CF (Carrying Food) on 4 August and FL (Fledgling) on 16 August—indicate late renesting or possible double-brooding.

BREEDING PHENOLOGY

CODE		# OF RECORDS	DATE RANGE
C	Courtship	1	19 Jun
NB	Nest Building	3	9 Jun–16 Jun
ON	Occupied Nest	2	9 Jun–11 Jun
NE	Nest with Eggs	3	28 May–4 Jul
FY	Feeding Young	49	11 Jun–4 Aug
FL	Fledged Young	42	16 Jun–16 Aug

DISTRIBUTION

Virginia's Warblers have a small breeding range, mainly in the Four Corners states of Colorado, Utah, New Mexico, and Arizona, with minor extensions into bordering states. A few occupy the Guadalupe Mountains of western Texas and the foothills just south of the Arizona–New Mexico border. They winter in the dense scrub of Mexico's semi-arid west-central highlands. Johnson (1976) mentioned that Virginia's Warblers have expanded westward into California, where they may come into contact with Nashville warblers; some authorities advocate lumping Virginia's, Nashville, and Colima warblers into a single superspecies (Phillips et al. 1964).

B&N maintained that Virginia's Warblers in Colorado occur between 5,000 and 7,000 feet (1525–2285 m) in elevation. Atlas data show that they nest from less than 6,000 feet (1829 m) to over 9,000 feet (2745 m). They concentrate in the western quarter of Colorado, along the Eastern Slope foothills from Wyoming to New Mexico, and parallel to the Upper Arkansas River drainage between 6,500 and 8,000 feet (1980–2440 m). Atlas results Confirmed breeding in latilongs suspected of hosting breeders, expanded breeding into places not expected to provide them with nesting habitat (Latilong Study), and showed a somewhat wider distribution throughout western Colorado than portrayed by A&R. Atlasers most often recorded abundance code A3 (53% of the reports); this corroborates Cooke's 1897 report that Virginia's Warblers are the most common warbler species in many parts of Colorado.

Atlas workers reported five incidents of Brown-headed Cowbird parasitism on Virginia's Warblers. The reports came from throughout Colorado, and all entailed nestling or fledged cowbirds with adult warblers. Friedmann's earlier account (1963) lacked any reports of parasitism on Virginia's Warblers; the Atlas and other reports (Chace and Cruz 1996; Appendix C) hint that the rate of parasitism on this warbler may be on the rise, or that the Atlas encouraged exploration of new habitats and study of comparatively unknown species.

BREEDING EVIDENCE

REPORTED IN 343 (20%) OF 1745 PRIORITY BLOCKS

Possible	148 blocks	(43%)
Probable	99 blocks	(29%)
Confirmed	96 blocks	(28%)

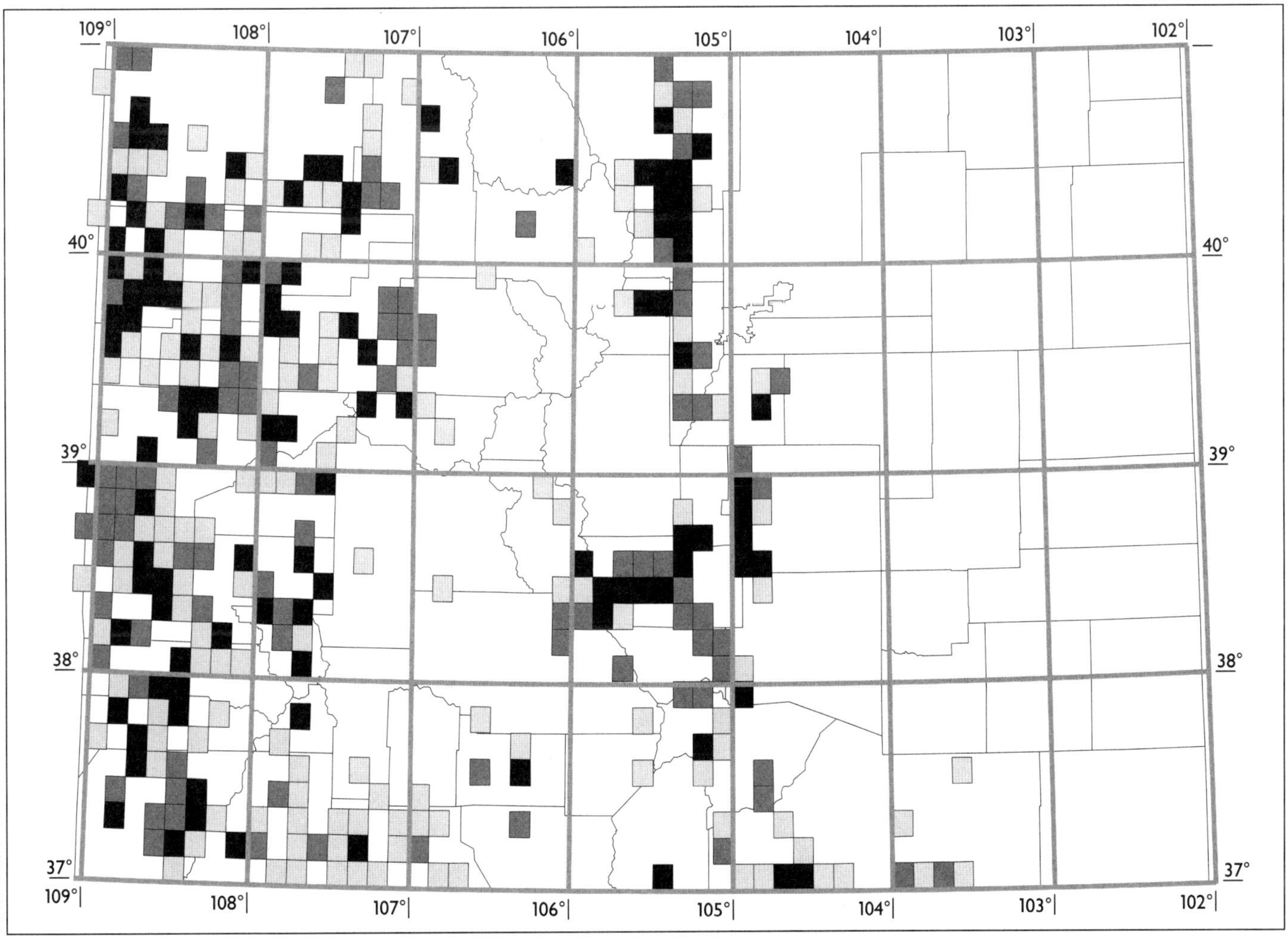
109°
108°
107°
106°
105°
104°
103°
102°
40°
39°
38°
37°

COLORADO DIVISION OF WILDLIFE / WILDLIFE RESOURCE INFORMATION SYSTEM

YELLOW WARBLER *Dendroica petechia*

How delightful to meet a common species, the Yellow Warbler, that unlike the starling and the House Sparrow cheers all who see it. Everyone appreciates the "summer yellow bird" or "wild canary" singing from the highest available perch around the house and garden.

HABITAT

Warbler species arrange themselves by habitat. Yellow Warblers come closest to the status of habitat generalists (Ehrlich et al. 1988), but even they require that the habitat remain deciduous. On the plains of eastern Colorado Atlas volunteers found Yellow Warblers mainly in cottonwood and willow-clogged stream courses and rural farmyards. They discovered them in farmhouse lilacs and shelterbelts with various trees. In the mountains, as on the plains, these warblers inhabit the stream bottoms—narrowleaf cottonwood copses and willow carrs—and some nest in aspen forests. Less than 2% occupied conifers. Second in use to the deciduous riparian habitat are plains and montane willow carrs.

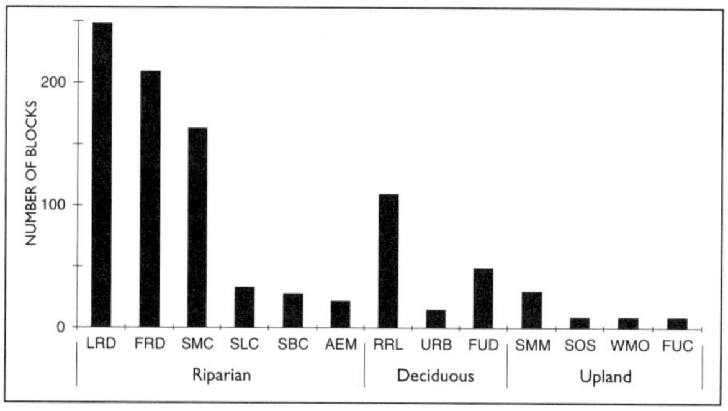

NUMBER OF BLOCKS
0
100
200
LRD
FRD
SMC
SLC
SBC
AEM
RRL
URB
FUD
SMM
SOS
WMO
FUC
Riparian
Deciduous
Upland

Years ago these sweet-singing warblers bred in the old trees of Front Range cities, but this has diminished in recent years, probably due to pesticides that eliminate the warbler's food supply (HEK).

Atlas habitat observations illustrate how Colorado warbler species partition habitats. Those breeding in riparian environments use different parts of the woodland. On the plains and western riparian systems Yellows use the trees, chats the thickets; in the foothills Yellows use the taller cottonwoods, willows, and alders. Virginia's and MacGillivray's occur respectively in hillside and streamside shrubs. Wilson's, usually at elevations higher than Yellows, use patches of low-growing willows up to timberline. Orange-crowneds favor aspens with understory; Yellow-rumped (Audubon's) and Black-throated Grays choose, respectively, mountain conifers and pinyon/juniper. Yellowthroats specialize in cattail marshes.

BREEDING

"Hurry up and do it" seems to be the motto of these efficient breeders. They often require as little as three months or less (May–July) in their nesting territory. In the Churchill, Manitoba, area these warblers arrive, nest, and depart within seven weeks (Harrison 1984).

Atlas work found that in Colorado these birds linger longer. Although some volunteers recorded NE (nest with eggs) as early as 6 May, some of these warblers were still feeding young as late as 19 August.

Males arrive before the females and claim territory by persistent song. Requirements for a desirable territory include suitable nest sites, concealing cover, tall singing-posts, and open space (Kendeigh 1941). These warblers feed almost entirely on insects, especially the larvae of many species that are terribly destructive to foliage. Forbush (1907) wrote, "It would be hard to find a summer bird more useful among the shade trees or to the orchards and small-fruit gardens than this species."

Increased singing by the males signals the arrival of the females (Bent 1953). After short courtships, females begin at once to build nests. They place their neat cups of grasses and other fine materials in upright crotches of trees or bushes and line them with cottony felted fibers. If the habitat is particularly choice, several pairs may nest semi-colonially, and females may steal nesting materials from each other (Harrison 1984). Exclusivity hardly exists: in Vermont six other species nested within 30 feet (9 m) of a Yellow Warbler nest (Smith 1943). At a spring-fed pond in an Atlas block on the plains, a pair nested near Red-winged Blackbirds, Common Yellowthroats, House Sparrows, and Bullock's Orioles (Walt Kuenning, RRK).

BREEDING PHENOLOGY

CODE		# OF RECORDS	DATE RANGE
C	Courtship	3	16 May–15 Jun
NB	Nest Building	29	12 May–28 Jun
ON	Occupied Nest	22	29 May–8 Jul
NE	Nest with Eggs	11	6 May–7 Jul
NY	Nest with Young	9	23 May–18 Jul
FY	Feeding Young	153	23 May–3 Aug
FL	Fledged Young	59	30 May–19 Aug

BY RUTH R. KUENNING

DISTRIBUTION

The various subspecies of Yellow Warblers breed from Alaska across Canada to Newfoundland, and south to Peru and even the Galápagos Islands. They have the greatest range of any wood-warbler (Terres 1980). Within the conterminous United States, only Common Yellowthroats have a greater breeding range, because Yellow Warblers do not breed in the deep Southeast (Harrison 1984). The most widespread warbler in Colorado (but second most numerous), they breed in every latilong, and atlasers found them in every county. Except for differing details of "primary range," the Atlas map resembles A&R's summer-and-migration map. Atlas abundance codes show them most abundant in north-central and west-central Colorado—essentially Latilongs 38107, 39107, and 40107, plus North Park. Atlas work attributes less importance to southwestern and southeastern pinyon/juniper woodlands, the northwestern sage-brush-covered plateaus, and the croplands of the High Plains.

Prime targets of Brown-headed Cowbirds (Friedmann 1963), Yellow Warblers work unremittingly, not always successfully, to foil these parasites. Numerous references report that these warblers roof over the cowbird eggs with additional nesting material so that they will not hatch. A nest with as many as six layers has been reported (Sprunt 1979), each layer with its complement of cowbird and warbler eggs. Nesting near a colony of Yellow-headed Blackbirds benefits these warblers because the blackbirds will drive away the cowbirds (Terres 1980). Twelve Atlas volunteers reported cowbird parasitism on this species in 12 different counties (Chace and Cruz 1996; Appendix C).

Even though Yellow Warblers appeared on the National Audubon Society Blue List and the Colorado Special Concern list (Webb 1985) because of population or range reductions and cowbird parasitism, BBS numbers show a modest increase continent-wide, of 0.6%/year. The survival potential of these widespread warblers, with their generalist, deciduous habitat preference, probably surpasses that of other warblers with narrower habitat requirements.

BREEDING EVIDENCE

REPORTED IN 735 (42%) OF 1745 PRIORITY BLOCKS

Possible	230 blocks	(31%)
Probable	225 blocks	(31%)
Confirmed	280 blocks	(38%)

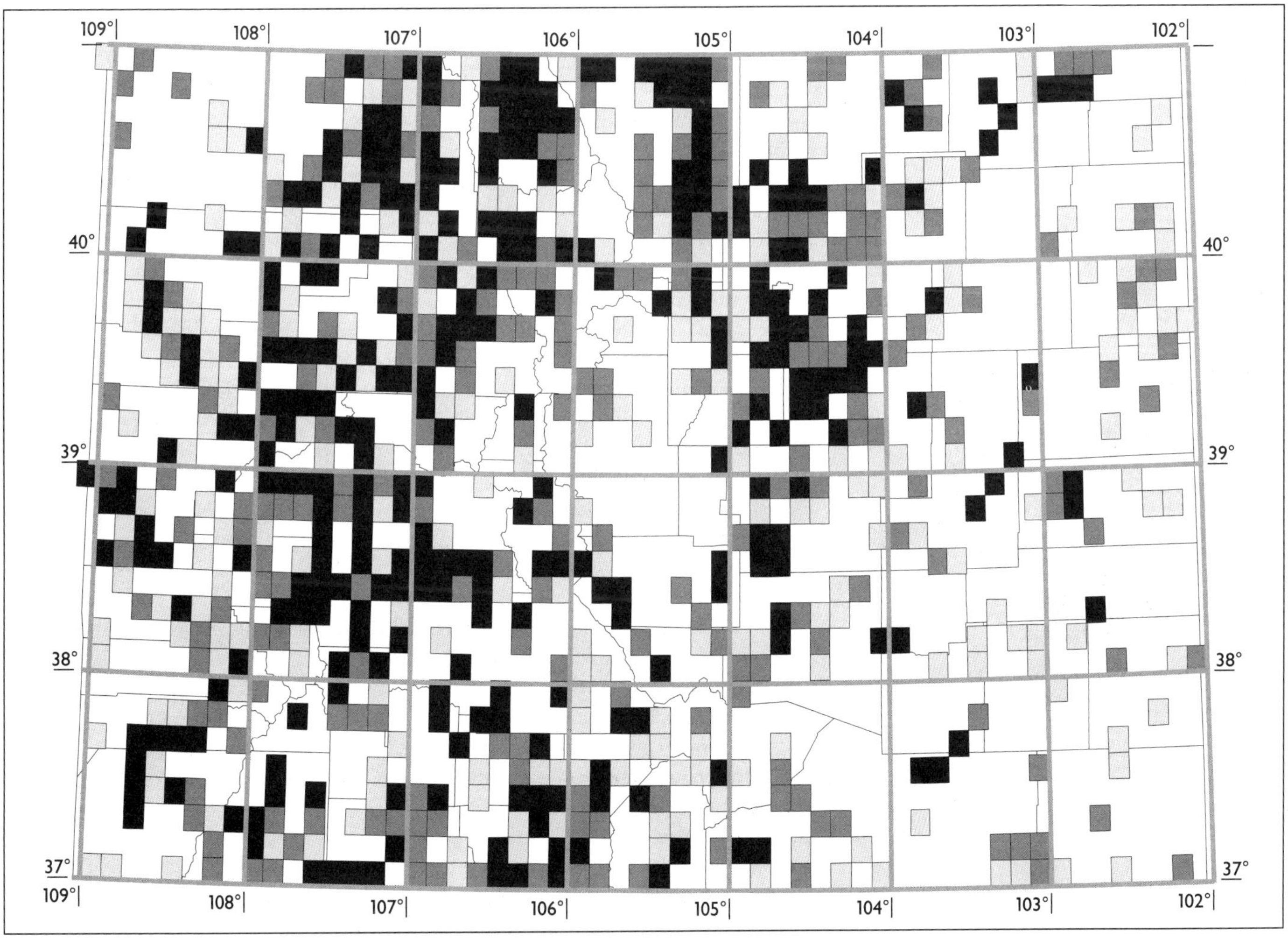

109°
108°
107°
106°
105°
104°
103°
102°
40°
39°
38°
37°

COLORADO DIVISION OF WILDLIFE / WILDLIFE RESOURCE INFORMATION SYSTEM

For many Atlas workers, Yellow-rumped Warblers served as an indicator species in blocks containing aspen or conifer forests. A field card with no check-mark beside "Yellow-rumped Warbler" meant the block needed more work.

HABITAT

Colorado Yellow-rumps—"Audubon's Warblers"—inhabit the conifer and aspen forests of the mountains at elevations between 7,000 and 11,500 feet (2135–3500 m), where they breed abundantly. Atlas workers observed Yellow-rumped Warblers more often in coniferous forests than in any other habitat type (67%). The warblers also breed in aspen forests (21%), especially those adjacent to or mixed with coniferous forests.

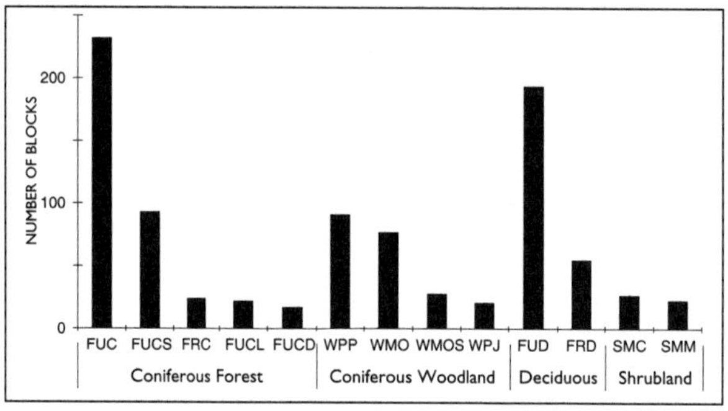

NUMBER OF BLOCKS
200
100
0
FUC
FUCS
FRC
FUCL
FUCD
WPP
WMO
WMOS
WPJ
FUD
FRD
SMC
SMM
Coniferous Forest
Coniferous Woodland
Deciduous
Shrubland

By extrapolating specific conifer codes (FUCS, WMOL) to the generic codes (FUC, WMO), Atlas habitat reports project Yellow-rumps as most frequent in spruce/fir —31% of habitat reports; ponderosa pine, lodgepole pine, and Douglas-fir each had 5–10% of the habitat designations. Proportionately, blocks with conifers had more breeding pairs—almost half (45%) reported A4 and A5 abundance codes; aspen had almost as many—38% A4 and A5 codes. (Many blocks had both habitats.)

Yellow-rumped Warblers consume a diet almost entirely of insects. They glean insects from the canopy foliage in aspens and concentrate on the lower branches in conifers (Scott and Crouch 1988). As they sally out from perches for insects they utter a diagnostic *whit* call note and fan white tail-patches upon retreat, their yellow rumps conspicuous all the while. They also feed on the ground. These insectivores sometimes glide, swallow-like, over water, while skimming insects from the surface (Terres 1980).

The other two most common conifer species, Ruby-crowned Kinglets and Mountain Chickadees, differ slightly in size, but, more significantly, in feeding styles. The tinier kinglets focus on the upper half of a tree, and their gleaning and hovering techniques enable them to pick insects from above and underneath the foliage. The larger Mountain Chickadees feed entirely by gleaning and show no preference as to the part of the tree in which they feed. Yellow-rumps spend 70% of their feeding efforts gleaning from foliage and twigs, and 15% in hawking insects like a flycatcher (Airola and Barrett 1985). Perhaps their flycatching techniques caught the attention of many Atlas workers: food-related codes accounted for 66% of all Confirmed breeding records.

BREEDING

Yellow-rumped Warblers usually build cup nests on horizontal branches, close to the tree trunk, 4–50 feet (1.2–15 m) high (Harrison 1979). Females lay four eggs and incubate them 12–14 days, often raising two broods. Both parents feed the young.

Of the 36 nests they found, atlasers submitted seven nest cards that documented nests in a variety of tree species. Three nests, positioned near the trunks of aspen trees, occupied the hollow of a broken branch, the base of a split, and a space behind dead bark. Four other nests rested on branches in conifers, three in Engelmann spruce and one in ponderosa pine.

Both male and female warblers feign injury at the approach of an observer during the nestling period (Jensen *in* Bent 1953). Atlasers recorded distraction behavior on four occasions corresponding to that period.

Brown-headed Cowbirds have long occupied the Colorado plains, but recently they have expanded their range above 8,200 feet (2500 m). The Atlas helped document this upward expansion into the mountains by providing the first Colorado records of Brown-headed Cowbird parasitism on Yellow-rumped Warblers—from five counties (Chace and Cruz 1996; Appendix C). All seven cases describe warblers feeding fledged cowbirds.

BREEDING PHENOLOGY

CODE		# OF RECORDS	DATE RANGE
C	Courtship	3	16 Jun–16 Jul
NB	Nest Building	17	26 May–6 Jul
ON	Occupied Nest	9	10 Jun–6 Jul
NE	Nest with Eggs	1	26 Jun
NY	Nest with Young	10	13 Jun–15 Jul
FY	Feeding Young	220	14 May–9 Aug
FL	Fledged Young	118	19 Jun–19 Aug

DISTRIBUTION

The breeding range of these warblers extends from the boreal forests of Alaska, northern Canada, and the north-eastern U.S. south through the western mountains to Guatemala. Yellow-rumped Warblers occur as "Audubon's," the western, yellow-throated subspecies, and the eastern, white-throated "Myrtle." Historically considered separate species because of differences in plumage, the two forms have breeding ranges that overlap in northern Canada, where they hybridize freely. Taxonomists now consider them subspecies of a single species, the Yellow-rumped Warbler.

Breeding widely throughout the coniferous parts of Colorado, "Audubon's" Warblers are the most abundant of Colorado warblers, with triple the number of Yellows (the next most common). They rank as the 11th most numerous Colorado bird species according to Atlas abundance estimates, with over 1 million breeding pairs.

Only 5–9% of the breeding distribution of this species occurs within the state (Priorities 1995). Atlas abundance codes show them more abundant in Latitudes 37 and 40, probably because more blocks in these areas have appropriate habitat.

Breeding Confirmations collected by atlasers in the Cold Spring and Douglas mountains clarified and extended the known distribution of Yellow-rumps in northwestern Colorado. Overall, the distribution and abundance shown on the Atlas map remain consistent with recent literature.

Gregarious in migration and often the last warblers to leave in the fall, Yellow-rumps move through Colorado in waves of 5–50 birds on the Eastern Slope and hundreds on the Western Slope, in late September and October. Although a few winter in the state at lower elevations in riparian habitats, most winter south to central Mexico and Panama.

Brood parasitism by Brown-headed Cowbirds poses a new threat to Yellow-rumped Warblers in Colorado. As a new host species Yellow-rumps lack evolutionary defense mechanisms that have developed among many species long exposed to brood parasitism. Overall, Yellow-rumped Warblers contend with minor threats on both breeding and wintering grounds.

BREEDING EVIDENCE

REPORTED IN 638 (37%) OF 1745 PRIORITY BLOCKS

Possible	126 blocks	(20%)
Probable	133 blocks	(21%)
Confirmed	379 blocks	(59%)

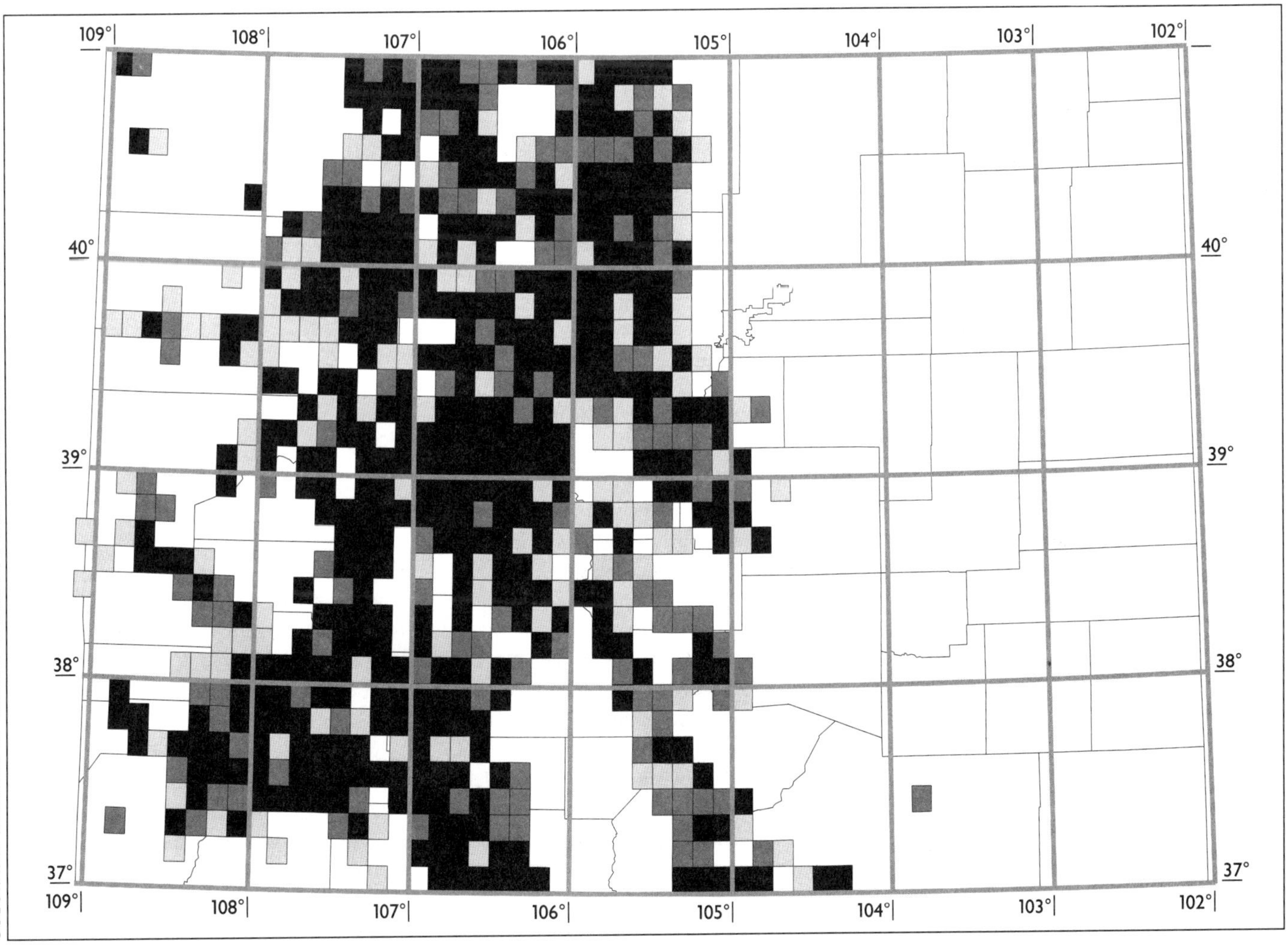
109°
108°
107°
106°
105°
104°
103°
102°
40°
39°
38°
37°

COLORADO DIVISION OF WILDLIFE / WILDLIFE RESOURCE INFORMATION SYSTEM

BLACK-THROATED GRAY WARBLER *Dendroica nigrescens*

The name of the Black-throated Gray Warbler does little justice to the fineness of its dress. In pinyon/juniper country, where shades of gray dominate the avian plumages, the Black-throated Gray Warbler's striking appearance sets it apart from its plainer neighbors.

HABITAT

Within Colorado, Black-throated Gray Warblers occupy, almost exclusively, mature pinyon/juniper habitats. Atlasers found that scrub oak, sagebrush, and other habitats lying adjacent to stands of pinyon/juniper will occasionally host these birds, but in Colorado rarely, if ever, do Black-throated Grays breed much beyond the boundaries of mature pinyon/juniper woodlands.

No other Colorado warbler exhibits such an exclusive dependency on pinyon/juniper habitats. Nevertheless scrub oaks that border on, or intrude into, pinyon/juniper woodlands often bring Virginia's Warblers into close proximity with Black-throated Grays. Sedgwick (1987) found these two species occupying similar habitats in northwestern Colorado. Virginia's Warblers, however, show a much stronger proclivity toward shrubby areas within the pinyon/juniper than do Black-throated Grays.

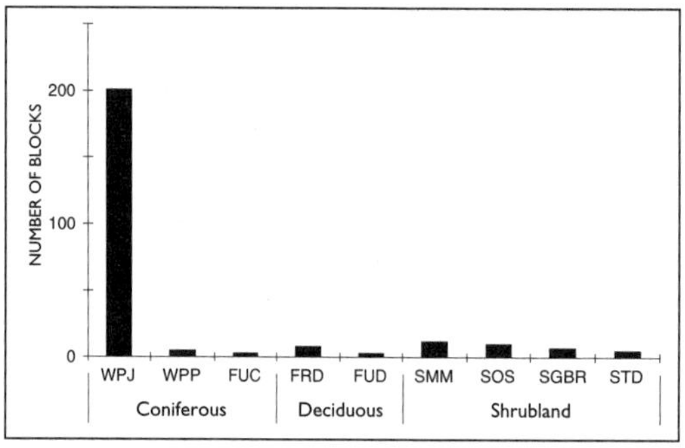
NUMBER OF BLOCKS
200
100
0
WPJ
WPP
FUC
FRD
FUD
SMM
SOS
SGBR
STD
Coniferous
Deciduous
Shrubland

Generally, Black-throated Gray Warblers build their nests 5–40 feet (2–12 m) above the ground and several feet out from the main trunk on a horizontal limb (Harrison 1984). In Colorado, the pinyons and junipers they prefer rarely grow much over 20 feet (6 m) in height, so that these warblers cannot place nests as high as in some other parts of their range.

BREEDING

Black-throated Gray Warblers arrive early on their territories. In Mesa Verde National Park, where they breed at elevations over 7,000 feet (2100 m), and in balmy Mesa County they begin arriving during the middle of April (Marilyn Colyer pers. comm., Dexter and Levad 1992, respectively). Atlasers detected nest-building activity as early as 8 May and fledglings by 16 June. A wide range of Confirmation dates, even at similar latitudes and elevations, hints at renesting or that this species might, on occasion, double-brood. It also implies that much early breeding activity escaped the notice of atlasers who did not begin surveying blocks until June.

The scaled-down height of nests in pinyon/juniper country would seem to make nest-finding relatively easy. Atlasers, however, located but one nest with eggs and four other nests with young. Apparently, the behavior of Black-throated Gray Warblers offers few clues to the locations of their nests.

Brown-headed Cowbirds must find it equally difficult to locate their nests. Although cowbirds regularly parasitize nests in pinyon/juniper country, Black-throated Gray Warblers seem little impacted. The only two Colorado records of cowbird parasitism on this species resulted from Atlas work, in Moffat and Garfield counties (Chace and Cruz 1996; Appendix C).

BREEDING PHENOLOGY

CODE		# OF RECORDS	DATE RANGE
C	Courtship	1	8 May
NB	Nest Building	5	8 May–22 Jun
ON	Occupied Nest	2	7 Jun–4 Jul
NE	Nest with Eggs	1	23 May
NY	Nest with Young	4	13 Jun–8 Jul
FY	Feeding Young	44	28 May–24 Jul
FL	Fledged Young	36	16 Jun–6 Aug

DISTRIBUTION

Black-throated Gray Warblers breed across much of the western United States and into British Columbia. Within this range, however, they occupy a puzzling combination of habitats. In the Southwest, they find their niche in arid pinyon/juniper, oak, and manzanita habitats. In the Northwest and British Columbia, however, they inhabit brushy areas and edges associated with the soggy fir forests of the region (Godfrey 1966). During winter, they retreat to western Mexico. Early observers, though uncertain of the distribution, thought them rare (Cooke 1897, Sclater 1912). By 1965 subsequent observers had assembled a more accurate picture of the range (B&N).

Few surprises emerge from the Atlas data regarding the Colorado range of this species.

The discovery of singing and territorial birds as far north as Buena Vista along the Arkansas River drainage, however, merits special attention. First Confirmations of breeding came from two western Colorado latilongs with prior records of probable breeding.

Black-throated Gray Warblers sustain a sizable population among the pinyon/juniper woodlands of western Colorado. Likewise, they thrive along the southern exposures off the Arkansas River corridor between Salida and Florence. On the other hand, atlasers failed to find them in the densities indicated by A&R in the foothills from Florence south to Trinidad and eastward across Las Animas County. Although this region boasts extensive stretches of pinyon/juniper habitat, they turned up here in only three blocks.

Much of the pinyon/juniper country favored by Black-throated Gray Warblers has escaped development. Chaining, the practice of uprooting small trees (especially pinyons and junipers) to make land more suitable for grazing, threatened their habitat but the heyday of chaining has passed (Ron Lambeth pers. comm.). Black-throated Gray Warblers do not appear on a sufficient number of BBS routes for the data to show statistically significant trends in their population. BBS routes along wide rights-of-way miss these warblers because their songs do not carry as far as those of many other pinyon/juniper birds (Ron Lambeth pers. comm.).

BREEDING EVIDENCE

REPORTED IN 200 (11%) OF 1745 PRIORITY BLOCKS

☐	Possible	57 blocks	(29%)
▩	Probable	53 blocks	(26%)
■	Confirmed	90 blocks	(45%)

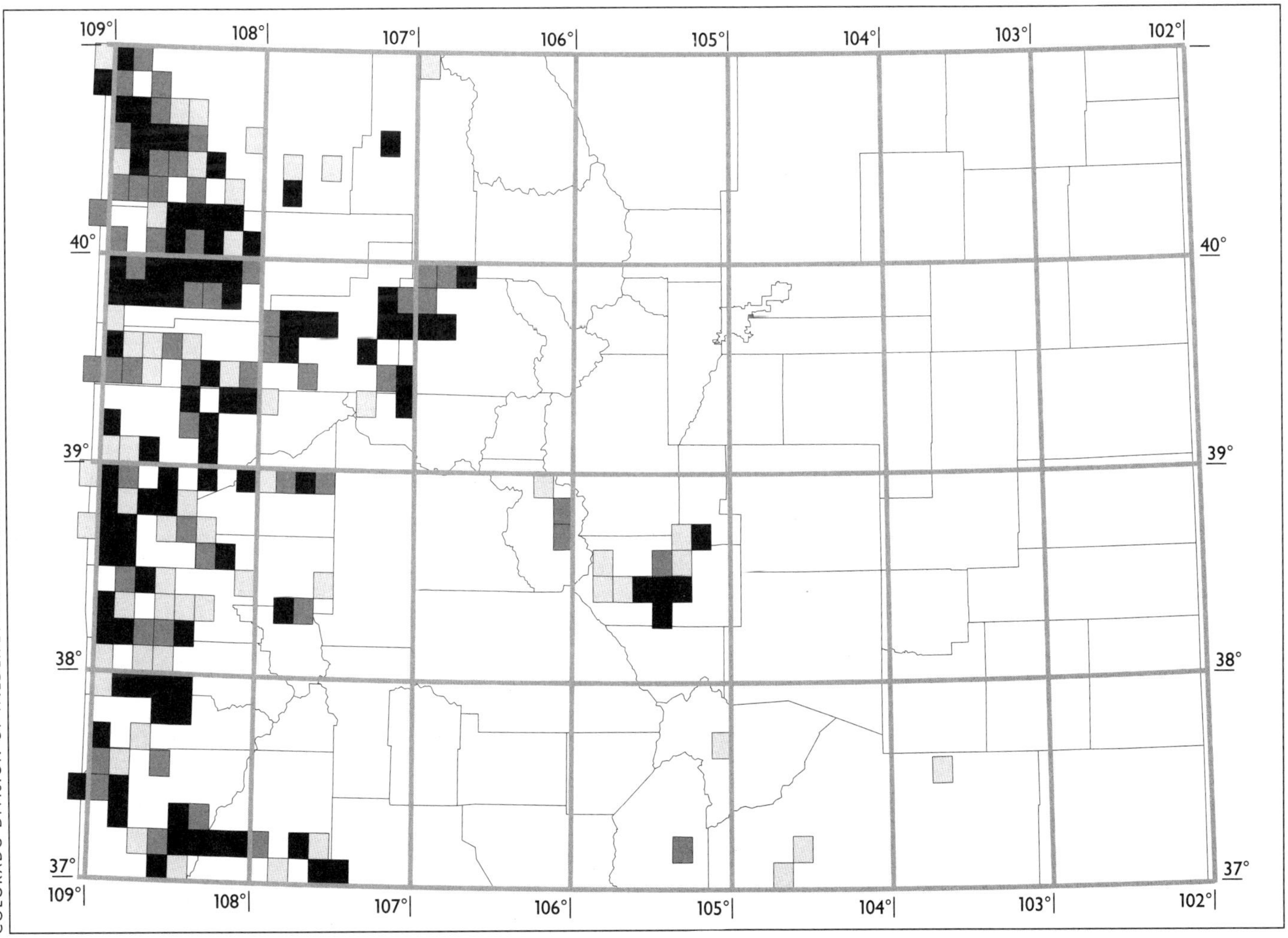
109°
108°
107°
106°
105°
104°
103°
102°
40°
39°
38°
37°

COLORADO DIVISION OF WILDLIFE / WILDLIFE RESOURCE INFORMATION SYSTEM

Only occasionally dropping down into the understory, Grace's Warblers live most of their lives high in tall ponderosa pines and build their nests near the tips of branches, often 50 feet or more above the ground. Atlasers, apparently willing to risk stiff necks but not broken ones, found the birds in several new areas but failed to locate a single nest.

HABITAT

Grace's Warblers require ponderosa pine and prefer a scrub oak understory (Terres 1980, Ehrlich et al. 1988), although they may occur in open ponderosa woodland without much oak (Kip Stransky pers. comm.). In addition to ponderosas, Atlas workers reported these warblers in three blocks with scrub oak mixed with ponderosa and in four other habitats, each of which either intermixes with or lies adjacent to ponderosa. Of these habitats, atlasers reported only one block in which the warblers occurred only in scrub oak. Some observers have noted a preference for open stands of mature ponderosas containing some very tall trees to serve as song posts (Bent 1953, Staicer 1989).

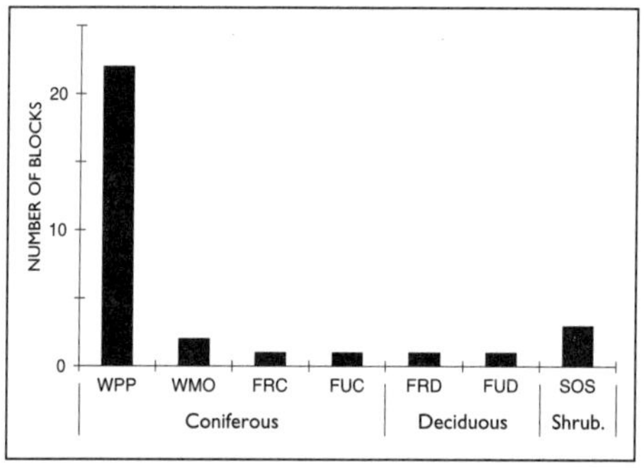
NUMBER OF BLOCKS
0
10
20
WPP
WMO
FRC
FUC
FRD
FUD
SOS
Coniferous
Deciduous
Shrub.

Orange-crowned, Virginia's, and Yellow-rumped warblers may inhabit the same areas as Grace's Warblers, but they exploit different niches. Orange-crowned and Virginia's nest and feed on or near the ground and use primarily oak rather than pine. The relationship between Grace's and Yellow-rumps may resemble the relationship between Cape May and Yellow-rumped warblers in northeastern coniferous forests (cf. MacArthur 1958, Ehrlich et al. 1988). Yellow-rumps spend most of their time in the lower parts of the trees or on the ground in the Northeast and may behave similarly in the pines of the Southwest, but the relationship needs more study. Grace's, like the Cape May Warbler, dwells near the top.

BREEDING

Most published dates on the breeding cycle of Grace's Warbler, one of the least known of all the warblers, come from Arizona (Webster 1961, Staicer 1989). Males arrive at northern Arizona breeding territories around 20 April (Staicer 1989) and begin singing from the tallest pines to attract a mate. Breeding in Colorado may begin a couple of weeks later; the earliest Colorado Atlas records carry mid May dates, perhaps marking the arrival of Atlas workers rather than warblers. Mating and nest-building take place in May (Bent 1953, Staicer 1989). Sources differ as to the height of nests—from 4 feet (1.2 m; Sclater 1912), to 20–40 feet (6–12 m; Terres 1980), and to 50–60 feet (15–18 m; Ligon 1961) They construct compact nests with vegetable fiber, hair, and oak catkins (Ligon 1961). In late May or early June, females begin incubation, which takes 14 days. Although no research has measured the fledgling period, it probably parallels the 10–12 days for most other *Dendroica* warblers. Young birds in northern Arizona leave the nest in late June (Staicer 1989). Atlas workers Confirmed breeding in Colorado only by observing feeding activity (CF, FY, FF), the earliest Confirmation on 26 June and all others in July.

BREEDING PHENOLOGY

CODE		# OF RECORDS	DATE RANGE
FY	Feeding Young	5	26 Jun–17 Jul
FL	Fledged Young	2	4 Jul–11 Jul

DISTRIBUTION

Grace's Warblers nest from southern Nevada, southern Utah, and southwestern Colorado south to Nicaragua. Knowledge of the species' northern and eastern limits has expanded in recent years. Bent (1953) described its habitat as extending north to Fort Lewis and Pagosa Springs. B&N indicated that Grace's Warblers "breed regularly in the ponderosa pine belt of Archuleta, La Plata and other southwestern counties."

Early Atlas records extended the range across the crest of the San Juan Mountains to the north side of the San Juans and to the Uncompahgre Plateau all the way north to Unaweep Canyon in southern Mesa County (Jacks Canyon, 38108G5). At least a few of these treetop birds nest all along the west side of the Uncompahgre, in ponderosa/scrub oak associations.

Singing males and juvenile birds seen in the Wet Mountains near Rye in 1977 through 1984 seemed to represent only a temporary outpost for this species east of the Continental Divide (Griffiths et al. 1978, 1980, Kingery 1984). An Atlas trip in 1994, however, turned up one or two singing males in the Wet Mountains near Wetmore, suggesting that these mountains still harbor a small, isolated population of these birds.

Atlasers seldom encountered large numbers of Grace's Warblers. They recorded only one instance of multiple singing males (7 or more) and reported A2 (2–10 pairs) most frequently as an abundance code. Although the range expansion established by Atlas work may indicate a growing population, it could just as easily represent the discovery of previously overlooked birds.

Although cowbird parasitism of Grace's Warblers nests rarely occurs (Terres 1980), atlasers watched an adult Grace's feed two fledgling warblers and a fledgling cowbird in the Barkelew Draw block (38108A4; Brenda Wright pers. comm; Appendix C.). Dependence upon mature ponderosa pines and the value of those trees as timber may combine to pose a significant threat to Grace's Warblers, although field workers found the birds in areas that had been selectively logged (Kip Stransky pers. comm.). Possible threats and profound ignorance of its status place it high on the three conservation priority lists.

BREEDING EVIDENCE

REPORTED IN 22 (1%) OF 1745 PRIORITY BLOCKS

▢	Possible	13 blocks	(59%)
■	Probable	3 blocks	(14%)
■	Confirmed	6 blocks	(27%)

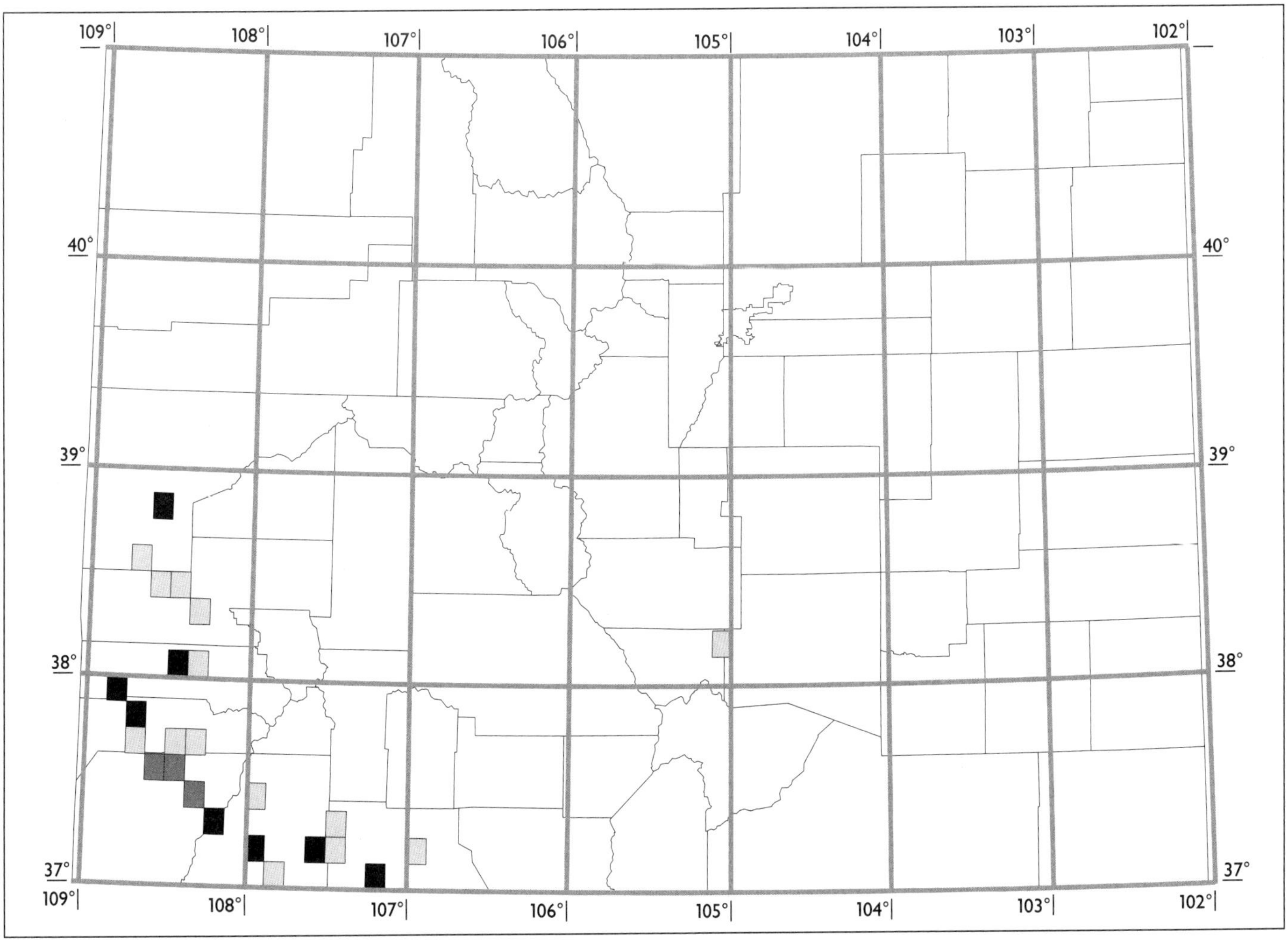
109°
108°
107°
106°
105°
104°
103°
102°
40°
39°
38°
37°

COLORADO DIVISION OF WILDLIFE / WILDLIFE RESOURCE INFORMATION SYSTEM

AMERICAN REDSTART *Setophaga ruticilla*

"With what dainty grace he spreads his tail, half opens his wings, and pirouettes from limb to limb like a village belle with coquettishly held skirts tripping the maze of a country dance!" (Chapman 1907). Latin Americans call the lively American Redstart "candelita" or "little torch" (Craves 1994).

HABITAT

Breeding habitats for American Redstarts include open, moist, deciduous woodlands with good undergrowth of shrubs and young trees, mixed coniferous/deciduous forests, and willow and alder thickets. Atlasers found redstarts in streamside habitats in all the blocks in which they reported them.

Redstarts eat insects almost exclusively, particularly those associated with forest trees (Sherry and Holmes 1997). The flashing of the pale-colored segment of the tail during feeding presumably dislodges insects from vegetation (Keast et al. 1995). Colorful accounts abound of redstarts flashing their contrasting orange-red and black wings and tail while feeding. Chapman (1907) perhaps most vividly describes the foraging technique of these vivacious, seemingly tireless warblers as "a mad series of darts and dives and whirls, of onward rushes and as sudden stops, which yield not one insect but many."

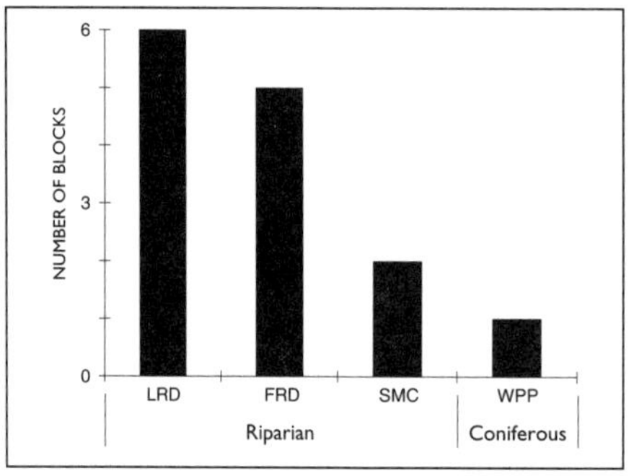
NUMBER OF BLOCKS
0
3
6
LRD
FRD
SMC
WPP
Riparian
Coniferous

The only true aerial feeders among wood-warblers, redstarts combine morphological and ecological features of warblers and flycatchers. They combine the long wings (which provide energetic advantages in long migration flights) and basic skeleton of warblers with the long flycatcher tail for greater mobility and control. With flycatchers they also share flattened bills and long rictal bristles, adaptations for aerial feeding (Sherry and Holmes 1997).

Exhibiting another trait unusual among wood-warblers, male redstarts do not acquire full adult plumage until the July of their first potential breeding season. The duller, female-like plumage of young males may protect them from predation as they gain experience at territory defense. Benefits of bright plumage in older males, such as obtaining higher-quality territories that increase chances of attracting females, may outweigh the costs of increased predation risk (Procter-Gray and Holmes 1981).

BREEDING

Redstarts have two song types. To attract a mate, they repeatedly deliver an advertising song. After pairing, they sing randomly a series of songs as notice of territorial possession. Most redstarts mate monogamously, but up to 30% of the males may mate polygynously, in an unusual way: 80–1640 feet (25–500 m) may separate polygynous territories, the more distant ones beyond earshot of each other. As the males shuttle between territories, they switch between song types, depending on the stage in the breeding cycle (Sherry and Holmes 1997).

Females build open cup nests, commonly 10–20 feet (3–6 m) above ground, and rarely as low as a few inches off the ground or as high as 70 feet (21 m; Griscom and Sprunt 1979). They weave the nest around three or more upright twigs of a tree or shrub (Sclater 1912, Bent 1953, Griscom and Sprunt 1979).

Redstarts reuse nests of several other species. Yezerinac (1993) observed a female redstart incubating three eggs, and two weeks later observed three redstart nestlings, in a Yellow Warbler nest after the warbler offspring had fledged. Reports of redstart use of five Red-eyed Vireo and one Yellow-throated Vireo nests in widely separated locations suggest that this behavior occurs occasionally (Bent 1953, Yezerinac 1993). By bypassing nest-building, normally single-brooded redstarts may replace a clutch lost to predation at a lower energy cost (Yezerinac 1993). In a separate curious incident, an adult male redstart fed a female Yellow Warbler and her nestlings and chased away the male warbler (Mannan 1979).

One of the most common victims of Brown-headed Cowbird parasitism, redstarts, like Yellow Warblers, sometimes build new nest floors over cowbird eggs (Bent 1953, Sherry and Holmes 1997). Colorado, with its few redstarts, has no records of cowbird parasitism.

Breeding redstarts spend 29–36 days from nest-building to fledging, and up to three weeks feeding young out of the nest (Sherry and Holmes 1997). Scant Atlas observations provide no Colorado nesting phenology.

BY BEVERLY K. BAKER

BREEDING PHENOLOGY

CODE		# OF RECORDS	DATE RANGE
FY	Feeding Young	1	25 Jun

DISTRIBUTION

Redstarts have a widespread breeding range, from southeastern Alaska and British Columbia across Canada, south to Oregon, Wyoming, eastern Texas, and the eastern U.S. except Florida. They winter from Baja California, central Florida, and central Mexico south through the West Indies to Ecuador and northwestern Brazil. In Colorado and other western states they occur in disjunct populations.

Sclater (1912) called the redstart infrequent along the foothills and rare or absent in western Colorado, a status that Atlas observations affirm. The Atlas map reflects a distribution similar to the A&R summer range map, with a few blocks along the edge of the Front Range and three in North and Middle parks. Atlasers Confirmed breeding at a banding station on Fort Carson, with observations of recently fledged young during July (a first latilong breeding record), and at Rye when a male carrying a dragonfly dove into dense shrubs next to a boxelder on 25 June 1994. Along the Yampa River near Craig a singing bird on 6 June 1987 was a Possible breeder. This record, within the window of Colorado breeding dates (Nelson 1993), and plausible considering breeding records in Utah and Wyoming, adds latilong 40106 to the possible breeding latilongs.

The American Redstart symbolizes Partners in Flight/Aves de las Americas (PIF), an international organization formed in 1990 for conservation of migratory birds throughout the Americas; it has a goal of keeping common birds common. BBS data showed American Redstarts declining at 1.2% per year from 1966 to 1979. From 1980 to 1994 the drop apparently tapered off, suggesting that previous declines may have leveled. Will redstarts, considered one of the three most abundant warblers in North America (Griscom and Sprunt 1979), stay common, fulfilling PIF's goal for its emblem species?

BREEDING EVIDENCE

REPORTED IN 9 (1%) OF 1745 PRIORITY BLOCKS

☐	Possible	7 blocks	(78%)
■	Probable	1 block	(11%)
■	Confirmed	1 block	(11%)

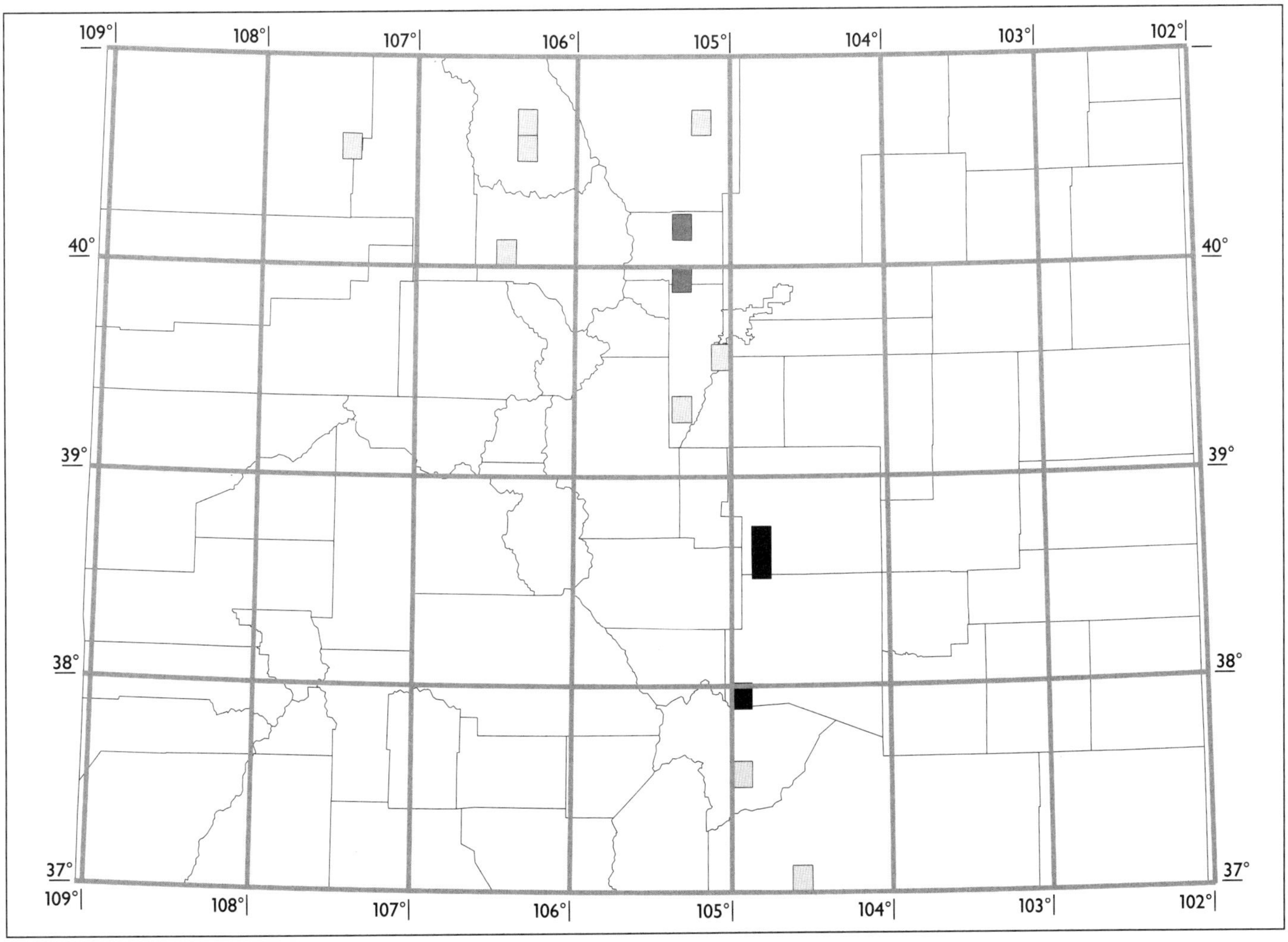
109°
108°
107°
106°
105°
104°
103°
102°
40°
39°
38°
37°

COLORADO DIVISION OF WILDLIFE / WILDLIFE RESOURCE INFORMATION SYSTEM

For birds wedded to the ground, Ovenbirds' strong preference for tall, closed-canopy, deciduous, forests seems anomalous. Colorado birds counter the anomaly by breeding in foothills ponderosa pine communities, none of which matches the canopy height, density, or deciduousness typical of eastern Ovenbird habitats.

HABITAT

Ovenbirds put their nests on the ground and feed on or near the ground. Yet they occur almost exclusively in large tracts of tall, all or almost all deciduous, forests. They manifest their dependence on the most important component of their breeding regime—large, contiguous, interior forests—through a signficant pattern of edge avoidance (Kroodsma 1984).

In the East Ovenbirds typically nest in climax deciduous forests, usually oaks or maples mixed with other trees. They show a broad tolerance for different plant communities

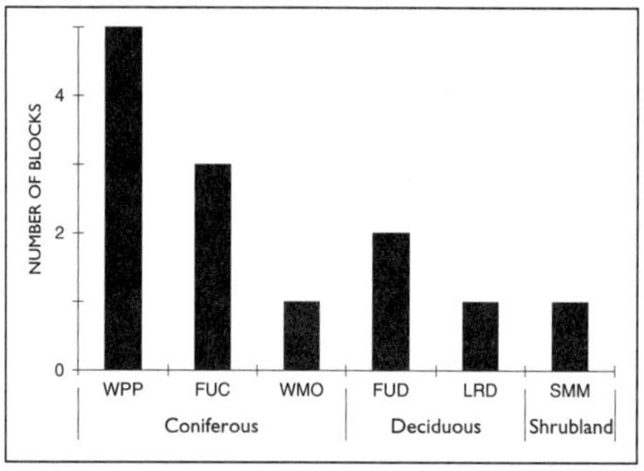
NUMBER OF BLOCKS
0
2
4
WPP
FUC
WMO
FUD
LRD
SMM
Coniferous
Deciduous
Shrubland

by breeding in forests with varied understories and ground covers as well as with different dominant tree species. These habitats share common structural characteristics: canopy heights of 50–75 feet (16–22 m) and relatively closed canopies—60–90% closed (Van Horn and Donovan 1994). Depending on forest type, they require minimum expanses of 250 acres (100 ha) to as much as 2,200 acres (885 ha; Van Horn and Donovan 1994).

Several studies have examined breeding Ovenbirds and their success in contiguous forest habitat and fragmented forests. In New Jersey, Quebec/Ontario, and Missouri the birds had far less success breeding in fragmented forests. In contiguous forests 75–87% of the males found mates, but in fragmented forests only 24% (Missouri) to 58% (Quebec/Ontario) found mates (Villard et al. 1993).

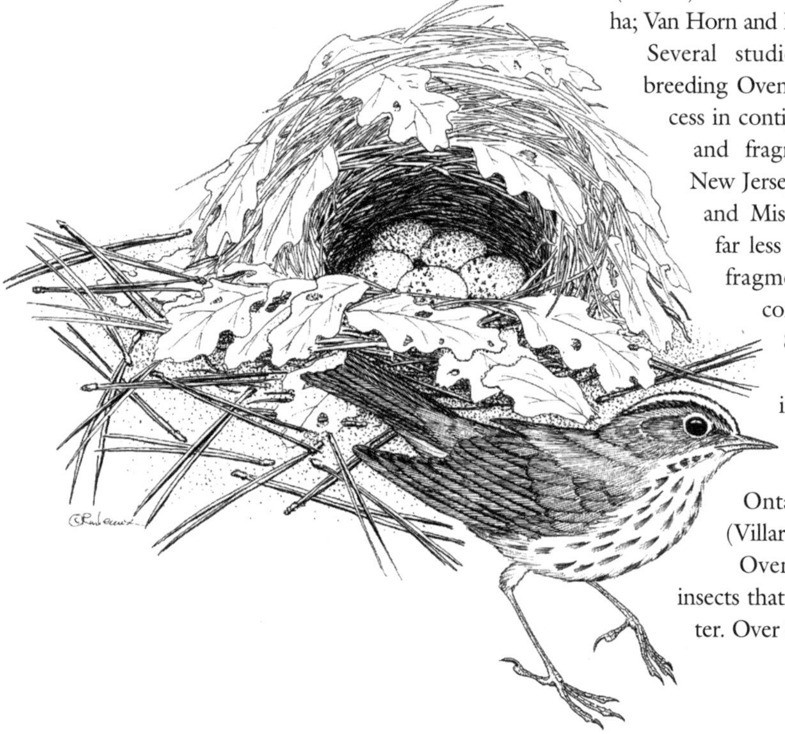

Ovenbirds eat forest insects that they find in leaf litter. Over 80% of their feeding occurs on the ground. Distance from the forest edge, through its effect on litter moisture, may influence prey abundance (Van Horn and Donovan 1994).

The habitats Ovenbirds use in Colorado do not meet ideal eastern standards. They do use large contiguous woodlands and avoid climax coniferous forests. All Atlas records come from conifer forests that vary from spindly to mature ponderosa pines, and from forests of mainly ponderosa to mixed conifer forests including pines and Douglas-firs. Most sites have a scrub oak understory, which may harbor one basic element of the habitat—one that the ground-dwelling birds detect and that researchers have not. The oakleaf litter may provide essential components for nest construction (Chan Robbins pers. comm.) or perhaps may provide a necessary dietary component.

BREEDING

Males arrive on the breeding grounds 9–14 days before the females, and stake out territories 0.5–4.5 acres in size (0.2–1.8 ha), with an average of 3 acres (1.2 ha) each. The more males within a forest section, the more likely they will find mates.

Each female picks a mate and his territory, and then picks a nest site with two characteristics: approachable from two directions and with "a certain amount of light" (Hann 1937). They most frequently build their nests close to an opening within the forest (Hann 1937). This permits easy access for the birds, but it may also help predators to find the ground-nesting females more easily—especially in fragmented forests.

The female builds the nest on the ground, and partly covers it with leaf litter. A sizable affair, the oven-shaped nest sphere has average measurements of 6 inches wide by 5 inches high (16x12 cm). By the end of the nesting cycle the nest sits in a depression probably created by the nest activity. In Colorado Ovenbirds build their nests in dense scrub oak thickets (B&N). With only one Confirmation, the Atlas merely adds one phenology date to the two breeding records cited in B&N (nests found on 1 June and 22 June), viz., a third June breeding record that could refer either to a nest or to fledglings. Atlasers recorded singing and territorial birds from 20 May to 7 July. According to Atlas estimates, fewer than 400 pairs of Ovenbirds (and perhaps fewer than 200 pairs) nest in Colorado.

The males use song for two purposes: establishment and maintenance of territory and as an "All's well" message to mate and young. The young, hushed by an alarm chirp from a parent, begin moving again when the males resumes singing (Hann 1937).

Ovenbirds fall prey to cowbirds more than most birds. They do not recognize cowbirds as a threat, and females will even work on a nest while, within a few feet, a female cowbird watches (Hahn 1937). Because Ovenbirds tend to build near open spaces, the cowbirds apparently have little trouble finding nests.

BREEDING PHENOLOGY

CODE		# OF RECORDS	DATE RANGE
FY	Feeding Young	1	21 Jun

DISTRIBUTION

Ovenbirds nest throughout deciduous forests from Labrador to northeastern British Columbia, south to the Carolinas, Arkansas, and central Wyoming. They winter from Florida and central Mexico to the West Indies and Panama. BBS show a modest but statistically significant increasing trend of 0.7% per year; the species may have declined earlier in the century and increased recently. In Colorado, Sclater (1912) cited four records and declared them rare. In 1939 Niedrach found nine singing males within one square mile of Perry Park (Douglas County), theorized that they had increased in numbers, and classified them as "rather common" within their foothills habitat (B&N).

The Atlas found that the small Colorado population, separated from the rest of the species' range, occupies disjunct patches along a narrow strip of the Front Range from Larimer County to the New Mexico line. Atlasers found Ovenbirds in the same pockets of the same dis-contiguous strip reported by B&N and A&R. Atlasers found two new breeding locations, in Pueblo and Las Animas counties. Suitable habitat in Jefferson and Douglas counties hosts the largest number. The Atlas block selection precluded investigating all appropriate habitats along the foothills; an intensive survey would probably elicit more Ovenbirds in bigger clumps.

BREEDING EVIDENCE

REPORTED IN 7 (0%) OF 1745 PRIORITY BLOCKS

Possible	5 blocks	(71%)
Probable	1 block	(14%)
Confirmed	1 block	(14%)

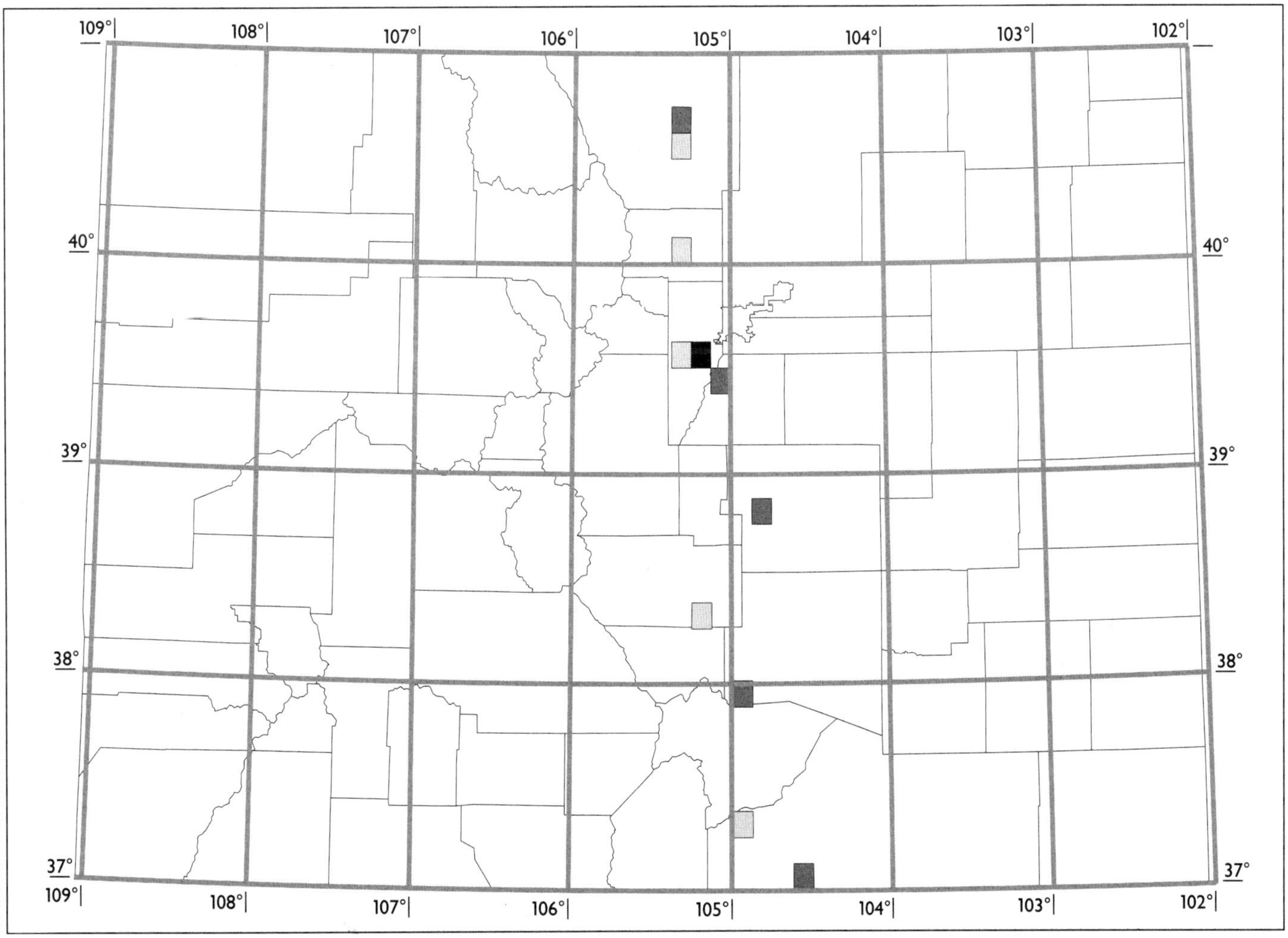
109°
108°
107°
106°
105°
104°
103°
102°
40°
39°
38°
37°

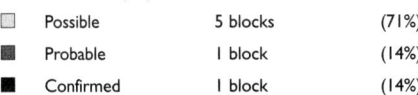

COLORADO DIVISION OF WILDLIFE / WILDLIFE RESOURCE INFORMATION SYSTEM

With a beak full of insects the bird made its way along the slim branch, its tail constantly bobbing up and down. Suddenly interrupted by an intruder, the bird launched into a scolding display of chips while flying from branch to branch. Failing to scare off the unwelcome visitor, the bird disappeared momentarily behind a willow before returning, minus the meal in its beak. It certainly didn't appreciate the significant contribution it had just made to Colorado ornithology.

HABITAT

Throughout their range Northern Waterthrushes show a penchant for breeding in shady, moist habitats that surround wooded ponds, streams, bogs, and swamps (Bent 1953, Griscom and Sprunt 1979). In Colorado, the birds have selected middle-elevation willow and alder thickets bordering stream courses around 8,000 feet (2450 m). Thick, nearly impenetrable cover characterizes the haunts of these elusive warblers. Searching for them requires waders, an ability to negotiate dense tangles of branches and unstable ground, and a keen sense to detect and to avoid surprising the many moose who share this watery domain.

Northern Waterthrushes breed along the Michigan River in the open expanses of North Park. Winding through the eastern half of Jackson County, the Michigan River embraces a maze of braided channels. Irrigated hay meadows of timothy flank the river, and these are bordered by extensive sagebrush uplands. Similar conditions occur along many other rivers in North Park and adjacent areas (e.g., Laramie River Valley). A thorough search of these rivers will probably yield more reports of breeding waterthrushes.

These terrestrial wood-warblers usually appeared in slower-moving or recently drained side-channels crisscrossed with many intertwining branches (Dillon 1995). These dark, moist micro-habitats afford the birds a place to conceal their nest among the overhanging banks, stumps, or tangled roots of upturned trees or bushes (Bent 1953, Harrison 1984).

BREEDING

The status of this migrant species changed abruptly when the author observed a pair of agitated birds carrying food on 27 June 1994 in the Gould NW Atlas block (40106F2). A search produced no nest or young but the adult's behavior indicated they had young nearby. A month later, two separate birds were observed in the same area, one of which was gathering food, possibly for young nearby. Two searches the following year yielded several birds including an adult feeding a fledgling on 26 July 1995 (Urling Kingery VF).

North Park has produced three other intriguing records, all within 10 miles of the Gould NW Atlas block. A search of the Michigan River northeast of Walden pro-duced one adult waterthrush, also on 26 July 1995 (MBD). Arapaho NWR personnel reported one on the refuge on 16 July 1989 (A&R).

Less than 6 miles (10 km) southeast of the original sighting, on the North Fork of the Canadian River in the Colorado State Forest, CDOW banded a female waterthrush. This bird, caught on 7 August 1995, had a receding brood patch (Jim Dennis pers. comm.). Whether this bird nested nearby or was an early migrant remains uncertain. In eastern parts of their range, waterthrushes begin their fall migration as early as mid July (Bent 1953, Eaton 1995). In Colorado fall migrants typically appear, on the plains at least, from mid August into October (A&R, Holt 1997).

BREEDING PHENOLOGY

CODE		# OF RECORDS	DATE RANGE
FY	Feeding Young	1	27 Jun
FL	Fledged Young	1	26 Jul

DISTRIBUTION

Northern Waterthrushes breed across Alaska and Canada south to the northeastern states, northern Minnesota, Wisconsin, and Michigan. In the West they range through the northern Rockies to Idaho and western Montana. Isolated populations breed in Oregon, North Dakota, and southeastern Saskatchewan (Eaton 1995). They spend the winter months in the Caribbean islands and from central Mexico to northern South America (Eaton 1995).

This species regularly occurs as a rare to uncommon spring and fall migrant in Colorado (A&R). The state has at least two previous summer records: one from Estes Park on 16 July 1910 (Widmann 1911) and one from Rocky Mountain National Park on 25 July 1962 (A&R). The two earlier records presumably account for the 1983 AOU checklist status of probable breeding in Colorado.

Elsewhere in the West, a disjunct population, first discovered in Oregon in 1977 (Contreras 1988), continues to expand its range there and in Washington (Eaton 1995). In northwestern Wyoming, south of the main range, waterthrushes probably breed in Grand Teton National Park (Kingery 1991b, Scott 1993).

Much of the habitat that hosts Colorado waterthrushes is on private land, difficult to access, and generally uninviting to birders. Atlas protocol required observers to search areas that they might not normally explore. The Confirmation of waterthrushes exemplifies the positive outcome of this strategy. An intriguing question then arises: have waterthrushes expanded their breeding range into Colorado or have they always been here? We will never know about their past, but it seems likely that the birds have bred in small numbers for some time, and their present range may expand even more. All it will take is for someone to venture into their tangled haunts to find them.

BREEDING EVIDENCE

REPORTED IN 2 (0%) OF 1745 PRIORITY BLOCKS

Possible	0 blocks	(0%)
Probable	0 blocks	(0%)
Confirmed	2 blocks	(100%)

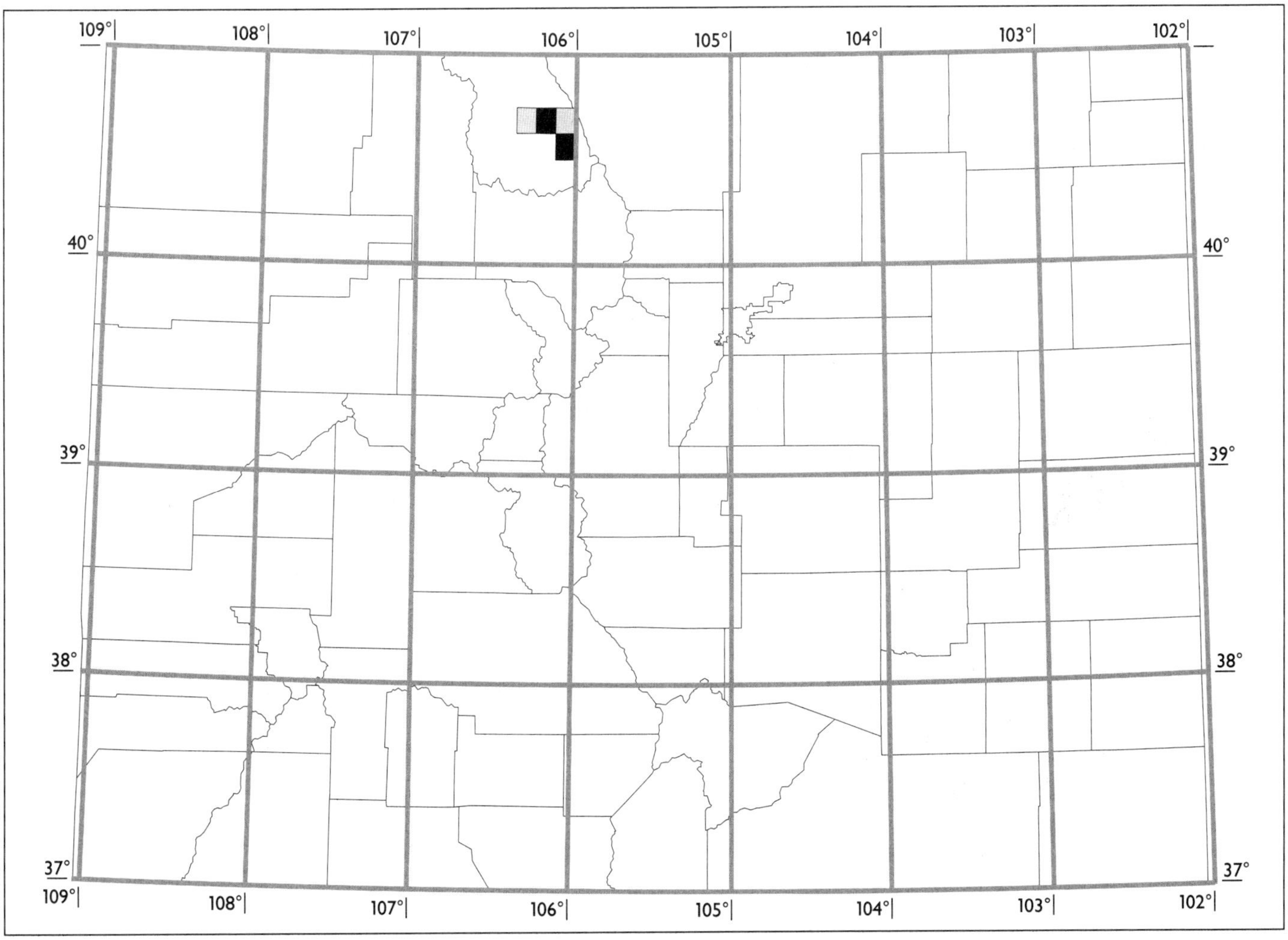
109°
108°
107°
106°
105°
104°
103°
102°
40°
39°
38°
37°

COLORADO DIVISION OF WILDLIFE / WILDLIFE RESOURCE INFORMATION SYSTEM

MACGILLIVRAY'S WARBLER *Oporornis tolmiei*

John Kirk Townsend first named this species "Tolmie's Warbler" to honor his ornithologist friend, Dr. William F. Tolmie, who spent most of his life in the Pacific Northwest with the Hudson Bay Company. Later, John James Audubon renamed the species "MacGillivray's Warbler" to honor his close friend, Dr. William MacGillivray. Although Audubon usurped the common name for his friend (an editor of his works), the scientific taxonomy rule of priority requires the scientific name to retain Tolmie's name. Some ornithologists think that the American Ornithologists' Union should restore the common name to Tolmie's Warbler, because Audubon's friend rarely visited North America (Ehrlich et al. 1988, Pitocchelli 1995).

HABITAT

If its name reflected its habitat, MacGillivray's Warbler would be called the "shrub" or "undergrowth" warbler. Even the earliest accounts of this bird in Colorado indicated that "any patch of shrubbery or tangled growth of bushes is sure to be selected as the summer abode of one or more pairs of these birds" (Henshaw 1875). Other historical references (Cooke 1897, 1900) gave no real clues about the species' abundance or distribution in the state. (Cooke, in his first paper [1897], used the scientific name, *macgillivrayi*, whereas later [1900] he switched to *tolmiei.*)

Eighty percent of all Atlas sightings occurred in four mountain shrub habitats—montane carrs, mountain shrublands, riparian deciduous, and scrub oaks. Another 10% occurred in aspen forests; the aspen forests where they nest most frequently have a shrubby understory. Atlas findings parallel those of A&R, who gave the primary habitats as moist ravines with scrub oak in the foothills life zone through shrubby understories of all forest types in the montane life zone. Within these areas, MacGillivray's Warblers seem to prefer large shrubs found at the interface between steep slopes and riparian or otherwise moist areas. Indeed, over half the Atlas reports came from riparian habitats (both shrubs and trees).

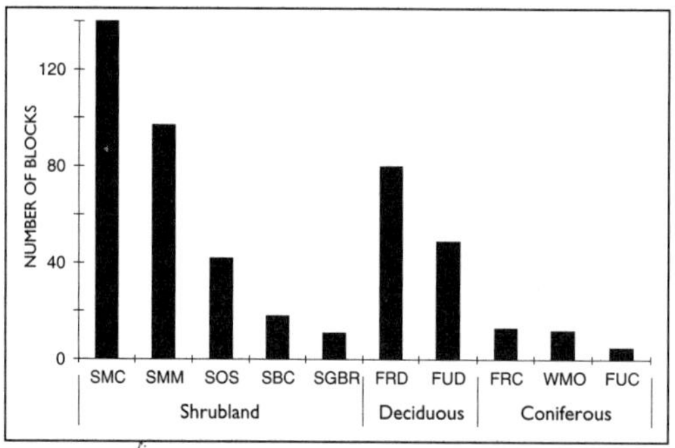

NUMBER OF BLOCKS
0
40
80
120
SMC
SMM
SOS
SBC
SGBR
FRD
FUD
FRC
WMO
FUC
Shrubland
Deciduous
Coniferous

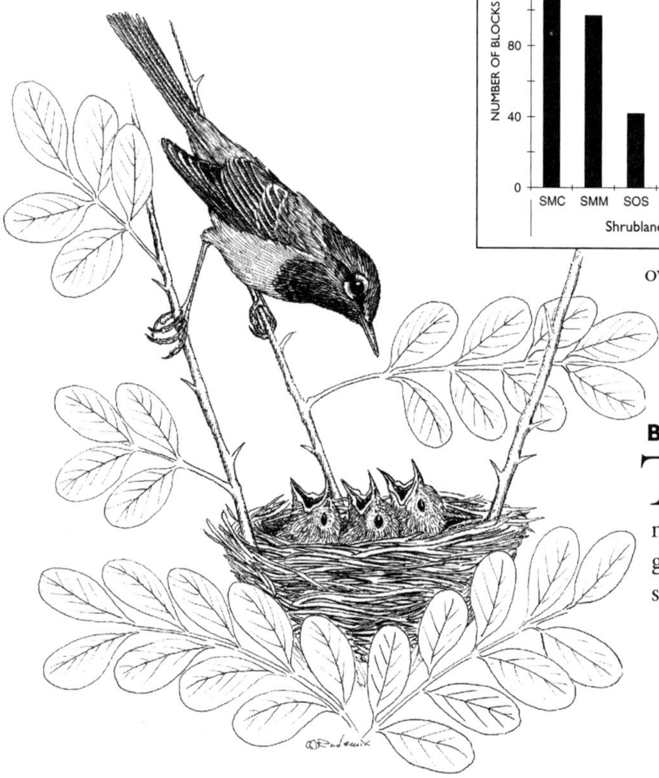

BREEDING

These warblers build poorly constructed, but well-hidden nests in clumps of grass on the ground or near the ground in shrubs. Coarse grasses, willow bark strips, and straw compose the outer cup of the nest, and the inner cup consists of fine grasses, root fibers, and hair (Pitocchelli 1995). Three to five (usually four) eggs are laid on successive days. Incubation lasts 11–13 days. The female alone broods the young, which fledge in 8–9 days (Pitocchelli 1995).

Atlas Confirmation dates fell within a narrow time frame, especially those with nests. Reflecting the well-concealed nature of the nests, Atlas workers found only four nests with eggs and only three nests with young. By far, the most common Confirmation codes were feeding young and fledged young. MacGillivray's Warblers are single-brooded (Pitocchelli 1995). The nesting process begins in early June, and nests with eggs begin to appear during the third week of June; dependent young being fed were noted most commonly the first week of July; and fledged young appear, on the average, shortly thereafter in mid July.

Atlasers had a low but respectable Confirmation percentage, but these furtive birds do not make it easy. Each male sings a slightly different song (Pitocchelli 1995)—which excuses those field workers who have trouble recognizing the species by song.

BREEDING PHENOLOGY

CODE		# OF RECORDS	DATE RANGE
NB	Nest Building	8	31 May–9 Jul
NE	Nest with Eggs	4	22 Jun–27 Jun
NY	Nest with Young	3	6 Jul–18 Jul
FY	Feeding Young	70	17 Jun–6 Aug
FL	Fledged Young	39	16 Jun–11 Aug

DISTRIBUTION

MacGillivray's is the western member of a superspecies that includes the Mourning Warbler but not the similar Connecticut Warbler. In Alberta and British Columbia MacGillivray's and Mourning come within 31–125 miles (50–200 km) of each other (Pitocchelli 1995). MacGillivray's breed north to southeastern Alaska and south to southeastern Coahuila and southern Nuevo Léon, Mexican states immediately south of Texas (AOU 1983).

MacGillivray's Warblers may have expanded their range since the 1880s due to large-scale logging of forests (Pitocchelli 1995). Overall, they have probably increased in logged areas (due to increased shrubby habitats created by logging), but they may have declined in areas in which livestock grazing disturbs shrubby habitats (Pitocchelli 1995).

Their occurrence in Colorado reflects the distinctly western continental distribution. Western Slope blocks contained the vast majority of breeding Confirmations, concentrated in mid-elevation mountainous areas. This distribution coincides with historical accounts listing MacGillivray's as one of the most common western Colorado warblers and as breeding from the base of the foothills to 9,000 feet (2750 m; Cooke 1897). The Front Range breeding records range from the foothills shrublands to blocks along the Continental Divide. The highest Confirmation came from the Mount Elbert block (39106A4) at 11,000 feet (3350 m). MacGillivray's were noticeably absent from North and South parks and the San Luis Valley. Historical accounts make no mention of these areas either.

BBS data, although not statistically significant, uniformly show declines by state, national, and physiographic regions. MacGillivray's Warblers need careful monitoring in Colorado because of the severe decline indicated by BBS data.

BREEDING EVIDENCE

REPORTED IN 380 (22%) OF 1745 PRIORITY BLOCKS

☐	Possible	126 blocks	(33%)
■	Probable	123 blocks	(32%)
■	Confirmed	131 blocks	(34%)

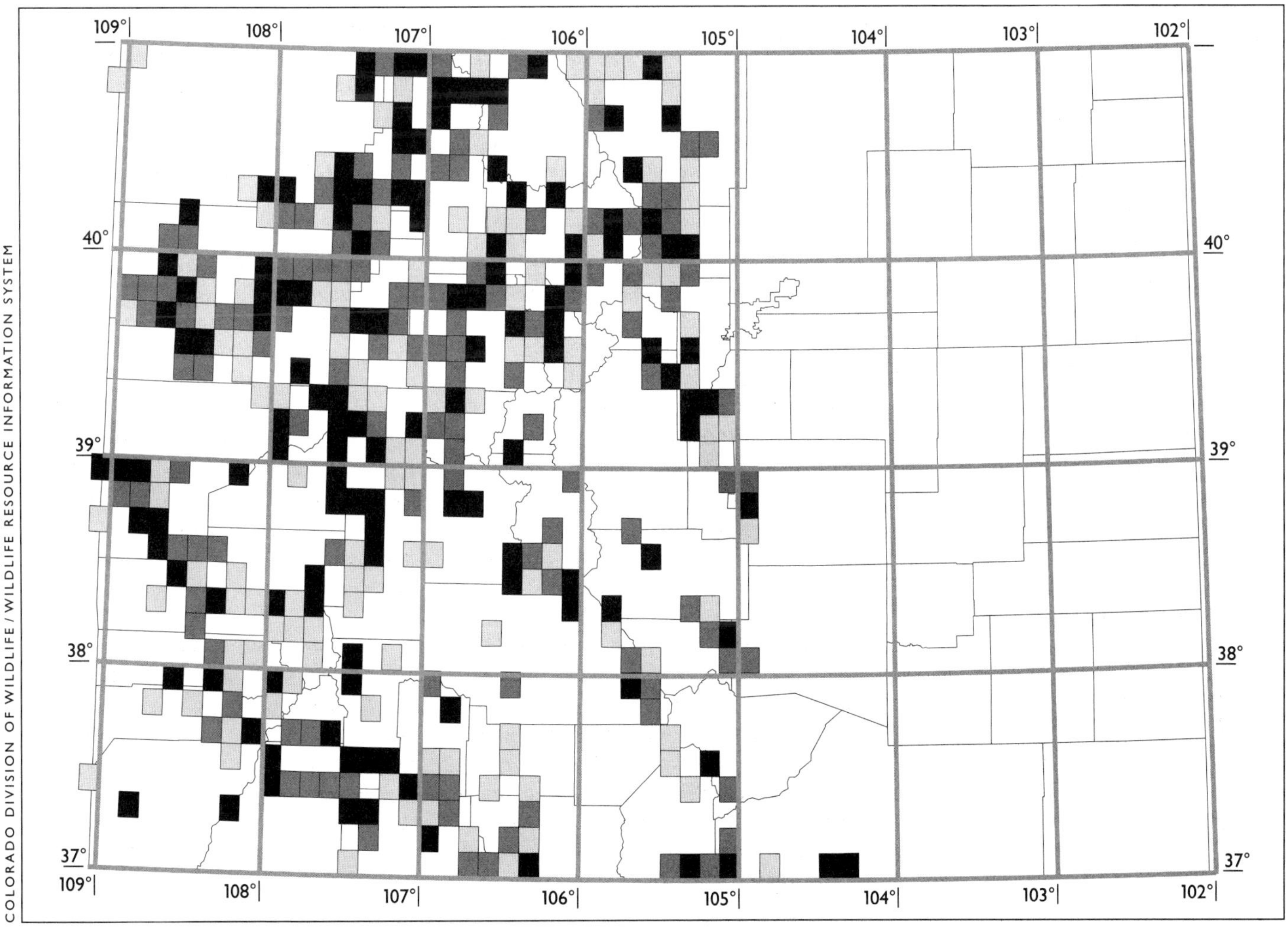
109°
108°
107°
106°
105°
104°
103°
102°
40°
39°
38°
37°
COLORADO DIVISION OF WILDLIFE / WILDLIFE RESOURCE INFORMATION SYSTEM

Common Yellowthroat

The "witchety" birds sing the same distinctive rhythm across their range, yet the songs have minute differences from place to place. Common Yellowthroats also display varying habitat divisions; Colorado birds have a more discriminating disposition than yellowthroats in the Northeast. The "Lone Ranger" birds enliven Colorado marshes with color, sound, and action.

HABITAT

Like many birds that span the continent, Common Yellowthroats use different habitats in the East and the West. The habitats have in common a component of wetness, but eastern Yellowthroats display more cosmopolitan tastes. They inhabit brush thickets and woodland edges, tangles, overgrown roadsides, old fields with clumps of shrubbery—usually (but not always) in moist or marshy areas (Arbib *in* Andrle and Carroll 1993). In upstate New York yellowthroats seem to sing from every shrubby roadside regardless of moisture (HEK), and an early New England author said that any bush in Vermont may contain a yellowthroat (Ellison *in* Laughlin and Kibbe 1985). In Michigan about half the atlas observations came from shrubby wetlands, 20% from open wetlands, and 10% from shrubby uplands (Brewer et al. 1991). In coastal California (Marin County) 80% occur in marshes and swamps (Shuford 1993).

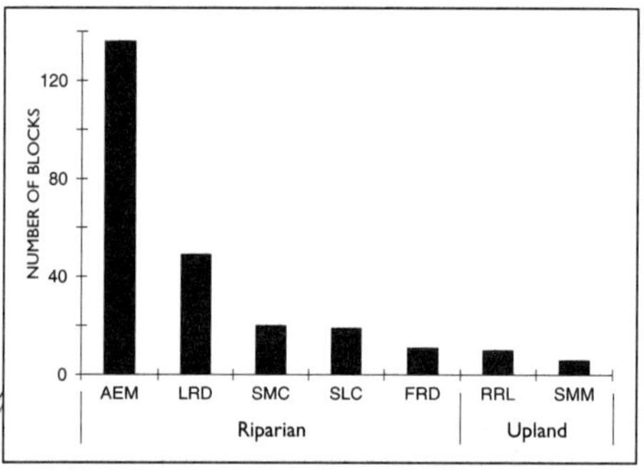
NUMBER OF BLOCKS
0
40
80
120
AEM
LRD
SMC
SLC
FRD
RRL
SMM
Riparian
Upland

Colorado yellowthroats behave more like their California relatives. Here they most often breed in cattail marshes, and some favor low thickets alongside streams, ponds, and gravel pits. Over half the Atlas habitat reports came from emergent wetlands—mostly cattail marshes. Streamside habitats—plains and foothills riparian areas, deciduous woodlands and shrubby willows—composed 40% of the sightings. This may understate the marsh connection because Atlas habitat codes lack the detail necessary to depict moisture conditions; many riparian habitats contain marshy micro-habitats. The narrower habitat preferences of western yellowthroats restrict the Colorado distribution, particularly on the Western Slope. Ridgway, in 1877, opined that, on the higher and drier hillsides of the Great Basin, MacGillivray's Warblers replace yellow-throats (Ridgway *in* Ryser 1985).

Common Yellowthroats sing a distinctive song known by most observers as *Witchety witchety witchety*. Although songs maintain an identifiable cadence and volume, they actually vary across the country (Mulvihill *in* Brauning 1992).

Yellowthroats consume a diet of 99% animal matter—principally Hymenoptera (bees and wasps), Hemiptera (leaf bugs, leafhoppers, lice, scales), and Coleoptera (beetles; Gross *in* Bent 1953). They catch their insect prey on or near the ground by gleaning from low vegetation, bushes, small trees, and even mud. During the breeding season they forage within a foot of the ground 90% of the time; after nesting they ascend as high as 18 feet (5.5 m) to feed (Shuford 1993).

BREEDING

Yellowthroats can nest very close together. A Vermont study showed 26 pairs/100 acres (64/100 ha; Ellison in Laughlin and Kibbe 1985). In Colorado Breeding Bird Censuses their densities varied from 6 to 21 pairs/100 acres (15–52 pairs/100 ha; *Am. Birds* various). A yellowthroat territory will, in California, typically include open water or damp ground (Shuford 1993); Colorado birds probably have similar requisites.

They build bulky nests—large for a small bird—of cattail leaves and other material, in cattails, shrubs, or small trees. Sometimes they place them quite close to the ground, backed against a shrub stem. Rarely do they place them higher than 20 inches (48 cm) above the ground (Arbib *in* Andrle and Carroll 1988, Shuford 1993).

Atlas dates show a long breeding season, from May nest-building to late August food-carrying; in other states yellowthroats have two broods, and the Colorado dates imply that yellowthroats also produce two broods in Colorado.

BREEDING PHENOLOGY

CODE		# OF RECORDS	DATE RANGE
NB	Nest Building	2	24 May–14 Jun
ON	Occupied Nest	3	28 May–19 Jul
NE	Nest with Eggs	2	31 May–9 Jun
FY	Feeding Young	10	1 Jun–23 Aug
FL	Fledged Young	4	10 Jun–22 Jul

DISTRIBUTION

Common Yellowthroats breed from southern Yukon to the Atlantic and through the U.S. to central Mexico. In the East, atlasers in Ohio, Kentucky, Maryland, New York, Pennsylvania, Vermont, and New Hampshire found them in 95–100% of their Atlas blocks. Yellowthroats winter from the southern U.S. to the Caribbean islands and Panama.

Their predilection for swamps and marshes constrains their Colorado distribution. The Atlas map shows them breeding in marshes at the eastern and western edges of the mountains and along the major stream corridors. They mostly avoid the central mountains and Middle and South parks, but find suitable marshes in North Park and the San Luis Valley. Colorado's low percentage of Atlas blocks with yellowthroats (compared to their ubiquitous presence in the East) reflects the drier and more varied nature of habitats in the state.

For the most part yellowthroats do not breed above 8,000 feet (2440 m). Perhaps the most surprising result of Atlas work is the large number of yellowthroats found in Rio Blanco County, on the Roan Plateau and westerly. The scattered marshes there, probably not explored before by birdwatchers, supported only a few pairs (2–10 breeding pairs estimated in most blocks).

Cattail marshes flourish on the outflow of settlement's sediments; sewage contains nutrients that nourish them. Cattails nurture yellowthroats, at least in Colorado. Human settlement shows little sign of abatement, and unless disaster strikes the birds elsewhere, the masked witchety-birds will savor safe prospects in Colorado.

BREEDING EVIDENCE

REPORTED IN 173 (10%) OF 1745 PRIORITY BLOCKS

□	Possible	76 blocks	(44%)
■	Probable	77 blocks	(44%)
■	Confirmed	20 blocks	(12%)

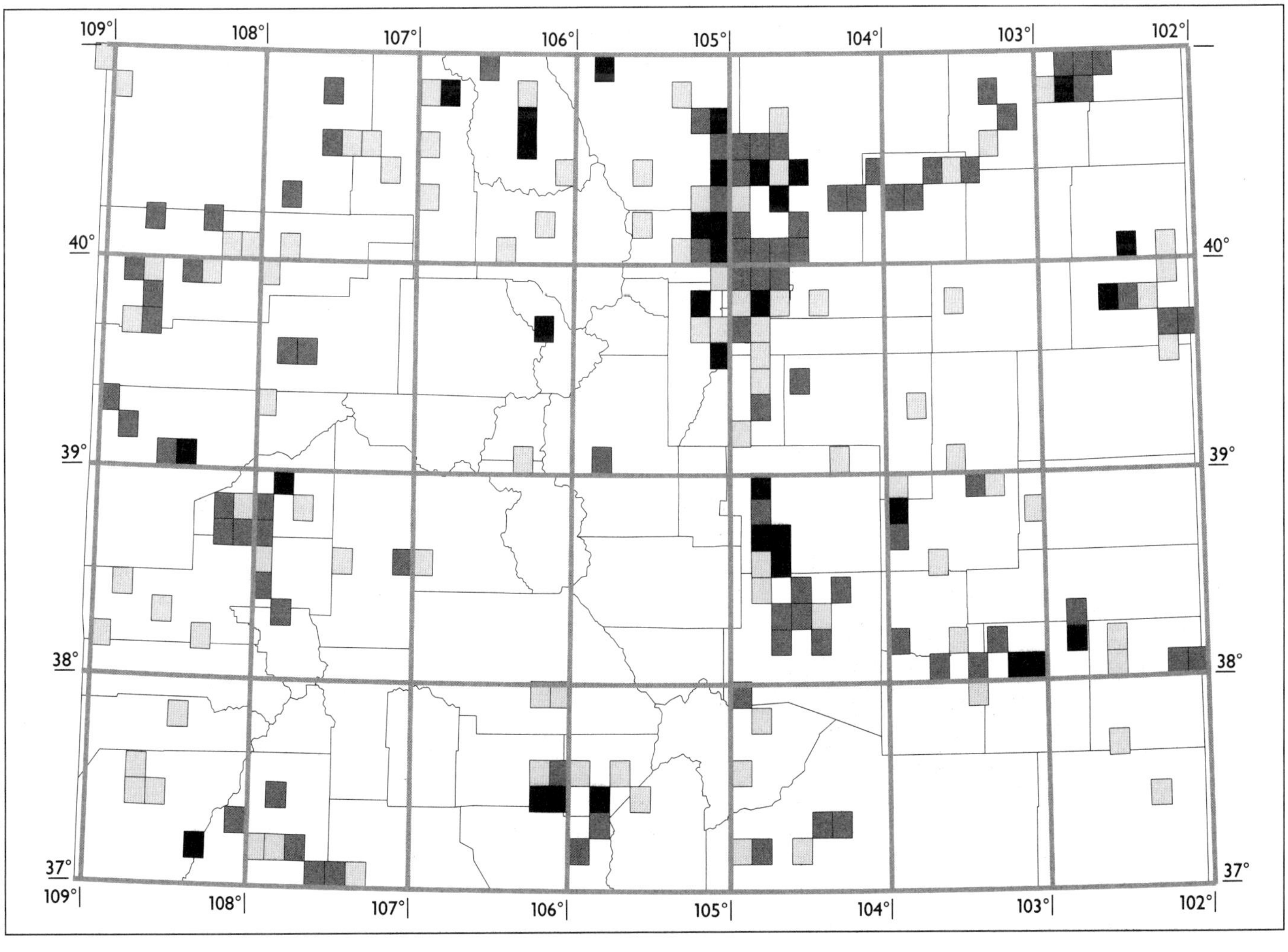
109°
108°
107°
106°
105°
104°
103°
102°
40°
39°
38°
37°

COLORADO DIVISION OF WILDLIFE / WILDLIFE RESOURCE INFORMATION SYSTEM

Wilson's Warblers breed in places popular with Colorado campers, hikers, and fishermen: along streams in high mountain meadows. Named after the "Father of Ornithology," Alexander Wilson, these birds breed in riparian thickets throughout the mountainous area of Colorado.

HABITAT

Widespread in their high-country breeding haunts but often elusive, Wilson's Warblers choose willow thickets and other brushy tangles that provide an abundance of insects, their primary food. Wilson's Warbler breeding areas are closely tied to willow or alder thickets at the edges of streams, sloughs, lakes, and beaver ponds. Montane carrs (52%) and subalpine willow carrs (39%) make up nine-tenths of the Atlas habitat codes. Other riparian corridors accounted for another 5% of Atlas breeding codes; half of the other codes that atlasers listed were second and third habitats in addition to willow carrs.

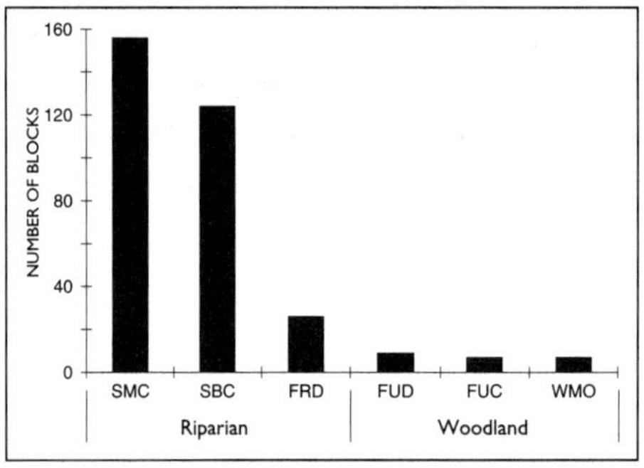
160
120
80
40
0
NUMBER OF BLOCKS
SMC
SBC
FRD
FUD
FUC
WMO
Riparian
Woodland

Wilson's start out at elevations where the other common riparian warbler, the Yellow, stops. The two species overlap to some degree along streams from 8,000 to 10,000 feet (2440–3048 m). In Boulder County, Wilson's Warbler and Lincoln's Sparrow are the most common breeding species in montane and subalpine willow carrs (Hallock 1984).

BREEDING

Wilson's Warblers breed from 6,000 to 12,000 feet (1830–3660 m), according to Cooke (1897), who asserted that their center of abundance during breeding season lay both just above and just below timberline. He deemed them, in July, the most numerous insect-eating bird above timberline (overlooking American Pipits). A&R gave summer elevations from 10,000 to 13,000 feet (3048–3960 m), with 10,000 to 11,500 feet (3048–3500 m) being most common. Many Atlas blocks with Wilson's Warbler have wide elevation ranges, more than 4,000 feet (1220 m) in some cases. Montane carr, the most common habitat code, consists of middle-elevation willow thickets, generally from 9,000 to 10,500 feet (2745–3100 m). The prevalence of montane carrs among Atlas habitat codes suggests, in contrast to Cooke's opinion and in agreement with A&R, that substantial numbers of Wilson's breed well below timberline. Perversely, the only nest card reported a nest at 8,200 feet (2500 m) in Jefferson County (Deckers 39105C2).

In the Sierra Nevadas, as opposed to coastal parts of the range, Wilson's Warblers have larger clutch sizes and higher rates of nest success, and regularly exhibit polygyny (Stewart et al. 1977). In Colorado data from the Bailey Nesting Area in Summit County suggest the same (Lane and Leukering 1994). Possibly polygyny is likely in areas when males hold territories that vary greatly in the quality of resources (Stewart et al. 1977, Ehrlich et al. 1988). Females tend to choose superior males—by inference those that have high-quality territories. For example, females mate with already mated males who have already established territories in high-quality habitats, despite the presence of bachelor males with inferior habitats. Nesting in better-quality habitat apparently allows a female to raise her young without help from a male already occupied with his first mate's brood.

In Colorado, the quality of riparian areas varies widely from one drainage to another, or even across a fence in the same drainage, due to differing grazing practices. Low-elevation riparian areas have degenerated in quality because of water diversions and grazing; this may explain why fewer breed at lower elevations than Cooke (1897) suggested, if he was indeed correct.

Cowbirds uncommonly victimize these warblers (Ehrlich et al. 1988). Atlasers, who found only one Wilson's Warbler nest, reported no incidents save for "many" at the Bailey nesting area, where the Colorado Bird Observatory conducts intensive bird studies. Chace and Cruz (1996) reported eight instances between 1985 and 1992 in Boulder, Clear Creek, and Summit counties.

BREEDING PHENOLOGY			
CODE		# OF RECORDS	DATE RANGE
C	Courtship	1	14 Jul
NB	Nest Building	6	27 May–18 Jun
ON	Occupied Nest	5	8 Jun–4 Aug
NE	Nest with Eggs	8	16 Jun–7 Jul
NY	Nest with Young	11	16 Jun–18 Jul
FY	Feeding Young	88	23 Jun–24 Aug
FL	Fledged Young	25	21 Jun–18 Aug

DISTRIBUTION

These bright yellow warblers breed from the treeline in northern Alaska and Canada south in the western mountains to southern California and northern New Mexico; also east across Canada and south into the mountains of New York and New England. In Michigan in 1988 atlasers confirmed breeding for the first time, in that state's Upper Peninsula (Brewer et al. 1991).

Probable and Confirmed Atlas records cover the higher mountains and plateaus of Colorado. Superimposing the A&R map of Wilson's Warbler distribution onto their map of Yellow Warbler fills in the holes, but Atlas maps show more extensive distribution of both species, with many blocks having both. The Atlas map shows gaps across northwestern Saguache County—the Cochetopa Pass country—and in the Sangre de Cristos. Wilson's Warbler habitat certainly occurs in these areas, but less extensively than in other mountainous areas. The Cochetopa Hills, drier than much of the high country, probably have less suitable willow carr habitat. In the Sangres many Priority blocks did not coincide with high-altitude riparian areas and Atlas coverage was less thorough. To the A&R map the Atlas adds breeding areas in the Front Range (in he Tarryalls) and Wet Mountains, and expands the breeding areas in the Flat Tops, Grand Mesa, and San Juans.

The high-altitude habitat would seem to face few threats other than grazing pressure, so the species, at least in Colorado, appears secure.

BREEDING EVIDENCE			
REPORTED IN 276 (16%) OF 1745 PRIORITY BLOCKS			
☐	Possible	64 blocks	(23%)
▩	Probable	70 blocks	(25%)
■	Confirmed	142 blocks	(51%)

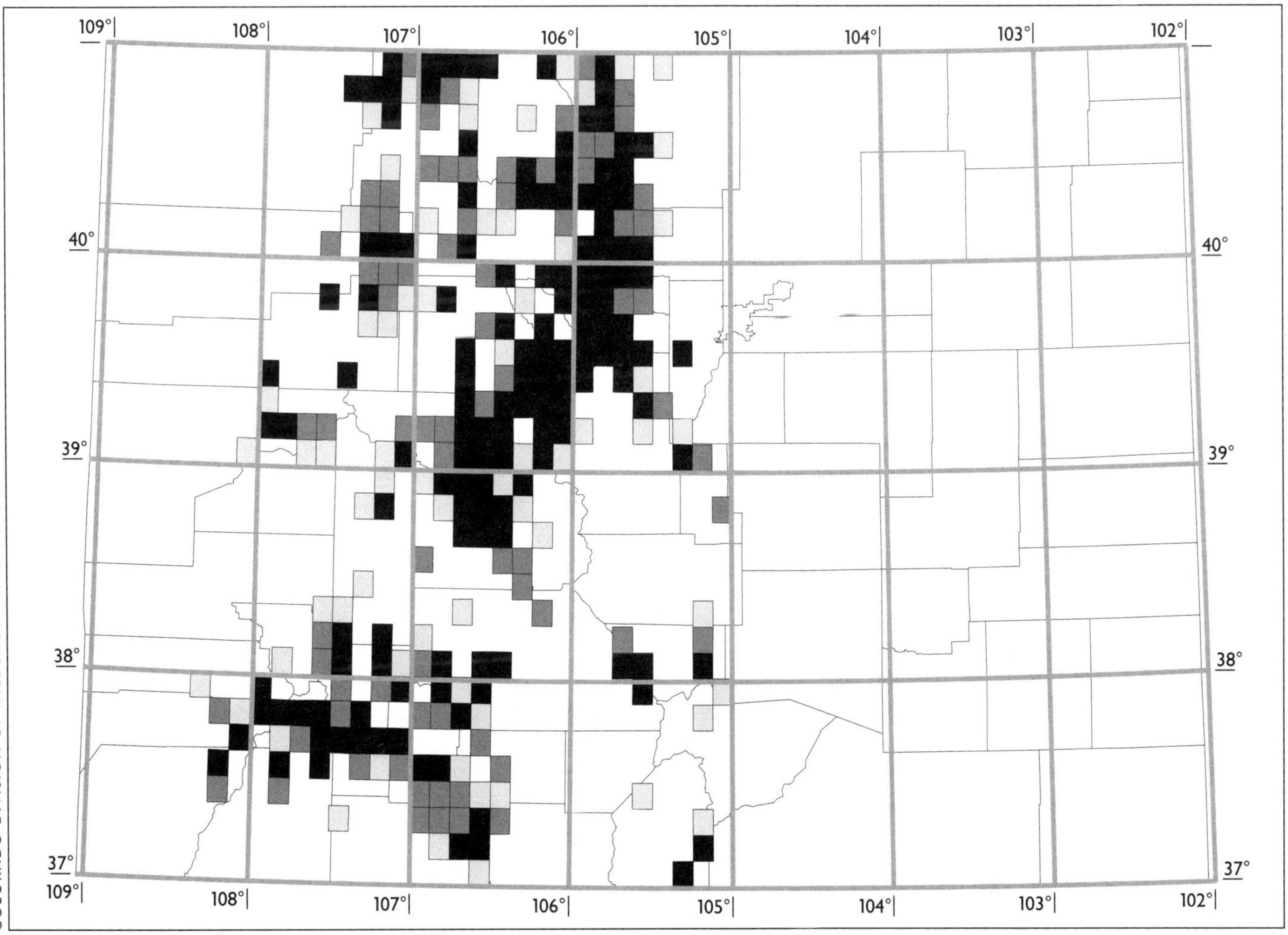
109°
108°
107°
106°
105°
104°
103°
102°
40°
39°
38°
37°

COLORADO DIVISION OF WILDLIFE / WILDLIFE RESOURCE INFORMATION SYSTEM

YELLOW-BREASTED CHAT

Icteria virens

Emanating from tangled thickets, Yellow-breasted Chat songs combine an astonishing array of calls, whistles, squawks, chatters, chugs, and gurgles. The source of this ventriloquial jumble of sounds may perch in the open one moment and dive out of sight the next. This display of vocal dexterity, which includes the lowest notes of any warbler, tantalizes observers during May and June. In July male chats, like their mates, retire to the midst of the tangles.

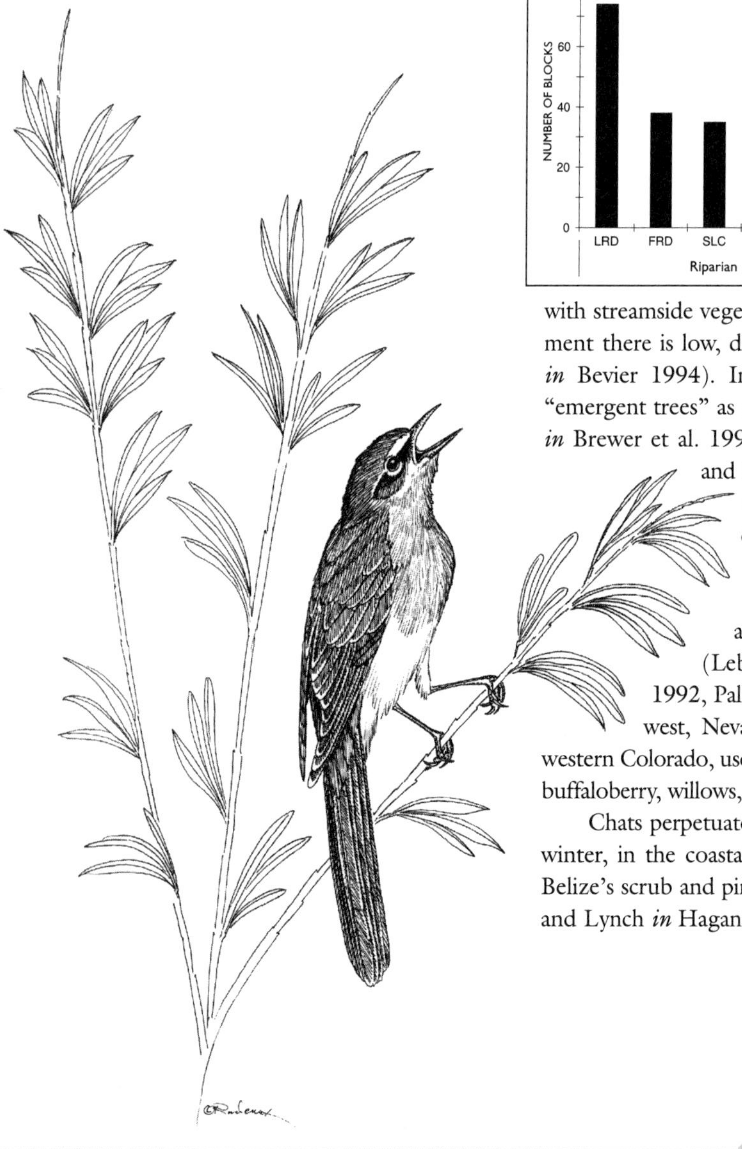

HABITAT

In Colorado Yellow-breasted Chats typically nest in dense riparian thickets along plains and western valley river systems and foothills streams—more than 94% of Atlas habitat reports specified riparian locations. Most nest below 7,000 feet (2135 m), although a few stray as high as 8,000 feet (2440 m) and spread out to mountain shrub communities, most often scrub oak and skunkbrush (A&R). Within the riparian community, atlasers found that they frequent mixed shrubby thickets; the bushes include hawthorn, willows, wild rose, skunkbrush, wild plum, and chokecherry, mixed with boxelders and young cottonwoods. Atlasers found them most frequently on the plains and western valleys (two-thirds of the Atlas reports).

In the East chats use a variety of shrubby habitats; although most commonly associated

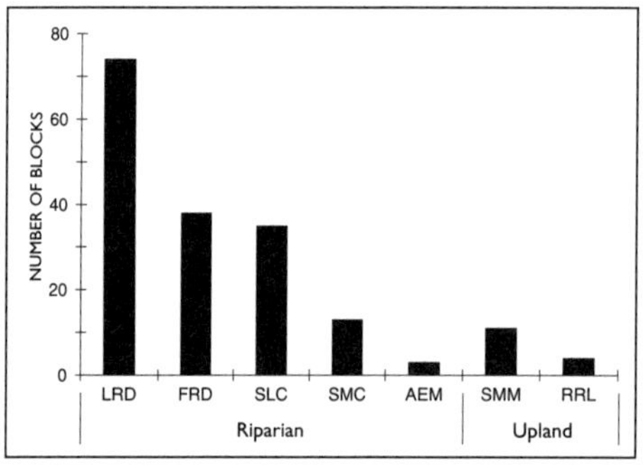

NUMBER OF BLOCKS
80
60
40
20
0
LRD
FRD
SLC
SMC
AEM
SMM
RRL
Riparian
Upland

with streamside vegetation, the essential element there is low, dense vegetation (Askins *in* Bevier 1994). In Michigan they favor "emergent trees" as song perches (Reinoehl *in* Brewer et al. 1991) and in Pennsylvania and Kentucky they use artificial habitats created by clear-cuts, abandoned fields, utility corridors, reclaimed strip mines, and pine plantations (Leberman *in* Brauning 1992, Palmer-Ball 1996). Farther west, Nevada birds, like some in western Colorado, use large, dense thickets of buffaloberry, willows, and roses (Bent 1953).

Chats perpetuate their shrubby ways in winter, in the coastal scrub of Mexico and Belize's scrub and pine savannas (Petit et al. and Lynch *in* Hagan and Johnson 1992).

BREEDING

Colorado chats nest within a narrow range of dates, from late May to the end of June, and probably raise but one brood. Atlasers found only two nests—B&N described hours-long sessions trying to locate nests of these skulkers—and atlasers Confirmed them in only 13% of the blocks with chats.

Starting in early May along the Front Range, earlier in the western valleys, male chats furtively sing their versatile babel of cackles, whistles, mews, and squeaks. A singing male "laughs dryly, gurgles derisively, whistles triumphantly, chatters provokingly, and chuckles complacently" (Taverner *in* Bent 1953).

Males often adhere to a daily pattern: one routinely followed a fixed route in which he used the same singing perches, in the same order, at about the same time of day. He repeated the pattern throughout the day (Bent 1953). Male chats also indulge in a hovering flight song that includes medleys similar to their ground-based concerts, and which they deliver with fluttering wings, dangling legs, and raised head.

Chats build their nests within the tangles, 1–4 feet from the ground. They have attached Colorado nests to scrub oak (in a bed of poison ivy), coralberry, and gooseberry (B&N). They build bulky nests of dead leaves, grasses, and plant stems, and small vines, with a lining of finer grasses, stems, bark, and even horsehair.

The female does most or all the incubation, of 3–5 eggs. The nestlings, for the first few days, receive sustenance by regurgitation, and then graduate to solid foods such as beetles, grasshoppers, butterflies, and especially hairless caterpillars (Wheelock *in* Bent 1953).

Cowbirds commonly deposit eggs in chat nests, but the eggs rarely hatch; of 100 parasitized nests, in only three did the eggs hatch (Friedman *in* Bent 1953). The Atlas had one observation of parasitism, a pair of chats attending a fledgling (Chace and Cruz 1996; Appendix C).

BREEDING PHENOLOGY

CODE		# OF RECORDS	DATE RANGE
C	Courtship	5	21 May–7 Jul
NB	Nest Building	3	22 May–1 Jun
ON	Occupied Nest	1	26 Jun
NE	Nest with Eggs	1	27 Jun
FY	Feeding Young	4	15 Jun–16 Jul
FL	Fledged Young	2	25 May–22 Jul

DISTRIBUTION

Yellow-breasted Chats nest throughout the U.S. into northern Mexico, and winter from Mexico to Panama. The Colorado distribution reflects the pattern of plains stream bottoms, western river corridors, and foothills shrub zones.

Although they may pack territories close together in one place—almost in loose colonies—they may spurn nearby apparently suitable habitat (B&N). Atlasers estimated fewer than 10 nesting pairs in over 60% of the blocks where they found chats. They provided A3 estimates (11–100 pairs) in 37%, and deemed only two blocks to have more than 100 nesting pairs.

The Atlas distribution map, in outline, resembles that of A&R, but atlasers found the "hurly burly" of this "garrulous" singer (Keyser 1902) only patchily within this range. Almost equal numbers of blocks on each side of the Continental Divide had them. In the San Luis Valley, with only one prior record, atlasers found chats in three blocks, all along streams with dense shrubby tangles (Rich Levad pers. comm., HEK).

Although Sclater (1912) regarded the chat as a common bird in the state, this does not seem applicable today. The spotty distribution reflects the spotty nature of thick brush that remains along Colorado streams. Very likely settlers, subdividers, and grazing animals have cleared away the dense tangles that this secretive bird needs.

BREEDING EVIDENCE

REPORTED IN 136 (8%) OF 1745 PRIORITY BLOCKS

☐	Possible	70 blocks	(51%)
■	Probable	48 blocks	(35%)
■	Confirmed	18 blocks	(13%)

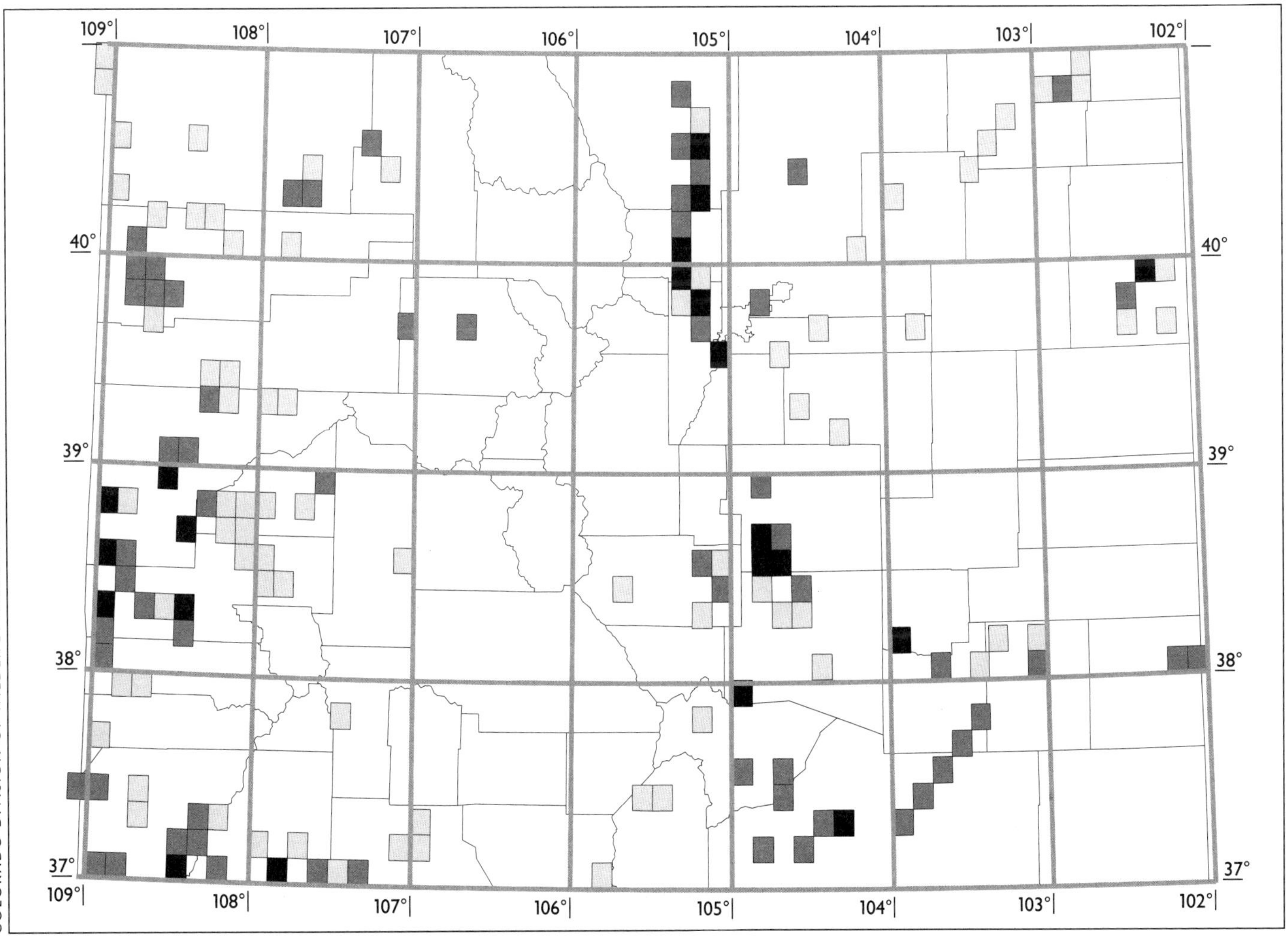
109°
108°
107°
106°
105°
104°
103°
102°
40°
39°
38°
37°

COLORADO DIVISION OF WILDLIFE / WILDLIFE RESOURCE INFORMATION SYSTEM

Seldom conspicuous, even in full song and despite their vivid plumage, Hepatic Tanagers move restlessly from tree to tree, gleaning insects and spiders as they move. Because few biologists have studied them, Colorado Atlas observations provide a modest amount of new information about the habitat requirements of these retiring birds of the Southwest.

HABITAT

Hepatic Tanagers, like Colorado's common Western Tanagers, nest in the pines. In the Southwest, the center of their U.S. abundance, they occur commonly in "dense oaks" (Phillips et al. 1964), and among ponderosa pines and Douglas-firs (Brandt 1951). They breed in the lower zone of pinyon pines in Los Alamos County, New Mexico, primarily in the lower canyon bottoms (Travis 1992), and in the pinyon pines of the Chisos Mountains, Texas (Wauer 1973).

Colorado's few Hepatic Tanagers demur from the traditional habitat used by their relatives in Arizona. Here they breed in

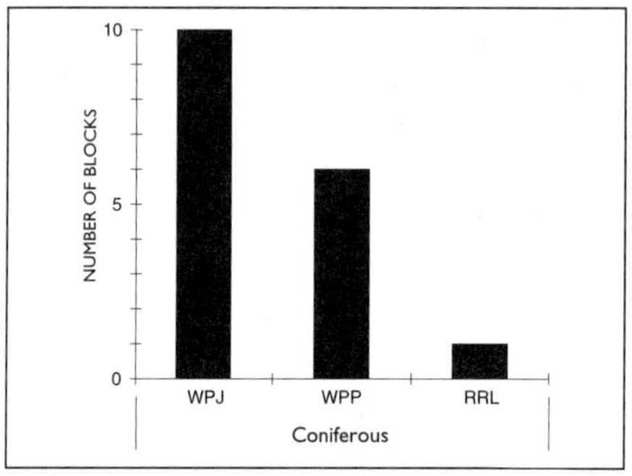
NUMBER OF BLOCKS
10
5
0
WPJ
WPP
RRL
Coniferous

the pinyon pine-covered mesas. They pick niches that include tall trees.

Two different mini-habitats fill this need. In some blocks tall, mature ponderosa pines grow in the deeper and moister draws that lead up to the canyon rims. Within the draws, ponderosas tower above the more abundant pinyons and junipers, which grow on the sides of the draws; the tanagers feed and sing from the ponderosas (e.g. Tobe, 37103B5, Pine Canyon, 37103A6).

These tanagers also use a sub-habitat without ponderosas—mesas covered on their tops and sides with pinyon pines. Here males use as song perches the taller pinyons that grow at the edge of the mesa. The birds cover a relatively large territory—several acres at least. One patrolled an area in the Brown Canyon block (37103E3) at least 300 by 50 yards (3 acres, 1.2 ha), all on the canyon rim; it also flew a half-mile across the canyon to the mesa rim on the other side. The male of a second pair sang from both the tall pinyons on the canyon rim and from trees below (HEK). It appears that the tanagers encompass the mesa top in their territories in order to include the tall pinyons for high singing perches.

Slowly and deliberately they move high in the pines gleaning insects from the foliage and smaller branches of the oaks and pines. Sometimes they sally out, flycatcher-like, to catch prey, but rarely do they descend to the ground.

BREEDING

Hepatic Tanagers place their nests in tall trees; one of the few nests reported so far (in Arizona) was 50 feet (15 m) up in a tall ponderosa, 12 feet (3 m) out from the trunk, in a fork near the end of the branch. The birds had suspended the nest between two prongs of the fork and made it of green grass, green and gray forb stems, and flower stalks and blossoms. They lined it neatly with finer dry and green grasses. Another nest was 25 feet (8 m) up in a sycamore tree (Bent 1958).

Atlaser Van Truan on 15 June 1990 found a nest in a side canyon of the Purgatoire River (Doss Canyon North, 37103D7). He saw the female drop onto the nest, located on a horizontal branch on the northwest side of a ponderosa pine 30 feet (9 m) up and 8 feet (2.5 m) out from the trunk. On 2 July the birds had abandoned the first nest and 200 yards (180 m) away built a second one, also 30 feet (9 m) up and on the north side of the pine. The male fed the female as she sat on the nest. The birds covered a territory of 600 by 450 feet (180 x 140 m), which spanned across to both sides of the canyon.

BREEDING PHENOLOGY

CODE		# OF RECORDS	DATE RANGE
C	Courtship	2	30 May–20 Jun
ON	Occupied Nest	1	15 Jun
NY	Nest with Young	1	24 Jun
FY	Feeding Young	1	20 Jun
FL	Fledged Young	1	10 Jul

DISTRIBUTION

Hepatic Tanagers breed from northern New Mexico and Arizona into Mexico and Central America at least as far as Costa Rica. The first Colorado observation, in July 1973, consisted of a pair near Parachute Creek—where subsequently no one has seen them. The Las Animas County population was discovered in 1978, with breeding confirmed in 1980 (A&R).

Until Atlas field work, Colorado breeding was known from only one or two sites in Las Animas County, where the Atlas documented them as breeders in several blocks. A fledgling in the Cedarwood block (37104H5), in July 1993, extended the breeding range into Pueblo County (Dave Johnson VF). Birdwatching tourists from Ohio saw a male at the Royal Gorge (38105D3) in 1989 (Juliet Howard, VF, photo); in 1990 another Ohio group found a pair in the same place; during three and a half hours of observation the male sang occasionally and carried a probable fecal sac in its bill (Norman Walker, VF).

The Atlas map documents a small contingent of breeding Hepatic Tanagers in southern Colorado; they occupy an area where Western Tanagers, for the most part, do not occur—the two species have in common only two of nine blocks in Las Animas County, plus the Royal Gorge block. Apparently the Hepatics breed almost entirely on private land; they may have a somewhat wider distribution than shown on the Atlas map, but not much more. A&R speculate that a series of observations of migrants in Cañon City and Pueblo involved birds en route to breeding sites not yet discovered (like the Royal Gorge area). Private land in Pueblo, Huerfano, Custer, and Fremont counties may harbor a few more breeding pairs.

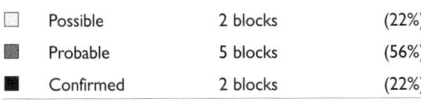

BREEDING EVIDENCE

REPORTED IN 9 (1%) OF 1745 PRIORITY BLOCKS

☐	Possible	2 blocks	(22%)
■	Probable	5 blocks	(56%)
■	Confirmed	2 blocks	(22%)

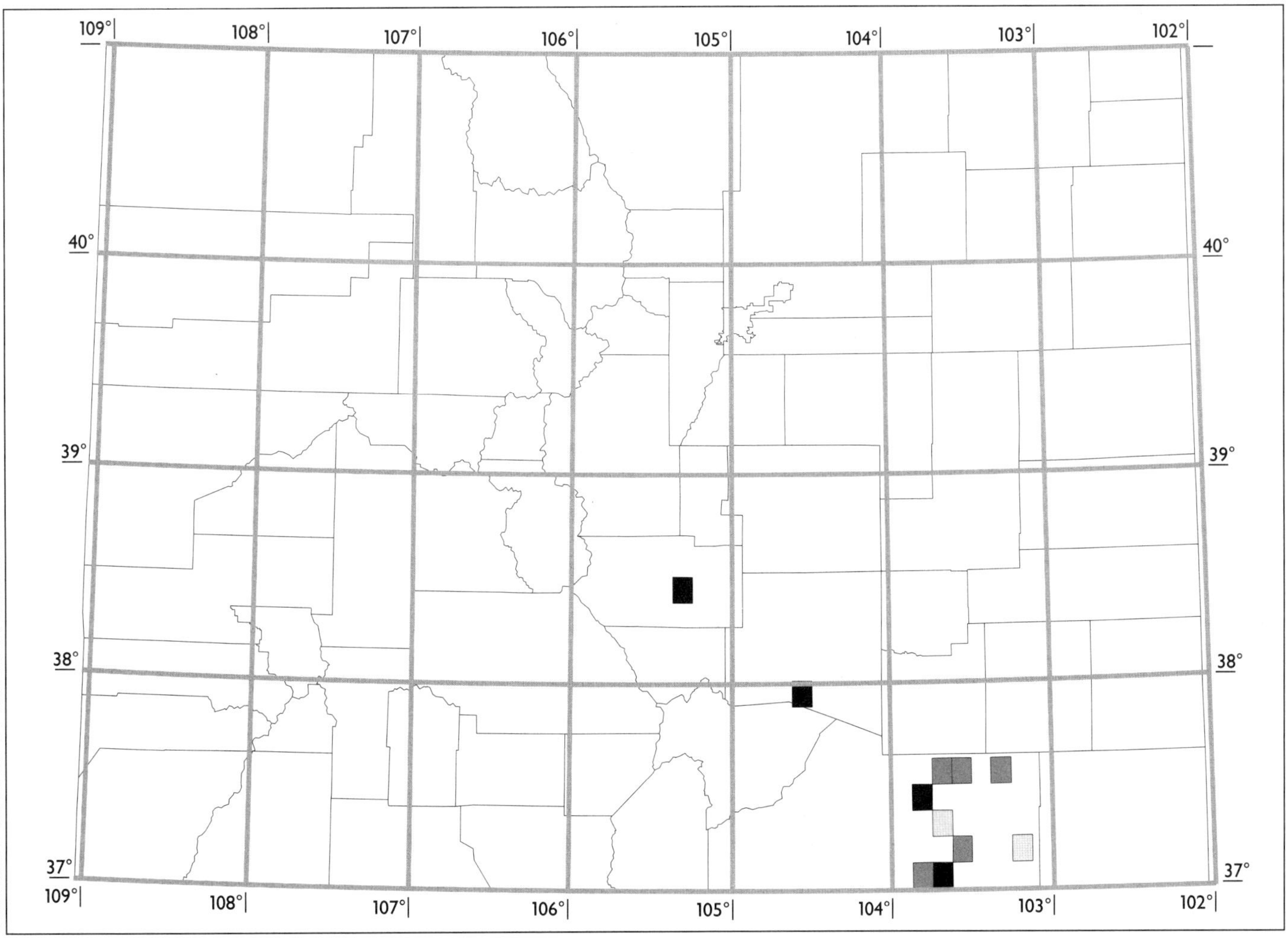
109°
108°
107°
106°
105°
104°
103°
102°
40°
39°
38°
37°

COLORADO DIVISION OF WILDLIFE / WILDLIFE RESOURCE INFORMATION SYSTEM

For a few months each year, Western Tanagers grace us with their distinguished presence. Although arguably the most dazzling of Colorado's Neotropical migrants, they frequently escape the notice of those who do not recognize the male's unremarkable song.

HABITAT

Almost as widespread as they are colorful, Western Tanagers occupy a variety of habitats from the foothills up to subalpine regions. Colorado tanagers, atlasers found, breed in ponderosa pine and aspen woodlands and a great variety of foothills and mid-elevation coniferous forests. Smaller numbers occur in pinyon/juniper woodlands and oak shrublands. Breeding elevations range almost entirely above 6,000 feet (1,800 m).

In California Western Tanagers typically use mixed-species forests with a canopy cover

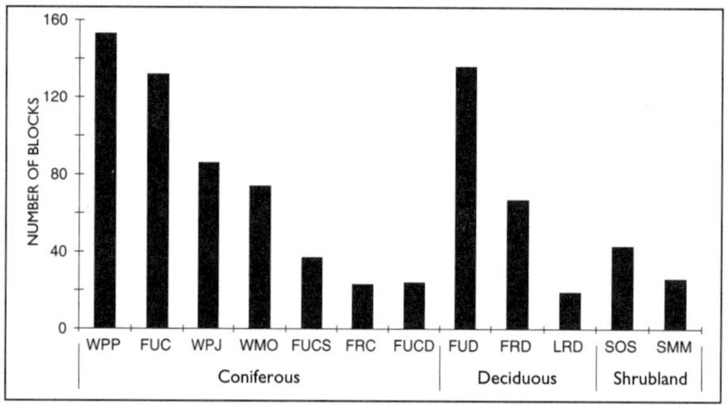
NUMBER OF BLOCKS
160
120
80
40
0
WPP
FUC
WPJ
WMO
FUCS
FRC
FUCD
FUD
FRD
LRD
SOS
SMM
Coniferous
Deciduous
Shrubland

of 66.6% (Shuford 1993). Although Rocky Mountain forests lack the high diversity of other North American forests, Atlas habitat codes imply that here also, tanagers use the more diverse woodlands. Non-diverse, open-canopy ponderosa and pinyon/juniper woodlands did compose 28% of the habitat reports, but in one-third the tanagers used them in combination with other conifers.

One-quarter of Atlas reports cited two or more habitat types. The combinations suggest that atlasers found tanagers in forests of mixed species and that, within an area the size of an Atlas block, the birds use several forest types. If atlasers reported more than one habitat, they most often combined FUC (which could indicate a number of different tree species) with ponderosa pine or pinyon/juniper. When they used more specific codes, they most frequently cited low- to mid-elevation forests—mixed woodlands and forests, conifer riparian forests, and Douglas-fir and lodgepole pine forests. Only a few cited spruce/fir forests (37 of 832 reports), which concurs with A&R, who regarded tanagers as "generally rare" in such forests. One-quarter of the habitat reports specified aspen forests, but two-thirds of those listed second and third habitats as well.

Western Tanagers typically place their nest in the fork of a horizontal branch of a coniferous tree (Project Tanager 1995). The four Atlas nest cards recorded nest sites in four different conifer species: white fir, Douglas-fir, ponderosa pine, and Engelmann spruce. Estimated nest heights ranged from 18 to 40 feet (5–12 m). Whatever the tree type used for nesting, Western Tanagers prefer fairly open areas in the forest for nesting (Johnsgard 1986). They consume a wide variety of insects typical of woodland areas, supplemented by berries and small fruits where available (Bent 1958).

BREEDING

Atlasers recorded adults carrying food over the entire breeding season, quite a broad range of dates. Scarlet and Summer tanagers engage in courtship feeding (Project Tanager 1995), which might account for the early dates, but such behavior apparently has gone unrecorded for Western Tanagers. Also, male Summer Tanagers bring food to incubating females on an occasional basis (Robinson 1996). If Western Tanagers share this trait, this and courtship feeding could explain some of the early observations of food carrying.

Reports of dependent fledglings clustered more tightly than did reports of carrying food. The overwhelming majority of dependent fledgling observations came during July, with nestlings reported only from mid June until early July. Project Tanager (1995) indicated a nestling period of approximately two weeks for this species. Collectively, these data build a compelling argument against the possibility of double-brooding in Colorado.

Clutch sizes typically vary from three to five eggs. In southern parts of their range, females typically lay a clutch of three eggs (Project Tanager 1995). Little documentation of cowbird parasitism exists for this species. Chace and Cruz (1996; Appendix C), however, reported two recent cases of cowbird parasitism on Western Tanagers in Colorado.

BREEDING PHENOLOGY

CODE		# OF RECORDS	DATE RANGE
C	Courtship	10	30 May–4 Jul
NB	Nest Building	19	21 May–12 Jul
ON	Occupied Nest	21	1 Jun–17 Jul
NY	Nest with Young	7	14 Jun–22 Jul
FY	Feeding Young	136	3 Jun–7 Aug
FL	Fledged Young	38	7 Jun–9 Aug

DISTRIBUTION

Only two tanager species, Western and Hepatic, breed in Colorado. Although Hepatic Tanagers must stretch their range to its very northernmost limits just to reach Colorado, Western Tanagers breed throughout the mountainous regions of western North America from southeastern Alaska and Mackenzie to the Mexican border. During the winter months, most retreat to central Mexico south to Costa Rica. A few winter along the coast of California (Kaufmann 1992).

All foothills and mountainous regions of Colorado support populations of Western Tanagers. Numbers peak, however, on both sides of the chain of mountain parks running south to north through the center of the state. The southern San Juan Mountains have the highest abundance according to Atlas abundance codes, in Latilongs 37106 and 37107; the northern Front Range and Park Range (Latilongs 40105 and 40106) and the Wet Mountains (38105) have other centers of abundance.

Atlas field workers contributed first Confirmations of breeding for Western Tanagers in four latilongs, three of them in the western half of the state. Two clusters of records provide notable additions to the breeding range: four blocks in eastern Las Animas County represent a notable eastern extension, and seven blocks in western Moffat County, in the ponderosa and pinyon pines, add breeding tanagers north of the Yampa River near Dinosaur National Monument.

These breeding records probably do not represent changes in range, which has no doubt remained stable since the arrival of the first settlers. These birds most frequently use trees standing in open areas for nest sites and thrive in relatively open ponderosa pine woodlands. Therefore, forest fragmentation and tree thinning exert a lesser impact on Western Tanagers than on Scarlet Tanagers and other species that require large, continuous stands of forest.

BREEDING EVIDENCE

REPORTED IN 591 (34%) OF 1745 PRIORITY BLOCKS

☐	Possible	202 blocks	(34%)
▩	Probable	174 blocks	(29%)
■	Confirmed	215 blocks	(36%)

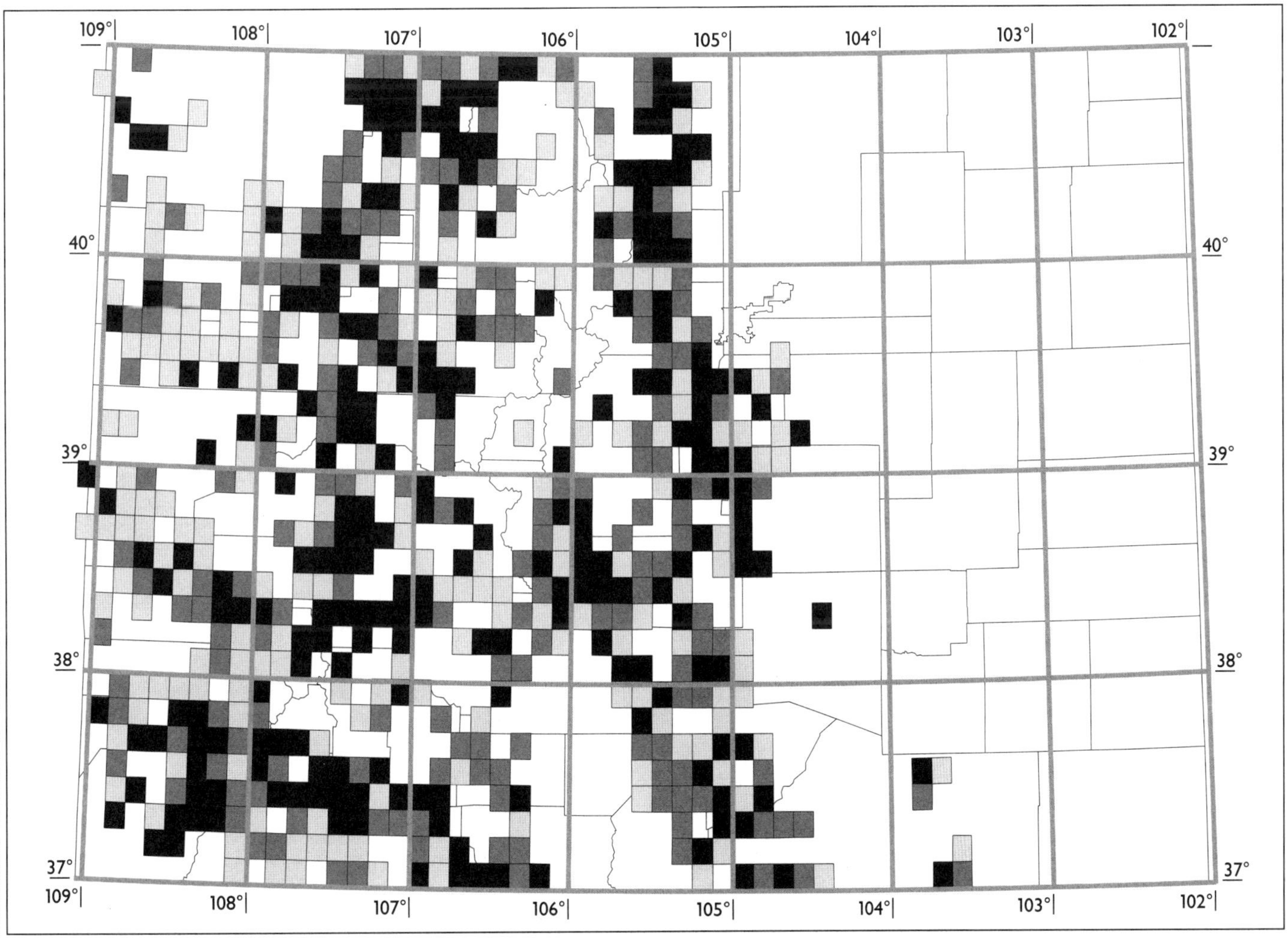
109°
108°
107°
106°
105°
104°
103°
102°
40°
39°
38°
37°

Green-tailed Towhees, true dandies of the hillside shrublands, wear an intriguing olive coat vividly punctuated and stunningly set off by a debonair reddish-brown chapeau. From their shrubby perches these ground-centered towhees take in some of the best views in the state—canyons, rivers, and white-peaked mountains.

HABITAT

Atlas workers found the majority of Green-tailed towhees on dry shrubby hillsides and sagebrush flats. The shrubs most commonly identified with montane shrublands (SMM—the most frequently used habitat code) are snowberry, serviceberry, chokecherry, bitterbrush, mountain mahogany, squawapple, scrub oak, and sagebrush. Additionally, these towhees reside on hillsides covered with pinyons and junipers, as well as in riparian shrublands. Green-tailed Towhees avoid dense forests except in openings and where conditions allow shrubs to form (Bent 1968). Atlas reports show that this species can follow the shrub habitat all the way to the edge of the tundra.

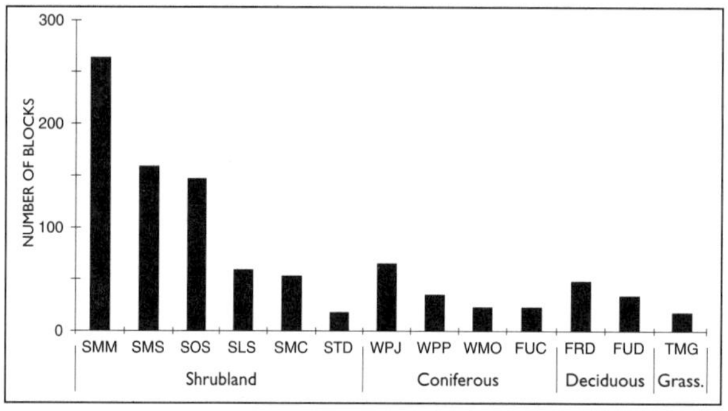
NUMBER OF BLOCKS
300
200
100
0
SMM SMS SOS SLS SMC STD
Shrubland
WPJ WPP WMO FUC
Coniferous
FRD FUD
Deciduous
TMG
Grass.

A combination of stout legs, short wings, long tail, and conical bill makes these towhees superbly suited for a life under shrubs. Their usual food-gathering technique includes taking a little jump forward and a little kick backwards in order to dislodge insects and to discover seeds.

Green-tailed Towhees nest in brushy areas with open spaces between the shrubs. If disturbed or perturbed, a Green-tail often scurries away in a "rodent run," which makes it resemble a scooting chipmunk; it drops to the ground without opening its wings and darts across the ground. A predator may have trouble distinguishing towhee from rodent and therefore not search for a nest (Ryser 1985).

BREEDING

The nest, situated either on the ground or very close to it, consists of twigs, stems, grass, bark, and, when available, hair. Structurally, the open nest is thick-walled and fairly deeply cupped (B&N). Of the 14 nest cards turned in by atlasers, the towhees had tucked their nests under sagebrush seven times, snowberries four times, and one each under bitterbrush, chokecherry, and scrub oak.

Atlasers made a significant contribution to the phenology for this abundant Colorado species; they achieved a high Confirmation rate of 49%. Previous data suggested a nesting season of 21 May to 10 August (Nelson 1993). Among the 351 Confirmations, the earliest nest-building came on 3 May and the last fledgling on 22 August. During this very long breeding season, two-thirds of the Confirmations occurred after 1 July. The high Confirmation percentage, combined with the late dates, does not prove that Green-tailed Towhees produce a second brood but does strongly suggest that this phenomenon occurs.

Brown-headed Cowbirds parasitize these towhees (Chace and Cruz 1996; Appendix C). Atlas field workers encountered this four times: one nest with eggs in Larimer County and dependent fledglings in Routt, Montrose, and Chaffee counties.

BREEDING PHENOLOGY

CODE		# OF RECORDS	DATE RANGE
C	Courtship	1	7 Jun
NB	Nest Building	12	3 May–6 Jul
ON	Occupied Nest	11	19 May–16 Jul
NE	Nest with Eggs	33	24 May–18 Jul
NY	Nest with Young	8	10 Jun–13 Jul
FY	Feeding Young	107	6 Jun–16 Aug
FL	Fledged Young	172	2 Jun–22 Aug

DISTRIBUTION

Birds of the southern Rockies and Great Basin, Green-tailed Towhees spend summers in the dry montane and plateau interior of the western United States. The towhees carry on their business from southwestern Montana to the east slope of the Sierras, south to northeastern Arizona and west-central New Mexico; to the east they reach only to the western edge of the shortgrass prairie at the foothills. In the entire breeding range the average elevation is 7,300 feet (2225 m); they breed at higher altitudes in the south than in the north.

The Colorado range of this quintessential western species starts at the mountains' edge; 92% of the Atlas records lie west of Longitude 105. Over 900 Atlas records verify that Green-tailed Towhees in Colorado are most at home in the foothills, low mountains, and mesas, and are absent only from the low western valleys and interiors of the intermountain parks where dryland shrubs do not grow. They flourish in greater abundance on the Western Slope, which has more suitable habitat than the Eastern Slope (Gregg 1963, A&R). In the broad brushstroke, Atlas results concur with A&R, except that atlasers did not detect Green-tails in the interior of the San Luis Valley. These elegant towhees of the dry hillside shrublands do not regularly inhabit the higher elevations, but a post-breeding dispersal to higher altitudes does occur (Packard 1946).

Atlas abundance calculations rank this towhee as the thirteenth most numerous species in Colorado, with almost a million breeding pairs. The estimates also demonstrate that northwestern Colorado has a high abundance and contains over half of the state's Green-tails. BBS data likewise depict this species as reaching some of the highest densities in its entire range in northwestern Colorado (Price et al. 1995, Dobbs et al. 1998). Two publications, using BBS data, claim that Colorado contains 20–40% of the species' entire breeding population (Price et al. 1995, Priorities 1995). However, given the small sample sizes in the Intermountain West, such statements deserve cautious treatment (Dobbs et al. 1998, RR).

Forty-nine Colorado BBS routes recorded the towhee during 1968–1996, with an average of over 10 birds per route. Despite the high counts, BBS data prove inadequate to provide population trends for either Colorado or North America. Regardless, the abundance of Green-tailed Towhees in western Colorado probably makes the state a very important part of the species' breeding world. The status and distribution of these rufous-chapeauxed towhees will wax and wane with the fortunes of the dry hillside shrub and stretches of sagebrush.

BREEDING EVIDENCE

REPORTED IN 687 (39%) OF 1745 PRIORITY BLOCKS

☐	Possible	178 blocks	(26%)
▦	Probable	169 blocks	(25%)
■	Confirmed	340 blocks	(49%)

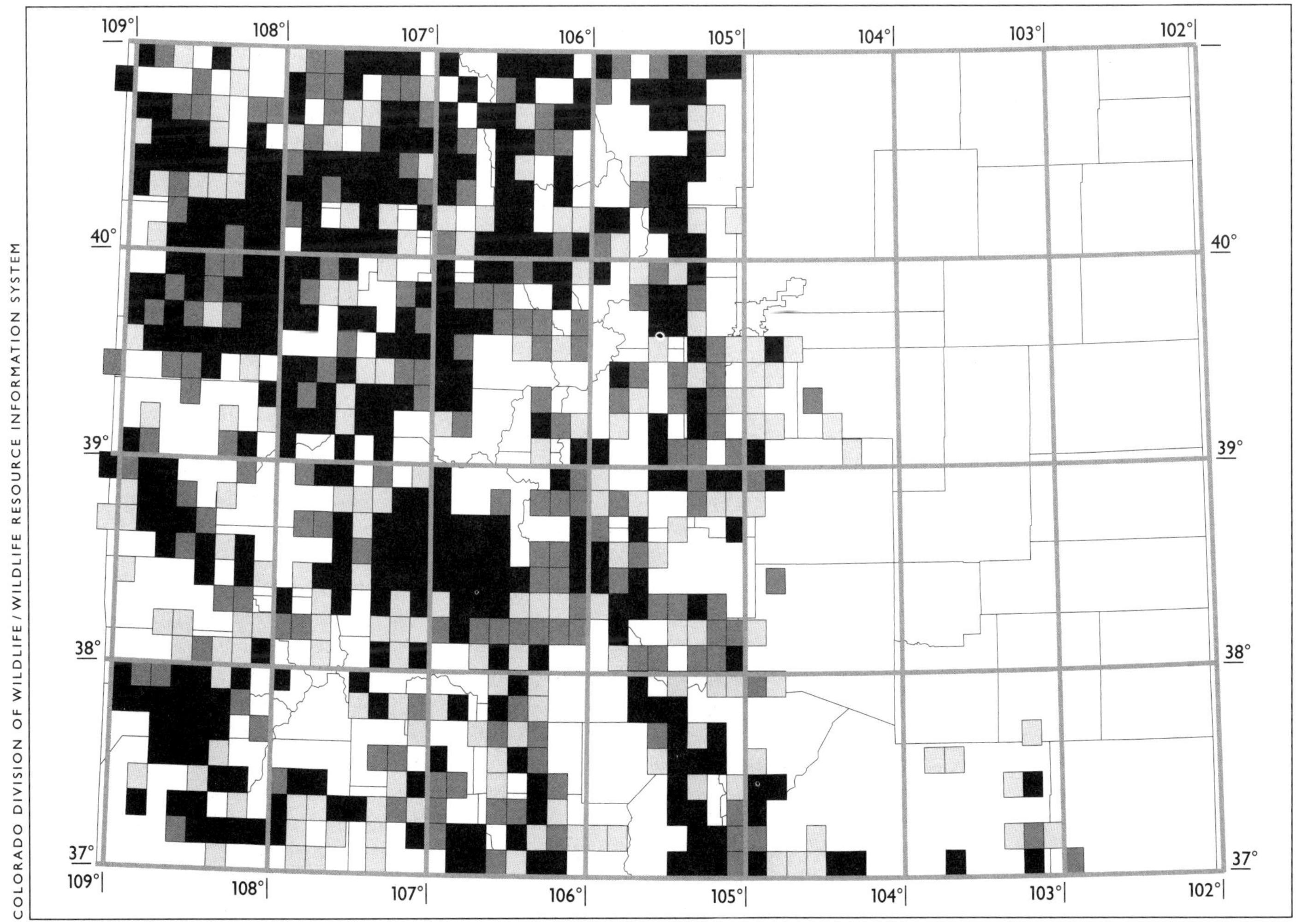
109°
108°
107°
106°
105°
104°
103°
102°
40°
39°
38°
37°
COLORADO DIVISION OF WILDLIFE / WILDLIFE RESOURCE INFORMATION SYSTEM

Because they nest and feed in tall, often impenetrable brush, one might expect to have difficulty finding Spotted Towhees. Atlas workers probably missed very few of these birds though, for they constantly emit cat-like calls and trilling songs, and their volume control stays set on high.

HABITAT

In 1995, after completion of Atlas field work, the Rufous-sided Towhee became two species, Spotted Towhee and Eastern Towhee (*Pipilo erythrophthlamus*; AOU 1995, Greenlaw 1996a). Both species may nest in Colorado; some atlasers distinguished putative Eastern Towhees by their distinct plumage and vocalizations. "Pure" Eastern Towhees, however, are unlikely although birds with genes of the eastern species occur in northeastern Colorado; the presence of

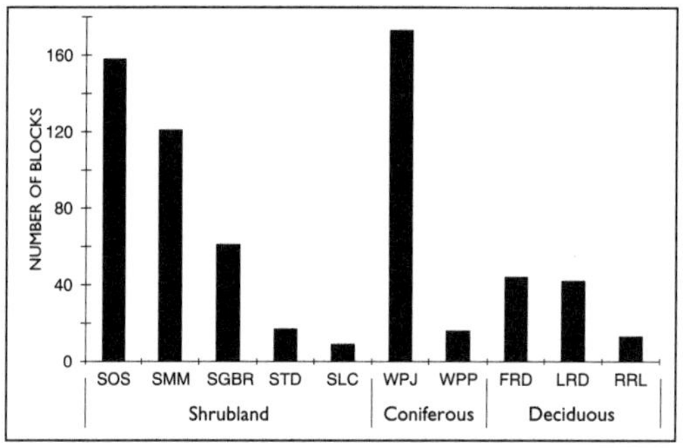
NUMBER OF BLOCKS
160
120
80
40
0
SOS
SMM
SGBR
STD
SLC
WPJ
WPP
FRD
LRD
RRL
Shrubland
Coniferous
Deciduous

Eastern Towhees as a breeding species needs documentation by specimens (Jon Greenlaw pers. comm.). The extensive collection of the two towhee types in the Denver Museum of Natural History does not include any birds from northeastern Colorado with eastern plumage affinities.

B&N located the center of abundance for Spotted Towhees in the scrub oak country where the Upper Sonoran joins the Transition zone. A&R also indicated the importance of scrub oak and added shrubby pinyon/juniper woodlands and riparian thickets as important habitats. The emphasis on scrub oak over pinyon/juniper derives more from the location of observers than from the preferences of the birds.

Atlas records confirm the significance of scrub oak, but they also suggest that earlier accounts have understated pinyon/juniper's significance. In Colorado, "shrubby pinyon/juniper" probably comes closest to defining the most frequently used habitat, although these birds also use scrub oak where pinyon/juniper does not grow. Atlasers also recorded sagebrush in 12% of the blocks, and Atlas work reveals that Spotted Towhees also make considerable use of lowland and foothills riparian habitats, another 12% of the total. The riparian habitats that they frequent have a strong component of shrubs.

The densest populations of Spotted Towhees reside between 5,000 and 7,000 feet (1525–2135 m; A&R). Around 7,000 feet (2135 m), where scrub oak stands closer to aspen than to pinyon/juniper, Green-tailed Towhees generally outnumber Spotted Towhees. Also, Spotted Towhees spurn broad homogeneous sagebrush stands where Green-tails frequently nest, although Spotted Towhees do use the taller Great Basin sagebrush growing in drainages and riparian zones (RL).

BREEDING

Spotted Towhees (in California) raise two (occasionally three) broods (Greenlaw 1996b). With birds on nests from mid May to late July, evidence of double-brooding in Colorado remains circumstantial (Jon Greenlaw pers. comm.). The clutch usually contains 3 or 4 eggs, incubation takes 12–14 days, and fledglings leave the nest after another 9–11 days (Greenlaw 1996b).

The birds build their nests on the ground beneath a bush, usually away from the taller shrubs where the male sings. When approached too closely, they sneak away on the ground in a manner suggesting a small mammal (Haws and Hayward *in* Bent 1968).

Atlasers found only 15 nests, the dates falling into a six-week range: 12 May–20 June. These dates appear to define a rather narrow window of nesting for birds that supposedly raise two or even three broods, but the dates recorded for fledglings suggest multiple broods. Juvenile towhees wear plumage easily distinguished from the adult plumage, and atlasers Confirmed breeding in 128 blocks by finding fledglings—60% of all Confirmations. Dates ranged from 28 May to 25 August, dates that could easily encompass more than one brood by a single pair. Spotted Towhee fledglings often beg noisily for food, and they respond readily to pishing; the ease of Confirming this species by finding fledglings made it unnecessary for atlasers to look for nests once youngsters became mobile.

The abundance codes reported for this species form a neat bell-shaped curve with 19 reports of A1, 94 of A2, 188 of A3, 98 of A4, and 16 of A5. Spotted Towhees have a low incidence of parasitism by Brown-headed

Cowbirds (Greenlaw 1996b), with only two Colorado records (Chace and Cruz 1996; Appendix C, Rich Bunn pers. comm.).

BREEDING PHENOLOGY

CODE		# OF RECORDS	DATE RANGE
C	Courtship	1	29 May
NB	Nest Building	7	27 May–24 Jun
ON	Occupied Nest	4	12 May–2 Jun
NE	Nest with Eggs	7	3 Jun–19 Jun
NY	Nest with Young	5	15 Jun–20 Jun
FY	Feeding Young	61	29 May–25 Aug
FL	Fledged Young	127	28 May–16 Aug

DISTRIBUTION

Spotted Towhees breed from southern Canada to the Mexican highlands, between the Pacific coast and the eastern flanks of the Rockies. The range extends east to the western Dakotas and western Nebraska (Greenlaw 1996b).

The Atlas map shows that in Colorado the range forms two north–south stripes, one along the foothills and a broader one along the western border. These areas coincide closely with the combined ranges of pinyon/juniper woodlands and scrub oak on Colorado's vegetation map (except north of Denver, where montane shrubs replace scrub oak).

Eastern Towhees breed from southeastern Manitoba and southern Maine to the Gulf coast and west to Iowa, eastern Kansas, and Louisiana. A "zone of introgression" occurs in the central Dakotas and throughout Nebraska, up the Platte to northeastern Colorado; in Colorado Spotted Towhees dominate the gene pool (Greenlaw 1996a).

Atlasers Confirmed breeding of presumed Eastern Towhees in four blocks in Logan, Sedgwick, and Phillips counties (Dave Hallock pers. comm.) and saw territorial behavior in at least one of the Yuma County blocks (HEK). Despite field marks and songs pointing to Eastern Towhees, confirmed breeding by that species awaits more definitive proof.

Along the Arkansas downstream from Pueblo, atlasers did not observe any towhees after mid May, and neither towhee breeds along the Arkansas in Kansas (Greenlaw 1996a, 1996b).

Plains rivers mostly lack shrubby components, which in part accounts for the distribution gap between the two towhees on the plains. A comparison of the plains distribution of Spotted and Eastern towhees with another shrub-obligate, the Lazuli Bunting, shows the buntings with a much greater presence on the plains. Either towhees and buntings have different habitat requirements or some other, unknown, factor inhibits a connection between the two towhee species.

BREEDING EVIDENCE

REPORTED IN 471 (27%) OF 1745 PRIORITY BLOCKS

☐	Possible	124 blocks	(26%)
▩	Probable	149 blocks	(32%)
■	Confirmed	198 blocks	(42%)

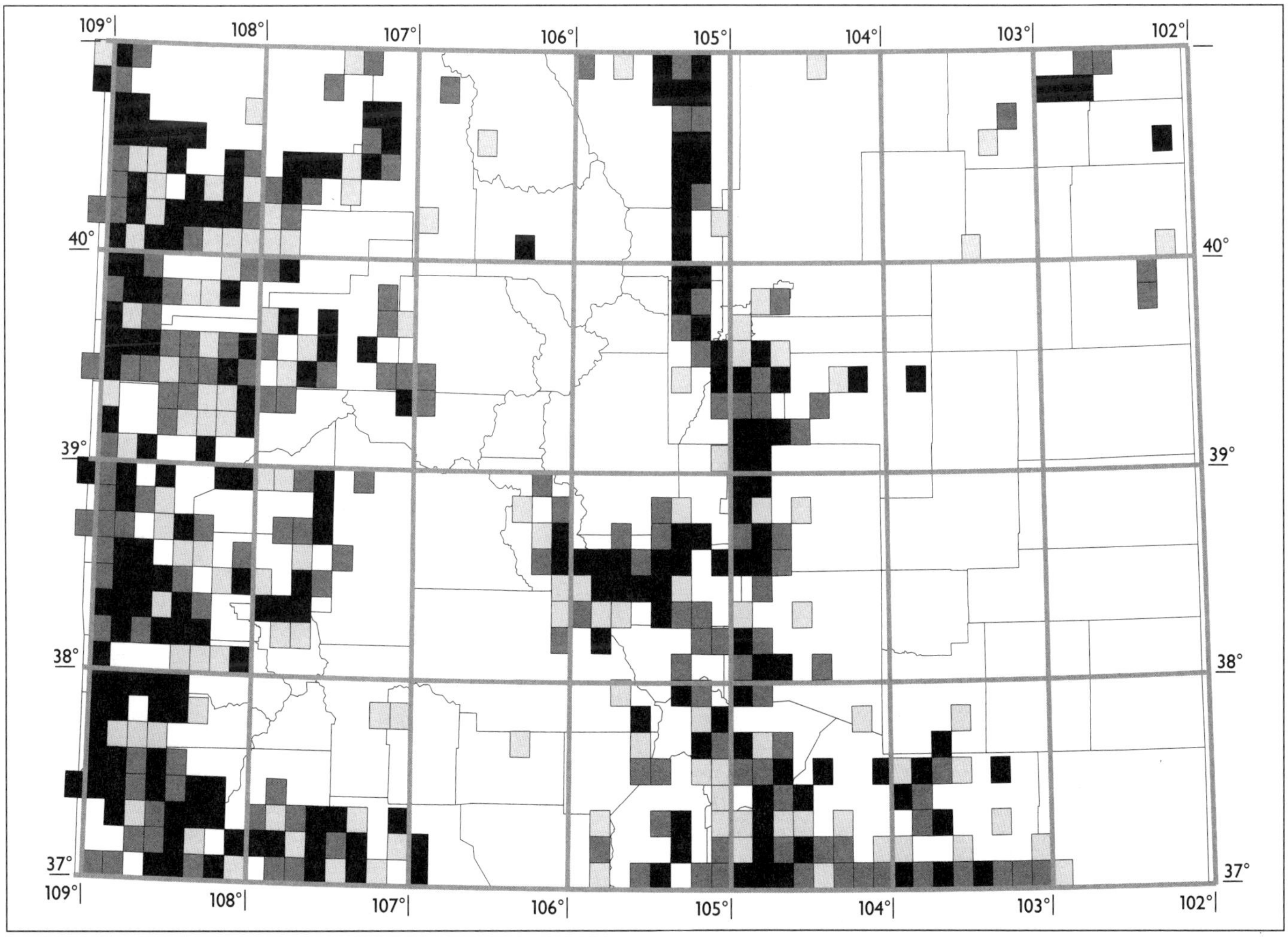
109°
108°
107°
106°
105°
104°
103°
102°
40°
39°
38°
37°

COLORADO DIVISION OF WILDLIFE / WILDLIFE RESOURCE INFORMATION SYSTEM

Some describe Canyon Towhees as quiet, unobtrusive, and secretive, yet others consider them the friendliest and most trusting of the towhees (Ligon 1961, Marshall 1960). In areas far from human influence towhees portray a shy and retiring manner; yet around human habitations they tame easily and venture boldly around yards and buildings.

HABITAT

One of Colorado's regional specialties, Canyon Towhees inhabit pinyon/juniper woodlands from the southern Front Range foothills to the canyons, arroyos, and mesas of southeastern Colorado. Although primarily associated with pinyon/juniper, they also occupy grasslands with cholla cactus, shrublands, sagebrush, and riparian woodland, and they occasionally frequent the vegetation around farm and ranch buildings (A&R, B&N, Sutton 1967). Many of these habitats flank rocky slopes or outcrops at the bases of cliffs or canyon walls. These terrestrial foragers prefer to feed in open areas adjacent to escape cover (Marshall and Johnson *in* Bent 1968).

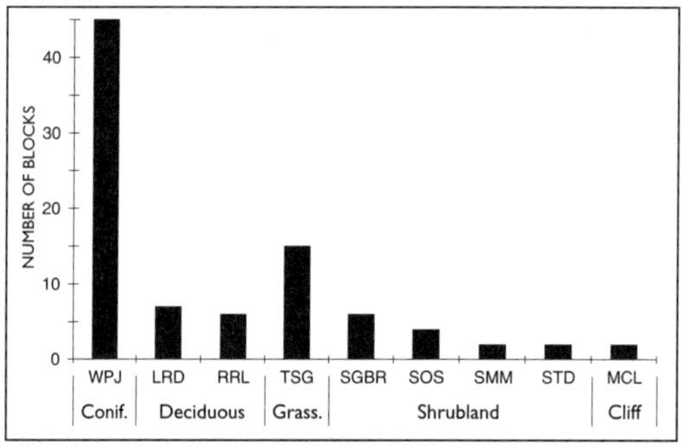
NUMBER OF BLOCKS
0
10
20
30
40
WPJ
LRD
RRL
TSG
SGBR
SOS
SMM
STD
MCL
Conif.
Deciduous
Grass.
Shrubland
Cliff

Atlas results reflect the affinity that Canyon Towhees show for pinyon/juniper. Nearly half of all the observations and over half of the breeding Confirmations occurred in pinyon/juniper. Cactus grasslands came in a distant second.

BREEDING

Spring arrives early in towhee country. As residents that maintain year-round territories, towhees get a head start on nesting—probably well before most atlasers departed for the field. The nesting season probably extends further than even the broad range of dates that Atlas records show. In Pueblo County, Percival (1992) reported a nest with three young from 20 to 30 April. In Oklahoma, Sutton (1967) discovered a recently completed nest on 2 September. This extended season suggests that towhees raise two and possibly even three broods (Bendire 1890, Ligon 1961, Sutton 1967).

Canyon Towhees often build their nests in the lower reaches of juniper trees (B&N). Other sites include pinyon pine, cholla cactus, sagebrush, yucca, ornamental plants, porches, and outbuildings (Bailey 1928, B&N, Ligon 1961, Sutton 1967). One Atlas record described a nest 6 feet (1.8 m) up on the rafter of an abandoned house. Towhees lay three to four eggs in a bulky nest of twigs, coarse grass, and forbs (Ligon 1961, Marshall and Johnson *in* Bent 1968).

This species suffers little cowbird parasitism. No reports surfaced during the Atlas nor in the research of Colorado records by Chace and Cruz (1996). In Oklahoma, Sutton (1967) examined over 30 nests, all free of cowbird eggs or chicks.

BREEDING PHENOLOGY

CODE		# OF RECORDS	DATE RANGE
NB	Nest Building	5	10 May–3 Jul
ON	Occupied Nest	1	15 May
NE	Nest with Eggs	7	23 May–27 Jun
NY	Nest with Young	1	7 Jun
FY	Feeding Young	11	29 May–14 Jul
FL	Fledged Young	6	1 Jun–7 Aug

DISTRIBUTION

Canyon Towhees live year round in dry semi-desert and scrub habitats from west-central Arizona to southeastern Colorado, and south through western Texas and Mexico. In Colorado their range closely aligns with the distribution of pinyon/juniper east of the Rocky Mountains (Davis 1951). Their range has changed little since early accounts when Cary (1909) described them as common south of the Arkansas River and east of the Sangre de Cristo Range except for the treeless stretches in Prowers and Baca counties. Canyon Towhees occurred sparingly north of Pueblo, Fremont, and El Paso counties (B&N).

With few exceptions, the distribution of Atlas records follows the range outlined in A&R. Records extend from western Baca County northwest into Fremont County, with the majority of sightings in central and eastern Las Animas County. The highest reported densities (11–100 pairs/block) occur in Las Animas, Otero, Bent, and Baca counties. To the northwest in Pueblo, Huerfano, and

Fremont counties, the distribution of towhees becomes patchy with lower average densities reported (2–10 pairs/block).

Canyon Towhee sightings occurred in two new locations: in two blocks on the southwestern flank of the Wet Mountains, and in one block near Antonito in the San Luis Valley. The only prior record from the west side of the Wet Mountains came from an historical account by Lowe (1894) who reported that the species occurred sparingly at 10,000 feet (3050 m). This report was eventually discounted (Davis 1951). In the San Luis Valley, Ryder (1965) reported on a pair of towhees 23 miles west of Antonito in August 1959. An out-of-range Atlas report came from north of Cortez in Montezuma County, not included on the Atlas map. Several undocumented sightings exist from the San Luis Valley and Western Slope (CFO, DFO, var.). These sightings suggest seasonal wandering or disjunct populations and warrant further examination. Atlasers did not locate Canyon Towhees in Boulder County, where during the 1960s to mid 1980s they appeared periodically near Lyons (A&R).

Few data exist for this species in Colorado, with much of our knowledge coming from historical studies in adjacent states. Further searching in areas with isolated sightings will help establish just where these birds occur.

BREEDING EVIDENCE

REPORTED IN 77 (4%) OF 1745 PRIORITY BLOCKS

Possible	27 blocks	(35%)
Probable	18 blocks	(23%)
Confirmed	32 blocks	(42%)

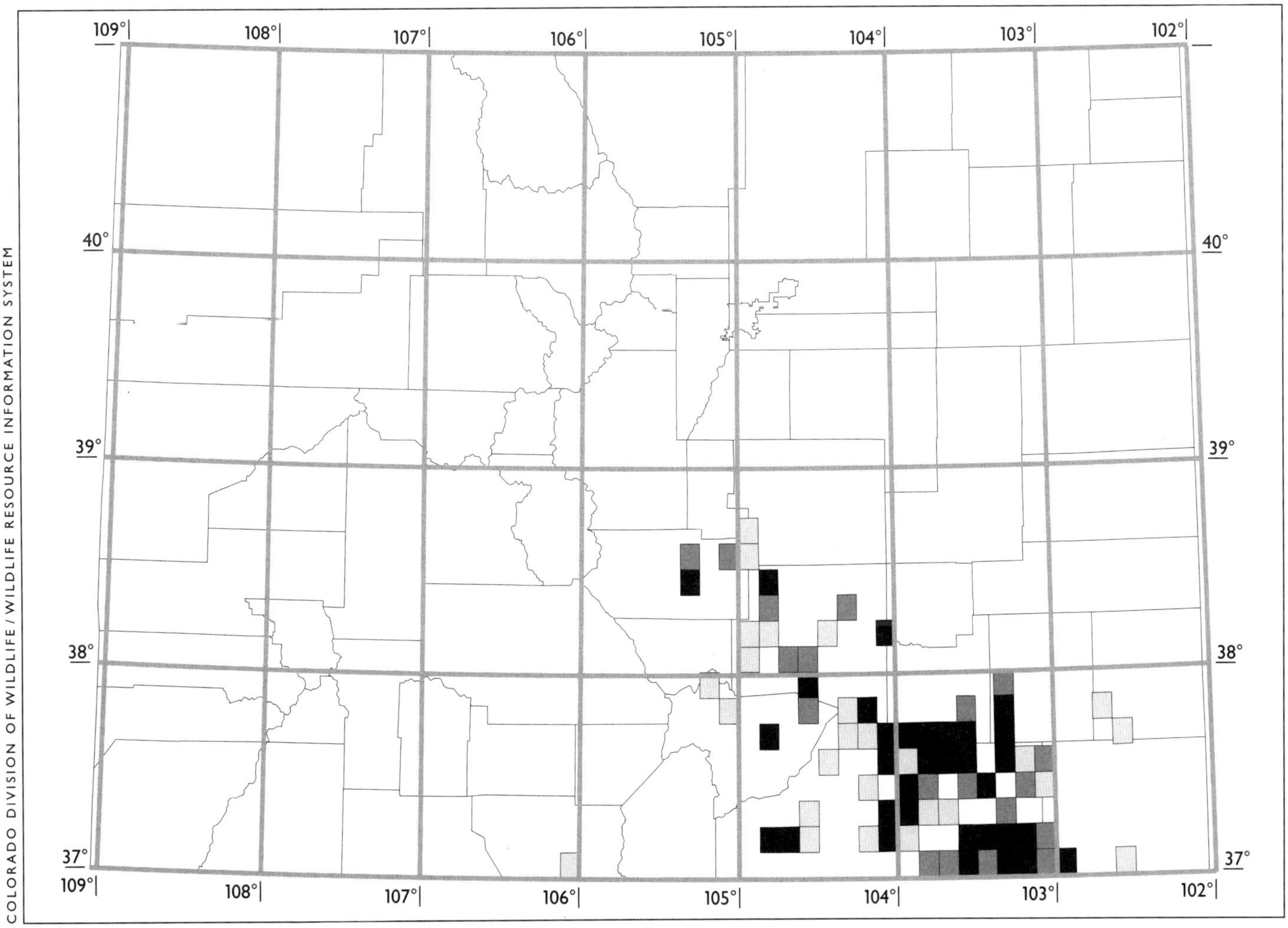
109°
108°
107°
106°
105°
104°
103°
102°
40°
39°
38°
37°

COLORADO DIVISION OF WILDLIFE / WILDLIFE RESOURCE INFORMATION SYSTEM

Cassin's Sparrows cloak themselves with a top-of-the-line camouflage, which makes them very difficult to see or to identify during most of the year. During the breeding season, however, everything changes. The territorial male will perch in full view on his lookout post, then launch himself into a conspicuous, fluttering flight while spilling out the sharps and flats of his haunting, unmistakable song.

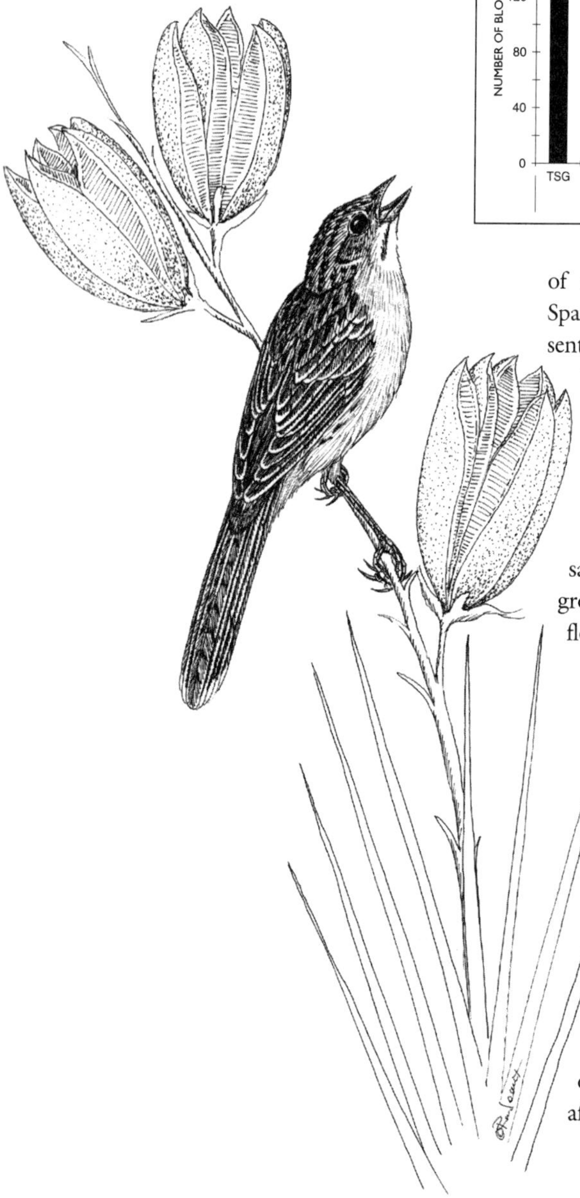

HABITAT

Cassin's Sparrows inhabit various open, arid grasslands and shortgrass prairies dotted with sandsage, yuccas, shrubs, cacti, and mesquites (Bent 1968, B&N, A&R). They use the taller plants as lookout posts and singing perches and avoid grasslands without them. Ironically their density generally corresponds negatively to the density of woody vegetation. When the right mix of habitat components and climatic conditions do come together, high local densities can give the impression of colonial nesting (Bent 1968). Atlas blocks, however, could not provide the level of resolution necessary either to corroborate or to refute this.

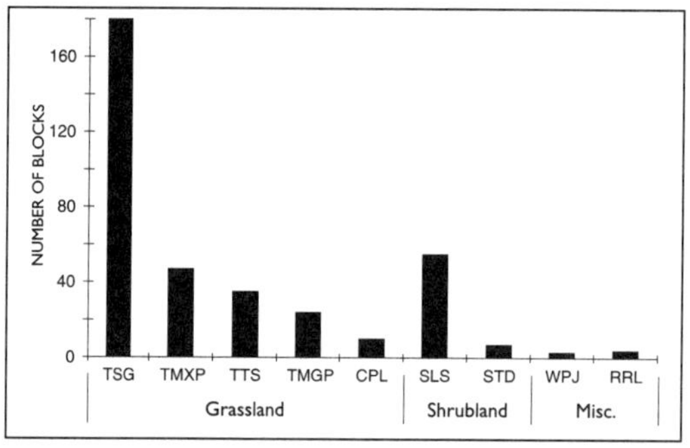
NUMBER OF BLOCKS
160
120
80
40
0
TSG
TMXP
TTS
TMGP
CPL
SLS
STD
WPJ
RRL
Grassland
Shrubland
Misc.

Shortgrass prairie composed nearly 50% of all Atlas habitat records for Cassin's Sparrows, and sandsage shrublands represented another 25%. Typically, the birds build their nests on the ground at the base of a shrub, yucca, cactus, clump of grass, or up to one foot (0.3 m) above the ground in a shrub or cactus (Bent 1968). The only Atlas nest card for this species reported a typical nest site in sandsage at 4.8 inches (12.2 cm) from the ground. Although nesting Cassin's often flee at the slightest sign of disturbance (Bent 1968), most observers still had trouble finding their well-concealed nests, precluding further nest-site details.

BREEDING

Cassin's Sparrows begin arriving in Colorado around mid April (A&R). With only two exceptions (17, 26 April), however, atlasers did not detect any until the second half of May. The tendency of Cassin's to delay singing and other courtship activities after arriving on the breeding grounds (Bent 1968) obscures their true arrival pattern. Once they do wind up, they sing incessantly, often all night. Courtship includes conspicuous larking flight displays and territorial disputes that often end with vigorous chasing and calling (Bent 1968). Despite these behaviors, only 20% of all Atlas records resulted in breeding Confirmations, most likely due to the birds' otherwise cryptic existence. Unmated males do not skylark, so that song flights usually evidence, at the least, paired birds (Bock and Scharf 1994).

Reported nesting dates span 1 March through August (Oberholser 1974, Maurer et al. 1989). Atlas data fell well within that range, the majority occurring from early June through mid July. Egg dates also fell between the extreme dates (16 May, 22 July) that others have reported for the western plains region (Kingery and Julian 1971, Johnsgard 1979). Although double-brooding in Cassin's Sparrows remains unconfirmed, Johnsgard (1979) surmised from nesting dates that they double-brood in Texas. The late Atlas fledgling dates of 4 and 10 August, and the observation of an apparently dependent fledgling on 15 August (Dorn and Dorn 1995), also suggest the possibility of second broods in Colorado and Wyoming respectively.

They have an opportunistic nesting strategy, very likely keyed to rainfall because it stimulates both insects and seeds in arid grasslands (Phillips 1944, Maurer et al. 1989, Carl Bock pers. comm.). A&R stated the reverse: that Cassin's do not nest when rainfall promotes growth of tall grass. Although well adapted to arid regions, Cassin's Sparrows do require insects (especially grasshoppers) to feed hungry young, and many insects emerge only when stimulated by moisture. In Arizona, the species' breeding season coincides with the July and August monsoons (Maurer et al. 1989).

Some Cassin's may also undergo a dispersal in the middle of a breeding season to raise a second brood elsewhere if conditions become unfavorably dry (Phillips 1944, Hubbard 1977). Other researchers believe instead that the birds simply do not become detectable until favorable breeding conditions develop (Carl Bock pers. comm.). Overall, they seem highly responsive to climatic variations, as reflected in their patchy, irregular breeding distribution.

BREEDING PHENOLOGY

CODE		# OF RECORDS	DATE RANGE
C	Courtship	11	22 May–17 Jul
NB	Nest Building	4	11 Jun–28 Jun
ON	Occupied Nest	2	26 May–8 Jun
NE	Nest with Eggs	5	27 May–22 Jun
NY	Nest with Young	5	8 Jun–18 Jul
FY	Feeding Young	29	2 Jun–15 Jul
FL	Fledged Young	19	21 May–10 Aug

DISTRIBUTION

Cassin's Sparrows range from southeastern Arizona, eastern Colorado, and southwestern Nebraska into northern Mexico. Cooke (1909) listed the first Colorado record near Springfield on 27 May 1895, and expressed surprise when they nested as far north as Barr Lake in 1907. In 1993, Dorn and Dorn (1995) confirmed breeding in southeastern Wyoming.

Normally, these birds stay well east of the mountains (B&N, A&R) at elevations of under 5,200 feet (1585 m). Atlas records of Cassin's concentrated in southeastern Colorado, but many also nested in the northeast.

Cassin's Sparrows belong to the ranks of infrequent host species for Brown-headed Cowbirds (Friedmann 1963, Kingery and Julian 1971, Chace and Cruz 1996). Instead, their conservation problems stem from agricultural and ranching practices, particularly "clean farming," removing brush from pasturelands (Oberholser 1974), and grazing (Bock and Bock 1988). Livestock grazing has highly negative impacts, possibly the "major reason" for the apparent national decline (Carl Bock pers. comm.).

BBS data indicate a long-term major decline in the U.S., but the nomadic and cyclic nature of Cassin's breeding may affect BBS results. Numbers *detected* fluctuate greatly from year to year; non-singing birds have such secretive habits that observers may miss them unless local climate conditions stimulate the larking song displays of these otherwise secretive sparrows. On one BBS route in New Mexico with large numbers, the count swung from 16 to 210 to 29 over six years, and on another, from 60 one year to zero the next (Sandy Williams pers. comm.).

Nonetheless, Colorado contains a possible 20% of the U.S. breeding range of this species, and with the alarming long-term BBS trend, Colorado conservation priorities list this sparrow high among Colorado species needing management attention.

BREEDING EVIDENCE

REPORTED IN 336 (19%) OF 1745 PRIORITY BLOCKS

☐	Possible	99 blocks	(29%)
▩	Probable	170 blocks	(51%)
■	Confirmed	67 blocks	(20%)

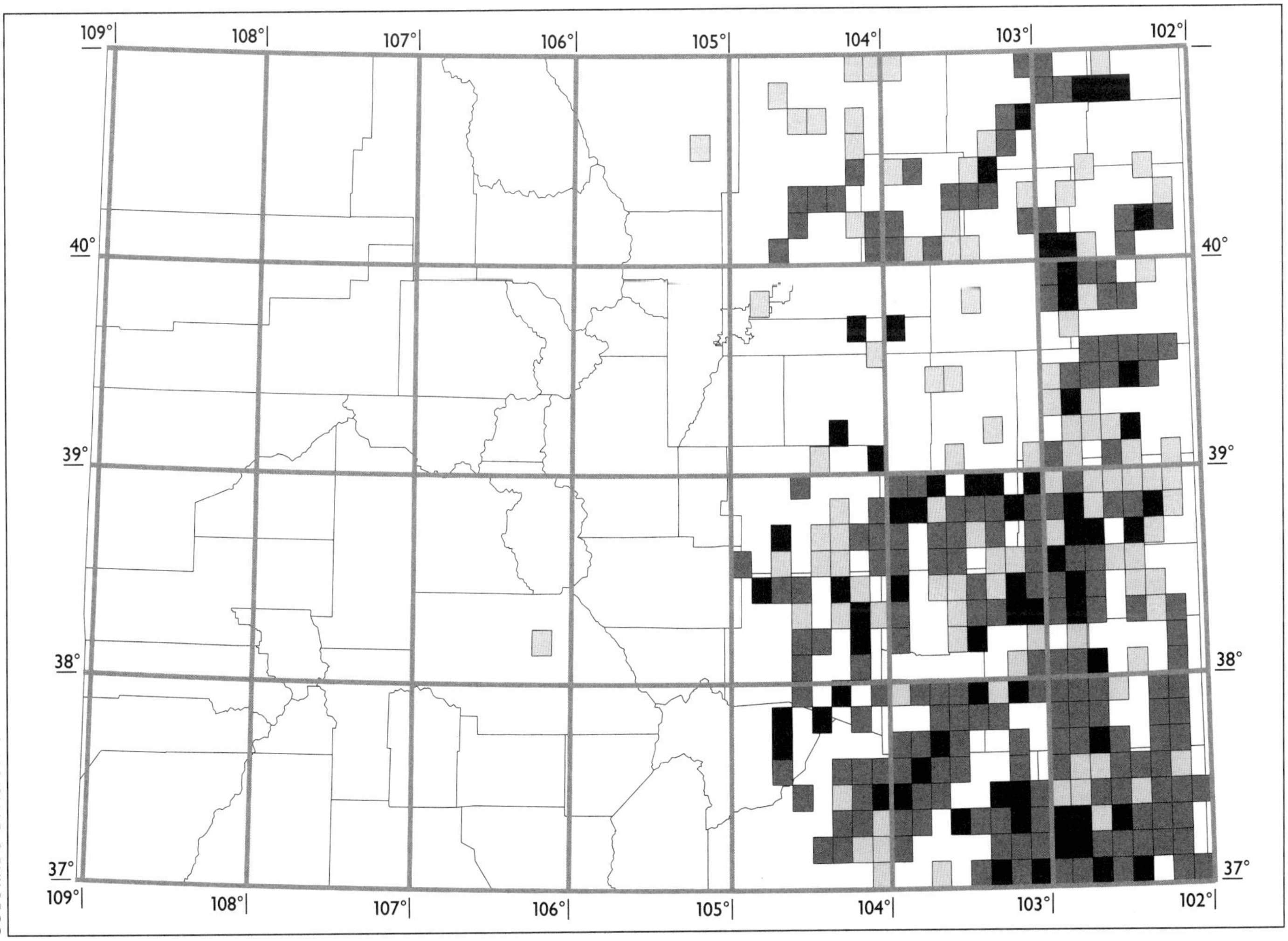
109°
108°
107°
106°
105°
104°
103°
102°
40°
39°
38°
37°

COLORADO DIVISION OF WILDLIFE / WILDLIFE RESOURCE INFORMATION SYSTEM

A parched, rocky hillside, where scrub oaks and skunkbrush grow knee-high and the grass turns brown by early June, must seem like nirvana to a Rufous-crowned Sparrow. In southeastern Colorado these secretive seed-eaters nest almost exclusively in shrubby, rocky sites on canyon slopes or on the margins of pinyon/juniper woodland.

HABITAT

Throughout their range, Rufous-crowned Sparrows nest in mixed shrub habitats, often on grassy and rocky hillsides (Ligon 1961, Pulich 1988, Shuford 1993). Colorado atlasers found them in similar habitats within pinyon/juniper woodland, scrub oak woodland, mixed shrubland, and shortgrass prairie. According to Atlas field worker estimates, nesting populations exceeded 100 pairs in only one block, Cobert Mesa North (37103A5), in southern Las Animas County. Here the sparrows nested in scrub oak and mixed shrub thickets on moderately grazed canyon slopes.

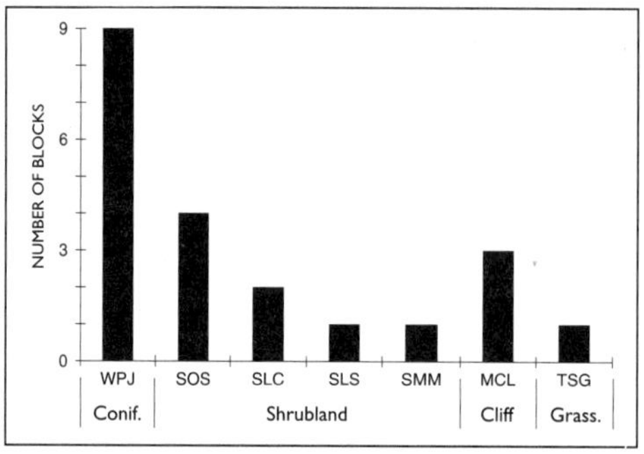
NUMBER OF BLOCKS
0
3
6
9
WPJ
SOS
SLC
SLS
SMM
MCL
TSG
Conif.
Shrubland
Cliff
Grass.

Shuford (1993) characterized Rufous-crowned Sparrows as "short distance colonizers" that may invade areas where fire or other disturbances have opened up the shrub cover. In southeastern Colorado, they frequently occupy heavily grazed canyons and hillsides. In some areas, grazing and trampling by cattle may benefit Rufous-crowned Sparrows by opening up existing shrub habitat; in others, grazing and trampling reduce the availability of shrubs. These sparrows' affinity for steep, rocky hillsides suggests that they gravitate toward areas intermediate between the heavily grazed canyon bottoms and the ungrazed rimrock.

BREEDING

Rufous-crowned Sparrows nest in a depression on the ground, often at the base of a small shrub or tuft of grass, or up to three feet off the ground in low shrubs (Bent 1968). They feed primarily on grass and forb seeds, fresh grass stems, and tender plant shoots. They supplement this vegetable diet with various arthropods, especially during the breeding season (Beal 1910).

Atlas field workers confirmed breeding in only four of 15 Priority blocks in which they found Rufous-crowned Sparrows. A nest with eggs in Dennis Canyon, Las Animas County (37103A3), found on 27 May, lay in a hollow on the ground on a rocky, south-facing slope covered with short grasses and small shrubs (SRJ). A nest in Tubs Springs (37102A7), Baca County, observed on 11 June, lay at the base of a 4-foot-high scrub oak on the edge of a relatively dense canyon-bottom thicket (SRJ). [The two nests found by the author are apparently the first actual nests discovered in Colorado—Ed.] Atlasers observed adults feeding young in Rock Canyon (37103F3) on 18 June and in Riley Canyon (37103F5) in July.

The low Confirmation rate for Rufous-crowned Sparrows reflects their secretiveness. Pulich (1988) described these sparrows as "quite wary and often heard but not seen," and noted that individuals will respond to squeaking or to owl calls. Ligon (1961), writing about Rufous-crowned Sparrow behavior in New Mexico, said, "Its pleasing and distinctive song . . . directs attention to a nearby bush, where one may glimpse the reddish-crowned songster, only to have him vanish and presently be heard some distance away." During Atlas surveys in southeastern Colorado, the skittishness of singing males frequently confounded efforts to locate and to Confirm active nests.

BREEDING PHENOLOGY			
CODE		# OF RECORDS	DATE RANGE
ON	Occupied Nest	1	11 Jun
NE	Nest with Eggs	1	27 May
FY	Feeding Young	1	18 Jun

DISTRIBUTION

Rufous-crowned Sparrows inhabit semi-desert country from coastal California inland along the Sierra foothills, eastward across Arizona and New Mexico to southeastern Colorado, and southward to Oaxaca in Mexico. Otero, Bent, Las Animas, and Baca counties lie at the northeastern tip of their North American breeding range. Their winter range closely parallels their breeding range.

Sclater (1912) did not include this species among Colorado's birds. Bailey and Niedrach (1965) noted that although no nest records existed for Colorado, several specimens had been collected in the canyons of Las Animas and Baca counties. A&R characterized the Rufous-crowned Sparrow as an "uncommon resident" of Baca and Las Animas counties, and the Latilong tudy (1988) reported breeding in these two counties. The string of Atlas records along the Purgatoire River north into Otero and Bent counties extends the known Colorado breeding range northward by about 50 miles (80 km).

Rufous-crowned Sparrow population dynamics have received little attention from biologists.

BREEDING EVIDENCE		
REPORTED IN 15 (1%) OF 1745 PRIORITY BLOCKS		
Possible	5 blocks	(36%)
Probable	6 blocks	(43%)
Confirmed	4 blocks	(21%)

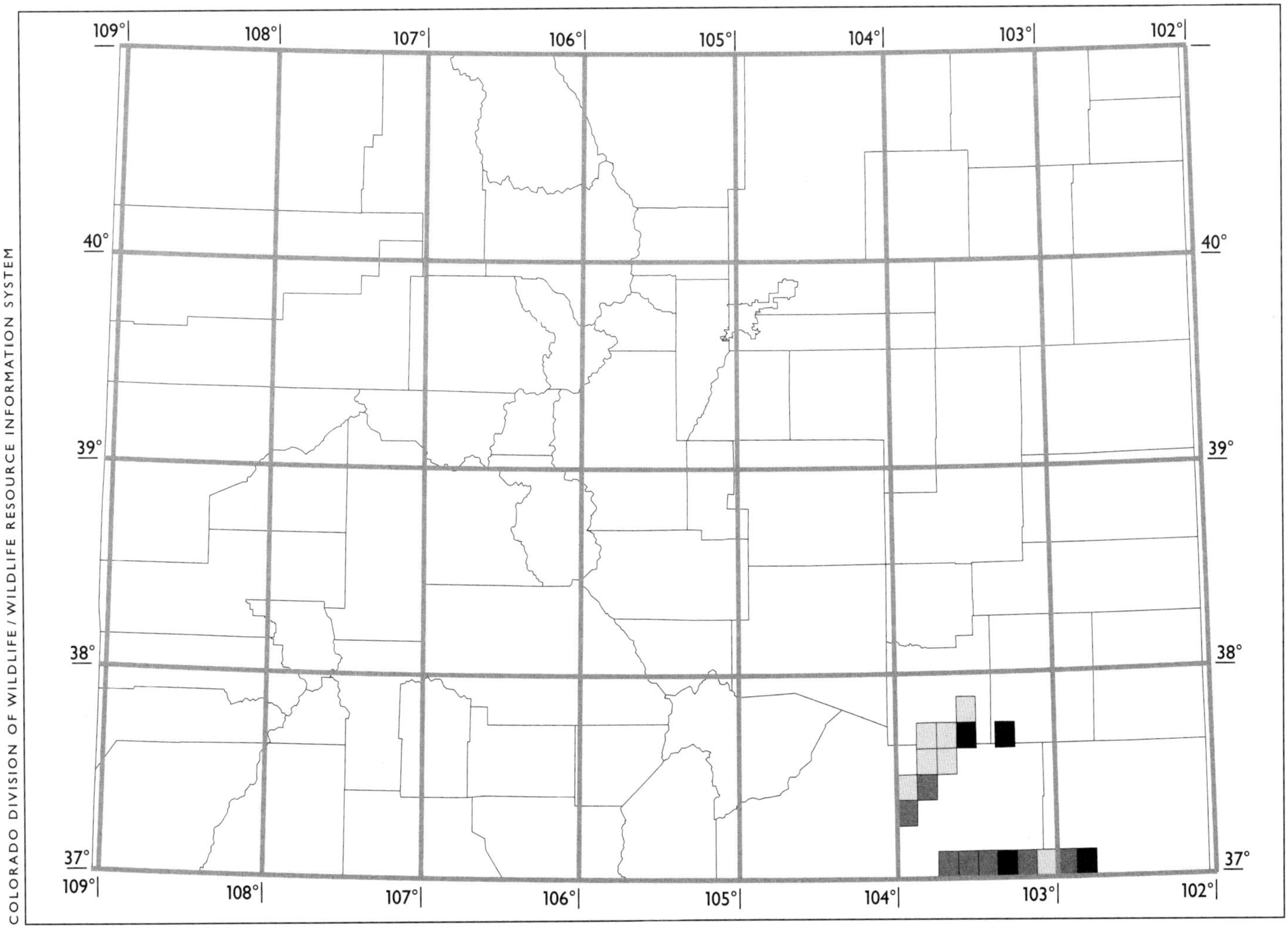
109°
108°
107°
106°
105°
104°
103°
102°
40°
39°
38°
37°
COLORADO DIVISION OF WILDLIFE / WILDLIFE RESOURCE INFORMATION SYSTEM

Hearty, if not melodious, singers, Chipping Sparrows enliven nearly all of Colorado's woodland and shrubland habitats. Among true sparrows, only Lark Sparrows can rival the distribution of Chipping Sparrows within our state.

HABITAT

Although breeding on neither prairie nor tundra, Chipping Sparrows make themselves at home in most naturally occurring habitats between these two extremes of Colorado's topography. Elsewhere in North America they readily adapt to parks, gardens, and residential areas, but Colorado birds show a marked preference for less disturbed locales. Coniferous woodlands, especially pinyon/juniper and ponderosa pine, support, by far, the greatest numbers. Even where especially dry conditions exist in pinyon/juniper woodlands, atlasers regularly recorded densities of 11–100 breeding pairs per block.

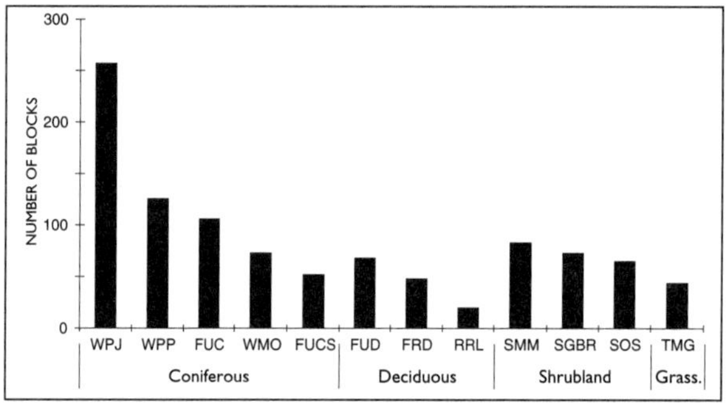
300
200
100
0
NUMBER OF BLOCKS
WPJ
WPP
FUC
WMO
FUCS
FUD
FRD
RRL
SMM
SGBR
SOS
TMG
Coniferous
Deciduous
Shrubland
Grass.

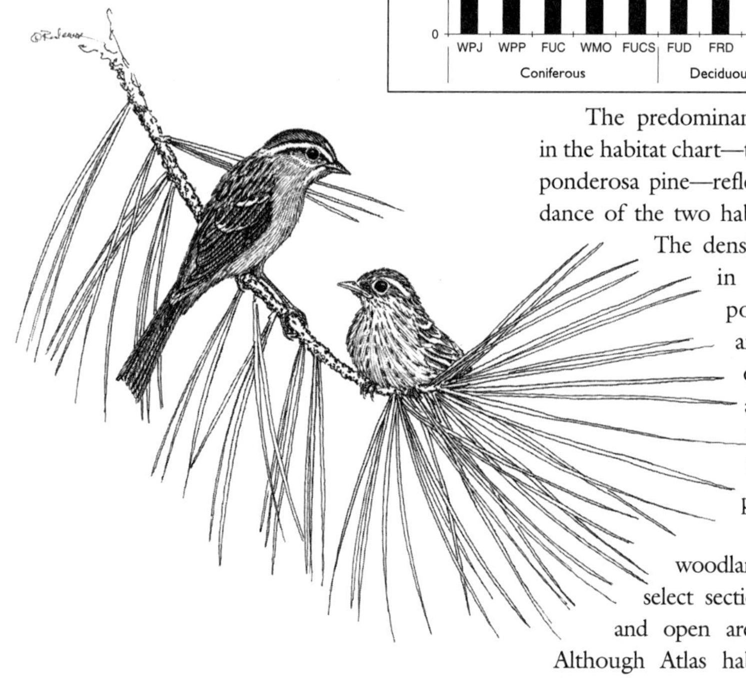

The predominance of pinyon/juniper in the habitat chart—twice as many blocks as ponderosa pine—reflects the relative abundance of the two habitats within the state. The densities of breeding pairs in pinyon/juniper and ponderosa woodlands are about equal; atlasers estimated A4 and A5 abundance codes in 36% of pinyon/juniper blocks and 28% of ponderosa blocks.

In pinyon/juniper woodlands Chipping Sparrows select sections with smaller trees and open areas (Sedgwick 1987). Although Atlas habitat codes make no reference to tree size, the four nest cards recorded by field workers—all in pinyon/juniper—provide anecdotal support for this thesis. In each, the birds situated the nests in small trees or low vegetation in or at the edge of an open woodland.

Shrubland habitats also produced several Atlas records. Scrub oak, mountain sagebrush, serviceberry, and a number of other taller shrubs provide breeding habitat. Most of these records come from mixed shrubland/woodland habitats (usually pinyon/juniper, ponderosa pine, or Douglas-fir).

A tantalizing question raised by Atlas records concerns whether Chipping Sparrows breed in krummholz. A&R speculated, but lacked proof, that these birds occasionally breed at high elevations. Twice, field workers found fledglings in krummholz habitat at suggestive dates (30 June and 14 July). Still, whether the adults nested in the krummholz or brought the young there after fledging remains uncertain (Versaw 1995).

Chipping Sparrows formerly nested in the Denver parks (Niedrach and Rockwell 1939) but no longer do so. In eastern Colorado the Atlas map shows no Confirmed breeding below the level of ponderosa and pinyon pines, with one notable exception. The first (and only) Confirmed breeding record on the plains away from the mountains came from Red Lion SWA: on 18 July 1992 two adults fed three or four juveniles, completely dependent on the adults (John and Bill Prather VF) .

BREEDING

Chipping Sparrows return to their Colorado breeding grounds earlier than most other summer residents; they begin to arrive across Colorado in mid April (A&R, Dexter and Levad 1992).

Atlasers reported nest-building and several nests during May and early June. Almost all fledgling observations came during July; the first five observations of fledglings, from 17 May to 9 June, came from Latilong 37108 in the southwestern corner. The responsibilities of nest-building fall entirely on the female. Likewise, she assumes primary responsibility for incubating the eggs—typically four, frequently three (Harrison 1979). Chipping Sparrows often host cowbirds, substantiated by five Atlas reports of parasitism (Chace and Cruz 1996; Appendix C).

Related species with which Chipping Sparrows could compete include Spotted and Green-tailed towhees, Vesper and Lark sparrows, and Dark-eyed Juncos. Each forages primarily on the ground and each has some habitat overlap with Chipping Sparrows. Of these, the towhees and Vesper Sparrows employ double scratching, unlike Chipping Sparrows, to expose food items on the ground. This difference should limit the extent of competition. The larger Lark Sparrows tend to feed on larger insects than do Chipping Sparrows, although both consume large numbers of seeds (Ryser 1985). Chipping Sparrows and Dark-eyed Juncos have similar diets and methods of foraging, but juncos tend to inhabit forests and Chipping Sparrows incline toward woodlands and meadows for feeding. They also select different nest sites—juncos on the ground, Chipping Sparrows in trees or shrubs.

BREEDING PHENOLOGY

CODE		# OF RECORDS	DATE RANGE
C	Courtship	12	28 Apr–27 Jun
NB	Nest Building	25	12 May–6 Jul
ON	Occupied Nest	6	17 May–2 Jul
NE	Nest with Eggs	16	7 May–23 Jul
NY	Nest with Young	14	15 May–27 Jul
FY	Feeding Young	180	31 May–25 Aug
FL	Fledged Young	179	17 May–25 Aug

DISTRIBUTION

Chipping Sparrows breed south of the treeline throughout Canada and the U.S. except in the southern Great Plains. Some winter in southern Arizona, New Mexico, and Texas, and many continue on to Mexico.

The Atlas map places their Colorado breeding range almost entirely west of Interstate 25 (Longitude 105). Only in the Black Forest and the pinyon/juniper country east of Walsenburg and Trinidad do they make large-scale advances east of the highway. Several records of Possible and Probable breeding birds in northern Las Animas County represent an extension of range shown by A&R. Likewise, several Confirmations extend the known range into Dinosaur National Monument in the northwest. The three Possible plains records (Latilongs 37102 and 40102) were all birds observed in June.

Prior and subsequent to breeding, Chipping Sparrows pass through almost every square foot of the state. During post-breeding dispersal they even visit alpine tundra.

Although logging poses a threat to many species, Chipping Sparrows benefit from the open areas produced by logging, which increases the available foraging surface (Franzreb and Ohmart 1978).

The abundance of Chipping Sparrows within Colorado makes any threat seem insignificant. However, unabated conversion of woodlands into suburban-type settings will diminish the available habitat for this attractive sparrow and eventually reduce its numbers unless they can adapt to suburban living in Colorado the way they have in the East.

BREEDING EVIDENCE

REPORTED IN 777 (45%) OF 1745 PRIORITY BLOCKS

☐	Possible	237 blocks	(31%)
▪	Probable	135 blocks	(17%)
■	Confirmed	405 blocks	(52%)

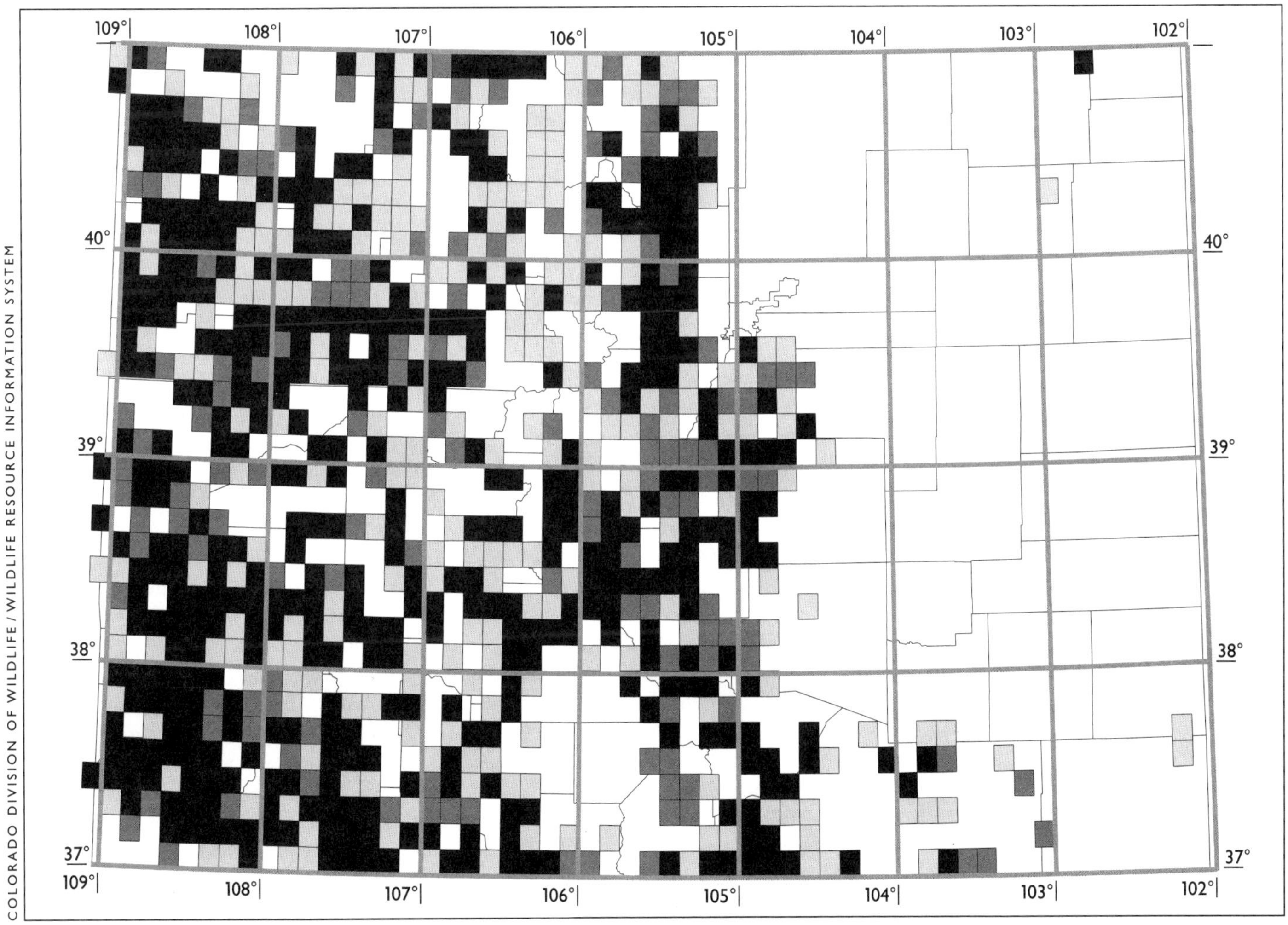
109°
108°
107°
106°
105°
104°
103°
102°
40°
39°
38°
37°
COLORADO DIVISION OF WILDLIFE / WILDLIFE RESOURCE INFORMATION SYSTEM

Physician and amateur ornithologist Thomas Mayo Brewer (1814–1880), having defended the introduction of the House Sparrow to North America, might not deserve the honor of having a *native* sparrow named after him. Nevertheless, the man surely earned the honor by liberally sharing data on bird biology and eggs with professional ornithologists such as John Cassin, who had the privilege in 1856 of tagging this little sparrow as Brewer's.

HABITAT

Brewer's Sparrows prefer sagebrush, primarily big sagebrush. Seventy-five percent of the Atlas Confirmations occurred in sagebrush, and many other habitats featured sagebrush. Ligon (1961) saw Brewer's Sparrows as "almost coextensive with rabbitbrush and big sage." Ignoring the timberline Brewer's Sparrow of Canada and Alaska, the breeding range roughly parallels the North American range of big sagebrush. Other shrub species that form similar stand characteristics, such as greasewood, hopsage, and saltbushes in the low country or mountain mahogany and snowberry higher up, may attract lesser densities of nesting Brewer's Sparrows (B&N, Bent 1968, Medin 1990, A&R).

At timberline in the Flat Tops Wilderness on 29 June 1988, Hugh and Urling Kingery detected two singing

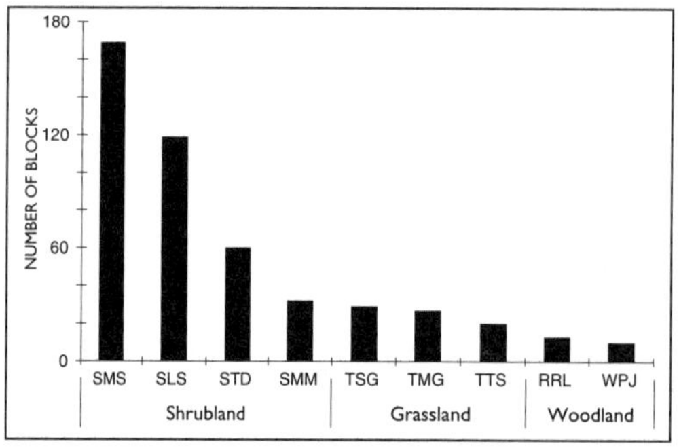
180
120
60
0
NUMBER OF BLOCKS
SMS
SLS
STD
SMM
TSG
TMG
TTS
RRL
WPJ
Shrubland
Grassland
Woodland

Brewer's Sparrows in 2–5-foot willows and on 3 August Confirmed breeding nearby by finding fledglings. Another Probable nester in willow habitat came from Taylor Park northeast of Gunnison, but which also has extensive high-country sagebrush. A proposed new species, the Timberline Sparrow (*Spizella taverneri*), occurs in southeastern Alaska and western Canada (Doyle 1997, AOU 1998). Conceivably these Colorado high-country birds could more closely resemble the Timberline Sparrow in vocalizations, ecology, and morphology and thereby belong to this potential species.

Habitat characteristics correlated with dense populations include a dominance of shrubs, little diversity in shrubs, more forbs, less grass, more bare ground, less duff, and relatively level ground with little rockiness (Short 1984). Atlas locations for Brewer's Sparrows indicate preference for sagebrush of middle heights. Moderate, incomplete burns in sagebrush, the kind typical of prescribed fires, do not harm important components of nesting habitat (Petersen and Best 1987).

In the sagebrush community Brewer's Sparrows take the foliage-feeding position, and Sage and Vesper Sparrows, their close associates, forage mostly on or from the ground (Wiens et al. 1990).

BREEDING

Brewer's Sparrows start to arrive in mid April, with full numbers at the end of the month (A&R, Atlas). They sing in migration and on the nesting ground. People describing their songs offer differing impressions: canary-like (Whetmore, 1964); cicada-like (Swarth *in* Bent 1968); a long, sorry-sweet song (Hyde 1979); having marked continuity and variety (Hartshorne in Bent 1968); and weak, monotonous, and not very noticeable (Pitelka *in* Bent 1968). It sounds to me like a speedy little handy man with a ratchet wrench. A field of them sounds like a busy shop of ratchet work; Bent refers to this as "flock trilling." These little workers in unbuttoned khaki overalls sing an astonishingly long spasm of trills and buzzes as they build and repair nesting territories.

After pairing they become quieter (Best and Petersen 1985). This variable, plus annual fluctuations in nesting schedules due to weather, complicates population monitoring of this species. Increased singing from delayed pairing or failed nesting might falsely indicate relatively high populations.

Brewer's Sparrows begin nesting late and nest only once a season, which would appear to put them at a competitive disadvantage (Reynolds 1981). However, of all the common shrub-steppe birds, Brewer's Sparrows seem best able to respond to conditions in an opportunistic fashion (Rotenberry and Wiens 1991; see also Dawson et al. 1979). In wetter years they have larger clutch sizes.

Each nest almost certainly will have hair as part of the lining, which typically protects 3–5 eggs (Harrison 1979). Brown-headed Cowbirds sometimes parasitize this sparrow's nesting attempts (Friedmann *in* Bent 1968), although Colorado has no records (Chace and Cruz 1996). An extended study in southwestern Idaho noted no cowbird parasitism, considerable predation, and that the previous winter's precipitation correlated most significantly with clutch and brood sizes and fledging success (Rotenberry and Wiens 1989).

After fledging, the Brewer's Sparrow family often moves to higher elevations, mixing with other species, especially Chipping Sparrows.

BREEDING PHENOLOGY

CODE		# OF RECORDS	DATE RANGE
C	Courtship	3	28 May–11 Jun
NB	Nest Building	7	17 May–23 Jun
ON	Occupied Nest	6	11 Jun–22 Jun
NE	Nest with Eggs	12	27 May–9 Jul
NY	Nest with Young	4	20 Jun–24 Jul
FY	Feeding Young	62	5 Jun–31 Jul
FL	Fledged Young	74	31 May–6 Aug

DISTRIBUTION

The winter range of Brewer's Sparrows spans from Death Valley to west-central Texas, and south through western Mexico to Jalisco and Guanajuato (AOU 1983). Immediately north of this, their sagebrush summer range reaches into the three western provinces of Canada.

Confirmations of this sagebrush-obligate concentrated in Colorado's greatest sagebrush counties, Moffat, Rio Blanco, Jackson, and Gunnison. Differences emerge east and west of the 106th longitude. East of this division only four of 29 nesting Confirmations occurred in sagebrush, suggesting marginal nesting habitat. These 29 Confirmations constitute only 21% of the blocks on the east side of the state with Brewer's Sparrows; by contrast, west of the line atlasers Confirmed nesting in 42% of the blocks. Apparently, low bird numbers played no part in the disparity in Confirmation rates; Atlas observers reported only slightly fewer birds in the occupied eastern blocks. The difficulty in Confirming nesting could have resulted from lower nesting success; eastern Colorado may function as a population sink for this species.

Brewer's Sparrows, usually the most abundant bird on their nesting grounds, reach densities of 370–740 birds/mi^2 (150–300 birds/km^2), and that number can exceed 1,200 (500) if they face reduced competition from Sage Thrashers or little threat from Loggerhead Shrikes (Reynolds 1981, Rotenberry and Wiens 1989). BBS data show a large declining trend over the whole Brewer's Sparrow breeding range—the annual drop of 3.7% converts to a 31-year drop since 1966 of over 60% of the total population. Nobody seems to know why, so we can hope that this sparrow simply has a naturally long population cycle.

BREEDING EVIDENCE

REPORTED IN 424 (24%) OF 1745 PRIORITY BLOCKS

☐	Possible	181 blocks	(43%)
■	Probable	78 blocks	(18%)
■	Confirmed	165 blocks	(39%)

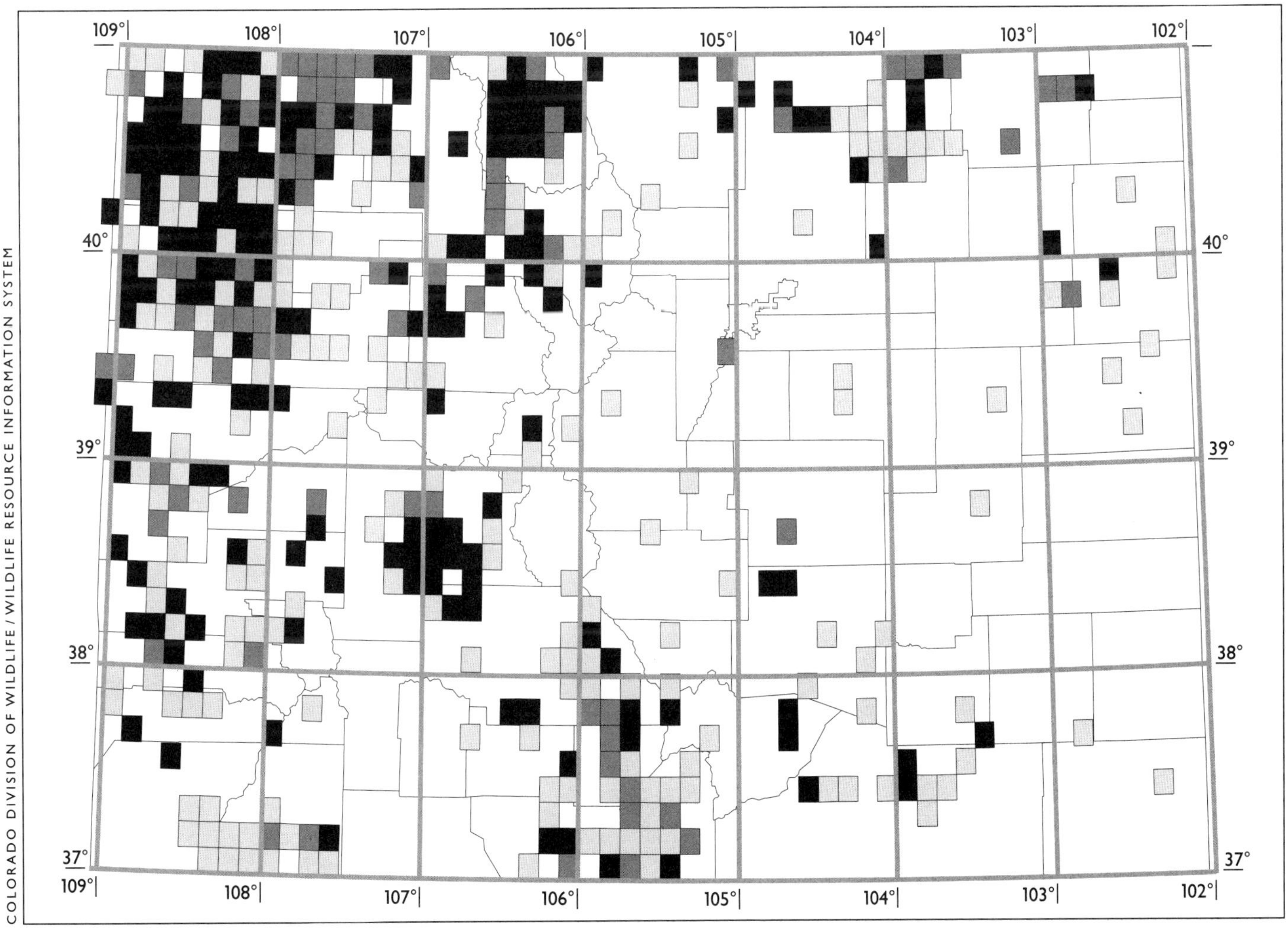
109°
108°
107°
106°
105°
104°
103°
102°
40°
39°
38°
37°

COLORADO DIVISION OF WILDLIFE / WILDLIFE RESOURCE INFORMATION SYSTEM

The haunting song of the Field Sparrow, with its clear notes accelerating and ascending slightly to finish in a trill, is rarely heard in Colorado. From the East, their fields turning to shrublands and then to forests, a few pairs have drifted westward along shrubby riparian corridors. As if to acknowledge their eastern roots, most Colorado Field Sparrows breed within sight of Kansas or Nebraska. Unlike other recent colonizers like Blue Jays, starlings, and Common Grackles, Colorado Field Sparrows show little affinity for city life. Instead they share their few haunts with critters almost unknown in the rest of Colorado—chiggers.

HABITAT

Field Sparrows require tallgrass meadows for nesting and foraging, and shrubs or other small trees and perches for singing. Their preferred grassland habitats typically contain uncommonly tall grass (Terres 1980). Tall grass and forbs usually indicate little to no sustained grazing by cattle.

In the eight years of the Atlas, field workers found Field Sparrows in only six Priority blocks—all within 20 miles of the state line. Atlasers used five different habitat codes: one grassland code, two for shrubs, one rural, and only LRD (which includes tamarisk thickets) more than once. This suggests that they use habitats both uncommon in Colorado and not easily categorized.

In northeastern Colorado, the birds use small cottonwoods and shrubby thickets composed of three-leaf sumac (skunkbrush) and wild plum, sometimes with scattered, skinny hackberry trees, as the preferred woody habitat. In one Priority block on the lower Arkansas (Holly West, 38102A2), all nesting activity occurred in tall tamarisks (which have invaded the disturbed river bottoms), a habitat mostly lacking in biological diversity.

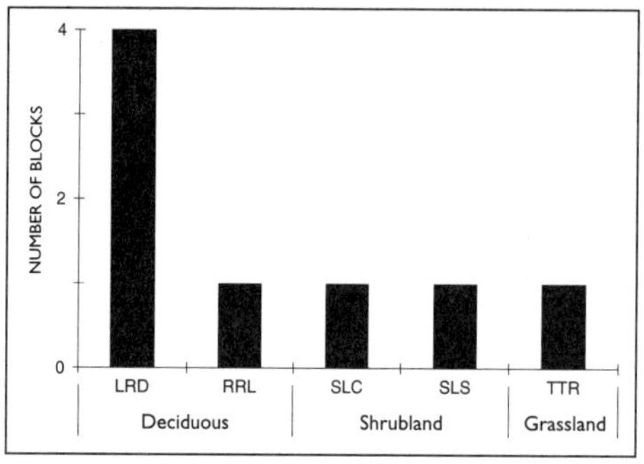
NUMBER OF BLOCKS
4
2
0
LRD
RRL
SLC
SLS
TTR
Deciduous
Shrubland
Grassland

BREEDING

Atlas field work produced the first confirmation of Colorado nesting on 6 June 1990, when Bill Prather found a nest with eggs in a small shrub at Tamarack Ranch SWA (40102G6), still the only confirmation of Colorado nesting. In 1991 Prather found a pair acting secretively in a brush patch in the same area; the female repeatedly returned to the same bush but he did not find nest or young (Bill Prather VF).

The sparrows place their nests either on the ground or up to 5 feet (1.5 m) high in a shrub or sapling. In the East, nest height above ground varies with time of the year. June nests average 6 inches (15 cm) above the ground, July nests 13 inches (34 cm), and August nests 17 inches (43 cm). This probably corresponds to the height and stage of maturation of the surrounding grass and forbs (Welty 1982).

Field Sparrows make several nesting attempts per season, depending on the brood success or failure. One female laid 28 eggs in ten nesting attempts in one season (Carey et al. 1994). Inasmuch as atlasers heard singing males and saw pairs in Colorado in July (DLN, HEK), they possibly could have been setting up territories for second or third nesting attempts.

BREEDING PHENOLOGY

CODE		# OF RECORDS	DATE RANGE
NE	Nest with Eggs	1	6 Jun

DISTRIBUTION

Field Sparrows breed in the central and eastern U.S., north into southern Ontario, west to eastern Montana, and south to central Oklahoma and Texas. Colorado Atlas reports came only from the easternmost parts of the South Platte, Republican, and Arkansas rivers.

The Holly West block had the most Field Sparrows. From the tamarisk thickets at least ten singing males proclaimed their presence. Six miles to the east, Hamet SWA, on the state line (Holly East, 38102A1), had almost as many (DLN). The sparrows may have a continuous population along the lower Arkansas River for at least 10 miles (16 km) west from the Kansas line.

BBS data show a large decrease of Field Sparrows across the continent (over 3% per year). Probably the transformation of old farm fields into forest has diminished available habitat. Also, as a frequent cowbird host, Field Sparrows suffer from nest parasitism. Although most of the colonizing Field Sparrows have moved into state wildlife areas, these areas have cattle grazing nearby (and cowbirds achieve abundance without the benefit of four-legged grazers). The glut of cowbirds could threaten their tenuous breeding foothold in Colorado.

BREEDING EVIDENCE

REPORTED IN 5 (0%) OF 1745 PRIORITY BLOCKS

Possible	1 block	(20%)
Probable	4 blocks	(80%)
Confirmed	0 blocks	(0%)

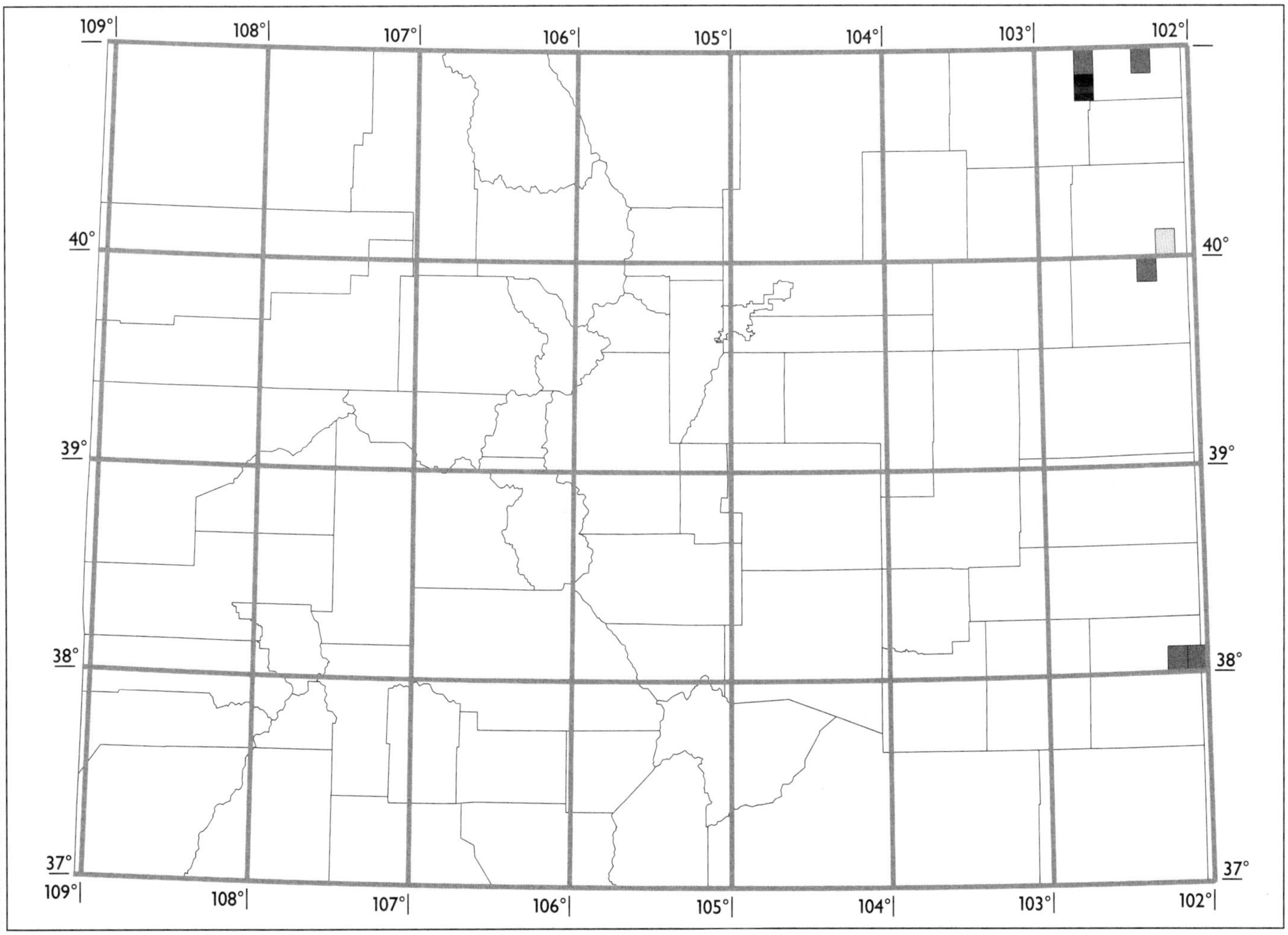
109°
108°
107°
106°
105°
104°
103°
102°
40°
39°
38°
37°

COLORADO DIVISION OF WILDLIFE / WILDLIFE RESOURCE INFORMATION SYSTEM

At some time of year Vesper Sparrows make an appearance in almost every county in the conterminous U.S. and also much of Canada and Mexico. They sing very well, and do so in most of those counties. The face, if seen often enough, becomes unique, not so plain or nondescript as it first seems. Some of birding's fondest acquaintances come this way, born of subtle looks rather than gorgeous plumage.

HABITAT

Despite their vast itinerary, Vesper Sparrows seek a narrow set of habitat conditions, especially within the nesting range. Both parts of the scientific name, the Greek *Pooecetes* and the Latin *gramineus,* refer to their grass-inhabiting nature. Before the 1880s people called them "Grass Finches" (Ligon 1961).

Atlas data show that in Colorado the densest populations occur in middle- to higher-elevation sagebrush. The Atlas also shows that montane grasslands support high population densities, as does lower-elevation sagebrush in northwestern Colorado. Sparsely or patchily distributed shrubs with a good cover of grasses make the best habitat (Schaid et al. 1983). Scattered pinyon or ponderosa pines, or junipers, or corn,

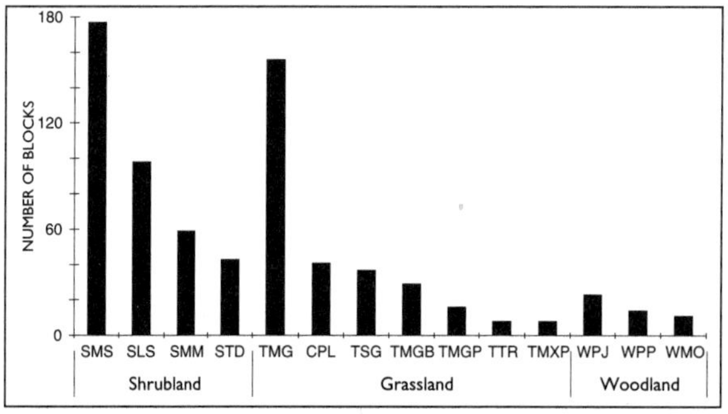

180
120
60
0
NUMBER OF BLOCKS
SMS SLS SMM STD TMG CPL TSG TMGB TMGP TTR TMXP WPJ WPP WMO
Shrubland
Grassland
Woodland

especially where volunteering in a fallow field, also provide requisite structural complexity for some nesting (Best and Rodenhouse 1984, Perritt and Best 1989).

The Atlas affirmed nesting activities in all these land-cover types, plus shortgrass, mountain shrub, riparian areas, seeded rangeland, aspen, clear-cuts, timberline, and alpine habitats. Recognize, however, that most Confirmations mean that Vesper Sparrows include these habitats in their nesting territories, but not necessarily as nest sites or, in the alpine, dispersal after fledging.

Tall woody plant structures provide strategic perches and merely incidental nest concealment (Castrale 1982). Petersen and Best (1987) report cases of Vesper Sparrows colonizing sagebrush sites a few years after fire and herbicide treatments. They speculate that these ground nesters arrive only after some critical accumulation of herbaceous cover. Conversely, too much ground cover can accumulate; on more lush, tallgrass sites, moderate livestock grazing improves Vesper Sparrow habitat (Kantrud and Kologiski 1983, Bock et al. 1992). Atlas workers and others (Bent 1968) observe that Vesper Sparrows seem especially fond of dust baths, which should make dirt roads and cow paths important habitat features. Perhaps European settlement expanded Vesper Sparrow habitat in Colorado.

BREEDING

Displaying male Vesper Sparrows strut and run with wings raised and tail feathers spread; they rise up in a short flight song (Bent 1968). This species reportedly can simultaneously sing separate nonharmonic notes from each side of its syrinx, an exhibit of the "two-voice phenomenon" (Ritchison 1980). A mnemonic depiction of the classic song goes, "too too, tee tee, *chup-chup-chup-chup*-cheat-a-little-eat-a-little *chip*." Frequently the singer omits the italicized part of the above transliteration, and some Vesper Sparrows completely reorganize the phrasing and change the emphasis.

The nest, its thick rim sunken to ground level, cups 3–6 whitish or pale greenish-white eggs variably dotted, blotched, and scrawled with browns and gray (Harrison 1979). Vesper Sparrows have two broods per nesting season (Bent 1968, Ehrlich et al. 1988). A food manipulation study of Vesper Sparrows showed that in most years food supply has no effect on nesting success (Adams et al. 1994). Failures most often occurred because of predation (see also Vickery et al. 1992).

BREEDING PHENOLOGY

CODE		# OF RECORDS	DATE RANGE
C	Courtship	3	11 May–28 May
NB	Nest Building	6	25 May–6 Jun
ON	Occupied Nest	7	12 May–22 Jun
NE	Nest with Eggs	26	24 May–23 Jul
NY	Nest with Young	13	30 May–19 Jul
FY	Feeding Young	94	25 May–14 Aug
FL	Fledged Young	118	4 Jun–24 Aug

DISTRIBUTION

Vesper Sparrows breed from southern Canada to the Appalachian Mountains, Ohio River, and much of the West. Colorado has a few winter records, but most Vesper Sparrows winter no closer than southern New Mexico (AOU 1983, A&R).

Migrant Vesper Sparrows spread all across the state (A&R). In Gunnison flocks often arrive in early spring snowstorms, mingling with Horned Larks and taking advantage of the dark strip along roads where snow melts as fast as it falls (Hyde 1979). On the plains during April and May atlasers found legions of Vesper Sparrows singing their trill-laden songs, but the sparrows do not stay there to breed. Most plains migrants probably head north, as others move directly into the mountains and western plateaus and valleys where Atlas data portray a sparrow widely distributed in grasslands and shrublands.

They have a greater presence in central Colorado than previously realized, with solid populations in South Park and the Gunnison Basin. Vacant spots on the map occur in the high country where conifer forests dominate, particularly in the San Juans, Elks, Sawatch, Gores, and Front Range.

The Atlas map, showing the grassy plains vacant for a species associated with grasslands, looks like a mistake of a reversed east-to-west image. A missing vegetative component in the plains doesn't easily explain the almost complete evacuation by this bird. The reason may lie in the solar heat budget of the nesting season. Vesper Sparrows seem to retreat from lower latitudes, elevations, and ground cover as the advancing season's temperature rises. Eastern Colorado temperatures may simply remain too warm too long. The lower deserts of west-central Colorado see the same retreat of Vesper Sparrows as the gaps in nesting Confirmations indicate. In North Dakota in 1988, hot weather caused the birds to abandon nests during incubation and cut short the nesting season (George et al. 1992).

Camp and Best (1995) calculated a nest density of one per 1.45 acres (1.5/ha) on Iowa roadside habitats. The High Atlas abundance estimates on several sites suggests comparable or greater Vesper Sparrow densities in Colorado. BBS data show a small decrease across the continent but greater drops in the East. Eastern atlases report diminishing numbers, due to woodlands supplanting abandoned farmlands (e.g., Cadman et al. 1987, Brauning 1992, Robbins 1996). Farm and natural vegetation relationships differ in the West. Although woodlands will not displace shrublands and grasslands here, other more subtle changes can impact these essential components of Vesper Sparrow breeding habitat. This should warn westerners that even these sparrows bear monitoring, simply because of their abundance in these habitats.

BREEDING EVIDENCE

REPORTED IN 599 (34%) OF 1745 PRIORITY BLOCKS

☐	Possible	215 blocks	(36%)
■	Probable	119 blocks	(20%)
■	Confirmed	265 blocks	(44%)

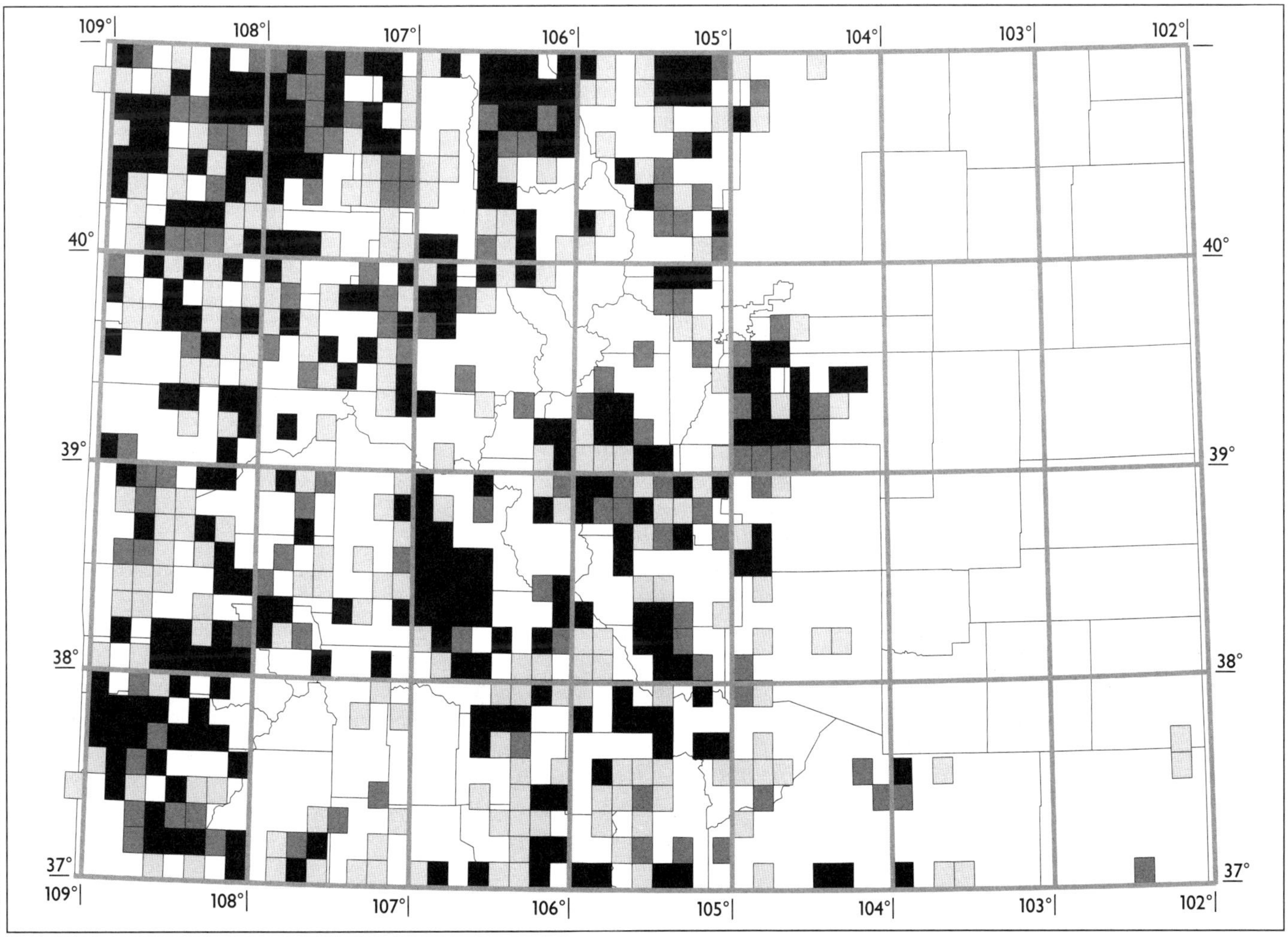
109°
108°
107°
106°
105°
104°
103°
102°
40°
39°
38°
37°

COLORADO DIVISION OF WILDLIFE / WILDLIFE RESOURCE INFORMATION SYSTEM

LARK SPARROW

Chondestes grammacus

The prince of sparrows sings a long medley of sweet warbles and trills, interspersing it every so often with rude and embarrassing snorts. Its namers chose to overlook this disquieting thing and called it "Lark Sparrow." The exuberant bird sometimes does break into a larking flight song.

HABITAT

Lark Sparrows demand two characteristics for nesting habitat: open views and a variety of vegetation heights. Scattered woody vegetation with herbs and grasses of low to modest density provide the required physiognomy (Renwald 1977). As the Atlas map shows, the birds mostly avoid the mountains. Suitable habitat where atlasers found the birds include grasslands with junipers, greasewood, yucca, galletagrass, and western wheatgrass, and on the plains, grasslands within cottonwood stands and the cholla-studded shortgrass prairie. Good habitat may include pastures with scattered trees and shrubs. Scattered rock outcrops projecting a meter or two high can also help to create the required physical arrangement (Rumble 1987).

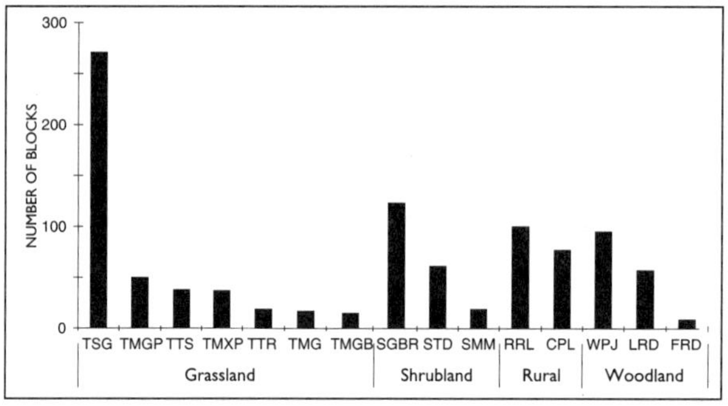
300
200
100
0
NUMBER OF BLOCKS
TSG TMGP TTS TMXP TTR TMG TMGB SGBR STD SMM RRL CPL WPJ LRD FRD
Grassland
Shrubland
Rural
Woodland

Lark Sparrow habitat usually improves with moderate-intensity fires and livestock grazing on more productive ground. The same events on drier lands can quickly diminish suitability of habitat (Bock et al. 1992, Renwald 1977).

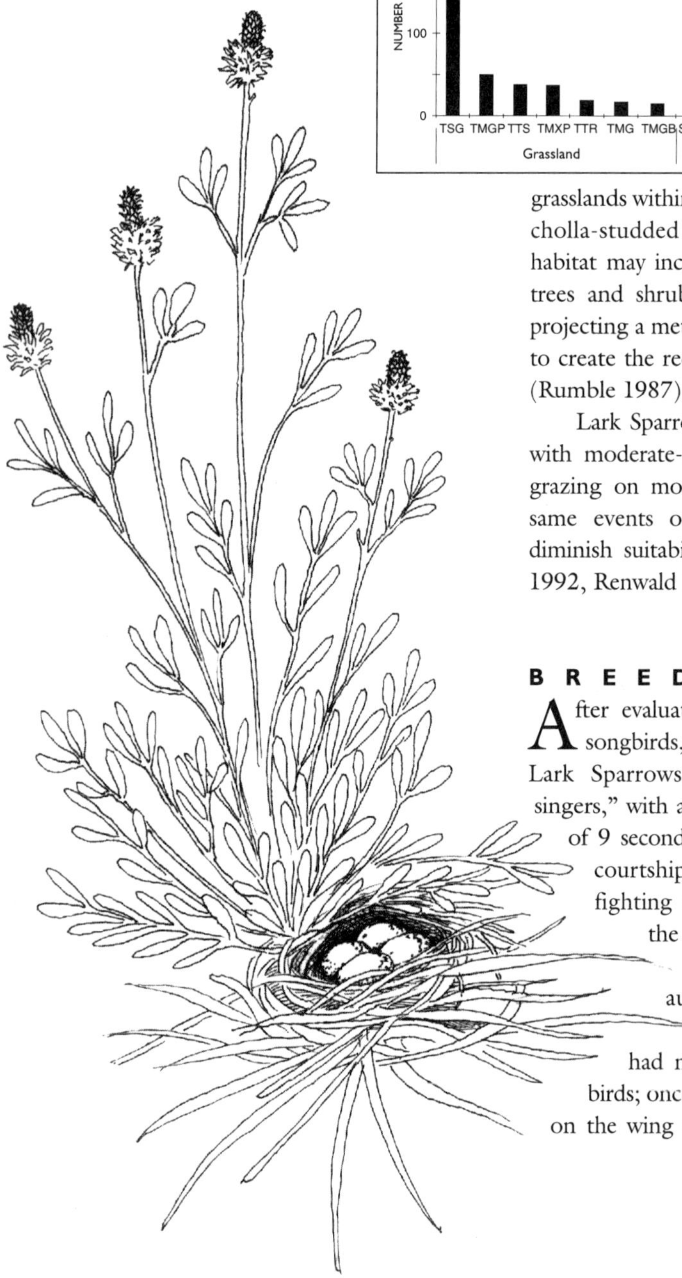

BREEDING

After evaluating several of the world's songbirds, Hartshorne (1973) placed Lark Sparrows among the "very good singers," with a sing-time/pause-time ratio of 9 seconds/10 seconds. Their songs, courtship strutting, chasing, and fighting make them conspicuous in the spring by late April or early May (B&N, A&R). One author (Cameron *in* Bent 1969) wrote that he had never seen such pugnacious birds; once five or six of them fighting on the wing almost hit him in the face.

Fall flocks, sometimes up to 20, may perch on a fence and suddenly burst into song (Wetmore 1964).

Although Lark Sparrows arrive on the nesting ground later than many migrant sparrows, Atlas Confirmations begin before mid May, slightly earlier than most sparrows. The lengthy nesting period continues into mid August; Atlas observers saw young birds being fed as late as mid August, which proved typical for sparrows.

A Lark Sparrow will most often build a nest in a ground depression. Eight of the nine nests described by Atlas observers were on the ground hidden by vegetation. The other nest sat in a depression below ground level under a large boulder. Atlas observers Confirmed nesting in 52% of the Priority blocks where they detected them. That 26% of the Confirmations were nest finds testifies to the relative ease of finding Lark Sparrow nests.

Lark Sparrows reuse their own nests and, at least in Oklahoma and Texas, will use nests of other birds and add their own linings (McNair 1984). To do this they accept a great variety of settings, including rock and tree cavities. In Oklahoma, Sutton (1967) reported a pair of Lark Sparrows feeding not only their own young, but also the young of mockingbirds in another nest not far away. Two earlier records from B&N and four Atlas records from Las Animas, Phillips, and Yuma counties show that Lark Sparrows also foster cowbirds (Chace and Cruz 1996; Appendix C).

The behavioral profile of this ruddy-faced sparrow—long, varied, persistent song with larking flight; migrant-group singing (Wetmore 1964); desperate fighting; ostentatious court-ship displays including ritual passing of a twig (Bent 1968); trespass use of other birds' nests and feeding others' nestlings—reveals a bird of extreme temperament.

BREEDING PHENOLOGY

CODE		# OF RECORDS	DATE RANGE
C	Courtship	23	26 Apr–16 Jul
NB	Nest Building	23	17 May–22 Jun
ON	Occupied Nest	9	29 May–9 Jul
NE	Nest with Eggs	36	10 May–15 Jul
NY	Nest with Young	13	31 May–10 Jul
FY	Feeding Young	181	27 May–10 Aug
FL	Fledged Young	146	19 May–19 Aug

DISTRIBUTION

Unlike many other Neotropical migrants, Lark Sparrows winter across a range as wide as their summer range, which stretches from the west side of the Appalachians to the east side of the coastal ranges and Sierra Nevadas. They retreat in the fall to the Atlantic, Pacific, and Gulf coasts southward through the Caribbean and Mexico to the northern Central American countries. The winter range looks as maritime as the summer range appears to avoid the coasts. The zone of compromise, where these sparrows occur year round in the interior and along the coasts, lies in California, northern Mexico, and Texas.

East of the 94th Meridian, in midwestern America, numbers of nesting Lark Sparrows have declined severely this century (Buckelew and Hall 1994, Brauning 1992, Peterjohn and Rice 1991, Brewer et al. 1991) probably due to the reduced need for horse pasture (Pearson 1936). More recently and for unknown reasons, populations have dipped in the arid West, where unused pasture often does not revert to dense shrubs and trees or become cropland (Dobkin 1994). BBS data on this sparrow depict a large decrease across the continent, a drop of 3.2%/year.

Still common, Lark Sparrows can hardly go unnoticed during migration anywhere in Colorado except in the higher and forested parts of the state. The blank sections of this sparrow's Atlas map approximate the outline of Colorado's high country. Atlas data verify and refine earlier publications in revealing the highest numbers and pervasiveness of Lark Sparrows in the southeastern part of the state (B&N, A&R). The Atlas discovered some localized populations such as those in southern Hinsdale and Gunnison counties. These singers are largely absent from the intermountain parks from North Park to the San Luis Valley. The Atlas changed the Latilong Study status of Lark Sparrows to "Breeder" in all seven latilongs that had not already obtained that status.

BREEDING EVIDENCE

REPORTED IN 771 (44%) OF 1745 PRIORITY BLOCKS

Possible	186 blocks	(24%)
Probable	163 blocks	(21%)
Confirmed	422 blocks	(55%)

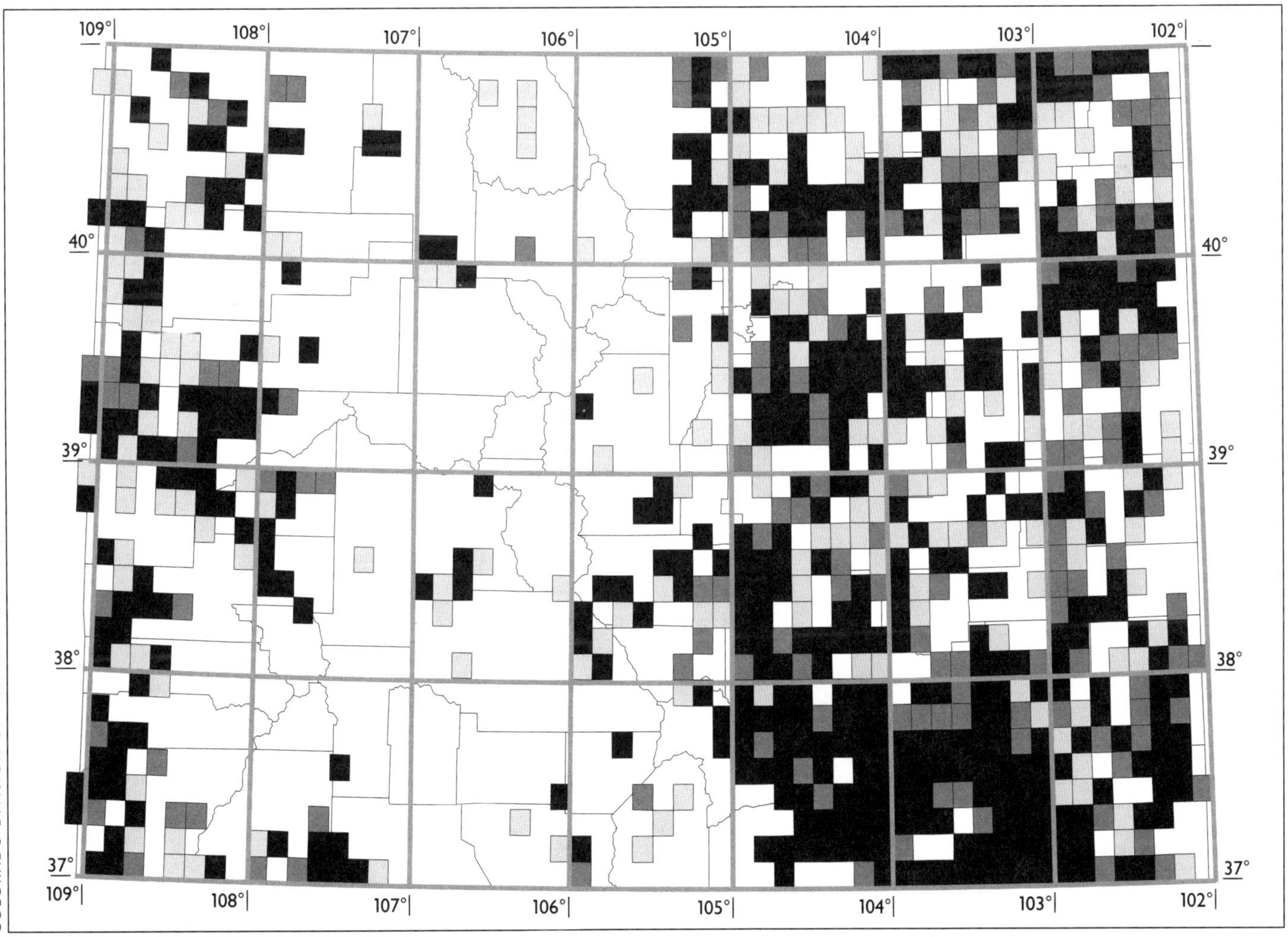
109°
108°
107°
106°
105°
104°
103°
102°
40°
39°
38°
37°

COLORADO DIVISION OF WILDLIFE / WILDLIFE RESOURCE INFORMATION SYSTEM

BLACK-THROATED SPARROW *Amphispiza bilineata*

For a whole month, a captive of this bibbed seed-eater drank no water and ate only dry food. Remaining healthy, the little champ never slowed down (Smyth and Bartholomew 1966). To do this its excretions became extremely concentrated. This kind of performance surely makes the Black-throated Sparrow, physiologically, the North American bird species best adapted to desert life.

HABITAT

Other desert birds survive the hot and dry periods by staying near surface water, reducing activity during the heat of the day, spending much time in shade, and eating juicy invertebrates. These "desert" sparrows do drink water when available, yet reportedly abandon surface water the moment new herbaceous shoots or significant insect hatches appear (Linsdale 1938, Hill 1980, Wingfield et al. 1992).

The subspecies *A. b. deserticola*, the desert-dwelling Black-throated Sparrow, occupies the greater part of the species' range in Colorado and elsewhere. These birds prefer arid, open vegetation types with occasional larger shrubs, such as greasewood, sagebrush, spiny hop-sage, and blackbrush (Fautin 1946). The Atlas found them predominantly in pinyon/juniper woodland (in 14 blocks), tall desert shrub types (9 blocks), lowland sagebrush (6 blocks), and low desert shrub (4 blocks). Widely spaced Utah junipers typified the woodland type. Rare and local in southeastern Colorado, the subspecies *A. b. opuntia* carries the generic name of the "cholla" cactus. Cane cholla serves these sparrows' arid shrub habitat requirement in southeastern Colorado.

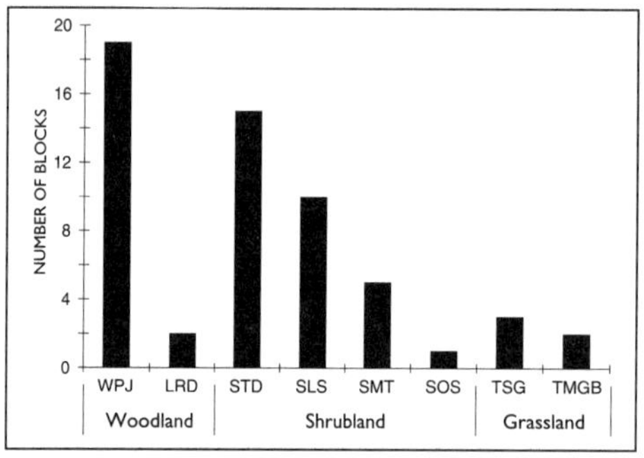
NUMBER OF BLOCKS
0
4
8
12
16
20
WPJ
LRD
STD
SLS
SMT
SOS
TSG
TMGB
Woodland
Shrubland
Grassland

The vegetative structure of the winter grounds resembles the Colorado summer habitat (Davis 1972, Repasky and Schluter 1994). In a survey of Mexican birds by forest types, Hutto labelled the Black-throated Sparrow a year-long thornforest resident, suggesting a broader habitat range in Mexico than in Colorado (*in* Hagan and Johnson 1989).

BREEDING

These well-dressed southwesterners sing their delicate towhee-like songs to equally handsome mates. Even though visually striking, Black-throated Sparrows depend more on song than on visual displays to maintain territory (Heckenlively 1967). Their more modestly plumed nearest relatives, the Sage Sparrows, humbly offer just one basic sweet song, but the highly repertoired Black-throated Sparrows excel in song duration and frequency range (Heckenlively 1970). One songster in the Juniata Atlas block (38108H3) sang a different song on at least five different song posts within an hour (RL). Variable song in individual birds probably discourages passing individuals of the species from attempting to settle in an apparently saturated habitat (Mike Baker pers. comm.).

Atlas observers detected Black-throated Sparrows in only 38 blocks (plus eight Non-priority blocks). They Confirmed nesting in 12 blocks: fledglings in five and adults carrying food or feeding young in the other seven. The lack of nest reports (only one, in a Non-priority block) testifies that these sparrows conceal their nests well. Black-throated Sparrows usually build their nests in the middle of a bush within 2 feet of the ground. The sturdy nest of grass and other plant fibers, and hair, has fine materials lining the cup (Bent 1968). These sparrows occasionally foster cowbirds (Chace and Cruz 1996) and one hosted a Sage Sparrow egg (Gustafson 1976). Harrison (1979) pronounced the female a "tight sitter; may not leave nest until almost touched." Black-throated Sparrows generally produce two broods per year (Ehrlich et al. 1988). Their stay in Colorado, at the north end of their range, appears long enough to raise a second brood, although second broods remain undocumented here. Atlas surveyors Confirmed nesting between 12 May and 16 July.

Breeding territories in New Mexico measured about 390–490 feet across ("about 120 to 150 m"; Heckenlively 1967). Atlas observers found isolated pairs, suggesting that this species can utilize small patches of suitable habitat. For this reason, the Black-throated Sparrow looks like an extinction-resistant species.

BREEDING PHENOLOGY

CODE		# OF RECORDS	DATE RANGE
C	Courtship	2	22 May–9 Jun
ON	Occupied Nest	1	12 May
FY	Feeding Young	7	5 Jun–13 Jul
FL	Fledged Young	6	9 Jun–16 Jul

DISTRIBUTION

The breeding range of Black-throated Sparrows spans 22 degrees of latitude from southeastern Oregon almost to Mexico City. In winter they retreat from Colorado to the southern two-thirds of their breeding range.

In Colorado, Black-throated Sparrows reach the northeastern edge of their range. Atlas workers discovered high population densities of these birds only south and west of Cortez. For the Rangely, Grand Junction, Gateway, and Naturita populations, the Atlas records post low to moderate numbers. In the five blocks in southeastern Colorado, the estimates dipped to 1–10 pairs per block. Atlas observations expanded the known breeding range to include parts of two latilongs in southeastern Colorado (37102, 37104) and one in the northwest (40108). The latter bridges a gap between west-central Colorado and a known population just over the border in Wyoming. Atlasers found them in only one, Non-priority, block in Pueblo and Fremont counties, where other authors have reported them as rare breeders (Latilong Study, A&R). Atlas block selection may have skipped over localized nest sites.

BBS data show a large declining trend over the continent of 3.9% per year, but the low numbers limit any conservation attention for these sparrows in Colorado.

Wildfire and cheatgrass unite as chief enemies of Black-throated Sparrows. Fire destroys the shrubs. By occupying the fire-disturbed sites, cheatgrass retards or prevents recovery of the shrubs by capturing the water and by fueling recurring fires. Livestock grazing has benign or mixed effects on these arid shrubland birds (Medin 1986). Grazing can affect grass and shrub cover and favors the spread of the flammable cheatgrass.

Most natural bird populations fluctuate, and a migrant reacts to factors local and distant. Even so, land managers should consider the Black-throated Sparrow a useful indicator of arid shrubland health because of the bird's detectability and the vulnerability of its habitat to alteration by common human uses.

BREEDING EVIDENCE

REPORTED IN 38 (2%) OF 1745 PRIORITY BLOCKS

Possible	18 blocks	(47%)
Probable	8 blocks	(21%)
Confirmed	12 blocks	(32%)

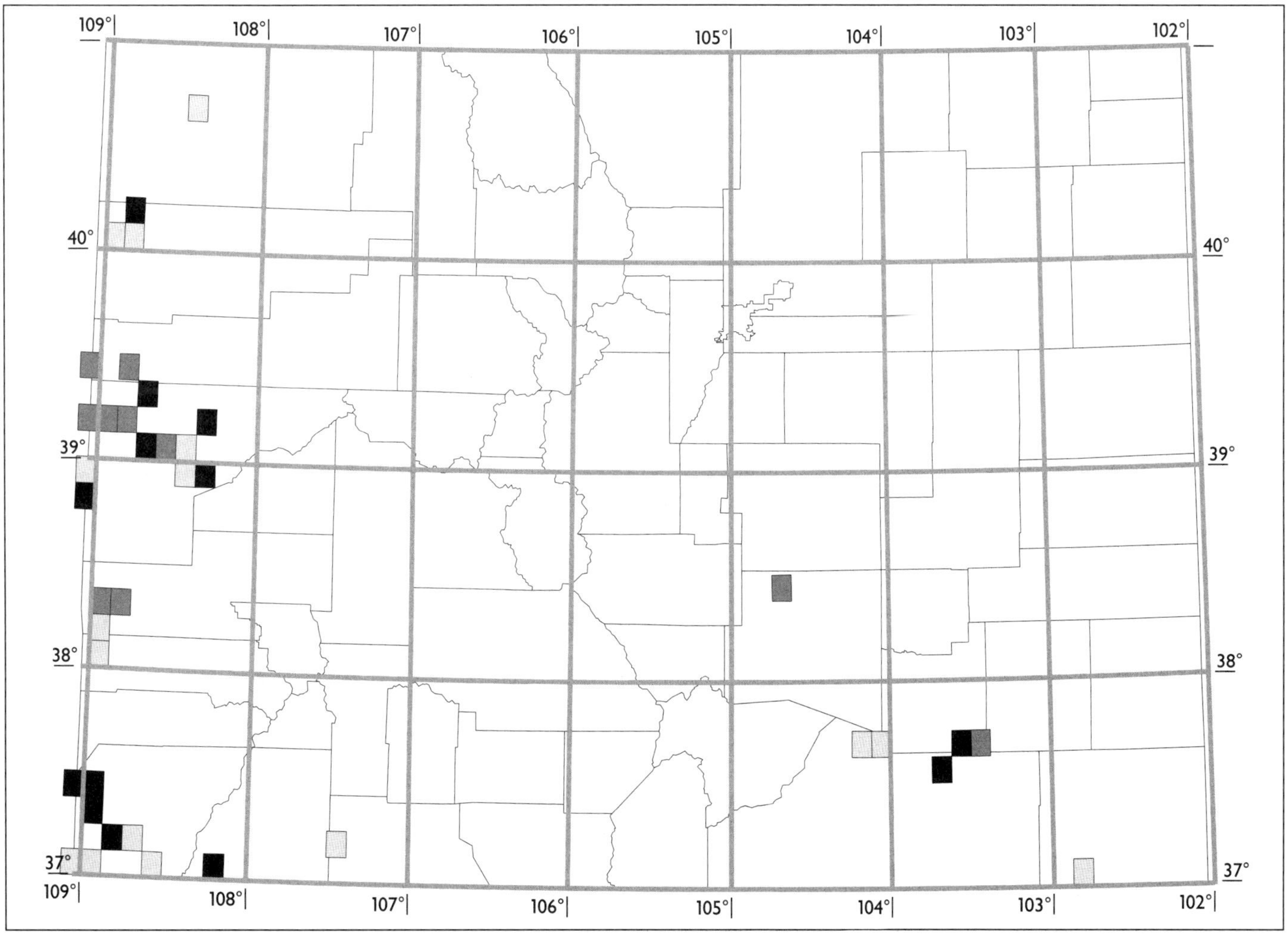
109°
108°
107°
106°
105°
104°
103°
102°
40°
39°
38°
37°

COLORADO DIVISION OF WILDLIFE / WILDLIFE RESOURCE INFORMATION SYSTEM

Ol' Slim hoisted his saddle onto the rail and clumped into the bunkhouse. "The Sage Sparruhs 'r back. I reckon it'll be safe t'move some calves onto the range now." ... This button-breasted bird is western, and returns to Colorado when many ranches begin calving.

HABITAT

Sage Sparrows cross any open country habitat in migration, pause to forage, and then, some time after arriving on the breeding terrain, begin to sing (Rich 1980b). For nesting they select only sizable, low-elevation stands of big sagebrush or mixed big sagebrush and greasewood.

Atlas records reveal that high-country sagebrush and plains sandsage, plentiful in Colorado, do not make suitable nesting habitat, nor do sagebrush parks of 30 acres or less—also common in the state (Knick and Rotenberry 1995).

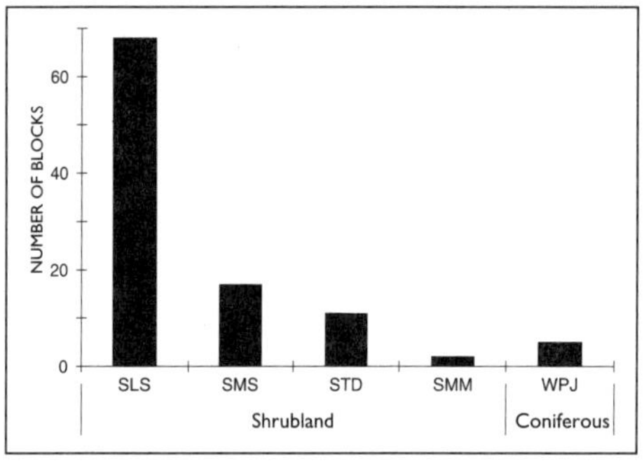
NUMBER OF BLOCKS
0
20
40
60
SLS
SMS
STD
SMM
WPJ
Shrubland
Coniferous

Within those narrow parameters of plant species, stand size, and elevation fits a generous amplitude of stand characteristics. Shrub heights can go from 18 to 40 inches (45–100 cm) and canopy cover of 10–60% (Rich 1980b). Bock et al. (1992), citing many authors, concluded that Sage Sparrows probably respond positively to grazing. This ground feeder prefers only a modest amount of understory vegetation. Moderate, incomplete burns in sagebrush, the kind typical of prescribed fires, do not harm important components of Sage Sparrow nesting habitat (Petersen and Best 1987a). Some of the tolerance these birds show for habitat disturbance, however, results from philopatry, the strong Sage Sparrow loyalty to territory (Wiens et al. 1986).

In the absence of Brewer's Sparrows, Sage Sparrows numerically dominate the sagebrush bird community (Rotenberry and Wiens 1989). Of Atlas blocks with Sage Sparrows, 55% also had Brewer's Sparrows. In these blocks both sparrows post significantly higher abundances than in blocks with only one, which emphasizes the similarity in habitat preferences of the two species. However, even on sites where Brewer's outnumber Sage Sparrows, evidence on competition suggests that Sage Sparrows are more likely to suppress Brewer's numbers than the reverse (Reynolds 1981).

BREEDING

These sagebrush sparrows begin to return to Colorado in February and reach full numbers in mid April (A&R). Unusual among songbirds, they arrive on the nesting territory in pairs (Rich 1980b, Best and Petersen 1985). The earliest Atlas date fell on 1 March with several singing males reported on a west-central Colorado site. The birds would have arrived well before this date because the first stages of territory definition occur with little or no singing.

Few birds have a song so difficult to describe, yet so easy to remember. To most human ears all Sage Sparrows sound the same; however, each individual bird keeps its own song version. Sound spectrograms can identify each bird year after year.

"Walking-in-line" displays define territory boundaries (Rich 1983). Neighboring males fly and run to the edges of their territories and march the boundary line side-by-side, no doubt daring a trespass. If the boundary walkers begin head-bobbing gestures, a fight always ensues. Researcher Best said that "walking-in-line" displays always defined precise boundaries. On Rich's study site, with lower density, Sage Sparrows did not often engage in "walking-in-line" and boundaries flexed more.

Males with mates have significantly larger territories than unmated males (Reynolds 1981, Petersen and Best 1987b). Estimated Sage Sparrow densities range from 130 to 518 birds/mi^2 (50–200 birds/km^2) but on the low-density site Rich mapped territories as large as 17 acres—38/mi^2 (15/km^2)—the largest known for any sparrow (Rich 1980b).

Sage Sparrows achieve greater nesting success in years of higher arthropod production; however, weather during the nesting season significantly influences a parent bird's ability to harvest the productivity. In contrast Brewer's Sparrows seem relatively unaffected by inclement weather. Predation, primarily by snakes, affected fledging success rates the most at one Idaho site (Rotenberry and Wiens 1989, 1991).

Sage Sparrows construct cup nests, usually at mid bush, with an ample mass of screening foliage above (Rich 1980a). The birds apparently build ground nests only following devastation of the shrubs in their territories or because of an early-season need for warmth (Winter and Best 1985).

Occasionally cowbirds deposit eggs in Sage Sparrow nests. Atlasers found a pair of Sage Sparrows attending a cowbird fledgling in Moffat County (Chace and Cruz 1996; Appendix C). In Nevada apparently at least one Sage Sparrow parasitized a Black-throated Sparrow nest (Moldenhauer and Wiens 1970).

BREEDING PHENOLOGY

CODE		# OF RECORDS	DATE RANGE
C	Courtship	1	5 Jun
NB	Nest Building	1	13 Jun
NE	Nest with Eggs	1	17 May
NY	Nest with Young	1	19 Jun
FY	Feeding Young	10	16 May–24 Jul
FL	Fledged Young	23	7 Jun–10 Aug

DISTRIBUTION

Sage Sparrows winter in the southwestern U.S. and adjacent Mexico, in creosote bush and saltbush country. They breed in the Great Basin from the Columbia and Snake rivers to southern Nevada, east to the Continental Divide and Four Corners.

The Atlas probably identified every significant population center in Colorado. It missed them in La Plata County, where A&R show summer records, but every population probably has blocks with nesting Sage Sparrows present but undetected. For example, the observed western Montrose County population probably still reaches Blue Mesa in southern Mesa County (pre-Atlas BLM records).

The Atlas map shows that Sage Sparrows do not nest as high as their obligate plant, sagebrush, grows. Extensive sagebrush in Middle Park, the Roan Plateau, and upper Glade Park does not support breeding Sage Sparrows. In the Colorado River drainage Atlas records from an enclave in Eagle County at 6,600 feet (2010 m) marked the upper limits, until the discovery in 1997 of territorial birds at 7,800–8,000 feet (2375–2440 m) in southern Gunnison County (Kim Potter pers. comm.). In Moffat County they nest up to 7,500 feet (2285 m) and in the San Luis Valley to 8,400 feet (2560 m).

Ornithologists may split this poorly known species into two or more species, but Colorado's form will remain the Sagebrush Sparrow (DeBenedictis 1995).

BREEDING EVIDENCE

REPORTED IN 93 (5%) OF 1745 PRIORITY BLOCKS

☐	Possible	35 blocks	(37%)
▩	Probable	22 blocks	(24%)
■	Confirmed	36 blocks	(39%)

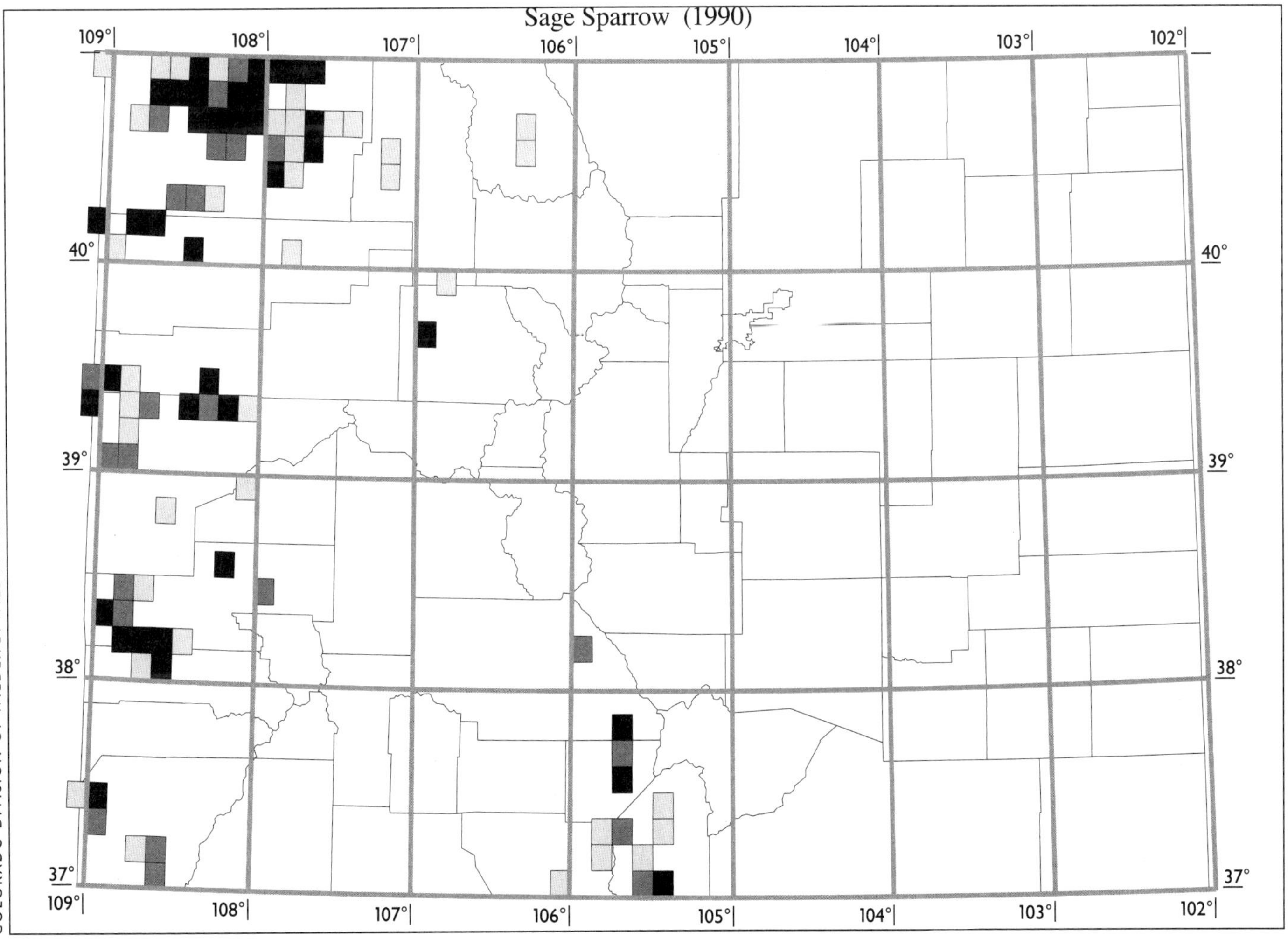
Sage Sparrow (1990)
109°
108°
107°
106°
105°
104°
103°
102°
40°
39°
38°
37°

COLORADO DIVISION OF WILDLIFE / WILDLIFE RESOURCE INFORMATION SYSTEM

LARK BUNTING

Calamospiza melanocorys

The legislature didn't plan it this way, but by designating the Colorado columbine as state flower and the Lark Bunting as state bird, they covered the whole state. In June and July, state flowers decorate the Colorado high country and state birds decorate the plains landscape—both visually and vocally. As the most gifted singers of the larking plains birds, Lark Buntings lack only a good press agent to rank them in the same league with Shelley's skylark, Keats' nightingale, Van Dyke's Hermit Thrush, and Muir's Water Ouzel.

HABITAT

Lark Buntings shoot up "as though propelled from guns, pouring out the most infectious and passionate song [of] any bird in the United States" (Whittle 1922). Males soar into the sky, stretch their wings into a V, and float down rocking like butterflies as they sing their "ecstatic" song. They combine whistled notes, trills, toots, chirrups, buzzes, and unmusical chugs into rhythmic, complex songs that ring across the prairie (Baumgarten *in* Bent 1968). Alas—most Coloradans have never looked at the state bird, unless they live on the plains.

Atlasers found more Lark Buntings in shortgrass prairie (42% of habitat codes) than in mid and tall grasses, but those, together with croplands, comprise another 42%. Buntings use shortgrass, if not heavily grazed and if interspersed with taller vegetation (sunflowers, ragweed, thistles). In the San Luis Valley buntings use farm fields and desert shrubs.

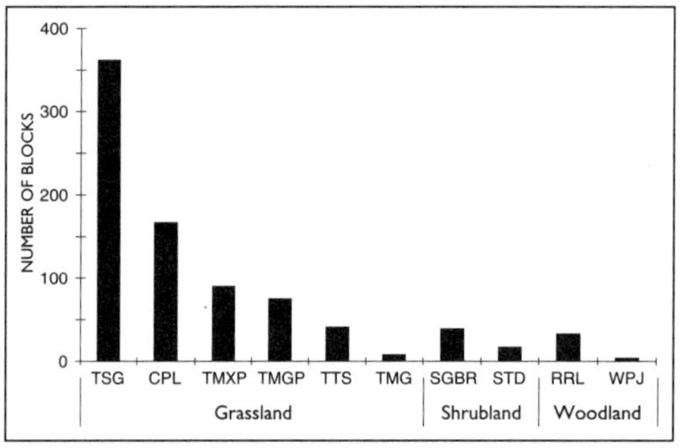

NUMBER OF BLOCKS
400
300
200
100
0
TSG
CPL
TMXP
TMGP
TTS
TMG
SGBR
STD
RRL
WPJ
Grassland
Shrubland
Woodland

Prairie birds partition their habitat by space, time, and food. Compared with Horned Larks and McCown's and Chestnut-collared longspurs, Lark Buntings start nesting later, occupy taller vegetation, and feed larger insect prey to their young. Larks and longspurs occupy barren sections of shortgrass prairie; Lark Buntings pick taller vegetation and moister situations (Creighton and Baldwin 1974). Atlas data show that Horned Larks start nesting a month before the buntings and continue two weeks longer. The two longspurs start, on average, four days before the buntings, with the last fledgling seen three weeks before the last bunting fledgling.

BREEDING

Hundreds of flocks of hundreds of migrating state birds roll across the plains in May; males sing their extraordinary songs even as they move north. The rolling flocks branch off to breeding sites, where groups of males enter a pre-territorial phase that involves communal flight displays (Taylor and Ashe 1976). In this display males perform sensational larking displays. In a few days males disperse and begin to fill the skies with primary song flights to attract females and to establish territories (Ervin 1981). Only at this time of mate-finding (less than a week) do they strongly defend a territory. Mated males seldom display because they help with incubation, brooding, and feeding young; during these phases they tolerate other buntings in their territory (Tom Shane pers. comm.).

The female examines nest sites as the male follows (Creighton 1971). She builds the nest primarily with grasses, and places it in a shallow depression on the ground under or next to a forb, shrub, or clump of grass. Nest placement must meet several criteria: warmth from the morning sun; shade, from midday and afternoon sun, provided by forbs, grass, or shrubs; ventilation and cooling from the prairie's south winds; and sometimes shelter from cold north winds (Shane 1974, With and Webb 1993). Lark Buntings measure the quality of a territory by its ability to protect the nest site from solar radiation, which increases nestling survival (Pleszczynska 1978).

Lark Buntings occasionally host cowbird eggs, yet no records exist of a Lark Bunting actually fledging a cowbird chick. In the only Colorado record, the cowbird hatched but probably did not fledge (Porter 1973).

An estimate using Atlas abundance codes suggests that Colorado has 1,600,000 breeding pairs of Lark Buntings—making them the fourth most numerous bird in Colorado.

BREEDING PHENOLOGY

CODE		# OF RECORDS	DATE RANGE
C	Courtship	30	6 May–30 Jun
NB	Nest Building	11	14 May–28 Jun
ON	Occupied Nest	10	22 May–13 Jul
NE	Nest with Eggs	16	24 May–12 Jul
NY	Nest with Young	15	20 May–20 Jul
FY	Feeding Young	125	1 Jun–2 Aug
FL	Fledged Young	130	28 May–5 Aug

DISTRIBUTION

Most Lark Buntings nest on prairies from southern Alberta and Manitoba to New Mexico and Texas, with Colorado hosting a major portion; a few breed in grassland basins in and west of the Rockies. They winter from the southwestern states to central Mexico. Since 1976 their winter range has moved north; small flocks now winter in Baca County and southwestern Kansas (Shane and Seltman 1995).

They nest in most Atlas blocks on the plains. A few breed in North and South parks and west of the mountains. Compared with A&R, atlasers found fewer in northwestern Colorado and a significant though small nesting population (2,100 pairs according to the Atlas estimates) in the San Luis Valley.

Lark Buntings have a nomadic nature; although the Atlas map shows them blanketing the plains, numbers in any one place vary annually. Wet years generate increases and drought causes declines (Ron Ryder pers. comm.). BBS statistics suggest a major decline, yet huge flocks still roll across the prairies. In its 31 years the BBS has probably produced insufficient data to reflect the long-term population cycles (Shane 1996).

The "prairie bobolinks" have waxed and waned for many years. In 1931, the year that the legislature designated the Lark Bunting as state bird, Bergtold reported that, ironically, they became uncommon in many places where they usually nested abundantly.

In 1912 Sclater regarded them as "one of the commonest summer birds on the dry eastern prairie," as did B&N 50 years later. Now, near the urban centers, only a few nest, their habitat eliminated by the march of suburbia and monoculture croplands. Farther east, despite declines shown by the BBS, the large Atlas numbers show that they maintain their abundance. Each spring over a million state birds grace the prairies with their extraordinary songs.

BREEDING EVIDENCE

REPORTED IN 650 (37%) OF 1745 PRIORITY BLOCKS

□	Possible	83 blocks	(13%)
▩	Probable	240 blocks	(37%)
■	Confirmed	327 blocks	(50%)

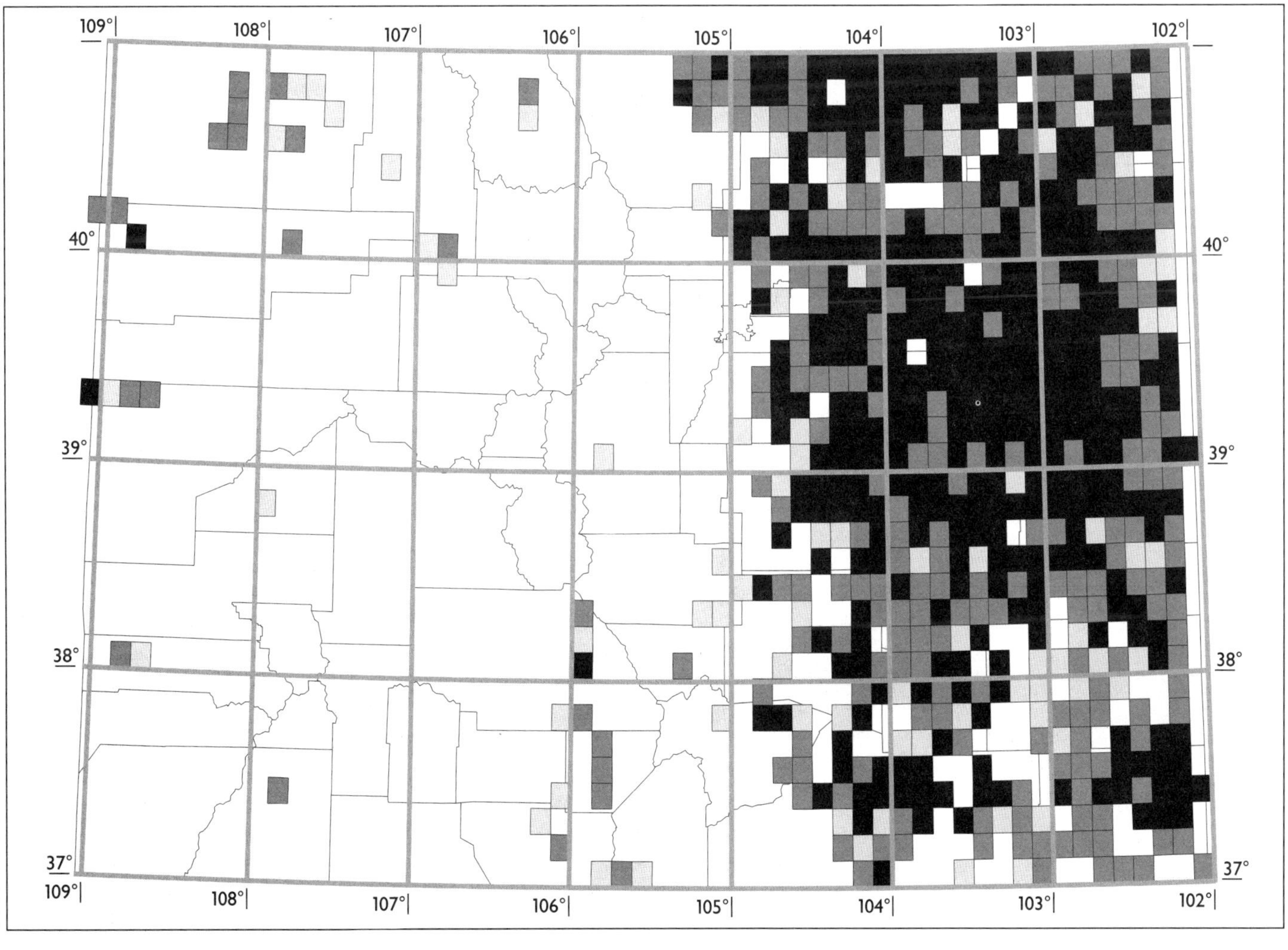
109°
108°
107°
106°
105°
104°
103°
102°
40°
39°
38°
37°

COLORADO DIVISION OF WILDLIFE / WILDLIFE RESOURCE INFORMATION SYSTEM

Though seldom seen due to their secretive habits, Savannah Sparrows range from 3,600 feet (1100 m) in northeastern Colorado to 12,000 feet (3660 m) in the mountains. When disturbed, these grassland birds run like mice or dart into the air and dive into a bush or clump of grass, where they vanish. This trait, and their penchant for moist meadows for breeding and nesting, meant wet feet for atlasers and helped the birds remain seldom seen.

HABITAT

Across their range throughout most of North America, Savannah Sparrows inhabit open country such as grassy meadows, cultivated fields (especially alfalfa), lightly grazed pastures, roadsides, coastal grasslands, sedge bogs, edges of salt marshes, and tundra (Wheelwright and Rising 1993). They favor dense ground vegetation, especially grasses, and moist micro-habitats (Wiens 1969).

Moist, grassy, mountain meadows provide Savannah Sparrows' favored habitats in Colorado. Nesting occurs in new growth, usually grasses or sedges at the edge of moist areas, or in irrigated fields of hay and alfalfa (B&N). Atlas data confirm this information, with croplands, montane grassland, and emergent marshes accounting for 58% of habitats reported. Salt meadows, mountain sagebrush, montane carr, and willow carr make up another 22%. Most of these seven habitats occur in close proximity to one another, so not surprisingly account for 80% of reported records.

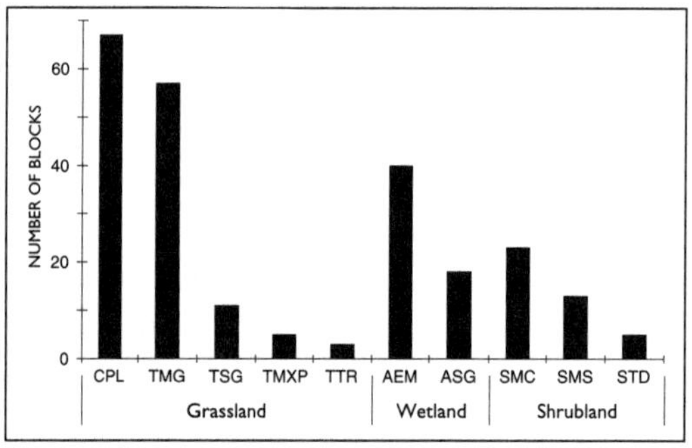
NUMBER OF BLOCKS
0
20
40
60
CPL
TMG
TSG
TMXP
TTR
AEM
ASG
SMC
SMS
STD
Grassland
Wetland
Shrubland

BREEDING

Savannah Sparrows observe a variation on customary territorial concepts. Although territories that the males defend have very small dimensions—0.26 acre (0.11 ha) in Michigan—the defended area often has unoccupied space around it (Ryser 1985). Males spend only about 66% of their time on territory; they feed in non-territorial common areas, yet females feed mostly within their mates' territories. Males define the territories from singing perches and by chases. Females also defend the territory and may force monogamy on some males, some of whom practice polygyny (9% in Nova Scotia, 40% in Michigan). Pairs whose members survive typically breed together in subsequent years (Wheelwright and Rising 1993).

The female alone builds the nest, well hidden on the ground. They conceal nests under grass clumps with a canopy of dead grasses and herbs, although sometimes they omit the canopy. Experienced nesters orient the nest away from prevailing winds. Some populations have second broods; this depends on climate and whether the male assumes post-fledging care of the first brood (Wheelwright and Rising 1993). Atlas dates provide sufficient time for second broods, although this is not confirmed for Colorado.

Although throughout its range the recorded frequency of brood parasitism appears low (Wheelwright and Rising 1993), four records of cowbird brood parasitism exist for Colorado, including two records from the Atlas (Chace and Cruz 1996; Appendix C). Studies of Savannah Sparrow life history have concentrated on only a few of the 17 subspecies, mainly those in Nova Scotia, Quebec, and the Upper Midwest (Wheelwright and Rising 1993).

BREEDING PHENOLOGY

CODE		# OF RECORDS	DATE RANGE
NB	Nest Building	1	22 Jul
ON	Occupied Nest	4	10 Jun–21 Jun
NE	Nest with Eggs	9	7 Jun–22 Jul
NY	Nest with Young	3	18 Jun–8 Jul
FY	Feeding Young	37	12 Jun–5 Aug
FL	Fledged Young	20	21 Jun–6 Aug

DISTRIBUTION

Savannah Sparrows breed across the continent from the Arctic Ocean south to the Great Basin, Rockies, and Appalachians. In Colorado historical accounts refer to Savannah Sparrows as "a common summer resident . . . not an uncommon breeder up to nearly 12,000 ft." (Cooke 1897); "occurs throughout the state in moist meadows to 9200 ft." (B&N). Atlas records report at least Possible breeding for all but 14 counties in Colorado, 11 of them on the eastern plains. Elevations ran from Possible breeding at 3,600 feet (1100 m), and Confirmed breeding from 4,600 feet (1400 m) to 11,100 feet (3385 m).

Although they breed throughout Colorado, Savannah Sparrows concentrate in wet streamside meadows and irrigated hayfields in the mountainous regions of the state. North, Middle, and South parks, the upper Yampa Valley, the upper Gunnison Valley, and the San Luis Valley contain the bulk of Atlas Confirmations and the largest aggregations of records. Breeding also occurs on higher plateaus with open grasslands and sedge meadows, such as the Flat Tops and Grand Mesa. Atlas results differ somewhat from A&R by showing patchy distribution in the San Luis Valley and only four blocks in the upper Arkansas; recent habitat losses may explain these differences.

Surprisingly, scattered pairs of Savannah Sparrows breed on the plains, with Confirmations in wet meadows from Larimer and southern Weld, Lincoln, and Baca counties. Plains records show no discernible pattern and appear localized, with few contiguous blocks, as one might expect for a bird requiring wet grass on the dry plains. The western valleys had similarly scattered reports.

Populations increased significantly in western North America from 1969 to 1989 (BBS unpubl. data, *in* Wheelwright and Rising 1993) while declining slightly across the continent (1966–1996). Forest clearing and more intensive agriculture may explain these increases (Wheelwright and Rising 1993). Irrigated hay meadows form an important component of Savannah Sparrow habitats in mountain parks and western valleys in Colorado. Prior to settlement these areas were likely already wet grassy meadows, so that loss of irrigated hay meadows to urban and resort development for uses such as golf courses, residences, and ski area base facilities may negatively affect Savannah Sparrow populations in the future.

BREEDING EVIDENCE

REPORTED IN 191 (11%) OF 1745 PRIORITY BLOCKS

☐	Possible	65 blocks	(34%)
▩	Probable	54 blocks	(28%)
■	Confirmed	72 blocks	(38%)

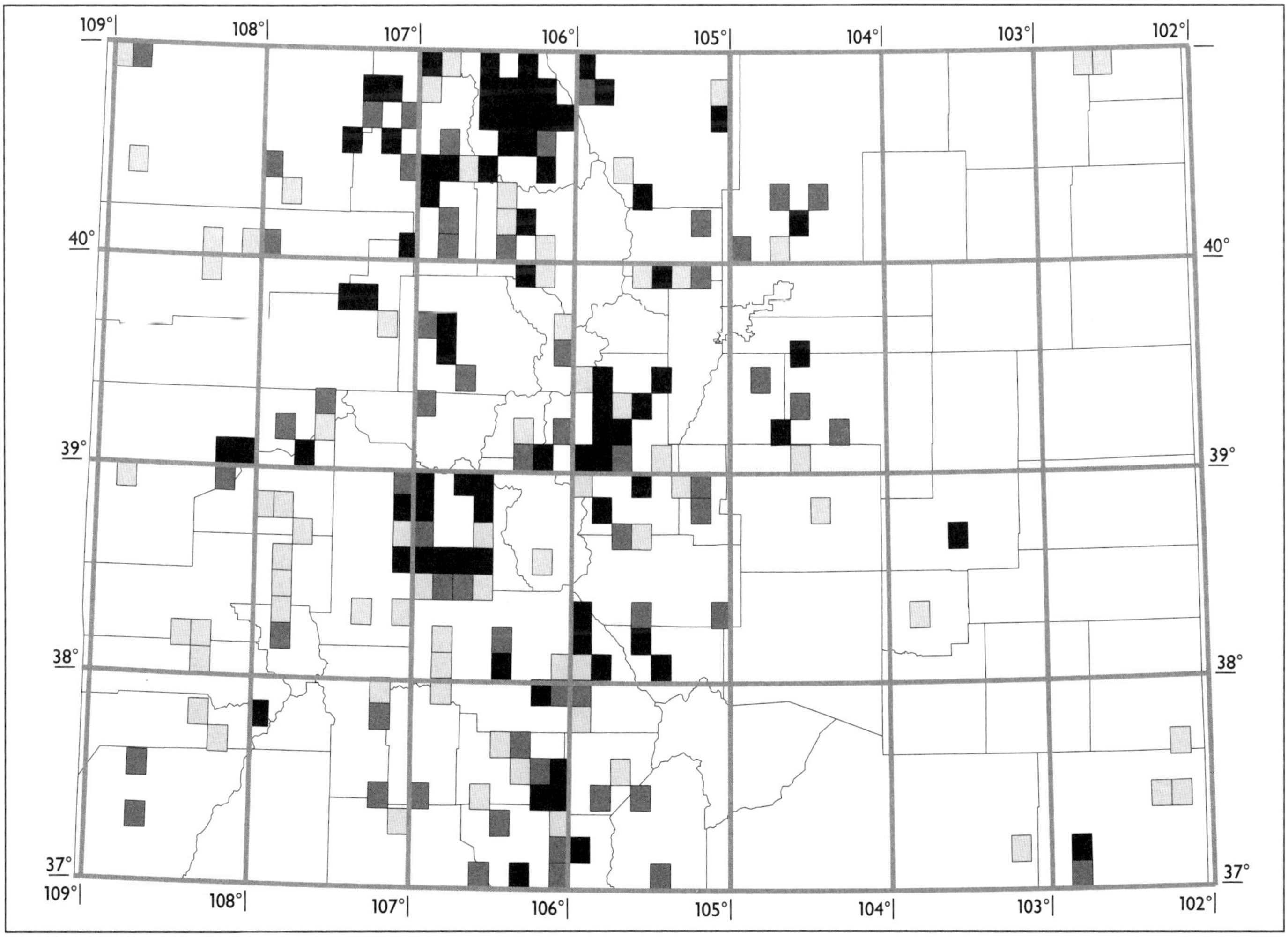
109°
108°
107°
106°
105°
104°
103°
102°
40°
39°
38°
37°

COLORADO DIVISION OF WILDLIFE / WILDLIFE RESOURCE INFORMATION SYSTEM

Piccolos of the grasslands orchestra, Grasshopper Sparrows practice obscurity. Only when the breeding season comes do the males throw back their heads to proclaim territory and woo their mates with thin, wiry, songs, almost inaudible to human ears. These small brown birds blend into the grass, and their exultant song loses itself among the real grasshopper voices of the prairie.

HABITAT

Whether the grass is tall or short, in pasture or prairie, Grasshopper Sparrows require only that the grass furnish the protection they need to live inconspicuously. Over 83% of the Atlas records are from various types of grasslands; these sparrows use more tall grass than other plains sparrows.

In different grassland ecosystems Grasshopper Sparrows select breeding sites with different components. In the East they may pick sparser vegetation than in the West. In Colorado they almost exclusively prefer the prairie grasses (Kendeigh 1941) or grasslands with rabbitbrush or saltbush (B&N). Although these hiders-in-the-grass avoid

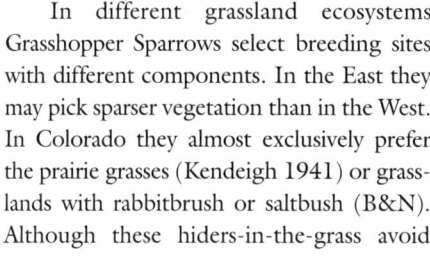

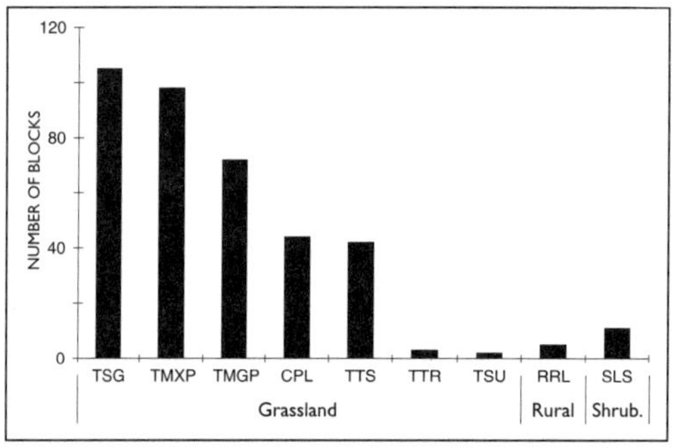

120
80
40
0
NUMBER OF BLOCKS
TSG
TMXP
TMGP
CPL
TTS
TTR
TSU
RRL
SLS
Grassland
Rural
Shrub.

extensive shrublands, they need shrubs, other tall plants (e.g., sunflowers), or fence posts as song posts. They forage exclusively on the ground, and need exposed bare ground for success. They require large grassland tracts for breeding (250 acres [100 ha] in Maine, 75 acres [30 ha] in Illinois) even though they maintain small territories (Vickery 1996).

Although named for their song (Pearson 1942), Grasshopper Sparrows live up to their name in another way: they consume large numbers of grasshoppers. The principal food is insects, in summer 60% grasshoppers. When feeding young they remove the "less digestible" body parts, particularly legs, by vigorous shaking (Vickery 1996).

BREEDING

Grasshopper Sparrows dribble into their breeding grounds in early May. Males arrive first and establish their 1–3-acre (0.4–1.2 ha) territories by singing, at first only in the morning. Two or three weeks later they sing all day and even into the night (Vickery 1996). Although they sometimes nest in loose colonies (A&R), they vigorously defend individual territories with songs and crouching flutter-wing displays to intimidate rival males. Females also sing, and answer the males' wooing songs with trills (Vickery 1996).

Nest-building reaches its peak in late May or early June, and a second nesting may occur in late June or July. Atlas dates strongly suggest that Grasshopper Sparrows in Colorado do not engage in a second nesting.

Females alone build the nests, domed structures of grass and lined with rootlets and horsehair, in a slight depression in the ground at the base of a protecting weed or grass clump (Vickery 1996). Females never go directly to or from the nest but drop to the ground some distance away and trickle, mouse-like, to the site. Females remain on the nest until an intruder or predator almost tramples it. Then they flutter helplessly, feigning injury to lure the intruder away (Vickery 1996).

Males sing while the females incubate, but they stop and drop into the grass if danger threatens. After the young hatch, both parents feed them. Although they try to fool approaching observers by swallowing the food (Smith *in* Bent 1968), slightly over half of the Atlas Confirmed breeding records involved adults feeding young, showing the diligence of atlasers.

Nonparental attendants carry food and brood young of unrelated nestlings. Both juveniles unrelated to parents and adults from neighboring territories who recently lost nests to predators do this. They make 9–50% of provisioning visits to certain nests. Parents facilitate nonparental attendants by moving out of the nest, but vigorously chase away unrelated birds that do not bring food (Vickery 1996).

BREEDING PHENOLOGY

CODE		# OF RECORDS	DATE RANGE
C	Courtship	1	6 Jun
NB	Nest Building	1	13 Jun
ON	Occupied Nest	2	8 Jun–23 Jun
NE	Nest with Eggs	3	22 Jun–25 Jun
NY	Nest with Young	1	13 Jul
FY	Feeding Young	47	1 Jun–23 Jul
FL	Fledged Young	24	17 Jun–29 Jul

DISTRIBUTION

Grasshopper Sparrows have a large range. They breed coast to coast from southern Canada to the southern U.S. but avoid vast stretches of western basins, deserts, mountains, and forests. They winter from the southeastern U.S. to the West Indies and Central America. As settlers cleared the eastern forests for pastures or hayfields, they expanded nesting habitat for this grassland-dependent species (Howell 1932, Vickery 1996) but the sparrows' penchant for grasslands ensures a spotty distribution.

In Colorado they inhabit the plains, most abundantly near the Kansas border. All Atlas observations came from east of 105° 20' longitude. Abundance peaks in the northeastern corner of the state, north of the 40th Parallel, in Phillips, Sedgwick, Logan, and northern Yuma and Washington counties. The distribution dwindles closer to the mountains, where expanses of grassland likewise dwindle; the one Confirmation near Denver occurred on the Rocky Mountain Arsenal NWR, which has extensive grasslands.

The various conservation priority schemes target this species for attention. BBS figures suggest a serious decline in North America: 3.5%/year. Their numbers, however, like those of many grassland breeders, fluctuate annually (B&N). In 1959 Knorr observed that in Colorado these sparrows exhibit sporadic influxes similar to Dickcissels. Habitat loss imposes severe impacts, particularly conversion of pasture to crops (Vickery 1996). In some parts of their range Savannah Sparrows could usurp their preferred habitat (Smith *in* Bent 1968) but the only observed interactions between the two involve countersinging (Vickery 1996). Displacement probably does not occur in Colorado because Savannah Sparrows breed chiefly in the western two-thirds of the state.

Hawks seldom take Grasshopper Sparrows because of the sparrows' ability to disappear into the grass, yet Loggerhead Shrikes capture them frequently. Impaled Grasshopper Sparrows are "quite common" in Oklahoma (Vickery 1996). Skunks, weasels, and cats successfully kill these little birds and rob their nests; the high level of predation influences the two-broods-per-season pattern (Vickery 1996). Their chief enemies are humans who, by converting grasslands to croplands and mowing the grass that protects nests, diminish these essential members of the grassland concertos.

BREEDING EVIDENCE

REPORTED IN 327 (19%) OF 1745 PRIORITY BLOCKS

☐	Possible	94 blocks	(29%)
▩	Probable	156 blocks	(48%)
■	Confirmed	77 blocks	(23%)

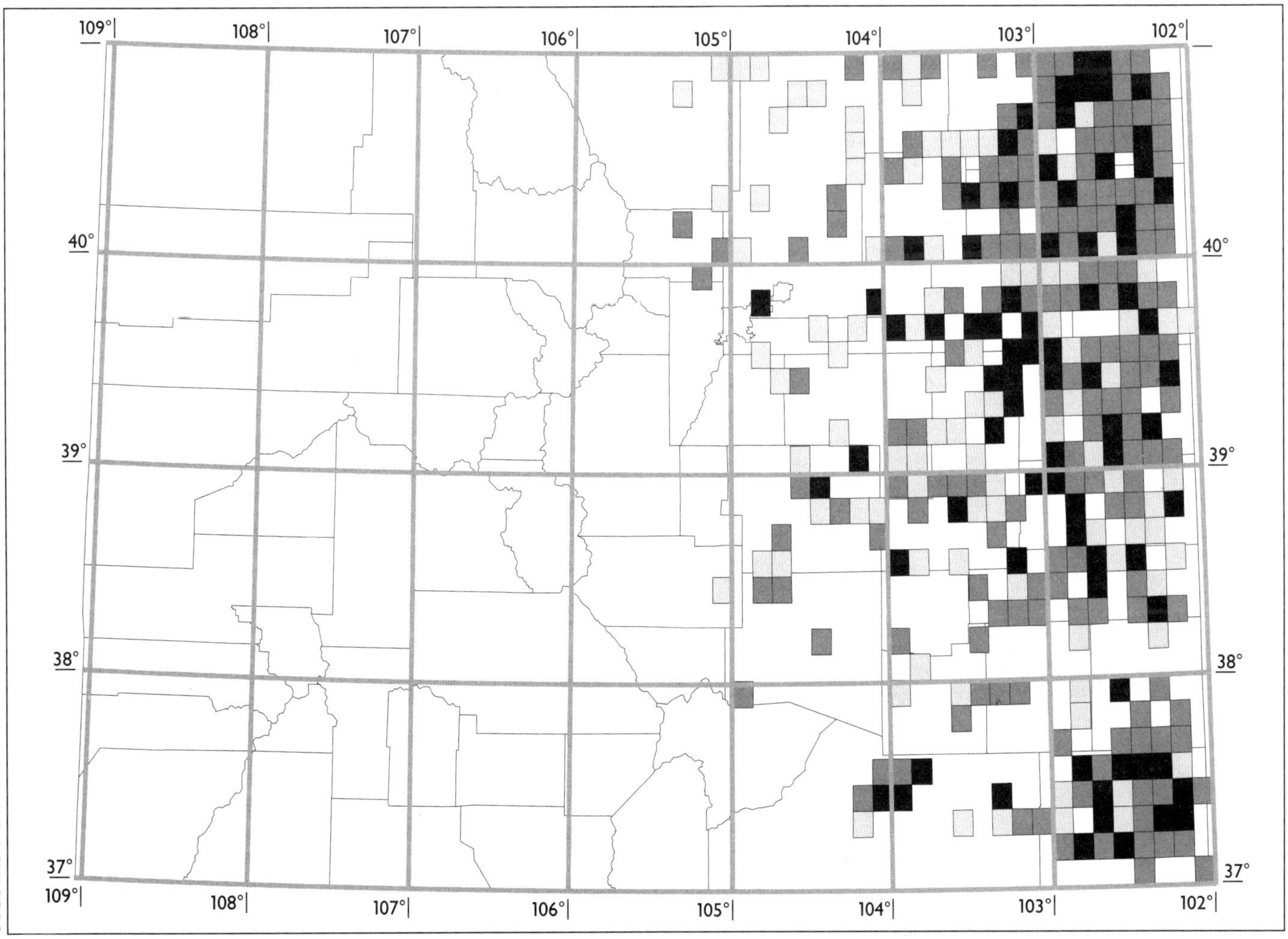
109°
108°
107°
106°
105°
104°
103°
102°
40°
39°
38°
37°

COLORADO DIVISION OF WILDLIFE / WILDLIFE RESOURCE INFORMATION SYSTEM

These ground-scratchers have little Colorado breeding history reported in Colorado because of their secretive nesting habits and inaccessible higher-elevation habitat haunts—and, perhaps, because of their pre-1950 rarity. Although atlasers gathered much new information concerning their Colorado distribution, these sparrows remained true to their name and "out-foxed" Atlas workers, who did not discover a single nest.

HABITAT

Fox Sparrows of the Rockies, once called Slate-colored Sparrows (and about to be again; Rising 1996), spend most of the breeding season in riparian willow shrublands and wet, willow-grown meadows from 7,500 to 11,000 feet (2285–3350 m). Over 80% of Atlas observations occurred in high- and mid-elevation willow carrs. In other parts of the range Fox Sparrows occupy a variety of brushlands, montane thickets, and moist, rank growths of willows, alders, and bottomland trees (Cadman et al. 1987, Roberson and Tenney 1993).

Whatever the plant community, Fox Sparrows inhabit dense shrubby understory associated with watercourses. Instead of choosing habitat based on species of plants present, these sparrows focus on more general habitat characteristics such as the presence of dense shrubby vegetation for concealment from predators (Burns and Hackett 1993). They forage and travel about their territories through leaf litter and brush-covered alleyways.

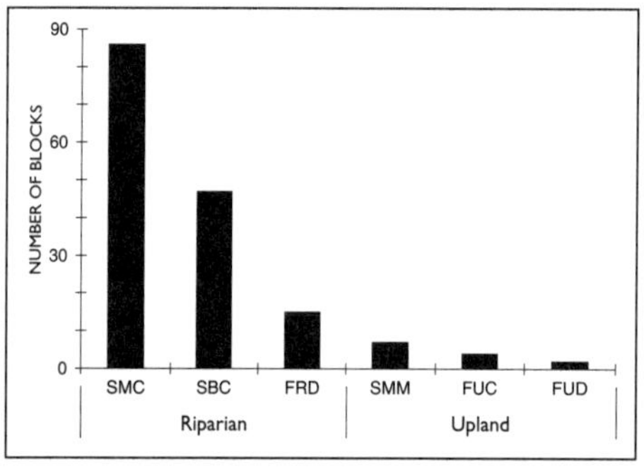
90
60
30
0
NUMBER OF BLOCKS
SMC
SBC
FRD
SMM
FUC
FUD
Riparian
Upland

Fox Sparrows have adapted themselves to the harsh weather of high elevations by specializing in selecting dense understory in which they place their nests. Regular grazing that reduces the density of the understory diminishes the number of available nest sites.

BREEDING

Males and females return to breeding grounds at the same time and establish pair bonds within a week of arrival, although after the males have established their territories. Both sexes sing. The songs sound similar except that females vocalize at a somewhat weaker volume (Saunders 1910).

Fox Sparrows construct bulky nests of sticks and bark, densely lined with fine grass. They most frequently nest on the ground, but sometimes higher up. Early in the season they have a tendency to place nests higher than nests built later (Philipp 1925). Snow cover probably influences nest height (Threlfall 1982). The birds situate above-ground nests in a variety of locations within a nest plant as long as they can conceal the nest well (Burns and Hackett 1993).

Egg-laying begins immediately after nest construction. While females incubate and brood the young, males defend territories and later assist in feeding. Some Fox Sparrow populations raise two broods (Rogers 1994). Atlasers first recorded territorial behavior on 7 May, and fledgling-associated activities ranged from mid June into August, suggesting two broods for Colorado birds also. Atlas workers saw Fox Sparrows use the broken-wing distraction display to coax away predators.

Of all the Atlas blocks with Fox Sparrow observations, atlaser efforts Confirmed breeding in only 19%; territorial behaviors comprised 62% of the observations. The birds exhibit secretive nesting and feeding habits (only two observations of nest-building and no nests discovered). Atlaser Randy Lentz documented the first case of Brown-headed Cowbird parasitism on Fox Sparrows in Colorado, at over 10,000 feet (3048 m) near Breckenridge (Chace and Cruz 1996; Appendix C). Cowbirds at higher elevations use watercourses as travel routes (KMP); as the cowbird elevation range continues to rise, Fox Sparrows may increasingly fall host to this nest parasite.

BREEDING PHENOLOGY

CODE		# OF RECORDS	DATE RANGE
C	Courtship	1	14 Jun
NB	Nest Building	2	27 May–28 May
FY	Feeding Young	13	24 Jun–23 Jul
FL	Fledged Young	11	15 Jun–8 Aug

DISTRIBUTION

These short-distance migrants have wintering and breeding ranges restricted to North America. They breed across Canada and the West, and winter south of Colorado along the Pacific coast and in the southern U.S. and northern Mexico.

In Colorado they breed regularly west of the Continental Divide, but have a somewhat spotty distribution. They show a localized distribution and lower numbers on the Eastern Slope, due to a lack of suitable willow carrs along the steep Front Range stream canyons. The low percentage of Atlas blocks with these sparrows may relate in part to the rather limited distribution of their primary habitat and in part to the inaccessible character of those habitats. Skeptically, Cooke (1897) described "Slate-colored Sparrows" as a "summer resident; rare. The status of this species as a Colorado bird is very unsatisfactory." He knew of no one who had confirmed Ridgway's 1873 statement that they bred in Colorado, but in 1876–1877 Carter had collected three breeding records in Grand County (B&N). The Denver/Boulder area had no records up to 1939 (Alexander 1937, Niedrach and Rockwell 1939). B&N reported few nesting records but thought them "evenly distributed in their preferred habitat." By 1992 A&R reported a comparatively wide distribution.

Understanding Fox Sparrow distribution in Colorado has come only recently and the Atlas made huge strides in this area. Atlasers Confirmed breeding in eight latilongs and added Probable breeding in Latilong 38105. The most notable range expansions include the Flat Tops and the White River and Uncompahgre plateaus.

That early observers found so few Fox Sparrows implies an expanding range—their loud sweet song certainly draws attention. It would seem that observers in Boulder County would have found existing populations before 1937 and that other observers would have found more on the Western Slope prior to B&N's opus.

A Fox Sparrow nest inspired designation of the Alfred M. Bailey Bird Nesting Area. George and Marie Shier in 1964 found the nest on Rock Creek in Summit County; Dr. Bailey (of B&N) took his camera to record activity at the nest. Because of this, at the time, rare, discovery, the Denver Field Ornithologists, Denver Museum of Natural History, and U.S. Forest Service designated the site as the Bailey Bird Nesting Area. The Colorado Bird Observatory now conducts extensive summer field work there (Willow Lakes, 39106F2).

BREEDING EVIDENCE

REPORTED IN 143 (8%) OF 1745 PRIORITY BLOCKS

Possible	68 blocks	(47%)
Probable	48 blocks	(34%)
Confirmed	27 blocks	(19%)

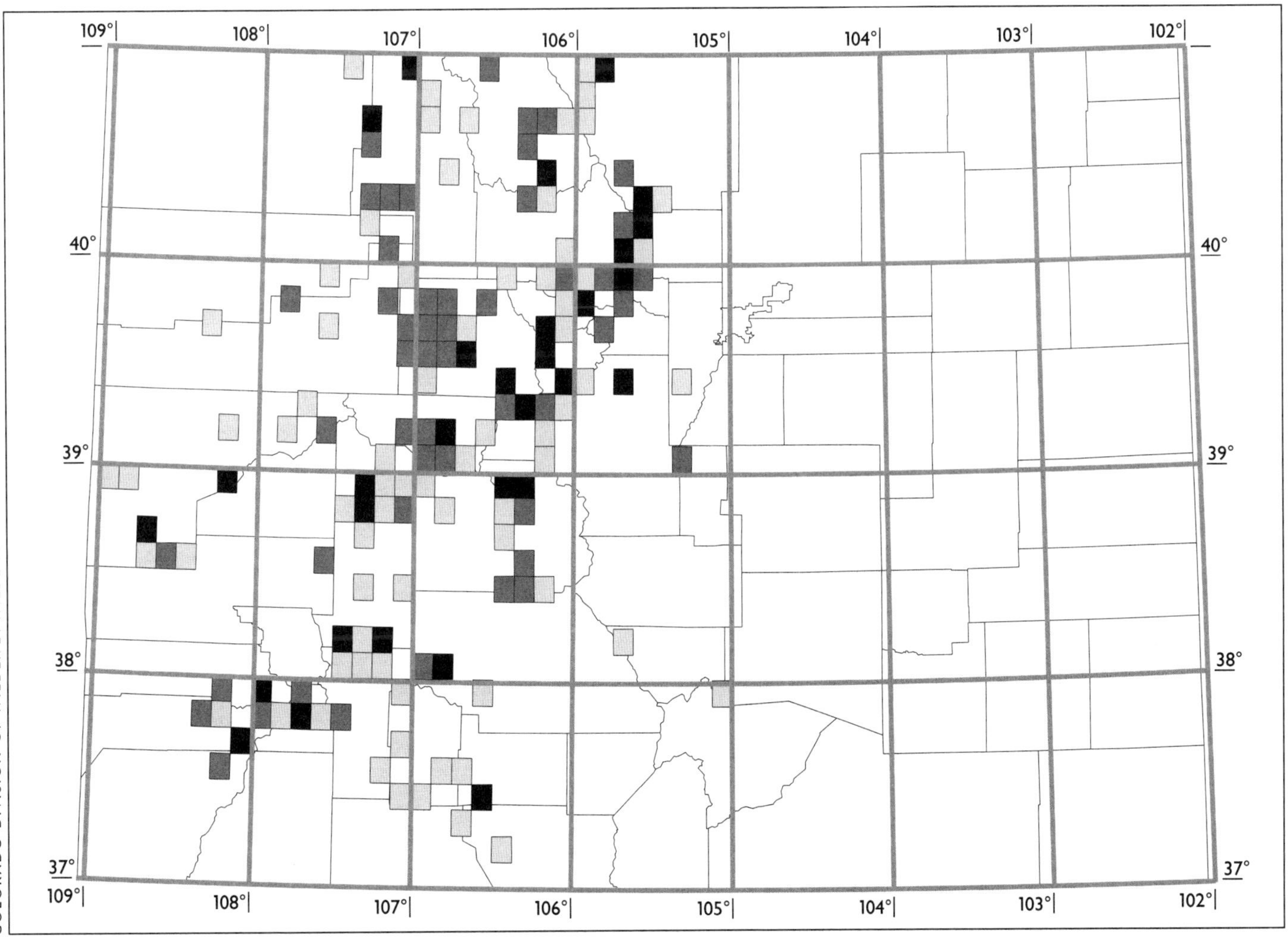
109°
108°
107°
106°
105°
104°
103°
102°
40°
39°
38°
37°

COLORADO DIVISION OF WILDLIFE / WILDLIFE RESOURCE INFORMATION SYSTEM

If we could read a Song Sparrow sonogram in the same way that a computer scans a bar code, we might instantly know the singer's birthplace and nesting habitat. No two adult males sound exactly alike, and young sparrows quickly learn to discriminate among the vocalizations of their parents, their neighbors, and strangers. The most variable North American species, Song Sparrows have evolved into 30 subspecies in the U.S. and Canada (AOU 1957), with subtle to distinct variations in plumage, size, habits, and song repertoires (Rising 1996).

HABITAT

Song Sparrows in the western U.S. inhabit shrub thickets near running or standing water, marshes, moist forest clearings, and moist forest edges. They appear to prefer clumpy vegetation containing dense pockets for nesting and more open areas for foraging (Marshall 1948).

In Colorado, atlasers found these sparrows in an astonishing 32 habitats, ranging from croplands on the plains to willow carrs in the high mountains. Nine-tenths of the sightings, however, occurred in three habitat categories: lowland and mountain willow carrs (37%), lowland and mountain riparian woodlands (33%), and emergent wetlands (10%). Highest nesting densities generally occurred in blocks containing one or more of these habitats.

In montane willow carrs, Song Sparrows compete for nest sites with Lincoln's and Fox sparrows. Lincoln's Sparrows usually nest on the ground at the base of willows or in meadows and bogs between willow clumps. Song and Fox sparrows often nest a few feet off the ground in the densest willow thickets (Finch 1987). Fox and Lincoln's sparrows range to higher elevations than Song Sparrows (Hallock 1984, Finch 1987).

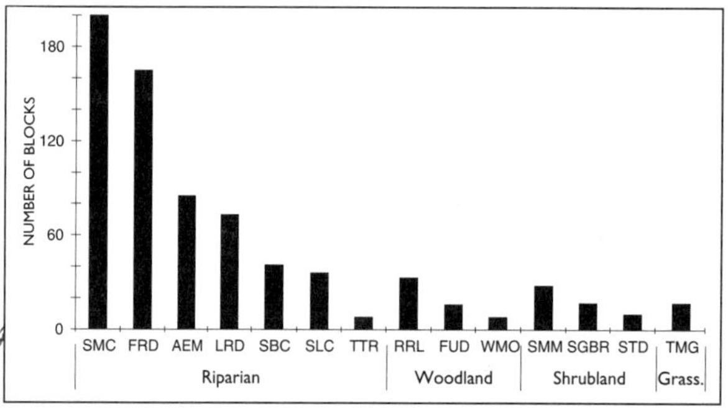
NUMBER OF BLOCKS
180
120
60
0
SMC FRD AEM LRD SBC SLC TTR
Riparian
RRL FUD WMO
Woodland
SMM SGBR STD
Shrubland
TMG
Grass.

When foraging, Song Sparrows hop along the ground and scratch out seeds and invertebrates from the mud, grass, and leaf litter. They also hop through low vegetation gleaning insects and nabbing seeds and small fruits, and occasionally they flycatch from a low perch (Marshall 1948, Bent 1968). Insects compose the bulk of their diet during the breeding season, with miscellaneous seeds predominating during other times of the year (Beal 1910).

BREEDING

Pairs begin nest-building in mid May. They often raise two clutches, with the second brood fledging as late as mid August (Bent 1968). Atlasers observed nest-building on the plains at the edge of the Front Range as late as 10 July. Nesting phenology showed little altitudinal variation. Nest dates (NB, ON, NE, and NY) ranged from 20 May to 10 July on the Front Range piedmont, from 30 May to 22 July in the central mountains, and from 27 May to 9 August in the Western Slope valleys and plateaus.

Song Sparrows experience a high rate of nest parasitism by Brown-headed Cowbirds, with seven Atlas and numerous other records. Atlasers found a cowbird egg in a Song Sparrow nest in Water Canyon (39108H7), Rio Blanco County, and observed cowbird fledglings with adult Song Sparrows in blocks in Moffat, Garfield, Jackson, Eagle, Teller, and Adams counties (Chace and Cruz 1996; Appendix C).

The relatively low Confirmation rate for Song Sparrows (35%) may reflect their penchant for nesting in shrub thickets. Dependent fledglings (FL) or adults carrying food to young (FY) accounted for nearly four-fifths of the Confirmations. Atlasers found actual nests (ON, NE, NY) in only 31 of the 589 blocks where they found Song Sparrows.

BREEDING PHENOLOGY

CODE		# OF RECORDS	DATE RANGE
C	Courtship	3	2 Jun–10 Jun
NB	Nest Building	12	30 May–13 Jul
ON	Occupied Nest	5	8 Jun–29 Jun
NE	Nest with Eggs	14	20 May–8 Jul
NY	Nest with Young	14	27 May–9 Aug
FY	Feeding Young	104	18 May–8 Aug
FL	Fledged Young	68	14 Jun–22 Aug

DISTRIBUTION

Song Sparrows breed from the Aleutian Islands to Newfoundland and south to the mid South and Baja California. In an area that arcs north from Missouri to South Dakota to the edge of the Rocky Mountains, the breeding range omits Nebraska, Kansas, Texas, most of New Mexico, and, judging by

Atlas results, eastern Colorado. A separate population breeds in the highlands of central Mexico. The winter range covers the entire U.S.

B&N and A&R described this species as a "common" summer resident in the western valleys, mountain parks, and eastern plains. The Latilong Study reports breeding in all latilongs except 40102 and 39103 in the northeast and 38102 in the southeast.

Atlas data show that high nesting densities do indeed occur in the western valleys, mountain parks, and along the northern Front Range, but that nesting density falls off rapidly east of the Front Range. Only on the west edge of the eight latilongs lying east of the 104th Meridian (39103 and 38103) did Atlas field workers Confirm nesting, and then in only two Non-priority blocks, one 50 miles south of the other, in eastern Elbert and western Lincoln counties. All the Possible records between Longitudes 102 and 104 on the Atlas map refer to singing birds observed 1–21 June (except for one 22 May and one 10 July). Atlasers coded the two Probables as P and T (Pair and Territorial); both of these and most of the Possibles involved multiple birds (A2 and A3 abundance codes). Despite these records, it seems likely that these sparrows do not breed in the eastern quarter of Colorado, particularly because Kansas has only one breeding record, that on its eastern border (Thompson and Ely 1992), and in Nebraska they breed only on the eastern edge (Johnsgard 1979).

An absence of shrub vegetation along heavily grazed stream courses on the plains may limit Song Sparrow nesting opportunities. In addition, Song Sparrows appear to avoid hot, humid summer climates, as in Kansas, Oklahoma, Texas, and the Deep South.

Loss of wetland habitat has reduced populations in coastal California (Shuford 1993). BBS data throughout North America show a slight decline. One Colorado population study (the Indian Peaks seasonal counts) shows that in the Arapaho and Roosevelt national forests, where wetlands receive substantial protection, breeding populations appear to have remained steady or increased slightly from 1982 to 1991 (Hallock 1994). Nevertheless the Atlas map displays a wide distribution and the conservation priority programs regard Song Sparrow populations as safe.

BREEDING EVIDENCE

REPORTED IN 582 (33%) OF 1745 PRIORITY BLOCKS

☐	Possible	206 blocks	(35%)
▩	Probable	168 blocks	(29%)
■	Confirmed	208 blocks	(36%)

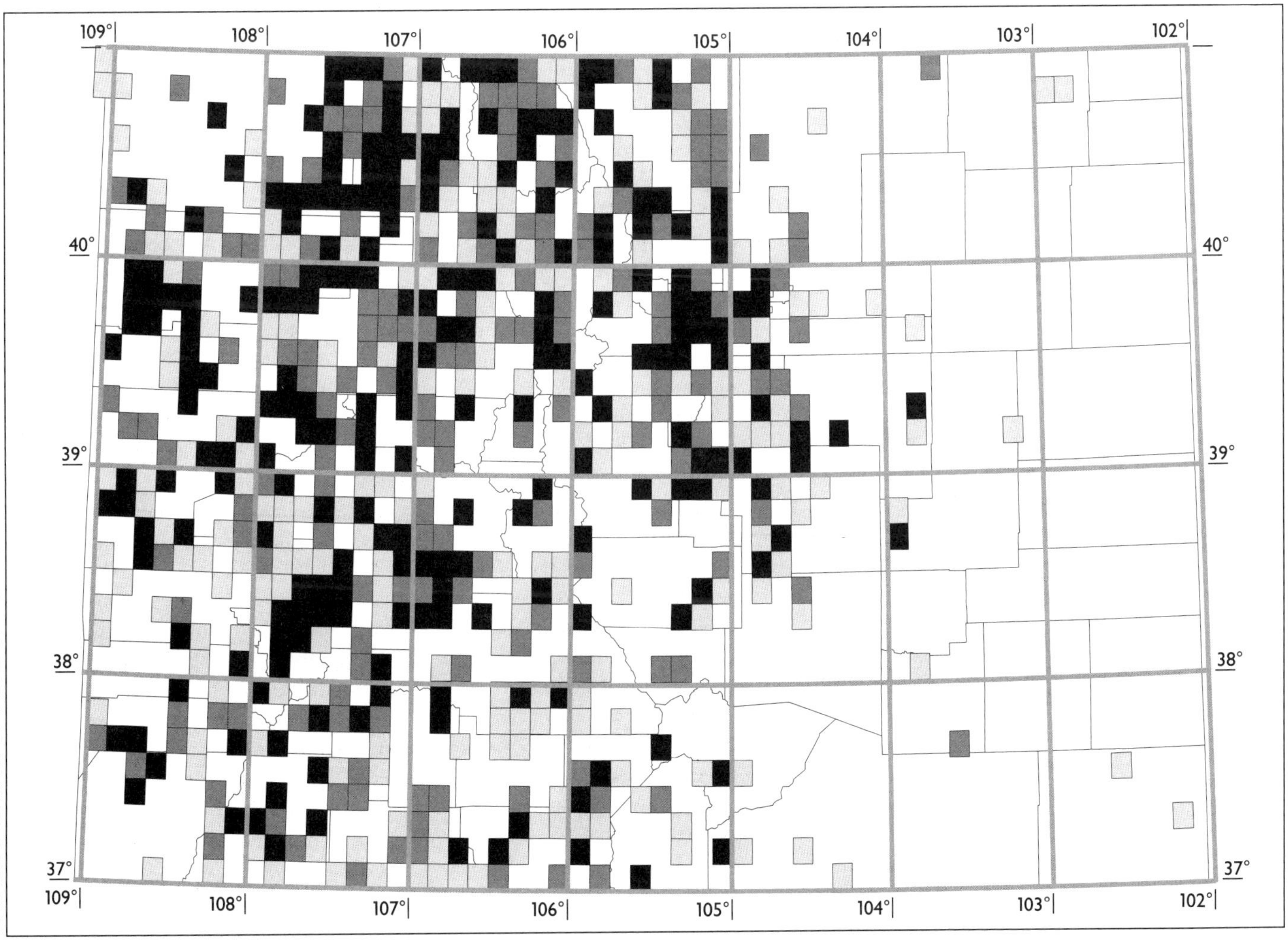
109°
108°
107°
106°
105°
104°
103°
102°
40°
39°
38°
37°

COLORADO DIVISION OF WILDLIFE / WILDLIFE RESOURCE INFORMATION SYSTEM

Although characterized by many as shy and elusive, Lincoln's Sparrows undergo a remarkable change in early summer. With the surge of hormones accompanying the breeding season, these birds become intensely curious when intruders enter their territories. They become, in fact, one of the easiest species to "pish" up onto an open perch. Best of all, they flood the mountain valleys with lively songs of rollicking trills.

HABITAT

Lincoln's Sparrows breed in boggy habitats across the northern third of the continent. These habitats range from sites dominated by tamarack to sites dominated by willows or sedges. With considerable frequency these sparrows also use aspen groves for breeding sites (Ammon 1995, Rising 1996).

Closer to home, middle- and high-elevation willow carrs yielded the overwhelming majority of Atlas records of nests with eggs or young. Aspen groves hosted most of the rest. Nevertheless, a handful of Atlas records reveal that mesic or wet meadows dominated by shrubby cinquefoil—a habitat type largely unrecognized in scientific articles—also provide suitable nesting habitat in Colorado. Regardless of the dominant

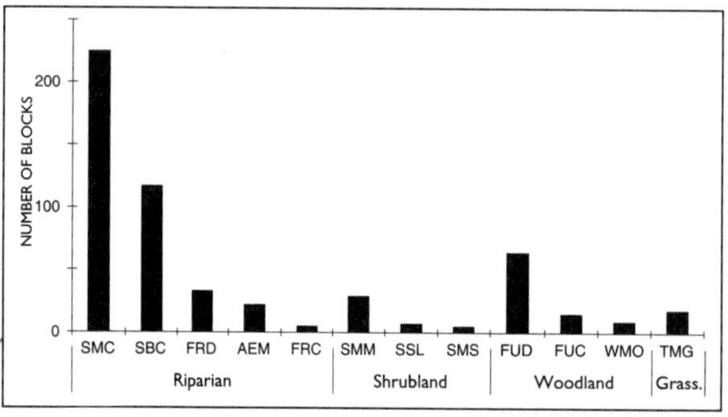
NUMBER OF BLOCKS
200
100
0
SMC
SBC
FRD
AEM
FRC
SMM
SSL
SMS
FUD
FUC
WMO
TMG
Riparian
Shrubland
Woodland
Grass.

plant species, these birds typically conceal their nests at the base of a shrub or in a thick growth of sedges (Rising 1996). Surrounding water levels may reach to the base of an active nest (AV).

Although a handful of Atlas records came from coniferous forests, these probably represent habitat used only for observation and singing. Foraging takes place primarily on wet ground and amid dense foliage where the birds breed (Ryser 1985, Ammon 1995).

Atlasers found several species occupying the same shrublands as Lincoln's Sparrows. At the lower end of their breeding range, these include Song Sparrows, Yellow Warblers, and Dusky Flycatchers, and at higher elevations, Fox and White-crowned sparrows and Wilson's Warblers. Lincoln's can breed in much smaller willow patches than the others. Of all these species, probably Wilson's Warblers most frequently live in close association with Lincoln's Sparrows. Although the prey items of these two species overlap somewhat, the active foraging style of Wilson's Warblers contrasts with the markedly slower foraging style of Lincoln's Sparrows. The diet of Wilson's Warblers thus tends more toward active arthropods, ones that fly in response to disturbance. The diet of Lincoln's Sparrows inclines toward slower and more hidden arthropods, including a much higher percentage of larvae (Raley and Anderson 1990).

BREEDING

Lincoln's Sparrows start to return to lower elevations by late April and in May arrive in the mountains to establish territories. Males start singing their cadences of trills as soon as they arrive on the breeding grounds; nesting soon follows. All Atlas records of nest-building fall before mid June. Records of eggs and nestlings span a much broader range of dates—from June into early August. Although these data hint at double-brooding, Ammon (1995) reports double-brooding as rare and occurring only during favorable years. Because weather-related factors at high elevations might easily disrupt the breeding cycle, later dates may reflect postponed nesting or second attempts.

A typical nest has 3–5 eggs, usually 4 or 5. Females incubate the eggs for about 12 days. In a majority of Colorado nests, one egg will hatch asynchronously, one day later than the others. Both parents participate in feeding nestlings. The young leave the nest about 10–12 days after hatching and spend the next week honing their flight skills and hiding amid dense cover (Ammon 1995, Rising 1996).

Brown-headed Cowbirds occasionally parasitize Lincoln's Sparrow nests (Chace and Cruz 1996; Appendix C). The cowbird nestlings, however, fare dismally. There exist no records of cowbirds fledging successfully from Lincoln's Sparrow nests. Predators of nests and young include Sharp-shinned Hawks, weasels, chipmunks, and, infrequently, Gray Jays. Not surprisingly, better concealed nests suffer lower rates of predation (Ammon 1995).

BREEDING PHENOLOGY

CODE		# OF RECORDS	DATE RANGE
NB	Nest Building	5	30 May–13 Jun
ON	Occupied Nest	5	15 Jun–6 Jul
NE	Nest with Eggs	19	11 Jun–30 Jul
NY	Nest with Young	8	29 Jun–8 Aug
FY	Feeding Young	144	20 Jun–8 Aug
FL	Fledged Young	75	20 Jun–24 Aug

DISTRIBUTION

Lincoln's Sparrows breed throughout Canada and southern Alaska. Within the lower 48 states, they nest only in the western mountain ranges and in northern parts of border states from Minnesota eastward. They winter along the Pacific coast, in the southern tier of states and in most of Mexico.

Cooke (1897) described Lincoln's Sparrows as common from 7,000 feet (2135 m) to timberline. And, despite our penchant for destroying willow-grown areas for roads, reservoirs, towns, subdivisions, and cattle grazing, Lincoln's Sparrows remain common within our borders.

Atlas data provide no great surprises regarding the distribution of these sparrows within the state. They occupy all of Colorado's mountain ranges although the center of abundance follows the Continental Divide. The map demonstrates a definite lower altitudinal limit at about 8,000 feet (2440 m), and abundant distribution from there to timberline. Field workers did substantiate breeding for the first time in three western Colorado latilongs, but this speaks far more about the habits of Colorado's birdwatchers than it does about the range of the Lincoln's Sparrow.

BBS data reflect a fairly high upward trend of almost 3% per year for this species, with no trend established for Colorado. Although this may appear to dispel the notion that loss of habitat implies loss of numbers, BBS routes (begun only recently, in 1968 in the West, and until 1992, too few routes in Colorado to track them) follow already established roads and never travel through reservoirs. We have no mechanism for turning back the clock and comparing current populations to those that existed prior to the roads and reservoirs, many of which displaced sizable chunks of breeding habitat.

BREEDING EVIDENCE

REPORTED IN 413 (24%) OF 1745 PRIORITY BLOCKS

☐	Possible	69 blocks	(17%)
▩	Probable	83 blocks	(20%)
■	Confirmed	261 blocks	(63%)

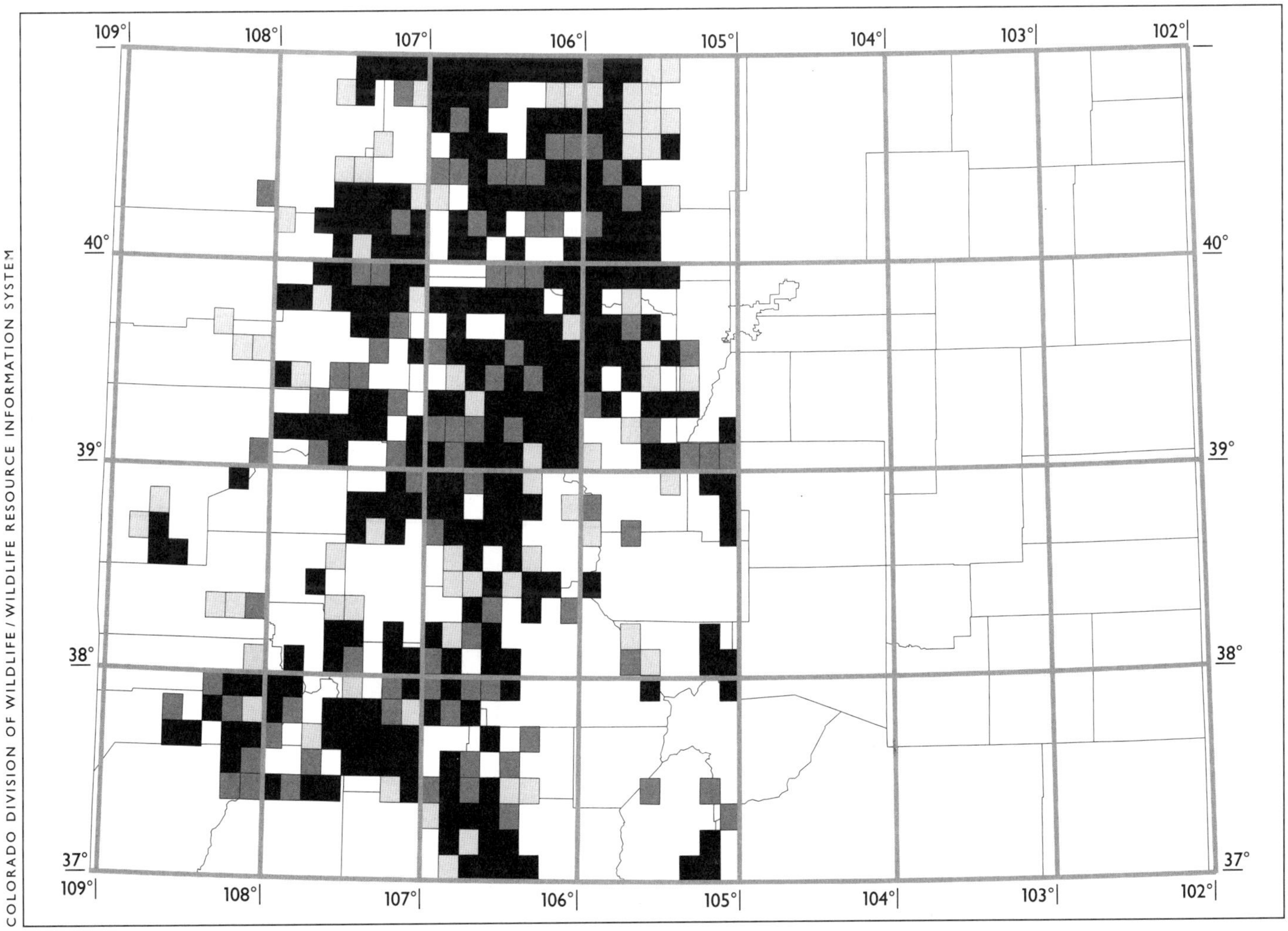
109°
108°
107°
106°
105°
104°
103°
102°
40°
39°
38°
37°
COLORADO DIVISION OF WILDLIFE / WILDLIFE RESOURCE INFORMATION SYSTEM

The haunting songs of White-crowned Sparrows add an unforgettable aura to the subalpine willows and timberline krummholz shrublands. Their captivating trills, much studied by ornithologists, differ from valley to valley and mountain range to mountain range. With discontinuous available habitat discontinuous populations have developed regional dialects. Although all North American *Zonotrichia* occur in Colorado, only this species breeds in the state—mainly at high elevations during June, July, and early August.

HABITAT

Atlas habitat reports support the typical view that White-crowned Sparrows typically nest in willow carrs: two-thirds of the nests found and 52% of the habitat reports came from mid- and high-elevation carrs; another 9% came from timberline krummholz thickets (scrubby, often prostrate, conifers at timberline). These sparrows also nest in other habitats that meet their ecological requirements—in Colorado these include aspen groves, conifer woodlands, montane shrublands, and sagebrush flats.

Three habitat essentials compose their minimum: grass, either pure or mixed with other plants; bare ground for foraging; and dense shrubs or small conifers thick enough to provide a roost and to conceal a nest. Often the habitat includes standing or running water on or near the territory and tall coniferous trees, generally on the periphery of the territory. The most suitable sites have patchy distributions of bare ground, grass, and shrubs (Chilton et al. 1995).

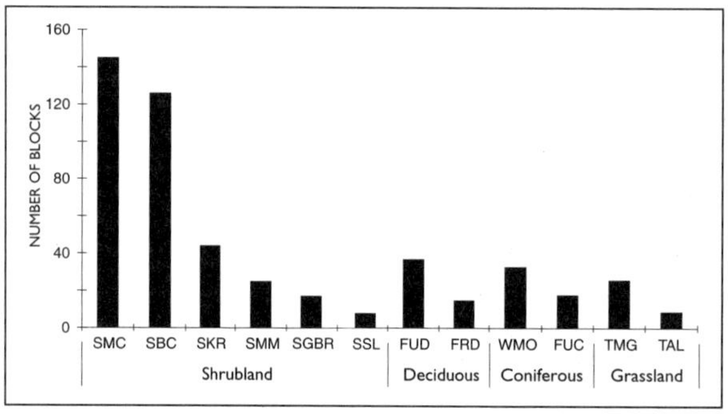
160
120
80
40
0
NUMBER OF BLOCKS
SMC
SBC
SKR
SMM
SGBR
SSL
FUD
FRD
WMO
FUC
TMG
TAL
Shrubland
Deciduous
Coniferous
Grassland

Individuals breeding in riparian habitat cannot tolerate changes in vegetation structure caused by summer cattle grazing (Knopf et al. 1988). In contrast, other human activities such as road construction, logging, and farming often create new habitat, albeit transitory, by providing bare ground and grasslands in previously wooded regions.

Males learn song at two to three months of age, and their sensitivity to learning lasts only three months. They learn songs characteristic of their population, although not necessarily those of their fathers. They develop regional dialects (Baptista 1975, Chilton et al. 1995).

Wilson's Warblers, Fox Sparrows, and Lincoln's Sparrows often nest in willow carrs with White-crowned Sparrows. White-crowned territories may overlap those of Fox Sparrows but White-crowneds displace Chipping Sparrows and Dark-eyed Juncos (Chilton et al. 1995).

BREEDING

Male White-crowneds usually arrive seven days or more before females in order to establish territories (King and Mewaldt 1987); they often return to their previous year's territory (Chilton et al. 1995). Singing perches define territories, and White-crowneds engage in aggressive territorial defense, especially at the boundaries (Petrinovich and Patterson 1982). Nesting and foraging by both sexes take place within the territory.

White-crowneds nest on the ground or close to it. Females usually place their nests under or within dense vegetation (Harrison 1979). Atlasers found most nests on the ground or on low branches of willows and conifers. In years of heavy snow accumulation White-crowneds may nest more frequently off the ground and at greater heights (Morton 1978). On Niwot Ridge in the Indian Peaks west of Boulder, they placed 48% of their nests on the ground at the base of willows and constructed most above-ground nests in spruce trees at the lower edges of meadows where runoff prevented ground nesting (King and Mewaldt 1987).

The female constructs the nest with fine twigs, rootlets, grasses, leaves, forb stems, and bark shreds, and she lines her nest with finer materials including mammal hair. Clutches, usually 4–5 eggs, can range from 3 to 6 (Harrison 1979).

Most Colorado White-crowneds breed in June, July, and early August. Nest-building begins in mid May and can last to early July (19 June was the median Atlas date) whereas nests with eggs were found from 21 May (unusually early) to 20 July (median 29 June). Half of the few nests found with young were before 8 July and half after that date, and fledglings appeared from 28 June to 22 August (22 July median).

The males' territorial defense does not succeed well. About one-third of the nestlings have different fathers than the male who feeds them (Chilton et al. 1995). Colorado has no records of cowbird parasitism (Chace and Cruz 1996), although parasitism of this subspecies has occurred elsewhere (Morton et al. 1993).

BREEDING PHENOLOGY			
CODE		# OF RECORDS	DATE RANGE
C	Courtship	5	28 May–1 Jul
NB	Nest Building	10	28 May–14 Jul
ON	Occupied Nest	4	14 Jun–21 Jul
NE	Nest with Eggs	16	21 May–20 Jul
NY	Nest with Young	7	6 Jul–23 Jul
FY	Feeding Young	106	7 Jun–8 Aug
FL	Fledged Young	68	28 Jun–22 Aug

DISTRIBUTION

White-crowned Sparrows breed throughout much of boreal North America, most often in montane willow carrs, but they breed at sea level along the Pacific coast (Chilton et al. 1995; Dunn et al. 1995). Those that visit Colorado during winter (mainly *Z. l. gambelii*) have different head markings and bill color than Colorado breeders. The Colorado breeding populations belong to subspecies *oriantha*, and they winter primarily in western Mexico.

Colorado birds breed in all montane areas of the western two-thirds of the state, from the Front Range and Sangre de Cristos west to the Park Range and south to the San Juans. Atlasers found the first Confirmed breeding on Cold Spring Mountain in northwestern Moffat County and a surprising contingent along shrubby streams in the San Luis Valley.

According to Atlas abundance estimates, White-crowned Sparrows rank as one of the most abundant breeding birds in Colorado; only 22 other species had greater numbers. Atlasers estimated 107 blocks with 101–1,000 pairs (of 355 with abundance codes), and in 23 blocks scattered over the state they estimated over 1,000 breeding pairs. Highest concentrations breed from the Front Range west to the Flat Tops and north in the Park Range (Latilongs 39105, 39106, 39107, and 40107).

BBS data report that White-crowned populations across the continent have decreased significantly. Dobkin (1994) also reports continent-wide declines. Colorado BBS routes undercontributed to these data because of a paucity of mountain routes, as explained under Lincoln's Sparrow. The apparent abundance of White-crowned Sparrows in Colorado, as shown by the Atlas, means that the Colorado conservation priority schemes give the species a low priority.

BREEDING EVIDENCE			
REPORTED IN 362 (21%) OF 1745 PRIORITY BLOCKS			
☐	Possible	70 blocks	(19%)
▩	Probable	78 blocks	(22%)
■	Confirmed	214 blocks	(59%)

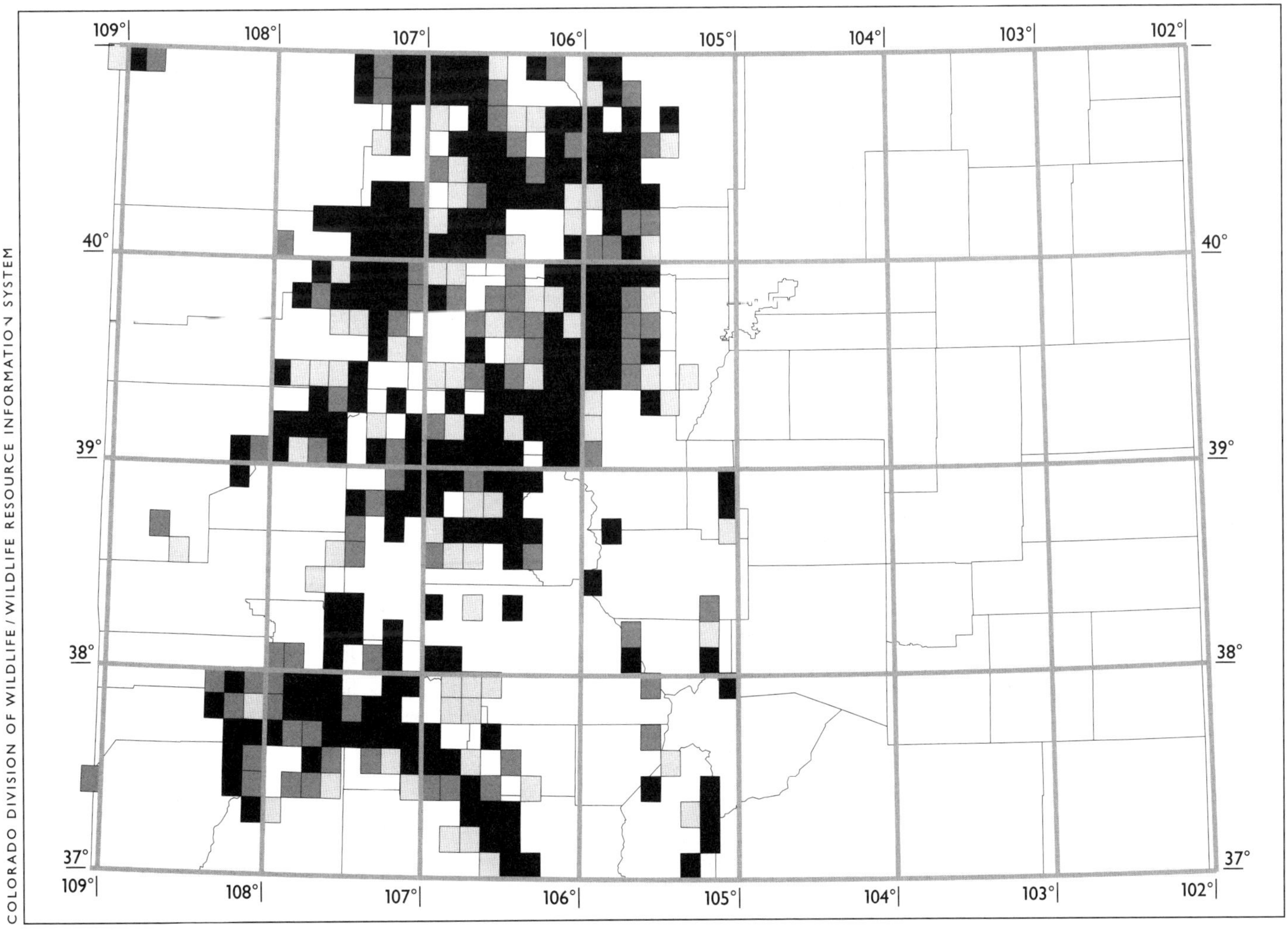

109°
108°
107°
106°
105°
104°
103°
102°
40°
39°
38°
37°

COLORADO DIVISION OF WILDLIFE / WILDLIFE RESOURCE INFORMATION SYSTEM

DARK-EYED (GRAY-HEADED) JUNCO *Junco hyemalis caniceps*

On the northern slopes of the Wasatch Mountains in Utah, Colorado's Gray-headed Juncos freely hybridize with juncos of another subspecies, the Pink-sided Junco (Ryser 1985). This miscegenation led taxonomists to lump these populations together as one species. Along with the Oregon, White-winged, and Slate-colored Juncos, they now form a single species, the Dark-eyed Junco, much to the disappointment of listers.

HABITAT

Gray-headed Juncos nest in a wide variety of wooded habitats (Atlasers reported 18 forest and woodland habitat codes and eight shrubland codes) from the pinyon/juniper zone to timberline (Ryser 1985). They most frequently nest in coniferous forests—70% of all Atlas reports—but also extensively use aspen (20%) and occasionally other deciduous and shrubby habitats. Unspecified upland coniferous forest led all habitat types reported by Atlas workers. If these reports include spruce/fir at the same rate that it occurs among specified conifer reports (FUCL, WMOS, etc.), then 33% of all reported codes indicate this habitat. This proportion would affirm earlier accounts of the importance of spruce/fir

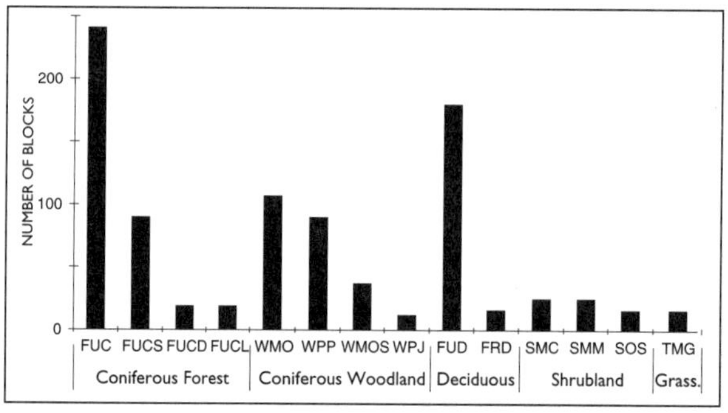
NUMBER OF BLOCKS
0
100
200
FUC
FUCS
FUCD
FUCL
WMO
WPP
WMOS
WPJ
FUD
FRD
SMC
SMM
SOS
TMG
Coniferous Forest
Coniferous Woodland
Deciduous
Shrubland
Grass.

(A&R, B&N). Aspen comes in a clear second with 20% of the reports. Pinyon/juniper appeared as a habitat for Possible or Probable breeding birds in a significant number of blocks; however, no nests were found in this habitat, and the only Confirmations involved mid July fledglings, birds that could have wandered quite far from their natal area.

The large number of habitats reported derives partially from this bird's preference for edges (Hadley 1969). Reports in grass and shrub habitat probably indicate birds on the edges of, or within clearings in, the expected woodland habitats.

BREEDING

In April and May, Gray-headed Juncos follow the receding snow line upward from their wintering grounds in the foothills and adjacent plains (Hadley 1969). Nests found by atlasers show a clear pattern of the density and timing of breeding: Atlas workers found a few nests in mixed conifer forests in late May and early June, many nests in aspen during the last two weeks of June, and many nests in spruce/fir forest from the middle of June to the middle of July. Two August nests in aspen support accounts of some Gray-headed Juncos raising two broods (Hadley 1969, Ehrlich et al. 1988). Juncos feed their young exclusively on insects; the mean hatching dates for first clutches falls approximately one week before the biomass of insects peaks in their breeding habitats (Smith and Anderson 1982).

Juncos typically build their nests of grasses and leaves on the ground beneath a sheltering shrub, log, or rock (Hadley 1969, Smith and Anderson 1982). Rarely, Dark-eyed Juncos will build a nest in a tree, shrub, or building (Ehrlich et al. 1988), and one of the 15 nest-record cards completed by Atlas workers describes a nest 6 feet from the ground on a Douglas-fir branch (Alan Wallace).

Atlasers found this species rather easy to Confirm because the streaked plumage of the fledglings clearly distinguishes them from the adults. The Atlas Confirmation rate for this species rivaled that of the ubiquitous American Robin and those of Cliff Swallows and Black-billed Magpies with their conspicuous nests.

BREEDING PHENOLOGY

CODE		# OF RECORDS	DATE RANGE
C	Courtship	2	12 Jun–20 Jun
NB	Nest Building	9	12 May–10 Jul
ON	Occupied Nest	16	19 May–15 Jul
NE	Nest with Eggs	50	18 May–13 Aug
NY	Nest with Young	20	26 May–22 Jul
FY	Feeding Young	150	2 Jun–24 Aug
FL	Fledged Young	216	10 Jun–24 Aug

DISTRIBUTION

Dark-eyed Juncos breed through the forests of Alaska and Canada, in the East south through the Appalachians, and in all the western mountain ranges. The Gray-headed form breeds in the southern Rocky Mountains, with its center of abundance in Colorado, where only Gray-headeds breed. Colorado has no reports of Gray-headed Juncos breeding with other subspecies. The

Atlas map closely resembles a map of Colorado's montane forests, the birds' absence indicating unforested parks and valleys within the mountainous region.

Atlasers usually reported dozens or hundreds of pairs per block (abundance codes A3 and A4 respectively), and 27 blocks came in with abundance codes of A5 (more than a thousand breeding pairs). Atlasers' abundance codes depict the highest populations in the northern Front Range, the Flat Tops, Gore, Park, and San Juan mountains, which together have over half the state's population (Latilongs 39107, 40106, 40105, 39106, 37106, and 37107 in order of abundance). The total estimate of over 1,100,000 pairs, places Gray-headed Junco as the seventh most numerous Colorado breeding species.

Although Atlas workers reported two sightings of Brown-headed Cowbird fledglings with Dark-eyed Junco adults (Chace and Cruz 1996; Appendix C), the ground-nesting habits of the Dark-eyed Junco largely protect it from cowbird parasitism (Terres 1980). Its preference for edges makes timber cutting a benefit for this species more than a threat. In Colorado's mountain forests, Gray-headed Juncos will continue to contend with Mountain Chickadees, Ruby-crowned Kinglets, and Pine Siskins for the title as "most numerous bird."

BREEDING EVIDENCE

REPORTED IN 613 (35%) OF 1745 PRIORITY BLOCKS

☐	Possible	75 blocks	(12%)
▩	Probable	87 blocks	(14%)
■	Confirmed	451 blocks	(74%)

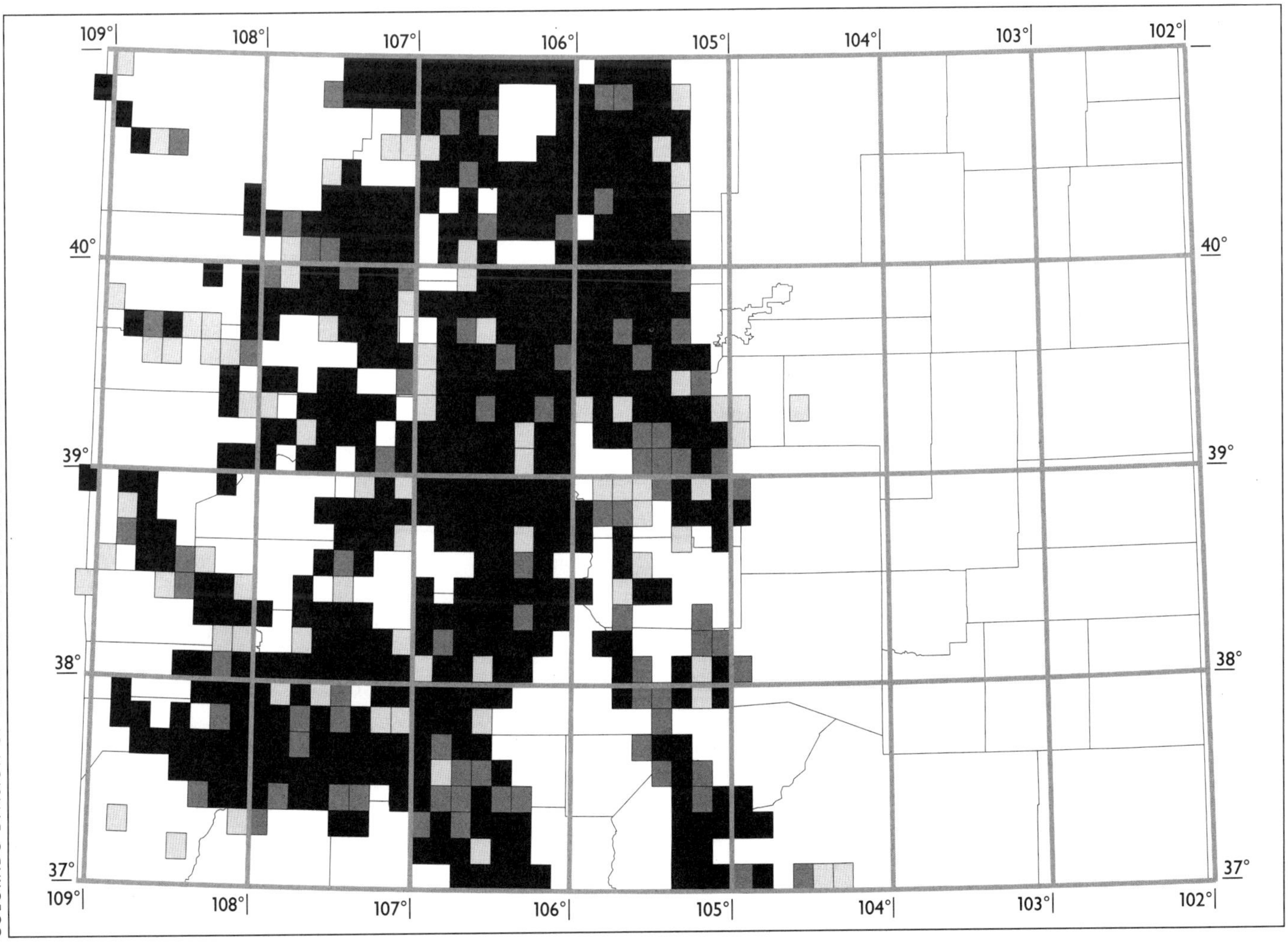
109°
108°
107°
106°
105°
104°
103°
102°
40°
39°
38°
37°

COLORADO DIVISION OF WILDLIFE / WILDLIFE RESOURCE INFORMATION SYSTEM

A May morning on the plains of eastern Colorado finds the air resounding with the songs of our "larking" birds. Distinctive voices in the choir come from McCown's Longspurs fluttering upward, hanging in the air, and parachuting to the earth. All the while their warbling voices ring out as they repeat the performance over and over again.

HABITAT

Male McCown's Longspur displays attract females to their territories on the shortgrass and grazed mixed-grass prairies of the Great Plains. Although, in the McCown's Longspurs' chosen habitat, the grass at times barely covers the ground, prickly pear, patches of lupine and locoweed, along with cow patties, form mini-habitats.

McCown's occupy areas of very sparse grass (buffalo grass and blue grama), large amounts of exposed soil, and low diversity of other plants. These longspurs may have evolved with the bison—they choose grazed pastures and nest geographically irregularly across the prairie.

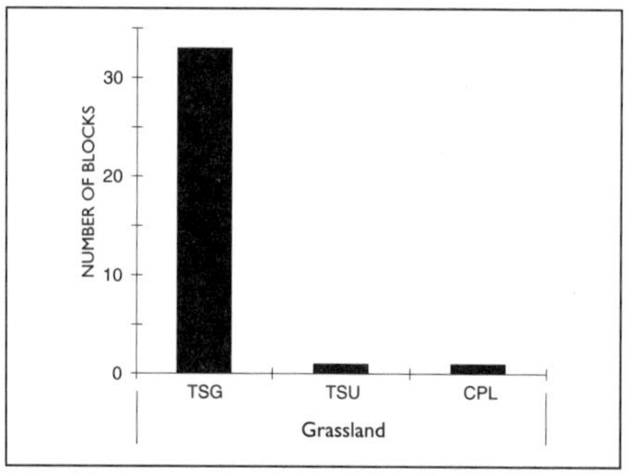
NUMBER OF BLOCKS
30
20
10
0
TSG
TSU
CPL
Grassland

Horned Larks tend to use similar habitats. The two species avoid direct competition for nest sites and food resources by staggering their nesting cycles: the McCown's nesting cycle lags two weeks behind that of Horned Larks. McCown's and Chestnut-collared longspurs eat similar food and have similar timing for their nest cycles, but the space they use differs significantly: Chestnut-collareds favor taller grass with more plant diversity (Creighton and Baldwin 1974) even though both longspurs may breed in close proximity (B&N). Lark Buntings, which also share the prairie with these species, choose habitats similar to those of Chestnut-collared Longspurs.

On the Pawnee National Grassland, McCown's have higher densities on heavily grazed pastures (17–31 pairs/100 acres, 41–76/100 ha) than on lightly grazed pastures (5.5–17 pairs/100 acres, 14–41/100 ha). Densities on six plots averaged only 5–6 pairs/100 acres (12–14/100 ha), demonstrating that they do not occupy all the available range on the Pawnee (With 1994).

These prairie habitats keep McCown's Longspurs well supplied with grass and forb seeds as well as the insects they eat. During breeding, seeds compose three-quarters of the adults' food. They feed nestlings a different diet, mostly insects: on the Pawnee, half grasshoppers, one-third beetles, plus butterflies and moths. Grasshoppers, which they capture by stalking, predominate in their summer diet. The proportion of grasshoppers increases as the season progresses, both for adults and for nestlings as they grow older (Mickey 1943, With 1994).

BREEDING

The April arrival of these longspurs on their prairie breeding grounds is an exciting event for their human observers. Males, which arrive some two weeks before the females (Mickey 1943), perform aerial displays as they flutter upward 30 feet (10 m), hang suspended for a second, then descend on V-shaped wings, showing the startling white linings and the white and black T-patterned tails. These displays establish territories and attract the females, which soon make compact nests on the ground, usually in the shelter of grass tufts, small cacti, or manure piles (With 1994).

Only females incubate the 2–5 eggs, but both parents feed and brood the nestlings. The males often continue their aerial display during incubation, even descending close to the nests, thus revealing their location. Atlas workers found McCown's Longspurs' nests and eggs as early as 18 May and feeding young only until 29 June, a relatively short period. Atlasers also found fledglings for an equally short period, only a month.

BREEDING PHENOLOGY

CODE		# OF RECORDS	DATE RANGE
C	Courtship	3	23 May–2 Jun
ON	Occupied Nest	1	16 Jun
NE	Nest with Eggs	6	18 May–15 Jun
NY	Nest with Young	1	18 Jun
FY	Feeding Young	5	28 May–29 Jun
FL	Fledged Young	8	14 Jun–15 Jul

DISTRIBUTION

McCown's Longspurs historically bred over a wide range in the prairie regions of the western Great Plains, from Oklahoma and Colorado north to western Minnesota, Manitoba, and Alberta (Bent 1968). The breeding range has shrunk drastically (Krause 1968, With 1994). They have not bred in Oklahoma, Minnesota, South Dakota, and Manitoba since 1915 (With 1994) although a South Dakota atlaser reported territorial behavior in 1993 (Peterson 1995).

Northern Weld County remains the center of breeding for Colorado. The solid zone of blocks there demonstrates the importance of that area. However, atlasers discovered previously unknown breeding locations in east-central Colorado 100 miles (160 km) south of their Weld County stronghold. Most of the known pre-Atlas breeding areas for these longspurs in Colorado have been on public lands, specifically Pawnee National Grassland. The Atlas method of surveying one block in each topographic map sent volunteers to work intensively in previously unvisited, privately owned pastures. As a result atlasers found McCown's Longspurs to the south, in Washington, Elbert, Lincoln, and Kit Carson counties. The Confirmed sites, on private ranches near Matheson and Hugo (RRK, David Pantle, VFs), moved the known breeding range south 115 miles (185 km). These same ranches also hosted Chestnut-collared Longspurs, whose breeding range also moved south.

The decrease in overall range and abundance has come about since 1900 (With 1994). Observers attribute the decrease to a variety of factors, including control of prairie wildfires, plowing, and "too rampant" use of pesticides (Oberholser 1974, With 1994). Some include overgrazing in this litany but, because these longspurs select overgrazed pastures in some circumstances, the practice most likely has varying effects on their abundance. The decline probably has a closer link to conversion of the shortgrass prairie to agriculture and urban development (With 1994).

BREEDING EVIDENCE

REPORTED IN 36 (2%) OF 1745 PRIORITY BLOCKS

□	Possible	4 blocks	(11%)
▩	Probable	11 blocks	(31%)
■	Confirmed	21 blocks	(58%)

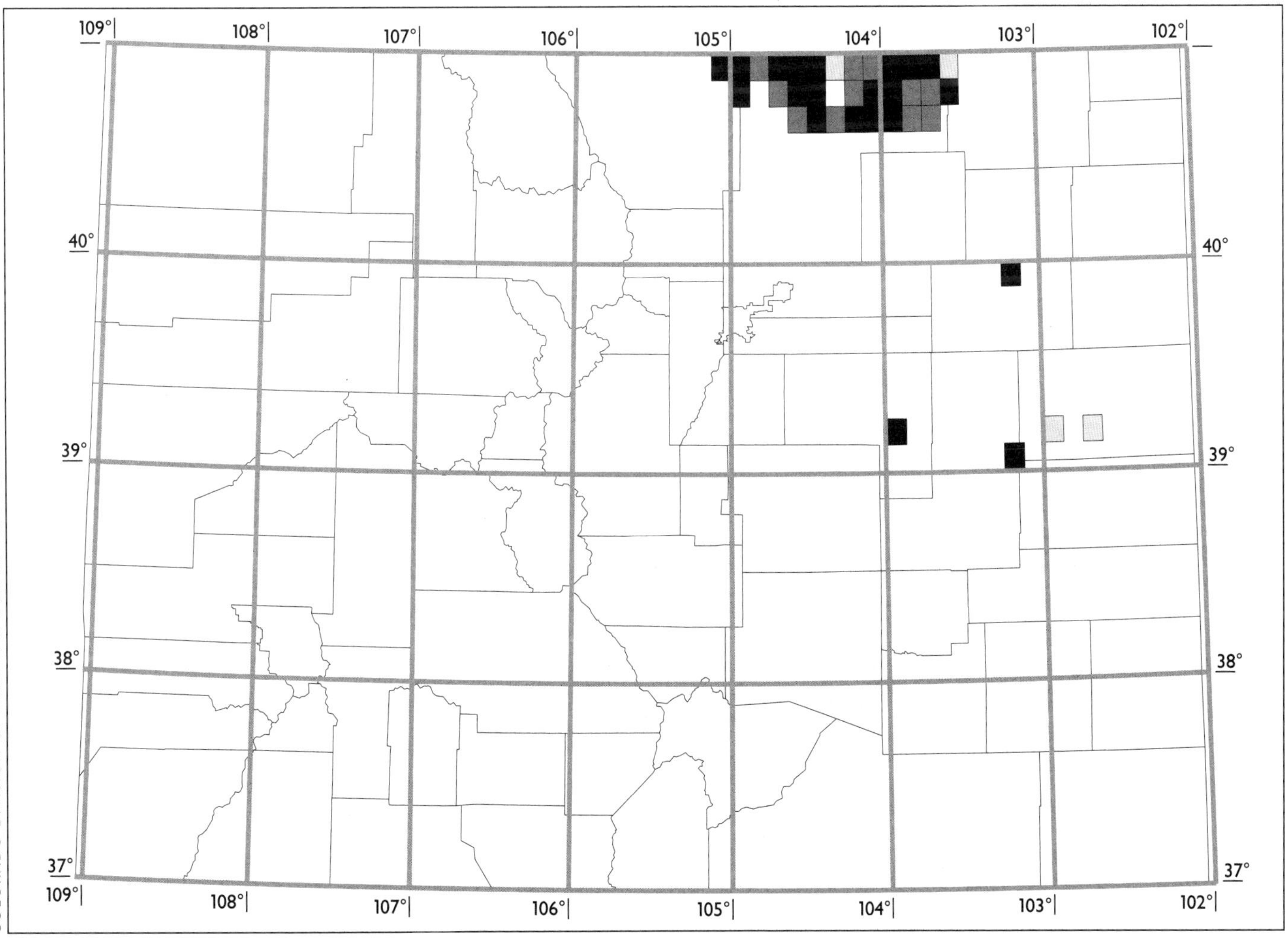
109°
108°
107°
106°
105°
104°
103°
102°
40°
39°
38°
37°

COLORADO DIVISION OF WILDLIFE / WILDLIFE RESOURCE INFORMATION SYSTEM

Many voices, though from only a few species, comprise an incomparable dawn chorus on the prairie. Lacking natural perches, their owners "lark" —sing on the wing—and present an extraordinary musical spectacle. Although scarce on the Colorado prairies, Chestnut-collared Longspurs, where they do occur, add a meadowlark-like song to the music of the grassland.

HABITAT

Atlasers located Chestnut-collared Longspurs only in plains grasslands; they reported principally shortgrass prairie, but also, in three blocks, taller mid-grass prairies. Atlas habitat codes lack the detail to differentiate grass species within the shortgrass prairie—important with Colorado's two nesting longspurs because they split the prairie according to grass height.

The less numerous Chestnut-collareds choose a scarcer habitat within the shortgrass—taller grass in damper locations—than do the more abundant McCown's Longspurs (DuBois 1937). On the Pawnee National Grassland, Chestnut-collareds occupy pastures typified by shortgrass sedge and two taller bunch grasses, the native western wheatgrass and the introduced crested wheatgrass. They pick areas with less bare ground than do McCown's Longspurs and use Canada thistle for song posts (Creighton and Baldwin 1974). They find this grass in slightly rolling country and along the valley floors (B&N).

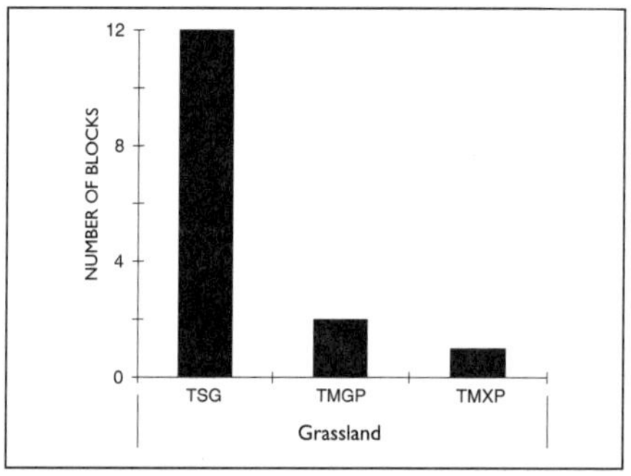
12
8
4
0
NUMBER OF BLOCKS
TSG
TMGP
TMXP
Grassland

Southeast of Hugo, on adjoining cattle ranches belonging to the Bledsoe family, atlasers found several male Chestnut-collared Longspurs in breeding plumage. They sang in isolated patches of taller grass in moist areas, near stock ponds carved out of the arid shortgrass prairie.

Chestnut-collareds share their habitat with the more cosmopolitan Lark Buntings, from which they lessen competition by extending their breeding over a longer time in the summer. They consume a diet similar to that of McCown's Longspurs—insects smaller than the ones Lark Buntings eat. Compared with Lark Buntings, they eat more Coleoptera and Orthoptera (beetles and grasshoppers) and less Diptera (flies, mosquitos, gnats; Creighton and Baldwin 1974).

In states farther north, they breed on the outskirts of towns where golf courses, airports, and idle lands provide cover, as well as on the remaining grassy plains (Bent 1968).

BREEDING

Early observers only suspected breeding in Colorado (Sclater 1912; Bergtold 1928). Not until 1931 was breeding finally confirmed, at Pawnee Buttes (B&N). Chestnut-collared Longspurs inhabit Colorado from late March to late November (A&R) and their breeding season lasts from 1 May to 31 July (Nelson 1993, Atlas).

The birds hide the nest under a clump or tuft of grass, the rim flush with the surface of the ground (Bob Dorn pers. comm.). They lay 3–5 eggs, sometimes 6, and the female, only, incubates for 12–13 days. Atlasers reported breeding from 20 May to 29 July. They located two nests, on 20 May and 18 June, both in shortgrass prairie, the micro-habitat not described.

Cowbirds occasionally victimize these longspurs (Ehrlich et al. 1988) although Colorado has no records (Chace and Cruz 1996).

BREEDING PHENOLOGY

CODE		# OF RECORDS	DATE RANGE
ON	Occupied Nest	2	20 May–18 Jun
FY	Feeding Young	3	20 Jun–29 Jul
FL	Fledged Young	1	14 Jul

DISTRIBUTION

Chestnut-collared Longspurs breed in the shortgrass prairie of the western Great Plains, from southern Canada to eastern Colorado. They winter from just south of the breeding range to the southern Great Plains and the Southwest into Mexico.

Their range has contracted. They formerly nested in western Kansas as far east as Hays (Thompson and Ely 1992). Historically, they may have nested widely throughout the shortgrass prairie of eastern Colorado until grazing and plowing transformed the native prairie.

BY DAVID PANTLE

Prior to the Atlas, breeding in Colorado had been observed only in northern Weld County. The Latilong Study reported breeding in only two of the 28 latilongs, both adjacent to the Wyoming border. Atlasers expanded this to one more latilong. First, atlasers on a BBS survey in southern Washington County discovered a small breeding population 60 miles (95 km) south of the Pawnee on one section (1 mi^2, 2.6 km^2) of grazed remnant prairie. Other atlasers then found two more groups, breeding even farther south (115 mi., 185 km), one near Matheson in Elbert County (Ruth and Walt Kuenning VF), and the Bledsoe ranch group, southeast of Hugo on both sides of the Lincoln/Cheyenne county line (DP).

Human activities have greatly reduced the abundance of bird species that inhabit only grasslands. More of the prairie has disappeared in recent years, due to another round of "sod busting" of marginal sandy slopes. Discovery by atlasers of several scattered breeding populations of Chestnut-collared (and McCown's) Longspurs on large private ranches emphasizes the ecological importance of the remaining prairie pastures.

BREEDING EVIDENCE

REPORTED IN 14 (1%) OF 1745 PRIORITY BLOCKS

Possible	5 blocks	(36%)
Probable	4 blocks	(28%)
Confirmed	5 blocks	(36%)

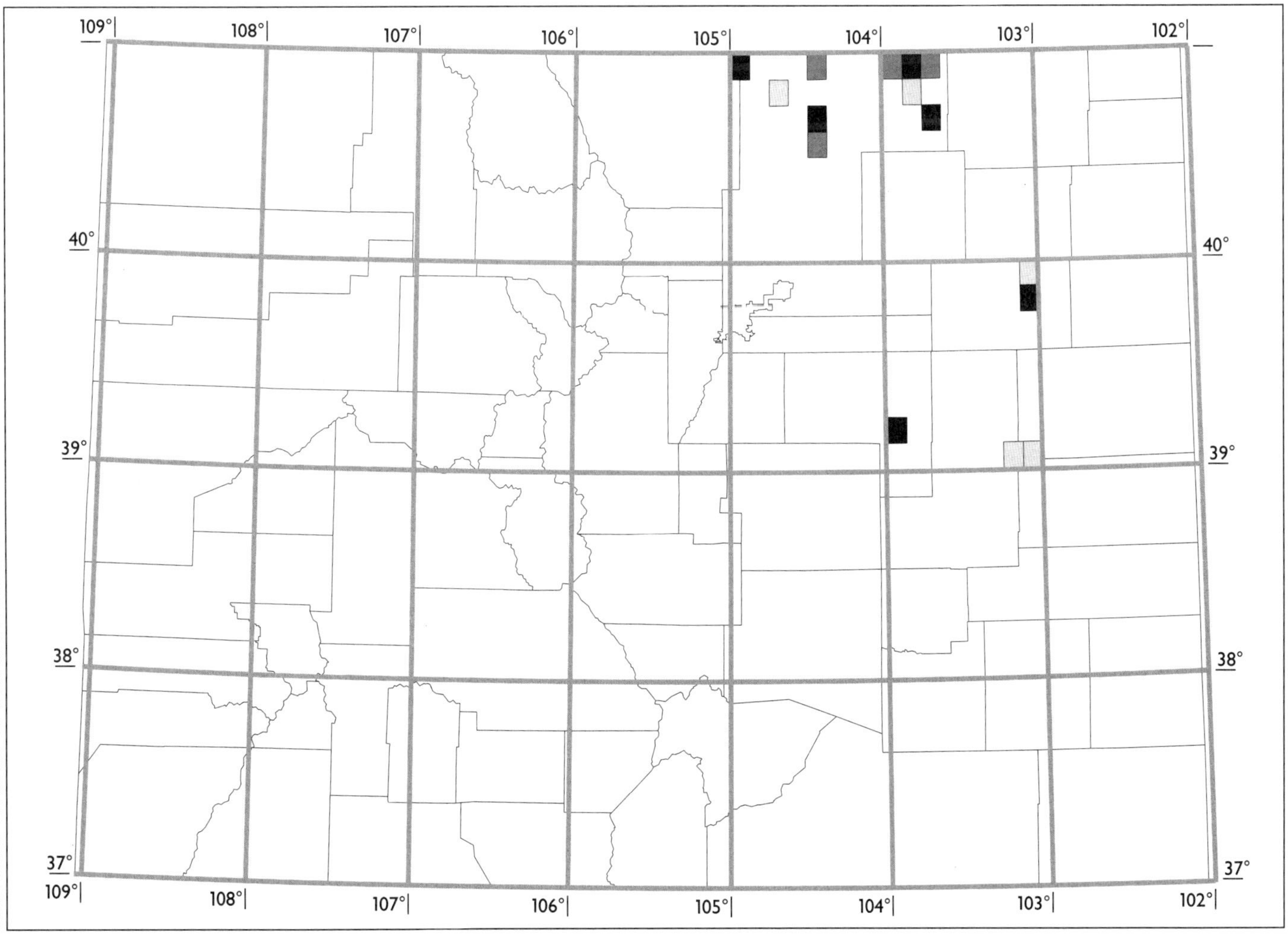
109°
108°
107°
106°
105°
104°
103°
102°
40°
39°
38°
37°

COLORADO DIVISION OF WILDLIFE / WILDLIFE RESOURCE INFORMATION SYSTEM

NORTHERN CARDINAL *Cardinalis cardinalis*

The cardinal serves as an American icon: seven states have adopted it as their state bird; numerous sports teams, from professional football to high schools, call themselves "cardinals;" and the red birds probably outnumber all others on Christmas greeting cards and rural mailboxes. Denverites, especially newcomers, often ask, "Why don't we see cardinals in Denver?"

HABITAT

In the East, cardinals inhabit early successional habitats: shrubby fields, brushy woodland edges, and dense tangles in deciduous woods and in urban and suburban landscapes. They occur abundantly in the shrub/sapling stage in the sequence from farm or meadows to forest (Meade *in* Andrle and Carroll 1988, Schutsky *in* Brauning 1992, Palmer-Ball 1996).

The tangles and thick brush that attract cardinals in the East do not extend into Colorado, at least not extensively. The few sites where atlasers found them, all in Yuma County, lie along low-elevation streams with a combination of trees and shrubs.

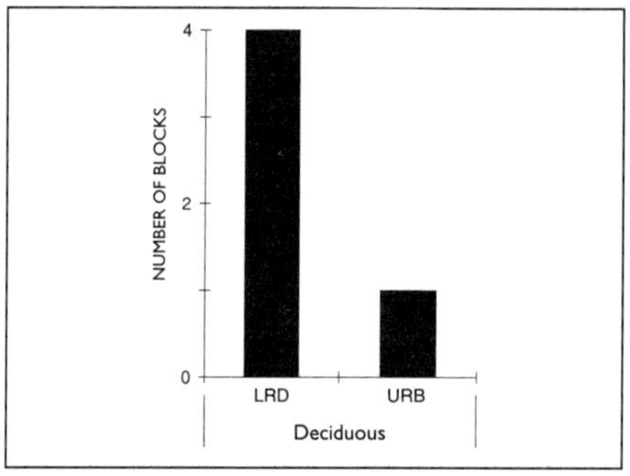
4
2
0
NUMBER OF BLOCKS
LRD
URB
Deciduous

Hackberries, boxelders, and cottonwoods form the tree component; skunkbrush and a variety of shrubby trees comprise the shrub component. Near Idalia, draws dissect the level plains landscape into the "Breaks," each gully filled with several acres of skunkbrush, hawthorns, hackberries, and tall cottonwoods. Most Colorado stream bottoms, plains towns, and Front Range urban centers lack the tangles of shrubs, vines, and trees essential to cardinal prosperity.

The cardinal diet consists mainly of fruits and seeds (70%); insects, spiders, centipedes, snails, and slugs compose the balance (Richards *in* Foss 1994). Nestlings eat almost entirely insects.

BREEDING

While trying to attract females, the flamboyant males stretch their necks and fight each other. Mated pairs sway together and both sing alternately and in unison during courtship. A male may perch near a female, extend his crest, neck, and body, and sidestep down to the female while singing rapidly. After mating they become secretive during the nesting process (Bent 1968).

Cardinals hide their nests well in the thickest part of the vegetation; they use both deciduous and coniferous sites (Peterjohn and Rice 1991). Most studies peg nests at an average of 4–5 feet (1.2–1.5 m) in height, mainly in thickets (Schutsky *in* Brauning 1992, Richards *in* Foss 1994, Palmer-Ball 1996); in Iowa the birds place their nests higher, at an average height of 15 feet (4.5 m; Jackson et al. 1996). The only two described Colorado nests, both in Littleton, were 40–50 feet high, in a maple and a cottonwood (B&N). The nest has four layers: an outer one of rough weedy branches, the next of dead leaves, then a layer of bark, and an inside liner of fine grass (Jackson et al. 1996).

Northern Cardinals nest over an extended period: nests with eggs from 3 April to 16 August in Pennsylvania (Brauning 1992), and breeding seasons in Michigan from mid April to mid September (Bent 1968) and Kansas from 1 April to 20 September (Thompson and Ely 1992). Cardinals may have as many as four broods (Johnsgard 1979). The Atlas nest-building record on 12 July may have represented a second or third brood.

BREEDING PHENOLOGY

CODE		# OF RECORDS	DATE RANGE
NB	Nest Building	1	12 Jul

DISTRIBUTION

Originally birds of the Southeast and Southwest, Northern Cardinals over the last century spread north to southern Canada, but not west into Colorado. One pair nested in Littleton in 1924, and two pairs in 1926 (Bergtold 1927, B&N). Until the Atlas, the state had no other breeding records, although B&N thought it a rare resident along the eastern edge.

In 1991 the Atlas sent a news release to newspapers in Wray and Julesburg, asking if any residents had nesting cardinals. The eight replies (all from Wray) showed two clusters of nesting Northern Cardinals: one in the town of Wray and the other in the Beecher Island Atlas block (39102G2). Atlasers found cardinals in only one other, adjacent, block. Breeding seems confined to Yuma County, along two branches of the Republican River and tributaries such as Black Wolf Creek.

Some Northern Cardinals occasionally stray west to the edge of the foothills (one each on two Denver Christmas Counts in 1996), but it seems unlikely that they will ever brighten very many backyards along the Front Range.

BREEDING EVIDENCE

REPORTED IN 2 (0%) OF 1745 PRIORITY BLOCKS

Possible	0 blocks	(0%)
Probable	1 block	(50%)
Confirmed	1 block	(50%)

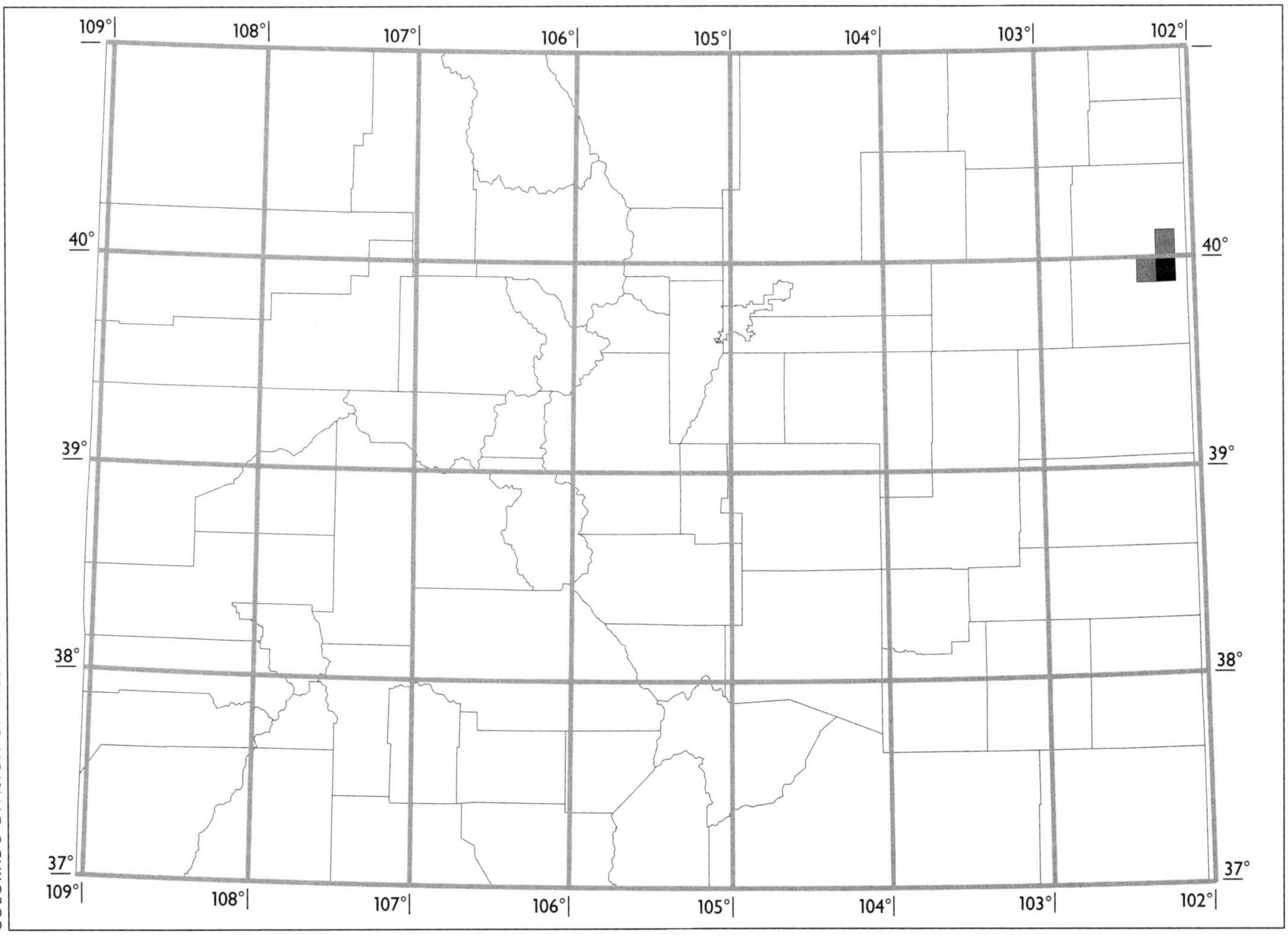
109°
108°
107°
106°
105°
104°
103°
102°
40°
39°
38°
37°

COLORADO DIVISION OF WILDLIFE / WILDLIFE RESOURCE INFORMATION SYSTEM

Its song sounds confusingly like that of the ubiquitous American Robin; its habits tend toward the furtive; and it seems to pop up in almost any habitat. These characteristics endow the Black-headed Grosbeak with a degree of unpredictability that led one Atlas worker to dub it the "Surprise Bird."

HABITAT

Atlas workers recorded 33 different habitat codes for the Black-headed Grosbeak, one of Colorado's most cosmopolitan birds. A&R stated that this species breeds primarily in ponderosa pine, aspen, foothill riparian forest, pinyon/juniper woodlands, and scrub oak shrublands. Atlas workers found them using all of these habitats extensively, each of them constituting 7–15% of the 716 habitat reports from 573 blocks (Priority and Non-priority). According to A&R, they reside much less frequently in lowland riparian forests than in several other habitats; in Atlas reports, however, lowland riparian forests (LRD) came in second in number of reports with 95, 13% of the total. By a slim margin, first place went to foothills riparian forests (FRD) with 108 reports (15%). Several other habitats had significant shares of the total: scrub oak (13%), pinyon/juniper (12%), aspen (11%), ponderosa pine (9%), and mountain scrub (7%). In addition, Atlas workers found these birds using willow carrs (SLC, SMC) in 32 blocks. The requirement for thick cover, either as canopy or undergrowth, seems to unify all this diversity (Ryser 1985), but this species may require the diversity itself (Hill 1995).

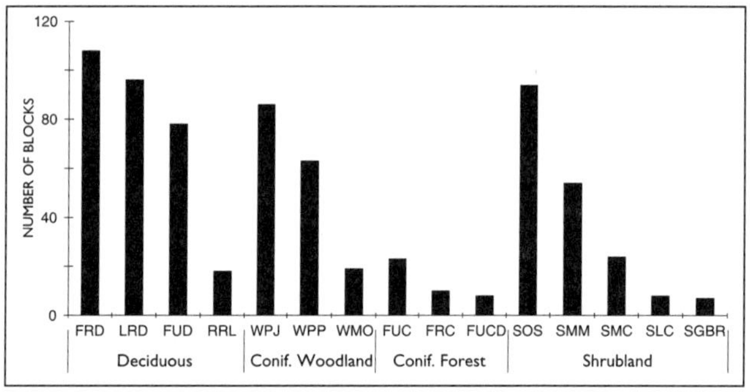
NUMBER OF BLOCKS
0
40
80
120
FRD
LRD
FUD
RRL
WPJ
WPP
WMO
FUC
FRC
FUCD
SOS
SMM
SMC
SLC
SGBR
Deciduous
Conif. Woodland
Conif. Forest
Shrubland

BREEDING

Males arrive and begin singing on the breeding territory in May, the earliest Atlas record falling on 8 May. Nest-building begins soon after. Bent (1968) reported nests in Colorado from 21 May to 17 July, and all nesting activity observed by Atlas workers except for one nest found on 19 July fell within those parameters. B&N noted egg-laying activity in Colorado only in June. No second broods have been recorded previously (Hill 1995), and Atlas dates cluster rather tightly into a pattern of incubation and hatching in June with young birds off the nest but still dependent on parents through July into early August. Nests initiated after early June probably represent renesting attempts after failure (Hill 1995). Failures may occur frequently, for in New Mexico at least, Black-headed Grosbeaks suffer significantly from nest predation by Steller's Jays and Western Scrub-Jays, with losses of 60% in one study (Hill 1995).

BREEDING PHENOLOGY

CODE		# OF RECORDS	DATE RANGE
C	Courtship	3	17 May–4 Jul
NB	Nest Building	14	28 May–9 Jul
ON	Occupied Nest	8	31 May–13 Jul
NE	Nest with Eggs	3	18 Jun–7 Jul
NY	Nest with Young	2	12 Jun–19 Jul
FY	Feeding Young	74	7 Jun–28 Aug
FL	Fledged Young	67	18 Jun–4 Aug

DISTRIBUTION

The Black-headed Grosbeak's breeding range extends from southwestern Canada across the western United States to southern Mexico; it winters in Mexico. Atlas work has clarified conflicting accounts in earlier descriptions of its Colorado distribution. B&N described it as a "common summer resident on the prairies," but A&R described it as rare on the eastern plains; Atlas work supports the latter; few reports came from prairie blocks or from riparian zones in prairie. A&R, however, called it rare in lower western valleys, but Atlas workers found the species common there.

In general, Black-headed Grosbeaks in Colorado breed between 5,000 and 8,000 feet (1525–2440 m). A few may nest higher or lower when appropriate habitat extends to those elevations. Their range extends, in two braod bands, along the Front Range foothills and Western Slope plateaus. They follow the Colorado and Arkansas river valleys into the central mountains and a few breed northeasterly along the South Platte.

Although Atlas workers found Black-headed Grosbeaks in many blocks, they seldom found many in any one block. They reported an abundance code of A4 in only 21 Priority blocks and reported no codes of

A5. They most commonly determined abundance codes A2 (209 reports) and A3 (196 reports). Hill (1995) notes that the world population certainly numbers in the millions. Atlas abundance codes suggest a Colorado population of about 100,000 pairs. The songs of this species resemble those of American Robins and Western Tanagers, birds that often frequent the same habitats, and even very experienced observers occasionally mistake the songs. This circumstance, combined with the Black-headed Grosbeak's rather inconspicuous habits, suggests that Atlas work may somewhat underrepresent both its distribution and its abundance.

BBS data indicate a large increase in population in Colorado (but with an average of only one bird/route) and in the pinyon/juniper region (3.7 birds/route); across the West more routes show increases than decreases. Yet these birds have a localized winter distribution in Mexico and that habitat is suffering moderate (11–25%) losses (Priorities 1995). Population trends bear watching to see whether or not the BBS increase continues.

BREEDING EVIDENCE

REPORTED IN 509 (29%) OF 1745 PRIORITY BLOCKS

▢	Possible	195 blocks	(38%)
▦	Probable	139 blocks	(27%)
■	Confirmed	175 blocks	(34%)

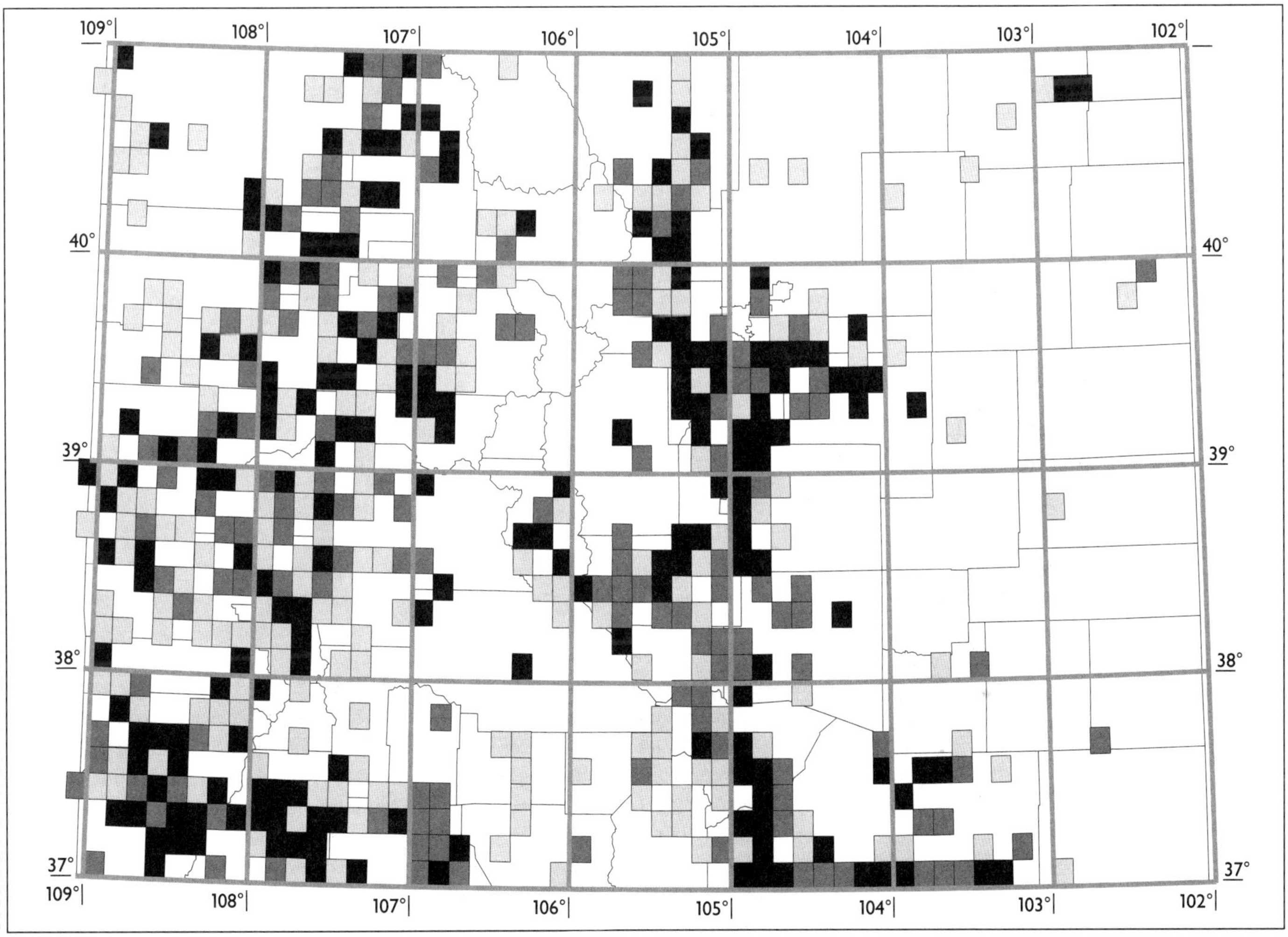
109°
108°
107°
106°
105°
104°
103°
102°
40°
39°
38°
37°

COLORADO DIVISION OF WILDLIFE / WILDLIFE RESOURCE INFORMATION SYSTEM

Although Blue Grosbeaks wear hues that dull in comparison to the vivid blues flaunted by some of Colorado's other blue birds, they certainly "take the blue ribbon" for recycling efforts! These resourceful birds embellish their nests with a potpourri of trash that may include string, cotton, insulation, shreds of fabric, snakeskin, cellophane, paper, and scraps of plastic bags (Ingold 1993, CPM). They also prize horsehair for lining their nests.

HABITAT

Atlas workers discovered Blue Grosbeaks in habitats typical of the western half of their range (Stabler 1959, Bent 1968, Ingold 1993). The birds associate strongly with the edges of lowland and foothills riparian habitats, which compose nearly one-half of all habitats reported. Rural areas, which provide shelterbelts, overgrown fields, garden shrubs, irrigated orchards, and shrubby corridors along irrigation ditches, also compose a significant part of habitats reported. The grosbeaks even use the dense and pervasive stands of tamarisk invading Colorado's lowland riparian corridors.

Blue Grosbeaks tend to place their nests in dense vegetation situated in otherwise open habitats. Supporting structures include small, low trees or shrubs, as well as tangles of briers, vines, or other thickets of vegetation

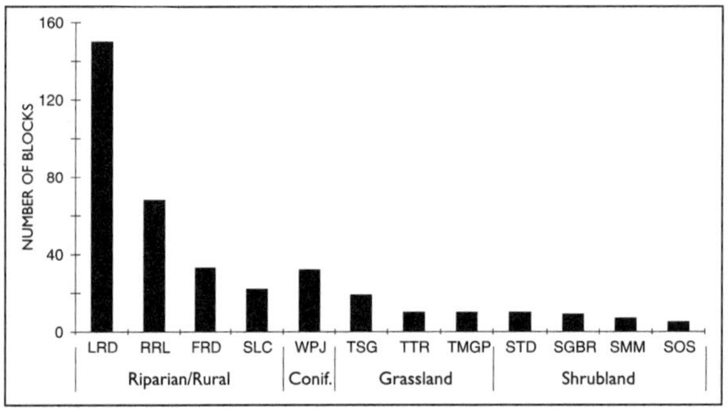
NUMBER OF BLOCKS
160
120
80
40
0
LRD
RRL
FRD
SLC
WPJ
TSG
TTR
TMGP
STD
SGBR
SMM
SOS
Riparian/Rural
Conif.
Grassland
Shrubland

(Ingold 1993). They usually build their nests within 3–8 feet (1–2.5 m) of the ground, although heights range from 6 inches (15 cm) to 21 feet (6.4 m) (Bent 1968).

BREEDING

Of the 300 blocks in which Atlas workers reported Blue Grosbeaks, only 22% had breeding Confirmations, of which few involved nests. Perhaps the somewhat impenetrable character of their preferred nest sites precluded observers from finding nests. The vast majority of block records resulted from Atlas workers detecting individual birds or pairs in suitable nesting habitat, probably a reflection of the unabashed habits of male Blue Grosbeaks guarding their females or proclaiming territory ownership. Typically, they broadcast their busy, conversation-like songs or sharp, metallic *tinks* from conspicuous perches. Also, whether nest-building, incubating, or feeding young, both females and males frequently carry on prolonged bouts of alarm-calling and tail-flicking before ducking into the dense cover concealing their nests (CPM). The birds' conspicuous behaviors and the openness of their habitats probably conspired to yield most Confirmations.

Blue Grosbeaks take their time while migrating north in spring. In Colorado, they begin to arrive after the first week in May (A&R). One Atlas worker caught a notable exception—a Probable nester in the far southeastern part of the state on 4 May. No other records came in until 15 May. The most exceptional nesting date came from a Non-priority block just north of Fort Collins, where the birds started their second clutch on or about 17 August. The three young fledged on 9 September and then left the area one day later (Melcher and Giesen 1996), 18 days after the penultimate Atlas record of a fledgling on 22 August. Perhaps their inclination to dine on grasshoppers (Ingold 1993), which become more prevalent as the summer progresses, encourages Blue Grosbeaks to nest relatively late in the season. The mid August nesting date followed a very wet spring, which probably delayed nesting even more.

Apparently, Blue Grosbeaks rarely fall victim to cowbird parasitism (Friedmann 1963, Friedmann et al. 1977); one atlaser reported an adult Blue Grosbeak with a fledgling cowbird (Chace and Cruz 1996; Appendix C).

BREEDING PHENOLOGY

CODE		# OF RECORDS	DATE RANGE
C	Courtship	3	18 Jun–23 Jun
NB	Nest Building	12	29 May–22 Jun
ON	Occupied Nest	4	28 May–28 Jun
NE	Nest with Eggs	6	12 Jun–24 Jun
NY	Nest with Young	3	23 Jun–31 Jul
FY	Feeding Young	31	15 May–7 Aug
FL	Fledged Young	11	6 Jun–22 Aug

DISTRIBUTION

Atlas results slightly expand the known breeding distribution (B&N, Latilong Study, A&R) of Blue Grosbeaks in Colorado. The majority concentrate in the southeastern quadrant of the state, where the birds find a rich variety of habitats suitable to them, and along most major and some minor riparian systems throughout the remaining part of Colorado. The Front Range corridor from Denver to Wyoming produced a surprising complex of records. The birds generally do not occupy areas above 6,000 feet (1830 m) in elevation, but the handful of exceptions includes a Possible breeder in North Park at about 8,200 feet (2500 m). The pairs found in the San Luis Valley, for which atlasers submitted verification forms, added another, higher-altitude dimension to their breeding territory. The northwestern quadrant of Colorado, which sits along the fringe of the species' northern range limits, yielded the lowest number of blocks reporting Blue Grosbeaks, and the birds altogether avoided the state's mountainous mid section. The most commonly used Atlas abundance code (A2) indicates that the species, though widespread, occurs in relatively low densities.

Historically, Blue Grosbeaks only ranged across the southern half of the U.S. During this century, however, they expanded their range northward, particularly in the Great Plains, where shelterbelts and abandoned farmsteads provided new habitat (Ingold 1993). In Colorado, Cooke (1897) described them as occurring only in the Arkansas Valley, Sclater (1912) found no records from the Western Slope, and B&N described them as "uncommon" in 1965, whereas the more current A&R listed their status as "uncommon to common," depending on location. While many species of Neotropical migrants undergo declines, Blue Grosbeak populations continue to increase (BBS data), possibly in response to man-induced habitat changes like fragmentation (Ingold 1993). The birds spend the non-breeding season in coastal regions of southern Mexico and Central America, however, where the effects of agriculture and deforestation and introduction of exotic plants on their populations remain uncertain (Ingold 1993).

BREEDING EVIDENCE

REPORTED IN 303 (17%) OF 1745 PRIORITY BLOCKS

□	Possible	117 blocks	(39%)
▩	Probable	119 blocks	(39%)
■	Confirmed	67 blocks	(22%)

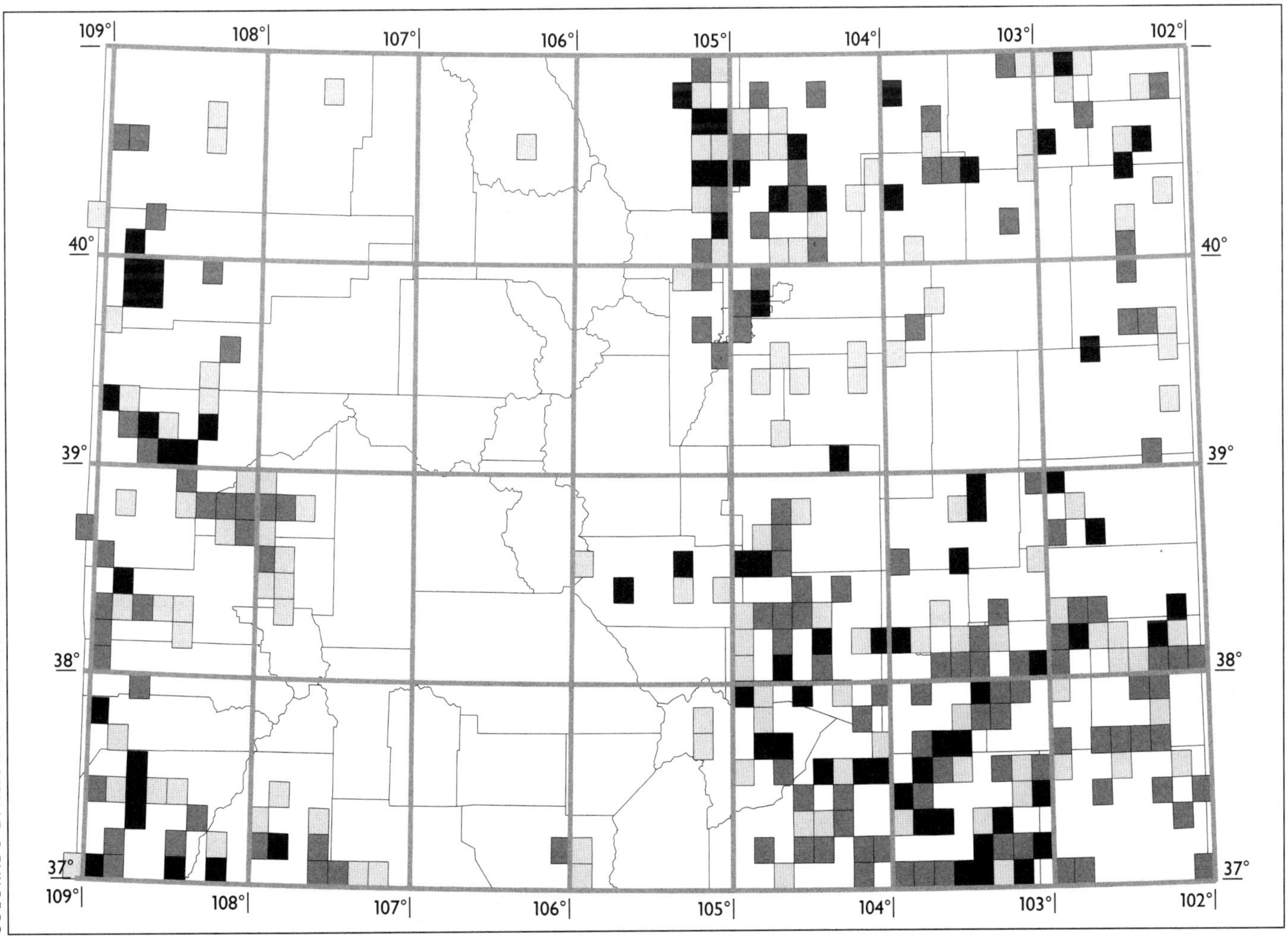
109°
108°
107°
106°
105°
104°
103°
102°
40°
39°
38°
37°

COLORADO DIVISION OF WILDLIFE / WILDLIFE RESOURCE INFORMATION SYSTEM

A warbling song persistently echoes off canyon walls; the ear leads the eye to a bundle of feathers in a kaleidoscope of turquoise, fawn, and white, perched atop a scrub oak, framed by a backdrop of cliffs that have taken 300 million years to reach the just-right patina of red. This habitat, in the Pennsylvanian Age formations of the Front Range, epitomizes a typical setting for Lazuli Buntings (Benedict 1990).

HABITAT

Atlasers found "The Lazuli" most at home in shrublands. The birds especially favor the shrubs associated with the Transition Zone, 5,500 to 7,000 feet (1675–2135 m), particularly riparian areas that allow rich mixes of Transition Zone shrubs with those of the eastern plains and western valleys. Atlasers also found mountain shrublands, comprised mostly of mountain mahogany and serviceberry, of almost equal importance. To a lesser degree these buntings use a variety of other shrub habitats, notably those associated with pinyon and juniper, as well as sagebrush. The fact that they usually build their nests within 4 feet of the ground, coupled with their combined granivorous and insectivorous diet, connects the birds to the shrublands.

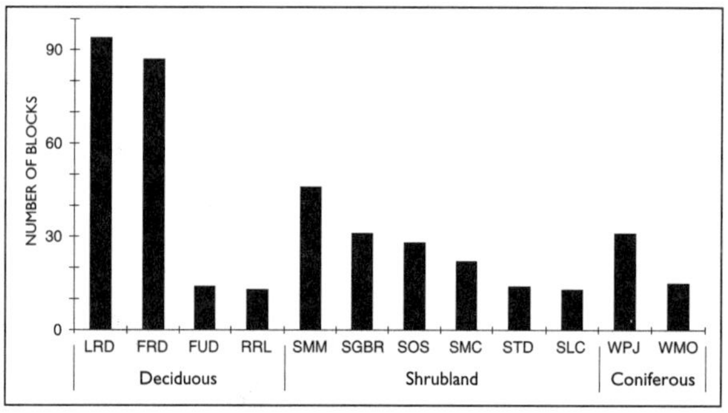

NUMBER OF BLOCKS
0
30
60
90
LRD
FRD
FUD
RRL
SMM
SGBR
SOS
SMC
STD
SLC
WPJ
WMO
Deciduous
Shrubland
Coniferous

Even though all available references limit Lazuli Buntings to shrubby habitats (e.g., Greene et al. 1996), 42% of Atlas habitat reports came from riparian woodlands. The generality of Atlas habitat codes precludes details about the understory where observers found the birds, but most riparian areas the birds use have those rich shrubland mixes.

Nevertheless, Lazuli Buntings do nest in cottonwood groves at Chatfield State Park; they use edge habitat that includes patches of willows, but also sing from the tall cottonwoods (Bottorff et al. 1971–1984). The buntings typically use taller shrubs and trees within their territory for song perches.

The birds feed on the ground, in shrubs and grasses, and up to 65 feet (20 m) in cottonwoods, aspen, willows, and chokecherries. They pick up seeds on the ground and they perch on stems of grasses and other plants to extract seeds. They glean insects from tree and shrub foliage and even pursue flying insects by sallying out from perches (Greene et al. 1996).

BREEDING

In 1841, John James Audubon supplied one of the first descriptions of the nest: "usually placed in willows along margins of streams [and] composed of sticks of fine grasses and cow or buffalo hair" (*in* Bent 1968). Atlasers found a protracted nest-building season, from 15 May to 2 July. Although the birds evinced no hard evidence of second nesting attempts, the circumstantial evidence of the extended Atlas breeding phenology points in that direction. The late dates also could involve renesting after a failure of the first attempt. Only 22% of the Atlas observations Confirmed nesting, which not only speaks of the stealth of Lazulis, but more importantly may provide insight into their success in breeding.

Atlasers found one nest containing eggs of Brown-headed Cowbirds and one pair feeding fledgling cowbirds (Appendix C). Other reports (Sclater 1912, B&N) also substantiate cowbird parasitism of the species in Colorado. At the Canadian northern edge of the range, essentially no cowbird parasitism occurs, but in a Montana study site, all of 44 nests suffered hits from cowbirds. A Utah survey found an intermediate rate—24–36% (two different years). The rate seems to depend on breeding chronology: in Montana cowbird density peaked during the start of the bunting nesting season (Greene et al. 1996).

Hybridization between Lazuli and Indigo buntings can and does occur to some degree where the two are sympatric (Payne 1992). However, atlasers made only four reports of hybrids, none documented with a written description.

BREEDING PHENOLOGY

CODE		# OF RECORDS	DATE RANGE
C	Courtship	6	4 May–24 Jun
NB	Nest Building	12	15 May–2 Jul
ON	Occupied Nest	1	23 Jun
NE	Nest with Eggs	1	15 Jun
NY	Nest with Young	2	23 Jun–28 Jul
FY	Feeding Young	35	2 Jun–5 Aug
FL	Fledged Young	19	13 Jun–20 Aug

BY ROBERT RIGHTER

DISTRIBUTION

Lazuli Buntings range westward from the 100th Meridian and south from southwestern Canada to southern California and south-central New Mexico. As Indigo Buntings have extended their range to the west, Lazuli Buntings have extended their range to the east (Bent 1968). Now an extensive zone of overlap exists, shifting west, from Oklahoma and Colorado to South Dakota and Montana. Lazulis are not extending their range east as fast as Indigos move west (Greene et al. 1996).

Prior to man's habitat interventions through the creation of water impoundments along the main east–west watercourses, the rivers on the plains flooded every spring, thereby curtailing the floristic growth that now occurs. Previously an expanse of shrubless grasslands geographically separated Lazuli and Indigo buntings from each other. By altering the annual regimen of the stream, irrigation runoff and controlled releases from upstream reservoirs have created a botanic bridge to the Rockies that induced range extensions for both buntings.

Atlas sightings have helped to establish that the main distribution of Lazulis follows the brush and shrub habitat that lies along foothills and low mountains. The map shows a spotty distribution, one linked to the presence of appropriate shrubby habitat. Occasionally and marginally this species ranges higher in elevation but seems restricted by the montane conifer forests when it does. The shortgrass prairie on the eastern plains also acts as a limiting factor. Lazulis nest easterly in shrubby habitats along the Arkansas and South Platte valleys and tributaries.

There seems little evidence that the status of this colorful bunting has changed since the turn of the century, at least in the main part of the range (Sclater 1912, B&N, A&R). Irrigation ditches north of the Arkansas in Otero and Bent counties now provide shrubby habitats that the birds exploit. No doubt the distribution has and will change subtly with the ebb and flow of brush and shrub.

BREEDING EVIDENCE

REPORTED IN 292 (17%) OF 1745 PRIORITY BLOCKS

☐	Possible	127 blocks	(43%)
▩	Probable	100 blocks	(34%)
■	Confirmed	65 blocks	(22%)

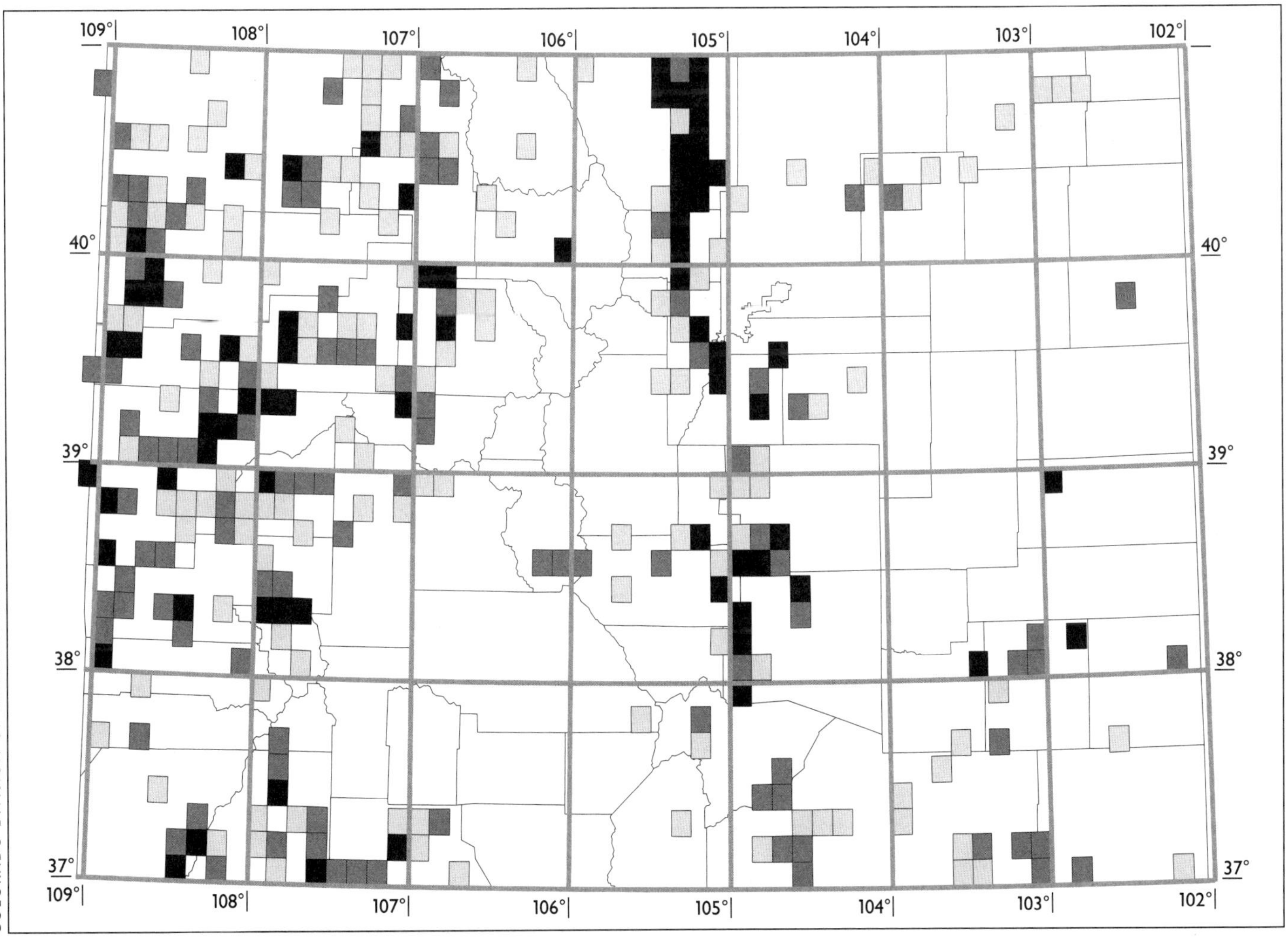

109°
108°
107°
106°
105°
104°
103°
102°
40°
39°
38°
37°

COLORADO DIVISION OF WILDLIFE / WILDLIFE RESOURCE INFORMATION SYSTEM

Colorado is being invaded. Who better than by Indigo Buntings? We can expect to see more of these five-and-a-half inch conquistadors in the years to come. The stunning indigo-blue color of the males, flashing through the shrubbery, will no doubt attract attention more often than the petite, mouse-brown females.

HABITAT

Indigo Buntings occur in a variety of brush and shrubs that grow along rivers, creeks, roads, shelterbelts, agricultural fields, forest and field edges, as well as clearings. This habitat can be stable for a period of time or tenuous, depending upon man's involvement and the successional stage of the habitat. Audubon (1861) perceptively described this bunting as "not a forest bird but [one that] prefers the skirts of the woods."

Over 60% of the Atlas sightings came from riparian areas. Tamarisk, another invader (not as welcome), now grows along many creeks and rivers as one of the dominant

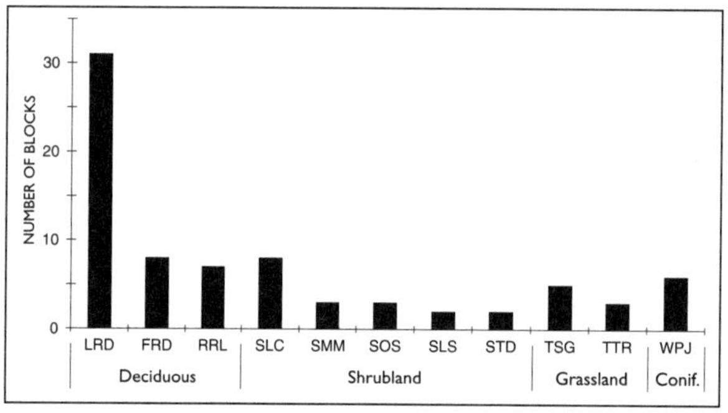
NUMBER OF BLOCKS
30
20
10
0
LRD
FRD
RRL
SLC
SMM
SOS
SLS
STD
TSG
TTR
WPJ
Deciduous
Shrubland
Grassland
Conif.

shrubs of the lower segments of this habitat. This rapidly spreading shrub may favorably influence the status and distribution of these buntings. The rest of the Atlas observations came from varied habitats, most associated with shrub and brush. Curiously, these snazzy all-blue buntings shy away from urban environments even when they could find suitable habitat (A&R, Payne 1992).

Lazuli Buntings, in many instances, use some of this same habitat, and Atlas field workers reported occasional hybridization (four sight reports) between the two species.

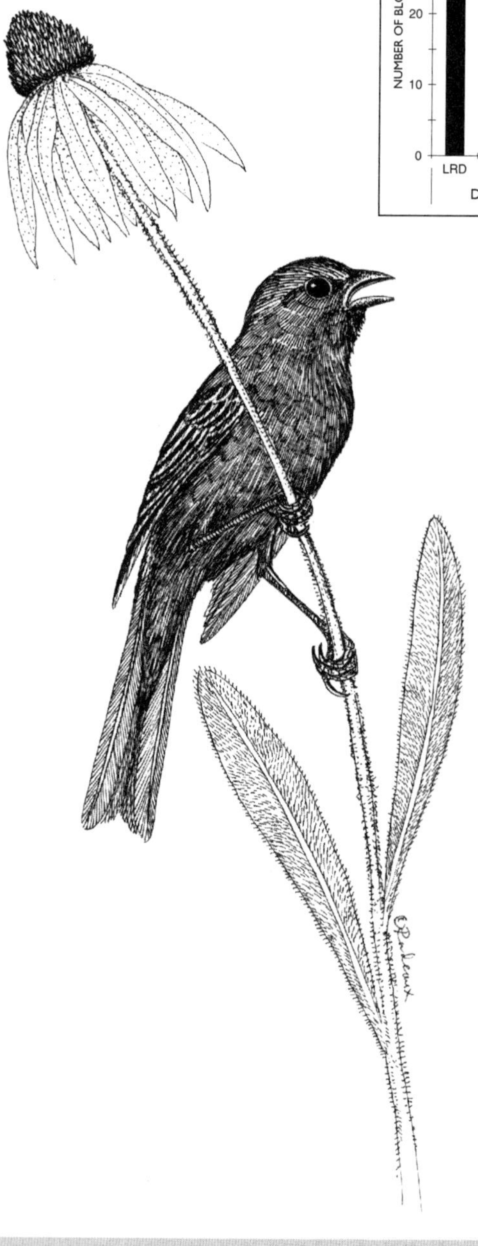

BREEDING

Indigo Buntings start arriving in the state early in May. The female, craftily working alone, builds an open nest about one yard from the ground, utilizing a multitude of vegetative materials. The first nest for the state was found in late summer, 8 August 1953. The earliest reported fledgling during the Atlas was on 13 July. The late dates of these nesting records suggest that the species produces at least two broods while in Colorado.

Detecting breeding Confirmations was frustrating for keen-eyed Atlas field workers, as they recorded only four in 77 sightings. This parallels the low breeding Confirmation percentage recorded for Lazuli Buntings. The cunning of these buntings at camouflaging their nests and fledglings from would-be predators may explain their success and recent expansion.

Brown-headed Cowbirds parasitize this species. A female Indigo Bunting will reject spotted cowbird eggs laid in the nest before she lays her unspotted eggs. If the cowbird lays her eggs in the nest after the bunting has started to lay, she accepts the eggs (B&N, Payne 1992).

BREEDING PHENOLOGY

CODE		# OF RECORDS	DATE RANGE
C	Courtship	1	4 Jun
FY	Feeding Young	2	10 Jul–28 Jul
FL	Fledged Young	2	13 Jul–22 Jul

DISTRIBUTION

The vast majority of Indigo Buntings summer in the U.S. east of the 100th Meridian from Minnesota and Maine south to the Gulf coast. They winter from Florida and southern Texas to Panama. Beginning in the 1940s they began to pioneer the Southwest (Payne 1992). This expansion coincided with man's manipulation of the western river systems with water impoundments that allowed willows and non-native tamarisks to expand (Rosenberg et al. 1991). For example, by 1964 Indigo Buntings had started to breed in Oak Creek and Prescott, Arizona (Phillips et al. 1964), and in the Grand Canyon region. Only 20 years later, they had become "uncommon" summer visitors to the Grand Canyon (Brown et al. 1987).

In Colorado, it took 80 years from collection of the first specimen in El Paso County on 8 May 1872 (Sclater 1912) until discovery of the first nest near Morrison in 1953. The Western Slope had no records until 1969 (Davis 1970). Now these buntings are frequent but local summer residents on the eastern plains and western valleys (A&R). Even if they arrived via the Platte and Arkansas, the Atlas map shows that most inhabit the southern half of the state, particularly southeastern Colorado. Half the records and three of the four Confirmations came from the southeast. Not only the river courses, but also shrubby hillsides, support them.

Prior to the Atlas, northwestern Colorado, with comparatively little field work, had no breeding records (A&R). Atlas field workers ferreted out six sightings for the region, including a Confirmed breeding in Garfield County (Horse Mtn., 38107F7), demonstrating that Indigo Buntings have more of a presence in this region than previously known.

Despite this western pioneering, the BBS for North America shows a modest drop of 0.7% per year, which computes to a 31-year decline in the total population of 20%. The gradual increase in Colorado—involving comparatively few birds—neither shows up on BBS routes nor has a significant effect on total numbers of this bird.

As water impoundments continue to curtail flooding on the major rivers of the eastern plains and western valleys, and as the newly emerging shrub habitat prospers and lures Indigo Buntings westward, we should expect to see more of these sapphire-blue opportunists adorning our shrubs.

BREEDING EVIDENCE

REPORTED IN 62 (4%) OF 1745 PRIORITY BLOCKS

Possible	34 blocks	(55%)
Probable	24 blocks	(39%)
Confirmed	4 blocks	(6%)

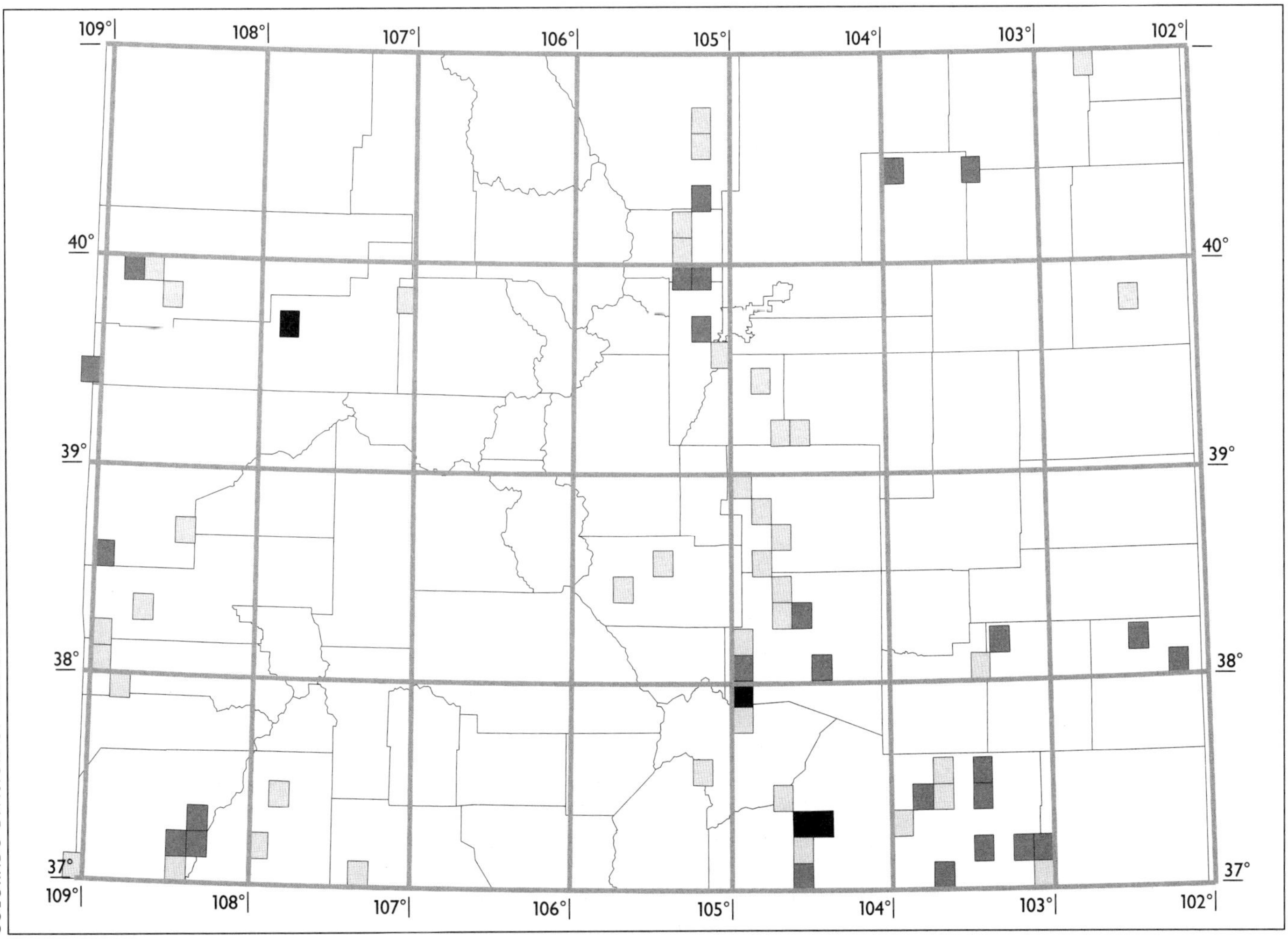
109°
108°
107°
106°
105°
104°
103°
102°
40°
39°
38°
37°

COLORADO DIVISION OF WILDLIFE / WILDLIFE RESOURCE INFORMATION SYSTEM

Singers of a simple, name-dropping (their own) song, Dickcissels play "hide-and-seek": where, if at all, will these nomadic breeders appear this summer? (Roberts 1932). The hoarse, incessant, unmusical *Dick-ciss-ciss-ciss* has a certain summery quality—the persistency of the notes matches the mood of summer heat (Pearson 1942).

HABITAT

The inconstancy of Dickcissels stems from their habitat requirements—a plains habitat that varies from year to year because precipitation varies so much. They seek disturbed agricultural fields; older fields with dense vegetative cover attract them. They have a second requirement: perches, either forb or woody, that poke above other vegetation in the field. They use fence posts and utility wires but still require natural song perches in their territories. Suitable perches average 19 inches (49 cm) above the rest of the field (Zimmerman 1971).

Dickcissels use mixed and tallgrass prairies, croplands with alfalfa, clover, and timothy, and abandoned or fallow crop-

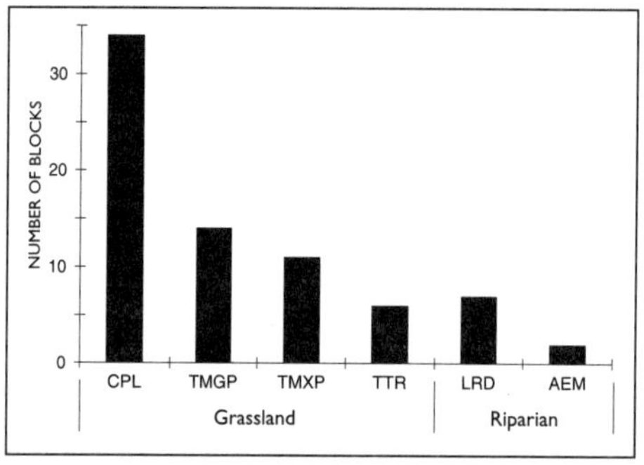
NUMBER OF BLOCKS
0
10
20
30
CPL
TMGP
TMXP
TTR
LRD
AEM
Grassland
Riparian

lands—ephemeral habitats that explain the annual hide-and-seek. Within their chosen fields, successful males pick territories that have the greatest volume of vegetation (Zimmerman 1971).

During the Atlas Dickcissels picked traditional disturbed sites—swales with tall dense growth, fence-rows with thick sunflowers—and a recent (since 1986) phenomenon: federally sponsored Conservation Reserve lands. Because government policies change, this boon to Dickcissels may disappear from their habitat options. This happened in 1996 when the government allowed farmers to harvest the C.R.P. fields before the end of the Dickcissel nesting season (Gordon East pers. comm.).

Colorado lies at the western fringe of the tall or midgrass prairie, and our Dickcissels have a particularly irregular annual schedule. From year to year, different sections of the plains receive different amounts of precipitation, and that affects plant growth dramatically, at least for Dickcissels. Dickcissels' irregularity, erratic mass invasions, and extra-limital excursions result from drought in the Great Plains (John Zimmerman pers. comm.).

BREEDING

During courtship and territory establishment, Dickcissels sing their unmusical chant throughout the day—as much as 50% of daylight hours (Schartz and Zimmerman 1971). They have a 1:1 sex ratio but practice polygyny, which leaves many singing males as bachelors (Zimmerman 1966).

Females pick their mates according to availability of suitable nest sites (Zimmerman 1982). After the young hatch, the females do not respect male territories and often forage beyond their mates' boundaries. The Dickcissel community accepts female wandering without aggression (Zimmerman 1966).

Males with good territories attract females; many with good song perches but sparse vegetation keep their space but remain bachelors (40%). Successful males may have two to five mates—although usually their mates have staggered nesting cycles (Zimmerman 1982). Polygynous males fledge more young, but monogamous males fledge a better percentage. Clutch size, which varies from 2 to 6 eggs, increases later in the season (Harmeson 1974).

The males act "very attentive to the female but not to the nest"; as the females build nests the males follow without helping (Zimmerman 1966). By themselves, females build nests, incubate, brood, and feed the young (Harmeson 1974). The process takes 6–7 weeks and females do not initiate second broods (Zimmerman 1982).

Atlas dates show a short two-month range of breeding dates, from late May to 1 August. In the peak year of 1994, dates progressed from south to north: observations in Latitudes 37 and 38 dominated before 20 June (10 of 13); Latitudes 39 and 40 dominated after that (11 of 15), with three of four Confirmations in July.

The female builds the nest on, or within a foot of, the ground. Grasses and forbs arching over the top conceal it well from human observers (hence we know little about Dickcissel breeding behavior) but apparently not from a plethora of predators. Dickcissels lose high percentages of nests and young to snakes, ground squirrels, and other predators.

Dickcissels service a substantial number of cowbirds; a Kansas study found cowbird eggs in 18 of 19 nests (Elliott 1978). The number of cowbird eggs averages 2.3–2.9/ parasitized nest and host nests produce an average 0.4–0.6 cowbirds each (Zimmerman 1982). Cowbirds impact most heavily on low-density nesting populations (Fretwell 1986).

BBS surveys show a drop of 1.6% per year (60% of the routes showed declines); this statistically significant and alarming trend suggests that the population has dropped by over a third in 31 years.

BREEDING PHENOLOGY

CODE		# OF RECORDS	DATE RANGE
NY	Nest with Young	1	23 Jul
FY	Feeding Young	5	21 Jun–1 Aug
FL	Fledged Young	1	20 Jul

DISTRIBUTION

Dickcissels nest mainly in the Great Plains, from southern Alberta and Michigan to northern New Mexico, Texas, and Georgia. Most winter in Venezuela.

Early ornithologists saw them rarely in Colorado. Observers found them near Colorado Springs in 1871, Sterling in 1892, and Wray in 1908, but few others found them at all (Cooke 1897, 1909, Sclater 1912). In the 1930s and 1940s they nested close to the foothills at Boulder, Golden, Littleton, and Colorado Springs. About that time observers began to encounter them more widely in northeastern Colorado (B&N).

Atlasers found them sparingly, in only 10–15% of High Plains blocks; of those they found 36% in one year, 1994. Also in 1994 three to five males sang along a farm road in the San Luis Valley near Antonito. They used song perches in fields with sparse bushy vegetation. On 16 July a female carrying food for her young dove into thick roadside willow brush. This constitutes the westernmost breeding record for the species. A brief foray to this site failed to find Dickcissels in 1995 (Urling Kingery, HEK).

Settlers, breaking the prairie into cropland, probably turned inhospitable shortgrass into the disturbed habitat that Dickcissels favor. That would explain the birds' rarity in the nineteenth century and their profusion in the past 75 years.

BREEDING EVIDENCE

REPORTED IN 59 (3%) OF 1745 PRIORITY BLOCKS

▢	Possible	28 blocks	(47%)
▩	Probable	25 blocks	(42%)
■	Confirmed	6 blocks	(10%)

COLORADO DIVISION OF WILDLIFE / WILDLIFE RESOURCE INFORMATION SYSTEM

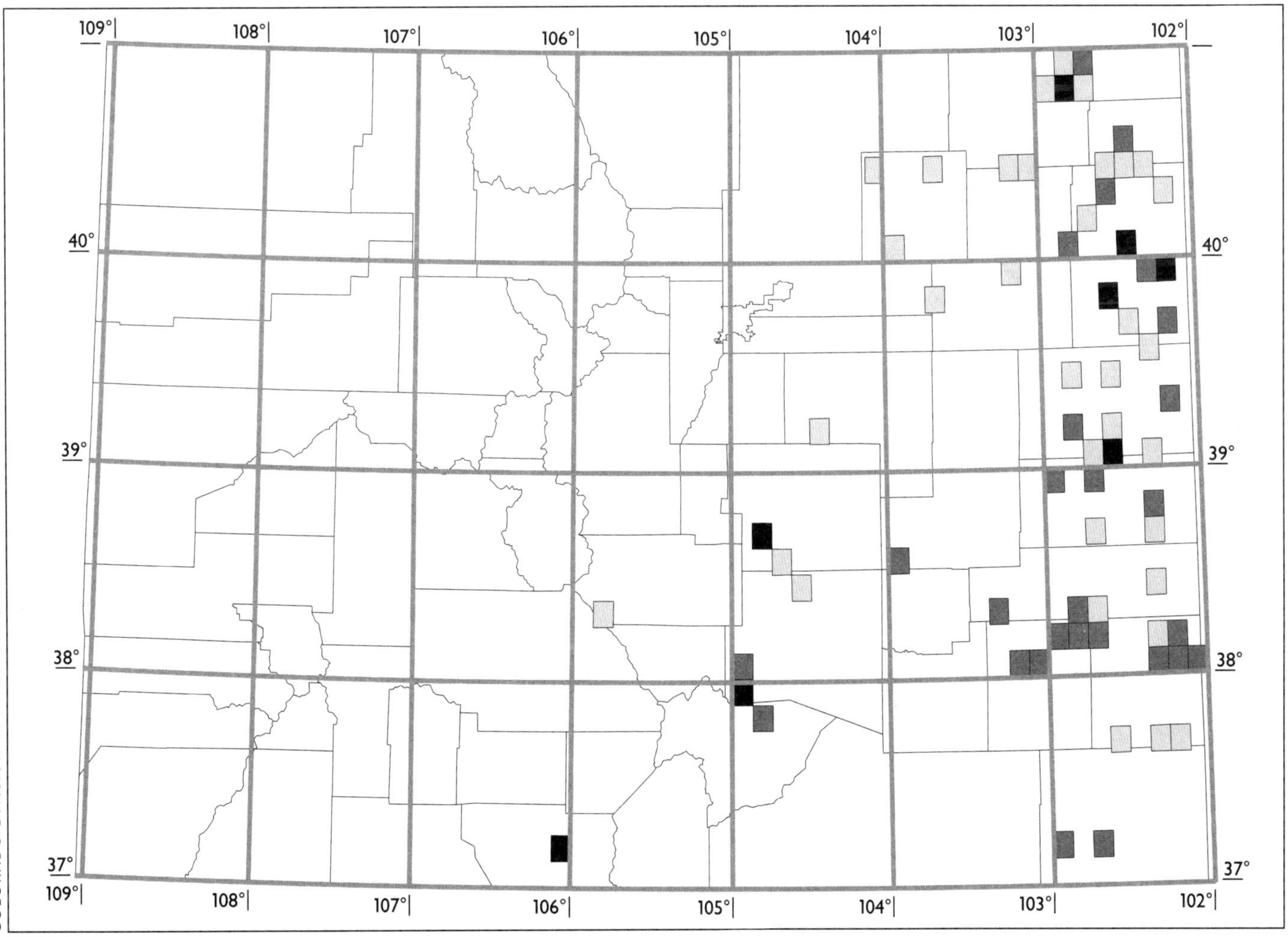

109°
108°
107°
106°
105°
104°
103°
102°
40°
39°
38°
37°

One of Colorado's early ornithologists, Norman Betts (1913), reported Bobolinks near the beginning of this century in a large meadow southeast of Boulder. Nearly 100 years later, in spite of encroaching urbanization, that site remains one of the largest and best documented colonies of the species in Colorado. The persistence of that colony, however, demonstrates the risks that Bobolinks endure in Colorado from changing agricultural practices and land use.

HABITAT

Bobolinks generally breed in grassy meadows and irrigated hay fields, but they require additional habitat features that limit their distribution. The birds nest in fairly wet, though not marshy, fields. Nest sites usually require forbs nearby as well as perching sites for territorial males. Isolated stalks of forbs (e.g., wild asparagus) in hay fields frequently serve as perches for males about to launch into their flight songs. Bobolinks show a strong preference for hay fields, more than eight years old, that have high litter cover (Martin and Gavin 1995). They prefer new growth to old, so grazing, haying, and burning before or after the nesting season all tend to improve habitat quality (Thompson and Strauch 1986).

Many Bobolink populations in the western U.S. seek actively cultivated hayfields in naturally occurring moist areas, but

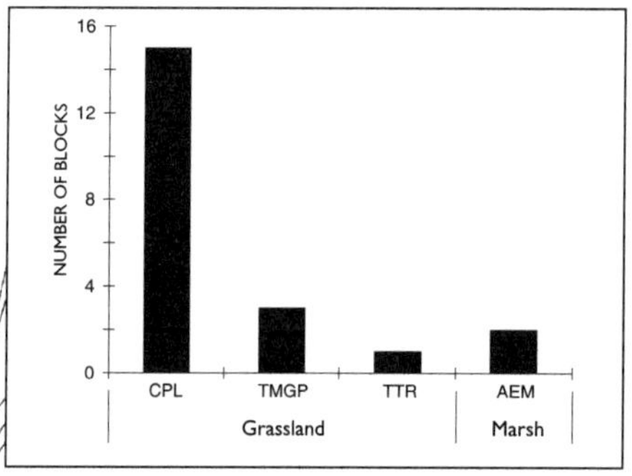
16
12
8
4
0
NUMBER OF BLOCKS
CPL
TMGP
TTR
AEM
Grassland
Marsh

Colorado birds use irrigated hay fields as primary habitat. This caused Hamilton (1962) to suggest that these populations are relics from historically wetter times, but a lack of genetic differentiation in western and eastern birds suggests recent range expansion from the east (Wittenberger 1978). Nevertheless, western populations tend to be isolated and inbred (Avery and Oring 1977).

Atlasers in Colorado found Bobolinks in appropriate habitat. Cropland, especially hayfields, composed 15 of 21 habitat reports. Other habitats included emergent wetlands and mixed grasslands.

BREEDING

Bobolinks build their nests on the ground from coarse and fine grasses. The tall grass and scattered herbaceous cover of their preferred wet fields not only conceal these ground nests. They also allow breeding birds to approach and to exit their nest sites by walking along the ground rather than by flying directly to or from the nest when intruders, including human observers, come within the nest vicinity. This behavior characteristic makes nests especially difficult to locate. They raise only one brood. Once the young fledge, Bobolinks usually leave nesting areas for secluded molting areas (Bent 1958).

In a study of breeding status on the historical breeding ground near Boulder, Thompson and Strauch (1986) found seven broods that fledged from 2 July to 16 July, with a mean fledging date of 7 July; however, all Atlas records come from May and June. The earliness of Atlas observations, the difficulty in locating nests, and the early exit of birds from nesting areas make it easy to explain the fairly low Confirmation rate for Bobolinks.

Atlas abundance estimates suggest that Colorado has fewer than 750 nesting pairs, probably around 500.

BREEDING PHENOLOGY

CODE		# OF RECORDS	DATE RANGE
C	Courtship	1	15 Jun
NY	Nest with Young	1	15 Jun
FY	Feeding Young	1	30 Jun

DISTRIBUTION

The breeding range for Bobolinks extends across the northern U.S. and southern Canada. In the West, they occur in scattered and fairly small colonies. Atlasers in Colorado found the species mainly in the northern quarter of the state, but observations stretched from the northeastern corner to the northwestern corner. In general, the Atlas probably understates the breeding presence of the species in Colorado because of the difficulty in Confirming nesting, the shortness of the nesting cycle, and Atlas block selection.

Supplementing the field cards from Atlas field workers, however, a researcher who studied Bobolinks during the Atlas period provided his records, six of them Confirmations from Non-priority blocks (Steve Martin pers. comm.). These data demonstrate breeding across northern Colorado from Fort Collins to eastern Moffat County.

Bobolinks are a species at some risk. They have small and isolated populations, and almost all nests are in actively mowed hay fields. Early harvests destroy nests and kill unfledged young. However, agricultural management practices that include mid July haying and winter grazing are compatible with population maintenance.

BREEDING EVIDENCE

REPORTED IN 11 (1%) OF 1745 PRIORITY BLOCKS

☐	Possible	4 blocks	(36%)
▪	Probable	4 blocks	(36%)
■	Confirmed	3 blocks	(27%)

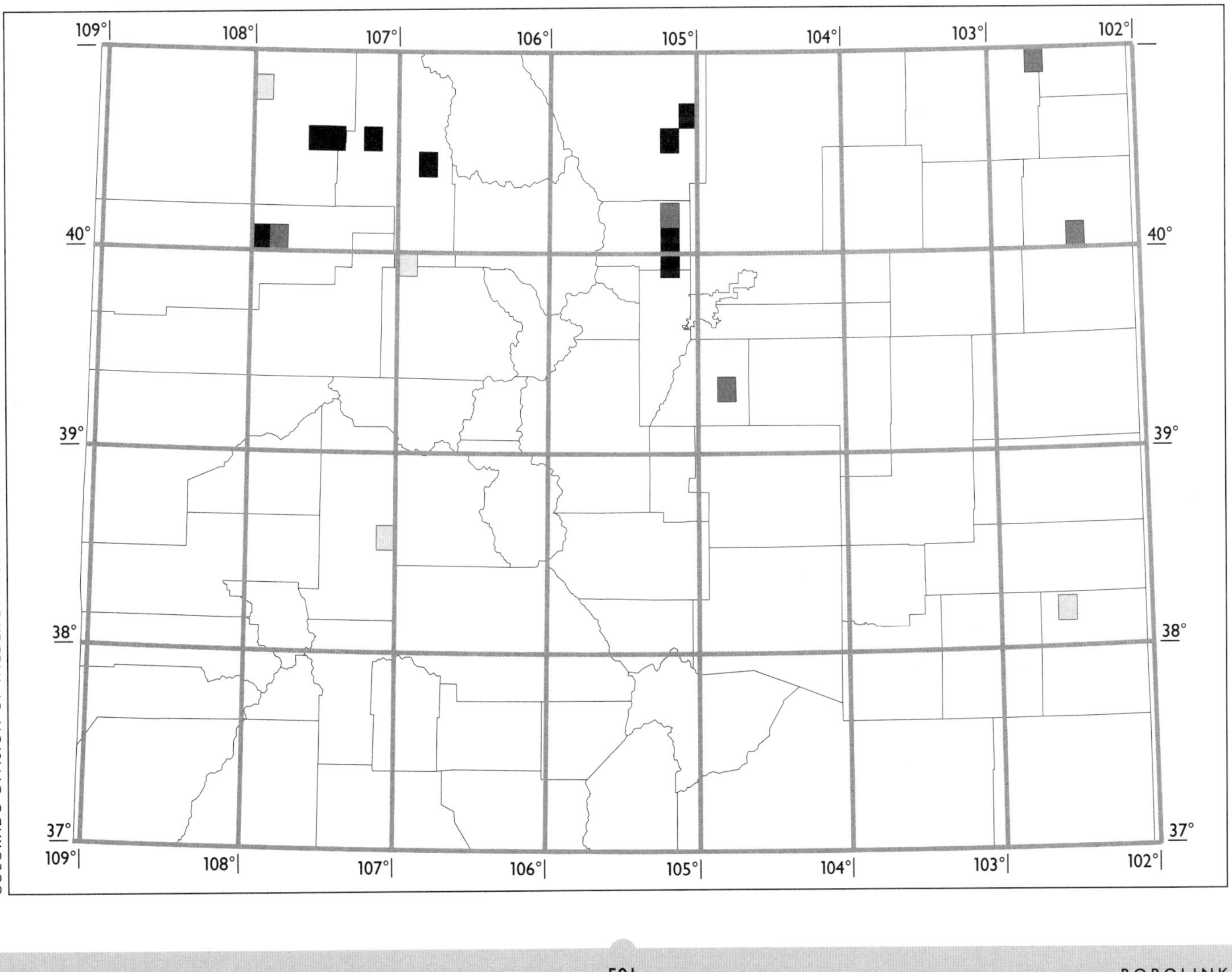
109°
108°
107°
106°
105°
104°
103°
102°
40°
39°
38°
37°

COLORADO DIVISION OF WILDLIFE / WILDLIFE RESOURCE INFORMATION SYSTEM

The *oka-ree* of the Red-winged Blackbirds sounding from wetlands heralds the early spring throughout much of Colorado. Each male flashes his red epaulets as he defends an extensive territory. A dominant male, with a good territory, may attract and mate with several females, perhaps even a dozen, in order to pass on his genes to as many offspring as possible.

HABITAT

We expect to find Red-winged Blackbirds in cattail marshes, but they can exploit a wide variety of wetland and upland habitats and prey types, making the species one of the most numerous in the U.S. (Yasukawa and Searcy 1995). Marshes and other aquatic and riparian areas accounted for only 57% of the breeding habitats noted by atlasers. Agricultural lands and croplands composed another 28%, and various other habitats, including woodlands, shrublands, grasslands, and urban areas, composed the rest. Red-winged Blackbirds can forage in trees, and thus use riparian territories near trees, whereas Yellow-headed Blackbirds will not breed in such areas (Orians 1985). Red-wings will also nest away from water.

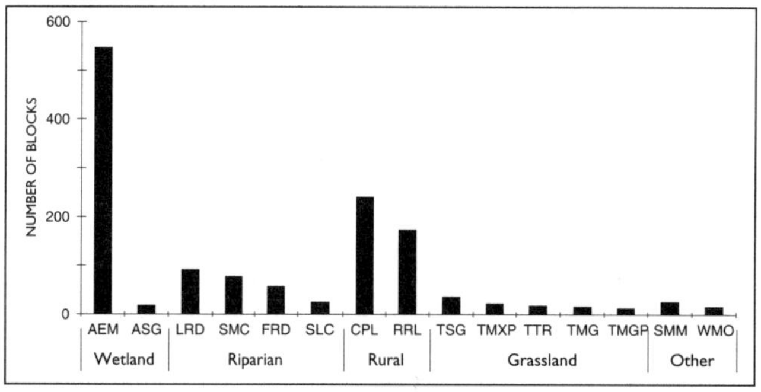
NUMBER OF BLOCKS
600
400
200
0
AEM ASG LRD SMC FRD SLC CPL RRL TSG TMXP TTR TMG TMGP SMM WMO
Wetland
Riparian
Rural
Grassland
Other

They employ a seasonal dietary switch: mainly insects in the summer and mostly seeds and crop residues in the winter. They also flock to bird feeders and congregate around feedlots (Yasukawa and Searcy 1995).

BREEDING

Male Red-winged Blackbirds stake out territories with calls and aggressive actions (B&N), often in large colonies. They are polygynous; up to 15 females may breed within one male's territory, although the mean size of a harem is five or less (it varies by quality of habitat). One study suggests that within a male's territory females defend smaller nesting territories, but other researchers discount female aggressiveness as an equivalent to territorial defense (Yasukawa and Searcy 1995). Many males, especially first-year birds, do not secure mates. Renesting following destruction of a nest and second nestings are common, thus accounting for the range of Atlas dates for nest-building.

Each female builds a nest, lays 3–5 eggs, incubates, and cares for the nestlings, in a total of 24–32 days, plus 2–3 weeks feeding fledged young (Yasukawa and Searcy 1995). The male defends the territory, repels predators, and guards the females and nests. A male may produce more young by spending his time courting and advertising for more females than by feeding his existing young (Orians 1985). To some degree males participate in feeding nestlings; this varies in different parts of the range, from 6% of the males in Washington to 90% in Wisconsin and New York (Yasukawa and Searcy 1995). DNA testing shows that, for 25% of the young in one study area, the social parents (the ones defending the nest) were not the biological parents (Westneat 1993). In spite of the guarding by the male, the Red-winged Blackbirds are a favorite host of cowbirds (Chace and Cruz 1996; Appendix C).

Females weave nests around supporting cattails or other vegetation. Nest record cards recorded several unusual supports for nests—a drift of dead tumbleweeds, thistles, clumps of grass, shrubs in rural shelterbelts with no water near, and the only small shrub along a reservoir bank. Red-wings now breed in agricultural fields, where they can find large numbers of insects (Orians 1985). They even display from the tops of alfalfa plants.

Atlas data show that the breeding season lasts for four and one-half months, from mid April to the end of August. During that period Red-wings produce at least two broods, possibly three. Atlasers reported Confirmed or Probable breeding in a very high 83% of the blocks.

BREEDING PHENOLOGY

CODE		# OF RECORDS	DATE RANGE
C	Courtship	11	28 Apr–27 Jun
NB	Nest Building	43	21 Apr–10 Jul
ON	Occupied Nest	76	22 Apr–30 Jul
NE	Nest with Eggs	63	8 Apr–7 Jul
NY	Nest with Young	34	16 Apr–16 Jul
FY	Feeding Young	205	5 May–12 Aug
FL	Fledged Young	149	11 May–26 Aug

DISTRIBUTION

Red-winged Blackbirds, among the most abundant North American birds, breed from the Yukon across Canada, through all of the United States, south to Central America and the Caribbean. In August the adults leave their breeding grounds for a nearby secluded area in which to molt (Stokes 1979). In the fall Red-wings form into flocks and migrate south from the northern part of the range. They form immense winter congregations in the southern states, where they can be destructive to crops.

They breed commonly throughout Colorado in suitable habitat, from the plains to mountain valleys over 9,000 feet (2745 m; B&N). One Atlas nest report gave an elevation of 10,100 feet (3080 m; Dan Bridges, nest card).

Colorado's Red-wings reach their greatest abundance in the triangle from Denver north along the Front Range and northeast along the South Platte, mainly in wetlands and croplands, where atlasers coded the preponderance of A4 and A5 abundance estimates. Other peaks of abundance occur in the wetlands of the San Luis Valley and the San Juan Basin.

Observers found them in 1,046 Priority blocks, the eighth highest for any species. Surprisingly, despite this broad Colorado occurrence, using Atlas abundance codes they rank only 18th in total nesting pairs. People may perceive Red-wings as more numerous than they are because of their conspicuousness and their clamorous winter roosts.

In North America Red-winged Blackbirds may have become the most abundant species today, but they were not always. These birds benefit from the way we change landscapes (Orians 1985). In 1892 ranchers fed corn to 200,000 sheep in the Big Thompson and Poudre valleys and Red-wings flocked to feed on the scattered grain (Cooke 1897). Across the country they have increased partly due to the greater productivity of lakes and marshes and masses of cattails fertilized by agricultureal run-off and treated sewage. In the South in recent years they mass in winter flocks of up to a million. Because the large winter flocks often cause considerable damage to agricultural crops, federal researchers spend substantial efforts at finding ways to control this damage and the birds that cause it.

BREEDING EVIDENCE

REPORTED IN 1046 (60%) OF 1745 PRIORITY BLOCKS

☐	Possible	168 blocks	(16%)
■ (grey)	Probable	281 blocks	(27%)
■	Confirmed	597 blocks	(57%)

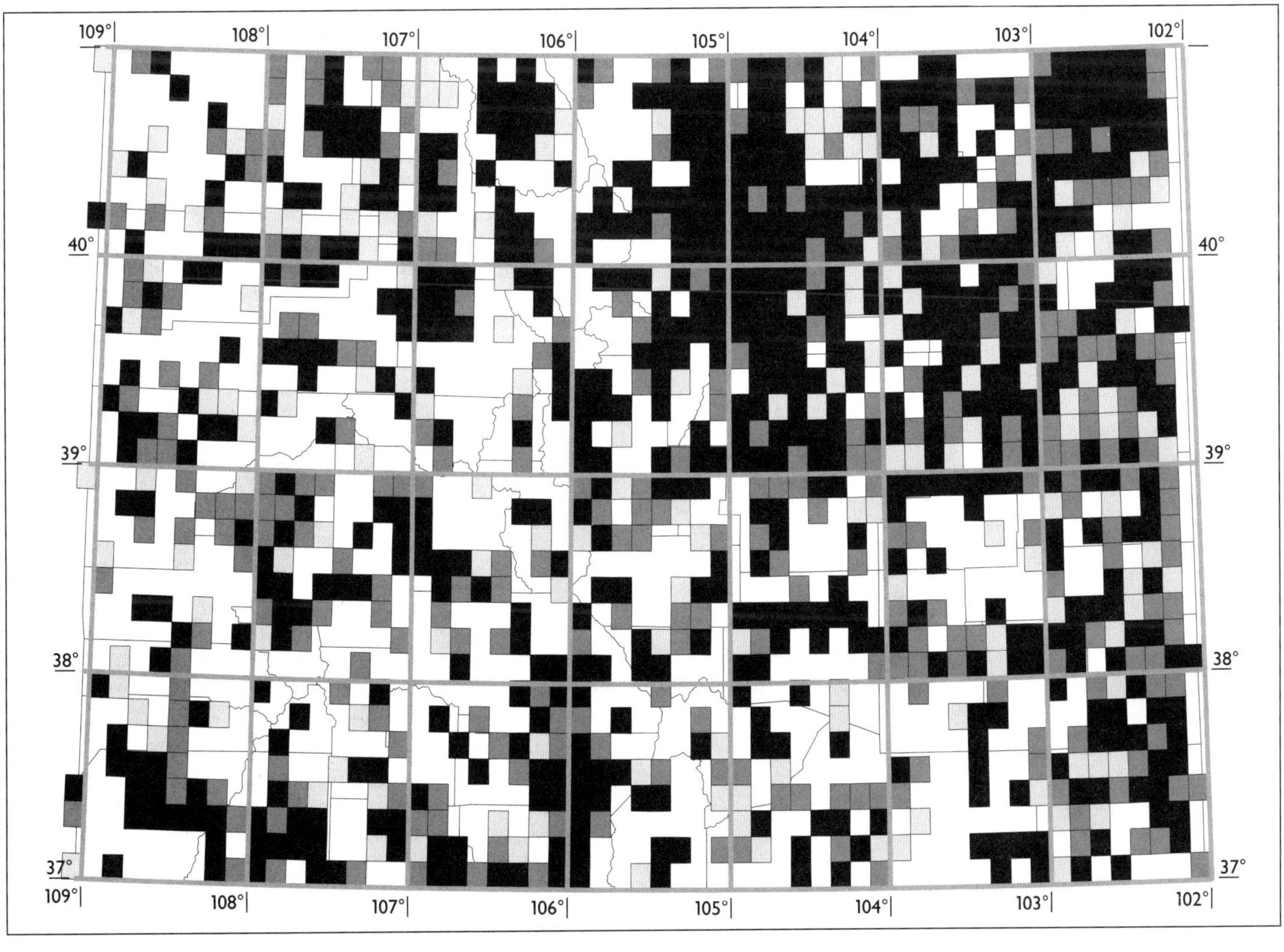
109°
108°
107°
106°
105°
104°
103°
102°
40°
39°
38°
37°

COLORADO DIVISION OF WILDLIFE / WILDLIFE RESOURCE INFORMATION SYSTEM

WESTERN MEADOWLARK *Sturnella neglecta*

For 30 years after Lewis and Clark encountered Western Meadowlarks, scientists ignored the birds; hence, when in 1844 John James Audubon bestowed on the species a scientific name, he called it "neglected" (Bent 1958). Today Western Meadowlarks have a well-deserved reputation for their sterling songs. Six states have designated the Western Meadowlark as their state bird—recognition not afforded to the musically challenged Eastern Meadowlark.

HABITAT

Western Meadowlarks, "the very spirit of the boundless prairie" (Bent 1958), breed in open country, in Colorado from the lowest point (3,340 feet) to 10,200 feet (1020–3110 m; Sclater 1912, B&N, Atlas data). They choose habitats with healthy grass and thick litter cover (Lanyon 1994). Atlasers found them in shortgrass and midgrass prairies, croplands, montane grasslands, desert shrublands, sagebrush, and cottonwood groves interspersed with grasslands.

Meadowlarks catch prey by poking their awl-like bills, closed, into the ground, then spreading them open; this works especially well on burrowing spiders. They also probe the bases of grass clumps, favorite daytime hiding places of nocturnal insects (Orians 1985).

Like Neotropical migrants, meadowlarks consume large quantities of insects that injure crops: beetles, weevils, wireworms,

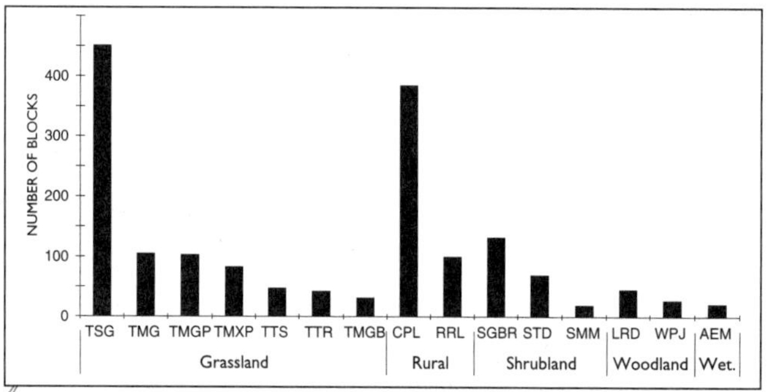
NUMBER OF BLOCKS
0
100
200
300
400
TSG TMG TMGP TMXP TTS TTR TMGB
Grassland
CPL RRL
Rural
SGBR STD SMM
Shrubland
LRD WPJ
Woodland
AEM
Wet.

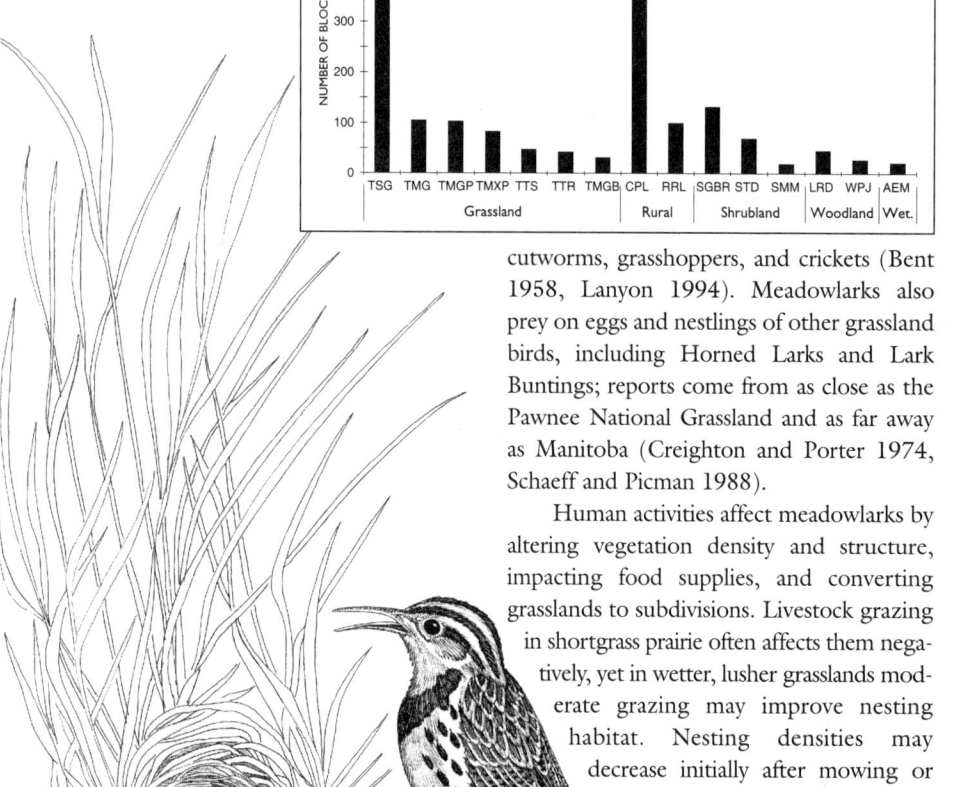

cutworms, grasshoppers, and crickets (Bent 1958, Lanyon 1994). Meadowlarks also prey on eggs and nestlings of other grassland birds, including Horned Larks and Lark Buntings; reports come from as close as the Pawnee National Grassland and as far away as Manitoba (Creighton and Porter 1974, Schaeff and Picman 1988).

Human activities affect meadowlarks by altering vegetation density and structure, impacting food supplies, and converting grasslands to subdivisions. Livestock grazing in shortgrass prairie often affects them negatively, yet in wetter, lusher grasslands moderate grazing may improve nesting habitat. Nesting densities may decrease initially after mowing or fire, then rebound as vegetation recovers (Ryder 1980, Pylypec 1991, Bock et al. 1993, Saab et al. 1995). Likewise insecticide applications that decrease food supplies can decrease breeding numbers (George et al. 1995).

BREEDING

"Chupp, rattle, and roll" describes just some of the variety of sounds made by meadowlarks. Each male's repertoire averages seven different liquid, flute-like songs, which he switches in response to intruding males (Lanyon 1994). Males become conspicuous with their songs and courtship posturing by late March, when they begin forming and then vigorously defending territories averaging 10 acres (4 ha). They mark the perimeters by singing from perches providing unobstructed views—fence posts, wires, trees, and shrubs (B&N, Schroeder 1974).

Nest construction begins by mid April. Most Atlas nest-building dates came between 13 May and 21 June. Females conceal the ground nest in a small depression. They use dried grasses for the outside and finer grasses for the inside lining. Usually a partial or full dome hides the nest. A runway, or tunnel, 1–5 feet long, leads to a side opening (Schroeder 1974, Lanyon 1994). Atlasers achieved a high Confirmation rate, but only 11% involved nests, verifying the birds' skills at nest concealment.

The nesting cycle from nest-building through fledgling independence lasts 46–55 days (Lanyon 1994). Females build second nests following successful first nestings and repeatedly attempt renesting after a failure. Atlasers observed adults carrying food or feeding young from mid May to late July—a span of over 70 days, enough time for double broods. Meadowlarks are polygynous; males usually have two mates (rarely three) concurrently (Lanyon 1994).

Certain agricultural methods—tillage with discs, plows, and treaders, and mowing for hay or weed control—destroy nests, eggs, young, and incubating and brooding females. Changed mowing and tillage practices and modified farm implements can reduce mortality (Rodgers 1983, Basore 1984, Hays and Farmer 1990, Rodenhouse et al. 1995).

Predation often leads to low nest success (Basore 1984, Rodenhouse et al. 1995). Falcons and eagles prey on meadowlarks, as do foxes, skunks, dogs, other mammals, and snakes. Domestic farm cats pose a particular threat to vulnerable ground-nesting meadowlarks (Harrison 1992, Lanyon 1994).

Sclater (1912) reported the only Colorado instance of Brown-headed Cowbird parasitism. Bent (1958) considered it unlikely that cowbird young could compete with the larger young meadowlarks; no definitive research has established the effects of cowbird parasitism (Lanyon 1994).

BREEDING PHENOLOGY

CODE		# OF RECORDS	DATE RANGE
C	Courtship	11	7 Apr–13 Jul
NB	Nest Building	12	14 Apr–4 Jul
ON	Occupied Nest	24	2 May–19 Jul
NE	Nest with Eggs	33	9 May–5 Jul
NY	Nest with Young	19	13 May–22 Jul
FY	Feeding Young	386	13 May–28 Jul
FL	Fledged Young	246	10 May–6 Aug

DISTRIBUTION

Western Meadowlarks nest across southern Canada to southern Ontario, through the western Great Plains to the Pacific coast and northern Mexico. In the fall they leave the northern parts of their range; many migrate to Mexico. During the twentieth century breeders have spread east to the Great Lakes, an expansion tied to forest clearing and agriculture (Lanyon 1994). Their midwestern range overlaps that of Eastern Meadowlarks, but interbreeding occurs rarely (Lanyon 1994).

Meadowlarks rank as the third most frequently reported Atlas species, substantiating their wide distribution and high numbers. They reach high densities in eastern Colorado where each year five to ten BBS routes record them on all 50 stops. Atlas field workers Confirmed breeding in 57 of the 60 counties where they found them. The highest-elevation Confirmation came at 10,200 feet (3110 m) in South Park (Alma, 39106C1). Atlasers verified breeding in the one latilong (39103) that had lacked that status—and almost covered that latilong with Confirmations.

Meadowlarks are extremely sensitive to any human presence in their nesting territories. They "invariably" desert a nest if disturbed during incubation (Lanyon 1994) and delay nest visits or desert if disturbed with young in the nest (B&N, Schroeder 1974). Recreational activities also have negative effects. In a Boulder Open Space unit with intense recreational pressure meadowlarks do not nest close to trails (Miller and Knight 1995). Even though birdwatching, photography, and field studies, as well as farming activities, disturb nesting meadowlarks, these songsters have managed to persist successfully away from heavy human activity. Nonetheless Partners in Flight has identified the meadowlarks' stronghold, shortgrass prairie—no longer "boundless"—as a top habitat conservation priority for Colorado breeding birds.

BREEDING EVIDENCE

REPORTED IN 1187 (68%) OF 1745 PRIORITY BLOCKS

☐	Possible	183 blocks	(15%)
▩	Probable	264 blocks	(22%)
■	Confirmed	740 blocks	(62%)

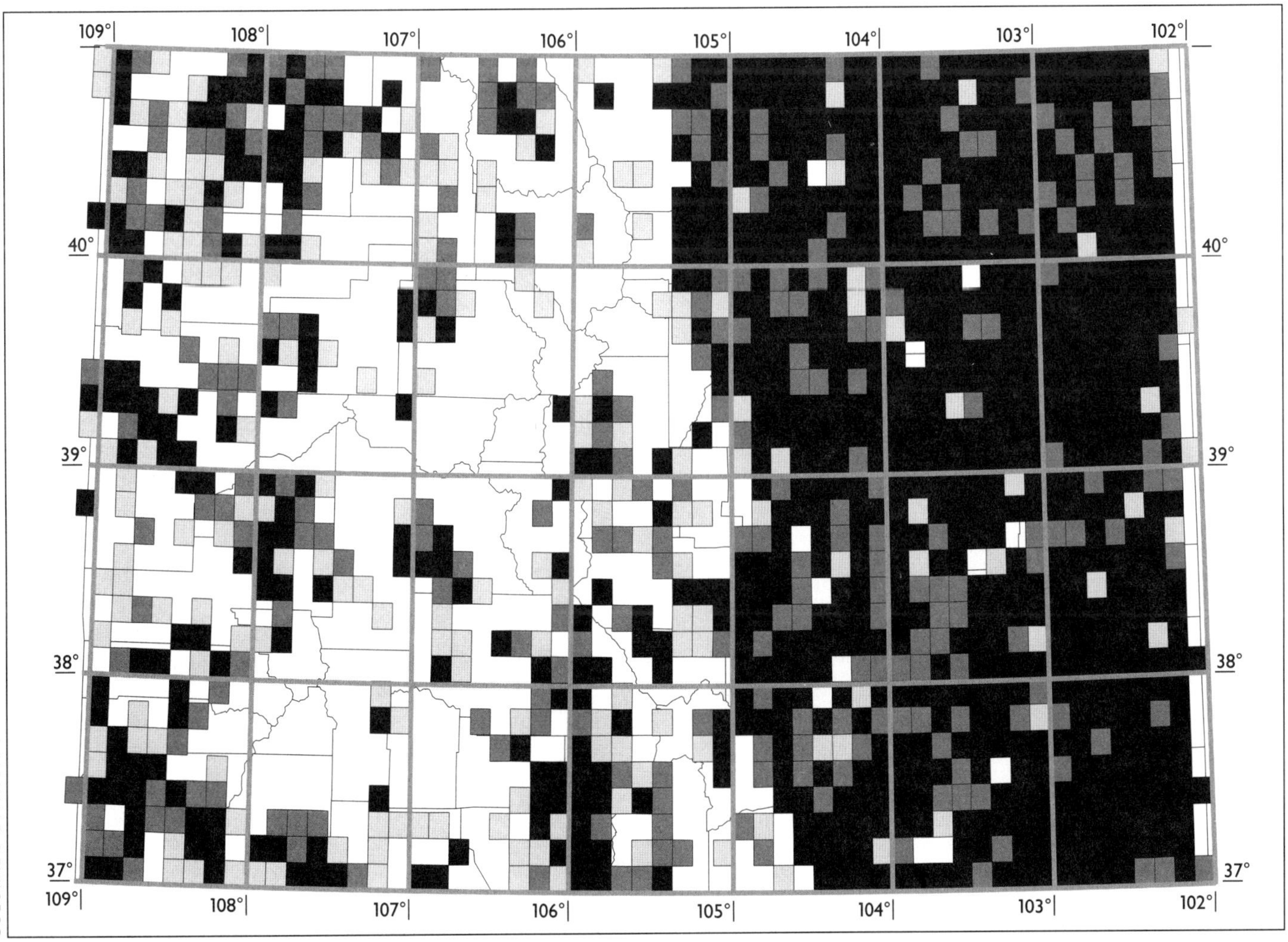
109°
108°
107°
106°
105°
104°
103°
102°
40°
39°
38°
37°

COLORADO DIVISION OF WILDLIFE / WILDLIFE RESOURCE INFORMATION SYSTEM

The bright yellow heads and white wing patches, contrasting with jet-black bodies, make for striking courtship displays by Yellow-headed Blackbirds. Males posture and employ their colors to guard territories for their harems, and females even use similar displays to protect their space from female trespassers.

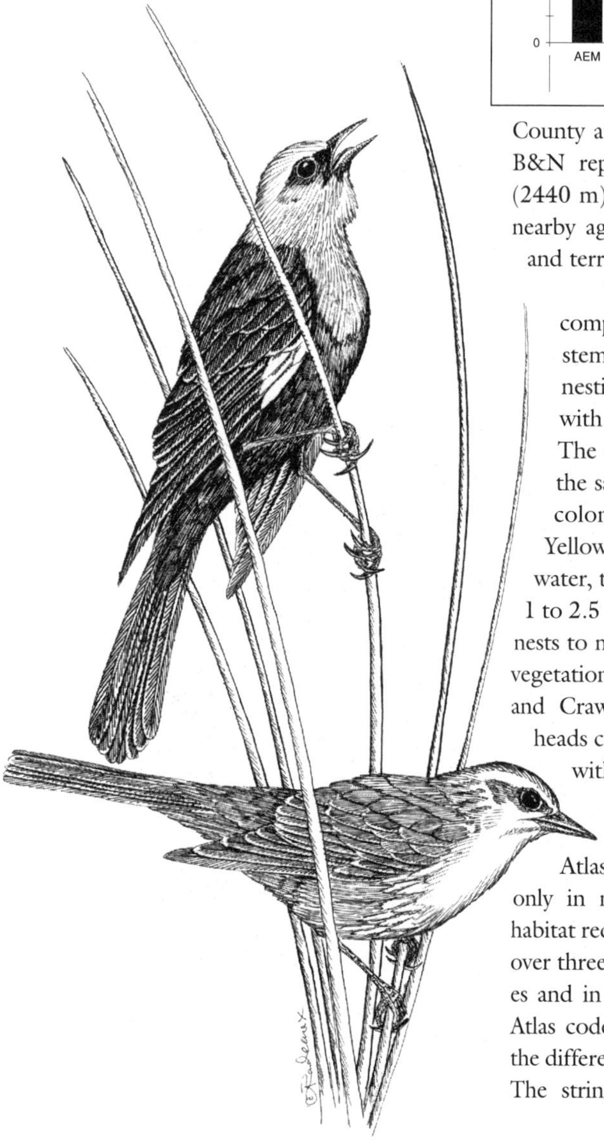

HABITAT

Yellow-headed Blackbirds require an aquatic habitat for breeding, with dense emergent vegetation such as cattails, bulrushes, and reeds, in water at least a foot deep. They nest in colonies, with numbers dependent on the size and vegetation of the marsh and the abundance of food within territories or on nearby upland habitats that the birds often use for foraging (Twedt and Crawford 1995). They breed on the plains, mountain parks, and western valleys, and into the foothills. Sclater (1912) reported them breeding at Twin Lakes in Lake

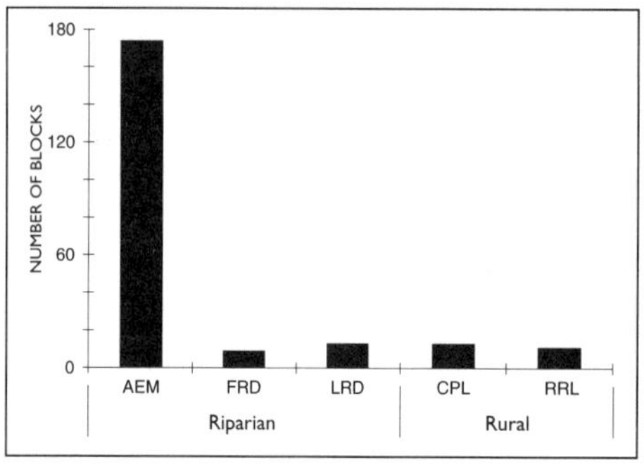

180
120
60
0
NUMBER OF BLOCKS
AEM
FRD
LRD
CPL
RRL
Riparian
Rural

County at about 9,000 feet (2745 m), and B&N reported nesting up to 8,000 feet (2440 m). They feed in the marshes and in nearby agricultural lands, gathering aquatic and terrestrial insects and seeds.

Their less expansive presence, compared to Red-winged Blackbirds, stems from the habitat requirement for nesting and feeding sites in marshes with standing water several feet deep. The two blackbird species may nest in the same marsh, but they separate their colonies according to water depth. Yellow-heads locate nests *only* over open water, the depth most often ranging from 1 to 2.5 feet (32–76 cm). They attach their nests to marsh plants, either last year's dead vegetation or this year's new growth (Twedt and Crawford 1995). The larger Yellow-heads can supplant Red-wings, even those with already established territories, and relegate them to the fringes (Ryser 1985).

Atlasers basically found Yellow-heads only in marshes, reflecting their exacting habitat requirements; Red-wings occurred in over three times as many blocks with marshes and in six times as many blocks overall. Atlas codes do not detail water depth, but the difference no doubt hangs on that factor. The stringent habitat requisites may also impact negatively on the statewide abundance estimate for Yellow-heads (see Appendix C).

BREEDING

Flocks of males arrive in late March or April (A&R). Colonies vary in size. Atlasers reported 63 Priority blocks (46% of those reporting abundance codes) with ten or fewer nests; 61 (45%) with 11–100, and only 12 (9%) with over 100. They reported Non-priority block abundance in exactly the same percentages. The Atlas field manual instructed field workers to report abundance of polygynous species by estimating the total number of nests, not territorial males.

Aggressive males link different songs with different displays. In "symmetrical song-spread posture" males spread their wings wide, fan and lower their tails, stretch up their heads, and (sometimes) arch their wings above their backs. The song given with this posture has clear introductory syllables and a short trill, and lasts 1–2 seconds. In "asymmetrical song-spread," they spread their wings only slightly, hold the right wing out alone or at least farther out than the left wing, and turn their heads so that the birds sing over their left shoulders. Songs last up to four seconds, with introductory notes and a prolonged trill (Ryser 1985).

Each male establishes and guards a territory to which he attracts one to six females (Twedt and Crawford 1995). A colony may have as many as 25 to 30 nests in "15 feet square" (225 ft^2, 21 m^2; Bent 1958). After the vegetation has grown, they construct nests just above the water level. Females weave the nest from wet, dead vegetation.

The female alone incubates the clutch, normally three or four eggs, for 10 to 12 days. The female feeds the nestlings principally aquatic insects and spiders at first, then adds grasshoppers, beetles, and other insects as they get older. The youngsters leave the nest at nine to ten days of age; they cling to dense vegetation until, at about three weeks, they can fly (Richter 1984). Brown-headed Cowbird parasitism appears to be only accidental (Ortega and Cruz 1991), and atlasers reported only one incident; in contrast several early Colorado observers found multiple instances (Chace and Cruz 1996; Appendix C).

Adults finish molting by mid September after which Yellow-heads gather in flocks as they migrate out of Colorado. During this period, they may join other blackbirds in enormous congregations that sometimes heavily damage crops. Bent (1958) reports: "On the whole, the bird is probably more beneficial than harmful, except in a few places where it is sufficiently numerous to cause appreciable damage to crops."

BREEDING PHENOLOGY

CODE		# OF RECORDS	DATE RANGE
C	Courtship	1	31 May
NB	Nest Building	4	21 Apr–30 May
ON	Occupied Nest	8	7 May–1 Aug
NE	Nest with Eggs	4	11 May–18 Jun
NY	Nest with Young	6	29 May–4 Jul
FY	Feeding Young	36	28 May–19 Jul
FL	Fledged Young	29	2 Jun–5 Aug

DISTRIBUTION

Yellow-headed Blackbirds are summer residents from coastal British Columbia east to the Prairie Provinces and Wisconsin, south to central California, New Mexico, and Kansas, wherever the birds find appropriate habitat. They winter in the southwestern U.S. and northern Mexico.

The Atlas map shows Yellow-heads in clusters of blocks all over the state, but without the ubiquity of Red-wings. In the mountains they favor the large parks but not small mountain marshes; on the plains they find breeding habitat at lakes, ranch ponds, reservoirs, or river marshes. The marshes created in connection with irrigation reservoirs north and east of Denver hold high concentrations, as do more natural wetlands in the San Luis Valley. Surprisingly, atlasers recorded no breeders around Bonny Reservoir and few along the Arkansas; Atlas blocks may have omitted the deep-water marshes favored by Yellow-heads.

Due to specialized habitat requirements and population declines current conservation priority schemes have listed the Yellow-headed Blackbird as meriting attention in Colorado and nationally. Of particular concern, water management of reservoirs, such as drawdowns and low capacity during drought years, can eradicate emergent vegetation. This species needs watching.

BREEDING EVIDENCE

REPORTED IN 169 (10%) OF 1745 PRIORITY BLOCKS

☐	Possible	47 blocks	(28%)
■	Probable	39 blocks	(23%)
■	Confirmed	83 blocks	(49%)

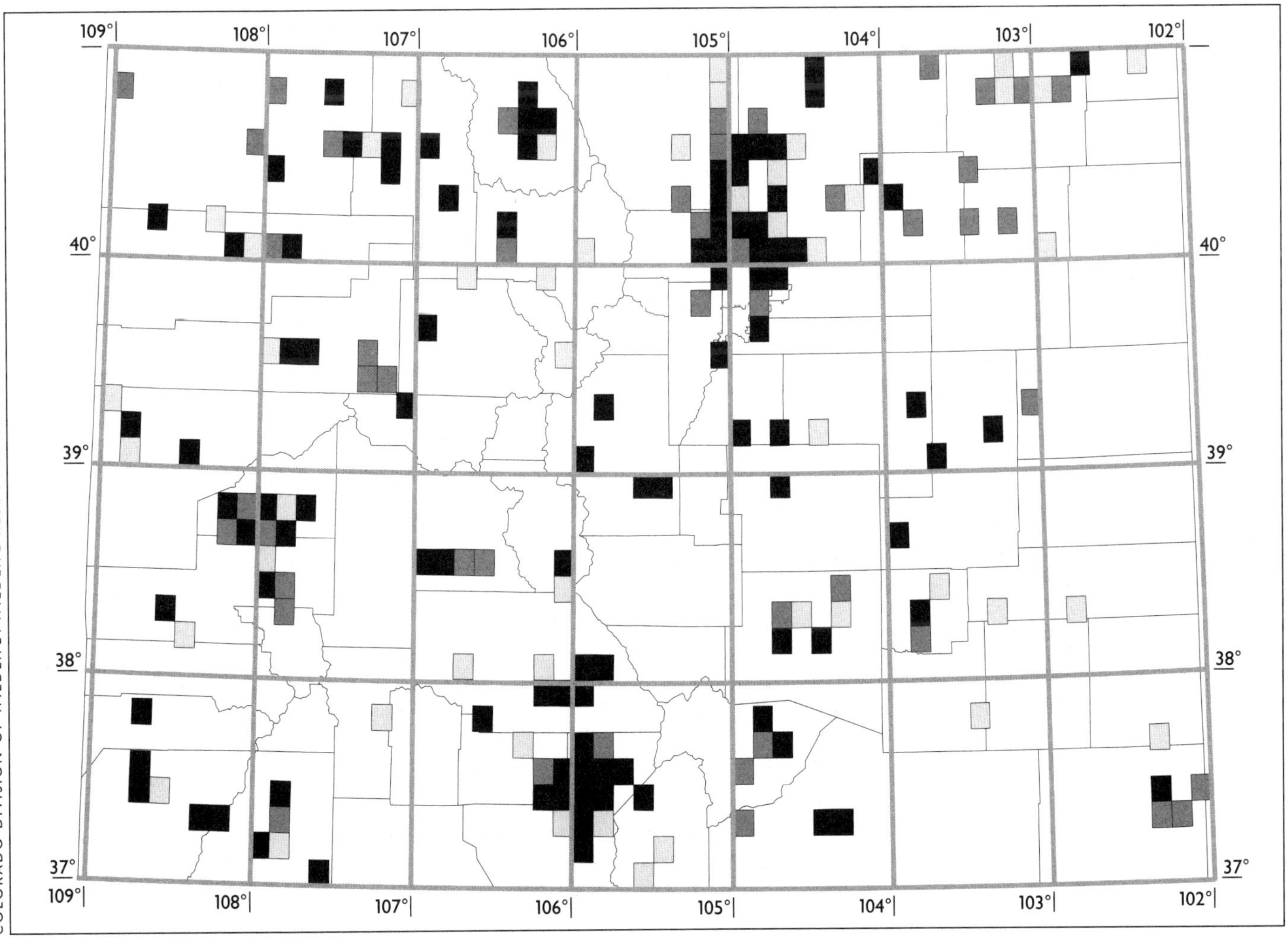
109°
108°
107°
106°
105°
104°
103°
102°
40°
39°
38°
37°

COLORADO DIVISION OF WILDLIFE / WILDLIFE RESOURCE INFORMATION SYSTEM

As with many black birds, Brewer's Blackbirds make up in behavior what they lack in color. Their repertoire of antics (Williams 1952) includes a threatening "head-up" display, during which the male points his bill straight upward, flattens his feathers, and draws himself up into an iridescent column of arrogance. When performing the flirtatious feather-fluffing and tail-spreading "ruff-out" display, males and females exaggerate themselves further by uttering *squee* or *schlr-r-r-r-up* sounds.

HABITAT

Brewer's Blackbirds use a variety of habitats ranging from wetlands to semi-arid areas (Bent 1958). Atlas results reflect their preference for open areas, including mountain parks, wet pastures, farmsteads, marshes, shrublands, and even desert scrub. Suitable nesting habitat must provide some form of vertical nesting cover (usually small trees or shrubs), a good foraging area (usually a wetland or stream margin with very short vegetation), and relatively high perches from which males can watch for predators and guard their mates (Williams 1952, Bent 1958, Orians and Horn 1969). In Colorado, this ideal juxtaposition of habitat components occurs most often in cropland/rural areas, which composed one-third of the Atlas habitats recorded, 80% of them from blocks in the foothills, intermountain valleys, and Western Slope. These blackbirds use shrublands in another one-quarter of the blocks, half of them sagebrush.

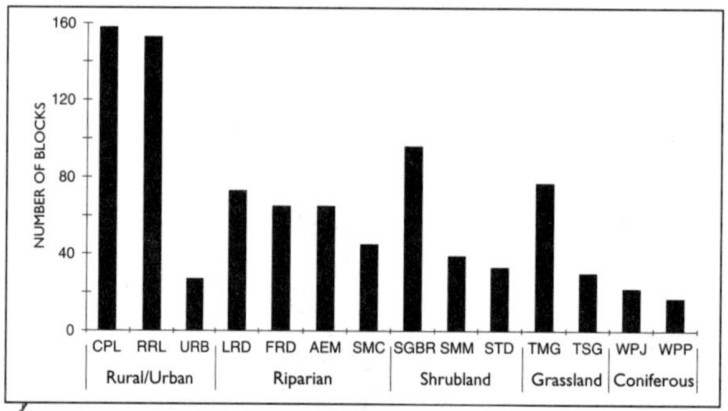
NUMBER OF BLOCKS
160
120
80
40
0
CPL RRL URB LRD FRD AEM SMC SGBR SMM STD TMG TSG WPJ WPP
Rural/Urban
Riparian
Shrubland
Grassland
Coniferous

The birds build their nests, most typically, on the ground at the base of a shrub or clump of grass, or in a shrub or small tree, although at times they nest in snags up to 150 feet (46 m) from the ground (Furrer 1975, Ritter and Purcell 1983). Occasionally, they settle in artificial structures such as birdhouses (Bancroft 1986) and haystacks (B&N). Because they prefer to nest in small colonies (Horn 1968), habitat opportunities for colony formation may influence their choice of nest sites.

BREEDING

By late March migrating Brewer's Blackbirds begin to appear in Colorado (Bent 1958, A&R). Atlas workers first recorded breeders on 27 April—multiple males singing in suitable nesting habitat. The sexes apparently segregate during migration (Bent 1958); males arrive on the breeding grounds first.

Brewer's Blackbird ranked high (30th) for its proportion of Confirmations, possibly owing to the open habitat they use and their colonial nesting habit. Atlas workers extended the latest egg date in Colorado (B&N) from 17 June to 8 July. Breeding phenology dates fell within the normal range.

Typically, 5–20 pairs of Brewer's Blackbirds form a breeding colony, but colonies may consist of up to 100 pairs. They rarely nest singly (Bent 1958), although Atlas nest cards never mentioned neighboring nests. These birds double-brood on occasion, but the synchronous breeding habitats of colonial nesters may preclude most opportunities for a second brood. Unsuccessful birds, however, will attempt to renest up to three times (Johnsgard 1979). Brewer's Blackbirds have a polygynous mating propensity (Verner and Willson 1969), which varies somewhat geographically: polygynous in Monterey County, Calif. (Williams 1952), monogamous in Washington (Horn 1968).

BREEDING PHENOLOGY

CODE		# OF RECORDS	DATE RANGE
C	Courtship	11	18 May–14 Jul
NB	Nest Building	9	10 May–1 Jul
ON	Occupied Nest	13	19 Apr–13 Jul
NE	Nest with Eggs	13	19 May–8 Jul
NY	Nest with Young	27	1 Jun–13 Jul
FY	Feeding Young	242	20 May–28 Jul
FL	Fledged Young	106	20 May–22 Aug

DISTRIBUTION

Brewer's Blackbirds breed from central British Columbia and Mackenzie east to Michigan and Ontario around Lake Huron, south to Texas and northern Baja California. They arrived in Michigan only in the 1930s (Brewer et al. 1991) and first bred in Ontario in 1953, prospered, and then declined (Cadman et al. 1987). They spend the winter from North Carolina west to Arizona and south to southern Mexico. A few birds stay year round in the Pacific coast states, Nevada, Utah, and parts of Colorado.

Atlas results show them nesting from urban areas to wilderness, and from the plains to about 10,000 feet (3048 m) in elevation, with most occurring between

5,000 and 8,000 feet (1525–2440 m). Atlasers found them concentrated on the Western Slope, especially in lowland riparian and irrigated valleys, mountain parks, and edges of mountain streams. On the Eastern Slope, they occurred more sporadically along mountain, foothills, and plains riparian systems, around farmsteads, and in the Black Forest.

At one time they nested in Denver's City Park (B&N), where their instinct to mob all potential predators made life difficult for themselves and for people attempting to use the area. B&N characterized the City Park birds as "pugnacious and cocky," but the same behavior took on a different quality when B&N described them driving away magpies, bullsnakes, and other predators from their nests. Since publication of B&N in 1965 Common Grackles have overrun urban parks and residential areas in Front Range cities. Now only a few Brewer's Blackbirds nest in the city, and only on the fringes (HEK). No one has documented a cause-and-effect relationship between the grackles' arrival and Brewer's departure. In fact the opposite occurred in an Ontario study area: Brewer's Blackbirds displaced Common Grackles (Stepney 1979), but apparently only on a small scale; across the province grackles greatly outnumber Brewer's in numbers and distribution (Cadman et al. 1987).

During the twentieth century, Brewer's Blackbird populations grew and their range expanded eastward as a result of ranching, agriculture, and clearing brush and forests (Stepney and Power 1973). Nonetheless Brewer's Blackbirds face restraints on their productivity. Brown-headed Cowbirds heavily parasitize Brewer's Blackbirds (Friedmann 1963, Friedmann et al. 1977, Chace and Cruz 1996; Appendix C). Only one Atlas worker, however, reported parasitism (cowbird egg in one nest). Common Grackles may out-compete them as they invade the Brewer's range (A&R). Other possible problems for the species include blackbird control programs in agricultural areas. Although Brewer's Blackbirds do eat some grain and fruit, the bulk of their diet consists of insects known to damage crops, including caterpillars, cutworms, grasshoppers, beetles, dragonflies, and diptera (Orians and Horn 1969).

BREEDING EVIDENCE

REPORTED IN 671 (38%) OF 1745 PRIORITY BLOCKS

☐	Possible	143 blocks	(21%)
▩	Probable	131 blocks	(20%)
■	Confirmed	397 blocks	(59%)

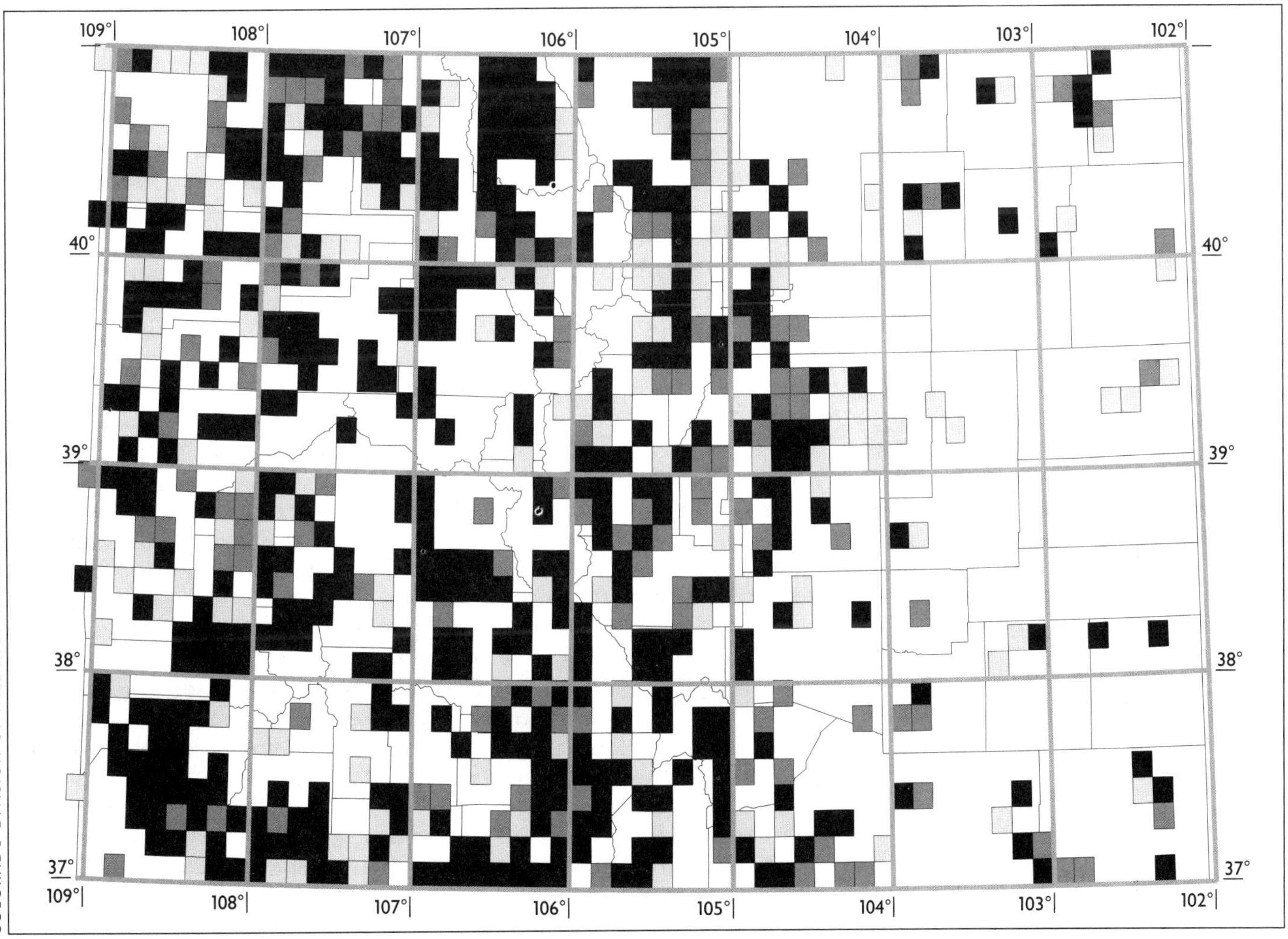
109°
108°
107°
106°
105°
104°
103°
102°
40°
39°
38°
37°

COLORADO DIVISION OF WILDLIFE / WILDLIFE RESOURCE INFORMATION SYSTEM

COMMON GRACKLE *Quiscalus quiscula*

Inflating to over twice their size, male Common Grackles, splendidly colored in purple and bronze, gather in competitive courting groups to woo one female. Noisy, large, and conspicuous, these grackles have recently expanded their range west as they adapt successfully to human-altered environments.

HABITAT

Two-thirds of the Common Grackles found by atlasers used habitats associated with humans; of those, 80% used rural habitats such as farmyards and shelterbelts, the rest used croplands and urban sites. A substantial number—21%—used riparian habitats, mainly broad-leaved but also narrow-leaf cottonwoods.

Colorado grackles flock to human-altered habitats that enhance foraging opportunities. They search for insects (20–25% of their diet) along the ground in open areas usually connected to the riparian drainages, shelterbelts, and urban "forests" where they nest. Although they also consume bird eggs, nestlings, and adult passerines, these compose only a small portion of their diet, and, taking the year-round average, less than 1% of the diet (Beal 1910). The bulk of the diets of these dietary generalists consists of fruit, grain, grass and forb seeds, and nuts (Davidson 1994). Their flocking behavior and grain consumption can have dramatic impacts on crops, which has earned them great human animosity. They often steal food from ground-foraging birds such as robins (Ehrlich et al. 1988), and they also monopolize feeding stations, which frustrates people trying to attract a diversity of bird species to their yards.

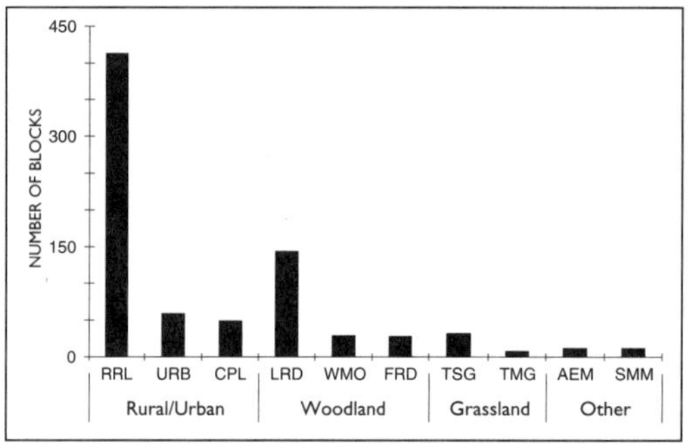
450
300
150
0
NUMBER OF BLOCKS
RRL
URB
CPL
LRD
WMO
FRD
TSG
TMG
AEM
SMM
Rural/Urban
Woodland
Grassland
Other

BREEDING

Common Grackles return to Colorado in late March and depart for their southern U.S. wintering areas in late October (A&R). Common Grackles usually defend only a small area around the nest—and they often form loose colonies of up to 100 nesting pairs. Adults, however, vigorously defend their nest from egg and nestling predators, which often include neighboring grackles. During courtship displays, a male fluffs his body feathers, spreads his tail and wings, and points his bill up, while vocalizing. Often groups of four or five males display around one female (Stokes 1979). They orient their bodies to the female so as to maximize their striking iridescent plumage. Grackles typically are monogamous, but polygyny occurs.

Females build a bulky cup-shaped nest most often in dense coniferous trees and shrubs (Maxwell et al. 1976), but also in deciduous trees and shrubs and sometimes cavities (Ehrlich et al. 1988). Six Atlas nest records illustrate the diversity of nest sites used by Common Grackles. These records describe nests built in a Russian-olive (nest 7 feet high), Siberian elm (18 feet), cottonwood (40 feet), barn (8 feet), tree cavity (4 feet), and cliff (5 feet). At Chatfield Reservoir from 1978 to at least 1984, Common Grackles built nests in the bottoms of the stick nests of Great Blue Herons; the number ranged from 4 to 17 pairs (Bottorff et al. 1971–1984).

Grackles occasionally have two broods. They lay four or five greenish-white to light brown eggs (Ehrlich et al. 1988). Eggs hatch after 13–14 days of incubation by the female and fledge 12–15 days later (Peer and Bollinger 1997). Brown-headed Cowbirds rarely parasitize Common Grackles, although Colorado has one record, derived from the Atlas (Chace and Cruz 1996; Appendix C). Even when they do host one, the cowbird rarely fledges (Peer and Bollinger 1997). Both sexes tend the young, primarily on insects and spiders—75% of their diet (Maxwell and Putnam 1972).

BREEDING PHENOLOGY

CODE		# OF RECORDS	DATE RANGE
C	Courtship	4	4 May–25 Jun
NB	Nest Building	33	14 Apr–16 Jun
ON	Occupied Nest	51	26 Apr–4 Jul
NE	Nest with Eggs	13	6 May–3 Jul
NY	Nest with Young	30	25 May–29 Jun
FY	Feeding Young	183	23 May–28 Jul
FL	Fledged Young	96	22 May–1 Aug

DISTRIBUTION

Common Grackles have their main distribution in eastern North America and historically did not breed in great numbers west of the central Great Plains (Sclater 1912, Peer and Bollinger 1997). Habitat alteration facilitated the western range expansion of the past 40 years, and now grackles nest commonly throughout eastern Colorado. The recent range expansion into Colorado has coincided with the disappearance of Brewer's Blackbirds from the majority of human-altered habitats that grackles prefer (A&R).

The Atlas map reflects the eastern origin of Colorado's grackles. East of Longitude 105, 71% of the Atlas blocks reported them, and 83% of the blocks east of 105 and north of the 39th Parallel reported them. West of 105, half the grackles use manmade habitats, a quarter use riparian woodlands. East of 105 they have a greater affinity for the human landscape: 72% use manmade habitats and 20% breed along the riparian corridors.

Thought of economically as one of the most significant "pest" bird species in North America, Common Grackles damage sprouting and ripening corn and other crops. They congregate in noisy roosts preliminary to migration and in winter (Peer and Bollinger 1997). Their urban/suburban roosts, relatively small in Colorado, can mount to 500 in residential Denver (HEK).

BBS data document a large increase in Colorado, of 6%/year, even though the continental trend shows a substantial decrease (1.6%/year). The population increase comes, in part, from planting shelterbelts and the growth of cottonwood corridors across the Great Plains (Robbins et al. 1986, Peer and Bollinger 1997), habitats in which atlasers found colonies of nesting grackles. Modern grain-harvesting techniques provide spillage on the wintering grounds, and so reduce overwinter mortality. These human-aided increases in the abundance and range of grackles place additional pressure on their breeding-ground competitors and victims. In addition, the expansion has led to huge post-breeding flocks, sometimes numbering in the millions, that cause grain crop damage and nuisances, which in turn has led to expensive control efforts. USF&WS has devoted many studies to the economic impacts (see Peer and Bollinger 1997). Now one of the most abundant species on the continent, Common Grackles face efforts by USF&WS to reduce their numbers (Peer and Bollinger 1997). These control measures may have brought about the drop in continental numbers.

BREEDING EVIDENCE

REPORTED IN 652 (37%) OF 1745 PRIORITY BLOCKS

☐	Possible	127 blocks	(19%)
■	Probable	97 blocks	(15%)
■	Confirmed	428 blocks	(66%)

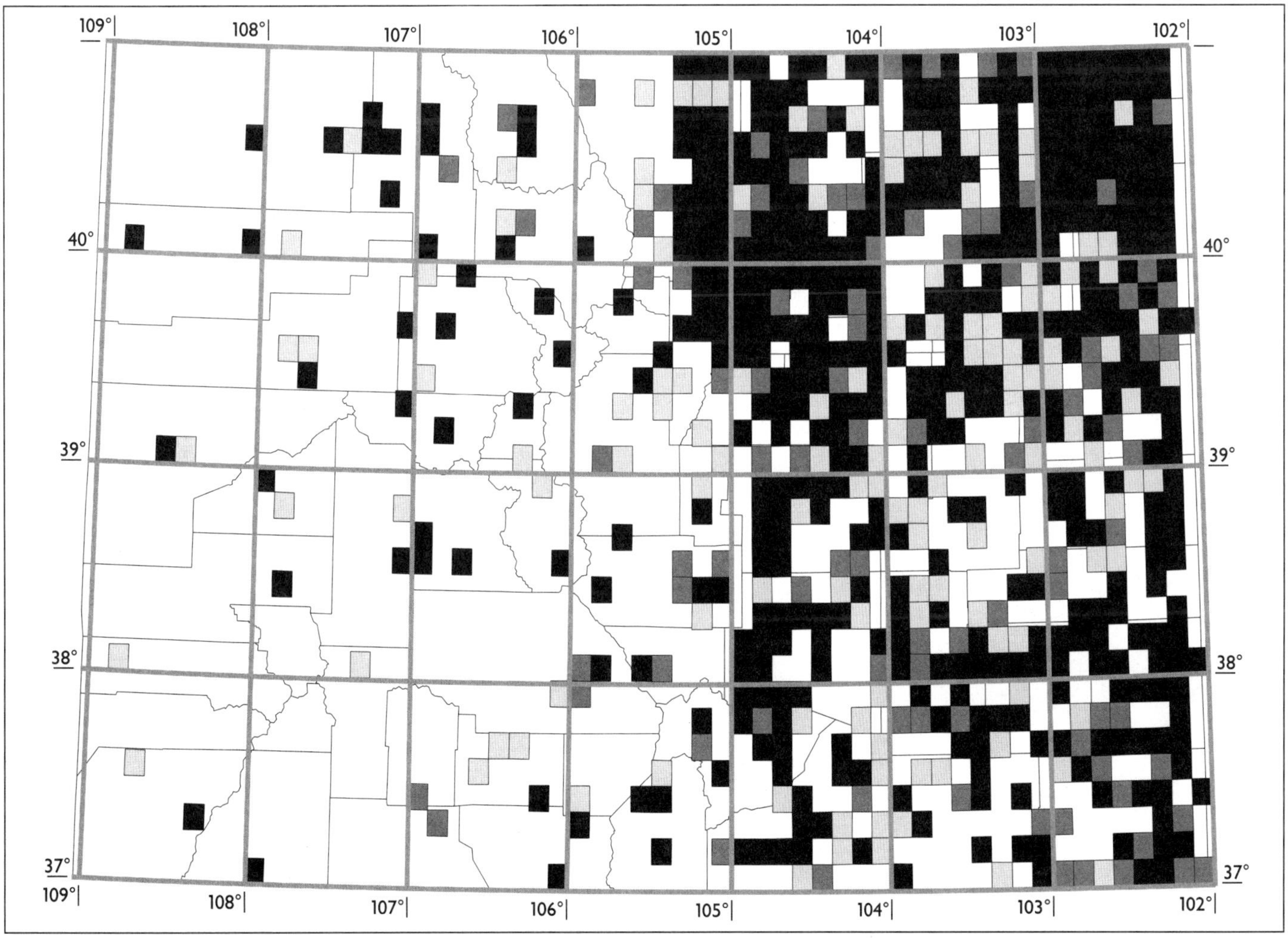
109°
108°
107°
106°
105°
104°
103°
102°
40°
39°
38°
37°

COLORADO DIVISION OF WILDLIFE / WILDLIFE RESOURCE INFORMATION SYSTEM

GREAT-TAILED GRACKLE *Quiscalus mexicanus*

With a few strategically placed white feathers, the jet-black male Great-tailed Grackle might easily pass for a magpie. Even in character the male Great-tailed Grackle resembles that spirited corvid; he patrols his territory with a swagger and bluster reminiscent of a barnyard magpie.

HABITAT

As Great-tailed Grackles expand their range in Colorado, one wonders what impacts these aggressive birds may have on species sharing their habitats. At the present time, however, the numbers and distribution of the grackles remain sparse enough to render such judgments speculative.

In Colorado, as elsewhere, these grackles have nested in marshes, evergreen trees, and windbreaks (A&R). More typically they nest in dense colonies in trees, and prefer open areas with scattered trees (Steve Martin pers. comm.). Near the edge of their range they often nest in marshes (Faanes and Norling 1981). However, these big grackles use marsh vegetation for nest sites "only when there are no suitable trees or tall shrubs in close proximity to a water source" (Selander and Giller 1961), a condition not reflected by atlas observations. Almost half of the nest sites atlasers described were in cattails (most with nearby trees and bushes) although rural types produced the greatest number of Atlas habitat reports. These habitats include both shelterbelts and areas adjacent to farm buildings. The latter hold abundant stores of food and nest sites, and areas with open water (natural and man-made) frequently lie nearby. Foraging areas frequented by these birds include grain fields, cattle pens, and pond margins (Selander and Giller 1961).

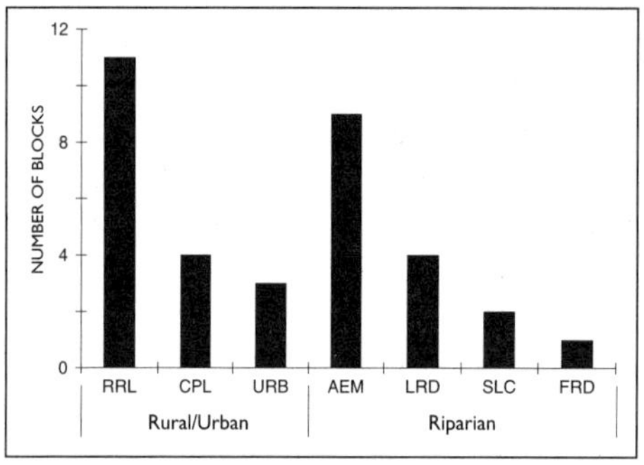
NUMBER OF BLOCKS
0
4
8
12
RRL
CPL
URB
AEM
LRD
SLC
FRD
Rural/Urban
Riparian

In east-central Colorado one breeding colony nested in a mature shelterbelt nearly 2 miles from the closest water source (HEK). This contrasts with Selander and Giller's statement (1961) that this species "is not found . . . at any great distance from the ocean, lakes, ponds, or streams."

BREEDING

The Atlas data provide a precariously small sample from which to draw conclusions about the breeding cycle of Great-tailed Grackles. The phenological dates recorded, however, indicate nothing extraordinary for a bird that maintains a small wintering population within the state. Fledglings appeared by late May; dates for dependent young extended only into mid July.

In Colorado, as elsewhere, males preside over a colony of smaller, plainer females. Males take no part in nest-building or incubation of the eggs. Rather, like other polygynous blackbirds, they spend their days seeking to attract additional females while guarding their colonies and females against interlopers (Orians 1985).

Females defend a tiny area around their own nest against the intrusions of other females. Once the young have hatched, females assume the entire responsibility for feeding the typical brood of three (Selander and Giller 1961). No Colorado records of double-brooding exist, although the species often double-broods in other parts of its range.

BREEDING PHENOLOGY

CODE		# OF RECORDS	DATE RANGE
C	Courtship	1	30 Apr
NB	Nest Building	2	14 May–15 Jun
ON	Occupied Nest	2	29 May–14 Jun
NY	Nest with Young	3	1 Jun–23 Jun
FY	Feeding Young	3	9 Jun–13 Jul
FL	Fledged Young	2	24 May–1 Jun

DISTRIBUTION

Breeding from Kansas to Arizona south to Peru, Great-tailed Grackles have a rapidly expanding range. BBS statistics show a huge increase, almost 5%/year.

As if to frustrate speculation regarding its origins, the first Colorado record of a Great-tailed Grackle came in 1970 from Gunnison (Kingery 1972), close to the center of the state. Certainly it arrived from points south, but by which route? Many early records, including the first breeding record, came from Monte Vista (Stepney 1975), suggesting the Rio Grande Valley as a port of entry. The Rio Grande hypothesis fares poorly, however, in accounting for Arkansas Valley and San Juan Basin birds.

By 1961, Great-tailed Grackles had reached both Clayton and Aztec, on opposite sides of New Mexico close to the Colorado border (Ligon 1961). The Arkansas Valley and San Juan Basin populations may well derive from these populations. In all probability, the species arrived via a general march northward across New Mexico, by crossing the border at multiple points.

Great-tailed Grackles now have a wide distribution, but in relatively small numbers, across most of southern Colorado up to an elevation of 8,000 feet (2400 m). Northern Colorado has few so far, most between Greeley and Denver. The period of Atlas field work (1987–1995) witnessed a nearly statewide expansion of both range and numbers of Great-tailed Grackles. New Confirmations of breeding occurred in seven latilongs, all in the northern half of the state.

Still, Atlas map may underrepresent them. Great-tailed Grackles fall well short of occupying all available habitat in the state and, as the map portrays, have a spotty distribution. They have colonized sites in certain Non-priority blocks (e.g., Mancos, Del Norte) and probably more.

Once in Colorado, Great-tailed Grackles wasted little time before showing up during the colder months. The first winter record came from Gunnison during the winter of 1973–1974 (A&R) and the birds have now established a winter range within the state. Here, as elsewhere along the northern end of their range, they retreat modestly from their summer breeding range.

Many attribute the northward advance of this species to growth of irrigation in the southwestern United States. The expansion of Great-tailed Grackles in the Great Basin and the Southwest closely followed the era of massive water projects in these arid regions. Selander and Giller (1961), however, suggested that intensive cattle grazing and the conversion of prairie lands into farms and settlements with shade trees and crops (assisted by irrigation) also had much to do with the species' movement northward across Texas earlier this century. In all likelihood, these factors also facilitated the spread of grackles farther west.

BREEDING EVIDENCE

REPORTED IN 27 (2%) OF 1745 PRIORITY BLOCKS

□	Possible	10 blocks	(37%)
▒	Probable	5 blocks	(19%)
■	Confirmed	12 blocks	(44%)

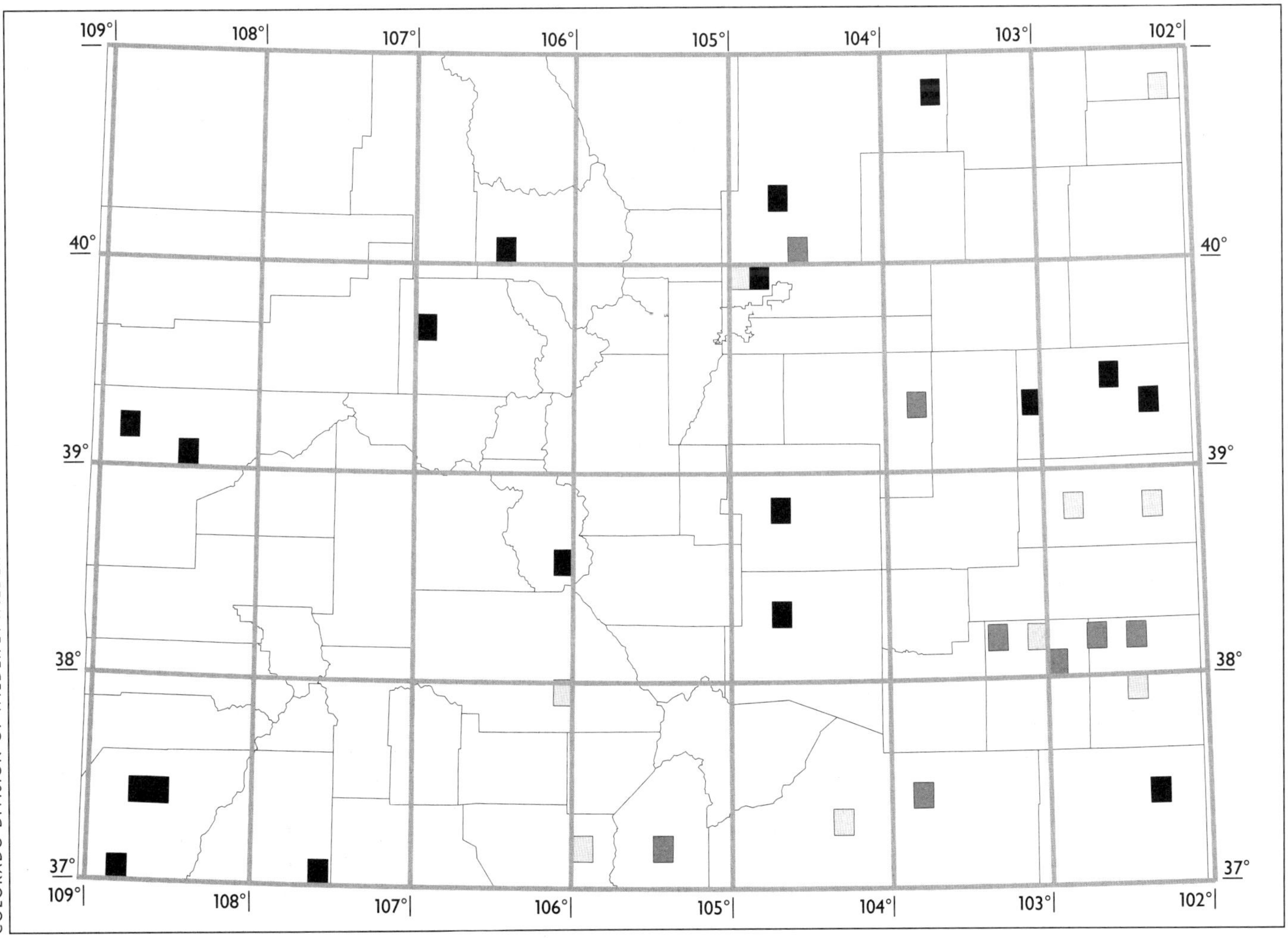
109°
108°
107°
106°
105°
104°
103°
102°
40°
39°
38°
37°

COLORADO DIVISION OF WILDLIFE / WILDLIFE RESOURCE INFORMATION SYSTEM

Brown-headed Cowbird

Originally limited to shortgrass prairie, "buffalo-birds" used to follow herds of bison and feed on insects stirred up by the animals. Cowbirds of the genus *Molothrus* do not build their own nests and most are brood parasites. Such behavior probably pre-adapted Brown-headed Cowbirds for a commensal relationship with the nomadic bison.

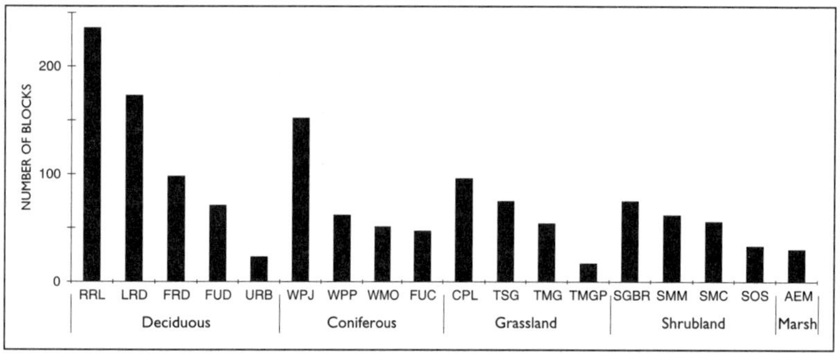
NUMBER OF BLOCKS
0
100
200
RRL LRD FRD FUD URB
Deciduous
WPJ WPP WMO FUC
Coniferous
CPL TSG TMG TMGP
Grassland
SGBR SMM SMC SOS
Shrubland
AEM
Marsh

HABITAT

Brown-headed Cowbirds, unlike most birds, breed and forage in different habitats. During the breeding season females search for nests in the morning, usually near the edges of woodlands or forests; in the afternoon they move to open areas to roost and to forage on the ground for seeds and arthropods, often in association with livestock (Rothstein et al. 1980). Their abundance increases when humans provide foraging opportunities (e.g., seeds, livestock). Rather than extensive forests or fields, cowbirds like grasslands with low or scattered trees and edge habitats with ample food sources, close to potential forest-dwelling hosts (Gates and Gysel 1978, Brittingham and Temple 1983, Johnson and Temple 1990). They typically avoid unlogged forests (Airola 1986).

Atlasers recorded these habitat generalists in 46 different habitats, an assortment that reflects the diversity of habitats used by host species. The top four, all below 7,000 feet (2745 m), were rural, riparian woodlands (plains and foothills), croplands, and pinyon/juniper woodlands. Distribution gaps occur in areas of extensive cropland, desert shrub, and backbones of the high mountains.

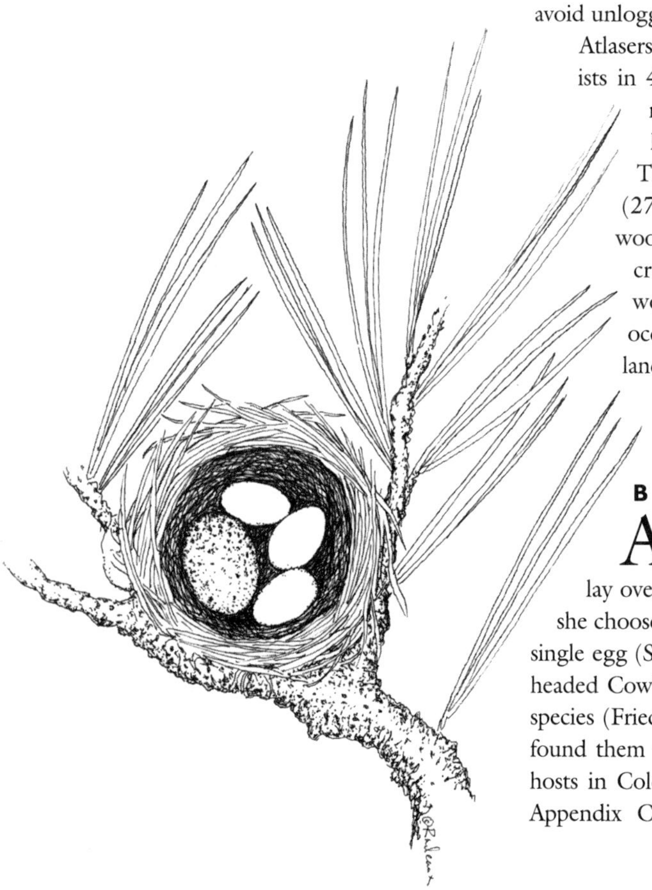

BREEDING

As brood parasites, cowbirds do not build nests. A female can lay over 40 eggs per year; almost daily she chooses another nest in which to lay a single egg (Scott and Akney 1980). Brown-headed Cowbirds have parasitized over 200 species (Friedmann and Kiff 1985). Atlasers found them with 51 of the 59 now-known hosts in Colorado (Chace and Cruz 1996; Appendix C), which includes 21 species added by Atlas field work: Gray and Cordilleran flycatchers, Eastern Phoebe, Western Kingbird, Bewick's and House wrens, Mountain Bluebird, Brown Thrasher, Virginia's, Yellow-rumped, Black-throated Gray, and Grace's warblers, Yellow-breasted Chat, Western Tanager, Blue Grosbeak, Sage and Fox sparrows, Common Grackle, Orchard Oriole, American Goldfinch, and Cassin's Finch.

Egg-laying females display certain preferences in host nests: in order, smaller host eggs; two host eggs, large closed nests (over 3 inches, 7.6 cm diameter) or small open nests (over 2 inches, 5.1 cm, diameter) rather than large open nests (Lowther 1993).

Some species with a long history of exposure to cowbird parasitism recognize the foreign egg and reject it or abandon the nest. Others, "acceptors," fail to recognize the intruding egg or physically cannot remove the large cowbird egg. The impact of cowbirds on acceptor hosts varies. Some, such as Red-winged Blackbirds, tolerate parasitism (Ortega and Cruz 1991) whereas some local populations of Plumbeous Vireos would disappear due to parasitism without the immigration of outsiders the following season to bolster the population (Chace 1995). Cowbirds have the most detrimental effect on hosts with localized or limited distributions.

Cowbirds impact reproductive success by removing host eggs, having eggs that hatch faster (11 days) than most hosts, and reducing host egg-hatching success. Young cowbirds also out-compete their nest mates for food and space in the nest.

BREEDING PHENOLOGY

CODE		# OF RECORDS	DATE RANGE
C	Courtship	137	26 Apr–18 Jul
NE	Nest with Eggs	39	10 May–29 Jul
NY	Nest with Young	13	24 May–10 Jul
FY	Feeding Young	5	22 Jun–7 Jul
FL	Fledged Young	148	8 May–23 Aug

DISTRIBUTION

Originally birds of the Great Plains grasslands, Brown-headed Cowbirds have spread from coast to coast, to northern Canada, and into central Mexico. In Colorado, following herds of bison, they originally ranged across the plains, the four mountain parks, and perhaps northwestern Colorado.

Edwin James reported that "cow buntings" followed the Long expedition "five or six miles, alighting on the ground . . . within a few paces of our horses' feet" (Marsh 1931). This earliest Colorado record, in 1820, preceded the bison slaughter; then they ranged into Middle and South parks and perhaps to higher elevations. They probably followed the bison—seasonal migrations on the plains and altitudinal migrations from the mountain parks up to the alpine (Meaney and VanVuren 1994, Chace and Cruz unpubl. data). The near extermination of bison in the late 1800s limited cowbirds to low-elevation grasslands and foothills in eastern and western Colorado as cattle replaced bison.

Now that large numbers of domestic livestock graze in the mountains, so once again do cowbirds roam there. When cowbirds followed nomadic bison, specific local bird populations faced sporadic parasitism; now with static cattle herds, and consequently static cowbirds, potential hosts face constant pressure. Atlasers recorded them in most montane blocks but not in high-altitude blocks.

Atlas map and abundance estimates show the same geographic concentrations. According to Atlas abundance estimates, one-third of the state's cowbirds occur in the six northwestern latilongs. More than one-fifth occupy the southwestern six latilongs, and, surprisingly, only about one-fifth occur in the nine plains latilongs—their presumed place of origin.

According to BBS results, cowbird populations have decreased since 1966, probably due to reforestation in the eastern U.S. They have, however, increased significantly in Colorado. High densities of cowbirds and severely impacted avian communities have brought appeals for cowbird-control programs. Such programs are expensive, labor-intensive, and non-permanent. They work best for host species that are endangered and have limited distribution (e.g., Kirtland's Warbler). Controls at large winter roosts cost relatively little but are largely ineffective because of wide breeding-ground dispersal from a single winter roost. Removal efforts also reduce selection pressure on hosts that allows the evolution of anti-cowbird defense behaviors (Sealy 1992).

Having destroyed bison herds and created landscape conditions that enable cowbirds to thrive, do we have an obligation to protect birds endangered by cowbirds?

BREEDING EVIDENCE

REPORTED IN 1119 (64%) OF 1745 PRIORITY BLOCKS

Possible	424 blocks	(38%)
Probable	489 blocks	(44%)
Confirmed	206 blocks	(18%)

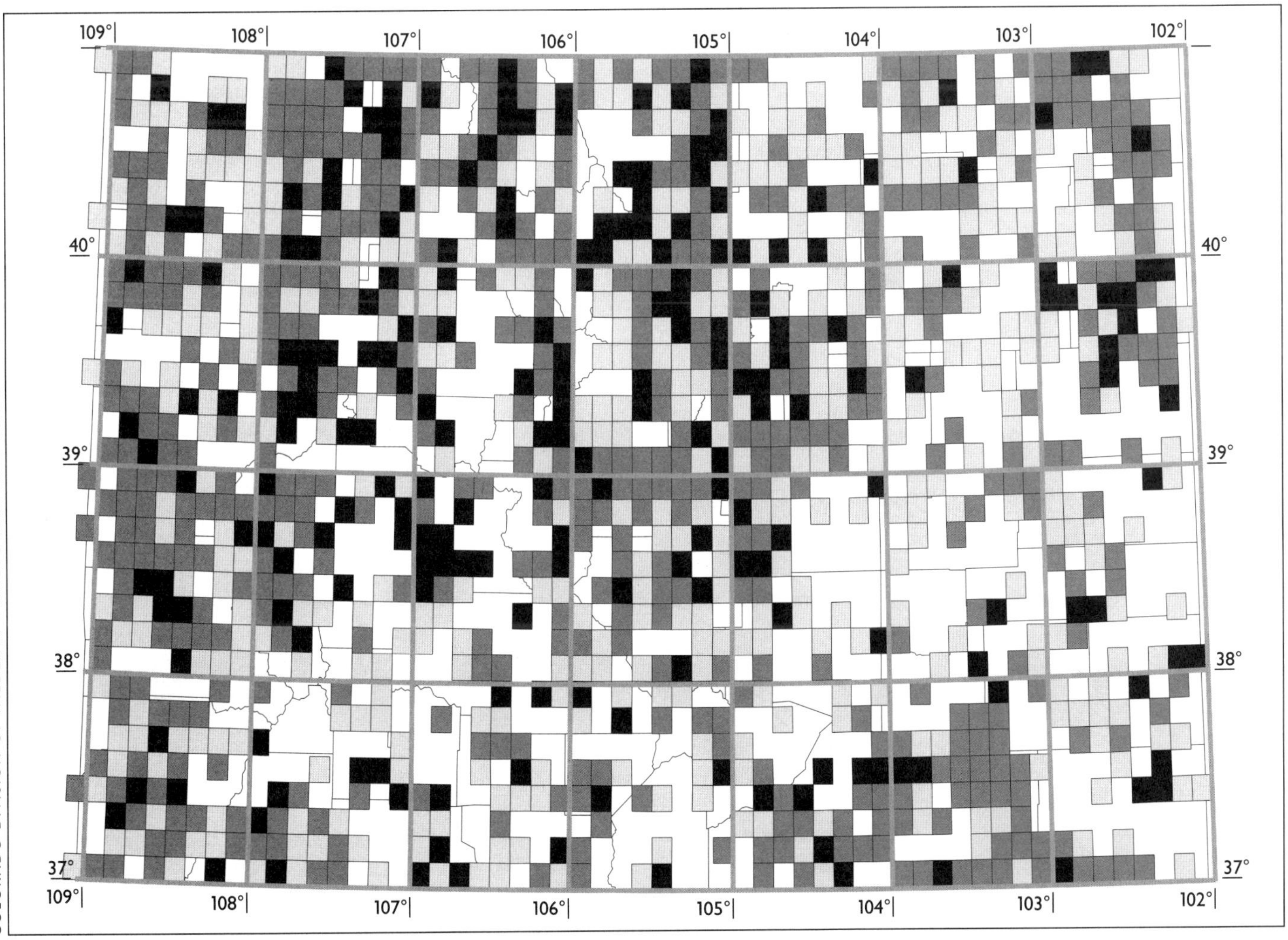
109°
108°
107°
106°
105°
104°
103°
102°
40°
39°
38°
37°

COLORADO DIVISION OF WILDLIFE / WILDLIFE RESOURCE INFORMATION SYSTEM

For many years the insulting name, "bastard Baltimore Oriole," clung to this unique chestnut-and-black oriole. Although it has lost the "bastard" designation, it continues to be saddled with *spurius* because an early ornithologist assumed a female Baltimore Oriole was a male Orchard Oriole (Wilson 1832). The common name gives a better picture of this species, whose robin-like song rings out in orchards and shade trees.

HABITAT

In the East, these birds live up to their name by preferring orchards, especially those of pears, peaches, and apples, near human habitations (Bent 1958). They also occur plentifully in shade trees in yards and along streets and roads. On the western edge of their range, as Atlas data show for Colorado, they reside mainly in riparian woodlands, shelterbelts, and farmyard trees on the plains. In all parts of their range, they avoid heavily wooded areas by choosing semi-open areas without dense cover.

Atlas work in Colorado reports them entirely in deciduous habitats, but reveals a habitat dichotomy not previously articulated. Atlasers found almost equal use of two major habitats: 47% riparian woodlands and 49%

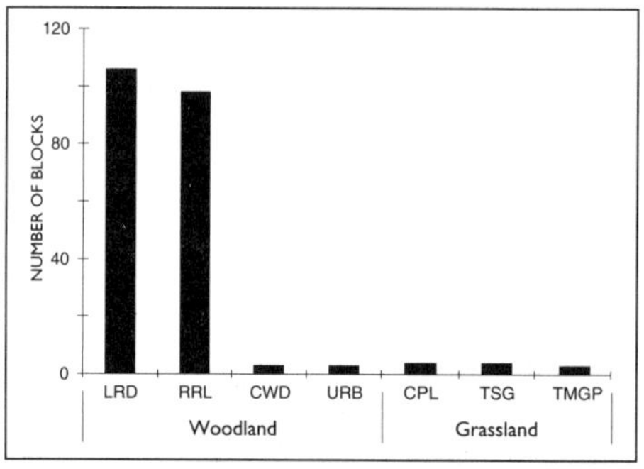
120
80
40
0
NUMBER OF BLOCKS
LRD
RRL
CWD
URB
CPL
TSG
TMGP
Woodland
Grassland

rural, which includes farmyards, shelterbelts, and croplands. The newly found, wide use of High Plains farmyards explains the substantial change to the previously described Colorado range that atlasers detected.

BREEDING

Like many of "our" breeding birds, Orchard Orioles spend most of their year in their winter home. They live in Central America a full nine months of the year. They leave in mid April and have returned as early as 20 July (Skutch *in* Bent 1958). Atlas work shows these birds building nests in late May and early June. Volunteers recorded two exceptionally late sightings of fledged young in early August.

Males, both those in full adult plumage and those in first-year plumage, arrive in their breeding areas about a week before the females (Sealy 1980). Courtship begins when the females arrive. Males display by rising from the treetops in full song and continue to sing as they descend to the shelter of the leaves (Bent 1958). Both partners build the nest, a hanging basket usually made of woven grasses. They build a less pendulous nest than that of Bullock's Orioles; usually the width exceeds the depth (Scharf and Kren 1996). Nests of western birds have larger openings than those in the East, possibly for heat dissipation. Orchard Orioles tend to use smaller-diameter, shorter trees than Bullock's, and within riparian forests use denser sections (Scharf and Kren 1996). Nevertheless the nests of the two species are too similar to distinguish with certainty without seeing the builders.

Unlike many species, male Orchard Orioles breed in first-year plumage, which resembles the yellow plumage of the females, except for a black bib. First-year breeders tend to nest in marginal habitat (Scharf and Kren 1996). In the first year of colonizing a new area, as many first-year males breed as do mature males (Sealy 1980). In Colorado, as in other parts of their range, these orioles nest sociably with other species of birds in the same tree. Neighbors commonly include both Eastern and Western kingbirds. Males do not incubate, but they feed the females and the hatchlings in the nests and remove fecal sacs; both parents feed the fledglings (Bent 1958).

Atlasers recorded two instances of adults feeding juvenile cowbirds, consistent with the thesis that Orchard Orioles, unlike Bullock's, are cowbird "acceptors" (Chace and Cruz 1996; Appendix C). Typically, the presence of cowbird young reduces oriole fledging success, although the orioles abandon nests when the cowbird lays before the host (Scharf and Kren 1996).

BREEDING PHENOLOGY

CODE		# OF RECORDS	DATE RANGE
C	Courtship	1	2 Jun
NB	Nest Building	8	27 May–14 Jun
ON	Occupied Nest	19	17 May–30 Jun
NE	Nest with Eggs	1	4 Jun
NY	Nest with Young	3	22 Jun–29 Jun
FY	Feeding Young	22	9 Jun–25 Jul
FL	Fledged Young	11	2 Jun–8 Aug

DISTRIBUTION

Mainly a species of the Great Plains and southeastern U.S., Orchard Orioles range from the Canadian border to central Mexico. They occur more commonly in the southern states than in the northern states, and only accidentally west of the Rocky Mountains. They winter from central Mexico to northern South America.

In 1897 only a single specimen, taken by Allen in Denver, existed for Colorado (Cooke 1897). By 1958 Bent mentioned only Yuma County and Denver as breeding areas, and in 1965 B&N reported these orioles only on the eastern edge of Colorado, and could cite confirmed nesting only in 1906 and 1962. The Latilong Study reported breeding across eastern Colorado but more recently, A&R showed them nesting almost exclusively along the South Platte to Denver and the Arkansas upstream to Florence, plus a group in Yuma County.

The Atlas map looks quite different. Atlas data show a species not confined to the river valleys. Their range now extends westward almost to the foothills, from Denver north. They do breed up the Platte to Denver, but in the Arkansas Valley atlasers found them only as far west as Crowley County. They also populate farms and towns across the High Plains where they broadcast their un-oriole-like songs while hidden in the leaves.

Across the continent these small orioles show a significant BBS decline of 1.8% per year, but too few Colorado routes traverse their habitat to establish a trend here.

Whether or not the expanded Atlas map depicts a recent expansion in range seems uncertain, as the Latilong Study map shows them breeding in essentially the same places as the Atlas. The Atlas map may portray breeding in High Plains rural habitats that existed before the BBS started in 1966. The spread of Orchard Orioles has probably occurred gradually during the twentieth century, the range expansion facilitated by post-settlement (since 1860) creation of new habitat suitable for these sweet singers.

BREEDING EVIDENCE

REPORTED IN 193 (11%) OF 1745 PRIORITY BLOCKS

☐	Possible	60 blocks	(31%)
▩	Probable	60 blocks	(31%)
■	Confirmed	73 blocks	(38%)

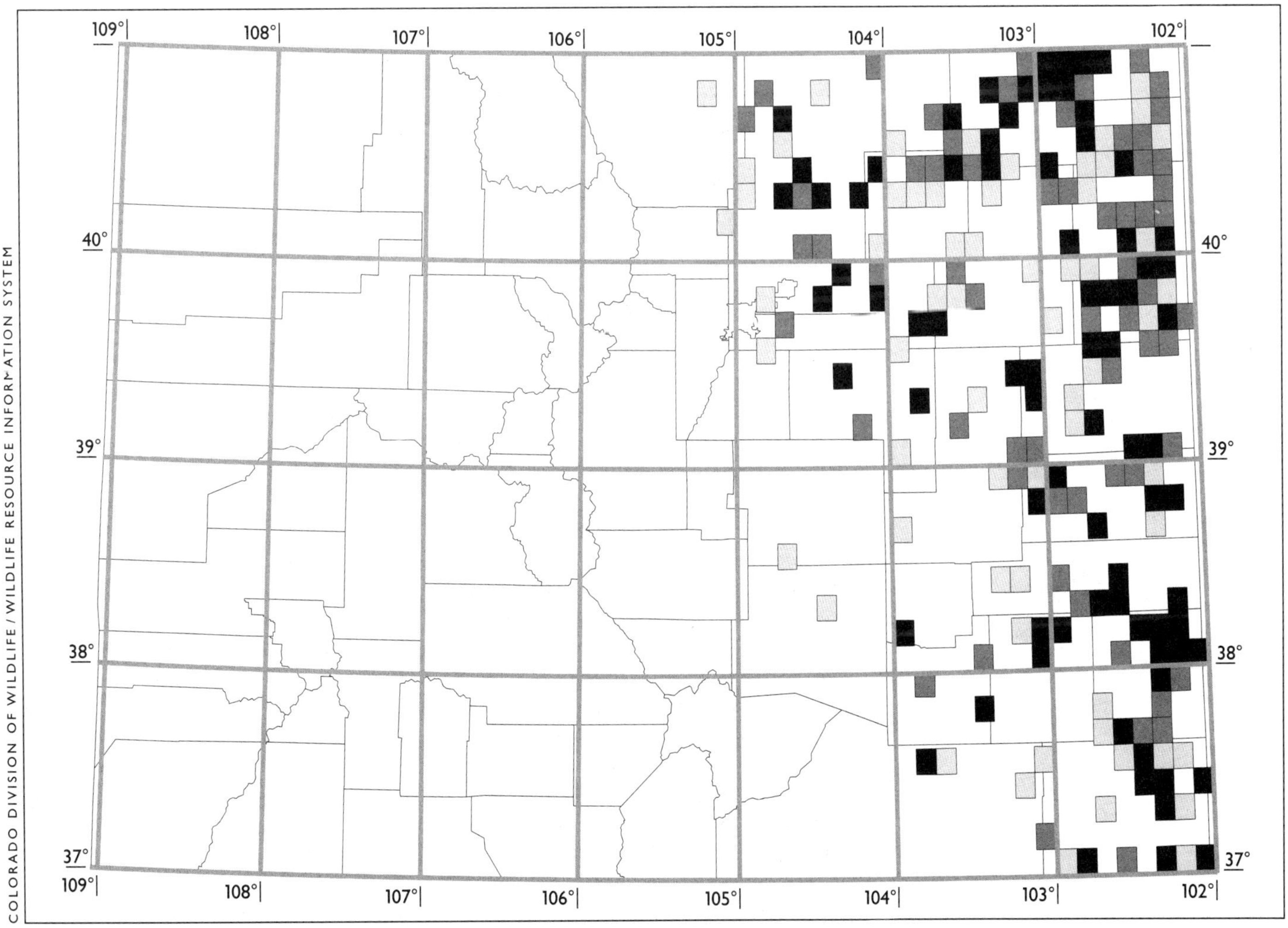
109°
108°
107°
106°
105°
104°
103°
102°
40°
39°
38°
37°
COLORADO DIVISION OF WILDLIFE / WILDLIFE RESOURCE INFORMATION SYSTEM

From 1983 to 1995 the American Ornithologists' Union considered the Northern Oriole a single species. In 1995 the AOU split the species into Bullock's and Baltimore species, reinstating the status of a decade earlier. Baltimore's solid black head and Bullock's orange-and-black-patterned head make them far easier to separate than small shorebirds, gulls, sparrows, and flycatchers, but the differences go beyond appearance. Different vocalizations, body size, molting patterns, nest site placement, and migration timing, plus genetic analysis, led to the decision to re-separate the species.

HABITAT

Because of their similar behavior and habitats and prior one-species status, this account addresses mainly Bullock's, with differences from Baltimore pointed out. The range map for Baltimore appears on page 544.

Both oriole species show a strong dependence on mature deciduous trees, in Colorado native cottonwoods. Cottonwoods and exotic trees used in landscaping dominate Atlas nest cards, although native junipers housed two nests found by atlaser Randy Lentz. The high proportion of rural habitat reports makes this tie to cottonwoods less obvious without examining the typical Colorado farm. Clusters of plains cottonwoods or Siberian elms shield

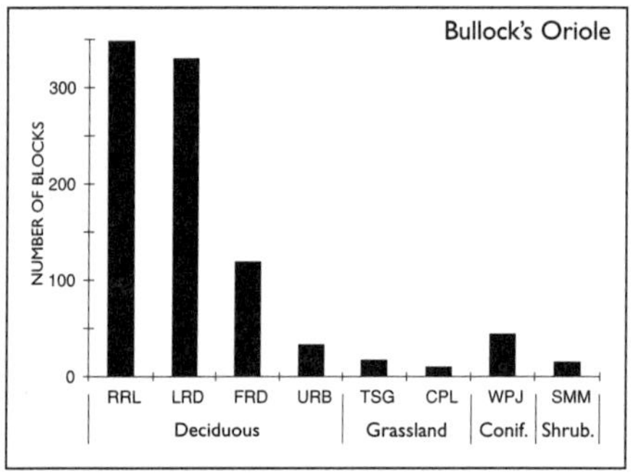
Bullock's Oriole
NUMBER OF BLOCKS
300
200
100
0
RRL
LRD
FRD
URB
TSG
CPL
WPJ
SMM
Deciduous
Grassland
Conif.
Shrub.

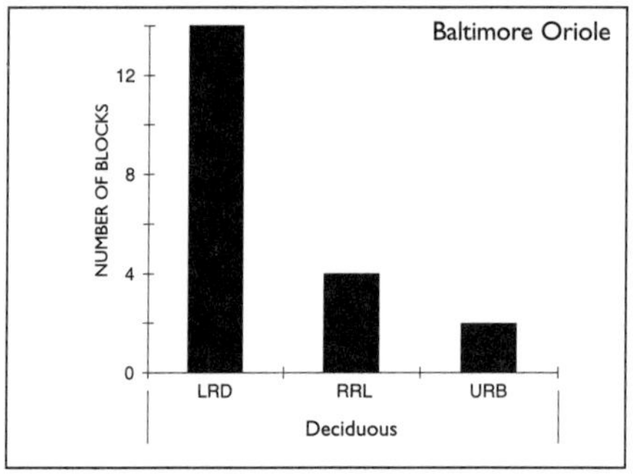
Baltimore Oriole
NUMBER OF BLOCKS
12
8
4
0
LRD
RRL
URB
Deciduous

plains farmyards, and planted clumps of narrowleaf cottonwoods or poplars surround mountain ranchhouses, which also frequently sit close to timbered riparian strips.

The close tie between orioles and deciduous habitat led to combining Bullock's and Baltimore as Northern Oriole. The treeless Great Plains originally formed a barrier between the two species. As Americans settled the plains they planted trees around farms, in towns, and as windbreaks; these eventually provided a bridge that allowed the two species to meet; where they met they interbred, leading ornithologists to question whether the oriole complex consisted of one or two species (AOU 1983, 1995).

Orioles feed on soft foods, mainly insects and nectar in summer and fruits in winter. Juniper berries may explain their presence in juniper habitats. Deciduous trees provide insects for food and sites for the long, hanging nests of these species. Once the leaves drop from a deciduous tree, that dangling nest jumps out; UN (used nests) accounted for 184 Bullock's Oriole confirmations. How many unseen nests occurred in junipers or other coniferous trees? Probably not many, but undoubtedly some. All records of Baltimore Orioles came from three habitats—mostly cottonwood stream bottoms, plus a few rural and urban habitats.

BREEDING

Although Atlas data for Baltimore Orioles contain few samples, the dates coincide with the Bullock's dates. Both orioles begin nesting from mid May into June and take about a week to construct their nests (Jackson 1993). Nests dangle from the forks of limbs, near the tips of the branches, generally about halfway up the tree. Multiple nests clustered in one tree may represent birds returning in subsequent years (Jackson 1993), but they can also represent semi-colonial nesting (Greg Butcher pers. comm.).

The degree of territorial behavior depends on the importance of the nest habitat as a foraging area. In riparian sites with an ample food supply, orioles enforce typical songbird territories. They nest semi-colonially when they have limited foraging opportunities in the riparian habitat (e.g., a narrow strip of cottonwoods that lacks shrub cover, such as an irrigation ditch). Then they may feed in fields next to the nest trees, and several pairs may nest in the same grove or even in the same tree (Pleasants 1979; Greg Butcher pers. comm.).

Nests consist of plant fibers and the inner bark of junipers and willows, with lining from those materials and down from plants, wool, and fine moss. They sometimes incorporate oddities such as yarn, thread, and, in an Atlas block southeast of Hayden, monofilament line (too loose to hold eggs and replaced by a more normal nest 5 feet away; NMB).

Brown-headed Cowbirds often lay eggs in oriole nests; orioles have developed mechanisms to thwart them. Orioles recognize and remove cowbird eggs from the nest by piercing them with their beaks and carrying them from the nest (Sealy and Neudorf 1995). A second, less common, defense consists of burying the cowbird eggs in nesting material (Hobson and Sealy 1978) so that they do not receive the same incubation and care as the orioles' eggs. Not all parents succeed in protecting their nests. For Colorado, one Atlas and two older reports of parasitism exist (Chace and Cruz 1996; Appendix C).

BREEDING PHENOLOGY

BULLOCK'S ORIOLE

CODE		# OF RECORDS	DATE RANGE
NB	Nest Building	39	13 May–27 Jun
ON	Occupied Nest	130	11 May–30 Jul
NE	Nest with Eggs	4	4 Jun–9 Jul
NY	Nest with Young	55	25 May–19 Jul
FY	Feeding Young	106	26 May–31 Jul
FL	Fledged Young	48	8 Jun–24 Jul

BALTIMORE ORIOLE

CODE		# OF RECORDS	DATE RANGE
NB	Nest Building	2	15 May–4 Jun
ON	Occupied Nest	1	23 Jun
NY	Nest with Young	2	8 Jun–3 Jul
FY	Feeding Young	1	22 Jun
FL	Fledged Young	1	23 Jul

DISTRIBUTION

Both species breed in North America only. They split the continent up the Great Plains: Bullock's Orioles on the west and Baltimore Orioles on the east. The ranges of both species reach north into southern Canada and Bullock's breed south into central Mexico.

The basic Colorado distribution of these species has changed little since 1897 (Cooke). Baltimore Orioles occur only on the eastern border, primarily along the South Platte and Republican river drainages. Although in 1889 a pair of Baltimore Orioles bred in Boulder County (Alexander 1937), that aberration has not recurred. Three Atlas reports of hybrids came from northeastern Colorado.

Bullock's breed throughout the state, including all but one block where atlasers found Baltimores. They occur thickly on the plains in both riparian and many urban/rural habitats. Similarly, on the Western Slope, they occupy the stream bottoms, towns, and ranchyards. Atlasers found fewer in North, Middle, and South parks than shown by A&R; otherwise the two maps coincide. The orioles seem to recognize an altitudinal boundary of about 7,500 feet (2285 m), but probably habitat rather than altitude confines them.

As the old plains cottonwoods age and die without the replenishment by younger trees, the river bottoms will become more open, and oriole habitat may diminish. Conceivably, flood control could partially recreate the Great Plains barrier and isolate these two species again.

BREEDING EVIDENCE

(BULLOCK'S ORIOLE)

REPORTED IN 762 (44%) OF 1745 PRIORITY BLOCKS

☐	Possible	128 blocks	(17%)
▩	Probable	81 blocks	(11%)
■	Confirmed	553 blocks	(72%)

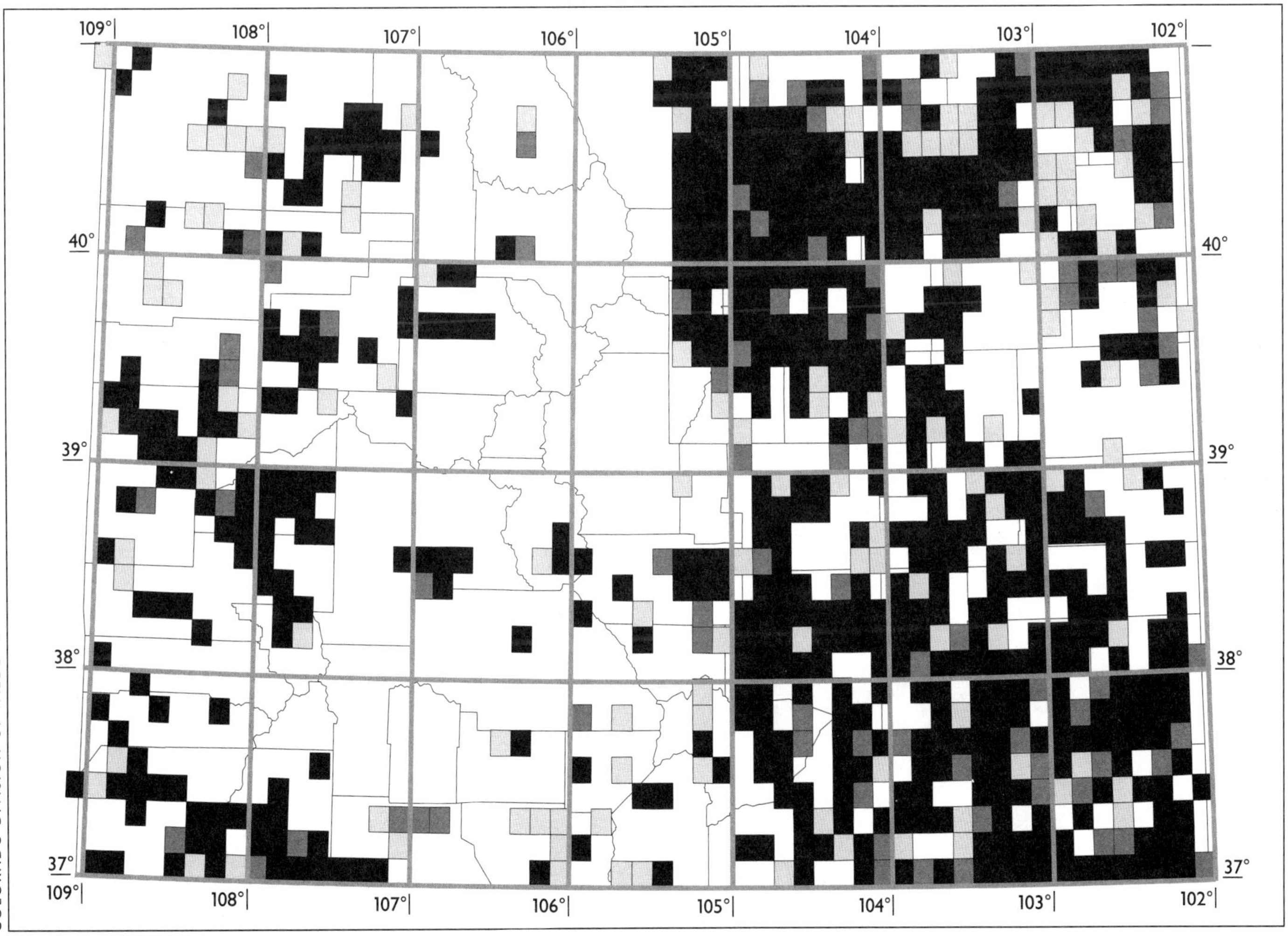
109°
108°
107°
106°
105°
104°
103°
102°
40°
39°
38°
37°

COLORADO DIVISION OF WILDLIFE / WILDLIFE RESOURCE INFORMATION SYSTEM

This desert jewel, adorned in brilliant lemon-yellow and black, did not even appear on the Colorado bird list until 1975 (A&R). Since that discovery of two separate, vagrant males in the same week near Denver, observers have found them regularly at a few sites—all near the Utah line.

HABITAT

Atlas results affirm that Colorado's few breeding pairs of Scott's Orioles live in an intermediate zone of the dry mesas, hills, canyons, and plateaus of far western Colorado. They inhabit groves of sparse juniper with widespread yuccas. They avoid the real desert below the juniper tree line, and during breeding they go no higher than the lower limits of Colorado's desert evergreens. The state has suitable habitat only below 5,500 feet (1700 m) and in only a few places (A&R).

To the south and west of Colorado they also use flat or rolling plains of hard, gravelly soil generally bare of vegetation and typified

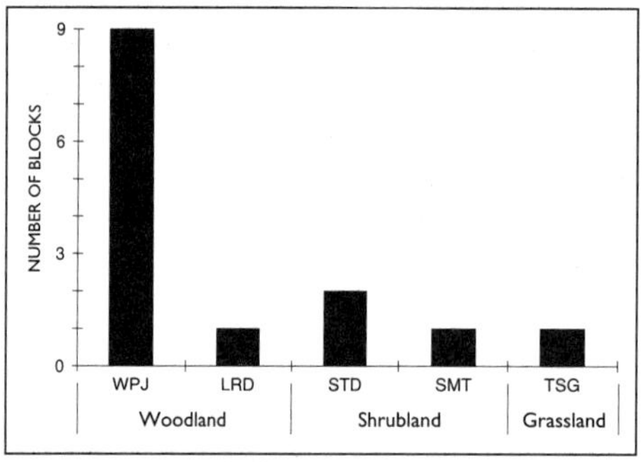
NUMBER OF BLOCKS
0
3
6
9
WPJ
LRD
STD
SMT
TSG
Woodland
Shrubland
Grassland

by low, scraggly, omnipresent creosote bushes and yucca. Attracting the orioles to this desolate region are widely scattered soapweed yuccas, picturesque plants in which the birds nest (Bent 1958). Scott's Orioles not only nest in yuccas but use the plant leaves to construct their nest. Even when orioles build nests in scrub pinyons or junipers, the nest-building materials come mainly from fiber obtained from the edges of yucca blades (Ligon 1961).

All Atlas observations came from pinyon/juniper and desert scrub. Atlas habitat codes did not distinguish between pure juniper and mixed woodlands, but the observations came from isolated juniper-only stands and adjoining desert shrublands.

BREEDING

Colorado's first breeding records involved fledglings seen in 1979 west of Rangely and close to a longtime Utah nesting area (Gent 1987, A&R). Colorado has records of four nests, two before the Atlas and two in 1997. A nest found near Dinosaur on 18 June 1981 held three eggs. Using strands of wheatgrass, the female had built it 6 feet up in a 9-foot juniper. She had started to build by 5 June. Although she concealed it well in the foliage, the nest, "decidedly loose and delicate," allowed diffused light to pass through the whole structure (Ed Hollowed, CFO records). A juniper tree also held the second nest, near Fruita: 4 feet high and empty on 31 May 1982, the nest contained three eggs on 2 June. By the time of a subsequent visit on 25 June an unknown predator had raided the nest, and one partially feathered dead nestling sprawled underneath it (Lonnie Renner nest card). In 1997 two pairs had nests in junipers west of Mack, 5 and 10 feet above the ground. One fledged three young (banded) and one, with two eggs, ultimately failed (CD, Rich Levad pers. comm.).

The two reports during the Atlas period of parents feeding young birds came on 24 June and 1 July. A single report of a fledged young bird occurred on 7 July. Four Atlas workers reported pairs of orioles from 28 May until 7 July.

In the Southwest, the birds place their nests in the dead drooping blades of the yuccas beneath the live crown, which provides ample protection from both weather and potential enemies. The yellowish cup is so securely laced that it may remain intact for several years. The female oriole lays 2–4 eggs and incubates for 14 days (Harrison 1984, Terres 1980). The young are fed regurgitated insects, fruit, and berries for the first five days and then receive whole food. After 14 days of attendance by both parents the young leave their nest (Ehrlich et al. 1988).

BREEDING PHENOLOGY

CODE		# OF RECORDS	DATE RANGE
FY	Feeding Young	2	24 Jun–1 Jul
FL	Fledged Young	1	13 Jun

DISTRIBUTION

Scott's Orioles breed from central Utah and southern California to western Texas, south to southern Mexico. Prior to the Atlas, the only Colorado breeding records came along the Utah line in western Mesa, Rio Blanco, and Moffat counties.

Atlasers found Scott's Orioles in only eight Priority and five Non-priority blocks—but in three new places. They found them in the known Grand Valley site but did not find them near either Rangely (western Rio Blanco) or Dinosaur (western Moffat) where Atlas blocks did not include the old sites.

Atlasers Confirmed breeding southwest of Cortez in Montezuma County, where orioles occurred in five blocks. A field worker found a pair of orioles northeast of Dinosaur in central Moffat County but did not Confirm breeding.

In southeastern Colorado, in Las Animas County 35 miles south of La Junta on the Purgatoire River, a pair including a singing male and three other observations hinted at another isolated breeding site (Aaron Ellingson VF, Rich Bunn pers. comm.). Scott's Orioles regularly occur near Logan in eastern New Mexico, only 100 miles (160 km) south of the Colorado line (Sandy Williams pers. comm.).

Scott's Orioles arrive in Colorado in May and depart by the middle of August (A&R). Breeding season in the southern part of their range, southern California and Arizona, extends from late April to mid May. Their Colorado dates no doubt lag behind; the few Atlas reports, combined with prior reports, suggest a mid May to mid July cycle.

A few additional breeding sites may likely be found in Colorado. In Utah, Scott's Orioles breed in three latilongs adjacent to Colorado and probably in the fourth (Walters 1983). Several potential areas lie so close to the Utah-Colorado line that Priority blocks did not include them.

BREEDING EVIDENCE

REPORTED IN 9 (1%) OF 1745 PRIORITY BLOCKS

Possible	4 blocks	(44%)
Probable	2 blocks	(22%)
Confirmed	3 blocks	(33%)

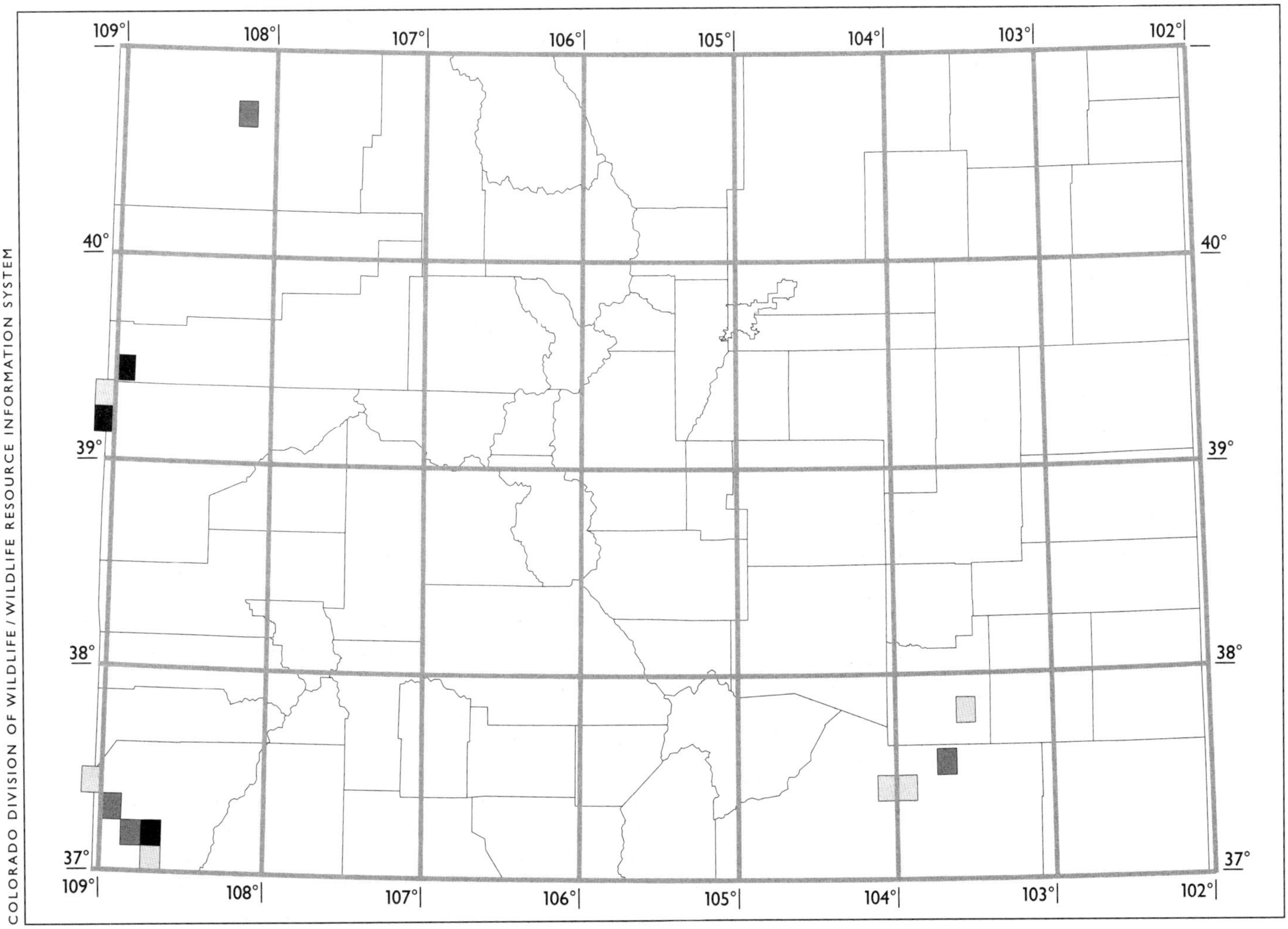
109°
108°
107°
106°
105°
104°
103°
102°
40°
39°
38°
37°

COLORADO DIVISION OF WILDLIFE / WILDLIFE RESOURCE INFORMATION SYSTEM

BROWN-CAPPED ROSY-FINCH *Leucosticte australis*

But for a few roads and many trails that allow access to cliffs above timberline, only serious mountain climbers would encounter nesting Brown-capped Rosy-Finches. Their loud, grating call notes sound very different from the soft and melodic calls of Horned Larks, American Pipits, and occasional Rock Wrens, the only other small passerines that normally nest above timberline.

HABITAT

Only in the alpine zone of the high mountains do Brown-capped Rosy-Finches nest, and only in the vertical cliffs and crags. No other bird species exploit this habitat. Rosy-finches place their nests either in deep crevices in solid rock or in cracks on looser rock faces. The spotty distribution of this habitat serves to concentrate nesting into loose colonies. On many precipitous peaks, groups of hundreds may nest, yet they may not occur at all on peaks of equal elevation blanketed with tundra vegetation of mat willows, grasses, sedges, and brilliant alpine flowers.

Additionally, they need suitable foraging grounds. Primarily insect-eaters in the summer, they frequent rocky ridge lines and snowfields. They take advantage of insects caught in updrafts or strong winds. When winds cross snowfields, updrafts are cut off and insects fall stunned on the snow surface, where these finches forage on an abundant food source.

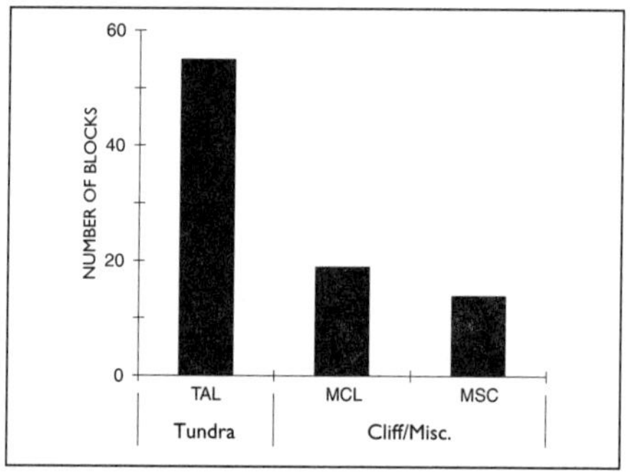

NUMBER OF BLOCKS
60
40
20
0
TAL
MCL
MSC
Tundra
Cliff/Misc.

Like other species inhabiting remote regions where human activity has traditionally been minimal, they show little fear of man.

BREEDING

Brown-capped Rosy-Finches (and other rosy-finch species) are one of the few passerines whose females determine territories and choose nest sites. Males have poorly developed songs because they do not use song to defend territories. Males outnumber females by a 6:1 ratio. Several males display in front of individual females; the female chooses one and they form a monogamous bond that lasts throughout the nesting season. Surplus males spend the breeding season fighting with other males (Johnson 1983, Shreeve 1980).

The female builds a bulky grass nest, but she may economize effort by maintaining the same territory and using the same nest cup several years in a row (Johnson 1983, Shreeve 1980). Single-brooded, she incubates the 4–5 eggs for 12–14 days until they hatch. The young fledge in about 18 more days. As soon as the young gain independence, they begin flocking with other recent fledglings. Flocks of adults quickly re-form after the breeding season.

The number of Atlas rosy-finch reports was surprisingly low given the large amount of suitable habitat in the state. Access to alpine cliff habitat in Atlas blocks often proved very difficult: it frequently required backpacking or mountaineering. Nevertheless, in the San Miguel Range one climber found a nest with young at 14,200 feet (4328 m) near the summit of Mount Wilson (Chuck LaRue, nest card).

Atlasers Confirmed breeding during a fairly narrow range of dates—from a nest on 22 June to the last fledgling seen on 22 August. These dates start earlier than B&N's earliest observations in mid July, but B&N observed a pair feeding nestlings on 30 August on Loveland Pass. The two-month range of Atlas dates reflects in part the difficulty of access to the alpine zone early in the summer; probably the nesting season extends over a longer period of time.

BREEDING PHENOLOGY

CODE		# OF RECORDS	DATE RANGE
C	Courtship	3	24 Jun–12 Jul
ON	Occupied Nest	1	22 Jun
NY	Nest with Young	2	16 Jul–18 Jul
FY	Feeding Young	12	4 Jul–13 Aug
FL	Fledged Young	10	25 Jul–22 Aug

DISTRIBUTION

Rosy-finches of the genus *Leucosticte* have a circumpolar distribution in Asia and North America. Ornithologists have debated the taxonomy of rosy-finches for decades. Classed as separate species until 1983, then lumped together (AOU 1983), the three kinds of North American rosy-finches regained species status in 1993 (AOU 1993). The three species nest in cliffs in western North America: Gray-crowned from Alaska to Montana and California; Black in Idaho, Montana, and Utah; and Brown-capped endemic to the southern

Rockies—mainly Colorado but across the Wyoming border in the Snowy Range and across the New Mexico border in the southern Sangre de Cristos.

Atlasers found Brown-capped Rosy-Finches breeding in all the Colorado mountain ranges rising above timberline. Rosy-finches swirl over cliffs and snowfields from the San Miguels and San Juans across to the Sangre de Cristos and the Spanish Peaks; from the West Elks and Elks up the Sawatch to the Mosquitos and Gores, from the Flat Tops Wilderness and Park Range over all of the Front Range from Pikes Peak to the Mummy Range, Never Summers, and Rawah Wilderness.

Atlasers did not find them in the continuity in which they occur in these high peaks, for two reasons. First, the selection of Atlas blocks missed many nesting cliffs: Priority blocks contained the summits of only three of the 54 14,000-foot peaks. Second, in some blocks with suitable habitat, atlasers did not undertake the climbs to rosy-finch cliffs. The map reflects both factors, especially in the Sangre de Cristos. Although they undoubtedly occur all along that precipitous but narrow cordillera, only four blocks recorded them there.

The nesting grounds of this species are probably among the most stable and secure of any Colorado bird. The federal government owns and protects most alpine habitat, and the logistics of reaching cliffs with rosy-finches means that they seldom feel the footprints of man.

The recent increase in rock and mountain climbing could pose threats to a few nests. Fortunately, most technical routes are either on cliffs below timberline or on a few highly popular peaks. The challenges of climbing above timberline—the dangerous nature of the high-altitude rocks and the physical trials posed by altitude—mean that few rosy-finch cliffs will have pitons pounded into them and climbing ropes draped across them.

BREEDING EVIDENCE

REPORTED IN 62 (4%) OF 1745 PRIORITY BLOCKS

☐	Possible	29 blocks	(47%)
▩	Probable	10 blocks	(16%)
■	Confirmed	23 blocks	(37%)

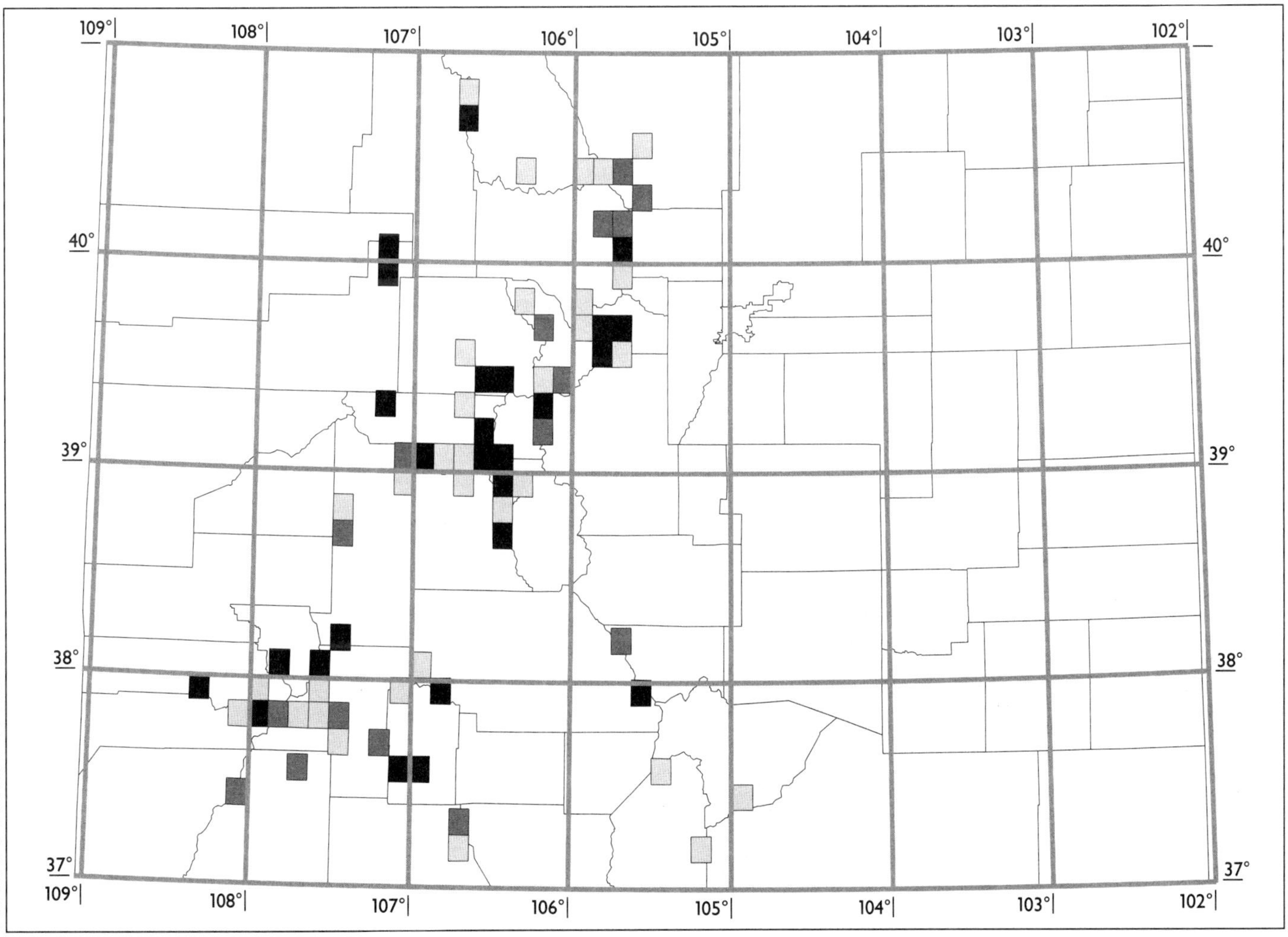
109°
108°
107°
106°
105°
104°
103°
102°
40°
39°
38°
37°

COLORADO DIVISION OF WILDLIFE / WILDLIFE RESOURCE INFORMATION SYSTEM

PINE GROSBEAK *Pinicola enucleator*

Whether foraging alongside a mud puddle or singing their cheery notes from atop an Engelmann spruce, Pine Grosbeaks pay little heed to hikers passing through the high country. Their homes fall amid spectacular terrain; 6-foot-tall featherless bipeds must seem rather insignificant in a land of such dimensions.

HABITAT

Perhaps more than any other bird of the family Fringillidae, Pine Grosbeaks depend on Engelmann spruce and subalpine fir. Unlike Pine Siskins, Red Crossbills, and Cassin's Finches, which sustain healthy populations in the forests below the subalpine zone, Pine Grosbeaks remain almost exclusively in boreal regions during the summer months.

The Atlas data wonderfully illustrate Pine Grosbeaks' interconnectedness with subalpine habitats. Aspen, lodgepole, and even ponderosa pine each produced a smattering of records, but nothing approaching the

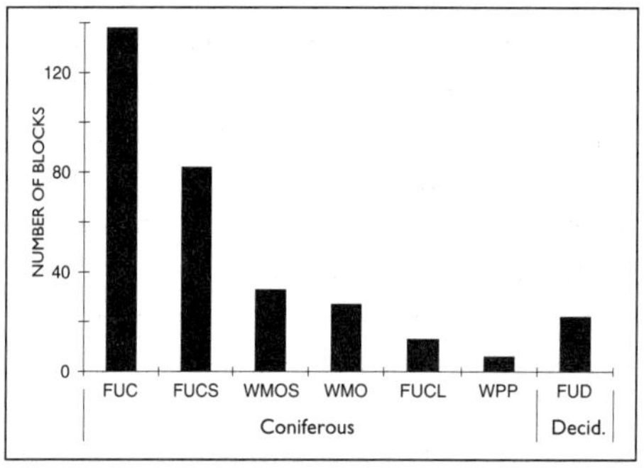

NUMBER OF BLOCKS
0
40
80
120
FUC
FUCS
WMOS
WMO
FUCL
WPP
FUD
Coniferous
Decid.

number drawn from spruce/fir forests and woodlands. Addition-ally, many of the records from aspen habitat emanate from aspen stands mixed with or in close proximity to spruce/fir.

Although Pine Grosbeaks frequent dense forests of spruce/fir, they more commonly dwell in open spruce/fir woodlands, sometimes mixed with aspen. They nest and spend much of their time foraging and drinking near shaded streams or in meadows. Seeds of the Engelmann spruce supply the major portion of their diet (French 1954). Other food items include the tips of spruce branches, small flying insects, berries, and plant and aspen seeds and foliage.

Pine Grosbeaks normally construct their nests at heights of 2–25 feet (1–8 m) in spruce or fir trees (Ehrlich et al. 1988). Although this does not seem very high, the dense growth on branches of trees used for nesting makes locating their nests unusually difficult. B&N reported 1942 as the date of the first nest found in Colorado.

BREEDING

Pine Grosbeaks breed across most subalpine regions of Colorado, but pieces of information regarding their breeding habits remain cloaked in mystery. The difficulty that attends finding their nests has hampered gathering breeding information.

In Utah's Uinta Mountains breeding normally begins by the first week in June and the young, in most cases, leave the nest by early July (French 1954). Atlas records show greater variability. Nest-building activity occurred as late as 19 July and observations of fledglings came as early as 22 June. Condition of the seed crop no doubt plays a role as it does with other seed-eating species. Males feed their mates throughout the courtship period until the eggs hatch (Adkisson 1981). The clutch normally contains four eggs (Ehrlich et al. 1988).

During breeding season, the parents develop gular sacs in their mouth that store food they gather to bring to their young. Parents often approach the nest together and feed the young over a relaxed interval lasting a few minutes (French 1954). Parents continue to feed begging fledglings well past the time they appear fully developed (B&N).

The late date at which fledglings first appear makes it highly improbable that Pine Grosbeaks double-brood in Colorado. There does exist, however, a record of an active nest that still held two hatchlings and one egg (all of which subsequently fledged) on 7 August 1951 (B&N).

BREEDING PHENOLOGY

CODE		# OF RECORDS	DATE RANGE
C	Courtship	1	25 Jul
NB	Nest Building	5	6 Jul–25 Jul
ON	Occupied Nest	2	1 Jun–7 Jul
NE	Nest with Eggs	1	28 Jun
NY	Nest with Young	2	6 Jul–18 Jul
FY	Feeding Young	24	8 Jun–5 Aug
FL	Fledged Young	54	22 Jun–24 Aug

DISTRIBUTION

Pine Grosbeaks enjoy a broad distribution across the boreal regions of Europe, Asia, and North America. Several subspecies breed in North America; the Colorado race breeds from the mountains of

British Columbia and Alberta as far south as the mountains of eastern Arizona and northern New Mexico.

Although Pine Grosbeaks inhabit nearly all subalpine regions of the state, their numbers increase markedly from east to west, especially in the southern half of the state. They achieve their highest densities, however, in the central and northern mountains. The Atlas shows only widely scattered reports from drier mountain areas such as eastern Park County, the Greenhorn Mountains, and the Sangre de Cristo Range and none from Pikes Peak although they occur on Christmas Counts there (HEK). The wetter San Juan Range, on the other hand, yielded them in almost every block.

Atlas field workers recorded first Confirmations of breeding in two western Colorado latilongs. Atlas records show a somewhat less restricted range statewide (usually reaching to lower elevations) than indicated by A&R. The one report of Pine Grosbeaks from the Greenhorn Mountains extends the known summer range of this species within the state.

During the winter months, Pine Grosbeaks tend to wander only short distances from their breeding grounds. The winter range varies greatly according to seed crop conditions. On rare occasions, they retreat to the lower foothills in search of food during winter months. Pinyon/juniper woodlands provide favored wintering grounds when the pinyons hold good crops of seeds (A&R).

BREEDING EVIDENCE

REPORTED IN 261 (15%) OF 1745 PRIORITY BLOCKS

Possible	76 blocks	(29%)
Probable	97 blocks	(37%)
Confirmed	88 blocks	(34%)

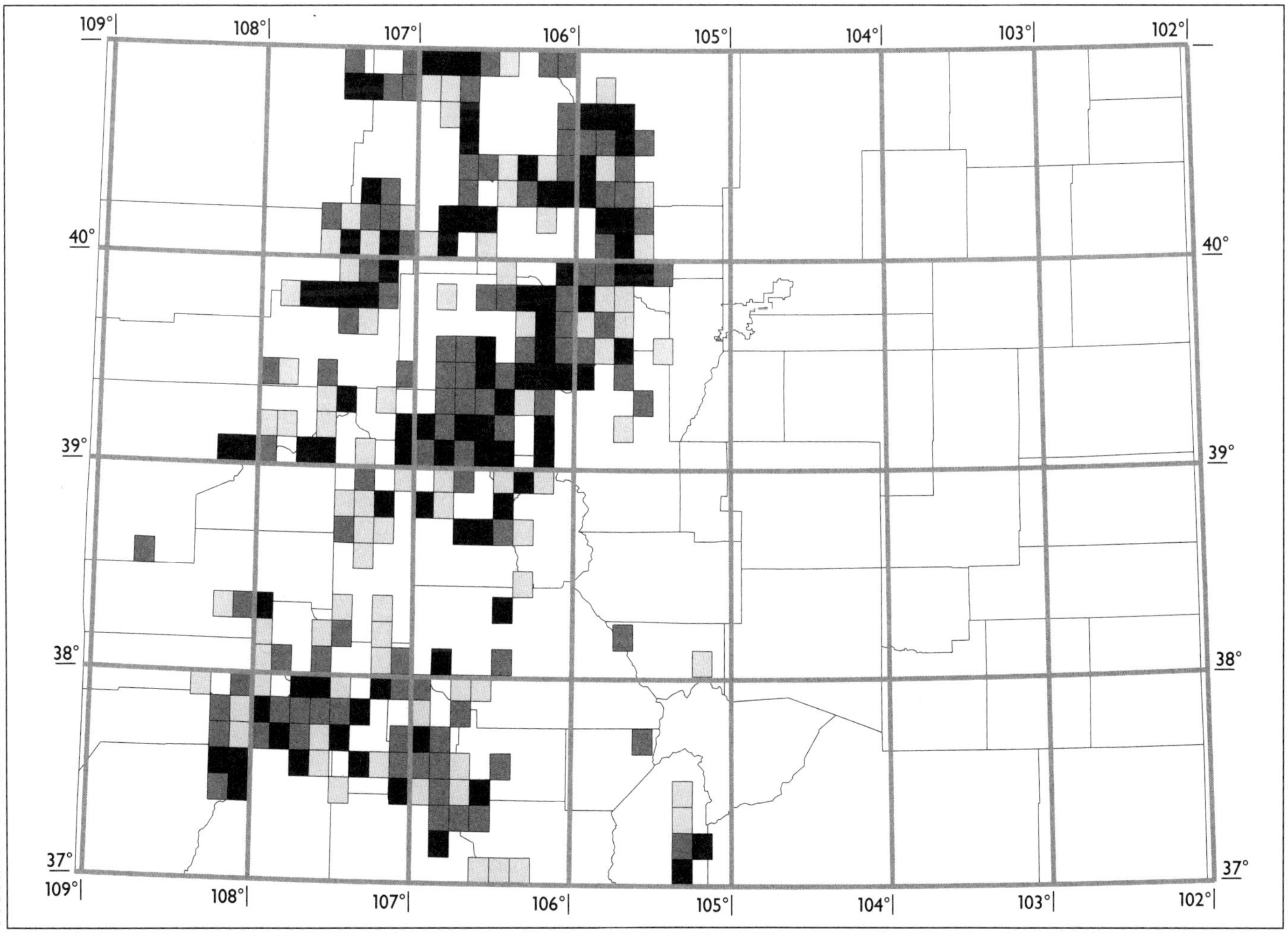
109°
108°
107°
106°
105°
104°
103°
102°
40°
39°
38°
37°

COLORADO DIVISION OF WILDLIFE / WILDLIFE RESOURCE INFORMATION SYSTEM

Standard references say that Cassin's Finches breed in subalpine forests and hide their nests in tall conifers. Atlas data divulge something else. In about 30% of the Atlas blocks the habitat did not conform to the "books"—atlasers found the finches in deciduous forests, riparian areas, pinyon/juniper, and urban and rural settings.

HABITAT

Most authors say that Cassin's Finches spend the summer in the high mountains of the West—the cool forests of the upper montane and subalpine, especially tall spruces, Douglas-firs, and lodgepole and ponderosa pines (see Hahn 1996). The main Colorado breeding range extends from 8,000 to 11,000 feet (2430–3350 m; A&R).

Atlas reports agree that substantial numbers breed in the high country. High-country conifers claimed 45%, mostly the expected spruce/fir forests; another 6% came from mid-elevation conifers, and 9% from ponderosa pine.

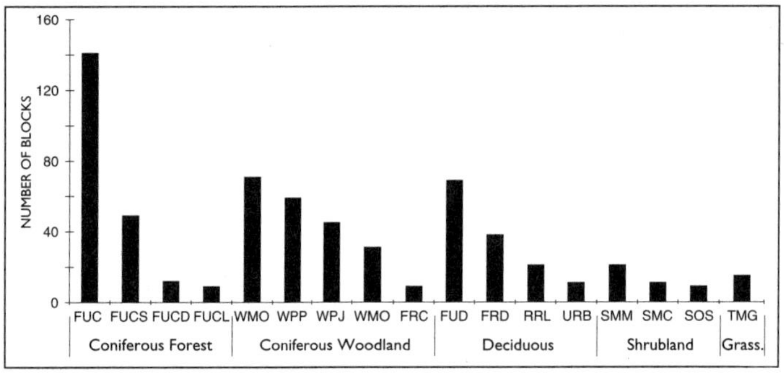
160
120
80
40
0
NUMBER OF BLOCKS
FUC FUCS FUCD FUCL WMO WPP WPJ WMO FRC FUD FRD RRL URB SMM SMC SOS TMG
Coniferous Forest
Coniferous Woodland
Deciduous
Shrubland
Grass.

Other habitats composed about 30% of habitats reported. Unexpectedly, pinyon/junipers had 7% (45 blocks)—including several Confirmations. Cassin's also breed in towns within their range (Balph 1978). Atlasers tallied them in 11 urban and 23 rural habitats—in high-country towns Gunnison, Cripple Creek, Dillon, and Kremmling and pinyon/juniper communities Dove Creek and Cedaredge.

Ample numbers use aspen groves and riparian trees for nesting: 107 blocks (17%). In the Axial block (40107C7), at the wrong elevation (6,300 feet, 1920 m) in the wrong habitat (narrowleaf cottonwoods), Judy Ward (1988) watched a breeding flock of at least four pairs building nests and feeding young. This grove, several miles from other woodlands, contained 50 trees within 4 acres (1.5 ha).

Mainly ground foragers, Cassin's Finches feed mostly on seeds. They also eat evergreen buds, aspen and willow catkins, seeds, berries, and insects (Hahn 1996).

BREEDING

Cassin's nest colonially; availability of food such as aspen staminate buds may determine the varying nesting locations (Salt 1952, Samson 1976). In Oregon, marked pairs returned to the same nest site a second year (Mewaldt and King 1985), suggesting that the pair bond persists from year to year. During nest-building and egg-laying, the male defends only an area around his mate, near the nest or when feeding (Samson 1976).

The male song, a rapid rushed warble, sometimes contains at the end a clear imitation of other birds. They incorporate the calls and songs of neighboring flickers, Steller's Jays, Western Tanagers, Red Crossbills, Pine Siskins, and Evening Grosbeaks (Hahn 1996). Juvenile males do not attain adult plumage until age 14 months; they sing throughout the breeding season, even though they do not breed unless an unusual opportunity arises (Samson 1976).

Atlasers found most nesting activity in late May and June, but they saw some nest-building in July. Cassin's Finches do not raise second broods; however, when June snowstorms destroyed nests, many of the Oregon birds attempted replacement nestings, most apparently abandoned when the flock departed to molt (Mewaldt and King 1985). Atlas cards show neither whether birds building nests in July had lost earlier nests nor if they completed breeding.

Females lay 4–5 eggs and incubate about 12 days; males feed females on or near the nest (Samson 1976, Ward 1990). Both parents feed the young. Brown-headed Cowbirds parasitize them (three Atlas records; Chace and Cruz 1996; Appendix C) and they confront nest predators such as Gray Jays (Mountain View Crest block, 37107E6; Kingery and Kingery 1995).

BREEDING PHENOLOGY

CODE		# OF RECORDS	DATE RANGE
C	Courtship	18	26 Mar–14 Jul
NB	Nest Building	27	8 May–24 Jul
ON	Occupied Nest	10	29 May–3 Aug
NE	Nest with Eggs	1	25 Jun
NY	Nest with Young	10	28 Apr–22 Jul
FY	Feeding Young	52	25 May–24 Aug
FL	Fledged Young	69	26 May–17 Aug

DISTRIBUTION

Cassin's Finches breed in the mountains from southwestern Canada to northern New Mexico and Arizona. B&N's surprisingly short account had little data; the two museum associates did not pursue Cassin's Finches as avidly as other mountain species. (The recent AOU monograph reported 37 topics with little or no data; Hahn 1996.) A&R provided the first comprehensive picture of Colorado distribution, but even they reported these finches as occurring mainly in the high mountains.

In fact, Cassin's Finches occur throughout the mountains and in much of the southern Colorado pinyon/juniper woodlands. Atlasers did not find them in Las Animas County (included by A&R as having a few breeders) but did find them in the Black Forest (especially Douglas County). Enigmatically, several pairs of probable breeders were in pinyon/junipers in two isolated Atlas blocks south of La Junta.

The Atlas map suggests a bottom elevation limit of 6,000–6,500 feet (1825–2000 m), inasmuch as the finches did not appear at the edge of the Front Range foothills or in the pinyon/juniper woodlands in southeastern Colorado.

Atlas abundance estimates show the greatest density in the north-central mountains—the Flat Tops, Elks, Gores, Park (Zirkel) Range, and Grand Mesa (in Latilongs 39106, 39107 and southern 40106 and 40107). Two-thirds (26 of 39) of the A4 estimates came from this area. The habitat preferences of the A4 estimates divided 3:1, conifers over aspen.

Do they compete with the House Finches, which have started to move into the mountains? In winter, both finches often use the same feeders in Front Range towns. A 30-year tally (1965–1994) of Christmas Counts for Colorado Springs shows a steady increase in House Finches from 683 to 1,342, yet erratic numbers for Cassin's Finches. Cassin's had a high count of 403 in 1987 (that year the compiler provided a finch identification packet; Cindy Lippincott pers. comm.) but an average of only 65 (Ben Sorensen pers. comm.). Although one count, mainly on the plains, cannot determine a trend, it reflects a typical Front Range pattern.

Atlas maps of the two finches show little overlap. Most House Finches stay below 6,000 feet (1830 m), except in the San Luis Valley. The two species do share several blocks in the Glenwood Springs–Eagle area where Cassin's Finches have their stronghold (Latilongs 39107 and 39106). House Finches occupy Las Animas County—could they have driven out the Cassin's Finches that previous observers reported?

BREEDING EVIDENCE

REPORTED IN 491 (28%) OF 1745 PRIORITY BLOCKS

Possible	181 blocks	(37%)
Probable	147 blocks	(30%)
Confirmed	163 blocks	(33%)

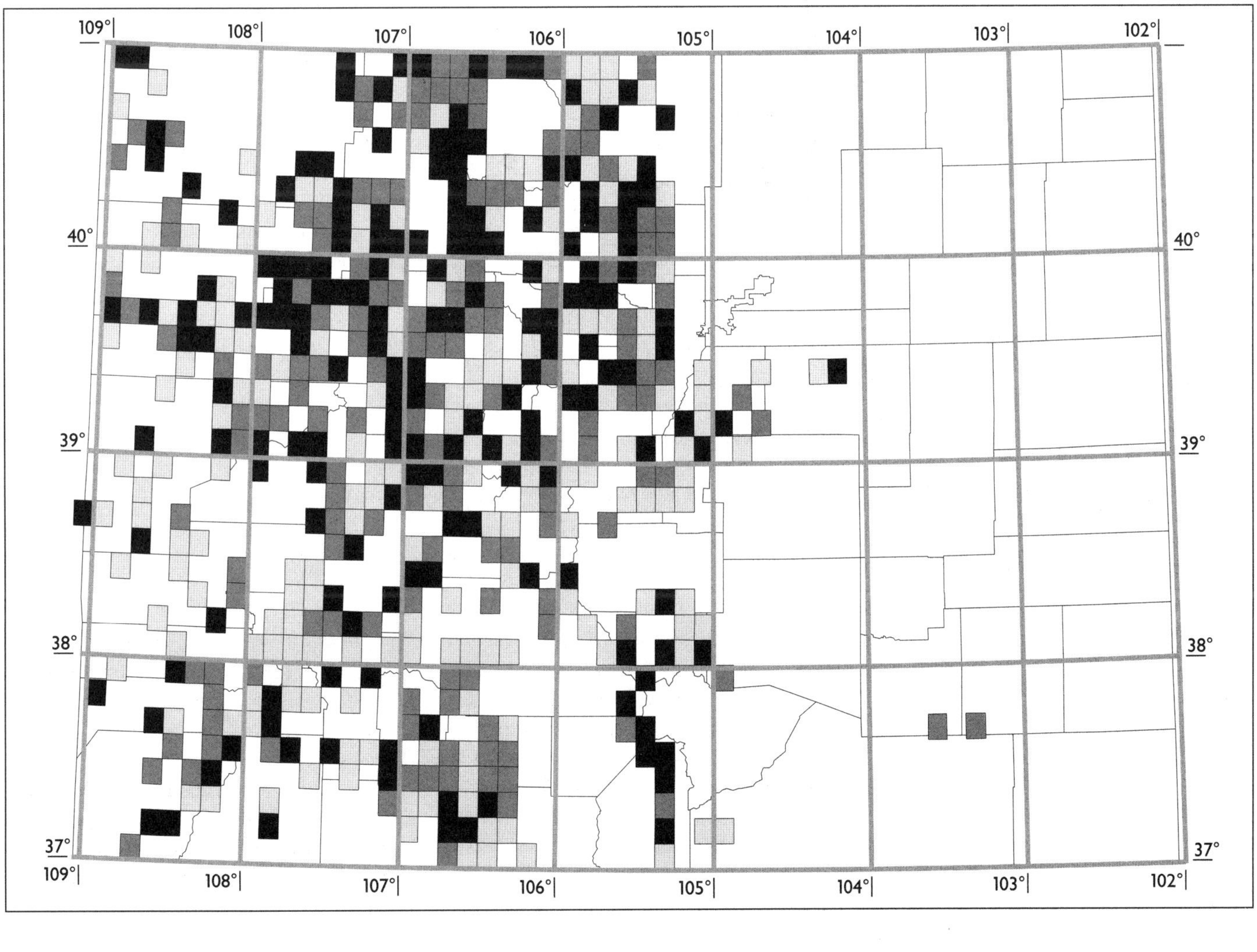

109°
108°
107°
106°
105°
104°
103°
102°
40°
39°
38°
37°

COLORADO DIVISION OF WILDLIFE / WILDLIFE RESOURCE INFORMATION SYSTEM

In early March, in the tips of spruce trees in the mountain valleys, the most urbanized of Colorado's native birds start to sing. In high-country communities House Finches replace the proverbial robins as the first sign of spring. With the first notes people lift their eyes, attitudes perk up, and folks call local birders to ask, "What was that bird I heard singing this morning?" Meanwhile, along streams of the Colorado plateau and in pinyon/juniper woodlands, these finches have started to reclaim the haunts of their ancestors, and in prairie towns, to colonize new territory.

HABITAT

House Finches demonstrate an adaptability to a wide range of environments. Nearly 270 Atlas reports (of 573) evidenced a strong attachment to human activities, ranging from industrial sites to rural shelterbelts and farmhouses. No other native bird species has urbanized better. Even early reports on the species noted this strong attachment to human habitation (Cooke 1897). When it comes to undomesticated habitats, pinyon/junipers dominated; dense, prickly branches provide suitable nest sites.

Prior to the arrival of European settlers in the West, House Finches nested in cholla cactus, thickets, and cottonwoods along streambeds, as well as in conifers (B&N). Thomas Say collected a male, the first of the species known to science, at the confluence of Cheyenne and Fountain creeks (now in Colorado Springs) on 12 July 1820 (Marsh 1931); the date and place imply breeding in a riparian stream bottom.

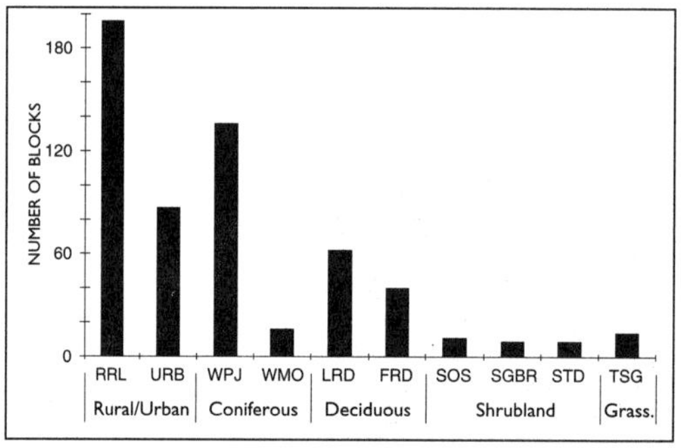
NUMBER OF BLOCKS
180
120
60
0
RRL
URB
WPJ
WMO
LRD
FRD
SOS
SGBR
STD
TSG
Rural/Urban
Coniferous
Deciduous
Shrubland
Grass.

Nesting habitat simply requires dense cover. The dense prickly branches of junipers provide excellent habitat. In urban Colorado exotic juniper plantings do equally well, as do blue spruce, ivy-covered walls, hanging plants, and house eaves.

House Finches consume an almost exclusively vegetable diet, especially seeds, all year; unlike most passerines they feed their nestlings few insects. In the wild, favorite food sources include buds, seeds of thistle, mustard, and knotweed, and fruits (Hill 1993). In breeding season males and non-breeding females form feeding flocks. They attack favorite food sources such as annual seeds; composite flowers gone to seed (dandelions, sunflowers, etc.) attract the flocks. Feeders with thistle and sunflower seeds also provide food sources. But House Finches can feed on a wide variety of foods, including hot dog buns at Coors Field.

BREEDING

Although Atlas reports of nest-building tie closely to latititude, Atlas Confirmations as a whole do not. Because of the Atlas protocol (which required listing of the "highest" code; NB was lowest and therefore most frequently replaced by higher codes) the NB reports alone cannot delineate a northward progression of nesting by date. The only NB report in August, on the 18th, probably represents a second or third brood. Groups of fledged young appear at feeders in Kremmling at two different times during the summer, indicating that even at this elevation (7,350 feet, 2240 m), House Finches can successfully raise two broods (NMB).

Like many finch species, these red-topped songsters weakly defend small territories—space around the nest with a radius of perhaps 14 feet (4 m). In preferred nesting locations, pairs will nest within 3 feet (1 m) of each other, forming loose colonies (Hill 1993). The male's defense of nest and mate abates when she starts to incubate. House Finches occasionally feed young of other species. One report of a robin nest had eggs and eventually young of both species, with both sets of parents feeding the young (B&N). The benefit for the House Finches is unclear.

BREEDING PHENOLOGY

CODE		# OF RECORDS	DATE RANGE
C	Courtship	7	12 Mar–13 Jul
NB	Nest Building	27	12 Mar–25 Jun
ON	Occupied Nest	30	12 May–29 Jul
NE	Nest with Eggs	8	25 Apr–2 Jul
NY	Nest with Young	19	1 May–16 Aug
FY	Feeding Young	56	13 May–16 Aug
FL	Fledged Young	63	13 May–6 Aug

DISTRIBUTION

Ignore the range maps in the field guides. House Finches now feed and breed from coast to coast. In Colorado the Atlas recorded them in all but three high mountain counties. The Continental Divide, which may pose an elevation challenge, separates two and trisects the third. In fact the entire mountain center of the state has only scattered Atlas records; the few Confirmations of breeding there occurred mostly in rural and urban settings. Instead the finch populations concentrate along the Front Range and in the western and southern border counties. At the start of the Atlas, they did not nest in plains towns where, by its conclusion, atlasers found them breeding (e.g., Joes, Arriba, Kit Carson, Wray, Haxtun, Brush, Fort Morgan, and Lamar).

Grinnell and Miller (1944) set an upper limit of 6,562 feet (2000 m) in California but in Colorado, atlasers found them in breeding season above 7,000 feet (2145 m) from Larimer to San Juan counties. They Confirmed breeding at 8,700 feet (2665 m) in Larimer and 7,900 feet (2420 m) in Gunnison counties and reported Possible breeding at 10,000 feet in Park County.

House Finches originally occurred from the Pacific coast to the Rockies, with some spilling over to the edge of the plains. As late as 1947 Peterson's *Field Guide to the Birds* (eastern version) did not even mention House Finches. The pet trade gets credit for introducing them to the East Coast, where stores sold them as cage birds in the first half of the century. The International Migratory Bird Treaty made trapping and selling without a federal permit illegal. With the threat of enforcement, shop owners simply turned their inventory loose (Jackson 1992). Prolific breeders, House Finches quickly established themselves along the Atlantic seaboard and spread inland, until now they span the continent.

The coincidence of their spread across the Great Plains with their arrival in Colorado plains towns raises a question as to the direction from which the plains colonizers came. This species does not appear to create major problems in the new Colorado range, unlike starlings, House Sparrows, and cowbirds. In any case we can no longer do anything more than sit back and enjoy the bubbly songs of bright red, singing House Finches.

BREEDING EVIDENCE

REPORTED IN 491 (28%) OF 1745 PRIORITY BLOCKS

◻	Possible	163 blocks	(33%)
▩	Probable	133 blocks	(27%)
■	Confirmed	195 blocks	(40%)

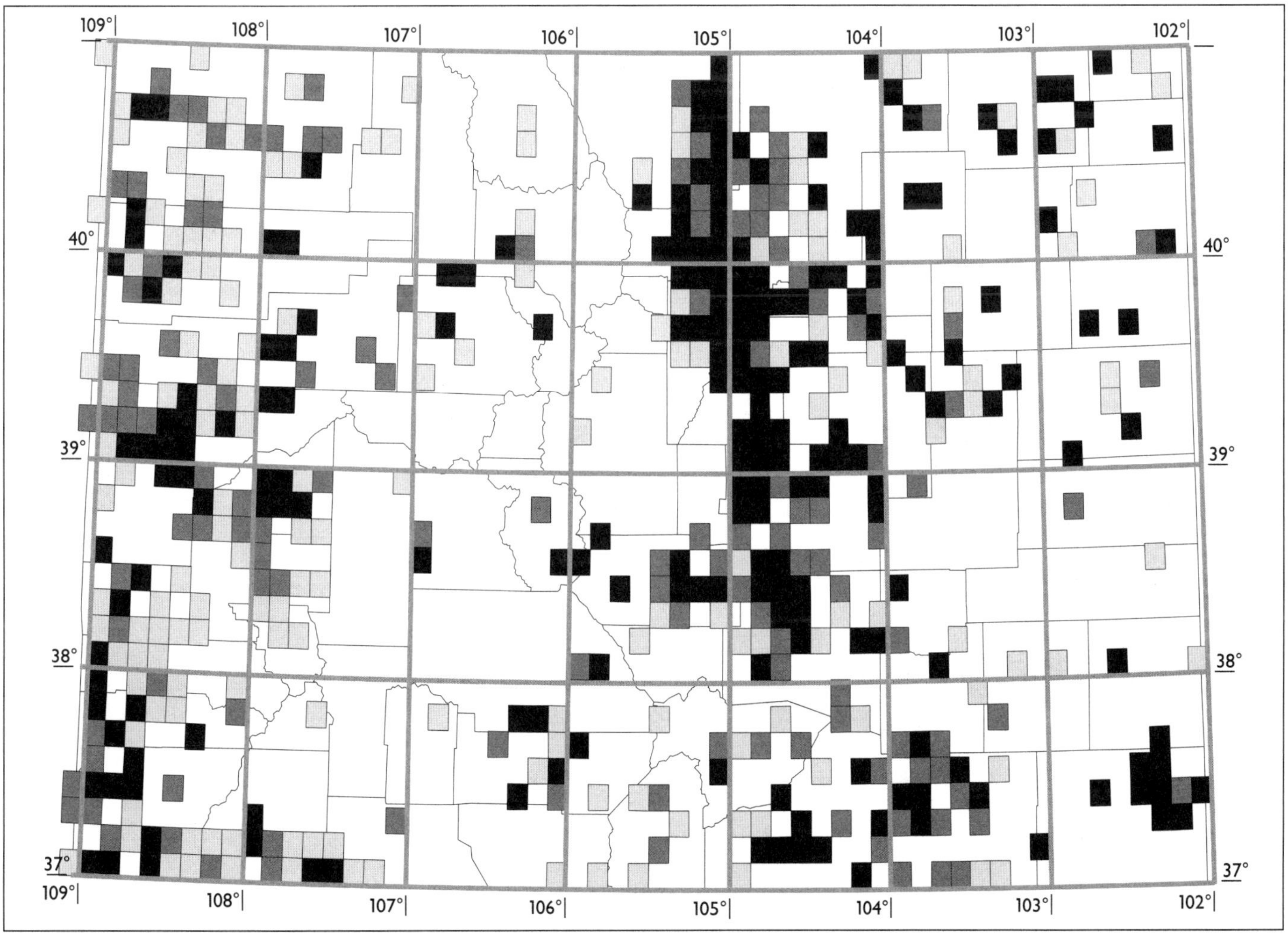
109°
108°
107°
106°
105°
104°
103°
102°
40°
39°
38°
37°

Descriptions of Red Crossbills invariably include the term "irregular." Recent research suggests that "complex" might suit them better. Once taxonomists have sorted out this complexity, the Red Crossbill may proliferate into as many as eight species. If this comes about, they might approach Darwin's Finches as an example of adaptive radiation, and as an identification challenge for birders they may surpass migrating *Empidonax* flycatchers.

HABITAT

Red Crossbills depend upon conifers for survival (Benkman 1989a). Dependence upon particular species of conifers has separated Red Crossbills into different "Types" (Groth 1988, Benkman 1993a, Knox 1992). Each crossbill Type uses a variety of conifers through the year, but during late winter and early spring, when survival depends upon feeding efficiency, each Type specializes in extracting and husking seeds from the cones of a single species of conifer (Benkman 1993a). These trees include Douglas-fir, ponderosa pine, and lodgepole pine; crossbill Types specially adapted to each occur in Colorado (Groth 1993).

Lodgepole pines and Engelmann spruce have similarly sized seeds; the crossbill Types that specialize in these two conifers have similar palates. However, the scales of lodgepole and Engelmann cones have different sizes and quality. For lodgepole cones—larger and tougher—the lodgepole-crossbill Type needs a bill with greater depth in order to extract seeds more efficiently (Benkman and Miller 1996).

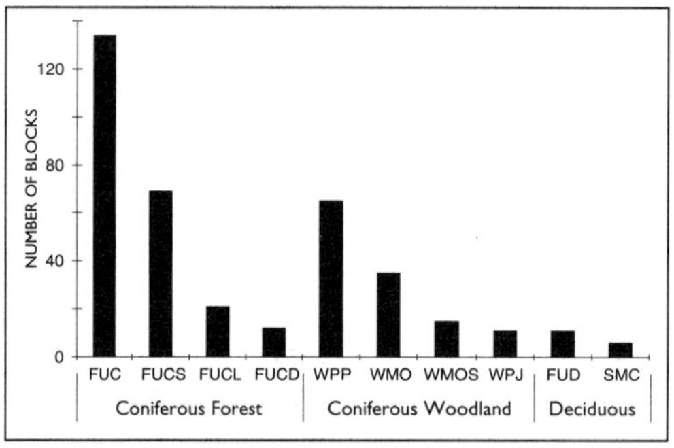
NUMBER OF BLOCKS
120
80
40
0
FUC
FUCS
FUCL
FUCD
WPP
WMO
WMOS
WPJ
FUD
SMC
Coniferous Forest
Coniferous Woodland
Deciduous

Because dietary overlap is highest during the summer (Benkman 1987), and because most Atlas records bear June, July, and August dates, Atlas data cannot separate the different Red Crossbill Types. In fact, in Colorado, spruce/fir forest appears more significant as a breeding habitat than any other conifer. Generic conifer codes (WMO, FUC) together make up 47% of all reported codes. If these codes include each conifer in the same proportion that atlasers specified specific trees (WMOS, FUCD, etc.), then 58%, 15%, and 7% of all habitat codes respectively indicate spruce/fir, lodgepole, and Douglas-fir. Ponderosa ran a distant second to spruce/fir with 19%. Atlas workers observed Red Crossbills in a number of other habitats, including deciduous trees, but crossbills' inefficiency at handling non-conifer seeds rules out these habitats for possible breeding (Benkman 1988b). Atlas workers reported no breeding Confirmations in them.

BREEDING

Red Crossbills breed whenever they find sufficient food for egg formation and the prospect of enough food to feed their nestlings (Benkman 1990). A&R reported "at least eight records of mid winter nesting in Colorado." During a major invasion of Red Crossbills in 1948, Colorado observers found 19 nests between 1 February and 18 April in the ponderosa pine hills near Parker, and 15 more nests in 1952 between 16 January and 30 June (Bailey et al. 1953). In the 1970s observers on Christmas Counts at Denver and Evergreen found nests (HEK). Observers have found few nests since. Atlas workers, who reported Red Crossbills in 341 Priority blocks, found only a single nest.

On 24 and 25 July 1990, Craig Benkman (pers. comm.) observed three nests being built in ponderosa pine (by Type 2, ponderosa-adapted birds) at Round Mountain Campground near Wilkerson Pass (Tarryall, 39105A4). One female was both building a nest and feeding two dependent young. Between 10 and 13 August he captured 15 (Type 5) crossbills in Engelmann spruce on Grand Mesa (Skyway, 39107A8). One female had a well-developed brood patch. On 18 and 19 September 1993, Benkman netted four female (Type 5, lodgepole-adapted) crossbills at Lincoln Gulch Campground near Independence Pass (New York Peak, 39106A6). All had brood patches.

Field workers can readily identify the heavily streaked juvenile crossbills, and Atlas workers Confirmed breeding in 53 blocks by finding fledglings. Immature birds, however, like adults, can wander over great distances, and many of the birds that produced Confirmations may have hatched elsewhere. Fledglings being fed (FF, 5 blocks) probably hatched nearby, and these codes bear more significance. The median dates for fledgling observation in ponderosa pine fell on 6

June, and in spruce/fir on 17 July, more than a month later. This may reflect either different times of cone ripening or Atlas workers visiting higher elevations later in the season. Crossbills feed their young by regurgitation (Terres 1980); therefore atlasers could not Confirm breeding by observing adults carrying food.

BREEDING PHENOLOGY

CODE		# OF RECORDS	DATE RANGE
C	Courtship	3	20 Jun–7 Aug
NB	Nest Building	9	25 Mar–30 Jul
FY	Feeding Young	1	30 Jun
FL	Fledged Young	66	24 Mar–21 Oct

DISTRIBUTION

Red Crossbills cruise the coniferous forests of the entire Northern Hemisphere, in North America from Alaska east to the Atlantic in Canada and New England, and south in the western mountains into Mexico. Specific nesting locations in Colorado shift from year to year with the availability of cones, but crossbills generally reside in coniferous forests throughout the state. A&R indicated scattered records, but no breeding, at low elevations on the eastern plains, and they could cite "no breeding records from the higher mountains."

Atlas data provide evidence that plentiful breeding takes place in Colorado's high-elevation spruce/fir forests. More than half of the Confirmations for this species came in spruce/fir habitats, and song, which occurs most frequently during the breeding cycle (Bailey et al. 1953), was reported from 22 Priority blocks, 19 of them spruce/fir.

The Colorado bird conservation and monitoring lists rank Red Crossbills very low, deeming them ecological generalists; however, the highly specialized winter requirements for each Type (Benkman 1993a) would seem to merit a higher degree of concern. The likely forthcoming split of Red Crossbills into six to eight species will raise major new conservation issues (Van Remsen pers. comm.). Introduction of a competitor, the red squirrel, has virtually extirpated the Newfoundland Red Crossbill (Benkman 1989, 1992a). Logging at short rotation ages could cause a decline in other crossbill Types "disproportionate to forest loss" (Benkman 1993b).

BREEDING EVIDENCE

REPORTED IN 339 (19%) OF 1745 PRIORITY BLOCKS

☐	Possible	195 blocks	(57%)
▩	Probable	70 blocks	(21%)
■	Confirmed	74 blocks	(22%)

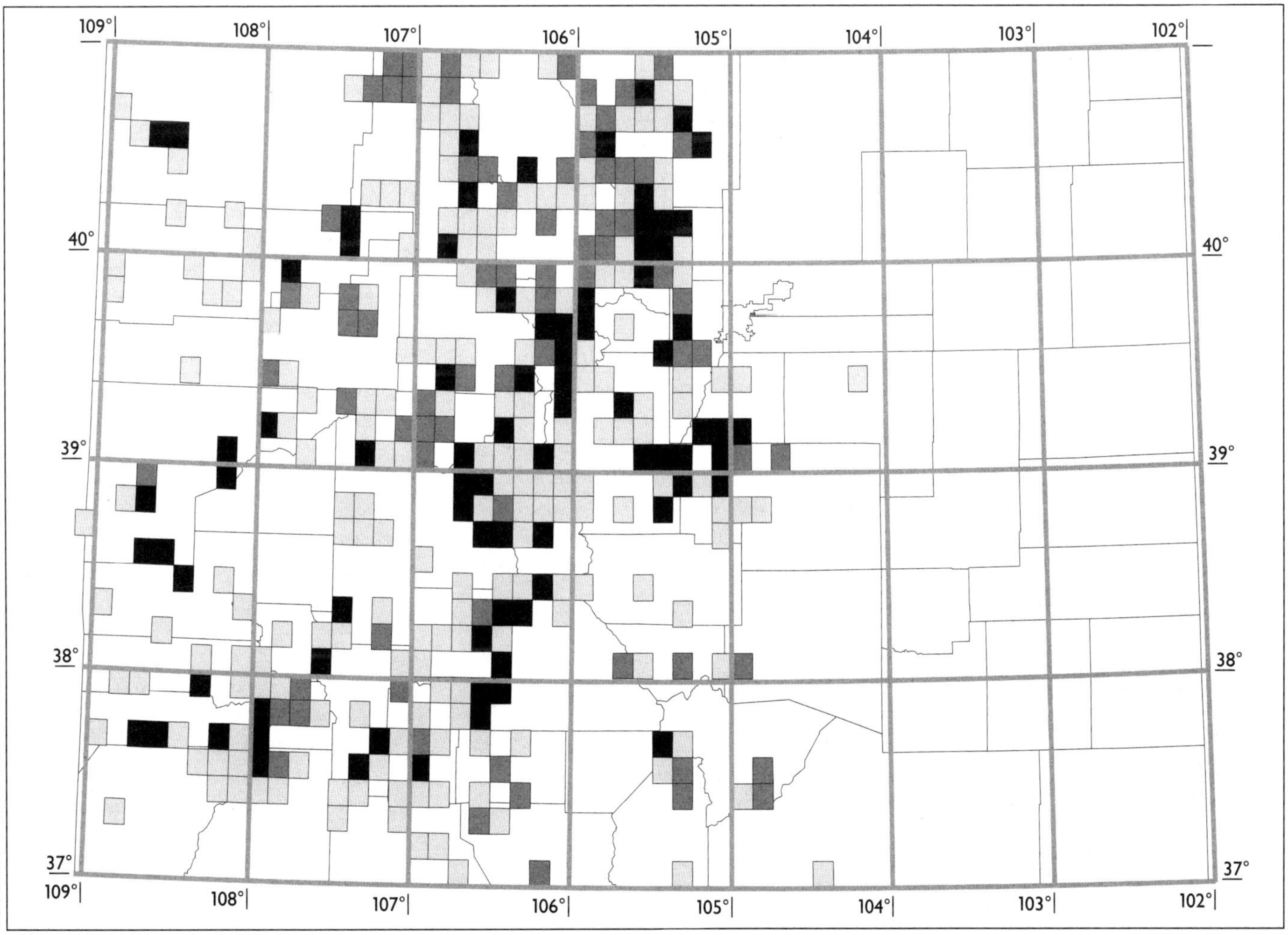
109°
108°
107°
106°
105°
104°
103°
102°
40°
39°
38°
37°

WHITE-WINGED CROSSBILL *Loxia leucoptera*

Clogged by deep snow through its long winters and teeming with mosquitoes during its brief summers, Colorado's beautiful but forbidding spruce/fir forest clings tightly to its secrets. In the past two decades, observers have finally wrested a rudimentary knowledge of the Boreal Owl from its grip. And only in this last decade have Atlas field work and other research gathered the first clues for resolving another of its mysteries: the status of the White-winged Crossbill.

HABITAT

Crossbills breed only when and where an appropriate conifer species produces adequate seed crops (Benkman 1988a, 1990). White-winged Crossbills breed in coniferous forests that produce large crops of small cones: black spruce, white spruce, and tamarack in the Far North, red spruce in the Northeast, and Engelmann spruce in the Rocky Mountains (Benkman 1992b). The specialized beak of this species enables it to feed efficiently only on small-seeded spruce cones, so it rarely visits any other habitat (Benkman 1988b, 1992b). Field workers reported five habitat codes for White-winged Crossbills: FUC and WMO—general coniferous forest and woodland codes that include Engelmann spruce; FUCS and WMOS—which specifically identify Engelmann spruce; and SKR—krummholz—dwarf Engelmann spruce growing at timberline.

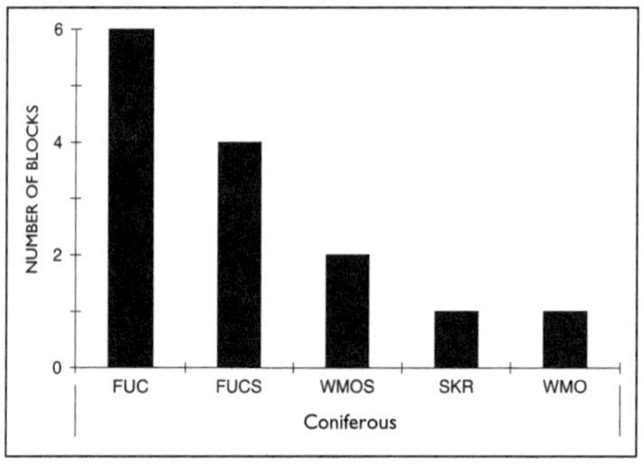
6
4
2
0
NUMBER OF BLOCKS
FUC
FUCS
WMOS
SKR
WMO
Coniferous

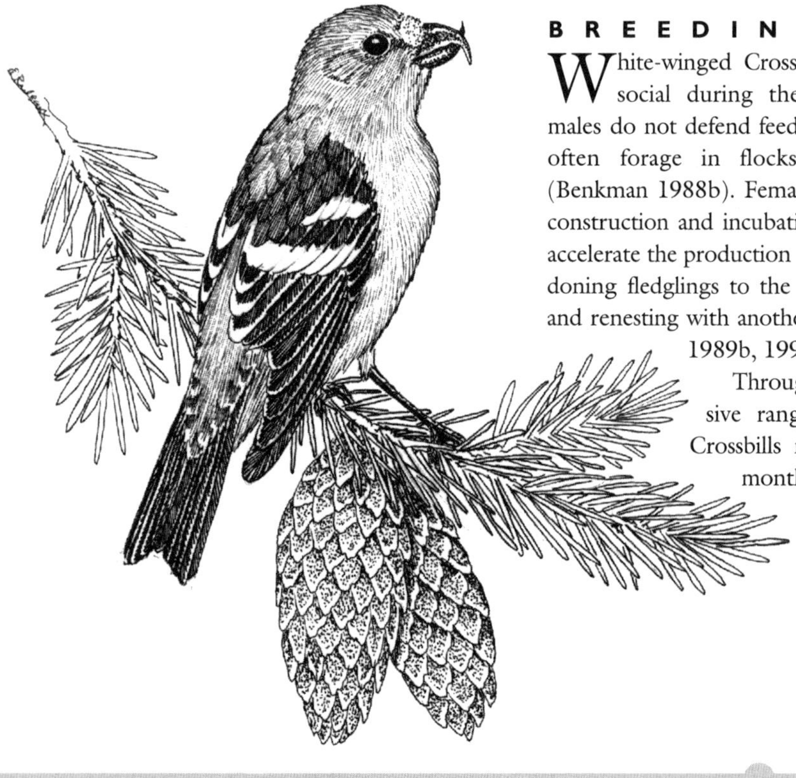

BREEDING

White-winged Crossbills remain quite social during the nesting period; males do not defend feeding territories and often forage in flocks during nesting (Benkman 1988b). Females do all the nest construction and incubating, and they may accelerate the production of young by abandoning fledglings to the care of the males and renesting with another mate (Benkman 1989b, 1992b).

Throughout their extensive range, White-winged Crossbills nest during every month of the year; peak nesting periods correspond to cone-ripening phenology (Benkman 1990, 1992b). Very scanty data suggest that nesting in Colorado occurs during two separate periods. During the most significant period, the summer months, breeding depends on the ripening of spruce seeds. A handful of Colorado records corresponds to this period. Atlas workers found dependent fledglings once each in July and August and independent juvenile birds in September. These observations correspond to developing spruce cones. Elsewhere, when spruce produce large cone crops, these crossbills commonly produce more than one brood during summer and early autumn (Benkman 1992b).

In Canada, breeding diminishes during the fall, and then a second peak occurs in January and February when white and red spruce produce large cone crops and many seeds remain in the cones through the winter months (Benkman 1992b). Atlas records indicate that White-winged Crossbills in Colorado's Engelmann spruce forests also breed during the winter, perhaps somewhat later than in Canada. Atlasers observed nest-building in February and March. Neither February nest produced young, and observers did not find the nests of the March crossbills. Near Leadville, however, an Atlas worker found adults feeding an awkward little fledgling on 4 April 1995 (Elaine Hill VF). An observation on 7 June 1987 of fledglings incapable of flight provided Colorado's first breeding record (Groth 1992). That date seems too early for foraging on developing cones, and these birds probably depended on the remains of the previous year's huge cone crop (Craig Benkman pers. comm.).

BREEDING PHENOLOGY

CODE		# OF RECORDS	DATE RANGE
NB	Nest Building	2	4 Feb–2 Mar
FL	Fledged Young	4	4 Apr–7 Aug

DISTRIBUTION

The circumpolar range of White-winged Crossbills spans the boreal forests of Canada and Alaska in North America, and the boreal forests of Scandinavia and Siberia in Eurasia. It extends south into northern

New England, the northern edge of the Great Lakes states, the northern Cascades, and the northern Rocky Mountains. Small, isolated patches of the range lie in the southern Rockies of Colorado and New Mexico (Benkman 1992b). Atlas workers found evidence of White-winged Crossbills breeding along the Continental Divide from Cameron Pass to Leadville, in the mountains north of Steamboat Springs, on Grand Mesa, and in the San Juan Mountains near Spring Creek Pass. White-wingeds also occur on the White River Plateau (Groth 1992). These areas encompass the most extensive spruce/fir forests in the state. It seems likely that observers will eventually find White-winged Crossbills in other spruce/fir areas, such as the Elk and West Elk mountains.

Colorado's White-winged Crossbills may represent a recent range expansion (Groth 1992), but more likely they represent a small, permanent population. The species appears in the earliest records of Colorado birds: Cooke (1897) cited a report from the San Juan Mountains dated 1881. B&N regarded this species as a "rare straggler," and reported several records, many from the same areas that Atlas workers found them. Atlas Confirmations and the recent work of other researchers suggest that Colorado hosts a small outpost population of perhaps a few hundred birds.

BREEDING EVIDENCE

REPORTED IN 7 (0%) OF 1745 PRIORITY BLOCKS

☐	Possible	5 blocks	(71%)
▪	Probable	0 blocks	(0%)
■	Confirmed	2 blocks	(29%)

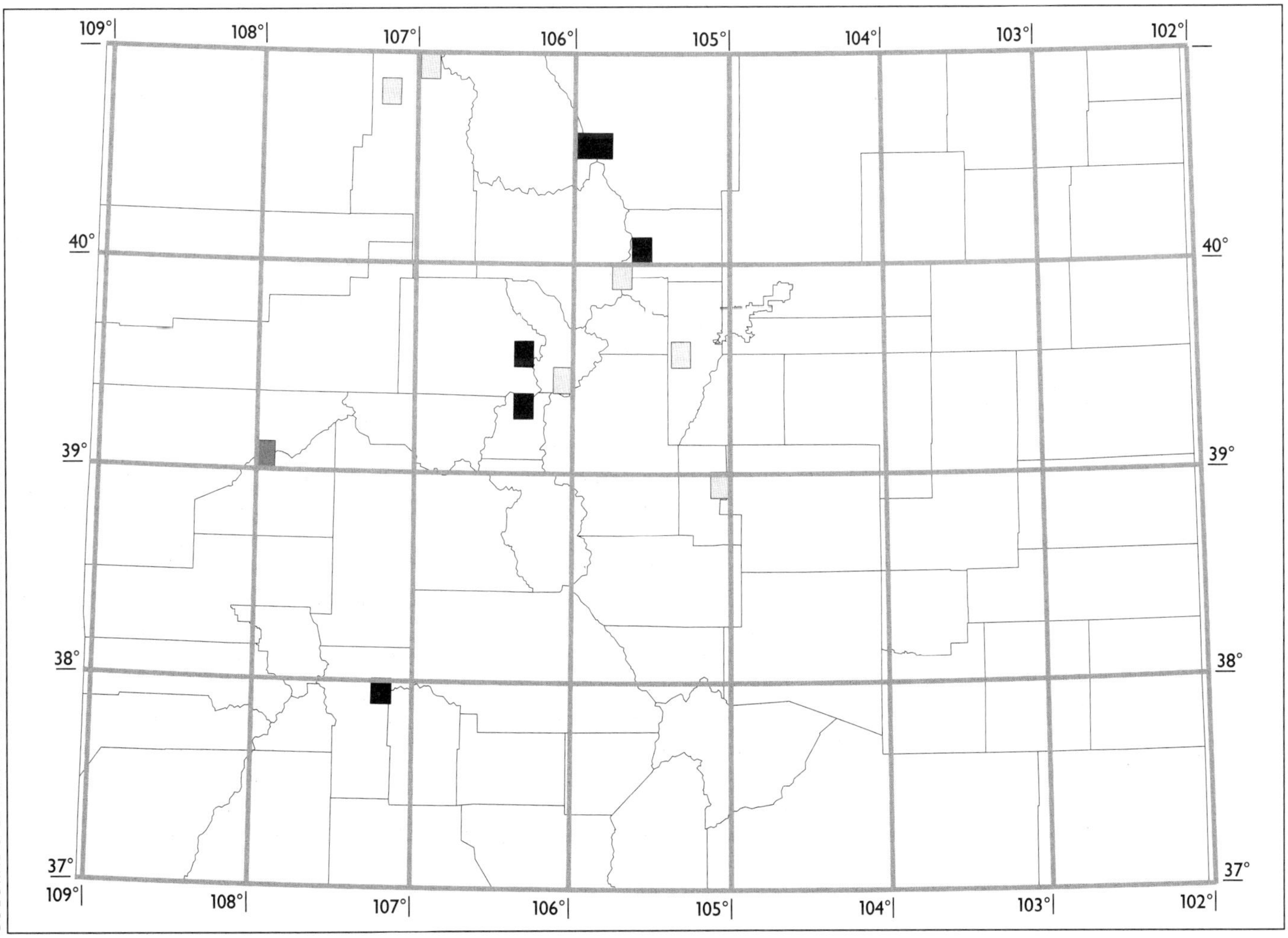
109°
108°
107°
106°
105°
104°
103°
102°
40°
39°
38°
37°

COLORADO DIVISION OF WILDLIFE / WILDLIFE RESOURCE INFORMATION SYSTEM

In Colorado, "Pine" Siskins use pines, but they use spruce/fir more. Seed specialists, they feed on pine seeds, but more so on spruce and dandelion seeds. They nest in pines—but also in spruce trees, willows, aspen, and even lilacs.

HABITAT

Atlas field work found that Pine Siskins have habitat preferences more generalized than the name suggests. Codes specifying pine trees—ponderosa, lodgepole, and pinyon—made up about 20% of the codes reported; however, siskins use spruce/fir much more frequently than pines. Projection of generic codes (WMO and FUC) to the ratio of comparable codes citing specific conifers (WMOS, FUCL, etc.) results in assigning half of all reports to spruce/fir. Conifers composed 70% of all habitats reported.

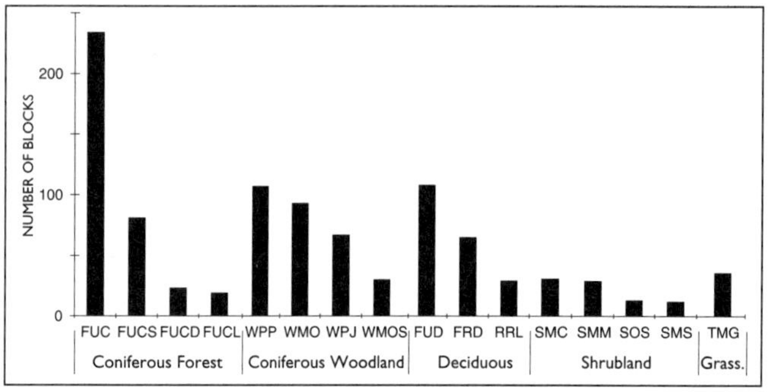
NUMBER OF BLOCKS
0
100
200
FUC FUCS FUCD FUCL WPP WMO WPJ WMOS FUD FRD RRL SMC SMM SOS SMS TMG
Coniferous Forest
Coniferous Woodland
Deciduous
Shrubland
Grass.

Siskins make significant use of deciduous habitats as well; aspen and narrowleaf cottonwoods, the two most abundant deciduous trees in the Colorado mountains, together accounted for about 20% of all habitat reports.

The remaining 10% covered nearly every other habitat in the state from western grassland to alpine tundra. Many of these reports described traveling and foraging birds, but published accounts (e.g., B&N) imply that siskins could nest in almost any habitat. Across their broad range, these eclectic birds have used hemlock, cedars, redwood, cypress, boxelders, maples, oaks, cottonwoods, lilacs, and eucalyptus (Palmer *in* Bent 1968). B&N mentioned ponderosa, blue spruce, and Engelmann spruce; A&R wrote, "primarily in coniferous forest (especially spruce/fir), and rarely in riparian areas, aspen forest, and shrublands." Atlas records agree, except that siskins use these last habitats more than rarely.

Given their feeding preferences, the siskins' strong tie to conifers seems surprising. Although they do feed on conifer seeds, they also feed regularly in meadows, sagebrush, and urban areas, on thistles, dandelions, and related plants that do not grow in conifer forests. Timber harvest and other activities that fragment conifer forests benefit siskins by creating openings where a variety of seed sources can grow (Keller 1987).

BREEDING

Siskins typically build their nests toward the ends of conifer branches where dense needles provide hiding cover. Blue spruce and other landscaping conifers create nesting habitat in rural and urban settings. Atlasers had trouble finding the well-concealed nests, which the females rarely leave during incubation and the early stages of brooding. The nests seemed easier to find in deciduous trees, as five of the eight Atlas nest cards came from these habitats even though the birds nest much more frequently in conifers.

Incubation lasts 13 days and the young remain in the nest for 15 days (Weaver and West 1943). Adults continue to feed the young for several days after fledging (Palmer *in* Bent 1968). Siskins feed both mates and young by regurgitation. During the breeding season siskins remain social—they usually nest in loose colonies and travel in flocks when away from the nest (Palmer *in* Bent 1968).

Their breeding schedules, like those of many finches, vary erratically. In New England and on the Pacific Coast, Pine Siskins generally breed early: courtship behavior (mate feeding) starts in late February and egg-laying peaks from late April to early May (Palmer *in* Bent 1968, Shuford 1993, MacLeod *in* Foss 1994). Colorado dates range from 21 April to August (B&N). Atlas dates also run later, with no breeding reported until May. Nest-building peaked in June, but lasted to 31 July; FY and FL dates peaked in July. Late summer breeding (exemplified by the late August fledglings) coincides with composites going to seed. The breeding season can extend even later—to late September 1989 during a large red spruce crop (Craig Benkman pers. comm.).

Whether or not siskins initiate a second broods is uncertain. Some probably do, some may move to a new locality for second attempt, and different parts of the population may breed at different times (Palmer *in* Bent 1968, MacLeod *in* Foss 1994).

BREEDING PHENOLOGY

CODE		# OF RECORDS	DATE RANGE
C	Courtship	29	2 May–26 Jul
NB	Nest Building	38	5 May–31 Jul
ON	Occupied Nest	7	26 May–4 Aug
NY	Nest with Young	2	26 Jun–10 Jul
FY	Feeding Young	69	6 Jun–24 Aug
FL	Fledged Young	108	28 May–21 Aug

DISTRIBUTION

Pine Siskins breed from Alaska to eastern Canada and across the northern U.S. south through the western mountains into Mexico. In Colorado the pattern of Atlas observations matches the distribution of forested lands across the western, mountainous portion. Although B&N reported nests in Denver's City Park and Limon, atlasers found no nesting on the plains. Siskins do not nest in treeless prairies, deserts, parks, or sagelands, but they visit many of these open areas to forage, especially during the winter.

Siskin populations fluctuate widely based on seed sources; "average" numbers do not typically occur. The birds may breed in large numbers at a site one year and then desert it for years afterward. The random chance of finding flocks renders population estimates based on abundance codes ambiguous. They rank among the ten most abundant birds in the state according to Atlas abundance figures, with over a million breeding pairs. Their erratic roving habits make long-term trends and patterns difficult to detect.

Despite this uncertainty, atlasers' reports show that siskin abundance increases from pinyon/juniper woodlands up to spruce/fir forests. Far fewer birds use pinyon/juniper (9% of the blocks reported A4 codes and none reported A5s) than use other coniferous habitats (43% reported A4 or A5 codes). More than 85% of the Atlas reports from pinyon/juniper indicated only Possible breeding, and siskins do not use this habitat in Las Animas County. The Confirmation rate in pinyon/juniper reached only 5%. In contrast, the Confirmation rate in Priority blocks that specified spruce/fir habitats approached 40%, and 90% of the abundance codes of A5 came from spruce/fir blocks.

Regardless of their erratic roving habits, these lively, stripey "pine" birds enliven mountain summers in the pines, firs, and spruces where they nest and the meadows, shrubs, and aspens where they feed.

BREEDING EVIDENCE

REPORTED IN 732 (42%) OF 1745 PRIORITY BLOCKS

☐	Possible	335 blocks	(46%)
▩	Probable	164 blocks	(22%)
■	Confirmed	233 blocks	(32%)

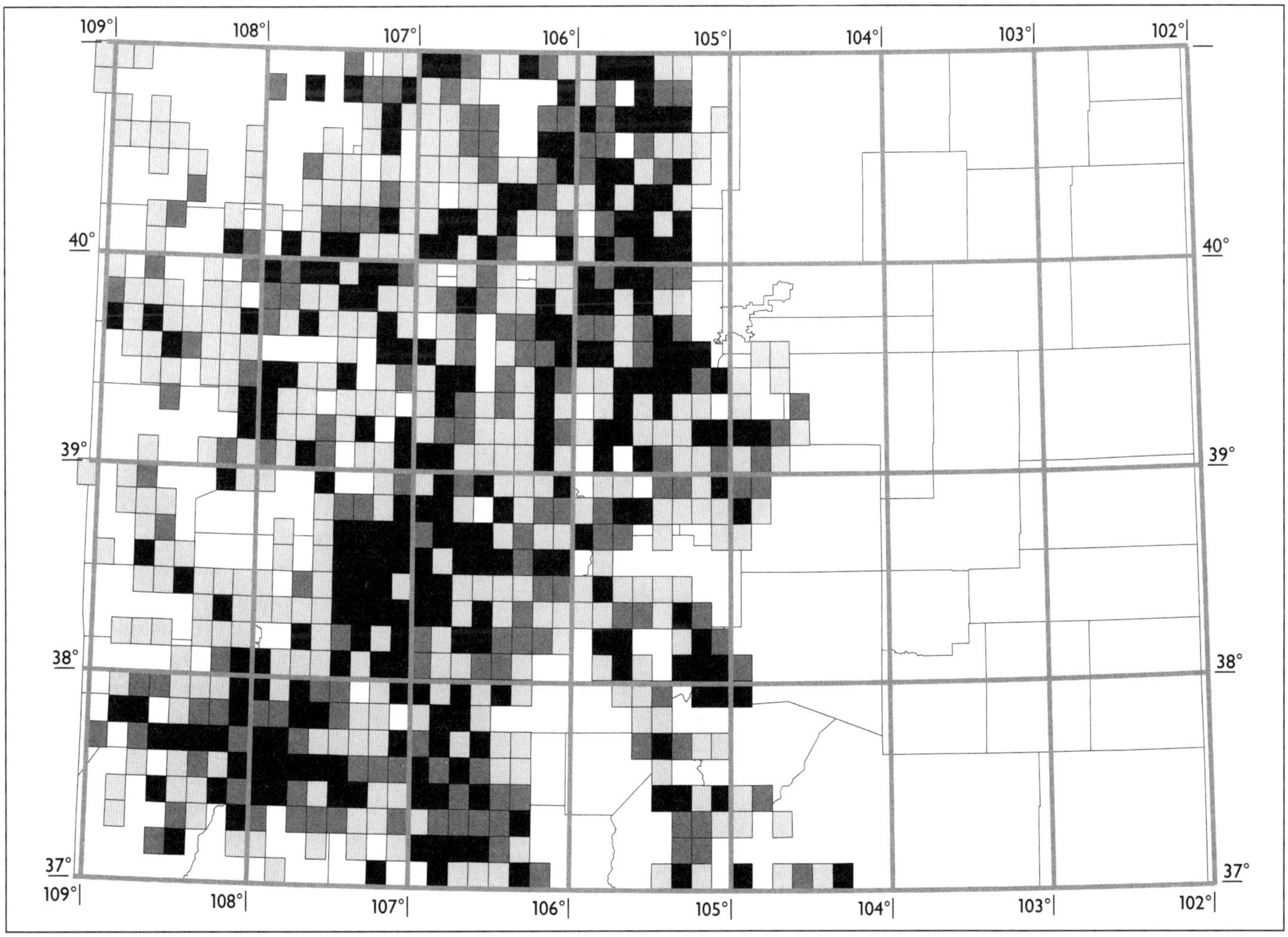
109°
108°
107°
106°
105°
104°
103°
102°
40°
39°
38°
37°

COLORADO DIVISION OF WILDLIFE / WILDLIFE RESOURCE INFORMATION SYSTEM

Less conspicuous, less studied, and less familiar than the other goldfinch, the Lesser Goldfinch confirms the paradox that sometimes less is more. For Atlas workers, this very combination made the bird more challenging and more interesting.

HABITAT

Prior to the Atlas, accounts of Lesser Goldfinches in Colorado indicated that they reside "where scrub oaks merge with the ponderosa pine" (B&N) and that they breed "in riparian forests, shrubland (mostly Gambel oak), and ponderosa pine forests" (A&R). Atlas data reveal that this description fits the birds' actual habitat use only in north-central Colorado, within an hour's drive of most of Colorado's human population. Elsewhere, this species much more frequently associates with pinyon/juniper than with either scrub oak or ponderosa. In fact, atlasers reported pinyon/juniper more frequently than any other single habitat; however, narrowleaf and broad-leaved cottonwoods came in a close second and third, and, combined, the number of riparian cottonwood forest reports surpasses that of pinyon/juniper.

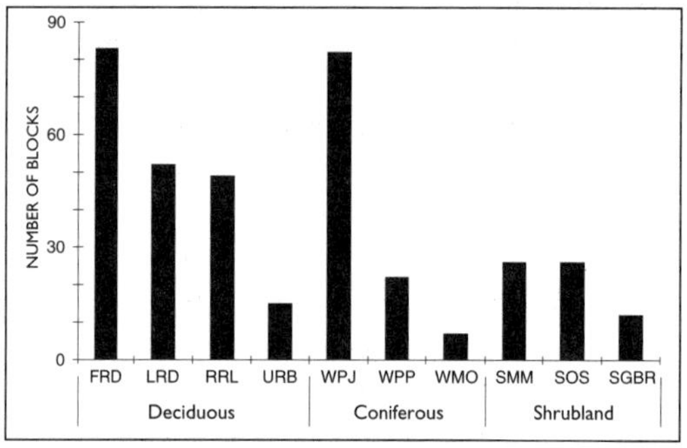
NUMBER OF BLOCKS
90
60
30
0
FRD LRD RRL URB WPJ WPP WMO SMM SOS SGBR
Deciduous
Coniferous
Shrubland

Several accounts note these birds' dependence upon water (B&N, Terres 1980, Ehrlich et al. 1988). This requirement, along with their need for the presence of herbaceous plants (especially thistle) that provide their diet, probably influences the habitats that Lesser Goldfinches use more than the distribution of tree and shrub species in which it nests. The specific location of nests varies greatly; five nest record cards named five different species of trees.

BREEDING

Lesser Goldfinches arrive in Colorado from their wintering grounds in mid May and begin nesting activities in June. Gregarious in the winter, this species supposedly does not maintain that habit during nesting (Gross *in* Bent 1968), but the birds' need for frequent drinking to digest their diet of seeds leads them to concentrate their nesting in prime habitat and makes them appear colonial.

Dates reported by Atlas workers indicate a rather uniform and relatively brief nesting season for this species, with nest-building occurring primarily in the last three weeks of June. Two reports of nest-building in mid July probably represent renesting attempts.

Females lay four or five eggs that hatch after 12 days of incubation (Gross *in* Bent 1968). Research has not yet determined the nestling period for this little-studied species, but it probably approximates the 11–14-day period of the other goldfinches (Linsdale *in* Bent 1968, Ehrlich et al. 1988). Adults feed the young primarily by regurgitation (Linsdale *in* Bent 1968). Courtship activities include mate feeding, behavior that closely resembles fledgling feeding, so that interpretation of early FY and FF dates requires caution. Atlasers began seeing fledglings in early July, exactly 30 days following the first reports of nest building. Fledgling observations continued regularly through July and August.

Like most other cardueline finches, these birds make poor cowbird hosts (Terres 1980), due to their relatively late nesting and their granivorous diet. Colorado has two reports of cowbird parasitism (Chace and Cruz 1996).

BREEDING PHENOLOGY

CODE		# OF RECORDS	DATE RANGE
C	Courtship	6	24 Jun–3 Aug
NB	Nest Building	20	6 Jun–12 Jul
ON	Occupied Nest	2	3 Jun–5 Jul
NE	Nest with Eggs	1	11 Jul
FY	Feeding Young	16	18 Jun–25 Aug
FL	Fledged Young	19	6 Jul–25 Aug

DISTRIBUTION

Lesser Goldfinches breed from coastal Oregon and the Great Basin to Panama. Colorado lies at the northern limit of the breeding range in the interior West, and their population in Colorado thins from south to north; Atlas workers reported birds in 105 blocks south of 38 degrees and in only 41 blocks north of 40 degrees. This pattern reverses that of the American Goldfinches—whose numbers concentrate in the northeast and thin to the south.

These two species both use riparian cottonwood and rural and urban plantings extensively, often in the same areas. Where the ranges of the two species overlap, Atlas workers reported both species in most of the blocks in which they found either. In Latilong 38107, for example, atlasers found American Goldfinches and Lesser Goldfinches each in 13 blocks. In ten of these blocks they found both species; in six of these ten blocks they reported both species in the same habitat; and in half of these six blocks they listed the same date for both species. In the northern part of the state, where American Goldfinches reach their highest density, the two species seem less compatible. There, Atlas workers found Lesser Goldfinches restricted to foothill habitats like narrowleaf cottonwood, ponderosa pine, and scrub oak and nearly absent from the irrigated and riparian areas inhabited by American Goldfinches.

Where Atlas workers found this species, they often found them relatively abundant; they reported A3 (11–100) most frequently, and reported A4 (101–1,000 pairs) in 5% of the blocks. This density enables the Lesser Goldfinches to maintain a population in Colorado that probably exceeds that of the more widely distributed American Goldfinches.

BREEDING EVIDENCE

REPORTED IN 307 (18%) OF 1745 PRIORITY BLOCKS

Possible	137 blocks	(45%)
Probable	115 blocks	(37%)
Confirmed	55 blocks	(18%)

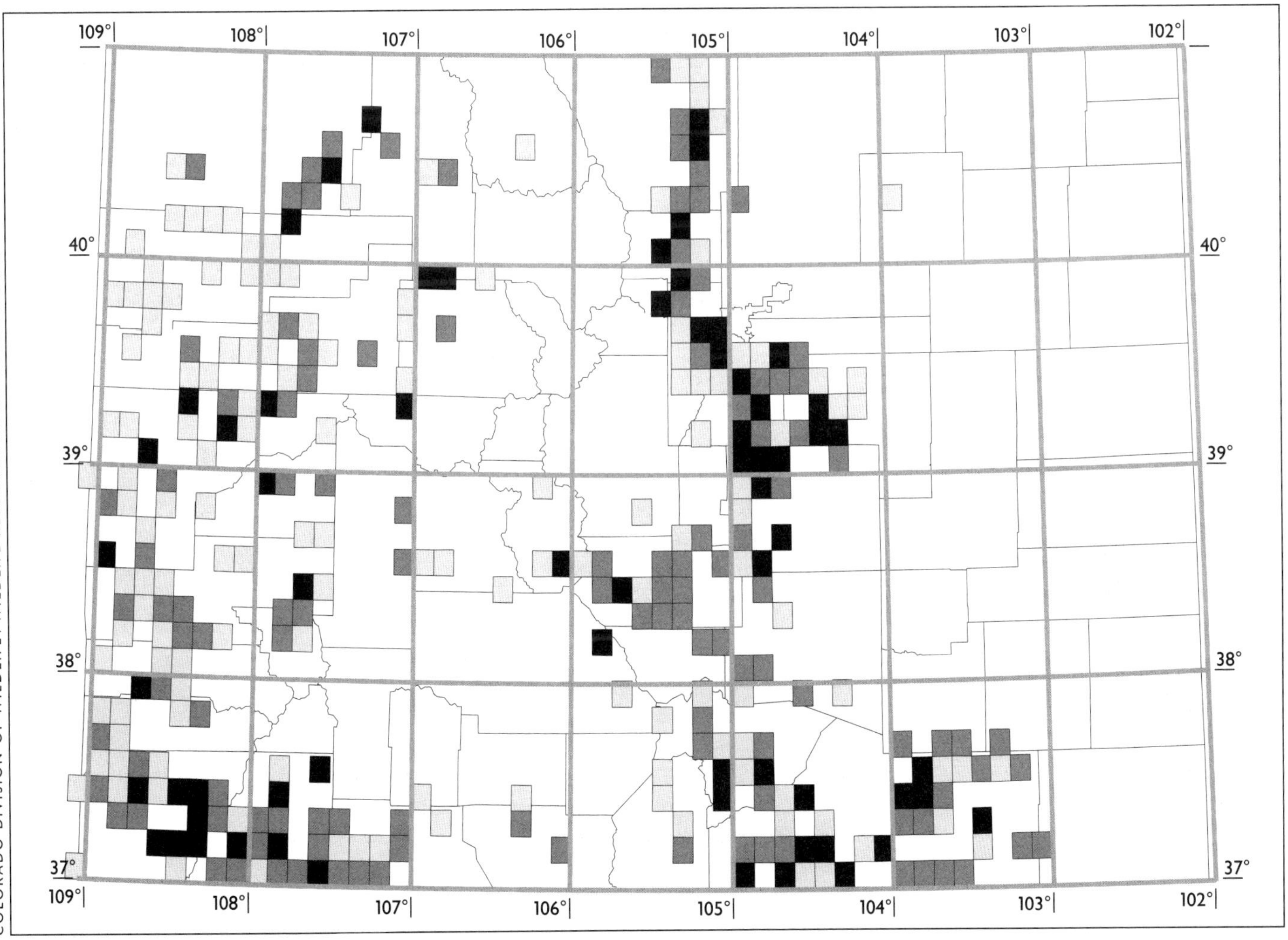
109°
108°
107°
106°
105°
104°
103°
102°
40°
39°
38°
37°

COLORADO DIVISION OF WILDLIFE / WILDLIFE RESOURCE INFORMATION SYSTEM

The lively, vocal, gregarious habits of the American Goldfinch, along with its tolerance of human presence, have led to a literature replete with descriptions of a joyous, friendly, and even affectionate bird. This penchant for ascribing human emotions to this bird has made it a sort of Bambi of the bird world.

HABITAT

American Goldfinches nest in a variety of deciduous shrubs and trees; they prefer small isolated stands or the edges of larger stands (Nickell 1951, Middleton 1993). Proximity to an abundant food supply (the seeds of composites, especially thistles) seems more important than any specific cover plants in their choice of nest sites. In Colorado, riparian habitats and the riparian-like habitats of irrigated farms and towns most frequently provide the conditions required for nesting. Although Atlas workers reported 26 different habitat codes for this species, more than 65% of all reported codes designated riparian forest (LRD, FRD) and rural woodlots and shelterbelts (RRL). Another 15% indicated grass, shrub, and other habitats found in association with these. All of the mountain shrub reports came from blocks north of 40 degrees, where this habitat often grows in riparian zones.

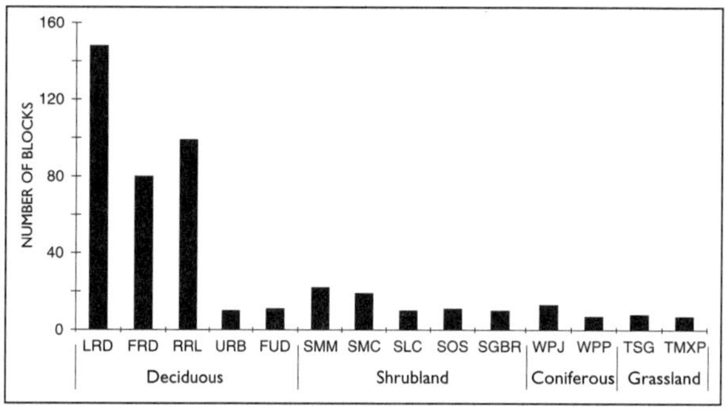
NUMBER OF BLOCKS
160
120
80
40
0
LRD FRD RRL URB FUD
SMM SMC SLC SOS SGBR
WPJ WPP
TSG TMXP
Deciduous
Shrubland
Coniferous
Grassland

Atlas workers reported this species using pinyon/juniper, ponderosa pine, and aspen, but most reports of these habitats also included other, more commonly used habitats. American Goldfinches probably nest in conifers and aspens only rarely, if at all.

BREEDING

Nesting dates for the American Goldfinch vary rather widely with subspecies (Middleton 1993). The much studied eastern populations (*C. t. tristis*) do not start to nest until, at the earliest, mid June (Middleton 1993). This pattern of late nesting extends as far west as eastern Kansas (Thompson and Ely 1992). In California on the Pacific Slope, *C. t. salicamans* begins much earlier, in April and May (Middleton 1993). Birds in the interior West (*C. t. pallidus*) begin breeding between these extremes, with nests reported in Idaho in early June (Burleigh 1971) and egg records in British Columbia in late May (Tyler *in* Bent 1968). Both *tristis* and *pallidus* nest in Colorado (Middleton 1993).

The different timing of the two subspecies' breeding cycles may account for dates scattered from May to July for the onset of breeding activities, but Atlas workers submitted a number of Possible breeding and Probable breeding codes bearing dates earlier than the rather extensive literature on this species supports. They reported numerous observations in April and May, mostly singing birds, pairs, or birds in suitable habitat, but also one observation of nest-building. Several factors render early season Atlas data ambiguous: American Goldfinches' winter habitat resembles their summer habitats; they sing, court, and may form pair bonds in their winter flocks (Middleton 1993); and they often build nests that they do not use (Tyler *in* Bent 1968). Early reports of these activities may indicate birds that have not yet reached their breeding grounds. Banders at Lykins Gulch near Lyons provided the earliest unambiguous breeding evidence when they netted a male on 19 May 1995 with partial cloacal protuberance (Virginia Dionigi pers. comm.), which would indicate nesting within the week or two following (Pyle et al. 1987, Mike Carter pers. comm.). In view of the dubious nature of early singing and pairing as on-site breeding behavior, the Atlas database dropped all Possible and Probable codes to Observed if they occurred prior to 1 May.

Cowbirds do not frequently parasitize eastern American Goldfinch nests because the breeding times for the two species barely overlap (Mariani et al. 1993). Young cowbirds do not thrive on the hosts' granivorous diet, so goldfinches rarely succeed in rearing them (Middleton 1993); however, two Atlas workers in two blocks did observe adult goldfinches feeding young cowbirds (Chace and Cruz 1996; Appendix C).

BREEDING PHENOLOGY

CODE		# OF RECORDS	DATE RANGE
C	Courtship	6	27 May–7 Jul
NB	Nest Building	14	8 Jun–12 Aug
ON	Occupied Nest	3	10 Jun–23 Jul
NE	Nest with Eggs	1	30 Jun
NY	Nest with Young	1	17 Jun
FY	Feeding Young	8	4 Jun–24 Aug
FL	Fledged Young	14	7 Jun–7 Sep

DISTRIBUTION

The coast-to-coast breeding range of the American Goldfinch lies from southern Canada to the mid-U.S., from Colorado north. The range map for this species in the Birds of North America series (Middleton 1993) shows it (inaccurately) extending only into the northern corners of the state. Indeed, the largest populations do nest in the northern half of the state, especially in the northeast, where atlasers reported this species in 145 Priority blocks, approximately doubling the number of northwest, southwest, and southeast quadrants—82, 67, and 74 blocks receptively. This pattern would appear even more marked if some early reports of Possible and Probable breeding described migrating birds.

Populations of this species seem quite stable (Priorities 1995). In some areas competition from House Sparrows has reduced goldfinch numbers (Tyler *in* Bent 1968), and the spread of House Finches in eastern North America may also affect its population (Middleton 1993). On the other hand European settlement of North America has fragmented forests and created openings and edges preferred by goldfinches as well as by thistles, their staple food. These modifications have probably expanded the range of this species and made it more numerous than before settlement (Middleton 1993).

Although atlasers found American Goldfinches scattered widely throughout Colorado, they seldom found them in large numbers, with A2 (2–10 pairs) constituting 60% of the codes reported. These birds' specialization on moist habitats in a dry state keeps its total population relatively low, and Lesser Goldfinches, although found in fewer blocks, probably outnumber them.

BREEDING EVIDENCE

REPORTED IN 365 (21%) OF 1745 PRIORITY BLOCKS

□	Possible	205 blocks	(56%)
▩	Probable	117 blocks	(32%)
■	Confirmed	43 blocks	(12%)

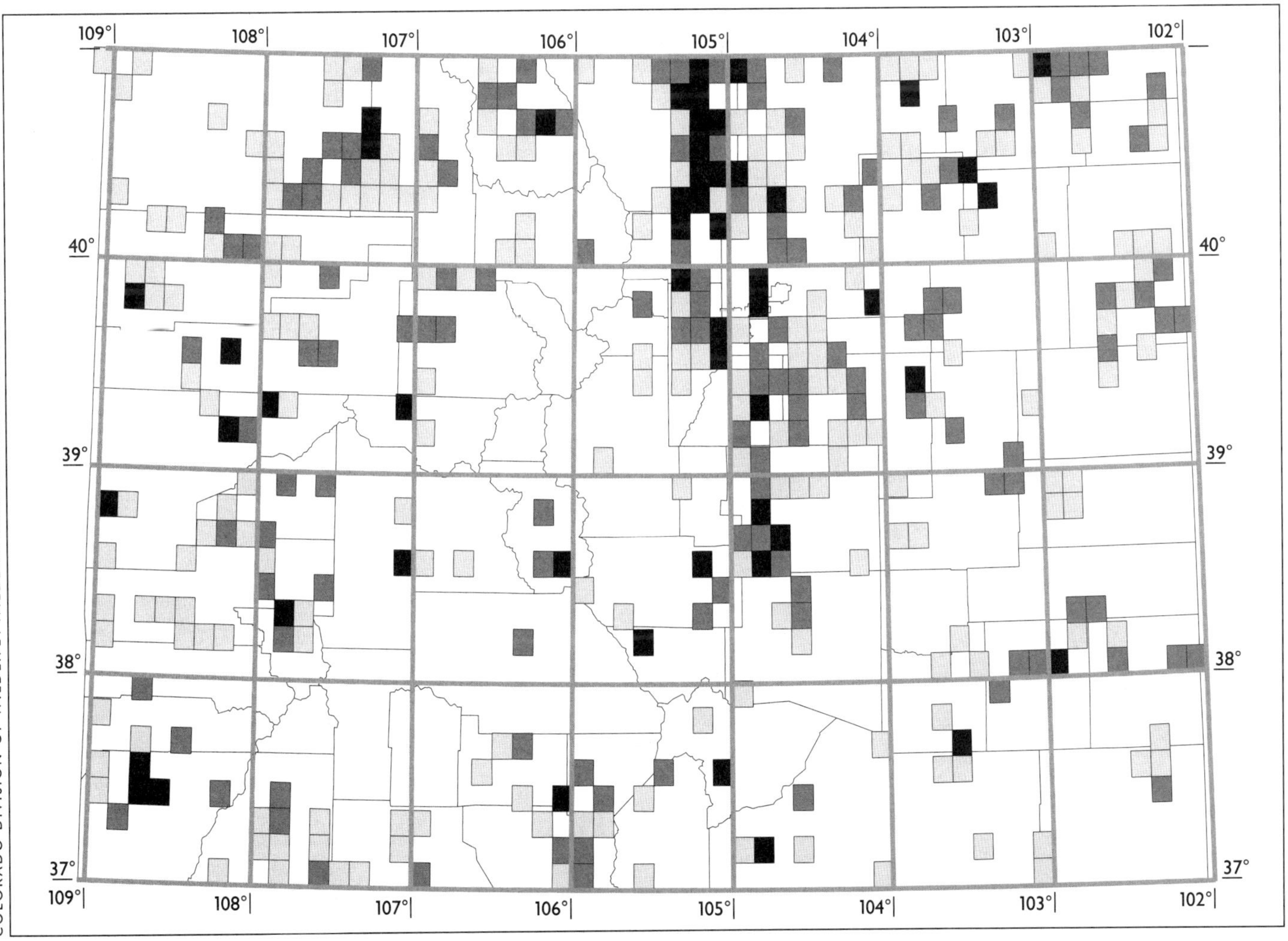
109°
108°
107°
106°
105°
104°
103°
102°
40°
39°
38°
37°

COLORADO DIVISION OF WILDLIFE / WILDLIFE RESOURCE INFORMATION SYSTEM

Evening Grosbeaks seem more conspicuous in winter than summer. Winter flocks throng feeders, chatter noisily, and devour enormous quantities of sunflower seeds. In spring, the flocks disperse, and the grosbeaks seem to disappear. As they begin the nesting cycle, they become secretive and carefully camouflage their nests, so that observers easily overlook them.

HABITAT

Evening Grosbeaks breed primarily in coniferous forests, especially ponderosa pines, but also use Douglas-firs and Engelmann spruces. Long ponderosa needles help camouflage and protect their nests. Evening Grosbeaks feed on insects only during breeding; they may select nest sites near high densities of spruce budworms—elsewhere breeding populations respond positively to budworm outbreaks. Because Douglas-fir is the principal host of spruce budworms in Colorado, numbers of nesting Evening Grosbeaks may vary according to the budworm cycle in forests containing large numbers of Douglas-firs (Bekoff et al. 1987). Atlasers reported them in Douglas-fir in only four blocks.

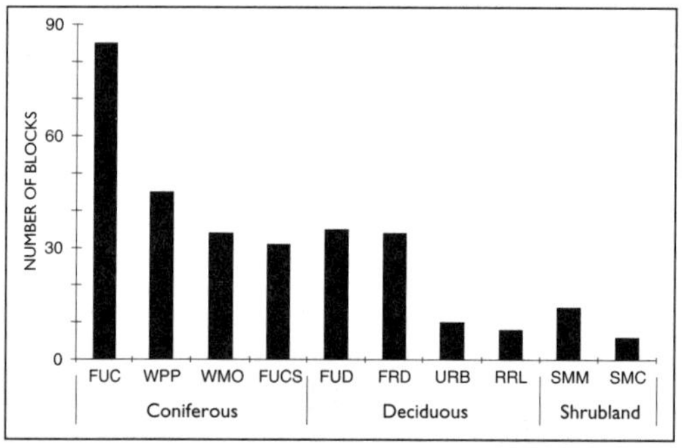
NUMBER OF BLOCKS
0
30
60
90
FUC
WPP
WMO
FUCS
FUD
FRD
URB
RRL
SMM
SMC
Coniferous
Deciduous
Shrubland

The grosbeaks select relatively open areas, and avoid dense deciduous stands (Bekoff et al. 1987). They nest irregularly from the plains near the foothills almost to timberline, and they occasionally place nests in evergreens near city residences (B&N).

In other states they nest in deciduous as well as in coniferous trees (Bent 1968). Atlasers reported 23% of their observations from deciduous forests and only two nests, one in an aspen, the other in a ponderosa.

After the young fledge, the parents soon lead them to adjacent areas, where atlasers often found them in mixed coniferous and aspen forests in mid July. About two-thirds of the habitat reports specified coniferous woods in the mountains and foothills.

BREEDING

In the Front Range Evening Grosbeaks breed from late May to at least mid July (Bekoff et al. 1987). They arrive as mated pairs at the breeding area in mid to late May, and do not attempt to establish a territory. They begin most nests in late May or early June. They are well adapted for breeding in habitats in Colorado that have severe weather and short growing seasons. These grosbeaks produce only one brood (Scott and Bekoff 1991).

The female gathers the nest material and builds the nest, although the male accompanies her during material collection. After the young hatch, the pair forages together; they return to the nest together with food for the nestlings, although the female does most of the feeding, by regurgitation of insects. The male removes the fecal sacs (Elkins *in* Andrle and Carroll 1988).

Early observers in Colorado first found Evening Grosbeaks only in winter (Cooke 1897), then had difficulty finding nests of this "irregular wanderer" (Cooke 1900, 1909), but they assumed that some Evening Grosbeaks breed in the mountain parks (Sclater 1912). One author reported them as an infrequent summer resident, with only two breeding records (Bergtold 1928). Later observers discovered nests with eggs from 13 June to 22 July, and adults with young from 22 June to 6 August (B&N).

More recent observers had greater success locating Evening Grosbeaks and documenting breeding. The Latilong Study reported breeding in 16 of the 28 latilongs. Atlasers Confirmed breeding in two additional latilongs in northwestern Colorado. Atlas workers compiled a relatively low Confirmation rate, only 23%, most Confirmations based on observing fledglings or adults feeding young.

Atlasers observed Evening Grosbeaks gathering nesting materials in at least eight blocks, but had difficulty finding the well-concealed nests; they found only two occupied nests, one in an aspen and one in a conifer. A nest card submitted for Teller County (Roberta Winn) reports a pair beside the road, the female gathering grass. Then both flew into an Engelmann spruce to a partially completed nest. Atlasers observed two pairs gathering nest material, one in aspen, the other in willows.

Atlasers reported a longer breeding season than earlier described, with adults feeding young as late as 28 August, and accompanying fledglings as late as 3 September. They also observed pair formation as early as 18 April and 1 May.

BY DAVID PANTLE

BREEDING PHENOLOGY

CODE		# OF RECORDS	DATE RANGE
C	Courtship	5	20 May–23 Jun
NB	Nest Building	8	30 May–5 Jul
ON	Occupied Nest	2	15 Jun–16 Jun
FY	Feeding Young	16	17 Jun–28 Aug
FL	Fledged Young	25	26 May–3 Sep

DISTRIBUTION

Evening Grosbeaks breed in a narrow band of the spruce belt across southern Canada, south into New England, and also down through the mountains of the West into Mexico. In winter, they wander widely and erratically through most of the United States.

Thought of as local and irregular breeders, Evening Grosbeaks had, until the Atlas, yielded little information about their Colorado habits (A&R). B&N reported only five nesting locations (Denver, Boulder, Longmont, Estes Park, and Routt County)—both plains and foothills sites. They depicted a species that converges on a place for one or a few years to breed (e.g., Boulder 1941–1944, Denver 1939, 1948, 1956, 1963) but which then vanishes.

The Atlas fleshes out a previously skeletal breeding map. Atlasers reported them throughout the wooded mountains and foothills, although widely scattered and in relatively small numbers. They found no plains nesting. The map confirms breeding in all the latilongs with expanses of conifers. The most solid breeding came from the ponderosas in the Black Forest between Denver and Colorado Springs and the hills east of the Arkansas River between Salida and Buena Vista.

Atlas data show no striking geographic patterns that would corroborate the erratic breeding reported by B&N. A faint cyclic pattern may emerge: in 1994 atlasers recorded twice as many Confirmations as in any other year, and they reported 37% of both the total reports and total Confirmations in 1993–1994.

Evening Grosbeaks presently face little threat on their breeding and wintering grounds. They flourish and have expanded their range to the east and the south in the U.S. Increasing utilization of sunflower seed in feeders in the winter has further affected the winter distribution of Evening Grosbeaks, although more in the eastern U.S. than in Colorado.

BREEDING EVIDENCE

REPORTED IN 254 (15%) OF 1745 PRIORITY BLOCKS

☐	Possible	136 blocks	(54%)
▩	Probable	59 blocks	(23%)
■	Confirmed	59 blocks	(23%)

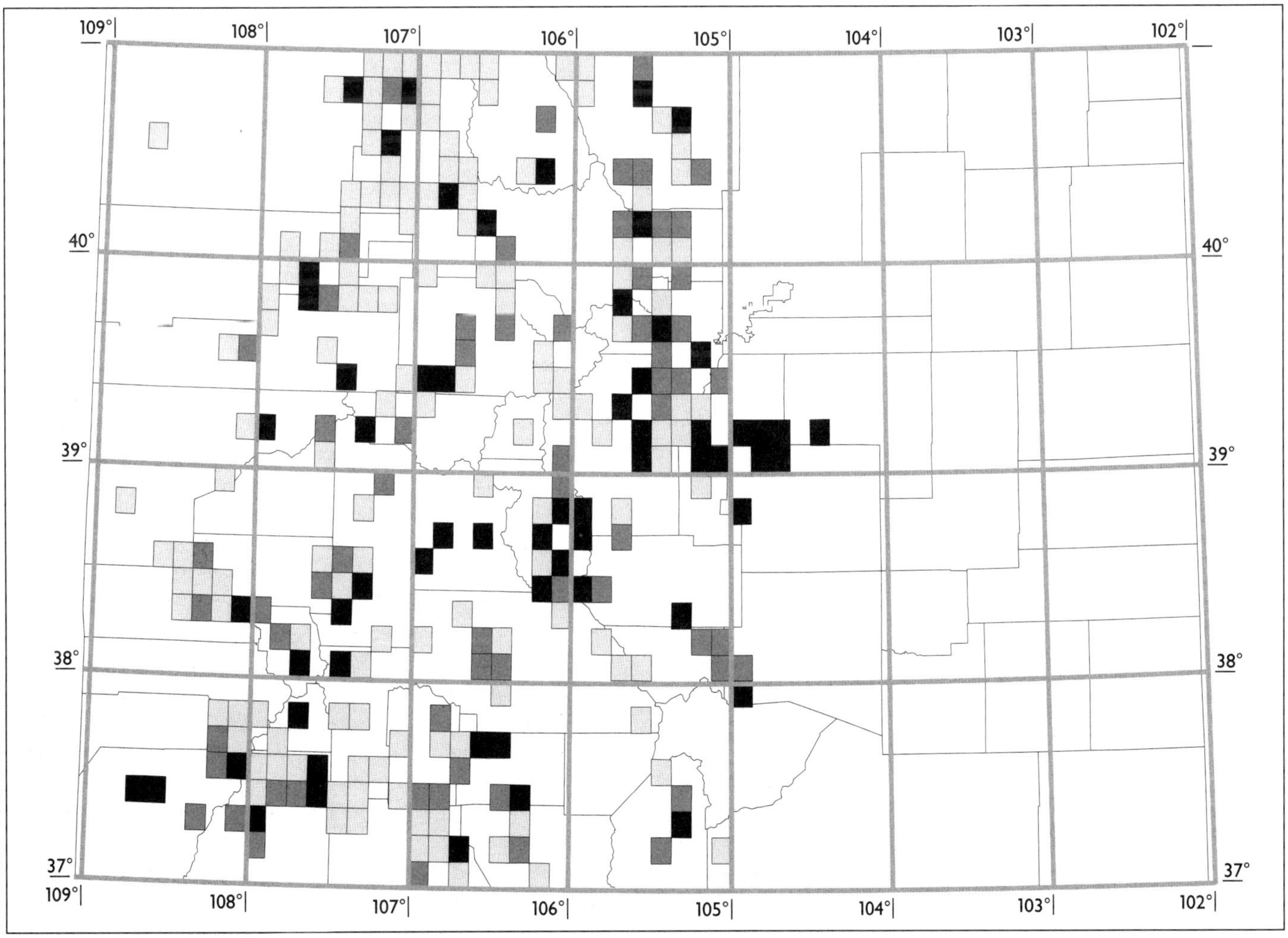

109°
108°
107°
106°
105°
104°
103°
102°
40°
39°
38°
37°

COLORADO DIVISION OF WILDLIFE / WILDLIFE RESOURCE INFORMATION SYSTEM

With virtual unanimity, birdwatchers regard the House Sparrow with scorn, a situation that has led to a number of misconceptions about them. These misconceptions include the beliefs that the birds are virtually ubiquitous, that they followed man into the state, and that they represent a significant threat to other species.

HABITAT

The perceived ubiquity of House Sparrows mirrors more the location of people rather than a universal abundance of the species. Their presence in Colorado reflects a range expansion from more easterly sites where introduction occurred midway through the nineteenth century (Lowther and Cink 1992). Furthermore, when they arrived in Colorado at around the beginning of the twentieth century, they did not so much follow man as they followed horses, cattle, and train-spilled grain. Competition between House Sparrows and native species

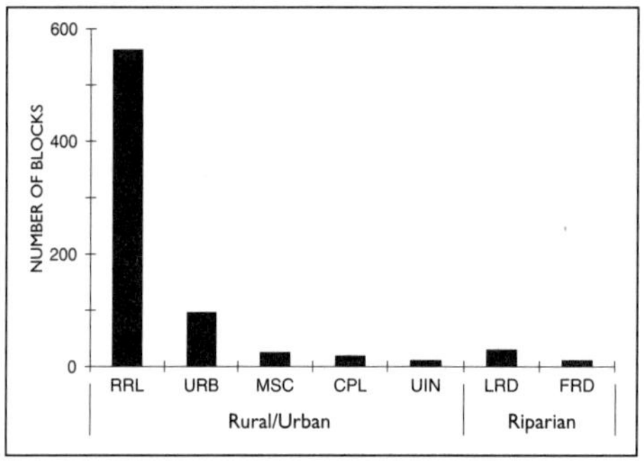
600
400
200
0
NUMBER OF BLOCKS
RRL
URB
MSC
CPL
UIN
LRD
FRD
Rural/Urban
Riparian

for nesting sites occurs only where the presence of man has already imposed other threats to natives.

Nevertheless, few people appreciate House Sparrows; as a result we tend to ignore the species. The scientific literature, however, does not reflect the tendency of field observers to disregard House Sparrows. Lowther and Cink (1992) report that the scientific bibliography relating to the species has more than 4,800 entries.

The association of House Sparrows with humans and human-modified environments narrowly limits their habitat selections. The primary habitat found by Atlas observers, rural farm and ranches including abandoned sites, composed 73% of Atlas observations. Ninety-one percent of all observations took place in a composite of human-modified environments, including rural and urban areas, croplands, and structures such as bridge abutments. During non-breeding seasons, House Sparrows, non-migratory in Colorado, stay in the same habitats and often use old nests as nighttime roosts (Lowther and Cink 1992).

House Sparrows feed almost entirely on cereal grain and seeds. In urban areas, the species also depends heavily on bird seed. During the late spring and summer months, invertebrates comprise up to 9% of a House Sparrow's diet, associated with feeding of fledglings (Lowther and Cink 1992).

BREEDING

House Sparrows build impressive nests, typically constructed of dried grasses with a lining of finer materials. Many nests, especially in urban areas, they conceal in the eaves of buildings or other suitable cavities, or stuff behind thick vines on the sides of houses. The size of the nests depends on the size of the space—the nest expands to fill the available volume (Lowther and Cink 1992). Although House Sparrows do compete with other cavity-nesters for nest sites, the impacts of human presence (such as habitat alteration and introduction of non-native predators like house cats) temper the sparrows' effect on native species.

House Sparrows do not nest exclusively in enclosed spaces. In trees they build globular structures shaped like a squashed ball, 10–15 inches (30–40 cm) in diameter, with side entrances (Lowther and Cink 1992). Atlasers found impressive structures in the dead portions of old poplar shelterbelts around farmhouses. Other common residents of the scattered farm building and shelterbelt associations in which House Sparrows are most prevalent include Western and Eastern kingbirds, House Wrens, and Common Grackles, all of which are capable of successful nest-site defense or nest in different sites.

Most observations of nest-building in Colorado occurred from mid April to late June. The earliest outlying observation, on 27 February, seems isolated only because few atlasers began field work so early in the nesting cycle. On the other hand, this particular observation may reflect gathering of nesting materials by House Sparrows in order to improve roosting sites (Lowther and Cink 1992). It takes about 40 days from laying of the first egg to independence of fledged young. The extended period during which observers in Colorado found House Sparrows on nests (27 February to 27 July) supports the widespread evidence from other studies for multiple clutches of eggs per season for some individuals. Colorado has one old (1909) record of cowbird parasitism, an egg under incubation in a nest built in an old magpie nest (Chace and Cruz 1996).

BREEDING PHENOLOGY			
CODE		# OF RECORDS	DATE RANGE
C	Courtship	8	15 Apr–13 Jul
NB	Nest Building	65	27 Feb–14 Jul
ON	Occupied Nest	180	27 Feb–22 Jul
NE	Nest with Eggs	9	27 Apr–12 Jul
NY	Nest with Young	42	15 Apr–27 Jul
FY	Feeding Young	134	8 May–6 Aug
FL	Fledged Young	52	13 May–22 Aug

DISTRIBUTION

Native to Eurasia from the British Isles to Siberia and northern Africa to Burma, House Sparrows now range almost worldwide, courtesy of introductions. In North America they range coast to coast, from Quebec to British Columbia and south.

Introduced into the U.S. in 12 states from 1854 to 1881, House Sparrows had arrived in Colorado by 1895 (B&N). They are considerably less common in Colorado than they are in other states in the East. Forest, grassland, mountain, and desert habitats in Colorado remain free of them, although urban areas in the mountains, such as Leadville (10,200 feet, 3110 m) and Fairplay (10,000 feet, 3048 m), host the highest-elevation populations in North America (Lowther and Cink 1992). BBS results show a significant downward trend in House Sparrow populations throughout the continent, a trend reflected in BBS figures for the High Plains and for Colorado.

In all probability, House Sparrow populations in Colorado will continue to decrease. Isolated, abandoned farm sites on the eastern plains currently harbor many colonies. The nesting habitat quality of these farm sites for House Sparrows deteriorates as buildings collapse and trees and bushes die off from a lack of irrigation. As the human population becomes more concentrated on the plains, House Sparrows may disappear from large empty areas that return to grassland status or turn into cropland.

BREEDING EVIDENCE		
REPORTED IN 692 (40%) OF 1745 PRIORITY BLOCKS		
Possible	81 blocks	(12%)
Probable	66 blocks	(9%)
Confirmed	545 blocks	(79%)

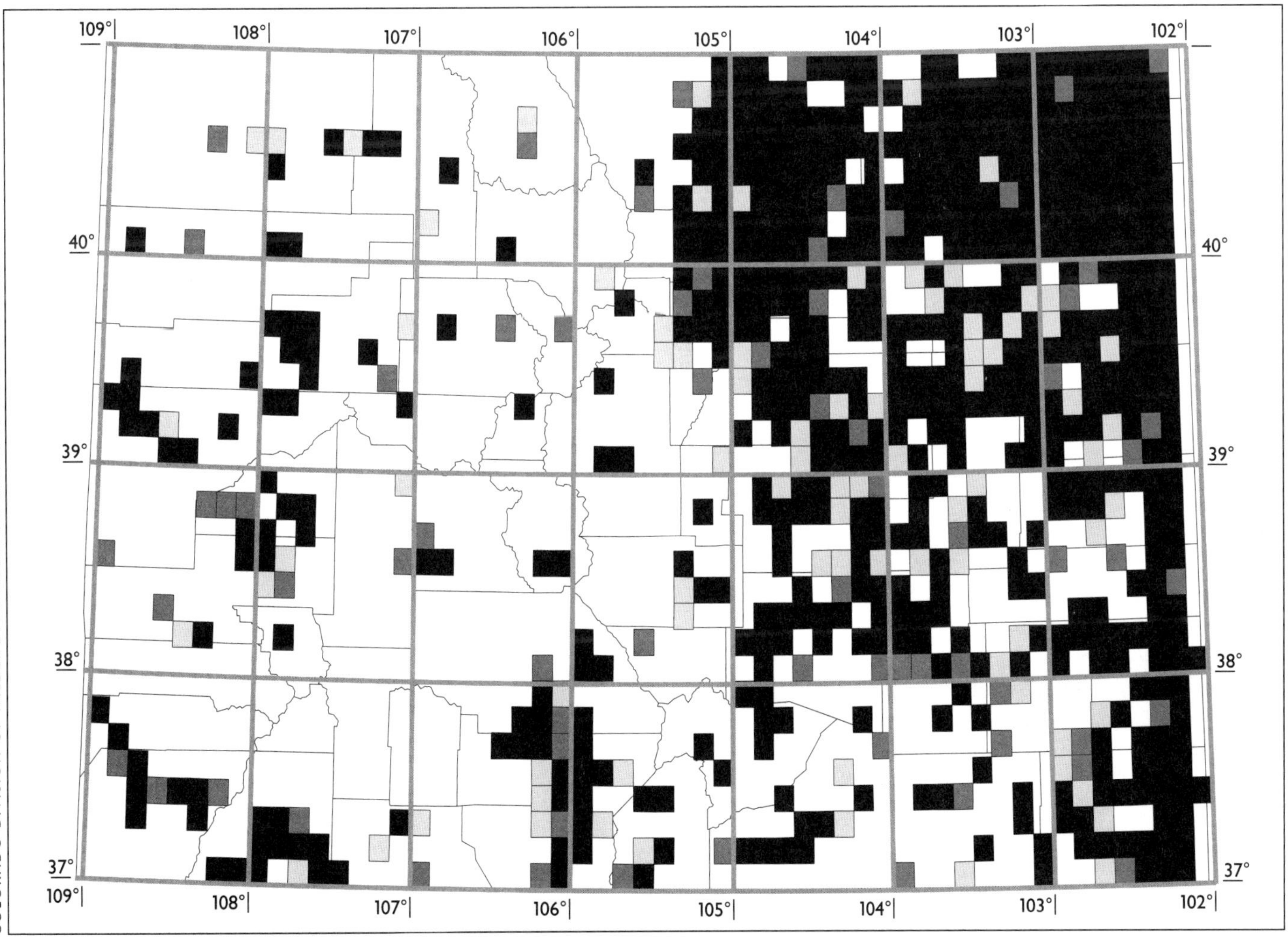
109°
108°
107°
106°
105°
104°
103°
102°
40°
39°
38°
37°

COLORADO DIVISION OF WILDLIFE / WILDLIFE RESOURCE INFORMATION SYSTEM

From page 266

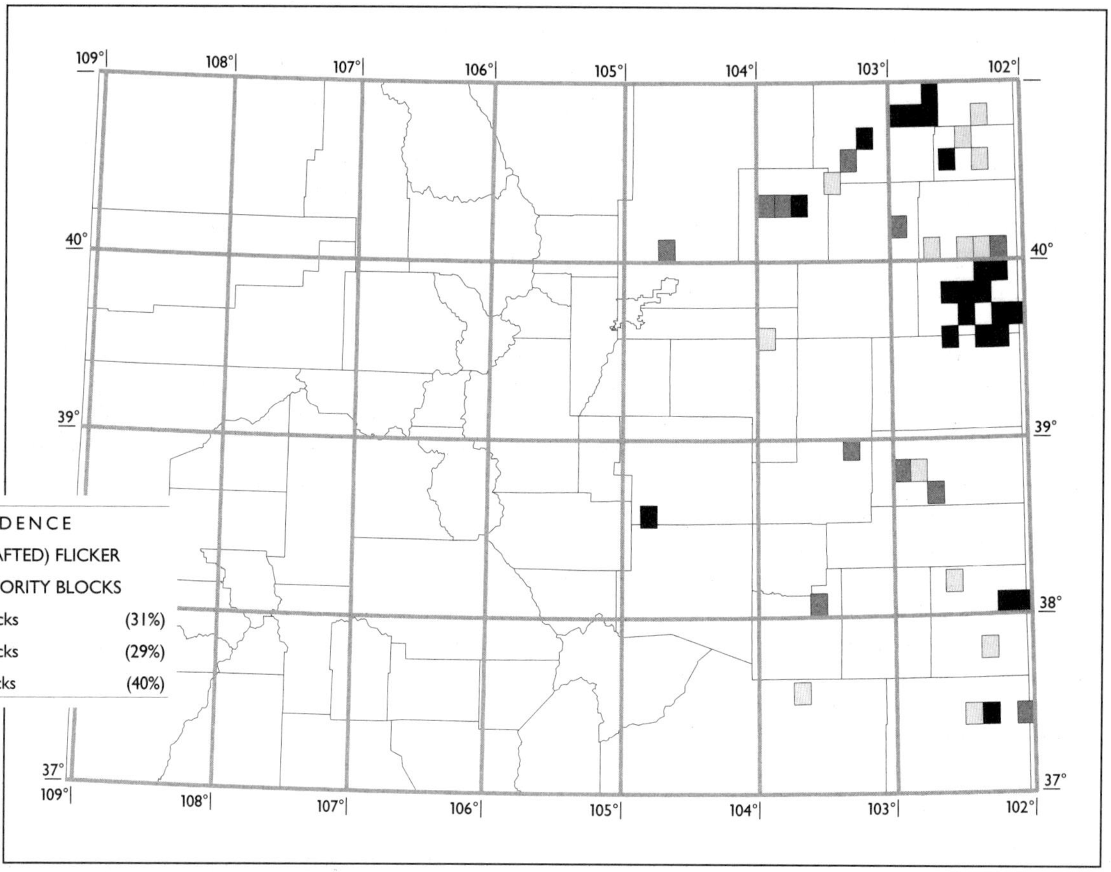
109° 108° 107° 106° 105° 104° 103° 102°
40° 39° 38° 37°

BREEDING EVIDENCE

NORTHERN (YELLOW-SHAFTED) FLICKER

REPORTED IN 35 (2%) OF 1745 PRIORITY BLOCKS

Possible	11 blocks	(31%)
Probable	10 blocks	(29%)
Confirmed	14 blocks	(40%)

From page 518

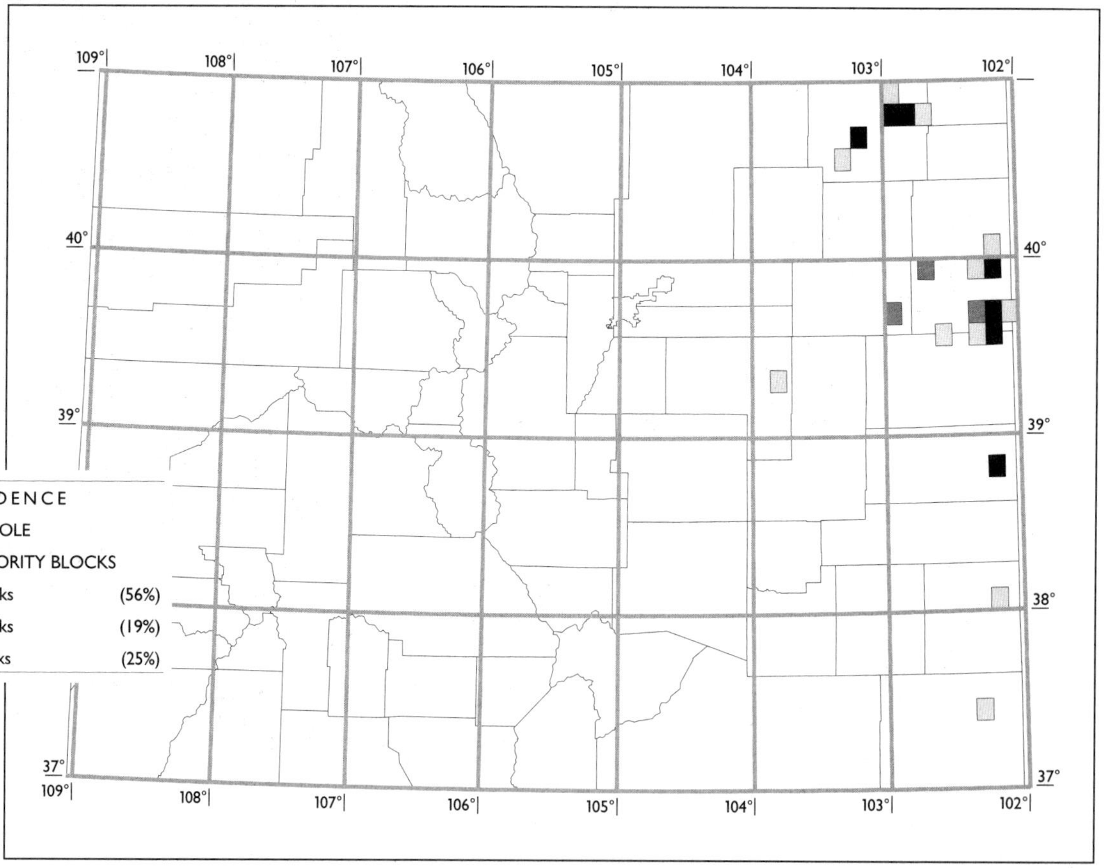
109° 108° 107° 106° 105° 104° 103° 102°
40° 39° 38° 37°

BREEDING EVIDENCE

BALTIMORE ORIOLE

REPORTED IN 16 (1%) OF 1745 PRIORITY BLOCKS

Possible	9 blocks	(56%)
Probable	3 blocks	(19%)
Confirmed	4 blocks	(25%)

SUPPLEMENTAL SPECIES ACCOUNTS

This chapter discusses three categories of Colorado breeding birds: species Confirmed during the Atlas but at less than three sites; species that displayed breeding behavior but not Confirmed; and species for which the Atlas has no confirming records but which have bred in the state before or after the Atlas time period. Three maps show the blocks where birds in the first two groups occurred.

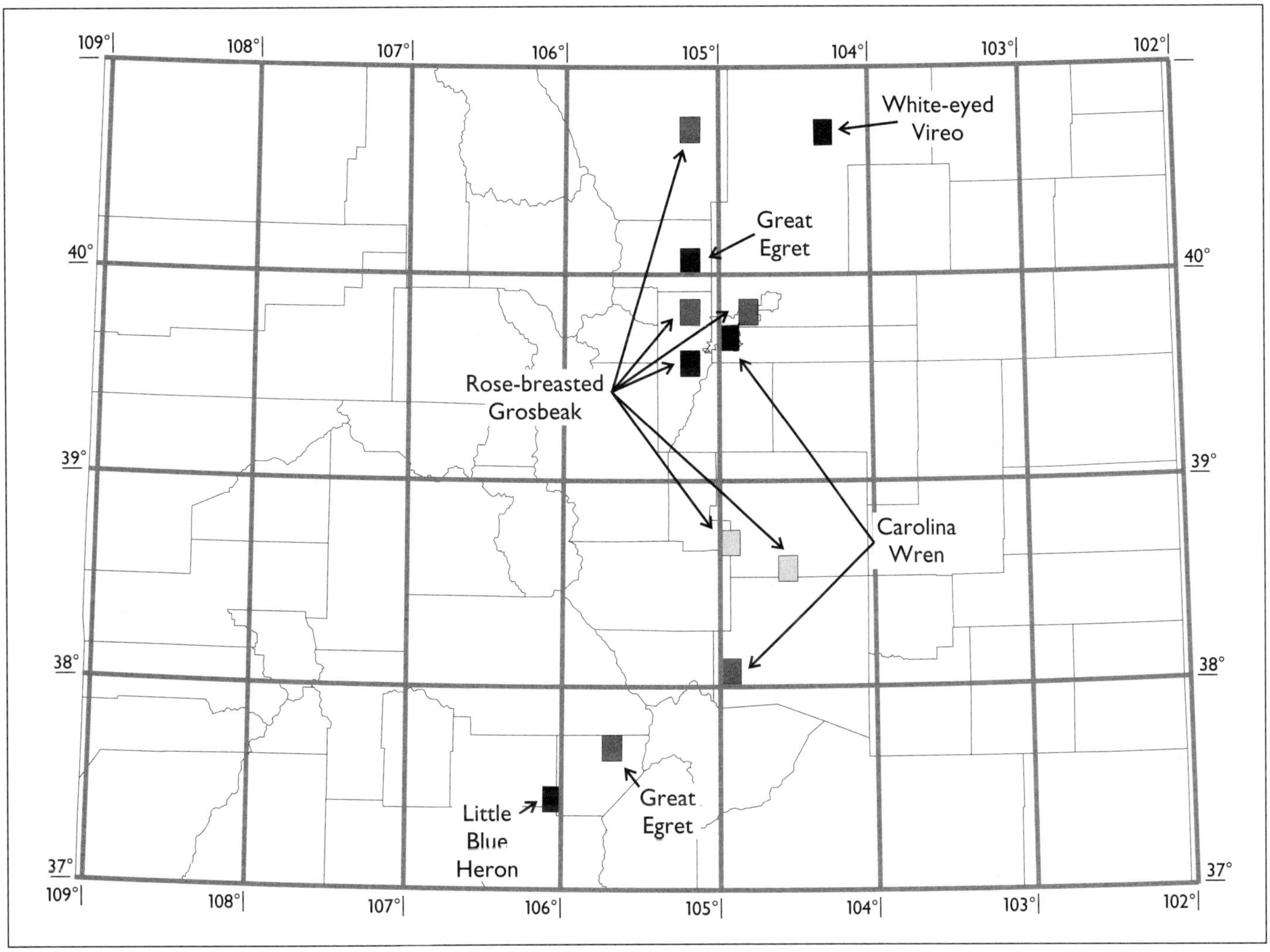
109°
108°
107°
106°
105°
104°
103°
102°
40°
39°
38°
37°
White-eyed Vireo
Great Egret
Rose-breasted Grosbeak
Carolina Wren
Great Egret
Little Blue Heron

RARE BREEDING SPECIES

Atlasers Confirmed breeding by all the species in this section during the Atlas time period, but recorded three or fewer Confirmations.

GREAT EGRET
Ardea alba

One or two pairs of Great Egrets have nested continuously in a heronry on Boulder Creek 8 miles (13 km) east of Boulder since at least 1972. They nest among 100 or more pairs of Great Blue Herons in a large cottonwood grove closed to public access. The state has one other breeding record, from Riverside Reservoir in 1983 (A&R). —H.E.K.

LITTLE BLUE HERON
Egretta caerulea

Although Little Blue Herons have been reported at least 86 times for Colorado (A&R), the only definite record of nesting is one in the San Luis Valley (Ryder et al. 1989).

At least one pair (possibly two) nested on Parker Pond, Monte Vista NWR. An adult attended a nest of two eggs and three young on 20 June 1988. On 5 July the nest contained four dead young and one cold egg. Two young were preserved as specimens (DMNH 39515 and 39519). Although observers have reported adult Little Blue Herons in the area since 1988, no other nests or young have been found.

—Ronald A. Ryder

ACORN WOODPECKER
Melanerpes formicivorus

Atlas field worker Mark Yaeger discovered the first state record of this spectacular woodpecker at Lake Dorothey SWA southeast of Trinidad (and accessible only through Raton, New Mexico; Barela, 37104A3). Although a family stayed in the vicinity for a year after the initial discovery on 20 July 1994, the birds did not nest after their discovery (Yaeger 1994).

After this find, observers from Durango belatedly reported that Acorn Woodpeckers had inhabited a ridge covered with scrub oak and ponderosa pines within a Durango subdivision (Durango West, 37107C8) for the two years before Yaeger found the birds at Lake Dorothey. The Durango birds brought young to feeders for at least three years during the Atlas period (Chris Hagen, Chad Ferry VF).

In both places the habitat consisted of scrub oak interspersed with ponderosa pines. At Lake Dorothey the woodpeckers had storage trees in dead ponderosas. At Durango they share their habitat with Lewis's Woodpeckers; the hillside has three towering ponderosas with plentiful snags. The Acorn Woodpeckers protect their acorn storage sites (one of which was a utility pole) by aggressively chasing away the Lewis's Woodpeckers (Chris Hagen, Chad Ferry pers. comm.). —H.E.K.

LEAST FLYCATCHER
Empidonax minimus

Suspected for over 70 years of nesting in Colorado, this species was finally verified as a breeder during the Atlas period. Bill Prather discovered the first Colorado nest on 4 July 1988 at a site near Lyons where he and his family (Inez and John) had heard singing birds for three years. The nest had three young. Prather watched two of them fledge on 17 July, and saw the three together on 19 to 24 July. He also saw a second family group 50 feet (15 m) away (Prather 1988).

Among the tall cottonwoods on the South Platte River in Chatfield State Park, Least Flycatchers maintained territories for the decade of the 1980s, with as many as six singing birds recorded in a day, from 14 May to 22 June (Joey Kellner pers. comm.). Then, still in the park but on Plum Creek in the Littleton Atlas block (39105E1), atlasers found nests in a grove of medium-height (40–50 feet, 12–15 m) young cottonwoods for three successive years.

Each year from 1990 to 1992 atlasers found one nest about 25 feet (8 m) high in a cottonwood at the edge of the same small opening in the woodland. Two adults fed two begging fledglings on 30 June 1992. In 1991, a half-mile away, a second pair held a territory in a similar cottonwood copse (J.B. Hayes, HEK, VF, nest cards). Atlas reports also include a singing bird in Washington County on 7 June 1994 (Beth Dillon VF).

In June 1996 Least Flycatchers nested near Livermore (north of Fort Collins) in a riparian/grassland mixture with cottonwoods and willows. The birds built their nest only 6 feet (2 m) off the ground in a cottonwood (Dave Hallock pers. comm.).—H.E.K.

VERMILION FLYCATCHER
Pyrocephalus rubinus

In 1981 a Wyoming ornithologist, Helen Downing, visiting relatives on a farm near Akron, confirmed Colorado's first breeding Vermilion Flycatchers. Her relatives had first noticed the bird in their shelterbelt on 29 April, and she observed a pair there on 1 May. By early June the flycatchers had built a nest 10 feet (3 m) up in a Siberian elm tree near the farmhouse. A hailstorm killed the female on the nest; DMNH has nest, nestlings, and specimen in its collection (Downing 1981).

During the Atlas, at the park below the John Martin Dam (Hasty, 38102A8), a pair of these flycatchers built a nest that Chris Wood found on 12 May 1994. Later in May high winds blew the nest out of the tree; the female disappeared but the male, singing vigorously, remained through 25 May (Kingery 1994). —H.E.K.

WHITE-EYED VIREO
Vireo griseus

A pair of White-eyed Vireos tried to nest at Crow Valley campground, in the Pawnee National Grassland. After the first observation on 15 May 1992, they began building a nest on 17 May. The vireos placed their nest about 2 feet from the ground in a boxelder growing in a small stand of shrubs.

The next weekend, Memorial Day, several large groups of people choked the campground, including one contingent of 200; the total campground population may have reached 500 people each day. Many birdwatchers, bird photographers, and casual visitors passed by, visited, examined, or photographed the birds and the nest site. Nighttime temperatures dropped to 34°F (1°C) on 27 May and snow fell north of Crow Valley that evening. By 27 May the vireos abandoned their nesting attempt (Prather 1993). —H.E.K.

CAROLINA WREN
Thryothorus ludovicianus

The Carolina Wrens that Marty Poole first discovered in her Englewood backyard in December 1988 remained there through the winter. In April and May Poole heard one wren singing on a daily basis, and on 21 May 1989 confirmed Colorado's first breeding record. The wrens had built a nest on a shelf 3 feet (1 m) above the floor of an unused playhouse, in an alcove formed by a sliding door. Two nestlings had fledged by 30 May and the adults tended to them in a yard across the street (Kingery 1992b).

In Beulah a singing wren stayed for four years, 1992–1995, but never displayed additional breeding behavior or attracted a mate (Pat Flynn pers. comm.). —H.E.K.

GOLDEN-WINGED WARBLER
Vermivora chrysoptera

A male Golden-winged Warbler attended a fledgling, apparently also a Golden-winged Warbler, at Roxborough State Park on 17 July 1993. Observers had seen the male through the spring, although they never observed a female. Virginia's Warblers occur commonly in the vicinity, a mature scrub oak community. Observers speculated about possible hybridization (J.B. Hayes pers. comm., VF). —H.E.K.

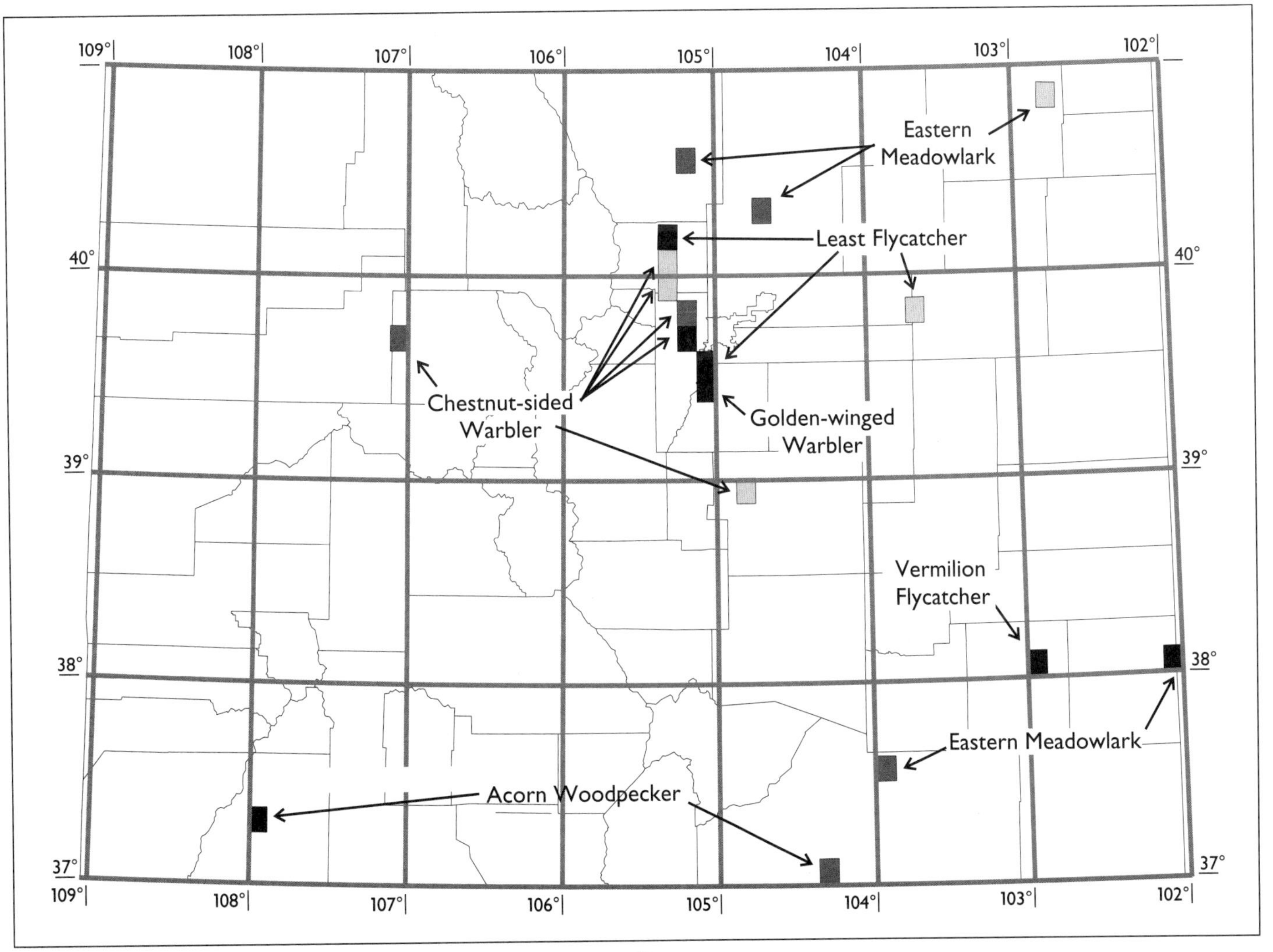
109°
108°
107°
106°
105°
104°
103°
102°
40°
39°
38°
37°
Eastern Meadowlark
Least Flycatcher
Chestnut-sided Warbler
Golden-winged Warbler
Vermilion Flycatcher
Eastern Meadowlark
Acorn Woodpecker

CHESTNUT-SIDED WARBLER
Dendroica pensylvanica

Colorado's first two breeding records of Chestnut-sided Warblers came in 1968 and 1975, from the edge of the foothills at Colorado Springs and Boulder. The Atlas database contains six records—one nest, three singing males, and two territorial males.

The one Atlas Confirmation occurred in Apex Gulch, a Jefferson County park 2 miles (3 km) south of Golden (Morrison, 39105F2). Clogged with alder and hawthorn thickets, the gulch cuts through the lower foothills. A female sat in a nest built in the fork of a small tree, with a male nearby, on 4 June 1991. Ten to 15 days later the nest was empty, the birds gone; the nest probably failed, although young birds might have hatched and fledged in the interim (David Gulbenkian pers. comm., nest card).

Five miles (8 km) to the northeast in the Wheat Ridge Greenbelt (Golden, 39105G2) a male sang on territory from 25 May to 30 June 1989 (Duane Nelson VF). Far to the west in an Eagle County block (Dotsero 3910761), Jack Merchant (VF) studied the behavior of a territorial male along the Eagle River from 15 June to 2 July 1994. The bird sang continuously during a ten-minute observation on 15 June. On 28 June and 2 July the Chestnut-sided followed a female Yellow Warbler that was feeding young in the nest, but Merchant never saw the Chestnut-sided feed the nestlings.

Atlasers recorded individual singing males in more typical foothills-edge blocks near Eldorado Springs, Boulder, and the Air Force Academy. —H.E.K.

ROSE-BREASTED GROSBEAK
Pheucticus ludovicianus

Rose-breasted Grosbeaks migrate regularly through the state. A pair nested in Longmont in 1894 (Cooke 1898). Another brought young to a feeder in Loveland in 1925 (B&N).

At her feeder in the foothills west of Littleton, Candy Vallardo (VF) observed Rose-breasted Grosbeaks in four of the years from 1988 to 1995. In 1992 a pair came regularly to the feeder. The male first appeared 3 July, the female on 5 August. Vallardo observed both feeding a fledgling; she last saw all three on 12 September 1992. The other Atlas records are pairs or singing males observed from 4 June to 4 July.

In 1996 (post-Atlas), along the Hanging Lake trail near Glenwood Springs, a "brilliantly attired adult male Rose-breasted Grosbeak," holding a green caterpillar in its bill, flew into a shrub with three fledgling grosbeaks. The fledglings perched unsteadily and begged noisily in a chokecherry bush. All of them then disappeared. Observers did not see a female (Kim Potter pers. comm., Potter 1997). —H.E.K.

EASTERN MEADOWLARK
Sturnella magna

Eastern Meadowlarks look almost identical to the abundant Western Meadowlarks, and each species may learn the song of the other (B&N, Lanyon 1995). Any identification and documentation should include the diagnostic *zert* call note of the Eastern Meadowlark. In Kansas they occur mainly in the eastern third of the state, whereas Western Meadowlarks, the Kansas state bird, predominate in the western two-thirds of Kansas (Thompson and Ely 1992). Even so, a few Easterns breed in eastern Colorado.

Eastern Meadowlarks prefer to breed in pastures, but also nest in fields, meadows, old orchards, and other open country. Where the ranges of Eastern and Western meadowlarks overlap, Easterns prefer river bottoms and more poorly drained grasslands, usually with taller grass, whereas Westerns prefer the uplands and better drained, prairie-type grasslands (Lanyon 1993). Excessive grazing reduces breeding density, because it reduces grassy cover needed for nesting (Roseberry and Klimstra 1970).

Colorado has had several records of territorial birds, but until 1992, no confirmations of breeding. Observers suspected but did not confirm breeding at Red Lion SWA, where a small population lived from 1975 or earlier until 1985. Then the fields were burned and the habitat altered, and the birds did not appear again until 1990 (A&R). Atlasers reported a territorial Eastern Meadowlark from one block west of Fort Collins from 12 June to 25 July, 1995 (Horsetooth Res., 40105E2).

Hamet SWA, on the Kansas line east of Holly, where the Arkansas River flows into Kansas. It has extensive tall grass fields that dip toward the tamarisk thickets and cottonwoods of the river. Observers reported Eastern Meadowlarks there starting in June 1992 (Duane Nelson, Mark Janos VF). On 5 July 1992, Joe Roller found two birds: an adult giving the *zert* call and a young bird, barely able to fly. The *zert*-ing bird flew after the young bird and Roller twice observed the adult feed an insect to the young one. One Eastern Meadowlark was at the site on 13 May 1993 (HEK). —David Pantle

PROBABLE BREEDING SPECIES

During the Atlas several out-of-range species maintained territories or remained at the same site without any Confirmation of nesting. Colorado has no confirmed breeding records of any for these species.

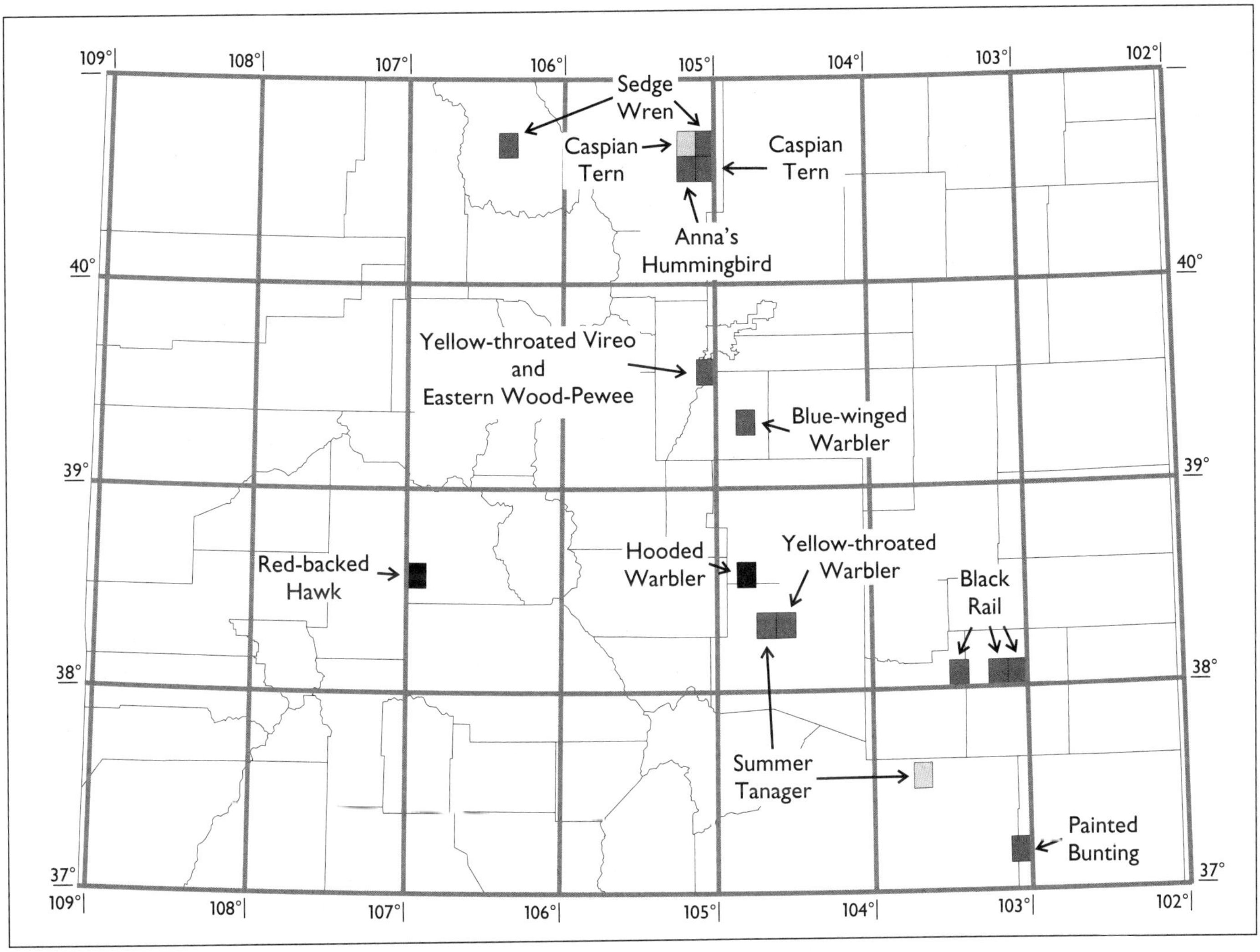

109°
108°
107°
106°
105°
104°
103°
102°
40°
39°
38°
37°
Sedge Wren
Caspian Tern
Caspian Tern
Anna's Hummingbird
Yellow-throated Vireo and Eastern Wood-Pewee
Blue-winged Warbler
Red-backed Hawk
Hooded Warbler
Yellow-throated Warbler
Black Rail
Summer Tanager
Painted Bunting

RED-BACKED HAWK

Buteo polyosoma

The presence of a Red-backed Hawk in Colorado over eight consecutive breeding seasons (1987–1994) presents an ecological and behavioral anomaly that we may never fully understand. The Red-backed Hawk typically resides from Cape Horn north to the central Colombian Andes. Meyer de Schaunesee (1970) described the species as non-migratory.

On 31 August 1987 Mark Daily discovered a Red-backed Hawk, seemingly paired with a Swainson's Hawk, on the Allen Ranch north of Gunnison. In 1988 a Red-backed Hawk was again observed on the ranch, this time nesting with a Swainson's Hawk. Observations of plumage and behavior made at this time support identification of the Red-backed Hawk as a female and the Swainson's as a male (Ron Meyer pers. comm.).

Biologists from the Peregrine Recovery Program banded the lone nestling on 13 August 1988 and the adult Swainson's Hawk on 14 August 1988. This nestling fledged, with its plumage described as "like a Swainson's" (Kingery 1988). This bird is widely regarded as a hybrid produced by the Swainson's/Red-backed Hawk pair, although no definitive evidence supports this conclusion. Observers saw what is probably the same pair in all subsequent years through 1994, although they did not confirm any additional nesting attempts (Ron Meyer pers. comm., Kingery 1988–1993).

This bird poses two breeding mysteries. First, did it mate with the Swainson's or did it pair up after the Swainson's male lost his first mate? Second, can it be classified as a wild bird (Wheeler 1988) or did it escape from captivity (Allen 1988)? The CFO records committee decided not to add the species to the state list because of the latter question (Nelson 1991b). —Frank J. Hein

RUFFED GROUSE
Bonasa umbellus

Colorado's first confirmed Ruffed Grouse record came on 24 October 1988: one killed during hunting season, between the Utah line and the west boundary of Dinosaur National Monument (Webb and Reddall 1989). A search of that area on 27 September 1991 by a CFO search party discovered only one Ruffed Grouse among several Blue Grouse (Nelson 1992). Subsequent observers have not reported any.

A juvenile was collected at this site in October 1990. The specimen, now at the Denver Museum of Natural History, has no accompanying data, except that published in Braun et al. (1994).

Ruffed Grouse juveniles disperse no more than a mile or so from their natal sites, but by October this juvenile could have moved into the state from Utah, only a half-mile away. The October date of the specimen and the location provide probable but not definite breeding in Colorado.

BLACK RAIL
Laterallus jamaicensis

At Bent's Old Fort National Historic Site near Las Animas, Dan Bridges on 7 May 1991 heard a Black Rail calling from a 40-acre (16-ha) cattail marsh. Through 22 June he and other observers heard and saw one rail and thought they heard two (Bridges 1992c). The calling behavior, however, purportedly corresponds with that of unmated males in the East (Bill Howe pers. comm.). Since 1991 observers have located Black Rails in this and several other marshes in the Arkansas Valley, mainly in Bent County.

Bridges spent several hours trying to confirm breeding, without success, in both 1991 and subsequent years. The Atlas therefore records this species as Territorial, but unconfirmed as a Colorado breeding species.

CASPIAN TERN
Sterna caspia

A pair, seen continuously along the Poudre River and adjacent wetlands in Fort Collins from 21 May to 9 July 1992, never revealed a nest site even though they displayed courtship behavior in May and carried sticks (Bill Howe, Dave Leatherman VFs; A&R).

ANNA'S HUMMINGBIRD
Calypte anna

An immature male came to David Steingraeber's feeder in the foothills west of Fort Collins from 9 July into August 1990; in August it began to molt into brighter plumage (Dave Leatherman VF, photograph). Since males acquire adult plumage three to four months after hatching, this was a hatch-year bird. How far first-year birds may wander is not known and the closest nesting site is several hundred miles away in southeastern Arizona (Russell 1996). Because Anna's breed in the winter in California, this bird was, probably, a vagrant (Van Remsen pers. comm.). The state has a handful of other records.

EASTERN WOOD-PEWEE
Contopus virens

In a cottonwood stream bottom along Plum Creek in the Littleton Atlas block in Chatfield State Park (39105E1), an Eastern Wood-Pewee sang its distinctive *Pee-a-Wee* for three weeks, 25 May to 18 June 1991. The pewee patrolled a circle of about 100 yards (90 m). Observers saw neither a second bird nor evidence of nesting (HEK, VF).

YELLOW-THROATED VIREO
Vireo flavifrons

Single Yellow-throated Vireos maintained territories for three years at Chatfield State Park near Denver. A male sang on territory from 22 May to 18 June 1990, 13 May into June 1991, and in 1992 (Kingery 1990, 1991a, 1992).

SEDGE WREN
Cistothorus platensis

The state has three records of territorial behavior by these wrens. Singing males displayed territorial behavior at Lower Latham Reservoir from 25 to 30 May 1969 (A&R), a behavior they exhibit in Louisiana prior to spring migration (Van Remsen pers. comm.).

At Arapaho NWR during the Atlas, Bill Howe (CFO RBF) observed a Sedge Wren that sang on territory from 15 June to 11 July 1988, and which even carried a blade of grass as if to a nest. Neither he nor other observers saw a mate and they did not confirm breeding (Kingery 1988c). Another Sedge Wren sang and fed in a wet meadow near Wellington, Larimer County, from 15 to 27 June 1993 (Steve Martin, Dave Leatherman VF, photo; Kingery 1993).

BENDIRE'S THRASHER
Toxostoma bendirei

Colorado birders have reported Bendire's Thrashers from the San Luis Valley for several years. Atlas observers studied possible Bendire's in several sites in the Valley, including two locations where Bendire's Thrashers had been reported previously. The thrashers that the atlasers found lacked the appropriate field marks of Bendire's Thrashers (see Kaufman 1990), because all had breast striping to some degree or another. The observers concluded that these birds better fit the description of Sage Thrashers than Bendire's (Rawinski et al. 1995, Bob Andrews, Bob Righter, Urling Kingery, HEK, pers. comm. and VF). The Atlas Technical Committee agrees.

Accordingly the Colorado Atlas does not accept Bendire's Thrashers as a verified breeding bird in the state. Possibly future observers will confirm their presence, but this will require either specimens or better field tools to establish the validity of this species as a Colorado breeding bird.

BLUE-WINGED WARBLER
Vermivora pinus

Blue-winged Warblers returned to the same site in Castlewood State Park from 1991 to 1993. The site features scattered thickets of scrub oak and willows and 1–3-foot forbs along both sides of Cherry Creek. A pair was present from 4 to 5 June 1991, and then a vigorously singing male stayed from 6 to 19 June. In 1992 and 1993 a male returned and sang at the same site. In that time no one substantiated an actual nesting attempt (Kingery 1991b, 1992, 1993).

YELLOW-THROATED WARBLER
Dendroica dominica

City Park in Pueblo attracted a Yellow-throated Warbler that sang from the same area from 9 to 28 May 1992; the same bird or another appeared 3 miles (5 km) away and sang regularly from 3 to 11 June (Mark Janos, CFO RBR).

HOODED WARBLER
Wilsonia citrina

Colorado observers have two enigmatic breeding records of Hooded Warblers: an unmated female with a brood patch, and a pair that laid eggs, abandoned the nest, but two weeks later fed fledglings.

The banding station on Fort Carson southwest of Colorado Springs (Cheyenne Mtn., 38104F7) repeatedly netted a female Hooded Warbler from 16 May to 17 June 1994. On 23 May she carried spiderwebs in her bill. As the weeks progressed, bander Susan Craig (VF) observed the warbler develop a brood patch. Craig saw no sign of a male, nor of any nest or young (Kingery 1994).

In post-Atlas 1998, a pair of these warblers nested in Gregory Canyon west of Boulder (40105A3). From 4 to 31 July the nest had two warbler eggs and a cowbird egg. The female abandoned the nest in mid July, yet on 31 July a female Hooded Warbler fed two dependent, recently fledged, warblers (Gent 1998, Peter Gent pers. comm, RBF).

SUMMER TANAGER
Piranga rubra

Observers found Summer Tanagers, exhibiting limited breeding behavior, in three locations during the Atlas period. At Olive Marsh in Pueblo a male sang and showed Agitated behavior on 6 June 1989 and a pair appeared briefly on 9 July 1989 (Dave Silverman VF). Others sang on territory in the canyon country of Las Animas County on 16 June 1993 and 10 July 1994 (Aaron Ellingson VF).

PAINTED BUNTING
Passerina ciris

Observers studied a singing male Painted Bunting from 28 May to 12 June 1994 in a shrubby draw in Cottonwood Canyon below Carrizo campground, Comanche National Grassland (Carrizo Mtn., 37103B1). Observers have found singing Painted Buntings in the same locality from 1995 to 1997.

HISTORICAL NESTING SPECIES

During the Atlas, observers did not Confirm breeding by the following species for which the state has nesting records. Three seem dubious or poorly documented. (See previous section for Hooded Warbler [breeding in Colorado verified in 1998].)

LEAST BITTERN
Ixobrychus exilis

Colorado has eight records of breeding for these secretive bitterns. B&N list six from the Denver-Boulder-Greeley area and one from Prewitt Reservoir near Sterling. In addition, near Boulder on 2 July 1979, Horst Droger discovered a nest with five eggs, perched in cattails over water 2 feet deep. A photograph taken on 21 July showed an adult with five young (Droger 1979). The Atlas has two Possible records, from LaSalle (40104C6) and Hygiene (40105B2).

YELLOW-CROWNED NIGHT-HERON
Nyctanassa violacea

One pair of Yellow-crowned Night-Herons nested for two to three years on an island in Denver's City Park lake, within a nesting colony of Black-crowned Night-Herons. In 1983 they produced one chick (Kingery 1983). In 1984 and 1985 observers saw one bird on a nest, but did not see a second one in the colony (Kingery 1984, 1985). The state also has a number of records of immature, but mobile, Yellow-crowneds. The Atlas has no records.

HARLEQUIN DUCK
Histrionicus histrionicus

Early reports had Harlequin Ducks breeding in Grand and Summit counties before 1900, and possibly in San Juan and La Plata counties. Three specimens taken by Edwin Carter bear dates of 15 May 1875 and 21 May 1876. B&N doubted the validity of this duck as a nesting species because Carter, an avid egg collector, did not have any sets from this flashy duck in his collection. The Atlas has no records.

HOODED MERGANSER
Lophodytes cucullatus

B&N did not regard the breeding of this species as documented adequately, and discounted the nineteenth-century reports of breeding. During the Atlas, however, ducks of this species summered in Englewood for several years. In 1996, on 29 April, Cat Anderson observed a female with five ducklings swimming on a pond next to the Highline Canal. The next winter, as she cleaned a nest box at the pond, she found four eggs—two of Wood Duck and two of Hooded Merganser. These two observations at last verify breeding by Hooded Mergansers in Colorado (Kingery in press).

BROAD-WINGED HAWK
Buteo platypterus

A pair reportedly nested in Grandview Cemetery in Fort Collins in 1978 (Gent 1987). No written information substantiates this report, which originated as a verbal report to the CFO Records Committee (Peter Gent pers. comm.). Without supporting documentation, the CFO Records Committee and the Atlas deem it best to categorize Broad-winged Hawk as a hypothetical Colorado breeding species (PG, MJ, HEK). The Atlas has one record in the Possible category, of a bird seen in June 1995 south of Colorado Springs.

MARBLED GODWIT
Limosa fedoa

Lois Webster found a nest with four eggs on the Pawnee Grassland on 28 May 1984; she followed its progress from two eggs to four eggs to broken eggshells evidencing predation by a fox or coyote (Kingery 1984). No one has reported possible breeding godwits before or since her discovery. The Atlas has no records other than migrants.

RING-BILLED GULL
Larus delawarensis

Cooke (1900) said he found these gulls "breeding quite commonly June 18, 1898, at the San Luis Lakes." This, and a report from F.M. Drew (1885), present a perplexing dilemma, inasmuch as no observers since then have mentioned Ring-billed Gulls as Colorado breeders. These gulls are now widespread in the state throughout the year. It seems remarkable that they would breed in great numbers yet only one observer would report them. Subsequent events have probably rendered the site unsuitable for breeding. A dam has augmented the size of San Luis Lake (and turned it into one lake), and may have flooded the nest site. The lake also hosted a fishing and hunting club in the early 1900s, possibly intensifying disturbances and its unsuitability for nesting gulls.

MAGNIFICENT HUMMINGBIRD
Eugenes fulgens

In 1965 the Denver Museum of Natural History collected a female and two eggs from a nest west of Boulder (B&N), the only Colorado breeding record. The Atlas has one report: 3-6 visited a feeder 15 July 1994 at Electra Lake north of Durango (Caye Geer pers. comm.).

NORTHERN PARULA
Parula americana

A female built a nest near Red Rocks Park west of Denver in 1982. Observed from 11 to 16 June, she attended the nest until a storm destroyed it; the nest had no eggs and no male was present (A&R). The Atlas has no records other than migrants.

BAY-BREASTED WARBLER
Dendroica castanea

An Aiken Audubon Society trip to Westcreek, Douglas County, on 24 June 1978 discovered a pair of Bay-breasted Warblers, and participants observed the female carrying nesting material into a blue spruce. A month later, on 24 July 1978, Roberta and Vincent Winn twice observed a male Bay-breasted Warbler collecting insects and flying off with them. They subsequently observed two fledglings, almost certainly Bay-breasteds, although they did not see adults feed them (Winn 1979). The Atlas has no records other than migrants.

APPENDIX A: BLOCK STATISTICS

EXPLANATION

Species per block: Number of species recorded by type of breeding code, plus total species in block. This list includes Border Blocks not included in Block Statistics (Map 1 and Figure 2). PO=Possible, PR=Probable, CF=Confirmed.

CMPL: An X indicates block completed.

Alternate: indicated sector substituted as Priority block (instead of SE sector).

BB: A paid field worker surveyed these blocks, wholly or partially.

Quadcode	Map Name	Observer	Species per Block PO	PR	CF	Total	CMPL	Alternate	BB
Q3710211	Midway S.E.	CFO	20	4	14	38			
Q3710212	Midway S.W.	Nelson	14	9	5	28		SW	x
Q3710213	Moore Draw S.E.	Giesen	2	7	10	19	X		
Q3710214	Moore Draw S.W.	Giesen	7	11	15	33	X		
Q3710215	Campo	Giesen/Versaw	7	12	11	30	X		x
Q3710216	Campo S.W.	Versaw	6	9	7	22	X		x
Q3710217	Tubs Springs	Stv Jones	7	19	21	47	X		
Q3710218	Big Hole Canyon	Stv Jones	12	18	27	57	X		
Q3710222	Midway	Nelson	2	7	12	21	X		x
Q3710223	Moore Draw N.E.	Nelson	1	6	10	17	X		x
Q3710224	Moore Draw N.W.	Nelson	0	9	14	23	X		x
Q3710225	Campo N.E.	Nelson	3	3	9	15	X		x
Q3710226	Campo N.W.	Nelson	2	9	11	22	X		x
Q3710227	Edler	Stv Jones	5	5	15	25	X		
Q3710228	Reader Lake	Stv Jones	4	6	14	24	X		
Q3710232	Stonington	Thompson	9	8	21	38	X		
Q3710233	Walsh S.E.	Thompson	3	10	22	35	X		
Q3710234	Vilas South	Thompson	7	2	7	16	X		
Q3710235	Bisonte	Versaw	8	5	9	22	X		x
Q3710236	Springfield S.W.	Versaw	9	6	5	20	X		x
Q3710237	Pritchett	Maguire	6	6	13	25	X		
Q3710238	Lone Rock	Leatherman/Yaeger	10	6	10	26	X		x
Q3710241	Saunders	Thompson	4	13	22	39	X		
Q3710242	Bartlett	Thompson	10	5	20	35	X		
Q3710243	Walsh	Thompson	16	22	33	71	X		
Q3710244	Vilas North	Thompson	6	6	24	36	X		
Q3710245	Springfield East	Yaeger	6	8	11	25	X		x
Q3710246	Springfield West	Thompson	10	7	22	39	X		
Q3710247	Harbord	Yaeger	7	9	7	23	X		x
Q3710248	Pritchett N.W.	Slater/Yaeger	10	3	8	21	X		x
Q3710252	Lycan	Rauch/Maguire	13	2	15	30	X		
Q3710253	Two Buttes S.E.	Maguire/Thompson	5	7	23	35	X		
Q3710254	Two Buttes	Thompson	10	6	19	35	X		
Q3710255	Horse Creek Springs	Lentz	15	9	13	37	X		
Q3710256	Big Rock Grange	Yaeger	4	6	7	17	X		x
Q3710257	McEndree Ranch	Yaeger	6	7	14	27	X		x
Q3710258	Deora	Dillon/Wilson	11	8	11	30	X		
Q3710262	Webb	Thompson/Maguire	12	7	13	32	X		
Q3710263	Plains Community	Lentz/Thompson	14	11	22	47	X		
Q3710264	Two Buttes N.W.	Nelson	1	11	14	26	X		x
Q3710265	Two Buttes Reservoir	Thompson/CFO	27	17	24	68	X		
Q3710266	Hasser Ranch	Nelson	8	7	21	36	X		x
Q3710267	Floating W Ranch	Maguire	6	14	11	31	X		
Q3710268	Pipe Spring	Yaeger	13	6	9	28	X		x
Q3710272	Two Butte Springs	Nelson	2	4	10	16	X		x
Q3710273	North Plum Creek S.E.	Righter/Pratt	14	17	12	43	X		
Q3710274	Barrel Spring	Nelson	2	4	6	12	X		x
Q3710275	Gobblers Knob	Nelson	1	5	11	17	X		x

Quadcode	Map Name	Observer	Species per Block PO	PR	CF	Total	CMPL	Alternate	BB
Q3710276	Cat Creek	Cairo/Lentz	8	7	16	31	X		
Q3710277	Dripping Spring	Yaeger	10	3	5	18	X		x
Q3710278	Hand Springs	Dillon/Yaeger	7	7	7	21	X		x
Q3710282	Durkee Creek N.W.	Sleeper/Nelson	4	8	13	25	X		x
Q3710283	North Plum Creek N.E.	Nelson	4	8	15	27	X		x
Q3710284	North Plum Creek N.W.	Nelson	1	3	16	20	X		x
Q3710285	Cat Creek N.E.	Wright	5	0	8	13	X		x
Q3710286	Cat Creek N.W.	Wright	9	8	8	25	X		x
Q3710287	Denny Lake	Wright/Kingery	4	1	3	8	X		x
Q3710288	High Rock	Kingery	10	6	13	29	X		
Q3710311	Furnish Canyon East	Stv Jones	19	14	26	59	X		
Q3710312	Furnish Canyon West	Stv Jones	13	14	21	48	X		
Q3710313	Dennis Canyon	Stv Jones	14	9	19	42	X		
Q3710314	Jesus Canyon	Stv Jones/Opler	18	18	19	55	X		
Q3710315	Cobert Mesa North	Stv Jones/Kingerys	22	20	12	54	X		
Q3710316	Pine Canyon	Bannon/Kingery	12	15	24	51	X		
Q3710317	Branson S.E.	Dillon/Pratt	16	13	22	51	X		
Q3710318	Branson	Lentz/Righter	21	8	19	48	X		
Q3710321	Carrizo Mountain	BCNA	14	26	38	78	X		
Q3710322	Pintada Creek	Stv Jones/BCNA	19	20	32	71	X		
Q3710323	Kim South	Stv Jones	11	16	17	44	X		
Q3710324	Dalerose Mesa	Graves/Dennis	16	18	19	53	X		
Q3710325	Tobe	CFO	10	4	11	25	X		
Q3710328	Trementina Canyon	Opler/Kingery	15	5	4	24			
Q3710331	Utleyville	Versaw	4	7	11	22	X		x
Q3710332	Andrix	Stv Jones	5	10	9	24	X		
Q3710333	Kim North	Bachant/ Stv Jones	13	12	22	47	X		
Q3710334	Cherry Canyon	Dick/Cushman/Jones	14	17	12	43	X		
Q3710335	Villegreen	Versaw	4	3	6	13	X		x
Q3710336	Humbar Spring	Dillon	25	8	20	53	X		x
Q3710337	Doss Canyon South	Graves/Dillon	16	9	22	47	X		x
Q3710338	Painted Canyon	Kingerys	16	15	17	48	X		
Q3710341	Table Mesa	Versaw/Wright	11	3	10	24	X		x
Q3710342	Buck Canyon	Munshi	14	13	29	56	X		x
Q3710343	Robbers Roost Canyon	Munshi	7	5	15	27	X		
Q3710344	Icehouse Canyon	Graves	1	13	13	27	X		
Q3710345	Plum Canyon	Finch	5	18	6	29	X		
Q3710346	Johnson Canyon	Graves	7	12	9	28	X		
Q3710347	Doss Canyon North	Graves	10	16	9	35	X		
Q3710348	Rock Crossing	Youkey/Truan	6	7	51	64	X		
Q3710351	Walker Canyon	Dillon	18	12	11	41	X		
Q3710352	Plug Hat Ranch	Dillon/Wright	12	5	8	25	X		x
Q3710353	Brown Canyon	Kingery	18	10	15	43	X		
Q3710354	Lost Canyon	Graves	0	17	16	33	X		
Q3710355	Beaty Canyon	Janos/Finch	22	17	18	57	X		
Q3710356	O V Mesa	Finch	15	22	18	55	X		
Q3710357	Stage Canyon	Youkey/Truan	12	9	49	70	X		
Q3710358	Lockwood Arroyo	Youkey/Truan	13	6	12	31	X		
Q3710361	Clay Ranch	Yaeger	4	8	10	22	X		x
Q3710362	Ninaview	Munshi	14	13	4	31	X		
Q3710363	Rock Canyon	Bachant	17	13	15	45	X		
Q3710364	Corbin Canyon	Graves	0	8	13	21	X		
Q3710365	Riley Canyon	Versaw/Jones	20	13	26	59	X	SW	x
Q3710366	Packers Gap	Youkey/Truan/Versaw	12	18	18	48	X		
Q3710367	Sheep Canyon	Truan/Versaw	13	5	14	32	X		x
Q3710368	Bloom	Truan/Kingery	6	12	10	28	X		
Q3710371	Toonerville S.E.	Graves	10	4	3	17			
Q3710372	Toonerville	Graves	3	4	6	13			

Quadcode	Map Name	Observer	Species per Block				CMPL	Alternate	BB
			PO	PR	CF	Total			
Q3710373	Turkey Canyon	Gustafson/Dillon	11	6	18	35	X		x
Q3710374	Higbee	Graves/Dillon	13	20	21	54	X		x
Q3710375	La Junta S.E.	Graves/Versaw/Dillon	30	12	5	47	X		x
Q3710376	La Junta S.W.	Graves	11	6	8	25	X		
Q3710377	Timpas	Graves	4	9	1	14			
Q3710378	Timpas S.W.	Graves	1	8	0	9			
Q3710381	Toonerville N.E.	Kingery/Pratt	7	5	4	16	X	SW	
Q3710382	Gilpin	Truan/Kingery/Pratt	8	9	7	24	X		
Q3710383	Hackamore Ranch	Kingerys	13	16	16	45	X		
Q3710384	Thompson Arroyo	Locke/Dillon	13	8	17	38	X		x
Q3710385	La Junta	Graves	10	3	6	19	X		
Q3710386	Hawley	Kingery/Pratt	2	4	8	14	X		
Q3710387	Timpas N.E.	Mollhoff/Graves	4	9	15	28	X		
Q3710388	Timpas N.W.	Kingery	5	0	2	7	X		
Q3710411	Trinchera	D Johnson	15	4	14	33	X		
Q3710412	Abeyta	Graves	5	13	16	34	X		
Q3710413	Barela	Yaeger	25	10	24	59	X		x
Q3710414	Fishers Peak	Yaeger	36	10	14	60	X		x
Q3710415	Starkville	Kingery/Wolff/CFO	23	22	15	60	X		
Q3710416	Valdez	Yaeger	29	10	19	58	X		x
Q3710417	Little Pine Canyon	Yaeger	27	7	19	53	X		x
Q3710418	Tercio	Yaeger	23	5	22	50	X		x
Q3710421	Trinchera Cave	D Johnson	20	14	20	54	X		
Q3710422	Patterson Crossing	Yaeger	12	6	23	41	X		x
Q3710423	Mooney Hills	Truan/Yaeger	11	12	10	33	X		x
Q3710424	Trinidad East	D Johnson	10	17	20	47	X		
Q3710425	Trinidad West	Graves/Yaeger	39	10	26	75	X		x
Q3710426	Madrid	Graves	12	9	15	36	X		
Q3710427	Weston	Yaegers	8	11	30	49	X		
Q3710428	Vigil	Yaegers	19	7	26	52	X		
Q3710431	Lambing Spring	Yaeger	19	7	20	46	X		x
Q3710432	Model	Yaeger	3	2	8	13	X		x
Q3710433	Earl	D Johnson	25	9	24	58	X		
Q3710434	Hoehne	D Johnson	18	8	23	49	X		
Q3710435	Ludlow	Yaeger	21	7	15	43	X		
Q3710436	Delagua	Yaeger	20	8	12	40	X		
Q3710437	Gulnare	Crosby/Kempf	24	4	24	52	X		
Q3710438	Herlick Canyon	Crosby/Johnson	36	9	17	62	X		
Q3710441	Brown Sheep Camp	Youkey/Truan	6	8	10	24	X		
Q3710442	Tyrone	Youkey/Truan	9	9	14	32	X		
Q3710443	Seven Lakes Res.	Yaeger	6	6	13	25	X		x
Q3710444	Vega Corral	Yaeger	9	4	9	22	X		x
Q3710445	The Hogback	Nelson	6	15	25	46	X		x
Q3710446	Aguilar	Schreier	13	21	19	53	X		
Q3710447	Santa Clara	Nelson	1	14	17	32	X		x
Q3710448	Spanish Peaks	Ketchen	21	8	11	40	X		
Q3710451	Thatcher	Youkey/Truan	8	6	32	46	X		
Q3710452	Bates Lake	Mollhoff/Truan	5	6	13	24	X		
Q3710453	Hidden Valley Ranch	Watts	10	10	12	32	X		
Q3710454	Little Dome	Sandstrom-Smith	9	8	21	38	X		
Q3710455	Pryor S.E.	Roth	15	13	8	36	X		
Q3710456	Pryor	Nelson	3	12	16	31	X		x
Q3710457	Walsenburg South	Nelson	6	14	23	43	X		x
Q3710458	Ritter Arroyo	Ketchen	17	2	4	23	X		
Q3710461	Delhi	Mollhoff	11	11	19	41	X		
Q3710462	Sun Valley Ranch	Stigen	12	7	8	27	X		x
Q3710463	Jones Lake Spring	Sandstrom-Smith	14	7	15	36	X		
Q3710464	So. Rattlesnake Butte	Yaeger	8	1	4	13	X		

Quadcode	Map Name	Observer	Species per Block PO	PR	CF	Total	CMPL	Alternate	BB
Q3710465	Cucharas Reservoir	Roth	14	16	9	39	X		
Q3710466	Maria Reservoir	Nelson	6	10	25	41	X		x
Q3710467	Walsenburg North	Tucey/Leathermn	12	16	16	44	X		
Q3710468	Black Hills	Yaeger	9	10	9	28	X		
Q3710471	Snowden Lake	Stigen	2	3	3	8	X		x
Q3710472	Sanford Hills	D Johnson	12	8	21	41	X		
Q3710473	Myers Canyon	Stigen	18	11	10	39	X		x
Q3710474	No. Rattlesnake Butte	Yaeger	3	1	3	7	X		
Q3710475	Capps Springs	Stigen	20	12	12	44	X		x
Q3710476	Lascar	Silverman	10	1	12	23	X		
Q3710477	Huerfano Butte	D Johnson	12	13	25	50	X		
Q3710478	Hayden Butte	Nelson	2	1	7	10	X		x
Q3710481	Apishapa Bridge	D Johnson	8	8	13	29	X		
Q3710482	Yellowbank Creek	Stigen	3	1	3	7	X		x
Q3710483	Red Top Ranch	Silverman	12	5	11	28	X		
Q3710485	Cedarwood	D Johnson	17	7	19	43	X		
Q3710486	Graneros Flats	Nelson	3	4	8	15	X	SW	x
Q3710487	Colorado City	F Harrison	12	6	7	25	X		
Q3710488	Rye	Silverman	9	18	43	70	X		
Q3710511	Torres	Bridges	10	7	17	34	X		
Q3710512	Culebra Peak	Bridges	6	9	18	33	X		
Q3710513	La Valley	Bridges	12	6	23	41	X		
Q3710514	Sanchez Reservoir	Bridges	14	8	26	48	X		
Q3710515	Garcia	Kingerys	27	16	11	54	X		
Q3710516	Sky Valley Ranch	Swift	5	7	16	28	X		
Q3710517	Kiowa Hill	Kingerys	9	4	15	28	X		
Q3710518	Lobatos	Pantles	6	10	10	26	X		
Q3710521	Stonewall	Gulbenkian	17	32	13	62	X		
Q3710522	El Valle Creek	Gulbenkian	8	8	7	23	X		
Q3710523	Taylor Ranch	Wallace	6	32	16	54	X		
Q3710524	San Luis	Graves	15	7	15	37	X		
Q3710525	San Acacio	Graves/Kingerys	13	5	7	25	X		
Q3710526	Mesito Reservoir	Kingerys	3	0	3	6	X		
Q3710527	Manassa N.E.	Levad	14	3	4	21	X		x
Q3710528	Manassa	Pantles	15	16	27	58	X		
Q3710531	Cucharas Pass	Ketchen	16	7	16	39	X		
Q3710532	Trinchera Peak	Ketchen	12	5	6	23	X		
Q3710533	Ojito Peak	Wallace	30	10	16	56	X		
Q3710534	Fort Garland S.W.	Graves	15	12	6	33	X		
Q3710535	Blanca S.E.	Levad	5	1	1	7	X		x
Q3710536	Lasauses	Levad	1	4	2	7	X		x
Q3710537	Pikes Stockade	Levad	20	12	9	41	X		x
Q3710538	La Jara	Darnell	12	2	13	27	X		
Q3710541	Cuchara	Yaeger	17	5	24	46	X		x
Q3710542	McCarty Park	Gulbenkian	11	16	6	33	X		
Q3710543	Trinchera Ranch	Wallace	17	24	10	51	X		
Q3710544	Fort Garland	Graves/Levad	26	12	21	59	X		x
Q3710545	Blanca	Levad	28	16	26	70	X		x
Q3710546	Baldy	Levad	5	2	1	8	X		x
Q3710547	Alamosa East	Morkill	16	8	45	69	X		
Q3710548	Alamosa West	Darnell	12	6	14	32	X		
Q3710551	La Veta	Yaeger	16	4	26	46	X		
Q3710552	La Veta Pass	Fisher	27	5	20	52	X		
Q3710553	Russell	Fisher	13	15	11	39	X		
Q3710554	Blanca Peak	Knapp/Plaska	32	18	12	62	X		
Q3710555	Twin Peaks	Levad	30	5	7	42	X		x
Q3710556	Dry Lakes	Navo	4	5	2	11			
Q3710557	Hooper S.E.	Brandt	1	9	32	42	X		

Quadcode	Map Name	Observer	Species per Block				CMPL	Alternate	BB
			PO	PR	CF	Total			
Q3710558	Mount Pleasant School	Levad	20	12	18	50	X		x
Q3710561	Farisita	Silverman	16	14	21	51	X		
Q3710562	Little Sheep Mtn.	Yaeger	20	8	25	53	X		x
Q3710563	Red Wing	Yaeger	15	4	6	25	X		x
Q3710564	Mosca Pass	Yaeger	20	7	23	50	X		x
Q3710565	Zapata Ranch	Rawinski	16	22	6	44	X		
Q3710566	Medano Ranch	Schrupp/Levad	2	7	6	15	X		
Q3710567	Hooper East	Ketchen	11	6	3	20	X		
Q3710568	Hooper West	Darnell/Levad	4	1	17	22	X		x
Q3710571	Badito Cone	D Johnson	25	11	7	43	X		
Q3710572	Gardner	Bright	23	11	9	43	X		
Q3710574	Medano Pass	Yaeger	28	12	13	53	X		x
Q3710575	Liberty	Ketchen/Kingerys	16	4	24	44	X		
Q3710576	Sand Camp	Ron Garcia	11	0	23	34	X		
Q3710577	Deadman Camp	Stigen	3	3	0	6	X		x
Q3710578	Deadman Camp S.W.	Kingerys	10	8	10	28	X		
Q3710581	San Isabel	Barber	12	9	6	27	X		
Q3710582	Bear Creek	Ketchen	17	7	13	37	X		
Q3710583	Devils Gulch	Ketchen	15	9	8	32	X		
Q3710584	Beck Mountain	Yaeger	22	7	15	44	X		x
Q3710585	Crestone Peak	Fisher	7	10	5	22			
Q3710586	Crestone	Levad	25	4	13	42	X		x
Q3710588	Moffat South	Gillihan	16	14	17	47	X	SW	
Q3710611	Antonito	Graves/Kingerys	22	11	15	48	X		
Q3710612	Fox Creek	Graves/Versaw	16	14	21	51	X		x
Q3710613	Osier	Bridges	15	5	26	46	X		
Q3710614	Cumbres	Staatz	21	2	17	40	X		
Q3710615	Archuleta Creek	Kingerys	24	8	19	51	X		
Q3710616	Chama Peak	Kingerys	24	11	30	65	X	CW	
Q3710617	Chromo	Kingerys/Boyle	22	4	31	57	X		x
Q3710618	Edith	Gables	17	13	19	49	X		
Q3710621	Goshawk Dam	Kingerys	11	7	24	42	X		
Q3710622	Vicente Canyon	Stigen/Kingerys	11	4	5	20	X		x
Q3710623	La Jara Canyon	Gllhan/Kingerys	35	9	6	50			
Q3710624	Spectacle Lake	Kingerys	25	11	25	61	X		
Q3710625	Victoria Lake	Ketchen	15	0	12	27	X		
Q3710626	Elephant Head Rock	Boyle	26	4	36	66	X		x
Q3710627	Harris Lake	Noss/Kingerys	22	10	29	61	X		
Q3710628	Serviceberry Mtn.	Versaw	26	17	24	67	X		x
Q3710631	Capulin	Stigen	16	6	4	26	X		x
Q3710632	Centro	Burkhart	25	1	10	36	X		
Q3710633	Terrace Reservoir	Burkhart/Kingerys	32	11	16	59	X		
Q3710634	Red Mountain	Rawinski	18	16	14	48	X		
Q3710635	Platoro	Woodwards	9	10	15	34	X		
Q3710636	Summit Peak	Kingerys	12	8	10	30	X		
Q3710637	Blackhead Peak	Roth	31	30	9	70	X		
Q3710638	Jackson Mountain	Roth	33	24	11	68	X		
Q3710641	Waverly	Darnell	12	3	16	31	X		
Q3710642	Fulcher Gulch	Burkhart	13	7	7	27	X		
Q3710643	Greenie Mountain	Burkhart/Levad	27	10	25	62	X		x
Q3710644	Jasper	Navo/Kingerys	20	15	7	42			
Q3710645	Summitville	Levad	24	3	15	42	X		x
Q3710646	Elwood Pass	Rawinski	8	9	5	22	X		
Q3710647	Wolf Creek Pass	Figgs/Versaw	19	28	23	70	X		x
Q3710648	Saddle Mountain	Figgs/Graves	25	39	28	92	X		
Q3710651	Homelake	Markley/Versaw	14	6	22	42	X		
Q3710652	Monte Vista	Versaw	9	3	9	21			x
Q3710653	Dog Mountain	Alves	19	17	11	47	X		

			Species per Block						
Quadcode	Map Name	Observer	PO	PR	CF	Total	CMPL	Alternate	BB
Q3710654	Horseshoe Mountain	Hallock	19	12	21	52	X		x
Q3710655	Del Norte Peak	Roth	29	8	7	44	X		
Q3710656	Beaver Creek Res.	Levad	20	5	9	34	X		x
Q3710657	Mount Hope	Figgs/Levad	15	9	14	38	X		x
Q3710658	South River Peak	Nelson	6	8	18	32	X		x
Q3710661	Center South	Stigen	6	3	2	11	X		x
Q3710662	Sevenmile Plaza	Stigen	9	3	5	17	X		x
Q3710663	Del Norte	Rawinski	32	20	19	71	X		
Q3710664	Indian Head	Levad	30	8	16	54	X		x
Q3710665	South Fork East	Versaw	20	7	22	49	X		x
Q3710666	South Fork West	Kenvin/Stigen	21	6	20	47	X		x
Q3710667	Lake Humphreys	Levad	18	3	5	26	X		x
Q3710668	Spar City	Nelson	3	8	14	25	X		x
Q3710671	Center North	Stigen	6	4	2	12	X		x
Q3710672	La Garita	Gillihan	5	3	9	17	X		
Q3710673	Twin Mountains S.E.	Kingery/Levad	7	1	14	22	X		x
Q3710674	Twin Mountains	Kingerys/Andrews/Righter	13	5	17	35	X		
Q3710675	Pine Cone Knob	Kingerys	26	15	29	70	X		
Q3710676	Pool Table Mtn.	Stigen	13	4	10	27	X		x
Q3710677	Wagon Wheel Gap	Stigen	19	9	15	43	X		x
Q3710678	Creede	Hallock	21	9	19	49	X		x
Q3710681	Harrence Lake	Kingerys	12	6	16	34	X		
Q3710682	Swede Corners	Kingerys	9	10	12	31	X		
Q3710683	Lime Creek	Yaeger	19	0	14	33	X		x
Q3710684	Lookout Mountain	Kingerys/Andrews/Righter	18	10	24	52	X		
Q3710685	Bowers Peak	Kingery/Andrews/Righter	16	15	7	38	X		
Q3710686	Mesa Mountain	Stigen	15	6	8	29	X	SW	x
Q3710687	Halfmoon Pass	Hernbrode	15	1	13	29	X		
Q3710688	San Luis Peak	Stigen	10	5	8	23	X		x
Q3710712	Pagosa Junction	Kingerys	31	7	28	66	X		
Q3710713	Carracas	Kingerys	26	16	25	67	X		
Q3710714	Allison	Kingerys/Leatherman	25	14	26	65	X		
Q3710715	Tiffany	Shryock/Levad	23	7	18	48	X		x
Q3710716	Ignacio	Kingerys	23	9	10	42		NW	
Q3710717	Bondad Hill	Kingerys	42	12	11	65	X	CW	
Q3710718	Long Mountain	Graves/Versaw/Kingerys	11	11	13	35			
Q3710721	Oak Brush Hill	Versaw/Kingerys/Levad	35	5	21	61	X		x
Q3710722	Lonetree Canyon	Kingerys/Levad	9	5	8	22			x
Q3710723	Chimney Rock	Versaw/Levad	21	4	17	42	X		x
Q3710724	Pargin Mountain	Levad	33	3	5	41	X		x
Q3710725	Bayfield	Kingerys	33	10	29	72	X		
Q3710726	Gem Village	Boyle	24	12	16	52	X		x
Q3710727	Loma Linda	Waters/Kingerys	35	9	29	73	X		
Q3710728	Basin Mountain	Levad	43	4	31	78	X		x
Q3710731	Pagosa Springs	Johnson/various	35	10	27	72	X		
Q3710732	Chris Mountain	Roth	22	15	12	49	X		
Q3710733	Devil Mountain	Kingerys	22	2	18	42	X		
Q3710734	Baldy Mountain	Kingerys	23	13	27	63	X		
Q3710735	Ludwig Mountain	Pantles	13	14	27	54	X		
Q3710736	Rules Hill	Graves/Levad	34	9	12	55	X		x
Q3710737	Durango East	Childress	13	15	20	48	X		
Q3710738	Durango West	Berry/Tomberlin/Gulbenkian	30	9	44	83	X		
Q3710741	Pagosa Peak	Kingerys/Clark	29	4	31	64	X		
Q3710742	Oakbrush Ridge	Levad	27	5	5	37	X		x
Q3710743	Bear Mountain	D Clark	26	8	4	38	X		
Q3710744	Granite Peak	D Clark	32	7	9	48	X		
Q3710745	Vallecito Reservoir	Rees	30	1	22	53	X		
Q3710746	Lemon Reservoir	Pantles	13	14	22	49	X		

Quadcode	Map Name	Observer	Species per Block PO	PR	CF	Total	CMPL	Alternate	BB
Q3710747	Hermosa	Noss/Levad	19	23	12	54	X		x
Q3710748	Monument Hill	Shields/Kingerys	14	8	21	43	X		
Q3710751	Palomino Mountain	Nelson	6	4	14	24	X		x
Q3710752	Cimarrona Peak	Figgs/Kingerys	32	7	24	63	X		
Q3710753	Granite Lake	Figgs/Kingerys/Reid	23	18	32	73	X		
Q3710754	Emerald Lake	Boyle	17	2	22	41	X		x
Q3710755	Columbine Pass	Parkers	5	3	25	33	X		
Q3710756	Mountain View Crest	Kingerys	17	1	18	36	X		
Q3710757	Electra Lake	Geer	11	15	38	64	X		
Q3710758	Elk Creek	K Potter	22	4	17	43	X		x
Q3710761	Workman Creek	Levad	25	6	15	46	X		x
Q3710762	Little Squaw Creek	Hill	11	5	10	26	X		
Q3710763	Weminuche Pass	Boyle	12	0	9	21	X		x
Q3710764	Rio Grande Pyramid	Boyle	13	0	16	29	X		x
Q3710765	Storm King Peak	Boyle	19	0	18	37	X		x
Q3710766	Snowdon Peak	Wuerthele	21	10	4	35	X		
Q3710767	Engineer Mountain	Geer	17	7	31	55	X		
Q3710768	Hermosa Peak	Axford/K Potter/Levad	19	3	27	49	X		
Q3710771	Bristol Head	Staatz	25	4	11	40	X		
Q3710772	Hermit Lakes	Staatz	34	12	17	63	X		
Q3710773	Finger Mesa	Hernbrode	39	10	26	75	X		
Q3710774	Pole Creek Mtn.	Blackburn/Gulbenkian	16	24	17	57	X		
Q3710775	Howardsville	Fagans/Simon	21	2	14	37	X		
Q3710776	Silverton	Parkers	6	8	29	43	X		
Q3710777	Ophir	Kingerys	10	5	17	32	X		
Q3710778	Mount Wilson	Hallock	17	14	26	57	X		
Q3710781	Baldy Cinco	Nelson	3	9	21	33	X		x
Q3710782	Slumgullion Pass	Hayes	27	4	24	55	X		
Q3710784	Redcloud Peak	Woodwards	6	11	24	41	X		
Q3710785	Handies Peak	Simon	8	5	8	21	X		
Q3710786	Ironton	Simon	12	7	9	28	X		
Q3710787	Telluride	Boyle	16	0	13	29	X		x
Q3710788	Gray Head	Hallock	18	16	33	67	X		
Q3710811	Pinkerton Mesa	Kingerys	26	6	12	44		CW	
Q3710812	Redmesa	Graves/Levad	25	11	26	62	X		x
Q3710813	Red Horse Gulch	Kingerys	21	10	0	31		NE	
Q3710814	Greasewood Canyon	Guadagno/Kingery	42	1	11	54		NW	
Q3710815	Moqui Canyon	Guadagno	12	7	13	32			
Q3710816	Tanner Mesa	Colyer	3	5	0	8			
Q3710817	Sentinel Peak S.E.	Kingery/Dexter	10	9	18	37	X		
Q3710818	Sentinel Peak S.W.	Kingery/Dexter	13	9	12	34	X		
Q3710821	Kline	Boyle	18	5	21	44	X		x
Q3710822	Mormon Reservoir	Boyle	19	9	25	53	X		x
Q3710823	Trail Canyon	Boyle	22	4	15	41	X		x
Q3710824	Moccasin Mesa	Guadagno	21	34	17	72	X		
Q3710825	Wetherill Mesa	Colyer	3	17	12	32			
Q3710826	Towaoc	Colyer/Guadagno	16	6	17	39			
Q3710827	Mariano Wash East	Guadagno	7	9	7	23			
Q3710828	Mariano Wash West	Colyer	0	5	0	5			
Q3710831	Hesperus	Parkers	16	11	35	62	X		
Q3710832	Thompson Park	Boyle	29	6	18	53	X		x
Q3710833	Mancos	Colyer	3	11	11	25			
Q3710834	Point Lookout	Colyer	7	8	31	46	X		
Q3710835	Cortez	Colyer/Atkinson	7	21	14	42	X		
Q3710836	Mud Creek	Fyler	6	8	24	38	X		
Q3710837	Battle Rock	Dexter	25	9	23	57	X		
Q3710838	Bowdish Canyon	Versaw	10	9	7	26			
Q3710841	La Plata	Versaw	13	5	8	26			

Quadcode	Map Name	Observer	Species per Block				CMPL	Alternate	BB
			PO	PR	CF	Total			
Q3710842	Rampart Hills	Colyer/Versaw/Fyler	13	5	21	39	X		
Q3710843	Millwood	Colyer	15	23	42	80	X		
Q3710844	Dolores East	Versaw/Fyler	17	12	37	66	X		
Q3710845	Dolores West	Wagner	24	17	32	73	X		
Q3710846	Arriola	Fyler	13	3	60	76	X		
Q3710847	Woods Canyon	Versaw	22	8	26	56	X		
Q3710848	Negro Canyon	Fyler	14	10	26	50	X		
Q3710851	Orphan Butte	Waters/Kingerys	13	6	18	37	X		
Q3710852	Wallace Ranch	Versaw	22	10	19	51	X		
Q3710853	Stoner	Fyler	18	1	23	42	X		
Q3710854	Boggy Draw	Fyler	14	12	27	53	X		
Q3710855	Trimble Point	Versaw/Fyler	25	10	36	71	X		
Q3710856	Yellow Jacket	Versaw/Fyler	21	12	33	66	X		
Q3710857	Pleasant View	Fyler	18	5	27	50	X		
Q3710858	Ruin Canyon	Harbaugh	8	7	18	33	X		
Q3710861	Rico	Hallock	12	14	30	56	X		
Q3710862	Clyde Lake	Kingery/Fyler	20	6	25	51	X		
Q3710863	Nipple Mountain	Fyler	16	10	28	54	X		
Q3710864	Willow Spring	Versaw	23	15	36	74	X		
Q3710865	Narraguinnep Mtn.	Bridges	23	8	33	64	X		
Q3710866	Doe Canyon	Robinson/Fyler	30	5	39	74	X		
Q3710867	Cahone	Versaw/Fyler	15	1	24	40	X		
Q3710868	Champagne Spring	Wright	15	15	18	48	X		x
Q3710871	Dolores Peak	Stigen	19	7	8	34	X		x
Q3710872	Groundhog Mountain	Roth	18	36	9	63	X		
Q3710873	Groundhog Reservoir	Pantles	13	17	25	55	X		
Q3710874	South Mountain	Wright	29	9	22	60	X	NE	x
Q3710875	Glade Mountain	Boyle	24	6	17	47	X		x
Q3710876	The Glade	Robinson	19	5	35	59	X		
Q3710877	Secret Canyon	Fyler	8	2	35	45	X		
Q3710878	Dove Creek	Harbaugh	18	2	28	48	X		
Q3710881	Little Cone	Stigen	14	9	15	38	X		x
Q3710882	Beaver Park	Pantles	12	10	23	45	X		
Q3710883	Lone Cone	Kingery/LaRue/Stigen	23	11	12	46	X		x
Q3710884	North Mountain	Wright	27	4	21	52	X	NE	x
Q3710885	McKenna Peak	Wright	18	10	21	49	X		x
Q3710886	Dawson Draw	Lambeth/Wright	16	23	26	65	X		
Q3710887	Joe Davis Hill	Wright	29	5	23	57	X		x
Q3710888	Egnar	Harbaugh	17	3	19	39	X		
Q3710911	Aneth S.E.-Yellow Rock Pt.	Righter	21	0	0	21			
Q3710931	Wickiup Canyon	Versaw	6	6	4	16			
Q3710941	Ruin Point	Kingerys/Versaw	18	8	21	47	X		
Q3810211	Holly East	Nelson/Kingery	12	15	23	50	X		x
Q3810212	Holly West	Nelson	5	24	27	56	X		x
Q3810213	Granada	Kuennings	9	8	20	37	X		
Q3810214	Carlton	Wright	7	1	4	12	X		x
Q3810215	Lamar East	Wright	4	3	7	14	X		x
Q3810216	Lamar West	Kingery/Pratt	6	1	8	15	X		
Q3810217	Prowers	Nelson	2	5	11	18	X		x
Q3810218	Hasty	Nelson	2	8	13	23	X		x
Q3810222	Holly N.W.	Kuennings	7	8	16	31	X		
Q3810223	Granada N.E.	Wright	17	1	16	34	X		x
Q3810224	Granada N.W.	Graves	3	3	15	21	X		
Q3810225	May Valley	Slater	19	4	14	37	X		
Q3810226	Wiley	Wright/Kingery	12	2	14	28	X		x
Q3810227	McClave	Kingery/Pratt	16	8	27	51	X		
Q3810228	Lubers	Kingery/Pratt	13	6	18	37	X		
Q3810232	Lake Devore	Sleeper/Nelson	5	9	14	28	X		x

Quadcode	Map Name	Observer	Species per Block PO	PR	CF	Total	CMPL	Alternate	BB
Q3810233	Sheridan Lake S.E.	Kingery	2	3	7	12	X		
Q3810234	Sheridan Lake S.W.	Stph Jones/Kingery	3	3	7	13	X		
Q3810235	Chivington S.E.	Graves	3	8	16	27	X		
Q3810236	Neenoshe Reservoir	Nelson	10	19	21	50	X		
Q3810237	Swede Lake	Nelson	17	26	20	63	X		x
Q3810238	Rose Ranch	Nelson	6	6	1	13	X		x
Q3810242	Stuart	Kingery	4	3	3	10	X		
Q3810243	Sheridan Lake	Kingery/Pratt	9	1	10	20	X		
Q3810244	Brandon	Kingery/Pratt	4	6	9	19	X		
Q3810245	Chivington	Graves	3	12	19	34	X		
Q3810246	Alkali Lake	Wings	1	4	1	6	X		
Q3810247	Eads	Pantles	1	6	6	13	X		
Q3810248	Hawkins	Wings	3	11	10	24	X		
Q3810252	Lake Albert	Kingery/K Potter	5	4	7	16	X		x
Q3810253	Cheyenne Wells 3 S.E.	Kingery/Pratt	5	3	10	18	X		
Q3810254	Cheyenne Wells 3 S.W.	Kingery/Pratt	6	3	5	14	X		
Q3810255	Kit Carson 4 S.E.	Hernbrode	19	8	11	38	X		
Q3810256	Oswald Ranch	Kingery/Hallock	7	1	5	13	X		x
Q3810257	Dunlap Ranch	Leatherman/Hallock	6	2	5	13	X		x
Q3810258	Arsenic Lake S.W.	Hallock	3	6	9	18	X		x
Q3810262	Lake Albert N.W.	K Potter	4	4	7	15	X		x
Q3810263	Cheyenne Wells 3 N.E.	Kingery	10	0	10	20	X		
Q3810264	Cheyenne Wells 3 N.W.	Kingery	6	1	5	12	X		
Q3810265	Kit Carson 4 N.E.	K Potter	3	4	9	16	X		x
Q3810266	Kit Carson 4 N.W.	Kingery	10	5	23	38	X		
Q3810267	Lewis Lake	Hallock/Leatherman	9	2	10	21			x
Q3810268	Arsenic Lake	Kingery	8	8	9	25	X		
Q3810272	Arapahoe	K Potter	10	2	17	29	X		x
Q3810273	Cheyenne Wells	Kingery	8	4	15	27	X		
Q3810274	Cheyenne Wells S.W.	Ketchen/Hallock	7	3	7	17			x
Q3810275	Firstview	Ketchen/Hallock	8	4	7	19			x
Q3810276	Eureka Creek South	Kingery/K Potter	10	5	7	22	X		x
Q3810277	Kit Carson	Pantles	13	9	25	47	X		
Q3810278	Sorrento	Kingery	9	11	25	45	X		
Q3810282	Arapahoe N.W.	Kingery	7	4	13	24	X		
Q3810283	Cheyenne Wells N.E.	Kingery	8	5	12	25	X		
Q3810284	Cheyenne Wells N.W.	Kingery	10	8	11	29	X		
Q3810285	Landsman Hill	Kingery	9	4	15	28	X		
Q3810286	Eureka Creek North	Kingery	7	5	9	21	X		
Q3810287	Big Spring	Kingery	5	8	22	35	X		
Q3810288	Kit Carson N.W.	Kingery	11	5	24	40	X	SW	
Q3810311	Kreybill	Kingery	17	8	10	35	X		
Q3810312	Las Animas	Graves	17	3	7	27	X		
Q3810313	Cornelia	Graves	9	4	10	23	X		
Q3810314	Hadley	Dillon/Nelson	13	9	27	49	X		x
Q3810315	Cheraw	King	5	21	9	35	X		
Q3810316	Rocky Ford	King/Versaw	7	17	21	45	X		x
Q3810317	Manzanola	King	4	20	3	27	X		
Q3810318	Elder	King	2	16	4	22	X		
Q3810321	Tree Top Ranch	Graves	21	10	14	45	X		
Q3810322	Bishop Ranch	Graves	17	3	3	23			
Q3810323	McIntosh Ranch	Graves/Kingerys	18	6	4	28			
Q3810324	Lewis Ranch	Janos/Dillon	19	9	22	50	X		x
Q3810325	Meredith Hill	K Martin/Versaw	22	11	16	49	X		
Q3810326	Sugar City	K Martin/Versaw	7	3	4	14	X		
Q3810327	Ordway	King	17	15	7	39	X		
Q3810328	Olney Springs	Truan	9	17	23	49	X		
Q3810331	Haswell S.E.	Nelson	5	3	6	14	X		x

Quadcode	Map Name	Observer	Species per Block				CMPL	Alternate	BB
			PO	PR	CF	Total			
Q3810332	Long Lake	Nelson	5	2	8	15	X		x
Q3810333	Arlington	Nelson	17	16	19	52	X		x
Q3810334	Houston Lakes	Nelson	3	4	9	16	X		x
Q3810335	Todd Point	F Harrison	6	5	3	14	X		
Q3810336	Lake Henry	King	10	10	11	31	X		
Q3810337	Nero Hill	F Harrison	14	5	11	30	X		
Q3810338	Antelope Mesa	F Harrison/Truan	11	7	15	33	X		
Q3810341	Haswell N.E.	Pratt	2	5	11	18	X		
Q3810342	Haswell	Nelson	10	3	11	24	X		x
Q3810343	Arlington N.E.	Versaw	4	1	10	15	X		x
Q3810344	Trimble Lake	Versaw	4	3	7	14			x
Q3810345	The Pinnacles	F Harrison	3	5	3	11	X		
Q3810346	Box Springs	F Harrison	9	8	6	23	X		
Q3810347	Windmill Lake	F Harrison	5	4	2	11	X		
Q3810348	Ninemile Spring	F Harrison/Truan	3	2	21	26	X		
Q3810351	Galatea	Kuennings	6	2	13	21	X		
Q3810352	Galatea S.W.	Versaw	7	2	11	20	X		x
Q3810353	Bluff Spring	Versaw	5	0	3	8		NW	x
Q3810354	Scott Draw	Versaw	5	0	0	5			x
Q3810355	Metz Springs	Versaw	9	2	11	22	X	SW	x
Q3810356	Sharp Lake	Dillon	1	8	17	26	X		x
Q3810357	Walker Point	Versaw/Kingerys	8	0	6	14			x
Q3810358	Cockleburr Springs	Versaw	3	6	8	17	X		x
Q3810361	Galatea N.E.	Kuennings	2	7	9	18	X		
Q3810363	Barrel Springs Draw	Leatherman	7	3	6	16	X		
Q3810364	Hubbard Lake	Versaw	8	2	11	21	X	SW	x
Q3810365	Karval	Brevillier/Kuennings	6	8	8	22	X		
Q3810366	Forder	Brevillier/Kuennings	7	5	6	18	X		
Q3810367	Peace Valley	Kuennings	9	6	19	34	X		
Q3810368	Sanborn Reservoir	Kuennings	6	6	12	24	X		
Q3810371	Wild Horse	Kuennings	2	4	15	21	X		
Q3810372	Aroya	Versaw	4	2	11	17	X		x
Q3810373	Rock Basin	Leatherman	7	2	7	16	X		
Q3810374	McKenzie Draw	Versaw	6	5	19	30	X		x
Q3810375	Stanley Gulch	Pantles	7	7	11	25	X		
Q3810376	Punkin Center	Kuennings	4	1	13	18	X		
Q3810377	Kutch S.E.	Versaw	7	4	6	17	X		x
Q3810378	Kutch S.W.	Kuennings	8	8	8	24	X		
Q3810381	Sanders Ranch	Pantles	7	4	13	24	X		
Q3810382	Schafer Reservoir	Leatherman	4	3	9	16	X		
Q3810383	Boyero	Pantles/Leatherman	11	10	18	39	X		
Q3810384	Kinney Lake	Dillon	4	10	12	26	X		x
Q3810385	Beckman Lake	K Potter	4	2	4	10	X		x
Q3810386	Punkin Center N.W.	Pantles	7	7	14	28	X		
Q3810387	Kutch	Kuennings	14	6	18	38	X		
Q3810388	Kutch N.W.	Versaw	10	6	7	23	X		x
Q3810411	Hardesty Reservoir	King	2	22	2	26	X		
Q3810412	Flying A Ranch	Ketchen	7	2	5	14	X		
Q3810413	Chicos Well	Ellington	5	1	3	9		CE	
Q3810414	Doyle Bridge	D Johnson	18	7	18	43	X		
Q3810415	Goat Butte	Nelson	3	9	16	28	X		x
Q3810416	Verde School	Nelson	3	8	13	24	X		x
Q3810417	Muldoon Hill	Nelson	4	15	18	37	X	SW	x
Q3810418	Beulah	Whitfield	27	30	16	73	X		
Q3810421	Fowler	Dickson	11	5	28	44	X		
Q3810422	Nepesta	Dickson	8	1	18	27	X		
Q3810423	Avondale	D Johnson	10	3	13	26	X		
Q3810424	Vineland	Silverman	29	16	24	69	X		

Quadcode	Map Name	Observer	Species per Block: PO	PR	CF	Total	CMPL	Alternate	BB
Q3810425	Southeast Pueblo	Whitfield	9	6	12	27	X		
Q3810426	Southwest Pueblo	D Johnson/Silverman	16	8	29	53	X		
Q3810427	Beulah N.E.	Whitfield	16	13	14	43	X		
Q3810428	Owl Canyon	Whitfield	28	9	18	55	X		
Q3810431	Grandview School S.E.	F Harrison	7	8	5	20	X		
Q3810432	Boone Hill	D Johnson	5	2	13	20	X		
Q3810433	North Avondale	Whitfield	12	8	20	40	X		
Q3810434	Devine	D Johnson	14	10	18	42	X		
Q3810435	Northeast Pueblo	D Johnson/Roth	36	15	18	69	X		
Q3810436	Northwest Pueblo	AVAS	13	14	38	65	X		
Q3810437	Swallows	Yaeger	8	11	23	42	X		
Q3810438	Hobson	D Johnson	14	5	15	34	X		
Q3810441	Grandview School	F Harrison	4	6	5	15	X		
Q3810442	Highlands Church	Yaeger	4	10	10	24	X		
Q3810443	North Avondale N.E.	Knight	17	15	7	39			
Q3810445	Piñon	Roth	20	21	19	60	X		
Q3810446	Steele Hollow	Ketchen	4	6	11	21	X		
Q3810447	Stone City	Truan	5	11	17	33			
Q3810448	Pierce Gulch	Monaco/Johnson	10	3	7	20	X		
Q3810451	Truckton S.E.	Kingerys	10	2	7	19		CW	
Q3810452	Edison School	Kuennings	8	4	11	23	X		
Q3810453	Hanover S.E.	Versaw	7	3	2	12		NW	x
Q3810454	Hanover	Versaw	7	1	5	13		CW	x
Q3810455	Fountain S.E.	Versaw	5	2	1	8		SW	x
Q3810456	Buttes	Conover/Versaw	15	6	14	35	X		x
Q3810457	Timber Mountain	Maynards	24	2	8	34	X		
Q3810458	Mount Pittsburg	Maynards	22	4	14	40	X		
Q3810461	Truckton N.E.	Kuennings/Sedgwick	5	7	17	29	X		
Q3810462	Truckton	Kuennings	3	6	17	26	X		
Q3810463	Hanover N.E.	Winternitz	6	3	1	10			
Q3810464	Hanover N.W.	Winternitz	8	0	3	11			
Q3810466	Fountain	Maynards	22	18	21	61	X		
Q3810467	Cheyenne Mountain	Maynard	18	13	15	46	X		
Q3810468	Mount Big Chief	Conover	32	15	21	68	X		
Q3810471	Rush	Kuennings/Freed	10	2	14	26	X		
Q3810472	Yoder	Kuennings/Freed	12	0	16	28	X		
Q3810473	Big Springs Ranch	Versaw	5	2	6	13	X		x
Q3810474	Ellicott	Kuennings/Freed	15	4	20	39	X		
Q3810475	Corral Bluffs	Kuennings/Conover	10	3	16	29	X		
Q3810476	Elsmere	Ferguson/Graves	15	7	15	37	X		
Q3810477	Colorado Springs	Williams	21	5	22	48	X		
Q3810478	Manitou Springs	Higgins	12	16	32	60	X		
Q3810481	Holtwold Store	Martin	10	6	12	28	X		
Q3810482	Rush N.W.	Kuennings	8	2	15	25	X		
Q3810483	Holcolm Hills	Kuennings	8	2	14	24	X		
Q3810484	Haegler Ranch	Sorenson	12	6	19	37	X		
Q3810485	Falcon	Kuennings	11	6	17	34	X		
Q3810486	Falcon N.W.	Winternitz	24	12	19	55	X		
Q3810487	Pikeview	Kuennings/Versaw	10	10	29	49	X		
Q3810488	Cascade	Maynards	18	21	21	60	X		
Q3810511	Saint Charles Peak	Silverman	18	13	21	52	X		
Q3810512	Deer Peak	Percival/Johnson	21	6	15	42	X		
Q3810513	Rosita	G Miller	8	18	20	46	X		
Q3810514	Aldrich Gulch	G Miller	1	13	14	28	X		
Q3810515	Horn Peak	Silverman	21	8	20	49	X		
Q3810516	Rito Alto Peak	Kingerys	16	14	11	41			
Q3810517	Mirage	Stv Jones	12	7	17	36	X		
Q3810518	Moffat North	Morkill	22	6	16	44	X		

Quadcode	Map Name	Observer	Species per Block				CMPL	Alternate	BB
			PO	PR	CF	Total			
Q3810521	Wetmore	Roth	19	23	13	55	X		
Q3810522	Hardscrabble Mtn.	Roth	21	25	18	64	X		
Q3810523	Mount Tyndall	G Miller/Kingerys	17	7	6	30			
Q3810524	Westcliffe	G Miller/Kingerys	7	3	6	16	X		
Q3810525	Beckwith Mountain	G Miller	12	11	17	40	X		
Q3810526	Electric Peak	Bright/Waddington	16	13	10	39	X		
Q3810527	Valley View Hot Spgs.	Stv Jones	17	10	23	50	X		
Q3810528	Villa Grove	Swies/Kingerys	12	9	14	35	X		
Q3810531	Florence S.E.	Monaco/Peterson	14	2	2	18			
Q3810532	Rockvale	Roth	26	32	8	66	X		
Q3810533	Curley Peak	Evans	15	27	30	72	X		
Q3810534	Iron Mountain	Kingerys	13	9	15	37		CW	
Q3810535	Hillside	G Miller	16	15	6	37			
Q3810536	Cotopaxi	Bright	17	17	9	43	X		
Q3810537	Coaldale	Kingerys	13	5	8	26		CE	
Q3810538	Bushnell Peak	Gulbenkian	19	11	22	52	X		
Q3810541	Florence	Watts/Graves	23	16	25	64	X		
Q3810542	Cañon City	Graves	5	5	19	29	X		
Q3810543	Royal Gorge	Truan	21	23	23	67	X		
Q3810544	McIntyre Hills	Lentz	23	6	28	57	X		
Q3810545	Echo	Lentz	18	8	23	49	X		
Q3810546	Arkansas Mountain	Lentz	17	5	29	51	X		
Q3810547	Howard	Dillon	14	18	28	60	X		x
Q3810548	Wellsville	Lentz/Brady	15	2	27	44	X		
Q3810551	Phantom Canyon	Ketchen	19	14	22	55	X		
Q3810552	Cooper Mountain	Leatherman	4	1	3	8			
Q3810553	Rice Mountain	Dillon	15	13	31	59	X		x
Q3810554	Gribble Mountain	Versaw	18	15	19	52	X		x
Q3810555	Hall Gulch	Kingerys	17	5	5	27		CE	
Q3810556	Waugh Mountain	Dillon	9	20	20	49	X		x
Q3810557	Jack Hall Mountain	Pantles	12	6	13	31	X		
Q3810558	Salida East	Pantles	14	4	15	33	X		
Q3810561	Big Bull Mountain	Versaw/Brekke	18	7	12	37	X		x
Q3810562	Cripple Creek South	Versaw/Brekke	17	8	27	52	X	SW	x
Q3810563	High Park	Versaw	26	3	17	46	X		x
Q3810564	Cover Mountain	Marchand	10	14	11	35	X		
Q3810565	Thirtyone Mile Mtn.	Versaw	12	1	2	15		SW	x
Q3810566	Black Mountain	Guthrie	13	32	23	68	X		
Q3810567	Gribbles Park	Anderson	11	16	17	44	X		
Q3810568	Cameron Mountain	Kuennings	12	9	18	39	X		
Q3810571	Pikes Peak	Sorensens	22	1	8	31			
Q3810572	Cripple Creek North	Winn	14	9	12	35	X		
Q3810573	Wrights Reservoir	Lane	20	7	10	37	X		
Q3810574	Witcher Mountain	Kuennings	15	9	19	43	X		
Q3810575	Thirtynine Mile Mtn.	Kuennings	15	6	20	41	X		
Q3810576	Dicks Peak	Kingerys	15	8	6	29			
Q3810577	Agate Mountain	Dillon	8	18	17	43	X		x
Q3810578	Castle Rock Gulch	Versaw	19	8	17	44	X		x
Q3810581	Woodland Park	Sorensens	19	7	34	60	X		
Q3810582	Divide	Winn	25	11	22	58	X		
Q3810583	Lake George	Sorensens/Kneuer	16	12	34	62	X		
Q3810584	Elevenmile Canyon	Taggart	23	11	24	58	X		
Q3810585	Spinney Mountain	Kneuer	27	9	17	53	X		
Q3810586	Guffey N.W.	Crowley/Versaw	10	8	11	29	X		x
Q3810587	Antero Res. N.E.	Kuennings	11	5	18	34	X		
Q3810588	Antero Res.	Versaw	13	6	12	31	X		x
Q3810611	Hickey Bridge	Wait	9	8	6	23	X		
Q3810612	Saguache	Yaeger	15	1	9	25	X		x

Quadcode	Map Name	Observer	Species per Block				CMPL	Alternate	BB
			PO	PR	CF	Total			
Q3810613	Laughlin Gulch	Yaeger	17	4	13	34	X		x
Q3810614	Lake Mountain	Dexter	15	10	18	43	X		x
Q3810615	Grouse Creek	Dexter	15	9	14	38	X		
Q3810616	Saguache Park	Wright	25	3	22	50	X		
Q3810617	Elk Park	Wright	16	0	20	36	X		
Q3810618	Stewart Peak	McGinley	12	8	4	24	X		
Q3810621	Graveyard Gulch	Kingerys/Pratt	15	14	16	45	X		
Q3810622	Klondike Mine	Pantles	10	12	12	34	X		
Q3810623	Lake Mountain N.E.	Pantles	4	20	24	48	X		
Q3810624	Trickle Mountain	Dexter	18	9	20	47	X		
Q3810625	North Pass	Dexter	22	8	13	43	X		
Q3810626	Cochetopa Park	Wright	19	9	15	43	X		
Q3810627	Cold Spring Park	Wright	26	2	14	42	X		
Q3810628	Rock Creek Park	Dexter	20	7	15	42	X		
Q3810631	Whale Hill	Kingerys	25	9	26	60	X		
Q3810632	Bonanza	Staatz	14	4	11	29	X		
Q3810633	Chester	Wait/Yaeger	22	6	20	48	X		x
Q3810634	Sargents Mesa	Dunmire/Kingerys	15	5	18	38	X		
Q3810635	West Baldy	Yaeger	15	6	13	34	X		x
Q3810636	Razor Creek Dome	Levad	19	2	21	42	X		
Q3810637	Sawtooth Mountain	Boyle/Funk	11	4	19	34	X		
Q3810638	Spring Hill Creek	Funk	16	1	18	35	X		
Q3810641	Poncha Pass	Kingerys	27	7	19	53	X		
Q3810642	Mount Ouray	Kingerys	15	18	26	59	X		
Q3810643	Pahlone Peak	Kingerys/Versaw	19	10	10	39	X		
Q3810644	Sargents	Versaw	25	10	21	56	X		x
Q3810645	Doyleville	Dexter	29	1	13	43	X		
Q3810646	Houston Gulch	Dexter/Boyle	24	12	25	61	X		x
Q3810647	Iris	Meyer	15	10	35	60	X		
Q3810648	Iris N.W.	Meyer	10	17	33	60	X		
Q3810651	Salida West	Ebright	14	6	45	65	X		
Q3810652	Maysville	Hilty/Versaw	33	12	25	70	X		x
Q3810653	Garfield	Roth	11	18	12	41	X		
Q3810654	Whitepine	Stph Jones	16	22	12	50	X		
Q3810655	Pitkin	Traver/Boyle	23	7	23	53	X		
Q3810656	Parlin	Meyer	8	7	41	56	X		
Q3810657	Signal Peak	Meyer	13	9	35	57	X		
Q3810658	Gunnison	Meyer	23	17	56	96	X		
Q3810661	Nathrop	Versaw	17	6	20	43	X		x
Q3810662	Mount Antero	Whipple/Versaw	14	12	22	48	X		x
Q3810663	St. Elmo	Kingerys	13	3	1	17			
Q3810664	Cumberland Pass	Dexter/Versaw/Stph Jones	12	7	16	35	X		
Q3810665	Fairview Peak	Meyer	10	3	25	38	X		
Q3810666	Crystal Creek	Meyer	7	3	21	31	X		
Q3810667	Almont	Meyer	9	3	29	41	X		
Q3810668	Flat Top	Meyer/Blank	14	20	17	51	X		
Q3810671	Buena Vista East	Kingery	25	12	21	58	X		
Q3810672	Buena Vista West	Gabel/Versaw	17	23	15	55	X		x
Q3810673	Mount Yale	Pantles	11	5	16	32	X		
Q3810674	Tincup	Dexter	10	7	14	31	X		
Q3810675	Taylor Park Res.	Wright	11	6	12	29	X		
Q3810676	Matchless Mountain	Theimer	15	6	19	40	X		
Q3810677	Cement Mountain	Dexter/Meyer	29	4	14	47	X		
Q3810678	Crested Butte	Meyer	6	14	33	53	X		
Q3810681	Marmot Peak	Moore/Versaw	15	14	24	53	X		x
Q3810682	Harvard Lakes	Fergusons/Kingerys	34	5	14	53	X		
Q3810683	Mount Harvard	Kramer/Versaw	11	4	13	28	X		x
Q3810684	Winfield	Wright	15	0	19	34	X		

Quadcode	Map Name	Observer	Species per Block PO	PR	CF	Total	CMPL	Alternate	BB
Q3810685	Pieplant	Meyer	15	5	30	50	X		
Q3810686	Italian Creek	Wright	13	19	6	38	X		
Q3810687	Pearl Pass	Meyer	9	6	23	38	X		
Q3810688	Gothic	Meyer	17	4	31	52	X		
Q3810711	Mineral Mountain	Wright/Stigen	20	4	15	39	X		x
Q3810712	Cannibal Plateau	Wright	20	2	27	49	X		
Q3810713	Lake City	Funk	24	5	16	45	X		
Q3810714	Uncompahgre Peak	Wright/Funk	18	1	19	38	X		x
Q3810715	Wetterhorn Peak	Wright	17	4	16	37	X		
Q3810716	Ouray	Dexter	23	4	19	46	X		
Q3810717	Mount Sneffels	Wright	16	12	8	36	X		
Q3810718	Sams	Wright/Facer	25	1	10	36	X		x
Q3810721	Rudolph Hill	Funk	12	1	18	31	X		
Q3810722	Powderhorn Lakes	Dexter	17	2	18	37	X		
Q3810723	Alpine Plateau	Funk	15	7	24	46	X		
Q3810724	Sheep Mountain	Wright	18	5	18	41	X		x
Q3810725	Courthouse Mountain	Wright	23	5	13	41	X		
Q3810726	Dallas	Dexter	39	9	23	71	X		
Q3810727	Ridgway	Dexter	33	14	30	77	X		
Q3810728	Horsefly Peak	Dexter/Levad	33	9	13	55	X		x
Q3810731	Powderhorn	Funk	19	0	18	37	X		
Q3810732	Gateview	Funk	20	5	18	43	X		
Q3810733	Poison Draw	Funk	13	5	19	37	X		
Q3810734	Lost Lake	Wright/Funk	24	0	24	48	X		x
Q3810735	Washboard Rock	Dexter/Funk	16	3	23	42	X		
Q3810736	Buckhorn Lakes	Durant/Guadagno	23	6	32	61	X	SW	
Q3810737	Colona	Dexter	26	22	25	73	X		
Q3810738	Government Springs	Guadagno	23	10	30	63	X		
Q3810741	Big Mesa	Funk	17	5	20	42	X		
Q3810742	Carpenter Ridge	Funk	18	2	18	38	X		
Q3810743	Sapinero	Funk	11	7	20	38	X		
Q3810744	Curecanti Needle	Wright/Funk	25	5	26	56	X		
Q3810745	Cimarron	Dexter	29	14	30	73	X		
Q3810746	Cerro Summit	Guadagno	18	8	24	50	X		
Q3810747	Montrose East	Wainwrights	21	15	9	45	X		
Q3810748	Montrose West	Dexter/Horn/Levad	21	27	15	63	X		
Q3810751	McIntosh Mountain	Alves	15	15	45	75	X		
Q3810752	West Elk Peak S.W.	Funk	18	1	16	35	X		
Q3810753	Little Soap Park	Funk	18	3	23	44	X		
Q3810754	X Lazy F Ranch	Funk/Shuster	8	23	16	47	X		
Q3810755	Cathedral Peak	Kingerys/Rowe	20	7	22	49	X		
Q3810756	Grizzly Ridge	Dexter	13	0	5	18			
Q3810757	Red Rock Canyon	Dexter/Horn	21	5	9	35	X		
Q3810758	Olathe	Dexter/Levad	15	11	18	44	X		x
Q3810761	Squirrel Creek	Funk	11	3	22	36	X		
Q3810762	West Elk Peak	Dexter/Funk	18	2	18	38	X		
Q3810763	Big Soap Park	Bridges	23	3	28	54	X		
Q3810764	Mount Guero	Kingerys	23	11	17	51	X		
Q3810765	Crawford	Rowe	20	9	22	51	X		
Q3810766	Grand View Mesa	Ohlheiser/Dexter	25	14	13	52	X		
Q3810767	Black Ridge	Woodwards/Carey	19	9	15	43	X		
Q3810768	Olathe N.W.	Dexter	9	2	12	23	X		
Q3810771	Mt. Axtell	Purrington/Meyer	18	10	36	64	X		
Q3810772	Anthracite Range	Levad	21	5	23	49	X		
Q3810773	West Beckwith Peak	Kingery	19	11	26	56	X		
Q3810774	Minnesota Pass	Shuster/Kngery	31	6	21	58	X		
Q3810775	Paonia	Guadagno	13	1	27	41	X		
Q3810776	Hotchkiss	Rowe/Wright/Levad	19	11	27	57	X		

Quadcode	Map Name	Observer	Species per Block: PO	PR	CF	Total	CMPL	Alternate	BB
Q3810777	Lazear	Rowe	18	14	20	52	X		
Q3810778	Orchard City	Dexter	25	7	12	44	X		
Q3810781	Oh-Be-Joyful	Purrington	32	18	10	60	X		
Q3810782	Marcellina Mountain	Kingery/Wang/Bogart	25	8	14	47	X		
Q3810783	Paonia Reservoir	Bailey	16	6	8	30			
Q3810784	Somerset	Wang	3	8	7	18			
Q3810785	Bowie	Guadagno	29	29	36	94	X		
Q3810786	Gray Reservoir	Carey	22	14	16	52	X		
Q3810787	Dry Creek	Guadagno	17	15	41	73	X		
Q3810788	Cedaredge	Woodwards/Mohan/Dexter/Levad	23	13	40	76	X		
Q3810811	Placerville	Wright/Gulbenkian/Boyle	23	10	35	68	X		x
Q3810812	Gurley Canyon	Boyle	34	10	15	59	X		x
Q3810813	Oak Hill	Wright	32	13	19	64	X		
Q3810814	Barkelew Draw	Wright	28	4	20	52	X		x
Q3810815	Basin	Wainwrights/Wright	20	3	15	38	X		x
Q3810816	Gypsum Gap	Lambeth	13	2	5	20			
Q3810817	Hamm Canyon	Wainwrights	9	4	7	20	X		
Q3810818	Horse Range Mesa	Fyler	16	3	8	27			
Q3810821	Hotchkiss Reservoir	Grother/Levad/Dexter	27	11	11	49	X		
Q3810822	Sanborn Park	Wright	37	8	12	57	X		
Q3810823	Norwood	Dexter	28	6	16	50	X		
Q3810824	Redvale	Wright	31	7	21	59	X		x
Q3810825	Naturita	Wright	16	8	18	42	X		x
Q3810826	Naturita N.W.	Kingery	15	10	8	33	X		
Q3810827	Bull Canyon	Kingery/Bridges	17	8	21	46	X		
Q3810828	Anderson Mesa	Versaw/Wright	29	13	13	55	X		x
Q3810831	Pryor Creek	Levad/Grother	26	5	22	53	X		x
Q3810832	Antone Spring	Boyle/Grother	28	11	20	59	X		
Q3810833	Ute	Wright/Dexter	13	5	19	37	X		
Q3810834	Big Bucktail Creek	Wright	24	14	22	60	X		x
Q3810835	Nucla	Jensen/Lambeth/Stigen	26	5	24	55	X		x
Q3810836	Uravan	Leatherman/Graves/Boyle	21	10	13	44	X		
Q3810837	Davis Mesa	Wright	19	7	16	42	X		x
Q3810838	Paradox	Lambeth/Wright	25	16	11	52	X		x
Q3810841	Dry Creek Basin	Boyle	29	13	12	54	X		
Q3810842	Davis Point	Lambeth	29	9	24	62	X		
Q3810843	Moore Mesa	Shuster/Barr/Stigen	28	9	9	46	X		
Q3810844	Starvation Point	Barr/Stigen	37	12	18	67	X		
Q3810845	Windy Point	Wright	27	5	20	52	X		x
Q3810846	Atkinson Creek	Boyle	26	4	19	49	X		x
Q3810847	Red Canyon	Wright/Wainwrights/Graves	31	11	18	60	X		x
Q3810848	Roc Creek	Ferguson	10	3	6	19			
Q3810851	Hoovers Corner	Kretzmeyer/Adamus/Dexter	26	5	21	52	X		
Q3810852	Camel Back	Dexter	17	7	17	41	X		
Q3810853	Cottonwood Basin	Dexter	32	15	7	54	X		
Q3810854	Kelso Point	Dexter	26	19	24	69	X		
Q3810855	Snipe Mountain	Levad	29	10	19	58	X		
Q3810856	Uncompahgre Butte	Frederick	21	12	45	78	X		
Q3810857	Calamity Mesa	Williams/Clements/Lambeth	27	10	7	44	X		x
Q3810858	Juanita Arch	Frederick/Hahn	27	15	32	74	X		
Q3810861	Delta	Dexter/Levad	18	20	20	58	X		x
Q3810862	Roubideau	Goulding/Dexter	23	8	7	38	X		
Q3810863	Good Point	Dexter	16	11	7	34	X		
Q3810864	Escalante Forks	Dexter	28	15	14	57	X		
Q3810865	Keith Creek	Hahn/Henwood	25	12	15	52	X		
Q3810866	Casto Reservoir	Levad/Henwood	20	13	22	55	X		
Q3810867	Pine Mountain	Levad/Henwood	19	10	23	52	X		x
Q3810868	Gateway	Cunningham/Levad	19	4	8	31	X		

Quadcode	Map Name	Observer	Species per Block PO	PR	CF	Total	CMPL	Alternate	BB
Q3810871	North Delta	Dexter/Levad	14	14	13	41	X		x
Q3810872	Point Creek	Janos/Adamus/Dexter	13	17	5	35	X		
Q3810873	Dominguez	Guadagno	20	24	19	63	X		
Q3810874	Triangle Mesa	Henwood	31	2	3	36	X		
Q3810875	Jacks Canyon	Grohs/Levad	23	11	13	47	X		x
Q3810876	Snyder Flats	Levad	30	15	35	80	X		x
Q3810877	Fish Creek	Dexter	33	14	30	77	X		
Q3810878	Two V Basin	Tindall/Massey	31	9	16	56	X		
Q3810881	Hells Kitchen	Dexter	22	11	14	47	X		
Q3810882	Indian Point	Dexter	28	14	30	72	X		
Q3810883	Juniata Reservoir	Lambeth	11	8	17	36	X		
Q3810884	Whitewater	Levad	14	10	22	46	X		
Q3810885	Island Mesa	Levad	20	12	26	58	X		
Q3810886	Glade Park	Wright	25	17	14	56	X		x
Q3810887	Payne Wash	Lambeth/Wright	36	9	23	68	X		x
Q3810888	Bieser Creek	Ireland	19	9	11	39			
Q3810961	Dolores Point North	Lambeth	25	8	13	46	X		
Q3810971	Steamboat Mesa	Lambeth	3	0	5	8			
Q3810981	Marble Canyon	Pearson	19	6	32	57	X		
Q3910211	Mount Sunflower	Kingery	3	0	1	4			
Q3910212	Mount Sunflower S.W.	Kingery	9	7	10	26	X		
Q3910213	Burlington 3 S.E.	Kingery	8	12	13	33	X		
Q3910214	Burlington 3 S.W.	Hallock	7	6	10	23	X		x
Q3910215	Alpine Ranch	Kingery/Hallock	7	3	6	16	X		x
Q3910216	Alpine Ranch S.W.	Kingery	11	1	6	18	X		
Q3910217	Stratton 3. S.E.	Hallock	2	8	6	16	X		x
Q3910218	Stratton 3 S.W.	Hallock	5	5	7	17	X		x
Q3910222	Mount Sunflower N.W.	Kingery	5	3	10	18	X		
Q3910223	Burlington 3 N.E.	Kingery/Hallock	6	4	9	19	X		x
Q3910224	Burlington 3 N.W.	Rauch/Hallock	7	2	11	20	X		x
Q3910225	Alpine Ranch N.E.	Kingery/Hallock	7	2	7	16	X		x
Q3910226	Alpine Ranch N.W.	Kingery	6	7	15	28	X		
Q3910227	Stratton 3 N.E.	Kingery	7	2	11	20	X		
Q3910228	Stratton 3 N.W.	Kingery	2	2	13	17	X		
Q3910232	Peconic	Hartman	3	4	13	20	X		
Q3910233	Burlington	Kingery	8	3	9	20	X		
Q3910234	Bethune	Rauch/Hallock	11	4	9	24	X		x
Q3910235	Stratton	Kingery/Hallock	6	1	6	13	X		x
Q3910236	Vona	Hallock	5	4	8	17	X		x
Q3910237	Seibert	Chace	10	6	9	25	X		
Q3910238	Flagler Reservoir	Kingery	6	2	14	22	X		
Q3910242	Kanorado N.W.	Kingery/Hallock	10	3	12	25	X		x
Q3910243	Burlington N.E.	Kingery/Mollhoff/Hallock	3	10	17	30	X		x
Q3910244	Settlement	Rauch/Hallock	6	5	9	20	X		x
Q3910245	Tuttle	Kingery	12	8	20	40	X		
Q3910246	Stratton N.W.	Flageolle	4	8	18	30	X		
Q3910247	Seibert N.E.	Chace	2	4	6	12	X		
Q3910248	Seibert N.W.	Chace	7	1	11	19	X		
Q3910252	Bonny Reservoir So.	Mollhoff	9	17	6	32	X		
Q3910253	Idalia S.E.	Kingery/Mollhoff/Dillon	13	13	21	47	X		x
Q3910254	Idalia S.W.	Dillon	3	12	16	31	X		x
Q3910255	Kirk	Dillon	9	14	26	49	X		x
Q3910256	Joes S.W.	Hayes/Chace	13	5	14	32	X		
Q3910257	Cope S.E.	Hayes/Chace	11	0	12	23	X		
Q3910258	Cope S.W.	Chace	8	3	6	17	X		
Q3910261	Hale Ponds	Dillon	12	12	12	36			x
Q3910262	Bonny Reservoir No.	Kingery	21	15	33	69	X		
Q3910263	Idalia	Kingery	7	3	12	22	X		

Quadcode	Map Name	Observer	Species per Block PO	PR	CF	Total	CMPL	Alternate	BB
Q3910264	Spring Canyon	Kingery	13	3	18	34	X		
Q3910265	Adler Creek	Maguire	5	2	17	24	X		
Q3910266	Joes	TenBrink	1	2	20	23	X		
Q3910267	Cope	Hayes	9	3	6	18			
Q3910268	Cope N.W.	Hayes	13	4	10	27	X		
Q3910272	Beecher Island	Hayes	7	7	8	22	X		
Q3910273	Wildcat Canyon	Kingery/Dillon	4	13	14	31	X		x
Q3910274	Vernon S.W.	Kingery	19	18	27	64	X		
Q3910275	Abarr S.E.	Graves/Chace	10	13	23	46	X		
Q3910276	Abarr	Graves/Maguire	6	5	19	30	X		
Q3910277	De Nova S.E.	Roller	4	10	6	20	X		
Q3910278	De Nova	Roller	15	10	11	36	X		
Q3910282	Beecher Island N.W.	Hayes/Chace	21	14	29	64	X		
Q3910283	Vernon	Kingery	10	18	29	57	X		
Q3910284	Vernon N.W.	Kingery/Dillon	11	14	17	42	X		x
Q3910285	Heartstrong	Dillon	3	9	13	25	X		x
Q3910286	Beverly Grove	Schwalbes/Dillon	9	7	14	30	X		x
Q3910287	De Nova N.E.	Schwalbes	8	5	9	22	X		
Q3910288	De Nova N.W.	Schwalbes	6	9	7	22	X		
Q3910311	Hugo 4 S.E.	Pantles	6	4	14	24	X		
Q3910312	Bledsoe Ranch	Pantles	6	10	19	35	X		
Q3910313	Clifford	Kuennings	1	1	10	12	X		
Q3910314	Hugo S.W.	K Potter	13	2	5	20	X		x
Q3910315	Lake S.E.	Kuennings	10	2	14	26	X		
Q3910316	Long Creek	K Potter	5	8	7	20	X		x
Q3910317	Matheson S.E.	K Potter	7	1	11	19	X		x
Q3910318	Matheson S.W.	Kuennings	12	1	5	18	X		
Q3910321	Hugo 4 N.E.	Kuennings	6	4	11	21	X		
Q3910322	Hugo 4 N.W.	Kuennings	6	5	8	19	X		
Q3910323	Sevenmile Ranch	Pantles	6	5	18	29	X		
Q3910324	Hugo	Kuennings	3	1	7	11	X		
Q3910325	Barron Creek	Kuennings	11	9	18	38	X		
Q3910326	Lake	Kuennings	5	2	16	23	X		
Q3910327	Matheson N.E.	Kuennings	5	4	12	21	X		
Q3910328	Matheson	Kuennings	8	2	19	29	X		
Q3910331	Flagler	Bleck	7	5	25	37	X		
Q3910332	Flagler S.W.	McDonald	6	0	12	18	X		
Q3910333	Arriba	Bleck	6	3	15	24	X		
Q3910334	Genoa East	Bleck	8	2	10	20	X		
Q3910335	Genoa West	Bleck	5	4	11	20	X		
Q3910336	Limon	Kuennings	10	5	15	30	X		
Q3910337	River Bend	Graves	5	13	16	34	X		
Q3910338	Beuck Draw	Graves	2	2	9	13	X		
Q3910341	Flagler N.E.	Bleck	9	1	13	23	X		
Q3910342	Flagler N.W.	McDonald	4	0	14	18	X		
Q3910343	Arriba N.E.	Bleck	8	0	12	20	X		
Q3910344	Arriba N.W.	Hayes	7	0	9	16	X		
Q3910345	Walks Camp Park	Hayes	11	1	10	22	X		
Q3910346	T Draw	Calkum/Flageolle	5	1	19	25	X		
Q3910347	Barking Dog Spring	Calkum	2	6	23	31	X		
Q3910348	Agate	Graves	4	1	11	16	X		
Q3910351	Anton S.E.	McDonald	6	1	12	19	X		
Q3910352	Thurman	McDonald	3	0	15	18	X		
Q3910353	Shaw	Hayes	13	0	10	23	X		
Q3910354	Lindon S.W.	Hayes	10	1	8	19	X		
Q3910355	Lusto Springs	Dillon	6	4	16	26	X		x
Q3910356	Last Chance S.W.	Graves	4	0	5	9	X		
Q3910358	Noonen Res. S.W.	Dillon/Versaw	15	3	25	43	X		x

Quadcode	Map Name	Observer	Species per Block: PO	PR	CF	Total	CMPL	Alternate	BB
Q3910361	Arickaree	McDonald	9	0	15	24	X		
Q3910362	Anton	A Kelly	6	0	8	14	X		
Q3910363	Lindon N.E.	Pimental	6	4	7	17	X		
Q3910364	Lindon	Pimental	5	2	7	14	X		
Q3910365	Last Chance	TenBrink	1	3	19	23	X		
Q3910366	Last Chance N.W.	TenBrink	4	6	24	34	X		
Q3910367	Cottonwood Val. N	Kingery/Dillon	11	9	18	38	X		x
Q3910368	Noonen Reservoir	Pimental	15	3	10	28	X		
Q3910371	Elba S.E.	Kingery	8	1	8	17	X		
Q3910372	Elba S.W.	Kingery	2	2	7	11	X		
Q3910373	Antelope Creek S.E.	Kingery/Dillon	8	4	15	27	X		x
Q3910374	Dry Gulch	Kingerys	5	7	21	33	X		
Q3910375	Woodlin School	Dillon	7	8	17	32	X		x
Q3910376	Wetzel Creek	Dillon	11	5	20	36	X		x
Q3910377	Shamrock S.E.	Malone/Hayes	3	2	5	10	X	CE	
Q3910378	Poison Springs	Malone	7	3	7	17	X	NW	
Q3910381	Elba N.E.	Kingery	8	2	11	21	X		
Q3910382	Elba	Kingery	6	3	12	21	X		
Q3910383	Antelope Creek East	McDonald	3	1	12	16	X		
Q3910385	Woodrow	Malone	7	3	15	25	X		
Q3910386	Woodrow N.W.	Malone	5	0	10	15	X		
Q3910387	Shamrock	Kingery/Hayes	4	2	5	11	X		
Q3910388	Potty Brown Creek	Hayes	3	2	5	10	X		
Q3910411	Alta Vista	Kuennings	7	5	13	25	X		
Q3910412	Ramah South	Kuennings	2	2	18	22	X		
Q3910413	Calhan	Kuennings	8	6	21	35	X		
Q3910414	Peyton	Brevillier	16	2	15	33	X		
Q3910415	Eastonville	Brevillier/Kuennings	16	6	14	36	X		
Q3910416	Black Forest	Dayhoff	9	3	25	37	X		
Q3910417	Monument	Dayhoff	15	17	31	63	X		
Q3910418	Palmer Lake	Kuennings	19	16	28	63	X		
Q3910421	Simla	Gallagher	9	6	12	27	X		
Q3910422	Ramah North	Gallagher	15	14	10	39	X		
Q3910423	Fondis	Dayhoff	11	2	34	47	X		
Q3910424	Bijou Basin	Dayhoff	6	13	36	55	X	SW	
Q3910425	Elbert	Kuennings	11	7	32	50	X		
Q3910426	Cherry Valley School	Dayhoff	14	9	38	61	X		
Q3910427	Greenland	Dayhoff	18	12	22	52	X		
Q3910428	Larkspur	Dayhoff	23	8	38	69	X		
Q3910431	Kuhns Crossing	Winternitz/Versaw	18	5	4	27			
Q3910432	Bijou S.W.	Winternitz/Graves	7	10	20	37	X		
Q3910433	Big Gulch	Winternitz/Graves	16	7	13	36	X		
Q3910434	Kiowa	Staatz	23	9	15	47	X		
Q3910435	Elizabeth	Herold	12	17	25	54	X		
Q3910436	Russellville Gulch	Kingerys/Graves	11	5	12	28	X		
Q3910437	Castle Rock South	Wilson/Graves	8	29	50	87	X		
Q3910438	Dawson Butte	Staatz	16	8	10	34			
Q3910441	Cattle Gulch	Graves	3	6	19	28	X		
Q3910442	Bijou	Besser	13	7	24	44	X		
Q3910443	Kiowa N.E.	Kingery	14	10	21	45	X		
Q3910444	Kiowa N.W.	Besser	11	7	16	34	X		
Q3910445	Cabin Gulch	Herold	10	19	17	46	X		
Q3910446	Ponderosa Park	DMNH/Graves	12	26	25	63	X		
Q3910447	Castle Rock North	Tuesday Birders/Leatherman	25	29	21	75	X		
Q3910448	Sedalia	Schock	22	3	25	50	X		
Q3910451	Deer Trail	Versaw	11	2	16	29	X		x
Q3910452	Byers S.W.	Besser	5	7	19	31	X		
Q3910453	Strasburg S.E.	Graves	6	7	18	31	X		

Quadcode	Map Name	Observer	Species per Block PO	PR	CF	Total	CMPL	Alternate	BB
Q3910454	Strasburg S.W.	Graves	11	6	22	39	X		
Q3910455	Watkins S.E.	Besser	10	11	16	37	X		
Q3910456	Piney Creek	Springston	23	2	28	53	X		
Q3910457	Parker	Schock	17	13	21	51	X		
Q3910458	Highlands Ranch	Allison/Besser	14	14	14	42	X		
Q3910461	Peoria	J King	2	5	8	15	X		
Q3910462	Byers	Mullineaux	10	7	20	37	X		
Q3910463	Strasburg	Mullineaux	3	7	13	23	X		
Q3910464	Strasburg N.W.	BLM - Minges	19	9	15	43	X		
Q3910465	Watkins	Bleck	5	9	25	39	X		
Q3910466	Coal Creek	Kingery	10	8	22	40	X		
Q3910467	Fitzsimons	Hale	12	5	20	37	X		
Q3910468	Englewood	Maguire/Kingery	12	9	24	45	X		
Q3910471	Leader S.E.	K Potter	5	9	22	36	X		
Q3910472	Leader S.W.	Pimental	10	2	8	20	X		
Q3910473	Roper School	Bleck	4	0	12	16	X		
Q3910474	Bennett	Bleck	13	8	28	49	X		
Q3910475	Manila	Valasek	10	3	10	23	X		
Q3910476	Box Elder School	Valasek	13	7	11	31	X		
Q3910477	Sable	Lentz/Hales	13	2	28	43	X		
Q3910478	Commerce City	Lentz	7	3	20	30	X		
Q3910481	Leader	Besser	7	10	11	28	X		
Q3910482	Leader N.W.	Pimental	16	0	9	25	X		
Q3910483	Living Springs	Bleck	7	1	20	28	X		
Q3910484	Sunnydale	Mullineaux	7	4	12	23	X		
Q3910485	Horse Creek	Mullineaux	3	8	12	23	X		
Q3910486	Mile High Lakes	Gordon	10	4	4	18	X		
Q3910487	Brighton	Gordon/Besser	11	9	17	37	X		
Q3910488	Eastlake	TenBrink/Carter	9	6	18	33	X		
Q3910511	Mount Deception	Winn/Baker	22	10	20	52	X		
Q3910512	Signal Butte	Winn	12	10	26	48	X		
Q3910513	Hackett Mountain	Heller/Winn/Kneuer/Versaw	14	19	14	47	X		
Q3910514	Tarryall	Taggart	19	12	19	50	X		
Q3910515	Glentivar	Guthrie	11	15	25	51	X		
Q3910516	Sulphur Mountain	Crowley/Versaw	17	8	16	41	X		x
Q3910517	Hartsel	Kuennings	14	10	21	45	X		
Q3910518	Garo	Kneuer	20	4	8	32	X		
Q3910521	Dakan Mountain	Winn/Heller/Versaw	20	8	20	48	X		
Q3910522	Westcreek	Winn	14	13	36	63	X		
Q3910523	Cheesman Lake	Winn	16	10	10	36	X		
Q3910524	McCurdy Mountain	Erthal	13	5	3	21	X		
Q3910525	Farnum Peak	Blaj/Versaw	18	12	19	49	X		x
Q3910526	Eagle Rock	Van Erp	28	4	25	57	X		
Q3910527	Elkhorn	Bridges	21	5	19	45	X		
Q3910528	Fairplay East	M Roberts	27	13	10	50	X		
Q3910531	Devils Head	Staatz	19	12	8	39	X		
Q3910532	Deckers	Besser	14	5	16	35	X		
Q3910533	Green Mountain	Lentz/Kingery	24	5	14	43	X		
Q3910534	Windy Peak	Simon	6	15	4	25	X		
Q3910535	Topaz Mountain	Lentz	17	6	18	41	X		
Q3910536	Observatory Rock	Van Erp/Taggart	26	3	27	56	X		
Q3910537	Milligan Lakes	Bridges	20	8	22	50	X		
Q3910538	Como	Pratt	22	8	11	41	X		
Q3910541	Kassler	Maguire/Lockett	30	12	25	67	X		
Q3910542	Platte Canyon	Vincent	8	16	11	35			
Q3910543	Pine	Kingery	47	6	7	60	X		
Q3910544	Bailey	Simon	12	20	21	53	X		
Q3910545	Shawnee	Maguire/Lockett	19	10	22	51	X		

Quadcode	Map Name	Observer	Species per Block PO	PR	CF	Total	CMPL	Alternate	BB
Q3910546	Mount Logan	Lentz	9	3	21	33	X		
Q3910547	Jefferson	Hannay	14	7	22	43	X		
Q3910548	Boreas Pass	Besser/Lentz	19	6	27	52	X		
Q3910551	Littleton	Kingerys	20	13	46	79	X		
Q3910552	Indian Hills	Hay/Vallado	11	25	28	64	X		
Q3910553	Conifer	M Brown	23	4	39	66	X		
Q3910554	Meridian Hill	Schock/Solomon	19	15	26	60	X		
Q3910555	Harris Park	Brockner	13	22	19	54	X		
Q3910556	Mt. Evans	Carter	11	8	16	35	X		
Q3910557	Montezuma	Lentz	11	2	25	38	X		
Q3910558	Keystone	Bailey/Nelson	8	3	18	29	X		x
Q3910561	Fort Logan	Oliver	5	6	36	47	X		
Q3910562	Morrison	Finch	15	13	34	62	X		
Q3910563	Evergreen	Simon/Brockner	29	20	25	74	X		
Q3910564	Squaw Pass	Gables/Roberts	19	11	22	52	X		
Q3910565	Idaho Springs	Hayden	12	16	25	53	X		
Q3910566	Georgetown	Murdock	26	14	3	43	X		
Q3910567	Grays Peak	Sperger	9	5	12	26	X		
Q3910568	Loveland Pass	Kingerys	10	4	18	32	X		
Q3910571	Arvada	Breckon	4	5	16	25	X		
Q3910572	Golden	Nelson	12	16	24	52	X		
Q3910573	Ralston Buttes	Gulbenkian	18	27	22	67	X		
Q3910574	Black Hawk	Phillips	27	5	21	53	X		
Q3910575	Central City	Van Erp	14	8	30	52	X		
Q3910576	Empire	Enright	16	9	15	40	X		
Q3910577	Berthoud Pass	Foster	17	2	13	32	X		
Q3910578	Byers Peak	Enright/Nelson	8	5	16	29	X		x
Q3910581	Lafayette	Besser	8	4	15	27	X		
Q3910582	Louisville	Kaempfer	14	16	24	54	X	CE	
Q3910583	Eldorado Springs	Nelson	20	32	45	97	X		
Q3910584	Tungsten	Hansley	20	12	24	56	X		
Q3910585	Nederland	Hallock	23	14	38	75	X		
Q3910586	East Portal	Hallock	25	11	30	66	X		
Q3910587	Fraser	Powell/Pantles	13	15	31	59	X		
Q3910588	Bottle Pass	Moody/Nelson	17	5	26	48	X		x
Q3910611	Jones Hill	Lentz	14	9	26	49	X		
Q3910612	South Peak	Lentz	12	4	22	38	X		
Q3910613	Granite	Schwalbes	34	6	9	49	X		
Q3910614	Mount Elbert	Lentz	10	1	25	36	X		
Q3910615	Independence Pass	Lentz	13	3	20	36	X		
Q3910616	New York Peak	Versaw	13	1	12	26	X		x
Q3910617	Hayden Peak	Vidal	14	18	15	47	X		
Q3910618	Maroon Bells	Nelson	6	7	16	29	X		x
Q3910621	Fairplay West	Nelson	7	10	21	38	X		x
Q3910622	Mount Sherman	Lentz	11	8	22	41	X		
Q3910623	Leadville South	Piombino/Versaw	28	8	20	56	X		x
Q3910624	Mount Massive	Hyden	17	6	8	31			
Q3910625	Mt. Champion	Lentz	15	0	18	33	X		
Q3910626	Thimble Rock	Dillon	12	6	10	28			x
Q3910627	Aspen	Vidal	14	23	33	70	X		
Q3910628	Highland Peak	Vidal	14	22	8	44	X		
Q3910631	Alma	McMenamy/Kuhlman/Pratt	19	8	18	45	X		
Q3910632	Climax	Nelson	5	7	15	27	X		x
Q3910633	Leadville North	D Ward	12	13	22	47	X		
Q3910634	Homestake Res.	D Ward/Schwalbe	10	10	16	36	X		
Q3910635	Nast	Merchant	22	7	11	40	X		
Q3910636	Meredith	Merchant	19	8	16	43	X		
Q3910637	Ruedi Reservoir	K Potter	16	5	20	41	X		x

Quadcode	Map Name	Observer	Species per Block PO	PR	CF	Total	CMPL	Alternate	BB
Q3910638	Woody Creek	Vidal	15	11	30	56	X		
Q3910641	Breckenridge	Black/Lentz	17	5	30	52	X		
Q3910642	Copper Mountain	Hallock	16	7	22	45	X		x
Q3910643	Pando	Besser/DMNH	12	9	10	31	X		
Q3910644	Mt. Holy Cross	Merchant/Kingery	14	16	10	40	X		
Q3910645	Mount Jackson	Merchant	7	2	6	15			
Q3910646	Crooked Creek Pass	Merchant	17	14	17	48	X		
Q3910647	Red Creek	K Potter	19	6	22	47	X		x
Q3910648	Toner Reservoir	Graves	29	8	33	70	X		
Q3910651	Frisco	McMenamy	7	20	30	57	X		
Q3910652	Vail Pass	Schenck/Kingery	16	12	30	58	X		
Q3910653	Red Cliff	Righter	14	12	14	40	X		
Q3910654	Minturn	J Potter	16	10	4	30			
Q3910655	Grouse Mountain	Daigneault/Smith	4	3	6	13			
Q3910656	Fulford	Claussner/Merchant	17	17	23	57	X		
Q3910657	The Seven Hermits	Merchant	19	13	15	47	X		
Q3910658	Suicide Mountain	Merchant	10	19	30	59	X		
Q3910661	Dillon	Black	15	20	22	57	X		
Q3910662	Willow Lakes	Kingerys	7	9	25	41	X		
Q3910663	Vail East	Righter	6	24	17	47	X		
Q3910664	Vail West	Righter	11	25	20	56	X		
Q3910665	Edwards	J Potter	21	14	9	44	X		
Q3910666	Wolcott	Kunkel	5	19	33	57	X		
Q3910667	Eagle	Merchant	18	19	53	90	X		
Q3910668	Gypsum	Merchant	15	15	42	72	X		
Q3910671	Ute Peak	Barrett	15	13	18	46	X		
Q3910672	Squaw Creek	Woodwards	11	13	36	60	X		
Q3910673	Mount Powell	Stv Jones	18	7	17	42	X		
Q3910674	Piney Peak	Hallock	20	7	19	46	X		x
Q3910675	Lava Creek	Merchant	12	17	19	48	X		
Q3910676	State Bridge	Merchant	11	20	26	57	X		
Q3910677	Castle Peak	Merchant	23	14	18	55	X		
Q3910678	Burns South	Merchant	21	16	23	60	X		
Q3910681	Sylvan Reservoir	Barrett	14	3	25	42	X		
Q3910682	Battle Mountain	Barrett	18	12	30	60	X		
Q3910683	King Creek	Phelps/Kingerys	19	8	28	55	X		
Q3910684	Sheephorn Mountain	Kingery	28	17	20	65	X		
Q3910685	Radium	Hallock	23	16	26	65	X		x
Q3910686	McCoy	Ewing	22	3	36	61	X		
Q3910687	Blue Hill	Ewing	29	5	36	70	X		
Q3910688	Burns North	Ewing	28	12	31	71	X		
Q3910711	Snowmass Mountain	L Roberts/Schofield	17	1	25	43	X		
Q3910712	Marble	Merchant/Vidal	18	10	9	37	X		
Q3910713	Chair Mountain	Wright/K Potter	28	2	20	50	X		x
Q3910715	Electric Mountain	Guadagno	16	6	26	48	X	SW	
Q3910716	Chalk Mountain	Levad	16	3	17	36	X		x
Q3910717	Leon Peak	Abramson	14	0	1	15	X		
Q3910718	Grand Mesa	Woodwards/Levad	10	12	32	54	X		x
Q3910721	Capitol Peak	Nelson	3	12	18	33	X		x
Q3910723	Placita	Clark/Zerbi	13	4	39	56	X		
Q3910724	Elk Knob	Kingerys	19	8	27	54	X		
Q3910725	Spruce Mountain	Kingerys	15	17	28	60	X		
Q3910726	Porter Mountain	Abramson	30	6	7	43	X		
Q3910727	Vega Reservoir	Lambeth/K Potter	20	7	29	56	X		x
Q3910728	Collbran	Goulding/K Potter	26	7	28	61	X		x
Q3910731	Basalt	Zerbi	10	4	59	73	X		
Q3910732	Mount Sopris	K Potter	31	2	12	45	X		x
Q3910733	Stony Ridge	K Potter	26	5	27	58	X		x

Quadcode	Map Name	Observer	Species per Block: PO	PR	CF	Total	CMPL	Alternate	BB
Q3910734	Quaker Mesa	Zerbi	12	4	23	39	X		
Q3910735	Flatiron Mountain	Grode	26	10	13	49	X		
Q3910736	Hightower Mountain	Kingerys	15	3	28	46	X		
Q3910737	South Mamm Peak	Radice	18	16	34	68	X		
Q3910738	Hawxhurst Creek	Radice et al	21	14	49	84	X		
Q3910741	Leon	Fuller/Levad/Vidal	22	14	34	70	X		x
Q3910742	Carbondale	Fuller/Johnson	27	29	24	80	X		
Q3910743	Cattle Creek	Clark	13	2	7	22			
Q3910744	Center Mountain	Zerbi	11	9	27	47	X		
Q3910745	Gibson Gulch	K Potter	30	9	21	60	X		x
Q3910746	Hunter Mesa	Graves/K Potter	21	22	28	71	X		x
Q3910747	North Mamm Peak	K Potter	25	7	29	61	X		x
Q3910748	Rulison	K Potter	21	7	19	47	X		x
Q3910751	Cottonwood Pass	Merchant	25	21	18	64	X		
Q3910752	Shoshone	Zerbi	13	13	25	51	X		
Q3910753	Glenwood Springs	Zerbi/K Potter	11	18	47	76	X		
Q3910754	Storm King Mtn.	Conway/K Potter	6	5	6	17			
Q3910755	New Castle	Ligon/Levad/K Potter	34	8	28	70	X		x
Q3910756	Silt	Graves/Levad/K Potter	21	8	32	61	X		
Q3910757	Rifle	Merchant/Zerbi/K Potter	17	8	42	67	X		
Q3910758	Anvil Points	Graves/K Potter	15	18	27	60	X		x
Q3910761	Dotsero	Merchant	24	16	30	70	X		
Q3910762	Broken Rib Creek	Merchant	24	15	20	59	X		
Q3910763	Carbonate	Grode	26	9	18	53	X		
Q3910764	Adams Lake	K Potter/Green	23	2	29	54	X		x
Q3910765	Deep Creek Point	Grode	26	17	16	59	X		
Q3910766	Rifle Falls	Zerbi	12	10	30	52	X		
Q3910767	Horse Mountain	Zerbi/K Potter	35	13	39	87	X		x
Q3910768	Rio Blanco	Graves/K Potter	27	18	34	79	X		x
Q3910771	Sugarloaf Mountain	K Potter	26	14	32	72	X		x
Q3910772	Sweetwater Lake	Merchant	17	25	27	69	X		
Q3910773	Deep Lake	Green/Hallock	25	8	23	56	X		
Q3910774	Blair Mountain	Hallock	19	8	24	51	X		x
Q3910775	Meadow Creek Lake	Merchant	24	10	19	53	X		
Q3910776	Triangle Park	Jordan	21	7	12	40	X		
Q3910777	Red Elephant Point	Grode	27	12	20	59	X		
Q3910778	Thirteenmile Creek	Wright	17	9	22	48	X		x
Q3910781	Dome Peak	Kingerys	16	8	18	42	X		
Q3910782	Trappers Lake	Kingerys	11	9	22	42	X		
Q3910783	Big Marvine Peak	Conway/Boyle	11	19	19	49	X		x
Q3910784	Oyster Lake	Cunningham	23	15	6	44			
Q3910785	Buford	Pantles	15	24	35	74	X		
Q3910786	Big Beaver Res.	Pantles	12	14	31	57	X		
Q3910787	Veatch Gulch	K Potter	30	15	22	67	X		x
Q3910788	Lo 7 Hill	K Potter	18	21	37	76	X		x
Q3910811	Skyway	Davis/Radice	20	6	5	31			
Q3910812	Lands End	McGinley	16	2	25	43	X		
Q3910813	Palisade	Lambeth	22	10	29	61	X		
Q3910814	Clifton	Levad	6	9	7	22	X		
Q3910815	Grand Junction	Levad	22	18	34	74	X		
Q3910816	Colorado Natl. Mon.	Hall/Woodward	19	13	31	63	X		
Q3910817	Battleship Rock	Kladder	19	15	29	63	X		
Q3910818	Sieber Canyon	Kladder	17	4	18	39	X		
Q3910821	Molina	Kladder/Levad	31	4	21	56	X		x
Q3910822	Mesa	Krehbiel/Levad	19	5	40	64	X		x
Q3910823	Cameo	McGinley	27	1	26	54	X		
Q3910824	Round Mountain	Lambeth/Wright	17	11	9	37	X		x
Q3910825	Corcoran Point	Lambeth/Levad	10	5	10	25	X		x

			Species per Block						
Quadcode	Map Name	Observer	PO	PR	CF	Total	CMPL	Alternate	BB
Q3910826	Fruita	Levad	9	5	23	37	X		x
Q3910827	Mack	Pearson	17	12	48	77	X		
Q3910828	Ruby Canyon	Dexter/Lambeth	8	14	18	40	X		
Q3910831	Housetop Mountain	Levad	27	3	20	50	X		x
Q3910832	De Beque	Hahn/Lambeth/Ireland	27	12	19	58	X		
Q3910833	Wagon Track Ridge	Williams/McVean/Levad	24	13	23	60	X		
Q3910834	Winter Flats	Williams	8	6	29	43	X		
Q3910835	Corcoran Peak	Tindall/Graham/Dexter/Wright	23	10	9	42	X		x
Q3910836	Ruby Lee Reservoir	Renner/Wright	8	9	12	29	X		
Q3910837	Highline Lake	Dexter/Levad	12	9	14	35	X		x
Q3910838	Badger Wash	Dexter	11	17	27	55	X		
Q3910841	Grand Valley	Graves/Stigen	17	20	10	47	X		x
Q3910842	Red Pinnacle	Kladder/Levad	30	9	14	53	X		x
Q3910843	Long Point	Kladder	20	15	13	48			
Q3910844	The Saddle	Moyer/Lambeth	28	4	11	43	X		
Q3910845	Middle Dry Fork	McVean/Lambeth	32	7	5	44	X		
Q3910846	Garvey Canyon	Wright/Stigen/Lambeth	27	11	4	42	X		x
Q3910847	Howard Canyon	Pearson	20	14	20	54	X		
Q3910848	Carbonera	Dexter/Lmbeth	15	19	13	47	X		
Q3910851	Forked Gulch	Levad	31	8	21	60	X		x
Q3910852	Circle Dot Gulch	Graves/Levad	36	12	23	71	X		x
Q3910853	Mount Blaine	Levad	24	1	19	44	X		x
Q3910854	Desert Gulch	Wright	19	22	22	63	X		x
Q3910855	Henderson Ridge	Heywood/Lambeth	28	12	14	54	X		
Q3910856	Calf Canyon	Brigham/Stigen	31	2	4	37	X		x
Q3910857	Douglas Pass	Hahn/Henwod/Levad	20	0	7	27			
Q3910858	Baxter Pass	Wlliams/Lambeth	24	6	13	43	X		
Q3910861	McCarthy Gulch	Levad	13	6	19	38	X		x
Q3910862	Cutoff Gulch	Wright	16	13	16	45	X		x
Q3910863	Bull Fork	Wright	25	8	9	42	X		x
Q3910864	Figure Four Spring	Wright	36	4	10	50	X		x
Q3910865	Razorback Ridge	Fagans/Wright	29	10	8	47	X		x
Q3910866	Brushy Point	Lambeth	16	4	15	35	X		
Q3910867	Big Foundation Creek	Lambeth	32	12	7	51	X		
Q3910868	East Evacuation Cr.	Pearson	22	1	21	44	X		
Q3910871	No Name Ridge	McGinley/Wright	31	6	15	52	X		x
Q3910872	Jessup Gulch	Wright	19	5	15	39	X		x
Q3910873	Rock School	Wright	22	8	11	41	X		x
Q3910874	Yankee Gulch	Toolen	20	4	11	35	X		
Q3910875	Black Cabin Gulch	Dexter	30	9	23	62	X		x
Q3910876	White Coyote Draw	Hawksworth	21	7	29	57	X		
Q3910877	Texas Mountain	Hawksworth	21	7	34	62			
Q3910878	Texas Creek	Wright	15	15	17	47	X		x
Q3910881	Segar Mountain	Wright	15	18	23	56	X		x
Q3910882	Greasewood Gulch	Wright	19	6	10	35	X		x
Q3910883	Square S Ranch	Wright	21	11	26	58	X		x
Q3910884	Wolf Ridge	Lambeths	14	15	12	41	X		
Q3910885	Sagebrush Hill	Dexter	15	6	16	37	X		
Q3910886	Philadelphia Creek	Lambeth	24	6	16	46	X		
Q3910887	Water Canyon	Boyle/Lambeth	15	4	9	28	X		x
Q3910888	Banta Ridge	Wright	14	11	20	45	X		x
Q3910921	Bitter Creek Well	Lambeth/Ferreira	10	4	10	24			
Q3910931	Bar X Wash	McGinley	3	1	7	11	X		
Q3910941	Jim Canyon	Creedon	13	20	2	35	X		
Q4010212	Wray	Stv Jones	11	10	13	34	X		
Q4010213	Robb	Kaempfer	11	8	11	30	X		
Q4010214	Eckley	Kaempfer	21	6	22	49	X		
Q4010215	Schramm	Soderberg	9	1	10	20	X		

			Species per Block						
Quadcode	Map Name	Observer	PO	PR	CF	Total	CMPL	Alternate	BB
Q4010216	Yuma South	Soderberg	8	1	12	21	X		
Q4010217	Otis S.E.	Stv Jones	9	7	14	30	X		
Q4010218	Snyder Lake	Stv Jones	18	2	17	37	X		
Q4010222	Wray N.W.	Kaempfer	7	11	13	31	X		
Q4010223	Eckley N.E.	Kaempfer	9	11	11	31	X		
Q4010224	Eckley N.W.	Kaempfer	12	7	14	33	X		
Q4010225	Yuma N.E.	Borichevsky	7	5	9	21	X		
Q4010226	Yuma North	Kaempfer	6	3	12	21	X		
Q4010227	Hyde	Stv Jones	8	3	16	27	X		
Q4010228	Otis	Kaempfer	8	7	17	32	X		
Q4010232	Alvin S.W.	Pusateri/Hallock	9	8	15	32	X		x
Q4010233	Wauneta	Kaempfer	6	10	8	24	X		
Q4010234	Old Baldy	Kaempfer	8	7	11	26	X		
Q4010235	Clarkville S.E.	Borichevsky/Kaempfer	9	13	9	31	X		
Q4010236	Clarkville S.W.	Kaempfer	10	9	11	30	X		
Q4010237	Lone Star	Kaempfer	12	4	12	28	X		
Q4010238	Burdett	Kaempfer	8	8	10	26	X		
Q4010242	Alvin N.W.	Mollhoff	3	14	12	29	X		
Q4010243	Wauneta N.E.	Opler	13	3	12	28	X		
Q4010244	Fiddler Peak	Opler	10	7	11	28	X		
Q4010245	Clarkville N.E.	Opler	8	6	15	29	X		
Q4010246	Clarkville	Opler	11	4	13	28	X		
Q4010247	New Haven	Opler	9	0	11	20	X		
Q4010248	Glacken Hill	Opler	10	4	17	31	X		
Q4010252	Amherst S.W.	Hallock	7	8	12	27	X		x
Q4010253	Holyoke	Hallock	9	6	10	25	X		x
Q4010254	Paoli	Whitney/Dillon	2	7	17	26	X		x
Q4010255	Haxtun S.E.	Hallock	8	7	14	29	X		x
Q4010256	Rockland	Whitney/Dillon	13	6	20	39	X		x
Q4010257	St. Petersburg	Opler	10	3	9	22	X		
Q4010258	Leroy	Opler	12	1	13	26	X		
Q4010262	Amherst	Mollhoff/Hallock	9	11	12	32	X		x
Q4010263	Holyoke N.E.	Hallock/Wieland	8	7	16	31	X		x
Q4010264	Holyoke N.W.	Pulliam/Wieland	7	4	14	25	X		
Q4010265	Haxtun East	Hallock	6	5	10	21	X		x
Q4010266	Haxtun West	Hallock/Wieland/Kreher	11	6	25	42	X		x
Q4010267	Fleming	Prather	7	4	12	23	X		
Q4010268	Uhler Ranch	Prather	9	2	14	25	X		
Q4010272	Venango S.W.	Mollhoff	5	14	14	33	X		
Q4010273	Julesburg S.E.	Opler/Pulliam	10	3	15	28	X		
Q4010274	Julesburg S.W.	Opler/Pulliam	9	3	15	27	X		
Q4010275	Marks Butte	Prather	5	3	14	22	X		
Q4010276	Tamarack Ranch	Pulliam	8	2	16	26	X		
Q4010277	Crook	Sedgwick/Hallock	10	4	16	30	X		x
Q4010278	Proctor	Prather	12	7	6	25	X		
Q4010282	Venango N.W.	Mollhoff	2	11	8	21	X		
Q4010283	Julesburg	Stv Jones	13	5	24	42	X		
Q4010284	Ovid	Stv Jones	7	5	17	29	X		
Q4010285	Sedgwick	Stv Jones	7	12	25	44	X		
Q4010286	Julesburg Reservoir	Prather	19	27	33	79	X		
Q4010287	Twin Buttes	Dillon	12	18	29	59	X		x
Q4010288	Haystack Butte	Dillon	7	9	24	40	X		x
Q4010311	Akron S.E.	Soderberg	2	8	11	21	X		
Q4010312	Akron S.W.	Soderberg	0	4	15	19	X		
Q4010313	Pinneo S.E.	Soderberg	3	8	14	25	X		
Q4010314	Rago	Kaempfer	10	4	18	32	X		
Q4010315	Gary	Kaempfer	10	3	15	28	X		
Q4010316	Huey Ranch	Kaempfer	3	5	8	16	X		

			Species per Block						
Quadcode	Map Name	Observer	PO	PR	CF	Total	CMPL	Alternate	BB
Q4010317	Vallery S.E.	BCNA	4	4	20	28	X		
Q4010318	Adena	Kaempfer	6	4	12	22	X		
Q4010321	Platner	Soderberg	3	8	16	27	X		
Q4010322	Akron	Soderberg	9	10	24	43	X		
Q4010323	Fremont Butte	Wainwrights/Kaempfer	8	9	10	27	X		
Q4010324	Pinneo	Wainwrights/Kaempfer	11	9	7	27	X		
Q4010325	Miller Ranch	Wainwrights/Kaempfer	9	5	11	25	X		
Q4010326	Round Top	Wainwrights/Kaempfer	8	4	7	19	X		
Q4010327	Lamb	Wainwrights	9	5	17	31	X		
Q4010328	Vallery	R Davis	9	14	7	30	X		
Q4010331	Buffalo Spgs Ranch S.E.	Kaempfer	5	4	13	22	X		
Q4010332	Buffalo Spgs Ranch	Soderberg	0	6	20	26	X		
Q4010333	Merino S.E.	Anderson/Kaempfer	11	8	12	31	X		
Q4010334	Merino S.W.	Litz/Kaempfer	6	6	13	25	X		
Q4010335	Brush East	Stv Jones	4	6	17	27	X		
Q4010336	Brush West	Litz	8	3	29	40	X		
Q4010337	Fort Morgan	Browns	15	14	25	54	X		
Q4010338	Weldona	Browns	15	20	21	56	X		
Q4010341	Buffalo Spgs Ranch N.E.	Opler	15	3	14	32	X		
Q4010342	Buffalo Spgs Ranch N.W.	Opler/Lentz	15	5	17	37	X		
Q4010343	Merino	Opler/Lentz	11	5	11	27	X		
Q4010344	Messex	Andrews	26	20	22	68	X		
Q4010345	Antelope Springs	Opler	23	5	16	44	X		
Q4010346	Dead Horse Springs	Opler	15	8	13	36	X		
Q4010347	Peace Valley School	Opler/Berry	14	9	20	43	X		
Q4010348	Judson Hills	Opler/Berry	6	4	13	23	X		
Q4010351	Reiradon Hill	Opler	8	6	15	29	X		
Q4010352	Sterling South	Opler/Dillon	6	9	15	30	X		x
Q4010353	Atwood	Erthal	16	14	17	47	X		
Q4010354	Willard	Opler	12	3	9	24	X		
Q4010355	Stoneham S.E.	Opler	18	1	16	35	X		
Q4010356	Stoneham	Opler	10	1	12	23	X		
Q4010357	Raymer	Opler	13	4	12	29	X		
Q4010358	Buckingham	Opler	22	1	15	38	X		
Q4010361	Galien	Mollhoff		1	10	11	X		
Q4010362	Sterling North	Maguire/Dillon	13	20	28	61	X		x
Q4010363	Atwood N.E.	Hawksworth	10	1	15	26	X		
Q4010364	Wild Horse Lake	Hawksworth	9	1	12	22	X		
Q4010365	Stoneham N.E.	Opler	15	3	15	33	X		
Q4010366	Stoneham N.W.	Barber	8	9	19	36	X		
Q4010367	Raymer N.E.	Barber	6	10	23	39	X		
Q4010368	Raymer N.W.	Barber	4	1	22	27	X		
Q4010371	Iliff	Maguire	4	7	16	27	X		
Q4010372	Padroni	Holt	12	5	20	37	X		
Q4010373	N. Sterling Res.	Holt/Dillon	14	9	29	52	X		x
Q4010374	Westplains	Dillon	6	5	4	15	X		x
Q4010375	Avalo S.E.	Mollhoff	0	2	17	19	X		
Q4010376	Avalo	Cheney/Dillon	4	2	21	27	X		x
Q4010377	Gatehook Spring	Hallock	12	13	17	42	X		x
Q4010378	Pawnee Buttes	Barber	9	4	25	38	X		
Q4010381	Peetz	Dillon/Leatherman	10	9	16	35	X		x
Q4010382	Padroni N.W.	Dillon/Leatherman	9	6	11	26	X		x
Q4010383	Kirchnavy Butte	Dillon	9	7	12	28	X		x
Q4010384	Chimney Canyons	Mollhoff	0	1	10	11	X		
Q4010385	Dipper Spring	Dillon	9	6	21	36	X		x
Q4010386	Battle Canyon	Hallock	7	16	20	43	X		x
Q4010387	Vim School	Hallock	12	7	17	36	X		x
Q4010388	Dolan Spring	Hallock	10	9	18	37	X		x

Quadcode	Map Name	Observer	Species per Block PO	PR	CF	Total	CMPL	Alternate	BB
Q4010411	Hoyt	Zipser/Kaempfer	20	8	13	41	X		
Q4010412	Wiggins S.W.	Wainwrights	8	3	14	25	X		
Q4010413	South Roggen	Wainwrights	4	5	14	23	X		
Q4010414	Prospect Valley	Woolf/Dirckx	18	14	12	44	X		
Q4010415	Keenesburg	Kaempfer	20	23	28	71	X		
Q4010416	Hudson	J Hill/Dirckx	17	16	28	61	X		
Q4010417	Fort Lupton	Browns	10	16	25	51	X		
Q4010418	Frederick	Arndt/Dirckx	10	11	23	44	X		
Q4010421	Wiggins	Wainwrights	8	5	16	29	X		
Q4010422	Omar	McConkey/Kaempfer	9	7	14	30	X		
Q4010423	Roggen	Hyden	3	6	4	13			
Q4010424	Tampa	Browns	7	7	13	27	X		
Q4010425	Klug Ranch	DW King	10	12	19	41	X		
Q4010426	Milton Reservoir	BBC	6	5	19	30	X		
Q4010427	Platteville	Arndt/Wainwrights	13	9	20	42	X		
Q4010428	Gowanda	Wainwrights	7	12	23	42	X		
Q4010431	Orchard	Browns	8	7	14	29	X		
Q4010432	Masters	Martin/Browns	15	21	19	55	X		
Q4010433	Dearfield	Piombino	12	15	26	53	X		
Q4010434	Hardin	Piombino/Swies	4	3	22	29	X		
Q4010435	Valley View School	Piombino/Swies	6	6	21	33	X		
Q4010436	La Salle	Swies	10	16	36	62	X		
Q4010437	Milliken	Arndt/Wainwrights	10	12	14	36	X		
Q4010438	Johnstown	Hyden	24	15	7	46	X		
Q4010441	Sunken Lake	Opler	17	9	28	54	X		
Q4010442	Greasewood Lake	Opler	8	2	4	14	X		
Q4010443	Point of Rocks	Opler	6	0	8	14		NE	
Q4010444	Barnesville	Opler	1	1	7	9		SW	
Q4010445	Kersey	Cairo	17	8	26	51	X		
Q4010446	Greeley	Holt/Kozan	16	5	22	43	X		
Q4010447	Bracewell	Holt/Cairo	9	7	21	37	X		
Q4010448	Windsor	Holt	17	7	26	50	X		
Q4010451	Keota S.E.	Opler	5	1	7	13	X		
Q4010452	Dutch Girl Lake	Opler	16	2	12	30	X		
Q4010453	Fosston	Opler	4	4	12	20	X		
Q4010454	Cornish	Hernandez	10	5	14	29	X		
Q4010455	Galeton	Holt	14	1	25	40	X		
Q4010456	Eaton	Opler	11	7	26	44	X		
Q4010457	Severance	Opler	17	11	17	45	X		
Q4010458	Timnath	Opler	15	8	20	43	X		
Q4010461	Keota	Barber	1	2	14	17	X		
Q4010462	Keota N.W.	Opler	15	1	15	31	X		
Q4010463	Briggsdale	Guertin	9	2	11	22	X		
Q4010464	Baker Draw	Erthal	11	7	14	32	X		
Q4010465	Purcell	Opler	13	4	17	34	X		
Q4010466	Antelope Reservoir	Opler	14	4	22	40	X		
Q4010467	Nunn	Opler	7	5	16	28	X		
Q4010468	Cobb Lake	Householder	7	6	19	32	X		
Q4010471	Grover S.E.	Graul	5	3	16	24	X		
Q4010472	Grover South	Dillon	3	3	9	15	X		x
Q4010473	Hereford S.E.	Leatherman	1	0	7	8			
Q4010474	Reno Reservoir	Van Erp/Dillon	11	9	20	40	X		x
Q4010475	Chalk Bluffs S.E.	Ryder	4	7	19	30	X		
Q4010476	Chalk Bluffs S.W.	Sedgwick/Dillon	9	2	18	29	X		x
Q4010477	Dover	Ryder	7	16	22	45	X		
Q4010478	Carr S.W.	Opler	8	4	14	26	X		
Q4010481	Grover N.E.	Dillon	4	5	18	27	X		x
Q4010482	Grover North	Hawksworth/Dillon	4	3	12	19	X		x

			Species per Block						
Quadcode	Map Name	Observer	PO	PR	CF	Total	CMPL	Alternate	BB
Q4010483	Hereford	Leatherman	12	5	7	24	X		
Q4010484	Hereford N.W.	Van Erp	12	13	23	48	X		
Q4010485	Chalk Bluffs East	Dillon	5	6	18	29	X		x
Q4010486	Chalk Bluffs West	Barber	0	0	9	9	X		
Q4010487	Carr East	Martins	7	7	16	30	X		
Q4010488	Carr West	Opler	13	4	22	39	X		
Q4010511	Erie	Swies	10	8	35	53	X		
Q4010512	Niwot	Browns	14	14	22	50	X		
Q4010513	Boulder	Stv Jones	23	21	57	101	X		
Q4010514	Gold Hill	BBC	13	7	40	60	X		
Q4010515	Ward	Kaempfer	20	5	21	46	X		
Q4010516	Monarch Lake	Hallock	18	10	26	54	X		
Q4010517	Strawberry Lake	Wells/Nelson	8	13	23	44	X		x
Q4010518	Granby	Collins	30	16	26	72	X		
Q4010521	Longmont	J Harrison	21	10	37	68	X		
Q4010522	Hygiene	Borichevsky	10	13	21	44	X		
Q4010523	Lyons	Browns	22	25	36	83	X		
Q4010524	Raymond	Simon/Nielsen	26	28	25	79	X		
Q4010525	Allens Park	Bouricius	13	19	55	87	X		
Q4010526	Isolation Peak	Bouricius	8	18	22	48	X		
Q4010527	Shadow Mountain	Connor/Nelson	14	11	36	61	X		
Q4010528	Trail Mountain	Jasper	9	16	22	47	X		
Q4010531	Berthoud	Means	5	8	36	49	X		
Q4010532	Carter Lake Res.	Kaempfer	25	17	24	66	X		
Q4010533	Pinewood Lake	BBC	35	9	40	84	X		
Q4010534	Panorama Peak	Stv Jones	24	20	32	76	X		
Q4010535	Longs Peak	Figgs/Nelson	15	23	25	63	X		
Q4010536	McHenrys Peak	Hill/Connor	19	10	14	43	X		
Q4010537	Grand Lake	Lewis/RMNP	16	12	20	48	X		
Q4010538	Bowen Mtn.	Lentz	19	0	23	42	X		
Q4010541	Loveland	Means	10	7	30	47	X		
Q4010542	Masonville	Kramer/Holt	16	13	38	67	X		
Q4010543	Drake	Hawksworth	22	10	26	58	X		
Q4010544	Glen Haven	Harden	14	8	24	46	X		
Q4010545	Estes Park	Opler/Connor/Pulliam	34	8	38	80	X		
Q4010546	Trail Ridge	Ryder	21	16	56	93	X		
Q4010547	Fall River Pass	Barber/Melcher/RMNP	18	8	12	38	X		
Q4010548	Mount Richthofen	Wilkinson/Lentz	14	2	17	33	X		
Q4010551	Fort Collins	Opler	11	10	25	46	X		
Q4010552	Horsetooth Reservoir	Opler/Leatherman	32	15	38	85	X		
Q4010553	Buckhorn Mountain	Wilkinson/Pulliam	22	26	26	74	X		
Q4010554	Crystal Mountain	Moody	17	5	8	30			
Q4010555	Pingree Park	Hein/RMNP	21	8	13	42	X		
Q4010556	Comanche Peak	Hein/Connor	11	5	9	25			
Q4010557	Chambers Lake	Ryder/Connor	11	17	15	43	X		
Q4010558	Clark Peak	Ryder	5	18	16	39	X		
Q4010561	Wellington	Rhoads	30	8	18	56	X		
Q4010562	Laporte	Leatherman	16	13	37	66	X		
Q4010563	Poudre Park	Hollngsworth/Lisk	22	7	34	63	X		
Q4010564	Big Narrows	Hawksworth	18	4	25	47	X		
Q4010565	Rustic	Barber/Hein	11	10	24	45	X		
Q4010566	Kinikinik	Barber	8	6	20	34	X		
Q4010567	Boston Peak	Anderson	17	13	14	44	X		
Q4010568	Rawah Lakes	Dillon	17	7	14	38	X		
Q4010571	Buckeye	Ryder	20	12	16	48	X		
Q4010572	Livermore	Rhoades	21	5	28	54	X		
Q4010573	Livermore Mountain	Cringan/Pulliam	32	22	13	67	X		
Q4010574	Haystack Gulch	Dillon	24	14	28	66	X		

Quadcode	Map Name	Observer	Species per Block				CMPL	Alternate	BB
			PO	PR	CF	Total			
Q4010575	Red Feather Lakes	Cringan	20	11	27	58	X		
Q4010576	South Bald Mountain	Barber	21	9	11	41	X		
Q4010577	Deadman	Dillon	11	8	11	30	X		x
Q4010578	Glendevey	Simmons/Hallock	30	15	25	70	X		x
Q4010581	Round Butte	Ryder	6	15	28	49	X		
Q4010582	Table Mountain	Martins	10	6	30	46	X		
Q4010583	Virginia Dale	Cringan/Hallock	16	16	21	53	X		x
Q4010584	Cherokee Park	Cringan/Hallock/Curdts	24	16	30	70	X		x
Q4010585	Diamond Peak	Dillon	25	11	14	50	X		
Q4010586	Eaton Reservoir	Martins	14	7	16	37	X		
Q4010587	Sand Creek Pass	Martins	14	8	20	42	X		
Q4010588	Crazy Mountain	Hallock/Berry/Opler	31	10	28	69	X		x
Q4010611	Hot Sulphur Springs	Pantles	12	16	23	51	X		
Q4010612	Parshall	Gable	7	13	7	27	X		
Q4010613	Junction Butte	Hallock	12	11	20	43	X		x
Q4010614	Kremmling	Barrett	26	24	35	85	X		
Q4010615	Gore Pass	Jasper/Levad	12	16	23	51	X		x
Q4010616	Lynx Pass	Ewing/K Nelson	21	7	19	47	X		
Q4010617	Toponas	Washburne/Ewing	16	14	38	68	X		
Q4010618	Trapper	Ewing	21	10	22	53	X		
Q4010621	Cabin Creek	Wells	21	14	2	37			
Q4010622	Corral Peaks	Gables/Barrett	19	9	17	45	X		
Q4010623	Gunsight Pass	Hallock	16	19	29	64	X		x
Q4010624	Hinman Reservoir	Liewer	35	13	32	80	X		
Q4010625	Tyler Mountain	Lieewer	22	18	26	66	X		
Q4010626	Gore Mountain	Ferner/Hallock	18	8	19	45	X		
Q4010627	Green Ridge	K Nelson	17	2	23	42	X		
Q4010628	Yampa	Fratt	29	5	4	38			
Q4010631	Radial Mountain	Lentz	10	5	19	34	X		
Q4010632	Parkview Mountain	Lentz	14	6	29	49	X		
Q4010633	Hyannis Peak	Hallock	18	9	19	46	X	NW	x
Q4010634	Whiteley Peak	Liewer	24	11	31	66	X		
Q4010635	Lake Agnes	Liewer	19	11	15	45	X		
Q4010636	Walton Peak	Hallock	13	11	21	45	X		x
Q4010637	Blacktail Mountain	Hoffmann	13	9	16	38	X		
Q4010638	Oak Creek	Levad	19	4	27	50	X		x
Q4010641	Jack Creek Ranch	Simmons/Pantle	11	10	23	44	X		
Q4010642	Rand	Barton/Dillon	22	10	23	55	X		x
Q4010643	Buffalo Peak	Hallock	17	10	17	44	X		x
Q4010644	Spicer Peak	Gillespie/Barrett	22	7	20	49	X		
Q4010645	Rabbit Ears Peak	Opler/Dillon	16	13	33	62	X		x
Q4010646	Mount Werner	NAS/Kingerys	19	12	16	47	X		
Q4010647	Steamboat Springs	Toolen/NAS/Hallock	29	22	26	77	X		x
Q4010648	Cow Creek	Paton/Hallock	20	19	22	61	X		x
Q4010651	Gould	Ryder	5	25	12	42	X		
Q4010652	Owl Ridge	Pantles	5	19	26	50	X		
Q4010653	MacFarlane Reservoir	Barton/Ryder/Dillon	12	10	24	46	X		x
Q4010654	Coalmont	Barton/Dillon	13	12	30	55	X		x
Q4010655	Teal Lake	Pulliam/Dillon	10	9	37	56	X		x
Q4010656	Buffalo Pass	Kingerys/NAS	17	6	24	47	X		
Q4010657	Rocky Peak	Simon	24	12	33	69	X		
Q4010658	Mad Creek	Simon	20	13	23	56	X		
Q4010661	Johnny Moore Mtn.	Dillon	15	15	26	56	X		x
Q4010662	Gould N.W.	Dillon	7	14	29	50	X		x
Q4010663	Walden	Arapaho NWR	16	12	47	75	X		
Q4010664	Delaney Butte	Dillon	13	16	26	55	X		x
Q4010665	Pitchpine Mountain	Dillon	8	22	30	60	X		x
Q4010666	Mount Ethel	Dillon	13	3	21	37	X		x

Quadcode	Map Name	Observer	Species per Block				CMPL	Alternate	BB
			PO	PR	CF	Total			
Q4010667	Floyd Peak	Litteral	22	6	9	37	X		
Q4010668	Clark	Hallock	23	15	27	65	X		x
Q4010671	Shipman Mtn.	Dillon	15	2	14	31			x
Q4010672	Eagle Hill	Dillon	10	6	21	37	X		x
Q4010673	Cowdrey	Dillon	8	8	26	42	X		x
Q4010674	Lake John	Dillon	4	10	28	42	X		x
Q4010675	Boettcher Lake	Guertin/Dillon	16	25	23	64	X		x
Q4010676	Mount Zirkel	Hallock	24	6	17	47	X		x
Q4010677	Farwell Mountain	Halverson/Mullineaux/Walton	25	16	25	66	X		
Q4010678	Hahns Peak	Kingerys/NAS	27	3	25	55	X		
Q4010681	Old Roach	Hallock	25	7	17	49	X		x
Q4010682	Kings Canyon	Dillon	17	6	12	35	X		x
Q4010683	Northgate	Dillon	16	16	29	61	X		x
Q4010684	Independence Mtn.	Dillon	15	6	30	51	X		x
Q4010685	Pearl	Hallock	23	17	24	64	X		x
Q4010686	Davis Peak	Rechel/Hallock	20	13	22	55	X		x
Q4010687	West Fork Lake	Hallock	19	9	22	50	X		x
Q4010688	Elkhorn Mountain	Righter/Barrett	23	22	31	76	X		
Q4010711	Orno Peak	Bridges	18	6	25	49	X		
Q4010712	Devils Causeway	Massey	21	8	25	54	X		
Q4010713	Ripple Creek	Boyle	18	9	22	49	X		x
Q4010714	Lost Park	Toolen/Reichert	22	4	30	56	X		
Q4010715	Fawn Creek	Boyle	24	7	24	55	X		x
Q4010716	Sawmill Mountain	Boyle	21	8	29	58	X		x
Q4010717	Rattlesnake Mesa	Reichert	34	5	32	71	X		
Q4010718	Meeker	Reichert	23	13	26	62	X		
Q4010721	Sand Point	K Nelson/Giffen	21	3	22	46	X		
Q4010722	Dunckley Pass	Gables/K Nelson	21	9	14	44	X		
Q4010723	Pagoda Peak	Massey	19	11	17	47	X		
Q4010724	Slide Creek	Falice	19	17	30	66	X		
Q4010725	Sleepy Cat Peak	Massey	22	10	20	52	X		
Q4010726	Thornburgh	Massey	19	9	9	37	X		
Q4010727	Ninemile Gap	Toolen	23	21	16	60	X		
Q4010728	Devils Hole Gulch	Levad	27	4	16	47	X		x
Q4010731	Rattlesnake Butte	Wright	27	14	17	58	X		x
Q4010732	Dunckley	Hallock	17	23	29	69	X		x
Q4010733	Hayden Gulch	Toolen/Hallock	18	21	26	65	X		
Q4010734	Pagoda	Bridges	30	2	32	64	X		
Q4010735	Hamilton	Bridges	19	8	28	55	X		
Q4010736	Monument Butte	J Ward	17	23	18	58	X		
Q4010737	Axial	J Ward	10	17	32	59	X		
Q4010738	Easton Gulch	Levad	19	4	17	40	X		x
Q4010741	Milner	Wright	13	4	9	26	X		x
Q4010742	Mount Harris	CFO	34	21	19	74	X		
Q4010743	Hayden	Wright/Kingerys	25	4	18	47	X		x
Q4010744	Breeze Mountain	Toolen	16	9	11	36			
Q4010745	Castor Gulch	Luke	15	9	35	59	X		
Q4010746	Round Bottom	Wright	14	18	29	61	X		x
Q4010747	Horse Gulch	J Ward	7	7	18	32	X		
Q4010748	Juniper Hot Springs	Levad	4	5	8	17	X		x
Q4010751	Wolf Mountain	Wright	18	9	21	48	X		x
Q4010752	Hooker Mountain	Roz Garcia	26	9	62	97	X		
Q4010753	Rock Spring Gulch	Hallock/Renner	23	23	46	92	X		x
Q4010754	Ralph White Lake	Toolen	24	19	27	70	X		
Q4010755	Craig	J Ward	12	23	30	65	X		
Q4010756	Pine Ridge	Boyle	10	7	10	27	X		x
Q4010757	Lay S.E.	Marsh	7	6	7	20	X		
Q4010758	Lay	Marsh	14	10	15	39	X		

Quadcode	Map Name	Observer	Species per Block: PO	PR	CF	Total	CMPL	Alternate	BB
Q4010761	Pilot Knob	Loughridge/Wright	26	15	24	65	X		x
Q4010762	Quaker Mountain	Roz Garcia	4	7	22	33			
Q4010763	Slide Mountain	Hallock	32	11	41	84	X		x
Q4010764	McInturf Mesa	Boyle	13	5	11	29	X		x
Q4010765	Craig N.E.	Boyle/J Ward	5	9	7	21	X		x
Q4010766	Craig N.W.	Boyle/J Ward	6	7	14	27	X		x
Q4010767	Iron Springs	Boyle/J Ward	3	1	9	13	X		x
Q4010768	Adobe Springs	Boyle	6	6	9	21	X		x
Q4010771	Meaden Peak	Sears/Barrett	30	8	21	59	X		
Q4010772	Bears Ears Peaks	Renner	7	15	34	56	X		
Q4010773	Buck Point	Skorkowsky/Barrett	19	14	19	52	X		
Q4010774	Freeman Reservoir	Kingerys	10	4	31	45	X		
Q4010775	Fortification	J Ward/CFO	13	32	22	67	X		
Q4010776	East Timberlake Cr	J Ward/Hollowed/Dexter	11	14	10	35	X		
Q4010777	Great Divide	J Ward/Hollowed/Boyle	12	14	7	33	X		
Q4010778	Mayberry Spring	Howes/Garcia	5	19	20	44	X		
Q4010781	Shield Mountain	Hallock	24	9	24	57	X		x
Q4010782	Tumble Mountain	Hallock	19	15	24	58	X		x
Q4010783	Fly Creek	Hallock	15	17	20	52	X		x
Q4010784	Bakers Peak	CFO	20	25	26	71	X		
Q4010785	Fortification N.E.	J Ward	11	8	18	37	X		
Q4010786	Pole Gulch	Boyle/J Ward	11	10	12	33	X		x
Q4010787	Thornburgh Gulch	J Ward	3	4	10	17	X		
Q4010788	Bighole Butte	Boyle/J Ward	14	10	11	35	X		x
Q4010811	Buckskin Point	Wright	30	12	36	78	X		x
Q4010812	White River City	Toolen	36	6	27	69	X		
Q4010813	Barcus Creek S.E.	Hollowed	29	15	13	57	X		
Q4010814	Barcus Creek	Hollowed/K Potter	17	9	16	42	X		x
Q4010815	Calamity Ridge	K Potter	13	13	22	48	X		x
Q4010816	Gillam Draw	Levad	18	7	19	44	X		x
Q4010817	Rangely	Hawksworth	25	10	7	42	X		
Q4010818	Banty Point	Wright	15	5	15	35	X		x
Q4010821	White Rock	Levad	29	4	10	43	X		
Q4010822	Indian Valley	Wright	23	4	18	45	X		x
Q4010823	Smizer Gulch	Dexter	26	17	19	62	X		x
Q4010824	Rough Gulch	Wright	23	8	29	60	X		x
Q4010825	Divide Creek	Wright	16	10	22	48	X		x
Q4010826	Cactus Reservoir	Wright	23	14	22	59	X		x
Q4010827	Rangely N.E.	K Potter	14	10	16	40	X		x
Q4010828	Mellen Hill	Levad	6	5	8	19	X		x
Q4010831	Price Creek	Levad	19	2	19	40	X		x
Q4010832	Citadel Plateau	Levad	28	1	11	40	X	SW	x
Q4010833	Wapiti Peak	Levad	17	3	9	29	X		x
Q4010834	Elk Springs	Wright	16	11	16	43	X		x
Q4010835	M F Mountain	Figgs/Levad	11	2	7	20	X		x
Q4010836	Skull Creek	Figgs	29	7	4	40			
Q4010837	Lazy Y Point	Figgs	21	14	17	52	X		
Q4010838	Plug Hat Rock	Figgs	20	25	10	55	X		
Q4010841	Juniper Mountain	Levad/Petch	16	9	22	47	X		x
Q4010842	Cedar Knob	Toolen/Lentz	14	5	21	40	X		
Q4010843	Cross Mtn. Canyon	Toolen/Lentz	14	2	6	22			
Q4010844	Twelvemile Mesa	Toolen/Lentz/Levad	22	1	10	33	X		
Q4010845	Indian Water Canyon	Wright	15	8	16	39	X		x
Q4010846	Haystack Rock	K Potter	17	8	16	41	X		x
Q4010847	Tanks Peak	K Potter	25	12	27	64	X		x
Q4010848	Hells Canyon	K Potter	19	9	15	43	X		x
Q4010851	Maybell	Peck/Toolen	28	10	11	49	X		
Q4010852	Sunbeam	Dexter	13	9	7	29	X		

			Species per Block						
Quadcode	Map Name	Observer	PO	PR	CF	Total	CMPL	Alternate	BB
Q4010853	Peck Mesa	Robinson/Dexter	19	19	7	45	X		
Q4010854	Lone Mountain	Dexter	20	15	7	42	X		
Q4010855	Limestone Hill	Wright	17	4	17	38	X		x
Q4010856	Greystone	Boyle	20	3	24	47	X		x
Q4010857	Zenobia Peak	Toolen/Dexter/Wright	16	3	19	38	X		
Q4010858	Canyon of Lodore South	Dexter	18	11	8	37	X		
Q4010861	Bald Mountain	Boyle	4	6	9	19	X		x
Q4010862	Ninemile Hill	Toolen	10	8	9	27	X		
Q4010863	Sevenmile Draw	Kingerys	18	7	24	49	X		
Q4010864	Clay Buttes	Kingery	19	8	9	36	X		
Q4010865	Sheephead Basin	Kingery	12	9	5	26			
Q4010866	Vermillion Mesa	Facer/Boyle	11	11	12	34	X		x
Q4010867	Jack Springs	Toolen	20	5	24	49	X		
Q4010868	Canyon of Lodore North	Toolen/Zinn/Dunnick	25	5	17	47	X		
Q4010871	The Nipple S.E.	Levad	20	6	9	35	X		x
Q4010872	The Nipple	Toolen/Kingery/Levad	14	6	16	36	X		x
Q4010873	Lang Spring	Roth/Brevillier/Boyle	7	8	10	25	X		x
Q4010874	Sheepherder Springs	Lambeth	4	1	7	12	X		
Q4010875	G Spring	Roth	0	23	17	40	X		
Q4010876	Irish Canyon	Toolen	11	3	6	20	X		
Q4010878	Lodore School	Kingery/Fellows	23	26	19	68	X		
Q4010881	The Nipple N.E.	Levad	4	5	10	19	X		x
Q4010882	Reservoir Draw	Boyle	11	4	8	23	X		x
Q4010883	Powder Wash	Brevillier/Boyle	9	4	9	22	X		x
Q4010884	Coffeepot Spring	Levad	8	3	8	19	X		x
Q4010885	Hiawatha	Toolen	11	0	6	17	X		
Q4010886	Sugarloaf Butte	Toolen	15	1	7	23	X		
Q4010887	Sparks	Luke	10	33	17	60	X		
Q4010888	Beaver Basin	Kingerys	22	8	27	57	X		
Q4010921	Dinosaur	Levad	12	10	21	43	X		x
Q4010951	Jones Hole	Bridges	0	0	1	1			
Q4010971	Swallow Canyon	Bridges	36	3	11	50			
Q4010981	Willow Creek Butte	Madsen	42	2	1	45	X		

APPENDIX B: ABUNDANCE ESTIMATES

These tables present estimates of the number of breeding pairs for most species that breed in the state. Derived from Atlas abundance codes (Priority blocks only), they delineate low, medium, and high calculations. Table B1 gives statewide totals and calculates the number of pairs per acre and its reciprocal, how many acres it takes to support one pair of birds. Table B2 provides species numbers. Table B3 explains the formula used to compute these figures.

The Results section in the front of the book discusses the varied ways that atlasers reported abundance codes. Because of this variability, users of the book should not construe these numbers as true population figures. The calculation for total breeding pairs in Colorado, even the High Estimate, portrays an extremely low number. Certainly Colorado has far more breeding birds than one pair per acre.

Statistically these abundance numbers present what statisticians term a "first approximation."Despite low precision, the numbers at least provide a starting point to evaluate the relative abundance of species that breed in Colorado.

The random selection of Atlas blocks distorts extrapolations for some species, those that breed in sites that are not evenly distributed across the landscape. The breeding requirements of certain species produce too patchy a distribution for valid extrapolations, which assume uniform availability of suitable nesting habitat throughout the extrapolated area (Van Remsen pers. comm.).

A natural continuum of nest-habitat patchiness moves from evenly distributed to highly clumped (Van Remsen pers. comm.). Atlas figures provide valid estimates for species that have a relatively "two-dimensional," or even, distribution of nesting habitat (those that nest, for example, in grasslands, shrublands, conifers, aspen, urban and rural areas, and, probably, tundra). These projections do not work well for highly colonial species (grebes, herons, ibis) or those which nest in non-randomly distributed habitat such as lakes, marshes, cliff ledges, and waterfalls. This concern affects some species of grebes, waterfowl, falcons, marsh birds, and shorebirds, both eagles, and Black Swifts. In addition, none of the handful of colonies of American White Pelicans or California Gulls fell within Priority blocks, so that the Atlas has no abundance estimate at all for those two species. To address this concern, Table B2 includes a column that evaluates which species may have unsatisfactory extrapolations and why. It also shows estimates from the Division of Wildlife and a few other sources for waterfowl, gamebirds, and certain raptors.

TABLE B1
ESTIMATES OF BREEDING PAIRS, STATEWIDE

	Low Estimate	Mid Estimate	High Estimate
Total breeding pairs	12,051,143	40,058,021	72,602,621
Averages			
Pairs/acre: i.e., each acre hosts this average fraction of a breeding pair	0.18	0.60	1.09
Acres/pair: i.e., each breeding pair occupies an average of this many acres	5.54	1.67	0.92

Numbers used for statewide calculations

Total square miles	104,247
Total acres	66,718,080

TABLE B2

ESTIMATES OF BREEDING PAIRS, BY SPECIES

Col. 1: NON-RANDOM
Denotes species for which Atlas extrapolations may be unsatisfactory.
Colonial: colonial nesting species.
Non-random: species breeds in non-randomly distributed habitat.

Col. 3–5: ATLAS ESTIMATES
Low, Mid, and High estimates derived from Atlas abundance codes.

Col. 6: RANK
Abundance rank, from most abundant to least, based on the Atlas Mid-estimate.

Col. 7: OTHER ESTIMATES—NUMBER
Population estimates from other sources.

Col. 8: OTHER ESTIMATES—SOURCE
Source for Column 7. A key at the end of the table identifies the source.

Non-Random		Atlas Estimates				Other Estimates	
Distribution	Species	Low	Mid	High	Rank	Number	Source
	Pied-billed Grebe	705	1,476	2,290	189		
Colonial	Eared Grebe	545	1,064	1,292	198		
Colonial	Western Grebe	321	677	962	207		
Colonial	Clark's Grebe	146	146	146	226		
Colonial	American White Pelican	-	-	-	-	2,075	Ry 3
Colonial	Double-crested Cormorant	520	1,108	1,569	196	1,000	Ry 1
Non-random	American Bittern	209	384	497	213		
Colonial	Great Blue Heron	2,262	4,619	6,136	154	1,512	M
Colonial	Snowy Egret	856	1,728	1,867	185	600	Ry 2
	Green Heron	48	48	48	239		
Colonial	Black-crowned Night-Heron	1,400	3,461	4,835	168		
Colonial	White-faced Ibis	1,283	3,482	4,896	167		
Non-random	Turkey Vulture	6,453	13,888	21,211	133		
	Canada Goose	6,991	27,065	46,782	114		
	Wood Duck	560	1,136	1,680	195	579	Ga
	Gadwall	3,616	10,736	16,963	140	10,982	Ga
Non-random	American Wigeon	1,070	2,331	3,424	173	2,214	Ga
	Mallard	21,165	79,102	138,748	85	30,026	Ga
	Blue-winged Teal	2,310	5,185	8,030	151	2,135	Ga
	Cinnamon Teal	3,500	11,893	20,335	139	6,277	Ga
	Northern Shoveler	2,218	7,224	11,726	146	3,708	Ga
	Northern Pintail	1,523	3,364	5,087	169	2,474	Ga
	Green-winged Teal	3,780	8,656	13,524	145	5,685	Ga
Non-random	Canvasback	67	67	67	237	21	Ga
Non-random	Redhead	736	1,640	2,432	186	3,383	Ga
	Ring-necked Duck	1,026	2,244	3,343	174	330	Ga
Non-random	Lesser Scaup	779	1,610	2,145	187	2,516	Ga
	Bufflehead	24	24	24	250	15	Ga
	Barrow's Goldeneye	42	42	42	240		
	Common Merganser	1,770	3,857	5,873	163	373	Ga
	Ruddy Duck	806	1,777	2,611	182	672	Ga
Non-random	Osprey	110	110	110	232		
	Mississippi Kite	101	101	101	233		
Non-random	Bald Eagle	114	114	114	231	26	C
	Northern Harrier	2,179	2,179	2,179	175		
	Sharp-shinned Hawk	1,945	4,403	7,477	157		
	Cooper's Hawk	2,649	2,649	2,649	171		
	Northern Goshawk	1,249	1,249	1,249	192		
	Swainson's Hawk	6,050	12,127	19,464	136		
	Red-tailed Hawk	10,927	23,297	38,015	118		
	Ferruginous Hawk	992	992	992	200		
Non-random	Golden Eagle	2,631	2,631	2,631	172	800–1200	C

Non-Random Distribution	Species	Atlas Estimates				Other Estimates	
		Low	Mid	High	Rank	Number	Source
	American Kestrel	14,869	33,810	52,535	111		
Non-random	Peregrine Falcon	236	236	236	222	83	C
Non-random	Prairie Falcon	989	989	989	201	400–600	C
	Chukar	1,253	3,780	5,688	164		
	Ring-necked Pheasant	18,160	63,977	104,715	91		
	Sage Grouse	4,227	17,857	31,262	125		
	White-tailed Ptarmigan	2,171	4,342	5,325	158		
	Blue Grouse	14,057	57,094	99,093	97		
	Sharp-tailed Grouse	220	473	671	210		
	Greater Prairie-Chicken	2,205	6,538	9,590	147	3,000–6,000	Gi
	Lesser Prairie-Chicken	594	1,080	1,080	197	1,000–2,000	Gi
	Wild Turkey	5,587	23,456	41,371	117		
	Scaled Quail	6,494	23,988	40,230	116		
	Gambel's Quail	544	1,147	1,532	194		
	Northern Bobwhite	2,128	4,200	5,108	159		
Non-random	Virginia Rail	894	1,861	2,592	178		
Non-random	Sora	2,073	4,461	6,337	156		
	American Coot	3,433	12,060	20,629	138	8,000	F
	Sandhill Crane	220	220	220	223		
Non-random	Snowy Plover	24	24	24	248	31	E&M
Non-random	Piping Plover	24	24	24	249	16	E&M
	Killdeer	44,164	169,323	289,160	59		
	Mountain Plover	1,826	4,026	6,084	160		
Non-random	Black-necked Stilt	113	210	240	224		
Non-random	American Avocet	1,683	3,492	4,908	166		
Non-random	Willet	36	36	36	242		
	Spotted Sandpiper	10,420	22,121	30,507	119		
	Upland Sandpiper	280	632	992	208		
	Long-billed Curlew	943	2,056	3,233	177		
	Common Snipe	11,983	41,312	68,068	108		
Non-random	Wilson's Phalarope	2,977	10,469	17,708	141		
Non-random	Forster's Tern	63	63	63	238		
Non-random	Least Tern	84	138	138	228	23	CDOW
Non-random	Black Tern	169	337	469	215		
	Rock Dove	29,683	98,450	155,076	74		
	Band-tailed Pigeon	3,731	12,123	19,324	137	70,000	B
	Mourning Dove	392,020	1,490,767	2,735,244	4		
	Yellow-billed Cuckoo	885	1,790	2,475	181		
	Greater Roadrunner	728	1,768	2,920	183		
	Barn Owl	691	1,374	2,164	191		
	Flammulated Owl	1,807	3,763	5,009	165		
	Western Screech-Owl	570	1,214	1,776	193		
	Eastern Screech-Owl	234	494	747	209		
	Great Horned Owl	7,522	16,517	27,603	128		
	Northern Pygmy-Owl	423	938	1,518	202		
	Burrowing Owl	5,727	20,885	36,646	121		
	Spotted Owl	6	6	6	255	12	C
	Long-eared Owl	291	291	291	217		
	Short-eared Owl	118	118	118	230		
	Boreal Owl	238	238	238	221		
	Northern Saw-whet Owl	1,404	3,125	4,579	170		
	Common Nighthawk	45,206	188,508	329,877	54		
	Common Poorwill	13,962	46,571	73,375	105		
Non-random	Black Swift	334	831	1,388	205	< 270	K
	Chimney Swift	899	1,831	2,375	179		
	White-throated Swift	20,678	73,693	119,916	87		

Non-Random Distribution	Species	Atlas Estimates Low	Mid	High	Rank	Other Estimates Number	Source
	Black-chinned Hummingbird	31,285	87,283	122,726	80		
	Broad-tailed Hummingbird	205,362	828,075	1,516,471	14		
	Belted Kingfisher	2,826	6,014	9,784	149		
	Lewis's Woodpecker	3,954	14,055	24,310	132		
	Red-headed Woodpecker	2,257	4,656	6,563	153		
	Acorn Woodpecker	18	18	18	253		
	Red-bellied Woodpecker	30	30	30	245		
	Williamson's Sapsucker	13,802	45,674	72,861	106		
	Red-naped Sapsucker	32,018	111,645	179,913	68		
	Ladder-backed Woodpecker	650	1,383	1,957	190		
	Downy Woodpecker	14,031	55,977	97,968	98		
	Hairy Woodpecker	27,838	98,000	160,571	75		
	Northern (Red-shafted) Flicker	97,139	325,437	511,726	37		
	Northern (Yellow-shafted) Flicker	1,509	4,882	7,777	152		
	Olive-sided Flycatcher	19,362	71,348	119,404	88		
	Western Wood-Pewee	106,939	529,392	963,514	25		
	Willow Flycatcher	2,572	5,204	6,651	150		
	Least Flycatcher	18	18	18	252		
	Hammond's Flycatcher	23,451	63,119	86,976	93		
	Gray Flycatcher	38,498	175,759	320,737	58		
	Dusky Flycatcher	64,867	310,786	566,231	39		
	Cordilleran Flycatcher	32,828	109,746	173,821	70		
	Black Phoebe	24	24	24	247		
	Eastern Phoebe	351	772	1,167	206		
	Say's Phoebe	15,790	62,536	112,010	94		
	Ash-throated Flycatcher	18,794	63,568	101,095	92		
	Great Crested Flycatcher	36	36	36	241		
	Cassin's Kingbird	4,936	16,502	26,718	129		
	Western Kingbird	81,515	260,185	398,596	43		
	Eastern Kingbird	14,507	47,448	76,850	104		
	Scissor-tailed Flycatcher	18	18	18	251		
	Loggerhead Shrike	7,318	16,700	26,109	127		
	Northern Three-toed Woodpecker	3,741	14,368	24,891	130		
	Bell's Vireo	150	282	306	218		
	Gray Vireo	4,201	12,420	18,255	135		
	Plumbeous Vireo	73,489	297,213	544,061	40		
	Warbling Vireo	345,820	842,146	1,572,584	13		
	Red-eyed Vireo	144	144	144	227		
	Gray Jay	23,395	85,929	140,973	81		
	Steller's Jay	46,757	207,178	377,832	51		
	Blue Jay	3,931	14,328	24,652	131		
	Western Scrub-Jay	28,400	96,864	153,820	76		
	Pinyon Jay	29,297	121,408	221,963	66		
	Clark's Nutcracker	22,406	88,626	152,656	78		
	Black-billed Magpie	45,830	167,532	278,548	60		
	American Crow	13,927	53,662	93,748	100		
	Chihuahuan Raven	848	1,795	2,860	180		
	Common Raven	13,959	47,685	84,768	103		
	Horned Lark	938,608	2,246,503	4,199,890	1		
	Purple Martin	788	1,733	2,495	184		
	Tree Swallow	93,345	432,585	788,923	31		
	Violet-green Swallow	132,049	639,093	1,164,358	20		
	Northern Rough-winged Swallow	17,211	60,111	98,608	96		
	Bank Swallow	13,902	34,415	44,174	110		
	Cliff Swallow	184,183	747,493	1,368,671	17		
	Barn Swallow	104,891	424,216	775,924	32		

Non-Random Distribution	Species	Atlas Estimates Low	Mid	High	Rank	Other Estimates Number	Source
	Black-capped Chickadee	48,226	217,343	396,353	49		
	Mountain Chickadee	326,633	1,173,645	2,157,763	6		
	Juniper Titmouse	27,751	85,001	126,473	83		
	Bushtit	17,483	61,724	99,196	95		
	Red-breasted Nuthatch	43,879	140,785	216,216	62		
	White-breasted Nuthatch	46,470	207,044	377,613	52		
	Pygmy Nuthatch	51,461	184,531	339,142	56		
	Brown Creeper	21,834	71,283	110,790	89		
	Rock Wren	55,020	255,692	465,916	45		
	Canyon Wren	7,096	29,254	51,013	113		
	Bewick's Wren	46,089	188,430	344,899	55		
	House Wren	327,548	1,032,685	1,906,956	9		
Non-random	Marsh Wren	483	905	1,053	203		
	American Dipper	4,066	9,111	13,855	143		
	Golden-crowned Kinglet	25,559	104,566	191,276	72		
	Ruby-crowned Kinglet	418,100	1,011,162	1,889,318	10		
	Blue-gray Gnatcatcher	104,875	473,560	864,641	30		
	Eastern Bluebird	180	180	180	225		
	Western Bluebird	16,826	52,785	97,179	102		
	Mountain Bluebird	114,527	481,178	879,572	29		
	Townsend's Solitaire	22,761	88,690	151,875	77		
	Veery	3,045	6,221	8,077	148		
	Swainson's Thrush	9,213	32,881	54,015	112		
	Hermit Thrush	154,432	555,209	1,020,385	24		
	American Robin	590,168	1,903,131	3,511,966	2		
	Gray Catbird	958	2,110	3,182	176		
	Northern Mockingbird	41,310	123,563	227,793	65		
	Sage Thrasher	55,976	182,171	335,954	57		
	Brown Thrasher	4,093	8,879	12,972	144		
	Curve-billed Thrasher	248	248	248	220		
	Dark-eyed (Gray-headed) Junco	343,084	1,103,625	2,037,118	7		
	European Starling	144,767	501,100	921,734	27		
	American Pipit	35,022	131,823	241,945	63		
	Cedar Waxwing	1,974	3,977	5,005	161		
	Orange-crowned Warbler	31,612	88,620	125,074	79		
	Virginia's Warbler	60,542	262,153	478,998	42		
	Yellow Warbler	79,125	349,196	637,169	34		
	Chestnut-sided Warbler	32	32	32	244		
	Yellow-rumped Warbler	354,152	971,564	1,804,674	11		
	Black-throated Gray Warbler	60,577	239,053	438,123	47		
	Grace's Warbler	1,338	4,542	7,266	155		
	American Redstart	162	259	259	219		
	Ovenbird	189	343	371	214		
	Northern Waterthrush	66	120	120	229		
	MacGillivray's Warbler	46,703	213,941	390,157	50		
	Common Yellowthroat	7,047	25,752	43,257	115		
	Wilson's Warbler	60,483	206,257	379,676	53		
	Yellow-breasted Chat	5,609	21,662	37,056	120		
	Hepatic Tanager	78	78	78	236		
	Western Tanager	82,133	341,005	623,575	36		
	Green-tailed Towhee	310,386	856,377	1,590,112	12		
	Spotted Towhee	208,626	588,180	1,091,107	22		
	Canyon Towhee	6,615	19,812	29,427	123		
	Cassin's Sparrow	86,177	315,681	579,829	38		
	Rufous-crowned Sparrow	1,215	3,947	6,114	162		
	Chipping Sparrow	198,874	827,502	1,514,120	15		

Non-Random Distribution	Species	Atlas Estimates Low	Mid	High	Rank	Other Estimates Number	Source
	Brewer's Sparrow	176,591	520,484	963,749	26		
	Field Sparrow	162	312	366	216		
	Vesper Sparrow	176,985	618,099	1,137,179	21		
	Lark Sparrow	190,752	755,420	1,384,157	16		
	Black-throated Sparrow	5,100	13,505	18,324	134		
	Sage Sparrow	25,535	108,854	199,047	71		
	Lark Bunting	540,470	1,308,072	2,444,009	5		
	Savannah Sparrow	41,120	128,379	237,046	64		
	Grasshopper Sparrow	92,201	289,039	533,763	41		
	Fox Sparrow	4,885	19,835	34,915	122		
	Song Sparrow	42,663	144,655	229,752	61		
	Lincoln's Sparrow	149,826	487,635	899,516	28		
	White-crowned Sparrow	232,371	580,760	1,083,311	23		
	McCown's Longspur	23,407	68,655	127,178	90		
	Chestnut-collared Longspur	565	1,061	1,161	199		
	Northern Cardinal	12	12	12	254		
	Rose-breasted Grosbeak	32	32	32	243		
	Black-headed Grosbeak	32,420	111,090	178,250	69		
	Blue Grosbeak	10,339	35,296	58,278	109		
	Lazuli Bunting	25,531	100,043	182,998	73		
	Indigo Bunting	715	1,545	2,447	188		
	Dickcissel	2,899	9,358	14,795	142		
	Bobolink	207	462	669	211		
	Red-winged Blackbird	171,429	671,944	1,231,090	18		
	Eastern Meadowlark	24	24	24	246		
	Western Meadowlark	528,006	1,900,006	3,493,099	3		
	Yellow-headed Blackbird	14,809	44,856	66,936	107		
	Brewer's Blackbird	86,727	392,813	716,631	33		
	Common Grackle	77,148	341,625	623,502	35		
	Great-tailed Grackle	413	875	1,272	204		
	Brown-headed Cowbird	64,310	242,673	406,349	46		
	Orchard Oriole	5,557	17,581	28,564	126		
	Baltimore Oriole	230	451	624	212		
	Bullock's Oriole	33,233	119,174	198,229	67		
	Scott's Oriole	90	90	90	234		
	Brown-capped Rosy-Finch	6,178	19,184	28,899	124		
	Pine Grosbeak	22,887	76,944	121,023	86		
	Cassin's Finch	51,787	237,524	433,041	48		
	House Finch	96,014	257,076	477,491	44		
	Red Crossbill	26,508	85,642	132,770	82		
	White-winged Crossbill	84	84	84	235		
	Pine Siskin	318,127	1,050,534	1,937,056	8		
	Lesser Goldfinch	24,471	80,501	125,737	84		
	American Goldfinch	15,185	55,559	93,437	99		
	Evening Grosbeak	14,293	52,820	87,844	101		
	House Sparrow	222,154	661,838	1,224,766	19		

Sources of Other Estimates

Code	Source	Year of estimate
B	Braun 1994	1972
C	Jerry Craig pers. comm.	1996
CDOW	Colorado Division of Wildlife	
E&M	Estelle and Mabeel 1994	1994
F	Frederickson et al.	1977
Ga	Jim Gammonley pers. comm. [DOW estimates for 1996, from six places only: South Platte River drainage, Poudre River drainage, North Park, Yampa Valley, Browns Park NWR, and San Luis Valley]	1996
Gi	Giesen 1994a	Late 1980s
K	Knorr 1961	1961
M	Miller and Graul 1987	1978–1983
Ry 1	Ryder 1996	1980s
Ry 2	Ron Ryder pers. comm.	1976
Ry 3	Ron Ryder pers. comm.	1994
V	Van Sant and Braun 1990	1981–1983

TABLE B3

FORMULA FOR CALCULATIONS

This table summarizes the formula used to compute the estimates in Tables B1 and B2.

Column A: Pertinent Abundance Code.

Column B: Number of pairs used in calculation for that Abundance Code.

Column C: Adjustment to B, if no A5 codes reported. (The formula makes similar adjustments if no A4 and A5 reported, etc.)

Column D: Sample calculations, assuming 5 blocks reported for each Abundance Code; multiplied by 6 to derive the statewide total, because Atlas surveyed only one out of every six blocks. (Tables B1 and B2 include an adjustment for blocks without abundance estimates.)

Column E: Sample calculations, assuming 5 blocks reported for each Abundance Code, but no A5 codes reported.

A Abundance Code	B # of pairs used for calculation	C Adjustment for no A5	D equals	E Assume no A5 codes
Low				
A1	1		30	30
A2	2		60	60
A3	11		330	330
A4	101		3,030	3,030
A5	1001		30,030	____
Total			33,480	3,450
Middle				
A1	1	1	30	30
A2	5.5	5.5	165	165
A3	55	55	1,650	1,650
A4	555	200	16,650	6,000
A5	2000		60,000	____
Total			78,495	7,845
High				
A1	1	1	30	30
A2	10	10	300	300
A3	100	100	3,000	3,000
A4	1000	200	30,000	6,000
A5	10000		300,000	____
Total			333,330	9,330

APPENDIX C: SUMMARY OF ATLAS COWBIRD OBSERVATIONS

* = New Colorado record of host species

Total for Species		Species	Breeding Code	Date	Habitat Codes			Quadcode	Map Name	County	Observer
1		Willow Flycatcher	NE	06/22/87				Q4010663	Walden	Jackson	Sedgwick, James
4		Dusky Flycatcher	NE	06/20/93	SMM			Q3710865	Narraguinep Mt.	Montezuma	Bridges, Dan
		Dusky Flycatcher	NY	07/01/95				Q3910667	Eagle	Eagle	Merchant, Jack
		Dusky Flycatcher	FL	07/16/94				Q4010588	Crazy Mtn.	Larimer	Hallock, Dave
		Dusky Flycatcher	FL	07/20/94	FRD			Q4010684	Independence Mt.	Jackson	Dillon, Beth
2	*	Gray Flycatcher	NY	06/06/92	WPJ			Q4010862	Ninemile Hill	Moffat	Toolen, John
		Gray Flycatcher	NE	07/03/94	WPJ			Q3910832	DeBeque	Mesa	Ireland, Terry
1	*	Cordilleran Flycatcher	FF	07/24/91	WMO			Q3810581	Woodland Park	Teller	Sorensen, Ben & Sally
1	*	Eastern Phoebe	NY	06/15/91	WPP			Q3710218	Big Hole Canyon	Baca	Jones, Steve
2	*	Western Kingbird	FL	07/07/94	RRL			Q4010416	Hudson	Weld	Dirckx, Jo & Ferd
		Western Kingbird	FL	07/04/94	CPL			Q4010418	Frederick	Weld	Dirckx, Jo & Ferd
1		Gray Vireo	NE		WPJ			Q3910816	Colo Natl Mon.	Mesa	Hutchings, Scott
4		Plumbeous Vireo	NE	06/05/91	WPP	SOS		Q3710844	Dolores E.	Montezuma	Versaw, Alan
		Plumbeous Vireo	NE	06/21/88	WPP			Q3910448	Sedalia	Douglas	Schock, Mary Jane
		Plumbeous Vireo	FF	07/02/93				Q4010573	Livermore Mt.	Larimer	Dillon, Beth
		Plumbeous Vireo	FF	07/16/92	WPP			Q4010523	Lyons	Boulder	King, D.W.
9		Warbling Vireo	NY	06/24/94	FRD			Q3810546	Arkansas Mt.	Fremont	Lentz, Randy
		Warbling Vireo	NE	06/24/88	RRL	CPL		Q3810657	Signal Pk.	Gunnison	Meyer, Ron
		Warbling Vireo	NE	06/26/89	WPP			Q3810814	Barkelew Draw	San Miguel	Dexter, Coen
		Warbling Vireo	FF	07/07/93	LRD			Q3810553	Rice Mtn.	Fremont	Dillon, Beth
		Warbling Vireo	NY	07/09/95				Q3910761	Dotsero	Eagle	Merchant, Jack
		Warbling Vireo	FL	07/13/94				Q4010671	Shipman Mtn.	Jackson	Dillon, Beth
		Warbling Vireo	FL	07/15/93	WMO			Q3710616	Chama Peak	Archuleta	Kingery, Hugh
		Warbling Vireo	FL	07/22/90	LRD			Q3710551	LaVeta	Huerfano	Yaeger, Mark
		Warbling Vireo	FF	07/30/93	FRD			Q3810834	Big Bucktail Cr.	Montrose	Dexter, Coen
1		Red-eyed Vireo	NE	05/15/89	LRD	URB		Q3810436	N.W. Pueblo	Pueblo	Truan, Van
2		Horned Lark	FF					Q4010253	Holyoke	Phillips	Wieland, Jack
		Horned Lark	FL	06/29/93	STD			Q3710588	Moffat S.	Saguache	Gillihan, Scott
3	*	Bewick's Wren	NE					Q3710358	Lockwood Arroyo	Las Animas	Youkey, Don
		Bewick's Wren	NY	05/27/88				Q3710451	Thatcher	Las Animas	Youkey, Don
		Bewick's Wren	NY	05/27/88				Q3710451	Thatcher	Las Animas	Youkey, Don
1	*	House Wren	FL	07/20/94	FRD	WPP		Q3710413	Barela	Las Animas	Yaeger, Mark
2		Ruby-crowned Kinglet	FL	07/06/94	CPL	FUCS	FUD	Q4010634	Whitely Pk.	Grand	Liewer, Jim
		Ruby-crowned Kinglet	NY	07/10/89	CPL			Q3810656	Parlin	Gunnison	Meyer, Ron
5		Blue-gray Gnatcatcher	NE	05/25/90	WPJ			Q3810765	Crawford	Delta	Rowe, Maurice
		Blue-gray Gnatcatcher	NY	06/19/94	WPJ			Q4010876	Irish Canyon	Moffat	Toolen, John
		Blue-gray Gnatcatcher	FY	06/22/93	WPJ			Q4010825	Divide Cr.	Rio Blanco	Dexter, Coen
		Blue-gray Gnatcatcher	NE	06/26/95	SOS			Q3710716	Ignacio	La Plata	Kingery, Hugh
		Blue-gray Gnatcatcher	NY	06/30/89	WPJ			Q3910816	Colo. Natl. Mon.	Mesa	Brevillier, Toni
3	*	Mountain Bluebird	NE					Q3710451	Thatcher	Las Animas	Youkey, Don
		Mountain Bluebird	NY	05/27/88	WPJ	TSG		Q3710451	Thatcher	Las Animas	Youkey, Don
		Mountain Bluebird	NY	05/27/88	WPJ	TSG		Q3710451	Thatcher	Las Animas	Youkey, Don
2		Hermit Thrush	FL	06/26/94	WMO			Q3710741	Pagosa Peak	Mineral	Kingery, Urling & Hugh
		Hermit Thrush	NE	07/29/90	WMOS			Q3910622	Mt. Sherman	Park	Lentz, Randy
1	*	Brown Thrasher	NY	06/14/94	LRD			Q3910255	Kirk	Kit Carson	Dillon, Beth
5	*	Virginia's Warbler	FL	07/00/95				Q3910816	Colo. Natl. Mon.	Mesa	Hutchings, Scott
		Virginia's Warbler	FL	07/07/94	WPJ			Q3810562	Cripple Cr. So.	Teller	Versaw, Alan
		Virginia's Warbler	FL	07/13/94	SLS			Q4010824	Rough Gulch	Rio Blanco	Wright, Brenda
		Virginia's Warbler	FF	07/13/89				Q3810478	Manitou Spgs.	El Paso	Higgins, Ann
		Virginia's Warbler	FF	07/20/88	WPP	SMM		Q3910574	Blackhawk	Gilpin	Phillips, Polly
12		Yellow Warbler	NE	05/20/87				Q3910456	Piney Creek	Douglas	Springston, Larry
		Yellow Warbler	NE	06/03/88	SLC	LRD		Q3910435	Elizabeth	Elbert	Herold, Jim & Verna

Total for Species		Species	Breeding Code	Date	Habitat Codes	Quadcode	Map Name	County	Observer
		Yellow Warbler	NE	06/12/93	STD	Q3710547	Alamosa E.	Alamosa	Gillihan, Scott
		Yellow Warbler	NE	06/16/93	SMC	Q4010664	Delaney Butte	Jackson	Dillon, Beth
		Yellow Warbler	NE	06/22/87		Q4010663	Walden	Jackson	Sedgwick, James
		Yellow Warbler	NE	06/24/94	SMC	Q4010518	Granby	Grand	Collins, Wally
		Yellow Warbler	FF	07/01/93	FRD	Q3910755	New Castle	Garfield	Levad, Rich
		Yellow Warbler	FL	07/10/89	LRD	Q3810421	Fowler	Otero	Dickson, Bob & Johnnie
		Yellow Warbler	FF	07/11/94	CPL SMC	Q3810771	Mt. Axtell	Gunnison	Meyer, Ron
		Yellow Warbler	NY	07/16/89	FRD	Q3710514	Sanchez Res.	Costilla	Bridges, Dan
		Yellow Warbler	FF	07/19/92		Q3810466	Fountain	El Paso	Brevillier, Toni
		Yellow Warbler	FY	08/04/90	WMO	Q3910723	Placita	Garfield	Zerbi, Vic
7	*	Yellow-rumped Warbler	FL	06/23/90	SMC WMO	Q3710747	Hermosa	La Plata	Parker, Jim & Marianne
		Yellow-rumped Warbler	FL	06/28/94	SMM FRD	Q4010763	Slide Mtn.	Routt	Hallock, Dave
		Yellow-rumped Warbler	FF	07/10/93	FRD	Q4010735	Hamilton	Moffat	Bridges, Dan
		Yellow-rumped Warbler	FL	07/20/94	FUD	Q3710768	Hermosa Peak	La Plata	Potter, Kim
		Yellow-rumped Warbler	FL	07/21/94	FUD	Q3910727	Vega Res.	Mesa	Potter, Kim
		Yellow-rumped Warbler	FL	07/21/91	WMOS WMOL	Q4010646	Mt. Werner	Routt	Kingery, Urling & Hugh
		Yellow-rumped Warbler	FF	07/23/93	FUD TMG	Q3710647	Wolf Creek Pass	Archuleta	Versaw, Alan
2	*	Black-throated Gray Warbler	FY	07/04/90	FRD	Q3910757	Rifle	Garfield	Merchant, Jack
		Black-throated Gray Warbler	FY	07/08/87	SMS FRD CPL	Q4010737	Axial	Moffat	Ward, Judy
1	*	Grace's Warbler	FF	06/26/89	WPP	Q3810814	Barkelew Draw	San Miguel	Dexter, Coen
1		MacGillivray's Warbler	FF	07/18/94	SMC	Q3810676	Matchless Mtn.	Gunnison	Ford, Fern
1	*	Yellow-breasted Chat	FF	06/29/89		Q3910283	Vernon	Yuma	Kingery, Hugh
1		Wilson's Warbler	FL	07/24/88		Q3910662.	Willow Lakes	Summit	CBO
1	*	Western Tanager	FY	07/13/89	FRD	Q3810478	Manitou Spgs.	El Paso	Higgins, Ann
4		Green-tailed Towhee	NE	05/30/91	FRD	Q4010762	Quaker Mt.	Routt	Garcia, Roz
		Green-tailed Towhee	FL	06/28/93	FRD	Q3810845	Windy Point	Montrose	Dexter, Coen
		Green-tailed Towhee	FF	07/11/91	LRD	Q4010582	Table Mtn.	Larimer	Martin, Kathy
		Green-tailed Towhee	FF	07/18/95	SMM WPJ	Q3810682	Harvard Lakes	Chaffee	Kingery, Urling & Hugh
1		Spotted Towhee	FL	06/29/91	LRD	Q3910551	Littleton	Douglas	Kingery, Hugh
5		Chipping Sparrow	FL	07/01/91	WPJ	Q3910834	Winter Flats	Mesa	Williams, E
		Chipping Sparrow	NY	07/04/88		Q4010535	Longs Peak	Larimer	Spahn, Robert
		Chipping Sparrow	FF	07/13/94	SLS TMGB	Q3910868	E. Evacuation Cr.	Garfield	Pearson, E.W.
		Chipping Sparrow	FL	07/21/92		Q3710684	Lookout Mtn.	Saguache	Kingery, Urling & Hugh
		Chipping Sparrow	NE	07/23/93	WPP SBC	Q4010545	Estes Park	Larimer	Opler, Paul
3		Vesper Sparrow	NY	06/07/87	SMS	Q3810668	Flattop	Gunnison	Meyer, Ron
		Vesper Sparrow	FY	07/10/93	SLS	Q4010717	Rattlesnake Mes.	Rio Blanco	Reichert, Chuck
		Vesper Sparrow	FF	07/11/92	TMG	Q3810587	Antero Res N.E.	Park	Kuenning, Walt & Ruth
4		Lark Sparrow	NE	06/03/90	WPJ	Q3710358	Lockwood Arroyo	Las Animas	Truan, Van
		Lark Sparrow	NE	06/20/92	TSG WPJ	Q3710316	Pine Cyn.	Las Animas	Bannon, Brett
		Lark Sparrow	FF	06/26/94	TSG	Q4010253	Holyoke	Phillips	Wieland, Jack
		Lark Sparrow	FF	07/28/89		Q3910274	Vernon S.W.	Yuma	Kingery, Hugh
1	*	Sage Sparrow	FL	07/08/92	LRD	Q4010863	Sevenmile Draw	Moffat	Kingery, Hugh
2		Savannah Sparrow	FF	07/21/93	AEM	Q3710682	Swedes Corner	Saguache	Kingery, Urling & Hugh
		Savannah Sparrow	FL	07/27/94	SMC	Q4010674	L. John	Jackson	Dillon, Beth
1	*	Fox Sparrow	FL	07/16/94	SBC	Q3910641	Breckenridge	Summit	Lentz, Randy
7		Song Sparrow	FF	06/22/95	SMC	Q3910511	Mt. Deception	Teller	Versaw, Alan
		Song Sparrow	NE	06/22/87		Q4010663	Walden	Jackson	Sedgwick, James
		Song Sparrow	NE	07/04/91	SLC	Q3910887	Water Canyon	Rio Blanco	Hawksworth, Dave
		Song Sparrow	FF	07/06/94		Q3910477	Sable	Adams	Hetrick, Mindy
		Song Sparrow	FL	07/08/92		Q4010863	Sevenmile Draw	Moffat	Kingery, Hugh
		Song Sparrow	FL	07/20/94	FUD SMC	Q4010517	Strawberry L.	Grand	Nelson, Duane
		Song Sparrow	FF	07/21/93	SOS SMM FRCD	Q3910741	Leon	Eagle	Levad, Rich
1		Lincoln's Sparrow	FL	08/01/94	TMG	Q3910535	Topaz Mtn.	Park	Lentz, Randy
9		Gray-headed Junco	FF			Q4010526	Isolation Peak	Boulder	Bouricius, Steve
		Gray-headed Junco	NE	06/06/92	FUC	Q4010575	Red Feather Lks.	Larimer	Cringan, Alex

Total for Species		Species	Breeding Code	Date	Habitat Codes			Quadcode	Map Name	County	Observer
		Gray-headed Junco	FY	07/04/90	WPP			Q3810513	Rosita	Custer	Miller, Gary
		Gray-headed Junco	FF	07/10/92	FUD			Q4010774	Freeman Res.	Moffat	Kingery, Hugh
		Gray-headed Junco	FL	07/13/94	FUC			Q4010671	Shipman Mtn.	Jackson	Dillon, Beth
		Gray-headed Junco	FL	07/23/92	FUD			Q3710753	Granite L.	Hinsdale	Kingery, Urling & Hugh
		Gray-headed Junco	FF	07/26/93	FUD			Q3810667	Almont	Gunnison	Meyer, Ron
		Gray-headed Junco	FL	07/28/94	FUD			Q4010655	Teal L.	Jackson	Dillon, Beth
		Gray-headed Junco	FY	08/06/91	FUD			Q4010634	Whiteley Pk.	Grand	Barrett, Norm
14		Red-winged Blackbird	NE					Q3910456	Piney Creek	Douglas	Springston, Larry
		Red-winged Blackbird	NE		WPP	RRL	SMM	Q3910583	Eldorado Spgs	Jefferson	Jones, Steve
		Red-winged Blackbird	NY					Q3810333	Arlington	Kiowa	Nelson, Duane
		Red-winged Blackbird	NE					Q4010561	Wellington	Larimer	Martin, Steve
		Red-winged Blackbird	NE	05/23/89	AEM	LRD		Q3910274	Vernon S.W.	Yuma	Kingery, Hugh
		Red-winged Blackbird	NE	06/02/87	LRD			Q3910466	Coal Cr.	Arapahoe	Kingery, Hugh
		Red-winged Blackbird	NE	06/02/87	SMC			Q4010575	Red Feather Lks.	Larimer	Cringan, Alex
		Red-winged Blackbird	FL	06/06/94	RRL			Q3710284	N. Plum Crk N.W.	Prowers	Nelson, Duane
		Red-winged Blackbird	NE	06/08/89	SLC			Q4010344	Messex	Washington	Andrews, Bob
		Red-winged Blackbird	NE	06/11/89	SLC	AEM		Q3910283	Vernon	Yuma	Kingery, Hugh
		Red-winged Blackbird	NE	06/13/92	RRL			Q4010268	Uhler Ranch	Logan	Prather, Bill
		Red-winged Blackbird	NE	06/21/90	AEM	LRD		Q4010531	Berthoud	Larimer	Means, Ann
		Red-winged Blackbird	NE	06/24/93	CWD			Q3910232	Peconic	Kit Carson	Hartman, Susan
		Red-winged Blackbird	FL	07/22/92	AEM			Q3710752	Cimarrona Pk.	Hinsdale	Kingery, Urling & Hugh
1		Western Meadowlark	NE		RRL	CPL		Q4010561	Wellington	Larimer	Martin, Steve
1		Yellow-headed Blackbird	NE	06/00/93	AEM			Q3810861	Delta	Delta	Adamus, Paul
1		Brewer's Blackbird	NE	06/05/91	AEM			Q3810658	Gunnison	Gunnison	Meyer, Ron
1	*	Common Grackle	NE	06/14/92	RRL			Q4010286	Julesburg Res.	Sedgwick	Prather, Bill
2	*	Orchard Oriole	FL	07/17/94	LRD	LRD		Q3910275	Abarr S.E.	Yuma	Chace, Jim
		Orchard Oriole	FF	07/23/93	LRD	RRL		Q3910282	Beecher I.. N.W.	Yuma	Chace, Jim
1		Bullock's Oriole	FF	07/23/92				Q3910282	Beecher I. N.W.	Yuma	Chace, Jim
1	*	Blue Grosbeak	FL	07/14/93	RRL	CPL		Q3710812	Redmesa	La Plata	Levad, Rich
2		Lazuli Bunting	NE	05/13/88	TSG			Q3910437	Castle Rock S.	Douglas	Wilson, Rob
		Lazuli Bunting	FL					Q3910435	Elizabeth	Elbert	Herold, Jim & Verna
1	*	Cassin's Finch	FL	06/22/94	FUCS			Q4010722	Dunckley Pass	Rio Blanco	Nelson, Kathleen
3		House Finch	NE	05/10/94	URB	FRD		Q3810788	Cedaredge	Delta	Woodward, Ronda
		House Finch	NE	07/02/89	LRD			Q3810884	Whitewater	Mesa	Levad, Rich
		House Finch	NE	07/14/90	RRL			Q3710243	Walsh	Baca	Thompson, Janeal
2	*	American Goldfinch	FL	07/01/91	RRL			Q4010285	Sedgwickk	Sedgwick	Jones, Steve
		American Goldfinch	FF	07/22/93	LRD			Q3910561	Fort Logan	Denver	Bridges, Barbara

APPENDIX D: SAMPLE FIELD CARD

Dates	Hab	Abu		Ob	Po	Pr	Cf
6/24	FRD	2	Vireo, Warbling				NY
			Red-eyed				
			Warbler, Or-crowned				
6/18	WPJ SOS	4	Virginia's				CF
6/24	FRD SMC	3	Yellow				ON
			Y-rmp (Audubon's)				
5/22	WPJ	3	Blk-throated gray		X		
			Grace's •				
			Redstart, American				
			Ovenbird •				
			Warbler, MacG'ray's				
			Yellowthroat, Common				
			Warbler, Wilson's				
6/18	SMC	2	Chat, Yellow-breasted		X		
6/18	WPJ	3	Tanager, Western				ON
6/24	WPJ SOS	4	Grosbeak, B-headed			A	
7/10	FRD	2	Blue				CF
6/18	FRD SMM	3	Bunting, Lazuli		X		
6/18	FRD SMM	2	Indigo		X		
			Dickcissel				
6/24	SSL SMM	2	Towhee, Green-tailed				CF
6/24	WPJ SMM	5	Rufous-sided				CF
			Brown •				
			Sparrow, Cassin's				
			Rufous-crowned •				
6/24	WPJ SMM	3	Chipping				NY
			Brewer's				
			Vesper				
6/24	TTR	2	Lark				CF
			Black-throated •				
			Sage •				
			Bunting, Lark				
			Sparrow, Savannah				
			Grasshopper				
			Fox •				
5/22	SMC	2	Song		X		
			Lincoln's				
			White-crowned				
			Junco, Gray-headed				
			Longspur, McCown's •				
			Chestnut-collared •				
			Bobolink				

Dates	Hab	Abu		Ob	Po	Pr	Cf
6/18	SMC	2	Blackbird, Red-winged				CF
			Meadowlark, Western				
			Blackbird, Yel-hd				
6/24	TTR	3	Brewer's				CF
			Grackle, Common				
6/24	FRD	3	**Cowbird, Brn-head'd**				NY
			Oriole, Orchard •				
6/24	FRD	2	N (Bullock's)				ON
			Scott's •				
			Finch, Rosy				
			Grosbeak, Pine				
			Finch, Cassin's				
6/18	FRD WPJ	2	House				NB
			Crossbill, Red				
			Siskin, Pine				
6/18	FRD WPJ	3	Goldfinch, Lesser				NB
			American				
			Grosbeak, Evening				
			Sparrow, House				
6/24			Calif. Gull	O			
Total	52			1	17	5	29

Notes and Comments:

1   51

Cowbird nestling in Warbling Vireo Nest

BREEDING CODES

OBSERVED (Ob)
- O Non-breeder or migrant observed during breeding season

POSSIBLE (Po)
- # Possible breeder observed or heard
- X Singing Male

PROBABLE (Pr)
- M 7 Singing Males
- P Pair
- T Territory
- C Courtship
- V Visiting nest site
- A Agitated behavior
- N Nest building (Wrens or Woodpeckers)

CONFIRMED (Cf)
- NB Nest building
- PE Physiological evidence
- DD Distraction display
- UN Used nest
- FL Fledged Young
- ON Occupied nest
- FY Feeding young
- FS Fecal sac
- NE Nest with eggs
- NY Nest with young

ABUNDANCE CODES

Total Pairs in Block
- A1 1 breeding pair
- A2 2–10 breeding pairs
- A3 11–100 breeding pairs
- A4 101–1000 breeding pairs
- A5 More than 1000 breeding pairs

For Explanation of Habitat Codes, please refer to your field worker's manual.

•Fill out a verification form if these species are found nesting outside their known ranges.

Boldface Species: Abundance data especially important.

OTHER OBSERVERS - Name	Address	City	Zip

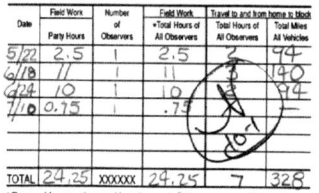

COLORADO BREEDING BIRD ATLAS
FIELD CARD 3810546

ATLAS BLOCK INFORMATION

Block Worked Previous Year	
No	X
Yes	

Atlas Year 19 94

Block Finished?	
No	
Yes	X

Map Name Arkansas Mtn.

Map Number 3810546

Atlas Region R

Block: (Circle Sector) NW NE CW CE SW (SE)

Elevation: Min. 6240 Max: 7680

ATLASER

Name Randy Lentz

Address 15903 East Rice PL. #A

City Aurora State CO Zip 80015

Telephone: (H) (303) 680-3381

(W) (______) ____________

BLOCK WORK THIS YEAR

Date	Field Work Party Hours	Number of Observers	Field Work •Total Hours of All Observers	Travel to and from home to block: Total Hours of All Observers	Travel to and from home to block: Total Miles All Vehicles
5/22	2.5	1	2.5	2	94
6/18	11	1	11	3	140
6/24	10	1	10	2	94
7/10	0.75	1	.75		—
TOTAL	24.25	XXXXXX	24.25	7	328

•Party Hours times Number of Observers = Total Hours

Flicker

Dates	Hab	Abu		Ob	Po	Pr	Cf
			Grebe, Pied-billed				
			Eared				
			Western				
			Clark's				
			Cormorant, D-crested				
			Bittern, American				
			Heron, Great Blue				
			Egret, Snowy				
			Heron, Bl-cr Night				
			Ibis, White-faced				
			Goose, Canada				
			Duck, Wood				
			Teal, Green-winged				
			Mallard				
			Pintail, Northern				
			Teal, Blue-winged				
			Cinnamon				
			Shoveler, Northern				
			Gadwall				
			Wigeon, American				
			Canvasback •				
			Redhead				
			Duck, Ring-necked •				
			Scaup, Lesser •				
			Merganser, Common •				
			Duck, Ruddy				
5/22	MCL	1	Vulture, Turkey		#		
			Kite, Mississippi •				
			Harrier, Northern				
			Hawk, Sharp-shinned				
			Cooper's				
			Goshawk, Northern •				
			Hawk, Swainson's				
6/18	MCL	1	Red-tailed			A	
			Ferruginous				
			Eagle, Golden				
			Kestrel, American				
			Falcon, Peregrine •				
			Prairie •				
			Chukar •				
			Pheasant, R-necked				
			Grouse, Blue				

Dates	Hab	Abu		Ob	Po	Pr	Cf
			Ptarmigan, W-tailed				
			Grouse, Sage •				
			Prairie-chicken, G •				
			Lesser •				
			Grouse, Sharp-tailed				
			Turkey, Wild				
			Bobwhite, Northern				
			Quail, Scaled				
			Gambel's •				
			Rail, Virginia				
			Sora				
			Coot, American				
			Crane, Sandhill •				
			Killdeer				
			Plover, Mountain				
			Stilt, Black-necked				
			Avocet, American				
6/24	OWS	2	Sandpiper, Spotted		#		
			Upland				
			Curlew, Long-billed •				
			Snipe, Common				
			Phalarope, Wilson's				
			Tern, Forster's				
			Black				
			Dove, Rock				
			Pigeon, Band-tailed				
6/18	WPJ FRD	3	Dove, Mourning		#		
			Cuckoo, Bl-billed				
			Yellow-billed				
			Roadrunner, Greater				
			Barn-Owl, Common				
			Owl, Flammulated				
			Screech-Owl, Eastern•				
			Western •				
			Owl, Great Horned				
			Pygmy-Owl, Northern				
			Owl, Burrowing				
			Long-eared				
			Short-eared				
			Boreal •				
			N. Saw-whet				
6/18	WPJ	3	**Nighthawk, Common**			T	

Dates	Hab	Abu		Ob	Po	Pr	Cf
6/24	WPJ	2	**Poorwill, Common**		X		
			Swift, Black •				
			Chimney				
5/22	MCL	3	White-throated		#		
			Hummingbird, Blk-ch				
6/18	FRD WPJ	3	Broad-tailed				NE
6/18	OWS	1	Kingfisher, Belted		#		
			Woodpecker, Lewis'				
			Red-headed				
			Sapsucker, Red-naped				
			Williamson's				
			Woodpecker, Lad-bck.•				
			Downy				
			Hairy				
			N. Three-toed •				
			Flicker, N (R-shaft)				
			N (Yellow-Shafted)				
			Flycatcher, Ol-sided				
6/18	FRD	2	Wood-Pewee, Western				NB
			Flycatcher, Willow				
			Hammond's				
			Dusky				
6/18	WPJ	4	Gray			M	
			Western				
			Phoebe, Eastern •				
			Say's				
6/24	WPJ	2	Flycatcher, Ash-thr				CF
			Kingbird, Cassin's				
			Western				
			Eastern				
			Lark, Horned				
			Swallow, Tree				
5/22	MCL FRD	3	Violet-green		#		
			Rough-winged				
			Bank				
5/22	MCL FRD	3	Cliff				ON
6/18	FRD MSC	2	Barn				ON
			Jay, Gray				
			Steller's				
			Blue				
6/24	WPJ	2	Scrub				FL
6/18	WPJ	2	Pinyon		#		

Dates	Hab	Abu		Ob	Po	Pr	Cf
6/18	WPJ	2	Nutcracker, Clark's		#		
			Magpie, Black-billed				
			Crow, American				
			Raven, Chihuahuan				
5/22	WPJ	1	Common		#		
			Chickadee, Blk-capped				
6/18	WPJ FRD	3	Mountain				NY
6/18	WPJ	3	Titmouse, Plain		#		
6/18	WPJ SOS	2	Bushtit		#		
			Nuthatch, Red-brstd				
6/18	WPJ	3	White-breasted				FL
			Pygmy				
			Creeper, Brown				
6/24	SMM	3	Wren, Rock				FL
6/24	MCL SMM	3	**Canyon**				CF
			Bewick's				
6/18	FRD	3	House				CF
			Marsh				
			Dipper, American				
			Kinglet, Golden-cr				
			Ruby-crowned				
5/22	WPJ SMM	3	Gnatcatcher, Blue-Gr				ON
			Bluebird, Eastern •				
			Western				
6/24	WPJ	2	**Mountain**			P	
			Solitaire, Townsend's				
			Veery •				
			Thrush, Swainson's				
			Hermit				
6/18	FRD	3	Robin, American				CF
			Catbird, Gray				
			Mockingbird, Northern				
			Thrasher, Sage				
			Brown				
			Curve-billed •				
			Pipit, Water				
			Waxwing, Cedar				
			Shrike, Loggerhead				
			Starling				
			Vireo, Bell's •				
			Gray				
6/18	WPJ	4	Solitary				NB

APPENDIX E: SCIENTIFIC NAMES OF OTHER SPECIES

Compiled by Beverly K. Baker

Scientific names from Harrington 1954, Weber 1976, and Carter 1988. Bracketed scientific names from Weber and Wittmann 1996a, 1996b.

TREES AND SHRUBS

Alder, mountain (thinleaf)	*Alnus tenuifolia* [*A. incana tenuifolia*]
Aspen, quaking	*Populus tremuloides*
Birch, river (Rocky Mtn.)	*Betula occidentalis*
Birch, bog (dwarf)	*Betula glandulosa*
Blackbrush	*Coleogyne ramosissima*
Box-elder	*Acer negundo* [*Negundo aceroides*]
Buffaloberry	*Shepherdia canadensis*
Buckbrush, Fendler	*Ceanothus herbaceous*
Cactus, cholla	*Opuntia imbricata* (*Opuntia arborescens*), [*Cylindropuntia imbricata*]
Chokecherry	*Prunus virginiana subsp. melanocarpa* [*Padus virginiana*]
Cholla — see cactus	
Cinquefoil, shrubby	*Potentilla fruticosa* [*Pentaphylloides floribunda*]
Coralberry	*Symphoricarpos* spp.
Cottonwood, broad-leaved	See Fremont and plains cottonwood
Cottonwood, Fremont	*Populus fremontii* [*P. deltoides ssp. wislizenii*]
Cottonwood, narrowleaf	*Populus angustifolia*
Cottonwood, plains	*Populus sargentii* [*P. deltoides ssp. monilifera*]
Creosote bush	*Larrea divaricata*
Currant, squaw (wax)	*Ribes cereum*
Dogwood, red osier	*Cornus stolonifera*
Douglas-fir	*Pseudotsuga menziesii (P. taxifolia)*
Elm, Siberian (Chinese)	*Ulmus pumila*
Fir, subalpine	*Abies lasiocarpa* [*A. bifolia*]
Fir, white	*Abies concolor*
Gooseberry	*Ribes* spp.
Greasewood	*Sarcobatus vermiculatus*
Hackberry	*Celtis occidentalis*
Hawthorn	*Crataegus* spp.
Hop-sage, spiny	*Grayia spinosa* [*Atriplex grayi*]
Juniper, common or low	*Juniperus communis*
Juniper, one-seed	*Juniperus monosperma* [*Sabina monosperma*]
Juniper, Rocky Mountain	*Juniperus scopulorum* [*Sabina scopulorum*]
Juniper, Utah	*Juniperus utahensis* [*Sabina osteosperma*]
Manzanita	*Arctostaphylos patula* subsp. *platyphylla*
Mountain mahogany	*Cercocarpus* spp.
Oak, scrub	*Quercus gambelii*
Oak, shinnery	*Quercus havardii*
Pine, lodgepole	*Pinus contorta*
Pine, limber	*Pinus flexilis*
Pine, pinyon	*Pinus edulis*
Pine, ponderosa	*Pinus ponderosa*
Pine, bristlecone	*Pinus aristata*
Pinyon/juniper	*Pinus edulis* and *Juniperus utahensis* (mostly), *J. scopulorum*, or *J. monosperma*
Plum, wild	*Prunus americana*
Rabbitbrush	*Chrysothamnus* spp.
Rose, wild	*Rosa* spp.
Russian-olive	*Eleagnus angustifolia*
Sage, silver	See sagebrush
Sagebrush, big & mountain	*Artemisia tridentata* & *Artemisia frigida* [*Seriphidium tridentatum, S. vaseyanum*]
Sagebrush, sand	*Artemisia filifolia* [*Oligosporus filifolius*]
Salt-cedar—see tamarisk	
Saltbush	*Atriplex* spp.
Saltbush, four-winged	*Atriplex canescens*
Sandsage—see sagebrush, sand	
Serviceberry	*Amelanchier* spp.
Shadscale	*Amelanchier oreophila*
Skunkbrush (Skunkbush)	*Rhus trilobata* [*R. aromatica*]
Snakeweed	*Gutierrezia* spp. [*Gutierrezia sarothrae*]
Snowberry	*Symphoricarpos* spp.
Soapweed — see yucca	
Spruce, Colorado blue	*Picea pungens*
Spruce, Engelmann	*Picea engelmannii*
Spruce, red	*Picea rubens*
Tamarisk; salt-cedar	*Tamarix gallica* [*T. ramosissima*]
Willow	*Salix* spp.
Willow, peach-leaf	*Salix amygdaloides*
Willow, sandbar	*Salix exigua*
Winterfat	*Ceratoides lanata (Eurotia lanata)* [*Krascheninnikovia lanata*]
Yucca	*Yucca* spp.

GRASSES

Bluestem, big	*Andropogon gerardi*
Bluestem, little	*Andropogon scoparius*
Bluestem, sand	*Andropogon hallii*
Buffalo grass	*Buchloe dactyloides*
Bulrush	*Scirpus* spp. [*Schoenoplectus* spp.]
Cattail	*Typha* spp.
Cheatgrass	*Bromus tectorum* [*Anisantha tectorum*]
Corn	*Zea mays*
Dropseed, sand	*Sporobolus* spp.
Galleta grass	*Hilaria jamesii*
Grama, blue	*Bouteloua gracilis*
Grama, side-oats	*Bouteloua curtipenndula*
Junegrass	*Koeleria* spp.
Muhly, mountain	*Muhlenbergia* spp.
Needle-and-thread	*Stipa comata*
Reeds; reed, common; carrizo	*Phragmites communis* [*P. australis*]

Ricegrass, Indian	*Oryzopsis hymenoides*
Switchgrass	*Panicum vergatum*
Three-awn, red	*Aristida* spp.
Timothy	*Phleum* spp.
Wheatgrass	*Agropyron* spp. [*Eremopyrum triticeum; Pseudoroegneria spicata; Thinopyrum* spp.]
Wheatgrass, crested	*Agropyron cristatum*
Wheatgrass, western	*Agropyron* [*Pascopyrum*] *smithii*

FORBS

Cocklebur	*Xanthium* spp.
Columbine, Colorado	*Aquillegia caerulea*
Dandelion	*Taxaxacum officinale*
Grape, wild	*Vitis* spp.
Knapweed	*Centaurea* spp. [*Acosta, Acroptilon*]
Kochia	*Kochia scoparia*
Locoweed	*Oxytropis* spp.
Loosestrife, Purple	*Lythrum salicaria*
Lupine	*Lupinus* spp.
Paintbrush spp.	*Castilleja*
Poison-ivy	*Toxicodendron rydbergii* [*Rhus rydbergii*]
Pondweed	*Potamogeton* spp.
Prickly pear	*Opuntia* spp.
Spurge, Leafty	*Euphoria esola* [*Tithymalus esula*]
Sweet-clover, white	*Melilotus albus*
Thistle	*Breea, Carduus, Cirsium, Onopordum*

ANIMALS

Bear, black	*Ursus americanus*
Beaver	*Castor canadensis*
Bullsnake	*Pituophis melanoleucus*
Coyote	*Canis latrans*
Ground squirrel	*Spermophilus* sp.
Bison	*Bison bison*
Elk	*Cervus elaphus*
Moose	*Alces alces*
Prairie dog	*Cyonomys* sp.
Vole, red-backed	*Clethrionomys gapperri*
Warbler, Kirtland's	*Dendroica kirtlandii*

WORKS CITED

Carter, J.L. 1988, Trees and shrubs of Colorado. Johnson Books, Boulder.

Fitzgerald, J.P., C.A. Meaney, and D.M. Armstrong. 1994. Mammals of Colorado. Denver Mus. of Nat. Hist., Denver.

Harrington, H. D. 1954. Manual of the plants of Colorado: for the identification of the ferns and flowering plants of the state. Sage Books, Denver.

Weber, W. A. 1976. Rocky Mountain Flora. Colorado Assoc. University press, Boulder.

Weber, W. A. and R. C. Wittmann. 1996a. Colorado flora: eastern slope. Revised ed. University Press of Colorado, Niwot.

Weber, W. A. and R. C. Wittmann. 1996b. Colorado flora: western slope. Revised ed. University Press of Colorado, Niwot.

APPENDIX F: GAZETTEER

Site Name	Atlas Block Number	Atlas Block Name	County
Adobe Creek Reservoir	38103C3	Arlington NE	Bent/Kiowa
Air Force Academy	3810487	Pikeview	El Paso
Akron	40103B2	Akron	Washington
Alamosa NWR	37105D7	Alamosa East	Alamosa
Antero Reservoir	38105H8	Antero Res.	Park
Arapaho NWR	40106E3	MacFarlane Res.	Jackson
	40106F3	Walden	Jackson
Arriba	39103C3	Arriba	Lincoln
Aspen	39106B7	Aspen	Pitkin
Axial	40107C7	Axial	Moffat
Bailey	39105D5	Shawnee	Park
Bailey Bird Nesting Area	39106F2	Willow Lakes	Summit
Banner Lakes SWA	40104A5	Keenesburg	Weld
Barr Lake St. Park	39104H7	Brighton	Adams
Basalt	39107C1	Basalt	Town—Eagle; Atlas block—Pitkin
Battlement Mesa	39107D8	Rulison	Garfield
Bents Fort Natl Mon.	38103A4	Hadley	Otero
Beulah	38104A8	Beulah	Pueblo
Big Creek Lakes	40106H5	Pearl	Jackson
Bledsoe ranch	39103A2	Bledsoe Ranch	Lincoln/Cheyenne/Kit Carson
Blue Lake	See Adobe Creek Res.		
Bowers Peak Atlas block	37106H5	Bowers Peak	Saguache
Brown Canyon	37103E3	Brown Canyon	Las Animas
Browns Park NWR	40108G7	Lodore School	Moffat
Brush	40103C6	Brush	Morgan
Bonny Reservoir SRA	39102F2	Bonny Res. South	Yuma
Boulder	40105A3	Boulder	Boulder
Buena Vista	38106G2	Buena Vista West	Chaffee
Burlington	39102C3	Burlington	Kit Carson
Cameron Pass	40105E8	Clark Peak	Larimer/Jackson
Canon City	38105D2	Canon City	Fremont
Carbondale	3910702	Carbondale	Garfield
Carrizo Canyon, campground	37103B1	Carrizo Mtn.	Baca
Castlewood St. Park	39104C7	Franktown	Douglas
Cedaredge	3810788	Cedaredge	Delta
Chatfield St. Park	39105E1	Littleton	Douglas/Jefferson
Cherry Creek Reservoir	39104F7	Fitzsimons	Arapahoe
Chester Atlas block	3810633	Saguache	Saguache
Colorado City	37104H7	Colorado City	Pueblo
Colorado Natl Mon.	39108A6	Colo. Natl Mon.	Mesa
Colorado Springs	38104G7	Colorado Springs	El Paso
Connected Lakes unit of Colorado River State Park	39108A5	Grand Junction	Mesa
Cortez	37108C5	Cortez	Montezuma
Craig	40107E5	Craig	Moffat
Cripple Creek	38105F2	Cripple Creek	Teller
Crow Valley campground	40104F3	Briggsdale	Weld
Deadman Hill	40105G7	Deadman	Larimer
Debeque	39108C2	Debeque	Mesa
Del Norte	37106F3	Del Norte	Rio Grande
Dillon	39106F1	Dillon	Summit
Dillon Reservoir	39106E1	Frisco	Summit
Dove Creek	37108G8	Dove Creek	Dolores
Durango	37107C8	Durango West	La Plata
Eagle	39106F7	Eagle	Eagle
Elbert	39104B5	Elbert	Elbert

Site Name	Atlas Block Number	Atlas Block Name	County
Electra Lake	37107E7	Electra Lake	La Plata
Elevenmile Reservoir	38105H5	Spinney Mtn.	Park
Englewood	39104F8	Englewood	Arapahoe
Evans ranch	39105F4	Squaw Pass	Clear Creek
Fairplay	39106B1	Fairplay	Park
Flatirons	39105H3	Eldorado Spgs.	Boulder
Florence	38105D1	Florence	Fremont
Fort Carson	38104E7	Timber Mtn.	El Paso
	38104F7	Cheyenne Mtn.	El Paso
Fort Collins	40105E1	Fort Collins	Larimer
Fort Lyon	38103A2	Las Animas	Bent
Fort Morgan	40103C7	Fort Morgan	Morgan
Fort Garland	37105D4	Fort Garland	Costilla
Garden of the Gods	38104G8	Manitou Spgs.	El Paso
Garfield	38106E3	Garfield	Chaffee
Gateway	38108F8	Gateway	Mesa
Gerald Fyler's backyard	3710846	Arriola	Montezuma
Glade Park	39108A7	Battleship Rock	Mesa
Golden Gate Canyon State Park	39105G4	Blackhawk	Gilpin
Glenwood Springs	39107E3	Glenwood Spgs.	Garfield
Gothic	38106H8	Gothic	Gunnison
Granby	40105A8	Granby	Grand
Granby, Lake	40105B7	Shadow Mtn.	Grand
Grand Junction	39108A5	Grand Junction	Mesa
Grand Lake (lake)	40105B7	Shadow Mtn.	Grand
Grand Lake (town)	40105C7	Grand Lake	Grand
Great Plains Reservoirs	38102C6/38102C7	Neenoshe Res./Swede Lake	Kiowa/Kiowa
Greeley	40104D6	Greeley	Weld
Gunnison	3810658	Gunnison	Gunnison
Hasser Ranch	37102F6	Hasser Ranch	Baca/Prowers
Haxtun	40102F6	Haxtun West	Phillips
Holly East	38102A1	Holly East	Prowers
Holly West	38102A2	Holly West	Prowers
Holyoke	40102E3	Holyoke	Phillips
Horsetooth Reservoir	40105E2	Horsetooth Res.	Larimer
Idalia	39102F3	Idalia	Yuma
Independence Pass	39106a5	Independence Pass	Pitkin/Lake
Jackson Reservoir	40104D1	Sunken L.	Morgan
Joes	39102F6	Joes	Yuma
John Martin Reservoir	38102A8/38103A1	Hasty/Kreybill	Bent/Bent
Juniata Reservoir	38108H3	Juniata Res.	Mesa
Kenosha Pass	39105D7	Jefferson	Park
Kim	37103B3	Kim South	Las Animas
Kirchnavy Buttes	40103H3	Kirchnavy Buttes	Logan
Kit Carson	38102G7	Kit Carson	Cheyenne
Kremmling	40106A4	Kremmling	Grand
La Veta	37105E1	La Veta	Huerfano
Lake City	38107A3	Lake City	Hinsdale
Lake Dorothey SWA	37104A3	Barela	Las Animas
Leadville	39106C3	Leadville	Lake
Lincoln Gulch campground	39106A6	New York Peak	Pitkin
Livermore	40105G2	Livermore	Larimer
Little Bear Peak	37105E4	Blanca Peak	Alamosa
Longmont	40105B1	Longmont	Boulder
Loveland	4010541	Loveland	Larimer
Loveland Pass	39105F8	Loveland Pass	Clear Creek/Summit
Lower Latham Reservoir	40104C6	La Salle	Weld
Lyons	40105B3	Lyons	Boulder

Site Name	Atlas Block Number	Atlas Block Name	County
Mancos	37108C3	Mancos	Montezuma
Meeker	40107A8	Meeker	Rio Blanco
Monte Vista NWR	37106D1/37106D2/37106E1	Waverly/Fulcher Gulch/Homelake	Rio Grande
Mount Evans	39105E6	Mt. Evans	Clear Creek
Mount Wilson	37107G8	Mt. Wilson	Dolores
Mummy Range	40105D6	Trail Ridge	Larimer
Navajo Reservoir	37107A4	Allison	Archuleta
Neenoshe Reservoir	38102C6	Neenoshe Res.	Kiowa
Neeskaw Reservoir	38102C6	Neenoshe Res.	Kiowa
Niwot Ridge	40105A5	Ward	Boulder
Ophir	37107G7	Ophir	San Juan
Ouray	38107A6	Ouray	Ouray
Pagosa Springs	37107C1	Pagosa Spgs.	Archuleta
Palmer Divide	39104A7	Monument	El Paso
Pawnee Buttes	40103G8	Pawnee Buttes	Weld
Perry Park	39104B9	Larkspur	Douglas
Phantom Canyon	38105E1	Phantom Canyon	Fremont
Pikes Peak	38105G1	Pikes Peak	El Paso
Pintada Canyon	37103B2	Pintada Creek	Las Animas
Prewitt Reservoir	40103D4	Messex	Washington
Rangely	40108A7	Rangely	Rio Blanco
Raton Mesa	37104A4	Fishers Peak	Las Animas
Red Lion SWA	40102H6	Julesburg Res.	Logan
Red Feather Lakes	40105G5	Red Feather Lakes	Larimer
Red Rocks Park	39105F2	Morrison	Jefferson
Riverside Reservoir	40104C3	Dearfield	Weld
Rocky Mountain Arsenal	39104G7	Commerce City	Adams
Round Mtn. campground	39105A4	Tarryall	Park
Roxborough St. Park	39105D1	Kassler	Douglas
Russell Lakes SWA	37106H1/37106H2	Saguache/Saguache	Harrence Lake/Swede Corners
Saguache	38106A2	Saguache	Saguache
San Luis Lake(s) (SWA)	37105F6	Medano Ranch	Alamosa
Sheep Canyon Atlas block	37103F7	Sheep Canyon	Las Animas/Otero
Silver Plume	39105F6	Georgetown	Clear Creek
Spanish Peaks	37104D8	Spanish Peaks	Huerfano
Springfield	37102D5	Springfield East	Baca
State Bridge	39106G6	State Bridge	Eagle
Steamboat Springs	40106D7	Steamboat Spgs.	Routt
Sweetwater Lake	39107H2	Sweetwater L.	Garfield
Tamarack SWA	40102G6/40102G7	Tamarack Ranch/Crook	Logan
Taylor Park	38106G5	Taylor Park Res.	Gunnison
Trail Ridge Road	40105D6/40105D7	Trail Ridge/Fall River Pass	Larimer/Larimer/Grand
Trinidad	37104B5	Trinidad West	Las Animas
Twin Lakes	39106A3	Granite	Lake
Two Buttes Reservoir	37102F5	Two Buttes Res.	Baca
Union Reservoir	40105B1	Longmont	Weld
Walden	40106F3	Walden	Jackson
Walsenburg	37104F7	Walsenburg North	Huerfano
Walsh	37102D3	Walsh	Baca
Waterton Canyon	39105D1	Kassler	Douglas/Jefferson
Westcreek	39105B2	Westcreek	Douglas
Winter Park	39105H7	Fraser	Grand
Wolcott	39106F6	Wolcott	Eagle
Woodland Park	38105H1	Woodland Park	Teller
Wray	40102A2	Wray	Yuma
Yankee Boy Basin	37107H7	Telluride	Ouray
Yuma	40102B6	Yuma North	Yuma

CITATIONS AND REFERENCES

CERTAIN STANDARD REFERENCES

The text uses shorthand for certain seminal references to Colorado birds:

A&R: Andrews, Robert A., and Robert Righter. 1992. Colorado birds. Denver Museum of Natural History, Denver. The most recent standard reference book on Colorado bird distribution; meticulously researched. Maps, elevation charts, seasonal occurrences.

B&N: Bailey, A.M., and R.J. Niedrach. 1965. Birds of Colorado. Denver Museum of Natural History, Denver. The basic reference on Colorado birds. Two volumes of descriptions, distribution, records, and discussion.

BBS: 1966–1996. Breeding Bird Survey Trend Analysis. Unpubl. data furnished by the Breeding Bird Survey Office, National Biological Service, Patuxent Wildlife Research Center, Laurel, MD.

Latilong Study: Kingery, H.E. 1988. Colorado bird distribution study. Colorado Field Ornithologists in cooperation with the Colorado Division of Wildlife, Denver. Maps occurrences seasonally by latilong from 1965 to 1987, derived from contributions by most Colorado birdwatchers.

Priorities 1995: Colorado Bird Observatory and Colorado Division of Wildlife, 1995. Priorities for bird conservation and monitoring in Colorado. Categorizes bird species using Partners in Flight criteria for developing priorities for bird conservation and monitoring.

CITATIONS AND REFERENCES

Abbott, C.G. 1929. Watching Long-crested Jays. Condor 31:124–125.

Adam, C.G. 1987. Status of the Eastern Screech Owl in Saskatchewan with reference to adjacent areas. Pp. 268–276 *in* Biology and conservation of northern forest owls. U.S. For. Serv. Gen. Tech. Rep. RM-142.

Adams, J.S., R.L. Knight, L.C. McEwen, and T.L. George. 1994. Survival and growth of nestling Vesper Sparrows exposed to experimental food reductions. Condor 96:739–748.

Adkisson, C.S. 1981. Geographic variation in vocalizations and evolution of North American Pine Grosbeaks. Condor 83:277–288.

Ahlquist, J.E., A.H. Bledsoe, J.T. Ratti, and C.G. Sibley. 1987. Divergence of the single-copy DNA sequences of the Western Grebe (*A. occidentalis*) and Clark's Grebe (*A. clarkii*) as indicated by DNA-DNA hybridization. Postilla No. 200.

Aiken C., and E. Warren. 1914. Birds of El Paso County. Colorado Coll. Publ. Gen. Ser. 74, Sci. Ser. No. 13, Colo. Springs, CO.

Airola, D.A. 1986. Brown-headed Cowbird parasitism and habitat disturbance in the Sierra Nevada. J. Wildl. Manage. 50:571–575.

Airola, D.A., and R.H. Barrett. 1985. Foraging and habitat relationships of insect-gleaning birds in Sierra Nevada mixed-conifer forest. Condor 87:205–206.

Alexander, G. 1937. The birds of Boulder County Colorado. Univ. Colo. Stud. 24:79–105.

Alisauskas, R.T., and T.W. Arnold. 1994. American Coot. Chapt. 9 *in* T.C. Tacha and C.E. Braun eds., Migratory shore and upland game bird management in North America. Int. Assoc. Fish and Wildl. Agencies. Allen Press, Lawrence, KS.

Allen, A.A. 1924. A contribution to the life history and economic status of the screech owl (*Otus asio*). Auk 41:1–16.

Allen, R.W., and M.M. Nice. 1952. A study of the breeding biology of the Purple Martin. Am. Midl. Nat. 47:606–665.

Allen, S. 1988. Some thoughts on the identification of Gunnison's Red-backed Hawk and why it's not a natural vagrant. J. Colo. Field Ornithol. 22:8–13.

Altman, R. 1997. Olive-sided Flycatcher in western North America: status review (draft). USFWS, Portland, OR.

American Ornithologists' Union. 1957. Check-list of North American birds, 5th ed. Lord Baltimore Press, Baltimore, MD.

American Ornithologists' Union. 1983. Check-list of North American birds, 6th ed. Allen Press, Lawrence, KS.

American Ornithologists' Union. 1985. Thirty-fifth supplement to the AOU check-list of North American birds. Auk 106:532–538.

American Ornithologists' Union. 1989. Thirty-seventh supplement to the AOU check-list of North American birds. Auk 106:532–538.

American Ornithologists' Union. 1993. Thirty-ninth supplement to the AOU check-list of North American birds. Auk 110:675–682.

American Ornithologists' Union. 1995. Fortieth supplement to the AOU check-list of North American birds. Auk 112:819–830.

American Ornithologists' Union. 1997. Forty-first supplement to the AOU check-list of North American birds. Auk 114:542–552.

American Ornithologists' Union. 1998. Check-list of North American birds, 7th ed. Allen Press, Lawrence, KS.

Ammon, E.M. 1995. Lincoln's Sparrow. *In* The birds of North America no. 191 (A. Poole and F. Gill eds.). Acad. Nat. Sci., Philadelphia, and Am. Ornithol. Union, Washington, DC.

Andersen, D.E. 1991. Management of North American grasslands for raptors. Pp. 203–210 *in* B.A.G. Pendleton and D.L. Krake, eds., Proc. of the midwest raptor management symposium and workshop. Natl. Wildl. Fed. Sci. Tech. Ser. No. 15, Washington, DC.

Anderson, A.H., and A. Anderson. 1973. The Cactus Wren. Univ. Arizona Press, Tucson.

Anderson, S.H. 1976. Comparative food habits of Oregon nuthatches. Northwest Sci. 50:213–221.

Andrews, B.J., M. Sullivan, and J.D. Hoerath. 1996. Nest site reuse in the Western Wood-Pewee. Wilson Bull. 108:378–380.

Andrews R. 1987. Distribution ecology of Brown Thrasher and Gray Catbird. J. Colo. Field Ornithol. 21:32.

Andrews, R.A., and R. Righter. 1992. Colorado birds. Denver Mus. Nat. Hist., Denver.

Andrle, R.F., and J.R. Carroll eds. 1988. The atlas of breeding birds in New York state. Cornell Univ. Press, Ithaca, NY.

Angell, T. 1969. Study of the Ferruginous Hawk: adult and brood behavior. Living Bird 8:225–241.

Angell, T. 1978. Ravens, crows, magpies, and jays. Univ. Washington Press, Seattle.

Anonymous. 1967. Colorado Spring Counts. J. Colo. Field Ornithol. 2:10–13.

Anthony, A.W. 1892. Birds of southwestern New Mexico. Auk 9:351–369.

Armbruster, J.S. 1982. Wood Duck displays and pairing chronology. Auk 99:116–122.

Armistead, H.T. 1983. Bank Swallow. Pp. 302–304 *in* J. Farrand ed. The Audubon Society master guide to birding. Vol. 2. Knopf, New York.

Armstrong, D.M. 1972. Distribution of mammals in Colorado. Monogr. Univ. Kansas Mus. Nat. Hist. 3:1–415.

Audubon, J.J. 1838. Ornithological biography, vol. 4.

Audubon, J.J. 1861. The birds of America. Reissue, Roe Lockwood & Son. New York.

Austin, G.T. 1976. Sexual and seasonal differences in foraging of Ladder-backed Woodpeckers. Condor 78:317–323.

Austin, J.E., and M.R. Miller. 1995. Northern Pintail. *In* The birds of North America. no. 163 (A. Poole and F. Gill eds.). Acad. Nat. Sci., Philadelphia, and Am. Ornithol. Union, Washington, DC.

Austin, O.L. Jr. 1961. Birds of the world. Golden Press, New York.

Austing, G.R., and J.B. Holt Jr. 1966. The world of the Great Horned Owl. J.B. Lippincott, Philadelphia.

Avery, M., and L.W. Oring. 1977. Song dialects in the Bobolink. Condor 79:113–118.

Bailey, A.M. 1974. Second nesting of Broad-tailed Hummingbirds. Condor 76:350.

Bailey, A.M., and F. Brandenburg. 1940. The Snowy Plover in Colorado. Condor 42:128.

Bailey, A.M., and F.G. Brandenburg. 1941. Colorado nesting records. Condor 43:73–74.

Bailey, A.M., and R.J. Niedrach. 1938. Nesting of Virginia's Warbler. Auk 55:176–178.

Bailey, A.M., and R.J. Niedrach. 1965. Birds of Colorado. Denver Mus. Nat. Hist., Denver.

Bailey, A.M., R.J. Niedrach, and A.L. Baily. 1953. Red Crossbills of Colorado. Mus. Pictorial No. 9, Denver Mus. Nat. Hist., Denver.

Bailey, F.M. 1928. Birds of New Mexico. Judd and Detweiler Press, Washington, DC.

Bailey, R.G. 1978. Ecosystems of the United States. U.S. Dept. Agric., For. Serv., RARE II, Map B, 1:7500000.

Baldassarre, G.A., and E.G. Bolen. 1994. Waterfowl ecology and management. John Wiley and Sons, New York.

Baldwin, P.H., and J.R. Koplin. 1966. The Boreal Owl as a Pleistocene relict in Colorado. Condor 68:299–300.

Balgooyen, R.G. 1976. Behavior and ecology of the American Kestrel in California. Univ. Calif. Pub. Zool. 103:1–83.

Balph, M.H. 1978. Some population trends among Cassin's Finches in northern Utah. North Am. Bird Banding 3(1):12–15.

Baltosser, W.H. 1986. Nesting success and productivity of hummingbirds in southwestern New Mexico and southeastern Arizona. Wilson Bull. 98:353–367.

Bancroft, J. 1986. Brewer's Blackbirds successfully nest in birdhouse. Blue Jay 44:188.

Baptista, L.F. 1975. Song characteristics and demes in sedentary populations of the White-crowned Sparrow *(Zonotrichia leucophrys nuttalli)*. Univ. Calif. Pub. in Zool. 105:1–52.

Barlow, J.C. 1962. Natural history of the Bell's Vireo *Vireo bellii* (*Audubon*). Univ. Kansas Publ. 12:241–296.

Barlow, J.C. 1980. Patterns of ecological interactions among migrant and resident vireos on the wintering grounds. Pp. 79–107 *in* A. Keast and E.S. Morton, eds., Migrant birds in the Neotropics: ecology, behavior, distribution, and conservation. Smithsonian Inst. Press, Washington DC.

Barrowclough, G.F., and R.J. Gutierrez. 1990. Genetic variation and differentiation in the Spotted Owl (*Strix occidentalis*). Auk 107:737–744.

Basore, N.S. 1984. Breeding ecology of upland birds in no-tillage and tilled cropland. M.S. Thesis, Iowa State Univ., Ames.

Bates, J.W., and M.O. Moretti. 1994. Golden Eagle (*Aquila chrysaetos*) population ecology in eastern Utah. Great Basin Nat. 54:248–255.

Baumgarten, H.E. 1968. Lark Bunting. *In* A.C. Bent, Life histories of North American cardinals, grosbeaks, buntings, towhees, finches, sparrows, and allies (part 2). U.S. Natl. Mus., Washington, DC.

BBS. 1966–1996. Breeding bird survey trend analysis. Unpubl. data furnished by the Breeding Bird Survey Office, Biological Research Division, U.S. Geological Survey, Patuxent Wildlife Research Center, Laurel, MD.

Beal, F.E.L. 1897. The blue jay and its food. Pp. 197–207 *in* Yearbook. U.S. Dept. Agric., Washington, DC.

Beal, F.E.L. 1910. Birds of California in relation to the fruit industry part II. U.S. Dept. Agric. Biol. Surv. Bull. 34.

Beal, F.E.L. 1911. Food of the woodpeckers of the U.S. Dept. Agric. Biol. Surv. Bull. 37.

Beal, F.E.L. 1912. Food of our more important flycatchers. U.S. Dept. Agric. Biol. Surv. Bull. 44.

Beason, R.A. 1995. Horned Lark. *In* The birds of North America no. 195 (A. Poole and F. Gill eds.). Acad. Nat. Sci., Philadelphia, and Am. Ornithol. Union,Washington, DC.

Beaver, D.L., and P.H. Baldwin. 1975. Ecological overlap and the problem of competition and sympatry in the Western and Hammond's flycatchers. Condor 77:1–13.

Bechard, M.J., and J.K. Schmutz. 1995. Ferruginous Hawk. *In* The birds of North America no. 172 (A. Poole and F. Gill eds.). Acad. Nat. Sci., Philadelphia, and Am. Ornithol. Union, Washington, DC.

Beidleman, R.G. 1949. Guide to the birds of prey of Colorado. Univ. Colo. Mus. Leaf. No. 6. Boulder.

Bekoff, M., A.C. Scott, and D.A. Conner. 1987. Nonrandom nest-site selection in Evening Grosbeaks. Condor 89:819–829.

Belanger, L., and R. Couture. 1988. Use of man-made ponds by dabbling duck broods. J. Wildl. Manage. 52:718–723.

Bellrose, F.C. 1976. Ducks, geese and swans of North America. 2nd ed. Stackpole Books, Harrisburg, PA.

Bellrose, F.C. 1980. Ducks, geese and swans of North America. 3rd ed. Stackpole Books, Harrisburg, PA.

Belthoff, J.R., and G. Ritchison. 1989. Natal dispersal of Eastern Screech-Owls. Condor 91:254–265.

Bendire, C.E. 1889. *Picicorvus columbianus* (Wils.) Clarke's Nutcracker, its nest and eggs, etc. Auk 6:226–236.

Bendire, C.E. 1890. Notes on *Pipilo fuscus mesoleucus* and *Pipilo aberti*, their habits, nests, and eggs. Auk 7:22–29.

Bendire, C. 1892. Life histories of North American birds: with special reference to their breeding habits and eggs. U.S. Natl. Mus. Spec. Bull. 1. Washington, DC.

Bendire, C. 1893. The cowbirds. U.S. Natl. Mus. Rep., Washington, DC.

Bene, F. 1947. The feeding and related behavior of hummingbirds with special reference to the black-chin *Archilochus alexandri (Bourcier and Mulsant)*. Mem. Boston Soc. Nat. Hist. 9:395–478.

Benedict, A.D. 1991. A Sierra Club naturalist's guide to the southern Rockies. Sierra Club Books, San Francisco.

Benkman, C.W. 1987. Food profitability and the foraging ecology of crossbills. Ecological Monogr. 57:251–267.

Benkman, C.W. 1988a. Why White-winged Crossbills do not defend feeding territories. Auk 105:370–371.

Benkman, C.W. 1988b. Seed handling ability, bill structure, and the cost of specialization for crossbills. Auk 105:715–719.

Benkman, C.W. 1988c. Flock size, food dispersion, and the feeding behavior of crossbills. Behavioral Ecology and Sociobiology 23:167–175.

Benkman, C.W. 1989a. On the evolution and ecology of island populations of crossbills. Evolution 43:1324–1330.

Benkman, C.W. 1989b. Breeding opportunities, foraging rates, and parental care in White-winged Crossbills. Auk 106:483–485.

Benkman, C.W. 1990. Intake rates and the timing of crossbill reproduction. Auk 107:376–386.

Benkman, C.W. 1992a. A crossbill's twist of fate. Natural History (Dec.):39–42.

Benkman, C.W. 1992b. White-winged Crossbill. *In* The birds of North America no. 27 (A. Poole P. Stettenheim and F. Gill eds.). Acad. Nat. Sci., Philadelphia, and Am. Ornithol. Union, Washington, DC.

Benkman, C.W. 1993a. Adaptation to single resources and the evolution of crossbill (Loxia) diversity. Ecological Monogr. 63:305–325.

Benkman, C.W. 1993b. Logging, conifers, and the conservation of crossbills. Conservation Biology 7:473–479.

Benkman, C.W., and R.E. Miller. 1996. Morphological evolution in response to fluctuating selection. Evolution 50:2499–2504.

Benson, L.A., C.E. Braun, and W.C. Leininger. 1993. Sage Grouse response to burning in the big sagebrush type. Proc. issues and technology *in* The management of impacted western wildlife. Thorne Ecol. Inst. 5:97–104.

Bent, A.C. 1919. Life histories of North American diving birds. U.S. Natl. Mus. Bull. 107, Washington, DC.

Bent, A.C. 1921. Life histories of North American gulls and terns. U.S. Natl. Mus. Bull. 113, Washington, DC.

Bent, A.C. 1922. Life histories of North American petrels and pelicans and their allies. U.S. Natl. Mus. Bull. 121, Washington, DC.

Bent, A.C. 1923. Life histories of North American wild fowl, part 1. U.S. Natl. Mus. Bull. 126, Washington, DC.

Bent, A.C. 1925. Life histories of North American wild fowl, part 2. U.S. Natl. Mus. Bull. 130, Washington, DC.

Bent, A.C. 1926. Life histories of North American marsh birds. U.S. Natl. Mus. Bull. 135, Washington, DC.

Bent, A.C. 1927. Life histories of North American shore birds, part 1. U.S. Natl. Mus. Bull. 142, Washington, DC.

Bent, A.C. 1929. Life histories of North American shore birds, part 2. U.S. Natl. Mus. Bull. 146, Washington, DC.

Bent, A.C. 1932. Life histories of North American gallinaceous birds. U.S. Natl. Mus. Bull. 162, Washington, DC.

Bent, A.C. 1937. Life histories of North American birds of prey, part 1. U.S. Natl. Mus. Bull. 167, Washington, DC.

Bent, A.C. 1938. Life histories of North American birds of prey, part 2. U.S. Natl. Mus. Bull. 170, Washington, DC.

Bent, A.C. 1939. Life histories of North American woodpeckers. U.S. Natl. Mus. Bull. 174, Washington, DC.

Bent, A.C. 1940. Life histories of North American cuckoos, goatsuckers, hummingbirds, and their allies. U.S. Natl. Mus. Bull. 176, Washington, DC.

Bent, A.C. 1942. Life histories of North American flycatchers, larks, swallows, and their allies. U.S. Natl. Mus. Bull. 179, Washington, DC.

Bent, A.C. 1946. Life histories of North American jays, crows, and titmice. U.S. Natl. Mus. Bull. 191, Washington, DC.

Bent, A.C. 1948. Life histories of North American nuthatches, wrens, thrashers, and their allies. U.S. Natl. Mus. Bull. 195, Washington, DC.

Bent, A.C. 1949. Life Histories of North American thrushes, kinglets, and their allies. U.S. Natl. Mus. Bull. 196, Washington, DC.

Bent, A.C. 1950. Life histories of North American wagtails, shrikes, vireos, and their allies. U.S. Natl. Mus. Bull. 197, Washington, DC.

Bent, A.C. 1953. Life histories of North American wood warblers. U.S. Natl. Mus. Bull. 203, Washington, DC.

Bent, A.C. 1958. Life histories of North American blackbirds, orioles, tanagers, and allies. U.S. Natl. Mus. Bull. 211, Washington, DC.

Bent, A.C. 1968. Life histories of North American cardinals, grosbeaks, buntings, towhees, finches, sparrows, and allies. U.S. Natl. Mus. Bull. 237, Washington, DC.

Bergin, T.M. 1992. Habitat selection by the Western Kingbird in western Nebraska. Condor 94:903–911.

Bergin, T.M. 1993. The influence of habitat structure, reproductive success, and nest predation on nest site selection of Western Kingbirds, Mourning Doves, and Common Grackles. Ph.D. Diss. Bowling Green State Univ., Bowling Green, KY.

Bergman, C.A. 1985. Invaders from the far north. Natl. Wildl. 23(6):34–39.

Bergman, R.D., P. Swain, and M.W. Weller. 1970. A comparative study of nesting Forster's and Black terns. Wilson Bull. 82:435–444.

Bergtold, W.H. 1927. The cardinal in Colorado. Auk 44:108.

Bergtold, W.H. 1928. A guide to Colorado birds. Smith-Brooks Printing Co., Denver.

Bergtold, W.H. 1931. The season (Denver region). Bird-Lore 33:275.

Best, L.B., and K.L. Petersen. 1985. Seasonal changes in detectability of Sage and Brewer's sparrows. Condor 87:556–558.

Best, L.B., and N.L. Rodenhouse. 1984. Territory preference of Vesper Sparrows in cropland. Wilson Bull. 96:72–82.

Betts, N. 1913. Birds of Boulder County Colorado. Univ. Colo. Stud. 10:177–232.

Bevier, L.R. 1994. The atlas of breeding birds of Connecticut. Natl Audubon Soc. and Audubon Council Conn., Hartford.

Bieniasz, K.A. 1978. Biology of the Greater Sandhill Crane in Routt County Colorado. M.A. Thesis. Univ. Northern Colo., Greeley.

Bierregaard, R.O. Jr. 1994. Family *Cathartidae* (New World Vultures). Pp. 23–41 *in* del Hoyo, J.A. Elliott, and J. Sargatal eds. Handbook of the birds of the world. Lynx Edicions, Barcelona.

Biodiversity Legal Foundation. 1995. Petition for a rule to list the Columbian Sharp-tailed Grouse *Tympanuchus phasianellus columbianus* as "threatened or endangered" in the conterminous United States under the Endangered Species Act. 16 U.S.C. sec. 1531 et seq. (1973) as amended. USFWS document.

Blancher, P.J. 1982. Food supply, predation, and potential for competition between Western and Cassin's kingbirds. Ph.D. Diss. Queen's Univ. Kingston Ontario.

Blancher, P.J., and R.J. Robertson. 1984. Resource use by sympatric kingbirds. Condor 86:305–313.

Bloom, P.H. 1980. The status of the Swainson's Hawk in California. 1979. Final Report II-8.0 U.S. Bur. Land Manage. and Fed. Aid in Restoration, Proj. W-54-R-12, Calif. Dept. Fish and Game, Sacramento.

Bock, C.E. 1970. Ecology and behavior of the Lewis' Woodpecker *(Asyndemus lewis)*. Univ. California Publ. Zool. Univ. California Press, Berkeley.

Bock, C.E., and J.H. Bock. 1974. On the geographical ecology and evolution of the Three-toed Woodpeckers *Picoides tridactylus* and *P. arcticus*. Am. Midl. Nat. 92:397–405.

Bock, C.E., and J.H. Bock. 1983. Responses of birds and deer mice to prescribed burning in ponderosa pine. J. Wildl. Manage. 47:836–840.

Bock, C.E., and J.H. Bock. 1988. Grassland birds in southeastern Arizona: impacts of fire grazing and alien vegetation. Pp. 43–58 *in* P.D. Goriup, ed. Ecology and conservation of grassland birds. ICBP Tech. Publ. No. 7. Int. Counc. Bird Preservation, Cambridge England.

Bock, C.E., and D.E. Fleck. 1995. Avian response to nest box addition in two forests of the Colorado Front Range. J. Field Ornithol. 3:352–362.

Bock, C.E., and W.C. Scharf. 1994. A nesting population of Cassin's Sparrows in the sandhills of Nebraska. J. Field Ornithol. 65:472–475.

Bock, C.E., H.H. Hadow, and P. Somers. 1971. Relations between Lewis' and Red-headed woodpeckers in southeastern Colorado. Wilson Bull. 83:237–248.

Bock, C.E., V.A. Saab, T.D. Rich, and D.S. Dobkin. 1993. Effects of livestock grazing on Neotropical migratory landbirds in western North America. Pp. 296–309 *in* D.M. Finch and P.W. Stangel, eds. Status and management of Neotropical migratory birds. U.S. For. Serv. Gen. Tech. Rep. RM-229.

Bohl, W. 1957. Chukars in New Mexico 1931–1957. New Mexico Dept. Game and Fish, Santa Fe.

Bollinger, E.K., and T.A. Gavin. 1992. Eastern Bobolink populations: ecology and conservation in an agricultural landscape. Pp. 497–506 *in* Ecology and conservation of Neotropical migrant landbirds (J.M. Hagan, III and D.W. Johnson, eds.): Smithsonian Inst. Press, Washington, DC.

Bonnett, M., and K. Zimmerman. 1991. Politics and preservation: the Endangered Species Act and the Northern Spotted Owl. Ecol. Law Q. 18:105–171.

Borland, H. 1956. High, wide and lonesome. J.B. Lippincott, Philadelphia.

Borland, H. 1970. Country editor's boy. J.B. Lippincott, Philadelphia.

Bosakowski, T. 1986. Short-eared Owl roosting strategies. Am. Birds 40:237–240.

Botero, J.E., and D.H. Rusch. 1994. Foods of Blue-winged Teal in two Neotropical wetlands. J. Wildl. Manage. 58:561.

Bottorff, R.L. 1972. *In* Van Velzen, W.T. Breeding bird censuses. Am. Birds 26:980.

Bottorff, R., N. Hurley, F. Justice, J. Justice, R. Kelley, H.E. Kingery, U. Kingery, D. Stotz, and J. Trainor. 1971–1984. *In* Van Velzen W.T. Breeding bird censuses. Am. Birds 25:966 26:980 27:996 28:1036 29:1125 31:71 32:83 33:81 34:73 35:87 37:84 38:98.

Boula, K. 1982. Food habits and roost sites of Northern Saw-whet Owls in northeastern Oregon. Murrelet 63:92–93.

Boulder County Audubon Society. 1975–1997. Monthly wildlife inventories. Boulder County Aud. Soc. newsletter, Boulder.

Bowen, R.V. 1997. Townsend's Solitaire. *In* The birds of North America no. 269 (A. Poole and F. Gill eds.). Acad. Nat. Sci Philadelphia and Am. Ornithol. Union,Washington, DC.

Boyd, R.L. 1976. Behavioral biology and energy expenditure in a Horned Lark population. Ph.D. Thesis Colo. State Univ., Fort Collins.

Bradbury, W.C. 1918a. Notes on the nesting habits of the White-throated Swift in Colorado. Condor 20:103–110.

Bradbury, W.C. 1918b. Notes on the nesting of the Mountain Plover. Condor 20:157–163.

Brandt, H. 1951. Arizona and its bird life. Bird Research Found. Cleveland, OH.

Braun, C.E. 1973. Distribution and habitats of Band-tailed Pigeons in Colorado. Proc. West. Assoc. State Game and Fish Comm. 53:336–344.

Braun, C.E. 1976. Methods for locating, trapping, and banding Band-tailed Pigeons in Colorado. Colo. Div. Wildl. Spec. Rep. 39.

Braun, C.E. 1980. Alpine bird communities of western North America: implications for management and research. Pp. 280–291 *in* R.M. DeGraaf tech. coord. Workshop Proc. Mgmt. of W. Forests and Grasslands for Nongame Birds U.S. For. Serv. Gen. Tech. Rep. INT-86.

Braun, C.E. 1994. Band-tailed Pigeon. Chapter 5 *in* T.C. Tacha and T.C. Braun eds. Migratory Shore and Upland Game Management In North America. Intl. Assoc. Fish Wildl. Agencies. Allen Press, Lawrence, KS.

Braun, C.E. 1995. Distribution and status of Sage Grouse in Colorado. Prairie Nat. 27:1–9.

Braun, C.E., and G.E. Rogers. 1971. White-tailed Ptarmigan in Colorado. Colo. Div. Game Fish and Parks. Tech. Publ. No. 27.

Braun, C.E., M.F. Baker, R.L. Eng, J.S. Gashwiler, and M.H. Schroeder. 1976a. Conservation committee report on effects of alteration of sagebrush communities on the associated avifauna. Wilson Bull. 88:165–171.

Braun, C.E., T. Britt, and R.O. Wallestad. 1977. Guidelines for maintenance of Sage Grouse habitat. Wildl. Soc. Bull. 5:99–106.

Braun, C.E., J.H. Enderson, M.R. Fuller, Y.B. Linhart, and C.D. Marti. 1996. Northern Goshawk and forest management in the southwestern United States. Wildl. Soc. Tech. Rev. 96-2.

Braun, C.E., K.M. Giesen, R.W. Hoffman, T.E. Remington, and W.D. Snyder. 1994. Upland bird management analysis guide, 1994–1998. Colo. Div. Wildl. Rep. 19.

Braun, C.E., R.B. Davies, J.R. Dennis, K.A. Green, and J.L. Sheppard. 1992. Plains Sharp-tailed Grouse recovery plan. Colo. Div. Wildl., Denver.

Braun, C.E., R.W. Hoffman, and G.E. Rogers. 1976b. Wintering areas and winter ecology of White-tailed Ptarmigan in Colorado. Colo. Div. Wildl. Spec. Rep. No. 38.

Braun, C.E., K. Martin, and L.A. Robb. 1993. White-tailed Ptarmigan. *In* The birds of North America, no. 68 (A. Poole and F. Gill, eds.). Acad. Nat. Sci., Philadelphia, and Am. Ornithol. Union, Washington, DC.

Brauning, D.W., ed. 1992. Atlas of breeding birds in Pennsylvania. Univ. Pittsburgh Press, Pittsburgh.

Brawn, J.D. 1991. Environmental effects on variation and covariation in reproductive traits of Western Bluebirds. Oecologia 86:193–201.

Breeding Bird Survey Trend Analysis. 1966–1985. Unpubl. data, Breeding Bird Surv. Off., Natl Biol. Serv., Patuxent Wildl. Res. Center, Laurel, MD.

Breiding, G.H. 1943. Starling nesting in Colorado. Wilson Bull. 55:247.

Breitwisch, R., P.G. Merritt, and G.H. Whitesides. 1986. Parental investment by the Northern Mockingbird: male and female roles in feeding nestlings. Auk 103:152–159.

Brewer, R., G.A. McPeek, and R.J. Adams. 1991. The atlas of breeding birds of Michigan. Michigan St. Univ. Press, East Lansing.

Brewer, T.M. 1872. Colorado birds reported by C.E.H. Aiken. Proc. Boston Soc. Nat. Hist. 15:193–210.

Bridges, D. 1992a. Relative abundance of owls in Colorado. J. Colo. Field Ornithol. 26:27–28.

Bridges, D. 1992b. Northern Saw-whet Owls vs. Boreal Owls above 10,000 feet in the Wet, Sangre de Cristo, and Culebra mountains of south-central Colorado: a preliminary report. J. Colo. Field Ornithol. 26:29–31.

Bridges, D. 1992c. Black Rail saga. J. Colo. Field Ornithol. 26:57–60.

Brittingham, M.C., and S.A. Temple. 1983. Have cowbirds caused forest songbirds to decline? BioSci. 33:31–35.

Brockner, W.W. 1984. Brown-headed Cowbird parasitizing Mountain Chickadee nest. J. Colo. Field Ornithol. 18:109–110.

Brodkorb, P. 1978. Catalogue of fossil birds, part 5. Bull. Florida State Mus. 23(3):1–228.

Brown, B.T. 1992. Nesting chronology, density and habitat use of Black-chinned Hummingbirds along the Colorado River, Arizona. J. Field Ornithol. 63:393–400.

Brown, B.T. 1993. Bell's Vireo. *In* The birds of North America, no. 35 (A. Poole and F. Gill, eds.). Acad. Nat. Sci., Philadelphia, and Am. Ornithol. Union, Washington, DC.

Brown, B.T., S.W. Carothers, and R.R. Johnson. 1987. Grand Canyon birds. Univ. Arizona Press, Tucson.

Brown, C.R., and M.B. Brown. 1995. Cliff Swallow. *In* The birds of North America, no. 149 (A. Poole and F. Gills, eds.). Acad. Nat. Sci., Philadelphia, and Am. Ornithol. Union, Washington, DC.

Brown, C.R., A.M. Knott, and E.J. Damrose. 1992. Violet-green Swallow. *In* The birds of North America, no. 14 (A. Poole, P. Stettenheim, F. Gill, eds.). Acad. Nat. Sci., Philadelphia, and Am. Ornithol. Union. Washington, DC.

Brown, J.L. 1963. Aggressiveness, dominance and social organization in the Steller Jay. Condor 65:460–484.

Brown, M., and J.J. Dinsmore. 1986. Implications of marsh size and isolation for marsh bird management. J. Wildl. Manage. 50:392–397.

Brugger, K.E., L.N. Arkin, and J.M. Gramlich. 1994. Migration patterns of Cedar Waxwings in the eastern United States. J. Field Ornithol. 65:381–387.

Buchanan, D. 1995. Grouse hunters finding birds in short supply. Grand Jct. Daily Sentinel. Aug. 20, page 7D.

Buckelew, A.R. Jr., and G.A. Hall. 1994. West Virginia breeding bird atlas. Univ. Pittsburgh Press, Pittsburgh.

Bump, S.R.1986. Yellow-headed Blackbird *(Xanthocephalus xanthocephalus)* nest defense: aggressive responses to Marsh Wrens (*Cistothorus palustris*). Condor 88:3.

Burget, M.L. 1957. The Wild Turkey in Colorado. Colo. Game Fish Dept., Fed. Aid Rep. W-30-R.

Burgoyne, P. 1980. Bird population changes and manipulation of a ponderosa pine forest on the Kaibab Plateau, Arizona. Ph.D. Diss., Brigham Young Univ., Provo, UT.

Burleigh, T.D. 1971. Birds of Idaho. Caxton Printers, Caldwell, ID.

Burns, K.J., and S.J. Hackett. 1993. Nest and nest-site characteristics of a western population of Fox Sparrow (*Passerella iliaca*). Southwest Nat. 38:277–279.

Cabe, P.R. 1993. European Starling. *In* The birds of North America, no. 48 (A. Poole and F. Gill, eds.). Acad. Nat. Sci., Philadelphia, and Am. Ornithol. Union, Washington, DC.

Cade, B.S., and R.W. Hoffman. 1990. Winter use of Douglas-fir forests by Blue Grouse in Colorado. J. Wildl. Manage. 54:471–479.

Cade, T.J. 1982. The falcons of the world. Cornell Univ. Press, Ithaca, NY.

Cade, T.J., et al., eds. 1988. Peregrine Falcon population: their management and recovery. Peregrine Fund, Boise, ID.

Cadman, M.D., P.F.J. Eagles, F.M. Helleiner. 1987. Atlas of the breeding birds of Ontario. Univ. of Waterloo Press. Waterloo.

Caffrey, C. 1992. Female-biased delayed dispersal and helping in American Crows. Auk 109:609–19.

Calder, W.A. 1973. Microhabitat selection during nesting of hummingbirds in the Rocky Mountains. Ecol. 54:127–134.

Calder, W.A. 1981. Heat exchange of nesting hummingbirds in the Rocky Mountains. Nat. Geog. Soc. Res. Rep. 13:145–169.

Calder, W.A. 1994. When do hummingbirds use torpor in nature? Physiol. Zool. 67:1051–1076.

Calder, W.A., and L.L. Calder. 1992. Broad-tailed Hummingbird. *In* The birds of North America, no. 16 (A. Poole and F. Gill, eds.). Acad. Nat. Sci., Philadelphia, and Am. Ornithol. Union, Washington, DC.

Camp, M., and L.B. Best. 1995. Nest density and nesting success of birds in roadsides adjacent to rowcrop fields. Am. Midl. Nat. 131:347–358.

Campbell, H., D.K. Martin, P.E. Ferkovich, and B.K. Harris. 1973. Effects of hunting and some other environmental factors on Scaled Quail in New Mexico. Wildl. Monogr. 34.

Campo, J.J., B.C. Thompson, J.C. Barron, R.C. Telfair II, P. Durocher and S. Gutreuter. 1993. Diet of Double-crested Cormorants wintering in Texas. J. Field Ornithol. 64:135–144.

Cannings, R.J. 1993. Northern Saw-whet Owl. *In* The birds of North America, no. 42 (A. Poole and F. Gills, eds.). Acad. Nat. Sci., Philadelphia, and Am. Ornithol. Union, Washington, DC.

Carey, A.B., M.M. Hardt, S.P. Horton, B.L. Biswell. 1991. Spring bird communities in the Oregon Coast Range. U.S. For. Serv. GTR-285.

Carson, R. 1962. Silent Spring. Houghton Mifflin, Boston.

Carter, J.L. 1988, Trees and shrubs of Colorado. Johnson Books, Boulder, CO.

Carter, M., K. Barker, and J. Reddall. 1992. Population trends of Great Blue Herons and Black-crowned Night-Herons in South Platte River valley. J. Colo. Field Ornithol. 26:131.

Carter, M., G. Fenwick, C.Hunter, D. Pashley, D. Petit, J. Price, and J. Trapp. 1996. For the future. Am. Birds 50:238–240.

Carter, M.D. 1986. The parasitic behavior of the Bronzed Cowbird (*Molothrus aeneus*) in south Texas. Condor 88:11–25.

Carty, D. 1986. Geese of the Front Range. Colo. Outdoors 35(6):3–5.

Cary, M. 1909. New records and important range extensions of Colorado birds. Auk 26:180–185.

Cary, M. 1911. A biological survey of Colorado. N. Amer. Fauna 33:1–256.

Case, N.A., and O.H. Hewitt. 1963. Nesting and productivity of the Red-winged Blackbird in relation to habitat. Living Bird 2:7–20.

Castrale, J.S. 1982. Effects of two sagebrush control methods on nongame birds. J. Wildl. Manage. 46:945–952.

Chace, J.F. 1995. The factors affecting the reproductive success of the Solitary Vireo (*Vireo solitarius plumbeus*) in Colorado. M.A. Thesis, Univ. Colo., Boulder.

Chace, J.F., and A. Cruz. 1996. Knowledge of the Colorado host relations of the parasitic Brown-headed Cowbird (*Molothrus ater*). J. Colo. Field Ornithol. 30:67–81.

Chace, J.F., A. Cruz, and A. Cruz, Jr. 1997. Nesting success of the Western Wood-Pewee in Colorado. Western Birds 28:110–112.

Chace, J.F., A. Cruz, and R.E. Marvil. In press. Reproductive interactions between Brown-headed Cowbirds and Solitary Vireos in Colorado. *In* T.L. Cooke, ed. The ecology and management of cowbirds. Univ. Texas Press, Austin.

Chadwick, D. 1989. Never snicker at a snipe. Natl Wildl. 27:12–15.

Chapin, E.A. 1925. Food habits of the vireos. U.S. Dept. Agric. Bull. 1355.

Chaplin, S.B. 1982. The energetic significance of huddling behavior in Common Bushtits (*Psaltriparus minimus*). Auk 99:424–430.

Chapman, F.M. 1907. The warblers of North America. D. Appleton & Co., New York. Reprinted 1968, Dover Publ., New York.

Chase, C.A., III. 1979. Breeding shorebirds in the Arkansas Valley. J. Colo. Field Ornithol. 13:3–6.

Chase, C.A., III. 1980. Summer report, June 1–August 31, 1979. J. Colo. Field Ornithol. 14:38–51.

Chase, C.A., III. 1984. Gull Hybridization: California x Herring. J. Colo. Field Ornithol. 18:62.

Chilton, G., M.C. Baker, C.D. Barrentine, and M.A. Cuningham. 1995. White-crowned Sparrow. *In* The birds of North America, no. 183 (A. Poole and F. Gills, eds.). Acad. Nat. Sci., Philadelphia, and Am. Ornithol. Union, Washington, DC.

Choate, E.A. 1985. The dictionary of American bird names. Rev. ed. Revised by Raymond A. Paynter, Jr. Harvard Common Press, Boston.

Christensen, G. 1970. The Chukar Partridge: its introduction, life history and management. Nevada Dept. of Fish and Game, Reno.

Chu, D.S., J.D. Nichols, J.B. Hestbeck, and J.E. Hines. 1995. Banding reference areas and survival rates of Green-winged Teal, 1950–89. J. Wildl. Manage. 59:487–498.

Church, K.E., J.R. Saver, and S. Droege. Population trends of quails in North America. Proc. Natl. Symp. Quail IV:44–54.

Cimprich, D.A., and F.R. Moore. 1995. Gray Catbird. *In* The birds of North America, no. 167 (A. Poole and F. Gill, eds.). Acad. Nat. Sci., Philadelphia, and Am. Ornithol. Union, Washington, DC.

Clark, L., and J.R. Mason. 1985. Use of nest material as insecticide and anti-pathogene agents by the European Starling. Oecologia 67:169–176.

Clark, R.J. 1975. A field study of the Short-eared Owl, (*Asio flammeus*) Pontopiddan, in North America. Wildl. Monogr. 47.

Clements, J.F. 1981. Birds of the world: a checklist. Facts on File, New York.

Cogswell, H.L. 1949. Alternate care of two nests in the Black-chinned Hummingbird. Condor 51:176–178.

Colorado Bird Observatory and Colorado Division of Wildlife. 1995. Priorities for bird conservation and monitoring in Colorado. CBO and CDOW, Denver.

Colorado Division of Wildlife. 1989. Colorado statewide waterfowl management plan, 1989–2003. CDOW, Fort Collins.

Colorado Field Ornithologists. 1965–1995. Seasonal reports. J. Colo. Field Ornithol.

Colorado Natural Heritage Program. 1995. Colorado's Natural Heritage: rare and imperiled animals, plants, and natural communities. Colo. State Univ., Fort Collins.

Colwell, M.A., and J.R. Jehl, Jr. 1994. Wilson's Phalarope. *In* The birds of North America, no. 83 (A. Poole and F. Gill, eds.) Acad. Nat. Sci, Philadelphia, and Am. Ornithol. Union, Washington, DC.

Conry, J.A. 1978. Resource utilization, breeding, biology and nestling development in an alpine passerine community. Ph.D. Diss., Univ. Colo., Boulder.

Contreras, A. 1988. Northern Waterthrush summer range in Oregon. West. Birds 19:41–42.

Conway, C.J. 1995. Virginia Rail. *In* The birds of North America, no. 173 (A. Poole and F. Gill, eds.). Acad. Nat. Sci., Philadelphia, and Am. Ornithol. Union, Washington, DC.

Conway, C.J., and W.R. Eddleman. 1994. Virginia Rail. Pp. 192–206 *in* T.C. Tacha and C.E. Braun eds. Migratory shore and upland game bird management in North America. Int. Assoc. of Fish and Wildl. Agencies, Washington, DC.

Cook, A.G. Birds of the desert region of Uintah County, Utah. Great Basin Nat. 44:584–620.

Cook, K.J. 1978. Field observations of two heron species in Gunnison County. Unpubl. paper. Western State College, Gunnison, CO.

Cooke, W.W. 1897. The birds of Colorado. State Agric. Coll. 37. Tech. Ser. 2. Smith-Brooks Printing Co., Denver.

Cooke, W.W. 1898. Further notes on the birds of Colorado. State Agric. Coll., Bull. 44. Smith-Brooks Printing Co., Fort Collins.

Cooke, W.W. 1900. The birds of Colorado. (Second appendix to Bull. no. 37). Agric. Exp. Sta. Bull. No. 56. Agric. Coll. Colo., Fort Collins.

Cooke, W.W. 1909. The birds of Colorado—third supplement. Auk 26:400–422.

Cooper, R.J., K.M. Dodge, P.J. Martinat, S.B. Donahoe, and R.C. Whitmore. 1990. Effects of diflubenzuron application on eastern deciduous forest birds. J. Wildl. Manage. 54:486–493.

Copelin, F.F. 1963. The Lesser Prairie Chicken in Oklahoma. Oklahoma Wildl. Cons. Dept. Tech. Bull. No.6.

Coues, E. 1874. Birds of the Colorado valley. U.S. Gov. Printing Off., Washington, DC.

Coues, E. 1895. The expeditions of Zebulon Montgomery Pike to the headwaters of the Mississippi River through Louisiana territory and in New Spain, etc. Francis P. Harper, NY.

Craig, G.R. 1991. Peregrine Falcon restoration program. Job Prog. Rep., Colo. Div. Wildl. Res. Rep.

Craig, G.R. 1993. Peregrine Falcon restoration program. Job Prog. Rep., Colo. Div. Wildl. Res. Rep.

Craig, G.R. 1994. Peregrine Falcon restoration program. Job Prog. Rep., Colo. Div. Wildl. Res. Rep.

Craighead, J.J., and F.C. Craighead, Jr. 1956. Hawks, owls, and wildlife. Wildl. Manage. Inst., Washington, DC., and Stackpole Books, Harrisburg, PA. Reprinted 1969, Dover Publ., New York.

Craves, J.A. 1994. Forest fire-tail: for nonstop action, follow the American Redstart. Birder's World. 8(3):12–16.

Creighton, P.D. 1971. Nesting of the Lark Bunting in north-central Colorado. Grassland Biome, U.S. Int. Biol. Prog. Tech. Rep. No. 29.

Creighton, P.D. 1974. Habitat exploitation by an avian ground-foraging guild. Ph.D. Thesis, Colo. State Univ., Fort Collins.

Creighton, P.D., and P.H. Baldwin. 1974. Habitat exploitation by an avian ground-foraging guild. Grassland Biome, U.S. Intl. Biological Program Tech. Rep. No. 263, Fort Collins, CO.

Creighton, P.D., and D.K. Porter. 1974. Nest predation and interference by Western Meadowlarks. Auk 91:177–178.

Crockett, A.B. 1975a. Ecology and behavior of the Williamson's Sapsucker in Colorado. Ph.D. Thesis, Univ. of Colo., Boulder.

Crockett, A.B. 1975b. Nest site selection by Williamson's and Red-naped sapsuckers. Condor 77:365–368.

Crockett, A.B., and P.L. Hansley. 1977. Coition, nesting, and post-fledging behavior of Williamson's Sapsucker in Colorado. Living Bird 16:7–19.

Crockett, A.B., and H.H. Hadow. 1975. Nest site selection by Williamson and Red-naped sapsuckers. Condor 77:365–368.

Cruz, A., and N.J. Sanders. 1996. Foraging behavior of the Western Wood Pewee in the Colorado Front Range. J. Colo. Field Ornithol. 30:169.

Csada, R.D., and R.M. Brigham. 1992. Common Poorwill. *In* The birds of North America, no. 32 (A. Poole, P. Stettenheim, and F. Gill, eds.). Acad. Nat. Sci., Philadelphia, Am. Ornithol. Union, Washington, DC.

Csada, R.D., and R.M. Brigham. 1994. Breeding biology of the Common Poorwill at the northern edge of its distribution. J. Field Ornithol. 65:186–193.

Culotta, J.M., and R.W. Hoffman. 1989. Winter distribution and diet of Merriam's Wild Turkey along the northern Front Range of Colorado. Colo. Div. Wildl., Fed. Aid Rep. W-152-R.

Curson, J., D. Quinn, and D. Beadle. 1994. Warblers of the Americas. Houghton Mifflin, Boston.

Cuthbert, F.J., and M. Louis. 1993. The Forster's Tern in Minnesota: status, distribution, and reproductive success. Wilson Bull. 105:184–187.

Darley, J.A., D.M. Scott, and N.K. Taylor. 1977. Effects of age, sex, and breeding success on site fidelity of Gray Catbirds. Bird-Banding 48:145–151.

Darling, J.L. 1970. New breeding records of *Toxostoma curvirostre* and *T. bendirei* in New Mexico. Condor 72:366–367.

Davidson, A.H. 1994. Common Grackle predation on adult passerines. Wilson Bull. 106:174–175.

Davis, C.M. 1978. A nesting study of the Brown Creeper. Living Bird 17:237–263.

Davis, J. 1951. Distribution and variation of the Brown Towhee. Univ. California Publ. Zool. 52:1–120.

Davis, J., G.F. Fisler, and B.S. Davis. 1963. The breeding biology of the Western Flycatcher. Condor 65:337–382.

Davis, L.I. 1972. A field guide to the birds of Mexico and Central America. Univ. Texas Press, Austin.

Davis, T.A., and Ackerman, R.A. 1985. Adaptations of Black Tern (*Chlidonias niger*) eggs for water loss in a moist nest. Auk 102:640–643.

Davis, W.A. 1969. Birds in western Colorado. Colo. Field Ornithol., Denver.

Davis, W.A. 1970. Additions and corrections to "Birds in Western Colorado." J. Colo. Field Ornithol. 8:30–32.

Davis, W.E., Jr. 1993. Black-crowned Night-Heron. *In* The birds of North America, no. 74 (A. Poole and F. Gill, eds.). Acad. Nat. Sci., Philadelphia, and Am. Ornithol. Union, Washington, DC.

Davis, W.E., Jr., and J.A. Kushlan. 1994. Green Heron. *In* The birds of North America, no. 129 (A. Poole and F. Gill, eds.). Acad. Nat. Sci., Philadelphia, and Am. Ornithol. Union, Washington, DC.

Davis, W.J. 1980. The Belted Kingfisher, *Megaceryle alcyon*: its ecology and territoriality. M.S. Thesis, Univ. Cincinnati, OH.

Davis, W.J. 1982. Territory size in *Megaceryle alcyon* along stream habitat. Auk 99:353–362.

Dawson, W.L. 1909. Birds of Washington. Occidental Publ. Co., Seattle.

Dawson, W.R., C. Carey, C.S. Adkisson, and R.D. Ohmart. 1979. Responses of Brewer's and Chipping sparrows to water restriction. Physiol. Zool. 52:529–541.

DeBenedictis, P.A. 1995. Sage Sparrow mysteries. Birding 27:134–137.

del Hoyo, J., A. Elliott, and J. Sargatal. 1992. Handbook of birds of the world. Vol. 1, Ostrich to ducks. Lynx Edicions, Barcelona.

del Hoyo, J., A. Elliott, and J. Sargatal. 1994. Handbook of birds of the world. Vol. 2. New world vultures to guinea fowl. Lynx Edicions. Barcelona.

del Hoyo, J., A. Elliott, and J. Sargatal. 1996. Handbook of birds of the world. Vol. 3, Hoatzins to auks. Lynx Edicions. Barcelona.

Delacour, J. 1956. Waterfowl of the world, vol. 2. Country Life Ltd., London.

DeLong, A.K, J.A. Crawford, and D.C. DeLong, Jr. 1995. Relationships between vegetational structure and predation of artificial sage grouse nests. J. Wildl. Mange. 59:88–92.

Derrickson, K.C., and R. Breitwisch. 1992. Northern Mockingbird. *In* The birds of North America, no. 167 (A. Poole and F. Gill, eds.). Acad. Nat. Sci., Philadelphia, and Am. Ornithol. Union, Washington, DC.

Derrickson, S.R. 1977. Aspects of breeding behavior in the Pintail (*Anas acuta*). Ph.D. Diss., Univ. Minnesota, St. Paul.

Dexter, C. 1995. Possible hybridization of Scissor-tailed Flycatcher and Western Kingbird. J. Colo. Field Ornithol. 29:20.

Dexter, C., and R. Levad. 1992. Bird check list for the Grand Valley and surrounding high country of Mesa County, Colorado. Grand Valley Audubon Soc., Grand Junction, CO.

Denver Field Ornithologists. 1965–1995. Lark Bunting. Vol. 1–30.

Denver Field Ornithologists. 1967. Summary of Colorado's 1967 spring counts. Denver Field Ornithologists, Denver.

Dickson, J.G., ed. 1992. The Wild Turkey: biology and management. Stackpole Books, Harrisburg, PA.

Dillon, M.B. 1995. Observations of Northern Waterthrushes exhibiting breeding behavior in Colorado. J. Colo. Field Ornithol. 29:84–87.

Dixon, K.L., and J.D. Gilbert. 1964. Altitudinal migration in the Mountain Chickadee and Black-capped Chickadee in Colorado. Condor 66:61–64.

Dobbs, R.C., P.R. Martin, and T.E. Martin. 1998. Green-tailed Towhee. *In* The birds of North American, no. 368 (A. Poole and F. Gill, eds.). The birds of North America, Inc. Philadelphia, PA.

Dobkin, D.S. 1994. Conservation and management of Neotropical migrant landbirds in the northern Rockies and Great Plains. Univ. Idaho Press, Moscow.

Dorn, J.L. 1972. The Common Raven in Jackson Hole, Wyoming. M.S. Thesis, Univ. Wyoming, Laramie.

Dorn, J.L., and R.D. Dorn. 1990. Wyoming birds. Mountain West Publ., Cheyenne, WY.

Dorn, R.D., and J.L Dorn. 1994. Further data on screech-owl distribution and habitat use in Wyoming. West. Birds 25:35–42.

Dorn, R.D., and J.L. Dorn. 1995. Cassin's Sparrow nesting in Wyoming. West. Birds 26:104–106.

Downing, H. 1981. The first nesting of Vermilion Flycatchers in Colorado. J. Colo. Field Ornithol. 15:75–76.

Doyle, T.L. 1997. The Timberline Sparrow, *Spizella (breweri) taverneri*, in Alaska, with notes on breeding habitat and vocalizations. West. Birds 28:1–12.

Drennan, S.R. 1992–1994. Ninety-third Christmas bird count, National Audubon Society. Am. Birds 47:879–880, 48:763, 49:734.

Droger, H.K. 1979. Female Least Bittern with young. J. Colo. Field Ornithol. 13:62–63.

DuBois, A.D. 1937. Notes on coloration and habits of the Chestnut-collared Longspur. Condor 39:104–107.

DuBowy, P.J. 1985. Feeding ecology and behavior of post-breeding Blue-winged Teal and Northern Shovelers. Can. J. Zool. 63:1292–1297.

DuBowy, P.J. 1996. Northern Shoveler. *In* The birds of North America, no. 217 (A. Poole and F. Gill, eds.). Acad. Nat. Sci., Philadelphia, and Am. Ornithol. Union, Washington, DC.

Duebbert, H.F. 1966. Island nesting of the Gadwall in North Dakota. Wilson Bull. 78:12–25.

Duncan, D.C. 1987. Nesting of Northern Pintails in Alberta: laying date, clutch size, and renesting. Can. J. Zool. 65:234–246.

Dundas, H. 1995. Burrowing Owl status and conservation programs. Bird Trends 4:21–22.

Dunkle, F.W. 1977. Swainson's Hawks in the Laramie Plains, Wyoming. Auk 94:65–71.

Dunn, E.H., and D.J. Agro. 1995. Black Tern. *In* The birds of North America, no. 147 (A. Poole and F. Gills, eds.). Acad. Nat. Sci., Philadelphia, and Am. Ornithol. Union, Washington, DC.

Dunn, J.L., K.L. Garrett, and J.K. Alderfer. 1995. White-crowned Sparrow subspecies: identification and distribution. Birding 27:182–200.

Dunning, J.B., Jr., and R.K. Bowers Jr. 1990. Lethal temperatures in Ash-throated Flycatchers nest located in metal fence poles. J. Field Ornithol. 61:98–103.

Ealey, D.M. 1977. Aspects of the ecology and behavior of a breeding population of dippers (*Cinclus mexicanus*: Passeriformes) in southern Alberta. M. Sc. Thesis, Univ. Alberta, Edmonton.

Eaton, S.W. 1992. Wild Turkey. *In* the birds of North America, no. 22 (A. Poole and F. Gill, eds.). Acad. Nat. Sci., Philadelphia, and Am. Ornithol. Union, Washington, DC.

Eaton, S.W. 1995. Northern Waterthrush. *In* The birds of North America, no. 182 (A. Poole and F. Gill, eds.). Acad. Nat. Sci., Philadelphia, and Am. Ornithol. Union, Washington, DC.

Eckhardt, R.C. 1976. Polygyny in the Western Wood-Pewee. Condor 78:561–562.

Eddleman, W.R., F.L. Knopf, B. Meanley, F.R. Reid, and R. Zembal. 1988. Conservation of North American rallids. Wilson Bull. 458–475.

Edwards, E.P. 1972. A field guide to the birds of Mexico. Sweet Briar, VA.

Ehrlich, P.R., and G.C. Daily. 1993. Birding for fun—sapsuckers, swallows, willow, aspen and rot. Am. Birds. 47:18–20.

Ehrlich, P.R., D.S. Dobkin, and D. Wheye. 1988. The birder's handbook: a field guide to the natural history of North American birds. Simon and Schuster, New York.

Ehrlich, P.R., D.S. Dobkin, and D. Wheye. 1992. Birds in jeopardy. Stanford Univ. Press, Stanford, CA.

Elliott, P.F. 1978. Cowbird parasitism in the Kansas tallgrass prairie. Auk 95:161–167.

Ellis, K.E., J.B. Parrish, J.R. Murphy, and G.H. Richins. 1989. Habitat use by breeding male sage grouse: a management approach. Great Basin Nat. 49:404–407.

Ellison, W.G. 1992. Blue-gray Gnatcatcher. *In* The birds of North America, no. 23 (A. Poole, P. Stettenheim, and F. Gill, eds.). Acad.Nat. Sci., Philadelphia, and Am. Ornithol. Union, Washington, DC .

Ely, D.C. 1996. News from the field: the winter report (December 1995, January & February 1996). J. Colo. Field Ornithol. 30:112–122.

Emlen, J.T. 1952. Social behavior in nesting Cliff Swallows. Condor. 56:177–199.

Emlen, J.T. 1954 Territory, nestbuilding, and pair formation in the Cliff Swallow. Auk. 71:16–35.

Enderson, J.H. 1964. A study of the Prairie Falcon in the central Rocky Mountain region. Auk 81:332–352.

Enderson, J.H. 1965. A breeding and migration survey of the Peregrine Falcon. Wilson Bull. 77:327–339.

Enderson, J.H., G.R. Craig, and W.A. Burnham. 1988. Status of peregrines in the Rocky Mountain and Colorado Plateau. Pp. 83–86 *in* T.J. Cade et al., ed. Peregrine Falcon population: their management and recovery. Peregrine Fund, Boise, ID.

Enriquez, R.E. 1979. Greater Sandhill Crane habitat management plan. U.S. For. Serv., Routt Natl For., CO.

Ensign, J.T. 1983. Nest site selection, productivity, and food habits of Ferruginous Hawk in southeastern Montana. M.S. Thesis. Montana State Univ., Bozeman.

Erickson, R.C. 1948. Life history and ecology of the Canvasback, *Nyroca valisineria* (Wilson), in southeastern Oregon. Ph.D. Diss., Iowa State Coll., Ames.

Erickson, K.A., and A.W. Smith. 1985. Atlas of Colorado. Colorado Assoc. Univ. Press, Boulder.

Erskine, A.J. 1992. Atlas of breeding birds of the Maritime Provinces. Nova Scotia Mus., Halifax.

Ervin, S. 1977. Flock size, composition, and behavior in a population of Bushtits. Bird-Banding 48:97–109.

Ervin, W.E. 1981. Environmental influences of song-flight in Lark Buntings: behavioral adaptations for communication. M.A. Thesis, Univ. Colo., Boulder.

Estelle, V., and T. Mabee. 1994. Breeding success of Least Terns, Piping Plovers and Snowy Plovers: evaluation of predator exclosures in southeast Colorado, 1994. Unpubl. rep. submitted to Colo. Div. Wildl. and Bur. Land Manage. Colo. Bird Obs., Brighton.

Evans, K.E. 1968. Characteristics and habitat requirements of the Greater Prairie Chicken and Sharp-tailed Grouse—a review of the literature. U.S. Dept. Agric. Conserv. Res. Rep. No. 12.

Evans, P.J., M.J.R. Miller, M.E. Barker, and S. Postupalsky. 1994. Birds breeding in or beneath Osprey nests in the Great Lakes basin. Wilson Bull. 106:743–749.

Evans, R.M., and F.L. Knopf. 1993. American White Pelican. *In* The birds of North America, no. 57 (A. Poole and F. Gill, eds.). Acad. Nat. Sci., Philadelphia, and Am. Ornithol. Union, Washington, DC.

Ewell, H., and A. Cruz. 1998. Foraging behavior of the Pygmy Nuthatch (*Sitta pygmaea*). West. Birds 29:169–173.

Faanes, C.A., and W. Norling. 1981. Nesting of the Great-tailed Grackle in Nebraska. Am. Birds 35:148–149.

Facemire, C.F. 1978. Ground nesting Kansas thrashers. Bull. Kansas Ornithol. Soc. 29:21–22.

Farrand J., Jr., ed. 1983. The Audubon Society master guide to birding. Vol. 1–3 . Alfred A. Knopf, New York.

Fautin, R.W. 1946. Biotic communities of the northern desert shrub biome in western Utah. Ecol. Monogr. 16:251–310.

Feare, C.J. 1984. The Starling. Oxford Univ. Press, New York.

Feare, C.J. 1994. Changes in the numbers of common starlings and farming practice in Lincolnshire. British Birds 87:200–204.

Feerer, J.L. 1977 Niche partitioning by Western Grebe polymorphs. Unpubl. M.S. Thesis. Humboldt State Univ., Arcata, CA.

Fergus, J. 1993. Upland encounters. Outdoor Life. Oct. 1993.

Fiester, M. 1973. Blasted beloved Breckenridge. Pruett, Boulder.

Figgins, J.D. 1913. The status of Gambel Quail in Colorado. Condor 10:158.

Figgins, J.D. 1914. The fallacy of the tendency toward ultraminute distinctions. Auk 26:272–291.

Figgs, M. 1984. Montane willow carr. *In* Van Velzen and Van Velzen, Forty-seventh breeding bird census. Am. Birds 38:113.

Finch, D.M. 1987. Bird-habitat relationships in riparian communities of southeastern Wyoming. Ph.D. Diss., Univ. Wyoming, Laramie.

Finch, D.R., and R.T. Reynolds. 1988. Bird response to understory variation and conifer succession in aspen forests. *In* Issues and technology in the management of impacted western wildlife, Symp. Proc. Thorne Ecological Inst. Boulder, CO.

Finlay, J.C. 1971. Breeding biology of Purple Martins at the northern limit of their range. Wilson Bull. 83:255–269.

Fischer, D.H. 1980. Breeding biology of Curve-billed Thrashers and Long-billed Thrashers in southern Texas. Condor 82:392–397.

Fischer, R.B. 1958. The breeding biology of the Chimney Swift *Chaetura pelagica* (Linnaeus). Univ. Press New York, Albany.

Fischl, J., and D.F. Caccamiso. 1987. Relationships of diet and roosting behaviour in the European Starling. Am. Midl. Nat. 117:395–404.

Fitch, H.S. 1963. Observations on the Mississippi Kite in southwestern Kansas. Univ. Kansas Publ. Mus. Nat. Hist. 12:503–519.

Fitch, H.S., F. Swenson, and D.F. Tillotson. 1946. Behavior and food habits of the Red-tailed Hawk. Condor 48:205–257.

Fite, M.K. 1984. Vocal behavior and interactions among parents and offspring in the American Dipper. M.S. Thesis, Utah State Univ., Logan.

Fitton, S. 1973. Screech-Owl distribution in Wyoming. West. Birds 24:182–188.

Fitzner, R.E., and R.H. Gray 1994. Winter diet and weights of Barrow's and Common goldeneye in southcentral Washington. Northwest Sci. 68:172–177.

Fitzner, R.E., D. Berry, L.L. Boyd, and C.A. Reick. 1977. Nesting of Ferruginous Hawks (*Buteo regalis*) in Washington. Condor 79:245–249.

Fitzgerald, J.P., C.A. Meaney, and D.M. Armstrong. 1994. Mammals of Colorado. Denver Mus. Nat. Hist., Denver, and Univ. Press of Colorado, Boulder.

Fitzpatrick, J.W. 1980. Wintering of North American tyrant flycatchers in the Neotropics. Pp. 67–78 *in* A. Keast and E.S. Morton, eds. Migrant birds in the Neotropics: ecology, behavior, distribution, and conservation. Smithsonian Inst. Press, Washington, DC.

Flack, J.A.D. 1976. Bird populations of aspen forests in western North America. Ornithol. Monogr. No 19. Am. Ornithol. Union, Washington, DC.

Fletcher, N., and G. Sheppard. 1994. The Northern Goshawk in the Southwestern region—1992 status report. U.S. For. Serv. internal document.

Forbes, M.R., and C.D. Ankney. 1988. Nest attendance by adult Pied-billed Grebes. Can. J. Zool. 66:2019–2023.

Forbush, E.H. 1907. Useful birds and their protection. Mass. Board Agric., Boston.

Foss, C.R., ed. 1994. Atlas of breeding birds in New Hampshire. Audubon Soc. of New Hampshire, Dover.

Fox, G.A. 1974. Changes in eggshell quality of Belted Kingfishers nesting in Ontario. Can. Field-Nat. 88:358–359.

Franzreb, K.E. 1984. Foraging habits of Ruby-crowned and Golden-crowned kinglets in an Arizona montane forest. Condor 86:139–145.

Franzreb, K.E. 1985. Foraging ecology of Brown Creepers in a mixed-coniferous forest. J. Field Ornithol. 56:9–16.

Franzreb, K.E., and R.D. Ohmart. 1978. The effects of timber harvesting on breeding birds in a mixed-coniferous forest. Condor 80:431–441.

Frary, L.G. 1954. Waterfowl production on White River Plateau, Colo. M.S. Thesis, Colo. A&M Coll., Fort Collins.

Fredrickson, L.H., J.M. Anderson, F.M. Kozlik, and R.A. Ryder. 1977. American coot. Pp. 123–147 *in* G.C. Sanderson, ed. Management of migratory shore and upland game birds in North America. Int. Assoc. Game and Fish Agencies, Washington, DC.

French, N.R. 1954. Notes on breeding activities and on gular sacs in the Pine Grosbeak. Condor. 56:83–85.

Fretwell, S.D. 1986. Distribution and abundance of the Dickcissel. Current Ornithol. 4:211–242.

Friedl, T.W.P. 1993. Intraclutch egg-mass variation in geese: a mechanism for brood reduction in precocial birds? Auk 110:129–132.

Friedmann, H. 1963. Host relations of the parasitic cowbirds. U.S. Natl. Mus. Bull. 233, Washington, DC.

Friedmann, H., and L.F. Kiff. 1985. The parasitic cowbirds and their hosts. Proc. West. Found. Vert. Zool. 2:225–304.

Friedmann, H., L.F. Kiff, and S.I. Rothstein. 1977. A further contribution to knowledge of the host relations of the parasitic cowbirds. Smithsonian Contrib. to Zool. 235:1–75.

Fuller, M.R., D. Bystrak, C.S. Robbins, and R.M. Patterson. 1987. Trends in American Kestrel counts from the North American breeding bird survey. Raptor Res. Rep. 6:22–27.

Furrer, R.K. 1975. Breeding success and nest site stereotypy in a population of Brewer's Blackbirds (*Euphagus cyanocephalus).* Oecologia 20:339–350.

Gabrielson, I.N., and F.C. Lincoln. 1959. Birds of Alaska. Stackpole Books, Harrisburg, PA, and Wildlife Management Institute, Washington, DC.

Gabrielson, I.N., and S.G. Jewett. 1940. The birds of Oregon. Oregon State College Monogr., Stud. Zool., No. 2.

Galbreath, D., and R. Moreland. 1953. The Chukar Partridge in Washington. Washington State Game Dept.

Gale, D. 1884–1888. Notes 1884–88. [Notebooks copied and annotated by Junius Henderson, with index.] In Univ. Colorado Libr., Boulder.

Gammonley, J.H. 1996. Cinnamon Teal. *In* The birds of North America, no. 209 (A. Poole and F. Gill, eds.). Acad. Nat. Sci., Philadelphia, and Am. Ornithol. Union, Washington, DC.

Ganey, J.L., and R.P. Balda. 1989. Home-range characteristics of Spotted Owls in northern Arizona. J. Wildl. Manage. 53:1159–1165.

Ganey, J.L., and R.P. Balda. 1994. Habitat selection by Mexican Spotted Owls in northern Arizona. Auk 111:162–169.

Ganey, J.L., R.P. Balda, and R.M. King. 1993. Metabolic rate and evaporative water loss of Mexican Spotted Owls and Great Horned Owls. Wilson Bull. 105:645–656.

Ganey, J.L., J.A. Johnson, R.P. Balda, and R.W. Skaggs. 1988. Mexican Spotted Owl. Pp. 145–150 *in* R.L. Glinski, et al., eds. Proc. southwest raptor management symposium and workshop. Natl. Wildl. Fed., Washington, DC.

Gates, J.E., and L.W. Gysel. 1978. Avian nest dispersion and fledgling success in field-forest ecotones. Ecol. 59:871–883.

Gauthier, G. 1988a. Philopatry, nest-site fidelity, and reproductive performance in Bufflehead. Auk 107:126–132.

Gauthier, J., and Y. Aubry. 1996. The breeding birds of Quebec. Prov. Quebec Soc. for the Protection of Birds and Can. Wildl. Serv., Montreal.

Gehlbach, F.R. 1994. The Eastern Screech-owl: life history, ecology, and behavior in the suburbs and countryside. Texas A&M Univ. Press, College Station.

Gehlbach, F.R. 1995. Eastern Screech-Owl. *In* The birds of North America, no. 165 (A. Poole and F. Gill, eds.). Acad. Nat. Sci., Philadelphia, and Am. Ornithol. Union, Washington, DC.

Gent, P. 1987. Colorado Field Ornithologists' Records Committee report for 1978–1985. West. Birds 18:97–108.

Gent, P. 1998. Hooded Warblers nest in Colorado. J. Colo. Field Ornithol. 32:230–233.

George, J. 1994. Central region colonial nesting waterbirds inventory. Colo. Div. Wildl.

George, J. 1995. Central region colonial nesting waterbirds inventory. Colo. Div. Wildl.

George, J. 1996. Central region colonial nesting waterbirds inventory. Colo. Div. Wildl.

George, J.L., C.E. Braun, R.A. Ryder, and E. Decker. 1991. Response of waterbirds to experimental disturbances. Proc. of issues and technology in the management of impacted western wildlife. Thorne Ecol. Inst. 5:52–59.

George, T.L., A.C. Fowler, R.L. Knight, and L.C. McEwen. 1992. Impacts of a severe drought on grassland birds in western North Dakota. Ecol. Appl. 2:275–284.

George, T.L., L.C. McEwen, and B.E. Petersen. 1995. Effects of grasshopper control programs on rangeland breeding bird populations. J. Range Manage. 48:336–342.
Gibbons, D.W., J.B. Reid, and R.A. Chapman. 1993. New atlas of breeding birds in Britain and Ireland: 1988–1991. T & AD Poyser. London.
Gibson, F. 1971. The breeding biology of the American Avocet (*Recurvirostra americana*) in central Oregon. Condor 73:444–454.
Giesen, K.M., and C.E. Braun. 1979. Nesting behavior of female White-tailed Ptarmigan in Colorado. Condor 81:215–217.
Giesen, K.M. 1994a. Breeding range and population status of Lesser Prairie-Chickens in Colorado. Prairie Nat. 26:175–182.
Giesen, K.M. 1994b. Movements and nesting habitat of Lesser Prairie-Chickens in Colorado. Southwest. Nat. 39:96–98.
Giesen, K.M., and C.E. Braun. 1993. Status and distribution of the Columbian Sharp-tailed Grouse in Colorado. Prairie Nat. 25:237–242.
Giesen, K.M., C.E. Braun, and T.A. May. 1980. Reproduction and nest-site selection by White-tailed Ptarmigan in Colorado. Wilson Bull. 92:188–199.
Giezentanner, J.B. 1970. Avian distribution and population fluctuations on the shortgrass prairie of north-central Colorado. Grassland Biome, U.S. Intl. Biol. Program Tech. Rep. No. 62, Fort Collins, CO.
Gilbert, D.W., D.R. Anderson, J.K. Ringelman, and M.R. Szymczak. 1996. Response of nesting ducks to habitat and management on the Monte Vista National Wildlife Refuge, Wildl. Monogr. 131. The Wildlife Soc.
Gilmer, D.S., and R.E. Stewart. 1984. Swainson Hawk nesting ecology in North Dakota. Condor 86:12-18.
Glahn, J.F. 1974. Study of breeding rails with recorded calls in north-central Colorado. Wilson Bull. 86:206-214.
Gleason, R.L., and T.H. Craig. 1979. Food habits of Burrowing Owls in southeastern Idaho. Great Basin Nat. 39:274–276.
Glinski, R., and S. Ambrose. 1990. Status report: return of the Peregrine Falcon in the southwest and Alaska from the brink of extinction. Eyas 13:4–5.
Glinski, R.L., and R.D. Ohmart. 1983. Breeding ecology of the Mississippi Kite in Arizona. Condor 85:200–207.
Godfrey, W.E. 1986. The birds of Canada. Rev. ed. Natl. Mus. Nat. Sci., Ottawa.
Goldberg, N.H. 1979. Behavioral flexibility and foraging strategies in Cassin's and Western kingbirds (*Tyrannus vociferans* and *T. verticalis*) breeding sympatrically in riparian habitats in central Arizona. Ph.D. Diss., Univ. Illinois, Champaign.
Goldstein, M.I., B. Woodbridge, M.E. Zaccagnini and S.B. Canavelli. 1996. An assessment of mortality of Swainson's Hawks on wintering grounds in Argentina. J. Raptor Res. 30:106–107.
Gooch, B. 1993. The much maligned snipe. Quail Unlimited 12:22–25.
Gooders, J., and T. Boyer. 1986. Ducks of North America and the northern hemisphere. Facts on File, New York.
Goodge, W.R. 1959. Locomotion and other behavior of the Dipper. Condor 61:4–17.
Goodwin, D. 1976. Crows of the world. Cornell Univ. Press, Ithaca, NY.
Goodwin, D. 1983. Pigeons and doves of the world. 3rd ed. Cornell Univ. Press, Ithaca, NY.
Gorenzel, W.P. 1979. Production, spatial, and temporal relationships of the American Coot in Colorado. M.S. Thesis, Colo. State Univ., Fort Collins.
Gorenzel, W.P., R.A. Ryder, and C.E. Braun. 1981. American Coot response to habitat changes on a Colorado marsh. Southwest. Nat. 26:59–65.
Gorenzel, W.P., R.A. Ryder, and C.E. Braun. 1982. Reproduction and nest site characteristics of American Coots at different altitudes in Colorado. Condor 84:59–65.
Gorenzel, W.P., R.A. Ryder, and C.E. Braun. 1983. American coot distribution and migration in Colorado. Wilson Bull. 93:115–118.
Gorsuch, D.M. 1934. Life history of the Gambel's Quail in Arizona. Univ. Arizona Bull. Vol. V, No.4.
Goss, B.F. 1883. Notes on the breeding habits of Maximilian's Jay (*Gymnocitta cyanocephala*) and Clarke's Crow (*Picicorvus columbianus*). Bull. Nuttall Ornithol. Club 8:43–45.
Graber, R.R., J.W. Graber, and E.L. Kirk. 1974. Illinois birds: *Tyrannidae*. Ill. Nat. Hist. Surv., Biol. Notes No. 86.
Graham, T.B., and A.W. Nettell. 1992. Pinyon/juniper scrub *in* Resident bird counts. J. Field Ornithol. 63:102–103.
Grant, G.S., and T.L. Quay. 1977. Breeding biology of Cliff Swallows in Virginia. Wilson Bull. 89:286–290.
Graul, W.D. 1973. Adaptive aspects of the Mountain Plover social system. Living Bird 12:69–94.
Graul, W.D. 1975. Breeding biology of the Mountain Plover. Wilson Bull. 87:6–31.
Graul, W.D., and L.E. Webster. 1976. Breeding status of the Mountain Plover. Condor 78:265–267.
Gray, J.A. 1945. Land birds at sea. Condor 47:215–216.
Gray, M.T. 1995. DOW Working for Wildlife: Peregrine Falcon. Colorado's Wildlife Company, Colo. Div. Wildlife, Denver.
Green, G.A., and R.G. Anthony. 1989. Nesting success and habitat relationships of Burrowing Owls in the Columbia River basin, Oregon. Condor 91:347–354.
Greene, E., V.R. Muehter, and W. Davison. 1996. Lazuli Bunting. *In* The birds of North America, no. 232 (A. Poole and F. Gill, eds.). Acad. Nat. Sci., Philadelphia, and Am. Ornithol. Union, Washington, DC.
Greenlaw, J.S. 1978. The relation of breeding schedule and clutch size to food supply in the Rufous-sided Towhee. Condor, 80:24–33.
Greenlaw, J.S. 1996a. Eastern Towhee. *In* The birds of North America, no. 262. (A. Poole and F. Gill, eds.) Acad. Nat. Sci., Philadelphia, and Am. Ornithol. Union, Washington, DC.
Greenlaw, J.S. 1996b. Spotted Towhee. *In* The birds of North America, no. 263. (A. Poole and F. Gill, eds.). Acad. Nat. Sci., Philadelphia, and Am. Ornithol. Union, Washington, DC.
Gregg, R.E. 1963. The ants of Colorado. Univ. Colo. Press, Boulder.
Griese, H.J., R.A. Ryder, C.E. Braun. 1980. Spatial and temporal distribution of rails in Colorado. Wilson Bull. 92:96–102.
Griess, J. 1995. Burrowing Owl progress report. Rocky Mt. Arsenal, Aurora, CO.

Griffiths, D.A., C. Griffiths, and D. Silverman. 1978. Suspected nesting of Grace's Warbler on the Eastern Slope. J. Colo. Field Ornithol. 12:41–42.

Griffiths, D.A., C. Griffiths, and D. Silverman. 1980. Grace's Warbler nesting on the Eastern Slope. J. Colo. Field Ornithol. 14:37.

Grinnell, J., and A.H. Miller. 1944. The distribution of birds of California. Pac. Coast Avifauna 27:1–608.

Grinnell, J., and T. Storer. 1924. Animal life in the Yosemite.

Grinnell, J., J. Dixon, and J.M. Linsdale. 1930. Vertebrate natural history of a section of northern California through the Lassen Peak region. Univ. California Publ. Zool. 35:273–280.

Griscom, L., and A. Sprunt, Jr. 1979. The warblers of America: a popular account of the wood warblers as they occur in the western hemisphere. Rev. E.M. Reilly, Jr. Doubleday, Garden City, NY.

Gross, A.O. 1921. The Dickcissel (*Spiza americana*) of the Illinois prairies. Auk 38:163–184.

Groth, J.G. 1988. Resolution of cryptic species in Appalachian Red Crossbills. Condor 90:745–760.

Groth, J.G. 1992. White-winged Crossbill breeding in southern Colorado, with notes on juveniles' calls. West. Birds 23:35–37.

Groth, J.G. 1993. Evolutionary differentiation in morphology, vocalizations, and allozymes among nomadic sibling species in the North American Red Crossbill (*Loxia curvirostra*) complex. Univ. Calif. Publ. Zoology 127:1–143.

Gruson, E.S. 1972. Words for birds. Quadrangle Books, New York.

Gullion, W.G. 1960. The ecology of Gambel's Quail in Nevada and the arid southwest. Ecol. 41:518–536.

Gustafson, G.R. 1976. Sage Sparrow egg in a Black-throated Sparrow nest. Auk 92:805–806.

Gutierrez, R. 1971. Observations on the breeding biology and behavior of Mourning Doves in Fort Collins, Colorado. J. Colo. Field Ornithol. 10:10–16.

Gutierrez, R.J., A.B. Franklin, and W.S. Lahaye. 1995. Spotted Owl. *In* The birds of North America, no. 179 (A. Poole and F. Gill, eds.). Acad. Nat. Sci., Philadelphia, and Am. Ornithol. Union, Washington, DC.

Hadley, N. 1969. Breeding biology of the Gray-headed Junco, *Junco caniceps* (Woodhouse) in the Colorado Front Range. J. Colo. Field Ornithol. 5:15–21.

Hadow, H.H. 1973. Winter ecology of migrant and resident Lewis' Woodpeckers in southeastern Colorado. Condor 75:210–224.

Hagan, J.M., III, 1993. Decline of the Rufous-sided Towhee in the eastern United States. Auk 110:863–874.

Hagan, J.W., and D.W. Johnson, eds. 1992. Ecology and conservation of Neotropical migrant landbirds. Smithsonian Press. Washington, DC.

Hahn, T.P. 1996. Cassin's Finch. *In* The birds of North American, no. 240 (A. Poole and F. Gill, eds.). Acad. Nat. Sci., Philadelphia, and Am. Ornithol. Union, Washington, DC.

Haig, S.M. 1992. Piping Plover. *In* The birds of North America, no. 2 (A. Poole, P.Stettenheim, and F.Gill, eds.). Acad. Nat. Sci., Philadelphia, and Am. Ornithol. Union, Washington, DC.

Haig, S.M., T. Eubanks, W. Harrison, R. Lock, L. Pfannmuller, M. Ryan, and J. Sidle. 1992. 1991 Piping Plover breeding census. Unpubl. letter to cooperating agencies.

Hainsworth, F.R., B.G. Collins, and L.L. Wolf. 1977. The function of torpor in hummingbirds. Physiol. Zool. 50:215–222.

Hall, G.A. 1983. West Virginia birds: distribution and ecology. Carnegie Mus. Nat. Hist. Spec. Publ. No. 7.

Hallock, D. 1984a. Status and avifauna of willow carrs in Boulder County. J. Colo. Field Ornithol. 18:100–105.

Hallock, D. 1984b. Montane willow carr. *In* Van Velzen and Van Velzen, Forty-seventh breeding bird census. American Birds 38:112.

Hallock, D. 1990. A study of breeding and winter birds in different age-classed lodgepole pine forests. J. Colo. Field Ornithol. 24:2–16.

Hallock, D. 1992. Montane mesic willow carr. J. Field Ornithol. 63:97–98 (suppl).

Hallock, D. 1994. A decade of Indian Peaks bird counts. Boulder County Nature Assoc., P.O. Box 673, Boulder, CO 80306.

Hamas, M.J. 1975. Ecological and physiological adaptations for breeding in the Belted Kingfisher (*Megaceryle alcyon*). Ph.D. Diss., Univ. Minnesota, Minneapolis.

Hamas, M.J. 1994. Belted Kingfisher. *In* The birds of North America, no. 84 (A. Poole and F. Gill, eds.) Acad. Nat. Sci., Philadelphia, and Am. Ornithol. Union, Washington, DC.

Hamerstrom, F., F.N. Hamerstrom, and C.J. Burke. 1985. Effect of voles on mating systems in a central Wisconsin population of Harriers. Wilson Bull. 97:332–346.

Hamil, H. 1976. Colorado without mountains. Lowell Press. Kansas City, MO.

Hamilton, R.C. 1975. Comparative behavior of the American Avocet and the Black-necked Stilt (*Recurvirostridae*). Am. Ornithol. Monogr., No. 17:1–98.

Hamilton, W.J., III. 1962. Bobolink migratory patterns and their experimental analysis under night skies. Auk 79:208–233.

Hancock, J., and J. Kushlan. 1984. The herons handbook. Harper and Row, New York.

Hancock, J.A., J.A. Kushlan, and M.P. Kahl. 1992. Storks, ibises, and spoonbills of the world. Academic Press, London.

Hann, H.W. 1937. Life history of the Oven-bird in southern Michigan. Wilson Bull. 49:145–237.

Hann, H.W. 1950. Nesting behavior of the American Dipper in Colorado. Condor 52:49–62.

Harlow, D.L., and P.H. Bloom. 1987. Buteos and the Golden Eagle. Pp. 102–110 *in* B.G. Pendleton, et al., eds. Proc. of the western raptor management symposium and workshop. Nat. Wildl. Fed. Sci. Tech. Ser. No. 12. Washington, DC.

Harlow, R.C. 1922. Breeding habits of the Northern Raven in Pennsylvania. Auk 39:399–410.

Harmeson, J.P. 1974. Breeding ecology of the Dickcissel. Auk 91:348–359.

Harrington, H.D. 1954. Manual of the plants of Colorado: for the identification of the ferns and flowering plants of the state. Sage Books, Denver.

Harrison, C. 1978. A field guide to nests, eggs, and nestlings of North American birds. William Collins Sons & Co., Glasgow.

Harrison, G.H. 1992. Is there a killer in your house? Natl. Wildl. 30(6):10–13.

Harrison, H. 1979. A field guide to western bird nests. Houghton Mifflin, Boston.

Harrison, H.H. 1984. Wood warblers world. Simon and Schuster, New York.

Hart, S.L., and A.B. Hulbert, eds. 1932. Zebulon Pike's Arkansaw Journal. Colo. Coll. and Denver Publ. Libr. No *loc. cit.*

Hartshorne, C. 1973. Born to sing: an interpretation and world survey of bird song. Indiana Univ. Press, Bloomington.

Haug, E.A., B.A. Millsap, and M.S. Martell. 1993. Burrowing Owl). *In* The birds of North America, no. 61 (A. Poole and F. Gill, eds.). Acad. Nat. Sci., Philadelphia, and Am. Ornithol. Union, Washington, DC.

Haws, T.G., and G.L. Hayward. 1968. Spotted Towhee. Pp. 583–590 *in* Bent, A.C. Life histories of North American cardinals, grosbeaks, buntings, towhees, finches, sparrows, and their allies, part 2. O.L. Austin Jr., ed. U.S. Natl. Mus. Bull. No. 237, Washington, DC.

Hayman, P., J. Marchant, and T. Prater. 1986. Shorebirds an identification guide to the waders of the world. Houghton Mifflin, Boston.

Hays, R.L., and A.H. Farmer. 1990. Effects of CRP on wildlife habitat: emergency haying in the midwest and pine plantings in the southeast. Trans. N. Am. Wildl. Nat. Resour. Conf. 55:30–39.

Hayward, C.L., C. Cottam, A.M. Woodbury, and H.H. Frost. 1976. Birds of Utah. Great Basin Nat. Memoirs No. 1. Brigham Young Univ., Provo, UT.

Hayward, G.D. 1997. Forest management and conservation of Boreal Owls in North America. J. Raptor Res. 31:114–124.

Hayward, G.D., and E.O. Garton. 1988. Resource partitioning among forest owls in the River of No Return Wilderness, Idaho. Oecologia 75:253–265.

Hayward, G.D., and P.H. Hayward. 1993. Boreal Owl. *In* The birds of North America, no. 63 (A. Poole and F. Gill, eds.). Acad. Nat. Sci., Philadelphia, and Am. Ornithol. Union, Washington, DC.

Hayward, G.D., and J. Verner, eds. 1994. Flammulated Owl locations and distribution of associated vegetative ecosystems in the United States. Accompanying Flammulated, Boreal, and Great Gray Owls in the United States: a technical conservation assessment. U.S. Dept. Agric. For. Serv., Gen. Tech. Rep. RM-253, Fort Collins, CO.

Hayworth, A.M., and W.W. Weathers 1984. Temperature regulation and climatic adaptation in Black-billed and Yellow-billed magpies. Condor 86:19–26.

Heckenlively, D.B. 1967. Role of song in territoriality of Black-throated Sparrows. Condor 69:429–430.

Heckenlively, D.B. 1970. Song in a population of Black-throated Sparrows. Condor 72:24–36.

Heindel, M. 1996. Field identification of the Solitary Vireo complex. Birding 28:458–471.

Heinrich, B. 1989. Ravens in winter. Summit Books. New York.

Heinrich, B. 1993. Kinglets realm of cold. Nat. Hist. 102:6–9.

Hejl, S.J., and R.E. Woods. 1991. Bird assemblages in old growth and rotation-aged Douglas-fir/ponderosa pine stands in the northern Rocky Mountains: a preliminary assessment. Pp. 285–292 *in* D.M. Baumgarter and J.E. Lotan, eds. Symp. proc., interior Douglas-fir: the species and its management. Washington State Univ., Pullman.

Hejl, S.J., R.L. Hutto, C.R. Preston, and D.M. Finch. 1995. Effects of silvicultural treatments in the Rocky Mountains Pp. 220–244 *in* Ecology and Management of Neotropical Migratory Birds (T.E. Martin and D.M. Finch, eds.). Oxford University Press, Oxford, England.

Hekstra, G.P. 1973. Scops and Screech Owls. *In* J.A. Burton, ed. Owls of the world. Dutton, New York.

Henderson, J. 1908. An annotated list of the birds of Boulder County, Colorado. Univ. Colo. Studies, 6:220–242.

Henshaw, H.W. 1875. Report upon the ornithological collections made in portions of Nevada, Utah, California, Colorado, New Mexico, and Arizona, during the years 1871, 1872, 1873 and 1874 in report upon geographical and geological explorations and surveys west of the one hundredth meridian, vol. 5. U.S. Gov. Printing Off., Washington, DC.

Herron, G.B., C.A. Mortimore, and M.R. Rawlings. 1985. Nevada raptors: their biology and management. Nevada Dept. Wildl. Biol. Bull. 8.

Hersey, L.J., and R.B. Rockwell. 1909. An annotated list of the birds of the Barr Lake District, Adams County, Colorado. Condor 11:109–122.

Hertz, P.E., J.V. Remsen, Jr., and S.I. Zones. 1976. Ecological complementarity of three sympatric parids in a California oak woodland. Condor 78:307–316.

Hester, F.E., and J. Dermid. 1973. The world of the Wood Duck. J.B. Lippincott, Philadelphia.

Hickey, J.J., and L.B. Hunt. 1960. Initial songbird mortality following a Dutch elm disease control program. J. Wildl. Manage. 24:259–265.

Higuchi, H. 1986. Bait-fishing by the Green-backed Heron *Ardeola striata* in Japan. Ibis 128:285–290.

Higuchi, H. 1988. Individual differences in bait-fishing by the Green-backed Heron *Ardeola striata,* associated with territory quality. Ibis 130:39–44.

Hill, G.E. 1993. House Finch. *In* The birds of North America, no. 46. (A. Poole and F. Gill, eds.). Acad. Nat. Sci., Philadelphia, and Am. Ornithol. Union, Washington, DC.

Hill, G.E. 1995. Black-headed Grosbeak. *In* The birds of North America, no. 143. (A. Poole and F. Gill, eds.). Acad. Nat. Sci., Philadelphia, and Am. Ornithol. Union, Washington, DC.

Hill, H.O. 1980. Breeding birds in a desert scrub community in southern Nevada USA. Southwest. Nat. 25:173–180.

Hirshman, S. 1998. Black Swifts (*Cypseloides niger*) in Box Canyon, Ouray, Colorado. J. Colo. Field Ornithol. 32:53–60.

Hobson, K.A., and S.G. Sealy. 1987. Cowbird egg buried by a Northern Oriole. J. Field Ornithol. 58:222–224.

Hochbaum, H.A. 1944. The Canvasback on a prairie marsh. Am. Wildl. Inst., Washington, DC.

Hoffman, R.W. 1962. The Wild Turkey in eastern Colorado. Colo. Dept. Game and Fish Tech. Publ. 12.

Hoffman, R.W. 1963. The Lesser Prairie Chicken in Colo. J. Wildl. Manage. 27:726–32.

Hoffman, R.W. 1965. The Scaled Quail in Colorado. Colo. Dept. Game, Fish, and Parks. Tech. Publ. No.18.

Hoffman, R.W. 1968. Roosting sites and habits of Merriam's turkeys in Colorado. J. Wildl. Manage. 32:859–866.

Hoffman, R.W. 1981. Population dynamics and habitat relationships of Blue Grouse. Colo. Div. Wildl. Fed. Aid in Wildl. Restor. Rep.W-37-R-34.

Hoffman, R.W. 1990. Chronology of gobbling and nesting activities of Merriam's Wild Turkey. Proc. Natl. Wild Turkey Symp. 6:25–31.

Hoffman, R.W. 1995. CDOW Memorandum, 8/10/95, to Colorado Div. Wildl. Regional Personnel.

Hoffman, R.W., and C.E. Braun. 1977. Characteristics of a wintering population of White-tailed Ptarmigan in Colorado. Wilson Bull. 89:107–115.

Hoffman, R.W., and K.M. Giesen. 1983. Demography of an introduced population of White-tailed Ptarmigan. Can. J. Zool. 61:1758–1764.

Hoffman, R.W., H.G. Shaw, M.A. Rumble, B.F. Wakeling, C.M. Mollohan, S.D. Shemnitz, R. Engel-Wilson, and D.A. Hengel. 1993. Management guidelines for Merriam's Wild Turkeys. Colo. Div. Wildl. Div. Rep. 18.

Hohman, W.L. 1986. Incubation rhythms of Ring-necked Ducks. Condor 88:290–296.

Holden, C.H., Jr., and C.E.H. Aiken. 1872. Notes on the birds of Wyoming and Colorado territories.

Holland, T.M., and C. Schultz. In press. First Boreal Owl nests documented in western Colorado. Southwest. Nat.

Holmes, R.T., and S.K. Robinson. 1988. Spatial patterns, foraging tactics and diets of ground-foraging birds in a northern hardwood forest. Wilson Bull. 100:377–394.

Holroyd, G.L. 1993. Dark secrets: discovering the unusual habits of the Black Swift. Birder's World 7(5):22–25.

Holt, D.W., and S.M. Leasure. 1993. Short-eared Owl. *In* The birds of North America, no. 62 (A. Poole and F. Gill, eds.). Acad. Nat. Sciences, Philadelphia, and Am. Ornithol. Union, Washington, DC.

Holt, H.R. 1997. A birder's guide to Colorado. American Birding Association, Colo. Springs.

Holt, H.R., and J.A. Lane. 1987. A birder's guide to Colorado. L & P Press, Denver.

Hooper, D.C. 1951. Waterfowl nesting at Minto Lakes, Alaska. Proc. 2nd Alaskan Sci. Conf.:318–321.

Horak, G.J. 1970. A comparative study of the foods of the Sora and Virginia Rail. Wilson Bull. 82:206–213.

Horak, G.J. 1984. Kansas Prairie-Chickens. Kansas Dept. Wildlife and Parks, Wildl. Bull. No. 3.

Horn, H.S. 1968. The adaptive significance of colonial nesting in the Brewer's Blackbird *(Euphagus cyanocephalus)*. Ecol. 49:682–694.

Houston, C.S., and M.J. Bechard. 1983. Red-tailed Hawk distribution in Saskatchewan. Blue Jay 41:99–109.

Houston, C.S., and M.J. Bechard. 1984. Decline of the Ferruginous Hawk in Saskatchewan. Am. Birds 38:166–170.

Houston, C.S., and J.K. Schmutz. 1995. Declining reproduction among Swainson's Hawks in prairie Canada. J. Raptor Res. 29:198–201.

Howell, A.H. 1924. Birds of Alabama. Brown Printing Co. Montgomery.

Howell, A.H. 1932. Florida bird life. Coward-McCann, New York.

Howell, S.N.G., and S. Webb. 1995. A guide to the birds of Mexico and northern Central America. Oxford Univ. Press, New York.

Howes-Jones, Daryl. 1985. Relationships among song activity, context, and social behavior in the Warbling Vireo. Wilson Bull. 97:4–22.

Hubbard, J.P. 1977. The status of Cassin's Sparrow in New Mexico and adjacent states. Am. Birds 31:933–941.

Hughes, A.J. 1993. Breeding density and habitat preference of the Burrowing Owl in northeastern Colorado. M.S. Thesis, Colo. State Univ., Fort Collins.

Hughes, J.M. 1996. Greater Roadrunner. *In* The birds of North America, no. 244 (A. Poole and F. Gill, eds.). Acad. Nat. Sci., Philadelphia, and Am. Ornithol. Union, Washington, DC.

Hupp J.W., and C.E. Braun. 1991. Geographic variation among Sage Grouse in Colorado. Wilson Bull. 103:255–261.

Hutto, R.L. 1995. USFS northern region songbird monitoring program: distribution and habitat relationships. Admin. rept., USFS Region 1. Missoula, MT.

Hyde, A.S. 1979. Birds of Colorado's Gunnison country. Western State Coll., Gunnison.

Ingold, D.J. 1987. Documented double-broodedness in Red-headed woodpeckers. J. Field Ornithol. 58:234–235.

Ingold, D.J. 1989. Nesting phenology and competition for nest sites among Red-headed and Red-bellied woodpeckers and European Starlings. Auk 106:209–217.

Ingold, J.L. 1993. Blue Grosbeak. *In* The birds of North America, no. 79 (A. Poole and F. Gill, eds.). Acad. Nat. Sci., Philadelphia, and Am. Ornithol. Union, Washington, DC.

Ingold, J.L., and G.E. Wallace. 1994. Ruby-crowned Kinglet. *In* The birds of North America, no. 119 (A. Poole and F. Gill., eds.). Acad. Nat. Sci., Philadelphia, and Am. Ornithol. Union, Washington, DC.

Jackson, J.A. 1976. A comparison of some aspects of the breeding ecology of Red-headed and Red-bellied woodpeckers in Kansas. Condor 78:67–76.

Jackson, J.A. 1992. From Hollywood to Broadway. Birder's World. 6(1):12–15.

Jackson, J.A. 1993. Lord Baltimore's oriole. Birder's World 7(2):12–15.

Jackson, L.S., C.A. Thompson, and J.J. Dinsmore. 1996. The Iowa breeding bird atlas. Univ. Iowa Press, Iowa City.

Jacobs, B. 1986. Birding on the Navajo and Hopi Reservations. Jacobs Publ. Co. Sycamore, MO.

James, P.C., and G.A. Fox. 1987. Effects of some insecticides on productivity of Burrowing Owls. Blue Jay 45:65–71.

Janes, S.W. 1987. Status and decline of Swainson's Hawks in Oregon: role of habitat and interspecific competition. Oregon Birds 13:165–179.

Jewett, S.G., W.P. Taylor, W.T. Shaw, and J.W. Aldrich. 1953. Birds of Washington State. Univ. Washington Press, Seattle.

Joern, W.T., and J.F. Jackson. 1983. Homogeneity of vegetational cover around the nest and avoidance of nest predation in mockingbirds. Auk 100:497–499.

Johnsgard, J. 1979. Birds of the Great Plains. Univ. Nebraska Press, Lincoln.

Johnsgard, P.A. 1973. Grouse and quails of North America. Univ. Nebraska Press, Lincoln.
Johnsgard, P.A. 1975. Waterfowl of North America. Indiana Univ. Press, Bloomington.
Johnsgard, P.A. 1978. Ducks, geese, and swans of the world. Univ. Nebraska Press, Lincoln.
Johnsgard, P.A. 1979. Birds of the Great Plains: breeding species and their distribution. Univ. Nebraska Press, Lincoln.
Johnsgard, P.A. 1981. The plovers, sandpipers, and snipes of the world. Univ. Nebraska Press, Lincoln.
Johnsgard, P.A. 1983. The hummingbirds of North America. Smithsonian Inst. Press, Washington, DC.
Johnsgard, P.A. 1986a. Birds of the Rocky Mountains. Colo. Assoc. Univ. Press, Boulder.
Johnsgard, P.A. 1986b. Birds of the Rocky Mountains: breeding species and their distribution. Univ. Nebraska Press, Lincoln.
Johnsgard, P.A. 1987. Diving birds of North America. Univ. Nebraska Press, Lincoln.
Johnsgard, P.A. 1988. Northern American owls: biology and natural history. Smithsonian Inst. Press, Washington, DC.
Johnsgard, P.A. 1990. Hawks, eagles, and falcons of North America. Smithsonian Inst. Press, Washington, DC.
Johnson, B.R., and R.A. Ryder. 1977. Breeding densities and migration periods of Common Snipe in Colorado. Wilson Bull. 89:116–121.
Johnson, D.H., and J.W. Grier. 1988. Determinants of breeding distributions of ducks. Wildl. Monogr. No. 100.
Johnson, E.J., and L.B. Best. 1982. Factors affecting feeding and brooding of Gray Catbird nestlings. Auk 99:148–156.
Johnson, K. 1995. Green-winged Teal. *In* The birds of North America, no. 193 (A. Poole and F. Gill, eds.). Acad. Nat. Sci., Philadelphia, and Am. Ornithol. Union, Washington, DC.
Johnson, L.S. 1993. The cost of polygyny in the House Wren. J. Anim. Ecol. 62:669.
Johnson, L.S., and W.A. Searcy. 1993. Nest site quality, female mate choice and polygyny in the House Wren. Ethology. 95:265.
Johnson, N.K. 1963. Biosystematics of sibling species of flycatchers in the *Empidonax hammondii-oberholseri-wrightii* complex. Univ. California Publ. Zool. 66:79–238.
Johnson, N.K. 1976. Breeding distribution of Nashville and Virginia's warblers. Auk 93:219–230.
Johnson, N.K. 1996. Speciation in vireos. I. Macrogeographic patterns of allozyme variation in the *Vireo solitarius* complex in the contiguous United States. Condor 97:903–919.
Johnson, N.K., and J.A. Marten. 1992. Macrogeographic patterns of morphometric and genetic variation in the Sage Sparrow complex. Condor 94:1–19.
Johnson, R.E. 1983. Nesting biology of the Rosy Finch on the Aleutian Islands, Alaska. Condor 85:447–452.
Johnson, R.G., and S.A. Temple. 1990. Nest predation and brood parasitism of tallgrass prairie birds. J. Wildl. Manage. 54:106–111.
Johnston, R.F. 1992. Rock Dove. *In* The birds of North America. no.13 (A. Poole and F. Gill, eds.). Acad. Nat. Sci., Philadelphia, and Am. Ornithol. Union, Washington, DC.
Jones, A.H. 1930. Pygmy Nuthatches and wrens. Bird-Lore 32:426–427.
Jones, P.W., and T.M. Donovan 1996. Hermit Thrush. *In* The birds of North America, no. 261 (A. Poole and F. Gill, eds.). Acad. Nat. Sci., Philadelphia, and Am. Ornithol. Union, Washington, DC.
Jones, S.L., and J.S. Dieni. 1995. Canyon Wren. *In* The birds of North America, no. 197 (A. Poole and F. Gill, eds.). Acad. Nat. Sci., Philadelphia, and Am. Ornithol. Union, Washington. DC.
Jones, S.R. 1991. Distribution of small forest owls in Boulder County, Colorado. J. Colo. Field Ornithol. 25:55–70.
Jones, S.R. In press. Boulder County's last Burrowing Owls? J. Colo. Field Ornithol.
Joy, S.M., R.T. Reynolds, R.L. Knight, and R.W. Hoffman. 1989. Feeding ecology of Sharp-shinned Hawks nesting in deciduous and coniferous forests in Colorado. Condor 96:455–467.
Joyner, D.E. 1977. Behavior of Ruddy Duck broods in Utah. Auk 94:343–349.
Joyner, D.E. 1983. Parasitic egg laying in Redheads and Ruddy Ducks in Utah: incidence and success. Auk 100:717–725.
Kalla, P.I., and F.J. Alsop III. 1983. The Mississippi Kite in Tennessee. Am. Birds 37:146–149.
Kalmbach, E.R., and I.N. Gabrielson. 1921. Economic value of the starling in the United States. U.S. Dept. Agric. Bull. 808.
Kaminski, R.M., and H.H. Prince. 1984. Dabbling duck-habitat associations during spring in Delta Marsh, Manitoba. J. Wildl. Manage. 48:37–50.
Kantrud, H.A., and R.L. Kologiski. 1982. Effects of soils and grazing on breeding birds of uncultivated upland grasslands of the northern Great Plains. Wildl. Res. Rep. 15. USDI, Fish and Wildl. Serv.
Kantrud, H.A., and R.L. Kologiski. 1983. Avian associations of the northern Great Plains grasslands. J. Biogeog. 10:331–350.
Kaufman, K. 1990. A field guide to advanced birding. Houghton Mifflin Co., Boston.
Kaufman, K. 1992. Western Tanager. Bird Watcher's Digest 14(5):33–38.
Keast, A., and E.S. Morton, eds. 1980. Migrant birds in the Neoptropics: ecology, behavior, distribution, and conservation. Smithsonian Inst. Press, Washington, DC.
Keast, A., and S. Saunders. 1991. Ecomorphology of the North American Ruby-crowned (*Regulus calendula*) and Golden-crowned (*R. satrapa*) kinglets. Auk 108:880–888.
Keast, A., L. Pearce, and S. Saunders. 1995. How convergent is the American redstart (*Setophaga ruticilla*), Parulinae) with flycatchers (Tyrannidae) in morphology and feeding behavior? Auk 112:310–325.
Keenan, W.J., III. 1981. Green heron fishing with mayflies. Chat 45:41.
Keith, L.B. 1961. A study of waterfowl ecology on small impoundments in southeastern Alberta. Wildl. Monogr. 6.
Keller M. 1987. The effect of forest fragmentation on birds in spruce/fir old growth forest. Ph.D. Diss., Univ. Wyoming, Laramie.
Keller, M., and S.H. Anderson. 1992. Avian use of habitat configurations created by forest cutting in southeastern Wyoming. Condor 94:55–65.

Kendeigh, S.C. 1941. Birds of a prairie community. Condor 43:165–174.

Kessel, B. 1957. A study of the breeding biology of the European Starling in North America. Condor 55:49–67.

Keyser, L.S. 1902. Birds of the Rockies. A.C. McClurg, Chicago.

Kilham, L. 1958. Territorial behavior of wintering Red-headed Woodpeckers. Wilson Bull. 70:347–358.

Kilham, L. 1959. Mutual tapping of the Red-headed Woodpecker. Auk 76:235–236.

Kilham, L. 1960. Courtship and territorial behavior of Hairy Woodpeckers. Auk 77:259–270.

Kilham, L. 1977a. Early breeding season behavior of Red-headed Woodpeckers. Auk 94:231–239.

Kilham, L. 1977b. Nest-site differences between Red-headed and Red-bellied woodpeckers in South Carolina. Wilson Bull. 89:164–165.

Kilham, L. 1984. Cooperative breeding of American Crows. J. Field Ornithol. 55:346–356.

King, J.R., and L.R. Mewaldt. 1987. The summer biology of an instable insular population of White-crowned Sparrows in Oregon. Condor 89:549–565.

Kingery, H.E. 1972. Great Basin-central Rocky Mountain region. Am. Birds 26:882–887.

Kingery, H.E. 1973. The Curve-billed Thrasher (*Toxostoma curvirostre*) - its status in Colorado and the first recorded nest for the state. J. Colo. Field Ornithol. 17:16–18.

Kingery, H.E. 1979–1993. Mountain West. Am. Birds. 1979—33:883-886. 1980a—34:293-296—1980b—34:914–918. 1982a—36:315-317. 1982b—36:1000-1003. 1983—37:1010-1013. 1984—38:1044-1047. 1985a—39:330. 1985b—39:942-945. 1988b—42:372,486. 1988c—42:1321-1324. 1990a—44:1161-1164. 1990b—44:1311. 1991b—45:475-479. 1991c—45:1141-1144. 1992—46:1157-1160. 1993a—46:1157-1160. 1993b—47:1130-1133.

Kingery, H.E. Mountain West. Natl. Audubon Soc. Field Notes. 1994a—48:322-324. 1994b—48:966-970.

Kingery, H.E., ed. 1987, 1990a. Field worker's handbook. Colo. Breeding Bird Atlas Partnership, Denver.

Kingery, H.E. 1988a. Colorado bird distribution study. Colo. Field Ornithologists, in cooperation with Colo. Division of Wildlife, Denver.

Kingery, H.E. 1989. Ages of Loggerhead Shrike young. J. Colo. Field Ornithol. 23:73.

Kingery, H.E. 1991a. Spotted Owl records in Colorado. J. Colo. Field Ornithol. 25:15–18.

Kingery, H.E. 1992b. First Colorado nest for Carolina Wren. Colo. Field Ornihtol. J. 26:67.

Kingery, H.E. 1996. American Dipper. *In* The birds of North America, no. 229 (A. Poole and F. Gill, eds.). Acad. Nat. Sci., Philadelphia, and Am. Ornithol. Union, Washington, DC.

Kingery, H.E., and U.C. Kingery. 1995. Gray Jay as predator on Cassin's Finch nestlings. J. Colo. Field Ornithol. 29:17.

Kingery, H.E. In press. Hooded Merganser nests in Colorado. Colo. Field Ornithol.

Kingery, H.E., and P.R. Julian. 1971. Cassin's Sparrow parasitized by cowbird. Wilson Bull. 83:439.

Kirsch, L.M. 1969. Waterfowl production in relation to grazing. J. Wildl. Manage. 33:821–828.

Klaas, E.E. 1970. A population study of the Eastern Phoebe and its social relationship with the Brown-headed Cowbird. Ph.D. Diss. Univ. Kansas, Lawrence.

Klimkiewicz, M.K., and A.G. Futcher. 1989. Longevity records of North American birds: suppl.1. J. Field Ornithol. 60:469–494.

Knick, S.T., and J.T. Rotenberry. 1995. Landscape characteristics of fragmented shrubsteppe habitats and breeding passerine birds. Conserv. Biol. 9:1059–1071.

Knight, R.L., and M.W. Call. 1980. The Common Raven. U.S. Bur. Land Manage. Tech. Note no. 344. Denver.

Knopf, F.L. 1991. Status and conservation of Mountain Plovers: the evolving regional effort: report of research activities. U.S Fish and Wildl. Natl Ecol. Res. Ctr, Fort Collins, CO.

Knopf, F.L. 1996. Mountain Plover. *In* The birds of North America, no. 211 (A. Poole and F. Gill, eds.). Acad. of Nat. Sci., Philadelphia, and Amer. Ornithol. Union, Washington, DC.

Knopf, F.L., and B.J. Miller. 1994. *Charadrius montanus*—montane, grassland, or bare-ground plover? Auk 111:504–506.

Knopf, F.L., and L. Webster. 1976. Breeding status of the Mountain Plover. Condor 78:265–267.

Knopf, F.L., and T.E. Olson. 1984. Naturalization of Russian-olive: implications to Rocky Mountain wildlife. Wildlife Society Bulletin 12:289–298.

Knopf, F.L. J.A. Sedgwick, and R.W. Cannon. 1988. Guild structure of a riparian avifauna relative to seasonal cattle grazing. J. Wildl. Manage. 52:280–290.

Knorr, O.A. 1959. Birds of El Paso County. Univ. Colo. Press, Boulder.

Knorr, O.A. 1961. The geographical and ecological distribution of the Black Swift in Colorado. Wilson Bull. 73:155–170.

Knox, A.G. 1992. Species and pseudospecies: the structure of crossbill populations. Biol. J. Linnean Soc. 47:325–335.

Kortright, F.H. 1942. Ducks, geese, and swans of North America. Stackpole Co., Harrisburg, PA.

Kortright, F.H. 1943. The ducks, geese, and swans of North America. Am. Wildl. Inst., Washington, DC.

Krause, H. 1968. McCown's Longspur. *In* A.C. Bent. Life histories of North American cardinals, grosbeaks, buntings, towhees, finches, sparrows, and allies, pt.3. O.L. Austin, ed. U.S. Natl. Mus. Bull. No. 237, Washington, DC.

Kroodsma, D.E. 1973. Coexistence of Bewick's Wrens and House Wrens in Oregon. Auk 90:341–352.

Kroodsma, D.E., and J. Verner. 1978. Complex singing behaviors among *Cistothurus* wrens. Auk 95:703–716.

Kroodsma, D.E., and J. Verner. 1997. Marsh Wren. *In* The birds of North America, no. 308 (A. Poole and F. Gill, eds.). Acad. of Nat. Sci., Philadelphia, and Amer. Ornithol. Union, Washington, DC.

Kroodsma, R.L. 1984. Effect of edge on breeding forest bird species. Wilson Bull. 96:426–433.

Kuenning, R.R. 1994. Observations on Blue Jay cooperative breeding, May through July 1993. J. Colo. Field Ornithol. 28:29–33.

Lack, D. 1956. A review of the genera and nesting habits of swifts. Auk 72:1–32.
Lane, J.A., and H.R. Holt. 1973. A birder's guide to Denver and eastern Colorado. L & P Photography, Sacramento.
Lane, L., and T. Leukering. 1994. Alfred M. Bailey bird nesting area project, 1994 report. Colo. Bird Observatory, Brighton.
Lanyon, W.E. 1993. A lark or two. Am. Birds. 47:1050–1057.
Lanyon, W.E. 1994. Western Meadowlark. *In* The birds of North America, no. 104 (A. Poole and F. Gill, eds.). Acad. Nat. Sci., Philadelphia, and Am. Ornithol. Union, Washington, DC.
Lanyon, W.E. 1995. Eastern Meadowlark. *In* The birds of North America, no. 160 (A. Poole and F. Gill, eds.). Acad. Nat.Sci., Philadelphia, and Am.Ornithol. Union, Washington, DC.
Laskey, A.R. 1962. Breeding biology of mockingbirds. Auk 79:596–606.
Laughlin, S.B., and D.P. Kibbe, eds. 1985. The atlas of breeding birds of Vermont. Univ. Press of New England, Hanover, NH.
Lawrence, L. deK. 1966. A comparative life-history study of four species of woodpeckers. Am. Ornithol. Union Ornithol. Monogr. No. 5.
Leatherman, D. 1995. A brief biography and interview with Dr. Ronald A. Ryder - April 1995. J. Colo. Field Ornithol. 29:44–50.
Lederer, N., and S. Armstead. 1995. Cliff-nesting raptors in Boulder County and vicinity: 1994 status report. Boulder County Nature Assoc., Boulder, CO.
Leonard, M.L., and J. Picman. 1987. The adaptive significance of multiple nest building of salt Marsh Wrens. Anim. Behav. 35:1.
Leopold, A. 1949. Sand County almanac. Oxford Univ. Press, New York.
Levad, R. 1989. Western Screech-Owls in the Grand Valley. J. Colo. Field Ornithol. 23:107–108.
Levad, R. 1993. Grand Junction owling trip report, March 13, 1993. J. Colo. Field Ornithol. 27:131–132.
Levad, R. 1994. Barn Owls in western Colorado. J. Colo. Field Ornithol. 28:174–175.
Li, P., and T.E. Martin. 1991. Nest-site selection and nesting success of cavity-nesting birds in high elevation forest drainages. Auk 108:405–418.
Ligon, J.C., and D.A. Griffiths. 1972. Black Phoebe nesting in Colorado. J. Colo. Field Ornithol. 13:3–4.
Ligon, J.S. 1946. History and management of Merriam's Wild Turkey. Univ. New Mexico Publ. Biol. No. 1.
Ligon, J.S. 1961. New Mexico birds and where to find them. Univ. New Mexico Press, Albuquerque.
Lindauer, I.E. 1983. A comparison of the plant communities of the South Platte and Arkansas River drainages in eastern Colorado. Southwestern Naturalist 28:249–259.
Linkhart, B.D. and R.T. Reynolds. 1987. Brood division and postnesting behavior in Flammulated Owls. Wilson Bull. 99:240–243.
Linsdale, J.M. 1938. Environmental responses of vertebrates in the Great Basin. Am. Midl. Nat. 19:1–206.
Logan, C.A. 1983. Reproductively dependent song cyclicity in mated male mockingbirds (*Mimus polyglottos*). Auk 100:404–413.
Lokemoen, J.T., H.F. Duebbert, and D.E. Sharp. 1990. Homing and reproductive habits of Mallards, Gadwalls, and Blue-winged Teal. Wildl. Monogr. 106.
Lovell, H.B. 1958. Baiting of fish by a Green Heron. Wilson Bull. 70:280–281.
Lowe, W.P. 1894. A list of the birds of the Wet Mountains, Huerfano County, Colorado. Auk 11:266–270.
Lowther, P.E. 1993. Brown-headed Cowbird. *In* The birds of North America, no. 47 (A. Poole and F. Gill, eds.). Acad. Nat. Sci., Philadelphia, Amer. Ornithol. Union, Washington, DC.
Lowther, P.E., and C.L. Cink. 1992. House Sparrow. *In* The birds of North America, no. 12 (A. Poole, P. Stettenheim, and F. Gill, eds.). Acad. Nat. Sci., Philadelphia, and Am. Ornithol. Union, Washington, DC.
Lunk, W.A. 1962. The Rough-winged Swallow: *Stelgidopteryx ruficollis* (Vieillot): a study based on its breeding biology in Michigan. Pub. of the Nuttall Ornithol. Club, No. 4.
Lymn, N., and S.A. Temple. 1991. Land-use changes in the Gulf Coast region: links to declines in midwestern Loggerhead Shrike populations. Passenger Pigeon 53:315–325.
Lynch, P.J., and D.G. Smith. 1984. Census of Eastern Screech-Owls (*Otus asio*) in urban open-space areas using tape-recorded song. Am. Birds 38:388–391.
MacArthur, R.H. 1958. Population ecology of some warblers of northeastern coniferous forests. Ecol. 39:599–619.
MacCracken, J.G., D.W. Uresk, and R.M. Hansen. 1985. Vegetation and soils of Burrowing Owl nest sites in Conata Basin, South Dakota. Condor 87:152–154.
Mack, G.D., and L.D. Flake. 1980. Habitat relationships of waterfowl broods on South Dakota stock ponds. J. Wildl. Manage. 44:695–700.
Madge, S., and H. Burn. 1994. Crows and Jays: a guide to the crows, jays and magpies of the world. Houghton Mifflin, New York.
Madson, C. 1990. Can the Can come back? Wyoming Wildl. 54 (3):12–17.
Maguire, J. 1990. Atlas people—the ones we meet. Colo. Bird Atlas Newsl. 10:11–12.
Manci, K. 1986. The American Coot. Colo. Outdoors 35(6):8–10.
Mannan, R.W. 1979. American Redstart assists at Yellow Warbler nest. Bird-Banding 50:263.
Mannan, R.W. 1984. Habitat use by Hammond's Flycatcher in old-growth forests, northeastern Oregon. Murrelet 65:84–86.
Manolis, T. 1977. Foraging relationships of Mountain Chickadees and Pygmy Nuthatches. West. Birds. 8:18–20.
Manuwal, D.A. 1991. Spring bird communities in the southern Washington Cascade Range. Pp. 161–174 *in* L.F. Ruggiero, K.B. Aubry, A.B. Carey, and M.H. Huff, tech. coords. Wildlife and vegetation of unmanaged Douglas-fir forests. U.S. For. Serv. Gen. Tech. Rep. PNW-GTR-285.
Mariani, C.L., C.G. Earley, and C McKinnon. 1993. Early nesting by the American Goldfinch, *Carduelis tristis*, and subsequent parasitism by the Brown-headed Cowbird, *Molothrus ater*, in Ontario. Can. Field-Nat. 107:349–350.

Marks, J.S. 1986. Nest-site characteristics and reproductive success of Long-eared Owls in southwestern Idaho. Wilson Bull. 98:547–560.

Marks, J.S., and J.H. Doremus. 1988. Breeding-season diet of Northern Saw-whet Owls in southwestern Idaho. Wilson Bull. 100:690–694.

Marks, J.S., D.L. Evans, and D.W. Holt. 1994. Long-eared Owl. *In* The birds of North America, no. 133 (A. Poole and F. Gill, eds.). Acad. Nat. Sci., Philadelphia, and Am. Ornithol. Union, Washington, DC.

Marsh, T.G. 1931. A history of the first records of all the birds reported to have been seen within the present boundaries of the state of Colorado prior to settlement. Master's Thesis, Univ. Denver.

Marshall, J.T. 1948. Ecological races of Song Sparrows in the San Francisco Bay region. Part I: Habitat and abundance. Part II: Geographic variation. Condor 50:193–215, 233–256.

Marshall, J.T. 1960. Interrelations of Abert and Brown towhees. Condor 62:49–64.

Marshall, J.T., Jr. 1967. Parallel variation in north and middle American screech-owls. West. Found. Vert. Zool. Monogr. 1:1–72.

Marti, C.D. 1974. Feeding ecology of four sympatric owls. Condor 76:45–61.

Marti, C.D. 1979. Status of owls in Utah. *In* Schaeffer and S.M. Ehlers, eds. Proc. of the National Audubon society symposium on owls of the west: their ecology and conservation. Natl. Audubon Soc.

Marti, C.D. 1992. Barn Owl. *In* The birds of North America, no. 1 (A. Poole, P. Stettenheim, and F. Gill, eds.). Acad. Nat. Sci., Philadelphia, and Am. Ornithol. Union, Washington, DC.

Marti, C.D. 1994. Barn Owl reproduction: patterns and variation near the limit of the species' distribution. Condor 96:468–484.

Marti, C.D., and C.E. Braun. 1975. Use of tundra habitats by Prairie Falcons in Colorado. Condor 77:213–214.

Marti, C.D., and J.S. Marks. 1989. Medium-sized owls. Pp. 124–133 *in* B.G. Pendleton, ed. Proc. western raptor management symposium and workshop. Natl. Wildl. Fed. Sci. Tech. Ser. No. 12. Washington, DC.

Martin, A., H.S. Zim, and A.L. Nelson. 1951. American wildlife and plants—a guide to wildlife food habits. McGraw-Hill. Republ. 1961, Dover Publ., New York.

Martin, D.J. 1973a. Selected aspects of Burrowing Owl ecology and behavior. Condor 75:446–456.

Martin, D.J. 1973b. Burrow digging by Barn Owls. Bird-Banding 44:59–60.

Martin, R.F. 1971. The Canyon Wren *(Catherpes mexicanus)* raiding food storage of a trypoxylid wasp. Auk 88:677.

Martin, S.G., and T.A. Gavin. 1995. Bobolink. *In* The birds of North America, no. 176 (A. Poole and F. Gill, eds.). Acad. Nat. Sci., Philadelphia, and Am. Ornithol. Union, Washington, DC.

Martin, S.G., and K. Martin. 1996. Hybridization between a Mountain Chickadee and Black-capped Chickadee in Colorado. J. Colo. Field Ornithol. 30:60–65.

Martin, S.G., P.H. Baldwin, and E.B. Reed. 1974 Recent records of birds from the Yampa Valley, northwestern Colorado. Condor 76:113–116.

Martin, T.E., and J. Roper. 1988. Nest predation and nest site-selection of a western population of the Hermit Thrush. Condor 90:51–57.

Martinez-Vilalta, A., and A. Motis. 1992. Family *Ardeidae* (Herons). Pp. 376–429 *in* del Hoyo, J., Elliott, A., and Sargatal, J. ed. Handbook of the birds of the world. Lynx Edicions, Barcelona.

Marvil, R.E., and A. Cruz. 1989. Impact of Brown-headed Cowbird parasitism on the reproductive success of the Solitary Vireo. Auk 106:476–480.

Massey, B.W. 1976. Vocal differences between American Least Terns and the European Little Tern. Auk 93:760–773.

Maurer, B.A., E.A. Webb, and R.K. Bowers. 1989. Nest site characteristics and nestling development of Cassin's and Botteri's sparrows in southeastern Arizona. Condor 91:736–738.

Maxwell, G.R., II, J.M. Nocilly, and R.I. Shearer. 1976. Observations at a cavity nest of the Common Grackle and an analysis of grackle nest sites. Wilson Bull. 88:505–507.

Maxwell, G.R., II, and L.S. Putnam. 1972. Incubation care of young, and nest success of the Common Grackle (*Quiscalus quiscula*) in northern Ohio. Auk 89:349–359.

McCallum, D.A. 1994a. Review of technical knowledge: Flammulated Owls. Pp. 14–46 *in* Flammulated, Boreal, and Great Gray Owls in the United States—a technical conservation assessment. U.S. For. Serv. Gen. Tech. Rep. RM-253.

McCallum, D.A. 1994b. Conservation status of Flammulated Owls in the United States. Pp. 74–79 *in* Flammulated, Boreal, and Great Gray Owls in the United States—a technical conservation assessment. U S. For. Serv. Gen. Tech. Rep. RM-253.

McCallum, D.A. 1994c. Flammulated Owl. *In* The birds of North America, no. 9 (A. Poole and F. Gill, eds.). Acad. Nat. Sci., Philadelphia, and Am. Ornithol. Union, Washington, DC.

McCallum, D.A., F.R. Gahlbach, and S.W. Webb. 1995. Life history and ecology of Flammulated Owls in a marginal New Mexico population. Wilson Bull. 107:530–537.

McCallum, D.A., W.D. Graul, and R. Zaccagnini. 1977. The breeding status of the Long-billed Curlew in northeastern Colorado. Auk 94:599–601.

McCrimmon, D.A., Jr. 1978. Nest site characteristics among five species of herons on the North Carolina coast. Auk 95:267–280.

McDonald, C.B., J. Anderson, J.C. Lewis, R. Mesta, A. Ratzlaff, T.J. Tibbitts, and S.O. Williams III. 1991. Mexican Spotted Owl (*Strix occidentalis*) status review. U.S. Fish Wildl. Serv. Endangered Spec. Rep. 20, Albuquerque.

McEllin, S.M. 1979a. Nest sites and population demographics of White-breasted and Pygmy nuthatches in Colorado. Condor 81:348–352.

McEllin, S.M. 1979b. Population demographics, spacing, and foraging behaviours of White-breasted and Pygmy nuthatches in ponderosa pine habitat. Pp. 301–329 *in* C.J.G. Dickson et al., eds. The role of insectivorous birds in forest ecosystems. Academic Press, New York.

McGovern, M., and J.M. McNurney. 1986. Densities of Red-tailed Hawk nests in aspen stands in the Piceance Basin, Colorado. J. Raptor Res. 20:43–45.

McHenry, M.G. 1971. Breeding and post-breeding movements of Blue-winged Teal (*Anas discors*) in southwestern Manitoba. Ph.D. Diss., Univ. Oklahoma, Norman.

McKinney, F. 1965. The displays of the American Green-winged Teal. Wilson Bull. 77:112–121.

McKinney, F. 1970. Displays of four species of blue-winged ducks. Living Bird 9:29–64.

McKinney, F., and P. Stolen. 1982. Extra-pair-bond courtship and forced copulation among captive Green-winged Teal (*Anas crecca carolinensis*). Anim. Behav. 30:461–474.

McKnight, D.E. 1974. Dry-land nesting by Redheads and Ruddy Ducks. J. Wildl. Manage. 38:112–119.

McKusick, C.R. 1980. Three groups of turkeys from southwestern archeological sites. Contrib. Sci. Nat. Hist. Mus., Los Angeles County 330:225–235.

McNair, D.B. 1948. Reuse of other species nests by Lark Sparrows. Southwest. Nat. 29:506–509.

McPherson, J.M. 1987. A field study of winter fruit preferences of Cedar Waxwings. Condor 89:293–306.

Meaney, C.A., and D. VanVuren. 1993. Recent distribution of bison in Colorado west of the Great Plains. Proc. Denver Mus. Nat. Hist. 3:1–10.

Means, B., and R. Means. 1992. Audubon to Xantus. Academic Press and Harcourt Brace Jovanovich, London.

Medin, D.E. 1986. Grazing and passerine breeding birds in a Great Basin low-shrub desert. Great Basin Nat. 46:567–572.

Medin, D.E. 1990. Birds of a shadscale (*Atriplex confertifolia*) habitat in east central Nevada. Great Basin Nat. 50:295–298.

Melcher, C.P. 1992. Avifauna responses to intensive browsing by elk in Rocky Mountain National Park. M.S. Thesis, Colo. State Univ., Fort Collins.

Melvin, S.M., and J.P. Gibbs. 1994. Sora. Pp. 209–217 *in* T.C. Tacha and C.E. Braun, eds. Migratory shore and upland game bird management in North America. Intl. Assoc. Fish and Wildl. Agencies, Washington, DC.

Merola, M. 1995. Observations on the nesting and breeding behavior of the Rock Wren. Condor 97:2.

Mewaldt, L.R. 1956. Nesting behavior of the Clark Nutcracker. Condor 58:3–23.

Mewaldt, L.R., and J.R. King. 1985. Breeding site faithfulness, reproductive biology, and adult survivorship in an isolated population of Cassin's Finches. Condor 87:494–510.

Meyer de Schauenesee, R. 1970. A guide to the birds of South America. Livingston Publ. Co., Wynnewood, PA.

Mickey, F.W. 1943. Breeding habits of McCown's Longspur. Auk 60:181–209.

Middleton, A.L.A. 1978. The annual cycle of the American Goldfinch. Condor 80:401–406.

Middleton, A.L.A. 1993. American Goldfinch. *In* The birds of North America, no. 80 (A. Poole and F. Gill, eds.). Acad. of Nat. Sci., Philadelphia, PA, and Amer. Ornithol. Union, Washington, DC.

Miller, G.C. 1978. Riverside Reservoir Colorado: 1977 nesting season report. J. Colo. Field Ornithol. 33:19–20.

Miller, G.C., and W.D. Graul. 1987. Inventories of Colorado's Great Blue Herons. J. Colo. Field Ornithol. 21:59–66.

Miller, G.C., and R.A. Ryder. 1979. Cattle Egret in Colorado. West. Birds 10:37–41.

Miller, J.H., and M.T. Green. 1987. Distribution, status, and origin of Water Pipits breeding in California. Condor 89:788–797.

Miller, M.R., J.P. Fleskes, D.L. Orthmeyer, and D.S. Gilmer. 1992. Survival and other observations of adult female Northern Pintails molting in California. J. Field Ornithol. 63:138–144.

Miller, S.G., and R.L. Knight. 1995. Recreational trails and bird communities. Unpubl. rep. submitted to City of Boulder Open Space and City of Boulder Mountain Parks.

Miller, S.G., and R.L. Knight. 1996. Notes on the reproductive success of the Western Wood-Pewee (*Contopus sordidulus*) along the Front Range of Colorado. Unpubl. data.

Miller, S.G., and R.L. Knight. In press. Influence of recreational trails on breeding bird communities. Ecolog. Applic.

Miller, S.J., and D.W. Inouye. 1983. Roles of the wing-whistle in the territorial behaviour of male Broad-tailed Hummingbirds (*Selasphorus platycercus*). Anim. Behav. 31:689–700.

Millsap, B.A. 1981. Distributional status of falconiformes in west central Arizona. U.S. Dept. Int., Bur. Land Manage., Tech. Note 355:1–102.

Millsap, B.A., and P.A. Millsap. 1987. Burrow nesting by Common Barn-Owls in north central Colorado. Condor 89:668–670.

Minor, W.F., M. Minor, and M.F. Ingraldi. 1993. Nesting of Red-tailed Hawks and Great Horned Owls in a central New York urban/suburban area. J. Field Ornithol. 64:433–439.

Mirsky, E.N. 1976. Ecology of co-existence in a wren-wrentit-warbler guild. Ph.D. Diss., Univ. California, Los Angeles.

Mitchell, P.A. 1968. The food of the Dipper (*Cinclus mexicanus* Swainson) on two western Montana streams. M.A. Thesis, Univ. Montana, Missoula.

Moldenhauer, R.R., and J.A. Wiens. 1970. The water economy of the Sage Sparrow, (*Amphispiza belli nevadensis*). Condor 72:265–275.

Molini, W.A. 1976. Chukar partridge species management plan. Nevada Dept. Fish and Game, Reno.

Moore, W.S. 1995. Northern Flicker. *In* The birds of North America, no. 166 (A. Poole and F. Gill, eds.). Acad. Nat. Sci., Philadelphia, and Am. Ornithol. Union, Washington, DC.

Morehouse, E.L., and R. Brewer. 1968. Feeding of nestling and fledgling Eastern Kingbirds. Auk 85:44–54.

Morrison, C.F. 1888. A list of some birds of La Plata County, Colo., with annotations. Ornithol. and Oologist 13:70–75, 107–108, 115–116, 139–140.

Morrison, M.L., and M.P. Yoder-Williams. 1984. Movements of Steller's Jays in western North America. North Am. Bird Bander 9(2):12–15.

Morton, E.S. 1992. What do we know about the future of migrant landbirds? Pp. 579–589 *in* J.M. Hagan, III and D.W. Johnston, eds., Ecology and conservation of Neotropical migrant landbirds. Smithsonian Inst. Press, Washington, DC.

Morton, M.L. 1978. Snow conditions and the onset of breeding in the Mountain White-crowned Sparrow. Condor 80:285–289.

Morton, M.L., K.W. Sockman, and L.E. Peterson. 1993. Nest predation in the Mountain White-crowned Sparrow. Condor 95:72–82.

Moskoff, W. 1995. Veery. *In* The birds of North America, no. 142 (A. Poole and F. Gill, eds.). Acad. Nat. Sci., Philadelphia, and Am. Ornithol. Union, Washington, DC.

Moulton, P., M. Moulton, and J. Sundine. 1976. Observations of nesting Long-eared Owls. J. Colo. Field Ornithol. 28:4–7.

Muir, J. 1894. The mountains of California. Century Co., New York.

Munro, J.A. 1949. Studies of waterfowl in British Columbia: Green-winged Teal. Can. J. Res. 27:149–178.

Murphey, D.A., and T.S. Baskett. 1952. Bobwhite mobility in central Missouri. J. Wildl. Manage. 16:498–510.

Murphy, M.T. 1983. Nest success and nesting habits of Eastern Kingbirds and other flycatchers. Condor 85:208–219.

Murphy, M.T. 1986. Temporal components of reproductive variability in Eastern Kingbirds. Ecol. 67:1483–1492.

Mutel, C.F., and J.C. Emerick. 1992. From grassland to glacier, the natural history of Colorado and the surrounding region, Johnson Books, Boulder.

Nelson, D.L. 1991. Breeding of Least Terns, Piping Plovers and Snowy Plovers in the Arkansas River Valley in 1991. Unpubl. rep. submitted to Colo. Div. Wildl. and Bur. Land Manage., Denver.

Nelson, D. 1991b. The C.F.O. Records committee report for 1989. J. Colo. Field Ornithol. 25:119–120.

Nelson, D. 1992a. Ruffed Grouse in Moffat County, Colorado: some thoughts about their status in the state. 26:1–3.

Nelson, D.L. 1992b. [1993] Breeding of Interior Least Terns and Piping Plovers in the Arkansas River valley in 1992. Unpubl. rep. submitted to Colo. Div. of Wildl. and Bur. Land Manage. Colo. Bird Obs., Brighton.

Nelson, D.L. 1993. Manual on use of breeding codes. Colo. Bird Atlas Partnership, Denver.

Nelson, D.L. 1993b. Piping plovers: A case for how they colonized southeast Colorado and implications based on that scenario for their future as a breeding species. J. Colo. Field Ornithol. 27:80–88.

Nelson, D.L. 1995. Habitat management for Least Terns and Piping Plovers in Colorado. Unpubl. report submitted to Colo. Div. of Wildlife, Denver.

Nelson, D.L., and C.S. Aid. 1993. Breeding of Piping Plovers and Interior Least Terns in Colorado in 1993, with notes on Snowy Plovers. Unpubl. rep. submitted to Colo. Div. Wildl. and Bur. Land Manage. Colo. Bird Obs., Brighton.

Nelson, D.L., and M.F. Carter. 1990. Nesting of Least Tern and Piping Plover, southeast Colorado. Unpubl. rep. submitted to Colo. Div. of Wildl. Colo. Bird Observatory, Brighton.

Nero, R.W., R.J. Clark, R.J. Knapton, and R.H. Hamre, eds. 1987. Biology and conservation of northern forest owls. U.S. For. Serv. Gen. Tech. Rep. RM-142. Fort Collins, CO.

Nettleship, D.N., and D.C. Duffy, eds. 1995. Colonial waterbirds Vol.18 (Spec. Publ. 1).

Neudorf, D.L. 1992. Reactions of four passerine species to threats of predation and cowbird parasitism: Enemy recognition or generalized responses?. Behav. 123:84–105.

Nice, M.M. 1937. Studies in the life history of the Song Sparrow. Trans. Linnean Soc. N.Y. 4:1–246.

Nice, N.M. 1922–23. A study of nesting Mourning Doves. Auk 39:457–474, 40:37–58.

Nickell, W. 1951. Studies of habitats, territory, and nests of the Eastern Goldfinch. Auk 68:447–470.

Nickell, W.P. 1965. Habitats, territory and nesting of the Gray Catbird. Am. Midl. Nat. 73:433–478.

Niedrach, R.J., and R.B. Rockwell. 1939. Birds of Denver and mountain parks. Denver Mus. Nat. Hist. Popular Ser. No. 5., Denver.

Nolan, V. 1960. Breeding behavior of the Bell Vireo in southern Indiana. Condor 62:225–244.

Nolte, K.R., and T.E. Fulbright. 1996. Nesting ecology of Scissor-tailed Flycatchers in south Texas. Wilson Bull. 108:302.

Oberholser, H.C. 1938. The bird life of Louisiana. Louisiana Dept. Conserv. Bull.

Oberholser, H.C. 1974. The bird life of Texas. Vol. 2. Univ. Texas, Austin.

Odum, E.P. 1941a. Annual cycle of the Black-capped Chickadee, 1. Auk 58:314–333.

Odum, E.P. 1941b. Annual cycle of the Black-capped Chickadee, 2. Auk 58:518–534.

Ohlendorf, H.M. 1974. Competitive relationships among kingbirds (*Tyrannus*) in Trans-Pecos, Texas. Wilson Bull. 86:357–373.

Ohlendorf, H.M. 1976. Comparative breeding ecology of phoebes in Trans-Pecos Texas. Wilson Bull. 88:255–271.

Olendorff, R.R. 1972. Large birds of prey of the Pawnee National Grassland: nesting habits and productivity 1969–1971. U.S. Intern. Biol. Prog.Tech. Rep. 151, Grassland Biome, Fort Collins, CO.

Olendorff, R.R. 1976. The food habits of North American Golden Eagles. Am. Midl. Nat. 95:231–236.

Olendorff, R.R. 1993. Status, biology, and management of Ferruginous Hawks: a review. Raptor Res. and Tech. Asst. Ctr., Spec. Rep. U.S. Bur. Land Manage., Boise, ID.

Orians, G.H. 1985. Blackbirds of the Americas. Univ. Washington Press, Seattle.

Orians, G.H., and H.S. Horn. 1969. Overlap in foods and foraging of four species of blackbirds in the potholes of central Washington. Ecology 50:930–938.

Oring, L.W. 1964. Behaviour and ecology of certain ducks during the post-breeding period. J. Wildl. Manage. 28:223–233.

Oring, L.W. 1968. Growth, molts and plumages of the Gadwall. Auk 85:355–380.

Oring, L.W. 1969. Summer biology of the Gadwall at Delta, Manitoba. Wilson Bull. 81:44–54.

Oring, L.W., D.B. Lank, and S.J. Maxson. 1983. Population studies of the polyandrous Spotted Sandpiper. Auk 100:272–285.

Ortega, C.P., and A. Cruz. 1991. A comparative study of cowbird parasitism in Yellow-headed Blackbirds and Red-winged Blackbirds. Auk 108:16–24.

Packard, F.M. 1946. Midsummer wanderings of certain Rocky Mountain birds. Auk 62:371–94.

Page, G.W., J.S. Warriner, J.C. Warriner, and P.W.C. Paton. 1995. Snowy Plover. *In* The birds of North America, no. 154 (A. Poole and F. Gill, eds.). Acad. Nat. Sci., Philadelphia, and Am. Ornithol. Union, Washington, DC.

Palmer, D.A. 1986. Habitat selection, movements and activity of Boreal and Saw-whet Owls. M.S. Thesis, Colo. State Univ., Fort Collins.

Palmer, D.A., and R.A. Ryder. 1984. The first documented breeding of the Boreal Owl in Colorado. Condor 86:215–217.

Palmer, R.S. 1962. Handbook of North American birds. Vol. 1. Yale Univ. Press. New Haven, CT.

Palmer, R.S. 1976a. Handbook of North American birds, Vol. 2. Yale Univ. Press. New Haven, CT.

Palmer, R.S. 1976b. Handbook of North American birds. Vol. 3. Yale Univ. Press. New Haven, CT.

Palmer, R.S. 1988a. Handbook of North American birds. Vol. 4. Yale Univ. Press. New Haven, CT.

Palmer, R.S. 1988b. Handbook of North American birds. Vol. 5. Yale Univ. Press. New Haven, CT.

Palmer-Ball, B., Jr. 1996. The Kentucky breeding bird atlas. Univ. Press Kentucky, Lexington.

Parker, J.W. 1974. The breeding biology of the Mississippi Kite in the Great Plains. Ph.D. Diss. Univ. Kansas, Lawrence.

Parker, J.W. 1988. The ace dive-bomber of the prairie is a terror on the green. Smithsonian (July):54.

Parker, J.W., and J.C. Ogden. 1979. The recent history and status of the Mississippi Kite. Am. Birds 33:119–129.

Paszowski, C.A. 1982. Vegetation, ground, and frugivorous foraging of the American Robin. Auk 99:701–709.

Paszowski, C.A. 1984. Macrohabitat use, microhabitat use, and foraging behavior of the Hermit Thrush and Veery in a northern Wisconsin forest. Wilson Bull. 96:286–292.

Payne, Robert B. 1992. Indigo Bunting. *In* The birds of North America, no. 4 (A. Poole, P. Stettenhein, F. Gill. eds.). Acad. Nat. Sci., Philadelphia, and Am. Ornithol. Union, Washington, DC.

Pearson, T.G. 1942. Birds of America. Garden City Publ. Co., Garden City, NY.

Pearson, T.G., ed.-in-chief. 1936. Birds of America. Doubleday & Co., New York.

Peer, B.D., and Bollinger, E.K. Common Grackle. 1997. *In* The birds of North America, no. 271 (A. Poole and F. Gill, eds.). Acad. Nat. Sci., Philadelphia, and Am. Ornithol. Union, Washington, DC.

Pendleton, B.A.G., and D.L. Krahe. 1991. Proceedings of the midwest raptor management symposium and workshop. Natl. Wildl. Fed. Sci. Tech. Ser. No. 15, Washington, DC.

Percival, B. 1992. *In* Monthly Report, Lark Bunting. Denver Field Ornithol. 27:80.

Percival, B., V. Truan, and L. Lilly. 1994. Colo. Field Ornithologists Rare Bird Report. Unpubl. rep.

Peregrine Falcon Restoration Program. Job Progress Report, Colo. Div. Wildl. Res. Rep.

Perrit, J.E., and L.B. Best. 1989. Effects of weather on the breeding ecology of Vesper Sparrows in Iowa crop fields. Am. Midl. Nat. 121:355–360.

Peterjohn, B.G., and D.L. Rice. 1991. The Ohio breeding bird atlas. Ohio Dept. Natural Resources, Columbus.

Petersen, A.J. 1955. The breeding cycle in the Bank Swallow. Wilson Bull. 67:235–286.

Petersen, K.L., and L.B. Best. 1987a. Effects of prescribed burning on nongame birds in a sagebrush community. Wildl. Soc. Bull. 15:317–329.

Petersen, K.L., and L.B. Best. 1987b. Territory dynamics in a Sage Sparrow population: are shifts in site use adaptive? Behav. Ecol. Sociobiol. 21:351–358.

Petersen, K.L., and L.B. Best. 1991. Nest-site selection by sage thrashers in southeastern Idaho. Great Basin Nat. 51:261–166.

Peterson, R.A. 1995. The South Dakota breeding bird atlas. South Dakota Ornithologists Union, Aberdeen.

Peterson, R.T. 1941. A field guide to western birds. Houghton Mifflin, Boston.

Peterson, R.T. 1941b. The Steller's Jay. Leaflet No. 146 (senior). Natl. Audubon Soc.

Peterson, R.T. 1990. A field guide to western birds. 3rd ed. Houghton Mifflin, Boston.

Petrinovich, L., and T.L. Patterson. 1982. The White-crowned Sparrow: stability, recruitment, and population structure in the Nuttall subspecies (1975–1980). Auk 99:1–14.

Pezzolesi, L.S. 1994. The western Burrowing Owl: Increasing prairie dog abundance, foraging theory, and nest site fidelity. M.S. Thesis, Texas Tech. Univ., Lubbock.

Philipp, P.B. 1925. Notes on some summer birds of Magdalen Islands. Can. Field-Nat. 39:75–78.

Phillips, A., J. Marshall, and G. Monson. 1964. The birds of Arizona. Univ. Arizona Press, Tucson.

Phillips, A.R. 1944. Status of Cassin's Sparrow in Arizona. Auk 61:409–412.

Phillips, A.R. 1951. Some observations of birds in southern Colorado. Condor 53:50–51.

Phillips, J.C. 1922. A natural history of the ducks. Houghton Mifflin, New York.

Picman, J. 1977a. Destruction of eggs by the Long-billed Marsh Wren (*Telmatodytes palustris palustris*). Can. J. Zool. 55:1914–1920.

Picman, J. 1977b. Intraspecific nest destruction in the Long-billed Marsh Wren (*Telmatodytes palustris palustris*). Can. J. Zool. 55:1997–2003.

Pitocchelli, F. J. 1995. MacGillivray's Warbler. *In* The birds of North America, no. 159 (A. Poole and F. Gill, eds.). Acad. Nat. Sci., Philadelphia, and Am. Ornithol. Union, Washington, DC.

Pleasants, B.T. 1979. Adaptive significance of the variable dispersion pattern of breeding Northern Orioles. Condor 81:38–34.

Pleszczynska, W.K. 1978. Microgeographic prediction of polygyny in the Lark Bunting. Sci. 201:935–937.

Pleszczynska, W.K., and R.I.C. Hansell. 1980. Polygyny and decision theory: testing of a model in Lark Bunting (*Calamospiza melanocorys*). Am. Nat. 116: 821–830.

Plumpton, D.L., and R.S. Lutz. 1993a. Prey selection and food habits of Burrowing Owls in Colorado. Great Basin Nat. 53:299–304.

Plumpton, D.L., and R.S. Lutz. 1993b. Nesting habitat use by Burrowing Owls in Colorado. J. Raptor Res. 27:175–179.

Pollock, R. 1991. A neighborhood helper. WildBird 5(11):32–33.

Pollock, R., and J. Pollock. 1993. A Mountain Bluebird summer. Colo. Outdoors (May/June):25–27.

Porter, D.K. 1973. First observation of cowbird parasitism on Lark Buntings in Colorado. J. Colo. Field Ornithol. 16:18.

Porter, D.K., and R.A. Ryder. 1974. Avian density and productivity studies and analysis on the Pawnee Site in 1972. Grassland Biome, U.S. Intl. Biol. Prog. Tech. Rep. 252, Fort Collins, CO.

Porter, D.K., M.S. Strong, J.B. Giezentanner, and R.A. Ryder. 1975. Nest ecology, productivity and growth of the Loggerhead Shrike on the shortgrass prairie. Southwest Nat. 19:429–436.

Porter, R.D., and C.M. White. 1973. The Perergrine Falcon in Utah, emphasizing ecology and competition with the Prairie Falcon. Brigham Young Univ. Sci. Bull. Biol. Series 18:1–74.

Poston, H.J. 1969. Relationships between the Shoveler and its breeding habitat at Strathmore, Alberta. Can. Wildl. Serv. Rep. Ser. 6:132–137.

Potter, K. 1997. Rose-breasted Grosbeak (*Pheucticus ludovicianus*) with young in Garfield County, Colorado. J. Colo. Field Ornithol. 31:15.

Pough, R.H. 1951. Audubon water bird guide: water, game and large land birds. Doubleday and Co., Garden City, NY.

Power, H.W., and M.P. Lombardo. 1996. Mountain Bluebird. *In* The birds of North America, no. 222 (A. Poole and F. Gill, eds.). Acad. Nat. Sci., Philadelphia, and Am. Ornithol. Union, Washington, DC.

Poysa, H., M. Rask, and P. Nummi. 1994. Acidification and ecological interactions at higher trophic levels in small forest lakes: perch and Common Goldeneye (abstract). J. Field Ornithol., 67:188–202.

Prather, B. 1988. Least Flycatcher nesting in Colorado. J. Colo. Field Ornithol. 22:134–136.

Prather, B. 1991. 1991 convention report. J. Colo. Field Ornithol. 25:96–100.

Prather, I. 1993. White-eyed Vireo nest: controversy in Colorado. J. Colo. Field Ornithol. 27:111–114.

Pravosudov, V.V., and T.C. Grubb, Jr. 1993. White-breasted Nuthatch. *In* The birds of North America, no. 54 (A. Poole and F. Gill, eds.). Acad. Nat. Sci., Philadelphia, and Am. Ornithol. Union, Washington, DC.

Preston, C.R., and R.D. Beane. 1993. Red-tailed Hawk. *In* The birds of North America, no. 52 (A. Poole and F. Gill, eds.). Acad. Nat. Sci., Philadelphia, and Am. Ornithol. Union, Washington, DC.

Preston, C.R., and R.D. Beane. 1996. Occurrence and distribution of diurnal raptors in relation to human activity and other factors at Rocky Mountain Arsenal, Colorado. Pp. 365–374 *in* D. Bird, et al., eds. Raptors in human landscapes. Academic Press, London.

Prestt, I., and R. Wagstaffe. 1973. Barn and Bay Owls. *In* J.A. Burton, ed. Owls of the world. E.P. Dutton, New York.

Price, F.E., and C.E. Bock. 1983. Population ecology of the Dipper (*Cinclus mexicanus*) in the Front Range of Colorado. Studies in avian biology no. 7. Cooper Ornithol. Soc. Allen Press, Lawrence, KS.

Price, J., S. Droege, and A. Price. 1995. The summer atlas of North American birds. Academic Press, London.

Procter-Gray, E., and R.T. Holmes. 1981. Adaptive significance of delayed attainment of plumage in male American Redstarts: tests of two hypotheses. Evol. 35:742–751.

Project tanager reference booklet. 1995. Cornell Laboratory of Ornithology. Cornell Univ., Ithaca, NY.

Prose, B.L. 1985. Habitat suitability index models: Belted Kingfisher. U.S. Fish Wildl. Serv. Biol. Rep. No. 82.

Provost, M.W. 1947. Nesting of birds in the marshes of northwest Iowa. Am. Midl. Nat. 38:485–503.

Pulich, W.M. 1988. The birds of north central Texas. Texas A&M Univ. Press, College Station.

Pulliam, W.M. 1995. How to find a Boreal Owl. Web page, http://lamar.colostate.edu/~bbill/borowl.html

Pyle, P., S.N.G. Howell, R.P. Yunick, and D.F. DeSante. 1987. Identification guide to North American passerines. Slate Creek Press, Bolinas, CA.

Pylypec, B. 1991. Effects of fire on bird populations in a fescue prairie. Can. Field-Nat. 105:346–349.

Rakestraw, J. 1995. A closer look: Lesser Prairie-Chicken. Birding 27:209–212.

Raley, C. M., and S H. Anderson. 1990. Availability and use of arthropod food resources by Wilson's Warblers and Lincoln's Sparrows in southeastern Wyoming. Condor 92:141–150.

Ratcliffe, D.A. 1962. Breeding density in the Peregrine Falcon, *Falco peregrinus*, and the Raven, *Corvus corax*. Ibis 104:13–39.

Rawinski, J.J., L. Rawinski, and J. Poe. 1995. Bendire's Thrasher field trip results. The controversy continues! J. Colo. Field Ornithol. 29:105–107.

Rawinski, J.J., R. Sell, P. Metzger, H. Kingery, and U. Kingery. 1993. Young Boreal Owls found in the San Juan Mountains, Colorado. J. Colo. Field Ornithol. 27:57–59.

Regosin, J.V. 1994. Scissor-tailed Flycatchers eject Brown-headed Cowbird eggs. J. Field Ornithol. 65:508–511.

Regosin, J.V. 1995. Aspects of breeding biology and social organization in Scissor-tailed Flycatcher. Condor 97:154–164.

Reller, A.W. 1972. Aspects of behavioral ecology of Red-headed and Red-bellied woodpeckers. Midl. Nat. 88:270–290.

Remington, T., and W. Snyder. 1991. Past and present pheasants. Colo. Outdoors 40:11–13.

Remington, T.E., and R.W. Hoffman. 1996. Food habits and preferences of Blue Grouse during winter. J. Wildl. Manage. 60:808–817.

Rennicke, J. Colorado Wildlife. 1990. Falcon Press, Helena, MT.

Renwald, J.D. 1977. Effect of fire on Lark Sparrow nesting densities. J. Range Manage. 30:283–285.

Repasky, R.R., and D. Schluter. 1994. Habitat distributions of wintering sparrows along an elevational gradient: tests of the food, predation and microhabitat structure hypotheses. J. Anim. Ecol. 63:569–582.

Reynolds, R.T. 1989. The status of accipiter populations in the western United States. Pp. 92–101 *in* B. Pendleton, K. Steenhof, and M.N. Kockert, eds. Proc. west. raptor manage. symp. workshop. Natl. Wildl. Fed., Washington, DC.

Reynolds, R.T., and B.D. Linkhart. 1987a. Fidelity to territory and mate in Flammulated Owls. Pp. 234–238 *in* Biology and conservation of northern forest owls. U.S. For. Serv. Gen. Tech. Rep. RM-142.

Reynolds, R.T., and B.D. Linkhart. 1987b. The nesting biology of Flammulated Owls in Colorado. Pp. 239–248 *in* Biology and conservation of northern forest owls. U.S. For. Serv. Gen. Tech. Rep. RM-142.

Reynolds, R.T., and B.D. Linkhart. 1992. Flammulated Owls in ponderosa pine: evidence of preference for old growth. Pp. 166–169 *in* Old-growth forests in the Southwest and Rocky Mountain regions, Proc. of a workshop. U.S. For. Serv. Gen. Tech. Rep. RM-213.

Reynolds, R.T., S. Joy, and T.B. Mears. 1990. Predation and observation records of Boreal Owls in western Colorado. J. Colo. Field Ornithol. 24:99–101.

Reynolds, R.T., D.P. Kane, and D.M. Finch. 1991. Tree-nesting habitat of Purple Martins in Colorado. Unpubl. report.

Reynolds, R.T., R.A. Ryder, and B.D. Linkhart. 1989. Small forest owls. Pp. 134–143 *in* Proc. western raptor management symposium and workshop. Natl. Wildl. Fed., Washington, DC.

Reynolds, T.D. 1981. Nesting of the Sage Thrasher, Sage Sparrow, and Brewer's Sparrow in southeastern Idaho. Condor 83:61–64.

Reynolds, T.D., and T.D. Rich. 1978. Reproductive ecology of the Sage Thrasher (*Oreoscoptes montanus*) on the Snake River Plain in southeastern Idaho. Auk 95:580–582.

Rice, J. 1978. Ecological relationships or two interspecifically territorial vireos. Ecol. 59:526–538.

Rice, W.R. 1982. Acoustical location of prey by the Marsh Hawk: adaptation to concealed prey. Auk 99:403–413.

Rich, T. 1980a. Nest placement in Sage Thrashers, Sage Sparrows, and Brewer's Sparrows. Wilson Bull. 92:362–368.

Rich, T. 1980b. Territorial behavior of the Sage Sparrow: spatial and random aspects. Wilson Bull. 92:425–562.

Rich, T. 1981. Microgeographic variation in the song of the Sage Sparrow. Condor 83:113–119.

Rich, T. 1983. "Walking-In-Line" behavior in Sage Sparrow territorial encounters. Condor 87:496–497.

Rich, T. 1984. Monitoring burrowing owl populations: implications of burrow reuse. Wildl. Soc. Bull. 12:178–180.

Rich, T. 1985a. The organization and structure of Sage Sparrow song: locatability, distance transmission, and contrast. Murrelet 66:1–10.

Rich, T. 1985b. A Sage Thrasher nest with constructed shading platform. Murrelet 66:18–19.

Rich, T., and S.I. Rothstein. 1985. Sage Thrashers reject cowbird eggs. Condor 87:561–562.

Richter, W. 1984. Nestling survival and growth in the Yellow-headed Blackbird, *Xanthocephalus xanthocephalus.* Ecol. 65:597–608.

Ridgway, R. 1877. Ornithology (Report of the United States geological exploration of the fortieth parallel. Vol. 4, part 3). U.S. Gov. Print. Off., Washington, DC.

Righter, B., H. Kingery, and R. Wilson. 1989. A House Wren at 11,800 feet. J. Colo. Field Ornithol. 23:106.

Rilling, R., and D. Falzone-Schrim. 1993. Nesting colony study in the San Luis Valley. Colo. Div. Wildl., Monte Vista.

Ringelman, J.K. 1991. Evaluating and managing waterfowl habitat. Colo. Div. Wildl., Div. Rep. No. 16, Denver.

Ringelman, J.K., and K.J. Kehmeier. 1990. Buffleheads breeding in Colorado. J. Colo. Field Ornithol. 24:46–48.

Rising, J.D. 1983. The Great Plains hybrid zones. Current Ornithology 1:131–157.

Rising, J.D. 1996. A guide to the identification and natural history of the sparrows of the United States and Canada. Academic Press, San Diego.

Ritchison, G. 1980. The two-voice phenomenon in Vesper Sparrows. Kentucky Warbler 56(2):27–30.

Ritter, L.V., and K. Purcell. 1983. Cavity-nesting Brewer's Blackbirds. West. Birds 14:205.

Robbins, C.S. 1996. Atlas of the breeding birds of Maryland and the District of Columbia. Univ. Pittsburgh Press, Pittsburgh.

Robbins, C.S., B. Bruun, and H.S. Zim. 1983. Birds of North America. Golden Press, New York.

Robbins, C.S., D. Bystrak, and P.H. Geissler. 1986. The breeding bird survey: its first fifteen years, 1965–1979. U.S. Fish and Wildl. Serv., Resour. Publ. 157.

Robbins, C.S., J.R. Sauer, R.S. Greenburg, and S. Droege. 1989. Population declines in North American birds that migrate to the Neotropics. Proc. Natl. Acad. Sci. 86:7658–7662.

Robbins, D.W., J.B. Reid, and R.A. Chapman. 1993. The new atlas of breeding birds in Britain and Ireland: 1988–1991. T & AD Poyser, London.

Roberson, D., and C. Tenney. 1993. Atlas of the breeding birds of Monterey County, California. Monterey Peninsula Audubon Soc., Carmel.

Roberts, T.S. 1932. Birds of Minnesota. Univ. Minnesota Press, Minneapolis.

Robertson, R.J., B.J. Stutchbury, and R.R. Cohen. 1992. Tree Swallow. *In* The birds of North America, no. 11 (A. Poole, P. Stettenheim, F. Gill, eds.). Acad. Nat. Sci., Philadelphia, and Am. Ornithol. Union, Washington, DC.

Robinson, J.A, L.W. Oring, J.P. Skorupa, and R. Boettcher. 1997. American Avocet. *In* The birds of North America, no. 275 (A. Poole and F. Gill, eds.). Acad. of Nat. Sci., Philadelphia, and Amer. Ornithol. Union, Washington, DC.

Robinson, W.A. 1986. Effect of fruit ingestion on Amelanchier seed germination. Bull. Torrey Bot. Club 113:131–134.

Rockwell, R.B. 1907. Some Colorado notes on the Rocky Mountain Screech-Owl. Condor 9:140–145.

Rockwell, R.B. 1908. An annotated list of the birds of Mesa County, Colorado. Condor 10:152–180.

Rockwell, R.B. 1909. The use of magpies' nests by other birds. Condor 11:90–92.

Rockwell, R.B. 1911. Nesting notes on the ducks of the Barr Lake region, Colorado. Condor 14:117–131.

Rocky Mountain/Southwestern Peregrine Falcon Recovery Team. 1977. American Peregrine Falcon, Rocky Mountain and Southwest population, recovery plan. U.S. Fish and Wildl. Serv., Washington, DC.

Rodenhouse, N.L., L.B. Best, R.J. O'Connor, and E.K. Bollinger. 1995. Effects of agricultural practices and farmland structures. Pp. 269–292 *in* T.E. Martin and D. M. Finch, eds. Ecology and management of Neotropical migratory birds: a synthesis and review of critical issues. Oxford Univ. Press, New York.

Rodgers, R.D. 1983. Reducing wildlife losses to tillage in fallow wheat fields. Wildl. Soc. Bull. 11(1):31–38.

Rogers, C.M. 1994. Avian nest success, brood parasitism and edge-independent reproduction in an Alaskan wetland. J. Field Ornithol. 64:433–586.

Rogers, G.E. 1964. Sage Grouse investigation in Colorado. Colo. Dept. Game, Fish and Parks, Tech. Publ. 16.

Rogers, G.E. 1968. The Blue Grouse in Colorado. Colo. Game, Fish and Parks Dept., Tech. Publ. No. 2.

Rogers, G.E. 1969. The Sharp-tailed Grouse in Colorado. Colo. Div. Game, Fish and Parks GF-R-T-23.

Root, R.B. 1969. The behavior and reproductive success of the Blue-gray Gnatcatcher. Condor 71:16–31.

Root, T. 1988. Atlas of wintering North American birds, an analysis of Christmas bird count data. Univ. Chicago Press, Chicago.

Roseberry, J.L., and W.D. Klimstra. 1970. The nesting ecology and reproductive performance of the Eastern Meadowlark. Wilson Bull. 82:243–267.

Rosenberg, K.V., R. Ohmart, and W.C. Hunter. 1991. Birds of the lower Colorado River valley. Univ. Arizona Press, Tucson.

Rosenfield, R.N., and J. Bielefeldt. 1993. Cooper's Hawk. *In* The birds of North America, no. 75 (A. Poole and F. Gill, eds.). Acad. Nat. Sci., Philadelphia, and Am. Ornithol. Union, Washington, DC.

Rotella, J.J., and J.T. Ratti. 1992. Mallard brood movements and wetland selection in southwestern Manitoba. J. Wildl. Manage. 56:508–515.

Rotenberry, J.T., and J.A. Wiens. 1989. Reproductive biology of shrubsteppe passerine birds: geographical and temporal variation in clutch size, brood size, and fledging success. Condor 91:1–14.

Rotenberry, J.T., and J.A. Wiens. 1991. Weather and reproductive variation in shrubsteppe sparrows: a hierarchical analysis. Ecol. 72:1325–1335.

Rothstein, S.I., J. Verner, and E. Stevens. 1980. Range expansion and diurnal changes in dispersion of the Brown-headed Cowbird in the Sierra Nevada. Auk 97:253–267.

Ruelas Inzunza, E. 1998. El Niño visits the river of raptors. Hawk Mountain News, Spring 1998:26–30.

Ruggiero, L.F. et al. 1991. Wildlife and vegetation of unmanaged Douglas-fir forests. U.S. For. Serv. Gen. Tech. Rep. PNW-GTR-285.

Rumble, M.A. 1987. Avian use of scoria rock outcrops. Great Basin Nat. 47:625–630.

Rumsey, R.L. 1970. Woodpecker nest failures in creosoted utility poles. Auk 87:367–369.

Russell, D.N. 1971. Food habits of the starling in eastern Texas. Condor 73:369–372.

Russell, S.M. 1996. Anna's Hummingbird, *In* The birds of North America, no. 226. (A. Poole and F. Gill, eds.). Acad. Nat. Sci., Philadelphia, and Am. Ornithol. Union, Washington, DC.

Rutherford, W.H., and C.R. Hayes. 1976. Stratification as a means for improving waterfowl surveys. Wildl. Soc. Bull. 4:74–78.

Ryder, R.A. 1950. New breeding records for Colorado. Condor 52:133–134.

Ryder, R.A. 1951. Waterfowl production in the San Luis Valley, Colorado. M.S. Thesis, Colo. A&M Coll., Fort Collins.

Ryder, R.A. 1959. Interspecific intolerance of the American Coot in Utah. Auk 76:424–442.

Ryder, R.A. 1961. Coot and duck productivity in Northern Utah. Trans. North Am. Wildl. Conf. 26:134–147.

Ryder, R.A. 1963. Migration and population dynamics of American Coots in western North America. Proc. XIII Intl. Ornithol. Congr. 441–453.

Ryder, R.A. 1964. California Gulls nesting in Colorado. Condor 66:440.

Ryder, R.A. 1965. A checklist of the birds of the Rio Grande drainage of southern Colorado. Preliminary Draft. Colo. State Univ. Fort Collins.

Ryder, R.A. 1967. Distribution, migration and mortality of the White-faced Ibis (*Plegadis chihi*) in North America. Bird-Banding 38:257–277.

Ryder, R.A. 1978. Breeding distribution, movement, and mortality of snowy egrets in North America. Pp. 197–205 *in* A. Sprunt IV, J.C. Ogden, and S. Winckler, eds. Wading Birds. Natl. Audubon Soc. Res. Rep. No. 7, New York.

Ryder, R.A. 1980. Effects of grazing on bird habitats. Pp. 51–66 *in* Management of western forests and grasslands for nongame birds. U.S. For. Serv. Gen. Tech. Rep. INT-86, Ogden, UT.

Ryder, R.A. 1981. Movements and mortality of White Pelicans fledged in Colorado. Colonial Waterbirds 4:72–76.

Ryder, R.A. 1991a. Distribution and status of the Boreal Owl in western North America: an update. J. Colo.-Wyo. Acad. Sci. 23:26.

Ryder, R.A. 1991b. Distribution, status, migration and harvest of Colorado Redheads (abstract). J. Colo. Field Ornithol. 25:102.

Ryder, R.A. 1993. The status of White-faced and Glossy ibises in Colorado. J. Colo. Field Ornithol. 27:102.

Ryder, R.A. 1994. Waterbirds of the San Luis Valley, past, present and future. J. Colo. Field Ornithol. 28:100.

Ryder, R.A. 1995. Double-crested Cormorant in Colorado: historical breeding status and migration (abstract). J. Colo. Field Ornithol. 29:168.

Ryder, R.A. 1996. Status and movements of Double-crested Cormorants in Colorado and Wyoming (abstract). J. Colo.-Wyo. Acad. Sci. 38:12.

Ryder, R.A., W.D. Graul, and G.C. Miller. 1979. Status, distribution, and movements of Ciconiiformes in Colorado. Proc. Colonial Waterbird Group Conf. 3:49–58.

Ryder, R.A., and D.E. Manry. 1994. White-faced Ibis. *In* The birds of North America, no. 130 (A. Poole, P. Stettenheim, and F. Gill, eds.). Acad. Nat. Sci., Philadelphia, and Am. Ornithol. Union, Washington, DC.

Ryder, R.A., D.A. Palmer, and J.J. Rawinski. 1987. Distribution and status of the Boreal Owl in Colorado. *In* Biology and conservation of northern forest owls: a symposium proceedings. U.S. For. Serv. Gen. Tech. Rep. RM-142.

Ryder, R.A., R.W. Schnaderbeck, and C.W. Jeske. 1989. The Little Blue Heron, a new breeding bird for Colorado. J. Colo. Field Ornithol. 23:102.

Ryser, F. 1985. Birds of the Great Basin. Univ. Nevada Press, Reno.

Saab, V.A., C.E. Bock, T.D. Rich, and D.S. Dobkin. 1995. Livestock grazing effects in western North America. Pp. 311–356 *in* T.E. Martin and D.M. Finch, eds. Ecology and management of Neotropical migratory birds: a synthesis and review of critical issues. Oxford Univ. Press, New York.

Sabo, S.R. 1980. Niche and habitat relationships in subalpine bird communities in White Mountains, New Hampshire. Ecol. Monogr. 50:241–259.

Sakai, H.F. 1987. Response of Hammond's and Western flycatchers to different aged douglas-fir stands in northwestern California. Master's Thesis. Humboldt State Univ., Arcata, CA.

Sakai, H.F., and B.R. Noon. 1991. Nest-site characteristics of Hammond's and Pacific-slope flycatchers in northwestern California. Condor 93:563–574.

Salata, L.R. 1983. Status of the Least Bell's Vireo on Camp Pendleton California: research done in 1983. Final Rep., U.S. Fish Wildl. Serv., Laguna Niguel.

Salomonsen, F. 1968. The moult migration. Wildfowl 19:5–24.

Salt, G.W. 1952. The relation of metabolism to climate and distribution in three finches of the genus *Carpodacus.* Ecol. Monogr. 22:121–151.

Salyer, J.C., and K.F. Lagler. 1946. The Eastern Belted Kingfisher, *Megaceryle alcyon alcyon* (Linnaeus) in relation to fish management. Trans. Amer. Fish. Soc. 76:97–117.

Samson, F.B. 1976. Territory, breeding density, and fall departure in Cassin's Finch. Auk 93:477–497.

Sandfort, W. 1952. Chukar Partridge. Colo. Conserv. (Mar.–Apr.):15–18.

Sandfort, W. 1967. A decade of Chukar hunting. Colo. Outdoors. Nov.–Dec. 20–23.

Sauer, J.R., and S. Droege. 1992. Geographic patterns in population trends of Neotropical migrants in North America. Pp. 26–42 *in* J.M. Hagan III and D.W. Johnston, eds. Ecology and conservation of Neotropical migrant landbirds. Smithsonian Inst. Press, Washington, DC.

Sauer, J.R., S. Schwartz, B.G. Peterjohn, and J.E. Hines. 1996. The North American breeding bird atlas home page. Version 94.3. Patuxent Wildlife Research Center. Laurel, MD.

Saunders, A.A. 1910. Singing female Slate-colored Fox Sparrow. Condor 12:80.

Savard, J.L., G.E. Smith, and J.N. Smith. 1991. Duckling mortality in Barrow's Goldeneye and Bufflehead broods. Auk 108:568–577.

Schaeff, C., and J. Picman. 1988. Destruction of eggs by Western Meadowlarks. Condor 90:935–937.

Schaid, T.A., D.W. Uresk, W.L. Tucker, and R.L. Linder. 1983. Effects of surface mining on the Vesper Sparrow in the northern Great Plains. J. Range Manage. 36:500–503.

Schaller, G.B. 1994. Breeding behavior of White Pelican at Yellowstone Lake, Wyoming. Condor 66:3–23.

Scharf, W.C., and J. Kren. 1996. Orchard Oriole. *In* The birds of North America, no. 255 (A. Poole and F. Gill, eds.). Acad. Nat. Sci., Philadelphia, and Am. Ornithol. Union, Washington, DC.

Schartz, R.L., and J.L. Zimmerman. 1971. The time and energy budget of the male Dickcissel. Condor 73:68–76.

Schemnitz, S.D. 1961. Ecology of the Scaled Quail in the Oklahoma Panhandle. Wildl. Monogr. No. 8. Wildl. Soc., Washington, DC.

Schemnitz, S.D. 1994 Scaled Quail. *In* The birds of North America, no.106 (A. Poole and F. Gill, eds.). Acad. Nat. Sci., Philadelphia, and Am. Ornithol. Union, Washington, DC.

Schmutz, J.A. 1988. Reproductive performance and habitat use of Rio Grande Wild Turkeys. M.S. Thesis, Colo. State Univ., Fort Collins.

Schmutz, J.K. 1984. Ferruginous Hawk and Swainson's Hawk abundance and distribution in relation to land use in southeastern Alberta. J. Wildl. Manage. 48:1180–1187.

Schmutz, J.K. 1987. Factors limiting the size of the breeding population of Ferruginous Hawks. Pp. 189–191 *in* G.L. Holroyd, et al., eds. Proc. of the workshop on endangered species in the prairie provinces. Prov. Mus. Alta. Nat. Hist. Occas. Pap. No. 9.

Schmutz, J.K., and D.J. Hungle. 1989. Population of Ferruginous and Swainson's hawks increase in synchrony with ground squirrels. Can. J. Zool. 67:2596–2601.

Schmutz, J.K., R.W. Fyfe, U. Banasch, and H. Armbruster. 1991. Routes and timing of migration of falcons nesting in Canada. Wilson Bull. 103:44–58.

Schmutz, J.K., S.H. Brechtel, K.D. De Smet, D.G. Hejertaas, G.L. Holroyd, C.S. Houston, and R.W. Nero. 1992. Recovery plan for the Ferruginous Hawk in Canada. Prep. for Recovery of Natl. Endangered Wildl. (RENEW), Ottawa.

Schmutz, J.K., S.M. Schmutz, and D.A. Boag. 1980. Coexistence of three species of hawks (*Buteo* spp.) in the prairie-parkland ecotone. Can. J. Zool. 58:1075–1089.

Schnell, J.H. 1958. Nesting behavior of goshawks. Condor 60:377–403.

Schorger, A.W. 1952. Introduction of the domestic pigeon. Auk 69:462–463.

Schreur, J.L. 1987. Ciconiiform reproductive success and public viewing in the San Luis Valley, Colorado. M.S. thesis, Colo. State Univ., Fort Collins.

Schroeder, M.A., and C.E. Braun. 1992. Seasonal movement and habitat use by Greater Prairie-Chickens in northeastern Colorado. Colo. Div. Wildl. Spec. Rep. 68.

Schroeder, M.A., and C.E. Braun. 1993a. Movement and philopatry of Band-tailed Pigeons captured in Colorado. J. Wildl. Manage. 57:103–112.

Schroeder, M.A., and C.E. Braun. 1993b. Partial migration in a population of Greater Prairie-Chickens in northeastern Colorado. Auk 110:21–28.

Schroeder, M.A., and L.A. Robb. 1993. Greater Prairie-Chicken. *In* The birds of North America, no. 36 (A. Poole, P. Stettenheim, and F. Gill, eds.). Acad. Nat. Sci., Philadelphia, and Am. Ornithol. Union, Washington, DC.

Schroeder, R.L., II. 1974. A study of the nesting behavior of Western Meadowlarks near Fort Collins, Colorado. J. Colo. Field Ornithol. 21/22:6–10.

Sclater, W.H. 1912. A history of the birds of Colorado. Witherby and Co., London.

Scott, A.C., and M. Bekoff. 1991. Breeding behavior of Evening Grosbeaks. Condor 93:71–81.

Scott, D.M. 1977. Cowbird parasitism on the Gray Catbird at London, Ontario. Auk 94:18–27.

Scott, D.M., and C.D. Ankney. 1980. Fecundity of the Brown-headed Cowbird in southern Ontario. Auk 97:677–683.

Scott, O.K. 1993. A birder's guide to Wyoming. Am. Birding Assoc., Colo. Springs, CO.

Scott, V.E., and G.L. Crouch. 1987. Response of breeding birds to commercial clearcutting of aspen in southwestern Colorado. U.S. For. Serv. Res. Note RM-475. Rocky Mt. For., and Range Exp. Stn., Fort Collins.

Scott, V.E., and G.L. Crouch. 1988. Summer birds and mammals of aspen-conifer forests in west-central Colorado. U.S. For. Serv. Gen. Tech. Rep. RM-280, Rocky Mt. For. and Range Exp. Stn., Fort Collins.

Scott, V.E., G.L. Crouch, and J.A. Whelan. 1982. Responses of birds and small mammals to clearcutting in a subalpine forest in central Colorado. Res. Note RM-422. U.S. For. Serv., Rocky Mt. For. and Range Exp. Stn., Fort Collins.

Sealy, S. 1978. Possible influence of food on egg-laying and clutch size in the Black-billed Cuckoo. Condor 80:103–104.

Sealy, S.G. 1980. Breeding biology of Orchard Orioles in a new population in Manitoba. Can. Field-Nat. 94:154–158.

Sealy, S.G. 1992. Removal of Yellow Warbler eggs in association with cowbird parasitism. Condor 94:40–54.

Sealy, S.G., and R.C. Bazin. 1995. Low frequency of observed cow-bird parasitism on Eastern Kingbirds. Behav. Ecol. 6:140–145.

Sealy, S.G., and D.L. Neudorf. 1995. Male Northern Orioles eject cowbird eggs: implications for the evolution of rejection behavior. Condor. 97:369–375.

Sedgwick, J.A. 1975. A comparative study of the breeding biology of Hammond's (*Empidonax hammondii)* and Dusky (*Empidonax oberholseri*) Flycatchers. M.A. Thesis. Univ. Montana, Missoula.

Sedgwick, J.A. 1987. Avian habitat relationships in pinyon/juniper woodland. Wilson Bull. 99:413–431.

Sedgwick, J.A. 1993a. Dusky Flycatcher. *In* The birds of North America, no. 78 (A. Poole and F. Gill, eds.). Acad. Nat. Sci., Philadelphia, and Am. Ornithol. Union, Washington, DC.

Sedgwick, J.A. 1993b. Reproductive ecology of the Dusky Flycatcher in western Montana. Wilson Bull. 105:84–92.

Sedgwick, J.A. 1994. Hammond's Flycatcher. *In* The birds of North America, no. 109 (A. Poole and F. Gill, eds.). Acad. Nat. Sci., Philadelphia, and Am. Ornithol. Union, Washington, DC.

Sedgwick, J.A., and F.L. Knopf. 1988. A high incidence of Brown-headed Cowbird parasitism of Willow Flycatchers. Condor 90:253–256.

Sedgwick, J.A., and F.L. Knopf. 1989. Regionwide polygyny in Willow Flycatchers. Condor 91:473–475.

Sedgwick, J.A., and R.A. Ryder. 1986. Effects of chaining pinyon/juniper on nongame wildlife. Pp. 541–551 *in* R.L. Everett, comp. Proc., pinyon/juniper conference. U.S. For. Serv. Intermt. Res. Sta. Gen. Tech. Rep. INT-215., Reno, NV.

Selander, R.K. 1954. A systematic review of the booming nighthawks of western North America. Condor 56:57–82.

Selander, R.K., and D.R. Giller. 1961. Analysis of sympatry of Great-tailed and Boat-tailed grackles. Condor 63:29–86.

Semel, B., and P.W. Sherman. 1986. Dynamics of nest parasitism in Wood Ducks. Auk 103:813–815.

Seymour, N.R. 1974. Site attachment in the Northern Shoveler. Auk 91:423–427.

Shane, T.G. 1972. The nest site selection behavior of the Lark Bunting, *Calamospiza melanocorys.* M.S. Thesis, Kansas State Univ., Manhattan.

Shane, T.G. 1974. Nest placement by Lark Buntings. Bird Watch 2(8):1–3. Bird Populations Institute, Manhattan, KS.

Shane, T.G. 1995. The historical development of wintering Lark Bunting populations north of the thirty seventh parallel in Colorado and Kansas. Kansas Ornithol. Soc. Bull. 46:36–39.

Shane, T.G. 1996. The Lark Bunting: in peril or making progress? J. Colo. Field Ornithol. 30:162–168

Shane, T.G., and S.S. Seltman. 1995. The historical development of wintering Lark Bunting populations north of the thirty-seventh parallel in Colorado and Kansas. Kansas Ornithol. Soc. Bull. 46:36–39.

Sharp, B.S. 1985. White-faced Ibis management guidelines: Great Basin population. U.S. Fish Wildl. Serv., Portland, OR.

Shelton, A.D. 1971. A population study of the breeding hawks in the genus *Buteo* in Marion County, Ohio. M.S. Thesis, Ohio State Univ., Columbus.

Sherrod, S.K. 1978. Diets of North American Falconiformes. Raptor Res. 12:49–121.

Sherry, T.W., and R.T. Holmes. 1997. American Redstart. *In* The birds of North America, no. 277 (A. Poole and F. Gill, eds.). Acad. Nat. Sci., Philadelphia, and Am. Ornithol. Union, Washington, DC.

Short, H.L. 1984. Habitat suitability index models: Brewer's Sparrow. U.S. Fish Wildl. Serv. FWS/OBS-82/10.83.

Short, L.L. 1982. Woodpeckers of the world. Delaware Mus. Nat. Hist, Monogr. Ser. No. 4:179–184.

Short, L.L., and J.J. Morony, Jr. 1970. A second hybrid Williamson's x Red-naped Sapsucker and an evolutionary history of sapsuckers. Condor 72:310–315.

Shreeve, D.F. 1980. Behaviour of the Aleutian Grey-crowned and Brown-capped Rosy Finches *Leucosticte tephrocotis.* Ibis 122:145–165.

Shuford, W.D. 1993. The Marin County breeding bird atlas. Bushtit Books, Bolinas, CA.

Shukman, J.M. 1993. Breeding biology and distribution limits of phoebes in western Kansas. Kansas Ornithol. Soc. Bull. 44:25–29.

Sidle, J.G., and W.F. Harrison. 1990. Recovery plan for the interior population of the Least Tern (*Sterna antillarum*) U.S. Fish Wildl. Serv, Twin Cities, MN.

Sidle, J.G., and E.M. Kirsch. 1993. Least Tern and Piping Plover nesting at sand pits in Nebraska. Colonial Waterbirds 16:139–148.

Siegfried, W.R. 1978. Habitat and the modern range expansion of the Cattle Egret. Pp. 315–324 *in* A. Sprunt IV, J.C. Ogden, and S. Winckler, eds. Wading birds. Natl. Audubon Soc. Res. Rep. No. 7, New York.

Skutch, A.F. 1985. Life of the woodpecker. Ibis Publ., Santa Monica, CA.

Skutch, A.F. 1987. Helpers at birds' nests. Univ. Iowa Press, Iowa City.

Slack, R.D. 1976. Nest guarding behavior by male Gray Catbirds. Auk 93:292–300.

Sloane, S.A. 1996. Incidence and origins of supernumeraries at Bushtit (*Psaltriparus minimus)* nests. Auk 113:757–770.

Smith, D.G., and C.D. Marti. 1976. Distributional status and ecology of Barn Owls in Utah. Raptor Res. 10:33–44.

Smith, D.G., and J.R. Murphy. 1973. Breeding ecology of raptors in the eastern Great Basin of Utah. Brigham Young Univ. Sci. Bull., Biol. Ser. 18.

Smith, D.G., A. Devine, and D. Walsh. 1987. Censusing Screech Owls in southern Connecticut. Pp. 255–267 *in* Biology and conservation of northern forest owls. U.S. For. Serv. Gen. Tech. Rep. RM–142.

Smith, D.G., C.R. Wilson, and H.H. Frost. 1972. The biology of the American Kestrel in central Utah. Southwest. Nat. 17 :73–8

Smith, K.C. 1978. Range extension of the Blue Jay into western North America. Bird-Banding 49:208–214.

Smith, K.G., and D.C. Andersen. 1982. Food, predation, and reproductive ecology of the Dark-eyed Junco in northern Utah. Auk 99:650–661.

Smith, R.L. 1968. Grasshopper Sparrow. *In* A.C. Bent. Life histories of North American cardinals, grosbeaks, buntings, towhees, sparrows and their allies, pt.2. O.L. Austin, Jr., ed. U.S. Natl. Mus. Bull. 237, Washington, DC.

Smith, S.M. 1993. Black-capped Chickadee. *In* The birds of North America, no. 39 (A. Poole, P. Stettenheim, and F. Gill, eds.). Acad.Nat. Sci., Philadelphia, and Am. Ornithol. Union, Washington, DC.

Smith, W.P. 1943. Some Yellow Warbler observations. Bird-Banding 14:57–63.

Smyth, M., and G.A. Bartholomew. 1966. The water economy of the Black-throated Sparrow and the Rock Wren. Condor 68:447–458.

Snapp, B.D. 1976. Colonial breeding in the Barn Swallow (*Hirundo rustica*) and its adaptive significance. Condor. 78:471–480.

Snow, C.S. 1994. A third Oklahoma Brown Thrasher nest on the ground. Bull. Okla. Ornithol. Soc. 27:16.

Snyder, W.D. 1967. Experimental habitat improvement for Scaled Quail. Colo. Dept. Game, Fish, Parks. Tech. Publ.19, Denver.

Sorensen, B. 1995. Colorado Springs Christmas count records, 1965–1994, Cassin's Finch, House Finch. Unpubl. data.

Sorenson, M.D. 1991. The functional significance of parasitic egg laying and typical nesting in Redhead ducks: an analysis of individual behaviour. Anim. Behav. 42:771+.

Sorenson, M.D. 1993. Parasitic egg-laying in Canvasbacks: frequency, success, and individual behavior. Auk 110:57–69.

Spencer, H.E., Jr. 1953. Cinnamon Teal (*Anas cyanoptera*)—its life history, ecology and management. M.S. Thesis, Utah State Agric. Coll., Logan.

Sprunt, A., Jr. 1979. Yellow Warbler. Pp. 77–80 *in* L. Griscom and A. Sprunt, Jr., eds. Warblers of America. Rev. E.M. Reilly, Jr. Doubleday, New York.

Stabler, R.M. 1959. Nesting of the Blue Grosbeak. Condor 61:46–48.

Stahlecker, D.W., and R.B. Duncan. 1996. The Boreal Owl at the southern terminus of the Rocky Mountains: undocumented longtime resident or recent arrival? Condor 98:156–163.

Stahlecker, D.W., and H.J. Griese. 1977. Evidence of double brooding by American Kestrels in the Colorado high plains. Wilson Bull. 89:618–619.

Staicer, C.A. 1989. Grace's Warbler singing behavior. Auk 106:49–63.

Stepney, P.G.L. 1979. Competitive and ecological overlap between Brewer's Blackbird and the Common Grackle, with consideration of associated foraging species. Ph.D. dissertation, Univ. of Toronto, Ontario.

Stepney, P.H.R. 1975. First recorded breeding of the Great-tailed Grackle in Colorado. Condor 77:208–210.

Stepney, P.H.R., and D.M. Power. 1973. Analysis of the eastward breeding expansion of Brewer's Blackbird plus general aspects of avian expansions. Wilson Bull. 85:452–464.

Stewart, A.C., R.W. Campbell, and S. Dickin. 1996. Use of dawn vocalizations for detecting breeding Cooper's Hawks in an urban environment. Wildl. Soc. Bull. 24:291–293.

Stewart, P.A. 1969. Movements, population fluctuations, and mortality among Great Horned Owls. Wilson Bull. 81:155–162.

Stewart, R.E., and H.A. Kantrud. 1973. Ecological distribution of breeding waterfowl populations in North Dakota. J. Wildl. Manage. 37:39–50.

Stewart, R.M. 1973. Breeding behavior and life history of the Wilson's Warbler. Wilson Bull. 85:21–30.

Stewart, R.M., R.P. Henderson, and K. Darling. 1977. Breeding ecology of the Wilson's Warbler in the high Sierra Nevada, California. Living Bird 16:83–102.

Stiehl, R.B. 1985. Brood chronology of the Common Raven. Wilson Bull. 97:78–87.

Stiles, F.G., and A.J. Negret. 1994. The nonbreeding distribution of the Black Swift: A clue from Colombia and unsolved problems. Condor 96:1091–1094.

Stillwell, J.E., and N.J. Stillwell. 1955. Notes on songs of Lark Buntings. Wilson Bull. 67:138–139.

Stockard, C.R. 1905. Nesting habits of birds in Mississippi. Auk 22:146–158.

Stokes, D.W. 1979. A guide to bird behavior. Vol.1. Little, Brown, Boston.
Stokes, D.W., and L.Q. Stokes. 1983. A guide to bird behavior. Vol. 2. Little, Brown, Boston.
Stokes, D.W., and L.Q. Stokes. 1989. A guide to bird behavior, Vol. 3. Little, Brown, Boston.
Stokes, D.W., and L.Q. Stokes. 1991a. The bluebird book: a complete guide to attracting bluebirds. Little, Brown, Boston.
Stokes, D.W., and L.Q. Stokes. 1991b. Eastern, Mountain, and Western bluebirds: America's Feathered Treasures. WildBird. Mar. 1991:26.
Storer, R.W., and G.L. Nuechterlein. 1992. Western and Clark's Grebe. *In* The birds of North America, no. 26 (A. Poole, P. Stettenheim, and F. Gill, eds.). Acad. Nat. Sci., Philadelphia, and Am. Ornithol. Union, Washington, DC.
Storer, S., and R.W. Storer. 1965. The color phases of the Western Grebe. Living Bird 4:59–63.
Stouffer, P., and D. Caccamise. 1991. Roosting and diurnal movements of radio-tagged American Crows. Wilson. Bull. 103:387–400.
Stout, G.D., ed. 1967. The shorebirds of North America. Viking Press, New York.
Strickland, D., and H. Ouellet, 1993. Gray Jay. *In* The birds of North America, no. 40 (A. Poole and F. Gill, eds.). Acad. Nat. Sci., Philadelphia, and Am. Ornithol. Union, Washington, DC.
Strong, M.A., and R.A. Ryder. 1971. Avian productivity on the Pawnee Site of north-central Colorado. Grassland Biome, U.S. Intl. Biological Program Tech. Rep. 82, Fort Collins, CO.
Sullivan, J.O. 1973. Ecology and behavior of the Dipper, adaptations of a passerine to an aquatic environment. Ph.D. Diss., Univ. Montana, Missoula.
Sutcliffe, S.M., R.E. Bonney, Jr., and J.D. Lowe. 1986. Proc. of the second northeastern breeding atlas conference, April 25–27, 1986. Cornell Univ. Lab. Ornithol., Ithaca, NY.
Sutton, G.M. 1967. Oklahoma birds: their ecology and distribution with comments on the avifauna of the southern Great Plains. Univ. Oklahoma Press, Norman.
Svoboda, P.L., K.E. Young, and V.E. Scott. 1980. Recent nesting records of Purple Martins in western Colorado. West. Birds 11:195–198.
Swanson, G.M. 1971. An unusual Townsend's Solitaire nest. J. Colo. Field Ornithol. 9:30.
Sydeman, W.J., M. Guentat, and R.P. Balda. 1988. Annual reproductive yield in the cooperative Pygmy Nuthatch (*Sitta pygmaea*). Auk 105:70–77.
Szymczak, M.R. 1975. Canada Goose restoration along the foothills of Colorado. Colo. Div. Wildl., Tech. Publ. 31.
Szymczak, M.R. 1986. Characteristics of duck populations in the intermountain parks of Colorado. Colo. Div. Wildl. Tech. Publ. No. 35, Denver.
Szymczak, M.R. 1995. Number and location of Canada Geese goslings planted in Colorado, 1963–1992. Unpubl. manuscript.
Tacha, T.C., and C.E. Braun, ed. 1994. Migratory shore and upland game bird management in North America. Intl. Assoc. Fish and Wildl. Agencies. Allen Press, Lawrence, KS.
Tacha, T.C., S.A. Nesbitt, and P.A. Vohs. 1992. Sandhill Crane. *In* The birds of North America, no. 31 (A. Poole, P. Stettenheim, eds.). Acad. Nat. Sci., Philadelphia, and Am. Ornithol. Union, Washington, DC.
Tate, J., Jr. 1986. The blue list for 1986. Am. Birds 40:227–236.
Taylor, D.L., and W.J. Barmore, Jr. 1980. Post-fire succession of avifauna in coniferous forests of Yellowstone and Grand Teton National Parks, Wyoming. Pp. 130–144 *in* Management of western forests and grasslands for nongame birds, U.S. For. Serv. Gen. Tech. Rep. INT-86.
Taylor, D.M., and C.D. Littlefield. 1986. Willow Flycatcher and Yellow Warbler response to cattle grazing. Am. Birds 40:1169–1173.
Taylor, M.A., and F.S. Guthery. 1980. Status, ecology, and management of the Lesser Prairie Chicken. U.S. For. Serv. Gen. Tech. Rep. RM-77.
Taylor, S.V., and V.M. Ashe. 1976. The flight display and other behaviors of male Lark Buntings (*Calamospiza melanochorys* (sic)). Bull. Psychonomic Soc. 7:527–529.
Teather, K. 1992. Foraging patterns of male and female Scissor-tailed Flycatchers. J. Field Ornithol. 63:318–323.
Tegetmeier, W.B. 1911. Pheasants, their natural history and practical management. 5th ed. H. Cox, London.
Telfair, R.C., II. 1983. The Cattle Egret: a Texas focus and world view. Kleberg Stud. Nat. Resour. Tex. Agric. Exp. Stn., Texas A&M Univ., College Station.
Telfair, R.C., II. 1994. Cattle Egret. *In* The birds of North America, no. 113 (A. Poole and F. Gill, eds.). Acad. Nat. Sci., Philadelphia, and Am. Ornithol. Union, Washington, DC.
TenBrink, J. 1995. Common Poorwills *(Phalaenoptilus nutalli)* nesting in Elbert County, Colorado. J. Colo. Field Ornithol. 29:185–186.
TenBrink, J. 1997. Tumbling (?) poorwills. J. Colo. Field Ornithol. 31:13–14.
Terborgh, J.W. 1980. The conservation status of Neotropical migrants: present and future. Pp. 21–34 *in* A. Keast and E.S. Morton, eds., Migrant birds in the Neotropics: ecology, behavior, distribution, and conservation. Smithsonian Inst. Press. Washington, DC.
Terres, J.K., ed. 1980. The Audubon society encyclopedia of North American birds. Knopf, New York.
Thompson, B.C., J.A. Jackson, J. Burger, L.A. Hill, E.M. Kirsch, and J.L. Atwood. 1997. Least Tern, *In* The birds of North America, no. 290 (A.Poole and F. Gill, eds.). Acad. Nat. Sci., Philadelphia, and Am. Ornithol. Union, Washington, DC.
Thompson, M.C., and C. Ely. 1989. Birds in Kansas. Vol. 1. Univ. Kansas, Lawrence.
Thompson, M.C., and C. Ely. 1992. Birds in Kansas. Vol. 2. Univ. Kansas, Lawrence.
Thompson, R.W., and J.G. Strauch. 1986. Habitat use by breeding birds on City of Boulder open space. Boulder County Nature Association, Boulder, CO.
Threlfall, W., and J.R. Blacquiere. 1982. Breeding biology of Fox Sparrow in Newfoundland. J. Field Ornithol. 53:235–239.

Tobalske, B.W. 1992. Evaluating habitat suitability using relative abundance and fledging success of Red-naped Sapsuckers. Condor 94:550–553.
Tomback, D.F. 1977. Foraging strategies of Clark's Nutcracker. Living Bird 16:123–161.
Tomback, D.F. 1980. How nutcrackers find their seed stores. Condor 82:10–19.
Townsend, J.K. 1839. Narrative of a journey across the Rocky Mountains to the Columbia River. H. Perkins, Philadelphia.
Tramontano, J.P. 1964. Comparative studies of the Rock Wren and Canyon Wren. Master's Thesis, Univ. Arizona, Tucson.
Trauger, D.L., and J.H. Stoudt. 1974. Looking out for the Canvasback. Part II. Ducks Unlimited 38:30–31, 42, 44, 45, 48, 60.
Travis, J.R. 1992. Atlas of the breeding birds of Los Alamos County, New Mexico. Pajarito Ornithol. Surv., Los Alamos National Laboratory, Los Alamos.
Truslow, F.K. 1966. Ground-nesting Great Horned Owl: a photographic study. Living Bird 5:177–186.
Tuck, L.M. 1972. The snipes: a study of the genus *Capella*. Can. Wildl. Serv. Monogr. Ser., No. 5. Ottawa.
Tufts, R.W. 1961. Birds of Nova Scotia. Nova Scotia Mus., Halifax.
Turner, A., and C. Rose. 1989. Swallows and martins an identification guide and handbook. Houghton Mifflin, Boston.
Twedt, D.J., and R.D. Crawford. 1995. Yellow-headed Blackbird. *In* The birds of North America, no. 192 (A. Poole and F. Gill, eds.). Acad. Nat. Sci., Philadelphia, and Am. Ornithol. Union, Washington, DC.
Tweit, R.C. 1996. Curve-billed Thrasher. *In* The birds of North America, no. 235 (A. Poole and F. Gill, eds.). Acad. Nat. Sci., Philadelphia, and Am. Ornithol. Union, Washington, DC.
Tyler, J.D. 1994. How often do Brown Thrashers nest on the ground in Oklahoma? Bull. Okla. Ornithol. Soc. 27:4–6.
Tyler, J.D., and K.C. Parkes. 1992. Hybrid Scissor-tailed Flycatcher x Western Kingbird specimen form southwestern Oklahoma. Wilson Bull. 104:178–181.
U.S. Dept. Agriculture, For. Serv. 1994. Neotropical migratory bird reference book. U.S. For. Serv., Pacific Southwest Region, Fisheries, Wildlife, and Rare Plants Staff.
U.S. Fish and Wildlife Service. 1988. Great Lakes and Northern Great Plains Piping Plover Recovery Plan. U.S. Fish Wildl. Serv., Twin Cities, MN.
U.S. Fish and Wildlife Service. 1991. Endangered and threatened wildlife and plants; animal candidate review for listing as endangered or threatened species. 50 CFR part 17.
U.S. Fish and Wildlife Service. 1992. Endangered and threatened wildlife and plants—notice of finding on petition to list the Ferruginous Hawk. Federal Register 57:37507–37513.
U.S. Fish and Wildlife Service. 1995. Migratory nongame birds of management concern in the United States: the 1995 list. Migratory bird management office, Washington, DC.
U.S. Fish and Wildlife Service and Canadian Wildlife Service. 1994. Status of waterfowl and fall flight forecast. U.S. Fish Wildl. Serv., Washington, DC.
Udvardy, M. 1977. The Audubon society field guide to North American birds. Knopf, New York.
Ure, J., P. Briggs, and S.W. Hoffman. 1991. Petition to list as endangered the Ferruginous Hawk (*Buteo regalis*), as provided by the Endangered Species Act of 1973, as amended in 1982. Ferruginous Hawk Project, Salt Lake City, UT.
VanCamp, L.F., and C.J. Henny. 1975. The screech-owl: its life history and population ecology in northern Ohio. U.S. Fish Wildl. Serv., North Am. Fauna No. 71.
Van Dyke, H. 1911. The poems of Henry Van Dyke. Charles Scribner's Sons, New York.
Van Horn, M.A., and T. Donovan. 1994. Ovenbird. *In* The birds of North America, no. 88 (A. Poole and F. Gill, eds.). Acad. Nat. Sci., Philadelphia, and Am. Ornithol. Union, Washington, DC.
Van Sant, B.F., and C.E. Braun. 1990. Distribution and status of Greater Prairie-Chickens in Colorado. Prairie Nat. 22:225–230.
Vance, D.R., and R.L. Westemeier. 1979. Interactions of pheasants and prairie chickens in Illinois. Wildl. Soc. Bull. 7:221–225.
Veblen, T.T., K.S. Hadley, E.M. Nel, T. Kitzberger, M. Reid, and R. Villalba. 1994. Disturbance regime and disturbance interactions in a Rocky Mountain subalpine forest. J. Ecology 82:125–135.
Verbeek, N.A.M. 1967. Breeding biology and ecology of the Horned Lark in alpine tundra. Wilson Bull. 79:208–218.
Verbeek, N.A.M. 1970. Breeding ecology of the Water Pipit. Auk 87:425–451.
Verbeek, N.A.M. 1975. Comparative feeding behavior of three coexisting tyrannid flycatchers. Wilson Bull. 87:231–240.
Verner, J. 1965. Breeding biology of the Long-billed Marsh Wren. Condor 67:6–30.
Verner, J. 1994. Current management situation: Flammulated Owls. Pp. 10–13 *in* Flammulated, Boreal, and Great Gray owls in the United States: a technical conservation assessment. U.S. For. Serv. Gen. Tech. Rep. RM-253.
Verner, J., and G.H. Engelsen. 1970. Territories, multiple nest building, and polygyny in the Long-billed Marsh Wren. Auk 87:557–567.
Verner, J., and M.F. Willson. 1969. Mating systems, sexual dimprphism, and the role of male North American passerine birds in the nesting cycle. Ornithol. Monogr. No. 9.
Versaw, A. 1995. Breeding Chipping Sparrows at 12,200 feet? J. Colo. Field Ornithol. 29:135–137.
Vickery, P.D. 1996. Grasshopper Sparrow. *In* The birds of North America, no. 239 (A. Poole and F. Gill, eds.). Acad. Nat. Sci., Philadelphia, and Am. Ornithol. Union, Washington, DC.
Vickery, P.D., M.L. Hunter Jr., and J.V. Wells. 1992. Evidence of incidental nest predation and its effects on nest of threatened grassland birds. Oikos 63:281–288.
Vierling, K.T. 1997. Habitat selection of Lewis' Woodpeckers in southeastern Colorado. Wilson Bull. 109:121–130.
Villard, M., P.R. Martin, and C.G. Drummond. 1993. Habitat fragmentation and pairing success in the Ovenbird (*Seiurus aurocapillus*). Auk 110:759–768.
Viverette, C.B., S. Struve, L. Goodrich, and K. Bildstein. 1996. Decreases in migrating Sharp-shinned Hawks (*Accipiter striatus*) at traditional raptor migration watch sites in eastern North America. Auk 113:32–40.
Voous, K.H. 1988. Owls of the northern hemisphere. MIT Press, Cambridge, MA.

Wallmo, O.C. 1956. Ecology of Scaled Quail in west Texas. Texas Game and Fish Comm., Austin.

Walters, R. 1983. Utah bird distribution: latilong study. Utah Dept. Nat. Resour. Div. Wildl. Resour.

Wampole, J.H. 1951. North Park: sageland and waterfowl. Colo. Conserv., October:7–9.

Ward, J. 1988. Cassin's Finch nests in cottonwoods. Colo. Bird Atlas Newsletter 4:6.

Ward, J. 1990. Cassin's Finch nesting in atypical habitat. J. Colo. Field Ornithol. 24:102–104.

Warkentin, I.G., and P.C. James. 1988. Trends in winter distribution and abundance of Ferruginous Hawks. J. Field Ornithol. 59:209–214.

Warren, E.R. 1908. Northwestern Colorado bird notes. Condor 10:18–26.

Warren, E.R. 1910. Bird notes from Salida, Chaffee County, Colorado. Auk 27:142–151.

Waser, N.M. 1976. Food supply and nest-timing of Broad-tailed Hummingbirds in the Rocky Mountains. Condor 78:133–135.

Wauer, R.H. 1973. Birds of Big Bend National Park and vicinity. Univ. Texas Press, Austin.

Weaver, R.L., and F.H. West. 1943. Notes on the breeding of the Pine Siskin. Auk 60:492–504.

Webb, B. 1982a. Distribution and nesting requirements of Montane forest owls in Colorado. J. Colo. Field. Ornithol. 16:26–31, 58–65, 76–82.

Webb, B. 1982b. Distribution and nesting requirements of montane forest owls in Colorado. M.A. Thesis, Univ. Colo., Boulder.

Webb, B. 1985. Birds subgroup report. Pp. 33–39 *in* B.L. Winternitz and D.W. Crumpacker, eds., Colorado wildlife workshop—species of special concern. Colo. Div. Wildl., Denver.

Webb, B., and J. Reddall. 1989. Recent state record specimens of birds at the Denver Museum of Natural History. J. Colo. Field Ornithol. 23:121.

Weber, W.A. 1976. Rocky Mountain Flora. Colorado Assoc. University Press, Boulder.

Weber, W.A., and R.C. Wittmann. 1996a. Colorado flora: eastern slope. Revised ed. University Press of Colorado, Niwot.

Weber, W.A., and R.C. Wittmann. 1996b. Colorado flora: western slope. Revised ed. University Press of Colorado, Niwot.

Webster, H., Jr. 1944. A survey of the Prairie Falcon in Colorado. Auk 61:609–616.

Webster, J.D. 1961. Revision of the Grace's Warbler. Auk 78:554–566.

Weeks, H.P., Jr. 1978. Clutch size variation in the Eastern Phoebe in southern Indiana. Auk 95:656–666.

Weeks, H.P., Jr. 1979. Nesting ecology of the Eastern Phoebe in southern Indiana. Wilson Bull. 91:441–454.

Weeks, H.P., Jr. 1994. Eastern Phoebe. *In* The birds of North America, no. 94 (A. Poole and F. Gill, eds.). Acad. Nat. Sci., Philadelphia, and Am. Ornithol. Union, Washington, DC.

Weidmann, U. 1990. Plumage quality and mate choice in Mallards (*Anas platyrhynchos*). Behav. 115:127+.

Welch, B.L., F.J Wagstaff, and J.A. Roberson. 1991. Preference of wintering sage grouse for big sagebrush. J. Range Manage. 44:462–465.

Welch, B.L., J.C. Pederson, and R.L Rodriguez. 1988. Selection of big sagebrush by Sage Grouse. Great Basin Naturalist 48:274–279.

Weller, M.W. 1959. Parasitic egg laying in the Redhead (*Aythya americana*) and other North American Anatidae. Ecol. Monogr. 29:333–365.

Weller, M.W. 1965. Chronology of pair formation in some of the Nearctic *Aythya* (Anatidae). Auk 82:227–235.

Welty, J.C. 1982. The life of birds. 2nd ed. Saunders Publ., Philadelphia.

West, N.E. 1984. Successional patterns and productivity potential of pinyon/juniper ecosystems. Pp. 1301–1322 *in* Developing strategies for rangeland management. Westview Press, Boulder, CO.

Westneat, D.F. 1993. Polygyny and extrapair fertilizations in eastern Red-winged Blackbirds (*Agelaius phoeniceus*). Behav. Ecol. 4:49–60.

Weston, F.M. 1949. Blue-gray Gnatcatcher. *In* A.C. Bent, Life histories of North American thrushes, kinglets, and their allies. U.S. Natl. Mus., Washington, DC.

Wetmore, A. 1920. Observations on the habits of birds at Lake Burford, New Mexico. Auk 37:221–247.

Wetmore, A. 1964. Song and garden birds of North America. National Geographic Society, New York.

Wheeler, B. 1988. A Red-backed Hawk in Colorado. J. Colo. Field Ornithol. 22:5–7.

Wheelock, I.G. 1904. Birds of California. A.C. McClurg, Chicago.

Wheelwright, N.T., and J.D. Rising. 1993. Savannah Sparrow. *In* The birds of North America, no. 45 (A. Poole and F. Gill, eds.). Acad. Nat. Sci., Philadelphia, and Am. Ornithol. Union, Washington, DC.

White, D.H., and A.J. Krynitsky. 1986. Wildlife in some areas of New Mexico and Texas accumulate elevated DDE residues. Arch. Environ. Contamination and Toxicol. 15:149–158.

White, H.C. 1953. The eastern Belted Kingfisher in the maritime provinces. Bull. Fish. Resour. Board Can., No. 97:1–44.

Whitman, P.L. 1988. Biology and conservation of the endangered Interior Least Tern: a literature review. U.S. Fish Wildl. Serv., Biol. Rep. 88(3).

Whittle, C.L. 1922. Miscellaneous bird notes from Montana. Condor 24:73–76.

Widmann, O. 1911. List of birds observed in Estes Park, Colorado, from June 10 to July 18, 1910. Auk 28:304–319.

Wiens, J.A. 1969. An approach to the study of ecological relationships among grassland birds. Ornithol. Monogr. 8:1–93.

Wiens, J.A. 1982. Song pattern variation in the Sage Sparrow (*Amphispiza belli*): dialects or epiphenomena? Auk 99:208–229.

Wiens, J.A., J.T. Rotenberry, B. Van Horne. 1986. A lesson in the limitations of field experiments: shrubsteppe birds and habitat alteration. Ecol. 67:365–376.

Wiens, J.A., B. Van Horne, J.T. Rotenberry. 1990. Comparisons of the behavior of Sage and Brewer's sparrows in shrubsteppe habitats. Condor 92:264–266.

Wiley, R.H., Jr. 1978. The lek mating system of the Sage Grouse. Sci. Am. 238:114–125.
Williams, C.S., and W.H. Marshall. 1938. Duck nesting studies, Bear River Migratory Bird Refuge, Utah, 1937. J. Wildl. Manage. 2:29–48.
Williams, L. 1952. Breeding behavior of the Brewer Blackbird. Condor 54:3–47.
Williams, O., and P. Wheat. 1971. Hybrid jays in Colorado. Wilson Bull. 83:343–47.
Williamson, P. 1971. Feeding ecology of the Red-eyed Vireo (*Vireo olivaceus*) and associated foliage-gleaning birds. Ecol. Monogr. 41:129–152.
Willoughby, E.M., and T.J. Cade. 1964. Breeding behavior of the American Kestrel (Sparrow Hawk). Living Bird 3:75-96.
Wilson, A. 1832. American Ornithology.
Wingfield, J.C., C.M. Vleck, and M.C. Moore. 1992. Seasonal changes of the adrenocortical response to stress in birds of the Sonoran Desert. J. Exp. Zool. 264:419–428.
Winkler, D.W. 1996. California Gull. *In* The birds of North America, no. 259 (A. Poole and F. Gill, eds.). Acad. Nat. Sci., Philadelphia, and Am. Ornithol. Union, Washington, DC.
Winkler, H., D.A. Christie, and D. Nurney. 1995. Woodpeckers of the world. Houghton Mifflin, Boston.
Winn, R. 1979. Bay-breasted Warblers summer in Colorado. J. Colo. Field Ornithol. 13:21–23.
Winter, B.M., and L.B. Best. 1985. Effect of prescribed burning on placement of Sage Sparrow nests. Condor 87:294–295.
Winternitz, B. 1976. Temporal change and habitat preference of some montane breeding birds. Condor 78:383–393.
With, K.A. 1994. McCown's Longspur. *In* The birds of North America, no. 96. (A. Poole and F. Gill, eds.). Acad. Nat. Sci., Philadelphia, and Am. Ornithol. Union, Washington, DC.
With, K.A., and Balda R.P. 1990. Intersexual variation and factors affecting parental care in Western Bluebirds: a comparison of nestling and fledgling periods. Can. J. Zool. 68:733–742.
With, K.A., and D.R. Webb. 1993. Microclimate of ground nests: the relative importance of radiative cover and wind breaks for three grassland species. Condor 95:401–413.
Wittenberger, J.F. 1978. The breeding biology of an isolated Bobolink population in Oregon. Condor 80:355–371.
Wolf, B.O. 1997. Black Phoebe. *In* The birds of North America, no. 268 (A. Poole and F. Gill, eds.). Acad. Nat. Sci., Philadelphia, and Am. Ornithol. Union, Washington, DC.
Woodbridge, B., K.K. Finley, and S.T. Seager. 1995. An investigation of the Swainson's Hawk in Argentina. J. Raptor Res. 29:202–204.
Woofinden, N.D., and J.R. Murphy. 1989. Decline of a Ferruginous Hawk population: a 20-year summary. J. Wildl. Manage. 53:1127–1132.
Yaeger, M. 1987. Second Colorado nesting site for Black Phoebe. Colo. Bird Atlas Newsl. 2:2.
Yaeger, M. 1994. First Colorado record of Acorn Woodpecker. J. Colo. Field Ornithol. 28:141–144.
Yasukawa, K., and W.A. Searcy. 1995. Red-winged Blackbird. *In* The birds of North America, no. 184. (A. Poole and F. Gill, eds.). Acad. Nat. Sci., Philadelphia, and Am. Ornithol. Union, Washington, DC.
Yezerinac, S.M. 1993. American Redstarts using Yellow Warblers' nests. Wilson Bull. 105:529–530.
Yosef, R. 1996. Loggerhead Shrike. *In* The birds of North America, no. 231 (A. Poole and F. Gill, eds.). Acad. Nat. Sci., Philadelphia, and Am. Ornithol. Union, Washington, DC.
Young, G. 1981. The troubled waters of Mono Lake. Natl Geog. 160:504–519.
Young, J.R., J.W. Hupp, J.W. Bradbury, and C.E. Braun. 1994. Phenotypic divergence of secondary sexual traits among Sage Grouse, *Centrocercus urophasianus,* populations. Anim. Behav., 47:1353–1362.
Zeleny, L. 1976. The bluebird. Indiana Univ. Press, Bloomington.
Zerbi, V. 1985. Purple Martins nesting on McClure Pass. J. Colo. Field Ornithol. 19:53–54.
Zeuner, F.E. 1963. History of domesticated animals. Hutchinson and Co., London.
Zimmerman, J.L. 1966. Polygyny in the Dickcissel. Auk 83: 534–546.
Zimmerman, J.L. 1971. The territory and its density-dependent effect in *Spiza americana.* Auk 88:591–612
Zimmerman, J.L. 1982. Nesting success of Dickcissels (*Spiza americana*) in preferred and less preferred habitats. Auk 99:292–298.
Zwank, P.J., K.W. Kroel, D.M. Levin, G.M. Southward, and R.C. Rommé. 1994. Habitat characteristics of Mexican Spotted Owls in southern New Mexico. J. Field Ornithol. 65:324–334.
Zwickel, F.C. 1992. Blue Grouse. *In* The birds of North America, no. 15 (A. Poole, P. Stettenheim, and F. Gill, eds.). Acad. Nat. Sci., Philadelphia, and Am. Ornithol. Union, Washington, DC.

INDEX

ATLAS AUTHORS

Atlas Editor **Hugh Kingery** directed the Colorado Atlas from its inception in 1987. His editing endeavors have ranged from a boy scout newsletter to CFO's *Journal* to *Trail and Timberline*, magazine of The Colorado Mountain Club. (He also wrote a book on the club's history.) Over the years the Mountain Club, Denver Audubon Society, and Denver Field Ornithologists each elected him as President. *American Birds* carried his byline as Regional Editor for the Mountain West from 1972 to 1996. For the AOU's *Birds of North America* he wrote the monograph on American Dipper.

After exploring 224 Colorado Atlas blocks (plus 20 in New York) Hugh and his wife Urling still watch birds in terms of Atlas codes. They backpacked into seven blocks and spent only one day mired in a mudhole and wondering how to get out.

A wildlife biologist for the Boulder District, Roosevelt National Forest, **Beverly Baker** conducts environmental analyses ranging from wildlife and plant habitat to noxious weed control. With Colorado Partners in Flight Bev developed *Wonders on the Wing*, a school video and curriculum about the state's migratory songbirds.

Norm Barrett, as a wildlife biologist for USFS in Kremmling, coordinated Atlas field work across the Routt and Arapaho forests and surveyed 20 blocks himself. On the two forests he monitored breeding Buffleheads, Northern Goshawks, and Boreal Owls. After completion of Atlas field work Norm transfered to Oregon's Rogue River National Forest.

Formerly a CDOW biologist and now an environmental consultant, **Steve Boyle** conducts biological inventories and assessments and arranges conservation easements for farmers and ranchers. Before his Atlas field work (57 Atlas blocks as a volunteer and paid field worker) he did Spotted Owl and small-owl searches for CDOW and USFS.

A Ph.D. candidate at the University of Colorado, **Jameson Chace** has studied vireos and cowbirds and now, in Arizona, interactions between Brown-headed and Bronzed cowbirds and their impacts on local birds. Jim's search of Atlas field cards for cowbird reports produced an article in the CFO *Journal* and Appendix C to this book.

In 1988 **Michael Carter** founded the CBO to further the conservation of Rocky Mountain and Great Plains birds. CBO sponsors bird banding at Barr Lake, Brighton, and Dillon (see Fox Sparrow account), educates students and adults, monitors birds for CDOW, and more. Mike served as President of CFO and Chairman of the West Working Group, Partners in Flight and sat on the Atlas Steering Committee.

Coen Dexter teaches chemistry at Central High School in Grand Junction. An Atlas regional coordinator, he and his wife, Brenda Wright (a paid field worker), worked on 127 Atlas blocks. A veteran river rafter, Coen in 1998 discovered 28 Black Phoebes on the San Miguel River.

M. Beth Dillon works as a wildlife biologist for BRD USGS (Biological Resources Division, U.S. Geological Survey) in Fort Collins. Beth's field work includes bird surveys and research in Colorado, Texas, Montana, and Mexico. Beth helped the Atlas as regional coordinator, rendezvous organizer, paid field worker, records reviewer, and had an Atlas triumph: in North Park she discovered Northern Waterthrushes—a new breeding species for the state.

During the Atlas **Frank J. Hein** worked at the Denver Museum of Natural History as a Boslog fellow and curatorial assistant in vertebrate zoology. His duties included supervision of the Dinosaur Ridge hawk watch.

Non-game Migratory Bird Coordinator for Region 6, USF&WS, in Denver, **Stephanie Jones** covered ten Atlas blocks in conjunction with BBS surveys. Her monograph on the Canyon Wren in The Birds of North America series inspired (actually necessitated) her to conduct original research on that little known species. Steph has helped lead Colorado Partners in Flight since 1992.

Steve Jones works as a teacher in Boulder and an environmental consultant specializing in breeding bird ecology. He led field trips to 16 Atlas blocks in the canyons of southeastern Colorado and covered nine plains blocks near Wray. He has co-authored three books: *Shortgrass Prairie, Boulder County Nature Almanac*, and *Colorado Nature Almanac.*

Professor of Economics at the University of Colorado, **Bill Kaempfer** coordinated the Atlas region that swept through Boulder from the Continental Divide to Nebraska. Bill completed 34 blocks, most on the High Plains. He co-founded *Cobirds*, an Email chat line for Colorado birders. To subscribe, send a message, "subscribe COBIRDS [your name]" to listproc@lists.colorado.edu.

Ruth Kuenning and her late husband Walt coordinated the Colorado Springs Atlas region and surveyed 59 Atlas blocks themselves. They particularly relished exploring ranch blocks between Colorado Springs and Punkin Center, where they found longspurs, Mountain Plovers, Long-billed Curlews, and a roadside Ferruginous Hawk nest that they monitored for several years.

Ron Lambeth, a celebrated wit and a biologist with the U.S. Bureau of Land Management in Grand Junction, served as a regional coordinator. Ron probably knows Colorado pinyon/juniper and desert habitats and birds better than anyone does. He is presently working on the Bird Conservation Plan of Colorado Partners in Flight.

An entomologist for the Colorado State Forest Service, **Dave Leatherman** travels the state to provide advice about insect control. He contributed data for 37 Atlas blocks and excellent photographs for the Atlas color folio.

Tony Leukering coordinates CBO's statewide bird monitoring project, from which he extrapolated data for the Atlas. He has worked for three other bird observatories and for *American Birds* as Christmas Bird Count editor.

Thirty years of teaching English at Grand Junction's Central High School honed the writing skills that grace **Rich Levad's** species accounts. He worked in 86 Atlas blocks as volunteer and paid field worker and has conducted point counts for CBO. He has a predilection for nighttime surveys: in the Grand Junction area he has located over 100 roost holes of Western Screech-Owls.

Cynthia Melcher works as an avian research biologist with BRD USGS. Her M.S. thesis on bird responses to overgrazing by elk led her to survey Atlas blocks in the high country of Rocky Mountain National Park. She edits the CFO *Journal*.

Duane Nelson completed 56 blocks as a volunteer and paid field worker. He initiated the Dinosaur Ridge hawk watch at Morrison. For eight years he has monitored endangered species in the Arkansas Valley: Piping Plovers and Least Terns. He believes in on-the-ground management: e.g., making safe nesting islands (to reduce egg losses, especially from foxes, coyotes, raccoons, and snakes).

A Professor of Entomology at Colorado State University and retired scientific editor with BRD USGS, **Paul Opler** of Fort Collins atlased 71 blocks, served as co-coordinator for the big northeastern Colorado region, and chaired the Atlas publication committee. In 1998 Paul completed work on a Peterson *Field Guide to Western Butterflies.*

David Pantle retired from law practice in Denver and moved to a home overlooking the Arkansas River in Canon City. With his wife Sherrill he worked on 35 Atlas blocks in Colorado and 23 in Kansas. He provided wise counsel on the Atlas Steering Committee and served as Secretary of the Colorado Field Ornithologists and Arkansas Valley Audubon.

Kim Potter conducts biological inventories for the White River National Forest and monitors Sage Grouse for CDOW. Atlas work (volunteer and paid) took her to 55 blocks. In 1998, surveying for Black Swifts, she found five new colonies and a sixth colony in a cave where, in 1962, a spelunker had reported "nesting birds."

Charles R. Preston contributed guidance to the Atlas as soon as he arrived at the Denver Museum of Natural History in 1990 as Chairman of Zoology and Curator of Ornithology. He directed ecological studies at Rocky Mountain Arsenal NWR and Comanche National Grassland. In 1998 he became Founding Curator, Draper Natural History Museum, Buffalo Bill Historical Center, in Cody, Wyoming.

Co-author of *Colorado Birds*, **Robert Righter** served as Atlas Treasurer for three years. Working with Cornell University's Laboratory of Ornithology, he authored an audio field guide, *Bird Songs of the Rocky Mountain States and Provinces*, out in 1999. In one of the 18 Atlas blocks Bob covered, he recorded one species for the audio: "booming" by a Common Nighthawk.

Richard Roth, as Wildlife Program Manager for the Rocky Mountain Region of USFS, facilitated Atlas field work in many ways. He served as a regional coordinator and helped with USFS access and contacts. He completed 15 Colorado Atlas blocks and four in Kansas. Dick initiated the surveys that located pockets of breeding Mexican Spotted Owls in Colorado's Wet Mountains.

This book's chapter on Colorado Ornithologists describes the extraordinary achievements of **Ronald A. Ryder**, emeritus Professor of Wildlife Biology at Colorado State University.

James Sedgwick, research biologist with BRD USGS, specializes in *Empidonax* flycatchers: their lifetime reproductive success, population dynamics, and habitats. In 1998 he took sound recordings in Colorado and Utah to examine vocal and genetic differentiation of Willow Flycatcher subspecies. He served as co-coordinator for the Fort Collins Atlas region.

John Toolen, habitat biologist for CDOW in northwestern Colorado and Atlas coordinator for the same region, surveyed 30 blocks. An active promoter of bird conservation, John sits on the boards of the Colorado Bird Observatory and the Colorado chapter of The Wildlife Society, and he co-founded the local chapter of the Society for Conservation Biology.

During the Atlas **Alan Versaw** taught mathematics in Mancos and Colorado Springs; he now works as a computer programmer in Colorado Springs. Besides volunteer and paid field work in 95 blocks, he helped the Atlas with data entry, database proofreading, and as regional coordinator. Alan co-founded *Cobirds*, the Email chat line about Colorado birds.

Roberta Winn worked on eight Atlas blocks around Woodland Park and helped Ruth and Walt Kuenning in 34 blocks. She retired 25 years ago from the Conservation Library, Denver Public Library, and moved to the mountains. There she became fascinated with Flammulated Owls and monitored nests near Woodland Park.

Professor of Zoology **Barbara Winternitz** took her Colorado College ornithology classes on field trips into three Atlas blocks. She co-chaired the 1985 CDOW workshop that produced the state's first Species of Special Concern list.

The art of conservationist, atlaser, and species account **author Mark Yaeger** sparkles from almost every page of this book (see page 38).

BLOCKS

PRIORITY BLOCK:

Blocks, roughly 3 by 3.5 miles on a side, in the South East sector of each topographic map.

NON-PRIORITY BLOCK:

Similarly-sized blocks in other sectors of a topographic map.

BLOCK CODES:

CBAP used the code system applied by the U.S. Geological Survey for topographic maps. This consists of three parts: two numbers define the latitude and three identify the longitude; these specify a latilong. Each latilong contains 64 maps—eight rows of eight maps. The last two units of the code define the location of the map within the latilong, first a letter (A–H corresponding to the numbers 1–8) which locates it along the north/south axis, and second a number (1–8) which places it east and west. The numbering system starts from the southeast corner (of the latilong or the state).

For example, the Mount Elbert quadrangle, with a code of 39106A4, lies in latilong 39106, i.e., north of Latitude 39 and west of Longitude 106. "A4" places it as the first map north and fourth west of the southeast corner of the latilong. In the computer listings used in the appendix, numbers replace the letters (A–H = 1–8). See Figure 1 on page 6.

CERTAIN STANDARD REFERENCES

The text uses shorthand for certain seminal references to Colorado birds:

A&R: Andrews, Robert A., and Robert Righter. 1992. *Colorado Birds.* Denver Museum of Natural History, Denver. The most recent, standard, reference book on Colorado bird distribution; meticulously researched. Maps, elevation charts, seasonal occurrences.

B&N: Bailey, A. M., and R. J. Niedrach. 1965. *Birds of Colorado.* Denver Museum of Natural History, Denver. The basic reference on Colorado birds. Two volumes of descriptions, distribution, records, and discussion.

BBS: 1966–1996. *Breeding Bird Survey Trend Analysis.* Unpublished data furnished by the Breeding Bird Survey Office, National Biological Service, Patuxent Wildlife Research Center, Laurel MD.

Latilong Study: Kingery, H. E. 1988. *Colorado Bird Distribution Study.* Colorado Field Ornithologists, in cooperation with the Colorado Division of Wildlife. Denver. Maps occurrences, seasonally, by latilong, from 1965–1987, derived from contributions by most Colorado bird-watchers.

Priorities 1995: Colorado Bird Observatory and Colorado Division of Wildlife, 1995. *Priorities for Bird Conservation and Monitoring in Colorado.* Categorizes bird species, using Partners in Flight criteria, for developing priorities for bird conservation and monitoring.

BREEDING CODES

The following codes apply to a species seen or heard during its breeding season:

OBSERVED

O Migrants and non-breeding species observed in block during breeding season.

POSSIBLE

SPECIES FOUND or breeding calls heard in suitable nesting habitat.

X SINGING MALE present in suitable nesting habitat during its breeding season.

PROBABLE

M MULTIPLE MALES: seven different singing males heard on one day in suitable nesting habitat.

P PAIR observed in suitable nesting habitat.

T permanent TERRITORY presumed through territorial behavior: defense, chasing other birds, or song in the same location on at least two occasions a week or more apart.

C COURTSHIP behavior between a male and a female or COPULATION. Includes display or food exchange.

V VISITING probable nest site, but no further evidence obtained.

A AGITATED behavior or anxiety calls of adult, indicating nest site or young in vicinity.

N NEST BUILDING by Bald Eagles or excavation of nest hole by wrens and woodpeckers, and construction of dummy nests.

CONFIRMED

NB NEST BUILDING or adult carrying nesting material.

PE PHYSIOLOGICAL EVIDENCE. Refers to cloacal protuberances and brood patches. For use only by banders.

DD DISTRACTION DISPLAY or injury feigning.

UN USED NEST or eggshells found.

FL recently FLEDGED young (or downy young of ducks, grebes, grouse, and sandpipers still dependent on adults), with limited mobility, including young incapable of sustained flight.

ON OCCUPIED NEST indicated by adult entering or leaving nest site in circumstances indicating occupied nest.

FY FEEDING YOUNG: adult seen carrying food for young, or feeding recently fledged young. Also subdivided into:
- **CF** Adult CARRYING FOOD for young.
- **FF** FEEDING FLEDGLING: adult feeding fledgling.

FS Adult carrying FECAL SAC away from nest for disposal.

NE NEST with EGGS.

NY NEST with YOUNG seen or heard.